SI PREFIXES

Factor	Prefix	Symbol	Origin
10^{24}	yotta	Y	Italian *ott(o)* — eight (3 × 8 = 24)
10^{21}	zetta	Z	Italian *etta* — seven (3 × 7 = 21)
10^{18}	exa	E	Greek *exa* — out of
10^{15}	peta	P	Greek *peta* — spread out
10^{12}	tera	T	Greek *teratos* — monster
10^{9}	giga	G	Greek *gigas* — giant
10^{6}	mega	M	Greek *mega* — great
10^{3}	kilo	k	Greek *khilioi* — thousand
10^{2}	hecto	h	Greek *hekaton* — hundred
10^{1}	deca[a]	da	Greek *deka* — ten
10^{-1}	deci	d	Latin *decimus* — tenth
10^{-2}	centi	c	Latin *centum* — hundred
10^{-3}	milli	m	Latin *mille* — thousand
10^{-6}	micro	μ	Greek *micros* — very small
10^{-9}	nano	n	Greek *nanos* — dwarf
10^{-12}	pico	p	Italian *piccolo* — small
10^{-15}	femto	f	Greek *femten* — fifteen
10^{-18}	atto	a	Danish *atten* — eighteen
10^{-21}	zepto	z	Greek *(h)epto* — seven (−3 × 7 = −21)
10^{-24}	yocto	y	Greek *octo* — eight (−3 × 8 = −24)

[a]Alternative spelling: deka.

Greek Alphabet

Greek Letter	Name	Equivalent	Sound When Spoken
Α α	Alpha	A	al-fah
Β β	Beta	B	bay-tah
Γ γ	Gamma	G	gam-ah
Δ δ	Delta	D	del-tah
Ε ε	Epsilon	E	ep-si-lon
Ζ ζ	Zeta	Z	zay-tah
Η η	Eta	E	ay-tay
Θ θ	Theta	Th	thay-tah
Ι ι	Iota	I	eye-o-tah
Κ κ	Kappa	K	cap-ah
Λ λ	Lambda	L	lamb-dah
Μ μ	Mu	M	mew
Ν ν	Nu	N	new
Ξ ξ	Xi	X	zzEye
Ο ο	Omicron	O	om-ah-cron
Π π	Pi	P	pie
Ρ ρ	Rho	R	row
Σ σ	Sigma	S	sig-ma
Τ τ	Tau	T	tawh
Υ υ	Upsilon	U	opp-si-lon
Φ φ	Phi	Ph	figh or fie
Χ χ	Chi	Ch	kigh
Ψ ψ	Psi	Ps	sigh
Ω ω	Omega	O	o-may-gah

ENHANCED
WebAssign
Increased Engagement.
Improved Outcomes.
Superior Service.
Exclusively from Nelson Education and Cengage Learning, Enhanced WebAssign® combines the exceptional content that you know and love with the most powerful and flexible online homework solution, WebAssign. Enhanced WebAssign engages students with immediate feedback, rich tutorial content, and interactive eBooks, helping students to develop a deeper conceptual understanding of subject matter. Instructors can build online assignments by selecting from thousands of text-specific problems or supplementing with problems from any Nelson Education or Cengage Learning textbook in our collection.
With Enhanced WebAssign, you can
• Help students stay on task with the class by requiring regularly scheduled assignments using problems from the textbook.
• Provide students with access to a personal study plan to help them identify areas of weakness and offer remediation.
• Focus on teaching your course and not on grading assignments. Use Enhanced WebAssign item analysis to easily identify the problems that students are struggling with.
• Make the textbook a destination in the course by customizing your Cengage YouBook to include sharable notes and highlights, links to media resources, and more.
• Design your course to meet the unique needs of traditional, lab-based, or distance learning environments.
• Easily share or collaborate on assignments with other faculty or create a master course to help ensure a consistent student experience across multiple sections.
• Minimize the risk of cheating by offering algorithmic versions of problems to each student or fix assignment values to encourage group collaboration.
Learn more at www.cengage.com/ewa
Iofoto/Shutterstock, Monkey Business Images/Shutterstock, Monkey Business Images/Shutterstock, Bikeriderlondon/Shutterstock, Goodluz/Shutterstock, StockLite/Shutterstock, Shutterstock, Andresr/Shutterstock, Andresr/Shutterstock, Dboystudio/Shutterstock, Pablo Cavlog/Shutterstock, WaveBreakMedia/Shutterstock

NELSON
EDUCATION

PHYSICS
AN ALGEBRA-BASED APPROACH

PHYSICS
AN ALGEBRA-BASED APPROACH

Ernie McFarland
University of Guelph

Alan J. Hirsch

Joanne M. O'Meara
University of Guelph

Physics: An Algebra-Based Approach

by Ernie McFarland, Alan J. Hirsch, and Joanne M. O'Meara

Vice President, Editorial Higher Education:
Anne Williams

Senior Publisher:
Paul Fam

Marketing Manager:
Leanne Newell

Senior Developmental Editor:
Roberta Osborne

Photo Researcher:
Derek Capitaine

Permissions Coordinator:
Derek Capitaine

Production Project Manager:
Christine Gilbert

Production Service:
Integra Software Services Pvt. Ltd.

Copy Editor:
Frances Robinson

Proofreader:
Erin Moore

Indexer:
Integra Software Services Pvt. Ltd.

Design Director:
Ken Phipps

Managing Designer:
Franca Amore

Interior Design:
Jerilyn Bockorick, Nesbitt

Cover Design:
Courtney Hellam

Cover Image:
David Clapp/Getty Images

Compositor:
Integra Software Services Pvt. Ltd.

Printed and bound in the United States of America
1 2 3 4 18 17 16 15

For more information contact Nelson Education Ltd., 1120 Birchmount Road, Toronto, Ontario, M1K 5G4. Or you can visit our Internet site at http://www.nelson.com

Library and Archives Canada Cataloguing in Publication

McFarland, Ernie, author
Physics: an algebra-based approach / Ernie McFarland (University of Guelph), Alan J. Hirsch, Joanne M. O'Meara (University of Guelph). — First edition.

Includes index.
ISBN 978-0-17-653186-7 (bound)

1. Mathematical physics—Textbooks. 2. Algebraic logic—Textbooks. I. Hirsch, Alan J., author II. O'Meara, Joanne M. (Joanne Michelle), author III. Title.

QC20.7.A4M34 2015
530.15 C2014-908371-8

ISBN-13: 978-0-17-653186-7
ISBN-10: 0-17-653186-6

BRIEF TABLE OF CONTENTS

CONTENTS

CHAPTER **5**
Newton's Laws of Motion 105

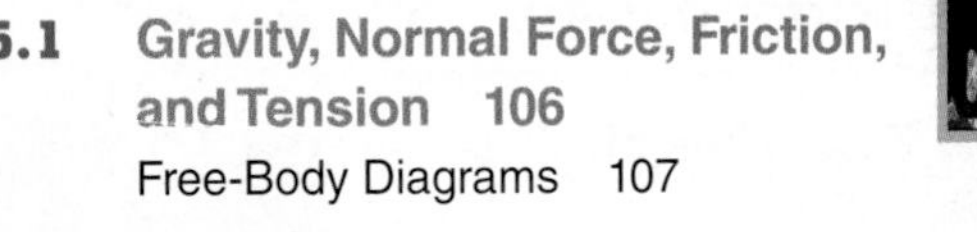

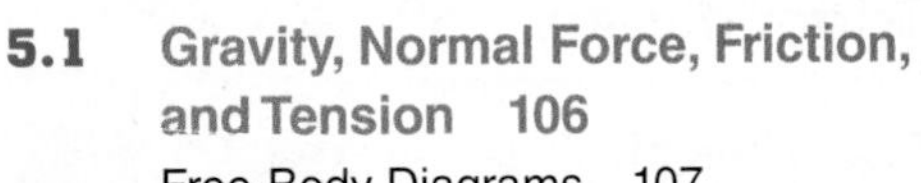

Stefan Schurr/Shutterstock

CHAPTER **6**
Applying Newton's Laws 139

Mitch Gunn/Shutterstock.com

CHAPTER **7**
Work, Energy, and Power 161

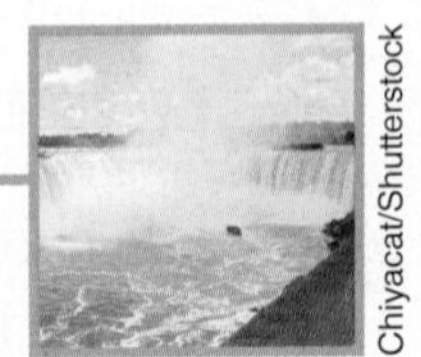

Chiyacat/Shutterstock

CHAPTER **8**
Momentum and Collisions 195

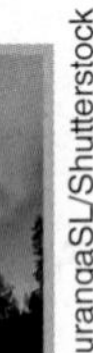

SurangaSL/Shutterstock

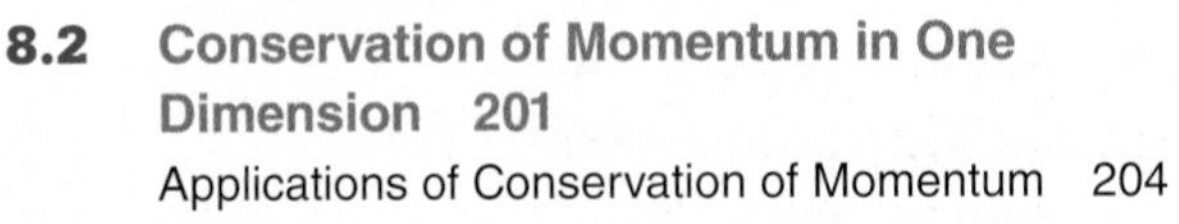

CHAPTER **9**
Gravitation 223

MarcelC/Thinkstock

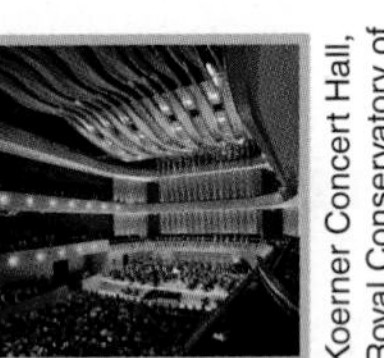
Koerner Concert Hall, Royal Conservatory of Music. Photo: Tom Arban

Gemasolar solar thermal plant, owned by Torresol Energy © SENER

© Prisma Bildagentur AG/Alamy

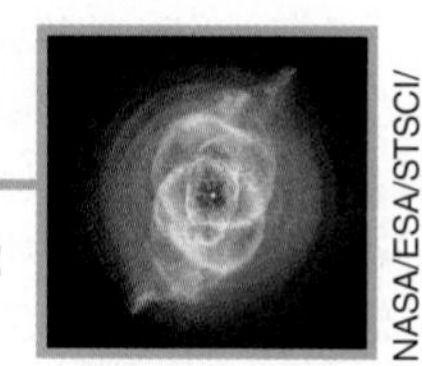
NASA/ESA/STSCI/ Science Photo Library

cowardlion/Shutterstock

Dave Massey/ Shutterstock

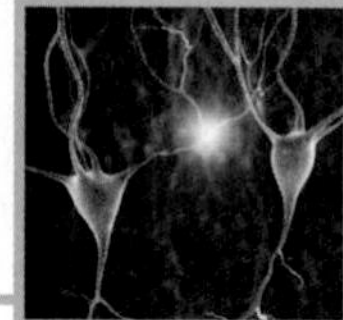
Alfred Pasieka/Science Photo Library

CHAPTER 21 Electrical Resistance and Circuits 677

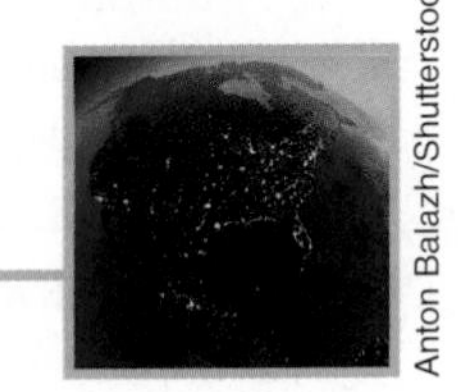
Anton Balazh/Shutterstock

CHAPTER 22 Magnetism 705

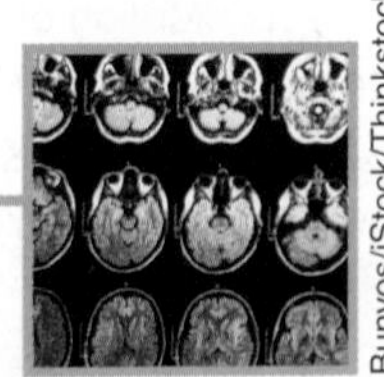
Bunyos/iStock/Thinkstock

CHAPTER 23 Electromagnetic Induction 735

Jesus Keller/Shutterstock

CHAPTER 24 Nuclear Physics: Theory and Medical Applications 763

Brandon Alms/Shutterstock

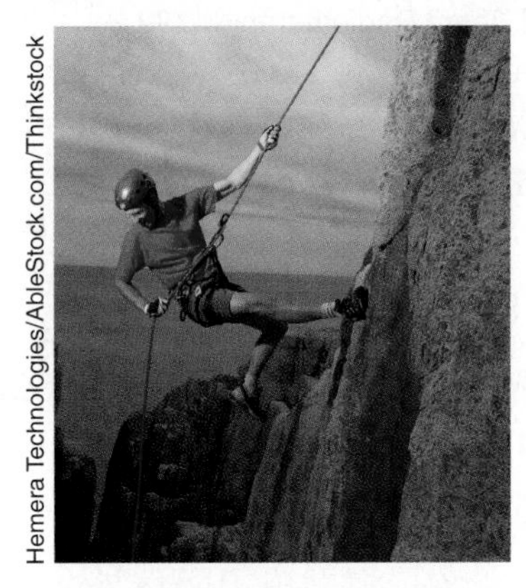
Hemera Technologies/AbleStock.com/Thinkstock

The forces involved in rock climbing are analyzed in Chapter 11.

Physics: An Algebra-Based Approach is informed by a combined 70+ years of teaching physics to first-year university and senior high school students in Ontario. We undertook this project because we have had great difficulty finding a text that approaches the subject at the appropriate level, is modern and engaging, and ensures mastery of the essentials required for success in subsequent studies in a variety of disciplines in science. We have used the core of this text in the classroom for over 20 years at the University of Guelph with more than 10 000 students. Over that period, we made updates and added features as we learned from our students what their challenges were.

This book is designed for science students who have a limited background in physics, which often leads them to approach the subject with trepidation. As a result, throughout the book we emphasize the relevance of physics to students' everyday lives in order to provide context to their learning. We are exceedingly careful about notation, terminology, and significant digits, as well as being quite deliberate in our mathematics as we work through the more than 200 sample problems in the book.

Papa Bravo/Shutterstock

Electrostatic forces are discussed in Chapter 19 in the context of the gecko's ability to walk on ceilings.

Features

There are a number of features in this text that are unique in the marketplace. First, we have included the answers to *all* of the 2200+ questions at the back of the book, not just the odd-numbered questions. We strongly believe that students should attempt as many problems as they can in order to develop their problem-solving skills, and, given the widespread use of online homework systems, we feel that the practice of withholding answers to the even-numbered questions in order for instructors to set assignments is outdated. Second, we have intentionally not included icons to indicate the level of difficulty of individual questions. We feel that such icons deter students from attempting some of the more challenging questions, which is counter to our philosophy of "practise, practise, practise." Third, we have also intentionally not included a summary of equations at the end of each chapter. As noted by one of the reviewers of this book, we strive to encourage critical thinking and problem solving rather than encouraging a "plug-and-chug" approach to questions. The equations are clearly indicated throughout the chapters and students will benefit from reading a sentence or two (or more!) of surrounding text as they find their way toward a solution to a problem.

Nicola Keegan/Shutterstock

Squid mobility is a biological example discussed in Chapter 8.

In developing *Physics: An Algebra-Based Approach* from the textbook used at the University of Guelph, we have enhanced the book significantly through many features designed to assist students in learning about physics and its numerous applications. These features also assist instructors as they share their knowledge and enthusiasm with their students. For example,

- Helpful **Tips** are provided in marginal text boxes with titles such as "Math Tip," "Notation Tip," "Units Tip," "Problem-Solving Tip," "Terminology Tip," and "Learning Tip." There are 80 Tip boxes throughout the text.
- Thirty "**Tackling Misconceptions**" boxes ensure that the most common misconceptions experienced by students are addressed.
- "**Problem-Solving Strategies**" boxes are provided in several chapters to discuss particular approaches to solving problems in a given context.
- "**Try This!**" boxes in every chapter suggest hands-on activities that give practical, concrete experience of physics concepts using simple, easily found materials. For example, in one of the more than a hundred activities in the text, a simple method for measuring the speed of light with a microwave oven is described.

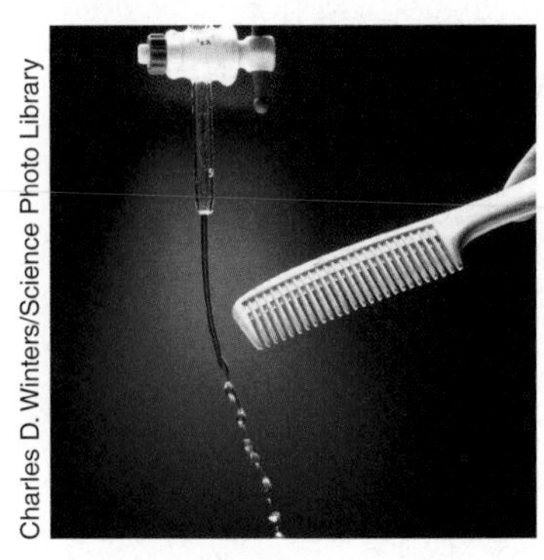
Charles D. Winters/Science Photo Library

A "Try This!" activity in Chapter 19.

- Over 90 "**Fermi Questions**," named after the famous physicist Enrico Fermi, give students practise in making real-life assumptions and calculations involving physics principles and applications.
- More than 200 "**Did You Know?**" boxes introduce interesting scientific and historical facts, as well as modern and futuristic applications of physics.
- "**Profiles in Physics**" boxes summarize brief biographies for nine interesting scientists, enhancing the historical context of physics discoveries.
- Useful **reference information**, often contained in appendices, is included in the inside front and back covers of our book for easy access.
- Examples of **physics applications** from countries around the world are included to emphasize the international nature of the discipline.

Oliver Hoffmann/Shutterstock

A "Did You Know?" box in Chapter 22 introduces students to ferrofluid and its many applications.

Evidence-Based Pedagogy

Our extensive collective experience in the classroom and lecture hall is clearly evident throughout the book. So too is our familiarity with the latest developments in Physics Education Research in active learning techniques, alternative conceptions in introductory physics, instructional material design, and refined approaches to problem-solving skills development. For example,

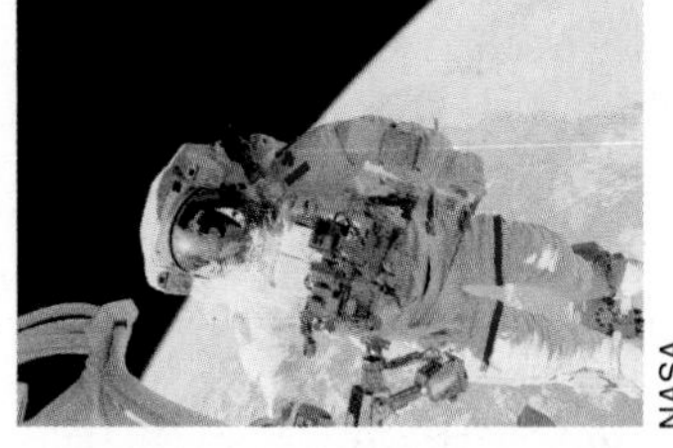

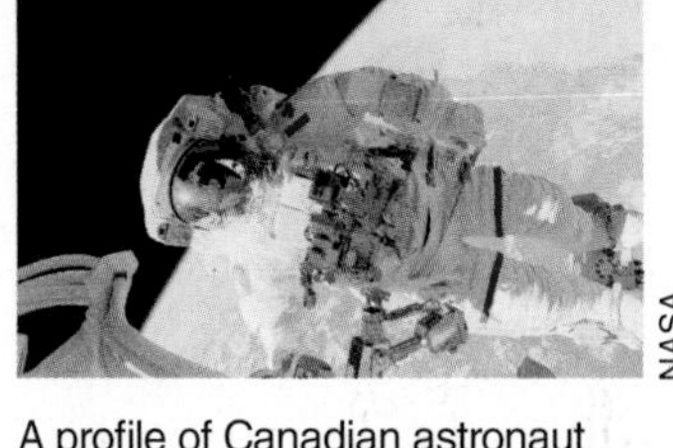

NASA

A profile of Canadian astronaut Chris Hadfield is included in our discussion of weightlessness in Chapter 9.

- Awareness of **common difficulties** in understanding certain concepts has led to a careful explanation of these concepts. For example, the introduction of centripetal force in some textbooks leads to the incorrect impression that centripetal force is a separate force of nature that produces circular motion. In this book, the phrase "centripetal force" appears only once in a "Tackling Misconceptions" box, and many examples are given to illustrate the variety of forces (gravity, normal force, etc.) that can produce the centripetal acceleration toward the centre of a circle. As well, we include conceptual questions in which the students are asked to identify the force(s) that provide(s) the centripetal acceleration in various situations.

dwphotos/Shutterstock

Our discussion of centripetal acceleration is just one example of the way in which our text is informed by Physics Education Research.

- **Exercises** for the student are placed at the end of each section. These exercises, designed to provide "just-in-time learning," are to be completed before the student continues on to the next section of the book.
- **Sample problems** (10 per chapter, on average) are presented in a clear and concise fashion.
- Consistency in the **use of symbols** is found throughout the text (e.g., variables such as d and v are denoted using italics, and vectors such as $\vec{\boldsymbol{F}}$ and $\vec{\boldsymbol{g}}$ are denoted using boldface italics with an arrow above).
- Care has been taken to ensure that rules of **significant digits** presented early in the text are followed consistently throughout.
- Since **free-body diagrams** are important in analyzing many situations in the topic of mechanics, they are introduced painstakingly. There are many introductory questions in which students are asked only to draw free-body diagrams for objects in various circumstances. The introduction of Newton's laws of motion is followed by more complex problems involving free-body diagrams.

muratart/Shutterstock.com

The "Fosbury flop" is the basis of a question in Chapter 11 involving centre of mass.

- Much more attention has been paid to **careful wording** than in many textbooks. As an example, consider the term "acceleration." Many books are careful in the chapters on motion to ensure that acceleration is considered (correctly) as a vector, with both magnitude and direction, but in the chapters on force the term "acceleration" is often used incorrectly, particularly in problems, to indicate only the magnitude of the acceleration. This can lead to confusion on the part of the students.

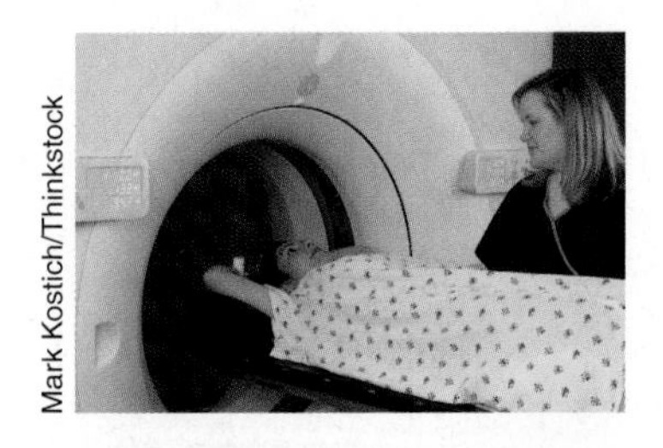

Mark Kostich/Thinkstock

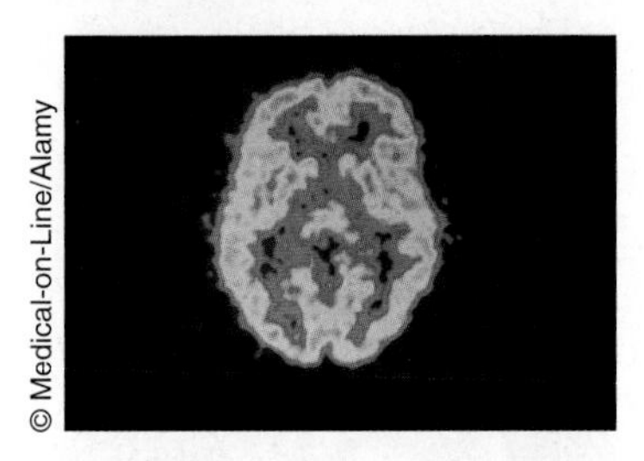

© Medical-on-Line/Alamy

Positron emission tomography is featured in the topic of momentum in Chapter 8.

- To add interest, **real-life examples** are included, not only in the narrative sections but also in the problems. Many of the examples are taken from the biological sciences. For example, the medical application involving positron emission tomography (PET) is featured in the topic of momentum in Chapter 8.
- Student understanding of many of the concepts is explored by **conceptual questions** as well as numerical problems. For example, in the section on conservation of momentum, students are given word descriptions of various collisions and asked to choose those in which the total momentum of a particular pair of objects is constant.
- On average, there are 93 questions per chapter, with a total of **2238 questions** in the text. These questions have been very carefully designed to build in complexity, ranging from straightforward confidence-building exercises to questions that require the integration of a number of different concepts in the text.

Ancillaries

Instructor Resources

The **Nelson Education Teaching Advantage (NETA)** program delivers research-based instructor resources that promote student engagement and higher-order thinking to enable the success of Canadian students and educators. Be sure to visit Nelson Education's **Inspired Instruction** website at http://www.nelson.com/inspired to find out more about NETA.

The following Instructor Resources have been created for *Physics: An Algebra-Based Approach*. Access these ultimate tools for customizing lectures and presentations at www.nelson.com/instructor.

NETA Test Bank: This resource, by Vahid Rezania (MacEwan University), includes multiple-choice questions written according to NETA guidelines for effective construction and development of higher-order questions.

The NETA Test Bank is available in a new, cloud-based platform. **Nelson Testing Powered by Cognero®** is a secure online testing system that allows you to author, edit, and manage test bank content from any place you have Internet access. No special installations or downloads are needed, and the desktop-inspired interface, with its drop-down menus and familiar, intuitive tools, allows you to create and manage tests with ease. You can create multiple test versions in an instant, and import or export content into other systems. Tests can be delivered from your learning management system, your classroom, or wherever you want. Nelson Testing Powered by Cognero for *Physics: An Algebra-Based Approach* can be accessed through www.nelson.com/instructor.

NETA PowerPoint: Microsoft® PowerPoint® lecture slides for every chapter have been created by Manuel Díaz-Avila (Mount Royal University). The slides include key figures, tables, and photographs from the textbook. NETA principles of clear design and engaging content have been incorporated throughout, making it simple for instructors to customize the deck for their courses.

Image Library: This resource consists of digital copies of figures, short tables, and photographs used in the book. Instructors may use these jpegs to customize the NETA PowerPoint slides or create their own PowerPoint presentations.

Instructor's Solutions Manual: This manual, prepared by Ernie McFarland, Alan J. Hirsch, Joanne M. O'Meara, Anna Kiefte (Acadia University), and Naeem Syed Ahmed (Laurentian University), contains complete solutions to all exercises and end-of-chapter questions and problems.

DayOne: Day One–Prof InClass is a PowerPoint presentation that instructors can customize to orient students to the class and their text at the beginning of the course.

Enhanced WebAssign

Nelson Education's **Enhanced WebAssign**®, the leading homework system for math and science, has been used by more than 2.2 million students. Created by instructors for instructors, EWA is easy to use and works with all major operating systems and browsers. EWA adds interactive features to go far beyond simply duplicating text problems online. Students can watch solution videos, see problems solved step-by-step, and receive feedback as they complete their homework.

Enhanced WebAssign allows instructors to easily assign, collect, grade, and record homework assignments via the Internet. This proven and reliable homework system uses pedagogy and content from Nelson Education's best-selling physics textbooks, then enhances it to help students visualize the problem-solving process and reinforce physics concepts more effectively. EWA encourages active learning and time on task, and respects diverse ways of learning.

Student Ancillaries

Enhanced WebAssign

Nelson Education's Enhanced WebAssign® is a groundbreaking homework management system that combines our exceptional physics content with the most flexible online homework solution. EWA will engage you with immediate feedback, rich tutorial content, and interactive features that go far beyond simply duplicating text problems online. You can watch solution videos, see problems solved step-by-step, and receive feedback as you complete your homework.

Visit http://www.cengage.com/ewa for more information and to access EWA today!

Student Solutions Manual

This manual, prepared by Ernie McFarland, Alan J. Hirsch, Joanne M. O'Meara, Anna Kiefte (Acadia University), and Naeem Sayed Ahmed (Laurentian University) contains solutions to all odd-numbered questions and problems.

Acknowledgments

The production of this physics textbook was a major undertaking and would not have been possible without the expertise and guidance of many people.

The authors thank the wonderful team at Nelson Education for their knowledge and direction: Paul Fam, Senior Publisher; Roberta Osborne, Senior Developmental Editor; Christine Gilbert, Production Project Manager; Leanne Newell, Marketing Manager; and Frances Robinson, Copy Editor. Many thanks go also to colleagues at the University of Guelph for feedback on earlier versions of this book: Martin Williams, Jim Hunt, and Elisabeth Nicol, as well as to the thousands of students who provided formal and informal comments about these versions.

We also wish to thank the faculty members at other institutions who reviewed the text and provided many useful suggestions for additions and changes:

Zaven Altounian, McGill University
Marco Bieri, Simon Fraser University
James Brewer, British Columbia Institute of Technology
Ron Cundall, Lethbridge College
Bertrand Dion, CÉGEP de Sainte-Foy
Michael Gericke, University of Manitoba
Ioulia Kvasnikova, Douglas College

Mark Laidlaw, University of Victoria
Edward Mathie, University of Regina
Donna McKenzie, Grant MacEwan University
Bruce Miller, Georgian College
Vesna Milosevic-Zdjelar, University of Winnipeg
Yoichi Miyahara, McGill University
Ted Monchesky, Dalhousie University
Pierre Ouimet, University of Regina
Vahid Rezania, Grant MacEwan University
Georg Rieger, University of British Columbia
Claudia Schubert, Conestoga College
Clinton Sheehan, Northern College
Zbigniew M. Stadnik, University of Ottawa
Anne Topper, Queen's University

Special thanks go to Stanford Downey and Jamie Rasor at Stanford Downey Architects Inc., Toronto, for sharing with the authors their expertise in the topic of controlling resonance oscillations (Section 13.4 in the text). Mr. Downey was the architect for the transformation of One King West, the 51-storey condominium/hotel constructed atop a 12-storey heritage building more than 100 years old. The high-rise building has a large array of water pools on the top floor that ensure stability in high winds.

Most importantly, we give our largest and most heartfelt thanks to our spouses: Mary McMartin, Judy Evans, and Carl Svensson. They provided love and support, and made countless sacrifices as we spent many long hours working on this project.

Photo by Philip Castleton Photography Inc. Courtesy of Stanford Downey Architects Inc.

One King West in Toronto has features that apply physics principles to reduce oscillations in high winds.

This book is dedicated to our children: Grant & Steve, Dan & Lianna, and Hannah & Mara, as representatives of our future generations of great thinkers and dreamers. Your endless curiosity and enthusiasm are inspiring—never stop asking questions. We can't wait to see where your passions take you from here!

**The most beautiful thing we can experience is the mysterious.
It is the source of all true art and science.**

– Albert Einstein

Help Us to Improve

This textbook has had extensive classroom-testing and revision over several years, with constructive criticism from students and instructors helping to improve the current version. We welcome your input as well: feel free to contact us with your suggestions for further enhancements.

Send feedback on *Physics: An Algebra-Based Approach* to paba@uoguelph.ca

Ernie McFarland
Alan J. Hirsch
Joanne M. O'Meara

ABOUT THE AUTHORS

Ernie McFarland

Photography by Rolly Meisel

Ernie McFarland is Emeritus Professor of Physics at the University of Guelph, where he recently retired after teaching for 35 years. He has received five teaching awards, including the Ontario Confederation of University Faculty Associations Teaching Award, the Canadian Association of Physicists (CAP) Medal for Excellence in Undergraduate Teaching, and a 3M National Teaching Fellowship. He is co-author of three books: *Energy, Physics, and the Environment* (3rd Ed., Thomson, 2007), *Physics for the Biological Sciences* (5th Ed., Nelson, 2012), and *Physics for Tomorrow's World* (with Alan Hirsch, Revised 2nd Ed., Nelson, 2012), which was the precursor to this current text. He is also sole author of *Einstein's Special Relativity—Discover it for Yourself* (Fitzhenry & Whiteside, 1998).

Ernie received a Bachelor of Science degree in Physics from the University of Western Ontario, and a Masters of Science in Theoretical Nuclear Physics from McMaster University. He has been actively involved in the Ontario Association of Physics Teachers (OAPT), of which he is founding President, and received honourary life memberships in the OAPT and in the Society for Teaching and Learning in Higher Education. He has given over 70 invited talks and workshops on teaching physics and about university teaching in general, and has written many papers about physics teaching.

Ernie is well-known for his lecture demonstrations, and edited "The Demonstration Corner" column in the *OAPT Newsletter* for 25 years. He and a colleague presented demonstration shows at elementary schools for many years, and appeared 19 times on CTV's *Good Morning Canada*.

Ernie has a strong interest in energy and the environment, and gave the Herzberg Memorial Public Keynote Lecture "Energy: Where on Earth Are We Going?" at the 2006 CAP Congress.

He has had significant international involvement. For the American Association of Physics Teachers, he has served on four committees, chaired the Apparatus Committee, and was a member of the Editorial Board of *The Physics Teacher* magazine. In 1999, he was elected to the International Commission on Physics Education, and was chosen as Secretary from 2002 to 2005. He was on the International Advisory Board for physics education conferences in India and Japan.

Ernie and his wife Mary enjoy hiking, travelling, and getting together with their four children, other family members, and friends. Ernie also likes playing his guitars, and is trying to improve his golf game.

Alan J. Hirsch

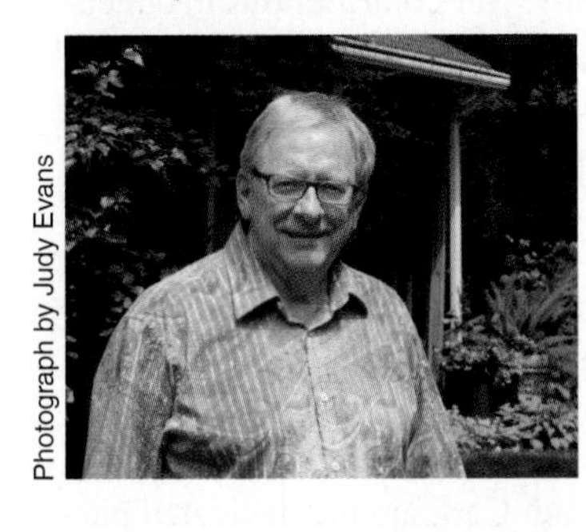
Photograph by Judy Evans

Alan Hirsch cherishes the experience of teaching high school physics, science, and mathematics in Ontario for 31 years. His classrooms became the inspiration and experimental incubator for many textbooks that bear his name. He developed and wrote Ontario's first textbook written exclusively for college-bound students, *Physics: A Practical Approach* (Wiley: 1st ed. 1981; 2nd ed. 1991). He was also the sole author of *Physics for a Modern World* (a Grade 11 academic text; Wiley: 1986), and *Physics 12 College Preparation* (Nelson: 2004). He was a co-author of *Nelson Physics 11* (2002), *Nelson Physics 12* (2003), as well as *Physics for Tomorrow's World* (with Ernie McFarland), which became the basis for this current text (Nelson: Revised 2nd ed.

2012). He has also co-authored 13 science texts for Grades 7 to 10 for use across Canada. Several of these physics and science texts have been translated into French, and *Physics for a Modern World* was adapted for use in Australia. For several years, he also wrote amusement-park workbooks and coordinated model roller coaster contests for the annual physics/math day at Canada's Wonderland, near Toronto, where he had no choice but to ride the coasters to verify data!

Alan received a Bachelor of Science with majors in physics and mathematics from McGill University and teaching certificates from Queen's University and the University of Toronto. He has been an active member of the Ontario Association of Physics Teachers (OAPT), and served as its Section Representative to the American Association of Physics Teachers for five years, during which time he and Ernie McFarland were invited to write a comprehensive article titled "Physics Teaching in Canada," published in *The Physics Teacher* magazine in April, 1992. In 1997, he was awarded a life membership to the OAPT in recognition of his contributions to physics education and the association.

Although officially retired from active teaching, Alan is as busy as ever. When he is not reading or writing, he indulges in a variety of passions, such as gardening or travelling to explore nature, history, architecture, and culture, while trying to improve his photography skills. He and his wife, Judy, are thankful to have visited more than 130 countries and consider travelling to be the best way to learn.

Joanne O'Meara

Photograph by Susan Bubak

Joanne O'Meara has been teaching physics at the post-secondary level since 1999. From the beginning, Joanne has tried to make the subject more accessible and relevant through strong links between theory and application, emphasizing the importance of unifying fundamental principles within the context of interdisciplinary learning. Joanne joined the Department of Physics at the University of Guelph in 2002 and has served as the Associate Chair of the department for several years. She has played a pivotal role in the ongoing evolution of the undergraduate physics curriculum at Guelph—for physics majors and non-physics majors alike. At the national level, Joanne has held executive positions with the Division of Physics Education within the Canadian Association of Physicists and chaired the National Task Force on Undergraduate Physics Revitalization. Joanne is committed to the betterment of physics education at all levels, as evidenced by her development of instructional videos, physics demonstration shows, and teacher workshops for elementary and secondary schools, as well as her repeated appearances on the Discovery Channel's *Daily Planet*.

Joanne earned a Bachelor of Science in applied physics in 1995 and a Doctorate in medical physics in 1999, both from McMaster University. Her contributions to education are recognized with a number of awards, including the Presidential Distinguished Professor Award (2004–2006, University of Guelph); the Special Merit Teaching Award (2009, University of Guelph Faculty Association); the Medal for Excellence in Teaching Undergraduate Physics (2011, Canadian Association of Physicists); and the John Bell Award (2014, the University of Guelph's highest honour for outstanding contributions to university education). Joanne's research applies radiation physics principles to the assessment of the elemental composition of unknown samples, ranging from embedded shrapnel in battlefield survivors to soils and minerals on the surface of Mars.

Outside the university setting, Joanne and her husband, Carl, are the dedicated parents of two incredibly inquisitive and adventurous girls. From skiing, snowshoeing, and tobogganing in the winter to kayaking, hiking, biking, and camping in the summer, they are happiest when enjoying the beauty of the great outdoors!

Measurement and Types of Quantities

1

Kameel4u/Shutterstock

Figure 1-1 The dials and screen in a helicopter cockpit provide measurements that are crucial to the safe operation of the helicopter.

Every dial in a helicopter cockpit involves a measurement (Figure 1-1). The safe operation of all vehicles, from cars to helicopters, depends on the accurate measurements of such quantities as speed, direction, pressure, and electric voltage. From mundane measurements such as the temperature and time to cook a pizza, to the life-saving measurements needed to track flying aircraft to prevent mid-air collisions, examples of the importance of accurate measurements are everywhere in our modern world. Diabetics monitor their health over the course of a day by routinely measuring the glucose level in their blood. Drivers travel to their destinations using dashboard navigation systems that rely on data from global positioning satellites orbiting Earth. Engineers continuously monitor power output and cooling systems at power stations to ensure safe operation. Construction workers repeatedly

CHAPTER **OUTLINE**

measure lengths and angles when building new homes, the importance of which is demonstrated by the common woodworking mantra: measure twice, cut once! Whether it is checking the time on your cell phone or consulting the weather forecast before heading out the door in the morning, you make use of countless measurements over the course of a day.

We begin this text with topics related to measurement and other mathematical concepts. You may have studied some of these topics before, but just as a musician needs to learn the rules and practise the skills required to become proficient, in physics you need to learn how to use accurate measurements and to master other mathematical skills before tackling the physics concepts presented in later chapters.

1.1 The Importance of Measurement

qualitative description a statement that indicates the quality of some object, event, or idea

quantitative description a statement that indicates a quantity

measurement a numerical quantity found by measuring with a device or an instrument

We use a great variety of ways to describe things: a cappuccino is robust; the results of a medical research experiment are encouraging; a marathoner's energy seems boundless, and so on. These statements are called **qualitative descriptions** because each one indicates the quality of some object, event, or idea. Such descriptions are subjective; in other words, their meanings depend on a person's previous experience and values, so they differ from one person to another. Thus, these descriptions do not have much value where comparisons must be made or communication must be exact.

Now consider these descriptions: the mass of phosphorus in a vitamin pill is 0.125 g; the average healthy body temperature is 37°C; a lecture room can accommodate 600 people; the ratio of the value of a quarter to a dollar is 1:4. These statements are called **quantitative descriptions** because each one indicates a quantity. They have the same meaning to everyone who understands them, thus they are objective and they form the basis of comparing and communicating about objects, events, and ideas. Many quantitative descriptions involve counting (600 people), some involve ratios with no units (1:4), while others involve measurements (0.125 g and 37°C).

A **measurement** is a numerical quantity found by measuring with a device or an instrument.[1] A measurement is expressed by a number followed by a unit. In the body temperature example, 37 is the number and °C (degrees Celsius) is the unit. To see how important the unit is, compare a temperature of 37 in the Celsius scale with 37 in the Fahrenheit scale. The numbers are the same, but the measurements are not! See also Figure 1-2.

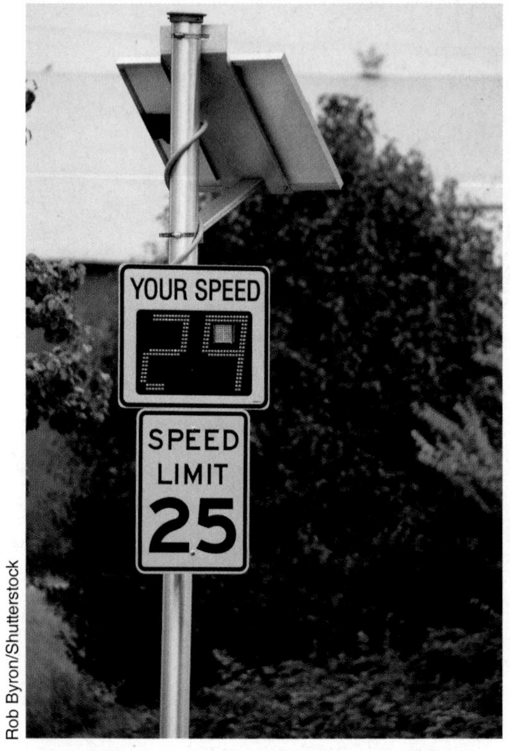

Figure 1-2 This sign shows poor communication because the measurement does not have a unit. The reader could argue that it means 25 km/h, 25 mi/h, or something else.

Measurement in Our Daily Lives

How do measured quantities affect our daily lives? Measurement of *time* is important to us: an alarm signals wake-up time; classes begin and end at scheduled times; a part-time job involves payment per unit of time. *Length* is also relevant: the stopping distance of a vehicle is greater on an icy road than on a dry road; the distance you walk, jog, or run each day affects your state of physical fitness. *Mass*, too, must be considered: groceries are often priced according to their mass in grams or kilograms (or pounds or ounces if the British Imperial measurement system is used); a common headache tablet contains 325 mg of acetylsalicylic acid. No doubt you can think of other examples of time, length, and mass, as well as measurements of other quantities that affect your life.

[1]The word "measurement" can also mean the act of measuring.

Measurement in the Past

In ancient civilizations, such as those of China, Egypt, Greece, and Rome, measured quantities made it easier to trade or barter; to build structures such as temples, pyramids, roads, and irrigation systems; and to schedule festivities, agricultural endeavours, and other events. Measurements at that time were very crude by today's standards. Consider, for example, the measurement of time. Large time intervals were measured using repeated events involving celestial bodies such as the Sun, Moon, and stars (Figure 1-3). Shorter time intervals could be measured approximately with dripping water or sand flowing through a narrow opening into a graduated container. More accurate timing devices began to appear during the Renaissance in Europe, especially after Galileo Galilei (1564–1642) of Italy discovered the principle of the pendulum clock.[2]

Along with improvements in measurement came developments in scientific experimentation. From about the late 1500s onward, science has continually advanced, in many ways due to the progress in measurement techniques. Evidence of this, as well as other examples of measurement in the past, will be presented later in the text.

DID YOU KNOW?

In 1602, Sanctorius, a medical friend of Galileo, invented the pulsilogium (a simple pendulum) to measure human pulse rates. The pendulum's length was adjusted until it matched the pulse rate, then the rate was recorded as the length of the pendulum.

Measurement in Science and Physics

What do scientists in general and physicists in particular do? This is not an easy question to answer; each physicist would likely give a slightly different response. However, a graphic representation of the scientific process, illustrated in Figure 1-4, shows elements common to all forms of scientific research: *curiosity, creativity,* and *collaboration.* You can see that the questions in curiosity lead to creative ideas, which in turn lead to research. This research is not conducted in isolation; rather it is performed with peers and through collaboration, often with scientists around the world.

When the creative ideas in the scientific process require experimentation and/or mathematical analysis for the research, then *measurements* are crucial. A physicist conducting an experiment analyzes the measured data and tries to make sense of the results within the framework of existing theories, in collaboration with his/her team. In some cases, this analysis may indicate the need for a new theory to explain the observations. Often this work will lead to new questions, which the research team will try to answer by designing new experiments and performing more measurements. Thus, measurement is a crucial step in the process of physics experimentation and discovery.

(a) Filip Fuxa/Shutterstock

(b) Kippy Spilker/Shutterstock

(c) Nick Reynolds Photography/Shutterstock

Figure 1-3 Examples of time measurement using repeated natural events. **(a)** It is believed that these structures at Stonehenge in England were used to predict eclipses and the changing of the seasons over 3500 years ago. **(b)** As Earth rotates on its axis once each 24 h, stars in the night sky appear to revolve around the North Star at a rate of 15° every hour, as shown in this time-lapse photograph. **(c)** Earth's rotation is also responsible for the moving shadow on the calibrated dial of this sundial in London, England.

[2]A profile of Galileo is found in Chapter 2, page 45.

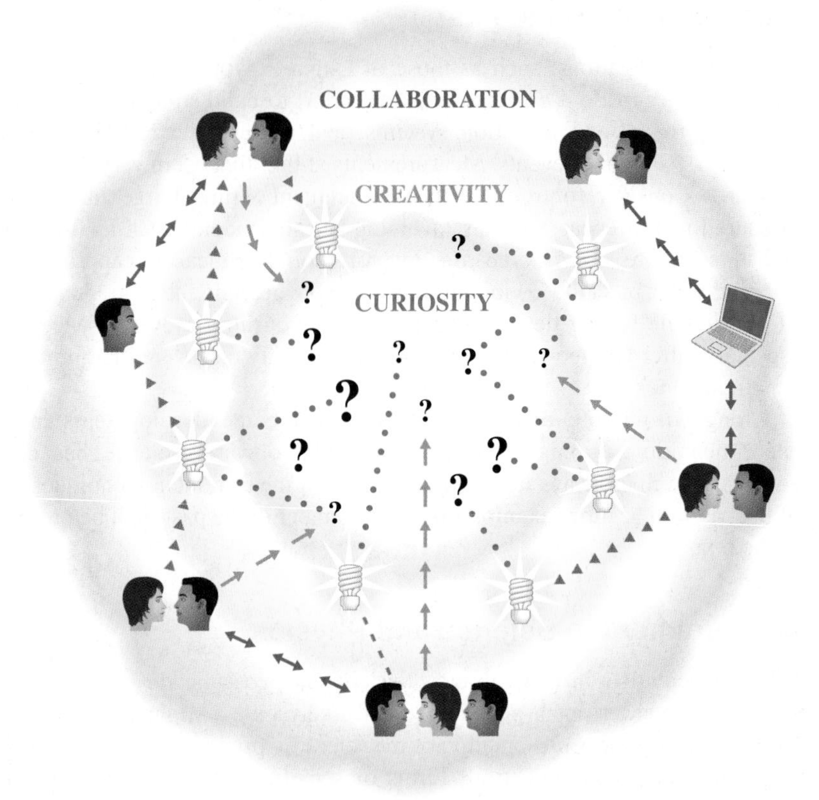

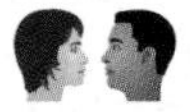

Figure 1-4 The process of scientific advancement has three main interrelated characteristics: curiosity, creativity, and collaboration. Results of collaboration lead back to asking more questions.

nanotechnology the application of the study of the properties of extremely tiny particles, some as small as an atom

Research in fundamental physics often leads to practical applications. For example, measurement plays a key role in biomedical engineering as theory is put into practice in the design and construction of everything from artificial heart valves and pacemakers to contact lenses, tooth braces, and artificial limbs. As well, many advances in modern medicine, from diagnostic imaging to radiation therapy for treating cancer, have arisen from applying knowledge gained from basic nuclear physics research. Even the GPS device in your smart phone makes use of physics research from the abstract world of Einstein's general relativity.

Another field in which measurement is crucial is the rapidly developing **nanotechnology**, the application of the study of the properties of extremely tiny particles, some as small as an atom. ("Nano" stems from nanometre, nm, which is 10^{-9} m or one-billionth of a metre. To get an idea of how small this is, a human hair is approximately 40 000 to 80 000 nm in diameter!) Nanotechnology is applied to gather vast numbers of measurements using electronic sensors that analyze data on airplanes, trains, power plants, furnaces, and many other technologies, aiding in their safe and efficient operation. Nanoparticles and nanobots (or nanorobots) can be used in a large variety of applications, such as creating efficient solar cells, making fire-resistant glass, increasing computer memory and speed, and treating cancerous tumours (Figure 1-5). More details of nanotechnology will be presented later in the text.

Since measurement is important in understanding physics and solving physics problems, it is the focus of the next three sections.

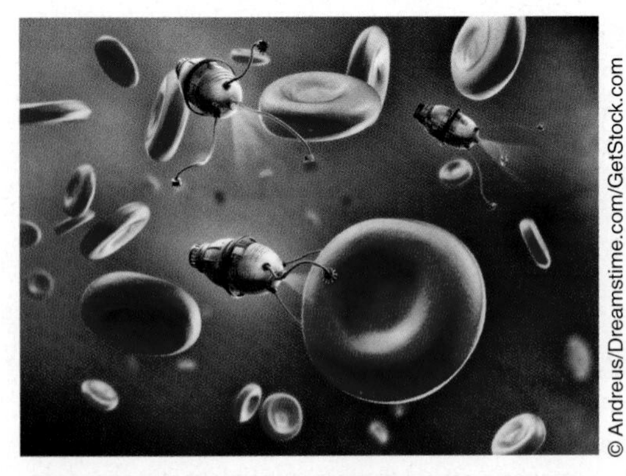

Figure 1-5 In this artist's conception of nanotechnology, nanobots move through the bloodstream repairing some cells.

EXERCISES

1-1 Give three examples of qualitative descriptions involving your sense of (a) taste and (b) smell.

1-2 State three qualitative and three quantitative descriptions of yourself and/or a friend.

1-3 Classify each of the following as a qualitative description or quantitative description:

(a) a dozen roses

(b) a deafening noise

(c) a slow diffusive motion in the bloodstream

(d) a 100 m dash

(e) a pulse rate of 56 min^{-1}

(f) a blood pressure of 120/80 mm Hg

1-4 For each of the following, cite two examples of measurements that affect your daily lives: time, length, mass, and volume. Use examples other than those given in the text.

1-5 Name and give examples of quantities other than time, length, mass, and volume that you might encounter.

1-6 List disadvantages of using celestial bodies as reference objects to measure time.

1-7 Describe, with examples, the difference(s) between theoretical aspects of physics and practical applications of physics research.

1.2 Metric Units, Prefixes, and Conversions

In order for measurements to be useful in science, industry, and commerce, they must be stated using a set of standard units. A **standard unit** is one that can be reproduced according to its definition and does not change; thus it means the same for everyone using it. An inch and a centimetre are examples of standard units. The most common set of standard units in the world is the metric system. Another standard system is the British Imperial system, which is used in only a few countries in the world, including the United States. This book will use the metric system almost exclusively. Scientists throughout North America use the metric system, even though certain industries are slow in making the change from the British to the metric system. Furthermore, measurements associated with physics are easier to understand and manipulate using the metric system.

standard unit a unit of measurement that can be reproduced according to its definition and does not change

DID YOU KNOW?

The word "mile" in the British system of measurement originated from the Latin word *mille*, which means one thousand. The original mile, used during the time of the Roman Empire, was called *milia passum*, or a thousand double paces.

Base Units in the Metric System

A **base unit**, also called a **fundamental unit**, is a standard unit of measurement from which other units may be derived. Although there are seven base units in the metric system, only the three used to measure length, mass, and time will be considered in this chapter. The base unit used to measure electric current will be introduced later in the text. (The inside front cover of the text includes a complete summary of the metric base units.)

base (or fundamental) unit a standard unit of measurement from which other units may be derived

DID YOU KNOW?

The word "metre" was chosen from the Greek word *metron*, which means "to measure."

The metric system has several possible units for each measured quantity, such as centimetres, metres, and kilometres for length. However, by international agreement beginning in 1960, the system that has become standard is the *Système International d'Unités*, or simply **SI**. SI units are used often in physics.

SI the international metric system, *Système International d'Unités*

Notation Tip

In science, numbers with several digits are written in sets of three digits separated by a space rather than a comma. For example, there are 86 400 s in one day and π is 3.141 592 654 (to 10 digits).

The metre (m) is the SI base unit of length.[3] It was originally defined in Paris in 1790, during the French Revolution, by the French Academy of Sciences as one ten-millionth (10^{-7}) of the assumed distance from the equator to the North Pole. Almost 100 years later, in 1889, the metre was redefined as the distance between two fine lines engraved near the ends of a metal bar. This standard metre was stored under controlled conditions near Paris, and copies were sent to countries around the world.

[3]The international spelling of metre will be used throughout this text. This also applies to the unit of liquid volume, the litre, symbol L.

Currently the standard metre is defined as the distance light travels in a vacuum in 1/(299 792 458) of a second. This quantity is believed to be constant and is reproducible anywhere, so it is an excellent standard.

The SI base unit of time is the second (s). Originally it was defined as 1/(86 400) of a mean solar day (1 day = 24 h = 1440 min = 86 400 s). A mean solar day is the length of a day from one high noon to the next, averaged over a period of one year. A tiny error was apparent with this definition because the length of a day depends on Earth's rotation, and that rotation is slowing down very gradually. Thus, an atomic standard was adopted for the second in 1967. One second is the time for 9 192 631 770 vibrations of light of a certain wavelength emitted from a cesium-133 atom, which is believed to be another constant quantity. It is estimated that the world's most accurate cesium-fountain atomic clock in London, England, will not gain or lose more than one second in 138 million years! This clock is part of a worldwide network of clocks with the accuracy needed to operate the Global Positioning System (GPS), which will be discussed further in Chapter 9. See Figure 1-6.

DID YOU KNOW?

Battery-powered watches and clocks with quartz crystals are common. The battery causes the crystal to vibrate at a high frequency, which causes an electric current at that frequency. A microchip changes the frequency to 1.0 Hz, or a period of 1.0 s, which is used to operate the clock.

DID YOU KNOW?

Time standards can be heard on "speaking telephones" or found on the Internet. For example, by dialing 613-745-1576 you can access a voice that states the time every 10 s, with a background click broadcast every second. The websites http://www.nrc-cnrc.gc.ca/eng/services/time/web_clock.html and http://tycho.usno.navy.mil/time.html also provide time checks.

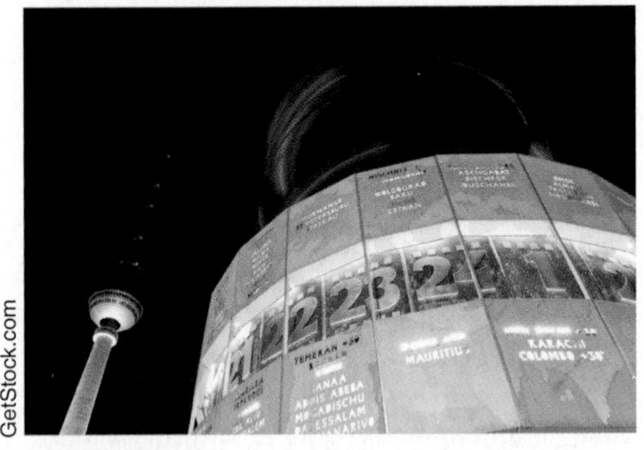

Figure 1-6 This accurate atomic clock in Alexander Platz in Berlin, Germany, doubles as a tourist attraction.

The SI base unit of mass is the kilogram (kg). The kilogram is the only base unit with a prefix (kilo), but it is a more convenient size than a gram. As well, the kilogram is the only standard unit not now defined in terms of natural events. Originally the kilogram was defined as the mass of a litre of water at 4°C at sea level, and a metal cylinder with the same mass was built as a standard and kept in France. Later, a small discrepancy was discovered in the measurement of the mass of the water, and the metal cylinder became the standard kilogram. Copies of this standard kilogram mass are kept in major centres around the world, including one in Ottawa (Figure 1-7).

Figure 1-7 The standard kilogram kept in France was used to make duplicate standards for other countries. Each standard is well-protected from the atmosphere. Shown here is Canada's standard, kept in Ottawa, Ontario.

In summary, the standard SI base units for length, time, and mass are, respectively, the metre, second, and kilogram.

Scientific Notation

Very small or very large numbers are awkward to write in long form or key into a calculator. This problem is solved by using powers of ten and expressing the numbers in **scientific notation**. Using scientific notation, also called "standard form," a non-zero digit is placed before the decimal point with other digits after it, then the number is multiplied by the appropriate power of 10. Huge ranges of large to small measurements, listed in Table 1-1, illustrate why scientific notation is more convenient than long form.

scientific notation the method of expressing very large or small numbers using a non-zero digit before the decimal point and other digits after it, then multiplying by the appropriate power of 10

SAMPLE PROBLEM 1-1

What is the ratio of the mass of the Sun to the mass of a uranium atom?

Solution

From Table 1-1, the ratio is $\dfrac{2 \times 10^{30}\ \text{kg}}{4 \times 10^{-25}\ \text{kg}} = 5 \times 10^{54}$

Thus, the ratio of the masses is $5 \times 10^{54} : 1$.

Notice that the units cancel out. Algebraically, units behave just like numbers.

Derived Units

The square metre (m^2) is a unit that denotes surface area. It is an example of a **derived unit**, a measurement unit that can be stated in terms of one or more of the base (SI) units. In this case, the unit is derived by multiplying length × length. Other examples of derived units are

- the cubic metre (m^3) for volume
- the metre per second (m/s) for speed
- the kilogram per cubic metre (kg/m^3) for density

Other derived units will be introduced throughout the text. The inside front cover of this book lists many of these units. Notice that several derived units are named after famous scientists. Notice also that common units such as watts, hertz, and volts are part of SI.

derived unit a measurement unit that can be stated in terms of one or more of the base (SI) units

Table 1-1

Using Scientific Notation to Indicate Extreme Measurements

Lengths	
Distance to most remote quasar	2×10^{26} m
Distance to Andromeda Galaxy (Figure 1-8)	2×10^{22} m
Distance from Sun to nearest star	4×10^{16} m
Average distance from Earth to Sun	1.5×10^{11} m
Diameter of Earth	1.3×10^{7} m
Diameter of red blood cell	8×10^{-6} m
Diameter of typical bacteria (Figure 1-9)	1×10^{-6} m
Diameter of hydrogen atom	1×10^{-10} m
Diameter of proton	1×10^{-15} m
Times	
Estimated age of universe	5×10^{17} s
Age of solar system	1.5×10^{17} s
Time since start of human existence	1×10^{13} s
Age of Egyptian pyramids	1.4×10^{11} s
One year (approximate value)	3.16×10^{7} s
Time interval between heartbeats	1 s
Duration of nerve impulse	1×10^{-3} s
Period of typical radio waves	1×10^{-6} s
Shortest pulse of light in laboratory	1×10^{-15} s
Lifetime of a W particle (a boson)	3×10^{-25} s
Masses	
Known universe (estimated)	1×10^{53} kg
Milky Way Galaxy	8×10^{41} kg
Sun	2×10^{30} kg
Earth	6×10^{24} kg
Small mountain	1×10^{12} kg
Ocean liner	8×10^{7} kg
Car	1×10^{3} kg
Mosquito	1×10^{-5} kg
Speck of dust	7×10^{-10} kg
Bacterium	1×10^{-15} kg
Uranium atom	4×10^{-25} kg
Electron	9×10^{-31} kg

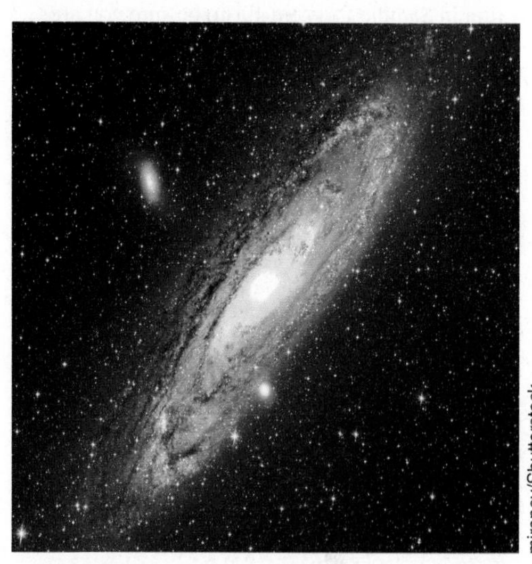

mironov/Shutterstock

Figure 1-8 The Andromeda Galaxy, which consists of millions of stars, is similar in shape to our own Milky Way Galaxy. The distance from our solar system to the Andromeda Galaxy is 2×10^{22} m. There are thousands of galaxies in the known universe.

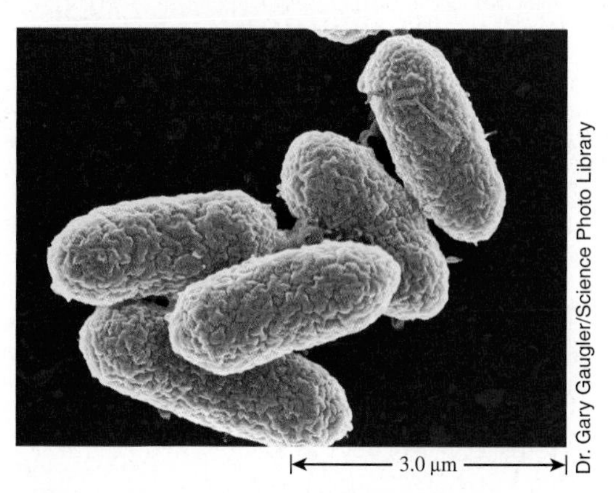

Dr. Gary Gaugler/Science Photo Library

Figure 1-9 This scanning electron microscope image shows rod-shaped Salmonella bacteria with a diameter of about 1 µm and a length of about 3 µm.

Math Tip

When manipulating numbers containing exponents, remember to apply the exponent laws where appropriate:

$$x^m \cdot x^n = x^{m+n}$$
$$x^m \div x^n = x^{m-n}$$
$$(x^m)^n = x^{mn}$$
$$(xy)^n = x^n y^n$$
$$\left[\frac{x}{y}\right]^n = \frac{x^n}{y^n}$$

Math Tip

Use your calculator to perform the calculation in Sample Problem 1-1 to be sure you are using the EE or EXP key correctly. Avoid the error made by some students who try to enter a number like 10^{24} and get 10^{25} instead because they enter 10 EE 24 instead of the correct entry, 1 EE 24.

Metric Prefixes

Although all lengths could be stated using the base unit of the metre, it is sometimes more convenient to state lengths in larger or smaller units, such as the kilometre, the centimetre, or the very tiny nanometre. The metric prefixes, such as "centi," "kilo," and "nano" are based on multiples of 10, a feature of the metric system that is a great advantage. The inside front cover of this book lists the 20 metric prefixes as well as their symbols, meanings, and origins.

Conversions within the Metric System

Measurements to be added or subtracted must have a common unit. For example, you cannot add centimetres to metres without first converting to a common unit. Skill in converting one metric unit to another will assist you in analyzing physical situations and solving physics problems.

In general, when converting from one metric unit to another, multiply by a conversion ratio such as (100 cm)/(1 m) in which the numerator and denominator are equivalent. Such a ratio equals, effectively, one; multiplying by one does not change the quantity. To determine the ratio either rely on your memory, which will improve with practice, or refer to the list of metric prefixes on the inside front cover.

SAMPLE PROBLEM 1-2

Convert 47.5 mm to metres.

Solution

The conversion ratio is either (1 m)/(1000 mm) or (10^{-3} m)/(1 mm). The "mm" must be in the denominator to cancel the "mm" in 47.5 mm.

$$\begin{aligned} 47.5 \text{ mm} &= 47.5 \text{ mm} \times \frac{1 \text{ m}}{1000 \text{ mm}} \\ &= 0.0475 \text{ m} \end{aligned}$$

Thus, the length is 0.0475 m.

SAMPLE PROBLEM 1-3

Add 34 cm and 4.20 m.

Solution

Before adding, the measurements must be expressed in the same units. In this case, we will use metres.

$$\begin{aligned} 34 \text{ cm} + 4.20 \text{ m} &= 0.34 \text{ m} + 4.20 \text{ m} \\ &= 4.54 \text{ m} \end{aligned}$$

Thus, the sum is 4.54 m (or 454 cm).

DID YOU KNOW?

Measurement errors can be very serious. In July, 1983, Air Canada Flight 143 from Montreal to Edmonton ran out of fuel before reaching Edmonton and was forced to crash land near Gimli, Manitoba, thus earning the nickname "the Gimli Glider." The crew responsible for fuel made an error in calculating the number of litres of fuel per kilogram required. In 1999, a space mission to Mars failed because the rocket manufacturer listed quantities in British Imperial units but the NASA scientists thought the quantities were in metric units.

SAMPLE PROBLEM 1-4

Convert 3.07×10^{12} ps to megaseconds.

Solution

From the inside front cover, pico = 10^{-12} and mega = 10^{6}. To avoid errors, the best way to perform this conversion is to proceed in two steps, first converting picoseconds to seconds, and then seconds to megaseconds. Both these steps can be carried out in one line:

$$3.07 \times 10^{12}\ \text{ps} = 3.07 \times 10^{12}\ \text{ps} \times \frac{10^{-12}\ \text{s}}{1\ \text{ps}} \times \frac{1\ \text{Ms}}{10^{6}\ \text{s}}$$
$$= 3.07 \times 10^{-6}\ \text{Ms}$$

Thus, the time is 3.07×10^{-6} Ms. Notice how the ratios were arranged so that "ps" cancelled "ps," and "s" cancelled "s," leaving Ms.

DID YOU KNOW?

The world's smallest ruler used for measuring lengths in an electron microscope was developed in Canada for the National Research Council. Its divisions are only 18 atoms apart.

SAMPLE PROBLEM 1-5

Convert the density measurement of 5.3 kg/cm³ to kilograms per cubic metre.

Solution

The conversion relating centimetres and metres is 100 cm = 1 m. Since the denominator contains cm³, which represents cm·cm·cm, we must apply the factor three times, that is, we must cube it: $(100\ \text{cm})^3 = (1\ \text{m})^3$. Hence, we have

$$5.3\ \frac{\text{kg}}{\text{cm}^3} = 5.3\ \frac{\text{kg}}{\text{cm}^3} \times \frac{(100\ \text{cm})^3}{1\ \text{m}^3}$$
$$= 5.3 \times 10^{6}\ \frac{\text{kg}}{\text{m}^3}$$

Thus, the answer is $5.3 \times 10^{6}\ \text{kg/m}^3$.

DID YOU KNOW?

For many people, their height and arm span are essentially the same value. Leonardo da Vinci and other Renaissance artists codified this observation.

TRY THIS!

Applying Estimation Skills

Using your hand span (Figure 1-10), arm span, shoe length, and average walking pace is a skill that will help you estimate a great variety of measurements. Use a metre stick or metric tape to determine the following measurements in the unit stated:

- your hand span (in centimetres)
- your arm span (in metres)
- the sole of your shoe (in centimetres)
- the length of your natural pace (in metres)

Practise using the quantities you determined above to calculate, as accurately as possible,

(a) the surface area of a desk or lab bench
(b) the surface area of the floor of a room
(c) the distance you travel in 1.0 min while walking at a comfortable speed

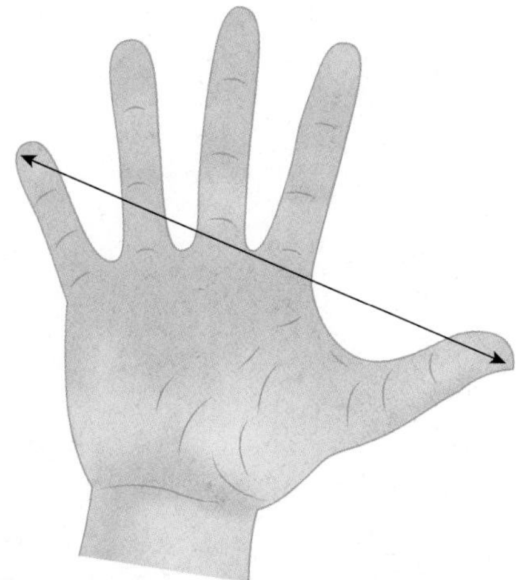

Figure 1-10 Measuring your hand span.

EXERCISES

1-8 List reasons why the original definitions of the metre and second were unsatisfactory.

1-9 Use the values from Table 1-1 to calculate the approximate value of these ratios:

(a) largest length to smallest length
(b) longest time to shortest time
(c) greatest mass to least mass

Which quantity has by far the greatest range of values?

1-10 If you could count one dollar each second, how long would it take to count to one billion (10^9) dollars? Express your answer in seconds and years.

1-11 Determine an estimate for the number of stars in our Milky Way Galaxy by assuming that the mass of the average star is the same as the Sun's mass, and the masses of other bodies, such as planets, are insignificant (Reference: Table 1-1).

1-12 The mass of a hydrogen atom is about 1.7×10^{-27} kg. If the Sun consists only of hydrogen atoms (actually, it is about 95% hydrogen), how many such atoms are in the Sun? (Use Table 1-1 as reference.)

1-13 Simplify

(a) $(2.1 \times 10^{-24})(4.0 \times 10^{9})$

(b) $(2 \times 10^{19})(2 \times 10^{-43})(2 \times 10^{-12})$

(c) $(6.4 \times 10^{21}) \div (8.0 \times 10^{12})$

(d) $(3.88 \times 10^{-2}\ \text{m}) \div (2.00 \times 10^{-7}\ \text{s})$

1-14 State and give examples of the difference between a derived unit and a base unit.

1-15 Many derived units are named after famous scientists. Name four of these and, in each case, state what the unit equals in terms of SI base units. (Reference: inside front cover.)

1-16 Describe what patterns can be seen in the list of metric prefixes on the inside front cover.

1-17 Write these measurements without using prefixes:

(a) The longest cells in the human body are motor neurons at a length of 1.3×10^{-1} dam.

(b) Each bristle of a carbon nanotube is 30 nm in diameter.

(c) At 1.23×10^{4} km, the Hudson Bay shoreline is the longest bay shoreline in the world.

(d) Each day, *Voyageur 1* spacecraft travels another 1.486×10^{-3} Tm away from Earth.

1-18 Perform the following unit conversions, showing your work. Express the final answer in scientific notation.

(a) 20 ms to seconds

(b) 8.6 cm to micrometres

(c) 3.28 g to megagrams

(d) 105 MHz to kilohertz

(e) 2.4×10^{-3} MW/m^2 to milliwatts per square metre

(f) 9.8 m/s^2 to metres per square microsecond

(g) 4.7 g/cm^3 to kilograms per cubic metre

(h) 53 people/km^2 to people per hectare (ha) (**Note:** 1 ha = 10^4 m^2)

1-19 Simplify

(a) 280 mm + 37 cm

(b) 9850 mm − 1.68 m

1-20 An American tourist driving in Canada decides to convert the speed limit of 100 km/h to miles per hour, with which she is more familiar. She knows that there are 5280 ft in 1 mi, that 1 ft contains 12 in, and that 1 in = 2.54 cm. Using this information and your knowledge of SI prefixes, determine the speed limit in miles per hour.

1.3 Dimensional Analysis

From earlier science courses, you are familiar with the definition of density as mass/volume or, using symbols, $\rho = m/V$, where ρ is the Greek letter rho. Rearranging this equation to solve for mass, we have mass = density × volume or $m = \rho V$. If this equation is used to solve a problem, the unit resulting from the product of density and volume must equal the unit of mass. The equals sign dictates that the units as well as the numbers are equal on both sides of the equation. An example will illustrate this. Let us find the mass of 0.20 m^3 of pure aluminum, which has a density of 2.7×10^3 kg/m^3.

$$\begin{aligned} m &= \rho V \\ &= 2.7 \times 10^3 \frac{\text{kg}}{\text{m}^3} \times 0.20\ \text{m}^3 \\ &= 5.4 \times 10^2\ \text{kg} \end{aligned}$$

Thus, the mass is 5.4×10^2 kg.

In this example, both sides of the equation have dimensions of mass. The symbols for the common dimensions of length, mass, and time are written L, M, and T, respectively. We use square brackets to indicate the dimensions of a quantity; for instance, $[V]$ means the dimensions of volume V, and since volume is length × length × length, we have $[V] = \text{L}^3$.

The process of using dimensions to analyze a problem or an equation is called **dimensional analysis**. In this process, the quantities are expressed in terms of dimensions, such as length, mass, and time, and then the expressions are simplified algebraically. If you develop an equation during the solution of a physics problem, dimensional analysis is a useful tool to ensure that both the left-hand and right-hand sides of your

dimensional analysis the process of using dimensions, such as length, mass, and time, to analyze a problem or an equation

equation have the same dimensions. If they do not, you must have made an error in developing the equation.

As an example, the dimensional analysis of the density equation, $m = \rho V$, involves determining the dimensions of both sides of the equation to ensure that they are equivalent:

$$\text{Dimensions of left-hand side (L.H.S.)} = [m] = \text{M}$$

$$\text{Dimensions of right-hand side (R.H.S.)} = [\rho][V] = \frac{\text{M}}{\text{L}^3} \times \text{L}^3 = \text{M}$$

Thus, the L.H.S. and R.H.S. of the equation $m = \rho V$ have the same dimensions.

SAMPLE PROBLEM 1-6

A student reads a test question which asks for a distance d, given a time t of 2.2 s and a constant acceleration a of 4.0 m/s^2. Being unaware of the equation to use, the student decides to try dimensional analysis and comes up with the equation $d = at^2$. Is the equation dimensionally correct?

Solution

The dimensions involved are L for distance, T for time, and, based on the units m/s^2, L/T^2 for acceleration. Now,

$$\text{Dimensions of L.H.S} = [d] = \text{L}$$

$$\text{Dimensions of R.H.S} = [a][t^2] = \frac{\text{L}}{\text{T}^2} \times \text{T}^2 = \text{L}$$

Thus, the dimensions are equal on both sides of the equation, and the equation is dimensionally correct. Unfortunately for this student, the equation is not correct from the point of view of physics. The correct equation is $d = \frac{1}{2}at^2$, but since the "$\frac{1}{2}$" has no dimensions, it does not change the dimensions of the R.H.S. This illustrates a limitation of dimensional analysis; that is, it cannot detect errors involving multiplication or division by dimensionless constants.

EXERCISES

1-21 Write the dimensions of the following measurements:

(a) a speed of 6.8 m/s

(b) a speed of 100 km/h

(c) an acceleration of 9.8 m/s^2

(d) a density of 1.2 g/cm^3

1-22 Determine the final type of quantity produced (e.g., length, speed, etc.) in the following dimensional operations:

(a) $\text{L} \times \text{T}^{-1}$

(b) $(\text{L}/\text{T}^2) \times \text{T} \times \text{T}$

(c) $\text{M} \times \text{L}^{-3}$

(d) $(\text{L}/\text{T}^3) \times \text{T}$

1-23 Use the information in the inside front cover to write the dimensions of these derived units:

(a) newton

(b) watt

(c) hertz

1-24 When a person is jogging at a constant speed, the distance travelled equals the product of speed and time, or $d = vt$. Show that the dimensions on both sides of this equation are equal.

1-25 Assume that for a certain type of motion, the distance travelled is related to time according to the equation $d = kt^3$, where k is a constant. Determine the dimensions of k.

1-26 Prove that the equation $d = v_0t + \frac{1}{2}at^2$ is dimensionally correct. (d is a distance, v_0 is initial velocity, t is time, and a is acceleration.)

1.4 Significant Digits

Quantities that are counted are exact, and there is no uncertainty about them. Two examples are found in these statements:

There are 27 bones in a human hand.

Ten dimes have the same value as one dollar.

uncertainty in measurement, the range of values in which the true value is expected to lie; also called possible error

Measured quantities are much different because they are never exact; every measurement has some uncertainty associated with it. Measurement **uncertainty** is the range of values in which the true value is expected to lie; it is also called possible error. Uncertainty should accompany a measured value; in fact, the reporting of a measured value is not complete without stating the uncertainty of the measurement. The uncertainty may arise from the measurement device, environmental conditions, properties of the item being measured, or even the skill of the person doing the measuring.

As an example, suppose you are measuring the mass of a friend in kilograms. You might try a bathroom scale and get 56 kg, then a scale in a physician's office and get 55.8 kg, and finally a sensitive electronic scale in a scientific laboratory and get 55.778 kg. Not one of these measurements is exact; even the most precise measurement, the last one, has an uncertainty. For the bathroom scale, the reading of 56 kg could mean that the mass is closer to 56 kg than either 55 kg or 57 kg, and the measurement could be written as 56 kg $\pm$ 1 kg. However, you may believe that you can read the scale to the closest 0.5 kg, in which case you would record the measurement as 56 kg $\pm$ 0.5 kg. In many instruments, the uncertainty is stated either on the scale or in the owner's manual. For example, the scale in the physician's office could have an uncertainty of $\pm$ 0.2 kg, so the measurement would be recorded as 55.8 kg $\pm$ 0.2 kg, and if the uncertainty of the electronic scale is $\pm$ 0.001 kg, the measurement is recorded as 55.778 kg $\pm$ 0.001 kg.

significant digits the digits in any measurement that are reliably known

In any measurement, the digits that are reliably known are called **significant digits**. These include the digits known for certain and the single last digit that is estimated. Thus, the mass of 56 kg $\pm$ 1 kg has two significant digits, and 55.778 kg $\pm$ 0.001 kg has five significant digits.

The number of significant digits can be determined by applying these rules:

- Zeroes placed before other digits are not significant: 0.089 kg has 2 significant digits.
- Zeroes placed between other digits are always significant: 4006 cm has 4 significant digits.
- Zeroes placed after other digits behind a decimal are significant: both 5.800 km and 703.0 N have 4 significant digits.
- Zeroes at the end of a number are significant only if they are indicated to be so using scientific notation to place the zeroes after a decimal. For example, the distance 5 800 000 km may have anywhere from 2 to 7 significant digits. By using scientific notation, we can judge which digit is the estimated one, so we can determine the number of significant digits:

  5.8×10^6 km has 2 significant digits

  5.800×10^6 km has 4 significant digits

  $5.800\,000 \times 10^6$ km has 7 significant digits

percent error the difference between the measured and accepted values of a measurement expressed as a percentage

If the accepted value of a measurement is known, the **percent error** of an experimental measurement can be found using

$$\text{percent error} = \frac{|\text{measured value} - \text{accepted value}|}{\text{accepted value}} \times 100\%$$

Calculations Based on Measurements

Measurements made in experiments or given in problems are often used in calculations. For example, you might be asked to find the average speed of a cyclist given the measurement of the time it takes to travel a certain distance. The final answer of the problem should take into consideration the number of significant digits of each measurement, and it may have to be rounded off.

When adding or subtracting measured quantities, the final answer should have no more than one estimated digit.

SAMPLE PROBLEM 1-7

Add 123 cm + 12.4 cm + 5.38 cm

Solution

123 cm (the "3" is estimated)

12.4 cm (the "4" is estimated)

+ 5.38 cm (the "8" is estimated)

140.78 cm (the "0," the "7," and the "8" are estimated)

Thus, the answer should be rounded off to one estimated digit, or 141 cm.

Math Tip

In Sample Problems 1-7 and 1-8, the calculations were done first then the answer was rounded off. When calculated answers are rounded off to the appropriate number of significant digits, the following rules apply:

- If the first digit to be dropped is 4 or less, the preceding digit is not changed; e.g., to three significant digits 3.814 becomes 3.81.
- If the first digit to be dropped is more than 5, or a 5 followed by digits other than zeroes, or a 5 alone or followed by zeroes, the preceding digit is raised by 1; e.g., to three significant digits 5.476 becomes 5.48; 9.265 221 becomes 9.27; and 1.265 becomes 1.27.[4]

When multiplying or dividing measured quantities, the final answer should have the same number of significant digits as the original measurement with the least number of significant digits.

SAMPLE PROBLEM 1-8

A cyclist travels 4.00×10^3 m on a racetrack in 292.4 s. Calculate the average speed of the cyclist.

Solution

Average speed (v) is distance (d) divided by time (t):

$$v = \frac{d}{t} = \frac{4.00 \times 10^3 \text{ m}}{2.924 \times 10^2 \text{ s}}$$

$$= 1.368.. \times 10^1 \text{ m/s}$$

This answer should be rounded off to three significant digits. Hence, we obtain an average speed of 1.37×10^1 m/s or 13.7 m/s.

Most quantities written in this book have either two or three significant digits, although some quantities require more significant digits. The answers to the numerical problems have been written to the correct number of significant digits after rounding off following the rules given in this section. You will be able to practise the rules of significant digits with every numerical problem.

[4] In some specialized cases in statistical calculations, the digit preceding a 5 is not changed if it is even, but it is raised by 1 if it is odd; e.g., 1.265 becomes 1.26 and 1.275 becomes 1.28. This rule exists to avoid the accumulated error that would occur if the 5 were always rounded up.

Round-off Error

Assume you are helping a friend choose floor tiles for a room with a length (l) of 5.17 dm and a width (w) of 3.41 dm. At a price of \$2.49/dm², how much will the tiles cost? You begin by finding the surface area (A) of the floor:

$$\begin{aligned} A &= l \times w \\ &= 5.17 \text{ dm} \times 3.41 \text{ dm} \\ &= 17.6297 \text{ dm}^2 \end{aligned}$$

If you follow the rules of rounding off answers, this area becomes 17.6 dm².

Next you calculate the cost:

$$\begin{aligned} \text{Cost} &= A \times \text{rate} \\ &= 17.6 \text{ dm}^2 \times \frac{\$2.49}{\text{dm}^2} \\ &= \$43.824, \text{ or } \$43.80 \text{ to three significant digits} \end{aligned}$$

But if you had simply left the area calculation in your calculator before multiplying by the rate, the calculated cost would have been:

$$\begin{aligned} \text{Cost} &= A \times \text{rate} \\ &= 17.6297 \text{ dm}^2 \times \frac{\$2.49}{\text{dm}^2} \\ &= \$43.89795, \text{ or } \$43.90 \text{ to three significant digits} \end{aligned}$$

You can see that, to obtain a final answer valid to three significant digits, it is necessary to keep *more* than three significant digits in intermediate answers. Rounding off too early introduces error, which is referred to as *round-off error.*

Hence, when doing calculations, remember to *keep all the digits in your calculator until the final answer is determined,* and then round off this answer to the correct number of significant digits. If it is necessary to write down an intermediate answer, use the correct number of significant digits, but keep the intermediate answer with all its digits in your calculator.

Order-of-Magnitude Estimations

When solving numerical problems in physics, as well as in everyday experiences, it is good to be able to estimate the value of a quantity. This skill applies when checking that a calculated answer makes sense, but it also can be applied when you are trying to decide if you can believe some advertising or reports in the media.

order-of-magnitude estimation a calculation based on reasonable assumptions to obtain a value expressed to a power of 10; also called a *Fermi question* after Enrico Fermi

An **order-of-magnitude estimation** is a calculation based on reasonable assumptions to obtain a value expressed to a power of 10. Often there is only one significant digit in the calculated value.

Consider, for example, an estimation of the volume of air in litres inhaled by a person in one year. We begin with two assumptions. First, a person would inhale about one litre (1 L) of air with each breath. (You might find a more accurate value by devising a water-displacement experiment.) Second, the number of breaths a person would take per minute would likely be between 10 and 20, let's say 15. Now we are ready for the calculation. Notice the cancellation of units in this calculation.

$$\begin{aligned} \text{volume of air} &= \frac{1 \text{ L}}{\text{breath}} \times \frac{15 \text{ breaths}}{\text{min}} \times \frac{60 \text{ min}}{\text{h}} \times \frac{24 \text{ h}}{\text{day}} \times \frac{365 \text{ d}}{\text{year}} \\ &= 8 \times 10^6 \text{ L / year (one significant digit)} \end{aligned}$$

Thus the volume of air inhaled by a person is approximately 8×10^6 L/year. Order-of-magnitude estimations are often rounded off to the nearest power of ten. In this example, we would then obtain 10^7 L/year.

This type of question will be asked from time to time in this text. Such a problem is called a *Fermi question,* named after Enrico Fermi, a famous physicist and professor who often asked his students to estimate quantities impossible to measure directly.

PROFILES IN PHYSICS

Enrico Fermi (1901–1954): Combining Education with Experimental and Theoretical Physics

Enrico Fermi was a brilliant physicist who was at the forefront of 20th century science (Figure 1-11). He is celebrated as one of the leaders in understanding nuclear reactions and learning how to control them. He is remembered as a great teacher who used estimation questions, along with many other methods, to inspire his students to "think physics." He was honoured by having an element named after him (Fermium, atomic number 100) as well as a class of particles called "fermions," which have a particular characteristic spin.

Born in Rome in 1901, Fermi entered higher education at an institute in Pisa, Italy, at age 17. By age 20 he was already publishing articles in scientific journals. He studied at various European universities and became a physics professor at the University of Rome when he was 24.

When Fermi began concentrating less on theoretical physics and more on experimental physics, he discovered how to cause neutrons to slow down during a nuclear reaction. This important discovery later led to the development of controlled nuclear fission reactions and nuclear fission bombs. For this discovery as well as other contributions, Fermi was awarded the Nobel Prize in Physics in 1938. Shortly after receiving this award in Sweden, he and his family moved to the U.S.A. where he later became a U.S. citizen.

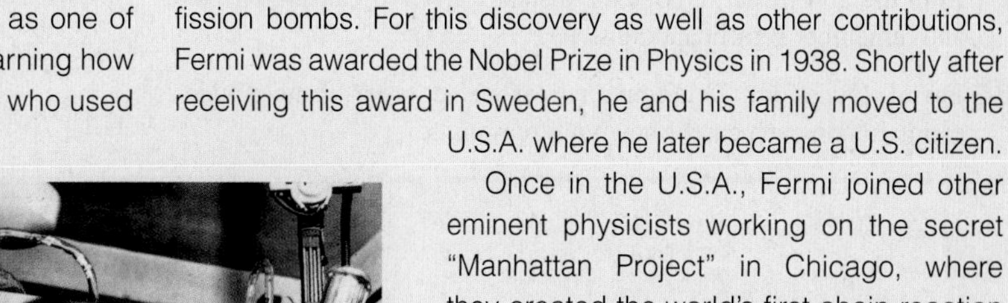

Figure 1-11 Enrico Fermi in the lab.

Once in the U.S.A., Fermi joined other eminent physicists working on the secret "Manhattan Project" in Chicago, where they created the world's first chain-reaction "atomic pile." In July 1945, while witnessing a major nuclear explosion test, he performed the type of calculation he often asked his students to do: he estimated the power of the nuclear blast. Not surprisingly, his estimation was "in the ballpark."

Enrico Fermi combined experimental and theoretical physics in a way that no other leading physicist did. He was highly respected and well-liked, and he preferred easier explanations to harder ones. He died at age 53 of cancer, possibly related to his close work with radioactive nuclear materials.

EXERCISES

1-27 Three different scales are for sale at three different prices. A potential buyer measures the mass of a box on the scales and obtains these values:

42.40 kg ± 0.005 kg

42.4 kg ± 0.05 kg

42.4 kg ± 100 g

Which scale do you think would be the most expensive? the least expensive? Why?

1-28 State the number of significant digits in each measurement:

(a) 0.04 Tm

(b) 400.20 pm

(c) 8.10×10^6 kg

(d) 0.008 200 μs

1-29 Round off each quantity to three significant digits, and write the answer in scientific notation.

(a) 38 510 Gm

(b) 0.000 940 488 MW

(c) 55.055 dam

(d) 876.50 kL

(e) 0.076 550 μg

1-30 A student performs measurements to determine the density of pure water at 4°C, and obtains an average result of 1.08×10^3 kg/m^3. The accepted value is 1.00×10^3 kg/m^3. What is the percent error in the experimental measurement?

1-31 A rectangular mirror measures 1.18 m by 0.378 m. Find the perimeter and the surface area of the mirror, expressing your answers to the appropriate number of significant digits.

1-32 Use the data on the inside back cover to determine the difference in masses between

(a) a neutron and a proton

(b) a proton and an electron

1-33 Assume that the Earth–Sun distance has a constant value of $1.495\ 988 \times 10^{11}$ m and the Earth–Moon distance is constant at 3.844×10^{8} m. Also assume that all three lie in the same plane. Determine the greatest and least Moon–Sun distances, taking into consideration significant digits.

1-34 Find the time it takes light, travelling at 3.00×10^{8} m/s, to travel from the Sun to Earth. (Use the distance given in Question 1-33, and remember significant digits.)

1-35 **Fermi Question:** Determine a reasonable order-of-magnitude estimation in each case. Show your reasoning.

(a) Assume that the mass of an average cell in your body is 1×10^{-12} kg. Calculate the number of cells in your body.

(b) What is the mass in kilograms of all the hamburger patties consumed in North America in one year?

(c) Assume that a Ferris wheel at an amusement park were to fall off its support and become a gigantic rolling wheel. How many rotations, travelling in a straight line, would it have to make to travel from Calgary to Winnipeg?

(d) Canada's coastline, including along its islands, is the longest in the world at about 2×10^{5} km. If Canada's entire population were lined up side-by-side with outstretched arms just touching, how many total populations would equal the coastline length?

1.5 Scalars and Vectors

scalar quantity a quantity with magnitude (or size) but no direction (also called a scalar)

vector quantity a quantity with both magnitude and direction (also called a vector)

A tourist, trying to find a museum in an unfamiliar city, asks how to get there and is told, "All you have to do is walk one-and-a-half kilometres from here." This information is obviously not very useful without a direction. A measurement of 1.5 km is an example of a **scalar quantity**, one with magnitude but no direction. (Magnitude means size.) A measurement such as 1.5 km west is an example of a **vector quantity**, one with both magnitude and direction.

Some common scalar quantities and typical examples are

Length or distance	A race track is 100 m long.
Mass	The mass of a newborn baby is approximately 3 kg.
Time	The time interval between heartbeats is about 1.0 s.

Vector quantities are common in physics. The ones used most frequently in the study of motion and forces are

Displacement	A crane is used to lift a steel beam 40 m upward.
Velocity	A military jet was travelling at 1600 km/h east.
Acceleration	The acceleration due to gravity on the Moon is 1.6 m/s^2 downward (toward the centre of the Moon).
Force	A person falling about 50 cm onto a hard surface and landing on a heel experiences a force of about 2×10^{4} N upward, enough to fracture a bone.

DID YOU KNOW?

Scalar is derived from the Latin word *scala*, which means "ladder" or "steps" and implies magnitude. Vector is a copy of the Latin word *vector*, which means "carrier" and implies something being carried from one place to another in a certain direction. In biology, a vector can be an organism that carries disease or an agent that transfers genetic material.

Scalar quantities, also called scalars, are easy to manipulate algebraically. The ordinary rules of addition, subtraction, multiplication, and division apply. Vector quantities, or vectors, are more complex. They require special symbols and rules of addition, subtraction, and so on. The general nature of scalars and vectors will be considered as you proceed through Chapter 2. However, details regarding vectors are presented in Chapter 3, which prepares you for vector analysis in two-dimensional motion in Chapter 4.

EXERCISES

1-36 List three scalar quantities other than those already given, and give an example of each.

1-37 For motion that you have experienced today, write two specific examples of displacement and two of velocity.

LOOKING BACK...LOOKING AHEAD

This chapter has emphasized the importance of measurement and basic mathematical skills in physics. Details of the SI units were presented; this system will be used throughout the text. The topics of scientific notation, uncertainties, significant digits, and order-of-magnitude estimations were featured. Scalar and vector quantities were introduced and compared.

In the remainder of the text, you will apply and extend your knowledge of measurement and skills in mathematical operations. The next chapter deals with motion in one dimension. Then Chapter 3 describes mathematical details of vectors and trigonometry required for the study of motion in two dimensions in Chapter 4. Displacement, velocity, and acceleration vectors are important there.

CONCEPTS AND SKILLS

Having completed this chapter, you should now be able to do the following:

- State the SI base units of length, mass, and time and give examples of derived units.
- Convert from one metric unit to another.
- Write numbers using scientific notation.
- Apply the exponent laws, especially for multiplication and division involving powers.
- Verify that the dimensions of an equation are correct by using dimensional analysis.
- Write measurements to the appropriate number of significant digits.
- Calculate the percent error of a measured value knowing the accepted value.
- Perform calculations ($+, -, \times$, and $\div$) involving measured quantities and round off the answer to the appropriate number of significant digits.
- Develop skill in estimating quantities and calculate order-of-magnitude estimations.

KEY TERMS

You should now be able to define or explain each of the following words or phrases:

qualitative description
quantitative description
measurement
nanotechnology
standard unit
base (or fundamental) unit
SI
scientific notation
derived unit
dimensional analysis
uncertainty
significant digits
percent error
order-of-magnitude estimations
scalar quantity (or scalar)
vector quantity (or vector)

Chapter Review

MULTIPLE-CHOICE QUESTIONS

1-38 The mass of a heavyweight male wrestler, in grams, is closest to

(a) 10^3
(b) 10^4
(c) 10^5
(d) 10^6
(e) 10^7

1-39 Which length is the smallest?

(a) 10^6 cm
(b) 10^8 mm
(c) 10^{10} mm
(d) 10^{12} nm
(e) 10^{14} pm

1-40 Order the following four masses in descending size: A: 10^{-4} Mg; B: 10^7 kg; C: 10^{-15} Yg; D: 10^2 Gg.

(a) D, B, C, A
(b) A, C, B, D
(c) B, C, D, A
(d) B, D, A, C
(e) none of these

1-41 How many significant digits are there in the sum 460.299 + 390.0008 + 6.123 + 5.07?

(a) 7
(b) 6
(c) 5
(d) 4
(e) 3

1-42 How many significant digits are there in the product 8.005 00 $\times$ 0.005 380 1?

(a) 7
(b) 6
(c) 5
(d) 4
(e) 3

1-43 An airline passenger of mass 78 kg has a 20.7 kg checked bag and a 4.19 kg carry-on bag. The total mass, to the correct number of significant digits, is

(a) 1.0289×10^2 kg
(b) 1.029×10^2 kg
(c) 1.03×10^2 kg
(d) 1.0×10^2 kg
(e) 1×10^2 kg

Review Questions and Problems

1-44 List reasons why measurement is important for use in (a) society and (b) physics.

1-45 List the SI base units of length, time, and mass.

1-46 Why is it not necessary to have a base unit for area in the SI?

1-47 A high-rise apartment building is 30 storeys high. Estimate the height of the building in metres and decametres, showing your reasoning.

1-48 Convert each measurement to the units indicated in parentheses. State the answers in scientific notation.

(a) Mt. Everest is 8.85 km high. (metres, decimetres, centimetres)

(b) The biggest animal known was a blue whale with a mass of 1.90×10^2 Mg. (kilograms, grams, centigrams)

(c) A "cosmic year" is the time our solar system takes to complete one revolution around the centre of the Milky Way Galaxy. It is about 2.2×10^8 years. (seconds, microseconds, and exaseconds)

1-49 Determine the ratio of each of the following masses to your own mass:

(a) The world's largest erratic boulder (i.e., one moved by a glacier), located in Alberta, has a mass of 16 Mg.

(b) The lightest human birth on record is 0.26 kg.

(c) The heaviest mass supported by a person's shoulders is 5.2×10^2 kg.

(d) A prize winning pumpkin had a mass of 6.7×10^6 dg (Figure 1-12).

(e) A recipe calls for 28 g of sea salt.

Figure 1-12 A typical prize-winning pumpkin (Question 1-49(d)).

1-50 The larva of a certain moth consumes 86 000 times its own mass at birth within its first 48 h. How much would a human with a mass at birth of 3.0 kg have to consume in the first 48 h to compete with this phenomenal eater? Express your answer in kilograms, then in a unit that requires a number between 1 and 1000.

1-51 Simplify

(a) 2.00×10^2 km $- 3.0 \times 10^3$ m

(b) $(4.4 \times 10^3$ m$) \div (2.0 \times 10^5$ s$^2)$

(c) $(4.4 \times 10^3$ m$) \div (2.0 \times 10^5$ s$)^2$

1-52 The equation for the area of a circle is $A = \pi r^2$, and the equation for the area of a right-angled triangle is $A = bh/2$.

(a) What do the symbols r, b, and h represent?

(b) Determine the dimensions of the right side of each equation. What do you conclude?

1-53 The list below shows the symbols and dimensions of some quantities used in physics.

Force (F)	$[F] = \text{M}\cdot\text{L/T}^2$
Energy (E)	$[E] = \text{M}\cdot\text{L}^2\text{/T}^2$
Power (P)	$[P] = [E]\text{/T}$
Pressure (p)	$[p] = [F]\text{/L}^2$

(a) What do the symbols M, L, and T mean?

(b) Make a similar list for these quantities: speed (v), acceleration (a), and area (A).

(c) Express the dimensions of power and pressure in terms of M, L, and T.

(d) How can the dimensions of energy be expressed in terms of mass and speed?

(e) How can the dimensions of power be expressed in terms of mass, speed, and acceleration?

1-54 If two measurements have different dimensions, can they be added? multiplied? Give an example to illustrate each answer.

1-55 Before a wooden metre stick is imprinted with millimetre and centimetre markings, it is important that the wood be cured (i.e., properly dried). Explain why.

1-56 An experiment is used to determine the speed of light in a transparent material. The measured value is 1.82×10^8 m/s and the accepted value is 1.86×10^8 m/s. What is the percent error in the measurement?

1-57 A customer buys three 120 cm pieces of rope from a new roll that is 500 m long. Assume both measurements have three significant digits. What length of rope is left on the roll? (Don't forget significant digits.)

1-58 The official diameter of the discus for women is 182 mm and for men it is 221 mm. If each discus is rolled along the ground like a wheel for 25.0 revolutions, how much farther does the men's discus travel?

1-59 A soccer goal is 7.32 m wide and 2.44 m high. Calculate the area of the goal opening.

1-60 The equation for the volume of a cylinder is $V = \pi r^2 h$, where r is the radius and h is the height. Find the volume in cubic centimetres of a solid cylinder of gold that is 8.4 cm in diameter and 22.8 cm in height. Write your answer in scientific notation with the appropriate number of significant digits.

1-61 **Fermi Question:** Determine an order-of-magnitude estimation for each quantity. (Show your reasoning.)

(a) the number of your normal paces it would take to walk 1.6 km

(b) the number of kernels of corn in a container that holds 1.0 L (**Hint:** It may help you to know that a cube 1.0 dm (= 10 cm) on a side has a capacity of 1.0 L.)

(c) volume of water (in litres) in a typical above-ground backyard swimming pool (See the hint in (b) above.)

(d) the number of times your heart has beat in your lifetime to date

(e) the number of people in the world who are sleeping at the time you are answering this question

1-62 Classify each of the following as either scalar or vector.

(a) the force exerted by your biceps on your forearm as you hold a weight in your hand

(b) the number of cars in a parking lot

(c) the reading on a car's odometer

(d) the reading on a car's speedometer

(e) the gravitational force of the Moon on Earth

(f) the age of the universe

Applying Your Knowledge

1-63 Do the prefixes kilo, mega, and giga when used in connection with computers (e,g., kilobyte) have the same meaning as in SI? Explain your answer.

1-64 **Fermi Question:** Determine an estimate for the number of cells in your index finger. Assume that the cells are spherical and have a diameter of about 10 μm.

1-65 **Fermi Question:** The heart of an average adult pumps about 5 L of blood per minute. Estimate the volume of blood your heart will pump from now until the end of your own life expectancy.

1-66 Determine the approximate time it takes a fingernail to grow 1 nm. State your assumptions and show your calculations.

DID YOU KNOW?

The answer to Question 1-66 indicates how quickly layers of atoms are assembled in the process of protein synthesis. It took many years to grow the longest fingernails in the world, a world record of 8.65 m in total for all ten fingernails.

1-67 At one time, Enrico Fermi stated that his 50 min lecture lasted one microcentury. Determine how close his estimate was.

1-68 If volume were a base dimension (V) in the SI, what would be the dimensions of length? of area?

1-69 (a) In Canada, the speed limit on many highways is 100 km/h. Convert this measurement to metres per second.

(b) The fastest measured speed of any animal is 97 m/s, recorded as a peregrine falcon was diving (Figure 1-13). Convert the measurement to kilometres per hour.

(c) Suggest a convenient way of changing metres per second to kilometres per hour and vice versa.

Figure 1-13 A peregrine falcon in flight (Question 1-69(b)).

1-70 Recent measurements have caused scientists to state that the length of a mean solar day increases by 1 ms each century.

(a) How much longer would a day be 3000 years from now?

(b) How long will it take from now for the day to be one minute longer?

1-71 Refer to Figure 1-3(b) on page 3.

Show mathematically why the stars in the photograph appear to move at the rate of 15° every hour.

Determine approximately how long the time-exposure photograph lasted.

1-72 **Fermi Question:** Imagine that you are able to cover a typical football field with $10 bills laid out flat. Determine an estimate of the height to which those bills would reach if stacked tightly flat on top of each other.

One-Dimensional Kinematics

2

Jay Nemeth/Red Bull Stratos (via AP Images)

Figure 2-1 Felix Baumgartner's historic jump was from an altitude much higher than the typical 11 000 m altitude of passenger aircraft. As he fell toward Earth, his pressurized suit protected him from temperatures lower than −50°C with, of course, a wind chill factor making it even more dangerous.

On October 14, 2012, Felix Baumgartner jumped out of a specially designed, high-tech capsule at an altitude of 39 000 m and fell freely downward toward Earth's surface (Figure 2-1). He reached a speed record for a human falling through air of about 1350 km/h. Analyzing this fall is part of the study of motion presented in this chapter.

Humans have always been concerned with motion. In prehistoric times, hunters tossed spears at moving animals, nomads crossed rivers in search of hospitable living areas, and everyone was influenced by the apparent motion of the Sun, Moon,

CHAPTER **OUTLINE**

and stars in the sky. Our current way of life involves obvious examples of motion, such as athletics, our vast networks of transportation, etc. It also involves less obvious motion, such as satellite tracking weather systems to warn us of hurricanes and other disasters, a radioactive tracer as it passes through a patient's circulatory system, or motion of subatomic particles studied by scientific researchers and nanotechnologists.

kinematics the study of motion

The study of motion is called **kinematics**, which stems from the Greek word for motion, *kinema*. Kinematics involves descriptions of motion using both words and mathematical equations. We will not study the causes of changes in motion (which involve forces) until Chapter 5, at which point you will have a thorough understanding of kinematics.

DID YOU KNOW?

A quick check of the dictionary reveals several words with the prefix "kine." For example, kinesiology is the study of muscles, especially the mechanics of human body motion, and kinesics is the study of non-linguistic body movements, such as gestures used as a method of communication.

In this chapter, kinematics is restricted to motion in a straight line, in other words, motion in one dimension. The concepts studied here can be applied to more complex motions, some of which will be seen in later chapters.

2.1 Distance and Speed

speed the time rate of change of distance

A person begins jogging and covers the first 30 m in 10 s (Figure 2-2). This means that, during each second, the distance travelled is, on average, 3.0 m; we say that the average speed is 3.0 m/s. **Speed** is the time rate of change of distance. The quantities distance, speed, and time are scalar quantities, which means they involve magnitude but not direction.

In introducing physics we try to keep the descriptions as simple as possible. Thus, in the case of the jogger, when we consider average speed we don't consider that the person travels less than 3 m in the first second and perhaps more than 3 m in the last second. Also, we are not concerned about the motion of the person's arms or legs. We simply treat the moving body as a whole. Furthermore, we do not consider whether the path of motion is straight or curved or otherwise; rather we consider the total distance travelled in a certain amount of time.

average speed the ratio of the total distance travelled to the elapsed time

The equation for the **average speed** of a moving object is

$$\text{average speed} = \frac{\text{total distance travelled}}{\text{elapsed time}}$$

or, using symbols, $$\text{speed}_{\text{av}} = \frac{d}{\Delta t} \qquad (2\text{-}1)$$

where the subscript "av" indicates "average" and the Greek letter Δ (delta) is used to denote the change in a quantity, in this case a change in time, that is, a time interval. Normally "t" represents the time at which a single event occurs, whereas Δt represents the time between two events, that is, the elapsed time. (Δt is *not* the product of Δ and t.)

Warren Goldswain/Shutterstock

Figure 2-2 When jogging, one of the variables that can be measured is the average speed.

The units of average speed can be any length unit divided by any time unit. Typical units are metres per second (m/s) and kilometres per hour (km/h). In Chapter 1 you learned how to convert from metres per second to kilometres per hour, and vice versa. Table 2-1 shows examples of typical speeds in various units.

Table 2-1

Some Typical Speeds (listed from slowest to fastest)

Motion	Speed in m/s	Speed in Alternate Unit
Continental drift	9.5×10^{-10} m/s	30 mm/yr
Growth of hair on human head	5×10^{-9} m/s	5 nm/s
Slow-moving glacier (Figure 2-3)	6×10^{-7} m/s	2 m/yr
Growth of wild grass in spring	1×10^{-7} m/s	25 cm/month
Running athlete	10 m/s	1.0 dam/s
Express elevator in skyscraper	17 m/s	61 km/h
Hockey puck after hard slapshot (Figure 2-4)	50 m/s	180 km/h
Electrical pulse along an axon	100 m/s	360 km/h
Sound in air	340 m/s	0.34 km/s
Supersonic jet	750 m/s	2.7 Mm/h
Earth in orbit around Sun	3×10^{4} m/s	30 km/s
Electron in hydrogen atom	2.2×10^{6} m/s	2.2 Mm/s
Lightning upward return stroke	1.4×10^{8} m/s	140 Mm/s
Light in a vacuum	3.0×10^{8} m/s	300 Mm/s

Figure 2-3 Pressure from higher up, as well as melting or calving along the edges, can cause a glacier, such as the Perito Moreno Glacier in Argentina, to move.

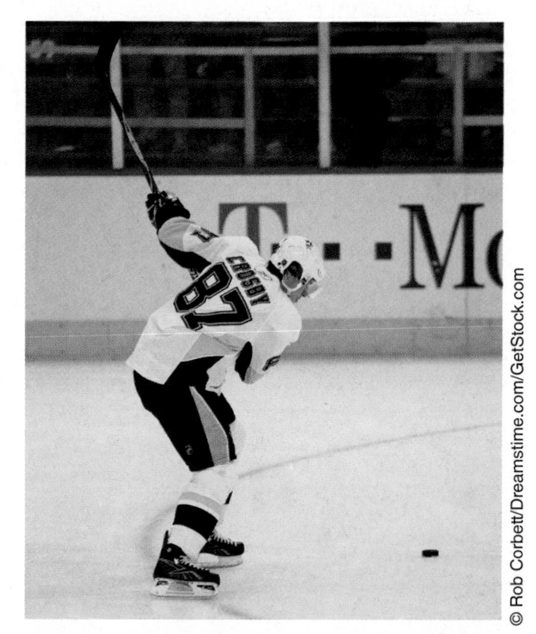

Figure 2-4 A puck can reach speeds approaching 200 km/h after a slapshot.

SAMPLE PROBLEM 2-1

A swimmer crosses a circular swimming pool of diameter 18 m in 22 s.

(a) Calculate the swimmer's average speed.

(b) At that same average speed, how long would the swimmer take to swim around the edge of the pool?

Solution

(a) We use Eqn. 2-1:

$$\text{speed}_{\text{av}} = \frac{d}{\Delta t} = \frac{18\,\text{m}}{22\,\text{s}}$$

$$= 0.82\,\text{m/s (rounded off to two significant digits)}$$

Thus, the average speed is 0.82 m/s.

(b) Eqn. 2-1 for average speed can be rearranged to solve for time. Thus,

$$\Delta t = \frac{d}{\text{speed}_{\text{av}}}$$

In this case, the distance travelled is the circumference of the pool, which is π times the diameter. Hence,

$$\Delta t = \frac{\pi \cdot \text{diameter}}{\text{speed}_{\text{av}}} = \frac{\pi\,(18\text{ m})}{0.82\,\text{m/s}}$$

$$= 69\,\text{s (rounded off to two significant digits)}$$

Thus, the time taken is 69 s.

Note: Significant digits and rounding of numbers were discussed in Section 1.4.

DID YOU KNOW?

A tiny geolocator placed harmlessly on the back of a migratory bird enables ornithologists to track the routes taken by the bird. One amazing discovery revealed that an American Golden-Plover (Figure 2-5) flew around the Pacific Ocean from Alaska to American Samoa, then returned to Alaska the next mating season, covering a total distance of 24 000 km at a record average speed during migration of 100 km/h!

Paul Reeves Photography/Shutterstock

Figure 2-5 American Golden-Plover.

Instantaneous Speed

instantaneous speed the speed at any given instant

uniform motion motion at a constant speed in a straight line (one dimension)

If you are travelling by car to a destination 100 km away and you have one hour to get there, your required average speed is 100 km/h. To achieve this average, parts of the trip would have to be at speeds greater than 100 km/h to make up for the starting and stopping times and other times when the speed is less than 100 km/h. There are two different types of speeds considered here: the average speed for the entire time elapsed, and the speed at any given instant, called the **instantaneous speed**, symbol v (from the word velocity). In a car the value indicated by the speedometer is the instantaneous speed.

From now on in the text whenever the word speed is used it will be assumed to be the instantaneous speed. If the average speed is required, it will be called, simply, average speed.

Litwin Photography/Shutterstock

Figure 2-6 During certain time intervals at an air show, the planes display uniform motion.

Uniform Motion

The simplest type of motion to study is called **uniform motion**, which is motion at a constant speed in a straight line (one dimension). One common example is a car travelling at constant speed along a straight, smooth highway, and another example is shown in Figure 2-6. With uniform motion for a specific time interval, the average speed and the instantaneous speed have the same value.

EXERCISES

2-1 Determine the average speed in metres per second in each case. Express the final answer in scientific notation using the correct number of significant digits.

(a) The current woman's world record for the 4.00×10^2 m track race is 43.2 s.

(b) Sound travels 18.8 cm through a human muscle in 0.119 ms.

(c) The Moon takes 27.3 days to complete one orbit around Earth. The average Earth–Moon distance is 3.84×10^5 km.

(d) Assume the combined length of the small and large intestines is 9.0 m and indigestible waste takes between 12 h and 36 h to pass through these organs. (In this case, find the range of the average speeds.)

2-2 In outer space, light travels at 3.00×10^8 m/s. Determine the time in seconds for each of the following:

(a) Light travels from the Sun to Earth. The average distance from Earth to the Sun is 1.50×10^{11} m.

(b) Laser light is sent from Earth, reflects off a mirror on the Moon, and returns to Earth. (See Question 2-1(c) for the distance between Earth and the Moon.)

2-3 (a) Write an equation for distance travelled in terms of average speed and time.

(b) Prove that your equation in (a) is dimensionally correct. (Dimensional analysis was discussed in Section 1.3.)

2-4 Determine the total distance travelled in each case.

(a) Sound travelling at 344 m/s crosses a room in 0.0350 s.

(b) Thirty-two scuba divers ride an underwater tricycle for 60.0 h at an average speed of 1.74 km/h. Express this answer in kilometres and metres.

(c) The women's world-record, 6-day (144 h), long-distance race was set at an average speed of 6.14 km/h.

(d) A high-speed train travels at 575 km/h for 75.0 min (Figure 2-7).

2-5 Under what conditions can each of the following situations occur?

(a) instantaneous speed equals average speed

(b) instantaneous speed is greater than average speed

(c) instantaneous speed is less than average speed

2-6 **Fermi Question:** In an international competitive walking event, walkers must ensure that the rear foot does not leave the ground before the advancing foot makes contact. If you were in this type of race, estimate your best time to complete a 10 km road walk. (Remember to state your assumptions, show your calculations, and round off your answer appropriately.)

© Li Dongliang/Dreamstime.com/GetStock.com

Figure 2-7 A high-speed train in China.

2.2 Position, Displacement, and Velocity

A tourist travelling by car through the Italian Alps enters a mountain pass like the one shown in Figure 2-8 (a). The road winds back and forth as it climbs to the top of the pass and back down. The tourist takes 90 min to travel 45 km, but the views are spectacular. Meanwhile a local driver, who has seen the views often, chooses to pay a toll and drive through a tunnel at the base of the mountain to get to the same destination as the tourist. It takes only 9.0 min to travel 15 km straight east through the tunnel. The two *distances* are different (15 km compared with 45 km), but the vehicles have the same displacement (15 km east) from the start of the motion to the end (Figure 2-8 (b)). **Displacement** is the change of position from the initial point to the final point.

displacement the change of position from the initial point to the final point

The direction of a displacement can be specified as east, west, north, up, down, and so on. However, at this stage in kinematics it is also convenient to use a simple reference line to define direction, because we are using mainly motion in one dimension. We will call this reference line the x-axis. This axis can be horizontal or any other direction, depending on the situation. Consider Figure 2-9 (a), which shows a car in two different positions, initial and final, along the x-axis. The same situation is shown in (b) with the initial position labelled x_1 and the final position labelled x_2, relative to a reference point called the origin (labelled 0). Then, the displacement is defined as the change in position:

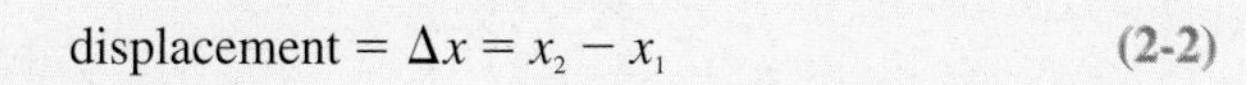

$$\text{displacement} = \Delta x = x_2 - x_1 \quad \text{(2-2)}$$

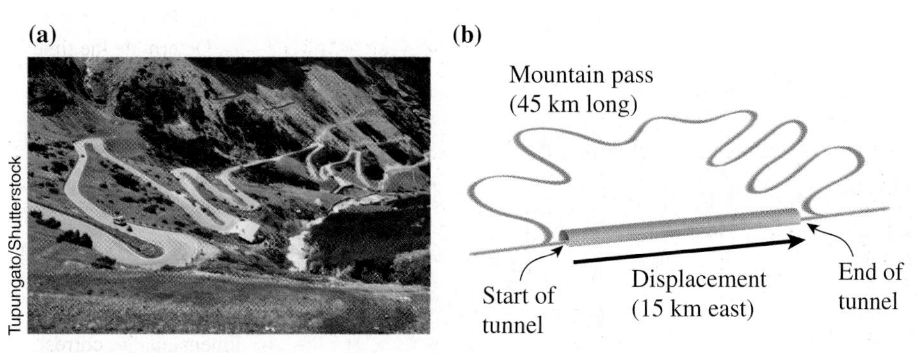

Figure 2-8 **(a)** A mountain pass in the Italian Alps. **(b)** Whether a car travels through the tunnel or across the mountain pass, the displacement for the trip is the same.

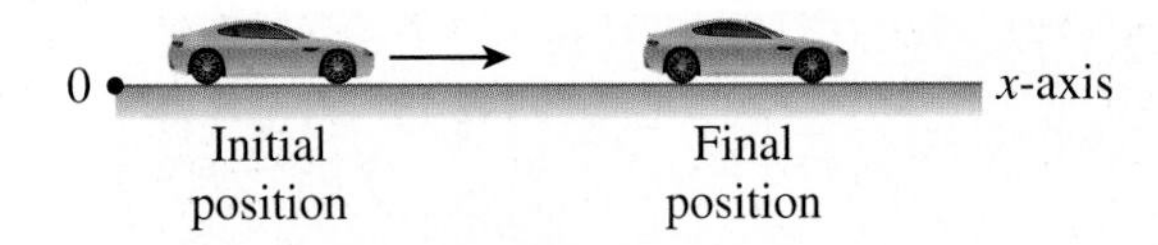

(a) A car is travelling to the right along an x-axis.

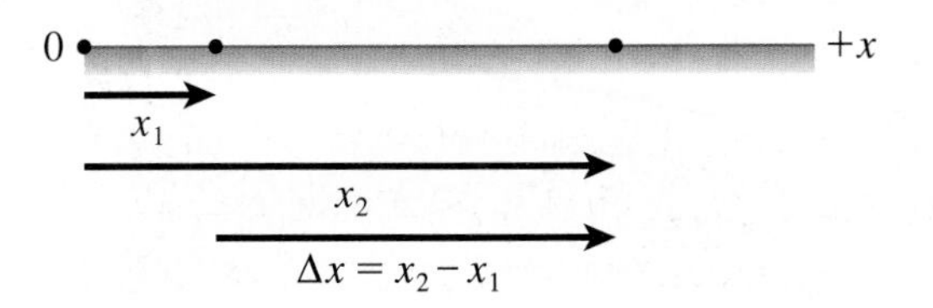

(b) Using points to represent the positions shown in (a)

Figure 2-9 Illustrating displacement $\Delta x = x_2 - x_1$.

Strictly speaking, displacement is a vector quantity. (Recall that scalar and vector quantities were discussed in Section 1.5.) However, we will not concern ourselves with details of the vector nature of displacement until Chapter 4 (Two-Dimensional Kinematics).

Average Velocity

Velocity is the vector counterpart of speed. Thus, **velocity** is the time rate of change of *position*. (Recall that speed is the time rate of change of *distance*.) To determine the **average velocity** over a period of time, we use the definition

$$\text{average velocity} = \frac{\text{change of position}}{\text{elapsed time}} = \frac{\text{displacement}}{\text{elapsed time}}$$

velocity the time rate of change of position

average velocity the ratio of the displacement to the elapsed time for that displacement

To define average velocity in terms of symbols, we consider the one-dimensional motion of an object such as a car that is at position x_1 at time t_1, and at position x_2 at time t_2. The change of position, that is, the displacement, is $\Delta x = x_2 - x_1$, and the elapsed time is $\Delta t = t_2 - t_1$. Hence, the average velocity (symbol v_{av}) is

$$v_{av} = \frac{\Delta x}{\Delta t} = \frac{x_2 - x_1}{t_2 - t_1} \quad (2\text{-}3)$$

Again, strictly speaking, average velocity is a vector, but its vector nature will not be described in detail until Chapter 4.

SAMPLE PROBLEM 2-2

Compare the average velocities and average speeds for the two cars travelling in the Italian Alps (Figure 2-8).

Solution We begin with the average velocities. In kinematics problems, it is useful to define the positive direction of our coordinate system. Since both cars end up travelling east, it is convenient to define the $+x$-direction as east. In writing up a problem solution, you can make this definition in words, or by a small label on your drawing: $+x$ (east). For both cars, the displacement is $\Delta x = +15$ km. (Note that this is a positive quantity, meaning that the displacement is eastward. A displacement of $\Delta x = -15$ km would be westward in this case.)

(a) For the car that went through the tunnel,

$$v_{av} = \frac{\Delta x}{\Delta t} \quad \text{where } \Delta t = 9.0\,\text{min} = 0.15\,\text{h}$$

$$= \frac{15\,\text{km}}{0.15\,\text{h}}$$

$$\therefore v_{av} = 1.0 \times 10^2\,\text{km/h}$$

Since the average velocity of this car is positive, the average velocity is eastward. Hence, the average velocity is 1.0×10^2 km/h eastward.

(b) For the car that went over the mountain pass,

$$v_{av} = \frac{\Delta x}{\Delta t} \quad \text{where } \Delta t = 90\,\text{min} = 1.5\,\text{h}$$

$$= \frac{15\,\text{km}}{1.5\,\text{h}}$$

$$= 1.0 \times 10^1\,\text{km/h}$$

Since this average velocity is positive, it too is eastward. Hence, the average velocity of this car is 1.0×10^1 km/h eastward. The two average velocities are very different even though the displacement is the same for both cars.

Next we compare the average speeds.

(c) For the car that went through the tunnel,

$$\text{speed}_{\text{av}} = \frac{d}{\Delta t} = \frac{15\,\text{km}}{0.15\,\text{h}} = 1.0 \times 10^2\,\text{km/h}$$

(d) For the car that went over the mountain pass,

$$\text{speed}_{\text{av}} = \frac{d}{\Delta t} = \frac{45\,\text{km}}{1.5\,\text{h}} = 3.0 \times 10^1\,\text{km/h}$$

Thus, the average speeds are 1.0×10^2 km/h and 3.0×10^1 km/h.

Notation Tip

Notice how we deal with directions by using positive and negative signs. We defined the $+x$-direction as east, and since (in this problem) the displacement is eastward, we represent it as a positive quantity (+15 km). The average velocities of both cars turned out to be positive, indicating that they are eastward. Notice also that we do not use boldface type for the symbols Δx and v_{av}. The reason for this will be explained in Chapter 4.

TACKLING MISCONCEPTIONS

Comparing Average Speed and Average Velocity

Average speed and average velocity are *not* the same. For the car that went over the mountain pass, the average speed was 30 km/h and the average velocity was 10 km/s east. For the car that went through the tunnel, the average speed of 100 km/h was equal to the magnitude of the average velocity.

Instantaneous Velocity

Average velocity is defined for a particular period of elapsed time. However, we can also consider the velocity at a particular instant of time, that is, the **instantaneous velocity**, represented by the symbol v. We will give a formal definition of instantaneous velocity later, but for now, you can think of it as being the same as instantaneous speed (Section 2.1), with the important addition that instantaneous velocity (being a vector) has a direction as well as a magnitude. Instantaneous speed is the magnitude of the instantaneous velocity. For example, if a car has an instantaneous velocity of 100 km/h west, its instantaneous speed is 100 km/h.

In the remainder of the text, the term velocity will indicate instantaneous velocity; when we are discussing average velocity, it will be clearly indicated.

instantaneous velocity the velocity at any particular instant of time

Graphing Uniform Motion

Graphing provides a useful way of studying motion. We will begin by studying graphs for uniform motion, which is motion with a constant velocity. (How does this definition compare with the definition for uniform motion given in Section 2.1?)

Suppose that a commuter train is travelling at a constant velocity of 20 m/s south for 3.0 min along a straight stretch of track. In other words, the velocity at any instant—the instantaneous velocity—is 20 m/s south. If we define the $+x$-direction as south, and choose the origin of our coordinate system (i.e., $x = 0$) to correspond to the train's position at some initial time $t = 0$, then the position–time data are as shown below and the corresponding graph of the motion is shown in Figure 2-10.

Time t (s):	0	60	120	180
Position x (m):	0	1.2×10^3	2.4×10^3	3.6×10^3

An important quantity to calculate on such a graph is the slope of the line. The equation to find the slope of a straight line on a graph is

$$\text{slope} = \frac{\text{change in quantity plotted on vertical axis}}{\text{corresponding change in quantity plotted on horizontal axis}}$$

or in this case, $\text{slope} = \dfrac{\text{change in position}}{\text{change in time}}$

In symbols, $m = \dfrac{\Delta x}{\Delta t}$

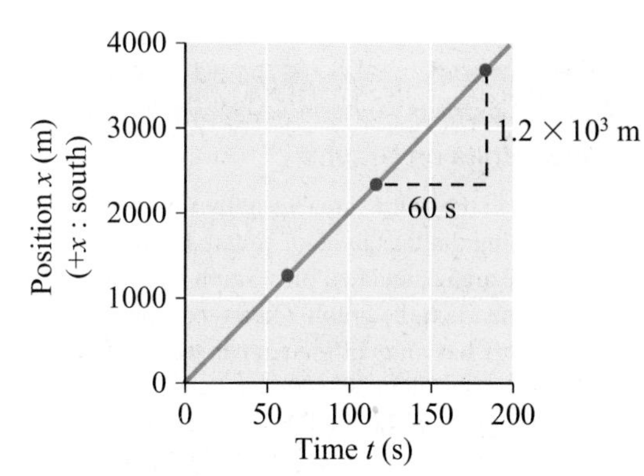

Figure 2-10 Position-versus-time graph for the commuter train.

Notice that the right-hand side of this expression, $\Delta x/\Delta t$, is the average velocity defined in Eqn. 2-3. Hence, the slope of a position–time graph gives the average velocity during the time interval Δt.

For example, let us calculate the slope between $t = 120$ s and $t = 180$ s.

$$m = \frac{\Delta x}{\Delta t} = \frac{(3.6 \times 10^3 - 2.4 \times 10^3)\,\text{m}}{(180 - 120)\,\text{s}}$$

$$\therefore\ m = \frac{1.2 \times 10^3\,\text{m}}{60\,\text{s}} = 20\,\text{m/s}$$

Thus, the slope is 20 m/s. Since this is a positive quantity, and the $+x$-direction is south, we can state that the average velocity of the train is 2.0×10^1 m/s south (assuming two significant digits) between the times 120 s and 180 s. The slope is constant, so the average velocity would be the same no matter where we calculated it along the line. It should be apparent that for uniform motion, the average velocity during any time interval is equal to the instantaneous velocity at any specific time.

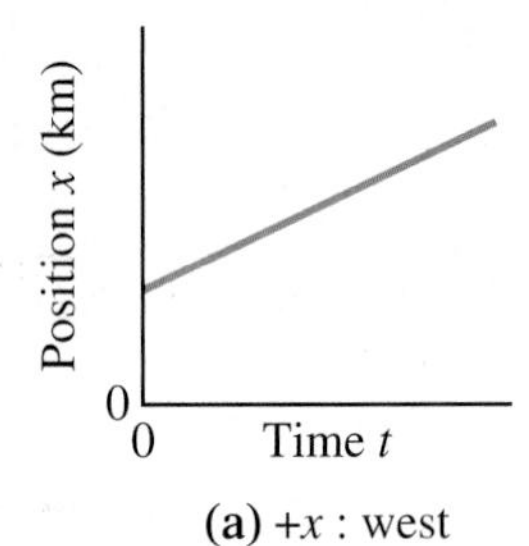

(a) $+x$: west

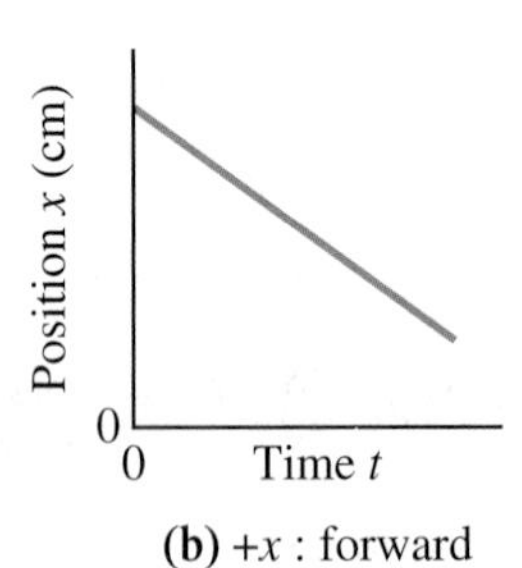

(b) $+x$: forward

Figure 2-11 Sample Problem 2-3.

SAMPLE PROBLEM 2-3

Describe the motion represented by each position–time graph shown in Figure 2-11.

Solution

(a) Since the slope of the line is constant, the velocity is constant. The slope is positive, and since $+x$ is toward the west, the velocity is in the westerly direction. The initial position is not at the origin, and the object is moving away from the origin.

(b) Again the slope of the line is constant; however, its value is negative. The $+x$-direction in this case is forward, and hence the velocity is constant in the backward direction. The initial position is not at the origin, and the moving object is travelling back toward the origin.

For uniform motion, the velocity is constant, and the velocity–time graph is a straight line parallel to the time axis (Figure 2-12). The area under the line (down to the time axis) on this graph can be calculated. In this case it is the area of a rectangle, $v \times \Delta t$. In calculating the area on a graph, it is important to include the units on the axes of the graph to determine what the area represents. For the graph shown, the units of the area are (m/s)(s) = m; in other words, the area gives the displacement, Δx, during the time interval for which the area is found.

TACKLING MISCONCEPTIONS

Area Units on Graphs

Although the word "area" usually implies something that has units of square metres, etc., the area calculated on a graph has units that depend on the graph. On a v-t graph, if the v-axis has units of metres per second and the t-axis has units of seconds, then the area "height times width" will have units of metres. Thus, units must be considered for area on graphs just as they must be for slopes of lines on graphs.

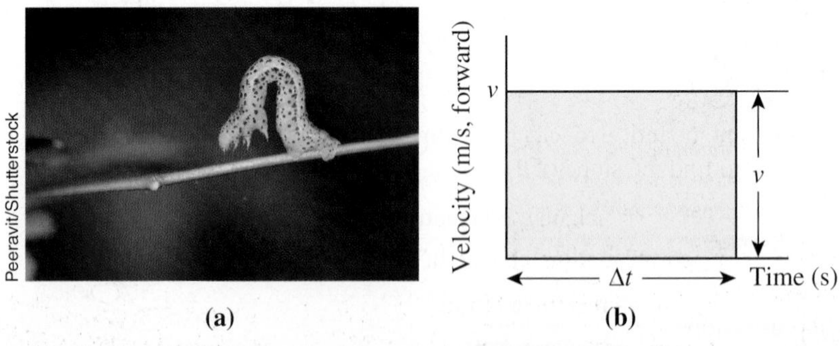

Figure 2-12 **(a)** A caterpillar crawling along a twig at a constant velocity is an example of uniform motion. **(b)** For uniform motion, the velocity–time graph is a straight line parallel to the time axis. The area of the rectangle, $v\Delta t$, is the displacement, Δx.

Graphing Non-Uniform Motion

non-uniform motion motion with a changing velocity, in other words a change in speed (the magnitude of velocity), or a change in direction, or both

tangent to a curve a straight line parallel to a curved line at a particular point

Although uniform motion is easy to study, it is not common. **Non-uniform motion**, in which the velocity changes, occurs for most examples of motion. Non-uniform motion involves a change in speed (the magnitude of velocity), or a change in direction, or both. In other words, the instantaneous velocity is changing.

Consider, for example, a car starting from rest and slowly speeding up. A possible set of position–time data for this motion is given below, and the corresponding graph of the motion is shown in Figure 2-13. The $+x$-direction in this case is simply "forward."

Time t (s):	0	2.0	4.0	6.0	8.0
Position x (m):	0	4.0	16	36	64

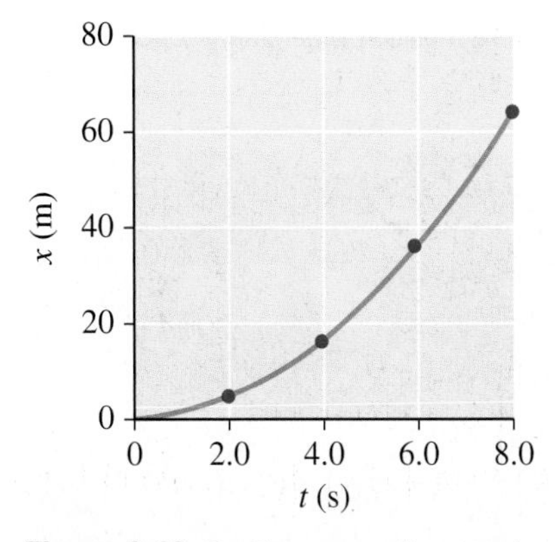

Figure 2-13 Position-versus-time graph for an example of non-uniform motion.

Since the position–time graph shown in Figure 2-13 is not a straight line, the velocity is obviously not constant. How can we determine the velocity at any instant? Recall that for uniform motion the slope of the line on a position–time graph indicates the velocity. In this case, the slope of the curve continually changes, which means that the instantaneous velocity continually changes. To find the slope of a curved line at a particular instant, we draw a straight line parallel to the curve at that instant. This straight line is called a **tangent to a curve**. *The slope of the tangent to a curve on a position–time graph gives the instantaneous velocity.*

Figure 2-14 shows the tangent drawn at $t = 2.0$ s for the same motion shown in the previous graph. In Figure 2-15 we compare this tangent with the average velocities between $t = 2.0$ s and later times. From $t = 2.0$ s to $t = 8.0$ s, the Δt is 6.0 s and the average velocity is the slope of line A in Figure 2-15. From $t = 2.0$ s to $t = 6.0$ s, Δt is 4.0 s and the average velocity is the slope of line B. Similarly, when $\Delta t = 2.0$ s, the average velocity is the slope of C. Notice that as Δt gets smaller, the slopes of the lines get nearer to the slope of the tangent that we want. In other words, as Δt becomes smaller, the average velocity ($\Delta x/\Delta t$) gets closer to the instantaneous velocity (v). Using symbols,

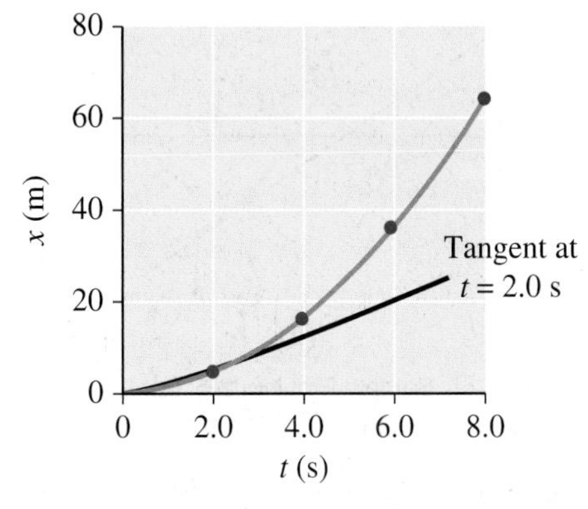

Figure 2-14 The instantaneous velocity equals the slope of the tangent at a particular time, in this case at $t = 2.0$ s.

$$v = \lim_{\Delta t \to 0} \frac{\Delta x}{\Delta t} \tag{2-4}$$

This is read as: "The instantaneous velocity v equals the limit, as Δt approaches zero, of Δx divided by Δt." (In the notation of calculus, the Δ symbols are replaced by "d" symbols to represent very small quantities. Thus, the instantaneous velocity is $v = dx/dt$.)

TRY THIS!

Drawing Tangents with a Mirror

You can use a small, unframed plane mirror to draw a tangent to a curved line on a graph. Place the mirror so it is to perpendicular to the line at the point chosen. Adjust the mirror so the curve on the paper lines up with its image, then draw a line perpendicular to the mirror to obtain the tangent.

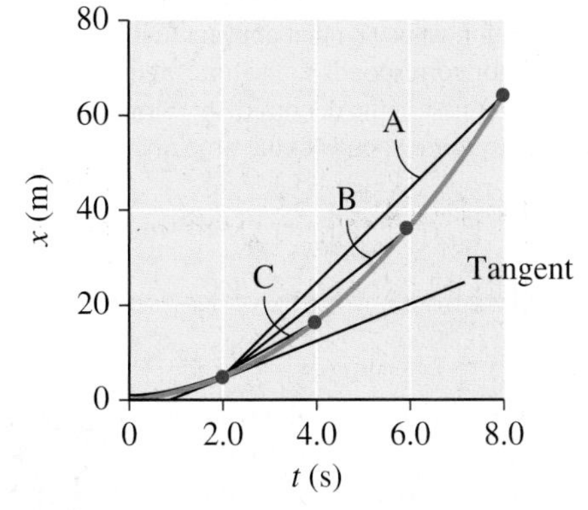

Figure 2-15 Comparing the slope of the tangent at $t = 2.0$ s with the average velocities represented by the slopes of lines A, B, and C. Notice that the slope of C is almost the same as the slope of the tangent.

To summarize, the *instantaneous velocity* is the velocity at a particular instant, and it equals the slope of the tangent on the position–time graph at that instant. The *instantaneous speed* of a moving object equals the magnitude of its instantaneous velocity.

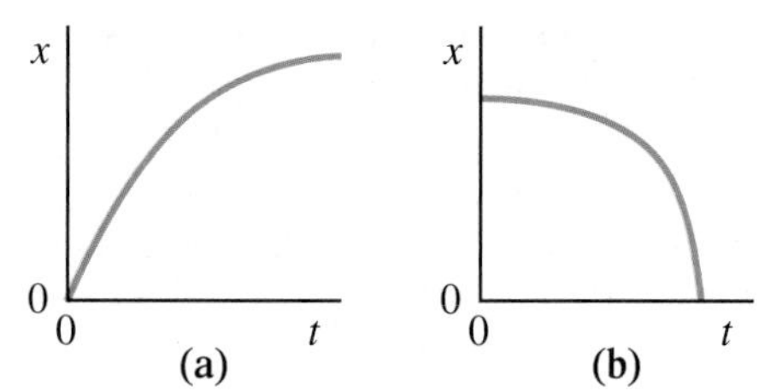

Figure 2-16 Position–time graphs for Sample Problem 2-4.

SAMPLE PROBLEM 2-4

Describe the motion represented by the curves in the position–time graphs of Figure 2-16. For both graphs, the $+x$-direction is east.

Solution

(a) The slope of the (tangent to the) curve is positive at all times and is relatively steep at the beginning. Thus, the velocity starts off high in the easterly direction, then gradually reduces to zero. The object starts at the origin and then moves away from the origin.

(b) The slope is zero at the beginning, then it becomes negative. Thus, the velocity starts off at zero, then gradually increases in magnitude in the westerly direction. The object starts at a position east of the origin, then moves westward until it reaches the origin.

(a)

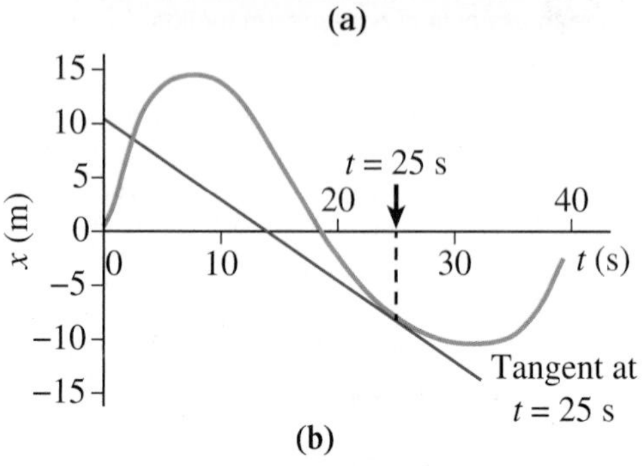

(b)

Figure 2-17 Sample Problem 2-5: **(a)** A soccer linesman. **(b)** An x-t graph of the linesman's motion.

SAMPLE PROBLEM 2-5

Figure 2-17 (a) shows a linesman at the side of a soccer field, and (b) shows a graph of position versus time for his back-and-forth motion in one dimension along the side line. The $+x$-direction is west. At time $t = 25$ s, what is the linesman's instantaneous velocity?

Solution

Since we are asked for the instantaneous velocity at $t = 25$ s, we draw a tangent to the curve at this point, and then calculate the slope of this tangent. We can use the position and time coordinates of *any* two points on this line to calculate the slope. It is easiest if we extrapolate the line back to intersect both the x-axis and the t-axis, giving us these two points on the line: $(t_1, x_1) = (0 \text{ s}, 10.5 \text{ m})$ and $(t_2, x_2) = (14 \text{ s}, 0 \text{ m})$.

Calculating the slope,

$$\begin{aligned} m &= \frac{\Delta x}{\Delta t} = \frac{x_2 - x_1}{t_2 - t_1} \\ &= \frac{(0 - 10.5)\,\text{m}}{(14 - 0)\,\text{s}} \\ &= -0.75\ \text{m/s} \end{aligned}$$

The negative sign indicates that the velocity is in the $-x$-direction, that is, eastward. Thus, the instantaneous velocity is approximately 0.75 m/s east. (We say "approximately" because of the uncertainty in our hand-drawing of the tangent.)

Math Tip

When calculating velocity by determining the slope of a line, it is important to "match up" the x- and t-values in the numerator and denominator. In the solution to the Sample Problem 2-5, the position value (x_2) that appears first in the numerator corresponds to the time value (t_2) appearing first in the denominator. Similarly, the position x_1 corresponds to the time t_1.

EXERCISES

2-7 The windsock shown in Figure 2-18 is used at a small airport. Does this device indicate a scalar quantity or a vector quantity? What is that quantity?

Figure 2-18 A windsock (Question 2-7).

2-8 (a) Under what conditions could the total distance travelled equal the magnitude of the displacement?

(b) Is it possible for the distance travelled to exceed the magnitude of the displacement? Explain, using an example.

(c) Is it possible for the magnitude of the displacement to exceed the distance travelled? Explain, using an example.

2-9 (a) State the difference between instantaneous speed and instantaneous velocity.

(b) Under what condition(s) could the average speed equal the magnitude of the average velocity?

2-10 A city bus travels a straight route that is 12 km from one end to the other. The bus takes 48 min to complete the route, including all stops *and the return* to the initial position.

(a) Calculate the average speed of the bus over the entire route.

(b) Calculate the average velocity of the bus over the entire route.

(c) Why are these answers different?

2-11 A billiard ball travels 0.46 m in the $+x$-direction—having started at the origin ($x = 0$)—bounces off another ball to travel 0.84 m in the opposite direction, then bounces from the edge of the billiard table finally coming to rest 0.12 m from that edge. The entire motion is one-dimensional and takes 2.5 s. Determine the billiard ball's

(a) average speed

(b) final position

(c) average velocity

2-12 The graph in Figure 2-19 represents the position–time graph[1] of an automobile travelling westward; that is, the $+x$-direction is west. Assuming two significant digits, determine the

(a) average speed between 0.0 s and 4.0 s; between 0.0 s and 8.0 s

(b) average velocity between 8.0 s and 10 s; between 12 s and 16 s; and between 0.0 s and 16 s

(c) instantaneous speed at 6.0 s; at 10 s

(d) instantaneous velocity at 14 s

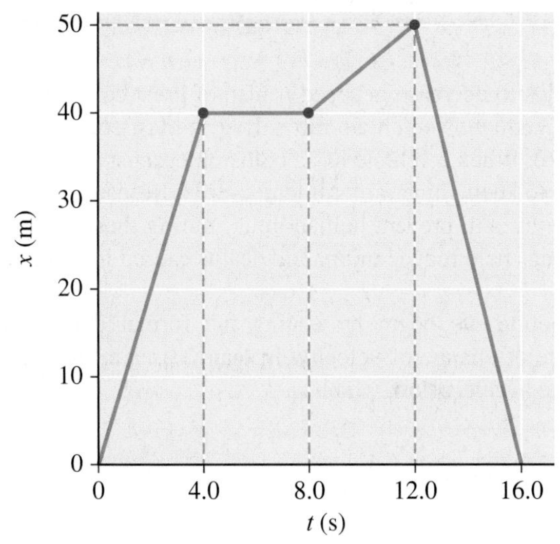

Figure 2-19 Question 2-12.

2-13 A car driver sees an emergency and reacts to put on the brakes. During the 0.20 s reaction time, the car maintains its uniform velocity of 28 m/s forward. What is the car's displacement during the time it takes the driver to react?

2-14 At an average velocity of 8.5 m/s straight down, how long does it take a parachutist (Figure 2-20) to descend from an altitude of 2.4×10^3 m above sea level to 7.2×10^2 m above sea level?

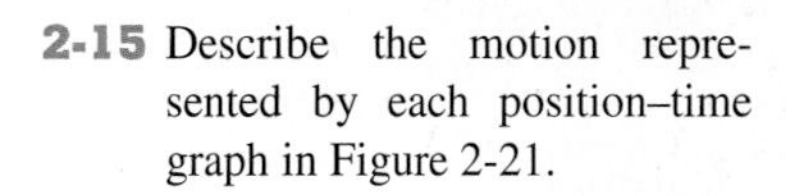

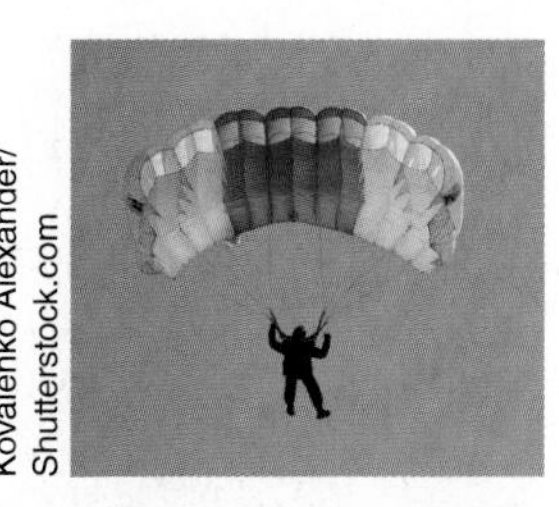

Kovalenko Alexander/ Shutterstock.com

Figure 2-20 A parachutist (Question 2-14).

2-15 Describe the motion represented by each position–time graph in Figure 2-21.

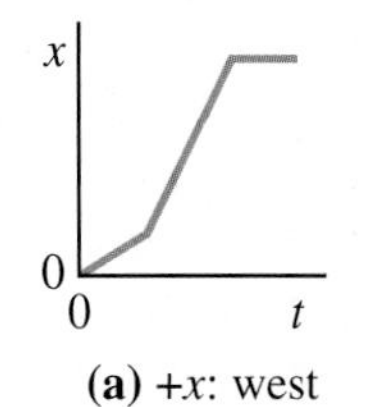

(a) $+x$: west

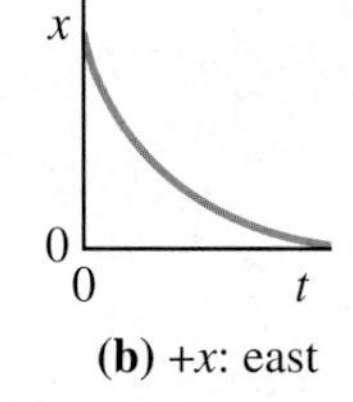

(b) $+x$: east

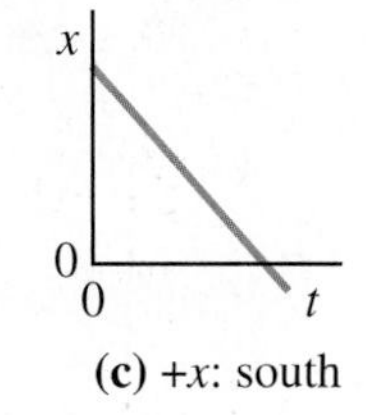

(c) $+x$: south

Figure 2-21 Question 2-15.

2-16 Determine the area, with units, between the line and the horizontal axis on the velocity–time graph in Figure 2-22. (The velocity is northward.) What does this area represent?

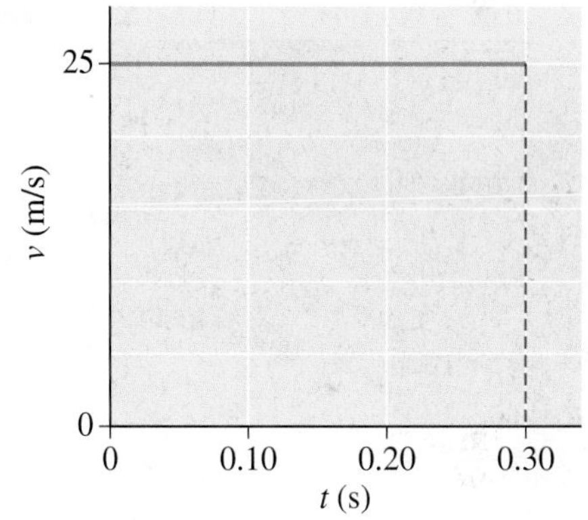

Figure 2-22 Question 2-16.

2-17 Determine the instantaneous velocity at 2.5 s, 4.5 s, and 6.0 s from the graph in Figure 2-23. The $+x$-direction is south.

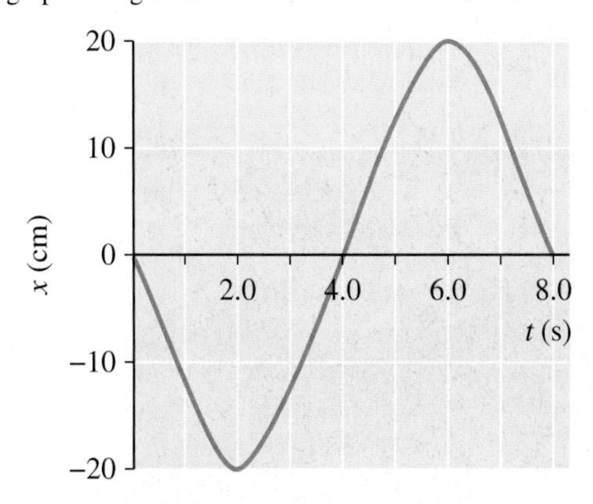

Figure 2-23 Question 2-17.

[1]The abrupt changes in the slopes of the lines in such graphs are idealized. The speed of an automobile cannot instantly change from 10 m/s to 0 m/s, for example; the sharp corners shown in graphs such as Figure 2-19 would be rounded for the motion of an actual automobile. However, the sharp corners make it easier for you to answer the questions.

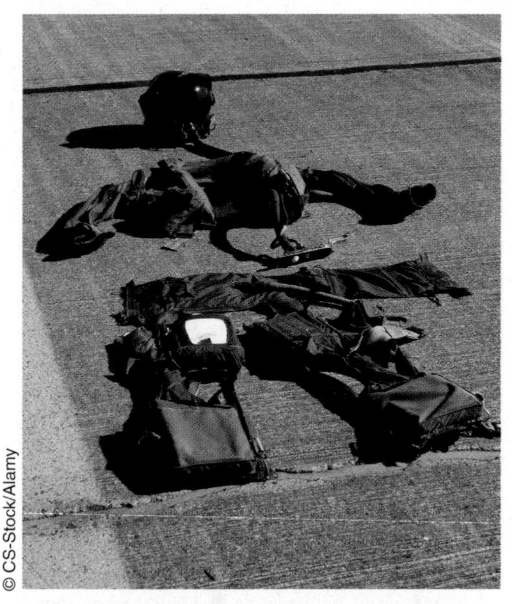

© CS-Stock/Alamy

Figure 2-24 Fighter-jet pilots and astronauts wear anti-gravity suits (also called *g*-suits) similar to this one to reduce loss of blood to the head.

2.3 Acceleration

In this chapter so far you have encountered both uniform motion and non-uniform motion. The latter, in which there is a change in the magnitude of the velocity, or the direction of the velocity, or both, is known formally as accelerated motion.

Acceleration is common in our daily lives, and some scientists devote their studies entirely to analyzing acceleration and its effects. One type of acceleration that we cannot escape is caused by the force of gravity pulling objects toward Earth. (We will study this acceleration in Section 2.5, as well as discuss gravitation in more detail in Chapter 9.) Acceleration in sports is also important. A sprinter with a greater acceleration has a better chance of winning a short race. Acceleration occurs in all types of transportation: automobiles, aircraft, bicycles, even walking. Pilots of military aircraft experience high acceleration, especially during combat operations or when the emergency escape system is activated. During extreme acceleration, a loss of blood to the head can reduce vision and cause unconsciousness. Special training and specially designed body suits (Figure 2-24) help reduce these effects. Astronauts also experience high acceleration, especially during take-off from Earth's surface. Astronauts must be placed in a lay-back position to prevent loss of blood to the head during upward acceleration (Figure 2-25).

DID YOU KNOW?

The first anti-gravity space suit, or *g*-suit, was designed in 1941 by a Canadian engineer, pilot, and inventor named Wilbur R. Franks at the University of Toronto, Ontario. His original suit had an outer layer that was filled with water that exerted pressure on the abdomen and legs to prevent blood from pooling in the lower parts of the body. The design helped save the lives of thousands of Allied flight pilots during World War II. Later designs used air pressure instead of water pressure. In his extreme jump in 2012, Felix Baumgartner, featured at the beginning of this chapter, tested the newest design of the anti-gravity suit.

NASA

Figure 2-25 Final launch preparations for two astronauts.

acceleration the time rate of change of velocity

average acceleration the ratio of the change in velocity to the elapsed time

Acceleration that corresponds to decrease in speed is also of great concern. When an aircraft travelling at high speed lands on an aircraft carrier, it must come to a stop extremely quickly (Figure 2-26). When a vehicle has a collision or must stop quickly, the driver and passengers tend to keep going, at least temporarily. Research is carried out to try to improve safety devices to prevent human injury during these situations. Headrests, seat belts, and air bags help reduce injury and deaths caused in automobile collisions.

To analyze accelerated motion we begin by stating the formal definition of **acceleration**: it is the time rate of change of velocity. In many cases in this text we will be concerned with **average acceleration**, which is

$$\text{average acceleration} = \frac{\text{change in velocity}}{\text{elapsed time}}$$

In symbols,
$$a_{av} = \frac{\Delta v}{\Delta t} = \frac{v_2 - v_1}{t_2 - t_1} \quad \textbf{(2-5)}$$

where v_1 is the velocity at time t_1, and v_2 is the velocity at time t_2.

Jason and Bonnie Grower/Shutterstock

Figure 2-26 The arrester cables on an aircraft carrier help the aircraft come to a stop very quickly.

Since velocity is a vector quantity, acceleration is also a vector quantity. Again because the motion studied in this chapter is in one dimension, details about the vector nature of acceleration are left for Chapter 4. The important thing to remember here is that vector quantities have a direction and a magnitude.

Various units can be used to denote acceleration, as long as they derive from a velocity unit divided by a time unit. Examples of units are found in the next two sample problems.

SAMPLE PROBLEM 2-6

In an outdoor lacrosse game, a player uses a lacrosse stick (Figure 2-27) to accelerate the ball from rest to 1.4×10^2 km/h west in 0.16 s. Determine the average acceleration of this motion.

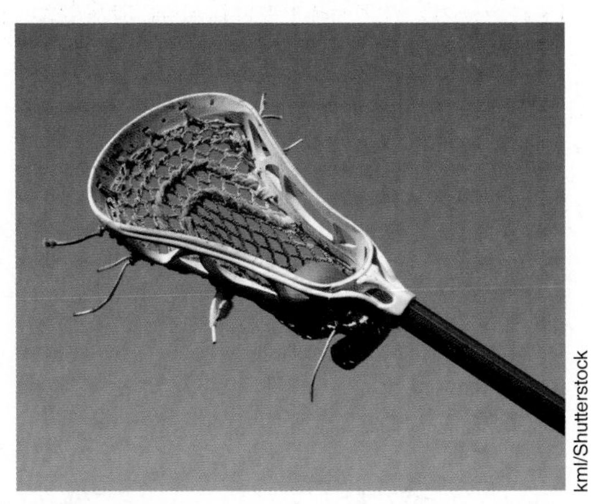

Figure 2-27 A lacrosse stick and ball.

kml/Shutterstock

Solution

Define the $+x$-direction to be west. We use Eqn. 2-5:

$$a_{av} = \frac{\Delta v}{\Delta t}$$

In this problem, $\Delta v = v_2 - v_1 = (1.4 \times 10^2 - 0)$ km/h, and $\Delta t = 0.16$ s.

Hence, $a_{av} = \dfrac{1.4 \times 10^2 \text{ km/h}}{0.16 \text{ s}} = 8.8 \times 10^2 \text{ (km/h)/s}$

Since the result of our calculation is a positive quantity, the average acceleration is in the $+x$-direction, that is, west. Thus, the average acceleration is 8.8×10^2 (km/h)/s west.

SAMPLE PROBLEM 2-7

A motorcycle rider is travelling at 23 m/s in the $+x$-direction. Suddenly the rider brakes to prevent an accident and comes to a stop in 3.2 s.

(a) Determine the magnitude of the rider's average acceleration.

(b) Use the units found in (a) to show that acceleration has the dimensions L/T^2.

Solution

(a) Using Eqn. 2-5,

$$a_{av} = \frac{\Delta v}{\Delta t}$$

In this problem, $\Delta v = v_2 - v_1 = (0 - 23)$ m/s $= -23$ m/s, and $\Delta t = 3.2$ s.

Notice that Δv is negative. It is important to remember that a change (Δ) in a quantity must always be calculated as the *final* value of the quantity minus the *initial* value.

Calculating a_{av}, $$a_{av} = \frac{-23 \text{ m/s}}{3.2 \text{ s}} = -7.2 \text{ (m/s)/s} = -7.2 \text{ m/s}^2$$

Thus, the average acceleration is -7.2 m/s^2, or 7.2 m/s^2 in the negative-x-direction.

(b) The units of average acceleration determined in part (a) are m/s^2. Since metre is a unit of length, and second is a unit of time, the dimensions of acceleration are L/T^2.

TACKLING MISCONCEPTIONS

Negative Acceleration

Notice in Sample Problem 2-7 that the acceleration is negative. If a moving object has an initial velocity that is positive, then negative acceleration means that the object slows down.[2] However, *if an object is already moving in the negative direction, then a negative acceleration means that the object is speeding up.* For example, if the $+x$-direction is east and a car is heading west, thus having a negative velocity, then a westward acceleration (which is also negative) would indicate that the car is speeding up in the westward direction.

[2]Slowing down is sometimes called deceleration, but this term will not be used in this book.

Constant Acceleration

In the special situation where the velocity changes uniformly with time, the acceleration is constant or uniform. Consider a snowboarder (Figure 2-28 (a)), for example, who starts from rest at the top of a ski slope and experiences a constant acceleration of 3.0 m/s^2 down the slope. The acceleration remains constant for the first 6.0 s. A table of data for this example is shown below and the corresponding velocity–time graph is shown in Figure 2-28 (b). The $+x$-direction is along the slope, downward.

Time (s)	0.0	1.0	2.0	3.0	4.0	5.0	6.0
Velocity v (m/s)	0.0	3.0	6.0	9.0	12	15	18

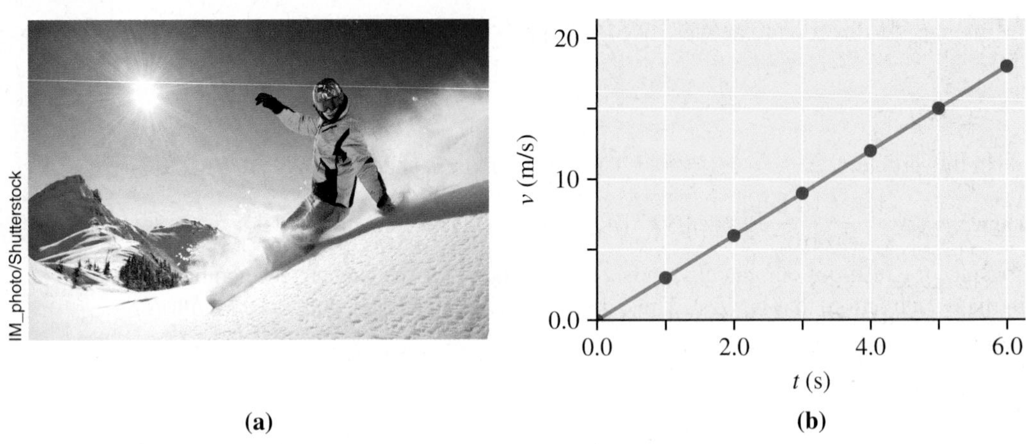

Figure 2-28 **(a)** A snowboarder can experience constant acceleration for short periods of time. **(b)** Velocity–time graph for the constant-acceleration example. The $+x$-direction is downward.

The slope of the line on the velocity–time graph in Figure 2-28 (b) is

$$\text{slope} = \frac{\Delta v}{\Delta t} = \frac{18 \text{ m/s}}{3.0 \text{ s}} = 3.0 \text{ m/s}^2$$

However, $\Delta v/\Delta t$ is just average acceleration. In this example, the slope is constant, which corresponds to an average acceleration that is constant. In addition, the acceleration at any instant—the instantaneous acceleration (symbol a)—is also constant. (We will define instantaneous acceleration mathematically shortly.) The acceleration–time graph for the above example is shown in Figure 2-29.

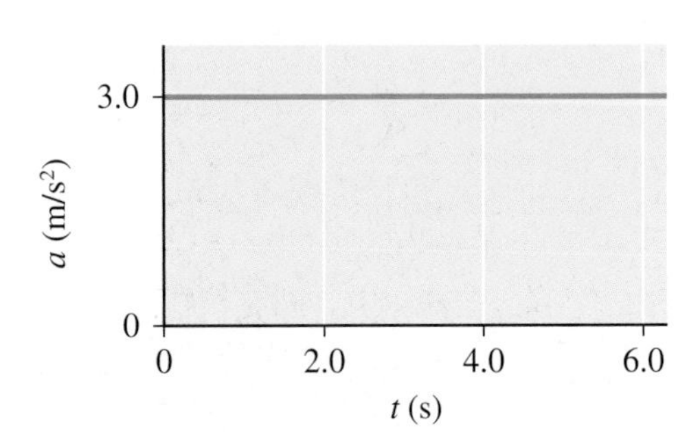

Figure 2-29 Acceleration–time graph for the constant-acceleration example. The $+x$-direction is along the slope, downward.

What additional information can be found from the acceleration–time graph? Recall from the graphing of uniform motion that the area under the line on a velocity–time graph indicates the displacement, that is, the change in position. Similarly, the area under the line on an acceleration–time graph indicates the change in velocity. This is verified in the next sample problem.

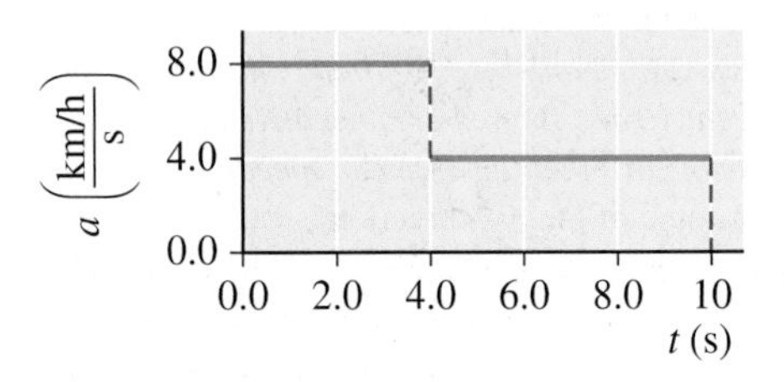

Figure 2-30 Sample Problem 2-8. The $+x$-direction is forward.

SAMPLE PROBLEM 2-8

A car, with a standard transmission, reaches 30 km/h in first gear, after which it undergoes acceleration for 10 s in second and third gears. The acceleration for this ten-second interval is indicated in Figure 2-30. (The $+x$-direction is "forward.") Determine the magnitude of the velocity at the end of 4.0 s of this acceleration and at the end of 10 s. Then draw a velocity–time graph corresponding to the acceleration–time graph.

Solution

We consider the first 4.0 s. The definition of average acceleration, $a_{av} = \Delta v/\Delta t$, can be rearranged to give the change in velocity: $\Delta v = a_{av}\Delta t$

The average acceleration is 8.0 (km/h)/s, and $\Delta t = 4.0$ s. Hence,

$$\Delta v = 8.0\,\frac{\text{km/h}}{\text{s}} \times 4.0\,\text{s} = 32\,\text{km/h}$$

Thus, the velocity increases by 32 km/h during the first 4.0 s. Since the velocity at the beginning of this time interval was 30 km/h, the velocity at the end of 4.0 s is (30 + 32) km/h = 62 km/h, in the forward direction.

Notice that the change in velocity equals the area under the line on the acceleration–time graph between $t = 0$ and $t = 4.0$ s. This region has a height of 8.0 (km/h)/s and a width of 4.0 s, and hence an area (height × width) of 32 km/h. This means that we can determine the change in velocity by finding the area under the line, which we will do to determine the change in velocity from $t = 4.0$ s to $t = 10$ s:

$$\begin{aligned}\text{area under line} &= \text{height} \times \text{width} \\ &= 4.0\,\frac{\text{km/h}}{\text{s}} \times 6.0\,\text{s} \\ &= 24\,\text{km/h}\end{aligned}$$

Therefore, at $t = 10$ s, the velocity is $v = 62$ km/h + 24 km/h = 86 km/h, forward.

The corresponding velocity–time graph is shown in Figure 2-31. Notice again that during each period of constant acceleration, the velocity changes linearly with time.

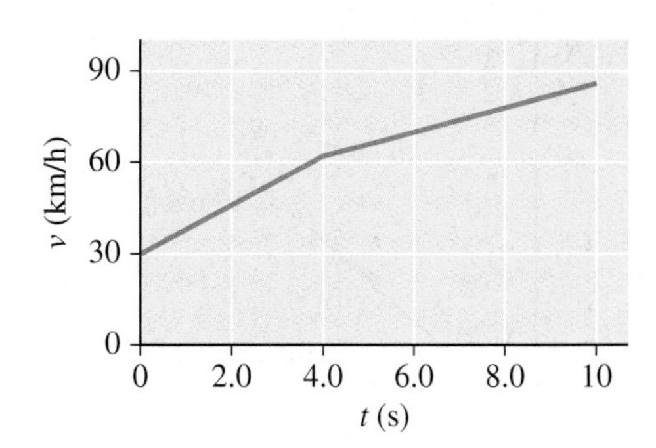

Figure 2-31 Sample Problem 2-8 (solution). The $+x$-direction is forward.

Math Tip

The lines on the a-t graph in Sample Problem 2-8 were above the t-axis, resulting in a positive area and thus an increase in velocity. If the line on the a-t graph is below the t-axis, the area is negative, which means a decrease in velocity. If the velocity is already negative, then a decrease in velocity would result in an even larger negative velocity.

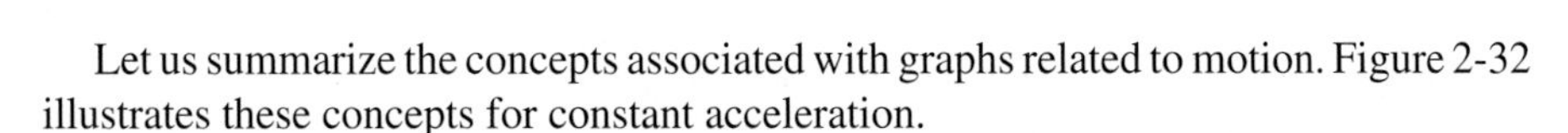

Let us summarize the concepts associated with graphs related to motion. Figure 2-32 illustrates these concepts for constant acceleration.

- The three graphs commonly drawn are position versus time (or displacement v. time), velocity versus time, and acceleration versus time.
- Two important quantities found from graphs are slope and area under the line (or, more generally, area under the curve).
- The slope of a position–time graph gives velocity.
- The slope of a velocity–time graph gives acceleration.
- The area under the line (curve) on an acceleration–time graph indicates the change in velocity.
- The area under the line (curve) on a velocity–time graph indicates the change in position, that is, the displacement.

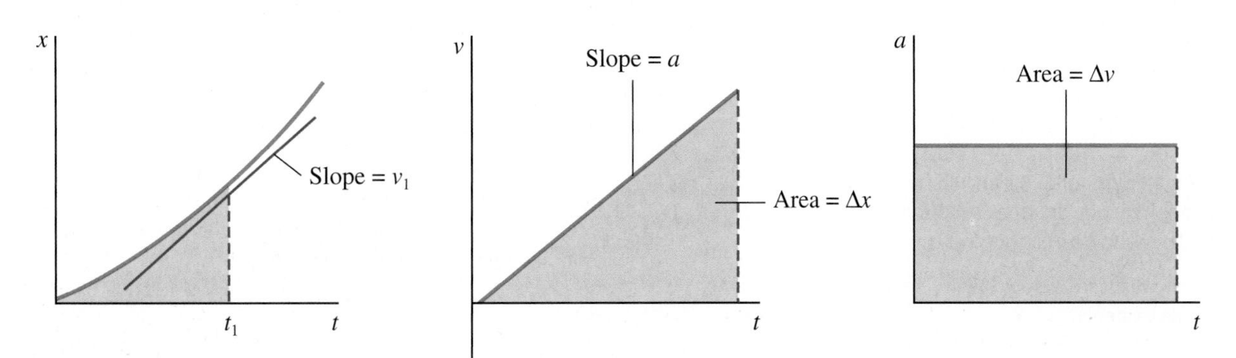

Figure 2-32 Position, velocity, and acceleration graphs for constant acceleration summarizing the usefulness of calculating slopes and areas.

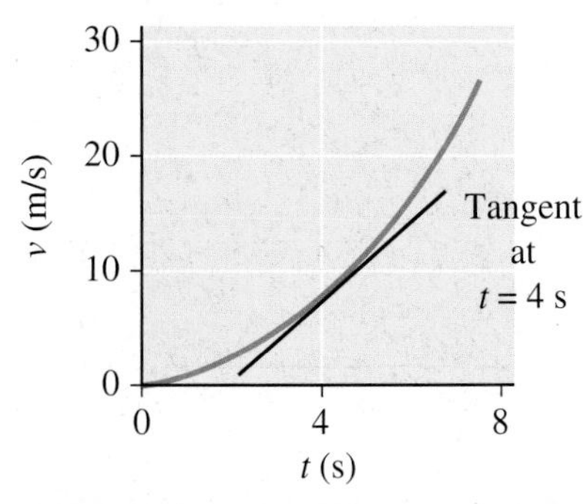

Figure 2-33 A velocity–time graph showing non-constant acceleration. The acceleration at time $t = 4$ s equals the slope of the tangent to the curve at this time. The $+x$-direction is forward. Since the slope is positive, the direction of the acceleration at $t = 4$ s is forward.

Variable (Non-Constant) Acceleration

The velocity–time graph for an object undergoing constant acceleration is a straight line. For non-constant acceleration, however, the velocity–time graph is curved—the slope changes as the acceleration changes. An example of a velocity–time graph showing increasing acceleration is shown in Figure 2-33, in which the $+x$-direction is forward. The acceleration at any instant, called the **instantaneous acceleration**, is the slope of the tangent to the curve at that instant. In the figure, a tangent is drawn at $t = 4$ s. Just as instantaneous velocity is defined as the limit, as $\Delta t \rightarrow 0$, of $\Delta x/\Delta t$, instantaneous acceleration (symbol a) is defined as the limit, as $\Delta t \rightarrow 0$, of $\Delta v/\Delta t$:

$$a = \lim_{\Delta t \rightarrow 0} \frac{\Delta v}{\Delta t} \qquad (2\text{-}6)$$

In other words, as Δt approaches zero, the average acceleration ($\Delta v/\Delta t$) approaches the instantaneous acceleration. (Using the notation of calculus, instantaneous acceleration is written as $a = dv/dt$.)

instantaneous acceleration the acceleration at any instant

From now on, whenever the word acceleration is used it will be assumed to be the instantaneous acceleration. If average acceleration is required, it will explicitly be called average acceleration (symbol a_{av}).

SAMPLE PROBLEM 2-9

Describe the motion represented by the graph in Figure 2-34. The $+x$-direction is "forward."

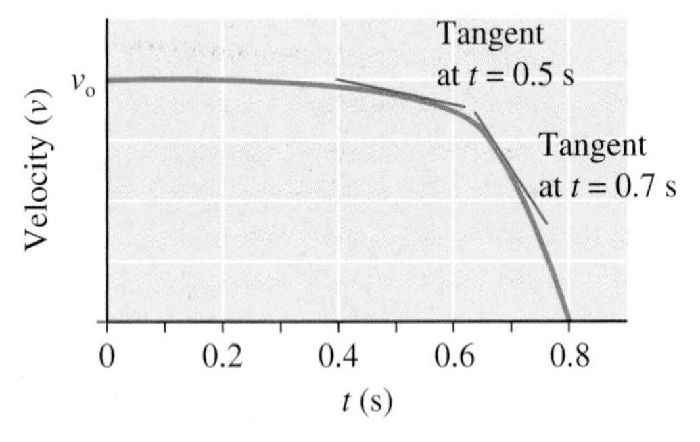

Figure 2-34 Sample Problem 2-9. The $+x$-direction is forward.

Solution

The motion begins with an initial velocity, v_0, and undergoes a slight slowing down or negative acceleration followed by a negative acceleration with increasing magnitude. (Tangents are drawn at 0.5 s and 0.7 s. Their slopes are negative, and the magnitude of the slope at 0.7 s is larger than that at 0.5 s.) The object, while always heading in the forward direction, slows down from an initial speed,[3] symbol v_0, at time $t = 0.0$ s, and comes to rest at $t = 0.8$ s.

EXERCISES

2-18 Which of the following could be units of acceleration?
- (a) km/min^2
- (b) cm·s^{-2}
- (c) (Mm/h)·h
- (d) (dam/s)/h

2-19 A flock of geese is travelling south during the fall migration. The $+x$-direction is south. Describe the flock's motion when its acceleration is (a) zero, (b) positive, (c) negative.

2-20 Is it possible to have westward velocity with eastward acceleration? Explain, with an example.

2-21 Calculate the magnitude of the average acceleration of the following two land-speed record holders, both of which started from rest:
- (a) In 1902, the racer reached 34.2 m/s in 58.5 s.
- (b) In 1997, the racer reached 341 m/s in 9.45 s.

2-22 In an archery tournament (Figure 2-35) an arrow, travelling at 42.8 m/s east, strikes a target and comes to a stop in 3.12×10^{-2} s. Determine the magnitude and direction of the arrow's average acceleration during this short period of time.

[3] The symbol v_0 is read "v zero" or "v sub zero" and means the velocity at time $t = 0$; it can also be called v_1 or the initial velocity.

Figure 2-35 An archery tournament (Question 2-22).

2-23 A supersonic jet travelling from New York City to London reduces its velocity from 1.65×10^3 km/h east to 1.12×10^3 km/h east before reaching the western shores of England.

(a) If this change of velocity takes 345 s, determine the jet's average acceleration in kilometres per hour per second ((km/h)/s).

(b) Determine this acceleration in metres per second squared.

2-24 Figure 2-36 (a) shows a bobcat kitten walking along a fallen tree, and (b) shows a graph of velocity versus time for the kitten's motion. The $+x$-direction is north. Draw the corresponding acceleration–time graph.

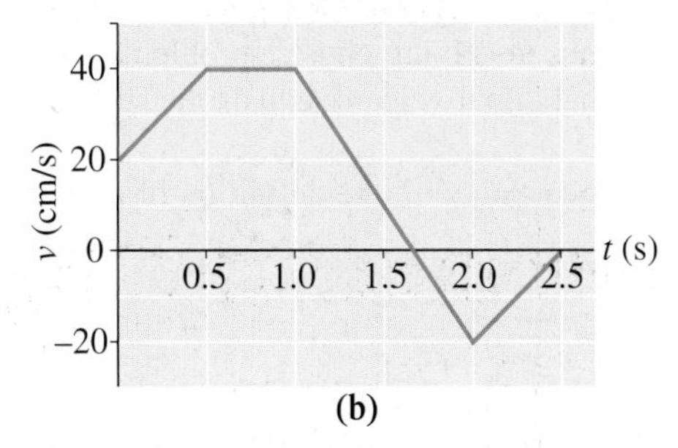

Figure 2-36 Question 2-24. **(a)** A bobcat kitten. **(b)** Velocity graph: the $+x$-direction is north.

2-25 The data shown in the table below were recorded during an experiment involving measurements of the velocity of a baby crawling on a floor. The $+x$-direction is "forward."

(a) Plot a velocity–time graph for this motion.

(b) Determine the slopes of the different line segments on the graph.

(c) Use the values found in (b) to plot an acceleration–time graph for this motion.

Time (s):	0.0	2.0	4.0	6.0	8.0	10	12
Velocity (cm/s):	10	15	20	15	10	5.0	0.0

2-26 The acceleration–time graph of a football lineman being pushed by other players is illustrated in Figure 2-37. The $+x$-direction is west. Plot the corresponding velocity–time graph, assuming the initial velocity is zero.

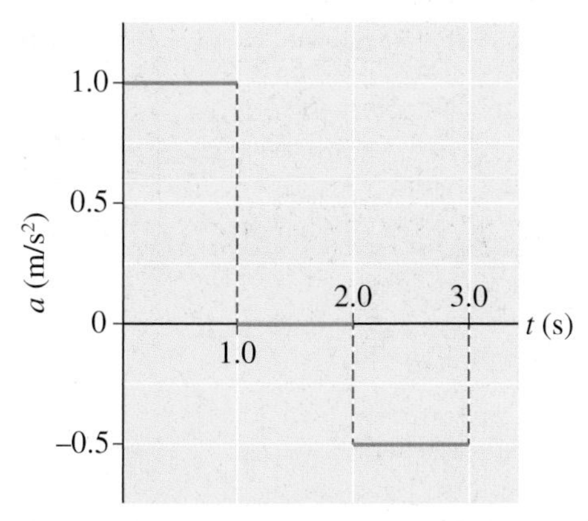

Figure 2-37 Question 2-26. The $+x$-direction is west.

2-27 Under what conditions would average acceleration and instantaneous acceleration be equal?

2-28 Describe the motion represented in each graph in Figure 2-38. Be sure to consider the axes of each graph.

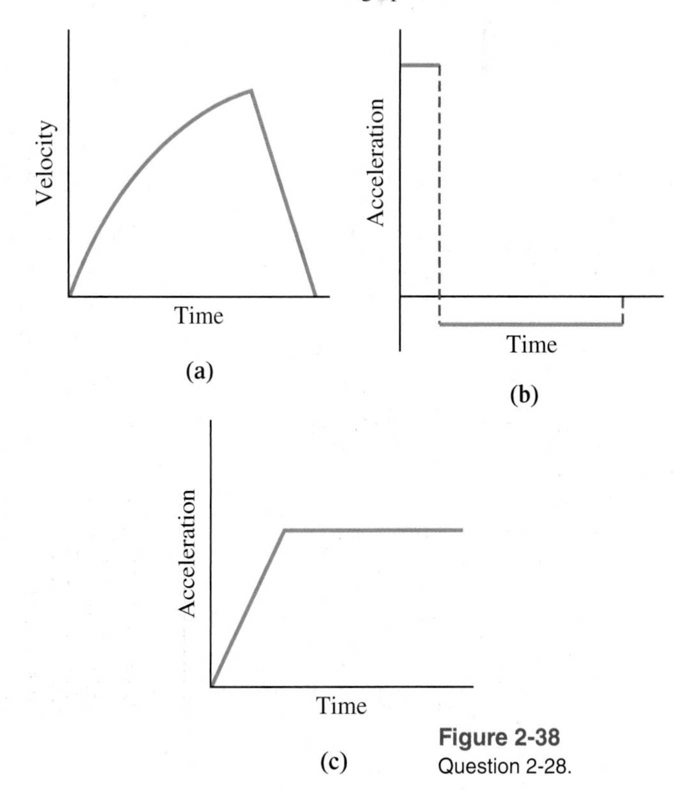

Figure 2-38 Question 2-28.

2.4 Solving Constant-Acceleration Problems

Some of the joys of studying physics relate to understanding things that go on around us and learning about applications of physics principles. But studying physics also involves the development of skills in solving problems. To experience success in this aspect of physics requires practice. In this section, you will gain knowledge in how to approach problem-solving in physics, and you will experience practice in solving kinematics problems logically.

We begin this section by deriving several equations that will be useful in a variety of circumstances, and then apply those equations in solving problems.

Equations of Kinematics for Constant Acceleration

A number of very useful equations can be derived by using substitution in equations already known. We start with the definition of average acceleration (Eqn. 2-5):

$$a_{av} = \frac{\Delta v}{\Delta t} = \frac{v_2 - v_1}{t_2 - t_1}$$

Since we are considering only constant acceleration in this section, the instantaneous acceleration, a, at any time equals the average acceleration, a_{av}, during any time interval, and we can replace a_{av} in the above equation with a. Hence, we can write

$$a = \frac{v_2 - v_1}{t_2 - t_1}$$

In dealing with constant-acceleration situations, it is usually convenient to set the initial time t_1 equal to zero, and to write the final time t_2 simply as t. In addition, the initial velocity (at $t = 0$) is normally written as v_0, and the final velocity (at time t) is written just as v. Making these changes,

$$a = \frac{v - v_0}{t - 0}$$

Rearranging this equation to solve for v,

$$v = v_0 + at \quad \text{or} \quad \Delta v = at \text{ (for constant acceleration)} \qquad \textbf{(2-7)}$$

Equation 2-7 will be used many times in solving physics problems involving constant acceleration. In words, it states that "final velocity equals the sum of initial velocity and acceleration times time."

We now develop further equations, starting with the definition of average velocity from Eqn. 2-3:

$$v_{av} = \frac{\Delta x}{\Delta t} = \frac{x_2 - x_1}{t_2 - t_1}$$

Since the acceleration is constant, the velocity is changing at a uniform rate (i.e., linearly), and therefore the average velocity is midway between the initial and final velocities:

$$v_{av} = \frac{v_0 + v}{2}$$

Substituting this expression for v_{av} into the previous equation,

$$\frac{v_0 + v}{2} = \frac{x_2 - x_1}{t_2 - t_1}$$

As we did before, we now write the final time t_2 as t, and set the initial time t_1 to be zero. In addition, we write the initial position x_1 as x_0 (at time $t = 0$), and the final position x_2 simply as x (at time t):

$$\frac{v_0 + v}{2} = \frac{x - x_0}{t - 0}$$

Solving for $x - x_0$ gives $x - x_0 = \frac{1}{2}(v_0 + v)t$

Rearranging to solve for final position x,

$$x = x_0 + \tfrac{1}{2}(v_0 + v)t \quad \text{or} \quad \Delta x = \tfrac{1}{2}(v_0 + v)t \quad \text{(for constant acceleration)} \qquad \textbf{(2-8)}$$

To obtain an equation for final position in terms of initial position, initial velocity, acceleration, and time, we substitute the expression for v from Eqn. 2-7 ($v = v_0 + at$) into Eqn. 2-8 to eliminate the final velocity:

$$x = x_0 + \tfrac{1}{2}(v_0 + (v_0 + at))t$$

Simplifying,

$$x = x_0 + v_0 t + \tfrac{1}{2}at^2 \quad \text{or} \quad \Delta x = v_0 t + \tfrac{1}{2}at^2 \quad \text{(for constant acceleration)} \qquad \textbf{(2-9)}$$

Finally, we derive an equation involving initial and final velocities, displacement, and acceleration. We begin with Eqn. 2-8,

$$x = x_0 + \tfrac{1}{2}(v_0 + v)t$$

From Eqn. 2-7, we can obtain the following expression for the time, t:

$$t = \frac{v - v_0}{a}$$

Substituting this expression for t into Eqn. 2-8,

$$x = x_0 + \tfrac{1}{2}(v_0 + v)\frac{(v - v_0)}{a}$$

Simplifying the right-hand-side, $x = x_0 + \dfrac{v^2 - v_0^2}{2a}$

$$\therefore\ v^2 - v_0^2 = 2a(x - x_0) \quad \text{or} \quad v^2 - v_0^2 = 2a\Delta x \quad \text{(for constant acceleration)} \qquad \textbf{(2-10)}$$

With equations 2-7, 2-8, 2-9, and 2-10, we are now in a position to solve any problem involving *constant acceleration* in one dimension. We list these equations below for handy reference:

$$v = v_0 + at \quad \text{or} \quad \Delta v = at \qquad \textbf{(2-7)}$$

$$x = x_0 + \tfrac{1}{2}(v_0 + v)t \quad \text{or} \quad \Delta x = \tfrac{1}{2}(v_0 + v)t \qquad \textbf{(2-8)}$$

$$x = x_0 + v_0 t + \tfrac{1}{2} at^2 \quad \text{or} \quad \Delta x = v_0 t + \tfrac{1}{2} at^2 \qquad \textbf{(2-9)}$$

$$v^2 - v_0^2 = 2a(x - x_0) \quad \text{or} \quad v^2 - v_0^2 = 2a\Delta x \qquad \textbf{(2-10)}$$

where a = acceleration (must be constant to use the above equations);

x_0 = (initial) position at time $t = 0$ x = (final) position at time t

v_0 = (initial) velocity at time $t = 0$ v = (final) velocity at time t

Problem-Solving Tip

Notice that in each of Eqns. 2-7 to 2-10 at least one variable is not present: Eqn. 2-7 does not contain the position variables x and x_0; Eqn. 2-8 does not contain acceleration, a; Eqn. 2-9 does not have final velocity v; and Eqn. 2-10 does not involve time t. Noting which variables are eliminated in the various equations makes problem solving easier.

When solving problems in kinematics, it is often handy to choose the initial position, x_0, to be zero. This produces simplifications in Eqns. 2-8 to 2-10.

Learning how and when to use these kinematics equations takes practice. Before proceeding to solve some sample problems, we list in Table 2-2 the important quantities we have encountered so far in kinematics.

Table 2-2

Important Quantities in Kinematics

Quantity	Type of Quantity	Symbol(s)
Distance	scalar	d
Average speed	scalar	speed_{av}
Speed	scalar	v
Time interval	scalar	Δt, $t_2 - t_1$, or $t - t_0$
Initial position	vector	x_0 or x_1
Final position	vector	x or x_2
Displacement (change in position)	vector	Δx, $x_2 - x_1$, or $x - x_0$
Initial velocity	vector	v_0 or v_1
Final velocity	vector	v or v_2
Average velocity	vector	v_{av}
Average acceleration	vector	a_{av} (or a if a = constant)
Acceleration	vector	a

Applying the Kinematics Equations

To develop skill in solving problems that may involve up to five or even six variables, it is wise to follow a logical set of steps. Some basic problem-solving steps are listed below.

Problem-Solving Strategies for Solving Kinematics Problems

- Read the problem then draw a sketch to help you understand the problem better.
- Define a positive direction to use in that particular problem, and then be consistent with your signs for the various quantities. (For example, if you decide that the positive direction is east, then the velocity of a westward-moving bicycle is negative.)
- Write down what is given, including words that translate into quantities, using the proper symbols and consistent

units. (For example, if an object "starts from rest," its initial velocity is zero, or $v_0 = 0$, and if acceleration is given in metres per second squared and time is given in minutes, change the time to seconds.)

- Determine which equations involve the quantities in the problem, and plan an attack involving these equations. Sometimes only one equation is needed.
- When using an individual equation, rearrange it algebraically to solve for the unknown quantity before substituting the numerical values into the equation.
- Write the answer with the correct units, the appropriate number of significant digits, and a direction if the answer is a vector.
- Check to be sure the answer makes sense. If it seems too large or too small, check your steps and arithmetic.

These steps will be illustrated in the sample problems that follow.

SAMPLE PROBLEM 2-10

A motorcycle rider shifts from second to third gear when travelling at 15 m/s. She then speeds up for 3.6 s with a constant acceleration of magnitude 5.2 m/s². How far did she travel during the acceleration?

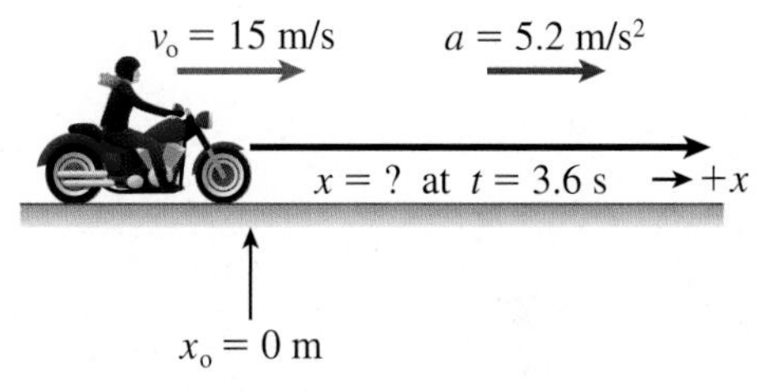

Figure 2-39 Sample Problem 2-10.

Solution

Define the $+x$-direction to be forward, and choose the initial position to be $x_0 = 0$ m (Figure 2-39).

The known quantities are: $v_0 = 15$ m/s

$$t = 3.6 \text{ s}$$
$$a = 5.2 \text{ m/s}^2$$
$$x_0 = 0 \text{ m}$$

The quantity required is x, and we note that the final velocity, v, is not provided.

We can use Eqn. 2-9 directly: $x = x_0 + v_0 t + \frac{1}{2}at^2$

$$\therefore \quad x = (0\,\text{m}) + (15\,\text{m/s})(3.6\,\text{s}) + \tfrac{1}{2}(5.2\,\text{m/s}^2)(3.6\,\text{s})^2$$
$$= 54\,\text{m} + 34\,\text{m}$$
$$= 88\,\text{m}\ (\text{to two significant digits})$$

Thus, the cyclist travelled 88 m.

(a)

SAMPLE PROBLEM 2-11

A rocket is launched vertically from rest (Figure 2-40 (a)), and when it reaches an altitude of 4.7 km above the launch pad, its speed is 630 m/s. What is the magnitude of the rocket's acceleration (assumed constant)?

Solution

Choose the $+x$-direction to be up, and choose the initial position (the launch pad) to be $x_0 = 0$ m (Figure 2-40 (b)).

The known quantities are

$$v_0 = 0 \text{ m/s (“launched vertically \textit{from rest}”)}$$
$$x = 4.7 \text{ km} = 4.7 \times 10^3 \text{ m}$$
$$v = 630 \text{ m/s}$$
$$x_0 = 0 \text{ m}$$

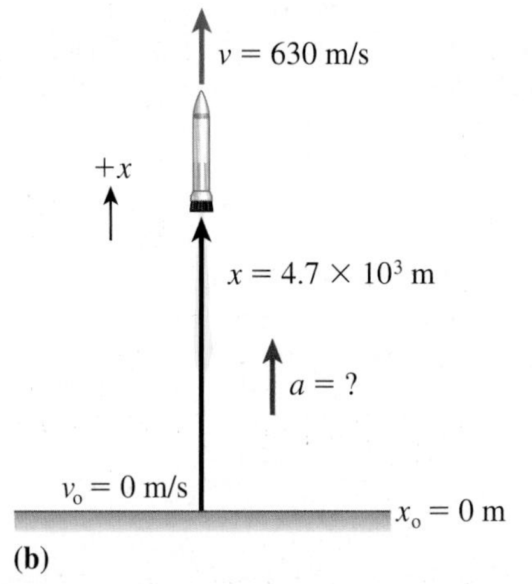

(b)

Figure 2-40 Sample Problem 2-11. **(a)** A rocket launch. **(b)** Data for the problem.

The required quantity is the magnitude of the acceleration, a, and we note that the time, t, is not provided. We can use Eqn. 2-10, which does not involve time, to determine the acceleration:

$$v^2 - v_0^2 = 2a\ (x - x_0)$$

We first simplify this equation by noting that both v_0 and x_0 are zero.
Hence, $v^2 = 2ax$.
We now rearrange this equation to solve for the required acceleration, a:

$$a = \frac{v^2}{2x}$$

Substituting values, $a = \dfrac{(630\text{ m/s})^2}{2(4.7 \times 10^3\text{ m})}$

$$\therefore\ a = 42\text{ m/s}^2$$

Thus, the rocket's acceleration is 42 m/s^2 in magnitude.

SAMPLE PROBLEM 2-12

In a game of curling, a rock (Figure 2-41 (a)), is sent along the ice where it experiences a constant acceleration of 26 cm/s^2 in a direction opposite to the initial velocity. The rock travels 28 m from release before coming to rest. Determine the rock's (a) initial velocity and (b) travel time.

(a)

(b)

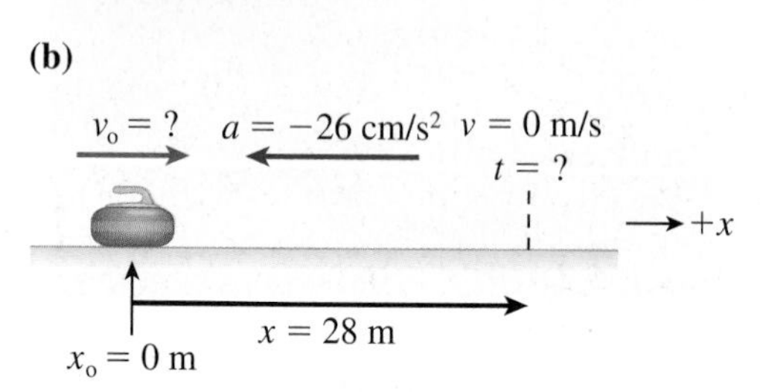

Figure 2-41 Sample Problem 2-12. **(a)** Curling. **(b)** Data for the problem.

Solution

We select the $+x$-direction to be in the direction of the rock's motion, and choose the point of release to be at $x_0 = 0$ m (Figure 2-41 (b)).
The given quantities are

$$x = 28\text{ m}$$
$$v = 0\text{ m/s}$$
$$a = -26\text{ cm/s}^2 = -0.26\text{ m/s}^2$$
$$x_0 = 0\text{ m}$$

Notice that the acceleration is negative because its direction is opposite to the $+x$-direction (rock's motion). If we had chosen the positive direction to be opposite to the rock's motion, the acceleration would have been positive, but the initial velocity and the final position would have been negative.

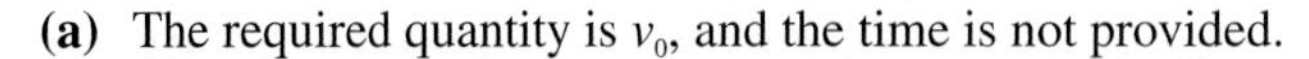

(a) The required quantity is v_0, and the time is not provided.
We can use Eqn. 2-10: $v^2 - v_0^2 = 2a\ (x - x_0)$
Substituting $v = 0$ m/s and $x_0 = 0$ m, and then solving for v_0^2,

$$v_0^2 = -2ax$$

Taking the square root,

$$v_0 = \pm\sqrt{-2ax}$$

Substituting values,

$$v_0 = \pm\sqrt{-2(-0.26\text{ m/s}^2)(28\text{ m})}$$
$$= \pm 3.8\text{ m/s}$$

Since the initial velocity must be positive (i.e., forward), we use only the positive square root. Thus, the initial velocity is 3.8 m/s forward.

(b) The required quantity is t. Since we know x, x_0, v, v_0, and a, we could use any of Eqns. 2-7, 2-8, or 2-9 to find t. The easiest is Eqn. 2-7: $v = v_0 + at$.

Rearranging to solve for t, $t = \dfrac{v - v_0}{a}$

Substituting values,

$$t = \frac{(0 - 3.8)\ \text{m/s}}{-0.26\ \text{m/s}^2}$$

$$= 15\ \text{s (to two significant digits)}$$

Hence, the time taken for the curling rock to slow down to a stop is 15 s.

Using an App to Measure Acceleration

Download an app that allows you to measure acceleration on a hand-held device. Learn how to measure the acceleration of the device as you produce various examples of horizontal straight-line motion. How close to constant acceleration (positive or negative) can you achieve?

EXERCISES

2-29 You are asked to find the constant acceleration needed by an object, having a known initial velocity, to travel a known distance in a known time. Which equation (2-7, 2-8, 2-9, or 2-10) is appropriate for this situation?

2-30 Prove that Eqn. 2-10 is dimensionally correct.

2-31 Rearrange Eqn. 2-8 to solve for the time, t.

2-32 The equations derived in this section could have been derived by starting with a velocity–time graph showing constant acceleration. An example of such a graph is shown in Figure 2-42. Use the fact that the area under the line on a velocity–time graph indicates the displacement $(x - x_0)$ to derive Eqn. 2-8.

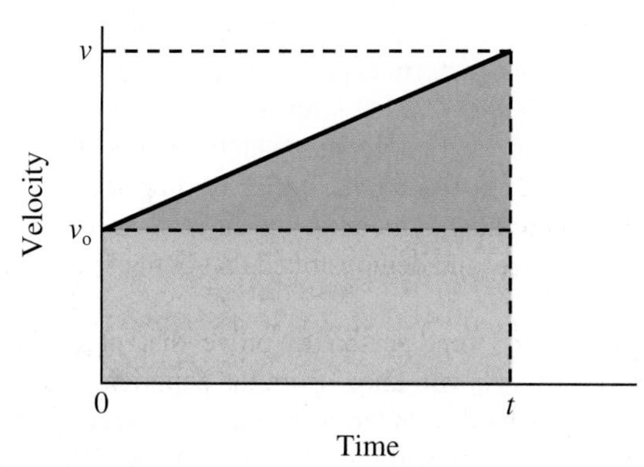

Figure 2-42 Question 2-32.

2-33 In a badminton game (Figure 2-43), a "bird" is struck so that its horizontal velocity is +12.3 m/s. Assuming that air resistance causes its acceleration to be $-2.6\ \text{m/s}^2$, what is the magnitude of its horizontal velocity after 1.5 s? Neglect its vertical motion.

photofriday/Shutterstock.com

Figure 2-43 Badminton (Question 2-33).

2-34 A baseball, travelling horizontally with a speed of 41 m/s (almost 150 km/h), is hit by a bat and has its velocity changed to 45 m/s in the opposite direction. The ball is in contact with the bat for 2.0 ms. What is the average acceleration experienced by the ball while being hit by the bat?

2-35 A sprinter leaves the starting block and, for a time of 3.3 s, undergoes a constant acceleration of magnitude $2.4\ \text{m/s}^2$.

(a) How far has the sprinter run in this time?

(b) How fast is the sprinter running at the end of this time?

2-36 A car travelling at 40 km/h on the entrance ramp of an expressway accelerates uniformly to 100 km/h in 36 s.

(a) Express 36 s in hours.

(b) How far (in kilometres) does the car travel in this time?

(c) What is the magnitude of the car's average acceleration during this time?

2-37 A rocket begins its third stage of launch at a velocity of 2.28×10^2 m/s forward. It undergoes a constant acceleration of $62.5\ \text{m/s}^2$ while travelling 1.86 km, all in the same direction. What is the rocket's speed at the end of this motion?

2-38 **Fermi Question:** If you slide your calculator or similar device along a smooth, flat surface, it will come to a stop in a measureable distance. Estimate the average acceleration of the device during its motion on a particular surface from the instant it leaves your hand until it stops.

2-39 A train accelerates as it enters a valley from the top of a hill. It accelerates uniformly at $0.040\ \text{m/s}^2$ for 225 s, and during this time travels 4.0 km. What was the magnitude of the velocity at the start of the acceleration?

2-40 An electron travelling at 6.74×10^7 m/s west enters a force field that reduces its velocity to 2.38×10^7 m/s west. While this (constant) acceleration is occurring, the electron experiences a displacement of 0.485 m in the same direction.

(a) How long did the force field take to cause this velocity change?

(b) What was the electron's acceleration during this time?

Digital Vision/Thinkstock

Figure 2-44 The high diving board.

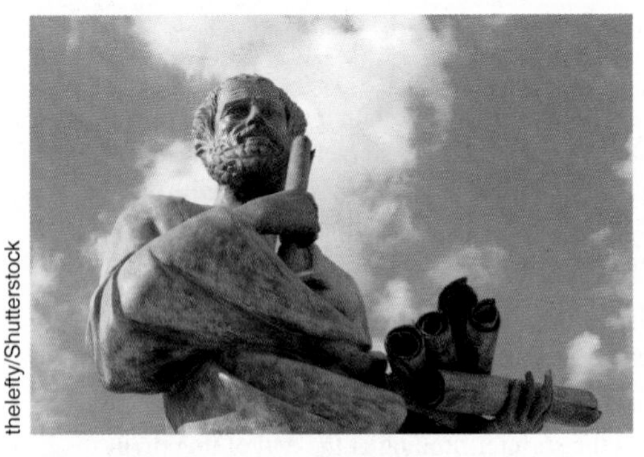

thelefty/Shutterstock

Figure 2-45 A statue of Aristotle at the Aristotle University, Thessaloniki, Greece.

acceleration due to gravity the rate of increase of velocity with time for an object falling with negligible air resistance

free fall the action of any object falling with negligible air resistance

DID YOU KNOW?

Aristotle and other Greek philosophers believed that objects beyond Earth, namely the stars, were composed of a fifth element which they called quintessence. This word stems from *quinte*, meaning fifth, and *essentia*, meaning essence.

2.5 Acceleration Due to Gravity

A person who dives from a 10 m high board (Figure 2-44) enters the water at a speed of about 50 km/h. Had the diver started from a lower height, the speed would have been less; and, of course, if the height had been larger, the speed would have been greater. The farther an object falls toward Earth, the faster is its speed (neglecting air resistance). The rate of increase of a falling object's velocity with time, or acceleration, is called the **acceleration due to gravity**.

Do all objects accelerate at the same rate near Earth's surface? If you drop an unfolded piece of paper and a pencil at the same instant, the paper will flutter downward and will land much later than the pencil. However, if you crumple the paper into a tight ball and try the experiment again, you will notice that the two objects land at about the same time. In this second case, the effect of air resistance has been reduced, and the accelerations are essentially the same. Thus, we observe that if air resistance is negligible, the acceleration due to gravity of objects at the same location is constant. Any object falling with negligible air resistance is said to be undergoing "**free fall**."

People have not always believed this observation. In fact, even great scientists of the past, such as the Greek philosopher Aristotle, 384 BCE to 322 BCE (Figure 2-45),[4] thought that heavier objects fell faster than lighter objects. He and his followers based this belief on the observation that a stone falls to Earth more quickly than a feather. To explain this observation, the ancient Greeks stated that all matter on Earth was composed of four elements, or pure substances: earth, water, air, and fire. All objects tend to reach their natural level, depending on their composition. For example, a stone is composed mainly of earth, so it falls rapidly toward its natural level, Earth. A feather is composed of more air than earth, so it tends to fall more slowly. Smoke from a fire rises because it is composed mainly of fire, which has a natural level above air, water, and earth.

Aristotle's theory of falling objects, as well as others of his theories, were accepted for nearly two thousand years. Finally, during the Renaissance era in Europe, scientists began to realize that experimentation was needed to test and verify theories. In a famous book titled *The Advancement of Learning*, published in 1605, an English philosopher and statesman, Francis Bacon, proposed that theories should be based on experimental facts. However, he did not perform experiments involving gravity. It was the renowned scientist from Italy, Galileo Galilei, who truly revolutionized science. He realized the importance of controlled experiments, which involve changing only one variable at a time to determine the effects on other variables. In performing motion experiments, for example, Galileo eliminated the effect of air resistance by dropping heavy and light objects of the same size, and demonstrated that both objects fell at the same rate toward Earth.

Galileo (Figure 2-46) was also the first person to prove that objects undergo acceleration as they fall toward Earth. As you know from the experience of dropping objects, they fall so quickly that it is hard to judge if they are accelerating or simply travelling really fast. Aristotle believed that a falling stone maintained a constant, high speed. He had no way of measuring short time intervals to prove otherwise. Galileo, however, devised ingenious ways of measuring time and was able to verify that acceleration does indeed occur.

Measuring the Acceleration Due to Gravity

With today's technological advancements, we have many ways of observing the acceleration of falling objects. One way is to use stroboscopic photography, with light

[4]BCE: Before the Common (or Current or Christian) Era, sometimes written as BC (Before Christ). CE: Common (or Current or Christian) Era, sometimes written as AD (*Anno Domini*, i.e., "in the year of the Lord").

PROFILES IN PHYSICS

Galileo Galilei (1594–1642): A Giant among Scientists

"I, Galileo Galilei, son of the late Vincenzo Galilei, of Florence, aged 70 years, being brought to judgment...abandon the false opinion that the Sun is the centre and immovable, and I will not hold, defend, or teach this false doctrine in any manner."

Chris Hill/Shutterstock

Figure 2-46 The tomb of Galileo Galilei in the Basilica of Santa Croce, Florence, Italy.

Galileo Galilei was a great and daring leader in the scientific world of the Renaissance (Figure 2-46). He was born in Pisa, Italy, in 1564, just 42 years after the first European ship sailed around the world (proving that Earth is round). That same year the great Italian artist, Michelangelo, died, and the famous English playwright, William Shakespeare, was born. When just 17, Galileo entered the University of Pisa to study medicine, but he soon became more interested in the physical sciences and mathematics. By the age of 26 he was appointed professor of mathematics at the University of Pisa, but three years later he moved to the University of Padua.

Even early in his career he performed numerous experiments, made many important discoveries, and wrote valuable scientific reports and books. He discovered that a pendulum of fixed length always swings with the same period of time for each vibration. This discovery led to the invention of pendulum clocks, which were valuable in scientific investigations. He used rolling balls on inclined planes to discover that the acceleration due to gravity does not depend on mass and that it remains constant. One legend, not necessarily true, is that he dropped two balls of different mass from the Leaning Tower of Pisa to verify that heavy and light objects fall at the same rate (Figure 2-47). He learned of the discovery that two lenses aligned properly could magnify distant objects, and he applied this breakthrough to the invention of the astronomical telescope. In 1609, with his newly made telescope, he made a number of significant findings, including the fact that the Moon has craters, the planet Venus goes through a full set of phases, the Sun has dark regions, now called sunspots, the Milky Way Galaxy is composed of individual stars, and the planet Jupiter has moons that revolve around it. The astronomical discoveries helped persuade Galileo that, contrary to the belief of Aristotle and the teachings of the Catholic Church, Earth was not the centre of the universe, with the Sun revolving around it. Rather Earth and other planets revolve around the Sun, a theory proposed earlier by a Polish astronomer, Nicholas Copernicus (1473–1543).

Galileo described many of his discoveries and theories in the Italian language rather than the traditional Latin for scientific documents. Thus, a larger number of people could read his works. His publications caused controversy with the Roman Catholic Church, and in 1616 the Inquisition, the judicial arm of the Catholic Church, required Galileo to stop teaching and writing about the Sun-centred theory.

However, Galileo continued writing and in 1632, when he was 68, he published a work entitled *Dialogue Concerning the Great World Systems,* in which he compared the Earth-centred and Sun-centred theories. His belief in the latter theory was obvious, and in his book he scorned the theories of the Pope and other leaders. The Roman Catholic Inquisition found him guilty of teaching false theories, and forced him to deny his own beliefs (see the quotation at the beginning of this profile.). Rather than sending the aging and unhealthy man to jail, the Inquisition put him under house arrest, denied him visitors, and forbade him from publishing his works outside of Italy.

Despite his persecution, Galileo continued to write in secrecy. In 1636 a book, *Dialogues of Two New Sciences,* was published in Holland after being smuggled out of Italy. This was his greatest work, dealing with his studies of motion.

In 1637 Galileo went blind but continued to write until he died in 1642. This was the same year that another great scientist, Isaac Newton, was born. Many of Galileo's discoveries and ideas helped Newton and other scientists develop new theories. The age of true science had begun.

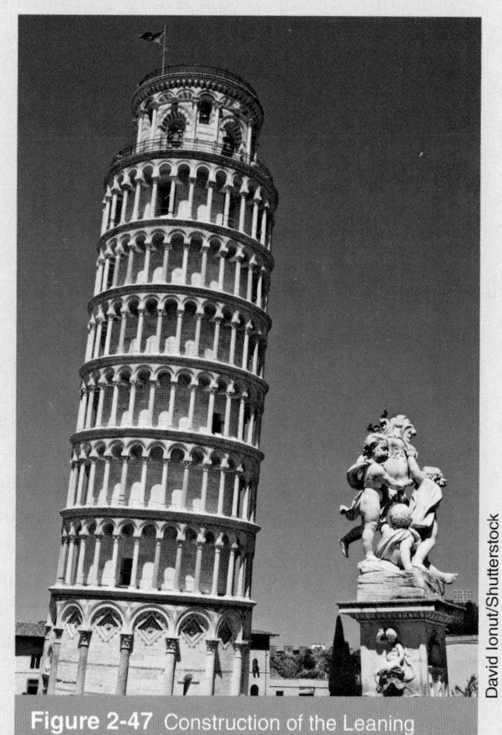

David Ionut/Shutterstock

Figure 2-47 Construction of the Leaning Tower of Pisa, Italy, began in the 12th century.

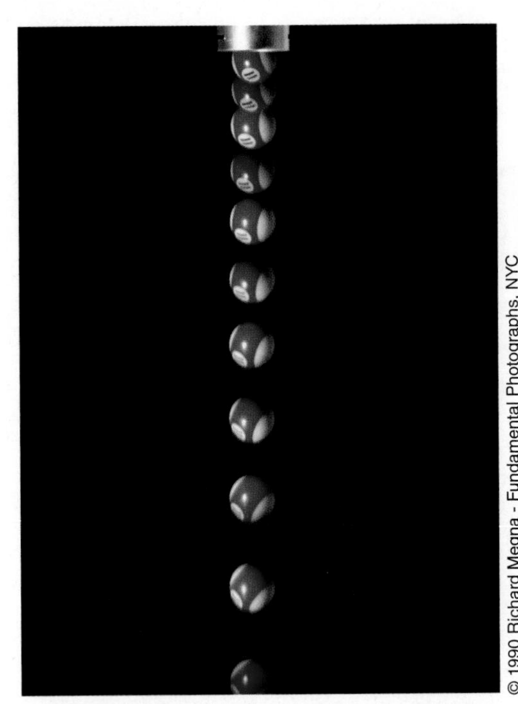

Figure 2-48 A stroboscopic photo of a ball falling vertically from rest.

$\vec{g}$ the symbol for the acceleration due to gravity

DID YOU KNOW?

Knowing the acceleration due to gravity to seven or more significant digits is of great interest to certain groups of people. Geologists and geophysicists can use the information to determine the structure of Earth's interior and help locate areas having high concentrations of mineral deposits. Military experts are concerned with variations in the acceleration due to gravity in the operation of such devices as cruise missiles. Space scientists use the data to help calculate the paths of artificial satellites.

flashing at regular intervals to reveal the motion of the falling object (Figure 2-48). Such a photograph can be analyzed to find the distance the object has fallen at the end of each known time interval. Then the kinematics equation $x = x_0 + v_0t + \frac{1}{2}at^2$ can be reduced to $x = \frac{1}{2}at^2$ because $v_0 = 0$ (and x_0 can be chosen to be zero), and the acceleration can be determined by rearrangement to find $a = 2x/t^2$.

With methods such as this, the acceleration of a falling object is found to be constant. Near Earth's surface, the acceleration, to two significant digits, is 9.8 m/s² downward. This acceleration is used so often that it is given a special symbol, $\vec{g}$, the acceleration due to gravity.[5] The magnitude of this acceleration is $g = 9.8$ m/s². (Do not confuse an italicized g with the non-italicized g that is the symbol for "gram.") The value of $g = 9.8$ m/s² applies to objects undergoing free fall, that is, objects that are not affected appreciably by air resistance.

Scientists throughout the world use more sophisticated techniques to measure the acceleration due to gravity at different locations. For example, at the International Bureau of Weights and Measures in France, investigations have been performed in a vacuum chamber in which an object is propelled upward using a special elastic band. At the top and bottom of the object are mirrors that reflect laser beams used to determine the time of flight. Using this technique, g at the bureau in Paris is found to be 9.809 260 m/s².

The acceleration due to gravity varies slightly from one location to another. In general, the value is greater near Earth's north and south poles than near the equator, and it is greater at lower altitudes than at higher altitudes. The average value, to two significant digits, is 9.8 m/s². Detailed reasons for the variations in g are discussed in Chapter 9, where Table 9-2 lists g at several locations. (If three significant digits for g are required when solving a problem in this textbook, use $g = 9.80$ m/s².)

So far we have considered falling objects only. But the acceleration due to gravity applies to objects that have been thrown upward and are being pulled back toward Earth. Neglecting air resistance and considering one dimensional vertical motion, if you throw a ball straight upward, once the ball leaves your hand its speed decreases at a uniform rate, with an acceleration of 9.8 m/s² downward for its entire flight; that is, if $+y$ is chosen to be upward, then $a = -g = -9.8$ m/s² when v is upward, when $v = 0$ for the instant at the top of the flight, and when v is downward. It is important to realize that the directions of the velocity and acceleration can be opposite (in this case when the ball is rising) or the directions can be the same (when the ball is falling). Notice that even for the brief instant when $v = 0$ as the ball changes directions, downward acceleration still exists. (The force of Earth's gravity causing that acceleration still exists; you will learn more about this concept in Chapter 5.)

Calculations Involving the Acceleration Due to Gravity

Because free fall involves constant acceleration, the kinematics equations developed in Section 2.4 can be applied. However, we will make one important change when writing these equations: we will write y rather than x for position (Table 2-3). This will prevent confusion in the following chapters where x is used for horizontal motion and y for vertical motion.

When applying these equations, remember that positive and negative signs are important. Begin each solution to a problem by choosing which direction is positive, then maintain this convention throughout the entire solution. It is important to note

[5]The symbol $\vec{g}$ is the first example in this text showing the formal notation for a vector: boldface type in italics, with a short arrow above. This notation will be used extensively in subsequent chapters.

Table 2-3

Equations of Kinematics Written for Vertical Motion

Variables Involved	Variable(s) Eliminated	Equation	
v, v_0, t	y, y_0	$v = v_0 + at$	**(2-7b)**
y, y_0, v_0, v, t	a	$y = y_0 + \frac{1}{2}(v_0 + v)t$	**(2-8b)**[6]
y, y_0, v_0, a, t	v	$y = y_0 + v_0 t + \frac{1}{2}at^2$	**(2-9b)**
y, y_0, v_0, v, a	t	$v^2 - v_0^2 = 2a(y - y_0)$	**(2-10b)**

that the symbol g represents $+9.8$ m/s². Therefore, if upward is chosen to be positive, then the acceleration is $a = -g = -9.8\ \text{m/s}^2$. However, if downward is chosen to be positive, $a = g = +9.8\ \text{m/s}^2$.

SAMPLE PROBLEM 2-13

A ball is thrown vertically with an initial velocity of 8.5 m/s upward. What maximum height will the ball reach above its starting point?

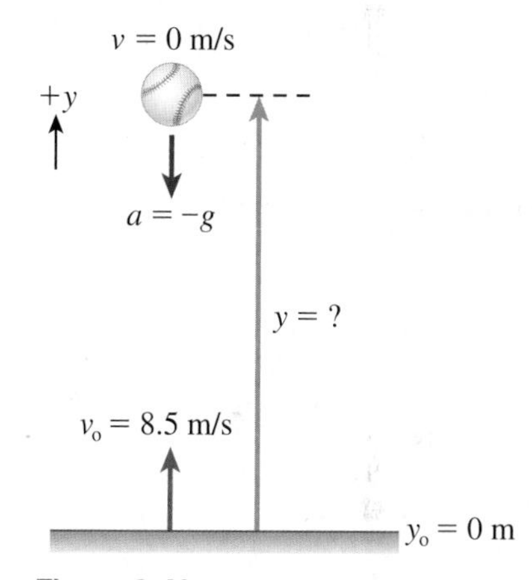

Figure 2-49 Sample Problem 2-13.

Solution

Let us define upward as positive. We also choose $y_0 = 0$ m as the starting position (Figure 2-49).

The given quantities are

$$v_0 = +8.5\ \text{m/s}$$

$$a = -g = -9.8\ \text{m/s}^2$$

$$v = 0\ \text{m/s}$$

$$y_0 = 0\ \text{m}$$

We are able to state that $v = 0$ m/s because at the top of its flight, the ball comes to rest instantaneously (then reverses its direction of motion). The required quantity is y, and the appropriate equation is

$$v^2 - v_0^2 = 2a(y - y_0)$$

Substituting $v = 0\ \text{m/s}, y_0 = 0\ \text{m}, \text{and}\ a = -g$,

$$-v_0^2 = 2(-g)y$$

Rearranging to solve for y, $y = \dfrac{v_0^2}{2g}$

Substituting numerical values,

$$y = \frac{(8.5\ \text{m/s})^2}{2(9.8\ \text{m/s}^2)}$$

$$= +3.7\ \text{m (to two significant digits)}$$

Thus, the ball rises to a height of 3.7 m above its original position.

Notation Tip

It is incorrect to write $g = -9.8$ m/s². The symbol "g" represents the magnitude of the vector $\vec{g}$, and the magnitude of a non-zero vector is always a positive quantity.

[6]When applying this equation as well as Eqns. 2-9b and 2-10b, you may prefer to use Δy rather than $y - y_0$.

TRY THIS!

Height of a Ball Throw

Design and carry out an experiment to determine the maximum height you and/or a friend can throw a ball, such as a tennis ball, straight upward. Besides a calculator and the ball, the only apparatus you will need is a stopwatch.

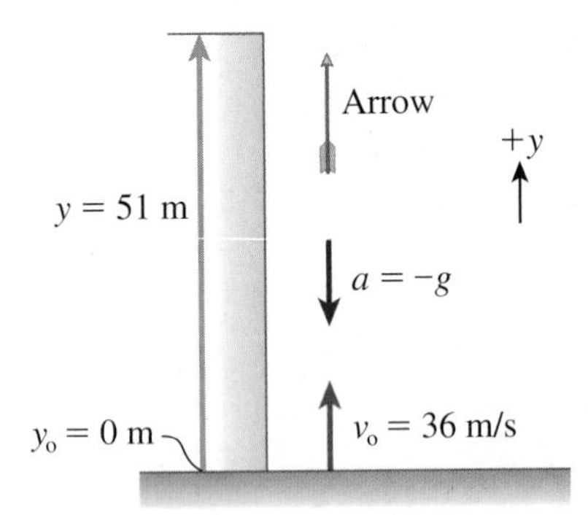

Figure 2-50 Sample Problem 2-14.

Math Tip

The quadratic formula is useful for finding the roots of a quadratic equation, that is, an equation involving a square function (quadratic is Latin for "square"). If the quadratic equation is written in this form,

$$ax^2 + bx + c = 0, \quad \text{where} \quad a \neq 0,$$

then its roots are

$$x = \frac{-b \pm \sqrt{b^2 - 4ac}}{2a}$$

SAMPLE PROBLEM 2-14

An arrow is shot straight upward beside a building that is 51 m high. The arrow's initial speed is 36 m/s. Neglecting air resistance, find

(a) the time(s) when the arrow passes the top of the building

(b) the total time of flight

Solution

Let us define upward as positive, and choose the initial position to be $y_0 = 0$ m (Figure 2-50).

(a) The given quantities are

$$v_0 = +36\text{ m/s}$$

$$a = -g = -9.8\text{ m/s}^2$$

$$y = 51\text{ m}$$

$$y_0 = 0\text{ m}$$

The required quantity is t, and one way to find the total time is to use

$$y = y_0 + v_0 t + \tfrac{1}{2}at^2$$

Taking care to use consistent units, we substitute the given quantities into the equation, but we omit the units for convenience. Thus,

$$51 = 0 + 36t - 4.9t^2$$

Rearranging this equation so that it has the standard form of a quadratic equation,

$$4.9t^2 - 36t + 51 = 0$$

Using the quadratic formula, $t = \dfrac{-b \pm \sqrt{b^2 - 4ac}}{2a}$

where $a = 4.9$, $b = -36$, and $c = 51$.

$$\text{Thus, } t = \frac{-(-36) \pm \sqrt{(-36)^2 - 4(4.9)(52)}}{2(4.9)}$$

$$= \frac{36 \pm 17.2}{9.8}$$

$$= 1.9\text{ s and } 5.4\text{ s}$$

Thus, there are two positive roots of this equation. This means that the arrow passes the top of the building at $t = 1.9$ s on the way up, and again at 5.4 s on the way down.

(b) There are various ways of finding the total time of flight. The given quantities are

$$v_0 = +36\text{ m/s}$$

$$a = -g = -9.8\text{ m/s}^2$$

$$y_0 = 0\text{ m}$$

$$y = 0\text{ m} \quad \text{(assuming the arrow returns to its initial position)}$$

The required quantity is t, and the appropriate equation is

$$y = y_0 + v_0 t + \frac{1}{2}at^2$$

Substituting $y_0 = y = 0\,\text{m}$, and $a = -g$,

$$0 = v_0 t - \frac{1}{2}gt^2$$

Factoring out a "t," $0 = t(v_0 - \frac{1}{2}gt)$

Hence, either $t = 0$, or $v_0 - \frac{1}{2}gt = 0$. The first solution ($t = 0$) corresponds to the initial condition: at time $t = 0$, the arrow is at position $y = 0$. The second solution is the one we want; solving for t gives

$$t = \frac{2v_0}{g} = \frac{2(36\,\text{m/s})}{9.8\,\text{m/s}^2} = 7.3\,\text{s}$$

Thus, the total time of flight is 7.3 s.

A faster solution would have been to realize that if the arrow took 1.9 s to rise to the top of the building, from (a) above, it would also take 1.9 s to travel from the top of the building down to the initial position. Add this to the 5.4 s it took to arrive at the top of the building on the way down to obtain the total time of 7.3 s.

A third technique in this case is to find the time for the arrow to rise to the top of its flight (where $v = 0$), then multiply that time by 2. The equation to use is $v = v_0 + at$, which can be solved for t when $v = 0$.

In solving problems involving constant acceleration, remember that there is often more than one way to find the solution, especially in problems with more than one part.

Terminal Speed

A skydiver jumps out of an aircraft and experiences free fall for a short period of time (Figure 2-51 (a)). The diver's speed increases rapidly, and the air resistance to the diver's motion is soon strong. Eventually this resistance becomes so high that it prevents any more increase in speed. At this stage, the diver has reached **terminal speed**,[7] which is the maximum speed reached by an object falling in a gas or liquid. At terminal speed, the object's speed (and velocity) is constant, and its acceleration has become zero. A graphical representation of the speed of a falling object as a function of time is shown in Figure 2-51 (b).

In general there are two situations in which an object can reach terminal speed in air:

- an object has a large surface area relative to its mass—examples include a parachute and a table-tennis ball;
- an object is falling with a high speed. This is what occurred when Felix Baumgartner made his record-breaking jump as described in the chapter introduction, although his speed was much higher where the atmosphere is less dense than it would have been at lower altitudes.

The terminal speeds of various falling objects in air are listed in Table 2-4. The values listed for the baseball, tennis ball, and table-tennis ball make sense when you think of the size and mass of each ball. Terminal speeds are also important in liquids and in gases other than air.

[7]Sometimes this is called terminal velocity, but since velocity is a vector and we are interested here only in the magnitude of the velocity (i.e., the speed), terminal speed is a more accurate term to use in this discussion.

Determining Your Reaction Time

Use a 30 cm ruler or a metre stick to determine the reaction times of you and your friends. Person A holds one end of the ruler and lets the ruler hang lengthwise vertically. Person B, whose reaction time is being determined, places a thumb and forefinger almost together just below the bottom of the ruler, so that when A releases the ruler, it will fall between B's thumb and forefinger. Without warning, person A drops the ruler and B tries to grasp it as soon as possible after it starts to fall freely.

(a) Find the average distance of several falls, then use a kinematics equation to determine reaction time.

(b) Repeat the procedure while B is being distracted (for instance, while texting or talking on the phone).

(c) Use your answer to (b) to determine how far a car, moving at 100 km/h, would travel before B reacts in an emergency and applies the brakes.

2happy/Shutterstock

(a)

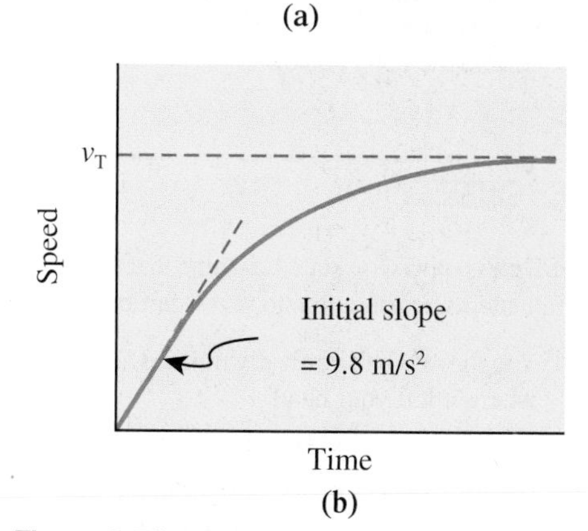

(b)

Figure 2-51 **(a)** Skydiving. **(b)** The general shape of a speed–time graph for a falling object that reaches terminal speed, v_T.

terminal speed the maximum speed reached by an object falling in a gas or liquid

DID YOU KNOW?

Red blood cells in a human blood sample in a test tube reach terminal speed almost instantaneously. The settling rate for an average male is about 3 mm/h, for an average female it is about 4 mm/h, but in diseased conditions, it can become as high as 100 mm/h.

Table 2-4

Approximate Terminal Speeds of Objects Falling in Air

Object	Terminal Speed	
	(m/s)	(km/h)
Human	53	190
Baseball	40	140
Tennis ball	30	110
Table-tennis ball	20	72
Raindrop	7	25
Human with open parachute	5 to 10	18 to 36
Dandelion seeds	0.5	1.8
Dust particle (typical)	0.02	0.07

danielkreissl/Shutterstock

(a)

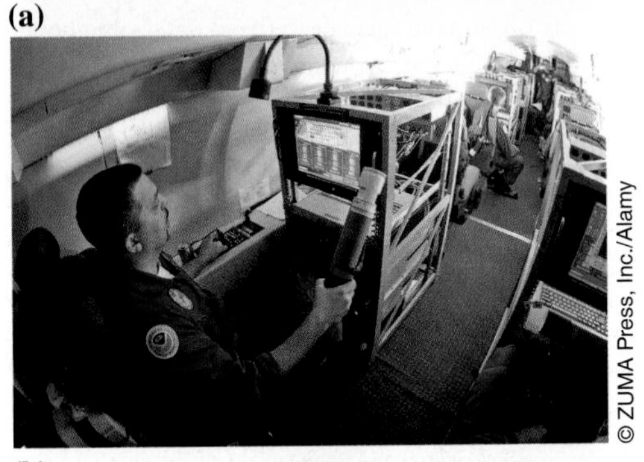

© ZUMA Press, Inc./Alamy

(b)

Figure 2-52 **(a)** Seed dispersal of dandelions. **(b)** The National Oceanic and Atmospheric Administration (NOAA) sends aircraft into storms to study their features. Here a technician prepares a dropsonde that will be deployed into the eye of a hurricane, sending back data as it falls at terminal speed toward the ocean below.

Notice the difference in the terminal speeds of the last two entries in Table 2-4. Dandelion seeds (Figure 2-52 (a)) use wind for dispersal, and if their terminal speed were lower they may travel farther in a breeze, possibly landing where the conditions for their survival might not be ideal. Dust particles have a much lower terminal speed, so they can travel a much greater distance horizontally before they land. This has important implications when a large volcanic eruption spews both large and small dust particles into the atmosphere: the large particles settle relatively quickly but the fine particles can travel thousands of kilometres, affecting global weather patterns sometimes for years.

DID YOU KNOW?

A dropsonde is an expendable weather research device containing a set of instruments used to measure weather-related variables such as temperature, humidity, and air pressure. The device is dropped from an aircraft (Figure 2-52 (b)) at an altitude of about 6500 m and within 10 s a small parachute opens, causing the device to reach a downward terminal speed of approximately 50 km/h. Dropsondes are valuable in studying hurricanes, tornadoes, and other extreme weather events.

EXERCISES

Note: Unless otherwise stated, assume that the falling object is undergoing free fall and its acceleration, to two significant digits, is 9.8 m/s^2 down.

2-41 You throw a ball vertically upward and catch it at the same position where it left your hand.

(a) How does its final velocity compare with its initial velocity?

(b) How does the rise time compare with the fall time?

(c) What is the ball's velocity at the top of its flight?

(d) What is the acceleration of the ball as it is rising? at the top of its flight? as it is falling?

2-42 In each situation described below, assume the object starts from rest.

(a) A steel ball is dropped from the top of the Leaning Tower of Pisa to the piazza below, a distance of 55 m. Calculate the maximum speed of the ball in metres per second just at it lands.

(b) To entertain tourists, divers in Acapulco, Mexico, dive from a cliff and hit the water at a speed of 27 m/s (Figure 2-53). How high is the cliff above the water level?

(c) A stone is dropped from a bridge and lands in the water 2.4 s later. Calculate the maximum speed of the stone.

2-43 Prove that a free-falling ball dropped vertically from rest travels three times as far in the second second as in the first second.

2-44 A baseball pitcher throws a ball vertically straight upward and catches the ball 4.2 s later.

(a) With what velocity did the ball leave the pitcher's hand?

(b) What was the maximum height reached by the ball?

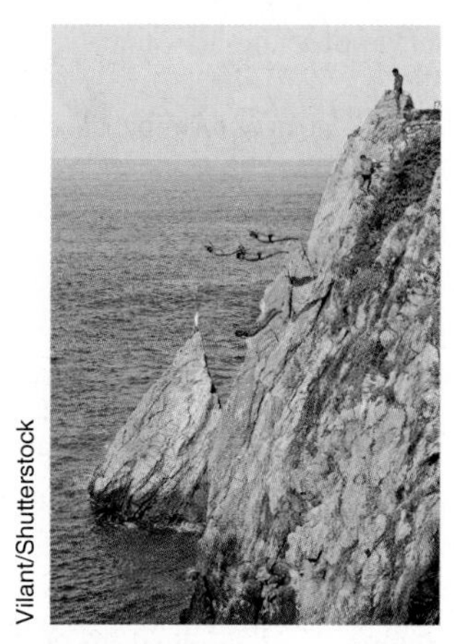
Vilant/Shutterstock

Figure 2-53 Cliff diving (Question 2-42(b)).

2-45 A hot-air balloon (Figure 2-54) is moving at a velocity of 2.1 m/s upward when the balloonist drops a ballast (a heavy mass used for control) over the edge. The ballast takes 3.8 s to land on the ground below.

(a) How high was the balloon when the ballast was dropped?

(b) With what speed did the ballast strike the ground?

LittleStocker/Shutterstock

Figure 2-54 Hot-air balloon.

2-46 Assume that a high jumper has a vertical velocity of 5.112 m/s upward when jumping.

(a) Calculate and compare the maximum increase in height achieved by the jumper in London (g_L = 9.823 m/s^2) and Denver (g_D = 9.796 m/s^2). Remember to use the correct number of significant digits.

(b) Why do you think g_L is larger than g_D?

2-47 An astronaut drops a camera from rest while exiting from a spacecraft on the Moon. The camera takes 1.7 s to fall 2.3 m.

(a) Determine the magnitude of the acceleration due to gravity on the Moon.

(b) What is the ratio of the magnitude of the acceleration due to gravity on Earth to that on the Moon?

2-48 Assume that, during the first minute of blast off, a space craft has an average acceleration of magnitude 5 g, that is, 5 times the acceleration due to gravity on the surface of Earth. Calculate the craft's speed in metres per second and kilometres per hour after the first minute. (These values are approximate because the craft does not experience constant acceleration in a straight line.)

2-49 A person throws a golf ball vertically upward on Earth and the ball remains in flight a total time of 2.6 s.

(a) How long did the ball rise?

(b) What was its initial velocity?

(c) With the same initial speed on Mars, how long would the ball remain in flight there? (Refer to the inside back cover for a table of data for the planets.)

2-50 In a laboratory experiment a motion detector interfaced to a computer is used to determine that the time for a falling steel ball to travel the final 0.80 m before striking the floor is 0.087 s. With what speed does the ball strike the floor?

2-51 A stone is thrown vertically from a bridge with a velocity of 14 m/s downward. At what time will the stone reach the water 21 m below the bridge? (**Hint:** Apply the quadratic formula to solve this problem. Explain the meaning of both roots of the solution.)

2-52 Sketch a speed–time graph for a skydiver who jumps from an aircraft, reaches terminal speed some time later, then opens the parachute and reaches another terminal speed. Refer to Table 2-4.

LOOKING BACK...LOOKING AHEAD

In this chapter several concepts from Chapter 1 have been extended and applied, including measurement, metric units, dimensional analysis, and scalar and vector quantities. The focus has been on the kinematics (motion) variables of speed, velocity, and acceleration, in each case distinguishing average and instantaneous values. Emphasis was placed on solving problems related to motion, mostly using equations, but also by analyzing graphs. The acceleration due to gravity provided the most common example of constant acceleration from our everyday experiences.

The motion studied in this chapter was one-dimensional. In Chapter 3, properties of vectors will be discussed in detail and then applied to the study of two-dimensional motion in Chapter 4, then to the forces causing acceleration in Chapters 5 and 6.

CONCEPTS AND SKILLS

Having completed this chapter you should now be able to do the following:

- Calculate the average speed, average velocity, and average acceleration of an object moving in one dimension.
- Recognize the conditions required for uniform motion.
- Determine an object's displacement given its initial and final positions.
- Plot position–time, velocity–time, and acceleration–time graphs involving uniform and non-uniform motion, and use slopes and areas on those graphs to determine unknown quantities.
- Describe the motion portrayed in the types of graphs named above.

- Apply the constant-acceleration equations involving initial and final positions, time, initial and final velocities, and acceleration to solve for any one of the variables, given at least three of the other variables.
- Apply the constant-acceleration equations to situations involving the acceleration due to gravity.
- Recognize the advantage of using a systematic strategy in problem solving.
- Describe the meaning and cause of terminal speed in air.

KEY TERMS

You should now be able to define or explain each of the following words or phrases:

kinematics
speed
average speed (defining equation)
instantaneous speed
uniform motion
displacement
velocity
average velocity (defining equation)
instantaneous velocity
non-uniform motion
tangent to a curve
acceleration
average acceleration (defining equation)
instantaneous acceleration
acceleration due to gravity
free fall
$\vec{g}$
terminal speed

Chapter Review

MULTIPLE-CHOICE QUESTIONS

Note: Unless otherwise stated, assume that any object that is falling is undergoing free fall and its acceleration, to two significant digits, is 9.8 m / s² down.

2-53 During the time interval that an object undergoes uniform motion,

(a) the instantaneous velocity and the average velocity are equal
(b) the direction of motion must remain constant
(c) the displacement is directly proportional to the time interval
(d) all of the above are true
(e) only (a) and (b) above are true

2-54 For any given object, which of the following pairs of quantities cannot be both constant and nonzero during the same time interval ?

(a) the acceleration and the velocity
(b) the magnitude of the acceleration and the acceleration
(c) the distance and the displacement
(d) the speed and the velocity

2-55 A cyclist, moving westward on a level, straight path, coasts to a stop. During this motion, the directions of the acceleration, instantaneous velocity, and displacement are, respectively

(a) west, west, west
(b) east, west, west
(c) east, east, west
(d) west, east, east

2-56 In a coin toss with upward defined as the positive direction, a quarter rises vertically (motion A), reaches maximum height (condition B), and then falls vertically (motion C). During A, B, and C, the quarter' s acceleration, in metres per second squared, is, respectively,

(a) −9.8; 0; +9.8
(b) +9.8; +9.8; +9.8
(c) −9.8; 0;−9.8
(d) −9.8; −9.8;−9.8

2-57 You step off a high diving board (height Δy above the water) with an initial velocity of zero. The magnitude of the velocity with which you hit the water is proportional to

(a) g and Δy
(b) $\sqrt{g}$ and $\sqrt{\Delta y}$
(c) g and $\sqrt{\Delta y}$
(d) $\sqrt{g}$ and Δy
(e) $g(\Delta y)^2$

Questions and Problems

2-58 In the Canadian Grand Prix auto race (Figure 2-55), the drivers travel a total distance of 304.29 km in 69 laps around the track. Assume the fastest lap time is 84.118 s. Determine the average speed in metres per second for this lap.

2-59 A driver's handbook states that, for any specific speed, the safe distance separating your car from the car ahead is the distance you would travel in 2.0 s at that speed.

(a) Determine this distance if the speed of your car is 25 m/s (90 km/h).
(b) Approximately how many car lengths is this distance?

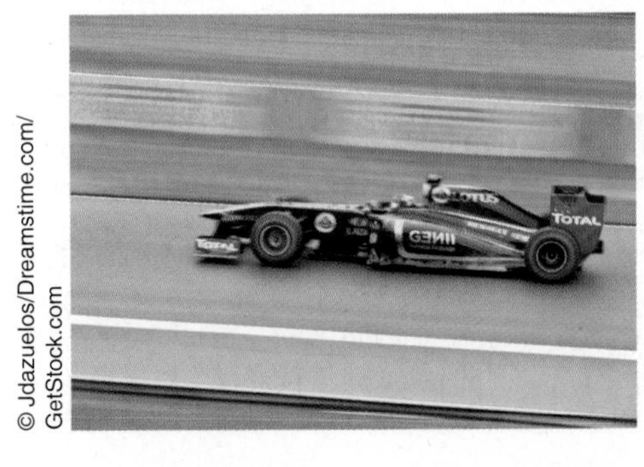

Figure 2-55 At the Canadian Grand Prix auto race (Question 2-58).

2-60 An impatient driver, wanting to save time, drives along a city bypass at an average speed of 115 km/h where the speed limit is 90.0 km/h. If the bypass is 18.0 km long, how much time (in minutes) does the driver save by breaking the speed limit? (By the way, the driver consumes about 20% more fuel at the higher speed.)

2-61 Two joggers, 1 and 2, move at constant speeds in the same direction along a straight park pathway, as illustrated in Figure 2-56. The joggers' positions at the same time are indicated by identical letters. Starting positions are not shown. Do the joggers ever have the same instantaneous speeds? If so, at which position(s)? If not, explain how you know.

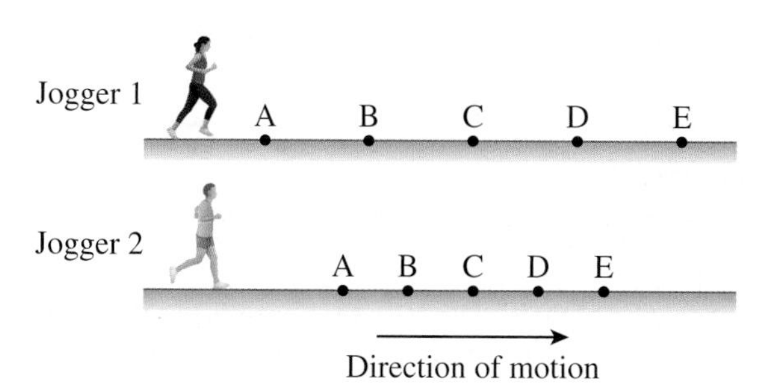

Figure 2-56 Question 2-61.

2-62 A migrating bird flies at 80 km/h for 30 min, then at 60 km/h for 1.5 h. Assuming two significant digits in the given numbers, calculate the bird's

(a) total distance travelled

(b) average speed

2-63 An eagle flies at 24 m/s for 1.6×10^3 m then glides at 18 m/s for 1.2×10^3 m.

(a) How long (in seconds) does this motion take?

(b) What is the eagle's average speed for this motion?

2-64 An electromagnetic signal, travelling at the speed of light (3.00×10^8 m/s), is sent from Earth to a satellite located 4.8×10^7 m away (Figure 2-57). The satellite receives the signal and after a delay of 0.55 s sends a return electromagnetic signal to Earth, still the same distance away. What total time elapses between sending the signal and receiving the return signal on Earth?

Figure 2-57 A communications satellite (Question 2-64).

2-65 A distance of 100 m can be run by one person in 10 s and by another person in 12 s, both at a constant speed. If the two runners pass by a reference position at the same instant, by what distance is the faster runner ahead of the slower runner after the faster runner has run 100 m from the reference position?

2-66 The planet Venus takes 2.1×10^7 s to complete its orbit around the Sun. The orbit of Venus can be approximated as a circle of radius 1.1×10^{11} m.

(a) Determine its average speed in metres per second and kilometres per hour.

(b) Find the magnitude of its average velocity after it has completed half a revolution around the Sun.

(c) Repeat (b) for one complete revolution around the Sun.

2-67 For uniform motion, compare

(a) instantaneous speed with average speed

(b) instantaneous velocity with average velocity

(c) instantaneous speed with instantaneous velocity

2-68 Copenhagen lies 1100 km straight east of Glasgow and 1700 km straight west of Moscow. An airplane flies from Copenhagen to Glasgow in 2.2 h, stays there for 1.0 h, then flies to Moscow in 3.1 h. Expressing the final answers in metres per second, determine the plane's

(a) average speed for the entire trip

(b) average speed while in the air

(c) average velocity for the entire trip

2-69 A professional baseball pitcher throws a fastball with a speed of 42 m/s, and 0.44 s later a batter bats the ball straight over the pitcher's head. The ball travels at 48 m/s horizontally until it is caught by a fielder 1.9 s after the hit. (We are neglecting any vertical motion.)

(a) What is the distance from the pitcher's mound to home plate?

(b) Determine the baseball's average speed for the entire motion.

(c) Calculate the magnitude of its average velocity for the entire motion.

2-70 Describe the motion in each time segment of the position–time graph shown in Figure 2-58.

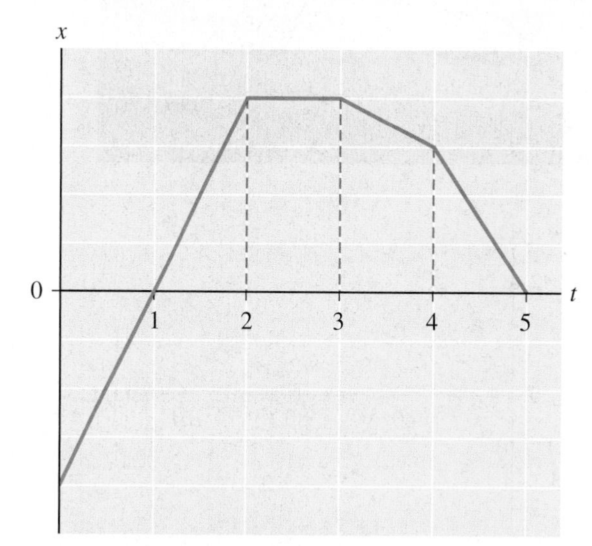

Figure 2-58 Question 2-70.

2-71 What quantity can be calculated from a position–time graph to give the velocity of an object? How can that quantity be found if the position–time data give a curve?

2-72 How can a velocity–time graph be used to determine

(a) displacement

(b) acceleration

2-73 Sketch the velocity–time graph that corresponds to the position–time graph shown in Figure 2-59. The positive direction is south.

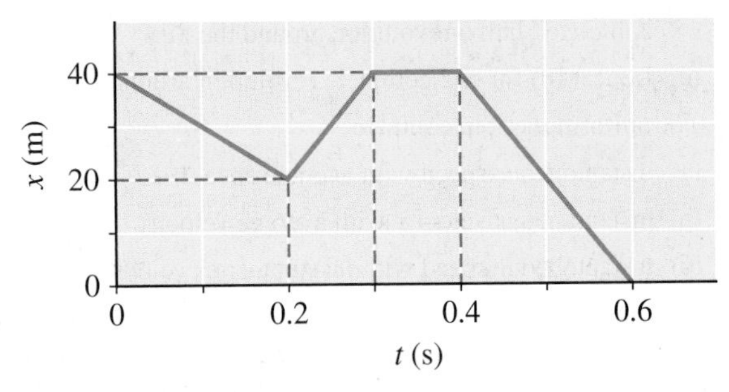

Figure 2-59 Question 2-73.

2-74 For motion in one dimension, state the difference between uniform motion and constant acceleration. Sketch displacement-time, velocity–time, and acceleration–time graphs of both types of motion.

2-75 Figure 2-60 (a) shows an insect crawling in a straight line along a twig and (b) shows a graph of its position as it crawls for 120 s.

(a) At approximately what times is the insect not moving?

(b) During which time intervals is the insect's velocity positive? negative?

(c) Find the insect's approximate velocity at 20 s and then at 40 s.

(a)

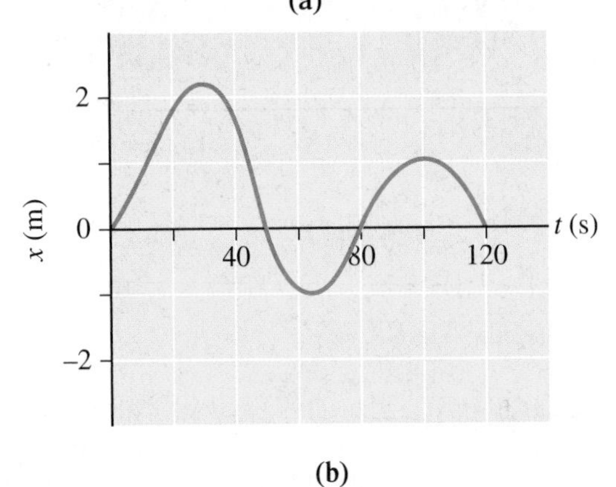

(b)

Figure 2-60 Question 2-75: **(a)** Insect crawling along a twig. **(b)** x-t graph of insect's motion.

2-76 Describe an example when an object has zero speed and a non-zero acceleration.

2-77 A dog is running east with an initial speed, v_0. Its acceleration, a, is in the westerly direction. Which diagram in Figure 2-61 best represents the position of the dog (the dots) and its velocity and acceleration vectors (the arrows)? Explain.

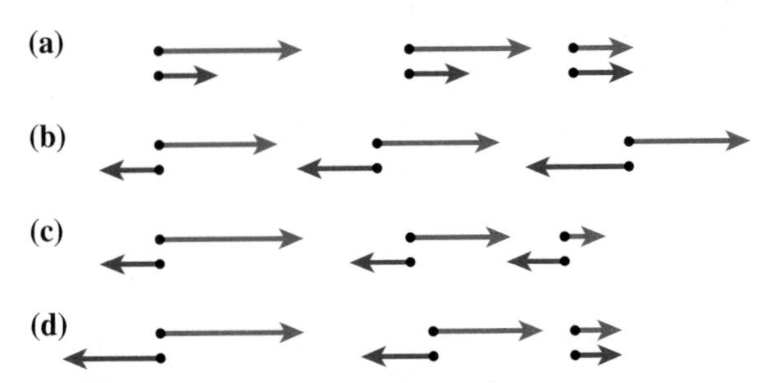

Figure 2-61 Question 2-77. The red arrows are velocity vectors and the purple arrows are acceleration vectors. East is toward the right.

2-78 Assume that the $+x$-direction for a one-dimensional recreational path is south. Describe the motion of a runner on the path if the runner has

(a) a positive velocity and a positive acceleration

(b) a positive velocity and a negative acceleration

(c) a negative velocity and a positive acceleration

(d) a negative velocity and a negative acceleration

2-79 As a squirrel is running forward along a branch, its acceleration is decreasing and its speed is increasing. Sketch a velocity–time graph for the squirrel.

2-80 (a) Sketch the acceleration–time graph that corresponds to the velocity–time graph shown in Figure 2-62.

(b) Assuming the initial position is zero, sketch the position–time graph of the same motion.

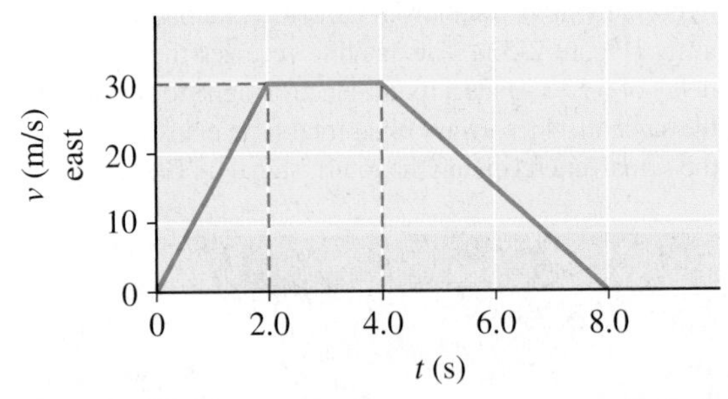

Figure 2-62 Question 2-80.

2-81 The table below gives the velocity–time data of the winning car in a drag race.

Time (s)	0.00	1.00	2.00	3.00	4.00
Velocity (m/s, forward)	0.0	14.4	38.8	59.1	74.2

(a) Plot a velocity–time graph of the motion.

(b) During which second was the car's average acceleration the greatest? the least? Determine the car's average acceleration for each of those seconds.

(c) Find the magnitude of the car's average acceleration for the entire motion.

2-82 A motorcycle is travelling at 22 m/s west. It then experiences the acceleration shown in Figure 2-63.

(a) Plot a velocity–time graph of the motion from time $t = 0.0$ s to $t = 10$ s.

(b) Determine the displacement of the motorcycle during the 10 s interval.

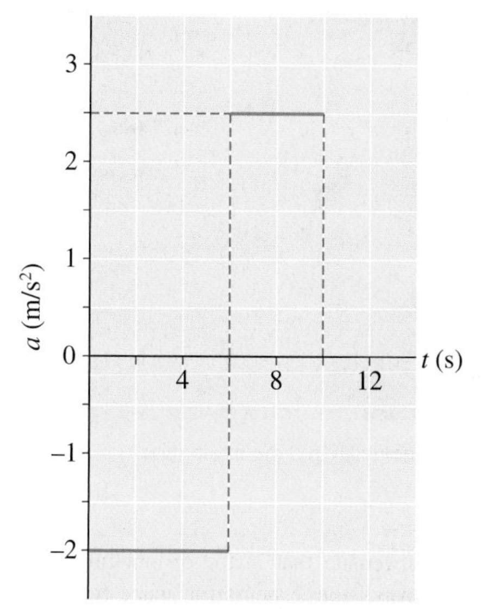

Figure 2-63 Question 2-82. The positive direction is west.

2-83 Figure 2-64 (a) shows a helicopter while the pilot is searching for a safe place to land, and (b) shows a v-t graph of the helicopter's motion.

(a)

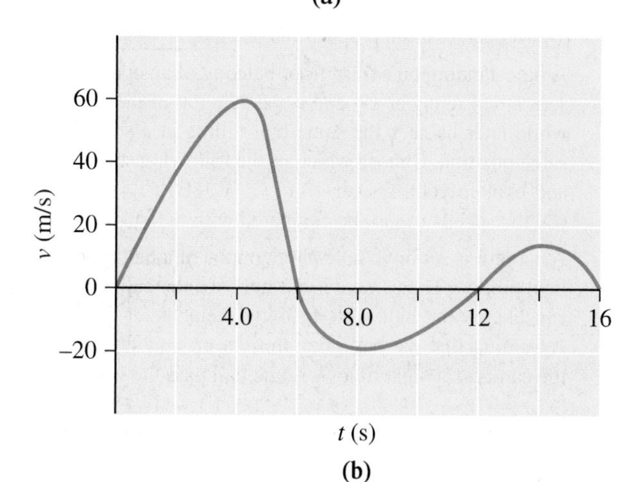

(b)

Figure 2-64 Question 2-83. **(a)** Helicopter. **(b)** v-t graph of the helicopter's motion, with the positive direction south.

(a) At what times is the helicopter stopped?

(b) When is the helicopter's acceleration zero?

(c) During which time intervals is the helicopter accelerating northward?

(d) Calculate the approximate instantaneous acceleration at $t =$ 6.0 s and $t =$ 12 s.

2-84 In a ride at an amusement park, the cars start from rest and accelerate rapidly forward, covering the first 15 m in 1.2 s.

(a) What is the magnitude of the cars' average acceleration (assumed constant)?

(b) What is the speed of the cars at 1.2 s?

(c) Express the magnitude of the average acceleration in terms of g.

2-85 Give an example to show that an object's velocity can reverse direction when the acceleration is constant.

2-86 A bicyclist, travelling at 4.0 km/h at the top of a hill, coasts downhill with constant acceleration, reaching a speed of 33 km/h in 33 s. What distance, in metres, does the cyclist travel in that time?

2-87 Describe the motion represented by each graph in Figure 2-65.

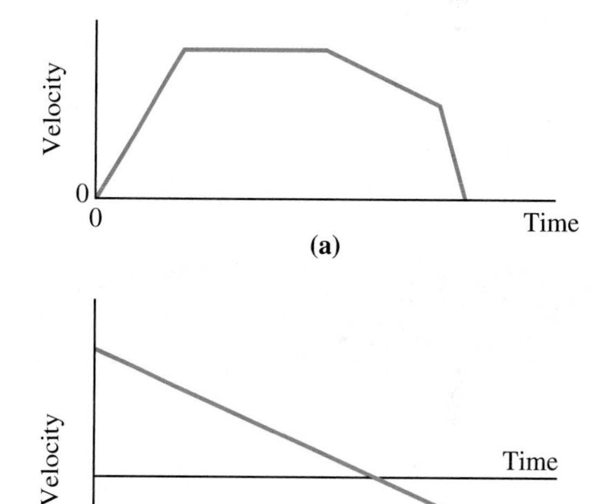

(b)

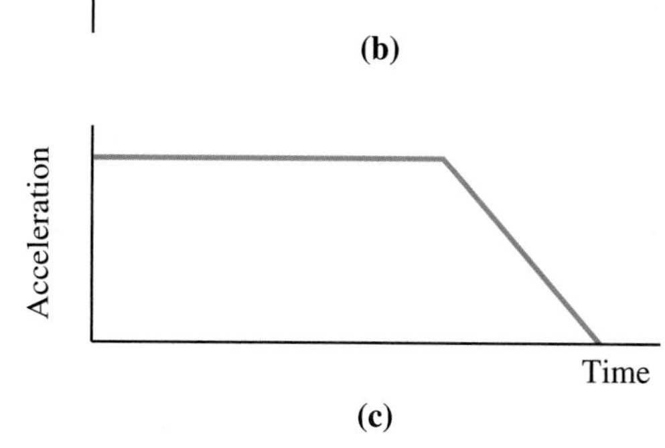

(c)

Figure 2-65 Question 2-87.

2-88 The heart's left ventricle accelerates blood from rest to a velocity of 25 cm/s forward during a displacement of 2.0 cm forward. Determine

(a) the blood's acceleration

(b) the time needed for the blood to reach its maximum velocity

2-89 (a) With a constant acceleration of magnitude 1.6 m/s^2, how long will it take a bus to travel 2.0×10^2 m, if the bus starts from rest?

(b) Repeat (a) if the bus starts with a speed of 8.0 m/s. (**Hint:** Use the quadratic formula.)

2-90 In track-and-field competition, the current fastest time for the woman's 100 m dash is 10.5 s, and the time for the women's 400 m relay is 41.4 s (Figure 2-66). It appears as if, on average, each of the four women in the relay can run 100 m in less than 10.5 s. Explain this apparent discrepancy. (**Hint:** Consider acceleration as well as the photo.)

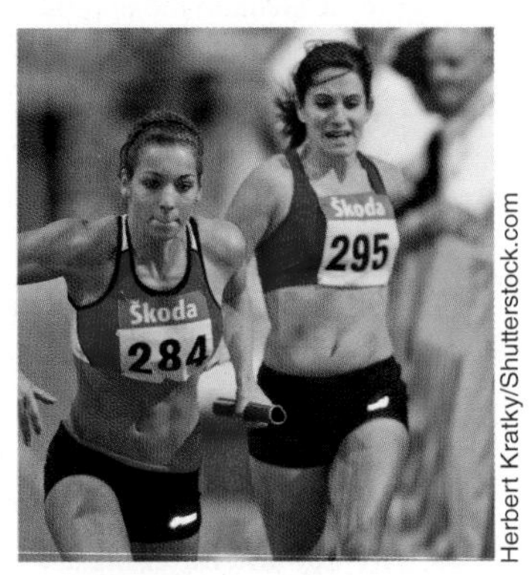

Figure 2-66 The baton must be transferred from one runner to another in a relay race (Question 2-90).

2-91 Research has shown that a driver with no distractions requires an average of 0.80 s to engage a car's brakes after seeing an emergency. This reaction time applies to people who have not been drinking alcoholic beverages. Approximate reaction times for beer consumers are shown in Figure 2-67. Use the data from this graph to determine the distance travelled while reacting (reaction distance) to complete the table below.

	Reaction distance		
Speed	**No alcohol**	**3 bottles**	**5 bottles**
50 km/h (14 m/s)	?	?	?
90 km/h (25 m/s)	?	?	?
120 km/h (33 m/s)	?	?	?

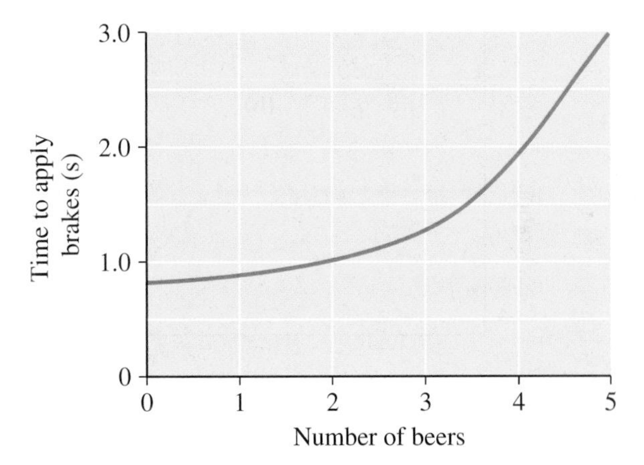

Figure 2-67 Reaction times for beer-drinking drivers (Question 2-91).

2-92 In its final trip upstream to its spawning territory, a salmon must jump to the top of a waterfall that is 1.9 m high (Figure 2-68). With what minimum initial vertical velocity must the salmon jump to get to the top of the waterfall?

Figure 2-68 Spawning fish in a mountain river (Question 2-92).

2-93 Some scientists speculate that in the distant future one of the more powerful propulsion systems for space travel will be an matter–antimatter system. Assuming this system could provide an acceleration with a magnitude of 30 g, how many hours would it take a rocket, starting from rest, to reach a maximum speed that is 10% of the speed of light? (We are neglecting the effects of special relativity, which would begin to play a role at such a high speed. The speed of light is 3.0×10^8 m/s.)

2-94 A boy throws a stone downward from a bridge with an initial speed v_0, and a girl throws a stone upward from the same bridge with the same initial speed, v_0. Compare the velocities of their stones upon reaching the water below.

2-95 A steel ball, S, and a table-tennis ball, T, are dropped from the top of a building. Assume that S experiences free fall, and T experiences air resistance and eventually reaches terminal speed.

(a) On a single set of axes, draw a velocity–time graph comparing the motions of the two balls. Assume downward is positive.

(b) Repeat, assuming upward is positive.

2-96 A boy standing on a third floor balcony of an apartment building sees a ball rising at a speed v_1 past the top of the railing. A short while later he sees the same ball falling at a speed v_2, past the same position. Choose which of the following statements is true, and explain your answer. (A) $v_1 < v_2$; (B) $v_1 > v_2$; (C) $v_1 = v_2$; (D) the speeds cannot be compared with the information given.

2-97 A camera is set up to take photographs of a ball undergoing vertical motion. The camera is 5.2 m above the ball launcher, which can launch the ball with an initial velocity of 17 m/s upward. Assuming that the ball goes straight up and straight down past the camera, at what times will the ball pass the camera?

Applying Your Knowledge

2-98 **Fermi Question:** A patient with a detached retina is warned by an eye specialist that if she slows down with an acceleration greater than 2 g in magnitude, the retina could pull away entirely from the sclera. Help the patient decide whether or not a vigorous game of tennis would be acceptable. Estimate values of running speeds and stopping distances or stopping times to determine your answer.

2-99 On a highway, two posts are set up 1.0 km apart to help motorists judge the accuracy of their speedometers. An information sign tells the motorists to drive at exactly 90 km/h between the posts and to measure the time to travel the 1.0 km. For a motorist who follows this procedure and requires a time of 38 s, determine the actual speed of the car.

2-100 A helicopter is travelling horizontally, straight toward a cliff. When 700 m from the cliff, it sends a sonar signal and 3.40 s later receives the reflected signal. Assuming the signal travels at 350 m/s, determine the speed of the helicopter.

2-101 Two cars, A and B, are stopped for a red light beside each other at an intersection. The light turns green and the cars accelerate. Their velocity–time graphs are shown in Figure 2-69. The positive direction is "forward."

(a) At what time(s) do A and B have the same velocity?

(b) When does B overtake A? (**Hint:** Their displacements must be equal at that time, and displacement can be found from a velocity–time graph.)

(c) How far have the cars travelled when B overtakes A?

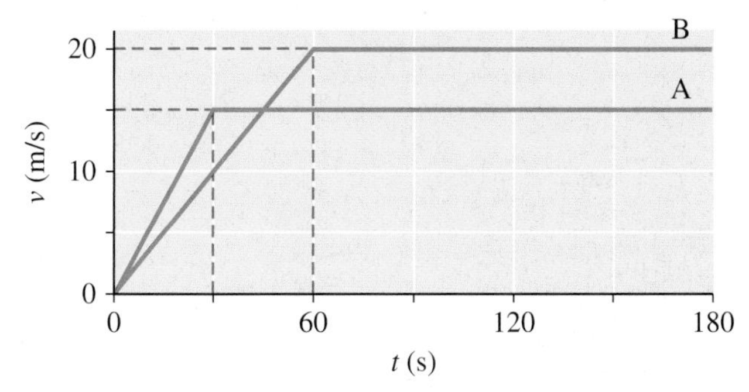

Figure 2-69 Question 2-101.

2-102 A fish swimming at a constant speed of 18 m/s spots a stationary barracuda. Just as the fish passes the barracuda, the predator begins swimming with a constant acceleration of magnitude 2.2 m/s^2. The fish and barracuda are moving in the same direction.

(a) How far does the barracuda swim before catching the fish?

(b) At what time will this occur? (**Hint:** Graphing may help you visualize this problem.)

2-103 The current world record for the men's pole vault is over 6.1 m (Figure 2-70).

(a) How long does the vaulter take to fall the first 50 cm on the way down?

(b) Repeat (a) for the last 50 cm, say from the 5.3 m mark to the 5.8 m mark on the way down before landing on the safety mat.

(c) Explain why the vaulter appears to be in "slow motion" near the top of the jump.

Figure 2-70 Pole vaulting (Question 2-103).

2-104 Scientists use specially designed helmets with embedded electronic sensors to study the effects of impacts on hockey players' heads during a game. The sensors transmit data to a computer, which records the data. In one test of 14-year-old players, the impacts ranged from an acceleration of magnitude 77 m/s^2 (an impact on the chin) to 2.2×10^2 m/s^2 (an impact to the upper head).

(a) Express the range of accelerations in terms of g.

(b) If the impact to the upper head lasts 12 ms, determine the stopping distance during the collision. What assumption is required to perform this calculation?

(c) Some major league football players experience head impacts resulting in 100 gs of acceleration, yet the concussions of the teen players may be more damaging. Give a reason for this apparent discrepancy.

2-105 A runner starts from rest and runs 100 m in 10.0 s. He runs with a constant acceleration for the first 4.00 s, and then with a constant speed for the remaining 6.00 s. What was the magnitude of his constant acceleration?

2-106 (a) Assume that somehow a tennis ball accelerates freely at 9.8 m/s^2 downward until it reaches its terminal speed of 3.0×10^1 m/s. Determine the height from which the ball must be dropped from rest so that it travels at its terminal speed for the last 1.5 s of its flight.

(b) What is wrong with the assumption in (a)? To illustrate your answer, sketch a v-t graph of the ball's motion for the situation in (a), and another graph of the true motion.

Vectors and Trigonometry 3

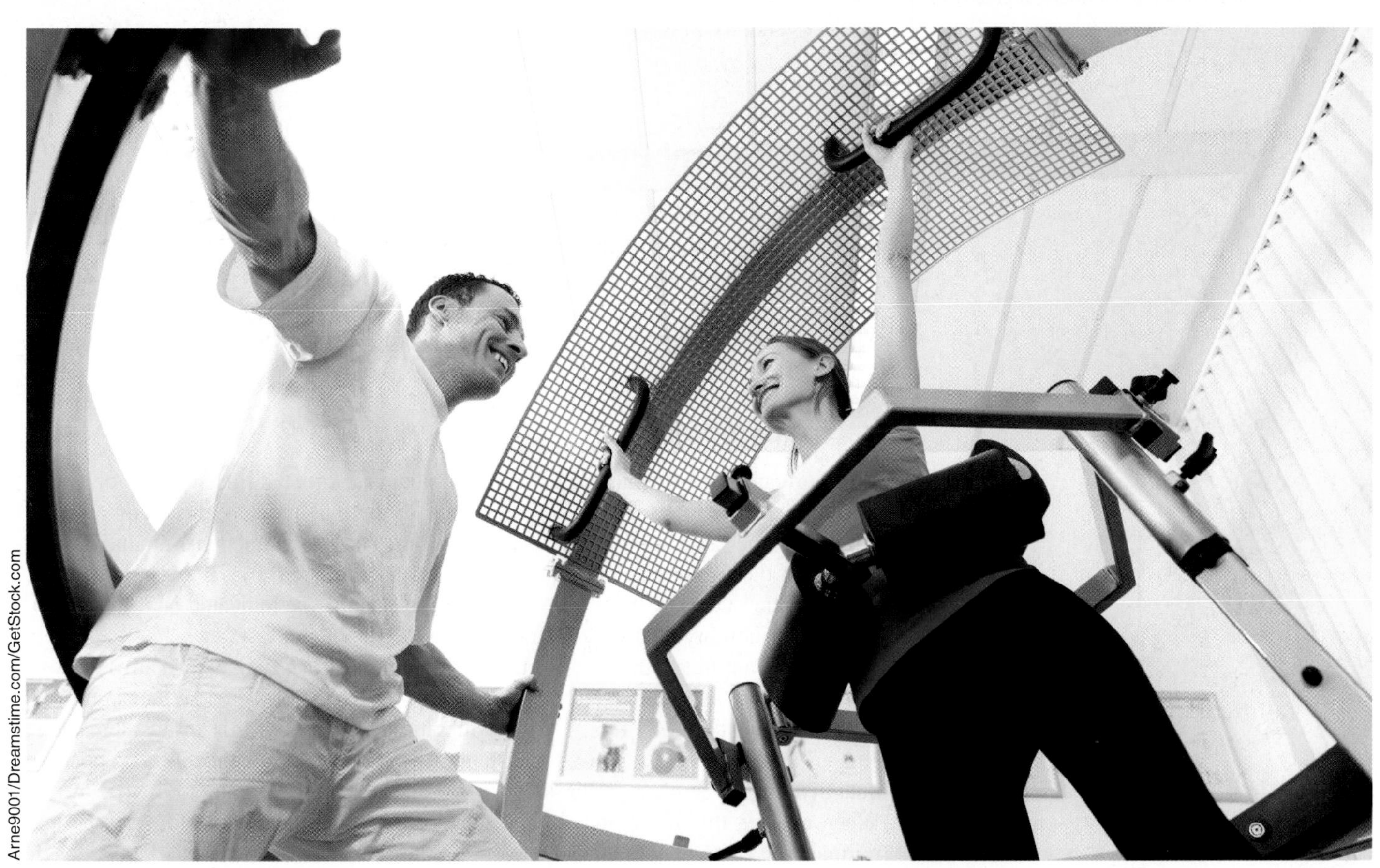
Arne9001/Dreamstime.com/GetStock.com

Figure 3-1 The world's first three-dimensional therapy system, called a "Space Curl," provides relief to patients with back problems, and helps improve coordination of sports enthusiasts and astronauts.

Many devices used for physiotherapy, rehabilitation, and exercise are designed by applying principles of forces and angles. For instance, the device shown in Figure 3-1 is an exercise and physiotherapy system invented by NASA that allows the user to stretch and move in three dimensions. In order to fit the body sizes, strengths, and needs of possible patients, the machine must be designed with adjustable tensions and angles. This chapter presents the mathematical skills required to understand how to calculate angles as well as the magnitudes and directions of forces and other physical quantities.

A force is an example of a vector quantity, or simply a vector. Recall from Section 1.5 that a vector has both magnitude and direction. This differs from a scalar quantity or scalar, which has only magnitude. Many more details about vectors combined with the mathematics of triangles, called trigonometry, are required in order to apply what you

CHAPTER OUTLINE

Learning Tip

Various protractor apps are available for phones and other personal devices. Protractor apps can be used to measure angles from the horizontal or vertical; some (including "Handy Tools") have other sets of tools, such as a surface level, a caliper, a plumb bob, and a ruler; and many of these apps are free to download.

learned about distance, speed, and acceleration in one dimension in Chapter 2, to motion in two dimensions in Chapter 4. If you have studied vectors and trigonometry in a mathematics course, some of the material in this chapter may be a review, although the notation might be different from what you have seen before. If the mathematical details of vectors and trigonometry are new to you, this chapter is an important introduction to Chapter 4 as well as many other parts of the text.

3.1 General Properties of Vectors

In this section, we look at some general properties of vectors. Scale diagrams with distance measurements, angle measurements, and directions are important here, so you are advised to have a metric ruler and a protractor handy as you read this section. A common example of a scale diagram is a map, which includes directions, normally with north toward the top of the page, east toward the right, etc.

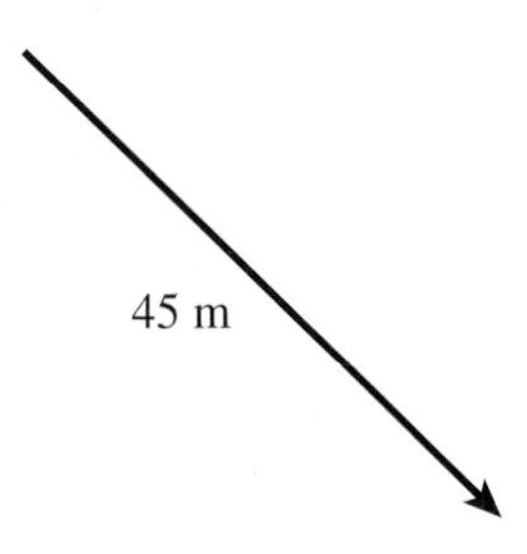

(a) Scale: 1.0 cm = 10 m

(b) Scale: 1.0 cm = 5.0 m/s²

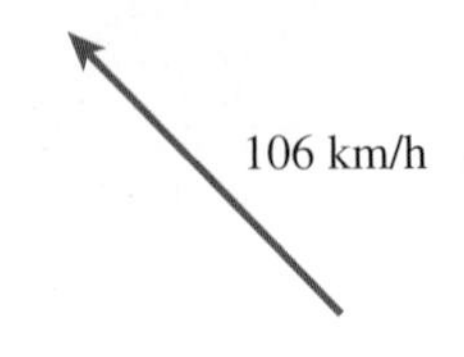

(c) Scale: 1.0 cm = 40 km/h

Figure 3-2 Examples of vector quantities.

Vector Symbols

A vector can be represented in a diagram by using a directed line segment or arrow. The length of the line segment is proportional to the magnitude of the vector, and the direction is the same as the direction of the vector. The line segment has an initial point (the tail of the arrow) and a terminal point (the head of the arrow). If a vector is drawn using a scale, such as 1.0 cm = 100 km, the scale should be indicated on the diagram. Figure 3-2 shows typical vectors drawn to scale.

Vector quantities in this book will be written in **boldface type** with an arrow above the symbol, for example, $\vec{\boldsymbol{A}}$. The magnitude of a vector is indicated either in normal (not boldface) type, such as A for the magnitude of $\vec{\boldsymbol{A}}$, or with an absolute value sign around the vector symbol, for instance $|\vec{\boldsymbol{A}}|$. This magnitude is always a positive quantity (unless zero), regardless of the direction of the vector; thus, $A = |\vec{\boldsymbol{A}}| \geq 0$. When reading this book, it is important to distinguish between boldface type and normal type since they indicate different quantities.

Directions of Vectors

Various ways can be used to denote vector directions. In this book directions are usually written after the magnitude and units of the measurement. For example, a velocity might be given as 100 km/h west. Typical examples of directions are listed below:

east	30° east of north
upward	22.5° south of west
forward	5.2° above the horizon

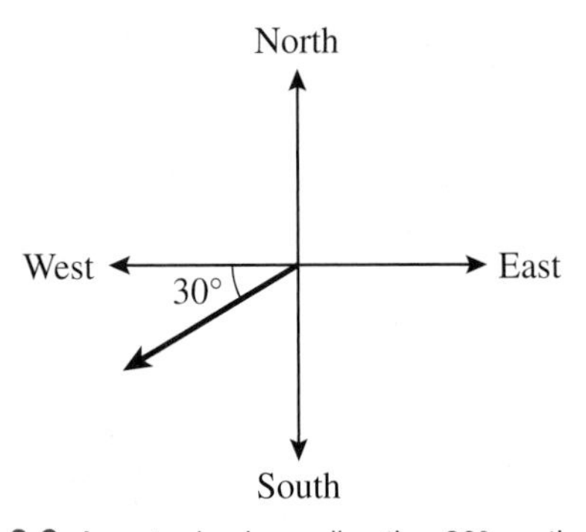

Figure 3-3 A vector having a direction 30° south of west.

Figure 3-3 shows a vector having a direction 30° south of west, which is the same as 60° west of south. (This direction is sometimes written [W 30° S] or [S 60° W].)

Multiplying a Vector by a Scalar

A vector can be multiplied by a positive scalar, producing a vector with the same direction as the original vector but a different magnitude (unless the scalar happens to be "1," in which case the multiplication produces no change). Thus, $3\vec{A}$ is a vector 3 times as long as $\vec{A}$ and in the same direction. Multiplication of a vector by a negative scalar quantity produces a vector in the opposite direction. For example, the vector $-2\vec{A}$ is twice as long as $\vec{A}$ and points in the opposite direction; if $\vec{A}$ points east, then $-2\vec{A}$ points west. Multiplication of a vector by a scalar is illustrated in Figure 3-4.

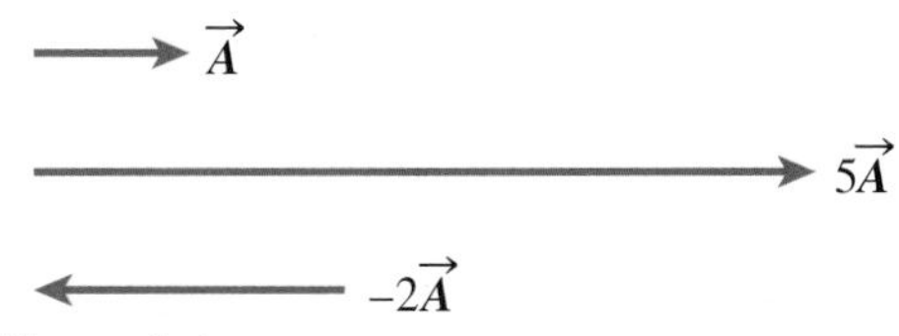

Figure 3-4 Multiplying a vector by a scalar.

EXERCISES

3-1 A velocity vector $\vec{v}$ is 22 m/s west. What are (a) $|\vec{v}|$? (b) v?

3-2 Use a ruler and a protractor to determine the magnitude (with units) and direction of each vector shown in Figure 3-5.

3-3 Draw a vector diagram to indicate each of the following vectors. In each case indicate the scale used to draw the diagram.

(a) $\vec{A} = 5.0 \times 10^3$ km 50° north of west

(b) $\vec{B} = 0.040$ m/s 10° south of east

(c) $\vec{C} = 38$ N at 25° above the horizontal (where N is the symbol for newton, the SI unit of force)

(d) $5\vec{C}$, where $\vec{C}$ is the vector given in (c) above

(e) $-3\vec{C}$, where $\vec{C}$ is the vector given in (c) above

3-4 An acceleration vector $\vec{a}$ is 1.5 m/s² downward. What is $-2\vec{a}$?

3-5 **Fermi Question:** Imagine two of your fingers walking straight forward with their largest pace, creating a vector $\vec{W}$. Estimate in kilometres the resultant (magnitude and direction) of $10^7\,\vec{W}$.

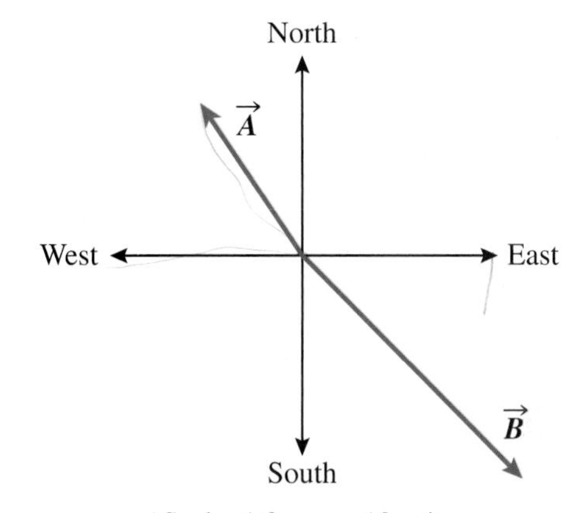

Figure 3-5 Vectors for Question 3-2.

3.2 Adding and Subtracting Vectors Using Scale Diagrams

If a person walks 6.0 m west, then turns and walks 8.0 m south, the total *distance* travelled is 14.0 m (a scalar quantity). However, the *displacement,* which is a vector that points in a straight line from the initial position directly to the final position, has a magnitude less than 14.0 m and a specific direction. One way to find this displacement is to add the vectors using a scale diagram. If we let $\vec{A} = 6.0$ m west and $\vec{B} = 8.0$ m south, these vectors can be added as shown in Figure 3-6. The result of this addition is the total displacement, which is called the resultant displacement. If we use $\vec{R}$ as the symbol for the resultant displacement, then $\vec{R} = \vec{A} + \vec{B}$, where the "+" sign here means a *vector* addition. The result of any vector addition is often called a **resultant**, but it can also be called a resultant vector, net vector, or vector sum. The word resultant will be used in this text, including specific cases such as resultant displacement, resultant force, and so on. (Resultant will also be applied to vector subtraction later in this section.)

resultant the result of a vector addition (or subtraction) (also called a resultant vector)

In Figure 3-6, the resultant displacement, $\vec{R} = \vec{A} + \vec{B}$, can be found by using the scale to determine the magnitude, and a protractor for the angle. In this example, $\vec{R} = 10.0$ m 53° south of west. (Because the vectors $\vec{A}$ and $\vec{B}$ are perpendicular to each other, the magnitude in this particular case could also be found using the Pythagorean relation $R = \sqrt{A^2 + B^2}$.) Using a scale diagram to determine the sum of vectors clearly has limited accuracy. In Section 3.4 we will introduce a more accurate way to add (and subtract) vectors using quantities called components of vectors.

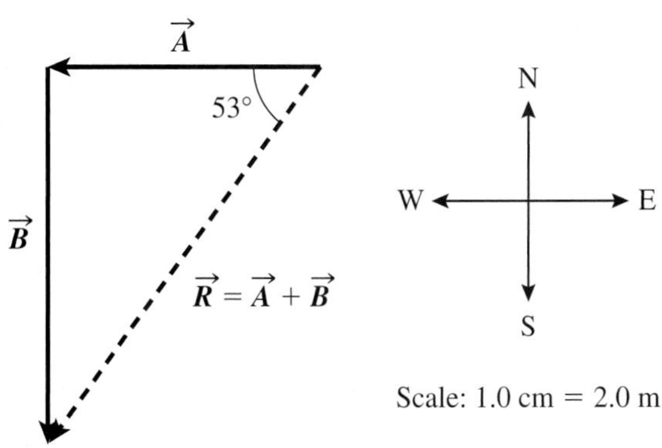

Figure 3-6 Adding vectors using a scale diagram.

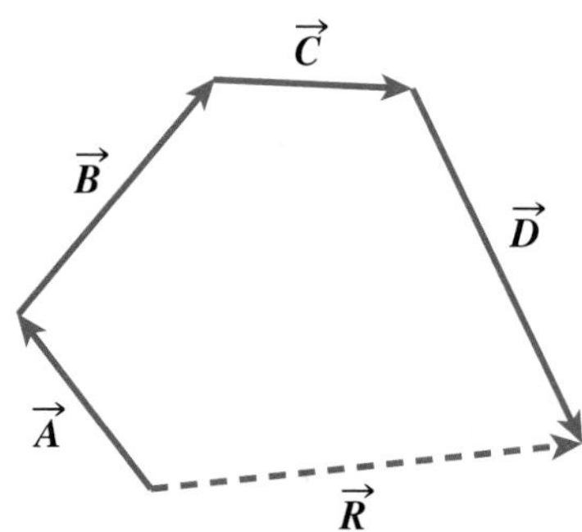

Figure 3-7 Placing four vectors head-to-tail to obtain the resultant $\vec{R}$, where $\vec{R} = \vec{A} + \vec{B} + \vec{C} + \vec{D}$.

Notice in Figure 3-6 that the two vectors, $\vec{A}$ and $\vec{B}$, are placed so that the tail of $\vec{B}$ touches the head of $\vec{A}$, and the resultant vector $\vec{R}$ is drawn from the start (tail) of $\vec{A}$ to the finish (head) of $\vec{B}$. The process of adding vectors by placing them tail to head to find the resultant vector can be extended to any number of vectors having the same units (Figure 3-7). Notice that the resultant vector $\vec{R}$ points from the *tail* of the *first* vector, $\vec{A}$, to the *head* of the *last* vector, $\vec{D}$.

In the previous two examples, the vectors were already placed conveniently tail to head, so the resultant was simply drawn from the tail of the first vector to the head of the last one. If, however, the vectors are not already placed tail to head in the diagram, they can be moved around so they become tail to head. This process is shown in Figure 3-8. When moving a vector on a diagram it is important that the redrawn vector has the same magnitude and direction as the initial vector, otherwise it is not the same vector.

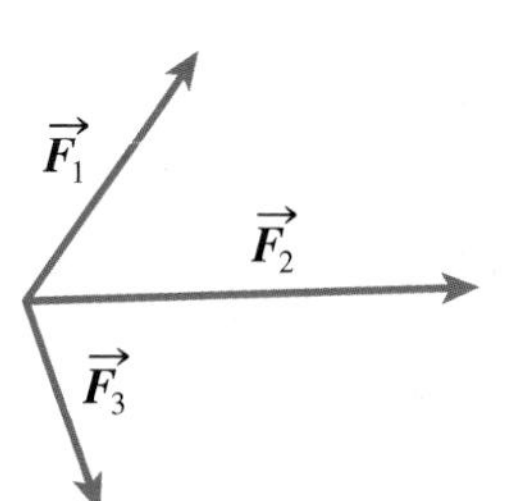

(a) Three forces act at a single point.

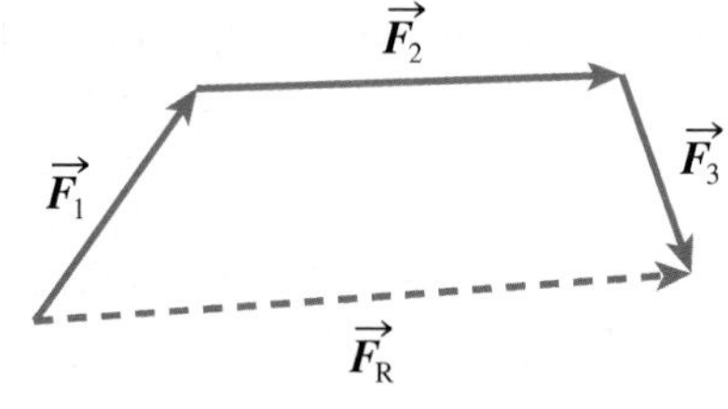

(b) $\vec{F}_2$ and $\vec{F}_3$ are moved so that the net force $\vec{F}_R$ can be found.

Figure 3-8 Adding vectors by moving them parallel to themselves.

TACKLING MISCONCEPTIONS

Are Vectors Like Scalars?

It's a misconception to think that if vector $\vec{A} = \vec{B} + \vec{C}$ then the magnitude of $\vec{A}$ is equal to the sum of the magnitudes of $\vec{B}$ and $\vec{C}$. That is true only if $\vec{B}$ and $\vec{C}$ are in the same direction. In all other cases, $A \neq B + C$. Remember to always treat vectors differently from scalars.

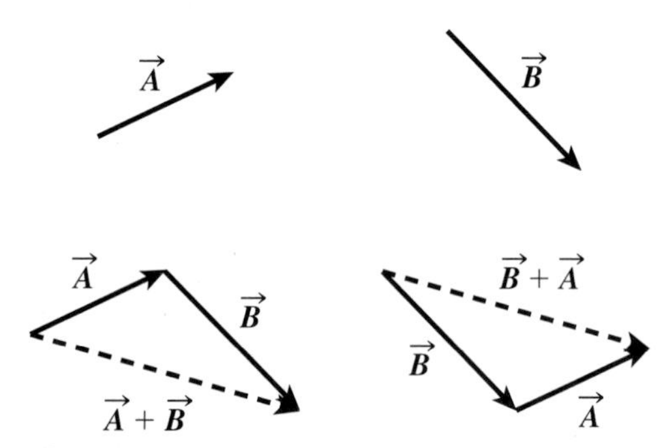

Figure 3-9 Illustrating the commutative law for vector addition: $\vec{A} + \vec{B} = \vec{B} + \vec{A}$.

Properties of Vector Addition

In mathematics courses you have learned of the commutative and associative laws. These laws can be applied to vector addition.

- Vector addition is commutative; in other words, the order of addition does not matter:

$$\vec{A} + \vec{B} = \vec{B} + \vec{A}$$

- Vector addition is associative; that is, if more than two vectors are added it does not matter how we group them:

$$(\vec{A} + \vec{B}) + \vec{C} = \vec{A} + (\vec{B} + \vec{C})$$

We will not present formal proofs of these laws in this text. However, Figure 3-9 illustrates the commutative law for vector addition. Notice in Figure 3-9 that $\vec{A} + \vec{B}$ has the same magnitude and direction as $\vec{B} + \vec{A}$.

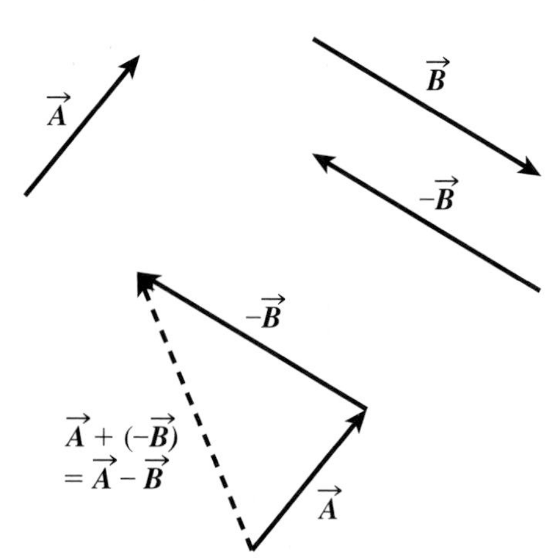

Figure 3-10 Subtracting vectors using a diagram.

Vector Subtraction

Vector subtraction occurs often in physics. We define the subtraction $\vec{A} - \vec{B}$ to be the sum of $\vec{A}$ and $-\vec{B}$, where $-\vec{B}$ has the same magnitude as $\vec{B}$ but is opposite in direction. Thus,

$$\vec{A} - \vec{B} = \vec{A} + (-\vec{B})$$

In other words, to subtract $\vec{B}$ from $\vec{A}$, add the opposite of $\vec{B}$ to $\vec{A}$. This is shown in Figure 3-10. It is left as an exercise (Question 3-7) for you to prove, using a diagram, that $\vec{A} - \vec{B}$ does not equal $\vec{B} - \vec{A}$.

EXERCISES

3-6 Add each set of vectors by using a scale diagram.

(a) $\vec{C}$ = 28 km 20° north of east; $\vec{D}$ = 34 km 30° west of north

(b) $\vec{E}$ = 300 N east; $\vec{F}$ = 250 N 55° south of east; $\vec{G}$ = 600 N 15° south of west

3-7 Assume $\vec{X}$ = 5.5 units 30° west of north, and $\vec{Y}$ = 4.5 units 45° west of south. Use a scale diagram to find

(a) $\vec{X} - \vec{Y}$

(b) $\vec{Y} - \vec{X}$

(c) How does $(\vec{X} - \vec{Y})$ compare to $(\vec{Y} - \vec{X})$? What conclusion can be made?

(d) Does your conclusion in (c) apply if $\vec{X}$ and $\vec{Y}$ are equal?

3-8 Find $\vec{A} + \vec{B} - \vec{C}$ diagrammatically, given $\vec{A}$ = 80 m 50° west of north, $\vec{B}$ = 60 m south, and $\vec{C}$ = 90 m 30° south of west.

3.3 Trigonometry

trigonometry the branch of mathematics that deals with the relationships between the angles and sides of triangles and the calculations based on them

Consider the situation in which town planners want to provide wheelchair access to a community swimming pool (Figure 3-11). If you draw a side-view diagram of the ramp, you will notice that the diagram consists of a triangle with sides and angles that can be measured and calculated. Analyzing such a triangle is an example of the application of **trigonometry**, which is the branch of mathematics that deals with the relationships between the angles and sides of triangles and the calculations based on them. It is important to develop skill in using your calculator to solve physics problems involving trigonometry.

Figure 3-11 For someone in a wheelchair who wants to swim for sport, exercise, or therapy, a ramp helps to make the pool accessible.

Sine, Cosine, and Tangent

The trigonometric functions, such as sine (sin), cosine (cos), and tangent (tan), are defined in terms of the right-angled triangle shown in Figure 3-12. In that triangle, the sides are labelled relative to the angle theta (θ): o for opposite side, a for adjacent side, and h for hypotenuse. Then the defining equations of $\sin\theta$, $\cos\theta$, and $\tan\theta$ are

$$\sin\theta = \frac{\text{opposite side}}{\text{hypotenuse}} = \frac{o}{h} \quad (3\text{-}1)$$

$$\cos\theta = \frac{\text{adjacent side}}{\text{hypotenuse}} = \frac{a}{h} \quad (3\text{-}2)$$

$$\tan\theta = \frac{\text{opposite side}}{\text{adjacent side}} = \frac{o}{a} \quad (3\text{-}3)$$

DID YOU KNOW?

Creating accessibility for wheelchairs is just one example of "universal design." In this type of design, the components of a structure or device are arranged in the most user-friendly way possible. The principle can be applied in homes, playgrounds, schools, office buildings, transportation vehicles, and so on. It takes into consideration the needs of people with disabilities, as well as children (especially for safety reasons) and senior citizens. More than 14% of Canadians and almost 50% of people over 65 years of age have some degree of disability.

Based on these defining equations, you should be able to prove that $\tan\theta = \sin\theta / \cos\theta$.

If θ is a very small angle, $\sin\theta \approx \tan\theta$, as you can verify in Figure 3-13. (You can also verify this on your calculator by comparing, for example, sin 0.5° with tan 0.5°.) Scientific calculators can express angles in two or three different units. When cleared or first turned on, the calculator usually indicates angles in degrees (DEG). Pushing the appropriate key (e.g., DGR) will change the units to radians (RAD) or grads (GRA, where 90° = 100 grads). Unless otherwise noted, only degrees will be used in this book; make sure your calculator is in the appropriate mode when doing calculations involving angles.

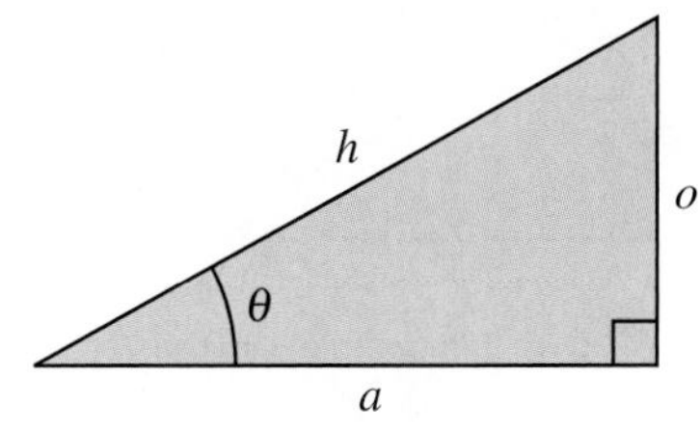

Figure 3-12 The right-angled triangle used to define $\sin\theta$, $\cos\theta$, and $\tan\theta$.

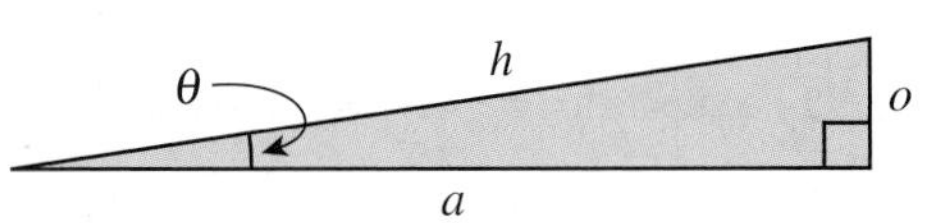

Figure 3-13 For very small values of θ, $\sin\theta$ is approximately equal to $\tan\theta$.

SAMPLE PROBLEM 3-1

In Figure 3-13, if o is 3.50 cm and a is 4.20 cm, apply trigonometry to determine θ and h.

Solution

We begin by rearranging Eqn. 3-3, $\tan\theta = \frac{o}{a}$, to find θ.

$$\theta = \tan^{-1}\left(\frac{o}{a}\right) = \tan^{-1}\left(\frac{3.50\,\text{cm}}{4.20\,\text{cm}}\right) = 39.8°$$

Next we rearrange Eqn. 3-1, $\sin\theta = \frac{o}{h}$, to find h, although Eqn. 3-2 works equally well:

$$h = \frac{o}{\sin\theta} = \frac{3.50\,\text{cm}}{\sin 39.8°} = 5.47\,\text{cm}$$

Thus, the length of h is 5.47 cm. (In this example, the solution can also be found by using the law of Pythagoras, $h^2 = o^2 + a^2$, which applies to right-angled triangles.)

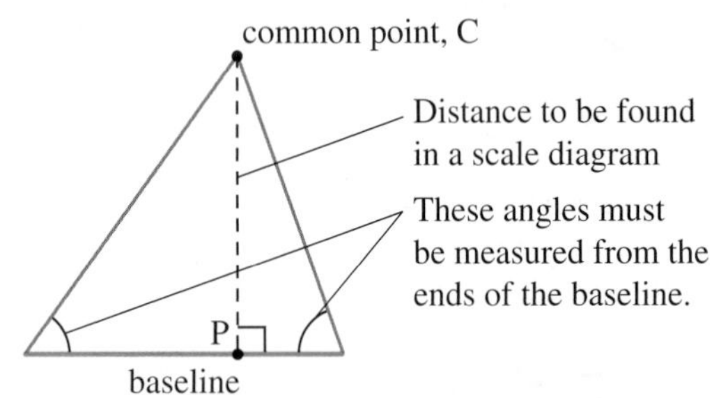

Figure 3-14 In triangulation, a baseline and two angles to a common point must be measured or estimated and then a scale diagram used to determine an unknown distance indirectly. Notice that the broken line representing the shortest distance from C to P is perpendicular to the baseline.

TRY THIS!

Using Triangulation to Estimate a Horizontal Distance

"Triangulation" is a method of measuring a distance indirectly by drawing a scale diagram of a triangle with one known side, called the baseline, and two angles from the ends of the baseline to a common point, C, as shown in Figure 3-14. You will need a protractor (either a plastic one or an app on a personal device) and a ruler to follow the steps below to practise triangulation.

- Choose at least one unknown distance to find indirectly. An outdoor example is the shortest distance from a position (P) on a sidewalk to a distant tree. An indoor example is the distance from where you are located (P) to a light switch.
- Choose as large a baseline as possible. Measure and record the baseline using your natural pace or shoe length for a longer distance or your hand span for a shorter distance.
- Measure and record the angle from each end of the baseline to the common point, C. Be sure to line up the protractor as exactly as possible along the baseline.
- Using an appropriate scale, draw a scale diagram to determine the perpendicular "unknown" distance from the common point to position P on the baseline.
- List ways in which you think you could increase the accuracy of your answer.

Sine Law

For any triangle, such as the one in Figure 3-15, the sine law applies:

$$\frac{\sin A}{a} = \frac{\sin B}{b} = \frac{\sin C}{c} \qquad \textbf{(3-4)}$$

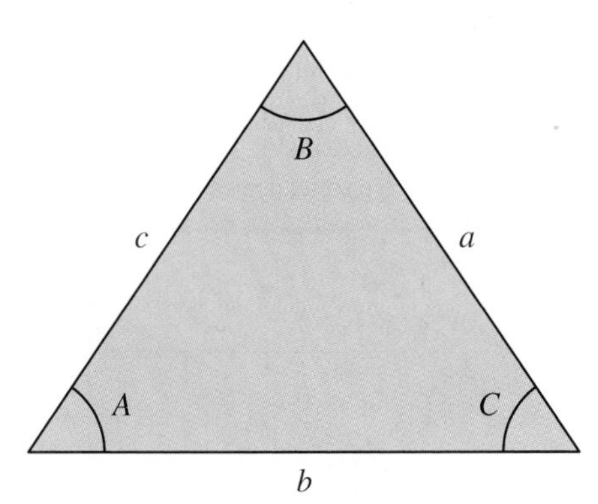

Figure 3-15 In this triangle, side a is opposite angle A, etc.

SAMPLE PROBLEM 3-2

In Figure 3-15, if $A = 31.5°$, $C = 40.2°$, and $c = 13.2$ cm, find the length of a.

Solution

Rearranging Eqn. 3-4 $\frac{\sin A}{a} = \frac{\sin C}{c}$ to determine a, we have

$$a = \frac{c \sin A}{\sin C}$$

$$= \frac{(13.2\ \text{cm})(\sin 31.5°)}{\sin 40.2°}$$

$$= 10.7\ \text{cm}$$

Thus, the length of a is 10.7 cm.

Cosine Law

For any triangle, such as the one in Figure 3-15, the cosine law applies:

$$a^2 = b^2 + c^2 - 2bc \cos A \qquad (3\text{-}5)$$

Depending on what is given and what is required to find, other useful forms of the cosine law are

$$b^2 = a^2 + c^2 - 2ac \cos B \qquad (3\text{-}6)$$

$$c^2 = a^2 + b^2 - 2ab \cos C \qquad (3\text{-}7)$$

SAMPLE PROBLEM 3-3

In Figure 3-15, if $a = 4.5$ cm, $b = 5.5$ cm, and $c = 6.5$ cm, find angle A.

Solution

From Eqn. 3-5, $a^2 = b^2 + c^2 - 2bc \cos A$,

$$2bc \cos A = b^2 + c^2 - a^2$$

$$\cos A = \frac{b^2 + c^2 - a^2}{2bc}$$

$$A = \cos^{-1}\left(\frac{b^2 + c^2 - a^2}{2bc}\right)$$

$$= \cos^{-1}\left(\frac{(5.5\ \text{cm})^2 + (6.5\ \text{cm})^2 - (4.5\ \text{cm})^2}{2(5.5\ \text{cm})(6.5\ \text{cm})}\right)$$

$$= 43°$$

Thus, angle A is 43°.

TRY THIS!

Estimating Vertical Height Using Triangulation

Design and carry out an investigation to determine the height of a tall tree or a building (such as a campus building). For clues, refer to Figure 3-16.

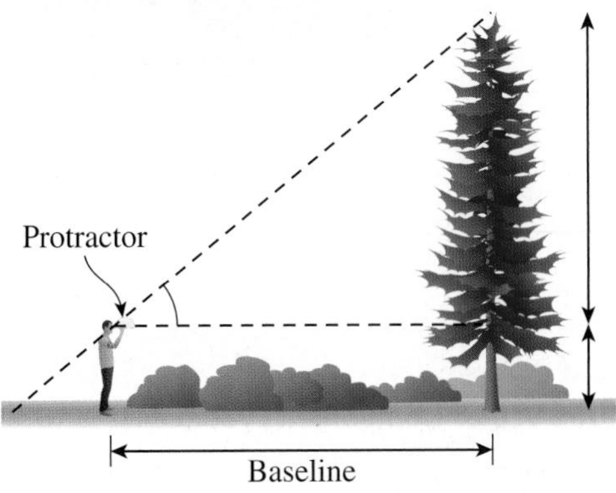

Figure 3-16 Estimating vertical heights.

Trigonometric Identities

The definitions of sine and cosine can be applied to any angles, whether they are in triangles or not. The following relationships or trigonometric identities are useful in some circumstances in this text:

$$\sin 2\theta = 2 \sin \theta \cos \theta \quad (3\text{-}8)$$

$$\sin^2 \theta + \cos^2 \theta = 1 \quad (3\text{-}9)$$

$$\cos(360° - \theta) = \cos \theta \quad (3\text{-}10)$$

EXERCISES

3-9 Refer to Figure 3-12 to determine the unknown quantities:

(a) If $\theta = 36°$ and $a = 1.4$ m, find h.

(b) If $\theta = 28.5°$ and $h = 65.8$ mm, find a.

(c) If $h = 0.85$ km and $a = 0.39$ km, apply trigonometry to find o.

(d) If $h = 55.8$ cm and $o = 17.1$ cm, find a. (Apply trigonometry, and then check your answer by applying the law of Pythagoras.)

3-10 A hiker uses triangulation to determine the approximate horizontal distance across a canyon (Figure 3-17). The baseline parallel to one side of the canyon is 130 paces long and the angles from the ends of the baseline to a common point are, respectively, 66° and 77°.

(a) Use a scale diagram to determine the width of the canyon in paces.

(b) Assuming the hiker's average pace is 75 cm, determine the canyon's width in metres.

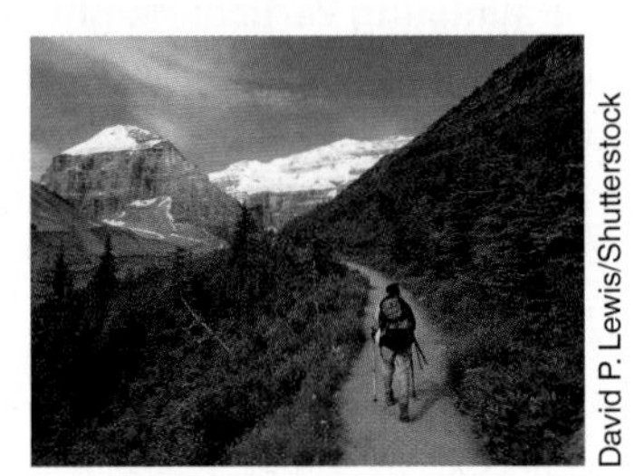

David P. Lewis/Shutterstock

Figure 3-17 Question 3-10.

3-11 Refer to Figure 3-15 to determine the unknown quantities:

(a) If $B = 41°$, $C = 17°$, and $b = 12$ cm, find c.

(b) If $A = 106°$, $a = 28.5$ mm, and $c = 10.2$ mm, find angle C.

(c) If $a = 8.2$ cm, $b = 7.1$ cm, and $c = 6.0$ cm, find angle B.

(d) If $C = 33.3°$, $a = 18.5$ m, and $b = 26.5$ m, find c.

3-12 A person's arm is injured and is placed in a sling, like that shown in Figure 3-18. Assume that the angle between the upper arm and the forearm is 62° and the distance from the tip of the shoulder (S) to the tip of the elbow (E) is SE = 37 cm. The distance from the tip of the shoulder to the tip of the end of the open hand (H) is SH = 44 cm. Determine the length (EH) of the forearm.

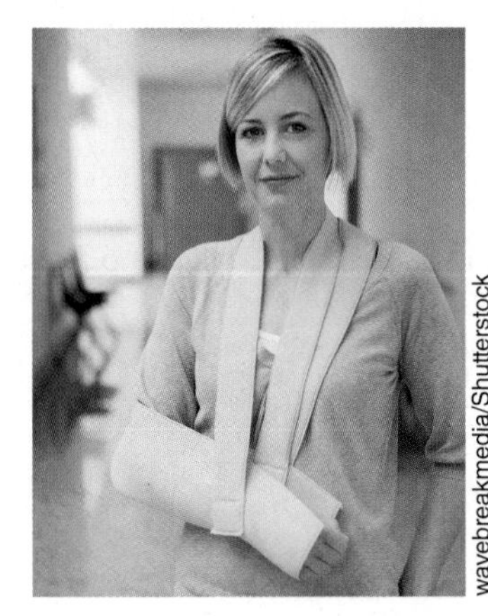

wavebreakmedia/Shutterstock

Figure 3-18 Question 3-12.

TACKLING MISCONCEPTIONS

Are Vector Components Vectors?

Because components of vectors are not themselves vectors, you should not place an arrow over the hand-written letters (such as A_x and A_y) representing the components.

3.4 Adding and Subtracting Vectors Using Components

Adding and subtracting vectors using diagrams, as presented in Section 3.2, is useful and instructive because the process helps to visualize the physical situation. However, the amount of uncertainty that the technique produces in the resultant vector may be rather large. To reduce the uncertainty, vectors can be added by a mathematical technique using the components of vectors. This technique is especially useful when adding three or more vectors.

The **component of a vector** is a projection of a vector along an axis, usually of a rectangular coordinate system. Any vector can be completely described by its rectangular components. Rectangular components, by definition, are perpendicular to each other. They are also known as orthogonal components. "Orthogonal" stems from the Greek *orthos* which means "right" and *gonia* which means "angle." In this book we will use two rectangular components because only two-dimensional situations will be considered. (In three dimensions, three components are needed.)

component of a vector a projection of a vector along an axis, usually of a rectangular coordinate system

Consider vector $\vec{A}$ in a rectangular coordinate system with a horizontal x-axis and a vertical y-axis, as seen in Figure 3-19. The projection of $\vec{A}$ along the x-axis is the x-component and has the symbol A_x. The projection of $\vec{A}$ along the y-axis is the y-component and has the symbol A_y. In this textbook, we will show components with lines that differ in colour from the vector colour. Notice that $\vec{A}$ is a vector, but the components A_x and A_y are not vectors; rather they are positive or negative numbers, having the same units as $\vec{A}$.

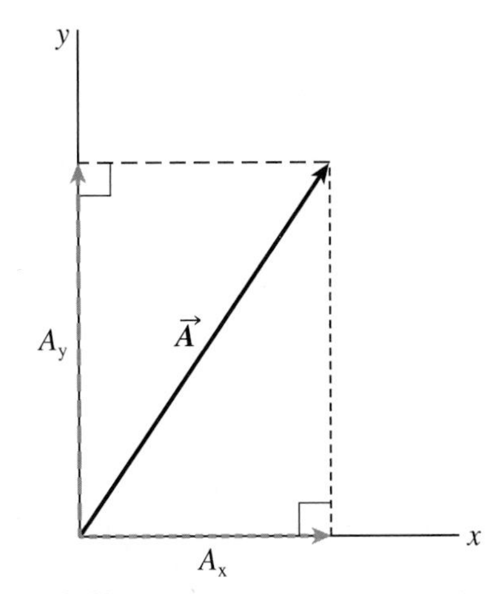

Figure 3-19 The components of a vector for horizontal–vertical axes.

Although many vector situations involve horizontal and vertical axes, sometimes the axes are tilted at an angle. In that circumstance, the components of the vector are still drawn perpendicular to each other. A typical example of a situation in which you might choose a set of axes at an angle to the horizontal and vertical is shown in Figure 3-20. In this case, $\vec{F}$ could represent the downward force of gravity on a person on a ramp that is inclined at an angle θ to the horizontal, with F_x parallel to the ramp and F_y perpendicular to the ramp.

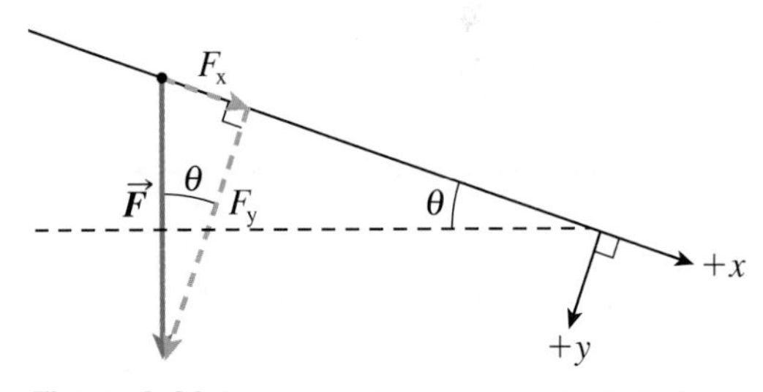

Figure 3-20 For a situation such as a person on a ramp, it is usually convenient to choose the x-y axes parallel and perpendicular to the ramp.

To calculate the components of a vector, the trigonometric ratios of sine, cosine, and tangent can be applied. (Components can also be found by using a scale diagram, but we are aiming for higher accuracy here.)

SAMPLE PROBLEM 3-4

Determine the x- and y-components of the vectors shown in Figure 3-21.

Solution

Notice that $+x$ has been chosen to the right and $+y$ has been chosen upward. Hence, if the x-component of a vector is directed toward the right, it is positive, and if directed toward the left, it is negative. Similarly, an upward y-component is positive, and a downward y-component is negative.

$$A_x = (+10\,\text{cm})\cos 30° = 8.7\,\text{cm} \qquad A_y = (+10\,\text{cm})\sin 30° = 5.0\,\text{cm}$$

$$B_x = (+15\,\text{cm})\cos 70° = 5.1\,\text{cm} \qquad B_y = (-15\,\text{cm})\sin 70° = -14\,\text{cm}$$

Thus, the components are $A_x = 8.7\,\text{cm}$, $A_y = 5.0\,\text{cm}$, $B_x = 5.1\,\text{cm}$, and $B_y = -14\,\text{cm}$.

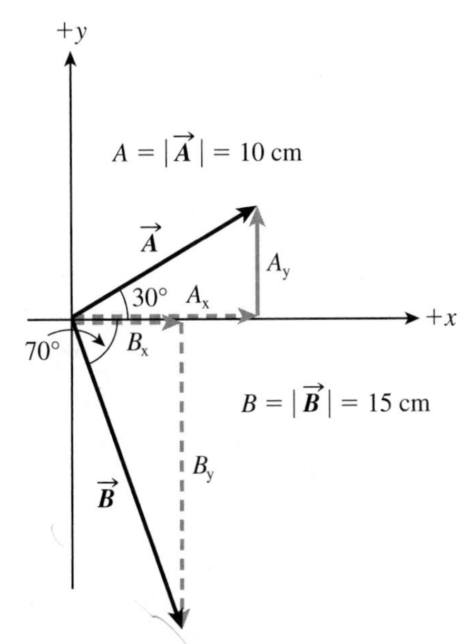

Figure 3-21 Sample Problem 3-4.

Adding Vectors

When two vectors are added to give a resultant vector, the x-components of the original vectors add to give the x-component of the resultant, and the y-components of the original vectors add to give the resultant y-component. In symbols, for the addition $\vec{R} = \vec{A} + \vec{B}$,

$$R_x = A_x + B_x$$
$$R_y = A_y + B_y$$

TACKLING MISCONCEPTIONS

Using Cosine and Sine for Vector Components

Some students think that the x-component of a vector is *always* found using cosine and the y-component is *always* found using sine. Although this was the situation in Sample Problem 3-4, it is not always true. For example, in Figure 3-21, if the components of vector $\vec{A}$ had been drawn above the vector such that the angle between $\vec{A}$ and the y-axis is 60°, then A_x would be $A \sin 60°$ and A_y would be $A \cos 60°$. See also Question 3-13 on page 69.

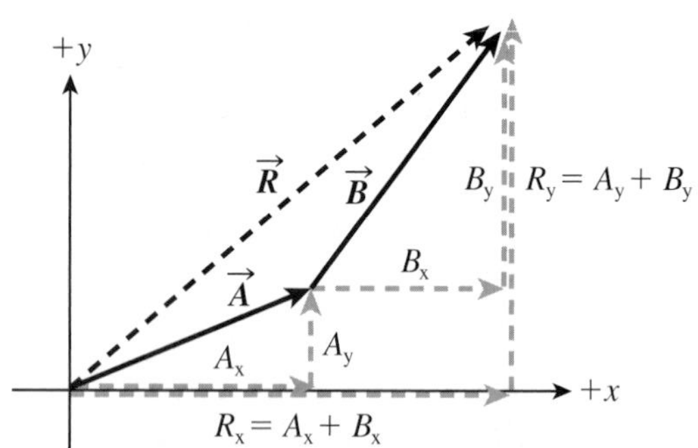

Figure 3-22 Adding components of vectors $\vec{A}$ and $\vec{B}$ to give the components of the resultant vector $\vec{R}$.

This addition of components is illustrated in Figure 3-22.

To add any number of vectors by components, the steps listed below are used. These steps are applied in Sample Problem 3-5 below.

- Define an x-y coordinate system, indicating which directions are positive. (Remember you are free to choose the most convenient directions for the x-y axes.)
- Determine the x- and y-components of all the vectors.
- Add all the x-components to find the resultant x-component.
- Add all the y-components to find the resultant y-component.
- Find the magnitude and direction of the resultant vector. (For this step, the Pythagorean relation, $R = \sqrt{R_x^2 + R_y^2}$, and/or trigonometric ratios can be used.)

SAMPLE PROBLEM 3-5

A person walks 50 m 30° north of east, then 60 m east, and finally 80 m 40° south of west. Use components to find the person's final displacement, $\vec{R}$. Assume two significant digits for each measurement.

Solution

1. In this situation, it is convenient to choose the x-y coordinate system to be the east-west, north-south directions. We will call the displacement vectors $\vec{A}$, $\vec{B}$, and $\vec{C}$ respectively, as shown in Figure 3-23 (a).
2. The x- and y-components of the vectors are shown in Figure 3-23 (b). They are

$$A_x = (50\,\text{m})\cos 30° = 43\,\text{m}$$

$$A_y = (50\,\text{m})\sin 30° = 25\,\text{m}$$

$$B_x = 60\,\text{m}$$

$$B_y = 0\,\text{m}$$

$$C_x = -(80\,\text{m})\cos 40° = -61\,\text{m}$$

$$C_y = -(80\,\text{m})\sin 40° = -51\,\text{m}$$

   Notice that since $\vec{B}$ points east, $B_y = 0$ m. Note also that the components of $\vec{C}$ are both negative, since $\vec{C}$ has components that are southward and westward.
3. The resultant x-component is

$$R_x = A_x + B_x + C_x$$

$$= (43 + 60 - 61)\,\text{m}$$

$$\therefore\ R_x = 42\,\text{m}$$

4. The resultant y-component is

$$R_y = A_y + B_y + C_y$$

$$= (25 + 0 - 51)\,\text{m}$$

$$R_y = -26\,\text{m}$$

5. The resultant components and the resultant displacement are shown in Figure 3-23 (c). R or $|\vec{R}|$ can be found using the Pythagorean relation:

$$R = \sqrt{(42\,\text{m})^2 + (26\,\text{m})^2}$$

$$= 49\,\text{m}$$

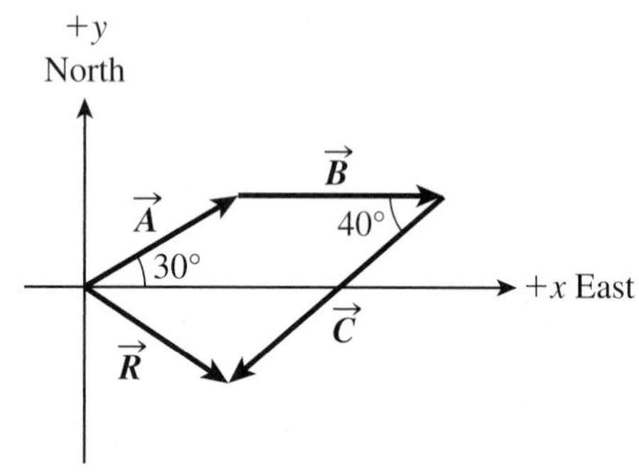

(a) Showing the various displacement vectors

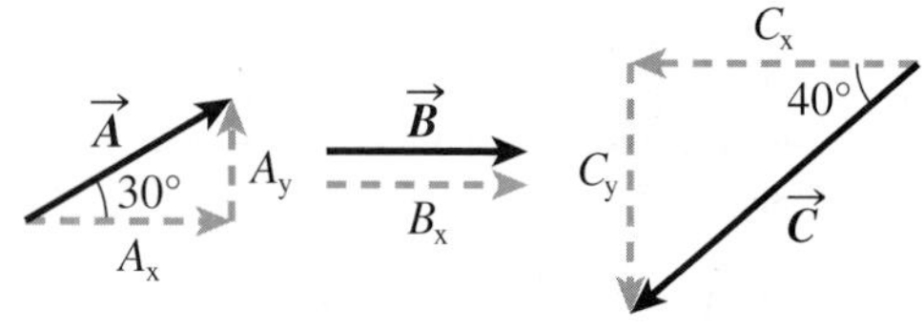

(b) Showing the components of the vectors

$R_x = 42$ m

θ

$R_y = -26$ m

$\vec{R}$

(c) The resultant vector

Figure 3-23 Sample Problem 3-5.

The direction of $\vec{R}$ can be found from $\tan\theta = \frac{26\,\text{m}}{42\,\text{m}}$, hence, $\theta = \tan^{-1}\frac{26\,\text{m}}{42\,\text{m}} = 32°$.

Thus, the resultant displacement is 49 m 32° south of east.

Although a calculator gives answers almost instantaneously, do not rely on the answers without thinking about them. Inverse trig functions, such as $\tan^{-1}$, provide an example of the limitations of calculators. In the range of 0° to 360°, there are always two angles with the same tangent. However, your calculator will give you only one angle, namely the smaller of the two possible angles. For example, $\tan 20° = \tan 200° = 0.364$, but when you find $\tan^{-1} 0.364$ on your calculator, you get only 20°. Thus, you must decide on the appropriate way to interpret answers given by your calculator.

Subtracting Vectors

Components can also be used to perform subtraction of vectors. If we want to subtract vector $\vec{B}$ from $\vec{A}$ to give $\vec{C}$ (i.e., $\vec{C} = \vec{A} - \vec{B}$), the components of $\vec{C}$ can each be determined by subtraction using the components of $\vec{A}$ and $\vec{B}$:

$$C_X = A_X - B_X$$

$$C_y = A_y - B_y$$

EXERCISES

3-13 Determine the x- and y-components of each vector shown in Figure 3-24. In each case, the $+x$-axis is to the right and the $+y$-axis is toward the top of the page.

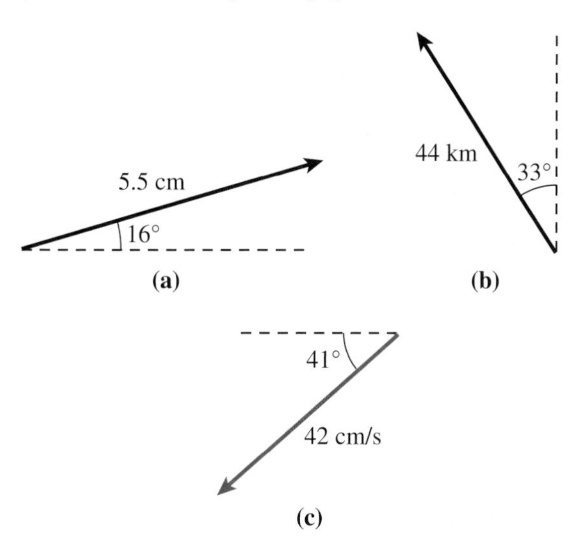

Figure 3-24 Question 3-13.

3-14 Figure 3-25 shows a ski hill inclined at an angle of 16° to the horizontal. The x-y coordinate system is chosen as shown, with the x-axis along the hill and the y-axis perpendicular to the hill. The acceleration due to gravity is the vector $\vec{g}$, which is 9.8 m/s² vertically downward, as illustrated. Determine the components of $\vec{g}$ along the axes shown.

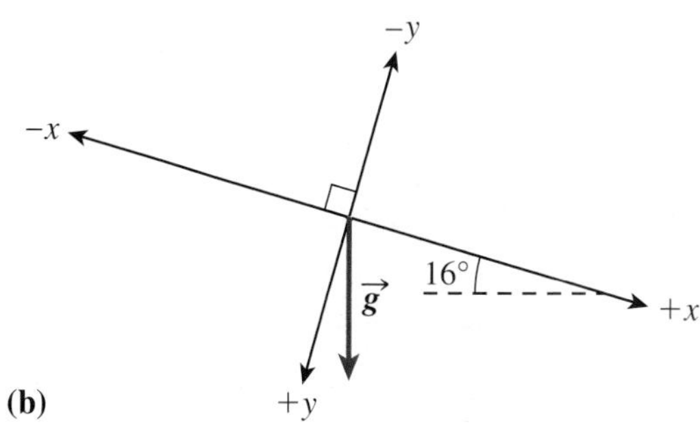

Figure 3-25 Question 3-14: **(a)** A typical ski hill. **(b)** Choosing $+x$- and $+y$-axes for the hill.

3-15 Find the magnitude and direction of the sum of vectors $\vec{A} + \vec{B} + \vec{C}$ using the component technique. Assume two significant digits. Show your steps clearly. $\vec{A} = 20\,\text{km}$ north; $\vec{B} = 30\,\text{km}$ east; $\vec{C} = 50\,\text{km}$ 55° west of north.

3-16 With the vectors $\vec{A}$, $\vec{B}$, and $\vec{C}$ from the previous question, use the component technique to determine the magnitude and direction of (a) $\vec{A} - \vec{B}$ and (b) $\vec{A} - \vec{B} - \vec{C}$.

3-17 Three people on a seaside dock are pulling on ropes attached to a yacht. The forces, $\vec{D}$, $\vec{E}$, and $\vec{F}$, are measured in newtons, N, and are shown in Figure 3-26. The magnitudes of the forces are $D = 281$ N, $E = 192$ N, and $F = 255$ N. Their directions are indicated in the figure. Find the magnitude and direction of the resultant of these three forces.

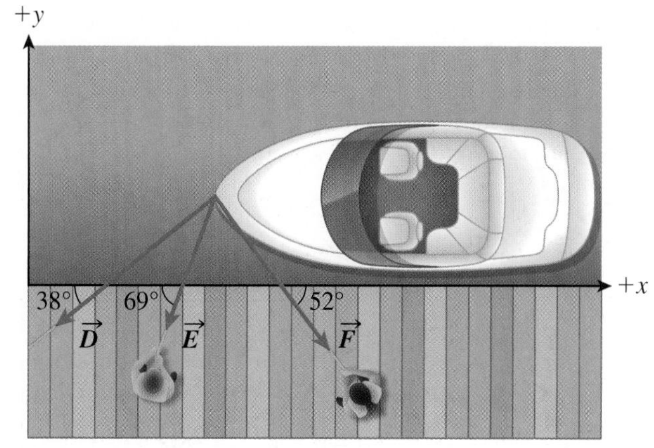

Figure 3-26 Question 3-17.

LOOKING BACK...LOOKING AHEAD

Chapters 1 and 2 introduced vectors briefly, and this chapter presented details of vector terminology and properties, and the addition and subtraction of vectors. Trigonometry was introduced and then applied to the important concept of components of vectors. These concepts will be applied to two-dimensional kinematics in Chapter 4 and throughout many parts of the book.

CONCEPTS AND SKILLS

Having completed this chapter you should now be able to do the following:

- Determine the magnitude and direction of a vector in a scale diagram.
- Use a scale diagram to find the resultant vector when vectors are added or subtracted.
- Apply the common trigonometric functions (sine, cosine, and tangent) as well as the sine and cosine laws to solve problems involving vectors.
- Choose an appropriate *x*-*y* coordinate system for any particular situation involving two or more vectors and their components.
- Perform vector addition or subtraction using components.

KEY TERMS

You should now be able to define or explain each of the following words or phrases:

resultant (or resultant vector)

trigonometry

component of a vector

Chapter Review

MULTIPLE-CHOICE QUESTIONS

3-18 When does $\vec{A} - \vec{B} = \vec{B} - \vec{A}$?
(a) never
(b) always
(c) only when $\vec{A} = \vec{B}$
(d) more information is needed to answer the question

3-19 If vector $\vec{A}$ is 10 m 45° south of west and vector $\vec{B}$ is 24 m 45° south of east, then $\vec{A} + \vec{B}$ has a magnitude of
(a) 26 m
(b) 34 m
(c) 14 m
(d) 0 m
(e) none of these

3-20 Vectors $\vec{A}$ and $\vec{B}$ face west, and vector $\vec{C}$ faces east. All three vectors are equal in magnitude. The resultant of $2\vec{A} - (2\vec{B} - 2\vec{C})$ is equal to
(a) $2\vec{A}$
(b) $\vec{A}$
(c) $2\vec{C}$
(d) $-\vec{A}$
(e) $2\vec{A} - 2\vec{C}$

3-21 In Figure 3-27, vector $\vec{A}$ is equal to
(a) $\vec{B} - \vec{D}$
(b) $\vec{D} - \vec{B}$
(c) $\vec{C} + \vec{E} - \vec{D}$
(d) $\vec{C} - \vec{D}$
(e) $\vec{D} - \vec{C}$

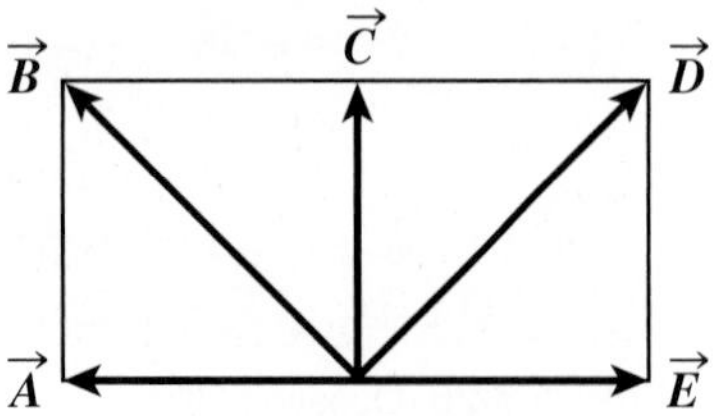

Figure 3-27 Question 3-21.

3-22 In Figure 3-28, if you are given the values of sides *a* and *c* and the angle between them, and you know there are 180° in a triangle, which of the following can be used to determine the other three variables?
(a) the cosine law, used twice
(b) the cosine law, followed by the sine law
(c) the sine law, used twice
(d) the sine law, followed by the cosine law
(e) Both (a) and (b) above are possible.

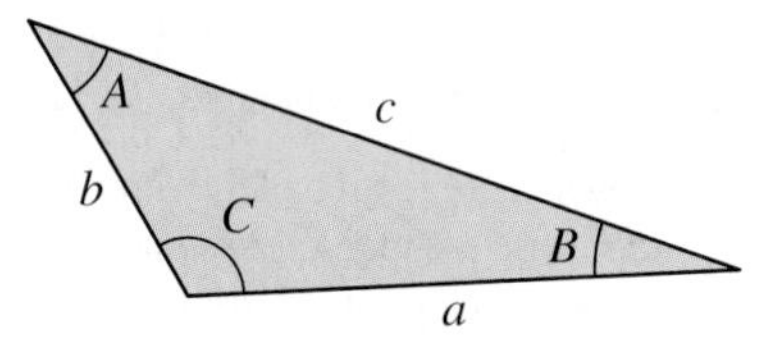

Figure 3-28 Question 3-22.

3-23 In Figure 3-29, the component of vector $\vec{A}$ along the inclined plane is
(a) $A \sin b$
(b) $A \sin a$
(c) A
(d) zero
(e) none of these

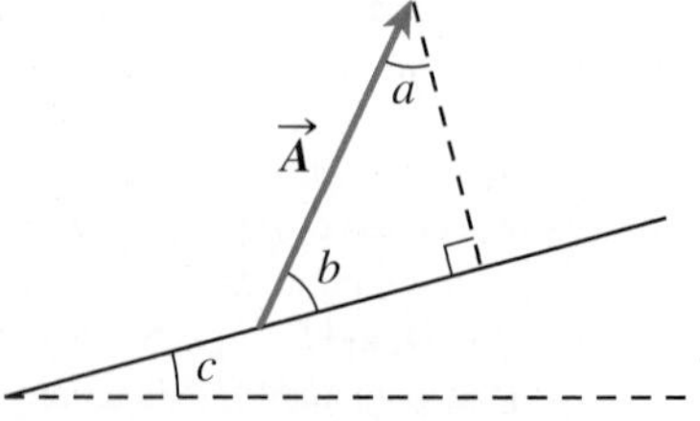

Figure 3-29 Question 3-23.

Review Questions and Problems

3-24 Assume that a ramp used for competitive water-skiing jumps is 6.70 m in length from the position where the ramp emerges from the water to the top of the ramp, which is 1.80 m above the water surface. Determine the angle the ramp makes with the water.

3-25 A student is standing 14 m from a campus building wall and measures an angle of 51° between a horizontal line and the top of the wall. If the student's eyes are 1.7 m above the ground, determine the height of the wall using a scale diagram or trigonometry.

3-26 On a mountain highway with switchbacks, a straight stretch of road rises uniformly between hairpin turns at an angle of 6.8° above the horizontal (Figure 3-30). If the elevation at the top of the straight stretch is 92 m above the bottom of the stretch, how long is that part of the highway?

Figure 3-30 Question 3-26.

3-27 Can a component of a vector have a magnitude greater than the vector's magnitude? Explain.

3-28 If vector $\vec{A}$ is perpendicular to vector $\vec{B}$, write an expression for the magnitude of $\vec{A} + \vec{B}$ in terms of the magnitudes of $\vec{A}$ and $\vec{B}$.

3-29 Can a vector have zero magnitude if one of its components is different from zero? Justify your answer.

3-30 (a) Can two vectors having the same magnitude be combined to give a zero resultant vector? If so, how?

(b) Can two vectors having different magnitudes be combined to give a zero resultant vector? What about three vectors? Explain.

3-31 If $\vec{A} = 36$ km east and $\vec{B} = 53$ km west, find $\vec{A} + \vec{B}$, $\vec{B} + \vec{A}$, $\vec{A} - \vec{B}$, and $\vec{B} - \vec{A}$.

3-32 Two muscles at the back of your leg, called the lateral and medial heads of the gastrocnemius muscle, exert forces $\vec{L}$ and $\vec{M}$, respectively, on the Achilles tendon (Figure 3-31). The forces are measured in newtons, N. Use components to determine the resultant force on the tendon.

DID YOU KNOW?

The Achilles tendon, behind the ankle, is the strongest and thickest tendon in the human body. It undergoes stress while walking and about double that stress while running. It is named after Achilles, a strong, heroic fighter in Greek mythology whose mother held him by the ankle as she dipped him into a magical river to ward off a forecast of early death. With the water unable to touch the ankle, Achilles' heel was vulnerable, and Achilles died after an arrow, likely poisoned, struck him there.

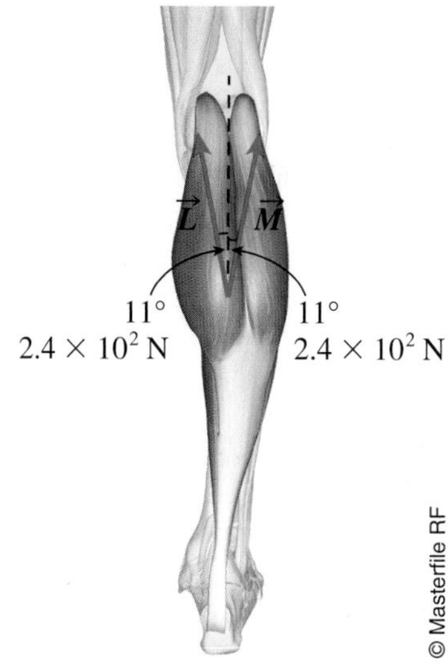

Figure 3-31 Question 3-32.

3-33 A golfer drives a ball 214 m due east of the tee, then hits it 96 m 30° north of east, and finally putts the ball 12 m 40° south of east. The ball sinks into the hole. Determine the magnitude and direction of the displacement needed from the tee to yield a hole-in-one. Use (a) a scale diagram and (b) components. Compare the two answers.

3-34 Use components to find $\vec{C} + \vec{D} - \vec{E}$ in Figure 3-32.

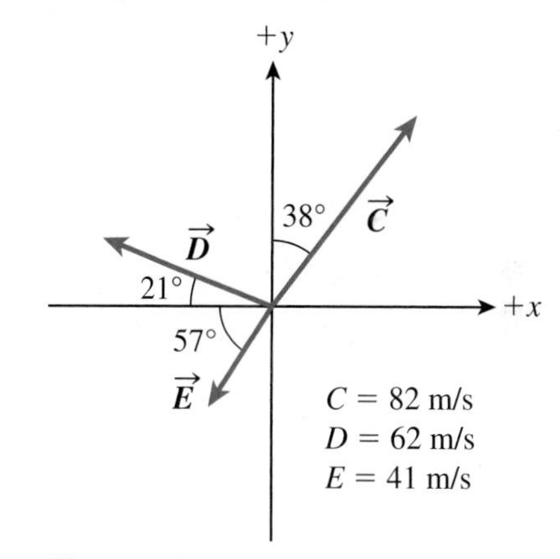

Figure 3-32 Question 3-34.

3-35 Determine the vector $\vec{C}$ that must be added to the sum of $\vec{A}$ and $\vec{B}$ in Figure 3-33 to give a net displacement of (a) zero and (b) 4.0 km W.

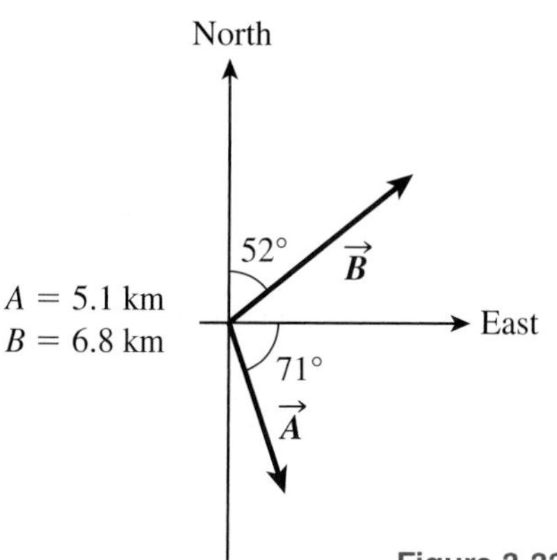

Figure 3-33 Question 3-35.

Applying Your Knowledge

3-36 If the component of a certain vector $\vec{B}$ along another vector $\vec{A}$ is zero, what conclusion(s) can be stated? (A diagram will help you visualize this question.)

3-37 **Fermi Question:** Assume that a displacement vector can be drawn from a person's nose to his or her toes. For a city of one million people, estimate the resultant displacement vector of all the nose-to-toes vectors at (a) 4 p.m. and (b) 4 a.m. Explain your reasoning.

3-38 Miami is 2050 km due south from Ottawa, and Chicago is 1780 km 30° west of north from Miami. Using components, determine the magnitude and direction of the displacement of an airline flight that goes directly from Ottawa to Chicago.

3-39 A roofer places a box of nails on a roof that is at an angle of 22.5° to the horizontal. The gravitational force acting on the box is 49.0 N vertically downward. Determine the component of the gravitational force on the box that is (a) parallel to the roof and (b) perpendicular to the roof.

3-40 Figure 3-34 (a) shows the tow bar and tow line used in competitive water skiing, and (b) shows the dimensions of the tow bar. Assuming the tow line segments attached to the bar are equal in length, determine that length and the angles in the triangle.

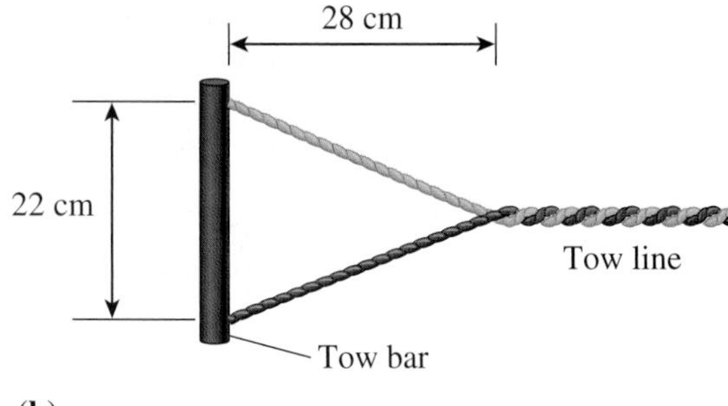

Figure 3-34 Question 3-40. **(a)** Using a tow bar. **(b)** Dimensions of the tow bar.

3-41 An amateur astronomer uses two points in Earth's orbit around the Sun as a baseline to estimate the radius of Saturn's orbit around the Sun. (Saturn is the sixth planet from the Sun.) As shown in Figure 3-35, the baseline is 3.0×10^8 km and the angles to Saturn, taken 6 months apart, are both 84°. Determine the distance from Saturn to the Sun. (When astronomers use triangulation to measure such large distances, they take into consideration the motion of the distant object.)

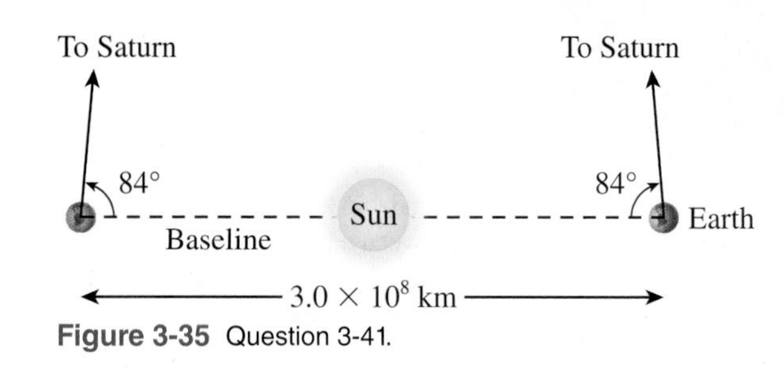

Figure 3-35 Question 3-41.

3-42 The photos in Figure 3-36 show a variety of stairways, from the Spanish Steps in Rome to stairs in a building interior and a ramp to an airplane. Notice that each step of a stairway has a "rise" and a "run."

(a) Express the angle of a stairway using one of the trigonometric functions in terms of rise and run.

(b) Simply by thinking about it, estimate the rise, run, and angle of a stairway that you use regularly.

(c) When convenient, measure the values of the rise and run (using, for example, a ruler or your hand span), and calculate the angle of the stairway.

(d) What would be the ideal rise, run, and angle for your own stride in ascending and descending a stairway efficiently and safely. Explain your answer.

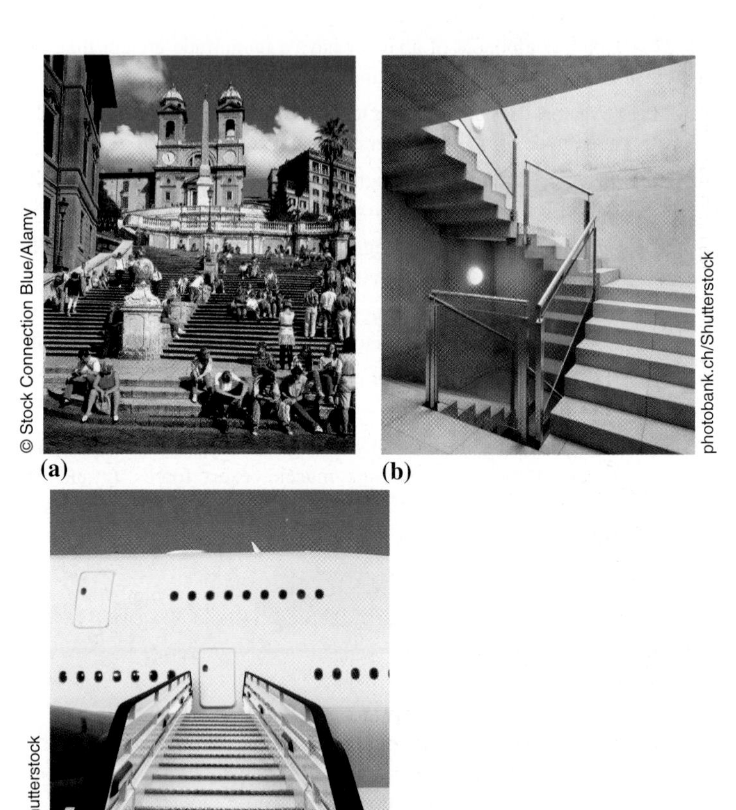

Figure 3-36 Question 3-42. A variety of stairways. **(a)** The Spanish Steps in Rome, Italy. **(b)** Stairs in a building interior. **(c)** Airplane exit stairway.

4 Two-Dimensional Kinematics

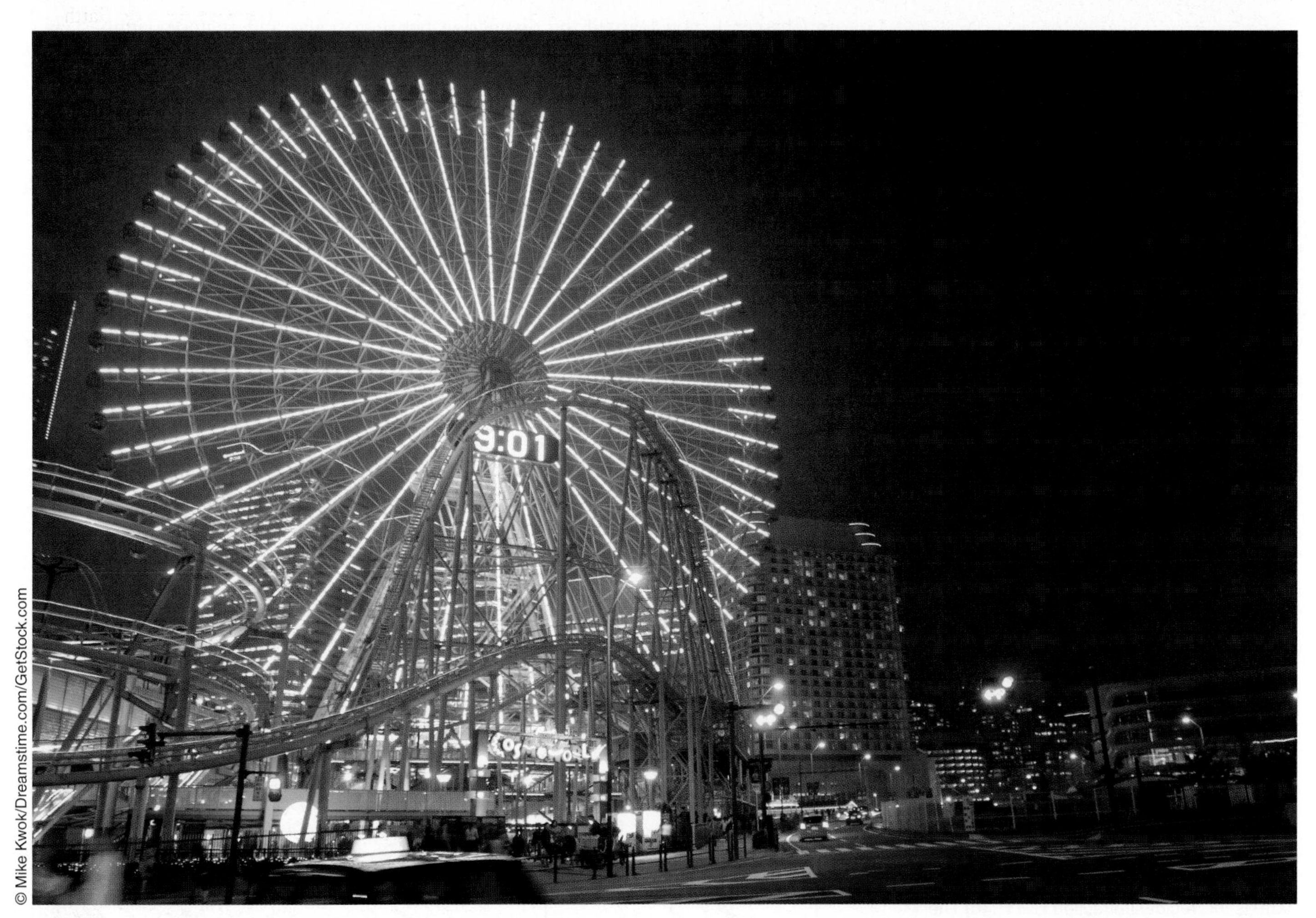

Figure 4-1 Passengers on the huge Ferris wheel at the Cosmo World Amusement Park in Yokohama, Japan, experience circular motion.

In Chapter 2, you studied motion in a straight line, in other words, in one dimension. However, most motion we experience occurs in either two or three dimensions. Fortunately, much of the information you learned in one-dimensional kinematics applies to two-dimensional kinematics, so only part of this chapter will appear new.

Two-dimensional motion occurs in many situations, such as the movement of cars racing around tracks or travelling around corners, or the circular motion of the Ferris wheel in Figure 4-1. It applies to huge objects such as planets that travel in elliptical paths around the Sun, and to tiny objects such as electrons that travel in regions around the nucleus of an atom. This type of motion is also common in sports where, for example, a ball travels in an arched path through the air. Two-dimensional motion will be analyzed in this chapter using some basic definitions and equations.

CHAPTER OUTLINE

The concepts of kinematics in two dimensions can be applied to more complex motions in three dimensions, a task beyond the scope of this book.

4.1 Displacement and Velocity in Two Dimensions

In Chapter 1, you learned the distinction between scalar and vector quantities, and in Chapter 3, you studied some properties of the addition and subtraction of vectors. Chapter 2 began with a description of distance and speed in one dimension. These scalar quantities have the same definitions and equations in two dimensions, so they will not be described here, although you will be asked to solve problems involving them. The majority of Chapter 2 was devoted to the vector quantities of displacement, velocity, and acceleration in one-dimensional motion. In this section, we will combine displacement and velocity with the mathematics of vectors from Chapter 3. One major difference between Chapter 2 and this chapter is that we omitted the vector notation in Chapter 2 because we were dealing with only one direction. Here, however, the vector notation will be used in most situations: directions are important in two-dimensional motion.

Displacement in Two Dimensions

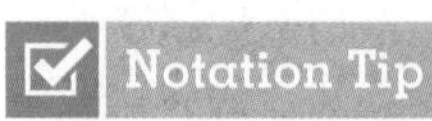

Notice that the $\vec{r}$ s are in boldface type with an arrow above, indicating vectors. From now on in this textbook, all vectors will have this format and, like all other variables, will be italicized.

Recall from Section 2.2 that displacement is the change of position, and has a direction from the initial position to the final position of a moving object. In one dimension, we used $\Delta x = x_2 - x_1$ for displacement (or $\Delta y = y_2 - y_1$ for vertical motion) because the motion was constrained to be along only one reference axis. In two dimensions, we consider motion in a reference x-y plane, and use $\vec{r}$ as the symbol for position, and $\Delta\vec{r}$ for displacement. Figure 4-2 shows how the defining equation for displacement, Eqn. 4-1, applies to two dimensions.

$$\Delta\vec{r} = \vec{r}_2 - \vec{r}_1 \qquad (4\text{-}1)$$

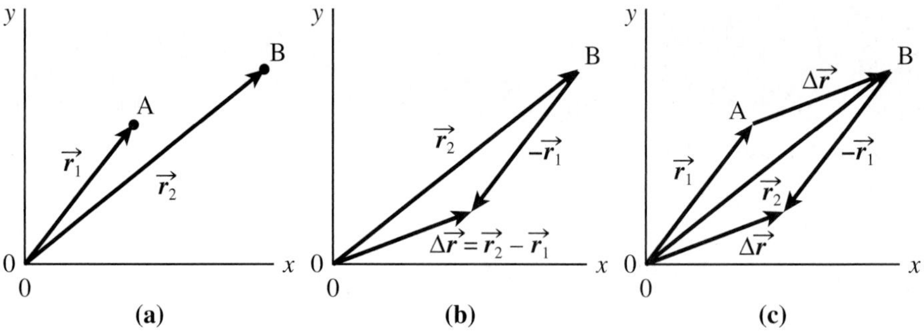

Figure 4-2 **(a)** An object moves from point A to point B. Its position vector, relative to an origin, (0, 0), changes from $\vec{r}_1$ to $\vec{r}_2$. **(b)** The displacement of the object is defined as $\Delta\vec{r} = \vec{r}_2 - \vec{r}_1$. It is shown here as the vector subtraction $\vec{r}_2 - \vec{r}_1$. **(c)** The displacement can also be drawn as the vector from the initial point, A, to the final point, B. Notice that this displacement has the same magnitude and direction as the displacement drawn in (b); hence, these two displacement vectors are in fact the same vector.

As a moving object goes from an initial position $\vec{r}_1$, having components x_1 and y_1 (Figure 4-3), to a final position $\vec{r}_2$ with components x_2 and y_2, the components of the displacement vector $\Delta\vec{r}$ are

$$\Delta x = x_2 - x_1 \text{ and } \Delta y = y_2 - y_1 \qquad (4\text{-}2)$$

Notice in Figure 4-3 that the object moves from point A, having position coordinates (x_1, y_1), to point B, with coordinates (x_2, y_2).

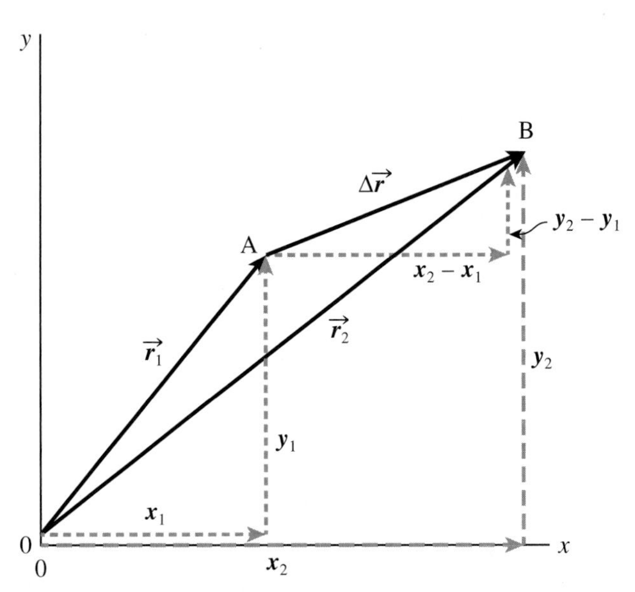

Figure 4-3 The components of the displacement vector, $\Delta\vec{r}$, are: $\Delta x = x_2 - x_1$ and $\Delta y = y_2 - y_1$.

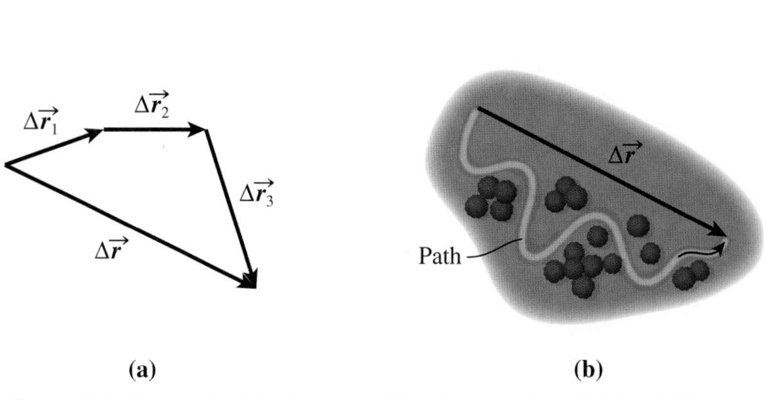

Figure 4-4 The resultant displacement, $\Delta\vec{r}$, is the same in both **(a)** and **(b)**, although the paths taken are different. Notice in (a) that the resultant displacement is the vector addition of the individual displacements.

The displacement, $\Delta\vec{r}$, defined by Eqn. 4-1 is sometimes called the **resultant displacement**. It is the vector addition of any individual displacements that an object has undergone, as in Figure 4-4 (a). Notice that the resultant displacement depends only on the initial and final positions, not on the path taken. Figure 4-4 shows two distinctly different paths that have the same resultant displacement.

resultant displacement the vector addition of the individual displacements that an object has undergone

SAMPLE PROBLEM 4-1

A kite surfer (Figure 4-5) on a frozen lake follows the path shown in Figure 4-6. Determine the surfer's

(a) total distance travelled

(b) (resultant) displacement

Solution

(a) The total distance travelled, being a scalar quantity, is 22 m + 34 m = 56 m.

(b) If we did not care about accuracy, a scale diagram could be used to find the displacement. We will use components to give a more accurate answer. Let us call the two displacements $\vec{A}$ and $\vec{B}$, and define the $+x$- and $+y$-directions to be east and north respectively (Figure 4-7 (a)). The components of $\vec{A}$ and $\vec{B}$ are

$$A_x = A\cos 36^\circ = (22\text{ m})\cos 36^\circ = 18\text{ m} \qquad A_y = A\sin 36^\circ = (22\text{ m})\sin 36^\circ = 13\text{ m}$$

$$B_x = B\cos 65^\circ = (34\text{ m})\cos 65^\circ = 14\text{ m} \qquad B_y = -B\sin 65^\circ = -(34\text{ m})\sin 65^\circ = -31\text{ m}$$

The x-component of the resultant displacement is

$$\Delta x = A_x + B_x = (18 + 14)\text{ m} = 32\text{ m}$$

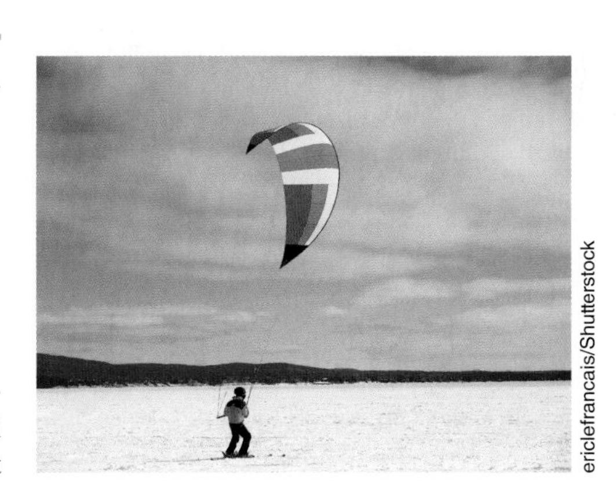

Figure 4-5 Kite surfing on a frozen lake.

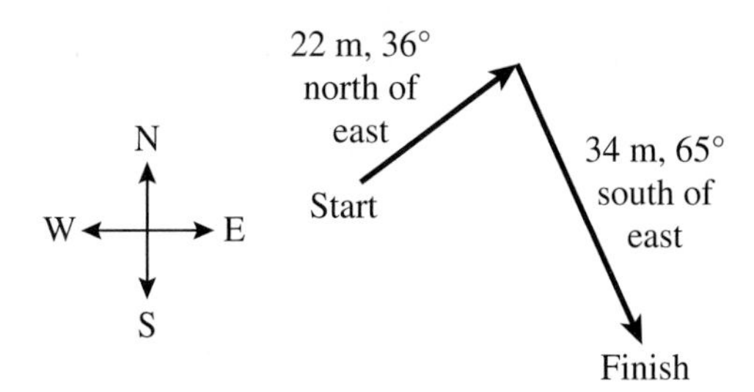

Figure 4-6 Sample Problem 4-1.

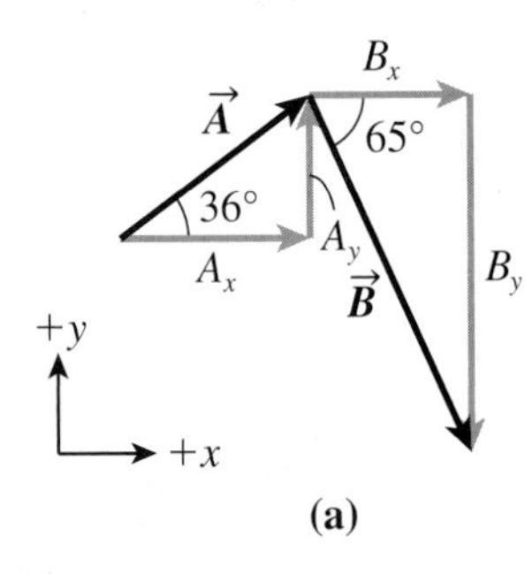

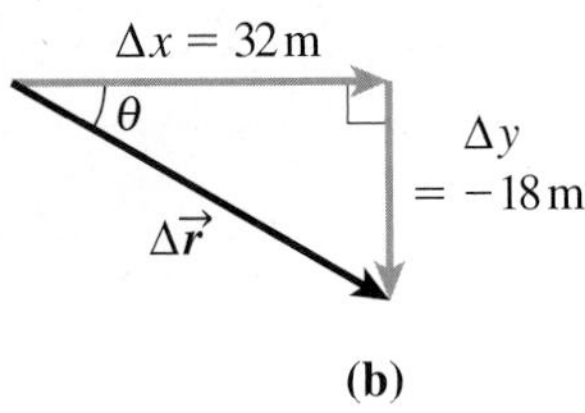

Figure 4-7 **(a)** Components of $\vec{A}$ and $\vec{B}$. **(b)** The resultant displacement $\Delta\vec{r}$.

The y-component of the resultant displacement is

$$\Delta y = A_y + B_y = (13 - 31)\,\text{m} = -18\,\text{m}$$

Thus, the magnitude of the resultant displacement (Figure 4-7 (b)) is

$$\Delta r = \sqrt{(32\,\text{m})^2 + (-18\,\text{m})^2} = 37\,\text{m}$$

The direction (angle θ in Figure 4-7 (b)) of the resultant displacement can be found using $\theta = \tan^{-1}\left(\frac{18\,\text{m}}{32\,\text{m}}\right) = 29°$.

Thus, the resultant displacement is 37 m 29° south of east.

Velocity in Two Dimensions

In Chapter 2, we found that the *average velocity*[1] is given by

$$\text{average velocity} = \frac{\text{displacement}}{\text{elapsed time}} \quad \text{or} \quad v_{av} = \frac{\Delta x}{\Delta t}$$

Using symbols appropriate for two dimensions, this equation becomes

$$\vec{v}_{av} = \frac{\Delta\vec{r}}{\Delta t} \tag{4-3}$$

The x- and y-components of this average velocity vector are given by

$$v_{av,\,x} = \frac{\Delta x}{\Delta t} \quad \text{and} \quad v_{av,\,y} = \frac{\Delta y}{\Delta t} \tag{4-4}$$

One type of problem in which you are asked to find average velocity is shown in the next sample problem. Other types of questions will be introduced later in the chapter.

SAMPLE PROBLEM 4-2

If the time for the kite surfer's motion in Sample Problem 4-1 took 37 s, what was the surfer's (a) average speed and (b) average velocity?

Solution

(a) $\text{speed}_{av} = \frac{d}{\Delta t} = \frac{56\,\text{m}}{37\,\text{s}} = 1.5\,\text{m/s}$

(b) $\vec{v}_{av} = \frac{\Delta\vec{r}}{\Delta t} = \frac{37\,\text{m}\ 29°\ \text{south of east}}{37\,\text{s}} = 1.0\,\text{m/s}\ 29°\ \text{south of east}$

Thus, the surfer's average speed is 1.5 m/s, and average velocity is 1.0 m/s 29° south of east.

Math Tip

In Chapter 3, you saw how to add vectors using either components or a scale diagram. An alternate technique uses the laws of sines and cosines from trigonometry, which was also presented in Chapter 3. These laws are especially convenient in problems involving two vectors that are not perpendicular to each other. They become less useful when more than two vectors are involved. If you have expertise in using these laws, you might wish to try them out on some vector-addition problems. You should have a thorough knowledge of how to use components, since many topics in later chapters involve this concept.

Terminology Tip

Notice in Sample Problem 4-2 that the magnitude of the average velocity does not equal the average speed. However, as you will see, the instantaneous speed is equal to the magnitude of the instantaneous velocity.

[1]See Chapter 2, page 26, for a formal definition of average velocity.

Instantaneous Velocity

Instantaneous velocity[2] can also be found for two-dimensional motion. The defining equation in this case is

$$\vec{v} = \lim_{\Delta t \to 0} \frac{\Delta \vec{r}}{\Delta t} \quad (4\text{-}5)$$

(a)

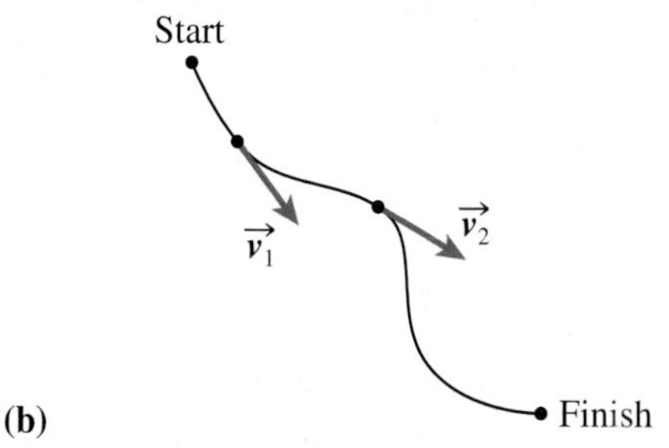

(b)

Figure 4-8 **(a)** A water strider. **(b)** The direction of the instantaneous velocity for the water strider moving along a curved path is tangent to the curve.

Remember that the symbol $\vec{v}$ with no subscript means instantaneous velocity. The *magnitude* of the instantaneous velocity equals the instantaneous speed of a moving object. Furthermore, for an object travelling in a curved path, such as the water strider in Figure 4-8 (a), the instantaneous velocity has a direction that is tangent to the curve at the instant involved (Figure 4-8 (b)).

The x- and y-components of the instantaneous velocity are

$$v_x = \lim_{\Delta t \to 0} \frac{\Delta x}{\Delta t} \text{ and } v_y = \lim_{\Delta t \to 0} \frac{\Delta y}{\Delta t} \quad (4\text{-}6)$$

SAMPLE PROBLEM 4-3

Figure 4-9 shows a circular track of radius 1.5×10^2 m. A runner takes 2.8×10^2 s to run around the track. Assuming that the runner's speed is constant, and that the runner is running clockwise around the track, determine the instantaneous velocity at A and at B.

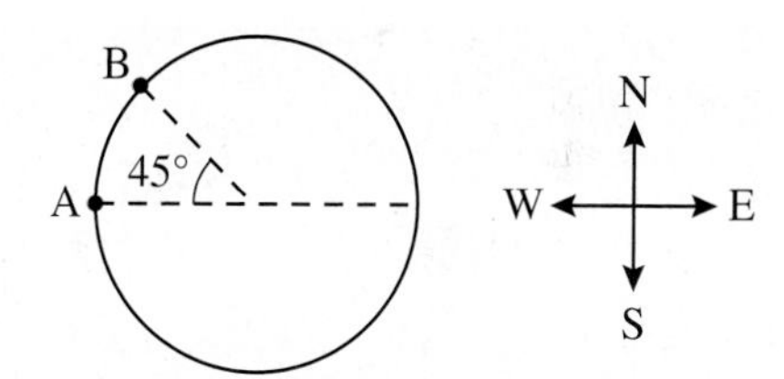

Figure 4-9 Sample Problem 4-3.

Solution

The runner's average speed is

$$\begin{aligned} \text{speed}_{av} &= \frac{d}{\Delta t} \\ &= \frac{2\pi r}{\Delta t} \\ &= \frac{2\pi\,(1.5 \times 10^2\,\text{m})}{2.8 \times 10^2\,\text{s}} \\ &= 3.4\,\text{m/s} \end{aligned}$$

Since the speed is constant, the instantaneous speed is 3.4 m/s at any time, and the magnitude of the instantaneous velocity is also 3.4 m/s at any time. Since the direction of the instantaneous velocity is always tangent to the curve, the velocity at position A is 3.4 m/s north, and at B, it is 3.4 m/s 45° east of north. These velocities are shown in Figure 4-10.

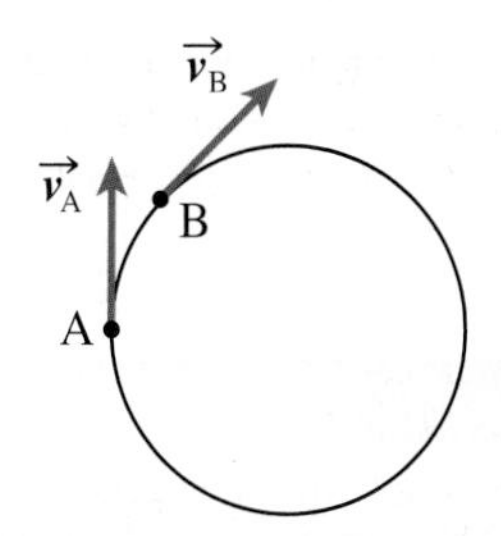

Figure 4-10 Sample Problem 4-3 (solution).

[2]See Chapter 2, page 27, for a formal definition of instantaneous velocity.

EXERCISES

4-1 Determine the displacement in each case.

(a) A person walks 38 m north, then 68 m west.

(b) A glider travels (horizontally) 550 m north, then 750 m east, and then 920 m south. (Assume two significant digits.)

(c) A boat on a lake follows the path shown on Figure 4-11, where the scale is indicated.

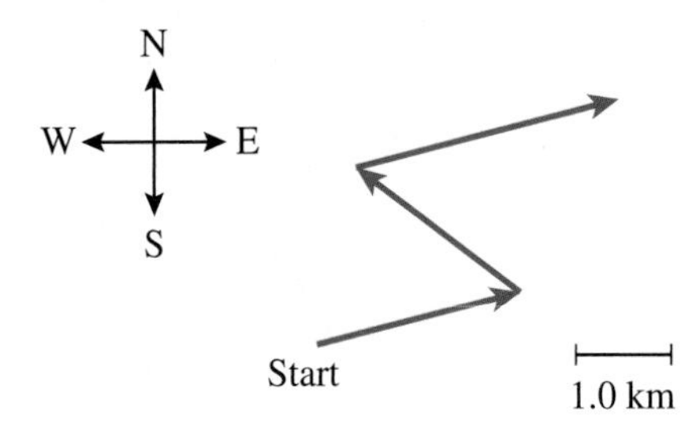

Figure 4-11 Question 4-1.

4-2 A drone (Figure 4-12) travels 12.0 km north, then 30.0 km west, then 21.0 km south, and finally 14.0 km east. The time for the entire trip is 1.25 h. Determine the drone's

(a) total distance travelled

(b) resultant displacement

(c) average speed in kilometres per hour for the entire trip

(d) average velocity for the entire trip

Oleg Yarko/Shutterstock

Figure 4-12 A remotely controlled drone, also known as an unmanned air vehicle (UAV), can be as small as a hummingbird or much larger, like the one shown here (Question 4-2).

DID YOU KNOW?

Drones may be best known for their use in military operations. However, thousands of drones in over 50 countries have many other uses, including rescue operations, traffic control, peering into hurricanes and volcanoes, monitoring inaccessible areas, such as archaeological sites and flooded regions, and monitoring livestock and agriculture.

4-3 A clock mounted on a vertical wall has a sweep second hand with a tip that is 14 cm from the centre of the clock.

(a) What is the average speed of the tip of the second hand?

(b) Determine the instantaneous velocity of the tip when it passes the 6:00 o'clock position; the 10:00 o'clock position.

(c) Find the tip's average velocity between the 1:00 o'clock position and the 5:00 o'clock position.

4-4 A skater on Ottawa's Rideau Canal (Figure 4-13) travels in a straight line 8.5×10^2 m 25° north of east, then 5.6×10^2 m in a straight line 21° east of north.

(a) Find the skater's resultant displacement.

(b) If the motion took 4.2 min, determine the skater's average speed and average velocity.

Howard Sandler/Shutterstock

Figure 4-13 Skating on Ottawa's Rideau Canal (Question 4-4).

4-5 State in words (such as "up and to the right") the direction of the instantaneous velocity at each labelled position in the diagrams of curved motion illustrated in Figure 4-14. In Figure 4-14 (a), the object is moving counterclockwise around the circle.

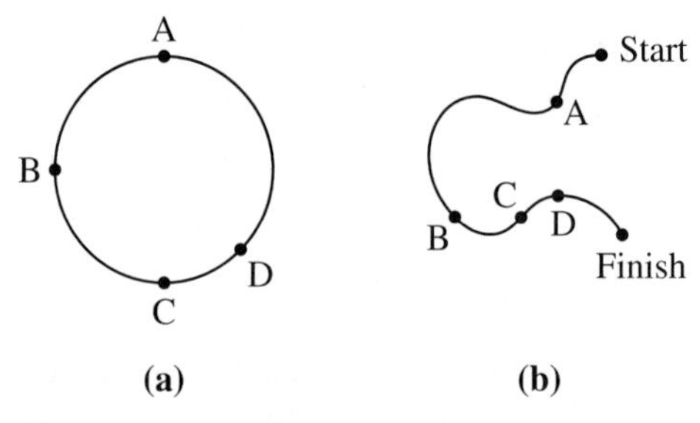

Figure 4-14 Question 4-5.

4-6 The average velocity of a car during a 14 min drive is 72 km/h 32° west of south.

(a) What is the car's displacement from its starting position?

(b) If the $+x$- and $+y$-directions are east and north respectively, what are the x- and y-components of the average velocity?

4.2 Acceleration in Two Dimensions

A race car is travelling around a circular track at a constant speed of 150 km/h. Is the car accelerating? The answer is yes, because the velocity of the car is changing in direction even though its magnitude is constant. Acceleration in two dimensions occurs when an object's velocity undergoes a change in magnitude, or direction, or both.

To analyze acceleration in two dimensions, recall from Chapter 2 that acceleration is the time rate of change of velocity. The equation introduced there for *average acceleration*[3] also applies to two-dimensional motion. Thus,

$$\vec{a}_{av} = \frac{\Delta \vec{v}}{\Delta t} = \frac{\vec{v}_2 - \vec{v}_1}{t_2 - t_1} \quad (4\text{-}7)$$

where $\vec{v}_2$ and $\vec{v}_1$ are the velocities at times t_2 and t_1 respectively. Notice that acceleration and velocity are vector quantities and $\vec{v}_2 - \vec{v}_1$ is a *vector* subtraction.

The components of the average acceleration are

$$a_{av,\,x} = \frac{\Delta v_x}{\Delta t} = \frac{v_{2x} - v_{1x}}{t_2 - t_1} \text{ and } a_{av,\,y} = \frac{\Delta v_y}{\Delta t} = \frac{v_{2y} - v_{1y}}{t_2 - t_1} \quad (4\text{-}8)$$

where, for example, v_{2x} represents the x-component of velocity $\vec{v}_2$.

SAMPLE PROBLEM 4-4

The straight stretches of a horse racing track are 250 m long and 160 m apart. The ends of the track are semicircles, as shown in Figure 4-15. A horse is galloping at a constant speed of 14 m/s in the direction shown. Determine the horse's average acceleration from (a) C to D, then (b) C to E.

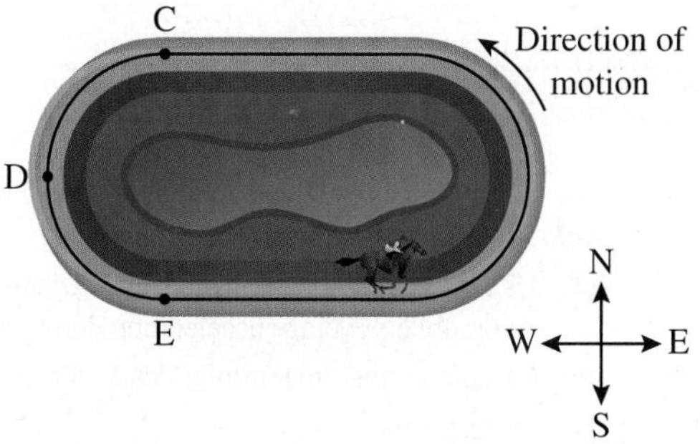

Figure 4-15 Sample Problem 4-4.

Solution

(a) The distance around the semicircle from C to E is $\pi r = \pi\,(80\text{ m}) = 251$ m. Hence, the distance from C to D is (251 m)/2 = 126 m. The time to gallop from C to D is

$$\Delta t = \frac{d}{v_{av}} = \frac{126\text{ m}}{14\text{ m/s}} = 9.0\text{ s}$$

Point C is at the beginning of the semicircle, so the horse's velocity there is $\vec{v}_C = 14$ m/s west. At D its velocity is $\vec{v}_D = 14$ m/s south. The vector subtraction of $\Delta\vec{v} = \vec{v}_D - \vec{v}_C$ is shown in Figure 4-16 (a). The magnitude of $\Delta\vec{v}$ is given by the Pythagorean relation:

$$\Delta v = \sqrt{v_D^2 + v_C^2} = \sqrt{(14\text{ m/s})^2 + (14\text{ m/s})^2} = 20\text{ m/s}$$

The direction of $\Delta\vec{v}$ (angle θ in Figure 4-16 (a)) is simply 45° east of south, since $|\vec{v}_D| = |\vec{v}_C|$.

The average acceleration is therefore

$$\vec{a}_{av} = \frac{\Delta\vec{v}}{\Delta t}$$

$$= \frac{20\text{ m/s } 45^\circ\text{ east of south}}{9.0\text{ s}}$$

$$= 2.2\text{ m/s}^2\ 45^\circ\text{ east of south}$$

Thus, the average acceleration from C to D is 2.2 m/s² 45° east of south.

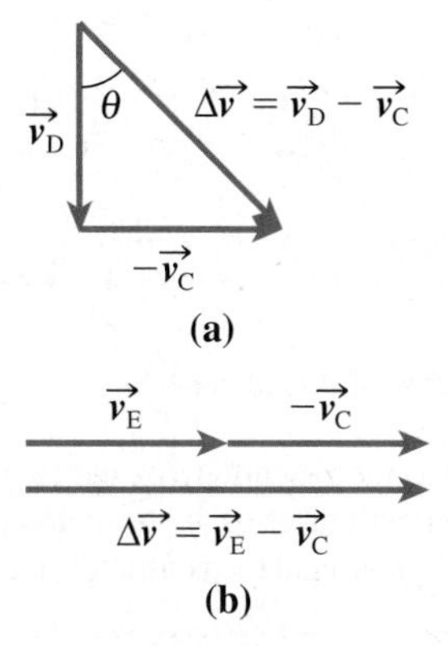

Figure 4-16 Sample Problem 4-4 (solution).

[3]See Chapter 2, page 32, for the definition of average acceleration.

(b) The time for the horse to gallop from C to E is double that in (a) above, or 18 s. The vector subtraction of $\Delta\vec{v} = \vec{v}_E - \vec{v}_C$ is shown in Figure 4-16 (b), giving $\Delta\vec{v} = 28$ m/s east.

$$\text{Thus, } \vec{a}_{av} = \frac{\Delta\vec{v}}{\Delta t} = \frac{28 \text{ m/s east}}{18 \text{ s}} = 1.6 \text{ m/s}^2 \text{ east}$$

Therefore, the average acceleration from C to E is 1.6 m/s^2 east.

Instantaneous Acceleration

The definition of *instantaneous acceleration* in two dimensions is the same as in one dimension.[4] Thus,

$$\vec{a} = \lim_{\Delta t \to 0} \frac{\Delta\vec{v}}{\Delta t} \quad \text{(4-9)}$$

Again, when no subscript accompanies the symbol "$\vec{a}$," the acceleration is assumed to be the instantaneous acceleration.

If the acceleration is constant, the constant-acceleration equations from Chapter 2 can also be used in two dimensions. For example, the expressions for $\vec{v}$ and $\vec{v}_0$ are

$$\vec{v} = \vec{v}_0 + \vec{a}t \text{ and } \vec{v}_0 = \vec{v} - \vec{a}t$$

The equation to find $\vec{v}$ involves a vector addition (see Question 4-10), and the equation to find $\vec{v}_0$ involves a vector subtraction (see Question 4-11).

EXERCISES

4-7 An automobile has a velocity of 25 m/s east, and 15 s later its velocity is 25 m/s south.

(a) Using a rough sketch, estimate the magnitude and direction of the automobile's average acceleration during the 15 s time interval.

(b) Calculate the magnitude and direction of the average acceleration.

4-8 Determine the average acceleration for the object shown in Figure 4-17 as it travels from A to B in 8.5 s. The two dashed lines are westward.

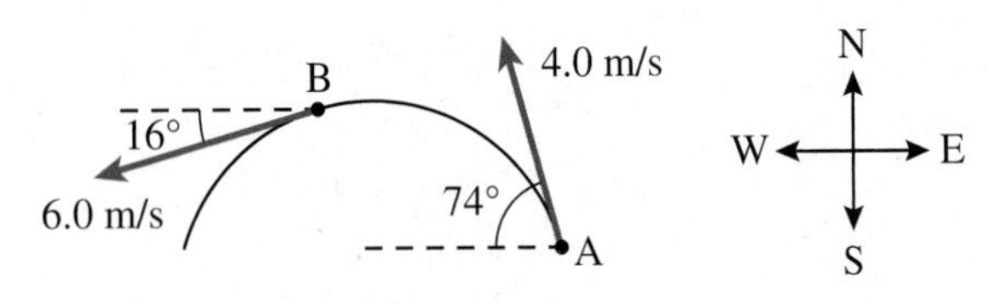

Figure 4-17 Question 4-8.

4-9 Determine the magnitude of the average acceleration during the time interval it takes each object described below to complete half a revolution around the central object.

(a) Planet Mercury travels once around the Sun in 7.60×10^6 s at an average speed of 47.8 km/s.

(b) Mars takes 687 (Earth) days to travel once around the Sun in its orbit of average radius 2.28×10^7 m.

(c) A military satellite takes 80.0 min to travel once around Earth in an orbit having a diameter of 1.29×10^7 m.

4-10 A shortfin mako, the world's fastest shark, initially swimming east at 6.4 m/s, undergoes an average acceleration of 2.0 m/s^2 south for 7.0 s. Determine the final velocity of the shark after this time interval.

4-11 A ball is thrown from a stationary hot-air balloon with an initial unknown velocity. The ball accelerates at 9.8 m/s^2 downward for 2.0 s, at which time its instantaneous velocity is 24 m/s at an angle of 45° below the horizontal. Determine the magnitude and direction of the initial velocity. (Neglect air resistance.)

4-12 A motor scooter is travelling along a winding highway at a constant speed of 82 km/h. At 3:00 p.m., the scooter is heading 38.2° east of north, and at 3:15 p.m. it is heading 12.7° south of east. Define the $+x$- and $+y$-directions to be east and north respectively. Use Eqn. 4-8 to determine the x- and y-components of the average acceleration during this time interval.

[4] See Chapter 2, page 36, for a formal definition of instantaneous acceleration.

4.3 Introduction to Projectile Motion

The following motions all have some common characteristics:

- A javelin is thrown through the air.
- A howler monkey jumps from one branch of a tree to another.
- The shark mentioned in Question 4-10 flies though the air after jumping out of the water.
- A ballet dancer executes a forward jump called a *jeté* (see Figure 4-66 at the end of the chapter).
- A skateboarder skates at top speed off the top end of a ramp (Figure 4-18).

Figure 4-18 Once the skateboarder leaves the ramp, he or she becomes a "projectile."

Each motion involves an object moving through the air without any propulsion system and following a curved path, such as the path of a tennis ball illustrated in Figure 4-19. An object moving through the air in this way is called a **projectile**, and the motion of such an object is called **projectile motion**. (A more specific definition of this type of motion will be given soon.)

projectile an object moving through the air without any propulsion system and following a curved path

projectile motion the curved motion of a projectile

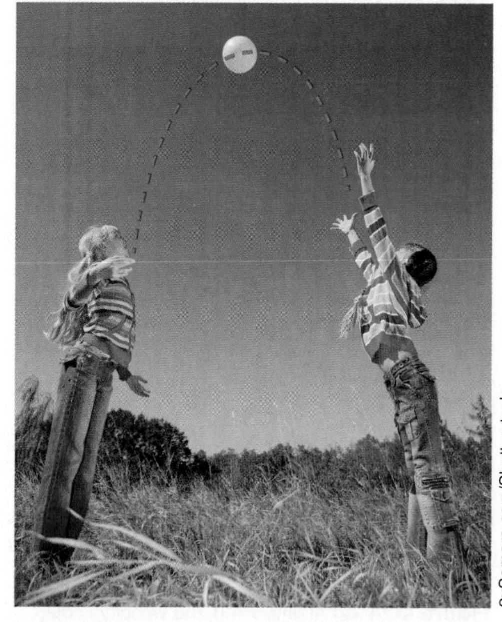

Figure 4-19 A typical path followed by a projectile, in this case a tennis ball.

Comparing Free Fall and Projectile Motion

Choose a thin, stiff piece of cardboard (for example, a file folder) to make the coin-launching device shown in Figure 4-20. Cut the cardboard to a strip about 24 cm long and 12 cm wide. Fold the cardboard lengthwise as illustrated, and tape or staple the ends to add strength to the launcher. Predict, with reasons, the order in which the coins will reach the floor after they leave the launcher as described below.

Holding the launcher level, place one coin on each side of the vertical barrier, then listen carefully as you flick the launcher sideways horizontally causing one coin to fall straight downward and the other to become a projectile. Compare the landing times by listening to the coins hit the floor. If necessary, repeat the experiment to ensure that the projected coin leaves at the same instant as the other coin.

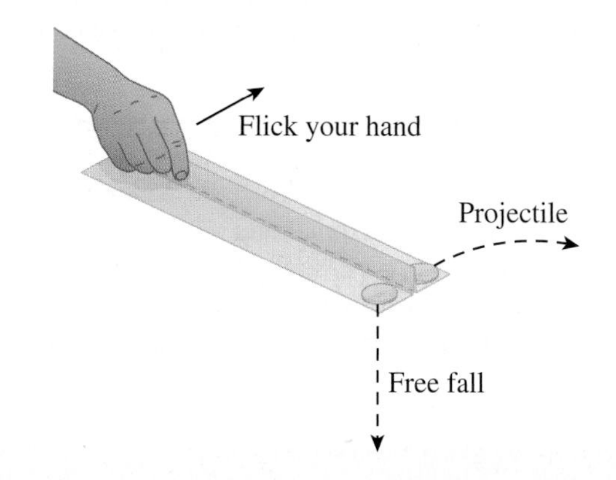

Figure 4-20 Comparing the landing times for free fall and projectile motion. The width of the platforms holding the coins should be about the same size as the coins.

To begin the study of projectile motion, consider the situation illustrated in Figure 4-21. Three daring individuals are leaving the edge of a cliff above water. One person is going to drop straight down into the water, while the other two are running, one faster than the other. If all three leave the edge of the cliff at the same instant following the paths shown in the diagram, which one will land first? By experimentation and close observation, we learn that all three land at the same time. This observation greatly affects our method of analyzing projectile motion.

Experiments and demonstrations, including the activity described in the Try This! box in this section, can be used to verify that this observation is true. Figure 4-22 shows a photograph of two balls released simultaneously from a device that projects

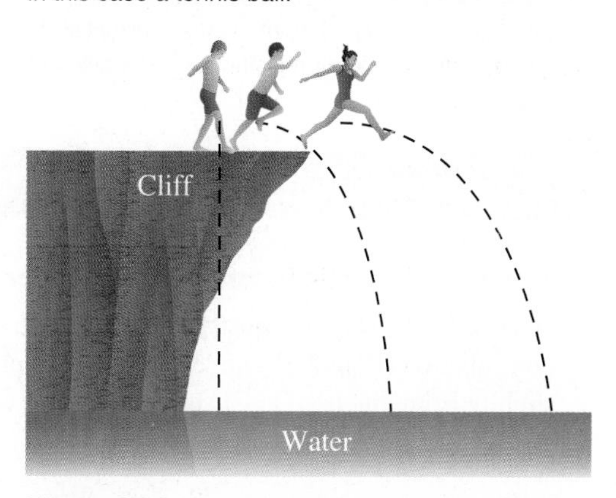

Figure 4-21 When three people leave the edge of a cliff at the same instant, one dropping straight down and two running with different speeds, which will land in the water first?

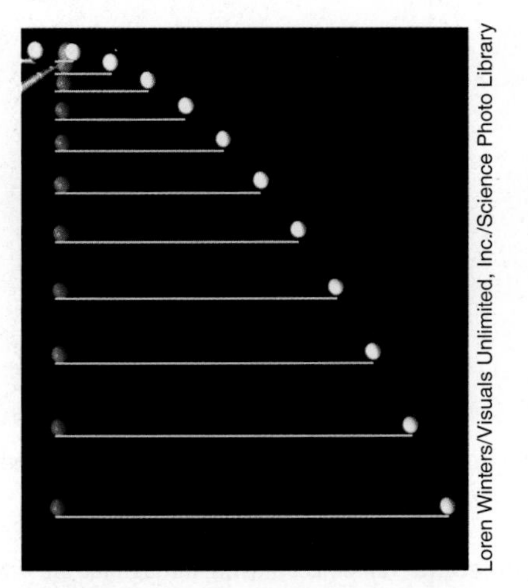

Loren Winters/Visuals Unlimited, Inc./Science Photo Library

Figure 4-22 This photograph, taken in a darkened room with a strobe light flashing at equal time intervals, shows the positions of two falling balls, one falling straight downward and the other projected sideways.

one ball horizontally while releasing the other straight down from rest—a more high-tech version of your homemade coin launcher. The time between flashes of the strobe light remains constant. Although one ball falls straight down, and the other moves horizontally while falling downward, both balls fall through the same vertical distance with each flash. Notice that the horizontal distance moved between flashes remains constant, but the vertical distance fallen increases during each succeeding time interval.

Based on the observation of the motion of the balls in Figure 4-22, we can make the following important conclusions about projectile motion:

- The horizontal velocity of a projectile is constant.
- In the vertical direction, the projectile undergoes constant acceleration downward, that is, the acceleration due to gravity.
- For a projectile, the horizontal and vertical motions are independent of each other except that they have a common time.

These conclusions are based on the assumption that air resistance can be neglected, an assumption also made when we analyzed one-dimensional motion in Chapter 2.

Using these facts, we can now write a formal definition of this type of motion. **Projectile motion** is motion with a constant horizontal velocity combined with constant vertical acceleration caused by gravity. The path of an object undergoing this type of motion is a *parabola*, and the curved path is called *parabolic*. Because the horizontal and vertical motions do not depend on each other, we can use separate sets of equations to analyze projectile motion. The uniform velocity equations from Section 2.2 apply to the horizontal motion and the constant-acceleration equations from Sections 2.4 and 2.5, using $g = 9.8\ \text{m/s}^2$ (or $9.80\ \text{m/s}^2$, if needed), apply to the vertical motion.

think4photop/Shutterstock

Figure 4-23 An archer fish.

DID YOU KNOW?

The archer fish (Figure 4-23) uses a deadly accurate jet of water from its mouth to knock a prey from a branch above the water. The fish presses its tongue against a groove in its mouth to form a tube, and then compresses its gills to spurt the stream of water in the air. The stream travels in a parabolic arc as a projectile, striking grasshoppers, spiders, or butterflies up to 2 m away!

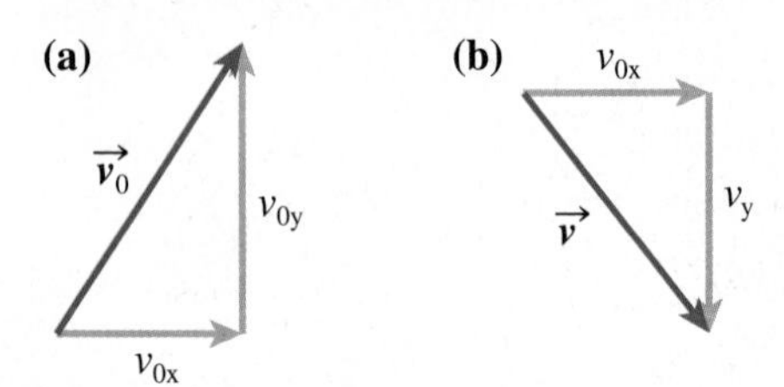

Figure 4-24 **(a)** At time $t = 0$, the velocity $\vec{v}_0$ of a projectile has a horizontal component, v_{0x}, and a vertical component, v_{0y}. **(b)** At some time t later, the projectile's velocity $\vec{v}$ has the same horizontal component (neglecting air resistance) and a different vertical component, v_y.

Refer to the text box titled Kinematics Equations for Projectile Motion, which summarizes the kinematics equations for both the horizontal and vertical portions of projectile motion. The symbols are not in boldface type, since they do not represent vectors. For example, v_{0x} represents the initial velocity's x-component (not a vector), and "t" represents time (a scalar quantity). See also Figure 4-24 for an explanation of the symbols v_{0x}, v_{0y}, and v_y.

Problem-Solving Strategies

Kinematics Equations for Projectile Motion

Horizontal (x) Motion: constant-velocity (zero acceleration); only one equation (Eqn. 2-3) applies, but we will change the notation slightly from that used in Chapter 2. We start with Eqn. 2-3:

$$v_{\text{av},x} = \frac{\Delta x}{\Delta t}$$

where we have explicitly indicated that this is the *x-component* of the average velocity. Since there is constant velocity in the x-direction, $v_{\text{av},x}$ always equals the initial x-component of velocity, which we write as v_{0x}. We write Δt as $t - t_0$, and since we will always choose the initial time t_0 to be zero, we are left simply with "t." For the displacement, Δx, we write $x - x_0$. Hence, we have $v_{0x} = \dfrac{x - x_0}{t}$, which can be rearranged to

$$x = x_0 + v_{0x}t \quad \text{or} \quad \Delta x = v_{0x}t \qquad \textbf{(4-10)}$$

(continued)

Vertical (y) Motion: downward constant acceleration of magnitude g. The four kinematics equations for free fall given in Table 2-3 apply here. We rewrite them below, including y-subscripts to clearly indicate the components of vectors:

$$v_y = v_{0y} + a_y t \quad \text{or} \quad \Delta v_y = a_y t \qquad (4\text{-}11)$$

$$y = y_0 + \tfrac{1}{2}(v_{0y} + v_y)t \quad \text{or} \quad \Delta y = \tfrac{1}{2}(v_{0y} + v_y)t \qquad (4\text{-}12)$$

$$y = y_0 + v_{0y}t + \tfrac{1}{2}a_y t^2 \quad \text{or} \quad \Delta y = v_{0y}t + \tfrac{1}{2}a_y t^2 \qquad (4\text{-}13)$$

$$v_y^2 - v_{0y}^2 = 2a_y(y - y_0) \quad \text{or} \quad v_y^2 - v_{0y}^2 = 2a_y(\Delta y) \qquad (4\text{-}14)$$

Equations 4-10 to 4-14 can be used to solve a large variety of problems involving projectile motion. When solving these problems, remember that time is the common variable between the horizontal motion and the vertical motion.

projectile motion (formal definition) motion with a constant horizontal velocity combined with constant vertical acceleration caused by gravity, resulting in a parabolic path

SAMPLE PROBLEM 4-5

A ball is thrown off a balcony with an initial velocity of 20 m/s horizontally.

(a) Determine the location of the ball at later times of 1.0 s, 2.0 s, 3.0 s, and 4.0 s.

(b) Show these positions in a diagram.

(c) Name the shape of the resulting curved path.

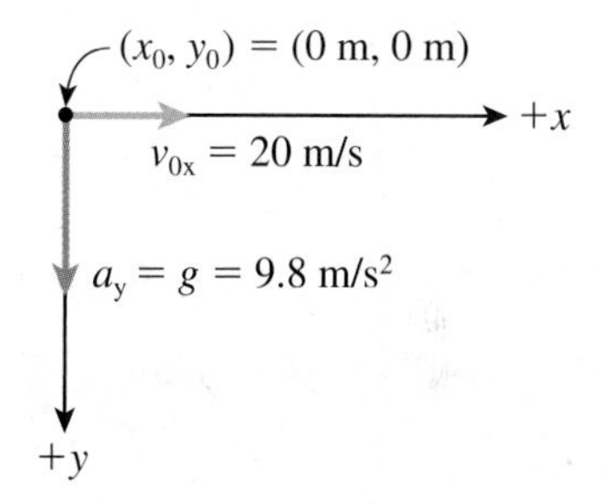

Figure 4-25 Sample Problem 4-5.

Solution

(a) We choose the $+x$-direction to be the horizontal direction in which the ball is thrown, and the $+y$-direction to be downward (Figure 4-25). Let the position as the ball leaves the balcony be $(x_0, y_0) = (0\text{ m}, 0\text{ m})$.

The horizontal component of the initial velocity, v_{0x}, is given as 20 m/s. This component remains constant; thus, during each second, the ball moves 20 m horizontally. Hence, after the first second, it has moved 20 m; after the second, 40 m, etc. The horizontal positions for the first four seconds are shown in the table below. More formally, these positions can be determined from Eqn. 4-10: $x = x_0 + v_{0x}t$.

In the vertical direction, the y-component of acceleration is $a_y = g = 9.8\text{ m/s}^2$ (a positive quantity, since downward is positive). The vertical position can be found at the end of each second using Eqn. 4-13: $y = y_0 + v_{0y}t + \frac{1}{2}gt^2$, where $y_0 = 0\text{ m}$, and $v_{0y} = 0\text{ m/s}$. Thus, $y = \frac{1}{2}gt^2$. Results of using this equation are presented in the table below.

Time, t (s):	0.0	1.0	2.0	3.0	4.0
x-component of position x (m):	0.0	20	40	60	80
y-component of position y (m):	0.0	4.9	20	44	78

(b) Figure 4-26 shows a diagram of the position of the ball at the required times. The positions are joined with a smooth curve.

(c) The curved path shown in Figure 4-26 is a parabola.

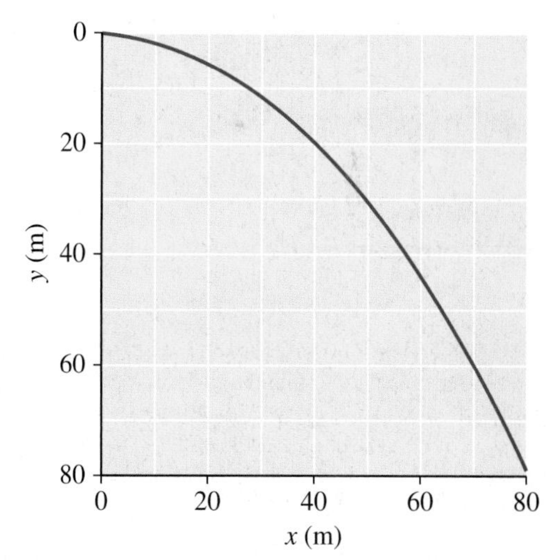

Figure 4-26 Sample Problem 4-5 (solution).

As seen in Sample Problem 4-5, once an object has become a projectile, one of the variables to consider is the horizontal position. The horizontal distance that a projectile travels from launch to landing is called its **horizontal range**. You should be able to predict at least one factor that the horizontal range depends on.

horizontal range the horizontal distance that a projectile travels from launch to landing

SAMPLE PROBLEM 4-6

A child travels down a water slide, leaving it with a velocity having only a horizontal component of 4.6 m/s. The child then undergoes projectile motion and lands in the swimming pool below, as illustrated in Figure 4-27. The end of the slide is 3.5 m above the pool. Determine

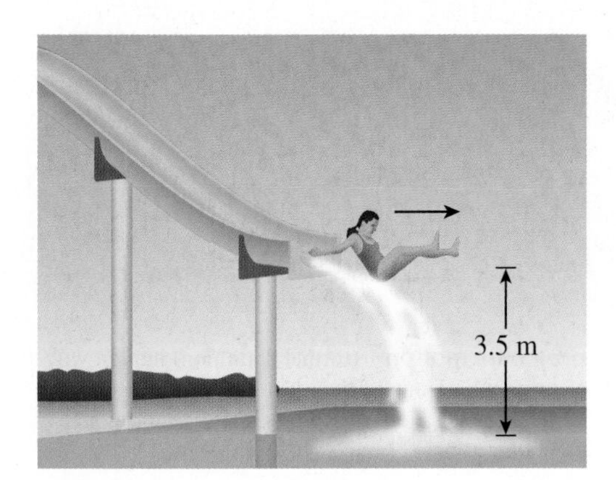

Figure 4-27 Sample Problem 4-6.

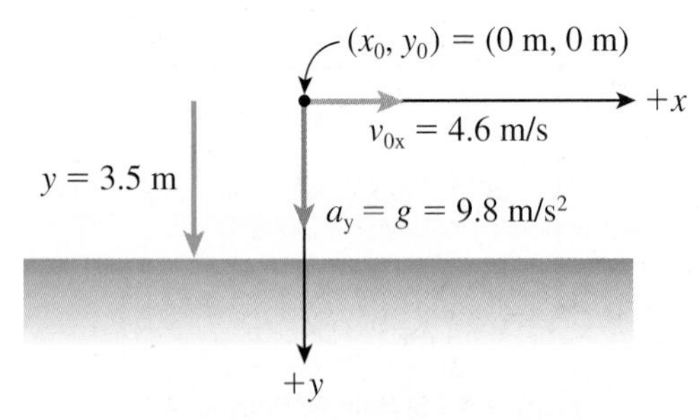

Figure 4-28 Given quantities. (Sample Problem 4-6.)

(a) the time the child remains airborne

(b) the horizontal range the child travels while in the air

(c) the velocity with which the child enters the water

Solution

(a) We define the $+y$-direction to be downward and the $+x$-direction to be the horizontal direction that the child leaves the slide (Figure 4-28). We choose the position as the child leaves the slide to be $(x_0, y_0) = (0\,\text{m}, 0\,\text{m})$.

It is useful in projectile-motion problems to list the various quantities such as velocity, position, etc., in two columns, one for horizontal motion and one for vertical motion:

Horizontal (constant velocity)	Vertical (constant acceleration)
$x_0 = 0$ m	$y_0 = 0$ m
$x = ?$ (required to find in (b))	$y = 3.5$ m
$v_{0x} = 4.6$ m/s	$v_{0y} = 0$ m/s
$t = ?$ (required to find in (a))	$v_y = ?$ (needed in (c))
	$a_y = g = 9.8$ m/s²
	$t = ?$ (required to find in (a))

The horizontal motion has two unknowns, so they cannot be determined until the vertical motion is analyzed. Thus, let us begin by using the vertical motion to find the time. The appropriate equation is the one with no v_y, that is, Eqn. 4-13:

$$y = y_0 + v_{0y}t + \tfrac{1}{2}a_y t^2$$

Substituting $y_0 = 0\,\text{m}$, $v_{0y} = 0\,\text{m/s}$, and $a_y = g$ yields $y = \tfrac{1}{2}gt^2$

Hence, $t^2 = \dfrac{2y}{g}$;

$$\therefore t = \pm\sqrt{\frac{2y}{g}} = \pm\sqrt{\frac{2(3.5\,\text{m})}{9.8\,\text{m/s}^2}} = \pm 0.85\,\text{s}$$

Only the positive root applies, so the time for the fall is 0.85 s.

(b) We now use this time to find the horizontal displacement, x.

$$x = x_0 + v_{0x}t = 0\,\text{m} + (4.6\,\text{m/s})(0.85\,\text{s}) = 3.9\,\text{m}$$

Thus, the child hits the water 3.9 m horizontally from the end of the slide. In other words, the horizontal range of the child's projectile motion is 3.9 m.

(c) To determine the magnitude and direction of the velocity as the child enters the water, we first find its x- and y-components. The x-component, v_{0x}, is constant at 4.6 m/s. The y-component can be found by using Eqn. 4-11, 4-12, or 4-14. We will use Eqn. 4-11:

$$v_y = v_{0y} + a_y t = 0\,\text{m/s} + (9.8\,\text{m/s}^2)(0.85\,\text{s}) = 8.3\,\text{m/s}$$

Knowing the x- and y-components of the velocity, we can determine the magnitude and direction (Figure 4-29) using the Pythagorean relation and trigonometry:

$$|\vec{v}| = \sqrt{(4.6\,\text{m/s})^2 + (8.3\,\text{m/s})^2} = 9.5\,\text{m/s}$$

$$\theta = \tan^{-1}\frac{8.3\,\text{m/s}}{4.6\,\text{m/s}} = 61°\ \text{below the horizontal}$$

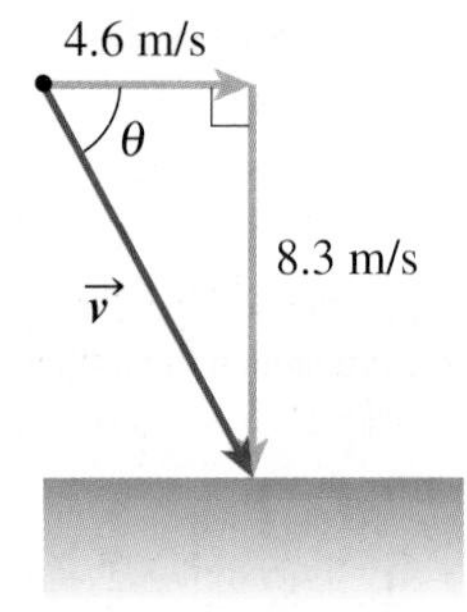

Figure 4-29 The magnitude of the velocity as the child enters the water is $v = 9.5$ m/s and $\theta = 61°$. (Sample Problem 4-6.)

Hence, the velocity is 9.5 m/s at an angle of 61° below the horizontal.

In some projectile-motion problems the initial velocity is unknown, but it can be found if enough information is provided, as seen in the next sample problem.

SAMPLE PROBLEM 4-7

A helicopter, travelling horizontally, is 80 m above the ground when a relief package is released and leaves the helicopter with the same horizontal velocity as the helicopter. The package subsequently lands on the ground 100 m horizontally ahead. What is the helicopter's velocity? (Assume two significant digits.)

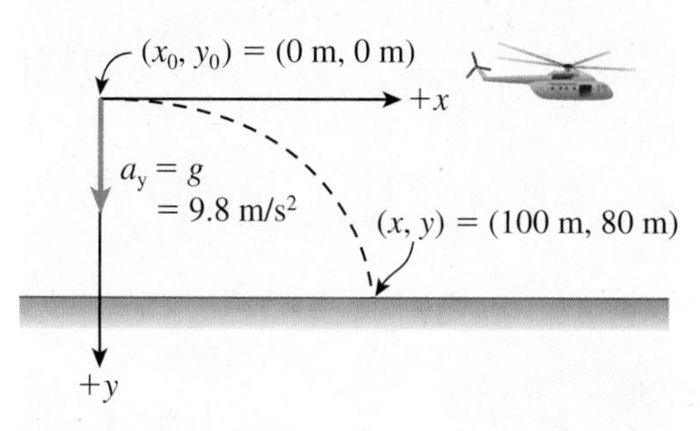

Figure 4-30 Given quantities in Sample Problem 4-7.

Solution

We select the $+x$-direction to be the horizontal direction that the helicopter is travelling, and the $+y$-direction to be downward (Figure 4-30). We choose the point from which the package is released to be at $(x_0, y_0) = (0\,\text{m}, 0\,\text{m})$. The quantities involved are shown in the table.

Horizontal motion	Vertical motion
$x_0 = 0$ m	$y_0 = 0$ m
$x = 100$ m	$y = 80$ m
$v_{0x} = ?$ (required)	$v_{0y} = 0$ m/s
$t = ?$	$v_y = ?$
	$a_y = g = 9.8$ m/s^2
	$t = ?$

The common variable, time, can be found by considering the vertical motion. We use Eqn. 4-13:

$$y = y_0 + v_{0y}t + \tfrac{1}{2}a_y t^2$$

Substituting $y_0 = 0\,\text{m}$, $v_{0y} = 0\,\text{m/s}$, and $a_y = g$, we get

$$y = \tfrac{1}{2}gt^2 \quad \text{and} \quad t^2 = \frac{2y}{g}$$

$$\therefore t = \pm\sqrt{\frac{2y}{g}} = \pm\sqrt{\frac{2\,(80\text{ m})}{9.8\text{ m/s}^2}} = \pm 4.0\,\text{s}$$

Only the positive root applies, so the time is 4.0 s. Using this time in Eqn. 4-10 rearranged to find v_{0x} for the horizontal motion,

$$v_{0x} = \frac{\Delta x}{t} = \frac{x - x_0}{t} = \frac{(100 - 0)\,\text{m}}{4.0\,\text{s}} = 25\,\text{m/s}$$

The initial velocity of the package is 25 m/s horizontally, so the helicopter must also have been travelling at 25 m/s horizontally when the package was released.

EXERCISES

Note: In all questions related to projectile motion, assume air resistance is negligible.

4-13 A squash ball is struck horizontally and moves freely through the air. What is its vertical acceleration? horizontal acceleration?

4-14 Why is an airplane flying through the air not an example of projectile motion?

4-15 A marble is rolled off a table at a horizontal velocity of 1.7 m/s. If the table top is 0.86 m above the floor,

(a) How long will the marble be in flight?

(b) Where will the marble land?

4-16 A tennis player serves a ball horizontally, giving it a speed of 24.0 m/s from a height of 2.50 m. The player is 12.0 m from the net and the top of the net is 0.900 m above the court surface.

(a) How long is the ball in the air, assuming it clears the net and lands in the other court?

(b) How far horizontally does the ball travel?

(c) With what velocity does the ball strike the court surface?

(d) By how much distance does the ball clear the net?

4-17 A stone is thrown horizontally from a high cliff with a speed of 5.0 m/s. Determine the stone's position and velocity at $t = 0.0$ s, 0.50 s, 1.0 s, 1.5 s, and 2.0 s. Then draw a diagram showing the path of the projectile (an x-y plot) and the velocity at each position plotted. Include the components of each velocity vector.

4-18 With what horizontal speed must a bowler pitch a cricket ball so that it falls 90 cm (two significant digits) as it travels 18.4 m toward the batsman?

Figure 4-31 A fly ball in baseball.

4.4 Solving Projectile Motion Problems

The sample problems concerning projectile motion in Section 4.3 all had an initial velocity that was horizontal, as this is the simplest case with which to start our discussions of projectile motion. The same kinematics equations apply if the initial velocity is at some angle to the horizontal, which means that v_{0y} is different from zero. However, when dealing with such situations, we need to be careful with the signs of the various quantities involved in the problem. As an example, suppose that the initial velocity has a vertical component that is upward, such as when a fly ball is hit in baseball (Figure 4-31). If the $+y$-direction is chosen to be upward, then v_{0y} will be positive, and the vertical position, y, will also be positive as the ball flies through the air (Figure 4-32 (a)). However, a_y will be negative, since the gravitational acceleration is downward. If, instead, we choose the $+y$-direction to be downward, then v_{0y} and y will be negative, and a_y will be positive (Figure 4-32 (b)). Of course, whether we choose the $+y$ direction to be up or down has no effect on the physics of the problem, that is, on how far the ball will travel, what its maximum height will be, how long it will be in the air, etc. As long as we are consistent in our signs with either approach, we will get the same answer at the end of our calculations.

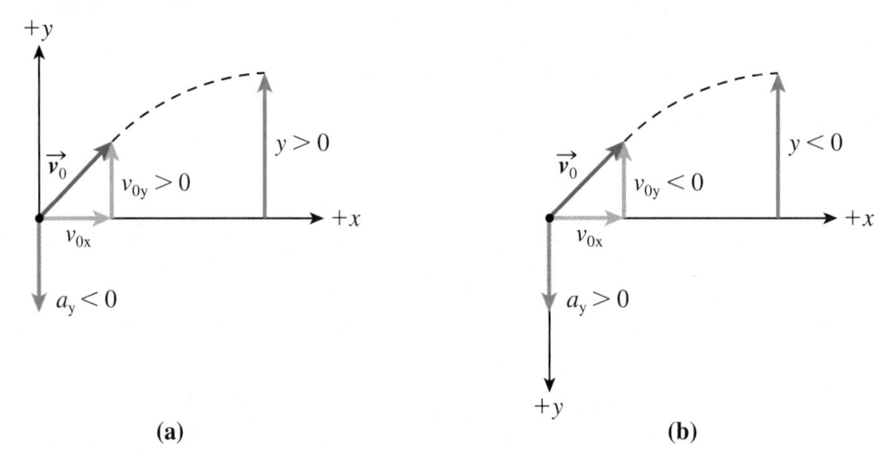

Figure 4-32 Choosing the $+y$-direction to be **(a)** upward or **(b)** downward affects the signs of y, v_{0y}, and a_y, but not the outcome of the analysis of the problem.

SAMPLE PROBLEM 4-8

A woman strikes a golf ball, which leaves the ground with an initial velocity of 42.0 m/s at an angle of 32.0° above the horizontal. Determine the ball's

(a) horizontal range of travel

(b) maximum height

(c) horizontal distance travelled when it is 15.0 m above the ground

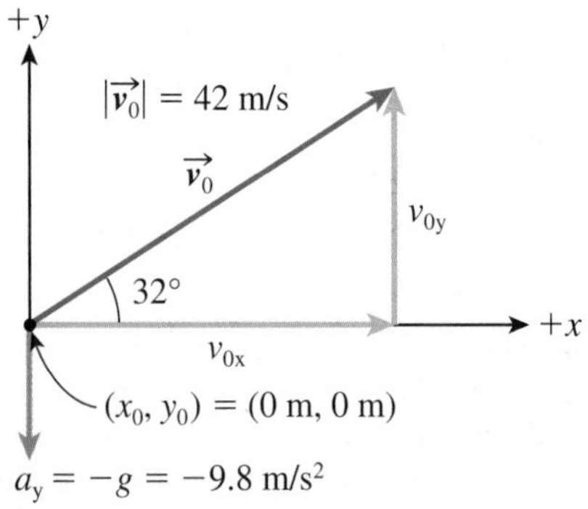

Figure 4-33 Set-up for Sample Problem 4-8.

Math Tip

To review trigonometric functions such as sine and cosine, refer to Chapter 3, Section 3.3.

Solution

(a) We choose the $+x$-direction to be the horizontal direction toward which the ball is hit, and the $+y$-direction to be vertically upward (Figure 4-33). We select the point where the ball is hit to be at $(x_0, y_0) = (0\text{ m}, 0\text{ m})$. We begin by finding the vertical and horizontal components of the initial velocity:

$$v_{0y} = (42.0\text{ m/s})(\sin 32.0°) = 22.257\text{ m/s} = 22.3\text{ m/s}$$

$$v_{0x} = (42.0\text{ m/s})(\cos 32.0°) = 35.618\text{ m/s} = 35.6\text{ m/s}$$

Next, we list the quantities of the horizontal and vertical motions separately.

Horizontal (v_x = constant)	Vertical (a_y = constant)
$x_0 = 0$ m	$y_0 = 0$ m
$x = ?$ (required to find in (a))	$y = 0$ m for (a)
$v_{0x} = 35.6$ m/s	$v_{0y} = 22.3$ m/s
$t = ?$	$v_y = -22.3$ m/s for (a), by symmetry
	$a_y = -g = -9.80$ m/s^2
	$t = ?$

Notice in the above list that as the ball hits the ground (assumed level), it returns to the position $y = 0$ m, and its y-component of velocity is reversed from its initial value (just as for a ball going straight up and down).

In order to determine the horizontal range, we will first use the vertical data to find the time, and then use the time in calculating the range. There is more than one way to calculate the time. We will use Eqn. 4-13:

$$y = y_0 + v_{0y}t + \tfrac{1}{2}a_y t^2$$

Substituting numerical values,

$$0 = 0 + (22.3\,\text{m/s})t - 4.90\,(\text{m/s}^2)t^2$$

Factoring out a "t": $0 = t(22.3\,\text{m/s} - 4.90\,\text{m/s}^2\,t)$

Hence, either $t = 0$ (when the ball is struck), or $22.3\text{ m/s} - 4.90\text{ m/s}^2\,t = 0$ (when the ball lands). Solving this latter equation for t (while keeping the extra digits in the calculator) gives $t = 4.54$ s.

Now, the horizontal range can be found from Eqn. 4-10:

$$x - x_0 = v_{0x}t = 35.6\,\text{m/s} \times 4.54\,\text{s} = 1.62 \times 10^2\,\text{m}$$

Thus, the horizontal range is 1.62×10^2 m.

(b) In order to find the ball's maximum height, we need to recognize that, at the peak of the trajectory, the ball's y-component of instantaneous velocity is zero, that is, $v_y = 0$ m/s. (This statement is similar to that for a ball travelling vertically straight up and down.) Knowing this, we can use Eqn. 4-14 to find the height:

$$v_y^2 - v_{0y}^2 = 2a_y(y - y_0)$$

Substituting with $v_y = 0\,\text{m/s}$, $a_y = -g$, *and* $y_0 = 0\,\text{m}$ gives

$$v_{0y}^2 = 2gy$$

Hence, $$y = \frac{v_{0y}^2}{2g} = \frac{(22.3\,\text{m/s})^2}{2(9.80\,\text{m/s}^2)} = 25.3\,\text{m}$$

Thus, the ball rises to a height of 25.3 m above the ground.

(c) We are required to find the horizontal distance travelled when the ball is 15.0 m above the ground. In other words, we are asked to determine "x," given that $y = 15.0$ m. We first use vertical information to find the time, and then use the time to determine "x." Eqn. 4-13 will give us the time:

$$y = y_0 + v_{0y}t + \tfrac{1}{2}a_y t^2$$

Substituting known quantities,

$$15.0\,\text{m} = 0\,\text{m} + (22.3\,\text{m/s})t - 4.90\,(\text{m/s}^2)t^2$$

This is a quadratic equation, which we rearrange into the standard form

$$(4.90\,\text{m/s}^2)t^2 - (22.3\,\text{m/s})t + 15.0\,\text{m} = 0$$

Using the quadratic formula (and keeping the extra digits in the calculator),

$$t = \frac{-b \pm \sqrt{b^2 - 4ac}}{2a} \quad \text{where } a = 4.90\,\text{m/s}^2, b = -22.3\,\text{m/s, and } c = 15\,.0\,\text{m}$$

$$= \frac{-(-22.3\,\text{m/s}) \pm \sqrt{(-22.3\,\text{m/s})^2 - 4(4.90\,\text{m/s}^2)(15.0\,\text{m})}}{2(4.90\,\text{m/s}^2)}$$

$$= 0.82\text{ s or } 3.72\text{ s}$$

Thus, the ball is 15.0 m above the ground twice during the flight, once on the way up and again on the way down. We can now use Eqn. 4-10 to determine the horizontal position, x, for each time:

$$x_1 = x_0 + v_{0x}t_1 = 0\,\text{m} + (35.6\,\text{m/s})(0.82\,\text{s}) = 29\,\text{m}$$

and

$$x_2 = x_0 + v_{0x}t_2 = 0\,\text{m} + (35.6\,\text{m/s})(3.72\,\text{s}) = 1.32 \times 10^2\,\text{m}$$

Thus, when the ball has an elevation of 15.0 m, the horizontal distance that it has travelled is either 29 m or 1.32×10^2 m.

Maximum Horizontal Range

As you learned in previous sample problems, the horizontal range of a projectile can be found by applying the kinematics equations one step at a time. However, it is possible to derive an equation to find the range of a projectile given its initial speed, v_0, and angle of projection, θ (Figure 4-34). This derivation is given here for the special case in which the projectile lands at the same elevation from which it began.

The horizontal range of a projectile is the displacement, Δx, or $x - x_0$. It is convenient to choose the initial x-component of position, x_0, to be zero. Hence, we can write the range simply as the x-position, x, which is equal to the product $v_{0x}t$ (from Eqn. 4-10). The time, t, is the same as the time for the projectile to rise and fall back to its original vertical position, so we can determine this time using the vertical motion equations. Thus, for the vertical motion, using Eqn. 4-13,

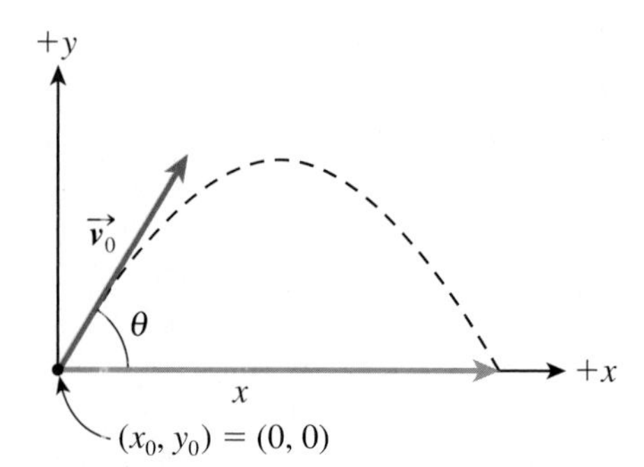

Figure 4-34 The horizontal range, x, of a projectile.

$$y = y_0 + v_{0y}t + \tfrac{1}{2}a_y t^2$$

Choosing the forward direction as $+x$ and up as $+y$, and substituting $y = 0$ m (for landing), $y_0 = 0\,\text{m}$, $v_{0y} = v_0 \sin\theta$, and $a_y = -g$:

$$0 = 0 + (v_0 \sin\theta)t - \tfrac{1}{2}gt^2$$

Thus, $\quad 0 = t(v_0 \sin\theta - \tfrac{1}{2}gt)$

Therefore, either $t = 0$ (takeoff), or $v_0 \sin\theta - \tfrac{1}{2}gt = 0$ (landing). Solving the latter equation for t gives

$$t = \frac{2v_0 \sin\theta}{g}$$

Going back to the horizontal motion,

$$x = v_{0x}t = (v_0 \cos\theta)(t) = v_0 \cos\theta \times \frac{2v_0 \sin\theta}{g} = \frac{v_0^2}{g}2\sin\theta\,\cos\theta$$

DID YOU KNOW?

Galileo (1564–1642) advanced the study of projectile motion by using a parabola to describe the motion of a projectile that is not influenced by air resistance. He proved that the maximum horizontal range occurs when the angle of projection is 45°.

However, $2\sin\theta\cos\theta = \sin 2\theta$. Hence, the horizontal range is given by

$$x = \frac{v_0^2}{g}\sin 2\theta$$

where v_0 is the magnitude of the initial velocity of the projectile at an angle θ above the horizontal. This applies only if the projectile lands at the same level from which it started. Do not use the equation for any other circumstance. (A general equation for range could be derived, but it is somewhat more complex.)

We can use the equation for range to determine the maximum range of a projectile. The maximum value of $\sin 2\theta$ is 1, and it occurs when $2\theta = 90°$ or $\theta = 45°$. Thus, the maximum range occurs when the projection angle is 45°, and the equation for range becomes

$$x_{MAX} = \frac{v_0^2}{g} \quad (\text{for } \theta = 45°)$$

The actual horizontal range would be somewhat less if air resistance were taken into consideration.

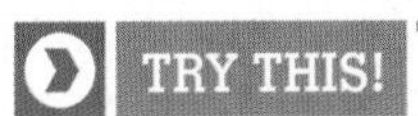

Speed of a Ball Throw

Choose a ball (a golf ball, tennis ball, baseball, etc.) that you can throw safely with a maximum horizontal range outdoors. Perform the measurement(s) and calculations needed to determine the magnitude of the initial velocity of the ball. (Possible measurements include distance, time, and launch angle.)

EXERCISES

Note: In all questions related to projectile motion, assume air resistance is negligible.

4-19 A soccer ball is kicked and it undergoes projectile motion, landing back on the field.

(a) How does its rise time compare to its fall time?

(b) What is the ball's acceleration at the top of its flight?

4-20 A cannon is set at an angle of 45° above the horizontal. It gives a cannonball a muzzle speed of 2.2×10^2 m/s. Determine the cannonball's

(a) maximum height

(b) time of flight (to the same vertical level)

(c) horizontal range (to the same vertical level)

4-21 A football is placed on a line 25 m from the goalpost. The placement kicker kicks the ball directly toward the goalpost, giving it a velocity of 21 m/s at an angle of 47° above the horizontal. The horizontal bar of the goalpost is 3.0 m above the field. How far above or below the horizontal bar of the goalpost will the ball travel?

4-22 Northern leopard frogs are native to many areas in Canada and the United States, and are known for their jumping ability (Figure 4-35). Suppose that a leopard frog is sitting on a log, 15 cm above the surface of a pond. Startled by a predator, the frog jumps from the log with an initial velocity of 5.7 m/s at an angle of 24° above the horizontal, and lands in the pond.

Roger de Montfort/Shutterstock

Figure 4-35 A Northern leopard frog (Question 4-22).

(a) How long does the frog remain airborne?

(b) How far does the frog travel horizontally before landing?

(c) Determine the frog's velocity just before landing.

4-23 A girl throws a ball onto the roof of a house, as illustrated in Figure 4-36, and prepares to catch the ball with a baseball glove held 1.0 m above the ground. The ball rolls off the roof with a speed of 3.2 m/s.

(a) After leaving the roof, how long will the ball take to land in her glove?

(b) How far from the horizontal edge of the roof should she hold her glove?

(c) What is the ball's velocity as it reaches her glove?

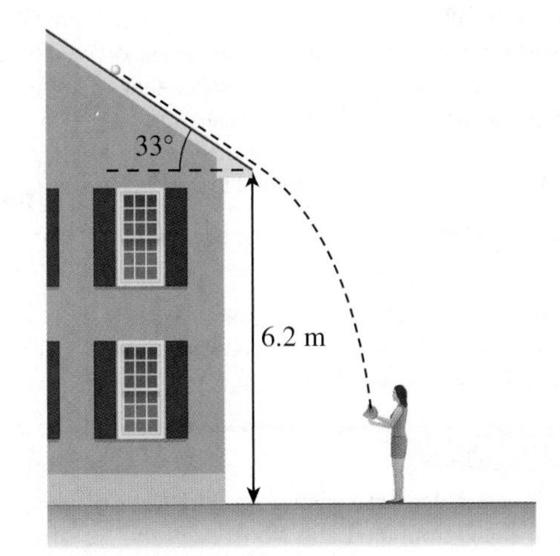

Figure 4-36 Question 4-23.

4-24 An arrow leaves a bow with an initial speed of 85 m/s.

(a) Assuming the arrow lands at the same elevation as its starting position, determine its horizontal range for these angles of projection above the horizontal: 20°; 70°; 40°; 50°; and 45°.

(b) Write conclusions based on your answers in (a) above.

4-25 What is the maximum range of a projectile if its initial speed is
(a) 2.2×10^2 m/s?
(b) 5.0×10^2 m/s?

4-26 **Fermi Question:** An experiment is set up to determine the maximum horizontal distance a rubber stopper tied to a string can be tossed when a person twirls it in a vertical circle and then releases the string at an appropriate time. Determine an estimate of that horizontal range. (For safety reasons, you should *not* perform this experiment.)

4.5 Uniform Circular Motion

Figure 4-37 Riders on this carousel undergo circular motion.

Imagine you are seated on an amusement park ride that rotates (Figure 4-37). If the radius of the path you are travelling remains the same and the speed of the ride remains constant, then this type of motion is called **uniform circular motion**. Consider the vector representation of this type of motion shown in Figure 4-38, in which an object is moving in a circle, centre C, with a radius r. Although the object's speed is constant, its velocity is constantly changing because its direction is constantly changing. At some initial instant, the velocity vector, $\vec{v}_1$, is perpendicular to the position vector, $\vec{r}_1$; at some time later, $\vec{v}_2$ is perpendicular to $\vec{r}_2$, and so on. Thus, in uniform circular motion, the velocity vector is perpendicular to the position vector.

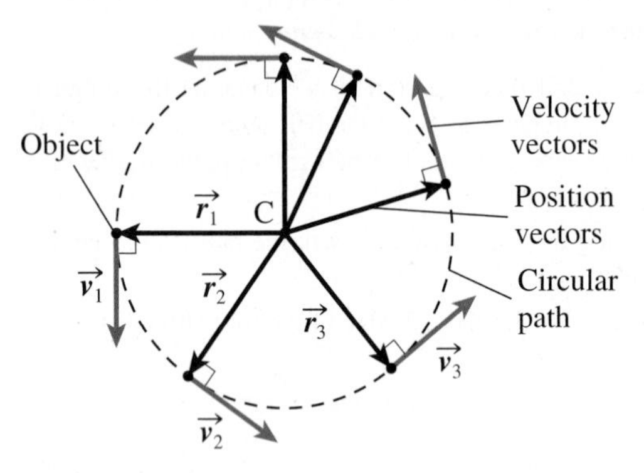

Figure 4-38 In uniform circular motion, the speed of the object remains constant but the velocity vector changes because its direction changes. Notice that at any instant the position vector (also called the radius vector) is perpendicular to the velocity vector, and the velocity vector is always tangent to the circle.

uniform circular motion motion at a constant speed in a circle (or arc) of constant radius

centripetal acceleration the acceleration that occurs in uniform circular motion; it is an instantaneous acceleration toward the centre of the circle

Uniform circular motion occurs for the individual "particles" of any object spinning at a constant rate. Examples of such objects are electric fans and motors, lawn mower blades, wheels (from the point of view of the centre of the wheel), and rotating rides at amusement parks. Circular, or almost circular, motion occurs also for objects or particles in orbits around other objects. For example, we often make the assumption that the motion of a planet around the Sun, or a satellite around Earth, or an electron around a nucleus is uniform circular motion. We will discuss these scenarios in more detail in subsequent chapters.

As you learned in Section 4.2, an object travelling at constant speed is undergoing acceleration if the direction of the velocity is changing. This is certainly true for uniform circular motion. The type of acceleration that occurs in uniform circular motion is called **centripetal acceleration** for reasons that will be clear shortly. Centripetal acceleration is an instantaneous acceleration (as opposed to an average acceleration). We will study centripetal acceleration in two stages: First we will consider its direction, and then we will determine its magnitude.

DID YOU KNOW?

The word "centripetal" was coined by Sir Isaac Newton. It stems from the Latin words *centrum* which means centre and *petere* which means seek. Therefore, centripetal means seeking the centre or toward the centre. Don't confuse centripetal with centrifugal. "Fugal" stems from the Latin *fugere* which means flee, so centrifugal means fleeing away from the centre. Centripetal motion is described further in Chapter 6.

The Direction of Centripetal Acceleration

Recall that the defining equation for instantaneous acceleration is Eqn. 4-9:

$$\vec{a} = \lim_{\Delta t \to 0} \frac{\Delta \vec{v}}{\Delta t}$$

To apply this definition to uniform circular motion, we will draw vector diagrams and perform vector subtractions. This is done in four stages in Figure 4-39. In each succeeding diagram, the time interval Δt between the initial velocity $\vec{v}_0$ and the final velocity $\vec{v}$ becomes smaller. In the final diagram it is clear that, as the time interval approaches zero, the direction of the change of velocity, $\Delta \vec{v}$, is close to being toward the centre of the circle. From Eqn. 4-9 above, the acceleration is in the same direction as the change of velocity, since division of $\Delta \vec{v}$ by the scalar Δt has no effect on direction. We conclude that *the direction of the centripetal acceleration is toward the centre of the circle.* Notice that the centripetal acceleration and the velocity are perpendicular to each other, a fact that is commonly misunderstood.

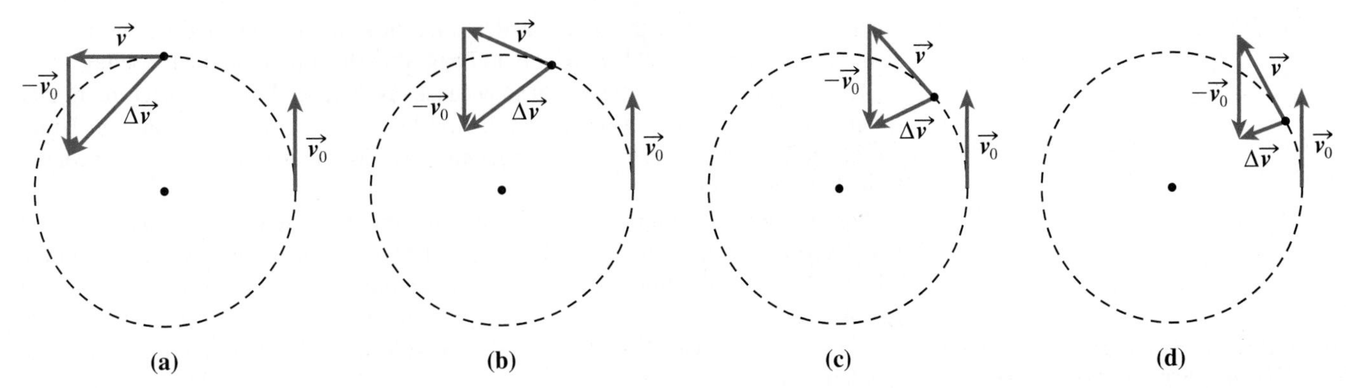

Figure 4-39 As the time interval between $\vec{v}_0$ and $\vec{v}$ decreases, the direction of $\vec{v}$ becomes closer and closer to facing the centre of the circle. In (d), Δt is small and the $\Delta\vec{v}$ vector is essentially perpendicular to the instantaneous velocity vector $\vec{v}$.

The Magnitude of Centripetal Acceleration

To derive an equation for the magnitude of the centripetal acceleration in terms of the instantaneous speed and the radius of the circle, consider the diagram in Figure 4-40 (a). It shows a particle in uniform circular motion as it moves from an initial position, $\vec{r}_0$, to a subsequent position, $\vec{r}$. Its corresponding velocities are $\vec{v}_0$ and $\vec{v}$. (Since we have *uniform* circular motion, $|\vec{v}_0| = |\vec{v}|$.) The change in position is $\Delta\vec{r}$ and the change in velocity is $\Delta\vec{v}$. Both of these quantities involve a vector subtraction, as shown in Figure 4-40 (b) and (c).

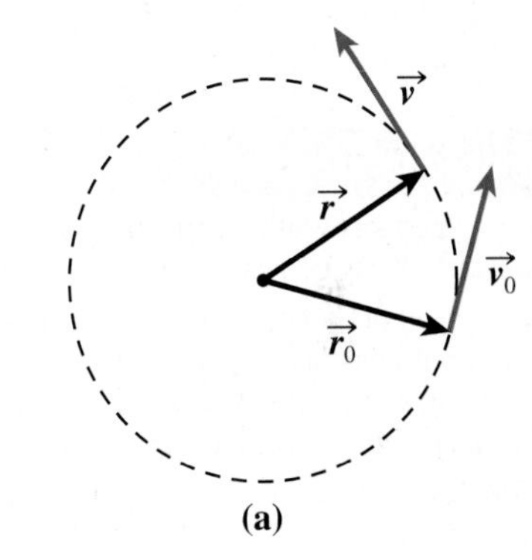

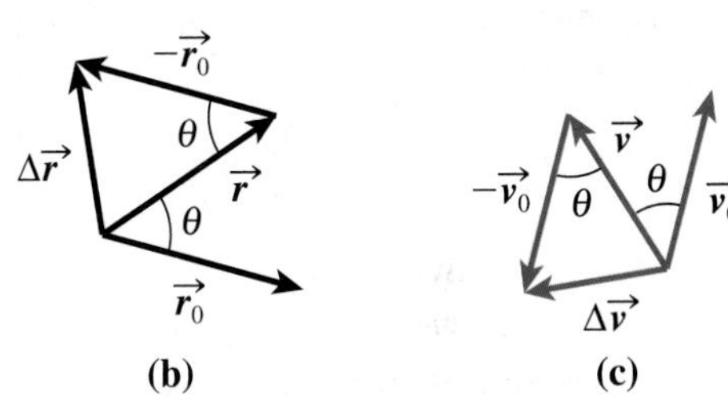

Figure 4-40 Diagrams for uniform circular motion used to derive the magnitude of centripetal acceleration.

The triangles in Figure 4-40 (b) and (c) are isosceles because $|\vec{r}_0| = |\vec{r}|$, and $|\vec{v}_0| = |\vec{v}|$. Since $\vec{v}_0 \perp \vec{r}_0$ and $\vec{v} \perp \vec{r}$, the two triangles are similar. Therefore, the following equation can be written:

$$\frac{|\Delta\vec{v}|}{|\vec{v}|} = \frac{|\Delta\vec{r}|}{|\vec{r}|} \quad \text{or} \quad |\Delta\vec{v}| = \frac{|\vec{v}| \times |\Delta\vec{r}|}{|\vec{r}|}$$

Now, the magnitude of the centripetal acceleration ($\vec{a}_c$) is

$$|\vec{a}_c| = \lim_{\Delta t \to 0} \frac{|\Delta\vec{v}|}{\Delta t}$$

We divide both sides of the equation for $|\Delta\vec{v}|$ by Δt to obtain

$$\frac{|\Delta\vec{v}|}{\Delta t} = \frac{|\vec{v}|}{|\vec{r}|} \times \frac{|\Delta\vec{r}|}{\Delta t}$$

$$\therefore \quad |\vec{a}_c| = \lim_{\Delta t \to 0} \frac{|\Delta\vec{v}|}{|\Delta t|} = \lim_{\Delta t \to 0}\left[\frac{|\vec{v}|}{|\vec{r}|} \times \frac{|\Delta\vec{r}|}{\Delta \mathrm{t}}\right]$$

Now, $\lim\limits_{\Delta t \to 0} \dfrac{|\Delta\vec{r}|}{\Delta t} = |\vec{v}|$ (magnitude of the instantaneous velocity)

$$\therefore \quad |\vec{a}_c| = \frac{|\vec{v}|}{|\vec{r}|} \times |\vec{v}|$$

Hence, the magnitude of the centripetal acceleration is

$$a_c = \frac{v^2}{r} \qquad (4\text{-}15)$$

where v is the speed of the object undergoing uniform circular motion, and r is the radius of the circle or arc. Eqn. 4-15 will be used frequently in problem-solving.

The equation $a_c = v^2/r$ makes sense because, as the speed of the object in circular motion increases, the direction of the velocity changes more quickly, requiring a larger acceleration; as the radius becomes larger, the direction changes more slowly, meaning a smaller acceleration.

For objects undergoing uniform circular motion, it is often the case that the speed is not known, but the radius and the period of time for one complete trip around the circle are known. To find the centripetal acceleration with this information we use the fact that the speed is constant and equals the distance travelled in one complete revolution ($2\pi r$) divided by the **period of revolution**, T, which is the time for one revolution. Thus,

period of revolution the time for one revolution of an object in uniform circular motion (symbol: T)

$$v = \frac{2\pi r}{T}$$

Substituting this expression for v into $a_c = v^2/r$, we obtain

$$a_c = \frac{4\pi^2 r}{T^2} \qquad \text{(4-16)}$$

where a_c is measured in metres per second squared when r is measured in metres and T in seconds.

Finally, the **frequency of revolution**, f, is the number of revolutions per second. Frequency is the reciprocal of period and is measured in cycles per second, or s^{-1}. This unit is given a special SI name and symbol, the hertz (Hz), named after the German physicist Heinrich Hertz, 1857–1894. Since $f = 1/T$, the equation for centripetal acceleration can then be written

frequency of revolution the number of revolutions per second of an object in uniform circular motion (symbol: f)

$$a_c = 4\pi^2 r f^2 \qquad \text{(4-17)}$$

Problem-Solving Tip

The following facts will be applied in analyzing situations involving centripetal acceleration:

- The speed of an object in uniform circular motion is constant and can often be found by using $v = 2\pi r/T$, where T is the period of one revolution of radius r.
- The centripetal acceleration is directed toward the centre of the circle and its magnitude can be found using one or more of these equations:

$$a_c = \frac{v^2}{r} = \frac{4\pi^2 r}{T^2} = 4\pi^2 r f^2$$

SAMPLE PROBLEM 4-9

A child on a merry-go-round is 4.4 m from the centre of the ride and is travelling at a constant speed of 1.8 m/s. What is the magnitude of the child's centripetal acceleration?

Solution

Using Eqn. 4-15, $a_c = \frac{v^2}{r} = \frac{(1.8\ \text{m/s})^2}{4.4\ \text{m}} = 0.74\ \text{m/s}^2$

Thus, the magnitude of the centripetal acceleration is $0.74\ \text{m/s}^2$.

SAMPLE PROBLEM 4-10

Find the magnitude and direction of the centripetal acceleration of a dust particle that is sitting on the outside edge of the blade of a ceiling fan that is 17.2 cm across. The fan is rotating clockwise when viewed from above at a rate of 45 rev/min, and at the instant in question, the dust particle is moving northward.

Solution

Converting 45 rev/min to hertz, $\frac{45\ \text{rev}}{\text{min}} \times \frac{1\ \text{min}}{60\ \text{s}} = 0.75\ \frac{\text{rev}}{\text{s}}$.

Thus, $f = 0.75$ Hz, or $0.75\ \text{s}^{-1}$, and $r = \frac{1}{2}$ (17.2 cm) = 8.6 cm. Using Eqn. 4-17,

$$\begin{aligned} a_c &= 4\pi^2 r f^2 \\ &= 4\pi^2 (8.6\ \text{cm})(0.75\ \text{Hz})^2 \\ &= 1.9 \times 10^2\ \text{cm/s}^2 \\ &= 1.9\ \text{m/s}^2 \end{aligned}$$

Since the fan is moving clockwise when viewed from above, the direction of the acceleration must be toward the east when the particle is travelling northward, as seen in Figure 4-41. Thus, the centripetal acceleration is $1.9\ \text{m/s}^2$ east.

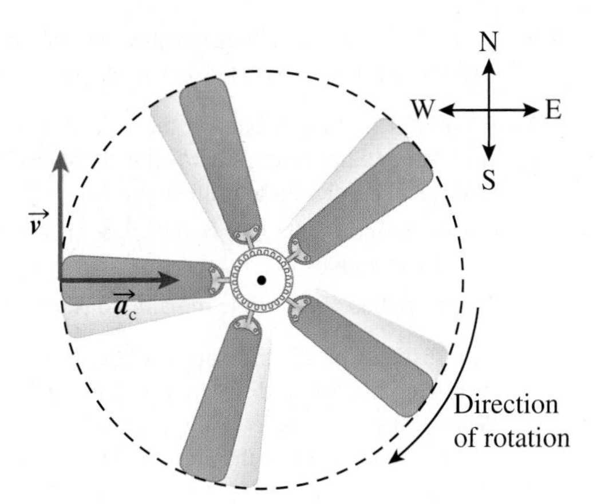

Figure 4-41 Sample Problem 4-10.

SAMPLE PROBLEM 4-11

Determine the frequency and period of revolution of an ice skater spinning such that the centripetal acceleration of her nose has a magnitude of $12\ \text{m/s}^2$. Assume the distance from her nose to the axis of rotation is 12 cm.

Solution

From $a_c = 4\pi^2 r f^2$, we have $f^2 = \dfrac{a_c}{4\pi^2 r}$

$$\therefore \quad f = \pm\sqrt{\frac{a_c}{4\pi^2 r}} = \pm\sqrt{\frac{12\ \text{m/s}^2}{4\pi^2(0.12\ \text{m})}} = \pm 1.6\ \text{s}^{-1} = \pm 1.6\ \text{Hz}$$

A negative frequency has no meaning, so the required frequency is 1.6 Hz.

$$T = \frac{1}{f} = \frac{1}{1.6\ \text{Hz}} = 0.63\ \text{s}$$

Thus, the period of revolution is 0.63 s.

TRY THIS!

Centripetal Acceleration during Exercise

Design an investigation and carry out the measurements needed to determine the magnitude of the centripetal acceleration of your fingertips as you twirl one arm in a full circle such that your hand undergoes uniform circular motion in the vertical plane.

EXERCISES

4-27 In the expression "uniform circular motion," what does the word "uniform" mean?

4-28 The motion of a particle undergoing uniform circular motion with a speed of 4.0 m/s is illustrated in Figure 4-42. For this motion,

(a) state the direction of the velocity vector, the acceleration vector, and the radius vector at the instant shown

(b) calculate the magnitude of the centripetal acceleration

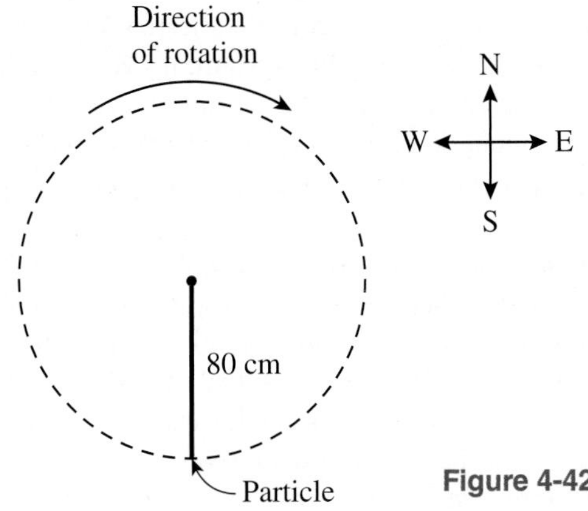

Figure 4-42 Question 4-28.

4-29 Determine the magnitude of the centripetal acceleration for each of the following motions, assuming that they involve uniform circular motion.

(a) A motorcycle travels at 25 m/s around a curve that has a radius of curvature of 1.2×10^2 m.

(b) In the death spiral in pairs figure skating (Figure 4-43), the speed of the woman's shoulder, which is 1.1 m from the centre of rotation, is 6.4 m/s.

© Duomo/CORBIS

Figure 4-43 The "death spiral" in pairs figure skating (Question 4-29(b)).

4-30 If the direction of an object moving with uniform circular motion is reversed, what happens to the direction of the centripetal acceleration?

4-31 A child is twirling a ball on the end of a string in a horizontal circle around her head. What is the effect on the magnitude of the centripetal acceleration of the ball if

(a) the speed of the ball remains constant but the radius of the circle doubles

(b) the radius of the circle remains constant but the speed doubles

4-32 At a distance of 25 km from the "eye" of a hurricane, the wind is moving at 180 km/h in a circle around the eye (Figure 4-44). What is the magnitude of the centripetal acceleration, in metres per second squared, of the air particles that make up the wind?

© Matthew Trommer/Dreamstime.com/GetStock.com

Figure 4-44 The eye of a hurricane (Question 4-32).

4-33 A cowhand is about to lasso a calf with a rope that is undergoing uniform circular motion. The time for one complete revolution of the rope is 1.2 s and the end of the rope is 4.3 m from the centre of the circle. Calculate the magnitude of the centripetal acceleration of the end of the rope.

4-34 Using an ultracentrifuge, which is a centrifuge operating at a very high rate of rotation, greatly reduces the sedimentation time of blood cells or other materials. (Chapter 6 explains the physics of centrifuges.)

(a) Calculate the magnitude of the centripetal acceleration of a point 7.4 cm from the centre of an ultracentrifuge rotating at 1.3 MHz.

(b) Compare your answer in (a) to the magnitude of the acceleration due to gravity on Earth.

4-35 **Fermi Question:** A circular bowl containing frozen berries is placed at the centre of the tray in a microwave oven. Estimate the magnitude of the centripetal acceleration of a berry along the inside of the bowl's edge as the tray is rotating. State all your assumptions.

4-36 **Fermi Question:** A land iguana is resting on the shoreline of a Galapagos Island in the Pacific Ocean.

(a) Assuming the island is located at the equator, estimate the magnitude of the centripetal acceleration experienced by the iguana due to Earth's rotation.

(b) What percentage of the magnitude of the acceleration due to gravity is your estimate in (a)?

Zoran Karapancev/Shutterstock.com

Figure 4-45 Motion can be described either with respect to the frame of reference of the moving parade float or with respect to the frame of reference of the stationary spectators, commonly called Earth's frame of reference.

frame of reference a coordinate system fixed to an object, such as Earth, relative to which a motion can be observed

relative velocity the velocity of an object relative to a specific frame of reference

4.6 Frames of Reference and Relative Velocity

The parade float with cartoon characters and passengers on it in Figure 4-45 is travelling at a constant velocity past parade enthusiasts by the side of the road. Relative to the spectators, both the float and everything on it are in motion. However, the cartoon characters and the passengers sitting on the float are not in motion relative to each other. The spectators by the side of the road are stationary relative to Earth's frame of reference, while everyone on the float is stationary relative to the float's frame of reference but moving relative to Earth's frame of reference. A **frame of reference** is a coordinate system fixed to an object, such as Earth, relative to which a motion can be observed. The motion observed depends on the frame of reference chosen.

Without thinking about it, we often choose Earth as the frame of reference to be used. For example, whenever we have mentioned the motion of a vehicle or an athlete, the motion has been assumed to be relative to Earth's frame of reference. Sometimes other frames of reference may be more convenient than Earth's frame. For example, if we study the motion of the planets relative to the solar system, the Sun's frame of reference is the logical choice. If we consider a particle on the rim of a rolling wheel, the (centre of the) wheel is the convenient frame of reference because the motion is then circular, whereas in Earth's frame of reference the motion is more complex (Figure 4-46).

The velocity of an object relative to a specific frame of reference is called **relative velocity**. Although all velocities are relative to some frame of reference, we have not bothered to use this expression before because we considered only motion relative to one frame of reference at a time. We are now going to study situations in which at least two frames of reference are involved. Such situations occur for travel in an airplane when there is a wind, for watercraft when there is a flowing river, or for passengers moving about in a train, to name just three examples.

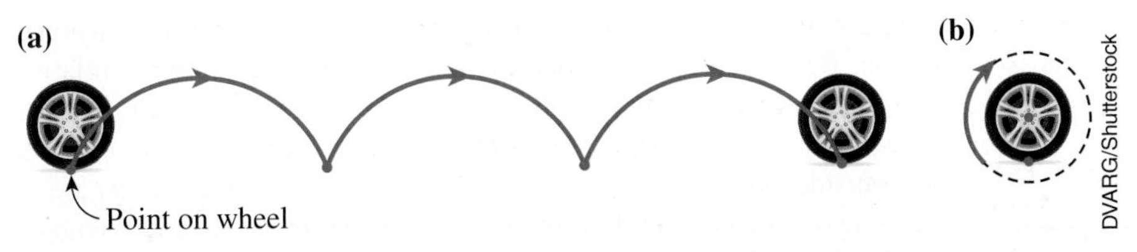

Figure 4-46 The motion of a point on a rolling wheel is more complex when observed from Earth's frame of reference **(a)**, but simple if viewed from the frame of reference of the wheel's centre **(b)**.

To analyze relative velocity in more than one frame of reference, we will introduce a new symbol and apply that symbol to the addition of relative velocity vectors. Consider the symbol $\vec{v}_{PE}$, the boldface vector symbol with a double subscript in capital letters. The first subscript represents the object whose velocity is stated relative to the second subscript, which is the frame of reference. Thus, the symbol $\vec{v}_{PE}$ is read "the velocity of object P relative to the frame of reference E." If P is a plane travelling at 510 km/h east in Earth's frame of reference, E, then

$$\vec{v}_{PE} = 510 \text{ km/h east}$$

Now let us add another frame of reference, A, which could be the air or wind affecting the plane's motion. Then $\vec{v}_{PA}$ is the plane's velocity relative to the air, and $\vec{v}_{AE}$ is the air's velocity relative to Earth. The three velocities mentioned so far are related through the *relative velocity equation*:

$$\vec{v}_{PE} = \vec{v}_{PA} + \vec{v}_{AE} \qquad \textbf{(4-18)}$$

Notice that the "+" sign in Eqn. 4-18 represents a vector addition to give the resultant relative velocity, $\vec{v}_{PE}$. This equation applies whether the motion is in one, two, or three dimensions. As an example, first consider one-dimensional motion in which a plane's velocity relative to the air, called the *air velocity*, is 450 km/h east, and the air's velocity relative to Earth is 60 km/h east (Figure 4-47). Using Eqn. 4-18,

$$\begin{aligned}\vec{v}_{PE} &= \vec{v}_{PA} + \vec{v}_{AE} \\ &= 450 \text{ km/h east} + 60 \text{ km/h east} \\ &= 510 \text{ km/h east}\end{aligned}$$

In this case the wind causes the velocity of the plane relative to Earth to increase, which makes sense since the velocities are in the same direction.

Figure 4-47 The velocities of an airplane and the air: $\vec{v}_{PA}$ is the velocity of the plane relative to air; $\vec{v}_{AE}$ is the velocity of the air relative to Earth; $\vec{v}_{PE}$ is the resultant velocity of the plane relative to Earth.

$$\vec{v}_{PE} = \vec{v}_{PA} + \vec{v}_{AE} \qquad \vec{v}_{XY} = \vec{v}_{XC} + \vec{v}_{CY}$$

$$\vec{v}_{KM} = \vec{v}_{KL} + \vec{v}_{LM} \qquad \vec{v}_{AE} = \vec{v}_{AB} + \vec{v}_{BC} + \vec{v}_{CD} + \vec{v}_{DE}$$

Figure 4-48 Understanding the pattern of the relative velocity equation will help you with its correct application.

The remaining examples in this section deal with two-dimensional motion. Before going on, notice the pattern of the equation for relative velocity. The left side of the equation has a single relative velocity and the right side has the vector addition of two relative velocities. Furthermore, the "outside" symbols on the right side (in this example, P and E) are in the same order as the symbols on the left side, while the "inside" symbols are the same. This pattern is illustrated in Figure 4-48 for the airplane example as well as other situations.

SAMPLE PROBLEM 4-12

An airplane is capable of travelling with an airspeed of 380 km/h. Determine its velocity relative to the ground if it is flying facing east and the wind velocity is

(a) 90 km/h north

(b) 90 km/h 35° south of east

Solution

(a) If we use G for ground, P for airplane, and A for air or wind, the relative velocity equation can be written

$$\vec{v}_{PG} = \vec{v}_{PA} + \vec{v}_{AG}$$
$$= 380\,\text{km/h east} + 90\,\text{km/h north (a vector addition)}$$

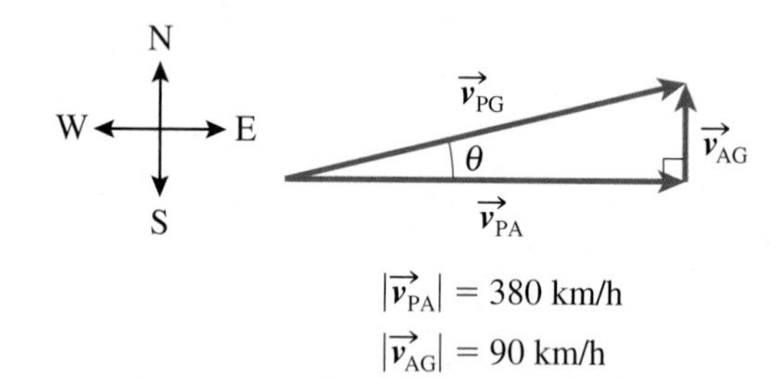

Figure 4-49 Sample Problem 4-12, part (a).

From Figure 4-49, we can see that the vector $\vec{v}_{PG}$ has a magnitude of

$$\sqrt{(380\text{ km/h})^2 + (90\text{ km/h})^2} = 3.9 \times 10^2\,\text{km/h}$$

and a direction given by

$$\theta = \tan^{-1}\left(\frac{90\,\text{km/h}}{380\,\text{km/h}}\right) = 13°\,\text{north of east}$$

Therefore, the airplane's velocity relative to the ground is 3.9×10^2 km/h 13° north of east.

(b) The diagram illustrating the vector addition $\vec{v}_{PG} = \vec{v}_{PA} + \vec{v}_{AG}$ is shown in Figure 4-50 (a). In order to determine the magnitude and direction of $\vec{v}_{PG}$, we could use components or the laws of cosines and sines, or we could find an approximate solution by using a scale diagram. We will utilize components. To avoid triple subscripts (such as $\vec{v}_{PGx}$), we will write $\vec{v}_1, \vec{v}_2,$ and $\vec{v}_R$ in place of $\vec{v}_{PA}, \vec{v}_{AG},$ and $\vec{v}_{PG}$, respectively. Hence, the vector addition becomes $\vec{v}_R = \vec{v}_1 + \vec{v}_2$, where the subscript "R" represents "resultant."

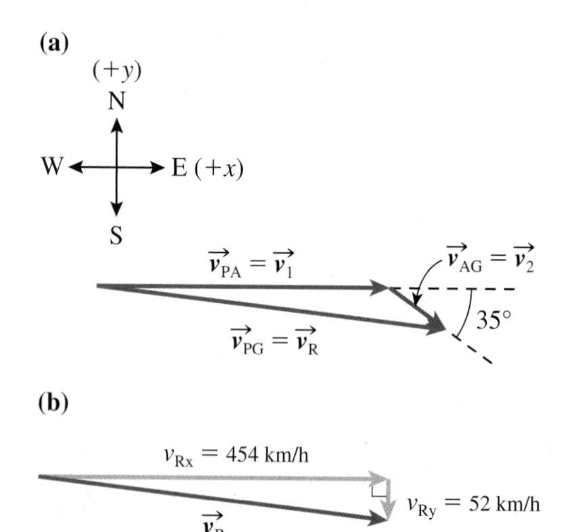

Figure 4-50 Sample Problem 4-12, part (b). **(a)** Adding the velocity vectors. **(b)** The components of the resultant velocity.

We now determine the x- and y-components of $\vec{v}_1$ and $\vec{v}_2$. We choose the $+x$- and $+y$-directions to be east and south respectively. Then $\vec{v}_1$ has only an x-component, 380 km/h. The components of $\vec{v}_2$ are

$$v_{2x} = (90\,\text{km/h})\cos 35° = 74\,\text{km/h}$$

$$v_{2y} = (90\,\text{km/h})\sin 35° = 52\,\text{km/h}$$

Adding components,

$$v_{Rx} = v_{1x} + v_{2x}$$
$$= (380 + 74)\,\text{km/h}$$
$$= 454\,\text{km/h}$$

$$\begin{aligned} v_{Ry} &= v_{1y} + v_{2y} \\ &= (0 + 52)\,\text{km/h} \\ &= 52\,\text{km/h} \end{aligned}$$

Figure 4-50 (b) shows the resultant velocity having these components. Its magnitude can be determined from the Pythagorean relation, and its direction from trigonometry. Hence, the resultant velocity, that is, the velocity of the plane relative to the ground, is 4.6×10^2 km/h at an angle of 6.5° south of east.

The next sample problem shows that quantities other than velocity can be found for relative motion.

SAMPLE PROBLEM 4-13

A kayaker, capable of paddling at a speed of 4.5 m/s in still water, is on a river that is flowing with a velocity of 3.2 m/s west. The river is 220 m wide.

(a) If the kayaker is paddling north, what is the kayak's velocity relative to the shore?

(b) How long will the kayak take to cross the river?

(c) Where will the kayak land relative to the starting position?

(d) If the kayaker had wanted to land straight across from the initial position, at what angle would he have had to aim?

Solution

(a) Figure 4-51 illustrates the situation described for (a), (b), and (c). Let S represent the shore, W the water, and K the kayak. The relative velocities are

$$\begin{aligned} \vec{v}_{KW} &= 4.5\,\text{m/s north} \\ \vec{v}_{WS} &= 3.2\,\text{m/s west} \\ \vec{v}_{KS} &= ? \end{aligned}$$

The relative velocity equation is

$$\begin{aligned} \vec{v}_{KS} &= \vec{v}_{KW} + \vec{v}_{WS} \\ &= 4.5\,\text{m/s north} + 3.2\,\text{m/s west} \end{aligned}$$

The magnitude of $\vec{v}_{KS}$ is $\sqrt{(4.5\,\text{m/s})^2 + (3.2\,\text{m/s})^2} = 5.5\,\text{m/s}$

and its direction (see Figure 4-51) is $\theta = \tan^{-1}\left(\dfrac{3.2\,\text{m/s}}{4.5\,\text{m/s}}\right) = 35°$ west of north.

Hence, the velocity of the kayak relative to the shore is 5.5 m/s at an angle of 35° west of north.

(b) We can determine the time by considering only the motion across the river. In this direction, the speed is constant ($|\vec{v}_{KW}| = 4.5\,\text{m/s}$) and the distance ($d$) to be covered is 220 m. Hence,

$$v_{KW} = \frac{d}{\Delta t} \quad \text{giving} \quad \Delta t = \frac{d}{v_{KW}} = \frac{220\,\text{m}}{4.5\,\text{m/s}} = 49\,\text{s}$$

Thus, the kayaker takes 49 s to cross the river.

(c) The current carries the kayak downstream (west) during the time it takes to cross the river. The distance travelled downstream is given by

$$v_{WS}\Delta t = (3.2\,\text{m/s})(49\,\text{s}) = 1.6 \times 10^2\,\text{m}$$

Thus, relative to the starting position, the kayak lands 1.6×10^2 m west and 220 m north.

Sheri Armstrong/Shutterstock

(a)

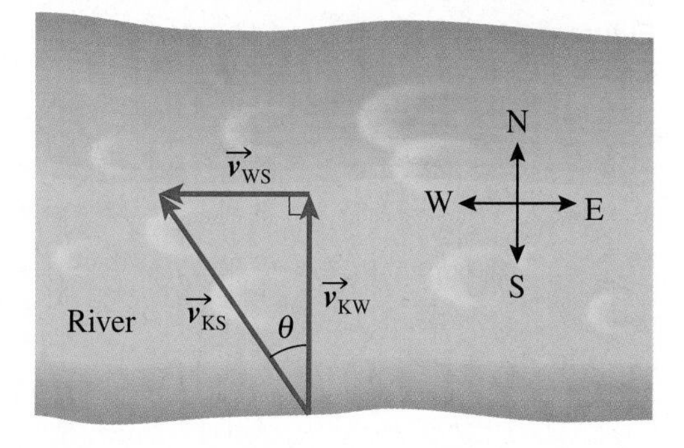

(b)

Figure 4-51 Sample Problem 4-13, parts (a), (b), and (c).

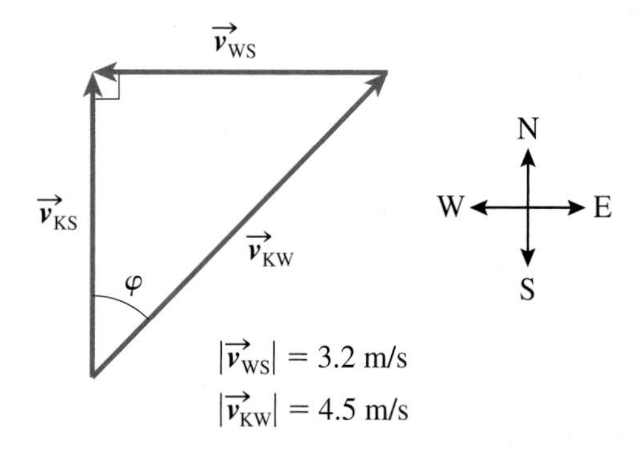

Figure 4-52 Sample Problem 4-13, part (d).

(d) Figure 4-52 shows the situation here. In this case, the velocity of the kayak relative to the water, $\vec{v}_{KW}$, is the hypotenuse of the triangle in (b). Notice that (the resultant velocity) $\vec{v}_{KS}$ points due north in order for the kayaker to travel straight across the river. Thus,

$$\sin\varphi = \frac{3.2\,\text{m/s}}{4.5\,\text{m/s}} \quad \text{giving} \quad \varphi = 45°$$

Thus, the kayak must head upstream at an angle of 45° to the shore, or 45° east of north.

Christophe Testi/Shutterstock

Figure 4-53 To a person in the frame of reference of a moving sidewalk, people that are standing beside the sidewalk appear to be moving in the opposite direction.

Sometimes it is useful to know that the velocity of object A relative to object B has the same magnitude as the velocity of B relative to A but is opposite in direction. In equation form this is written

$$\vec{v}_{AB} = -\vec{v}_{BA}$$

Consider, for example, the situation at an airport in Figure 4-53, in which a passenger, M, on a moving sidewalk goes past a person, S, standing by the side of the moving sidewalk. If $\vec{v}_{MS} = 1.5$ m/s north, then S is viewing M moving north at 1.5 m/s. However, in the frame of reference of the moving sidewalk, M observes that S appears to be moving at 1.5 m/s south, that is, $\vec{v}_{SM} = -\vec{v}_{MS} = -1.5$ m/s north or 1.5 m/s south. This is applied in a vector subtraction in the next sample problem.

SAMPLE PROBLEM 4-14

A small motorboat, M, is travelling at 8.4 m/s east on a lake, and a sailboat, S, is travelling nearby at 5.8 m/s north. Both velocities are stated in the frame of reference of the water, W. What is the velocity of the motorboat relative to the sailboat?

Solution

The relative velocity equation is

$$\vec{v}_{MS} = \vec{v}_{MW} + \vec{v}_{WS}$$

We are given $\vec{v}_{SW}$ rather than $\vec{v}_{WS}$, so we must use the fact that $\vec{v}_{WS} = -\vec{v}_{SW}$ to solve the problem. Hence, $\vec{v}_{MS} = \vec{v}_{MW} + (-\vec{v}_{SW})$.

This vector subtraction is shown in Figure 4-54. The velocity of the motorboat relative to the sailboat can now be determined using the Pythagorean relation and trigonometry. It is 1.0×10^1 m/s at an angle $\theta = 35°$ south of east.

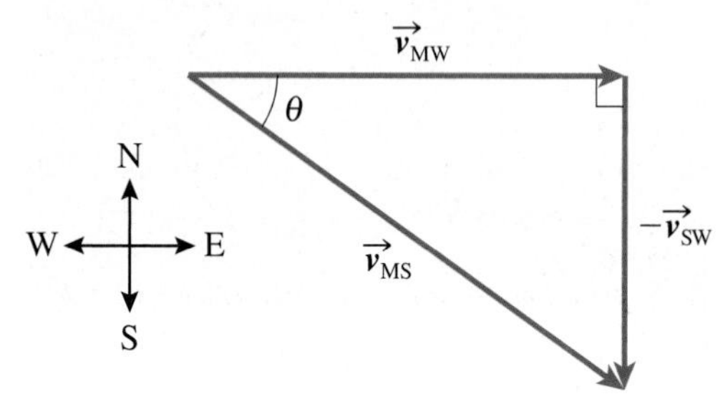

Figure 4-54 Sample Problem 4-14 (solution). $\vec{v}_{MS} = \vec{v}_{MW} + (-\vec{v}_{SW})$.

The analysis of relative velocity given here applies to velocities we commonly experience. However, it does not apply when the velocity becomes extremely high, that is, at or near the speed of light, 3.0×10^8 m/s. For example, if airplane A, travelling at 250 m/s east relative to Earth, is approaching airplane B, travelling at 200 m/s west relative to Earth, the velocity of B relative to A is 450 m/s west. However, if airplane A is approaching a beam of light that, relative to Earth, has a velocity of 3.0×10^8 m/s west, the velocity of the beam relative to A is still 3.0×10^8 m/s west, not the sum, 3.0×10^8 m/s + 250 m/s, west. The idea that the speed of light is constant relative to all frames of reference—that is, it does not depend on the motion of the observer—was first introduced by Albert Einstein in his Special Theory of Relativity in 1905.

EXERCISES

4-37 Something is wrong with each equation below. Rewrite each one to show the correct form.

(a) $\vec{v}_{AC} = \vec{v}_{AB} - \vec{v}_{BC}$

(b) $\vec{v}_{DE} = \vec{v}_{LD} + \vec{v}_{LE}$

(c) $\vec{v}_{MN} = \vec{v}_{NT} + \vec{v}_{TM}$ (show two ways of correcting this)

(d) $\vec{v}_{LP} = \vec{v}_{ML} + \vec{v}_{MN} + \vec{v}_{NO} + \vec{v}_{OP}$

4-38 A tourist on board a cruise ship (Figure 4-55) walks with a speed of 1.1 m/s relative to the ship. The ship is moving forward at a speed of 2.8 m/s relative to the water. Determine the tourist's velocity relative to the water if he is walking (a) toward the bow, (b) toward the stern, and (c) toward the port side.

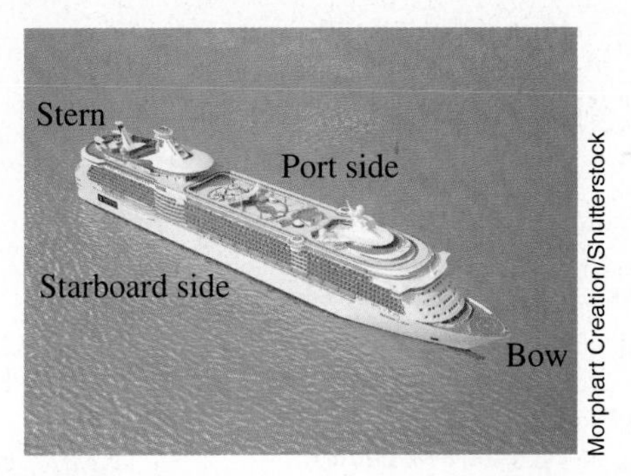

Figure 4-55 A cruise ship (Question 4-38).

4-39 A plane travelling with an air velocity of 320 km/h 30° south of west passes over Vancouver. The wind velocity is from the north at 80 km/h. Determine the location of the plane 2.0 h later. Assume two significant digits.

4-40 A helicopter can travel with an airspeed of 65 m/s. It is aimed in a direction 40° north of west. Assuming two significant digits, determine the velocity of the helicopter relative to the ground if the wind velocity is (a) 22 m/s 40° south of east, (b) 22 m/s from the west, (c) 22 m/s 60° north of east.

4-41 A swimmer, who can move at 0.80 m/s in calm water, starts swimming across a river that is 120 m wide. The river is flowing at 0.60 m/s. Assume two significant digits.

(a) Determine the swimmer's velocity relative to the shore.

(b) Find the location where the swimmer will land.

(c) At what angle would the swimmer have to aim upstream in order to land directly across from the starting point?

4-42 Two canoeists can move at the same speed in calm water. One begins canoeing straight across a river and the other aims her canoe at an angle upstream in the same river in order to land straight across from the starting position. The canoeists' speed is greater than the river-current speed. Which canoeist reaches the far side first? Explain your answer.

4-43 Assume that Montreal is 3100 km 15° north of east from San Francisco. An air flight leaves San Francisco at 7:30 a.m. and the pilot wants to land in Montreal 5.0 h later. Assume two significant digits.

(a) If there is no wind, what should be the pilot's velocity relative to the air?

(b) If the wind is blowing at 75 km/h 60° south of east, what should be the pilot's velocity relative to the air?

LOOKING BACK...LOOKING AHEAD

The quantities of distance, displacement, speed, velocity, and acceleration studied in one-dimensional motion in Chapter 2 have been extended to two dimensions in this chapter. The skills of adding and subtracting vectors, introduced in Chapter 3, have been reviewed. Distinctions between average and instantaneous velocities, and average and instantaneous accelerations, have been made for two-dimensional motion. Projectile motion was analyzed by separating the motion into horizontal and vertical components and applying the kinematics equations from Chapter 2. Uniform circular motion was analyzed using equations for centripetal acceleration in terms of speed, radius, period, and frequency. Finally, the concept of frames of reference was introduced so that relative velocity in different frames could be analyzed.

This completes our study of kinematics, and the foundation is now set for studying the forces that cause acceleration, presented in Chapters 2 and 4. Thus, Chapters 5 and 6 and parts of later chapters will apply the knowledge of kinematics in more detail.

CONCEPTS AND SKILLS

Having completed this chapter you should now be able to do the following for two-dimensional motion:

- Use the relationship involving average velocity, displacement, and time to determine one of these quantities, given the other two.
- Determine the instantaneous velocity of an object travelling in any path, including a curved path if the instantaneous speed is known.
- Use the vector subtraction of velocities to find the average acceleration of an object.
- Determine the unknown quantity in the following list, given the other quantities: average acceleration, time, initial velocity, and final velocity.
- Analyze projectile motion problems by applying the kinematics equations for constant velocity (horizontal component) and constant acceleration (vertical component).
- Solve problems related to the horizontal range of a projectile.
- Solve centripetal acceleration problems involving any or all of these variables: radius, speed, period of revolution, and frequency of revolution.
- Apply the relative velocity equation to solve problems involving two frames of reference.

KEY TERMS

You should be able to define or explain each of the following words or phrases for two-dimensional motion. In cases marked with an asterisk*, the original definition is found in Chapter 2, where one-dimensional motion was studied.

resultant displacement
average velocity*
instantaneous velocity*
average acceleration*
instantaneous acceleration*
projectile
projectile motion
horizontal range
uniform circular motion
centripetal acceleration
period of revolution
frequency of revolution
frame of reference
relative velocity

Chapter Review

Note: In all questions related to projectile motion, assume that air resistance is negligible.

MULTIPLE-CHOICE QUESTIONS

4-44 Assume $\vec{A}$ is the shortest displacement vector between your eyes and the top of the front wall of your lecture hall, and $\vec{B}$ is the shortest displacement vector between your eyes and the bottom of the same wall. These vectors make an angle of θ (above) and β (below) the horizontal. The height of the wall is

(a) $A \sin\theta - B \sin\beta$
(b) $A \sin\theta + B \sin\beta$
(c) $(A + B) \sin(\theta + \beta)$
(d) $\sqrt{|A|^2 + |B|^2}$
(e) $A \tan\theta + B \tan\beta$

4-45 Two balls, C and D, are thrown horizontally from the roof of a building at the same instant, with the initial speed of C double that of D. Assuming air resistance can be neglected,

(a) C and D hit the ground at the same time and at the same speed.
(b) C reaches the ground first.
(c) D reaches the ground first.
(d) C and D hit the ground at the same time, but C's final speed is double D's final speed.
(e) none of the above

4-46 A bottle-nosed dolphin (Figure 4-56) launches itself from the water through the air and follows a parabolic path. At the highest point in the path in the air,

(a) Neither the velocity nor the acceleration is zero.
(b) The velocity is not zero but the acceleration is zero.
(c) Both the velocity and the acceleration are zero.
(d) The velocity is zero but the acceleration is not zero.

4-47 An Olympic athlete launches a shot (Figure 4-57), and the shot follows a parabolic path. Point E is a position along the path as the shot is still rising, and point F is the highest position in the path. The relationships between the speed and the magnitudes of the accelerations of the shot at E and F are

(a) $v_E < v_F$ and $a_E < a_F$
(b) $v_E < v_F$ and $a_E = a_F$
(c) $v_E > v_F$ and $a_E = a_F$
(d) $v_E = v_F$ and $a_E \neq a_F$
(e) $v_E > v_F$ and $a_E < a_F$

4-48 The minute hand of a 12-hour clock rotates clockwise (as expected) around the centre of the clock, C. During the time the tip of the hand moves from the 12:00 o'clock position to the 6:00 o'clock position, the directions of the average acceleration and the centripetal acceleration are, respectively,

(a) horizontally to the right; horizontally to the left
(b) toward C; toward C
(c) horizontally to the right; toward C
(d) horizontally to the left; toward C
(e) vertically upward; vertically downward

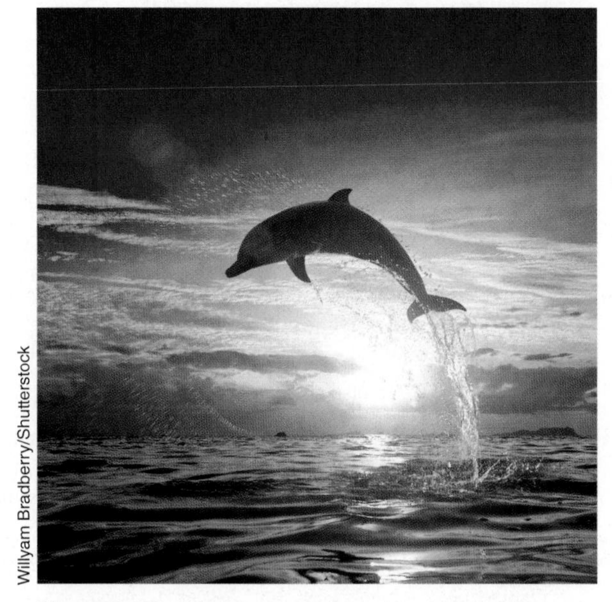
Willyam Bradberry/Shutterstock

Figure 4-56 A bottle-nosed dolphin (Question 4-46).

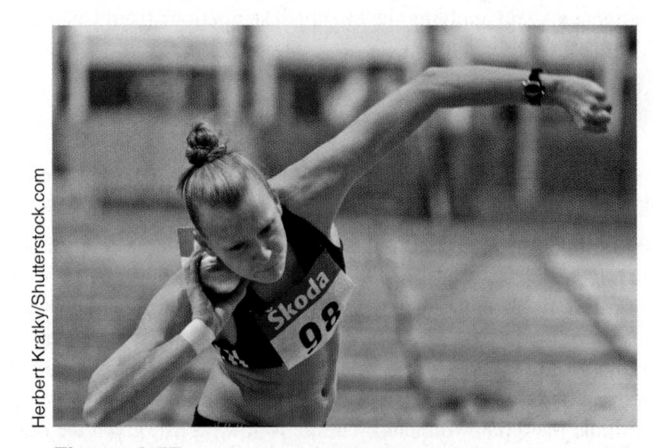

Herbert Kratky/Shutterstock.com

Figure 4-57 Preparing to launch the shot (Question 4-47).

Review Questions and Problems

4-49 A boy throws a ball into the air at a steep angle to the horizontal, then runs and catches the ball. Compare the displacement of the boy with the displacement of the ball at the end of the motion.

4-50 A resultant displacement, $\Delta\vec{r}$, is the sum of displacements $\vec{A}$ and $\vec{B}$. Is $|\Delta\vec{r}|$ necessarily as large as $|\vec{A}| + |\vec{B}|$? Explain.

4-51 The magnitudes of two displacements are 25 m and 42 m. What are the maximum and minimum values of the magnitude of their vector sum?

4-52 A square is inscribed in a circle of radius 12 m, as shown in Figure 4-58. One person walks from A to B along the edges of the square, and a second person walks along the circumference. Each person reaches B after 48 s. Calculate each person's

(a) average speed

(b) average velocity

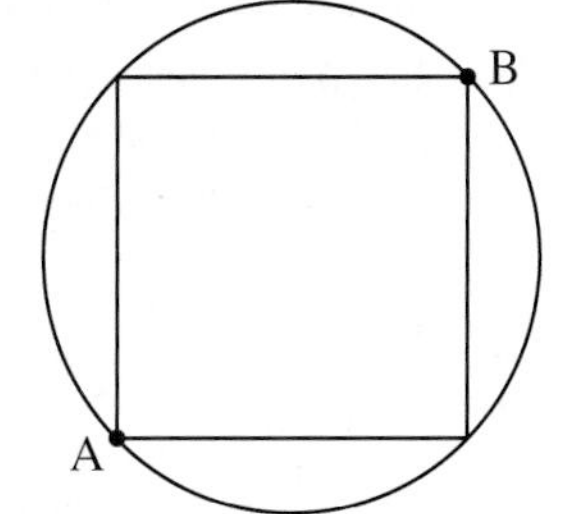

Figure 4-58 Question 4-52.

4-53 In 5.0 s, a firefighter slides 4.5 m down a pole and runs 6.8 m straight to a fire truck. Determine the firefighter's

(a) total distance moved

(b) average speed

(c) resultant displacement

(d) average velocity

4-54 A field hockey player (Figure 4-59) runs 16.0 m 35.0° south of west, then 22.0 m 15.0° south of east, all in 6.40 s.

(a) Determine the magnitude and direction of the resultant displacement.

(b) Find the average velocity of the player.

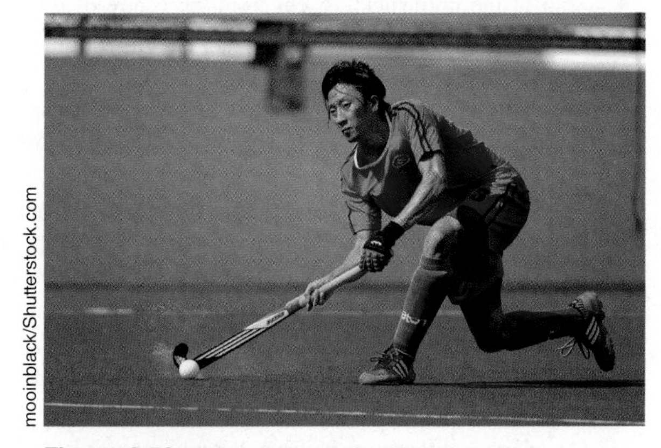

Figure 4-59 Field hockey player (Question 4-54).

4-55 The cheetah, probably the world's fastest land animal, can run up to 100 km/h over short distances (Figure 4-60 (a)). The motion of a cheetah chasing its prey at top speed is illustrated in Figure 4-60 (b), with north facing toward the top of the diagram. State the cheetah's instantaneous velocity, including the approximate direction, at positions A, B, and C.

(a)

(b)

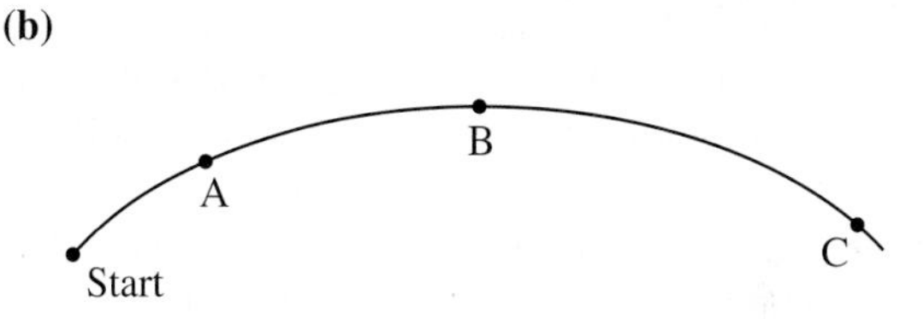

Figure 4-60 Question 4-55. **(a)** A running cheetah. **(b)** Path of the cheetah's motion.

DID YOU KNOW?

Unlike lions and leopards, cheetahs have claws that are only semi-retractable. The claws act like a sprinter's track shoes, digging in to aid in acceleration.

4-56 A bug is crawling along a checkered tablecloth on a picnic table. It starts at a position given by (x_0, y_0) = (42 cm, 28 cm), moves to position (x_1, y_1) = (29 cm, 36 cm), then moves to position (x_2, y_2) = (11 cm, 12 cm). Determine the magnitude and direction of the bug's displacement.

4-57 The driver of an automatic-transmission car has three different controls for producing acceleration. What are they?

4-58 An airplane is travelling at 140 m/s 28° south of east, then 94 s later at 150 m/s 28° east of south. Determine the airplane's average acceleration during this time.

4-59 A gasoline-powered go-kart is travelling at constant speed around a circular track, as illustrated in Figure 4-61. The diameter of the track is 1.2×10^2 m and the go-kart takes 28 s to complete one revolution. Determine the kart's

(a) speed

(b) instantaneous velocity at positions A, B, and C

(c) average velocity between positions A and D

(d) average acceleration between positions B and D, then between A and D

4-60 A train is travelling east at 23 m/s when it enters a curved portion of the track and experiences an average acceleration of 0.15 m/s^2 south for 95 s. What is the train's velocity at the end of this acceleration?

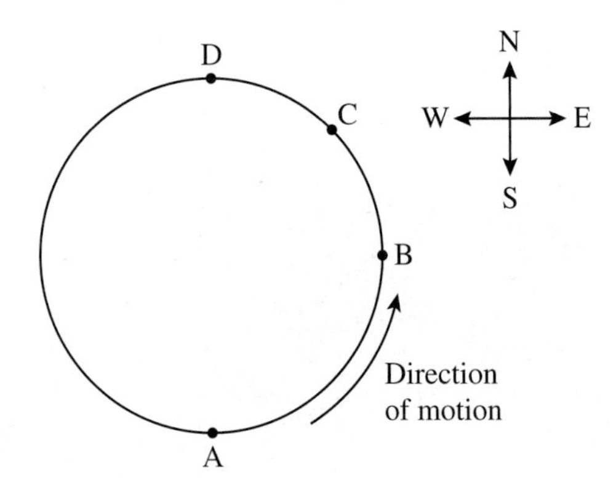

Figure 4-61 Question 4-59.

4-61 A race car driver wants to leave a curve in the track with a velocity of 54 m/s north. With what velocity should the car enter the curve if the average acceleration in the curve is 19 m/s² 45° north of west for 4.0 s?

4-62 A basketball tossed through the air is caught at the same elevation from which it was thrown. At what position(s) in its path is the speed of the ball the least? the greatest?

4-63 What are the horizontal and vertical accelerations of a projectile?

4-64 A child throws a snowball straight toward a tree with a horizontal velocity of 18 m/s. The tree is 9.0 m away and the snowball starts from a height of 1.5 m above the ground.

(a) How long will the snowball take to reach the tree?

(b) At what height above the ground will the snowball hit the tree?

(c) What is the snowball's velocity as it hits the tree?

4-65 A tennis ball is struck so that its initial velocity is horizontal. If it falls 1.5 m while moving 16 m horizontally, what was its initial velocity?

4-66 During World War I, the German army used a huge gun to bombard Paris with shells from a large distance away. (The gun was named "Big Bertha" by the Allied forces.) The gun fired shells with an initial speed of 1.1×10^3 m/s. The gun was set at an angle of 45° above the horizontal. Assume the shell landed at the same level as the gun.

(a) How long would each shell stay in the air?

(b) What was the maximum horizontal distance, in kilometres, the gun could be from Paris to try to hit it?

(c) What maximum height, in kilometres, did the shells reach?

4-67 An astronaut strikes a golf ball on the Moon where the magnitude of the acceleration due to gravity is 1.6 m/s². The ball takes off with a velocity of 32 m/s at an angle 35° above the Moon's horizontal and lands in a valley 15 m below the level where it started. Determine the golf ball's (a) maximum height, (b) time of flight, (c) horizontal distance travelled.

4-68 Neglecting air resistance, state another angle of projection above the horizontal that would result in the same range of a projectile fired at an angle of 38°; 15°; 44.5°. (Assume the projectile lands at the same elevation from which it starts.)

4-69 Is centripetal acceleration an average acceleration, an instantaneous acceleration, both, or neither?

4-70 Do all points along the second hand of a dial clock have the same centripetal acceleration? Explain your answer.

4-71 An engineer has calculated that the maximum centripetal acceleration of a car on a certain curve is 4.4 m/s² in magnitude. For a car travelling at 25 m/s, what is the minimum radius of curvature of this curve?

4-72 **Fermi Question:** Assume you are exercising on a stationary exercise bike. Estimate the frequency of rotation of the pedals in order for the pedals to experience a centripetal acceleration equal in magnitude to the acceleration due to gravity.

4-73 Determine the magnitude of the centripetal acceleration of the tip of the second hand, the minute hand, and the hour hand on the clock shown in Figure 4-62.

Figure 4-62 Question 4-73.

4-74 A turkey is centred on a rotating turntable in a microwave oven. The end of the drumstick, 16 cm from the centre of rotation, experiences a centripetal acceleration of magnitude 0.22 m/s². What is the period of revolution of the plate?

4-75 Various types of centrifuges are used to train astronauts and to research human space travel. Two examples are shown in Figure 4-63.

(a) An astronaut trainee in the Texas centrifuge is 15 m from the centre as the device rotates at a constant frequency of 21 rpm (revolutions per minute). What is the magnitude of the acceleration experienced by the trainee?

(b) Express the answer in terms of "*g*," the magnitude of the acceleration due to gravity. (**Note:** Humans may lose consciousness if exposed to accelerations of more than about 7 *g*'s.)

(c) Describe the differences between the two centrifuge designs shown, and speculate on the advantages of the newer, German design.

4-76 Assume you are standing on a train that is moving (relative to Earth's frame of reference), and you drop a ball to the floor.

(a) Describe the path of the ball from your frame of reference.

(b) Describe the path from the frame of reference of an observer standing on the ground beside the train.

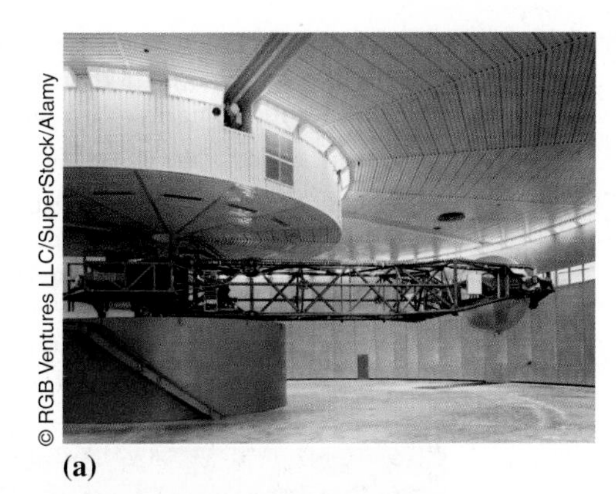

© RGB Ventures LLC/SuperStock/Alamy

Hermann J. Knippertz/AP Photo/The Canadian Press

(a) (b)

Figure 4-63 Question 4-75. **(a)** The centrifuge formerly used for astronaut training at the Johnson Space Center in Houston, Texas. **(b)** The Short-Arm Human Centrifuge in Cologne, Germany, can hold four human test subjects—two seated and two lying down—as well as cellular and biological samples. Its radius is 2.8 m and it can operate with a range of revolution frequencies from 5.7 rpm to 45 rpm.

4-77 State the direction of the instantaneous velocity and instantaneous acceleration of object A in each case from your own frame of reference.

(a) You are looking east while you are twirling a ball (A) on the end of a string around your head. The ball travels in a horizontal circle clockwise when viewed from above. The ball is directly in front of you at this instant.

(b) You are facing south when suddenly a snowball (A) passes in front of your eyes from left to right. The snowball had been released earlier with an initial velocity such that its current instantaneous velocity is horizontal.

4-78 A plane is headed southeast (i.e., 45° south of east) with an airspeed of 280 km/h. A wind is blowing at 80 km/h 20° east of north. What is the velocity of the plane relative to the ground? Assume two significant digits.

4-79 A swimmer who can swim at a speed of 0.80 m/s in still water heads directly across a river that is 80 m wide. She lands at a position that is 50 m downstream from where she started, but on the opposite side. Assume two significant digits.

(a) Determine the speed of the river current.

(b) Calculate the swimmer's velocity relative to the shore.

(c) In what direction should she have aimed to have landed directly across from her starting position?

4-80 Assume that the distance from London, England, to Rome, Italy, is 1.4×10^3 km, in the direction 47° south of east. A wind is blowing at 75 km/h from the west. The pilot of an airplane wants to fly directly from London to Rome in 3.5 h. What velocity of the airplane relative to the air should the pilot maintain?

Applying Your Knowledge

4-81 On a detailed map of Mount Everest, the following data are observed. The elevation of the top of the mountain is 8848 m. At a horizontal distance of 1500 m due north of the peak, in Tibet, the elevation falls to 7600 m. To the southwest, at a horizontal distance of 1800 m in Nepal, the elevation is 6800 m. Determine the average slope of each mountain side. Which side is steeper?

4-82 A bicyclist, travelling initially at 18.0 km/h east, experiences an acceleration for a certain time and reaches a velocity of 29.0 km/h 45.0° south of east. The magnitude of the acceleration is 1.40 (km/h)/s. Determine the (a) direction of the acceleration and (b) time for the acceleration.

4-83 A baseball player wants to determine the approximate speed given to a ball by a bat when the ball has its maximum range. How could this be done using only a measuring tape or metre stick and an equation?

4-84 **Fermi Question:** The women's world record for the hammer throw is about 79 m (Figure 4-64). Stating any assumptions and approximations, determine approximate values for

(a) the magnitude of the initial velocity of the throw

(b) the time of flight

(c) the maximum height reached by the hammer

© CTK/Alamy

Figure 4-64 Building up speed before releasing the "hammer" (Question 4-84).

4-85 If the speed of a particle in circular motion is increasing, is the acceleration still toward the centre of the circle? Use a diagram to prove your answer.

4-86 In a basketball game, a ball leaves a player's hand 6.10 m from the basket from a height of 1.20 m below the level of the basket. If the initial velocity of the ball is 7.80 m/s at an angle of 55.0° above the horizontal and in line with the basket, will the player score a basket? If not, by how much will the ball miss the basket?

4-87 To produce artificial gravity on a space colony, it is proposed that the colony should rotate (Figure 4-65). At the outside edge of the colony, suppose that the centripetal acceleration associated with the rotation is equal in magnitude to the acceleration due to gravity on Earth. For a colony that is 1.0 km in diameter, determine the frequency of rotation, the period of rotation, as well as the speed of a person at the edge of the colony (relative to the centre of the colony).

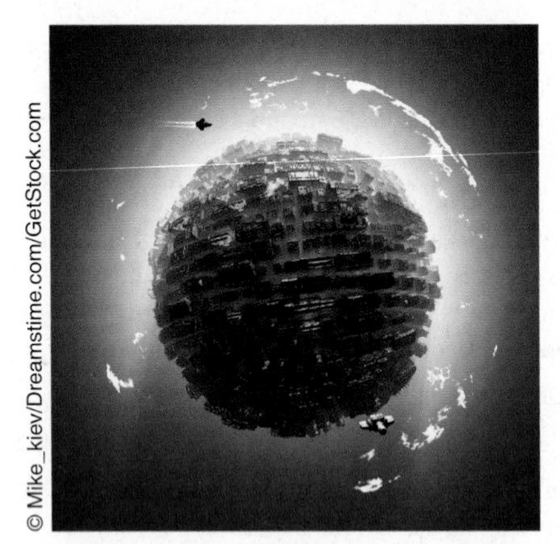
© Mike_kiev/Dreamstime.com/GetStock.com

Figure 4-65 An artist's conception of a futuristic space colony (Question 4-87).

4-88 Suppose a ballet dancer must execute a series of jumps (such as *grand jetés*, Figure 4-66), each timed at 0.70 s to be in rhythm with the music.

(a) How high would the dancer have to jump to accomplish the set of jumps?

(b) If the dancer leaves the stage floor at an angle of 35° above the horizontal, with what initial velocity did the jump occur?

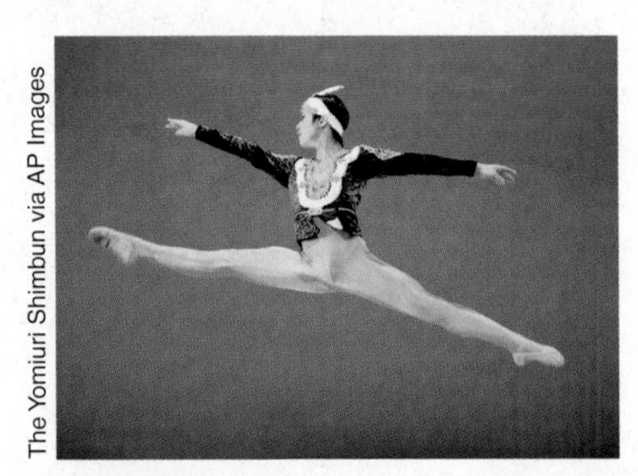
The Yomiuri Shimbun via AP Images

Figure 4-66 A *grand jeté* in a ballet competition (Question 4-88).

DID YOU KNOW?

A skillful ballet dancer can create a floating illusion in the middle of a *grand jeté* by changing the body configuration as the jump progresses. The illusion occurs as the dancer lifts his or her legs as high as possible during the mid-portion of the jump. (You can learn how this concept relates to centre of mass in Chapter 11.)

4-89 Greyhounds, the world's fastest dogs, can reach a speed of 70 km/h in only 6 strides covering a distance of 30 m. Some greyhounds are trained for racing, as seen in Figure 4-67.

(a) Calculate the magnitude of the greyhound's acceleration.

(b) **Fermi Question:** Imagine you are designing a circular greyhound racetrack so the lure used to entice the dogs to run in the correct direction moves in a large circle. Use reasonable assumptions to determine the magnitude of the lure's centripetal acceleration once it is moving at a constant speed.

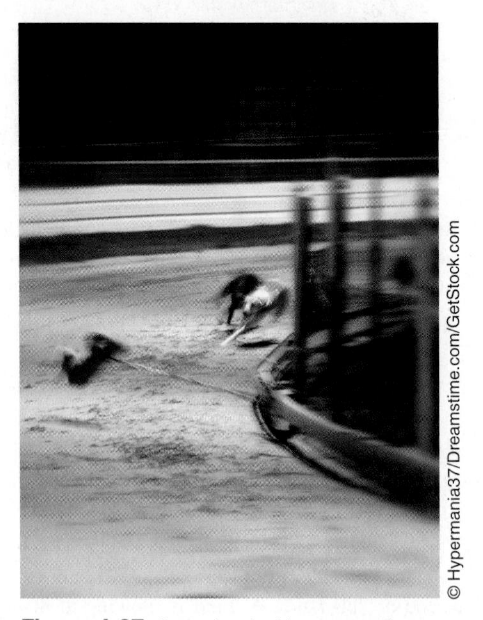
© Hypermania37/Dreamstime.com/GetStock.com

Figure 4-67 A mechanical lure, sometimes resembling a rabbit or other animal, must keep just ahead of the racing greyhounds (Question 4-89(b)).

4-90 To test the quality of a new glue product, the glue is used to attach a metal washer to the surface of a variable-speed propeller. The washer is 14 cm from the centre of the propeller, and when the propeller's frequency reaches 15 Hz, the washer flies off at an angle of 38° above the horizontal. What is the horizontal range of the washer, assuming it lands at the same height as when it left the propeller?

4-91 A helicopter flies at a speed of 120 km/h in still air. The wind is blowing at 65 km/h in a direction 38° south of east.

(a) Determine the direction the helicopter should head in order to fly straight west.

(b) What is the speed of the helicopter relative to the ground?

4-92 One method of estimating the speed of a falling raindrop can be used in a moving vehicle. Assume you are in a car or train travelling at 50 km/h. You observe raindrops on a vertical window making an angle of 30° to the vertical. Assuming the raindrops are falling straight down (relative to Earth's frame of reference), determine the speed of the drops relative to Earth in kilometres per hour.

Newton's Laws of Motion 5

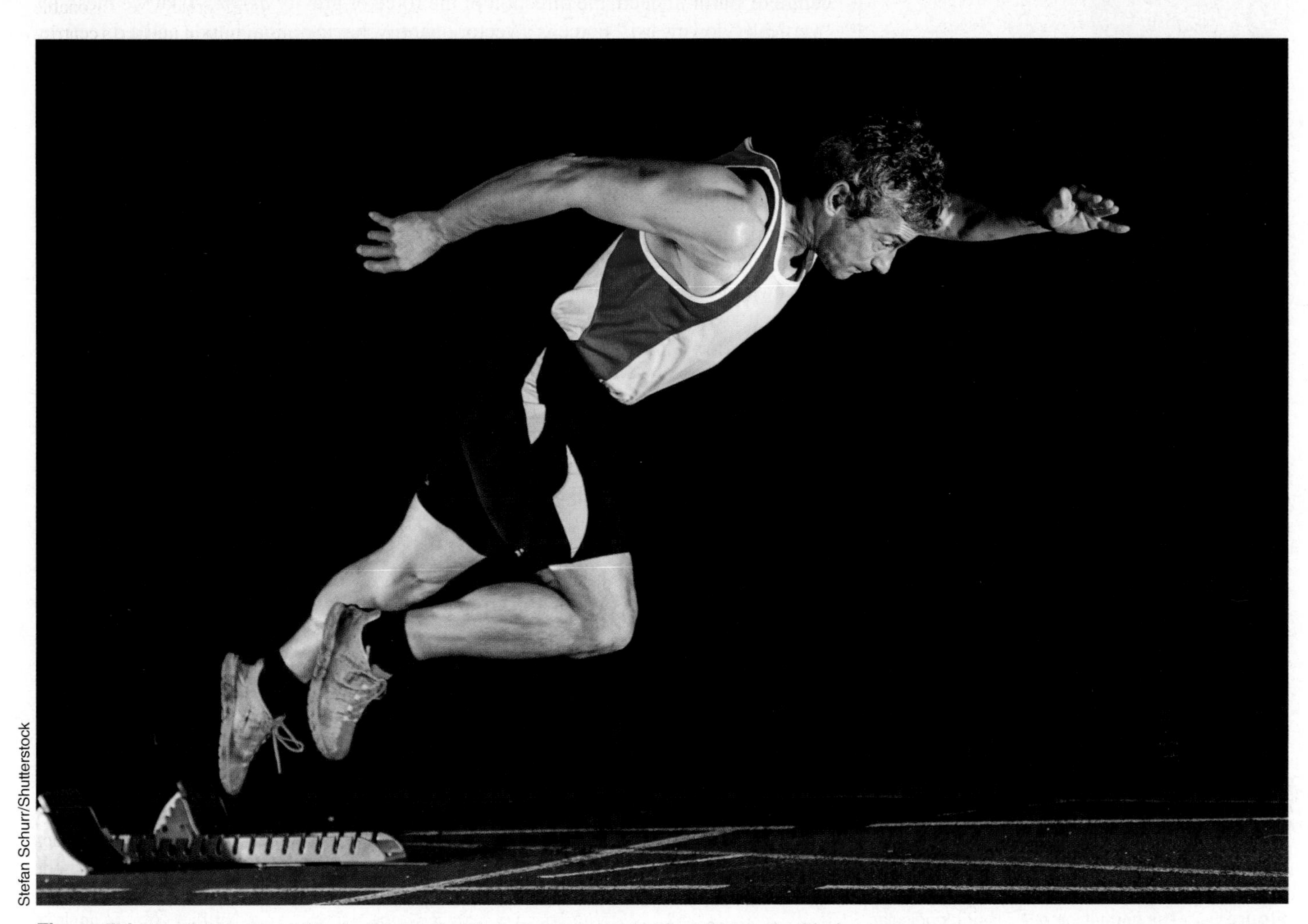

Stefan Schurr/Shutterstock

Figure 5-1 A sprinter relies on Newton's laws of motion when accelerating from the starting blocks.

In Chapters 2 to 4, we discussed kinematics (the study of motion) using concepts such as displacement, velocity, and acceleration. However, we did not consider the forces that cause objects to move the way they do. The movements of objects as large as stars or as small as red blood cells are determined by the forces acting on them. A knowledge of forces is essential for an understanding of motions as diverse as the graceful movements of a ballet dancer, the takeoff of an airplane, or the biting action of an animal's jaw. The three basic laws of motion, known as Newton's laws, are used in analyzing a wide variety of real-world situations in biomechanics, engineering, space science, and a host of other fields (Figure 5-1). The study of forces and their effects on the motion of objects is called **dynamics**.

CHAPTER OUTLINE

5.1 Gravity, Normal Force, Friction, and Tension

dynamics the study of forces and their effects on the motion of objects

gravity an attractive force between all objects

You have known since childhood that forces are pushes and pulls, and there are a number of forces that are important to you in your everyday activities. Perhaps the most important is the force of **gravity**, which is a force of attraction between all objects. For an object close to Earth, gravity pulls vertically downward toward the centre of Earth. Indeed, the direction of the force of gravity *defines* what we mean by "vertically downward." Force is a vector quantity, having a magnitude and a direction; the force of gravity is represented by the symbol $\vec{F}_G$.

Recall that in this book a vector is indicated in boldface type with an arrow above it, for example, $\vec{F}$ for force. The magnitude of a vector is indicated either in normal (not boldface) type, such as F for the magnitude of $\vec{F}$, or with an absolute sign around the vector symbol, for example, $|\vec{F}|$. The magnitude of a vector can only be positive or zero, never negative ($F = |\vec{F}| \geq 0$).

normal force the force perpendicular to surfaces in contact with each other

The downward force of gravity is acting on this book right now. However, the book does not actually move downward (unless you drop it). If the book is sitting, say on a table, the table exerts an upward support force that prevents the book from moving downward. This force is called the **normal force**, symbolized as $\vec{F}_N$. In this context, the word "normal" means "perpendicular." The essential feature of the normal force is that it is perpendicular to the surfaces in contact with each other, such as the lower surface of the book and the upper surface of the table. Its direction is not necessarily vertical, as shown in Figure 5-2 for a person standing on sloping ground.

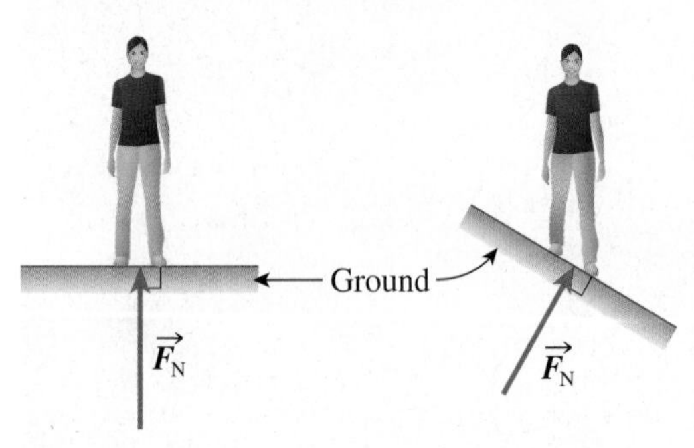

Figure 5-2 The normal force, $\vec{F}_N$, is perpendicular to the surfaces in contact.

force of friction the force parallel to surfaces in contact with each other, due to attractive electrical forces between molecules in the surfaces

If you give a box a quick hard push so that it slides along the floor, the box eventually slows down and stops. The force that causes the box to slow down is the **force of friction**, $\vec{F}_f$, exerted on the box by the floor. Friction is *parallel* to the surfaces in contact. At the molecular level, this force is due to attractive electrical forces between molecules in the box and molecules in the floor.

kinetic friction the friction force exerted between surfaces moving relative to each other

static friction the friction force exerted between surfaces that are not moving relative to each other

When a box slides along a floor, the surfaces in contact are moving relative to each other, and the friction is referred to as **kinetic friction**. In situations where the surfaces are not in motion relative to each other, the friction is called **static friction**. For instance, if a book is at rest on a table, and you exert a small horizontal force on the book and it remains at rest, it is static friction between the table and the book that is acting in a direction opposite to your force and preventing the book from moving. As another example, if you hold a book horizontally on the palm of your hand and you slowly increase its speed without the book sliding on your hand, the force that is accelerating the book is static friction between your hand and the book. Note that although the book is moving, the friction force is *static* friction because the book and your hand are not in motion relative to each other. Kinetic and static friction will be discussed in more detail in Section 6.1. For the remainder of this chapter, we will simply refer generally to the force of friction.

In most situations described in this book, we will be discussing the friction between solid objects in contact with each other. However, there is also friction when an object moves through a fluid, such as air or water. This is usually referred to as viscous friction, or air resistance (if the fluid is air). Viscous friction depends on the shape and speed of the object, and on the type of fluid. This type of friction will be discussed in more detail in Chapter 12 (Fluid Statics and Dynamics).

tension force exerted by ropes, strings, cables, tendons, etc.

Often objects are pulled or supported by ropes, strings, or cables. The force exerted by ropes, etc., is called **tension**, represented as $\vec{T}$. Figure 5-3 shows two people pulling on a rope. From the point of view of person A, the rope is pulling to the right; that is, the tension is to the right. However, from the point of view of person B, the tension is to the left. A rope, string, or cable always has a tension that pulls *away* from the object that we are considering. The magnitude of the tension is usually the same at each end. Since the tensions exerted on the people in Figure 5-3 have the same magnitude but opposite directions, one tension is shown as $\vec{T}$ and the other as $-\vec{T}$.

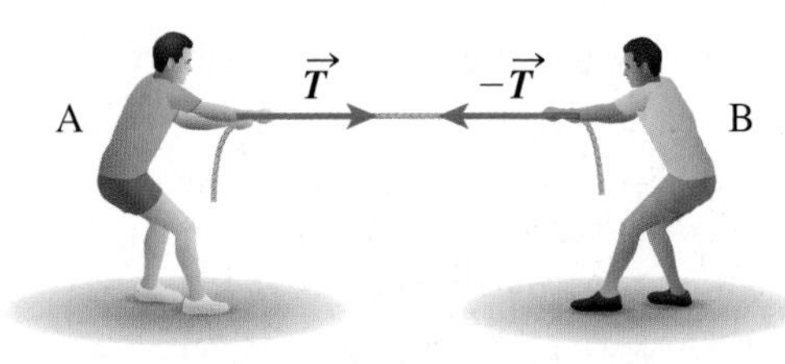

Figure 5-3 Tension pulls away from the object being considered.

DID YOU KNOW?

There is an old saying in engineering: "You can't push a rope." In other words, ropes can only pull objects—they cannot be used to push things.

Notice that in order for an object to experience a normal, friction, or tension force, it must be in direct contact with something (floor, rope, etc.) that exerts this force. Pushes and pulls exerted by people also are the result of direct contact. Such forces are often referred to as **contact forces**. However, in order for the force of gravity to act, contact between objects is not required, and hence gravity is referred to as an **action-at-a-distance force**. Can you think of other action-at-a-distance forces?

Four forces have now been introduced in this section of the book: gravity, normal force, friction, and tension. Table 5-1 lists the symbols used for these forces.

Table 5-1

Symbols Used for Common Forces in This Book

$\vec{F}_G$	gravity
$\vec{F}_N$	normal force
$\vec{F}_f$	friction
$\vec{T}$	tension

Free-Body Diagrams

In order to analyze situations and problems involving forces, it is handy to draw a diagram in which only the forces acting on a particular object are shown. Not shown are the object itself, nor other objects that are exerting the forces. The object itself is represented by a large dot. Such a diagram is called a **free-body diagram** or a **force diagram**. Usually the magnitudes of the forces are drawn roughly to scale if the magnitudes of the forces are known. Often the words "free-body diagram" are abbreviated as "**FBD**." Free-body diagrams are illustrated in the following two sample problems.

contact forces forces such as friction and normal force that, in order to be exerted, require contact between objects

action-at-a-distance forces forces such as gravity that do not require contact between objects in order to be exerted

free-body diagram (FBD, or **force diagram)** a diagram in which only the forces acting on a particular object are shown

SAMPLE PROBLEM 5-1

Draw a free-body diagram (FBD) for a crate that is being dragged across the floor by a person pulling horizontally on a rope.

Solution

Before drawing the FBD, it is often useful to make a sketch such as Figure 5-4 (a) to illustrate the situation being described, and to help in identifying the forces involved. We can see in the sketch that the crate is being pulled horizontally by a rope, which means that there is a horizontal tension, $\vec{T}$, acting on the crate. This tension is directed away from the crate. Notice that the person pulls on the rope, but does not touch the crate directly. Thus, the person does not exert a force on the crate—the rope does.

The crate is also in contact with the floor. The floor exerts a normal force $\vec{F}_N$ upward on the crate, and a friction force $\vec{F}_f$ horizontally, in the direction opposite to the tension in the rope. The final force acting on the crate is the force of gravity $\vec{F}_G$, acting vertically downward. The four forces acting on the crate are shown on the FBD in Figure 5-4 (b).

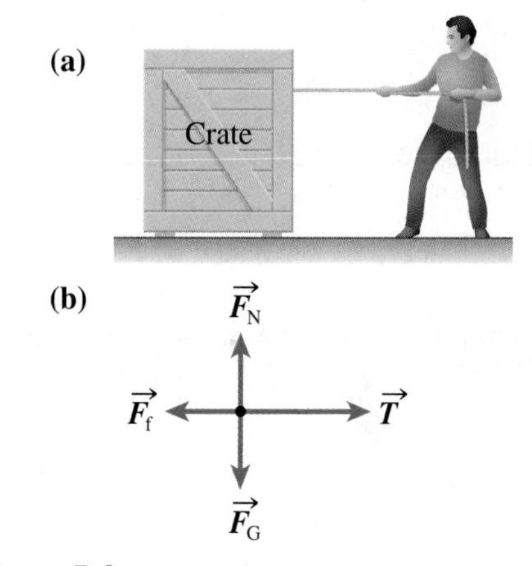

Figure 5-4 **(a)** A sketch for Sample Problem 5-1, showing the crate being pulled. **(b)** The free-body diagram for the crate.

SAMPLE PROBLEM 5-2

Draw a FBD for a woman who is standing in an elevator that is moving upward.

Solution

A sketch is shown in Figure 5-5 (a), and the FBD in Figure 5-5 (b). The downward force of gravity $\vec{F}_G$ is acting on the woman, and the only object that is in contact with the woman is the floor of the elevator, which exerts an upward normal force $\vec{F}_N$. Since there is nothing in this situation to imply that the woman is being pushed horizontally, there is no friction force exerted by the floor on the woman. Notice that the upward motion of the elevator is not important in determining which forces are acting. Whether the elevator is moving up, standing still, or moving down, the FBD is the same.

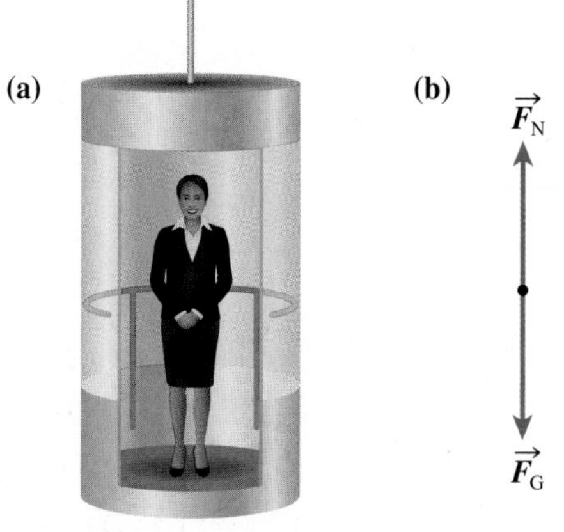

Figure 5-5 **(a)** A sketch for Sample Problem 5-2. **(b)** The FBD for the woman.

EXERCISES

In Questions 5-1 to 5-9 below, make a sketch of the situation, and draw a free-body diagram (FBD) for object A.

5-1 A ball (A) hangs at the end of a vertical string.

5-2 A cheeseburger (A) sits on a table.

5-3 A man (A) is standing in an elevator that is moving downward.

5-4 A red blood cell (A), or erythrocyte, is settling downward in blood plasma. (In addition to the force of gravity, there is an upward buoyant force and an upward fluid friction force.)

5-5 A metal girder (A) is being lifted upward by a cable connected to a crane.

5-6 A hockey puck (A) is sliding on ice.

5-7 A pen (A) is falling, having been knocked off a table. Neglect air resistance.

5-8 A box (A) full of car parts is being hauled across a floor by a person who is pulling on a rope that makes an angle of 30° upward from the horizontal.

5-9 A stove (A) is being pulled up a ramp into a truck by a cable that is parallel to the ramp. The ramp makes an angle of 20° with the horizontal.

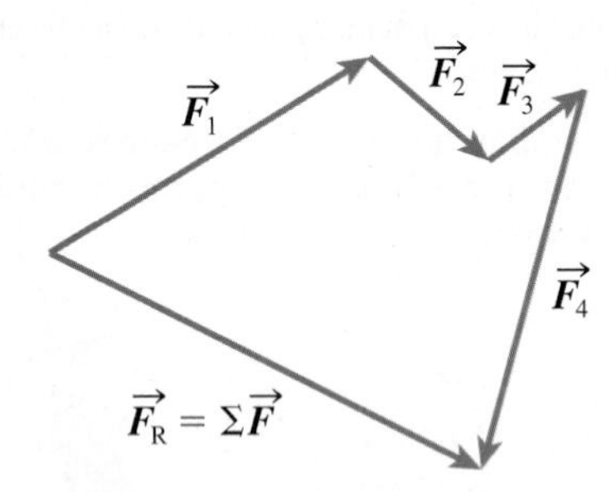

Figure 5-6 Using a diagram to add forces.

5.2 Adding Forces

In order to analyze how forces affect the motion of objects, we first need to review how to add vectors together. Since forces are vectors, they must be added as vectors. If one child pulls on a wagon with a force of 5 N eastward,[1] and another child pulls with a force of 5 N westward, then the magnitude of the total force is not ten newtons, but rather zero newtons.

The addition of vectors in one dimension, such as the forces on the wagon mentioned above, is straightforward. For vectors in two dimensions, we need to use more complex methods. One technique of adding vectors (discussed in Chapter 3) is to use a scale diagram in which the vectors are placed head-to-tail in sequence, as shown in Figure 5-6. The tail of force $\vec{F}_2$ is placed at the head of $\vec{F}_1$, the tail of force $\vec{F}_3$ at the head of $\vec{F}_2$, etc. The total force is a vector drawn from the tail of the first force $\vec{F}_1$ to the head of the last force $\vec{F}_4$ in the sequence.

resultant force (or **net force, total force, sum of the forces**) total of the forces acting on an object

The total of the forces acting on an object is given a variety of names: **resultant force**, **net force**, **total force**, or **sum of the forces**. In this book, we will use the term **resultant force**, and will symbolize it as $\vec{F}_R$ or $\Sigma\vec{F}$. The Greek capital letter sigma, Σ, means "sum of"; thus, $\Sigma\vec{F}$ means "sum of the forces."

Although a scale diagram is easy to draw, it does not give very accurate results, and components are usually used to add forces accurately. You might wish to review the addition of vectors before proceeding. A reminder: components of a vector are not themselves vectors, but rather are positive or negative numbers with units.

SAMPLE PROBLEM 5-3

At the instant shown in Figure 5-7 (a), two forces are being exerted on a baseball: the force of gravity ($\vec{F}_G$ = 1.5 N downward), and the force of air resistance ($\vec{F}_A$ = 0.50 N at an angle of 58° from the vertical). Determine the magnitude and direction of the resultant force on the ball.

Solution

We will use components to find the resultant force. We define the $+x$-axis to be to the left and the $+y$-axis to be downward, as shown in Figure 5-7 (b). Hence, $\vec{F}_G$ has only a y-component. The x- and y-components of $\vec{F}_A$ are

$$F_{Ax} = (0.50\,\text{N})\sin 58° = 0.42\,\text{N}$$

$$F_{Ay} = -(0.50\,\text{N})\cos 58° = -0.26\,\text{N}$$

[1]N is the symbol for the SI unit of force, the newton, which will be discussed in detail in Section 5.3.

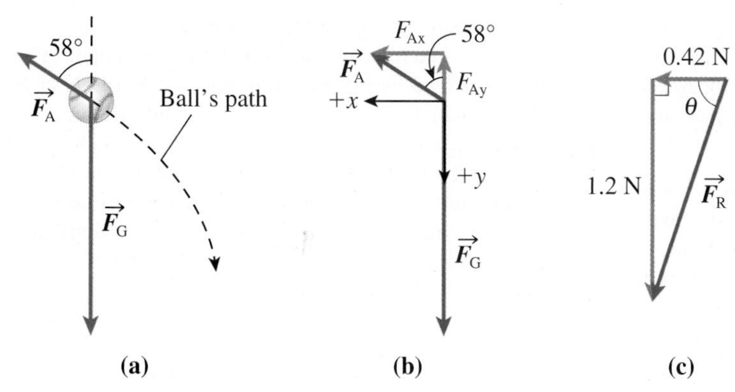

Figure 5-7 **(a)** Forces acting on the baseball in Sample Problem 5-3. **(b)** Components of the forces. **(c)** The resultant force.

Since $\vec{F}_G$ has no x-component, the x-component of the resultant force, F_{Rx}, is just F_{Ax}:

$$F_{Rx} = \Sigma F_x = F_{Ax} = 0.42\,\text{N}$$

Noting that $F_{Gy} = F_G$ (since $F_{Gx} = 0$), then the y-component of the resultant force is

$$F_{Ry} = \Sigma F_y = F_G + F_{Ay} = (1.5 - 0.26)\text{N} = 1.2\,\text{N}$$

Figure 5-7 (c) shows the components of the resultant force, which has a magnitude of

$$F_R\,(\text{or}\,|\vec{F}_R|) = \sqrt{(0.42\,\text{N})^2 + (1.2\,\text{N})^2} = 1.3\,\text{N}$$

The direction of $\vec{F}_R$ is given by angle θ in Figure 5-7 (c): $\theta = \tan^{-1}(1.2/0.42) = 71°$. Thus, the resultant force is 1.3 N at an angle of 71° from the horizontal.

EXERCISES

5-10 The force of gravity on a book is 19 N downward. When the book is being held stationary by a person's hand, the hand is providing a force of 19 N upward on the book.

(a) What is the magnitude of the resultant force on the book?

(b) If the hand is suddenly removed, what are the magnitude and direction of the resultant force on the book? (Neglect air resistance.)

In Questions 5-11 to 5-15, determine the magnitude and direction of the resultant force on object A.

5-11 A skydiver (A) in free-fall experiences

(a) a force of gravity of 586 N downward, and a force of air resistance of 492 N upward

(b) a force of gravity of 586 N downward, and air resistance of 586 N upward.

5-12 A toboggan (A) is being pulled along a horizontal snow-covered field by a person exerting a force of 72 N horizontally. In the opposite direction, there is a horizontal friction force on the toboggan of magnitude 69 N. The force of gravity on the toboggan is 153 N in magnitude, and the normal force exerted by the snow on the toboggan has a magnitude of 153 N.

5-13 A soaring bird (A) is subject to three forces at a particular moment: the downward force of gravity (3.27 N), a horizontal air resistance drag force (0.354 N), and an upward lift force (3.74 N).

5-14 A long-jumper (A) at landing (Figure 5-8) experiences two forces: the force of gravity (F_G = 538 N) and the force exerted by the ground on her feet (F_{feet} = 6382 N). Air resistance is negligible in comparison with these forces. (Note that the actual force exerted by the ground on the feet of a world-class jumper would be several times the value used in this problem.)

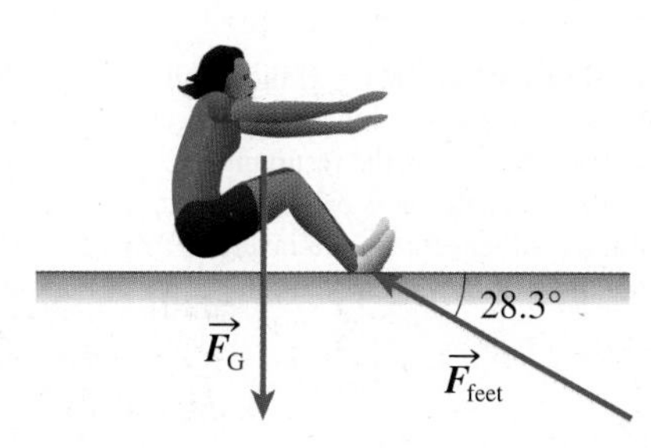

Figure 5-8 Question 5-14. (Forces not to scale.)

5-15 A quarterback (A) is hit simultaneously by two linebackers. The forces exerted by each linebacker are: horizontal, 412 N, 27.0° west of north; and horizontal, 478 N, 54.0° east of north. The force of gravity and the normal force exerted on the quarterback are equal in magnitude. Neglect friction.

5-16 A refrigerator is being moved along a horizontal floor by one person pulling on a rope that exerts a force of 252 N east, and by another person pushing east with a force of unknown magnitude. There is a friction force of 412 N west, and the force of gravity on the refrigerator and the normal force each have a magnitude of 1127 N. If the resultant force on the refrigerator is 2 N east, what is the magnitude of the unknown force?

5-17 A large crate is pulled across an icy sidewalk by two people pulling horizontally on ropes (Figure 5-9). The tension $\vec{T}_1$ in one of the ropes is 27 N east, and the resultant horizontal force on the crate due to the two rope tensions is 56 N at 16° south of east. Neglect friction. Determine the magnitude and direction of the tension $\vec{T}_2$ in the other rope.

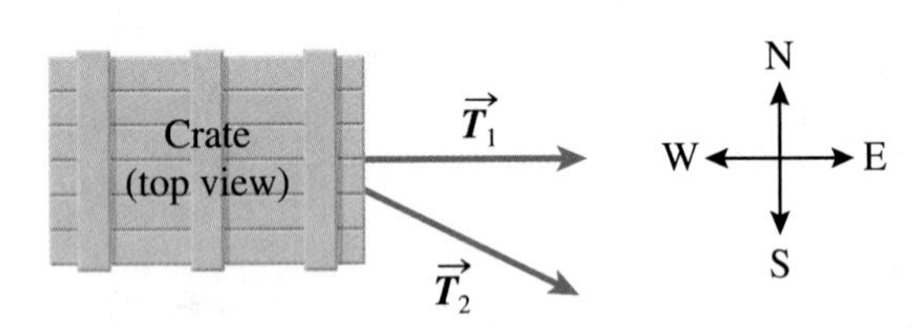

Figure 5-9 Question 5-17.

5-18 Figure 5-10 shows the three main forces acting on a horizontal forearm. The force of gravity, $\vec{F}_G$, has a magnitude of 11.6 N. The force exerted by the biceps muscle, $\vec{F}_B$, is 129 N at an angle of 72.0° relative to the horizontal. The contact force exerted by the humerus (the bone in the upper arm), $\vec{F}_H$, is 111 N at 70.1° relative to the horizontal. Define the $+x$-direction to be horizontal (either left or right), and the $+y$-direction to be vertical (either up or down). Using components in these directions, determine the magnitude of the resultant force on the forearm. (Be careful with significant digits.)

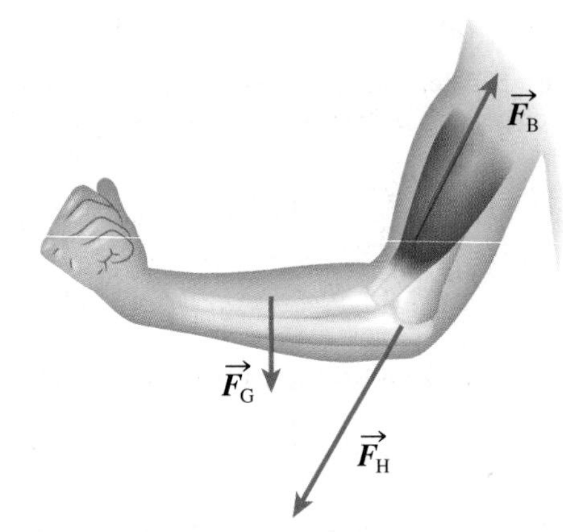

Figure 5-10 Forces acting on the forearm in Question 5-18. (Not to scale.)

5.3 Newton's Second Law of Motion

In 1687, Sir Isaac Newton (1642–1727) published one of the greatest scientific works of all time, the *Philosophiae Naturalis Principia Mathematica (Mathematical Principles of Natural Philosophy)*, usually referred to simply as the *Principia*. In this book, the results of the work of Newton and other scientists (called natural philosophers at that time) were compiled, including Newton's famous three laws of motion and his law of universal gravitation. The *Principia* was the culmination of centuries of thought concerning the motion of objects, from earthly ones such as falling apples to astronomical ones such as orbiting planets, and represented a huge step forward in our understanding of the workings of the universe.

The remainder of this chapter deals with Newton's laws of motion. We discuss his second law first. It might seem odd that Newton's first law is not treated right away, but as will be seen in the next section, Newton's first law is just a special case of his second law.

Newton's second law of motion If the resultant force on an object is not zero, the object experiences an acceleration in the direction of the resultant force. The magnitude of the acceleration is proportional to the magnitude of the resultant force, and inversely proportional to the mass of the object.

Newton's second law of motion relates the resultant force exerted on an object to the acceleration of the object. This law states

If the resultant force on an object is not zero, the object experiences an acceleration in the direction of the resultant force. The magnitude of the acceleration is proportional to the magnitude of the resultant force, and inversely proportional to the mass of the object.

This law should make intuitive sense to you. For a given object, if a larger resultant force is applied, then a larger acceleration results. For example, if you push this book across a table, then the harder you push, the larger is the acceleration of the book. For a given resultant force applied to a variety of objects then, the larger the mass of an object, the smaller is its acceleration. A given resultant force applied to a box of books produces a smaller acceleration than the same resultant force applied to a single book. A truck needs a bigger engine than a small car to get the same acceleration. With the bigger engine, a larger force can be applied to accelerate the truck, which has a large mass.

To write this law mathematically, we start with the proportionalities stated in the law:

$$a \propto F_{\mathrm{R}} \quad \text{and} \quad a \propto \frac{1}{m}$$

Combining these two proportionalities,

$$a \propto \frac{F_{\mathrm{R}}}{m}$$

The constant of proportionality is conventionally chosen to be one, so that we can now write

$$a = \frac{F_{\mathrm{R}}}{m}$$

Including the directions of the vectors, that is, writing the symbols for force and acceleration in boldface type, we have

$$\vec{a} = \frac{\vec{F}_{\mathrm{R}}}{m}$$

This is usually arranged with the force on the left-hand side:

$$\vec{F}_{\mathrm{R}} = m\vec{a} \quad \text{or} \quad \Sigma\vec{F} = m\vec{a} \qquad \textbf{(5-1)}$$

which is the customary way in which Newton's second law is written. For problem-solving, this relationship is most useful in terms of components:

$$\Sigma F_{\mathrm{x}} = ma_{\mathrm{x}} \quad \text{and} \quad \Sigma F_{\mathrm{y}} = ma_{\mathrm{y}} \qquad \textbf{(5-2)}$$

This law governs the motion of all macroscopic objects with which you are familiar: airplanes, planets, rivers, people, etc. It also works at the microscopic level of biological cells and small organisms such as amoebas. (It generally does not work at the level of molecules, atoms, nuclei, and subatomic particles. In this realm, a different set of rules called *quantum mechanics* is needed instead of the Newtonian mechanics that we are discussing.)

We are now in a position to define the SI unit of force: the newton. **A newton (N) is the magnitude of resultant force that, if applied to an object of mass 1 kg, will produce an acceleration of magnitude 1 m/s^2.** By substituting units into both sides of $F_{\mathrm{R}} = ma$, we see that 1 N is the same as 1 kg·m/s^2.

newton (N) the magnitude of resultant force that, if applied to an object of mass 1 kg, will produce an acceleration of magnitude 1 m/s^2

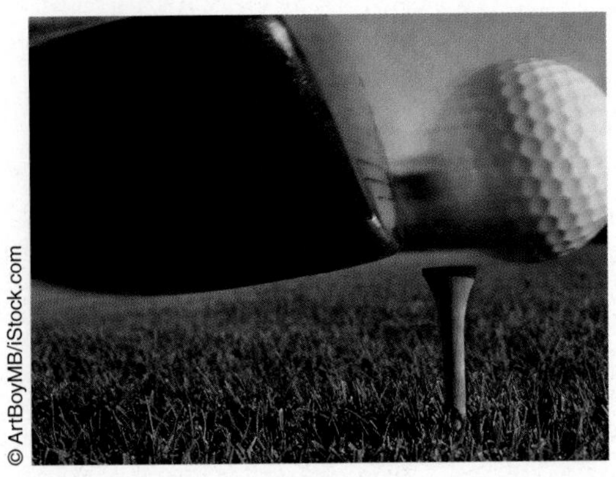
© ArtBoyMB/iStock.com

Figure 5-11 Sample Problem 5-4.

SAMPLE PROBLEM 5-4

A golf club hits a golf ball (mass 45.9 g) with a force of magnitude 5.0×10^3 N (Figure 5-11). What is the magnitude of the acceleration of the ball?

Solution

First, convert the mass unit to kilograms: $45.9\,\text{g} \times \dfrac{1\,\text{kg}}{1000\,\text{g}} = 4.59 \times 10^{-2}\,\text{kg}$

Since only magnitudes are important in this question, we rewrite Eqn. 5-1 in terms of magnitudes ($F_R = ma$), rearrange to solve for a, and substitute known quantities:

$$a = \frac{F_R}{m} = \frac{5.0 \times 10^3\,\text{N}}{4.59 \times 10^{-2}\,\text{kg}} = \frac{5.0 \times 10^3\,\text{kg}\cdot\text{m/s}^2}{4.59 \times 10^{-2}\,\text{kg}} = 1.1 \times 10^5\ \text{m/s}^2$$

Thus, the ball has an acceleration of magnitude 1.1×10^5 m/s^2.

Gravitational Acceleration and Force

If you drop a pencil, a comb, or any object near the surface of Earth, it falls with a downward acceleration of approximately 9.8 m/s^2 (if air resistance is neglected). This acceleration is due solely to the force of gravity acting on the object. If an apple of mass 0.10 kg is free-falling with an acceleration of 9.8 m/s^2 downward, then by Newton's second law, the force of gravity (in this case, the resultant force) on the apple is

$$\begin{aligned}\vec{F}_G = \vec{F}_R &= m\vec{a}\\ &= (0.10\,\text{kg})(9.8\,\text{m/s}^2\,\text{downward})\\ &= 0.98\,\text{kg}\cdot\text{m/s}^2\,\text{downward}\\ &= 0.98\,\text{N downward}\end{aligned}$$

This force of gravity is the same whether the apple is falling or not. If the apple is hanging on a tree, the force of gravity on it is still 0.98 N downward. Newton's second law gives us a method for determining the force of gravity on an object no matter what the object is doing, as long as the object is near the surface of Earth. All we need to do is multiply the object's mass by the acceleration due to gravity, $\vec{g}$, and we have the force of gravity on the object:

$$\vec{F}_G = m\vec{g} \quad \text{(force of gravity)} \qquad (5\text{-}3)$$

Notice that Eqn. 5-3 is a vector equation; the force of gravity is downward, in the same direction as the acceleration due to gravity. The magnitude of this force is simply mg.

For objects such as satellites or spaceships that are not near Earth's surface, the force of gravity cannot be calculated in this way. Gravity far from Earth's surface, and gravity due to other planets and the Sun, are discussed in Chapter 9.

The magnitude of the acceleration due to gravity is not a constant value of 9.8 m/s^2 at all points on Earth's surface. (The variability is due to Earth's non-spherical shape and non-uniform density. More details are provided in Chapter 9.) Thus, the force of gravity on a given object is not quite constant over Earth's surface. However, when solving problems in this book, assume that the gravitational acceleration at Earth's surface has a constant magnitude of 9.80 m/s^2, unless indicated otherwise.

The gravitational force on an object is sometimes referred to as the object's **weight**. Unfortunately, the term "weight" is often used in everyday conversation in place of "mass." For example, some people might say that the weight of an apple is 0.10 kg, when in fact this is the apple's mass. In this book, we will normally use the unambiguous "force of gravity" or "gravitational force," instead of "weight."

weight a term sometimes used to refer to the gravitational force on an object

Contrary to popular belief, it is easy to lose or gain weight. To lose weight, all you need to do is go to a place such as a mountain top, where the force of gravity is smaller, and your weight will be smaller! To gain weight, go to sea-level, where gravity is larger. Regrettably, these manoeuvres have no effect on your mass.

EXERCISES

5-19 A horizontal force is applied to a hockey puck of mass 0.15 kg at rest on ice. (Neglect friction.) If the magnitude of the resulting acceleration of the puck is 32 m/s^2, what is the magnitude of the force?

5-20 Magnetic forces are used to alter the direction of motion of the electron beam(s) in older TVs that have picture tubes. If a magnetic force of magnitude 3.20×10^{-15} N is exerted on an electron (mass $m_e = 9.11 \times 10^{-31}$ kg), what is the magnitude of the resulting acceleration?

5-21 A ball is thrown vertically upward. Neglect air resistance. On the way up, after the ball has left the thrower's hand,

(a) what is the direction of the acceleration of the ball?

(b) what is the magnitude of this acceleration?

(c) what is the direction of the resultant force on the ball?

(d) Repeat parts (a) to (c) when the ball is at the top of its path.

(e) Repeat parts (a) to (c) when the ball is on the way down.

5-22 A baseball is thrown from an outfielder to an infielder. Neglect air resistance. (Therefore, the trajectory will be parabolic.) Before the ball has reached its maximum height,

(a) what is the direction of the acceleration of the ball?

(b) what is the magnitude of this acceleration?

(c) what is the direction of the resultant force on the ball?

(d) Repeat parts (a) to (c) when the ball is at its maximum height.

(e) Repeat parts (a) to (c) after the ball has reached its maximum height.

5-23 (a) A constant resultant force is applied to various masses. Sketch the shape of a graph of the magnitude of the resulting acceleration versus mass.

(b) A variable resultant force is applied to a constant mass. Sketch the shape of the graph of the magnitude of the resulting acceleration versus the magnitude of the resultant force.

5-24 What is the magnitude of the force of gravity on a book of mass 2.45 kg?

5-25 Determine the mass (in grams) of a pencil if the magnitude of the force of gravity on it is 0.118 N.

5-26 The molar mass of Infectious Pancreatic Necrosis Virus (IPNV), which is a trout virus, is 5.4×10^7 g/mol. Determine the magnitude of the force of gravity on one IPNV particle. Avogadro's number is 6.02×10^{23} mol^{-1}.

5-27 The force of gravity on a 3.5 kg rock on Mars is 13 N in magnitude. What is the magnitude of the gravitational acceleration on Mars?

5-28 On Venus, the magnitude of the acceleration due to gravity is 8.9 m/s^2. If you were to travel to Venus, by what percentage would the magnitude of the force of gravity on you change?

5-29 (a) As a woodpecker's beak hits a tree (Figure 5-12), the speed of the bird's head decreases from 6.6 m/s to 0 m/s in a distance of 2.0 mm. Determine the magnitude of the acceleration. Assume horizontal motion and constant acceleration.

(b) How much time does it take for the head to slow down?

(c) What is the ratio of the magnitude of the force exerted by the tree on the head to the magnitude of the force of gravity on the head?

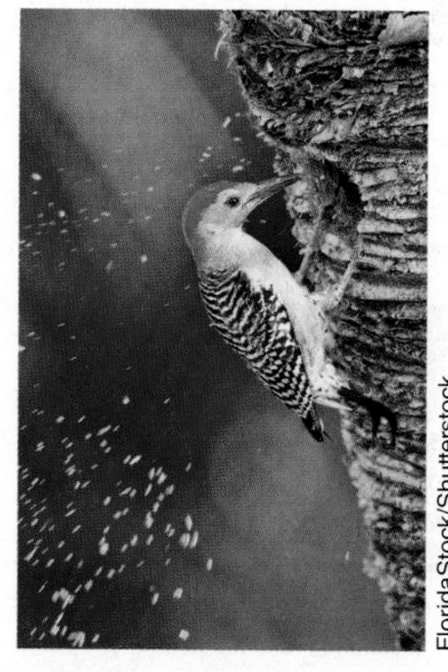

Figure 5-12 Question 5-29. The force on a woodpecker's head when drilling a hole in a tree is huge. Woodpeckers' heads have several unique features to reduce the shock to the brain, including a thin layer of fluid between the skull and brain.

5-30 **Fermi Question:** Determine the magnitude of the force required to accelerate an automobile from rest to a standard in-city driving speed in a typical time.

5.4 Newton's First Law of Motion

In Section 5.3, Newton's second law of motion was presented: an object on which there is a resultant force experiences an acceleration in the direction of the resultant force. The magnitude of this acceleration is proportional to the magnitude of the resultant force and inversely proportional to the object's mass. What if the resultant force is zero? According to Newton's second law, the object's acceleration is zero. Hence, the object's velocity must be constant both in magnitude and direction; that is, the object travels in a straight line at constant speed. (A special case of this situation is a constant speed of zero, in other words, the object is at rest.)

This particular circumstance of zero resultant force, zero acceleration, and constant velocity, discussed above using Newton's second law, is usually presented in a separate law, namely **Newton's first law of motion**, which deals with objects that experience no resultant force. It states

Newton's first law of motion If the resultant force on an object is zero, the object's velocity is constant; that is, the object travels in a straight line at constant speed (which could be zero).

If the resultant force on an object is zero, the object's velocity is constant; that is, the object travels in a straight line at constant speed (which could be zero).

At first glance, this law might seem surprising to you. You know that if you stop exerting a force on something, say a table that you have been pushing across the floor, it does *not* continue moving in a straight line at constant speed. It slows down and comes to rest. However, it comes to rest only because there has been a resultant force acting on it, namely the force of friction. If you push an object across a very smooth surface such as ice, and then stop pushing, it takes a long time for the object to come to rest because the force of friction is small. If you could reduce the force of friction to zero, so that the resultant force would be zero, then the object would continue travelling indefinitely in a straight line at constant speed. We are more accustomed to objects being at rest (i.e., having a constant speed of zero) if they are subject to no resultant force. Being at rest is just a special case of the more general situation in which objects have a constant non-zero speed in a straight line.

You have experienced many examples of Newton's first law in your everyday life. For example, when you are standing on a moving bus or subway train and the brakes are suddenly applied, you lurch forward. The brakes slow down the bus or train, but your body tends to continue to move forward at constant speed. Normally this forward motion is soon stopped by the force of friction between your feet and the floor, or by a force exerted because you are holding on to a post or handle. Imagine what would happen if you wore rollerblades on a bus! Because the skates provide very little friction, when the bus slows down, you would roll forward down the aisle until you collided with something (the windshield, for instance) that would stop you. In a car, seatbelts are used to stop your forward motion when the brakes are applied.

DID YOU KNOW?

93% of Canadians wear seatbelts. The 7% of Canadians who are not wearing seatbelts account for almost 40% of fatalities in vehicle collisions. (Source: Transport Canada)

The tendency of objects to continue moving with constant velocity or to stay at rest (if there is no resultant force) is sometimes called **inertia**, and Newton's first law is sometimes called the **law of inertia**. In a sense, this law should really be called Galileo's Law because it was discovered by Galileo Galilei (1564–1642), who performed a number of experiments on the motion of objects. Galileo was the first true experimental scientist. He studied nature not by just thinking about it, as people had done before him, but rather by doing careful experiments and analyzing the results mathematically.

inertia the tendency of objects to continue moving with constant velocity or to stay at rest (if there is no resultant force)

law of inertia another name for Newton's first law of motion

SAMPLE PROBLEM 5-5

A fish of mass 1.23 kg is swimming with a constant velocity of 0.56 m/s at an angle of 37° up from the horizontal. What is the resultant force on the fish?

Solution

The key phrase in this problem is "constant velocity." If the fish is swimming at constant velocity, then by Newton's first law the resultant force on the fish must be zero. This conclusion is true regardless of the detailed nature of the many forces acting on the fish: gravity, viscous friction, buoyant force, etc. The vector sum of all these forces must be zero in order for the fish to swim at constant velocity (Figure 5-13).

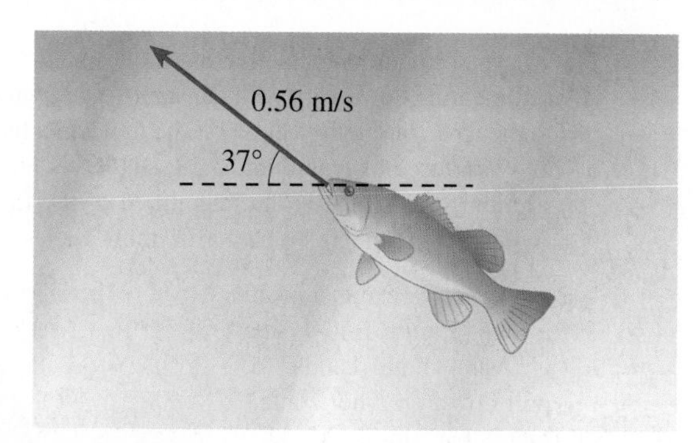

Figure 5-13 The resultant force is zero on a fish swimming at constant velocity.

TRY THIS!

Demonstrating Newton's First Law of Motion

You can easily demonstrate Newton's first law yourself with a coin and playing card. If you place the coin on the card which is projecting somewhat over the edge of a table, and then flick the card horizontally with your finger, the card will move across the table and the coin will stay essentially in place. As the card moves, it exerts only a small force on the coin for a very short time. Thus the resultant force on the coin is virtually zero and it stays at rest. (However, if you measure the position of the coin carefully, you will find that it has indeed moved somewhat because of the force of friction exerted by the card.) The magician's trick of pulling a tablecloth out from under a set of dishes on a table is similar. The dishes remain essentially at rest because the force exerted by the tablecloth is small and acts for only a short time. By the way, this trick is best left to magicians; don't try it with breakable dishes!

SAMPLE PROBLEM 5-6

A woman exerts a force of 7.5 N upward on her purse, and as a result it moves upward at constant velocity. What is the magnitude of the force of gravity on the purse?

Solution

Figure 5-14 shows the two forces exerted on the purse: $\vec{F}_{hand}$ exerted by the woman, and the downward force of gravity $\vec{F}_G$. Because we are told that the purse moves at constant velocity, we know that the resultant force must be zero. Therefore, the upward force and the downward force must have equal magnitudes. Since the upward force has a given magnitude of 7.5 N, then the downward force of gravity also has a magnitude of 7.5 N.

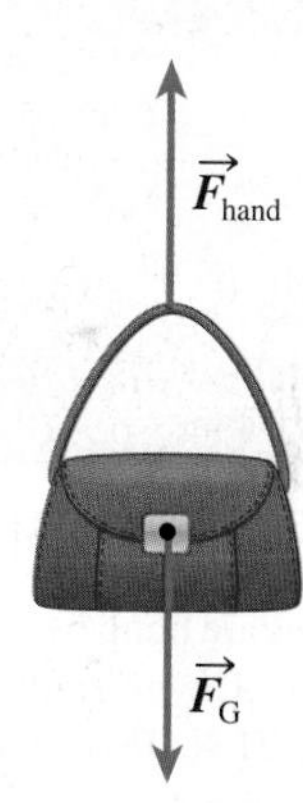

Figure 5-14 The forces on the purse in Sample Problem 5-6.

EXERCISES

In Questions 5-31 to 5-38 below, is the resultant force on object A zero?

5-31 A hamburger (A) is sitting on a table.

5-32 A red blood cell (A) is settling at a constant velocity of 4 mm/h downward in blood plasma.

5-33 An egg (A) is falling toward the ground, having fallen from a table.

5-34 A skydiver (A) is falling at "terminal speed," that is, is falling downward at a constant speed.

5-35 A hawk (A) of mass 2.12 kg is diving with a constant velocity of 10.6 m/s at an angle of 25.3° below the horizontal.

5-36 A marble (A) tied to the end of a string swings back and forth as a pendulum.

5-37 A planet (A) travels at constant speed in a circular orbit around a star.

5-38 A girl (A) is standing on a moving escalator that is half-way between the floors in a department store. The escalator is moving at constant velocity.

5-39 A boy is trying to push a table across a carpeted floor. He is exerting a horizontal force of 35 N, but the table is not moving. What are the magnitude and direction of the friction force exerted by the floor on the table?

5-40 A metal beam is hanging at rest from a vertical cable on a crane.

(a) If the magnitude of the tension in the cable is 2.57×10^4 N, what is the mass of the beam?

(b) If the same beam is being moved upward by the cable with a constant speed of 0.78 m/s, what is the magnitude of the tension in the cable?

5-41 In Figure 5-15, one person is lying flat, holding a rope that passes horizontally to a tree branch, then over the branch and downward to a friend who is hanging onto the lower end of the rope. There is negligible friction on the tree branch, and hence the branch only changes the direction of the rope, but does not affect the magnitude of its tension. The force of gravity on the hanging person has a magnitude of 708 N, and the people and rope are stationary.

(a) Draw a FBD for the hanging person, and determine the magnitude of the tension in the rope.

(b) Draw a FBD for the other person, and determine the magnitude of the friction force acting on this person.

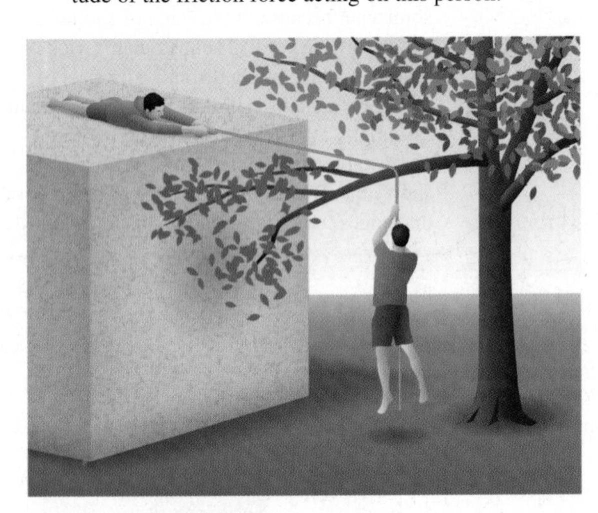

Figure 5-15 Question 5-41.

5-42 Three spheres are hung by light threads to form a holiday decoration (Figure 5-16). The magnitudes of the force of gravity on spheres A, B, and C are 21 N, 15 N, and 11 N, respectively.

(a) Draw a FBD for sphere C, and determine the magnitude of the tension in the lowest thread.

(b) Repeat for sphere B and the tension in the middle thread.

(c) Repeat for A and the highest thread.

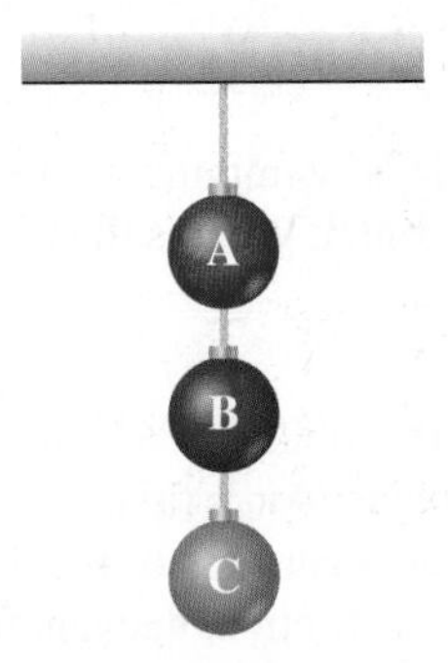

Figure 5-16 Question 5-42.

5-43 A large protein molecule having molar mass 4.1×10^{13} g/mol is sedimenting downward with constant velocity in water. In addition to the force of gravity acting on the molecule, there are an upward buoyant force of magnitude 5.13×10^{-13} N and an upward fluid friction force. What is the magnitude of the fluid friction force? Avogadro's number is 6.02×10^{23} mol^{-1}.

5-44 A woman is exercising in a gym (Figure 5-17). She is using both hands to pull down on a horizontal metal bar (mass 2.3 kg) attached at its centre to a cable that is vertically upward. The cable goes up over a pulley and then hangs down and is attached at its other end to a heavy weight of mass 13.6 kg. The woman is pulling the metal bar downward with a constant speed of 4.0 cm/s, and therefore the heavy weight is moving upward with the same speed.

Figure 5-17 Question 5-44.

(a) Draw a FBD for the heavy weight, and determine the magnitude of the tension in the cable.

(b) Draw a FBD for the metal bar, and determine the magnitude of the force exerted vertically downward on the bar by each of the woman's hands.

(c) Repeat (a) and (b) if the woman is pulling the metal bar with a downward acceleration of magnitude 0.50 m/s^2.

5.5 Solving Problems Using Newton's First and Second Laws

Newton's first and second laws of motion can be used in solving a wide variety of real-world problems in engineering, biomechanics, aerodynamics, astronomy, sports, space missions to other planets, etc. The simplest problems to deal with are those in which the forces are constant during the times of interest. Most of the problems that you will be asked to solve in this book are of the constant-force type. However, in research and industry today, many problems requiring Newton's second law involve complicated forces that are not constant over time. Computers can be used to analyze

Problem-Solving Strategy For Two-Dimensional Dynamics

1. Read the problem carefully. Check the definitions of words about which you are uncertain.
2. Make a diagram on which you include all relevant information, including any numerical values given.
3. Identify the object (or group of objects) whose motion is important.
4. Make a free-body diagram (FBD) for that object (or group). Be careful to use *only the forces acting on that object* (group). For simple problems, you can sometimes skip the diagram in step 2 and proceed directly to the FBD.
5. On your FBD, label an *x*-*y* co-ordinate system in which either the $+x$-direction or $+y$-direction is in the *direction that the object (group) is actually accelerating*. The other positive direction (if needed) is chosen at right angles to this direction. This choice of co-ordinate system means that the acceleration will have a positive component along one axis and a component of zero along the other axis, thus making the solution *much* easier.
6. Determine the *x*- and *y*-components of all known forces on your FBD.
7. Write down $\Sigma F_x = ma_x$ and/or $\Sigma F_y = ma_y$. On the next line of your solution, replace ΣF_x and/or ΣF_y with a list summing the appropriate *x*- and *y*-components of all the forces acting. On the right-hand side, replace m with the appropriate mass symbol for the problem (for example, m_1). If the object is not accelerating in the x (or y) direction, then write $a_x = 0$ (or $a_y = 0$); that is, use Newton's first law in this direction.
8. Repeat steps 3–7 for other objects (or groups) as necessary.
9. Use algebra to solve the resulting equations for the unknown(s). It is easier to do the algebra first and then substitute known numbers, rather than substituting numbers first, especially if you have to redo your work after making a mistake.
10. Substitute the known quantities and calculate the unknowns.
11. Check your answers. Are they reasonable? (For example, if you calculate the mass of a bird to be 757 kg, you know there must be a mistake somewhere.) Have you included the appropriate units in your answers?

these problems by considering the motion in a series of extremely short time intervals, during each of which the forces can be considered to be approximately constant. The motion of the object (spaceprobe, long-jumper, etc.) can then be determined for each interval and the complete motion worked out.

Because of the enormous variety of possible situations that can be analyzed using Newton's first and second laws, it is useful to employ a systematic approach that can be used to solve problems in this area. By using the approach outlined below, you should be able to tackle new problems with confidence.

SAMPLE PROBLEM 5-7

During the takeoff of a space vehicle, the acceleration is about 0.50 g upward. If the mass of the vehicle (including fuel, etc.) is 2.0×10^6 kg, determine the magnitude of the upward force on the vehicle, that is, the thrust due to the rockets.

Solution

After reading the problem carefully, we draw a diagram such as Figure 5-18 (a), transposing all the information from the problem to the diagram. Since it is the motion of the vehicle that is important, we then draw a FBD for the vehicle (Figure 5-18 (b)). At takeoff, since air resistance is negligible at low speeds, the only forces are the force of gravity and the upward rocket thrust.

The acceleration of the vehicle is upward; therefore, we choose the $+y$-direction to be upward. This is simply a convenience so that the acceleration of the vehicle will have a positive value. There is a smaller chance of mistakes if negative numbers can be avoided.

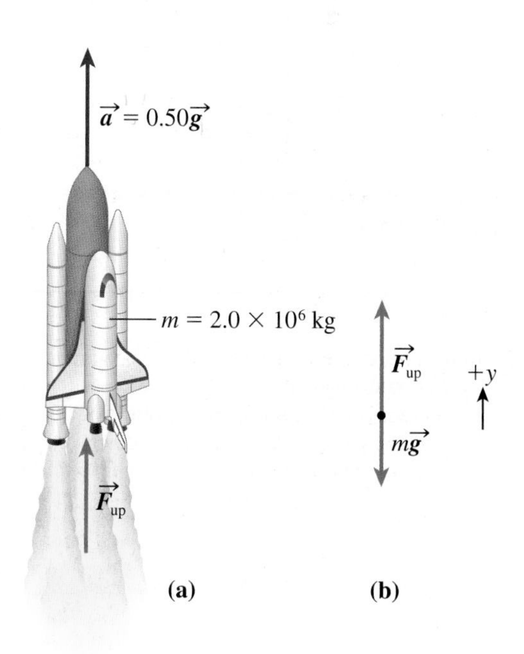

Figure 5-18 **(a)** Diagram for Sample Problem 5-7. **(b)** FBD for vehicle.

We now write Newton's second law in the y-direction: $\Sigma F_y = ma_y$

On the left-hand side of this equation, add all the forces: $F_{up} + (-mg) = ma_y$

Rearrange to solve for the unknown F_{up}:

$$F_{up} = mg + ma_y$$
$$= m(g + a_y)$$

Substitute known values (recall that the upward acceleration is given: $a_y = 0.50\ g = 4.9\ \text{m/s}^2$):

$$F_{up} = (2.0 \times 10^6\ \text{kg})((9.80 + 4.9)\ \text{m/s}^2)$$
$$= 2.9 \times 10^7\ \text{N}$$

Thus, the upward thrust is approximately 2.9×10^7 N in magnitude.

DID YOU KNOW?

At takeoff, more than 85% of a space vehicle's mass consists of fuel. This means that the maximum acceleration does not occur at takeoff, when the mass is very large, but later when most of the fuel has been used and the mass is much smaller.

Units Tip

Mass in kg multiplied by acceleration in m/s² gives force in kg·m/s², which is a newton (N).

SAMPLE PROBLEM 5-8

A box is being dragged at constant velocity across a horizontal floor by a person pulling on a horizontal rope. The magnitudes of the tension in the rope and the force of gravity on the box are 60.5 N and 213 N, respectively. Determine the magnitudes of: (a) the normal force on the box, and (b) the friction force on the box.

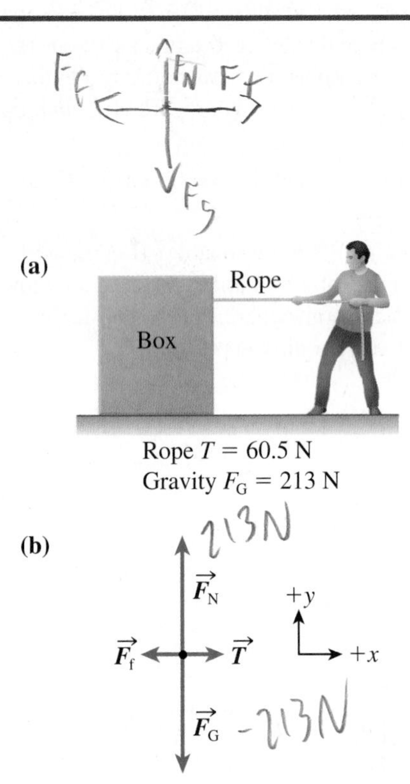

Figure 5-19 (a) Diagram for Sample Problem 5-8. **(b)** FBD for box.

Solution

(a) Figure 5-19 (a) shows a sketch of the situation, and Figure 5-19 (b) presents a FBD for the box.

Since the box is moving at constant horizontal velocity, the horizontal acceleration and horizontal resultant force must be zero.

In the vertical direction, the box has no motion; that is, its vertical velocity is constant at zero, and hence the vertical acceleration and resultant force are both zero.

Define an x-y co-ordinate system as shown in Figure 5-19 (b), and write

$$\Sigma F_x = 0 \quad \text{and} \quad \Sigma F_y = 0$$

In order to use these equations, we need to resolve the forces in the problem into x- and y-components. In this particular problem, tension and friction each have only an x-component, and gravity and the normal force each have only a y-component.

In part (a), we are asked to find the magnitude of the normal force, which has only a y-component, so we use $\Sigma F_y = 0$.

This gives $F_N + (-F_G) = 0$

Therefore, $F_N = F_G = 213$ N

Thus, the magnitude of the normal force is 213 N.

(b) Now use $\Sigma F_x = 0$ to find the friction force.

$$T + (-F_f) = 0$$

and therefore, $$F_f = T = 60.5\ \text{N}$$

Therefore, the magnitude of the friction force is 60.5 N.

Notation Tip

Remember that F_G (which could also be written as $|\vec{F}_G|$) is the magnitude of the force $\vec{F}_G$ and therefore is positive. Since the $+y$-direction is upward, then the y-component of the downward force $\vec{F}_G$ is negative and is written as $-F_G$.

SAMPLE PROBLEM 5-9

Two sleighs are connected by a horizontal rope, and the front sleigh is pulled by another horizontal rope with tension of magnitude 5.39 N. The sleigh masses are: front, 5.72 kg; rear, 6.89 kg. If the sleighs move together on a horizontal frozen lake on which friction is negligible, determine the magnitudes of: (a) the acceleration of the sleighs, and (b) the tension in the connecting rope.

Solution

(a) This problem illustrates how important the selection of object(s) for FBDs can be. Figure 5-20 (a) shows a diagram for the problem. Since we are asked first to determine the acceleration of the sleighs, we choose our FBD object to be the two sleighs together. The fact that the sleighs have the same acceleration (magnitude and direction) allows us to consider the sleighs as one object. The FBD is shown in Figure 5-20 (b). It shows the total gravitational force $(m_1 + m_2)\vec{g}$, the horizontal tension $\vec{T}_1$, and the total normal force $\vec{F}_N$ acting on the two sleighs.

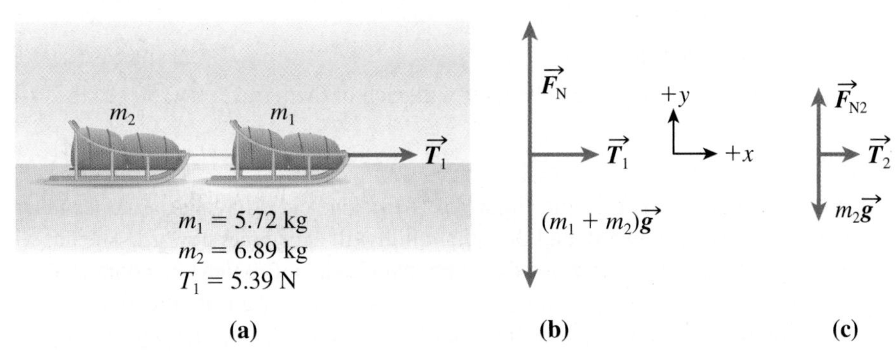

Figure 5-20 **(a)** Diagram for Sample Problem 5-9. **(b)** FBD for the two sleighs together. **(c)** FBD for rear sleigh.

Notice that the tension in the connecting rope is not shown. It pulls to the right on the rear sleigh and to the left on the front sleigh, and thus for the two sleighs considered together, there is no net contribution from this tension. Recall that tension pulls away from whatever object we are considering.

Since the sleighs accelerate to the right, the $+x$-direction is chosen to the right. We know that there will be no acceleration in the y-direction (the sleighs do not fly into the air, nor sink into the ground), and therefore we need to use only

$$\Sigma F_x = ma_x$$

The only force with an x-component is $\vec{T}_1$, and the total mass is $m = m_1 + m_2$.

Therefore, $$T_1 = (m_1 + m_2)a_x$$

Solving for the unknown a_x, $a_x = \dfrac{T_1}{m_1 + m_2}$

Substituting known values,

$$a_x = \frac{5.39\,\text{N}}{(5.72 + 6.89)\,\text{kg}} = 0.427\,\text{m/s}^2$$

Thus, the magnitude of the acceleration of the sleighs is 0.427 m/s^2.

(b) In order to determine the tension in the connecting rope, we need to use Newton's second law for one of the sleighs by itself. The only horizontal force acting on the rear sleigh is the tension in the connecting rope to the right, which we designate as $\vec{T}_2$. For the front sleigh, there are two horizontal forces, $\vec{T}_1$ and $-\vec{T}_2$, and therefore it will be easier to choose the rear sleigh for our FBD (Figure 5-20 (c)). Use Newton's second law:

$$\Sigma F_x = ma_x$$

Since $\vec{T}_2$ is the only force with an x-component, and $m = m_2$,

$$T_2 = m_2 a_x = (6.89 \text{ kg})(0.427 \text{ m/s}^2) = 2.94 \text{ N}$$

Therefore, the magnitude of the tension in the connecting rope is 2.94 N. You should check that you can also arrive at this answer by using a FBD for the front sleigh.

SAMPLE PROBLEM 5-10

Two children, of masses 35 kg and 45 kg, are hanging vertically from opposite ends of a rope that passes over a fixed horizontal metal bar in a playground. If friction between the rope and bar is neglected, what are the magnitudes of

(a) the acceleration of the children?

(b) the tension in the rope?

(c) If the children start from rest, how far will each of them travel in 0.50 s?

Solution

(a) A general diagram is given in Figure 5-21 (a). The child having the larger mass will accelerate downward, and the other child will accelerate upward. The accelerations of the two masses have the same magnitude (since they are connected by a single rope), but opposite directions. Because of the opposite directions of the accelerations, it is not appropriate to consider the masses as one object, and we draw a FBD for each mass (Figures 5-21 (b) and (c)).

Since the smaller mass accelerates upward, we choose its $+y$-direction upward (Figure 5-21 (b)). For the larger mass, its acceleration is downward and hence the $+y$-direction is chosen downward for this mass (Figure 5-21 (c)). This approach might seem unusual to you, but it ensures that the y-component of acceleration of *each* child will be positive, and we will be able to set these y-components equal as the solution proceeds.

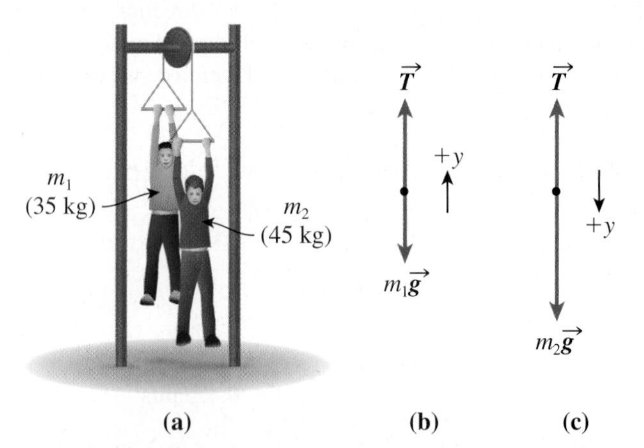

Figure 5-21 **(a)** Diagram for Sample Problem 5-10. **(b)** FBD for smaller mass. **(c)** FBD for larger mass.

We now use Newton's second law for the smaller child, using Figure 5-21 (b),

$$\Sigma F_y = ma_y$$

Thus,

$$T + (-m_1 g) = m_1 a_y \qquad \text{Eqn. [1]}$$

For the larger mass, using $\Sigma F_y = ma_y$ and Figure 5-21 (c) gives

$$m_2 g + (-T) = m_2 a_y \qquad \text{Eqn. [2]}$$

Since m_1 and m_2 are known, we now have two equations in two unknowns (a_y and T) that can be solved.

To solve the equations, first add them to eliminate T and then rearrange to find a_y. Adding and rearranging,

$$m_2 g - m_1 g = m_1 a_y + m_2 a_y$$

$$(m_2 - m_1)g = (m_1 + m_2)a_y$$

This gives

$$a_y = \frac{(m_2 - m_1)g}{m_1 + m_2}$$

(Alternatively, this expression for a_y could be determined by solving Eqn. [1] for T, substituting the resulting expression for T into Eqn. [2], and then rearranging to solve for a_y.)

Substituting known values,

$$a_y = \frac{((45 - 35)\text{ kg})(9.8\text{ m/s}^2)}{(45 + 35)\text{ kg}} = 1.2\text{ m/s}^2$$

Thus, the acceleration has a magnitude of 1.2 m/s².

(b) The value of a_y can now be substituted back into either of the original equations to solve for T, which turns out to be 3.9×10^2 N.

Notice that the rope tension is not equal in magnitude to the force of gravity on either of the two masses (3.4×10^2 N and 4.4×10^2 N). The tension must be greater than the force of gravity on the smaller mass in order to accelerate it upward, and smaller than gravity on the larger mass for a downward acceleration.

(c) Since we know the acceleration, we can use one of the constant-acceleration equations of kinematics to find how far each child moves in 0.50 s. The appropriate equation is Eqn. 2-9, which we write with y-subscripts to designate the vertical motion:

$$y = y_0 + v_{0y}t + \tfrac{1}{2}a_y t^2$$

We are told that the children start from rest; hence, $v_{0y} = 0$ m/s. We choose $y_0 = 0$ m for either child. Thus,

$$y = \tfrac{1}{2}a_y t^2 = \tfrac{1}{2}(1.2\text{ m/s}^2)(0.50\text{ s})^2 = 0.15\text{ m}$$

Hence, each child moves 0.15 m in the first 0.50 s after being at rest.

EXERCISES

5-45 Soon after jumping out of an airplane, a skydiver of mass 67 kg experiences an upward air resistance force of 567 N. What is the acceleration (magnitude and direction) of the skydiver?

5-46 A forklift truck is lifting a large crate of books vertically upward at a constant speed of 0.12 m/s. The force of gravity on the crate is 2.07×10^3 N in magnitude.

(a) What are the magnitude and direction of the force exerted by the forklift on the crate?

(b) Would you call this a normal force, friction force, or tension?

5-47 A hot-air balloon (Figure 5-22) has an acceleration of 0.24 m/s² downward. The total mass of the balloon, basket, and contents is 315 kg.

(a) Determine the magnitude of the upward buoyant force on the balloon, basket, and contents.

(b) The balloonists want to reduce the acceleration to zero, but there is no fuel left to heat the air in the balloon. How much ballast must be thrown overboard? Neglect air resistance.

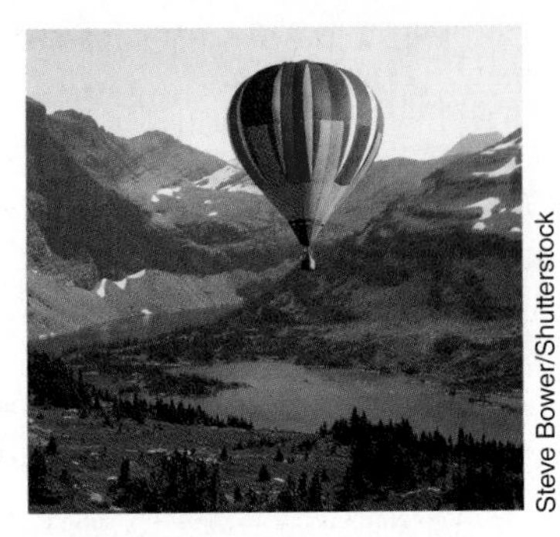

Steve Bower/Shutterstock

Figure 5-22 Question 5-47.

5-48 A woman having mass 55.3 kg is standing in an elevator that has an upward acceleration of 1.08 m/s^2. What is the magnitude of the normal force exerted by the floor on the woman? (This force is sometimes called the "apparent weight" of the woman. See the following question.)

5-49 (a) If the woman in the previous question were standing on bathroom scales in the elevator, what would the scales read in newtons? (Bathroom scales show the magnitude of the upward force exerted by the scales.)

(b) Repeat (a) if the elevator has a downward acceleration of 1.08 m/s^2.

(c) Repeat (a) if the elevator and woman are in free fall as a result of a break in the elevator cable.

(d) A person in free-fall is sometimes said to be "weightless." Why is this term used? Is the force of gravity on such a person zero?

(e) Repeat (a) if the elevator has a constant velocity of 1.08 m/s downward; and finally, a constant velocity of 1.08 m/s upward.

TRY THIS!

Elevator Physics

Here is something you can easily try in an elevator. Go into an elevator with a mass and a spring scale, suspend the mass from the scale and note the reading on the scale when the elevator is stationary. For each of the situations (a) to (f) listed below, predict whether the scale reading will greater than, less than, or the same as the stationary value. Holding the scale firmly against the elevator wall for stable readings, test your predictions when the elevator is

(a) just starting to move upward
(b) just starting to move downward
(c) moving at constant velocity upward
(d) moving at constant velocity downward
(e) slowing down while moving upward
(f) slowing down while moving downward

5-50 **Fermi Question:** Determine the sum of the magnitudes of the normal forces exerted on all the people on Earth.

5-51 When a karate expert breaks a brick (Figure 5-23), the velocity of her hand changes from (typically) 13 m/s downward to essentially 0 m/s in a time of only 3.0 ms.

(a) What is the acceleration (assumed constant) of the hand?

(b) If the mass of the hand is 0.65 kg, what are the magnitude and direction of the resultant force on it? Which object exerts almost all of this force on the hand?

(c) Compare (make a ratio of) the magnitude of the force in (b) to the magnitude of the gravitational force on the entire woman, if her mass is 56 kg.

Andrey Ushakov/Shutterstock

Figure 5-23 Question 5-51.

5-52 The oriental rat flea, a major carrier of bubonic plague, can jump a vertical distance more than 100 times its body length. (Imagine if you could do that!) During the initial phase of the jump, the flea is in contact with the ground for a typical time of 1.2 ms, and has an average upward acceleration of 1.0×10^3 m/s^2 (about 100 g). The mass of an average oriental rat flea is 2.1×10^{-7} kg. Neglect air resistance.

(a) During the initial phase of the jump, what is the magnitude of the resultant force on the flea?

(b) Draw a FBD for the flea in (a). What object exerts the upward force on the flea?

(c) After the flea has lost contact with the ground, what is the resultant force (magnitude and direction) on it?

(d) How high does the flea jump?

DID YOU KNOW?

The maximum acceleration during the oriental rat flea's takeoff has a magnitude of about 140 g. For a human doing a standing vertical jump, the maximum acceleration is only about 1 to 2 g in magnitude.

5-53 A man is pushing a stove straight across a kitchen floor at a constant speed of 25 cm/s. He is exerting a horizontal force of 85 N, and the force of gravity on the stove has a magnitude of 447 N.

(a) What are the magnitudes of the friction force and normal force exerted by the floor on the stove?

(b) What are the magnitude and direction of the total force exerted by the floor on the stove?

5-54 In Figure 5-24, two boxes are held in place by three ropes connected at point A, and by the force of friction (1.8 N) on box 2. The magnitudes of the force of gravity on boxes 1 and 2 are 2.5 N and 6.7 N, respectively.

(a) Draw a FBD for box 1, and determine the magnitude of the tension in the vertical rope.

(b) Draw a FBD for box 2, and determine the magnitudes of the tension in the horizontal rope and the normal force acting on box 2.

(c) Draw a FBD for point A, and determine the magnitude and the direction (angle θ) of tension $\vec{T}_3$.

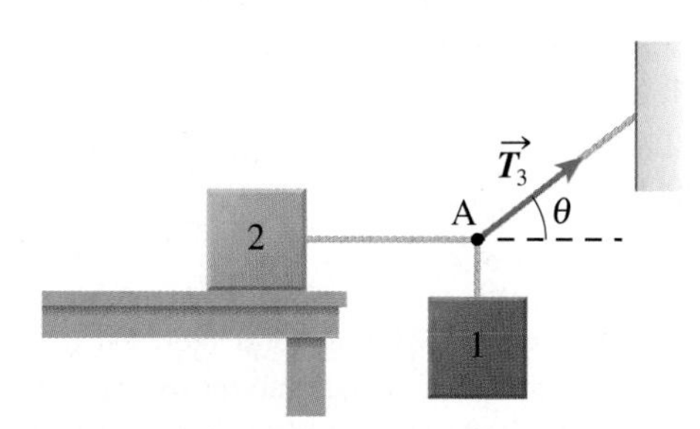

Figure 5-24 Question 5-54.

5-55 A sign outside a hair stylist's shop is suspended by two wires. The force of gravity on the sign has a magnitude of 55.7 N. If the angles between the wires and the horizontal are as shown in Figure 5-25, determine the magnitude of the tensions in the two wires.

Figure 5-25 Question 5-55.

5-56 Two sleds on level ground are connected by a horizontal rope, and the front sled is pulled by another horizontal rope with tension of magnitude 29 N. The sled masses are: front, 6.7 kg; rear, 5.6 kg. The magnitude of the force of friction on the front sled is 9 N; on the rear sled, 8 N. Determine the magnitudes of

(a) the acceleration of the sleds

(b) the tension in the connecting rope

(c) the normal force on the front sled

5-57 A large family has solved its food-shopping problem by connecting three shopping carts together with two horizontal ropes (Figure 5-26). The masses of the loaded carts are: m_1 = 15.0 kg, m_2 = 13.2 kg, and m_3 = 16.1 kg, Neglect friction. A third rope, which pulls the first cart, makes an angle of θ = 21.0° above the horizontal and has a tension of magnitude 35.3 N. Determine the magnitudes of: (a) the acceleration of the carts, (b) the tensions in the connecting ropes.

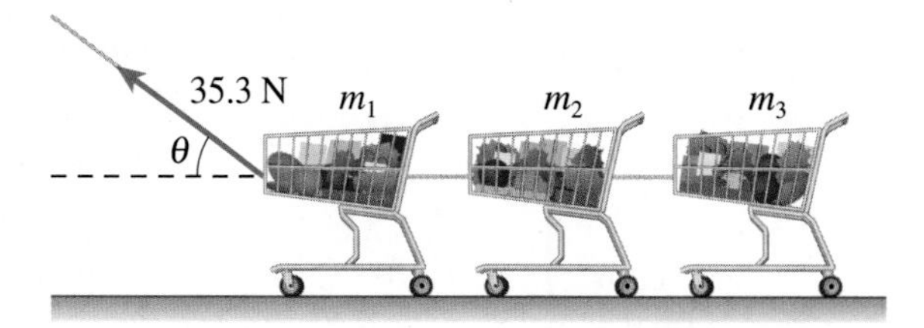

Figure 5-26 Question 5-57.

5-58 Two small boxes are attached to opposite ends of a string that passes over a fixed horizontal metal rod. The boxes hang vertically from the ends of the string. Friction between the rope and rod is negligible. When released, the boxes have an acceleration of magnitude 0.38 m/s^2. The mass of the smaller box is 0.75 kg.

(a) What is the mass of the other box?

(b) What is the magnitude of the tension in the string?

5-59 Two blocks are connected by a string passing over a stationary pulley (Figure 5-27). Block 2 sits on a horizontal surface. Neglect friction and the rotation of the pulley. If the masses are m_1 = 3.7 kg and m_2 = 2.7 kg, determine the magnitudes of: (a) the acceleration of the blocks (b) the tension in the string.

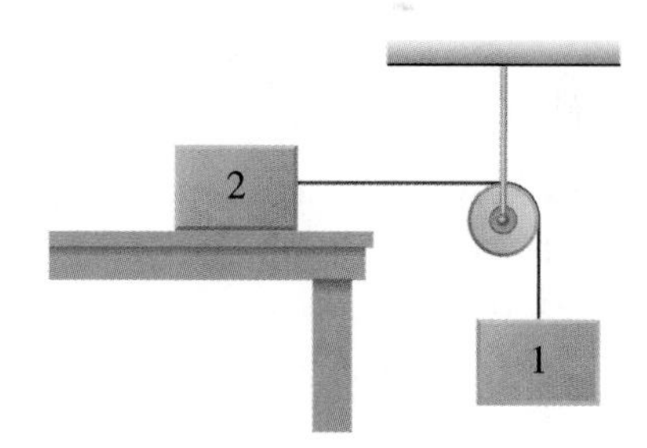

Figure 5-27 Questions 5-59 and 5-60.

5-60 Repeat the previous question, but include a friction force of magnitude 5.7 N between block 2 and the surface beneath it.

5.6 More Problem-Solving

Now that you have gained some experience in solving problems using Newton's second law, you have some idea of the wide variety of physical situations in which this fundamental law of nature is applicable, from launching a space vehicle to breaking a brick with a karate blow. You have seen many problems that involved a normal force, which was often equal in magnitude to the force of gravity on an object. In this section, we look at situations in which the normal force and the force of gravity do not have equal magnitudes.

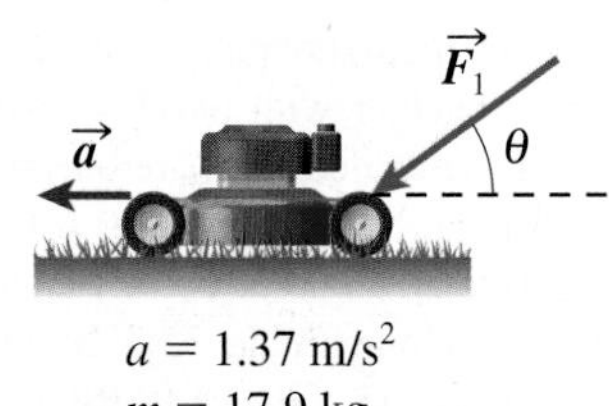

$a = 1.37$ m/s^2
$m = 17.9$ kg
$\theta = 35.1°$
$F_1 = 32.9$ N

(a)

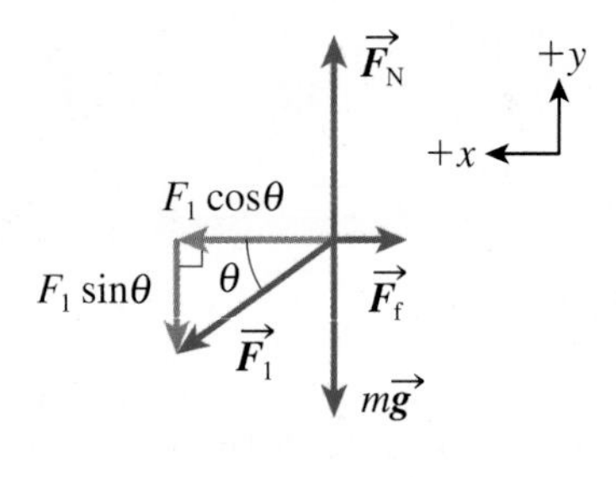

(b)

Figure 5-28 **(a)** Sketch for Sample Problem 5-11. **(b)** FBD for the lawnmower.

SAMPLE PROBLEM 5-11

A girl pushes a lawnmower of mass 17.9 kg from rest across a horizontal lawn by exerting a force of 32.9 N straight along the handle, which is inclined at an angle of 35.1° above the horizontal. If the magnitude of the acceleration of the mower is 1.37 m/s^2, determine the magnitudes of

(a) the normal force on the mower

(b) the friction force on the mower

If the acceleration lasts for only 0.58 s, and then the mower moves at constant velocity, find

(c) the final speed of the mower

(d) the magnitude of the force that the girl must then exert along the handle in order to maintain the constant speed

Solution

(a) After reading the problem carefully, we draw the general sketch (Figure 5-28 (a)) and the FBD for the mower (Figure 5-28 (b)). Since the mower accelerates to the left, the $+x$-direction is chosen to the left. The $+y$-direction is at right angles to this, either upward or downward (upward in this example). The only force that has both x- and y-components is the force exerted by the girl ($\vec{F}_1$), and these components are shown on the FBD.

We are first asked to find the magnitude of the normal force, which acts vertically. Therefore, we use Newton's second law in the y-direction:

$$\Sigma F_y = ma_y$$

Since there is no vertical acceleration, $a_y = 0$. Therefore, $\Sigma F_y = 0$. Writing the y-components of all the forces, and adding them to give zero,

$$F_N + (-mg) + (-F_1 \sin\theta) = 0$$

Hence,
$$\begin{aligned} F_N &= mg + F_1 \sin\theta \\ &= (17.9 \text{ kg})(9.80 \text{ m/s}^2) + (32.9 \text{ N})(\sin 35.1°) \\ &= 194 \text{ N} \end{aligned}$$

Therefore, the magnitude of the normal force is 194 N.

(b) To find the friction force, we use Newton's second law in the x-direction:

$$\Sigma F_x = ma_x$$

Summing the x-components of the forces, $F_1 \cos\theta + (-F_f) = ma_x$

Hence,
$$\begin{aligned} F_f &= F_1 \cos\theta - ma_x \\ &= (32.9 \text{ N})(\cos 35.1°) - (17.9 \text{ kg})(1.37 \text{ m/s}^2) \\ &= (26.9 - 24.5) \text{ N} \\ &= 2.4 \text{ N} \end{aligned}$$

The magnitude of the friction force is 2.4 N.

(c) In order to find the final speed of the mower, we can use one of the constant-acceleration equations of kinematics:

$$\begin{aligned} v_x &= v_{0x} + a_x t \\ &= 0 \text{ m/s} + (1.37 \text{ m/s}^2)(0.58 \text{ s}) \\ &= 0.79 \text{ m/s} \end{aligned}$$

Thus, the final speed of the lawnmower is 0.79 m/s.

(d) We now need to find the magnitude of the force (call it $\vec{F}_2$) that the girl needs to exert in order to maintain the constant speed of the mower. Figure 5-29 shows the FBD in this case. Force $\vec{F}_2$ is not as large as the original $\vec{F}_1$, and the new FBD has been drawn so that $F_2 \cos\theta$ has the same magnitude as F_f, so that there will be zero acceleration. Since the normal force is no longer the same magnitude as in parts (a), (b), and (c) of this problem, it is now labelled as $\vec{F}_{N2}$ instead of $\vec{F}_N$. In the x-direction, we use Newton's second law (or first law, since $a_x = 0$):

$$\Sigma F_x = ma_x = 0$$

$$F_2 \cos\theta + (-F_f) = 0$$

This gives

$$F_2 \cos\theta = F_f$$

$$F_2 = \frac{F_f}{\cos\theta} = \frac{2.4 \text{ N}}{\cos 35.1°} = 2.9 \text{ N}$$

Therefore, the girl needs to exert a force of magnitude 2.9 N.

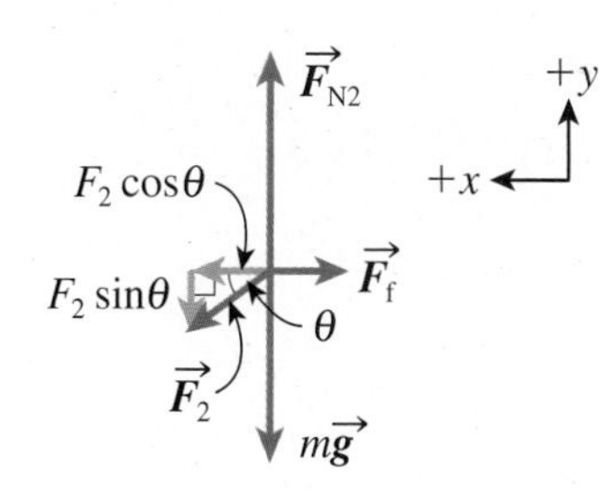

Figure 5-29 Sample Problem 5-11, part (d).

SAMPLE PROBLEM 5-12

A skier of mass 65 kg slides down a snow-covered hill inclined at 12° from the horizontal. If friction is neglected, what are the magnitudes of

(a) the normal force on the skier?

(b) the acceleration of the skier?

Solution

This problem illustrates the use of an x-y coordinate system that is not horizontal–vertical. Figures 5-30 (a) and (b) show a general diagram and the FBD for the skier. Since the skier's acceleration is along the incline, we choose the $+x$-axis in this direction. The $+y$-axis is perpendicular to this. The force of gravity has both x- and y-components, shown in Figure 5-30 (b) as $(mg)_x$ and $(mg)_y$, but the normal force has only a y-component.

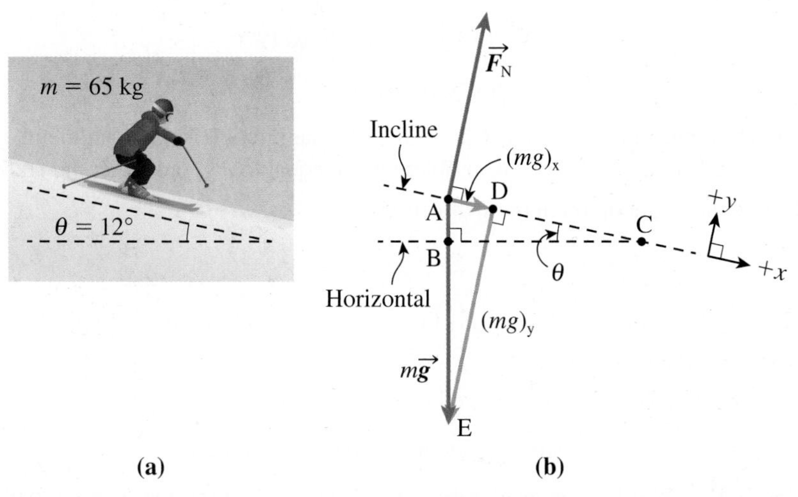

Figure 5-30 **(a)** Diagram for Sample Problem 5-12. **(b)** FBD for the skier.

In order to solve this problem, we will need to express the components of the force of gravity in terms of angle θ, which is the 12° angle between the incline and the horizontal. To do this, consider triangles ABC and ADE in Figure 5-30 (b).

In triangle ABC, angle ABC is 90°, since it lies between a horizontal line and the vertical force $m\vec{g}$. Angle ACB is θ. In order for the angles in triangle ABC to add up to 180°, angle BAC must be $(90° - \theta)$. In triangle ADE, side AE is the force $m\vec{g}$, and sides AD and DE are its x- and y-components, respectively. Angle EAD is the same as angle BAC; that is, $(90° - \theta)$. Angle ADE is 90°, since it is the angle between the x- and y-components. Since the angles in triangle ADE must add up to 180°, angle AED must be equal to θ. We now can use the following information about triangle ADE:

- angle ADE is 90°
- angle AED is θ
- the hypotenuse has a magnitude of mg
- side AD represents the x-component of $m\vec{g}$, that is, $(mg)_x$
- side DE represents the y-component of $m\vec{g}$, that is, $(mg)_y$

Since triangle ADE has a right angle (ADE), the components of $m\vec{g}$ can be written in terms of $\sin\theta$ and $\cos\theta$:

$$(mg)_x = mg\sin\theta$$

$$(mg)_y = -mg\cos\theta$$

Notice the negative y-component of $m\vec{g}$. (In Figure 5-30 (b), the y-component of $m\vec{g}$ and the $+y$-axis are in opposite directions.)

(a) To find the magnitude of the normal force, which is in the y-direction, we use Newton's second law:

$$\Sigma F_y = ma_y$$

The acceleration $a_y = 0$, because the skier does not accelerate perpendicular to the plane. (The skier does not sink into the plane, nor fly into the air.)

Thus,

$$\Sigma F_y = 0$$

$$F_N + (-mg\cos\theta) = 0$$

$$F_N = mg\cos\theta$$

$$= (65\ \text{kg})(9.8\ \text{m/s}^2)(\cos 12°)$$

$$= 6.2 \times 10^2\ \text{N}$$

Therefore, the normal force is 6.2×10^2 N in magnitude. (Notice that the magnitude of the normal force in this problem is not equal to mg, but rather $mg\cos\theta$.)

(b) To find the acceleration in the x-direction,

$$\Sigma F_x = ma_x$$

The only force in the x-direction is the x-component of gravity.

Therefore, $mg\sin\theta = ma_x$

Dividing both sides by mass m, and rearranging with a_x on the left,

$$a_x = g\sin\theta = (9.8\ \text{m/s}^2)(\sin 12°) = 2.0\ \text{m/s}^2$$

Hence, the skier has an acceleration of 2.0 m/s^2 in magnitude. Notice that the skier's mass was not used in the final calculation of the acceleration. Therefore, an object of *any mass* will slide down a 12° incline with the same acceleration (neglecting friction).

Problem-Solving Tip

When you first looked at this problem, you might have been tempted to choose a horizontal–vertical x-y co-ordinate system. If you had done this, both the normal force and the acceleration would have had x- and y-components, which would have made the solution much more difficult. *Remember always to choose the $+x$- or $+y$-axis in the direction of the acceleration of the object.* In particular, if a problem involves an object that moves up or down an incline, choose one co-ordinate axis parallel to the incline, and the other axis perpendicular to this.

EXERCISES

5-61 A woman is leaning on a table (Figure 5-31). The table has a mass of 25 kg and is stationary. If the force exerted by the woman on the table is 35 N at 37° below the horizontal, determine the magnitudes of

(a) the normal force exerted by the floor on the table
(b) the friction force exerted by the floor on the table

Figure 5-31 Question 5-61.

5-62 A newspaper carrier pulls her papers in a small sleigh during the winter. The rope with which she pulls the sleigh makes an angle of 25.0° above the horizontal, and the magnitude of the tension in the rope is 6.73 N. The force of gravity on the sleigh and papers has a magnitude of 35.6 N, and the sleigh is moving horizontally at constant velocity. Determine the magnitude and direction of the (a) normal force, and (b) friction force, exerted on the sleigh.

5-63 One day the rope used by the girl in the previous question broke, and she had to push her sleigh. She pushed slightly downward, with a force of 8.25 N at an angle of 15.2° below the horizontal. The force of gravity on the sleigh and papers was still 35.6 N in magnitude, and the sleigh still moved horizontally at constant velocity. Determine the magnitude and direction of

(a) the normal force
(b) the friction force
(c) the total force exerted on the sleigh by the ground

5-64 A man starts to pull a large chair across a horizontal floor by pulling at an angle of 15° above the horizontal (Figure 5-32). The force that he exerts is 91 N in magnitude, and there is a normal force of 221 N upward on the chair. If the chair's horizontal acceleration is 0.076 m/s^2 in magnitude, what are

Figure 5-32 Question 5-64.

(a) the mass of the chair?
(b) the magnitude of the friction force on the chair?

5-65 A girl's soapbox derby car is pushed along a horizontal road with a broom handle (Figure 5-33). The total mass of the car and girl is 36.7 kg, and a force of 72.7 N is applied along the direction of the broom handle, making an angle of 69.3° with the vertical. There is a friction force of magnitude 12.7 N on the car. Determine the magnitudes of

(a) the acceleration of the girl and car
(b) the normal force exerted by the ground on the girl and car
(c) the distance the car will travel in 1.50 s, assuming it started from rest

Figure 5-33 Question 5-65.

5-66 A box of mass m is pulled at constant velocity across a horizontal floor by a rope exerting a force of magnitude F applied at an angle of θ above the horizontal (Figure 5-34). In terms of given parameters and "g," determine expressions for the magnitudes of

(a) the normal force on the box
(b) the friction force on the box

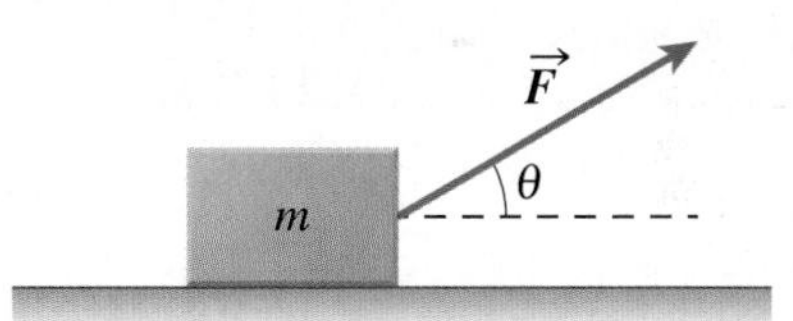

Figure 5-34 Question 5-66.

5-67 On the roller coaster shown in Figure 5-35, the cars and riders are starting to slide down an incline that makes an angle of 36° with the horizontal (that's steep!). Neglecting friction, what is the magnitude of the acceleration of the cars when they go down the incline?

Figure 5-35 Question 5-67.

5-68 A girl on a toboggan slides with an acceleration of magnitude 1.5 m/s^2 down a hill. If friction is neglected, what is the angle between the hill and the horizontal?

5-69 An otter of mass m slides down a muddy incline into a river. The incline makes an angle of ϕ with the horizontal. Determine an algebraic expression for the magnitude of the otter's acceleration in terms of given parameters and "g." Neglect friction.

Newton's third law of motion If object A exerts a force on object B, an equal but opposite force is simultaneously exerted by B on A.

5.7 Newton's Third Law of Motion

Take a look at Figure 5-36. What do you think happened to the lines painted on the road? Were the bends caused by an earthquake? Did the road-painters just have an "off-day"? The crooked lines are actually the result of **Newton's third law of motion**, which is his famous law of action and reaction. This law states that

If object A exerts a force on object B, an equal but opposite force is simultaneously exerted by B on A.

Ernie McFarland, author

Figure 5-36 Why aren't the lines straight on this road?

Another way of saying this is that for every action, there is an equal and opposite reaction.

The road lines in Figure 5-36 are at a busy intersection, where a large number of vehicles have to stop for traffic lights. When the light turns green, the vehicles accelerate over the lines on the road (Figure 5-37). In order for a vehicle to accelerate forward, its tires exert a backward force on the road (Figure 5-38). By Newton's third law, the road exerts a forward force on the tires, and hence on the vehicle. (This forward force that accelerates the vehicle is just the friction force between the road and the tires. Without friction, a car would simply sit and spin its wheels.) At the intersection shown in Figures 5-36 and 5-37, the backward force exerted by the vehicles on the road has caused the road surface to slip backward, carrying the painted lines with it. The upper layer of asphalt must have been poorly bonded to the underlying layers.

Ernie McFarland, author

Figure 5-37 A car accelerating over the lines.

Newton's third law comes into play whenever a force is exerted on an object. To illustrate the importance of this law, we consider some other forms of transportation (Figure 5-39):

(a) When we walk, our feet push backward on the ground; the ground pushes forward on our feet, thus moving us forward.

(b) Boat propellers push backward on the water; the water pushes forward on the propellers.

(c) A rocket expels hot gases downward; the gases push upward on the rocket.

(d) Airplane wings are shaped so that they deflect air downward; that is, they exert a downward force on the air; the air pushes upward on the wings.

In each of the transportation examples, note that there are only two objects involved, and that we are considering only one force exerted on each object. People often become confused about Newton's third law because they try to include more than two objects at a time, or because they introduce additional forces.

Figure 5-38 Forces when a car accelerates to the right.

For example, consider the situation of you standing on the ground. Earth exerts a downward force of gravity (the "action") on you. There is a common misconception that the "reaction" force in this instance is the upward normal force exerted by Earth on you. If this were so, the action-reaction pair would be two forces exerted on you, not one force exerted on you and one on Earth, as required in Newton's third law. The correct "reaction" force is the upward force of gravity exerted by you on Earth

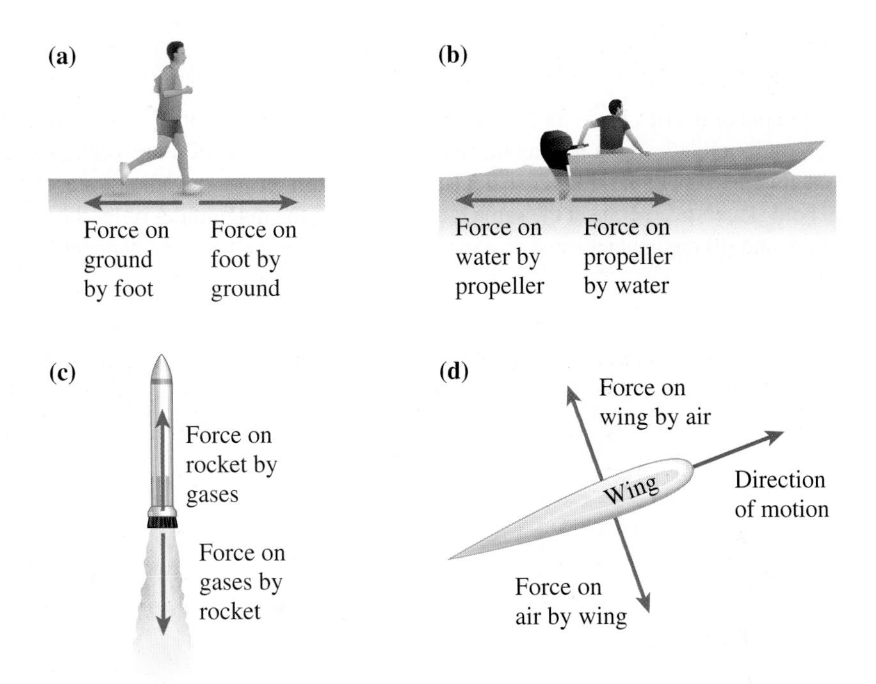

Figure 5-39 Newton's third law in transportation.

(Figure 5-40 (a)). It seems unusual to think of an *upward* force of gravity; however, because gravity is an attractive force between objects, the force of gravity exerted on Earth by you is indeed upward.

What about the upward normal force exerted by Earth on you? What action-reaction pair is it a part of? By Newton's third law, the other force in the pair must be a *downward* force exerted by you on Earth (Figure 5-40 (b)). This is the force exerted on Earth because you are in contact with it. This force is also a normal force because it is perpendicular to the surfaces in contact, namely the soles of your shoes and the ground. (Recall that "normal" means "perpendicular.") Note that this force is not the downward force of gravity, because that force is exerted downward on you, not on Earth. The downward force of gravity acts on you whether you are in contact with Earth or in mid-air during a jump, but the downward normal force that you exert on Earth acts only when you are touching it.

Notice that the two forces in an action-reaction pair are *always* the same type of force. They are both normal forces, or both gravitational forces, etc.

It does not matter which of the two forces you consider to be the "action," and which the "reaction." Cause and effect are not implied by these terms; the forces are applied at the same time.

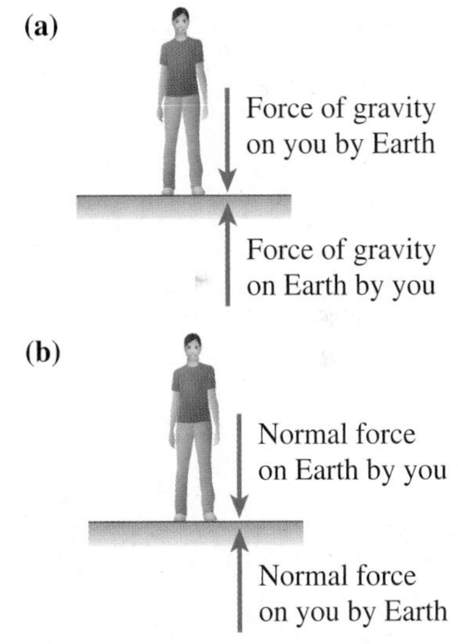

Figure 5-40 Two action–reaction pairs.

SAMPLE PROBLEM 5-13

A baseball player sliding into second base (Figure 5-41) experiences a friction force. What is the other force in the action-reaction pair? What is its name? What is its direction? What object exerts it? On what object is it exerted?

Solution

Let us call the friction force exerted by the ground on the player the "action." Its direction is opposite to the player's slide. By Newton's third law, the "reaction" is exerted by the player on the ground, in the same direction as the player's slide. It is also a friction force.

Figure 5-41 Sample Problem 5-13.

SAMPLE PROBLEM 5-14

Two people of masses $m_1 = 55$ kg and $m_2 = 65$ kg are standing, one just in front of the other, on rollerblades on a horizontal floor. A third person applies a horizontal force $\vec{F}$ of magnitude 84 N to the smaller skater (#1), who in turn pushes the other skater (#2), so that both skaters accelerate at the same rate. Determine the magnitudes of (a) their acceleration, and (b) the contact force $\vec{F}_C$ exerted by one skater on the other. Neglect friction.

Solution

(a) This problem is similar to Sample Problem 5-9 in which two sleighs connected by a rope were pulled by another rope. We first draw a diagram for the problem (Figure 5-42 (a)), and then a FBD for the two skaters together (Figure 5-42 (b)).

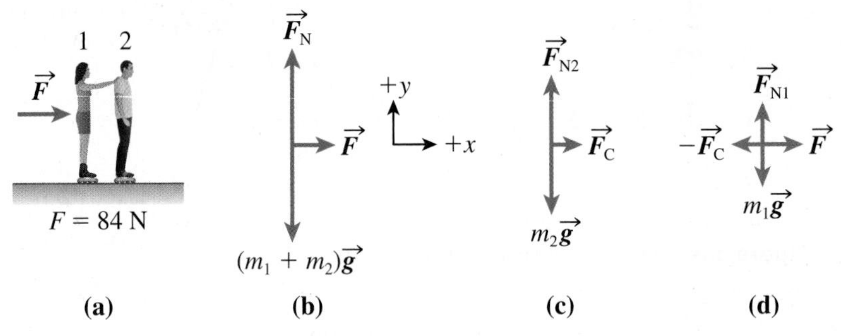

Figure 5-42 Sample Problem 5-14. **(a)** Diagram of the system. **(b)** FBD for the two skaters together. **(c)** FBD for skater 2. **(d)** FBD for skater 1.

We can solve for the acceleration by using Newton's second law in the x-direction:

$$\Sigma F_x = ma_x$$

$$F = (m_1 + m_2)a_x$$

Therefore, $a_x = \dfrac{F}{m_1 + m_2} = \dfrac{84 \text{ N}}{(55 + 65) \text{ kg}} = 0.70 \text{ m/s}^2$

Thus, the skaters' acceleration has a magnitude of 0.70 m/s^2.

(b) In order to determine the force exerted between the two skaters, we need to draw a FBD for only one skater. The force $\vec{F}_C$ exerted on skater 2 by skater 1 is to the right; it is this force that accelerates skater 2. By Newton's third law, skater 2 exerts a force on skater 1, of equal magnitude but to the left, that is, $-\vec{F}_C$.

Figure 5-42 (c) shows the FBD for skater 2. Using Newton's second law

$$\Sigma F_x = ma_x \quad \text{which gives} \quad F_C = m_2 a_x$$

Substituting known values, $F_C = (65 \text{ kg})(0.70 \text{ m/s}^2) = 46 \text{ N}$

Hence, the force exerted by one skater on the other has a magnitude of 46 N.

Alternatively, we could use a FBD for skater 1 (Figure 5-42 (d)), and use Newton's second law:

$$\Sigma F_x = ma_x \quad \text{giving} \quad F + (-F_C) = m_2 a_x$$

Thus, $\qquad F_C = F - m_1 a_x = 84 \text{ N} - (55 \text{ kg})(0.70 \text{ m/s}^2) = 46 \text{ N}$

This is the same answer as we determined with the FBD for skater 2, but using the FBD for skater 2 required fewer steps.

PROFILES IN PHYSICS

Sir Isaac Newton (1642–1727): Understanding Forces and Gravity

"Nature and Nature's laws lay hid in night;
God said, *Let Newton Be!* and all was light."

These lines, written by the English poet Alexander Pope, appear as the epitaph on Newton's tomb in Westminster Abbey, where Newton is buried along with other great Britons. Although he was undoubtedly one of the greatest scientists ever known, he had psychological problems and was not an easy person to get along with. He was very solitary and became involved in a number of personal controversies with other distinguished people of his time.

Figure 5-43 Sir Isaac Newton.

Isaac Newton (Figure 5-43) was born on December 25, 1642. His father had died three months before Isaac's birth, and when his mother remarried three years later, Isaac was sent to live with his grandmother. When he was eleven, his stepfather died and Isaac returned to his mother, half-brother, and two half-sisters. Some psychologists have suggested that these moves during his youth may have had negative effects on his personality.

In elementary school, Newton was a good, but not exceptional, student and his mother thought that perhaps he should become a farmer. However, a clergyman suggested that he should attend Cambridge University, which he entered in 1661. His academic talents were recognized at Cambridge by his mathematics professor, Isaac Barrow, and four years later, Newton received the degree of Bachelor of Arts. Soon after, the plague struck England, and the university was closed for two years to avoid the spread of the disease. Newton returned home to his mother, and in the quiet surroundings there, made a number of rapid and important advances in science and mathematics.

In a period of some 18 months, Newton invented calculus, discovered the law of universal gravitation, developed his three laws of motion, and made a number of discoveries in optics related to light, colour, and interference phenomena. His work in optics led him to build the first reflecting telescope. This list of achievements is enough to fill a normal person's lifetime; to accomplish all this in only 18 months is amazing! The famous story of an apple falling from a tree and hitting Newton on the head during this period, thus inspiring his development of the gravitation law, may or may not be true. However, the story is a reminder that his gravitation law is indeed universal, determining the motion of apples on Earth, and also the orbits of the moon, planets, stars, and galaxies.

When the plague was over, Newton returned to Cambridge for further studies, and was named Lucasian Professor of Mathematics in 1669. His first scientific paper, in 1672, was on light and colour, along with a description of his newly invented telescope. The paper was criticized by some other scientists, and Newton's reaction was withdrawal. He decided not to publish any more of his discoveries except when strongly encouraged by colleagues. As a result, he was later involved in a number of conflicts with other scientists and mathematicians concerning the credit for certain discoveries. Normally the first person to publish a result is acknowledged as the discoverer, but since Newton was publishing very little of his work, he wanted credit to be given to the person who first had the idea, even if it was kept secret.

Fortunately, Newton was persuaded by Edmund Halley (after whom Halley's comet is named) to publish his *Philosophiae Naturalis Principia Mathematica* in 1687, in which his laws of motion and universal gravitation were presented.

In 1692, Newton fell into a serious depression, with delusions of persecution. He recovered reasonably well, but having tired of the academic life of a professor, was made Warden of the Mint in London in 1691, where he busied himself with the development of new coinage. In 1699, he was promoted to Master of the Mint, a post he held until his death.

Newton became increasingly difficult to deal with in his later years. One of his most famous controversies was with the German philosopher and mathematician Gottfried Leibnitz over the discovery of calculus. Newton and Leibnitz had developed calculus independently, and initially there was a period of goodwill and mutual recognition. But by 1699 there was a great battle involving national pride and a number of supporters on both sides. The story is very complicated and the issue was never resolved to the satisfaction of the parties involved.

In 1703, Newton became President of the Royal Society, a group of distinguished British scientists, and he reigned over science in England with a firm hand until his death. He was ennobled in 1705 by Queen Anne, and thus became Sir Isaac Newton.

Newton spent only a small part of his adult life involved in physics and mathematics. He devoted a great deal of time to the study of theology, writing entire books on religious subjects, and he also worked on alchemy. He often professed disgust for his advances in physical science—an odd attitude for a man who had accomplished so much.

In 1727, Sir Isaac Newton died and was buried with state honours in Westminster Abbey.

EXERCISES

5-70 In each case below, one force is given. What is the other force in the action-reaction pair? Give its name (if possible) and direction, and indicate which object exerts it, and on which object it is exerted.

(a) The Sun exerts a gravitational force on Jupiter.

(b) A book on a table is supported by an upward normal force.

(c) A chef exerts a force on a baking pan to pull it out of an oven.

(d) A canoeist uses a paddle to exert a backward force on the water.

(e) Earth exerts a gravitational force on a falling apple.

(f) The upward force of air resistance (a type of friction) is exerted on a falling hailstone.

5-71 A train consisting of an engine and two cars is speeding up on a horizontal track with an acceleration of magnitude 0.33 m/s^2. Neglect friction on the cars. If each car has a mass of 3.1×10^4 kg, determine the magnitude and direction of

(a) the force exerted by the first car on the second

(b) the force exerted by the second car on the first

(c) the force exerted by the engine on the first car

5-72 Using Newton's third law of motion, explain how the water jets on a flyboard (Figure 5-44) enable a person to fly and hover in the air.

Figure 5-44 Question 5-72.

5-73 Two books are sitting side by side (touching each other) on a table. A horizontal applied force of magnitude 0.58 N causes the books to move together with an acceleration of 0.21 m/s^2 horizontally. The mass of the book to which the force is directly applied is 1.0 kg. Neglecting friction, determine

(a) the mass of the other book

(b) the magnitude of the force exerted by one book on the other

5-74 A figure skater of mass 56 kg pushes horizontally against the boards at the side of a skating rink for 0.75 s. Having started at rest, she leaves the boards with a speed of 75 cm/s. Neglect friction. What were the magnitudes of

(a) her acceleration (assumed constant)

(b) the force exerted on her by the boards

(c) the force that she exerted on the boards

5.8 The Fundamental Forces

There appears to be a wide variety of forces in the world around us—gravity, friction, normal force, air resistance, rope tension, muscle forces, and so on—but all forces can be traced to just four fundamental types. One of these is *gravity* (discussed earlier in this chapter), which is a force of attraction between all objects, and is responsible for the attraction of objects toward our own Earth. Gravity is the most important force at the astronomical level, and determines the motion of planets, stars, and galaxies.

electrical force a force that exists between charged objects or particles, and is responsible for holding together the atoms and molecules in all objects

electromagnetic interaction a name given to the combination of electrical and magnetic forces, which are very closely related

Most of the other forces with which we are directly familiar are fundamentally **electrical forces**.[2] Electrical forces exist only between charged objects or particles, such as negatively charged electrons and positively charged nuclei in atoms. The forces holding together the atoms and molecules in all the objects around us—cars, tables, people, alligators—are electrical. If a book is sitting on a table, the upward normal force exerted by the table on the book is due to the electrical forces between the molecules in the table. Muscle forces, friction forces, and forces exerted by objects (such as springs) which are being stretched or compressed are all fundamentally electrical forces. Magnetic forces have been shown to be very closely related to electrical forces, and often electrical and magnetic forces are referred to together as the **electromagnetic interaction**.

nuclear strong force the attractive force that holds together the protons and neutrons in nuclei

nuclear weak force a nuclear force related to processes such as beta-decay, and which has been shown to be essentially the same as the electromagnetic interaction

The other two fundamental forces are less noticeable in everyday life, since they are both related to the tiny nuclei of atoms. The **nuclear strong force** is the attractive force that holds together the protons and neutrons in nuclei. The **nuclear weak force** is related to processes such as beta-decay, in which nuclei emit high-speed electrons called beta-particles. A theory has shown that the weak force is essentially the same

[2]Electrical forces are considered in detail in Chapter 19.

as the electromagnetic interaction, and thus it is now more correct to say that there are three fundamental forces: *gravity, nuclear strong,* and *electroweak.* A number of physicists around the world are attempting to develop a theory in which the electroweak and nuclear strong forces are shown to be essentially the same force. Such a theory would be a "grand unified theory" (GUT). It might be possible eventually to show that all forces are simply different manifestations of the same single fundamental force, in a "theory of everything" (TOE).

DID YOU KNOW?

The 1979 Nobel Prize in Physics was awarded to Sheldon Glashow, Abdus Salam, and Steven Weinberg for their "unified model of the action of the weak and electro-magnetic forces."

EXERCISES

5-75 In each case below, identify the fundamental force that is responsible for the interaction given.

(a) The molecules in this book are bound to each other.

(b) A well-hit baseball returns toward the ground.

(c) Someone pushes you.

(d) Jupiter travels in an elliptical orbit around the Sun.

(e) The floor supports you.

(f) A moving boat slows down because of water resistance after the engine has been turned off.

(g) There are 92 protons and 143 neutrons contained in a small volume in a nucleus of uranium-235.

LOOKING BACK...LOOKING AHEAD

In previous chapters, we discussed kinematics, that is, the study of motion described by quantities such as position, velocity, and acceleration. We paid little attention to what caused the motion.

In the present chapter, the main emphasis has been on force, the physical quantity responsible for motion. More correctly, *force* is the quantity responsible for *acceleration*. Considerable space has been devoted to Newton's three laws of motion, and in the second law, we have seen the relation between (resultant) force and the acceleration studied in kinematics. Newton's laws have been applied to a wide variety of physical situations, from pushing a lawn mower to launching a space vehicle. We have discussed how all forces can be traced to just three fundamental types: gravity, nuclear strong, and electroweak.

The next chapter will continue the focus on forces with a detailed discussion of friction, and forces involved in circular motion.

CONCEPTS AND SKILLS

Having completed this chapter, you should now be able to do the following:

- Identify the forces acting on objects in a wide variety of physical settings.
- Draw free-body diagrams for objects in various situations.
- Add forces as vectors.
- Define a newton.
- Relate the gravitational force on an object to the gravitational acceleration.
- State Newton's three laws of motion, and use them in solving problems in one and two dimensions.
- Correctly identify action-reaction pairs of forces.
- List the fundamental forces and state how other forces are related to them.

KEY TERMS

You should be able to define or explain each of the following words or phrases:

dynamics
gravity
normal force
force of friction
kinetic friction
static friction
tension
contact forces
action-at-a-distance forces
free-body diagram (FBD, or force diagram)
resultant force (or net force, total force, sum of forces)
net force
total force
sum of forces
Newton's second law of motion
newton (N)
weight
Newton's first law of motion
inertia
law of inertia
Newton's third law of motion
electrical force
electromagnetic interaction
nuclear strong force
nuclear weak force

Chapter Review

MULTIPLE-CHOICE QUESTIONS

5-76 A ball is thrown vertically upward. As it travels upward after leaving the thrower's hand, which force(s) act(s) on it? Neglect air resistance.

(a) the downward force of gravity $\vec{F}_G$ only
(b) $\vec{F}_G$ and a decreasing upward force
(c) $\vec{F}_G$ and an increasing downward force
(d) $\vec{F}_G$ and a constant upward force
(e) $\vec{F}_G$ and the force due to the motion of the ball

5-77 A very large protein molecule is sedimenting downward in water at a constant velocity. There are three forces acting on the molecule: the force of gravity $\vec{F}_G$ downward, an upward buoyant force $\vec{F}_B$, and an upward fluid friction force $\vec{F}_{ff}$. Which of the following expresses the correct relation between the magnitudes of these three forces?

(a) $F_G > F_B + F_{ff}$
(b) $F_G = F_B + F_{ff}$
(c) $F_G < F_B + F_{ff}$

5-78 A baseball is hit hard by a bat toward the outfield. Which of the following is the correct list of force(s) acting on the ball as it is travelling through the air after being hit?

(a) the force of gravity and the force of the "hit"
(b) the force of gravity
(c) the force of gravity, the force of the "hit," and the force of air resistance
(d) the force of the "hit" and the force of air resistance
(e) the force of gravity and the force of air resistance

5-79 What is the approximate magnitude of the force of gravity (in newtons) on a typical adult person?

(a) 7
(b) 7×10^1
(c) 7×10^2
(d) 7×10^3

5-80 A person is pulling a toboggan at constant velocity across a rough horizontal surface by applying a force $\vec{F}$ at an angle of 20° up from the horizontal. Also acting on the toboggan are the force of gravity $\vec{F}_G$, the force of friction $\vec{F}_f$, and the normal force $\vec{F}_N$. Which of the following shows the correct relationships between the magnitudes of the four forces acting on the toboggan?

(a) $F = F_f$ and $F_N = F_G$
(b) $F = F_f$ and $F_N > F_G$
(c) $F > F_f$ and $F_N = F_G$
(d) $F > F_f$ and $F_N < F_G$
(e) $F < F_f$ and $F_N > F_G$

5-81 A soft-drink can is sitting on a table. If we consider the force of gravity on the soft-drink can to be the "action," then the "reaction" is

(a) the upward normal force on the can by the table
(b) the downward force of gravity on the table
(c) the downward normal force on the table by the can
(d) the upward force of gravity on Earth by the can

5-82 A small car has become stuck in deep snow, and the driver of a large truck offers to use his truck to push the car out of the snow. As the truck is pushing the car,

(a) the force exerted by the truck on the car is smaller than the force exerted by the car on the truck
(b) the force exerted by the truck on the car is larger than the force exerted by the car on the truck
(c) the force exerted by the truck on the car has the same magnitude as the force exerted by the car on the truck

5-83 A golf club is used to hit a golf ball. The mass of the golf club is much greater than that of the ball. During the collision between the golf club and the ball,

(a) the force exerted by the club on the ball is smaller than the force exerted by the ball on the club
(b) the force exerted by the club on the ball has the same magnitude as the force exerted by the ball on the club
(c) the force exerted by the club on the ball is larger than the force exerted by the ball on the club

Review Questions and Problems

5-84 A canoe is being paddled across a lake (Figure 5-45). There is a 75.3 N force exerted forward (north) by the water on the paddles, and a 67.2 N force of water resistance backward. In addition, there is a force on the canoe due to the wind: 7.8 N, *from* a direction 25.7° west of north. Determine the magnitude and direction of the resultant force on the canoe.

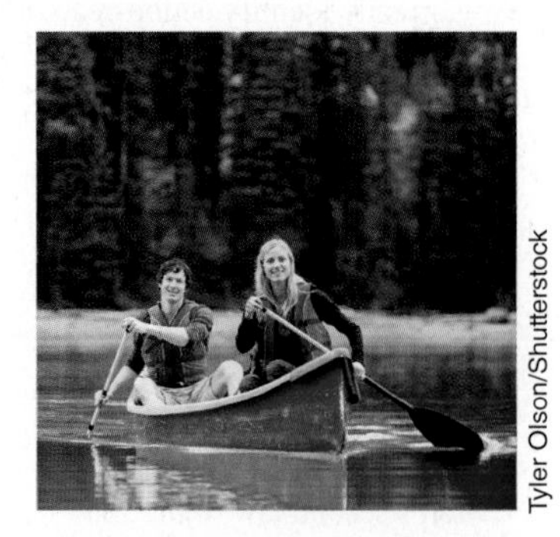

Tyler Olson/Shutterstock

Figure 5-45 Question 5-84.

5-85 The data below were obtained from an experiment in which a varying resultant force, $\vec{F}_R$, was applied to an object, and the resulting accelerations, $\vec{a}$, were measured. Using these data, plot a graph and determine the mass of the object.

F_R (N)	0	1.0	2.0	3.0	4.0	5.0	6.0	7.0
a (m/s²)	0	0.29	0.54	0.83	1.10	1.42	1.69	1.89

5-86 A certain fishing line can withstand a maximum tension of 63 N before breaking. A woman is using the line vertically to hold a fish of mass 0.92 kg. What is the magnitude of the maximum upward acceleration that she can give to the fish without breaking the line?

5-87 In running a 100 m sprint in 10 s, a world-class runner accelerates to a speed of about 8.0 m/s in the first 2.0 s. Determine the magnitude of the average horizontal force on a 63 kg runner during this interval.

5-88 Three salamis, P, Q, and R, having masses 1.1 kg, 1.5 kg, and 1.3 kg, respectively, are hung in a deli by strings (Figure 5-46). Determine the magnitude of the tension in the string joining P and Q.

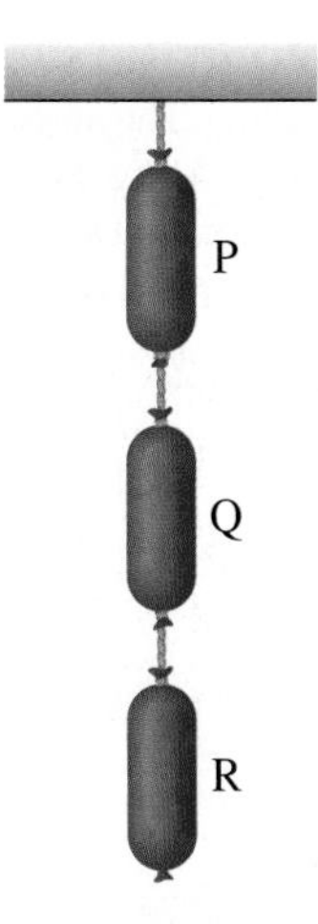

Figure 5-46 Question 5-88.

5-89 If you hang an object from a string inside a car, you can use it as an accelerometer to determine the car's acceleration by measuring the angle between the string and the vertical. Suppose the object you use is a metal washer that has a mass of 0.015 kg. If the string is 45 cm long, making an angle of 5.1° with the vertical, what is the magnitude of the car's acceleration? What information given in this question is not necessary for the solution?

5-90 A delivery girl is running with constant velocity, carrying a pizza toward a house. The mass of the pizza is 0.42 kg, and the speed of the girl (and pizza) is 2.0 m/s. Determine the magnitude and direction of the resultant force on the pizza.

5-91 Figure 5-47 shows the femur, tibia, and knee cap (patella) in a person's leg. The quadriceps tendon connects the patella to the quadriceps muscle in the thigh. This tendon exerts a tension $\vec{T}_Q$ on the patella, and this tension makes an angle $\alpha = 23°$ relative to the positive x-axis, as shown. The patella is connected to the tibia by the patella tendon, which exerts a tension $\vec{T}_P$ on the patella. The direction of this tension relative to the negative x-direction is

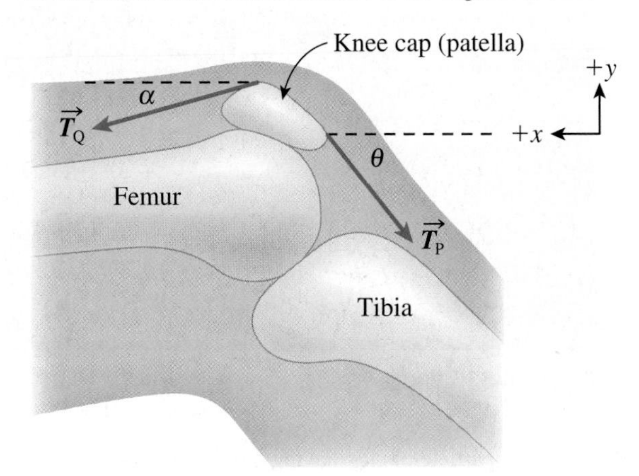

Figure 5-47 Question 5-91.

$\theta = 68°$. If the magnitudes of the tensions are $T_Q = 1.5 \times 10^3$ N and $T_P = 1.3 \times 10^3$ N, determine the magnitude and direction of the total force exerted on the patella by these two tensions.

5-92 A man is pushing an upholstered chair of mass 22.3 kg across a horizontal floor by exerting a force of 157 N at 71.0° from the vertical (Figure 5-48). The force of friction on the chair is 128 N in magnitude. Determine the magnitude of the acceleration of the chair.

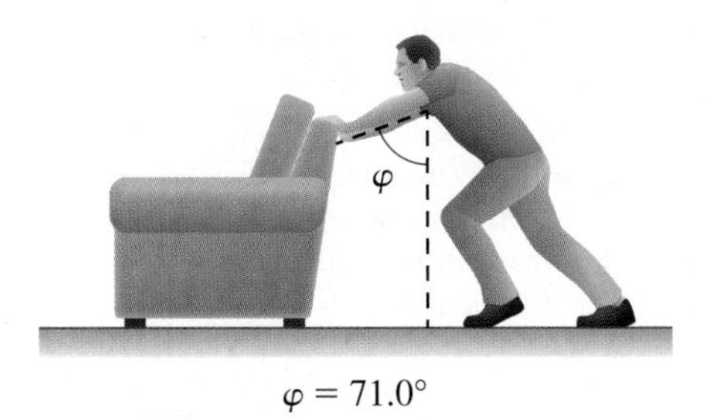

$\varphi = 71.0°$

Figure 5-48 Question 5-92.

5-93 The motion of a person who is ziplining (Figure 5-49) can be approximated as sliding down an incline. If a zipliner has a mass of 84 kg and is effectively sliding down an incline at an angle of 14° from the horizontal at a constant speed of 2.8 m/s, determine the magnitude of the friction force acting on the zipliner.

Figure 5-49 Question 5-93.

5-94 Two hamburgers that are touching each other are on a horizontal grill. The masses of the hamburgers are 113 g and 139 g, and friction is negligible. A horizontal force is applied to the 139 g hamburger, which in turn pushes the 113 g hamburger so that the hamburgers move together. If the applied force on the 139 g hamburger has a magnitude of 5.38×10^{-2} N, determine the magnitudes of

(a) the acceleration of the hamburgers

(b) the force exerted by one hamburger on the other

5-95 What is the fundamental force that is responsible for both the force of friction and the normal force?

Applying Your Knowledge

5-96 By using high-speed movies, it is possible to determine that the impact time during which a golf club is in contact with a golf ball is typically 1.0 ms, and that the speed of the ball when it leaves the club is about 65 m/s (2.3×10^2 km/h!). The mass of a golf ball is 0.045 kg. Determine the magnitude of the average force exerted by the club on the ball. (You can neglect the gravitational force on the ball because it is extremely small in comparison with the force from the club.)

5-97 **Fermi Question:** Determine the magnitude of the average force exerted by a seatbelt on an adult passenger sitting in a car, as the car does a panic stop from 100 km/h.

5-98 A treehouse has a vertical metal "firepole" for quick exits. A boy of mass 35.7 kg slides down the pole, starting from rest. The pole is 3.10 m high, and the boy wants to slide to the bottom in 2.00 s.

(a) What constant downward acceleration is required?

(b) What is the magnitude of the upward force of friction exerted on the boy by the pole?

5-99 If the mass of the leg and cast shown in Figure 5-50 is 17 kg, and the force exerted by the sling is 145 N upward and to the left at an angle of 83° above the horizontal, determine the magnitude and direction of the force exerted on the leg at the hip joint. Assume that the leg is stationary.

Richard Lyons/Shutterstock

Figure 5-50 Question 5-99.

5-100 When a resultant force is applied to a particular mass, m, an acceleration of magnitude a results. When the mass is increased by 2.0 kg, and the same resultant force is applied, the acceleration is halved. Determine mass m.

5-101 Repeat the previous question, but use a value of 0.37 for the ratio of the new acceleration to the old acceleration, instead of 0.50.

5-102 Below are experimental data obtained by allowing a constant, but unknown, resultant force to act on a variety of masses, m. The magnitudes of the accelerations, a, of the masses were then determined. Use the data to plot graphs of

(a) a versus m

(b) a versus $1/m$, and use this graph to determine the magnitude of the resultant force

m (kg)	0.57	1.1	1.6	2.0	2.5	3.2
a (m/s^2)	5.0	2.4	1.9	1.3	1.2	0.9

5-103 Tiny Tompfind, a man of mass 72 kg, steps from a table 92 cm to the floor below. Assume that his velocity is zero as he leaves the table. As he lands, he bends his knees so that he decelerates to a final speed of zero over a distance of 35 cm.

(a) What is Tiny's speed just as he starts to hit the floor?

(b) As he decelerates, what is the magnitude of the force exerted on him by the floor? Assume constant acceleration.

5-104 Three boxes are connected by two cables (Figure 5-51). Each cable passes over a pulley attached to a ceiling. The masses of the boxes are m_1 = 26 kg, m_2 = 38 kg, and m_3 = 41 kg. Friction and the rotational motion of the pulleys are neglected. Determine the magnitudes of

(a) the acceleration of the blocks

(b) the tensions in the connecting cables

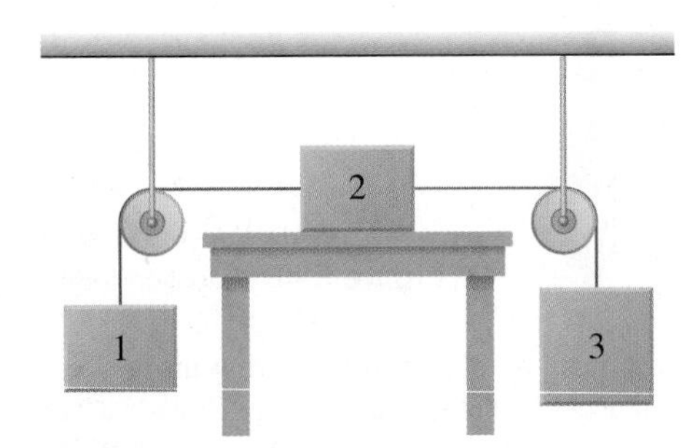

Figure 5-51 Question 5-104.

5-105 If the mass of the slackliner in Figure 5-52 is 75 kg, and (because the cable sags because of gravity acting on the person) the cable both in front of him and behind him is at an angle of 15° above the horizontal, determine the magnitude of the tension in the cable. Assume the person is stationary. (**Hint:** Consider the tension in the cable in front of the person separately from the tension in the cable behind him. These tensions have the same magnitude but different directions. Both pull away from the person (one forward, one backward), and both have an upward component.)

zapatisthack/iStock/Thinkstock

Figure 5-52 Question 5-105.

5-106 A small particle of clay having a mass of 1.06×10^{-14} kg is settling downward in water. At a certain time the particle's speed is 1.2 mm/h. The upward buoyant force on the particle has a magnitude of 4.0×10^{-14} N. There is also an upward fluid friction force on the particle; this force is proportional to the particle's speed v, and therefore its magnitude can be written as fv, where f is a constant called the friction factor. The numerical value of this constant depends on the size and shape of the particle, and on the viscosity of water. For this particular clay particle, f = 1.88×10^{-8} N·s/m. Determine the acceleration (magnitude and direction) of the particle.

5-107 A rocket ship is coasting in space from A to B (Figure 5-53). Between A and B, no forces act on the ship. At B, the rockets fire as shown and provide a constant acceleration until the ship reaches a point C in space. Make a sketch of the rocket's path between B and C.

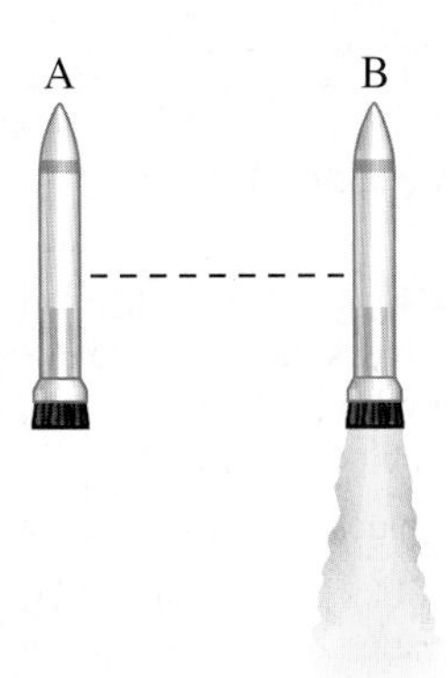

Figure 5-53 Question 5-107.

5-108 A mountain climber of mass 67.5 kg is using a rope to "stand" horizontally against a vertical cliff (Figure 5-54). The tension in the rope is 634 N at 27.0° from the vertical. Determine the magnitude and direction of the force exerted by the cliff on the climber's feet.

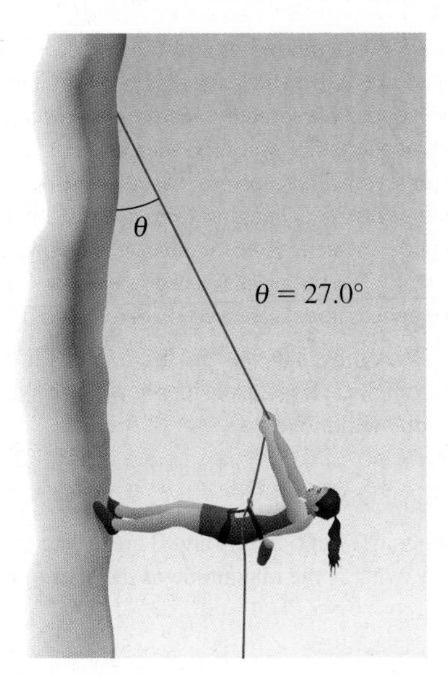

Figure 5-54 Question 5-108.

5-109 A girl on a toboggan slides down a hill inclined at 18° to the horizontal. Friction is negligible. If she starts from rest, what is her speed after sliding for 2.6 s?

5-110 A man is pushing a stroller up a ramp in a shopping mall. The mass of the baby and stroller is 23 kg, and the maximum force that the man can exert (parallel to the ramp) is 88 N. What is the maximum angle that the ramp can make with the horizontal, if the stroller is not to roll down the ramp? Neglect friction between the stroller wheels and the ramp.

5-111 Two boxes are connected by a string passing over a pulley (Figure 5-55). The incline makes an angle of 35.7° with the horizontal, and the masses of the boxes are m_1 = 5.12 kg and m_2 = 3.22 kg. Neglect friction and the rotational motion of the pulley.

(a) If the boxes start from rest, will box 1 slide up or down the incline?

What are the magnitudes of

(b) the acceleration of the boxes?

(c) the tension in the string?

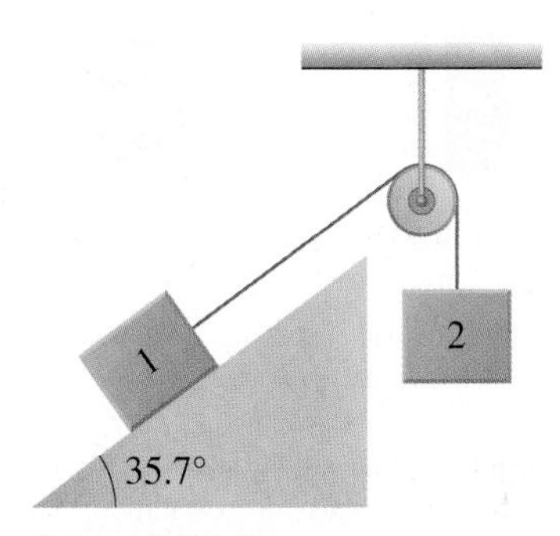

Figure 5-55 Question 5-111

5-112 A girl is pulling a wagon of mass 7.38 kg up a hill inclined at 75.7° to the *vertical* by applying a force parallel to the hill. If the acceleration of the wagon is 0.0645 m/s^2 up the hill, and if friction acting on the wagon is neglected, determine the magnitudes of

(a) the force applied by the girl, and

(b) the normal force on the wagon

5-113 Sometimes a little education can cause problems, as shown in the following classic story. A donkey is told by its owner to pull a sled, but it objects by citing Newton's third law: "If I pull on the sled, it pulls back on me with an equal but opposite force, and I will never be able to get going." Draw separate free-body diagrams for the donkey and sled, and explain why the donkey's argument is incorrect. Do not neglect friction.

5-114 A boy sitting in a boat pushes horizontally against a dock with a paddle for 1.2 s. During the push, the boy and boat move 0.55 m east. Neglect friction.

(a) If the boat was initially at rest, what is its final speed? Assume constant acceleration.

(b) If the total mass of the boy and boat is 95 kg, determine the magnitude and direction of the force exerted by the dock on the paddle.

(c) What were the magnitude and direction of the force exerted by the paddle on the dock?

5-115 (a) A chocolate bar of mass 0.055 kg is suspended from a spring scale that reads in newtons (Figure 5-56 (a)). What is the reading on the scale?

(b) The same chocolate bar is suspended from the same scale in the orientation shown in Figure 5-56 (b). What is the reading?

(c) Two identical bars of mass 0.055 kg are suspended from the scale as in Figure 5-56 (c). What is the reading?

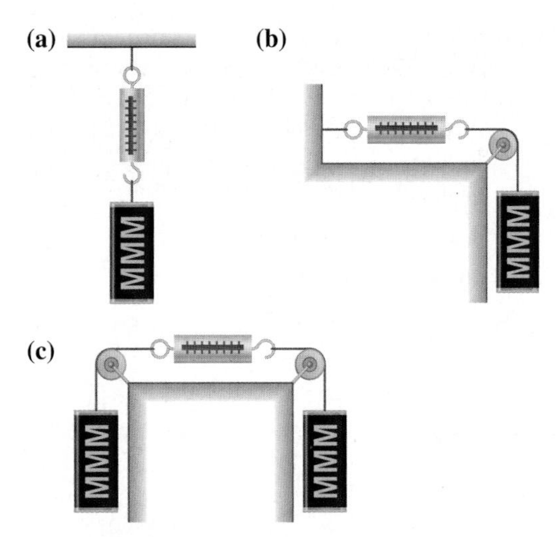

Figure 5-56 Question 5-115.

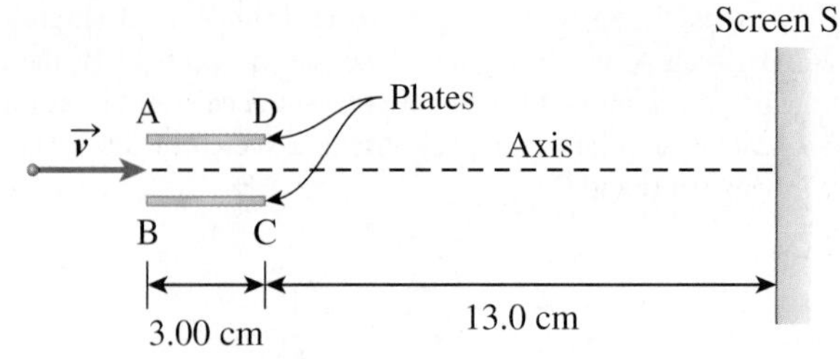

Figure 5-58 Question 5-117.

5-116 A long jumper just after takeoff (Figure 5-57) experiences two forces: gravity and air resistance. The magnitude of the air resistance can be calculated from: $F_{air} = kv^2$, where v is the jumper's speed, and the quantity k depends on the air density and the jumper's shape and size. For a typical jumper, k is about 0.25 kg/m. Reasonable takeoff velocity components are: horizontal, 9.2 m/s; vertical, 3.5 m/s.

(a) Use the takeoff velocity components to determine v^2 at takeoff.

(b) Determine F_{air} at takeoff. By considering units of k and v, check that you are getting the correct SI units for F_{air}.

(c) The air resistance is in the opposite direction to the jumper's velocity. Calculate the horizontal and vertical components of the takeoff air resistance.

(d) Draw a free-body diagram for the jumper (mass 55 kg) just after leaving the ground, and determine the magnitude and direction of the net force on her.

justasc/Shutterstock.com

Figure 5-57 Question 5-116.

5-117 In an oscilloscope, an electron beam is deflected by an electric force produced by charged metal plates (AD and BC in Figure 5-58). In the region ABCD, there is a uniform downward electric force of 3.20×10^{-15} N on each electron. The electrons enter halfway between A and B with a velocity of 2.25×10^7 m/s parallel to the plates. Outside ABCD, the electric force is zero, and the gravitational force can be neglected during the short time that the electrons take to reach the fluorescent screen S. When an electron hits S, how far below the axis will it be? The electron mass is 9.11×10^{-31} kg.

5-118 Figure 5-59 shows two blocks connected by a string that passes over a pulley. The blocks are sitting on inclines that make angles of 37° and 53° relative to the horizontal, as shown. Friction is negligible. The mass of block #2 is 2.7 kg.

(a) What mass of block #1 will allow the system to stay stationary?

(b) If the mass of block #1 is also 2.7 kg, what is the magnitude of the acceleration of the two blocks?

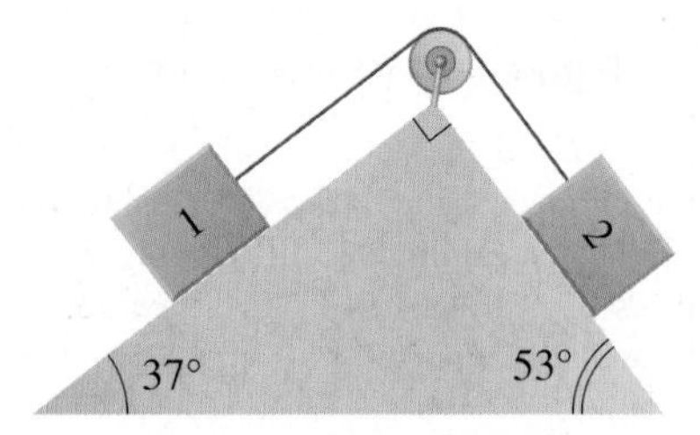

Figure 5-59 Question 5-118.

5-119 In Figure 5-60, two boxes A and B are shown. One end of a rope is attached to box B, which sits on a horizontal table. From box B, the rope extends horizontally and passes over a pulley attached to a corner of the table. The rope then extends downward, under a second pulley, and then upward; the other end of the rope is attached to a ceiling. Box A is hanging from a vertical wire attached to the second pulley, which is free to move up and down along with box A and the wire. The pulleys have negligible mass, and friction in the system can be ignored. The system is released from rest.

(a) As box A moves downward 1 cm, how far does box B move horizontally? (**Hint:** Consider how far the various portions of the rope will move.)

(b) What is the ratio of the magnitude of the acceleration of box B to that of box A?

(c) If the masses of boxes A and B are 2.0 kg and 1.0 kg, respectively, what is the magnitude of the acceleration of box A?

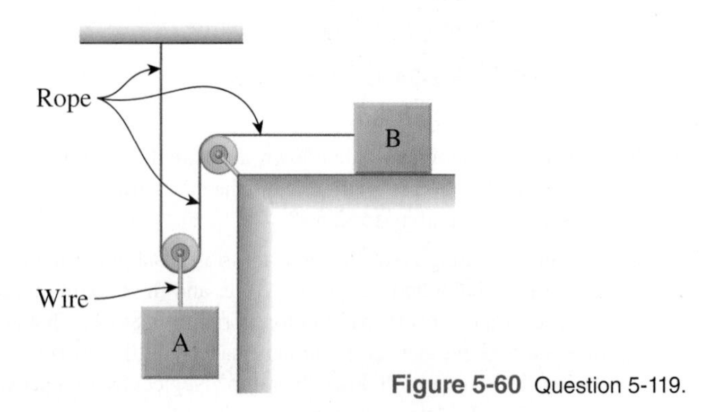

Figure 5-60 Question 5-119.

Applying Newton's Laws 6

Mitch Gunn/Shutterstock.com

Figure 6-1 The motion of a skier involves several forces, including kinetic friction.

Applications of Newton's laws are in evidence all around us, from the graceful movement of a flying bird, to the aerodynamic design of modern automobiles, to the motion of a skier coming down a hill (Figure 6-1). This chapter concentrates on applying Newton's laws in situations involving friction and circular motion. We will be exploring answers to questions such as, What is special about Teflon? Why are highway curves banked? and How does a centrifuge work?

CHAPTER **OUTLINE**

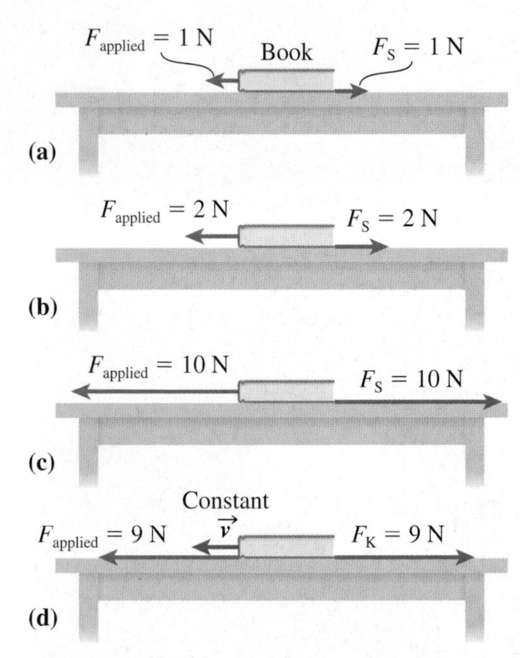

Figure 6-2 (a), (b), and **(c)** The book is stationary, as a horizontal applied force is increased from (a) 1 N, to (b) 2 N, to (c) 10 N. Static friction ($\vec{F}_S$) matches the applied force up to a maximum value (10 N in this example). (d) Once there is relative motion between the surfaces in contact, friction usually decreases. In this example, an applied force of magnitude 9 N moves the book at constant velocity because the kinetic friction force ($\vec{F}_K$) is only 9 N in magnitude.

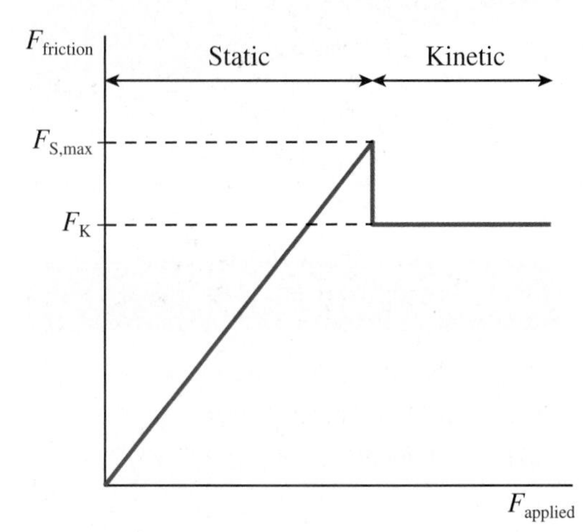

Figure 6-3 A graph of friction force versus applied force (magnitudes). The magnitude of kinetic friction (F_K) is usually less than the maximum magnitude of static friction ($F_{S,max}$).

coefficient of static friction the constant of proportionality (which depends on the types of materials in contact) between the maximum magnitude of static friction and the magnitude of the normal force

6.1 Static and Kinetic Friction

Imagine how different our world would be without friction. Most modes of transportation would be impossible—walking, using cars and trains, etc. We would be unable to hold a pen, or more importantly, a fork. Since friction is due to attractive electrical forces between molecules, a lack of friction would occur only if the electrical force did not exist. Without this force to hold atoms and molecules together, there would be no cars, trains, pens, forks, or even people to talk about these things.

Friction was discussed briefly in Chapter 5. Recall that the force of friction is parallel to the surfaces of objects in contact with each other. If the surfaces are moving (i.e., sliding) relative to each other, the friction is referred to as kinetic friction. If the surfaces are not in motion relative to each other, the friction is called static friction.

In this section, we will deal with friction more completely by introducing the concepts of coefficients of static and kinetic friction. Our discussion will apply to *rigid* objects that *slide* over each other. You know from everyday experience that *rolling* friction is considerably less than sliding friction—imagine trying to move an automobile by sliding it!—but we will not be dealing quantitatively with rolling friction in this book.

Figure 6-2 illustrates a demonstration that you can do yourself if you have a spring scale available (or even an elastic band). The scale is used to apply measurable horizontal forces on an object such as a book sitting on a table. For small applied forces, the book does not move because the force of friction is equal to the applied force (in magnitude). Since the surfaces in contact between the book and table are not moving relative to each other, the friction in this case is static friction ($\vec{F}_S$). As the force applied to the book increases, the static friction force does not increase in size indefinitely. It has a maximum magnitude ($F_{S,max}$) that is dependent on the types of materials in contact. The maximum magnitude also depends on the size of the normal force between the objects. In the example shown in Figure 6-2, $F_{S,max} = 10$ N.

Once an object begins to slide, we refer to the friction as kinetic friction ($\vec{F}_K$), that is, the friction force between two objects with surfaces moving relative to each other. The magnitude of the kinetic friction force is usually smaller than the maximum magnitude of static friction (Figures 6-2 (c) and (d), and Figure 6-3). Although kinetic friction usually decreases somewhat as the relative speed of the surfaces increases, in this book we will treat kinetic friction as being independent of speed. As is the case with the maximum value of static friction, the kinetic friction force depends on the kinds of materials in contact and on the magnitude of the normal force between the objects.

Coefficients of Friction

It has been found experimentally that the maximum magnitude of static friction between two objects is proportional to the magnitude of the normal force exerted between them, that is,

$$F_{S,max} \propto F_N \tag{6-1}$$

The constant of proportionality between $F_{S,max}$ and F_N depends on the types of materials in contact. The constant is higher for rough or sticky materials than for slippery materials. It is called the **coefficient of static friction** and is represented by μ_S (μ is the lower case Greek letter mu). Writing the relation between $F_{S,max}$ and F_N as an equation, we have

$$F_{S,max} = \mu_S F_N \qquad (6\text{-}2)$$

Remember that this equation deals only with the magnitudes of the forces. The directions of $\vec{F}_{S,max}$ and $\vec{F}_N$ are at right angles to each other ($\vec{F}_{S,max}$ is parallel to the surfaces in contact, and $\vec{F}_N$ is perpendicular to the surfaces). Values of μ_S range typically from 0.1 to 1 (Table 6-1). Notice that since μ_S is the ratio of the magnitudes of two forces, $F_{S,max}/F_N$, it has no units.

Eqn. 6-2 is sometimes expressed as an inequality involving F_S (not $F_{S,max}$) and F_N:

$$F_S \leq \mu_S F_N \qquad (6\text{-}3)$$

It has also been determined experimentally that the magnitude of kinetic friction is proportional to the magnitude of the normal force. The proportionality constant is called the **coefficient of kinetic friction**, represented by μ_K. Hence,

coefficient of kinetic friction the constant of proportionality (which depends on the types of materials in contact) between the magnitude of kinetic friction and the magnitude of the normal force

$$F_K = \mu_K F_N \qquad (6\text{-}4)$$

Since F_K is usually less than $F_{S,max}$, then $\mu_K \leq \mu_S$. Table 6-1 gives some representative values of the coefficients μ_S and μ_K. Note the low values of μ_S and μ_K for steel on Teflon. Teflon is a compound of carbon and fluorine that experiences extremely weak electrical forces with virtually all other types of molecules. Since the strength

DID YOU KNOW?

The science and technology of friction and lubrication is known as "tribology." The production of light by friction is called "triboluminescence"; you can demonstrate it in a darkened room by crushing (or biting) certain types of hard candy, such as Wintergreen mints. As you crush, the friction between the moving layers of candy causes a visible glow. "Nanotribology," which is the study of atomic friction, such as at a liquid–solid interface, may one day result in comfortable prosthetic joints and smoother-running machines.

Table 6-1

Approximate Coefficients of Static (μ_S) and Kinetic (μ_K) Friction

Materials	μ_S	μ_K
Copper on cast iron	1.05	0.29
Oak on oak (parallel to grain)	0.62	0.48
Oak on oak (cross grain)	0.54	0.32
Hard steel on hard steel (dry)	0.78	0.42
Hard steel on hard steel (greasy) (values depend on lubricant used)	0.11 – 0.23	0.03 – 0.11
Steel on Teflon	0.04	0.04
Leather on oak (parallel to grain)	0.61	0.52
Glass on glass	0.9 – 1.0	0.4
Rubber on asphalt, dry	1.1	1.0
Rubber on asphalt, wet	1.0	0.95
Rubber on ice	uncertain	0.005
Dry bone on bone	0.3	uncertain
Cartilage on synovial fluid (in human joints)	0.01	0.003

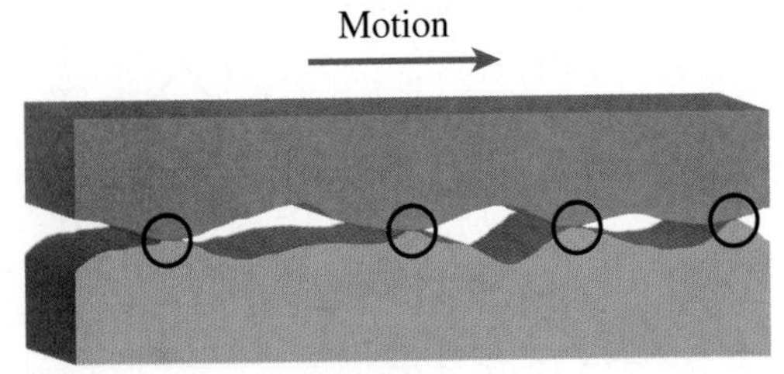

Figure 6-4 The molecular cause of friction. As one object slides over another, molecules bind together at points of close contact (circled). A force must be applied to break the bonds and maintain the motion.

of friction forces is determined by the strength of the intermolecular electrical forces, friction involving Teflon is very small. Teflon, which was created by chance by two research chemists in 1938, is used as a coating on some cookware because of its weak bonding to other materials (eggs, potatoes, etc.). As well, it is used in surgical implants because it does not interact with body fluids, and the space industry is also a large user of this inert material.

It must be emphasized here that coefficients of friction are poorly known, and poorly reproducible in experiments. They depend strongly on the state of the surfaces being used—the cleanliness, roughness, degree of oxidation, temperature, wear, etc. Therefore, in problems in this book, coefficients of friction will be given to two significant digits at most.

Figure 6-4 shows a highly magnified view of one object sliding over another. Even highly polished surfaces appear rough when magnified, and in regions where the molecules of the two surfaces come into close contact, electrical attractive forces cause the molecules to bind together; that is, molecular bonds are formed. This is the cause of friction. In order to produce and maintain a sliding motion, a force must be applied to break these bonds. As the sliding continues, new bonds are being constantly formed and broken. The molecules, which are stretched somewhat just before the bonds are broken, "snap back" into place as the bonds break, and vibrate with more energy than they had previously. We say that the thermal energy (or internal energy) of the molecules increases, which means that the objects get hotter. Whenever kinetic friction is acting, thermal energy is produced, and this effect is referred to as **frictional heating**. Sometimes an increase in thermal energy due to friction can produce melting of a material. For example, the motion of skis on snow melts some of the snow to produce a thin layer of water, which aids the sliding of the skis.

frictional heating the production of thermal energy as the result of kinetic friction acting between two sliding objects

Frictional heating is not sufficient to explain the existence of a thin layer of water under skate blades gliding on ice (Figure 6-5). Without this water, it would be very difficult to skate. For many years it was commonly thought that the pressure due to the weight of the skater on the blades is so large that melting occurs, or that a combination of this pressure and frictional heating produces the melting. However, it turns out that these mechanisms are not enough to generate melting. Rather, the thin layer of water is due to a phenomenon known as surface melting, which means that some solids at a temperature *below* the melting point naturally have a thin liquid layer all over the surface, even without anything pressing or moving against the surface. This was first suggested by the physicist Michael Faraday in 1842, and has recently been well-established experimentally for many solids. In the case of ice at −10°C, the surface water film has a thickness of only 2 nm, which is only about seven water molecules thick. It is this thin layer of water that permits ice-skating and accounts for the general slipperiness of ice.

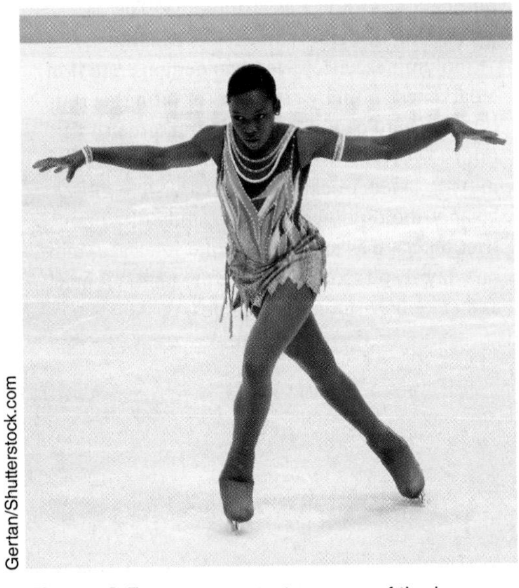

Figure 6-5 Until recently the cause of the low friction between ice and objects such as skates was not understood.

If two extremely smooth surfaces are placed in contact, a large number of molecular bonds can be formed, and it is possible for the two surfaces to weld together. You might have encountered this phenomenon yourself with flat microscope slides, which can stick together if they come into close contact. Another example occurs in machine shops, where small steel blocks (which are not magnetic) used for precision measurements can be temporarily bonded together by pushing the very flat edges of the blocks together. Thus a single block, say three units long, can be created from two blocks of lengths one unit and two units. The blocks can later be pulled apart by hand. These blocks are commonly known as "Jo blocks" (Figure 6-6), a short form for Johansson blocks, named after C.E. Johansson, a Swedish engineer.

A *permanent* weld can be achieved, say between two metal surfaces, if the surfaces are carefully polished and cleaned, and placed together in a chamber with very little air. (Otherwise, surface oxide films interfere with the strong bonding.) With a permanent weld, friction becomes essentially infinite.

TRY THIS!

Hot Hands

You can demonstrate frictional heating yourself by rubbing the palms of your hands together vigorously; they get quite hot!

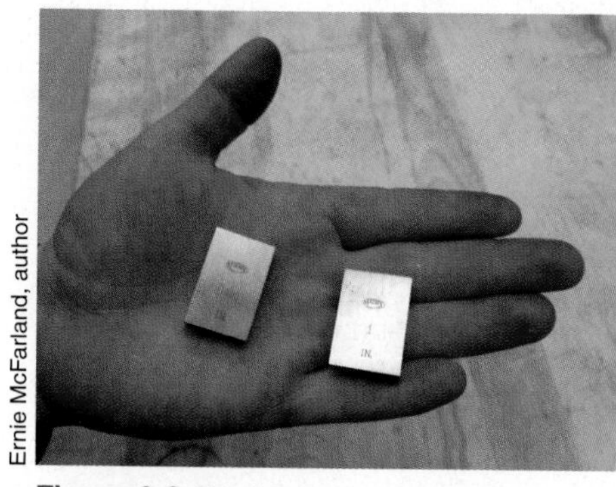

Ernie McFarland, author

Ernie McFarland, author

Figure 6-6 Two "Jo blocks," which are not magnetic, shown separated (left) and bonded together (right).

SAMPLE PROBLEM 6-1

A box of mass 7.52 kg is at rest on a horizontal wooden floor. The coefficients of friction between the box and the floor are static, 0.57, and kinetic, 0.45. A horizontal force is applied to the box. What are the magnitudes of the friction force and the acceleration of the box if the applied force has a magnitude of (a) 21 N? (b) 42 N? (c) 63 N?

Solution

(a) Figure 6-7 shows a diagram for the problem and the free-body diagram (FBD) for the box. We choose $+y$ upward and $+x$ in the direction of the applied force, as shown. In order to determine whether the box will have any acceleration at all, we need to calculate the maximum magnitude of static friction:

$$F_{\text{S,max}} = \mu_S F_N$$

The coefficient μ_S is known, but F_N is unknown. We can use Newton's first (or second) law in the y-direction to determine F_N:

$$\Sigma F_y = ma_y = 0$$

This gives

$$F_N + (-mg) = 0$$

$$F_N = mg = (7.52\,\text{kg})(9.80\,\text{m/s}^2) = 73.7\,\text{N}$$

Then

$$F_{\text{S,max}} = \mu_S F_N = (0.57)(73.7\,\text{N}) = 42\,\text{N}$$

Thus, the maximum static friction force possible is 42 N in magnitude.

In the x-direction, the only forces acting are the applied force and the friction force. Since the applied force has a magnitude of 21 N, which is less than $F_{\text{S,max}}$, then the static friction force will also have a magnitude of 21 N, and the box will have an acceleration of zero.

(b) The magnitude of the applied force (F_{applied}) is now 42 N, which is equal to $F_{\text{S,max}}$, and the box will be on the verge of slipping. The friction force will be at its maximum value of 42 N, and the box will still have an acceleration of zero.

(c) Now $F_{\text{applied}} = 63\text{ N} > F_{\text{S,max}}$, and the box will accelerate in the direction of the applied force. The box will move almost instantaneously and therefore we need to consider the kinetic friction force:

$$F_K = \mu_K F_N = (0.45)(73.7\,\text{N}) = 33\,\text{N}$$

To find the acceleration, we use Newton's second law:

$$\Sigma F_x = ma_x$$

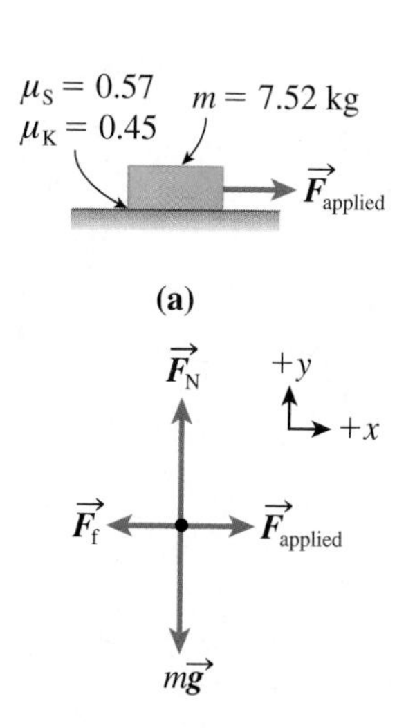

Figure 6-7 **(a)** Diagram for Sample Problem 6-1. **(b)** Free-body diagram for the box.

Therefore,

$$F_{\text{applied}} + (-F_K) = ma_x$$

$$a_x = \frac{F_{\text{applied}} - F_K}{m} = \frac{(63 - 33)\,\text{N}}{7.52\,\text{kg}} = 4.0\,\text{m/s}^2$$

Thus, when $F_{\text{applied}} = 63$ N, the box has an acceleration of magnitude 4.0 m/s².

SAMPLE PROBLEM 6-2

Toolbox
$m = 9.2$ kg
$\mu_S = 0.48$
$\varphi = 34°$

(a)

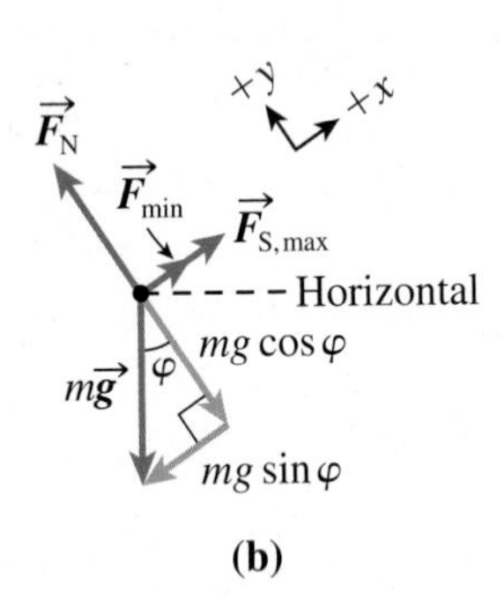

(b)

Figure 6-8 **(a)** Diagram for Sample Problem 6-2. **(b)** Free-body diagram for the toolbox.

A roofer is working on a roof inclined at 34° to the horizontal. Friction between the worker's boots and the roof is sufficient to prevent him from sliding, but his metal toolbox does slide. The coefficient of static friction between the roof and toolbox is 0.48, and the mass of the toolbox is 9.2 kg. What is the magnitude of the minimum force that the roofer must exert parallel to the slope of the roof to keep the box from sliding?

Solution

Figure 6-8 shows the general diagram and the FBD for the toolbox. If the worker is exerting the minimum possible force $\vec{F}_{\text{min}}$ (Figure 6-8 (b)), the toolbox must be on the verge of slipping, and static friction must be at its maximum possible value $F_{\text{S,max}}$. Since the toolbox would slip down the roof if released, the direction of friction must be upward parallel to the roof. As in situations in Chapter 5 involving motion along an incline, the x-y axes are chosen parallel and perpendicular to the incline. The x- and y-components of $m\vec{g}$ are shown in Figure 6-8 (b). Since the toolbox is at rest and remains at rest, the x- and y-components of its acceleration are zero.

We need to determine F_{min}. Using Newton's first or second law in the x-direction,

$$\Sigma F_x = ma_x = 0$$

Therefore, $F_{\text{min}} + F_{\text{S,max}} + (-mg\sin\varphi) = 0$

Hence, $F_{\text{min}} = mg\sin\varphi - F_{\text{S,max}}$ Eqn. [1]

We know m, g, and φ, but $F_{\text{S,max}}$ is unknown. However, $F_{\text{S,max}} = \mu_S F_N$, and we are given that $\mu_S = 0.48$, and we can obtain an expression for F_N by considering forces in the y-direction: $\Sigma F_y = ma_y = 0$

This gives $F_N + (-mg\cos\varphi) = 0$

$F_N = mg\cos\varphi$

Now we use $F_{\text{S,max}} = \mu_S F_N$, which gives

$F_{\text{S,max}} = \mu_S mg\cos\varphi$ Eqn. [2]

Using Eqn. [1],

$$\begin{aligned} F_{\text{min}} &= mg\sin\varphi - F_{\text{S,max}} \\ &= mg\sin\varphi - \mu_S mg\cos\varphi \quad \text{from Eqn. [2]} \\ &= mg(\sin\varphi - \mu_S\cos\varphi) \end{aligned}$$

Substituting known values, $F_{\text{min}} = (9.2\,\text{kg})(9.8\,\text{m/s}^2)(\sin 34° - 0.48\cos 34°) = 15\,\text{N}$. Thus, the minimum force required to prevent sliding is 15 N in magnitude.

Controlling Friction

From the coefficient of static friction of 1.05 for copper on cast iron, to 0.005 for the kinetic friction coefficient for rubber on ice, the scale of friction covers a huge range (Table 6-1). In some circumstances large friction is useful, such as in automobile braking, and in other cases small friction is wanted, as in moving parts in most types of machinery. Sometimes the size of friction cannot be altered substantially, but in many cases, unwanted friction can be reduced and desirable friction can be increased.

From your everyday experiences you already know some of the ways in which friction can be decreased. Ball bearings are frequently used in machines that have rotating parts, and lubricants such as oils are commonly employed in a variety of situations. Such lubricants fill in the depressions and pits that exist even on objects that appear smooth (Figure 6-4) and allow surfaces in contact to move easily past each other. Human joints have a wonderful lubricant, synovial fluid, that has static and kinetic friction coefficients with bone cartilage of only 0.01 and 0.003 respectively. This fluid allows human joints to operate easily without bones rubbing directly against each other.

For cars and airplanes, air friction is reduced by using streamlined designs that have been developed by using wind tunnels to test the flow of air around the objects. A vehicle that uses yet another way to reduce friction in travel over oceans or lakes is the hovercraft (Figure 6-9), which floats on a cushion of air as it travels across the water.

On the other hand, large friction is wanted between car tires and the road. Tires designed for a specific season (e.g., winter) use rubber that is developed to stay flexible in typical seasonal temperatures and therefore maintains good contact with the road. As a tire's temperature falls below its ideal range, the rubber stiffens and adheres less well to the road. Tire treads (Figure 6-10) are designed with grooves to allow water to be moved away from the tire, permitting the rubber to maintain contact with the asphalt underneath the water. If there is a large amount of water on the road, aquaplaning (or hydroplaning) can occur. This means that the tire grooves are unable to cope with the volume of water and the rubber itself is not able to touch the road. The result is that the tires start to slide and the car becomes more like a sled and is difficult to control.

Of course, driving on snow or ice is also difficult and dangerous. Especially treacherous is "black ice," which is an extremely thin layer of ice so transparent that the black asphalt is visible through the ice. Hence, the ice appears black and gives the visual impression of being asphalt. Some new devices can detect black ice by beaming infrared radiation onto a roadway and monitoring the reflected radiation, and might eventually be inexpensive enough to be used to provide warnings in areas where black ice is common.

Some types of catfish use friction in an unusual way to deter predators. These catfish have a bony dorsal fin that normally lies flat against the body. However, when the fish detects a predator, it uses muscles to move the dorsal fin upright (Figure 6-11) and push it against the bones in its spine. The fish takes advantage of the high coefficient of bone-on-bone friction to lock the fin securely in place. With this large bony fin in view, the fish becomes a much less attractive prey.

DID YOU KNOW?

Synovial fluid has the consistency of egg yolk, and the word "synovial" is partly derived from *ovum*, which is Latin for "egg."

Dave Turner/Shutterstock

Figure 6-9 A hovercraft moves with less friction than a boat as it floats on air over the water.

Paul Broadbent/Shutterstock

Figure 6-10 The grooves in a tire tread allow the rubber to stay in contact with a wet road.

dabjola/Shutterstock

Figure 6-11 When a catfish detects a predator, it moves its bony dorsal fin high on its spine and locks it in place with high bone-on-bone friction.

EXERCISES

6-1 If friction did not exist, what would happen if you tried to walk?

6-2 If a person slides a plate eastward across a table, in which direction is the force of friction on the plate? Is this friction static or kinetic?

6-3 A box of hamburger buns of mass 11 kg is at rest on a horizontal floor in a fast-food restaurant. The coefficients of friction between the box and the floor are static, 0.55, and kinetic, 0.47. A horizontal force is applied to the box. What are the magnitudes of the friction force and the acceleration of the box, if the applied force has a magnitude of (a) 55 N? (b) 61 N?

6-4 The coefficients of friction between a stove of mass 65 kg and the horizontal floor beneath it are static, 0.71, and kinetic, 0.64.

(a) What is the minimum horizontal force that must be applied to the stove to set it in motion? (b) Once the stove is moving, what magnitude of horizontal force will keep it moving at constant velocity?

6-5 A woman of mass 51 kg is standing on a wet floor. The coefficients of friction between her shoes and the floor are: static, 0.25, and kinetic, 0.17. In order to open a large door, she exerts a horizontal force of 135 N. Does she slide? Explain, using calculations.

6-6 A girl applies a horizontal force of magnitude 31 N to a chair of mass 5.8 kg. As a result, the chair slides across a horizontal floor with an acceleration of magnitude 0.32 m/s^2. What is the coefficient of kinetic friction between the chair and the floor?

6-7 A hockey puck travels 33 m along an ice rink before coming to rest. Its initial speed was 11 m/s.

(a) What is the magnitude of the acceleration of the puck?

(b) What is the coefficient of kinetic friction between the puck and the ice?

6-8 A small book is sitting on top of a large book, which is sitting on a horizontal tabletop. A student exerts a horizontal force on the large book and both books are accelerated together; that is, the small book does not slip on the large one.

(a) Draw a FBD for the small book when it is accelerating.

(b) What is the name of the force that produces the horizontal acceleration of the small book?

(c) If the acceleration of the books has a magnitude of 2.5 m/s^2, what is the smallest coefficient of static friction between the books that will prevent slipping?

6-9 Draw a FBD for the large book in the previous question when the book is accelerating. (This is tricky. Remember Newton's third law and do not neglect friction between the large book and the tabletop.)

6-10 A sleigh with two children (total mass 55.2 kg) is pulled along a horizontal snow-covered field by a rope that is at an angle of 35° above the horizontal (Figure 6-12). The coefficients of friction between the sleigh and the snow beneath it are static, 0.15, and kinetic, 0.095. What magnitude of rope tension is required to keep the sleigh moving at constant velocity? (**Hint:** $F_N \neq mg$.)

Figure 6-12 Question 6-10.

6-11 A child is sliding down a water slide inclined at 25° to the horizontal. The coefficient of kinetic friction between the child and the slide is 0.10. What is the magnitude of the acceleration of the child? (**Hint:** $F_N \neq mg$.)

6-12 In an experiment to determine the coefficient of static friction, μ_S, between two materials, a small block of one material rests on an incline made of the other material. The angle between the incline and the horizontal is increased to a value φ where the block is on the verge of slipping. (a) Determine an expression for μ_S in terms of φ. (b) Describe a similar experiment to determine μ_K.

6-13 **Fermi Question:** An automobile accelerates from rest to 100 km/h as quickly as possible on a horizontal road. Assuming that the tires are on the verge of slipping during this acceleration, estimate the coefficient of static friction between the tires and the road.

TRY THIS!

Tilting for Friction

You can easily determine the coefficient of static friction between your cell phone and the cover of this book (or any other two objects that are handy) by placing the phone on the horizontal book and gradually tilting the book so that the phone is on the verge of slipping. Measure the angle φ between the book and the horizontal. (You can use a protractor, use an app on a smartphone, or use a ruler and apply trigonometry to determine the angle.) The coefficient of static friction is given by $\mu_S = \tan\varphi$ (Question 6-12(a)). How could you determine the coefficient of kinetic friction? (Question 6-12(b)).

6.2 Forces in Circular Motion

There are many examples of circular motion in nature and in human-made devices. Many planets follow orbits that are approximately circular, each point on Earth travels in a daily circle because of Earth's rotation, and rotating shafts (whether in CD-players, food processors, or drills) move in circles. We saw in Chapter 4 that an object travelling at constant speed in a circle, or on a path that is a portion of a circle, is undergoing an acceleration directed toward the centre of the circle. By Newton's second law of motion, this centripetal acceleration must be due to a resultant force acting toward the centre. For instance, in the case of a planet in a circular orbit about a star, the force acting on the planet toward the centre of the circle is the force of gravity exerted by the star, as we will discuss in more detail in Chapter 9. In other situations, the force causing the centripetal acceleration might be a force such as tension, friction, etc., or a combination of such forces. The sample problems in this section illustrate a number of different forces providing the centripetal acceleration.

In Chapter 4, the magnitude of the centripetal acceleration of an object in circular motion was expressed as $a_c = v^2/r$, where v is the object's speed and r is the circle's radius. In dealing with problems involving *forces* on objects in circular motion, Newton's second law, $\Sigma\vec{F} = m\vec{a}$, is typically used along with $|\vec{a}| = a_c = v^2/r$. Since the acceleration of magnitude v^2/r is directed toward the centre of the circle, then the force represented by $\Sigma\vec{F}$ must be the resultant force directed toward the centre; this force can normally be easily determined from a free-body diagram.

Problem-Solving Strategy for Circular Motion

The general approach to solving problems involving forces was outlined in Section 5.5, and this approach still applies when dealing with circular motion. The only new step to be aware of is that since you know there is an acceleration of magnitude v^2/r directed toward the centre of the circle, then you should choose either $+x$ or $+y$ in this direction. Then, for example, if you choose $+x$ toward the centre, you would use $\Sigma F_x = ma_x = ma_c = mv^2/r$, where ΣF_x is the sum of the x-components (toward the centre) of all the forces acting on the object moving in a circle.

SAMPLE PROBLEM 6-3

The orbit of Uranus around the Sun is approximately a circle of radius 2.9×10^{12} m. The (approximately constant) speed of Uranus is 6.8×10^3 m/s, and its mass is 8.73×10^{25} kg.

(a) What is the name of the force that causes the centripetal acceleration in this case?

(b) What is the magnitude of this force?

(c) What is the orbital period of Uranus?

Solution

(a) A diagram is given in Figure 6-13 (a). The only force that acts on Uranus is the gravitational force of attraction toward the Sun. This is the force causing the centripetal acceleration.

(b) A free-body diagram for Uranus is shown in Figure 6-13 (b). The acceleration is toward the Sun, and the $+x$-axis has been chosen in this direction. Writing Newton's second law,

$$\Sigma F_x = ma_x \quad \text{which gives} \quad F_G = ma_x$$

Since a_x is the centripetal acceleration, we write $a_x = a_c = v^2/r$, giving $F_G = mv^2/r$. Substituting known values,

$$F_G = (8.73 \times 10^{25}\,\text{kg}) \times \frac{(6.8 \times 10^3\,\text{m/s})^2}{(2.9 \times 10^{12}\,\text{m})} = 1.4 \times 10^{21}\,\text{N}$$

Thus, the magnitude of the gravitational force is 1.4×10^{21} N.

(c) The orbital period is the time for one revolution, that is, the time to travel one circumference ($2\pi r$) of the circle. Since the speed is constant, we can write

$$\text{time} = \text{distance/speed.}$$

$$\text{Thus, } t = \frac{2\pi r}{v} = \frac{2\pi(2.9 \times 10^{12}\,\text{m})}{6.8 \times 10^3\,\text{m/s}} = 2.7 \times 10^9\,\text{s (or 84 yr)}$$

Therefore, the orbital period of Uranus is 2.7×10^9 s.

TACKLING MISCONCEPTIONS

Centripetal Force

Some books, especially those written several years ago, use the name "centripetal force" to refer to the resultant force causing centripetal acceleration. However, this can lead to a mistaken impression that this centripetal force is another force in nature that should be included on free-body diagrams along with other forces when dealing with circular motion. Since this approach is not correct, the term "centripetal force" is gradually falling out of favour and will not be used in this book.

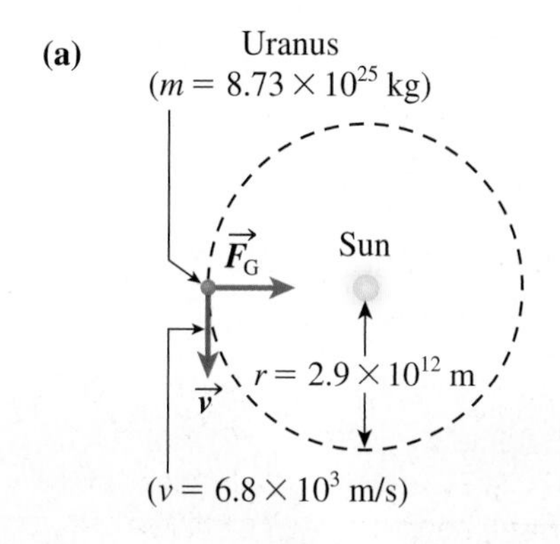

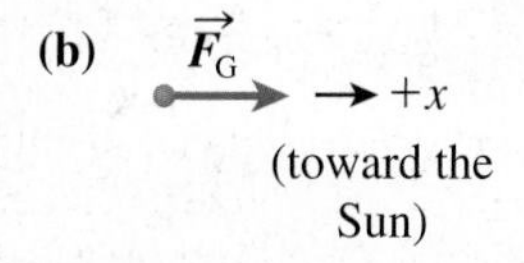

Figure 6-13 **(a)** Sample Problem 6-3. **(b)** Free-body diagram for Uranus.

DID YOU KNOW?

For something such as a spinning fan, the force providing the centripetal acceleration on each atom and molecule in the fan blades is an electrical force toward the centre of rotation, exerted by surrounding atoms and molecules. If the fan spins very rapidly, the electrical force might not be large enough to provide the required centripetal force, and the blades could fly apart; that is, the molecules would become separated. The strength of materials, which depends on the intermolecular electrical forces, is important in designing anything that spins rapidly, from food processors to jet engines. Materials that will be subject to high rotation speeds are usually tested for defects to ensure that they will not fly apart.

SAMPLE PROBLEM 6-4

A bird of mass 0.211 kg pulls out of a dive, the bottom portion of which can be considered as a circular arc of radius 25.6 m. At the bottom of the arc, if the bird's speed is 21.7 m/s, what is the magnitude of the upward lift on the bird's wings?

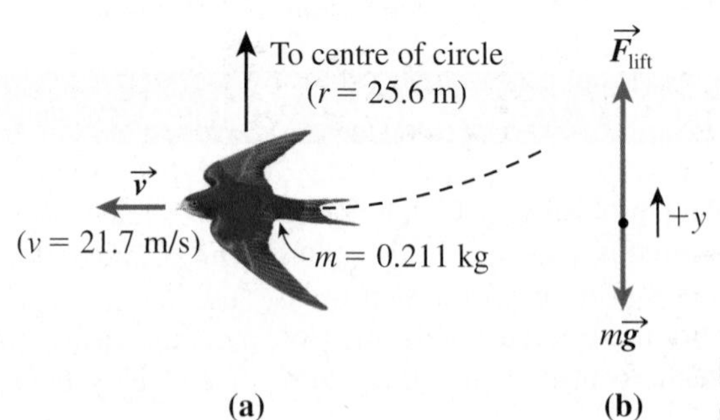

Figure 6-14 **(a)** Sample Problem 6-4. **(b)** Free-body diagram for the bird.

Solution

Figure 6-14 shows a diagram of the problem, and a free-body diagram for the bird. The forces acting on the bird are an upward lift $\vec{F}_{\text{lift}}$ exerted by the air, and the downward force of gravity $m\vec{g}$. As can be seen from Figure 6-14 (b) the resultant force toward the centre—that is, the force providing the centripetal acceleration—is the vector sum of $\vec{F}_{\text{lift}}$ and $m\vec{g}$. (We neglect the horizontal air-drag force since this force is not toward the centre of the circle, and therefore would play no part in our calculations for this problem.) Since the centripetal acceleration is upward, we choose the $+y$-axis in this direction. Using Newton's second law,

$$\Sigma F_y = ma_y$$

$$F_{\text{lift}} + (-mg) = ma_y$$

Substituting $a_y = a_c = v^2/r$, and solving for F_{lift},

$$\begin{aligned} F_{\text{lift}} &= mg + m\frac{v^2}{r} \\ &= m\left(g + \frac{v^2}{r}\right) \\ &= (0.211\,\text{kg})\left(9.80\,\text{m/s}^2 + \frac{(21.7\,\text{m/s})^2}{25.6\,\text{m}}\right) \\ &= 5.95\,\text{N} \end{aligned}$$

Therefore, the upward lift force has a magnitude of 5.95 N.

Note that the solution does not require that the bird travel at constant speed. The algebraic expression for F_{lift} is valid whether or not the speed v is changing. Of course, it was essential to substitute the correct numerical value for the bird's speed at the bottom of the dive in order to determine F_{lift} at that particular position.

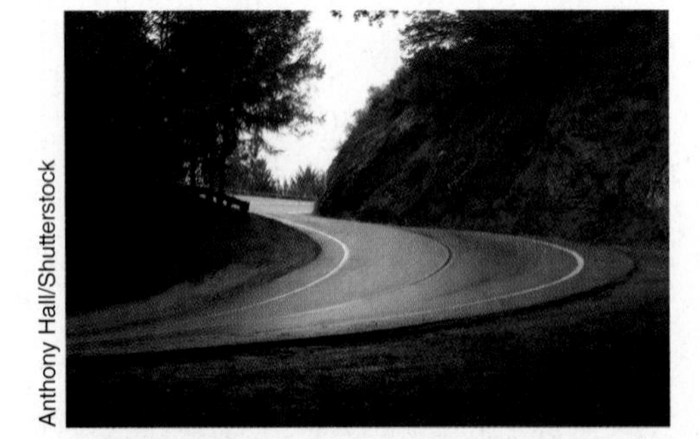

Figure 6-15 Banked curves on a road. (Sample Problem 6-5.)

SAMPLE PROBLEM 6-5

Highway curves are banked (Figure 6-15) so that even if the road is extremely slippery, cars will be able to negotiate the curve without sliding. **(a)** If friction is neglected, which force(s) provide(s) the centripetal acceleration of the car? **(b)** What is the proper banking angle (φ in Figure 6-16 (a)) for a car travelling at 90 km/h around a curve of radius 500 m? (Assume two significant digits.)

Solution

(a) The free-body diagram for the car is shown in Figure 6-16 (b). The car is travelling along part of a horizontal circle, and therefore the direction from the car to the centre of the circle is horizontal. Hence, the centripetal acceleration is horizontal, and the $+x$-axis has been chosen in this direction (with the $+y$-axis at right angles to it). We assume that the road is very slippery (i.e., no friction), and thus there are only two forces acting on the car, gravity $m\vec{g}$ and the normal force $\vec{F}_N$. The only force acting horizontally—that is, toward the centre of the circle—is a component of the normal force. Hence, this horizontal component of $\vec{F}_N$ is the force providing the centripetal acceleration.

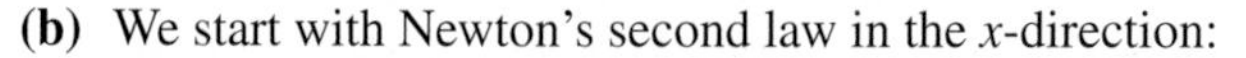

(b) We start with Newton's second law in the x-direction:

$$\Sigma F_x = ma_x \qquad \text{Eqn. [1]}$$

Since the only force with an x-component is $\vec{F}_N$, Eqn. [1] becomes $F_{Nx} = ma_x$, and we need an algebraic expression for F_{Nx} in order to continue. Writing the angle between the inclined curve and the horizontal as φ (as in Figure 6-16 (b)), and noting that $\vec{F}_N$ is at 90° to the incline, then the angle between $\vec{F}_N$ and the horizontal must be $(90° - \varphi)$, and the angle between $\vec{F}_N$ and the vertical is φ. The x- and y-components of $\vec{F}_N$ are then $F_{Nx} = F_N \sin\varphi$ and $F_{Ny} = F_N \cos\varphi$. Therefore,

$$F_{Nx} = F_N \sin\varphi = ma_x$$

Since a_x is the centripetal acceleration, $a_x = a_c = v^2/r$, which gives

$$F_N \sin\varphi = m\frac{v^2}{r} \qquad \text{Eqn. [2]}$$

Because F_N, φ, and m are unknown, we need another equation in order to solve for φ. In the y-direction, there is no acceleration since the car does not slide up or down the curve if the banking angle is correct. Therefore, using Newton's first or second law,

$$\Sigma F_y = ma_y = 0$$

$$F_N \cos\varphi + (-mg) = 0$$

$$F_N \cos\varphi = mg \qquad \text{Eqn. [3]}$$

We can eliminate F_N and m in one step by dividing Eqn. [2] by Eqn. [3]:

$$\frac{F_N \sin\varphi}{F_N \cos\varphi} = m\frac{v^2}{r}\frac{1}{mg} \quad \text{giving} \quad \tan\varphi = \frac{v^2}{rg}$$

$$\text{and therefore} \quad \varphi = \tan^{-1}\left(\frac{v^2}{rg}\right)$$

(Alternatively, we could have arrived at this same result by solving Eqn. [3] for F_N and substituting the result into Eqn. [2].)

Before substituting known values, we need to convert the units of the speed to metres per second:

$$90\frac{\text{km}}{\text{h}} \times \frac{1000\,\text{m}}{1\,\text{km}} \times \frac{1\,\text{h}}{3600\,\text{s}} = 25\,\text{m/s}$$

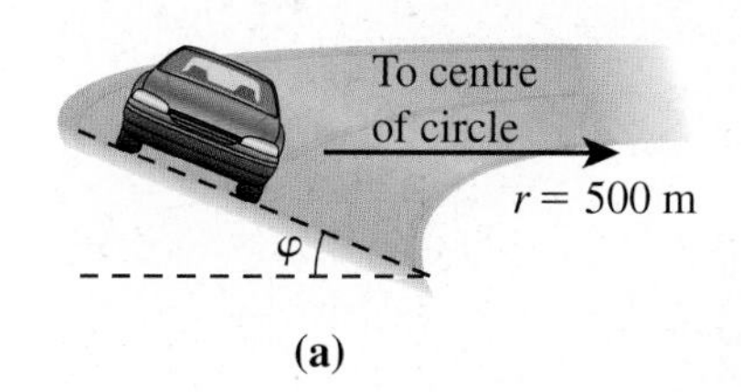

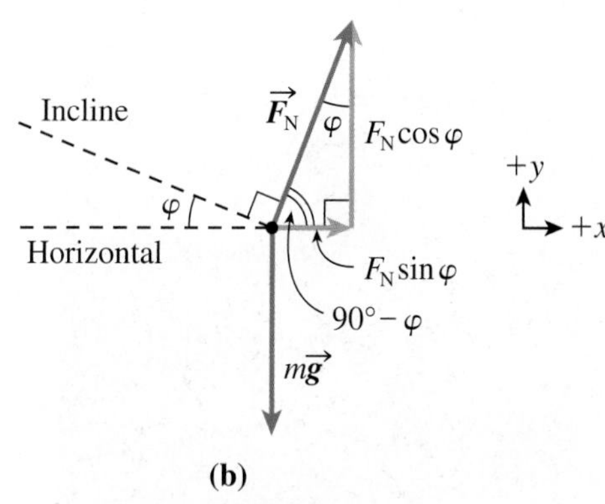

Figure 6-16 **(a)** Sample Problem 6-5. **(b)** Free-body diagram for the car.

TRY THIS!

Circular Motion Accelerometer

If you have a smartphone with accelerometers, here's an easy way to determine the magnitude of the centripetal acceleration of the phone as it undergoes circular motion. Just hold it flat in your hand at the end of an outstretched arm, rotate your arm and body, and your phone will provide information about the centripetal acceleration. (You will first need to install an app on your phone that gives data and/or a graph of acceleration versus time.)

Problem-Solving Tip

Forces and Centripetal Acceleration

In the three sample problems in this section, note that the force providing the centripetal acceleration has been

(i) gravity
(ii) the vector sum of aerodynamic lift and gravity
(iii) a component of the normal force

When solving problems, remember that the force (or combination of forces) causing the centripetal acceleration can be found automatically on the FBD as the resultant force acting toward the centre of the circle.

Perspectives - Jeff Smith/ Shutterstock.com

Figure 6-17 Bobsled tracks often have very large banking angles.

© CJ Crooks Design, www.cjcrooks.com

Figure 6-18 A shark with an accelerometer attached to its dorsal fin.

Substituting values, $\varphi = \tan^{-1}\left(\frac{v^2}{rg}\right) = \tan^{-1}\left[\frac{(25\,\text{m/s})^2}{(500\,\text{m})(9.8\,\text{m/s}^2)}\right] = 7.3°$

Thus, the correct banking angle is 7.3° (to two significant digits). Notice that the mass m "cancelled out" during the solution. This angle would be smaller if we were also to include the effect of friction.

Highway engineers must design roads with the proper banking angles. The next time you travel on a highway, notice that the banking angles for on- and off-ramps are larger than the angles for the more gradual curves on the main highway. Why is this so? Why is the posted speed for a ramp lower than the speed limit on the main highway? Having banked curves is also important for bobsled tracks (Figure 6-17), short-track bicycle racing, etc.

DID YOU KNOW?

Accelerometers such as those found in many smartphones are now routinely attached by biologists to fish, birds, amphibians, and mammals to monitor animal behaviour, movement, and bioenergetics (Figure 6-18). Often a tri-axial accelerometer unit is used; this consists of three accelerometers, one for each of the x-, y-, and z-directions. Whereas previous types of tracking monitors showed only the location of an animal, accelerometers provide data that indicate in detail how the animal is moving; for example, sharks have been shown to do barrel rolls and headstands. It is possible to determine whether an animal is feeding, resting, migrating, mating, etc., and energy expenditure can be estimated.

EXERCISES

6-14 In each case below, an object "A" is in circular motion. Draw a free-body diagram for object "A," and name the force(s) that provide(s) the resultant acceleration toward the centre of the circle.

(a) A communications satellite (A) is in a circular orbit around Earth.

(b) Earth (A) is in a circular orbit (approximately) around the Sun.

(c) A skier (A) slides over the top of a hump that has the shape of a circular arc.

(d) A car (A) travels around an *unbanked* curve on a highway without sliding. (**Hint:** The phrase "without sliding" is important.)

(e) A car (A) travels around a *banked* curve at the correct speed for the banking angle (i.e., friction is not required for the car to go around the curve).

(f) An electron (A) travels in a circular orbit around a nucleus (in a simple model of the atom).

6-15 A wet pair of jeans travels in a horizontal circle in the spin cycle of a washing machine.

(a) What object exerts the force on the jeans toward the centre of the machine?

(b) What is the name of the force exerted in part (a)?

(c) The radius of the circle in which the jeans travel is 26 cm. If the machine has a frequency of 8.5 rev/s, how far do the jeans travel in 1.0 s?

(d) What is the speed of the jeans?

(e) The mass of the wet jeans is 1.7 kg. What is the magnitude of the resultant force on the jeans?

6-16 Neptune follows an essentially circular orbit of radius 4.5×10^{12} m around the Sun. The mass of Neptune is 1.0×10^{26} kg, and the gravitational force exerted on Neptune by the Sun has a magnitude of 6.8×10^{20} N.

(a) What is Neptune's speed?

(b) How long does it take Neptune to complete one orbit around the Sun? Express your answer in seconds and years.

6-17 In an early model of the hydrogen atom, the electron (mass 9.11×10^{-31} kg) in its lowest energy state was thought of as travelling around the nucleus in a circle of radius 5.3×10^{-11} m. If this model is used, the speed of the electron is 2.2×10^{6} m/s.

(a) What is the name of the force that provides the centripetal acceleration?

(b) What are the magnitude and direction of this force?

6-18 A ball attached to the end of a string is whirled in a vertical circle (Figure 6-19). Neglect air resistance.

(a) Draw a free-body diagram for the ball at each of the points A, B, C, and D.

(b) In each free-body diagram, which force(s) make(s) up the resultant force component toward the centre of the circle?

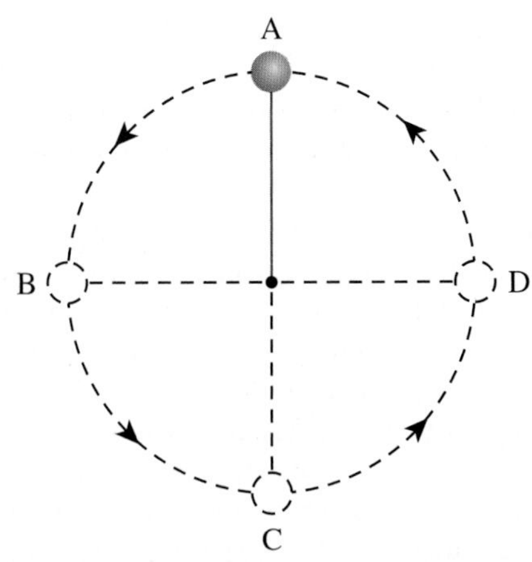

Figure 6-19 Question 6-18.

6-19 A young woman of mass 51.8 kg is on a swing (Figure 6-20). The radius of the circular arc that she follows is 3.80 m. When the woman is at the lowest point, where her speed is 2.78 m/s, what is the magnitude of the tension in each of the two supporting chains (assumed to be vertical)?

Figure 6-20 Question 6-19.

6-20 A car of mass 1.12×10^3 kg goes around an *unbanked* curve of radius 312 m at a speed of 28 m/s without sliding.

(a) What is the name of the force that provides the centripetal acceleration?

(b) What is the minimum coefficient of static friction between the tires and the road that will permit the car to go around the curve without sliding?

(c) If the radius of the curve were smaller, making the curve "sharper," would your answer to (b) increase or decrease? Explain.

(d) If the mass of the car were larger, how would your answer to (b) be affected?

6-21 Railroad tracks are banked at curves to reduce wear and stress on the wheel flanges and the rails, and to prevent the train from tipping over. If tracks are banked at an angle of 5.7° from the horizontal, and follow a curve of radius 5.5×10^2 m, what is the ideal speed for a train rounding the curve?

6-22 A child has tied a piece of wood of mass 0.47 kg to a string and is swinging it in a vertical circle of radius 83.0 cm.

(a) Draw a free-body diagram for the wood at the top of the circle.

(b) Determine the magnitude of the tension in the string, if the speed of the wood is 3.70 m/s when at the top of the circle.

6-23 In the previous problem, the wood will not "make it over the top" if its speed is too low. Determine the minimum speed of the wood at the top of its path if it is to follow a complete circle. Use the following steps:

(a) For a piece of wood of mass m travelling in a vertical circle of radius r, and having speed v at the top, determine the magnitude of the string tension T (at the top) in terms of r, v, m, and g (similar to part (b) of the previous problem).

(b) According to your expression in (a), what happens to T as v decreases?

(c) If the wood has a speed at the top such that it just *barely* completes the circle, what happens to T? (**Hint:** The string will go slack momentarily.)

(d) Use your results from (a) and (c) to determine an expression for the minimum speed v_{min} at the top for the wood to be barely able to complete the circle, in terms of g and r.

(e) Calculate v_{min} for $r = 83$ cm. Is the speed given in part (b) of the previous problem larger than v_{min}?

6.3 Centrifugal "Force"

You have probably heard of centrifugal "force" and might be wondering what it is. Centrifugal "force" is not a true force in the way that gravity, tension, and friction are. It is often referred to as a fictitious force or a pseudoforce, and hence quotation marks have been placed around the word "force" in the previous lines. In order to explain centrifugal "force," we need to begin with a discussion of inertial and noninertial frames of reference.

Inertial and Noninertial Frames of Reference

Imagine that you are travelling in a bus at constant speed along a straight, smooth highway. If you place a ball on the floor of the bus, it stays at rest relative to you and the bus, just as it would if you placed it on a floor in your home. The ball remains stationary because there is no resultant force acting on it: that is, the law of inertia (Newton's first law, presented in Section 5.4) is obeyed. Therefore, we call a constant-velocity bus an inertial frame of reference. A **frame of reference** is just

frame of reference any object (Earth, bus, particle, etc.) relative to which the positions, velocities, etc., of other objects can be measured

an object such as a bus, a room, or even a subatomic particle, relative to which the positions, velocities, etc., of other objects such as a ball can be measured.[1] In an **inertial frame of reference**, the law of inertia and other laws of physics are valid. (Any frame moving at constant velocity relative to an inertial frame is also an inertial frame.)

inertial frame of reference a frame of reference in which the law of inertia and other laws of physics are obeyed

Now consider what happens if the bus driver applies the brakes, thus giving the bus an acceleration in a direction opposite to its velocity. Relative to the bus, the ball appears to accelerate toward the front, even though there is still no resultant force acting on it! (There is nothing actually pushing the ball forward.) The law of inertia does not appear to be obeyed in the bus as it slows down. Therefore, the accelerating bus is referred to as a **noninertial frame of reference**. How can the ball accelerate toward the front of the braking bus if there is no resultant force on the ball? The answer is that we are looking at the motion of the ball from the "wrong" frame of reference, the noninertial frame of the accelerating bus. The situation is much simpler if we consider it from an inertial frame such as the ground. Relative to the ground, when the brakes are applied to the bus, the ball tends to continue moving forward at constant velocity (by Newton's first law).[2] Since the bus is slowing down, and the ball is not, then relative to the bus, the ball accelerates toward the front (Figure 6-21).

noninertial frame of reference a frame of reference in which the law of inertia and other laws of physics do not appear to be obeyed

fictitious force (pseudoforce or inertial force) a force that must be invented to explain the motion of an object relative to a noninertial frame of reference

If we try to analyze the motion of the ball in the noninertial frame of the accelerating bus, we have to invent a force toward the front of the bus. This **fictitious force**, also called a **pseudoforce** or an **inertial force**, would explain the acceleration of the ball. The analysis of motion in any noninertial frame involves fictitious forces, but no such forces are necessary in inertial frames. If a noninertial frame such as the bus accelerates in a straight line relative to an inertial frame, the fictitious force (on the ball) is in the opposite direction to the acceleration of the noninertial frame itself.

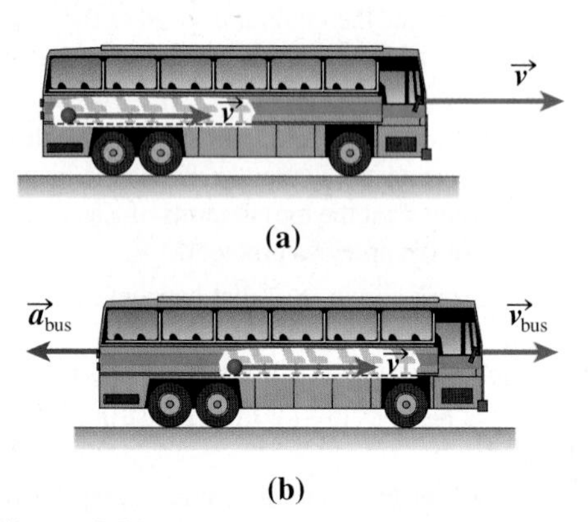

Figure 6-21 **(a)** A bus and a ball moving at constant velocity relative to the ground. Relative to the bus, the ball is at rest. **(b)** When the brakes are applied to the bus, it slows down. However, the ball tends to continue moving forward at its former (constant) velocity, relative to the ground (see footnote 2). Thus, relative to the bus, the ball accelerates toward the front.

We often assume that the surface of Earth is an inertial frame of reference. In actual fact, it is an accelerated (noninertial) frame. Because of Earth's daily rotation, the surface of Earth has a centripetal acceleration toward Earth's centre. In addition, Earth's annual motion around the Sun, and the motion of the Sun in the galaxy, etc., involve accelerations. However, because the accelerations involved in all these motions are small, Earth's surface is very close to being an inertial frame for everyday purposes.

All the laws of physics, such as Newton's three laws of motion, are valid in any inertial frame (bus moving at constant velocity, Earth, Sun, galaxy, or galactic cluster), and therefore there is no inertial frame that is somehow preferred over the others. However, the velocity of any object is different relative to different inertial frames. For example, a ball having zero velocity relative to a moving bus has a non-zero velocity relative to Earth. Consequently, there is no such thing as an absolute velocity, relative to some special absolute inertial frame.

Centrifuges and Centrifugal "Force"

In the bus example above, the motions were linear. Rotating objects provide examples of more complicated noninertial frames. Each point on a rotating object travels in a circle, and hence experiences a centripetal acceleration. Therefore, a rotating object is an accelerated, or noninertial, frame. One practical example is a centrifuge (Figure 6-22), which is a rapidly rotating device that separates substances in solution

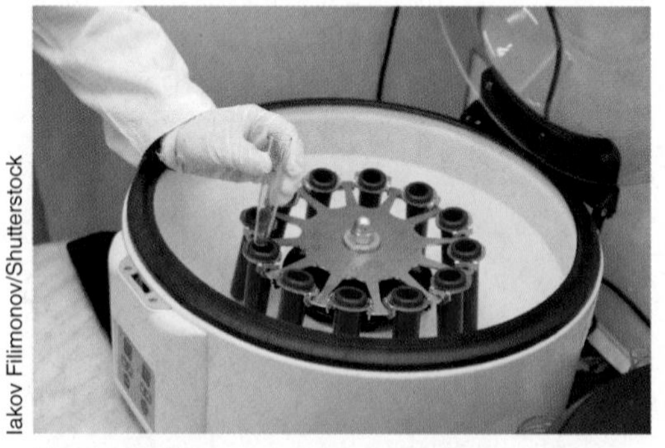

Figure 6-22 A centrifuge. The tops of the blue containers containing test tubes are hinged so that when the centrifuge rotates at high speed, the test tubes swing outward and become almost horizontal.

[1] More formally, a frame of reference (Section 4.6) is defined as an x-y-z coordinate system attached to an object such as a bus, room, etc. Without this coordinate system, it would be impossible to perform measurements of quantities such as position and velocity.

[2] The ball actually does not continue moving forward quite at constant velocity. However, if the floor of the bus had negligible friction, then any object sitting on the floor would indeed continue to slide forward at constant velocity relative to the ground as the bus is braking.

according to their densities. For example, if a sample of blood is allowed to stand in a test tube, the red blood cells settle to the bottom of the less dense plasma. Without a centrifuge, this settling is very slow, at a rate of about 3 mm/h for blood from males, and 4 mm/h for females. Using a centrifuge can speed up the settling by a factor of over a million.

Figure 6-23 shows how a centrifuge works. The test tubes containing samples are rotated in a circle that is almost horizontal. The rotation is very rapid—some centrifuges rotate with frequencies of millions of revolutions per minute. A dense molecule or cell near the "top" of a tube at A tends to continue moving at constant speed in a straight line, relative to Earth. This straight line motion carries it toward the "bottom" of the tube, as shown. Relative to the tube, the dense molecule is moving toward the bottom, and is settling out. Relative to Earth, the molecule is simply moving in a straight line, and the tube has an acceleration toward the centre of the centrifuge. The detailed motion of the molecule is affected by the viscous and buoyant forces exerted on it by the surrounding liquid, but the above discussion outlines the general principle involved in centrifugation.

If you could be stationary relative to the test tube—that is, rotating with it relative to Earth—then in order to explain the motion of the molecule outward from the centre of the circle, you would have to invent a fictitious force called the **centrifugal "force."** ("Centrifugal" means "fleeing from the centre.") The centrifugal "force" is directed outward from the centre of rotation in a rotating noninertial frame of reference. To explain the motion of the molecule relative to Earth (an inertial frame), the centrifugal "force" is not needed.

A detailed analysis of the motion of the molecule relative to the test tube or centrifuge would show that another fictitious force is also required, one that is perpendicular to the velocity of the molecule in the rotating frame. This force is called the **Coriolis "force,"** named after the French mathematician Gaspard Gustave de Coriolis (1792–1843). It is exerted on objects that are *moving* relative to rotating frames. Some effects of the Coriolis "force" relative to the rotating reference frame of Earth are discussed in the next subsection.

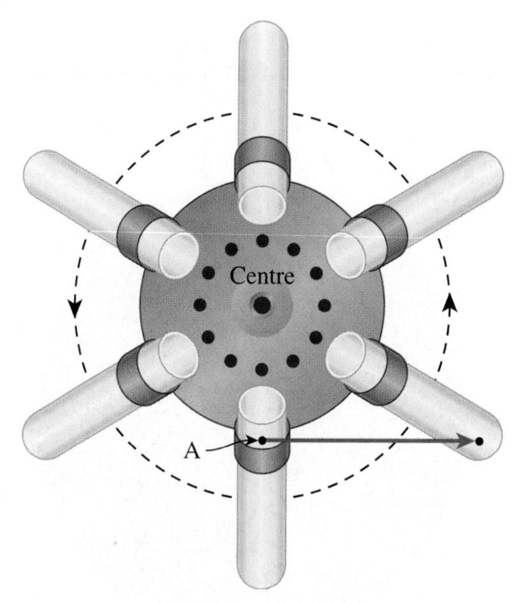

Figure 6-23 Operation of a centrifuge. A dense molecule in a liquid at A tends to continue in motion at constant velocity relative to Earth, and thus settles toward the bottom of the tube.

centrifugal "force" a fictitious force that is needed to explain the motion of an object relative to a rotating noninertial frame of reference; this "force" is directed outward from the centre of rotation

Coriolis "force" a fictitious force that is needed to explain the motion of an object relative to a rotating noninertial frame of reference; this "force" has a direction perpendicular to the object's velocity relative to the rotating frame

SAMPLE PROBLEM 6-6

A child is standing on a rotating playground merry-go-round. Draw a free-body diagram for the child and explain her motion in **(a)** the inertial frame of Earth and **(b)** the rotating frame of the merry-go-round.

Solution

(a) Figure 6-24 (a) shows a diagram for the problem. Relative to Earth, the child is travelling in a circle, and therefore there must be a force causing the child's centripetal acceleration toward the centre of the merry-go-round. Since we are told that the child is *standing,* the only possible horizontal force is the force of static friction, $\vec{F}_S$. This is the force providing the centripetal acceleration, and is shown in the free-body diagram (Figure 6-24 (b)).

(b) In the rotating frame of the merry-go-round, the child is stationary. On the free-body diagram (Figure 6-24 (c)), we have the three forces found in the inertial frame, and also a centrifugal "force" outward from the centre. (There is no Coriolis "force," because the child is not moving relative to the merry-go-round.) The child is stationary in the frame of the merry-go-round because the static friction force and the centrifugal "force" are equal in magnitude, but opposite in direction.

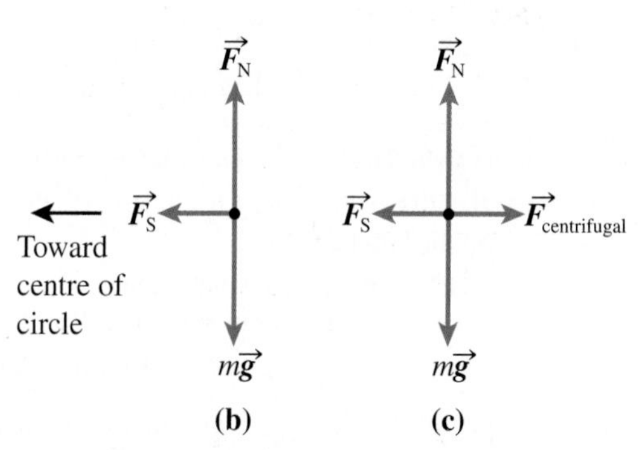

Figure 6-24 **(a)** Sample Problem 6-6. **(b)** Free-body diagram for girl, relative to Earth. **(c)** Free-body diagram for girl, relative to merry-go-round.

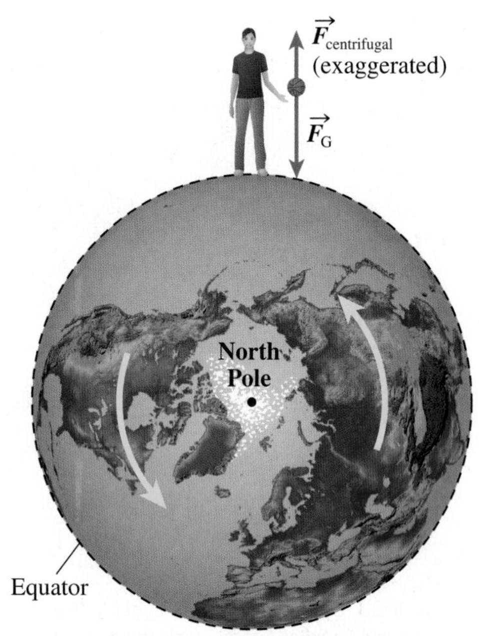

Figure 6-25 Relative to the rotating Earth (a noninertial frame), both a gravitational force and a centrifugal "force" are exerted on a falling ball. Thus, the ball's acceleration relative to Earth is less than the acceleration due to gravity alone.

Earth's Surface as a Noninertial Frame of Reference

Because Earth spins about its axis, it is a rotating noninertial frame of reference, although the effects due to the rotation are small. If you were to stand at the equator (Figure 6-25) and drop a ball, it would fall toward the centre of Earth because of the force of gravity. However, relative to the rotating Earth, there is also a centrifugal "force" on the ball directed outward from the centre of Earth. Therefore, the resultant force in the frame of Earth is less than the gravitational force, and the measured acceleration of the ball relative to Earth is less than the acceleration due to gravity alone. This is a real effect, albeit a small one; at the equator, the magnitude of the acceleration relative to Earth (what we normally think of as g) is only 0.35% less than that due to gravity alone. This centrifugal decrease of acceleration depends on latitude; the effect is maximum at the equator and zero at the poles.

The effect of the Coriolis "force" on an object moving relative to Earth is small unless the object is moving very fast or for a very long time. For example, unless a dropped ball falls for a long distance, the effect of the Coriolis "force" is negligible. However, the Coriolis "force" is responsible for the spiral shape of many weather patterns such as cyclones and hurricanes (Figure 6-26), and is important in the design of the sights of large military guns, for example. During a naval battle near the Falkland Islands in World War I, British gunners were surprised to see their shells landing 100 m to the left of German ships. The gunsights had been adjusted for the Coriolis "force" at 50° N latitude, but the battle was in the southern hemisphere, where this "force" produces a deflection in the opposite direction.

Photobank gallery/Shutterstock

Figure 6-26 A cyclone viewed from space. The spiral pattern is the result of the Coriolis "force."

TACKLING MISCONCEPTIONS

Toilets, Sinks and Coriolis "Force"

You might have heard that the Coriolis "force" causes water to drain from sinks, toilets, etc., in a clockwise direction in the northern hemisphere and a counterclockwise direction in the southern hemisphere. In fact, any tiny rotation initiated by the opening of the sink drain or the flushing of the toilet overwhelms any Coriolis effect. However, careful experiments do show the different directions of rotation in the two hemispheres, as long as precautions are taken to ensure that the opening of the drain hole causes no initial rotation.

EXERCISES

6-24 Suppose that you are riding in a train moving at constant velocity relative to Earth, and that you place a ball on the floor in the centre of the aisle. The ball is initially at rest relative to the train.

(a) Relative to the train, what happens to the ball if the brakes are applied to the train?

(b) Relative to the frame of reference of the train, what is the direction of the acceleration of the ball as the brakes are applied?

(c) Is the force causing this acceleration a real one?

6-25 Imagine that you are in a bus that is speeding up, and that you suspend a small ball from a string (Figure 6-27).

(a) In which direction (A, B, C) does the string hang?

(b) Draw a free-body diagram for the ball relative to the frame of Earth (assumed inertial). Is the ball accelerating? If so, what force is providing the acceleration?

(c) Draw a free-body diagram for the ball relative to the noninertial frame of the accelerating bus. Is the ball accelerating relative to this frame? Use your diagram to explain the ball's motion, or lack of it, relative to the bus.

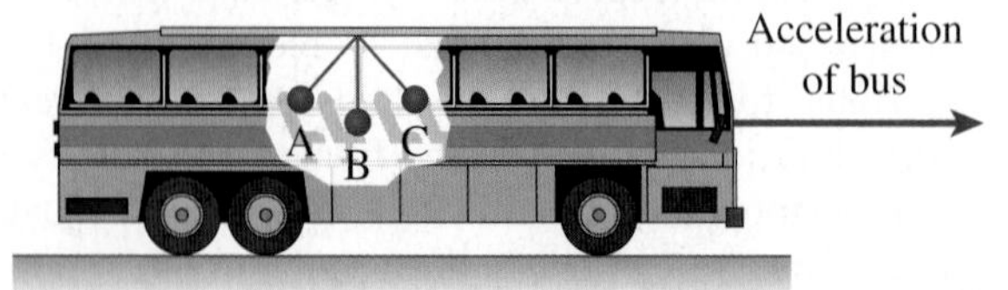

Figure 6-27 Question 6-25.

6-26 If you are sitting in a car that rounds a sharp curve to the right, you tend to move to the left, relative to the car. Explain your motion, first relative to Earth (assumed inertial), and then relative to the (noninertial) car. Diagrams might be helpful.

6-27 "Gravitropism" is the tendency of plants to align their stems and roots along the direction of the force of gravity. This tendency is caused by differential concentrations of auxins (plant hormones). If you had designed an experiment with seedlings placed in pots near the outer edge of a rotating platform that has a centre point C, in which direction would you expect to observe the stems to tilt: toward C, vertically upward, or away from C? Explain your answer.

LOOKING BACK...LOOKING AHEAD

Thus far in the book, we have been considering mechanics—the study of motion and forces. We started by studying motion itself through quantities such as displacement, velocity, and acceleration, and then considered the connection between motion and forces by discussing Newton's famous three laws of motion.

The application of Newton's laws has been the focus of this chapter. We have discussed static and kinetic friction, forces involved in circular motion, and the fictitious forces that are involved in noninertial frames of reference. The concepts that we developed have been illustrated through a variety of examples, such as "Jo blocks," the banking of highway curves, planetary motion, and centrifuges.

The next chapter will continue the development of mechanics by introducing work, energy, and power. A knowledge of these concepts is important in understanding many aspects of modern society—electrical power generation, automobile efficiency, etc.

CONCEPTS AND SKILLS

Having completed this chapter, you should now be able to do the following:

- Discuss the source of friction.
- List reasonable values of coefficients of friction between various materials.
- Solve problems involving coefficients of friction.
- Be able to identify the force(s) that provide(s) the centripetal acceleration in various examples of circular motion.
- Solve force problems involving circular motion.
- Explain how fictitious forces arise in noninertial frames.
- Discuss basic properties of the centrifugal and Coriolis "forces."
- Draw free-body diagrams for objects in simple situations involving noninertial frames.

KEY TERMS

You should be able to define or explain each of the following words or phrases:

coefficient of static friction
coefficient of kinetic friction
frictional heating
frame of reference
inertial frame of reference
noninertial frame of reference
fictitious force (or pseudoforce or inertial force)
pseudoforce
inertial force
centrifugal "force"
Coriolis "force"

Chapter Review

MULTIPLE-CHOICE QUESTIONS

6-28 The maximum magnitude of the static friction force acting on an object is usually

(a) less than the magnitude of the kinetic friction force
(b) equal to the magnitude of the kinetic friction force
(c) greater than the magnitude of the kinetic friction force

6-29 A student is trying to push a large box of books across a carpeted floor in a university residence room. The student is exerting a horizontal force of magnitude 95 N, but the box is not moving. The magnitude of the friction force exerted by the carpeted floor on the box is

(a) 0 N
(b) 95 N
(c) less than 95 N
(d) greater than 95 N

6-30 A woman is standing on a moving walkway in an airport (Figure 6-28). The walkway is moving at constant velocity, and since the woman is standing (not walking) her velocity is the same

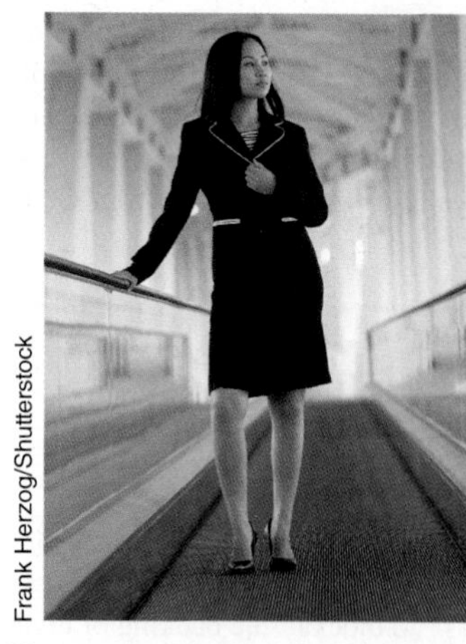

Figure 6-28 A woman on a moving walkway (Question 6-30).

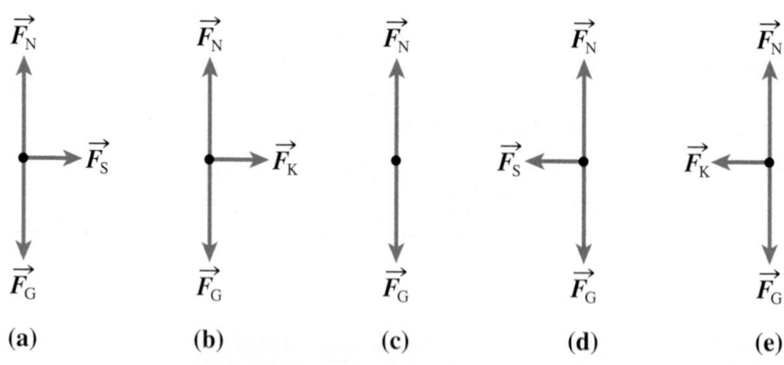

Figure 6-29 Question 6-30.

as that of the walkway. In Figure 6-29, which drawing shows the correct free-body diagram for the woman? The symbols have their usual meanings. The direction of the woman's velocity is toward the right in the diagrams.

6-31 Relative to Earth (assumed to be an inertial frame), which of the following statements is true for a test tube that is undergoing uniform circular motion in a centrifuge?

(a) The resultant force on the test tube is zero, and the velocity of the test tube is constant.
(b) The resultant force on the test tube is zero, and the speed of the test tube is constant.
(c) The resultant force on the test tube is not zero, and the speed of the test tube is changing.
(d) The resultant force on the test tube is not zero, and the velocity of the test tube is changing.
(e) The resultant force on the test tube is zero, and the velocity of the test tube is changing.

6-32 Which of the following is a correct list of all the real forces that act on a planet in a circular orbit around a star?

(a) the gravitational attractive force $\vec{F}_G$ toward the star, the centrifugal "force," and the force due to the motion of the planet
(b) $\vec{F}_G$ and the force due to the motion of the planet
(c) $\vec{F}_G$ and the centrifugal "force"
(d) $\vec{F}_G$

Review Questions and Problems

6-33 A horizontal force is applied to an object sitting at rest on a horizontal surface. The force is gradually increased and the object eventually moves. Sketch the shape of the graph of (the magnitude of) the friction force on the object versus (the magnitude of) the applied force.

6-34 A woman exerts a horizontal force of magnitude 512 N on a stationary refrigerator of mass 117 kg that is sitting on a horizontal floor. The coefficients of friction between the refrigerator and the floor are static, 0.69, and kinetic, 0.60. What happens? Show your reasoning.

6-35 A man is pushing horizontally on a table of mass 16 kg to move it across a horizontal floor. The coefficient of kinetic friction between the table and floor is 0.61. What magnitude of applied force is necessary to keep the table moving at constant velocity?

6-36 If the actual applied force in the previous question has a magnitude of 121 N, and the table starts from rest, how long will it take to travel 0.75 m?

6-37 A bird of mass 78 g is making a *horizontal* banked turn of radius 9.7 m at a speed of 17 m/s.

(a) What exerts the force that provides the centripetal acceleration of the bird?
(b) What is the magnitude of this force?

6-38 When an electrically charged particle enters a uniform magnetic field that is perpendicular to the particle's velocity, a magnetic force is produced that causes the particle to move in a circle at constant speed. (This will be discussed in detail in Chapter 22.)

(a) What is the radius of the resulting circle when an electron (mass 9.11×10^{-31} kg) travelling with speed 6.5×10^6 m/s is acted on by a magnetic force that has a magnitude of 1.2×10^{-15} N?
(b) How long does it take the electron to make one revolution?

6-39 A girl is standing on a rotating merry-go-round at a carnival. Her father makes the statement that, relative to the (inertial) frame of reference of Earth, the girl remains at the same distance from the centre of the merry-go-round because there is no resultant force on her. Comment on his statement.

Applying Your Knowledge

6-40 James Bond is being dragged along a horizontal surface by a rope attached to a large container of diamonds that is falling vertically. The rope passes over a pulley as shown in Figure 6-30. James has a mass of 83 kg, and the container of diamonds has a mass of 45 kg. If the magnitude of the acceleration of James and the diamonds is 1.2 m/s², determine the coefficient of kinetic friction between James and the surface beneath him. Neglect the rotational motion of the pulley.

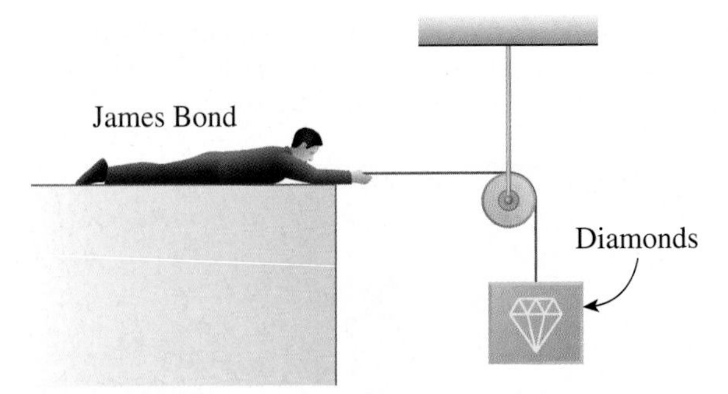

Figure 6-30 Question 6-40.

6-41 A car is speeding up on a level road; the tires are not slipping.

(a) What is the name of the force that provides the acceleration?

(b) If the coefficient of static friction between the tires and the road is 0.60, what is the maximum magnitude of acceleration the car can have without the tires slipping? Neglect air resistance.

6-42 A boy is pushing horizontally on a Teflon block to hold it against a vertical wall.

(a) If the block has a mass of 0.12 kg, what minimum magnitude of force must the boy apply if the block is not to slide down the wall? Assume a value of 0.050 for the coefficient of static friction between Teflon and the wall.

(b) Compare your answer in (a) to the magnitude of upward force required to simply hold the block in the boy's hand; that is, determine the ratio of your answer in (a) to the upward force magnitude.

Feeling Forces

This activity is related to problem 6-42. Using a bathroom scale, determine the weight of a very heavy book, preferably one with a slippery glossy cover. Then, holding the scale vertically, use it to press the book against a vertical wall, and determine how much force is required to barely hold the book in place. How does that force compare to the actual weight of the book?

6-43 A young man is standing on one foot, supported by crutches (Figure 6-31). The man's mass is 66 kg, and the force of gravity acting on him is balanced by three upward normal forces exerted by the floor: one force on his foot, and one force on each crutch. Assume that 60% of the total normal force is exerted on his foot, and that the remaining 40% is equally shared by the two crutches.

(a) Determine the magnitude of the normal force exerted on each crutch. (Neglect the mass of each crutch and the force of gravity acting on it.)

(b) The floor exerts a horizontal static friction force on each crutch as well; this force prevents the crutch from slipping sideways. If each crutch makes an angle of 85° relative to the horizontal (as seen from the front), and if the resultant force exerted by the floor on a crutch has a direction along the crutch, determine the magnitude of the static friction force on each crutch.

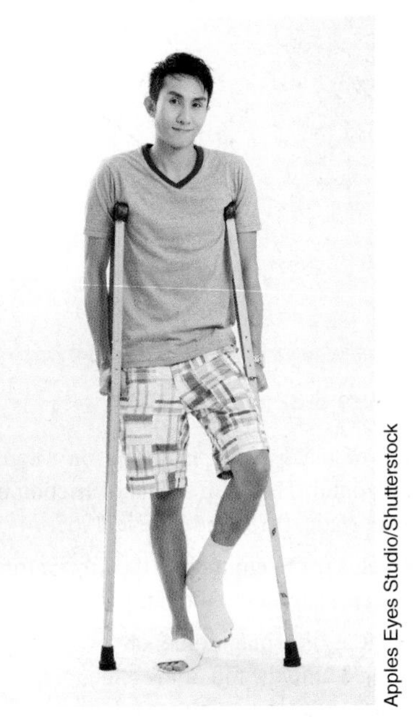

Apples Eyes Studio/Shutterstock

Figure 6-31 Question 6-43.

(c) If the coefficient of static friction between the floor and the tip of the crutch is 0.30, determine the minimum angle that each crutch can make with the floor without the crutch slipping.

Hint for Questions 6-44 to 6-49: $F_N \neq mg$.

6-44 A boy drags a huge garbage can of mass 27 kg along a horizontal driveway at a constant speed of 1.8 m/s. He exerts a force of 112 N at an angle of 27° above the horizontal. Determine the coefficient of kinetic friction between the can and the driveway.

6-45 A rope is used to exert a force of magnitude 21 N, at an angle of 31° above the horizontal, on a box at rest on a horizontal floor. The coefficients of friction between the box and the floor are static, 0.55, and kinetic, 0.50. Determine the smallest possible mass of the box, if it remains at rest.

6-46 A girl on a toboggan slides down a hill inclined at 32° to the horizontal. Starting from rest, she travels 12 m in 2.5 s.

(a) What is the magnitude of her acceleration?

(b) What is the coefficient of kinetic friction between the toboggan and the snow?

6-47 A skier is on a gradual slope inclined at 4.7° to the horizontal. She gives herself a push and starts down the slope with an initial speed of 2.7 m/s. If the coefficient of kinetic friction between her skis and the snow is 0.11, how far will she slide before coming to rest?

6-48 Two boxes are connected by a rope that passes over a pulley (Figure 6-32). Box #1 is on a ramp inclined at 35° to the horizontal, and the coefficient of kinetic friction between the box and ramp is 0.54. The masses of the boxes are $m_1 = 2.5$ kg and $m_2 = 5.5$ kg. Neglecting the motion of the pulley, and assuming that the velocity of each box is in the same direction as its acceleration, what is the magnitude of the acceleration of the boxes? Is the acceleration of Box #2 upward or downward?

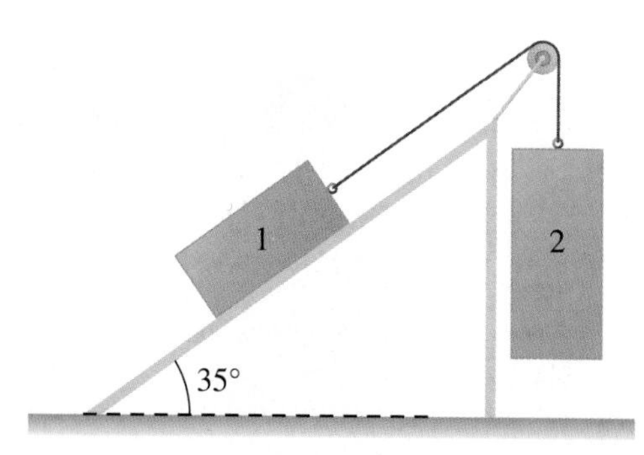

Figure 6-32 Question 6-48.

6-49 A crate of mass 22 kg is sitting on a ramp inclined at 45° to the horizontal. The coefficients of friction between the crate and ramp are static, 0.78, and kinetic, 0.65.

(a) What is the magnitude of the *largest* force that can be applied upward, parallel to the ramp, if the crate is to remain at rest?

(b) What is the magnitude of the *smallest* force that can be applied onto the top of the crate, perpendicular to the ramp, if the crate is to remain at rest?

6-50 Two blocks connected by a string passing over a pulley are on ramps inclined at 32° and 25°, as shown (Figure 6-33). The masses of block #1 and block #2 are 1.8 kg and 1.0 kg respectively. There is negligible friction between block #2 and the ramp on which it is sitting, but between block #1 and its ramp, the coefficients of static friction and kinetic friction are 0.15 and 0.10 respectively. The system is released from rest. What is the magnitude of the acceleration of the blocks?

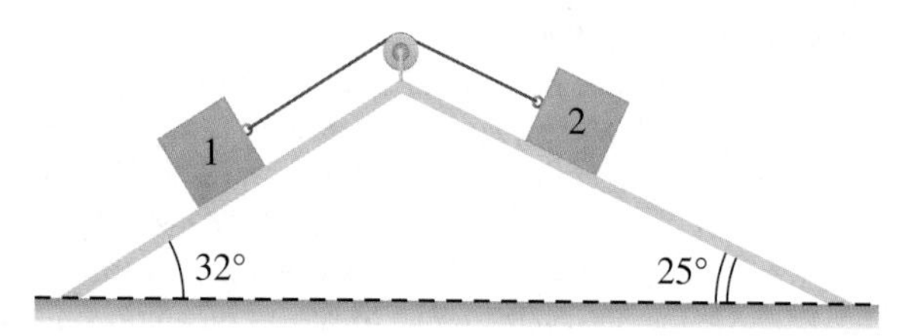

Figure 6-33 Question 6-50.

6-51 A *conical* pendulum (Figure 6-34) consists of a mass m (the bob) on the end of a string. The top end of the string is held fixed, and as the bob travels in a horizontal circle at constant speed, the string traces out a *cone*. For the pendulum shown, mass $m = 1.95$ kg, length $\ell = 1.03$ m, and angle $\theta = 36.9°$.

(a) Draw a free-body diagram for the bob at the instant shown.

(b) What force provides the centripetal acceleration?

(c) Determine the speed of the bob.

(d) How long does it take the bob to make one revolution?

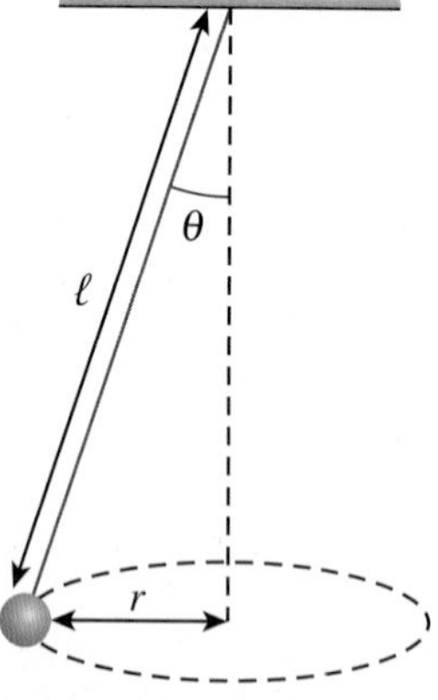

Figure 6-34 Question 6-51.

6-52 A skier slides over a hump that has the shape of a circular arc of radius 5.72 m (Figure 6-35). Determine his speed at the top if his skis just barely lose contact with the snow.

Figure 6-35 Question 6-52.

6-53 On a loop-the-loop roller coaster (Figure 6-36), passengers at any instant follow a path that can be considered part of a circle, and have an acceleration with a component toward the centre of the circle. The passengers' safety is assured by seatbelts and/or shoulder braces. However, the coaster is designed so that it will normally be travelling fast enough that the passengers will be able to loop-the-loop without being secured in their seats. Follow the steps below to determine the minimum speed, v_{min}, of the coaster at the top, if the passengers just make it around the loop without being fastened in the seats.

(a) Draw a free-body diagram for an unsecured passenger at the top of the loop if the speed is larger than v_{min}. Neglect friction.

(b) From (a), which forces together provide the centripetal acceleration?

(c) If the coaster is travelling just barely fast enough at the top so that the passenger makes it around the loop, what happens to one of the forces given in the answer to (b)? (**Hint:** The passenger momentarily loses contact with the seat, and falls ever so slightly downward away from it.)

(d) Using your result from (c), determine an expression for v_{min} in terms of r (the loop's radius) and g.

(e) Use a typical value of r to determine an estimate of v_{min}.

Figure 6-36 Question 6-53.

DID YOU KNOW?

In the early 1900s the first upside-down roller coasters had circular loops, which proved to be dangerous and uncomfortable. To make the rides safer and smoother, loops are now in the form of a Klothoid (sometimes spelled Clothoid) curve, as shown in Figure 6-36. Notice that the radius of the arc at the top of the loop is smaller than the radius lower down. As you can show in solving problem 6-53, the minimum speed needed for the cars and passengers to go around the top of the loop is given by $v_{min} = \sqrt{gr}$. Having a small radius at the top means that the minimum required speed is less than it would be with a large radius, and the ride is therefore safer.

6-54 A skateboarder is riding down a circular-arc ramp of radius 6.3 m (Figure 6-37). The board is inclined at an angle of 19° relative to the horizontal. If the skateboarder's speed is 2.8 m/s, and the mass of the skateboarder and board is 38 kg, determine the magnitude of the normal force exerted by the ramp on the boarder and board.

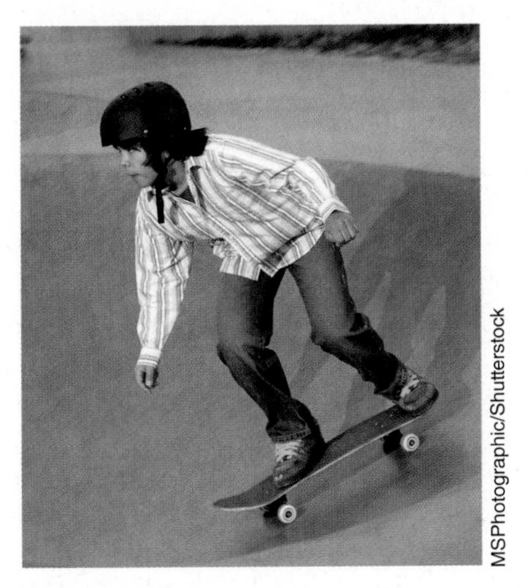

Figure 6-37 Question 6-54.

6-55 **Fermi Question:** Associated with Earth's daily rotation is a centripetal acceleration at each point on Earth. Estimate the magnitude of the resultant force acting on a typical adult person to provide this acceleration, if the person is standing at the equator. The equatorial radius of Earth is 6378 km.

6-56 A woman is standing without slipping in a subway car that is accelerating forward. She is not holding on to anything with her hands. Draw a free-body diagram for the woman in (a) the (inertial) frame of reference of Earth, and (b) the frame of reference of the subway car. (c) If the coefficient of static friction between the woman's shoes and the floor is 0.47, what is the maximum acceleration of the subway car if the woman is not to slip?

6-57 In an amusement park ride, people stand with their backs against the inside wall of a large cylindrical shell, which rotates in a horizontal circle (Figure 6-38).

(a) Relative to Earth (assumed inertial), each person is travelling in a circle. Draw a free-body diagram for one person, and identify the force that provides the centripetal acceleration. (Neglect friction between the person's feet and the floor.)

(b) Relative to the rotating noninertial frame of the ride, each person is stationary. Draw a free-body diagram for a person relative to this frame, and explain the lack of motion.

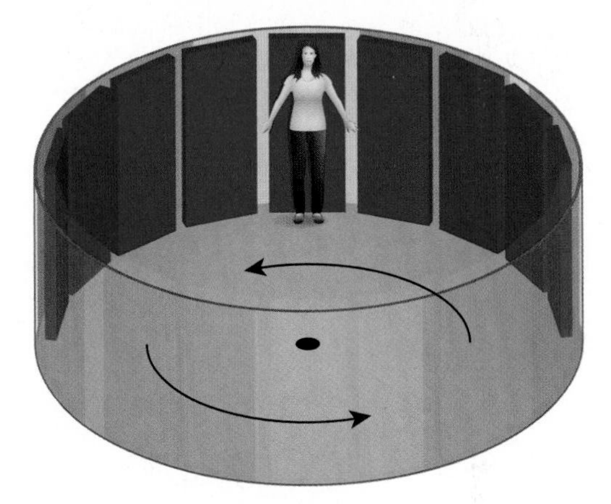

Figure 6-38 Question 6-57.

Work, Energy, and Power 7

Chiyacat/Shutterstock

Figure 7-1 Falling water, such as at Niagara Falls, contains a huge amount of energy. More than half of Canada's electricity is produced at hydroelectric plants where falling water spins electric turbines.

Although energy is hard to define precisely, it is familiar to all of us, and is probably the single most important "resource" that the world has (Figure 7-1). Without a continuous input of energy from the Sun, all life on Earth would cease to exist. Reliable sources of energy are required to operate most of the material objects associated with the high standard of living in North America—automobiles, televisions, computers, microwave ovens, airplanes, refrigerators, etc.—and large amounts of energy are used in their manufacture. An understanding of energy is necessary for a full appreciation of many major issues facing society today: use of finite resources (oil, gas, uranium, etc.), environmental pollution, climate change, and so on.

As we proceed through this chapter, we will see that there are various forms of energy, and that energy can be converted from one form into another. The most important

CHAPTER OUTLINE

concept in this chapter is the law of conservation of energy: a universal law that states that, although energy can be converted into different forms, it cannot be created or destroyed.

7.1 Work Done by a Constant Force

Many words have different meanings in different situations. For example, if a friend says to you, "Look at all the fans," the meaning of *fans* depends on whether you are in a soccer stadium or in an electrical appliance store. Similarly, the word *work* has a scientific meaning that is very different from its everyday meaning. If you hold a heavy box stationary for a long time, with your muscles aching and sweat dripping from your brow, most people would say that you are "working," but in the scientific sense you are doing no work at all. Before we begin discussions of energy, we need to introduce the closely related concept of work.

Suppose that you apply a constant horizontal force to push a shopping cart in a store (Figure 7-2). If the force has a magnitude F, and the displacement of the cart has a magnitude Δr, then the work (W) done by you on the cart is given by $W = F\,\Delta r$.

The force vector and the displacement vector are in the same direction in the above example. However, in general, the force and the displacement can be in different directions. If a lawnmower is pushed across level ground by a person applying a constant non-horizontal force (Figure 7-3), the work done by the person (or, we often say, by the force) on the lawnmower is

$$W = F\,\Delta r\,\cos\theta \qquad (7\text{-}1)$$

where

F = *magnitude* of the constant force (hence, $F \geq 0$)

Δr = *magnitude* of the displacement (hence, $\Delta r \geq 0$)

θ = angle between the vectors $\vec{F}$ and $\Delta\vec{r}$ (with these two vectors drawn with their tails touching)

work when a constant force $\vec{F}$ is applied to an object while the object undergoes a displacement $\Delta\vec{r}$, the work done on the object is given by $W = F\,\Delta r\,\cos\theta$, where θ is the angle between $\vec{F}$ and $\Delta\vec{r}$

joule the SI unit of work and energy, equivalent to a newton·metre (N·m) or kilogram·metre²/second² (kg·m²/s²)

Eqn.7-1 *defines* the **work** done by a constant force. Note that work is a *scalar* quantity: it has a magnitude, but no direction.

If $\vec{F}$ and $\Delta\vec{r}$ are in the same direction, the angle θ between them is zero. Hence, $\cos\theta = 1$ and we have $W = F\,\Delta r$ as before.

In SI units, F is in newtons, Δr in metres, and $\cos\theta$ is unitless. Therefore, we would expect that the unit of work would be newton·metre (N·m). However, this is conventionally called a **joule** (J), named after James Prescott Joule (1818–1889), an

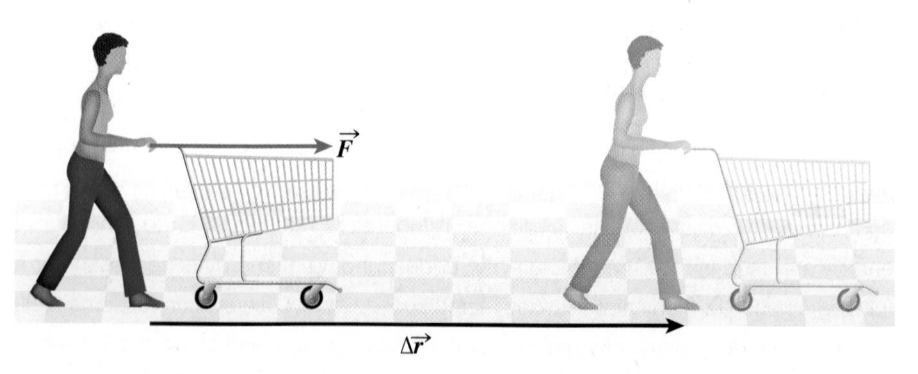

Figure 7-2 The work (W) done on an object when a force ($\vec{F}$) is applied and the object's displacement ($\Delta\vec{r}$) is in the same direction as the force is $W = F\Delta r$.

Figure 7-3 The work done on the lawnmower is $W = F\Delta r \cos\theta$, where θ is the angle between the constant force ($\vec{F}$) and the displacement ($\Delta\vec{r}$).

outstanding British experimental physicist who was one of the leaders in the development of the law of conservation of energy (Section 7.4).

We can see from the equation $W = F\,\Delta r\cos\theta$ that, in order for work to be done, three important conditions must be met:

1. A force must be applied to an object.
2. The object must undergo a displacement; that is, it must move.
3. The force and displacement must not be at right angles to each other. If they are, $\theta = 90°$, and therefore $\cos\theta = 0$ and $W = 0$.

The second and third conditions are discussed in more detail below.

Suppose that a lawnmower is pushed against a tree (Figure 7-4). If the person continues to push, with the lawnmower firmly at rest against the tree, the person is doing no work on the lawnmower because its displacement is zero. Notice how the scientific and everyday concepts of work are different. In normal conversation we would say that someone exerting a force on a stationary lawnmower is working, or doing work. However, in the scientific sense, no work is being done on the lawnmower.

Figure 7-4 No work is done on the lawnmower if it remains at rest ($\Delta\vec{r} = 0$).

When the lawnmower is being pushed across level ground, the force of gravity is not doing work on the lawnmower (Figure 7-5). Gravity is acting vertically downward and the lawnmower's displacement is horizontal. Since the force and displacement are perpendicular to each other, $\cos\theta = \cos 90° = 0$ and the work done is zero. Can you think of another force that does no work on the lawnmower in this situation because the force is perpendicular to the displacement?

In the equation $W = F\,\Delta r\cos\theta$, the quantities F and Δr represent magnitudes of vectors, and hence they are positive quantities (or possibly zero). However, $\cos\theta$ may

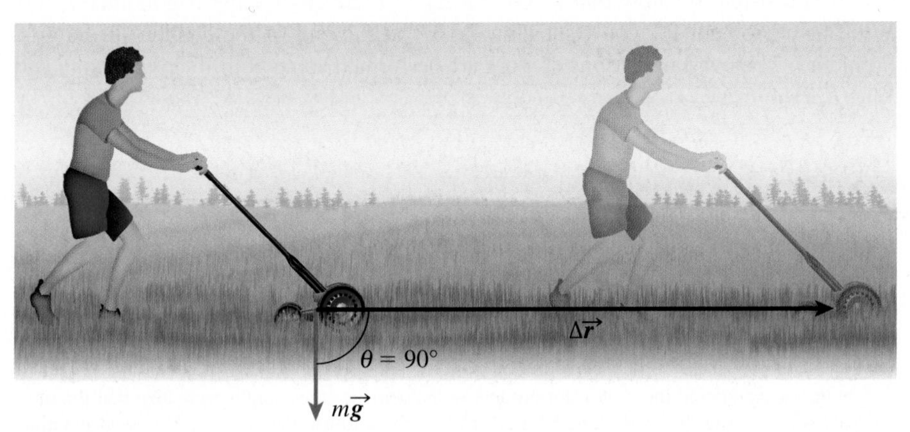

Figure 7-5 As the lawnmower moves across the ground, the force of gravity does no work on the lawnmower because $\theta = 90°$. Therefore, $\cos\theta = 0$ and $W = 0$.

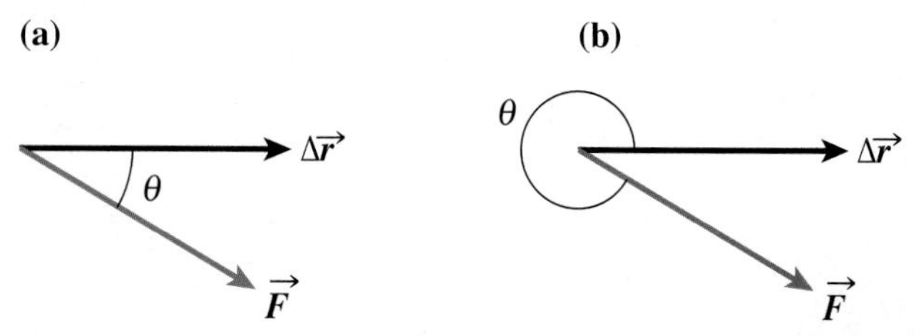

Figure 7-6 Either **(a)** the smaller angle or **(b)** the larger angle between $\vec{F}$ and $\Delta\vec{r}$ can be used as angle θ in calculating work using $W = F\Delta r \cos\theta$. Typically the smaller one is used.

be either positive or negative, depending on the value of θ. The angle θ may be chosen as either the larger or smaller angle between the vectors $\vec{F}$ and $\Delta\vec{r}$ (Figure 7-6) since $\cos(360° - \theta) = \cos\theta$. (You might like to use your calculator to confirm this for a few angles.) Normally the smaller angle is used.

Depending on the sign of $\cos\theta$, the work done on an object can be either positive or negative. As discussed further in Section 7.2, positive work indicates that the force is tending to increase the speed of the object. Negative work means that the force is tending to decrease the speed of the object on which the work is being done.

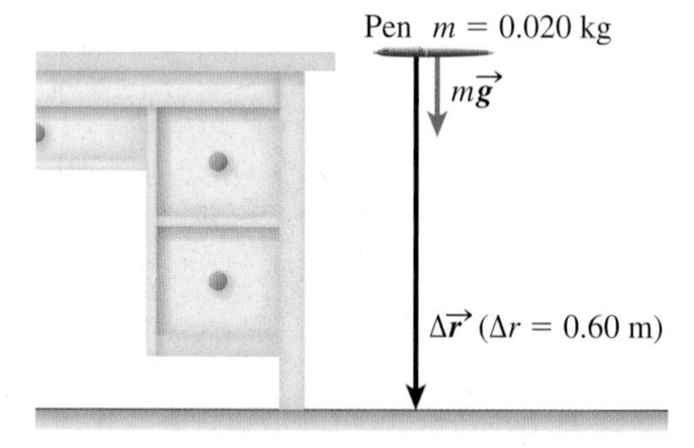

Figure 7-7 Sample Problem 7-1.

Units Tip

The units of mg are kilogram·metre/second^2, which is a newton (N). Multiplying this by the units of Δr (metre) gives newton·metre, equivalent to a joule (J). Since $\cos\theta$ is unitless, multiplication by $\cos\theta$ produces no change in units.

SAMPLE PROBLEM 7-1

A pen of mass 0.020 kg falls from a desktop to the floor. If the desktop is 0.60 m above the floor, what is the work done on the pen by gravity during the fall?

Solution

Figure 7-7 provides a diagram. Since the force of gravity ($m\vec{g}$) and the displacement ($\Delta\vec{r}$) are in the same direction, the angle (θ) between these two vectors is zero. To calculate the work done, we use Eqn. 7-1:

$$\begin{aligned} W &= F\,\Delta r\,\cos\theta \\ &= mg\,\Delta r\,\cos\theta \\ &= (0.020\,\text{kg})(9.8\,\text{m/s}^2)(0.60\,\text{m})(\cos 0°) \\ &= 0.12\,\text{J} \end{aligned}$$

Thus, the work done is 0.12 J.

Using Vector Components to Determine Work

There is an alternative way to define work (in two dimensions):

$$W = F_x\Delta x + F_y\Delta y \qquad (7\text{-}2)$$

where F_x, F_y, Δx, and Δy are the x- and y-components of the applied force and the displacement respectively. Each of these four quantities can be either positive or negative. It is possible to show that $F_x\Delta x + F_y\Delta y = F\,\Delta r\cos\theta$ using trigonometry; this is left as a problem for you (Question 7-107) at the end of the chapter. In Sample Problem 7-2 below, calculations of work are performed using both $W = F\,\Delta r\cos\theta$ and $W = F_x\Delta x + F_y\Delta y$.[1]

SAMPLE PROBLEM 7-2

A woman pushes a laser printer of mass 2.5 kg at constant velocity across a horizontal tabletop by exerting a force of 22 N at 25° below the horizontal. Determine the work done by each force acting on the printer if its displacement is 0.25 m in magnitude.

[1] If you have encountered the vector dot product in mathematics, you might recognize that the most compact way to write the work done by a constant force is to use the vector dot product notation, $W = \vec{F}\cdot\Delta\vec{r}$, but this notation will not be discussed further in this book. When solving problems, it is more convenient to use either $W = F\,\Delta r\cos\theta$ or $W = F_x\Delta x + F_y\Delta y$.

Solution using $W = F\,\Delta r \cos\theta$

We need first to identify the forces. Figures 7-8 (a) and (b) show a diagram for the problem and a free-body diagram for the printer. The forces are gravity $m\vec{g}$, normal force $\vec{F}_N$, kinetic friction $\vec{F}_K$, and the force $\vec{F}_W$ exerted by the woman.

It is now useful to check if any of the forces are at right angles to the displacement; if so, the work done by such a force is zero (from $W = F\,\Delta r \cos\theta$, with $\theta = 90°$). Figure 7-8 (c) shows the relative directions of the various forces and the displacement $\Delta\vec{r}$. Since both gravity and the normal force make a 90° angle with $\Delta\vec{r}$, then the work done by each of these forces is zero:

$$\text{work done by gravity: } W = mg\,\Delta r \cos 90° = 0\,\text{J} \text{ (since } \cos 90° = 0)$$

$$\text{work done by normal force: } W = F_N\,\Delta r \cos 90° = 0\,\text{J}$$

We now calculate the work done by $\vec{F}_W$, which makes an angle of 25° with $\Delta\vec{r}$:

$$\text{work done by force } \vec{F}_W\text{: } W = F_W\,\Delta r \cos\theta = (22\,\text{N})(0.25\,\text{m})\cos 25° = 5.0\,\text{J}$$

Before calculating the work done by friction $\vec{F}_K$, we need to determine the magnitude of this force, which is not given in the problem. This unknown force magnitude can be determined by considering horizontal forces and using Newton's first law (since the printer is not accelerating horizontally):

$$\sum F_x = ma_x = 0$$

On the left-hand side of this equation, we need the x-components of all the forces. The normal force and gravity have zero x-components. From Figure 7-8 (b), where the $+x$-direction has been chosen toward the right, the x-component of $\vec{F}_K$ is simply F_K, and the x-component of $\vec{F}_W$ is

$$F_{Wx} = -F_W \cos 25° = -(22\,\text{N}) \cos 25° = -20\,\text{N}$$

Substituting x-components of forces into Newton's first law,

$$F_{Wx} + F_K = 0$$

Thus, $(-20\,\text{N}) + F_K = 0$, which gives $F_K = 20\,\text{N}$.

We can now calculate the work done by the friction force $\vec{F}_K$, noting from Figure 7-8 (c) that the angle between $\vec{F}_K$ and $\Delta\vec{r}$ is 180°:

$$\text{work done by friction: } W = F_K\,\Delta r \cos 180° = (20\,\text{N})(0.25\,\text{m})(-1) = -5.0\,\text{J}$$

(The negative work done by $\vec{F}_K$ indicates that friction, if acting alone, would decrease the printer's speed, and the positive work done by $\vec{F}_W$ shows that this force alone would increase the speed. The total work done by all the forces is zero, showing that the speed of the printer is not changing. These statements will be justified formally in Section 7.2.)

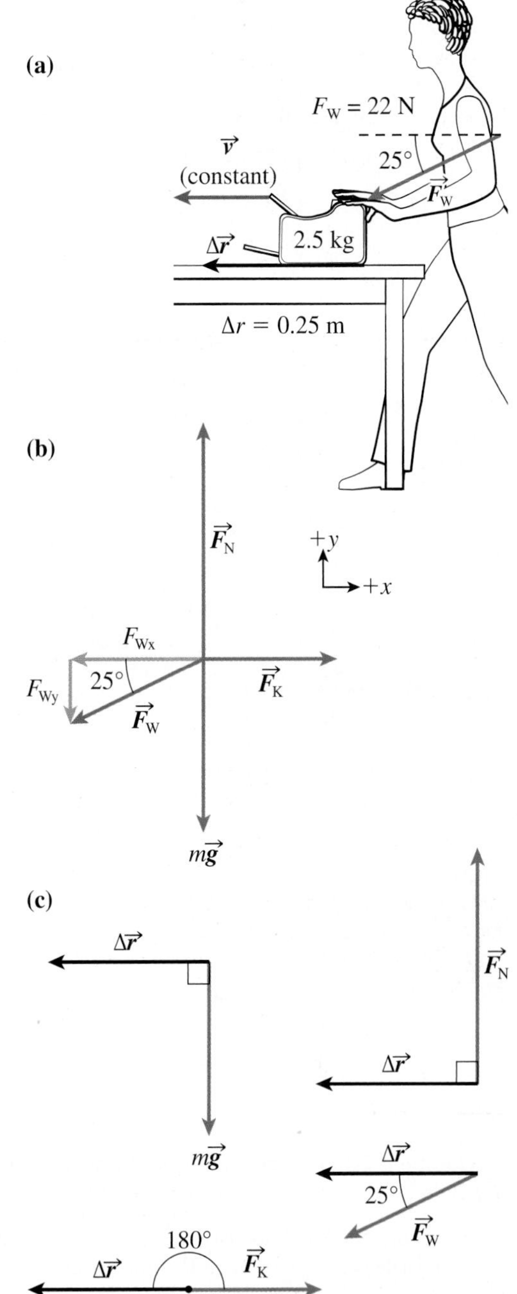

Figure 7-8 (**a**) Sample Problem 7-2. (**b**) Free-body diagram for the printer. (**c**) Angles between $\Delta\vec{r}$ and the four forces.

Alternative Solution using $W = F_x\Delta x + F_y\Delta y$

To use $W = F_x\Delta x + F_y\Delta y$, we need to determine the x- and y-components of the displacement and the forces. The x-y axes are shown in Figure 7-8 (b). The displacement vector $\Delta\vec{r}$ has components of $\Delta x = -0.25$ m and $\Delta y = 0$. Since Δy is zero, the product $F_y\Delta y$ will be zero for each force. Therefore, to calculate the work done by each force, we need to calculate only $F_x\Delta x$. The x-components of each force are

$$mg_x = 0\,\text{N} \qquad F_{Nx} = 0\,\text{N}$$

$$F_{Wx} = -20\,\text{N} \qquad F_{Kx} = 20\,\text{N}$$

(To see how F_{Wx} and F_{Kx} were determined, look at the above solution that uses $W = F\,\Delta r \cos\theta$.)

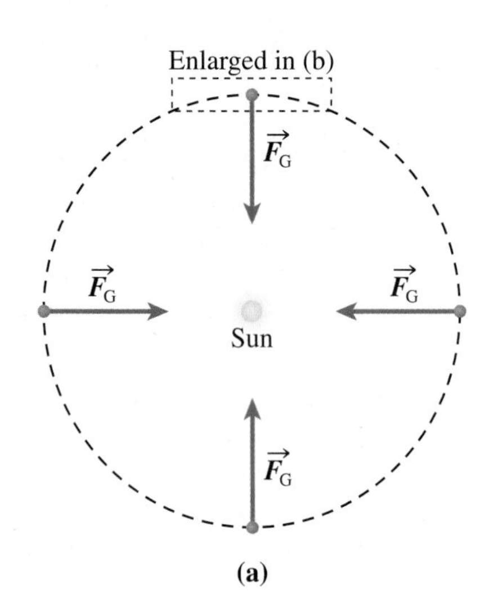

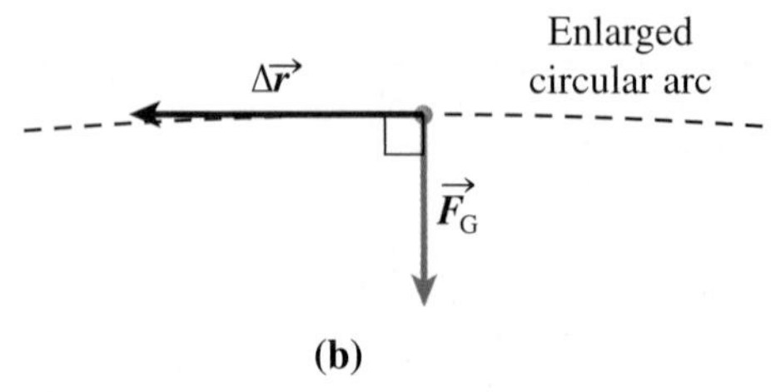

Figure 7-9 **(a)** As Venus moves in its orbit, which is almost a perfect circle, the gravitational force on it changes in direction. **(b)** For any small portion of the orbit, $\Delta\vec{r} \perp \vec{F}_G$; therefore, $\theta = 90°$, and $W = F_G\, \Delta r \cos\theta = 0$.

Calculating the work done by each force,

Gravity: $W = mg_x \Delta x = (0\,\text{N})(-0.25\,\text{m}) = 0\,\text{J}$

Normal: $W = F_{Nx} \Delta x = (0\,\text{N})(-0.25\,\text{m}) = 0\,\text{J}$

Force $\vec{F}_W$: $W = F_{Wx} \Delta x = (-20\,\text{N})(-0.25\,\text{m}) = 5.0\,\text{J}$

Friction: $W = F_{Kx} \Delta x = (20\,\text{N})(-0.25\,\text{m}) = -5.0\,\text{J}$

All the answers are of course the same as those calculated previously using

$$W = F\Delta r \cos\theta.$$

Work Done in Circular Motion

An important special case to consider is the work done by the force providing the centripetal acceleration of an object undergoing circular motion (or moving on a path that is a circular arc). Strictly speaking, the equation $W = F\, \Delta r \cos\theta$ is valid only if $\vec{F}$ is constant in both magnitude and direction. Nonetheless, we can use it to show that the work done by a force providing centripetal acceleration is zero, even though such a force is continually changing in direction.

As an example, we use the motion of Venus about the Sun. The orbit of Venus is very close to being a circle, and the force providing the centripetal acceleration in this situation is the gravitational force $\vec{F}_G$ on Venus; this force changes in direction as the planet moves (Figure 7-9 (a)). However, if we consider any small portion of the orbit, where $\Delta\vec{r}$ is very small (Figure 7-9 (b)), the force is essentially constant in magnitude and direction. Hence, the work done as Venus moves along this small portion can be determined using $W = F\, \Delta r \cos\theta$. The displacement $\Delta\vec{r}$ is perpendicular to the force; that is, $\theta = 90°$. Therefore, $\cos\theta = 0$, and the work $W = 0$. Since the work done is zero for any small piece of the orbit, then the work done is zero over the entire orbit (which can be considered to be a series of small pieces). This analysis can be applied to any force providing a centripetal acceleration, and so we can state that *the work done by any force providing the centripetal acceleration of an object in circular motion is zero.*[2]

EXERCISES

In Questions 7-1 to 7-9 below, is work being done on object A by force $\vec{F}$? If yes, is the work positive or negative?

7-1 A forearm A is stationary in a horizontal orientation. The biceps muscle is exerting a force $\vec{F}$ vertically upward on the forearm.

7-2 A woman, walking at constant velocity, carries a golf bag A by exerting an upward force $\vec{F}$ on it (Figure 7-10). The golf bag is moving horizontally as the woman walks.

bikeriderlondon/Shutterstock

Figure 7-10 Question 7-2.

7-3 A woman exerts an upward force $\vec{F}$ on a briefcase A to lift it to her desk.

7-4 A man exerts an upward force $\vec{F}$ on a backpack A as he lowers it at constant velocity toward the floor.

7-5 A baby stroller A is pushed along a horizontal sidewalk by a man who exerts a horizontal force $\vec{F}$.

7-6 The force of gravity $\vec{F}$ acts on a baseball A moving vertically upward.

7-7 A normal force $\vec{F}$ is exerted by a road on a car A accelerating away from an intersection.

7-8 An electron A travels in a circle as a result of a magnetic force $\vec{F}$ toward the centre of the circle.

7-9 A string tension $\vec{F}$ acts on a ball A attached to the string. The ball is swinging as a pendulum.

7-10 A force is applied to an object, which is undergoing a displacement. If the work done on the object by the force is zero, what can you conclude?

[2] In Chapter 10 we will see that if an object is travelling in a circle and a force is applied *tangentially* to the circle, then the work done by this tangential force is not zero, and the speed of the object changes.

7-11 A magnetic force on a moving charged particle is always perpendicular to the particle's velocity. What can you conclude about the work done on a charged particle by a magnetic force? (**Hint:** For a small time interval, how are the directions of the velocity and the displacement related?)

7-12 The work done by a force providing centripetal acceleration is always ___.

7-13 Express a joule in terms of SI base units: kilogram (kg), metre (m), and second (s).

7-14 A soccer ball of mass 0.425 kg is kicked straight up, and it reaches a maximum height of 11.8 m.

(a) What is the work done on the ball by gravity on the way up?

(b) On the way down?

(c) What is the total work done on the ball by gravity during the entire flight?

7-15 A soccer ball of mass 0.425 kg is kicked on a parabolic trajectory such that its maximum height is 11.8 m. What is the work done on the ball by gravity on the way up? on the way down? (**Hint:** Use $W = F_x\Delta x + F_y\Delta y$.)

7-16 While pushing a shopping cart down the aisle in a store, a person exerted a constant horizontal force of magnitude 2.7 N and did 17.0 J of work on the cart. How far did the person push the cart?

7-17 As the shopping cart in the previous question was pushed along the aisle, what was the work done on the cart by (a) the normal force and (b) the force of gravity?

7-18 A boy carries a hamburger of mass 0.25 kg on a tray at a constant velocity of 0.45 m/s horizontally, for a horizontal displacement of magnitude 6.7 m. What is the total work done on the hamburger?

7-19 (a) The weightlifter shown in Figure 7-11 is lifting a mass of 115 kg. What magnitude of force is required to lift this mass at a constant velocity of 0.330 m/s upward?

(b) As the mass is lifted a distance of 0.64 m, what is the work done by the upward force? by gravity?

Figure 7-11 Question 7-19.

7-20 A woman pushes a lawnmower by exerting a constant force of magnitude 9.30 N. If she does 87 J of work on the lawnmower while pushing it 11 m across level ground, what is the angle between the horizontal and the force that she applies to the mower?

7-21 A girl pulls a sleigh of mass 4.81 kg up a hill inclined at 12.7° to the horizontal (Figure 7-12). The vertical height of the hill is 27.3 m. Neglecting friction between the sleigh and the snow, determine how much work the girl must do on the sleigh to pull it at constant velocity up the hill.

Figure 7-12 Question 7-21.

7-22 Repeat the preceding question, if the vertical height is still 27.3 m but the angle is 14.1°. What general conclusion does this suggest?

7-23 The girl in the previous two questions now slides down the hill on the sleigh. If the girl's mass is 25.6 kg, determine the total work done on the girl and sleigh during the slide. Neglect friction.

7.2 Kinetic Energy and the Work-Energy Theorem

Think about a golfer hitting a golf ball. The club exerts a force on the ball and the ball moves parallel to the force; hence, the club does work on the ball. The speed of the ball increases and we show in this section that there is a close connection between the total work done on an object (such as the golf ball) and a special quantity that depends on the object's mass and speed. This quantity is the object's **kinetic energy**.

kinetic energy energy of motion, equal to ½ mv^2, where m and v represent an object's mass and speed respectively

What is kinetic energy? In words, we describe it as "energy of motion." We shall see that we write kinetic energy mathematically as $\frac{1}{2}mv^2$, where m and v represent an object's mass and speed respectively. The connection between work and kinetic energy is known as the **work-energy theorem**, which states that the total work done on an object by all the forces acting on it equals the change in the object's kinetic energy.

work-energy theorem the total work done on an object by all the forces acting on it equals the change in the object's kinetic energy

We first prove the work-energy theorem for a one-dimensional constant-force situation. Consider the total work done by all the (constant) forces acting on an object,

$$W_{TOT} = \Sigma(F_x\Delta x)$$

where the subscript "TOT" is an abbreviation for "total," and "Σ" indicates a sum including all the forces. Since the displacement Δx of an object in a given case is the same for all the forces, we can remove Δx from the summation:

$$W_{TOT} = (\Sigma F_x)\Delta x$$

Now $\Sigma F_x = ma_x$ from Newton's second law.

Thus, $$W_{TOT} = ma_x\,\Delta x$$

For constant forces, the resulting a_x is constant, and we can relate Δx to the initial and final velocities by using Eqn. 2-10 for constant-acceleration kinematics:

$$v_x^2 - v_{0x}^2 = 2a_x(x - x_0)$$

The difference between the final and initial positions, $x - x_0$, is just the displacement, Δx. Hence, the previous equation can be written as

$$v_x^2 - v_{0x}^2 = 2a_x\Delta x$$

Therefore, $\Delta x = \dfrac{v_x^2 - v_{0x}^2}{2a_x}$

Substituting for Δx in the expression for W_{TOT} gives

$$W_{TOT} = ma_x\left(\frac{v_x^2 - v_{0x}^2}{2a_x}\right)$$

$$= \tfrac{1}{2}mv_x^2 - \tfrac{1}{2}mv_{0x}^2$$

The quantity $\frac{1}{2}mv_x^2$ is the kinetic energy of the object in this one-dimensional situation.

Thus, $\frac{1}{2}mv_x^2 - \frac{1}{2}mv_{0x}^2$ is the change in the object's kinetic energy.

We have just proven the work-energy theorem:

The total work done on an object equals the change in the object's kinetic energy.

It is important to note that the change in kinetic energy equals the total work done by *all* the forces, or equivalently, the work done by the *resultant* force (ΣF_x). If we consider the work done by a single force (out of many acting), we can say that if the work done is positive, then the force is tending to increase the object's kinetic energy; if the work done is negative, the force tends to decrease the kinetic energy. Of course, an increase in kinetic energy means also an increase in speed. Similarly, a decrease in kinetic energy means a decrease in speed.

Although the theorem has been proven only for one-dimensional constant forces, it is valid also in two and three dimensions, and for varying forces. Thus, we can write the theorem in a more general form without the x-subscripts:

$$W_{TOT} = \tfrac{1}{2}mv^2 - \tfrac{1}{2}mv_0^2 \quad \text{(work-energy theorem)} \qquad \textbf{(7-3)}$$

where v and v_0 represent the object's final and initial speeds respectively.

In this book, we will represent the kinetic energy of an object by E_K:

$$E_K = \tfrac{1}{2}mv^2 \quad \text{(kinetic energy)} \qquad \textbf{(7-4)}$$

Thus, the work-energy theorem can be written most compactly as

$$W_{TOT} = \Delta E_K \quad \text{(work-energy theorem)} \qquad \textbf{(7-5)}$$

Note that kinetic energy is a scalar quantity, as is work.

It is useful at this point to verify that the units of work are the same as the units of kinetic energy. To do this, we determine the units of work and kinetic energy in terms of SI base units: kilogram, metre, and second.

Units of work: The fundamental SI unit of work is the joule (J), which is equivalent to a newton·metre (N·m). By remembering Newton's second law ($\Sigma\vec{F} = m\vec{a}$), we can rewrite N as kg·m/s^2. Thus N·m is (kg·m/s^2)·m, that is, kg·m^2/s^2.

Units of kinetic energy: Consider the expression $\frac{1}{2}mv^2$ for kinetic energy. The "$\frac{1}{2}$" is unitless; the unit for mass m is kg; for v^2, (m/s)^2. Therefore, the basic SI unit for kinetic energy is kg·m^2/s^2, the same as the unit that we determined for work. Neither kg·m^2/s^2 nor N·m is commonly used, being replaced instead by J (joule).

Notice that kinetic energy depends on the *square* of an object's speed. This means that relatively modest increases in the speed of an object can lead to rather large increases in its kinetic energy. Traffic accidents involving vehicles moving at high speeds are much more destructive than accidents at lower speeds because of the much higher kinetic energies involved. As an example, consider a car of mass 1.2×10^3 kg moving at a speed of 25 m/s (90 km/h). The car's kinetic energy is

$$E_K = \frac{1}{2}mv^2 = \frac{1}{2}(1.2 \times 10^3\,\text{kg})(25\,\text{m/s})^2 = 3.8 \times 10^5\,\text{J}$$

Suppose now that the car increases its speed by 20%, to 30 m/s (108 km/h). Calculation of the new kinetic energy gives 5.4×10^5 J. This represents a 44% increase in the kinetic energy, from only a 20% increase in the speed! In an accident, all this energy must be dissipated somehow, much of it in deformation of the car and its occupants. You might have heard the phrase "Speed kills," but perhaps a better one would be "Kinetic energy kills."

DID YOU KNOW?

Cars are designed with "crumple zones" which deform during a collision so that much of the kinetic energy of the car(s) is dissipated in producing the deformation (Figure 7-13). This lessens the possibility of serious injuries to the driver and passengers. The most common place for crumple zones is at the front of the vehicle, but sometimes they are in other locations as well.

Figure 7-13 The crumpling of the front of this car during a collision absorbed a large amount of kinetic energy, thus helping to reduce injuries of the occupants.

SAMPLE PROBLEM 7-3

A baseball of mass 0.15 kg is travelling with a velocity of 36 m/s horizontally just before it is caught in a glove.

(a) What is the kinetic energy of the ball before being caught?

(b) During the catch, what is the change in the ball's kinetic energy?

(c) The ball moves 0.047 m horizontally during the catch. Use work and energy to determine the magnitude and direction of the force (assumed constant) exerted on the ball by the glove.

Solution

(a) Figure 7-14 shows a diagram for this problem. The initial kinetic energy of the ball is

$$E_K = \frac{1}{2}mv_0^2 = \frac{1}{2}(0.15\,\text{kg})(36\,\text{m/s})^2 = 97\,\text{J}$$

(b) As a result of the catch, the ball comes to rest and its kinetic energy decreases to 0 J. The change in kinetic energy is

$$\Delta E_K = \frac{1}{2}mv^2 - \frac{1}{2}mv_0^2 = (0 - 97)\,\text{J} = -97\,\text{J}$$

Notice that ΔE_K is always the *final* kinetic energy minus the *initial* kinetic energy. The negative value for ΔE_K indicates that the kinetic energy has decreased (by 97 J in this problem).

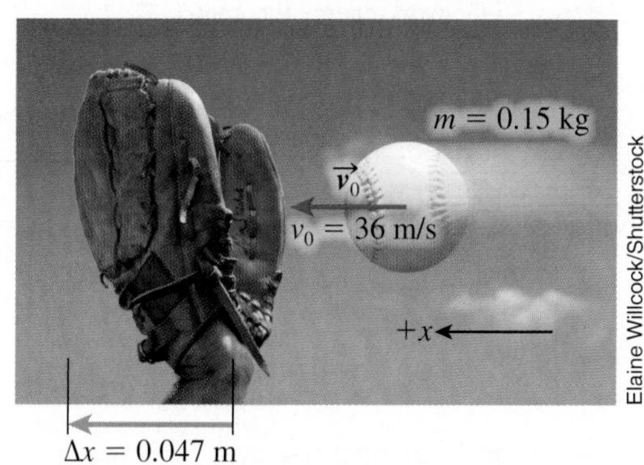

Figure 7-14 Sample Problem 7-3.

(c) We can use the work-energy theorem to find the force exerted by the glove, which is the only horizontal force:

$$W_{TOT} = \Delta E_K$$

$$F_x \Delta x = \Delta E_K$$

$$F_x = \frac{\Delta E_K}{\Delta x} = \frac{-97\,\text{J}}{0.047\,\text{m}} = -2.1 \times 10^3\,\text{N}$$

Hence, the force is 2.1×10^3 N in the $-x$-direction, that is, in the direction opposite to the ball's original velocity.

Problem-Solving Tip

Always label the $+x$-direction (and $+y$ if needed) on your diagrams. Notice in Figure 7-14 that the $+x$-direction has been chosen to be in the direction of the original velocity of the ball.

EXERCISES

7-24 Is kinetic energy a scalar or vector quantity?

7-25 (a) If the speed of a cyclist doubles, by what factor does the cyclist's kinetic energy increase?
(b) Repeat, if the cyclist's speed triples.

7-26 What is the kinetic energy of a car of mass 1.50×10^3 kg that is moving with a velocity of 18.0 m/s eastward?

7-27 (a) If the speed of the car in the previous question increases by 15.0%, what is the car's new kinetic energy?
(b) By what percentage has the car's kinetic energy increased?
(c) How much work was done on the car to increase its kinetic energy?

7-28 A sprinter of mass 55 kg has a kinetic energy of 3.3×10^3 J. What is her speed?

7-29 A red blood cell moving with a speed of 3.2 cm/s in an artery has a kinetic energy of 6.1×10^{-17} J. What is the mass of the red blood cell?

7-30 A plate of mass 0.353 kg falls (essentially from rest) from a table to the floor 89.3 cm below.
(a) What is the work done by gravity on the plate during the fall?
(b) Use the work-energy theorem to determine the speed of the plate just before it hit the floor.

7-31 An executive throws her pen straight up into the air after signing a lucrative contract. If the pen has a mass of 0.090 kg and an initial velocity of 5.6 m/s upward,
(a) what is the work done by gravity on the pen as it travels 0.75 m upward?
(b) what is the pen's speed after travelling 0.75 m upward? (Use the work-energy theorem.)

7-32 A nurse is injecting a child with 3.0×10^{-3} kg of vaccine (Figure 7-15). With what magnitude of force (assumed constant) must the nurse push on the plunger of the syringe, if the speed of the vaccine coming out is 1.4 m/s and the length of the plunger's motion is 3.4 cm? Neglect friction.

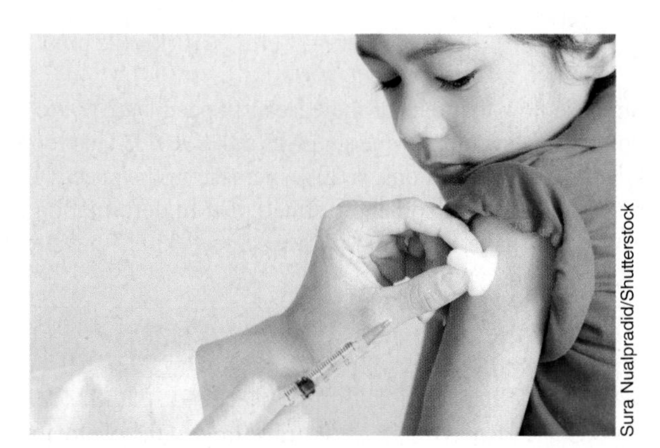
Sura Nualpradid/Shutterstock

Figure 7-15 Question 7-32.

7-33 A girl pulls a large box of mass 20.8 kg across a horizontal floor. She is exerting a force on the box of 95.6 N inclined at 35.0° above the horizontal. The kinetic friction force on the box has a magnitude of 75.5 N. Use the work-energy theorem to determine the speed of the box after being dragged 0.750 m, assuming that it starts from rest.

7-34 **Fermi Question:** Estimate the kinetic energy of a world-class female sprinter moving at top speed.

7.3 Gravitational Potential Energy

We have seen that an object has kinetic energy when it is moving. However, even when something is at rest, it can have energy. To illustrate this, suppose that you hold a ball in your hand and then drop it. Gravity pulls it down toward Earth's surface, increasing the ball's kinetic energy. When you were holding the ball in your hand, the ball had the *potential* to develop kinetic energy; the ball needed only to be released. We say that the ball had **gravitational potential energy**, which we define in words as energy due to elevation. A mathematical expression for gravitational potential energy is developed below.

gravitational potential energy energy due to elevation; $E_P = mgy$ close to Earth's surface

Figure 7-16 shows a ball falling from rest. Notice the vertical coordinate system, with $+y$ being upward. The initial position of the ball is represented by "y," and its final position by $y = 0$. We will use the work-energy theorem to equate the work done by gravity to the increase in the ball's kinetic energy during the fall. The initial kinetic energy is zero, and the final kinetic energy is $\frac{1}{2}mv^2$.

$$W_{\mathrm{TOT}} = \Delta E_{\mathrm{K}}$$

$$F_y \Delta y = \tfrac{1}{2} mv^2 - 0$$

Now, $F_y = -mg$, and $\Delta y = 0 - y = -y$. Therefore $(-mg)(-y) = \frac{1}{2} mv^2$, which gives $mgy = \frac{1}{2} mv^2$. Hence, the work done by gravity is mgy, and this equals the increase in the kinetic energy of the ball.

We now interpret the quantity mgy in another way. The gravitational potential energy of an object of mass m at an elevation y is *defined* to be mgy. This is the gravitational potential energy *relative* to the position where $y = 0$. (Sample Problem 7-4 below will provide further clarification on this point.) We represent gravitational potential energy by E_{P} ("P" for potential). The SI unit for gravitational potential energy, and for all types of energy, is the joule (J).

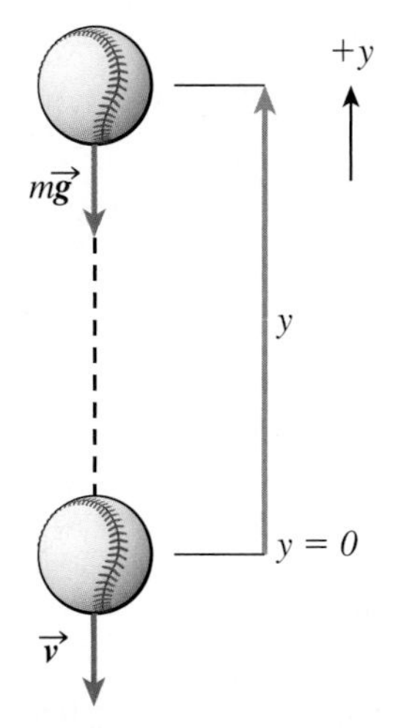

Figure 7-16 A ball falling from rest.

$$E_{\mathrm{P}} = mgy \text{ (gravitational potential energy)} \qquad (7\text{-}6)$$

Notice that gravitational potential energy depends on vertical position (y), but has no dependence on horizontal position.

The equation that we developed above, $mgy = \frac{1}{2} mv^2$, can now be given the following interpretation. Prior to falling, the ball had maximum gravitational potential energy (mgy) and no kinetic energy (since it was at rest). When the ball falls to $y = 0$, its gravitational potential energy is converted into an equal amount of kinetic energy ($\frac{1}{2} mv^2$). At the lowest point ($y = 0$), the ball has no gravitational potential energy but maximum kinetic energy.

Eqn. 7-6 for gravitational potential energy is valid only for small elevations (y) above the surface of Earth, that is, for elevations over which the magnitude of the gravitational acceleration (g) is approximately constant. In Chapter 9, a more general expression for gravitational potential energy will be introduced. As we will see in Chapter 9, the more general expression is consistent with the simplified expression given in Eqn. 7-6 for an object near Earth's surface, so we will focus on Eqn. 7-6 for the remainder of this chapter.

Notice that gravitational potential energy depends on mass. Anyone who has ever tried to hammer a large nail or spike into a piece of wood knows that the job is easier if a large hammer is used. This is simply because a large hammer has more mass and thus more gravitational potential energy when lifted; therefore, it is able to deliver more energy to the nail. Similarly, a heavy axe is much easier to use than a small hatchet for splitting wood, or in the case of the firefighter shown in Figure 7-17, breaking through the roof of a house on fire.

Patricia Marks/Shutterstock

Figure 7-17 Because the head of this firefighter's axe has a large mass, it has a large gravitational potential energy that makes it more useful than a small hatchet in providing energy to break through a roof of a house on fire.

SAMPLE PROBLEM 7-4

A ruler of mass 0.12 kg falls from a table to the floor 0.60 m below.

(a) Relative to the floor, what is the ruler's gravitational potential energy when on the table? on the floor?

(b) Relative to the table, what is the ruler's gravitational potential energy when on the table? on the floor?

(c) What is the change in the ruler's gravitational potential energy during the fall, relative to the floor? relative to the table?

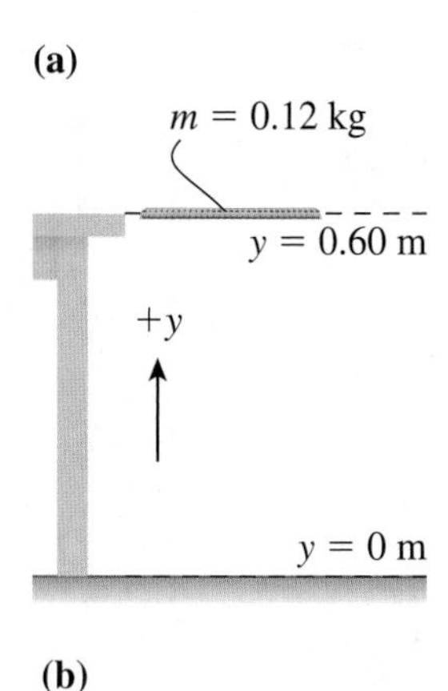

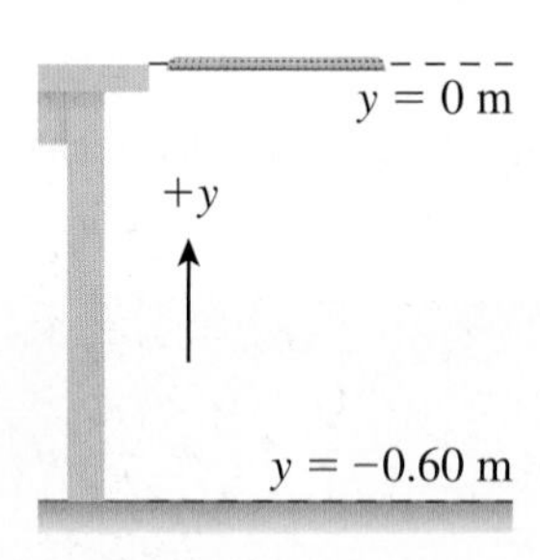

Figure 7-18 (**a**) Sample Problem 7-4, part (a). (**b**) Sample Problem 7-4, part (b).

Units Tip

Notice the units in the calculation of mgy. We have (kg)(m/s^2)(m) = kg·m^2/s^2, which is equivalent to a joule (J).

Solution

(a) A diagram is given in Figure 7-18 (a). Since the question asks for gravitational potential energy relative to the floor, we have chosen the $y = 0$ position to be at the floor. *The* $+y$*-direction is upward, as it must be whenever we use* $E_P = mgy$. When the ruler is on the table, it has an elevation of $y = 0.60$ m, and its gravitational potential energy is

$$\begin{aligned} E_P &= mgy \\ &= (0.12\,\text{kg})(9.8\,\text{m/s}^2)(0.60\,\text{m}) \\ &= 0.71\,\text{J (relative to the floor)} \end{aligned}$$

When the ruler is on the floor, $y = 0$, and therefore its gravitational potential energy is 0 J (relative to the floor).

(b) Since we are now asked for the gravitational potential energy relative to the table, we choose $y = 0$ to be at the level of the table (Figure 7-18 (b)). Therefore, when the ruler is on the table, its gravitational potential energy is 0 J (relative to the table). On the floor, since $+y$ must still be upward, $y = -0.60$ m. (Notice the negative sign.) Thus, the ruler's gravitational potential energy on the floor is

$$E_P = mgy = (0.12\,\text{kg})(9.8\,\text{m/s}^2)(-0.60\,\text{m}) = -0.71\,\text{J} \quad \text{(relative to the table)}$$

(c) To determine the change in the ruler's gravitational potential energy, we subtract the initial value from the final value.
Relative to the *floor*, this change is

$$\Delta E_P = (0 - 0.71)\,\text{J} = -0.71\,\text{J}$$

Relative to the *table*, the change is

$$\Delta E_P = (-0.71 - 0)\,\text{J} = -0.71\,\text{J}$$

Thus, the change in gravitational potential energy is -0.71 J, regardless of our choice of position for $y = 0$.

Choosing the $y = 0$ Position

A few comments should be made here about the selection of the $y = 0$ position. Gravitational potential energy is a relative quantity, always measured with respect to some arbitrary position where $y = 0$. In the sample problem above, the wording of the questions essentially told us which positions to choose for $y = 0$, but as this chapter continues, you will often have to make your own choice of $y = 0$ position in problem solutions. In most cases, it does not matter where $y = 0$ is chosen, but *it is usually easiest to choose* $y = 0$ *at the lowest possible position in the diagram* (e.g., at floor level in the preceding sample problem). This ensures that all y-values and all gravitational potential energies will be positive (or zero) and will be easy to work with.

EXERCISES

7-35 Use $E_P = mgy$ to write the SI unit of gravitational potential energy in terms of the base units of kilogram (kg), metre (m), and second (s). Similarly, determine the SI unit of kinetic energy. Are the units the same? What is the conventional SI unit used for all types of energy?

7-36 A girl of mass 55 kg climbs a set of stairs with a total vertical rise of 3.4 m. Relative to the bottom of the stairs, what is her gravitational potential energy at the top?

7-37 An orange of mass 125 g falls from a branch to the ground 3.50 m below.

(a) Relative to the ground, what is the gravitational potential energy of the orange on the branch? on the ground?

(b) Relative to the branch, what is the gravitational potential energy of the orange on the branch? on the ground?

(c) During the orange's fall, what is its change in gravitational potential energy relative to the ground? relative to the branch?

7-38 After being hit by a bat, a baseball of mass 0.15 kg reaches a maximum height where its gravitational potential energy has increased by 22 J from the point where the ball was hit. What is the ball's maximum height above the point where it was hit?

7-39 A figure skater lifts another figure skater (mass 56 kg) above his head (Figure 7-19). The height that the skater is lifted is 1.7 m.

(a) How much work is done by gravity on the skater who is lifted?

(b) Assuming that the lifting is done at constant speed, how much work is done by the skater doing the lifting?

(c) By how much does the gravitational potential energy of the lifted skater increase?

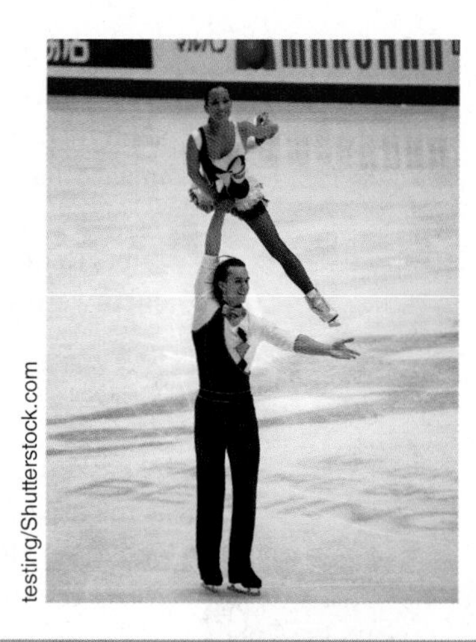

testing/Shutterstock.com

Figure 7-19 Question 7-39.

7.4 Law of Conservation of Energy

What does a person who is running have in common with a television? What does the production of starlight have in common with waterfalls? What does a nuclear power station have in common with a moving car? They all involve the conversion of one form of energy into other forms, and therefore are all governed by an extremely important and pervasive law of nature: the **law of conservation of energy**. As far as we know, this law cannot be violated, and it is one of the fundamental principles at work in the universe. It states

law of conservation of energy energy can be converted into different forms, but it cannot be created or destroyed

Energy can be converted into different forms, but it cannot be created or destroyed.

As an example, consider a falling ball. As already shown in Section 7.3, as the ball falls, its gravitational potential energy decreases and its kinetic energy increases by an equal amount. Thus, as the gravitational potential energy is converted into kinetic energy, the total amount of energy is unchanged. Figure 7-20 shows the various energies of a falling ball of mass 0.204 kg at a few points during its fall. Notice that, although both the ball's gravitational potential energy and kinetic energy change, the total energy (i.e., the sum of gravitational potential energy and kinetic energy) remains constant.

	E_P (J)	E_K (J)	E_P (J) + E_K (J)
Ball — $y = 3.0$ m	6.0	0.0	6.0
— $y = 2.0$ m	4.0	2.0	6.0
— $y = 1.0$ m	2.0	4.0	6.0
— $y = 0.0$ m	0.0	6.0	6.0

Figure 7-20 Gravitational potential energy, kinetic energy, and total energy of a falling ball.

So far we have discussed only two forms of energy, but there are many other types—thermal energy, electrical energy, sound energy, and so on. In this book, we will be applying conservation of energy to many different physical situations involving various forms of energy. Conservation of energy applies to all processes involving energy—conversion of food energy into kinetic energy of your body, production of sunlight, commercial generation of electrical energy, conversion of chemical energy in gasoline to kinetic energy of an automobile, interactions of subatomic particles, etc.

Note that "conservation of energy" differs from "energy conservation," which refers to the careful use of energy resources. Conservation of energy is a law of nature; energy conservation is something that we should all practise.

SAMPLE PROBLEM 7-5

A basketball player makes a free-throw shot (Figure 7-21) at the basket. As the ball is released from the player's hand, the ball's speed is 7.2 m/s and its height is 2.21 m above the floor. What is the ball's speed as it goes through the hoop, which is 3.05 m above the floor? Neglect air resistance.

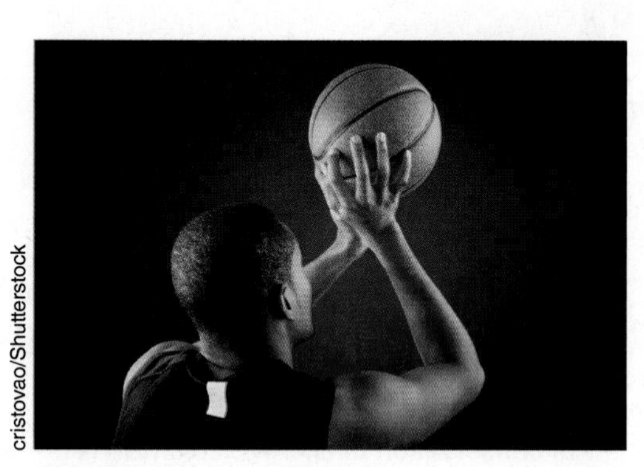

Figure 7-21 Sample Problem 7-5.

Solution

A diagram is given in Figure 7-22. By the law of conservation of energy, the total energy of the ball is constant as it travels through the air. In this problem, the two forms of energy involved are kinetic energy and gravitational potential energy. Referring to the release point as position 1, and to the point where the ball goes through the hoop as position 2, we can write

$$\text{Total energy at } 1 = \text{Total energy at } 2$$

$$\frac{1}{2} mv_1^2 + mgy_1 = \frac{1}{2} mv_2^2 + mgy_2$$

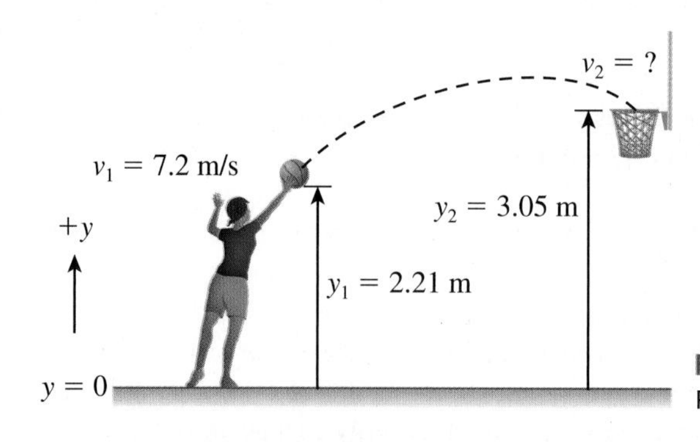

Figure 7-22 Drawing for Sample Problem 7-5.

Notice that the details of the trajectory are unimportant in this equation for conservation of energy. The direction of the initial velocity and the particular parabolic path that the ball follows do not affect the kinetic and potential energies, which are scalar quantities that do not depend on direction.

Note also in Figure 7-22 that we have chosen the $y = 0$ position to be at the lowest position of interest in the diagram, in this case the floor. This choice ensures that all gravitational potential energies in this problem are positive quantities.

We are asked to determine v_2. Dividing the conservation-of-energy equation by m and moving the gy_2 term to the left-hand side,

$$\frac{1}{2} v_1^2 + g(y_1 - y_2) = \frac{1}{2} v_2^2$$

Multiplying by 2,

$$v_1^2 + 2g(y_1 - y_2) = v_2^2$$

Solving for v_2,

$$\begin{aligned} v_2 &= \sqrt{v_1^2 + 2g(y_1 - y_2)} \\ &= \sqrt{(7.2\,\text{m/s})^2 + 2(9.8\,\text{m/s}^2)(2.21\,\text{m} - 3.05\,\text{m})} \\ &= 5.9\,\text{m/s} \end{aligned}$$

Thus, the speed of the ball as it passes through the hoop is 5.9 m/s. Notice that this is less than the speed when the ball was released: because the ball is at a higher elevation when it passes through the hoop, its gravitational potential energy is greater, meaning that its kinetic energy and speed must be less.

Mechanical Energy

Often the sum of kinetic energy and potential energy is called **mechanical energy**. Notice that we did not specify *gravitational* potential energy. Another kind of potential energy that can be included in mechanical energy is elastic potential energy, which is discussed later in this chapter.

mechanical energy the sum of kinetic energy, gravitational potential energy, and elastic potential energy

EXERCISES

7-40 Why do rollercoaster rides always start by going up a hill?

7-41 An egg is cracked and the contents (mass 0.052 kg) are dropped from rest 11 cm above a frying pan. Choose $y = 0$ at the position of the frying pan. Determine

(a) the initial gravitational potential energy of the egg contents

(b) the final gravitational potential energy of the egg contents

(c) the change in gravitational potential energy as the egg falls

(d) the kinetic energy and speed of the egg contents just before hitting the pan

7-42 A stone (mass 50 g) is dropped from rest from a roof.

(a) Neglecting air resistance, what is the stone's speed when it has fallen 15 m? Use conservation of energy.

(b) If air resistance is included, would your answer to (a) increase, decrease, or remain the same?

7-43 Two skiers start from rest at the same point at the top of a slope. They take different routes to the bottom, but end at the same point. Neglecting friction and air resistance, how do their final speeds compare?

7-44 A grasshopper (Figure 7-23) of mass 0.024 kg jumps vertically upward with an initial speed of 2.9 m/s. Neglecting air resistance, use conservation of energy to determine

(a) its speed on the way up, 0.20 m above the takeoff point

(b) its maximum height

(c) its speed on the way down, 0.20 m above the takeoff point

Figure 7-23 Question 7-44.

7-45 A woman throws a ball, which hits a vertical wall at a height of 1.2 m above the ball's release point. Just as it hits the wall, the ball's speed is 9.9 m/s. With what speed did the ball leave the woman's hand? Use conservation of energy and neglect air resistance.

7-46 A river is flowing with a speed of 3.74 m/s just upstream from a waterfall of vertical height 8.74 m. During each second, 7.12×10^4 kg of water pass over the fall.

(a) Relative to the bottom of the fall, what is the gravitational potential energy of this mass of water?

(b) If there is complete conversion of gravitational potential energy to kinetic energy, what is the speed of the water at the bottom of the fall?

7-47 A stick is thrown from a cliff 27 m high with an initial velocity of 18 m/s at 37° above the horizontal.

(a) Use conservation of energy to determine the speed of the stick just before it hits the ground below the cliff. Neglect air resistance.

(b) Repeat using the angle 37° below the horizontal.

7-48 A small ice cube is released from rest at the top edge of a hemispherical bowl (Figure 7-24). When it reaches the bottom of the bowl, its speed is 1.5 m/s. Use conservation of energy to find the radius of the bowl in centimetres. Neglect friction.

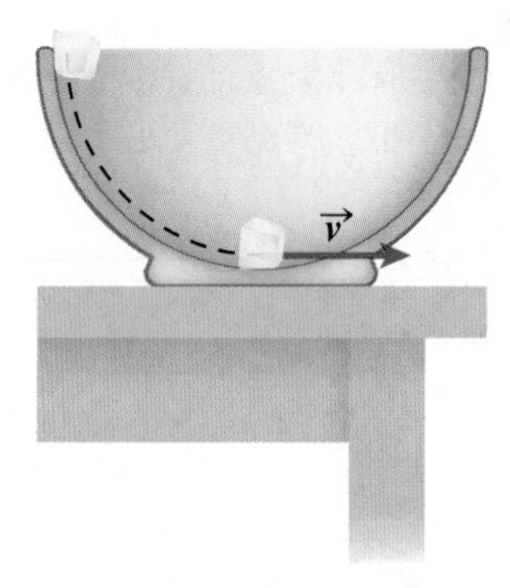

Figure 7-24 Question 7-48.

7.5 Work Done By Friction

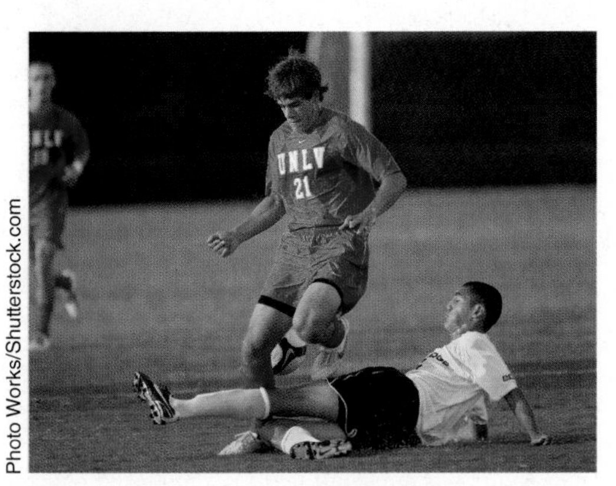

Figure 7-25 As the soccer player slides, his kinetic energy is decreasing.

Think about a soccer player sliding along the ground (Figure 7-25). As he is sliding, his speed and kinetic energy are decreasing, and when the slide is over, there is no kinetic energy left. Where did it go? While he was sliding, there was no change in elevation, so there has been no change in gravitational potential energy. We shall see in this section that the kinetic energy has been converted into a different form, thermal energy, and that the amount of thermal energy produced is closely related to the work done by friction.

During the soccer player's slide, he undergoes a horizontal displacement $\Delta\vec{r}$ (Figure 7-26). The only force that does work on him is the kinetic friction force $\vec{F}_K$. (Gravity and the normal force are both perpendicular to the displacement and hence do no work.) The kinetic friction force is in the opposite direction to the displacement, and therefore the angle θ between this force and the displacement is 180°. The work done by kinetic friction is $W = F_K\,\Delta r \cos 180°$, and since $\cos 180° = -1$, this gives

$$W = -F_K \Delta r$$

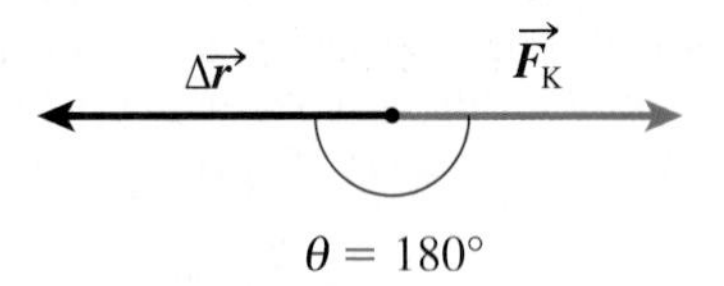

Figure 7-26 As the soccer player slides, kinetic friction is the only force that does work on the player.

Since F_K and Δr are both positive quantities (being magnitudes of vectors), the work done by friction is negative. This means that friction is removing kinetic energy from the player. Since energy is always conserved, this energy must be converted into another form, in this case, into **thermal energy** (also called **internal energy**) of the player and the ground. As the player slides, the atoms in the player and the ground move around with more energy; that is, the player and the ground get hotter.

thermal energy (or internal energy) the energy associated with the haphazard motion (including vibrations and rotations) of atoms and molecules; when kinetic friction acts on an object, thermal energy is produced

Whenever kinetic friction does negative work to slow something down, the magnitude of this work done equals the thermal energy produced. For example, if the work done by kinetic friction is −5 J, then the thermal energy produced is +5 J. Since kinetic friction is always opposite to the direction of the displacement, the work done by kinetic friction can always be written as $-F_K\Delta r$ (as above), and the magnitude of this work is just $F_K\Delta r$. Thus, we can write the thermal energy produced by kinetic friction as $F_K\Delta r$, which is a positive quantity. In this book, we will use E_{th} as the symbol for thermal energy. Thus,

$$E_{th} = F_K\Delta r \text{ (thermal energy produced by friction)} \qquad (7\text{-}7)$$

where F_K is the magnitude of the kinetic friction force, and Δr is the magnitude of the displacement.

Thermal energy is the energy associated with the haphazard motion (including vibrations and rotations) of atoms and molecules. As the thermal energy of an object increases, the atoms and molecules move more quickly. For a monatomic gas such as helium, the thermal energy is just the total kinetic energy of all the atoms. For more complicated molecules, or for atoms in solids, the thermal energy is partly kinetic and partly in a form called electric potential energy. More details of the production of thermal energy by kinetic friction were discussed in Section 6.1.

heat energy transferred between objects as a result of a difference in temperature between them

The terms "heat" and "thermal energy" are often confused. **Heat** refers to energy *transferred* between objects as a result of a difference in temperature between them, whereas thermal energy refers to the energy contained in the objects themselves. Thus, we say that heat flows from a cup of hot coffee, decreasing the thermal energy of the coffee and increasing the thermal energy of the air around it.

SAMPLE PROBLEM 7-6

A baseball player makes a horizontal slide into home plate.

(a) If he slides for 3.0 m before stopping, and the force of friction on the player has a constant magnitude of 3.3×10^2 N, how much thermal energy is produced during the slide?

(b) If the player's mass is 55 kg, what was his speed before sliding?

Solution

(a) Figure 7-27 shows a diagram for the problem. We can write the thermal energy (E_{th}) as $F_K \Delta r$, and since we know F_K and Δr, we can easily calculate E_{th}:

$$E_{th} = F_K \Delta r$$
$$= (3.3 \times 10^2\,N)(3.0\,m)$$
$$= 9.9 \times 10^2\,J$$

Thus, the thermal energy produced is 9.9×10^2 J.

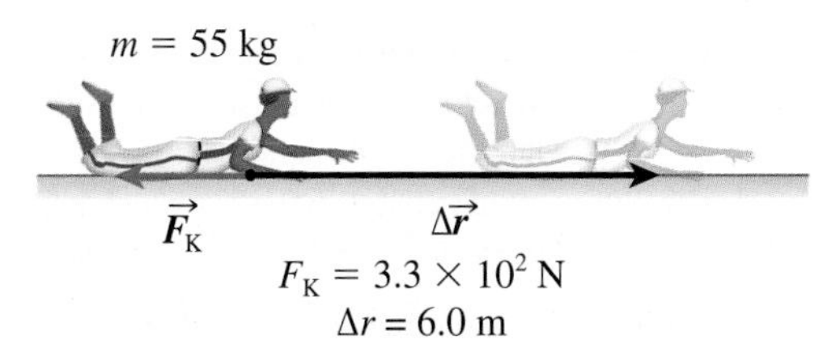

Figure 7-27 Sample Problem 7-6.

(b) By the law of conservation of energy, the player's initial kinetic energy must equal the thermal energy produced during the slide since there is no kinetic energy remaining at the end of the slide. (Since the slide is horizontal, there is no change in gravitational potential energy.) Thus,

$$\frac{1}{2}mv_0^2 = E_{th}$$

Rearranging to solve for the initial speed v_0,

$$v_0 = \sqrt{\frac{2E_{th}}{m}} = \sqrt{\frac{2(9.9 \times 10^2\,J)}{55\,kg}} = 6.0\,m/s$$

Therefore, the player's initial speed was 6.0 m/s.

SAMPLE PROBLEM 7-7

A small boy is pushing a large pot containing a plant across a floor by exerting a constant horizontal force. The pot and plant have a combined mass of 27 kg. The kinetic friction force on the pot has a magnitude of 93 N. If the pot starts from rest, and has a speed of 0.74 m/s after being pushed for 2.0 m, what is the magnitude of the force exerted by the boy?

Solution

There are many ways to solve this problem; we will use conservation of energy. Figure 7-28 shows a diagram.

The boy provides the pot with energy, which equals the work done by him. If he pushes with a horizontal force of magnitude F, the energy supplied by him is just $F\Delta r$ (where $\Delta r = 2.0$ m) since $\vec{F}$ and $\Delta\vec{r}$ are in the same direction.

By conservation of energy, the energy provided ($F\Delta r$) equals the sum of the change in the kinetic energy of the pot and the thermal energy produced in the floor and the bottom surface of the pot:

Energy provided = (change in the kinetic energy) + thermal energy

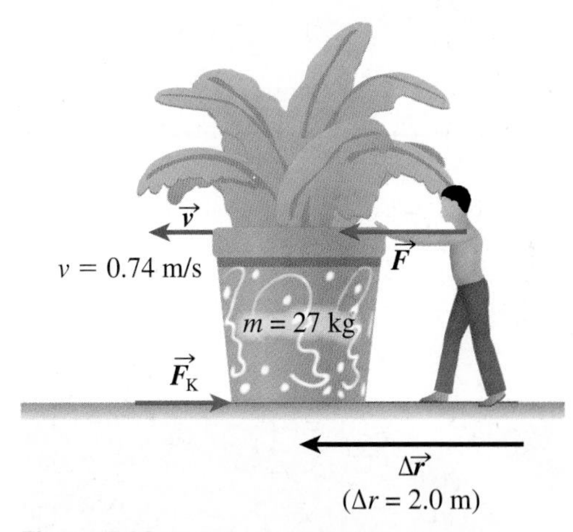

Figure 7-28 Sample Problem 7-7.

The kinetic energy can be determined from the expression $\frac{1}{2}mv^2$, since both m and v are known and the pot starts from rest ($v_0 = 0$). The thermal energy is just $F_K\Delta r$, and both F_K and Δr are given.

Thus,

$$F\Delta r = \frac{1}{2}mv^2 + F_K\Delta r$$

Solving for the unknown F,

$$F = \frac{mv^2}{2\Delta r} + F_K = \frac{(27\,kg)(0.74\,m/s)^2}{2(2.0\,m)} + 93\,N = 97\,N$$

Therefore, the boy exerts a force of magnitude 97 N.

Problem-Solving Tip

Sample Problem 7-7 could also have been solved by determining the acceleration of the pot using a one-dimensional kinematics equation, and then finding the magnitude of the force from Newton's second law of motion. However, using conservation of energy is often the simpler approach in such situations, since it involves only scalar quantities (work and energy) rather than vectors (force, velocity, and displacement).

EXERCISES

7-49 What is the SI unit of thermal energy?

7-50 (a) Suppose that you push this book across a horizontal desk at constant velocity. You are supplying some energy to the book. Into what form(s) does this energy go?

(b) Suppose that you push the book with a larger force so that the book speeds up. Into what form(s) does the energy now go?

7-51 A kinetic friction force of magnitude 67 N acts on a box as it slides across the floor. If the displacement of the box along the floor is 3.5 m in magnitude,

(a) What is the work done by friction on the box?

(b) How much thermal energy is produced?

7-52 As a penguin (Figure 7-29) slides across snow (assumed horizontal), 1.4×10^2 J of thermal energy is produced. If the kinetic friction force acting on the penguin is 19 N in magnitude, how far did the penguin slide?

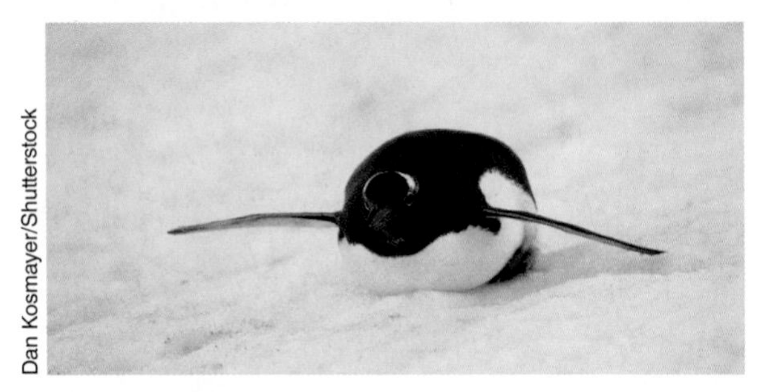

Dan Kosmayer/Shutterstock

Figure 7-29 Question 7-52.

7-53 A whiteboard eraser of mass 55 g slides along the ledge at the bottom of a whiteboard. Its initial speed is 1.9 m/s, and, after sliding for 54 cm, it comes to rest.

(a) Determine the eraser's initial and final kinetic energies.

(b) Into what form of energy is the kinetic energy converted?

(c) Use conservation of energy to determine the magnitude of the friction force acting on the eraser.

7-54 A pen of mass 0.057 kg is sliding across a horizontal desk. In sliding 25 cm, its speed decreases to 5.7 cm/s. What was its initial speed if the force of kinetic friction exerted on the pen by the desk is 0.15 N in magnitude? Use conservation of energy.

7-55 A clerk pushes a file cabinet (mass 22.0 kg) across the floor by exerting a horizontal force of magnitude 98 N. The force of kinetic friction acting on the cabinet is 87 N in magnitude. Assuming that the cabinet starts from rest, what is its speed after moving 1.2 m? Use conservation of energy.

7-56 The brakes are applied to a car travelling at 85 km/h along a horizontal road. The wheels lock and the car skids to a halt in 48 m. The magnitude of the kinetic friction force between the skidding car and the road is 7.5×10^3 N.

(a) How much thermal energy is produced during the skid?

(b) In what form was this thermal energy before the skid?

(c) Use conservation of energy to determine the mass of the car.

(d) Determine the coefficient of kinetic friction between the tires and the road.

7.6 Other Types of Energy

So far in this chapter we have discussed kinetic energy, gravitational potential energy, and thermal energy. But there are many forms of energy with which you are familiar that we have not yet introduced—chemical energy in gasoline, electrical energy, food energy, and so on. Although we do not have space to discuss all the possible forms of energy, Table 7-1 provides a brief description of some of them.

Denis Kuvaev/Shutterstock.com

Figure 7-30 When a pole-vaulter bends the pole, it stores elastic potential energy.

Elastic Potential Energy in Sports

Elastic potential energy is important in many sports. For example, the poles used by pole-vaulters store elastic potential energy when bent (Figure 7-30). Pole-vaulting is essentially a conversion of the vaulter's initial kinetic energy into gravitational potential energy at the top of the vault, with an intermediate conversion into elastic potential energy. As the pole bends early in the vault, kinetic energy is converted into elastic potential energy. As it straightens, the elastic potential energy changes to gravitational potential energy (Figure 7-31).

We can use conservation of energy to estimate the height attainable by pole-vaulters. We start by assuming that the initial kinetic energy is completely converted into gravitational potential energy, ignoring the small kinetic energy as the vaulter goes over the bar:

$$\frac{1}{2}mv^2 = mgy$$

Table 7-1

Various Forms of Energy

Light energy	Light energy is carried by travelling oscillations called electromagnetic waves (or their particle equivalent, photons). Visible electromagnetic waves are referred to as light, but there are invisible kinds such as gamma rays, x rays, ultraviolet waves, infrared waves, microwaves, and radiowaves.
Electrical energy	Electrical energy usually refers to the passage of electrons along wires, such as in appliances in your home.
Electric potential energy	Just as there is gravitational potential energy associated with the gravitational force, there is electric potential energy connected with the electric force. Gravitational potential energy changes as masses are moved relative to each other, and electric potential energy changes as charges are moved (more about this topic in Chapters 20 and 21).
Chemical energy	Chemical energy is a combination of kinetic energy of electrons in atoms and molecules and electric potential energy of the charged electrons and nuclei. For example, gasoline has chemical energy that is released as thermal energy when it is burned. During this burning, the electrons and nuclei are put into different arrangements with a resulting decrease in chemical energy and production of an equivalent amount of thermal energy.
Food energy	This is a type of chemical energy.
Nuclear energy	Nuclear energy is a potential energy associated with the nuclear strong force. It can be converted to other forms by rearrangements of the particles inside a nucleus, by fusing certain small nuclei together, or by breaking some types of large nuclei apart.
Sound energy	Sound energy is associated with the movement of compression waves through materials such as air and water. It is a combination of electric potential energy and kinetic energy of molecules.
Elastic potential energy	This type of energy is stored in objects that are stretched (or compressed), such as elastic bands, springs, and various types of sports equipment (as discussed in this section).

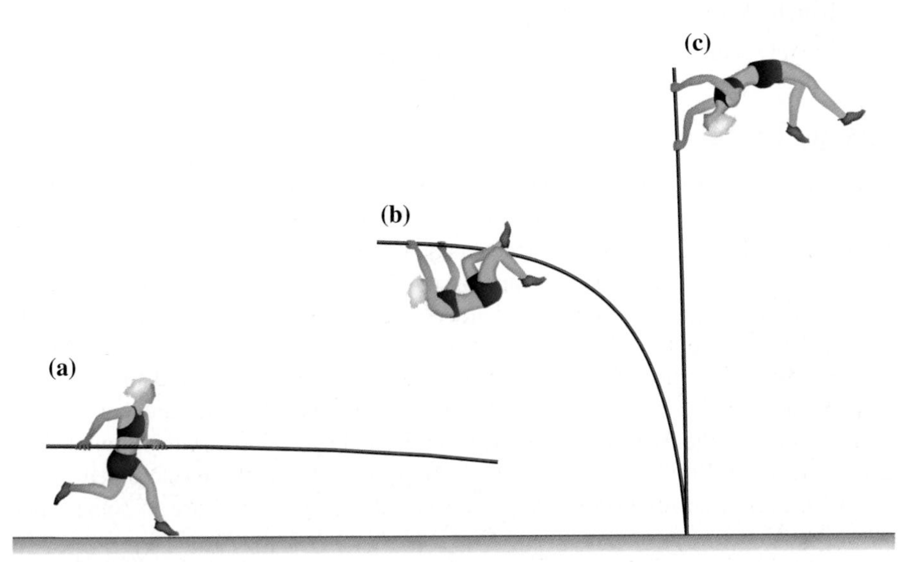

Figure 7-31 In the pole-vault, the vaulter starts with (**a**) kinetic energy, and finishes with (**c**) gravitational potential energy. (**b**) In between, much of the energy is stored in the pole as elastic potential energy.

Solving for height y,

$$y = \frac{v^2}{2g}$$

A typical running speed (v) for a pole-vaulter is about 10 m/s. (People can run faster, up to roughly 12.5 m/s, but not carrying a pole.) Since we are making only an estimate, we will use 10 m/s² for g.

Thus,

$$y \approx \frac{(10\,\text{m/s})^2}{2(10\,\text{m/s}^2)} \approx 5\,\text{m}$$

However, some energy is lost when the pole hits the ground, and thermal energy is generated in the pole (and thus lost) when it is bent and then straightened. With modern fibreglass poles, the energy lost is very small, roughly 10% of the total. A 10% decrease in the final gravitational potential energy means that the height should be decreased by 10%, to 4.5 m.

There are two refinements still to be made. When we use the equation $E_P = mgy$ for gravitational potential energy, y represents the height of the *centre of mass* (CM) of the object, in this case the vaulter. The CM of an object is a point at the average position of the object's mass at any instant.[3] Our y-value of 4.5 m indicates that the height of the vaulter's CM increases by 4.5 m during the vault. The CM of a typical vaulter is about 1 m from the ground when running, and just clears the bar (or can even go under it!) at the top of the vault. Therefore, in order to determine the height of the CM above the *ground* as the vaulter clears the bar, we need to add 1 m, giving a value of about 5.5 m.

The last adjustment arises because, at the top of the vault, the athlete essentially does a handstand, thus providing more energy and increasing the height by about 0.3 m. We now have a final estimate of about 5.8 m, which corresponds closely to the heights attained by world-class athletes. (As of January 2015, the world record was 6.16 m for men and 5.06 m for women.)

There are numerous other examples of storage and recovery of elastic potential energy in sports equipment:

- tennis racquets (frames and strings)
- many types of balls, which become compressed on impact (in golf, tennis, baseball, basketball, football, etc.)
- flexible shafts in golf clubs and hockey sticks.

DID YOU KNOW?

A high-tech application of elastic potential energy in sports is the design of tuned running tracks. When a runner's foot strikes the track, the track is deformed as it absorbs energy from the runner. Some of this energy is stored as elastic potential energy in the track material, and partially returned to the runner during the rebound of the track as the runner's foot leaves the track. Running tracks can be designed with the right springiness for effective return of energy to the runner, with a 2% to 3% reduction in running times.

Energy Units

Although the SI unit of energy is the joule, there are a number of other units in common use. Some are used scientifically, and some in the everyday world. Table 7-2 lists the names, abbreviations, and joule equivalents of a number of these units.

The electron volt (eV) and its common multiples, keV (10^3 eV), MeV (10^6 eV), etc., are used widely in atomic and subatomic physics. We will see this unit again in Chapter 24 when nuclear physics is discussed. The various "calories" and the British thermal unit arose as units for heat and thermal energy in the 1700s and 1800s, when it was not apparent that what we now call thermal energy is indeed just another form of energy. Note that the food calorie, the kilocalorie, and the Calorie (with a capital "C") are all the same. In North America, food energy is still commonly measured in the archaic Calories, but civilization is proceeding at a more rapid pace in some other countries, where joules are being used (Figure 7-32).

Figure 7-32 SI units are taken seriously in Australia.

Table 7-2

Non-SI Energy Units

Name	Abbreviation	Joule Equivalent
Electron volt	eV	1.60×10^{-19} J
calorie	cal	4.184 J
Kilocalorie	kcal	4184 J
Calorie	Cal	4184 J
Food calorie	Cal	4184 J
British thermal unit	Btu	1054 J

[3]Centre of mass is discussed in detail in Section 11.1.

EXERCISES

7-57 When an insect such as a flea jumps, the energy is provided not only by muscle, but also by an elastic protein that has been compressed like a spring. If a flea of mass 210 μg jumps vertically to a height of 65 mm, and 75% of the energy comes from elastic potential energy stored in the protein, determine the initial quantity of stored elastic potential energy. Neglect energy losses due to air resistance.

7-58 Use the conversion factors in Table 7-2 to perform the following unit conversions:

(a) 13.6 eV to joules

(b) 1.3 MeV to joules

(c) 2.3 cal to joules

(d) 2.8 J to Calories

(e) 3.7 J to British thermal units

(f) 4.57 cal to British thermal units

7-59 Convert a food energy of 87 Cal (a typical value for an apple) to kilojoules.

7.7 Power

In everyday speech, the terms force, energy, and power are often used interchangeably, but the scientific meanings of these words are different. We have already studied force and energy, and now turn our attention to power. The SI unit of power, the watt, is one with which you are undoubtedly familiar; for example, hairdryers and light bulbs are labelled according to their power consumption in watts.

Power is the rate at which energy is produced or used; that is, it is energy divided by time. For instance, if a car's kinetic energy increases, the power input to the car is the increase in the kinetic energy divided by the time taken.

power rate at which energy is produced or used, that is, energy divided by time

watt SI unit of power, abbreviated as W; equivalent to joule per second (J/s)

Mathematically, we write power (P) as

$$P = \frac{E}{t} \text{ (power)} \qquad (7\text{-}8)$$

where E is energy and t is time.

With energy in joules (J) and time in seconds (s), we get a unit of joule per second (J/s) for power. This unit is conventionally referred to as a **watt** (W), in honour of the Scottish physicist James Watt (1736–1819), who modified the steam engine to greatly improve its efficiency.

$$\text{W (watt)} = \frac{\text{J}}{\text{s}} \text{ (SI unit of power)}$$

Watts and kilowatts (kW) are often used to indicate the power consumption of home appliances such as light bulbs and electric heaters, and megawatts (MW) are used in describing power output of electrical plants. A 60 watt light bulb uses 60 J of electrical energy each second. A typical electrical generating station produces about 1000 MW of electrical power, enough for about 500 000 homes. As you sit reading this book, you generate about 100 W of thermal power, which is then radiated away. The solar power striking Earth's surface is approximately 1.78×10^{17} W or 178 PW. (SI prefixes are given on the inside front cover.)

A unit of power that is often used for engines is the *horsepower* (hp), which is equivalent to 746 W. This term was created in the late 1700s by James Watt to compare the power available from steam engines to that of draft horses.

DID YOU KNOW?

The average basal metabolic rate (BMR) for a typical person is approximately 100 W. This is the rate at which energy is being processed by the person and largely dissipated as heat to the surrounding environment. If there are 10 people together in a room, for example, then their total heating power is 1000 W, which is the same as a typical electric baseboard heater!

SAMPLE PROBLEM 7-8

In a 100 m race, a certain sprinter has a power output of 1.12 kW for 10.3 s (Figure 7-33). Determine the energy output during this time.

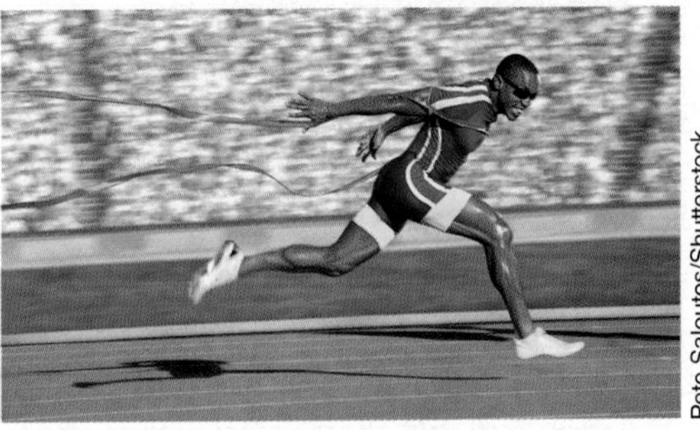

Pete Saloutos/Shutterstock

Figure 7-33 Sample Problem 7-8.

Solution

We first convert the power to watts:

$$1.12\,\text{kW} \times \frac{1000\,\text{W}}{1\,\text{kW}} = 1.12 \times 10^3\,\text{W} \,(\text{or } 1.12 \times 10^3\,\text{J/s})$$

Since $P = \frac{E}{t}$, we have

$$E = Pt = (1.12 \times 10^3\,\text{W})(10.3\,\text{s}) = (1.12 \times 10^3\,\text{J/s})(10.3\,\text{s}) = 1.15 \times 10^4\,\text{J}$$

Therefore, the sprinter uses 1.15×10^4 J of energy.

Units Tip

Since a watt is equivalent to a joule per second, then watts (W) multiplied by seconds (s) gives joules (J).

Kilowatt·Hour: An Energy Unit

One of the most confusing energy units is the kilowatt·hour (kW·h). Because part of the unit is kilowatt, which is a unit of power, many people believe that the kilowatt·hour is a power unit. However, since the kilowatt·hour involves multiplying a kilowatt by a time unit (hour), the kilowatt·hour is actually a unit of energy, as shown in more detail below.

Since power is energy divided by time ($P = E/t$), then we can write

$$Pt = E$$

This equation indicates that the product of power and time is energy (regardless of the particular units used). If SI units are used, then power in watts multiplied by time in seconds gives energy in watt·seconds, but since a watt is defined as a joule per second, then a watt·second is just a joule. But we can express power and time in other units; if power has units of kilowatts and time has units of hours, then the product—still an energy—has units of kilowatt·hours (kW·h).

We can express 1 kW·h in terms of joules by performing a unit conversion, remembering that 1 kW = 10^3 W, 1 W = 1 J/s, and 1 h = 3600 s:

$$1\,\text{kW}\cdot\text{h} \times \frac{10^3\,\text{W}}{1\,\text{kW}} \times \frac{1\,\text{J/s}}{1\,\text{W}} \times \frac{3600\,\text{s}}{1\,\text{h}} = 3.6 \times 10^6\,\text{J} = 3.6\,\text{MJ}$$

Hence, 1 kW·h is equivalent to 3.6 MJ.

The kilowatt·hour is a handy unit of electrical energy consumption, since the total electrical power requirement of a house is often in the range of kilowatts and time can easily be measured in hours. Using joules for electrical energy would result in extremely large numbers: a typical home consumes 500 to 1000 kW·h of electrical energy per month, which is of the order of 10^9 J! For electrical energy consumption at national and world levels, gigawatt·hours (GW·h) and terawatt·hours (TW·h) are often used. For example, electrical energy production in Canada in 2010 was 589 TW·h.[4]

TRY THIS!

Electric Power

Some modern electric meters mounted on homes show both the energy consumption in kilowatt·hours since the meter was installed and the current power usage in kilowatts (Figure 7-34). If your home has one of these meters, you can determine the power consumption of an appliance by first noting the power reading without an appliance running and then turning the appliance on and checking the new reading. How much power do you think is consumed by an electric oven? clothes dryer? vacuum cleaner? As well, some devices such as televisions and computer printers consume electrical energy even when they are turned off. You might want to unplug all such "sleeping" electrical users and see how much the power usage drops.

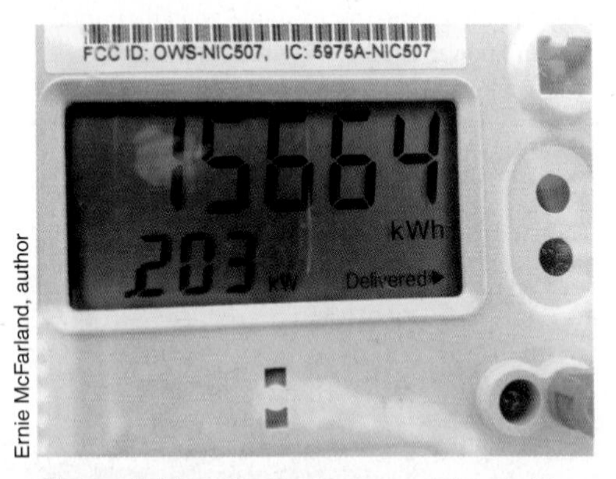

Figure 7-34 This electric meter on the outside of a house shows an energy consumption of 15 664 kW·h since the meter was installed, and a current power usage of 0.203 kW.

SAMPLE PROBLEM 7-9

A 100 W light bulb is turned on for 12 h. What is the energy consumption in kilowatt·hours?

Solution

To determine the energy in kilowatt·hours, we multiply the power in kilowatts by the time in hours. Since the power is given in watts, we first convert this to kilowatts:

$$100\,\text{W} \times \frac{1\,\text{kW}}{10^3\,\text{W}} = 0.10\,\text{kW} \text{ (assuming 2 significant digits)}$$

Then, $E = Pt = (0.10\text{ kW})(12\text{ h}) = 1.2\text{ kW}\cdot\text{h}$

Thus, the electrical energy consumed is 1.2 kW·h.

[4]Source: Natural Resources Canada.

EXERCISES

7-60 Determine

(a) the power input to a ball if its kinetic energy increases from 11 J to 35 J in a time of 0.86 s

(b) the electrical energy (in joules) used by a 100 W light bulb that is turned on for 65.0 min

(c) the electrical energy (in kilowatt·hours) consumed by a 1.10 kW hairdryer that is used for 5.7 min

(d) the power input (in kilowatts) to a woman of mass 66 kg standing in an elevator that lifts her through a vertical distance of 15 m in 2.4 s (without increasing her speed)

7-61 The metabolic rate in humans is proportional to the volume of oxygen supplied by the blood. For each litre of oxygen supplied, the energy produced is 2.0×10^4 J. Determine the metabolic power generated by a person completely at rest who consumes 15 L of oxygen per hour.

7-62 If you eat a 250 Calorie candy bar, for how many minutes could you cycle from this energy input? Assume a power output of 400 W while cycling and two significant digits in given data.

7-63 The Sun's power output is about 4×10^{26} W. How much energy is released by the Sun in 1 yr?

DID YOU KNOW?

A supernova (a massive exploding star) releases about 100 times as much energy in 1 s as does our Sun in its entire lifetime.

7-64 What is the cost of using a 100 W light bulb for 24 h if the average price of electrical energy is 9.7¢ per kilowatt·hour?

7-65 A typical small car moving at 80 km/h can travel 14 km using 1.0 L of gasoline. What is the power consumption (in watts) of such a car? The energy content per unit volume for gasoline is about 35 MJ/L.

7-66 A 2.65 kW motor is used to lift crates vertically upward in a factory. What is the maximum speed at which a crate of mass 147 kg can be lifted by this motor? Assume that all the motor's power goes into lifting. (**Hint:** Determine the height that the crate can be lifted in 1.00 s.)

7.8 Efficiency of Energy Conversions

Often we are interested in converting one form of energy into another. For instance, we use light bulbs to convert electrical energy into light, and plants convert light energy into chemical energy via photosynthesis. Since energy is always conserved, the total energy before any conversion always equals the total energy after. However, the energy might not be going into forms that we consider useful. For example, only 5% of the electrical energy used in an incandescent light bulb is converted into light. The remaining 95% goes into thermal energy. Since the purpose of light bulbs is the generation of light, we say that they are only 5% efficient. The **efficiency** of an energy conversion is the ratio of useful energy out to total energy in:

efficiency ratio of useful energy out to total energy in, during an energy conversion

$$\text{efficiency} = \frac{\text{useful energy out}}{\text{total energy in}} \qquad (7\text{-}9a)$$

Writing this with symbols,

$$\eta = \frac{\text{useful } E_{\text{out}}}{\text{total } E_{\text{in}}} \qquad (7\text{-}9b)$$

where efficiency is represented by η (the lower-case Greek letter "eta"). Efficiencies are very often quoted as percentages:

$$\eta = \frac{\text{useful } E_{\text{out}}}{\text{total } E_{\text{in}}} \times 100\% \qquad (7\text{-}9c)$$

Since power is just the rate of energy use ($P = E/t$), power can be inserted into the definition of efficiency in place of energy:

$$\eta = \frac{\text{useful } P_{out}}{\text{total } P_{in}} \times 100\% \qquad \textbf{(7-9d)}$$

By the law of conservation of energy, the useful energy (or power) out can never be greater than the total energy (or power) in, and thus efficiencies are never greater than 100%.

Efficiencies for various energy conversions are listed in Table 7-3. Notice the low efficiencies of incandescent lights and automobile engines. Electrical generators have a very high efficiency (about 98%); these are the machines that spin in electrical generating plants and generate the electrical currents. However, most electrical plants have a much lower *overall* efficiency. In order to spin the generators, steam is often used that has been produced by burning a fossil fuel (coal, oil, or natural gas) or by using a nuclear reactor to provide heat. Only about 30% to 40% of the thermal energy in the steam can be converted into kinetic energy of the spinning generators; the rest is wasted and simply goes into heating the surrounding air (Figure 7-35) or water in a nearby lake or river. Hydroelectric plants use falling water to spin the generators and have a much higher efficiency (about 95%). In some countries such as Sweden, the waste heat from steam-fired generators is often used in neighbouring industries that require heat for industrial processes; the waste heat from the industries can then be used to heat homes, etc. This production of both electricity and useful heat is referred to as cogeneration.

Kodda/Shutterstock

Figure 7-35 These towers at a nuclear electrical generating station in France eject waste heat into the surrounding air.

People have an efficiency of only about 25%. If you lift a 1.0 kg mass through a vertical distance of 1.0 m, you increase its gravitational potential energy by

$$mgy = (1.0\,\text{kg})(9.8\,\text{m/s}^2)(1.0\,\text{m}) = 9.8\,\text{J}$$

but the actual energy used by your body to do this is roughly four times as great (about 40 J). Approximately 30 J goes into thermal energy of your body. When you are producing a great deal of useful energy, such as when running, you generate a lot of wasted thermal energy and become very hot.

Table 7-3

Approximate Efficiencies of Energy Conversions

Device	Type of Energy In	Type of Useful Energy Out	Efficiency
Incandescent light	electrical	light	5%
Fluorescent light	electrical	light	20%
Light-emitting diodes (LEDs)	electrical	light	30–50%
Automobile engine	chemical	kinetic	14%
Dry-cell battery	chemical	electrical	90%
Electrical generator	kinetic	electrical	98%
Steam–electric power plant	chemical or nuclear	electrical	30–40%
Hydroelectric plant	kinetic	electrical	95%
People	chemical (food)	any useful energy	25%

SAMPLE PROBLEM 7-10

A person of mass 70 kg (two significant digits) requires about 5.0×10^5 J of food energy to walk 2.0 km on level ground. How much *more* energy is required if the walk is done at 24° above the horizontal? The efficiency for conversion of food energy to usable energy is 25% (Table 7-3). Express the answer in joules and food calories.

Solution

In terms of energy, the only difference between the level walk and the inclined walk is that there is an increase in gravitational potential energy, mgy, in the inclined walk (Figure 7-36), where $y = (2.0 \times 10^3 \text{ m})(\sin 24°)$:

$$\begin{aligned} E_P &= mgy \\ &= (70 \text{ kg})(9.8 \text{ m/s}^2)(2.0 \times 10^3 \text{ m})(\sin 24°) \\ &= 5.6 \times 10^5 \text{ J} \end{aligned}$$

Then, using Eqn. 7-9b,

$$\eta = \frac{\text{useful } E_{out}}{\text{total } E_{in}}$$

Therefore,

$$\text{total } E_{in} = \frac{\text{useful } E_{out}}{\eta} = \frac{5.6 \times 10^5 \text{ J}}{0.25} = 2.2 \times 10^6 \text{ J}$$

Thus, the total additional (food) energy required is 2.2×10^6 J.
To convert to food calories, we use the conversion factor provided in Table 7-2 (1 food calorie = 1 Cal = 4184 J):

$$2.2 \times 10^6 \text{ J} \times \frac{1 \text{ Cal}}{4184 \text{ J}} = 5.3 \times 10^2 \text{ Cal}$$

Hence, the extra energy needed is 5.3×10^2 food calories (Cal or kcal).

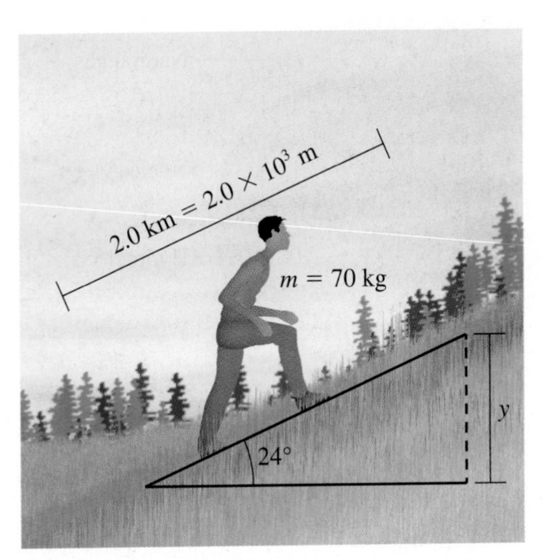

Figure 7-36 Sample Problem 7-10.

SAMPLE PROBLEM 7-11

What mass of coal is required per year in an electrical power plant that produces 1.0×10^3 MW of electrical power? The plant has an efficiency of 36% in converting the coal's chemical energy into electrical energy, and the energy content per unit mass of coal is 26 MJ/kg.

Solution

First convert all units to base SI units for convenience:
Since 1 MW = 10^6 W, then $1.0 \times 10^3 \text{ MW} = (1.0 \times 10^3) \times 10^6 \text{ W} = 1.0 \times 10^9$ W.
Similarly, 1 MJ = 10^6 J, and hence 26 MJ/kg = 26×10^6 J/kg = 2.6×10^7 J/kg.

Our plan will be to use the given efficiency and useful power out to calculate the required total power input to the plant, then to find the total energy input required for a time of one year, and finally to calculate the mass of coal based on its energy content per unit mass.

Start with Eqn. 7-9d:

$$\eta = \frac{\text{useful } P_{out}}{\text{total } P_{in}} \times 100\%$$

Rearrange to find total P_{in}:

$$\text{total } P_{in} = \frac{\text{useful } P_{out}}{\eta} = \frac{1.0 \times 10^9 \text{ W}}{0.36} = 2.8 \times 10^9 \text{ W or } 2.8 \times 10^9 \text{ J/s}$$

Using power = energy/time:

$$\text{total } P_{in} = \frac{\text{total } E_{in}}{t} \quad \therefore E_{in} = (\text{total } P_{in})t$$

Substituting numbers,

$$\text{total } E_{in} = (2.8 \times 10^9 \text{ J/s})\left(1 \text{ yr} \times \frac{365 \text{ d}}{1 \text{ yr}} \times \frac{24 \text{ h}}{\text{d}} \times \frac{60 \text{ min}}{\text{h}} \times \frac{60 \text{ s}}{\text{min}}\right) = 8.8 \times 10^{16} \text{ J}$$

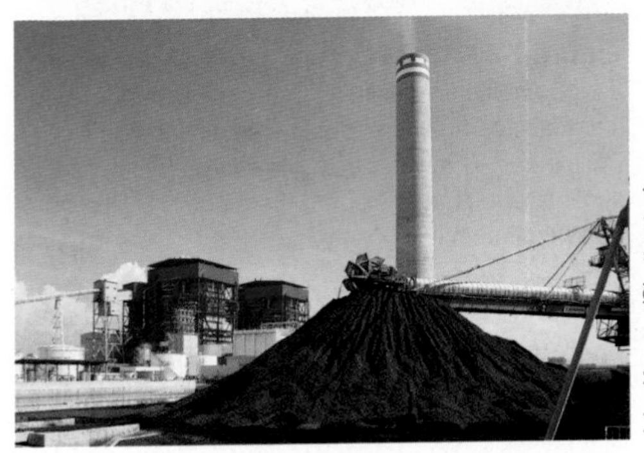

Figure 7-37 Coal-fired electrical power plants use huge amounts of coal every day.

Thus, 8.8×10^{16} J of energy must be provided by the coal. The mass of coal required is determined by dividing the required energy by the energy content per unit mass:

$$\text{mass } m = \frac{8.8 \times 10^{16}\,\text{J}}{2.6 \times 10^{7}\,\text{J/kg}} = 3.4 \times 10^{9}\,\text{kg}$$

Thus, the annual coal requirement is 3.4×10^{9} kg (3.4 billion kg!). This represents about 100 railroad cars of coal per day (Figure 7-37).

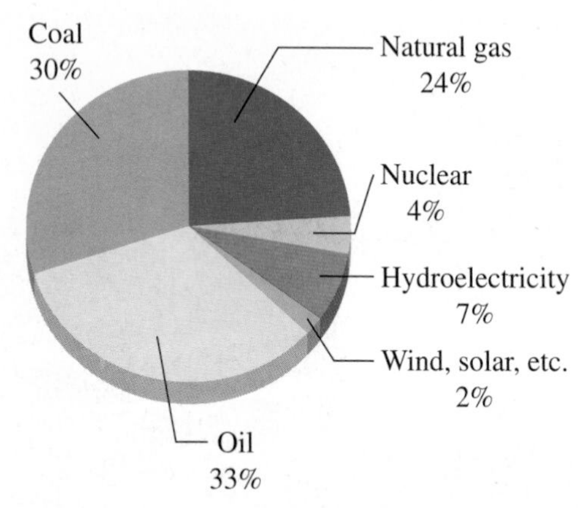

Figure 7-38 World energy consumption by source in 2013. (Source: Data from *BP Statistical Review of World Energy*, June 2014)

Energy Sources and Conservation

If energy is always conserved, why should we try to conserve it; that is, why should we use as little energy as possible in transportation, industry, etc.? Part of the answer is that, in all conversions and uses of energy, there is generation of some thermal energy, which is a form of energy that is not very useful to us. It is impossible to convert thermal energy to other forms with 100% efficiency. Thus, for example, when thermal energy is produced in an automobile engine, most of it is essentially lost to us forever. Once we have burned the gasoline, its energy has been used and is not retrievable. As well, as shown in Figure 7-38, fossil fuels (oil, coal, and natural gas) account for 87% of annual world energy consumption. Fossil-fuel reserves are finite, and so it makes sense to consume these resources wisely since we can use them only once.

greenhouse effect warming of Earth's atmosphere due to the absorption of radiation (especially infrared radiation) by atmospheric gases such as carbon dioxide

Using fossil fuels also creates environmental problems such as climate change due to the **greenhouse effect**. This refers to the warming of the atmosphere due to the absorption of radiation by atmospheric gases such as carbon dioxide produced by combustion of fossil fuels. As more carbon dioxide is put into the atmosphere, more infrared radiation[5] emitted by Earth can be absorbed in the atmosphere. The energy in this radiation is thus trapped in the atmosphere, which gets warmer. This warming of the atmosphere is leading to changes in climate, including an increase in extreme weather events in many parts of the world, including Canada.

Renewable energy sources such as the wind (Figure 7-39), Sun, geothermal energy, waves, and tides are being developed worldwide, but as seen in Figure 7-38, they contribute only a small fraction of the total energy used at present. The choice of energy sources in any specific country depends on many factors: local availability, cost, environmental impact, availability of qualified personnel to develop the technology, and so on.

TebNad/Shutterstock

Figure 7-39 The number of wind turbines generating electricity worldwide is increasing rapidly.

DID YOU KNOW?

In developing countries there are often people living in shantytowns who find it difficult to afford electricity for basic needs such as lighting. In Brazil and the Philippines, daytime lighting is now being provided for many such people by "pop-bottle light bulbs" (Figure 7-40). Each light bulb is made from a used 2 L pop bottle, which is filled with water (and a small amount of bleach to prevent growth of bacteria and algae), and then the bottle is sealed. A hole is cut in the roof of the dwelling, the bottle is placed in the hole with a few centimetres of the bottle above the roofline, and the slight gap between the bottle and the edges of the hole is caulked to prevent rain leakage. When the sun shines, the water in the bottle acts like a large lens and glows as brightly as a 50 W bulb!

Jay Directo/AFP/Getty Images

Figure 7-40 The "light bulb" shown here is a used 2 L pop bottle filled with water that has been placed securely in a hole in a roof in a small home in the Philippines. The water acts like a large lens when sunlight enters the hole and the result is a glowing 50 W light bulb.

Energy Return on Investment

energy return on investment (EROI) the energy obtained from a source per unit of energy required (or spent) to obtain it

A parameter that is somewhat related to efficiency is **energy return on investment (EROI),** which is the energy obtained from a source per unit of energy required (or spent) to obtain it. For example, suppose that a wind turbine is being constructed.

[5]Infrared radiation is a form of electromagnetic radiation (Chapter 16)—like light, but invisible.

Energy has to be used to manufacture its components, including the base, tower, and rotor, and also to construct and maintain the structure. If the turbine produces 20 units of electrical energy during its lifetime for every unit of energy required to construct and maintain it, then its EROI would be 20. Clearly a high EROI is a great advantage for an energy source.

Table 7-4 shows the EROI for various liquid fuels and for sources used for electricity production. An EROI of about 5–9 is considered the minimum for an energy source to be truly useful. You can see that hydroelectricity has an extremely high EROI, and Canada is fortunate to have a large number of hydroelectric sites. Using wind for electricity generation also has a high EROI. At the other end of the scale, the production of ethanol fuel from corn consumes so much energy in farming, fertilizing, and refining that it is barely an energy-breakeven technology, and its EROI comes in well below the accepted minimum of 5. It should be mentioned here that neither efficiency nor EROI takes environmental effects into consideration.

Table 7-4

Energy Return on Investment (EROI) for Various Energy Sources[6]

Liquid Fuels	EROI	Electricity Sources	EROI
Conventional oil	16	Hydroelectricity	40+
Ethanol from sugarcane	9	Wind	20
Biodiesel from soy	5.5	Coal	18
Tar sands	5	Natural gas	7
California heavy oil	4	Solar photovoltaic	6
Ethanol from corn	1.4	Nuclear	5

DID YOU KNOW?

Canada's annual hydroelectricity production is third in the world, behind only China and Brazil.

EXERCISES

7-67 How much food energy (in joules) is required for a boy of mass 47 kg to climb a 12 m high tree, over and above that required for him to rest for an equal time? Use the appropriate efficiency from Table 7-3.

7-68 How many food calories are required for a person to lift a box of books (mass 15 kg) through a vertical distance of 0.30 m? Use the appropriate efficiency from Table 7-3.

7-69 A 52 kg woman is doing push-ups. During each one, she raises the centre of mass of her body by 0.25 m. How many push-ups must she do in order to use at least 1.0 Cal of food energy, over and above the energy required for her to rest for an equal time? Neglect the energy used in lowering her body, and use the appropriate efficiency from Table 7-3.

7-70 **Fermi Question:** Estimate the total power you produce in running as fast as you can up a flight of stairs (Figure 7-41). Show your reasoning. Use the appropriate efficiency from Table 7-3.

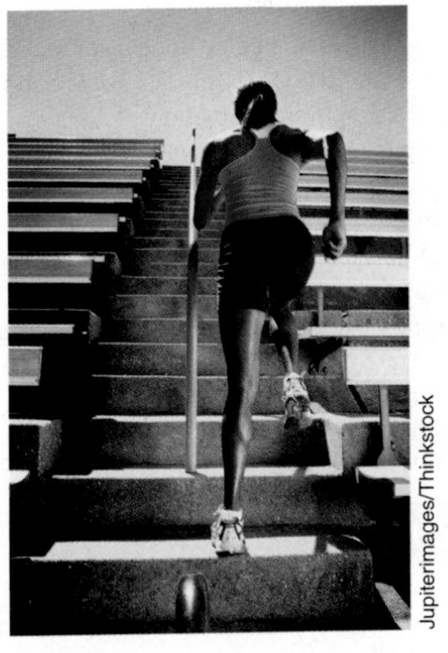

Jupiterimages/Thinkstock

Figure 7-41 Question 7-70.

[6]Source: M. Inman, "The True Cost of Fossil Fuels," *Scientific American*, Vol. 308, No. 4 (2013), pp. 58–61.

7-71 A car of mass 1.2×10^3 kg is driving up a hill 2.3 km long, inclined at 5.2° above the horizontal. How much more chemical energy (in megajoules) in the gasoline is required, in comparison with that used on a level road? Use the appropriate efficiency from Table 7-3.

7-72 Suppose that the efficiency of conversion of wind energy into electrical energy in a particular wind turbine is 35%.

(a) In what form is wind energy? (Choose from kinetic energy, gravitational potential energy, or elastic potential energy.)

(b) If the wind energy passing through the turbine each second is 7.5×10^6 J, how much electrical energy is generated per second?

(c) If the turbine is producing 2.2×10^6 J of electrical energy each second, how much wind energy is passing through it per second?

7-73 Electric heat is sometimes claimed to be a good way to heat homes because there is 100% conversion of electrical energy to thermal energy in electric heaters. While this statement is true, there are energy conversions in the *production* of the electrical energy that have efficiencies less than 100%. Assuming that the electrical energy is produced in a coal-fired plant, use the following efficiencies to calculate the *overall* efficiency of conversion of chemical energy in coal to thermal energy in a home.

Extraction of coal	96%
Transportation of coal	97%
Generation of electricity	33%
Transmission of electricity	90%
Electrical heating	100%

7-74 Calculate the overall efficiency of a fluorescent light powered by electricity from a coal-fired power plant. Use data from the previous question and from Table 7-3.

LOOKING BACK...LOOKING AHEAD

In earlier chapters, we concentrated on the motion of objects and on the forces causing the motion.

In this chapter, motion has been discussed from a different point of view by considering work and energy. We have seen how work and energy are related, and have introduced a number of types of energy in detail: kinetic energy, gravitational potential energy, and thermal energy. The important law of conservation of energy has been discussed and applied in situations from pole-vaulting to generation of electricity by burning coal. We have considered the concept of power and its relation to energy and time. Finally, the efficiency of energy conversions has been illustrated with a number of real-world examples: light bulbs, automobiles, etc.

The next chapter deals with another important law of mechanics: the law of conservation of momentum, particularly in its application to rocket propulsion and collisions of objects.

CONCEPTS AND SKILLS

Having completed this chapter, you should now be able to do the following:

- Identify whether work is being done in various physical situations.
- Define the work done by a constant force and use this definition in analyzing related physical situations.
- Explain why the work done by any force that is providing centripetal acceleration is zero.
- State the work-energy theorem and use it to solve problems.
- State the SI unit for work and energy, and relate it to the SI base units of kilogram, metre, and second.
- Solve problems related to kinetic energy and gravitational potential energy.
- Use the law of conservation of energy to analyze physical systems.
- Calculate the work done by friction in problems, and relate this work to the thermal energy produced.
- Give a qualitative description of various types of energy such as light energy, electrical energy, etc.
- Perform unit conversions involving energy units such as joules, electron-volts, Calories, etc.
- State the relation between power, energy, and time, and use this relation in problems.
- Use kilowatt·hour as an energy unit.
- Solve problems related to efficiency.
- State approximate efficiencies for devices such as automobile engines and light bulbs.
- Discuss why energy conservation is important.
- Do calculations involving energy, power, and efficiency in commercial production of electrical energy.

KEY TERMS

You should now be able to define or explain each of the following words or phrases:

work
joule (SI unit)
kinetic energy
work-energy theorem
gravitational potential energy
law of conservation of energy
mechanical energy
thermal energy (or internal energy)
heat
power
watt (unit)
efficiency
greenhouse effect
energy return on investment (EROI)

Chapter Review

MULTIPLE-CHOICE QUESTIONS

7-75 A joule is equivalent to

(a) kg·m/s^3
(b) kg·m^2/s^3
(c) kg^2·m^2/s^3
(d) kg·m/s
(e) kg·m^2/s^2

7-76 A car of mass 1200 kg skids to a stop in a distance of 25 m along a horizontal road. During the skid how much work is done on the car by the normal force?

(a) 0 J
(b) 2.9×10^5 J
(c) 3.0×10^4 J
(d) 1.2×10^4 J
(e) -2.9×10^5 J

7-77 A hockey puck of mass 0.20 kg is sliding across a horizontal ice surface. The coefficient of kinetic friction between the puck and the ice is 0.070. As the puck slides 2.0 m, how much work is done by the friction force on the puck?

(a) 0 J
(b) −0.27 J
(c) 0.29 J
(d) −4.0 J
(e) 4.0 J

7-78 The planet Venus follows an orbit that is approximately circular around the Sun. If the radius of this orbit is r, and the gravitational attractive force exerted on Venus by the Sun is $\vec{F}_G$, how much work is done on Venus by this gravitational force as Venus travels around half its orbit?

(a) 0
(b) $2F_G\pi r$
(c) $F_G\pi r$
(d) $2F_Gr$
(e) $F_G\pi r/2$

7-79 During a portion of the movement in a biceps curl, you are lifting a weight vertically upward. During this upward movement, which of the following expresses the correct mathematical statement about the change in the gravitational potential energy E_P of the weight?

(a) $\Delta E_P = 0$
(b) $\Delta E_P > 0$
(c) $\Delta E_P < 0$

7-80 A pencil falls from a table to the floor. As the pencil falls, ΔE_P is the change in the pencil's gravitational potential energy, and ΔE_K is the change in its kinetic energy. Neglect air resistance. Which of the following is correct?

(a) $\Delta E_P > 0$ and $\Delta E_K > 0$
(b) $\Delta E_P > 0$ and $\Delta E_K < 0$
(c) $\Delta E_P < 0$ and $\Delta E_K < 0$
(d) $\Delta E_P < 0$ and $\Delta E_K > 0$
(e) $\Delta E_P > 0$ and $\Delta E_K = 0$

7-81 A skateboarder is initially at rest at the top of a ramp (Figure 7-42). She then slides down the ramp, and attains a speed v at the bottom. To achieve a speed $2v$ at the bottom, how many times as high must a new ramp be? Neglect friction and air resistance.

(a) 2
(b) 3
(c) 4
(d) 6
(e) 8

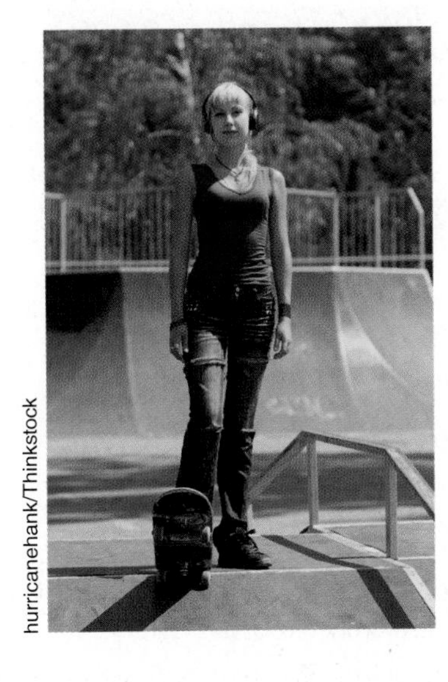
hurricanehank/Thinkstock

Figure 7-42 Question 7-81.

7-82 Two balls are dropped from the same height above the floor. One ball has twice the mass of the other one. If v_H represents the speed of the heavy ball just before it hits the floor, and v_L represents the speed of the light ball just before it hits the floor, which of the following is true? Neglect air resistance.

(a) $v_H = v_L$
(b) $v_H > v_L$
(c) $v_H < v_L$

7-83 A watt is equivalent to

(a) kg·m/s^3
(b) kg·m^2/s^3
(c) kg^2·m^2/s
(d) kg·m/s
(e) kg·m^2/s^2

7-84 The kilowatt·hour is a unit of

(a) power
(b) force
(c) energy
(d) time
(e) speed

Review Questions and Problems

7-85 Identify each quantity below as a vector quantity or a scalar quantity.

(a) work
(b) kinetic energy
(c) force
(d) velocity
(e) speed
(f) gravitational potential energy
(g) distance
(h) displacement
(i) thermal energy

7-86 Work can be expressed as $W = F\,\Delta r\cos\theta$. For constant F and constant Δr, sketch the shape of the graph of W versus θ for $0° \leq \theta \leq 360°$.

7-87 Biological muscle cells (Figure 7-43), which can be considered tiny nanomotors, transform chemical potential energy into mechanical work. If an active protein such as actin within a muscle cell produces a force of magnitude 7.2 pN over a displacement of 8.5 nm in the same direction as the force, how much work is done by the protein?

Designua/Shutterstock

Figure 7-43 Question 7-87. Actin, shown at the right in this figure, is a protein that plays an important role in muscle contraction.

7-88 What would happen to Earth's speed if the gravitational force toward the Sun did positive work on Earth?

7-89 Sketch the shape of the graph of kinetic energy E_K versus speed v for a given object.

7-90 A boy throws a stone vertically downward from a cliff with an initial speed of 4.03 m/s. Use the work-energy theorem to determine the height of the cliff if the speed of the stone at the bottom is 22.8 m/s. Neglect air resistance.

7-91 In throwing a 0.46 kg football (which is initially at rest), a young man applies a horizontal force over a horizontal displacement of 36 cm (Figure 7-44). The speed of the ball at release is 8.9 m/s. Use work and energy to determine the average magnitude of the force.

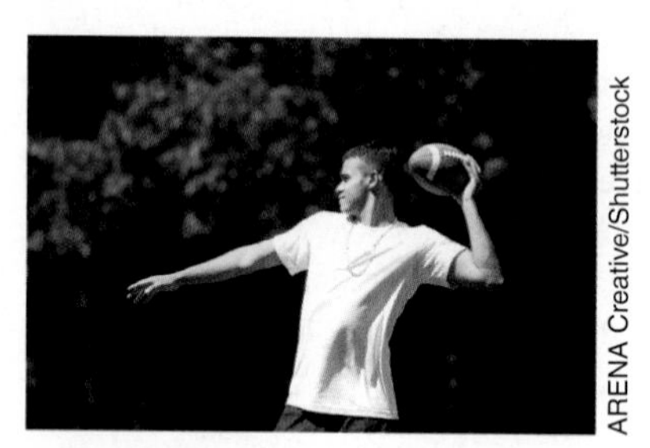

ARENA Creative/Shutterstock

Figure 7-44 Question 7-91.

7-92 A child drops a fork (from rest). If the speed of the fork is 4.3 m/s just before it hits the floor, how high was the fork above the floor when it was released? Use conservation of energy.

7-93 A hockey puck of mass 0.17 kg is sliding across a horizontal sheet of ice. The coefficient of kinetic friction between the puck and ice is 0.13. The initial speed of the puck is 3.7 m/s.

(a) What is the initial kinetic energy of the puck?

(b) Use analysis of forces to determine the magnitude of the friction force on the puck.

(c) As the puck slides 0.50 m, what is the work done by friction on the puck, and how much thermal energy is generated?

(d) What is the speed of the puck after sliding 0.50 m? (Use energy methods.)

7-94 Two men are dragging a boat of mass 57 kg straight across a level beach. Together, they are exerting a force of 3.40×10^2 N at 21° above the horizontal. The magnitude of the kinetic friction force on the boat is 2.90×10^2 N. Use work and energy to determine the speed of the boat after it has been dragged 2.3 m if it starts from rest.

7-95 The energy required to produce a metric tonne (1000 kg) of aluminum from aluminum ore is about 83 MBtu. Convert this value to megajoules. (Refer to Table 7-2.)

DID YOU KNOW?

The energy required to recycle aluminum is only about 25% of that needed to produce it from the ore.

7-96 Alpha particles (each of mass 6.6×10^{-27} kg) emitted by nuclei generally have kinetic energies in the range of 4–9 MeV (see Table 7-2 for an appropriate unit conversion factor). If an alpha particle has a kinetic energy of 4.0 MeV, what is its speed?

7-97 Suppose that you use a 60 W incandescent light bulb for 1.5 h.

(a) How much electrical energy is used (in kilowatt·hours and joules)?

(b) How much of this energy (in joules) goes into light? (Use Table 7-3.)

7-98 Table 7-5 shows the energy required per person·km for passenger transportation. Approximately how much energy is required for

(a) a person to ride a bicycle (Figure 7-45) for 17 km?

(b) a person to take an urban bus for the same distance?

Table 7-5

Approximate Energy Required per Person·km (Passenger Transportation)

Mode of Travel	Energy (in units of 10^5 J per person·km)
Bicycle	1.3
Walking	2.0
Bus (intercity)	6.3
Railroad	9.8–16
Bus (urban)	18
Automobile	23–40
Airplane	40

Figure 7-45 Question 7-98(a). A parking lot at a railway station in the Netherlands. In some countries, cycling is a very common mode of transportation.

7-99 Running at 16 km/h requires a total power output of about 0.56 kW. Determine the energy output for a person running 35 min at this speed.

7-100 A typical person of mass 70 kg (assume two significant digits) requires about 7.5×10^5 J of food energy to walk 3.0 km on level ground. How much more food energy is required if the walk is done at 21° above the horizontal? Express your answer in joules and Calories. Use Tables 7-3 and 7-2.

7-101 A colour television consumes about 145 W of electrical power. If the average price of electrical energy is 9.7¢/(kW·h), how much does it cost to watch television for 2.5 h?

7-102 Many appliances such as televisions have an "instant-on" feature, which means that the appliance is actually consuming electrical energy even when it is turned "off," in order that it will turn "on" immediately when needed. The typical power consumed by such devices when "off" ranges from about 0.5 W to 2.0 W. Suppose that a particular television consumes 2.0 W when "off," and that it is turned off when the owners are on vacation for 30 days.

(a) During the month, how much more electrical energy (in kilowatt·hours) does this television consume than one that consumes only 0.5 W when "off?"

(b) If the price of electrical energy is 9.4¢/kw·h, how much more does it cost for the month to have the 2.0 W television than a 0.5 W television?

DID YOU KNOW?

In some parts of Canada, the price of household electrical energy depends on when the energy is used. If energy is used when general electrical demand is high, the price is high; if demand is low, the price is low. For example, in Guelph, Ontario, in the summer of 2014, the price per kilowatt·hour was

- 13.5¢/(kW·h) between 11 a.m. and 5 p.m.
- 11.2¢/(kW·h) from 7 a.m. to 11 a.m. and from 5 p.m. to 7 p.m.
- 7.5¢/(kW·h) from 7 p.m. to 7 a.m. and on weekends and holidays

What do you think are advantages and disadvantages of this time-of-use pricing?

DID YOU KNOW?

Some countries such as Australia have a regulation that the maximum power allowed for appliances when turned "off" is 0.5 W.

Applying Your Knowledge

7-103 A man is pulling a vacuum cleaner of mass 5.7 kg straight across a horizontal rug by exerting a force of 19 N at an angle of 42° above the horizontal. There is a friction force on the vacuum of magnitude 12 N. If the vacuum is pulled for 1.5 m, what is the work done on it by each force? Use $W = F\,\Delta r \cos\theta$.

7-104 Solve the previous problem using $W = F_x\Delta x + F_y\Delta y$.

7-105 A baseball of mass 0.15 kg is thrown with an initial velocity of 29 m/s at an angle of 37° above the horizontal. Neglect air resistance.

(a) When it reaches its maximum height, how much work has been done on it during its flight?

(b) When it returns to its initial height, how much work has been done on it during its flight?

7-106 A waiter carrying a tray above his head accidentally inclines it at 19° to the horizontal, and a plate of mass 0.48 kg starts to slide. If the coefficient of kinetic friction between the plate and tray is 0.27, determine the total work done on the plate as it slides 41 cm along the tray.

7-107 Use trigonometry to show that $F\Delta r\cos\theta = F_x\Delta x + F_y\Delta y$.

7-108 A snowboarder of mass 73 kg (including the board) coasts up a hill inclined at 9.3° to the horizontal. Neglect friction. Use the work-energy theorem to determine how far along the hill he slides before stopping, if his initial speed at the bottom is 4.2 m/s.

7-109 A skater of mass 55.2 kg falls and then slides along the ice, undergoing a horizontal displacement of magnitude 4.18 m before stopping. If the coefficient of kinetic friction between the skater and the ice is 0.27, what was her speed when she started to slide? Use the work-energy theorem.

7-110 What is the original speed of an object if its kinetic energy increases by 50% when its speed increases by 2.00 m/s?

7-111 A boy is playing with a rope tied to a tree near his favourite swimming hole (Figure 7-46). Initially the boy is stationary and the rope (of length 3.7 m) makes an angle of 48° with the vertical. He then lifts his feet slightly and starts to swing freely. If air resistance is neglected, use conservation of energy to determine

(a) his speed at the bottom of the swing

(b) the maximum height, relative to his initial position, to which he can swing

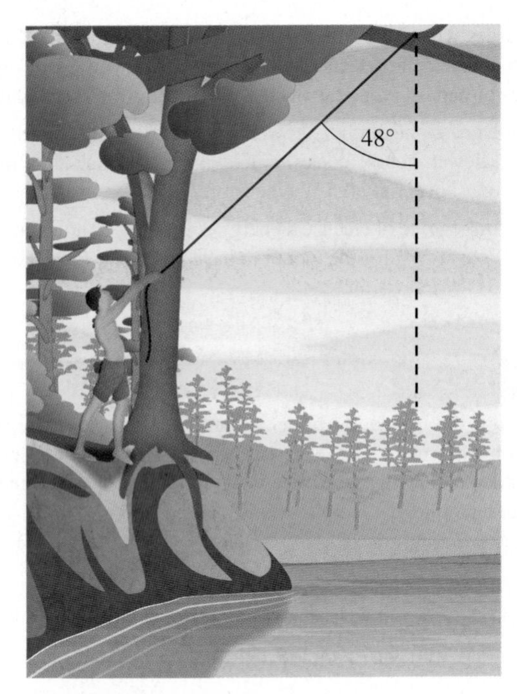

Figure 7-46 Question 7-111.

7-112 A skier slides down a slope of length 11.7 m inclined at an angle φ to the horizontal. If her initial speed is 65.7 cm/s, and her speed at the bottom of the slope is 7.19 m/s, determine φ. Use conservation of energy, neglecting friction and air resistance.

7-113 Two men hang onto opposite ends of a rope that passes over a pulley (Figure 7-47). The masses are $m_1 = 105$ kg and $m_2 = 67$ kg. Initially the men are at rest, with man #2 standing on the floor and man #1 supported with his feet 1.5 m above the floor. The support is then removed, and the men and rope are free to move. Use conservation of energy to determine the speed of the men just as man #1 touches the floor.

7-114 In a garden nursery a large box of flower seeds (mass 18 kg) is pushed 1.6 m along a horizontal floor by a woman who is exerting a force of 1.5×10^2 N at 22° below the horizontal. The coefficient of kinetic friction between the box and the floor is 0.55.

(a) Use Newton's laws of motion to determine the magnitudes of the normal force and friction force on the box.

(b) Use conservation of energy to determine the final speed of the box if it starts from rest.

(c) How much thermal energy is produced?

Figure 7-47 Question 7-113.

7-115 A snowboarder (mass 68 kg) slides down a hill inclined at 15° to the horizontal (Figure 7-48). The coefficient of kinetic friction between the snow and snowboard is 0.19. If the snowboarder starts from rest and slides 25 m down the hill, determine

(a) the magnitudes of the normal force and friction force on the snowboarder, using Newton's laws of motion

(b) the final speed of the snowboarder, using conservation of energy

(c) the thermal energy produced

Figure 7-48 Question 7-115.

7-116 In one type of tidal power plant, the incoming tide fills up a catchment area enclosed by concrete walls, and then at low tide, this water is allowed to fall through openings at the bottom of the concrete walls to spin turbines. In a particular tidal plant, the square catchment area has sides of 1.5 km, and the tide rises by 4.2 m. Assume that the process of emptying takes 2.0 h, and that all the energy of the water is converted to energy of the turbines. Determine the power (assumed constant) in megawatts available from the plant during emptying. The density of salt water is 1.03×10^3 kg/m^3. (**Hint:** When $E_P = mgy$ is used, "y" represents the elevation of the centre of mass.)

DID YOU KNOW?

Nova Scotia has one of the few tidal power plants in the world and the only one in North America (Figure 7-49). This plant takes advantage of the very high tides in the Bay of Fundy, and has a daily electrical output of 80–100 MW·h.

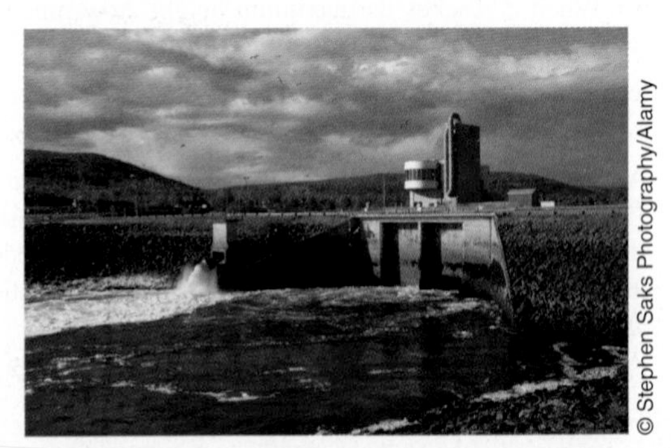

Figure 7-49 The Annapolis Tidal Power Plant in Nova Scotia.

7-117 Fermi Question:

(a) Assuming your brain operates at a total power of 20 W, estimate how much energy your brain has consumed since you were born.

(b) If your current average daily intake of energy is 10 MJ, how many days' worth of energy could make up for your brain's energy consumption that you determined in (a)?

7-118 The intensity of radiation from a distant source such as the Sun is inversely proportional to the square of the distance from the source. (We will discuss this "inverse square law" in more detail in Chapter 9.) At the top of Earth's atmosphere, the solar intensity (power per unit area) is 1.35 kW/m^2. If a spaceship is located halfway between the Sun and Earth, and has a solar panel of area 1.00×10^2 m^2, determine the total solar energy incident on the panel in a day.

7-119 An important force acting on cyclists (Figure 7-50) is air resistance, or drag. The magnitude of the drag force on a particular cyclist travelling at 6.5 m/s (or 23 km/h) is measured to be 8.9 N.

(a) Determine the work by the drag force on the cyclist in 1.0 s.

(b) In order to travel at constant velocity, how much energy must the cyclist provide in 1.0 s just to balance the work done by the drag force?

(c) What power must the cyclist provide to balance the power of the drag force?

(d) The drag force is proportional to the square of the speed of a moving object. If the speed of the cyclist is doubled to 13 m/s, determine the new magnitude of the drag force. What is the work done now by the drag force in 1.0 s? What is the power that the cyclist must provide because of drag? What is the ratio of this power to the power needed at 6.5 m/s? If the power P required to balance the drag power is written as $P \propto v^n$, where v is speed, what is the value of the exponent n?

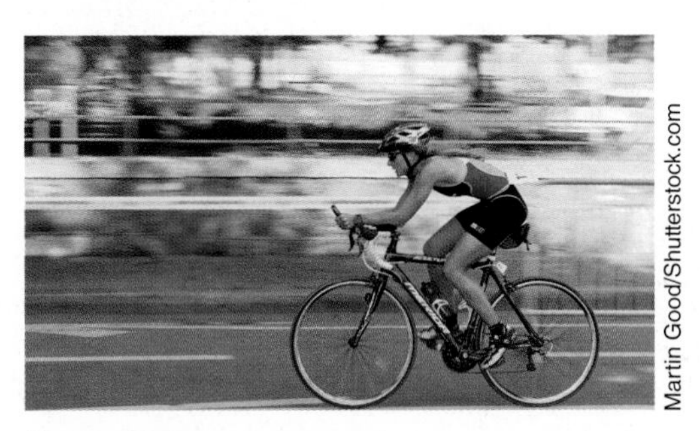

Figure 7-50 Question 7-119.

7-120 A woman is pushing a box straight across a horizontal floor with a constant speed of 0.74 m/s. The mass of the box is 24 kg, and the magnitude of the friction force on it is 98 N.

(a) During a time of 2.5 s, how much work does the woman do on the box?

(b) What is the power output of the woman (for the pushing)?

(c) If the woman has an efficiency of 25% for conversion of food energy to useful work, what is the internal power requirement of the woman when pushing (over and above the power required when resting)?

7-121 (a) Calculate the water flow-rate (in kilograms per second) over a waterfall producing electrical power of 97 MW from a turbine generator that is 93% efficient. The waterfall has a vertical height of 47 m. Neglect the kinetic energy of the water before it goes over the fall, and after it leaves the turbine.

(b) In a time of one week, the electrical energy consumption in a typical Canadian home is 150 kW·h. What is the average power requirement (in watts) for a typical home?

(c) How many typical homes could be supplied with electricity from the waterfall in (a)?

7-122 In a nuclear reactor, nuclear energy is converted to thermal energy by the fission (breakup) of nuclei of uranium-235. The energy content per unit mass of uranium-235 is 8.4×10^7 MJ/kg. What mass of uranium-235 is needed per year in a nuclear electrical generating plant that produces 1.0×10^3 MW of electrical power with 32% efficiency?

7-123 A stretched elastic band of mass 0.55 g is released so that its initial velocity is horizontal and its initial position is 95 cm above the floor. When it lands it has a horizontal displacement of 3.7 m from the initial point. What was the elastic potential energy stored in the stretched band? Neglect air resistance.

Momentum and Collisions 8

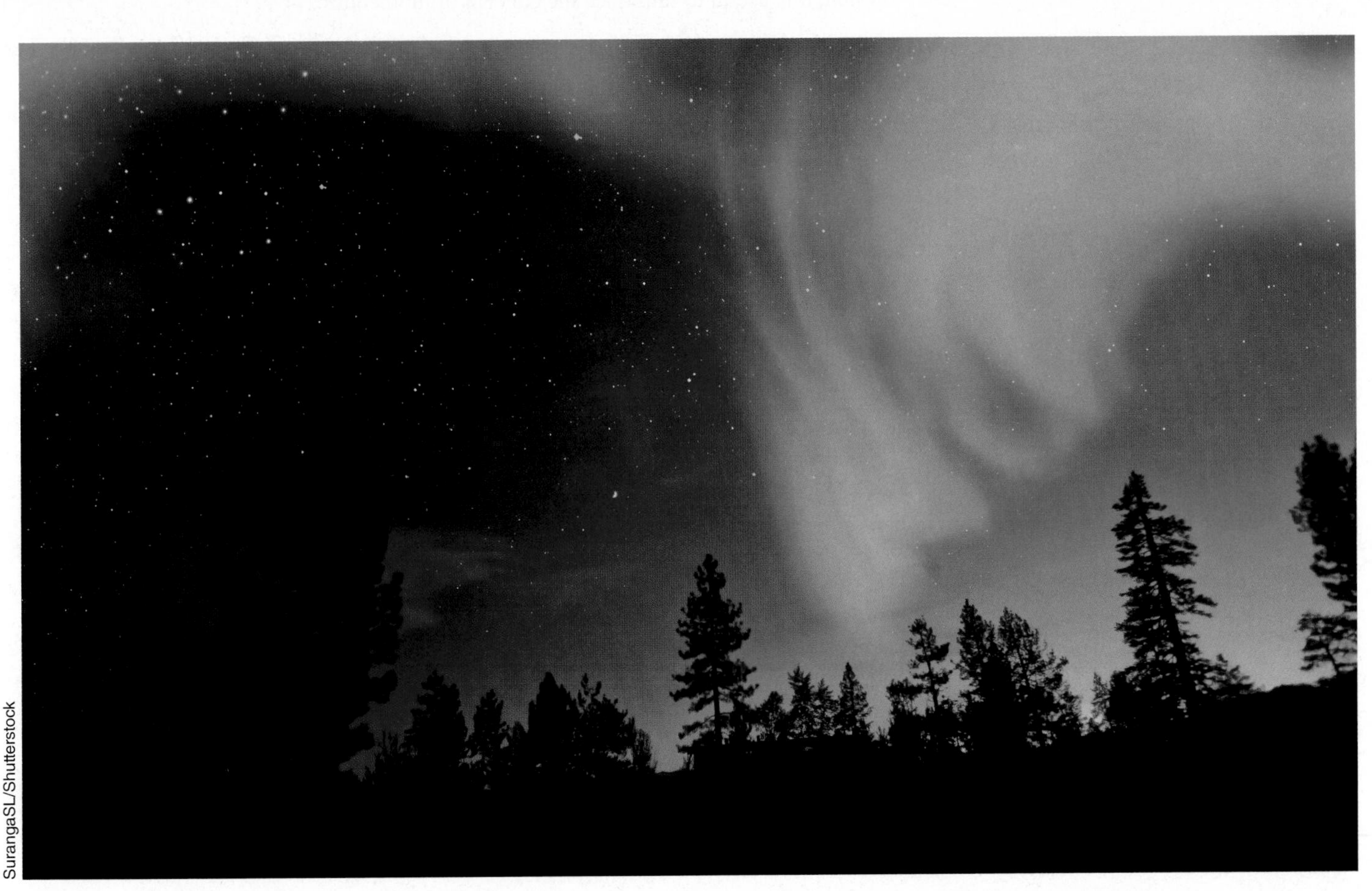
SurangaSL/Shutterstock

Figure 8-1 The spectacular display of the Northern Lights arises from collisions of cosmic rays from the Sun with our atmosphere.

Collisions are occurring around you all the time. Molecules in the air are continually colliding with each other and with objects (such as you) near them. Balls collide with tennis racquets, baseball bats, and golf clubs. Automobiles collide. The diffusion of molecules across biological cell membranes is the result of random molecular collisions. Energetic particles from the Sun undergo collisions with molecules in Earth's atmosphere to produce the northern and southern lights (*aurora borealis* and *aurora australis*; Figure 8-1). Even entire galaxies collide with each other. In order to understand collisions, and thereby the universe around us, we first need to study momentum and the law of conservation of momentum. We shall see that momentum is also useful in explaining processes such as rocket propulsion.

CHAPTER **OUTLINE**

8.1 Momentum

At one time or another, you have probably hit a baseball with a bat, a tennis ball with a racquet, or a golf ball with a club. In previous chapters the motion of objects such as baseballs, tennis balls, and golf balls has been described in terms of quantities such as velocity, acceleration, and kinetic energy. In order to discuss what happens in a collision, it is useful to introduce the concept of momentum.

momentum the product of the object's mass and velocity, usually represented by the symbol $\vec{p}$

The **momentum** of an object is defined as the product of the object's mass and velocity, and is usually represented by the symbol $\vec{p}$. Writing mass as m and velocity as $\vec{v}$, we have

$$\vec{p} = m\vec{v} \tag{8-1a}$$

Momentum depends on both the mass and velocity of an object. A large truck has more momentum than a small car travelling at the same speed. But it is possible for a rapidly moving small car to have the same momentum as a slow large truck.

Note that momentum is a vector, being the product of a scalar (mass) and a vector (velocity). Based on the definition given in Eqn. (8-1a), momentum will always be in the same direction as that of the velocity of the object. We will often be working with components of momentum:

$$p_x = mv_x \quad \text{and} \quad p_y = mv_y \tag{8-1b}$$

Since the SI units of mass and velocity are kilogram and metre/second respectively, the SI unit of momentum is kilogram·metre/second (kg·m/s). This unit does not have any special name (such as a newton or a joule) and is the standard unit of momentum.

Strictly speaking, what we are referring to as momentum is more correctly called linear momentum. However, *linear* is often omitted, except when there could be confusion with another quantity called angular momentum, which is discussed in detail in Chapter 10.

Momentum, Force, and Time

In Chapter 7, we saw how change in kinetic energy is related to force and displacement. We now investigate how change in momentum is associated with force and time. We begin by considering a one-dimensional situation; our example is a collision between a baseball and a bat to produce a line-drive, with the ball travelling horizontally just before and just after the collision (Figure 8-2).

© Chuck Savage/Corbis

Figure 8-2 A baseball being hit by a bat is an excellent example of the physics of collisions.

Suppose that during the collision with the bat, the baseball undergoes an acceleration (a_x) in a time interval (Δt) and its velocity changes from v_{0x} to v_x (Figure 8-3). These quantities are related by Eqn. 2-7 for constant-acceleration kinematics:

$$v_x = v_{0x} + a_x \Delta t$$

(The original form of Eqn. 2-7 used t instead of Δt, assuming that $t_0 = 0$. However, in dealing with momentum, it is customary to use Δt.)

For now, we assume that a_x is constant. The more general case of variable a_x, as would be the situation in an actual collision of a ball with a bat, is discussed later.

If we multiply the above equation by the mass (m) of the baseball and then re-arrange the equation slightly, we get

$$mv_x = mv_{0x} + ma_x\Delta t$$

$$mv_x - mv_{0x} = ma_x\Delta t$$

On the left-hand side of this equation, we now have the *change* in the baseball's momentum, that is, the final momentum (mv_x) minus the initial momentum (mv_{0x}). We can write this simply as Δp_x. Hence,

$$\Delta p_x = ma_x\Delta t$$

On the right-hand side, Newton's second law can be used to replace ma_x with the resultant force (ΣF_x or F_{Rx}) exerted by the bat on the ball:

$$\Delta p_x = (\Sigma F_x)\Delta t \quad (8\text{-}2)$$

This equation states that the *change in momentum equals the product of the resultant force and the time interval*. It is valid whenever any object's momentum changes. To change the momentum of, say, a car by a given amount, there is an infinite number of combinations of force and time intervals that could be used. For example, if we want to reduce a car's momentum (and thus its speed), we can apply the brakes strongly for a short time interval, or apply them gently for a longer time interval.

The relationship between momentum, force, and time interval could also be developed in the y- and z-directions:

$$\Delta p_y = (\Sigma F_y)\Delta t \quad \text{and} \quad \Delta p_z = (\Sigma F_z)\Delta t$$

Therefore, we can write a vector equation relating the change in the momentum vector to the (resultant) force vector and the time interval:

$$\Delta \vec{p} = (\Sigma \vec{F})\Delta t \quad \text{or} \quad \Delta \vec{p} = \vec{F}_R\Delta t \quad (8\text{-}3)$$

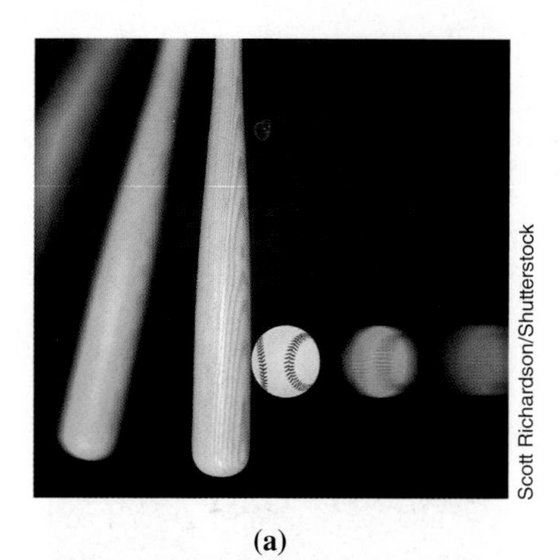

(a)

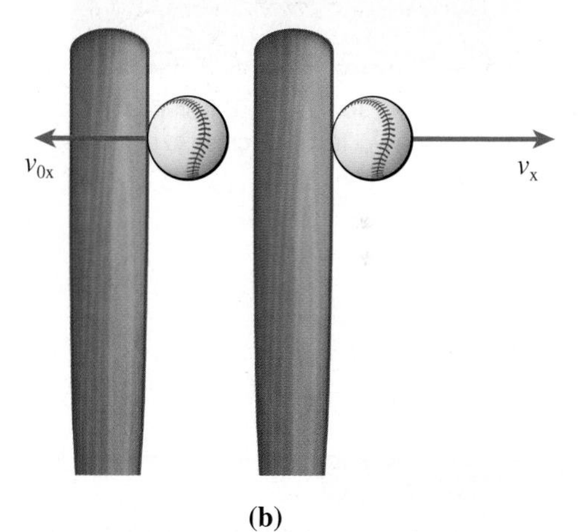

(b)

Figure 8-3 **(a)** Initial contact of ball and bat. **(b)** Final contact, after a time interval Δt.

This vector equation is very compact and shows that the change in momentum and the resultant force have the same direction. However, when solving problems it is usually easiest to work with individual components.

Sometimes the change in momentum is called **impulse**. However, since impulse is just the same as change in momentum, this additional term will not typically be used in this book.

impulse the change in an object's momentum ($\Delta \vec{p}$)

You are accustomed to seeing Newton's second law of motion written as $\Sigma \vec{F} = m\vec{a}$, but by re-arranging $\Delta \vec{p} = (\Sigma \vec{F})\Delta t$, we can express the second law in a different way:

$$\Sigma \vec{F} = \frac{\Delta \vec{p}}{\Delta t} \text{ (Newton's second law restated)} \quad (8\text{-}4a)$$

In other words, the resultant force on an object equals the time rate of change of the object's momentum. This is actually the way in which Newton originally stated his second law, although he referred to momentum as "quantity of motion." In terms of x-, y-, and z-components, we can write

$$\Sigma F_x = \frac{\Delta p_x}{\Delta t} \quad \Sigma F_y = \frac{\Delta p_y}{\Delta t} \quad \Sigma F_z = \frac{\Delta p_z}{\Delta t} \quad (8\text{-}4b)$$

TRY THIS!

Catching a Baseball

You can make use of Eqn. 8-3 when you catch a baseball. In order to stop the ball, you apply a force. Eqn. 8-3 tells us that we can either use a small force over a longer period of time or a large force over a shorter period of time for the same change in momentum of the ball. If you allow your hands to travel backward with the ball as it slows down, the time interval of the collision increases and the force you apply decreases. This means that the force applied by the ball on your hand also decreases—it hurts a lot less this way!

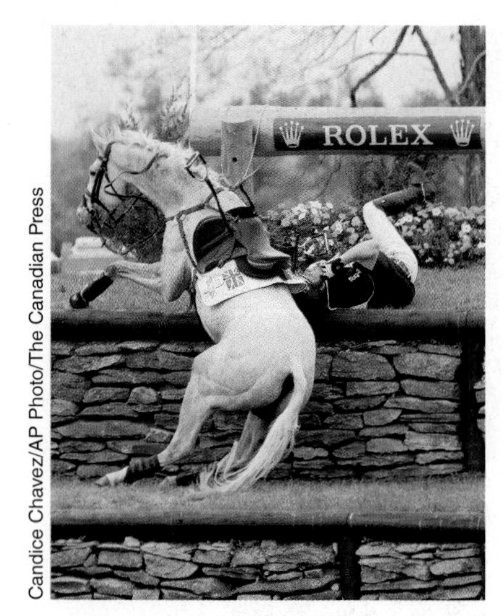

Candice Chavez/AP Photo/The Canadian Press

Figure 8-4 Oliver Townend, a British rider, attributes his survival from this crash to the airbag inside his safety vest.

Contact Time during a Collision

If we are trying to stop an object in motion with an applied force, Eqn. 8-3 tells us that we can use a small force over a longer time interval or a large force over a shorter time interval to achieve our goal. In many situations, such as catching a baseball or using an air bag to prevent injuries during a car crash, it is important to minimize the force needed to stop the moving object. This principle is at the forefront of many safety devices. Safety equipment for sports has extensive padding to increase the collision time due to the "give" of the material, thereby decreasing the necessary force applied. Nylon rope is used for safety tethers in rock climbing, as it retains elasticity even when extended to full length, thereby increasing the time interval over which the falling climber's momentum is stopped. The cables used on aircraft carriers to stop the aircraft as they land are designed to increase the time over which the planes are being slowed down, thereby minimizing the force exerted on the plane and its occupants. A recent development in safety equipment for equestrian sports has been attributed to saving the life of a rider in a 2010 accident in a Kentucky event (Figure 8-4). Oliver Townend was wearing a vest that contains an airbag that is triggered to deploy as soon as the rider leaves the saddle. This vest serves the same purpose as an airbag in a vehicle (see Did You Know?) and is designed to minimize the risk of injury from a fall from the horse.

Think about the opposite scenario of trying to get something moving with lots of speed. By examining Eqn. 8-3, we can see that increasing the contact time between the two objects will serve this purpose. Again returning to the sports world, in many cases athletes are encouraged to swing with "follow through," for example, in baseball, tennis, golf. When an athlete follows through in his/her swing, the contact time with the ball increases and therefore increases the change in momentum of the object for a given applied force.

DID YOU KNOW?

Airbags in motor vehicles are designed to reduce the net force exerted on a person by increasing the time over which the momentum of that person decreases when the vehicle stops suddenly. Since their wide-spread introduction to vehicles in the late 1990s, statistics show that airbags reduce the risk of a fatal injury in a head-on collision by about 30%. In addition, automobile design employs "crumple zones" to absorb much of the energy of collision through controlled deformation. These crumple zones are typically at the front of the vehicle to protect occupants in the case of head-on collisions, although they can be found at the rear and sides in some makes and models.

Average Force

So far, we have assumed that the acceleration of the object being considered (a baseball) is constant; this implies, of course, that the resultant force on the object is constant. In reality, in many circumstances involving change of momentum, the resultant force and the acceleration are not constant. Use of Eqn. 8-3 remains valid in such circumstances, as long as $\Sigma\vec{F}$ is interpreted as the *average* force acting on the object during the time interval Δt.

What exactly do we mean by average force? To answer this, consider Figure 8-5 (a), which shows the typical behaviour of the magnitude of the resultant force on an object as a function of time during a collision of duration Δt. It could represent, for instance, the magnitude of the resultant force acting on a male elk smashing antlers with another male during the rut. The area under the graph has been shaded. The average force is defined as the constant force, which, if applied for the same time interval Δt, would give the same amount of shaded area (Figure 8-5 (b)).

Since the resultant force on the elk during the collision is not constant over time, it is not obvious how we would use Eqn. 8-3 to determine the change in momentum. Which

TRY THIS!

Concussion Prevention

There has been a lot of discussion in the news media in recent years concerning concussions suffered by amateur and professional athletes in a number of sports. Through research online, investigate the key features necessary in the design of an effective helmet to protect hockey players from this very serious injury.

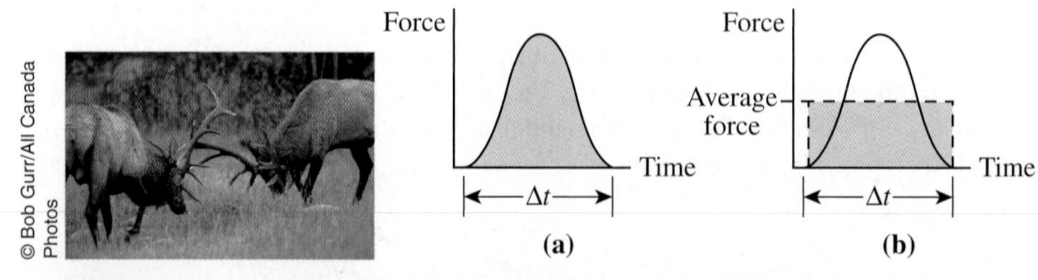

Figure 8-5 **(a)** Magnitude of force acting on an object during a typical collision. **(b)** The average force, acting for Δt, gives the same shaded area as in **(a)**.

value of $\Sigma\vec{F}$ should you use? Over what time interval? Integral calculus tells us that the change in the elk's momentum is the shaded area under the curve in Figure 8-5 (a), since it represents the product of the resultant force and time throughout the collision. So, by defining the average force as shown in Figure 8-5 (b), we can use Eqn. 8-3 with the value of the average force on the elk, since this gives rise to the same shaded area under the curve and therefore the same value of change in momentum.

SAMPLE PROBLEM 8-1

A tennis ball of mass 0.15 kg, travelling horizontally with a speed of 38 m/s, hits a racquet. Immediately after the collision, of duration 1.1 ms, the ball travels in the opposite direction horizontally with a speed of 51 m/s.

(a) What are the magnitude and direction of the initial momentum of the ball?

(b) What is the average force (magnitude and direction) exerted on the ball by the racquet?

(c) What is the ratio of the magnitude of this force to the magnitude of the force of gravity on the ball?

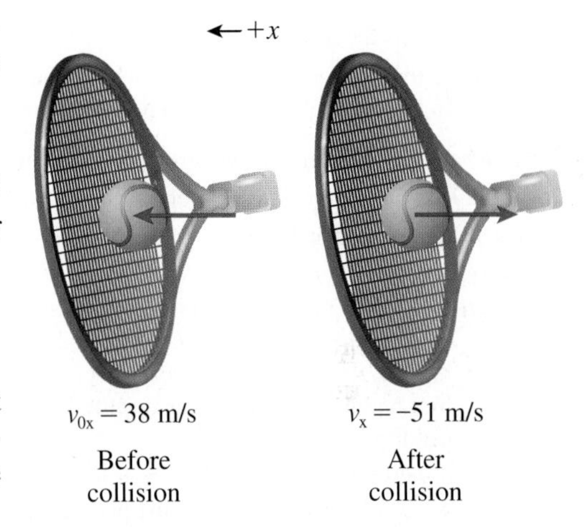

Figure 8-6 Sample Problem 8-1.

Solution

(a) We start by choosing a coordinate system. The problem is one-dimensional, and it matters little in this problem which direction we choose for $+x$. We choose it arbitrarily to be in the direction of the initial velocity of the ball (Figure 8-6). Thus, the x component of the initial velocity, v_{0x}, equals $+38$ m/s.

The initial momentum of the ball in the x-direction is

$$mv_{0x} = (0.15\,\text{kg})(38\,\text{m/s}) = 5.7\,\text{kg}\cdot\text{m/s}$$

Thus, the ball has an initial momentum of 5.7 kg·m/s in the same (horizontal) direction as the initial velocity.

(b) In order to determine the average force on the ball, we use Eqn. 8-4b:

$$\Sigma F_x = \frac{\Delta p_x}{\Delta t} = \frac{m(v_x - v_{0x})}{\Delta t}$$

Before substituting known quantities, it is important to note that v_x is a *negative* quantity (-51 m/s) since the ball's velocity after the collision is in the $-x$ direction. As well, we must remember to express Δt in seconds ($\Delta t = 1.1\text{ ms} = 1.1 \times 10^{-3}\text{ s}$). Now we substitute numbers

$$\Sigma F_x = \frac{(0.15\,\text{kg})(-51\,\text{m/s} - 38\,\text{m/s})}{1.1 \times 10^{-3}\,\text{s}} = -1.2 \times 10^4\,\text{N}$$

The average force exerted by the racquet on the ball (i.e., the resultant force on the ball) has a magnitude of 1.2×10^4 N and a direction (indicated by the negative sign) opposite to the ball's initial velocity (in the $-x$-direction).

(c) The force of gravity on the ball has a magnitude of

$$\begin{aligned} F_G &= mg \\ &= (0.15\,\text{kg})(9.8\,\text{m/s}^2) \\ &= 1.5\,\text{N} \end{aligned}$$

Therefore, the required ratio of the force exerted by the racquet to the force of gravity is

$$\frac{|\vec{F}_{\text{racquet}}|}{|\vec{F}_G|} = \frac{1.21 \times 10^4\,\text{N}}{1.47\,\text{N}} = 8.3 \times 10^3$$

Units Tip

Note the SI unit of momentum: mass in kg multiplied by velocity in m/s gives momentum in kg·m/s.

The average force exerted by the racquet is more than eight thousand times larger than the force of gravity. The forces between objects involved in collisions are often much larger than other forces such as gravity, and thus we typically can neglect other forces when analyzing collisions.

EXERCISES

8-1 (a) What is the SI unit of momentum?

(b) In the relation $\Delta\vec{p} = (\Sigma\vec{F})\Delta t$, what is the SI unit of $\Delta\vec{p}$ (from (a))? of $(\Sigma\vec{F})\Delta t$? Are these units the same?

8-2 What are the magnitude and direction of the momentum of

(a) a sprinter of mass 65 kg running east at 9.8 m/s?

(b) a red blood cell of mass 1.2×10^{-13} kg moving downward at a speed of 4.2 cm/s in an artery?

(c) an English bulldog of mass 24 kg travelling west at constant velocity on a skateboard, covering a distance of 1.0×10^2 m in 19.678 s? (This is the World Record, by the way, achieved by a bulldog named Tillman on July 30, 2009 in Los Angeles, California.)

(d) a 950 g diving Peregrine falcon moving at 320 km/h downward? (The fastest bird on the planet, a Peregrine falcon can reach diving speeds of about 350 km/h.)

8-3 **Fermi Question:** Compare the magnitude of your own momentum when you are running at top speed with that of an automobile travelling at the speed limit on a major highway.

8-4 An automobile speeds up as it leaves an intersection. Sketch the shape of the graph of the magnitude of the automobile's momentum as a function of its speed.

8-5 The momentum of a ball thrown at the target on a dunk tank has a magnitude of 4.8 kg·m/s. If the ball has a mass of 0.15 kg, what is the ball's speed?

8-6 What is impulse?

8-7 An egg of mass 59 g is dropped onto the floor, where it breaks and comes to rest. Its speed just before hitting the floor is 4.3 m/s. What are the magnitude and direction of

(a) its momentum before hitting the floor?

(b) its momentum after hitting the floor?

(c) its change in momentum?

8-8 The magnitude of the momentum of a car is 2.5×10^4 kg·m/s. A force of magnitude 2.1×10^3 N increases the car's momentum for 3.6 s. What is the magnitude of the car's new momentum?

8-9 A tennis ball of mass 59 g, having an initial velocity of 26 m/s horizontally, hits a net and comes to rest (then drops to the ground). If the force exerted on the ball by the net (the "net" force) is 1.3×10^2 N horizontally, how long is the ball in contact with the net during the collision? Express your answer in milliseconds.

8-10 A rubber ball of mass 95 g is dropped onto the floor. Just before it hits the floor, its speed is 4.4 m/s. It then rebounds upward, having a speed of 3.9 m/s just as it leaves the floor. What are the magnitude and direction of

(a) its momentum just before hitting the floor?

(b) its momentum just after hitting the floor?

(c) its change in momentum?

(d) the average force exerted on the ball by the floor, if they are in contact for a time interval of 4.2×10^{-3} s? Use momentum methods.

8-11 In the previous question, why can the force of gravity be neglected during the collision between the ball and floor?

8-12 A bird of mass 9.87×10^{-2} kg is flying with a velocity of 25.7 m/s at 11.7° above the horizontal. Determine its horizontal and vertical components of momentum.

8-13 A nitrogen molecule (mass 4.7×10^{-26} kg) in the air bounces off a wall (Figure 8-7). Its speed just before and after the collision is the same, 5.1×10^2 m/s.

(a) Draw a vector scale diagram, showing the initial momentum and final momentum of the nitrogen molecule.

(b) Use your diagram to perform a vector subtraction, thus giving the change in momentum (magnitude and direction) of the molecule.

(c) Use your result in (b) to determine the average resultant force (magnitude and direction) on the molecule during the collision if the molecule is in contact with the wall for 2.0×10^{-13} s.

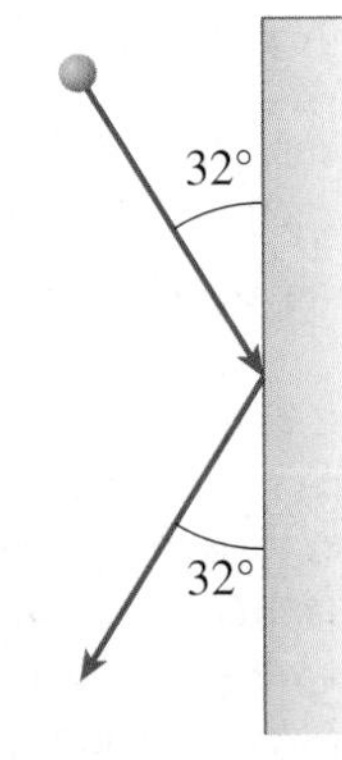

Figure 8-7 Question 8-13.

8-14 A stationary golf ball of mass 0.045 kg is hit by a club and leaves with a velocity of 2.4×10^2 km/h at 19° above the horizontal. The ball and club are in contact for 0.97 ms.

(a) Use momentum to determine the magnitude and direction of the average force exerted by the club on the ball. (Choose your $+x$ axis to be in the forward horizontal direction.)

(b) What is the average force (magnitude and direction) exerted by the ball on the club?

8.2 Conservation of Momentum in One Dimension

Have you ever had the experience of walking in a small untethered boat toward the dock and having the boat moving backward under you? It can be awkward to discover that, when you reach the end of the boat, it is no longer close enough to the dock for you to get out. This phenomenon can be explained by Newton's third law (you exert a backward force on the boat, and it exerts a forward force on you), but it also can be explained by a closely related law of nature: the **law of conservation of momentum**. This law is useful in a wide range of circumstances, such as walking in a boat, positron emission tomography (PET) scans used in medical diagnoses, rocket propulsion, and in many collisions of objects.

law of conservation of momentum If the resultant force acting on a system is zero, then the momentum of the system is constant.

To understand when the law of conservation of momentum is useful, we start with Eqn. 8-3, which relates the change in momentum, resultant force, and time interval:

$$\Delta\vec{p} = \vec{F}_R\Delta t$$

If the resultant force ($\vec{F}_R$) acting on an object is zero, then the change in momentum ($\Delta\vec{p}$) of the object is zero. Therefore, the momentum is constant or conserved. (This is really just another way of phrasing Newton's first law, which states that if the resultant force on an object is zero, its acceleration is zero, which means that its velocity is constant and therefore its momentum is also constant.)

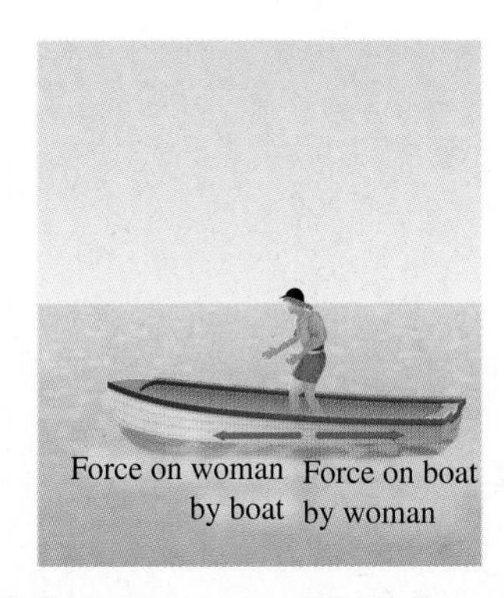

Figure 8-8 As the woman walks in the boat, the force exerted on the boat is equal and opposite to the force exerted on the woman. Hence, the resultant force on the boat–woman system is zero, and the momentum of this system is constant.

Conservation of momentum is equally valid for systems (collections) of objects. If the resultant force acting on a system is zero, then the momentum of the system is constant. This is the law of conservation of momentum.

To illustrate this law, we apply it to the case of a woman walking in a boat (Figure 8-8). In order to walk, the woman exerts a backward force on the boat, and by Newton's third law, the boat exerts an equal but opposite force on the woman. If we consider the system of the boat and woman together, these two forces will add to give a resultant force of zero. (We are neglecting the viscous friction between the boat and the water, and since there is no vertical acceleration, there must be a resultant force of zero vertically.) Therefore, the total momentum of the boat and woman is constant.

Suppose that both the boat and woman are initially stationary. The momentum of each is zero, and the total momentum of the boat–woman system is therefore zero (Figure 8-9 (a)). If the woman now starts walking, the total momentum of the system is constant; that is, it must remain zero (Figure 8-9 (b)). Mathematically,

$$m_1\vec{v}_1 + m_2\vec{v}_2 = 0$$

where "1" refers to the woman and "2" to the boat. The momentum of the woman in one direction is equal and opposite to the momentum of the boat in the other direction, and the total momentum is zero. (Remember that momentum is a vector.) Notice that the momentum of just the woman is *not* constant: initially she has zero momentum, and when walking, she has non-zero momentum. Similarly, the momentum of just the boat is not constant. When dealing with conservation of momentum, it is important to identify the system on which the resultant force is zero, and which therefore has constant momentum.

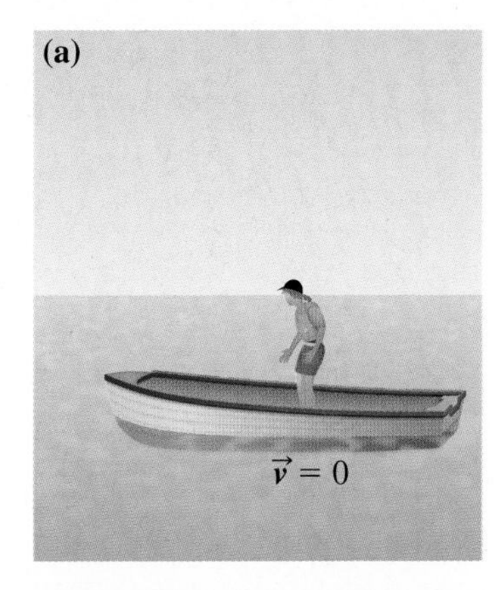

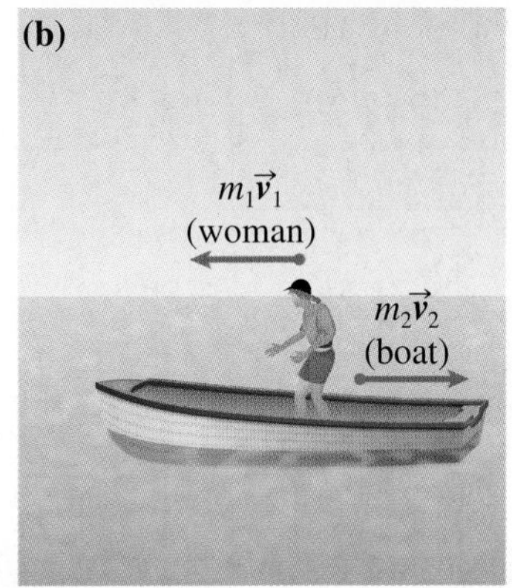

Figure 8-9 **(a)** If the boat and woman are stationary, the total momentum of the system is zero. **(b)** If the woman now walks, the total momentum is still zero; that is, $m_1\vec{v}_1 + m_2\vec{v}_2 = 0$.

SAMPLE PROBLEM 8-2

A man of mass 75 kg is standing in a stationary boat of mass 55 kg. He then starts walking toward one end of the boat at a speed of 2.3 m/s relative to the water. What is the resulting velocity (magnitude and direction) of the boat relative to the water? Neglect viscous friction between the boat and water.

Solution

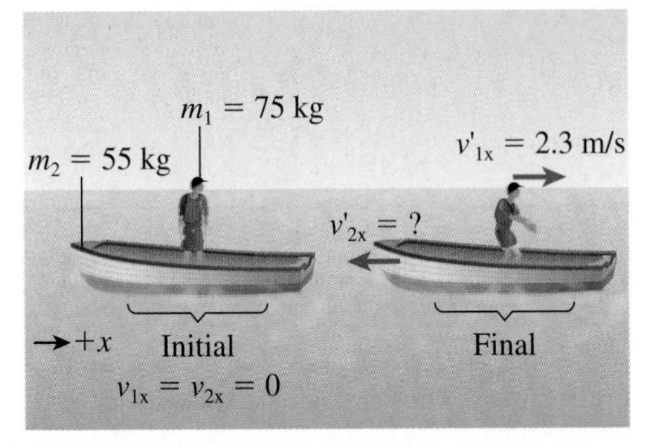

Figure 8-10 Sample Problem 8-2.

In problems involving conservation of momentum, it is extremely useful to draw diagrams showing the initial and final situations (Figure 8-10). A coordinate system is then chosen; in this problem, we arbitrarily select the $+x$ direction to lie in the direction of the man's final velocity. By the law of conservation of momentum, we know that, in the boat–man system, the initial momentum equals the final momentum, since there is no resultant force acting on this system. Mathematically,

$$m_1v_{1x} + m_2v_{2x} = m_1v'_{1x} + m_2v'_{2x}$$

where subscripts "1" and "2" refer to the man and boat respectively. The terms on the left-hand side of the equation are the initial conditions, while the terms on the right-hand side, with the primed superscript (′), are the final conditions.

In this problem, $v_{1x} = v_{2x} = 0$, since the man and boat are initially stationary. Thus, we can write

$$0 = m_1v'_{1x} + m_2v'_{2x}$$

Solving for the magnitude of the final velocity of the boat (v'_{2x}) and substituting numbers,

$$v'_{2x} = \frac{-m_1v'_{1x}}{m_2} = \frac{-(75\text{ kg})(2.3\text{ m/s})}{55\text{ kg}} = -3.1\text{ m/s}$$

Therefore, the boat's final velocity is 3.1 m/s in the opposite direction (indicated by the negative sign) to the man's final velocity.

SAMPLE PROBLEM 8-3

At the amusement park, two friends are having fun in the bumper cars. Hannah and her bumper car, with a combined mass of 108 kg, are travelling at a speed of 9.1 m/s. Mara and her bumper car, with a combined mass of 91 kg, are travelling directly toward Hannah with a speed of 6.3 m/s. After the collision, the bumper cars are stuck together. What is the speed of this pair just after the collision?

Solution

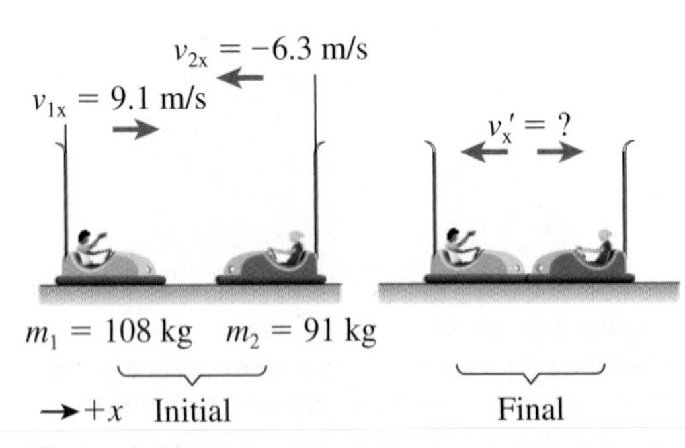

Figure 8-11 Sample Problem 8-3.

The initial and final diagrams are given in Figure 8-11. The $+x$-axis has been chosen in the direction of the initial velocity of Hannah. Therefore, the initial velocity of Mara is negative.

During the collision, there is no resultant force on the two-bumper car system. (The horizontal forces exerted between the cars are much larger than friction, which can therefore be neglected. In the vertical direction, there is no acceleration, which reflects no vertical resultant force.)

Therefore, the momentum of this system is constant. As in the previous problem we write

$$m_1v_{1x} + m_2v_{2x} = m_1v'_{1x} + m_2v'_{2x}$$

where "1" and "2" refer to Hannah and Mara respectively. The conditions before the collision have no superscript; those after the collision have a primed superscript. In this problem the two girls have the same final velocity since their bumper cars are stuck together, and so we write the x-components v'_{1x} and v'_{2x} each as v'_x:

$$m_1 v_{1x} + m_2 v_{2x} = (m_1 + m_2) v'_x$$

Solving for the unknown v'_x and substituting known numbers:

$$\begin{aligned} v'_x &= \frac{m_1 v_{1x} + m_2 v_{2x}}{m_1 + m_2} \\ &= \frac{(108\ \text{kg})(9.1\ \text{m/s}) + (91\ \text{kg})(-6.3\ \text{m/s})}{(108 + 91)\ \text{kg}} \\ &= +2.1\ \text{m/s} \end{aligned}$$

Therefore, the final velocity of the girls is 2.1 m/s in the direction (indicated by the positive sign) of the initial velocity of Hannah.

? TACKLING MISCONCEPTIONS

Is Momentum Constant in All Collisions?

Students often have the mistaken impression that momentum of a system is constant in *all* collisions. However, there are many collisions in which the resultant force on the particular colliding objects being considered is not zero, and therefore the momentum is not constant. For instance, when a person jumps from a ladder to a wooden deck, the momentum of the person–deck system is not constant because there is a huge normal force exerted by the deck supports (and the ground) during the collision. Stated another way, the deck is not free to move and cannot take up the momentum of the person. However, if we were to consider the system of the person, the deck, and Earth, momentum is conserved, because Earth is free to take up the person's momentum. Because the mass of Earth is very large compared to that of the person, the change in Earth's velocity when the person lands is, of course, very small.

Another type of situation in which the momentum of a particular pair of colliding objects is not constant occurs when one of the objects is constrained to rotate about hinges or an axle. For example, suppose that a snowball is thrown against an open gate (Figure 8-12), thus causing it to rotate. During the collision, there is a large force exerted on the gate by the hinges to prevent the gate from simply being forced backward by the ball without rotating; hence, the momentum of the ball–gate system is not constant. However, if the system being considered is the snowball, gate, and Earth, then the momentum of this system *is* conserved. In other words, momentum is always conserved when a collision occurs, but the system has to be chosen to include all the relevant objects that have momentum before and after the collision.

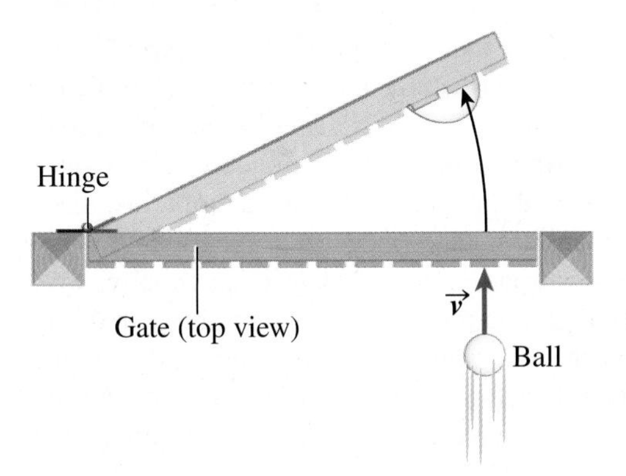

Figure 8-12 During the collision between a snowball and a hinged gate, the momentum of the ball–gate system is not constant because the hinges exert a non-negligible force on the gate.

DID YOU KNOW?

When rocket propulsion was first being discussed in the early 1900s, some scientists thought that rockets would never work outside Earth's atmosphere because they would have nothing to push against.

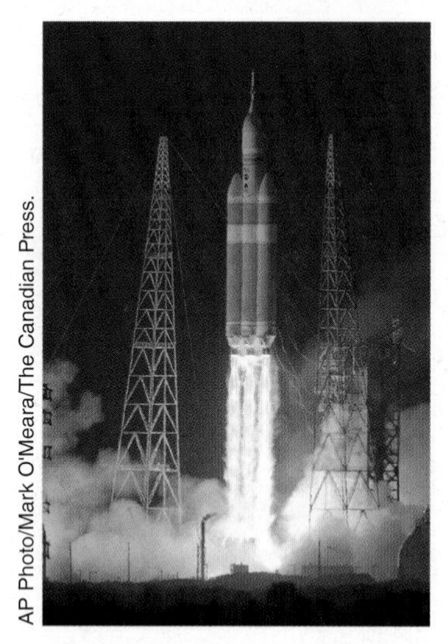

Figure 8-13 In this launch in December, 2014, a Delta IV Heavy rocket lifts off from Cape Canaveral, Florida, carrying an Orion spacecraft, which is being tested for future use in transporting humans to exciting destinations such as Mars or an asteroid.

Applications of Conservation of Momentum

1) Rocket Propulsion

An interesting application of conservation of momentum is rocket propulsion (Figure 8-13). Imagine a rocket ship at rest somewhere in space (Figure 8-14 (a)). It fires its engines—what this means is that hot gases are ejected from them. To eject these gases, the rocket ship must exert a force on them, and the gases exert an equal but opposite force on the rocket ship. The resultant force on the system of the rocket ship and gases together is zero, and thus the momentum of the system consisting of the rocket ship and gases together is constant. Since the rocket ship and its fuel (which provides the gases) are initially stationary, the momentum of the system is initially zero, and must remain zero. Thus, as the gases are ejected in one direction, the rocket ship moves in the opposite direction (Figure 8-14 (b)). In the same way, the shuttle launch in Figure 8-13 shows the gases being pushed down by the engines, leading to the shuttle being pushed up.

As a rocket ship continues to fire its engines, its fuel supply becomes smaller. Thus, there is a decrease in the mass of the rocket ship and remaining fuel. As a consequence of this variable mass, the detailed mathematics of rocket propulsion, which we will not present here, is more complicated than situations in which the mass of the system remains constant. However, the basic principle—conservation of momentum—is the same.

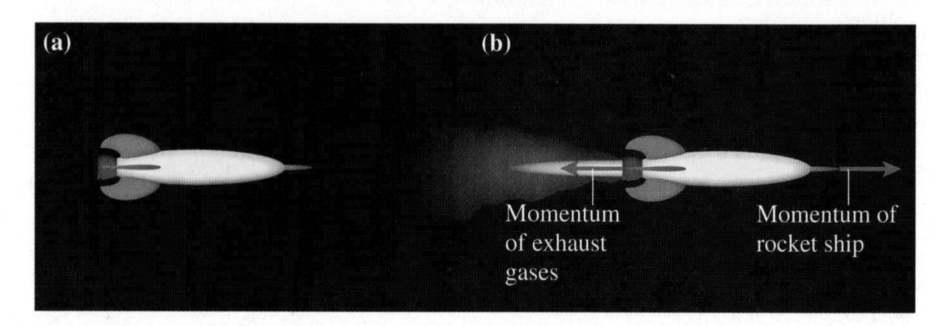

Figure 8-14 **(a)** Rocket ship at rest; momentum equals zero. **(b)** Rocket ship with engines firing; total momentum still equals zero.

DID YOU KNOW?

A squid's propulsion system also involves the principle of conservation of momentum (Figure 8-15). Much like a rocket ship ejecting hot gases in order to move in a particular direction, a squid can expel jets of water at high velocity. By conservation of momentum, the resulting momentum of the squid is equal in magnitude and opposite in direction to the momentum of the jet of water.

Figure 8-15 The principle of conservation of momentum allows the squid to move quickly out of harm's way.

2) Positron Emission Tomography (PET) Imaging

A very useful medical imaging technique that applies the principle of conservation of momentum is positron emission tomography (PET). A positron is a subatomic particle that has the same mass as an electron, but the positron has a positive electric charge. Positrons and electrons are antiparticles of each other, and if a positron and electron meet, they annihilate each other and emit two high-energy photons[1] called gamma (γ) rays.

To create a PET scan, a radioactive substance that emits positrons is injected into a patient and accumulates in the organ or region to be studied. For example, if a glucose solution is injected, it will accumulate preferentially at cancer sites because glucose is metabolized rapidly in cancerous tumours. Any positron that is emitted will encounter an electron in a surrounding molecule and annihilate with it. Before annihilation, these positrons and electrons have velocities that are small in magnitude and randomly oriented, so the initial momentum of the positron–electron system is essentially zero. Since momentum is conserved in the annihilation process, the total momentum of the two γ rays must also be zero. Therefore, the γ rays must move in opposite directions. The γ rays are detected in a ring-shaped detector (Figure 8-16 (a)) surrounding the patient, and with the aid of a computer, the source of the γ rays can be pinpointed

[1]The energy of electromagnetic radiation, such as visible light and γ rays, is quantized in packets called photons. The electromagnetic spectrum will be discussed in further detail in Chapter 16.

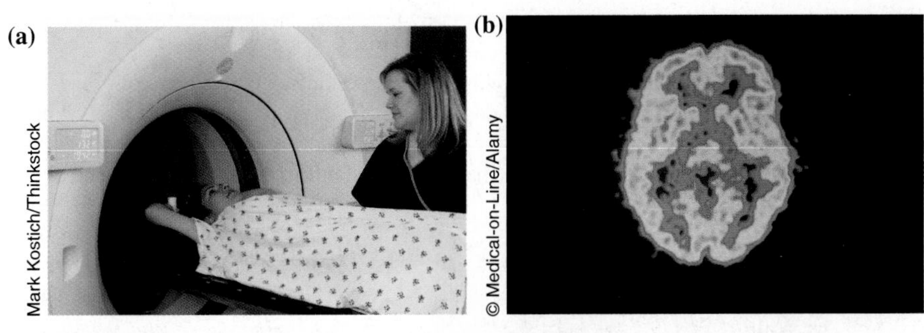

Figure 8-16 **(a)** A PET scanner in which the patient is surrounded by γ-ray detectors. **(b)** The image of a normal brain acquired by PET imaging. The bright locations are points of high metabolic activity where the positron-emitting glucose is being used.

DID YOU KNOW?

PET imaging can detect changes in the biological function of cells before they exhibit anatomical changes. This makes PET incredibly useful in detecting early stages of cancer, even before a tumour has developed.

and an image is created showing where the radioactive material has accumulated (Figure 8-16 (b)).

PET scans are commonly used to detect cancer and to create images during cancer therapy to determine if the therapy is effective. They can also be used to scan the heart to look for indications of coronary artery disease, or to evaluate damage from a heart attack. PET scans of the brain are useful in examining patients who have memory problems that might be the result of Alzheimer's disease, brain tumours, or other disorders. PET imaging provides information that is very different from the images acquired by other techniques such as computed tomography (CT) because it gives insight into tissue *function*, not just its structure. CT can only provide information on the anatomy of the location of interest.

3) Conservation of Momentum in Collisions

Conservation of momentum can be applied to many collisions. For example, in the collision of two billiard balls on a table (Figure 8-17), the force exerted by one ball is equal but opposite to the force exerted by the other, and hence the resultant force exerted on the system of the two balls together is zero. Therefore, the momentum of the two-ball system is constant; that is, the total momentum before the collision equals the total momentum after the collision. (We can safely neglect the friction between the balls and the table during the collision because it is very small compared to the forces exerted by the balls on each other. Vertically, there is no acceleration and thus there must be no resultant vertical force.)

Similarly, in the collision of two bumper cars at an amusement park, the force exerted by one car is equal but opposite to the force exerted by the other car (Figure 8-18). Neglecting other external forces such as friction, the resultant force on the system of two bumper cars is zero and the momentum of the system is therefore constant as discussed in Sample Problem 8-3. Next time you have the opportunity, conduct some qualitative experiments on conservation of momentum with your friends in the bumper cars!

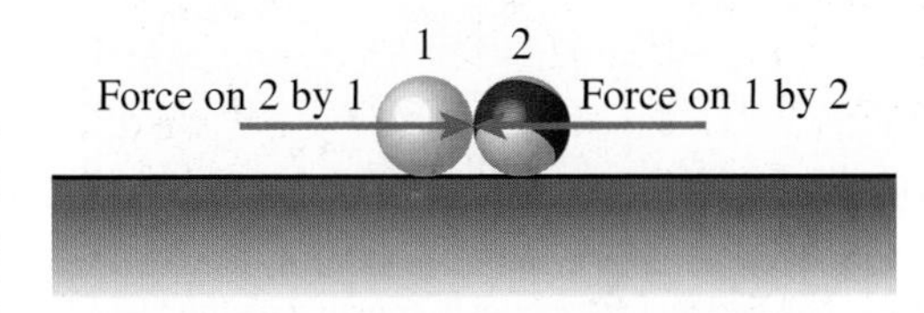

Figure 8-17 Two billiard balls colliding. The total momentum of the two-ball system is constant.

Figure 8-18 Conducting some qualitative experiments on conservation of momentum at the local amusement park.

TRY THIS!

Newton's Cradle

Figure 8-19 shows a popular science toy called a "Newton's Cradle." It consists of a row of metallic spheres suspended by strings from a support. If one of the end spheres is lifted and allowed to swing down and collide with the line of spheres, the sphere at the other end rises up, and none of the other spheres moves appreciably. Why does the toy behave in this way? Try to find one of these toys and play with it. See what happens when you start with two spheres lifted at one end.

Figure 8-19 Newton's Cradle.

DID YOU KNOW?

Nuclear and particle physicists make their living from controlling and observing collisions. From the first atomic structure experiments conducted by Ernest Rutherford to the international effort to discover the Higgs boson at the Large Hadron Collider (Figure 8-20), the observations made when subatomic particles collide have had a profound influence on our understanding of the universe.

Vincent Moncorge/Look At Sciences/ Science Photo Library

Figure 8-20 The Large Hadron Collider (LHC) under construction in Switzerland. The LHC is a 27-km-long high-energy particle accelerator. Since 'first beam' in 2008, the LHC has been probing the inner structure of matter in order to explain its origins.

4) Conservation of Momentum in Sports

When a tennis player hits a ball, is the momentum of the ball–racquet system constant (Figure 8-21)? The answer is not obvious. Perhaps the situation is similar to the snowball and gate described in the Tackling Misconceptions box on page 203; that is, the force exerted by the player's arms on the racquet during the collision might not be negligible in comparison with the forces between the racquet and ball. In this case, the momentum of the ball–racquet system would not be constant. We showed in Sample Problem 8-1, using realistic numbers, that the average force exerted by the racquet on the ball is about 10^4 N. This is approximately the same magnitude as the force of gravity on an object having a mass of 10^3 kg. Could a tennis player use her arms to hold up a thousand kilogram object? Obviously not. The player's arms can exert only a small force in comparison with the racquet–ball force, and hence the momentum of the ball–racquet system is indeed constant. This result has been confirmed in experiments.

A similar conclusion would be reached for baseball, cricket, and golf: the momentum of the system of the ball and bat or club is constant. In the case of golf, the head of the club is the important object that strikes the ball (because the head is massive relative to the club shaft), and the momentum of the head–ball system is constant. Momentum is also constant in collisions between players in sports such as hockey, rugby, and football, as considered in Question 8-84.

Yashkin Dmitry/Thinkstock

Figure 8-21 The momentum of the system of the ball and tennis racquet is constant as the two objects collide.

DID YOU KNOW?

The law of conservation of momentum in collisions was discovered by the Dutch physicist Christian Huygens (1629–1695).

EXERCISES

8-15 In which of the following situations is the momentum constant for the system consisting of objects A and B?

(a) A boy (A) stands in a stationary boat (B); the boy then walks in the boat. Neglect viscous friction.

(b) A bowling ball (A) collides with a bowling pin (B).

(c) A dart (A) is thrown against a door (B), which rotates as a result of the collision.

(d) A rocket (A) on a spaceship ejects hot gases (B) created by the burning of fuel.

(e) A moving billiard ball (A) strikes a stationary billiard ball (B).

(f) A woman uses a golf club (A) to strike a golf ball (B).

(g) A free-rolling railroad car (A) strikes a stationary railroad car (B).

(h) A hamburger (A), dropped vertically into a frying pan (B), comes to rest.

8-16 A boy of mass 45 kg stands on a stationary boat of mass 33 kg. He then walks with a speed of 1.9 m/s relative to the water. What is the resulting velocity (magnitude and direction) of the boat relative to the water? Neglect viscous friction.

8-17 Two ice-skaters, initially stationary, push each other so that they move in opposite directions. One skater (of mass 56.9 kg) has a speed of 3.28 m/s. What is the mass of the other skater if her speed is 3.69 m/s? Neglect friction.

8-18 A stationary artillery shell of mass 35 kg accidentally explodes, sending two fragments of mass 11 kg and 24 kg in opposite directions. The speed of the 11 kg fragment is 95 m/s. What is the speed of the other fragment?

8-19 Suppose that you drop a hairbrush and it falls downward.

(a) What is the direction of the gravitational force exerted by Earth on the hairbrush?

(b) What is the direction of the gravitational force exerted by the hairbrush on Earth?

(c) How do the forces in (a) and (b) compare in magnitude?

(d) What is the resultant force on the system consisting of Earth and the hairbrush?

(e) What can you conclude about the momentum of this system?

(f) If we consider Earth and the hairbrush to be initially stationary, how does Earth move as the hairbrush falls down?

8-20 A factory worker of mass 57 kg takes a ride on a large freely rolling cart having a mass of 26 kg. Initially the worker is standing on the cart and they both have a speed of 3.2 m/s relative to the floor. The worker then walks on the cart in the same direction as the cart is moving. Her speed is now 3.8 m/s relative to the floor. What is the new velocity (magnitude and direction) of the cart?

8-21 Two automobiles collide. One has a mass of 1.13×10^3 kg and is initially travelling east with a speed of 25.7 m/s. The other (mass 1.25×10^3 kg) has an initial velocity of 13.8 m/s west. The autos attach together during the collision.

(a) What is their common velocity (magnitude and direction) immediately after the collision?

(b) Determine the change in momentum (magnitude and direction) for each automobile.

(c) How are the quantities of the magnitude and direction for each automobile (found in part (b)) related?

(d) What is the total change in momentum of the two-automobile system?

8-22 An empty shopping cart, with a mass of 25 kg travelling with an initial speed of 2.2 m/s, collides with a stationary shopping cart full of groceries in the grocery store parking lot. Immediately after the collision, the carts move together with a speed of 0.75 m/s. What is the mass of the cart full of groceries?

8-23 A railroad car of mass 1.37×10^4 kg rolling north at 20.0 km/h collides with another railroad car of mass 1.12×10^4 kg, also initially rolling north but more slowly. After the collision, the coupled cars have a speed of 18.3 km/h. What was the initial speed of the other car?

8-24 A golf ball (mass 0.045 kg) is hit with a driver. The head of the driver has a mass of 0.15 kg and a speed of 56 m/s before the collision. The ball has a speed of 67 m/s as it leaves the clubface. What is the speed of the head immediately after the collision?

8-25 Two balls are rolling directly toward each other. One, of mass 0.25 kg, has a speed of 1.7 m/s; the other has a mass of 0.18 kg and a speed of 2.5 m/s. After the collision, the 0.25-kg ball has reversed its direction and has a speed of 0.10 m/s. What is the velocity (magnitude and direction) of the other ball?

8.3 Elastic and Inelastic Collisions

Imagine that two balls are thrown toward each other so that they collide: first two super balls (made from highly elastic material that bounces extremely well), then two tennis balls, and finally two balls made of soft putty. The collisions are very different. When the super balls collide, they bounce off each other with high speed; at the other extreme, the putty balls stick together and have only a small speed after the collision. The tennis ball collision is intermediate between the other two; the tennis balls bounce off each other with moderate speed. In each of these collisions, momentum is constant; in order to explain the differences between the collisions, we have to consider something other than momentum. That "something" is kinetic energy.

In the super ball collision, the total kinetic energy of the two balls after the collision is equal to the total kinetic energy before the collision. Such a collision is called an **elastic collision**. When the tennis balls collide, the total kinetic energy after the collision is not equal to the total kinetic energy before. This is an **inelastic collision**. For a tennis ball collision, the total final kinetic energy is *less* than the total initial kinetic energy. However, there can also be inelastic collisions in which kinetic energy increases, such as a collision that initiates an explosion, giving a total final kinetic energy that is *greater* than the initial kinetic energy.

elastic collision the total final kinetic energy is equal to the total initial kinetic energy in such a collision

inelastic collision the total final kinetic energy is not equal to the total initial kinetic energy in such a collision

When two objects *stick together* during a collision, as in the case of the putty balls, we have a **completely inelastic collision**. The decrease in total kinetic energy in a completely inelastic collision is the maximum possible. You might think that, if there is a maximum decrease in kinetic energy, the objects must be at rest after the collision,

completely inelastic collision the total final kinetic energy is less than the total initial kinetic energy. After such a collision, the objects stick together and move with the same velocity. The decrease in the total kinetic energy is the maximum that is possible.

but by the law of conservation of momentum, the objects usually must be moving. Visually, there is a bounciness or springiness in an elastic collision, whereas a completely inelastic collision has no springiness to it at all. The three categories of collision are summarized below in Table 8-1.

Table 8-1

Types of Collisions

Elastic	Total final kinetic energy = Total initial kinetic energy
Inelastic	Total final kinetic energy ≠ Total initial kinetic energy
Completely Inelastic	Total final kinetic energy < Total initial kinetic energy Objects stick together. Decrease in total kinetic energy is maximum possible.

Where does the kinetic energy go that is "lost" in inelastic and completely inelastic collisions? We know that energy is conserved, and so this lost kinetic energy must be transformed into other types of energy. For example, if two putty balls collide, they become warmer: kinetic energy has been converted into thermal energy. Depending on the objects that collide, kinetic energy could be converted into elastic potential energy, acoustical energy (sound), thermal energy, etc.

In practice, it is almost impossible to have a truly elastic collision between two macroscopic objects such as super balls. There is always some small amount of kinetic energy transformed into other forms. For instance, when such balls collide, thermal energy and acoustical energy are produced. (We can hear the collision.) Nonetheless, in this book, we include problems in which we treat certain collisions between

Problem-Solving Strategies for Collisions

When solving problems involving collisions, it is important to distinguish between inelastic, completely inelastic, and elastic collisions. Although we will point out differences below, in all collision problems that we consider in this book, momentum is constant:

$$m_1\vec{v}_1 + m_2\vec{v}_2 = m_1\vec{v}_1' + m_2\vec{v}_2' \qquad (8\text{-}5)$$

where m_1 and m_2 are the masses of the colliding objects, $\vec{v}_1$ and $\vec{v}_2$ are their velocities before the collision, and $\vec{v}_1'$ and $\vec{v}_2'$ are their velocities after the collision.

Inelastic Collisions

If the collision is inelastic, Eqn. 8-5 is the only one that can be used to solve the problem.

Completely Inelastic Collisions

Again, only the equation of conservation of momentum can be used here (Eqn. 8-5). However, since the objects stick together after this type of collision, we automatically know that their final velocities are equal ($\vec{v}_1' = \vec{v}_2' = \vec{v}'$). Eqn. 8-5 now becomes

$$m_1\vec{v}_1 + m_2\vec{v}_2 = (m_1 + m_2)\vec{v}'$$

where $\vec{v}'$ is the common final velocity of the objects after the collision. Sample Problem 8-3 in Section 8.2 gave an example of solving a completely inelastic collision problem.

Elastic Collisions

If a collision is known to be elastic (this is usually stated clearly in the problem), then the total kinetic energy before the collision equals the total kinetic energy after the collision. This gives a second equation to work with in solving a problem involving an elastic collision:

$$\tfrac{1}{2}m_1v_1^2 + \tfrac{1}{2}m_2v_2^2 = \tfrac{1}{2}m_1v_1'^2 + \tfrac{1}{2}m_2v_2'^2 \qquad (8\text{-}6)$$

This says that the total kinetic energy of the system before the collision (on the left-hand side of the equation) is the same as the total kinetic energy of the system after the collision (on the right-hand side of the equation). Remember that kinetic energy is a scalar that is calculated based on the speed of the object. The following two sample problems, Sample Problems 8-4 and 8-5, give examples of solving elastic collision problems using Eqns. 8-5 and 8-6 together.

macroscopic objects as being elastic; that is, we neglect the small kinetic energy that is lost. Collisions between molecules, atoms, and subatomic particles are often (but not always) perfectly elastic.

In the early history of the solar system, there were many completely inelastic collisions between relatively large objects such as the Moon (Figure 8-22 (a)) and smaller chunks of rock. Such collisions were the source of the craters on the Moon. Most similar craters that were formed on Earth have been destroyed by erosion due to rain and wind. However, some relatively recent craters still exist, such as the famous Barringer Crater in Arizona, which is believed to be about 25 000 years old. This crater, shown in Figure 8-22 (b), has a diameter of 1.3 km and a depth of 180 m and resulted from the collision between a large meteorite and Earth.

(a)

(b)

Figure 8-22 **(a)** Craters on the Moon are the result of completely inelastic collisions with large rocks. **(b)** Barringer Crater in Arizona was formed as a result of a completely inelastic collision.

SAMPLE PROBLEM 8-4

A billiard ball having mass m and initial speed v makes a head-on (one-dimensional) elastic collision with another billiard ball (initially stationary) having the same mass m. What are the final speeds of the two balls?

Solution

Initial and final diagrams are shown in Figure 8-23. The $+x$-axis has been chosen in the direction of motion of the initially moving ball. Since we are told that the collision is *elastic*, we know that the total initial kinetic energy equals the total final kinetic energy and we can use Eqn. 8-6. If the problem did not state that the collision is elastic, then we could not have made this assumption. Momentum is also constant in this collision. We can thus write two equations, one for kinetic energy (8-6) and one for momentum (8-5).

$$\frac{1}{2}mv_1^2 + \frac{1}{2}mv_2^2 = \frac{1}{2}mv_1'^2 + \frac{1}{2}mv_2'^2$$

$$m\vec{v}_1 + m\vec{v}_2 = m\vec{v}_1' + m\vec{v}_2'$$

where "1" and "2" refer to the (initially) moving and stationary balls respectively. The final speeds and velocities are primed, and the initial speeds and velocities are unprimed. These equations can be simplified quickly by

- dividing both sides of each of them by m
- setting $\vec{v}_2 = 0$ (and $v_2 = 0$) since ball #2 is initially stationary
- multiplying the kinetic energy equation by two to eliminate the fractions

This gives

$$v_1^2 = v_1'^2 + v_2'^2$$

$$\vec{v}_1 = \vec{v}_1' + \vec{v}_2'$$

Also since the motion is one-dimensional, in the first equation we can write $v_1^2 = v_{1x}^2$, $v_1'^2 = v_{1x}'^2$ and $v_2'^2 = v_{2x}'^2$. (If the motion had been two-dimensional, we would have written, for example, $v_1^2 = v_{1x}^2 + v_{1y}^2$.) We now have

$$v_{1x}^2 = v_{1x}'^2 + v_{2x}'^2 \qquad \text{Eqn. [1a]}$$

$$v_{1x} = v_{1x}' + v_{2x}' \qquad \text{Eqn. [2a]}$$

We are given that $v_{1x} = +v$ for the ball that is initially moving. Substitution of this into the two equations above gives

$$v^2 = v_{1x}'^2 + v_{2x}'^2 \qquad \text{Eqn. [1b]}$$

$$v = v_{1x}' + v_{2x}' \qquad \text{Eqn. [2b]}$$

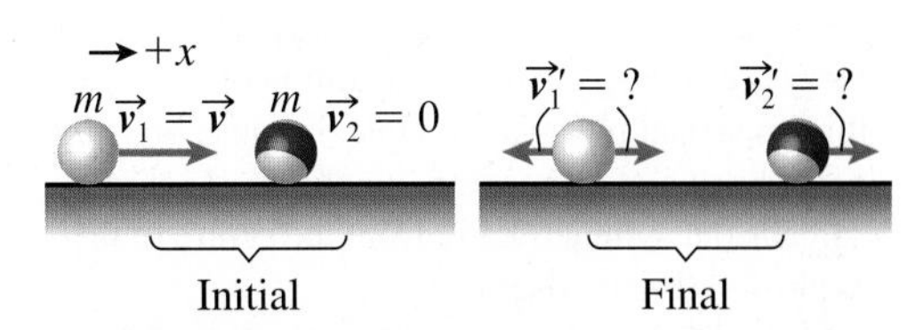

Figure 8-23 Sample Problem 8-4.

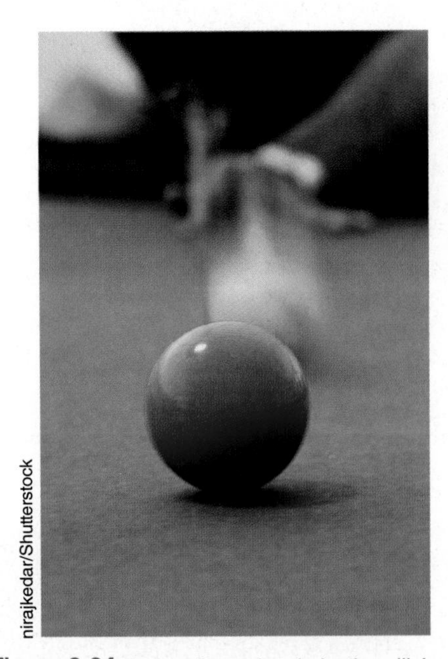

Figure 8-24 A one-dimensional elastic collision of two billiard balls.

Since the speed v is a given quantity, we now have two equations with two unknowns $(v'_{1x} \text{ and } v'_{2x})$. Our plan of attack will be to rearrange Eqn. [2b] to get v'_{1x} in terms of v'_{2x}, substitute this expression for v'_{1x} into Eqn. [1b], and solve the resulting equation for the unknown v'_{2x}. We then can use Eqn. [2b] to determine the other unknown, v'_{1x}.

First, rearranging Eqn. [2b]

$$v'_{1x} = v - v'_{2x}$$

Substituting for v'_{1x} in Eqn. [1b],

$$v^2 = (v - v'_{2x})^2 + v'^{\,2}_{2x}$$

Expanding and simplifying,

$$v^2 = v^2 - 2vv'_{2x} + v'^2_{2x} + v'^2_{2x}$$

$$0 = -2vv'_{2x} + 2v'^2_{2x}$$

$$0 = -v'_{2x}(v - v'_{2x})$$

Therefore, either $v'_{2x} = 0$ (which means that no collision occurred and hence this is not an appropriate solution) or $v - v'_{2x} = 0$, from which we conclude that $v'_{2x} = v$. Thus, $v'_{2x} = v$.

Substituting $v'_{2x} = v$ in Eqn. [2b] gives $v = v'_{1x} + v$

$$v'_{1x} = 0$$

Therefore, the ball (#1) that is initially moving is at rest after the collision ($v'_{1x} = 0$), and the ball (#2) that is initially stationary has the same speed after the collision that the other ball had before the collision ($v'_{2x} = v$). If you have had any experience with billiard balls, you know that this result is correct for head-on (one-dimensional) collisions (Figure 8-24). However, this conclusion is not valid for all one dimensional elastic collisions in which one object is initially stationary—the colliding objects must have the same mass. In the next sample problem, we investigate a one-dimensional elastic collision in which the objects have different masses.

DID YOU KNOW?

Momentum is also conserved in a system in which an object breaks apart or explodes. Fireworks are an exciting illustration of this important physics principle: because of conservation of momentum, the fireworks display spreads out symmetrically from the centre of the explosion (Figure 8-25).

Figure 8-25 The symmetry in the outward spread of the fireworks display arises from the principle of conservation of momentum.

SAMPLE PROBLEM 8-5

In the operation of a PET scanner, when a positron is emitted from a radioactive nucleus inside the patient, it typically travels a short distance (at most a few millimetres) before it undergoes annihilation. Suppose that while it is travelling a positron undergoes an elastic head-on (one-dimensional) collision with a stationary proton, that is, the nucleus of a hydrogen atom. If the speed of the positron just prior to the collision is 923 m/s, determine the velocity (magnitude and direction) of each particle after the collision. The mass of a positron is 9.109×10^{-31} kg, and the mass of a proton is 1.673×10^{-27} kg.

Solution

Figure 8-26 provides diagrams before and after the collision. The $+x$-axis has been chosen in the direction of the initial velocity of the positron. The basic setup of the solution is similar to that of the previous sample problem, but the analysis is more complicated because the particles have different masses. Since the collision is elastic, both kinetic energy and momentum are constant:

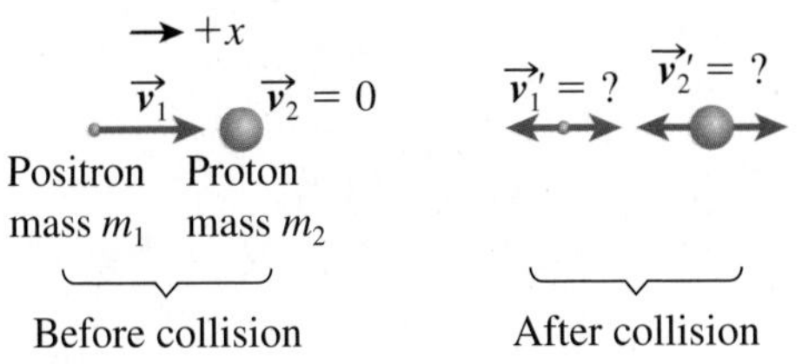

$$\tfrac{1}{2}\,m_1 v_1^2 + \tfrac{1}{2}\,m_2 v_2^2 = \tfrac{1}{2}\,m_1 v_1'^2 + \tfrac{1}{2}\,m_2 v_2'^2$$

$$m_1\vec{v}_1 + m_2\vec{v}_2 = m_1\vec{v}_1' + m_2\vec{v}_2'$$

where "1" and "2" refer to the positron and proton respectively. Since the motion is all in one dimension (along the x-axis), we can replace the velocity terms in the second equation with their corresponding x-components:

$$mv_{1x} + mv_{2x} = mv_{1x}' + mv_{2x}'$$

Since the proton is initially stationary, $v_{2x} = 0$ and there is no contribution to the initial kinetic energy of the system from the proton. Making this substitution in the equations above and multiplying the kinetic energy equation by two to eliminate the fractions, we have

$$m_1 v_1^2 = m_1 v_1'^2 + m_2 v_2'^2 \qquad \text{Eqn. [1a]}$$

$$m_1 v_{1x} = m_1 v_{1x}' + m_2 v_{2x}' \qquad \text{Eqn. [2a]}$$

As in Sample Problem 8-4, since there is no motion in the y-direction, $v_1^2 = v_{1x}^2$, $v_1'^2 = v_{1x}'^2$, and $v_2'^2 = v_{2x}'^2$.

Rewriting Eqns. [1a] and [2a] to express everything in terms of our unknown variables $(v_{1x}' \text{ and } v_{2x}')$, we get

$$m_1 v_{1x}^2 = m_1 v_{1x}'^2 + m_2 v_{2x}'^2 \qquad \text{Eqn. [1b]}$$

$$m_1 v_{1x} = m_1 v_{1x}' + m_2 v_{2x}' \qquad \text{Eqn. [2b]}$$

Since we are given values for m_1, m_2, and v_{1x}, we now have two equations with two unknowns $(v_{1x}' \text{ and } v_{2x}')$.

As in the previous problem, our plan will be to rearrange Eqn. [2b] to get v_{1x}' in terms of v_{2x}', substitute the expression for v_{1x}' into Eqn. [1b], and solve the resulting equation for v_{2x}'. We then can determine v_{1x}'. First we divide through by m_1:

$$v_{1x} = v_{1x}' + \frac{m_2}{m_1}\cdot v_{2x}'$$

Rearrange to get v_{1x}' in terms of v_{2x}',

$$v_{1x}' = v_{1x} - \frac{m_2}{m_1}\cdot v_{2x}' \qquad \text{Eqn. [3]}$$

We now substitute this expression into Eqn. [1b]:

$$m_1 v_{1x}^2 = m_1\left(v_{1x} - \frac{m_2}{m_1}\cdot v_{2x}'\right)^2 + m_2 v_{2x}'^2$$

As before, divide through by m_1 to give:

$$v_{1x}^2 = \left(v_{1x} - \frac{m_2}{m_1}\cdot v_{2x}'\right)^2 + \frac{m_2}{m_1}\cdot v_{2x}'^2 \qquad \text{Eqn. [4]}$$

Expanding the $\left(v_{1x} - \frac{m_2}{m_1}\cdot v_{2x}'\right)^2$ term on the right-hand side, Eqn. [4] becomes

$$v_{1x}^2 = v_{1x}^2 - 2v_{1x}\cdot\frac{m_2}{m_1}\cdot v_{2x}' + \left(\frac{m_2}{m_1}\cdot v_{2x}'\right)^2 + \frac{m_2}{m_1}\cdot v_{2x}'^2$$

Subtracting v_{1x}^2 from both sides of the equation:

$$0 = -2v_{1x}\cdot\frac{m_2}{m_1}\cdot v_{2x}' + \left(\frac{m_2}{m_1}\cdot v_{2x}'\right)^2 + \frac{m_2}{m_1}\cdot v_{2x}'^2$$

Figure 8-26 Sample Problem 8-5.

and then multiplying through by m_1 and dividing by m_2:

$$0 = -2v_{1x} \cdot v'_{2x} + \frac{m_2}{m_1} \cdot (v'_{2x})^2 + v'^{\,2}_{2x}$$

Almost there! We now need to divide through by the common factor of v'_{2x}:

$$0 = -2v_{1x} + \frac{m_2}{m_1} \cdot (v'_{2x}) + v'_{2x}$$

or

$$2v_{1x} = v'_{2x}\left(\frac{m_2}{m_1} + 1\right)$$

Therefore,

$$v'_{2x} = \frac{2v_{1x}}{\left(\frac{m_2}{m_1} + 1\right)}$$

$$v'_{2x} = \frac{2(923\ \text{m/s})}{\left(\frac{1.673 \times 10^{-27}\ \text{kg}}{9.109 \times 10^{-31}\ \text{kg}} + 1\right)}$$

which gives $v'_{2x} = 1.00$ m/s to the correct number of significant digits for our final answer.

We now substitute v'_{2x} into Eqn. [3] to determine v'_{1x}:

$$v'_{1x} = v_{1x} - \frac{m_2}{m_1} \cdot v'_{2x}$$

$$v'_{1x} = 923\ \text{m/s} - \frac{1.673 \times 10^{-27}\ \text{kg}}{9.109 \times 10^{-31}\ \text{kg}} \cdot 1.004\ \text{m/s}$$

$$v'_{1x} = -921\ \text{m/s}$$

Thus, the final results are $v'_{2x} = 1.00$ m/s and $v'_{1x} = -921$ m/s.

These values indicate that, after the collision, the proton has a velocity of 1.00 m/s in the direction in which the positron was originally moving, and the positron has a velocity of 921 m/s in the opposite direction (indicated by the negative value for v'_{1x}). This result makes physical sense: the low-mass positron rebounds from the heavy proton with very little reduction in speed, and the massive proton (compared to a positron!) moves very slowly after the collision.

TRY THIS!

A Ping Pong Ball as a Deadly Weapon

Hold a ping pong ball above a super ball and drop them together (touching) from waist height onto a hard floor. Watch out! Write down what you observe, and explain what happens. Assume that the collisions are one dimensional and elastic and that the mass of the ping pong ball is very much less than the mass of the super ball. (Another pair of balls that works well is a baseball above a basketball, *but safety precautions must definitely be taken in this case.*) Assuming elastic collisions, a very large mass difference between the balls, and negligible air resistance, show that the ping pong ball will go upward to nine times its release height above the floor.

EXERCISES

8-26 (a) In which type of collision do the objects stick together afterward?

(b) In an elastic collision, is the total final kinetic energy greater than, less than, or equal to the initial kinetic energy of the system?

(c) In which type of collision is the decrease in kinetic energy the maximum possible?

(d) In which type of collision is the total final kinetic energy not equal to the initial kinetic energy of the system?

8-27 During a snowball fight, two snowballs (each of mass 145 g) collide in mid-air in a completely inelastic collision. Just before the collision both balls are travelling horizontally, one ball with a velocity of 22 m/s north, and the other, 22 m/s south. What is the velocity (magnitude and direction) of each ball after the collision? Assume that the motion is in the horizontal plane both before and after the collision.

8-28 A proton travelling with an initial speed of 815 m/s collides head-on elastically with a stationary proton. What is the velocity (magnitude and direction) of each proton after the collision? Show your work.

8-29 A truck of mass 1.3×10^4 kg travelling at 9.0×10^1 km/h north collides with a car of mass 1.1×10^3 kg travelling at 3.0×10^1 km/h north.

(a) If the collision is completely inelastic, what is the velocity (magnitude and direction) of the vehicles immediately after the collision?

(b) Calculate the total kinetic energy before and after the collision, and the decrease in kinetic energy during the collision.

8-30 A bullet of mass 3.5 g travelling at 6.2×10^2 m/s collides horizontally with a stationary block of wood on a table. The bullet passes through the wood and emerges with a speed of 1.6×10^2 m/s. The wood has a speed of 6.6 m/s immediately after the collision.

(a) What is the mass of the wood?

(b) Is this collision elastic, inelastic, or completely inelastic?

8-31 A super ball (mass $m_1 = 22$ g) rolls with a speed of 3.5 m/s toward another (stationary) super ball (mass $m_2 = 27$ g). If the balls have a head-on (one-dimensional) elastic collision, what are the velocities (magnitude and direction) of the balls after the collision?

8-32 A nitrogen molecule (mass 4.65×10^{-26} kg) in the air undergoes a head-on elastic collision with a stationary oxygen molecule (mass 5.31×10^{-26} kg). After the collision, the nitrogen molecule has reversed its direction and has a speed of 34.1 m/s, and the oxygen is travelling at 481 m/s in the original direction of motion of the nitrogen. What was the initial speed of the nitrogen molecule?

8-33 An object of mass m makes an elastic collision with another object initially at rest and continues to move in the original direction but with one-third its original speed. What is the mass of the other object in terms of m?

8.4 Conservation of Momentum in Two Dimensions

If you think about it, life would be rather dull if the universe were one-dimensional. Fortunately we live in a three-dimensional universe; four-dimensional if you consider time as an additional dimension. We now discuss how conservation of momentum applies in more than one dimension.

In this chapter, we have already applied conservation of momentum to a number of one-dimensional phenomena, such as the head-on collision of bumper cars or the recoil of a boat when a person walks in it. Momentum is conserved because the forces exerted by the objects on each other are equal in magnitude but opposite in direction, and thus the resultant force on the system is zero (assuming that there are no other significant forces acting on the objects). If the resultant force on a system is zero, then the total momentum of the system is conserved as already discussed in Section 8.2.

In two- and three-dimensional situations, the same reasoning applies. The forces exerted between parts of a system are equal but opposite, the resultant force is thus zero, and the total momentum of the system is conserved. Both the resultant force and momentum are vector quantities. Thus, when we say that momentum is conserved, we mean that both the magnitude and direction of the momentum vector are unchanging. Alternatively, we can state that the x- and y-components of the momentum are each unchanging (if we are dealing with a two-dimensional collision). When solving problems involving two-dimensional conservation of momentum, we usually write two equations, one involving x-components and another involving y-components.

The great value of the law of conservation of momentum is its generality. It can be used in *any* situation in which a system is subject to a resultant force of zero. It applies to collisions between all sorts of objects: subatomic particles, tennis balls, and even galaxies consisting of billions of stars (Figure 8-27). It can be applied to systems that do not involve collisions. For example, in the Earth–Sun system, the gravitational force exerted by Earth on the Sun is equal in magnitude but opposite in direction to the force exerted by the Sun on Earth, and hence the momentum of the system is conserved. As Earth travels in its orbit around the Sun, the Sun wobbles a bit (i.e., it does not stay stationary) so that the momentum of the system is constant.

Bill Frische/Shutterstock

Figure 8-27 A collision on a truly massive scale when two galaxies collide.

SAMPLE PROBLEM 8-6

A boy (mass 47 kg) is standing on a raft of mass 53 kg that is drifting with a velocity of 1.2 m/s north. The boy then starts walking on the raft such that his net velocity is 0.75 m/s east. What is the resulting net velocity (magnitude and direction) of the raft? Neglect viscous friction between the raft and water.

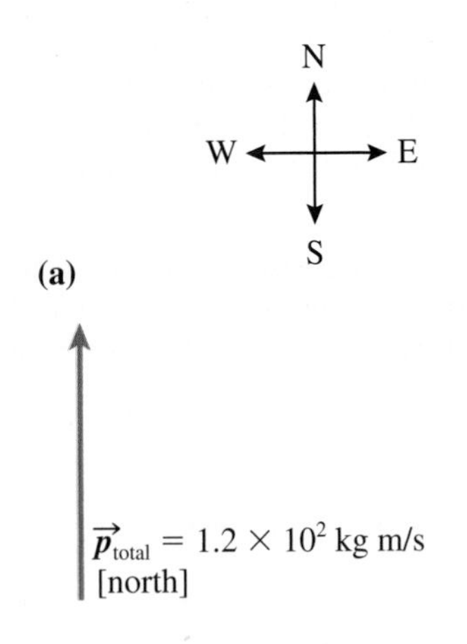

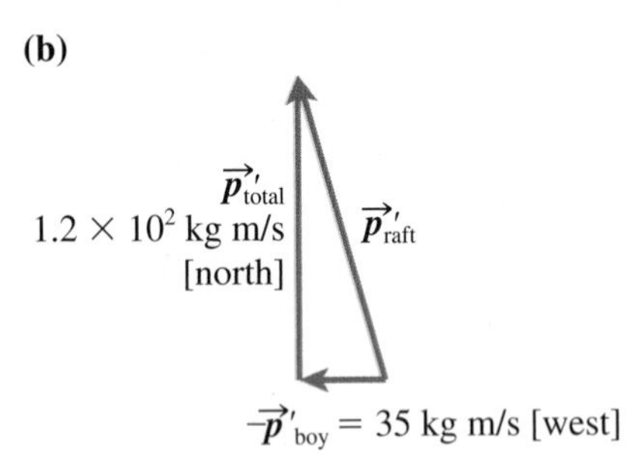

Figure 8-28 Sample Problem 8-6 (vector scale diagram solution). Scale: 1 cm = 45 kg·m/s. **(a)** Initial momentum. **(b)** Final momentum.

Approximate Solution (using a vector scale diagram)

Figure 8-28 shows an approximate solution to this problem, using a vector scale diagram. Since the boy and raft have the same initial velocity, the initial momentum is the total mass (100 kg) multiplied by the velocity (1.2 m/s north), giving 1.2×10^2 kg·m/s north (Figure 8-28 (a)).

When the boy walks east, the force exerted by him on the raft is equal and opposite to the force exerted on him by the raft, and therefore the momentum of the boy–raft system is constant. Thus, the total momentum must still be 1.2×10^2 kg·m/s north. The total momentum ($\vec{p}'_{total}$) of the system is the *vector* sum of the boy's momentum ($\vec{p}'_{boy}$) and that of the raft ($\vec{p}'_{raft}$), or $\vec{p}'_{total} = \vec{p}'_{boy} + \vec{p}'_{raft}$. By rearranging this expression, we can see that the unknown momentum of the raft can be written as $\vec{p}'_{raft} = \vec{p}'_{total} - \vec{p}'_{boy}$ or $\vec{p}'_{raft} = \vec{p}'_{total} + (-\vec{p}'_{boy})$. In order to draw this as a vector scale diagram, we draw the total momentum vector (1.2×10^2 kg m/s [north]) and the negative of the boy's momentum, that is, (47 kg)(0.75 m/s [west]), or 35 kg·m/s [west]. Figure 8-28 (b) shows the raft's final momentum as the vector sum of the total final momentum and the negative of the boy's final momentum. The momentum of the raft can be determined by direct measurement with a ruler and protractor: 1.3×10^2 kg·m/s at 73° north of west. To determine the net velocity of the raft, we divide its momentum by its mass. The result is 2.5 m/s at 73° north of west.

Probably the most difficult part of the solution is the drawing of Figure 8-28 (b). Look carefully at this diagram and think about how you would draw it, given only the total momentum northward and the boy's momentum eastward. Remember that the boy's momentum and the raft's momentum add as vectors to give the total momentum.

Exact Solution (using components)

A more accurate solution can be obtained by using vector components (Figure 8-29). We write equations expressing conservation of momentum in the x- and y-directions,

$$m_1 v_{1x} + m_2 v_{2x} = m_1 v'_{1x} + m_2 v'_{2x} \qquad \text{Eqn. [1]}$$

$$m_1 v_{1y} + m_2 v_{2y} = m_1 v'_{1y} + m_2 v'_{2y} \qquad \text{Eqn. [2]}$$

where "1" represents the boy and "2" the raft. The final velocities have a primed superscript, and the initial velocities have no superscript. We define the $+x$- and $+y$-directions to be east and north, respectively. This simplifies the solution, since the boy and raft initially each have a zero x-component of velocity ($v_{1x} = v_{2x} = 0$) and a common y-component ($v_{1y} = v_{2y} = 1.2$ m/s), as shown in Figure 8-29 (a).

In the final situation with the boy having a net velocity that is directed east (Figure 8-29 (b)), the boy's velocity has no y-component ($v'_{1y} = 0$) and a positive x-component ($v'_{1x} = 0.75$ m/s). In order that the total momentum remains constant, notice that the raft's final net velocity must have a positive y-component and a negative x-component. Since the masses are known (m_1 = 47 kg; m_2 = 53 kg), the only unknown quantities in Eqns. [1] and [2] are the components of the raft's final net velocity v'_{2x} and v'_{2y}.

We start by substituting $v_{1x} = v_{2x} = 0$ into Eqn. [1]:

$$0 = m_1 v'_{1x} + m_2 v'_{2x}$$

Rearranging and substituting other known values to solve for the unknown v'_{2x},

$$v'_{2x} = \frac{-m_1 v'_{1x}}{m_2} = \frac{-(47\text{ kg})(0.75\text{ m/s})}{53\text{ kg}} = -0.67\text{ m/s}$$

Similarly, substituting $v'_{1y} = 0$ into Eqn. [2], and then solving for v'_{2y},

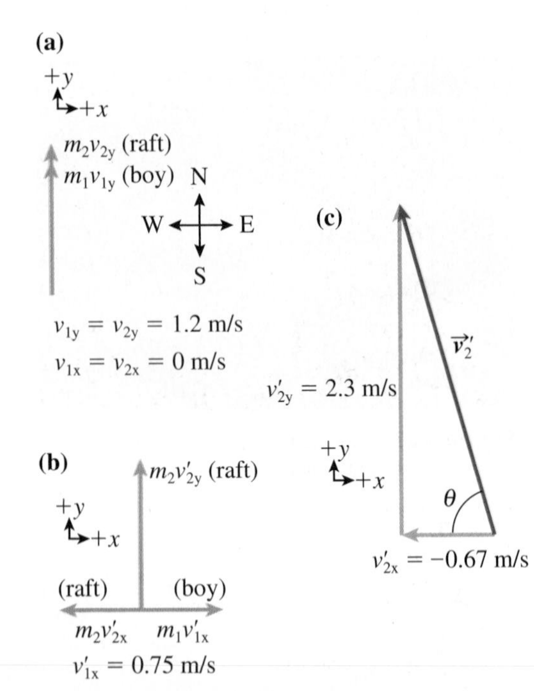

Figure 8-29 Solution to Sample Problem 8-6 using components. **(a)** Initial momentum. **(b)** Final momentum. **(c)** Final velocity of raft.

$$m_1v_{1y} + m_2v_{2y} = m_2v'_{2y}$$

$$v'_{2y} = \frac{m_1v_{1y} + m_2v_{2y}}{m_2}$$

$$= \frac{(47\,\text{kg})(1.2\,\text{m/s}) + (53\,\text{kg})(1.2\,\text{m/s})}{53\,\text{kg}}$$

$$= 2.3\,\text{m/s}$$

The magnitude and direction of the raft's final net velocity $\vec{v}'_2$ can now be determined from its components (Figure 8-29 (c)):

$$|\vec{v}'_2| = \sqrt{(v'_{2x})^2 + (v'_{2y})^2} = \sqrt{(-0.665\,\text{m/s})^2 + (2.26\,\text{m/s})^2} = 2.4\,\text{m/s}$$

and $\tan\theta = \dfrac{2.26\,\text{m/s}}{0.665\,\text{m/s}}$

Hence, $\theta = \tan^{-1}(2.26/0.665) = 74°$

Thus, the final velocity of the raft is 2.4 m/s at 74° north of west. Our scale diagram answer of 2.5 m/s at 73° north of west agrees well with this more accurate result.

SAMPLE PROBLEM 8-7

Two children are playing marbles. A collision occurs between two marbles of equal mass m. Marble #1 is initially moving while marble #2 is initially at rest. After the collision, marble #2 has a velocity of 1.10 m/s at an angle of 40.0° from the original direction of motion of marble #1. Marble #1 has a speed of 1.36 m/s after the collision. What was the initial speed of marble #1?

Solution

In order to solve this problem, it will help tremendously to start by drawing a diagram. Figure 8-30 shows the problem broken into two parts: the velocities of the marbles before the collision and the velocities of the marbles after the collision. We choose the coordinate system here in such a way as to simplify our solution since there is no y-component of velocity (and hence momentum) before the collision.

We are given the direction of the velocity for marble #2 after the collision. How do we know that marble #1 has a velocity directed at an angle θ below the $+x$-axis? Since there is no y-component of momentum before the collision, there must be no y-component of total momentum after the collision, so marble #1 must have a $-y$-component to its velocity in order to cancel out the $+y$-component to the velocity of marble #2 after the collision.

Conservation of momentum in both x- and y-directions can be written as

$$m_1v_{1x} + m_2v_{2x} = m_1v'_{1x} + m_2v'_{2x} \qquad \text{Eqn. [1]}$$

$$m_1v_{1y} + m_2v_{2y} = m_1v'_{1y} + m_2v'_{2y} \qquad \text{Eqn. [2]}$$

Before the collision, marble #1 has velocity components $v_{1x} > 0$, $v_{1y} = 0$, and the velocity components of marble #2 are $v_{2x} = v_{2y} = 0$ since it is at rest; v_{1x} is the unknown quantity we are trying to find in this problem. After the collision, both marbles have non-zero x- and y-components of velocity, as shown. Substituting into Eqns. [1] and [2] that $v_{1y} = v_{2x} = v_{2y} = 0$ and also that $m_1 = m_2 = m$,

$$mv_{1x} = mv'_{1x} + mv'_{2x} \qquad \text{Eqn. [3]}$$

$$0 = mv'_{1y} + mv'_{2y} \qquad \text{Eqn. [4]}$$

TRY THIS!

Scrambled Eggs

Design and build a container that protects an egg from breaking when it is dropped from a height of 2 m. Use items from around the house to cushion the egg at the end of its fall. Keep in mind that you want the change in momentum at impact to happen as slowly as possible in order to minimize the average force exerted on the egg. **Note:** you will probably need a few eggs in perfecting your design and it will likely get a little messy!

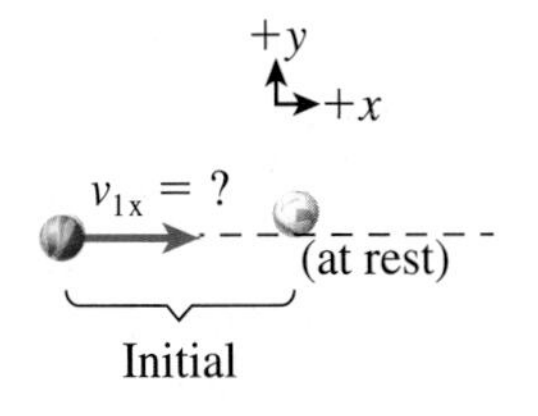

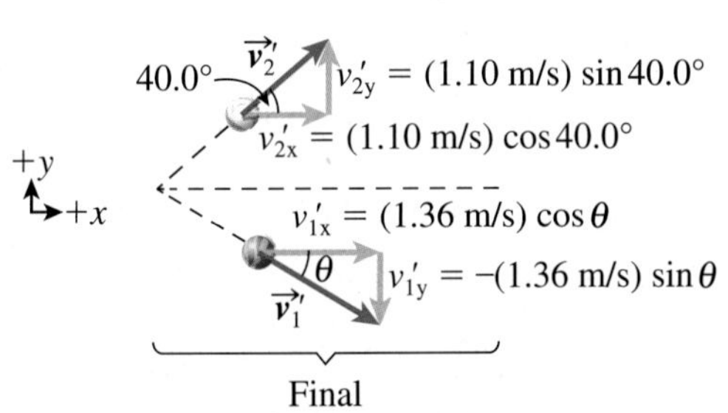

Figure 8-30 Sample Problem 8-7.

We can see from Eqns. [3] and [4] that the initial x-component of momentum of marble #1 is shared after the collision by both marbles, and that the two final y-components of momentum must add to zero. We simplify Eqns. [3] and [4] by dividing all terms by m:

$$v_{1x} = v'_{1x} + v'_{2x} \qquad \text{Eqn. [5]}$$

$$0 = v'_{1y} + v'_{2y} \qquad \text{Eqn. [6]}$$

Writing the velocities of the marbles before and after the collision in terms of their components using Figure 8-30 as our guide, we have

$$v'_{2x} = (1.10\,\text{m/s})\cos 40.0° \quad \text{and} \quad v'_{2y} = (1.10\,\text{m/s})\sin 40.0°$$

and

$$v'_{1x} = (1.36\,\text{m/s})\cos\theta \text{ and } v'_{1y} = -(1.36\,\text{m/s})\sin\theta$$

As already discussed, v'_{1y} must be negative so that the total y-component of the momentum after the collision is zero, equal to the initial value. Substituting the four final velocity components into Eqns. [5] and [6]:

$$v_{1x} = (1.36\,\text{m/s})\cos\theta + (1.10\,\text{m/s})\cos 40.0° \qquad \text{Eqn. [7]}$$

$$0 = -(1.36\,\text{m/s})\sin\theta + (1.10\,\text{m/s})\sin 40.0° \qquad \text{Eqn. [8]}$$

We have two equations here and two unknown variables: v_{1x} and θ. Since we are trying to determine v_{1x}, we will use Eqn. [8] to determine θ and then substitute this value into Eqn. [7] in order to solve the problem. From Eqn. [8],

$$\sin\theta = \frac{(1.10\,\text{m/s})\sin 40.0°}{1.36\,\text{m/s}}$$

$$\theta = 31.33°$$

Substituting this value of θ into Eqn. [7]:

$$v_{1x} = (1.36\,\text{m/s})\cos 31.33° + (1.10\,\text{m/s})\cos 40.0°$$

$$= 2.00\,\text{m/s}$$

Therefore, the initial speed of the moving marble is 2.00 m/s.

DID YOU KNOW?

The subatomic particle called the neutrino was first predicted to exist in 1930 by Wolfgang Pauli (1900–1958) in order to explain "missing" energy and momentum in nuclear reactions. It wasn't until almost 30 years later, in 1956, that neutrinos were first detected experimentally.

EXERCISES

8-34 A girl of mass 52 kg is standing on a cart of mass 26 kg that is free to move in any direction. Initially the cart is moving with a velocity of 1.2 m/s south relative to the floor. The girl then walks on the cart, and has a net velocity of 1.0 m/s west relative to the floor.

(a) Use a vector scale diagram to determine the approximate final velocity (magnitude and direction) of the cart.

(b) Solve this problem using components.

8-35 Two automobiles collide at an intersection. One car (mass 1.4×10^3 kg) is travelling south at 45 km/h and the other (mass 1.3×10^3 kg) is going east at 39 km/h. If the cars have a completely inelastic collision, what is their velocity (magnitude and direction) just after the collision?

8-36 Two spaceships from different nations have linked up in space and are coasting with their engines off, heading directly toward Mars.

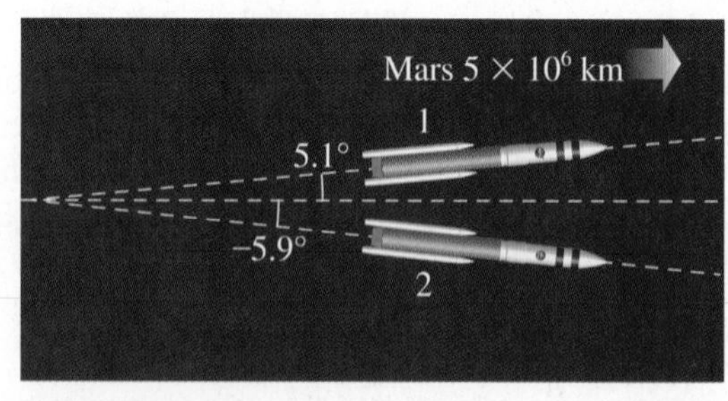

Figure 8-31 Question 8-36.

The spaceships are then thrust apart by the use of large springs. Spaceship #1 (mass $m_1 = 1.9 \times 10^4$ kg) then has a velocity of 3.5×10^3 km/h at 5.1° to its original direction, and ship #2 ($m_2 = 1.7 \times 10^4$ kg) has a velocity of 3.4×10^3 km/h at 5.9° to its original direction (Figure 8-31).

(a) Use vector components to find the original speed of the ships when they were together.

(b) Use a vector scale diagram to find the original speed of the ships when they were together.

8-37 Two balls of equal mass m undergo a collision. One ball is initially stationary. After the collision, the velocities of the balls make angles of 31.1° and 48.9° relative to the original direction of motion of the moving ball.

(a) Draw a diagram showing the initial and final situations. If you are uncertain about the final directions of motion, remember that momentum is conserved.

(b) If the initial speed of the moving ball is 2.25 m/s, what are the speeds of the balls after the collision?

(c) Use a vector scale diagram to determine approximate answers for part (b).

(d) Is this collision elastic? Justify your answer.

8-38 A neutron (mass 1.7×10^{-27} kg) travelling at 2.7 km/s collides with a stationary lithium nucleus (mass 1.2×10^{-26} kg). Afterward, the velocity of the lithium nucleus is 0.40 km/s at 54° to the original direction of motion of the neutron. If the speed of the neutron after the collision is 2.51 km/s, in what direction is it travelling?

LOOKING BACK...LOOKING AHEAD

Thus far in the book we have considered various aspects of mechanics: kinematics, Newton's laws of motion, energy, and momentum.

In this chapter we have defined (linear) momentum and have seen that the change in an object's momentum equals the product of the resultant force on the object and the time interval during which the resultant force acts. In addition, the law of conservation of momentum has been explained and used in discussions of various physical phenomena, such as collisions and rocket propulsion. Various types of collisions have been considered—elastic, inelastic, and completely inelastic—depending on the relationship between the initial kinetic energy and the final kinetic energy. Collisions in one and two dimensions have been discussed in detail.

Gravitation is the topic of the next chapter, as we continue our study of mechanics. The force of gravity is responsible for many of the large-scale properties of the universe, for example, the existence and motion of planets, stars, galaxies, and black holes.

CONCEPTS AND SKILLS

Having completed this chapter, you should now be able to do the following:

- Define (linear) momentum.
- State the relation between momentum, resultant force, and time interval, and use it in solving problems.
- State the law of conservation of momentum.
- Use conservation of momentum to solve problems (particularly involving collisions) in one and two dimensions using either vector scale diagrams or vector components.
- Distinguish situations in which momentum is not constant from those in which momentum is constant.
- State the essential features of elastic, inelastic, and completely inelastic collisions.

KEY TERMS

You should be able to define or explain each of the following words or phrases:

momentum	law of conservation of momentum	inelastic collision
impulse	elastic collision	completely inelastic collision

Chapter Review

MULTIPLE CHOICE QUESTIONS

8-39 The SI unit for momentum is

(a) newton
(b) joule
(c) watt
(d) kg·m/s
(e) kg·m²/s

8-40 One way that an octopus propels itself is by ejecting a jet of water from its gill chamber. Suppose that an octopus of mass 3.8 kg is initially at rest and ejects 0.80 kg of water. The speed of the water is 1.2 m/s. What is the resulting speed of the octopus? Neglect fluid friction and assume that the 0.80 kg of water is not included in the 3.8 kg mass of the octopus.

(a) 0.30 m/s
(b) 1.2 m/s
(c) 5.7 m/s
(d) 0.20 m/s
(e) 0.25 m/s

8-41 The contact times in collisions between various balls used in sports and the objects (bats, racquets, etc.) used to hit them are typically a few

(a) nanoseconds
(b) microseconds
(c) milliseconds
(d) picoseconds
(e) kiloseconds

8-42 Which of the following statements describe properties of completely inelastic collisions between two objects?

1. The objects stick together.
2. There is a maximum loss of kinetic energy.
3. Both objects are at rest after the collision.
4. Momentum is conserved.

(a) 1, 2, and 3
(b) 1 and 4
(c) 1, 2, 3, and 4
(d) 1, 2, and 4
(e) 3 and 4

8-43 In a collision between a small car and a large truck, how does the magnitude of the momentum change of the car (Δp_c) compare with the magnitude of the momentum change of the truck (Δp_t)?

(a) $(\Delta p_c) = (\Delta p_t)$
(b) $(\Delta p_c) < (\Delta p_t)$
(c) $(\Delta p_c) > (\Delta p_t)$
(d) impossible to say unless the masses and velocities are known

8-44 A billiard ball moving with speed v collides head-on with a stationary ball of the same mass. After the collision, the ball that was initially moving is at rest. What is the speed of the other ball after the collision?

(a) $v/2$
(b) 0
(c) v
(d) 2
(e) $3v/4$

8-45 A helium atom in the air collides with a wall, as shown in Figure 8-32. The speed of the atom is unchanged as a result of the collision. Which of the following diagrams shows the correct direction of the change of momentum of the helium atom?

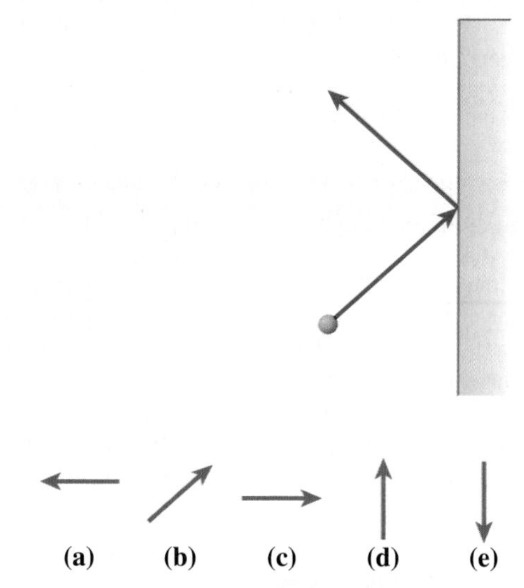

Figure 8-32 Question 8-45.

Review Questions and Problems

8-46 A number of objects of different masses have the same speed. Sketch the shape of the graph of the magnitude of momentum versus mass for these objects.

8-47 A car of mass 1.1×10^3 kg is travelling in a direction 22° north of east. Its eastward component of momentum is 2.6×10^4 kg·m/s. What is the speed of the car?

8-48 A car of mass 1.2×10^3 kg travelling west at 53 km/h collides with a telephone pole and comes to rest. The duration of the collision is 55 ms. What are the magnitude and direction of the average force exerted on the car by the pole?

8-49 A tennis ball of mass 59 g is thrown upward and then hit (served) just as it comes to rest at the top of its motion. It is in contact with the racquet for 5.1 ms and the average force exerted on the ball by the racquet is 324 N horizontally.

(a) What is the velocity (magnitude and direction) of the ball just as it leaves the racquet?

(b) How does the answer in part (a) change if the tennis player hits the ball with greater follow-through?

8-50 Each second, 1.3 kg of water leaves a hose at a speed of 24 m/s. The water, travelling horizontally, hits a vertical wall and is stopped. Neglecting any splash-back, what is the magnitude of the average force exerted by the water on the wall?

8-51 A girl of mass 38 kg is standing on a stationary wagon. She then walks forward on the wagon with a speed of 1.2 m/s, and the wagon moves backward with a speed of 3.0 m/s (both speeds relative to the ground). What is the mass of the wagon? Neglect friction.

8-52 A boy places a spring of negligible mass between two toy cars (Figure 8-33) of masses 112 g and 154 g. He compresses the spring and ties the cars together with a string. He then cuts the

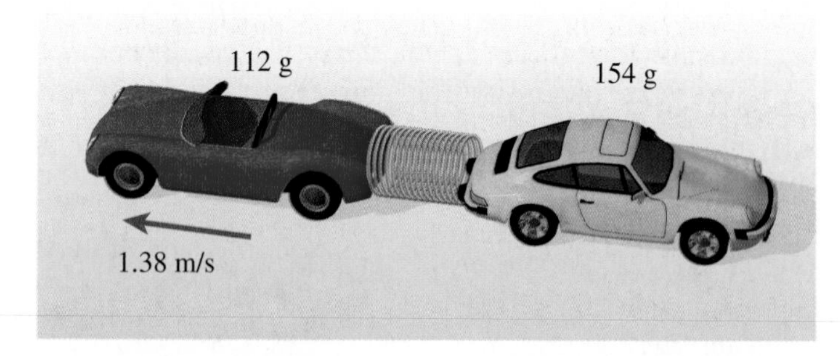

Figure 8-33 Question 8-52.

string, thus releasing the spring, and the cars move in opposite directions. The 112 g car has a speed of 1.38 m/s. What is the speed of the other car?

8-53 Give an example of a collision in which the momentum of the system of the colliding objects is not constant. Explain why momentum is not constant in this collision, but is constant in many other collisions.

8-54 A proton (mass 1.67×10^{-27} kg) travelling with an initial speed of 1.57 km/s collides head-on with a stationary alpha particle (mass 6.64×10^{-27} kg). The proton rebounds with a speed of 0.893 km/s. What is the speed of the alpha particle?

8-55 Two children each roll a marble on the ground so that there is a head-on collision. One marble has a mass of 2.5 g and a speed of 2.6 m/s before colliding, and the other has a mass of 2.9 g and a speed of 1.8 m/s. After the collision, the less massive marble has reversed direction and has a speed of 2.0 m/s. What is the velocity (magnitude and direction) of the other marble?

8-56 Two rocks in space undergo a collision. One has a mass of 2.67 kg and has an initial velocity of 1.70×10^2 m/s toward Jupiter; the other has a mass of 5.83 kg. After the collision, the rocks are both moving toward Jupiter, with speeds of 185 m/s and 183 m/s for the 2.67 kg and 5.83 kg rocks respectively. What was the initial velocity (magnitude and direction) of the more massive rock?

8-57 Two objects having masses $m_1 = 2.0$ kg and $m_2 = 4.0$ kg undergo head-on collisions. For the following initial ("i") and final ("f") velocities, identify the collisions as elastic, inelastic, or completely inelastic.

(a) $v_{1ix} = 6.0\,\text{m/s};\ v_{2ix} = 0;\ v_{1fx} = v_{2fx} = 2.0\,\text{m/s}$

(b) $v_{1ix} = 24\,\text{m/s};\ v_{2ix} = 0;\ v_{1fx} = -4.0\,\text{m/s};\ v_{2fx} = 14\,\text{m/s}$

(c) $v_{1ix} = 12\,\text{m/s};\ v_{2ix} = 0;\ v_{1fx} = -4.0\,\text{m/s};\ v_{2fx} = 8.0\,\text{m/s}$

8-58 Imagine that you are standing still and someone is about to run into you, making a completely inelastic collision. In which situation would your speed after the collision be greater: 1) If the runner has a large mass and small speed, or 2) if the runner has a small mass and large speed (but the same magnitude of momentum)? Show your work.

8-59 Two pucks on an air hockey table are equipped with spring bumpers and make an elastic head-on collision. One puck (mass 253 g) has an initial speed of 1.80 m/s. The other puck has a mass of 232 g and is initially stationary. What is the velocity (magnitude and direction) of each puck after the collision?

8-60 Two birds are flying horizontally and make a completely inelastic collision. Initially, one bird (mass 2.3 kg) is flying east at 18 m/s, and the other is flying west at 19 m/s. Afterward, the two birds tumble together at 3.1 m/s east. What is the mass of the second bird?

8-61 A girl is standing on a raft of mass 48 kg moving at 1.5 m/s west. The girl then walks on the raft with a net velocity of 1.0 m/s at 30° north of west. As the girl walks, the net velocity of the raft is 2.0 m/s at 10.5° south of west.

(a) Use a vector scale diagram to determine the approximate mass of the girl. Neglect viscous friction between the water and raft.

(b) Solve using components.

8-62 Two hockey pucks of equal mass undergo a collision on a hockey rink. One puck is initially at rest, and the other is moving with a speed of 5.4 m/s. After the collision, the velocities of the pucks make angles of 33° and 46° relative to the original velocity of the moving puck.

(a) Make a diagram showing the initial and final situations. Make sure that the geometry of your diagram is such that momentum is constant.

(b) Determine the speeds of the pucks after the collision.

(c) Find an approximate solution to this problem using a vector scale diagram.

8-63 Two trucks have a completely inelastic collision. After the collision, they have a velocity of 11 m/s at 35° south of west. Before the collision, one truck (mass 2.3×10^4 kg) has a velocity of 15 m/s at 51° south of west.

(a) Determine the initial velocity (magnitude and direction) of the other truck (mass 1.2×10^4 kg).

(b) What percentage of the initial kinetic energy is "lost" in the collision?

Applying Your Knowledge

8-64 The momentum of a runner triples as her speed increases by 3.00 m/s. What is her initial speed?

8-65 A man of mass 65 kg is sitting in a stationary go-kart of mass 35 kg that has broken down. He is holding a toolbox of mass 19 kg. He then throws the toolbox horizontally. If the resulting velocity of the man and go-kart is 1.1 m/s, what is the velocity (magnitude and direction) with which the man threw the toolbox? (Neglect frictional effects.)

8-66 Starting from rest, a skier of mass 66 kg slides down a hill that is 25 m high, and then makes a completely inelastic collision with a skier of mass 72 kg standing still at the bottom. Immediately after the collision, what is the speed of each skier? Neglect friction.

8-67 Liz and Aidan are playing on their family's zipline one afternoon and decide to recreate a "Tarzan and Jane" manoeuvre. Liz (mass 52 kg) starts at the highest platform from rest. Aidan (mass 41 kg) is standing on the intermediate platform that is 5.0 m lower than the highest point (Figure 8-34) and is scooped up by

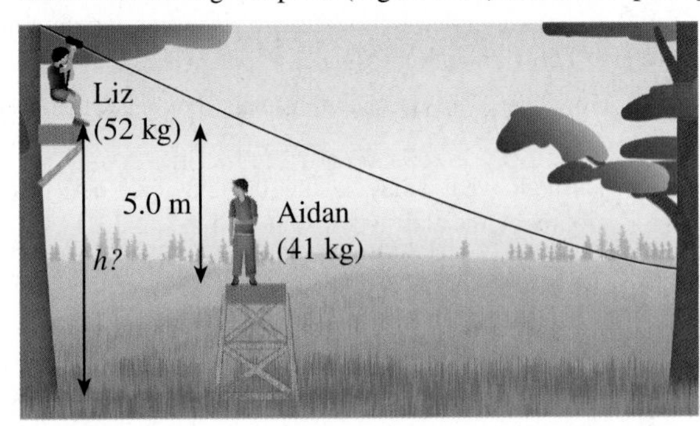

Figure 8-34 Question 8-67.

Liz as she reaches his level. They arrive together near ground level with a speed of 8.8 m/s. How high above ground level was Liz's starting point? (Neglect frictional losses throughout this problem.)

8-68 When fireworks explode, the fragments spread out symmetrically due to conservation of momentum. For simplicity, let's consider a firework of mass M that breaks up into three fragments, each of the same mass ($M/3$) (Figure 8-35). One fragment travels directly upward ($+y$-direction) after the explosion at a speed of 14 m/s and the second fragment moves with an initial velocity of 24 m/s, 22° below the horizontal to the right, as shown. Determine the velocity of the third fragment immediately after the explosion.

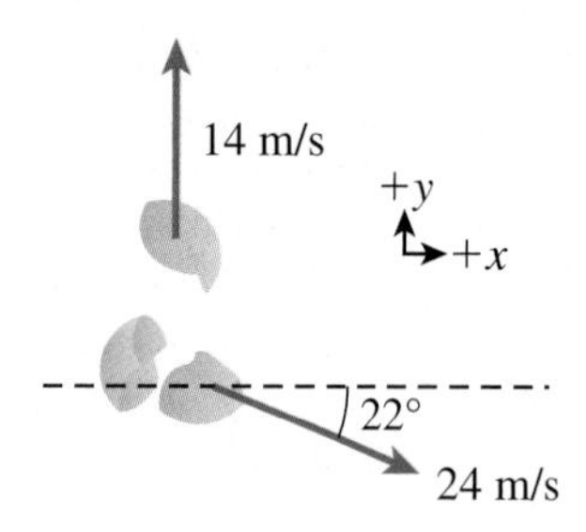

Figure 8-35 Question 8-68.

8-69 Figure 8-36 shows a water jet-pack in use, a means of locomotion much like rocket propulsion as discussed in this chapter. When the water is pushed directly downward, the person wearing the jet-pack is pushed directly upward. If we assume that the water leaves the nozzles at 11 m/s and the person and the jet-pack have a combined mass of 90 kg, determine the flow rate (in kg/s) required for this person to levitate above the water. Assume two significant digits.

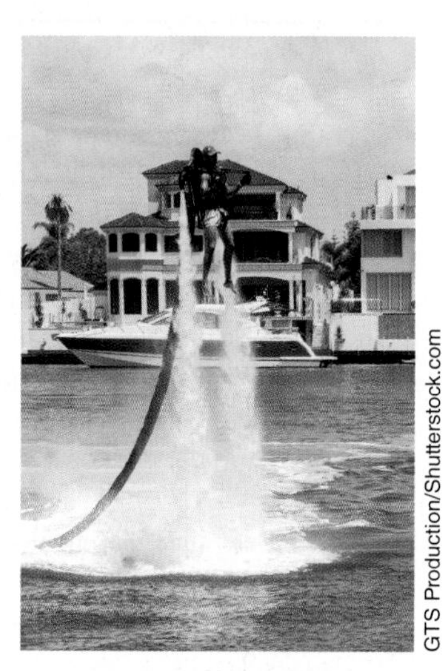

GTS Production/Shutterstock.com

Figure 8-36 Question 8-69.

8-70 After your first ball at the bowling alley, two pins remain standing. Your second attempt results in the ball hitting the pin in the centre such that it then hits the pin to the side—also known as getting a spare. The centre pin (mass = 1.5 kg) travels at 3.4 m/s, 73° to the initial direction of the ball in order to take out the pin to the side (Figure 8-37). The ball (mass = 6.8 kg) travels at 6.9 m/s at an angle of β relative to its initial direction after the elastic collision with the pin.

(a) How fast was the bowling ball moving when it hit the pin?

(b) Through what angle is the bowling ball deflected in its motion by the elastic collision with the pin?

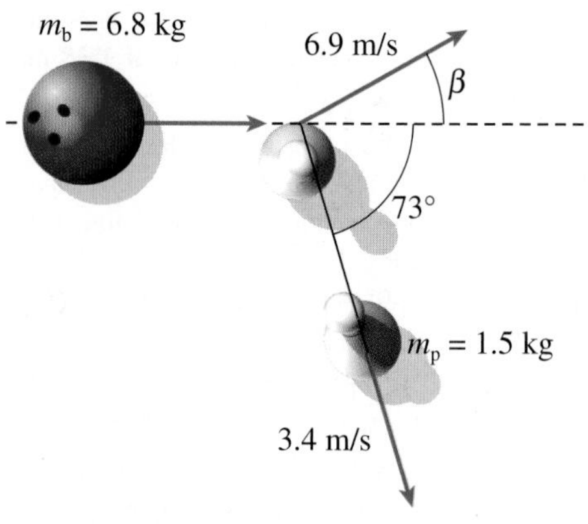

Figure 8-37 Question 8-70.

8-71 In a close curling game, your opponent's stone is in prime position—right in the centre of the house—and you need a strong throw to clear it out. Your stone hits the opposing stone with a speed of 1.50 m/s. After the elastic collision, your stone travels with a speed of 1.13 m/s at an angle α to its initial direction. Assume that the two stones have equal masses of 18.0 kg. (The house is the curling term used to describe the concentric, coloured circle region in which points are scored, as seen in Figure 8-38.)

(a) What is the velocity of the opponent's stone after the collision?

(b) If the coefficient of kinetic friction between the stone and the ice is 0.0168, does your opponent's stone clear the house?

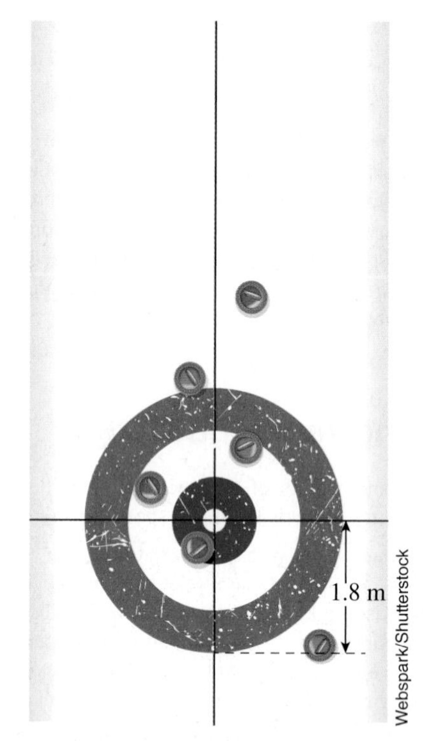

Webspark/Shutterstock

Figure 8-38 Question 8-71.

8-72 Two ice-skaters undergo a collision, after which their arms are intertwined and they have a common velocity of 0.85 m/s at 27° south of east. Before the collision, one skater (mass 71 kg) had a velocity of 2.3 m/s at 12° north of east, and the other skater's velocity was 1.9 m/s at 52° south of west. What is the mass of the second skater?

8-73 A space vehicle and its fuel tank are moving with a velocity of 3.50×10^2 m/s toward a space station. The fuel tank is then jettisoned to the rear, and the velocity of the tank is then only 2.50×10^2 m/s toward the station. What is the new speed of the space vehicle? The mass of the vehicle is 3.25 times that of the tank.

8-74 Two nuclei make a head-on elastic collision. One nucleus (mass m) is initially stationary. The other nucleus has an initial velocity $\vec{v}$, and a final velocity of $-\left(\frac{1}{5}\right)\vec{v}$. What is the mass of this nucleus?

8-75 Two subatomic particles have a collision. Initially the more massive particle (A) is at rest and the less massive particle (B) is moving. After the collision, the velocities of A and B make angles of 67.8° and 30.0°, respectively, to the original direction of B's motion. The ratio of the final speeds of the particles $\left(v'_B/v'_A\right)$ is 3.30. What is the ratio of the particle masses (m_A/m_B)?

8-76 A variety of objects of different masses are moving at different speeds, but the momentum of each object has the same magnitude. Sketch the shape of the graph of speed versus mass for these objects.

8-77 A piece of putty is dropped vertically onto a floor where it sticks. A rubber ball of the same mass is dropped from the same height onto the floor and rebounds to almost its initial height. For which of these objects is the magnitude of the change of momentum greater during the collision with the floor? Explain your answer.

8-78 Suppose that a small moving object hits a larger stationary object. In which situation is the force exerted by the small object (on the large object) of greater magnitude: if the small object bounces off the large object, or if it sticks to it? (Assume that the collision takes the same amount of time in either case.) Relate this to the use of rubber bullets by riot police.

8-79 A football of mass 0.41 kg is kicked from rest and given a velocity of 24 m/s at 29° above the horizontal. The ball is in contact with the kicker's foot for 8.2 ms. Determine

(a) the horizontal and vertical components of the ball's momentum,

(b) the horizontal and vertical components of the average force exerted by the foot on the ball (using momentum methods),

(c) the magnitude and direction of the force in (b).

8-80 Two rolling golf balls of the same mass make a collision. Initially, the velocity of one ball is 2.70 m/s east. After the collision, the velocities of the balls are 2.49 m/s at 62.8° north of west, and 2.37 m/s at 69.2° south of east. What is the unknown initial velocity (magnitude and direction)?

8-81 A ball (mass 0.25 kg) is attached to a string of length 26 cm (Figure 8-39). The ball is raised so that the string is taut and horizontal, and the ball is released so that at the bottom of its swing it undergoes an elastic head-on collision with another ball (mass 0.21 kg) that is free to roll along a horizontal table.

(a) Just before the collision, what is the speed of the swinging ball?

(b) Just after the collision, what is the speed of the 0.21 kg ball?

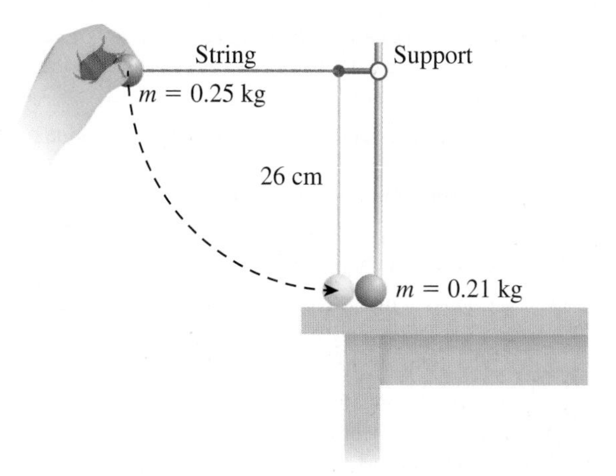

Figure 8-39 Question 8-81.

8-82 Figure 8-40 shows a device called a ballistic pendulum, ((a) schematically, (b) realistically) which was used to determine speeds of bullets before the advent of modern electronic timing. A bullet is shot horizontally into a block of wood suspended by two strings. The bullet remains embedded in the wood and the wood and bullet together swing upward.

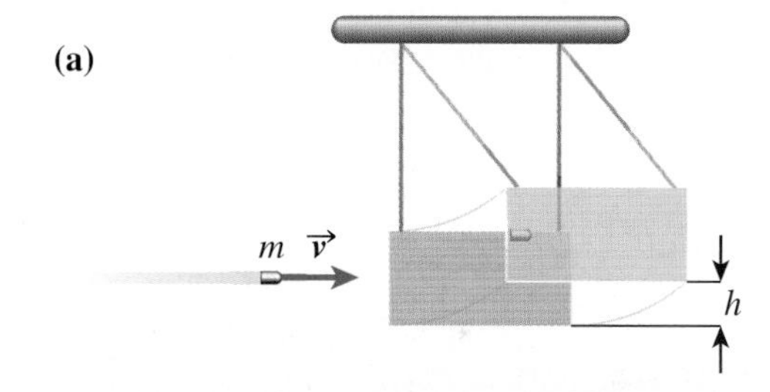

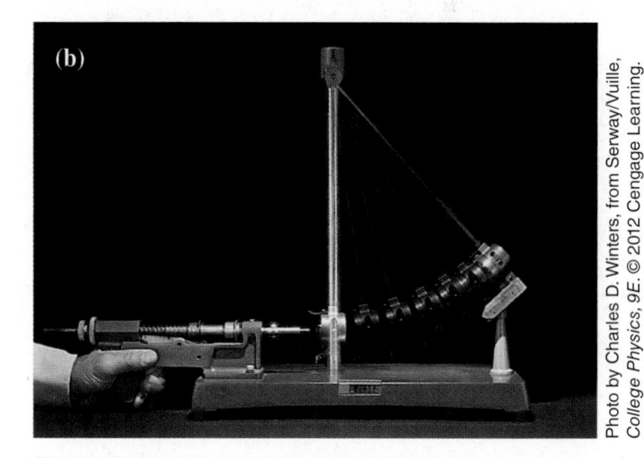

Photo by Charles D. Winters, from Serway/Vuille, *College Physics, 9E.* © 2012 Cengage Learning.

Figure 8-40 Question 8-82: Ballistic pendulum.

(a) Explain why the *horizontal* momentum of the bullet–wood system is constant during the collision, even though the strings exert tension forces on the wood.

(b) If the bullet and wood have masses m and M, respectively, and the bullet has an initial speed of v, determine an algebraic expression for the speed of the bullet and wood immediately after the collision (before they swing upward) in terms of m, M, and v.

(c) As the bullet and wood swing up, what law of nature can be used to relate the maximum vertical height to the speed just after the collision?

(d) Use your answers from (b) and (c) to determine an expression for the maximum vertical height h in terms of m, M, v, and g (magnitude of gravitational acceleration).

(e) Now rearrange your expression in (d) so that, if h is a known quantity, v can be calculated.

(f) If a bullet of mass 8.7 g hits a block of wood of mass 5.212 kg, and the bullet and wood swing up to a maximum height of 6.2 cm, what is the initial speed of the bullet?

8-83 A large ball of modelling clay (mass 4.5×10^2 g) is rolled on a tabletop so that it collides with a stationary small wooden box (mass 7.9×10^2 g). The collision is completely inelastic and the ball and box then slide on the table for a distance of 5.1 cm. If the speed of the ball is 2.2 m/s just before the collision, determine:

(a) the speed of the ball and box just after the collision, and

(b) the magnitude of the friction force acting on the ball and box.

8-84 Two rugby players simultaneously hit a stationary third player of mass 85 kg. Just before the completely inelastic collision, one player (mass 82 kg) is running with a velocity of 8.3 m/s north, and the other (mass 95 kg) has a velocity of 6.7 m/s at 25° west of north. Just after the collision, what is the velocity (magnitude and direction) of the three players combined?

8-85 An object of mass m_1 and initial velocity v_{1ix} undergoes a head-on elastic collision with a stationary object of mass m_2.

(a) In terms of m_1, m_2, and v_{1ix}, determine the final velocity of each mass.

What do your answers in (a) become if

(b) $m_1 = m_2$? (two billiard balls collide)

(c) $m_1 \gg m_2$? (a bowling ball hits a super ball)

(d) $m_1 \ll m_2$? (a super ball hits a bowling ball)

8-86 **Fermi Question:** A large cruise liner having a mass of 100 000 tonnes is arriving in a port. It is slowing down and is only 100 m from the dock. Estimate the magnitude of the liner's momentum and the force that is required to bring the liner to a halt just as it reaches the dock.

8-87 **Fermi Question:** The continents of Earth are in motion relative to the underlying core. If North America drifts about 1 to 2 cm per year, what is the magnitude of the momentum of this continent? How does this compare with the magnitude of the momentum of the large cruise ship in the previous problem?

8-88 **Fermi Question:** Two brothers, Cameron and Lucas, are stuck in the middle of a perfectly frictionless icy lake, about 100 m from shore. If there was friction they could just walk to shore; instead they will have to use their knowledge of conservation of momentum to return to safety. Cameron shoves his little brother Lucas away as hard as he can (without hurting him!). How long will it take to reach the shore? Who will reach safety first?

Gravitation 9

MarcelC/Thinkstock

Figure 9-1 A starry night away from city lights.

It is a wonderful experience to gaze up at the stars and planets on a warm summer night, away from bright city lights (Figure 9-1). People have been fascinated for millennia with the night sky, not only with its beauty, but also with its puzzles. Why do the Moon and planets move relative to the background stars? Why are different regions of the sky visible at different times of the year? What keeps stars and planets together? And so on. Many of the dominant features of the universe are due to gravitation. The force of gravity is responsible for the very existence of planets and stars, as well as for the patterns of their motion. Even the intricate beauty of the rings of Saturn (Figure 9-18) can be explained through an understanding of gravitation.

CHAPTER OUTLINE

9.1 Law of Universal Gravitation

Many early peoples were interested in astronomy, especially in the movements of the Sun, Moon, and planets. The Chinese and the Babylonians kept careful records of astronomical events for several centuries before the common era (BCE).[1] There is also evidence of astronomical interest in the Assyrians and Egyptians, the Mayans in Central America, and Bronze Age people in northwestern Europe, especially in the British Isles, as illustrated in Figure 9-2.

There was a great deal of study in astronomy in the ancient Greek civilization from 600 BCE to 400 CE. In the second century CE, the Greco-Egyptian astronomer Claudius Ptolemy created a complex geometrical representation of the solar system that explained and predicted the motion of the planets very well. In this model, Earth was at the centre of the solar system. This geocentric (Earth-centred) hypothesis and Ptolemy's general representation were accepted throughout the Middle Ages, until the advent of the heliocentric (Sun-centred) hypothesis of the Polish scientist, Nicholas Copernicus (1473–1543).

In 1543, Copernicus published *De Revolutionibus Orbium Coelestium* ("On the Revolutions of the Heavenly Spheres"), in which planetary motion was described in terms of orbits around the Sun. The Copernican model of the solar system was a significant departure from the thinking of the time and did not gain general acceptance for over a century. During this period, the German scientist, Johannes Kepler (1571–1630) was studying the details of the planets' orbits and published his three laws of planetary motion early in the 1600s. His laws seemed to indicate that the Sun occupied a special place in planetary orbits, consistent with the Copernican model. We will discuss Kepler's laws in more detail in Section 9.4.

DID YOU KNOW?

The word "planet" is derived from the Greek word *planets,* which means "wanderer." The planets wander in the sky relative to the stars, which appear essentially fixed relative to each other.

Kepler recognized that a force pulled on the planets toward the Sun, but he did not know the mathematical details of this force. It remained for Isaac Newton to unravel the force's secrets. (A short biography of Newton (1642–1727) is given in Section 5.7.)

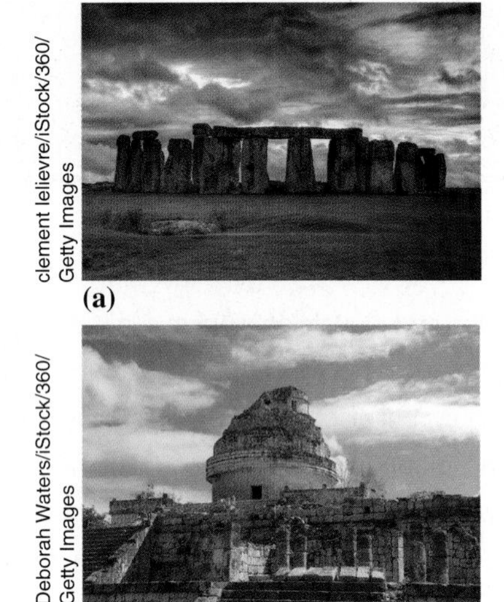

clement lelievre/iStock/360/Getty Images

Deborah Waters/iStock/360/Getty Images

© Henry Westheim Photography/Alamy

Figure 9-2 Archeological evidence of the interest of many ancient civilizations in studying the stars. **(a)** Stonehenge in the United Kingdom dates back to approximately 3100 BCE. **(b)** The Mayan astronomical observatory at Chichén Itzá in Mexico was built in approximately 1000 CE. **(c)** An astronomical sextant, built in 1673 at the Ancient Observatory of China, was used for measuring the distance between celestial bodies and the diameter of the Moon and Sun before the invention of the telescope.

[1]BCE: Before the Common (or Current or Christian) Era, sometimes written as BC (Before Christ). CE: Common (or Current or Christian) Era, sometimes written as AD (*Anno Domini*, i.e., "in the year of the Lord").

It is said that an apple falling from a tree (Figure 9-3) triggered Newton's thought that the same fundamental force attracting the apple toward Earth also attracts the Moon toward Earth, but this story might not in fact be true. His **law of universal gravitation** was published in the book called the *Principia* in 1687.

Figure 9-3 Woolsthorpe Manor, in Lincolnshire, UK, is the birthplace of Sir Isaac Newton. The home is seen here through the branches of the famous apple tree.

The law of universal gravitation states that two particles are attracted to each other by a force directed along the line between them; the magnitude of the force is proportional to the product of the particles' masses, and inversely proportional to the square of the distance between the particles. Mathematically, we write

$$F \propto \frac{m_1 m_2}{r^2}$$

where F is the magnitude of the force, m_1 and m_2 are the two masses, and r is the distance between the particles (Figure 9-4). We can write this law as an equation by introducing a constant of proportionality, conventionally given the symbol G:

$$F = \frac{Gm_1 m_2}{r^2} \quad (9\text{-}1)$$

law of universal gravitation two particles are attracted to each other by a force directed along the line between them; the magnitude of the force is proportional to the product of the particles' masses and inversely proportional to the square of the distance between the particles

The constant, G, is known as the **universal gravitation constant**. In SI units, it has the numerical value of

$$G = 6.67 \times 10^{-11}\,\text{N}\cdot\text{m}^2/\text{kg}^2$$

universal gravitation constant given the symbol G, a constant used in the law of universal gravitation having a value of 6.67×10^{-11} N·m²/kg²

The gravitational force is a mutual force of attraction between the two objects; each object experiences a force that is equal in magnitude and is directed toward the other. As we can see in Figure 9-4, the law of universal gravitation is in complete agreement with Newton's third law of motion.

Newton was able to *derive* Kepler's laws of planetary motion from his law of universal gravitation, and thus was able to explain all the details of planetary orbits—shapes, periods, etc.—from one basic law. The voyage of discovery and explanation started by Ptolemy and other early Greeks had been completed by Newton.

In the following sample problem, we explore the gravitational force for two rather different scenarios in order to get a sense of the strength of this interaction.

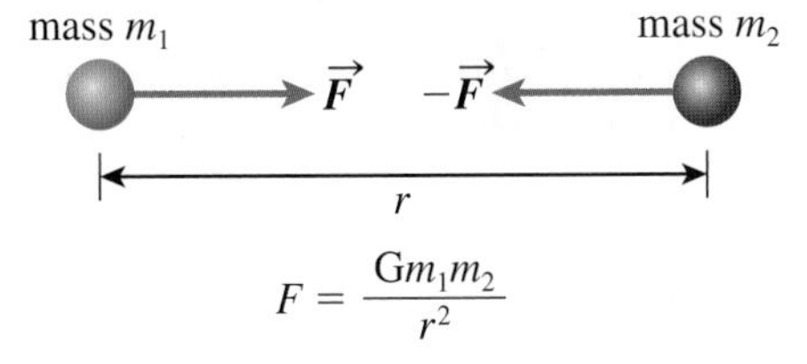

Figure 9-4 Newton's law of universal gravitation.

SAMPLE PROBLEM 9-1

Determine the magnitude of the gravitational force between

(a) two 1.0 kg masses separated by a distance of 1.0 m.

(b) Earth (mass 5.98×10^{24} kg) and the Sun (mass 1.99×10^{30} kg) when they are separated by 1.50×10^{8} km.

Solution

(a) This problem involves only straightforward substitution into Eqn. 9-1 for the gravitational force.

$$F = \frac{Gm_1 m_2}{r^2} = \frac{(6.67 \times 10^{-11}\,\text{N}\cdot\text{m}^2/\text{kg}^2)(1.0\,\text{kg})(1.0\,\text{kg})}{(1.0\,\text{m})^2} = 6.7 \times 10^{-11}\,\text{N}$$

Hence, the gravitational force between the 1.0 kg masses is only 6.7×10^{-11} N in magnitude.

DID YOU KNOW?

There is a conjecture in the world of theoretical physics that is called the Universal Limit on Maximum Luminosity. This idea proposes that no celestial object can emit more power than this upper limit, which is calculated as c^5/G, or the speed of light (3.0×10^8 m/s) to the power 5, divided by the gravitational constant. This gives a numerical value of 3.6×10^{52} W, which is a truly staggering value. For comparison, the total power output of our Sun is approximately 4×10^{26} W.

DID YOU KNOW?

The expression *standing on the shoulders of giants* is typically attributed to Sir Isaac Newton, although a variation was used in the 12th century by the theologian and author John of Salisbury. In 1675, Newton wrote: "If I have seen further, it is by standing upon the shoulders of giants" in a letter to Robert Hooke, a fellow scientist with whom Newton had a bitter rivalry. Since the expression has come to mean that intellectual progress is incremental and builds upon the work of predecessors, it is perhaps somewhat ironic that Newton may have been using the expression with a hint of sarcasm to his rival who was purportedly short in stature, rather than purely in humility.

(b) For the Earth–Sun system, we first convert the separation from kilometres to metres:

$$1.50 \times 10^8\,\text{km} \times \frac{10^3\,\text{m}}{1\,\text{km}} = 1.50 \times 10^{11}\,\text{m}$$

Then, substituting numbers into Eqn. 9-1,

$$F = \frac{Gm_1m_2}{r^2}$$
$$= \frac{(6.67 \times 10^{-11}\,\text{N}\cdot\text{m}^2/\text{kg}^2)(5.98 \times 10^{24}\,\text{kg})(1.99 \times 10^{30}\,\text{kg})}{(1.50 \times 10^{11}\,\text{m})^2}$$
$$= 3.53 \times 10^{22}\,\text{N}$$

Thus, the gravitational force between the Sun and Earth has a magnitude of 3.53×10^{22} N. This force is very large because of the huge masses involved, which more than compensate for the large Earth–Sun distance, even when this distance is squared as required in Eqn. 9-1.

Problem-Solving Tip

Very common errors arise in problem solving with gravitational forces and acceleration when students forget two important requirements of the variable r in Eqn. 9-1:

- r is always the distance from the centre of an object, not the surface
- r is always needed in metres in order for your final answer to be in the appropriate SI unit (of course this is not unique to Eqn. 9-1, but students often forget to make this change when distances between celestial masses are given in kilometres)

You may have noticed that in our earlier definition of the law of universal gravitation, we talked about the gravitational force between "particles," and Earth and the Sun are obviously not particles. However, since the Earth–Sun distance is very large compared to the size of the Sun and Earth, they can be considered to be particles and we can therefore freely apply Eqn. 9-1 in this scenario as we just did in the preceding sample problem. Later in this section, we discuss how to apply the law of universal gravitation to extended objects that cannot be treated as particles.

SAMPLE PROBLEM 9-2

Two objects separated by a distance, d, exert a mutual gravitational force. If the same objects are separated by a distance of $5d$, by what factor will the force have increased or decreased?

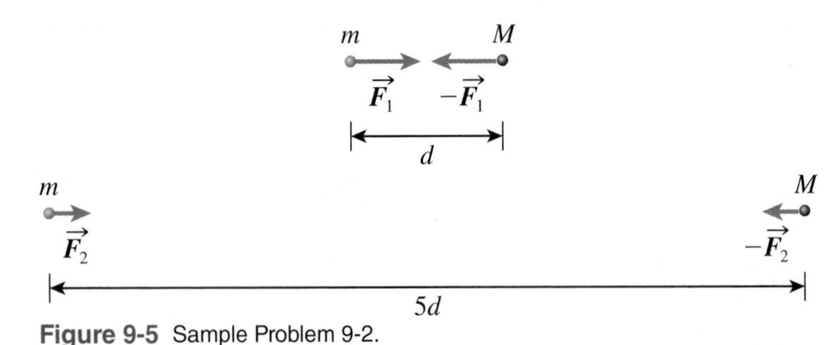

Figure 9-5 Sample Problem 9-2.

Solution

Let the two masses be m and M, and let the magnitude of the force be F_1 at separation d and F_2 at separation $5d$ (Figure 9-5). We are asked to find F_2 relative to F_1. When the separation is d, we have

$$F_1 = \frac{GmM}{d^2} \qquad \text{Eqn. [1]}$$

When the separation is $5d$,

$$F_2 = \frac{GmM}{(5d)^2} \qquad \text{Eqn. [2]}$$

Dividing Eqn. [2] by [1],

$$\frac{F_2}{F_1} = \frac{GmM}{(5d)^2} \div \frac{GmM}{d^2}$$
$$= \frac{GmM}{25d^2} \times \frac{d^2}{GmM}$$
$$= \frac{1}{25}$$

Therefore, as the separation increases from d to $5d$, the force decreases by a factor of 25.

Note that when we say that a quantity increases or decreases by a certain factor, the factor is conventionally a number greater than 1. Thus, we normally would say that the force decreases by a factor of 25, not by a factor of 1/25.

Determination of G

Because the gravitational force between non-astronomical objects (such as you and your best friend) is so small, the determination of the numerical value of the gravitation constant G is not easy. The first experiment that gave an accurate value of G was performed by the English scientist Henry Cavendish (1731–1810) in 1798, more than a century after Newton published his gravitation law.

Cavendish's device to measure G experimentally, called a torsional-balance apparatus, is sketched in Figure 9-6 (a). Two small lead spheres are fastened to the ends of a light horizontal rod, which is suspended at its centre by a fine fibre. Two larger lead spheres are positioned close to the small spheres and attract them gravitationally. The small spheres move slightly toward the large ones, causing the fibre to twist very slightly. If it has already been determined how much force is required to twist the fibre through various angles, then the gravitational attractive force of the spheres, and hence G, can be calculated. Figure 9-6 (b) shows a commercial device based on Cavendish's experiment. The smaller spheres are suspended inside the box and the larger spheres are clearly visible on the outside. A laser can be used to measure the degree of twist arising from the presence of the larger masses with great precision.

Because of the challenges in measuring G precisely, the experimental uncertainty in its value is rather large in comparison with uncertainties in other fundamental constants. As of October 2013, the best measured value of G was $(6.67545 \pm 0.00018) \times 10^{-11}\ \text{N}\cdot\text{m}^2/\text{kg}^2$, or $6.67545 \times 10^{-11}\ \text{N}\cdot\text{m}^2/\text{kg}^2 \pm 0.0027\%$, as determined by scientists at the International Bureau of Weights and Measures (BIPM) in France. While this percentage uncertainty might seem small to you, the uncertainty in the mass of an electron is much smaller, only $\pm 4.4 \times 10^{-6}\%$, and the charge on an electron is known to within $\pm 2.2 \times 10^{-6}\%$. If you are interested in the most recent values of any of the fundamental physical constants and their uncertainties, see http://physics.nist.gov/cuu/Constants/.

(a)

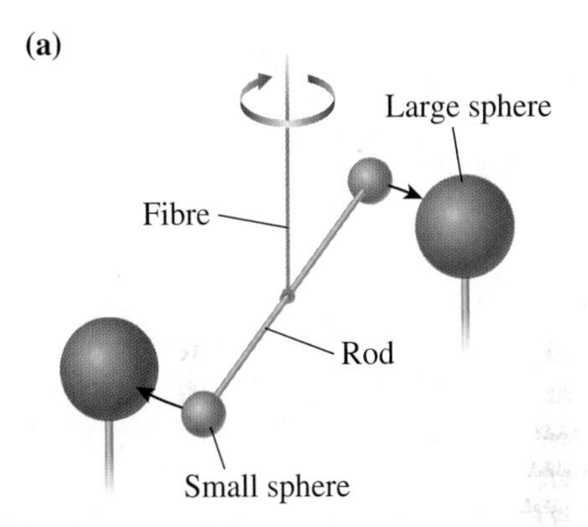

(b)

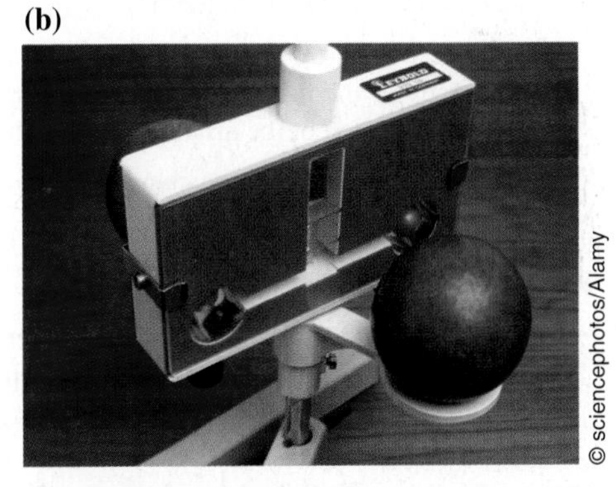

© sciencephotos/Alamy

Figure 9-6 **(a)** A sketch of the apparatus used by Henry Cavendish in 1798 to determine G. **(b)** Commercial apparatus for measuring G. The small spheres are suspended from a thin fibre within the box, with the larger spheres clearly visible on the outside.

Gravitational Force between Extended Objects

As we discussed briefly after Sample Problem 9-1, the law of universal gravitation refers to two "particles," that is, two objects small in size compared to the distance between them as illustrated in Figure 9-7 (a). If we want to determine the

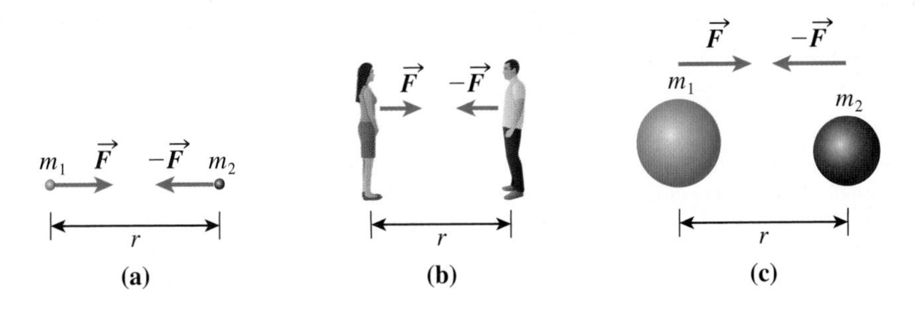

Figure 9-7 **(a)** For particles, Eqn. 9-1 can be applied directly. **(b)** For objects that are large compared to "r," Eqn. 9-1 must be used in conjunction with integral calculus. **(c)** A special case: objects with spherical symmetry. In this scenario, Eqn. 9-1 can still be used directly.

gravitational force between two objects that are large relative to the distance between them (for example, two people standing close together as shown in Figure 9-7 (b)), we cannot simply use the gravitation law as stated in Eqn. 9-1. That is, we cannot just use $F = Gm_1m_2/r^2$, with r being the distance between the centres of the objects. The underlying physics described in Eqn. 9-1 still applies as it is the law of *universal* gravitation, but we would have to use this expression in conjunction with integral calculus to add up (vectorially) all the gravitational forces between all the small particles that make up the objects—not an easy task! Such an exercise is beyond the scope of this textbook.

However, there is one exception to this: if the objects are spheres that have spherical symmetry,[2] then we can use Eqn. 9-1 directly no matter how close together the spheres might be. In this scenario, the distance r used in Eqn. 9-1 is the distance between the centres of the spheres, as shown in Figure 9-7 (c). For a person standing on Earth, the gravitation law gives a very good value for the force of gravity because (i) Earth's shape is very close to being a sphere with spherical symmetry, and (ii) the size of the person is small relative to the distance from the person to the centre of Earth.

EXERCISES

9-1 (a) What is the gravitational force (magnitude and direction) exerted by Earth (mass 5.98×10^{24} kg) on the Moon (mass 7.35×10^{22} kg)? Their average separation distance (centre-to-centre) is 3.8×10^8 m.

(b) What is the gravitational force (magnitude and direction) exerted by the Moon on Earth?

9-2 Two objects are moved farther and farther apart. Sketch the shape of the graph of the magnitude of the gravitational force on each object as a function of the distance between the objects.

9-3 In which of the following situations could we determine the magnitude of the gravitational force between the objects A and B by straightforward substitution of numbers into Eqn. 9-1 (assuming that values for m_A, m_B, and r were provided)?

(a) Jupiter (A) and Europa (B), one of Jupiter's moons

(b) a man (A) and a woman (B) hugging each other

(c) a baseball (A), travelling from a pitcher to a catcher, and Earth (B)

(d) two basketballs (A and B), sitting on a gym floor

(e) two books (A and B) beside each other on a bookshelf

9-4 As a meteor travels toward Earth, the Earth–meteor separation is halved. By what factor does the Earth–meteor gravitational force increase?

9-5 As a spaceship moves away from Earth, the gravitational force between them decreases by a factor of 9. By what factor has the spaceship–Earth separation increased?

9-6 At a certain distance above Earth's surface, the gravitational force on an object is 36 times smaller than its value at Earth's surface. What is this distance as a multiple of Earth's radius, r_E?

9-7 (a) What is the magnitude of the gravitational force between two oranges, each having mass 145 g, whose centres are separated by 35.0 cm?

(b) How does this compare to the magnitude of the gravitational force between one orange and Earth? Assume the orange is on Earth's surface and that the Earth has a mass of 5.98×10^{24} kg and an average radius of 6.37×10^6 m.

9-8 **Fermi Question:** What is the magnitude of the gravitational force between two scoops of ice cream stacked on top of each other on a cone? (Assume that the scoops are spherically symmetric!)

9.2 Gravity Due to Planets and Stars

Imagine trying to "weigh the Earth"—we cannot just pick it up and place it on a balance or a scale (Figure 9-8). So how do we know the mass of Earth? Cavendish's determination of the universal gravitation constant, G, in 1798 was important not only because it established a value for a fundamental constant of nature, but also because it permitted the measurement of the mass of Earth.

[2]Spherical symmetry means that at all points located the same distance from the centre of the sphere, there is the same composition, density, etc. These properties must not depend on the *direction* from the centre to a point, only on the *distance*. The following objects have spherical symmetry (or close to it): baseballs, planets, and gumballs. Any sphere with constant density also has spherical symmetry. If you were to form a sphere by gluing together half a baseball (simply sliced through the centre) and a hemispherical rock of the same radius, it would not have spherical symmetry because, at a given distance from the centre, you could find either rock or baseball stuffing, depending on the direction from the centre.

Figure 9-9 shows a person of mass m standing on Earth, which has a mass M and radius r. From the law of universal gravitation, Eqn. 9-1, the magnitude of the force of gravity on the person (and on the Earth) is

$$F = \frac{GmM}{r^2}$$

In Section 5.3, we expressed the gravitational force on an object of mass m on the surface of Earth as $F = mg$, where g is the magnitude of the acceleration due to gravity (9.80 m/s²). Therefore

$$mg = \frac{GmM}{r^2}$$

Cancelling out the mass of the person (m) that appears on each side of the equation gives us:

$$g = \frac{GM}{r^2} \qquad (9\text{-}2)$$

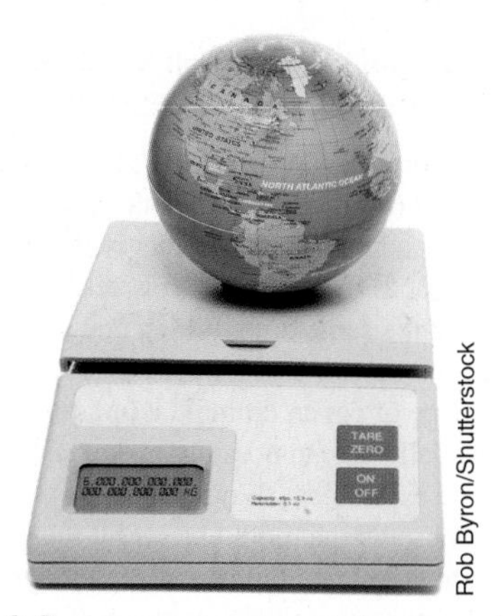

Rob Byron/Shutterstock

Figure 9-8 How can we measure the mass of Earth?

Thus, the gravitational acceleration at the surface of Earth (g) depends on the universal gravitation constant (G), as well as on the mass (M) and radius (r) of Earth. It does not depend on the mass of the person (or any other object). If g, G, and r are known, then M can be determined. The acceleration g is easy to measure, G is known from Cavendish's experiment (and from more recent ones), and r has been known reasonably well since the time of the early Greeks. Rearranging the above equation to solve for M, and substituting in values for g, G, and r,

$$M = \frac{gr^2}{G} = \frac{(9.80\,\text{m/s}^2)(6.37 \times 10^6\,\text{m})^2}{6.67 \times 10^{-11}\,\text{N}\cdot\text{m}^2/\text{kg}^2} = 5.96 \times 10^{24}\,\text{kg}$$

Our result for Earth's mass differs somewhat from the accepted value of 5.98×10^{24} kg because we assumed that Earth is a sphere and that g is constant at all points on Earth's surface. Actually, Earth is somewhat non-spherical and g is not constant, as discussed in more detail later in this section.

Notice that the relation $g = GM/r^2$ is valid not only for objects on Earth's surface, but also for objects *above* Earth's surface. Since r represents the distance from the object to Earth's centre, the numerical value of g will vary as an object moves away from the Earth's surface, such as when a satellite is launched. The symbol g still represents the magnitude of the gravitational acceleration, which decreases from a value of 9.80 m/s² at the surface with increasing distance r from the centre of Earth, according to $g = GM/r^2$ (Eqn. 9-2). We can also use Eqn. 9-2 for other planets, stars, etc., by simply substituting the appropriate mass M. This truly is the law of *universal* gravitation!

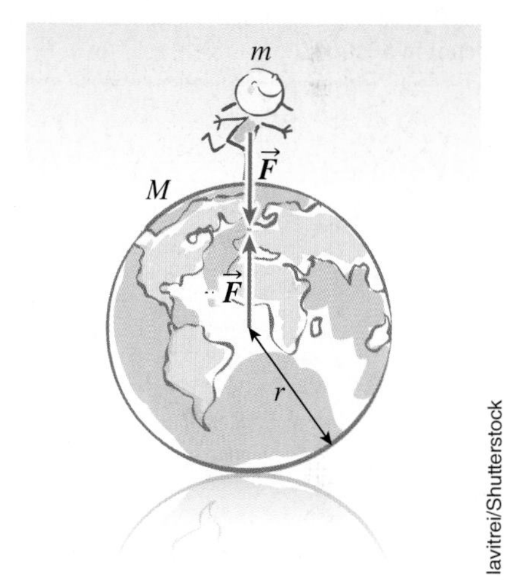

lavitrei/Shutterstock

Figure 9-9 The force of gravity on a person standing on Earth.

SAMPLE PROBLEM 9-3

For a satellite of mass 225 kg located 1.2×10^7 m above Earth's surface (Figure 9-10), determine the magnitudes and directions of:

(a) the gravitational force, and

(b) the resulting acceleration. Earth has a mass of 5.98×10^{24} kg and an average radius of 6.37×10^6 m.

Solution

(a) The magnitude of the force can be calculated from the law of universal gravitation:

$$F = \frac{Gm_1m_2}{r^2}$$

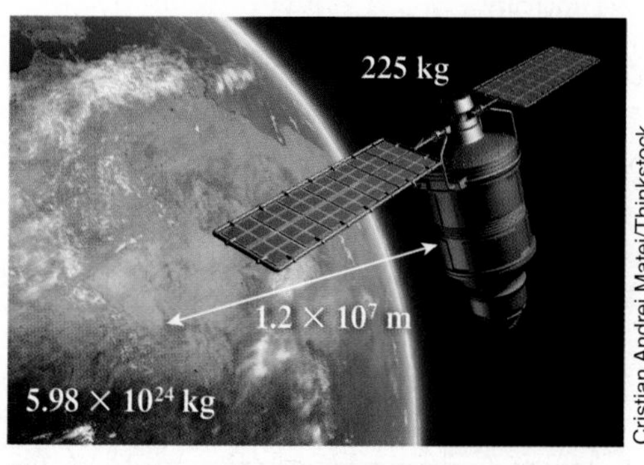

Cristian Andrei Matei/Thinkstock

Figure 9-10 Sample Problem 9-3.

Do All Objects Really Fall at the Same Rate near the Surface of Earth?

Standing on a table to get a little higher from the ground (safely!), hold a piece of paper in one hand and a shoe in the other with your arms straight out in front of you. Now release both objects at the same time. Which one reaches the ground first? Why? Now, take the exact same piece of paper, crumple it up into the tightest ball you can possibly make, and repeat the same experiment. What happens now? You should observe that both objects hit the ground at roughly the same moment, even though the shoe has more mass than the piece of paper. Once we minimize the effect of air resistance on the paper we can see more easily that, in the absence of other forces, objects of any mass will fall with an acceleration of g, as discussed in detail in Section 2.5.

Let m_1 and m_2 represent the mass of the satellite and Earth respectively. The separation r is the distance from the satellite to the *centre* (not the surface) of Earth, and therefore is the sum of the given height above Earth's surface and Earth's radius. Therefore

$$r = (6.37 \times 10^6 + 1.2 \times 10^7)\,\text{m}$$
$$= (0.637 \times 10^7 + 1.2 \times 10^7)\,\text{m}$$
$$= 1.84 \times 10^7\,\text{m}$$

(keeping an extra digit for carrying through to the next step in the calculation). We can now substitute numbers to calculate F:

$$F = \frac{(6.67 \times 10^{-11}\,\text{N}\cdot\text{m}^2/\text{kg}^2)(225\,\text{kg})(5.98 \times 10^{24}\,\text{kg})}{(1.84 \times 10^7\,\text{m})^2}$$
$$= 2.65 \times 10^2\,\text{N}$$

Thus, the gravitational force on the satellite is 2.7×10^2 N toward the centre of Earth.

(b) To find the gravitational acceleration of the satellite, we will use two different approaches. First, we will use Newton's second law ($F_R = m_1 a$), since we already know the magnitude of the resultant force, F_R, and the mass. Afterward, we will use the expression $g = GM/r^2$ that we derived above, where M is the mass of Earth and r is the distance from the satellite to Earth's centre. Using Newton's second law,

$$F_R = m_1 a$$

Hence,

$$a = \frac{F_R}{m_1} = \frac{2.65 \times 10^2\,\text{N}}{225\,\text{kg}} = 1.2\,\text{m/s}^2$$

Now using the second approach,

$$g = \frac{GM}{r^2} = \frac{(6.67 \times 10^{-11}\,\text{N}\cdot\text{m}^2/\text{kg}^2)(5.98 \times 10^{24}\,\text{kg})}{(1.84 \times 10^7\,\text{m})^2} = 1.2\,\text{m/s}^2$$

In both cases, we conclude that the acceleration of the satellite is 1.2 m/s^2 toward the centre of Earth. This confirms that we have not made an algebraic error in either case.

Table 9-1

Magnitudes of Surface Gravitational Accelerations (Relative to Earth g = 1)

Planet	Surface gravity (Earth g = 1)
Mercury	0.39
Venus	0.90
Earth	1.00
Mars	0.38
Jupiter	2.58
Saturn	1.11
Uranus	1.07
Neptune	1.40

SAMPLE PROBLEM 9-4

Calculate the magnitude of the gravitational acceleration on the surface of Mars, which has a mass of 6.4×10^{23} kg and a radius of 3.4×10^6 m.

Solution

This problem involves only a direct substitution into Eqn. 9-2:

$$g = \frac{GM}{r^2}$$
$$= \frac{(6.67 \times 10^{-11}\,\text{N}\cdot\text{m}^2/\text{kg}^2)(6.4 \times 10^{23}\,\text{kg})}{(3.4 \times 10^6\,\text{m})^2}$$
$$= 3.7\,\text{m/s}^2$$

Thus, the gravitational acceleration on the surface of Mars has a magnitude of 3.7 m/s^2, roughly one-third of that on the surface of Earth. Table 9-1 gives the magnitudes of the surface gravitational accelerations (relative to that on Earth) for all the planets. Further planetary data, such as mass, radius, etc., can be found in the inside back cover of the book.

Variations in g on Earth

Because Earth is rotating about its axis once per day, its shape is not perfectly spherical: it bulges slightly at the equator and is slightly "squashed" at the poles. The equatorial radius is 6378 km, the polar radius 6357 km. As a consequence, a person standing at the equator is farther away from the dense core near Earth's centre than a person standing at a pole, and the gravitational force and acceleration are therefore less at the equator. Because of this non-sphericity, the difference between the equatorial and polar g-values is 0.18%, and g-values in general increase with increasing latitude (i.e., as one moves toward a pole). Most other planets are also non-spherical, some such as Saturn visibly so (Figure 9-11).

An additional effect due to Earth's rotation is a centrifugal decrease in the measured g-value relative to Earth (Section 6.3). This effect is strongest at the equator and zero at the poles, and produces an additional increase in g-values as latitude increases. The polar g-value is 0.35% larger than the equatorial value because of the centrifugal effect. Therefore, Earth's equatorial bulge and its rotation both result in increasing values of g as we move from the equator to either pole, giving rise to a change of approximately 0.5% in total.

Gravitational force and acceleration on Earth's surface also depend on elevation. If you climb a mountain, you are moving farther from Earth's dense core and gravity becomes weaker. The decrease in g is roughly 2.0×10^{-3} m/s^2 for every kilometre of land elevation. Based on this value, we expect g to decrease by approximately 0.2% from sea level to the top of Mount Everest, which is not as large an effect as that seen due to the Earth's rotation and non-spherical shape.

In addition to variations in g-value due to Earth's shape, the centrifugal effect, and elevation, there are also variations due to the local density of Earth. For example, mineral deposits have high densities that produce local increases in g. Geologists use local variations to assist them in searching for minerals, oil, and gas, as discussed in more detail in Chapter 2.

Table 9-2 gives the latitude, altitude, and g-values for a number of different locations on Earth. Notice that for places close to sea level listed in the upper portion of

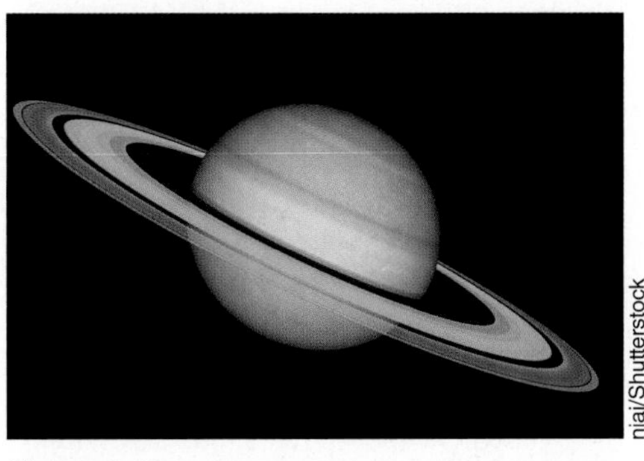
Figure 9-11 Notice the difference in the polar and equatorial radii of Saturn. Measure the radii on the image if you are not convinced.

DID YOU KNOW?

Precise measurements of g, conducted by NASA's Galileo mission during 34 orbits around Jupiter, have provided scientists with detailed information about the composition and structure of this gas giant in our solar system. Galileo, launched in 1989, was the first to measure Jupiter's atmosphere with a descent probe and the first to conduct long-term observations of the Jovian system from orbit (Figure 9-12). Galileo plunged into Jupiter's crushing atmosphere on Sept. 21, 2003.

Figure 9-12 A computer-generated image that superimposes NASA's Galileo orbiter with Jupiter and its moon Io. The orbiter is not to scale.

Table 9-2

Values of g at Various Locations

The upper portion of the table lists values at sea level for a variety of different locations on Earth, as indicated by their relevant latitude. The lower portion demonstrates the effect of altitude for a number of locations at similar latitudes. Notice that the effect from latitude is stronger than that of changing altitude, as discussed in the text.

Location	Latitude	Altitude (m)	g (m/s^2)
Jakarta, Indonesia	6.2° S	5	9.777
Hamilton, Bermuda	32.3° N	13	9.795
Victoria, British Columbia	48.4° N	8	9.813
Amsterdam, Holland	52.4° N	−2	9.817
Tuktoyaktuk, Northwest Territories	69.4° N	1	9.833
Hamilton, Bermuda	32.3° N	13	9.795
Tenerife, Canary Islands	28.3° N	1900	9.785
Picacho del Diablo, Mexico	31.0° N	3100	9.784
Mount Everest, Nepal	28.0° N	8800	9.763

TRY THIS!

Variations in *g* around the World

The Wolfram Alpha website (www.wolframalpha.com) is an excellent resource for using well-established data sets and algorithms to find answers to a wide variety of questions. For example, you can determine the acceleration due to gravity at a variety of locations on Earth through this site by typing in "acceleration due to gravity" and your particular location of interest.

the table, the g-values increase as the latitude increases due to the effects of the shape and rotation of Earth. We also see the effect of altitude, as the values of g decrease with increasing altitude for locations at similar latitude, as seen in the lower portion of Table 9-2.

This variation in g has implications for the world of sports. In Mexico City, which is close to the equator and also has a high altitude, there is a relatively low value of g of 9.776 m/s^2. As a result of low g and low air density at Mexico City, athletic performances there are improved in events such as the long jump and triple jump. For example, a world-class triple-jumper can jump about 13 cm farther at Mexico City than at Moscow. About 35% of the 13 cm increase is due to reduced gravity, and 65% to decreased air resistance. It is not surprising that a large number of world and Olympic records in track and field have been set at Mexico City.

EXERCISES

9-9 The mass of Earth is 5.98×10^{24} kg. Determine the magnitude and direction of the gravitational acceleration at a point in space located 7.4×10^7 m from the centre of Earth.

9-10 If we represent the magnitude of Earth's surface gravitational acceleration as 1 g, then what are the magnitudes of the accelerations (as fractions or multiples of g) at the following distances above Earth's surface?

(a) 1 Earth radius

(b) 2 Earth radii

(c) 9 Earth radii

9-11 A satellite is in a circular orbit 655 km above the surface of Earth, which has an average radius of 6.37×10^6 m and a mass of 5.98×10^{24} kg. Determine the magnitude of the gravitational acceleration at this height.

9-12 A satellite above Earth experiences a gravitational acceleration of 4.5 m/s^2 in magnitude.

(a) How far above Earth's surface is the satellite? The average radius of Earth is 6.37×10^6 m. Do not use the mass of Earth in your calculations, but *do* use surface $g = 9.8$ m/s^2.

(b) If the mass of the satellite is 5.6×10^2 kg, what is the magnitude of the gravitational force on the satellite?

9-13 If a planet of the same mass as Earth had a radius 2 times smaller than that of Earth, what would be the magnitude of the planet's surface gravitational acceleration, as a multiple of Earth's surface g?

9-14 (a) Calculate the magnitude of Mercury's surface gravitational acceleration. Its mass is 3.3×10^{23} kg and its diameter is 4878 km.

(b) Compare your answer to the value given in Table 9-1.

9-15 The Moon has a radius of 1.74×10^3 km, and a surface gravitational acceleration of magnitude 1.6 m/s^2.

(a) What is the mass of the Moon?

(b) What would be the magnitude of the gravitational force on *you* if you were on the Moon?

9-16 Does the gravitational acceleration on Earth's surface increase, decrease, or stay the same

(a) as latitude increases (at constant elevation)?

(b) as elevation increases (at constant latitude)?

9-17 **Fermi Question:** Estimate the percentage change in the acceleration due to gravity from the base to the summit of Mount Logan, the highest peak in Canada.

9.3 Gravitational Field

field a physical quantity that has a specific value at each point in space and time

gravitational field the gravitational force per unit mass at a particular location relative to a massive object such as Earth, the Sun, etc.

Physicists often like to think of forces in terms of quantities called fields. For example, there are gravitational fields, electric fields, magnetic fields—you have probably heard of Earth's magnetic field. But what is a field in the physics sense? This can be a difficult concept for students in introductory physics. The most straightforward definition is that a **field** can be any physical quantity that has a specific value at each point in space and time. Some fields are scalar quantities, such as the temperature profile in a given room. Some fields are vector quantities, such as the velocity of water in a river, since this quantity has both a magnitude and direction at each point in space and time in that river. We often make use of the field concept in physics when working with forces that act on objects at a distance, such as gravity.

Because Earth has mass, it creates in the space around it a gravitational field. If another object with mass (say the Moon) encounters this field, it experiences a gravitational force. So we can think of the force as a two-step phenomenon: Earth creates the field, and the field interacts with the Moon to produce the force. Fields are more than just abstract concepts; in advanced physics courses, students learn that fields are real physical entities that can carry energy and momentum.

To understand gravitational fields in a quantitative way, let's start by thinking about the gravitational force on a mass placed at a point P in space at a distance of 9×10^6 m from the centre of Earth. (This point is about 2600 km above Earth's surface.) The magnitude of the gravitational force on a mass at P can be calculated from Newton's law of universal gravitation: $F = Gm_1m_2/r^2$ where m_1 is the mass at P, m_2 is the mass of Earth, $r = 9 \times 10^6$ m, and G is the universal gravitational constant. For example, if a 1 kg mass is placed at P, the gravitational force turns out to be 5 N in magnitude. For a 2 kg mass, the force magnitude is 10 N; for a 3 kg mass, it is 15 N, and so on. Notice that the magnitude of the force is proportional to the amount of mass placed at P, that is, $F \propto m_1$. This follows directly from $F = Gm_1m_2/r^2$ of course.

Suppose that we now want to appreciate the strength of the gravitational force per each of these masses; we can divide the force magnitudes determined above by the corresponding masses (m_1). For example, divide 15 N by 3 kg, giving 5 N/kg, or divide 10 N by 2 kg, again giving 5 N/kg. The result is a constant, 5 N/kg in this particular instance. This quantity (force magnitude per unit mass) is *defined* to be Earth's **gravitational field** magnitude at point P. It does not depend on how much mass is placed at P, but it does depend on the mass of Earth, and on the distance between P and Earth's centre, as these values were used in the original calculations of the force magnitudes. In other words, the strength of the gravitational field of Earth depends only on its mass and how far we are from its centre.

This definition of gravitational field can be generalized now to any object with mass and at any location in space. If we imagine an object of mass M, it will exert a gravitational force on a second object with mass m that has a magnitude of GMm/r^2, where r is the distance from our source object M. If we define the gravitational field of this source object to be the force it exerts on other objects per unit mass of these other objects (m), we find that the gravitational field of our source has a magnitude of GM/r^2, where r is the distance from our source to the particular point of interest. Just as we discussed for the particular case of the gravitational field near Earth, the magnitude of this field depends only on the size of the source object and how far we are away from its centre.

The gravitational field is a *vector* quantity, having a magnitude and direction. Since the gravitational force is attractive, the gravitational force exerted on other objects points toward the source mass M; dividing this force by mass m gives a vector, the gravitational field, which also points toward M. We think of the gravitational field existing at point P even when there is no actual mass (m) placed there; the field exists in the space around the source object M.

Based on what we have learned so far then, the magnitude of the gravitational field associated with a source object of mass M at a distance r from its centre is written as

$$\text{the magnitude of the gravitational field} = \frac{GM}{r^2} \qquad (9\text{-}3)$$

Notice that the magnitude of the gravitational field is the same as the magnitude of the gravitational acceleration given in Eqn. 9-2. The gravitational field and the gravitational acceleration have the same direction as well—toward the centre of the source. Hence, gravitational acceleration and gravitational field represent essentially the same vector, and the same symbol is used for both: $\vec{g}$. Therefore, Eqn. 9-3 can be rewritten as

$$g = \frac{GM}{r^2} \quad \text{(magnitude of gravitational field)} \qquad (9\text{-}4)$$

Illuminating the Inverse Square Law

You will need the following materials for this experiment:

- A flashlight (or similar bright point-like light source)
- Some cardboard
- Ruler or measuring tape
- Light meter app for a smart phone or tablet

Remove the front cover of the flashlight to expose the bulb. Cut a square hole in the cardboard that is approximately 1.5 cm by 1.5 cm and mount the cardboard approximately 2.5 cm from the light bulb. Measure the light intensity from the bulb with your app at several (5 or more) distances (r) from the bulb. If you have a lot of environmental light that you cannot remove, you should do a background measurement (with your flashlight turned off) and subtract this value from your readings at each distance. Plot your data as intensity versus $1/r^2$ and describe what you see.

Even though this chapter is all about gravity, this experiment is relevant since the law of universal gravitation describes the force between two masses as having $1/r^2$ dependence, and the gravitational field of a mass varies as $1/r^2$ away from its centre. This is called the *inverse square law* and applies to many fields in physics: optics, nuclear physics, electricity, sound, etc. If you imagine the flashlight to be a massive celestial object, the spatial dependence observed with the light intensity is identical to the spatial dependence of the gravitational field of the massive celestial object.

DID YOU KNOW?

In order to run, you need to exert a downward force on the ground that is greater than the force of gravity acting on you. Astronauts on the Moon, where the gravitational field has a magnitude of only 1.6 N/kg instead of 9.8 N/kg as we have on Earth, found that they were running at very low speeds as a result, Figure 9-13.

Figure 9-13 An Apollo 16 astronaut walking on the Moon surface in 1972.

Table 9-3

Magnitudes of Surface Gravitational Field in our Solar System

Planet	Gravitational field strength (N/kg)
Mercury	3.8
Venus	8.8
Earth	9.8
Mars	3.7
Jupiter	25
Saturn	11
Uranus	10.5
Neptune	14

TACKLING MISCONCEPTIONS

Gravitational Field and Gravitational Force Are Not the Same Thing!

While they are closely related, the concepts of field and force are not exactly the same thing. When discussing gravity, there need to be *two objects* with mass interacting with each other for there to be a *force* acting; the magnitude of this force is given by Eqn. 9-1. However, whenever there is a *single object* with mass, there is automatically a gravitational *field* present around this object that has a magnitude given by Eqn. 9-4. While it could be argued that it is not strictly necessary to introduce the concept of fields when discussing gravity, you will find it tremendously useful in later chapters when we are discussing electric and magnetic interactions.

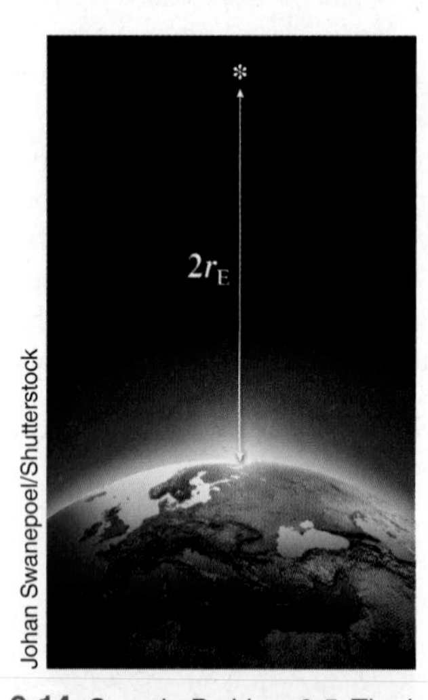

Figure 9-14 Sample Problem 9-5. The location of interest at which you are to determine the gravitational field, a distance of $2r_E$ above the surface, is indicated with an asterisk.

We now have two different interpretations for $g = GM/r^2$. The quantity g can represent the magnitude of either the gravitational acceleration or the gravitational field. To distinguish between the two concepts, the unit of newtons per kilogram (N/kg), that is, force/mass, is customarily used for field, and metres per second squared (m/s^2) is used for acceleration. Thus, we say that at Earth's surface the gravitational field has a magnitude of 9.80 N/kg, and the gravitational acceleration has a magnitude of 9.80 m/s^2. These units are actually equivalent, of course; since N = kg·m/s^2, then N/kg = (kg·m/s^2)/kg = m/s^2. Table 9-3 revisits the data presented in Table 9-1 but in the context now of the magnitude of the gravitational field at the surface of the planets in our solar system.

Some students find the following comparison helpful in appreciating the meaning of gravitational fields. If we know that silver costs approximately \$0.70 per gram and gold costs about \$50 per gram, it is immediately clear that gold has more inherent value. Similarly, by comparing gravitational field strengths at the surface of each planet (Table 9-3), it is immediately clear that Jupiter has the greatest gravitational effect on objects on its surface of all the planets in our solar system. Also, by knowing that silver costs \$0.70/gram, we can easily determine how expensive it is to buy 100 grams, 200 grams, or 4000 grams. Similarly, with the gravitational field strength on the surface of Mars having a value of 3.7 N/kg, we can easily determine how much gravitational force Mars will exert on a 100 kg, 200 kg, or 4000 kg object placed on its surface, as we will see in the next sample problem.

Calculating Gravitational Force from Gravitational Field

If the gravitational field $\vec{g}$ at some position is known, then the gravitational force $\vec{F}$ exerted on an object of mass m placed at that position can easily be determined. Since field is force per unit mass, then the force is just the product of the mass and the field:

$$\vec{F} = m\vec{g} \quad \text{(gravitational force)} \tag{9-5}$$

SAMPLE PROBLEM 9-5

(a) What is the gravitational field (magnitude and direction) due to Earth at a height of 2.00 Earth radii above Earth's surface (Figure 9-14)? The average radius of Earth is 6.37×10^6 m and its mass is 5.98×10^{24} kg.

(b) What is the gravitational force on a satellite of mass 178 kg at this height?

Solution

(a) The magnitude of the gravitational field is given by Eqn. 9-4:

$$g = \frac{GM}{r^2}$$

where M is the mass of Earth and r is the distance from the centre of Earth. In this particular problem, r = 3.00 (Earth radii) = 3.00 (6.37×10^6 m) = 1.91×10^7 m. Substituting numbers,

$$g = \frac{(6.67 \times 10^{-11}\,\text{N}\cdot\text{m}^2/\text{kg}^2)(5.98 \times 10^{24}\,\text{kg})}{(1.91 \times 10^7\,\text{m})^2}$$
$$= 1.09\,\text{N/kg}$$

The direction of the field is the same as the direction of the gravitational force exerted by Earth on any mass at this position, that is, toward the centre of Earth. Thus, the gravitational field is 1.09 N/kg toward the centre of Earth.

(b) The magnitude of the gravitational force on the satellite is the product of the satellite's mass and the magnitude of the gravitational field (Eqn. 9-5):

$$F = mg = (178\,\text{kg})(1.093\,\text{N/kg}) = 195\,\text{N}$$

Thus, the force on the satellite is 195 N toward the centre of Earth.

EXERCISES

9-18 What is the gravitational field (magnitude and direction) of the Sun at the position of Earth? The mass of the Sun is 1.99×10^{30} kg and the average Sun–Earth distance is 1.50×10^{11} m.

9-19 If the total gravitational field at a point in interstellar space is 5.42×10^{-9} N/kg in magnitude, what is the magnitude of the gravitational force at this point on an object of mass:
(a) 1.00 kg?
(b) 2.00 kg?

9-20 (a) Titan, a moon of Saturn, has an inherent gravitational field that has a magnitude of 1.3 N/kg on its surface. Titan's mass is 1.3×10^{23} kg. What is its radius in kilometres?
(b) What is the magnitude of the force of gravity that Titan exerts on a rock (mass 0.181 kg) resting on its surface?

9-21 What is the gravitational field (magnitude and direction) due to a golf ball of mass 0.045 kg at a distance of 15 cm from the centre of the ball?

9-22 What is the gravitational field (magnitude and direction) on the surface of a comet that is roughly spherical, with a radius of 7.0 km and a uniform density of 2700 kg/m^3?

9-23 **Fermi Question:** Estimate the magnitude of the gravitational field of the Sun at the farthest reaches of our solar system.

9.4 Orbits, Kepler's Laws, and Weightlessness

Here are some common questions asked by students: If satellites are attracted gravitationally toward Earth, why do they not just fall into Earth? If planets are attracted toward the Sun, why do they not plummet into the Sun and meet a fiery end? What does "weightlessness" mean—is the force of gravity zero on astronauts working on the International Space Station? These are good questions. We first deal with the case of satellites orbiting Earth.

Suppose that you drop a hamburger from rest. We can calculate how far it falls in 1.0 s by using one of the equations for constant-acceleration kinematics given in Chapter 4:

$$y = y_0 + v_{0y}t + \tfrac{1}{2}a_y t^2$$

The initial vertical velocity, v_{0y}, is zero, and assuming that the hamburger starts at $y_0 = 0$ m we have

$$y = \tfrac{1}{2}a_y t^2$$

Choosing $+y$ to be downward means that $a_y = +9.80$ m/s^2, and using the given time of 1.0 s,

$$y = \tfrac{1}{2}a_y t^2 = \tfrac{1}{2}(9.80\,\text{m/s}^2)(1.0\,\text{s})^2 = 4.9\,\text{m}$$

Thus, the hamburger falls 4.9 m in 1.0 s. Now suppose that instead of dropping the hamburger, you throw it horizontally. Since the initial motion is horizontal, the initial vertical component of velocity is zero, just as it was for the

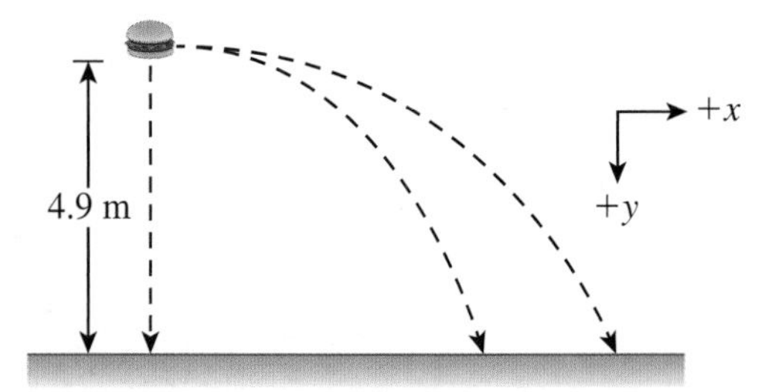

Figure 9-15 In 1.0 s, an object falls 4.9 m vertically regardless of its initial horizontal velocity.

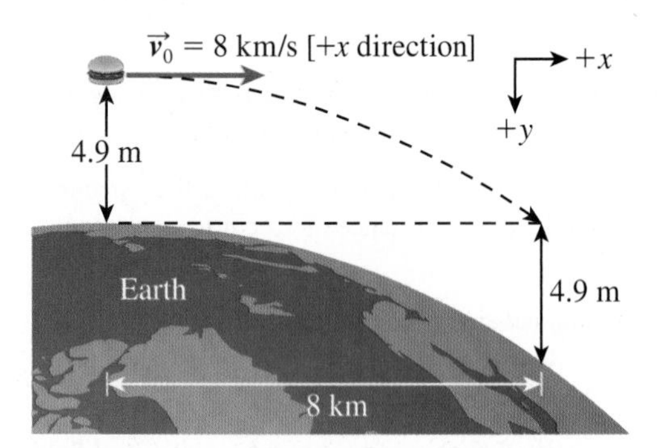

Figure 9-16 If an object is projected horizontally with a speed of 8 km/s, then in 1.0 s, it falls 4.9 m and travels 8 km horizontally. Due to the curvature of Earth, the object is no closer to the surface after 1.0 s and the object has achieved near-Earth orbit! (Horizontal and vertical distances are not to scale in this diagram.)

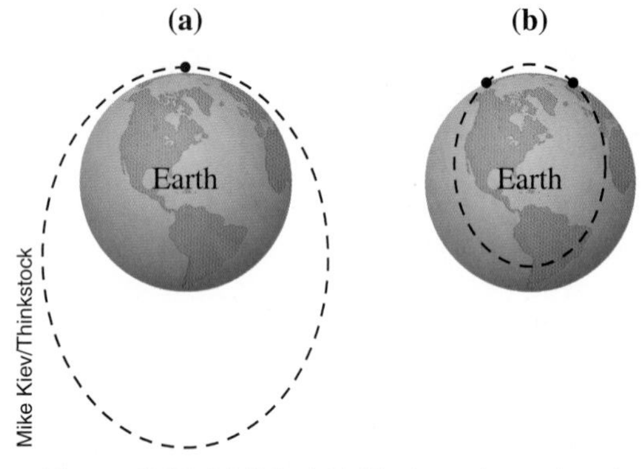

Mike Kiev/Thinkstock

Figure 9-17 (a) If the initial horizontal speed, v_{0x}, is somewhat greater than 8 km/s, the object follows an elliptical orbit. (b) If $v_{0x} < 8$ km/s, the object tries to follow an elliptical orbit, but crashes into Earth.

dropped hamburger. Therefore, the hamburger still falls 4.9 m vertically in 1.0 s, but in addition moves horizontally at constant velocity. If the initial horizontal velocity is, say, 2.0 m/s, then in 1.0 s, the hamburger moves 2.0 m horizontally as well as 4.9 m vertically. The vertical motion of the hamburger is unaffected by the horizontal motion, so it does not matter how fast you throw the hamburger, it still falls 4.9 m in 1.0 s (Figure 9-15). Increasing the initial horizontal velocity just increases how far the hamburger travels horizontally.[3] Air resistance has been neglected.

What if you could throw the hamburger horizontally at 8 km/s? In 1.0 s, the hamburger would fall 4.9 m and move 8 km horizontally. However, Earth's curvature is such that, in 8 km, the surface of Earth drops 4.9 m (Figure 9-16), and thus the hamburger gets no closer to the surface of Earth. You have put the hamburger into a "close-Earth" orbit! It is "falling" toward Earth but Earth keeps curving away from it, and the hamburger just goes round and round in a circular orbit. Because the gravitational force is always perpendicular to the hamburger's circular motion, this force does zero work on the hamburger, which therefore travels at constant speed.

So the answer to the question "Why do satellites not fall into Earth?" is twofold: (i) the satellites go too fast, and (ii) Earth is curved. At higher altitudes, the speed required for a circular orbit decreases, as shown in Sample Problem 9-6 below.

Of course, in reality, the "close-Earth" orbit that we have described would be strongly affected by air resistance. Actual satellites must be at least 150 km above Earth's surface for a long-term orbit. Even at 150 km, which is relatively low for a satellite, air resistance is large enough that satellites eventually slow down and burn up as they enter lower (and thus denser) parts of the atmosphere.

What would happen if a hamburger were thrown horizontally near Earth with a speed somewhat greater than 8 km/s? This speed is too large for a circular orbit, and the hamburger follows a curve called an ellipse, such as shown in Figure 9-17 (a). Similarly, if the initial speed is somewhat less than 8 km/s, the hamburger follows an ellipse but crashes into Earth before it travels very far (Figure 9-17 (b)). We will explore elliptical orbits in greater detail through the study of Kepler's laws of planetary motion.

If the initial horizontal speed is great enough, the hamburger will have sufficient energy to leave Earth's gravitational field and never return. We will discuss the concept of "escape velocity" further in Section 9.5.

The above discussion can be applied to planetary motion around the Sun, except that the numbers are different. Essentially, planets do not fall into the Sun because they go too fast and because the Sun is curved. Similarly, moons do not fall into planets.

SAMPLE PROBLEM 9-6

Determine the speeds of particles in Saturn's rings:

(a) for a particle at the innermost boundary, 6.7×10^7 m from the centre of Saturn, and

(b) for a particle at the outer edge of the easily visible rings, 13.9×10^7 m from Saturn's centre (Figure 9-18). The mass of Saturn is 5.64×10^{26} kg. Assume that the particles in Saturn's rings are travelling in circular orbits around the planet.

Solution

(a) The only appreciable force acting on an individual particle is the gravitational force toward Saturn, which gives rise to the circular motion. The gravitational force ($\vec{F}$) on a particle at the outer edge of the rings is shown; the $+x$-direction is toward Saturn, that is, in the direction of the centripetal acceleration of the particle. We use Newton's second law in the x-direction,

[3]Projectile motion is covered in detail in Section 4.3 in Chapter 4.

$$F_x = ma_x$$

where F_x is the gravitational force and m is the mass of the particle. We can write F_x as GmM/r^2, where M is Saturn's mass, and r is the distance from the particle to the centre of Saturn. Recall from Chapter 4 that the magnitude of the centripetal acceleration of an object travelling with speed v in a circular path of radius r is v^2/r. Therefore, we can also write $a_x = v^2/r$.

Thus, we have $\dfrac{GmM}{r^2} = m\dfrac{v^2}{r}$

Dividing by m, and multiplying by r, $\dfrac{GM}{r} = v^2$

Rearranging for the speed, $v = \sqrt{\dfrac{GM}{r}}$

Substituting numbers,

$$v = \sqrt{\frac{(6.67 \times 10^{-11}\,\text{N}\cdot\text{m}^2/\text{kg}^2)(5.64 \times 10^{26}\,\text{kg})}{6.7 \times 10^7\,\text{m}}}$$

$$= 2.4 \times 10^4\,\text{m/s}$$

Thus, a particle in circular orbit at the innermost edge of Saturn's rings has a speed of 2.4×10^4 m/s.

(b) For a particle at $r = 13.9 \times 10^7$ m, we have

$$v = \sqrt{\frac{GM}{r}}$$

$$= \sqrt{\frac{(6.67 \times 10^{-11}\,\text{N}\cdot\text{m}^2/\text{kg}^2)(5.64 \times 10^{26}\,\text{kg})}{13.9 \times 10^7\,\text{m}}}$$

$$= 1.65 \times 10^4\,\text{m/s}$$

The speed of a particle in circular orbit at the outer edge of the easily visible rings is 1.65×10^4 m/s. Notice that the speed required to maintain a stable circular orbit decreases as the distance from the planet's centre increases.

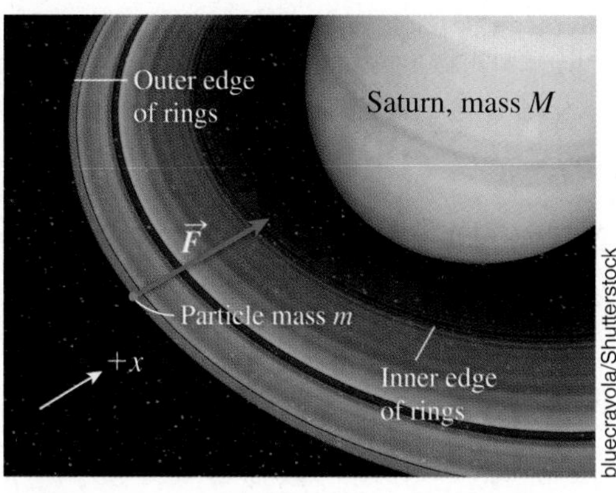

Figure 9-18 Sample Problem 9-6; the rings of Saturn.

DID YOU KNOW?

Saturn is not the only planet that has a ring system, but it is the only one that has rings visible from Earth. Jupiter and Uranus also have rings, and Neptune appears to have a series of arcs instead of continuous rings.

The relationship $v = \sqrt{GM/r}$ developed in the solution to the above sample problem is valid not just for the rings of Saturn; it works for any object travelling with speed v in a circular gravitational orbit of radius r about a central object having mass M.

Kepler's Laws

In the late 1500s the geocentric model of the solar system was generally accepted, even though Copernicus had published his heliocentric hypothesis in 1543. In the early 1600s, the German mathematician-astronomer Johannes Kepler (1571–1630) published three laws of planetary motion, based primarily on his analysis of accurate observations of planets made by the Danish astronomer Tycho Brahe (1546–1601). These laws indicated that the Sun occupies a special place in the orbits of the planets, and eventually the heliocentric theory gained acceptance.

Tycho Brahe (often referred to simply as Tycho) made extremely careful observations, using instruments that he designed himself, of the positions of the stars and planets from the early 1570s until his death in 1601. His work was exceptional, improving the precision of previous observations significantly, all without the benefit of a telescope, which was not invented until after his death. Most of these observations were made at his observatory in Denmark under the patronage

Kepler's first law the orbit of each planet is an ellipse with the Sun at one focus

Kepler's second law the line joining the Sun and the planet sweeps out equal areas in equal time intervals

Kepler's third law the square of the time (T^2) in which a planet completes one revolution about the Sun is proportional to the cube of its average distance (r^3) from the Sun

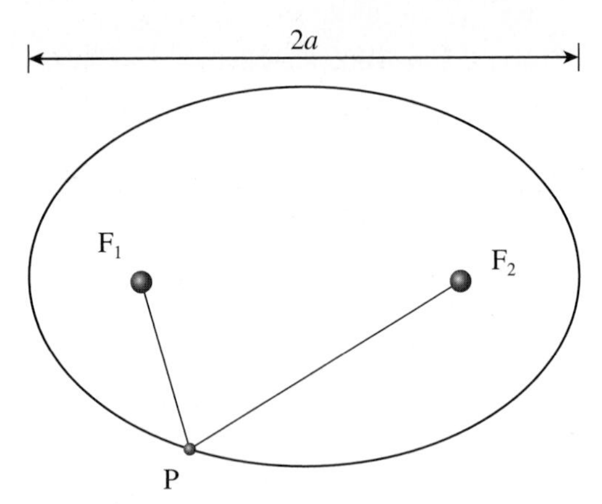

Figure 9-19 An ellipse. For any point P, $PF_1 + PF_2$ = constant. The symbol a denotes the semi-major axis of the ellipse.

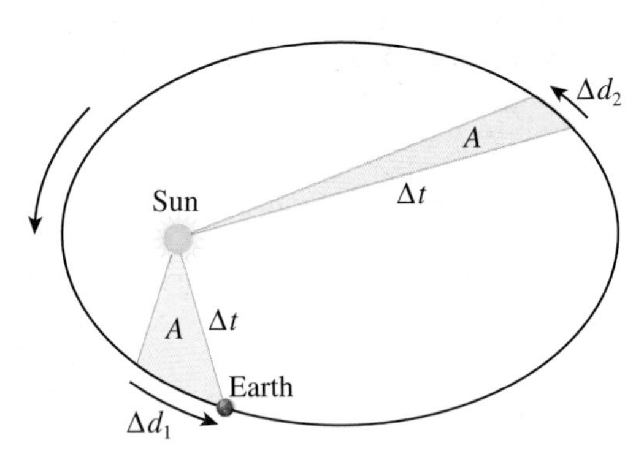

Figure 9-20 Kepler's second law: the line joining the Sun and the planet sweeps out equal areas, A (shaded regions), in equal time intervals, Δt. When Earth is further away, the arc length Δd_2 swept out in time Δt is shorter than the arc length swept out when the planet is closer in order for the area to be the same.

of the Danish king. Unfortunately, although he was an excellent astronomer, he was also arrogant and extravagant, and in 1597 he lost the king's support and was compelled to leave Denmark. After travelling around Europe for two years he ended up in Prague, where he became court astronomer for Emperor Rudolph II of Bohemia. In 1600, a year before his death, he invited Johannes Kepler to join his group, as Kepler was looking for a new institution due to political pressures. It was not a good match. Tycho was secretive and did not want his data to be used to support the Copernican model, as he was working on his own model of the solar system. Kepler, in turn, did not have a great deal of respect for Tycho, considering him to be merely an aristocrat dabbling in astronomy. When Tycho died in 1601, by his own later account Kepler admits that he "*... quickly took advantage of the absence, or lack of circumspection, of the heirs, by taking the observations under my care, or perhaps usurping them ...*".[4]

Kepler spent a great deal of time analyzing the orbit of Mars, for which the observational data were most complete. He tried various mathematical curves to fit the orbit, but was unsuccessful until he used an ellipse. An ellipse (Figure 9-19) is a closed figure such that the sum of the distances from any point P on it to two other fixed points (the foci F_1 and F_2) is a constant, that is, $PF_1 + PF_2 =$ constant. A circle is a special case of an ellipse for which the two foci are at the same position at the centre of the circle. **Kepler's first law** of planetary motion states that the orbit of each planet is an ellipse with the Sun at one focus.

Although Kepler's first law states correctly that the planetary orbits are ellipses, for most of the planets the ellipses are not very elongated. In fact, if you were to draw a scale diagram of the orbits of the planets (except for Mercury), they would look very much like circles. For example, the distance from Earth to the Sun varies only by about 3% during its annual motion about the Sun.

Even before Kepler had established that the orbit of Mars is an ellipse, he had determined that Mars speeds up as it approaches the Sun and slows down as it moves away from it. This was stated in a precise way in **Kepler's second law** of planetary motion: the line joining the Sun and the planet sweeps out equal areas in equal time intervals. Kepler's second law is illustrated in Figure 9-20. The statement that equal areas are swept out in equal times is equivalent to saying that each planet is moving most rapidly when closest to the Sun and least rapidly when farthest from the Sun. Earth is closest to the Sun around January 4th each year and farthest around July 5th.

Kepler's third law gives the relationship between the period (T) of a planet's orbit, which is the time taken to make one revolution, and its average distance (r) from the Sun. Table 9-4 lists the periods and distances for the planets. Can you see a mathematical relationship between the period and distance? Clearly the period increases as the distance increases, but not in a linear fashion. Kepler, using the data for the planets known at the time (Mercury to Saturn), was able to conclude that $T^2 \propto r^3$ or $T \propto r^{3/2}$. In Question 9-32 you are asked to confirm this proportionality.

In words, **Kepler's third law** of planetary motion states that the square of the time (T^2) in which a planet completes one revolution about the Sun is proportional to the cube of its average distance (r^3) from the Sun. Kepler's findings were highly controversial because they contradicted the widely held geocentric model of the solar system.

Kepler's third law can be easily proven for circular orbits. In Sample Problem 9-6 we showed that the speed v of an object in a circular orbit of radius r around a central object of mass M is given by $v = \sqrt{GM/r}$. The time T required for one complete orbit is just the total distance travelled ($2\pi r$) divided by the orbital speed. Therefore

[4]Arthur Koestler, *The Sleepwalkers*, Penguin Books, 1990.

$$T = \frac{2\pi r}{v} = \frac{2\pi r}{\sqrt{GM/r}}$$

Squaring both sides of the equation gives

$$T^2 = \frac{4\pi^2 r^2}{GM/r} = \frac{4\pi^2 r^3}{GM} = Cr^3 \qquad (9\text{-}6)$$

where $C = \frac{4\pi^2}{GM}$

C is a constant that depends on the mass of the Sun (or another central object). Therefore, from Newton's law of universal gravitation and centripetal acceleration, we have proven Kepler's third law, that is $T^2 \propto r^3$, in the case of circular orbits. The proof involving elliptical orbits is somewhat more complicated and we will not go into the details in this book; we will merely note that Eqn. 9-6 holds for elliptical orbits as well, where the semi-major axis of the ellipse is used (a) (Figure 9-19) instead of the radius of a circular path (r). Eqn. 9-6 is very useful in astronomy because it allows the determination of an unknown mass (M) of an object through observations of the period (T) and the radius (r) of the orbit of a second object around it.

Table 9-4

Periods and Average Distances from the Sun (1 AU = 1 Astronomical Unit = Average Earth–Sun Distance)

Planet	Period (Earth years)	Distance (AU)
Mercury	0.24	0.39
Venus	0.62	0.72
Earth	1.00	1.00
Mars	1.88	1.52
Jupiter	11.9	5.20
Saturn	29.5	9.59
Uranus	84.1	19.2
Neptune	165	30.1

DID YOU KNOW?

The slowly changing distance between Earth and the Sun throughout the year as it travels along its elliptical path is not the cause of the seasons. In the northern hemisphere we are closer to the Sun in the winter, with our closest distance in the early days of January each year. The seasons arise instead because the axis of Earth's daily rotation is not perpendicular to the plane of Earth's orbit around the Sun (Figure 9-21). As a result of this "tilted" axis of rotation, the North Pole faces somewhat away from the Sun during months close to December, resulting in winter in the northern hemisphere. In months close to June, the North Pole points slightly toward the Sun and the South Pole points away from it, reversing the seasons.

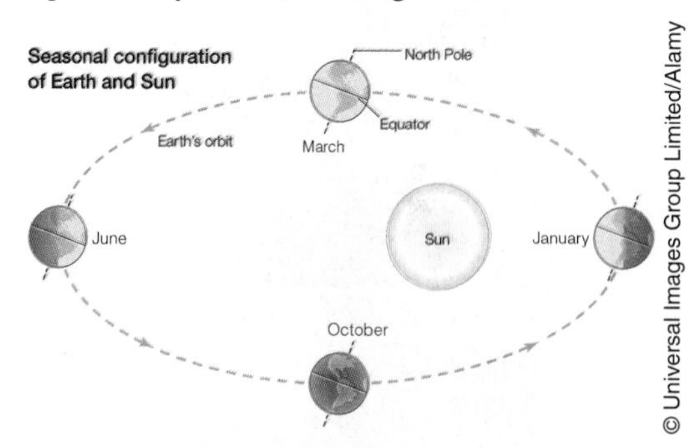

Figure 9-21 The tilted axis of rotation is the cause of the seasons on Earth, not our elliptical orbit.

Weightlessness

You have probably seen photographs and videos of astronauts who are floating in orbiting space shuttles and the International Space Station (ISS), such as the photograph of Julie Payette shown in Figure 9-22. You even may have experienced weightlessness yourself right here on Earth if you have ever been on a free-fall ride such as the Drop Tower at Canada's Wonderland (Figure 9-23). What exactly does "weightless" mean? It cannot mean that the force of gravity on the astronauts is zero. The astronauts are close enough to Earth that the force of gravity is only a little smaller than gravity on Earth's surface (see Question 9-30), and it is gravity that causes the astronauts and their ship to orbit Earth.

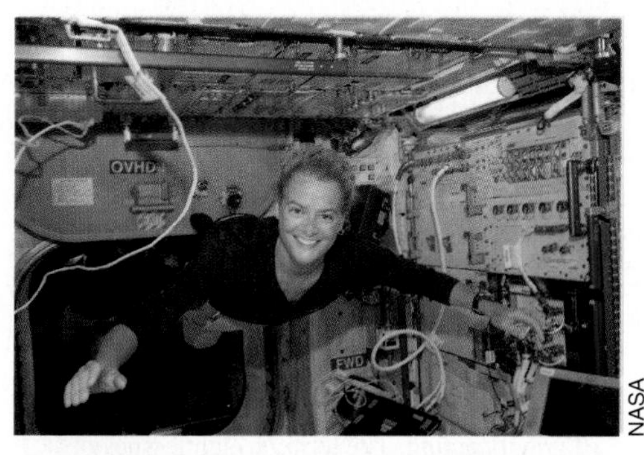

Figure 9-22 Canadian astronaut Julie Payette demonstrating "weightlessness" on the International Space Station during Shuttle mission STS-127 in July 2009.

TRY THIS!

Controlled Free Fall near the Surface of Earth

The next time you are at a major amusement park, such as Canada's Wonderland, you can experience the sensation of weightlessness for yourself. The ride called "Drop Tower" at Canada's Wonderland is approximately 70 m tall and reaches speeds of up to 100 km/h in a matter of seconds (Figure 9-23). For a portion of the descent, riders are in free fall with only the force of gravity acting upon them, before the braking mechanism is engaged to slow everyone down safely before landing.

sahua d/Shutterstock

Figure 9-23 Thrill seekers momentarily experiencing weightlessness on an amusement park free-fall ride.

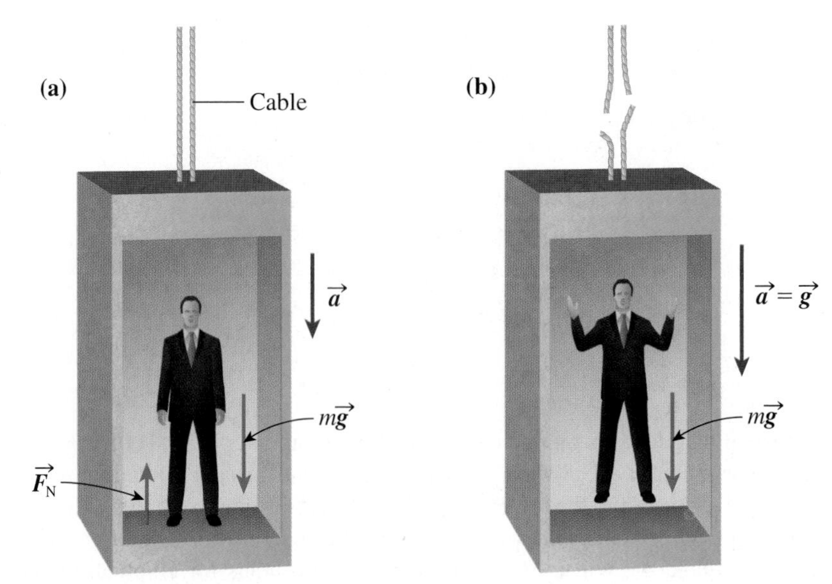

Figure 9-24 **(a)** A person in an elevator accelerating downward. **(b)** A person and elevator in free-fall: $\vec{a} = \vec{g}$ and $\vec{F}_N = 0$. The person is "weightless."

To understand weightlessness, we start with a down-to-Earth example, a person standing in a moving elevator (Figure 9-24 (a)). We assume for the moment that the elevator has a downward acceleration of magnitude a. The only forces acting on the person are gravity ($m\vec{g}$) downward and the upward normal force ($\vec{F}_N$) exerted by the floor. We now use Newton's second law:

$$\Sigma F_y = ma_y$$

Choosing $+y$ to be downward, we have

$$mg + (-F_N) = ma_y$$

Solving for $F_N = mg - ma_y = m(g - a_y)$.

Thus, the magnitude of the normal force depends on the downward acceleration. Larger values of a_y are associated with smaller values of F_N. If the person is standing on bathroom scales on the elevator floor, the scales provide the upward normal force and the scale reading displays the magnitude of this force.

What if the elevator cable is cut and the elevator and person go into free-fall (Figure 9-24 (b)), that is, they both have a downward acceleration of $\vec{a} = \vec{g}$? Then $F_N = m(g - g) = 0$; this means that the person, who is "floating" in the elevator, experiences no force from the elevator floor (or scales). It is the force F_N that gives the usual sensation of "weight," and in the absence of this force, the person feels "weightless." The bathroom scales would read zero.

weightlessness the sensation associated with being in free fall, in which an object is in motion with only the force of gravity acting

Thus, what **weightlessness** really means is *being in free-fall;* that is, being in motion with only the force of gravity acting. There are no forces exerted by a person's surroundings such as floors, walls, etc. Astronauts are weightless, not because they are in space but because they are in free-fall (along with their ship). If you jump from a chair or table to the floor, you are weightless during your short period of free-fall. Some people have the misconception that weightlessness is somehow associated with the lack of air in space, but the foregoing discussion shows that this is not the case.

TACKLING MISCONCEPTIONS

"Weightlessness" Does Not Mean Zero Gravity!

As we have seen here in our discussion, astronauts in orbit around Earth in the International Space Station definitely have a strong force of gravity acting on them; the gravitational field of Earth at this altitude is about 90% that of the field strength on the surface of Earth. Astronauts float in the ISS, experiencing the sensation of "weightlessness," because they and their surroundings are all in free fall with only the force of gravity acting. The term "weightlessness" is an unfortunate misnomer, as is another term used in this scenario: "micro-gravity."

Weightlessness has advantages and disadvantages for astronauts. It is easy for them to move around the spacecraft and easy to carry massive objects. However,

PROFILES IN PHYSICS

Canadian Astronaut, and Social-Media Celebrity, Colonel Chris Hadfield (1959–)

Chris Hadfield was born in Sarnia, Ontario in 1959 and joined the Canadian Armed Forces in 1978. After completing his bachelor's degree in mechanical engineering at Royal Military College in Kingston, Ontario in 1982, he continued his training as a pilot in Portage La Prairie, Manitoba, Moose Jaw, Saskatchewan, and Cold Lake, Alberta. He then went on to fly CF-5s and CF-18s for the North American Defence Command (NORAD) as well as working as an exchange officer with the US Navy. Overall, Hadfield has flown over 70 different types of aircraft in his career.

In 1992, Chris Hadfield and three others were selected to join the Canadian Space Agency (CSA) as astronauts from a field of over 5000 applicants. He has gone to space three times since then: in 1995 he served as Mission Specialist 1 on Russia's space station Mir and in 2001 he flew again as Mission Specialist 1 to the International Space Station (ISS). During the 2001 mission, Hadfield became the first Canadian to perform a spacewalk, spending enough time outside the space station over two such outings that he orbited Earth 10 times (Figure 9-25). His final mission, in 2012/2013, involved another visit to the ISS. During this mission, Commander Hadfield again made Canadian history by becoming the first from Canada to command a spacecraft, from March 13, 2013 to his departure in May of the same year. While at the ISS in 2012/2013, Chris Hadfield and his crew orbited Earth 2336 times!

Figure 9-25 Chris Hadfield in 2001, on the first spacewalk by a Canadian astronaut, as part of a mission to build the Canadarm2 robot on the International Space Station.

It was during his last mission that Commander Hadfield became an international celebrity through his use of social media platforms to bring space exploration to millions in a way that had not been done before. His YouTube videos on brushing your teeth or preparing meals while experiencing "weightlessness" and his gorgeous photos of our planet have been viewed by millions around the world. During his mission, he gained an average of 11 000 followers each day across all his social media platforms (Twitter, Facebook, YouTube, etc.), engaging so many of us with his lessons on science in space. His signoff from the ISS, a video of his version of the classic David Bowie tune *Space Oddity*, had over 22 million views when it was taken down in May 2014. Through his passion for space exploration, millions of us were taken along for the ride, likely inspiring a whole new generation of young Canadians to reach for the stars.

an accidental bump from someone will send an astronaut to the other end of the cabin. Think about the problem of opening a sticky drawer in free-fall (remember Newton's third law). Weightless astronauts experience a number of physiological effects. They temporarily grow 2 to 3 cm in height because of the decompression of the spongy discs in their spines. Their faces become puffy because the body fluids are also in free-fall and can more easily be in the facial region; when the astronauts are on the ground, the fluids tend to be more concentrated in the lower parts of the body. The cardiovascular system becomes deconditioned; normally it is harder to pump the blood up toward the head than down toward the feet, but in a weightless environment it is easy to pump the blood anywhere in the body. If astronauts are in space for an extended period of time, they must exercise to keep the cardiovascular system in condition so that they will be able to adapt when they return to Earth.

TRY THIS!

Weightlessness on Earth

You can demonstrate weightlessness yourself without travelling to an amusement park. Make a hole in a Styrofoam cup about 2 cm above the bottom, and fill the cup with water. Not surprisingly, the water will flow out of the hole. (Have a pail ready!) Now stand on a chair or table and fill the cup with water, keeping a finger over the hole. Drop the cup of water into a pail on the floor. Describe and explain what happens. If you have a partner to work with, taking a video of your experiment that you can replay in slow motion may help in observing the results.

EXERCISES

9-24 (a) A ball is thrown horizontally with an initial speed of 3.0 m/s. After 2.0 s, how far has it travelled horizontally? vertically? Neglect air resistance.

(b) What would your answers be for an initial speed of 5.0 m/s?

9-25 Earth is attracted by gravity toward the Sun; why does it not fall into the Sun?

9-26 Why does the gravitational force acting on a satellite in a circular orbit around Earth not change the speed of the satellite?

9-27 (a) What is the speed of a satellite in a circular orbit 525 km above the surface of Earth? Earth has an average radius of 6368 km and a mass of 5.98×10^{24} kg.
(b) How long does it take the satellite to complete one orbit?

9-28 (a) If a satellite is to have a circular orbit about Earth (mass 5.98×10^{24} kg) with a period of 4.0 h, how far (in kilometres) above the centre of Earth must it be?
(b) What must be its speed?

9-29 A satellite could be placed in a circular orbit very close to the surface of the Moon because there would be no air resistance. What would be the required speed for such a satellite? The Moon's mass and radius are 7.36×10^{22} kg and 1738 km.

9-30 The typical orbital height for the International Space Station is 370 km above Earth's surface.
(a) Determine the magnitude of the gravitational acceleration at this height. Earth's (average) radius is 6.37×10^6 m and its mass is 5.98×10^{24} kg.
(b) Express your answer to (a) as a percentage of surface-level g.
(c) Determine the orbital speed (in km/s) required to keep the ISS on a circular path.
(d) Calculate the period of the orbit of the ISS around Earth.

9-31 Earth is closest to the Sun around January 4th each year and farthest around July 5th. On which of these dates is Earth travelling most rapidly? least rapidly? Use Kepler's second law.

9-32 Using the planetary data in Table 9-4, calculate the ratio T^2/r^3 for each planet. Does this verify Kepler's third law?

9-33 (a) Based on your calculations in the previous question, what is the average value of the constant of proportionality, C, in Eqn. 9-6, in SI units? (**Note:** 1 AU is equal to 1.50×10^{11} m.)
(b) From your answer in part (a), what is the mass of the Sun in kilograms?

9-34 Deimos, a moon of Mars, has a period of 1.26 days and an average distance from the centre of Mars of 2.35×10^4 km. What is the mass of Mars?

9-35 An astronaut is in a rocket ship that is blasting off. The ship is 10 km above the surface of Earth and the rockets are still firing. Is the astronaut weightless? Explain your answer.

9-36 An astronaut in the ISS in orbit around Earth has been placed in the centre of the cabin by her astronaut "friends." She is stationary relative to the cabin and is not touching the walls, ceiling, or floor.
(a) Why is this a problem for the astronaut?
(b) How could she solve this problem?

9-37 (a) Can weightless astronauts walk in the usual way?
(b) How do you think weightless astronauts drink?

9-38 If riders were experiencing free fall the entire time they were on Drop Tower, what would be their maximum speed? How does this compare with the known top speed at Canada's Wonderland? Does this make sense? (See the Try This! text box on page 240 for the necessary details for this calculation.)

9-39 **Fermi Question:** Estimate how long it takes Mercury, the planet closest to the Sun in our solar system, to complete one orbit. How old in "Mercury years" are you?

9.5 Gravitational Potential Energy in General

What are black holes? Why does Earth's atmosphere not contain hydrogen or helium? Why does the Moon have no atmosphere at all? How fast would a rocket have to be going in order to escape completely from Earth? from Mercury? In order to answer these questions, we need to know how to calculate gravitational potential energy for objects far from the surface of Earth (or another planet, or the Moon, etc.).

In Chapter 7, we saw that the gravitational potential energy (E_P) of a mass m near Earth's surface is mgy, where g is the magnitude of the gravitational acceleration (9.80 m/s^2), and y is the elevation of the mass relative to some arbitrary position ($y = 0$). The expression mgy is valid only near Earth's surface, that is, where g is approximately a constant value. As an object moves upward in the atmosphere, g changes, and the equation $E_p = mgy$ cannot be used to determine gravitational potential energy.

The more general expression for gravitational potential energy that is used when an object of mass m, such as a rocket ship, is far from Earth's surface is

$$E_P = -\frac{GmM}{r} \quad (9\text{-}7)$$

where G is the universal gravitation constant, M is the mass of Earth, and r is the distance from the centre of Earth to the rocket ship. This expression is valid also near Earth's surface, as we shall see; we merely use the expression $E_P = mgy$ in the special case to simplify our calculations.

Math Tip

If you are learning calculus, notice the relationship between gravitational force and gravitational potential energy:

(i) $F_y = -\frac{d}{dy}(mgy) = -mg$

(ii) $F_r = -\frac{d}{dr}\left(-\frac{GmM}{r}\right) = -\frac{GmM}{r^2}$

The force, expressed as a component in the $+y$ (upward) or $+r$ (radial, also upward) direction, is the negative of the derivative of the potential energy.

Notice that as $r \rightarrow \infty$, that is, as the rocket ship moves very far from Earth, $E_P \rightarrow 0$. When using the equation $E_P = mgy$, we were free to choose the position where $E_P = 0$. However, for the equation $E_P = -GmM/r$, the position where $E_P = 0$ has been preselected at $r \rightarrow \infty$.

The equation $E_P = -GmM/r$ is valid not just for Earth and an object such as a rocket ship; this expression gives the gravitational potential energy of any two objects of masses m and M, separated by a distance r. Notice that, strictly speaking, this is the potential energy *of the two-mass system*, but we will often say that it is the potential energy associated with only one of the objects. As was the case for the equation $F = Gm_1m_2/r^2$, the objects either have to be far enough apart that they can be considered to be particles, or they must have spherical symmetry for Eqn. 9-7 to be directly applicable.

The negative sign in $-GmM/r$ often troubles students. Because G, m, M, and r are positive quantities, gravitational potential energy is always negative (unless zero). However, as mentioned in Chapter 7, the important quantity is not potential energy itself but rather the *change* in potential energy from one position to another. We illustrate this in the following calculation, and show that the negative sign should not be a cause for concern.

Figure 9-26 shows a satellite of mass 1.00×10^2 kg located 1.50×10^2 km above Earth's surface. The drawing is repeated three times: (a), (b), and (c). We will calculate in three different ways the gravitational potential energy of the satellite *relative* to Earth's surface, that is, the difference between its potential energy in its present position and its potential energy if it were at Earth's surface. We know, from our discussions in Chapter 7, that if the satellite was dropped from rest from its height of 150 km, its potential energy would decrease by this amount as it falls to Earth, and its kinetic energy would increase equivalently (if we neglect air resistance).

In Figure 9-26 (a), we choose $y = 0$ at Earth's surface, and use $E_P \approx mgy$. This is only an approximation because g is not constant up to a height of 150 km. The $+y$-direction is upward, as always when we use $E_P = mgy$. The calculation of the required difference in potential energy (ΔE_P) gives

$$\begin{aligned}\Delta E_P &\approx mgy_2 - mgy_1 \\ &\approx mg(y_2 - y_1) \\ &\approx (100\,\text{kg})(9.80\,\text{m/s}^2)(1.50 \times 10^5\,\text{m} - 0\,\text{m}) \\ &\approx 1.5 \times 10^8\,\text{J}\end{aligned}$$

Thus, the satellite has a gravitational potential energy of about 1.5×10^8 J relative to Earth's surface.

For Figure 9-26 (b), we again use $E_P \approx mgy$, but arbitrarily choose $y = 0$ at a height of 50 km above the satellite, that is, 200 km above Earth's surface. This is the position where $E_P = 0$. Therefore, the potential energy of the satellite 150 km above Earth is negative, and its potential energy at the surface would be even more negative. The difference in potential energy is

$$\begin{aligned}\Delta E_P &\approx mgy_2 - mgy_1 \\ &\approx mg(y_2 - y_1) \\ &\approx (100\,\text{kg})(9.80\,\text{m/s}^2)(-5.0 \times 10^4\,\text{m} - (-2.00 \times 10^5\,\text{m})) \\ &\approx 1.5 \times 10^8\,\text{J}\end{aligned}$$

This change of potential energy is just the same as in (a). The fact that the two potential energies, mgy_2 and mgy_1, are negative in this particular calculation is irrelevant; the potential energy of the satellite at a height of 150 km above the surface is still about 1.5×10^8 J larger than that at the surface.

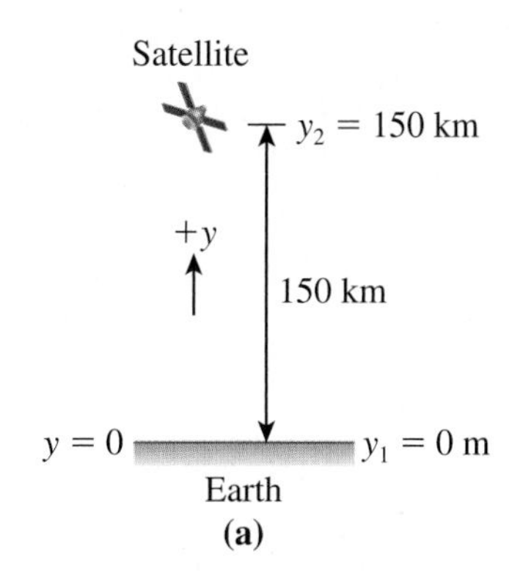

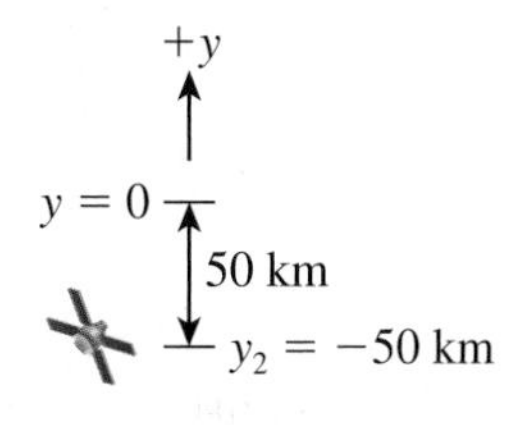

$y_1 = -200$ km
Earth
(b)

as $r \rightarrow \infty$
$E_P \rightarrow 0$
$r_2 = 6518$ km
$r_1 = 6368$ km
Earth
(c)

Figure 9-26 Calculating the gravitational potential energy of a satellite relative to Earth's surface. For **(a)** and **(b)**, $E_P = mgy$ will be used in our calculations. In (a), $y = 0$ and $E_P = 0$ is chosen to be at Earth's surface. In (b), $y = 0$ and $E_P = 0$ is chosen to be at a height of 50 km above the satellite. For **(c)**, the more general expression given by Eqn. 9-7 will be used in our calculations; with this expression, the reference location, where $E_P \rightarrow 0$, always occurs when $r \rightarrow \infty$.

Finally, for Figure 9-26 (c), we use the more appropriate expression for this scenario given by Eqn. 9-7. At an infinitely large distance from the centre of Earth ($r \rightarrow \infty$), $E_P \rightarrow 0$. The potential energy of the satellite 150 km above Earth is negative, and its potential energy at the surface is even more negative. At the surface, the distance from Earth's centre is $r_1 = 6368$ km (average value of Earth's radius), and at an elevation of 150 km, $r_2 = (6368 + 150)$ km $= 6518$ km. The calculation of ΔE_P gives

$$\begin{aligned}\Delta E_p &= -\frac{GmM}{r_2} - \left[-\frac{GmM}{r_1}\right]\\ &= -GmM\left[\frac{1}{r_2} - \frac{1}{r_1}\right]\\ &= -(6.67 \times 10^{-11}\,\text{N}\cdot\text{m}^2/\text{kg}^2)(100\,\text{kg})(5.98 \times 10^{24}\,\text{kg})\\ &\quad\left[\frac{1}{6.518 \times 10^6\,\text{m}} - \frac{1}{6.368 \times 10^6\,\text{m}}\right]\\ &= 1.44 \times 10^8\,\text{J}\end{aligned}$$

Thus, the difference in gravitational potential energy is 1.44×10^8 J. Our approximate answer of 1.5×10^8 J in (a) and (b) was not too bad! (Why not?)

Again, it does not matter that both potential energies are negative; the satellite's potential energy at the higher elevation is 1.44×10^8 J larger than at the surface. As you can see from this example, how much the potential energy *changes* from one position to another is not affected by whether the potential energy at a particular point is positive or negative. An object falls through a decrease in gravitational potential energy, regardless of the choice of the reference position for zero potential energy.

SAMPLE PROBLEM 9-7

What is the gravitational potential energy of the Earth–Moon system? The masses of Earth and the Moon are 5.98×10^{24} kg and 7.36×10^{22} kg, respectively, and their centre-to-centre distance is 3.84×10^8 m.

Solution

This problem involves just a straightforward substitution into Eqn. 9-7:

$$\begin{aligned}E_P &= -\frac{GmM}{r}\\ &= -\frac{(6.67 \times 10^{-11}\,\text{N}\cdot\text{m}^2/\text{kg}^2)(7.36 \times 10^{22}\,\text{kg})(5.98 \times 10^{24}\,\text{kg})}{3.84 \times 10^8\,\text{m}}\\ &= -7.64 \times 10^{28}\,\text{J}\end{aligned}$$

Thus, the potential energy of Earth and the Moon is -7.64×10^{28} J. It is not uncommon to refer to this just as the potential energy of the Moon, but it is more correct to say that it is the potential energy of the Earth–Moon system.

In the answer to Sample Problem 9-7 above, what does the potential energy of -7.64×10^{28} J actually mean? Recall that as $r \rightarrow \infty$, $E_P \rightarrow 0$; in other words, if Earth and the Moon were separated by a very large distance, their potential energy would be zero. Therefore, at their present separation, their potential energy is 7.64×10^{28} J *less* than the potential energy they would have at a very large (infinite) separation. This negative potential energy is also equal to the work that would be done by gravity if the Moon were to be moved from its present position to a distance infinitely far away.

This is similar to the situation of lifting an object that is close to the surface of Earth; as the object is lifted, the work done by gravity is negative.

SAMPLE PROBLEM 9-8

Suppose that an asteroid is moving directly toward Earth. If the asteroid's speed when it is 24 000 km from the centre of Earth is 4.2 km/s, what will its speed be when it is 12 000 km from Earth's centre? (Assume 2 significant digits in the given distances.) The mass of Earth is 5.98×10^{24} kg.

Solution

This problem involves using conservation of energy. The energies involved are the asteroid's kinetic energy and gravitational potential energy. Using subscript "1" for the energies when the asteroid is 24 000 km away, and "2" when it is 12 000 km away, we can express conservation of energy as

$$E_{K1} + E_{P1} = E_{K2} + E_{P2} \qquad \text{Eqn. [1]}$$

Because the asteroid is not close to Earth's surface, $E_P = -GmM/r$ (not $E_P = mgy$) must be used, where m and M represent the masses of the asteroid and Earth respectively. Substituting the expression for E_P into Eqn. [1], along with $E_K = \frac{1}{2}mv^2$ gives

$$\frac{1}{2}mv_1^2 - \frac{GmM}{r_1} = \frac{1}{2}mv_2^2 - \frac{GmM}{r_2} \qquad \text{Eqn. [2]}$$

We need to solve for the speed v_2. We first simplify Eqn. [2] by dividing all terms by m and multiplying by 2:

$$v_1^2 - \frac{2GM}{r_1} = v_2^2 - \frac{2GM}{r_2} \qquad \text{Eqn. [3]}$$

Notice that the mass of the asteroid has now been eliminated. This means that the asteroid's speed does not depend on its mass. Rearranging Eqn. [3] to solve for v_2,

$$v_2^2 = v_1^2 + 2GM\left(\frac{1}{r_2} - \frac{1}{r_1}\right)$$

$$v_2 = \sqrt{v_1^2 + 2GM\left(\frac{1}{r_2} - \frac{1}{r_1}\right)}$$

Substituting known quantities, remembering first to convert kilometres to metres,

$$v_2 = \sqrt{\left(4.2 \times 10^3\,\frac{\text{m}}{\text{s}}\right)^2 + 2\left(6.67 \times 10^{-11}\,\frac{\text{N}\cdot\text{m}^2}{\text{kg}^2}\right)(5.98 \times 10^{24}\,\text{kg})\left(\frac{1}{1.2 \times 10^7\,\text{m}} - \frac{1}{2.4 \times 10^7\,\text{m}}\right)}$$

$$= 7.1 \times 10^3\ \text{m/s}$$

$$= 7.1\ \text{km/s}$$

Thus, the speed of the asteroid when it is 12 000 km from Earth's centre will be 7.1 km/s. The asteroid is speeding up as it moves closer to Earth because of the gravitational force exerted on it by Earth.

DID YOU KNOW?

There is strong evidence indicating that the extinction of the dinosaurs 65 million years ago was caused by a huge collision between an asteroid or comet and Earth. It is believed that more than 50% of all plant and animal species alive on Earth at the time were destroyed in this catastrophic event.

Escape Speed

You know that if you throw something upward, it eventually falls back down. But if you could throw it fast enough, it would keep on going forever, gradually slowing down because of the pull of Earth's gravity on it but never returning. What goes up

escape speed minimum speed needed to escape from the gravitational pull of a celestial object

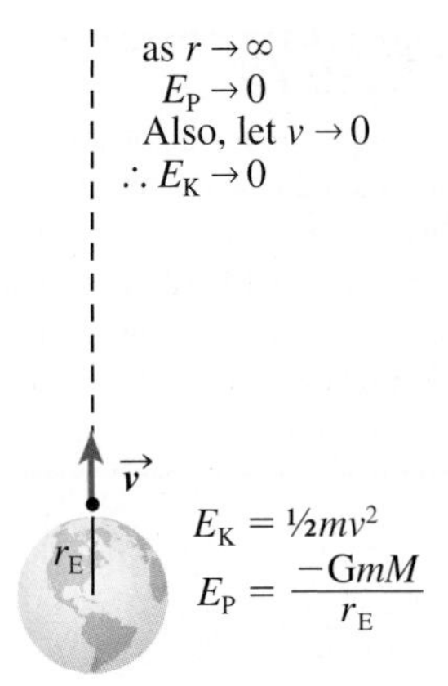

Figure 9-27 A space probe escaping from Earth.

does not always come down! How fast does an object (space probe, satellite, gas molecule) have to be travelling near Earth's surface in order not to return? The minimum speed required to escape from Earth (or another planet or star) is called the **escape speed**. We will now determine this speed for Earth.

Figure 9-27 shows an object—say a space probe—travelling with speed v near Earth's surface. Initially, the distance r that separates the probe and the centre of Earth is Earth's radius, r_E. As the probe moves away from Earth, Earth's gravity slows it down, but when the probe is very far away (i.e., as the separation distance $r \to \infty$), the gravitational force is effectively zero. If we neglect other forces acting on the probe, it would then travel at constant velocity (by Newton's first law of motion). In order to determine the *minimum* speed required to escape from Earth, we let the probe approach a constant speed of zero as $r \to \infty$. That is, the probe eventually tends to come to rest, but, since the gravitational force is approaching zero, it stays at rest.

We now consider the energies involved. When the probe (mass m) is close to Earth (mass M), it has kinetic energy, $E_K = \frac{1}{2}mv^2$, and gravitational potential energy, $E_P = -GmM/r_E$. When the probe is very far away, its kinetic energy approaches zero (since we are letting $v \to 0$), and its potential energy also approaches zero since, in Eqn. 9-7, $-GmM/r \to 0$, as $r \to \infty$. By the law of conservation of energy, the probe's total energy must remain constant, that is, the sum of kinetic and potential energies stays the same. Hence, total energy $(E_K + E_P)$ at Earth's surface = total energy $(E_K + E_P)$ as $r \to \infty$. Therefore,

$$\frac{1}{2}mv^2 + \left(-\frac{GmM}{r_E}\right) = 0 + 0$$

We now solve for the speed at Earth's surface, v. Dividing by m (which disappears), multiplying by 2, and rearranging,

$$v^2 = \frac{2GM}{r_E} \text{ and thus, } v = \sqrt{\frac{2GM}{r_E}}$$

Substituting numbers,

$$v = \sqrt{\frac{2(6.67 \times 10^{-11}\,\text{N}\cdot\text{m}^2/\text{kg}^2)(5.98 \times 10^{24}\,\text{kg})}{6.37 \times 10^6\,\text{m}}}$$
$$= 1.12 \times 10^4\,\text{m/s}$$
$$= 11.2\,\text{km/s}$$

Thus the escape speed from Earth is 11.2 km/s. If an object moves at this speed (or greater), it has sufficient kinetic energy to escape the gravitational pull of Earth. Note that the escape speed does not depend on the mass of the moving object: it is the same for space probes and gas molecules.

You probably know that Earth's atmosphere consists mainly of nitrogen and oxygen. These molecules are much more massive than atoms of hydrogen and helium, which are also gases. The speeds of gas molecules and atoms depend on the gas temperature and on the masses of the individual molecules or atoms. For a given temperature, the more massive molecules and atoms move more slowly than lighter ones. Therefore, nitrogen and oxygen molecules in the atmosphere move more slowly than hydrogen and helium atoms at the same temperature. It is this difference in speed that accounts for the composition of our atmosphere. Hydrogen and helium atoms move so quickly that they can escape from Earth's gravitational pull, whereas nitrogen and oxygen cannot escape because of their slower speeds (a lucky break for us oxygen-breathers).

The method that we used to find the escape speed from Earth can be applied to any planet, moon, or star. We simply use $v = \sqrt{2GM/r}$, where M and r represent the mass and radius of the object from which we wish to escape. Table 9-5 gives escape speeds

Table 9-5

Escape Speeds

Object	Mass (kg)	Radius (m)	Escape Speed (km/s)
Mercury	3.3×10^{23}	2.4×10^6	4.3
Earth	6.0×10^{24}	6.4×10^6	11.2
Moon	7.4×10^{22}	1.7×10^6	2.4
Mars	6.4×10^{23}	3.4×10^6	5.0
Jupiter	1.9×10^{27}	7.1×10^7	60
Sun	2.0×10^{30}	7.0×10^8	618

for various celestial objects. Note the low escape speed for our Moon; even nitrogen and oxygen (and other heavy gases) can escape from the Moon, which therefore has no atmosphere.

Black Holes

One of the most exotic objects believed to exist in the universe is the **black hole**: a region around a collapsed star where gravity is so strong that not even light can escape. When a massive star has had a long and energetic life its core implodes, releasing a great deal of energy and blowing off the outer layers of the star. This event is called a *supernova*. The star's brightness increases by roughly a factor of a billion (10^9) and then gradually decreases over a few months.

black hole a region around a collapsed star where gravity is so strong that not even light can escape

What the star turns into depends on what its original mass was. If the mass of its core was between about 1.5 and 3 solar masses,[5] then it becomes a *neutron star* consisting almost entirely of neutrons and having a radius of only about 10 to 20 km. The density of a neutron star is huge: approximately 10^{17} kg/m^3. A piece of neutron star the size of a sugar cube has a mass of about one hundred million tonnes! Objects have been observed in space that emit radiation in pulses; these *pulsars* are believed to be rotating neutron stars.

If the core mass of a star is above 3 solar masses (approximately), then there is no known force that can stop gravity from continuing the implosion of the core indefinitely until all the mass is concentrated at a single point in space that mathematicians call a singularity. In a fairly small region around the singularity, gravity is so strong that not even light can escape. It is this region that is called a black hole because neither light nor material particles can be emitted. The outer limit of a black hole is called the event horizon; we can never learn anything of events that occur inside this "horizon." The radius of a black hole is often called the Schwarzschild radius in honour of the German astrophysicist Karl Schwarzschild (1873–1916) who first calculated this critical size.

In order to calculate the Schwarzschild radius, we should use Einstein's general theory of relativity (which is *very* complicated). However, it turns out that the final answer is just the same as using Newtonian gravity. We can just use the equation for escape speed, $v = \sqrt{2GM/r}$, that we worked out earlier. At the Schwarzschild radius, by definition the escape speed is the speed of light, 3.00×10^8 m/s. The mass M is the mass of the collapsed star; we will use 4 solar masses, that is, 7.96×10^{30} kg. We want to solve for the radius r. First squaring the equation and rearranging for r,

$$v^2 = \frac{2GM}{r} \text{ and hence } r = \frac{2GM}{v^2}$$

Substituting numbers,

$$r = \frac{2(6.67 \times 10^{-11}\,\text{N}\cdot\text{m}^2/\text{kg}^2)(7.96 \times 10^{30}\,\text{kg})}{(3.00 \times 10^8\,\text{m/s})^2}$$
$$= 1.18 \times 10^4\,\text{m}$$
$$= 11.8\text{ km}$$

Thus, if a star with a core of 4 solar masses collapses to a singularity, then there is a black hole of radius 11.8 km around it. A star with a larger mass would have a larger Schwarzschild radius.

DID YOU KNOW?

In 1987, a supernova visible to the naked eye was observed by a Canadian, Ian Shelton. At the time, Shelton was a student research assistant in the University of Toronto's astronomy department. It was the first naked-eye supernova observed in 383 years (Figure 9-28). Ian Shelton subsequently went on to complete a Master's and a Doctorate at the University of Toronto.

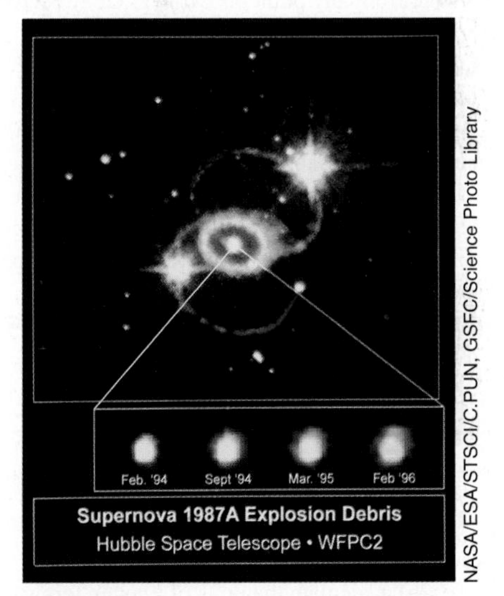

Figure 9-28 The remnants of the supernova observed in 1987 (referred to as SN1987a), as observed by the Hubble Space Telescope several years later.

[5] Masses of stars are often expressed in solar masses, that is, in terms of the mass of our Sun. A star of 2 solar masses has a mass which is twice that of our Sun (1 solar mass = 1.99×10^{30} kg).

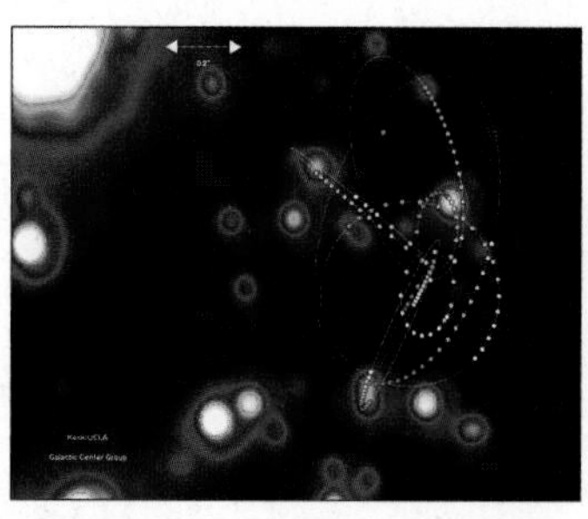

This image was created by Prof. Andrea Ghez and her research team at UCLA and are from data sets obtained with the W. M. Keck Telescopes.

Figure 9-29 The elliptical orbits of eight different stars near the centre of our galaxy, observed between 1995 and 2012, indicate the presence of a supermassive black hole.

DID YOU KNOW?

The global positioning system (GPS) consists of a network of satellites in orbit around the Earth at an altitude of about 20 000 km. The orbits are set up such that at least 4 satellites are always visible from any point on Earth. A GPS receiver is then able to determine its current position by triangulating on the known position of each satellite that is in view at that moment. The system has incredible precision, allowing you to determine your absolute position on the surface of the Earth to within 5 to 10 metres in only a few seconds! To achieve this level of precision, Einstein's theory of General Relativity must be included in the calculations.

Have black holes been detected? Perhaps a better question is: how could we detect a black hole? They are invisible and rather small as far as astronomical sizes are concerned, and hence they are rather difficult to find, so we have to look for them indirectly. Outside the event horizon, the gravity due to the collapsed star is quite strong, and we can observe events in this region. Any material there will be accelerated by the strong gravitational force, and will heat up to such a high temperature that x rays will be emitted. There are a number of x-ray sources in space that astronomers are confident are associated with black holes. One of the most popular candidates is Cygnus X-1, a powerful source of x rays in the constellation Cygnus. There is also evidence of a supermassive black hole at the centre of our own galaxy called Sagittarius A*, or Sgr A* for short; the trajectories of stars near this area (Figure 9-29) indicate the presence of a very massive, but unseen, object. To explain the observed stellar paths, the mass of this black hole must be about 4 *million* solar masses.

Einstein's Theory of Gravity

Earlier in this chapter, we wrote that "the voyage of discovery and explanation started by Ptolemy and other early Greeks had been completed by Newton." Newton's law of universal gravitation explained the motion of the planets and other astronomical objects, as well as objects here on Earth. However, it turns out that Newton's gravitation law is not quite correct in all cases. For example, the details of the orbit of Mercury do not quite fit with Newtonian predictions. In 1916, Albert Einstein published his general theory of relativity, which is very complicated mathematically. His theory deals with gravity in a more complete way and is able to explain Mercury's orbit and other subtle gravitational effects. So far, Einstein's theory has been able to handle everything required of it, but perhaps in time a new theory will be required to explain newly discovered phenomena.

EXERCISES

9-40 (a) Calculate the units of $-GmM/r$ in terms of kilograms, metres, and seconds.

(b) Calculate the units of $\frac{1}{2}mv^2$ in terms of kilograms, metres, and seconds.

(c) Are the units in (a) and (b) the same?

(d) What is the SI unit of energy, and how is it related to your answers above?

9-41 (a) Determine the gravitational potential energy of the Earth–Sun system. The masses of Earth and the Sun are 5.98×10^{24} kg and 1.99×10^{30} kg, respectively, and the average Earth–Sun separation is 1.50×10^{11} m.

(b) If Earth could be moved so that the Earth–Sun distance were extremely large (essentially infinite), what would be the gravitational potential energy of the system?

9-42 A satellite of mass 757 kg is in orbit 3.10×10^2 km above Earth's surface. Earth's mass and radius are 5.98×10^{24} kg and 6368 km. Determine the difference between the satellite's gravitational potential energy at its present elevation and its potential energy at ground level (a) approximately by using $E_P = mgy$, and (b) exactly by using $E_P = -GmM/r$. (c) If the satellite could be brought to rest and allowed to fall toward Earth, what would be its kinetic energy just before it hit Earth? Neglect air resistance.

9-43 If a rock of mass 1.50×10^2 kg is released from rest 761 km above the surface of Mercury, determine its speed just before it hits the surface. Mercury's mass and radius are 3.3×10^{23} kg and 2439 km.

9-44 Craters on the Moon (as shown in Fig. 8-22 (a)) are evidence of collisions with asteroids, comets, and other "rocks in space." Suppose that a large rock is travelling on a head-on collision course with the Moon. When the rock is 8500 km from the centre of the Moon, which has a radius of 1740 km, its speed is 6.3 km/s. The mass of the Moon is 7.36×10^{22} kg.

(a) As the rock moves from 8500 km to 1740 km from the Moon's centre, how is the change in its gravitational potential energy, ΔE_P, related to the change in its kinetic energy, ΔE_K?

(b) What will be the rock's speed when it is just about to hit the Moon's surface?

9-45 An atmospheric probe is fired from Earth and reaches an elevation of 452 km, where it comes to rest momentarily. Determine the probe's speed near Earth's surface. Neglect air resistance. The mass and radius of Earth are 5.98×10^{24} kg and 6368 km respectively.

9-46 Determine the escape speed from Uranus, which has a mass of 8.7×10^{25} kg and a diameter of 5.1×10^{4} km.

9-47 (a) What is the escape speed from a neutron star of mass 3.2×10^{30} kg (1.6 solar masses) and radius 16 km?

(b) What percentage of the speed of light (3.00×10^{8} m/s) is this?

9-48 Neptune has an escape speed of 24 km/s and a radius of 2.53×10^{4} km. What is its mass?

9-49 Calculate the Schwarzschild radius (in kilometres) for a collapsed star having a mass of 7.0×10^{30} kg, or 3.5 solar masses.

9-50 A black hole has a radius of 18 km. What is the mass of the object at its centre? Express your answer in (a) kilograms, and (b) solar masses. (1 solar mass = 1.99×10^{30} kg)

9-51 **Fermi Question:** Estimate the speed of an asteroid as it reaches the Sun if it starts from rest at the edge of the solar system.

LOOKING BACK...LOOKING AHEAD

The topics of Chapters 1 to 8 were related to mechanics: measurement, motion, forces, energy, and momentum. In this chapter, we have concentrated on the dominant force on the astronomical scale: gravitation. We have discussed Newton's law of universal gravitation and related it to the gravitational acceleration (and gravitational field) near planets and other astronomical objects. We have seen why satellites do not crash even though they are attracted by gravitation toward Earth, and have calculated speeds and radii of circular orbits around objects such as Saturn. Kepler's laws of planetary motion have been presented, and a careful discussion has been given of the often-misunderstood concept of weightlessness. The topic of gravitational potential energy has led us to the idea of escape speed and into the exotic realm of black holes. In the next chapter we will continue our discussion of mechanics as it applies to objects that are rotating.

CONCEPTS AND SKILLS

Having completed this chapter, you should now be able to do the following:

- State the contributions of Ptolemy, Copernicus, Kepler, and Newton in the development of our understanding of planetary motion.
- State Newton's law of universal gravitation and apply it in situations involving the force of gravity between masses.
- Discuss the scientific significance of Cavendish's experiment.
- Recognize situations in which $F = Gm_1m_2/r^2$ cannot be applied directly.
- Relate the gravitational acceleration on (or near) a planet or other object to the law of universal gravitation.
- Discuss reasons why gravitational acceleration is not constant over Earth's surface.
- Define gravitational field and be able to calculate its magnitude (near a planet, for example).
- Explain why satellites (and other objects in gravitational orbits) do not crash.
- Use the law of universal gravitation and the expression v^2/r for centripetal acceleration to solve problems involving circular orbits.
- State the general expression for gravitational potential energy.
- Use gravitational potential energy in problems involving speeds (particularly escape speeds) of objects subject to gravitational forces.
- Explain why different planets have different atmospheric compositions.
- Discuss the origin of a black hole.

KEY TERMS

You should be able to define or explain each of the following words or phrases:

law of universal gravitation
universal gravitation constant
field
gravitational field
Kepler's laws of planetary motion:
  Kepler's first law
Kepler's second law
Kepler's third law
weightlessness
escape speed
black hole

Chapter Review

MULTIPLE-CHOICE QUESTIONS

9-52 Two objects each of mass m, separated by a distance d, attract each other with a certain gravitational force. How many times larger is the gravitational force between two objects each of mass $2m$, separated by $d/2$?

(a) 16
(b) 8
(c) 4
(d) 2
(e) 32

9-53 If the distance between a spacecraft and Jupiter increases by a factor of 4, the magnitude of Jupiter's gravitational field at the position of the spacecraft

(a) increases by a factor of 4
(b) decreases by a factor of 4
(c) increases by a factor of 16
(d) decreases by a factor of 16
(e) decreases by a factor of 2

9-54 A 10-kg bag of flour and a 5-kg bag of sugar sit side-by-side on the floor in your kitchen. Which object experiences a greater gravitational field from Earth?

(a) the bag of flour
(b) the bag of sugar
(c) both experience the same gravitational field

9-55 A 10-kg bag of flour and a 5-kg bag of sugar sit side-by-side on the floor in your kitchen. Which object experiences a greater gravitational force from Earth?

(a) the bag of flour
(b) the bag of sugar
(c) both experience the same gravitational force

9-56 A space probe is travelling outward from Earth. At a particular time, the probe is at a distance of 2 Earth radii ($2r_E$) from Earth's centre. At this position, the gravitational force exerted by Earth on the probe has a magnitude F_1 and the gravitational potential energy of the probe is E_{P1}. A few hours later the probe is at a distance of $4r_E$ from Earth's centre, where the gravitational force has a magnitude F_2 and the gravitational potential energy of the probe is E_{P2}. Which of the following is correct?

(a) $F_2 = \frac{1}{2}F_1$ and $E_{P2} = \frac{1}{2}E_{P1}$
(b) $F_2 = \frac{1}{2}F_1$ and $E_{P2} = \frac{1}{4}E_{P1}$
(c) $F_2 = \frac{1}{4}F_1$ and $E_{P2} = \frac{1}{4}E_{P1}$
(d) $F_2 = \frac{1}{8}F_1$ and $E_{P2} = \frac{1}{2}E_{P1}$
(e) $F_2 = \frac{1}{4}F_1$ and $E_{P2} = \frac{1}{2}E_{P1}$

9-57 Which of the following is one of Kepler's Laws?

(a) A planet moves faster when it is nearer the Sun in its orbit and more slowly when it is further away.
(b) For every action there is an equal and opposite reaction.
(c) The gravitational force obeys the inverse square law.
(d) Earth is the centre of the solar system.

9-58 Our understanding of planetary motion in the solar system evolved over many years and included contributions from many scientists. Which scientist is credited with the collection of the data necessary to support the planet's elliptical motion?

(a) Ptolemy
(b) Johannes Kepler
(c) Tycho Brahe
(d) Isaac Newton

9-59 Our understanding of planetary motion in the solar system evolved over many years and included contributions from many scientists. Which scientist is credited with the long and difficult task of analyzing the data?

(a) Ptolemy
(b) Johannes Kepler
(c) Tycho Brahe
(d) Isaac Newton

9-60 Our understanding of planetary motion in the solar system evolved over many years and included contributions from many scientists. Which scientist is credited with the accurate explanation of the data?

(a) Ptolemy
(b) Johannes Kepler
(c) Tycho Brahe
(d) Isaac Newton

9-61 Staring through your telescope one night, you discover a new planet orbiting the Sun at a distance of 4.0 AU. What is its orbital period? (1 AU is the average distance between Earth and the Sun.)

(a) 2.5 Earth years
(b) 8.0 Earth years
(c) 16 Earth years
(d) 64 Earth years

9-62 (This question appeared in an international test of science achievement.) An aircraft flies in a vertical circular path of radius R at a constant speed. When the aircraft is at the top of the circular path the passengers feel "weightless." What is the speed of the aircraft? (g = magnitude of acceleration due to gravity)

(a) gR
(b) $(gR)^{1/2}$
(c) g/R
(d) $(g/R)^{1/2}$
(e) $2gR$

9-63 Two moons around a newly discovered planet have similar size but differ significantly in mass. Moon A has twice the mass of Moon B. How do the escape velocities compare for these two moons?

(a) $v_A \approx 2v_B$
(b) $v_A \approx \dfrac{v_B}{2}$
(c) $v_A \approx \sqrt{2}\,v_B$
(d) $v_A \approx \dfrac{v_B}{\sqrt{2}}$

Review Questions and Problems

9-64 The orbit of Venus is almost a perfect circle around the Sun. What is the acceleration (magnitude and direction) of Venus? The average Venus–Sun distance is 1.1×10^8 km. The mass of Venus is 4.9×10^{24} kg; of the Sun, 1.99×10^{30} kg.

9-65 What is the magnitude of the gravitational force between the Sun (mass 1.99×10^{30} kg) and Jupiter (mass 1.90×10^{27} kg) separated by 7.8×10^8 km?

9-66 (a) What are the magnitude and direction of the gravitational acceleration 3.10×10^6 m above Earth's surface? Earth's mass and radius are 5.98×10^{24} kg and 6.37×10^6 m.

(b) Using your answer from (a), determine the gravitational force (magnitude and direction) on a spacecraft of mass 1.25×10^4 kg at the elevation given in (a).

9-67 Suppose that a woman throws a golf ball straight up with a speed of 17.30 m/s. Using data from Table 9-2, determine how much higher the ball will travel in Jakarta than in Tuktoyaktuk. Neglect air resistance.

9-68 A space probe is at the position shown in Figure 9-30. Use components to determine the gravitational acceleration (magnitude and direction) of the probe due to the effects of Earth and the Moon. The masses of Earth and the Moon are 6.0×10^{24} kg and 7.4×10^{22} kg, respectively.

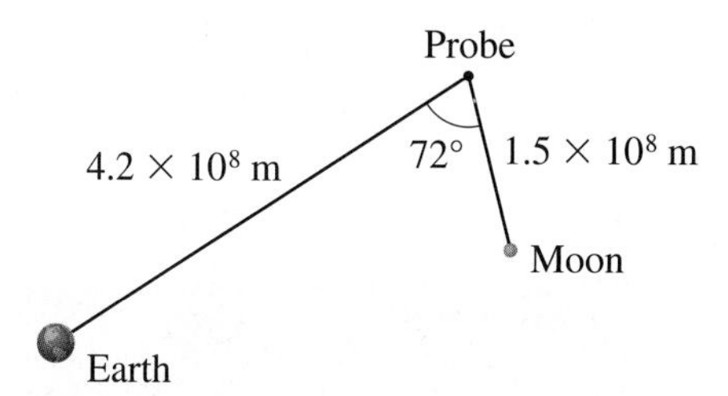

Figure 9-30 Question 9-68.

9-69 A newly-discovered exoplanet has a surface gravitational field strength that is 1.18 times stronger than that of Earth. Its diameter is 5.05×10^4 km. What is the mass of this planet?

9-70 The gravitational field at a certain height above Earth is 4.9 N/kg in magnitude. What is the gravitational force (magnitude and direction) on a spaceball of mass 0.14 kg at that height? (A spaceball is a baseball lost in space.)

9-71 What is the magnitude of Europa's gravitational field at a point in space 4.0×10^3 km from the centre of Europa? Europa, a moon of Jupiter, has a mass of 4.8×10^{22} kg.

9-72 At what elevation (in kilometres) above the surface of Uranus does its gravitational field have a magnitude of 1.0 N/kg? Its mass and radius are 8.73×10^{25} kg and 2.54×10^4 km.

9-73 Since a satellite in orbit around Earth is attracted gravitationally toward Earth, why does it not fall into Earth?

9-74 Explain in a few sentences what weightlessness means.

9-75 A satellite in a circular orbit around Earth has a speed of 6.42×10^3 m/s. The mass of Earth is 5.98×10^{24} kg.

(a) How far is the satellite from the centre of Earth?

(b) How long (in hours) does it take to complete one orbit?

9-76 Comets usually follow very elongated elliptical orbits with the Sun at one focus, with the speed of the comet varying along its path. If a comet is travelling at its highest speed, what can you conclude about its position relative to the Sun? Which one of Kepler's laws helps in answering this question?

9-77 Io (Figure 9-31), a moon of Jupiter first discovered by Galileo, completes one revolution around Jupiter in 1.77 days. Determine the mass of Jupiter, assuming that the orbit of Io is circular with a radius of 4.22×10^5 km.

9-78 Sketch the shapes of graphs of

(a) T versus r and

(b) T^2 versus r^3, where T is the orbital period of a planet and r is its average distance from the Sun. By including the data from Table 9-4 on the appropriate graph, verify Kepler's third law.

9-79 Tethys is a moon of Saturn that has a circular orbit. Its speed is 1.1×10^4 m/s and the mass of Saturn is 5.6×10^{26} kg. Determine the radius and period of the orbit of Tethys.

9-80 The gravitational potential energy of the Sun and Jupiter is -3.2×10^{35} J. Determine the distance between the Sun and Jupiter. The masses of the Sun and Jupiter are 1.99×10^{30} kg and 1.90×10^{27} kg respectively.

9-81 What speed is required to escape from Mimas, a moon of Saturn? The mass and diameter of Mimas are 3.7×10^{19} kg and 3.9×10^2 km.

9-82 If you are trying to escape from a planet, is the direction of your velocity important? For example, does it matter whether you move straight up or horizontally? Explain your answer. (Neglect atmospheric resistance.)

9-83 A probe is fired straight up from Earth with an initial speed of 9.0 km/s. How high above Earth's surface does it go? Earth's mass and radius are 5.98×10^{24} kg and 6368 km. Neglect air resistance.

DID YOU KNOW?

Io, one of Jupiter's moons, is the most volcanically active object in our solar system. Two sulfurous eruptions are visible in Figure 9-31, a colour composite image taken by NASA's Galileo spacecraft that orbited Jupiter from 1995 to 2003. At the left edge of Io in the image, a bluish plume rises about 140 km above the horizon (upper inset). Near the middle of Io, the ring-shaped Prometheus plume is seen rising about 75 km (lower inset). The Prometheus plume is visible in every image ever made of the region dating back to the Voyager mission of 1979, indicating that this plume may have been continuously active for decades.

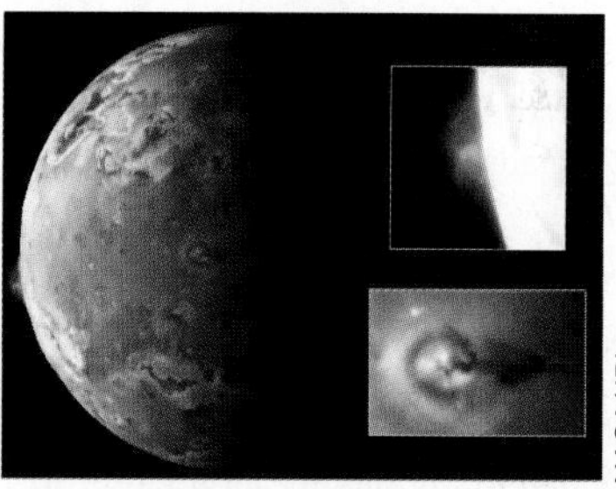

Figure 9-31 Io (Question 9-77), a moon of Jupiter, is the only moon in the solar system known to have active volcanoes. The insets shown on the right are closer views of two such volcanic eruptions underway when the image was taken in 1997.

9-84 A proton (mass 1.67×10^{-27} kg) is moving away from the Sun (mass 1.99×10^{30} kg). At a point in space 1.4×10^9 m from the Sun's centre, the proton's speed is 3.5×10^2 km/s.

(a) What is the proton's speed when it is 2.8×10^9 m from the Sun's centre?

(b) Will the proton escape from the Sun?

9-85 What is the radius (in kilometres) of a black hole surrounding a collapsed star of 4.5 solar masses (9.0×10^{30} kg)?

Applying Your Knowledge

9-86 Determine the resultant gravitational force (magnitude and direction) on the Moon (mass 7.35×10^{22} kg) due to the attraction of the Sun (mass 1.99×10^{30} kg) and Earth (mass 5.98×10^{24} kg). The Earth–Moon and Sun–Moon separations are 3.84×10^5 km and 1.50×10^8 km respectively. Assume that the Earth, Moon, and Sun have the positions shown in Figure 9-32.

to Earth
to Sun
Moon

Figure 9-32 Question 9-86.

9-87 What would be the magnitude of a planet's surface gravitational acceleration (as a multiple of Earth's surface g) if it had a mass that is 4 times smaller than that of Earth, and a radius 5/3 times smaller than that of Earth?

9-88 A satellite of mass 367 kg in a circular orbit around Earth has a speed of 3.9 km/s and an elevation of 2.5×10^7 m from the centre of Earth. Using only the numbers provided, determine the magnitude and direction of

(a) the acceleration of the satellite, and,

(b) the gravitational force on the satellite.

9-89 A small rock is heading toward Jupiter with a speed of 5.1 m/s. Its distance to the centre of Jupiter (mass 1.90×10^{27} kg) is 5.7×10^5 km. In a time of 3.0 s,

(a) how far will the rock travel, and

(b) what will be its new speed? (In such a short time, the rock will travel only a small fraction of the total distance to Jupiter, and thus the gravitational force on the rock will be essentially constant.)

9-90 Given only that Earth's surface gravitational field has a magnitude of 9.80 N/kg, determine the distance (as a multiple of Earth's radius, r_E) above Earth's surface at which the field is 3.20 N/kg.

9-91 The Moon and Earth have masses of 7.35×10^{22} kg and 5.98×10^{24} kg respectively, and are separated by a centre-to-centre distance of 3.84×10^8 m. At what distance from the centre of Earth, on the Earth–Moon line, is the total gravitational field due to Earth and the Moon zero? Is this location closer to Earth or the Moon? Does this make sense?

9-92 Two baseballs are released in space so that they are at rest relative to each other. They then move toward each other as a result of their mutual gravitational attraction. As they move, what happens to the total momentum of the two-ball system?

9-93 Venus has a surface gravity 0.91 times that of Earth and a diameter 0.95 times that of Earth. Using no numbers other than these, what is the mass of Venus relative to the mass of Earth, that is, what is the ratio of m_{Venus} to m_{Earth}?

9-94 Assume that the horizontal and vertical velocity components of a long-jumper at takeoff are 9.20 m/s and 3.50 m/s respectively. Neglecting air resistance, determine the length of jump at

(a) Moscow, where $g = 9.816$ m/s^2; and

(b) Mexico City, where $g = 9.779$ m/s^2. Assume that the jumper is at the same horizontal level at takeoff and landing. (More detailed calculations taking air resistance into account show that a jumper can travel 5 cm farther at Mexico City than at Moscow.)

9-95 A piece of space garbage is ejected from a spacecraft. It is at rest relative to the Sun (mass 1.99×10^{30} kg) and is 1.25×10^6 km from its centre. It starts to fall toward the Sun. How long does it take to travel the first 1.00×10^2 m? (Since 100 m $\ll 1.25 \times 10^6$ km, the gravitational force and acceleration will be roughly constant over this short distance.) Figure 9-33 illustrates just how big a problem we have with space garbage or debris around our own planet.

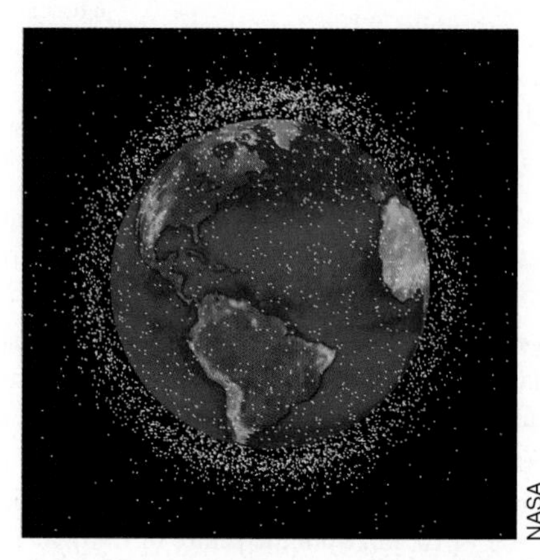

Figure 9-33 A computer-generated image from NASA's Orbital Debris Program. The image depicts all of the objects in low Earth orbit (within 2000 km of the surface) that are currently being tracked by NASA. Approximately 95% of the objects in this illustration are not functional satellites. The dots are scaled according to the image size of the graphic to optimize their visibility and are not scaled to Earth.

9-96 Some telecommunications satellites are in *geosynchronous orbits* above Earth, that is, they have periods of 24 h. As a result, since Earth turns on its axis once in 24 h and each satellite goes around Earth once in 24 h, any individual satellite stays positioned above a particular point on Earth.

(a) How far above Earth's surface must a geosynchronous satellite be? Earth's mass and average radius are 5.98×10^{24} kg and 6368 km.

(b) What is the satellite's speed?

9-97 Using the law of universal gravitation and the expression v^2/r for centripetal acceleration, prove Kepler's third law for circular orbits, that is, prove that $T^2 \propto r^3$, where T is the orbital period of the planet and r is the radius of the orbit.

9-98 Table 9-6 provides data concerning the moons of Uranus. Use these data to determine:

(a) the average value of the constant of proportionality, C, in Eqn. 9-6, and

(b) the mass of Uranus.

Table 9-6

Orbital Data for the Moons of Uranus (Question 9-98)

Moon	Average distance from Uranus ($\times 10^5$ km)	Period (Earth days)
Miranda	1.30	1.41
Ariel	1.92	2.52
Umbriel	2.67	4.14
Titania	4.39	8.71
Oberon	5.87	13.46

9-99 The Moon circles Earth once every 27.3 days at an average distance of 3.84×10^5 km from Earth's centre.

(a) Use this information to find the constant of proportionality, *C*, in Eqn. 9-6.

(b) Use this constant and Kepler's third law to determine the height above Earth's surface at which a satellite must travel to have a period of 12 hours. The average radius of Earth is 6.37×10^6 m.

9-100 When we say that the escape speed from Earth is 11.2 km/s, we mean the speed relative to the centre of Earth, not the surface. Any speed that a rocket has by virtue of starting on a rotating Earth can be used as part of the required 11.2 km/s for escape.

(a) Where on Earth is the speed due to rotation greatest?

(b) Determine this greatest rotation speed in kilometres per second. Use 6378 km for Earth's radius.

(c) In order to take best advantage of this starting speed, in what direction (east, south, north, west) should a rocket be launched? Explain your answer.

(d) Why is Cape Canaveral in Florida a good choice for rocket launches?

9-101 A typical neutron star has a mass that is twice that of our Sun with a radius of only 15 km. If an apple falls (from rest) from a height of 3.0 m above its surface, what is the speed of the apple when it reaches the "ground"?

9-102 **Fermi Question:** If you were standing on the surface of a typical neutron star, what is the approximate difference between the acceleration due to gravity at your feet and the acceleration due to gravity at your head? What would happen to you?

9-103 The maximum rate of rotation of a celestial object can be determined by thinking about the resultant force required for material at the surface to remain in place as it moves along a circular path. Assume a spherical object with uniform mass density ρ (mass per unit volume). Determine the maximum frequency of rotation of the sphere such that the gravitational force on the material at the surface is just sufficient to provide the necessary circular motion.

9-104 When travelling from one planet to another within a solar system, it can be shown that such a trip can be made with minimal energy required if the spacecraft leaves planet A when it is closest to the central star and is timed to arrive at planet B when it is furthest from the central star. Figure 9-34 illustrates one such trip of a spacecraft from Mars to Jupiter, leaving Mars at its closest point and arriving at Jupiter at its furthest point. Our Sun is at one focus of this elliptical path. Determine the time required to make the one-way trip from Mars to Jupiter along this path.

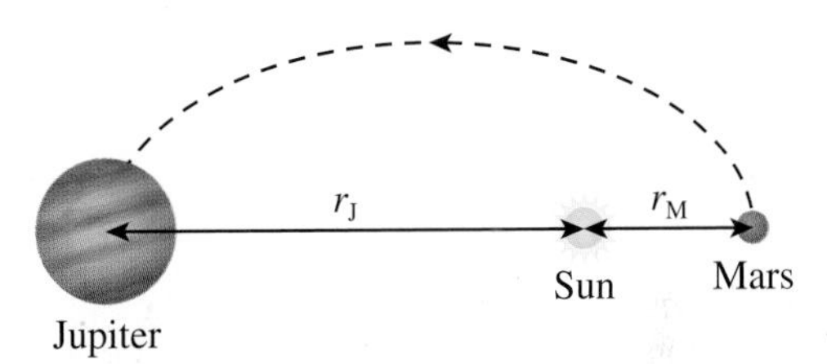

Figure 9-34 Question 9-104.

Rotational Motion 10

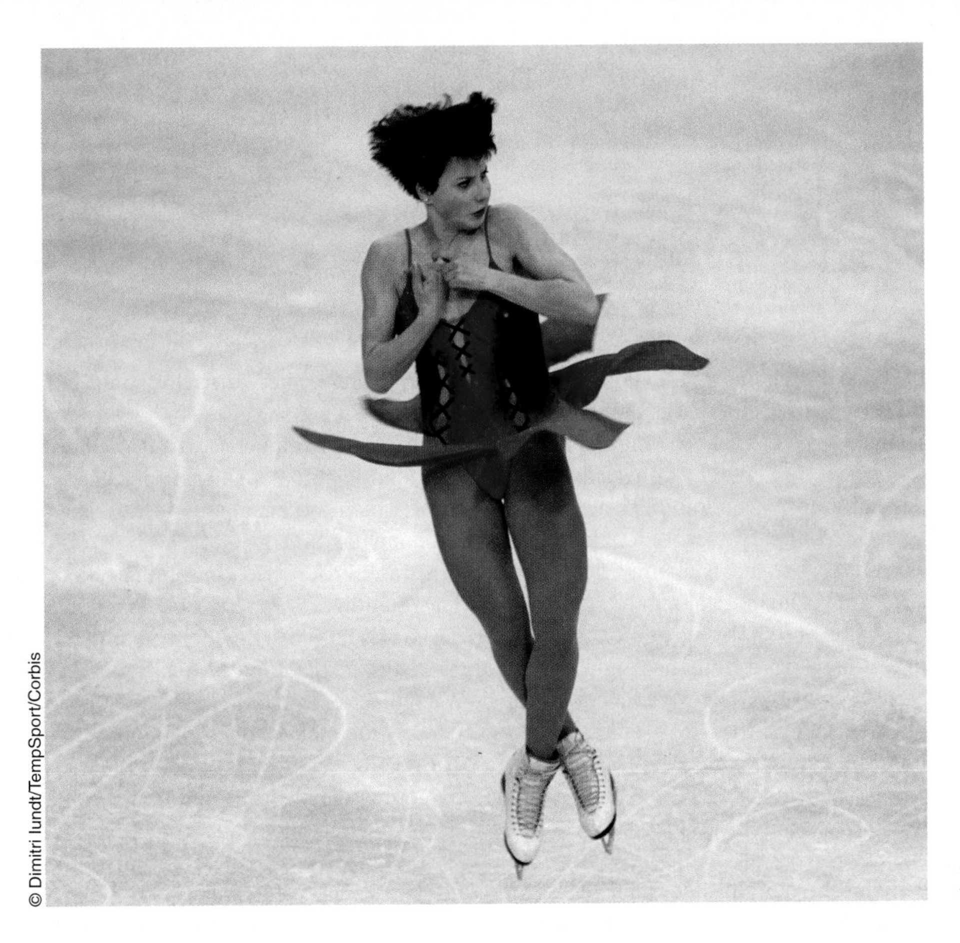

© Dimitri Iundt/TempSport/Corbis

Figure 10-1 A figure skater's performance incorporates beauty and athletics, but also a lot of interesting physics. When a skater goes into a spin and pulls her arms close to her body, she spins faster. Why? The answer is in Section 10.5.

Think of all the different things that rotate: Earth, Ferris wheels, propellers, figure skaters (Figure 10-1), food processors, drills, and wheels and motors of all kinds. An understanding of the motion of all these objects requires a study of rotation in general. We will see that there are a lot of similarities between rotational motion and the kinematics and dynamics introduced in Chapters 2 to 8.

CHAPTER **OUTLINE**

10.1 Rotational Kinematics

You probably have had occasion to watch (or ride) a rotating Ferris wheel (Figure 10-2 on page 256). The wheel's motion is rather interesting: every point on the structure has the same rotation rate (usually measured in revolutions per minute, or rpm), but the speed (in metres per second, say) of a point P_1 near the outer edge is much greater than the speed of a point P_2 closer to the centre. Every point on the wheel travels in a circle with its centre lying on a fixed line called the **rotation axis**. We say that the wheel rotates about (or around) its rotation axis. For a Ferris wheel the rotation axis is stationary, but for a wheel on a moving bicycle, the rotation axis is moving; relative to

rotation axis a line about which an object rotates; relative to this axis, each point on the rotating object travels on a circle with its centre on the axis

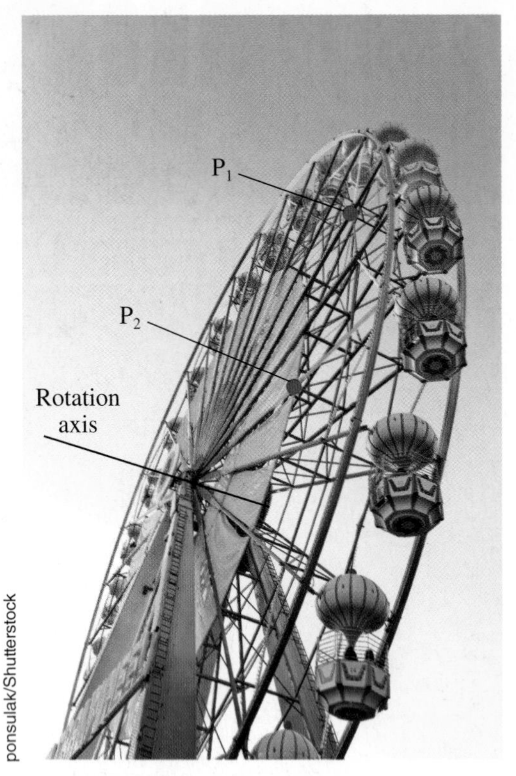

Figure 10-2 Every point (such as P_1 and P_2) on this Ferris wheel travels in a circle with its centre on the rotation axis. All points have the same rotation rate, but their speeds (in metres per second, for example) depend on the distance from the centre.

angular position the position of a point on a rotating object, measured as an angle between an arbitrary axis (drawn outward from the centre of rotation) and the line from the centre to the chosen point

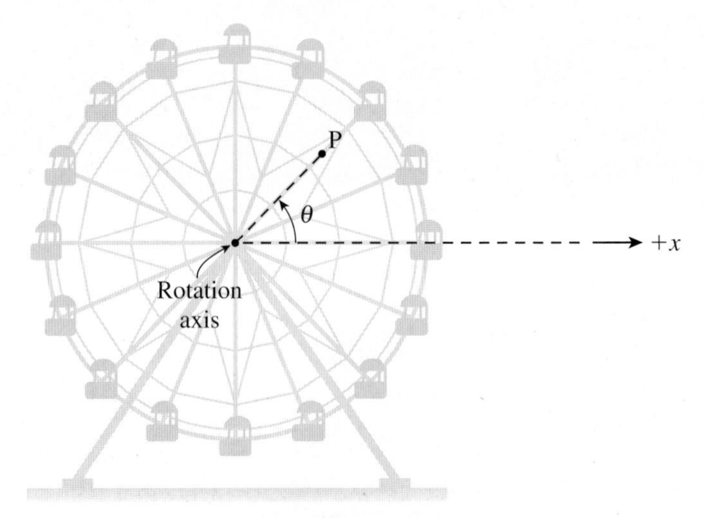

Figure 10-3 Angular position (θ) of point P.

the axis, each point on the bicycle wheel travels in a circle, but relative to the ground the motion of each point is more complicated.

In order to describe the motion of a rotating object such as a Ferris wheel, we need to introduce a new set of physical quantities: angular position, angular displacement, angular velocity, and angular acceleration, all of which are analogous to quantities such as position, velocity, etc., that you have encountered already.

Figure 10-3 shows a simplified view of a rotating Ferris wheel, or similar object, seen from a direction along the rotation axis. The rotation axis appears only as a dot in the centre since it is seen end-on, and a reference x-axis has been drawn in an arbitrary direction outward from the centre of the wheel. We define the **angular position** (θ) of a given point P on the wheel as the angle between the x-axis and the line drawn from the rotation axis to P. Note that θ is the lower case Greek letter "theta." Angular position is measured in degrees (°), radians (rad), or revolutions (rev).

The angular unit **radian** (rad) is defined by comparing the arc length on a circle to the circle's radius. Specifically, one radian is the angle at the centre of a circle that subtends an arc of length one radius (Figure 10-4). Therefore, there are 2π rad in one complete revolution, since an angle of 2π rad subtends an arc of length 2π radii ($2\pi r$), that is, one circumference. Any subtended arc length on a circle divided by the radius gives the central angle in radians.

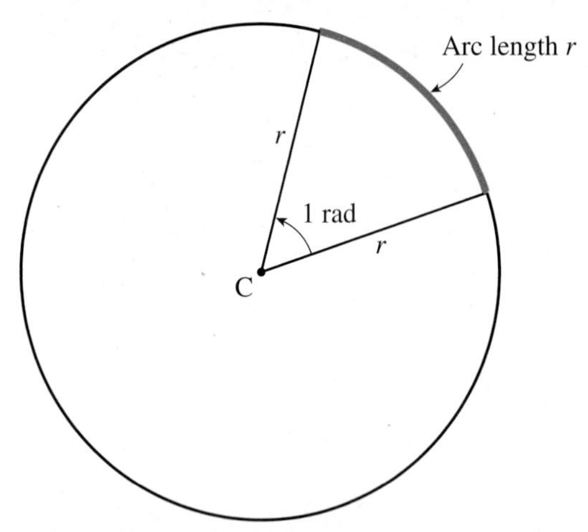

Figure 10-4 An angle of 1 rad at the circle's centre (C) subtends an arc of length one radius. In general, subtended arc length divided by radius gives the central angle in radians.

$$\frac{\text{subtended arc length on circle}}{\text{radius of circle}} = \text{central angle in radians} \qquad (10\text{-}1)$$

Since an angle in radians is defined as a length divided by another length (radius), then a radian is a dimensionless unit; that is, it is not a unit of length, time, etc. Since the radian is a unit of angle, the other angular units (degrees and revolutions) are also dimensionless, and any angle is a dimensionless quantity. Since angles can be measured in any of the three units mentioned so far, it is useful to know the conversion factors between them:

$$1 \text{ revolution} = 360° = 2\pi \text{ rad} \qquad (10\text{-}2)$$

$$\text{Since } 2\pi \text{ rad} = 360°, \text{ then } 1 \text{ rad} = \frac{360°}{2\pi} \approx 57.3°$$

Figure 10-5 shows the path of point P (from Figure 10-3) as the object rotates. The angular position of P is θ_1 at time t_1, and θ_2 at time t_2. Notice that the object is rotating counterclockwise in this case. In the time interval Δt (i.e., $t_2 - t_1$), point P has undergone an angular displacement $\Delta\theta = \theta_2 - \theta_1$. **Angular displacement** is the change in angular position. Notice that the angular displacement of *any* point on the wheel in a particular time interval is the same. If one point on the wheel rotates through 30°, all points on the wheel rotate through 30°. Therefore, *we usually refer to the angular displacement of the rotating object as a whole*, rather than to the angular displacement of a particular point. Angular displacement of a rotating object is often measured in revolutions rather than radians or degrees, since rotating objects are often making many revolutions per minute.

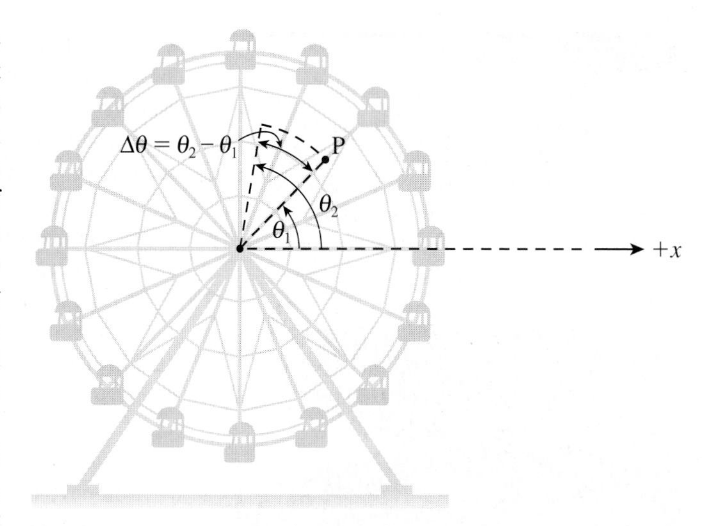

Figure 10-5 Angular displacement $\Delta\theta = \theta_2 - \theta_1$.

We can now discuss how fast the wheel is rotating. If the wheel has rotated through an angular displacement $\Delta\theta$ in a time interval Δt, it has an **average angular velocity** ω_{av} defined as the angular displacement divided by the time interval:

$$\omega_{av} = \frac{\Delta\theta}{\Delta t} \qquad (10\text{-}3)$$

where ω is the lower case Greek letter "omega." Average angular velocity can be measured in a variety of units; most commonly we will use radians per second (rad/s), or revolutions per minute (rpm).

The **instantaneous angular velocity** ω, also called simply **angular velocity**, is the limiting value of the average angular velocity as $\Delta t \to 0$.

$$\omega = \lim_{\Delta t \to 0} \frac{\Delta\theta}{\Delta t} \qquad (10\text{-}4)$$

The distinction between instantaneous angular velocity and average angular velocity is similar to that between instantaneous velocity and average velocity (Chapter 2). The angular velocity ω tells us how fast an object is rotating at any instant. In the case of, say, a rotating ceiling fan, its angular velocity is usually constant, and hence the fan's angular velocity and *average* angular velocity have the same value.

Just as we defined average acceleration and instantaneous acceleration in Chapter 2, we can now define average angular acceleration α_{av}, and instantaneous angular acceleration (or just angular acceleration) α, where α is the lower case Greek "alpha." The **average angular acceleration** of a rotating object is the change in angular velocity ($\Delta\omega$) divided by the time interval (Δt). **Instantaneous angular acceleration**, or just **angular acceleration**, α, is the limiting value of $\Delta\omega/\Delta t$ as $\Delta t \to 0$.

$$\alpha_{av} = \frac{\Delta\omega}{\Delta t} \qquad (10\text{-}5)$$

$$\alpha = \lim_{\Delta t \to 0} \frac{\Delta\omega}{\Delta t} \qquad (10\text{-}6)$$

Angular acceleration α is a measure of how rapidly the angular velocity is changing. If α is large, then ω is changing rapidly. When a CD is inserted into a computer's disc drive, the CD starts to spin and α is initially large as ω increases quickly, and then α

radian a unit of angular measure; one radian (approximately 57.3°) is the central angle in a circle that subtends an arc of length one radius

angular displacement change in angular position of a rotating object

average angular velocity average rotation rate of a rotating object during a time interval; angular displacement $\Delta\theta$ divided by the time interval Δt taken

DID YOU KNOW?

Escherichia coli bacteria swim by rotating thin flagella, each in the shape of a helix. These flagella (usually four of them) act as propellers and rotate with an average angular velocity of several hundred rpm. An *E. coli* bacterium is rod-shaped, about 1 μm in diameter and 2 μm long, and an individual flagellum extends several body lengths from the cell's body. The biological motor that turns each flagellum is tiny, only about 45 nm in diameter.

angular velocity (or instantaneous angular velocity) instantaneous rotation rate of a rotating object; the limit as $\Delta t \to 0$ of angular displacement $\Delta\theta$ divided by the time interval Δt taken

average angular acceleration a measure of how rapidly, on average, the rotation rate of a rotating object is changing during a time interval; change in angular velocity ($\Delta\omega$) divided by the time interval (Δt) taken

instantaneous angular acceleration (or angular acceleration) a measure of how rapidly the rotation rate of a rotating object is changing at a given instant; the limit as $\Delta t \to 0$ of change in angular velocity $\Delta\omega$ divided by the time interval Δt taken

DID YOU KNOW?

When a CD (Figure 10-6) starts to be read, its angular velocity is about 500 rpm, but as information is extracted, the CD undergoes a small angular acceleration that gradually decreases the angular velocity to about 200 rpm at the end of the disc. Why do you think this happens? (See Section 10.2 for the answer.)

Brian A Jackson/Shutterstock

Figure 10-6 As a CD is being played, its angular velocity decreases.

decreases to zero as ω becomes constant for a while. Angular acceleration is most commonly measured in units of radians per second squared, but other units (revolutions per minute squared, etc.) are possible.

You might be wondering whether rotational variables such as angular velocity are vectors or scalars. For the purposes of this book we will treat them as scalars, that is, positive or negative quantities that have no direction associated with them. If you take further courses in physics, you will learn that rotational variables are in fact vectors, but their vector nature is rather complicated and will not be discussed here.

It is customary in problems involving rotation to use positive values of ω as much as possible. Then, if α is positive, ω is increasing and the object is speeding up rotationally. If α is negative, ω is decreasing, and the object's rotation is slowing down. In many situations the angular acceleration α is constant; that is, the angular velocity is changing at a constant rate, and hence $\alpha_{av} = \alpha$. The Sample Problems provide a wide variety of helpful examples involving angular velocity and angular acceleration.

You have undoubtedly noticed that each rotational variable is an exact analogue of a linear variable introduced in Chapter 2: angular velocity corresponds to velocity, and so on. When acceleration, say a_x, is constant, there are equations that relate velocity, acceleration, etc., for example, $v_x = v_{0x} + a_x t$. When *angular* acceleration (α) is constant, there is a corresponding set of equations, given below:

Linear Equations (Constant a_x)	Rotational Equations (Constant α)	
$v_x = v_{0x} + a_x t$	$\omega = \omega_0 + \alpha t$	(10-7)
$x = x_0 + \frac{1}{2}(v_0 + v)t$	$\theta = \theta_0 + \frac{1}{2}(\omega_0 + \omega)t$	(10-8)
$x = x_0 + v_{0x}t + \frac{1}{2}a_x t^2$	$\theta = \theta_0 + \omega_0 t + \frac{1}{2}\alpha t^2$	(10-9)
$v^2 - v_{0x}^2 = 2a_x(x - x_0)$	$\omega^2 - \omega_0^2 = 2\alpha(\theta - \theta_0)$	(10-10)

Notice that each variable (e.g., velocity v_x) in the equations in the left-hand column has been replaced with its corresponding rotational variable (e.g., angular velocity ω) in the right-hand equations. The rotational equations relate angular acceleration α (which must be constant), time t, initial angular position θ_0 and initial angular velocity ω_0 (both at $t = 0$), final angular position θ and final angular velocity ω (both at time t). You will find them very useful in analyzing situations involving rotational motion and solving problems involving rotation.

Units Tip

Be careful with units when using the equations involving constant angular acceleration. The angular part of the units for θ, θ_0, ω, ω_0, and α must all be the same. For example, if θ_0 and θ are measured in *radians*, then ω and ω_0 must be in *radians* per second (or minute, etc.), and α must be in *radians* per second squared (or per minute squared, etc.). As well, the units of the time t must be the same as the time part of the units of ω, ω_0, and α. For example, if t is in *seconds*, then ω cannot be in revolutions per *minute*.

SAMPLE PROBLEM 10-1

The maximum angular velocity for a propeller on an airplane (Figure 10-7) is about 2500 revolutions per minute (rpm). Suppose that a propeller, starting with zero angular velocity, speeds up to this maximum with a constant angular acceleration of 17 rad/s^2.

(a) Convert 2500 rpm to radians per second. Assume two significant digits.

(b) How long does it take the propeller to reach its top angular velocity?

(c) How many revolutions does the propeller make in this time?

(d) When the propeller is turned off, it slows from 2500 rpm to rest in 23 s. What is the angular acceleration?

Solution

(a) Using the conversion factor given in Eqn. 10-2,

$$2500 \text{ rpm} = \frac{2500 \text{ rev}}{\text{min}} \times \frac{2\pi \text{ rad}}{1 \text{ rev}} \times \frac{1 \text{ min}}{60 \text{ s}} = 2.6 \times 10^2 \frac{\text{rad}}{\text{s}}$$

Therefore, the maximum angular velocity is 2.6×10^2 rad/s.

(b) We are asked to find how long the acceleration takes, that is, the time t. Given quantities are the initial angular velocity ($\omega_0 = 0$ rad/s), the final angular velocity ($\omega = 2.6 \times 10^2$ rad/s), and the angular acceleration ($\alpha = 17$ rad/s^2). We look for one of the constant-angular-acceleration equations that involves only the four quantities t, ω_0, ω, and α. Eqn. 10-7 is the appropriate equation:

$$\omega = \omega_0 + \alpha t$$

Rearranging to solve for t, and substituting numbers (using the value of ω in radians per second, not revolutions per second, since α is given in radians per second squared),

$$t = \frac{\omega - \omega_0}{\alpha} = \frac{(2.6 \times 10^2 - 0)\text{ rad/s}}{17\text{ rad/s}^2} = 15\text{ s}$$

Notice how the units work out: $\dfrac{\text{rad/s}}{\text{rad/s}^2} = \left(\dfrac{\text{rad}}{\text{s}}\right) \times \left(\dfrac{\text{s}^2}{\text{rad}}\right) = \text{s}$

Thus, it takes 15 s to reach the maximum angular velocity starting from rest.

(c) We need to find how many revolutions the propeller makes, that is, the angular displacement $\Delta\theta$, or $\theta - \theta_0$. (Alternatively, we could choose $\theta_0 = 0$ and simply find θ.) We know ω_0, ω, α, and t, and so any one of Eqns. 10-8, 10-9, or 10-10 will work. Eqn. 10-8 is probably the easiest to use

$$\theta = \theta_0 + \tfrac{1}{2}(\omega_0 + \omega)t, \text{ which gives } \Delta\theta = \theta - \theta_0 = \tfrac{1}{2}(\omega_0 + \omega)t$$

Substituting numbers, $\Delta\theta = \tfrac{1}{2}((0 + 2.6 \times 10^2)\text{ rad/s})(15\text{ s}) = 2.0 \times 10^3\text{ rad}$

Converting to revolutions, $2.0 \times 10^3\text{ rad} \times \dfrac{1\text{ rev}}{2\pi\text{ rad}} = 3.2 \times 10^2\text{ rev}$

Thus, the propeller makes 3.2×10^2 rev while speeding up to maximum angular velocity from rest.

(d) When the propeller slows down, its *initial* angular velocity ω_0 is now 2.6×10^2 rad/s and its *final* angular velocity ω is zero. The time t is 23 s, and we are asked to find the angular acceleration α. Since we are given ω_0, ω, and t, and we need α, we can use Eqn. 10-7: $\omega = \omega_0 + \alpha t$.

Solving for α, and substituting numbers,

$$\alpha = \frac{\omega - \omega_0}{t} = \frac{(0 - 2.6 \times 10^2)\text{ rad/s}}{23\text{ s}} = -11\text{ rad/s}^2$$

Hence, the propeller's angular acceleration is -11 rad/s^2. Notice the negative sign, which tells us that the propeller is slowing down; that is, the angular velocity is decreasing.

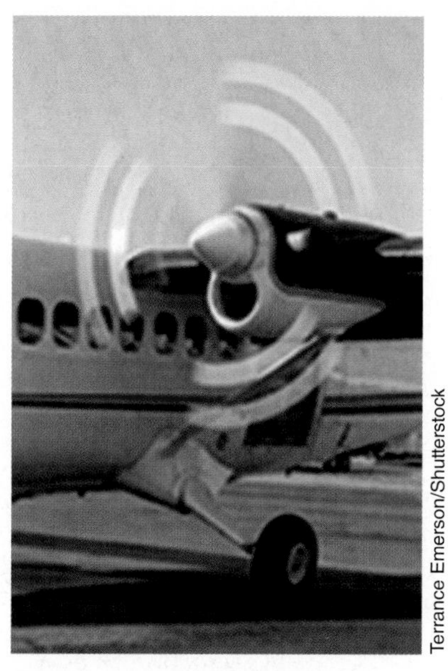
Terrance Emerson/Shutterstock

Figure 10-7 Sample Problem 10-1.

SAMPLE PROBLEM 10-2

A pottery wheel (Figure 10-8) is turned on and speeds up to 2.4×10^2 rpm in a time of 1.5 s.

(a) Assuming constant angular acceleration, how far does the wheel rotate in this time? Express your answer in revolutions and radians.

(b) What is the angular acceleration of the wheel in radians per second squared?

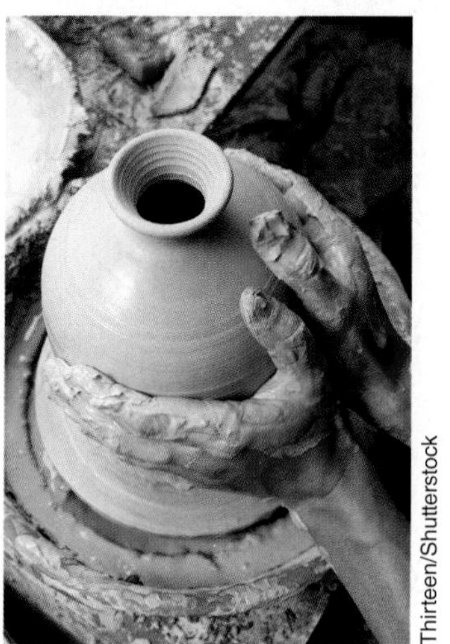
Thirteen/Shutterstock

Figure 10-8 Sample Problem 10-2.

Solution

(a) We are asked to find how far the wheel rotates, that is, to determine its angular displacement $\Delta\theta$, or $\theta - \theta_0$. Since we are told that the wheel is "turned on," we can assume that its initial angular velocity (ω_0) is zero. Its final angular velocity (ω) is 2.4×10^2 rpm, and the time t is 1.5 s. We then choose the equation that relates θ, θ_0, ω_0, ω, and t. The appropriate equation is Eqn. 10-8:

$$\theta = \theta_0 + \tfrac{1}{2}(\omega_0 + \omega)t, \text{ which gives } \Delta\theta = \theta - \theta_0 = \tfrac{1}{2}(\omega_0 + \omega)t$$

Before substituting numbers, we must check our units. In the angular part of the units, ω is in revolutions per minute (and $\omega_0 = 0$ rpm), and therefore $\Delta\theta$ will come out in revolutions, which is fine. However, in the time part of the units, ω and t do not match: ω is in revolutions per *minute*, and t is in *seconds*. We will convert the units of ω to revolutions per second (or we could have converted t to minutes). Doing this conversion,

$$2.4 \times 10^2 \frac{\text{rev}}{\text{min}} \times \frac{1 \text{ min}}{60 \text{ s}} = 4.0 \frac{\text{rev}}{\text{s}}$$

Now substituting numbers,

$$\Delta\theta = \tfrac{1}{2}(\omega_0 + \omega)t = \tfrac{1}{2}\left((0 + 4.0)\frac{\text{rev}}{\text{s}}\right)(1.5 \text{ s}) = 3.0 \text{ rev}$$

Thus, the wheel rotates 3.0 rev. To convert this to radians,

$$3.0 \text{ rev} \times \frac{2\pi \text{ rad}}{1 \text{ rev}} = 19 \text{ rad}$$

The angular displacement of the wheel is 3.0 rev, or 19 rad.

(b) We need to find the angular acceleration α in radians per second squared. We now know the angular displacement $\Delta\theta$ in revolutions and in radians, the final angular velocity ω in revolutions per minute and in revolutions per second, and the time t in seconds. The initial angular velocity ω_0 is zero, regardless of choice of units. In order to minimize our work, we should try to avoid a unit conversion if possible. Therefore, we choose $\Delta\theta$ in radians and t in seconds, and use Eqn. 10-9:

$$\theta = \theta_0 + \omega_0 t + \tfrac{1}{2}\alpha t^2 \quad \text{or} \quad \Delta\theta = \theta - \theta_0 = \omega_0 t + \tfrac{1}{2}\alpha t^2$$

Substituting $\omega_0 = 0$ rad/s, rearranging to solve for α, and substituting numbers,

$$\alpha = \frac{2\Delta\theta}{t^2} = \frac{2(19 \text{ rad})}{(1.5 \text{ s})^2} = 17 \text{ rad/s}^2$$

Hence, the angular acceleration of the wheel is 17 rad/s^2.

EXERCISES

10-1 In a circle of radius 11.7 cm, what angle in radians subtends an arc of length

(a) 11.7 cm?

(b) 23.4 cm?

(c) 73.5 cm?

10-2 In the previous question express the answers in degrees and revolutions.

10-3 Convert an angular velocity of 3.05 rad/s to

(a) radians per minute

(b) degrees per second

(c) revolutions per minute

10-4 What is the angular velocity of Earth about its axis, in radians per second?

10-5 Convert an angular acceleration of 6.70 rev/min^2 to
(a) radians per minute squared
(b) degrees per second squared
(c) radians per second squared

10-6 A springboard diver (Figure 10-9) has an angular displacement of 12 rad in a time interval of 0.85 s. What is his average angular velocity?

Figure 10-9 Question 10-6.

10-7 The angular velocity of an electric drill increases from 0 rad/s to 105 rad/s in a time interval of 0.550 s. What is its average angular acceleration?

10-8 An electric fan has a constant angular velocity of 2.00×10^3 rpm.
(a) How many revolutions will it make in 12 s?
(b) Through how many radians will it rotate in 15 s?
(c) When the fan is turned off, it takes 6.5 s for it to stop. What is its angular acceleration (assumed constant) during this time?

10-9 A Ferris wheel speeds up from rest with a constant angular acceleration of 0.105 rad/s^2 for a time of 3.75 s. Determine its final angular velocity and angular displacement.

10-10 Figure 10-10 shows an Osprey tilt-rotor aircraft, which has propellers that can rotate about a vertical axis (as shown) for vertical takeoff and landing. Once the craft is airborne, the mounting for each propeller can rotate 90° forward and the plane then looks like a normal propeller-driven plane as it flies. Suppose that a propeller of an Osprey speeds up from rest to an angular velocity of 1.5×10^3 rpm in a time of 6.28 s.
(a) What is its angular acceleration (assumed constant) in radians per second squared?
(b) Through how many revolutions does it turn during this time?

Figure 10-10 Question 10-10. An Osprey tilt-rotor aircraft lifting a vehicle from a ship.

10-11 The blades of a blender slow from 3.5 rad/s to 1.8 rad/s while rotating through 1.2 rad.
(a) What is the angular acceleration (assumed constant)?
(b) How long does this slowing-down take?
(c) Assuming that the angular acceleration stays constant, how much *longer* would it take for the blades to come to rest?

10-12 A pitched baseball (Figure 10-11) can have a rotation rate as high as 1.8×10^3 rpm. If this rotation rate remains constant, how many revolutions does the ball make on its trip to the plate? Assume that the ball is travelling with a constant horizontal speed of 38 m/s, and that the horizontal distance from the pitcher's mound to the plate is 18 m.

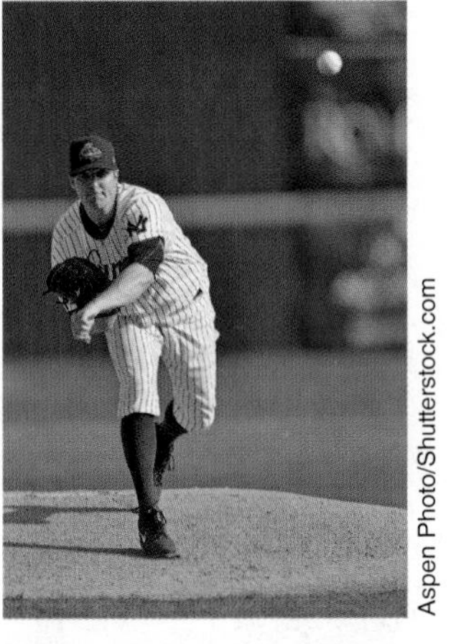

Figure 10-11 Question 10-12.

DID YOU KNOW?

When a baseball pitcher throws a knuckleball, it makes only $1\frac{1}{2}$ to 2 rotations as it travels to the plate. The lack of spin, combined with the contrast between the seams and the smooth parts of the ball, produces air pockets around the ball that cause its unpredictable movements.

10-13 **Fermi Question:** Estimate the average angular velocity (in radians per second) of the gymnast shown in Figure 10-12 while she is in the air above the balance beam.

Figure 10-12 Question 10-13.

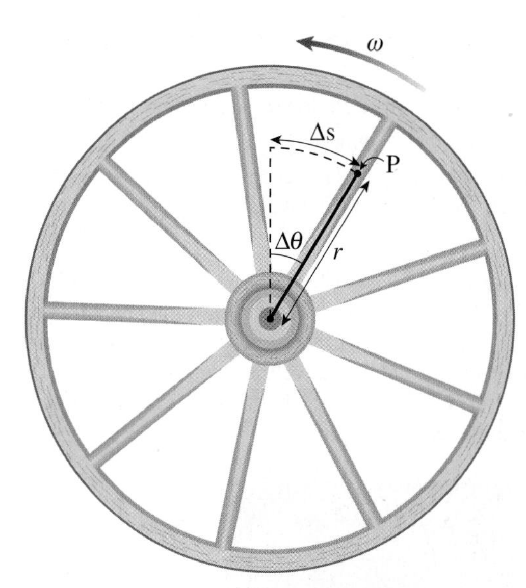

Figure 10-13 $\Delta s = r\Delta\theta$, if $\Delta\theta$ is in radians.

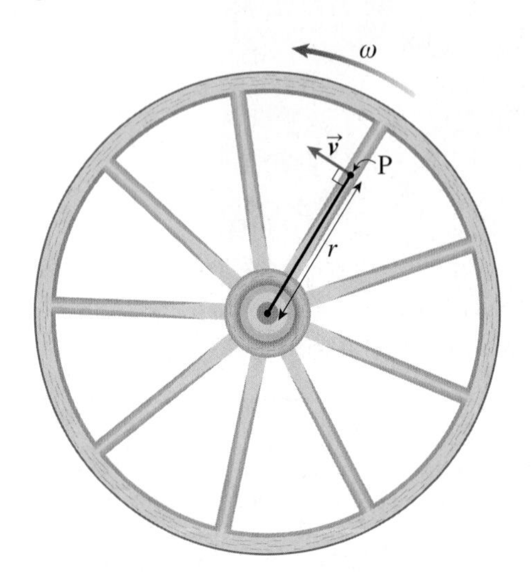

Figure 10-14 $v = \omega r$, if ω is in *radians* per second, or *radians* per minute, etc.

10.2 Relations between Rotational and Linear Quantities

One of the chief advantages of using a rotational variable such as angular velocity is that it applies to the rotating object as a whole; that is, each and every point on the object has the same angular velocity. However, the speed (in metres per second, for example) is different for points at different distances from the rotation axis. We now explore the relations between rotational quantities and the corresponding *linear* quantities: displacement, velocity, acceleration, etc.

Figure 10-13 shows the path of a point P at a distance r from the centre of a wheel rotating about a fixed axis. During a time interval Δt, the wheel undergoes an angular displacement $\Delta\theta$, and P travels a corresponding distance Δs along a circular arc. (Arc length is often represented by "s" or "Δs.") Recalling that subtended arc length divided by radius gives the angle in radians (Section 10.1), $\Delta s/r = \Delta\theta$ or

$$\Delta s = r\Delta\theta \qquad (10\text{-}11)$$

$\Delta\theta$ *must* be measured in *radians* for this relation to be true. We now divide by the time interval Δt:

$$\frac{\Delta s}{\Delta t} = r\frac{\Delta\theta}{\Delta t}$$

Letting $\Delta t \to 0$,

$$\lim_{\Delta t\to 0}\frac{\Delta s}{\Delta t} = r\lim_{\Delta t\to 0}\frac{\Delta\theta}{\Delta t}$$

The left-hand side of this equation is the (instantaneous) speed v of point P, and the right-hand side is the product of the radius r and the (instantaneous) angular velocity ω. Thus $v = r\omega$, usually written as

$$v = \omega r \qquad (10\text{-}12)$$

This expression gives the relationship between the linear speed v of a point at a distance r from the centre of an object having an angular velocity ω (Figure 10-14). Not surprisingly, the linear speed increases with increasing distance from the centre, and increases also with increasing angular velocity. We have implicitly assumed that the rotation axis is fixed. (For a *rolling* automobile wheel, which has a moving rotation axis, $v = \omega r$ gives the speed of any point on the wheel relative to the rotation axis, but not relative to the ground.)

Units Tip

Since $\Delta s = r\Delta\theta$ is valid only if $\Delta\theta$ is in *radians*, then $v = \omega r$ is valid only if ω has units of *radians* per second, or *radians* per minute, etc. As well, when using $v = \omega r$ there is another complication. Suppose you are given that $\omega = 3$ rad/s and $r = 2$ m, and are asked to calculate v. Substituting numbers into $v = \omega r$ gives $v = (3\text{ rad/s})(2\text{ m}) = 6$ rad·m/s. The unit of rad·m/s for speed seems odd. However, since radian is a *dimensionless* unit (Section 10.1), it can simply be discarded, and we write 6 rad·m/s = 6 m/s to give the usual SI unit for speed. Radians can also be inserted in units if needed. For example, $v = 6$ m/s and $r = 2$ m, and you need to calculate ω. Rearranging $v = \omega r$ to give $\omega = v/r$, and substituting numbers, $\omega = (6\text{ m/s})/(2\text{ m}) = 3\text{ s}^{-1}$. Now, insert (i.e., multiply) radians into the unit s^{-1} to give $\omega = 3$ rad/s.

DID YOU KNOW?

In Section 10.1 it was mentioned that the angular velocity of a CD decreases from about 500 rpm to 200 rpm during the time that the CD is being read. The data on a CD are stored as small indentations, or pits, encoded in a track spiralling around the centre. An infrared laser is shone on the pits as they rotate under it, and the reflected laser light hits a detector that decodes the signal. As the CD spins, the laser gradually moves outward, reading the part of the track closest to the centre first and the part farthest from the centre last. The code on the track must be read at a constant linear speed v of about 1.3 m/s. Since $v = \omega r$, when the track closest to the centre (where r is small) is being read, ω will be large (500 rpm), and when the track farthest from the centre (with large r) is being read, ω will be small (200 rpm).

Tangential Acceleration

Starting with $v = \omega r$, we can develop an expression for acceleration. Suppose that a rotating object is speeding up or slowing down. For a point P at a distance r from the rotation axis, if v and ω undergo changes of Δv and $\Delta\omega$ in a time interval Δt, then $\Delta v = \Delta\omega \cdot r$. Dividing by Δt and letting $\Delta t \to 0$,

$$\lim_{\Delta t \to 0} \frac{\Delta v}{\Delta t} = \lim_{\Delta t \to 0} \frac{\Delta \omega}{\Delta t} \cdot r$$

The left-hand side of this equation is the time rate of change of the speed of point P, and represents an (instantaneous) acceleration of P that is *tangential* to P's circular path. If this **tangential acceleration** is non-zero, P is speeding up or slowing down. We represent this acceleration by a_t (subscript "t" for "tangential"). The right-hand side of the equation is the product of the (instantaneous) angular acceleration α and the distance r from P to the rotation axis. Thus,

tangential acceleration for any point on an object rotating about a fixed axis, the tangential acceleration for the point is an acceleration component that increases or decreases the speed at that point; this component is tangential to the circle on which the point travels, and is perpendicular to the centripetal component of acceleration, which is directed toward the centre of the circle

$$a_t = \alpha r \qquad \textbf{(10-13)}$$

In order for this equation to be valid, *α must have radians in its units*. A fixed rotation axis has again been assumed.

This tangential acceleration a_t is not the centripetal acceleration acting toward the centre of the circle, discussed in Section 4.5. If P is on a wheel rotating about a fixed axis with *constant* angular velocity ω (i.e., angular acceleration $\alpha = 0$), the speed v of P is constant (from $v = \omega r$) and the acceleration of P has only a centripetal component ($a_c = v^2/r$). However, if the wheel is speeding up or slowing down (i.e., $\alpha \neq 0$), then the acceleration $\vec{a}$ of P has two components: a centripetal component ($a_c = v^2/r$ still, where v is now changing in time) and a tangential component ($a_t = \alpha r$). These components are perpendicular to each other (Figure 10-15), as were a_x and a_y in earlier chapters.

The tangential component a_t changes only the magnitude of the velocity $\vec{v}$, whereas the centripetal component a_c changes only the direction of $\vec{v}$. If $a_t > 0$, then the speed v of point P is increasing, and the acceleration component a_t is in the direction of velocity $\vec{v}$ at point P. Also, if $a_t > 0$, then $\alpha > 0$ (from $a_t = \alpha r$), and the angular velocity ω is increasing. If $a_t < 0$, the speed at point P is decreasing, and the acceleration component a_t is in the opposite direction to the velocity at P; as well, $\alpha < 0$, and ω is decreasing.

In Section 4.5 we wrote $a_c = v^2/r$ (Eqn. 4-15) for the magnitude of centripetal acceleration. We now know that $v = \omega r$, and so we can write $a_c = v^2/r = (\omega r)^2/r = \omega^2 r$. Thus, we now have an alternate expression for a_c:

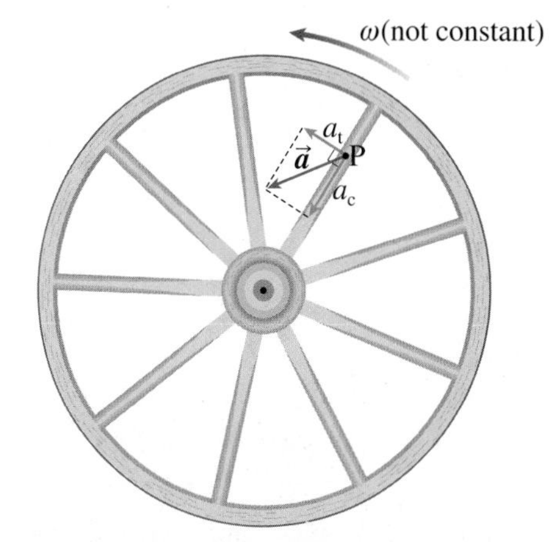

Figure 10-15 If the angular acceleration α is non-zero for a wheel rotating about a fixed axis, then the angular velocity ω is not constant, and the acceleration $\vec{a}$ of point P has a centripetal component ($a_c = v^2/r$) and a tangential component ($a_t = \alpha r$).

$$a_c = \omega^2 r \qquad \textbf{(10-14)}$$

This expression for a_c is useful when we know ω instead of v. (Of course, we could always calculate v from $v = \omega r$ and use $a_c = v^2/r$.) The expression $\omega^2 r$ for the magnitude of centripetal acceleration can be used *only* if ω has *radians* in its units.

SAMPLE PROBLEM 10-3

Just after being turned on, a circular saw blade (Figure 10-16) has an angular velocity of 2.4×10^2 rpm and an angular acceleration of 5.5×10^2 rad/s^2.

(a) What is the speed (in metres per second) at a point P on the blade 0.10 m from the centre?

(b) What is the magnitude of the acceleration at this point?

Kobets Dmitry/Shutterstock

Figure 10-16 Sample Problem 10-3.

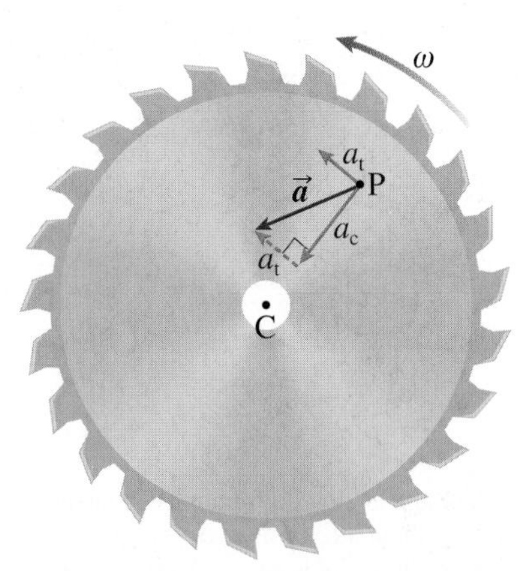

Figure 10-17 Sample Problem 10-3 (b). The acceleration $\vec{a}$ at point P has two components: a centripetal component a_c toward the centre C of the circle, and a tangential component a_t at right angles to the centripetal component.

Solution

(a) Since we are given the angular velocity ω and the distance (radius) r of the point from the centre, we can use $v = \omega r$ (Eqn. 10-12) to calculate speed. However, ω *must* have radians in its units for this equation to be used, and the given value is in rpm. Therefore, we first convert rpm to radians per second:

$$\omega = 2.4 \times 10^2 \frac{\text{rev}}{\text{min}} \times \frac{2\pi \text{ rad}}{1 \text{ rev}} \times \frac{1 \text{ min}}{60 \text{ s}} = 25 \frac{\text{rad}}{\text{s}}$$

Then, $v = \omega r = (25 \text{ rad/s})(0.10 \text{ m}) = 2.5 \text{ m/s}$

Therefore, the speed of the saw blade at point P (0.10 m from the centre) is 2.5 m/s. Notice that the "rad" has been omitted in the final unit for speed, as discussed in the Units Tip earlier in this section.

(b) Since the given angular acceleration is positive, the angular velocity is increasing, and the speed at point P is also increasing. Therefore, the tangential component of acceleration, a_t, is in the same direction as the velocity $\vec{v}$ at P. The (total) acceleration $\vec{a}$ at P has both a centripetal and a tangential component (Figure 10-17), and their magnitudes are given by

$$\text{Eqn. 10-14:} \quad a_c = \omega^2 r = (25 \text{ rad/s})^2 (0.10 \text{ m}) = 63 \text{ m/s}^2$$

$$\text{Eqn. 10-13:} \quad a_t = \alpha r = (5.5 \times 10^2 \text{ rad/s}^2)(0.10 \text{ m}) = 55 \text{ m/s}^2$$

Notice that "rad" has again been omitted in the units for a_c and a_t. Since these components are at right angles to each other, the magnitude of $\vec{a}$ can be determined from the Pythagorean relation:

$$a = \sqrt{a_c^2 + a_t^2} = \sqrt{(63)^2 + (55)^2} \text{ m/s}^2 = 84 \text{ m/s}^2$$

Thus, the acceleration at point P has a magnitude of 84 m/s^2.

SAMPLE PROBLEM 10-4

A ballet dancer is spinning (Figure 10-18) with an angular velocity of 5.2 rad/s and an angular acceleration of −0.90 rad/s^2.

(a) What is the speed at a point on her elbow, 0.37 m from the rotational axis?

(b) What will be the speed at the same point 1.0 s later?

Felix Mizioznikov/Thinkstock

Figure 10-18 Sample Problem 10-4.

Solution

(a) Use Eqn. 10-12: $v = \omega r = (5.2 \text{ rad/s})(0.37 \text{ m}) = 1.9 \text{ m/s}$

Therefore, the speed of the point on her elbow is 1.9 m/s.

(b) The tangential acceleration is given by Eqn. 10-13:

$$a_t = \alpha r = (-0.90 \text{ rad/s}^2)(0.37 \text{ m}) = -0.33 \text{ m/s}^2$$

Since the point on the dancer's elbow is travelling in a circle, its velocity $\vec{v}$ is tangential to the circle, and the tangential acceleration component changes the magnitude of this velocity $|\vec{v}|$, that is, the speed v. Since the tangential acceleration is constant (−0.33 m/s^2), we can use one of the equations of constant-acceleration kinematics (Eqn. 2-7 or Eqn. 4-11) in the tangential direction to determine the speed:

$$v = v_0 + a_t t = (1.9 \text{ m/s}) + (-0.33 \text{ m/s}^2)(1.0 \text{ s}) = 1.6 \text{ m/s}$$

Thus, the speed of the point on the elbow 1.0 s later is 1.6 m/s.

Alternate Solution for Part (b)

First use Eqn. 10-7 to find the angular velocity 1.0 s later:

$\omega = \omega_0 + \alpha t = (5.2 \text{ rad/s}) + (-0.90 \text{ rad/s}^2)(1.0 \text{ s}) = 4.3 \text{ rad/s}$

Then use Eqn. 10-12 to determine the speed:

$$v = \omega r = (4.3\ \text{rad/s})(0.37\ \text{m}) = 1.6\ \text{m/s}$$

Again, the speed of the point on the elbow is 1.6 m/s.

Rolling Objects

You have undoubtedly noticed that for a rolling ball or wheel, the more rotations it makes per second, the faster it moves across the ground. As well, a large rolling object moves across the ground faster than a small object having the same angular velocity. It can be shown that the linear speed V of the centre of a *rolling* (non-skidding) object is related to the object's angular velocity, ω, and radius, R, by

$$V = \omega R \qquad (10\text{-}15)$$

where ω must have *radians* in its units.

This relation $V = \omega R$ might at first glance appear to be the same as the $v = \omega r$ that we discussed earlier in this section. Even though these equations have the same form, the quantities are different (Figure 10-19). Upper-case V represents the linear speed of the *centre* of a *rolling* object; lower-case v is the linear speed of any *point P* on an object rotating about a *fixed axis*. R is the (outer) *radius* of the rolling object; r is the *distance from P to the rotation axis*; ω in either case is the object's angular velocity.

The linear acceleration of the centre of a rolling object (Figure 10-20) is related to its angular acceleration. Starting with Eqn. 10-15, $V = \omega R$, if V and ω undergo changes of ΔV and $\Delta\omega$ in a time interval Δt, then $\Delta V = \Delta\omega \cdot R$. Dividing by Δt and letting $\Delta t \rightarrow 0$,

$$\lim_{\Delta t \to 0} \frac{\Delta V}{\Delta t} = \lim_{\Delta t \to 0} \frac{\Delta \omega}{\Delta t} \cdot R$$

The left-hand side of this equation is the time rate of change of the speed of the centre of the object, which we write as the acceleration A of the centre. The right-hand side is the product of the object's angular acceleration α and the object's radius R. Thus,

$$A = \alpha R \qquad (10\text{-}16)$$

In order for this equation to be valid, *α must have radians in its units.*

Note that since α can be positive or negative, then A can also be positive (if the object's speed is increasing) or negative (if the speed is decreasing).

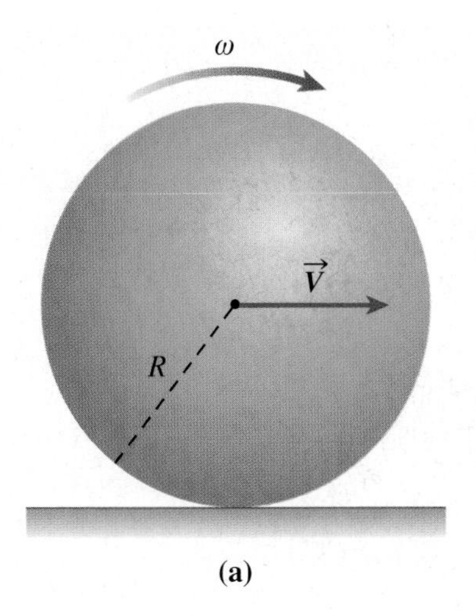

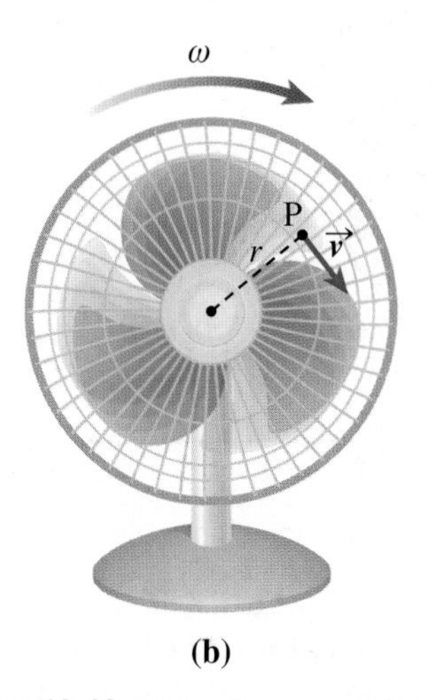

Figure 10-19 **(a)** A rolling object. The speed V of the centre is given by $V = \omega R$. **(b)** An object rotating about a fixed axis. The speed v of any point P at a distance r from the centre is given by $v = \omega r$.

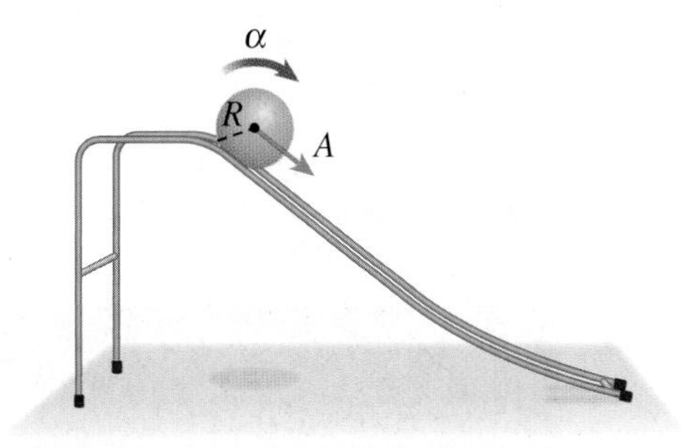

Figure 10-20 An accelerating rolling object. The acceleration of the centre is $A = \alpha R$.

SAMPLE PROBLEM 10-5

A golf ball (radius 2.1 cm) is putted uphill (Figure 10-21) with an initial speed of 0.63 m/s. As it rolls, it undergoes an acceleration of $-0.080\ \text{m/s}^2$. Determine

(a) the ball's angular acceleration

(b) the ball's initial angular velocity

(c) the ball's angular velocity after 2.0 s

(d) the ball's speed after 2.0 s

(e) the time required for the ball to come to rest

Terminology Tip

In Eqn. 10-15 ($V = \omega R$) and Eqn. 10-16 ($A = \alpha R$), the V and A represent the speed and acceleration, respectively, of the *centre* of the rolling object. However, it is common to refer to V and A as the speed and acceleration of the object (as a whole).

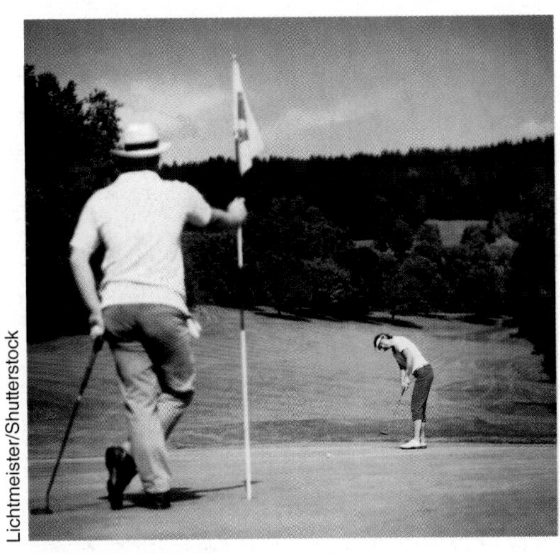

Figure 10-21 Sample Problem 10-5.

Units Tip

When the equation $\alpha = A/R$ is being used, then A in metres per second squared, divided by R in metres, gives second^{-2}. Since α is *angular* acceleration, we then insert radians to give radians per second squared. Remember, radians are included or removed as necessary.

Math Tip

In order to avoid round-off error (Section 1.4) when working through a problem with many steps, it is important to keep at least one extra digit in intermediate answers (or store intermediate answers with all their digits in your calculator).

Solution

(a) The angular acceleration can be determined from Eqn. 10-16: $A = \alpha R$.

First convert the units of the radius to metres: $2.1 \text{ cm} \times \dfrac{1 \text{ m}}{100 \text{ cm}} = 2.1 \times 10^{-2} \text{ m}$

Then, $\alpha = \dfrac{A}{R} = \dfrac{-0.080 \text{ m/s}^2}{2.1 \times 10^{-2} \text{ m}} = -3.81 \text{ rad/s}^2$

The angular acceleration is -3.8 rad/s^2 (to two significant digits).

(b) Since the radius and the initial speed are given, the initial angular velocity can be calculated from Eqn. 10-15: $V = \omega R$. Including subscripts "zero" for initial velocity and initial angular velocity, and solving for ω_0,

$$\omega_0 = \frac{V_0}{R} = \frac{0.63 \text{ m/s}}{2.1 \times 10^{-2} \text{ m}} = 30.0 \text{ rad/s}$$

Therefore, the initial angular velocity is 3.0×10^1 rad/s (to two significant digits).

(c) To find the angular velocity at $t = 2.0$ s, use Eqn. 10-7:

$$\omega = \omega_0 + \alpha t = (3.00 \times 10^1 \text{ rad/s}) + (-3.81 \text{ rad/s}^2)(2.0 \text{ s}) = 22.4 \text{ rad/s}$$

The ball's angular velocity after 2.0 s is 22 rad/s.

(d) Using Eqn. 10-15 to determine speed,

$$V = \omega R = (22.4 \text{ rad/s})(2.1 \times 10^{-2} \text{ m}) = 0.47 \text{ m/s}$$

Therefore, the ball's speed after 2.0 s is 0.47 m/s.

(e) When the ball comes to rest, $\omega = 0$ rad/s. Eqn. 10-7 can be used to find the time at which this happens:

$$\omega = \omega_0 + \alpha t, \text{ giving } t = \frac{\omega - \omega_0}{\alpha} = \frac{(0 - 30.0) \text{ rad/s}}{-3.81 \text{ rad/s}^2} = 7.9 \text{ s}$$

Thus, the ball comes to rest in a time of 7.9 s.

EXERCISES

10-14 A water sprinkler (Figure 10-22) is rotating at 55 rad/s. What is the speed in metres per second of a point on the sprinkler 5.2 cm from the rotation axis?

Figure 10-22 Question 10-14.

10-15 A point on the blades of a food processor has a speed of 3.1 m/s. If the point is 2.1 cm from the rotation axis, what is the processor's angular velocity in revolutions per minute?

10-16 Just after starting, a wind turbine has an angular acceleration of 0.78 rad/s^2. What is the tangential component of acceleration of a point on a turbine blade 12 m from the rotation axis?

10-17 A point on a helicopter rotor has a tangential component of acceleration of 78 m/s^2. How far from the rotation axis is this point if the rotor's angular acceleration is 95 rad/s^2?

10-18 Consider a point on a blade of a spinning electric fan.

(a) What is the direction of its centripetal component of acceleration?

(b) If the angular velocity of the blade is constant, what can you conclude about the tangential component of acceleration?

(c) If the speed of the point is increasing, what can you conclude about the angular acceleration?

10-19 A rotating merry-go-round in a playground (Figure 10-23) has an angular velocity of 35.2 rpm and an angular acceleration of 0.310 rev/s². For a point 2.00 m from the centre of the merry-go-round, determine the magnitude of the

(a) centripetal component of acceleration
(b) tangential component of acceleration
(c) acceleration

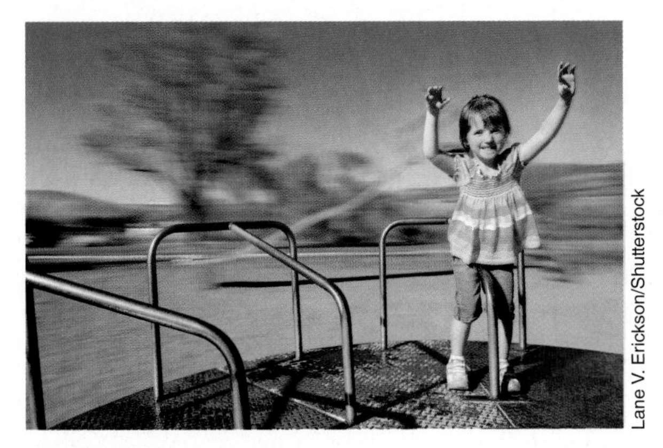

Lane V. Erickson/Shutterstock

Figure 10-23 Question 10-19.

10-20 The tub in a washing machine is starting to slow down after completing the spin cycle. It has an angular velocity of 18 rad/s, and an angular acceleration of −1.2 rad/s². After 5.0 s,

(a) What is the tub's angular velocity?
(b) How many revolutions has it made as it slows?
(c) What is the speed of a point 0.30 m from the tub's centre?
(d) What are the centripetal and tangential components of acceleration of this point?

10-21 A baseball of radius 3.7 cm is rolling with a speed of 3.4 m/s. What is its angular velocity?

10-22 If the wheels of the bicycle shown in Figure 10-24 have an angular acceleration of 0.35 rad/s², and the tires have a diameter of 66 cm, what is the acceleration of the bicycle and rider in metres per second squared?

Warren Goldswain/Shutterstock

Figure 10-24 Question 10-22.

10-23 Suppose that a dragster (Figure 10-25) starts from rest and proceeds with an acceleration of magnitude 22 m/s² for the first 4.5 s. At this time,

(a) What is the dragster's speed?
(b) How far has it travelled?
(c) If the rear tire radius is 0.34 m, what is the angular acceleration of the rear wheels?
(d) What is the angular velocity of the rear wheels after 4.5 s?
(e) How many rotations have the rear wheels made in 4.5 s?

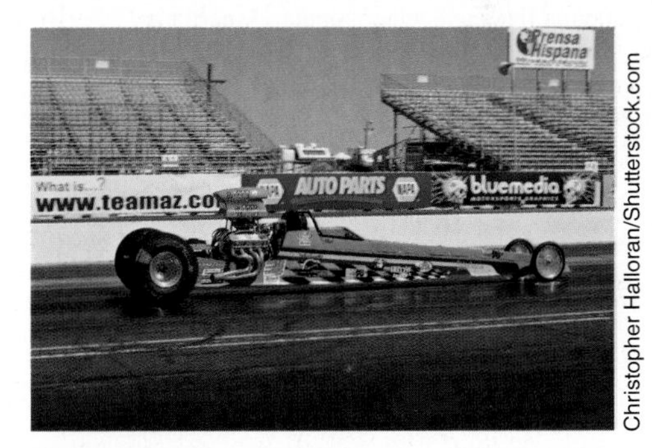

Christopher Halloran/Shutterstock.com

Figure 10-25 Question 10-23.

10-24 A soccer ball (diameter 21.8 cm) has been kicked and is rolling up a hill. Just after the kick, the ball has an angular velocity of 12.3 rad/s and an angular acceleration of −2.60 rad/s². Assuming constant angular acceleration, how far along the hill will the ball roll before stopping?

10.3 Rotational Kinetic Energy and Moment of Inertia

Think about the motion of a rotating fan (Figure 10-26). Because the fan is moving, it certainly has kinetic energy, but since the fan as a whole does not have one particular linear speed v that we can associate with it, we cannot just use $\frac{1}{2}mv^2$ to calculate its kinetic energy. However, since each particle on the fan has a speed at which it is moving and therefore a kinetic energy, we can write the kinetic energy (E_K) of the entire fan as being the sum of the kinetic energies of all the individual particles:

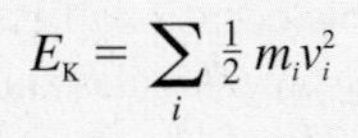

$$E_K = \sum_i \frac{1}{2} m_i v_i^2$$

Monkey Business Images/Shutterstock

Figure 10-26 A rotating fan.

where the Greek sigma (Σ) indicates that we are taking a sum, in this case the sum over all particles $i = 1, 2,\ldots$ on the rotating fan. As it stands, the expression $\sum_i \frac{1}{2} m_i v_i^2$ is not very useful, since the speeds (v_i) are different in general for the different particles. However, the angular velocity ω is the same for all the particles, and we can introduce ω (with no subscript i) into the expression for E_K by substituting $v_i = \omega r_i$ (Eqn. 10-12), where r_i is the distance of particle i from the rotation axis:

$$E_K = \sum_i \frac{1}{2} m_i \omega^2 r_i^2$$

Factoring out the $\frac{1}{2}$ and the ω^2, since they are the same for all the particles,

$$E_K = \frac{1}{2}\left(\sum_i m_i r_i^2\right)\omega^2 \qquad \textbf{(10-17)}$$

moment of inertia (or **rotational inertia**) rotational analogue of mass; for an object rotating about a fixed axis, the moment of inertia is the sum, for all particles on the object, of the product of each particle's mass and the square of its distance from the rotation axis

The quantity in parentheses, $\sum_i m_i r_i^2$, is called the **moment of inertia**, or **rotational inertia**, of the fan (or other rotating object). It is the rotational analogue of mass, and is usually symbolized by "I." Since the moment of inertia is calculated as (the sum of) mass times distance squared, its SI unit is kilogram·metre^2 (kg·m^2).

$$\text{Moment of inertia:} \quad I = \sum_i m_i r_i^2 \qquad \textbf{(10-18)}$$

rotational kinetic energy the kinetic energy associated with an object's rotation about an axis

Replacing $\sum_i m_i r_i^2$ in Eqn. 10-17 with I, we obtain a final expression for the **rotational kinetic energy** of an object rotating about a fixed axis:

$$\text{Rotational kinetic energy:} \quad E_K = \frac{1}{2} I\omega^2 \qquad \textbf{(10-19)}$$

translational kinetic energy an object's kinetic energy associated with the motion of the centre (of mass) of an object

The expression $\frac{1}{2} I\omega^2$ for rotational kinetic energy is analogous to $\frac{1}{2} mv^2$ for **translational kinetic energy**, that is, the kinetic energy associated with the motion of the centre (of mass) of an object. In developing the expression $E_K = \frac{1}{2} I\omega^2$, we have implicitly assumed that the object is rotating about a *fixed axis*. The angular velocity ω must have *radians* in its units in order for $E_K = \frac{1}{2} I\omega^2$ to be valid; normally ω is expressed in radians per second when $E_K = \frac{1}{2} I\omega^2$ is used.

How do we deal with the moment of inertia $I = \sum_i m_i r_i^2$? In words, $\sum_i m_i r_i^2$ means that we have to add up, for every particle in the rotating object, the product of mass and the square of the distance from the rotation axis. How would we do this? For a complicated shape such as a rotating fan blade, it would be difficult, requiring a computer to do calculus integrations. For objects with simpler shapes such as spheres, cylinders, and rods, the moments of inertia can be determined using calculus but without a computer. Table 10-1 provides algebraic expressions for the moments of inertia for some common shapes, assuming that each object has uniform density of material. When solving problems you will need to refer to this table occasionally to obtain moments of inertia. Each moment of inertia is the product of a dimensionless number (such as $\frac{1}{2}$, $\frac{2}{5}$, etc.), the total mass M of the object, and the square of a distance (radius R *or* length L) on the object.

Table 10-1

Moments of Inertia of Objects of Uniform Composition with Total Mass *M*

Object	Axis		Moment of Inertia
Thin ring or hoop	through centre	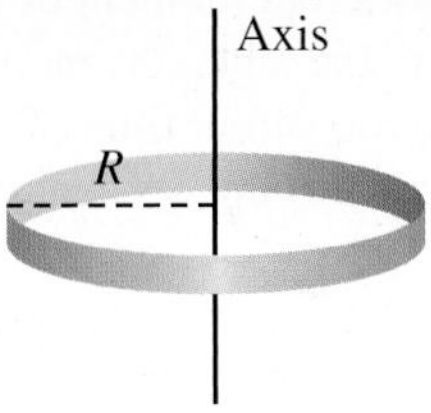 Axis, R	MR^2
Solid cylinder or disk	through centre	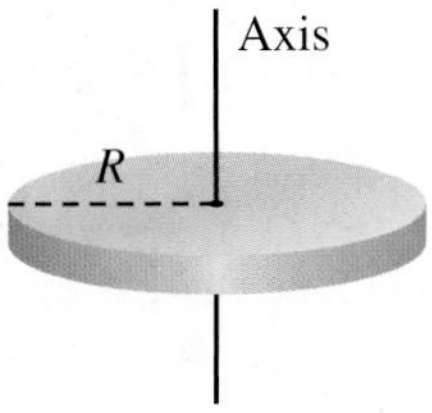 Axis, R	$\frac{1}{2}MR^2$
Solid sphere	through centre	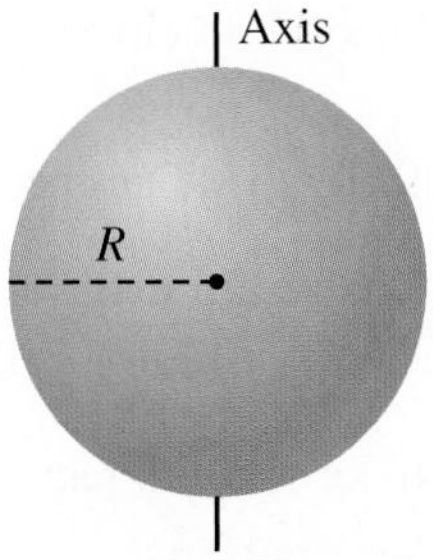 Axis, R	$\frac{2}{5}MR^2$
Thin-walled hollow sphere	through centre	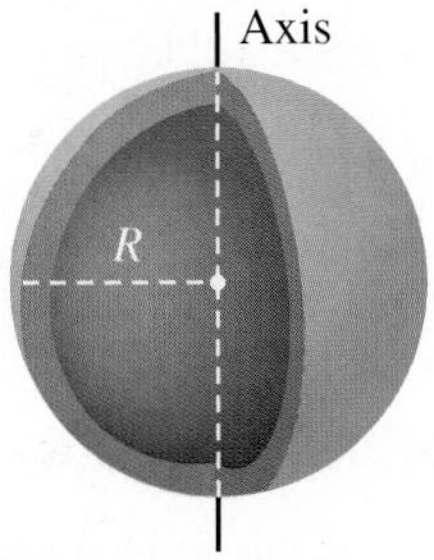 Axis, R	$\frac{2}{3}MR^2$
Thin rod	through centre	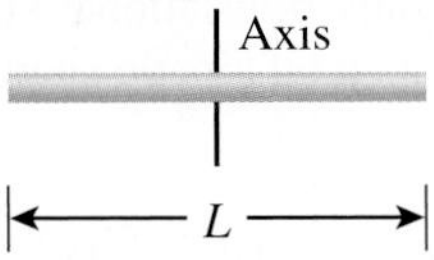 Axis, L	$\frac{1}{12}ML^2$
Thin rod	through end	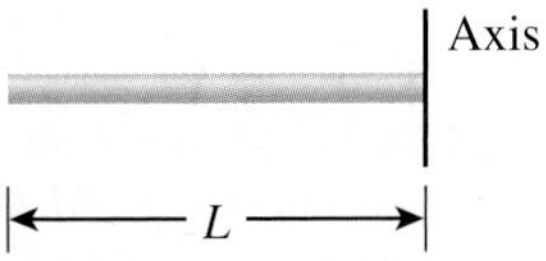 Axis, L	$\frac{1}{3}ML^2$

Notice that moment of inertia depends not only on the mass of the object, but also on its size. From $I = \sum_i m_i r_i^2$, we can see that I depends strongly on the distances, r_i, of the various particles from the rotation axis. *If the particles are distributed far from the rotation axis, the moment of inertia is large. If the particles are close to the axis, the moment of inertia is small.* Note in Table 10-1 that the position of the rotation axis has been specified in each case. The same object (such as a rod) rotating about a different rotation axis has a different moment of inertia since the distances, r_i, change for different axes. As shown in the table, the moment of inertia of a rod rotating about an axis through the end of the rod is four times larger than the moment of inertia of the same rod rotating about an axis through its centre.

Figure 10-27 Sample Problem 10-6.

SAMPLE PROBLEM 10-6

Determine the rotational kinetic energy of a disc-shaped grindstone (Figure 10-27) of mass 7.6 kg and radius 11 cm, rotating at 1.00×10^3 rpm about an axis through its centre.

Solution

Use Eqn. 10-19 for rotational kinetic energy: $E_K = \frac{1}{2} I\omega^2$. From Table 10-1 the moment of inertia of the disc is

$$I = \frac{1}{2} MR^2 = \frac{1}{2}(7.6 \text{ kg})\left(11 \text{ cm} \times \frac{1 \text{ m}}{100 \text{ cm}}\right)^2 = 0.046 \text{ kg}\cdot\text{m}^2$$

Converting the angular velocity to radians per second,

$$1.00 \times 10^3 \frac{\text{rev}}{\text{min}} \times \frac{2\pi \text{ rad}}{1 \text{ rev}} \times \frac{1 \text{ min}}{60 \text{ s}} = 105 \text{ rad/s}$$

Then, $E_K = \frac{1}{2} I\omega^2 = \frac{1}{2}(0.046 \text{ kg}\cdot\text{m}^2)(105 \text{ rad/s})^2 = 2.5 \times 10^2 \text{ kg}\cdot\text{m}^2\cdot\text{rad/s}^2$

Discarding the dimensionless "rad" in the units we have

$$E_K = 2.5 \times 10^2 \text{ kg}\cdot\text{m}^2/\text{s}^2 = 2.5 \times 10^2 \text{ J}$$

Thus, the kinetic energy of the grindstone is 2.5×10^2 J.

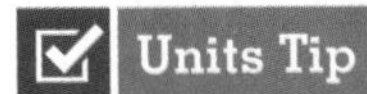

Units Tip

Notice in the solution to Sample Problem 10-6 that in order to obtain the kinetic energy in joules (equivalent to kilogram·metre²/second²), we needed first to perform unit conversions: the disc's radius to metres, and the angular velocity to radians per second.

TRY THIS!

Rolling Speeds

The answer to Sample Problem 10-7 on pg. 271 indicates that any uniform solid ball should roll down a ramp at the same speed, regardless of the ball's mass or radius. Try rolling solid balls of different masses and radii down an incline to confirm this. Following the method shown in Sample Problem 10-7, determine the speed for other shapes of rolling objects (thin ring or hoop, solid cylinder, etc.), and try rolling various objects down your incline to confirm the results. One particularly interesting pair of objects would be two cylindrical soup cans, one containing soup having a thin consistency (e.g., chicken broth) and the other having soup with a much thicker consistency (e.g., pea soup). Which one rolls faster? Why do you think this is so?

Kinetic Energy of Rolling Objects

A fan rotating about a fixed axis has only rotational kinetic energy, whereas a box sliding with no rotation has only translational kinetic energy. A *rolling* ball has both forms of kinetic energy, and its total kinetic energy is

$$E_K = \frac{1}{2} MV^2 + \frac{1}{2} I\omega^2 \quad \text{(rolling)} \qquad \textbf{(10-20)}$$

where M is the total mass of the ball, and V is the speed of its centre (strictly speaking, centre of mass). We have used upper case M and V to distinguish these quantities from the mass m and speed v of an individual particle in the ball. As discussed in Section 10.2, if the rolling ball is not slipping, we can write $V = \omega R$, where R is the ball's radius.

For an object such as a Frisbee flying through the air, its total kinetic energy is also the sum of translational kinetic energy and rotational kinetic energy, but $V \neq \omega R$ in this case because the Frisbee is not rolling.

SAMPLE PROBLEM 10-7

A uniform solid ball of mass 0.253 kg and radius 3.8 cm is released from rest at the top of a ramp of vertical elevation $y = 1.2$ m. It rolls down the ramp without slipping. Determine the ball's speed at the bottom of the ramp. Neglect energy losses due to friction. (This is a reasonable approach since rolling friction removes very little energy from the system.)

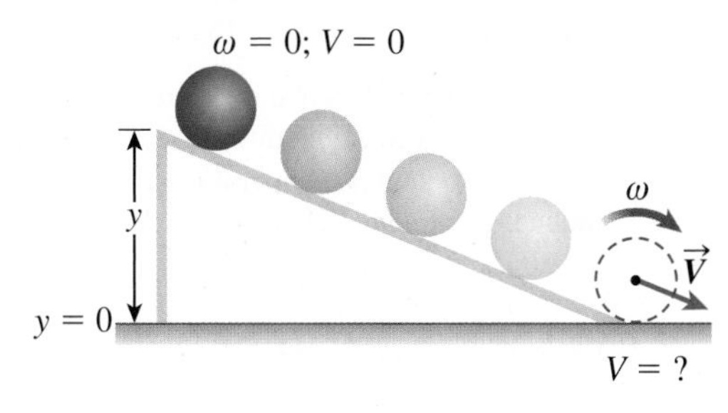

Figure 10-28 Sample Problem 10-7.

Solution

This problem involves conservation of energy. Gravitational potential energy of the ball at the top of the ramp is converted to translational kinetic energy and rotational kinetic energy at the bottom (Figure 10-28). The initial gravitational potential energy is Mgy, relative to the ball's position at the bottom of the ramp.

Therefore, $Mgy = \frac{1}{2}MV^2 + \frac{1}{2}I\omega^2$

Substituting $I = \frac{2}{5}MR^2$ for the moment of inertia of a solid ball (Table 10-1), and $\omega = V/R$ for a rolling object (since we want to find V, we eliminate ω),

$$Mgy = \frac{1}{2}MV^2 + \frac{1}{2}\left(\frac{2}{5}MR^2\right)\left(\frac{V}{R}\right)^2$$

This gives $Mgy = \frac{1}{2}MV^2 + \frac{1}{5}MV^2$

Dividing by M, and adding the two terms on the right-hand side,

$$gy = \frac{7}{10}V^2$$

Therefore,

$$V = \sqrt{\frac{10gy}{7}} = \sqrt{\frac{10(9.8\,\text{m/s}^2)(1.2\,\text{m})}{7}} = 4.1\,\text{m/s}$$

Therefore, the speed of the ball at the bottom of the ramp is 4.1 m/s. (This compares with $\sqrt{2gy} = 4.8$ m/s for an object *sliding* down an incline on which friction is negligible.)

Problem-Solving Tip

Note that the final answer to Sample Problem 10-7 does not depend on the ball's mass or radius, although numerical values were provided for these quantities. It might have been tempting to use these values to calculate the moment of inertia of the ball and work with that value in the solution, but that method would have involved more calculations than were needed. When solving problems, as a general rule it is a good idea to do algebra first (if possible), and substitute numerical values only at the end of a solution. The solution usually turns out to be simpler, and there is less chance of calculation errors.

EXERCISES

10-25 By substituting SI units for I and ω into the right-hand side of $E_K = \frac{1}{2}I\omega^2$, confirm that the resulting unit is equivalent to a joule.

10-26 By approximating each of the three wind turbine blades shown in Figure 10-29 as a thin rod with its rotation axis at one end, determine the total moment of inertia of the three blades, if each blade has a mass of 1.0×10^4 kg and a length of 42 m.

Jesus Keller/Shutterstock

Figure 10-29 Question 10-26.

10-27 A basketball player is spinning a basketball (which is a hollow sphere) on his fingertip at 5.0 rev/s (Figure 10-30). If the ball has a mass of 0.60 kg and a radius of 12 cm, determine its rotational kinetic energy.

XiXinXing/Shutterstock

Figure 10-30 Question 10-27.

10-28 A large cylindrical log falls from a log pile (Figure 10-31) and rolls along the ground. What is the log's kinetic energy if it has a mass of 127 kg, a radius of 15 cm, and a speed of 0.50 m/s?

10-29 A thin hoop of radius R starts from rest and rolls down an incline that has a vertical height y. Determine an algebraic expression for the angular velocity of the hoop at the bottom of the incline, in terms of R, y, and g. Neglect energy losses due to friction.

Figure 10-31 Question 10-28.

10.4 Torque and Newton's Second Law for Rotation

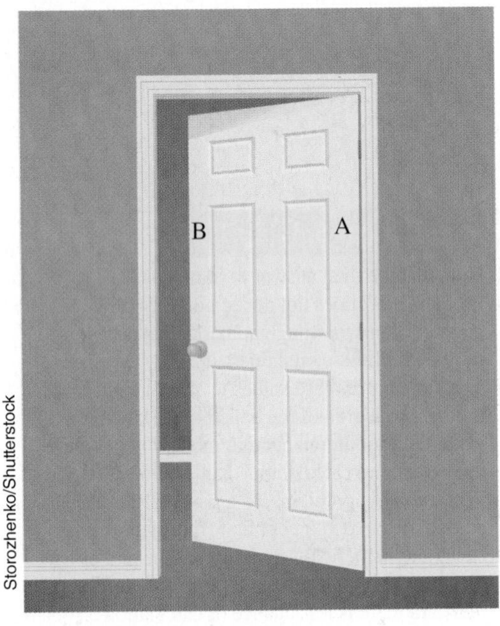

Figure 10-32 In order to open the door easily, would you push at A or at B?

Suppose that you want to push open a door, such as the partly open door in Figure 10-32. Where would you choose to push in order to open the door easily: at A, which is close to the hinges (where the rotation axis is), or at B, which is far from the hinges and rotation axis? From your personal experience, you know that it is easier to rotate the door open if you push at a point far from the rotation axis, that is, at point B. A smaller force is required to rotate the door if the force is exerted farther from the rotation axis. This leads us to the concept of **torque**, which is the rotational effect of a force. The word torque stems from the Latin word *torquere*, which means twist.

Based on your experience opening doors, using wrenches, etc., you know that the rotational effect provided by a force—that is, the torque that the force provides—depends not only on how large the force is, but on where it is applied. As an example, Figure 10-33 shows an overhead drawing of the door in Figure 10-32. A force $\vec{F}$ is being applied perpendicular to the door. The torque due to this force is given by

$$\tau = \pm Fd \qquad (10\text{-}21)$$

where the lowercase Greek letter tau (τ) represents torque, F is the magnitude of the force, and d is the perpendicular distance from the rotation axis to the line of action of the force. The perpendicular distance d is usually called the **moment arm** of the force. The $\pm$ sign will be discussed in a later paragraph. Eqn. 10-21 gives the general mathematical definition of torque. Since torque is the product of a force magnitude and a distance, its SI unit is newton·metre (N·m). It is important not to confuse this unit with the unit for work, the joule, in which the force and displacement were parallel to each other. For torque, the force and moment arm are perpendicular.

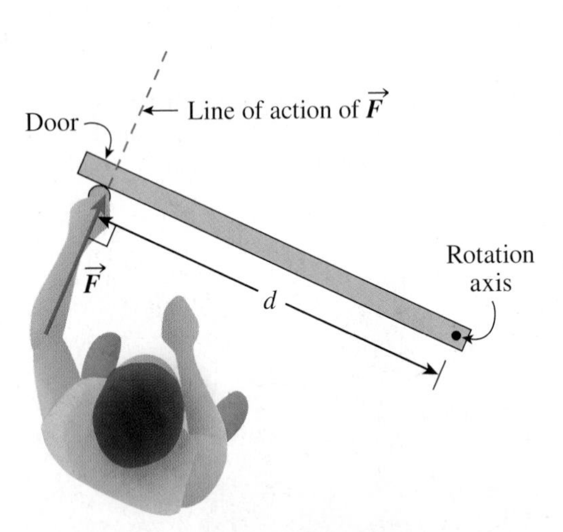

Figure 10-33 The torque τ due to $\vec{F}$ is $\tau = \pm Fd$, where d (the moment arm) is the perpendicular distance from the rotation axis to the line of action of $\vec{F}$.

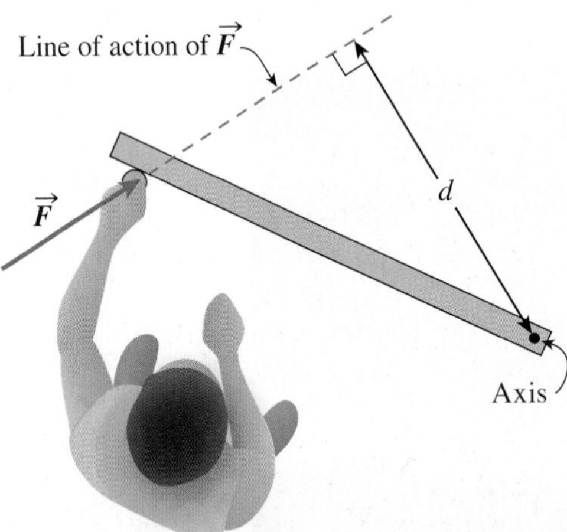

Figure 10-34 The force $\vec{F}$ is not perpendicular to the door. The force has the same magnitude as in Figure 10-33 and is applied at the same point on the door, but the torque is smaller because d is smaller.

In Figure 10-33 the force was perpendicular to the door. However, it could have been applied in any direction relative to the door. Figure 10-34 shows the force being applied to the door in a non-perpendicular direction. As a result, the line of action of the force had to be extended in order to draw a moment arm (starting at the rotation axis) perpendicular to the force's direction. This moment arm d is shorter than the one in Figure 10-33, even though the magnitude of the force and the point on the door where it is applied have not changed. Since d is shorter, the torque is smaller. The direction of $\vec{F}$ as well as the point where $\vec{F}$ is applied are both important in determining the length of the moment arm and the magnitude of the torque.

torque rotational effect of a force; torque equals $\pm$ the product of the magnitude of the force and the moment arm

moment arm perpendicular distance from an axis of rotation to the line of action of a force

Before working through a sample problem, we need to discuss the $\pm$ sign in Eqn. 10-21. Just as in earlier chapters on kinematics and dynamics where we were free to choose $+x$ and $+y$ directions to make problem-solving more straightforward, so also in rotational mechanics we can choose positive and negative rotational directions that are convenient. For example, we could choose counterclockwise rotations to be positive, and clockwise rotations to be negative, or vice versa. Since torques produce rotations, the signs of torques will be related to the directions of resulting rotations. In sample problems presented in the remainder of this section, suggestions and guidelines will be provided to aid you in the choice of positive and negative rotational directions.

SAMPLE PROBLEM 10-8

Calculate the torque (in newton·metres) exerted on the wrench by the 85 N force in each situation shown in Figure 10-35. The rotation axis is at the centre of the nut at the left-hand end of the wrench. Assume that 50° has two significant digits.

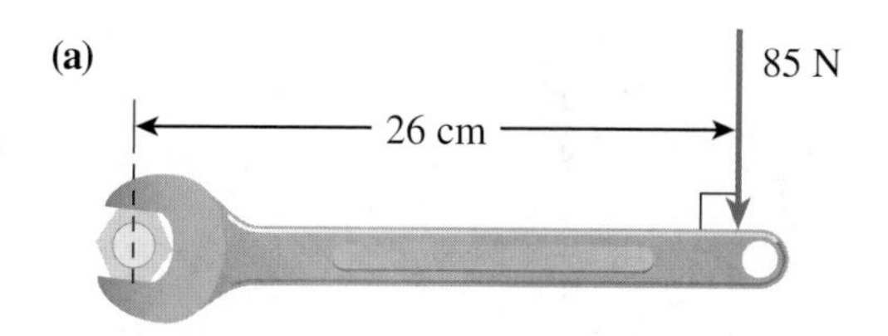

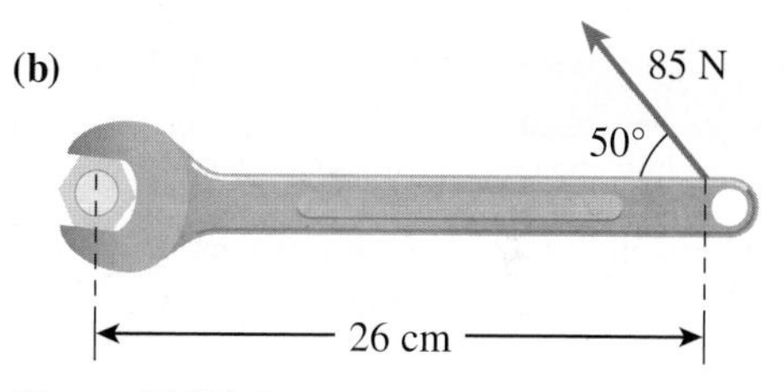

Figure 10-35 Sample Problem 10-8.

Solution

(a) In most cases in which there is only one torque acting, the positive rotational direction is chosen to be the direction in which the torque would rotate the object. In Figure 10-35 (a) the torque would produce a clockwise rotation, and this is chosen as the positive direction. The moment arm has a length of 26 cm since this distance is perpendicular to the line of action of the force, and the torque is

$$\tau = +Fd = +(85\text{ N})\left(26\text{ cm} \times \frac{1\text{ m}}{100\text{ cm}}\right) = +22\text{ N}\cdot\text{m}$$

Thus, the torque exerted on the wrench by the 85 N force is 22 N·m in the clockwise sense.

(b) Again, there is only one torque, and since it would produce a counterclockwise rotation, counterclockwise is chosen as the positive direction. In order to determine the moment arm, the line of action of the force has to be extended (Figure 10-36). The moment arm, of length d, is perpendicular to the force's line of action and forms one side of a right-angled triangle that has a hypotenuse of length 26 cm. To find d,

$$\sin 50° = \frac{d}{26\text{ cm}} \quad \therefore d = (26\text{ cm})\sin 50° = 2.0 \times 10^{1}\text{ cm} \times \frac{1\text{ m}}{100\text{ cm}} = 0.20\text{ m}$$

The torque is $\tau = +Fd = +(85\text{ N})(0.20\text{ m}) = +17\text{ N}\cdot\text{m}$

Hence, the torque exerted on the wrench by the 85 N force is 17 N·m in the counterclockwise sense in this case.

Note that the torque exerted by the same magnitude of force at the same point on the wrench is smaller than in part (a) because of the different direction (and thus different moment arm) of the force.

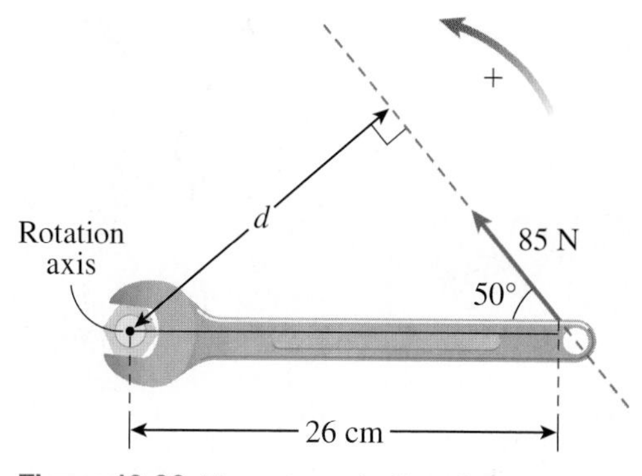

Figure 10-36 Moment arm for Sample Problem 10-8(b).

Math Tip

If there were any forces acting in part (a) of Sample Problem 10-8 to create a counterclockwise rotation, we would have used $\tau = -Fd$ for the torque exerted by it.

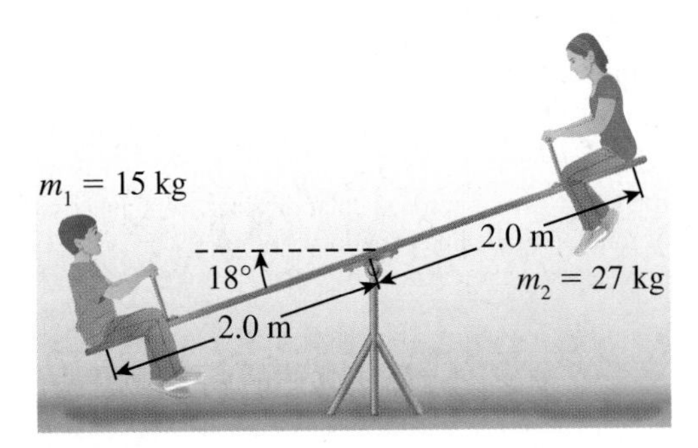

Figure 10-37 Sample Problem 10-9.

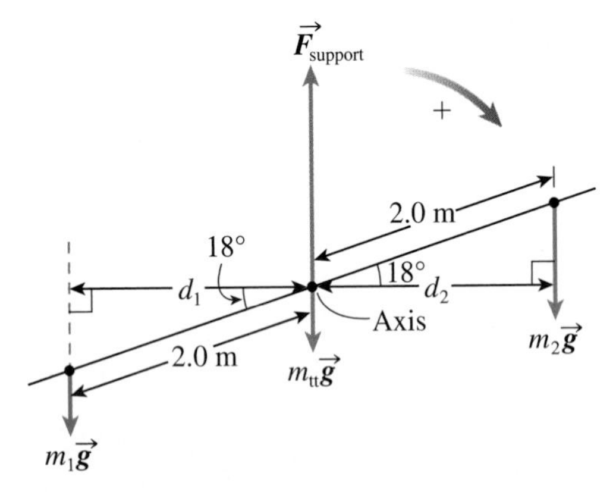

Figure 10-38 Forces and moment arms for Sample Problem 10-9.

SAMPLE PROBLEM 10-9

Two children are sitting on a teeter-totter consisting of a uniform wooden board resting on a support (Figure 10-37). The mass of the board (and handles for people to hold) is 16 kg. The children's masses are $m_1 = 15$ kg and $m_2 = 27$ kg. Each child is sitting 2.0 m away from the pivot. At the instant shown, neither child is touching the ground and the teeter-totter makes an angle of 18° relative to the horizontal. Considering the teeter-totter and children as a single system that can rotate, what is the resultant (total) torque acting on this system?

Solution

Figure 10-38 shows a free-body diagram for the system. Since moment arms will be important in the torque calculations, the locations of the forces are shown. Four external forces act on the system: the forces of gravity on the children and teeter-totter ($m_1\vec{g}$, $m_2\vec{g}$, and $m_{tt}\vec{g}$) and the upward force provided by the support ($\vec{F}_{support}$). Since $m_{tt}\vec{g}$ and $\vec{F}_{support}$ act directly at the rotation axis, their moment arms are zero and the torques due to these forces are zero.

Determining the moment arm d_1 for $m_1\vec{g}$,

$$\cos 18° = \frac{d_1}{2.0 \text{ m}} \quad \therefore d_1 = (2.0 \text{ m}) \cos 18° = 1.9 \text{ m}$$

The calculation for the moment arm d_2 for $m_2\vec{g}$ is the same as for d_1, giving $d_2 = 1.9$ m.

Since $m_2 > m_1$ and $d_1 = d_2$, then the clockwise torque due to $m_2\vec{g}$ will be greater than the counterclockwise torque due to $m_1\vec{g}$, and the positive rotational direction is therefore chosen to be clockwise. The resultant torque τ_R is

$$\tau_R = -m_1gd_1 + m_2gd_2 + m_{tt}g(0) + F_{support}(0)$$

Writing $d_1 = d_2 = d$, and factoring out g and d,

$$\tau_R = (-m_1 + m_2)gd = ((-15 + 27) \text{ kg})(9.8 \text{ m/s}^2)(1.9 \text{ m}) = 2.2 \times 10^2 \text{ N·m}$$

Thus, the resultant torque on the system at the instant shown is 2.2×10^2 N·m in the clockwise sense.

Remember to choose your positive rotational directional before beginning your calculation and be consistent with signs throughout. We could have chosen counter-clockwise as the positive direction in solving this problem. Then our expression for the resultant torque have been $\tau_R = +m_1gd_1 - m_2gd_2$, and the final answer would have been $\tau_R = -2.2 \times 10^2$ N·m, or 2.2×10^2 N·m in the clockwise sense as before.

Units Tip

Notice how the units have worked out:

$(\text{kg})(\text{m/s}^2)(\text{m}) = (\text{kg·m/s}^2)(\text{m}) = \text{N·m}$, which is the standard SI unit for torque

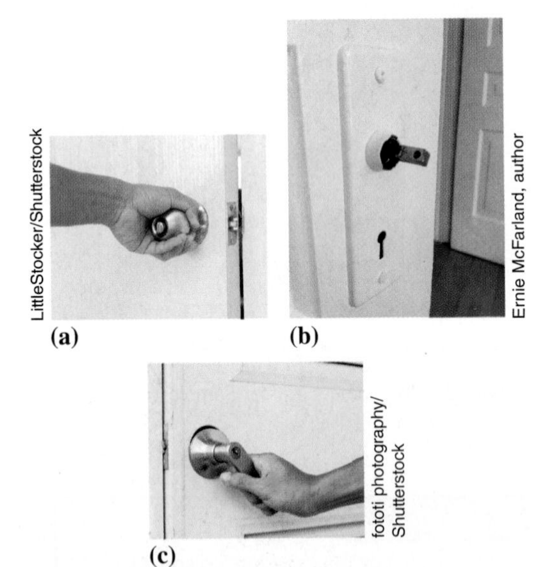

Figure 10-39 Providing enough torque to open a door latch is easy with **(a)** a door knob or **(c)** a door handle, but not with **(b)** the spindle inside a door knob.

LittleStocker/Shutterstock; Ernie McFarland, author; fototi photography/Shutterstock

TRY THIS!

What's Inside a Door Knob?

One type of object to which you frequently apply a torque is a door knob (Figure 10-39 (a)). Its size provides a moment arm of a few centimetres and most people can easily provide enough torque to open the latch. Try removing the door knob (a screwdriver or other tool might be needed) to reveal the spindle (Figure 10-39 (b)), which has a width typically less than a centimetre. Attempt to turn the spindle to open the latch. Because the spindle provides you with a much smaller moment arm, you will find it difficult (perhaps impossible) to provide the required torque. A moment arm considerably larger than that of a door knob is provided by a door handle (Figure 10-39 (c)), which gives a large torque for a small applied force. A handle is much easier to use for everyone, but especially for people with disabilities affecting their hands or wrists. The increasing use of handles is another example of universal design, in which devices are developed in the most user-friendly way, taking into account the needs of many types of people.

Newton's Second Law for Rotation

Many times in this chapter we have seen the similarities between rotational motion and translational motion: angular velocity corresponds to velocity, rotational kinetic energy to translational kinetic energy, etc. Not surprisingly, there is a rotational analogue to Newton's second law of motion. Recall that this law states that the resultant force on an object equals the product of the object's mass and acceleration ($\vec{F}_R = \Sigma\vec{F} = m\vec{a}$). You probably remember that the rotational analogue of mass is moment of inertia, and angular acceleration corresponds to acceleration, but what corresponds to force? The answer is torque, the rotational effect of a force. In order to give something an acceleration, a force is applied. To produce an *angular* acceleration, a torque is required. **Newton's second law for rotation** is

$$\tau_R = \Sigma\tau = I\alpha \qquad (10\text{-}22)$$

Newton's second law for rotation the resultant torque acting on an object equals the product of the object's moment of inertia and the object's angular acceleration

where τ_R, or $\Sigma\tau$, is the resultant torque, or the sum of the torques, on an object, I is the object's moment of inertia, and α is the resulting angular acceleration. In order for Eqn. 10-22 to be valid, α must have *radians* in its units; normally α is in radians per second squared. The relation $\tau_R = \Sigma\tau = I\alpha$ holds for rotation of a rigid object about a *fixed axis*. It is also valid in some special circumstances discussed in more advanced textbooks.

We can see from $\tau_R = \Sigma\tau = I\alpha$ how important moment of inertia is in determining the rotation of an object. For a given torque, an object with a large moment of inertia has a small angular acceleration.

SAMPLE PROBLEM 10-10

A pulley, constrained to rotate about a fixed horizontal axle, consists of a uniform disk of radius 7.5×10^{-2} m and mass 0.48 kg. A string is wrapped around the pulley and is used to provide a constant downward tension of magnitude 1.2 N on the disk. What is the resulting angular acceleration of the pulley? Neglect friction.

Solution

The forces acting on the pulley are shown in Figure 10-40. The downward force of gravity and the upward support force provided by the axle have lines of action that go through the rotation axis, and therefore the torque due to each of these forces is zero. The only non-zero torque is due to the string tension $\vec{T}$, and the moment arm for this force is just the pulley's radius R. Since the torque due to $\vec{T}$ causes a counterclockwise rotation of the pulley, we choose counterclockwise as the positive direction, and the torque due to $\vec{T}$ is TR. Newton's second law for rotation ($\Sigma\tau = I\alpha$) gives

$$+TR + (Mg)(0) + (F_{support})(0) = I\alpha$$

Substituting $I = \frac{1}{2}MR^2$ (from Table 10-1) for the disk,

$$TR = \frac{1}{2}MR^2\alpha$$

Dividing both sides by R, and rearranging to find α,

$$\alpha = \frac{2T}{MR} = \frac{2(1.2\text{ N})}{(0.48\text{ kg})(7.5 \times 10^{-2}\text{ m})} = 67\text{ rad/s}^2$$

Thus, the angular acceleration of the disk is 67 rad/s^2.

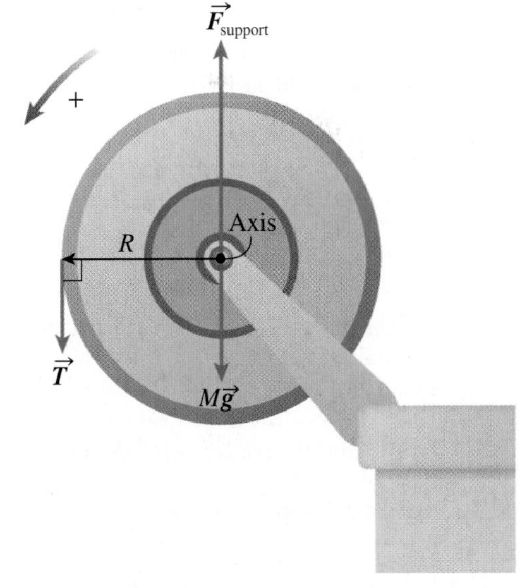

Figure 10-40 Sample Problem 10-10.

Units Tip

Notice the units in the calculation of α. The "2" is unitless, and so we have

$$\frac{\text{N}}{\text{kg}\cdot\text{m}},\text{ but N} = \text{kg}\frac{\text{m}}{\text{s}^2},\text{ which gives }\frac{\text{kg}\cdot\text{m}}{\text{s}^2(\text{kg}\cdot\text{m})} = \frac{1}{\text{s}^2}$$

Since α is *angular* acceleration, we insert radians to get rad/s^2.

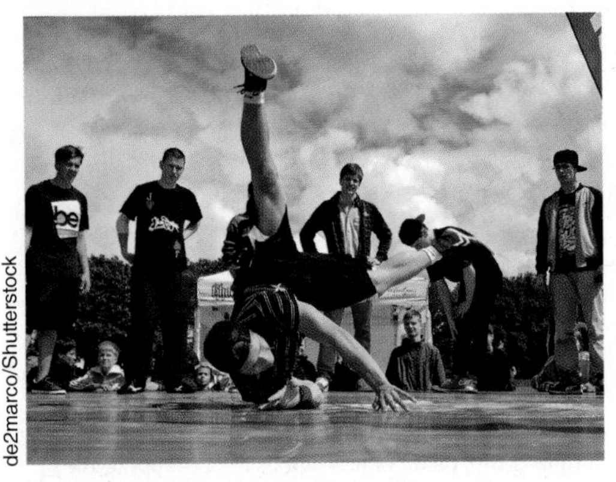

Figure 10-41 Sample Problem 10-11.

SAMPLE PROBLEM 10-11

The breakdancer shown in Figure 10-41 is rotating about a vertical axis. Suppose that he is slowing down because of a frictional torque exerted by the surface underneath him, and that this torque produces a change in his angular velocity from 5.0 rad/s to 4.0 rad/s in a time of 0.50 s. What is the magnitude of the frictional torque if the breakdancer's moment of inertia about the vertical axis is 4.2 kg·m²?

Solution

Since we are asked about only the frictional torque, Newton's second law for rotation becomes

$$\Sigma\tau = \tau_{\text{friction}} = I\alpha$$

To determine τ_{friction}, we need first to calculate the angular acceleration α using the information about the change in angular velocity. Use Eqn. 10-7:

$$\omega = \omega_0 + \alpha t, \text{ which gives } \alpha = \frac{\omega - \omega_0}{t} = \frac{(4.0 - 5.0)\text{ rad/s}}{0.50\text{ s}} = -2.0\text{ rad/s}^2$$

Now, $\tau_{\text{friction}} = I\alpha = (4.2\text{ kg}\cdot\text{m}^2)(-2.0\text{ rad/s}^2) = -8.4\text{ N}\cdot\text{m}$.

Notice that because α is negative, indicating that the angular velocity is decreasing, the resultant torque must also be negative.

Since magnitudes of quantities are positive, the magnitude of the frictional torque is 8.4 N·m.

EXERCISES

10-30 A farmer pushes on a farm gate with a force of 1.4×10^2 N perpendicular to the gate at a distance of 1.6 m from the rotation axis that passes through the hinges.

(a) Calculate the magnitude of the torque that the farmer exerts on the gate.

(b) Determine the magnitude of force required to cause the same torque if the distance is reduced to 0.80 m.

10-31 Each of the five drawings in Figure 10-42 shows a force vector of magnitude 48 N being applied to a rod of length 1.5 m that is constrained to rotate about a fixed axis (the black dot). In each case, determine the magnitude of the torque applied by the force, and indicate whether the rotation (if any) caused by the torque will be clockwise or counterclockwise.

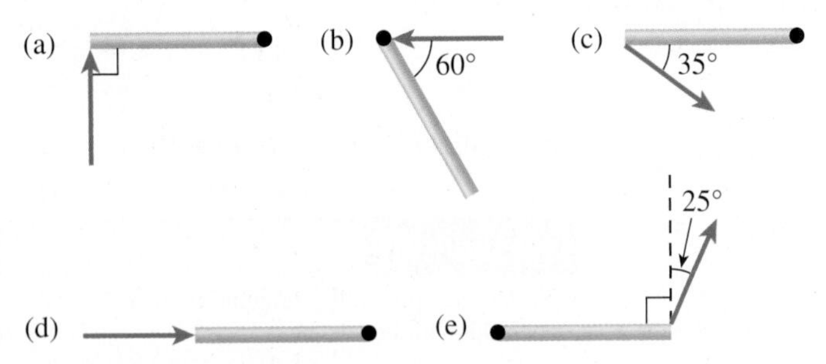

Figure 10-42 Question 10-31.

10-32 There is a relationship between the size of a vehicle and the size of its steering wheel. (Have you noticed the size of the steering wheel on a large bus?) Explain why.

10-33 In ballet, a spin called a pirouette can be performed with the male partner exerting two forces on the ballerina, as shown in Figure 10-43. One force is a push on the waist, and the other is a pull on the hand. Determine the sum of the two resulting torques on the ballerina in this case.

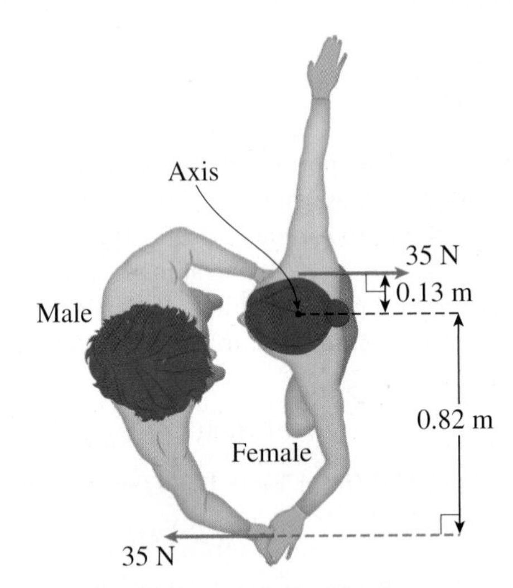

Figure 10-43 Question 10-33.

10-34 An electric fan motor exerts a torque of 0.83 N·m on the fan blades, which have a moment of inertia of 8.9×10^{-3} kg·m². What is the angular acceleration of the blades?

10-35 A wheel rotating about a fixed axis is subject to a torque of 2.50 N·m and an angular acceleration of 3.45 rad/s^2 results. Determine the wheel's moment of inertia.

10-36 The forearm of a certain person has a moment of inertia of 0.080 kg·m^2 about the elbow joint. The forearm is held vertically (Figure 10-44) and the biceps muscle exerts a horizontal force on the forearm at a perpendicular distance of 2.1 cm from the elbow's rotation axis. If the forearm has a resulting angular acceleration of 1.2 rad/s^2, what is the magnitude of the biceps muscle's force?

10-37 A pulley of radius 12.3 cm consists of a thin rim of mass 757 g and spokes that connect the thin rim to the pulley's centre. The spokes have negligible mass and hence negligible moment of inertia. The pulley is constrained to rotate about a fixed horizontal axle. A string is wrapped around the rim, and is used to exert a horizontal tension of magnitude 0.360 N on the pulley. What is the resulting angular acceleration? Neglect friction.

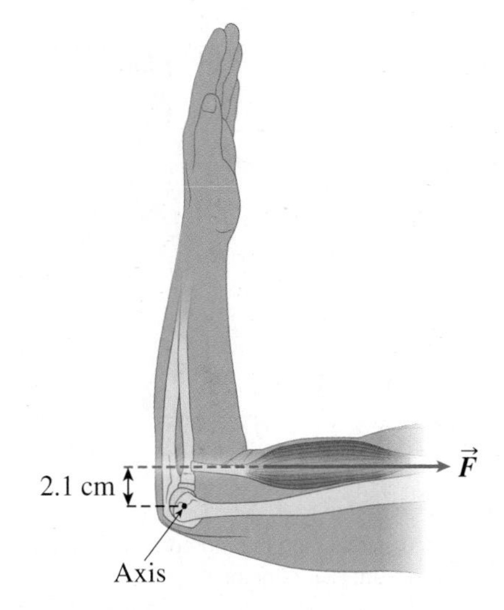

Figure 10-44 Question 10-36.

10.5 Angular Momentum

We now return to the question posed at the beginning of this chapter about the rotating figure skater whose angular velocity increases when she pulls her arms in. Why does this happen? By pulling in her arms, the skater is decreasing her moment of inertia, since some of her mass is being moved closer to the rotation axis. But what does this change in moment of inertia have to do with angular velocity? In order to answer this question, we have to understand the concept of angular momentum.

Recall that linear momentum (or just momentum, as we called it earlier) is the product of mass and velocity (Chapter 8). **Angular momentum** is the product of the rotational analogues of mass and velocity, that is, moment of inertia and angular velocity. Using the symbol L for angular momentum, we have

$$L = I\omega \tag{10-23}$$

This expression for angular momentum is valid for an object (such as a figure skater) rotating about a fixed axis (Figure 10-45). In SI units, I is in kilogram·metre^2 and ω is in radians per second. Therefore, you might expect that the unit for L would be kilogram·metre^2·radian per second, but the radian is dropped and the standard unit is kilogram·metre^2 per second.

Now we can begin to see that there might be some connection between moment of inertia and angular velocity, since their product is angular momentum. The remaining piece in the figure-skater puzzle is the **law of conservation of angular momentum**, which states that if the net torque acting on an object (or system of objects) is zero, then the angular momentum of the object (or system) is conserved. This is similar to the law of conservation of (linear) momentum: if the net force on a system is zero, the (linear) momentum is conserved.

We can now explain why skaters spin faster when they pull in their arms. Figure 10-46 (a) shows the forces acting on a spinning skater (neglecting friction, which is small). There are only the force of gravity and the normal force, both of which act directly along the rotation axis. Hence, the torque due to each of these forces is zero, and the resultant torque is zero. Therefore, the angular momentum of the skater is conserved; that is, it is constant. When he pulls his arms

angular momentum the product of moment of inertia and angular velocity (for an object rotating about a fixed axis)

law of conservation of angular momentum if the net torque acting on an object (or system of objects) is zero, then the angular momentum of the object (or system) is conserved

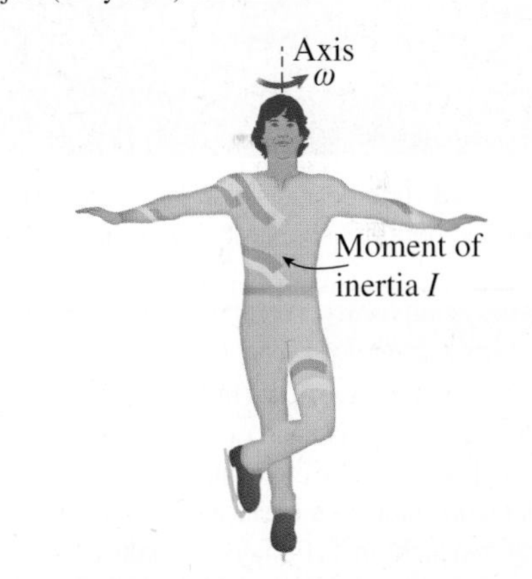

Figure 10-45 The angular momentum L of an object rotating about a fixed axis is given by $L = I\omega$.

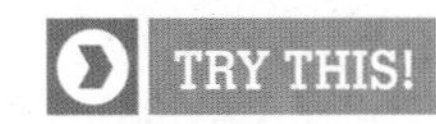

Spinning Physics

You can have some fun conserving angular momentum yourself if you have access to a chair or stool that can rotate. Sit on the chair with your feet off the floor and your arms stretched out to your sides. Ask a friend to push one of your arms to provide a torque so that you rotate slowly. Then fold your arms in on your chest. The effect is accentuated if you have a heavy book in each hand.

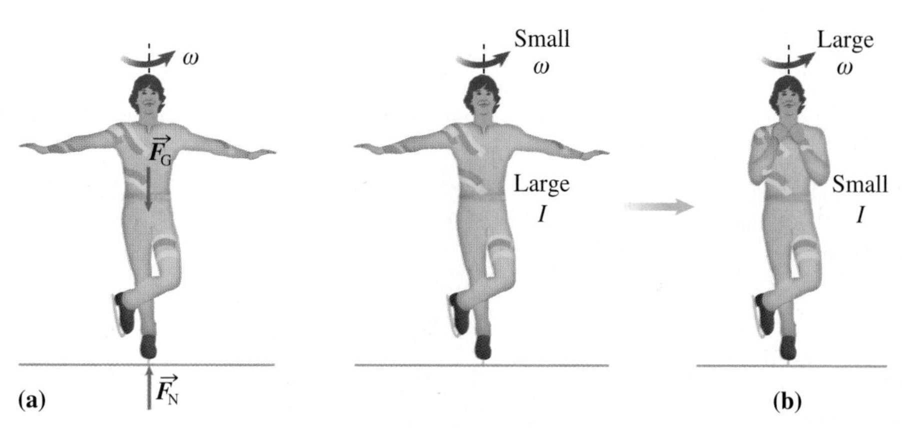

Figure 10-46 **(a)** The torque due to $\vec{F}_G$ and $\vec{F}_N$ is zero; therefore, the skater's angular momentum is conserved. **(b)** Reducing the moment of inertia (I) causes the angular velocity (ω) to increase. Angular momentum $(I\omega)$ remains constant.

Figure 10-47 The moment of inertia when doing a camel spin is about six times larger than that with arms hugging the body.

in (Figure 10-46 (b)), thus decreasing his moment of inertia, his angular velocity increases in order that his angular momentum remains the same. If he pushes his arms out, his moment of inertia increases, and his angular velocity decreases. (Of course, if the skater goes out of the spin and uses his skates to propel him into another manoeuvre or to stop, then his angular momentum is not conserved, since a torque has been exerted on him by the ice via his skates.)

For a typical skater, the moment of inertia with arms outstretched is about two times larger than that with arms hugging the body. Thus, by bringing the arms in, a skater can increase the angular velocity by roughly a factor of two. The moment of inertia in a camel spin (Figure 10-47) is about six times that with arms hugging the body.

Athletes other than skaters make use of the law of conservation of angular momentum. For example, divers (Figure 10-9 on pg. 261) often go into a tuck position to increase their angular velocity by decreasing the moment of inertia. They usually end their dives by extending their bodies, thus increasing the moment of inertia and decreasing their angular velocity. Gymnasts and trampolinists (Figure 10-48) also vary their moments of inertia during their routines.

Figure 10-48 Trampolinists alter their moments of inertia during performances, with dramatic results.

DID YOU KNOW?

When a cat falls upside down from a height, it almost always can land on its feet by using conservation of angular momentum. Its initial angular momentum is zero since it is not rotating. The cat first tucks in its front legs to reduce the moment of inertia of the front half of its body and rotates its front half in one direction by roughly 180°. Simultaneously, it extends its back legs to increase the moment of inertia of the back half of its body, and rotates the back half in the opposite direction, but only by as little as 10°. Since the rotations are in opposite directions, the total angular momentum is still zero. The cat then extends its front legs, rotates the front of its body a few degrees in the opposite direction that it originally rotated, while tucking in its rear legs and rotating them by roughly 180° in the original rotation direction of the front half. The result is that the cat has rotated both halves of its body by about 180° and lands on its feet. If you go online, you can probably find a slow-motion video that shows this.

SAMPLE PROBLEM 10-12

A skater, spinning with arms outstretched, has an angular velocity of 1.5 rad/s. Her moment of inertia is 1.8 kg·m². She then pulls in her arms so that her moment of inertia becomes 0.85 kg·m². Determine her angular momentum and final angular velocity.

Solution

Since we know the skater's initial moment of inertia (I_1) and initial angular velocity (ω_1), we can calculate her angular momentum (L) using Eqn. 10-23:

$$L = I_1\omega_1 = (1.8\ \text{kg}\cdot\text{m}^2)(1.5\ \text{rad/s}) = 2.7\ \text{kg}\cdot\text{m}^2\text{/s}$$

As discussed earlier in this section, the angular momentum of a skater is conserved while spinning. Therefore, $L = I_1\omega_1 = I_2\omega_2$. We need to find ω_2. Solving for ω_2 from $L = I_2\omega_2$,

$$\omega_2 = \frac{L}{I_2} = \frac{2.7\ \text{kg}\cdot\text{m}^2\text{/s}}{0.85\ \text{kg}\cdot\text{m}^2} = 3.2\ \text{rad/s}$$

Thus, the skater's angular momentum is 2.7 kg·m²/s, and her final angular velocity is 3.2 rad/s.

Angular Momentum and Kinetic Energy

We have described a number of situations in which the angular momentum of an object is conserved as its moment of inertia is changed. Is the object's kinetic energy conserved as well? The following calculations show that the answer is "no." Suppose that a figure skater with arms extended has a moment of inertia of $I_1 = 2\ \text{kg}\cdot\text{m}^2$ and an angular velocity of $\omega_1 = 1$ rad/s. Her kinetic energy is $E_{K1} = \frac{1}{2} I_1\omega_1^2 = 1$ J. She then pulls her arms in, reducing her moment of inertia to $I_2 = 1.0\ \text{kg}\cdot\text{m}^2$ and increasing her angular velocity to $\omega_2 = 2.0$ rad/s. Her kinetic energy is now $E_{K2} = \frac{1}{2} I_2\omega_2^2 = 2$ J. Clearly, her kinetic energy has increased. Where did the extra energy come from? While pulling her arms in, the skater used muscles that provided the energy. If she reverses the arm motion by extending her arms, her kinetic energy decreases from 2 J to 1 J, and the "lost" energy goes into heating her muscles.

EXERCISES

10-38 What is the angular momentum of a pulley with a moment of inertia $1.1 \times 10^{-2}\ \text{kg}\cdot\text{m}^2$ rotating at 12 rad/s?

10-39 A basketball player is spinning a basketball (a hollow sphere) on a fingertip. What is the angular momentum of the ball (mass 0.595 kg, radius 12.0 cm) if it is rotating at 283 rpm?

10-40 A platform diver decreases her moment of inertia by a factor of 1.8 in going from a layout (extended) position to a pike position (Figure 10-49).

(a) Does her angular velocity increase or decrease?

(b) By what factor?

Diego Barbieri/Shutterstock.com

Figure 10-49 Question 10-40.

10-41 A figure skater increases her moment of inertia from $0.90\ \text{kg}\cdot\text{m}^2$ to $1.90\ \text{kg}\cdot\text{m}^2$ by extending her arms. Her initial angular velocity is 9.5 rad/s. Determine her

(a) initial kinetic energy

(b) final angular velocity

(c) final kinetic energy

LOOKING BACK...LOOKING AHEAD

Everything discussed in this book so far has been in the area of mechanics: the study of motion and forces. We have presented concepts such as velocity, acceleration, forces, momentum, energy, and gravitation.

In this chapter, covering rotational motion, a number of new quantities have been introduced to describe rotation quantitatively: angular velocity, angular acceleration, moment of inertia, rotational kinetic energy, angular momentum, and so on. We have seen how rotational variables such as angular velocity are analogous to linear variables such as velocity, and have discussed rolling, which is a combination of rotational and linear motions. The accelerations, both centripetal and tangential, of points on rotating objects such as wheels have been dealt with in detail. We have employed Newton's second law for rotation in analyzing physical situations involving torque and angular acceleration. The important law of conservation of angular momentum has been presented and used to explain phenomena such as the increase in angular velocity of a figure skater pulling in his or her arms.

The next chapter will build on the concepts of forces and torques in a discussion of statics, stability, and elasticity. In particular, analysis of situations involving static equilibrium will be presented in detail.

CONCEPTS AND SKILLS

Having completed this chapter, you should now be able to do the following:

- Use the concepts of angular position, angular displacement, angular velocity (average and instantaneous), and angular acceleration (average and instantaneous) in analyzing physical situations involving rotation.
- State typical units for the above quantities and be able to perform unit conversions (e.g., from revolutions per minute to radians per second).
- Use the equations of constant-angular-acceleration kinematics in solving related problems.
- Analyze physical situations involving rotational and linear quantities for objects rotating about fixed axes and for rolling objects, and know the restrictions on units that can be used.

- Describe tangential acceleration, and explain how it is different from centripetal acceleration.
- Calculate the tangential and centripetal acceleration components in various circumstances, as well as the magnitude of the total acceleration.
- Solve problems related to moment of inertia and rotational kinetic energy, including problems involving rolling objects.
- Discuss how moment of inertia depends on the distribution of mass of an object.
- State in words the meaning of torque and how torque is calculated.
- State Newton's second law for rotation, and use it in analyzing problems.
- Calculate angular momentum in simple situations.
- State the law of conservation of angular momentum, and use it to explain changes in objects' angular velocities due to changes in their moments of inertia.
- Use the law of conservation of angular momentum in solving relevant problems.

KEY TERMS

You should be able to define or explain each of the following words or phrases:

rotation axis
angular position
radian (unit)
angular displacement
average angular velocity
instantaneous angular velocity
angular velocity
average angular acceleration
instantaneous angular acceleration
angular acceleration
tangential acceleration
moment of inertia
rotational inertia
rotational kinetic energy
translational kinetic energy
torque
moment arm
Newton's second law for rotation
angular momentum
law of conservation of angular momentum

Chapter Review

MULTIPLE-CHOICE QUESTIONS

10-42 The angular velocity of a wheel mounted on a fixed axle is decreasing. Which of the diagrams in Figure 10-50 best shows the direction for the acceleration of point P?

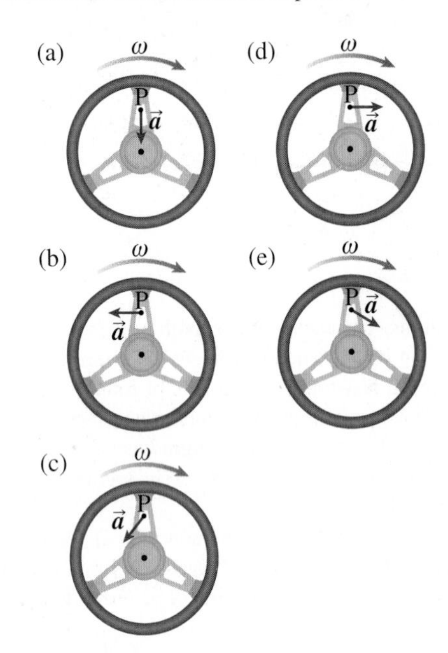

Figure 10-50 Question 10-42.

10-43 A point P on an airplane propeller has a centripetal acceleration of 3.0 m/s^2 and a tangential acceleration of 2.0 m/s^2. In a time of 1.0 s, by how much does the speed of P change?

(a) 1.0 m/s
(b) 2.0 m/s
(c) 3.0 m/s
(d) 4.0 m/s
(e) 5.0 m/s

10-44 Figure 10-51 shows four objects of equal mass. The first object is a solid sphere, the second a thin ring, the third a solid box of square cross-section, and the fourth a hollow thin-shelled box. Each object has an axis perpendicular to the page and through the centre of the object. Which of the following is a correct list of the objects in order of their moments of inertia, from largest to smallest? No calculations are required (just consider how the mass of each object is distributed about the rotation axis).

(a) 2, 4, 1, 3
(b) 1, 2, 3, 4
(c) 4, 3, 2, 1
(d) 4, 3, 1, 2
(e) 4, 2, 3, 1

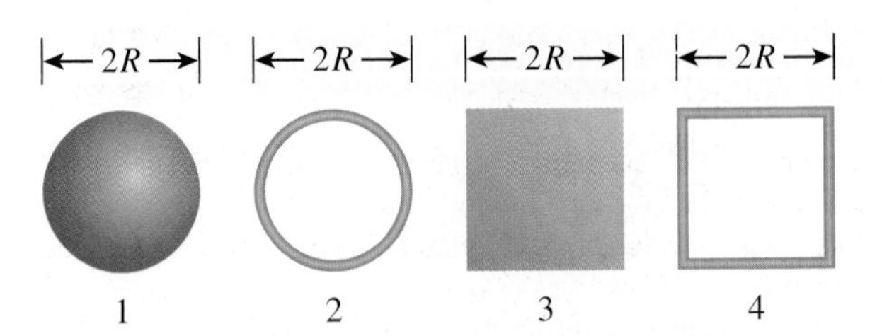

Figure 10-51 Question 10-44.

10-45 It is expected that the Sun will eventually shrink to a much smaller size, becoming what is known as a white dwarf star, about the size of Earth. Its mass will not change appreciably, but its density will increase considerably. As the Sun shrinks, its moment of inertia

(a) increases (b) decreases (c) remains the same

10-46 Refer to the previous question. As the Sun shrinks, its angular momentum

(a) increases (b) decreases (c) remains the same

10-47 Refer to the previous two questions. As the Sun shrinks, its angular velocity

(a) increases (b) decreases (c) remains the same

Review Questions and Problems

10-48 (a) In a circle of radius 10.5 cm, what angle in radians subtends an arc of length 15.7 cm?

(b) Express this angle in degrees and revolutions.

10-49 An amusement park ride (Figure 10-52) rotates through 45 rev in 93 s. What is its average angular velocity in

(a) revolutions per minute?

(b) radians per second?

Figure 10-52 Question 10-49.

dwphotos/Shutterstock

10-50 For the ride in the previous question what is the average period of rotation; that is, what is the average time for one revolution?

10-51 When an electric fan is turned on from rest, its angular acceleration is 157 rad/s^2 during a time interval of 0.85 s. What is its angular velocity at the end of this time interval in

(a) radians per second?

(b) revolutions per minute?

10-52 A car engine speeds up from 1.50×10^3 rpm to 2.33×10^3 rpm in a time of 5.65 s.

(a) What is its angular acceleration (assumed constant) in revolutions per minute squared?

(b) How many revolutions does it make in this time?

(c) If it continues to speed up with the same angular acceleration for another 3.50 s, what will be its final angular velocity?

10-53 A food processor (Figure 10-53) rotating at 377 rad/s is switched off and rotates through 473 rad while coming to rest.

(a) How long does it take to come to rest?

(b) What is its angular acceleration (assumed constant)?

Alan Poulson Photography/Shutterstock

Figure 10-53 Question 10-53.

10-54 By pulling her arms in, a figure skater increases her angular velocity while rotating through 1.7 rev in a time of 1.1 s. She has a constant angular acceleration of 8.1 rad/s^2.

(a) What are her initial and final angular velocities in radians per second?

(b) By what factor has her moment of inertia changed in the process of pulling her arms in?

10-55 For points on a rotating electric fan blade, sketch the shape of a graph of speed versus distance from the rotation axis (for a fixed angular velocity).

10-56 For a single point on a rotating electric fan blade, sketch the shape of a graph of speed versus the blade's angular velocity.

10-57 A tornado (Figure 10-54) is not a solid object rotating with a constant angular velocity at all points. Nevertheless, at any specific elevation above the ground and distance from the rotation axis, a given tornado will have a certain angular velocity. At a certain height above the ground, a particular tornado has a speed of 1.5×10^2 km/h at distance of 65 m from its central axis. What is the angular velocity (in revolutions per minute) at this location?

Minerva Studio/Shutterstock

Figure 10-54 Question 10-57.

10-58 The angular velocity of a drill (Figure 10-55) increases from 27 rad/s to 107 rad/s in 0.10 s, with a constant angular acceleration. For a point on the outside edge of the drill bit of diameter 0.62 cm, what are the magnitudes of the

(a) tangential acceleration?

(b) initial centripetal acceleration?

(c) initial total acceleration?

Dmitry Kalinovsky/Shutterstock

Figure 10-55 Question 10-58.

10-59 A car with tires of radius 33 cm has a speed of 90 km/h (assume two significant digits).

(a) What is the angular velocity of the wheels in radians per second?

(b) The speed 5.0 s later is 45 km/h. What is the angular acceleration (assumed constant) of the wheels?

10-60 The gymnast shown in Figure 10-56 is rotating around his supporting arm, which acts as the rotation axis. If the gymnast has an angular velocity of 3.1 rad/s and a rotational kinetic energy of 1.4×10^2 J, what is his moment of inertia about this axis?

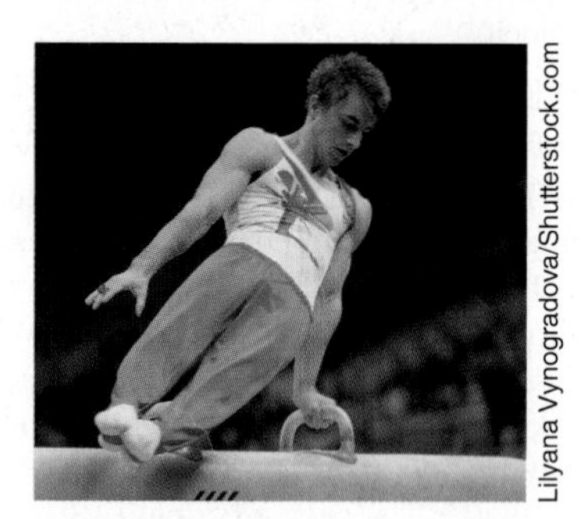

Figure 10-56 A gymnast rotating on a pommel horse (Question 10-60).

10-61 A uniform solid ball is rolling with a speed of 3.4 m/s and has a total kinetic energy of 0.36 J. What is its mass?

10-62 A wheel of radius 0.33 m and mass 1.3 kg rolls from rest down an incline with a vertical height of 1.6 m. Its speed at the bottom is 4.0 m/s. What is the moment of inertia of the wheel?

10-63 When you are pedalling a bicycle, the amount of torque you can produce on the pedals varies. At which position(s) of the left pedal is the torque produced by the left leg the greatest? the least? Explain.

10-64 A steel beam (Figure 10-57) is pivoted about an axis at P. The force of gravity on the beam has a magnitude of 3.6×10^2 N, and a person is using a rope to provide a tension of magnitude 2.2×10^2 N. Initially the beam is stationary.

(a) Find the magnitude of the torque about P caused by each of the two forces mentioned.

(b) In which direction will the beam rotate?

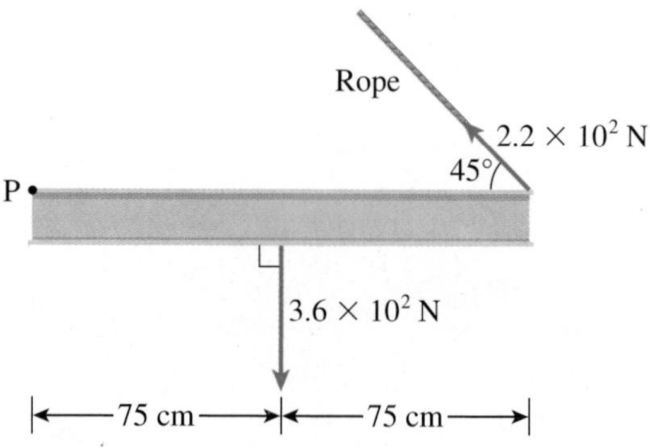

Figure 10-57 Question 10-64.

10-65 The blade assembly for a particular food processor has a moment of inertia of 8.3×10^{-4} kg·m². If the blades have an angular acceleration of 4.2×10^2 rad/s² when the processor is turned on, what is the torque applied to them?

10-66 A certain springboard diver has a moment of inertia of 6.1 kg·m² in the tuck position and 19.2 kg·m² in the layout (extended) position. When in the tuck position, the diver is rotating at 2.5 rev/s.

(a) What is the diver's angular momentum?

(b) If the diver now goes into the layout position, what is his angular velocity?

(c) Calculate the diver's initial and final kinetic energies.

(d) Account for the difference between the energies calculated in (c). Is the law of conservation of energy being violated?

Applying Your Knowledge

10-67 The crankshaft in an automobile engine undergoes a constant angular acceleration of 73.0 rad/s² while rotating through 84.4 rev in 1.15 s. What are its initial and final angular velocities in radians per second?

10-68 A spinning wheel (Figure 10-58) slows down to 14 rpm while rotating through 1.47 rev in 4.0 s. Determine its

Figure 10-58 Question 10-68.

(a) initial angular velocity in revolutions per minute

(b) angular acceleration (assumed constant) in revolutions per minute squared

10-69 An old 78.0 rpm record has a playing time of 2.57 min. Before playing, the turntable speeds up from rest with constant angular acceleration in a time of 1.83 s. After playing, it slows down to rest with constant angular acceleration in a time of 2.30 s. How many revolutions does the record make in total?

10-70 Two electric motors are started from rest. One (A) has a constant angular acceleration of 73.3 rad/s², and the other (B) has a constant angular acceleration of 98.7 rad/s². Motor B is started 0.550 s after A.

(a) How long after B starts will it have the same angular velocity as A?

(b) What is this angular velocity?

(c) How many revolutions will each motor have undergone at this time?

10-71 Figure 10-59 shows the acceleration of a point P that is 8.5 cm from the centre C of a grindstone rotating about a fixed axis. The acceleration is 85 m/s² in magnitude and makes an angle of 21° with the line joining P to the centre, as shown. Determine the grindstone's

(a) angular velocity

(b) angular acceleration

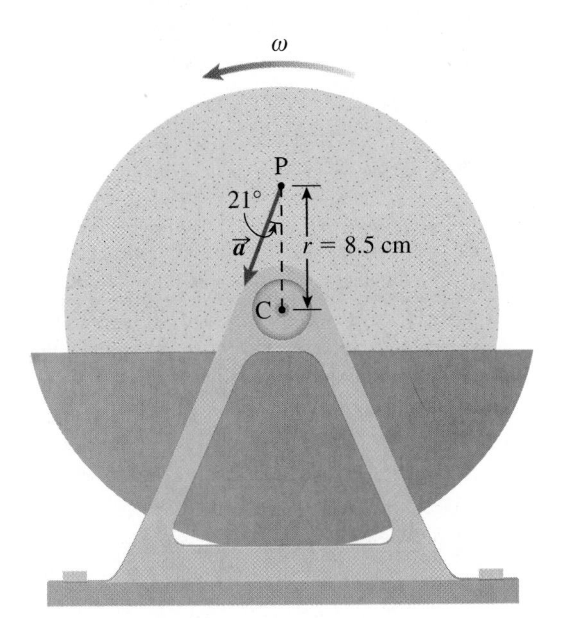

Figure 10-59 Question 10-71.

10-72 A ball is slowing down as it rolls on grass in a playground. Its angular acceleration is −1.8 rad/s² and its linear acceleration is −0.045 m/s².

(a) What is the ball's diameter in centimetres?

(b) When the ball's speed is 0.14 m/s, what is its angular velocity?

(c) If the ball's initial angular velocity was 9.0 rad/s, how long does it take to come to rest?

10-73 When a wheel of radius R is rolling along the ground, the total velocity vector of any point on the wheel relative to the *ground* is the vector sum of the velocity (magnitude $V = \omega R$) of the centre of the wheel and the velocity (magnitude $v = \omega r$) of the point due to the rotation around the centre. In terms of ω and R, determine the *speed* of each of the following points on the wheel:

(a) the top of the wheel

(b) the centre

(c) the bottom

(d) a point on the rim at the same horizontal level as the centre

10-74 A pulley of moment of inertia 9.36×10^{-2} kg·m², rotating about a fixed axle with an initial angular velocity of 23.4 rad/s, undergoes a constant angular acceleration of 5.12 rad/s² for 3.40 s. By how much does the kinetic energy of the pulley increase?

10-75 **Fermi Question:** By approximating a human forearm as a thin rod of mass M and length L, estimate its moment of inertia about the elbow joint.

10-76 A thin stick 1.5 m long is pivoted about one end (Figure 10-60). The stick is held in a horizontal position and released from rest. Neglecting friction, use conservation of energy to determine the stick's angular velocity when it is vertical. (**Hint:** When determining the change in gravitational potential energy of the stick, you must consider the change in position of the stick's centre of mass.)

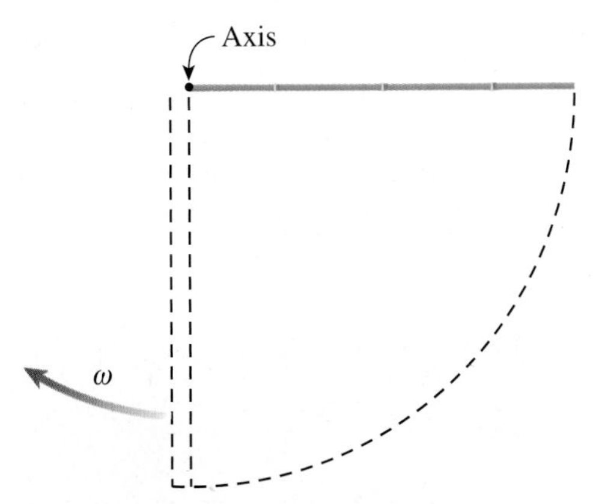

Figure 10-60 Question 10-76.

10-77 A bag of supplies is hung vertically from one end of a rope that is wound around a pulley on a ship. The pulley has a fixed rotation axis (Figure 10-61), a moment of inertia of 0.47 kg·m², and a radius of 0.25 m. The bag (mass 9.5 kg) is released from rest and falls toward a small boat below, causing the pulley to rotate. After the bag has fallen 2.5 m, what is the angular velocity of the pulley? (**Hint:** The speed of the bag is equal to the speed of any point on the rim of the pulley because of the rope connecting them.)

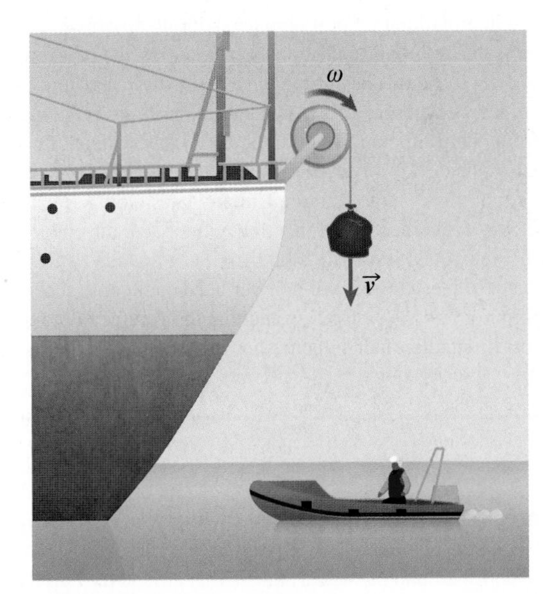

Figure 10-61 Question 10-77.

10-78 A centrifuge of moment of inertia 2.25×10^{-2} kg·m² rotating at 2.50×10^{4} rpm is brought to rest by a constant frictional torque of magnitude 2.31 N·m.

(a) What is the angular acceleration of the centrifuge?

(b) How long does it take to come to rest?

(c) How many revolutions does it make in coming to rest?

10-79 When a golf ball is hit by a golf club such as a driver (Figure 10-62), a large force is exerted by the club on the ball, and by Newton's third law the same magnitude of force is exerted by the ball on the club. If the ball–club collision occurs to the right or left of the club head's centre, then the force of the ball on the club exerts a torque on the club head that causes it to rotate about a vertical axis through its centre. This rotated club head propels the ball to the right or left of the intended trajectory. Suppose that a driver club head with a moment of inertia of 5.0×10^{-4} kg·m^2 strikes a ball off-centre and the club head rotates 11° during the collision, which occurs in a time of 1.0 ms. What was the magnitude of the torque (assumed constant) on the club head? (Assume that the club head had zero angular velocity about the vertical axis just prior to the collision.)

Figure 10-62 Question 10-79.

DID YOU KNOW?

To reduce the rotation of a golf club head during an off-centre collision with a golf ball (Question 10-79), during the past few decades golf club heads have been designed with increasingly large moments of inertia. This was done primarily by changing the heads from being solid and small to being hollow with all the mass around the outside. However, the result was that the club heads were becoming very large, and the United States Golf Association (USGA) was concerned that if this trend continued, the clubs would look ridiculous. As a result, in 2006 the USGA created a rule that the moment of inertia (MOI) of a driver head could not exceed 5900 g·cm^2 (or 5.9×10^{-4} kg·m^2), and that the volume of the driver head must be smaller than 460 cm^3. For more details, go online and search for "USGA MOI."

10-80 A wheel of radius R_1 (33.4 cm) with a wide hub of radius R_2 (14.3 cm) is mounted on a fixed axle (Figure 10-63). A rope is wrapped around the wheel and another around the hub. The ropes are pulled in opposite directions, as shown, with tensions of magnitude $T_1 = 3.35$ N and $T_2 = 5.51$ N.

(a) Which way will the wheel turn? Neglect friction.

(b) What is the resulting angular acceleration if the moment of inertia of the wheel is 0.270 kg·m^2?

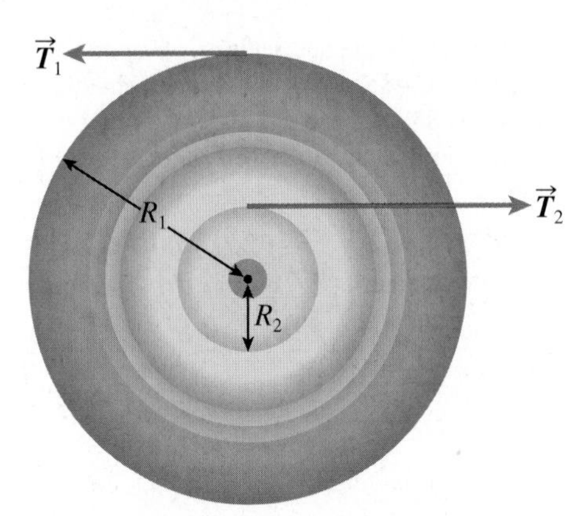

Figure 10-63 Question 10-80.

10-81 A thin rod of mass M and length L can rotate about a fixed horizontal axis through one end. It is initially held at an angle of 45° from the horizontal (Figure 10-64), and then released from rest.

(a) In terms of given parameters and "g," what is its angular acceleration at release, and what is the magnitude of the linear acceleration of its midpoint at this time?

(b) As the rod rotates downward, is its angular acceleration constant?

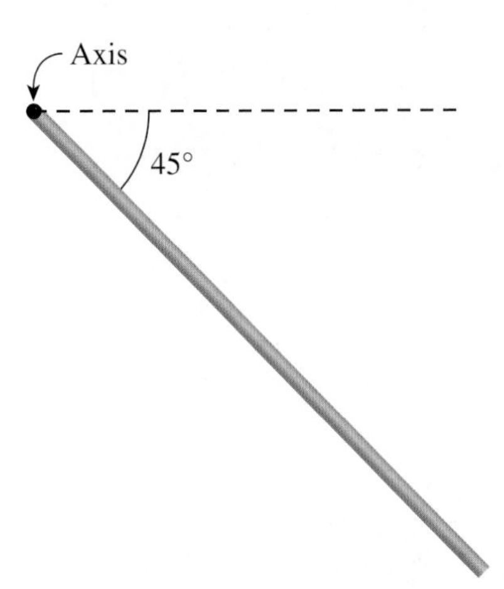

Figure 10-64 Question 10-81.

10-82 In Section 10.5 we stated that a figure skater can increase her moment of inertia by about a factor of two by simply extending her arms. How can this factor be as large as two when only a small fraction of the skater's mass is being redistributed?

10-83 A gymnast decreases her moment of inertia from 12 kg·m^2 to 6.0 kg·m^2 in a time of 1.1 s. If her initial angular velocity is 4.0 rad/s, what is her angular acceleration?

10-84 The Crab Nebula (Figure 10-65) is the remnant of a star that exploded in 1054 CE. The light from the explosion was visible to the naked eye in the night sky for almost 22 months. Much of the star's mass was ejected during the explosion and formed the nebula, but some of the mass was crushed into a neutron star having a radius of only 10 km (assume two significant digits) but a mass 1.4 times that of the Sun. The neutron star is spinning and once each revolution it emits a radio pulse that can be detected at Earth, and hence the star is called a pulsar. During the collapse of material to form the pulsar, angular momentum was conserved, and since the pulsar is very small it rotates very quickly, once every 0.033 s.

Figure 10-65 Question 10-84.

(a) What is the angular velocity of the pulsar in radians per second?

(b) The mass of the Sun is 1.99×10^{30} kg. Determine the moment of inertia of the Crab Nebula pulsar, assuming that it is a uniform solid sphere.

(c) What is the speed of a point on the surface of the pulsar?

(d) Determine the magnitude of the centripetal acceleration at the pulsar's surface.

10-85 A trampolinist is rotating in the tuck position with a rotational kinetic energy of 288 J. He then extends his body into a layout position and his moment of inertia increases by a factor of 4.5. Determine his new rotational kinetic energy.

10-86 **Fermi Question:** In the photograph of the dragster shown in Figure 10-25 in Section 10.2, notice that the front wheels are much smaller than the rear wheels. As the car is moving, the small wheels spin more rapidly than the large ones, but the total kinetic energy (translational plus rotational) of one of the small wheels is less than that of one of the large rear wheels. Estimate the ratio of the total kinetic energy of a large rear wheel to that of a small front wheel. You should find that this ratio is independent of the speed V of the dragster.

Statics, Stability, and Elasticity

11

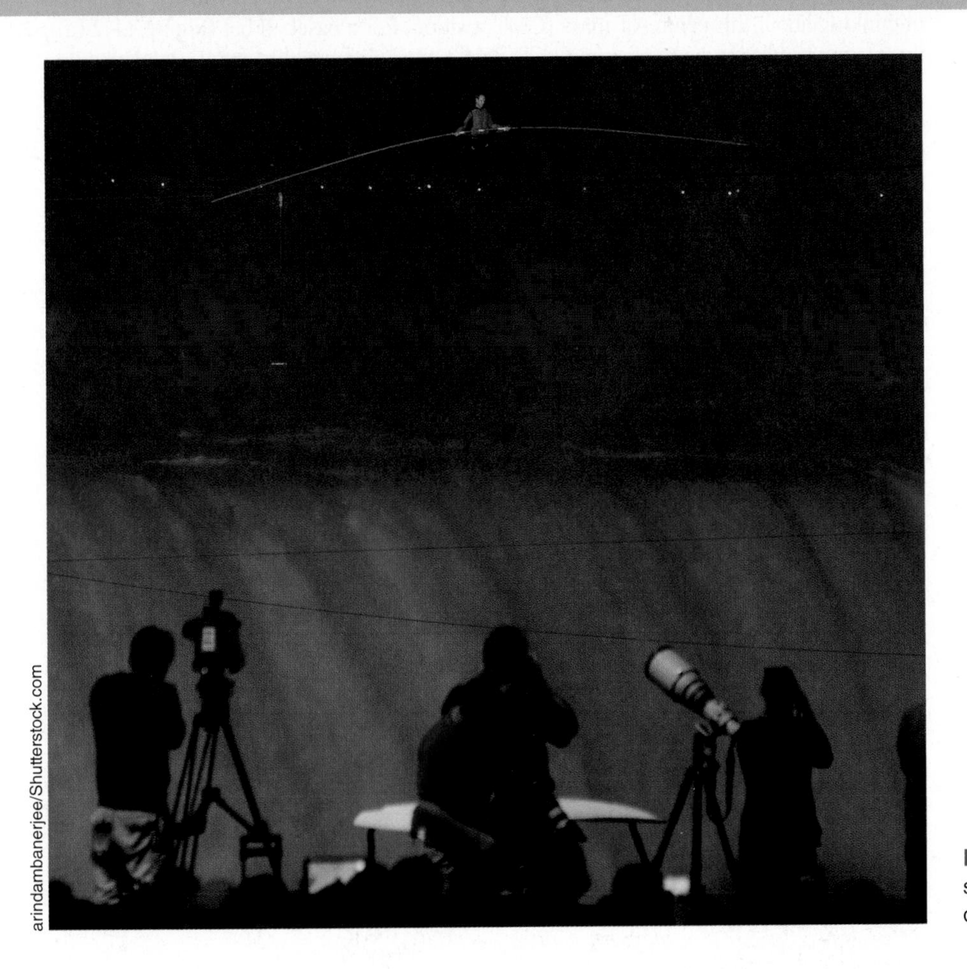

arindambanerjee/Shutterstock.com

Figure 11-1 On June 6, 2012, Nik Wallenda showed great stability and balance in making a daring tightrope walk across Niagara Falls.

The study of any balancing act such as tightrope-walking (Figure 11-1) involves the concept of stability. How can a swan rest peacefully while on one leg, or a ballerina stand motionless on the toes of one foot? Why can a calf or colt stand erect shortly after birth, while a human baby must wait many months before ever attempting to stand? These questions on stability will be addressed after the topic of statics is presented. As its name implies, **statics** is the study of forces and torques acting on structures at rest. The structures vary in size from large to small, from bridges to human tendons and bones. The study of these practical structures requires us to consider the concepts of stress and strain, and the important elastic properties of materials that allow them to deform without breaking. The effects of forces and torques in producing deformations in materials such as concrete and steel, bones and blood vessels, as well as structures such as arches will be analyzed in detail. You will find that there are countless applications of these topics in the fields of sports, biology and health sciences, architecture, engineering, toy manufacturing, fashion design, and others.

CHAPTER **OUTLINE**

11.1 Centre of Mass

statics study of forces and torques acting on structures at rest

centre of mass (CM) point located at the average position of the mass of an object

We start this chapter by considering the concept of **centre of mass (CM)**, which will be useful in our discussion of statics and stability. The centre of mass of an object is the point located at the average position of the mass of the object; we will see that in many situations the entire mass can be assumed to be concentrated there. Your intuition will serve you well in thinking about this centre-of-mass (CM) position. For a baseball bat (Figure 11-2 (a)), there is more mass at the thicker end and so the CM is closer to that end. For the wedding ring shown in Figure 11-2 (b), the average position of the mass is at the geometric centre of the ring, and therefore the CM is located there even though there is no mass actually at that point. The CM of a person depends on the specific distribution of mass in the individual but, generally speaking, the CM for men is located higher in the body than for women, since men tend to have more mass concentrated in the upper chest and shoulders.

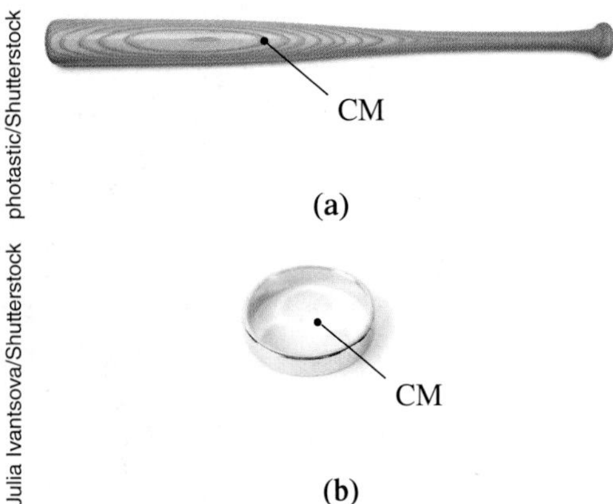
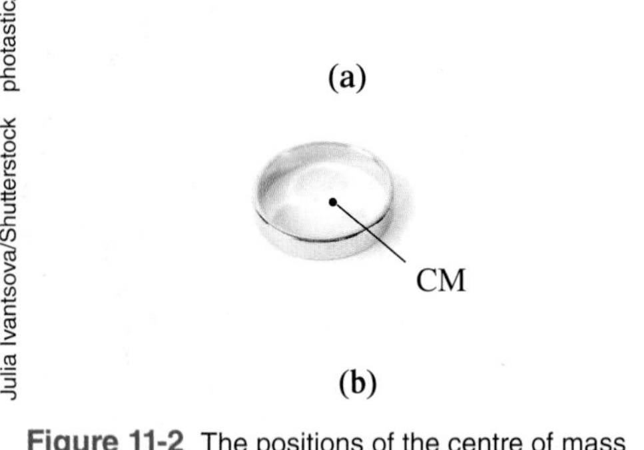

Julia Ivantsova/Shutterstock photastic/Shutterstock

Figure 11-2 The positions of the centre of mass (CM) of **(a)** a baseball bat, and **(b)** a wedding ring.

The position of the centre of mass can be expressed mathematically. For a collection of particles (which might make up an object) with masses m_1, m_2,... and x-components of position x_1, x_2,..., the x-component of the position of the centre of mass (CM) is given by

$$x_{CM} = \frac{m_1x_1 + m_2x_2 + m_3x_3 + \cdots}{m_1 + m_2 + m_3 + \cdots} = \frac{\sum_i m_ix_i}{\sum_i m_i} = \frac{\sum_i m_ix_i}{M} \qquad \textbf{(11-1a)}$$

where $\sum_i$ indicates a summation over all subscripts i, and M is the total mass of all the particles.

The y- and z-components of the CM position are defined in a similar way:

$$y_{CM} = \frac{m_1y_1 + m_2y_2 + m_3y_3 + \cdots}{m_1 + m_2 + m_3 + \cdots} = \frac{\sum_i m_iy_i}{\sum_i m_i} = \frac{\sum_i m_iy_i}{M} \qquad \textbf{(11-1b)}$$

$$z_{CM} = \frac{m_1z_1 + m_2z_2 + m_3z_3 + \cdots}{m_1 + m_2 + m_3 + \cdots} = \frac{\sum_i m_iz_i}{\sum_i m_i} = \frac{\sum_i m_iz_i}{M} \qquad \textbf{(11-1c)}$$

SAMPLE PROBLEM 11-1

Four "point particles" are distributed along an x-axis. Two particles each have mass m_1 = 1.0 kg, and the other two each have mass m_2 = 2.0 kg. Determine the position of the centre of mass of the particles for the mass distributions shown in Figures 11-3 (a) and (b). The 1.0 m spacing between the particles is the same in the two drawings.

(a) m_2 m_1 m_1 m_2 $+x$; $x = 0$; 1.0 m 1.0 m 1.0 m 1.0 m

(b) m_1 m_1 m_2 m_2 $+x$; $x = 0$

Figure 11-3 Sample Problem 11-1.

Solution

Using Eqn. 11-1a for the mass distribution in Figure 11-3 (a),

$$x_{CM} = \frac{\sum_i m_ix_i}{\sum_i m_i} = \frac{(2.0\,\text{kg})(1.0\,\text{m}) + (1.0\,\text{kg})(2.0\,\text{m}) + (1.0\,\text{kg})(3.0\,\text{m}) + (2.0\,\text{kg})(4.0\,\text{m})}{(2.0 + 1.0 + 1.0 + 2.0)\,\text{kg}} = 2.5\,\text{m}$$

Thus, the centre of mass is 2.5 m from the origin, which is at the geometric centre of the mass distribution. This is not surprising because the masses are distributed symmetrically about this point. Note that there is no actual mass at the CM of this particular distribution.

For the mass distribution in Figure 11-3 (b),

$$x_{CM} = \frac{\sum_i m_i x_i}{\sum_i m_i} = \frac{(1.0\,\text{kg})(1.0\,\text{m})+(1.0\,\text{kg})(2.0\,\text{m})+(2.0\,\text{kg})(3.0\,\text{m})+(2.0\,\text{kg})(4.0\,\text{m})}{(1.0+1.0+2.0+2.0)\,\text{kg}} = 2.8\,\text{m}$$

The centre of mass is 2.8 m from the origin. The CM is not at the geometric centre of the distribution, but is somewhat to the right of the centre because there is more mass to the right of centre. Notice again that there is no actual mass at the CM.

The result in Sample Problem 11-1 (a) that the CM of a symmetrical distribution of mass is at the geometric centre is a general result that is valid for any symmetrical distribution. Figures 11-4 (a), (b), and (c) show a uniform solid sphere, a hollow sphere, and a uniform rectangular wooden board, all of which have the CM at the geometric centre. A solid sphere does not actually need to be uniform in order to have the CM at the centre, but it must be spherically symmetric. This means that, although the detailed composition of the sphere can vary depending on the distance from the centre, at any one particular distance in any direction from the centre the composition of the sphere must be the same. For example, in a golf ball there are many layers of different materials, but at 1 cm from the centre in any direction there is always the same material. Figure 11-4 (d) shows an object made of uniform material, but since it is wider at one end than the other, its CM is closer to the wider (more massive) end, similar to the situation in Sample Problem 11-1(b). A boomerang is shown in Figure 11-4 (e) to illustrate a solid object that has an external CM.

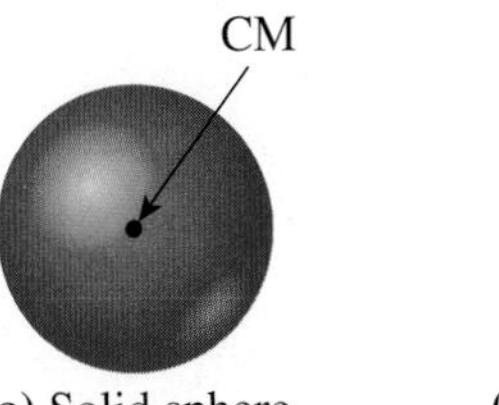

(a) Solid sphere

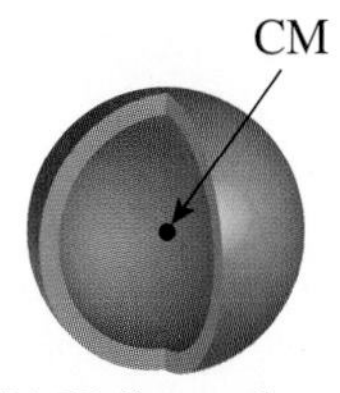

(b) Hollow sphere

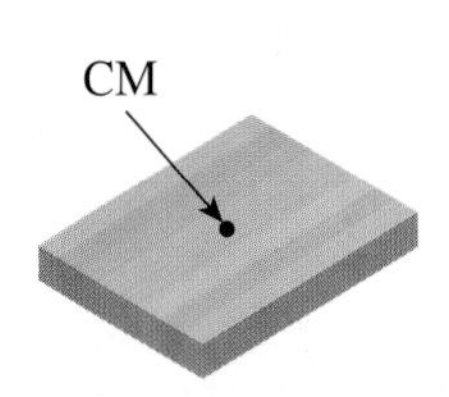

(c) Rectangular board

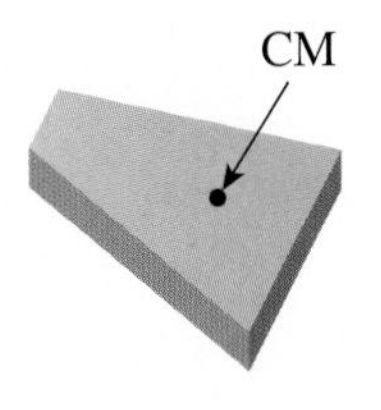

(d) Irregular object

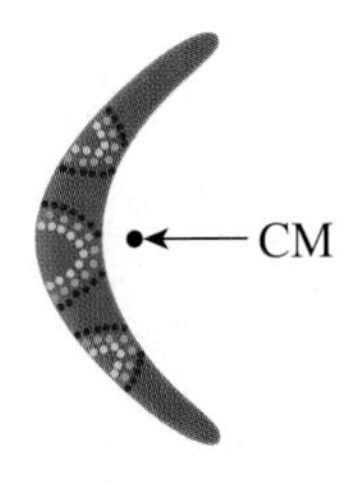

(e) Boomerang

Figure 11-4 The centre of mass of various objects: **(a)** solid sphere, **(b)** hollow sphere, **(c)** rectangular board, **(d)** irregular object, **(e)** boomerang.

SAMPLE PROBLEM 11-2

Figure 11-5 shows a carpenter's set square, having a mass of 240 g (three significant digits), and dimensions as follows: AB = EF = 4.00 cm, AO = 30.0 cm, and DE = 20.0 cm. Determine the position of the CM of the set square, expressed as x- and y-components of position relative to point O, with $+x$- and $+y$-directions as shown in the drawing. Assume that the set square is uniform in composition and thickness.

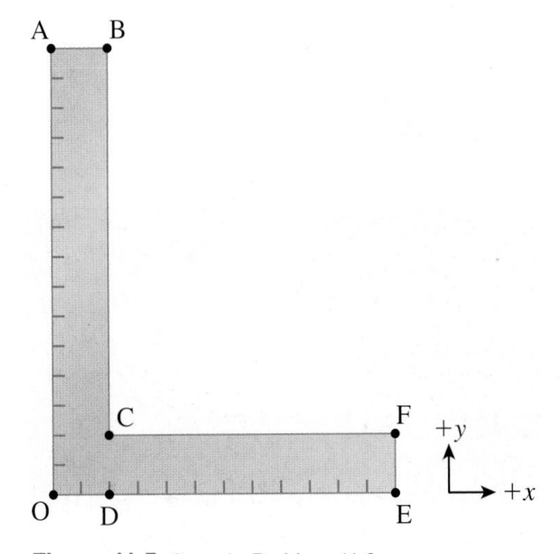

Figure 11-5 Sample Problem 11-2.

Solution

Our approach will be to consider the two arms of the set square as separate objects. First determine the position of the CM of each arm. Then we consider the mass of each arm to be located at the CM of the arm, and determine the CM position of these two masses.

Using the labels in Figure 11-5, we use OABD as one arm, and CDEF as the other arm.

Arm OABD is a rectangle of length 30.0 cm and width 4.00 cm. Because of this symmetric shape, the CM is at the geometric centre, which has coordinates relative to point O of

$$x_1 = (4.00\,\text{cm})/2 = 2.00\,\text{cm} \text{ (using subscript "1" for this arm)}$$

$$y_1 = (30.0\,\text{cm})/2 = 15.0\,\text{cm}$$

Similarly, arm CDEF is a symmetric rectangle of length 20.0 cm and width 4.00 cm, with its CM also at the geometric centre. Since this entire rectangle is displaced 4.00 cm (distance OD) from the origin O, the x-position coordinate of its CM is

$$x_2 = 4.00\,\text{cm} + (20.0\,\text{cm})/2 = 14.0\,\text{cm} \text{ (using subscript "2" for this arm)}$$

Its CM y-position is $y_2 = (4.00\text{ cm})/2 = 2.00$ cm.

We now consider the entire mass m_1 of arm #1 to be located at its CM, and the entire mass m_2 of arm #2 to be at its CM, and determine the overall CM of these two "point masses." But first we need to know the mass of each arm. Since the set square has a uniform thickness (i.e., in the $+z$-direction perpendicular to the drawing), the mass of each arm will be proportional to its area in the x-y plane.

$$\text{area of OABD} = A_1 = 30.0\,\text{cm} \times 4.00\,\text{cm} = 120\,\text{cm}^2\,\text{(three significant digits)}$$

$$\text{area of CDEF} = A_2 = 20.0\,\text{cm} \times 4.00\,\text{cm} = 80.0\,\text{cm}^2$$

$$A_{\text{total}} = A_1 + A_2 = 200\,\text{cm}^2\,\text{(three significant digits)}$$

The total mass is given: $m_{\text{total}} = 240$ g.

$$\therefore\ m_1 = \frac{A_1}{A_{\text{total}}} \times m_{\text{total}} = \frac{120\ \text{cm}^2}{200\ \text{cm}^2} \times 240\ \text{g} = 144\ \text{g}$$

$$m_2 = \frac{A_2}{A_{\text{total}}} \times m_{\text{total}} = \frac{80.0\ \text{cm}^2}{200\ \text{cm}^2} \times 240\ \text{g} = 96.0\ \text{g}$$

Using Eqns. 11-1a and 11-1b to determine the overall CM position,

$$x_{\text{CM}} = \frac{\sum_i m_i x_i}{\sum_i m_i} = \frac{m_1 x_1 + m_2 x_2}{m_{\text{total}}} = \frac{(144\,\text{g})(2.00\ \text{cm}) + (96.0\,\text{g})(14.0\ \text{cm})}{240\,\text{g}} = 6.80\,\text{cm}$$

$$y_{\text{CM}} = \frac{\sum_i m_i y_i}{\sum_i m_i} = \frac{m_1 y_1 + m_2 y_2}{m_{\text{total}}} = \frac{(144\,\text{g})(15.0\ \text{cm}) + (96.0\,\text{g})(2.00\ \text{cm})}{240\,\text{g}} = 9.80\,\text{cm}$$

Therefore, the CM of the set square is located at $x_{\text{CM}} = 6.80$ cm and $y_{\text{CM}} = 9.80$ cm, that is, 6.80 cm to the right of point O, and 9.80 cm above point O. Note that there is no mass at the CM position.

Centre of Mass and Newton's Laws of Motion

Ventura/Shutterstock.com

Figure 11-6 As this motorcycle and rider fly through the air, the only point following a parabolic projectile-motion trajectory is the CM of the rider and motorcycle together.

Although the formal mathematical definition of centre of mass (Eqn. 11-1) was not presented until this chapter, the concept of centre of mass is relevant to a number of topics in previous chapters. For example, in Chapter 5, Newton's second law of motion was stated as, "If the resultant force on an object is not zero, the object experiences an acceleration in the direction of the resultant force." More completely, this law should state "...the *centre of mass of* the object experiences an acceleration in the direction of the resultant force." As well, it is the centre-of-mass acceleration $\vec{a}$ that is given by $\vec{a} = \vec{F}_R/m$, where $\vec{F}_R$ is the resultant force acting on the object of mass m. If you think of the motion of an object such as the motorcycle and rider shown flying through the air in Figure 11-6, the only point that is actually following a parabolic projectile-motion trajectory with a constant acceleration of $\vec{g}$ is the CM of the motorcycle and rider considered as one object. All the other points on both the person and motorcycle have more complicated motions.

Newton's first law of motion should also be stated in terms of the CM: "If the resultant force on an object is zero, the velocity of *the centre of mass of* the object is constant; that is, the object's *centre of mass* travels in a straight line at constant speed (which could be zero)."

There have been references to CM earlier in this book. It was explicitly pointed out in Section 7.6 that, "when we use the equation $E_P = mgy$ for gravitational potential energy, y represents the height of the centre of mass (CM) of the object." As well, in Section 10.3 it was noted that the speed V in the mathematical expression for a rolling object's translational kinetic energy, $E_K = \frac{1}{2}MV^2$, is the speed of the CM of the object of mass M.

As we proceed through this chapter, we will see that CM is particularly important in the study of statics and stability.

Centre of Gravity

A concept closely related to centre of mass is **centre of gravity (CG)**, defined as the point in an object of mass m where the force of gravity $m\vec{g}$ can be assumed to act. It can be shown that if the gravitational field (or acceleration) $\vec{g}$ is constant for all points on the object, then the centre of gravity and centre of mass are at the same point. For the physical situations described in this chapter, it can be safely assumed that the centre of gravity and centre of mass coincide. Indeed, many times when drawing free-body diagrams in previous chapters, we implicitly assumed that the force of gravity was acting at the centre of mass of an object.

centre of gravity (CG) the point in an object where the force of gravity can be assumed to act; same as the centre of mass in most practical situations

Finding Centre of Mass Experimentally

The location of the centre of mass of an object with a simple shape such as the set square in Sample Problem 11-2 can be determined mathematically using Eqn. 11-1. However, using this equation for an irregularly shaped object would be difficult,[1] but it is possible to determine the approximate CM location for such an object experimentally. As an example, consider the situation shown in Figure 11-7. The pen can be balanced on a finger only if the CM is directly above the contact point between the finger and pen. Since the force of gravity, $m\vec{g}$, acts downward at the CM, the moment arm for $m\vec{g}$ relative to that contact point is zero and so the torque due to $m\vec{g}$ is zero. Therefore, the pen will remain stationary. If either the pen or the finger is moved so that the contact point is not directly below the CM, the torque due to $m\vec{g}$ will not be zero, and the pen will rotate and fall. Thus, balancing an object allows for a rough determination of the location of the CM.

Alternatively, if an object is suspended vertically from a string, for example, and allowed to rotate and become stationary, its final resting position will have the CM directly below the point of suspension. At this position, the torque due to $m\vec{g}$ will be zero. If this procedure is repeated with a couple of suspension points, the approximate position of the CM can be found.

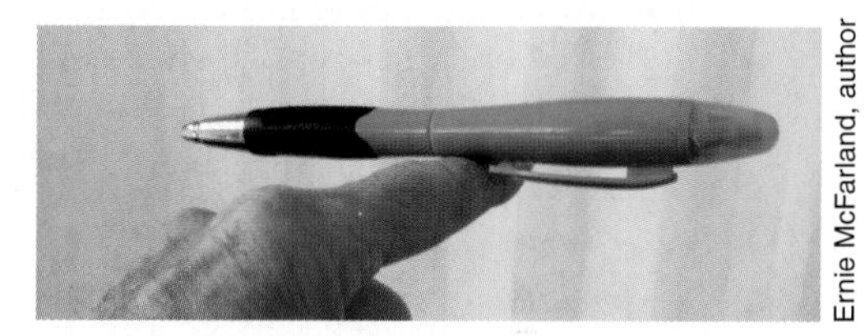

Ernie McFarland, author

Figure 11-7 When a pen is balanced on a finger, the CM of the pen must be located directly above the contact point between the finger and the pen.

TRY THIS!

Finding an Object's CM

Using one of the techniques described in the adjacent paragraphs, determine the approximate position of the CM of an irregularly shaped object, such as a spoon, hockey stick, soup ladle, banana, golf club, wrench, broom, coffee mug, etc.

EXERCISES

11-1 Four "point particles" have masses m_i and x- and y-components of position x_i and y_i as shown in Table 11-1.

(a) Determine the x- and y-components of the position of the CM of the system of particles.

(b) How far is the CM from the origin ($x = y = 0$ cm)?

11-2 A basketball of mass 0.60 kg and radius 0.120 m is sitting on a horizontal uniform wooden board of mass 1.45 kg, length 0.96 m, and thickness 0.018 m (Figure 11-8). The origin O of an x-y coordinate system and the $+x$- and $+y$-directions are shown in the drawing. The horizontal distance d between O and the point where the basketball touches the board is 0.22 m. What are the x- and y-components of the position of the CM of the system of the basketball and board?

Table 11-1

Data for Question 11-1

Particle Number i	m_i (g)	x_i (cm)	y_i (cm)
1	2.4	2.0	1.0
2	1.2	−3.2	4.1
3	3.7	4.6	−4.8
4	9.5	2.0	−6.1

[1]However, Eqn. 11-1 can be modified to give an expression involving integral calculus that could be used.

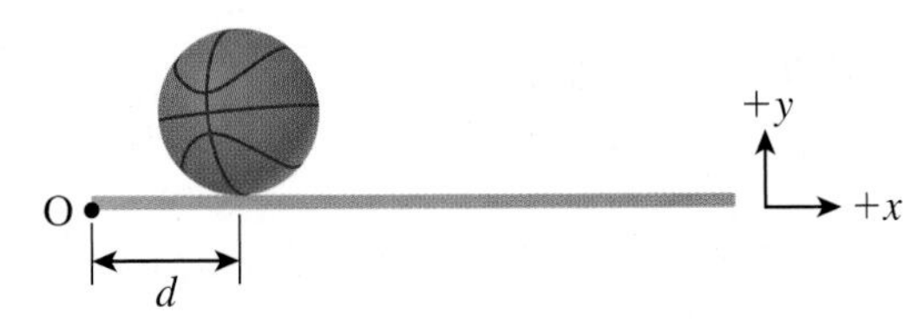

Figure 11-8 Question 11-2.

11-3 A horizontal rectangular wooden tabletop of uniform composition and mass 6.40 kg has a length of 0.840 m and a width of 0.680 m. Figure 11-9 shows the tabletop as seen from above, along with the origin O of an x-y coordinate system and the $+x$- and $+y$-directions.

(a) What are the x- and y-components of the position of the CM of the tabletop?

(b) A box of groceries of mass 2.40 kg is placed on the table. The CM of the tabletop and box together has coordinates of $x_{\text{CM}} = 0.380$ m and $y_{\text{CM}} = 0.350$ m. What are the x- and y-components of the position of the CM of the box?

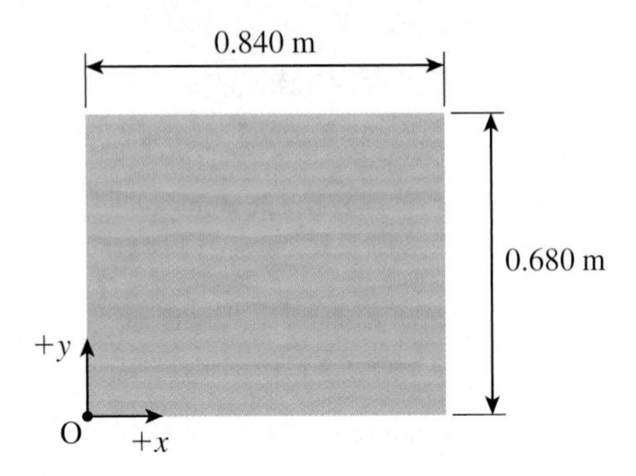

Figure 11-9 Question 11-3.

11-4 Describe the path of the CM of object "A" in each case, neglecting air resistance.

(a) A diver (A) runs off the end of a diving board, does a double somersault, then straightens completely to enter the water vertically downward.

(b) A firecracker (A) is travelling through the air and explodes.

11-5 The Fosbury flop is a style of high jumping, developed by Richard Fosbury in 1968, in which the jumper twists backward over the bar (Figure 11-10). Prior to 1968, jumpers went over the bar with their backs upward. (If you search online for "high jump evolution," you can see how jumping technique has changed over the past century.) When the Fosbury flop was first introduced it helped shatter the world high-jumping record by over 6 cm. The advantage of this technique is that the jumper's CM can remain basically below the bar for the entire jump. Explain why this would be an advantage to the jumper. (**Hint:** Think of energy.)

Figure 11-10 A high-jumper using the Fosbury flop (Question 11-5).

11.2 Static Equilibrium

The word equilibrium stems from the Latin words *aequus*, equal, and *libra*, balance. Thus, an object in equilibrium is in a state of balance. This means that the vector sum of all the forces acting on the object must be zero:

$$\Sigma \vec{F} = 0 \tag{11-2}$$

In addition, the sum of all the torques acting on the object must be zero:

$$\Sigma \tau = 0 \tag{11-3}$$

If these two conditions are met and the object is at rest, the object is in a state of **static equilibrium**; that is, it is stationary (with zero translational velocity $\vec{v}$ and zero angular velocity ω) and remains stationary. If the object is moving and the two conditions are met, the object is in **dynamic equilibrium**; that is, it has constant non-zero translational velocity or constant non-zero angular velocity, or both.

static equilibrium the state of a stationary object on which the sum of forces and sum of torques are both zero

dynamic equilibrium the state of a moving object on which the sum of forces and sum of torques are both zero

In this chapter, we are concerned with static equilibrium only. Applying the two conditions for static equilibrium allows us to analyze a wide variety of common physical situations.

Problem-Solving Strategy for Static Equilibrium

1. Draw a diagram on which you include all given information, including any numerical values.
2. Identify the object (or group of objects) on which the relevant forces and torques are acting.
3. Make a free-body diagram (FBD) for that object (or group).
4. Choose $+x$- and/or $+y$-directions and a pivot point relative to which torques will be calculated. Often there will be an obvious pivot point (for example, an elbow about which a forearm can rotate), and the solution will usually be easiest if that pivot point is chosen. Choose a positive direction (clockwise or counterclockwise) for torque calculations.
5. It is usually best to use $\Sigma\tau = 0$ first to write an equation involving the relevant forces and moment arms. Often one or more of the moment arms will be zero, and this will make subsequent steps easier.
6. Then use $\Sigma\vec{F} = 0$. For a typical two-dimensional problem, use $\Sigma F_x = 0$ and $\Sigma F_y = 0$ with the appropriate x- and y-components of force.
7. You now have three equations (one for torque, and two for components of forces), which can be used to solve for the unknown quantities.

SAMPLE PROBLEM 11-3

This problem shows one method that could be used to determine the position of a person's CM. Figure 11-11 shows a man lying horizontally on a uniform wooden board of mass $m_W = 5.0$ kg. The man has a mass of $m_M = 75$ kg and a height of $L = 1.80$ m, which is the same as the length of the board. The board is supported at the "foot end" by a block that exerts an upward force $\vec{F}_B$. At the "head end" of the board, a weigh scale exerts an upward force $\vec{F}_S$. If the scale reads 452 N (i.e., $F_S = 452$ N), determine the position of the man's CM (x_{CM} in the drawing) relative to the origin O at the bottom of his feet. Assume that the forces exerted by the block and by the scale are located at the ends of the board.

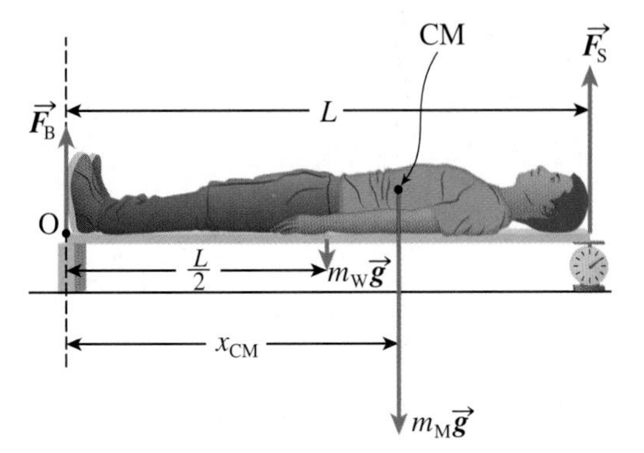

Figure 11-11 Sample Problem 11-3.

Solution

We are asked to find the distance x_{CM}, which is the moment arm relative to O for the force of gravity $m_M\vec{g}$ acting on the man, and so we will use $\Sigma\tau = 0$ about O since this equation will involve the unknown distance. Choosing point O as the pivot point for torque calculations also has the advantage that the moment arm and torque due to $\vec{F}_B$ will be zero; this is useful since we do not know the magnitude of $\vec{F}_B$. Let us choose clockwise to be the positive direction for the torques (although counterclockwise would work just as well).

All the distances shown in Figure 11-11 are perpendicular to the forces, and so no trigonometric calculations are needed to determine moment arms. The static equilibrium condition $\Sigma\tau = 0$ gives $m_W g(L/2) + m_M g(x_{CM}) - F_S(L) = 0$

Rearranging to solve for x_{CM},

$$x_{CM} = \frac{F_S(L) - m_W g(L/2)}{m_M g}$$

$$= \frac{(452\,\text{N})(1.80\text{ m}) - (5.0\,\text{kg})(9.8\text{ m/s}^2)(0.900\,\text{m})}{(75\,\text{kg})(9.8\text{ m/s}^2)}$$

$$= 1.0\,\text{m}$$

Thus, the man's CM is 1.0 m from the bottom of his feet. Of course, this distance is only one of the three components needed to specify the position of the CM completely, but the other two components could be estimated fairly well, based on the symmetry of the body.

TRY THIS!

Finding Your CM

Using the procedure discussed in Sample Problem 11-3, determine the position of your CM.

SAMPLE PROBLEM 11-4

Figure 11-12 (a) shows a human arm holding a dumbbell of mass $m = 12.0$ kg, with the horizontal forearm making an angle of 90.0° with the upper arm. Figure 11-12 (b) shows the forces exerted on the forearm: the downward force of gravity, $m\vec{g}$, on the dumbbell, an upward force, $\vec{F}_B$, exerted by the biceps muscle, and a downward force, $\vec{F}_H$, exerted by the humerus at the elbow joint, O. (For this problem, neglect the downward force of gravity acting on the forearm itself.) What are the magnitudes of $\vec{F}_B$ and $\vec{F}_H$? Useful distances are shown in Figure 11-12 (a).

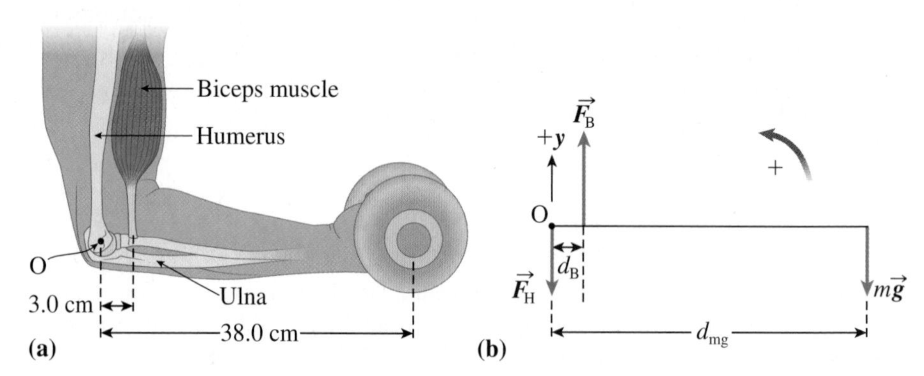

Figure 11-12 Sample Problem 11-4. **(a)** Arm holding a dumbbell. **(b)** Forces acting on the forearm.

Solution

There are two unknown quantities to be determined. If we choose the elbow joint O as the pivot point for torque calculations ($\Sigma\tau = 0$), then the moment arm for $\vec{F}_H$ will be zero and $\vec{F}_H$ will be eliminated from the calculations, thus allowing us to determine the magnitude of $\vec{F}_B$. Then we can use $\Sigma F_y = 0$ to determine the magnitude of $\vec{F}_H$.

Starting with $\Sigma\tau = 0$ with pivot point O and counterclockwise chosen arbitrarily as the positive direction for rotation (shown in Figure 11-12 (b)),

$$(F_H)(0) + F_B d_B - mgd_{mg} = 0$$

$$\therefore F_B = \frac{mgd_{mg}}{d_B} = \frac{(12.0 \text{ kg})(9.80 \text{ m/s}^2)(38.0 \text{ cm})}{3.00 \text{ cm}} = 1.49 \times 10^3 \text{ N}$$

(Notice that we did not need to convert the distances to metres, since only the ratio of the distances is involved.)

Now use $\Sigma F_y = 0$ with $+y$ upward:

$$F_B - F_H - mg = 0, \text{from which:}$$

$$F_H = F_B - mg = 1.49 \times 10^3 \text{N} - (12.0 \text{kg})(9.80 \text{ m/s}^2) = 1.37 \times 10^3 \text{N}$$

Therefore, the magnitudes of the force exerted by the biceps muscle and the force exerted by the humerus are 1.49×10^3 N and 1.37×10^3 N respectively. Note that the magnitudes of these forces are considerably larger than the magnitude of the gravitational force, $m\vec{g}$, acting on the dumbbell, which is only $(12.0 \text{ kg}) \times (9.80 \text{ m/s}^2) = 118$ N.

SAMPLE PROBLEM 11-5

A sign is hung from the centre of a uniform horizontal beam of length 2.4 m that is supported by a rope at an angle of 35° to the horizontal, as shown in Figure 11-13 (a). The beam is attached to a vertical wall. The force of gravity on the sign and beam has a magnitude of 4.5×10^2 N. Determine the magnitude of the tension in the rope and the magnitude and direction of the force exerted by the wall on the beam.

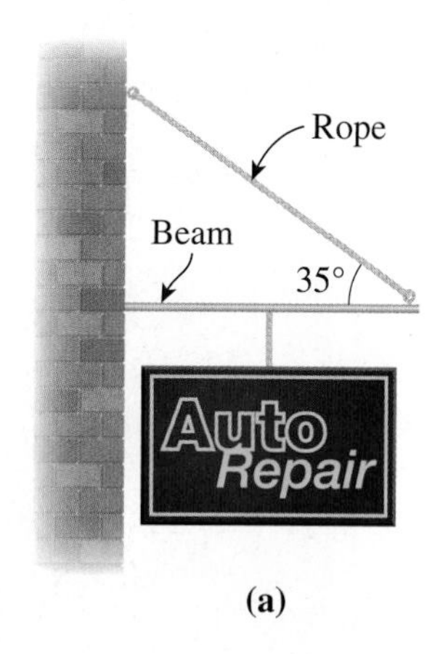

(a)

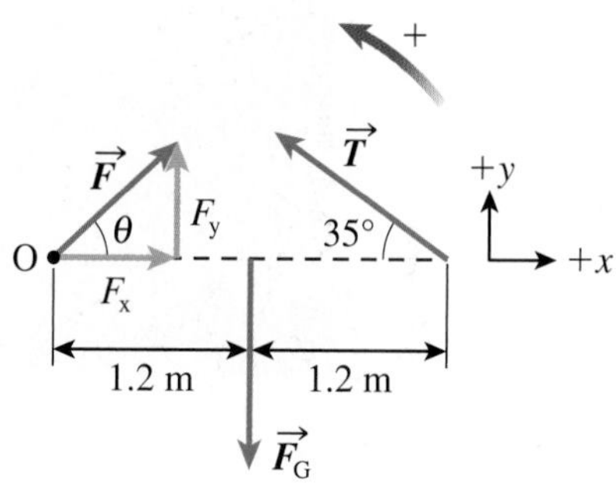

(b)

Figure 11-13 Sample Problem 11-5.
(a) A sign hanging from a beam supported by a rope. **(b)** Free-body diagram for the beam.

Solution

The FBD for the beam is shown in Figure 11-13 (b). The forces acting on the beam are the tension, $\vec{T}$, exerted by the rope, the force of gravity, $\vec{F}_G$, on the beam and sign, and the force $\vec{F}$ exerted by the wall. Since we do not know the direction of $\vec{F}$, we cannot determine a moment arm for this force unless we choose the pivot point carefully. However, we know the direction of $\vec{T}$, and so we can calculate its moment arm relative to any pivot point. We choose a pivot point O where the beam meets the wall so that the moment arm for $\vec{F}$ is zero, and this force will be eliminated from the torque calculations.

Use $\Sigma\tau = 0$ with pivot point O and the positive direction for rotation chosen arbitrarily as counterclockwise.

The moment arms for $\vec{F}_G$ and $\vec{T}$ are 1.2 m and (2.4 m)(sin 35°), respectively.

Therefore, $(F)(0) + (T)(2.4\text{ m})(\sin 35°) - (F_G)(1.2\text{ m}) = 0$.

$$T = \frac{(F_G)(1.2\text{ m})}{(2.4\text{ m})(\sin 35°)} = \frac{(4.5 \times 10^2\text{ N})(1.2\text{ m})}{(2.4\text{ m})(\sin 35°)} = 392\text{ N} = 3.9 \times 10^2\text{ N (to two significant digits)}$$

Since we now know the magnitudes and directions of $\vec{F}_G$ and $\vec{T}$, we use $\Sigma F_x = 0$ and $\Sigma F_y = 0$ to determine the x- and y-components of the unknown force $\vec{F}$. From these components we can find the magnitude and direction of $\vec{F}$. Choosing $+x$ to the right and $+y$ upward,

$$\Sigma F_x = 0 \text{ gives } F_x - T\cos 35° = 0$$

$$\therefore F_x = T\cos 35° = (392\text{ N})\cos 35° = 321\text{ N}$$

$$\Sigma F_y = 0 \text{ gives } F_y + T\sin 35° - F_G = 0$$

$$\therefore F_y = F_G - T\sin 35° = (4.5 \times 10^2\text{ N}) - (392\text{ N})\sin 35° = 225\text{ N}$$

The magnitude of $\vec{F}$ is $F = \sqrt{F_x^2 + F_y^2} = \sqrt{(321\text{ N})^2 + (225\text{ N})^2} = 3.9 \times 10^2\text{ N}$.
The direction of $\vec{F}$ can be specified by angle θ shown in Figure 11-13 (b):

$$\tan\theta = \frac{F_y}{F_x} \quad \therefore \theta = \tan^{-1}\left(\frac{F_y}{F_x}\right) = \tan^{-1}\left(\frac{225\text{ N}}{321\text{ N}}\right) = 35°$$

Therefore, the tension in the rope has a magnitude of 3.9×10^2 N, and the force exerted on the beam by the wall also has a magnitude of 3.9×10^2 N and a direction of 35° above the horizontal.

SAMPLE PROBLEM 11-6

Figure 11-14 (a) shows a rock-climber in static equilibrium. There are three forces acting on him: gravity downward, rope tension upward to the right, and a force exerted by the rock touching his boot. Figure 11-14 (b) shows the FBD for the climber. Since the force exerted by the rock has an unknown direction, the force vector $\vec{F}_{rock}$ has been

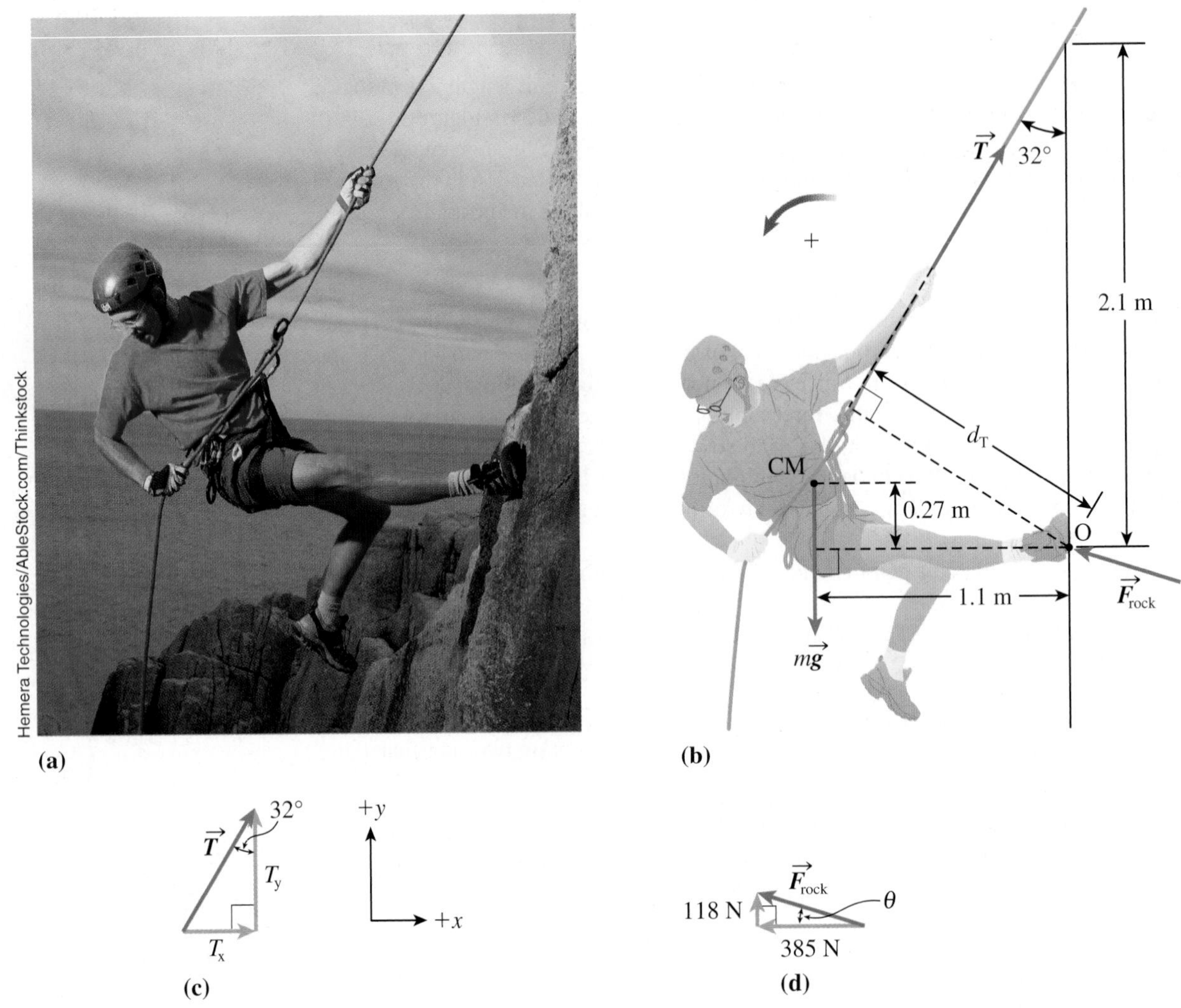

Figure 11-14 Sample Problem 11-6. **(a)** A rock-climber in static equilibrium. **(b)** Free-body diagram for the climber. **(c)** Components of $\vec{T}$. **(d)** Components of $\vec{F}_{rock}$.

drawn in an arbitrary direction. Figure 11-14 (b) also shows data, summarized in the next sentences. The rope makes an angle of 32° to the vertical. Relative to the point O where the boot touches the rock, the climber's CM is 1.1 m to the left and 0.27 m upward, and the vertical distance between O and the rope is 2.1 m. If the climber has a mass of 75 kg, determine

(a) the magnitude of the tension in the rope

(b) the magnitude and direction of the force $\vec{F}_{rock}$ exerted by the rock

Solution

(a) We will start by using $\Sigma\tau = 0$ with pivot point O. The moment arm and torque due to $\vec{F}_{rock}$ will be zero about O, and so by using $\Sigma\tau = 0$ we will be able to determine the magnitude of $\vec{T}$. Then for part (b) we can use $\Sigma F_x = 0$ and $\Sigma F_y = 0$ to determine the x- and y-components of $\vec{F}_{rock}$, and then its magnitude and direction.

As can be seen in Figure 11-14 (b), the moment arms about O for $m\vec{g}$ and $\vec{T}$ are 1.1 m and $d_T = (2.1 \text{ m})(\sin 32°)$ respectively. (Notice that the 0.27 m vertical distance related to the CM position has turned out to be irrelevant.) Using $\Sigma\tau = 0$ with the positive direction for rotation chosen as counterclockwise,

$$(F_{rock})(0) - (T)(2.1 \text{ m})(\sin 32°) + (mg)(1.1 \text{ m}) = 0$$

$$\therefore T = \frac{(mg)(1.1\ \text{m})}{(2.1\ \text{m})(\sin 32°)} = \frac{(75\ \text{kg})(9.8\ \text{m/s}^2)(1.1\ \text{m})}{(2.1\ \text{m})(\sin 32°)} = 727\ \text{N} = 7.3 \times 10^2\ \text{N (to two significant digits)}$$

Thus, the magnitude of the tension in the rope is 7.3×10^2 N.

(b) Figure 11-14 (c) shows the choice of $+x$- and $+y$-directions and the x- and y-components of $\vec{T}$, using the 32° angle that $\vec{T}$ makes with the vertical.
Using $\Sigma F_x = 0$,

$$T_x + F_{\text{rock,x}} = 0, \text{giving } F_{\text{rock,x}} = -T_x = -(727\ \text{N})\sin 32° = -385\ \text{N}$$

Using $\Sigma F_y = 0$,

$$T_y - mg + F_{\text{rock,y}} = 0, \text{giving } F_{\text{rock,y}} = mg - T_y = (75\ \text{kg})(9.8\ \text{m/s}^2) - (727\ \text{N})\cos 32° = 118\ \text{N}$$

The magnitude of $\vec{F}_{\text{rock}}$ is

$$F_{\text{rock}} = \sqrt{F_{\text{rock,x}}^2 + F_{\text{rock,y}}^2} = \sqrt{(-385\ \text{N})^2 + (118\ \text{N})^2} = 4.0 \times 10^2\ \text{N}$$

Figure 11-14 (d) shows the angle θ that $\vec{F}_{\text{rock}}$ makes with the horizontal, with θ given by

$$\theta = \tan^{-1}\left(\frac{118\ \text{N}}{385\ \text{N}}\right) = 17°$$

Therefore, the force exerted by the rock is 4.0×10^2 N at an angle of 17° above the horizontal.

SAMPLE PROBLEM 11-7

A woman of mass 62 kg is standing on a ladder of mass 11 kg and length 3.4 m. The ladder is leaning against a vertical wall that provides negligible friction. The angle between the floor and ladder is 59°. The woman is standing 2.1 m from the bottom of the ladder, as measured along the ladder. Determine

(a) the magnitude and direction of the force exerted by the wall on the ladder

(b) the magnitudes of the normal force and the static friction force exerted by the floor on the ladder

(c) the coefficient of static friction between the floor and ladder, if the ladder is on the verge of slipping

Solution

(a) Figure 11-15 (a) shows a FBD for the ladder and woman. The forces of gravity acting on the ladder and woman are $m_L\vec{g}$ and $m_W\vec{g}$, respectively. (Notice that $m_L\vec{g}$ acts at the CM of the ladder.) Since there is negligible friction between the ladder and wall, the wall exerts only a normal force $\vec{F}_{\text{N,wall}}$ on the ladder.[2] The floor exerts a normal force $\vec{F}_{\text{N,floor}}$ and a static friction force $\vec{F}_S$ on the ladder.
To determine the magnitude of $\vec{F}_{\text{N,wall}}$, we use $\Sigma\tau = 0$ with a pivot point O where the ladder touches the floor. Using this pivot point eliminates torques from both $\vec{F}_{\text{N,floor}}$ and $\vec{F}_S$ since these forces have a moment arm of zero relative to this point. For the three other forces, Figure 11-15 (b) shows the locations of the moment arms, given by

$$\text{for } m_L\vec{g},\ d_L = (1.7\ \text{m})\cos 59° = 0.876\ \text{m}$$

[2]Recall that friction is parallel to surfaces in contact, and normal force is perpendicular to the surfaces.

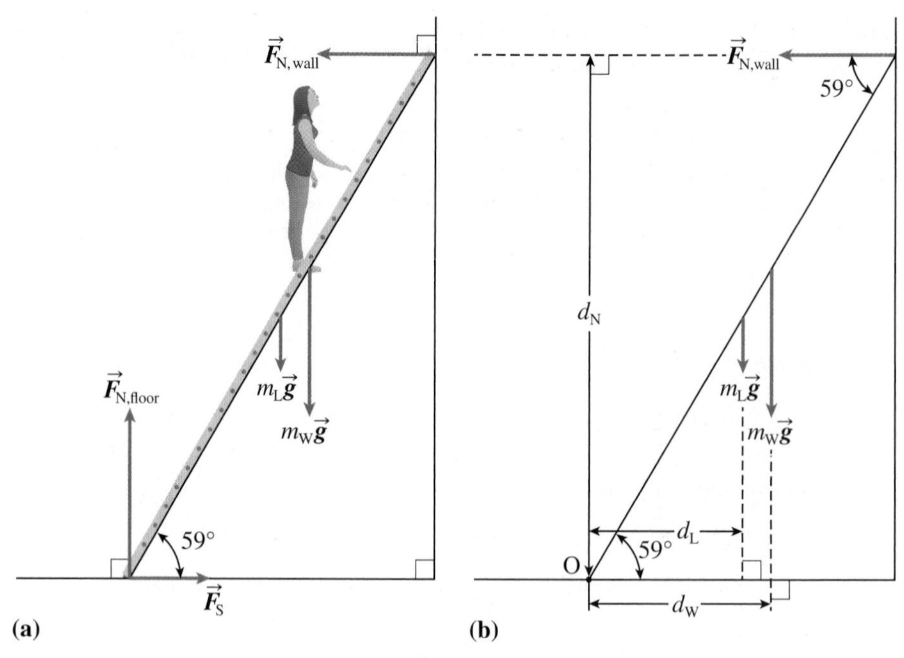

Figure 11-15 Sample Problem 11-7. **(a)** FBD for ladder and woman. **(b)** Moment arms relative to O for three forces.

$$\text{for } m_W\vec{g},\ d_W = (2.1\text{ m})\cos 59° = 1.08\text{ m}$$

$$\text{for } \vec{F}_{N,wall},\ d_N = (3.4\text{ m})\sin 59° = 2.91\text{ m}$$

Using $\Sigma\tau = 0$ with the positive rotation direction chosen as counterclockwise,

$$F_{N,wall}(2.91\text{ m}) - (m_W g)(1.08\text{ m}) - (m_L g)(0.876\text{ m}) = 0$$

$$F_{N,wall} = \frac{(m_W g)(1.08\text{ m}) + (m_L g)(0.876\text{ m})}{2.91\text{ m}}$$

$$= \frac{(62\text{ kg})(9.8\text{ m/s}^2)(1.08\text{ m}) + (11\text{ kg})(9.8\text{ m/s}^2)(0.876\text{ m})}{2.91\text{ m}}$$

$$= 2.6 \times 10^2\text{ N}$$

Thus, the force exerted by the wall has a magnitude of 2.6×10^2 N and a direction perpendicular to the wall.

(b) The magnitude of the normal force $\vec{F}_{N,floor}$ exerted by the floor on the ladder can be determined from $\Sigma F_y = 0$. With $+y$ chosen upward, $F_{N,floor} - m_L g - m_W g = 0$, giving

$$F_{N,floor} = (m_L + m_W)g = [(11 + 62)\text{ kg}](9.80\text{ m/s}^2) = 7.2 \times 10^2\text{ N}$$

The magnitude of the static friction force $\vec{F}_S$ can be calculated using $\Sigma F_x = 0$. With $+x$ chosen toward the right,

$$F_S - F_{N,wall} = 0, \text{ which gives } F_S = F_{N,wall} = 2.6 \times 10^2\text{ N}$$

Therefore, the magnitudes of the normal force and the static friction force exerted by the floor on the ladder are 7.2×10^2 N and 2.6×10^2 N respectively.

(c) If the ladder is on the verge of slipping, the static friction force has its maximum value, given by Eqn. 6-2: $F_{S,max} = \mu_S F_N$, where $F_N = F_{N,floor}$ in this problem.

$$\therefore \mu_S = \frac{F_{S,\,max}}{F_{N,floor}} = \frac{2.6 \times 10^2\ \text{N}}{7.2 \times 10^2\ \text{N}} = 0.36$$

Thus, if the ladder is on the verge of slipping, the coefficient of static friction between the floor and the ladder is 0.36.

EXERCISES

11-6 Can an object be in static equilibrium if only one force acts on it? Explain.

11-7 Give an example in which the net torque is zero but the net force is not zero.

11-8 Give an example in which the net force is zero but the net torque is not zero.

11-9 Two children are sitting on a teeter-totter that consists of a uniform wooden plank of length 6.0 m that rests on a pivot at its midpoint. One child of mass 28 kg sits at a distance of 2.4 m from the pivot. Where should a child of mass 32 kg sit in order that the teeter-totter be in static equilibrium?

11-10 **Fermi Question:** Suppose that you are just finishing the tightening of an automobile wheel nut that is now in static equilibrium. Estimate the maximum torque that you are applying to the wheel nut if you are using a tire wrench that is 40 cm long.

11-11 Two men are holding a uniform horizontal ladder on their shoulders (Figure 11-16). The ladder is 4.90 m long and has a mass of 9.1 kg. One man (#1) is supporting the ladder at 0.30 m from one end, and the other man (#2) is providing support at 1.30 m from the other end. Determine the magnitude of the supporting force exerted by each man.

bikeriderlondon/Shutterstock

Figure 11-16 Question 11-11.

11-12 A 1.6 kg fish is hanging from a horizontal fishing rod of length 1.7 m (Figure 11-17). A girl is holding onto the rod using her hands, as it rests against her stomach. The girl's hands are holding the rod at a point that is 45 cm from her stomach, and the force exerted by her hands makes an angle of 60° (two significant digits) relative to the vertical. Assume that the rod is rigid and has negligible mass.

(a) What is the magnitude of the force exerted on the rod by the girl's hands?

(b) What are the magnitude and direction of the force exerted on the rod by the girl's stomach?

(c) What are the magnitude and direction of the force exerted by the rod on the girl's stomach?

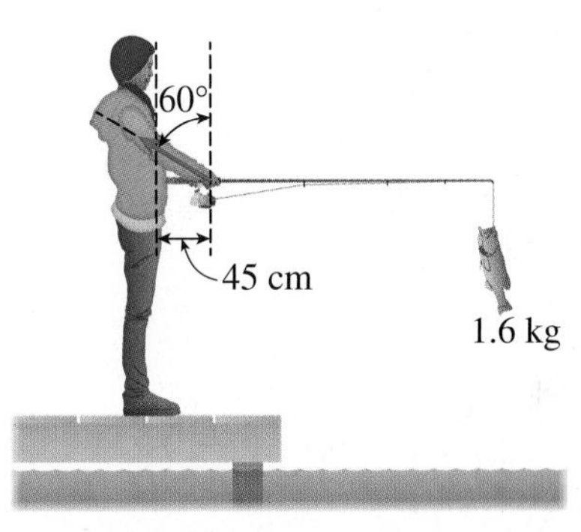

Figure 11-17 Question 11-12.

11-13 A uniform ladder having a length of 4.8 m and a mass of 15 kg is leaning against a vertical wall at an angle of 61° above the horizontal. Assume that there is friction at the floor but not at the wall. Determine the magnitude and direction of the force exerted on the ladder by (a) the wall and (b) the floor.

11-14 Repeat Question 11-13 if there is a 56 kg person standing 3.0 m from the bottom of the ladder (as measured along the ladder).

11-15 For Question 11-14 what is the minimum coefficient of static friction needed to prevent the ladder from slipping along the floor?

11-16 Figure 11-18 shows a person's arm held extended away from the body at an angle $\theta = 46°$ relative to the vertical. The arm is held in this position by the deltoid muscle, which is attached to the upper arm bone midway between the joint O and the centre of mass of the arm at C. If the mass of the arm is 3.8 kg and angle $\phi = 31°$, determine the magnitude of the tension T in the deltoid muscle.

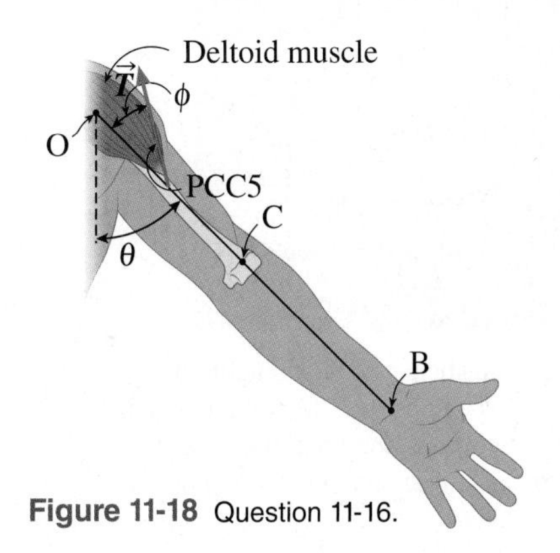

Figure 11-18 Question 11-16.

(a)

Jane McLoughlin/Shutterstock

(b)

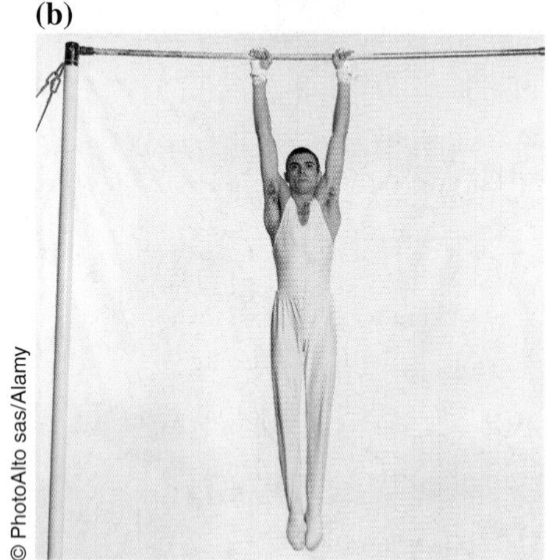

© PhotoAlto sas/Alamy

Figure 11-19 **(a)** Unstable equilibrium; CM above small support base. **(b)** Stable equilibrium; CM below support.

unstable equilibrium a situation in which an object in static equilibrium tends to fall or keep moving away from the equilibrium position after being disturbed

stable equilibrium a situation in which an object will return to equilibrium after being disturbed

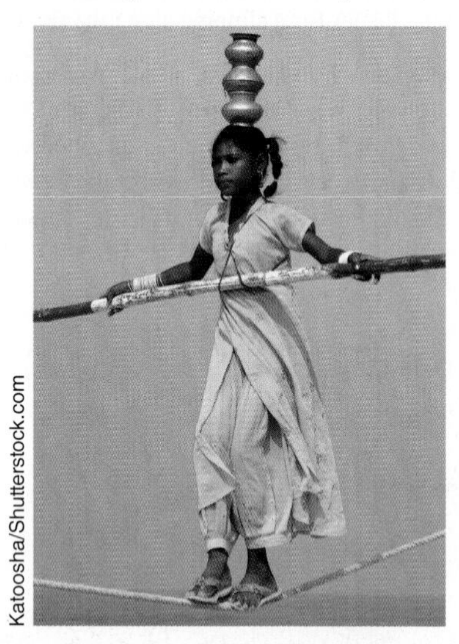

Katoosha/Shutterstock.com

Figure 11-20 Stability is greater if a tightrope-walker carries a pole, since the CM of the pole and person together is lower than that of the person alone. The pole also increases the moment of inertia, reducing rotation.

11.3 Stability

Consider the photograph shown in Figure 11-19 (a). The top gymnast is in static equilibrium with his CM above his supporting hand, but he is in a precarious position. If he were nudged only a little, his CM would move so that it would no longer be above his hand. The force of gravity acting at his CM would produce a torque about his hand, causing him to rotate and fall. He is said to be in **unstable equilibrium**, which refers to a situation in which an object in static equilibrium tends to fall or keep moving away from the equilibrium position after being disturbed.

In contrast, the CM of the gymnast in Figure 11-19 (b) is below the supporting rod. If this gymnast is nudged, the torque caused by the force of gravity acting at his CM will rotate him back toward the equilibrium position, and he will oscillate back and forth about that position, eventually coming to rest there. He is in **stable equilibrium**: a situation in which an object will return to equilibrium after being disturbed. An object can also be in stable equilibrium if its CM is above the area of support, if this area is large. For example, an automobile is normally in stable equilibrium because the supporting area is the large rectangle formed between the positions of the four tires. A small disturbance will not cause the automobile to topple. Similarly, a structure in the shape of a pyramid is generally in stable equilibrium.

From these definitions and examples, we can see that an object will be in unstable equilibrium if it has a small base and a CM high above the base. If an object has its CM below the support point or area, or if its CM is only a short distance above a large support base, it will be in stable equilibrium.

Stability and the Human Body

If you are standing on both feet, your support base consists of the rectangle enclosed by your feet (including your feet). As long as your CM is above this rectangle, you can easily maintain static equilibrium. If you stand on one foot or on tiptoes, you must move your CM to be above the smaller support base.

 TRY THIS!

Keeping Your Body Stable

Start by standing on two feet, and think about the location of your CM relative to the area enclosed by your feet. Now try to keep your CM in the same location and raise one foot. What happens? Adjust your CM position so that you can stand easily on one foot. Next, stand perpendicular to a wall with the right side of your body against the wall. Your right shoulder, hip, knee, and ankle should be touching the wall. Try to stand on your right foot. What happens? Why? Now move away from the wall and stand on your tiptoes. In which direction do you have to move your CM to do this? Now stand facing the wall and move toward it until your toes and your nose are touching it. Put your hands behind your back and try standing on your tiptoes. Again, what happens and why? Finally, move away from the wall and bend over, trying to touch your toes without bending your knees. While you are doing this, your CM actually moves outside your body to a position between your chest and your arms. Now try to bend forward to touch your toes with your back and legs against a wall.

An example of the application of stability to the human body occurs in tightrope walking (Figures 11-1 and 11-20). Have you ever noticed that the acrobats who walk on tightropes often carry a long pole that is heavily weighted at the two ends? The pole bends downward at the ends with the result that the CM of the person and pole together is lower than the CM of the person alone, therefore increasing stability. As well, the pole has a large moment of inertia, thus reducing the amount of rotation arising from any external torques.

Other Examples of Stability

The designs of various "balancing toys" are based on the main criterion of stable equilibrium: the CM of the object is below the support. Figure 11-21 shows examples of two such toys.

As a final example of stability, consider the famous Leaning Tower of Pisa (Figure 11-23), which is located in Italy in a city where Galileo once lived. In spite of the lean, the support base of the tower is still under the tower's CM, so the tower is in equilibrium.

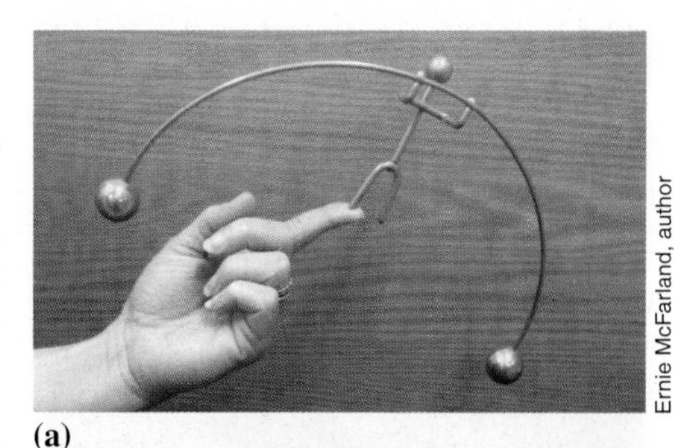

(a)

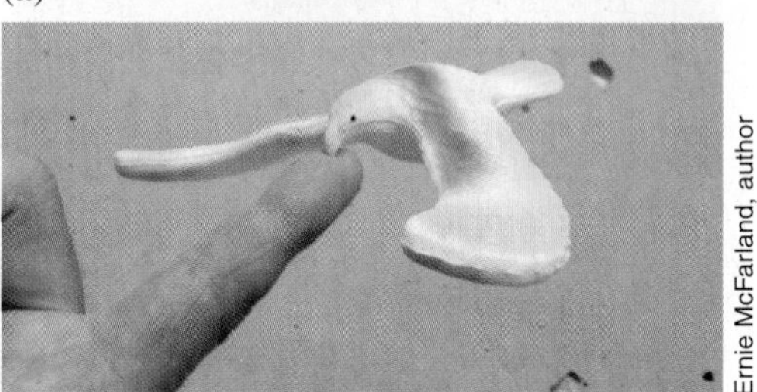

(b)

Figure 11-21 These toys are designed so that the CM is below the support, which is the finger in each case. (The bird has metal weights inside its wingtips.)

Strange Balancing in the Kitchen

Figure 11-22 shows an unusual balancing demonstration that you can create. Attach a spoon to a fork by sliding the tines of the fork around the round end of the spoon. Then insert a toothpick between the tines of the fork and balance the fork–spoon combination on the toothpick on the rim of a glass, as shown. You might have to try various sizes of forks and spoons in order to achieve the balancing. Because the handles of the fork and spoon point down and inward toward the glass, the CM of the fork–spoon–toothpick combination is located below the contact point between the toothpick and the glass. You can add a little drama by setting the ends of the toothpick on fire; the flames will travel along the toothpick and stop at the points where the toothpick touches the glass rim and the fork and spoon. Why do the flames stop there?

Figure 11-22 The CM of the fork–spoon–toothpick system is below the support point where the toothpick touches the glass.

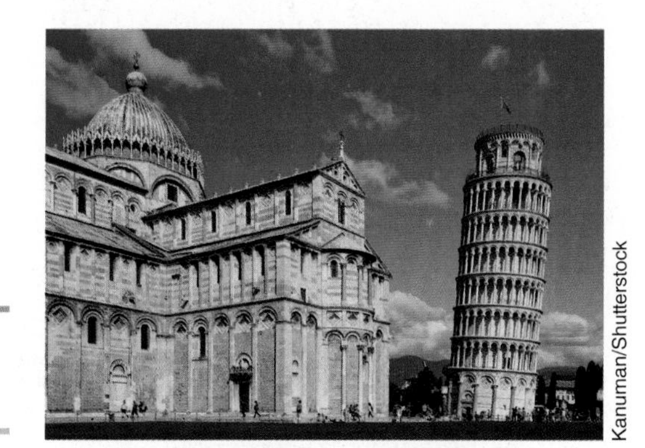

Figure 11-23 The Leaning Tower of Pisa.

SAMPLE PROBLEM 11-8

A heavy crate is sitting on the ramp at the rear of a truck. The ramp is inclined at an angle φ relative to the horizontal. The crate has a height of 1.4 m, and its length and width are each 1.0 m. Its CM is at its geometric centre.

(a) What is the maximum angle φ if the crate is not to tip over? Assume there is sufficient friction to prevent the crate from sliding.

(b) Repeat the question with the CM at a higher point, 1.0 m away from the lower edge of the crate.

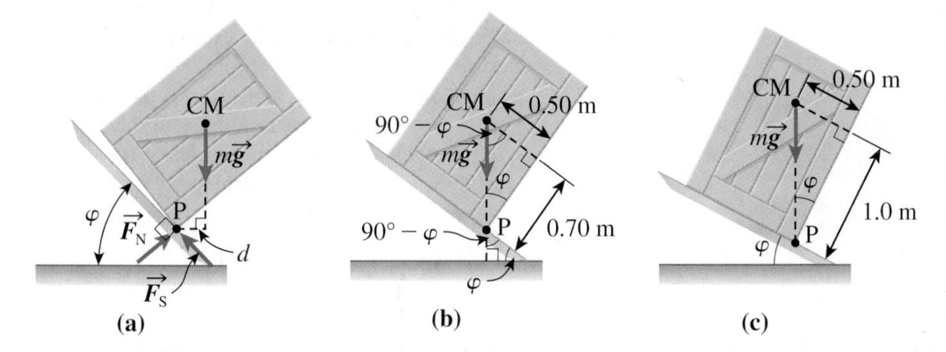

Figure 11-24 Sample Problem 11-8. **(a)** The crate is tipping about P because angle φ is large. Force $m\vec{g}$ points to the right of P. **(b)** Angle φ is at the maximum value for which the box does not tip, and $m\vec{g}$ points directly through P. **(c)** With the CM higher in the crate, φ becomes smaller.

DID YOU KNOW?

The loon (Figure 11-25) is a beautiful bird that is well adapted to water, where its feet help it dive and swim quickly. However, on land the loon is very awkward because its CM lies ahead of its feet, so it has to try to lean backward when it walks.

Brian Lasenby/Shutterstock

Figure 11-25 Loons are awkward walkers because of the position of their CM.

Solution

(a) Figure 11-24 (a) shows the forces that would be acting if angle φ is somewhat greater than the maximum value that we are asked to find. The crate has already tipped a small amount about an axis at the bottom front edge of the crate because of a torque due to the force of gravity, $m\vec{g}$, which has a moment arm d about this axis. The axis is seen end-on, indicated by point P. Because the crate has already tipped somewhat, the only contact between the crate and the ramp is at the bottom front edge, and the normal force, $\vec{F}_N$, and the static friction force, $\vec{F}_S$, act along this edge. These two forces each have a moment arm of zero relative to the rotation axis, and therefore provide no torque.

Figure 11-24 (b) shows the situation at the maximum angle φ for the crate not to tip, that is, when the crate is on the verge of tipping. The line of action of $m\vec{g}$ now passes right through the rotation axis and produces no torque. Forces $\vec{F}_N$ and $\vec{F}_S$ act along the front edge as in Figure 11-24 (a) and exert zero torque as well. Two similar triangles are shown, each with interior angles of 90°, φ, and 90° − φ. In the upper triangle, the CM is shown at its position halfway (0.70 m) from the base of the crate and halfway (0.50 m) from the side. In this triangle,

$$\varphi = \tan^{-1}\left(\frac{0.50\text{ m}}{0.70\text{ m}}\right) = 36°.$$

Therefore, the maximum angle is 36° if the crate is not to tip.

(b) Figure 11-24 (c) shows the CM now 1.0 m from the lower edge of the crate, instead of 0.70 m as in Figure-24 (b). The basic geometry is similar to that in Figure 11-24 (b), and the maximum angle φ is now given by $\varphi = \tan^{-1}\left(\frac{0.50\text{ m}}{1.0\text{ m}}\right) = 27°$.

Thus, for a higher CM, the maximum angle for no tipping has been reduced to only 27°. Raising the CM has clearly decreased the stability of the crate.

EXERCISES

11-17 Why is a large mass placed on the short end of a construction crane (Figure 11-26)?

Snvv/Shutterstock

Figure 11-26 Question 11-17.

11-18 Rank the filing cabinets shown in Figure 11-27 in order of stability, from least stable to most stable. All the drawers are equally filled.

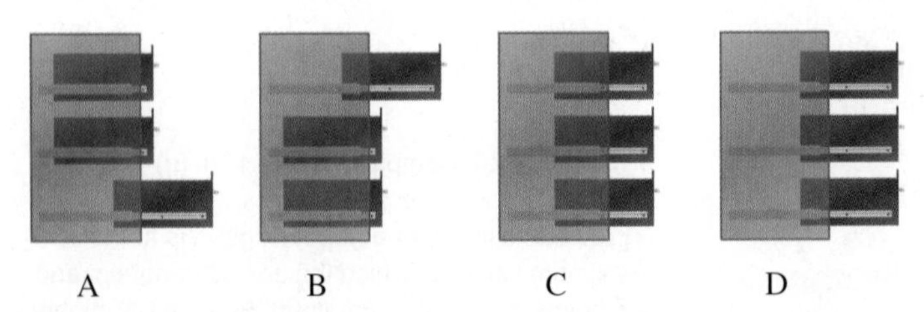

Figure 11-27 Question 11-18.

11-19 Describe how you would shift your body, or part of it, to maintain equilibrium in each case:

(a) You place a heavy backpack on your shoulders.

(b) You lift a heavy basket filled with apples from a tabletop using both hands.

(c) You lift a heavy suitcase with your left arm.

11-20 Is the long pole used by a tightrope walker more beneficial if it is straight, or if it is drooping? Explain.

11-21 A top-loading washing machine of mass 29 kg is sitting on a horizontal floor. Its CM is located along the central axis of the machine at a height of 62 cm above the floor, and 46 cm from each of the front, back, and sides. A person wants to tip the machine along its lower rear edge by exerting a horizontal force at its top front edge, which is 98 cm above the floor. This force is directed from the front to the rear of the machine. Assuming that friction between the machine and the floor is large enough to prevent sliding, what magnitude of force must the person exert in order to have the machine be on the verge of tipping?

11.4 Stress and Strain; Elasticity

In Section 11.2 we studied the forces and torques on objects in equilibrium, but we treated the objects as if they were totally rigid. All objects deform when forces and torques are applied; even extremely tough materials deform slightly (Figure 11-28). In this section we will study the deformation effects of forces and torques on materials and structures.

Figure 11-28 Even steel girders deform as a result of forces and torques exerted on them.

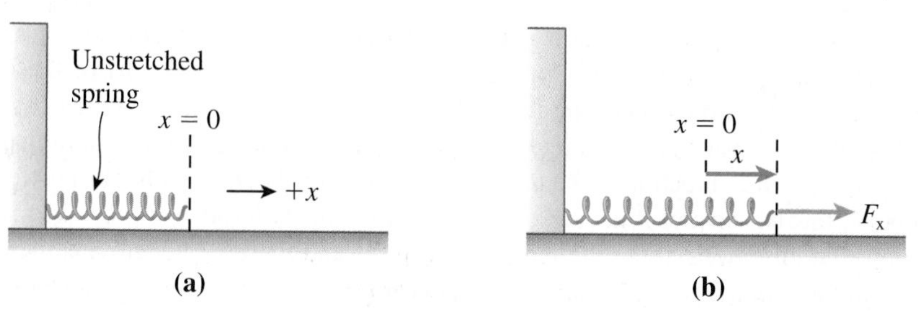

Figure 11-29 **(a)** An unstretched horizontal spring. **(b)** The spring stretches a distance x when pulled by force (component) F_x.

We begin by considering the stretching of a spring that is fixed at one end and can stretch horizontally as a result of an applied force (component) F_x (Figure 11-29). We can neglect friction in our analysis. Experiments show that the elongation, x, of the spring is proportional to F_x, as long as the force is not too large; limitations will be discussed later in this section. The proportionality between F_x and x is usually written as an equation, $F_x = kx$, where k is the **spring constant** for a given spring. The SI unit of k is newton per metre (N/m), and the spring constant represents the force per unit elongation of the spring, or the stiffness of the spring. A stiff spring has a large spring constant, and a soft spring has a small spring constant. If k is large, a large force is required to produce a given elongation. Sometimes the above equation is written as $F_x = -kx$, where F_x now represents the force exerted *by* the spring on whatever is stretching it, rather than the force exerted *on* the spring. In either case, the relation $F_x = \pm kx$ is referred to as **Hooke's law**,[3] named after the English scientist Robert Hooke (1635–1703). In summary,

spring constant force required per unit elongation of a spring; a measure of stiffness of a spring; symbol k, SI unit: newton per metre (N/m)

Hooke's law the proportional relation between the amount of elongation and the force causing the elongation; written as $F_x = \pm kx$

$$\text{Hooke's law: force exerted } on \text{ spring } F_x = kx \quad (11\text{-}4a)$$

$$\text{force exerted } by \text{ spring } F_x = -kx \quad (11\text{-}4b)$$

We now generalize Hooke's law, which is specifically for springs, to wide-ranging situations involving deformations caused by forces or torques. Think about the variety of ways you could deform a rod-shaped piece of rubber. You could pull on both ends to lengthen it a little, or push on both ends to shorten it, or you could twist it. You could also let it sink to the bottom of the ocean, where it would be squeezed by the high pressure there and shortened in all dimensions at once. For each type of deformation we define a quantity called **stress**, which is related to the force causing the deformation on a force-per-unit-area basis. We define another quantity, **strain**, which is a measure of the amount of deformation that occurs as a result of the stress. If the stress is small enough, these two quantities are related by a proportional relation with the constant of proportionality being the **elastic modulus**:

stress a quantity related to the force causing a deformation, on a force-per-unit-area basis

strain a measure of the amount of deformation that occurs as a result of a stress

elastic modulus the constant of proportionality between stress and strain (if the stress is sufficiently small)

$$\text{stress} = \text{elastic modulus} \times \text{strain} \quad (11\text{-}5)$$

[3] Although the relation $F_x = \pm kx$ is called "Hooke's law," it is really an experimental observation rather than a fundamental law of nature.

Notice that Eqn. 11-5 has the same general form as Hooke's law (Eqn. 11-4). The elastic modulus is analogous to the spring constant in Hooke's law. If an object has a large elastic modulus, it is difficult to deform. For each different type of deformation there is an equation of the same general form of Eqn. 11-5, but with differences depending on how the deformation is produced.

Tensile Stress and Strain; Young's Modulus

The simplest type of deformation is the stretching of an object as a result of a force pulling on it. Of course, if only one force acts on the object, the CM of the object accelerates in the direction of the force. Therefore, in order to have the object stretch without the CM moving, we need to have two forces of equal magnitude pulling on the object in opposite directions. The top part of Figure 11-30 shows a bar of unstretched length L_0 that has an area A at each end. Below this, forces $\vec{F}$ and $-\vec{F}$ are shown acting in opposite directions along the length of the bar, and the bar's length has increased by an amount ΔL. The bar is said to be under tension; this is similar to the tension in ropes, wires, etc., introduced in Chapter 5.

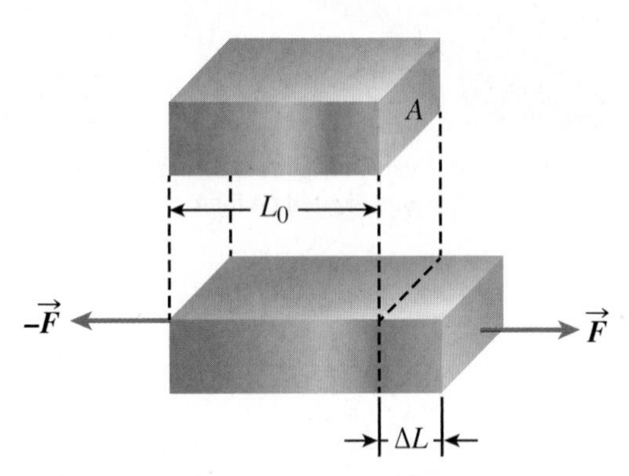

Figure 11-30 A bar of unstretched length L_0 and cross-sectional area A is stretched by forces $\vec{F}$ and $-\vec{F}$, producing an increase in length, ΔL.

tensile stress (stress) ratio of the magnitude of the external force $\vec{F}$ stretching an object to the cross-sectional area A of the object

tensile strain (strain) for an object that has been stretched, the ratio of the increase in length, ΔL, to the original unstretched length, L_0

Young's modulus ratio of tensile stress to tensile strain, for stresses small enough that strain and stress are proportional

The **tensile stress** (often just called **stress**) acting on the bar is defined as the ratio of the magnitude of the external force $\vec{F}$ to the area A of the surface to which the force is applied. The SI unit of tensile stress is newton per square metre (N/m^2), which is given the name of pascal (Pa), in honour of Blaise Pascal (1623–1662), a French physicist. As you will see in Section 11.5, the pascal is also the unit of pressure.

$$\text{tensile stress} = \frac{F}{A} \quad \textbf{(11-6)}$$

The **tensile strain** (often just called **strain**) is the ratio of the increase in length, ΔL, to the original unstretched length, L_0. Since tensile strain is a ratio of two lengths, always measured in the same units, it is dimensionless and has no units.

$$\text{tensile strain} = \frac{\Delta L}{L_0} \quad \textbf{(11-7)}$$

Experiments show that for sufficiently small tensile stress, the stress and strain are proportional to each other. The corresponding elastic modulus in this case is called **Young's modulus**,[4] symbolized as Y, and Eqn. 11-5 is written as

$$\frac{F}{A} = Y\frac{\Delta L}{L_0} \quad \textbf{(11-8)}$$

Since strain ($\Delta L/L_0$) is dimensionless, then Young's modulus (Y) has the same unit as stress (F/A), that is, newton per square metre, or pascal. A large value of Y indicates that a material is difficult to stretch, and a small value means the material is easy to stretch. Table 11-2 provides representative values of Young's modulus for various materials.

Table 11-2

Approximate Values of Young's Modulus Y

Material	Y (Pa)
Cartilage	$(1–4) \times 10^7$
Rubber	$(1–10) \times 10^7$
Tendon	$(2–10) \times 10^7$
Wood	$(9–11) \times 10^9$
Bone	$(1–2) \times 10^{10}$
Glass	$(5–9) \times 10^{10}$
Copper	1.1×10^{11}
Steel	2×10^{11}
Tungsten	4.1×10^{11}

[4]Young's modulus is named after Thomas Young, an English physician and physicist (1773–1829). A profile of Young and his achievements is given in Section 16.2.

Eqn.11-8 can be rearranged to give $\Delta L = \frac{FL_0}{AY}$. This shows that the distance ΔL that an object stretches is proportional to the unstretched length L_0. This means that an object twice as long will stretch twice as far for the same applied force. The distance stretched is inversely proportional to the cross-sectional area A of the object. Not surprisingly, the amount of stretch is proportional to the magnitude F of the applied force, and inversely proportional to Young's modulus Y.

The same definitions of stress and strain apply if the two forces push inward on the object, rather than pull outward. The object is then shortened, or compressed, and is said to be under **compression**. In this case, the stress and strain are called **compressive stress** and **compressive strain**. Since the change in length is negative, and since all quantities in Eqn. 11-8 are considered as being positive, ΔL in this case is interpreted as the *magnitude* of the change in length.

For many materials, Young's modulus has the same value for tension and compression. However, bone has a larger Young's modulus for tension than compression and, in contrast, concrete and stone have larger Y-values for compression than for tension. In other words, bone deforms more under compression than under tension, whereas concrete and stone deform more under tension.

compression the shortening of an object as a result of forces pushing inward on it

compressive stress ratio of the magnitude of the external force $\vec{F}$ compressing an object to the cross-sectional area, A, of the object

compressive strain for an object that has been compressed, the ratio of the magnitude of the decrease in length, ΔL, to the original unstretched length, L_0

DID YOU KNOW?

When fashion designers of the 19th century wanted to emphasize small female waists, they used tightly tied lace patterns surrounded above and below by great puffs of material. In the 20th century, with fewer maids to help tie the lace dresses for other women, women's fashions changed, and dresses made simply of hanging cloth were in style. Dresses made of material with fibres running vertically and horizontally experienced stresses that caused the dresses to wrinkle without clinging to the waist. This problem was overcome with the *bias cut*, made by a fashion designer in Paris, Madeleine Vionnet, in the 1920s. With her design, dresses were made with the fibres of the cloth at a 45° angle to the vertical (Figure 11-31), and the tensile stress due to the weight of the dress itself caused the material to stretch lengthwise and contract widthwise. This resulted in a garment that clings to the waist, for example, emphasizing the fashion lines desired.

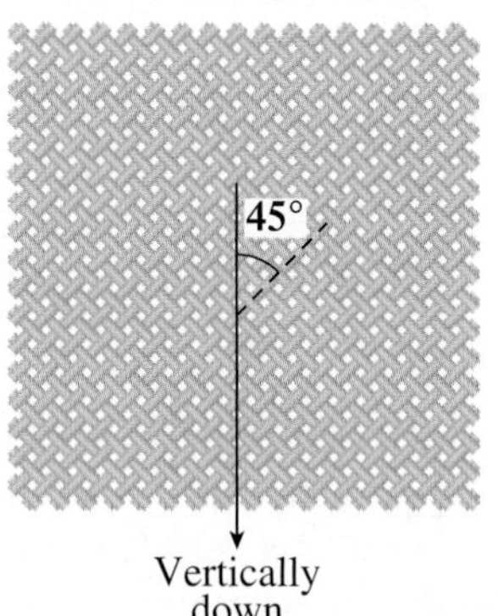

Figure 11-31 Garments made with bias-cut cloth have the fibres at 45° to the vertical, resulting in a garment that drapes softly and accentuates body lines.

SAMPLE PROBLEM 11-9

A 65 kg chandelier hangs from the ceiling of an auditorium on a single aluminum wire that is 8.5 m long and 0.16 cm^2 in cross-sectional area. Young's modulus for aluminum is 7.0×10^{10} Pa.

(a) Determine the stress applied to the wire.

(b) What is the resulting strain in the wire?

(c) By how much will the wire stretch, in millimetres?

Solution

(a) From Eqn. 11-6, stress $= \frac{F}{A}$. In this case, the relevant force is the force of gravity acting on the chandelier, which has a magnitude of

$$F = mg = (65\,\text{kg})(9.8\ \text{m/s}^2) = 637\,\text{N}$$

The area is $A = 0.16\,\text{cm}^2 \times \left(\frac{1\,\text{m}}{100\,\text{cm}}\right)^2 = 1.6 \times 10^{-5}\,\text{m}^2$.

$$\text{Stress}\ \frac{F}{A} = \frac{637\,\text{N}}{1.6 \times 10^{-5}\,\text{m}^2} = 4.0 \times 10^7\,\frac{\text{N}}{\text{m}^2} = 4.0 \times 10^7\,\text{Pa}$$

Thus, the stress is 4.0×10^7 Pa.

(b) Starting with Eqn. 11-8, $\frac{F}{A} = Y\frac{\Delta L}{L_0}$, the resulting strain is given by

$$\frac{\Delta L}{L_0} = \frac{1}{Y}\frac{F}{A} = \frac{1}{7.0 \times 10^{10}\,\text{Pa}}(4.0 \times 10^7\,\text{Pa}) = 5.7 \times 10^{-4}$$

The strain is 5.7×10^{-4}.

(c) From $\frac{\Delta L}{L_0} = 5.7 \times 10^{-4}$,

$$\Delta L = (5.7 \times 10^{-4})L_0 = (5.7 \times 10^{-4})(8.5\,\text{m}) = 0.0048\,\text{m} = 4.8\,\text{mm}$$

Therefore, the wire will stretch by 4.8 mm as a result of the hanging chandelier.

SAMPLE PROBLEM 11-10

Determine the amount of compression of a femur (thighbone) in millimetres for a person who is standing vertically on both legs, and whose upper-body mass is 56 kg. The inner and outer radii of the bony part of the femur are r = 1.3 cm and R = 2.0 cm respectively (Figure 11-32), and the length of the femur is 45 cm. Young's modulus for the bone is 2.0×10^{10} Pa.

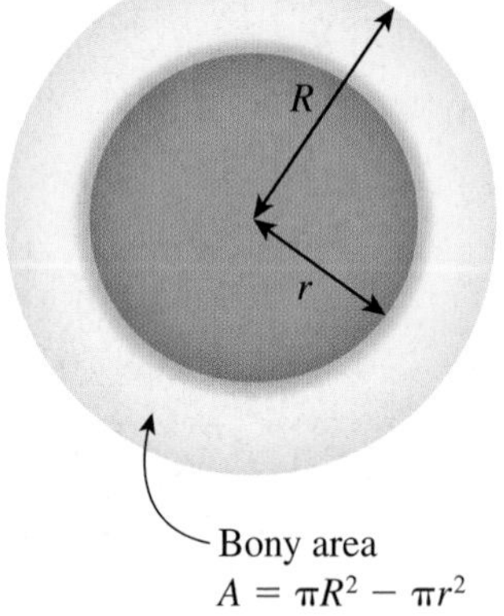

R

r

Bony area

$A = \pi R^2 - \pi r^2$

Figure 11-32 Cross-section of femur (Sample Problem 11-10).

Solution

We will use Eqn. 11-8, $\frac{F}{A} = Y\frac{\Delta L}{L_0}$, to find ΔL.

The central part of the femur is largely bone marrow, which provides effectively no support. Therefore, the area A needed is the cross-sectional area of the bony part of the femur. As can be seen in Figure 11-32, this area is determined by subtracting the area πr^2 of the inner marrow from the total area πR^2 of the femur: $A = \pi R^2 - \pi r^2 = \pi(R^2 - r^2)$

Since the person is standing on two legs, the force acting on one femur is one-half the force of gravity on the upper body: $F = \frac{1}{2}mg$.

From Eqn. 11-8, $\Delta L = \frac{FL_0}{YA} = \frac{(\frac{1}{2}mg)L_0}{Y\pi(R^2 - r^2)}$

$$\therefore \Delta L = \frac{\frac{1}{2}(56\ \text{kg})(9.8\ \text{m/s}^2)(0.45\ \text{m})}{(2.0 \times 10^{10}\ \text{Pa})\pi((0.020\ \text{m})^2 - (0.013\ \text{m})^2)} = 8.5 \times 10^{-6}\ \text{m} = 8.5 \times 10^{-3}\ \text{mm}$$

Thus, the femur is compressed by 8.5×10^{-3} mm.

Elasticity and Plasticity

We have stated that if stress is small enough, then strain is proportional to it, and a graph of stress versus strain is linear. But what happens when stress is not "small enough"? As stress continues to increase, eventually there is a limit to the proportionality. Figure 11-33 shows two typical stress–strain graphs; each graph starts with a straight line with slope equal to Young's modulus (consistent with Eqn. 11-8). However, at point A in each graph the linearity ends, and the stress at this point is known as the **proportional limit** for the material. Beyond this point, the graph can display a variety of shapes, as shown.

As the applied stress continues to increase beyond the proportional limit, the strain of the material increases and eventually point B on the graph is reached. The stress at this point is the **elastic limit**, and the region of the graph between the origin and B is the **elastic range**. The term **elasticity** refers to the ability of a material, having been stressed, to return to its original shape once the stress is removed. As long as the stress is below the elastic limit, the material remains in the elastic range and exhibits elasticity. When the stress is removed, many materials retrace the same stress–strain curve back to the origin of the graph as they relax. However, some materials, especially biomaterials and polymers, follow a different graph back to their original size and shape.

If the stress continues to increase beyond point B, and then the stress is removed, the material will not return to its original size. It will return to a zero-stress point by following a completely different stress–strain path, and will end with a positive strain when the stress has returned to zero. The material is said to exhibit **plasticity**, which means that it does not return to its initial size and shape, and the region of the stress–strain graph is the **plastic range**.

As stress increases even more, eventually the material will break, and the point C on the stress–strain graph at which this occurs is called, not surprisingly, the **breaking point** or the **fracture point**, which occurs at the **ultimate strength** of the material, that is, the maximum stress it can withstand without breaking. For some materials, there is a large increase in strain between the elastic limit and the breaking point, as in Figure 11-33 (a), and the material is said to be **ductile**, which means that it stretches a lot before breaking. Gold, silver, copper, and lead are examples of ductile materials. If there is only a small increase in strain (i.e., little deformation) between the elastic limit and fracture point (Figure 11-33 (b)), the object is **brittle**. Examples include most glasses and ceramics. Brittle objects often make a cracking sound when they break, and usually the broken pieces fit well together (Figure 11-34).

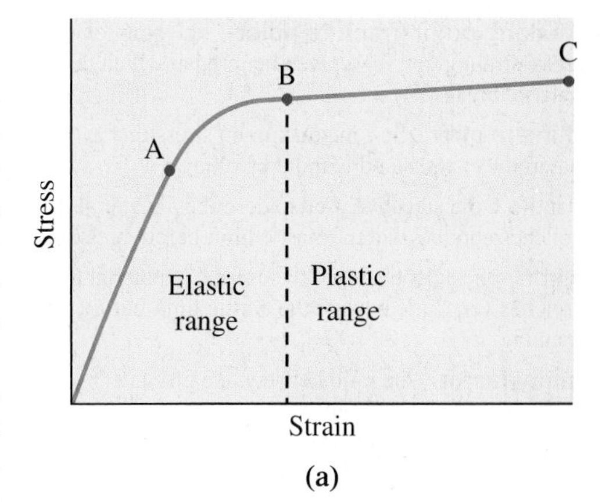

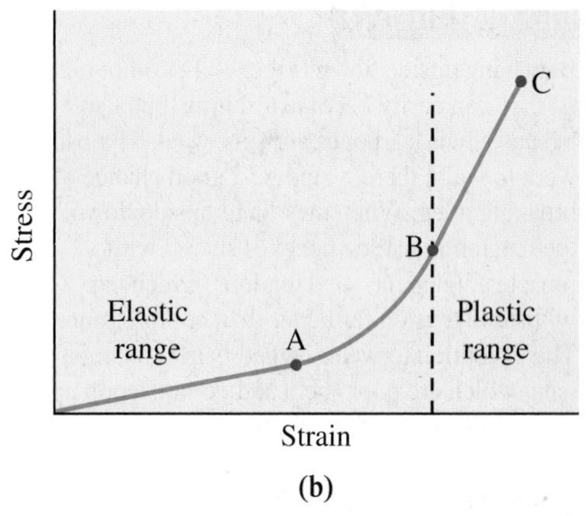

Figure 11-33 Two possible stress–strain relationships.

Figure 11-34 Glass is brittle, and broken pieces fit together easily.

DID YOU KNOW?

Samson Tech 12 is a type of fibre rope that has been engineered for exceptional strength as well as heat resistance. It is made from 12 strands of Technora fibre with a urethane coating. Tech 12 rope that is only 5 mm in diameter can support a mass of just over 2500 kg before breaking! It is used extensively in the entertainment industry, as it is stronger than metal wire and much friendlier to the bodies of stunt people and actors. It is also marketed for use in rescue operations such as the one depicted in Figure 11-35.

lidian/Shutterstock

Figure 11-35 Ropes used in rescue missions must be able to provide strong support in dangerous situations.

proportional limit for a given material, the maximum stress at which the stress and strain are proportional

elastic limit for a given material, the maximum stress at which the material exhibits elasticity

elastic range the region of the stress–strain graph in which a given material exhibits elasticity

elasticity the ability of a material, having been stressed, to return to its original shape once the stress is removed

plasticity the failure of a material to return to its initial size and shape when stress is removed

plastic range the region of the stress–strain graph in which a given material exhibits plasticity

breaking point (fracture point) the point on the stress–strain graph for a given material at which the material breaks

ultimate strength the maximum stress that a material can withstand without breaking

ductile the adjective used to describe a material that stretches a lot beyond the elastic limit before breaking

brittle the adjective used to describe a material that stretches very little beyond the elastic limit before breaking

safety factor the ratio between the ultimate strength and the maximum allowable stress for safety; a typical value is about 10

DID YOU KNOW?

Bats hang upside down (Figure 11-36) not only so they can easily become airborne, but also because their leg bones are very weak. If a bat were to walk, there would be a good chance of breaking a leg. When they hang upside down, the tension in the leg bones is shared with muscles, ligaments, and tendons, which can withstand tension far better than compression. The exception to weak-legged bats are vampire bats, which are poor fliers and do not scoop up bugs in the air. They have strong femurs that allow them to walk and hop along the ground as they approach their prey in search of blood.

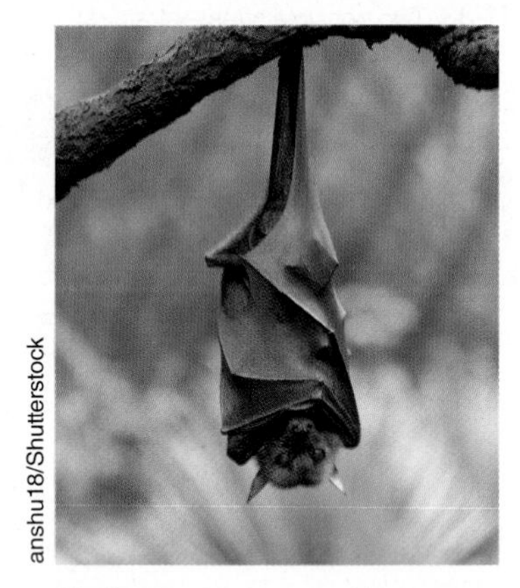

Figure 11-36 The leg bones of bats are very weak, and when bats hang upside down, the tension in their legs is then shared among muscles, ligaments, tendons, and the weak bones.

SAMPLE PROBLEM 11-11

The ultimate strength of bone under compression is 1.5×10^8 Pa. What is the maximum magnitude of weight that can be supported by a healthy human tibia (shinbone) of solid cross-sectional area 4.7 cm^2?

Solution

Since the ultimate strength is the maximum possible stress without breaking, then the maximum weight (of magnitude F_{max}) is given by $\frac{F_{max}}{A} = 1.5 \times 10^8$ Pa.

$$\therefore F_{max} = (1.5 \times 10^8 \text{ Pa})A = (1.5 \times 10^8 \text{ Pa})(4.7 \times 10^{-4} \text{ m}^2) = 7.1 \times 10^4 \text{ N}$$

Thus, the maximum magnitude of weight that can be supported is 7.1×10^4 N, which is the approximate weight of a 7000 kg mass.

The solution to the above sample problem reveals a rather high weight. Most engineers and designers try to build in a **safety factor** of about 10 for structures; in other words, the maximum *allowable* stress should be about 1/10 of the ultimate strength. Thus, it would be recommended that the maximum mass supported by a tibia be about 700 kg, which is still quite a lot!

Most biological materials consist of two or more molecular species. This allows the material to respond to a wide variety of mechanical stresses. For example, the walls of large blood vessels consist of two structural proteins: elastin and collagen. Figure 11-37 shows the stress–strain graph of a large blood vessel wall. The overall curve for the wall is the sum of the curves for the two components. For small stresses, it is mainly elastin that contributes to the stress–strain curve. The collagen consists of helical molecules that are essentially slack for small stresses. For larger stresses, the helices stretch and tighten up, and the collagen then is the main contributor to the shape of the stress–strain curve. Having these two proteins in the vessel walls means that for small increases in blood pressure (i.e., small increases in stress), the stretch of the walls is fairly large, but for larger increases, the walls stiffen up and stretch much less to prevent the vessels from swelling to very large sizes.

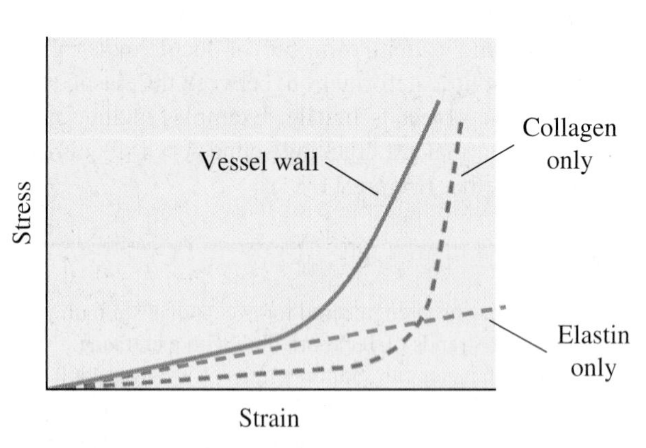

Figure 11-37 Stress–strain relationship for a large blood vessel wall.

EXERCISES

11-22 What are the SI units of (a) tensile stress, (b) Young's modulus, (c) tensile strain, (d) compressive stress, and (e) compressive strain?

11-23 A bird feeder of mass 1.5 kg is suspended from the branch of a tree by a string of diameter 1.8 mm. Calculate the stress in the string.

11-24 A mountain climber, with a mass of 68 kg including gear, is hanging from a vertical rope. What is the diameter of the rope, if the stress in the rope is 1.8×10^6 Pa?

11-25 A fish is held vertically by a fishing line that has a cross-sectional area of 3.0×10^{-6} m^2. If the stress in the line is 6.0×10^6 Pa, what is the fish's mass in kilograms?

11-26 A 40.0 kg child sits on a swing with two vertical ropes that are each initially 4.18 m long. Each rope increases in length by 2.0 cm.

(a) Determine the strain on each rope in both decimal and percentage notation.

(b) If the cross-sectional area of each rope is 4.5×10^{-4} m^2, what is the stress on each rope?

(c) Determine Young's modulus of the rope.

11-27 For large blood vessels, if the blood pressure increases by a small amount, the strain of the vessel walls increases considerably. However, if the pressure continues to increase, although the strain increases, its rate of increase is smaller. By considering what is happening at the molecular level, explain how this occurs.

11-28 A cylindrical bone of length 18 cm and solid cross-sectional area 2.5 cm^2 is subjected to stress tests, and the ultimate strength under compression is found to be 1.6×10^8 Pa.

(a) Assuming a safety factor of 8, what is the maximum magnitude of force that should be exerted on this bone?

(b) By how much will the length of the bone change (in centimetres) under this force? Young's modulus for the bone is 1.8×10^{10} N/m^2.

11.5 Shear and Bulk Stress and Strain; Architecture

In Section 11.4 we considered deformations caused by forces that either stretch or compress an object. However, another way in which forces can be applied to objects is shown in Figure 11-38: Two forces are being applied to an object in directions *parallel to* opposite surfaces. As a result, the top surface of area A moves relative to the bottom one, and adjacent molecules in the object are forced to slide past each other. This is similar to pages sliding in a book if opposite forces are applied to the book covers (Figure 11-39). In this type of situation, the stress is referred to as **shear stress**, which has essentially the same definition as tensile stress, except that the forces are now parallel to surfaces of area A, instead of being perpendicular to them.

$$\text{shear stress} = \frac{F}{A} \qquad (11\text{-}9)$$

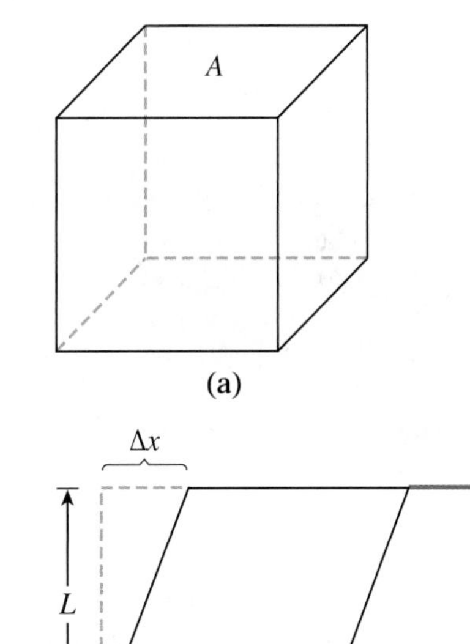

Figure 11-38 Deformation of an object under shear stress: (a) before deformation, (b) during deformation.

Notice that the SI unit of shear stress is the same as that of tensile stress: newton per square metre, or pascal.

In Figure 11-38 (b) the top surface has moved a distance Δx relative to the bottom one. The **shear strain** is defined as the ratio of Δx to the distance L between the surfaces. Since shear strain is a ratio of two lengths measured in the same units, it is dimensionless and has no units, as was the case for tensile strain.

$$\text{shear strain} = \frac{\Delta x}{L} \qquad (11\text{-}10)$$

For sufficiently small shear stresses, the shear stress and shear strain are related by a form of Hooke's law similar to that for tensile stress and tensile strain. In the shear case, the constant of proportionality is the **shear modulus** S, which, like Young's modulus, has an SI unit of newton per square metre, or pascal.

$$\frac{F}{A} = S\frac{\Delta x}{L} \qquad (11\text{-}11)$$

For a given material, the value of the shear modulus is typically about one half to one third of Young's modulus. This indicates that most materials resist tensile or compressive stresses better than shear stresses. Some representative values of shear modulus are given in Table 11-3.

(a)

Ernie McFarland, author

(b)

Ernie McFarland, author

Figure 11-39 (a) Book with no horizontal forces exerted on it. (b) Pages slide over each other as a rightward force is exerted on the top surface and a leftward force on the bottom surface.

shear stress ratio of magnitude of force F to area A, when a force $\vec{F}$ is applied parallel to a surface of area A on an object

shear strain for an object subject to a shear stress, the ratio of the distance Δx that one surface has moved (relative to the opposite surface) to the distance L between the surfaces

shear modulus ratio of shear stress to shear strain, for shear strains small enough that shear strain and shear stress are proportional

Table 11-3

Approximate Values of Shear Modulus S

Material	S (Pa)
Bone	$(3–8) \times 10^9$
Glass	3×10^{10}
Copper	4.5×10^{10}
Steel	8×10^{10}
Tungsten	1.6×10^{11}

SAMPLE PROBLEM 11-12

The shear modulus of a new polymer is to be determined. A sample disc of the material has a radius of 1.2 cm and a height of 0.50 cm. The base of the disc is fastened securely, and a shearing force of magnitude 58 N is applied parallel to the top surface of the disc (Figure 11-40). The top surface is observed to move a distance of 2.0 μm. What is the shear modulus of the material?

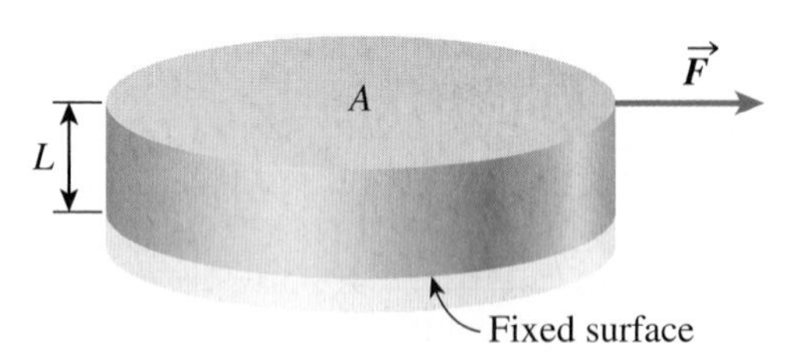

Figure 11-40 Sample Problem 11-12.

Solution

Rearranging Eqn. 11-11, $\frac{F}{A} = S\frac{\Delta x}{L}$, to solve for S gives $S = \frac{FL}{A\Delta x}$.

Area A is the circular cross-sectional area $A = \pi r^2$.

$$\therefore S = \frac{FL}{\pi r^2 \Delta x} = \frac{(58\ \text{N})(5.0 \times 10^{-3}\ \text{m})}{\pi(1.2 \times 10^{-2}\ \text{m})^2(2.0 \times 10^{-6}\ \text{m})} = 3.2 \times 10^8\ \text{Pa}$$

Thus, the shear modulus of this new polymer is 3.2×10^8 Pa.

torsion twisting of an object due to an applied torque

Torsion of Cylinders

A common type of shear deformation involves **torsion**, which is the twisting of an object due to an applied torque. We will consider the torsion of cylinders, both solid and hollow. Figure 11-41 shows a *solid* cylinder of radius r and height L with a fixed base. The cylinder is made of a material having a shear modulus S. A force $\vec{F}$ is applied tangentially to the circumference of the top surface, producing a torque $\tau = Fr$ that rotates the top surface through an angle θ relative to the bottom. Note that in this process, horizontal layers of the material are being forced to rotate and slide over each other. In a derivation beyond the scope of this book, it can be shown that the relationship between the applied torque τ and the angle θ (*in radians*) is given by

$$\tau = \frac{\pi r^4 S\theta}{2L} \quad (11\text{-}12)$$

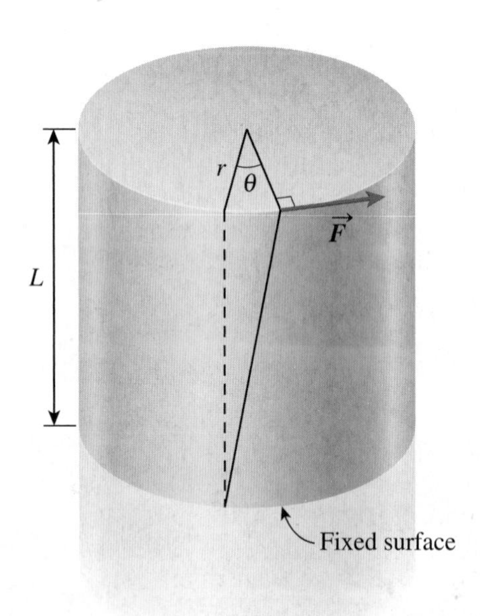

Figure 11-41 Torsional deformation of a solid cylinder due to a tangential force at the top when the base is fixed.

Figure 11-42 shows the corresponding situation for a *hollow* cylinder of height L, with inner and outer radii of r and R respectively. The average radius is $r_{av} = (r + R)/2$, and the thickness of the cylinder is $t = R - r$. The relationship between the torque and the angle of rotation θ (again *in radians*) is

$$\tau = \frac{2\pi r_{av}^3 tS\theta}{L} \quad (11\text{-}13)$$

SAMPLE PROBLEM 11-13

A solid copper cylinder has radius 3.0 cm and height 45 cm; its base is fixed to a horizontal surface. A force of magnitude 757 N is applied tangentially to the circumference of the top surface. Determine the angle in degrees that the top surface rotates relative to the bottom surface. (Reference: Table 11-3)

Solution

Start with Eqn. 11-12, $\tau = \frac{\pi r^4 S\theta}{2L}$, from which $\theta = \frac{2L\tau}{\pi r^4 S}$.

Since the force is applied tangentially, its moment arm is the radius r of the cylinder, and the torque is $\tau = Fr$.

$$\therefore \theta = \frac{2LFr}{\pi r^4 S} = \frac{2LF}{\pi r^3 S}$$

Then, using the shear modulus of copper given in Table 11-3,

$$\theta = \frac{2(0.45\,\text{m})(757\,\text{N})}{\pi(3.0 \times 10^{-2}\,\text{m})^3(4.5 \times 10^{10}\,\text{Pa})} = 1.78 \times 10^{-4}\,\text{rad} \times \frac{360°}{2\pi\,\text{rad}} = 0.010°$$

Therefore, the top surface rotates 0.010° relative to the bottom.

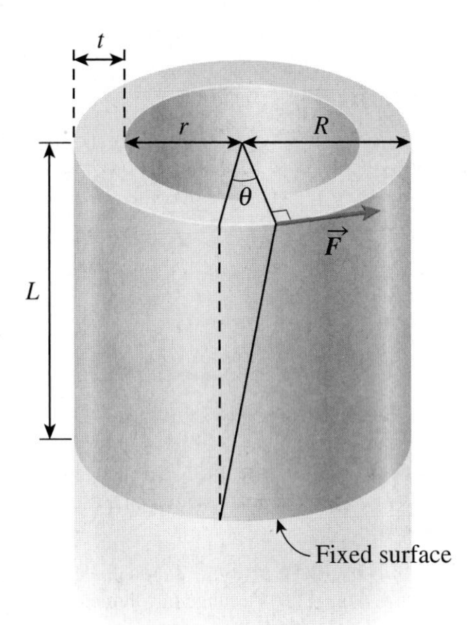

Figure 11-42 Torsional deformation of a hollow cylinder due to a tangential force at the top when the base is fixed.

SAMPLE PROBLEM 11-14

Two cylinders of the same length are made of the same material. One cylinder is solid with a radius of 3.0 cm. The other is hollow with inner and outer radii of 2.5 cm and 3.9 cm respectively. With these particular dimensions, the volumes of the two cylinders are the same, and since they are made of the same material, their masses and shear moduli are also the same. If the same torque is applied to each cylinder, determine the ratio of the twist angle of the solid cylinder to that of the hollow cylinder.

Solution

Use Eqn. 11-12 for the solid cylinder, with subscripts "S" (for solid) added to the radius and angle: $\tau = \frac{\pi r_S^4 S\theta_S}{2L}$. Rearrange to solve for the twist angle: $\theta_S = \frac{2L\tau}{\pi r_S^4 S}$

For the hollow cylinder, use Eqn. 11-13 with subscripts "H" (for hollow) where appropriate:

$$\tau = \frac{2\pi r_{H,av}^3 t_H S\theta_H}{L}$$

The angle θ_H is then given by $\theta_H = \frac{L\tau}{2\pi r_{H,av}^3 t_H S}$,

where average radius $r_{H,av} = \frac{(2.5 + 3.9)\,\text{cm}}{2} = 3.2\,\text{cm}$,

and thickness $t_H = (3.9 - 2.5)\,\text{cm} = 1.4$ cm

The ratio θ_S/θ_H is

$$\frac{\theta_S}{\theta_H} = \frac{2L\tau}{\pi r_S^4 S} \div \frac{L\tau}{2\pi r_{H,av}^3 t_H S} = \frac{2L\tau}{\pi r_S^4 S} \times \frac{2\pi r_{H,av}^3 t_H S}{L\tau} = \frac{4 r_{H,av}^3 t_H}{r_S^4}$$

DID YOU KNOW?

Approximately 40% of fractures of shin bones are the result of torsion.

Substituting numbers, $\frac{\theta_S}{\theta_H} = \frac{4r^3_{H,av}t_H}{r^4_S} = \frac{4(3.2\,\text{cm})^3(1.4\,\text{cm})}{(3.0\,\text{cm})^4} = 2.3$

Thus, the ratio of the twist angle of the solid cylinder to that of the hollow one is 2.3.

The answer to Sample Problem 11-14 indicates that it is advantageous to have hollow cylinders, pipes, bones, etc., instead of solid ones, since for the same mass of material, the amount of rotation for the same applied torque is less for a hollow cylinder. This might be one reason why animals have evolved hollow rather than solid leg bones. It certainly is an advantage for birds, whose bones are thin hollow tubes with a high "strength to weight" ratio.

bulk stress (volume stress) the change in pressure (force per unit area) exerted on all surfaces of an object

bulk strain (volume strain) for an object subject to bulk stress, the ratio of change in volume to original volume

pressure ratio of magnitude of force to surface area, when a perpendicular force is exerted on the surface of an object

bulk modulus ratio of bulk stress to bulk strain

compressibility reciprocal of bulk modulus

Bulk Stress and Bulk Strain

When a whale dives deep into the ocean, the force per unit area that the water exerts on its body increases with depth and is applied on all the surfaces of its body. As a result, the whale is squeezed and its volume decreases. This is a different type of stress from those that have been discussed so far (tensile, compressive, and shear), since this stress is applied to all surfaces of the whale at the same time. This kind of stress is referred to as **bulk stress** (or **volume stress**), and it produces **bulk strain** (or **volume strain**). In particular, this type of stress and strain affects objects immersed in fluids, that is, liquids and gases. As discussed in more detail in Chapter 12, a fluid exerts compressive forces on all the surfaces of an object immersed in it. For a given surface, the force acts in a direction perpendicular to the surface, and the magnitude of this force per unit area (F/A) of surface is referred to as **pressure** P.

Figure 11-43 shows a cube immersed in a fluid. The cube has a volume V_0 when the surrounding pressure is P_0. If the pressure increases by an amount ΔP, the cube's volume changes by an amount ΔV, where $\Delta V < 0$ as the material is being compressed. The bulk stress and bulk strain are defined as

$$\text{bulk (volume) stress} = \Delta\left(\frac{F}{A}\right) = \Delta P \quad \textbf{(11-14)}$$

$$\text{bulk (volume) strain} = \frac{\Delta V}{V_0} \quad \textbf{(11-15)}$$

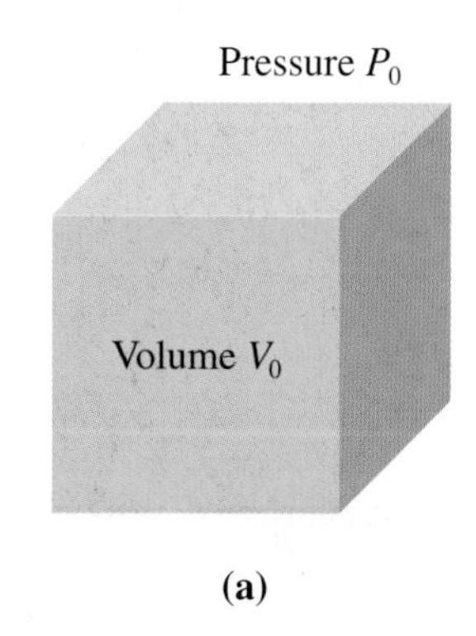

(a)

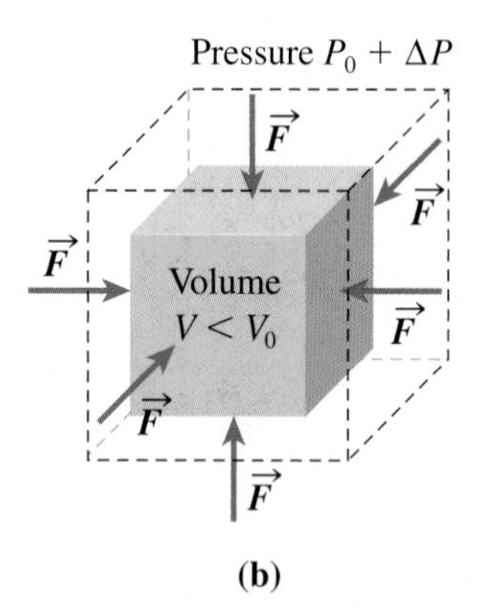

(b)

Figure 11-43 **(a)** A cube having volume V_0 when fluid pressure is P_0. **(b)** The pressure has increased by ΔP, and the cube has undergone a volume change $\Delta V < 0$. (The volume change has been exaggerated.)

For relatively small bulk stresses, the bulk strain is proportional to the bulk stress, and the elastic modulus that is the proportionality constant is the **bulk modulus** B. Hooke's law then takes the form

$$\Delta\left(\frac{F}{A}\right) = \Delta P = -B\frac{\Delta V}{V_0} \quad \textbf{(11-16)}$$

The negative sign is included in Eqn. 11-16 because ΔV is negative when the pressure increases ($\Delta P > 0$). Values of bulk modulus for several substances are given in Table 11-4. Notice that some liquids are included in Table 11-4, but there were no liquids included in previous tables of Young's modulus and shear modulus because liquids (and gases) simply flow when subjected to tensile or shear stresses. Values of bulk modulus for gases have not been listed in Table 11-4 because the values depend strongly on the initial pressure P_0. In some other reference sources, you might encounter the term **compressibility**, which is simply the reciprocal of the bulk modulus.

Note that the units of all the elastic moduli (tensile, shear, and bulk) are the same (pascal), as are the units of the various types of stress (pascal), and units of strain (no units).

Table 11-4

Approximate Values of Bulk Modulus *B*

Substance	*B* (Pa)
Ethyl alcohol	1.1×10^9
Water	2.2×10^9
Bone	1.5×10^{10}
Mercury	2.9×10^{10}
Glass	4×10^{10}
Copper	1.4×10^{11}
Steel	1.6×10^{11}
Tungsten	3.1×10^{11}

SAMPLE PROBLEM 11-15

The greatest ocean depth is in the Challenger Deep, which is a valley in the Mariana Trench in the western Pacific Ocean. The pressure at that depth is 1.1×10^8 Pa greater than the pressure at the ocean surface. If a ship carrying a load of steel were to sink from the surface into the Challenger Deep, what volume change would each cubic metre of steel undergo? Express the answer in cubic centimetres and as a percentage change. (Reference: Table 11-4)

Solution

Use Eqn. 11-16, $\Delta P = -B\dfrac{\Delta V}{V_0}$, and solve for ΔV, giving $\Delta V = -\dfrac{V_0 \Delta P}{B}$.

Substituting numbers, with B from Table 11-4,

$$\Delta V = -\frac{V_0 \Delta P}{B} = -\frac{(1.0\,\text{m}^3)(1.1 \times 10^8\,\text{Pa})}{1.6 \times 10^{11}\,\text{Pa}} = -6.9 \times 10^{-4}\,\text{m}^3 = -6.9 \times 10^2\,\text{cm}^3$$

As a percentage, $\dfrac{\Delta V}{V_0} = -\dfrac{6.9 \times 10^{-4}\,\text{m}^3}{1.0\,\text{m}^3} \times 100\% = -0.069\%$

Thus, the volume of one cubic metre of steel would decrease by $6.9 \times 10^2\,\text{cm}^3$, or 0.069%.

Stress and Strain in Architecture

One of the areas in which science and the humanities overlap is in architecture, or the design of structures. Although this topic could fill many books, we will look briefly at how forces and elasticity may be applied to architecture.

In comparing the strain under tensile, compressive, or shear stresses, many materials are much stronger under one type of stress than the other. Concrete is a good example of this. It has an elastic modulus that is about 10 times lower under tension than under compression. Thus, it is useful as a vertical support material, but it is poor when used to span a horizontal space. A horizontal slab (of any material) supported only at its ends sags under its own weight and becomes compressed along the top, but experiences tension on the bottom (Figure 11-44). With a concrete slab, this tension can result in cracking or breaking of the slab.

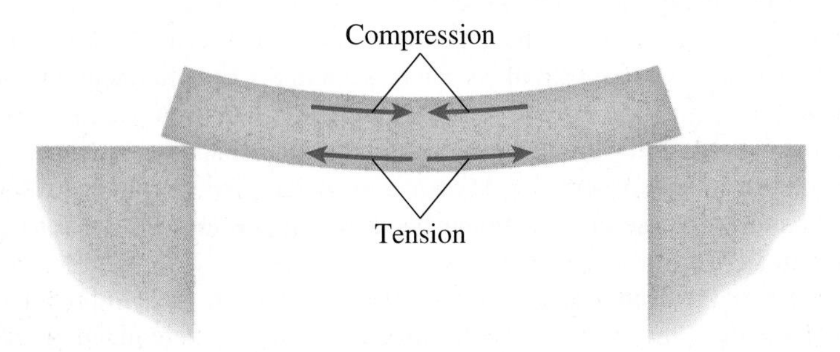

Figure 11-44 A horizontal slab sags under its own weight, and experiences both compression (top) and tension (bottom).

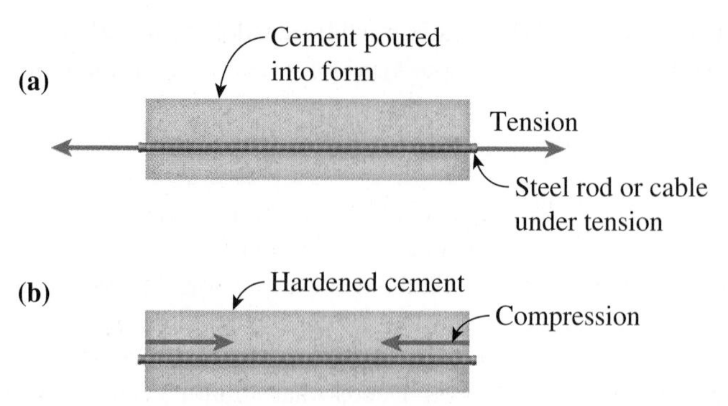

Figure 11-45 Prestressing concrete. **(a)** Cement poured onto steel under tension. **(b)** When the cement has hardened, tension is removed and the steel tries to contract.

Figure 11-46 In this beautiful Egyptian temple in Luxor, built about 3400 years ago, the horizontal spaces spanned are narrow because the stone slabs are not strong when used horizontally.

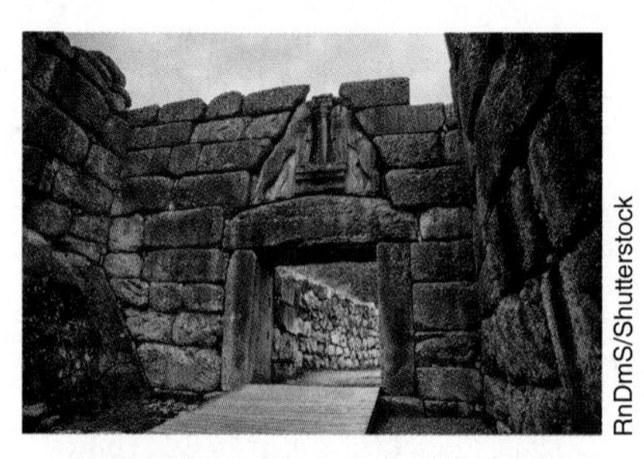

Figure 11-47 Using a triangular block to spread load to the sides in Mycenae, Greece.

Figure 11-48 A Roman arch using wedge-shaped stones in the arch itself to produce compressive stress and to transfer load to the strong side columns.

Various ways can be used to help make concrete stronger. The simplest way is to reinforce the concrete by placing metal rods where the concrete is to be poured. In this process, called *reinforcing concrete*, the metal helps make the concrete stronger under tension. The most common metal used is steel, which has the same rate of thermal expansion as concrete.

Another technique for strengthening concrete, which has been used for over 100 years, is to create a compression stress in the concrete *before* it has a load applied. The process is termed *prestressing concrete*. To accomplish this, steel rods or cables are placed under tension, then the cement is poured over them (Figure 11-45 (a)). Once the cement has hardened to become concrete, the tension on the rods or cables is removed and they try to contract, producing compression in the concrete (Figure 11-45 (b)). Loads can then be placed on the concrete.

A third way to strengthen concrete is to create a compression in the steel rods or cables *after* the concrete has set. This is referred to as *poststressing the concrete*. The rods or cables are compressed by tightening nuts on their ends, and the concrete is compressed as well. This is used for large structures, such as concrete bridges.

Other materials can be made stronger by adding additional components or by changing the way in which they are manufactured.

Spanning a Space

A doorframe is a simple example of a structure that spans a space. The frame consists of two vertical support posts and a horizontal beam across the top. This post-and-beam construction, using either stone or wood, was probably the first design used to span a space. Examples of this are found in ancient Egyptian temples that used heavy stone slabs for the horizontal span. The problem with using a single piece of stone or wood for this span is that it experiences tensile stress across the bottom, and these materials are weak under tension, similar to the situation for concrete (Figure 11-44). Thus, the width of the span had to remain small, as can be seen in the Egyptian temple in Luxor (Figure 11-46).

Improvements in design were made as builders tried to span larger widths. One of the earliest improvements occurred in Mycenae in ancient Greece (about 1200 BCE). Figure 11-47 shows the use of a triangular block which was able to transfer load via compression to angled blocks beside it.

The architects of the Roman Era, about 2000 years ago, made great improvements in the design of arches to span a space. Their semicircular arches were much more stable than the earlier spans because the wedge-shaped rocks or bricks pressed against each other, resulting in mostly compressive stress rather than tensile stress (Figure 11-48).

These arches could span large spaces, but because they produced large horizontal forces, they had to be surrounded by strong sides to prevent collapse.

After the Romans, the next great improvement in arches came with the pointed arch used in Gothic Europe (from the late 12th century to the 15th century). A pointed arch can span the same space as a semicircular arch, but needs only half the horizontal support force from the sides. To achieve this, the pointed arch needs to have a height that is twice that of the semicircular arch. This allowed a dramatic change in structures in Gothic times because the support buttresses could now be much smaller. Figure 11-49 shows Notre Dame Cathedral in Paris with its rather airy "flying buttresses."

These principles of spanning a space in two dimensions using arches also apply in three dimensions for domes. As seen in Figure 11-50, domes constructed today can be built to span a much larger space than in previous centuries. There are two main reasons for this: first, the mathematical analysis of structures is much more advanced; second, the materials used in construction are much lighter and stronger.

One other design that has had a great influence on building construction is the truss. A **truss** is any structure built with triangular shapes, which provide a very rigid structure and allow a load to be transferred over a wide area. The simplest form of truss is the A-frame, which probably was developed during the Medieval Ages (Figure 11-51 (a)). In bridges and other structures (Figures 11-51 (b) and (c)), trusses are much more complex, but the basic design principles of rigidity and load transfer are the same.

Mark III Photonics/Shutterstock

Figure 11-49 Notre Dame Cathedral in Paris, with its "flying buttresses" extending far from the arches.

Iurii Osadchi/Shutterstock.com

Figure 11-50 The Bolshoy Ice Dome used during the 2014 Winter Olympics in Sochi, Russia.

truss any structure built with triangular shapes

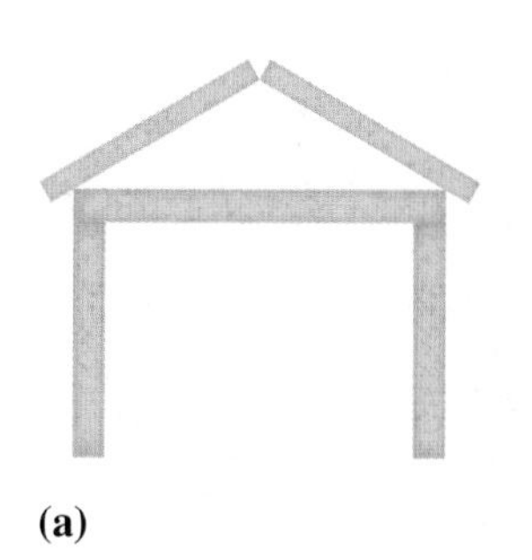
(a)

jessicakirsh/Shutterstock

(b)

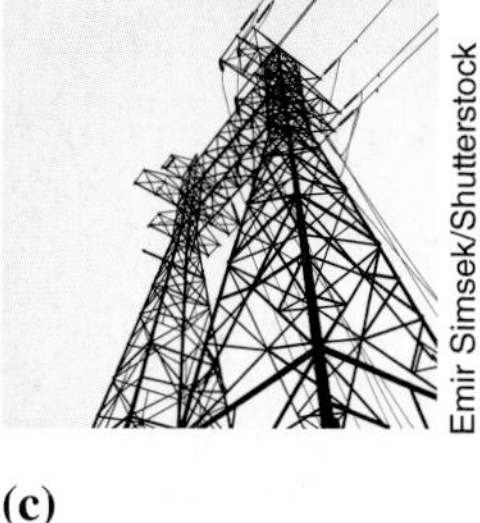
Emir Simsek/Shutterstock

(c)

Figure 11-51 **(a)** An A-frame truss. Trusses are commonly used in **(b)** bridges and **(c)** electric transmission towers, and many other structures.

EXERCISES

11-29 What are the SI units of (a) shear stress, (b) shear modulus, (c) shear strain, (d) pressure, (e) bulk stress, (f) bulk modulus, and (g) bulk strain?

11-30 In Figure 11-52, which lamp is being supported by (a) tension, (b) compression, and (c) shear?

11-31 A cylindrical bone of solid cross-sectional area 2.5 cm^2 and length 18 cm is found to have an ultimate strength under shear of 1.6×10^7 Pa. Assuming a safety factor of 10, determine the maximum shear force recommended for this bone if the force is applied parallel to one end of the bone and directed toward the centre of the end.

11-32 A football player is tackled and experiences a transverse (shearing) force on the femur (the main bone of the thigh) of

Figure 11-52 Question 11-30.

568 N at a part of the bone where the radius is 1.3 cm. If the shear modulus of the bone is 4.0×10^9 N/m^2, what is the strain on the bone?

11-33 A cylindrical rod of length 51 cm and diameter 6.0 mm is subjected to a horizontal shearing force parallel to its top surface (Figure 11-53), and the surface is displaced by 0.55 mm. The bottom end of the rod is fixed. If the force has a magnitude of 9.2×10^2 N, determine the shear modulus of the material in the rod.

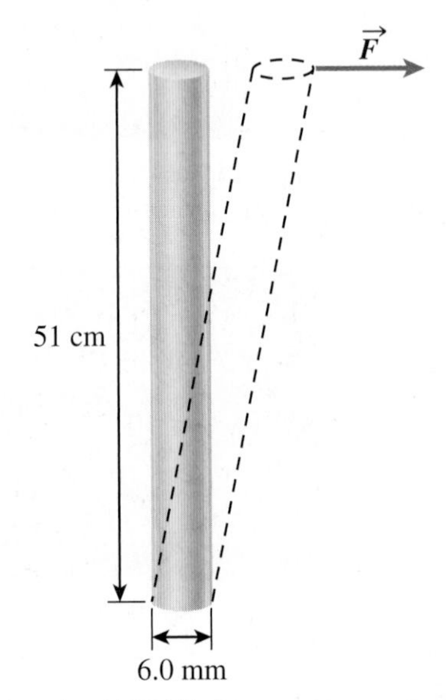

Figure 11-53 Question 11-33 (displacement is exaggerated).

11-34 A torque of magnitude 1.7×10^2 N·m is applied at one end of a solid cylindrical steel shaft that is 3.5 m long and 2.6 cm in radius. The other end of the shaft is held fixed, and the shaft undergoes torsion. The shear modulus for the steel is 8.0×10^{10} Pa. Through what angle in degrees does the shaft twist?

11-35 One end of a person's lower leg bone is twisted by 8.0° relative to the other end during a skiing accident. The bone is a hollow cylinder of average radius 2.5 cm, thickness 0.60 cm, and length 0.35 m. Its shear modulus is 4.2×10^9 Pa. How large is the torque acting on the bone?

11-36 Oil, with a bulk modulus of 1.7×10^9 Pa, is used in a hydraulic press to hoist cars in a repair garage. Determine the change of volume for each cubic metre of oil when the pressure on it increases by 2.0×10^3 Pa. Express your answer in cubic centimetres.

11-37 A certain liquid is subjected to pressure in order to measure its bulk modulus. A gauge shows that, for a pressure increase of 8.65×10^6 Pa, the volume decreases by 0.41%. Determine the bulk modulus of the liquid.

11-38 The graph in Figure 11-54 shows the relationship between the magnitudes of the stress and the strain for a certain material for bulk, tensile, and shear stresses. Use the data to determine the bulk modulus, Young's modulus, and shear modulus. What is the material? (See the tables in this chapter.)

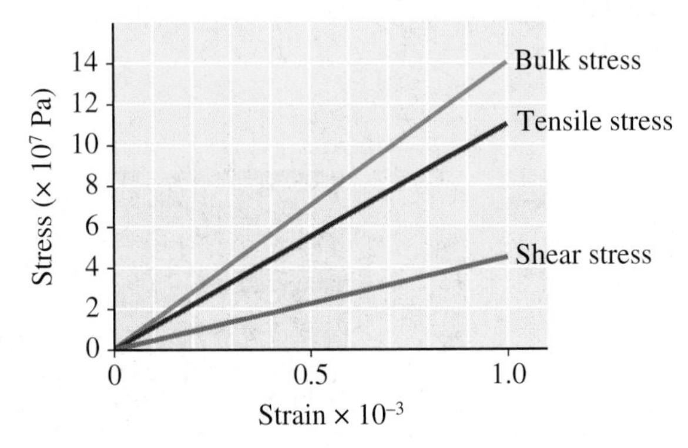

Figure 11-54 Question 11-38.

11-39 Describe why a material that is weak under tension cannot be used to span a space without strengthening the material.

LOOKING BACK...LOOKING AHEAD

This chapter builds on the foundations of mechanics that have been presented in previous chapters: acceleration, force, translational motion, torque, and rotational motion.

We began with the concept of centre of mass (CM) and calculations to determine the position of the CM in systems consisting of "point particles" and objects having common geometrical shapes. There was a discussion of experimental determination of the CM position for irregularly shaped objects. The important concept of static equilibrium was introduced, and many examples were presented that used forces and torques to analyze systems at rest. This led to the topic of stability, combining our knowledge of centre of mass and static equilibrium.

The deformation of objects by tension, compression, shearing, and bulk pressure (in fluids) was presented in detail, with discussions and examples involving the various types of stress, strain, and elastic modulus. Many biological examples were included, from the role of elastin and collagen in the stretching of blood vessel walls to a discussion of why bats hang upside down when resting. Finally, a brief examination of some changes in architecture over the centuries was given.

The next chapter continues the development of concepts in mechanics, with an emphasis on fluids. Topics include pressure, buoyancy, surface tension, and the movement of fluids such as water and blood.

CONCEPTS AND SKILLS

Having completed this chapter, you should now be able to do the following:

- Calculate the position of the centre of mass (CM) for "point–particle" systems and for objects having common geometrical shapes.
- State how centre of mass is related to Newton's laws of motion and to the concept of gravitational potential energy.
- State the condition required for the centre of gravity to be at the same position as the centre of mass.
- Describe how to determine the CM position experimentally for an irregularly shaped object.
- State the two conditions required for static equilibrium and apply them in solving numerical problems in a wide variety of physical situations involving forces and torques on objects at rest.
- Describe the conditions required for an object to be in stable or unstable equilibrium, and give examples of each type of equilibrium.
- Cite examples in which a lower CM results in greater stability.
- State Hooke's law in general for any elastic object.
- Describe how the basic types of stress—tensile, compressive, shear, and bulk—can be applied to an object.
- State the name of each of the three types of modulus.
- Apply the general equation "stress = elastic modulus × strain" in numerical analyses of specific situations involving tensile, compressive, shear, and bulk stresses.
- Recognize the advantages and limitations of the physical designs of the bones and arteries of the human body.
- Appreciate the improvements over the centuries in the design of structures needed to span a space.

KEY TERMS

You should be able to define or explain each of the following words or phrases:

statics
centre of mass (CM)
centre of gravity (CG)
static equilibrium
dynamic equilibrium
unstable equilibrium
stable equilibrium
spring constant
Hooke's law
stress
strain
elastic modulus
tensile stress
tensile strain
Young's modulus
compression
compressive stress
compressive strain
proportional limit
elastic limit
elastic range
elasticity
plasticity
plastic range
breaking point
fracture point
ultimate strength
ductile
brittle
safety factor
shear stress
shear strain
shear modulus
torsion
bulk stress
volume stress
bulk strain
volume strain
pressure
bulk modulus
compressibility
truss

Chapter Review

MULTIPLE-CHOICE QUESTIONS

11-40 Three "point particles" have masses m_1 = 12 g, m_2 = 24 g, and m_3 = 36 g, and x-components of position of x_1 = 1.2 cm, x_2 = 2.4 cm, and x_3 = −3.6 cm. The x-component of position of the centre of mass is

(a) 0.0 cm
(b) −0.80 cm
(c) 2.80 cm
(d) −1.2 cm
(e) 1.2 cm

11-41 Figure 11-55 shows a uniform horizontal beam of length L being supported by two vertical wires. One wire is at a distance of $L/3$ from the left-hand end of the beam, and has a tension $\vec{T}_1$. The other wire is $L/4$ from the right-hand end of the beam, and has a tension $\vec{T}_2$. The ratio T_1/T_2 is

(a) 3/4
(b) 4/3
(c) 3/2
(d) 2/3
(e) 2/1

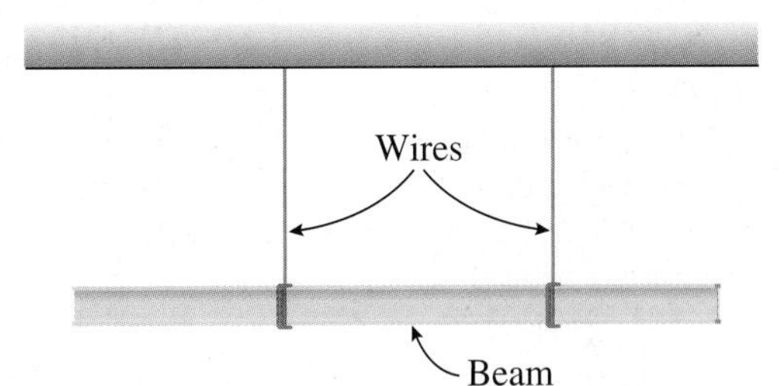

Figure 11-55 Question 11-41.

11-42 Figure 11-56 shows a uniform bar of mass m attached to a pivot at its upper end. The bar is being held at an angle θ relative to the vertical by a force $\vec{F}$ exerted at the lower end of the bar. This force makes an angle φ relative to the bar. What is the magnitude of $\vec{F}$?

(a) $\dfrac{mg \sin\theta}{2 \sin\varphi}$
(b) $\dfrac{mg \cos\theta}{\sin\varphi}$
(c) $\dfrac{mg \sin\theta}{2 \cos\varphi}$
(d) $\dfrac{2mg \sin\theta}{\sin\varphi}$
(e) $\dfrac{2mg \cos\theta}{\sin\varphi}$

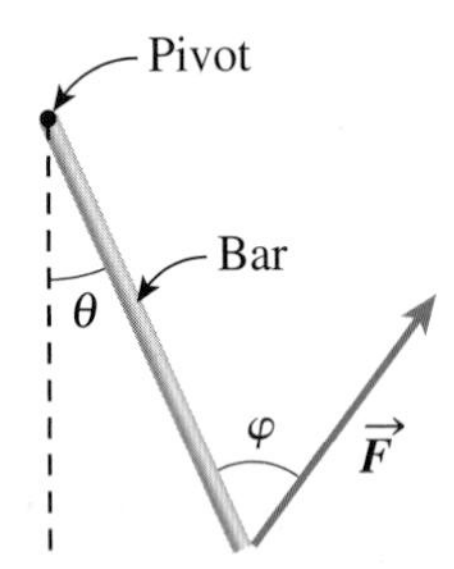

Figure 11-56 Question 11-42.

11-43 A heavy box is sitting on an incline that makes an angle θ to the horizontal. The box has a height of 1.2 m, and its length and width are each 0.80 m, and its CM is at its geometric centre. What is the maximum angle θ for the box not to tip over? Assume there is sufficient friction to prevent the box from sliding.

(a) 56°
(b) 42°
(c) 48°
(d) 34°
(e) 28°

11-44 Which of the physical quantities listed below cannot be expressed in a unit of newton per square metre?

(a) Young's modulus
(b) tensile stress
(c) shear strain
(d) shear stress
(e) shear modulus

11-45 A muscle tendon of radius r and length L_0 is stretched an amount ΔL by a constant force of magnitude F. How much would a similar tendon of radius $2r$ and length $2L_0$ be stretched by the same force?

(a) $\Delta L/4$
(b) $\Delta L/8$
(c) $2\Delta L$
(d) $4\Delta L$
(e) $\Delta L/2$

11-46 A piece of aluminum rod is stretched to its fracture point by a force of magnitude 7000 N. The magnitude of force required to stretch a longer length of the same diameter of aluminum rod to its fracture point is

(a) equal to 7000 N
(b) greater than 7000 N
(c) less than 7000 N

11-47 The shear modulus of steel is 8.0×10^{10} Pa, of copper 4.5×10^{10} Pa, and of tungsten 1.6×10^{11} Pa. For identical pieces (same dimensions) of these materials, rank them in order from easiest to most difficult to twist.

(a) copper, tungsten, steel
(b) copper, steel, tungsten
(c) steel, tungsten, copper
(d) tungsten, steel, copper
(e) steel, copper, tungsten

11-48 A solid cylindrical steel rod and a hollow cylindrical steel rod have the same mass and same length. The same magnitude of compressive force is applied to each one. How do the compression distances of the two rods compare?

(a) The compression distance of the solid rod is greater.
(b) The compression distance of the hollow rod is greater.
(c) The two rods have the same compression distance.

11-49 The solid rod and hollow rod in the previous question are now subjected to the same torque that causes each rod to twist. How do the twist angles of the two rods compare?

(a) The twist angle of the solid rod is greater.
(b) The twist angle of the hollow rod is greater.
(c) The two rods have the same twist angle.

11-50 An archaeologist discovers a cylinder of cross-sectional area 1.60 cm^2 and length 5.0 cm. When the cylinder is loaded with a 12.0 kg mass, careful microscopic analysis reveals a compression of 0.0750%. What is Young's modulus for the material of which the cylinder is made?

(a) 1.8×10^7 Pa
(b) 1.6×10^8 Pa
(c) 2.3×10^9 Pa
(d) 9.8×10^8 Pa
(e) 3.8×10^{10} Pa

11-51 A biomedical engineer (Figure 11-57) wishes to design a prosthesis of aluminum (shear modulus 2.6×10^{10} Pa) to replace a section of long bone (shear modulus 3.1×10^{10} Pa). If the average radius of the hollow aluminum tube she uses is the same as the average radius of the bone it replaces, and if the aluminum tube is to have exactly the same length and twist properties as the bone, then what should be the ratio of the thickness of the aluminum tube wall to the thickness of the bone "tube" wall?

(a) 0.84
(b) 0.59
(c) 1.7
(d) 1.0
(e) 1.2

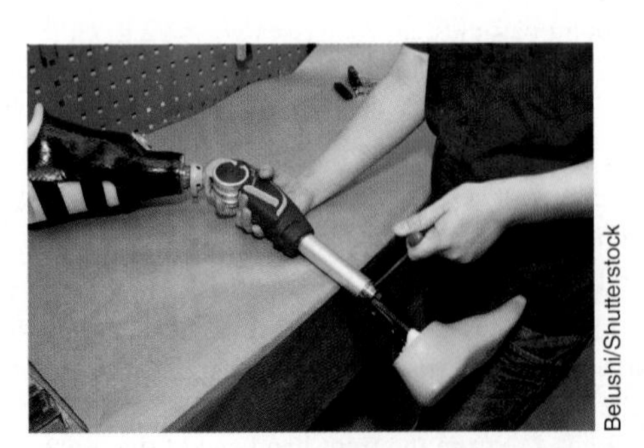

Figure 11-57 Question 11-51.

Review Questions and Problems

11-52 The masses of Earth and the Moon are 5.98×10^{24} kg and 7.35×10^{22} kg respectively, and the average centre-to-centre distance between them is 3.84×10^{5} km.

(a) Determine the distance in kilometres from the centre of Earth to the CM of the Earth–Moon system.

(b) The average radius of Earth is 6368 km. How far above or below Earth's surface is the CM of the Earth–Moon system?

11-53 A flag of mass 1.23 kg is hung from a horizontal pole of mass 4.50 kg attached to a vertical wall (Figure 11-58). The length and width of the flag are 1.20 m and 0.600 m respectively, and the length and diameter of the pole are 1.80 m and 0.024 m respectively. With the origin O of a coordinate system chosen at the top of the left end of the pole, and $+x$- and $+y$-directions to the right and downward respectively, determine the x- and y-components of position of the CM of the flag-plus-pole system. Assume that the flag and pole are each of uniform composition.

Figure 11-58 Question 11-53.

11-54 A support rope is tied to a tree at a height of 8.6 m above the ground (Figure 11-59). With a wind blowing from the left, the magnitude of the tension required in the rope to keep the tree stationary is 2.80×10^{2} N, and the torque about point P due to the tension has a magnitude of 1.80×10^{3} N·m. Determine the angle θ between the rope and the tree trunk.

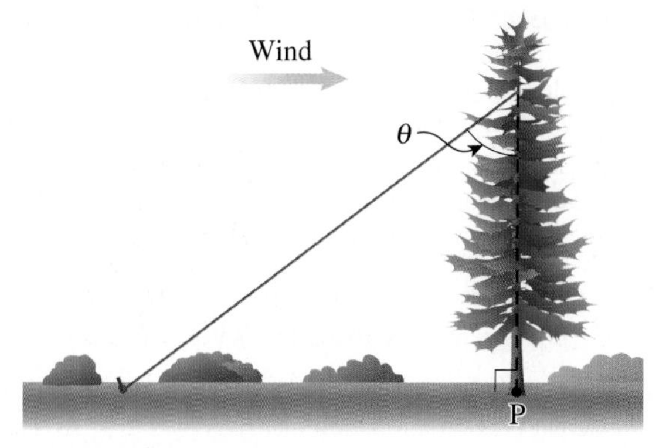

Figure 11-59 Question 11-54.

11-55 A boy is being transported in a wheelbarrow (Figure 11-60) by a man holding the handles at a horizontal distance of 92 cm from the wheel's centre. If the boy has a mass of 25 kg and his CM is at a horizontal distance of 42 cm from the wheel's centre, determine the magnitude of the total vertical force that the man is exerting. Assume static equilibrium.

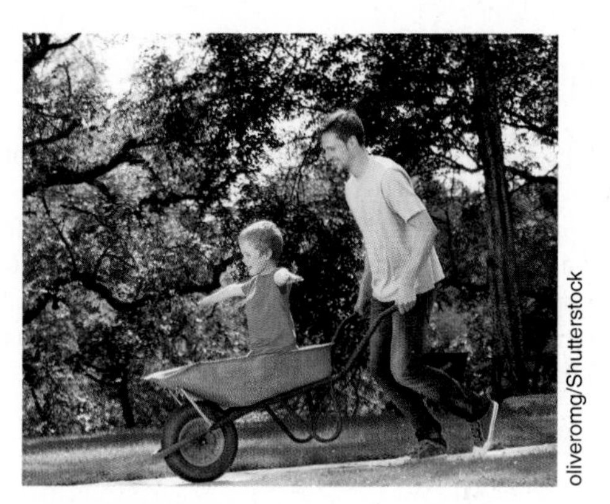

Figure 11-60 Question 11-55.

11-56 Figure 11-61 shows a mobile consisting of a horizontal bar of length 70 cm and negligible mass from which two decorations are hung by light strings. The bar is suspended by a vertical wire. Assume two significant digits in given data.

(a) How far from the left-hand end of the bar is the vertical wire attached?

(b) What is the magnitude of the tension in the wire?

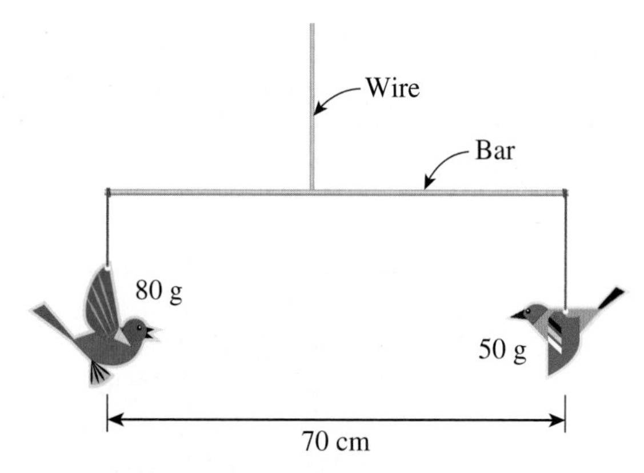

Figure 11-61 Question 11-56.

11-57 A woman of mass 62 kg is standing at the end of a uniform diving board of length 4.60 m and mass 22 kg (Figure 11-62 on pg. 320). The diving board is supported in two places. One support, A, is located 4.30 m from the end of the board, and the other support, B, is 3.60 m from the end. Determine the magnitude and direction of the support forces $\vec{F}_A$ and $\vec{F}_B$. Assume that the board remains horizontal with the woman standing on it.

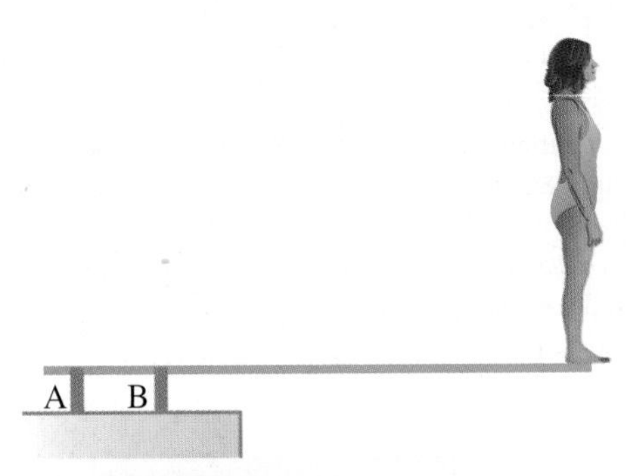

Figure 11-62 Question 11-57.

11-58 A boy of mass 35.0 kg is standing on a uniform plank of mass 23.5 kg and length 6.00 m (Figure 11-63). The plank sits on two supports, one at the left end and the other 1.50 m from the right end. What is the maximum distance x at which the boy can stand on the overhanging part of the plank without the plank starting to tip? (**Hint:** When the plank begins to tip, what happens to the force exerted by the left-end support?)

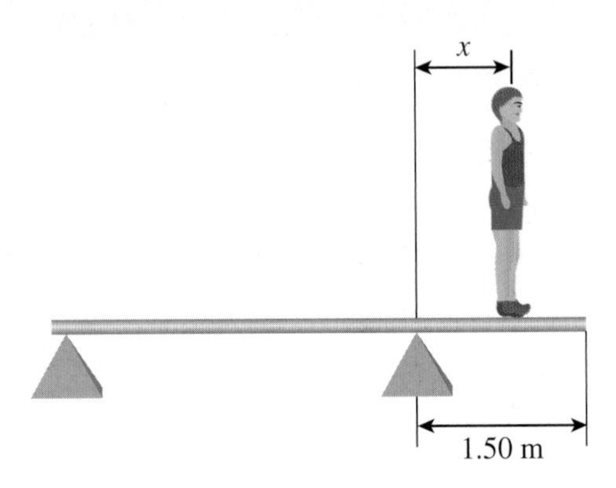

Figure 11-63 Question 11-58.

11-59 A uniform gate of mass 5.0 kg is held by two hinges (Figure 11-64). The upper hinge exerts only a horizontal force on the gate. Determine the magnitude and direction of the force exerted by each hinge.

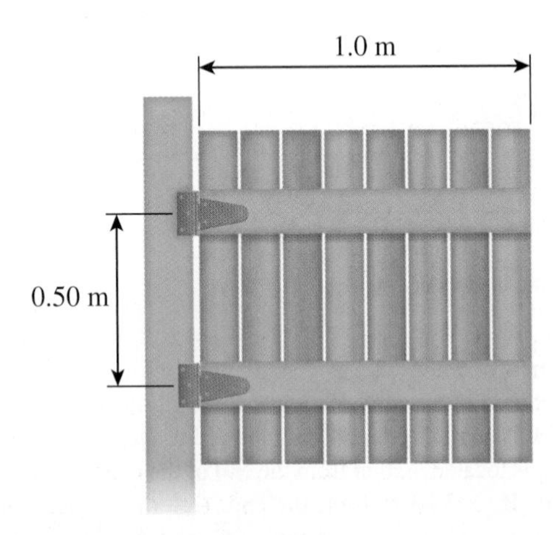

Figure 11-64 Question 11-59.

11-60 Suppose the arm in Sample Problem 11-4 (Figure 11-12 on pg. 294) is thrust forward so that the upper arm makes an angle of 25.0° with the vertical, while the forearm and hand holding the 12.0 kg mass stay horizontal. Assume that the force $\vec{F}_B$ exerted by the biceps muscle acts parallel to the upper arm. Determine the new values of the magnitude of $\vec{F}_B$ and the magnitude and direction of the force $\vec{F}_H$ exerted by the humerus at the elbow joint O. (As in Sample Problem 11-4, neglect the downward force of gravity acting on the forearm itself.)

11-61 Why does a four-legged animal find it much easier to stand and run after birth than a two-legged one?

11-62 An elevator in a gold mine is supported by a steel cable with a diameter of 3.2 cm and a Young's modulus of 2.0×10^{10} Pa. If the elevator and its passengers have a total mass of 850 kg, by how much does the cable stretch when the distance from the elevator cage to the elevator motor is

(a) 15 m?

(b) 2.2 km (neglect the mass of the cable)?

11-63 The elastic limit of the cable in the previous question is 2.5×10^8 Pa.

(a) Neglecting the mass of the cable, determine the stress in the cable. What percentage of the elastic limit is this stress?

(b) If the cable has a mass per unit length of 6.3 kg/m, and the length of hanging cable is 2.2 km, repeat (a) for the stress at the top of the cable.

11-64 A vertical steel column in a building under construction must be able to support a load of 5.5×10^3 kg, and the maximum amount of compression permitted is 1.0 mm. The column is 8.0 m high and Young's modulus for the steel is 2.0×10^{11} Pa. What is the minimum cross-sectional area (in square centimetres) allowed for the column?

11-65 The humerus is the upper arm bone between the elbow and the shoulder joint. It can be approximated as a hollow cylinder of length 33 cm, with outer radius 1.05 cm and inner radius 0.43 cm. Suppose a gymnast whose arm bone has these dimensions does a one-arm handstand. The mass of the gymnast, excluding the arm, is 65 kg.

(a) If Young's modulus for the humerus is 1.4×10^{10} Pa, what is the compressive strain of the humerus?

(b) By how much is the humerus compressed?

11-66 A 64 kg swimmer decides to build a diving board that is 2.4 cm thick and 23 cm wide. The length of the board is 5.5 m. One end of the board projects 4.8 m out over the water, and the remaining 0.7 m is clamped securely to a dock. When the swimmer stands on the end of the board over the water, what is the downward shear displacement of the board where she is standing if the shear modulus of the board is 4.5×10^6 Pa?

11-67 Two bones, one hollow and one solid, have the same shear modulus and the same length. The hollow bone has an outer radius of 2.00 cm and an inner radius of 1.40 cm. The radius of the solid bone is 1.8 cm. Determine the ratio

$$\frac{\text{torque required to twist the hollow bone by } 2.0°}{\text{torque required to twist the solid bone by } 2.0°}$$

11-68 A sample of oil has a bulk modulus of 1.8×10^9 Pa. If the external pressure on the oil increases by 1.2×10^6 Pa, what is the percentage change in the volume of the oil?

Applying Your Knowledge

11-69 Figure 11-65 shows a human upper arm and forearm. The CM of the upper arm is labelled CM_U, and the CM of the lower arm and hand is labelled CM_L. The origin O of an x-y coordinate system and $+x$- and $+y$-directions are shown, as well as the angle θ between the upper and lower arms. Use the following data to determine the x- and y-components of the position of the CM of the upper and lower arm and hand considered together as one system.

distance from CM_U to O = 0.15 m

distance from CM_L to O = 0.12 m

mass of upper arm = 2.2 kg

mass of lower arm and hand = 1.6 kg

θ = 120° (three significant digits)

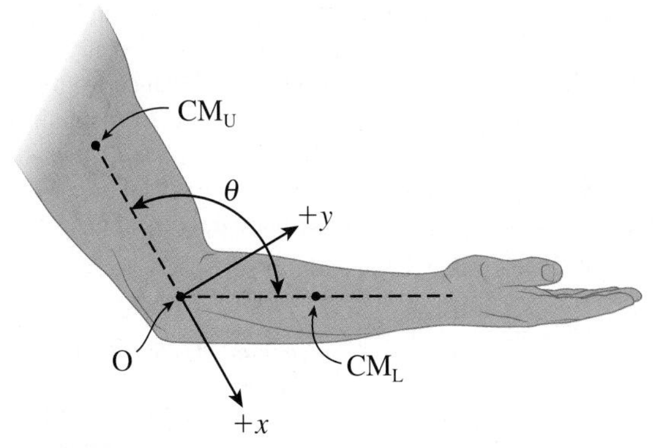

Figure 11-65 Question 11-69.

11-70 If you were constructing a hanging mobile like the one shown in Figure 11-66, would you start from the top or bottom? Explain why.

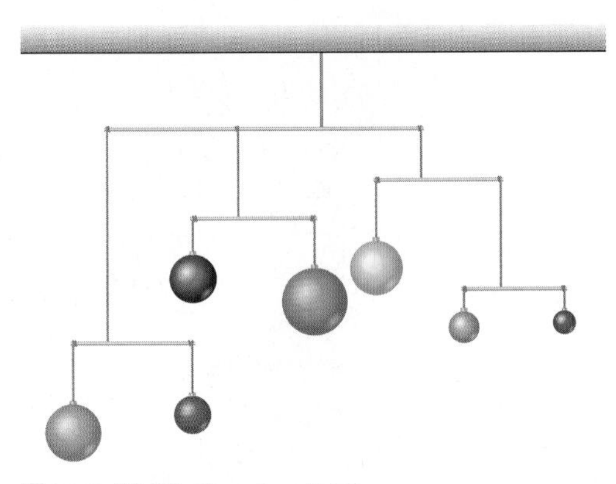

Figure 11-66 Question 11-70.

11-71 A uniform beam of length L and mass 225 kg is attached to a pivot at point P (Figure 11-67). The beam makes an angle of 52° relative to the horizontal and is held by a cable at 36° to the horizontal. The cable is attached to the beam at a distance of 0.20 L from the upper end of the beam. A mass m is suspended from the top of the beam.

(a) If the maximum possible magnitude of tension in the cable is 2.5×10^4 N, what is the maximum mass m?

(b) With m as determined in (a), what are the magnitude and direction of the force exerted on the beam at the pivot P?

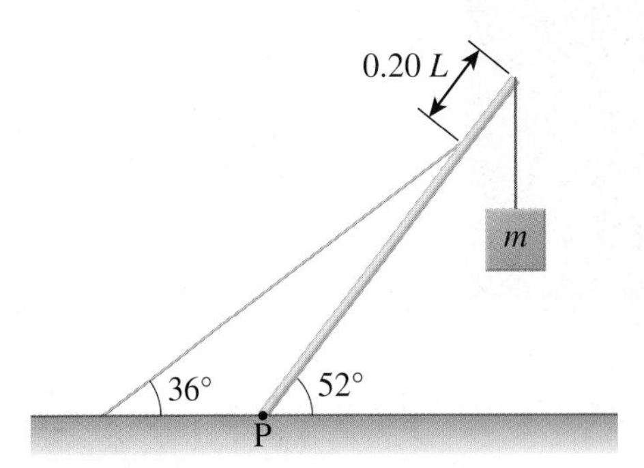

Figure 11-67 Question 11-71.

11-72 Figure 11-68 (a) shows a person's foot and lower leg when standing on tiptoe on a horizontal floor. The forces acting on the foot, represented as a bar ABC in Figure 11-68 (b), are a vertical force $\vec{F}_{floor}$ exerted by the floor, and forces $\vec{F}_{tibia}$ and $\vec{F}_{tendon}$ exerted by the tibia and Achilles tendon respectively. Distance AB = 18.0 cm, and AC = 25.0 cm. Angles are as shown. Assuming that the person is standing on two feet, $F_{floor} = mg/2$, that is, half the magnitude of the person's weight. The person's mass is 75 kg.

(a) What is the magnitude of $\vec{F}_{tendon}$?

(b) What are the magnitude and direction of $\vec{F}_{tibia}$?

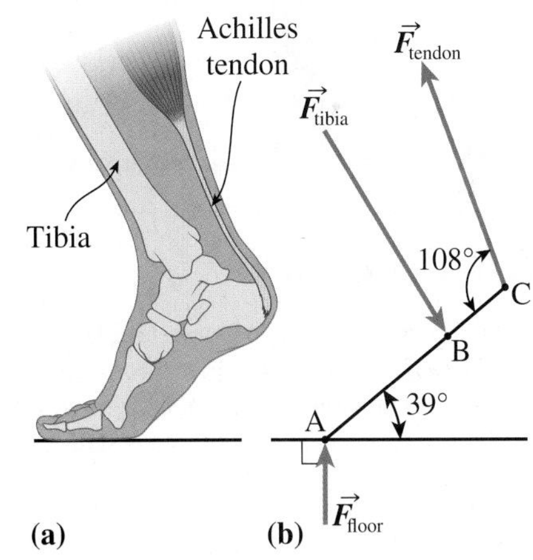

Figure 11-68 Question 11-72. **(a)** Standing on tiptoe. **(b)** Forces acting on the foot.

11-73 Figure 11-69 (a) on pg. 322 shows an athlete of mass 82 kg in the "iron cross" position, and Figure 11-69 (b) indicates the forces acting on one of the athlete's arms. The main muscles involved in supporting the athlete are the pectoralis major ("pecs") and the latissimus dorsi ("lats"). The total force exerted by the muscles on the arm, shown as $\vec{F}_M$, is exerted at a point 3.8 cm from the shoulder joint and has a direction 41° below the horizontal. There is also a force $\vec{F}_S$ exerted by the shoulder joint itself, and an upward force $\vec{F}_R$ exerted by the ring. Force $\vec{F}_R$ is exerted vertically upward at a perpendicular distance of 71 cm from the shoulder joint, and supports half the athlete's weight. Neglecting the weight of the arm itself, determine the magnitude of $\vec{F}_M$.

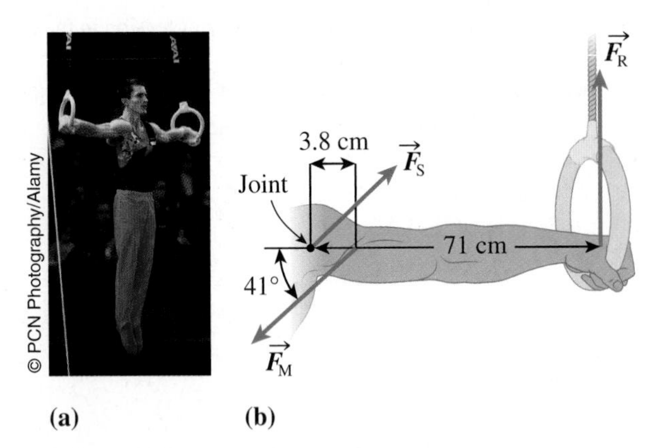

Figure 11-69 Question 11-73.

11-74 A man of mass 87 kg is standing on a uniform horizontal beam of mass 326 kg and length L (Figure 11-70). The right end of the beam is attached at a vertical wall to a pin P about which the beam can pivot. The man's position is $0.80L$ from P. The left end of the beam is supported by a cable attached to another vertical wall. The cable makes an angle of 131° to the horizontal.

(a) Determine the magnitude of the tension in the cable.

(b) What are the magnitude and direction of the force exerted on the beam at P?

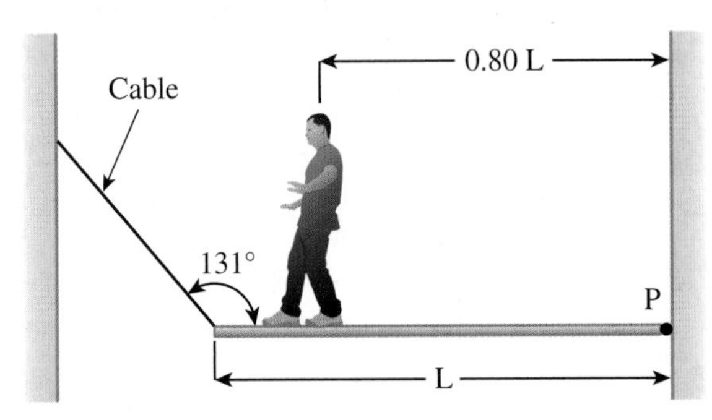

Figure 11-70 Question 11-74.

11-75 The top of a uniform ladder of mass 11.2 kg and length 4.00 m rests against a vertical wall and the bottom is on a horizontal floor, making an angle of 48.0° relative to the floor (Figure 11-71). A girl of mass 56.1 kg is standing on the ladder, 3.10 m from the bottom as measured along the ladder. Friction on the wall is negligible, but the coefficient of static friction between the ladder and the floor is 0.410. Under these conditions, the ladder will slip. Determine the minimum magnitude of horizontal force ($\vec{F}_{min}$) applied at the base of the ladder in order to prevent the ladder from slipping.

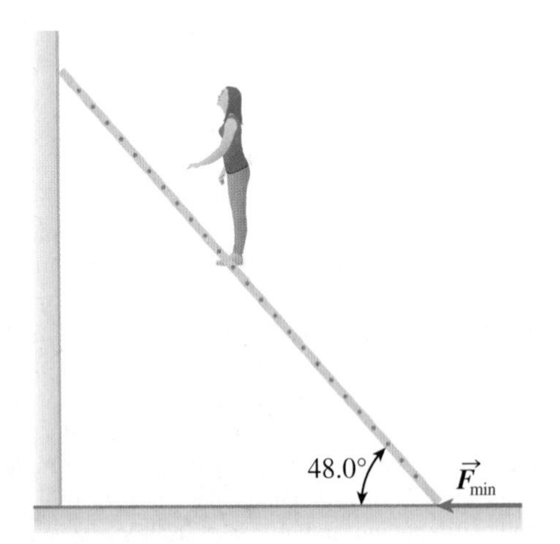

Figure 11-71 Question 11-75.

11-76 A person is standing halfway up a ladder that rests against a vertical wall; the bottom of the ladder is on a horizontal floor. The wall has negligible friction, but there is friction between the floor and the ladder, and the coefficient of static friction is μ_S. The ladder makes an angle of θ relative to the floor. In terms of μ_S, determine the minimum angle θ at which the ladder can rest without slipping.

11-77 This problem shows that a large force must be exerted by back muscles when a person bends over. Suppose that a woman is bending over while holding a box of mass $m_B = 15$ kg stationary above the floor (Figure 11-72). Her upper body has a mass $m_W = 44$ kg and a centre of mass located at C, which is 0.31 m from the vertebral joint O. The line OC is at an angle of 31° above the horizontal. The perpendicular distance from O to the line of action of $m_B\vec{g}$ is 0.37 m, as shown. To maintain the woman's posture, her back muscles exert a force $\vec{F}_M$ that has a moment arm $d_M = 1.5$ cm and acts parallel to OC. Force $\vec{F}_O$ represents the force exerted by the lower body directly at O.

(a) Determine the magnitude of $\vec{F}_M$.

(b) Calculate the ratio $\dfrac{F_M}{m_W g + m_B g}$.

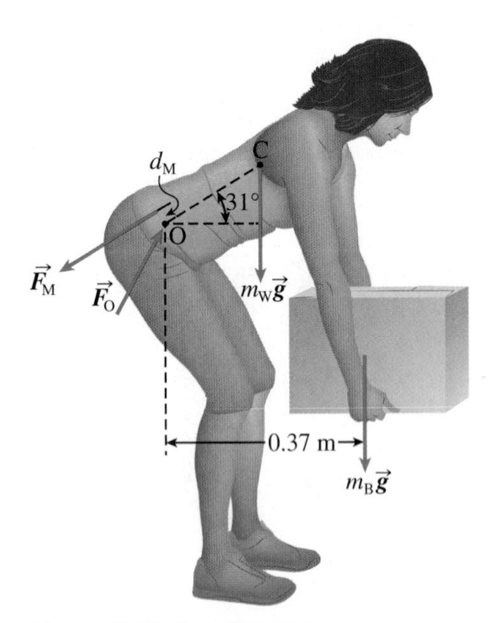

Figure 11-72 Question 11-77.

11-78 A rock-climber rests in a horizontal position while descending a vertical wall (Figure 11-73). The mass of the climber is 81.6 kg and his arm muscles are relaxed. The rope is fastened 20.0 cm above and 20.0 cm to the right of his centre of mass, which is 1.00 m from the wall. The rope makes an angle of 25.0° relative to the wall. Determine the magnitude and direction of the force exerted by the wall on the climber's feet. Assume this force acts at A, at the same horizontal level as the CM.

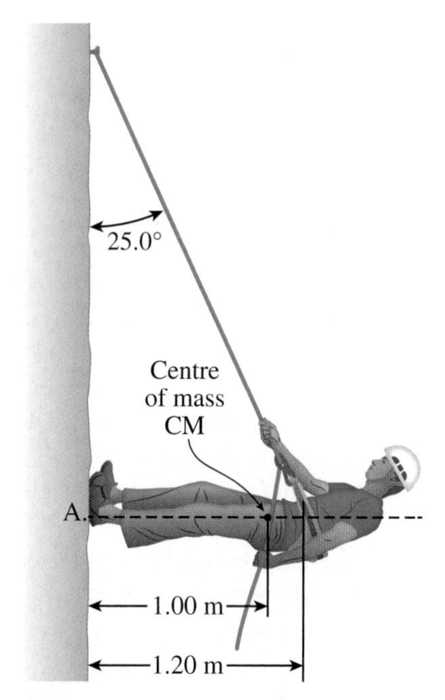

Figure 11-73 Question 11-78.

11-79 The rock-climber in the previous question is now using his hands to hold on to the end of a protruding horizontal rock surface and hang vertically from it. The rock is 22 cm wide, 15 cm high, and protrudes 2.0 m from a vertical rock face.

(a) What is the shear stress on the rock?

(b) If the shear strain of the rock is 9.0×10^{-7}, what is the shear modulus of the rock?

11-80 Figure 11-74 shows a person holding a 7.0 kg mass, which is attached to the end of a rope that makes an angle of 60° to the vertical. The forearm is held at 50° to the vertical upper arm. The force of gravity on the forearm and hand together has a magnitude of 30 N. Assume two significant digits in all data. If the CM of the forearm and hand is located halfway along the 40 cm length, and the biceps muscle exerts only a vertical force, determine

(a) the magnitude of the force exerted on the forearm by the biceps muscle

(b) the magnitude and direction of the force exerted by the humerus (the bone in the upper arm) on the forearm at the elbow

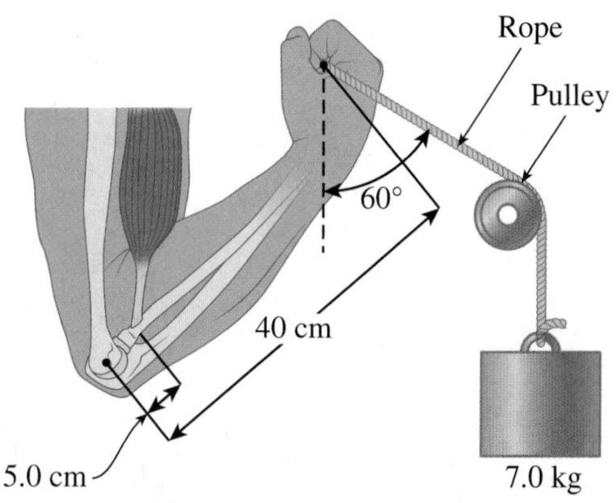

Figure 11-74 Question 11-80.

11-81 **Fermi Question:** Estimate the magnitude of force required to just barely tip a mid-size car sideways. Assume the force is exerted horizontally at the top edge of the car, and that there is sufficient friction to prevent the car from sliding.

11-82 Consider the motorized scooter and woman shown in Figure 11-75. Suppose that the horizontal distance between the centre of the rear wheels and the centre of the front wheels is 60 cm, and that the CM of the woman and scooter is located 20 cm forward of the centre of the rear wheels and 60 cm above the ground.

(a) If the total mass of the woman and scooter is 90 kg, determine the magnitude of the total normal force exerted by the ground at the two rear wheels, and the magnitude of the total normal force exerted by the ground at the front wheels. Assume two significant digits in given data.

(b) If the woman and scooter were to go forward up a ramp, what would be the maximum angle the ramp could make with the horizontal if the woman and scooter are not to tip backward?

Figure 11-75 Question 11-82.

11-83 A ramp is inclined at 36.0° to the horizontal. A box of mass 24.0 kg is to be placed on the incline, but the 36.0° angle is too large and the box will tip if left on its own. The box is 36.0 cm high, and its length and width are each 16.0 cm. The CM of the box is at its geometric centre. What minimum magnitude of force applied to the corner A of the box (Figure 11-76) will hold the base of the box on the ramp without tipping? The force is applied parallel to the ramp. Assume there is sufficient friction to prevent the box from sliding.

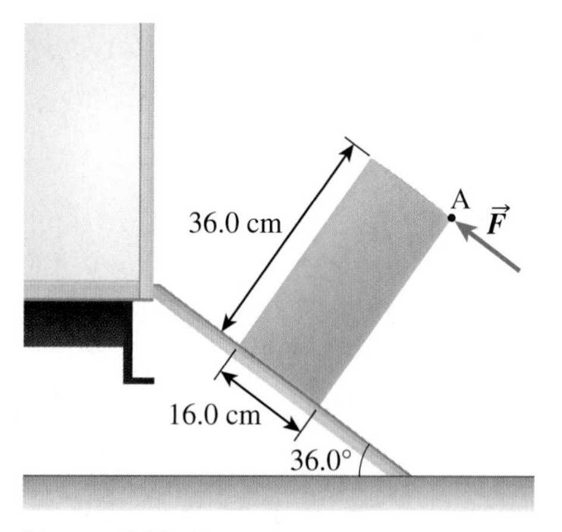

Figure 11-76 Question 11-83.

11-84 Two cylindrical rods, each of radius 0.22 cm, are placed end-to-end (Figure 11-77). One rod is made of copper of length 1.0 m and Young's modulus 1.1×10^{11} Pa, and the other is aluminum with a length of 2.0 m and Young's modulus 7.0×10^{10} Pa. Forces $\vec{F}$ and $-\vec{F}$ of equal magnitudes 6.0×10^3 N are applied at the ends of the rods, compressing them. What is the change in length of each rod in millimetres?

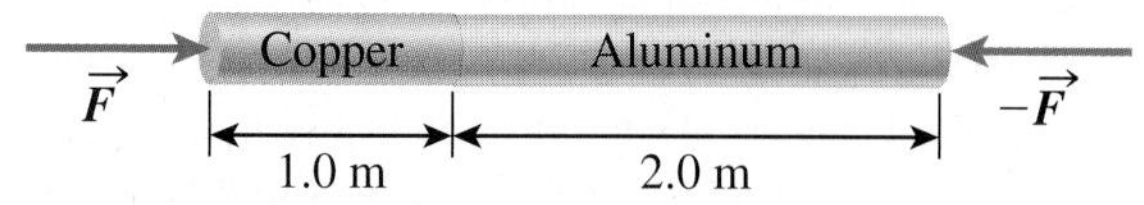

Figure 11-77 Question 11-84.

11-85 A car of mass 1857 kg is being towed with a steel cable of length 8.0 m, diameter 8.0 mm, and Young's modulus 2.1×10^{11} Pa. If the car is speeding up with an acceleration of magnitude 1.7 m/s^2, how much will the cable stretch in millimetres?

11-86 A solid cylinder of nickel has a radius of 2.0 cm and a length of 1.0 m. A hollow cylinder of nickel of the same length has an inner radius of 1.0 cm.

(a) In order for the hollow cylinder to have the same mass as the solid one, what must its outer radius be?

(b) Both cylinders are set vertically with their bases fixed to a horizontal surface. The same magnitude of force is applied tangentially to the circumference of the top surface of each cylinder. Determine the ratio of the twist angle of the solid cylinder to that of the hollow cylinder.

11-87 A uniform horizontal rod of length ℓ and mass m is supported at its ends by two vertical wires of equal lengths (Figure 11-78). The cross-sectional area of wire B is twice that of wire A.

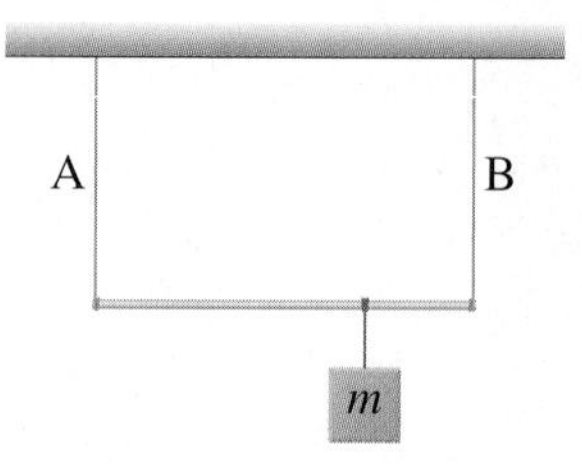

Figure 11-78 Questions 11-87 and 11-88.

Where should a mass m (same as the mass of the rod) be hung along the rod in order to produce equal stresses in A and B? Express your answer as a fraction of ℓ, measured from wire A.

11-88 Repeat the previous question if equal *strains* are to be produced in wires A and B. Additional information: Young's modulus for wire A is 1.5 times that for wire B.

11-89 A man of mass 75 kg falls downward from rest through a vertical height of 3.0 m and lands with his legs straight.

(a) What is his speed at impact?

(b) During the collision, his speed decreases to zero during a time of 0.030 s. What is the magnitude of his acceleration during this time?

(c) What is the magnitude of the force exerted on *each leg* during the collision?

(d) If this force is exerted on the tibia (shinbone), which has a solid cross-sectional area of 2.9 cm^2, what is the compressive stress on the tibia?

(e) The ultimate compressive strength for bone is 1.5×10^8 Pa. Express your answer from part (d) as a percentage of this value.

Fluid Statics and Dynamics 12

endopack/iStock/Thinkstock

Figure 12-1 This swimmer is relying on many fluids as he swims. Some are external, such as water and air, but many are internal, for example, blood, lung surfactant, and synovial fluid.

Fluids are crucial for the very existence of humans, other animals, and plants, as well as for the operation of devices such as automobiles, airplanes, etc. As a swimmer moves through the water (Figure 12-1), the water exerts an upward buoyant force, and as a result the swimmer does not sink. In addition to this force, the water exerts a viscous drag force on the swimmer. In order to generate energy, the swimmer is continually inhaling and exhaling air in order to extract oxygen and expel carbon dioxide. These gases (as well as other molecules) are carried throughout his body by blood being pumped by his heart. Lung surfactant (Section 12.3) makes it easy for his lungs to expand, and synovial fluid (Section 12.6) lubricates his joints.

This chapter explores the nature of fluids at rest and in motion. Topics include pressure in liquids and gases (especially the atmosphere), buoyancy, surface tension, and quantities such as volume flow rate and viscosity that are important in analyzing the movement of fluids such as blood.

CHAPTER **OUTLINE**

12.1 Properties of Fluids

fluid a substance that flows and takes the shape of its container

fluid statics the study of fluids at rest

Water is a common example of a fluid; air is another. A **fluid** is a substance that *flows* and takes the shape of its container. Thus, all liquids and gases are fluids. We begin with a study of fluids at rest, that is, **fluid statics**, and the first section deals with two important properties of fluids: density and pressure.

You know that a stone sinks in water and a cork floats on water. The stone is denser than water because its atoms are arranged in such a way that its mass is more concentrated than water. Conversely, the cork is less dense than water because its atoms are arranged in a less concentrated way than water molecules. Quantitatively we express this difference in terms of the **density** (ρ) of the substance, defined as the mass per unit volume. In terms of symbols,

density mass per unit volume of a substance; symbol ρ

$$\rho = \frac{m}{V} \tag{12-1}$$

where ρ = density

and m = the mass of a substance contained in volume V.

Note that ρ is the lowercase Greek letter rho.

The SI unit for density is kilogram per cubic metre (kg/m^3), but grams per cubic centimetre (g/cm^3) are often used as well:

$$1\,\frac{\text{g}}{\text{cm}^3} = 1 \times \frac{10^{-3}\,\text{kg}}{(10^{-2}\,\text{m})^3} = 1 \times 10^3\,\frac{\text{kg}}{\text{m}^3} \quad \text{or} \quad 1\,\frac{\text{kg}}{\text{m}^3} = 1 \times 10^{-3}\,\frac{\text{g}}{\text{cm}^3}$$

For an unconfined gas, the density is temperature dependent; for example, as temperature increases, the volume of the gas increases, and therefore the density decreases. When stating the density of a gas, we will assume a temperature of 0°C unless otherwise stated. For solids and liquids, there is also a temperature dependence but it is much smaller than for gases. Table 12-1 gives densities for various liquids, gases, and solids. Notice that the density of water at 4°C is a very convenient value (1.000×10^3 kg/m^3), and at 20°C its value to two significant digits is still 1.0×10^3 kg/m^3.

Table 12-1

Densities of Common Substances (at 20°C unless stated otherwise)

Substance	Density (kg/m^3)
Gases	
Helium	0.179
Air	1.29
Nitrogen	1.25
Oxygen	1.43
Liquids	
Ethyl alcohol	806
Water (at 4°C)	1.000×10^3
Water (at 20°C)	998.2
Blood (at 37°C)	1.06×10^3
Mercury	13.6×10^3
Solids	
Wood (balsa)	160
Wood (maple)	600–750
Ice (at 0°C)	917
Aluminum	2.70×10^3
Iron	7.86×10^3
Copper	8.96×10^3
Lead	11.3×10^3
Gold	19.3×10^3

DID YOU KNOW?

The density of most substances increases as temperature decreases, but water in a particular temperature range is unusual in this regard. The density of water has its maximum value of 1.00×10^3 kg/m^3 at 4°C, and as the temperature drops from 4°C to 0°C (where the water begins to turn to ice), its density decreases as the water molecules start to orient themselves into the less dense ice configuration. This means that in winter as lakes freeze from the top because of cold air, the ice remains floating at the surface, ensuring that fish and other aquatic life can survive in the water underneath at a temperature above 0°C.

SAMPLE PROBLEM 12-1

A measuring cup is placed on a kitchen scale and the cup's mass is displayed as 151 g. A volume of 125 mL of olive oil is then added to the cup, and the cup and oil are placed on the scale, giving a reading of 266 g. Determine the density of the oil in kilograms per cubic metre.

Solution

The mass of the oil is $m = (266 - 151)\text{ g} = 115\text{ g}$, and its volume is given as $V = 125\text{ mL} = 125\text{ cm}^3$. Using Eqn. 12-1, the oil's density is

$$\rho = \frac{m}{V} = \frac{115\text{ g}}{125\text{ cm}^3} = 0.920\frac{\text{g}}{\text{cm}^3} \times \frac{1\text{ kg}}{1000\text{ g}} \times \frac{(100\text{ cm})^3}{(1\text{ m})^3} = 9.20 \times 10^2\text{ kg/m}^3$$

Thus, the density of the olive oil is $9.20 \times 10^2\text{ kg/m}^3$.

specific gravity (s.g.) ratio of the density of an object to that of water at 4°C

TRY THIS!

Supercooled Water

It is possible to decrease the temperature of water to well below 0°C without the water turning to ice. Place an unopened, sealed plastic bottle of very pure water in a freezer for about 2–3 hours and then carefully remove it. With luck the water will still be liquid. Then shake the bottle or bang it on a solid surface, and the water will probably crystallize (i.e., freeze) within seconds. The explanation for this phenomenon is that the freezing of water at 0°C usually begins at a place in the water where there is an impurity or a surface imperfection in the container. If the water is very pure and the container is very smooth, the water molecules can stay in the liquid form below 0°C. Shaking or banging the bottle causes a disturbance that orients some of the water molecules into the ice formation and then the rest of the molecules quickly follow suit. If you are unsuccessful in your attempts, just search online for "supercooled water" and you can find videos showing this effect.

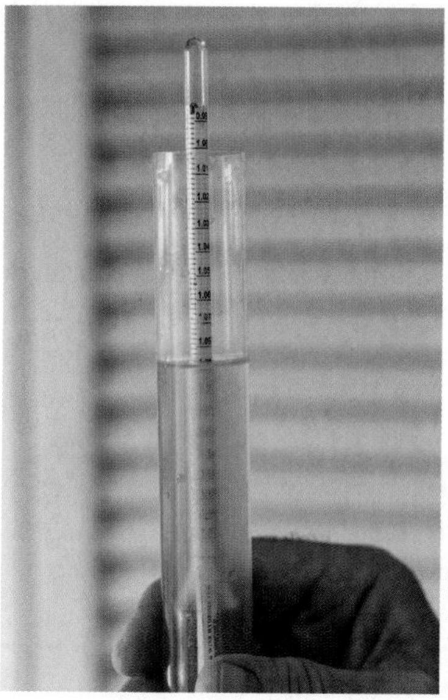

Specific Gravity

It is often convenient to express the density of a substance as the ratio of its density to that of water at 4°C ($1.00 \times 10^3\text{ kg/m}^3$). This ratio is called the **specific gravity (s.g.)**, or relative density of the substance. Since specific gravity is a ratio of two densities, it is a dimensionless quantity.

$$\text{s.g.} = \frac{\rho_{\text{substance}}}{\rho_{\text{water}}} \qquad (12\text{-}2)$$

In Table 12-1 the density of blood is given as $1.06 \times 10^3\text{ kg/m}^3$, and so its s.g. is

$$\text{s.g.} = \frac{\rho_{\text{blood}}}{\rho_{\text{water}}} = \frac{1.06 \times 10^3\text{ kg/m}^3}{1.00 \times 10^3\text{ kg/m}^3} = 1.06$$

Figure 12-2 Using a hydrometer (the floating cylinder) to measure the specific gravity of a liquid.

Specific gravities are measured by devices called hydrometers. The simplest hydrometer is a narrow glass or plastic cylinder weighted at one end. When placed in a liquid, the hydrometer floats vertically with the weighted end down (Figure 12-2). If the liquid has a small s.g., the hydrometer floats low in the liquid, and if the liquid has a high s.g., the hydrometer floats higher.[1] There is a vertical pre-calibrated scale along the hydrometer, and where the surface of the liquid meets the scale, the liquid's s.g. is simply given as the scale value there.

Hydrometers are regularly used in automobile service garages to measure the s.g. of the liquid in car cooling systems to ensure that there is the correct proportion of antifreeze and water. Hydrometers are also employed in breweries, distilleries, and wineries (Figure 12-3) to check the progress of fermentation. As sugars are converted

Figure 12-3 Hydrometers are used in wineries to monitor the fermentation progress.

[1]As discussed in detail in Section 12.4, the magnitude of the upward buoyant force on the hydrometer is proportional to the liquid's density. If the density of the liquid is large, then the buoyant force is large, and the hydrometer rides high in the liquid.

to alcohol, the s.g. of the fermenting mixture decreases. For example, prior to fermentation a typical beer wort has a s.g. of 1.050, and when fermentation is complete, the s.g. has dropped to 1.010.

Roberto Sitzia/iStock/Thinkstock

Figure 12-4 Snowshoes spread the force over a larger area than boots do, and therefore exert less pressure on the snow.

Pressure

A person wearing snowshoes (Figure 12-4) stays on top of the snow, while a person of equal mass wearing boots sinks into the snow. They both exert an equal force on the snow. Why does only one sink? The answer is the difference in pressure: snowshoes spread the force over a larger surface area, so the pressure is less.

If a force of magnitude F is applied in a direction perpendicular to a surface of area A, the **pressure** P is the force magnitude divided by the area:

pressure the ratio of force magnitude F to area A, when a force of magnitude F is applied in a direction perpendicular to a surface of area A; symbol P

$$P = \frac{F}{A} \tag{12-3}$$

In SI units with F in newtons and A in square metres, the unit of pressure is newtons per square metre (N/m^2), and this unit is given the name of **pascal** (Pa), in honour of Blaise Pascal (1623–1662), a French physicist, mathematician, inventor, writer, and philosopher who in his short life contributed greatly to our understanding of fluids.

pascal SI unit of pressure, equivalent to newton per square metre; symbol Pa

DID YOU KNOW?

Blaise Pascal was the first person to measure altitude above sea level using a barometer, a device that measures air pressure.

$$1 \text{ Pa} = 1 \frac{\text{N}}{\text{m}^2} \tag{12-4}$$

When you stand on one foot, the pressure that you exert on the floor is twice what it is when you stand on two feet because the force is exerted on half the area. Manufacturers of some rugs and other flooring products warn that people should not walk on flooring with high-heel shoes because the high pressure exerted by the small surface area of stiletto heels can cause damage.

Another example that illustrates how pressure varies with area is a "bed of nails." No one would be foolish enough to lie on a single nail protruding upward from a horizontal board. The pressure on the very small surface area would be so great that the nail would easily pierce the skin. However, a person can safely lie on a bed of nails if the nails are spread appropriately, about 2 cm apart (Figure 12-5). The force of the nails against the person is then spread out over a total surface area provided by hundreds of nails, so that any one nail does not puncture the skin.

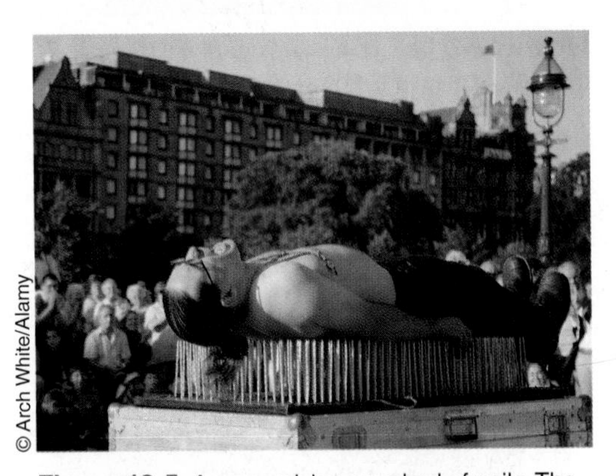

© Arch White/Alamy

Figure 12-5 A person lying on a bed of nails. The total contact area between the nails and skin is large enough that the resulting pressure will not pierce the skin.

TRY THIS!

Feeling Pressure

Place your left forearm flat on a desk or table. With the palm of your right hand press down hard on your left forearm near the elbow. Now try to exert the same force downward at the same location using only the tip of the index finger of your right hand. Explain what you feel and why.

SAMPLE PROBLEM 12-2

Determine the pressure in pascals and kilopascals exerted on the horizontal ground by a person of mass 82 kg if the total area that his shoes have in contact with the ground is 274 cm^2.

Solution

The magnitude of the downward force exerted on the ground by the person is

$$F_G = mg = (82 \text{ kg})(9.8 \text{ m/s}^2) = 8.0 \times 10^2 \text{ N}$$

The area is $264 \text{ cm}^2 \times \dfrac{(1 \text{ m})^2}{(100 \text{ cm})^2} = 2.64 \times 10^{-2} \text{ m}^2$

Using Eqn. 12-3,

$$P = \frac{F}{A} = \frac{8.0 \times 10^2\ \text{N}}{2.74 \times 10^{-2}\ \text{m}^2} = 2.9 \times 10^4 \frac{\text{N}}{\text{m}^2} = 2.9 \times 10^4\ \text{Pa} \times \frac{1\ \text{kPa}}{1000\ \text{Pa}} = 29\ \text{kPa}$$

Therefore, the pressure exerted on the ground is 2.9×10^4 Pa, or 29 kPa.

standard atmospheric pressure (or standard atmosphere) 1 atm = 101.3 kPa; typical pressure exerted by atmosphere at ground level

gauge pressure the difference between the pressure in one location and the pressure in another location (usually the atmosphere)

absolute pressure total pressure taking into account all sources of pressure

The pascal is a rather small unit of pressure. As a result, the pressure in pascals exerted by the person in Sample Problem 12-2 is huge: 2.9×10^4 Pa; whereas in kilopascals, which is a more convenient and commonly used unit, the pressure is 29 kPa. The pressure exerted at ground level by the atmosphere is approximately 100 kPa, but changes from place to place and from day to day. **Standard atmospheric pressure** (usually abbreviated as **standard atmosphere** or 1 atm) is 101.3 kPa (correct to four significant digits):

$$\text{standard atmosphere} = 1\ \text{atm} = 101.3\ \text{kPa} \quad (12\text{-}5)$$

Although we are not generally aware of atmospheric pressure, it is extremely large. The standard atmosphere of 101.3 kPa means that a force of 101.3×10^3 N is being exerted by the atmosphere on only 1 m^2 of area. If the 1 m^2 is on a horizontal surface, the atmospheric pressure is equivalent to that exerted by about 150 people, each of mass 70 kg, standing on that area!

Joe Belanger/Shutterstock

Figure 12-6 A tire pressure gauge measures the difference in pressure between the two ends of the gauge, that is, the pressure difference between the air inside the tire and the air in the atmosphere.

Gauge Pressure and Absolute Pressure

You probably know from experience that the pressure inside an automobile tire is larger than the pressure outside, and that a tire pressure gauge (Figure 12-6) can be used to measure this pressure. More specifically, what the gauge measures is the pressure *difference* between the two ends of the gauge. One end is open to the atmosphere and the other is connected to the air inside the tire via the valve stem that the gauge is pressed against. Suppose that a tire gauge is being used and gives a reading of 225 kPa. This means that the air pressure inside the tire is 225 kPa greater than the atmospheric air pressure outside. This 225 kPa is the gauge pressure of the air inside the tire, where **gauge pressure** is defined as the difference between the pressure in one location (e.g., inside the tire) and the pressure in another location (usually the atmosphere). We will assume that the atmospheric pressure is 1 atm, that is, 101 kPa (unless otherwise stated), and hence the actual total pressure of the air inside the tire is (101 + 225) kPa = 326 kPa. This is the **absolute pressure**, that is, the total pressure taking into account all sources of pressure. Since gauge pressure (P_{gauge}) is usually defined in terms of the pressure difference relative to atmospheric pressure, the absolute pressure (P_{abs}) is frequently written as

$$P_{abs} = P_{gauge} + 1\ \text{atm} \quad (12\text{-}6)$$

Pressure in Liquids

The molecules or atoms in liquids and gases move around in random motion (Figure 12-7), unlike molecules and atoms in solids, which are constrained to vibrate about fixed positions. Since atoms and molecules in liquids attract each other more strongly than atoms and molecules in gases, liquids are denser and the frequency of collisions is much higher. As fluid molecules and atoms move, they are constantly colliding with each other and with any surfaces in which they are in contact.

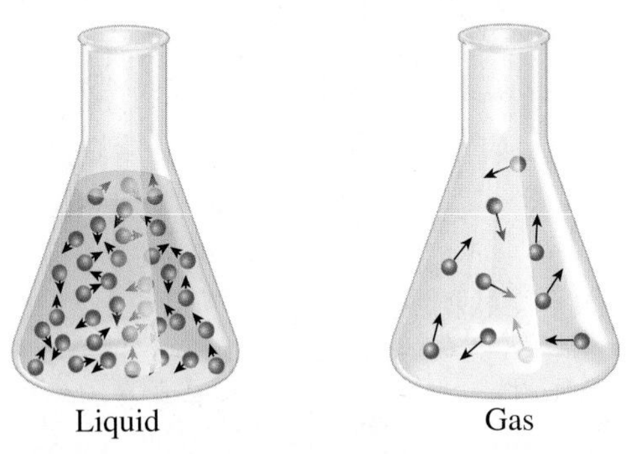

Figure 12-7 Atoms and molecules in liquids and gases are in random motion and are constantly undergoing collisions. These collisions are the source of pressure exerted within a fluid and on any surface that the fluid is touching.

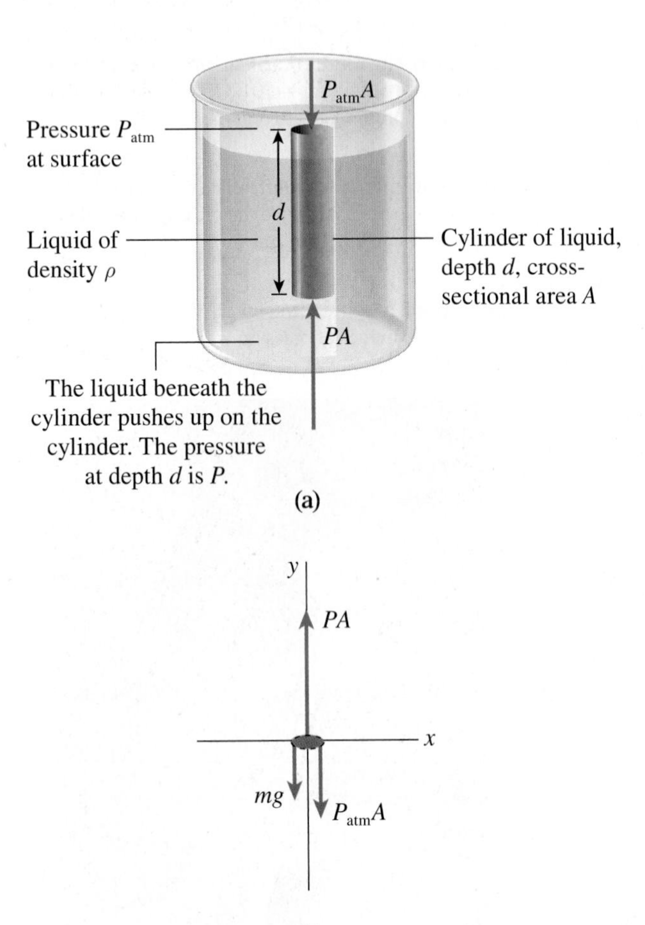

Figure 12-8 **(a)** Magnitudes of forces due to pressures P_{atm} and P acting on a cylinder of liquid. **(b)** FBD for the cylinder.

These collisions are the source of pressure in fluids, and because the motion and collisions are random, the pressure at any particular point in a fluid acts equally in all directions.

You probably know that pressure increases with depth in a liquid. We now develop an equation that expresses this increase quantitatively. Figure 12-8 (a) shows a liquid in a container, and a particular cylindrical portion of this liquid has been selected for analysis. This cylinder has a cross-sectional area A and a length d. The top of the cylinder is at the liquid's top surface and is open to the atmosphere, which has pressure P_{atm}. From Eqn. 12-3 ($P = F/A$, or $F = PA$), the magnitude of the downward force on the cylinder exerted by the atmosphere is $P_{atm}A$. The bottom of the cylinder contacts the liquid below it, which exerts an upward pressure P on the cylinder. The magnitude of the upward force due to this pressure is PA. In addition to these two pressure-based forces, the cylinder is subject to a gravitational force of magnitude mg, where m is the cylinder's mass. The three forces acting on the cylinder are shown in a FBD in Figure 12-8 (b). Although water molecules are constantly moving into and out of the cylinder, the cylinder as a whole is at rest, and the resultant force on it is zero. Writing Newton's first law of motion for the cylinder, with $+y$ chosen upward,

$$\Sigma F_y = 0, \quad \text{which gives} \quad PA - P_{atm}A - mg = 0 \qquad \text{Eqn. [1]}$$

From Eqn. 12-1, $m = \rho V$. The cylinder's volume V can be expressed as the product of cross-sectional area A and depth d, that is, $V = Ad$, giving $m = \rho Ad$. Substituting for m in Eqn. [1],

$$PA - P_{atm}A - \rho A dg = 0$$

Dividing by A and rearranging to solve for $P - P_{atm}$,

$$P - P_{atm} = \rho gd$$

Writing $P - P_{atm}$ as ΔP, we have $\Delta P = \rho gd$.
Therefore, the increase in pressure (ΔP) at a depth d in a liquid having density ρ is

$$\Delta P = \rho gd \qquad (12\text{-}7)$$

Although Eqn. 12-7 was developed with the assumption that the top of the cylinder of liquid was at the liquid's surface, it applies more generally and gives the pressure increase ΔP in a liquid of density ρ as the depth increases by an amount d. Note also that the area A does not appear in this equation, which is not dependent on any geometric detail of the liquid except for the depth.

SAMPLE PROBLEM 12-3

A diver is at sea level where the atmospheric pressure is 1 atm = 101 kPa. The diver then descends to a depth of 14.5 m. Determine the absolute and gauge pressures in kilopascals at that depth, given that the density of sea water is 1.03×10^3 kg/m^3.

Solution

Use Eqn. 12-7 to determine the increase in pressure at a depth of 14.5 m:

$$\Delta P = \rho gd = (1.03 \times 10^3 \text{ kg/m}^3)(9.80 \text{ m/s}^2)(14.5 \text{ m})$$

$$= 1.46 \times 10^5 \text{ Pa} \times \frac{1 \text{ kPa}}{10^3 \text{ Pa}} = 146 \text{ kPa}$$

At the water's surface, the only pressure is atmospheric and therefore the absolute pressure there is 101 kPa. Applying the pressure increase of 146 kPa at the depth of 14.5 m gives the absolute pressure there: $P_{abs} = (101 + 146)$ kPa = 247 kPa.

Recall that gauge pressure (P_{gauge}) is the pressure difference relative to atmospheric pressure; that is, $P_{gauge} = P_{abs} - 1$ atm (from Eqn. 12-6). Therefore, at the depth of 14.5 m, where $P_{abs} = 247$ kPa, we obtain

$$P_{gauge} = 247 \text{ kPa} - 1 \text{ atm} = (247 - 101) \text{ kPa} = 146 \text{ kPa}.$$

Another way to find the gauge pressure at the 14.5 m depth is to note first that at the water's surface where the only pressure is atmospheric, $P_{gauge} = P_{abs} - 1$ atm = 1 atm − 1 atm = 0 kPa. At the depth of 14.5 m, P_{gauge} increases by our calculated amount, $\Delta P = 146$ kPa, and thus has a value of $P_{gauge} = (0 + 146)$ kPa = 146 kPa.

Thus, the absolute pressure on the submerged diver is 247 kPa and the gauge pressure is 146 kPa.

Problem-Solving Tip

Notice in the solution to Sample Problem 12-3 that the pressure increase of 146 kPa applies to absolute pressure and also to gauge pressure. The increase in pressure given in Eqn. 12-7 ($\Delta P = \rho g d$) is valid for either absolute pressure or gauge pressure in a liquid as the depth increases by a distance d.

From the relationship $\Delta P = \rho g d$, it is evident that the pressure beneath the surface of a liquid depends on the depth, not the shape of a container. Thus, a swimmer who is 2 m below the water surface in a swimming pool experiences the same pressure as a swimmer 2 m below the surface of a large lake (assuming the water densities are the same).

Furthermore, a liquid in connected containers open to the atmosphere reaches the same level in all the containers. Figure 12-9 illustrates this principle for a liquid in four containers of different size and shape. The liquid levels are the same because the pressure is constant at a constant depth. If the pressure at the bottom of one container were larger than at the bottom of another, then the extra pressure would force liquid from the high-pressure region to the low-pressure one, and the result would be different liquid levels in the two containers, but this does not occur. The observation that "water seeks its own level" has been applied for centuries by people designing structures that must be horizontally level.

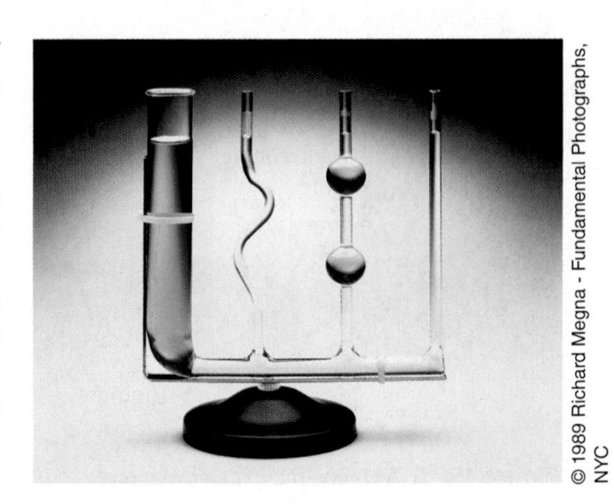

Figure 12-9 Despite the different shapes and diameters of the connected containers open to the atmosphere, the liquid is at the same level in all the containers.

Equation 12-7 was derived for a liquid of constant density. Liquids are generally quite incompressible, and the assumption of constant density as pressure increases with depth is quite reasonable. However, this is not the case for gases, since they are easily compressible and the density of a gas is in fact proportional to its pressure. An analysis of the relation between the pressure in Earth's atmosphere and height above Earth's surface will be given in Section 12.2. In principle, $\Delta P = \rho g d$ could be used to find the difference in pressure for a gas as long as the distance d is small, but this is not commonly done since the pressure differences are very small because of the low density of gases. For example, for air in a room 2.5 m high, $\Delta P = \rho g d$ gives a pressure at the floor only 0.03 kPa more than that at the ceiling.

Other Units of Pressure

There are a number of units other than pascal, kilopascal, and atm that can be used for pressure. A common one in meteorology is the bar, with 1 bar = 10^5 Pa = 100 kPa ≈ 1 atm. In some countries, pounds per square inch (psi) are used, with 14.7 psi = 1 atm. Most automobile mechanics in North America refer to tire pressures in psi.

From Eqn. 12-7 ($\Delta P = \rho g d$), we can see that a distance d of liquid can be used as a pressure indicator or unit. For example, blood pressure is commonly measured in millimetres of mercury (mm Hg). If someone's blood pressure is 120 mm Hg (to two significant digits), this is equivalent to the pressure at the bottom of a column of mercury 120 mm high. Using Eqn. 12-7 and the density of mercury given in Table 12-1, we can convert 120 mm Hg to pascals:

$$\Delta P = \rho g d = \left(13.6 \times 10^3 \frac{\text{kg}}{\text{m}^3}\right)\left(9.80 \frac{\text{m}}{\text{s}^2}\right)\left(120 \text{ mm} \times \frac{1 \text{ m}}{1000 \text{ mm}}\right) = 1.6 \times 10^4 \text{ Pa}$$

DID YOU KNOW?

Although humans need specially designed submersible vehicles to explore the depths of the oceans where the pressures can be hundreds of atmospheres, deep-sea organisms are well adapted to the crushing environment there. They lack compressible spaces in their bodies that would collapse under the extreme water pressure. Unlike land animals, they do not have lungs, ear canals, or other passageways.

DID YOU KNOW?

The standard atmosphere is defined as the gauge pressure at the bottom of exactly 0.76 m of mercury with $g = 9.806\,65 \text{ m/s}^2$ and a temperature of 0°C (at this temperature, $\rho_{Hg} = 13.595 \times 10^3 \text{ kg/m}^3$). This gives

$$1 \text{ atm} = \Delta P = \rho g d = \left(13.595 \times 10^3 \frac{\text{kg}}{\text{m}^3}\right)\left(9.806\,65 \frac{\text{m}}{\text{s}^2}\right)(0.760\,00 \text{ m}) = 1.0132 \times 10^5 \text{ Pa}$$

U-Tube Manometer

manometer a U-shaped transparent tube partly filled with a liquid, used to measure gauge pressure

A **manometer** is a device used to measure gauge pressure. It consists of a hollow U-shaped transparent tube partly filled with a liquid (mercury, water, etc.) and oriented vertically (Figure 12-10). One end of the tube is connected to a gas-filled vessel and the other end is open to the atmosphere. If the pressure in the vessel is the same as atmospheric pressure, the top of the liquid in the two parts of the tube will be at the same level. If the pressure in the vessel is larger than atmospheric pressure, the larger pressure will push liquid down in the vessel side of the tube (as in Figure 12-10). If the vessel pressure is less than atmospheric, the larger atmospheric pressure will push liquid down on the atmosphere side of the tube.

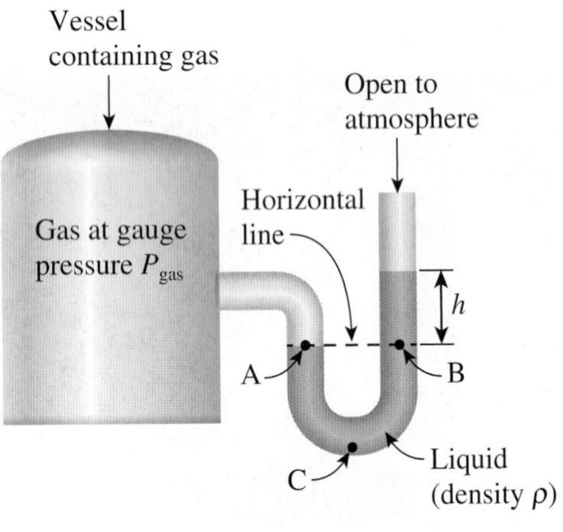

Figure 12-10 A U-tube manometer with one end connected to a gas at a pressure greater than atmospheric pressure.

We now determine an algebraic expression for the gauge pressure in the situation shown in Figure 12-10. Point A is the surface of the liquid (having density ρ) in the left-hand part of the U-tube, and point B is at the same horizontal level as A, as shown by the horizontal dashed line. Since these two points are at the same level in a connected fluid, the pressures at these points are equal: $P_A = P_B$. One way to think about this is to imagine starting at point C in the liquid near the bottom of the tube, where the pressure is P_C. Points A and B are at an equal distance above C. Going up the left-hand side of the tube to A, or going up the same distance in the right-hand side of the tube to B, the pressure will decrease from P_C by the same amount. Thus, the pressures at A and B must be the same. Now, since point A is at the surface of the liquid in contact with the gas in the vessel, the pressure there is the same as the pressure in the gas, $P_A = P_{gas}$. Point B is in the liquid at a depth h below the atmosphere, and the gauge pressure in the atmosphere is 0. Therefore, from Eqn. 12-7, the gauge pressure at B will be $P_B = 0 + \rho g h = \rho g h$. Since $P_A = P_B$, we have $P_{gas} = \rho g h$; that is, the gauge pressure of the gas is given simply by the pressure difference calculated using the height difference between the liquid levels in the U-tube. If the absolute pressure is needed, then Eqn. 12-6 would be used: $P_{abs} = P_{gauge} + 1$ atm.

If the liquid levels in Figure 12-10 were reversed, that is, if the level in the side open to the atmosphere was lower than that in the side open to the vessel, then the pressure in the atmosphere must be larger than the pressure in the gas in the vessel, and the gauge pressure of the gas would be $P_{gas} = -\rho g h$. Rather than having two separate equations depending on which liquid level is lower, we write just one equation for the difference in pressure indicated by a manometer:

$$\Delta P_{manometer} = \rho g h \tag{12-8}$$

When using this equation, note that *the pressure is always larger on the side of the manometer that has the lower liquid surface level, since the higher pressure there pushes the liquid down.*

SAMPLE PROBLEM 12-4

Figure 12-11 shows a large aquarium containing aquatic plants. The top portion of the aquarium contains air connected to a U-tube manometer containing mercury. The other side of the manometer is open to the atmosphere, and the mercury level on the atmospheric side of the manometer is 0.30 m lower than on the aquarium side.

The bottom portion of the aquarium contains water of depth 1.6 m. Determine the absolute and gauge pressures in kilopascals at points A, B, and C.

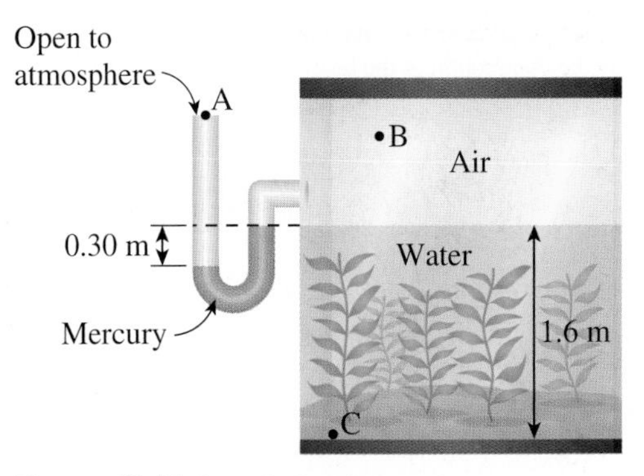

Figure 12-11 Sample Problem 12-4 .

Solution

Point A is in the atmosphere, and unless stated otherwise we assume the absolute pressure there is 1 atm, which is 101 kPa (correct to three significant digits). Since $P_{abs} = P_{gauge} + 1$ atm (Eqn. 12-6), then $P_{A,gauge} = 1$ atm $- 1$ atm $= 0$ kPa.

Thus, at A the absolute pressure is 101 kPa, and the gauge pressure is 0 kPa.

Point B is located in the air above the water in the aquarium. Since the mercury level on the atmospheric side of the manometer is lower than on the aquarium side, the pressure in the atmosphere must be greater than the pressure in the air (and at point B) in the aquarium . Using Eqn. 12-8 and the density of mercury from Table 12-1,

$$\Delta P_{manometer} = \rho_{Hg} g h_{Hg} = \left(13.6 \times 10^3 \frac{kg}{m^3}\right)\left(9.8 \frac{m}{s^2}\right)(0.30 \text{ m}) = 4.0 \times 10^4 \text{ Pa}$$

$$4.0 \times 10^4 \text{ Pa} \times \frac{1 \text{ kPa}}{10^3 \text{ Pa}} = 40 \text{ kPa (correct to two significant digits)}$$

Therefore, the atmospheric pressure outside the aquarium is 40 kPa larger than the air pressure at point B inside the aquarium. This 40 kPa difference applies to both absolute pressure and gauge pressure. The absolute pressure at B is then $P_{B,abs} = P_{atm} - 40$ kPa $= 61$ kPa, and the gauge pressure is $P_{B,gauge} = 0$ kPa $- 40$ kPa $= -40$ kPa.

(Alternatively, Eqn. 12-6 could be used to give the gauge pressure at B:

$$P_{B,gauge} = P_{B,abs} - 1 \text{ atm} = 61 \text{ kPa} - 101 \text{ kPa} = -40 \text{ kPa})$$

Hence, at B the absolute pressure is 61 kPa and the gauge pressure is –40 kPa (correct to two significant digits).

Point C is at the bottom of 1.6 m of water, and therefore the pressure there will be greater than at the surface by an amount given by Eqn. 12-7: $\Delta P = \rho_{water} g d_{water}$. Substituting numbers,

$$\Delta P = \left(1.0 \times 10^3 \frac{kg}{m^3}\right)\left(9.8 \frac{m}{s^2}\right)(1.6 \text{ m}) = 1.6 \times 10^4 \text{ Pa} \times \frac{1 \text{ kPa}}{10^3 \text{ Pa}} = 16 \text{ kPa}$$

Adding 16 kPa to the absolute air pressure of 61 kPa at the surface of the water gives 77 kPa for the absolute pressure at C.

For the gauge pressure at C, add 16 kPa to the gauge pressure of –40 kPa at the water's surface, giving −24 kPa.

(Alternatively, use Eqn. 12-6: $P_{C,gauge} = P_{C,abs} - 1$ atm $= 77$ kPa $- 101$ kPa $= -24$ kPa.)

Thus, at C the absolute pressure is 77 kPa and the gauge pressure is −24 kPa.

Figure 12-12 A Bourdon pressure gauge.

irman/iStock/Thinkstock

sphygmomanometer a device used to measure blood pressure, consisting of an inflatable cuff and a mercury manometer or other pressure meter

It is often more convenient to use a mechanical pressure gauge, rather than one that uses a liquid. A common type of gauge (Figure 12-12) uses a coiled metal tube that straightens when the pressure in it increases. This gauge is known as a Bourdon gauge, named after Eugene Bourdon who patented it in France in 1849. The end of the tube is connected to a gear assembly that rotates a pointer on a calibrated scale. Alternatively, the gauge can be connected to an electronic display. This gauge is often used to measure pressures in compressed-gas tanks, boilers, tires, etc.

Measuring Blood Pressure

A specialized device called a **sphygmomanometer** is used to measure blood pressure (Figure 12-13). It consists of an inflatable cuff that is wrapped around an arm and a U-tube mercury manometer or mechanical pressure meter used to measure the gauge pressure of the air in the cuff. After the cuff is placed around the arm, it is inflated

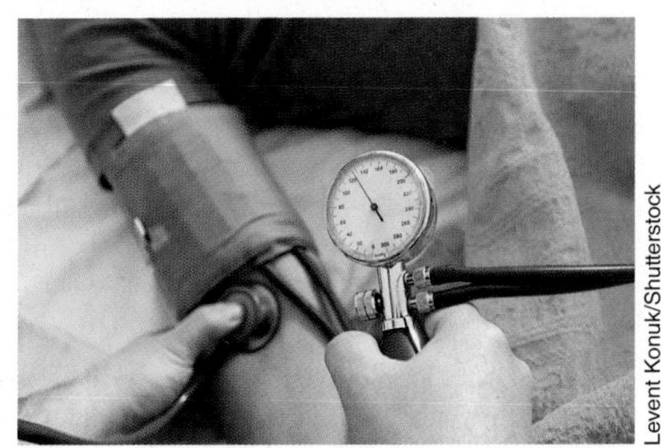

Figure 12-13 Measuring blood pressure using a sphygmomanometer.

Levent Konuk/Shutterstock

systolic pressure maximum blood pressure exerted by the contraction of the heart in a person's cardiac cycle

to a gauge pressure of 150–180 mm Hg, which is normally greater than the **systolic pressure** in the blood, which is the maximum pressure during contraction of the heart during a person's cardiac cycle. With the cuff pressure greater than the systolic pressure, the blood flow is stopped in the brachial artery in the arm. The cuff pressure is then gradually reduced as the nurse or physician listens with a stethoscope placed under the cuff; alternatively a sound sensor built into the cuff will monitor sound in the artery. With no blood flow, there is no sound generated, but just when the cuff pressure is reduced to the systolic pressure value, the blood starts to spurt turbulently through the artery during systole. This turbulent flow is noisy and can be detected by the stethoscope or other sound sensor, and the pressure at which it starts to occur is recorded as the systolic pressure.

diastolic pressure minimum blood pressure exerted by the heart during a person's cardiac cycle

The cuff pressure continues to be reduced until the noise disappears. At this point, the cuff pressure has been reduced just below the **diastolic pressure** in the blood, which is the minimum pressure during the cardiac cycle. With the cuff pressure below the diastolic pressure, blood is able to flow through the artery in a non-turbulent way during the entire cardiac cycle, and the cuff pressure at which the noise ceases is recorded as the diastolic pressure. Normal systolic pressure is usually about 120 mm Hg, and diastolic pressure around 80 mm Hg. The pressures are typically recorded as 120/80 ("120 over 80").

Notice in Figure 12-13 that the sphygmomanometer cuff is wrapped around the upper arm, which is roughly at the same elevation as the person's heart. As a result, the sphygmomanometer measures the blood pressure at the heart. What would happen to the measured blood pressure if the sphygmomanometer were wrapped around an ankle? (**Hint:** Remember Eqn. 12-7 ($\Delta P = \rho g d$).) For a giraffe, how would the blood pressure measured at the top of its neck differ from the blood pressure measured at the level of its heart?

DID YOU KNOW?

The word "sphygmomanometer" comes from the Greek words *sphygmos* meaning "pulse," *manos* meaning "thin or fine," and *metron* meaning "measure." Thus, a sphygmomanometer is an instrument that measures blood pulse (pressure, more correctly) in a fine (i.e., accurate) way.

EXERCISES

12-1 In each case determine the density in kilograms per cubic metre.

(a) A rock has a volume of 0.047 m^3 and a mass of 350 kg.

(b) The mass of 0.0450 L of glycerol is 56.7 g.

12-2 What mass of mercury would fill a cylindrical vessel that has a diameter of 18.4 cm and a height of 6.50 cm? (Reference: Table 12-1)

12-3 **Fermi Question:** Design and perform an experiment to determine an approximate value for your own density. Is your result reasonable? Why or why not?

12-4 What is the specific gravity of (a) lead, (b) ice (at 0°C)? (Reference: Table 12-1)

12-5 Explain each of the following statements, taking into consideration the difference between force and pressure.

(a) Small boxing gloves are more dangerous to the opponent than larger ones made of the same material.

(b) If trying to break a walnut by hand, it is much easier to use two walnuts pressed against one another than a single walnut in the hand.

12-6 Assume that a 55 kg woman places all her weight for a brief instant onto one heel of a high-heel shoe. If the circular heel has a radius of 5.0 mm, determine the pressure in kilopascals exerted by the heel on the floor.

12-7 **Fermi Question:** Estimate the average pressure (in kilopascals) you would exert on the floor if you were standing (a) in your running shoes, (b) on only your head (for an instant), and (c) in high heels.

12-8 The air pressure at the top surface of the water in a rectangular pool is 101 kPa.

(a) If the inside of the pool is 6.5 m wide and 9.4 m long, what total force is applied by the air on the surface of the water?

(b) If the pool has a maximum depth of 2.0 m, what is the maximum pressure at the bottom?

12-9 A tire pressure gauge measures the pressure in a bicycle tire to be 413 kPa. The atmospheric pressure is 101 kPa. What is the absolute pressure in the tire?

12-10 A gasoline storage tank, 13 m high, is vented to the atmosphere, so the pressure at the top of the tank is atmospheric pressure. Assuming the tank is full, what are the gauge and absolute pressures at the bottom of the tank? The density of gasoline is 6.8×10^2 kg/m^3.

12-11 Convert a pressure of 61.2 kPa to millimetres of mercury. Assume that $g = 9.80$ m/s^2.

12-12 Water must be pumped from a mine that is 95 m beneath ground level. Assuming atmospheric pressure remains approximately constant for this height difference, determine the minimum gauge pressure (in kilopascals) required for a pump placed in the water. (At this pressure, the pump must be able to support a water column of height 95 m.)

12-13 One end of a U-tube mercury manometer is open to the atmosphere and the other end is connected to the interior of a pipe carrying steam to heat a building. The mercury in the pipe side of the manometer is 35 cm lower than on the atmosphere side. What is the absolute pressure in kilopascals of the steam in the pipe?

12.2 Pascal's Principle; Pressure in the Atmosphere

Perhaps you have seen a car raised in a garage (Figure 12-14) and wondered how such a heavy object can be lifted so easily. The answer is that the pressure exerted by a liquid does the lifting. The liquid is enclosed in a system in which there is a movable piston at each of the two ends (Figure 12-15). The cross-sectional areas of the pistons, and of the cylinders in which they move, are different. An external force $\vec{F}_S$ is applied to the smaller piston, which has cross-sectional area A_S. As a result of $\vec{F}_S$ being applied, the pressure on this piston increases by an amount $P = F_S/A_S$. As the piston is pushed downward, the liquid molecules in contact with it experience the same increase in pressure, and since the molecules are constantly colliding with other liquid molecules, this pressure increase is transferred very quickly to every part of the liquid and to the walls of the container. This transfer of pressure was realized by Pascal, and is expressed as **Pascal's principle:**

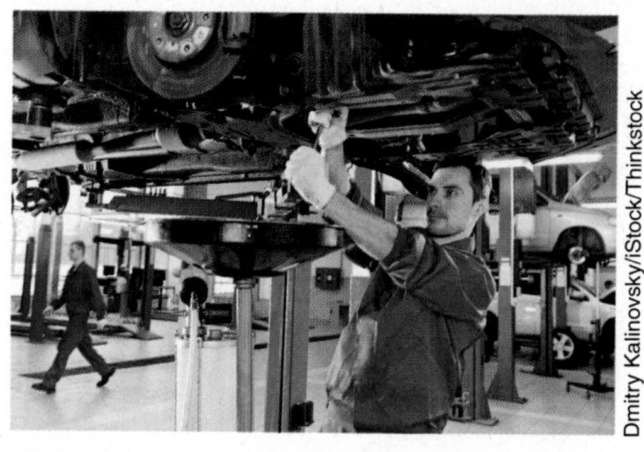

Figure 12-14 This car has been lifted by pressure in a liquid.

Pascal's principle A change in pressure applied to an enclosed fluid at rest is transferred undiminished to all parts of the fluid and to its container.

A change in pressure applied to an enclosed fluid at rest is transferred undiminished to all parts of the fluid and to its container.

Since the pressure P is transmitted to the other piston, which has a larger surface area A_L, we can write $P = F_L/A_L$, where F_L is the magnitude of the resulting force on this piston. Thus,

$$P = \frac{F_S}{A_S} = \frac{F_L}{A_L}, \text{ from which}$$

$$F_L = \frac{A_L}{A_S} F_S \quad (12\text{-}9)$$

Hence, the force on the larger piston is larger by a factor of A_L/A_S than the force on the small piston. This system is referred to as a **hydraulic lift** or a **hydraulic press**, where *hydraulic* means *operated by a fluid under pressure*. The result given in Eqn. 12-9 assumes that friction and the masses of the pistons are negligible, as are any differences in the fluid heights under the pistons that would create a pressure difference.

hydraulic lift (or **hydraulic press**) a mechanical device to lift objects, operated by a fluid under pressure

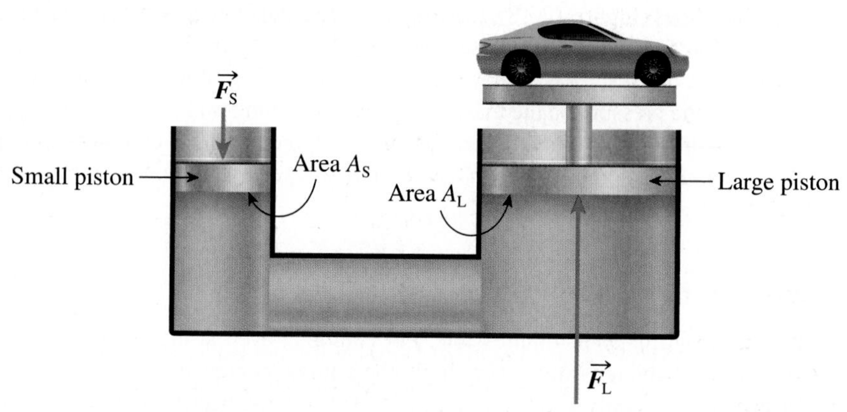

Figure 12-15 A hydraulic lift. The small downward force on a piston of small area results in a large upward force on a larger piston, which might be used to lift a car.

SAMPLE PROBLEM 12-5

A hydraulic lift is being used to hoist a car of mass 1100 kg. If the diameter of the large piston is 28 cm, and the diameter of the small piston is 6.0 cm, what is the minimum magnitude of force that could be applied to the small piston to lift the car?

Solution

Rearranging Eqn. 12-9 to solve for F_S gives $F_S = \frac{A_S}{A_L} F_L$

The force on the large piston is the gravitational force acting on the car, which has a magnitude of $F_L = mg = (1100 \text{ kg})(9.8 \text{ m/s}^2) = 1.08 \times 10^4 \text{ N}$.

The ratio of piston areas is $\frac{A_S}{A_L} = \frac{\pi r_S^2}{\pi r_L^2} = \frac{r_S^2}{r_L^2} = \frac{(3.0 \text{ cm})^2}{(14 \text{ cm})^2} = 0.0459$.

$$\therefore F_S = \frac{A_S}{A_L} F_L = (0.0459)(1.08 \times 10^4 \text{ N}) = 5.0 \times 10^2 \text{ N}$$

Thus, the minimum magnitude of force required on the small piston is only 5.0×10^2 N in order to lift this car.

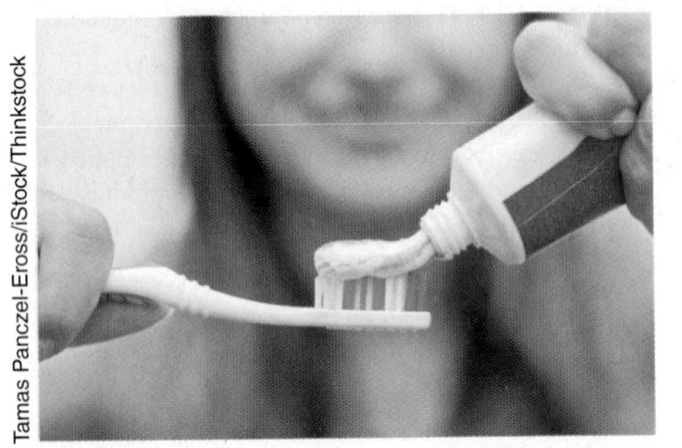

Tamas Panczel-Eross/iStock/Thinkstock

Figure 12-16 When a tube of toothpaste is squeezed, the pressure is exerted throughout all parts of the toothpaste, forcing some from the opening.

By supplying an increased force on the large piston, is nature giving us something for nothing? Since energy must be conserved, the answer is no. The work done by the small force in moving the small piston must equal the work done by the large force in moving the large piston (neglecting friction). This equality is achieved because the distance the small piston moves is larger than the distance moved by the large piston.

Pascal's principle has many applications. When we squeeze an open toothpaste tube (Figure 12-16), we are applying an external pressure that is transferred to all parts of the toothpaste, forcing some of it out of the open end. An interesting example in nature is the earthworm, which has a *hydrostatic skeleton* that consists of fluids held under pressure in a closed body compartment surrounded by a muscular structure. The earthworm propels itself forward by applying a series of muscular contractions along its body. Many other soft-bodied invertebrates, such as jellyfish (Figure 12-17), nematodes, starfish, sea urchins, and snails use a hydrostatic skeleton for propulsion or changing shape.

Vilainecrevette/Shutterstock

Figure 12-17 A jellyfish uses Pascal's principle for propulsion. It has a hydrostatic skeleton consisting of fluids under pressure surrounded by a muscular structure that exerts a sequence of contractions to move the jellyfish.

The principles of hydraulic lifts are used in many applications. A typical chair used by a barber or hairdresser rests on the large piston of a hydraulic lift. A small force applied on the small piston, usually by a hand lever or foot pedal, raises or lowers the chair. On any construction site there is hydraulically operated machinery in the form of bulldozers, backhoes (Figure 12-18), loaders, fork lifts, and cranes. All these machines operate using forces applied to enclosed liquids. The brakes on automobiles operate hydraulically, but the system is somewhat more complex. When a force is applied to the brake pedal, the pressure on the master cylinder containing brake fluid is increased. This pressure is transmitted equally through the fluid to eight other cylinders, two on each wheel. The increased pressure on the brake pads exerts a frictional resistance on the wheel disks, slowing down the car.

DID YOU KNOW?

The most common cause of glaucoma in the eye is an increase of liquid pressure in the aqueous humour, near the front of the eye. This increase is the result of fluid in the aqueous humour not draining away even though new fluid is entering. The increased pressure at the front of the eye is transmitted to the retina at the back of the eye, and blindness can result.

Pressure in the Atmosphere

Most liquids are virtually incompressible. For example, even at the deepest location in the ocean where the pressure is very high, the volume occupied by a specific mass of water decreases by only about 5% relative to the volume occupied at atmospheric pressure. In contrast, gases are easy to compress, and the volume of a given mass of gas decreases as the pressure exerted on it increases.

At temperatures and pressures near Earth's surface, most gases and gas mixtures obey the **ideal gas law**:

$$PV = nRT \qquad (12\text{-}10)$$

where P is the pressure exerted on the gas
V is the volume of the gas
n is the number of moles of gas
R is the molar gas constant (8.314 J·mol^{-1}·K^{-1})
and T is the absolute temperature, that is, the temperature in kelvins[2]

Rearranging the ideal gas law as $V = \frac{nRT}{P}$, we see that the volume of n moles of gas is inversely proportional to the pressure exerted on the gas. Another important feature of gases is that the density of a gas is proportional to the pressure on it. This follows directly from the ideal gas law, as shown in the next few lines.

The number of moles of gas, n, can be written as the mass m of gas divided by the molar mass M of the gas molecules: $n = m/M$. Substituting m/M for n in the ideal gas law,

$$PV = \frac{m}{M}RT, \quad \text{which gives} \quad P = \frac{m}{V}\frac{RT}{M}$$

Now m/V is the density ρ of the gas, and thus

$$P = \rho\frac{RT}{M} \quad \text{or} \quad \rho = \frac{M}{RT}P$$

Since R is a constant, and the molar mass M is constant for a given type of gas, the above result shows that for a constant temperature T, the density ρ of a gas is proportional to the pressure P exerted on it; that is, $\rho \propto P$.

The ideal gas law can be written with the molar gas constant R replaced with the product of two related constants, $R = k_B N_A$, where

$$\text{Boltzmann's constant } k_B = 1.381 \times 10^{-23} \text{ J/K}$$

$$\text{Avogadro's number } N_A = 6.022 \times 10^{23} \text{ mol}^{-1}$$

The ideal gas law then becomes $PV = nk_BN_AT$. The combination nN_A (number of moles multiplied by Avogadro's number) that appears in this expression is usually written as the number of gas molecules, N. This gives another form of the ideal gas law that is often useful:

$$PV = Nk_BT \qquad (12\text{-}11)$$

Starting with the ideal gas law as in Eqn. 12-11, it is possible to use calculus to derive an expression for the pressure P in the atmosphere as a function of elevation h above sea level. The result is the **barometric equation**:

$$P = P_0 e^{-mgh/(k_BT)} \qquad (12\text{-}12)$$

[2]Temperature in kelvins = (temperature in degrees Celsius) + 273.15.

Dmitry Kalinovsky/iStock/Thinkstock

Figure 12-18 A backhoe operates hydraulically; that is, it uses forces applied to enclosed liquids.

ideal gas law the relationship $PV = nRT$, where P is the pressure exerted on a gas, V is the volume of the gas, n is the number of moles of gas, R is the molar gas constant, and T is the temperature in kelvins

Units Tip

Note that the unit for absolute temperature is kelvin (K), not degrees kelvin, whereas the common everyday unit for temperature is *degrees* Celsius (°C).

DID YOU KNOW?

The unit "kelvin" is in honour of William Thomson (1824–1907), a Scottish mathematical physicist and engineer, also known as Lord Kelvin. He discovered that there is an absolute zero of temperature at approximately −273.15°C, or 0 K. He was the first U.K. scientist to be elevated to the House of Lords. Boltzmann's constant is named after Ludwig Boltzmann (1844–1906), an Austrian physicist who made many advances in the area of statistical mechanics, which explains how the properties of atoms and molecules determine the properties of matter as a whole.

The naming of Avogadro's number recognizes the contributions of Amedeo Avogadro (1776–1856). He was an Italian scientist who had a great influence in the development of molecular theory.

barometric equation the relationship expressing the exponential decrease of atmospheric pressure as a function of elevation above sea level

where P_0 is the sea-level air pressure, m is the average mass of an air molecule in kilograms, g is the magnitude of the gravitational acceleration, and T is the absolute temperature. In order to simplify the derivation, g and T have been assumed to be constant for elevations close to Earth's surface. Since $g = 9.8\ \text{m/s}^2$ (correct to two significant digits) up to an elevation of about 28 km, this assumption is quite reasonable. The temperature T also does not change appreciably (in kelvins) over elevations of physiological interest.

Since air density, ρ, is proportional to pressure, the exponential form of the barometric equation applies also to density:

$$\rho = \rho_0 e^{-mgh/(k_B T)} \qquad (12\text{-}13)$$

For an air molecule's average mass, m, in the barometric equation, the composition of air needs to be considered. Analysis of the molar fractions of dry air shows that it consists of 78% nitrogen (N_2), 20% oxygen (O_2), 0.9% argon (Ar), 0.03% carbon dioxide (CO_2), and small amounts of other atoms and molecules. With the molar masses of the two predominant gases being 28 g/mol for N_2 and 32 g/mol for O_2, the average molar mass is 29 g/mol. The calculation of m in kilograms per molecule is shown in the solution to Sample Problem 12-6.

SAMPLE PROBLEM 12-6

(a) Determine the absolute pressure in Earth's atmosphere at a height of 1000 m, as a percentage of the value at sea level. Assume a constant temperature of 20°C and two significant digits in given data.

(b) What is the density in Earth's atmosphere at 1000 m, as a percentage of the value at sea level?

Solution

(a) Eqn. 12-12 will be used, but since we are asked to find the pressure P as a percentage of sea-level pressure P_0, we first rearrange the equation to give the fraction P/P_0:

$$\frac{P}{P_0} = e^{-mgh/(k_B T)}$$

The mass m is the average mass of an air molecule in kilograms. Starting with the average molar mass of 29 g/mol,

$$29\,\frac{\text{g}}{\text{mol}} \times \frac{1\ \text{kg}}{10^3\ \text{g}} \times \frac{1\ \text{mol}}{6.022 \times 10^{23}\ \text{molecules}} = 4.8 \times 10^{-26}\ \text{kg/molecule}$$

Since "molecule" is dimensionless, we use $m = 4.8 \times 10^{-26}$ kg. As stated earlier, using $g = 9.8\ \text{m/s}^2$ is reasonable up to an elevation of 28 km, and the elevation h in this question is only 1000 m (or 1 km). Boltzmann's constant is $k_B = 1.381 \times 10^{-23}$ J/K, and the absolute temperature $T = (273 + 20)\ \text{K} = 293$ K. Substituting numbers,

$$\frac{P}{P_0} = e^{-mgh/(k_B T)} = e^{-[(4.8 \times 10^{-26}\ \text{kg})(9.8\ \text{m/s}^2)(1000\ \text{m})]\,/\,[(1.381 \times 10^{-23}\ \text{J/K})(293\ \text{K})]}$$

$$\therefore\ \frac{P_0}{P} = 0.89 = 89\%$$

Thus, the pressure at a height of 1000 m above sea level is 89% of the value at sea level.

(b) The density, ρ, of a gas is proportional to pressure, P. Therefore,

$$\frac{\rho}{\rho_0} = \frac{P}{P_0} = 89\%$$

Thus, the air density at 1000 m above sea level is 89% of that at sea level.

SAMPLE PROBLEM 12-7

(a) If sea-level atmospheric pressure is 100 kPa, at what height (in kilometres) will it have been reduced by 50% to 50 kPa? Assume a temperature of 22°C, and that given data have two significant digits.

(b) What *additional* height would be needed to reduce the pressure to 25 kPa, that is, to reduce it by an additional 50% of 50 kPa?

Solution

(a) Start with Eqn. 12-12, $P = P_0 e^{-mgh/(k_B T)}$. Since we need to find the height h, rearrange the equation to isolate h. Begin by dividing both sides of the equation by P_0 and taking the natural logarithm of both sides:

$$\ln\left(\frac{P}{P_0}\right) = \ln\left(e^{-mgh/(k_B T)}\right) = -\frac{mgh}{k_B T} \quad \text{(recalling that } \ln e^x = x\text{)}$$

Solving for h, $h = -\dfrac{k_B T}{mg} \ln\left(\dfrac{P}{P_0}\right)$

The temperature is $T = (273 + 22)\ \text{K} = 295\ \text{K}$.

As in Sample Problem 12-6, we have $m = 4.8 \times 10^{-26}$ kg (having started with $m = 29$ g/mol). Substituting numbers,

$$h = -\frac{(1.381 \times 10^{-23}\ \text{J/K})(295\ \text{K})}{(4.8 \times 10^{-26}\ \text{kg})(9.8\ \text{m/s}^2)} \ln\left(\frac{50\ \text{kPa}}{100\ \text{kPa}}\right) = 6.0 \times 10^3\ \text{m} = 6.0\ \text{km}$$

Notice that since ln(50/100) is negative, the value of h automatically turns out to be positive.

Thus, the height at which the pressure has been reduced by 50% is 6.0 km.

(b) Redoing the calculation with a reduction of 50% of 50 kPa, to give a pressure $P = 25$ kPa,

$$h = -\frac{(1.381 \times 10^{-23}\ \text{J/K})(295\ \text{K})}{(4.8 \times 10^{-26}\ \text{kg})(9.8\ \text{m/s}^2)} \ln\left(\frac{25\ \text{kPa}}{100\ \text{kPa}}\right) = 12 \times 10^3\ \text{m} = 12\ \text{km}$$

The *additional* height to achieve this 50% reduction is (12 – 6) km = 6 km.

Math Tip

Notice in Sample Problem 12-7 that the first 6 km of height reduced the original pressure of 100 kPa to half its value (50 kPa), and then the next 6 km reduced the pressure to 25 kPa, that is, half the 50 kPa. This result illustrates a general feature of exponential functions. For a given change in the independent variable (a change of $h = 6$ km in this situation), there is a constant fractional or percentage reduction (50% in this case) in the dependent variable (pressure).

Two follow-up questions for you to consider:

(a) At what height above sea level would the pressure be reduced by 50% again, to 12.5 kPa?

(b) Starting at an arbitrary height above sea level where the pressure is P_1, what additional height would reduce the pressure to 50% of P_1?

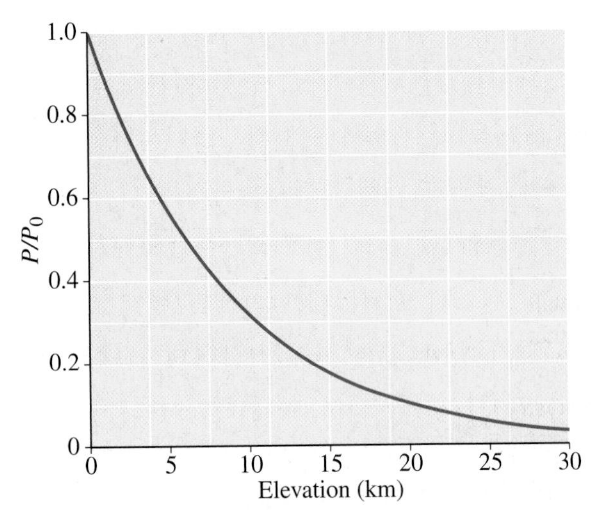

Figure 12-19 Pressure P in the atmosphere relative to sea-level pressure P_0, as a function of elevation.

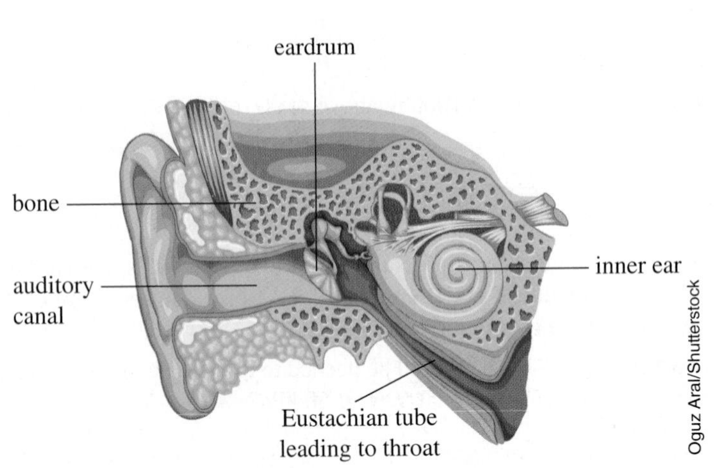

Figure 12-20 The Eustachian tube connects the air-filled region behind the eardrum to the air in the throat.

Figure 12-19 shows a graph of pressure versus elevation in the atmosphere. Notice how quickly the pressure (and therefore air density) decreases. The pressure falls to 30% of sea-level pressure at an elevation of only about 10 km, which is roughly the height at which jet aircraft fly. The cabins of airplanes need to be pressurized since the outside air is too thin to breathe at that height. Typical cabin pressures are about 80% of the pressure at sea level.

Under normal circumstances we do not notice atmospheric pressure because the pressure inside our bodies balances the pressure outside. However, it is possible to notice the effects of atmospheric pressure changing as a result of relatively small changes in elevation—while ascending or descending in an elevator, an aircraft, or in a car on a mountain highway, there can be a "popping" sensation in the ears. A duct called the Eustachian tube (Figure 12-20) connects the air behind the eardrum to the throat, where the air is at the same pressure as the surrounding environment. Normally, this tube is *closed*, so, for example, as the external pressure decreases while ascending in an aircraft, the pressure in the throat becomes less than the pressure behind the eardrum. This pressure difference causes the Eustachian tube to open, allowing air to flow from the area behind the eardrum to the throat, and the opening of the tube often results in a "pop." The popping effect can be reduced by yawning, swallowing, or gently blowing with the mouth and nose closed.

In many cases we take advantage of atmospheric pressure without thinking about it. The use of a medicine dropper or eye dropper is one example (Figure 12-21). When the bulb is squeezed, with no liquid in the dropper, air is forced out of the dropper. Then, when the end of the dropper is placed in a liquid and the bulb is released, the pressure inside the dropper becomes reduced, allowing the atmospheric pressure on the liquid to force the liquid up into the dropper.

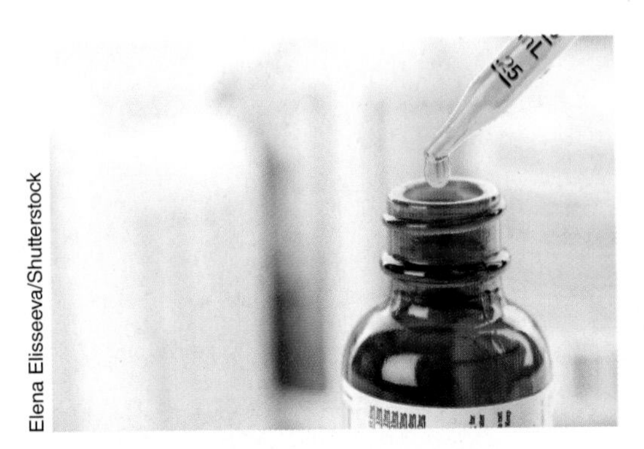

Figure 12-21 Taking advantage of atmospheric pressure.

barometer an instrument that measures atmospheric pressure

Measuring Atmospheric Pressure

A common instrument that measures atmospheric pressure directly is called a **barometer**. It was invented by the Italian scientist Evangelista Torricelli (1608–1647), who was a student of Galileo Galilei. Although Torricelli experimented with water, he discovered that mercury was a more convenient liquid to use in the barometer. To make a barometer, he took a glass tube closed at one end and filled it with mercury, and he then inverted it into a container also containing mercury (Figure 12-22). The mercury in the tube dropped a small amount, but

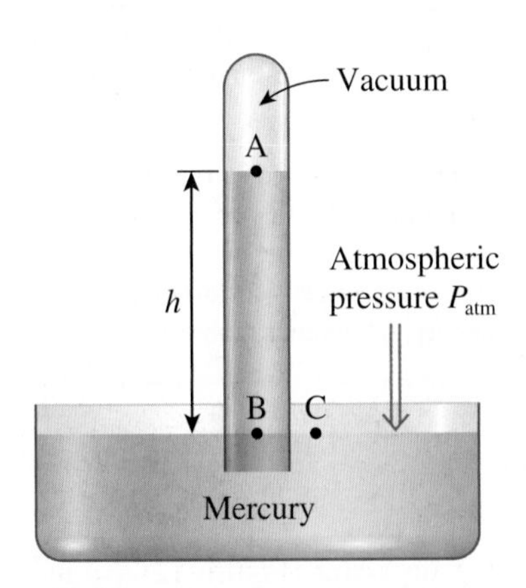

Figure 12-22 A mercury barometer. The atmospheric pressure P_{atm} can be calculated as ρgh, where ρ is the density of mercury.

Brad Calkins/iStock/Thinkstock

Figure 12-23 An aneroid barometer for home use.

then came to rest because the atmospheric pressure on the open mercury surface in the container was equal to the pressure caused by the mercury in the tube. The space in the tube above the mercury is essentially a vacuum, where the pressure is zero. Going down a distance h in the mercury column from point A in the vacuum to a point B that is level with the open mercury surface at C, the pressure increases by an amount given by Eqn. 12-7: $\Delta P = P_B - P_A = \rho gh$. Since A is in a vacuum, $P_A = 0$, and therefore $P_B = \rho gh$. Because points B and C are at the same level in a continuous liquid, the pressures there are equal; that is, $P_B = P_C$, which gives $\rho gh = P_{atm}$. Thus, the atmospheric pressure acting at C can be calculated simply as ρgh.

Although mercury barometers are still in use, a more convenient and safe type is the aneroid barometer ("aneroid" means "without liquid"). An aneroid barometer (Figure 12-23) consists of a flexible metal container with a partial vacuum inside. Changes in atmospheric pressure cause the top of the container to flex, and small levers transmit this movement to a pointer that moves over a graduated scale. Aneroid barometers with altitude scales are often used as altimeters in aircraft. Barometers for home use are normally of the aneroid type, and are sometimes referred to as "weather gauges" since low atmospheric pressure usually corresponds to wet weather and high pressure to fair weather.

EXERCISES

12-14 A large piece of equipment is being raised by a hydraulic lift in a factory. The cross-sectional area of the large piston is 0.091 m^2. The magnitudes of the forces exerted on the small and large pistons are 7.5×10^2 N and 1.2×10^4 N respectively. What is the radius of the small piston in millimetres?

12-15 **Fermi Question:** If you were able to stand on top of the small piston of a hydraulic lift whose diameter is equal to your shoe length, what diameter would the large piston have to be to lift a small SUV?

12-16 Sulphur Mountain (Figure 12-24) in Banff National Park, Alberta, has an elevation of 2450 m above sea level. Determine the atmospheric pressure at the top of Sulphur Mountain, assuming a sea-level pressure of 101 kPa and a temperature of 11°C.

Bernhard Richter/Shutterstock

Figure 12-24 Sulphur Mountain, Alberta (Question 12-16).

12-17 At what elevation in kilometres above sea level is the density of air 75% of that at sea level? Assume a temperature of 15°C.

12-18 Explain how you take advantage of atmospheric pressure when you use a straw to drink a liquid.

DID YOU KNOW?

The drinking straw (Question 12-18) was patented in 1888.

12-19 (a) If the height of mercury in a mercury barometer is 758 mm, what is the atmospheric pressure in kilopascals? Assume $g = 9.80\ \text{m/s}^2$. (Reference: Table 12-1)

(b) If the barometer had water instead of mercury, what would the water height be? Use your answer to explain why Torricelli used mercury.

12-20 Notice in the photograph of the aneroid barometer (Figure 12-23) that a "30" appears on one scale; this scale is indicating pressure in yet another unit: inches of mercury (1 inch = 2.54 cm). Determine the pressure in kilopascals corresponding to a reading of 30.0 inches of mercury. (Reference: Table 12-1)

BMJ/Shutterstock

Figure 12-25 A water strider standing on water.

12.3 Surface Tension

Perhaps you have seen a water strider standing on water (Figure 12-25) and wondered how this is possible, since the strider is standing right on top of the water, not immersed in it at all. If you look carefully at Figure 12-25, you will see that each of the strider's legs makes a small depression in the water without breaking through the surface.

The reason for the strider to be able to walk on water is that the surface of a liquid is different from the interior of a liquid. Figure 12-26 shows a representation of a liquid, showing the molecular forces of attraction on a molecule in the main body of the liquid and also on a molecule in the surface layer. The molecule in the main body experiences attractive forces in all directions, whereas the molecule at the surface feels negligible attractive forces upward because there is no liquid above it. As a result, molecules at the surface are pulled tightly to the main liquid below, and the surface acts somewhat like an elastic membrane under tension (think of a stretched sheet of plastic food wrap), and we say that there is a surface tension that lies in the plane of the surface and acts along the surface.

Figure 12-27 shows a view from above of molecules along a line (AB) of length ℓ on a liquid surface. These molecules experience forces of molecular attraction to the left and right along the surface. With the magnitude of the total rightward (or leftward) force on the line of molecules written as F, the **surface tension** of the liquid is defined as the intermolecular force magnitude F per unit length along the line along the surface. Denoting surface tension by the lowercase Greek letter gamma (γ),

surface tension magnitude of intermolecular force per unit length (symbol γ) in a direction perpendicular to a line of molecules on a liquid's surface

$$\text{surface tension } \gamma = \frac{F}{\ell} \qquad \text{(12-14)}$$

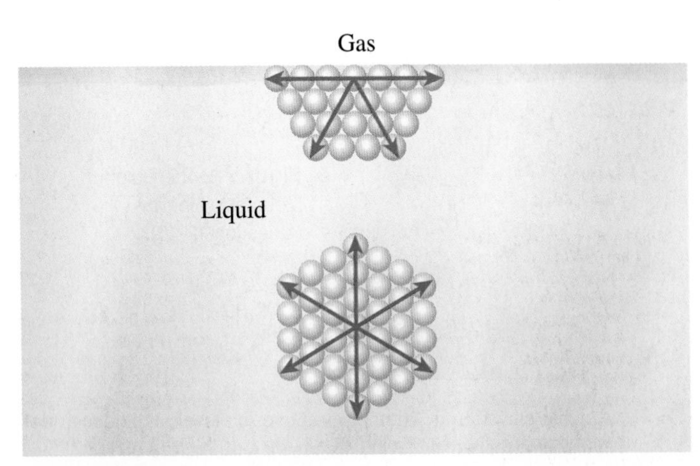

Figure 12-26 A molecule on the surface of a liquid experiences negligible molecular attractive forces upward, whereas a molecule in the liquid's interior experiences attractive forces in all directions.

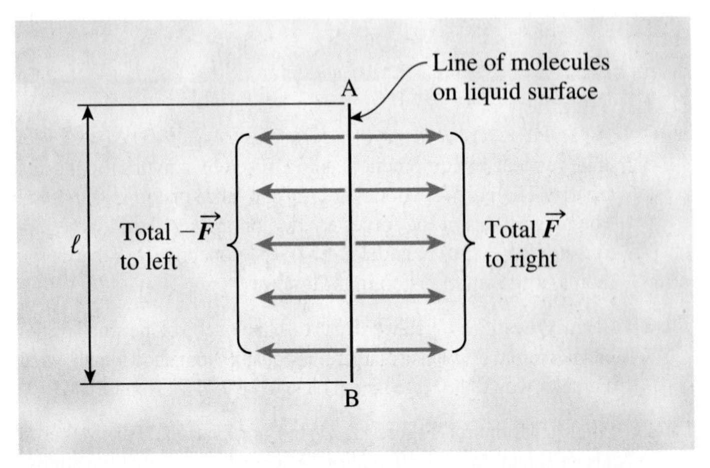

Figure 12-27 View from above of a line (AB) of molecules of length ℓ on the surface of a liquid. These molecules experience a total attractive force $\vec{F}$ to the right, and a total attractive force $-\vec{F}$ to the left.

The SI unit of surface tension is newton per metre (N/m).

Values of surface tension for various liquids are given in Table 12-2. Note that the surface tension of water at 100°C is less than that at 20°C. As the temperature of a liquid increases, the molecules are not bound as strongly because of the faster thermal motion.

We can now explain in more detail how a water strider can stand and walk on water. Figure 12-28 shows a magnified view of one of the strider's legs, which is making a small depression in the surface (as seen in Figure 12-25). The surface tension forces are exerted parallel to the water's surface, and since the surface in the depression is curved as shown, the surface tension forces for this surface have upward components. The directions of these forces are shown for two points in the depression created by the leg. Of course, the forces act all around the surface of the depression, not just at the two points shown. The upward components balance the force of gravity acting downward on the leg. (Since the strider has four legs, the downward force on each leg is $\frac{1}{4}$ of the total force of gravity on the strider.)

Molecules that lower the surface tension of liquids are called **surfactants** (an abbreviation for surface active agents). Detergents and soaps are surfactants; when they are added to water they reduce the surface tension and allow the water-soap combination to enter small crevices in clothing for better washing. Some insects not only stand on water, but use a surfactant to propel themselves. The rear of the insect's abdomen secretes a surfactant onto the water, thus reducing the surface tension behind the insect. The higher surface tension in the water ahead of the insect then pulls the insect forward.

Table 12-2

Surface Tensions of Various Liquids (at 20°C unless stated otherwise)

Liquid	Surface Tension (N/m)
Blood plasma	0.050
Ethyl alcohol	0.022
Lung surfactant (at 37°C)	0.001
Mercury	0.47
Soapy water	0.025
Water (at 20°C)	0.073
Water (at 100°C)	0.059

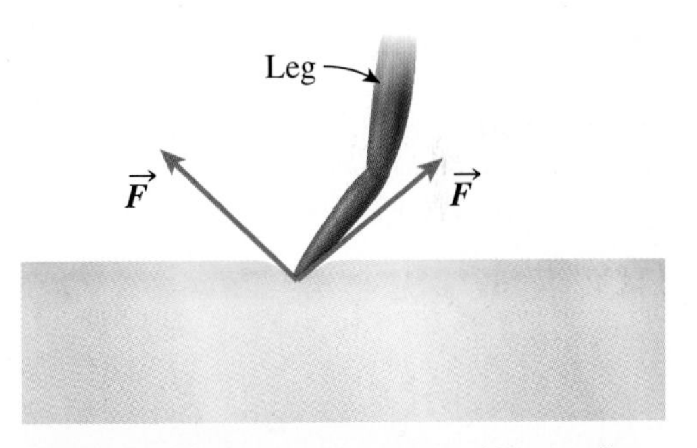

Figure 12-28 One leg of a water strider standing on water. Surface tension forces $\vec{F}$ parallel to the surface in the depression created by the leg have upward components that balance the downward gravitational force acting on the leg.

TRY THIS!

Surface-Tension Propulsion

Apply a small amount of soap to one end of a wooden toothpick, then place the toothpick on water in a sink or bathtub. The toothpick will move in a direction away from the soapy end. The soap reduces the surface tension where it contacts the water and the larger surface tension at the other end of the toothpick moves the toothpick forward.

surfactants molecules that lower the surface tension of liquids

Surface tension has an important connection to breathing. When we breathe in, air passes into our lungs through the trachea, which branches into smaller and smaller passages, eventually terminating in tiny pockets called alveoli (Figure 12-29). There are about 300 million alveoli in an adult human's lungs, having diameters ranging from 0.1 to 0.3 mm and a total surface area of about 100 m² (about 50 times larger than the skin's surface area). It is through this large alveolar area that gases are exchanged into and out of the blood in capillaries. When we inhale, the alveoli expand to expose as much surface area as possible to the incoming air. The interior surfaces of the alveoli are coated with a thin liquid, which must easily accommodate the alveolar expansion during inhalation. If the liquid was water with its high surface tension, the surface would be "tight" as a result of the large surface tension forces, and the necessary expansion could not occur, making normal breathing impossible. Instead of water, the coating of the alveolar surfaces is a liquid known as lung surfactant, which is a mixture of lipids and proteins. The surface tension of lung surfactant is about 70 times smaller than that of water (Table 12-2), and makes the required expansion of the alveoli possible.

When babies are born prematurely, they often are not capable of making lung surfactant themselves. This life-threatening situation, called Infant Respiratory Distress Syndrome (IRDS), frequently led to deaths prior to the 1980s, when artificial surfactants were developed that could be sprayed into the babies' lungs. These surfactants allow the infants to breathe until they begin to produce their own surfactants.

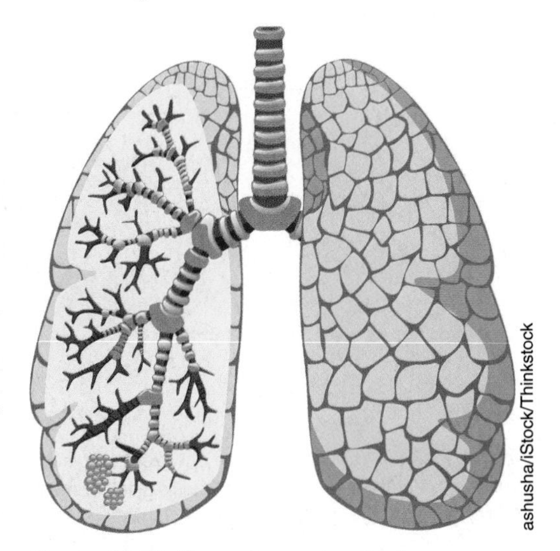

Figure 12-29 Human lungs, showing the branching of the trachea into smaller and smaller tubes ending in tiny pockets called alveoli (bottom left of figure, size exaggerated).

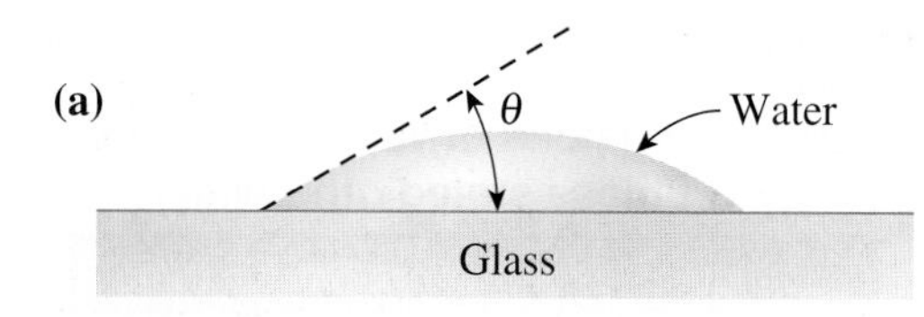

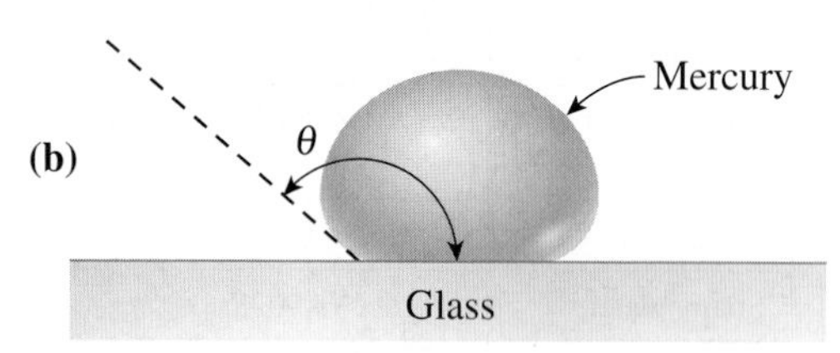

Figure 12-30 **(a)** A drop of water on glass with a contact angle $\theta = 30°$. **(b)** Mercury on glass with $\theta = 140°$.

cohesive forces forces exerted between like molecules

adhesive forces forces exerted between unlike molecules

contact angle for a liquid on a solid, the angle between the solid surface under the liquid and the edge of the liquid surface where it touches the solid

Ventin/Shutterstock

Figure 12-31 Drops of mercury "not wetting" a solid surface.

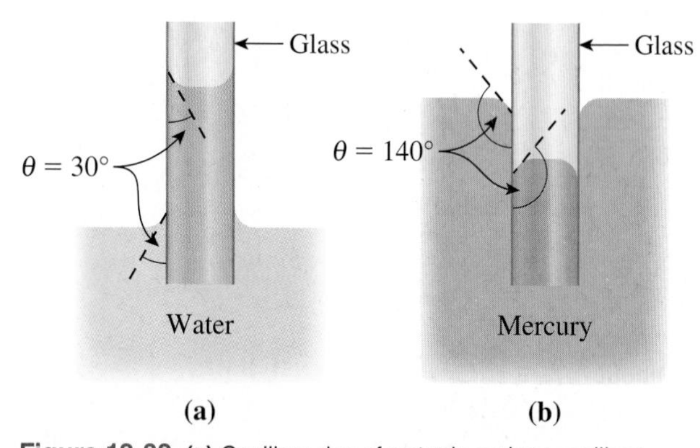

Figure 12-32 **(a)** Capillary rise of water in a glass capillary tube; contact angle $\theta = 30°$. **(b)** Capillary fall of mercury in a glass capillary tube; contact angle $\theta = 140°$.

Capillary Action

When drops of liquids are placed on horizontal solid surfaces, some spread out and become somewhat flattened, whereas others become almost spherical beads, depending on the liquid and the type of solid. When water is placed on glass, it flattens considerably (Figure 12-30 (a)), whereas mercury beads up on glass (Figure 12-30 (b)). In the case of water on glass, the attractive forces between water molecules themselves are less than the forces between water molecules and the glass molecules and atoms. Forces between like molecules (such as water) are known as **cohesive forces**, and forces between unlike molecules (water and glass) are **adhesive forces**. For water and glass, the adhesive forces are greater than the cohesive forces, whereas for mercury and glass, the cohesive forces are stronger.

The behaviour of liquids on solid surfaces is often characterized numerically by the **contact angle**, which is the angle measured between the portion of the solid surface under the liquid and the edge of the liquid surface where it touches the solid. The contact angles (θ) for water on glass[3] and mercury on glass are shown in Figure 12-30. For water on glass, $\theta = 30°$, and for mercury on glass, $\theta = 140°$. If a contact angle is between 0° and 90°, the liquid is said to "wet the surface," and for contact angles between 90° and 180°, the liquid "does not wet the surface." Figure 12-31 shows a photograph of drops of mercury "not wetting" a solid surface, with a contact angle greater than 90°.

This wetting (or not wetting) effect leads to the phenomenon of capillary rise (or fall). The word capillary means "hairlike," and capillary rise and fall occur in capillary tubes—that is, tubes with very small internal diameters—and in other situations where a liquid is in contact with small openings. If a glass capillary tube is placed vertically in water, the strong adhesive forces between water and glass cause the water to climb up inside the tube (Figure 12-32 (a)), wetting the glass. The contact angle between the water and glass where they meet at the top of the water column is $\theta = 30°$, which is the same value mentioned previously. Notice also that there is a contact angle of 30° where the water and glass meet outside the tube. If a glass capillary tube is placed in mercury, the mercury moves down inside the tube because of strong cohesive forces (Figure 12-32 (b)), and the contact angle is $\theta = 140°$, as before.

A general algebraic expression for the height that a liquid climbs in a tube can be determined by considering the forces acting on the cylindrical liquid column that has risen inside the tube. Figure 12-33 shows a liquid of density ρ and surface tension γ that has gone up a height h in a tube of radius r. The contact angle between the liquid and the tube is θ. The forces acting on the cylinder of liquid in the tube are the force of gravity $\vec{F}_G$ downward, and surface tension forces $\vec{F}$ all around the perimeter of the liquid surface at the top of the column. These surface tension forces are parallel to the liquid surface where it contacts the tube, and are therefore at the contact angle θ relative to the tube. We can use Eqn. 12-14 ($\gamma = F/\ell$) to determine the total magnitude of the surface tension forces acting around the circular circumference (length $\ell = 2\pi r$) where the liquid surface touches the tube: $F = \ell\gamma = 2\pi r\gamma$. The upward component of this total force is

$$F\cos\theta = 2\pi r\gamma\cos\theta$$

This is the magnitude of the upward force acting on the liquid column. The magnitude of the downward gravitational force on the liquid column

[3]If water is placed on glass that has been cleaned extremely carefully, the water spreads with a contact angle that is essentially zero instead of the 30° angle for water on glass that has not been specially prepared.

(of mass m) is mg. Writing $m = \rho V$ (from Eqn. 12-1), and noting that the volume of the liquid cylinder is the product of its cross-sectional area πr^2 and height h, the magnitude of the downward force is

$$mg = \rho \pi r^2 hg$$

Choosing $+y$ upward and using Newton's first law of motion $(\Sigma F_y = 0)$ gives

$$2\pi r \gamma \cos\theta - \rho \pi r^2 hg = 0, \text{ from which}$$

$$h = \frac{2\gamma \cos\theta}{\rho g r} \qquad (12\text{-}15)$$

Eqn. 12-15 shows that the height h of capillary rise is inversely proportional to the radius r of the capillary tube, and directly proportional to the surface tension of the liquid. Thus, for example, a decrease in radius by a factor of 10 results in an increase in height by a factor of 10. Although this equation was derived for capillary rise, it turns out that it applies equally well for capillary fall because the contact angle is then $\theta > 90°$, giving $\cos\theta < 0$ and a negative value of h, indicating a downward movement of the liquid.

Capillary rise is important in the upward movement of water and nutrients in the capillary tubes (xylem) of plants, and capillary tubes are often used in collecting small samples of blood from patients who first receive a needle prick on a finger (Figure 12-34). Capillary action can sometimes be problematic, for example, in concrete-block walls where water can seep through tiny pores in the mortar or concrete and cause damage.

Figure 12-33 Surface tension forces $\vec{F}$ and the gravitational force $\vec{F}_G$ acting on the cylindrical column of liquid that has risen a height h above the main body of liquid.

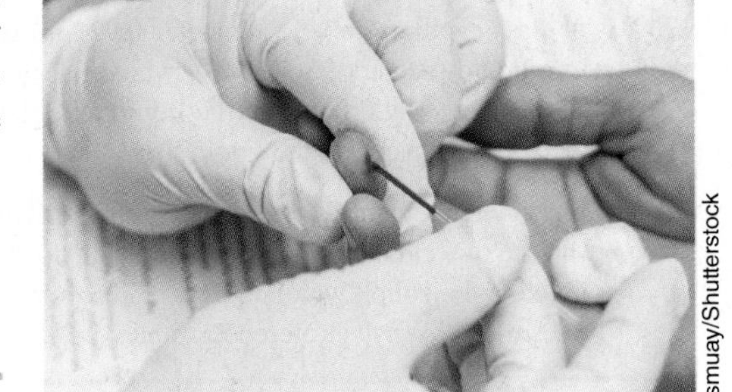

smuay/Shutterstock

Figure 12-34 Using a capillary tube to collect a small blood sample.

SAMPLE PROBLEM 12-8

A glass tube of inner diameter 1.0 mm is inserted into **(a)** water and **(b)** mercury. What is the height (in centimetres) of the resulting capillary rise or fall? Assume a contact angle of 30° (two significant digits) for water on glass and 140° for mercury on glass, and a temperature of 20°C. **(c)** Determine the height (in centimetres) of capillary rise for water if a tube of diameter 0.72 mm is used instead.

Solution

(a) Using Eqn. 12-15 with the density and surface tension of water from Tables 12-1 and 12-2,

$$h = \frac{2\gamma\cos\theta}{\rho g r} = \frac{2(0.073\,\text{N/m})\cos 30°}{(1.0\times10^3\,\text{kg/m}^3)(9.8\,\text{m/s}^2)\left(\dfrac{1.0\times10^{-3}\,\text{m}}{2}\right)} = 0.026\,\text{m} = 2.6\,\text{cm}$$

Thus, the water rises 2.6 cm in the tube.

(b) For mercury in the tube,

$$h = \frac{2\gamma\cos\theta}{\rho g r} = \frac{2(0.47\,\text{N/m})\cos 140°}{(13.6\times10^3\,\text{kg/m}^3)(9.8\,\text{m/s}^2)\left(\dfrac{1.0\times10^{-3}\,\text{m}}{2}\right)} = -0.011\,\text{m} = -1.1\,\text{cm}$$

The mercury falls 1.1 cm in the tube.

(c) The entire calculation for water could be redone for the new tube as in (a), but since the only change is the diameter (and hence radius) of the tube, we just note that height is inversely proportional to radius, $h \propto 1/r$, and therefore

$$\frac{h_2}{h_1} = \frac{r_1}{r_2}, \text{ which gives } h_2 = h_1 \frac{r_1}{r_2}$$

The numerical value of the ratio $\frac{r_1}{r_2}$ does not depend on the units used (except that the units for r_1 must be the same as that for r_2), and so we can just use millimetres as the unit for these radii. As well, the ratio of diameters is the same as the ratio of radii $\left(\frac{d_1}{d_2} = \frac{2r_1}{2r_2} = \frac{r_1}{r_2}\right)$, and since diameters are the given quantities, we have

$$h_2 = h_1 \frac{d_1}{d_2} = 2.6 \text{ cm} \times \frac{1.0 \text{ mm}}{0.72 \text{ mm}} = 3.6 \text{ cm}$$

Thus, the water rises 3.6 cm in the smaller tube.

EXERCISES

12-21 A glass capillary tube of diameter 0.92 mm is placed vertically in a liquid (density 986 kg/m^3), which rises 4.2 cm in the tube. The contact angle between the liquid and glass is 15°. What is the surface tension of the liquid?

12-22 If you were to gently lower a paper clip horizontally onto a calm surface of water at room temperature and then another paper clip onto ethyl alcohol, which clip would be more likely to remain at the surface? Why? (Reference: Table 12-2)

12-23 A plastic capillary tube of radius 0.48 mm is placed vertically in mercury, which goes down 0.80 cm in the tube. Determine the contact angle between the mercury and the plastic. Express your answer to the nearest degree. (Reference: Tables 12-1 and 12-2)

12-24 How would you expect the surface tension of lung surfactant to be affected by air pollution or cigarette smoke? How would that affect breathing?

12.4 Buoyancy and Archimedes' Principle

buoyant force an upward force exerted on an object in a fluid

If you try to hold a friend horizontally, you would find this task difficult—unless the friend is lying in water. Water and other fluids exert an upward force that causes objects in them to appear to be lighter. The force that pushes upward on an object in a fluid is called the **buoyant force**. This force helps hold a swimmer up in water and a hot-air balloon up in air.

The buoyant force can be understood and expressed mathematically by considering the forces exerted by a fluid on an object immersed in it. Figure 12-35 shows a rectangular block of height h immersed in a liquid. Since the pressure in a liquid increases with depth, the pressure on the block increases from the top to the bottom of the block. Writing the pressures at the top and bottom of the block as P_1 and P_2, respectively, Eqn. 12-7 gives $P_2 - P_1 = \rho_L gh$, where ρ_L is the density of the liquid. Since force magnitude equals pressure times area, the forces exerted at various places on the block also increase from top to bottom on the block, as shown in the drawing. Writing the area of the block's top and bottom surfaces as A, the magnitude of the downward force on the top surface is $F_1 = P_1 A$, and the magnitude of the upward force on the bottom surface is $F_2 = P_2 A$. Therefore, the magnitude of the net upward force (i.e., the buoyant force, $\vec{F}_B$) is

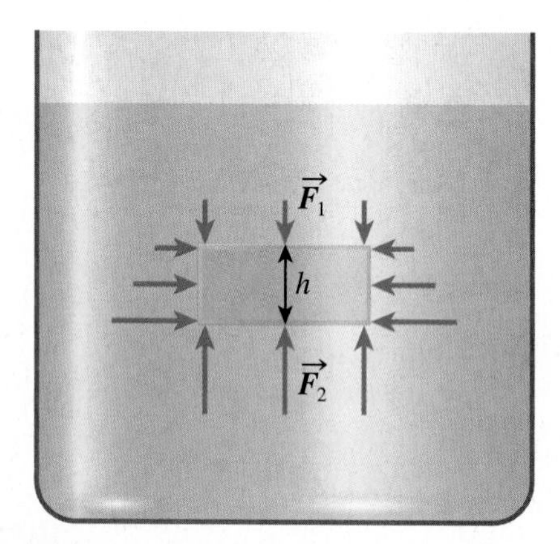

Figure 12-35 A block immersed in a liquid, showing forces exerted on the block by the liquid.

$$F_B = P_2 A - P_1 A = (P_2 - P_1)A = \rho_L ghA$$

The quantity hA equals the block's volume V, or equivalently, the volume of the liquid displaced by the block. Writing $hA = V$, we have $F_B = \rho_L Vg$. Now $\rho_L V$ is the mass m_L of liquid displaced by the block, and so $F_B = m_L g$. In words, what we have shown is that the buoyant force is equal in magnitude to the weight of the liquid displaced by the block. Although we developed this result for a block in a liquid, *it is valid for an object of any shape immersed wholly or partially in any fluid (liquid or gas)*. This relationship was discovered by a brilliant Greek scientist named Archimedes, who lived from about 287 BCE to 212 BCE. It is summarized in the following statement:

Archimedes' principle: The upward buoyant force on an object immersed wholly or partially in a fluid is equal in magnitude to the weight of the fluid displaced by the object.

Archimedes' principle The upward buoyant force on an object immersed wholly or partially in a fluid is equal in magnitude to the weight of the fluid displaced by the object.

Expressed as an equation,

$$F_B = m_f g = \rho_f V_f g \qquad (12\text{-}16)$$

where F_B is the magnitude of the buoyant force, and m_f, ρ_f, and V_f are the mass, density, and volume, respectively, of the fluid displaced by the object.

In addition to the upward buoyant force acting on an object in a fluid, there is of course the downward force of gravity. Thus, for an object momentarily at rest in a fluid, the magnitude of the resultant force is $F_G - F_B = mg - m_f g$ (assuming that $mg > m_f g$).[4] This difference between the true weight and the buoyant force is sometimes called the **apparent weight** of the object, since the object "appears" to weigh less in the fluid.

apparent weight difference between the force of gravity and the buoyant force on an object

Some fish and other aquatic animals have air-filled swim bladders that allow them to have an overall density which is the same as the water. As a result, they have an apparent weight of zero and can float without expending energy. As well, by changing the amount of gas in the bladder, an animal can become denser or less dense than water, and easily sink or rise to search for food. Some other aquatic animals have bones that are very porous to provide an overall density that is the same as that of water (see Question 12-92 about a cuttlefish).

SAMPLE PROBLEM 12-9

A piece of aluminum of mass 1.6 kg is completely immersed in water, and released from rest. What are the magnitudes of the **(a)** resultant force on the aluminum, **(b)** apparent weight of the aluminum, and **(c)** aluminum's acceleration immediately after release?

Solution

(a) From Table 12-1, the density ρ_{Al} of aluminum is 2.70×10^3 kg/m^3. The forces acting on the aluminum are gravity downward (magnitude $m_{Al} g$) and the buoyant force upward (magnitude $F_B = m_f g = \rho_f V_f g$, Eqn. 12-16). Since the aluminum is completely immersed, the volume V_f of water displaced equals the volume V_{Al}

[4]We will see in Section 12.6 that if an object is moving in a fluid, there is also a viscous drag force that must be taken into account.

TRY THIS!

Buoyancy at Home

Partly fill a glass with water, place it on a kitchen scale, and note the mass that is indicated. Now insert two or three fingers vertically into the water, and again note the mass that is measured. Do you expect it to be smaller, larger, or the same as the original measurement? Explain the actual result. (**Hint:** You might want to think about Newton's third law of motion.)

of the aluminum, which is $V_{Al} = \frac{m_{Al}}{\rho_{Al}}$. Therefore, we can write $F_B = \rho_f \frac{m_{Al}}{\rho_{Al}} g$. The density of aluminum is greater than that of water, and hence the force of gravity downward is larger than the buoyant force upward, and so we choose $+y$ downward. The magnitude of the resultant force is

$$\begin{aligned}\Sigma F_y &= m_{Al} g - \rho_f \frac{m_{Al}}{\rho_{Al}} g \\ &= m_{Al}\left(1 - \frac{\rho_f}{\rho_{Al}}\right) g \\ &= (1.6 \text{ kg})\left(1 - \frac{1.0 \times 10^3 \text{ kg/m}^3}{2.70 \times 10^3 \text{ kg/m}^3}\right)(9.8 \text{ m/s}^2) \\ &= 9.9 \text{ N}\end{aligned}$$

Thus, the magnitude of the resultant force on the aluminum is 9.9 N.

(b) The apparent weight is the difference in magnitude between the true weight and the buoyant force, and this quantity is just the magnitude of the resultant force, calculated to be 9.9 N in part (a).

(c) To find the magnitude of the acceleration, use Newton's second law of motion:

$$\Sigma F_y = ma_y, \text{ which gives } a_y = \frac{\Sigma F_y}{m} = \frac{9.9 \text{ N}}{1.6 \text{ kg}} = 6.2 \text{ m/s}^2$$

The acceleration of the aluminum is 6.2 m/s^2 in magnitude.

Rolf Weschke/iStock/Thinkstock

Figure 12-36 Floating logs (Sample Problem 12-10).

SAMPLE PROBLEM 12-10

A log having density 5.5×10^2 kg/m^3 is floating in water (Figure 12-36). What percentage of the log's volume is below the water line?

Solution

Let f represent the fraction of the log's volume below the water line. Using V_{log} for the volume of the entire log, then the volume of the log under water—that is, the volume V_f of fluid displaced—is $f V_{log}$. Since the log is floating, it is at rest, and the resultant force on it is zero. The two forces acting are gravity downward and buoyant force upward. Choosing $+y$ upward, $\Sigma F_y = m_{log} a_y = 0$ gives

$$F_B - m_{log} g = 0, \text{ or } \rho_f V_f g - \rho_{log} V_{log} g = 0$$

Substituting $f V_{log}$ for V_f gives $\rho_f f V_{log} g - \rho_{log} V_{log} g = 0$. Dividing by $V_{log} g$ and solving for f,

$$\rho_f f - \rho_{log} = 0 \therefore f = \frac{\rho_{log}}{\rho_f} = \frac{5.5 \times 10^2 \text{ kg/m}^3}{1.0 \times 10^3 \text{ kg/m}^3} = 0.55 = 55\%$$

Thus, the percentage of the log's volume under the water line is 55%. Notice that this percentage depends only on the densities of the log and water.

DID YOU KNOW?

Crocodiles can increase their density and thus remain lower in the water by swallowing stones. Some crocodiles have been found with as much as 5 kg of stones in their stomachs. This is obviously an advantage to the crocodile, but not to its prey.

EXERCISES

12-25 An iceberg having s.g. = 0.92 is floating in salt water with s.g. = 1.03 (Figure 12-37). What percentage of the iceberg's volume is above water?

12-26 A copper sphere of radius 8.00 cm is placed in liquid mercury.

(a) What fraction of the sphere's volume will be below the mercury's surface? (Reference: Table 12-1)

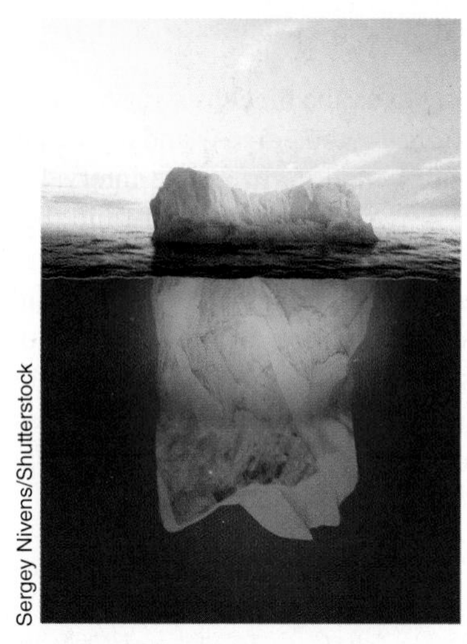
Sergey Nivens/Shutterstock

Figure 12-37 Most of an iceberg is under water (Question 12-25).

(b) What is the magnitude of the buoyant force exerted by the mercury on the sphere?

(c) What is the magnitude of the apparent weight of the sphere?

12-27 A helium-filled blimp (Figure 12-38) is sailing along at constant velocity at an elevation where the air density is 1.22 kg/m^3. The helium in the blimp has a volume of 5.3×10^3 m^3 and a density of 0.18 kg/m^3.

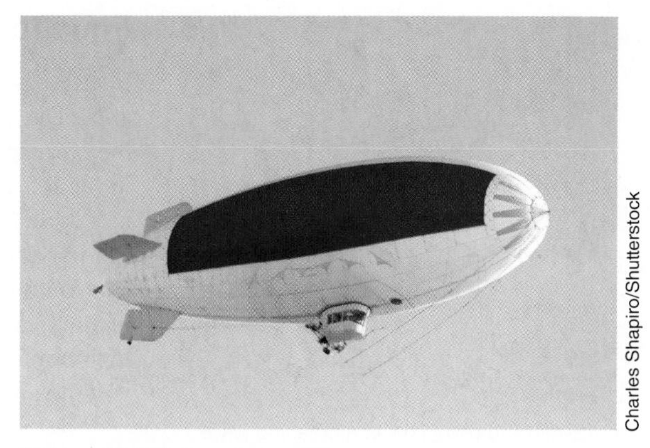
Charles Shapiro/Shutterstock

Figure 12-38 Question 12-27.

(a) What is the magnitude of the buoyant force exerted by the air on the volume of helium?

(b) What is the mass of the blimp and its contents (including the helium)?

12-28 **Fermi Question:** Estimate the magnitude of the buoyant force exerted by the air on a typical adult person.

12-29 Some children have built a flat raft made of wooden planks of density 5.5×10^2 kg/m^3. The raft is 2.0 m long, 1.7 m wide, and 4.0 cm thick. In order for the children not to get wet on a fresh-water lake (density 1.00×10^3 kg/m^3), what is the maximum mass that the raft can carry?

12.5 Fluid Dynamics

So far in this chapter the discussion has centred on stationary fluids, and now this section and the next one will focus on **fluid dynamics**, that is, the study of fluids in motion and objects moving in fluids. There are many important biological examples of moving fluids, including the circulation of blood and the breathing of air. As well, the movement of water in pipes to homes, businesses, and factories is crucial in modern society, as is the distribution of oil and natural gas in pipelines. Cars and airplanes moving through air and aquatic animals swimming in water involve other aspects of fluid dynamics.

© culture-images GmbH/Alamy

Figure 12-39 Streamlines of smoke particles passing over a car in a wind tunnel.

Streamline Flow and Ideal Fluids

The flow of a fluid can be described in one of two ways. The flow is called **streamline flow**, or **laminar flow**, if every particle of fluid that passes through a particular point always follows the same smooth path, known as a **streamline**. Streamline flow is smooth and steady, and streamlines do not cross each other. Figure 12-39 shows streamlines of smoke passing over a car in a wind tunnel. The other type of flow is not smooth and steady, but rather is irregular and chaotic, and is referred to as **turbulent flow** (Figure 12-40). A flow that is streamline can change to turbulent if the flow speed is increased above a certain value, or if the fluid encounters situations such as a rough surface over which it must flow.

Our discussion of fluid dynamics begins with the flow of **ideal fluids**, which have the following properties:

- The flow is streamline.
- Internal fluid friction (**viscosity**) between adjacent layers of fluid is negligible.
- The fluid is incompressible; that is, its density is constant.

fluid dynamics the study of fluids in motion and objects moving in fluids

streamline flow (or **laminar flow**) fluid flow that is smooth and steady; each fluid particle passing through a particular point always follows the same smooth path

streamline the smooth path followed by a particle in streamline flow

turbulent flow irregular and chaotic fluid flow

ideal fluids incompressible fluids in which flow is streamline and viscosity is negligible

viscosity internal friction between adjacent layers of moving fluid

Figure 12-40 Turbulent flow of a crashing wave.

Equation of Continuity

Figure 12-41 shows an ideal fluid flowing in a tube that has a varying cross-sectional area. The areas in the tube's left and right regions are A_1 and A_2, respectively, and the corresponding speeds of the fluid are v_1 and v_2. In a time interval Δt, the fluid in the left region moves a distance $\Delta x_1 = v_1 \Delta t$. The volume of fluid that has moved is $V_1 = A_1 \Delta x_1 = A_1 v_1 \Delta t$, and its mass is $m_1 = \rho A_1 v_1 \Delta t$, where ρ is the fluid's density. Similarly, in the right region during the same interval, the portion of fluid that moves has volume $V_2 = A_2 \Delta x_2 = A_2 v_2 \Delta t$ and mass $m_2 = \rho A_2 v_2 \Delta t$. Since the fluid is ideal and incompressible, the volume and mass of fluid that comes into the tube in the left region must equal the volume and mass leaving the tube in the right region; that is, volume and mass are conserved. Thus, $m_1 = m_2$, or $\rho A_1 v_1 \Delta t = \rho A_2 v_2 \Delta t$, which gives

$$A_1 v_1 = A_2 v_2 \qquad \text{(12-17)}$$

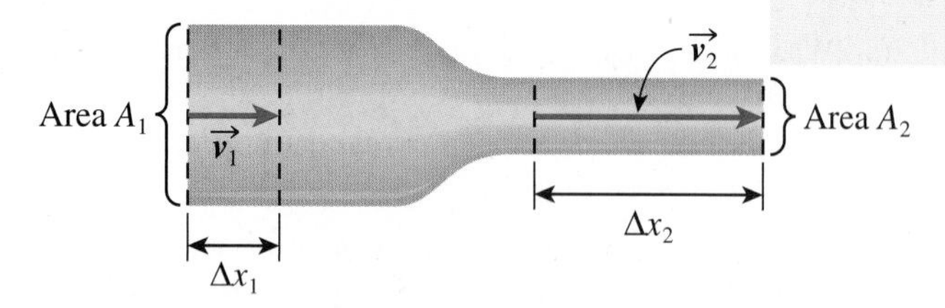

Figure 12-41 An ideal fluid flowing through a tube with a varying cross-sectional area.

This equation, known as the **equation of continuity**, states that for an ideal fluid flowing in a tube the product of cross-sectional area and speed is constant. This means that the speed is low in a region of high cross-sectional area, and high in a region of low cross-sectional area. (Most people have demonstrated this with a water hose at some time in their lives without realizing it.) The product of area and speed (Av) has units of metre^2 × metre/second, or metre^3/second. This unit indicates that the product Av gives volume per unit time—that is, **volume flow rate** (or **flow rate**), symbolized as Q—and so another way to state the equation of continuity is that volume flow rate Q is constant.

equation of continuity for an ideal fluid flowing in a tube the product of cross-sectional area and speed is constant; i.e., volume flow rate is constant

volume flow rate (or flow rate) volume of fluid flowing per unit time, or product of cross-sectional area and speed

$$\text{(volume) flow rate } Q = Av \qquad \text{(12-18)}$$

Terminology Tip

Be careful not to confuse *volume flow rate* (or *flow rate*) Q, which has units of cubic metres per second, and *flow speed* v, which has units of metres per second.

The equation of continuity applies in some circumstances where there is no actual tube containing the fluid. Figure 12-42 shows water flowing from a faucet. As the water moves downward, the force of gravity increases the water's speed. Since the volume flow rate must be the same along the column of water, then as the water speeds up, its cross-sectional area must decrease as seen in the photograph. The equation of continuity can also be used in situations where there is branching of tubes, as occurs in the human blood circulatory system (see Sample Problem 12-11). When analyzing such flow systems, it is important to remember that volume flow rate is constant, or that volume in equals volume out.

Figure 12-42 As the water flowing from this faucet speeds up because of gravity, its cross-sectional area decreases so that the volume flow rate remains constant.

SAMPLE PROBLEM 12-11

A volume of 5.0 L of blood is flowing per minute in the aorta. This blood eventually flows with a speed of 0.33 mm/s in capillaries that have an average diameter of 0.0080 mm. How many capillaries are there?

Solution

The volume flow rate in the aorta (subscript "a") is

$$Q_a = \frac{5.0\ \text{L}}{\text{min}} \times \frac{1\ \text{min}}{60\ \text{s}} \times \frac{1000\ \text{cm}^3}{1\ \text{L}} \times \frac{(1\ \text{m})^3}{(100\ \text{cm})^3} = 8.3 \times 10^{-5}\ \text{m}^3/\text{s}$$

This aortic volume flow rate equals the sum of the volume flow rates in the capillaries. Representing the number of capillaries as n, and using subscript "c" for capillary,

$$Q_a = nA_c v_c \therefore n = \frac{Q_a}{A_c v_c} = \frac{Q_a}{\pi r_c^2 v_c}$$

$$v_c = 0.33\ \frac{\text{mm}}{\text{s}} \times \frac{1\ \text{m}}{10^3\ \text{mm}} = 3.3 \times 10^{-4}\ \text{m/s}$$

$$r_c = \frac{0.0080\ \text{mm}}{2} \times \frac{1\ \text{m}}{10^3\ \text{mm}} = 4.0 \times 10^{-6}\ \text{m}$$

$$\therefore n = \frac{Q_a}{\pi r_c^2 v_c} = \frac{8.3 \times 10^{-5}\ \text{m}^3/\text{s}}{\pi(4.0 \times 10^{-6}\ \text{m})^2(3.3 \times 10^{-4}\ \text{m/s})} = 5.0 \times 10^9$$

Thus, there are 5.0×10^9 capillaries.

Bernoulli's Equation

Many fluid-flow situations can be analyzed using the equation of continuity and a second equation that involves the pressure, speed, and elevation of the fluid. This equation is named Bernoulli's equation in honour of the Swiss physicist Daniel Bernoulli (1700–1782) who derived it in 1738. As we will soon see, this equation is essentially a statement of conservation of energy in an ideal fluid.

Consider an ideal fluid of density ρ flowing steadily in a tube of varying cross-sectional area A and varying elevation y (Figure 12-43). The lower part of the tube has area A_1 and elevation y_1 relative to a horizontal reference level, and the upper part has area $A_2 < A_1$ and elevation $y_2 > y_1$. The pressure in the fluid is P_1 and P_2 in the lower and upper regions respectively. Consider the portion of fluid initially contained between points A and B. As it flows during a time interval Δt, its boundaries move to points C and D. This portion of fluid experiences a forward force, exerted by the fluid behind it, of magnitude P_1A_1 as its back edge moves a distance Δx_1 between A and C. As well, it experiences a backward force, exerted by the fluid ahead of it, of magnitude

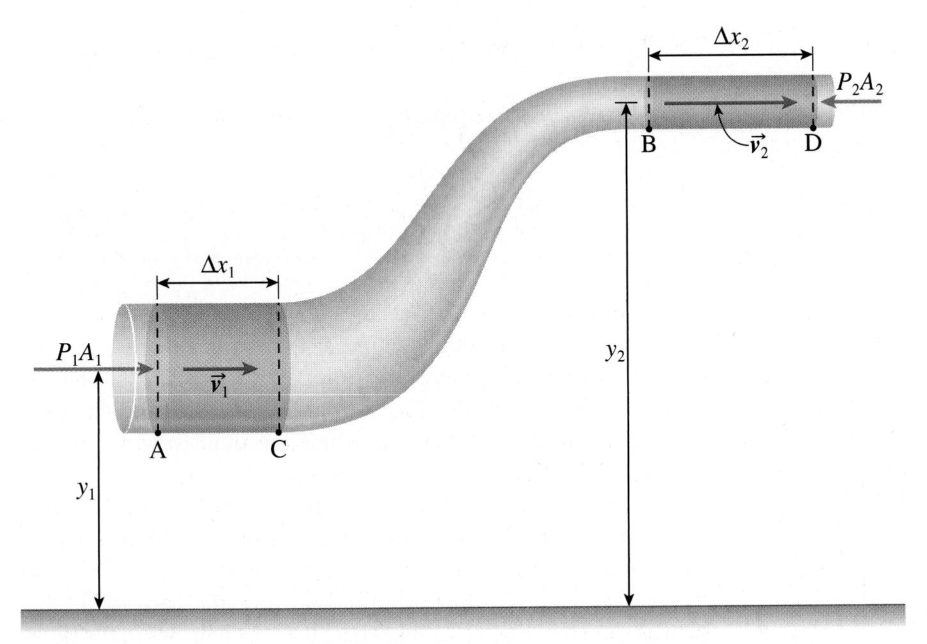

Figure 12-43 An ideal fluid flowing through a tube with a varying cross-sectional area and varying elevation.

P_2A_2 as its front edge moves a distance Δx_2 between B and D. The work done by the forward force at the back end is

$$W_1 = F_1\Delta x_1 = P_1A_1\Delta x_1 = P_1V$$

where V is the volume of fluid contained between A and C (shaded blue). The work done by the backward force at the front end has a similar expression but is negative because the force and displacement are in opposite directions:

$$W_2 = -F_2\Delta x_2 = -P_2A_2\Delta x_2 = -P_2V$$

The volume V that moved at the front end (shaded blue between B and D) is the same as at the lower end because the ideal fluid is incompressible. The total work done is

$$W_{tot} = P_1V - P_2V$$

This work changes the fluid's kinetic energy, E_K, and gravitational potential energy, E_P; that is, $W = \Delta E_K + \Delta E_P$. Comparing the fluid bounded by A and B with the fluid later bounded by C and D, there is a common volume between C and B (shaded grey). Therefore, the only *change* in energies involves only the fluid originally between A and C and the fluid later between B and D. These two regions of fluid have the same volume V and the same mass m. However, they have different cross-sectional areas (and therefore different flow speeds) and also different elevations, so the kinetic energies and the gravitational potential energies of the mass m will be different in the two regions. Writing the total work done $(P_1V - P_2V)$ as the changes in the energies,

$$P_1V - P_2V = \tfrac{1}{2}mv_2^2 - \tfrac{1}{2}mv_1^2 + mgy_2 - mgy_1$$

Rearranging the terms,

$$P_1V + \tfrac{1}{2}mv_1^2 + mgy_1 = P_2V + \tfrac{1}{2}mv_2^2 + mgy_2 \qquad \textbf{(12-19)}$$

This equation is a statement of conservation of energy. As an ideal fluid flows, the sum of kinetic energy, gravitational potential energy, and the energy (PV) available because the fluid is under pressure is constant. Although the above equation is a perfectly valid form of Bernoulli's equation, it is normally written in another way. Dividing by V and recalling that $m/V = \rho$ gives

$$P_1 + \tfrac{1}{2}\rho v_1^2 + \rho g y_1 = P_2 + \tfrac{1}{2}\rho v_2^2 + \rho g y_2 \qquad \textbf{(12-20)}$$

This is the usual form of **Bernoulli's equation**, often written as

$$P + \tfrac{1}{2}\rho v^2 + \rho g y = \text{constant} \qquad \textbf{(12-21)}$$

Bernoulli's equation at any point in the flow of an ideal fluid, the sum of the pressure, P, the kinetic energy per unit volume, $\frac{1}{2}\rho v^2$, and the gravitational potential energy per unit volume, $\rho g y$, is constant

In words, Bernoulli's equation states that, at any point in the flow of an ideal fluid, the sum of the pressure P, the kinetic energy per unit volume, $\frac{1}{2}\rho v^2$, and the gravitational potential energy per unit volume, $\rho g y$, is constant.

It is useful to consider two special cases of Bernoulli's equation. First consider the situation where the fluid is *static*; that is, $v_1 = v_2 = 0$. In this case, Eqn. 12-20 becomes $P_1 + \rho g y_1 = P_2 + \rho g y_2$, which gives $P_1 - P_2 = \rho g(y_2 - y_1)$. This is equivalent to $\Delta P = \rho g d$, which is Eqn. 12-7 that we developed earlier for the increase in pressure ΔP with depth d in a stationary liquid.

The second case is that of a fluid flowing in a horizontal tube, such as shown in Figure 12-44, where the tube narrows in the right-hand section. From the equation of continuity, the fluid's speed must increase in this region, that is $v_2 > v_1$. Since the tube is horizontal, $y_1 = y_2$, and Bernoulli's equation (Eqn. 12-20) becomes

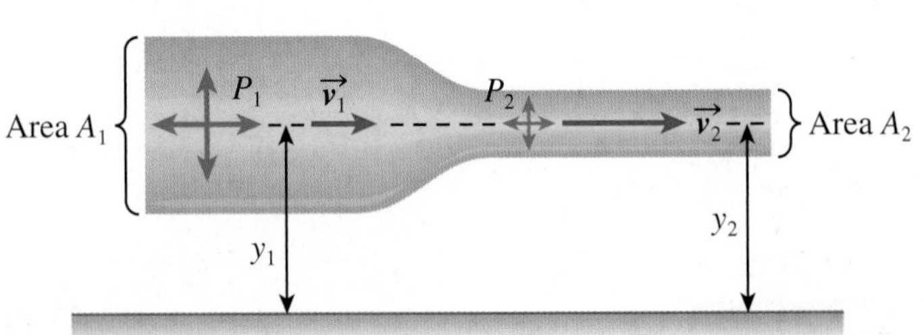

Figure 12-44 An ideal fluid flowing in a tube of constant elevation. Where the cross-sectional area decreases, the fluid's speed increases, and the pressure decreases.

$P_1 + \frac{1}{2}\rho v_1^2 = P_2 + \frac{1}{2}\rho v_2^2$. Since $v_2 > v_1$, then pressure P_2 must be less than pressure P_1. This rather surprising result indicates that *where the fluid is travelling faster, the pressure is less*. If you have ever held your hand in front of rapidly moving water coming from a hose, you might doubt this result. You certainly feel a large pressure, but the pressure in the water was not high until you placed your hand as a barrier in front of it and exerted a large force on the water. By Newton's third law of motion, the water exerted a large force back on your hand.

? TACKLING MISCONCEPTIONS

Pressure in a Moving Fluid

Sometimes people think that pressure in a moving fluid is only in the direction of motion, but because of molecular collisions, the pressure is exerted in all directions just as in the case of static fluids. Notice in Figure 12-44 that the pressures P_1 and P_2 are shown acting in a number of directions.

SAMPLE PROBLEM 12-12

If a pressure gauge is added to each of the two tube sections shown in Figure 12-44, the result is a device used to measure flow speeds, known as a Venturi meter, named after Giovanni Venturi, an Italian physicist (1746–1822). Suppose that oil of density 850 kg/m^3 is flowing through the horizontal tube, which narrows from a radius of 3.0 cm to 2.0 cm. The pressure gauges indicate a pressure drop from 62 kPa to 28 kPa as the oil goes from the wide area to the narrow area. Determine **(a)** the flow speed in the wide area, and **(b)** the flow rate.

Solution

(a) Using subscripts "1" and "2" for the wide and narrow regions respectively, the equation of continuity (Eqn. 12-17) is $A_1v_1 = A_2v_2$, and since the tube radii are given, $\pi r_1^2 v_1 = \pi r_2^2 v_2$, or

$$r_1^2 v_1 = r_2^2 v_2 \qquad \text{Eqn. [1]}$$

Bernoulli's equation is $P_1 + \frac{1}{2}\rho v_1^2 + \rho g y_1 = P_2 + \frac{1}{2}\rho v_2^2 + \rho g y_2$ (Eqn. 12-20). Since the tube is horizontal, $y_1 = y_2$, and hence

$$P_1 + \frac{1}{2}\rho v_1^2 = P_2 + \frac{1}{2}\rho v_2^2 \qquad \text{Eqn. [2]}$$

The only unknowns in Eqns. [1] and [2] are v_1 and v_2. Since we are asked to find v_1, first eliminate v_2 by rearranging Eqn. [1] to write v_2 in terms of v_1, and then substitute the expression for v_2 into Eqn. [2].

From Eqn. [1], $v_2 = (r_1/r_2)^2 v_1$

Substituting for v_2 in Eqn. [2], $P_1 + \frac{1}{2}\rho v_1^2 = P_2 + \frac{1}{2}\rho (r_1/r_2)^4 v_1^2$

Solving for v_1 and substituting numbers,

$$P_1 - P_2 = \frac{1}{2}\rho v_1^2[(r_1/r_2)^4 - 1]$$

$$v_1 = \sqrt{\frac{2(P_1 - P_2)}{\rho[(r_1/r_2)^4 - 1]}}$$

$$= \sqrt{\frac{2[(62 - 28) \times 10^3 \text{ Pa}]}{(850 \text{ kg/m}^3)[((3.0 \text{ cm})/(2.0 \text{ cm}))^4 - 1]}}$$

$$= 4.44 \text{ m/s}$$

Thus, the flow speed in the wide region is 4.4 m/s (correct to two significant digits).

Note that we did not need to change the units of r_1 and r_2 to metres since we were calculating their ratio.

(b) Since the speed v_1 is now known, the (volume) flow rate can be calculated using $Q = A_1v_1 = \pi r_1^2 v_1$ (Eqn. 12-18).

This gives $Q = \pi(3.0 \times 10^{-2} \text{ m})^2(4.44 \text{ m/s}) = 0.013 \text{ m}^3/\text{s}$.

Therefore, the flow rate is 0.013 m^3/s.

SAMPLE PROBLEM 12-13

Water enters a house in a horizontal pipe through a basement wall. The water has a speed of 1.4 m/s and an absolute pressure of 4.2×10^2 kPa in this inlet pipe, which has an inside radius of 1.0 cm. The water then travels upstairs in a vertical pipe, with an inside radius of 0.50 cm and a height of 6.0 m above the inlet pipe. This pipe branches to a sink, bathtub, etc. At the top of the 6.0 m pipe, what are the **(a)** flow speed, **(b)** absolute pressure in kilopascals, and **(c)** flow rate?

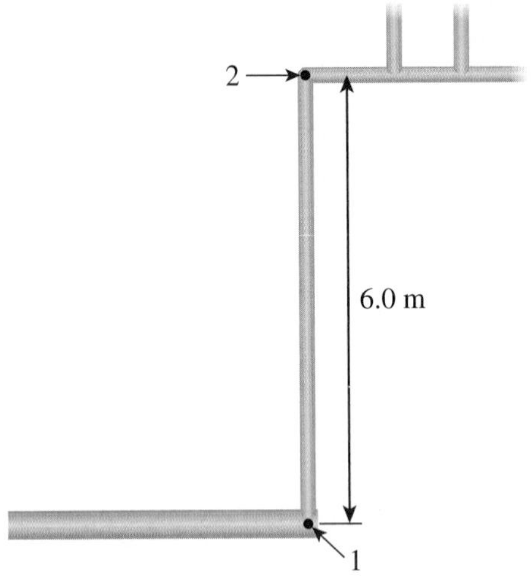

Solution

2

6.0 m

1

Figure 12-45 Sample Problem 12-13.

(a) Figure 12-45 shows a sketch of the situation. Choose point 1 to be in the inlet pipe and point 2 to be in the vertical pipe, 6.0 m above the inlet pipe. Choose $y_1 = 0$ m, and therefore $y_2 = 6.0$ m. The following information is given:

$$v_1 = 1.4 \text{ m/s}$$

$$P_1 = 4.2 \times 10^2 \text{ kPa} = 4.2 \times 10^5 \text{ Pa}$$

$$r_1 = 1.0 \text{ cm} = 1.0 \times 10^{-2} \text{ m}$$

$$\rho = 1.0 \times 10^3 \text{ kg/m}^3 \text{ (water)}$$

$$r_2 = 0.50 \text{ cm} = 5.0 \times 10^{-3} \text{ m}$$

Use the equation of continuity (Eqn. 12-17) to find the required flow speed v_2.

$$A_1v_1 = A_2v_2 \therefore \pi r_1^2 v_1 = \pi r_2^2 v_2 \text{ giving } v_2 = \left(\frac{r_1}{r_2}\right)^2 v_1 = \left(\frac{1.0 \text{ cm}}{0.50 \text{ cm}}\right)^2 (1.4 \text{ m/s}) = 5.6 \text{ m/s}$$

Thus, the flow speed at the top of the vertical pipe is 5.6 m/s.

(b) Bernoulli's equation (Eqn. 12-20) can be used to determine the pressure P_2.

$P_1 + \frac{1}{2}\rho v_1^2 + \rho g y_1 = P_2 + \frac{1}{2}\rho v_2^2 + \rho g y_2$, from which

$$\begin{aligned} P_2 &= P_1 + \tfrac{1}{2}\rho(v_1^2 - v_2^2) + \rho g(y_1 - y_2) \\ &= (4.2 \times 10^5 \text{ Pa}) + \tfrac{1}{2}(1.0 \times 10^3 \text{ kg/m}^3)((1.4 \text{ m/s})^2 - (5.6 \text{ m/s})^2) \\ &\quad + (1.0 \times 10^3 \text{ kg/m}^3)(9.8 \text{ m/s}^2)((0 - 6.0) \text{ m}) \\ &= (4.2 \times 10^5 \text{ Pa}) - (0.15 \times 10^5 \text{ Pa}) - (0.59 \times 10^5 \text{ Pa}) \\ &= 3.5 \times 10^5 \text{ Pa} \\ &= 3.5 \times 10^2 \text{ kPa} \end{aligned}$$

Therefore, the absolute pressure at the top of the pipe is 3.5×10^2 kPa.

(c) The (volume) flow rate is given by (Eqn. 12-18):

$$Q = A_2v_2 = \pi r_2^2 v_2 = \pi(5.0 \times 10^{-3} \text{ m})^2(5.6 \text{ m/s}) = 4.4 \times 10^{-4} \text{ m}^3\text{/s}$$

(The same value would be obtained using $Q = A_1v_1 = \pi r_1^2 v_1$.)

Thus, the volume flow rate is 4.4×10^{-4} m³/s.

Wendy Griffiths/iStock/Thinkstock

Figure 12-46 Sample Problem 12-14.

SAMPLE PROBLEM 12-14

A rain barrel full of water has a spigot near the bottom (Figure 12-46). **(a)** If the height of the water surface is 1.23 m above the spigot, what would be the speed of the water as it leaves the spigot? **(b)** If the spigot is oriented horizontally and is 42 cm above the ground, how far would the water travel horizontally before it hits the ground?

Solution

(a) Choose point 1 to be at the water surface in the barrel and point 2 to be at the spigot. The water at point 1 is open to the atmosphere, and is therefore at atmospheric pressure. The water emerging from the spigot is also open to the atmosphere, and hence is also at atmospheric pressure. Since $P_1 = P_2$, Bernoulli's equation becomes $\frac{1}{2}\rho v_1^2 + \rho g y_1 = \frac{1}{2}\rho v_2^2 + \rho g y_2$, and dividing each term by ρ gives $\frac{1}{2}v_1^2 + g y_1 = \frac{1}{2}v_2^2 + g y_2$. The equation of continuity $A_1 v_1 = A_2 v_2$ relates the areas and speeds at points 1 and 2. Since the cross-sectional area A_1 of the water surface is extremely large compared to a spigot's area A_2, we can assume that v_1 is negligibly small compared to v_2. Setting $v_1 = 0$, we now have $g y_1 = \frac{1}{2}v_2^2 + g y_2$. Solving for v_2,

$$v_2 = \sqrt{2g(y_1 - y_2)} = \sqrt{2(9.8\text{ m/s}^2)(1.23\text{ m})} = 4.9\text{ m/s}$$

Thus, the speed of the water as it leaves the spigot is 4.9 m/s.

(b) Once the water is moving through the air, it is a projectile experiencing a downward acceleration of magnitude 9.8 m/s^2, and can be treated the same as other projectiles (Section 4.3). We choose $+y$ to be downward, and first use Eqn. 4-13 to find the time for the water to fall downward 42 cm:

$$y = y_0 + v_{0y}t + \frac{1}{2}a_y t^2$$

We have $v_{0y} = 0$ m/s (since the spigot is horizontal), and we choose $y_0 = 0$ m, giving

$$y = \frac{1}{2}a_y t^2, \text{ from which } t = \sqrt{\frac{2y}{a_y}} = \sqrt{\frac{2(0.42\text{ m})}{9.8\text{ m/s}^2}} = 0.29\text{ s}$$

Now use Eqn. 4-10b, $x = x_0 + v_{0x}t$, with $x_0 = 0$ m:

$$x = v_{0x}t = (4.9\text{ m/s})(0.29\text{ s}) = 1.4\text{ m}$$

The water travels 1.4 m horizontally before it hits the ground.

Bernoulli's principle fluid pressure decreases (increases) when fluid speed increases (decreases)

The speed determined for the water leaving the spigot at the bottom of 1.23 m of water in Sample Problem 12-14(a) is the same speed that would be attained by a freely falling object dropped from rest at a height of 1.23 m. This should be expected since Bernoulli's equation is just an expression of conservation of energy. The equation $v_2 = \sqrt{2g(y_1 - y_2)}$ that we developed in our problem solution to relate fluid speed coming from an opening to the fluid height above the opening is sometimes called *Torricelli's law*, since it was first derived by Evangelista Torricelli about 100 years before Bernoulli developed his equation.

Applications of Bernoulli's Equation

Bernoulli's equation has many practical applications, often related to gases even though gases are not ideal fluids because they are compressible. Nevertheless, the general statement called **Bernoulli's principle** that fluid pressure decreases (increases) when fluid speed increases (decreases) is still valid for gases, although accurate numerical results cannot be obtained using Bernoulli's equation.

A common example is the use of a sprayer, such as a perfume atomizer (Figure 12-47). When the bulb is squeezed, a high-speed horizontal jet of air passes above a vertical tube leading to the perfume. The rapidly moving air is at low

Figure 12-47 A perfume atomizer uses Bernoulli's principle.

TRY THIS!

Demonstrating Bernoulli's Principle

Cut or tear a strip of paper (about 3–5 cm × 15–20 cm) and hold one end of it just touching your lower lip. The other end of the paper will hang down vertically. Now blow over the top of the paper, so that the air you are blowing just touches the paper. What happens to the paper? Why does this happen?

pressure, and the higher atmospheric pressure in the perfume container pushes some perfume upward into the lower pressure air and the air–perfume mixture is sprayed out. This basic type of apparatus in which atmospheric pressure pushes one fluid into another high-speed low-pressure fluid is used in paint sprayers, hose-end fertilizer sprayers, etc.

Burrowing animals such as prairie dogs (Figure 12-48) use Bernoulli's principle for ventilation of their burrows. There are normally two entrances to the burrows, one at a higher elevation than the other. When there is a wind, it speeds up to go over the constricted air space above the elevated entrance, and the air pressure decreases there. Some of the higher pressure air flowing over the lower entrance then flows through the burrow to the lower pressure area above the higher entrance. Marine worms also use a two-entrance system to provide water flow through their underwater burrows, and Bernoulli's principle is used by many other marine animals such as sponges to induce flow of nutrient-rich water.

Figure 12-48 The burrows of prairie dogs are ventilated as a result of Bernoulli's principle.

In people who have heavy plaque deposits on the walls of their arteries (a condition known as atherosclerosis), the Bernoulli effect can produce a phenomenon called *vascular flutter*. Since the internal area of an atherosclerotic artery is reduced by the plaque, the blood flow speed increases and the pressure drops. If it drops far enough, the walls of the artery can collapse inward under external pressure and stop the blood flow altogether. When the vessel is collapsed, there is no Bernoulli effect; the artery opens again and the blood flow resumes, only to be soon stopped again. This on-and-off blood flow can be detected with a stethoscope.

Another Bernoulli effect having to do with arterial walls is related to *aneurysms*, which are weakened areas of the walls that balloon outward, increasing the radius of the artery. Since the arterial area is large in the region of an aneurysm, the blood flow speed is low and the pressure is large. This large pressure can cause the ballooning to increase even further, with the possible risk of rupturing the artery.

Bernoulli's principle can be used to explain the aerodynamic lift on airplane wings. In Figure 12-49 showing the streamlines passing around an airplane wing in a wind tunnel, the streamlines above the wing are closer together than those below the wing, by design, indicating a higher flow speed above than below. From Bernoulli's principle, the pressure above the wing is less than that below and the wing experiences an upward force. Notice too that all the streamlines have been deflected downward as they pass over or under the wing. This indicates that the wing is exerting a downward force on the air and, by Newton's third law of motion, the air must exert an upward force on the wing.

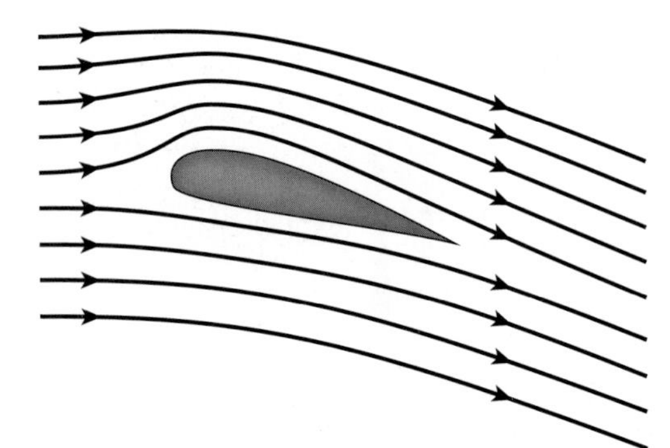

Figure 12-49 Streamlines around an airplane wing in a wind tunnel.

TACKLING MISCONCEPTIONS

Airspeed around Airplane Wings

Sometimes it is stated that the airspeed above an airplane wing is larger than that below because the air above the wing has to travel farther (because of the curve of the upper part of the wing) and must arrive at the rear of the wing at the same time as the air that travelled below the wing. This would mean that two adjacent air molecules that happen to part company at the wing's leading edge, with one going above the wing and the other going below it, would meet again at the rear of the wing. In fact, wind tunnel experiments show that the air above the wing travels so fast that it arrives at the rear of the wing earlier than the air below the wing.

Another connection to airplanes is that a common airspeed sensor on airplanes is a Bernoulli's-principle device known as a Pitot (pronounced "Pee-toe") tube, named after the French engineer Henri Pitot who developed the tube to measure water flow around bridge pilings in the Seine River in Paris in the early 18th century. Pitot tubes on airplanes are usually attached to the underside of wings (Figure 12-50), and point directly into the airflow. The tubes are heated to prevent them from becoming clogged with ice; if the heating element fails, the consequences can be disastrous. In 2009, an Air France flight from Rio de Janeiro to Paris crashed into the Atlantic Ocean, killing all 228 people aboard. The final report on the accident indicated that there were problems with the airspeed measurements likely due to ice-clogged Pitot tubes. As a result, the autopilot disconnected, the crew reacted incorrectly, and the plane went down.

Figure 12-50 A Pitot tube attached to an airplane wing.

EXERCISES

12-30 A portion of an artery that has been narrowed by atherosclerotic plaque has a radius of 2.2 mm, and the part of the artery leading up to it has a radius of 3.4 mm. What is the ratio of the blood's speed in the narrowed artery to that in the normal artery?

12-31 A fire truck pumps water at a rate of 75 kg/s onto a fire.

(a) Determine the volume flow rate of the water in cubic metres per second.

(b) If the inside diameter of the fire hose is 11 cm, what is the speed of the water in the hose?

12-32 A large artery (radius r_1) branches into three smaller arteries of equal radii r_2. The branching occurs in such a way that the flow speed of the blood is the same in each of the small arteries as in the large one. Determine the ratio r_2/r_1 to two significant digits.

12-33 A spoiler on the rear of a car (Figure 12-51) has a cross-sectional shape that somewhat resembles an airplane wing.

(a) Does the spoiler shown result in an upward or a downward force on the rear of the moving car? Explain your answer.

(b) What is the main advantage of using a spoiler?

Figure 12-51 A car with a rear spoiler (Question 12-33).

12-34 On a windy day, between tall buildings close together in the centre of a city it is common that the wind speed is larger (Figure 12-52) than in the surrounding countryside.

(a) Explain this phenomenon.

(b) Is the barometric pressure between the tall buildings larger than, smaller than, or the same as the pressure in the surrounding countryside? Explain your answer.

(c) Sometimes on very windy days, glass windows in buildings in a city centre will blow out of their frames and fall to the sidewalks and roads below. Why does this happen?

Figure 12-52 Question 12-34.

12-35 Figure 12-53 shows two plumbing designs for a kitchen drain.

(a) What is the purpose of the curved portion of the U-shaped drain?

(b) Which design would you recommend for a plumbing system? Explain your answer, using Bernoulli's principle.

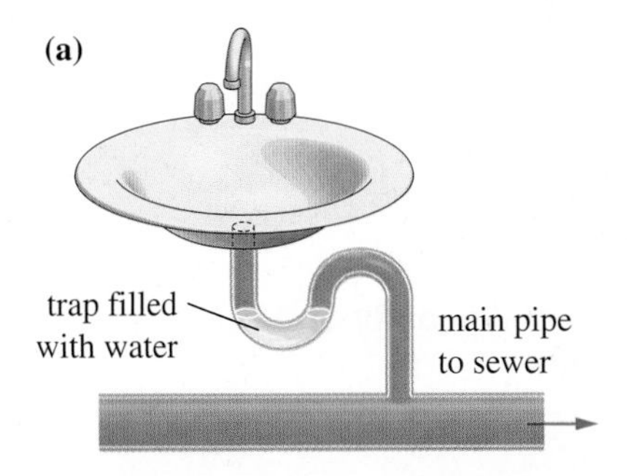

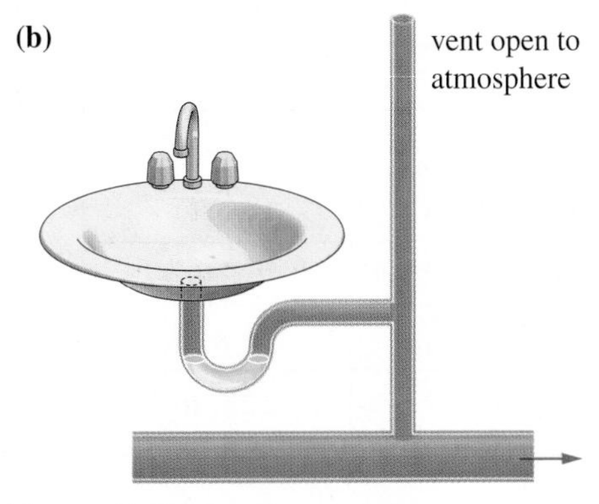

Figure 12-53 Question 12-35.

12-36 Oil of density 790 kg/m^3 is piped through a horizontal tube. At position A in the tube the pressure is 150 kPa and the speed of the oil is 3.10 m/s. Determine the speed of the oil at position B where the pressure is 130 kPa. Assume three significant digits in given data.

12-37 The faucet at the bottom of a large tank holding gasoline is 1.2 m above the ground. The top surface of the gasoline in the tank is 3.6 m above the faucet. The pressure at the top of the tank is atmospheric pressure. Assuming that the area of the tank is much greater than the area of the faucet, determine the speed with which the gasoline leaves the faucet.

12.6 Viscosity

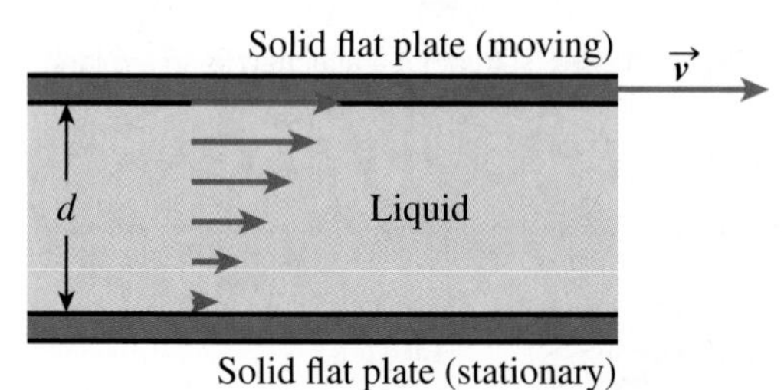

Figure 12-54 Measuring the viscosity of a liquid.

Our discussion of fluid dynamics in Section 12.5 considered ideal fluids, that is, fluids that are incompressible, have negligible viscosity, and are undergoing streamline flow. We now introduce viscosity, which is friction between adjacent layers of fluid as a result of cohesive forces between the molecules. Everyone knows that water pours more easily than honey; this is a result of water having a lower viscosity than honey.

Figure 12-54 shows one method that can be used to measure viscosity. Two horizontal solid flat plates are separated by a thin layer of liquid of thickness d. The bottom plate is stationary, and the top plate is being pulled at constant velocity $\vec{v}$ by a horizontal force $\vec{F}$. Each plate has a surface area A in contact with the liquid. A thin layer of liquid just below the top plate adheres to this plate and moves with the same velocity as the plate. Similarly, a thin liquid layer just above the bottom plate adheres to this plate and is stationary. In the main body of liquid, there is a gradual increase in speed from bottom to top, as shown. Experimentally, it has been shown that the magnitude of force F required to pull the top plate with speed v is proportional to v and to A, and inversely proportional to d:

$$F \propto \frac{vA}{d}$$

The constant of proportionality is the *coefficient of viscosity* of the liquid, represented by η (the lowercase Greek letter *eta*). The larger the liquid's coefficient of viscosity, the larger the force required to move the top plate and the layers of liquid sliding underneath it. Inserting η into the proportionality,

$$F = \eta \frac{vA}{d} \qquad (12\text{-}22)$$

Rearranging Eqn. 12-22 to solve for η gives $\eta = \frac{Fd}{vA}$. From this expression, we can determine the SI unit of viscosity:[5]

$$\frac{\text{N}\cdot\text{m}}{\text{m}\cdot\text{s}^{-1}\cdot\text{m}^2} = \left(\frac{\text{N}}{\text{m}^2}\right)\text{s} = \text{Pa}\cdot\text{s}$$

Thus, the SI unit for viscosity is the pascal·second. Many reference books list viscosities in units of poise (P), named after the French physician Jean Poiseuille (1799–1869):

$$1\ \text{P} = 1\ \text{g}\cdot\text{cm}^{-1}\cdot\text{s}^{-1} = 0.1\ \text{kg}\cdot\text{m}^{-1}\cdot\text{s}^{-1} = 0.1\ \text{Pa}\cdot\text{s}$$

The unit of centipoise (cP) is also still in use: 1 cP = 10^{-2} P = 10^{-3} Pa·s. Conveniently, the viscosity of water at 20°C is 1.00 cP.

Viscosities of several fluids are listed in Table 12-3. Notice that the viscosity of a liquid (e.g., water) decreases with increasing temperature, whereas that of a gas (e.g., air) increases. The viscosity of gases results from collisions due to random thermal

Table 12-3

Coefficients of Viscosity of Various Fluids

Fluid	Viscosity η (Pa·s)
Water (at 20°C)	1.00×10^{-3}
Water (at 100°C)	2.8×10^{-4}
Blood (at 37°C)	$3\text{–}4 \times 10^{-3}$
Ethanol (at 30°C)	1.0×10^{-3}
Liquid honey (at 20°C)	2–10
Air (at 20°C)	1.8×10^{-5}
Air (at 100°C)	2.2×10^{-5}

[5] Coefficient of viscosity is often referred to simply as viscosity.

motion. Molecules in gases are relatively far apart most of the time, and as the temperature increases they move more quickly and have more collisions with other molecules, thus increasing the frequency of interactions and the viscosity. Molecules in liquids are much closer together than in gases, and the viscosity is due to cohesive intermolecular forces. As temperature increases and the molecular kinetic energy increases, the fraction of molecules having enough energy to overcome the intermolecular forces increases, and the viscosity decreases, as you may have already observed in honey or maple syrup that has been warmed.

Newtonian and Non-Newtonian Fluids

A **Newtonian fluid** is one for which the viscosity varies only with temperature; water, air, and ethanol are Newtonian fluids. For a **non-Newtonian fluid**, the viscosity depends on other parameters such as flow speed, applied force, etc. A common example is latex paint, which has high viscosity when sitting in a paint can, but lower viscosity when being spread by a brush or roller, then higher viscosity again when it is stationary after being applied to a surface. Another example is a mixture of cornstarch and water, which has a low viscosity if a force is applied slowly, but a high viscosity when a force is applied rapidly. Search online for "running on cornstarch and water." You will find videos of people running on tanks or pools containing liquid mixtures of cornstarch and water. If they walk or stand on the mixtures, they sink.

Newtonian fluid a fluid that has a viscosity that varies only with temperature

non-Newtonian fluid a fluid that has a viscosity that depends on parameters other than temperature, such as flow speed, applied force, etc.

An important biological non-Newtonian fluid is synovial fluid, which fills the cavities in the synovial joints, such as the knee, of mammals. The viscosity of synovial fluid decreases as it is stressed during motion, and therefore it provides a smooth low-viscosity lubrication. When there is no motion, its viscosity increases, and therefore it remains in contact with the bones and does not simply flow toward the bottom of the joint.

Poiseuille's Law

In Section 12.5 when we used the equation of continuity and Bernoulli's equation to analyze the flow of an ideal fluid, it was implicit that, at a given cross-section of the flow tube, the velocity was the same at the centre of the tube as at the walls and at all points in between. In other words, the velocity vectors across the tube are all equal (Figure 12-55 (a)). This is a reasonable assumption for a low-viscosity fluid such as water flowing in a wide pipe. However, real fluids have viscosity and, when flowing in a tube, the layer of fluid adjacent to the tube walls adheres to the walls and has effectively zero velocity. The magnitude of the velocity gradually increases to a maximum in the centre of the tube. A detailed analysis shows that the velocity vectors have a parabolic profile; that is, the tips of the velocity vectors form a parabola (Figure 12-55 (b)).

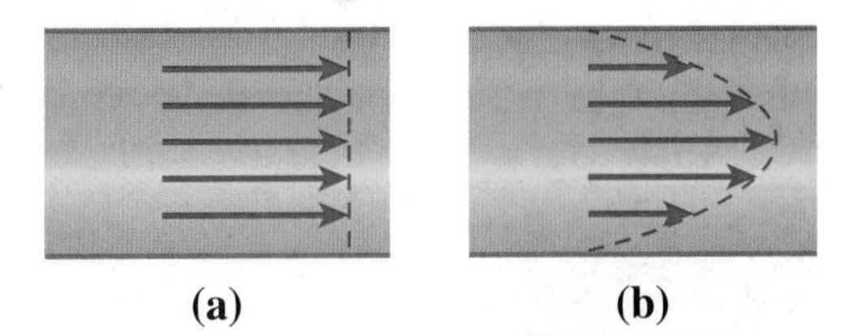

Figure 12-55 **(a)** The flow of an ideal fluid. The velocity is constant across the tube. **(b)** The flow of a viscous fluid. The velocity is zero at the walls of the tube and maximum at the centre; the velocity profile is parabolic.

In our earlier discussion of ideal fluid flow, we had $Q = Av$ for volume flow rate (Eqn. 12-18). For the flow of a viscous fluid, the volume flow rate is still the product of cross-sectional area and speed, but since the speed v is not constant across the area, the average speed v_{av} across the tube must be used. Hence, for viscous flow,

$$Q = Av_{av} \tag{12-23}$$

It can be shown that the speed v_c at the centre of the tube is twice the average speed: $v_c = 2v_{av}$.

Because viscosity is a friction force, it removes energy from the fluid as it flows. The energy that is removed appears as thermal energy; that is, the fluid increases in temperature. This energy must have been in another form before it became thermal energy, but which form? Earlier we had a form of Bernoulli's equation (Eqn. 12-19) that explicitly related the types of energy of a mass m of fluid as it moves:

$$P_1V + \tfrac{1}{2}mv_1^2 + mgy_1 = P_2V + \tfrac{1}{2}mv_2^2 + mgy_2$$

Consider a viscous fluid that is flowing from region 1 to region 2 in a tube of constant cross-sectional area and constant elevation. Since viscosity is removing energy as the fluid moves, then the sum of the terms on the right-hand side of the above equation must be less than the sum on the left. Which terms are changing? For a tube of constant area, the average speeds will be the same in the two regions and thus the kinetic energy in regions 1 and 2 will be the same (although the speeds are now average speeds). Similarly, for a tube of constant elevation, the potential energy terms will be the same. Therefore, it must be the PV terms that are different. The volume V of the mass of fluid is the same, and so it must be that *the downstream pressure P_2 is less than the upstream pressure P_1*. This conclusion is a crucial difference between the flow of a viscous fluid and the flow of an ideal fluid, in which the pressure would not change as the fluid flows in a tube of constant area and elevation.

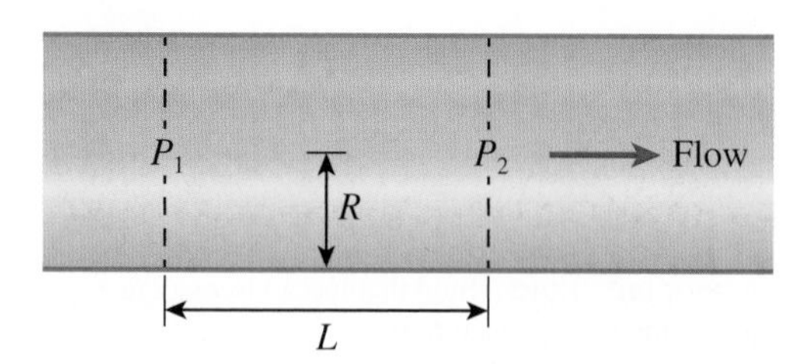

Figure 12-56 The flow of a viscous fluid. Upstream pressure P_1 is greater than downstream pressure P_2.

Figure 12-56 shows a fluid having viscosity η flowing in a tube of radius R. Two regions of the tube are separated by a distance L, and the pressures in the upstream and downstream regions are P_1 and P_2, respectively, where $P_1 > P_2$. In 1838 Jean Poiseuille, while studying flow in blood vessels, derived an equation relating the volume flow rate to the above quantities. This equation is now known as **Poiseuille's law**:

Poiseuille's law the relationship between volume flow rate, pressure change, viscosity of a fluid, and flow tube length and radius

pressure gradient the ratio of pressure change to flow tube length in the flow of a viscous fluid

$$Q = \frac{\pi R^4(P_1 - P_2)}{8\eta L} = \frac{\pi R^4 \Delta P}{8\eta L} \tag{12-24}$$

In this equation, the ratio $(P_1 - P_2)/L$ or $\Delta P/L$ is often called the **pressure gradient**.

It likely does not come as much of a surprise that the flow rate varies with pressure gradient: the bigger the pressure difference between upstream and downstream, the larger the flow rate. However, it is unusual in physical systems for one variable (Q in this case) to be dependent on a fourth power of another variable (R in this case). This means that if other variables remain constant, a reduction of the radius by only a factor of two will result in a reduction in the flow rate by a factor of sixteen.

SAMPLE PROBLEM 12-15

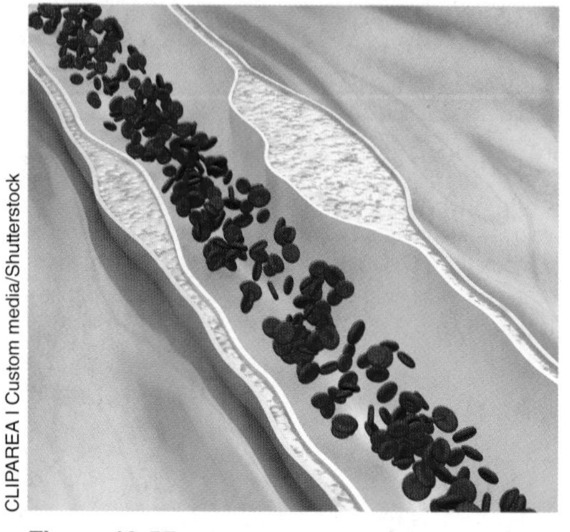
CLIPAREA I Custom media/Shutterstock

Figure 12-57 Artery with narrow diseased region (Sample Problem 12-15).

Suppose that a diseased portion of a certain artery (Figure 12-57) has a cross-sectional area only 0.50 times that of a healthy one. Determine the ratio of the pressure gradient in the diseased portion to that in the healthy one, if the same volume of blood is to be carried by the arteries per unit time.

Solution

We will use subscripts "D" for diseased and "H" for healthy. It is given that $A_D = 0.50A_H$ and $Q_D = Q_H$, and we are asked to determine $\frac{(\Delta P/L)_D}{(\Delta P/L)_H}$.

Poiseuille's law for the two arteries gives

$$Q_D = \frac{\pi R_D^4}{8\eta}\left(\frac{\Delta P}{L}\right)_D \text{ and } Q_H = \frac{\pi R_H^4}{8\eta}\left(\frac{\Delta P}{L}\right)_H$$

where it has been assumed that the viscosity η of the blood is the same in the two cases. Since $Q_D = Q_H$, we can write

$$\frac{\pi R_D^4}{8\eta}\left(\frac{\Delta P}{L}\right)_D = \frac{\pi R_H^4}{8\eta}\left(\frac{\Delta P}{L}\right)_H \text{ which gives } R_D^4\left(\frac{\Delta P}{L}\right)_D = R_H^4\left(\frac{\Delta P}{L}\right)_H$$

Algebraic rearrangement gives the required quantity $\frac{(\Delta P/L)_D}{(\Delta P/L)_H}$ in terms of the radii:

$$\frac{(\Delta P/L)_D}{(\Delta P/L)_H} = \frac{R_H^4}{R_D^4}$$

Now use $A_D = 0.50\,A_H$ to determine R_H^4/R_D^4, assuming that the arteries have circular cross-sections. Writing πr^2 for the areas, $\pi r_D^2 = 0.50\pi r_H^2$, then dividing by π and squaring the result, $r_D^4 = (0.50)^2 r_H^4 = 0.25\, r_H^4$.

Rearranging, $\dfrac{r_H^4}{r_D^4} = \dfrac{1}{0.25} = 4.0.$

The required quantity is then $\dfrac{(\Delta P/L)_D}{(\Delta P/L)_H} = \dfrac{R_H^4}{R_D^4} = 4.0.$

Therefore, the ratio of the pressure gradient in the diseased artery to that in the healthy artery is 4.0.

The result of Sample Problem 12-15 indicates why narrowing of an artery results in increased blood pressure that has to be produced by the heart. For the two arteries in the problem suppose that the required downstream pressure to push the blood through the rest of the circulatory system is 65 mm Hg, and that the pressure drop in the healthy artery is 10 mm Hg. This means that the pressure at the upstream end of the healthy artery is 75 mm Hg. For the diseased artery the pressure drop is 4.0 times that in the healthy one, that is, 40 mm Hg. The upstream pressure for the diseased artery must therefore be 105 mm Hg. This huge increase in pressure must be provided by the heart.

SAMPLE PROBLEM 12-16

A tank with a large cross-sectional area contains water at 20°C and a depth of 3.0 m (Figure 12-58). The top of the tank is open to the atmosphere. A narrow horizontal pipe of radius 1.0 mm and length 2.0 m leaves the tank at the bottom, as shown. Determine the volume flow rate and average speed of the water flowing in the pipe.

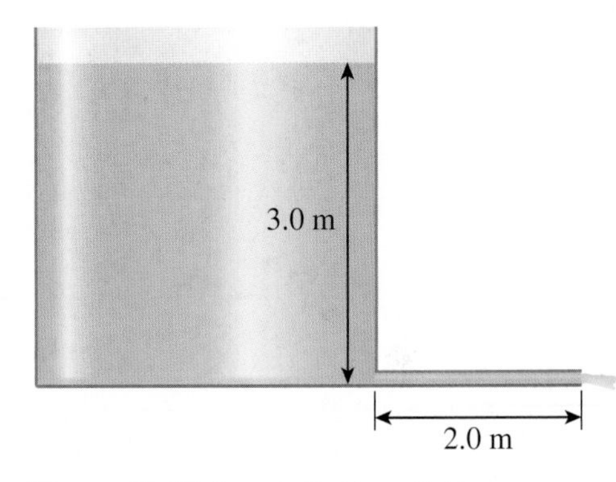

Figure 12-58 Sample Problem 12-16.

Solution

Label the point at the entrance to the pipe as "1," and at the exit as "2." The gauge pressure P_1 at point 1 is due to the depth of the water in the tank. Using Eqn. 12-7,

$$P_1 = \rho g d = (1.0 \times 10^3\ \text{kg/m}^3)(9.8\ \text{m/s}^2)(3.0\ \text{m}) = 2.94 \times 10^4\ \text{Pa}$$

At point 2 the water is open to the atmosphere, and the gauge pressure is zero: $P_2 = 0$ Pa.

Using Poiseuille's law (Eqn. 12-24) with the viscosity of water from Table 12-3,

$$Q = \frac{\pi R^4 (P_1 - P_2)}{8\eta L} = \frac{\pi(1.0 \times 10^{-3}\ \text{m})^4((2.94 \times 10^4 - 0)\ \text{Pa})}{8(1.00 \times 10^{-3}\ \text{Pa}\cdot\text{s})(2.0\ \text{m})} = 5.8 \times 10^{-6}\ \text{m}^3/\text{s}$$

Use Eqn. 12-23 to determine the average speed:

$$Q = Av_{av} \text{ or } v_{av} = \frac{Q}{A} = \frac{Q}{\pi R^2} = \frac{5.8 \times 10^{-6}\ \text{m}^3/\text{s}}{\pi(1.0 \times 10^{-3}\ \text{m})^2} = 1.8\ \text{m/s}$$

Thus, the volume flow rate is 5.8×10^{-6} m³/s and the average speed is 1.8 m/s for the water flowing in the pipe.

If Bernoulli's equation had been used as in Sample Problem 12-13, the calculated speed would be 7.7 m/s. The much lower actual average speed of 1.8 m/s indicates how large the viscous effects are for a fluid flowing through a small pipe, even for water which has a relatively low viscosity.

Problem-Solving Tip

Notice that Poiseuille's law is concerned only with fluid flowing in a cylindrical tube, pipe, etc., of constant radius and elevation. If there is constriction or dilation in the tube or a change in elevation, then only Bernoulli's equation can be used even though some error might be introduced by neglecting viscous effects. Such error will be small for a low-viscosity fluid flowing in a wide tube, but will be significant for a high-viscosity fluid, and/or narrow tubes.

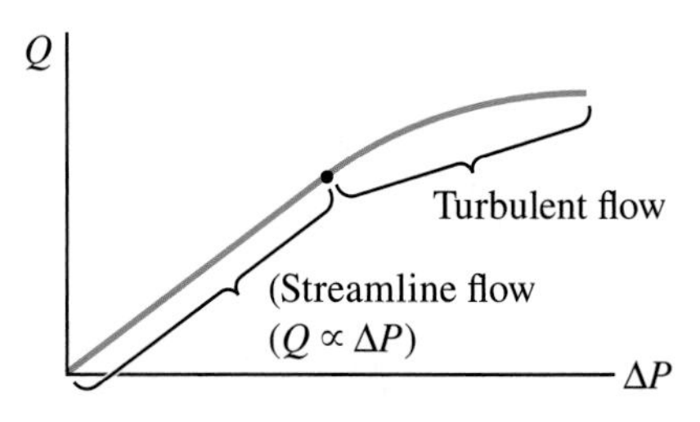

Figure 12-59 At low flow rates, the flow of a viscous fluid is streamline and $Q \propto \Delta P$, in accordance with Poiseuille's law. At higher flow rates, the flow becomes turbulent.

Turbulent Flow

When Poiseuille's law was introduced, we noted that it is fairly intuitive that volume flow rate Q increases as the pressure difference ΔP between upstream and downstream increases. Eqn. 12-24 expresses this linear relationship: $Q \propto \Delta P$. However, as ΔP continues to increase, this linearity holds only to a certain point (Figure 12-59), beyond which further increases in ΔP do not give rise to proportional increases in Q. Beyond this limit, larger and larger increases in ΔP are needed in order to give equal increases in Q, because streamline flow has ceased and turbulent flow (or turbulence) has begun. Turbulent flow is noisy and energy-wasting, and is not efficient in transporting fluid.

When flow is turbulent, Poiseuille's law is no longer applicable since Q is not proportional to ΔP. Bernoulli's equation also is not valid, since it requires that the fluid be ideal, and we will see shortly that turbulent flow is related to viscosity. However, it is still valid to define volume flow rate using Eqn. 12-23, $Q = Av_{av}$.

It is possible to determine the conditions at which flow will become turbulent. Osborne Reynolds, an English engineer and physicist (1842–1912), determined a dimensionless parameter now called the **Reynolds number**, R_e, defined by

Reynolds number a dimensionless parameter involving a fluid's density, average speed, and viscosity, as well as the diameter of the tube in which the fluid is flowing; if the Reynolds number is greater than about 2000, the flow is turbulent

$$R_e = \frac{\rho v_{av} D}{\eta} \qquad (12\text{-}25)$$

where ρ is the fluid's density, v_{av} is its average speed, D is the diameter of the tube in which the fluid is flowing, and η is the fluid's viscosity. If the Reynolds number is greater than about 2000, then the flow normally will be turbulent. However, if special precautions are taken, such as polishing the inside of the tube so that it is very smooth and increasing the fluid speed very gradually, then the Reynolds number can be as high as 16 000 before turbulence begins.

SAMPLE PROBLEM 12-17

At what minimum average flow speed would turbulence be expected in the aorta? Assume that the viscosity and density of blood are 3.5×10^{-3} Pa·s and 1.06×10^3 kg/m^3, respectively, and that the diameter of the aorta is 0.020 m.

Solution

Starting with Eqn. 12-25, $R_e = \dfrac{\rho v_{av} D}{\eta}$, solve for v_{av} with $R_e \approx 2000$:

$$v_{av} = \frac{\eta R_e}{\rho D} \approx \frac{(3.5 \times 10^{-3}\ \text{Pa}\cdot\text{s})(2000)}{(1.06 \times 10^3\ \text{kg/m}^3)(0.020\ \text{m})} \approx 0.3\ \text{m/s}$$

Thus, turbulence in the aorta would be expected for average flow speeds greater than about 0.3 m/s. Actual peak speeds are typically about this value. Experiments indicate negligible turbulence even during exercise except near heart valves during systole, but the turbulence quickly dies out in the aorta. Nowhere else in the human circulatory system does turbulence occur, except a very small amount at bifurcations.

It is important to remember that if flow is turbulent, then neither Bernoulli's equation nor Poiseuille's law can be used to analyze the flow. Turbulent flow is related to viscosity, and Bernoulli's equation requires that the fluid be ideal, that is, non-viscous. Poiseuille's law (in which $Q \propto \Delta P$) cannot be applied since Q is not proportional to ΔP when flow is turbulent.

Sedimentation

Any object moving through a fluid experiences a **drag force**, $\vec{F}_{\text{drag}}$, in the opposite direction to the object's velocity. For small slowly moving objects such as biological cells, dirt particles, large molecules, etc., the magnitude of this force is proportional to the object's speed; that is, $F_{\text{drag}} \propto v$, or

$$F_{\text{drag}} = fv \qquad (12\text{-}26)$$

where the constant of proportionality, f, is called the **friction factor**. In 1845 George Stokes, a scientist at the University of Cambridge in England, found that for a small spherical object the friction factor is given by

$$f = 6\pi\eta r \qquad (12\text{-}27)$$

where η is the coefficient of viscosity of the fluid and r is the radius of the object. Therefore, the magnitude of the drag force on a small spherical object is

$$F_{\text{drag}} = 6\pi\eta rv \qquad (12\text{-}28)$$

Equation 12-28 is known as **Stokes's law**.

Figure 12-60 shows a small spherical particle of radius r in a fluid. Assuming that the particle's density, ρ, is greater than the fluid's density, ρ_f, the downward force of gravity, $\vec{F}_G$, on the particle will be larger than the upward buoyant force, $\vec{F}_B$, and the particle will start to sink downward. As it moves, it will experience a drag force, $\vec{F}_{\text{drag}}$, upward that will increase as the particle's speed increases. Eventually the two upward forces together will be equal in magnitude to the downward force, the particle's acceleration will become zero, and the particle will continue downward at a constant speed called the **terminal speed** v_t.

An equation for terminal speed can be developed by first writing expressions for the magnitudes of the three forces acting on the particle. The volume of the spherical particle is $\frac{4}{3}\pi r^3$, and therefore its mass is $\rho\left(\frac{4}{3}\pi r^3\right)$. The magnitude of the downward gravitational force is then

$$F_G = \rho\left(\frac{4}{3}\pi r^3\right)g$$

The mass of fluid displaced by the particle is $\rho_f\left(\frac{4}{3}\pi r^3\right)$, and hence the magnitude of the upward buoyant force is

$$F_B = \rho_f\left(\frac{4}{3}\pi r^3\right)g$$

Using Eqn. 12-28, the magnitude of the upward drag force when the particle is moving at its terminal speed v_t is

$$F_{\text{drag}} = 6\pi\eta rv_t$$

Equating the magnitude of the downward force to the sum of the two upward forces ($F_G = F_B + F_{\text{drag}}$) gives

$$\rho\left(\frac{4}{3}\pi r^3\right)g = \rho_f\left(\frac{4}{3}\pi r^3\right)g + 6\pi\eta rv_t$$

Solving this equation for v_t,

$$v_t = \frac{2}{9}\frac{r^2 g}{\eta}(\rho - \rho_f) \qquad (12\text{-}29)$$

In laboratories where sedimentation in test tubes is used to collect biological cells and molecules for analysis, centrifuges are very often used since sedimentation is

drag force for an object moving in a fluid, the force exerted on the object by the fluid in a direction opposite to the object's velocity

friction factor the constant of proportionality between the magnitude of the drag force exerted by a fluid on a small object moving in the fluid and the speed of the object

Stokes's law the relationship between the magnitude of the drag force on a small spherical object moving in a fluid, the object's radius and speed, and the coefficient of viscosity of the fluid

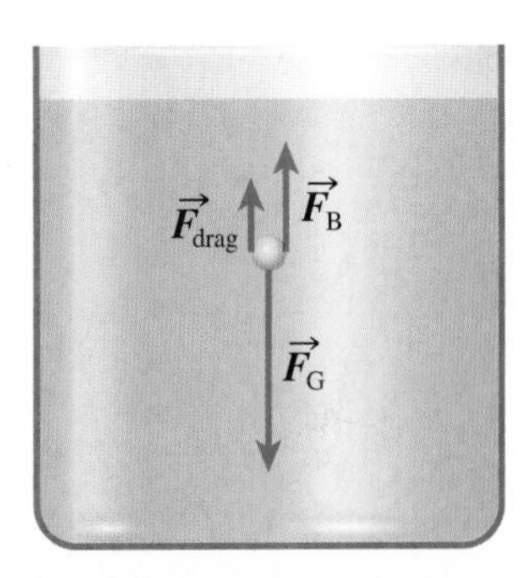

Figure 12-60 As a particle sediments downward in a fluid, three forces act on it: gravity $\vec{F}_G$ downward, buoyant force $\vec{F}_B$ upward, and drag force $\vec{F}_{\text{drag}}$ upward.

terminal speed constant downward speed as a particle settles in a fluid

TRY THIS!

Measuring Terminal Speed

You can determine the terminal speed of an object such as a coffee filter or a cupcake paper simply by dropping it and timing how long it takes it to fall a measured height. The higher the fall the better—you could do the drops while standing on a chair, or drop the object down a stairwell. You could also experiment with various numbers of nested filters or papers.

slow otherwise. The general operation of a centrifuge was discussed in Section 6.3. Relative to the rotating frame of reference of the centrifuge, there is a fictitious centrifugal "force" acting on the test tube contents. This "force" is directed outward from the axis of rotation. Relative to the laboratory, it is the test tubes that are experiencing a force inward toward the axis of rotation, providing a centripetal acceleration of magnitude $a_c = \omega^2 R$ (Eqn. 10-14), where ω is the angular velocity of the centrifuge and R is the radius of the circle in which the test tubes are rotating. (Recall from Chapter 10 that Eqn. 10-14 is valid only if ω has radians in its units.) The apparent centrifugal acceleration also has a magnitude $\omega^2 R$ directed outward from the rotation axis, that is, toward the bottom of the test tubes. Essentially, instead of the gravitational force pulling each cell or molecule (of mass m) toward the bottom of the test tubes, there is a much larger centrifugal "force" of magnitude $m\omega^2 R$ in that direction; that is, "g" has been replaced by $\omega^2 R$. This replacement of "g" by $\omega^2 R$ also occurs in the expression for the buoyant force. Thus, when a centrifuge is being used for sedimentation, Eqn. 12-29 can be used to determine the terminal speed by just replacing "g" with $\omega^2 R$:

$$v_t = \frac{2}{9}\frac{r^2\omega^2 R}{\eta}(\rho - \rho_f) \tag{12-30}$$

SAMPLE PROBLEM 12-18

A sample of large spherical protein molecules having molar mass of 2.0×10^6 g/mol and density 1.30×10^3 kg/m^3 is in water at 20°C. How long would it take these molecules to settle 3.0 mm in the water

(a) using gravity alone?

(b) using a centrifuge rotating at 5.0×10^3 rpm with a radius of rotation of 0.45 m for the test tubes?

Solution

(a) Use Eqn. 12-29 $v_t = \frac{2}{9}\frac{r^2 g}{\eta}(\rho - \rho_f)$ to calculate v_t, and then find time t from $v_t = d/t$ with $d = 3.0$ mm $= 3.0 \times 10^{-3}$ m. All quantities in the right-hand side of Eqn. 12-29 are known except for the radius r of each molecule, so this must first be determined. The mass m of one molecule is

$$m = \frac{M}{N_A} \qquad \text{Eqn. [1]}$$

where M is the molar mass and N_A is Avogadro's number.

Mass m can also be expressed as the product of density ρ and volume V, where V is the volume of a sphere of radius r:

$$m = \rho\left(\tfrac{4}{3}\pi r^3\right) \qquad \text{Eqn. [2]}$$

Equating the right-hand sides of Eqns. [1] and [2], and solving for r,

$$\rho\left(\tfrac{4}{3}\pi r^3\right) = \frac{M}{N_A}$$

$$r^3 = \frac{3M}{4\pi\rho N_A}$$

$$r = \sqrt[3]{\frac{3M}{4\pi\rho N_A}}$$

$$\therefore r = \sqrt[3]{\frac{3(2.0 \times 10^3\ \text{kg}\cdot\text{mol}^{-1})}{4\pi(1.3 \times 10^3\ \text{kg}\cdot\text{m}^{-3})(6.02 \times 10^{23}\ \text{mol}^{-1})}}$$

$$= 8.48 \times 10^{-9}\ \text{m}$$

The terminal speed is

$$v_t = \frac{2}{9}\frac{r^2 g}{\eta}(\rho - \rho_f)$$

$$= \frac{2}{9}\frac{(8.48 \times 10^{-9}\ \text{m})^2(9.8\ \text{m/s}^2)}{1.0 \times 10^{-3}\ \text{Pa}\cdot\text{s}}((1.30 - 1.00) \times 10^3\ \text{kg/m}^3)$$

$$= 4.70 \times 10^{-11}\ \text{m/s}$$

Then $v_t = d/t$ gives $t = \dfrac{d}{v_t} = \dfrac{3.0 \times 10^{-3}\ \text{m}}{4.70 \times 10^{-11}\ \text{m/s}} = 6.4 \times 10^7\text{s} = 2.0\ \text{yr}$

Therefore, the time taken for sedimentation by gravity alone is 2.0 yr, which is unreasonably long!

(b) The only difference when a centrifuge is used is that "g" is replaced by $\omega^2 R$. Therefore, the terminal speed is increased by a factor of $\dfrac{\omega^2 R}{g}$, and the time taken is decreased by this same factor. Remembering to convert the units of ω to radians per second, calculation of this factor gives

$$\frac{\omega^2 R}{g} = \frac{\left(5.0 \times 10^3\ \dfrac{\text{rev}}{\text{min}} \times \dfrac{2\pi\ \text{rad}}{1\ \text{rev}} \times \dfrac{1\ \text{min}}{60\ \text{s}}\right)^2 (0.45\ \text{m})}{9.8\ \text{m/s}^2} = 1.26 \times 10^4$$

Dividing the time from part (a) by this factor gives

$$\frac{6.4 \times 10^7\ \text{s}}{1.26 \times 10^4} = 5.1 \times 10^3\ \text{s} = 1.4\ \text{h}$$

Thus, the time taken for sedimentation by the centrifuge is only 1.4 h.

Drag Force in General

The discussion of drag force in the preceding subsection focused on the drag force on small particles moving slowly in a fluid. In this situation the relation $F_{drag} = fv$ (Eqn. 12-26) works well for calculations of F_{drag}. More completely, F_{drag} can be proportional to v^2 or it can be the sum of two parts, one of them proportional to v and the other proportional to v^2:

$$F_{drag} = fv + Cv^2 \qquad (12\text{-}31)$$

The constants f and C depend on the type of fluid, the size and shape of the moving object, and the orientation of the object as it moves. Typically the values of the constants must be determined experimentally for any specific situation. For small objects moving slowly, the linear term fv is usually the larger one, and for larger objects, especially ones moving quickly, the quadratic term Cv^2 is larger. For a baseball moving in air, for example, the Cv^2 term dominates for speeds greater than 1 cm/s, and the fv term is the larger one for speeds less than this.

Since the drag force removes energy from a moving object, whether the object is a car, bird, fish, etc., having a small drag force is often beneficial. Compare the shape of an antique car with that of a modern car (Figure 12-61). For these two cars travelling at

Rob Byron/Shutterstock.com

(a)

Maksim Toome/Shutterstock.com

(b)

Figure 12-61 For a given speed, the drag force on **(a)** an antique car is roughly twice that of the force on **(b)** a modern streamlined car.

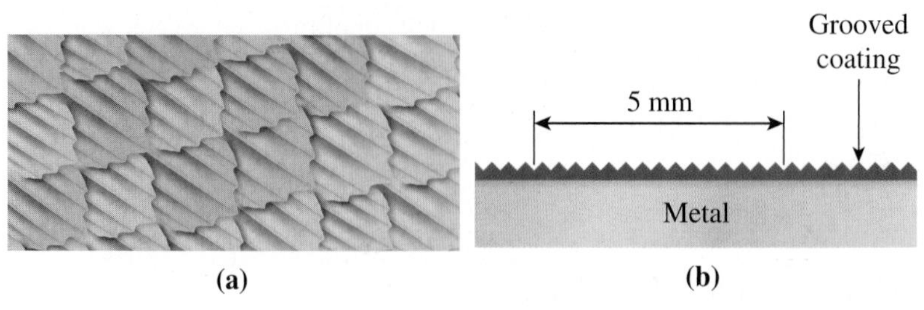

Figure 12-62 (a) Sharks have grooves in their skin (magnified here about 3000 times) that improve streamlining and reduce drag. (b) Some submarines have been developed that have a surface coating that mimics the sharks' grooves.

the same speed, the drag force on the modern one is about half that on the antique one. The antique is shaped like a box and provides a large impediment to the movement of air around it, but the modern one was developed from experiments in wind tunnels and designed to allow the air to flow smoothly around it with little turbulence. There are a number of improvements in the modern car to make it more streamlined: a low hood, aerodynamic external mirrors, low-profile windshield, concealed windshield wipers, wheel openings flush with the body, tapered rear, and so on.

Some animals have unique drag-reducing properties. Fast-swimming sharks have skin that has microscopic grooves parallel to the flow of the water (Figure 12-62 (a)) that reduce turbulence and increase speed. This method for drag reduction has now been applied to some submarines by the use of a thin plastic coating with fine grooves on the external surfaces (Figure 12-62 (b)). When penguins are swimming and need to increase speed to catch prey or escape predators, they reduce drag by releasing bubbles (Figure 12-63) that have been trapped among their feathers when they are above water. The water–air envelope that results around the penguin has a lower density and viscosity than the water alone, and therefore a lower drag. Penguins have been observed doubling and sometimes tripling their speed using this method. Modelling this approach, in 2010 commercial production of containerships and supertankers began with hulls constantly being "lubricated" with bubbles to reduce the energy expended by drag.

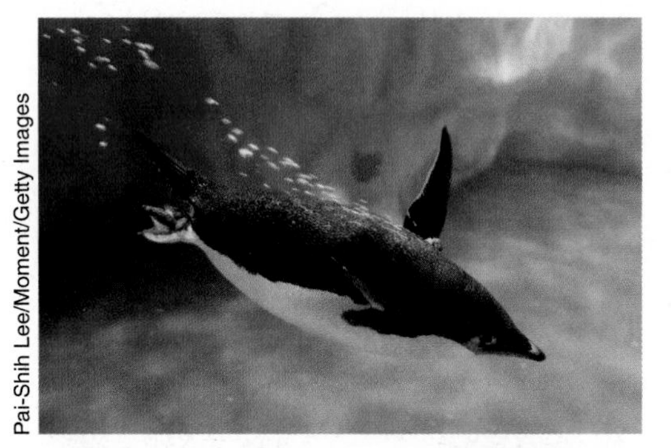

Figure 12-63 Penguins can reduce drag when needed by releasing air bubbles trapped in their feathers.

EXERCISES

12-38 A thin coating of oil having viscosity 0.25 Pa·s and thickness 0.50 mm has been placed between two horizontal microscope slides, each of length 4.0 cm and width 1.0 cm. What magnitude of horizontal force is required to pull one slide at a constant speed of 0.20 m/s relative to the other slide?

12-39 The volume flow rate of blood of viscosity 3.5×10^{-3} Pa·s in an artery of radius 0.50 cm is 6.0×10^{-5} m^3/s. What is the pressure gradient in the artery?

12-40 Why are compressor stations required at regular intervals along the cross-Canada natural gas pipeline? (These stations are located approximately 200 km apart; they do not add any gas to the pipeline.)

12-41 (a) Airports located at high altitude require longer runways for takeoff than airports at lower altitude. Partly this is because there is less oxygen available for fuel-burning. What is another reason for the longer runways for takeoff?

(b) At high altitude, runways also need to be longer for landing. Why?

12-42 A tank with a large cross-sectional area contains motor oil of density 889 kg/m^3 and a depth of 1.2 m. The top of the tank is open to the atmosphere. A narrow horizontal pipe of radius 1.6 mm and length 1.0 m leaves the tank at the bottom. Determine the average speed of the oil flowing in the pipe if

(a) the viscosity of the oil is neglected

(b) the viscosity (0.17 Pa·s) is included

12-43 Blood having density 1.06×10^3 kg/m^3 and viscosity 3.5×10^{-3} Pa·s is flowing with a speed of 11 cm/s and a Reynolds number of 558 in an artery. What is the radius of the artery in millimetres?

12-44 Spherical clay particles of diameter 3.8 μm and density 1.60×10^3 kg/m^3 are settling in water at 10°C. The water's density and viscosity are 1.00×10^3 kg/m^3 and 1.31×10^{-3} Pa·s respectively.

(a) What is the particles' terminal speed?

(b) Determine the magnitude of the drag force exerted on each particle.

12-45 Determine the terminal speed of a baseball of mass 145 g falling in air. The drag coefficients in Eqn. 12-31 for a baseball in air are $f = 0$ kg/s and $C = 1.3 \times 10^{-3}$ kg/m.

12-46 Molecules of a particular species are settling in water in a laboratory test tube. By what factor is the terminal speed increased if a centrifuge is used with an angular velocity of 5.2×10^{3} rpm and radius of rotation of 0.38 m, compared to the terminal speed if only gravity is used?

LOOKING BACK...LOOKING AHEAD

In previous chapters we discussed kinematics and dynamics, mainly of solid objects. The focus of the present chapter has been the study of fluids, both at rest and in motion. Many fundamental concepts and topics have been introduced: pressure, surface tension, buoyant force, Bernoulli's principle, viscosity, Poiseuille's law, drag force, sedimentation, and terminal speed.

The next chapter will continue with another topic in mechanics: oscillations and waves. This will serve as a prelude to chapters that follow with discussions of sound, music, and light.

CONCEPTS AND SKILLS

Having completed this chapter, you should now be able to do the following:

- Distinguish between absolute and gauge pressures, and analyze situations for liquids and gases involving pressure, density, specific gravity, and liquid depth.
- Describe how blood pressure is measured.
- Explain the operation of a hydraulic lift, and do calculations involving Pascal's principle.
- Use the barometric equation to solve problems involving pressures and heights in the atmosphere.
- Describe how surface tension allows small insects to stand and walk on water.
- Distinguish between adhesive forces and cohesive forces, and discuss how they are related to contact angle and capillary action.
- Perform calculations involving surface tension and capillary rise and fall.
- Use Archimedes' principle to solve problems involving buoyant force acting on objects wholly or partially immersed in a fluid.
- Know the properties of streamline flow and turbulent flow, and the characteristics of an ideal fluid.
- Solve numerical problems about ideal fluids, using the equation of continuity and Bernoulli's equation.
- Discuss a few commonplace examples of applications of Bernoulli's principle.
- Describe the difference between Newtonian and non-Newtonian fluids, and name examples of each type.
- Use Poiseuille's law to analyze situations involving the flow of viscous fluids.
- Explain how narrowing of blood vessels leads to increased blood pressure.
- Know the Reynolds number value at which turbulence is likely to begin, and do calculations involving the Reynolds number.
- Solve numerical problems involving the forces acting on a particle settling in a viscous fluid, and problems related to sedimentation due to gravity or to centrifugation.

KEY TERMS

You should be able to define or explain each of the following words or phrases:

fluid
fluid statics
density
specific gravity (s.g.)
pressure
pascal
standard atmospheric pressure
standard atmosphere
gauge pressure
absolute pressure
manometer
sphygmomanometer
systolic pressure
diastolic pressure
Pascal's principle
hydraulic lift
hydraulic press
ideal gas law
barometric equation
barometer
surface tension
surfactants
cohesive forces
adhesive forces
contact angle
buoyant force
Archimedes' principle
apparent weight
fluid dynamics
streamline flow
laminar flow
streamline
turbulent flow
ideal fluids
viscosity
equation of continuity
volume flow rate
flow rate
Bernoulli's equation
Bernoulli's principle
Newtonian fluid
non-Newtonian fluid
Poiseuille's law
pressure gradient
Reynolds number
drag force
friction factor
Stokes's law
terminal speed

Chapter Review

MULTIPLE-CHOICE QUESTIONS

12-47 Three pieces are cut from a copper pipe. The lengths of the pieces are 30 cm, 60 cm, and 90 cm. Which piece has the largest density?

(a) the 30 cm piece
(b) the 60 cm piece
(c) the 90 cm piece
(d) They all have the same density.

12-48 What is the SI unit of specific gravity?

(a) kilograms per cubic metre (kg/m^3)
(b) pascals (Pa)
(c) metres per square second (m/s^2)
(d) atmospheres (atm)
(e) specific gravity has no units

12-49 A diver is swimming downward in a lake (density 1.0×10^3 kg/m^3). As he moves down a distance of 10 m (two significant digits), which of the following is true about the pressure in the water around him?

(a) The gauge pressure increases by 9.8×10^4 Pa, and the absolute pressure does not change.
(b) The absolute pressure increases by 9.8×10^4 Pa, and the gauge pressure does not change.
(c) Both the gauge pressure and the absolute pressure increase by 9.8×10^4 Pa.
(d) The gauge pressure increases by 9.8×10^4 Pa, and the absolute pressure increases by 2.0×10^5 Pa.

12-50 A hydraulic lift is being used to hoist a large air conditioner of mass 458 kg at a construction site. The radii of the large and small pistons in the lift are 0.30 m and 0.050 m respectively. In order to hoist the air conditioner, what minimum magnitude of force must be exerted on the small piston?

(a) 1.2×10^2 N
(b) 13 N
(c) 7.5×10^2 N
(d) 76 N

12-51 On another planet it is found that the atmospheric pressure at ground level is 90 kPa, and the pressure 2.0 km above ground level is 60 kPa. What will the pressure be at 4.0 km above ground level?

(a) 30 kPa
(b) 40 kPa
(c) 45 kPa
(d) 50 kPa
(e) 33 kPa

12-52 A capillary tube is used to collect a small sample of blood. If a capillary tube of a radius that is half the radius of the original tube is then used, the height that the blood will go up in the second tube will be

(a) twice the original height
(b) half the original height
(c) the same as the original height

12-53 A stone is sinking slowly in a lake. At a particular time the stone is at point A, which is 2 m below the surface of the lake. A short time later, the stone is at point B, which is 4 m below the surface. If the density of the water is the same at A and B, the buoyant force on the stone at B is

(a) twice that at A
(b) greater than that at A, but not twice that at A
(c) equal to that at A

12-54 A cargo ship leaves the fresh water (s.g. = 1.00) of the St. Lawrence River and enters the salt water (s.g. = 1.03) of the Atlantic Ocean. When the ship is in the ocean, it rides

(a) higher in the water than when it is in the river
(b) lower in the water than when it is in the river
(c) at the same depth in the water as when it is in the river

12-55 A glass partially filled with water has an ice cube floating in it. The water level in the glass is at a particular level. After the ice cube has melted, the water level will be

(a) higher than the original level
(b) lower than the original level
(c) the same as the original level

12-56 A glass partially filled with water has an ice cube floating in it. The ice cube has a stone frozen inside it. The water level in the glass is at a particular level. After the ice cube has melted and the stone has fallen to the bottom of the glass, the water level will be

(a) higher than the original level
(b) lower than the original level
(c) the same as the original level

12-57 An artery of radius r divides into three arterioles, each of radius $r/6$. If the blood speed in the artery is v, what is the speed in each arteriole?

(a) $2v$
(b) $6v$
(c) $v/2$
(d) $12v$
(e) $v/6$

12-58 Water is flowing downward in a vertical pipe of constant radius. If viscosity is negligible, as the water flows downward, its pressure will

(a) increase (b) decrease (c) stay constant

12-59 Blood is flowing in a horizontal artery. As the blood moves into a region where plaque reduces the cross-sectional area of the artery, what happens to the blood's speed v and its pressure P? Neglect viscosity effects.

(a) v decreases and P decreases
(b) v increases and P increases
(c) v decreases and P increases
(d) v increases and P decreases
(e) v decreases and P remains constant

12-60 Which of the following statements is true for the viscosity, η_{gas}, of a gas and the viscosity, η_{liquid}, of a liquid? As temperature increases,

(a) η_{gas} increases and η_{liquid} decreases
(b) η_{gas} decreases and η_{liquid} decreases
(c) η_{gas} decreases and η_{liquid} increases
(d) η_{gas} increases and η_{liquid} increases

12-61 One artery (#1) carrying blood has atherosclerotic plaque and another (#2) does not. Artery #2 has a radius twice that of artery #1. If the volume flow rates and the blood's viscosity are the same in the two arteries, then the pressure gradients in the two arteries are related by

(a) $(\Delta P/L)_1 = \frac{1}{4}(\Delta P/L)_2$
(b) $(\Delta P/L)_1 = \frac{1}{2}(\Delta P/L)_2$
(c) $(\Delta P/L)_1 = 2\ \Delta P/L)_2$
(d) $(\Delta P/L)_1 = \frac{1}{16}(\Delta P/L)_2$
(e) $(\Delta P/L)_1 = 16\ (\Delta P/L)_2$

12-62 A viscous fluid is flowing in a tube with an average speed of 24 cm/s and a Reynolds number of 242. If another viscous fluid having the twice the density and twice the viscosity of the first fluid flows in the same tube, and also has a Reynolds number of 242, what is the average speed of the second fluid?

(a) 48 cm/s
(b) 12 cm/s
(c) 24 cm/s
(d) 96 cm/s
(e) 6.0 cm/s

Review Questions and Problems

12-63 A child on cross-country skis (Figure 12-64) will sink deeply in soft snow if the pressure on the snow exceeds 2.2 kPa. What is the minimum total surface area of skis that will prevent a 24 kg child from sinking deeply in the snow?

Levranii/Shutterstock

Figure 12-64 Question 12-63.

12-64 Determine the difference in blood pressure in kilopascals between the bottom of the feet and the top of the head of a person who is 1.74 m tall. (Reference: Table 12-1)

12-65 At what depth beneath the surface of water (at 4°C) is the pressure due to the water equal to an atmospheric pressure of 101 kPa? (Reference: Table 12-1)

12-66 A Canadian freediver, Mandy-Rae Cruickshank, set a women's world record by diving 88 m on one breath in the ocean near the Cayman Islands in 2007. What is the gauge pressure at that depth in atmospheres? Assume that the s.g. of saltwater is 1.03.

12-67 The difference in pressure between the top of a 13 cm column of a certain liquid and the bottom of the liquid is 1.6 kPa. Determine the density of the liquid.

12-68 A nurse measures a patient's blood pressure and records it as 134/95.

(a) According to this measurement, what is the patient's diastolic blood pressure (with appropriate units)?

(b) Convert your answer in (a) to kilopascals.

12-69 A client is sitting in a barber's chair that uses a small hydraulic lift. The barber applies a force of magnitude 99 N on the small piston and raises the chair and client at constant velocity. The cross-sectional area of the small piston is 28 cm^2, and the radius of the large piston is 0.14 m. What is the combined mass of the chair and client?

12-70 (a) What is the air density at the top of Mount Everest, which has an elevation of 8848 m above sea level? Assume a temperature of 0°C, and a sea-level air density of 1.3 kg/m^3.

(b) Explain why water boils at 66°C at the top of Mount Everest.

12-71 Sketch a graph showing the height of capillary rise as a function of capillary-tube radius, assuming that other relevant parameters remain constant.

12-72 A solution of sodium sulphate fertilizer has a specific gravity of 1.15 and a surface tension of 7.3×10^{-2} N/m. Calculate the maximum diameter (in millimetres) of xylem that would raise the solution to the top of a plant 0.50 m tall. Assume a contact angle of zero.

12-73 A helium-filled balloon is released from a weather station. The balloon has a volume of 1.50 m^3, and the total mass of the balloon and the attached package of instruments for measuring temperature, pressure, etc., is 1.62 kg. The volume of the instruments package is negligible. Determine the

(a) magnitude of the buoyant force on the balloon (Reference: Table 12-1)

(b) initial acceleration of the balloon and package

12-74 A small ferryboat that crosses a river is 8.3 m long and 5.3 m wide (Figure 12-65). If the boat settles an additional 6.5 cm into the river when a loaded truck pulls onto the boat, what is the mass of the truck?

Marcel Pelletier/iStock/Thinkstock

Figure 12-65 Question 12-74.

12-75 A basketball has an average density of 82 kg/m^3 and a radius of 12 cm. What magnitude of force is required to hold it stationary completely under water of density 1.000×10^3 kg/m^3? (The volume of a sphere of radius r is $\frac{4}{3}\pi r^3$.)

12-76 The air in a hot-air balloon has a temperature of 50°C and a density of 0.85 kg/m^3. The mass of the balloon and its basket and contents is 485 kg. If the balloon has an acceleration of 0.20 m/s^2 upward in air of density 1.28 kg/m^3, what is the volume of air in the balloon? Neglect air friction acting on the balloon.

12-77 Water is flowing in a pipe that branches at one point into three smaller pipes, each of which has a radius that is 0.25 times the radius of the large pipe. The speed of the water in each small pipe is 1.0 cm/s. What is the speed of the water in the large pipe, in metres per second?

12-78 Water is flowing with a volume flow rate of 8.0 L/min in a circular pipe that has an inside diameter of 12 mm.

(a) Determine the speed of the water in metres per second.

(b) What area (in square millimetres) of the opening at the end of the pipe would allow the water to exit at 5.5 m/s?

12-79 The water flow speed is 0.40 m/s in a section of garden hose lying on the ground. The inside radius of the hose is 5.0 mm. The end of the hose has a nozzle that is being held 0.90 m above the ground; the opening in the nozzle has an area of 0.30 cm^2. Assume that viscosity effects are negligible.

(a) With what speed does the water exit the nozzle?

(b) What is the gauge pressure of the water in the hose lying on the ground?

12-80 Water enters a house via a horizontal pipe having an inside radius of 0.80 cm. The water then goes up a vertical pipe of length 8.0 m that ends at a shower head having 36 circular openings each of radius 0.60 mm. The speed of the water as it leaves the shower head is 6.0 m/s. Assuming that viscosity effects are negligible, determine

(a) the flow rate

(b) the speed of the water in the horizontal pipe

(c) the gauge pressure of the water in the horizontal pipe

12-81 Blood of s.g. 1.05 is flowing in a horizontal artery of diameter 3.0 mm. At a smooth constriction, the blood pressure drops by 4.0×10^1 Pa. The flow rate is 1.0×10^{-6} m^3/s. What is the diameter (in millimetres) of the artery in the constricted area? Neglect viscosity.

12-82 Figure 12-66 shows a tank full of a viscous liquid that is flowing from the tank through two pipes of equal length. The pipes are at different elevations as shown, and the upper pipe has a radius twice that of the lower pipe. Determine the ratio of the flow rate in the upper pipe to that in the lower pipe. The top of the tank is open to the atmosphere.

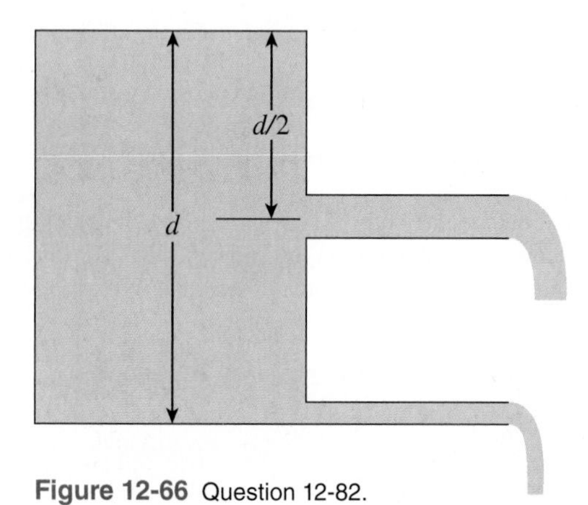

Figure 12-66 Question 12-82.

12-83 A rabbit's abdominal aorta has a radius of 1.5 mm. If the average blood speed is 50 cm/s, what is the Reynolds number? Use 0.0045 Pa·s for the viscosity of blood and 1.05 for its s.g. Is the flow turbulent?

12-84 Tomato bushy stunt virus is a spherical virus that can infect tomato plants and affect fruit production. Each virus particle has a diameter of 33 nm and a density of 1.30×10^3 kg/m^3. These virus particles are settling in water in a centrifuge of radius 32 cm spinning at 8.0×10^3 rpm. The water has a temperature of 10°C (two significant digits), and a density and viscosity of 1.00×10^3 kg/m^3 and 1.31×10^{-3} Pa·s respectively.

(a) Determine the terminal speed of the particles.

(b) How long (in minutes) will it take for the particles to settle 1.0 cm?

Applying Your Knowledge

12-85 **Fermi Question:** Estimate the pressure exerted on the pavement by each tire of a typical automobile.

12-86 Figure 12-67 shows an aquarium that has a length and width both equal to 1.00 m, and contains water having a depth of 30.0 cm. A U-tube manometer, open to the atmosphere at the right-hand end, is used to determine the pressure in the air above the water in the aquarium. The manometer fluid is water, and the water column height difference is 91 cm. Determine the gauge pressure and absolute pressure in kilopascals at the bottom of the water in the aquarium, correct to three significant digits.

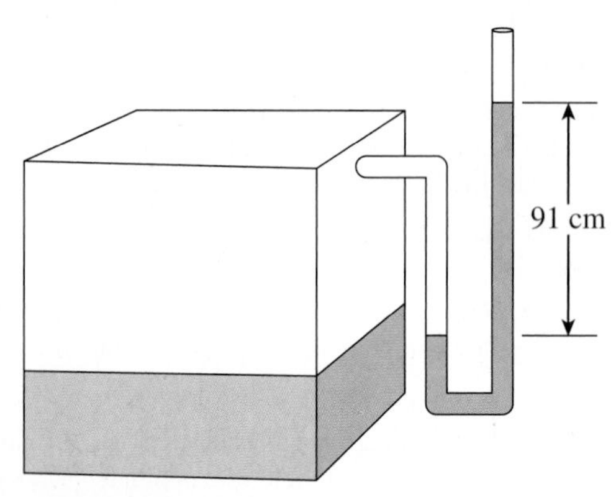

Figure 12-67 Question 12-86.

12-87 A planet in orbit around a distant star has an atmosphere many kilometres high. The atmosphere has a composition somewhat different from that of Earth's atmosphere, and the gravitational acceleration, g, on this planet has a different magnitude than Earth's g. At a height of 1.2 km above the planet's surface, the pressure is 0.82 P_0, where P_0 represents the pressure at the surface. At a height of 2.7 km above the surface, what will the pressure be in terms of P_0?

12-88 At the top of a mountain, the height of the column of mercury in a mercury barometer is 592 mm when the atmospheric pressure is 78.7 kPa. What is the magnitude of the gravitational field at this location?

12-89 A medical researcher wishes to measure the surface tension of a sample of experimental artificial lung surfactant. She determines that the specific gravity of the sample is 0.70, and that the sample's contact angle with glass is zero. She places a glass capillary tube vertically in a sample of water at 20°C, and measures a capillary rise of 15 cm. The water–glass contact angle is also zero. She then inserts the same tube in surfactant and observes a rise of 1.9 cm. What is the surface tension of the surfactant? (Reference: Table 12-2)

12-90 A blood sample is collected in a capillary tube of radius 0.80 mm. The contact angle between the blood and the tube is 11°, and the surface tension and density of the blood are 0.058 N/m and 1.06×10^3 kg/m^3 respectively. What mass (in grams) of blood rises in the tube?

12-91 The average density of an object can be conveniently determined by first weighing the object in air and then weighing it immersed in a fluid of density ρ_f. If the object's weight in air is W_a and the apparent weight in the fluid is W_f, show that the average density ρ of the object is given by $\rho = \rho_f \frac{W_a}{W_a - W_f}$.

12-92 (a) A cuttlefish (Figure 12-68) has porous bones that have a density of 6.20×10^2 kg/m^3, whereas the rest of its body has a density of 1067 kg/m^3. It lives in saltwater of density 1026 kg/m^3. In order for the cuttlefish to have the same overall density as the water, what percentage of the volume of its body consists of porous bones?

(b) The cuttlefish typically lives at a depth where the absolute pressure is 16 atm. Determine the depth.

Figure 12-68 A cuttlefish (Question 12-92).

12-93 A cylindrical buoy of mass 943 kg and diameter 1.28 m is floating in salt water of density 1.03×10^3 kg/m^3. A 78 kg man steps from a boat onto the buoy.

(a) How much will the buoy have moved down when it settles to its new equilibrium position?

(b) Later the man steps off to the buoy to the boat. Immediately after he has left the buoy, what is the resulting magnitude of the buoy's acceleration?

12-94 A hydrometer consisting of a spherical bulb and cylindrical stem floats first in water at 4°C (density 1.000×10^3 kg/m^3) and then in alcohol (Figure 12-69). When in water, 9.0 cm of the hydrometer stem is above the water surface, and when in alcohol, 2.0 cm of the stem is above the surface. The total volume of the hydrometer stem and bulb is 13.6 cm^3, and the cross-sectional area of the stem is 0.40 cm^2.

(a) What is the mass of the hydrometer?

(b) What is the density of the alcohol?

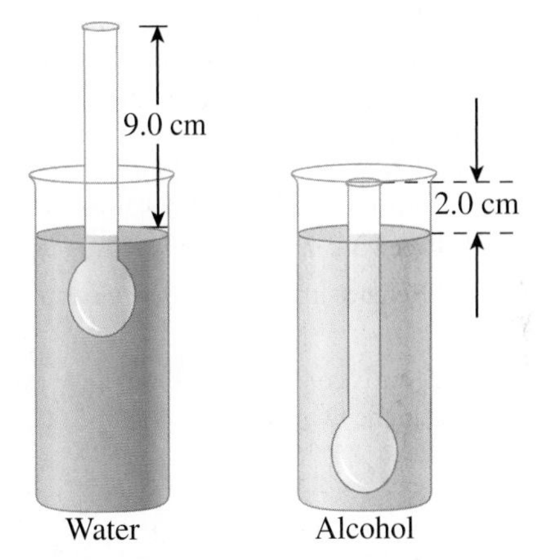

Figure 12-69 A hydrometer (Question 12-94).

12-95 A rubber ball of uniform density 942 kg/m^3 is floating in a container of water (density 1.000×10^3 kg/m^3). Oil of density 918 kg/m^3 is poured slowly into the container. The oil floats on top of the water and completely covers the ball. What percentage of the ball's volume is now under water?

12-96 A glass partially full of water is placed on a scale (S), and the mass of the glass and water is determined to be 0.151 kg. A cube of copper (s.g. 8.96), 2.0 cm on each side, is tied to a light string and suspended from a spring scale so that the copper is completely submerged in the water (Figure 12-70). What are the readings in kilograms on

(a) the spring scale?

(b) the scale S?

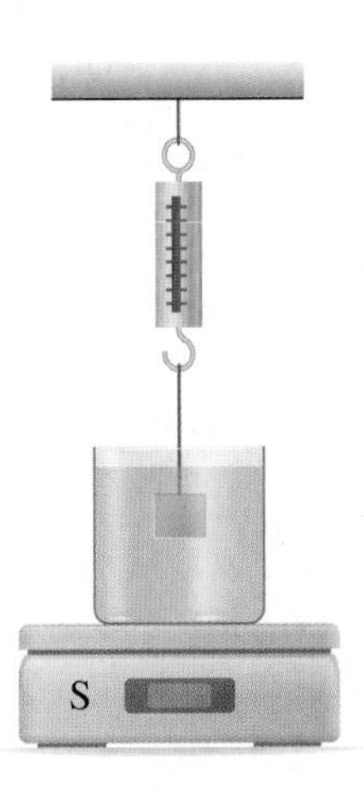

Figure 12-70 Question 12-96.

12-97 Air of density 1.3 kg/m^3 is flowing with a speed of 73 m/s over the top of the wing of a small airplane, and under the bottom of the wing with a speed of 62 m/s. The wing's surface area is 17 m^2 on the top and on the bottom. Treating air as an ideal fluid, and neglecting the small elevation difference between the top and bottom of the wing, determine the magnitude of the lift force on the wing.

12-98 In the normal region of a particular artery, the blood pressure is 100 mm Hg (correct to two significant digits), and the flow speed is 0.12 m/s. By what percentage would the pressure drop in this artery as the blood (density 1.05×10^3 kg/m^3) enters a region that has been narrowed by atherosclerotic plaque to a cross-sectional area one-fifth of normal? Assume that there is no change in elevation of the artery, and that viscosity is negligible.

12-99 Air (density 1.3 kg/m^3) is flowing through a horizontal circular pipe that narrows (Figure 12-71). The inside diameters of the wide and narrow regions of the pipe are 1.8 cm and 0.50 cm respectively. A U-tube containing mercury (s.g. 13.6) is connected to the two regions of the pipe, and the height difference, h, between the mercury levels is 1.0 cm.

(a) What is the difference in air pressure between the two regions of the pipe?

(b) Assuming that air is considered to be an ideal fluid, what are the air speeds in the two regions of the pipe?

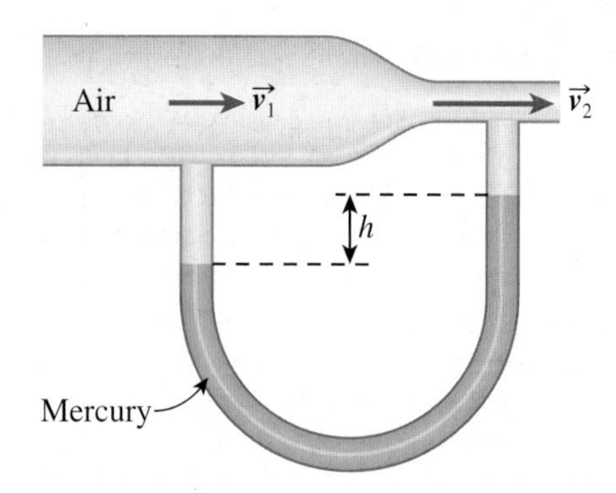

Figure 12-71 Question 12-99.

12-100 A horizontal hypodermic syringe (Figure 12-72) contains a vaccine with a s.g. of 1.12. The wide part of the syringe has an inner cross-sectional area of 1.7 cm^2, and the needle has an inside radius of 0.18 mm. A force is applied along the plunger and vaccine leaves the needle with a speed of 1.5 mm/s. Neglecting viscosity effects, determine the magnitude of the force applied to the plunger. Assume that prior to the application of the force, all pressures were at 1.0 atm, and that the pressure of the liquid in the needle remains at 1.0 atm as the force is applied.

Figure 12-72 Question 12-100.

12-101 Water (s.g. = 1.00) is flowing through a horizontal circular tube (Figure 12-73). The inside diameter of the wide portion of the tube is 24.0 cm. The water in the two vertical tubes (open to the atmosphere) is stationary at the levels shown. The volume flow rate is 1.70×10^{-2} m^3/s. What is the inside diameter in centimetres in the narrow portion of the tube? Neglect viscosity.

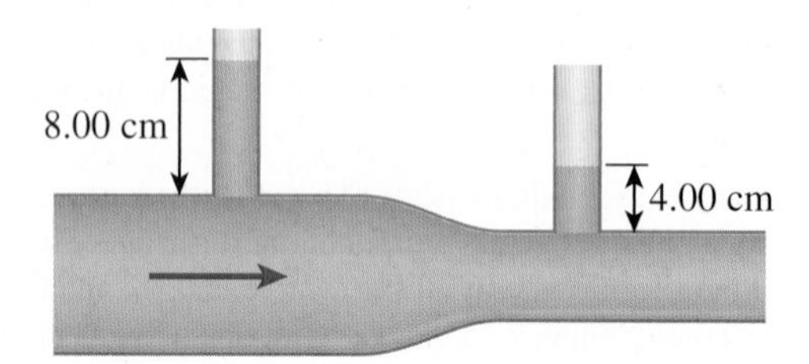

Figure 12-73 Question 12-101.

12-102 A large tank of water open to the atmosphere at the top has a spigot on its side at a height of 0.35 m above the bottom of the tank, which is sitting on the ground. The spigot is opened and water flows from it at angle of 32° above the horizontal, hitting the ground at a horizontal distance of 1.8 m from the spigot. Neglecting viscosity, and assuming that the cross-sectional area of the tank is much larger than that of the spigot, determine the height of the water in the tank above the spigot.

12-103 A patient is being given a blood transfusion intravenously (Figure 12-74). The bag containing blood is 0.97 m above the needle inserted into the patient's vein. The needle has a length of 3.0 cm and an inside radius of 0.22 mm. The patient's blood pressure in the vein is 12 mm Hg, and the blood's density and viscosity are 1.05×10^3 kg/m^3 and 4.5×10^{-3} Pa·s respectively. What is the flow rate through the needle in cubic centimetres per minute?

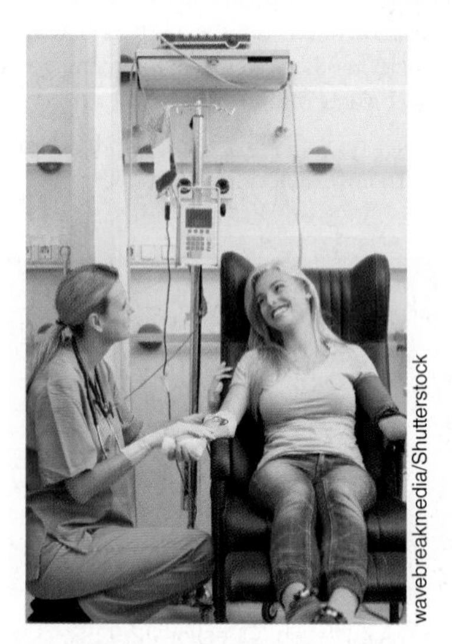

Figure 12-74 Question 12-103.

12-104 Determine the molar mass (in grams per mole) of spherical protein molecules of density 1.30×10^3 kg/m^3 that settle 2.0 cm in 360 min in a centrifuge spinning at 9.0×10^3 rpm. The centrifuge has a radius of 0.23 m, and the protein molecules are settling in water at 12°C, which has a density and viscosity of 1.00×10^3 kg/m^3 and 1.24×10^{-3} Pa·s respectively.

Oscillations and Waves 13

© StockShot/Alamy

Figure 13-1 This surfer captures some of the vast amount of energy of a wave at Teahupoo, Tahiti, in the South Pacific Ocean.

Oscillations are common: a child on a playground swing oscillates to and fro; a guitar string oscillates producing sound; electrons in an alternating current (AC) in North American oscillate sixty times each second; the wings of an insect also oscillate, causing a buzzing sound. Waves are also common; some are visible, such as ripples of water waves when we toss a stone into a calm pond, or the more complex waves on an ocean (Figure 13-1). We also can see waves on a rope or a string. But some waves, such as sound waves, microwaves, and radio waves, are invisible.

Waves are important in physics because certain types of energy, such as sound and light, travel from one place to another by some sort of wave action. Waves are caused by oscillations, so this chapter begins with a study of various types of oscillations and their properties. Then, several properties of waves we can see, such as waves on ropes, coiled springs, water, and even bridges, are presented. Many examples of

CHAPTER OUTLINE

wave properties, especially those related to sound, will already be familiar to you. This chapter leads directly to Chapters 14 and 16, which deal with the wave characteristics of sound and light respectively.

13.1 Oscillations and Simple Harmonic Motion

periodic motion (or harmonic motion) motion that is repeated at regular intervals of time

oscillation the periodic or harmonic motion of a particle or mechanical system (also called a vibration)

longitudinal oscillation a vibration in which the motion is parallel to the longitudinal axis

transverse oscillation a vibration in which the basic motion is perpendicular to the rest axis

As you are walking, if you take one step forward each second, your motion is called periodic or harmonic. **Periodic motion**, also called **harmonic motion**, is motion that is repeated at regular intervals of time. An important example is an **oscillation** or vibration, which is periodic motion of a particle or mechanical system.

We will examine the features of three common types of oscillation. First, consider the motion of a bungee-jumper or a mass on a coil spring vibrating up and down, as illustrated in Figure 13-2. This **longitudinal oscillation** is a vibration in which the motion is parallel to the longitudinal axis. The springs connected to wheels on vehicles to provide a smooth ride are another example of this type of oscillation. Notice that the final diagram in Figure 13-2 shows one way of labelling a complete cycle.

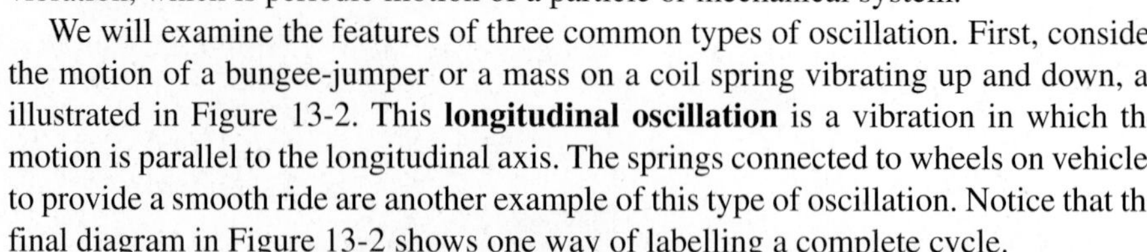

Next think of a child on a playground swing. Before any motion occurs, the swing is at its rest position, and the swing supports are along the vertical rest axis. When the child is set swinging to and fro, as shown in Figure 13-3 (a), the motion at the bottom of the oscillation is perpendicular to the rest axis. This motion, called a **transverse oscillation**,

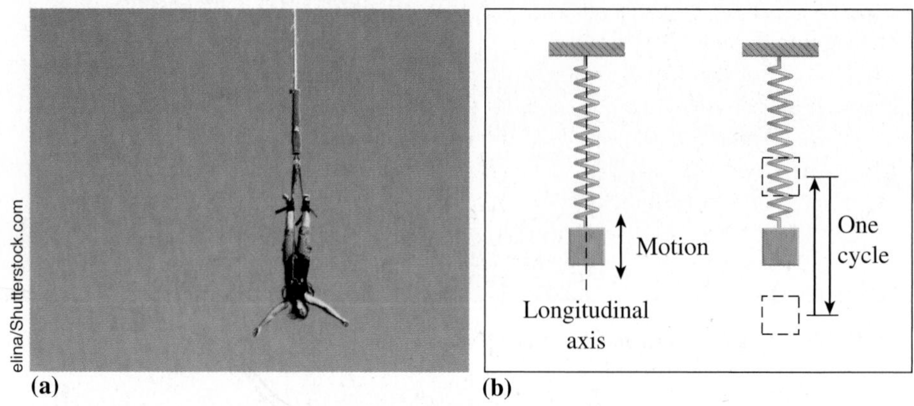

Figure 13-2 Longitudinal oscillations. **(a)** Bungee-jumping from a crane from a height of 150 m, reaching a maximum speed of 120 km/h! **(b)** Using a mass attached to a vertical spring to illustrate the motion and one cycle.

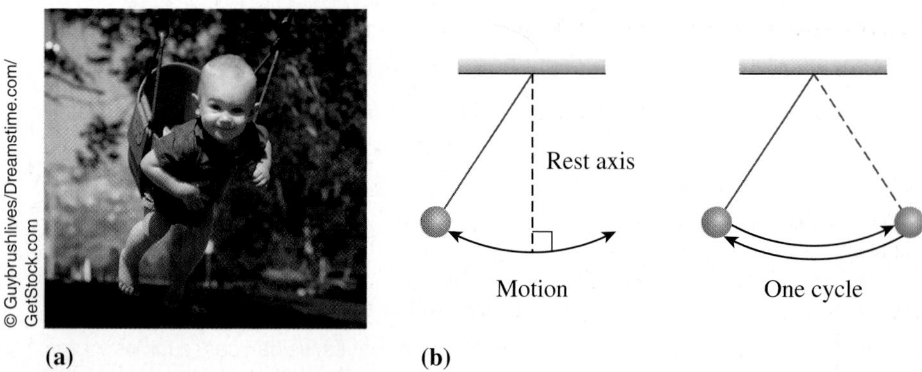

Figure 13-3 Transverse oscillations. **(a)** Toddler on a playground swing. **(b)** Using a simple pendulum to illustrate the motion and one cycle.

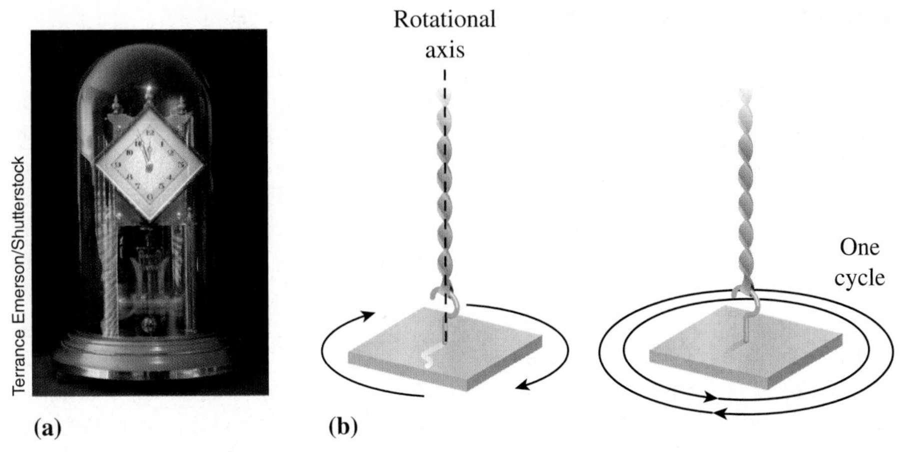

Figure 13-4 Torsional oscillations. **(a)** Ornate domed mantle clock with rotating components in blurred motion. **(b)** Illustrating the motion and one cycle.

torsional oscillation is a twisting around a rotational axis

amplitude (of oscillation) the maximum displacement of an oscillating object from its equilibrium position; symbol A

simple harmonic motion a periodic vibratory motion such that the force (and hence the acceleration) is directly proportional to the displacement; symbol SHM

is a vibration in which the basic motion is perpendicular to the rest axis. Figure 13-3 (b) illustrates a simple pendulum undergoing a transverse oscillation, as well as one example of how to label a complete cycle.

Finally, consider Figure 13-4 (a), which shows a mantle clock with components that oscillate around a rotational axis (in this case vertical). It provides one example of a **torsional oscillation**, a twisting around a rotational axis. Figure 13-4 (b) illustrates a mass rotating around and back as well as one complete cycle.

We will analyze longitudinal oscillations in this section, and study the other types of oscillation in more detail in later sections.

If we neglect friction and air resistance, the three types of oscillation described above are all harmonic. Let us consider the longitudinal oscillation of a mass on a flat surface connected to the end of a horizontal spring, as shown in Figure 13-5. The spring can be stretched or compressed along the horizontal axis, which we will call the x-axis.[1] Initially, the mass is at its equilibrium or rest position, with a displacement $x = 0$. A force is then applied, as shown in (b), to pull the mass to maximum displacement. This maximum displacement from the equilibrium position is called the **amplitude** of oscillation, A. If the mass is now released, the restoring force of the spring accelerates it to the left, as in (c). The force of the spring varies with the displacement, x, according to Hooke's law, $F_x = -kx$. Recall from Section 11.4 that the negative sign is included because the force exerted by the spring is opposite in direction to the displacement, x.

After the mass in Figure 13-5 (c) is released, it accelerates until it reaches maximum speed as it passes the original equilibrium position. Then the mass begins to compress the spring so that the displacement, x, is to the left. However, the restoring force of the spring is now to the right, so the acceleration is also to the right. Thus, again the acceleration and displacement are in opposite directions. The mass slows down and comes to a momentary stop at $x = -A$, as in (d), then begins to move toward the right past the equilibrium position at maximum speed, then on to $x = A$ again.

Since we are neglecting friction both in the spring and between the mass and the surface, this back-and-forth motion will continue. Such motion is called **simple harmonic motion**, which is a periodic vibratory motion such that the force (and hence the acceleration) is directly proportional to the displacement. The symbol used in this text for simple harmonic motion is SHM.

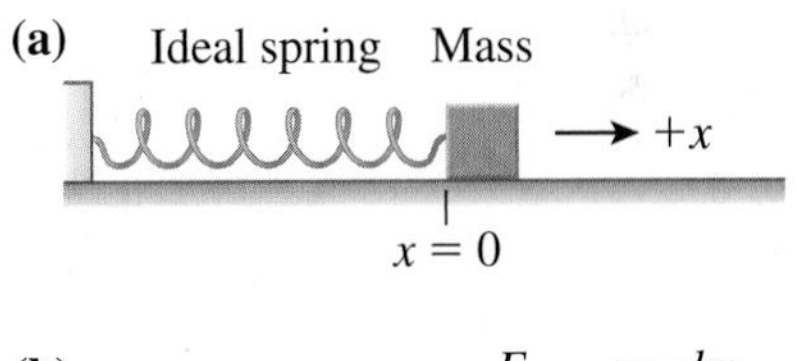

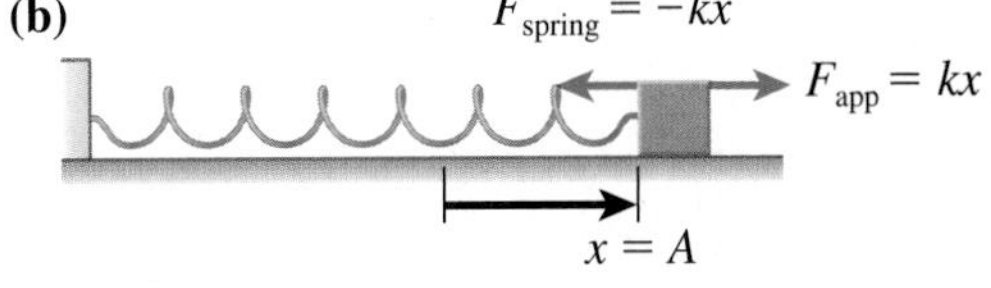

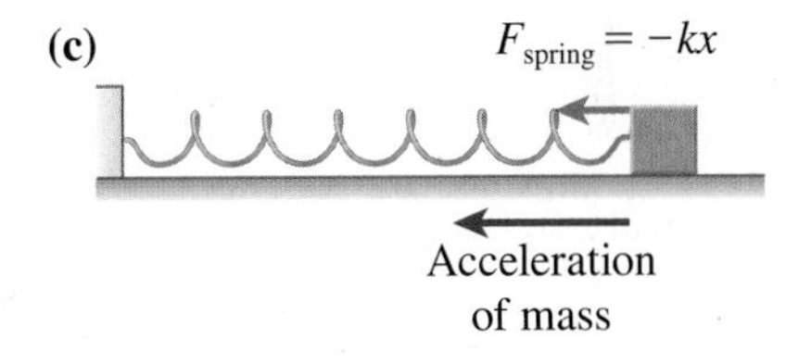

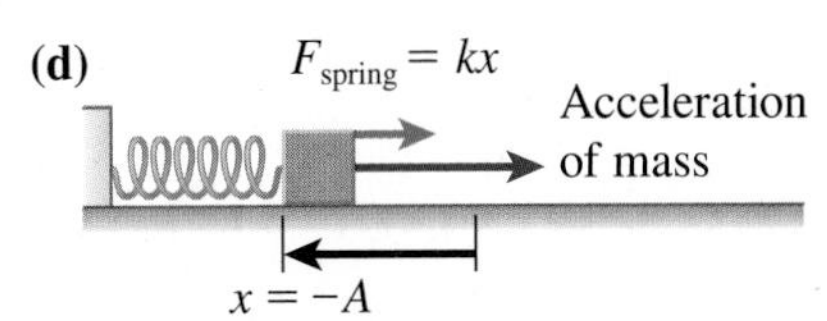

Figure 13-5 Using a longitudinal oscillation example to arrive at a definition of simple harmonic motion (SHM). **(a)** The mass attached to the spring is in the equilibrium position, $x = 0$. **(b)** An applied force on the mass stretches the spring so that $x = A$. The spring exerts a restoring force on the mass. **(c)** The applied force is removed and the restoring force causes the mass to accelerate to the left. **(d)** When the spring is compressed, the restoring force and the acceleration of the mass are to the right.

[1]Since the x-axis involves one dimension, quantities such as displacement, velocity, acceleration, and force can be treated as components of vectors, so vector symbols will not be necessary in this discussion.

Period, Frequency, and Phase

Period and frequency were introduced in the topic of circular motion (Section 4.5). The same definitions and equations apply to SHM, although in this case, the motion is generally linear. To review, *period* (T) is the amount of time for one complete cycle; it is measured in seconds (or seconds per cycle). *Frequency* (f) is the number of cycles per second; thus, it is the reciprocal of period. The official SI unit of frequency is the hertz (Hz), although it is sometimes expressed in cycles per second, or simply s^{-1}. Since period and frequency are reciprocals, the following equations can be used:

$$f = \frac{1}{T} \text{ and } T = \frac{1}{f} \qquad (13\text{-}1)$$

TACKLING MISCONCEPTIONS

Understanding Simple Harmonic Motion

Don't confuse SHM with other back and forth motions. For example, if basketball players are running back and forth across a gym during practice, their motion is not SHM even though the time taken for each trip may be constant, because at any instant their acceleration does not depend on their displacement. Also, since the acceleration of an object undergoing SHM varies with the displacement, the constant acceleration equations from Chapters 2 and 4 cannot be applied.

SAMPLE PROBLEM 13-1

Immediately after vigorous exercise, a person's heart beats 132 times in one minute. Calculate **(a)** the period and **(b)** the frequency of the heartbeat to two significant digits.

Solution

(a) By definition, period $= \frac{\text{time}}{\text{\# of cycles}} = \frac{60\text{ s}}{132\text{ cycles}} = 0.45\text{ s}$

Thus, the period is 0.45s.

(b) Using Eqn. 13-1, $f = \frac{1}{T} = \frac{1}{0.45\text{ s}} = 2.2\text{ Hz}$

Thus, the frequency is 2.2 Hz. This could have been found by applying the definition of frequency as the number of cycles per unit time.

phase the part of a cycle at which an object with SHM is found

The term **phase** refers to the part of a cycle at which an object with SHM or periodic motion is found. There are various ways of expressing phase, but for now, we will simply use fractions of a cycle. For example, if the mass in Figure 13-6 starts oscillating from the position of maximum displacement to the right, then at a phase of $\frac{1}{4}$ cycle the mass is at the equilibrium position, at $\frac{1}{2}$ cycle it is at the maximum displacement to the left, at 1 cycle it is back where it started, and so on. If two oscillating bodies have the same frequency, and start off at the same position, they will remain in phase with each other.

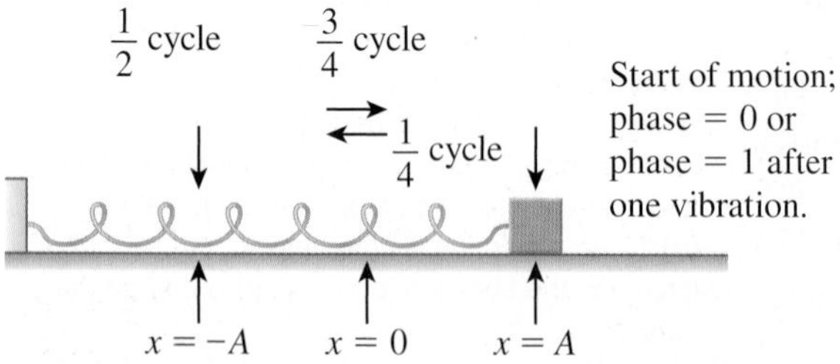

Figure 13-6 Using fractions of a cycle to indicate the phase of an oscillation.

The Reference Circle

To analyze SHM mathematically, we combine Hooke's law and Newton's second law with what is called a reference circle. Imagine that a mass attached to a horizontal spring is oscillating back and forth with SHM. At the same time, a handle pointing upward from a rotating disk is revolving with uniform circular motion. The circular motion of the handle provides the "reference circle." The frequency of revolution of the circular motion is adjusted to equal the frequency of oscillation of the SHM, and the two motions are adjusted to be in phase. Furthermore, the radius of the circle equals the amplitude of the SHM. A bright light source can be aimed from the side of the disk in such a way that it casts a shadow of the upright handle onto the mass in SHM, and this shadow appears to have the same motion as the mass (Figure 13-7). This verifies that we can use equations from uniform circular motion to derive equations for SHM.

In Chapter 4 you learned that an object in uniform circular motion with a radius r and a period T has a centripetal acceleration (a_c) with a magnitude given by Eqn. 4-16:

$$a_c = \frac{4\pi^2 r}{T^2}$$

from which we can write $T^2 = \frac{4\pi^2 r}{a_c}$ or $T = 2\pi\sqrt{\frac{r}{a_c}}$

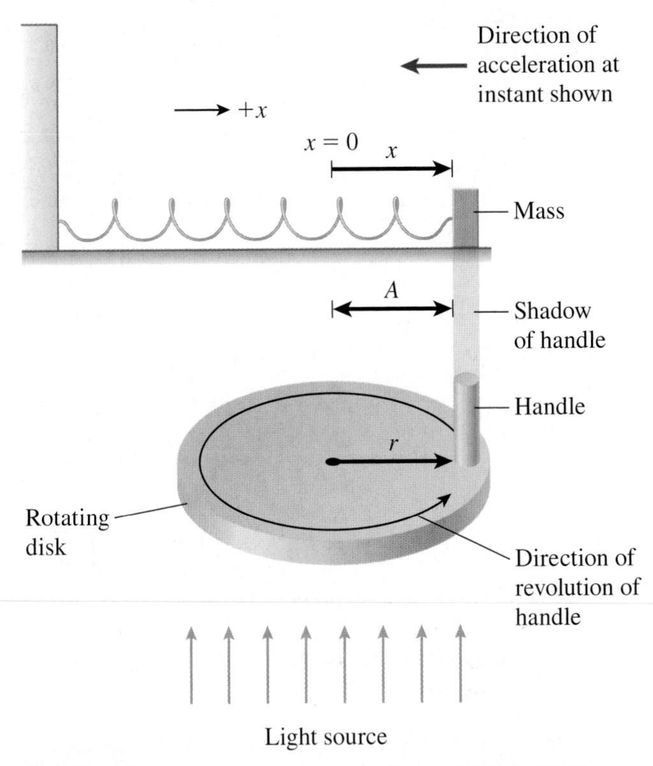

Figure 13-7 The reference circle: A handle on a disk is revolving with uniform circular motion at the same frequency as the mass on the end of a spring is undergoing SHM. The two motions are in phase. From the side, the motions appear identical, as illustrated by having a light source cast a shadow onto the oscillating mass. The motion of the handle of the reference circle is used to derive equations for SHM.

Now although the handle on the reference circle is undergoing constant acceleration or centripetal acceleration of magnitude a_c, its shadow is undergoing the same acceleration as the mass attached to the spring. The acceleration of the mass is not constant, as can be shown by Hooke's law ($F_x = -kx$) and Newton's second law ($F_x = ma_x$): equating the right-hand sides of these equations gives $-kx = ma_x$, from which $a_x = -(k/m)x$. Thus, since k and m are constants for this situation, the acceleration of a mass (or the shadow) undergoing SHM is proportional to the displacement, x, from the equilibrium position. Furthermore, the acceleration is opposite to the direction of the displacement, as indicated by the negative sign.

The above relation between the displacement and acceleration can be rewritten as $-x/a_x = m/k$; that is, the ratio of the displacement to the acceleration is constant. We now substitute this relation into the equation for period from Chapter 4 to obtain a general equation for the period of SHM:

$$T = 2\pi\sqrt{\frac{-x}{a_x}} \qquad \text{(13-2)}$$

Another equation results from substituting $-x/a_x = m/k$:

$$T = 2\pi\sqrt{\frac{m}{k}} \qquad \text{(13-3a)}$$

where m is the mass in kilograms oscillating on a spring having a spring constant k (in newtons per metre).

Finally, since frequency is the reciprocal of period,

$$f = \frac{1}{2\pi}\sqrt{\frac{k}{m}} \qquad \text{(13-3b)}$$

These equations for the SHM of a mass–spring system apply whether the motion is horizontal or vertical. Horizontal motion was used for the derivation because we then did not have to consider gravity, although the end result is the same. More details regarding SHM, including variables such as the speed and energy of the mass, are presented later in the chapter.

SAMPLE PROBLEM 13-2

A spring with a spring constant of 140 N/m has a 0.45 kg mass attached to it. The mass–spring system is placed horizontally, and the mass is displaced 15 cm and then released. Neglecting the mass of the spring and assuming the system undergoes SHM, determine the period and frequency of that motion.

Solution

Using Eqn. 13-3a,

$$T = 2\pi\sqrt{\frac{m}{k}} = 2\pi\sqrt{\frac{0.45\,\text{kg}}{140\,\text{N/m}}} = 0.36\,\text{s}$$

Now,

$$f = \frac{1}{T} = \frac{1}{0.36\,\text{s}} = 2.8\,\text{Hz}$$

Thus, the period is 0.36 s and the frequency is 2.8 Hz.

EXERCISES

13-1 For each situation shown in Figure 13-8, identify the oscillation as transverse or longitudinal.

(a) A daffodil oscillates back and forth in the wind.

(b) A child bounces up and down on a “pogo” stick.

(c) A pile driver pounds a metal post into the ground.

13-2 In each case, calculate the period of oscillation in seconds to two significant digits.

(a) A person’s eye blinks 15 times in 60 s.

(b) During part of its motion a compact disk rotates at a rate of 210 times per minute.

(c) The range of audible frequencies heard by a young, healthy human is 25 Hz to 20 kHz.

13-3 Calculate the frequency of oscillation in hertz in each case.

(a) The current world record for speed drumming (as of July 15, 2013), as measured by an electronic drumometer, was achieved with a count of 1208 strokes in 60.0 s.

(b) The needle of a sewing machine (Figure 13-9 (a)) oscillates up and down 720 times per minute.

(c) The period of oscillation of a metronome is 0.66 s (Figure 13-9 (b)).

images72/Shutterstock

(a)

© Julie Eydman/Fotolia.com

(b)

© Alarich/iStock.com

(c)

Figure 13-8 Question 13-1. **(a)** Daffodils in a garden. **(b)** Using a pogo stick. **(c)** A pile driver.

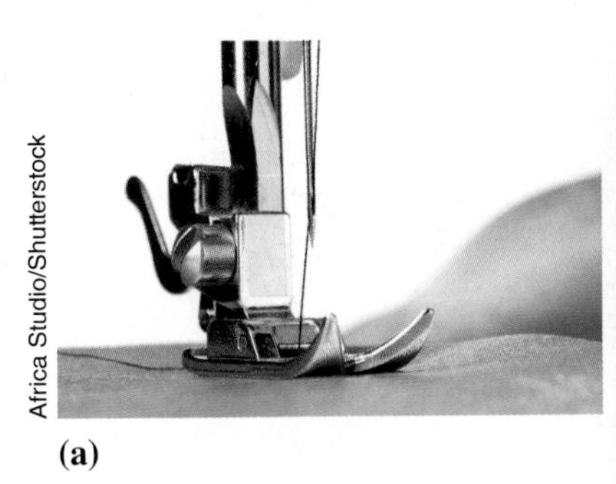

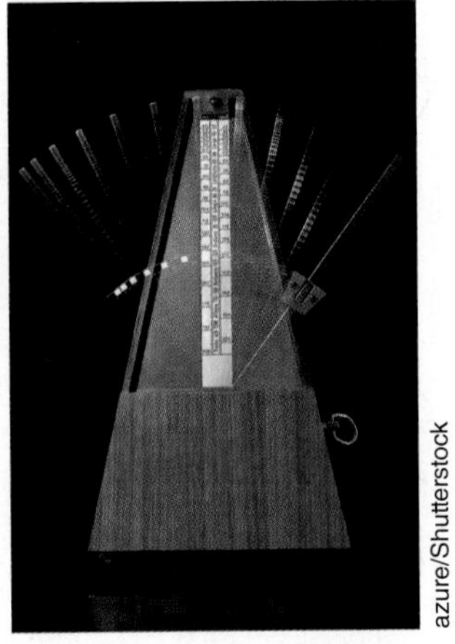

Figure 13-9 Question 13-3. **(a)** Sewing machine needle. **(b)** Metronome.

13-4 A 0.25-kg mass is attached to the end of a spring connected horizontally to a wall. The mass is displaced 8.5 cm, then released, and it undergoes SHM. The spring constant is 1.4×10^2 N/m. Assume the amplitude of oscillation remains constant.

(a) How far does the mass move in the first five cycles?

(b) Compare the phase after 2.5 cycles with the initial phase.

(c) What is the period of oscillation of the mass–spring system?

13-5 Show that $\sqrt{x/a}$ and $\sqrt{m/k}$ are dimensionally equivalent.

13-6 A vertical mass–spring system is bouncing up and down with SHM. In terms of the amplitude A, state the location or locations where

(a) the magnitude of the displacement from equilibrium position is maximum

(b) the speed is maximum

(c) the speed is minimum

(d) the magnitude of the acceleration is maximum

(e) the magnitude of the acceleration is minimum

13-7 A 0.10 kg mass is attached to a spring and set into vibratory motion. The frequency of oscillation is found to be 2.5 Hz. What is the spring constant of the spring in newtons per metre?

13-8 What mass hung from a spring of spring constant 1.4×10^2 N/m will give the mass–spring system a period of oscillation of 0.85 s?

13.2 The Simple Pendulum

A simple pendulum was given as an example of a transverse oscillation in Figure 13-3 (b). More formally, a simple pendulum consists of a bob or mass at the end of a string of negligible mass that is attached to a rigid support. Figure 13-10 illustrates several features needed to derive equations for the period and frequency of an oscillating simple pendulum. The path of the bob is an arc that has a radius r equal to the length of the pendulum, L. As defined earlier, the amplitude of oscillation (symbol A) is the maximum displacement of the bob from the equilibrium (or rest) position.

Based on Figure 13-10 as well as your experience, how do you think the period (T) of oscillation of a simple pendulum depends on

- the length (L) of the pendulum
- the amplitude (A) of the pendulum
- the mass (m) of the bob
- the magnitude of the acceleration due to gravity (g) (For example, how would the period on the Moon compare to that on Earth for pendulums of equal length and mass?)

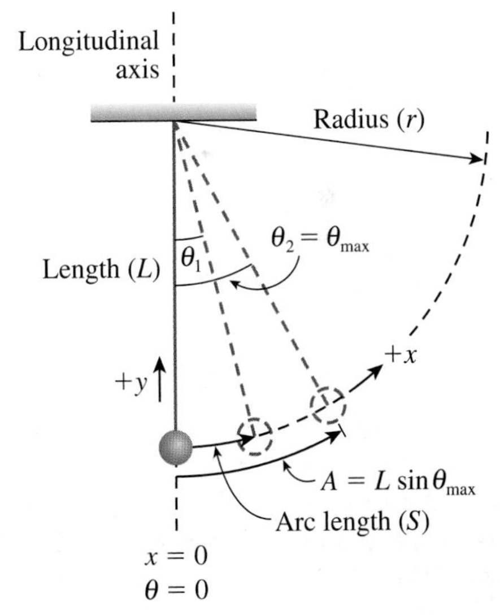

Figure 13-10 The bob of a simple pendulum moves in an arc that is part of a circle of radius r. The amplitude (A) occurs at the maximum angle, θ_{max}, the bob is displaced from the equilibrium (or rest) position along the longitudinal axis.

To explore these variables, we combine an analysis of rotational motion with an analysis of the forces acting on the pendulum bob. Figure 13-11 (a) shows the pendulum bob displaced to the right from equilibrium with the two forces acting on the bob included: gravity ($\vec{F}_G$) and tension ($\vec{T}$) in the string. The FBD in Figure 13-11 (b) shows that the restoring force causing the bob to accelerate back toward the equilibrium position is the x-component of $\vec{F}_G$. Since $+x$ is chosen to be in the direction of the arc to the right, this force is opposite in direction to the displacement. Thus

$$F_x = -mg\sin\theta$$
$$ma_x = -mg\sin\theta$$
$$a_x = -g\sin\theta$$

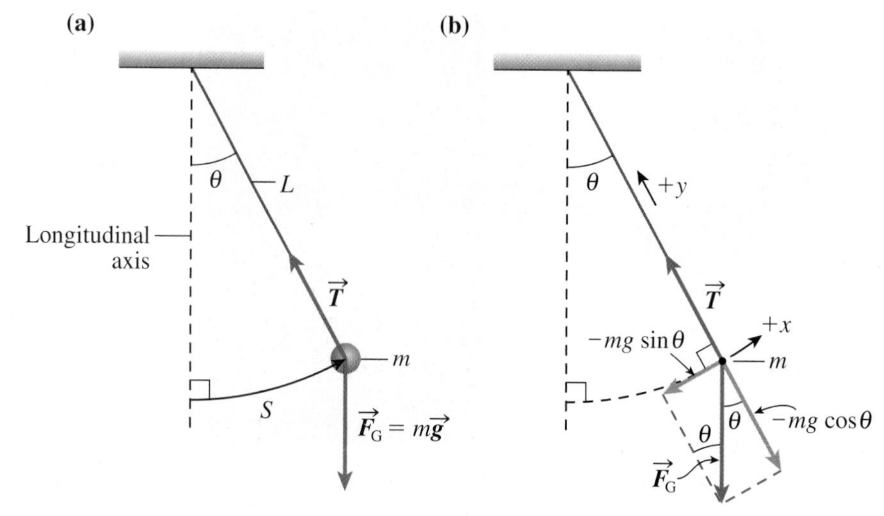

Figure 13-11 Forces acting on a simple pendulum: The forces of gravity and tension act on the pendulum bob, as shown in **(a)**. The component of the force of gravity perpendicular to the tension causes the pendulum to accelerate, as the FBD in **(b)** shows. The magnitude of this component is proportional to $\sin\theta$.

But for small angles of θ, $\sin\theta \approx \theta$ when θ is measured in radians. Thus, $a_x \approx -g\theta$. Furthermore, from Figure 13-11, $\theta = s/L$ when θ is in radians.

$$\therefore a_x = \frac{-gs}{L}$$

Now if we compare the bob's motion to the SHM described in Section 13.1, we see that for the small angles considered, the bob's motion is approximately one-dimensional, so the SHM equations can be applied. Thus, substituting the above equation into Eqn. 13-2, $T = 2\pi\sqrt{\frac{-x}{a_x}}$ and using the arc length s rather than x (which is valid for small angles of θ), we have

$$T = 2\pi\sqrt{\frac{L}{g}} \quad \text{(13-4a)}$$

$$\text{and } f = \frac{1}{2\pi}\sqrt{\frac{g}{L}} \text{ (since } f = 1/T) \quad \text{(13-4b)}$$

From these relations, we can conclude that, at a location where g = constant, $T \propto \sqrt{L}$ and $f \propto \frac{1}{\sqrt{L}}$ for a simple pendulum. Thus, at a particular location, the period or frequency of a pendulum depends only on the length, not on the mass of the pendulum bob or the amplitude of oscillation, assuming the amplitude is not large. We also conclude that we could determine the magnitude of the gravitational field, g, by measuring the period or frequency of known length.

TRY THIS!

Using a Pendulum to Determine the Local Value of *g*

Set up and carry out an experiment using a simple pendulum to determine the magnitude of the local gravitational field, g. Compare your answer to the accepted average value near Earth's surface.

SAMPLE PROBLEM 13-3

A microphone is suspended by a cable from a ceiling of an auditorium and is located 3.0 m above the floor of the stage. A former physics student notices the microphone moving back and forth in pendulum fashion, and determines that each minute it completes 8.0 cycles. How high is the auditorium ceiling above the floor?

Solution

The period of oscillation of the pendulum is the number of seconds per cycle, which in this case is $T = (60\ \text{s})/(8.0\ \text{cycles}) = 7.5\ \text{s}$.

$$\text{From} \quad T = 2\pi\sqrt{\frac{L}{g}}, \text{ we have } T^2 = \frac{4\pi^2 L}{g}$$

$$\therefore L = \frac{T^2 g}{4\pi^2} = \frac{(7.5\ \text{s})^2(9.8\ \text{m/s}^2)}{4\pi^2} = 14\ \text{m}$$

Thus, the height of the ceiling above the floor is 14 m + 3.0 m = 17 m.

Alexander Sakhatovsky/Shutterstock

Figure 13-12 A "grandfather," or pendulum, clock.

Galileo (1564–1642) was supposedly the first person to realize that a pendulum of fixed length has a regular period of oscillation.[2] He noticed this while watching a chandelier oscillate slowly back and forth in a cathedral in Pisa. Documents reveal that he understood how the principle of this type of motion could be applied to keeping time, but we don't know for certain whether he invented the pendulum clock. Certainly, such a clock was invented shortly after Galileo discovered the basic principle. The period of a pendulum clock (Figure 13-12) can be adjusted by raising or lowering the mass of the pendulum.

EXERCISES

Note: Unless otherwise stated, use $g = 9.8\ \text{m/s}^2$.

13-9 A pendulum clock is running somewhat fast. What should be done to correct the problem?

13-10 A 0.15 kg mass is suspended from a string such that the length of the resulting pendulum is 1.5 m. The mass is displaced so the arc length from the equilibrium position is 8.0 cm, and then it is released.

(a) Determine how far the mass travels in 3 complete cycles. Neglect friction effects.

(b) Calculate the period and frequency of the pendulum.

13-11 A monkey is swinging on a vine with a frequency of 0.34 Hz. How long is the vine–monkey pendulum?

13-12 An experiment is performed on Planet X to determine the acceleration due to gravity there. A pendulum of length 1.00 m is set up and its period of oscillation is measured to be 3.26 s. What is the magnitude of the acceleration due to gravity on Planet X?

13.3 The Sinusoidal Nature of Simple Harmonic Motion

In the topic of kinematics (Chapters 2 and 4), various equations were developed relating the variables displacement, position, velocity, acceleration, and time. Those equations applied to situations in which the acceleration was constant, so they cannot be used with SHM because the acceleration varies with the displacement. To find how the displacement, velocity, and acceleration depend on time, we use

[2] A profile of Galileo is featured in Chapter 2, page 45.

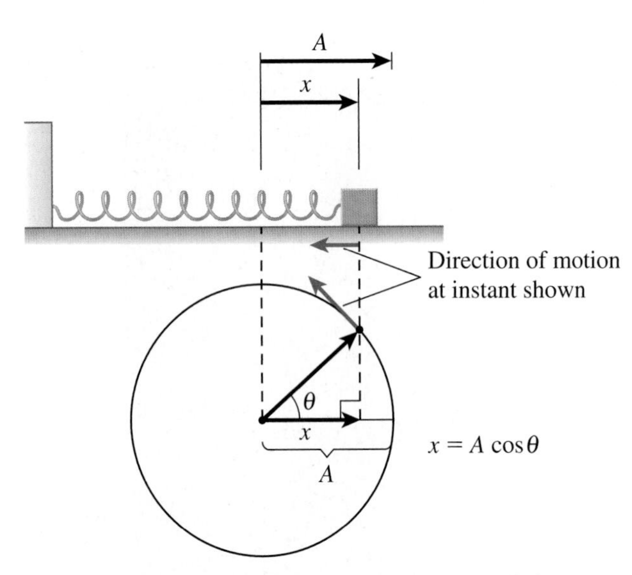

Figure 13-13 Using the reference circle to develop SHM equations involving displacement, velocity, and acceleration.

the reference circle described in Section 13.1. The reference circle is shown again in Figure 13-13 along with a mass–spring system undergoing SHM. We will use this set of diagrams to determine the displacement of the mass at any time.

In Figure 13-13, the period of rotation (T) and the amplitude (A) of the particle in circular motion are equal to the period of oscillation and amplitude, respectively, of the mass undergoing SHM. At the instant shown, the mass attached to the spring is moving to the left, and the corresponding particle on the reference circle is moving counterclockwise. The displacement (x) of the mass from its equilibrium position is given by the horizontal component of the particle's position in circular motion; thus, $x = A\cos\theta$ where θ is the angle measured counterclockwise from the $+x$-axis in the reference circle. One complete revolution of the particle takes a period of time, T, during which the angle θ increases by 2π radians. Thus, we can write the angle at any time, t, as $\theta = (2\pi t)/T$ (assuming that $\theta = 0$ at $t = 0$). Alternatively, since $f = 1/T$, we can write $\theta = 2\pi ft$. Therefore,

$$x = A\cos 2\pi ft \tag{13-5}$$

This is the equation for the displacement of a particle in SHM from its equilibrium position in terms of the amplitude A, the frequency f, and the time t after the motion began. Notice that at $t = 0$, $x = A$. In this equation, the angle ($2\pi ft$) must be measured in radians. It provides another way of indicating phase. (Fractions of a cycle were used in Section 13.1.)

Now imagine that a pen is attached to the mass as it oscillates back and forth, and that a piece of paper is pulled across the pen, as shown in Figure 13-14. The resulting line drawn by the pen forms a smooth shape, called a *sinusoidal curve.* This type of curve results any time a cosine or sine function is plotted on a graph of displacement versus time. It is a useful curve because it helps in the analysis of waves, which are caused by oscillations in the first place.

Next consider the instantaneous velocity vector, $\vec{v}$, of the particle in circular motion (Figure 13-15). The x-component of this velocity ($-v\sin\theta$) is the same as the instantaneous velocity of the mass in SHM. Now since the magnitude of the instantaneous

Math Tip

The equation developed here for the displacement of an object in SHM is a simplification of the general situation. The general equation for the displacement of an object in SHM is $x = A\cos(2\pi ft + \varphi)$. The angle ($2\pi ft + \varphi$) is called the *phase angle* or simply the *phase*, and the part φ is called the phase constant. This constant depends on where the motion of the SHM begins. In other words, the phase constant depends on the displacement of the object at $t = 0$ relative to the starting position, $x = A$ at $t = 0$, used in the derivation of the sinusoidal equations. In the equations in this text, we keep $\varphi = 0$, yielding the curve in Figure 13-14. One more fact may help reduce possible confusion. You may find in some reference material that the general equation for the displacement of a body undergoing SHM is $x = A\sin(2\pi ft + \varphi)$ rather than $x = A\cos(2\pi ft + \varphi)$. How does this happen? It is a matter of starting the motion when $t = 0$ at a position 90° out of phase with the derivation used in this text. The resulting curve is still a sinusoidal one.

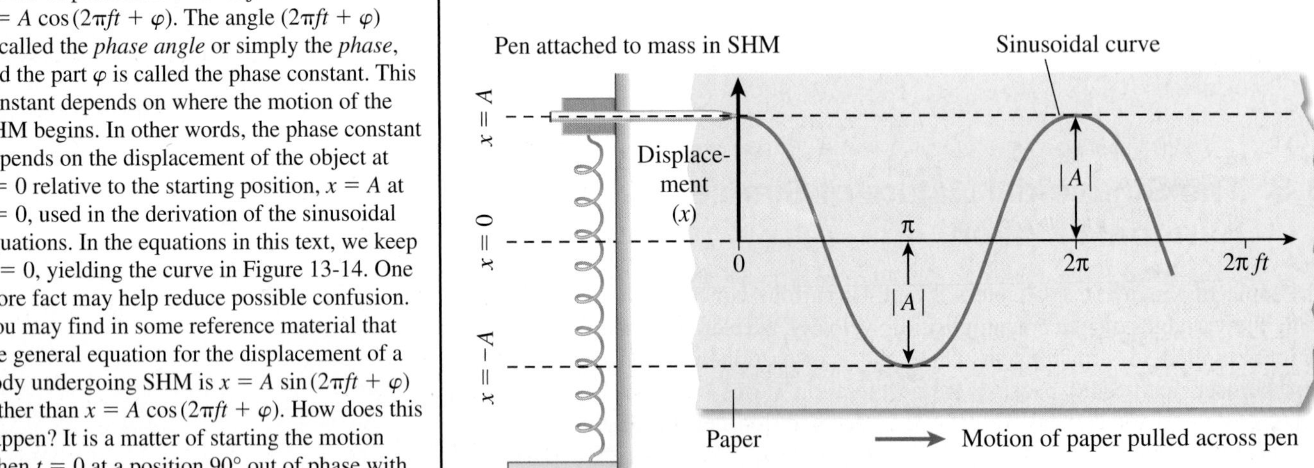

Figure 13-14 A pen attached to a mass in SHM produces a sinusoidal curve on a piece of paper pulled across the pen. This curve corresponds to the relation $x = A\cos 2\pi ft$.

velocity of the particle in circular motion is equal to its average speed ($v = 2\pi A/T$ or $v = 2\pi Af$), and since $\theta = 2\pi ft$ as in Eqn. 13-5, we can write the equation for the velocity of the mass in SHM as

$$v_x = -2\pi Af \sin 2\pi ft \quad (13\text{-}6)$$

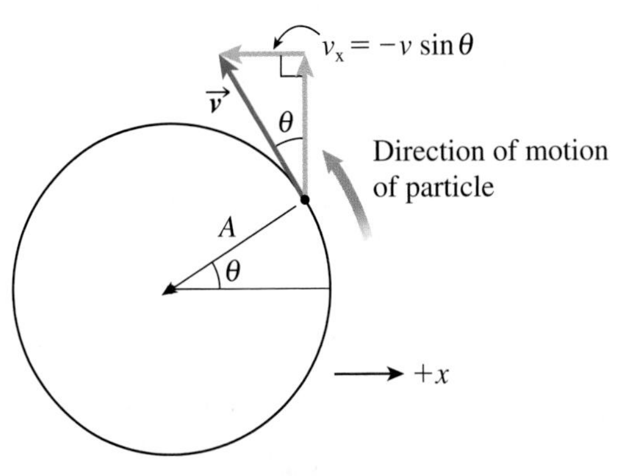

Figure 13-15 Determining the instantaneous velocity of a particle in SHM.

This is the equation for the magnitude of the instantaneous velocity of a particle in SHM in terms of the amplitude, frequency, and time after the motion began at $x = A$ at $t = 0$. Furthermore, the maximum magnitude of v_x occurs when $\sin 2\pi ft = -1$, that is, when the particle passes the $x = 0$ position.

Finally, consider the x-component of the centripetal acceleration of the particle on the reference circle, as shown in Figure 13-16. This component is given by $a_x = -a_c \cos\theta$, where $a_c = 4\pi^2 Af^2$. (This comes from $a_c = v^2/r = v^2/A = (2\pi Af)^2/A$.) Thus

$$a_x = -4\pi^2 f^2 A \cos 2\pi ft \quad (13\text{-}7a)$$

$$\text{or } a_x = -(2\pi f)^2 A \cos 2\pi ft \quad (13\text{-}7b)$$

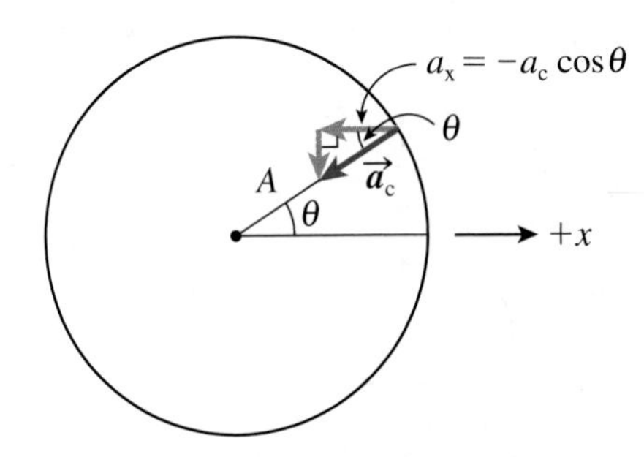

Figure 13-16 Determining the instantaneous acceleration of a particle in SHM.

These are the equations for the acceleration of a particle in SHM in terms of the amplitude, frequency, and time. Now since $x = A \cos 2\pi ft$, we can write the acceleration in terms of x and f. Thus

$$a_x = -(2\pi f)^2 x \quad (13\text{-}7c)$$

Notice in this equation that $a_x \propto x$, which is indeed the defining condition of SHM.

Here is a summary of the equations for the displacement, velocity, and acceleration of a particle undergoing SHM, assuming that at $t = 0$ the particle is at position $x = A$.

$$x = A \cos 2\pi ft \quad (13\text{-}5)$$

$$v_x = -2\pi Af \sin 2\pi ft \quad (13\text{-}6)$$

$$a_x = -4\pi^2 f^2 A \cos 2\pi ft \quad (13\text{-}7a)$$

Notice that in any of these equations, the frequency can be replaced by the reciprocal of the period because $f = 1/T$. The same equations apply to a mass on a vertical spring if the mass is at $x = A$ when $t = 0$. (See Sample Problems 13-4 and 13-5, where the role of gravity will be investigated.) If the mass were to start at any other position, the equations would be different by a phase constant φ, as described in the Math Tip text box on page 382.

Math Tip

It is important to remember in these SHM equations involving displacement, velocity, and acceleration, that we use radians, not degrees, for the angles. Thus, for example, in $\cos 2\pi ft$, if $t = T$, then $\cos 2\pi ft = \cos 2\pi = 1$, and so on.

Another consideration applies to anyone who has studied calculus. The equations for the velocity and acceleration in SHM can be found by using the first and second derivatives of the displacement with respect to time. Thus, since we have used cosine for displacement, velocity is sine and acceleration is cosine. Had we begun with sine for displacement, velocity would have been cosine and acceleration would have been sine.

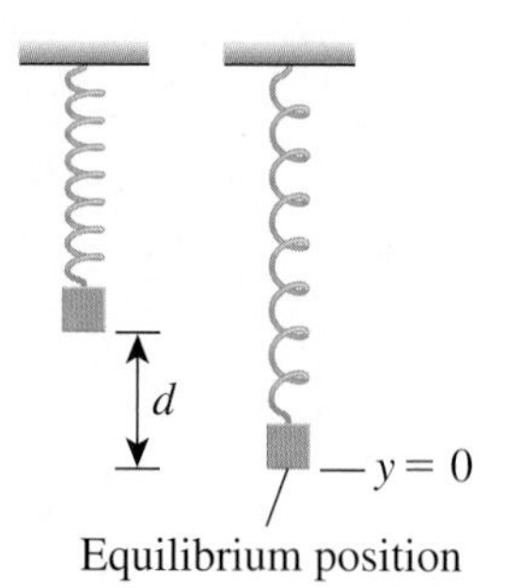

(b) FBD when system is in equilibrium.

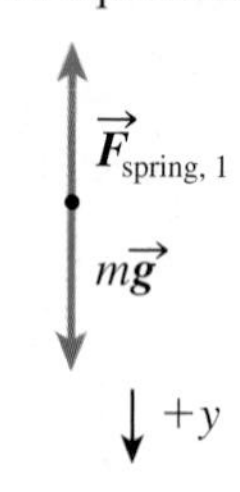

(c)

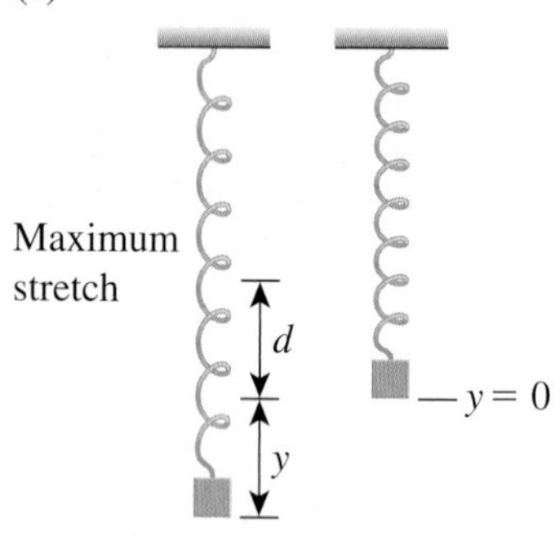

(d) FBD when mass is below equilibrium.

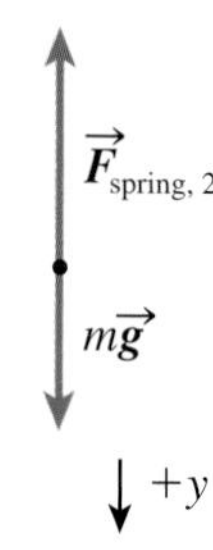

Figure 13-17 Sample Problem 13-4.

SAMPLE PROBLEM 13-4

Prove that the equation $T = 2\pi\sqrt{\frac{m}{k}}$ applies to a mass bouncing up and down on the end of a vertical spring that obeys Hooke's law. (Neglect the mass of the spring.)

Solution

When the mass is suspended on the vertical spring its equilibrium position $y = 0$ is at a location d below the unstretched position, as shown in Figure 13-17 (a). Notice in the diagram that we are using y for the displacement rather than x and $+y$ has been chosen downward. Using the FBD drawn in (b) for the mass at the equilibrium position,

$$\Sigma F_y = ma_y = 0$$

$$\therefore -F_{\text{spring},1} + mg = 0$$

where F is the magnitude of the restoring force of the spring ($F > 0$).

Thus, $F_{\text{spring},1} = mg$, or $kd = mg$.

Now if the spring is stretched so the mass is at the maximum displacement below the equilibrium position by a displacement of magnitude y, as shown in Figure 13-17 (c), then the FBD in (d) applies:

$$\Sigma F_y = ma_y \text{ where } a_y \neq 0$$

$$\therefore -F_{\text{spring},2} + mg = ma_y$$

$$-k(d + y) + mg = ma_y$$

$$-kd - ky + mg = ma_y$$

But since $kd = mg$,

$$-mg - ky + mg = ma_y$$

Hence, $-ky = ma_y$ and $\frac{-y}{a_y} = \frac{m}{k}$

This is the same relationship we proved for a horizontal spring, although we used y rather than x. Thus, the equation for the period of oscillation is the same.

SAMPLE PROBLEM 13-5

A 0.63 kg mass is suspended vertically from a spring of spring constant 24 N/m. The mass is raised 14 cm above the equilibrium position, as shown in Figure 13-18, and then allowed to drop.

(a) Determine the frequency of the resulting SHM.

(b) Calculate the velocity and acceleration 1.5 s after the mass is released.

Solution

(a) Using Eqn. 13-3b,

$$f = \frac{1}{2\pi}\sqrt{\frac{k}{m}} = \frac{1}{2\pi}\sqrt{\frac{24\ \text{N/m}}{0.63\ \text{kg}}} = 0.98\ \text{Hz}$$

Thus, the frequency is 0.98 Hz.

(b) Using upward as positive and y rather than x, and remembering to use radians to determine the sine and cosine values,

$$v_y = -2\pi A f \sin 2\pi f t$$
$$= -2\pi(0.14\,\text{m})(0.98\,\text{Hz})[\sin(2\pi \times 0.98\,\text{Hz} \times 1.5\,\text{s})]$$
$$= -0.14\,\text{m/s up or } 0.14\,\text{m/s down}$$

$$a_y = -(2\pi f)^2 A \cos 2\pi f t$$
$$= -(2\pi \times 0.98\,\text{Hz})^2(0.14\,\text{m})[\cos(2\pi \times 0.98\,\text{Hz} \times 1.5\,\text{s})]$$
$$= +0.85\,\text{m/s}^2 \text{ or } 0.85\,\text{m/s}^2 \text{ upward}$$

Thus, at 1.5 s the velocity is 0.14 m/s downward and the acceleration is 0.85 m/s² upward.

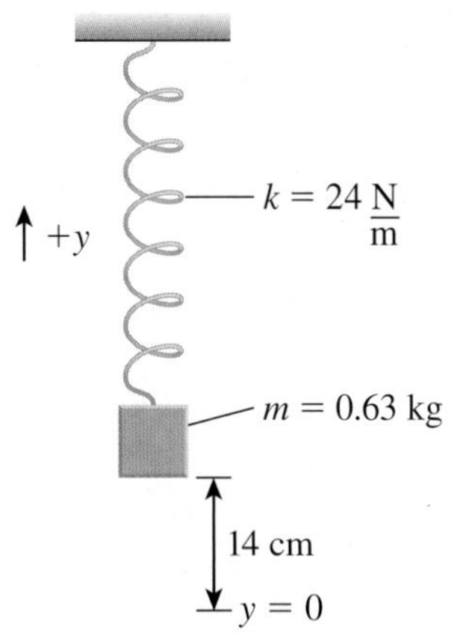

Figure 13-18 Sample Problem 13-5.

Plotting velocity–time and acceleration–time graphs of a mass in SHM yields sinusoidal curves, as seen in Figure 13-19. These graphs make sense because when the displacement is maximum in magnitude the mass has zero velocity but maximum magnitude of acceleration; when the displacement is zero the mass has maximum magnitude of velocity but zero acceleration.

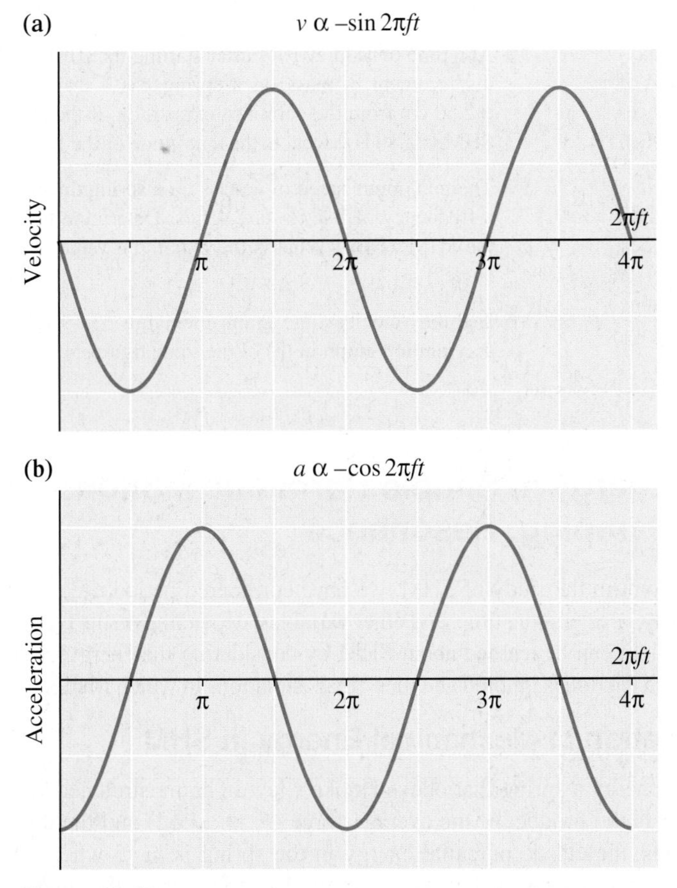

Figure 13-19 Graphs of velocity and acceleration for SHM.

SAMPLE PROBLEM 13-6

Use the displacement–time graph shown in Figure 13-14 on page 382 to verify the velocity–time graph in Figure 13-19.

Solution

From kinematics we know that the slope of the line on a displacement–time graph indicates the velocity. Thus we will look at the slope of the curved line in Figure 13-14 at three different times and compare them to the physical motion of the mass at the end of the horizontal spring. At time $t = 0$, the mass is at its maximum displacement $x = A$ and the slope of the line on the x-t graph is zero. This is true physically, as the mass is instantaneously at rest and just about to begin moving to the left. At $t = \pi/2$ radians (or one-quarter of the way through a complete cycle), the displacement is zero, and the velocity is maximum in the negative direction. Then, when $t = \pi$ radians (half way through the cycle), x is again maximum in magnitude, although it is negative, and the slope is again zero, which means that the velocity is zero. These situations correspond exactly with the velocity–time graph.

EXERCISES

Note: Remember to set your calculator to radians for SHM problems.

13-13 A ball is attached to a vertical unstretched spring. The ball is released from the spring's unstretched position at time $t = 0$ and drops a total distance of 16.0 cm (i.e., 8.0 cm below the SHM equilibrium position) before the spring reaches maximum stretch. The ball then undergoes SHM with a period of 0.36 s. Assuming upward is positive, calculate the displacement of the ball relative to the equilibrium position when $t = 0.25$ s.

13-14 A block attached to a horizontal spring is pulled to the right a distance of 18.0 cm from the equilibrium position. The block is released and the block–spring system undergoes SHM at $f = 1.25$ Hz. Assuming that positive is to the right, determine at 0.100 s after release the block's (a) displacement, (b) velocity, and (c) acceleration. Neglect friction.

13-15 For the block in the previous question, determine the displacement at times 0.000 s, 0.100 s, 0.200 s,...0.800 s. Then plot a graph of displacement as a function of time. What is the shape of the graph?

13-16 At a time of 1.00×10^2 s after starting its SHM from maximum displacement, a mass on a spring has a displacement that is +2.50 cm from the equilibrium position. If the frequency of the SHM is 2.80 Hz, what is the amplitude of the SHM?

13-17 The maximum speed of a mass on a spring undergoing SHM at a frequency of 8.4 Hz is 2.4 m/s. Determine the amplitude of the SHM. (**Hint:** What is the maximum value of the sine of an angle?)

13-18 Use the velocity–time graph in Figure 13-19 (a) to verify the acceleration graph in (b) of the same figure.

13.4 Energy in Simple Harmonic Motion; Damping; Resonance

Up to this point in the study of SHM, we have considered period, frequency, displacement, velocity, acceleration, time, and other variables, depending on the type of oscillation occurring. More can be learned about SHM by considering the energy of "ideal" situations in which friction is ignored and then "real" situations in which friction is considered.

Conservation of Mechanical Energy in SHM

The work done by a spring that obeys Hooke's law in being stretched or compressed a distance x is the product of the average force ($F = -kx/2$) and the distance (x), or $-kx^2/2$. Thus, the elastic potential energy in the spring is $kx^2/2$ when it is stretched or compressed a distance x. Let us apply this fact to a spring that undergoes SHM, as shown in Figure 13-20 (a). An "ideal" spring is stretched to the right a distance $x = A$ from the equilibrium position. The spring is called ideal because it experiences no friction, either internal or external. When released, the mass–spring system thus undergoes constant SHM. The elastic potential energy in the spring, for which we will use the symbol E_S, is maximum at the maximum stretch (i.e., when $x = A$, $E_S = kA^2/2$). According to the law of conservation of mechanical energy, when the mass is released, the total energy, E_T, of the system is shared between the elastic potential energy in the spring and the kinetic energy, E_K, of the mass. Thus

$$E_T = E_S + E_K$$

$$\text{or} \quad E_T = \tfrac{1}{2}kx^2 + \tfrac{1}{2}mv^2$$

where k is the spring constant of the spring,

x is the displacement of the mass from the equilibrium position,

m is the mass at the end of the spring,

and v is the instantaneous speed of the mass.

In this equation, m and k are constant, and the quadratic functions (x^2 and v^2) yield parabolas on a graph, so a plot of the energy of a mass–spring system has the features shown in Figure 13-20 (b). To understand the lines on this graph, begin with the mass at $x = -A$. Here, the speed is zero so the kinetic energy is zero but the compression of the spring is maximum, so the elastic potential energy is maximum and is equal to the total energy. Now the mass accelerates and when it reaches the equilibrium position, $x = 0$, it has maximum speed and thus maximum kinetic energy, but the spring has zero potential energy. Thus, the kinetic energy is the total energy. The situation at $x = A$ where the stretch is maximum is similar to that at $x = -A$. Between $x = -A$ and $x = A$, the kinetic energy and elastic potential energy are parabolic curves, and the total energy is constant. Numerical calculations involving this type of energy graph are found in Sample Problem 13-7 (c).

Before the mass is released its speed is zero and $x = A$, so the total energy of the system can be given by $E_T = \tfrac{1}{2}kA^2$. Then, at any position x we have

$$\tfrac{1}{2}kA^2 = \tfrac{1}{2}kx^2 + \tfrac{1}{2}mv^2$$

This expression can be used to find the instantaneous speed at any point x, as shown in the following derivation. Multiplying by 2 gives

$$kA^2 = kx^2 + mv^2$$

$$\therefore mv^2 = k(A^2 - x^2)$$

$$v^2 = \frac{k}{m}(A^2 - x^2)$$

$$v = \pm\sqrt{\frac{k}{m}(A^2 - x^2)} \text{ but since speed } v > 0,$$

$$v = \sqrt{\frac{k}{m}(A^2 - x^2)} \qquad \textbf{(13-8)}$$

From this equation, we can verify that the speed of the mass is zero when the displacement equals the amplitude (i.e., when $x = A$), and the speed is maximum when $x = 0$; thus

$$v_{max} = A\sqrt{\frac{k}{m}} \qquad \textbf{(13-9)}$$

(a)

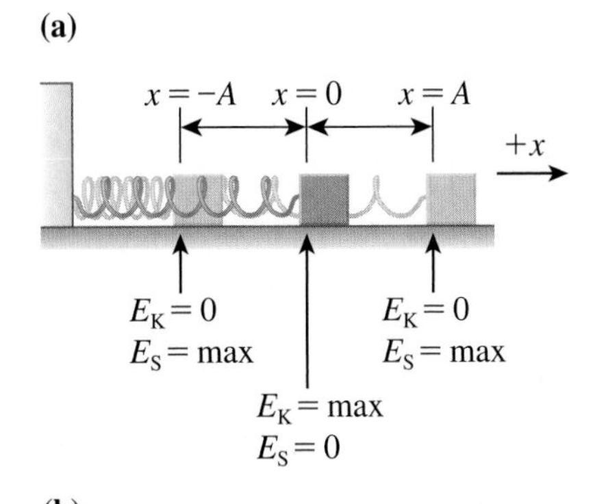

(b)

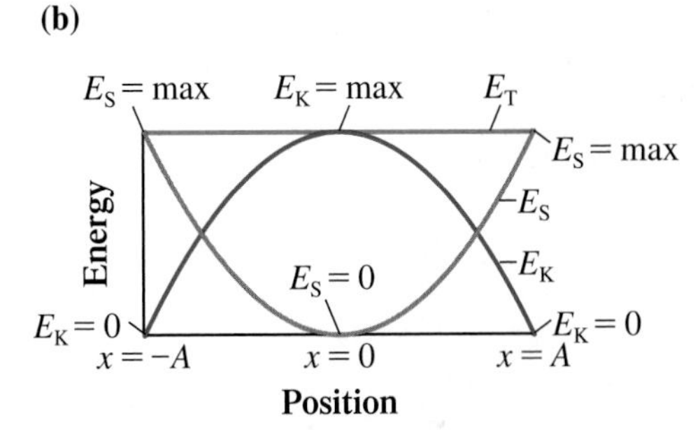

Figure 13-20 Mechanical energy in a mass–spring system. **(a)** An "ideal" spring undergoes SHM. **(b)** A typical energy graph of an ideal spring in which the total energy, E_T, is the sum of the elastic potential energy of the spring, E_S, and the kinetic energy of the mass, E_K.

SAMPLE PROBLEM 13-7

A 55 g box is attached to a horizontal spring that is then compressed 8.6 cm to the left. The spring constant is 24 N/m. The box is released and undergoes SHM.

(a) What is the maximum speed of the box?

(b) What is the speed of the box when it is 5.1 cm from the equilibrium position?

(c) Create a table of data to verify energy conservation; in other words, show that the sum of the elastic potential energy and kinetic energy of the mass–spring system is constant. Relate the data to the graph in Figure 13-20 (b).

Solution

(a) The given values in base SI units are $m = 0.055$ kg, $A = 0.086$ m, and $k = 24$ N/m. Applying the law of conservation of energy, $E_T = E_S + E_K$, we have

$$\tfrac{1}{2}kA^2 = \tfrac{1}{2}kx^2 + \tfrac{1}{2}mv^2$$

But at maximum speed, $x = 0$, so $\frac{1}{2}kA^2 = \frac{1}{2}mv^2$ or $kA^2 = mv^2$.

$$v_{max} = \sqrt{\frac{kA^2}{m}} = \sqrt{\frac{24\,\text{N/m}\,(0.086\,\text{m})^2}{0.055\,\text{kg}}} = 1.8\,\text{m/s}$$

Thus, the maximum speed of the box is 1.8 m/s.

(b) In this case $x = 0.051$ m. From $\frac{1}{2}kA^2 = \frac{1}{2}kx^2 + \frac{1}{2}mv^2$, $\frac{1}{2}mv^2 = \frac{1}{2}kA^2 - \frac{1}{2}kx^2$.

$$\therefore v^2 = \frac{kA^2 - kx^2}{m} = \frac{k(A^2 - x^2)}{m}$$

$$v = \sqrt{\frac{k(A^2 - x^2)}{m}} = \sqrt{\frac{24\,\text{N/m}((0.086\,\text{m})^2 - (0.051\,\text{m})^2)}{0.055\,\text{kg}}} = 1.4\,\text{m/s}$$

Thus, the speed is 1.4 m/s when the box is 5.1 cm from the equilibrium position.

(c) We apply the equations $E_S = \frac{1}{2}kx^2$ and $E_K = \frac{1}{2}mv^2$ where $v = \sqrt{\frac{k}{m}(A^2 - x^2)}$ for five values of x. One set of calculations is shown, and the results of all sets of calculations are shown in the table. When $x = 0.5A = 4.3$ cm or 0.043 m, the elastic potential energy is

$$E_S = \tfrac{1}{2}kx^2 = 0.5(24\,\text{N/m})(0.043\,\text{m})^2 = 2.2 \times 10^{-2}\,\text{J}$$

The speed is

$$v = \sqrt{\frac{k}{m}(A^2 - x^2)} = \sqrt{\left(\frac{24\,\text{N/m}}{0.055\,\text{kg}}\right)((0.086\,\text{m})^2 - (0.043\,\text{m})^2)} = 1.56\,\text{m/s}$$

The kinetic energy is

$$E_K = \tfrac{1}{2}mv^2 = 0.5(0.055\,\text{kg})(1.56\,\text{m/s})^2 = 6.7 \times 10^{-2}\,\text{J}$$

The total energy is $E_S + E_K = 2.2 \times 10^{-2}\,\text{J} + 6.7 \times 10^{-2}\,\text{J} = 8.9 \times 10^{-2}\,\text{J}$.

Position	$x = A$	$x = 0.5A$	$x = 0$	$x = -0.5A$	$x = -A$
x (m)	0.086	0.043	0	−0.043	−0.086
E_S (J)	8.9×10^{-2}	2.2×10^{-2}	0	2.2×10^{-2}	8.9×10^{-2}
v (m/s)	0	1.56	1.8	1.56	0
E_K (J)	0	6.7×10^{-2}	8.9×10^{-2}	6.7×10^{-2}	0
E_T (J)	8.9×10^{-2}	8.9×10^{-2}	8.9×10^{-2}	8.9×10^{-2}	8.9×10^{-2}

If plotted on an energy–position graph, the data in the table would yield the same type of graph shown in Figure 13-20 (b).

SAMPLE PROBLEM 13-8

Prove that the maximum speed of a mass on a spring in SHM is given by $2\pi fA$.

Solution

The maximum speed occurs when $x = 0$. As shown in Sample Problem 13-7, when applying the law of conservation of energy,

$$v_{max} = \sqrt{\frac{kA^2}{m}} = A\sqrt{\frac{k}{m}}$$

From Eqn. 13-3b in Section 13.1,

$$f = \frac{1}{2\pi}\sqrt{\frac{k}{m}} \text{ or } \sqrt{\frac{k}{m}} = 2\pi f$$

Combining these equations, we have $v_{max} = A\sqrt{\frac{k}{m}} = A(2\pi f)$.

Thus, the maximum speed is $2\pi fA$. This agrees with the results found in the previous section, where the velocity $v_x = -2\pi fA \sin 2\pi ft$ is maximum when $\sin 2\pi ft = -1$.

The law of conservation of mechanical energy can also be applied to other examples of SHM such as a mass on a vertical spring or a simple pendulum. In these cases, the gravitational potential energy must also be considered. However, we will restrict the Exercise questions related to this topic to mass–spring systems that are horizontal.

Damped Harmonic Motion

If you are in a car that goes over a bump in the road, you would not want the springs in the car to undergo continual SHM. You would expect the springs to settle down quickly and come to rest so you could ride in smooth comfort. This settling down is called damping. Thus, **damped harmonic motion** is periodic or repeated motion

damped harmonic motion periodic or repeated motion in which the amplitude of oscillation and the energy decrease with time

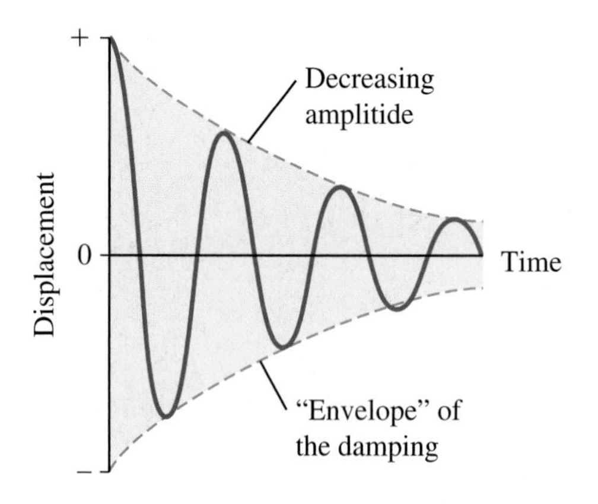

Figure 13-21 Example of a displacement–time curve representing damped harmonic motion.

in which the amplitude of oscillation and the energy decrease with time. A typical displacement–time curve representing damped harmonic motion with a fairly long damping time is shown in Figure 13-21.

Various mechanisms are designed specifically to have damping. A common example is the system of springs and shock absorbers used in automobiles (Figure 13-22). When a wheel goes over a bump in the road the spring and shock absorber are compressed quickly, but, if they are in good condition, they will soon stop bouncing up and down. The energy given to the spring and shock absorber is dissipated (i.e., is used up) through friction, which causes thermal energy and a small amount of sound energy.

The damping characteristics of an oscillating system depend on the time needed for the energy to be reduced to a negligible amount. Figure 13-23 illustrates four different damping situations. The damping in (a) is much slower than the damping in (b). The curve in (c) represents the fastest possible damping without the oscillating object moving past the equilibrium position. This oscillation is called *critically damped.* Springs in automobiles are approximately critically damped. The curve in (d) represents a very slow return to the equilibrium position after the initial displacement. This oscillation is called *overdamped.*

DID YOU KNOW?

"Damp" in the context used here means to check or deaden something. High-rise buildings must display damping characteristics in strong winds; a fireplace damper controls air motion through a chimney; and the damper pedal on a piano controls the time the sound remains audible. To illustrate this last example, try playing some notes on a piano without the damper pedal (the one on the right) suppressed. Then try pressing the pedal and again playing some notes. This time the dampers are removed from the keys and the sounds last much longer.

Resonance

Harmonic oscillations, such as pendulums and masses on springs, will oscillate at their natural frequencies once they are set into motion. For example, the natural frequency of a short pendulum is higher than that of a long pendulum.

Figure 13-22 A spring (in blue) and a shock absorber attached to a car wheel.

(a) Very slow damping.

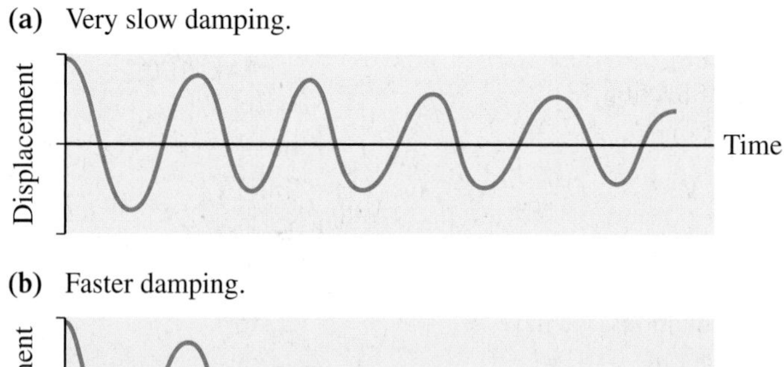

(b) Faster damping.

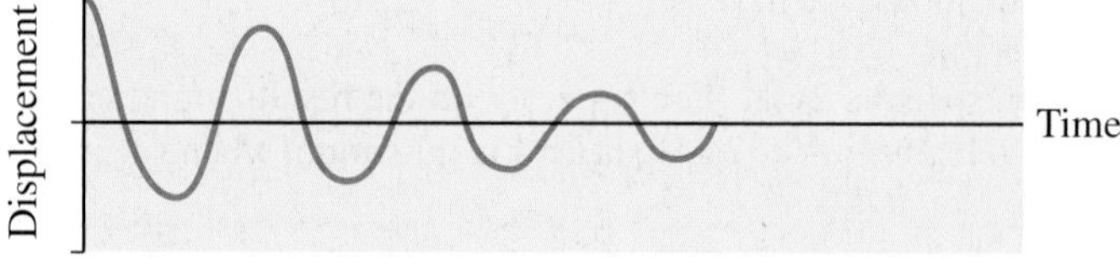

(c) Critical damping: the motion returns to equilibrium without going beyond it.

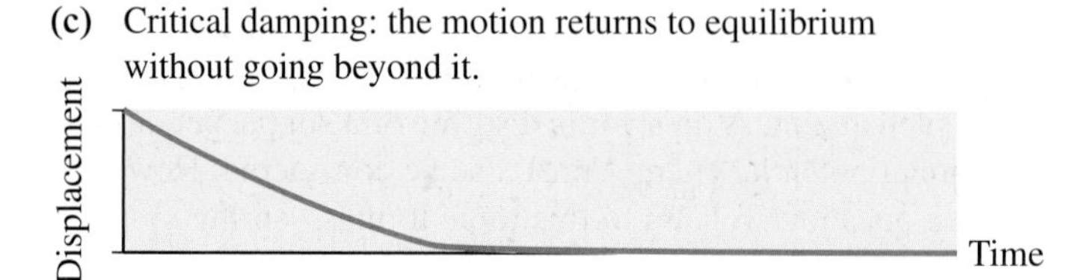

(d) Overdamping: the motion takes a long time to approach equilibrium.

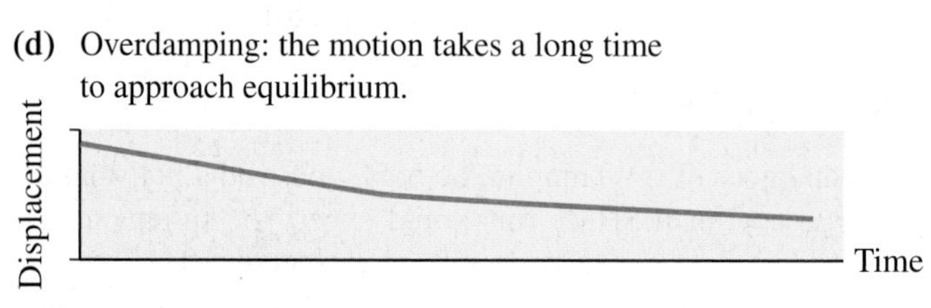

Figure 13-23 Using displacement–time graphs to illustrate four different examples of damping.

The natural frequency at which the oscillation occurs most easily is called the **resonant frequency**. When a system oscillates at its resonant frequency, large amplitudes are often observed, and these amplitudes can be sustained even with a small input force exerted at the same frequency. This phenomenon, in which an input force at the resonant frequency causes a large amplitude of oscillation, is referred to in physics as **resonance**.

resonant frequency the natural frequency at which an oscillation occurs most easily

resonance a natural oscillation caused by an input force at the resonant frequency that results in a large amplitude

forced oscillation a vibration in which the driving frequency is different from the resonant frequency, resulting in a small amplitude

Because the motion of most oscillating systems tends to dampen or decrease, energy must be given to the system to sustain the oscillation. Imagine, for example, pushing a child on a swing. To build up a large amplitude, you must exert a force on the swing seat, in other words impart energy, with the same frequency as the resonant frequency of the system. Thus, the energy transferred is maximum when the driving frequency equals the resonant frequency, which is the condition needed for resonance. If the driving frequency is different from the resonant frequency, the resulting oscillation is forced to occur at the driving frequency, and is thus called a **forced oscillation**. In this case, the oscillation energy that is developed is less than maximum and a small amplitude is achieved.

DID YOU KNOW?

When a crowd in a mosh pit at a concert jumps up and down in unison to the beat of the music at about 2 Hz, resonance can cause a large amplitude of oscillation of the floor. To reduce the chance that the oscillations could cause the floor to collapse, susceptible floors are designed to have a resonant frequency of at least 5 Hz.

 **TRY THIS!**

Resonance and Forced Oscillations

Use thin string or strong thread to set up a double pendulum system between two chairs or similar stable supports, as shown in Figure 13-24. For pendulum bobs use two equal-sized coins (quarters or larger) taped to the string. Tie the vertical strings tightly to the horizontal string so that the two pendulums have the same length, for example, 50 cm.

(a) Predict what you will observe when you pull one pendulum outward perpendicular to the horizontal string, and then release it. Try it, and watch carefully for several minutes. Use the law of conservation of energy and the concept of resonance to help explain what you observe.

(b) Shorten one of the pendulums to about half its original length, then repeat step (a). Use the concept of forced oscillations to help you explain what you observe.

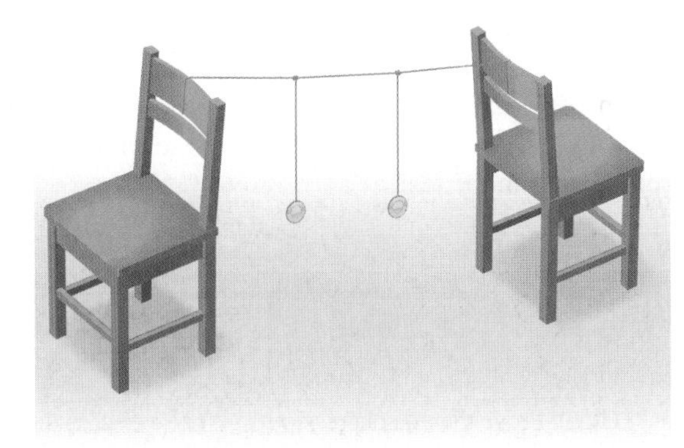

Figure 13-24 A double pendulum system.

(a)

AP Photo

(b)

Gregory Olsen/E+/Getty Images

Figure 13-25 **(a)** A "perfect storm" of conditions led to the collapse of the centre span of the Tacoma Narrows Bridge in Puget Sound, Washington, USA, on November 7, 1940. The main central span of the bridge is seen oscillating in the torsional mode as the amplitude of oscillation builds up. **(b)** Compare the rebuilt Tacoma Narrows Bridges, both of which are suspension bridges, with the original bridge. The most recent bridge is the one on the left.

A spectacular example in which a mechanical system began oscillating with a large amplitude and eventually broke apart is the collapse of a large suspension bridge in the State of Washington in 1940 (Figure 13-25 (a)). The Tacoma Narrows Bridge was suspended by large cables across a valley. Shortly after its completion, the bridge was observed to be unstable, especially in windy conditions when the kinetic energy of the air moving down the valley was transferred to the bridge. Although the explanation of what happened to cause the collapse is complex and the oscillations were not simple resonance, it is still valuable to explore the situation to learn how to prevent such disasters in the future.

TRY THIS!

Tacoma Narrows Bridge Video

Type "Tacoma Narrows Bridge Video" into your favourite search engine for some spectacular views of the oscillations and collapse.

DID YOU KNOW?

Parts of the human body have certain resonant frequencies. Research has shown that the eyes have resonant frequencies between 35 Hz and 75 Hz, the head between about 13 Hz and 30 Hz, and the entire body at about 6 Hz. Large amplitude oscillations at any of these frequencies could irritate or even damage parts of the body. In occupations such as road construction, efforts have been made to reduce the effects of mechanical oscillations on the human body.

Figure 13-26 One King Street West, Toronto, Ontario, is a very narrow, wide structure built atop a heritage building. The skyscraper would oscillate more in high winds if it did not have a set of large pools in the top interior.

One windy day, about four months after its official opening, the bridge began oscillating in the transverse mode, then one of the suspension cables came loose at a tower, and the entire 850 m centre span of the bridge began to oscillate in the torsional mode. The oscillation built up to such a large amplitude that the bridge collapsed. Nowadays, engineers test models of bridges and other structures in wind tunnels before beginning construction. Figure 13-25 (b) shows features of the new Tacoma Narrows Bridges that ensure that they are much stronger than the original bridge. For example, the stiffening girders or side supports of the road are flexible and only about 3 m high on the original bridge, but the new girders are deeper and are made with a triangular design that increases strength dramatically. What other changes can you observe?

Modern skyscrapers provide another example of the need to control mechanical resonance. In windy conditions, tall structures can oscillate, and the amplitude of oscillation can build up due to resonance, affecting the occupants of the buildings by causing nausea and motion sickness. Engineers install a *mass damper* near the top of a skyscraper to reduce the amplitude by absorbing some of the vibrational energy. One type of mass damper is a huge block of concrete or metal that is supported on a strong platform linked to large pistons and shock absorbers that can oscillate out of phase with the building's oscillation. Another type of damper uses huge tanks designed so that water sloshes back and forth in opposition to the oscillation of the building, absorbing energy and reducing the amplitude of oscillation. This method is used in the unique skyscraper at One King Street West in Toronto, Ontario, a building that has one of the narrowest thickness-to-width ratios of any building in the world (Figure 13-26). Such a design is sensitive to high winds, but the 10 large water tanks near the top of the building can be controlled to absorb the maximum amount of energy possible.

Resonance is important in more than just mechanical systems. Musical instruments produce sounds at resonant frequencies, the waves in microwave ovens have the resonant frequency of water molecules, and laser light is produced at specific resonant frequencies. Also, modern research into molecular structure is performed using a technique known as nuclear magnetic resonance. For example, hydrogen nuclei have a resonant frequency at which they will absorb electromagnetic radiation, so they can be detected in cancerous cells, brain matter, and other tissues in humans that have a large portion of hydrogen (in the form of water). Resonance is also important in the sending and receiving of radio and television signals. Details of several of these examples of resonance are presented later in the text.

EXERCISES

13-19 Figure 13-27 shows a mass–spring system at maximum compression (a) and maximum extension (b). The system is undergoing SHM.

(a) At what length(s) of the spring is the speed of the mass minimum?

(b) At what length(s) of the spring is the speed of the mass maximum?

(c) What is the amplitude of oscillation of the SHM?

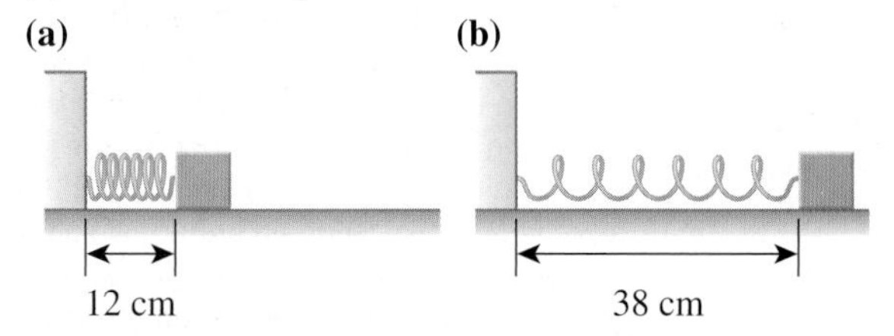

Figure 13-27 Question 13-19.

13-20 Prove that the expression $\sqrt{\frac{k}{m}(A^2 - x^2)}$ has dimensions of speed.

13-21 A toy car is attached to a horizontal spring and a force of magnitude 8.6 N exerted on the car causes the spring to stretch 9.4 cm.

(a) What is the spring constant of the spring?

(b) What is the energy of the toy–spring system in this position?

13-22 The energy of a mass–spring system undergoing SHM is 5.64 J. The mass is 0.128 kg and the spring constant is 244 N/m.

(a) What is the amplitude of the oscillation?

(b) Use two different approaches to find the maximum speed of the mass.

(c) Find the speed when the mass is 15.5 cm from the equilibrium position.

13-23 The amplitude of oscillation of a mass on a spring experiencing SHM is 0.18 m. The mass is 58 g and the spring constant is 36 N/m.

(a) Find the energy of the system and the maximum speed of the mass.

(b) What amplitude of oscillation would be required to double the energy?

(c) What is the maximum speed of the mass at this new energy?

13-24 The maximum amplitude of oscillation that a healthy human eardrum can withstand is approximately 10^{-7} m. At this amplitude the energy stored in the eardrum membrane is about 10^{-13} J. Determine the spring constant of the eardrum.

13-25 The graph in Figure 13-28 shows the energy relationships of a 120 g mass undergoing SHM on a horizontal spring. The variable x is the displacement from the equilibrium position.

(a) Which curve or line on the graph represents the total energy? the kinetic energy? the elastic potential energy?

(b) What is the amplitude of oscillation of the SHM?

(c) What is the spring constant of the spring?

(d) What is the maximum speed of the mass?

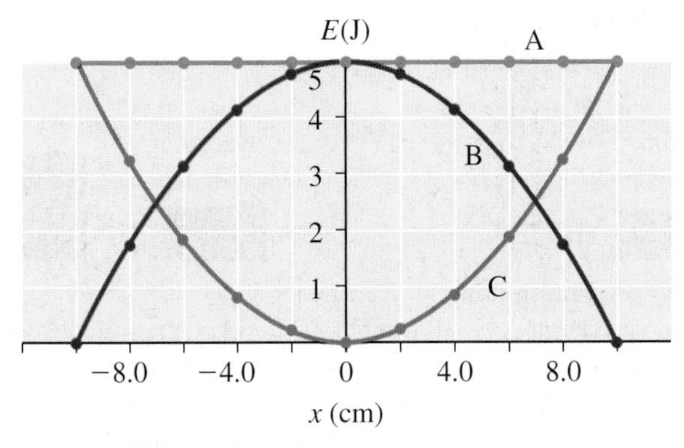

Figure 13-28 Question 13-25.

13-26 (a) Determine an equation for the speed of a mass on a spring in terms of the position of the mass and the frequency and amplitude of oscillation.

(b) A mass on a spring is pulled so the spring stretches 0.17 m. The mass is released and experiences SHM with a frequency of 1.8 Hz. What is the speed of the mass when it is 0.11 m from the equilibrium position?

13-27 State whether each device described should have fast, medium, or slow damping characteristics.

(a) a guitar string

(b) a metronome (used to practise music lessons)

(c) your eardrums

(d) saloon doors (Figure 13-29)

(e) the string on an archer's bow after the arrow leaves

Figure 13-29 Question 13-27 (d).

13-28 What happens to the resonant frequency of a simple pendulum as the length decreases? Use your answer to explain why you walk with your arms stretched out but you run with your arms bent at the elbow.

13-29 Describe how an athlete competing on a trampoline (Figure 13-30) would apply the concept of resonance to

(a) try to jump as high as possible

(b) to reduce the oscillation amplitude to a minimum before getting off the trampoline

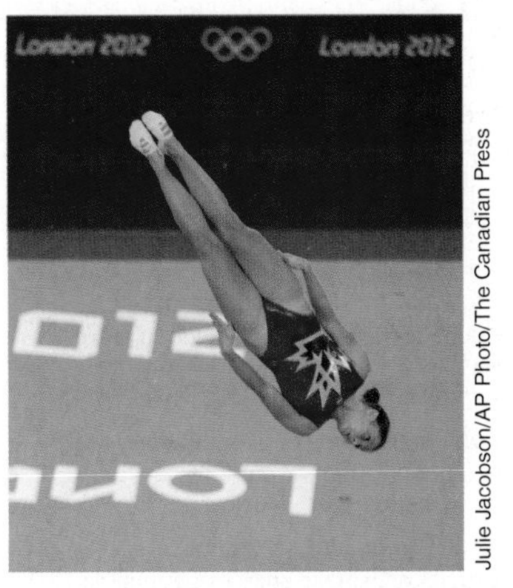

Figure 13-30 Canadian gymnast, Rosannagh MacLennan, performing at the 2012 Summer Olympics in London, England, where she won the gold medal in trampoline (Question 13-29).

13.5 Energy Transfer With Waves

You have seen various ways of transferring energy. One way is to do work on an object by applying a force to it—this is the way in which a driving force helps maintain the swinging of a child on a playground swing. Other ways involve such processes as conduction, convection, and radiation, three ways of transferring heat. The last method, radiation, is unlike the others because no particles are required for this form of energy transmission. Rather, it involves a type of wave.

wave a disturbance that transmits energy and/or information, but not matter

A **wave** is a disturbance that transmits energy and/or information, but does not transfer matter. The source of energy of a wave is an oscillation or a vibration. The energy may be transmitted by the wave to an object that changes the wave energy into some other form of energy. The operation of a radio is an example of transferring energy by waves called electromagnetic waves (Figure 13-31).

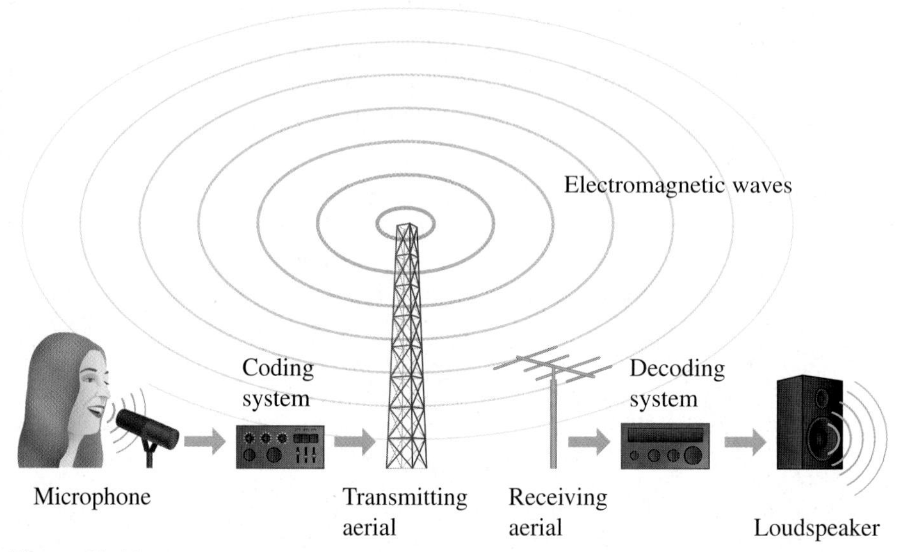

Figure 13-31 A simplified illustration of the operation of a radio: Sound energy is transformed into electrical energy, which is then transformed into electromagnetic waves. These waves carry the information to the receiving antenna where the reverse transformations occur.

Transverse and Longitudinal Waves

longitudinal wave (or compression wave) a wave in which the particles of the medium move parallel to the direction of wave propagation

compression the portion of a longitudinal wave where the particles are compressed

rarefaction the portion of a longitudinal wave where the particles are spread apart or rarefied

transverse wave (or shear wave) a wave in which the particles of the medium move perpendicular to the direction of wave propagation

crest the upper part of a transverse wave

trough the lower part of a transverse wave

Waves in a material medium are usually classified as either transverse or longitudinal, or a combination of these, depending on how the particles of the medium move relative to the direction of wave propagation. These motions correspond to the definitions of transverse and longitudinal oscillations. Thus, a **longitudinal wave** is one in which the particles of the medium move parallel to the direction of wave propagation. Longitudinal waves can propagate in all types of matter: solids, liquids, and gases. Figure 13-32 shows a longitudinal wave travelling along a stretched spring. The part of the spring where the coils are pressed together is called a **compression**, and a longitudinal wave is sometimes called a **compression wave**. Following the compression is part of the medium where the particles are spread out, or rarefied, called a **rarefaction**.

A **transverse wave**, also called a **shear wave**, is one in which the particles of the medium move perpendicular to the direction of wave propagation. Transverse or shear waves can propagate most easily in solids where a shifting plane can cause a set of molecules moving sideways to draw molecules in an adjacent plane along with it. Figure 13-33 shows that a transverse wave travelling along a rope has two main parts called a **crest**, the upper part of the wave, and a **trough**, the lower part of the wave.

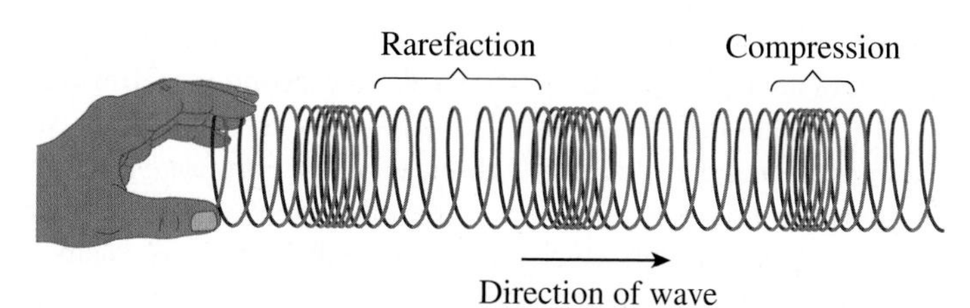

Figure 13-32 A longitudinal or compression wave, shown travelling along a stretched spring, consists of a compression and a rarefaction.

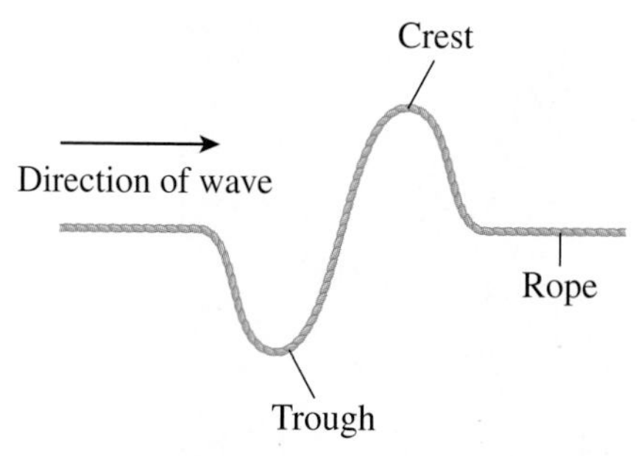

Figure 13-33 A transverse wave, shown travelling along a rope, consists of a crest and a trough.

Notice that these descriptions of longitudinal and transverse waves involve particle motion in materials. Waves that do not require a material in which to travel, namely electromagnetic waves such as light, have different properties. (Electromagnetic waves will be discussed in greater detail in Chapter 16.)

Oceanic and Seismic Waves

You have often observed waves on oceans or lakes, which are known as surface waves. A spectacular example of an ocean surface wave is shown in Figure 13-1. Waves can also occur beneath the surface of the water as well as through the solid and molten parts of Earth's interior.

The original source of energy of surface waves on oceans and lakes is the Sun. Solar energy heats air particles in the atmosphere, causing convection currents (wind) to be set up. The wind gives energy to the surface of the water, causing waves. The water waves transmit energy, but they do not transmit water molecules, at least not very far. For example, consider the situation in which a bottle is floating a few hundred metres from the shore of a lake where waves are approaching the shore. The bottle and the water molecules near it move in a complex pattern, but their basic motion is up and down as the waves pass beneath them; they do not move to the shore the way the waves do. Figure 13-34 (a) illustrates the action of the bottle, and (b) shows a simplified, almost circular path of the water molecules near the surface.

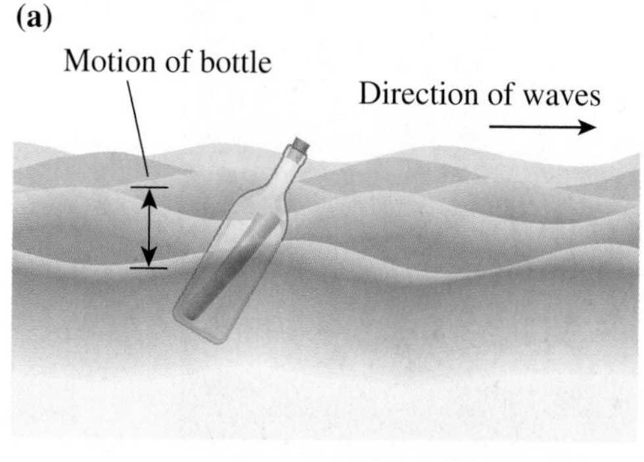

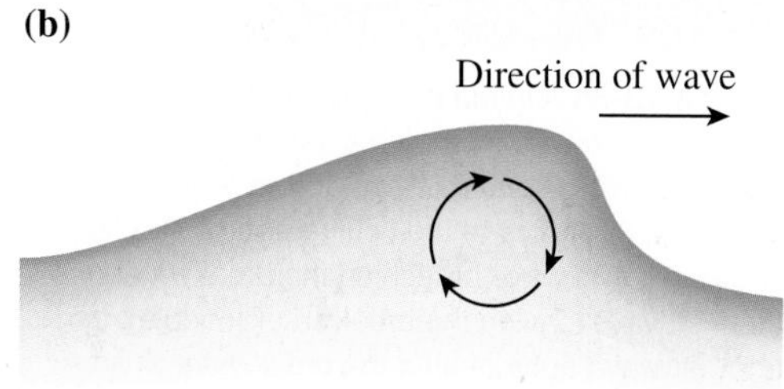

Figure 13-34 Surface waves on water. In **(a)**, a bottle floating on a wavy lake bobs up and down but does not advance appreciably with the waves. In **(b)**, the complex path of an individual water molecule of a surface wave is nearly circular.

An underwater earthquake or volcanic eruption can be the source of energy of a fast moving surface wave called a *tsunami*, which is set up on an ocean. In the deep part of the ocean, the tsunami may have a length of over 250 km, a height of about 5 m, and a speed of up to 800 km/h. Such a wave may pass unnoticed beneath a ship, despite the huge volume of water involved because the wave would take almost 20 min to pass beneath the ship. However, when a tsunami approaches a shoreline where the normal water depth is low and the wave meets with resistance, the height can build up to perhaps 30 m, causing great damage along the shore. Figure 13-35 on page 396 shows a sequence of photos taken after a severe earthquake in Japan in 2011.[3]

TACKLING MISCONCEPTIONS

Comparing Tsunamis and Tidal Waves

Some people mistakenly call a tsunami a "tidal wave." As its name suggests, a tidal wave relates to tides that are caused by the gravitational forces of the Moon and Sun on water on Earth. Tides do not cause tsunamis.

[3]Refer to the photograph and Did You Know? feature in Chapter 24 (page 784) describing the damage to the Fukushima Daiichi Nuclear Power Plant in Japan in this same earthquake.

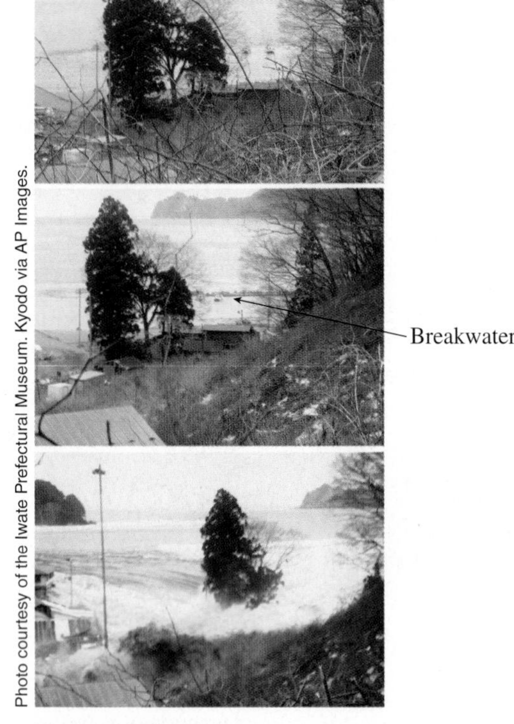

Figure 13-35 On March 11, 2011, a huge earthquake in Japan created a set of tsunami waves, causing great damage and loss of life. This sequence of photos at the fishing port of Miyoko shows the first wave breaching the breakwater (top), the first wave leaving the breakwater (middle), and another wave approaching the breakwater at a speed of 115 km/h (bottom).

Besides producing surface waves (tsunamis), the energy of earthquakes can produce **body waves** or **seismic waves** that move through the depths of Earth. The two types of body waves are called **primary (P) waves** and **secondary (S) waves**. P waves are longitudinal (compression waves), so they travel through both solids and liquids. S waves are transverse and they travel only through solids. Furthermore, P waves travel at a speed of about 4 to 8 km/s, while S waves travel more slowly, at about 2 to 5 km/s, depending on the material through which the waves are travelling. (The name "primary" is used because such waves arrive at a destination before the secondary waves.) Following an earthquake, oscillations are observed at seismological stations around the world, revealing the location of the earthquake as well as information about Earth's core. Figure 13-36 shows that only P waves are observed where Earth's liquid outer core is located between the earthquake and the seismological station. Studying the waves has helped scientists to better understand Earth's interior and its ongoing evolution.

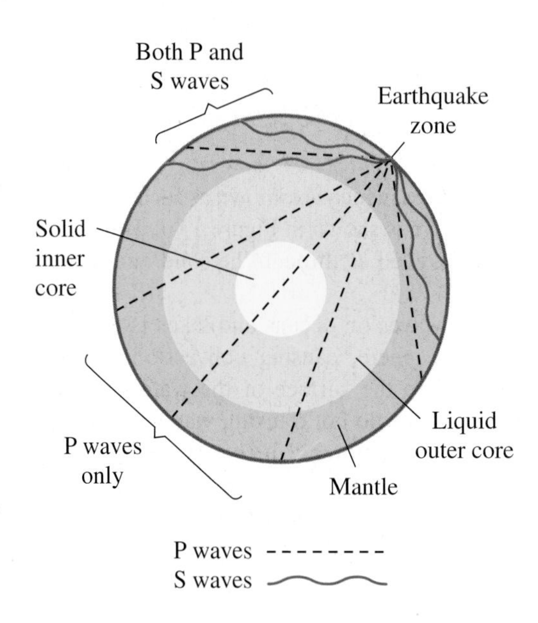

Figure 13-36 Earthquakes produce body waves (also called seismic waves) that travel through Earth as P and S waves. The waves in the diagram are approximations because local situations may cause them to undergo bending (not shown).

EXERCISES

13-30 Summarize the differences between longitudinal and transverse waves in a material medium by comparing
(a) the motion of the individual particles
(b) the type(s) of material in which they can travel
(c) the two main parts of each type of wave

13-31 Why should a tsunami *not* be called a tidal wave?

13-32 How long in minutes would it take a P wave to travel through the entire Earth, assuming the speed is constant at an average value of 8.0 km/s? (Earth's radius is about 6.4×10^3 km.)

13.6 Periodic Waves

A **periodic wave** is a travelling wave produced by a source oscillating at some constant frequency. To study periodic waves and learn the related terminology, consider the transverse wave on the rope shown in Figure 13-37 (a), which shows that the **axis of equilibrium** is the line along which the medium (in this case the rope) lies when not disturbed by the wave. The amplitude of a periodic wave, A, is the maximum displacement from the axis of equilibrium. The transverse wave consists of a series of crests and troughs above and below the equilibrium position. The hand at the left is the source of energy of the wave. In this case, the hand starts at a position along

body waves (or **seismic waves)** waves that move through Earth's interior; consist of primary and secondary waves

primary (P) waves longitudinal or compression waves that travel through both solid and liquid parts of Earth's interior at a speed of about 4–8 km/s

secondary (S) waves transverse waves that travel only through the solid part of Earth's interior at a speed of about 2–5 km/s

the equilibrium axis and then moves down, all the way up, and then back down to the starting position. The hand continues oscillating up and down with (we assume) a constant amplitude, A, and a constant frequency, f. Of course, if the frequency is constant, the period, T, is also constant because $T = 1/f$. An observer at any position along the wave will see the wave passing by with the same frequency, f. In an "ideal" situation in which no energy is absorbed by the medium, the amplitude also remains constant along the length of the rope.

periodic wave a travelling wave produced by a source oscillating at some constant frequency

axis of equilibrium the line along which a medium lies when not disturbed by a wave

Remember that the particles of the medium do not move along with the wave. In the case of the transverse wave shown, the motion of the individual particles of the medium is perpendicular to the propagation of the wave. To help visualize this, we draw a second wave that would occur a short time after the first wave, as seen in Figure 13-37 (b). The velocities of the individual particles of the medium are illustrated with the vertical arrows. Thus, the energy, in the form of a transverse wave, is moving to the right but the particles of the medium are moving up and down.

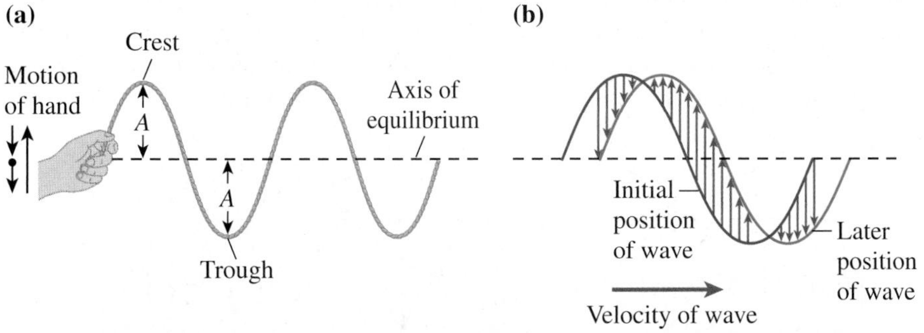

(a)
Crest
Motion of hand
A
A
Axis of equilibrium
Trough

(b)
Initial position of wave
Later position of wave
Velocity of wave

Figure 13-37 **(a)** Transverse periodic wave. The hand acts as a source of energy of the wave. It oscillates perpendicular to the axis of equilibrium at a frequency f and period $T = 1/f$, and with an amplitude A. **(b)** Illustrating the motion of the individual particles of a medium, in this case a rope, by drawing a second wave a short time after the first wave to observe where the particles move. The short vertical arrows represent particle velocities, whereas the long horizontal arrow represents the wave velocity.

Two positions along a wave are said to be *in phase* if they are at the same part of the wave. In Figure 13-38, the points B, E, and G are in phase with each other. They all occur on the crest of a wave in a part of the medium that is rising. B and D are not in phase with each other, even though they have the same displacement from the equilibrium position, because D is falling when B is rising. Can you see two other points that are in phase with each other?

wavelength (of the wave) the shortest distance between any two in-phase points on a wave; symbol λ

The shortest distance between any two in-phase points on a wave is called the **wavelength** of the wave. The wavelength is symbolized by lambda (λ), which is the lower case L in the Greek alphabet. Examples of wavelength are shown in Figure 13-38.

Similar terminology applies to a longitudinal wave. A mechanical longitudinal wave can be set up on a stretched coil or spring, such as a "Slinky" toy (Figure 13-39). Again, the hand at the left is the source of energy of the wave. Starting from an equilibrium position, the hand moves forward then all the way back, and then forward again to the starting position. It moves back and forth with motion at a constant amplitude, A, and frequency, f. An observer at any position along the spring will observe a wave passing by with the same frequency and amplitude as the source.

With a longitudinal wave, the particles of the medium (in this case, the spring) move back and forth parallel to the direction of wave

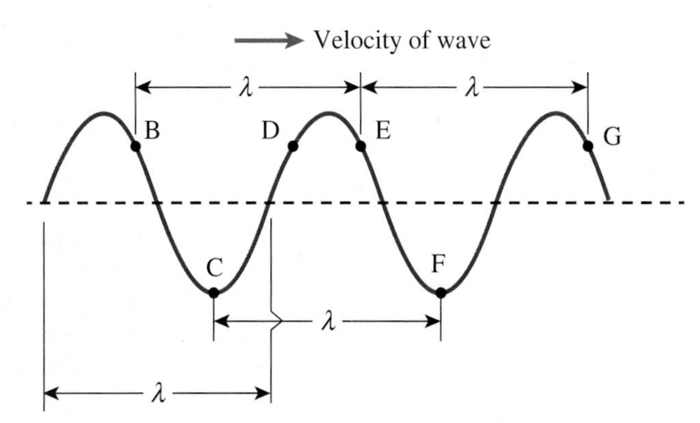

Figure 13-38 Illustrating in-phase positions and wavelength (λ) for a transverse periodic wave.

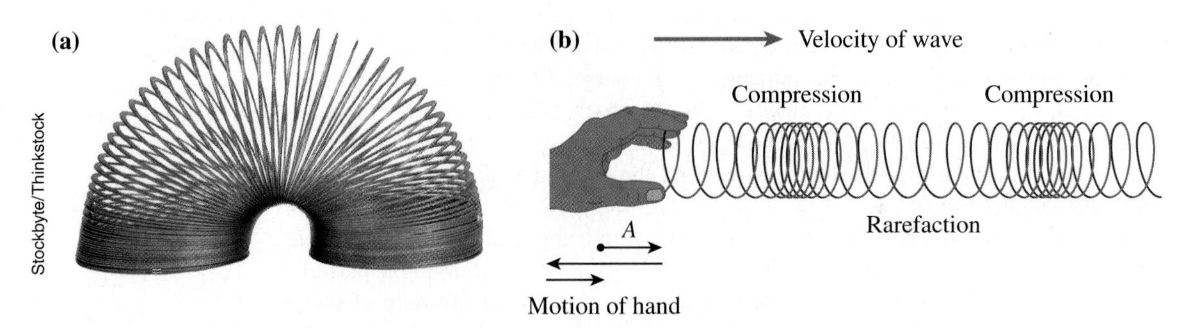

Figure 13-39 **(a)** A Slinky toy. **(b)** Longitudinal periodic wave: The hand is the source of energy. It oscillates parallel to the direction of motion of the wave at a frequency f and with amplitude A.

propagation. Again, the wavelength, λ, is the shortest distance between two in-phase points on the wave. In Figure 13-40 (a), points B, D, and F are all in phase, and C and E are in phase with each other. One wavelength is shown as the distance from B to D, from D to F, and from C to E. Figures 13-40 (b) and (c) show alternate ways of representing the longitudinal wave in (a).

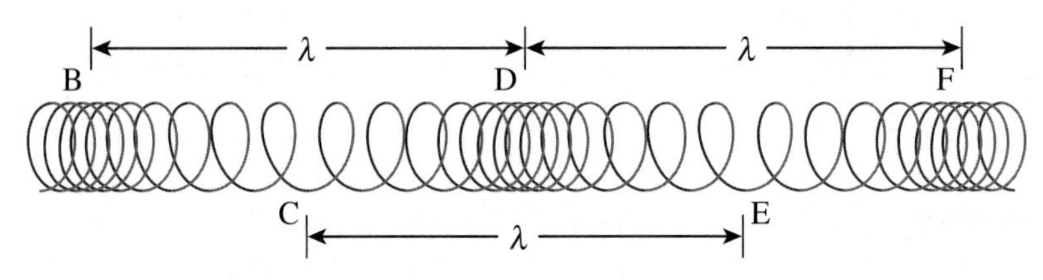

(a) Wavelengths and in-phase points.

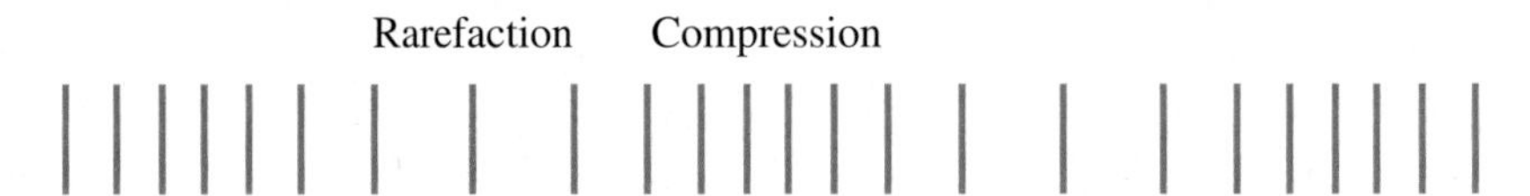

(b) A simplified diagram of longitudinal waves.

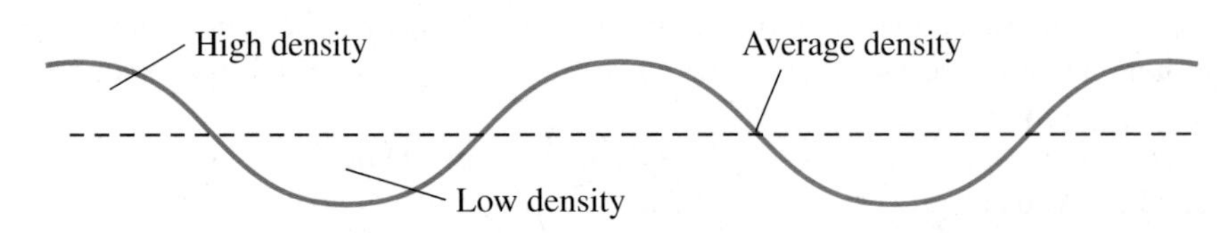

(c) Crests represent compressions where the particles of the medium have high density.

Figure 13-40 Characteristics of longitudinal periodic waves.

The Universal Wave Equation

As a wave performs its function of transferring energy from one place to another, it does so at a certain average speed. One way to find that average speed is to determine how long the wave takes to travel a known distance, then apply the equation speed = distance/time. With periodic waves travelling at a constant speed, each wavelength, λ, of the wave passes by a specific position in a length of time equal to the period, T, of the wave. Thus,

$$\text{since speed} = \frac{\text{distance}}{\text{time}}, \quad \text{speed} = \frac{\text{wavelength}}{\text{period}} \quad \text{or } v = \frac{\lambda}{T}$$

Now, since $f = \frac{1}{T}$, the equation $v = \frac{\lambda}{T}$ can be written

$$v = f\lambda \qquad \text{(13-10)}$$

This equation, called the **universal wave equation**, indicates the speed of a periodic wave in terms of the frequency and wavelength of the wave.

universal wave equation $v = f\lambda$ where v is the speed of a periodic wave and f and λ are the frequency and wavelength, respectively, of the periodic wave

SAMPLE PROBLEM 13-9

A machine in a wave pool (Figure 13-41) oscillates with a frequency of 1.4 Hz, producing periodic waves that are 6.5 m from crest to crest. What is the speed of the waves in the pool?

Solution

Since the source of the oscillation has a frequency of 1.4 Hz, the resulting waves will also oscillate at this frequency. We are given the wavelength, therefore this is a straightforward substitution into Eqn. 13-10:

$$\begin{aligned} v &= f\lambda \\ &= 1.4\,\text{Hz} \times 6.5\,\text{m} \\ &= 1.4\frac{\text{cycles}}{\text{s}} \times 6.5\frac{\text{m}}{\text{cycle}} \quad \text{(See the note below.)} \\ &= 9.1\,\text{m/s} \end{aligned}$$

Thus, the speed of the waves in the pool is 9.1 m/s. Note that the unit of frequency, hertz or Hz, was changed to the equivalent cycles/s and the unit of wavelength, metre or m, was changed to m/cycle to show how the cycles cancel, leaving metres per second, or m/s, the correct unit of speed.

TACKLING MISCONCEPTIONS

Wave and Particle Speeds

Eqn. 13-10 tells us the speed at which the wave propagates through a given medium, *not* the speed of the particles in the medium as they oscillate.

© Danita Delimont/Alamy

Figure 13-41 Indoor wave pool at the West Edmonton Mall, Alberta.

The speed of a periodic wave in a material medium depends on the medium through which the wave is travelling and the condition of that medium. For example, the speed of a transverse wave along a rope is likely different from the speed along a stretched Slinky coil under the same tension. Also, the speed of a wave along a Slinky coil with high tension is greater than the speed along the same coil with low tension.

TACKLING MISCONCEPTIONS

Rewriting the Universal Wave Equation

The universal wave equation is most commonly written $v = f\lambda$, which suggests that the speed of the wave depends on the frequency and the wavelength. However, in reality, the speed of a wave depends on the medium, and the frequency depends on the source of oscillation. So the true dependent variable is the wavelength, which means that the equation should be written $\lambda = \frac{v}{f}$. (You can relate this to Newton's second law of motion equation, commonly written $\vec{F}_{net} = m\vec{a}$ but more properly written $\vec{a} = \frac{\vec{F}_{net}}{m}$.)

Reflection and Transmission of Waves

You have studied the characteristics of waves travelling in one medium at a time. What happens when a travelling wave encounters the end of that medium? Let us look at how the answer depends on what, if anything, the wave encounters at the end of that medium. Consider a transverse pulse (a short portion of a wave) moving along a rope attached to a rigid barrier, such as a wall (Figure 13-42 (a)). When the pulse reaches

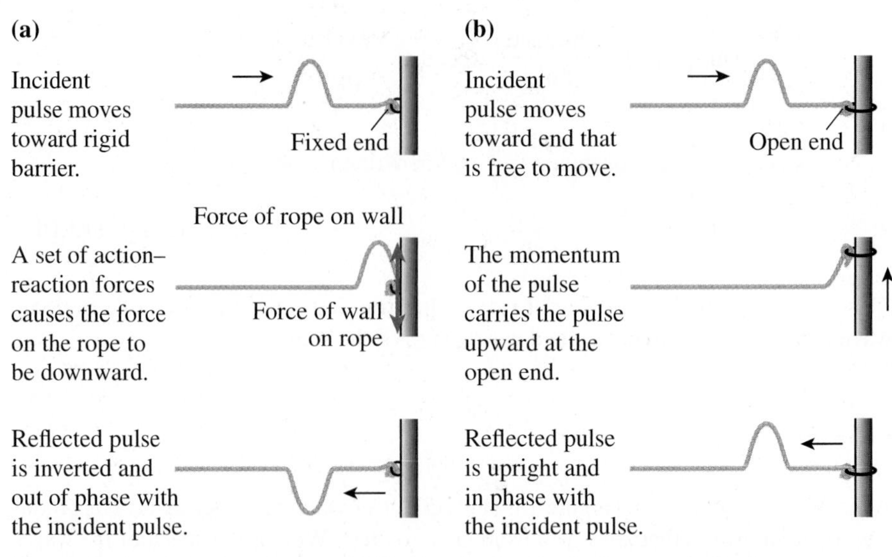

Figure 13-42 Reflection of pulses. **(a)** Fixed-end reflection. **(b)** Open-end reflection.

the barrier, it exerts an upward force on the barrier, and, according to Newton's third law of motion, the wall exerts an equal force downward on the rope. This downward force causes the reflected pulse to be on the opposite side of the rope. The reflection is called *negative;* in other words, the reflected pulse is out of phase with the incident pulse. A similar argument could be used to prove that an incident trough would reflect as a crest. Thus, we conclude that

When a pulse or wave undergoes fixed-end reflection, the reflected pulse or wave is inverted and out of phase with the incident one.

Next consider a transverse pulse reaching the end of a medium that is free to move. In Figure 13-42 (b) the free end for a rope is shown as a loop that can easily move up and down. When the pulse reaches the free end or "open end," it has nothing else to exert a force on, so its momentum keeps it rising above its original amplitude. This produces a return or reflected pulse on the same side of the equilibrium location. This reflection is called *positive;* in other words, the reflected pulse is in phase with the incident pulse. A wave acts in a similar manner. Thus, we conclude that

When a pulse or wave undergoes open-end reflection, the reflected pulse or wave is upright and in phase with the incident one.

Learning Tip

Technically, since a pulse is only part of a wave, it does not have a wavelength. That is why we refer to the pulse's length rather than wavelength.

Rather than striking a completely rigid or completely free end, a pulse[4] might strike another medium such that some of the energy is transmitted and some is reflected. The transmitted pulse is always positive or in phase. However, the phase of the reflected pulse depends on the two media involved. For a pulse trying to go from a "lighter" medium to a "heavier" one, the situation resembles fixed-end reflection, so the reflected pulse is negative, or out of phase, and the speed and length of the transmitted pulse are reduced (Figure 13-43 (a)). For a pulse trying to travel from a "heavier" medium to a "lighter" one, the reflected pulse acts as if it has struck a free or open end, so the reflected pulse is positive, or in phase, and the speed and length of the transmitted pulse

[4]Although the discussion here uses pulses, the same arguments apply to waves.

are increased (Figure 13-43 (b)). Fixed-end and open-end reflection are important in analyzing how some musical instruments create sound (Chapter 14) and how soap bubbles are observed to have rainbow colours (and other topics in Chapter 16).

(a) A pulse in a lighter medium undergoes in-phase transmission and out-of-phase reflection when it reaches a heavier medium.

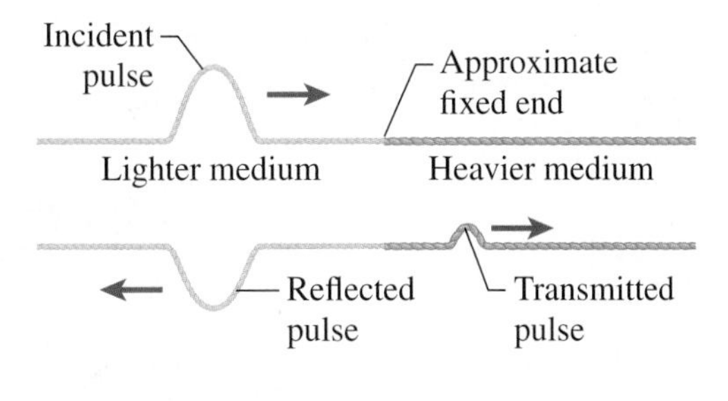

(b) When a pulse in a heavier medium reaches a lighter medium, both the transmitted and reflected pulses remain in phase.

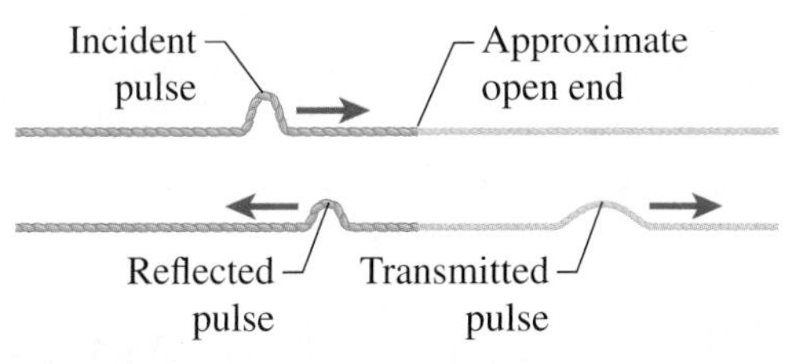

Figure 13-43 Reflection and transmission at the interface between two media.

EXERCISES

13-33 Describe the motion of the individual particles relative to the equilibrium position for (a) a transverse periodic wave and (b) a longitudinal periodic wave.

13-34 Describe the direction of propagation of the energy transmitted by (a) transverse waves and (b) longitudinal waves.

13-35 The wave in Figure 13-44 is travelling to the left in a medium. State the direction of the velocity of the particles of the medium at points B, C, D, and E.

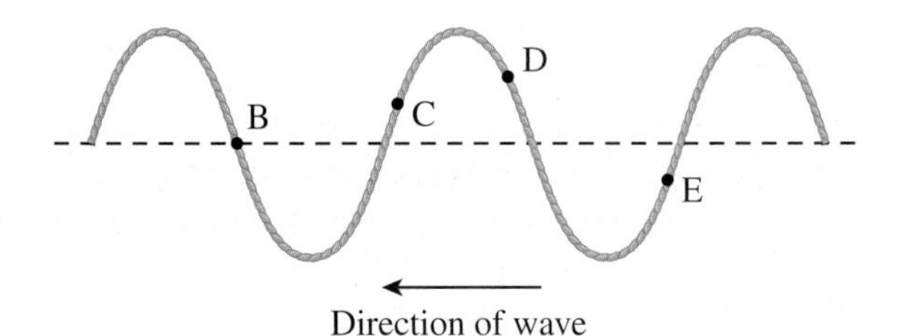

Figure 13-44 Question 13-35.

13-36 A guitar string is producing sound waves that have a wavelength of 0.391 m and a speed in air of 344 m/s. Determine the frequency of the string producing the sound.

13-37 Figure 13-45 shows two travelling waves with their dimensions labelled. Both waves are being generated at a frequency of 4.5 Hz. Determine the

(a) amplitude of the transverse wave

(b) wavelength of the transverse wave

(c) wavelength of the longitudinal wave

(d) speed of the transverse wave

(e) speed of the longitudinal wave

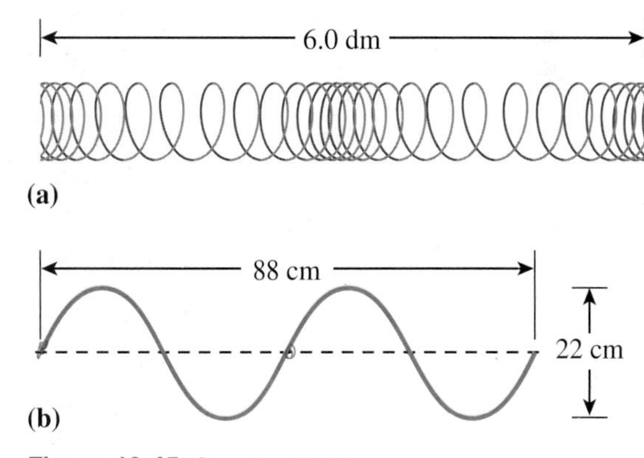

Figure 13-45 Question 13-37.

13-38 A buoy is anchored offshore to warn boaters in a lake of a shallow area. The buoy bobs up and down 6 times in 20 s as waves travel beneath it at a speed of 5.4 m/s. Assuming two significant digits, calculate the

(a) frequency of oscillation of the buoy

(b) wavelength of the water waves

13-39 Sound waves used to determine the depth of water in a lake are emitted from a boat with a period of 2.5×10^{-5} s. The wavelength of the waves in the water is 3.8×10^{-2} m and the reflected pulse is received 9.6×10^{-2} s after the pulse is emitted.

(a) Determine the speed of the pulse in the water.
(b) How deep is the lake at that location?

13-40 A bat searches for prey by emitting ultrasound with a wavelength of 3.73×10^{-3} m. If the ultrasound travels in air at 336 m/s, what is the period of the ultrasound waves?

13-41 In the therapeutic application of ultrasound, called *ultrasound diathermy,* high-intensity ultrasound waves are converted into thermal energy for deep heat treatments. If a transducer emits ultrasound waves with a period of 1.25 μs that travel at 1530 m/s in the body, determine
(a) the frequency of the ultrasound waves in megahertz
(b) the wavelength of the ultrasound waves in the body

13-42 Is reflection positive or negative for (a) fixed-end reflection and (b) open-end reflection?

13-43 A light rope is attached to a heavy rope, as illustrated in Figure 13-46 (a). Assume that the speed of a pulse or wave in the light rope is 2.0 cm/s and in the heavy rope it is 1.0 cm/s. An irregularly shaped trough is travelling from left to right, as shown. Draw a diagram showing what is happening 4.0 s after the instant shown.

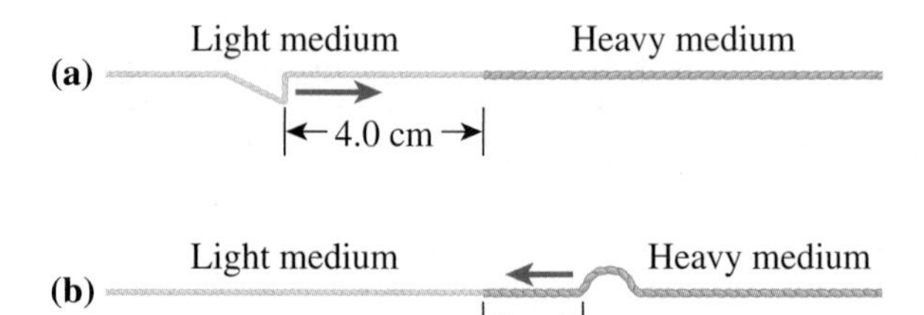

Figure 13-46 **(a)** Question 13-43. **(b)** Question 13-44.

13-44 Repeat Question 13-43 for a crest travelling from right to left, as shown in Figure 13-46 (b).

constructive interference an interference of waves in which the resulting displacement is larger than the displacements of the individual waves

destructive interference an interference of waves in which the resulting displacement is smaller than the displacements of the individual waves

13.7 Interference of Waves

We have observed the characteristics of waves when they travel in a single medium, when they reflect off a barrier, and when they travel into a second medium. But what happens when two waves in the same medium meet each other? They interfere with each other in one of two basic ways. When waves meet and the resulting displacement is larger than the individual displacements, the type of interference is called **constructive interference**. If the resulting displacement of the interfering waves is smaller, the interference is called **destructive interference**. In this section, we will study both types of interference and related phenomena for waves travelling in one dimension. The effects of interference are important in both sound and light, presented in Chapters 14 and 16 respectively.

Interference in One Dimension

Constructive and destructive interference can occur for transverse, longitudinal, and torsional pulses and waves, but it is most easily demonstrated with transverse pulses or waves in one dimension, that is, on a rope or coil.

Consider Figure 13-47, which shows both constructive and destructive interference of transverse pulses in one dimension.[5] For transverse pulses, constructive interference occurs when two positive pulses travelling toward each other meet, resulting in a larger positive pulse, or two negative pulses travelling toward each other meet, causing a larger negative pulse (not shown). Destructive interference of transverse pulses occurs when a positive pulse meets a negative pulse travelling in the opposite direction, resulting in a pulse of lower amplitude. If the initial positive and negative pulses have equal amplitude and shape as shown in (b), their displacements cancel each other for an instant, producing a position of zero amplitude, called a **node**. Then the positive and negative pulses continue travelling in their original directions.

If we observe any particular point in the medium where the pulses are interfering, the displacement that results is simply the algebraic addition of the displacements of the individual interfering pulses. The displacements are either positive or negative,

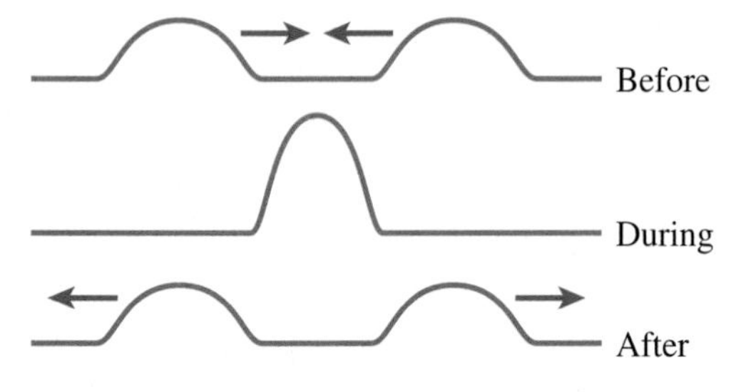

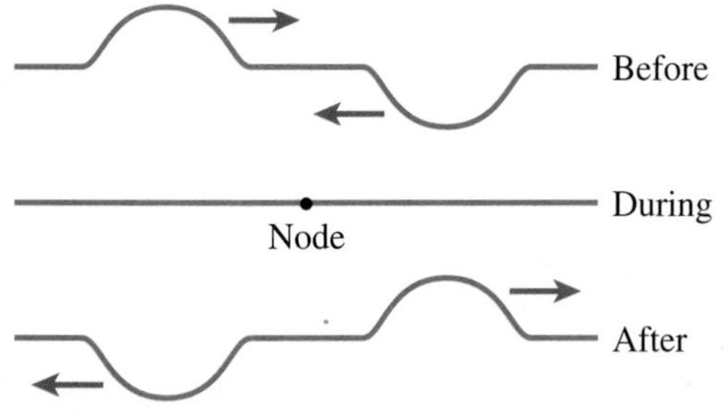

Figure 13-47 Interference of transverse pulses in one dimension. **(a)** Two positive pulses travelling in opposite directions produce a large positive pulse for an instant. **(b)** If a positive pulse meets a negative pulse of the same size and shape, a node (or zero amplitude) is produced for an instant.

[5]Interference of waves occurs in two dimensions, but discussion of this concept will be reserved for the topic of light where two-dimensional interference has important applications.

depending on whether they are above or below the equilibrium position respectively. Thus, for example, a displacement of +8.4 mm added to a displacement of −6.1 mm produces a displacement of +2.3 mm. The concept of displacement addition is summarized in the **principle of superposition**, which states

The resulting displacement of two interfering pulses at a point is the algebraic sum of the displacements of the individual pulses at that point.

(The word "superposition" stems from the verbs superimpose or superpose, both of which mean "to put on top of something else.")

The principle of superposition is useful for finding the resultant wave pattern when pulses or waves of unequal size or shape interfere with one another. It applies only to displacements that are reasonable in size because overly large pulses may cause distortion of the medium through which they are travelling.

This principle can be applied in diagrams, as shown for one-dimensional pulses in Figure 13-48. The arrows show the displacements of the individual pulses in the first two diagrams, then the resulting displacement in the third diagram.

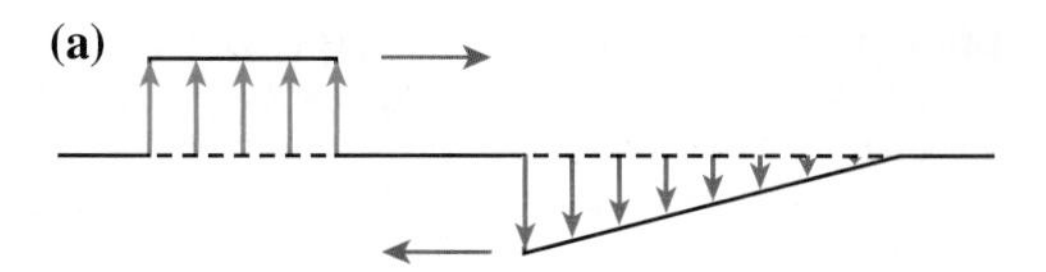

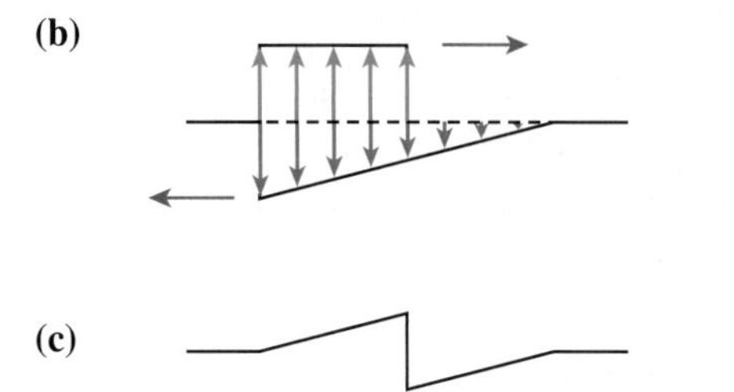

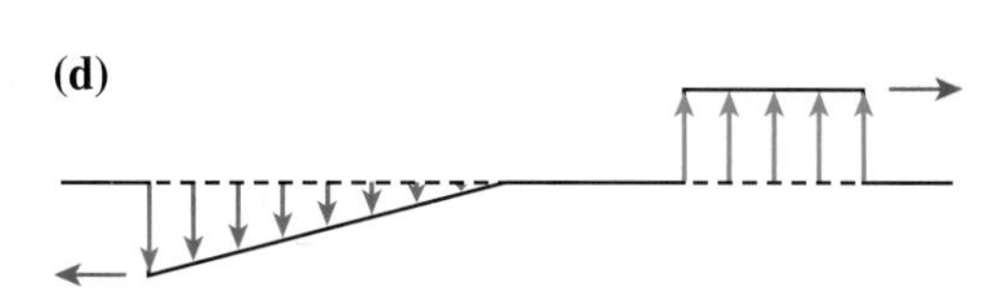

Figure 13-48 Applying the principle of superposition for pulses in one dimension.

Standing Waves in One Dimension

The principle of superposition can be applied not only to pulses, but also to periodic waves travelling in opposite directions. If periodic waves of equal wavelength and amplitude travel in opposite directions in the same medium, an interference pattern called a **standing wave** is set up such as those depicted in Figure 13-49. It has positions of zero displacement, or *nodes,* and positions of maximum displacement called **antinodes**. These nodes and antinodes remain in the same locations, which explains why the formation is called "standing."

node the resultant when a crest meets a trough of equal amplitude and length during destructive interference of transverse pulses or waves; a position of zero amplitude in an interference pattern

To learn how a standing wave pattern is set up on a rope, let us consider Figure 13-49 in detail. In each of (a) to (e), the orange wave is travelling to the right, the blue wave is travelling to the left, and the red line is the resultant wave found by applying the principle of superposition. In (a), the waves are seen at an instant when destructive interference occurs, leaving zero displacement. Then (b) is $\frac{1}{4}$ period later, producing constructive interference; (c) is $\frac{1}{2}$ period after (a), producing destructive interference; and (d) is $\frac{3}{4}$ period after (a), again producing constructive interference. The resulting standing wave pattern is seen in (e). Notice that the lines representing the antinodes show only the envelope or limit of the pattern. In fact, the particles of the medium oscillate between these limits.

If two people were asked to produce a standing wave pattern on a rope, they would stretch the rope and generate periodic waves starting at the two ends. To produce a good standing wave pattern, the waves would have to have the same amplitude (a somewhat difficult task) and the same wavelength, which means the same frequency or period (a very difficult task, especially at higher frequencies). To overcome this problem, one end of the rope can be tied to a support, and one person can send waves toward the fixed end where the waves will reflect to give waves travelling in both directions with the same amplitude and wavelength. (Fixed-end reflection was described in Section 13.6.) Let us now use this setup to describe various modes of oscillation of a standing wave.

Our discussion is based on the assumption that a node can occur at both ends of a rope or string. This is certainly true of strings on stringed instruments, such as a guitar, where various modes of oscillation can be set up. It is also true of the fixed end of a rope on which a person is demonstrating a standing wave pattern, although, at the end where the waves are generated, the displacement is slightly greater than zero. However, a rope is easier to observe than a guitar string, so we will consider it here.

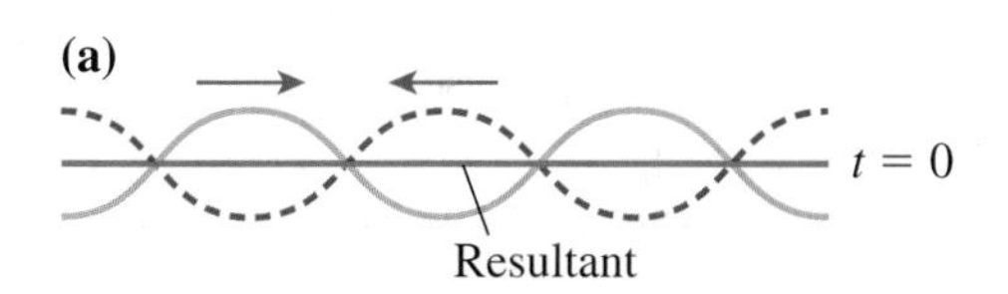

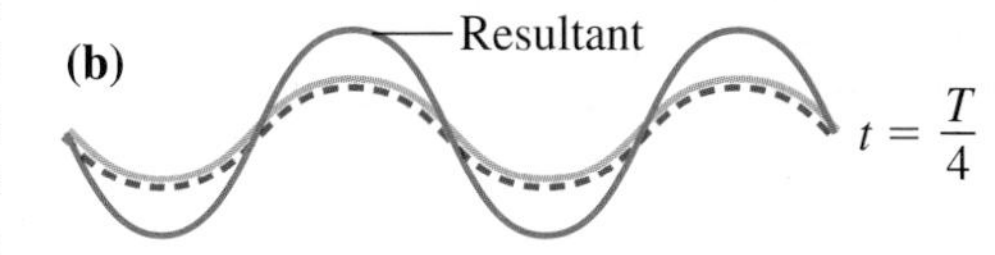

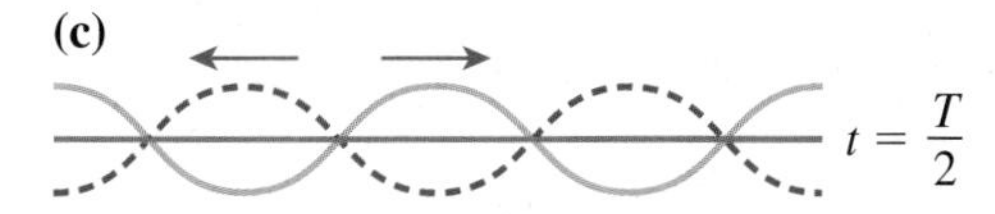

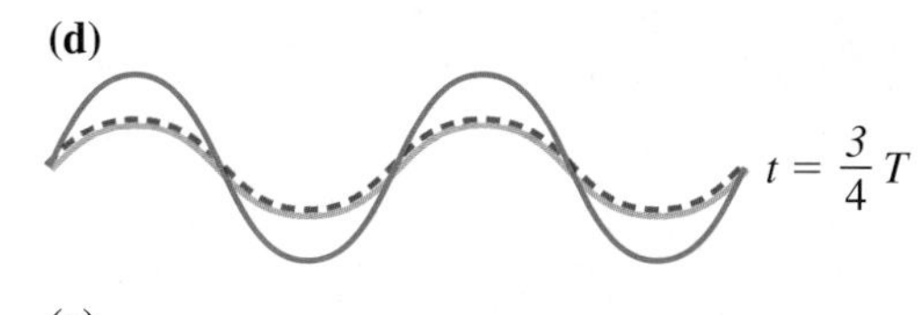

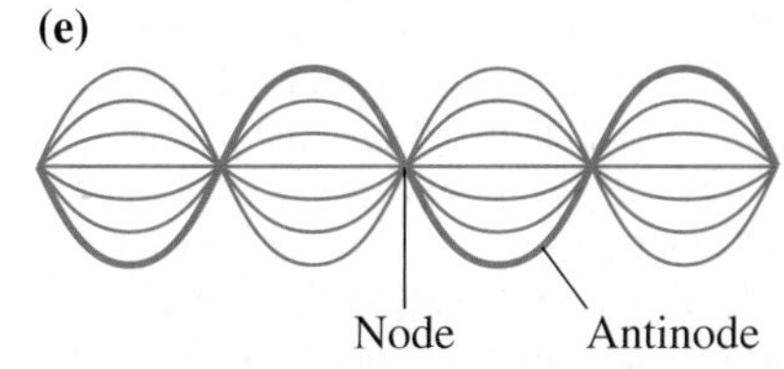

Figure 13-49 A standing wave pattern on a rope.

principle of superposition The resulting displacement of two interfering pulses at a point is the algebraic sum of the displacements of the individual pulses at that point.

standing wave the interference pattern that results when periodic waves of equal wavelength and amplitude travelling in opposite directions in the same medium meet

antinode a position of maximum displacement in an interference pattern

fundamental mode the mode with the longest wavelength on a standing wave; also called the first mode or the first harmonic

fundamental frequency the lowest resonant frequency in a standing wave pattern; corresponds to the fundamental mode

DID YOU KNOW?

Tidal action in the world's oceans and seas can set up standing wave patterns with very long wavelengths. A few places on Earth, including Tahiti in the South Pacific Ocean, have no noticeable tide because they lie on a tidal node. In contrast, some places, such as the Bay of Fundy in Eastern Canada, have very high tides due in part to the antinodes of standing waves.

Whenever a standing wave can be set up, the **fundamental mode** is the one with the longest wavelength. It can also be called the first mode or the first harmonic. On a rope or a string, the fundamental mode occurs when there is one antinode between the ends, as illustrated in Figure 13-50. In this case, the length of the rope is half the wavelength of the waves produced. To produce this pattern, the waves must be generated at the natural or resonant frequency which is the lowest frequency for the rope. The lowest resonant frequency is called the **fundamental frequency**, corresponding to the fundamental mode.

If the frequency of the generated wave is now doubled, the wavelength becomes half as long, as we know by applying the universal wave equation, $v = f\lambda$, where the speed v is constant. A new resonance is developed and the second mode of oscillation, or second harmonic, occurs (Figure 13-51 (a)). In this case, the length of the rope is one wavelength and there is one node between the two ends. The third and fourth modes, or harmonics, occur at higher resonant frequencies, producing two nodes then three nodes between the ends, as shown in Figures 13-51 (b) and (c). If you try producing these modes yourself as suggested in Try This! Producing Standing Waves below, you will appreciate why a standing wave pattern is possible only at specific resonant frequencies.

In observing all the standing waves presented in this section, we can draw the following important conclusion:

The distance between adjacent nodes in a standing wave pattern is half the wavelength ($\frac{1}{2}\lambda$) of the waves that produce the pattern.

The same relationship applies to the distance between adjacent antinodes.

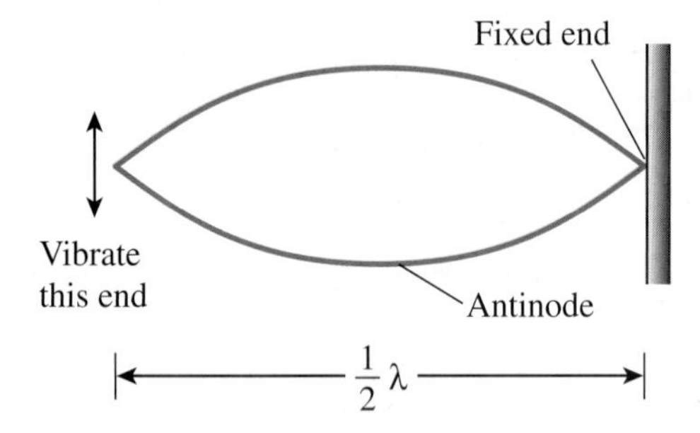

Figure 13-50 The fundamental mode of a transverse standing wave. This may also be called the first mode or first harmonic.

TRY THIS!

Producing Standing Waves

With a rope, such as a skipping rope, tied securely to a strong support about 1 m off the floor, try to create the standing wave patterns depicted in Figures 13-50 and 13-51. Using a small amplitude of oscillation will help you aim for patterns with even shorter wavelengths.

(a) Prove to yourself that standing waves occur at specific frequencies only.

(b) **Fermi Question:** Apply the universal wave equation to determine an estimate of the speed of the wave on the rope at one particular tension.

TRY THIS!

Wave Simulations

Enter the phrase "Wave on a String PhET" in a search engine and set up the simulation there so you can control the variables. For example, set the damping to zero, choose either fixed end or loose end, then try the manual setting so you can create a pulse or wave of whatever amplitude and length you want. Change the settings so you can learn how to simulate other wave features, such as standing waves.

SAMPLE PROBLEM 13-10

A 6.4 Hz source produces a standing wave pattern in the second mode on an 8.0 m rope.

(a) What is the speed of the waves producing the pattern?

(b) What is the fundamental frequency of the rope?

Solution

(a) In the second mode, the length of the rope equals the wavelength: $\lambda = 8.0$ m.

$$v = f\lambda = 6.4\,\text{Hz} \times 8.0\,\text{m} = 51\,\text{m/s}$$

Thus, the speed of the waves is 51 m/s.

(b) The fundamental frequency occurs when the length of the rope is $\frac{1}{2}\lambda$, so $\lambda = 2(8.0\text{ m}) = 16$ m. Now $f = \dfrac{v}{\lambda} = \dfrac{51\,\text{m/s}}{16\,\text{m}} = 3.2\,\text{Hz}$

Thus, the fundamental frequency is 3.2 Hz.

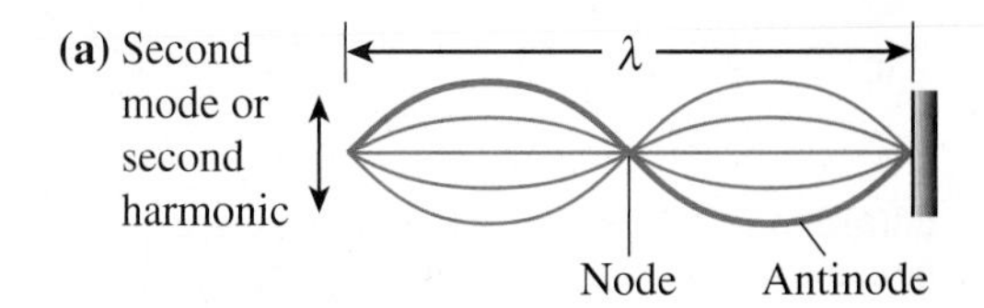

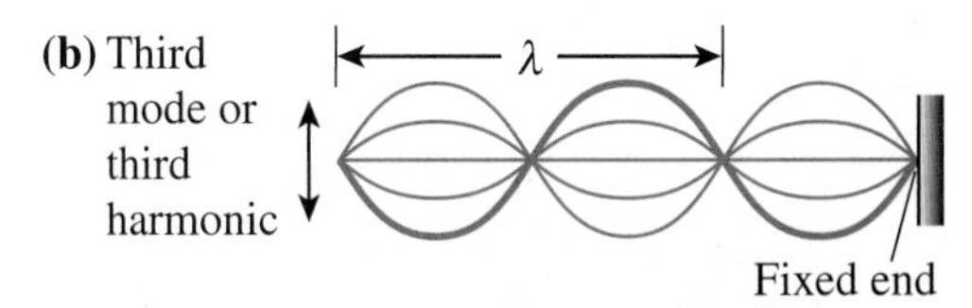

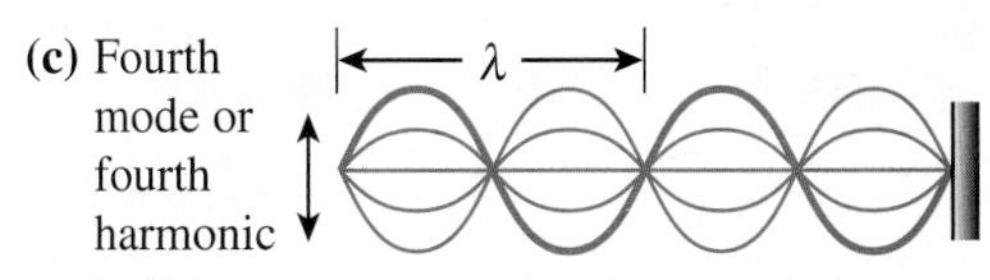

Figure 13-51 Modes of oscillation above the fundamental.

We have considered standing waves on ropes and strings only, that is, with transverse waves in one dimension. Standing waves can also be longitudinal and they can occur in two and three dimensions in a variety of media. For example, they can occur in air and in solid metal rods. In many cases, the modes may not be as simple as those on a string. For instance, instead of the modes occurring at 1*f*, 2*f*, 3*f*, etc., they may occur at 1*f*, 3*f*, 5*f*, etc. Standing waves are important in music (Chapter 14).

EXERCISES

13-45 State whether the type of interference is constructive or destructive in each case.

(a) A large trough meets a small trough.

(b) A large crest meets a small trough.

(c) A small rarefaction meets a large rarefaction.

(d) A large compression meets a small rarefaction.

13-46 What two conditions are necessary for two waves travelling in opposite directions on a rope to produce a node along the entire rope for an instant?

13-47 Apply the principle of superposition to draw the resulting pulses when the pulses shown in Figure 13-52 interfere. The point of superposition should be at the horizontal midpoints of the pulses.

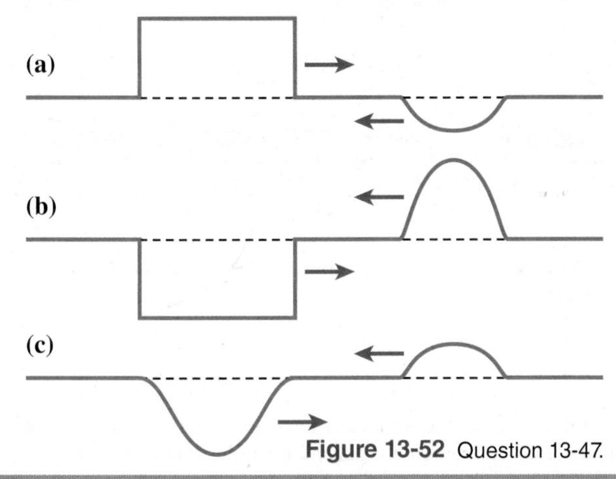

Figure 13-52 Question 13-47.

13-48 For a rope of fixed length attached rigidly at one end and oscillated at the other end, can a standing wave be set up at any arbitrary frequency? Why or why not?

13-49 Draw a scale diagram of a standing wave pattern that has four antinodes between the ends of a rope that is 6.0 m long. What is the wavelength of the waves producing this pattern?

13-50 The speed of a wave on a 6.2 m long rope is 7.6 m/s. Determine the frequency of oscillation required to produce a standing wave pattern having (a) 1 antinode, (b) 2 antinodes, and (c) 3 antinodes.

13-51 The fundamental frequency of a certain rope of length 3.8 m is 1.2 Hz.

(a) Determine the wavelength in the fundamental mode *and* the speed of the wave on the rope.

(b) Determine the frequency of the fourth and fifth modes or harmonics.

13-52 The distance between nodes on a standing wave pattern on a certain string is 16 cm and there are 5 nodes between the ends of the string. A single wave takes 1.2 s to travel to one end of the rope and back again.

(a) What is the wavelength of the waves producing the pattern?

(b) How long is the rope?

(c) What is the speed of the waves?

(d) What is the frequency of the waves?

(e) What mode of oscillation is occurring (i.e., which harmonic)?

13.8 Diffraction of Waves

diffraction the bending of a wave in a single medium as it passes by a barrier or through an opening

It is commonly known that we can hear around corners. The property of waves that allows this to occur is called diffraction. **Diffraction** is the bending of a wave in a single medium as it passes by a barrier or through an opening.[6] Diffraction of water waves can be observed on lakes and oceans as well as in water tanks. If straight periodic water waves meet the edge of a long straight barrier, they bend around the edge of it. The amount of bending, or diffraction, depends on the wavelength of the waves. At short wavelengths, the amount of diffraction is slight, but as the wavelength increases, so does the amount of diffraction (Figure 13-53).

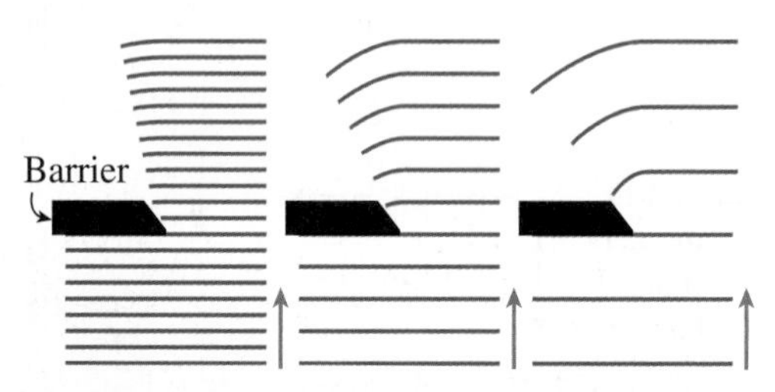

Figure 13-53 Diffraction of water waves around the edge of a barrier: as the wavelength increases, the amount of diffraction increases.

If straight periodic waves strike a straight barrier that is approximately the same size as the wavelength, diffraction occurs around both edges of the barrier. Behind the barrier, a sort of "shadow" is observed, then the waves meet and continue travelling in the medium. With short wavelengths, the shadow is relatively long; as the wavelength increases, the shadow decreases because the waves diffract more. Using the symbol w for the width of the barrier and λ for the wavelength, the amount of diffraction does not depend on w by itself or λ by itself; rather, it depends on the ratio of λ to w. If $\lambda/w \ll 1$, then little diffraction is noticed and a large shadow is observed, whereas, if $\lambda/w \geq 1$, the amount of diffraction is large and a small shadow is observed (Figure 13-54).

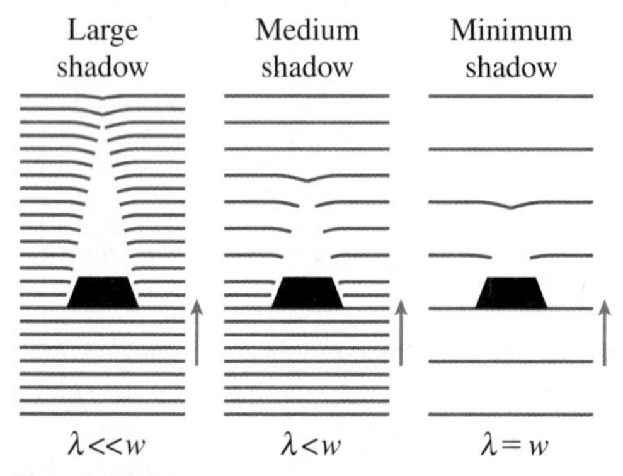

Figure 13-54 Diffraction of waves of wavelength λ around a small barrier of width w. For $\lambda \geq w$, the diffraction increases to a large amount, resulting in a small shadow.

Diffraction can also occur through a narrow opening or aperture. This situation is important, especially in the study of the wave nature of light (Chapter 16). If the width of the aperture, w, is much larger than the wavelength, which is comparable to saying $\lambda/w \ll 1$, little diffraction occurs (Figure 13-55 (a)). As the size of the aperture approaches the wavelength, the ratio λ/w approaches 1 and the amount of diffraction becomes greater (Figure 13-55 (b)). Thus, diffraction through an aperture becomes large when $\lambda/w \geq 1$, in other words, when the aperture width is less than or equal to the wavelength of the waves in the medium. In this case, the emerging waves appear almost as if they had come from a single source of circular waves (c). (Notice the faded regions in Figure 13-55 (c). The theoretical explanation and mathematical analysis of this will be dealt with in the study of the diffraction of light, Section 16.3.)

Although we have looked at diffraction for two-dimensional transverse water waves only, diffraction can occur in three dimensions for both transverse waves and longitudinal waves. In fact, the diffraction of sound waves explains why we can hear around corners. You will learn why we cannot *see* around corners in Chapter 16. (Try hypothesizing why the diffraction of light does not allow us to see around corners.) You will also learn what happens when two, three, or even thousands of tiny apertures are used rather than just a single one.

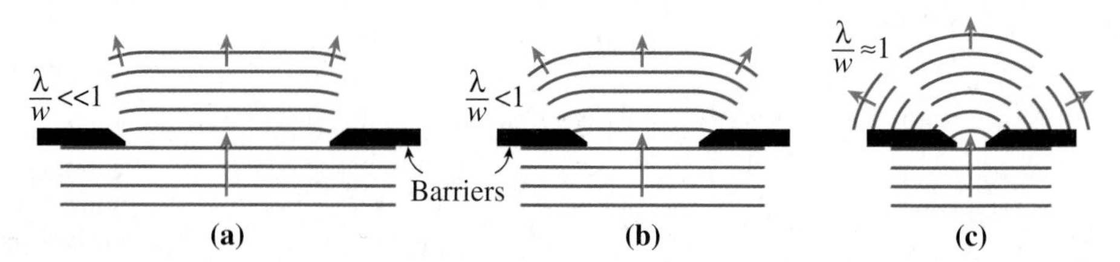

Figure 13-55 Diffraction through apertures: As the aperture width approaches the size of the wavelength, the diffraction increases.

[6]Be careful not to confuse diffraction, which occurs in one medium, with refraction, which occurs when waves enter a second medium. Refraction of light will be discussed in Chapter 15.

EXERCISES

13-53 Straight periodic waves with an initial wavelength of 12 mm are approaching a straight barrier that is 18 mm wide, and a shadow is produced behind the barrier. As the frequency of the source of waves decreases, what happens to the size of the shadow? Why?

13-54 What condition is necessary for maximum diffraction of waves of wavelength λ travelling around a sharp-edged barrier of width w?

13-55 Repeat Question 13-54 if the waves are travelling through an aperture of width w.

13-56 Speculate on the reason why the diffraction of light waves is something we usually do not observe.

LOOKING BACK...LOOKING AHEAD

The two main themes of this chapter, oscillations and waves, were presented in terms of mechanical systems. The mechanics principles applied included circular motion, Newton's second law of motion, Hooke's law for springs, and the law of conservation of mechanical energy.

Longitudinal, transverse, and torsional oscillations and their characteristics were introduced. Detailed mathematical analysis of simple harmonic motion (SHM) provided relationships involving period, frequency, displacement, velocity, acceleration, and amplitude. These relationships were derived using the reference circle and the law of conservation of energy. The transverse oscillation of a simple pendulum was also analyzed because it closely resembles SHM. Damped harmonic motion and resonance rounded off the topic of oscillations in a practical way.

The study of waves began with the function of waves, the types and characteristics of waves, and wave terminology for periodic waves. Then the transmission, reflection, interference (including the principle of superposition and standing waves), and diffraction were presented in detail.

The characteristics of waves presented in this chapter will be studied specifically for sound in Chapter 14 and for light in Chapter 16. Not all wave characteristics have been presented in this chapter. For example, the interference of waves in two dimensions as well as the polarization and scattering of light will be described in Chapter 16.

CONCEPTS AND SKILLS

Having completed this chapter, you should now be able to do the following:

- Describe and give examples of the three main types of oscillation.
- State the conditions required for simple harmonic motion (SHM) and cite examples of this type of motion.
- Determine the period and frequency of an oscillation by applying the defining equations or by knowing the relationship between period and frequency.
- Perform calculations involving the mass of an object on the end of a spring, the spring constant of the spring, and the period and/or frequency of the SHM of the system.
- Perform calculations involving the length of a simple pendulum with a small amplitude of oscillation, the magnitude of the acceleration due to gravity, and the period and/or frequency of the pendulum.
- For a body undergoing SHM, perform calculations involving the displacement, velocity, acceleration, amplitude, frequency, and time from the start of the motion.
- Apply the law of conservation of energy to a mass–spring system undergoing SHM to perform calculations related to the spring constant of the spring, and the mass, amplitude, and displacement of the oscillating body.
- Describe the motion and parts of longitudinal and transverse waves, including the wavelength and points that are in phase.
- Distinguish between surface waves, such as tsunamis, and body waves, such as primary and secondary seismic waves.
- Apply the universal wave equation to find any one of the speed, frequency, period, or wavelength of a periodic wave.
- Recognize and describe constructive and destructive interference in one dimension.
- Apply the principle of superposition to determine the results of the interference of pulses or waves.
- Recognize how different modes of oscillation of standing waves are produced, and apply the universal wave equation to calculate the speed, frequency, and wavelength of the waves that produce them.
- Describe how diffraction through apertures or around barriers depends on the wavelength of the waves and the width of the aperture or barrier.

KEY TERMS

You should be able to define each of the following words or phrases:

periodic motion
harmonic motion
oscillation
longitudinal oscillation
transverse oscillation
torsional oscillation
amplitude (of oscillation)
simple harmonic motion (SHM)
period
frequency[7]
phase
damped harmonic motion
resonant frequency
resonance

[7]Period and frequency were defined in Chapter 4, Section 4.5.

forced oscillation
wave
longitudinal (or compression) wave
compression
rarefaction
transverse (or shear) wave
crest
trough
body waves
seismic waves
primary (P) waves
secondary (S) waves
periodic waves
axis of equilibrium
wavelength
universal wave equation
constructive interference
destructive interference
node
principle of superposition
standing wave
antinode
fundamental mode
fundamental frequency
diffraction

Chapter Review

MULTIPLE-CHOICE QUESTIONS

13-57 The speed of a pulse on a string depends on the

(a) magnitude of the gravitational field
(b) amplitude of the pulse
(c) tension in the string
(d) length of the pulse

13-58 The mass attached to the spring in Figure 13-56 is undergoing SHM and at the instant shown is at its maximum displacement from the equilibrium position. Positive is chosen to be to the right. If v is the instantaneous speed and a is the instantaneous acceleration, then at the instant shown, which statement is true?

(a) $v = 0$; $a =$ maximum and positive
(b) $v =$ maximum; $a = 0$
(c) $v =$ maximum; $|a| =$ maximum, $a < 0$
(d) $v = 0$; $|a| =$ maximum, $a < 0$

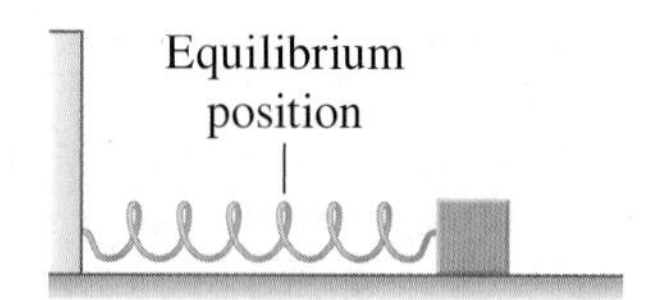

Figure 13-56 Question 13-58.

13-59 In a longitudinal wave, the individual particles of the medium

(a) move parallel to the direction of the wave's motion
(b) move perpendicular to the direction of the wave's motion
(c) move in ellipses
(d) move in circles

13-60 As the period of a periodic wave in a particular medium increases,

(a) the frequency remains constant and the wavelength increases
(b) the speed increases and the wavelength decreases
(c) the amplitude increases
(d) the frequency decreases and the wavelength increases
(e) none of the above is true

13-61 The superposition of two waves results in a standing wave if the waves have

(a) the same amplitude, different frequencies, and opposite directions of motion
(b) the same amplitude and direction of motion, but different frequencies
(c) the same frequency and amplitude, but opposite directions of motion
(d) the same frequency, amplitude, and direction of motion

13-62 A source of periodic waves, oscillating at an initial frequency f_1, results in waves of wavelength λ_1 travelling at a speed v_1 in a certain medium. If the frequency of the source changes to $2f_1$, then the wavelength and speed in the same medium become, respectively,

(a) $2\lambda_1$ and $2v_1$
(b) $2\lambda_1$ and v_1
(c) $0.5\lambda_1$ and $0.5v_1$
(d) $0.5\lambda_1$ and $2v_1$
(e) $0.5\lambda_1$ and v_1

Review Questions and Problems

13-63 State the relationship, if any exists, between the sets of variables listed below. Where possible, write an equation or mathematical variation (proportionality) statement based on the appropriate equation.

(a) period and frequency
(b) acceleration and displacement in SHM
(c) period and spring constant for a mass on a spring in SHM
(d) frequency and mass of a body in SHM on a spring
(e) period and length of a simple pendulum
(f) the frequency of a simple pendulum and the acceleration due to gravity
(g) hertz and cycles per second
(h) the maximum speed of a body in SHM and the amplitude of the motion
(i) the maximum elastic potential energy stored in a spring and the maximum kinetic energy of a mass undergoing SHM on a spring
(j) the damping time of a system undergoing damped harmonic motion and the amount of friction experienced by the system
(k) the wavelength of a periodic wave and the frequency of the source causing the wave (in a medium where the speed of the wave is constant)
(l) the amount of diffraction of waves passing through an aperture and the ratio of the wavelength of the waves to the aperture width

13-64 The horned sungem from South Africa has the fastest wing beat of any bird in the world at 1350 beats in only 15 s. Determine the period and frequency of this motion.

13-65 In the abbreviation SHM, what does the H stand for? What does the word mean?

13-66 A spider web (Figure 13-57) acts as a sort of elastic membrane that vibrates when an insect flies into it and becomes trapped. If the spring constant of the web is 0.75 N/m, what is the frequency of oscillation if a 0.40 g insect is caught?

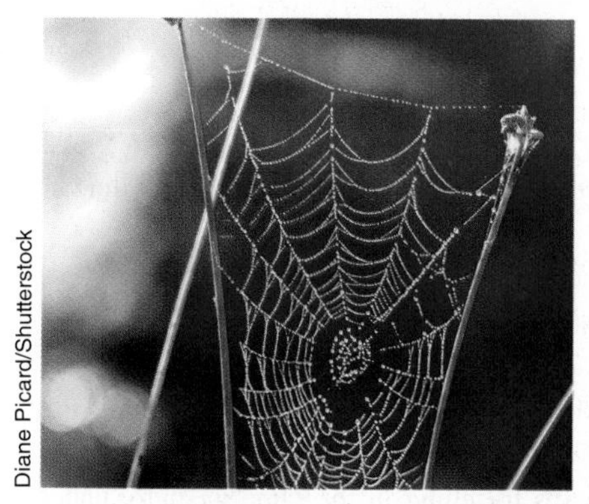
Diane Picard/Shutterstock

Figure 13-57 Question 13-66.

13-67 A 44 kg child bounces up and down on a pogo stick with a SHM of period 0.42 s.

(a) What is the spring constant of the pogo stick's spring? (Neglect the mass of the stick.)

(b) What would happen to the period of oscillation if a person with greater mass were to try bouncing on the stick?

13-68 What would happen to the resonant frequency of a playground swing if the person on the swing went from the sitting position to the standing position? Why?

13-69 A person buys a pendulum clock at the seacoast where the clock is in excellent adjustment, then moves to a village at a high altitude in the Rocky Mountains. What will happen to the clock's ability to keep time? Why? How can this problem be solved?

13-70 A simple pendulum is oscillating with a small amplitude, $x = A$.

(a) How far does the pendulum bob move in one cycle?

(b) At what position(s) of the cycle is the speed of the bob maximum? minimum?

(c) At what position(s) of the cycle is the magnitude of the acceleration of the bob maximum? minimum?

13-71 Assume that a demolition wrecking ball acts like a simple pendulum. If the length of the pendulum is 18 m, what are its period and frequency?

13-72 A geologist is using a sensitive simple pendulum of mass 95.00 g and length 42.50 cm to determine the gravitational field at a location above an assumed oil field. The period of the pendulum is found to be 1.307 s. Calculate the magnitude of the gravitational field (i.e., the acceleration due to gravity).

13-73 By what factor must a pendulum length be increased in order to increase its period by a factor of two?

13-74 The acceleration due to gravity at one location on the Moon is 1.63 m/s^2. What is the frequency of oscillation of a simple pendulum of length 0.854 m?

13-75 The diaphragm of a loudspeaker (Figure 13-58) is oscillating at a frequency of 420 Hz. At time $t = 0$, the oscillation is at its maximum displacement, $A = +0.82$ mm. At time $t = 0.010$ s, find the diaphragm's (a) displacement, (b) velocity, and (c) acceleration.

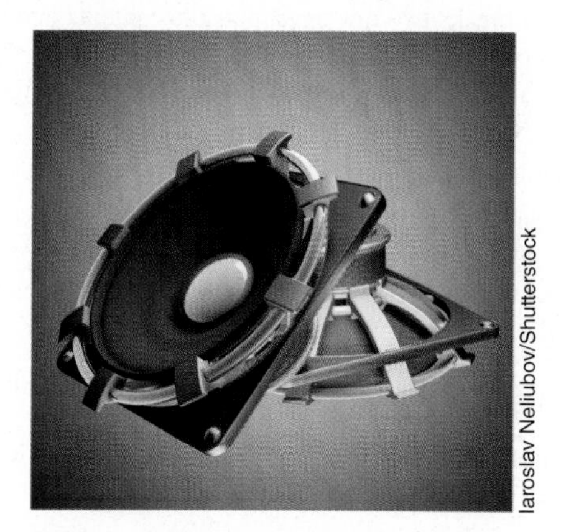
Iaroslav Neliubov/Shutterstock

Figure 13-58 Question 13-75.

13-76 The magnitude of the maximum velocity of a mass attached to a spring in SHM is 6.65 m/s. The period of the motion is 0.424 s.

(a) What is the amplitude of the oscillation?

(b) Assuming at $t = 0$, $x = A$, what is the velocity of the mass 2.50 s after its motion began?

13-77 A mass of 0.42 kg, attached to a horizontal spring of spring constant 38 N/m, undergoes SHM without friction at an amplitude of 5.3 cm.

(a) What is the energy of the mass–spring system?

(b) What is the maximum speed of the mass? (Use energy concepts.)

(c) What is the speed of the mass when the displacement is 4.0 cm?

(d) When the displacement is 4.0 cm, determine the kinetic energy and elastic potential energy. How does the sum of these energies compare to your answer in (a)?

13-78 A block attached to a horizontal spring of spring constant 75 N/m undergoes SHM with an amplitude of 0.150 m. If the speed of the mass is 1.7 m/s when its displacement is 0.120 m from the equilibrium position, what is the mass of the block?

13-79 Give examples in which damping of oscillations is (a) useful and (b) not useful.

13-80 Do you think it is possible to have any mechanical isolated harmonic motion that is not damped? ("Isolated" in this case means there is no external influence.) If so, give at least one example; if not, explain why not.

13-81 The end of a diving board exhibits damped harmonic motion after the diver jumps off. Assume that at the instant the diver leaves the board ($t = 0$), the end of the board is at its highest position, $x = 30$ cm, above the equilibrium position. The period of the oscillation is 0.50 s and the amplitude reduces to a negligible amount after 5 cycles. Draw a graph showing the approximate displacement as a function of time for this motion.

13-82 How could the concept of resonance be applied to try to free a car that is stuck in snow or mud?

13-83 If the resonant frequency of a child on a swing is f, at which, if any, of the following frequencies would it be possible to push the child to maintain a large amplitude of oscillation: $2f$; $3f$; $4f$; $\frac{1}{2}f$; f; $\frac{1}{4}f$?

13-84 A chef breaks two raw eggs, puts the contents into a hemispherical bowl, and proceeds to beat the eggs with a fork at a frequency *other* than the resonant frequency of the eggs in the bowl. Why "other"?

13-85 What is the main function of a wave?

13-86 Imagine a long, thin, solid metal rod suspended so it rests horizontally. (This imaginary rod may be 100 m long or more.) One end of the rod is struck, first parallel to the rod, then later perpendicular to the rod, as illustrated in Figure 13-59.

(a) Would energy from the blows travel at the same speed or at different speeds? Explain.

(b) Relate your answer in (a) to the speeds with which P and S waves travel through Earth.

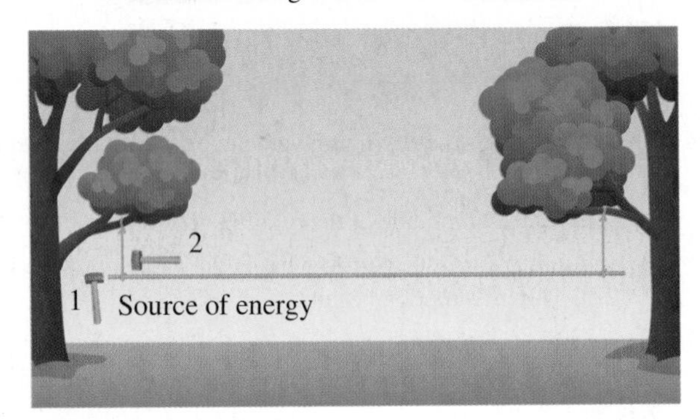

Figure 13-59 Question 13-86.

13-87 An earthquake occurs 7.2×10^3 km from a seismological station. How long after receiving the initial primary wave does the station receive the initial secondary wave? (Assume the P and S waves travel at 8.0 km/s and 4.5 km/s, respectively.)

13-88 Ultrasonic waves used to shatter a brain tumour operate at a frequency of 23.0 kHz; the speed of the waves in the body is 1.53 km/s. What is the wavelength of the ultrasound?

DID YOU KNOW?

The ultrasonic device used for destructive purposes is called a cavitron ultrasonic surgical aspirator (CUSA). After the tumour is shattered, the fragments are removed using a saline solution.

13-89 Determine the range of frequencies of FM radio stations in megahertz (MHz) if their range of wavelengths is from 2.78 m to 3.43 m. The speed of the waves in air is 3.00×10^8 m/s.

13-90 Determine the range of wavelengths of AM radio stations whose frequencies range from 530 kHz to 1610 kHz. The speed of the radio waves in air is 3.00×10^8 m/s. (Assume three significant digits.)

13-91 Draw the reflected pulse in each case in Figure 13-60.

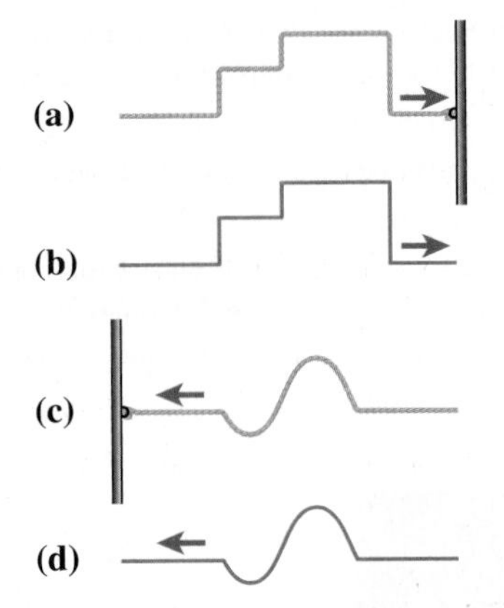

Figure 13-60 Question 13-91.

13-92 Two ropes, X and Y, are attached. The speed of waves in X is v and the speed in Y is $2v$. A wave of wavelength λ is travelling in X toward Y.

(a) For the wave that enters Y, what is the wavelength?

(b) Is the transmission from X to Y positive or negative?

(c) Is the reflection from the interface between X and Y positive or negative? How do the wavelength and amplitude of the reflected wave compare to the initial waves?

13-93 For longitudinal waves travelling in the same medium, under what condition(s) will (a) constructive interference and (b) destructive interference occur?

13-94 Two troughs, C and D, are travelling on a rope toward end E (Figure 13-61). Each trough is 10 mm wide, C is 7.0 mm deep, and D is 9.0 mm deep. After D reflects from E, it meets C and interferes for an instant.

(a) If E is a fixed end, what are the width and height (or depth) of the resulting pulse at maximum interference?

(b) Repeat (a) if E is an open end.

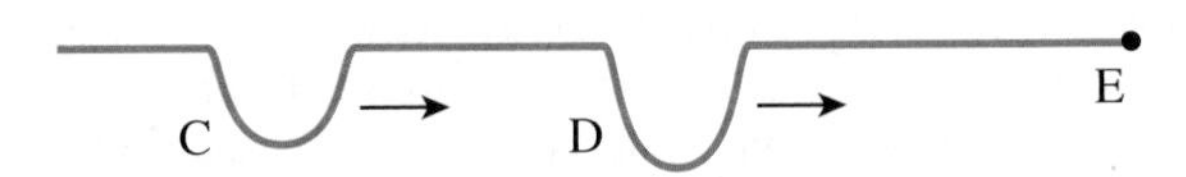

Figure 13-61 Question 13-94.

13-95 A standing wave pattern is set up on a string of length 54 cm fixed at both ends. The wavelength of the waves is 36 cm, the speed is 25 m/s, and the amplitude of the standing waves is 8.0 cm.

(a) What is the distance between adjacent nodes on the string?

(b) What is the frequency of oscillation of the source of the waves?

(c) Draw a scale diagram of this standing wave pattern.

13-96 The frequency of oscillation of the third mode or harmonic of a string with nodes at both ends is 660 Hz. What is the frequency of the fundamental mode? the fourth harmonic?

13-97 A set of water waves of wavelength 1.0 cm is approaching a barrier. Draw two diagrams that show the diffraction pattern that results if the width of the barrier is

(a) 5.0 cm (b) 1.0 cm.

Applying Your Knowledge

13-98 During flight, a sparrow's heart beats 800 times per minute and a pigeon breathes 400 times per minute. (Assume two significant digits.)

(a) Determine the frequencies of each of these activities in hertz.

(b) **Fermi Question:** Determine a reasonable estimate of the maximum frequency of your own heartbeat and breathing during or just after vigorous exercise.

(c) Speculate, with reasons, whether or not humans will ever be able to compete in the flying world with these high-revving "engines."

13-99 A car with four passengers has a total mass of 1800 kg. The spring constant for each of the four springs of the car is 1.2×10^5 N/m.

(a) Assuming the weight of the car is distributed evenly to the four springs, determine the compression of each spring.

(b) Neglecting damping, what would be the frequency of each spring after striking a bump?

(c) If the car were travelling at 25 m/s on a bumpy road, how far apart would the bumps have to be to set up a form of resonance of the front wheels?

13-100 To find the mass of an astronaut in the "weightless" conditions of space, a special oscillation device is used. The mass of the part that oscillates is 12 kg.

(a) If the device with nobody in it oscillates at a frequency of 1.1 Hz, what is the spring constant of the spring?

(b) When an astronaut sits in the device and the system oscillates with SHM, the new frequency of oscillation is 0.44 Hz. What is the mass of the astronaut?

13-101 After the telescope was invented, the famous Italian scientist, Galileo, first observed some moons of Jupiter in 1610. From the point of view of Earth, the moons appeared to be travelling with what we refer to as SHM. The moon Callisto was found to have a period of revolution around Jupiter of 16.8 (Earth) days. We now know that its radius of motion (or amplitude of the apparent SHM) is about 2.6×10^6 km.

(a) How does this situation relate to the reference circle used in the derivations in Sections 13.1 and 13.3?

(b) Plot a graph of Callisto's displacement (as observed from Earth) from the middle of Jupiter as a function of time for two complete cycles.

13-102 One tine of a 256 Hz tuning fork (Figure 13-62) is struck, causing both tines to move toward each other and then undergo SHM in opposite phases. In other words, the tines move toward each other, then away from each other, and so on. Assume that for a short while the amplitude of the SHM at the end of each tine remains constant at 3.2 mm. Determine the magnitude of the maximum velocity and the magnitude of the maximum acceleration at the end of each tine.

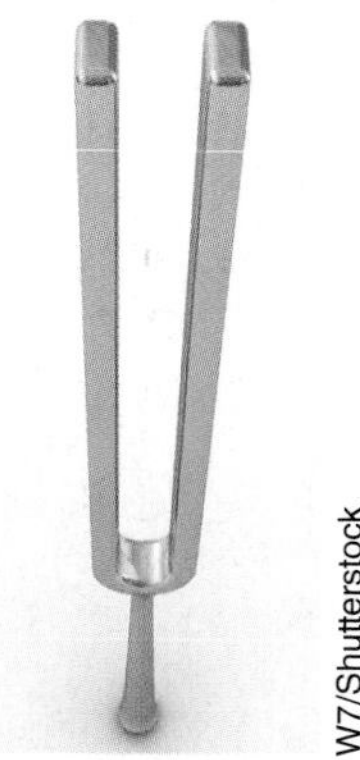
W7/Shutterstock

Figure 13-62 Question 13-102.

13-103 Fermi Question: An amusement park owner wants to design some exciting, new active-participation devices suitable for big kids (i.e., teenagers and older). One idea is to have a large vertical spring with a crossbar at the bottom that a person could grab onto and try bouncing up and down with a large amplitude. As a consultant, you are asked to help decide if such a device is feasible and, more importantly, safe.

(a) Choose a maximum amplitude of oscillation (in metres) that you would expect such a device would need to make it appeal to the average thrill seeker.

(b) Using the amplitude in (a) as well as the magnitude of your own weight (*mg*), calculate what spring constant would be required by the spring.

(c) Using your answer in (b), find the frequency of oscillation when you are on it.

(d) Determine your maximum speed and magnitude of acceleration when playing on the device. Is the device feasible? Is it safe? Explain.

13-104 Farm implements often produce oscillations that are irritating to a person riding in them. Assume you are hired by a manufacturer to design tractor seats that reduce oscillations transmitted to the person sitting in them to a minimum. Describe ways of accomplishing this goal. (Include the concepts of resonance and damping in your answer.)

DID YOU KNOW?

The study of designing devices to provide maximum human comfort is called ergonomics, or human engineering. Ergonomics is important in the airline and automobile industries, among others.

13-105 In 1831 in England, an army was marching in-step across a bridge suspended over a river. Unfortunately, the frequency of the steps matched a resonant frequency of the bridge, and the bridge collapsed. What could have been done to march the troops across the bridge at the same speed without causing the collapse? Explain your answer.

13-106 The Bay of Fundy on Canada's Atlantic Coast has among the highest tides in the world, up to a maximum of about 15 m. Such high tides result not only from the general shape of the bay (Figure 13-63), but also because of resonance effects. The length of the bay is such that the water in it oscillates back and forth with a period of about 13 h. This is just slightly more than the period of time between successive high tides or successive low tides, which is 12 h and 25 min.

(a) Why does "resonance" help increase the size of the tides in the Bay of Fundy?

(b) Proposals have been made to put a large dam across the mouth of the bay to harness tidal energy. Such a dam would reduce somewhat the effective length of the bay. How would this affect the resonant frequency of the water in the bay? Would this result in higher or lower maximum tides? Explain why you think so.

Figure 13-63 Satellite image of the Bay of Fundy separating the provinces of New Brunswick (on the left) and Nova Scotia (Question 13-106).

13-107 Timing is crucial when one trapeze acrobat must catch another (Figure 13-64). Use concepts presented in this chapter to explain how this timing is achieved.

Figure 13-64 Question 13-107. How can these acrobats achieve perfect timing?

13-108 Figure 13-65 shows a test being performed in a tsunami simulator tank at the Port and Research Institute in Yokosuka, Japan. The tank is 184 m long and 3.5 m wide, and can generate waves up to 3.5 m in height. Describe, with reasons, a possible experiment you think would be useful to test in this tank.

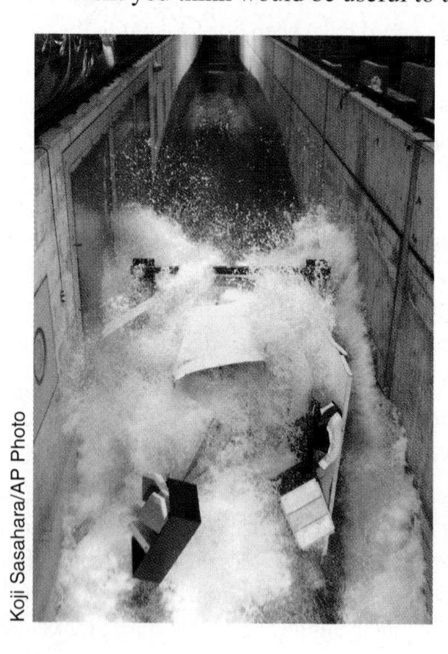

Figure 13-65 Question 13-108. A tsunami simulator tank with a wave striking a mock house with furniture and a mannequin.

13-109 On June 23, 2013, Nik Wallenda made a daring yet carefully planned walk across the Grand Canyon where the gorge was more than 460 m deep. (Wallenda was shown walking across Niagara Falls in the Chapter 11 introduction.) As shown in Figure 13-66, the steel cable, which was 5.1 cm in diameter, had several masses of various lengths suspended from it. On the day of the walk, there was a strong wind blowing along the canyon.

(a) Based on what you learned in this chapter, explain why the masses are suspended from the horizontal cable.

(b) Twice during the 22-minute crossing, Wallenda was bothered so much by the wind and the swaying of the cable that he stopped to kneel on one knee on the cable. Explain the cause(s) of the swaying of the cable, and explain the physics of the advantage of stopping to kneel temporarily.

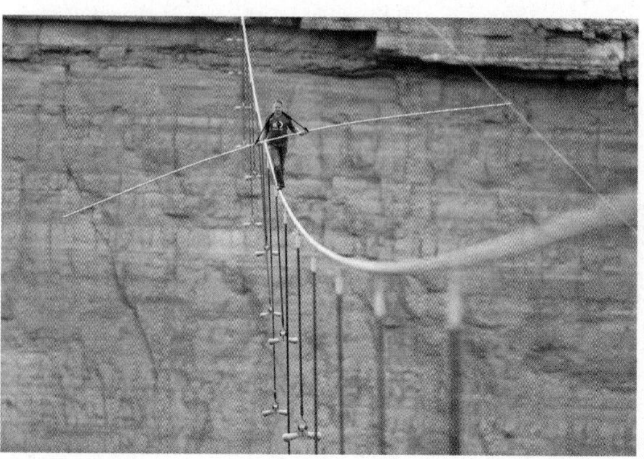

Figure 13-66 Question 13-109. The vertical pendulums suspended from the main cable are an important feature of the planning for Nik Wallenda's historic walk across the Grand Canyon.

13-110 A factory worker drops a 15 kg steel block from rest. The block falls 40 cm onto a vertical spring standing on the floor. The spring constant of the spring is 500 N/m. Once the block has made contact with the spring, it continues to move downward, compressing the spring until it comes to rest. (The spring then decompresses, firing the block upward.) Determine the distance that the spring was compressed by the block. Assume two significant digits in given data.

Sound and Music 14

Koerner Concert Hall, Royal Conservatory of Music. Photo: Tom Arban

Figure 14-1 Koerner Hall in Toronto, Ontario, has earned a reputation among musicians and patrons for its superb sound qualities that showcase pop, jazz, classical, and world music. The 1135-seat hall embraces the natural beauty of curved and varied wood surfaces, including the twisting ribbons of wood suspended from the ceiling. But beauty is secondary to the main function of the wood and other surfaces—namely, the stunning acoustics.

The sound and music industry is huge and complex, but it is based on physics principles that form a major portion of this chapter. Music venues, such as Toronto's Koerner Hall (Figure 14-1), have both visible and hidden architectural features that provide a balance of sound absorption and reflection to minimize unwanted sound and maximize desired sound, a concept discussed in the topic of acoustics. The architects and sound engineers who designed the renovations of the original historic hall, home of Canada's Royal Conservatory of Music, solved various sound challenges in unique ways. For example, sounds from nearby practice studios and a subway line are absorbed by rubber pads, and sounds from the musicians are spread throughout the hall by reflections from the undulating wood and other surfaces. Since its opening in 2009, Koerner Hall has become renowned as a concert hall to which musicians want to return.

Of course, the topic of sound relates to our sense of hearing, a sense many of us take for granted. Think for a moment how you would describe sound and hearing to someone

CHAPTER OUTLINE

who has been unable to hear since birth. Perhaps you could describe in written words or visual images the sounds a baby hears, such as a mother's soothing voice, or sounds of speech, singing, a piano, thunder, a car engine, a robin, fireworks, and so on. The visual images might include oscillations and waves presented in Chapter 13. You could describe the physical functions of the parts of the human ear, and you could have the person touch a loudspeaker or a musical instrument as it emanates sound. After completing this chapter, you will be able to understand and describe the characteristics of sound and music in much greater detail.

14.1 The Nature of Sound

What is sound? To answer this question, let us consider three aspects of sound: its production and transmission described in this section, and its detection presented in Section 14.2.

All sounds are produced by oscillations. The sound of rustling leaves originates when a wind causes the leaves to oscillate. The origin of the sound from a musical instrument depends on the instrument: the skin of a drum, the air in a trumpet, and the strings of a guitar all oscillate to produce sounds (Figure 14-2).

Daniel-life/Shutterstock

Figure 14-2 The fret board and vibrating strings of an acoustic guitar.

The energy that originates from an oscillation can be transmitted as a sound wave. Sound waves require a material medium through which to travel. A simple demonstration to verify this consists of an electric bell in a transparent jar from which the air can be evacuated. When the air has been removed by a vacuum pump, the gong of the bell can be seen oscillating but no sound can be heard (Figure 14-3).

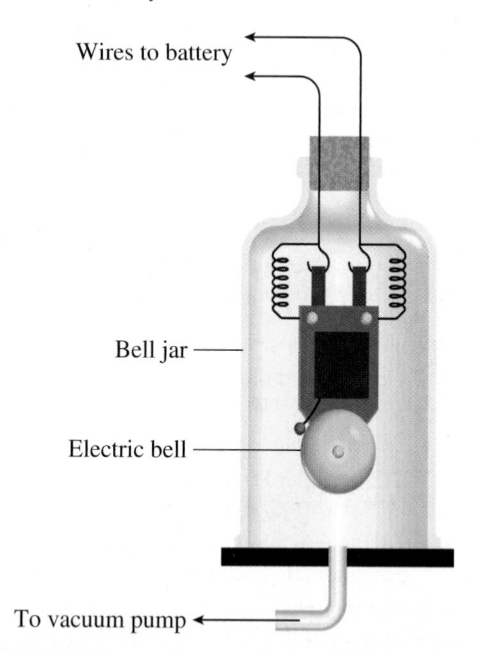

Figure 14-3 A bell in a vacuum cannot be heard, which provides evidence that sound requires a material medium to be transferred.

The human ear is sensitive to sound waves, so it is a common detector. However, humans are not alone in their ability to detect sound: animals and electronic sound receivers can also detect sound waves. Even if our ears cannot hear some of the sound waves that some animals and receivers can detect, such waves can still be considered to be sound. Thus, **sound** is a wave disturbance that originates with an oscillation, travels through a material medium, and can be detected by a listener or receiver.

sound a wave disturbance that originates with an oscillation, travels through a material medium, and can be detected by a listener or receiver

Sound Transmission in Air

Air is the most common medium through which sound waves we detect are transmitted. Air, like other gases, consists of large numbers of molecules that are constantly in motion, colliding often with nearby molecules. Between collisions, the molecules travel extremely quickly, approximately 1600 km/h, depending mainly on the temperature. At about 20°C and at sea level, each molecule of air has more than one billion collisions every second!

To visualize how these high-speed molecules act when a sound is produced nearby, consider the action of a drum skin (Figure 14-4 (a)). When the membrane is first struck it is depressed and the air molecules nearby become rarefied, or less densely packed. Then the membrane bounces outward, causing the air molecules to become compressed, or more densely packed. As the membrane continues to oscillate, the air molecules continue to experience rarefactions and compressions that travel as a sound wave away from the drum skin. You will recall that a wave consisting of a rarefaction and a compression is called a compression wave or a longitudinal wave. Thus, *in air sound is transmitted as a longitudinal wave.*

The rarefactions and compressions of these sound waves can be represented by a sinusoidal wave in which the troughs represent a low pressure of air molecules (rarefactions) and the crests represent a high pressure of air molecules (compressions),

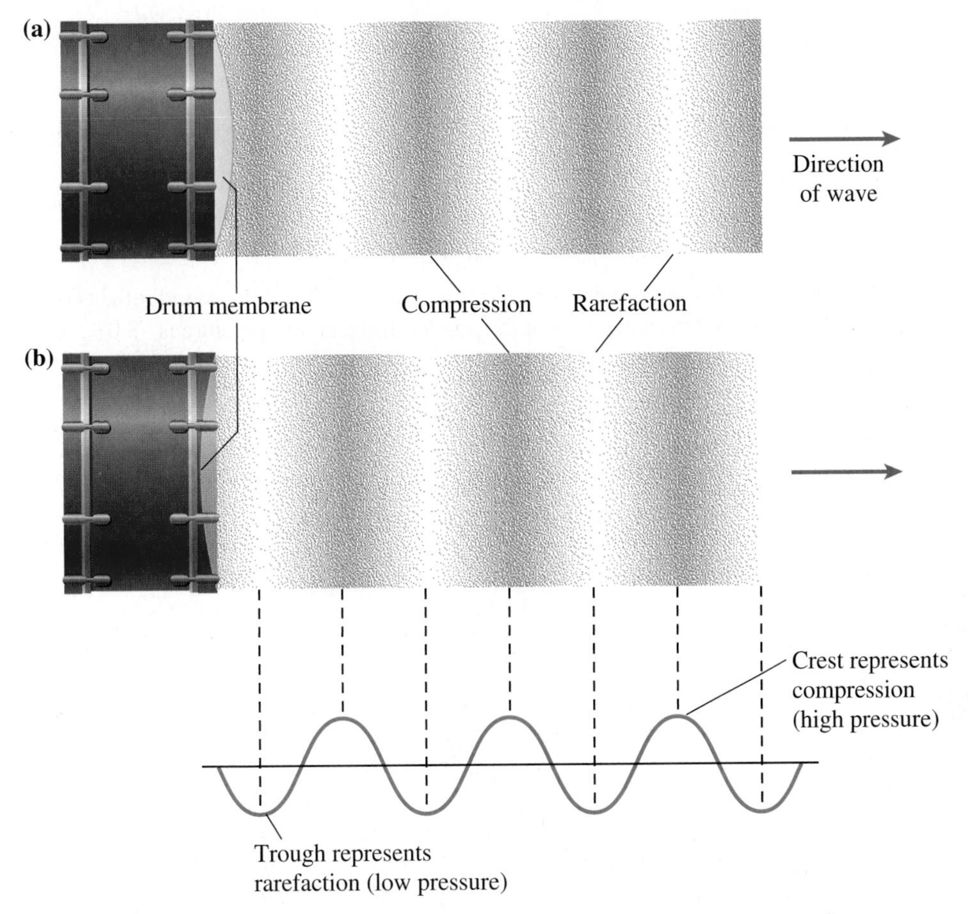

Figure 14-4 The production and transmission of a sound wave. **(a)** The vibrating membrane of a drum causes rarefactions and compressions of air molecules that travel outward in a longitudinal wave. **(b)** A sinusoidal wave can be used to represent a longitudinal wave. In this case the amplitude of the wave represents the pressure of the air molecules.

as shown in Figure 14-4 (b). It is important to remember that this sinusoidal wave is a representation of the pressure variations of the longitudinal wave. A sinusoidal wave could also be used to represent the displacement of air molecules from their average equilibrium position, but such a wave would not be in phase with the pressure wave representation.

The Speed of Sound in Air

You have had the experience of seeing a flash of lightning in the distance then hearing the thunder a short while later. The light from the flash travels at a speed of about 300 000 km/s in air, so it reaches you almost instantaneously. Sound, however, travels much more slowly in the air, so at large distances we notice a time delay between seeing lightning and hearing the sound it creates.

Experimental results show that the speed of sound in air changes when the temperature changes. For temperatures in the regions that we normally experience, the relationship is approximately linear. Figure 14-5 shows a graph of the speed of sound in air as a function of temperature. Notice that at 0°C the speed is 331 m/s, and the speed increases by about 6 m/s for every 10°C rise in temperature. In the limited temperature range we consider here, the increase, to three significant digits, is 0.606 (m/s)/°C. Thus, the equation of the line on the graph yields the speed of sound in air:

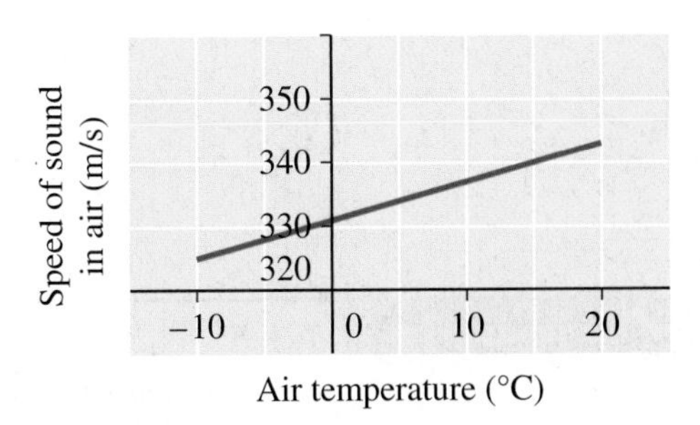

Figure 14-5 The speed of sound in air plotted as a function of temperature, valid for temperatures near 0°C.

Notation Tip

This text uses the symbol T for a variety of variables, namely T for period of oscillation, T for temperature in degrees Celsius or absolute temperature in kelvins, ΔT for temperature change, T_x or T_y for the components of the tension, $\vec{T}$, as well as T for the magnitude of tension. Thus, it is important to consider the context of the symbols in equations.

$$v = 331\ \text{m/s} + 0.606\frac{(\text{m/s})}{^\circ\text{C}}T \qquad (14\text{-}1)$$

where T is the temperature of the air in degrees Celsius (°C).

SAMPLE PROBLEM 14-1

The fog horn of a fishing trawler is sounded (Figure 14-6), and the sound reflects off a cliff and is received 6.4 s later back at the trawler. If the air temperature is –8.0°C, how far is the trawler from the cliff?

2009fotofriends/Shutterstock

Figure 14-6 A fog horn (Sample Problem 14-1).

Solution

First we find the speed of the sound in air.

$$\begin{aligned} v &= 331\ \text{m/s} + 0.606\,(\text{m/s})/^\circ\text{C} \times T \\ &= 331\ \text{m/s} + \left(0.606\ \frac{\text{m/s}}{^\circ\text{C}}\right)(-8.0^\circ\text{C}) \\ &= 331\ \text{m/s} - 4.8\ \text{m/s} \\ &= 326\ \text{m/s} \end{aligned}$$

$$\begin{aligned} \text{Now, } d &= vt \text{ where } t = 3.2\ \text{s} \\ &= 326\ \text{m/s} \times 3.2\ \text{s} \\ &= 1.0 \times 10^3\ \text{m (to two significant digits)} \end{aligned}$$

Thus, the trawler is 1.0 km from the cliff.

At room temperature (about 20°C) the speed of sound in air is about 343 m/s, which is more than 1200 km/h. As was stated earlier, the average speed of air molecules between collisions is about 1600 km/h, so the speed of a longitudinal wave travelling through air is about 3/4 as fast as the speed of the individual molecules. It should be evident why the speed of a wave changes as the temperature of the air changes. As the temperature rises, the speed of the individual air molecules increases, and the collisions can occur more rapidly, allowing the waves to travel through the air more quickly.

The Speed of Sound in General

You know from experience that sound can travel through walls, along metal fences, and through water. The speed of a wave in such media can be expressed in terms of the frequency and wavelength ($v = f\lambda$) or the period and wavelength ($v = \lambda/T$), as was presented with the universal wave equation in Section 13.6. However, the speed of a wave is found to vary from one medium to another, and it varies within a single medium depending on the condition of the medium. Thus, we can find an expression for the speed of sound waves in various media in terms of the properties of those media.

To discover how the speed of a wave depends on the properties of the medium, we will use an analogy with longitudinal waves travelling along springs. Consider a series of compressed springs, with each pair separated by a small block. One set of springs is very stiff (Figure 14-7 (a)), while a second set is very loose (as in (b)). A disturbance (a pulse or a wave) that begins at one end of each set will travel much

Learning Tip

The frequency (or period) of a sound depends on the source of the sound, so it remains constant even if the speed changes. Thus, from $v = f\lambda$, with a constant frequency if the speed increases the wavelength also increases.

faster in the stiff set than in the loose set of springs. The stiffness of a spring is a measure of its elasticity or elastic property. The relationship found experimentally is that the speed of the wave is proportional to the square root of the elastic property:

$$v \propto \sqrt{\text{elastic property}}$$

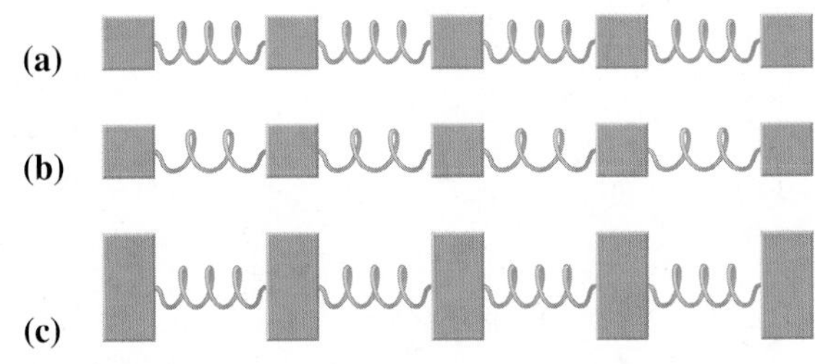

Figure 14-7 Determining the relationship between the speed of a longitudinal wave in a medium and the properties of the medium. **(a)** Stiff springs separated by low-mass blocks. **(b)** Loose springs separated by the same low-mass blocks. **(c)** Stiff springs separated by high-mass blocks.

To determine another factor that influences the speed of a wave, compare parts (a) and (c) of Figure 14-7. The same stiff springs used in (a) are used in (c), but in (c) the masses between the springs are much larger than in (a). As you might predict, the speed is slower with the larger masses because of their larger inertia. By experimentation, it is found that the speed varies inversely as the square root of the inertial property of the medium:

$$v \propto \frac{1}{\sqrt{\text{inertial property}}}$$

When these two relationships are combined, we obtain

$$v \propto \sqrt{\frac{\text{elastic property}}{\text{inertial property}}}$$

Thus, $$v = \text{constant} \times \sqrt{\frac{\text{elastic property}}{\text{inertial property}}}$$

Using experiments and derivations beyond the level we need to discuss here, this constant is found to have a value of one. Therefore, the speed of a wave travelling in a material medium is

$$v = \sqrt{\frac{\text{elastic property}}{\text{inertial property}}}$$

To carry the analogy of waves travelling in springs further, consider the structure of a solid medium. The molecules of the solid are arranged in a rather rigid formation. Imagine they are represented by the blocks. The molecules are held together by intermolecular forces, which are represented by the springs. Thus, a longitudinal wave can pass from molecule to molecule at a speed determined by the elasticity caused by the intermolecular forces and the inertia of the molecules. Our analogy could also apply fairly well to longitudinal waves travelling in liquids, although liquid molecules are free to move around one another. The analogy is an oversimplification for transverse waves travelling in solids, as well as longitudinal waves in gases where the molecular interaction is accomplished by collisions rather than a springiness of attraction and repulsion. However, the end results are the same, and the equation for speed in general can be adapted to all three states of matter: gases, liquids, and solids.

We can now use the general equation for speed to determine speed of a wave in specific situations. Consider first a string, such as a guitar string. The elastic property is the magnitude of the tension (T) in the string and the inertial property is the mass per unit length, called the **linear density** of the string (symbol μ, the lower case Greek letter mu). Thus,

linear density the mass per unit length or inertial property of a string; symbol μ (the Greek letter mu)

$$v = \sqrt{\frac{T}{\mu}} \qquad \textbf{(14-2)}$$

where v is the speed of a wave in a string measured in metres per second (m/s), T is the tension in the string measured in newtons (N), and μ is the linear density measured in kilograms per metre (kg/m).

This equation is logical for anyone with experience in plucking a stringed instrument because the speed increases as the string is pulled more tightly and the speed is lower in thicker strings.

SAMPLE PROBLEM 14-2

Prove that the equation $v = \sqrt{\frac{T}{\mu}}$ is dimensionally correct.

Solution

Since T is measured in newtons (N = kg·m·s^{-2} with dimensions M·L·T^{-2}) and μ is measured in kilograms/metre (kg·m^{-1} with dimensions M·L^{-1}), we can convert the right side of the equation to dimensions:

$$\sqrt{\frac{\text{M}\cdot\text{L}\cdot\text{T}^{-2}}{\text{M}\cdot\text{L}^{-1}}} = \sqrt{\text{L}^2\cdot\text{T}^{-2}} = \text{L}\cdot\text{T}^{-1}$$

Since speed has dimensions of length/time (L·T^{-1}), the equation is dimensionally correct.

SAMPLE PROBLEM 14-3

A piano string is under a tension of 2.5×10^2 N. If it is 1.2 m long and has a mass of 3.4 g, how fast does a wave travel along it?

Solution

In base SI units, $m = 3.4\,\text{g} \times \frac{1\,\text{kg}}{1000\,\text{g}} = 3.4 \times 10^{-3}\,\text{kg}$.

The linear density is $\mu = \frac{m}{L} = \frac{3.4 \times 10^{-3}\,\text{kg}}{1.2\,\text{m}} = 2.8 \times 10^{-3}\,\text{kg/m}$

Now,

$$v = \sqrt{\frac{T}{\mu}} = \sqrt{\frac{2.5 \times 10^2\,\text{N}}{2.8 \times 10^{-3}\,\text{kg/m}}} = 3.0 \times 10^2\,\text{m/s}$$

Thus, the speed of the wave in the string is 3.0×10^2 m/s.

density the inertial property of gases, liquids, and solids; symbol ρ (the Greek letter rho)

modulus the elastic property of gases, liquids, and solids; symbol B

For all other materials, including gases, liquids, and solids, the inertial property is the **density** (symbol ρ, the lower case Greek letter rho) and the elastic property is called the **modulus**. (Density was defined in Chapter 12 and the moduli of various materials were described in Chapter 11.) Every gas and liquid has its own characteristic bulk modulus (symbol B), while every solid has a characteristic modulus that depends on whether the travelling wave is transverse (shear) or longitudinal (compression). Table 14-1 summarizes the equations and symbols for the speed of a wave in various types of materials.

Notice in Table 14-1 that the speed of a wave in a material medium does not depend on the frequency or amplitude of the wave. Table 14-2 lists the speed of sound waves in various materials.

Table 14-1

Specific Equations for Finding the Speed of Waves in Different Media

Type of Material	Type of Wave	Equation for Speed	Meaning of Symbols (SI Units)
String	transverse	$v = \sqrt{\frac{T}{\mu}}$	T = tension (N) μ = linear density (kg/m)
Fluid (gas or liquid)	longitudinal	$v = \sqrt{\frac{B}{\rho}}$	B = bulk modulus (N/m^2) ρ = density (kg/m^3)
Thin rod or bar	longitudinal	$v = \sqrt{\frac{Y}{\rho}}$	Y = Young's modulus (N/m^2)
Solid	longitudinal or primary wave[1]	$v = \sqrt{\frac{B + S}{\rho}}$	S = shear modulus (N/m^2)
Solid	transverse or secondary wave	$v = \sqrt{\frac{S}{\rho}}$	

Table 14-2

The Speed of Sound in Various Media (at 0°C unless otherwise stated)

State	Material	Speed (m/s)	
Solid	aluminum	5100	
	glass	5030	
	steel	5030	
	maple wood	4110	
	bone (human)	4040	
	copper	3560	
	pine wood	3320	
	lead	1320	
	rubber	54	
Solid/liquid (at 37°C)	brain	1530	
	fat	1450	
	muscle	1580	
Liquid (at 25°C)	fresh water	1493	
	sea water	1533	(depends on salt content)
	methyl alcohol	1143	
Gas	hydrogen	1286	
	helium (at 20°C)	927	
	nitrogen (at 20°C)	350	
	air (at 0°C) (at 100°C)	331 386	
	oxygen	317	
	carbon dioxide	258	

[1]Primary and secondary waves were discussed in Section 13.5.

SAMPLE PROBLEM 14-4

Determine the speed of sound in a fresh-water lake where the density of the water is 1.0×10^3 kg/m^3 and the bulk modulus of the water is 2.2 GN/m^2.

Solution

First, convert the bulk modulus to base SI units:

$$2.2\frac{\text{GN}}{\text{m}^2} = 2.2\frac{\text{GN}}{\text{m}^2} \times \frac{1 \times 10^9\,\text{N}}{1\,\text{GN}} = 2.2 \times 10^9\,\text{N/m}^2$$

Now, $v = \sqrt{\dfrac{B}{\rho}} = \sqrt{\dfrac{2.2 \times 10^9\,\text{N/m}^2}{1.0 \times 10^3\,\text{kg/m}^3}} = 1.5 \times 10^3\,\text{m/s}$

Thus, the speed of sound in the water is 1.5×10^3 m/s.

EXERCISES

14-1 What oscillates to cause each of these sounds?

(a) notes played on a piano

(b) the buzzing of a mosquito

(c) thunder

14-2 Describe what happens to the density of air molecules as a sound wave passes through the air.

14-3 Explain why the speed of sound in air decreases as the temperature decreases.

14-4 A 440 Hz tuning fork (Figure 14-8) is struck in a room where the temperature is 22.2°C.

(a) Is the sound wave produced by the tuning fork transverse, torsional, or longitudinal?

(b) What is the wavelength of the sound wave? (Use three significant digits.)

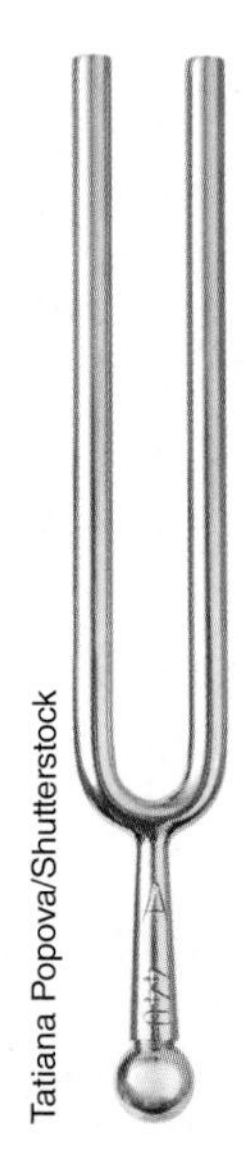

Tatiana Popova/Shutterstock

Figure 14-8 A 440 Hz tuning fork (Question 14-4).

14-5 A camper notices a lightning flash, then estimates the time before the thunder is heard to be about 3 s.

(a) Approximately how far away, in kilometres, did the lightning occur?

(b) Make up a useful "3-second rule" for thunder and lightning.

14-6 A starter's pistol is used to start a 200 m race on a straight track (Figure 14-9). A timer at the finish line sees a bright flash from the pistol at the starting line. If the air temperature is 15.0°C, how much later does the sound reach the timer? (Assume three significant digits.)

© age fotostock Spain, S.L./Alamy

Figure 14-9 A starter's pistol (Question 14-6).

14-7 A mechanic working on a steel pipeline drops a wrench onto the pipeline, and 1.10 s later another mechanic hears the resulting sound that passes along the pipeline. If the air temperature is 10.0°C, how long after the first sound will the second mechanic hear the sound that travelled through the air?

DID YOU KNOW?

The microphone beside the starter's pistol in Figure 14-9 sends a signal at the speed of light to speakers located at each starting block. This allows all runners to hear the pistol at the same instant, no matter how far they are from the pistol.

14-8 Two blocks of wood are banged together and the sound reflects off a building wall that is 242 m away. The sound takes 1.40 s to travel both ways. Determine the temperature of the air.

14-9 By what factor must the tension in a string increase to increase the speed of a wave in the string by a factor of 2?

14-10 In which type of medium (gas, liquid, or solid) is the speed of sound generally the greatest? Explain why this is so.

14-11 Prove that the equation $v = \sqrt{\frac{S}{\rho}}$ is dimensionally correct.

14-12 The bulk modulus of sea water is 2.23×10^9 N/m^2 and its density is 1.03×10^3 kg/m^3. Determine the speed of sound in this water.

14-13 Calculate the speed of sound in a cylindrical bone having a density of 1.6×10^3 kg/m^3 and a Young's modulus of 1.5×10^{10} N/m^2. (Treat the bone as a long, thin rod.)

14-14 The bulk modulus of a certain liquid is 1.2×10^9 N/m^2 and the speed of sound in the liquid is 1.2×10^3 m/s. Determine the density of the liquid.

14-15 The average density of Earth is 2.4×10^3 kg/m^3. Assume the speed of S waves through Earth is 4.5 km/s. Determine the average shear modulus of Earth.

14-16 Determine Young's modulus of aluminum, given that its density is 2.7×10^3 kg/m^3. (The speed of sound in aluminum is given in Table 14-2.)

14.2 Hearing and the Intensity of Sound

The human ear is sensitive to a wide range of frequencies and an extremely wide range of intensities of sounds. In this section we explore the mechanism by which we hear, and then we look at the frequency response of the ear and the difference between the intensity of sound and the more subjective quantity, the loudness of sound.

The Human Ear

The human ear consists of three main parts, the outer ear, the middle ear, and the inner ear (Figure 14-10 on page 422). The outer ear consists of the *pinna,* the external portion that we can see, and the *auditory canal.* The pinna directs the longitudinal sound waves into the auditory canal, allowing the waves to strike the *eardrum,* which connects the outer ear and the middle ear. The eardrum is an elastic membrane about the thickness of a human hair that oscillates with an amplitude as large as about 10^{-7} m to as small as 10^{-11} m, less than the diameter of an atom!

The oscillating eardrum transfers its energy to the middle ear, which consists of three tiny bones collectively called the *ossicles* and named the *malleus, incus,* and *stapes* (which is the smallest bone in the human body, at 3.6 mm). These bones are commonly called the *hammer, anvil,* and *stirrup* because of their shapes. They act like a system of levers to transfer energy to the *oval window,* which joins the air-filled middle ear to the liquid-filled inner ear.

The inner ear consists of a coiled, liquid-filled tube called the *cochlea* (from the Greek word *kocklias,* which means "snail"). The cochlea is shown in its correct position in Figure 14-10 (a) and uncoiled in (b). Incoming signals from the oval window cause the part of the cochlea called the *basilar membrane* to oscillate. (Basilar means "at the base.") Connected to the basilar membrane are thousands of tiny hairs and nerve endings that transform the energy of oscillation into electrical impulses that are sent via the *auditory nerve* to the brain for interpretation. The basilar membrane is tighter and thinner near the oval window, so that is where reactions to high frequency sounds occur. The other end of the membrane is looser and thicker, responding to low frequency sounds. Considering that the basilar membrane is only about 1.0 cm long, it is certainly a sensitive part of the hearing mechanism.

Two other parts of the ear shown in Figure 14-10 (a) are worth mentioning here. The *Eustachian tube* joins the middle ear to the throat. This tube allows air pressure to

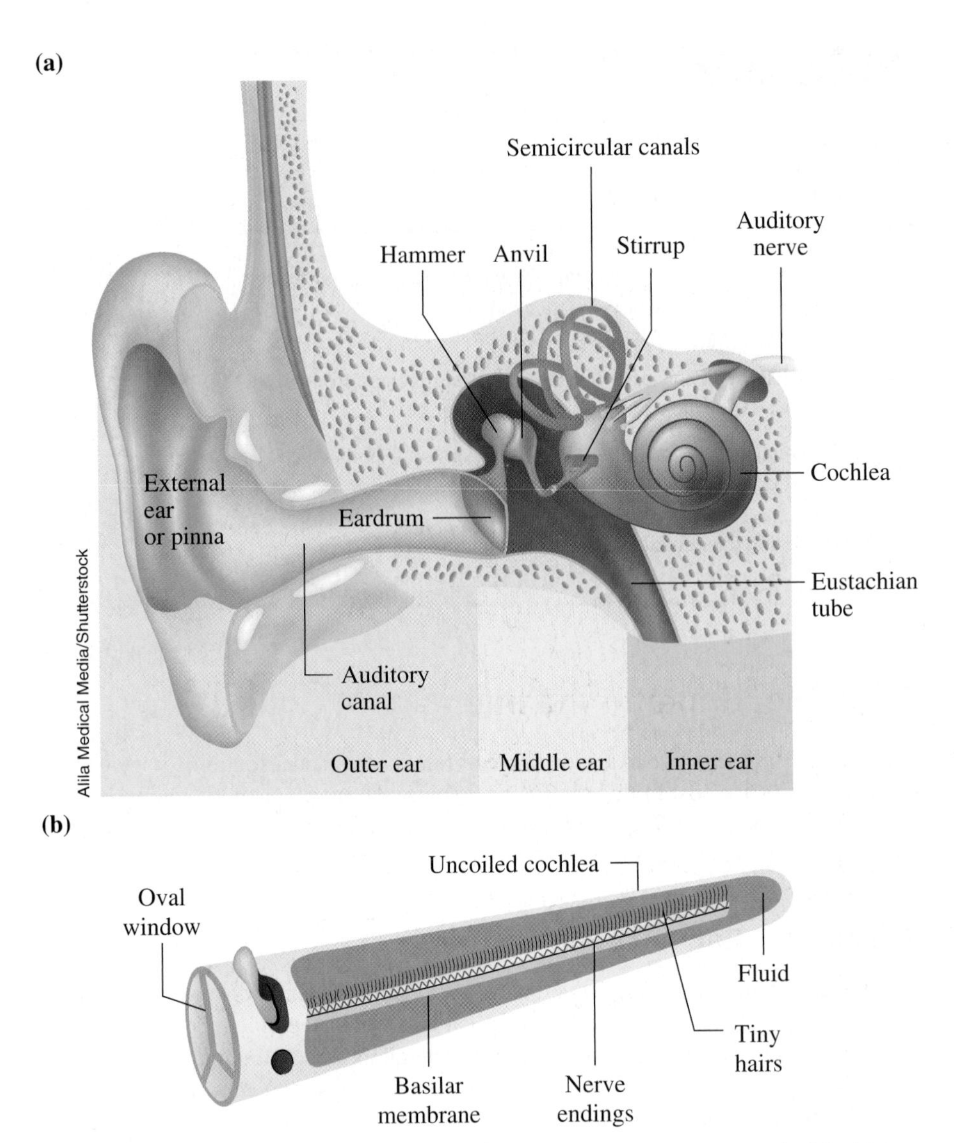

Figure 14-10 The human ear. **(a)** The three main parts (outer, middle, and inner) of the human ear. **(b)** The uncoiled inner ear.

DID YOU KNOW?

Hearing in animals is similar, although simpler, than human hearing. A cat has about 16 000 fibres in its basilar membrane, just over half the number in the human ear. Birds have only about 3000 fibres in the same membrane. The ears of insects are sometimes located in strange places: the thorax, the abdomen, or even on the forelegs.

equalize on both sides of the eardrum, causing the "popping" sensation during rapid ascent or descent in aircraft or elevators. The *semicircular canals* (above the stirrup) in the inner ear are important for controlling balance.

Approximately 5% of the population of North America suffers from hearing loss or deafness. If deafness results because impulses cannot travel through the auditory nerve to the brain, there is no possible treatment. However, if loss of hearing or deafness results from aging and/or damage to the eardrum, various types of hearing aids are available to improve hearing. Figure 14-11 (a) illustrates examples of these aids. One type has a case that fits behind the ear and is connected to a dome that fits into the ear canal. Another type fits directly into the ear canal and is thus hidden. In both devices, complex battery-operated electronics include a microphone and a loudspeaker that can be controlled to amplify or suppress chosen sound frequencies before the sound reaches the inner ear. Once the sound reaches the eardrum, energy is transmitted in the normal way to the brain. If deafness is caused by an impaired cochlea, a cochlear implant system can improve hearing. As illustrated in Figure 14-11 (b), a behind-the-ear microphone (in green) picks up the sound oscillations that are analyzed, converted into an electric code, and linked to a transmitting antenna (in grey). Signals then go to a small receiver just under the skin (also in grey), and from there to an array of electrodes in a fine coiled wire in the cochlea. The signals then transfer from the cochlea to the brain for interpretation.

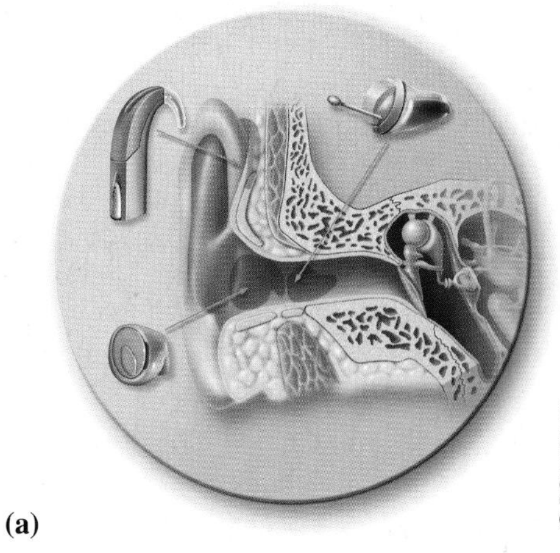

(a)

© BSIP SA/Alamy

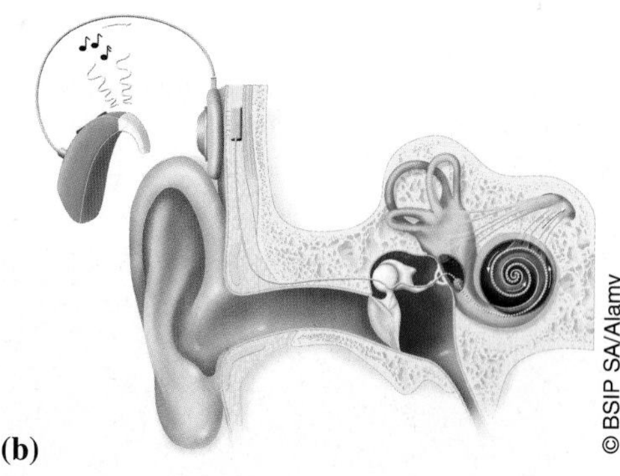

(b)

© BSIP SA/Alamy

Figure 14-11 Solutions to hearing impairment. **(a)** Illustrating examples of hearing aids, one type a behind-the-ear aid (shown on the left) and another type an in-the-ear-canal aid. **(b)** A cochlear implant system (described in the text).

The Human Audible Range

The response of the human ear to frequencies of sound varies greatly from person to person. The range of frequencies that an ear can detect is called the **human audible range**. The healthy ear of a young person may have an audible range from about 25 Hz to about 20 kHz. This range generally becomes smaller as a person becomes older. It may also be reduced by prolonged exposure to loud sounds.

human audible range the range of frequencies that a human ear can detect

Intensity and Loudness of Sound

The loudness of sounds we hear is relatively subjective, which means that it depends largely on the listener. One factor that influences loudness, however, can be analyzed objectively, a factor called intensity.

Sound intensity is a measure of the power of a sound per unit area. Since power is the rate at which energy is received, intensity could also be defined as (energy/time)/(unit area). The SI unit of sound intensity is watts per square metre, or W/m^2. From experience, we know that as we get farther from the source of a sound, the loudness we perceive becomes less. To relate this to sound intensity, consider an "ideal" situation in which a point source of sound is sending out sound waves in all directions and there are no barriers from which these waves can reflect. At a certain distance, r, from the source, the energy is spread out over a sphere of surface area $4\pi r^2$. If this distance is doubled to $2r$, the surface area of the sphere becomes $4\pi(2r)^2$, which is 4 times as large. Thus, the power per unit area reaching the larger sphere is 1/4 as much as at the smaller sphere. Similarly, if the distance becomes $3r$, the area is $4\pi(3r)^2$, which is 9 times as large. Thus, the power per unit area reaching the larger sphere is 1/9 as much as at the smaller sphere. This typical inverse square relationship can be expressed mathematically:

sound intensity a measure of the power of a sound per unit area, measured in watts per square metre, or W/m^2

$$I \propto \frac{1}{r^2}$$

where I is the intensity in watts per square metre and r is the distance in metres from the source of the sound.[2]

SAMPLE PROBLEM 14-5

At a distance of 59 m from a jet taking off, the intensity of the sound is 0.10 W/m^2. At what distance will the intensity be 1/10 of this value? (Neglect the effects of the reflection of sound off the ground.)

Solution

Since $I \propto \frac{1}{r^2}, \frac{I_2}{I_1} = \frac{r_1^2}{r_2^2}$

$$\therefore r_2 = r_1\sqrt{\frac{I_1}{I_2}} = (59\,\text{m})\sqrt{\frac{0.10\,\text{W/m}^2}{0.010\,\text{W/m}^2}} = 1.9 \times 10^2\,\text{m}$$

Thus, the distance is 1.9×10^2 m.

[2]The inverse square law is featured in a Try This! activity in Chapter 9, page 233.

Although the specific range varies from person to person, it is typically found that a healthy ear can hear intensities as low as 1.0×10^{-12} W/m^2, with intensities at the high end beginning to cause discomfort. The lowest sound intensity that a healthy ear can detect at 1000 Hz, called the **threshold of hearing**, is 1.0×10^{-12} W/m^2. The **threshold of pain** is the sound intensity at which a normal ear begins to feel pain; its value is estimated to be 1.0×10^0 W/m^2 or simply 1.0 W/m^2 at 1000 Hz. An example of this intensity is a jet taking off from a distance of about 20 m; this sound would be painful and without ear protection it could cause permanent damage to a person's ears. Both the threshold of hearing and the threshold of pain are subjective values. Notice that the threshold of pain is greater than the threshold of hearing by a factor of $10^0/10^{-12} = 10^{12}$, or 12 orders of magnitude. This means than a healthy human ear is sensitive to a range of intensities that vary up to a trillion times!

threshold of hearing the lowest sound intensity that a human can hear; estimated to be about 1.0×10^{-12} W/m^2 for a young child

threshold of pain the minimum sound intensity at which a human begins to feel pain; estimated to be 1.0 W/m^2 at 1000 Hz

This tremendous range of intensities is awkward to manipulate mathematically. This is one of the reasons that a different scale, based on logarithms, is used. This scale does *not* measure intensities; rather it measures the ratio of intensities using the value of 1.0×10^{-12} W/m^2 as reference. We now define **intensity level** (symbol β, the Greek letter beta) as the logarithm to the base 10 of the ratio of the intensity of a sound to the reference intensity.

intensity level the logarithm to the base 10 of the ratio of the intensity of a sound to the reference intensity, 1.0×10^{-12} W/m^2; symbol β; measured in bels (B) or decibels (dB)

Thus, $\beta = \log_{10} \frac{I}{I_0}$ where $I_0 = 1.0 \times 10^{-12}$ W/m^2

The unit of intensity level is the bel (B), named after Alexander Graham Bell (Figure 14-12) who invented the telephone.[3] The decibel (dB) is a more convenient size of intensity level, and it will be used often in this text (1.0 B = 10 dB). Thus,

$$\beta \text{(in decibels)} = 10 \log \frac{I}{I_0} \quad \text{(14-3)}$$

where I_0 is the reference intensity, 1.0×10^{-12} W/m^2. (Notice that the base "10" has been omitted from this notation. We will use only base 10 in this equation, so it can be omitted. See the Math Tip box for a further explanation of the logarithmic function.)

© Science Source

Figure 14-12 Alexander Graham Bell (1847–1922) is seen making his famous phone call inaugurating the line linking New York City and Chicago in 1892.

Math Tip

If you understand logarithms, especially the rules of logarithms, you will find intensity level easier to understand.

The logarithmic function is the inverse of the exponential function. Thus, if $x = a^y$, then $\log_a x = y$, which means that the logarithm of "x" to the base "a" is "y." For example, if the base is 10,

$$\log_{10} 10 = 1$$

$$\log_{10} 100 = 2$$

$$\log_{10} 10^n = n$$

It is also useful to be able to apply the rules of logarithms in solving intensity level problems. These rules are

(a) the product rule: $\log (a \times b) = \log a + \log b$

(b) the quotient rule: $\log \left(\frac{a}{b}\right) = \log a - \log b$

(c) the power rule: $\log a^b = b(\log a)$

The bel and decibel are ratios of sound intensities expressed in the logarithmic scale with base 10. When the base 10 is used the logarithm is called common. (We are not concerned here with the natural logarithm, which uses the base e.)

[3]Although Bell is credited with the invention of the telephone, a contemporary inventor named Elisha Gray filed a patent in Washington, D.C., for his version of a telephone on the same morning that Bell filed for his patent. Evidently that situation still causes controversy.

SAMPLE PROBLEM 14-6

Calculate the intensity level of the threshold of pain at 1000 Hz.

Solution

Recall that the threshold of pain is 1.0 W/m^2 at 1000 Hz.

$$\beta(\text{in dB}) = 10 \log \frac{I}{I_0}$$

$$= 10 \log \frac{1.0\,\text{W/m}^2}{1.0 \times 10^{-12}\,\text{W/m}^2}$$

$$= 10 \log 10^{12}$$

$$= 10 \times 12$$

$$= 120\,\text{dB}$$

Thus, the intensity level of the threshold of pain is 120 dB.

Similarly, the reference sound intensity (1.0×10^{-12} W/m^2) could be shown to have an intensity level of 0 dB. This tremendous range of intensities, from 10^{-12} W/m^2 to 10^0 W/m^2, has a range of intensity level from 0 dB to 120 dB. This illustrates the usefulness of a logarithmic scale in compressing large numbers.

Table 14-3 lists the intensities and the corresponding intensity levels of several sounds. The values stated are highly subjective and depend on several factors, so they are at best only approximate.

Table 14-3

Intensities and Intensity Levels of Sounds

Intensity (W/m^2)	Intensity Level (dB)	Ratio of Intensity to the Intensity at 0 dB	Example
10^{-12}	0	1	threshold of hearing
10^{-10}	20	10^2	average whisper at 1 m
10^{-8}	40	10^4	inside a car with engine on
10^{-6}	60	10^6	conversation at 1 m
10^{-4}	80	10^8	noisy street corner
10^{-2}	100	10^{10}	rock concert
10^0	120	10^{12}	threshold of pain

The intensities and intensity levels mentioned so far have been applicable to sounds at 1000 Hz only. The intensity levels perceived by the human ear are somewhat dependent on the frequency of the sound. Figure 14-13 on page 426 shows a plot of audible intensity level as a function of frequency. From the graph, it is evident that the average human ear is most sensitive to sound frequencies from just above 1000 Hz to about 5000 Hz. Lower frequencies must have higher intensity levels in order to be heard equally well.

DID YOU KNOW?

The intensity of sound, measured in watts per square metre (W/m^2), seems to be the exception to the variety of non-uniform units suggested for measurements. Consider sound intensity level. In North America it is normally measured in decibels (dB), but it may also be stated in bels (B).
At least five alternative names have been proposed for the decibel, including logit, decilit, decilog, decomlog, and decilu. In continental Europe, the unit of intensity level is the neper (1.0 Np = 8.686 dB). And there are still other units used in specific industries and applications!

DID YOU KNOW?

The logarithmic function was invented by John Napier, a Scottish lord, in 1594. Logarithms are applied not only to sound intensity levels, but also in radioactive decay, biological growth, concentration of chemicals, compound interest, and the intensity of earthquakes (the Richter scale).

DID YOU KNOW?

Have you ever noticed that television commercials are, in general, louder than the regular programs? The measured intensity of the sound from commercials may be no greater than regular programs. However, the frequencies used in the audio part of commercials correspond to those frequencies human ears are most sensitive to, so the perceived loudness is greater.

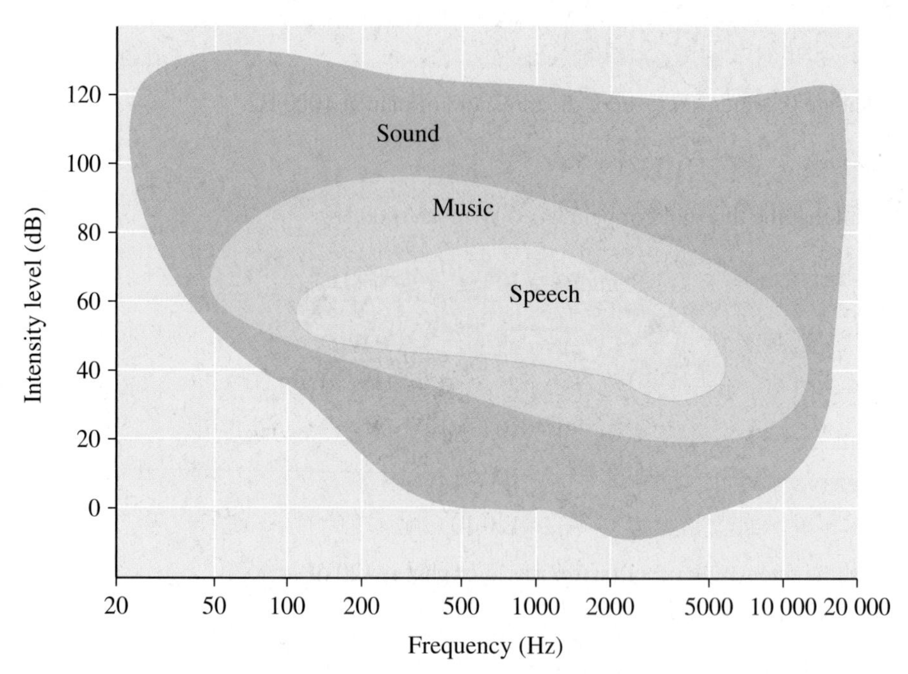

Figure 14-13 Sensitivity of the average human ear to different frequencies.

SAMPLE PROBLEM 14-7

By how much does the intensity level increase when the sound intensity doubles?

Solution

Let I be the initial intensity and I_0 be the reference intensity, 1.0×10^{-12} W/m^2.

$$\begin{aligned}\Delta\beta(\text{in dB}) &= 10 \log \frac{2I}{I_0} - 10 \log \frac{I}{I_0} \\ &= 10 \log \left(\frac{2I}{I_0} \div \frac{I}{I_0}\right) \\ &= 10 \log 2 \\ &= 10(0.301)\,\text{dB} \\ &= 3.0\,\text{dB}\end{aligned}$$

Thus, the intensity level increases by only 3.0 dB.

SAMPLE PROBLEM 14-8

Determine the intensity of an 84 dB sound from a stereo speaker.

Solution

Since $\beta = 84$ dB, $10 \log \frac{I}{I_0} = 84$, so $\log \frac{I}{I_0} = 8.4$

from which $\frac{I}{I_0} = 10^{8.4} = 2.5 \times 10^8$

$$\therefore I = 2.5 \times 10^8 (1.0 \times 10^{-12}\ \text{W/m}^2) = 2.5 \times 10^{-4}\ \text{W/m}^2$$

Thus, the intensity of the 84 dB sound is 2.5×10^{-4} W/m^2.

SAMPLE PROBLEM 14-9

The intensity level of the music at a concert changes from 51 dB to 63 dB. By what factor does the intensity increase?

Solution

Let $\beta_1 = 51$ dB and $\beta_2 = 63$ dB.

$$\Delta\beta = \beta_2 - \beta_1 = 63\ \text{dB} - 51\ \text{dB} = 12\ \text{dB}$$

$$\Delta\beta = 12\ \text{dB} = 10\left(\log\frac{I_2}{I_0} - \log\frac{I_1}{I_0}\right)$$

$$1.2\ \text{B} = \log\left(\frac{I_2}{I_0} \div \frac{I_1}{I_0}\right)$$

$$1.2\ \text{B} = \log\frac{I_2}{I_1}$$

$$15.8 = \frac{I_2}{I_1} \text{ (Use the } 10^x \text{ key on your calculator.)}$$

Thus, the intensity has been increased by a factor of about 16 times. An alternative solution is to find the intensities (in watts per square metre) of the 51 dB and 63 dB sounds, and then calculate their ratio.

EXERCISES

14-17 Starting with sound energy, describe the energy transformations that occur in human hearing.

14-18 State the factor by which the intensity of the first sound in each pair listed exceeds the second sound: (a) 110 dB, 80 dB; (b) 5 B, 3 B.

14-19 If a 30 dB sound is increased in intensity by a factor of 10^4, what is its new intensity level?

14-20 Refer to the plot in Figure 14-13. State the approximate threshold of hearing at (a) 100 Hz and (b) 10 000 Hz.

14-21 Three saxophones are producing sounds each of intensity level 75.0 dB (Figure 14-14). What is the intensity level of the resulting sound?

14-22 Determine the intensity level in decibels of a sound whose intensity is 6.4×10^{-4} W/m^2.

14-23 Find the intensity of a sound whose intensity level is (a) 100 dB and (b) 55 dB.

Stocksnapper/Shutterstock

Figure 14-14 A saxophone (Question 14-21).

14-24 Determine the change in intensity level (in decibels) when a sound intensity is increased by a factor of (a) 4.0, (b) 40, (c) 400. What pattern do you see in the answers?

14-25 The maximum intensity level of a clarinet is 86 dB (Figure 14-15). What is the intensity level if 5 clarinets are each producing sound at that level?

14-26 If 12 violins are producing sound of total intensity level 66 dB, what is the intensity level produced, on average, by each violin?

Andrey_Popov/Shutterstock

Figure 14-15 A clarinet (Question 14-25).

14.3 Infrasonics, Ultrasonics, and Echolocation

infrasonic sounds sound frequencies below 25 Hz, the low end of the human audible range (also called infrasound)

DID YOU KNOW?

The amplitude of the sound produced by human muscles is proportional to the load the muscle is lifting. Scientists use this observation to research muscle action in various activities, including sports.

It was stated earlier that the average maximum human audible range is from about 25 Hz to about 20 kHz. Sound frequencies lower than 25 Hz are called **infrasonic sounds** or *infrasound.* (The prefix *infra* is Latin for "lower than.") Such sounds can occur naturally during earthquakes or atmospheric pressure changes, or even when muscles in the body are activated. Some animals and birds, such as elephants and pigeons, can hear sounds with frequencies as low as 1 Hz. Infrasonic sounds can also be produced by devices such as the blowers in ventilating systems or large oscillating machinery in factories. Prolonged exposure to large amplitude infrasonic sounds can be irritating or, more seriously, can cause headaches, psychological problems, or even physical damage to body parts that resonate with the same frequency.

ultrasonic sounds sound frequencies above 20 kHz, the high end of the human audible range (also called ultrasound)

echolocation the finding of objects using reflected ultrasonic sounds

sonar an acronym for sound navigation and ranging

transducer a device that converts electrical signals into mechanical oscillations and then sound oscillations

Sound frequencies higher than 20 kHz are called **ultrasonic sounds** or *ultrasound.* (The prefix *ultra* is Latin for "higher than.") Many animals have ears that are sensitive to ultrasonic sounds, which explains why dog whistles are designed to attract the attention of dogs without bothering humans. Table 14-4 lists the approximate audible ranges of several animals.

Table 14-4

Approximate Audible Ranges

Animal	Audible Range (Hz)
Human	25 to 20 000
Dog	15 to 50 000
Cat	60 to 65 000
Bat	1000 to 120 000
Porpoise	150 to 150 000
Dolphin	250 to 220 000

The applications and uses of ultrasound are numerous. In nature, some animals, such as porpoises and certain bats, are able to navigate or hunt for prey by using ultrasound. Even in darkness, a bat that emits high frequency sounds can judge what is nearby when it receives the reflected sounds (Figure 14-16). There is evidence that some insects, such as moths, can hear the bats' sounds and thus try to evade their predator.

Humans have learned to mime the ability of these animals in a process called **echolocation**, which is the finding of objects using reflected ultrasonic sounds. This process is based on a technique developed during World War II called **sonar**, an acronym for sound navigation and ranging. In water, echolocation is used to determine the depth of an ocean or lake, or to locate submarines, sunken ships, or even schools of fish. A device called a **transducer** changes electrical signals into mechanical oscillations and then into sound oscillations at frequencies between 20 kHz and 100 kHz. The signals travel through the water, reflect off some object, and return to the transducer, which now acts as a receiver changing the sound waves back to

DID YOU KNOW?

A device called the "SonicScreen" or "Mosquito alarm" is used to deter groups of loitering teenagers by emitting a 17.4 kHz squeal that is ultrasonic to older patrons. According to the manufacturer, the maximum intensity level is 108 dB.

dean bertoncelj/Shutterstock

(a)

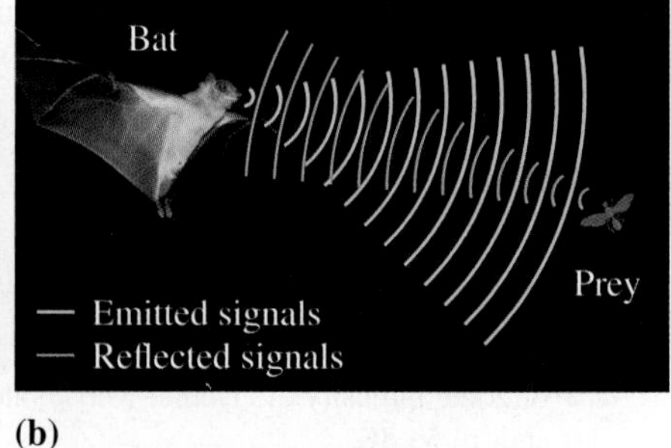

(b)

Figure 14-16 **(a)** A bat. **(b)** A bat emits high-frequency, low-wavelength sounds that reflect off objects such as prey. The bat uses the reflected signals to judge what is in its path.

mechanical oscillations and electrical signals. The ultrasound is sent out in pulses, so that the time between sending a pulse and receiving a reflected pulse can be measured (Figure 14-17). Then knowing the speed of sound in the water, the distance to the object can be found (distance = speed × time). Similar processes are used in air for some cameras that use ultrasonic sound for automatic focusing and for spectacles that help blind people avoid obstacles.

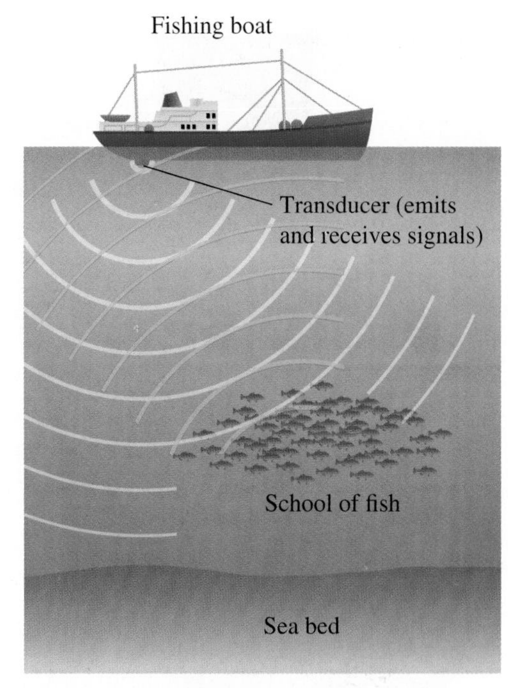

Figure 14-17 Using echolocation in water: the time for ultrasonic sound pulses to travel to and from an object and the speed of the sound in water are used to determine the distance to the object.

SAMPLE PROBLEM 14-10

A transducer on a submarine emits an ultrasonic signal and receives a reflected signal from a second submarine in 868 ms. If the speed of the ultrasound in the water is 1.46×10^3 m/s, how far away is the second submarine?

Solution

The time for the signal to travel the required distance is (868 ms)/2 = 434 ms = 4.34×10^{-1} s.

$$d = vt$$
$$= 1.46 \times 10^3 \text{ m/s} \times 4.34 \times 10^{-1} \text{ s}$$
$$= 6.34 \times 10^2 \text{ m}$$

Thus, the second submarine is 634 m away.

fotoknips/Shutterstock

Figure 14-18 Placing jewellery into an ultrasonic cleaner.

Ultrasonic sounds can be used as a cleaning agent (Figure 14-18). The object to be cleaned is placed in a liquid bath where ultrasonic frequencies in the range of 40 kHz can strike not only the exposed area of the object but also the inaccessible areas. The high frequency sounds have short wavelengths, which helps to jar loose unwanted particles. This technique is used to clean medical instruments, electronic components of watches, rings, etc.

In industry, ultrasound is used to detect otherwise invisible flaws in welded joints and other metal components. Flaws are detected by observing the ultrasonic waves that reflect off the interior of the metal part (Figure 14-19). Ultrasound can also be used to drill holes in hard materials such as glass, gemstones, aluminum, and steel.

DID YOU KNOW?

Daniel Kish, born with retinal cancer in California in 1966, had both eyes removed but has taught himself how to use echolocation to navigate public spaces. He set up an organization called *World Access for the Blind,* which has taught about 1000 students from over 30 countries to use human echolocation to help move around safely. Daniel and his students use up to two distinct clicks per second to detect buildings from about 50 m, cars and bushes from about 5 m, and many other objects. Research has shown that brain structures that process echo information are the same as those that process visual information in sighted people.

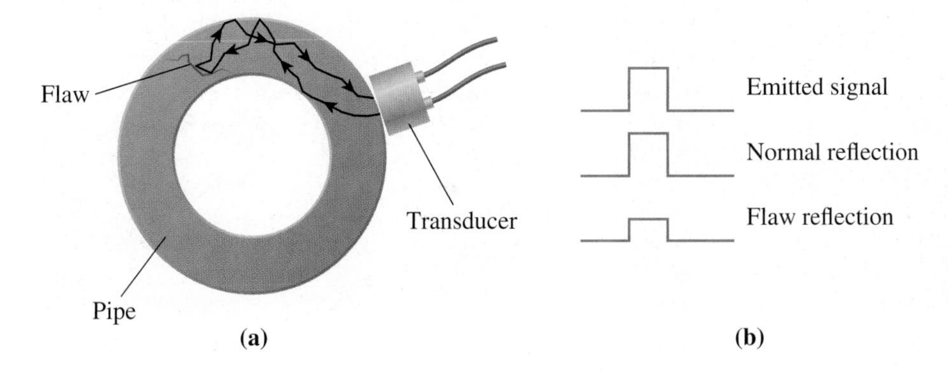

Figure 14-19 Using ultrasound to detect flaws. **(a)** A transducer sends signals that reflect off the metal components of a pipe as well as any flaws in the metal. **(b)** The shape of the reflected signals reveals any flaws.

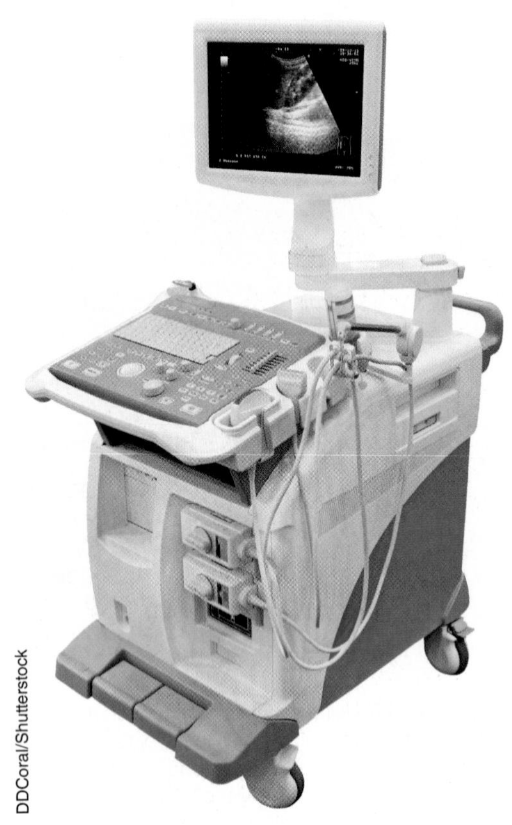
DDCoral/Shutterstock

Figure 14-20 A typical machine that uses ultrasound in diagnostic medicine.

DID YOU KNOW?

The process of diagnosing the brain with ultrasonic sound is called *echoencephalography*. Breaking this word into three parts helps make its meaning clear. "Echo" refers to the reflection of sound waves, "encephalo" comes from the Greek word for brain, *enkephalos*, and "graphy" comes from the Greek *graphos*, which means to write. What does echocardiography mean?

Applications of ultrasonics to the field of diagnostic and therapeutic medicine are both numerous and important. *Diagnostic medicine* is the process of discovering the source of a problem or an abnormality, whereas *therapeutic medicine* is the process of trying to heal or eliminate the problem or abnormality.

Consider first diagnostic medicine. A transducer/receiver is placed snugly against the surface of the patient's body and aimed toward the region to be diagnosed. Pulses of ultrasonic sounds enter the body, with some reflection of the waves occurring at interfaces between organs or tissues within the body. The reflected signals are picked up by the receiver and are then analyzed by computer (Figure 14-20).

The frequency of the emitted pulse depends on the part of the body being diagnosed. Ultrasonic sounds commonly used to test abnormalities of the eye have frequencies about 20 MHz, while those used in prenatal imaging have frequencies about 1.2 MHz. The lower frequency waves have longer wavelengths, allowing them to penetrate the tissue more readily. Higher frequencies have shorter wavelengths, which allow for fine detail on the resulting image.

There are two main types of diagnostic procedures that utilize ultrasound. Two-dimensional views, called B scans, use an array of transducers/receivers to scan across a region of the body. These scans provide information about the eye, liver, kidney, pancreas, heart, breast, and fetus. Three-dimensional scans are becoming more common, especially for investigating the health and development of a fetus in utero. The view is very detailed (Figure 14-21). Using ultrasound has some important advantages over other techniques. Based on how the sound waves interact with tissue, this is a much safer technique than using x rays, for example. In addition, ultrasound can provide better contrast between different types of tissue in some cases.

Consider also the therapeutic nature of ultrasound. It has long been known that heat helps relieve certain aches and pains. But energy from a hot-water bath or an infrared heat lamp cannot penetrate to the inner muscles, joints, and bones, where pain can be severe. Ultrasound can provide therapy to such regions. A transducer sends signals in the 1 MHz region at a relatively high intensity toward the painful region. The signals are absorbed and cause increased thermal activity (i.e., increased temperature), which helps body tissue repair itself (Figure 14-22). Ultrasonic sounds can also be used to destroy unwanted tissues, such as some cancerous tumours. For example, a device called a cavitron ultrasonic surgical aspirator is used by neurosurgeons to remove brain tumours with little or no damage to the surrounding tissue.

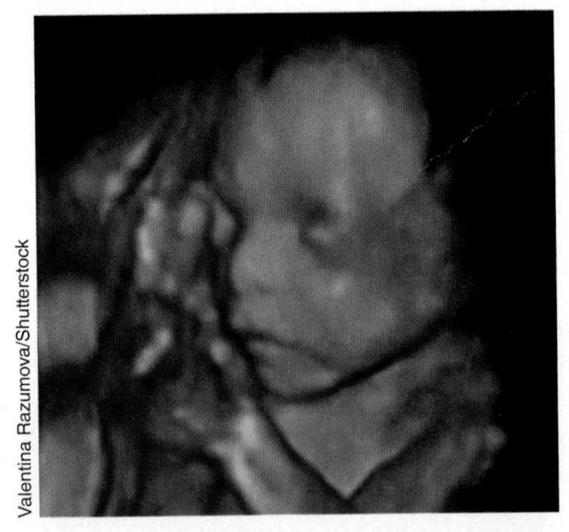
Valentina Razumova/Shutterstock

Figure 14-21 A 3-D ultrasound image of a fetus.

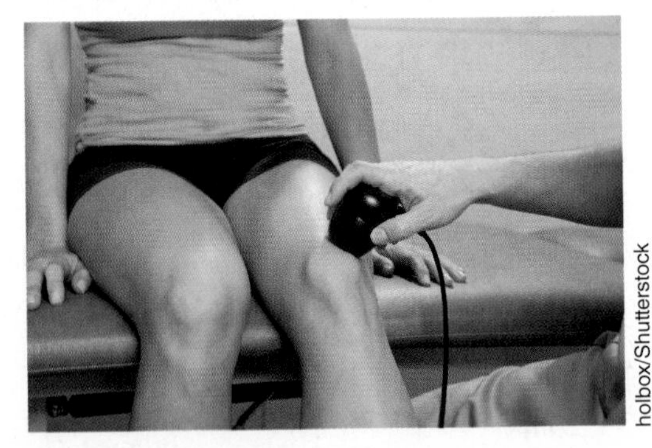
holbox/Shutterstock

Figure 14-22 Ultrasonic therapy.

Ultrasonic sounds have many other useful applications, including dental cleaning, cosmetic therapy, adding humidity to the air, and back-up warning features on cars, all depicted in Figure 14-23. However, not all applications are positive. For example, research has shown that marine life can be seriously affected by ultrasonic signals used for submarine communication and especially by explosive tests carried out to search for mineral deposits in oceans. (As was shown in Table 14-4, marine life such as porpoises and dolphins have very high frequency ranges of hearing.)

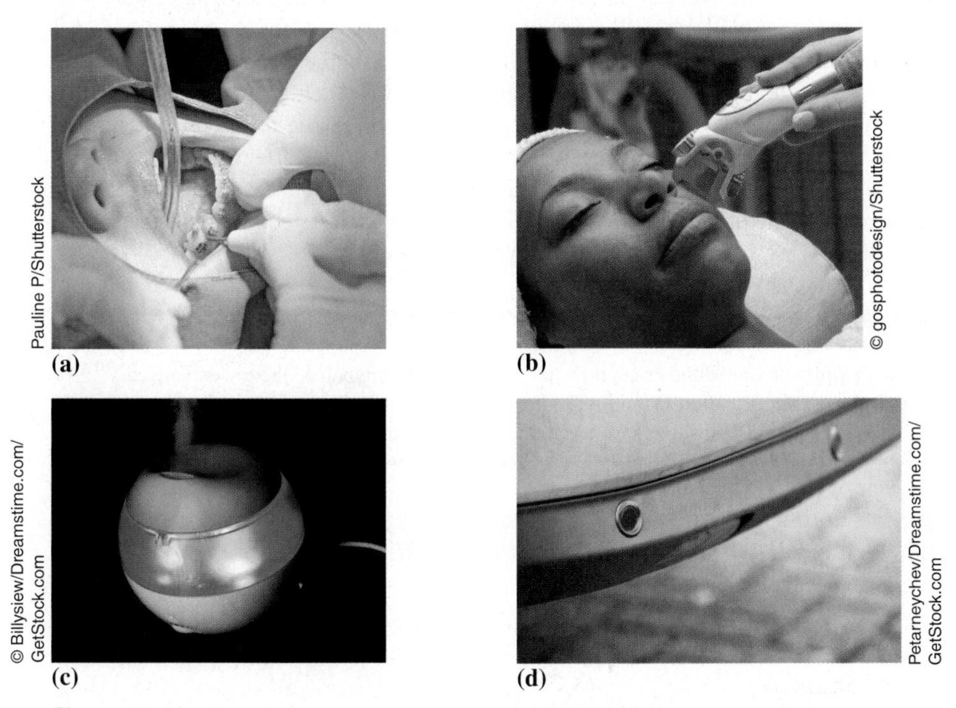

Figure 14-23 More applications of ultrasonic sounds. **(a)** Dental cleaning. **(b)** Cosmetic therapy. **(c)** Adding water vapour to the air. **(d)** Back-up warning system on a car.

EXERCISES

14-27 The speed of sound in fresh water is 1.5×10^3 m/s. Ultrasound sent from a boat on a fresh water lake strikes a school of fish and is received back at the surface after a total elapsed time of 120 ms. How far are the fish from the boat?

14-28 A camera with automatic ultrasonic focusing (Figure 14-24) is located 14.2 m from the subject to be photographed. If the air temperature is 25.0°C, what time interval, in milliseconds, elapses between sending the signal and receiving the reflected signal?

14-29 A bat is emitting ultrasonic sounds of wavelength 4.2 mm. If the air temperature is 5.0°C, determine the frequency of these sounds.

14-30 Describe the energy changes associated with an electric transducer.

14-31 Assume the speed of sound in human soft tissue is 1500 m/s. Determine the *range* of wavelengths, in millimetres, used in ultrasonic medical applications if the frequency ranges from 1.0 MHz to 20 MHz.

Figure 14-24 This lens system uses ultrasound for automatic focusing (Question 14-28).

14-32 What are two medical applications of ultrasonic sound?

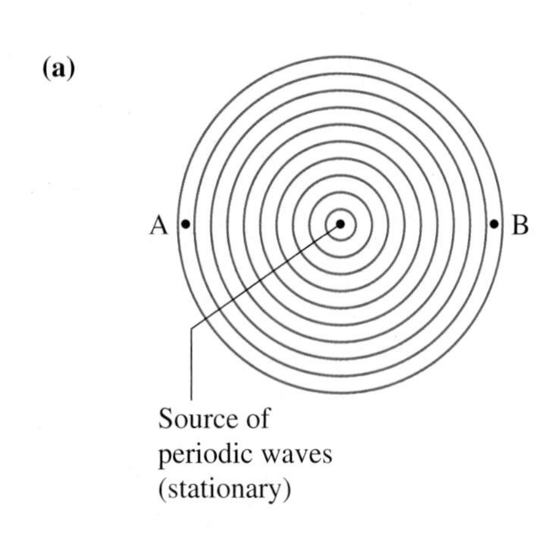

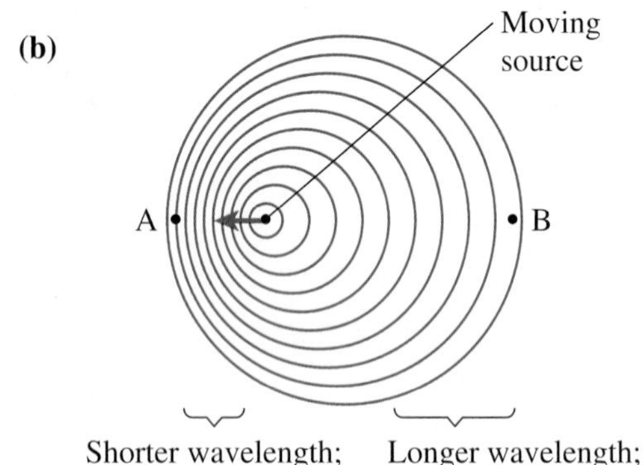

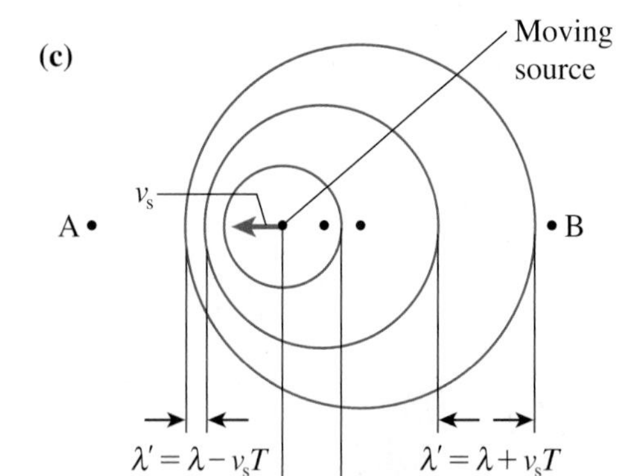

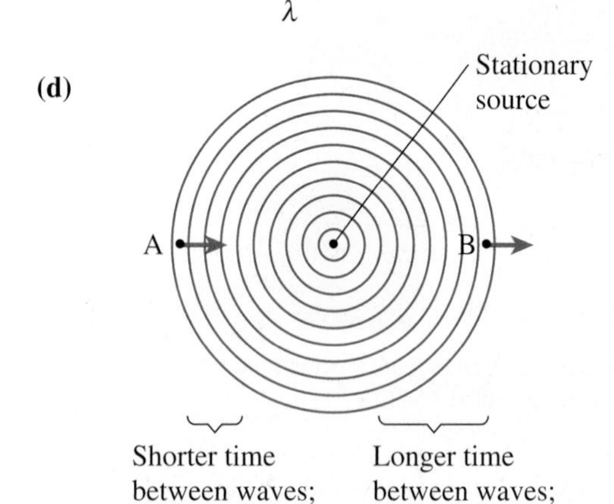

14.4 The Doppler Effect and Subsonic and Supersonic Speeds in Sound

As a noisy race car streaks past a person by the side of a racetrack, the person detects changes in pitch of the sound from the car. As the car approaches, the pitch becomes higher, and at the instant the car passes the observed pitch drops noticeably. Likely you recall hearing this effect from train whistles, car horns, or sirens on ambulances, fire trucks, or police cruisers. Similarly, a person sitting in a fast-moving vehicle with the window open will notice the sound from a stationary object change pitch as the vehicle goes past the sound source. Whether the sound source is moving or the listener is moving, or both, the observer interprets a change in pitch as an observed change in frequency. The observed changing frequency of waves due to the relative motion of the source of waves and the receiver is called the **Doppler effect**. It is named after Christian Johann Doppler, 1803–1853, an Austrian physicist and mathematician who first predicted the phenomenon in 1842. The Doppler effect is commonly experienced for sound waves, but it occurs for other types of waves as well.

DID YOU KNOW?

C. J. Doppler predicted the effect now named after him long before there were any cars or airplanes. Shortly after his initial prediction, in 1845, the effect was tested by having musicians play musical sounds from the open-air car of a moving train in Europe.

TACKLING MISCONCEPTIONS

Comparing Frequency and Pitch

Frequency and pitch are related, but they are *not* the same thing. Frequency is the number of cycles per second of a source of sound, whereas pitch is the interpretation of the sound heard by the human ear.

To understand why the Doppler effect occurs, consider Figure 14-25. In (a), sound waves are travelling outward from a stationary source, and stationary listeners A and B both hear the same frequency. In (b) as well as in the magnified view in (c), the source is moving to the left, causing the waves to be more closely spaced in the direction of motion and less closely spaced in the opposite direction. As the waves approach the stationary listener, A, they have a shorter wavelength than they would if the source were stationary, so the frequency is higher. (Recall that $v = f\lambda$.) For the stationary listener, B, however, the sound has a lower frequency because as the source is moving away the wavelength of the sound is longer. In (d), the source is once again stationary, while listener A is moving toward the source in the opposite direction to the wavefronts, thus experiencing a higher frequency. Listener B is moving away from the source in the same direction as the wavefronts, thus experiencing a lower frequency.

Calculating the Observed Frequency

The observed frequency heard by the listener can be found in terms of the frequency of the source, the speed of sound in air, and the speed of the source or the speed of the listener, or both. The equations derived here apply only if the relative motion of

Figure 14-25 The Doppler effect. **(a)** The source is stationary. Both listeners A and B hear the same frequency of sound. **(b)** As the source is moving to the left, listener A hears a higher frequency and listener B hears a lower frequency than in (a). **(c)** A magnified view of the situation in (b) showing the wavelengths detected by the observers. **(d)** With the source stationary, listener A moving toward the source hears a higher frequency than listener B moving away from the source.

Doppler effect the observed changing frequency of waves due to the relative motion of the source of waves and the receiver

Table 14-5

Symbols Used in Analyzing the Doppler Effect

f'	=	observed frequency heard by listener
f	=	frequency of sound emitted by source
λ'	=	observed wavelength heard by listener
λ	=	wavelength of sound emitted by source
T'	=	observed period heard by listener
T	=	period of sound emitted by source
v	=	speed of sound in air (or other medium)
v_S	=	speed of source
v_L	=	speed of listener

the source and listener is along an imaginary line joining them. (The Doppler effect equations for light are somewhat different because light has a maximum speed.) The symbols listed in Table 14-5 will be used in finding the relationships.

Both v_S and v_L are measured relative to the air (or other medium), which is assumed to be at rest relative to Earth's frame of reference.[4]

Consider the situation in which the listener is stationary and the source is moving at a speed v_S, as in Figure 14-25 (b) and (c). The sound from the source has a period T, a frequency f, a speed in air of v, and a wavelength in air of $\lambda = v/f$. As the source approaches the listener, the wavelength decreases because, by the time the one wave has travelled a wavelength λ, the source has moved a distance $d = v_S T$. Thus, the new wavelength is

$$\lambda' = \lambda - v_S T$$

$$\lambda' = \frac{v}{f} - v_S T$$

$$\lambda' = \frac{v}{f} - \frac{v_S}{f}$$

$$\lambda' = \frac{v - v_S}{f}$$

Finally, the observed frequency, f', heard by the listener is

$$f' = \frac{v}{\lambda'}$$

$$f' = \frac{v}{(v - v_S)/f} \quad \text{for a moving source } \textit{approaching} \text{ a stationary listener}$$

$$\text{or } f' = f\left(\frac{v}{v - v_S}\right)$$

Notice that this equation will always yield an observed frequency, f', higher than the source frequency, f.

Similarly, if the source is moving away from the listener, the wavelength increases to $\lambda' = \lambda + v_S T$. In Question 33 you are asked to prove that the resulting observed frequency heard by the listener is

[4]Frames of reference were discussed in Section 4.6.

$f' = f\left(\frac{v}{v + v_S}\right)$ for a moving source *receding* from a stationary listener.

Notice that this equation will always yield an observed frequency lower than the source frequency.

We can combine these two equations in one convenient equation:

$$f' = f\left(\frac{v}{v \mp v_S}\right) \text{ for a moving source and a stationary listener} \quad (14\text{-}4)$$

where "–" is used in the denominator when the source moves toward the listener and "+" is used when the source is moving away from the listener.

SAMPLE PROBLEM 14-11

On a cold day, when the speed of sound in air is 326 m/s, a truck horn is blaring a 348 Hz sound near a dangerous intersection.

(a) If the truck is travelling at 22.4 m/s, what observed frequency does a stationary listener at the intersection hear as the truck is approaching? receding?

(b) What is the observed wavelength of the sound as the truck is approaching?

Solution

(a) When the truck is approaching,

$$f' = f\left(\frac{v}{v - v_S}\right) = (348 \text{ Hz})\left(\frac{326 \text{ m/s}}{326 \text{ m/s} - 22.4 \text{ m/s}}\right) = 374 \text{ Hz}$$

When the truck is receding,

$$f' = f\left(\frac{v}{v + v_S}\right) = (348 \text{ Hz})\left(\frac{326 \text{ m/s}}{326 \text{ m/s} + 22.4 \text{ m/s}}\right) = 326 \text{ Hz}$$

Thus, the observed frequencies are 374 Hz and 326 Hz when the truck is approaching and receding respectively.

(b) From (a), the approaching frequency is 374 Hz.

$$\lambda' = \frac{v}{f'} = \frac{326 \text{ m/s}}{374 \text{ Hz}} = 0.872 \text{ m}$$

Thus, the observed wavelength is 0.872 m when the truck is approaching the intersection.

Next consider what happens when the source is stationary and the listener is moving. As illustrated in Figure 14-26, the approaching listener has a greater number of wavefronts striking the ears per second than the stationary listener. Let us compare the frequency heard by the two listeners. If the listener is travelling with a speed v_L, then in any time t the listener travels a distance $d = v_L t$ toward the source. The number of wavelengths received by this listener will exceed the number received by the stationary listener by d/λ. (For example, the moving listener in the diagram travels 2.0 cm and the wavelength is 0.5 cm, giving an extra 4 wavelengths encountered in time t.) Now since $d = v_L t$, the total number of extra wavelengths received by the listener in time t is $v_L\, t/\lambda$. Dividing by time t to translate this to the number of extra cycles *per second,* we have v_L/λ. Thus,

$$f' = f + \frac{v_L}{\lambda}$$

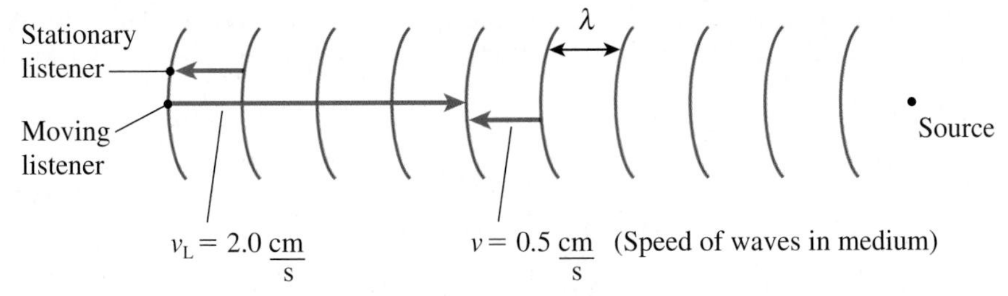

Figure 14-26 Deriving an equation for the Doppler effect when the listener is moving. A stationary listener would encounter one wavelength in the same time that the moving listener encounters 4 + 1 = 5 wavelengths.

$$f' = f + \frac{v_L}{v/f}$$

$$f' = f + \frac{fv_L}{v}$$

$$f' = f\left(1 + \frac{v_L}{v}\right)$$

$$f' = f\left(\frac{v + v_L}{v}\right)$$

for a listener approaching a stationary source.

For a listener moving away from the source, the number of wavefronts per second intercepted is less than for a stationary listener. It can be shown (Question 34) that in this case

$f' = f\left(\frac{v - v_L}{v}\right)$ for a listener receding from a stationary source.

Combining these two equations into one convenient equation for a stationary source and a moving listener, we have

$$f' = f\left(\frac{v \pm v_L}{v}\right) \text{ for a moving listener and a stationary source} \qquad (14\text{-}5)$$

where "+" is used in the numerator when the listener approaches the source and "−" is used when the listener moves away from the source.

Notice that when the listener is approaching the source, this equation gives a higher frequency and when the listener is receding, it gives a lower frequency.

SAMPLE PROBLEM 14-12

A motorcycle rider is approaching a stationary truck at a speed of 22.4 m/s. The truck's horn is sounding at 348 Hz, and the speed of the sound in air is 326 m/s. What observed frequency does the rider hear? (Notice that these values correspond to those given in Sample Problem 14-11, but in this case the source is stationary and the listener is moving.)

Solution

$$f' = f\left(\frac{v + v_L}{v}\right) = (348 \text{ Hz})\left(\frac{(326 \text{ m/s} + 22.4 \text{ m/s})}{326 \text{ m/s}}\right) = 372 \text{ Hz}$$

This value is slightly different than what was found in Sample Problem 14-11, although our ears may be unable to detect the difference.

What happens if *both* the source and the listener are in motion relative to the medium? In this case both factors are taken into consideration, and the observed frequency is

Math Tip

Be careful with the signs in Eqn. 14-6! A "+" in the numerator means the listener is moving toward the source, whereas a "+" in the denominator means the source is moving away from the listener.

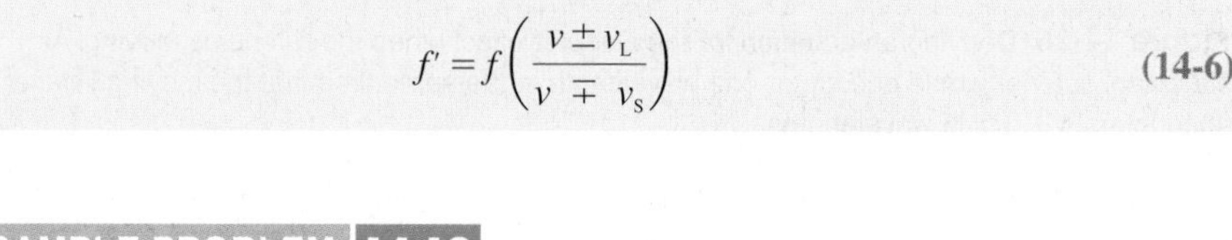

$$f' = f\left(\frac{v}{v \mp v_S}\right)\left(\frac{v \pm v_L}{v}\right)$$

$$f' = f\left(\frac{v \pm v_L}{v \mp v_S}\right) \quad \textbf{(14-6)}$$

SAMPLE PROBLEM 14-13

A fire truck, with a siren screeching at 1.00×10^3 Hz, is moving at 21.2 m/s in the same direction as a cyclist who is moving at 8.34 m/s. The speed of the sound in air is 341 m/s. What observed frequency does the cyclist hear?

Solution

Here the source (the fire truck) travels at v_S = 21.1 m/s *toward* the listener and the listener travels at v_L = 8.34 m/s *away from* the source. The general expression in Eqn. 14-6 therefore becomes $f' = f\left(\frac{v - v_L}{v - v_S}\right)$ given the relative motions involved here.

$$f' = f\left(\frac{v - v_L}{v - v_s}\right) = (1.00 \times 10^3\,\text{Hz})\left(\frac{341\,\text{m/s} - 8.34\,\text{m/s}}{341\,\text{m/s} - 21.2\,\text{m/s}}\right) = 1.04 \times 10^3\,\text{Hz}$$

Thus, the observed frequency is 1.04×10^3 Hz. Prove for yourself that if the cyclist were travelling in the opposite direction, toward the source, the observed frequency would be 1.09×10^3 Hz.

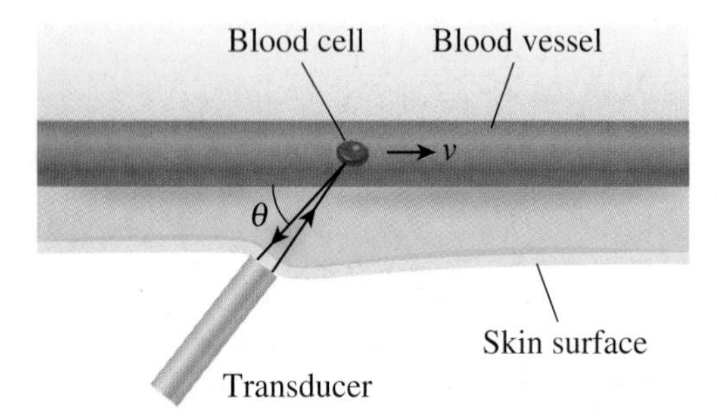

Figure 14-27 Using the Doppler effect to measure the speed of blood. A continuous ultrasonic sound wave is reflected off the moving blood cells. The transducer contains both a transmitter and a receiver.

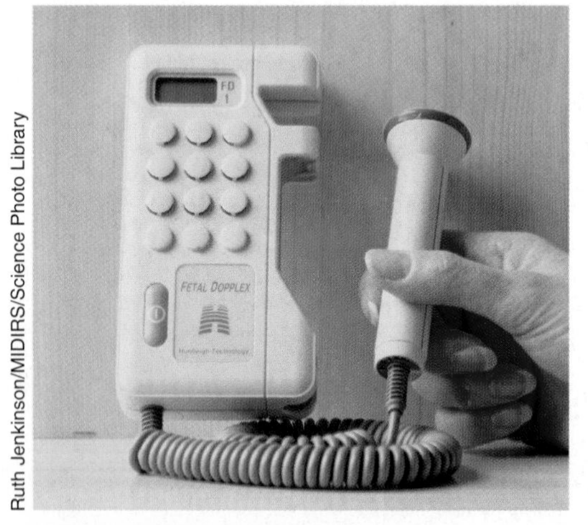

Ruth Jenkinson/MIDIRS/Science Photo Library

Figure 14-28 A Doppler-effect unit for monitoring the heartbeat of a fetus.

Applications of the Doppler Effect

Although the Doppler effect was first studied in 1842, practical applications did not evolve until the twentieth century. Anyone who has thrown a baseball to determine its speed at an amusement park or science centre has taken advantage of the Doppler effect. The device that measures the speed of the ball emits ultrasonic waves that reflect off the moving ball back toward a receiver. The change in frequency of the ultrasound is used to determine the speed of the ball.

A similar technique is used to measure the speed of blood cells in veins and arteries (Figure 14-27), although the complexity of the calculation is somewhat greater because the ultrasonic signals are usually at an angle to the motion of the blood cells. Information about the speed of blood helps locate constrictions in blood vessels because the blood flows more quickly in narrower passages.

The Doppler effect is also applied to monitor the growth of an unborn child between 12 and 20 weeks after conception, when other techniques are unreliable. In this case, ultrasound is reflected from the heart of the fetus to hear the heartbeat in real time using a simple handheld unit available in every doctor's office (Figure 14-28). In fact, the technology is so straightforward and cost-effective that expectant parents can rent their own handheld Doppler ultrasound unit for use at home throughout the pregnancy!

Subsonic and Supersonic Speeds

So far in our analysis of the Doppler effect in sound we have considered only speeds of the source or listener that are *subsonic speeds,* in other words, speeds that are less than the speed of sound in air. Now we will consider what happens when the speed of a source of sound approaches the speed of sound in air at that location. Figure 14-29 (a) shows the dramatic vapour cone that forms around a jet, and (b) illustrates the jet's sound waves when its speed reaches the speed of sound. The sound, in effect, is constantly catching up to the wavefronts in the direction it is moving.

When the speed of the source equals the speed of sound in air at that location, the speed of the source is given the name Mach 1. In general, the **Mach number** of a sound source is the ratio of the speed of the source to the speed of sound in air at that location. Thus, Mach number $= v_S/v$. Mach numbers above 1 indicate *supersonic speeds,* in other words, speeds that exceed the speed of sound. Thus, at 0°C near the surface of Earth, Mach 2.30 $= 2.30 \times 331$ m/s $= 761$ m/s. Mach number is named in honour of Ernst Mach, 1838–1916, an Austrian physicist and philosopher. (Do not confuse supersonic with ultrasonic. Supersonic refers to speeds; ultrasonic refers to frequencies.) The record high speed of a fixed-wing aircraft is approximately Mach 25. It is held by the space shuttle *Columbia* and was achieved where the speed of sound was about 300 m/s.

As an aircraft or other object approaches the speed of sound, the air compressions it produces are extremely closely spaced, creating a high pressure region called the **sound barrier**. To exceed the speed of sound, extra thrust is needed until the aircraft "breaks through" the sound barrier. Aircraft must be specially constructed to withstand the oscillations caused in breaking through the sound barrier to reach supersonic speeds.

When an object travels at supersonic speeds, the source of sound actually overtakes its own sound waves, causing the compressions to overlap each other. As you can see in Figure 14-30 (a), there are many instances where constructive interference occurs as compressions meet compressions.[5] In three dimensions, the overlapping compressions form what is called a **shock wave cone** that spreads outward from the source. In two dimensions, a similar effect produces a **bow wave**, commonly seen on water when the speed of a boat exceeds the speed of the waves in the water. The half-angle, θ, that a shock wave makes with the direction of travel can be found by using $\sin\theta = v/v_S$, as shown in Figure 14-30 (b). Thus, $1/\sin\theta = v_S/v =$ Mach number, or $\sin\theta = 1/$Mach number. The photo in (c) shows an important step in testing an aircraft's design in a wind tunnel before a high-speed craft is built.[6]

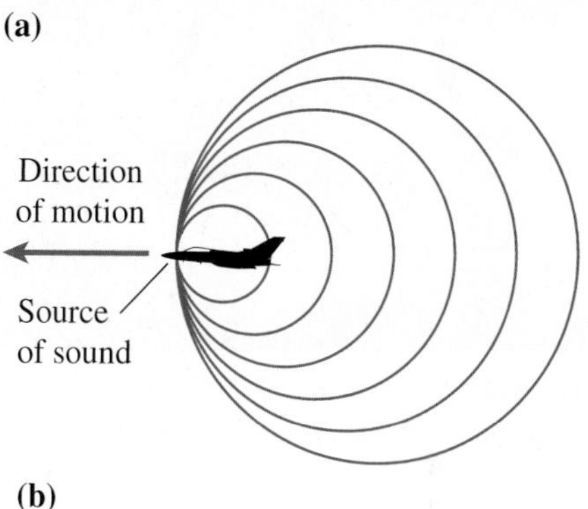

Figure 14-29 **(a)** As a jet's speed nears the speed of sound, a distinct vapour cone forms around it. **(b)** The circular waves in this diagram represent the jet's sound waves when its speed reaches the speed of sound in the surrounding air.

Mach number the ratio of the speed of the source to the speed of sound in air at that location

sound barrier a high pressure region in the air as an object approaches the speed of sound

shock wave cone an overlapping series of compressions in the air where an aircraft is travelling at a supersonic speed

bow wave a two-dimensional overlapping of water waves created where the speed of a boat exceeds the speed of the waves

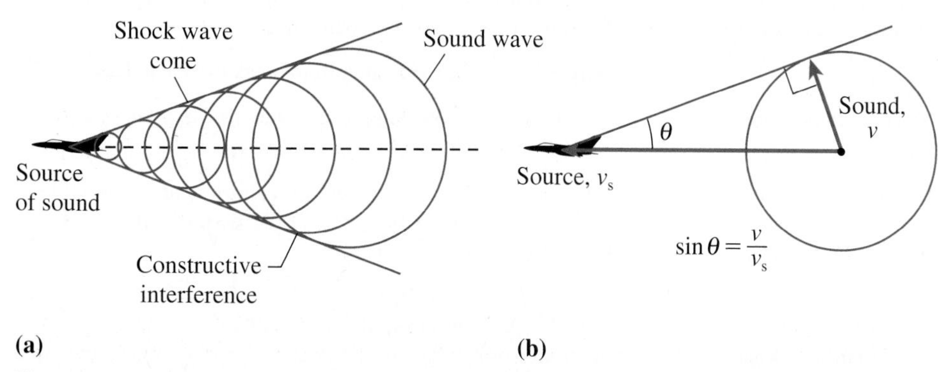

Figure 14-30 The source of sound is travelling faster than the sound. **(a)** In three dimensions, a shock wave cone is produced. **(b)** The Mach number can be expressed in terms of the half angle of a shock wave cone. **(c)** To help design high-speed aircraft, a model is tested in a wind tunnel where special photography reveals wave patterns as the air moves past the model at very high speeds.

[5]Constructive and destructive interference were described in Section 13.7.
[6]The specialized photography used to obtain the wind-tunnel image is called *Schlieren photography*.

sonic boom a very loud noise created near Earth's surface as a supersonic aircraft above creates a shock wave cone

If a shock wave from a supersonic aircraft approaches the ground before its energy is dispersed, it generates a **sonic boom**, a thunder-like noise that can damage buildings and ears, sometimes causing psychological problems to people exposed to them. To prevent serious sonic booms, supersonic aircraft are usually required to travel at very high altitudes. Military aircraft (Figure 14-31) commonly travel at supersonic speeds, but passenger airlines generally do not.

Sonic booms can also be produced by objects that are not emitting actual sound waves. A supersonic bullet, for example, causes compressions of air that produce a loud crack as the bullet passes by. The crack of a lion-tamer's whip or a bullwhip is another example (Figure 14-32) in which the sound produced is the mini-sonic boom arising from the tip of the whip travelling faster than the speed of sound in air.

James Steidl/Shutterstock

Figure 14-31 This supersonic SR-71 Blackbird military aircraft can fly at speeds up to Mach 3.

Maksym Bondarchuk/Shutterstock

Figure 14-32 A typical bullwhip.

EXERCISES

14-33 Prove that for a moving sound source that is receding from a stationary listener, the observed frequency is $f' = f\left(\frac{v}{v + v_S}\right)$. Be sure to indicate the meaning of each symbol. (**Hint:** Begin by writing an equation for the observed wavelength.)

14-34 Prove that for a moving listener who is receding from a stationary sound source, the observed frequency is $f' = f\left(\frac{v - v_L}{v}\right)$. Indicate the meaning of the symbols used. (**Hint:** What happens to the number of wavefronts heard by the listener moving in the same direction as the sound waves?)

14-35 A hollow, perforated plastic tube is twirled around, producing a sound with a frequency of 922 Hz. A listener across the room hears higher and lower frequencies as the end of the tube approaches and recedes at a speed of 25.5 m/s. If the speed of sound in air is 344 m/s, determine the two observed frequencies heard.

14-36 A bat, moving at a speed v_S, emits ultrasonic sounds of frequency 8.00×10^4 Hz. The bat's stationary prey "hears" an observed frequency of 8.55×10^4 Hz. If the speed of the ultrasound in air is 3.40×10^2 m/s, how fast is the bat flying?

14-37 A stationary submarine, A, is emitting ultrasonic sounds at 50 000 Hz in water, where the sound travels at 1450 m/s. At what observed frequency does submarine B, moving at 8.500 m/s toward A, receive the ultrasound? (Assume four significant digits.)

14-38 An ambulance with a siren at 824 Hz is travelling east at 90.0 km/h. A motorcycle is travelling at 72.0 km/h in the opposite direction. If the speed of the sound in air is 337 m/s, determine the observed frequency heard by the rider when the ambulance is (a) approaching and (b) receding.

14-39 Determine each of the following speeds in metres per second. The assumed speed of sound at the location of the sound is given in parentheses.

(a) The highest speed of air in a wind tunnel is approximately Mach 27. (340 m/s)

(b) The military aircraft shown in Figure 14-31 is travelling at Mach 3.0. (310 m/s)

(c) A hockey puck is found to be travelling at Mach 0.084. (330 m/s)

14-40 An aircraft travelling at 254 m/s has a speed of Mach 0.784. What is the local speed of sound in air?

14-41 Why is it difficult for an aircraft to break the sound barrier?

14-42 Can an aircraft travelling at subsonic speeds produce a shock cone? Why or why not?

14-43 What happens to the sharpness of a shock cone produced by a supersonic aircraft as the speed of the aircraft increases? Use a sketch to illustrate your answer.

14-44 The complete angle from one side of a shock cone to another is 36.4° when the speed of sound is 302 m/s. Find the speed of the source producing this shock cone. Express your answer as a Mach number.

14-45 Discuss the advantages and disadvantages of supersonic air travel (a) near large cities, (b) over rural areas, and (c) over the ocean.

14.5 Diffraction, Interference, and Resonance in Sound

Interference of waves in one dimension and diffraction of waves in two dimensions were presented in Chapter 13. Do sound waves in air, which are longitudinal waves, display these important characteristics? Furthermore, do sound waves exhibit resonance in a manner similar to mechanical systems, also studied in Chapter 13? These questions are explored in this section.

Diffraction of Sound

We can hear sound coming through open doorways or around corners very easily. The bending effect of waves in a single medium through an aperture or around a barrier is called *diffraction*. As described in Section 13.8, the amount of diffraction depends on the ratio of the wavelength of the waves to the width of the aperture or barrier. When this ratio is greater than or equal to about one, a large amount of diffraction occurs. Since the wavelength of sound waves in air varies from a few centimetres to about 14 m, sound can diffract easily around most objects in our daily experience.

SAMPLE PROBLEM 14-14

Fermi Question: Estimate the frequency of a sound that diffracts easily through an open doorway.

Solution

From $v = f\lambda$, we can estimate f if we can estimate v and λ. Assume that the temperature of the air in the room is about 20°C, so the speed of sound in the air is about 340 m/s. Next, assume that the width of a doorway is about 80 cm, or 0.8 m. Thus,

$$f = \frac{v}{\lambda} = \frac{340\ \text{m/s}}{0.8\ \text{m}} = 4 \times 10^2\ \text{Hz}$$

Therefore, the sound will diffract easily at a frequency of about 400 Hz or less. (A lower frequency means a longer wavelength and thus a good chance of diffraction.)

The answer to this sample problem reveals a clue about the design of audio loudspeaker systems. A typical loudspeaker setup (Figure 14-33) consists of two, three, or more separate loudspeakers of different sizes. The long wavelength sounds from the largest speaker are easily diffracted around obstacles and through doorways. The short wavelength sounds from the smaller speakers diffract much less around obstacles greater than about 10 cm, so the sound is more directional.

Figure 14-33 A typical loudspeaker system, with three sizes of speakers.

Beats and Beat Frequency

A special type of interference that is useful in music occurs when two sound sources have nearly the same frequency. The sound waves interfere constructively when they are in phase and then destructively when they are out of phase. The resulting periodic sounds of increasing and decreasing loudness are called **beats**. Once you have heard beats (as suggested in the Try This! activity) you will find them easy to recognize.

beats the interference pattern of periodic sounds of increasing and decreasing loudness caused by sounds of nearly equal frequency

Figure 14-34 on page 440 illustrates how the addition of two waves of slightly different wavelength (and thus slightly different frequency) can be added together. In this case, transverse waves are used to represent sound waves, and the principle of

DID YOU KNOW?

Loudspeakers of the future may be designed using nanotechnology.[7] Currently, researchers are embedding carbon nanotubes into an extremely thin membrane. Periodic electric current sent to the membrane causes the membrane to undergo rapid heating (when the current begins) and cooling (when the current stops). The resulting rapid expansions and contractions cause sound waves to emanate from the membrane. The membrane is flexible and inexpensive, and may result in walls and even clothing becoming loudspeakers!

TRY THIS!

Listening to Beats

Use a search engine to find at least one example of "acoustical beats" on YouTube or other source. If you can control variables such as frequency or amplitude, determine the effects of increasing or decreasing them.

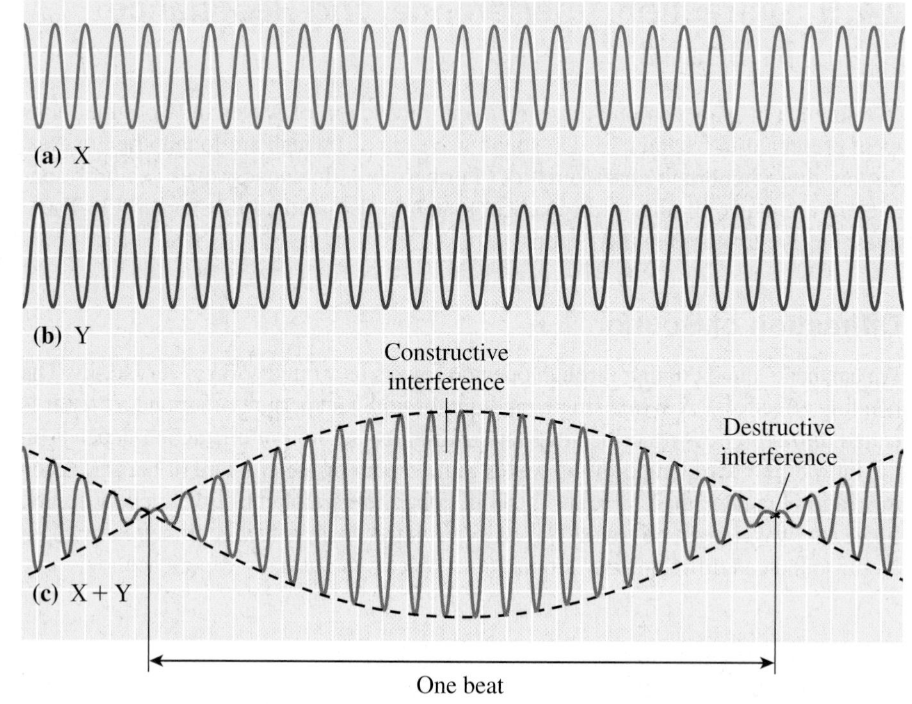

Figure 14-34 Waves X and Y have slightly different wavelengths and thus slightly different frequencies. When the principle of superposition is applied to add the two waves, the resultant pattern illustrates why beats have periodic increasing and decreasing loudness.

superposition (presented in Section 13.7) is applied to add the two waves together. The resulting wave at the bottom reveals a periodically changing overall amplitude, which is what we hear as beats.

If the sources have frequencies of 256 Hz and 258 Hz, there will be 2 beats heard per second, although the overall sound has a frequency of 257 Hz, which is the *average* of the two frequencies. Similarly, if the 256 Hz source is sounded with a 254 Hz source, there will be 2 beats heard per second with an overall frequency of 255 Hz. The number of beats per second is called the **beat frequency** (symbol f_{beat}); it is measured in hertz. Numerically it is the difference in the two frequencies. Thus,

beat frequency the number of beats per second; symbol f_{beat}

$$f_{beat} = |f_1 - f_2|$$

Therefore, the beat frequency in the examples above is 2 Hz.

For most sounds, a beat frequency up to about 7 Hz sounds relatively pleasing. However, with beat frequencies of about 15 Hz we begin to hear the two different notes that cause the beats, a sound that is discordant and unpleasant.

Although electronic tuning devices are convenient to use to tune musical instruments, the principle of beats can be applied to tune various instruments, including a piano. Most notes on a piano are produced by a set of two or three strings that must have the same frequency. Consider, for example, note A above middle C. Its frequency is 440 Hz, so a tuning fork of that frequency is sounded simultaneously with one string while the other two strings are kept silent. (An electronic tuning fork is often used for convenience.) If beats are heard, the tension in the string is adjusted until the beats are eliminated. Then the first string is sounded simultaneously with each of the other strings in turn; again, beats are used to help get the strings tuned.

DID YOU KNOW?

Pipe organs are able to use beats to produce pleasant vibrato sounds. If the organ register is set at *voix céleste*, for example, the sound that results from pressing a single key is made by two pipes almost equal in size. The pipes produce a sound with a beat frequency of about 2 to 3 Hz. These undulating sounds resemble the string section of an orchestra playing in unison. ("Voix céleste" means heavenly voice.)

[7]Nanotechnology, nanoparticles, and nanotubes were introduced in Chapter 1, page 4.

TRY THIS!

Investigating Beats

Using a stringed instrument such as a guitar, pluck two adjacent strings, one freely oscillating and the thicker one suppressed with one finger such that they produce the same note. Now detune one string slightly and pluck both strings again, listening for beats. Try various combinations of notes, and then describe what you observe about the production of beats on a single instrument, and how you could use beats to tune the instrument.

Resonance in Sound

Mechanical resonance (see Section 13.4) is a natural mechanical oscillation that results in a large buildup of amplitude. Similarly, resonance in sound is a natural sound oscillation of large amplitude; it can be produced by a small driving force. Likely you have heard the high-pitched squeal produced when a person runs a moist finger around the rim of a long-stemmed glass. If a nearby sound having the same frequency as the resonant frequency of the glass has a large enough amplitude, the glass may shatter (Figure 14-35).

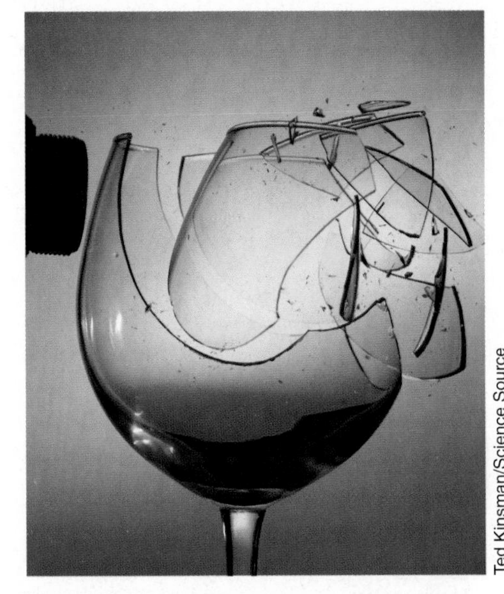

Ted Kinsman/Science Source

Figure 14-35 The energy of amplified sound waves having the resonant frequency of this wine glass causes the glass to shatter.

TRY THIS!

Wine Glass Resonance

Choose a long-stem wine glass (but *not* one made of fine crystal that may shatter) and add water to a depth of about 2 cm. Holding the base of the glass firmly on a table, glide a moistened finger around the edge of the glass. With an appropriate amount of friction between your finger and the rim, the glass will oscillate at the resonant frequency, producing a loud sound. (You may need practice to achieve a satisfactory sound.) Predict what effect changing the water level will have on the pitch of the sound produced. Try it. Then for added interest, search for "wine glass music" or the "glass armonica" on the Internet to see (and hear) wine-glass resonance in action!

Resonance is important in the production of sound by musical instruments. Each note from a musical instrument consists of at least one resonant frequency, called the fundamental frequency, combined with other resonant frequencies that are multiples of the fundamental. Details about how the main families of musical instruments produce resonance are found in Section 14.8.

A *sympathetic oscillation* can be set up in a sound source initially silent when it receives energy from another sound source that has the same resonant frequency. For example, if one tuning fork is struck and held close to a second, identical fork, the second fork soon begins oscillating at the same frequency, in other words, in sympathy. Energy is transferred through the air molecules by waves of the same frequency as the forks. The energy transferred is more noticeable if tuning forks of equal frequency are mounted on resonance boxes, shown in Figure 14-36.

Lonesome_tiger/Shutterstock

Figure 14-36 Mounted tuning forks.

DID YOU KNOW?

In some cases it is necessary to prevent an object from oscillating with maximum amplitude at its own resonant frequency or frequencies. A good loudspeaker, for example, is designed in such a way that its resonant frequencies, which are many, are controlled so that none is very dominant. Otherwise, the sound of some frequencies would be louder than that of other frequencies.

TRY THIS!

Observing Sympathetic Oscillations

Use a piano and/or a set of two acoustic guitars to observe sympathetic oscillations.

- On the piano, suppress the right (sustaining) pedal to free all the strings of the piano. Using your best singing voice, direct a loud, single note into the resonance box of the piano, and listen for some piano strings to oscillate in sympathy. Try various other notes.
- Using the set of two acoustic guitars that are in tune with each other, arrange them to face one another about 30 cm apart. Pluck the middle of one string on one guitar, wait a couple of seconds, and then stop the oscillations while listening for the effect on the second guitar. Try various notes, loudness, and waiting times, and describe what you observe.

EXERCISES

14-46 What evidence verifies that sound energy travels by means of waves?

14-47 The frequency range of a woman's soprano voice is from 262 Hz to 1047 Hz.

(a) Assuming the air temperature is 25.0°C, determine the wavelengths of these sounds.

(b) For each of the two frequencies named, what aperture width would allow the waves passing through to diffract a maximum amount?

14-48 Why are high-frequency loudspeakers more "directional" than low-frequency speakers?

14-49 Four tuning forks, with frequencies of 384 Hz, 380 Hz, 379 Hz, and 398 Hz, are available. What are all the possible beat frequencies when any two forks are sounded simultaneously? Which beat frequencies would sound relatively pleasant?

14-50 An electronic tuning fork set at 440 Hz is used to tune the A string of a guitar, and a beat frequency of 4 Hz is heard. Then the tension in the string is increased very slightly and a new beat frequency of 5 Hz is heard.

(a) What was the initial frequency of the string?

(b) What should be done to tune the string to 440 Hz?

14-51 How does the resonant frequency of a tuning fork relate to the size of the tuning fork?

14.6 Tonal Quality

quality (of a musical tone) a subjective measure of how pleasing a tone is to the human ear (also called tonal quality or timbre)

tone a sound produced by a musical instrument or a singing voice

harmonics the collection of frequencies of a musical tone

Assume that you hear a 440 Hz tone equally loudly from a piano, then from a violin, and finally from a saxophone. The three tones are the same frequency and loudness, but they sound different because each has a unique quality. The **quality** of a musical tone (or tonal quality) refers to how pleasing the tone is to the human ear. In this context, a **tone** is a sound produced by a musical instrument or a singing voice. Quality is also known by its French name, *timbre*.

Obviously the quality of a tone as interpreted by an individual listener depends on many subjective factors, including the types of music and the types of musical instruments the listener enjoys. However, we are not concerned with those subjective factors here. Rather we will study objectively how quality depends on the harmonic structure of the tone.

Most tones are complex; that is, they are composed of more than one individual frequency. (In musical terminology, each frequency that makes up a complex tone is called a partial tone, or simply a partial.) Together, the collected frequencies of a tone are called **harmonics**. As stated in Chapter 13, these harmonics are called first, second, third, and so on. The first harmonic is called the fundamental frequency, f_1, and the remaining harmonics are all whole-number multiples of f_1. Thus, the second harmonic has a frequency of $2f_1$, the fifth harmonic has a frequency of $5f_1$, etc.

Tones of varying complexity, starting with simple, are compared in Figure 14-37. Any of the three simple, pure tones in (a), each consisting of only one frequency, is said to have low quality; the sound is boring. Such a sound may be produced by a tuning fork or electrical signal generator, or sometimes by a person whistling. When the fundamental frequency, f_1, is combined with the second harmonic, $2f_1$, as in (b), the waveform is more complex and the quality is slightly higher than that of a pure tone.

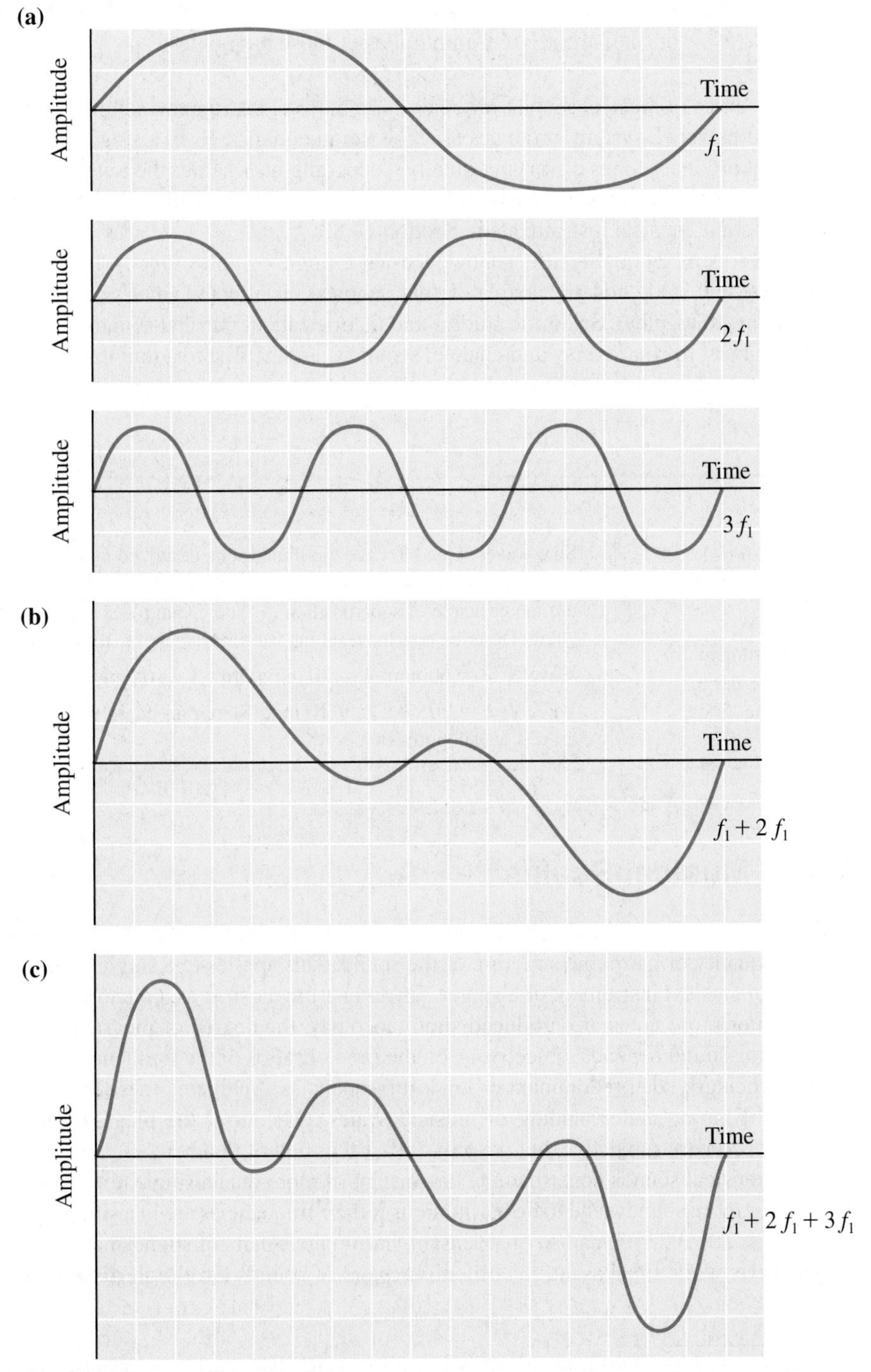

Figure 14-37 Comparing the quality of sounds of different waveforms. **(a)** Any one of the pure tones shown has a low quality. **(b)** Adding the first and second harmonics produces a more complex sound with a somewhat higher quality. **(c)** Adding the first, second, and third harmonics produces a sound of even higher quality.

The exact waveform depends on the amplitude of each harmonic and can be found by applying the principle of superposition; in this case, the second harmonic has an amplitude that is 80% that of the fundamental frequency. In (c), three harmonics are added to produce a complex, high-quality sound, with a cycle that repeats at the fundamental frequency. Such a sound may be produced by a musical instrument.

Thus, we conclude that

> Objectively, tonal quality depends on the number of harmonics that make up the tone and the relative amplitudes of those harmonics.

Every source of musical sounds, whether it is a musical instrument, a human voice, or a sound playback system, produces tones of unique quality. Even a single musical instrument produces tones of varying quality, depending on whether the tones are low pitched or high pitched and on the abilities of the performer. We will look at the tonal quality of some musical instruments in Section 14.8.

We have seen that quality affects how pleasant a single complex tone is. In the next section, we will study how the degree of "pleasantness" is affected when two or more tones are heard together. Scientific studies and theories about musical sounds are continually subject to controversy and change; scientists are the first to admit that there is much to learn about this exciting field.[8]

EXERCISES

14-52 The note G above middle C is struck on the piano ($f_1 = 392.0$ Hz). Determine the frequency of the second, third, and sixth harmonics.

14-53 The note C ($f_1 = 261.6$ Hz) is played on an instrument that produces only odd-numbered harmonics. Determine the frequency of the first three partial tones above the first harmonic.

14-54 Sine waves X and Y have the dimensions described below. Draw these waves neatly, one below the other so they begin in phase. Then use the principle of superposition to "add" X and Y at intervals of 2 mm. Describe how the resulting waveform relates to tonal quality.

Wave X: $\lambda = 60$ mm; $A = 20$ mm (Draw 1 wavelength.)

Wave Y: $\lambda = 20$ mm; $A = 10$ mm (Start Y in phase with X and draw 3 wavelengths of Y.)

14.7 Musical Scales

In the previous section the degree of pleasantness of a sound was seen to depend on the tonal quality or harmonic structure of the sound. This applies to a single tone produced by a musical instrument or a singer. On an instrument such as a piano or guitar, when two or more notes are produced simultaneously, the degree of pleasantness of the resulting sound depends, objectively, on the ratio or ratios of the frequencies of the notes, particularly the predominant or loudest frequencies. Analyzing these ratios will lead directly to an understanding of musical scales. Of course, the pleasantness of sounds is also very subjective, but we will omit discussion of that here.

consonance an indication of pleasant, harmonious pairs of sounds

dissonance an indication of unpleasant, inharmonious pairs of sounds

Many musical sounds consist at any one instant of more than a single note. In general, two or more sounds tend to be harmonious if their frequencies are in a simple ratio such as 1:1, 2:1, 3:2, 4:3, and so on. Pleasant, harmonious pairs of sounds have a high **consonance**, while unpleasant, inharmonious pairs of sounds have high **dissonance**,

[8] Some books use the term overtones when discussing harmonics above the fundamental frequency. For instruments that produce both odd- and even-numbered harmonics, the first overtone is the second harmonic, the second overtone is the third harmonic, and so on. For some instruments, the first overtone may be the third harmonic and the second overtone may be the fifth harmonic, etc. To avoid possible confusion, we have not used the term overtones.

or low consonance. Table 14-6 lists several ratios of frequencies in order of decreasing consonance. (The ratio of 1:1, called *unison*, is not listed in the table because it is not an interval, but its consonance is high.) The interval called an **octave** has one sound with double the frequency of another sound. For example, starting with a sound of 300 Hz, the next octaves are 600 Hz, then 1200 Hz, then 2400 Hz, and so on.

Let us look at how frequency ratios relate to the musical scale called the **equitempered scale**, which is the scale commonly used for musical compositions and in tuning most musical instruments, although there are many other ways of tuning instruments. (Other names for this scale are the equal scale, the tempered scale, and the equal tempered scale.) The standard frequency of this scale is 440 Hz on fixed-frequency instruments, such as the piano. This frequency is the note A above middle C on the piano. An octave below has a frequency of 220 Hz and an octave above has a frequency of 880 Hz. Figure 14-38 shows parts of the equitempered scale on a piano, including the notes and frequencies, as well as their staff notations (for reference only).

The calculated frequencies of the equitempered scale are based on two main facts:

- A note one octave above another has double the frequency.
- There are exactly 12 intervals per octave. On a piano keyboard, for example, there are five black keys and seven white keys per octave.

Thus, the equitempered scale uses a number (1.059 463) that when multiplied by itself 12 times gives the number 2. In other words, $\sqrt[12]{2} = 1.059463$ (to 7 significant digits). Using the symbol "a" to denote this special number and starting with the standard frequency of 440 Hz, we multiply and divide 440 by a, then by a^2, then by a^3, and so on. This process gives new notes and is repeated until the entire equitempered scale is calculated. The fact that each subsequent note is found by multiplying or dividing by the same ratio explains the name "equal tempered." Calculations for one octave of the equitempered scale are shown in Table 14-7. Notice in the table that a *semitone* is an interval of $\sqrt[12]{2}$ (or a), and a whole tone is a^2.

Table 14-6

Consonance and Some Frequency Ratios

	Interval	Ratio of Frequencies
	octave	2/1
Decreasing	fifth	3/2
consonance	fourth	4/3
or	major third	5/4
Increasing	major sixth	5/3
dissonance	minor third	6/5
	minor sixth	8/5

octave a sound interval with one sound having double the frequency of another sound

equitempered scale the musical scale commonly used for musical compositions, with a standard frequency of 440 Hz and 12 intervals per octave

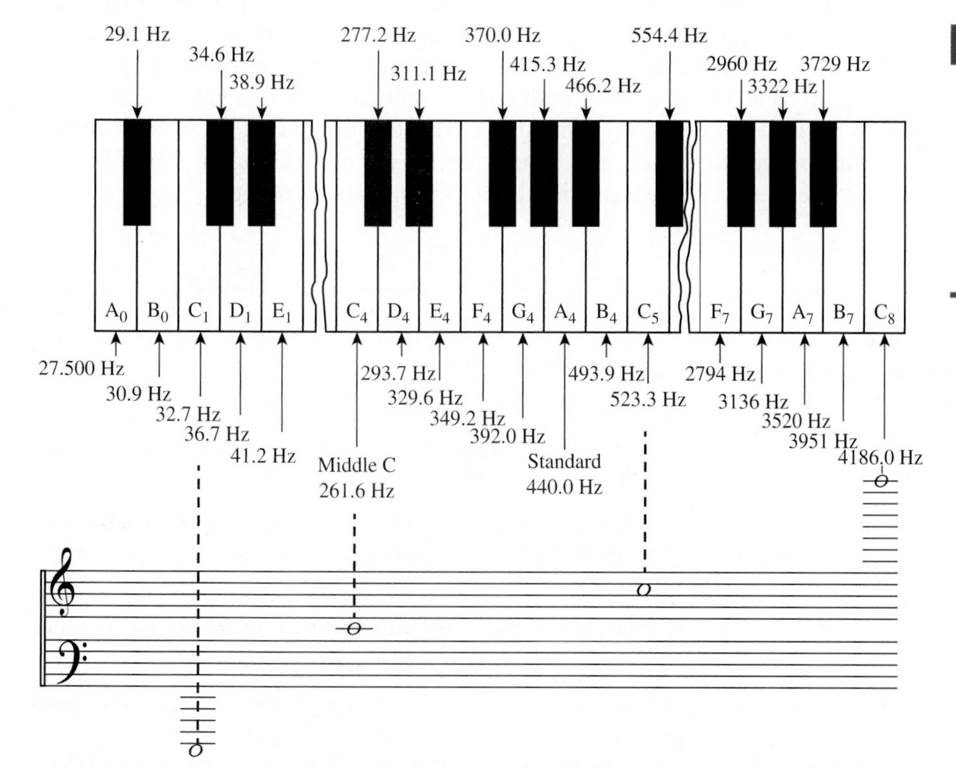

Figure 14-38 The equitempered scale, illustrated on a piano keyboard.

DID YOU KNOW?

The lowest note in classical music, D_2 or 73.4 Hz, is found in a composition by Mozart. Some singers have been able to sing sounds with frequencies lower than the lowest note on the piano, A_0, while other singers could sing above the highest note, C_8. However, these extreme frequencies have little musical value.

Table 14-7

The Equitempered Musical Scale

Note	Calculation (Using a = 1.059 463)	Frequency (Hz)	Approx. Ratio	Interval
A_4	standard	440.00	1/1	interval standard
A#, Bb[9]	$440 \times a^1$	466.16	16/15	semitone
B_4	$440 \times a^2$	493.88	9/8	major whole tone
C_5	$440 \times a^3$	523.25	6/5	minor third
C#, Db	$440 \times a^4$	554.37	5/4	major third
D_5	$440 \times a^5$	587.33	4/3	fourth
D#, Eb	$440 \times a^6$	622.25	45/32	augmented fourth
E_5	$440 \times a^7$	659.26	3/2	fifth
F_5	$440 \times a^8$	698.46	8/5	minor sixth
F#,Gb	$440 \times a^9$	739.99	5/3	major sixth
G_5	$440 \times a^{10}$	783.99	9/5	minor seventh
G#,Ab	$440 \times a^{11}$	830.61	15/8	major seventh
A_5	$440 \times a^{12}$	880.00	2/1	octave (eighth)

The equitempered scale was standardized relatively recently, in 1953. It is still not accepted everywhere in the Western world. One advantage of this scale is that it allows musicians to play a variable-frequency instrument such as a trombone along with a fixed-frequency instrument such as a piano. This flexibility is possible because all whole tones are equal in size and two semitones are exactly equal to one whole tone.

The disadvantage of this scale is that the resulting intervals or ratios are not exact. (Notice in Table 14-7 that the fourth column indicates "approximate" ratios.) This causes the intervals to sound slightly flat or slightly sharp. When variable-frequency instruments, such as violins, play alone, they are not restricted to this scale and can thus produce intervals of exact ratios that are of highest consonance.

EXERCISES

14-55 State the frequency of a note one octave above and one octave below a note with a frequency of (a) 392.0 Hz and (b) 1286 Hz.

14-56 State the frequency of a note two octaves above a note with a frequency of (a) 288.0 Hz and (b) 341.3 Hz.

14-57 For each pair of notes listed, determine their ratio of frequencies as a fraction, and state whether their consonance when sounded together is high or low.

(a) 600 Hz; 400 Hz

(b) 1200 Hz; 400 Hz

(c) 1500 Hz; 750 Hz

(d) 205 Hz; 200 Hz

14-58 Calculate the frequency of F_6 and E_3 on the equitempered scale. (Refer to Table 14-7.)

14-59 If B_5 on the equitempered scale is 987.77 Hz, find the frequencies of the notes just above and below it on a piano. (Use "a.")

14-60 Derive a two-octave equitempered scale based on six intervals per octave and a standard frequency of 600.00 Hz. Give an original name or symbol to each note. (**Hint:** $\sqrt[6]{2} = 1.122462$)

[9] In simple terms, the symbol "#" means sharp and the symbol "b" means flat, so A# has a higher frequency than A, and Bb has a lower frequency than B.

14.8 Musical Instruments and Resonance in Stringed and Wind Instruments

Each musical instrument is classified into one of four main types, depending on how oscillations are set up by the instrument. In this section we look at the basic principles of sound production in these types of instruments: stringed, wind, percussion, and electric and electronic. Some instruments, such as the piano and the human voice, may be classified into more than one type, but only the most common classification is used here.

Each classification type consists of several instruments (e.g., some wind instruments are the saxophone, the trumpet, and the oboe), and in some cases one instrument may be further subclassified according to the pitch range it produces (e.g., saxophones can be alto, tenor, baritone, and bass). In either case, a general characteristic within each set or subset of a class type is that *larger instruments have a lower range of pitches and smaller instruments have a higher range of pitches*. This important property is depicted visually in Figure 14-39 on page 448, which shows how the pitch ranges and notes of several instruments compare to the pitch ranges and notes on a full piano scale. Refer to this diagram as you read about the four main types of musical instruments and human singing voices.

In the discussion that follows we will examine not only the frequencies produced by the instruments, but also how the instruments are played and the quality of sound they produce.[10]

Stringed Instruments

Stringed instruments consist of two main parts: the oscillator and the resonator. The *oscillator* is the string, and the *resonator* is the box, case, or sounding board that the string is mounted on. A string by itself does not produce a loud or, necessarily, even a pleasing sound. It must be attached to a resonator, through which oscillations help improve the loudness and quality of the sound.

The frequency produced by an oscillating string depends on both the mode of oscillation and the physical properties of the string. The modes of oscillation (described in Section 13.7) consist of the fundamental frequency, called the first harmonic, and other harmonics that are whole-number multiples of the fundamental frequency. Plucking a string in the middle that is fixed at both ends produces a predominant fundamental frequency, while plucking the string at other locations produces more complex oscillations, often of higher quality. Figure 14-41 on page 449 shows some simplified examples of modes of oscillation of a string.

To calculate the frequencies of the various modes of oscillation of a string, we begin by finding the wavelengths of the transverse waves on the string in terms of the length of the string. Figure 14-42 on page 450 shows the standing wave pattern of the first three harmonics of a string. From the pattern, we can write an equation for the wavelength of the transverse waves on a string in terms of the length of the string, L, and the harmonic number, n:

$$\lambda_n = \frac{2L}{n} \quad \text{where } n = 1, 2, 3, \ldots$$

Now, using the universal wave equation, $v = f\lambda$, we have $f_n = v/\lambda_n$, and $f_n = \dfrac{v}{2L/n}$

Thus,

$$f_n = \frac{nv}{2L}$$

This equation can be used to find the frequency of the nth harmonic of a string in terms of the length of the string and the speed of a transverse wave in the string.

[10]In this chapter we have discussed the pitch, loudness, and quality (or timbre) of musical sounds, and have related those variables to physics principles. Other variables that affect music enjoyment and appreciation, such as rhythm, meter, tempo, and harmony, are left for books that deal specifically with music.

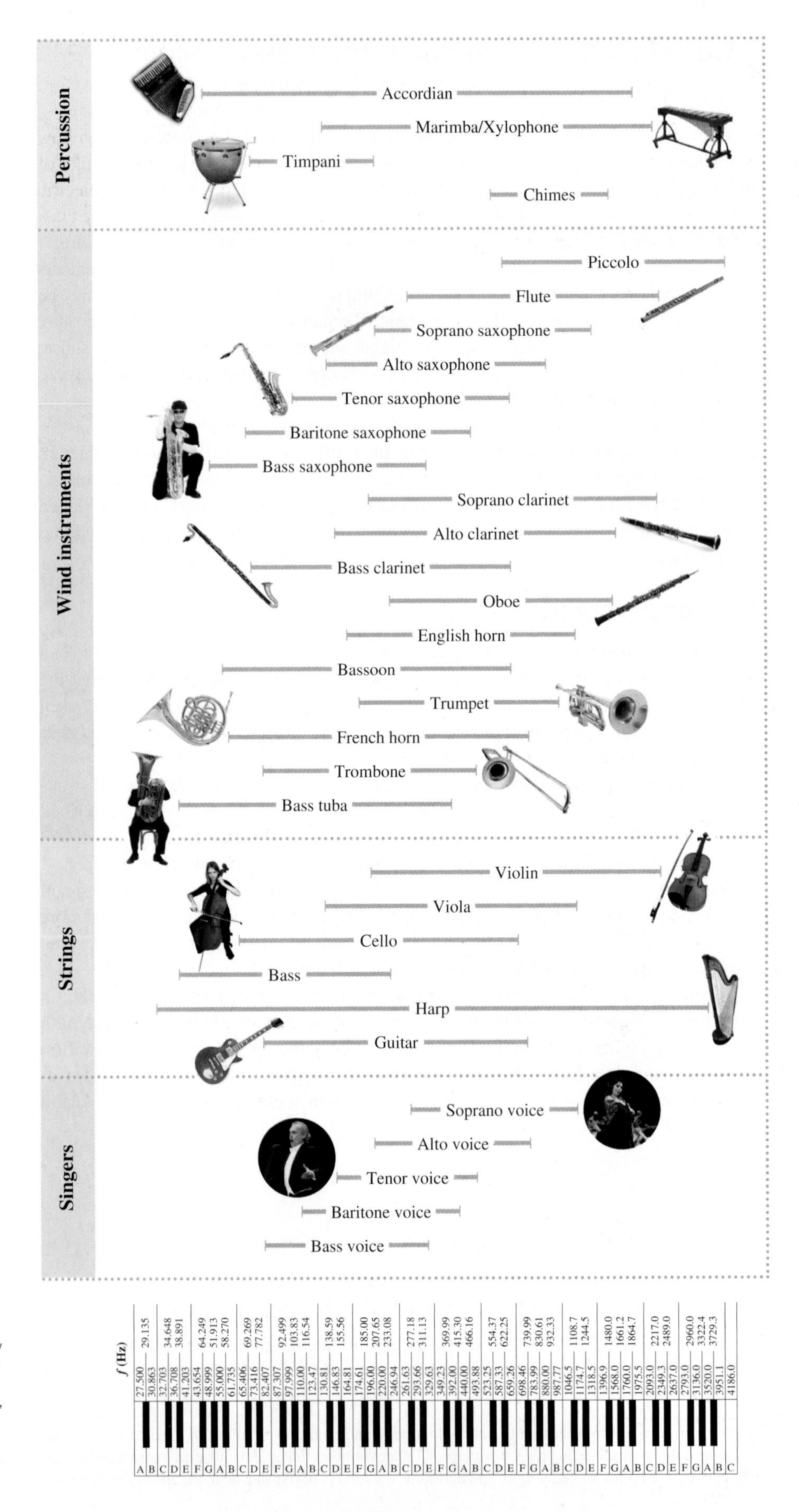

Figure 14-39 Because the piano has a large pitch range and its keyboard is so common, it is used here to illustrate the pitch ranges and notes of several instruments, as well as human singing voices.

Source: sbarabu/Shutterstock, nahariyani/Shutterstock, Vereshchagin Dmitry/Shutterstock, Andresr/Shutterstock, holbox/Shutterstock, moonsh8/Shutterstock, Bombaert Patrick/Shutterstock, Matthias G. Ziegler/Shutterstock, Andrey_Popov/Shutterstock, Matthias G. Ziegler//Shutterstock, Furtseff/Shutterstock, Furtseff/Shutterstock, Dmitry Skutin/Shutterstock, Matthias G. Ziegler//Shutterstock, mphot/Shutterstock, Serjio/Shutterstock, Vereshchagin Dmitry/Shutterstock, optimarc/Shutterstock, Ivica Drusany/Shutterstock.com, criben/Shutterstock.com

PROFILES IN PHYSICS

Daniel Levitin (1957–): Musician, Neuroscientist, and Author

For anyone, whether a musician or non-musician, who wants to learn about the relationship between music and the human brain, an ideal resource is Daniel Levitin's (Figure 14-40) best-selling book, *This Is Your Brain on Music* (Dutton/Penguin, 2006). Many of the physics principles related to music presented in this chapter are featured in Levitin's book, with added depth as he explains the brain's role in the creation and hearing of music.

At only 4 years of age, Daniel received his first tape recorder and, soon after, his experimentations with sound and music began. After studying music and psychology at university, Levitin took a break to work for 10 years as a musician, recording and sound engineer, and record producer for numerous rock groups. Levitin returned to university and earned undergraduate degrees in Cognitive Psychology and Cognitive Science, and a Ph.D. in Psychology. Currently, he is Professor of Psychology, Neuroscience, and Music at McGill University in Montreal, a leading research facility into music/brain interactions.

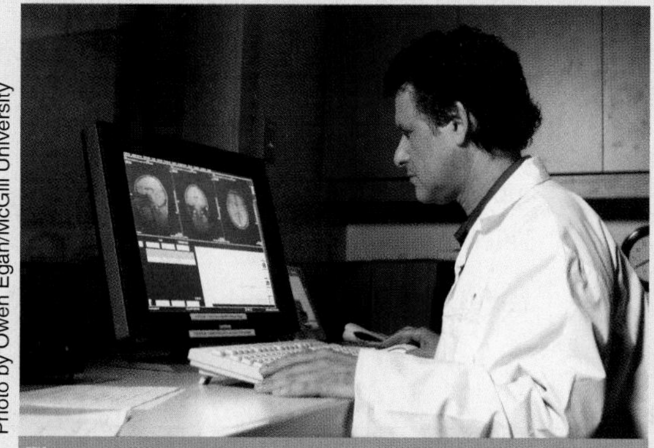

Figure 14-40 Daniel Levitin researches how music affects the brain.

Levitin has written numerous articles in audio magazines and scientific journals, as well as a second best-selling book related to music, *The World in Six Songs* (Dutton/Penguin, 2008). His books feature websites that provide samples of the songs that he describes in detail in the books. More recently, his research on the information age has led to the book *The Organized Mind* (Dutton, 2014).

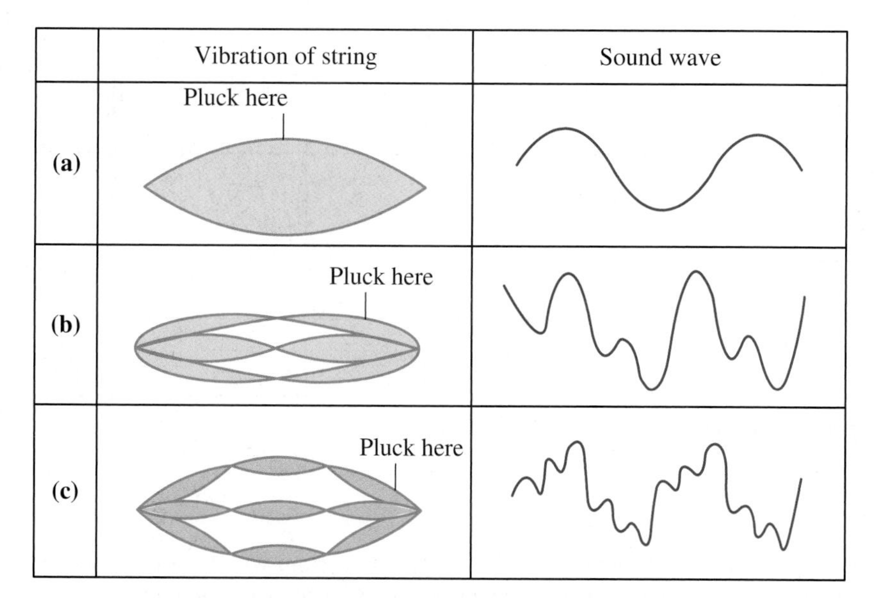

Figure 14-41 Changing the modes of oscillation of a string. **(a)** Relatively low quality. **(b)** Somewhat higher quality. **(c)** An even higher quality.

Finally, from Section 14.1, the speed of a wave in a string is $v = \sqrt{T/\mu}$, where T is the tension in the string in newtons (N) and μ is the linear density or mass per unit length of the string in kilograms per metre (kg/m). Thus, in general,

$$f_n = \frac{n}{2L}\sqrt{\frac{T}{\mu}} \quad (14\text{-}7)$$

and, specifically for the fundamental frequency,

$$f_1 = \frac{1}{2L}\sqrt{\frac{T}{\mu}}$$

These equations make sense not only to musicians who play stringed instruments, but also to anyone observing those musicians. Consider three important conclusions:

- Under constant tension, higher pitch notes are produced by pressing down on the string to make the length shorter. In fact, if the length is halved, the frequency doubles, producing a note one octave above the original. Thus, $f \propto \frac{1}{L}$.
- On strings of constant length, higher pitched notes are produced by increasing the tension in the string. This relationship is not linear because $f \propto \sqrt{T}$. This is applied when tuning stringed instruments.
- Thicker strings produce lower notes than thin strings of the same length. This agrees with the relationship $f \propto \frac{1}{\sqrt{\mu}}$ where μ is the mass per unit length.

N = node A = antinode			$f = \frac{v}{\lambda}$	$v = \sqrt{\frac{T}{\mu}}$
$n = 1$ $\left\{L = \frac{1}{2}\lambda_1\right\}$	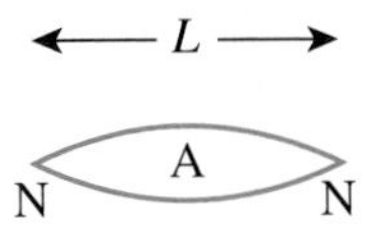 ← L → N A N	$\lambda_1 = \frac{2L}{1}$	$f_1 = \frac{v}{2L}$	$f_1 = \frac{1}{2L}\sqrt{\frac{T}{\mu}}$
$n = 2$ $\{L = \lambda_2\}$	 N N N A A	$\lambda_2 = \frac{2L}{2}$	$f_2 = \frac{2v}{2L}$	$f_2 = \frac{1}{L}\sqrt{\frac{T}{\mu}}$
$n = 3$ $\left\{L = \frac{3}{2}\lambda_3\right\}$	 N N N N A A A	$\lambda_3 = \frac{2L}{3}$	$f_3 = \frac{3v}{2L}$	$f_3 = \frac{3}{2L}\sqrt{\frac{T}{\mu}}$
↓	↓	↓	↓	↓
$n = 1, 2, 3, \ldots$		$\lambda_n = \frac{2L}{n}$	$f_n = \frac{nv}{2L}$	$f_n = \frac{n}{2L}\sqrt{\frac{T}{\mu}}$

Figure 14-42 Deriving an equation for the frequency of a string.

SAMPLE PROBLEM 14-15

The length of a B note guitar string is 68 cm. For a string with the same tension and linear density, what length would be required to produce a note **(a)** one octave higher and **(b)** two octaves higher than the original B?

Solution

(a) One octave higher means double the frequency and half the length, which is 34 cm.

(b) To achieve the next octave, the length must be reduced by another factor of 2; therefore the new length is 17 cm.

This problem could have been solved by using the relationship $f \propto \frac{1}{L}$ so that $\frac{f_2}{f_1} = \frac{L_1}{L_2}$, etc.

SAMPLE PROBLEM 14-16

A 2.00 m long piano string has a tension of 5.00×10^2 N and a linear density of 4.10×10^{-2} kg/m. Determine the

(a) fundamental frequency of the string

(b) frequency of the third harmonic of the string

Solution

(a) $f_1 = \dfrac{1}{2L}\sqrt{\dfrac{T}{\mu}} = \dfrac{1}{2 \times 2.00\,\text{m}}\sqrt{\dfrac{5.00 \times 10^2\,\text{N}}{4.10 \times 10^{-2}\,\text{kg/m}}} = 27.6\,\text{Hz}$

Thus, the fundamental frequency is 27.6 Hz.

(b) $f_n = nf_1 = 3 \times 27.6\,\text{Hz} = 82.8\,\text{Hz}$

Thus, the third harmonic frequency is 82.8 Hz.

Stringed instruments can be played by plucking, striking, or bowing, with each method producing a different quality of sound. Stringed instruments that are usually *plucked* include the guitar, banjo, mandolin, ukulele, and harp. Except for the harp, these instruments have from four to eight strings as well as frets to guide the placement of the fingers. The harp is more complex, with a hollow soundboard (the resonator), a vertical pillar, a curved neck, and 46 strings stretched between the pins on the soundboard and the pegs on the curved neck (Figure 14-43). The harpist can adjust the frequency of the string by using a set of seven foot pedals, each of which has three possible positions. This allows tuning in various major and minor keys.

Stringed instruments that are usually *bowed* belong to the violin family, which consists of the violin, viola, cello, and bass. The bow is made with dozens of fine fibres that are rubbed with rosin to increase friction when stroked perpendicular to the string. Each instrument has four strings and there are wooden soundboards at the front and back of the body. These instruments have no frets, so the frequency can be changed continuously, not necessarily in steps as in instruments with frets, such as the guitar. (The only other instrument capable of changing frequencies continuously is the trombone.) Stringed instruments emit a small amount of acoustic power. The maximum power from a piano is 0.44 W, while that from a bass drum is 25 W. You can appreciate why an orchestra needs many more violins than drums or trumpets.

Guitar Harmonics

If you have access to a guitar, here's a way to hear various harmonics. Place a finger gently on a string at the 12th fret, which is halfway along the string. Don't push the string down on the fret; just touch the string. Then use your other hand to pluck the string with a pick or a finger. Because of your finger halfway along the string, a node is created there and the first harmonic is suppressed, as well as all other odd-numbered harmonics. The rest of the length of the string is free to oscillate, and what is sounded is primarily the second harmonic but with other even-numbered harmonics sounding as well. The technique can also be used to hear the third harmonic by placing a finger gently on a string at the 7th fret (one-third of the way along the string), or the fourth harmonic at the 5th fret (one-quarter of the string length). You'll probably find the sounds of the harmonics to be rather pleasant.

Gorgev/Shutterstock

(a)

MIMOHE/Shutterstock

(b)

Figure 14-43 **(a)** A typical classical harp. **(b)** Details showing the strings attached to the pins on the soundboard.

DID YOU KNOW?

The squawks and squeals produced by inexperienced players bowing such stringed instruments as the violin originate when longitudinal waves are set up in the string. Such oscillations begin when the bow rubs a string lengthwise rather than exactly perpendicularly. The resulting longitudinal waves travel much faster than the transverse waves and produce a high, annoying pitch.

Wind Instruments

All wind instruments contain columns of oscillating air molecules. The frequency of oscillation of the air molecules, and thus the fundamental frequency and other harmonics of the sound produced, depend on whether the column is open at both ends or open at one end and closed at the other end. Standing waves are set up in the air columns. These standing waves differ from the standing waves produced on strings because, first, they are longitudinal rather than transverse, and, second, the pattern of nodes and antinodes is different. Strings always have displacement nodes at each end, whereas air columns have displacement antinodes at open ends and displacement nodes at closed ends.

The word displacement is important in standing waves in air columns: in this context it refers to the displacement of the air molecules from their equilibrium position. At an open end, the air molecules are free to oscillate, so their displacement can be maximum, producing an antinode. At a closed end, the air molecules have difficulty in oscillating far, resulting in a displacement node.

For our discussion of sound in air columns, we will focus on two possible designs, one with both ends of the column open and the other with one end closed. (The analysis of sound waves in actual wind instruments is complex, and will be left for texts devoted to the physics of music.)

First, consider a column of air of length L, open at both ends. A standing wave pattern can be set up in the column only at specific resonant frequencies, with the fundamental frequency, f_1, being the lowest possible frequency. The standing wave pattern has a displacement antinode at each end and, for the fundamental frequency or first harmonic, it has one displacement node between the ends. The next resonant frequency that causes a large, natural oscillation of air molecules is the second harmonic, and when it occurs there are two nodes between the ends. This pattern continues, as shown in Figure 14-44, with transverse waves being used to represent the displacements of longitudinal waves. As seen in the diagram, for an air column open at both ends, called an open-air column,

$$f_n = \frac{nv}{2L}, \text{ where } n = 1, 2, 3, ... \quad \textbf{(14-8)}$$

Comparing these results to strings, we see that in both cases the equations for the wavelength and frequency are identical, although the strings have displacement nodes at the ends and the open-air columns have displacement antinodes at the ends.

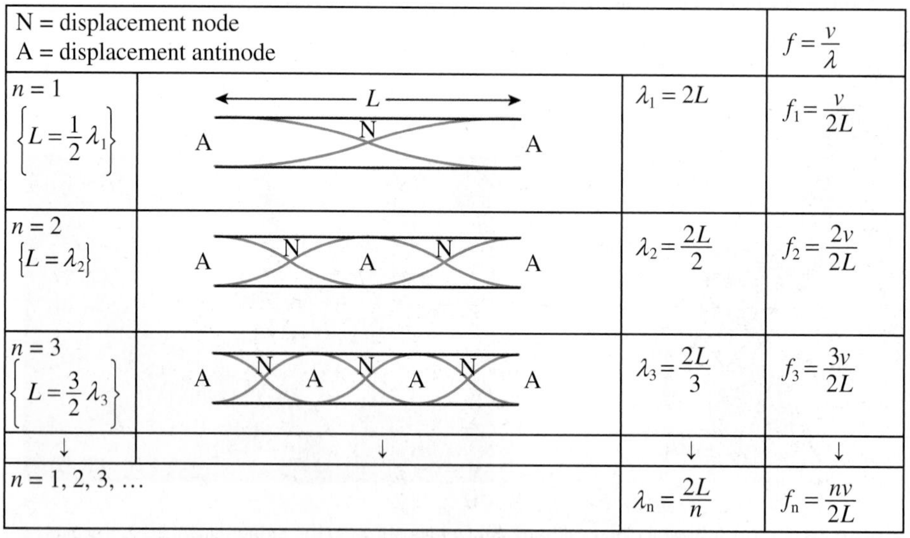

N = displacement node A = displacement antinode			$f = \frac{v}{\lambda}$
$n = 1$ $\left\{L = \frac{1}{2}\lambda_1\right\}$	← L → A N A	$\lambda_1 = 2L$	$f_1 = \frac{v}{2L}$
$n = 2$ $\{L = \lambda_2\}$	A N A N A	$\lambda_2 = \frac{2L}{2}$	$f_2 = \frac{2v}{2L}$
$n = 3$ $\left\{L = \frac{3}{2}\lambda_3\right\}$	A N A N A N A	$\lambda_3 = \frac{2L}{3}$	$f_3 = \frac{3v}{2L}$
↓	↓	↓	↓
$n = 1, 2, 3, \ldots$		$\lambda_n = \frac{2L}{n}$	$f_n = \frac{nv}{2L}$

Figure 14-44 Deriving an equation for the frequency of sounds in an open air column.

Next consider a column of air that is closed at one end, where a displacement node exists at the closed end. Figure 14-45 shows an air column with standing waves starting with the longest wavelength such that only $\frac{1}{4}\lambda$ fits into the length of the air column. Thus, $L = \frac{1}{4}\lambda_1$ or $\lambda_1 = 4L$. This produces the lowest resonant frequency, called, as usual, the fundamental frequency, or first harmonic, f_1. The next resonance occurs when $\frac{3}{4}\lambda$ fits into the length; that is, $\lambda_3 = 4L/3$. This produces the third harmonic because the frequency is 3 times that of f_1. Thus, there is no second harmonic produced in an air column closed at one end, which explains why the symbol f_3 is used rather than f_2. In general, there are no even numbered harmonics produced, just odd numbered harmonics. Therefore, for an air column closed at one end, commonly called a closed-air column,

$$f_n = \frac{nv}{4L}, \text{ where } n = 1, 3, 5, \ldots \qquad (14\text{-}9)$$

Whether an air column is open at both ends or only one end, there are four variables in the equation for the resonant frequency within the column. Knowing any three of these variables allows the fourth to be found.

N = displacement node A = displacement antinode			$f = \frac{v}{\lambda}$
$n = 1$ $\left\{L = \frac{1}{4}\lambda_1\right\}$	← L → N A	$\lambda_1 = \frac{4L}{1}$	$f_1 = \frac{v}{4L}$
$n = 3$ $\left\{L = \frac{3}{4}\lambda_3\right\}$	N A N A	$\lambda_3 = \frac{4L}{3}$	$f_3 = \frac{3v}{4L}$
$n = 5$ $\left\{L = \frac{5}{4}\lambda_5\right\}$	N A N A N A	$\lambda_5 = \frac{4L}{5}$	$f_5 = \frac{5v}{4L}$
↓	↓	↓	↓
$n = 1, 3, 5, \ldots$		$\lambda_n = \frac{4L}{n}$	$f_n = \frac{nv}{4L}$

Figure 14-45 Deriving an equation for the frequency of sounds in a column of air closed at one end.

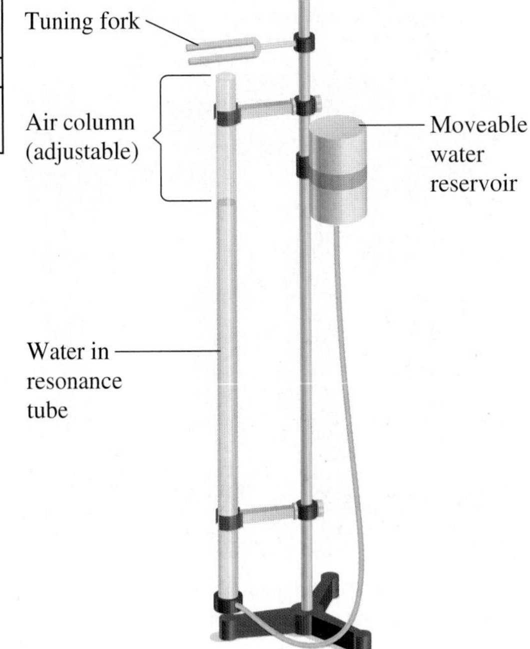

Figure 14-46 Resonance apparatus using water to control the length of the air column.

SAMPLE PROBLEM 14-17

A 512 Hz tuning fork is struck and held at the mouth of a vertical air column whose length can be adjusted by allowing water to move up and down in it, as shown in Figure 14-46. The water level starts at the top of the column and is decreased continuously. A loud resonant sound is heard when the air column is 16.4 cm long, then another loud resonant sound is heard when the air column is 49.5 cm long. Determine the speed of sound in the air.

Solution

The tuning fork has a fixed frequency and thus a fixed wavelength. The first resonance occurs when the air column is $\frac{1}{4}\lambda$ long and the second when the air column is $\frac{3}{4}\lambda$ long. Thus, the distance from one resonance to the next is $\frac{1}{2}\lambda$.

$$\tfrac{1}{2}\lambda = 49.5\,\text{cm} - 16.4\,\text{cm}$$
$$\therefore \lambda = 2(33.1\,\text{cm}) = 66.2\,\text{cm} = 0.662\,\text{m}$$

Now, $v = f\lambda = (512\,\text{Hz})(0.662\,\text{m}) = 339\,\text{m/s}$

Thus, the speed of sound in this apparatus is 339 m/s.

In the above example, we could have used the first resonance alone to find the wavelength, in which case $\frac{1}{4}\lambda = 16.4$ cm or $\lambda = 4 \times 16.4$ cm $= 65.6$ cm. This value does not agree with the answer in the sample problem, a situation that occurs in actual experiments. The discrepancy is caused by the end effects of an air column, and it is overcome by using the distance from one resonant loudness to another. In most wind instruments, end effects are taken into consideration by flaring the end of the air column or tube. This also helps the sound energy transfer effectively from the instrument to the room.

XiXinXing/Shutterstock

Figure 14-47 The flute is an example of an air reed instrument.

In some wind instruments, such as the pipe organ, the length of each air column is fixed. However, in most wind instruments, such as the trombone and flute, the length of the air column can be varied to allow for the production of tones of different pitch.

To cause the air molecules in a wind instrument to oscillate, something else must oscillate first. Following are descriptions of the four general mechanisms to force air molecules to oscillate.

First, in *air reed instruments*, air is blown across or through an opening, and the moving air sets up a turbulence inside the column of the instrument. Examples of such instruments are the pipe organ, flute, piccolo, recorder, and fife. Each pipe of a pipe organ has a fixed length, but the end of the pipe can be either open or closed. The flute and piccolo have keys that are pressed to change the length of the air column. The recorder and fife have side holes that must be covered with fingers to change the length of the air column and thus control the pitch. (See Figure 14-47.)

johnbraid/Shutterstock.com

Figure 14-48 The bagpipe is an example of a single-membrane reed instrument.

TRY THIS!

A Bottle Symphony

In a group, produce a "bottle symphony" by adding different levels of water to a set of pop or water bottles. Each bottle will act as a wind instrument when you blow air across its mouth. The length of the air column in each bottle determines the pitch of the sound. Tune each bottle by using a piano, or another instrument, or a frequency generator[11] as reference. Make up some simple tunes and play them.

In *single-membrane reed instruments,* moving air sets a single reed oscillating. Examples of these instruments include the saxophone, clarinet, and bagpipe. In the bagpipe, the reeds are located in four drone pipes attached to the bag, not to the mouth pipe. Again, the length of the air column is changed by holding down keys or covering side holes. (See Figure 14-48.)

Next, in *double-mechanical reed instruments*, moving air forces a set of two reeds to oscillate against each other, which in turn causes air in the instrument to oscillate. Examples include the oboe, English horn, and bassoon, in each of which keys are pressed to alter the length of the air column. (Refer to Figure 14-49.)

The final mechanism is found in *lip reed instruments*, also called *brass instruments*. In this type of instrument, the player's lips function as a double reed by oscillating, thus causing the air in the instrument to oscillate. None of the air escapes

[11]Free frequency generator software and other musical apps are available; examples are *Seventh String Tuning Fork*, *Raven Lite 1.0*, and *Audio Kit*.

though the side holes, as in other wind instruments; rather, the sound waves must travel all the way through the instrument. Examples of such instruments are the bugle, trumpet, trombone, French horn, and tuba. The length of the air column is changed either by pressing valves or keys that add extra tubing to the instrument, or, in the case of the trombone, by moving the U-shaped tube, called the slide (Figure 14-50).

The quality of sound from wind instruments is determined by such factors as the construction of the instrument and its components, the experience of the player, and the harmonics structure of the note played. Columns that are open at both ends, such as flutes and recorders, can produce both even- and odd-numbered harmonics, so the quality differs from columns that are closed at one end, which only produce odd-numbered harmonics. Furthermore, the waves produced in instruments that are cone-shaped are complex, which also affects the quality of the sound.

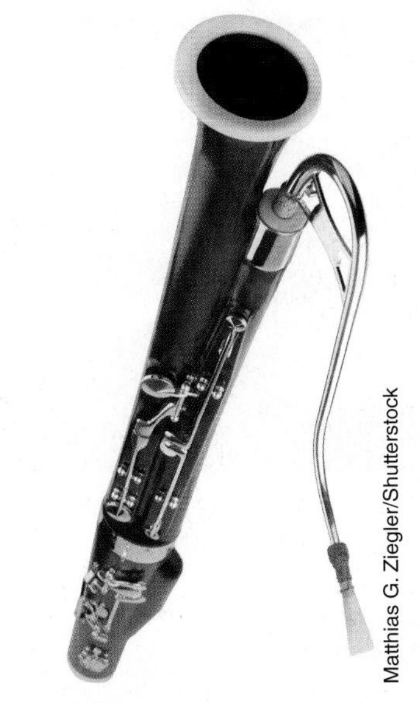

Figure 14-49 The classical bassoon is an example of a double-mechanical reed instrument.

Percussion Instruments

Percussion is the striking of one object against another. Percussion instruments are usually struck by a firm object such as a hammer, bar, or stick. These musical instruments were likely the first invented because they are relatively easy to make. Percussion instruments can be divided into three main categories.

Single indefinite pitch instruments are used for special effects or for keeping the beat of the music. Examples include the bass drum, triangle, and castanets. *Multiple definite pitch instruments* have strings, bars, or bells of different sizes that, when struck, produce their own resonant frequencies and harmonics. The best known of these instruments is the piano, which has strings but is commonly classified as a percussion instrument. A piano key is connected by a system of levers to a hammer that strikes the string or strings to produce a certain note. A modern piano has 88 notes, with a frequency range from 27.5 Hz to 4186 Hz. The lowest-pitch notes have one string, higher notes have two strings, and the 60 highest notes have three strings to increase the intensity of the sounds. The sounds from the strings are increased in loudness and quality by the wooden soundboard of the piano. Other examples are the tuning fork, orchestra bells, marimba, xylophone, and carillon.[12] *Variable pitch instruments* have a device used to rapidly change the pitch to a limited choice of frequencies. An example is the timpani, or kettle drum, which has a foot pedal for quick tuning (Figure 14-51).

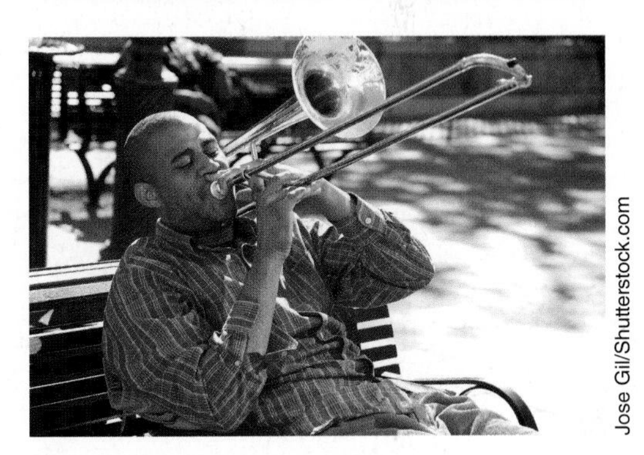

Figure 14-50 The trombone is an example of a lip reed instrument.

percussion the striking of one object against another

Some instruments, such as the accordion and harmonica, are difficult to classify as a single type of instrument. The accordion and harmonica use moving air to set reeds oscillating, but they do not have resonating air columns, so they are not usually called wind instruments. They are better classified as percussion instruments in which air knocks against reeds, causing them to oscillate (Figure 14-52 on page 456).

The Human Voice

The human voice (Figure 14-53 on page 456) is a fascinating instrument, consisting of a source of air (the lungs), oscillators (the vocal folds or vocal cords), and resonators (the lower throat or pharynx, mouth, and nasal cavity). To create most sounds, air from the lungs is forced through the trachea, or windpipe, and passes by the vocal folds, causing them to oscillate. The vocal folds are two bands of skin that act like a double reed. Loudness is controlled by the amount of air forced over the vocal folds. The pitch is controlled by muscular tension as well as by the size of the oscillating

Figure 14-51 The classical timpani or kettle drum is a variable pitch percussion instrument.

[12]Some compositions are much more complex than the music made by traditional percussion instruments. For example, Joseph Bertolozzi, a percussionist, composed "Tower Music" by recording about 400 sounds on the Eiffel Tower in Paris. You can check out this as well as his "Bridge Music" on the Internet to learn the numerous ways he controlled the sounds he wanted.

grynold/Shutterstock

Figure 14-52 The harmonica can be classified as a percussion instrument.

DID YOU KNOW?

Human infants cannot produce most sounds we understand as articulate speech until after they are about a year old when the larynx drops down in the throat. For infants born without a trachea, an extremely rare condition, the use of stem cells now makes the success rate of implanting an artificial trachea higher than previously possible. For one 2-year-old child, the surgeon implanted a 13 mm diameter tube made of plastic fibres surrounded by the child's own stem cells. For the first time in her life, the child was able to breathe without a tube down her throat. (For unrelated reasons, the breathing tube had been placed through her mouth rather than her nose.)

DID YOU KNOW?

Physicians use percussion when they tap a patient's chest or back and listen for sounds that indicate either clear or congested lungs.

parts. Since larger instruments have lower resonant frequencies, in general, male voices tend to have lower frequency ranges than female voices. The approximate frequency ranges of singers are shown in Figure 14-39 on page 448.

The quality of sound from the voice is controlled by the resonating cavities and the parts in them, such as the tongue and lips. Of course the quality of sound may also be improved by proper training. Trained singers can control such effects as *vibrato* (a slight, periodic changing of frequency) and *tremolo* (a slight periodic changing of amplitude).

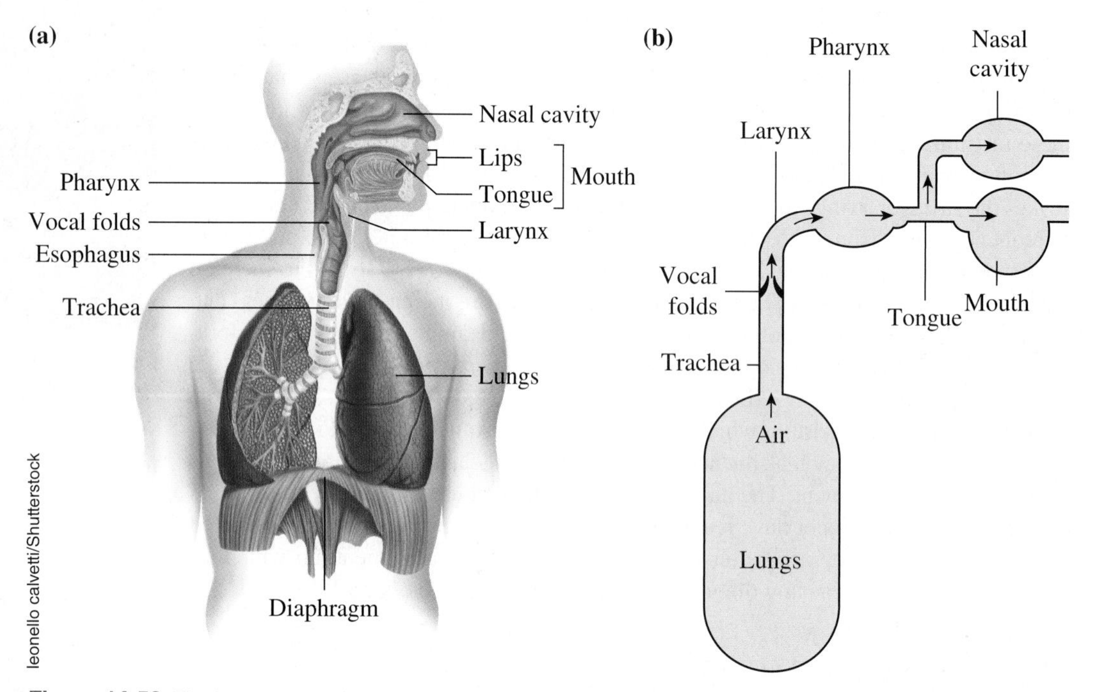

Figure 14-53 The human voice. **(a)** Basic structure. **(b)** Schematic diagram showing air flow.

Electrical and Electronic Instruments

Electrical instruments are made of three main parts, a *source* of sound, a *microphone*, and a *loudspeaker*. Many of the musical instruments discussed earlier in this section can be made into electrical instruments simply by adding a microphone and a loudspeaker. This is often done with stringed instruments that normally give out low amounts of power by attaching a microphone directly to the body of the instrument. In some cases the design of the instrument is altered. An electric guitar, for example, may have a solid body rather than the hollow body of acoustic guitars (Figure 14-54).

DID YOU KNOW?

Human voice waveforms are complex to analyze or to reproduce accurately. However, the use of computers makes it possible to recognize or reproduce many sounds. One application of this is the manufacture of robots that obey the oral commands of people who can speak but who cannot get around easily.

Loudspeakers are important in determining the quality of sound from an electrical instrument. A single loudspeaker does not have the same frequency range as our ears, so a set of two or three must be used to give both quality and frequency range, as described on page 439.

Figure 14-54 Example of an electric guitar.

DID YOU KNOW?

An interesting phenomenon occurs with the use of headphones used with portable MP3 players, laser disc players, etc. Some of the headphones do not create the low frequencies usually associated with base sounds, yet the listener actually "hears" the low-frequency sounds. The higher harmonics produced by the headphones cause a sensation that makes us believe we are hearing the absent first and second harmonics.

Electric Guitar Sounds

Listen to the sounds produced by an electric guitar with the electricity on and then off. Describe the differences in the sounds, especially in the quality.

Electronic instruments are much different from electrical and other instruments (Figure 14-55). Synthesizers and electronic pianos and organs are common electronic instruments that consist of four main parts: the *oscillator*, which causes the oscillations, the *filter circuit*, which selects frequencies that are sent to a *mixing circuit*, and the *amplifier and speaker system*, which makes the sound loud enough to be heard.

The sound produced by electronic instruments can be controlled by altering the waveforms of the emitted sound. Various basic waveforms, such as sine waves, square waves, sawtooth waves, triangular waves, and pulses (Figure 14-56) can be superimposed to create sounds that resemble the sound of almost any musical instrument. Further control is achieved by varying the attack and decay properties of the sound. The *attack*, which occurs at the beginning of a sound, can be sudden, delayed, or overshot. The *decay*, which occurs as the sound comes to an end, can be slow, fast, or irregular (Figure 14-57).

Figure 14-55 An example of an electronic synthesizer.

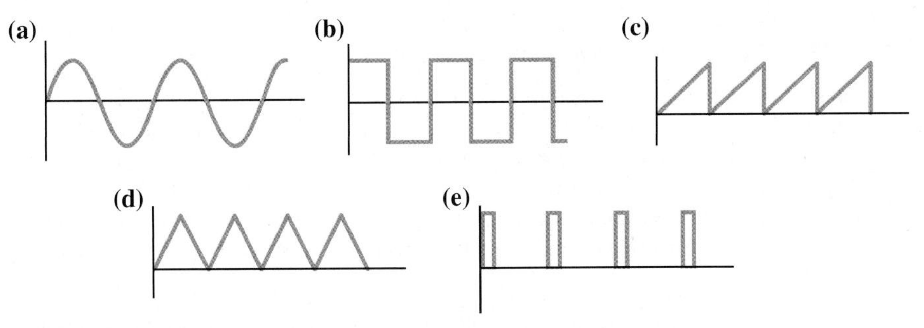

Figure 14-56 Examples of waveforms of electronic instruments. **(a)** Sine wave. **(b)** Square wave. **(c)** Sawtooth wave. **(d)** Triangular wave. **(e)** Periodic pulses.

Virtual Music

You have seen that electronic instruments use oscillators to generate music, electric circuits to amplify the sounds, and loudspeakers. The next advancement commonly used to generate and control music is called *Musical Instrument Digital Interface*, or MIDI. This system, which was standardized initially in 1983, provides a way of linking several devices together, all under the control of a computer. The end result is music that sounds like an individual musician, a band, or even a full orchestra.

The composer gives instructions to the computer using input devices such as a computer keyboard, a mouse, or a MIDI keyboard. These instructions include what notes to play, time interval and intensity of the notes, and properties of attack and decay. In this way, the composer can create virtual music of an entire orchestra, controlling the keys, tempo, and other details, one instrument at a time. After synthesizing and mixing the digital signals, the system sends them to an amplifier and speaker system.

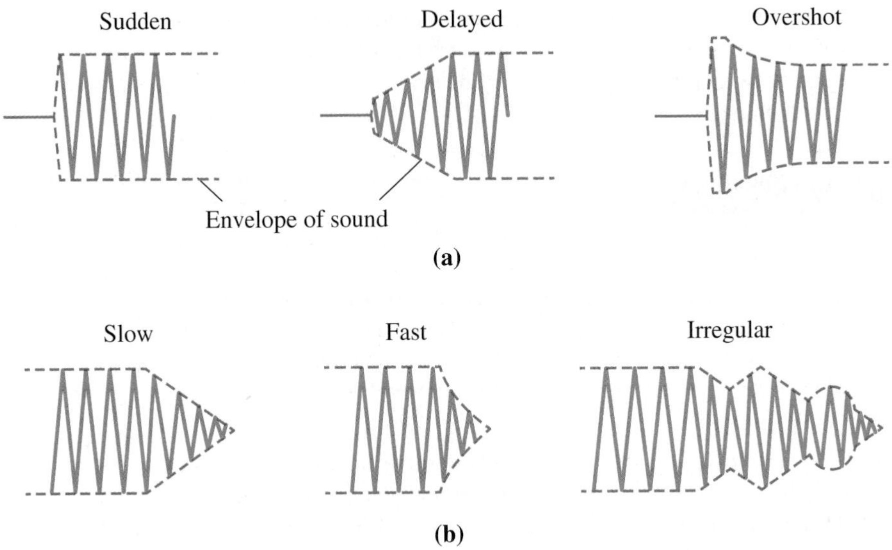

Sudden Delayed Overshot

Envelope of sound

(a)

Slow Fast Irregular

(b)

Figure 14-57 Examples of control patterns of electronic instruments. **(a)** Attack patterns. **(b)** Decay patterns.

The MIDI system is standardized so that software programs, hardware devices, and interfaces can be shared or interchanged. MIDI can also create sounds other than music, including special effects such as animal noises, screams, explosions, thunder, and so on, for movies and other purposes.

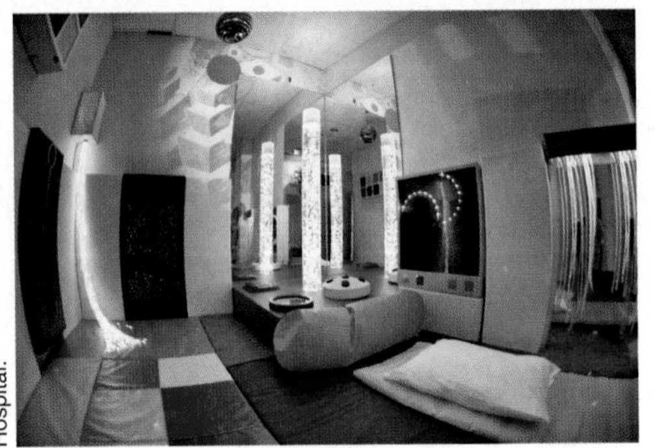

Photo by Richard Johnson, www.richardjohnson.ca. Courtesy of Holland Bloorview Kids Rehabilitation Hospital.

Figure 14-58 An example of a Snoezelen room. In this type of room people experience sensations related to the human senses, especially hearing, sight, and touch. Here a child is able to experience sound by touching a device that oscillates.

TRY THIS!

Controlled Multisensory Environments

An application that incorporates many of the concepts presented in this chapter is a *Snoezelen room*, one example of which is shown in Figure 14-58. It is a specially designed room with sensory-perception devices that simulate sounds and sights to enable individuals with hearing or sight challenges to experience what most of us take for granted. This type of room was developed in Holland in the 1970s and, currently, examples are found around the world. (The word Snoezelen comes from two Dutch words meaning "to explore" and "to relax.")

(a) Research "Snoezelen rooms" or "multisensory environments" to discover various ways that are used to simulate sounds.

(b) Describe how you would design a device for use in a Snoezelen room that would enable a deaf person with sight to experience your favourite type of music.

(c) Repeat (b) for a person that is both deaf and blind.

EXERCISES

14-61 Two strings used in a stringed instrument have linear densities of 40 g/m and 130 g/m. Which one would be used to produce lower pitches? Explain.

14-62 Determine which harmonics of the highest note of a piano ($f = 4186$ Hz) are audible to the average human ear.

14-63 A string has a third harmonic of 600 Hz. What are the first and second harmonics?

14-64 A piano string with a linear density of 4.0×10^{-4} kg/m has a fundamental frequency of 220 Hz. If the string is 0.92 m long, determine the

(a) speed of the transverse wave in the string

(b) tension in the string

14-65 A string 45 cm long has a mass of 12 g and is under a tension of 510 N. Determine the speed of transverse waves in the string.

14-66 Determine the frequency of the fifth harmonic of a 65 cm-long string with a linear density of 2.8×10^{-3} kg/m and under a tension of 150 N.

14-67 Determine the length of a string if the fourth harmonic is 880 Hz, the linear density is 6.4×10^{-4} kg/m, and the tension in the string is 120 N.

14-68 String X has half the linear density of string Y, but is under twice as much tension. How do the speeds of the transverse waves in the strings compare?

14-69 (a) What type of standing wave (transverse or longitudinal) is produced on a violin? What is the distance between adjacent nodes of this standing wave in terms of wavelength?

(b) Repeat (a) for standing waves in a trombone.

14-70 The keys of a saxophone, which acts like an open-air column, are pressed to produce an air column of length 45.0 cm.

(a) If the initial air temperature is 21.0°C, what is the fundamental frequency of a note produced in this column?

(b) What happens to the speed of the sound as the temperature increases? What happens to the fundamental frequency of a note produced in this column?

(c) After a brief period of time the temperature of the air in the saxophone goes up to 35.0°C. What is the fundamental frequency now? Was the original note sharp (higher frequency) or flat (lower frequency) compared with this note?

14-71 Assume you are asked to build a wooden resonating box on which to mount a 512 Hz tuning fork. It is to be used in a room where the air temperature is 20.0°C. Determine the length of the box (i.e., the length of the air column) if the box is (a) open at both ends and (b) closed at one end.

14-72 A pipe organ closed at one end has a length of 1.32 m and the speed of the sound is 345 m/s. Determine the frequencies of the three lowest sounds produced in this pipe.

14-73 A synthesizer adds a sawtooth wave, X ($\lambda = 4.0$ cm, $A = 1.0$ cm) and a square wave, Y ($\lambda = 4.0$ cm, $A = 1.5$ cm) to obtain a new sound. Draw one complete wavelength of X directly above one complete wavelength of Y, then use the principle of superposition to find their addition.

14.9 Acoustics

Suppose you are at a concert where the music has a full, rich sound, just what you hoped it would have. Then the music stops and someone on the stage speaks into a microphone, and you find the speech difficult to distinguish clearly. Likely you are in an auditorium or concert hall with properties designed for music but not for speech. As you will soon see, most rooms cannot provide good properties for both purposes at the same time.

The qualities of a room that determine how well sound is heard are called **acoustics**. The acoustics of a room depend on many factors, making the challenge of designing an auditorium or concert hall very complex.

acoustics the qualities of a room that determine how well sound is heard

Consider, first, all the sounds in a concert hall that are not produced by the musicians on the stage. Such unwanted sounds must be kept to a minimum. Obvious sounds, such as street noises, people's voices just outside the room, and doors as they are opened or closed, must be minimized. One single feature that helps achieve this is a double-door system with hall space between the doors. Less obvious sounds, such as sound caused by the air in heating or cooling systems, must also be minimized. Ducts for circulating air should have the fewest turns possible to prevent noise caused by air turbulence.

Next consider a physical quantity of a room called its reverberation time (to be defined shortly). Just as harmonic structure helps determine the quality of a tone, reverberation time helps determine the acoustics of a room. The sound produced by an instrument travels in all directions in a room. The listener hears the direct sound first, but soon hears many reflected sounds that are called reverberations. Very soon the sound builds up to a maximum intensity that depends on the initial intensity of the sound as well as the reflection and absorption properties of the room. As soon as the source of sound is cut off, the intensity of the sound drops exponentially. The amount of time for a sound in a room to decrease in intensity by a factor of one million from the maximum value is called the **reverberation time** of the room. A room with a reverberation time that is short has too little sound reflection, and the resulting sound is considered thin or shallow. A room with a reverberation time that is too long has too much sound reflection, causing the sounds that follow one another to clash.

reverberation time the amount of time for a sound in a room to decrease in intensity by a factor of one million from the maximum value

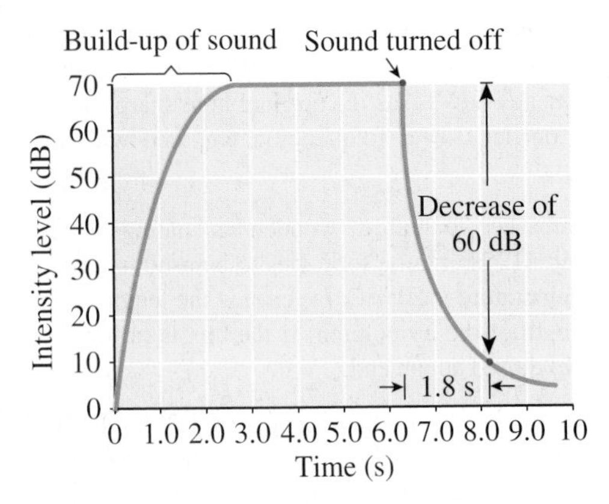

Figure 14-59 Graph of an experimental determination of reverberation time.

(a)

(b)

Figure 14-60 Examples of concert halls. **(a)** Moscow. **(b)** Budapest.

High-quality acoustics are possible with a reverberation time ranging from about 0.5 s in a small room used for speech, to over 3 s in a large hall used for orchestral performances, although these times vary for different frequencies of sounds. You can likely appreciate the difficulty of designing an auditorium with excellent acoustics, especially if more than one purpose is intended.

The reverberation time in a hall can be determined experimentally. A graph illustrating the process is shown in Figure 14-59, using intensity level (in decibels) rather than intensity (in watts per square metre). A drop in intensity of a factor of one million corresponds to a drop in intensity level of 60 dB. In the example shown, the sound level builds up to a maximum of 70 dB, and maintains this value until the source of the sound is stopped. The intensity level then drops by 60 dB to 10 dB in 1.8 s (i.e., the intensity drops to 10^{-6} of the maximum level). Thus, the reverberation time is 1.8 s.

The concert halls in the world that are liked most by musicians and audiences are found to have reverberation times between about 1.7 s and 2.1 s for orchestral music. What features of these halls help to provide good acoustics? The following is a list of some of these features; as you read the list, check the concert halls shown in Figure 14-60 and Figure 14-1 to see how their features would enhance or reduce the acoustics.

- The room should not be square or even rectangular; in such a room reflections could help set up standing waves for certain wavelengths of the sounds, causing loud and soft areas.
- The room should not be too large; the larger the room, the longer is the reverberation time.
- The audience density should be approximately one person for each 0.8 m^2 of floor area.
- The walls should not have areas that are concave (curved inward); such walls could cause reflected sound waves to concentrate to a region, producing unwanted loudness.
- The walls and ceilings should have features that diffuse the reflected sounds in many directions. These features should have various sizes (to reflect various wavelengths), shapes, and angles to the stage.
- Walls, ceilings, floors, and chairs should have appropriate absorbing properties. For example, an auditorium seat can be designed with upholstery that absorbs about the same amount of sound energy as a human would.
- The reverberation time should be about the same in all parts of the room.
- The musicians should be able to hear each other, at least to a certain extent.
- Some components can be altered to accommodate different purposes; for example, curtains can be moved, and reflecting discs suspended from the ceiling can be moved higher or lower.

Room acoustics are in no way exact, and they tend to change as technology advances and music likes and dislikes change. The current use of electronic equipment, for example, must be considered when designing new halls or auditoriums.

DID YOU KNOW?

The rooms available for the production and listening of music and the actual music played in them have greatly influenced each other. For example, medieval cathedrals in Europe were very large, with reverberation times around 5 s, sometimes as long as 10 s. The music composed to be heard in these cathedrals was very slow, and chanting was often used instead of speaking. Even now, speech in these cathedrals is difficult to hear clearly. When smaller chapels with much shorter reverberation times (about 1.5 s) were built, the music also changed. What we call Baroque music typifies the compositions written for smaller rooms. Music composed in later eras, such as the classical era typified by Mozart, requires a different reverberation time (about 1.6 s or 1.7 s) to result in the most pleasing sound.

EXERCISES

14-74 Discuss the acoustics of a room or an auditorium that you know well. Consider both what has been done to provide good acoustics and what could be done to improve the acoustics.

14-75 What is the intensity level of a sound in a room after reverberation time if the original intensity level is 85 dB?

14-76 What would happen to the reverberation time if a concert hall, designed for patrons without coats, is filled in the winter with people holding their winter coats in their laps?

14-77 An *anechoic chamber*, like the one in Figure 14-61, is a room designed to absorb as much sound as possible, isolating external sounds from sound testing or recording. From microphones to loudspeakers, radios to earphones, and musical instruments to materials used in constructing theatre walls, anechoic chambers play an important role in researching how to improve our enjoyment of sound and music. The reverberation time in an anechoic chamber is typically less than 0.05 s. Describe the features of the anechoic chamber shown that help it achieve such a short reverberation time.

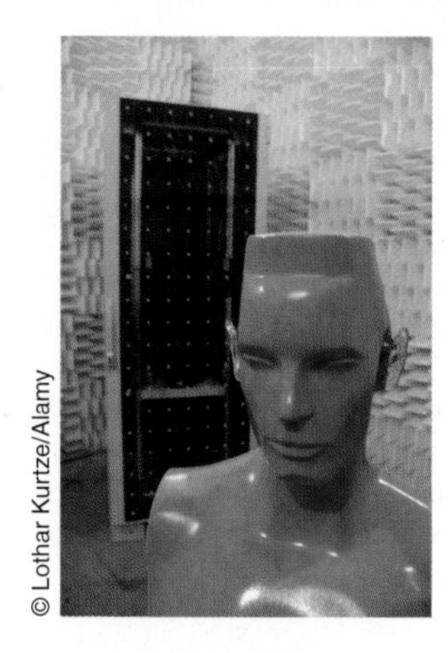

Figure 14-61 One use of an anechoic chamber is to test for acoustic control, in this case measuring noise with an artificial head (Question 14-77).

14-78 The human ear can distinguish an echo from the sound that caused the echo if the two sounds are separated by an interval of about 0.10 s or more. At an air temperature of 22.0°C, what is the shortest length for an auditorium that could produce a distinguishable echo of a sound emanating from one end?

LOOKING BACK...LOOKING AHEAD

Sound is produced by oscillations and transmitted by waves, so several concepts of mechanical oscillations and waves introduced in Chapter 13 were extended here for sound.

After discussing in general the production of sound and its transmission in air by longitudinal waves, we considered the speed of sound in air and other media. A brief look at the human ear was followed by a mathematical description of the intensity and intensity level of sound. We then concentrated on frequencies of sound beyond the human audible range, namely infrasonic and ultrasonic sounds and their applications. The Doppler effect, which occurs when there is relative motion of the source of sound and receiver, was described both qualitatively and quantitatively. This wave phenomenon was followed by the wave effects of diffraction, interference (including beats), and resonance.

After studying sound, we turned our attention to music, which included the topics of tonal quality, musical scales, and musical instruments. The chapter ended with an introduction to the complex field of acoustic design.

Sound waves are of the order of size associated with our everyday experiences, so the explanations and theories of sound are somewhat familiar to us. In the next three chapters we will explore the more abstract properties of light and discuss how light has both particle and wave characteristics.

CONCEPTS AND SKILLS

Having completed this chapter, you should now be able to do the following:

- Describe how sound is transmitted through air.
- Given either the air temperature or the speed of sound in air, determine the other quantity.
- Given any two of the elastic property, the inertial property, and the speed of sound in a material medium, determine the third quantity.
- Describe how the human ear receives sound and transmits it to the brain.
- State the average human audible range of frequencies.
- Recognize how the perceived loudness of a sound depends on the intensity of a sound, the distance from the source, and the frequency of the sound.
- Perform calculations involving sound intensity and intensity level.
- Describe how ultrasonic frequencies can be used in echolocation, and calculate unknown quantities involving speed, distance, and time.
- Describe applications of ultrasound.
- Describe the Doppler effect and explain its cause.
- Solve problems related to the observed frequency heard by a listener, the frequency of a source, the speed of sound, and the speeds of the listener and the source.
- Given any two of Mach number, speed of the source, and speed of the sound in air, find the third quantity.
- Explain the causes of the sound barrier, a shock wave cone, and a sonic boom.
- Recognize the condition required for large diffraction of sound waves.
- Explain the conditions required to produce beats, and use the principle of superposition to illustrate your explanation.

- Solve problems involving beat frequency.
- Describe how beats can be used to tune a musical instrument.
- Describe examples of resonance in sound.
- Determine the frequencies of notes one or more octaves above or below the given frequency of a musical note.
- State the standard frequency of the equitempered musical scale.
- Given the *n*th root of the number 2, calculate the frequencies in any *n*-interval equitempered scale based on any standard frequency.
- Given any four of the length of a vibrating string, the mode of oscillation, the tension, the mass per unit length, and the frequency, find the fifth quantity.
- Describe how resonance of sound waves is created in oscillating columns of air.
- Given any three of the length of a column of air, the speed of sound, the mode of oscillation, and the frequency, find the fourth quantity for either open-air columns or closed-air columns.
- Given the wavelength of a sound resonating in an air column or the distance between the resonance sounds, determine the other quantity.
- Describe the functions of the body parts that produce and control the sound of the human voice.
- Compare electrical and electronic musical instruments.
- Describe factors that affect the quality of the sound of musical instruments.
- Describe factors that affect the acoustics of an auditorium or a music hall.
- Describe factors that affect the reverberation time of an auditorium.
- Perform calculations involving the initial intensity level of a sound in an auditorium, reverberation time, and intensity level after the reverberation time.

KEY TERMS

You should be able to define each of the following words or phrases:

sound
linear density
density
modulus
human audible range
sound intensity
threshold of hearing
threshold of pain
intensity level
infrasonic sounds
ultrasonic sounds
echolocation
sonar
transducer
Doppler effect
Mach number
sound barrier
shock wave cone
bow wave
sonic boom
beats
beat frequency
quality
tone
harmonics
consonance
dissonance
octave
equitempered scale
percussion
acoustics
reverberation time

Chapter Review

MULTIPLE-CHOICE QUESTIONS

14-79 Blue whales communicate with low-frequency sounds of intensity 6.30×10^6 W/m^2. The intensity level of this sound is
(a) 63.0 dB
(b) 630 dB
(c) 18.8 dB
(d) 188 dB

14-80 We are unable to hear ultrasonic sounds because
(a) their wavelengths are too long
(b) their frequencies are too high
(c) their intensity level is too low
(d) their amplitude is too low

14-81 As a fire truck is approaching a listener, the pitch of the siren the listener hears increases because
(a) the observed wavelength of the sound waves increases
(b) the observed speed of the sound waves decreases
(c) the observed wavelength of the sound waves decreases
(d) the observed speed of the sound waves increases

14-82 An airplane is travelling at 1.10×10^3 km/h where the air temperature is 10.0°C. The plane is
(a) travelling at a supersonic speed
(b) travelling at a subsonic speed
(c) creating a shock wave cone in the air
(d) both (a) and (c) are true

14-83 Two guitar strings, X and Y, are plucked at the same instant, and a beat frequency of 4 Hz is heard. Then the tension in Y is increased and the new beat frequency is 1 Hz. If X has a frequency of 220 Hz, the original frequency of Y was
(a) 216 Hz
(b) 219 Hz
(c) 221 Hz
(d) 224 Hz

14-84 Four pure synthesizer notes have the following frequencies: A: 200.0 Hz; B: 400.0 Hz; C: 600.0 Hz; D: 800.0 Hz. They are sounded in these combinations: B with A; C with A; C with B; and D with C. The pair of sounds with the highest dissonance is
(a) C–B
(b) D–C
(c) C–A
(d) B–A

14-85 The fundamental frequency of an air column closed at one end is f_1. Between the ends of the column, the number of nodes present in a standing wave of frequency $7f_1$ is
(a) 3
(b) 4
(c) 5
(d) 6
(e) 7

14-86 An orchestra in a concert hall ends a song abruptly and in 1.2 s the sound level changes from 78 dB to 28 dB. The reverberation time in the hall is
(a) less than 1.2 s
(b) 1.2 s
(c) greater than 1.2 s

Review Questions and Problems

14-87 A puff of steam is seen rising from a ship's whistle, and 4.20 s later the sound from the whistle is heard. The air temperature is 18.0°C. How far away is the ship?

14-88 An orca whale (Figure 14-62), initially located 7.45 km from a hydrophone, emits a sound that is picked up by the hydrophone. If the orca is travelling at 24.0 km/h directly toward the hydrophone and the speed of the sound in the water is 1.47×10^3 m/s, how long after the sound is received will the whale arrive at the hydrophone?

Miles Away Photography/Shutterstock

Figure 14-62 An orca whale (Question 14-88).

14-89 A metal pipe, 7.00×10^2 m long, is filled with a liquid. One end of the pipe is struck with a hammer and the sound travels through the metal and the liquid at different speeds. The pipe has a Young's modulus of 1.10×10^{11} N/m^2 and a density of 8.93×10^3 kg/m^3. The bulk modulus of the liquid is 2.00×10^9 N/m^2 and its density is 1.05×10^3 kg/m^3. Determine which sound will reach the end of the pipe sooner and by how much time. (Treat the pipe as a thin bar.)

14-90 State the mathematical relationship between

(a) sound intensity and the distance from a (point) source of sound

(b) sound intensity level and sound intensity

14-91 If the initial intensity of a sound is I_1, what will be the intensity (in terms of I_1) when the distance from the source (a) doubles, (b) quadruples, and (c) halves?

14-92 The world's loudest insect, the African cicada, has a calling sound with an intensity level of 107 dB at a distance of 0.5 m. Determine the intensity of this sound.

14-93 Determine the intensity level of a stereo speaker that produces an intensity of 1.8×10^{-3} W/m^2 at a distance of 2.5 m.

14-94 The maximum intensity levels of a trumpet, a trombone, and a bass drum, each at a distance of 3.0 m, are 94.0 dB, 107 dB, and 113 dB respectively.

(a) What is the intensity level of three such trumpets playing simultaneously at the same distance of 3.0 m?

(b) What is the intensity level of the three different instruments played simultaneously, again at 3.0 m?

14-95 An ultrasonic pulse leaves the transducer aboard a submarine and the signal that reflects off a ship is received 2.28 s later.

(a) If the speed of the ultrasound in the water is 1.46 km/s, how far away is the ship?

(b) If the density of the water is 1.06×10^3 kg/m^3, what is its bulk modulus?

14-96 At a particular instant, the siren of an ambulance has an emission frequency of 1.15×10^3 Hz, the velocity of the ambulance is 26.0 m/s east, and the speed of sound in air is 337 m/s. Determine the observed frequency heard in each case:

(a) The ambulance is approaching a stationary listener.

(b) The ambulance is going away from a stationary listener.

(c) The listener is approaching the ambulance on a bicycle travelling at 8.00 m/s west.

14-97 Under what conditions could the following statement be true? "A subsonic speed at one location could be a supersonic speed at another location." Give an example of this situation.

14-98 A supersonic jet, travelling at Mach 1.80 where the speed of sound is 3.05×10^2 m/s, produces a shock wave cone.

(a) What is the jet's speed in metres per second and kilometres per hour?

(b) What is the total angle of the shock wave cone?

14-99 At a certain altitude the speed of sound in the air is 1087 km/h. Jet X is travelling at Mach 1.38 (south) and jet Y is travelling at Mach 1.89 (north). Determine the velocity of jet X relative to jet Y in kilometres per hour.

14-100 (a) What factor(s) determine the amount of diffraction of sound waves through an opening or around a barrier?

(b) **Fermi Question:** What is the approximate frequency of the sound waves that would diffract most easily around your head?

14-101 A 440 Hz tuning fork is sounded with a banjo string (Figure 14-63), and a beat frequency of 4 Hz is heard. Then a small metal mass is attached to one prong of the tuning fork, thus reducing the fork's frequency, and a new beat frequency of 3 Hz is produced during sounding. What is the frequency of the banjo string?

Jose Gil/Shutterstock

Figure 14-63 A banjo (Question 14-101).

14-102 Assume that a note of frequency 2640 Hz is the third harmonic of a sound. What are the frequencies of the first, second, and fourth harmonics?

14-103 Sound waveforms are viewed on a computer monitor at the same time the sounds are heard. The resulting waveforms are shown in Figure 14-64. Describe the sounds; in particular, compare the qualities.

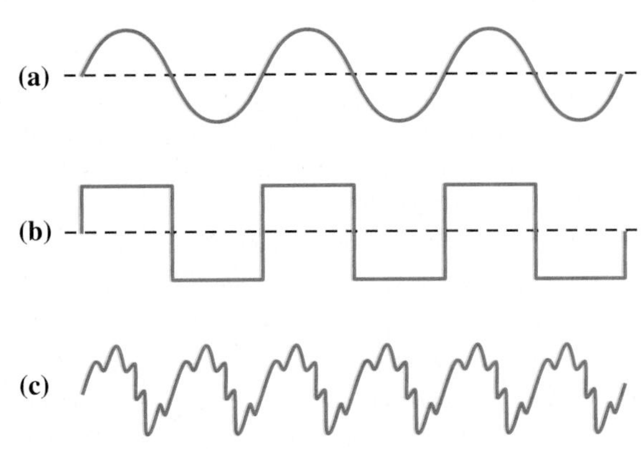

Figure 14-64 Waveforms (Question 14-103).

14-104 What is the frequency of a note (a) two octaves above E_4 and (b) one octave below E_4 on the equitempered scale? (Reference: Figure 14-38.)

14-105 Arrange the following ratios of frequencies in increasing order of dissonance: 2:1; 17:16; 3:2; 5:4; 97:96.

14-106 Determine the first three harmonic frequencies of a piano string that has a length of 88.0 cm and a linear density of 6.09×10^{-4} kg/m, and is under a tension of 2.90×10^{2} N.

14-107 A string of length 1.2 m has a mass of 6.0 g and is under a tension of 36 N. What tension is required in the string to produce a fundamental frequency one octave higher?

14-108 Adjustable air columns are used to study resonance in a room where the air temperature is 22.0°C. If the note G (784.0 Hz) is played at one end of an air column, calculate the lengths of the air column for the first two resonant frequencies if the column is (a) open at both ends and (b) closed at one end.

14-109 How can testing the reverberation time of an auditorium reveal the acoustic properties of the auditorium? What can be done to increase the reverberation time? What can be done to decrease it?

14-110 A sound level meter (Figure 14-65) registers 96 dB at a certain location in a concert hall during a drum roll. At what reading has the reverberation time elapsed? If that time is 0.95 s, describe the acoustics of the hall.

pwrmc/Shutterstock

Figure 14-65 A sound level meter (Question 14-110).

Applying Your Knowledge

14-111 A person's vocal chords oscillate with a wide range of frequencies, and the frequencies heard most strongly are those that produce resonances in the person's resonant cavities. For a given cavity (assumed to be closed at one end and open at the other), the simplest resonance occurs for a wavelength λ of the sound that is four times the length L of the cavity, $\lambda = 4L$ (or $L = \lambda/4$). At this particular wavelength, the sound will have a high intensity.

(a) If the length of the vocal cavity in a particular person is 22 cm, what wavelength of sound will produce the simplest resonance?

(b) If the air temperature in the vocal cavity is 36°C, what is the speed of sound there, and what frequency of sound will produce the simplest resonance?

(c) Suppose that the person inhales helium (Figure 14-66) and then speaks or sings. What wavelength of sound will produce the simplest resonance?

(d) Using the speed of sound in helium from Table 14-2, determine the frequency of sound that will produce the simplest resonance. Is this frequency higher or lower than the frequency determined in (b)? Describe how the person's voice will sound after inhaling helium.

14-112 Fermi Question: The outer ear canal (the auditory meatus) is open at one end.

(a) Estimate the length of the ear canal in millimetres and metres.

(b) Using your estimate in (a), determine an estimate for the fundamental frequency of the resonant sound in this tube.

(c) Use your answer in (b) to explain the lowest point in the graph of sensitivity of the human ear to different frequencies (Figure 14-13).

Rob Marmion/Shutterstock

Figure 14-66 Filling a balloon with helium (Question 14-111).

DID YOU KNOW?

The inert gas sulfur hexafluoride (SF_6), which has a higher density than air and is used as an insulator in particle accelerators, causes the voice to deepen when inhaled, just the opposite of the effect of helium. Demonstrations of this phenomenon can be found by searching "sulfur hexafluoride" on the Internet.

14-113 A worker wants to check the calibration of a tension meter that reads 1.20×10^{3} N when attached to a power cable. The cable is 100.0 m long and has a mass of 37.0 kg. The worker applies physics principles by tapping one end of the cable and measuring the time for the reflected sound to return. If the measured time is 3.40 s, how well calibrated is the tension meter?

14-114 Some of the terminology associated with stereo equipment is found in each part of this question.

(a) A recording system has a signal-to-noise ratio of 58 dB. Calculate the ratio of the intensity of the signal to intensity of the background noise.

(b) A stereo speaker has a flat response of 4 dB, which means that any two signals of equal power sent to the speaker result in sound intensity levels coming from the speaker that differ by 4 dB or less. Determine the maximum variation as a ratio of sound intensities produced by this speaker.

14-115 The wavelength of the waves in an ultrasonic cleaning bath correspond closely to the particles to be loosened and washed away. If these particles are 3.0×10^{-2} mm in diameter, what frequency (in megahertz) should be used in a bath containing alcohol (in which the speed of sound is about 1.2×10^{3} m/s).

14-116 **Fermi Question:** A bullwhip, which was shown in Figure 14-32, has a heavy handle and, when skillfully whipped, it transfers energy to the very narrow "popper" end such that the tip breaks the sound barrier. Estimate the minimum speed at which the tip must travel to create a mini-sonic boom at a summer rodeo.

14-117 (a) How do the acoustic features of a lecture hall with which you are familiar compare to the acoustic features of the hall shown in Figure 14-67?

photobank.ch/Shutterstock

Figure 14-67 A lecture hall (Question 14-117).

(b) Describe how the acoustics of the lecture hall with which you are familiar could be improved.

14-118 A mountain climber cannot stretch far enough to see how far a drop over a ledge is. To estimate the drop, she tosses a small stone horizontally over the ledge and listens for the sound of the stone hitting somewhere below. The total time between the toss and the return sound is 8.0 s and the air temperature is 20.0°C. How high is the drop? (**Hint:** Consider both kinematics and sound. The quadratic formula may be needed.)

Reflection, Refraction, and Dispersion of Light

15

Gemasolar solar thermal plant, owned by Torresol Energy © SENER

Figure 15-1 In this solar energy facility near Seville in Spain, 2,650 mirrors on the ground reflect light from the Sun to the tall solar tower, where the energy is focussed onto a central receiver containing molten compound. This molten compound is circulated to a heat exchanger at the bottom of the tower where it gives thermal energy to a circulating system of water. The heated water turns to steam, which, under pressure, is directed to turbines to generate clean electrical energy. Even after sunset, electrical energy can be produced for up to 15 h due to the thermal energy stored in the molten salt compound. The mirrors are heliostatic, which means they follow the Sun throughout the day, maximizing the energy collected at the tower.[1]

Without light from the Sun, life on Earth would not be possible. Light energy is needed for photosynthesis in plants and for maintaining the temperature of Earth's oceans and atmosphere needed to support life. A modern way of utilizing light from the Sun is shown in Figure 15-1. Numerous plane mirrors are placed in fields near the tall tower. The mirrors can be controlled to send the reflected light to a concentrated area near the top of the tower. As Earth rotates and the Sun's position in the sky changes, the mirrors are adjusted by computer to change their direction to aim the light in the desired direction. In this case, light from the Sun is transformed into thermal energy directly, and mechanical and electrical energies indirectly. If light can be transformed into

CHAPTER OUTLINE

[1]Sometimes not all the light is allowed to reflect from the mirrors to the tower. The bright area away from the tower results when the reflected light is so concentrated and intense it vaporizes specks of dust in the air, causing bright flashes, puffs of smoke, and audible crackles.

other forms of energy, can other forms of energy be transformed into light energy? In this chapter, we will explore answers to this and many other questions.

The light that reflects off the mirrors in Figure 15-1 is just one example of evidence of the straight-line motion of light. Other common examples are laser beams, shadow formation, and the fact that we are unable to see around solid objects. It is only natural that humans have learned to use thin, straight lines to represent "rays" of light. In this chapter, such rays will be used in diagrams to predict and explain what occurs when light strikes mirrors, prisms, and lenses. These "ray diagrams" will help us understand why the images we see in mirrors and lenses can vary in size relative to the object's size. This topic has many practical applications, some of which are described in this chapter, and some in later chapters.

As you study the fascinating topic of light, consider one more question: What evidence do we have that light acts like a particle, a wave, or a combination of both? This chapter will provide historical and experimental evidence to begin to answer this question, but more complete answers will be presented in Chapters 16 and 17.

15.1 Sources and Propagation of Light

We begin the topic of light by exploring answers to these questions: Why is light considered to be a form of energy? What are the basic properties of the transmission of light? How is the speed of light measured?

Light as Energy

To see evidence that light is a form of energy, we start by considering sources of light. Any object that is a source of light energy is called a **luminous object**. Luminous objects are classified according to the reason they emit light.

The most common luminous objects, called **incandescent sources**, emit light because they are at a high temperature. The Sun, in which nuclear reactions produce extremely high temperatures, is our most important incandescent light source. Other incandescent sources include incandescent light bulbs, which gain a high temperature from electrical energy, and fires, which gain a high temperature from chemical potential energy (Figure 15-2).

Many other luminous objects emit light in a spontaneous fashion, in other words they emit light as soon as they receive an input of energy.[2] **Gas discharge sources** (Figure 15-3 (a)), such as neon signs and mercury-vapour lamps, have a gas in a glass tube that becomes luminous when an electric current passes through the gas. A **fluorescent source**, such as a fluorescent tube or bulb, is also filled with a gas, such as mercury gas, but it has a coating of fluorescent material on the inside of the glass tube. Electric current passing through the gas causes the gas to emit energy that in turn strikes the fluorescent coating; the coating then emits light of many different energies, seen as white light.[3] **Light-emitting diodes (LEDs)** are efficient sources that emit light when an electric current travels through a semiconductor. **Phosphorescent materials**

luminous object any object that is a source of light energy

incandescent source an object that emits light due to a high temperature

gas discharge source a gas contained in a glass tube that becomes luminous when an electric current passes through it

fluorescent source an object that emits light when struck by high-energy waves or particles

light-emitting diode (LED) an efficient source that emits light when an electric current travels through a semiconductor

phosphorescent material a material that becomes luminous when struck by high-energy waves or particles, and remains luminous for a period of time

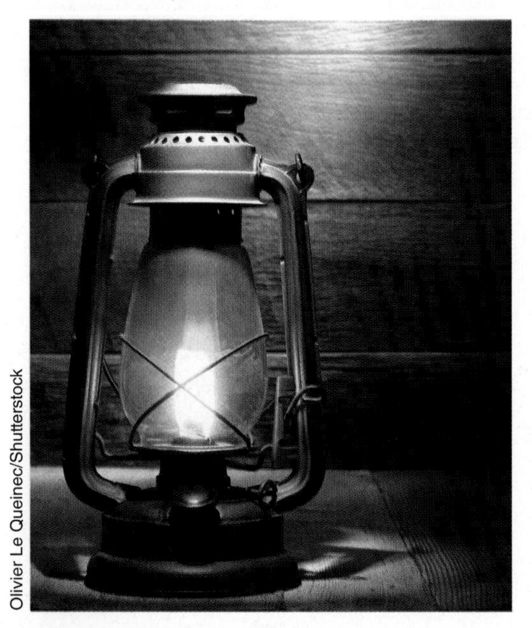

Figure 15-2 In this antique kerosene lamp, chemical potential energy is converted into light and thermal energies.

[2]Light sources that are classified as stimulated rather than spontaneous include the laser, described in Section 17.7. Another light source, triboluminescence, was described in Chapter 6, page 141.

[3]To further complicate the classification of light sources, the energized gases in neon signs and fluorescent lights are in the state of matter called *plasma*, which is created when a strong electric field strips electrons away from atoms. Other plasma sources include lightning, plasma television screens, and the Sun and all other stars in the universe.

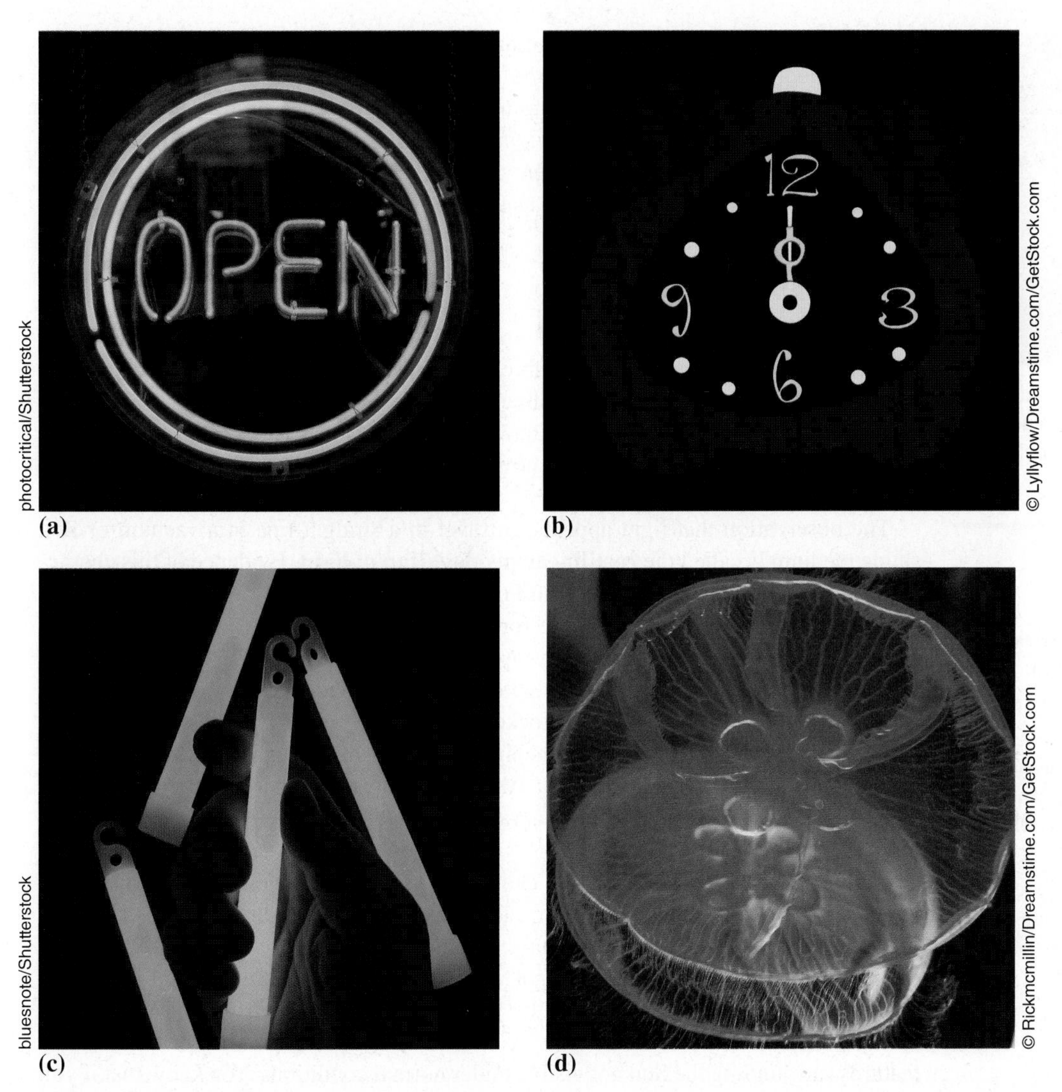

Figure 15-3 Other examples of luminous objects: **(a)** a neon sign, **(b)** a phosphorescent dial on an alarm clock, **(c)** chemiluminescent glow sticks, **(d)** a bioluminescent moon jellyfish (*Aurelia aurita*).

(Figure 15-3 (b)) become luminous when struck by high-energy waves or particles, and remain luminous for some time: the luminous dials on some clocks and watches emit light for several hours after they absorb energy. **Chemiluminescent materials** (Figure 15-3 (c)) react chemically to produce light without a noticeable increase in temperature. Some safety lights emit light energy in this manner. **Bioluminescent animals** (Figure 15-3 (d)), such as fireflies and certain fish and jellyfish, can emit light energy because of chemical reactions.

chemiluminescent material a material that produces light through a chemical reaction without a noticeable increase in temperature

bioluminescent animal an animal that can emit light energy because of chemical reactions

Certain light sources serve specific functions. For example, an electronic stroboscope, which can be used for entertainment or to study moving objects, emits short bursts of light at regular intervals. The photograph in Figure 15-4, on page 470 shows how stroboscopes can be used to study the mechanics of bird wings flapping at a high frequency. Another special type of light source, the laser, will be described in Section 17.7.

DID YOU KNOW?

Battery-powered hand-held lights, invented in the late 1800s, initially were called electrical hand torches. But with weak batteries and faulty contacts the light flickered frequently, resulting in the name *flashlight*, the name used to this day in North America.

No matter what the source is, light always originates from some form of energy. Since energy is conserved, we can surmise that light is a form of energy that has been transformed from some other form of energy. Similarly, when light strikes an object we observe that some light may be transmitted through the object and some may be reflected, but light that is neither transmitted nor reflected is absorbed. This absorbed

Hubert Raquet/Look At Sciences/Science Photo Library

Figure 15-4 Stroboscopic lights, flashing at a fairly high frequency, allow researchers to study the aerodynamic flight of a mechanical bird flapping its wings in a wind tunnel.

light is manifest in some other form of energy. For example, the light can change into thermal energy, increasing an object's temperature, or it can change into electrical energy in solar cells. Again, energy has not disappeared; it has simply changed forms.

Indeed, experiments confirm that *light is a form of energy that is visible to the eye*. As you will learn in greater detail later, visible light is only a very small part of all the energies that make up what is known as the electromagnetic spectrum.

A Modern Light Source

Use a search engine to look up FIPEL (field-induced polymer electroluminescent) foil.

- How does nanotechnology relate to this light source?
- Describe the advertised advantage(s) of this polymer foil over incandescent and fluorescent sources, LEDs, and OLEDs (organic light-emitting diodes).
- In your opinion, will this technology replace incandescent and fluorescent lighting?

Transmission of Light

Light can travel through air and glass and various other transparent media. However, light can also travel where no medium exists, in other words, through a vacuum. We know this because light that comes directly from the Sun or other stars or reflects off the Moon and planets reaches us through the vacuum of outer space. Thus, light energy is different from sound energy, which requires a material medium through which to travel.

The observation that light appears to travel in a straight line in a vacuum or in a single medium is called the **rectilinear propagation of light**. Evidence of this characteristic is shown in the photographs in Figure 15-5.

We now look more closely at the formation of shadows. If a source of light is a point source that emits light equally in all directions, any opaque object placed in the way of the light from this source will have a dark shadow behind it. The part of a shadow where light is completely blocked is called an **umbra**. If the source of light is an extended source, the shadow behind an object in the light's path may consist of lighter parts as well as the umbra. The lighter part of a shadow, where some light falls, is called a **penumbra**. Ray diagrams illustrating these concepts are shown in Figure 15-6.

Nature provides grand shadows called eclipses. To people on Earth, the most important types of eclipses are solar and lunar eclipses. A **solar eclipse**, or eclipse of the Sun, is a blocking of the Sun's light by the Moon as Earth passes through the Moon's shadow. As shown in Figure 15-7, observers on Earth where the Moon's umbra reaches see a *total eclipse* of the Sun, while observers where the Moon's penumbra reaches see a *partial eclipse*. A total solar eclipse allows observers to view features surrounding the Sun not visible under normal conditions. You know that if you view a bright light you can't see what is adjacent to the light without blocking the light. Similarly, the Sun's outer atmosphere, called the *corona*, flashes into view during a total solar eclipse (Figure 15-8 (a)). The corona consists of low-density gases that flare outward for millions of kilometres, showing how active and powerful the Sun is.

A **lunar eclipse**, or eclipse of the Moon, is a blocking of the Sun's light by Earth as the Moon passes through Earth's shadow. Both solar and lunar eclipses occur only if the Sun, Earth, and Moon are colinear, a situation that does not happen frequently. Other types of eclipses occur elsewhere in the solar system and beyond.

Terminology Tip

A *transparent* medium or material, such as clear glass, allows light to travel through it and images to be seen. A *translucent* material or object, such as frosted glass, allows some light to pass through it but does not allow clear images to be seen. An *opaque* material or object, such as a calculator, does not allow light to pass through it.

hxdyl/Shutterstock

(a)

David Varga/Shutterstock

(b)

Figure 15-5 Evidence of the rectilinear propagation of light. **(a)** Beams of light at a concert spread out in narrow, straight cones in the smoky mist. **(b)** With the Sun low in the sky, the shadows of the camels and their riders in a caravan provide evidence of the straight-line motion of light.

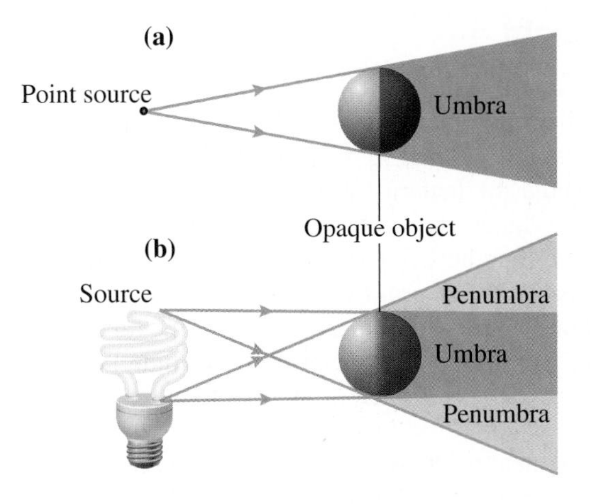

Figure 15-6 Shadow formation. **(a)** Although light from a point source spreads out in all directions, only two light rays from the source are needed in a diagram to locate the shadow behind the opaque object. **(b)** Light from an extended source casts a shadow with both an umbra and a penumbra. In this case, a minimum of four light rays are required to locate the umbra and penumbra behind the opaque object.

rectilinear propagation of light the observation that light appears to travel in a straight line in a vacuum or in a single medium

umbra the part of a shadow where light is completely blocked

penumbra the lighter part of a shadow where some light falls

solar eclipse a situation in which light from the Sun toward Earth is blocked by the Moon

lunar eclipse a situation in which light from the Sun toward the Moon is blocked by Earth

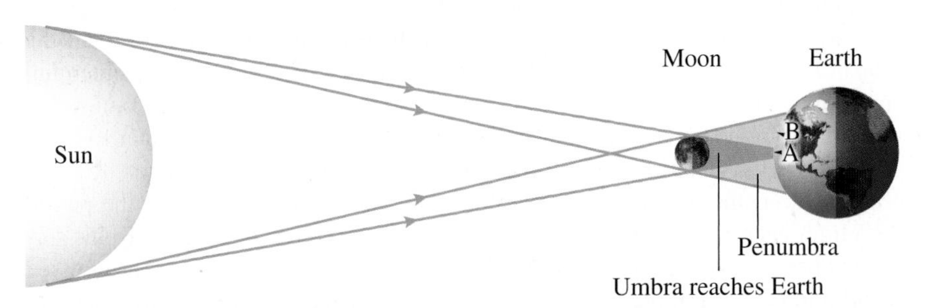

Figure 15-7 Illustrating a solar eclipse: Notice that only four straight light rays are needed to draw an eclipse diagram. The rays originate from the top and the bottom of the Sun and pass by the top and bottom of the object whose shadow we are locating, in this case, the Moon. An observer at position A on Earth would observe a total solar eclipse, and an observer at B would see a partial solar eclipse. Sizes and distances are not to scale.

TACKLING MISCONCEPTIONS

Using Ray Diagrams

Ray diagrams tend to be an oversimplification of the physical situation. For example, the Sun is a mass of violent gases, so it has no distinct surface. Thus, the rays in Figure 15-7 and other ray diagrams are simply representations to help explain the observations.

DID YOU KNOW?

It happens that the Moon's umbra is approximately the same length as the distance between the Moon and Earth. However, the Moon's orbit around Earth is not circular, so the Earth–Moon distance varies. If during a solar eclipse the Earth–Moon distance is greater than the Moon's umbra, the umbra does not reach Earth. Such an eclipse is called an *annular eclipse*, from the Latin *annulus*, which means ring. An observer on Earth directly in line with the Sun and the Moon would see a ring of light around the Moon (Figure 15-8 (b)).

(a)

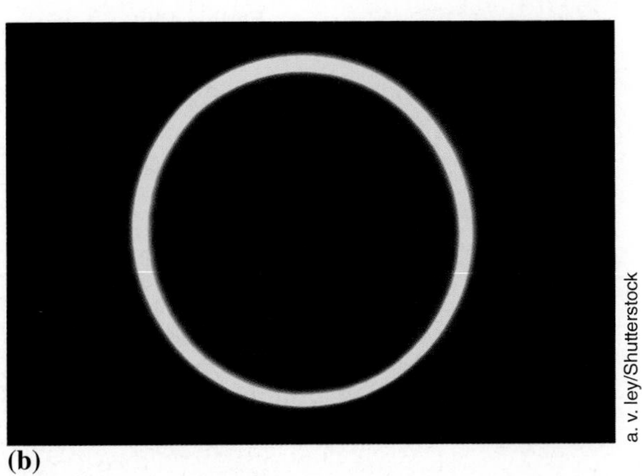

(b)

Figure 15-8 Solar eclipses. **(a)** This series of photos during a 2001 solar eclipse shows that the corona is visible when the eclipse is total. **(b)** During an annular eclipse, the ring of light surrounding the Moon is bright enough to prevent the corona from being seen.

Measuring the Speed of Light

The first recorded attempt to measure the speed of light in air was made by the great Italian scientist Galileo Galilei approximately 400 years ago. Both Galileo and his assistant had a lantern and the two men were separated by a distance of several kilometres. Galileo uncovered his lantern for an instant, and as soon as his assistant saw the light from that lantern he uncovered his own lantern for an instant. Galileo tried to measure the time between his initial sending of the light and observing the light from his assistant's lantern. He could not obtain any meaningful results. (Human reaction times would be far longer than the time for light to travel such a short distance. As well, timing devices of that era were crude.) Galileo concluded that the speed of light was either infinite or too great to measure using his technique.

The first fairly successful measurement of the speed of light resulted from a discovery made in 1676 by a Danish astronomer, Ole Roemer (1644–1710).[4] Roemer observed the planet Jupiter and its moons through a telescope. Jupiter is a large planet located much farther from the Sun than Earth. In particular, Roemer observed the motion of Jupiter's innermost moon, called Io. It is shown in Figure 15-9 just emerging from Jupiter's umbra. With Earth in position A, Roemer made note of the time when Io emerged from the umbra, then he observed that approximately 42.5 h later Io again emerged from the umbra after completing one revolution around Jupiter. Roemer expected that, as the months passed and more measurements were taken, the period of 42.5 h for each revolution of Io around Jupiter would remain constant. To his surprise, and because of his careful observations, he discovered that as Earth progressed in its orbit to position B, the observed period of time actually increased. Roemer reasoned, correctly, that the time difference resulted from the fact that the light travelling from Io to Earth had to cover an extra distance from A to B. Using his data, Roemer determined that the amount of time for light to travel a distance equal to the diameter of Earth's orbit (from C to D) was almost 22 min, or about 1300 s. Earth's orbital diameter, a measurement attributed to another scientist from Roemer's time, Christian Huygens, was believed to be about 3.0×10^{11} m, so using speed = distance/time, the speed of light was calculated to be about

$$v = \frac{d}{t} = \frac{3.0 \times 10^{11}\,\text{m}}{1.3 \times 10^{3}\,\text{s}} = 2.3 \times 10^{8}\,\text{m/s}$$

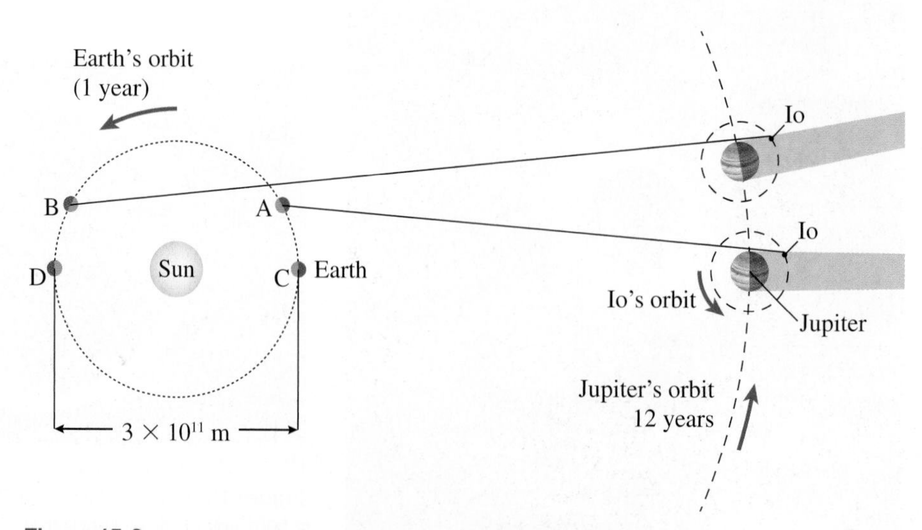

Figure 15-9 Roemer's observations of the moon Io orbiting Jupiter.

[4]Some references use Olaf for Roemer's first name.

This high speed (which is 230 million metres per second!) is lower than the currently accepted value of 3.0×10^8 m/s.

Nearly 200 years later, in 1849, a French physicist named Armand Fizeau measured the speed of light directly using a rather complicated technique, shown simplified in Figure 15-10. He sent light through the teeth of a wheel that could be rotated at varying speeds. The light reflected off a mirror on a hill 8.63 km away. At low speeds of rotation, the reflected light was blocked by the teeth, but as the rotation speed was increased, eventually light could be observed coming through the next space between the teeth. By determining the time for the wheel to advance by one space between the teeth, Fizeau was able to calculate the speed of light in air to be 3.13×10^8 m/s. His method lacked accuracy because it was difficult to judge exactly when the light coming through the space was the brightest. (See Question 15-6.)

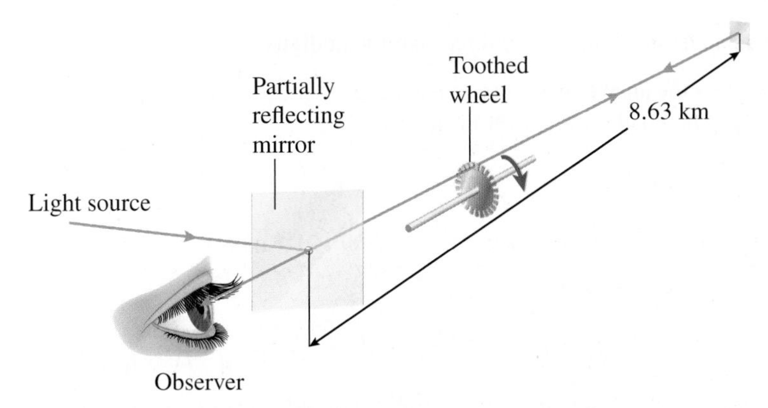

Figure 15-10 Fizeau's method of measuring the speed of light using a rotating wheel with teeth.

Fizeau's technique was improved upon slightly by another French scientist, Jean Foucault (1819–1868), who replaced the toothed wheel with an octagonal (8-sided) mirror, illustrated in Figure 15-11. Foucault used this technique to measure the speed of light both in air and in water. He found the speed of light in water to be only about 3/4 of the speed of light in air, a discovery that had a lasting effect on the proposed theories of light.

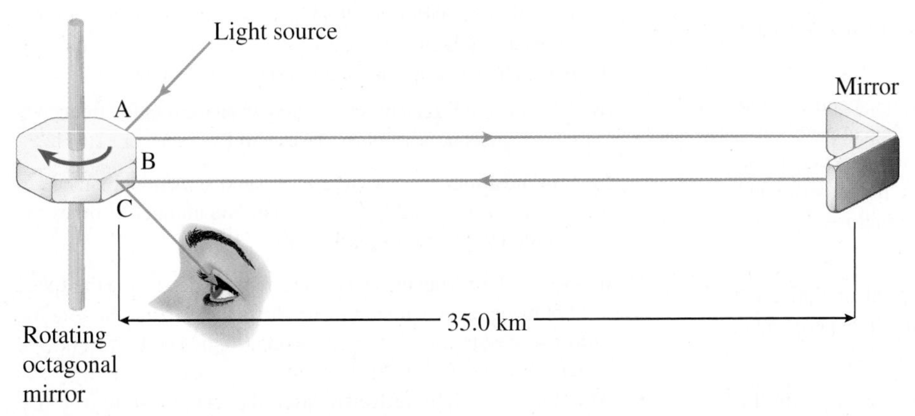

Figure 15-11 The rotating octagonal mirror used to measure the speed of light: this technique was used by Foucault and later by Michelson.

Twentieth century measurements of the speed of light have steadily increased in accuracy. An American Nobel-prize winning scientist, Albert Michelson (1852–1931), used the rotating octagonal mirror technique with a path length of 70.0 km. (See Question 15-8.) Nowadays, using a laser, the speed of light has been measured to nine significant digits: its value is 299 792 458 m/s. For most problems in this text, we will use three significant digits for this speed. Using the symbol c for the speed of light in air or a vacuum,

$$c = 3.00 \times 10^8 \text{ m/s} \qquad (15\text{-}1)$$

light year the distance light travels in a vacuum in one Earth year

This speed is used to determine the distance astronomers call a **light year**, which is the distance light travels in a vacuum in one Earth year. (See Sample Problem 15-1.)

SAMPLE PROBLEM 15-1

(a) Calculate one light year to three significant digits.

(b) Express the answer in (a) using a unit such that no power of 10 is required. (Reference: The inside front cover of the text)

Solution

(a) From $v = d/t$, $d = vt$, where $v = c = 3.00 \times 10^8$ m/s.

$$\begin{aligned} d &= ct \\ &= 3.00 \times 10^8 \frac{\text{m}}{\text{s}} (1 \text{ yr})\left(\frac{365 \text{ d}}{\text{yr}}\right)\left(\frac{24 \text{ h}}{\text{d}}\right)\left(\frac{3600 \text{ s}}{\text{h}}\right) \\ &= 9.46 \times 10^{15} \text{ m (to three significant digits)} \end{aligned}$$

Thus, one light year is 9.46×10^{15} m.

(b) From the list of metric prefixes, 1 Pm = 10^{15} m. Thus, one light year is 9.46 Pm.

TACKLING MISCONCEPTIONS

What Does a Light Year Mean?

A statement such as, "She is light years ahead of her time" is misleading. A light year is a measure of distance, not time.

EXERCISES

15-1 How can a vacuum jar (shown previously in Figure 14-3, page 414) be used to compare the transmission of sound and light without a medium?

15-2 Draw a ray diagram of (a) a lunar eclipse and (b) an annular eclipse.

15-3 Assume that when Galileo tried to measure the speed of light he was located 5 km from his assistant. Using the currently accepted speed of light (3.00×10^8 m/s), how long would the light have taken to travel the return trip in his experiment? Use your answer to explain why his experimental results were doomed to failure.

15-4 Using the currently accepted value of the speed of light in a vacuum, calculate the percent error in Roemer's experimental value of the speed of light.

15-5 Assuming that light takes 5.00×10^2 s to reach Earth from the Sun and that Earth's orbit is circular, what is the diameter of Earth's orbit?

15-6 (a) In Fizeau's method to determine the speed of light, the wheel had 720 teeth and the reflected light appeared bright when the wheel's rotation frequency was 25.2 Hz. The distance from the wheel to the reflecting mirror was 8.63×10^3 m. Assuming the reflected light passed through the space next to the space where the incident light was sent, what was the speed of the light?

(b) What would Fizeau have observed when the rotation frequency of the wheel described in (a) was half the value, or 12.6 Hz?

15-7 **Fermi Question:** Assume that light could somehow be made to travel in circles around Earth. Estimate the number of times the light could travel around Earth in 1 s.

15-8 In one of Albert Michelson's experiments to determine the speed of light using an octagonal mirror, the distance from the rotating mirror to the stationary mirror was 35.0 km. The light reflected to the observer when the frequency of rotation was 535 Hz. Assuming the light reflected after the octagonal mirror had rotated 1/8 of a revolution, determine Michelson's measurement of the speed of light.

15-9 What would Michelson have observed in his experiment if he had doubled the frequency of rotation of the octagonal mirror?

15-10 (a) The distance from Earth to the Andromeda Galaxy is 2.00×10^6 light years. Express this distance in metres.

(b) Current estimates of the radius of the universe as observed by our most powerful telescopes average about 1.4×10^{26} m. Express this enormous distance in light years.

15-11 Communication between the control centre on Earth and a robotic rover on Mars is carried out using electromagnetic waves that travel at the speed of light. If the distance between Earth and Mars at the time of the communication is 9.72×10^{10} m, what minimum time delay occurs between sending a signal and receiving a return signal?

15.2 Mirrors and Their Images

Probably the first mirror in which an image was observed was a smooth surface of water (Figure 15-12). Reflection of light off a smooth, regular surface is called **regular** or **specular reflection**; it is common in mirrors. If ripples or waves develop on the water, a clear image cannot be seen because the light scatters in many directions when it reflects. Reflection of light off an irregular surface is called **irregular** or **diffuse reflection**. It occurs for most objects we see, including the paper in this book. Figures 15-13 (a) and (b) compare the two types of reflection using light rays. (Waves of light could be used to show the same effects, but rays are more convenient.) Part (c) shows why paper, in this case filter paper, produces diffuse reflection.

In ancient civilizations, smooth polished metal was used to make mirrors. Nowadays, most common mirrors are made of glass with a silvered reflecting surface at the back of the glass. Some special mirrors are made with the reflecting surface at the front of the glass.

Images produced by mirrors or any other optical device can be described by using three main characteristics: attitude, type, and location.

- The *attitude* is either upright or inverted when compared to the original object.
- The *type* is either real or virtual. A **real image** is one that can be placed onto a screen; the light observed actually comes from the image. When you watch a movie you don't look at the movie projector; rather you look at the real image(s) on a screen. A **virtual image** is one that cannot be placed onto a screen; the light does not come from the image, so the observer must look at the optical device to see the image. For example, if you want to see the image produced by a plane mirror, you have to look at the mirror because the image is virtual. (To prove this, all you have to do is try to project an image from a plane mirror onto a screen or a wall.)
- Image location can be described in various terms, depending on the detail required. For example, stating that the image in a plane mirror is behind the mirror gives minimum detail, whereas stating the image-to-mirror distance (symbol d_i) is more detailed.

specular reflection (or **regular reflection**) reflection of light off a smooth, regular surface

diffuse reflection (or **irregular reflection**) reflection of light off an irregular surface

real image an image that can be placed onto a screen so the observer does not have to look at the optical device to see the image; for a mirror, it is in front of the mirror

virtual image an image that cannot be placed onto a screen so the observer can see the image only by looking at the optical device; for a mirror, it is behind the mirror

S.R. Lee Photo Traveller/Shutterstock

Figure 15-12 The smooth surface of water can act like a mirror, reflecting light and allowing an image of Japan's Mount Fuji to be seen.

(a) Smooth surface

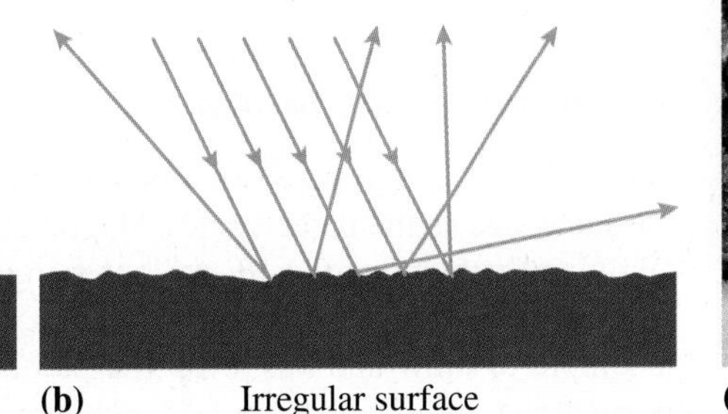

(b) Irregular surface

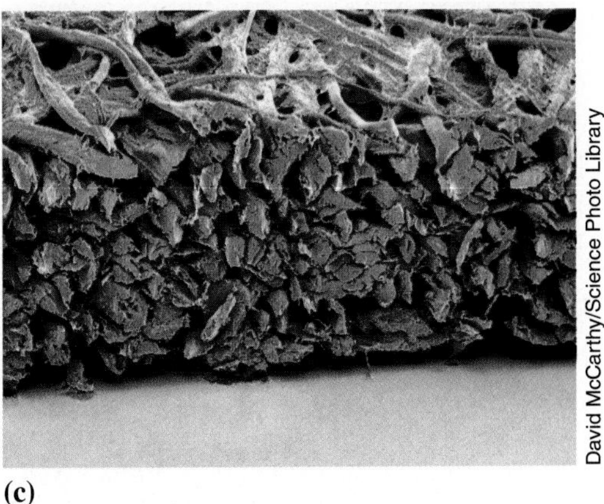

David McCarthy/Science Photo Library

(c)

Figure 15-13 Comparing specular and diffuse reflection: **(a)** specular reflection on a regular surface, **(b)** diffuse reflection on an irregular surface. **(c)** This highly magnified view of filter paper reveals why it is an irregular surface, thus producing diffuse reflection.

DID YOU KNOW?

Besides specular and diffuse reflection, there is a third type of reflection called *retro* or *reflex reflection*. It occurs when light strikes a specially made surface that reflects light back toward the source. Safety garments and traffic signs are made with retro-reflecting materials that bounce light back toward vehicle headlights. Such materials consist of an outer transparent layer, a thin layer of tiny glass beads, and an inner reflecting layer (Figure 15-14).

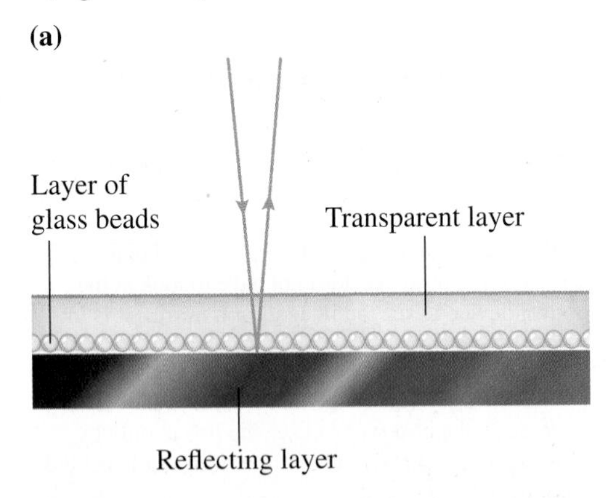

Figure 15-14 **(a)** One possible design of a retro-reflecting material. **(b)** Retro-reflective safety gear.

Quantitatively, it is useful to calculate the **linear magnification** (M) of an optical device, which is the ratio of the image height (h_i) to the object height (h_o).

Using symbols,

$$M = \frac{h_i}{h_o} \quad \text{(15-2)}$$

In this equation, the units cancel out, so linear magnification has no unit. However, the linear magnification may be either positive or negative, according to the following rules for a single optical device:

h_i is positive when the image is upright relative to the object

h_i is negative when the image is inverted relative to the object

The advantages of using rays rather than waves become apparent when ray diagrams are drawn to locate images in mirrors. Incident rays that originate from an object reflect off the mirror, and to locate the image of the object all we must do is find where the reflected rays intersect, or appear to intersect. After discussing how the light rays reflect, we will look at ray diagrams for both plane and curved mirrors.

The Laws of Reflection

Just as a non-spinning basketball that strikes a gymnasium floor bounces off at the same angle it struck, so also a light ray reflects off a mirror at the same angle it struck. In considering light rays that strike any optical device, we define all angles as measured from a line called a normal rather than from the surface of the optical device. The *normal*, symbol N, is a line perpendicular to the surface of the mirror, lens, or other optical device (like the normal drawn in free-body diagrams). The **angle of incidence** (θ_i) is the angle between the incident ray and the normal, and the **angle of reflection** (θ_r) is the angle between the reflected ray and the normal.

It can be shown experimentally that the following two *laws of reflection* apply to any light rays that undergo regular reflection.

- The incident ray, the normal, and the reflected ray all lie in the same plane.
- The angle of incidence equals the angle of reflection ($\theta_i = \theta_r$).

These laws apply whether the reflecting surface is plane or curved. When measuring angles of incidence or reflection at curved surfaces, it becomes clear why a normal is used to define angles (Figure 15-15).

linear magnification the ratio of the image height to the object height in an optical device

angle of incidence the angle between the incident ray and the normal; symbol θ_i

angle of reflection the angle between the reflected ray and the normal; symbol θ_r

Images in Plane Mirrors

When you look at yourself in a plane mirror, you see an image that is upright and the same size as the object. A ray diagram can be used to verify this observation by applying the fact that $\theta_i = \theta_r$, as shown in Figure 15-16 (a). There are several noteworthy features of this ray diagram. First, the object is a 1.3 cm-high arrow, which is simple to draw and has a top that is distinguishable from the bottom. Second, when drawing a ray diagram for a plane mirror, the easiest ray to draw from a point on the object is the one in which $\theta_i = 0°$, that is, the one that is along the normal, because it reflects straight back along the normal. For example, the ray from the top of the arrow along the normal to the right reflects back onto itself and returns to the top of the arrow. Another ray originating at the top of the arrow was chosen to travel up and to the right, and it reflects from the mirror with $\theta_i = \theta_r$. Third, the reflected rays that started from the top of the object do not intersect; in fact they spread apart from one another, so they must be extended straight back behind the mirror to find their (apparent) point of intersection. The same applies to the rays that started from the bottom of the object. The intersection of the extended rays is used to locate the image. Since the image is behind the mirror, it cannot be placed onto a screen, so it is virtual. Fourth, many more incident rays and their corresponding reflected rays could be drawn, but they are not needed. Lastly, the image is 1.3 cm high, the same size as the object, and it is located the same distance behind the mirror as the object is in front. Figure 15-16 (b) shows how an observer is able to see an image.

We can now summarize the three main characteristics of the image in a plane mirror and calculate the linear magnification of the mirror. The attitude is upright because the image arrow is facing the same way as the object arrow. The type is virtual because any image behind the mirror cannot be projected onto a screen. The location is behind the mirror such that the image distance equals the object distance. The linear magnification in the example in Figure 15-16 (a) is

$$M = \frac{h_i}{h_o} = \frac{+1.3\,\text{cm}}{+1.3\,\text{cm}} = +1.0$$

Plane mirrors have several uses besides checking our appearances. A town in Norway named Rjukan, surrounded by mountains, lived without direct sunlight striking it for six months each winter until 2013 when three large, computer-controlled, solar-powered plane mirrors were erected to reflect sunlight to the town's central square. Large mirrors are sometimes used in restaurants and hallways to try to make the rooms look bigger. Optometrists use a two-way plane mirror in a device called an ophthalmoscope to direct light into a patient's eye to examine the interior structure of the eye,

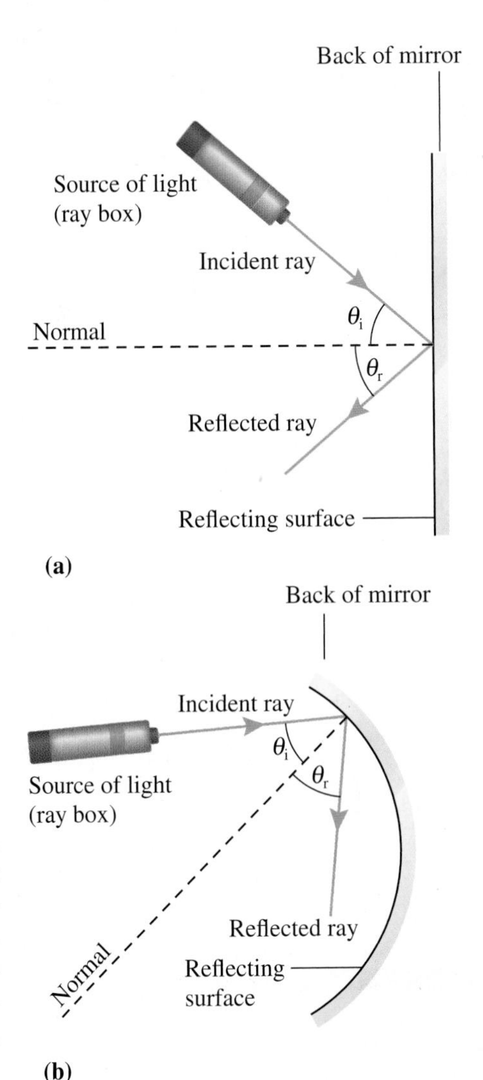

Figure 15-15 In mirrors, the angle of incidence equals the angle of reflection. In both the plane mirror **(a)** and the curved mirror **(b)** the reflecting surface shown is at the front of the mirror.

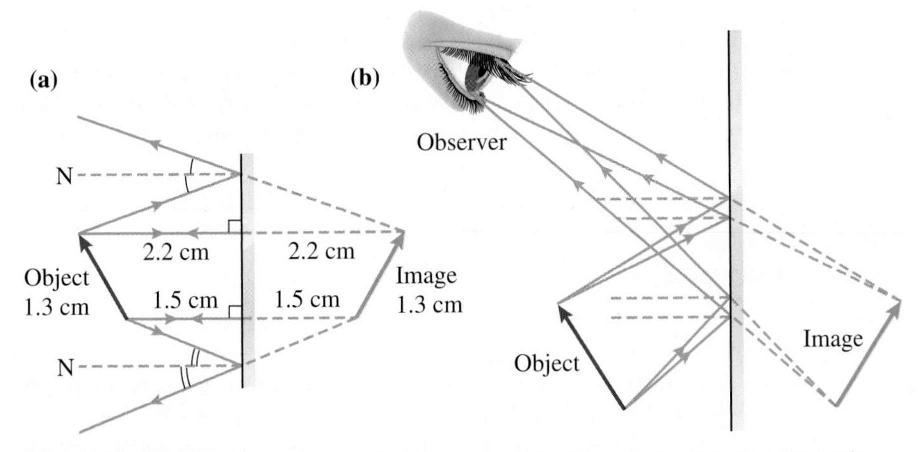

Figure 15-16 Using ray diagrams to locate an image of a 1.3 cm-high object in a plane mirror. **(a)** This type of diagram can be used to locate the image accurately. **(b)** This type of diagram shows how the observer sees an image.

especially the retina (Figure 15-17 (a)). One-way mirrors are used by observers such as researchers or security personnel to monitor activities in an adjacent room without being seen. The room under observation is well-lit and the glass mirror reflects much of the light back into the room, while the observer looks on from a dimly lit room; light from the bright room can enter the darker room, but very little light travels the other way. One design of automobile headlamps utilizes a plane mirror, allowing engineers to reduce slightly the size and aerodynamic resistance of automobile components (Figure 15-17 (b)). What other applications of plane mirrors have you seen?

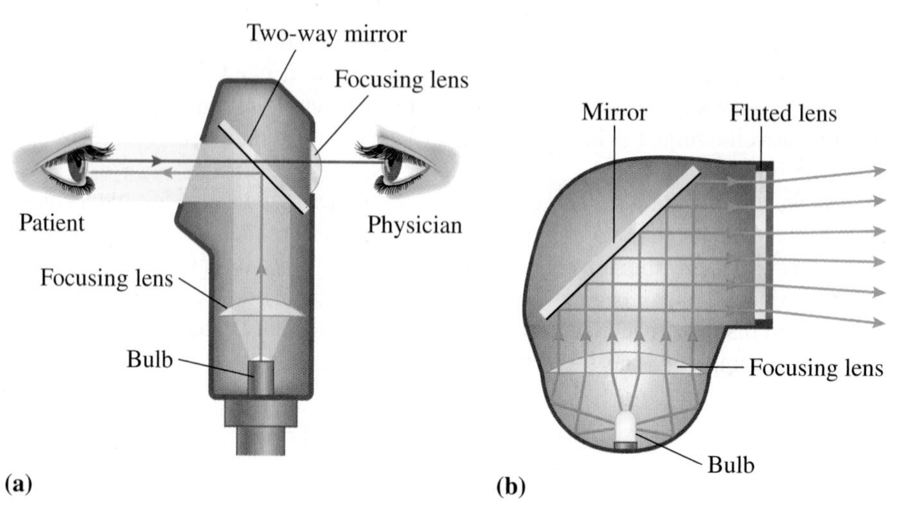

Figure 15-17 **(a)** The basic design of an ophthalmoscope. **(b)** A compact headlamp that utilizes a plane mirror.

Figure 15-18 Testing for lateral inversion.

TRY THIS!

Testing for Lateral Inversion

It is sometimes said that the image in a plane mirror is inverted horizontally or laterally (left-to-right) even though it is not inverted vertically. Don't be fooled by this misconception. A plane mirror does not invert an image either vertically or horizontally. The horizontal inversion is caused by the observer. To verify this, try the following activity. Print the word IMAGE on a piece of clear plastic and hold the plastic up to a mirror with the word IMAGE facing you, as shown in Figure 15-18. Next place a sheet of paper behind the plastic and face the word IMAGE toward the mirror. This time *you* caused the lateral inversion of the word. Describe what you observe. (**Note:** If you don't have plastic, you can use a dark marking pen and ordinary white paper, but you'll have to darken the backward printing on the reverse side.)

TRY THIS!

A Simple Kaleidoscope

Multiple images in two plane mirrors can be obtained by placing the mirrors at various angles to each other.

(a) Use two plane mirrors to determine experimentally the maximum number of images you can observe when the angle between the mirrors is 90°, 60°, and 45°.

(b) A kaleidoscope (Figure 15-19) is a device that applies the principle of multiple images by using two mirrors at an angle. If a kaleidoscope produces five images, what is the angle between the mirrors?

(c) Verify your answers in (a) and (b) by applying the equation $N = (360°/\theta) - 1$, where N is the number of images and θ is the angle between the plane mirrors.

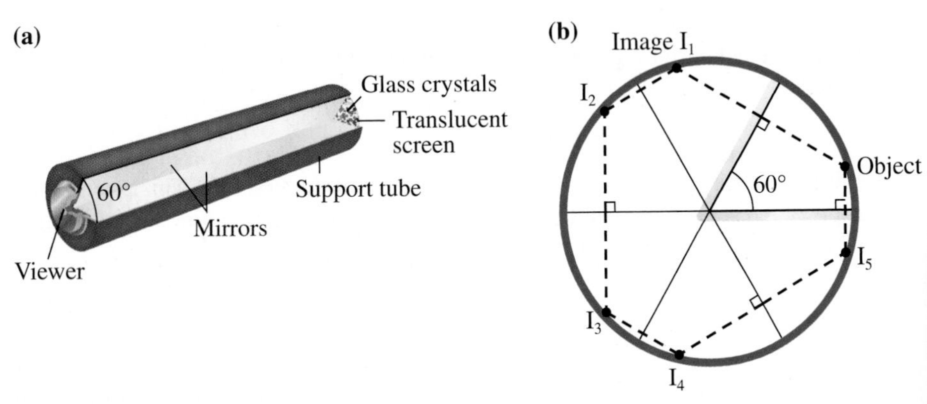

Figure 15-19 **(a)** One design of a kaleidoscope. **(b)** Locating five images of a sample "object."

Figure 15-20 The Cloud Gate Sculpture (affectionately called "The Bean" by locals) in Chicago's Millennium Park has curved reflecting surfaces that create a variety of unique images.

Curved Mirrors

Although plane mirrors are very common, they are not the only type of mirror used. Curved mirrors also have many applications, such as the popular sculpture shown in Figure 15-20 and others described shortly. We begin our discussion of curved mirrors with circular and spherical mirrors, and later we feature parabolic mirrors.

Curved mirrors are classified according to how they reflect light from a distant source. A **converging mirror** causes parallel light rays to converge or come together; its reflecting surface is concave (caved inward like the inside of a spoon). A **diverging mirror** causes parallel light rays to diverge or spread apart; its reflecting surface is convex (curved outward like the outside of a spoon). Figure 15-21 compares the shapes of these two types of mirrors and illustrates several definitions related to curved mirrors that have a cross-section that is part of a sphere. The *centre of curvature,* C, is the centre of the sphere. The *radius of curvature*, *r*, is the distance from C to the reflecting surface. The *principal axis* is a line drawn through C which strikes the middle, or vertex, V, of the mirror. The **focal point**, F, also called the principal focus, is the position where incident rays parallel and close to the principal axis meet or appear to meet after they are reflected. The **focal length**, *f*, is the distance from the focal point to the vertex of the mirror, that is, the distance FV. In diagrams involving the types of curved mirrors described here, it is useful to know that the focal length is half the radius of curvature, a fact that can be verified experimentally. For reasons that will be explained later, focal length is positive for converging mirrors and negative for diverging mirrors.

In the discussions that follow, the ray diagrams used to locate images in curved mirrors and in analyzing images mathematically have a circular cross-section; in other words, they are two-dimensional. To locate the image of an object in curved mirrors we draw incident rays from the object and locate their point of intersection after they reflect. Certainly the law of reflection that states $\theta_i = \theta_r$ holds for all incident rays striking a smooth curved surface. However, it is inconvenient to draw normals and measure angles in a diagram, so "rules" found experimentally or by applying $\theta_i = \theta_r$ provide a faster way of locating images.

converging mirror a mirror with a concave surface that causes parallel light rays to converge

diverging mirror a mirror with a convex shape that causes parallel light rays to diverge

focal point the position where the incident rays parallel and close to the principal axis of an optical device meet, or appear to meet, after they strike the device (also called the *principal focus*); symbol F

focal length (of a curved mirror) the distance from the focal point of a curved mirror to the vertex; symbol *f*

Starting with a circular converging mirror, the three rules used for ray diagrams are

- An incident ray parallel to the principal axis and relatively close to it reflects through the focal point.
- An incident ray through the focal point at a small angle to the principal axis reflects parallel to the principal axis.

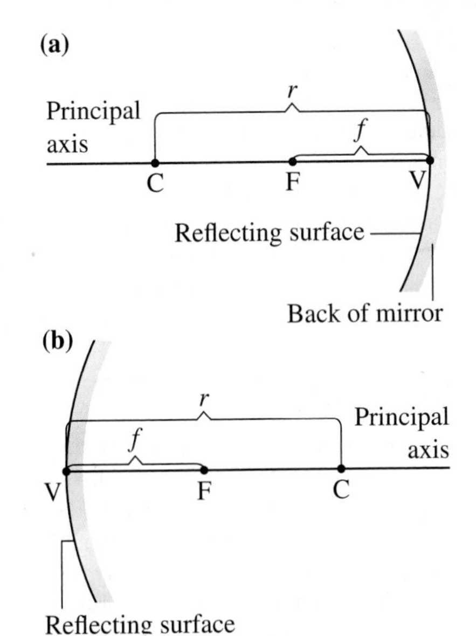

Figure 15-21 Curved mirrors with related symbols and ray reflections. **(a)** A converging mirror. **(b)** A diverging mirror.

- An incident ray through the centre of curvature reflects back onto itself. (This ray is actually a normal.)

reversibility of light rays the property of light that allows a path of light to be traced by reversing the direction of a known path

Notice that the second rule is simply the reverse of the first rule. This provides an example of the **reversibility of light rays**, a property of light that allows a path of light to be traced by reversing the direction of a known path.

SAMPLE PROBLEM 15-2

Apply the rules for a converging mirror to locate the image of a 13 cm high object located 72 cm from a converging mirror of focal length 17 cm. State the attitude and type of the image, measure the image distance (d_i), and calculate the linear magnification of the mirror.

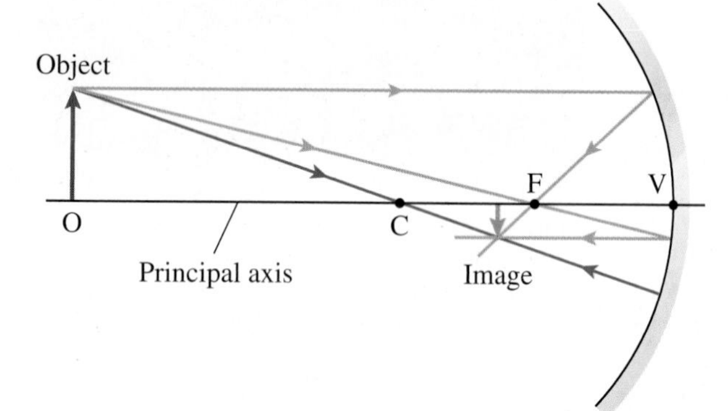

Figure 15-22 Sample Problem 15-2.

Solution

If $f = 17$ cm then $r = 34$ cm. Using the scale 1.0 cm = 10 cm, a curve of radius 3.4 cm is drawn with a compass, and C and F are located along the principal axis (Figure 15-22). As was shown in Figure 15-16 (a), the object is drawn as an arrow, which helps distinguish the top from the bottom, and it is placed so the bottom touches the principal axis. Three incident rays from the top of the object and their corresponding reflected rays are drawn according to the rules. The image of the top of the object is located at the intersection of the reflected rays, and the entire image can be drawn from there to the principal axis. In this case the resulting image is inverted, real because it is in front of the mirror, and located between the focal point and the centre of curvature 21 cm from the vertex.

The linear magnification is $M = \dfrac{h_i}{h_o} = \dfrac{-4\text{ cm}}{+13\text{ cm}} = -0.3$

The negative magnification corresponds to an inverted image.

Learning Tip

You can envision that a shape like a cosmetic mirror is a small portion of a larger sphere, but curved mirrors can have other shapes. For example, a cylindrical concave reflecting surface can be placed beneath a water-filled hose to focus solar energy to heat the water. Furthermore, on paper concave and convex mirrors are easiest to draw if they are simply part of a larger circle.

In the above sample problem, three rules were used, but any two of the three would have been adequate to locate the image. Notice that the image is in front of the mirror and if a screen were placed at the location of the image, the image would be in focus. Not all images produced by converging mirrors are real, as you will discover in Question 15-17.

The rules used to draw rays diagrams for diverging mirrors correspond to the rules for converging mirrors, but the wording is slightly different because the focal point and centre of curvature are located behind the mirror.

DID YOU KNOW?

In the Canary Wharf district of London, England, the concave front of a new 160 m, 37-storey skyscraper was observed to concentrate reflected light from the Sun to the street below for up to 2 h on clear days. The energy at the focal point of this converging "mirror" caused parts of a car to melt and other dangerous situations. The building has been called a "Fryscraper" and its nickname went from "Walkie-Talkie" (due to its shape) to "Walkie-Scorchie."

The rules for a circular diverging mirror are

- An incident ray parallel to the principal axis and relatively close to it reflects in line with the focal point.
- An incident ray toward the focal point at a small angle to the principal axis reflects parallel to the principal axis.
- An incident ray toward the centre of curvature reflects back onto itself.

SAMPLE PROBLEM 15-3

Repeat Sample Problem 15-2 using a circular diverging mirror, an object distance of 32 cm, and a focal length of −17 cm (because the focal point is behind the mirror).

Solution

The scale used to draw the diagram is 1.0 cm = 10 cm (Figure 15-23). The reflected rays—found by applying the rules—diverge and never meet, so they are extended behind the mirror to locate the image. The image is upright and virtual because it is located behind the mirror, between the vertex and the focal point, with d_i = 13 cm. Its linear magnification is $M = \frac{h_i}{h_o} = \frac{+4\,\text{cm}}{+13\,\text{cm}} = +0.3$.

The positive linear magnification corresponds to an upright image.

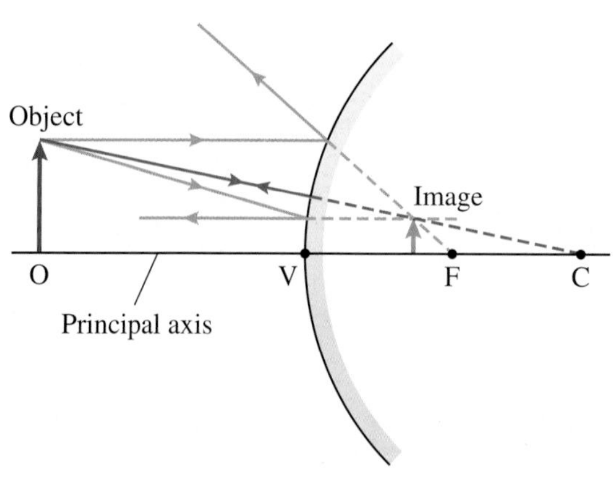

Figure 15-23 Sample Problem 15-3.

In both ray diagrams drawn for curved mirrors, the incident rays struck the mirror relatively close to the vertex. Incident rays parallel to the principal axis that strike a converging mirror far from the vertex do not reflect through the focal point (even though $\theta_i = \theta_r$). Since the reflected rays do not all meet at one place, the image is out of focus or blurry, especially near the outer edge of the mirror, a problem called **spherical aberration** (Figure 15-24 (a)). As long as the mirror is spherical this problem cannot be avoided. However, the problem is overcome by using a parabolic mirror, which has a focal point but no centre of curvature (Figure 15-24 (b)).

spherical aberration a problem with spherical curved mirrors in which the reflected rays from an object do not meet at the same location, resulting in a blurred image

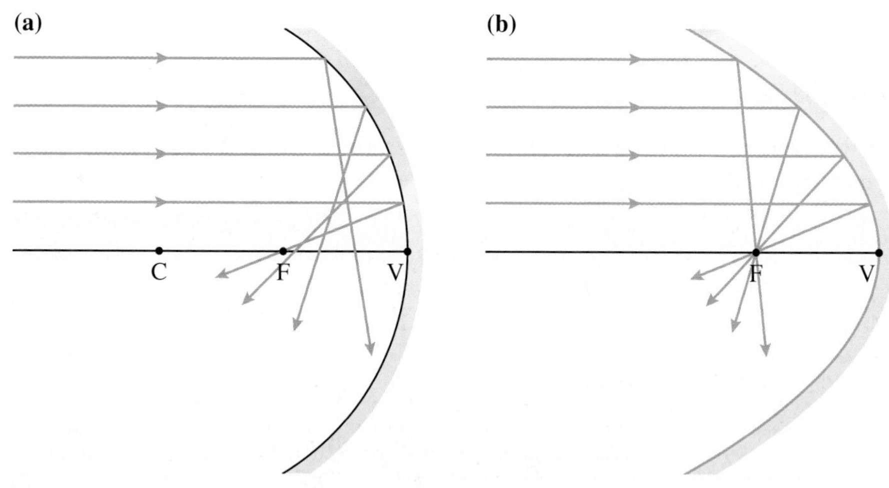

Figure 15-24 Spherical aberration and its correction. **(a)** Only the rays close to the principal axis reflect through F. **(b)** A parabolic mirror corrects for this aberration.

Learning Tip

Drawing or even sketching ray diagrams will help you visualize and remember the rules and images observed in mirrors. To make such diagrams as quick and easy as possible, questions in this chapter will use a radius that can be approximated with anything circular, such as a small drinking cup about 8 cm to 10 cm in diameter.

The Mirror Equation

A ray diagram can be used to derive an equation involving the variables focal length, f, object distance, d_o, and image distance, d_i. The derivation of the equation will be carried out using similar triangles in a ray diagram for a converging mirror (Figure 15-25).

In the diagram, the two reddish-shaded triangles are similar, so $\frac{h_i}{h_o} = \frac{f}{d_o - f}$.

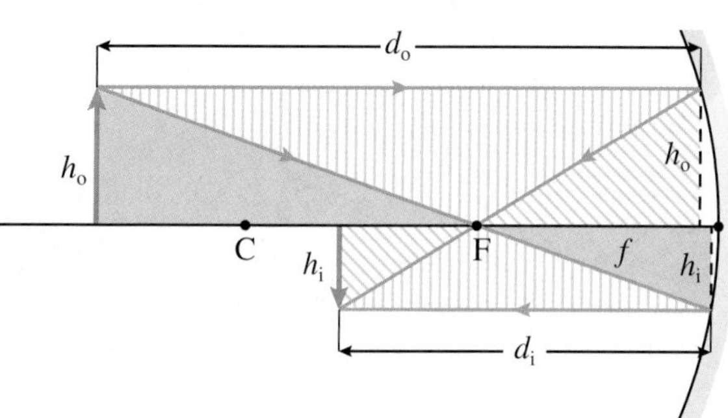

Figure 15-25 Using a ray diagram to determine the mirror equation.

Also, the two triangles with slanted blue are similar, so $\frac{h_i}{h_o} = \frac{d_i - f}{f}$.

From these equalities, $\frac{f}{d_o - f} = \frac{d_i - f}{f}$.

Thus, $f^2 = (d_i - f)(d_o - f)$ and $f^2 = d_o d_i - d_i f - d_o f + f^2$ or $d_o d_i = d_i f + d_o f$.
Dividing by $d_i\, d_o f$, we obtain

$$\frac{1}{f} = \frac{1}{d_o} + \frac{1}{d_i} \qquad \textbf{(15-3)}$$

This equation is called the *mirror equation*. If any two of the variables are given, the third can be found.

Another equation, this time involving linear magnification, can be derived from the ray diagram in Figure 15-25. The third set of similar triangles, those with the vertical green stripes, are similar, so we have $\frac{h_i}{h_o} = -\frac{d_i}{d_o}$

This allows another way of determining the linear magnification of a mirror.

$$M = \frac{h_i}{h_o} \text{ or } M = -\frac{d_i}{d_o} \qquad \textbf{(15-4)}$$

The mirror equation and these equations for linear magnification apply to both converging and diverging mirrors; however, rules regarding signs must be considered. Refer to Table 15-1.

Table 15-1

Sign Conventions for Curved Mirrors (using an object positioned upright on the principal axis)

Variable	Positive (+) Value	Negative (−) Value
Object distance (d_o)	always (for a real object)[a]	never
Object height (h_o)	always (real object)	never
Focal length (f)	converging mirror	diverging mirror
Image distance (d_i)	real image (in front of the mirror; inverted; below the principal axis)	virtual image (behind the mirror; upright; above the principal axis)
Image height (h_i)	virtual image (behind the mirror; upright; above the principal axis)	real image (in front of the mirror; inverted; below the principal axis)
Magnification (M)	virtual image (behind the mirror; upright; above the principal axis)	real image (in front of the mirror; inverted; below the principal axis)

[a]The variable d_o could be negative in a situation involving more than one optical device. For example, if a lens in front of a mirror produces an image behind the mirror, that image becomes the negative object for the mirror.

SAMPLE PROBLEM 15-4

Use the mirror equation to determine the image distance for the situation given in Sample Problem 15-2 and calculate the linear magnification using the ratio of distances.

Solution

$$\frac{1}{f} = \frac{1}{d_o} + \frac{1}{d_i}$$
$$\frac{1}{d_i} = \frac{1}{f} - \frac{1}{d_o}$$
$$\frac{1}{d_i} = \frac{1}{17\,\text{cm}} - \frac{1}{72\,\text{cm}}$$
$$d_i = 22\,\text{cm}$$

Thus, the image distance corresponds closely with the results in the ray diagram.

$$M = -\frac{d_i}{d_o} = -\frac{22\,\text{cm}}{72\,\text{cm}} = -0.31$$

Therefore, the linear magnification corresponds closely to the value found using the ratio h_i/h_o in Sample Problem 15-2.

Applications of Curved Mirrors

Curved mirrors have many useful applications. Whether the mirrors are converging or diverging, they are parabolic in shape to prevent the aberrations caused by circular or spherical mirrors, as was illustrated in Figure 15-24, on page 481.

Converging mirrors are often used as cosmetic or shaving mirrors. Various shapes are also seen in a "house of mirrors" at amusement parks. Converging reflectors are the basis of devices that send out light in a beam, such as flashlights and vehicle headlights. The filament of the light bulb is located near the focal point of the mirror, which allows the reflected rays to be nearly parallel, as shown in Figure 15-26 (a). Converging mirrors are also used to concentrate light in devices such as solar energy facilities and solar cookers (Figures 15-26 (b) and (c)). (You will discover other uses of converging mirrors in reflecting telescopes in Chapter 17.)

Images in diverging mirrors are always upright, virtual, and smaller than the object. The advantage of such mirrors is that they provide a much wider field of view than any other type of mirror, as shown in Figure 15-27 (a). Because of this wide-angled view, diverging mirrors are used as surveillance devices in stores to prevent theft. They are used at the front of a school bus to allow the driver to see if there is a small child in front of the bus (Figure 15-27 (b)), and they are used as rear-view mirrors on trucks, buses, and motorcycles. When using such a mirror, the driver must be aware that the image is smaller than the object, which causes the object to appear farther away than it actually is.

Figure 15-26 Applications of converging mirrors. **(a)** A flashlight produces a beam with nearly parallel edges by reflecting the light from the filament off the curved surface. **(b)** The black tubing containing circulating water absorbs light energy reflected from the converging reflectors on this solar energy farm. **(c)** A solar energy cooker is a valuable asset in remote regions where electricity is difficult or impossible to obtain.

(a)

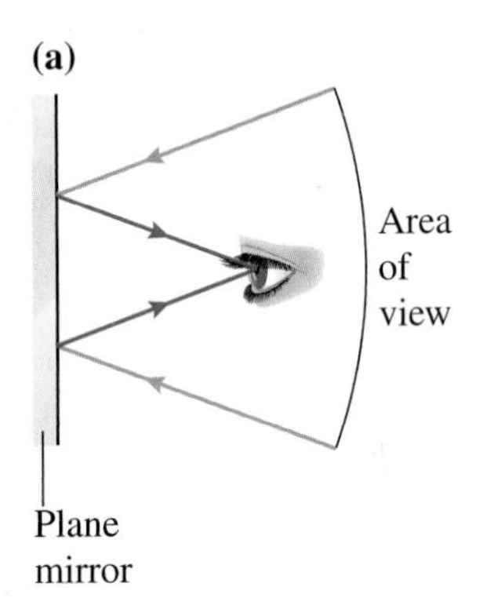

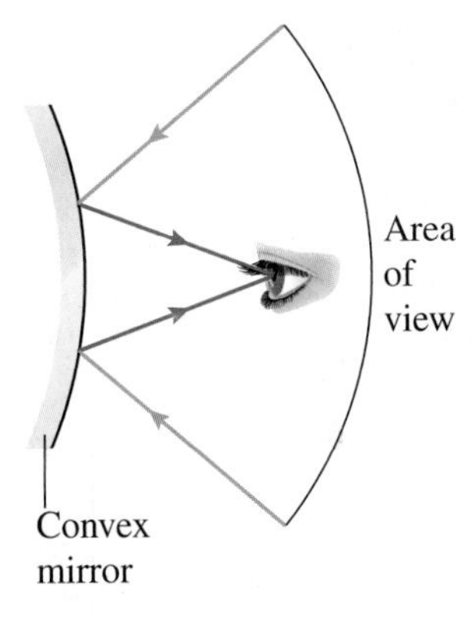

(b)

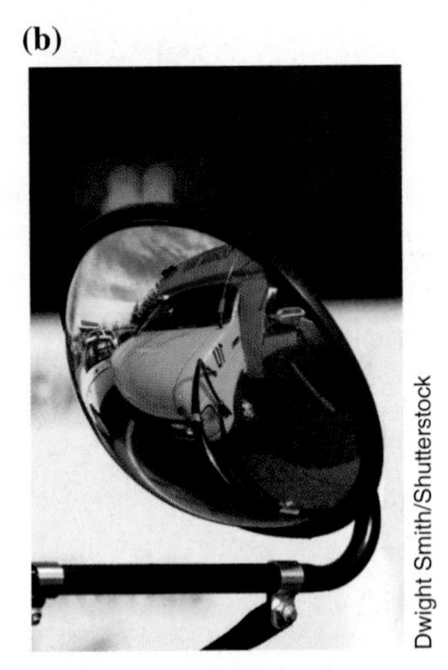

Figure 15-27 Diverging mirrors. **(a)** The diverging mirror produces a larger field of view than the plane mirror. **(b)** This school bus utilizes a diverging mirror for greater safety.

EXERCISES

15-12 Describe the difference between diffuse and specular reflection.

15-13 Assume you view yourself in a plane mirror from a distance, d_o, of 45 cm.

(a) What are the attitude and type of the image?

(b) How far is the image from the mirror?

(c) How far are you from your image?

(d) If you begin moving toward your image at a speed of 5 cm/s, how fast does your image approach you?

15-14 (a) Light rays strike a converging mirror such that the angle of incidence is 17°. What is the angle of reflection?

(b) A light ray strikes a diverging mirror along a normal. What are the angles of incidence and reflection?

15-15 Draw a ray diagram to locate the image of each object shown in Figure 15-28.

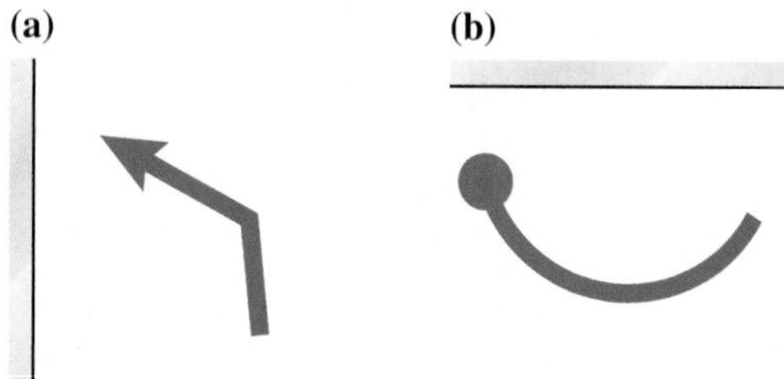

Figure 15-28 Question 15-15.

15-16 Figure 15-29 shows a point object (O) in front of two plane mirrors that are perpendicular to each other. Draw a detailed ray diagram to locate all the images produced. Check your diagram by placing two mirrors onto your diagram.

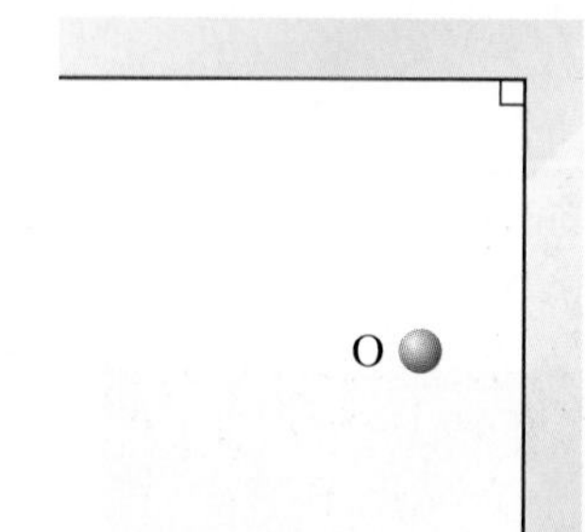

Figure 15-29 Question 15-16.

15-17 Use equations to determine the image distance, d_i, and the linear magnification, M, for each situation described below. Be careful with the positive and negative signs.

(a) converging mirror: $f = +2.5$ cm; $h_o = 1.0$ cm; $d_o = 7.5$ cm

(b) converging mirror: $f = +2.5$ cm; $h_o = 1.0$ cm; $d_o = 3.7$ cm

(c) converging mirror: $f = +2.5$ cm; $h_o = 1.0$ cm; $d_o = 1.2$ cm

(d) diverging mirror: $f = -2.5$ cm; $h_o = 1.0$ cm; $d_o = 7.5$ cm

15-18 To visualize each situation described in Question 15-17, sketch a ray diagram to locate the image of the object. In each sketch use two rules, and for each image found state its attitude and type, and estimate d_i and M. Compare your answers to what you found using equations.

15-19 A diverging mirror forms an image 1.0 cm from the vertex of the mirror. If the focal length of the mirror is −2.0 cm, determine the object distance and the magnification.

15-20 A converging mirror ($r = 17$ cm) is used to obtain the image of the Sun on a piece of paper. Estimate and then use an equation to calculate the image distance.

15-21 What is the "theoretical" focal length of a plane mirror? Use this fact combined with the mirror equation to prove that the image and the object are equidistant from the mirror.

15-22 Determine the focal length and state the type of mirror in each case. Assume two significant digits.

(a) $d_o = +30$ cm; $d_i = +30$ cm

(b) $d_o = +20$ cm; $d_i = -30$ cm

(c) $d_o = +20$ cm; $d_i = -10$ cm

(d) $d_o = +10$ cm; $d_i = -10$ cm

15-23 A man is viewing his face, which is 23 cm long, in a shaving mirror of focal length +40 cm from a distance of 24 cm. Use equations to determine the image distance and size, as well as the linear magnification of the mirror for this situation. State the attitude of the image. Assume two significant digits.

15-24 The radius of curvature of a rear-view mirror on a motorcycle is −50 cm. The driver, while stopped at an intersection, views a 1.6 m person standing 8.0 m away from the mirror. Determine the distance, size, and attitude of the image, and the linear magnification of the mirror. Assume two significant digits.

15.3 Refraction and Dispersion of Light

Magnifying glasses, binoculars, cameras, eyeglasses, and our eyes all rely on the bending of light as it travels from one medium to another. Other effects of the bending of light you have probably noticed are the apparent twinkling of stars, the shortened view of your legs as you walk through clear, waist-deep water, or the effect when viewing straws in water (Figure 15-30). The bending of light as it travels from one transparent medium to another at an angle is called the **refraction of light**.

refraction of light the bending of light as it travels from one transparent medium to another at an angle

Consider a beam of light travelling through air and striking the surface of a glass block at an angle of incidence, θ_i, different from 0°, as depicted in Figure 15-31. At the interface between the two media, some of the light is reflected such that the angle of reflection, θ_r, equals the angle of incidence. Some of the light enters the glass, whereupon it refracts toward the normal and travels in a straight line in the glass. The angle between the refracted ray and the normal is called the **angle of refraction**, symbol θ_R.

The reason the refraction occurs is that the speed of light in the glass is less than the speed of the light in air. More than a century ago, Jean Foucault discovered that light travels more slowly in glass and water than it does in air. An example that illustrates how a change in speed causes a change in direction is shown in Figure 15-32. A set of wheels travelling at a high speed on pavement reaches sand at an angle. The first wheel to reach the sand slows down first and the other wheels momentarily continue travelling at a greater speed on the pavement, causing the set of wheels to change its direction of motion. This simple model corresponds to the observation that waves refract when they travel at an angle from one medium into another where the speed is different.

What happens when the light travels from the glass back into the air, again at an angle of incidence in the glass other than 0°? Some of the light reflects at the surface between the media, and some light refracts as it emerges into the air, but this time, it refracts away from the normal because the speed of light increases when it emerges back into the air from the glass. Furthermore, it can be verified experimentally that the path of the light in the reverse direction is exactly the same as the initial path. This provides another example of the reversibility of light (Figure 15-33).

When light is refracted upon entering one medium from another, the amount of refraction depends on the relative speeds of light in the two media. For example, light is refracted more when entering glass from air at a given angle than when entering water from air at the same angle. Thus, light must travel more slowly in glass than in water. In general, then, we can state the following conclusions:

- When light travels at an angle from a medium in which the speed is faster to one in which the speed is slower, it is refracted toward the normal.
- When light travels at an angle from a medium in which the speed is slower to one in which the speed is faster, it is refracted away from the normal.
- At an angle of incidence of 0° the speed of light still changes in the second medium, but no refraction occurs.

Figure 15-30 From various angles, straws in water appear to be bent due to the refraction of light.

angle of refraction the angle between a refracted ray and the normal; symbol θ_R

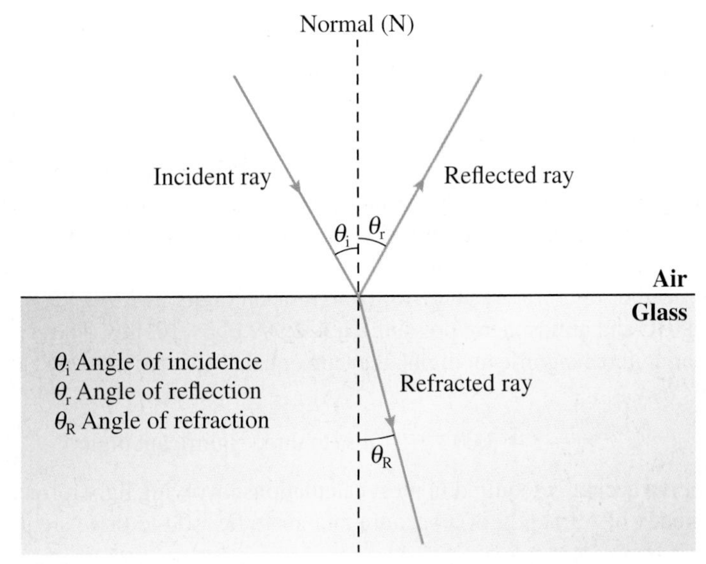

Figure 15-31 A beam of light in air striking glass undergoes partial reflection and partial refraction.

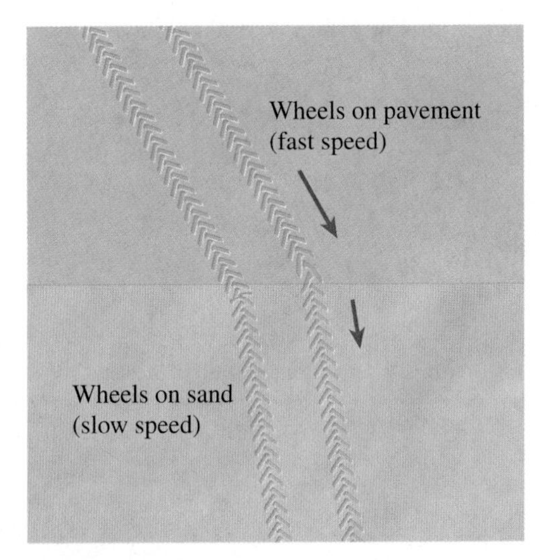

Figure 15-32 Illustrating how a change of speed causes refraction.

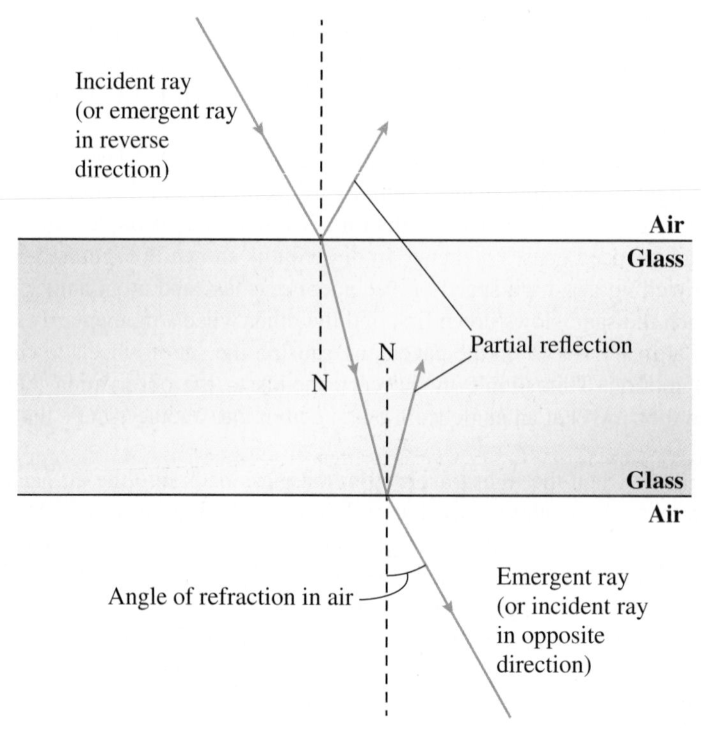

Figure 15-33 The refraction and reversibility of light.

A Rising Image

Place a coin in the middle of the base of an opaque container, such as a shallow dish or cup. Position your eyes at a level so you just miss seeing the coin (Figure 15-34). Slowly add water from another container to the dish or cup without disturbing the coin. Describe your observations and draw a ray diagram to explain them.

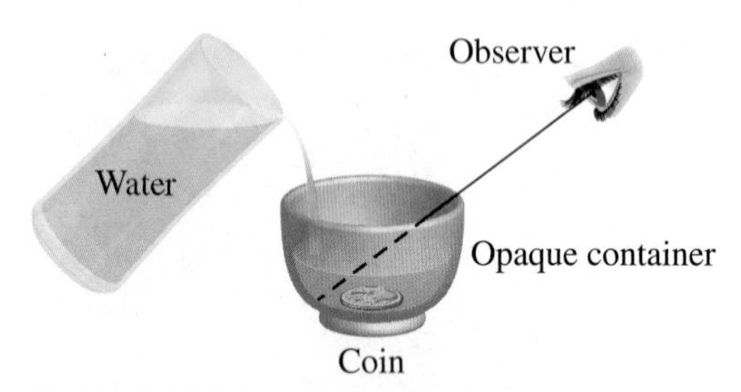

Figure 15-34 Setup for the Try This! activity.

Index of Refraction

Experiments show that light travels in a vacuum at a constant speed of 3.0×10^8 m/s, and it travels more slowly in other media. The ratio of the speed of light in a vacuum to the speed of light in another medium is defined as the **index of refraction** for the medium. Thus,

$$n = \frac{c}{v} \qquad \textbf{(15-5)}$$

where n is the index of refraction for light travelling from a vacuum, where the speed of light is c, into a medium where the speed of light is v. The units for this ratio of speeds cancel out, so index of refraction has no units.

Since the speed of light in a vacuum is greater than the speed of light in any other medium, the index of refraction of any other medium is always greater than one. Table 15-2 lists the speed of light and the index of refraction for several media; most data are given to three significant digits.

The speed of light in a vacuum, to six significant digits, is $2.997\,92 \times 10^8$ m/s, and in air (at 0°C and atmospheric pressure), it is $2.997\,05 \times 10^8$ m/s. These speeds are so close that, to three significant digits, they are equal. Thus, for most instances referred to in this text,

$$v_{air} = c = 3.00 \times 10^8 \text{ m/s (to three significant digits)}$$

Also, for the accuracy required in most calculations involving light refraction, we will use the index of refraction of a vacuum and air to be 1.00 to three significant digits. Thus,

$$n_{vacuum} = n_{air} = 1.00 \text{ (to three significant digits)}$$

index of refraction the ratio of the speed of light in a vacuum to the speed of light in another medium

Snell's Law of Refraction

As early as the second century CE, scientists studied the refraction of light, including measuring the angle of refraction. A Greco-Egyptian scientist named Ptolemy predicted that the amount of refraction that occurs when light travels from air into another medium could be determined by assuming that the ratio of the angle of incidence (θ_i) to the angle of refraction (θ_R) remained constant for that medium. This ratio of (θ_i/θ_R) applies reasonably well to small angles, but it does not apply to all angles.

Much later, around 1621, a Dutch mathematician named Willebrord Snell (1591–1626) discovered by experiment what is now called *Snell's law of refraction*:

The ratio of the sine of the angle of incidence to the sine of the angle of refraction remains constant for light travelling from air into another transparent medium.

In simplified form, the law is written $\dfrac{\sin\theta_i}{\sin\theta_R} = \text{constant}$

Snell did not relate the ratio of the sines of the angles to the ratio of the speeds in the two media, but later it was discovered that these ratios are equal. Thus, when light travels from a vacuum into a transparent medium at an angle, the index of refraction can be given by

$$n = \frac{\sin\theta_{\text{vacuum}}}{\sin\theta_{\text{medium}}}$$

This is one way of stating Snell's law of refraction in equation form. A more common way can be derived as follows: For two media, 1 and 2,

$$n_1 = \frac{\sin\theta_v}{\sin\theta_1} \quad \text{and} \quad n_2 = \frac{\sin\theta_v}{\sin\theta_2} \quad \text{where "v" stands for vacuum}$$

$$\therefore \quad n_1 \sin\theta_1 = \sin\theta_v \quad \text{and} \quad n_2 \sin\theta_2 = \sin\theta_v$$

Hence,

$$n_1 \sin\theta_1 = n_2 \sin\theta_2 \qquad (15\text{-}6)$$

This is the most common and probably the most useful way of writing Snell's law in equation form. When applying this equation, it is customary to use medium 1 as the medium in which the incident light is travelling. If either medium is air or a vacuum, then the index of refraction used, to three significant digits, is 1.00.

Table 15-2

Speed of Light and Indexes of Refraction for Various Media

(**Note:** The index of refraction of a material depends on the wavelength of the light. The data listed here are based on a wavelength of 600 nm.)

Medium	v ($\times 10^8$ m/s)	n
Air	2.997 05	1.000 29
Ice	2.31	1.30
Water	2.25	1.33
Glycerine	2.04	1.47
Olive oil	2.04	1.47
Benzene	2.00	1.50
Ruby	1.95	1.54
Fused quartz	2.05	1.46
Crown glass	1.97	1.52
Light flint	1.90	1.58
Heavy flint	1.82	1.65
Zircon	1.55	1.93
Diamond	1.24	2.42

TRY THIS!

Comparable Indexes of Refraction

The index of refraction of crown glass is approximately the same (about 1.5) as that of various liquids, such as olive oil. Predict what you would observe if you were to pour olive oil into a glass bowl or tumbler in which a small glass candle holder, a marble, or a similar object were resting on the bottom (Figure 15-35). Verify your prediction experimentally, and explain what you observe.

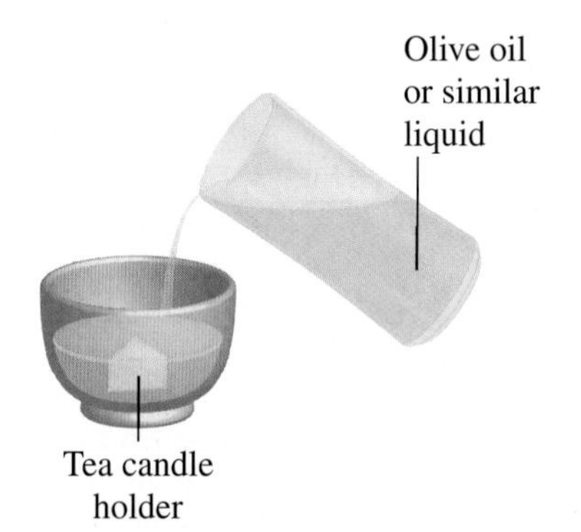

Figure 15-35 Viewing one material as a liquid is poured over it.

SAMPLE PROBLEM 15-5

Light in air strikes a transparent plastic rectangular block (Figure 15-36) at an angle, θ_1, of 65.4° and refracts at an angle, θ_2, of 37.9° in the block.

(a) Determine the speed of light in the block.

(b) Calculate the angle, θ_3, at which the light emerges from the block.

Solution

(a) Let air be medium 1, then $n_1 = 1.00$. From Snell's law ($n_1 \sin\theta_1 = n_2 \sin\theta_2$), the index of refraction of the plastic is given by

$$n_2 = \frac{n_1 \sin\theta_1}{\sin\theta_2} = \frac{1.00 \sin 65.4°}{\sin 37.9°} = 1.48$$

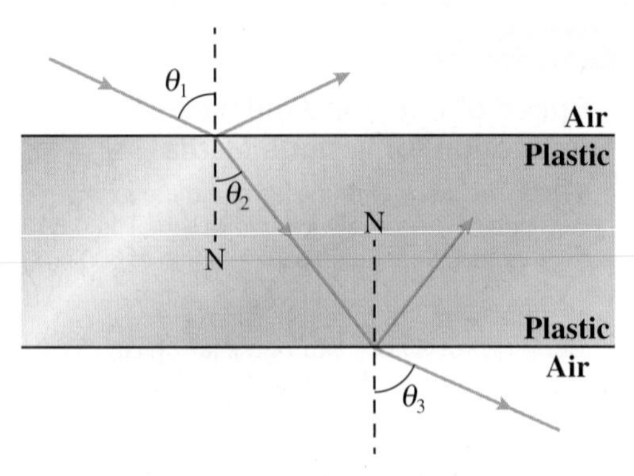

Figure 15-36 Sample Problem 15-5.

$$\text{From } n = \frac{c}{v},\ v_2 = \frac{c}{n_2} = \frac{3.00 \times 10^8\ \text{m/s}}{1.48} = 2.03 \times 10^8\ \text{m/s}$$

Thus, the speed of light in this medium is 2.03×10^8 m/s.

(b) From $n_2 \sin\theta_2 = n_3 \sin\theta_3$,

$$\sin\theta_3 = \frac{n_2 \sin\theta_2}{n_3} \quad \text{where } n_3 = 1.00 \text{ for air}$$

$$\therefore \quad \theta_3 = \sin^{-1}(1.48 \sin 37.9°) = 65.4°$$

Thus, the angle at which the light emerges is 65.4°, which is the same as the original angle of incidence in the air, as expected given the reversibility of light.

lateral displacement the sideways displacement of a light ray that strikes a transparent rectangular block (with parallel sides) at a non-zero angle of incidence

The final solution to the sample problem above illustrates an important feature of the refraction of light in a transparent block with parallel sides. The angle of refraction from the block to the air is equal to the original angle of incidence, so that the emergent ray is parallel to the incident ray. However, the light ray has undergone a sideways change of position, called a **lateral displacement**. As illustrated in Figures 15-37 (a) and (b), this lateral displacement is larger for larger angles of incidence and refraction back into the air. You can notice the effect of lateral displacement by viewing a printed page through a block of plastic or glass. However, when looking through a thin pane of glass the observed view appears normal because lateral displacement does not cause distortion.

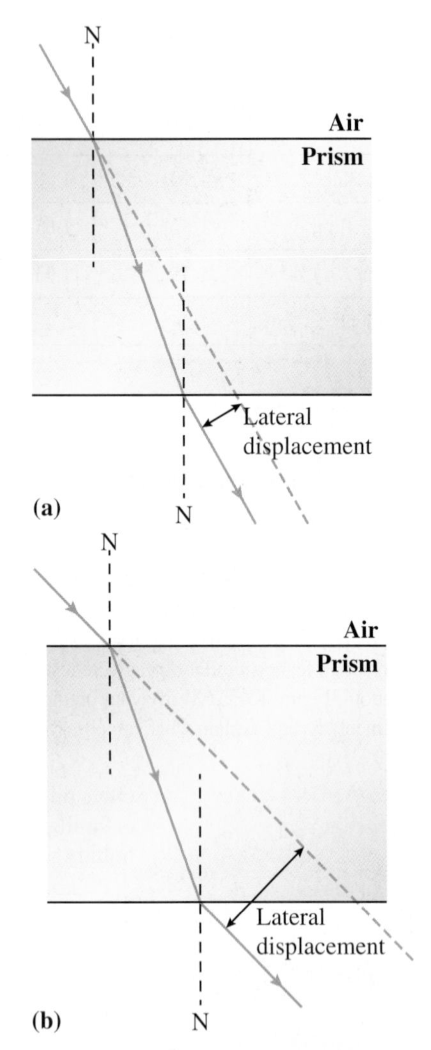

Figure 15-37 Lateral displacement in a transparent block with parallel sides. **(a)** When the angle of incidence and the angle of refraction in air are small, the lateral displacement is small. **(b)** When the angle of incidence and the angle of refraction in air are larger, the lateral displacement is larger.

Dispersion of Light

For thousands of years it was known that precious jewels such as diamonds cause sparkling colours to appear when viewed in white light. Scientists and philosophers alike thought that the colours were created by something within the jewels. Then, in 1666, when he was just 23 years old, Isaac Newton performed an important experiment verifying his hypothesis that the colours are a property of light. Newton passed a beam of sunlight through a triangular glass prism and discovered that white light can be separated into many colours, a process called **dispersion** (Figure 15-38 (a)). The resulting band of colours, called the **visible spectrum**, is composed of colours that blend into each other and are observed to be in the order red, orange, yellow, green, blue, and violet. A ray that emerges from the prism has a direction that deviates from the initial direction by an angle called the **angle of deviation**, shown in Figure 15-38 (b). As seen from Figure 15-38 (a), different colours have different angles of deviation.

It should be evident in Figure 15-38 that dispersion occurs because the different colours of light making up the visible spectrum are refracted at different angles upon

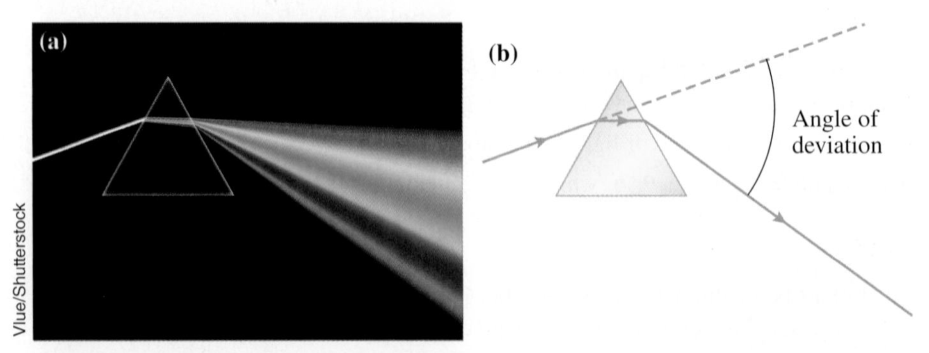

Figure 15-38 The dispersion of white light. **(a)** As a beam of white light enters a triangular glass prism it undergoes dispersion, which causes the light to split into its spectral colours. **(b)** Using a single ray to illustrate the angle of deviation.

travelling from one medium to another. The amount of refraction of each colour depends on the speed of light upon entering the second medium. (This speed, in turn, depends on the wavelength of the colour of light, an observation that supports the wave model of light, to be discussed later.) The index of refraction of a medium is related to the speed of light in the medium ($n = c/v$), so the index of refraction of the medium differs for different colours. Table 15-3 lists the index of refraction for the six basic colours of the visible spectrum in crown glass, a clear type of glass used in making optical instruments. Each index listed is an average of the range of indexes of refraction for the colour named.

Notice in Table 15-3 that red has the lowest index of refraction of the colours in crown glass. This indicates that red is the colour that slows down the least and thus is refracted the least in the glass.

Dispersion does not occur in rectangular blocks in which the normal for the ray that refracts into the air is parallel to the normal for the incident ray. However, it is commonly observed in transparent objects with edges that make sharp angles with each other, such as crystal chandeliers, jewellery, and bevelled mirrors. Dispersion also occurs in water droplets, allowing the formation of rainbows as described in the next section.

dispersion the splitting of white light into its spectral colours as it passes through a transparent medium

visible spectrum the components of white light visible to the human eye

angle of deviation the angle between the incident ray and the refracted ray beyond the prism for a prism with non-parallel sides

mirage an optical illusion caused by the refraction of light in layers of air of varying density

Table 15-3

The Index of Refraction for the Spectral Colours Travelling from Air into Crown Glass

Colour	Index of Refraction, n
Red	1.514
Orange	1.516
Yellow	1.517
Green	1.520
Blue	1.524
Violet	1.531

Mirages

On a hot summer day, a common sight along a paved road is an apparent shimmering pool of water that constantly recedes as you drive along. The image of the pool of water is an example of a **mirage**, an optical illusion caused by the refraction of light in layers of air of varying density. Another example of the same type of mirage is seen in a desert, often by weary travellers hoping to find water at an oasis (Figure 15-39).

How is this type of mirage created? The main condition is a very hot surface, such as pavement or sand, that causes the air near it to become hot and thus less dense than the air higher up. Figure 15-40 shows that as light from higher in the air approaches the region just above the hot surface where the density gradient is greatest, the light refracts more and more away from the normal, and thus curves back upward toward the cooler, more dense air. On a day with a clear blue sky, the observer sees this refracted light as a shining blue pool of water. If a tree near the horizon is viewed, its image appears to be below the actual tree, so this type of mirage is called an *inferior mirage*.

Under different conditions, for example, above a calm body of cold water, the most dense air is just above the surface, setting up the conditions required for a *superior mirage* in which the image appears above the actual object. In this case, the light from the object refracts back toward the surface and the image is seen floating in the sky (Figure 15-41). This optical illusion is often responsible for apparent sightings of UFOs and wild monsters.

Mirages provide an example in nature of the study of *gradient-index optics*, a study that has applications in lenses and is featured in Section 15.5.

© Robert Harding Picture Library Ltd/Alamy

Figure 15-39 A mirage on a dry salt pan in Etosha National Park in Namibia, with oryx gemsbok in the foreground.

Figure 15-40 The creation of a mirage over a hot surface.

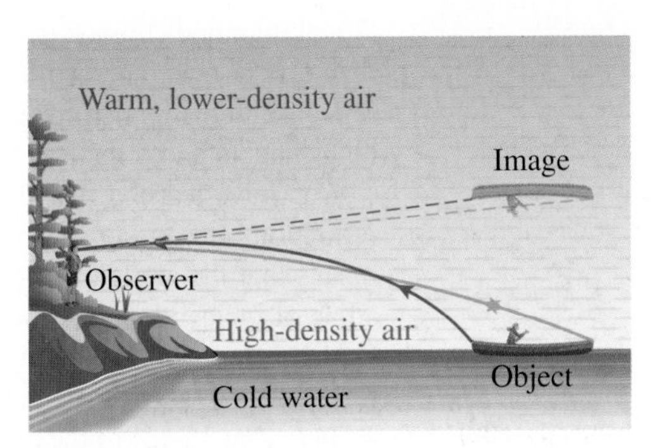

Figure 15-41 The creation of a superior mirage over a cold surface.

EXERCISES

15-25 When window shopping, have you ever noticed that some store windows are not vertical but are slanted to reduce glare from the Sun or street lights? To be effective, should such a window be slanted inward or outward at the top? Explain. (A diagram would help.)

15-26 The speed of light in a blood sample is 2.22×10^8 m/s. Find the index of refraction of the blood.

15-27 What is the speed of light in a clear soup that has an index of refraction of 1.37?

15-28 How long does it take light to travel 10.0 cm in water? (**Note:** $n_{water} = 1.33$.)

15-29 Assume the index of refraction for radio waves in ice is the same as for light; that is, $n_{ice} = 1.30$. Radio waves are sent downward from the ice surface through a layer of ice in the Arctic Ocean, and a signal that reflects from the bottom of the layer is received back at the source in 1.80×10^{-7} s. How thick is the layer of ice?

15-30 In each situation described below, light travels from medium A into medium B with a non-zero angle of incidence. State whether the light bends toward or away from the normal as it enters B.

(a) The speed of light in A is greater than the speed of light in B.

(b) The angle of incidence in A is greater than the angle of refraction in B.

(c) $n_A > n_B$

15-31 At what angle of incidence in air would a ray of light have to enter diamond ($n = 2.42$) in order to have an angle of refraction of 21.8°?

15-32 A ray of light is used to determine if a certain crystal is made of the gemstone, zircon. The angle of incidence in air is 48.5° and the angle of refraction in the crystal is 27.0°. What is the composition of the crystal? (Indexes of refraction are found in Table 15-2.)

15-33 Assume you are given a sample of an unknown transparent liquid. How would you determine the speed of light in the liquid?

15-34 A ray of light in air is aimed toward the surface of carbon tetrachloride ($n = 1.46$) at an angle of incidence of 71.5°. Determine the angle between the refracted ray in the liquid and the reflected ray in the air.

15-35 Light in air strikes a flint glass block ($n = 1.58$) at an angle of incidence of $a = 30.0°$, as shown in Figure 15-42. Determine all the angles shown in the diagram by applying the laws of reflection and refraction.

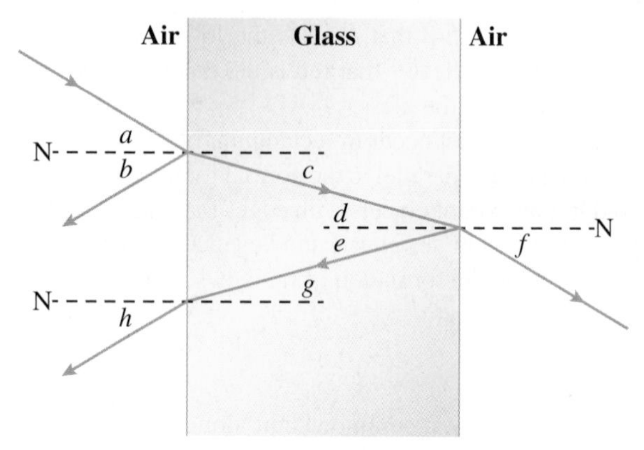

Figure 15-42 Question 15-35. (The diagram is not drawn to scale.)

15-36 A red laser beam enters a crown glass prism in the shape of an equilateral triangle. The angle of incidence is 60.0° and the index of refraction for the red light in the prism is 1.51. Determine the angle of refraction in the glass and the angle of refraction back into the air. (**Hint:** A diagram will help, especially in finding the final angle of refraction.)

15-37 For a ray of light in air striking the surface of a glass block ($n = 1.50$), what is the maximum angle of refraction?

15-38 White light in air approaches an equilateral prism made of crown glass at an angle of incidence of 45.6°. Which colour refracts more, red or blue? How much greater is the one angle of refraction within the prism than the other? (Reference: Table 15-3)

15-39 What conditions are required to observe a mirage in which the image of an object appears above the object?

FreshPaint/Shutterstock

Figure 15-43 The sparkle of a diamond results from the way in which light in the diamond reflects internally.

15.4 Total Internal Reflection

A real diamond (Figure 15-43) sparkles much more brilliantly and is more expensive than a fake one. The brilliance is caused by the way in which light in the diamond reflects internally before emerging into air. Learning about this internal reflection helps understand other effects of light, such as rainbows.

You have learned that the light that emerges from a medium of higher index of refraction into one of lower index of refraction refracts away from the normal. This means that the angle of refraction (θ_R) in the lower-index medium is greater than the angle of incidence (θ_i) in the higher-index medium. This is shown in Figure 15-44 (a) for light travelling from glass into air. Notice in the diagram that some of the light is reflected, but most of it emerges into the air.

In Figure 15-44 (b) the angle of incidence in the glass has increased and the angle of refraction in the air is almost 90°. (When this is observed experimentally, the colours of the visible spectrum are noticed.) In this case a relatively large portion of the light reflects back into the glass. Then when the angle in the glass is increased to a certain special angle, the emerging light just disappears along the surface of the glass, as in Figure 15-44 (c). This special angle, called the **critical angle** (θ_c), is the angle of incidence in a medium of greater index of refraction that allows the light in a medium of lesser index of refraction to disappear. Basically all the light is internally reflected. Any angle of incidence greater than the critical angle results in **total internal reflection** in the medium of higher index of refraction, as in Figure 15-44 (d). Total internal reflection can occur in any transparent medium that is adjacent to a medium with a lower index of refraction.

The fact that the critical angle in a medium results in the disappearance of the emergent light along the surface of the medium allows the use of Snell's law of refraction to calculate related quantities. For these calculations we assume that the theoretical value of the angle of refraction in the medium of lower index of refraction is 90°.

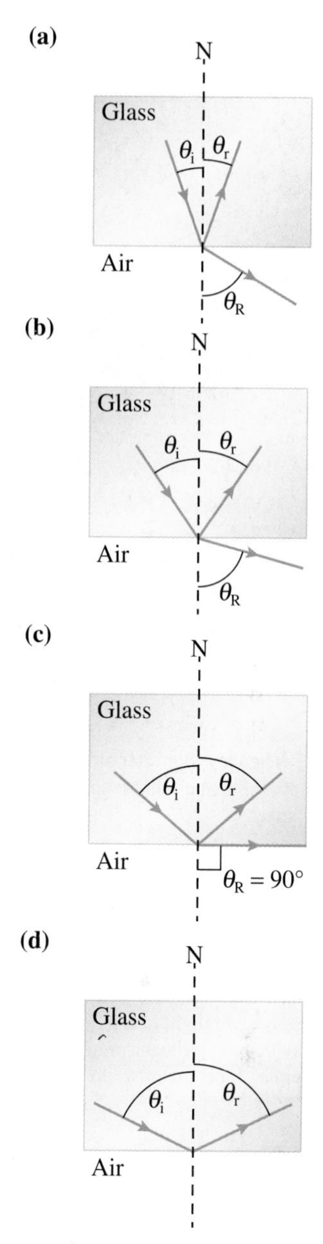

Figure 15-44 Refraction leading to total internal reflection. **(a)** Light travelling from a higher-index medium (glass) to a lower-index medium (air) at a small angle of incidence. **(b)** The angle in the glass increases. **(c)** When the light in the air just disappears along the surface of the glass, the angle *in the glass* is called the critical angle. **(d)** When the angle of incidence in the glass is greater than the critical angle, total internal reflection occurs in the glass.

A convenient form of Snell's law to solve critical angle problems is

$$n_1 \sin\theta_1 = n_2 \sin\theta_2$$

$$n_1 \sin\theta_c = n_2 \sin 90°$$

$$\therefore \ \sin\theta_c = \frac{n_2}{n_1} \qquad (15\text{-}7)$$

where "1" represents the medium of higher index of refraction and "2" represents the medium of lower index of refraction. In other words, Eqn. 15-7 applies when $n_1 > n_2$.

SAMPLE PROBLEM 15-6

Determine the critical angle for light travelling from water ($n = 1.33$) to air ($n = 1.00$).

Solution

Applying Eqn. 15-7 with 1 representing water and 2 representing air,

$$\sin\theta_c = \frac{n_2}{n_1} \quad \therefore \ \theta_c = \sin^{-1}\left(\frac{n_2}{n_1}\right) = \sin^{-1}\left(\frac{1.00}{1.33}\right) = 48.8°$$

Thus, the critical angle in the water is 48.8°.

What does this critical angle in water of approximately 49° mean to a swimmer in a waveless swimming pool? As shown in Figure 15-45, the swimmer beneath the surface looking upward has a view restricted to a circle that has an angular radius of about 49°. The view vertically above would be clearest, with distortion becoming more evident as the angle in the water approaches 49°.

Total internal reflection produces beautiful effects in nature and has found applications in many useful devices, as you will see shortly.

critical angle the angle of incidence in a medium of greater index of refraction that allows the light in a medium of lower index of refraction to disappear; symbol θ_c

total internal reflection the situation in which light in a medium of higher index of refraction strikes the interface with a medium of lower index of refraction at an angle greater than the critical angle, causing the light to reflect

Rainbows

The spectral colours of a rainbow are often written in the order red, orange, yellow, green, blue, and violet (R, O, Y, G, B, V). They are caused by both dispersion and total internal reflection within individual raindrops. To see a rainbow in the sky, the observer must be located between the raindrops and the Sun, as illustrated in Figure 15-46 (a). The Sun's rays travel to the water droplets where some of them totally internally reflect, then emerge from the droplets and travel to the observer. The rainbow caused

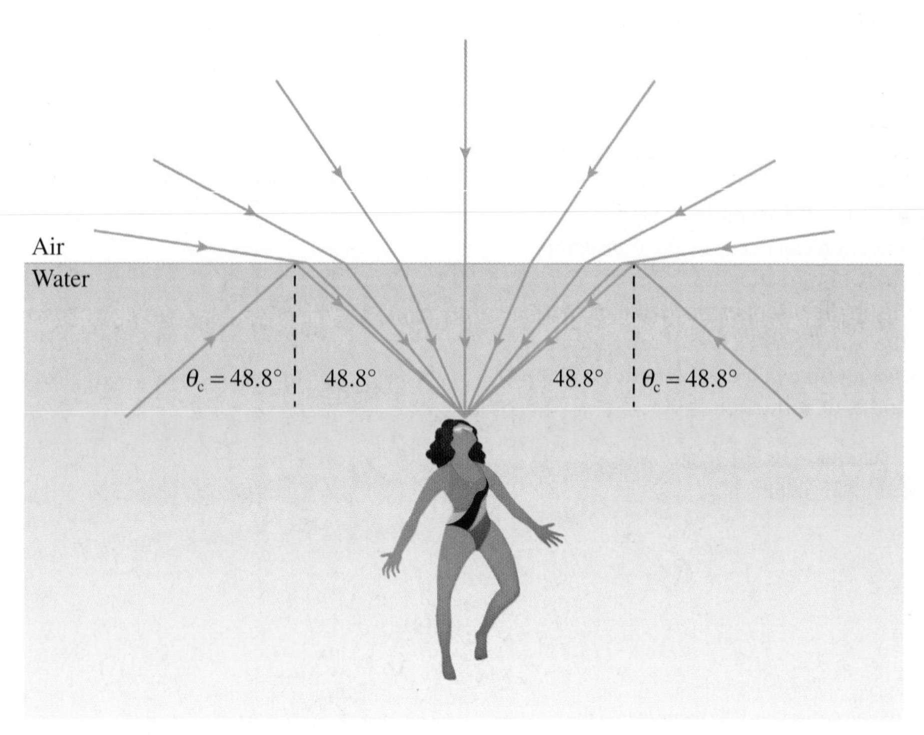

Figure 15-45 The view from beneath a smooth surface of water is restricted to a cone determined by the critical angle for light in the water.

(a)

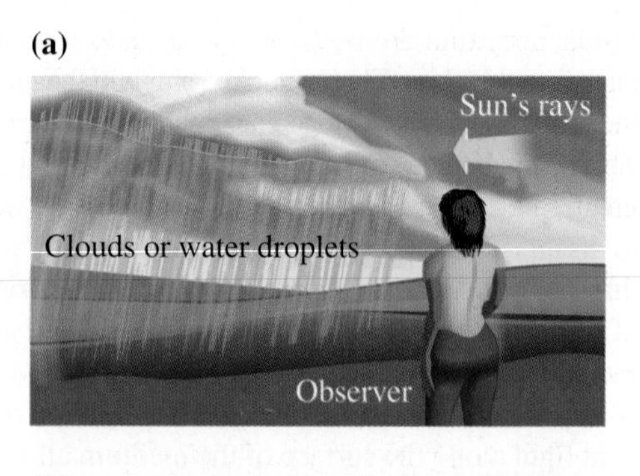

(b)

Figure 15-46 (a) The physical situation required to observe a rainbow. (b) The primary rainbow is brighter and lower in the sky than the secondary rainbow. Notice the order of the colours in each case.

primary rainbow a rainbow caused by the single internal reflections of sunlight in raindrops, resulting in the colours red, orange, yellow, green, blue, and violet

secondary rainbow a rainbow caused by two internal reflections of sunlight in raindrops, observed higher in the sky than the primary rainbow, resulting in the colours violet, blue, green, yellow, orange, and red

by single internal reflections is called the **primary rainbow**. Other orders of rainbows, which are much more difficult to observe than a primary rainbow, occur when the light is internally reflected more than once. For example, a **secondary rainbow**, which is observed higher in the sky than the primary rainbow, occurs after two internal reflections. Figure 15-46 (b) shows both a primary and a secondary rainbow.

To understand how the spectral colours are produced in a rainbow, consider Figure 15-47 (a). It shows a beam of white light entering a single drop of water, where it is refracted and dispersed into the colours of the visible spectrum. The various colours are reflected off the inner surface of the water (total internal reflection). As the

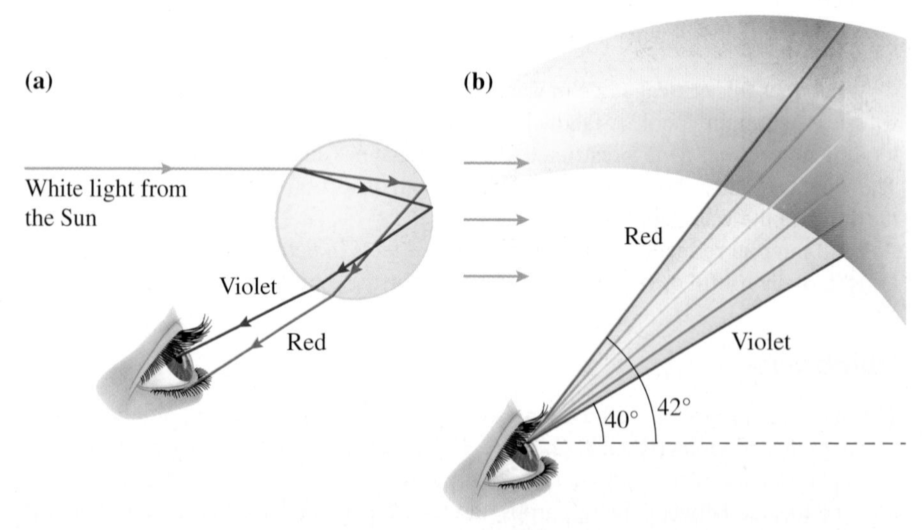

Figure 15-47 The formation of a primary rainbow. (a) White light is dispersed in each raindrop and undergoes a single total internal reflection. (b) Arrangement of the colours of a primary rainbow.

light emerges from the droplet, it is refracted again, but remains split into the spectral colours. In Figure 15-47 (a) violet light is entering the observer's eye from a region of the sky approximately 40° above the horizontal. The observer would have to look slightly higher in the sky to see the colours blue through red from nearby droplets. The resulting arrangement of colours of the primary rainbow is shown in Figure 15-47 (b).

To understand how a secondary rainbow is caused by a double internal reflection, refer to Figure 15-48. In this case the rainbow appears in an arc at an angle of about 52° above the horizontal. Notice that the colours of the secondary rainbow are in the reverse order compared to the colours of the primary rainbow.

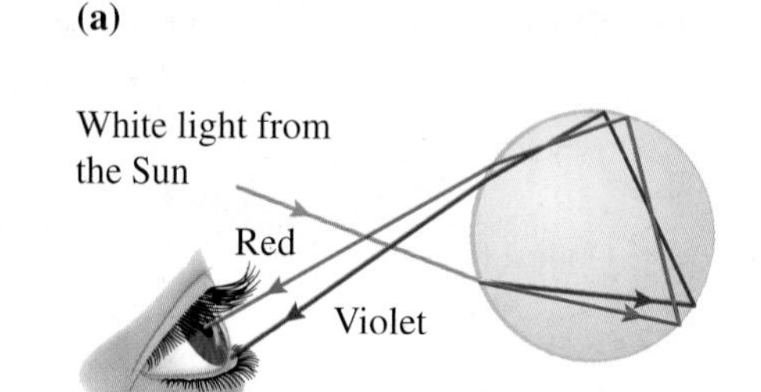

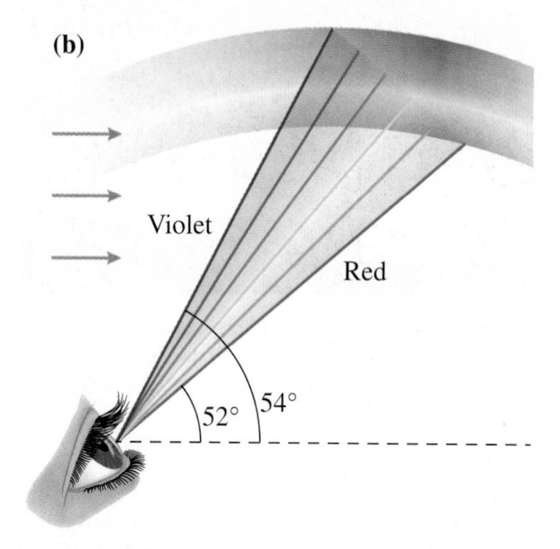

Figure 15-48 The formation of a secondary rainbow. **(a)** White light is dispersed and undergoes two total internal reflections. **(b)** Arrangement of the colours of a secondary rainbow.

Diamonds and Other Jewels

Why do real diamonds, like the one shown in Figure 15-43 at the beginning of this section, sparkle more than counterfeit ones? To explore this question, we begin by calculating the critical angle of diamond in air. From Table 15-2, the index of refraction of diamond is 2.42. Now using $\sin\theta_c = n_2/n_1$ with 1 representing diamond and 2 representing air,

$$\theta_c = \sin^{-1}\frac{n_2}{n_1} = \sin^{-1}\frac{1.00}{2.42} = 24.4°$$

Therefore, light rays that enter a real diamond will be totally internally reflected if they strike a surface within the diamond at any angle greater than 24.4°. A diamond is cut in such a way that when its motion relative to an observer is even very slight, the light that enters and exits will do so from a different surface. Thus, the chances are very high that light entering a diamond will be totally internally reflected many times before it exits. The result is a sparkling effect as the diamond is moved. Dispersion of white light can also occur, causing spectral colours along with sparkling.

Rare or artificial gems used for jewellery have indexes of refraction, and thus critical angles, that differ from those of diamond. Three examples are zircon (zirconium silicate, $n \approx 1.93$), cubic zirconia (zirconium dioxide, $n \approx 2.15$), and moissanite (silicon carbide, $n \approx 3.22$). With such a high index of refraction, moissanite has high sparkle possibilities.

Fibre Optics

fibre optics the applications of the transmission of light in transparent fibres

Thin, solid strands of transparent plastic or glass, called fibres, can be used to trap light through either total internal reflection or using a gradient index.[5] The study of the applications of total internal reflection in fibres is called **fibre optics**, which has important uses in the fields of communication, medicine, and industry.

Light that enters a solid fibre of glass surrounded by a medium of lower index of refraction totally internally reflects whenever it strikes the inside surface of the glass, and thus becomes trapped in the glass even if the fibre is curved (Figure 15-49 (a)). Fibres made for this purpose are composed of extremely pure glass, thus allowing light to be transmitted with little energy loss. When numerous glass fibres, each surrounded by a lower index of refraction layer called cladding, are bundled together, they form a light pipe (Figure 15-49 (b)).

Light pipes, which are used with lasers or light-emitting diodes (LEDs) as the light sources, have revolutionized communication systems. Telephone messages, television signals, and computer data can all be transmitted more quickly and accurately through these fibres than through traditional copper wires. Information is coded onto a light beam (from the laser or LED) and the light carries the information through the fibres to a receiver that decodes the information. There are two general ways of coding light for this purpose. One way, called *analogue modulation*, varies the intensity of the light

DID YOU KNOW?

The element carbon, atomic number 6, has more allotropes (crystalline forms) than any other element. One allotrope is diamond, the hardest known naturally occurring substance. Under great pressure beneath Earth's surface, it has formed into crystals over a period of millions of years. The largest known diamond, discovered in a South African mine in 1905, had an uncut mass of 621 g and was cut into 105 smaller diamonds. Scientists have learned how to make diamonds from another allotrope of carbon called graphite. These diamonds are used industrially for cutting and grinding, but they have limited use as jewellery because their maximum size is quite small. Many new uses are being explored for carbon nanoparticles, yet another allotrope.

[5] Discussion of gradient-index optical fibres is featured in Section 15.5 on lenses.

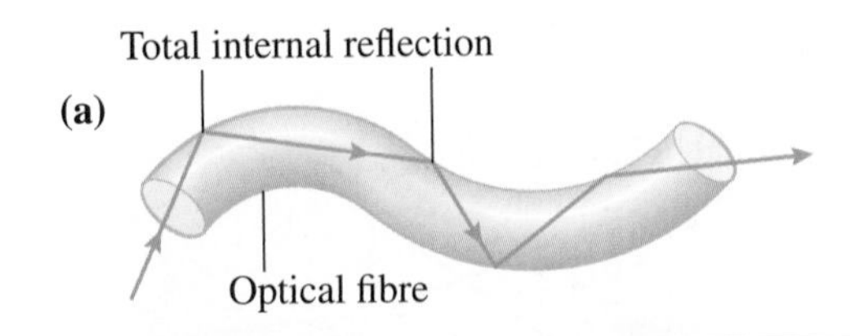

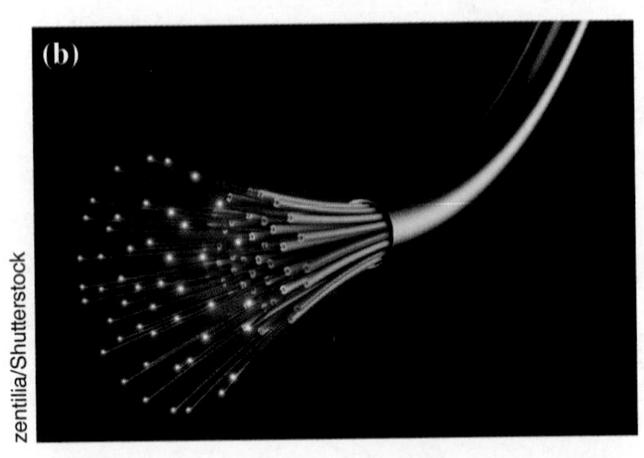

zentilia/Shutterstock

Figure 15-49 The basis of fibre optics. **(a)** Light in a solid glass fibre becomes trapped due to total internal reflection. **(b)** A light pipe or bundle of optical fibres.

DID YOU KNOW?

Glass fibres were made as long ago as 1880 by a most unusual method. An English physicist named Charles Boys attached the tail of an arrow to molten quartz and shot the arrow from a bow. The resulting fibres were as fine as a hair. Today's fibres used in fibre optics are about 1/10 the diameter of a hair and are made of fused silica.

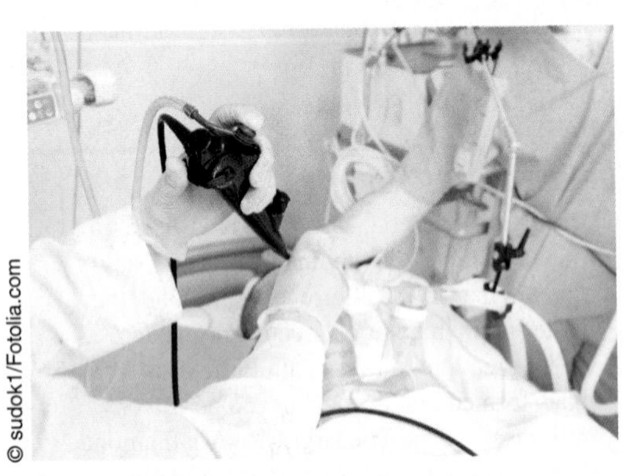

© sudok1/Fotolia.com

Figure 15-50 Using fibre optics in medicine: a surgeon manipulates an optical fibre endoscope to guide in tissue extraction in a biopsy.

beam. The other method, called *digital modulation*, uses on-off signals, similar to the binary code of a computer. Because lasers or LEDs can be pulsed on and off trillions of times each second, a single light beam can carry numerous amounts of data. For example, an under-water intercontinental cable can transmit more than 2.5 Tbits/s, which is 2.5×10^{12} bits of information per second, equivalent to more than 6×10^5 books each second, or 3×10^7 telephone conversations at once!

In the field of medicine, light pipes can be used to explore regions of the body that are normally hard to access. A flexible tube arrangement called an endoscope or a fibrescope consists of separate bundles of fibres to transmit light from a light source as well as transmit reflected light from the area viewed back to a camera or television screen (Figure 15-50). The endoscope may also be equipped to take tissue samples, thus preventing the need to cut into the body directly. Endoscopes, often called by specific names such as colonscopes, are used to explore such organs as the colon, stomach, and heart.

Light pipes similar to the ones used for medical applications are also used in industry to view normally inaccessible areas such as pipe valves, by plumbers in households for a similar purpose, and on some cars to send light from the signal lights back around to the driver, allowing the driver to judge whether the signal lights or parking lights are functioning properly.

Prism Applications

With the appropriate shape, a prism can act like a plane mirror. Plane mirrors reflect light directly, but prisms reflect light through total internal reflection. Plane mirrors have disadvantages in that they may become scratched or tarnished and do not last as long as prisms; as well, they absorb more light energy than prisms. Thus, the capacity of prisms to reflect light internally is useful when mirrors are either inconvenient or unsatisfactory for reflection.

Two devices that use this principle are binoculars and periscopes. Both have glass prisms that have a critical angle relative to air of less than 45°. Figure 15-51 (a) shows a simplified path of light passing through two prisms in a pair of binoculars. Without the prisms, the binoculars would have to be made longer to provide the same clear magnification and an upright image. The simplified view of the periscope, shown in Figure 15-51 (b), illustrates two prisms reflecting light internally to the observer. It is left as an exercise (Question 15-49) to verify that the image in this case is upright.

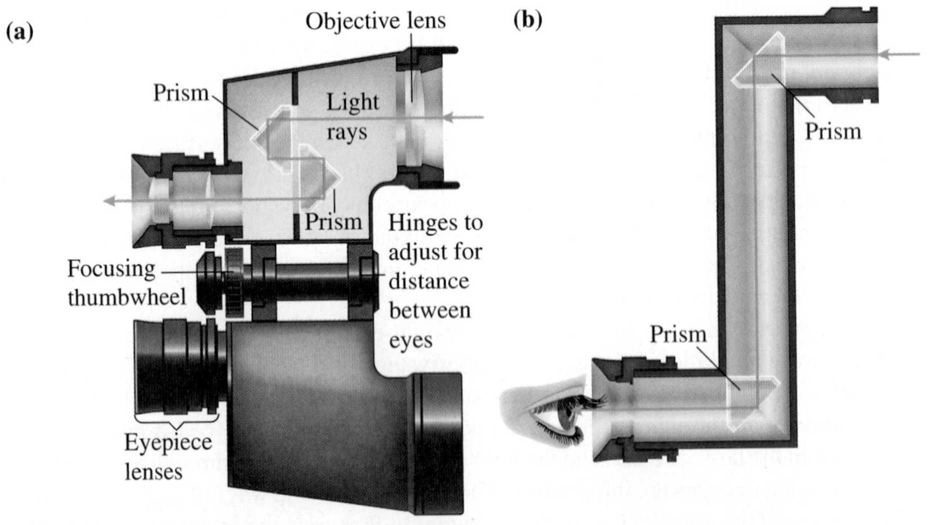

Figure 15-51 Applying total internal reflection in prisms. **(a)** The design of prism binoculars. **(b)** A simplified view of the design of a periscope.

EXERCISES

15-40 State two conditions necessary for total internal reflection.

15-41 Determine the critical angle for light travelling from zircon toward air.

15-42 The critical angle for a certain medium in a vacuum is 37.3°. Determine the index of refraction of the medium and use Table 15-2 to identify it.

15-43 In which medium, A or B, would total internal reflection be possible?

(a) $v_A > v_B$ where v = speed of light

(b) $n_A > n_B$ where n = index of refraction

(c) $\theta_A > \theta_B$ where θ = the angle in each medium

15-44 The speed of light in medium M is 2.04×10^8 m/s. What is the critical angle in M (relative to air)?

15-45 The critical angle of medium X surrounded by air is 45.8°. Determine the speed of light in X.

15-46 Determine the critical angle for light travelling from zircon (n = 1.93) toward benzene (n = 1.50).

15-47 Use the data from Table 15-2 to determine the critical angle of (a) heavy flint glass and (b) light flint glass surrounded by air. Assuming their edges are similarly cut, which medium would sparkle more if used as a gem? Explain why.

15-48 The core fibres used in light pipes are encased in thin films (the cladding), which are further encased in a protective plastic coating. Should the cladding have a lower or higher index of refraction than the core fibres? Why?

15-49 Use a diagram to prove that the image in a periscope (Figure 15-51 (b)) is upright.

15-50 Would a periscope with prisms made of crown glass (n = 1.52) function properly in an optical sense if it were submerged in water (n = 1.33)? Justify your answer.

15-51 Would a diamond (n = 2.42) submerged in water (n = 1.33) sparkle more or less than in air (n = 1.00)? Explain.

15.5 Lenses and Their Images

Anyone who uses a camera to take photographs and anyone who wears eyeglasses makes use of the refraction of light in lenses. Of course, the most important lens is the one in each of our eyes.

A **lens** is a transparent device that produces images by altering the direction of light. Light passing through a lens obeys Snell's law of refraction ($n_1 \sin\theta_1 = n_2 \sin\theta_2$). Thus, light that enters a glass lens from the air at an angle of incidence other than 0° refracts toward the normal, and when it emerges from the lens into the air it refracts away from the normal (Figure 15-52).

Like a curved mirror, a lens has a principal axis and may be either converging or diverging. To describe how light acts in lenses we will use rays of light for convenience. A **converging lens** causes light rays from a distant source to converge to a focal point. Such a lens is generally thicker in the middle than at the perimeter. A **diverging lens** causes light rays from a distant source to diverge as if they originated from a virtual focal point. Such a lens is generally thicker at the perimeter than in the middle. Figure 15-53 shows three converging lenses and three diverging lenses.

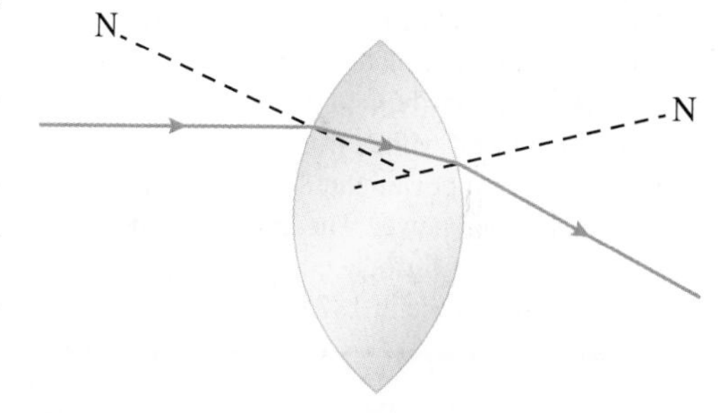

Figure 15-52 A lens alters the direction of light. Light entering the lens refracts toward the normal, and light emerging into the air refracts away from the normal. In this case, the lens is surrounded by a medium of lower index of refraction.

lens a transparent device that produces images by altering the direction of light

converging lens a lens that causes light rays from a distant source to converge to a focal point

diverging lens a lens that causes light rays from a distant source to diverge as if they originated from a virtual focal point

(a) Single convex lens

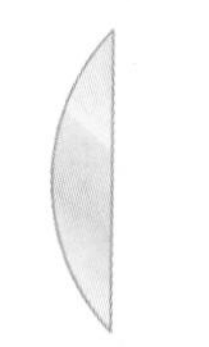

(b) Double convex lens

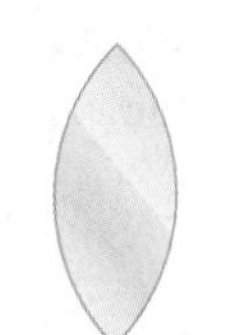

(c) Concavo-convex lens

(d) Double concave lens

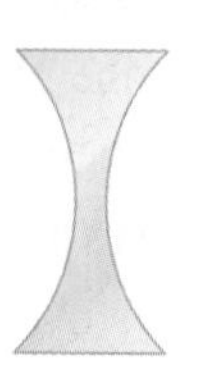

(e) Single concave lens

(f) Convexo-concave lens

Figure 15-53 The design of converging lenses **(a)**, **(b)**, and **(c)**, and diverging lenses **(d)**, **(e)**, and **(f)**.

primary focal point the position where rays of light parallel to the principal axis of a lens come together (for a converging lens) or appear to spread out from (for a diverging lens); symbol PF

secondary focal point the position on the side of a lens opposite to the primary focal point and equidistant from the lens; symbol SF

Because light can travel in either direction through a lens, each lens has two focal points, not just one as a curved mirror does. For convenience, we will call these focal points primary and secondary. The **primary focal point** (PF) results for rays that are parallel to the principal axis. Thus, for a converging lens the PF is located on the side of the lens where the light emerges; for a diverging lens the PF is located on the same side of the lens as the incident rays. In each case, the **secondary focal point** (SF) is located on the side of the lens opposite to the PF and equidistant from the lens. Refer to Figure 15-54, which also shows that the *focal length* of a lens is the distance from the middle of the lens to either focal point. (Notice that we have not defined the centre of curvature or radius of curvature for the curved part of a lens. A lens maker is concerned with them, but we do not need to know them for our discussion here.)

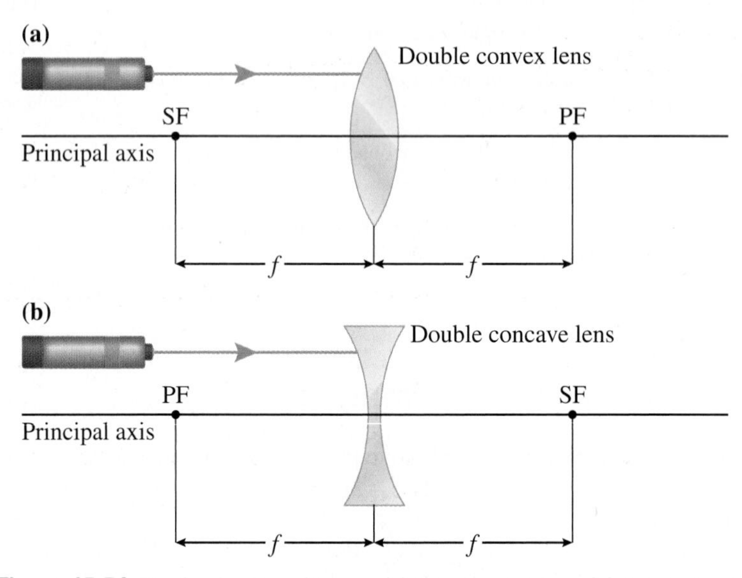

Figure 15-54 The focal points of lenses. **(a)** A converging lens. **(b)** A diverging lens.

DID YOU KNOW?

Strengths of eyeglass lenses and contact lenses could be compared by using focal lengths. However, optometrists and ophthalmologists use the power (P) of a lens, which is defined as the reciprocal of the focal length; that is, $P = 1/f$. The unit of this power is the dioptre, which is equivalent to 1/metre or m^{-1}. Thus, a lens prescription is stated in dioptres, with higher power lenses being used for eyes with poorer vision, and with converging lenses having a positive power and diverging lenses having a negative power. The dioptre has been in use since it was introduced at a medical conference in Europe in 1875.

Ray Diagrams for Lenses

One way to locate an image produced by a lens is to use a ray diagram. Light passing through a lens can be refracted twice, first when entering the lens and again when leaving. The path of this double refraction is tedious to determine by applying Snell's law of refraction, and it is not easy to draw accurately. So a convenient shortcut can be used; namely, a set of single refractions can be drawn at the middle of the lens. These refractions are determined experimentally and are the "rules" we apply to draw ray diagrams. The rules provide relatively accurate results, especially for lenses that are thin. The rules for converging and diverging lenses are summarized below and drawn in Figure 15-55. In each case, the third rule is an approximation because the light that travels through the middle of the lens may be displaced slightly rather than travelling exactly straight. However, the accuracy is still acceptable.

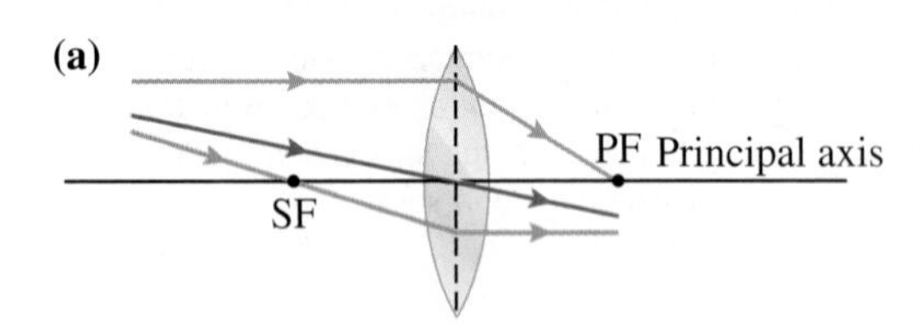

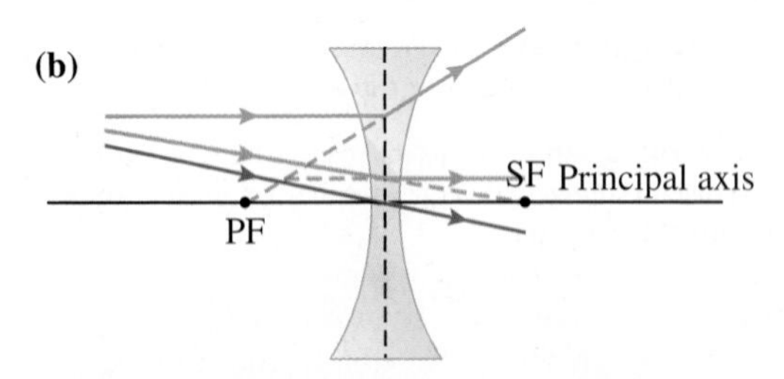

Figure 15-55 Illustrating the rules for drawing ray diagrams for lenses. **(a)** Converging lenses. **(b)** Diverging lenses.

Rules for drawing ray diagrams for lenses.

(a) Converging lenses:
- An incident ray parallel to the principal axis refracts through the PF.
- An incident ray through the SF refracts parallel to the principal axis.
- An incident ray through the middle of the lens at a small angle to the principal axis passes straight through the lens.

(b) Diverging lenses:

- An incident ray parallel to the principal axis refracts in line with the PF.
- An incident ray toward the SF refracts parallel to the principal axis.
- An incident ray through the middle of the lens at a small angle to the principal axis passes straight through the lens.

Traditionally, ray diagrams for lenses are drawn with the object located to the left of the lens and resting on the principal axis. Then rays are drawn according to any two of the three rules, and the image is drawn where the refracted rays intersect (for a real image) or appear to intersect when extended (for a virtual image). After the image is located, its attitude, type, and image distance can be stated and the linear magnification of the lens ($M = h_i/h_o$) can be calculated. Recall from Table 15-1 that h_o and h_i are positive when above the principal axis and negative when below the principal axis. This means that the linear magnification is negative whenever the image is inverted relative to the object.

SAMPLE PROBLEM 15-7

A 12 cm-high object is located 45 cm from a double convex lens of focal length 14 cm. Draw a ray diagram to locate the image of the object, state the characteristics of the image, and calculate the linear magnification of the lens.

Solution

An appropriate scale for this diagram is 1.0 cm = 10 cm. The principal axis and lens are drawn, and the primary and secondary focal points and the object are located using the scale. Then the incident and refracted rays are drawn according to the rules. (Two rules locate the image; the third rule is optional, but acts as a check.) The resulting image, shown in the ray diagram in Figure 15-56, is inverted, real, and located on the side of the lens opposite to the object, with $d_i \approx 20$ cm. The linear magnification of the lens is $M = h_i/h_o \approx -5\text{ cm}/12\text{ cm} \approx -0.4$.

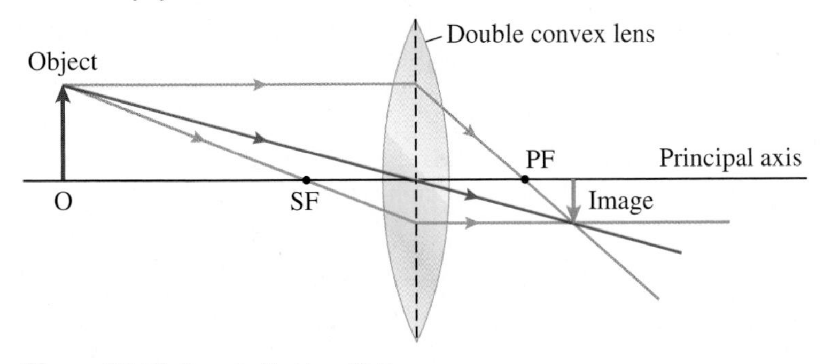

Figure 15-56 Sample Problem 15-7.

SAMPLE PROBLEM 15-8

Repeat Sample Problem 15-7 using a diverging double concave lens. (In this case the focal length is −14 cm.)

Solution

Again the diagram is drawn to scale (1.0 cm = 10 cm) and three incident and refracted rays are drawn using the rules. The refracted rays diverge so they must be extended straight back to where they intersect on the side of the lens where the incident light

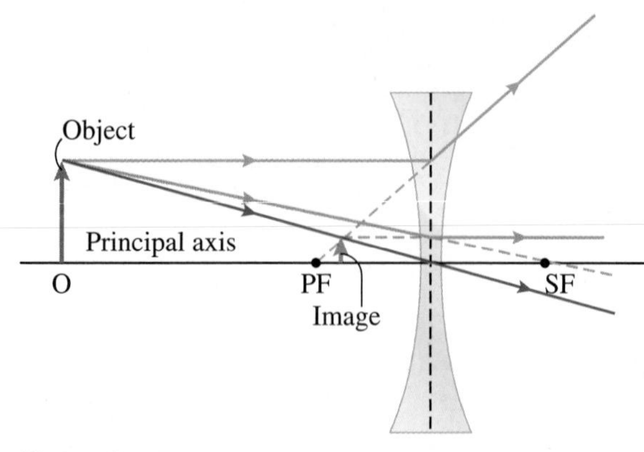

Figure 15-57 Sample Problem 15-8.

rays are located. The resulting image, shown in Figure 15-57, is upright, virtual, and located between the primary focal point and the lens, with $d_i \approx$ 11 cm. The linear magnification of the lens is $M = h_i/h_o \approx 3\text{ cm}/12\text{ cm} \approx +0.2$.

TRY THIS!

Creating Real Images

View and describe the formation of a real image by placing a converging lens (a simple magnifying glass will do) between a source of light and a screen. A good "source of light" is daylight passing through a window. Adjust the relative position of the lens and the screen to obtain a clear image. Try using a cosmetic mirror or other converging mirror to obtain a real image, and compare the results to what is produced by a lens. In each case, determine the approximate focal length of the device.

The Thin-Lens Equation

Images in lenses can be analyzed mathematically as well as with ray diagrams. To derive an equation that expresses the relation involving the focal length, object distance, and image distance, a ray diagram using a converging lens can be used. We begin by analyzing the triangles formed using the rays that utilize the focal points (Figure 15-58). In the first set of similar triangles (vertical green lines) to the left of the lens,

$$\frac{h_i}{h_o} = \frac{f}{d_o - f}$$

In the second set of similar triangles (slanted blue lines) to the right of the lens,

$$\frac{h_i}{h_o} = \frac{d_i - f}{f}$$

From these equalities, $\frac{f}{d_o - f} = \frac{d_i - f}{f}$.

Thus, $f^2 = (d_i - f)(d_o - f)$ and $f^2 = d_od_i - d_if - d_of + f^2$ or $d_od_i = d_if + d_of$. Dividing by $d_i d_o f$, we obtain

$$\frac{1}{f} = \frac{1}{d_o} + \frac{1}{d_i} \qquad \textbf{(15-8)}$$

where f = the focal length of the lens, d_o = the distance from the object to the lens, and d_i = the distance from the image to the lens. This equation is called the *thin-lens equation*. If any two of the variables are given, the third can be found.

The thin-lens equation is identical to the equation derived for curved mirrors. Similar sign conventions apply here, as summarized in Table 15-4.

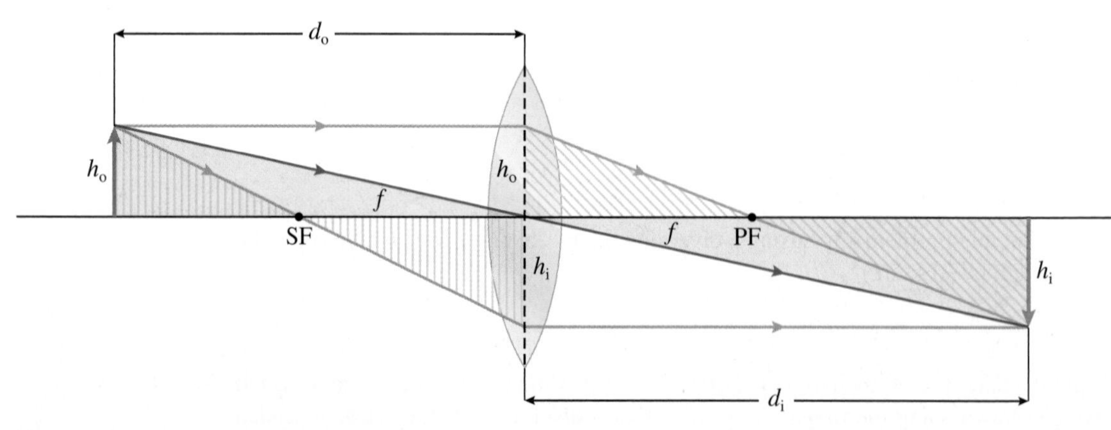

Figure 15-58 Deriving equations for a thin lens.

Table 15-4

Sign Conventions for Lenses (using an object positioned upright on the principal axis)

Variable	Positive (+) Value	Negative (−) Value
Object distance (d_o)	always (for a single lens)[a]	never
Object height (h_o)	always (for a single lens)	never
Focal length (f)	converging lens	diverging lens
Image distance (d_i)	real image (beyond the lens; inverted; below the principal axis)	virtual image (on the same side of the lens as the object; upright; above the principal axis)
Image height (h_i)	virtual image (on the same side of the lens as the object; upright; above the principal axis)	real image (beyond the lens; inverted; below the principal axis)
Magnification (M)	virtual image (on the same side of the lens as the object; upright; above the principal axis)	real image (beyond the lens; inverted; below the principal axis)

[a]The variable d_o is always positive for a single lens but can be negative if two or more lenses are used, as described in Chapter 17.

Notice also in Figure 15-58 that a ray that passes through the middle of the lens provides a third set of similar triangles (shaded red). In this set,

$$\frac{h_i}{h_o} = -\frac{d_i}{d_o} \tag{15-9}$$

Thus, the magnification of a lens can be found by using either the ratio h_i/h_o or the ratio $-d_i/d_o$.

SAMPLE PROBLEM 15-9

Use the thin-lens equation to calculate the image distance in **(a)** Sample Problem 15-7 and **(b)** Sample Problem 15-8. Also, use the object and image distances to check the linear magnification in each case.

Solution

(a) For the converging lens in Sample Problem 15-7, $f = 14$ cm and $d_o = 45$ cm.

$$\frac{1}{f} = \frac{1}{d_o} + \frac{1}{d_i}$$

$$\frac{1}{d_i} = \frac{1}{f} - \frac{1}{d_o}$$

$$\frac{1}{d_i} = \frac{1}{14\text{ cm}} - \frac{1}{45\text{ cm}}$$

$$d_i = +20\text{ cm}$$

A positive image distance means a real image on the far side of the lens, which in the scale diagram should be 2.0 cm from the lens. These answers check out correctly.

The linear magnification is $M = -d_i/d_o = -20\text{ cm}/45\text{ cm} = -0.45$. This value matches approximately the magnification found using the diagram, −0.4. The negative magnification means the image is inverted.

TRY THIS!

Reflection in Lenses

In an otherwise dark room with just one bright lamp on, hold a pair of eyeglasses about 30 cm from your eye so you can see an image of the lamp reflecting from the surface of the lens. What are the characteristics of the image you observe? Now turn the glasses around so the light reflects off the other side of the lens. Again, describe the image. Explain any differences between the two images observed.

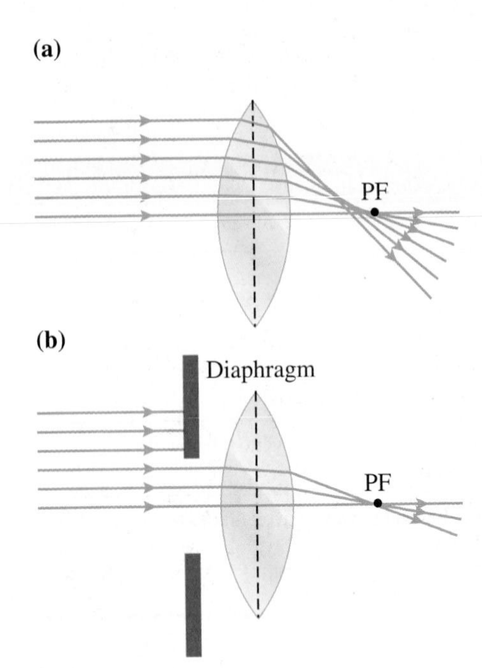

Figure 15-59 Spherical aberration in a converging lens. **(a)** The cause of the aberration. **(b)** Correcting the aberration.

(b) For the diverging lens in Sample Problem 15-8, $f = -14$ cm and $d_o = 45$ cm.

$$\frac{1}{d_i} = \frac{1}{f} - \frac{1}{d_o}$$

$$\frac{1}{d_i} = -\frac{1}{14 \text{ cm}} - \frac{1}{45 \text{ cm}}$$

$$d_i = -11 \text{ cm}$$

A negative image distance means the image is on the object side of the lens and it is a virtual image, which in the scale diagram should be located 1.1 cm from the lens. This checks out closely.

The linear magnification is $M = -d_i/d_o = -(-11 \text{ cm})/45 \text{ cm} = +0.24$. This value is close to the approximate value of 0.2 found using the scale diagram. The positive magnifications means the image is upright.

Lens Aberrations

Spherical lenses produce aberrations that result in imperfect images. These aberrations are not the fault of the lenses. Even well-made spherical lenses do not produce perfect images. The problems exist because the light passing through a lens obeys the law of refraction and not all rays come to an exact focus. We will examine two of the most common types of aberration here.

Spherical aberration, described previously for converging mirrors, is one type of aberration that occurs in lenses. For a converging lens, rays parallel to the principal axis but relatively far from it do not refract through the PF (Figure 15-59 (a)), resulting in an image that is not entirely in focus. In other words, at least part of the image is blurry. This aberration can be corrected by placing a diaphragm in front of the lens so that only the central portion of the lens is used (Figure 15-59 (b)).

Another type of problem, which occurs for lenses but not for mirrors, involves colours created around the edge of a lens. This problem, called **chromatic aberration**, occurs when white light undergoes dispersion into its spectral colours around the perimeter of a lens, as illustrated in Figure 15-60. The prefix chroma stems from the Greek *khroma,* which means colour. You may have noticed chromatic aberration of white light from a spotlight projected onto a screen or curtain at a concert or stage play. A similar effect, though perhaps less obvious, may be obtained with the light from a lecture hall projector striking a screen, especially if the lens is an inexpensive one. In glass, colours near the red end of the visible spectrum have a lower index of refraction, so they refract less than the colours at the violet end of the spectrum, as shown in Figure 15-61 (a). This type of aberration can be corrected by using a combination of a converging lens and a diverging lens, each made of a material with a different index of refraction, for example, flint glass and crown glass. One lens counteracts the aberration of the other, as illustrated in Figure 15-61 (b).

DID YOU KNOW?

Sometimes spherical aberration is desirable. For example, some beautiful photographic effects can be obtained by having the main subject in focus and the surroundings blurry. This effect is achieved by using a large lens aperture and a short exposure time. (Photography is described in further detail in Chapter 17.)

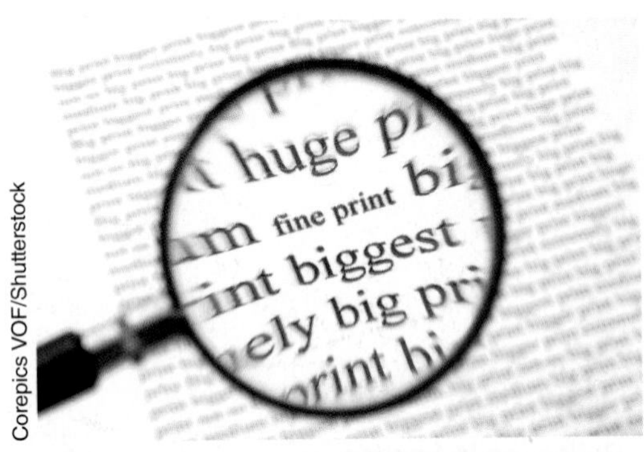

Corepics VOF/Shutterstock

Figure 15-60 An example of chromatic aberration in a converging lens: notice the colours around the perimeter of the lens. (The photograph also illustrates spherical aberration.)

Learning Tip

To help visualize the effects of spherical and chromatic aberration, look up the terms on the Internet. The photos and diagrams will help you understand the problems and their solutions.

Gradient-Index Lenses

You have seen (in Section 15.3) that a mirage is an example of gradient-index optics because it is caused by the changing index of refraction of layers of air due to temperature gradients. The same principle can be applied to lenses to help reduce aberration or to create lens shapes that are convenient for certain applications. For example, lenses used in some scanners, photocopiers, and overhead projectors are flat and are made with the index of refraction increasing from the centre to the outside edge, causing light from the source to converge as required (Figure 15-62).

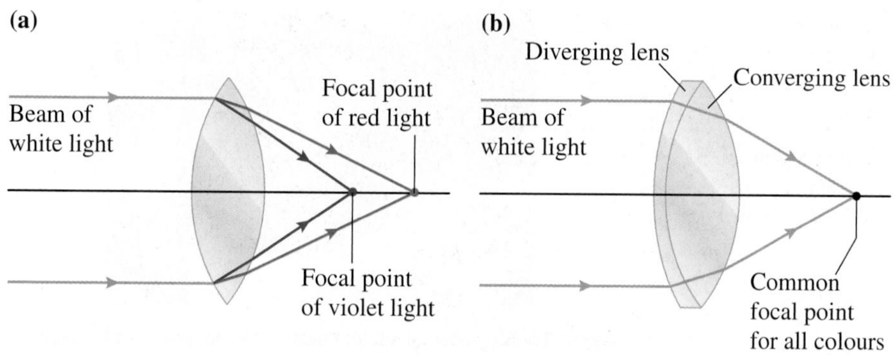

Figure 15-61 Chromatic aberration in a converging lens. **(a)** The cause of the aberration. **(b)** Correcting the aberration.

chromatic aberration a problem with lenses in which colours caused by the dispersion of light occur around the perimeter of the lens

Light pipes made with gradient-index optical fibres do not require the cladding used in conventional fibre optics (Section 15.4). Rather, the index of refraction of the fibre decreases from the centre of the fibre toward the outside, causing a cross-section of the fibre to act like a converging lens. This ensures that the transmitted light remains trapped in the fibre, even if the fibre goes around corners. Just as with a mirage that is an inferior image, the light that travels toward the lower-index of refraction part of the fibre refracts back toward the more dense part. This helps the light maintain a constant speed and increases the data transmission rate of the communication. Gradient-index optical fibres are also very flexible.

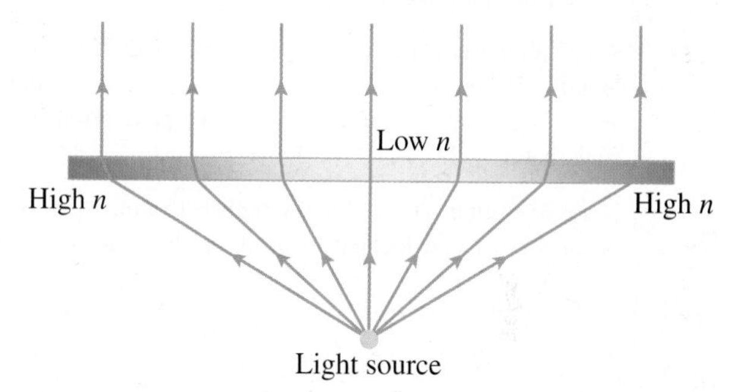

Figure 15-62 A simplified drawing of a flat gradient-index lens.

EXERCISES

15-52 The diagram in Figure 15-63 shows light rays in air approaching glass lenses. Draw large diagrams, similar in shape, and draw the approximate direction of each light ray as it travels into the glass and back into the air. Include a normal wherever a ray strikes a surface.

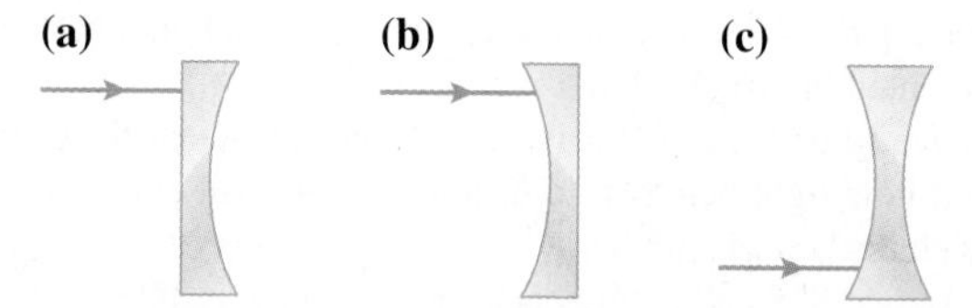

Figure 15-63 Question 15-52.

15-53 For each pair of lenses shown in Figure 15-64, which lens would you expect to have the greater focal length? Why? (Each lens is in air, and the index of refraction of each lens is given.)

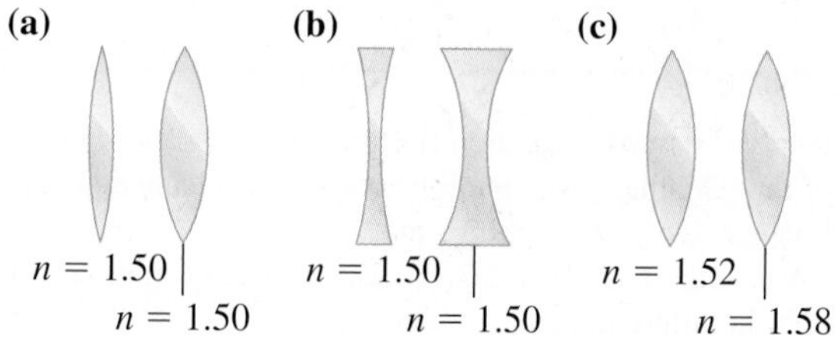

Figure 15-64 Question 15-53.

15-54 How would you find the approximate focal length of a converging lens experimentally?

15-55 You are given a lens from a discarded pair of eyeglasses. What experimental observations would help you determine whether the lens is a converging lens or a diverging lens?

15-56 Can an image in a double concave lens be placed onto a screen? Explain.

15-57 A piece of glass has a spherical bubble trapped inside. As light passes through the glass and the air bubble, does the bubble act like a converging lens or a diverging lens?

15-58 Use equations to determine the image distance, d_i, and the linear magnification, M, for each situation described below. Be careful with the positive and negative signs.

(a) converging lens: $f = +3.0$ cm; $h_o = 1.5$ cm; $d_o = 4.5$ cm

(b) converging lens: $f = +3.0$ cm; $h_o = 1.5$ cm; $d_o = 1.5$ cm

(c) diverging lens: $f = -3.0$ cm; $h_o = 1.5$ cm; $d_o = 3.0$ cm

15-59 To visualize each situation described in Question 15-58, sketch a ray diagram to locate the image of the object. In each sketch use two rules, and for each image found, state its attitude and type, and estimate d_i and M. Compare your answers to what you found using equations.

15-60 Using Questions 15-58 and 15-17 as reference, compare lenses with curved mirrors. Specifically, compare the properties of the images produced for various object locations (beyond the focal point, at the focal point, or closer than the focal point).

15-61 At a certain time the distance between Earth and the Moon is 3.85×10^8 m. A lens of focal length +0.154 m is used to place an image of the Moon onto a screen on Earth. How far should the screen be from the lens?

15-62 A lens is used to produce a concentrated beam of light starting from a light bulb.

(a) What type of lens should be used?

(b) What should be the approximate location of the light bulb in terms of the focal length of the lens?

15-63 A technician is using magnifying lenses called loupes of focal length +22 cm to work on a dentition mould (Figure 15-65). Where would the mould have to be located to produce an image such that $d_i = 2.0\ d_o$?

15-64 For a diverging lens of focal length −15 cm, where would the object have to be located to produce an image such that $d_i = -d_o/3.0$?

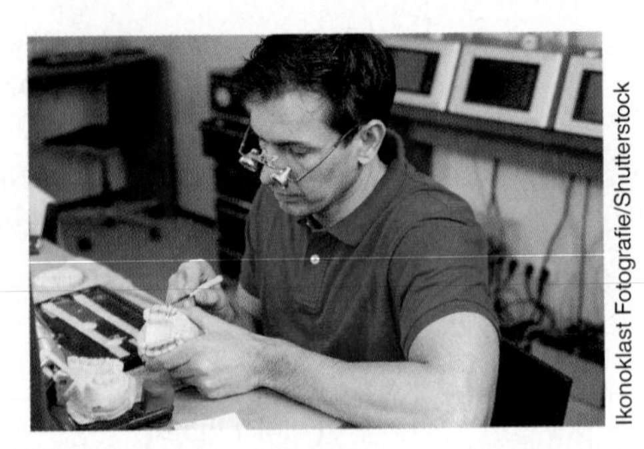

Figure 15-65 A dental lab technician with surgical loupes applies porcelain to a dentition mould (Question 15-63).

15-65 A lens is needed to provide an image with a height of −6.00 cm. The object is a person who is 1.50 m tall standing 8.0 m away from the lens.

(a) What type of lens is required?

(b) What is the image distance?

(c) What is the focal length of the lens?

15-66 Use a diagram to help explain how you would design a flat gradient-index lens that is diverging.

LOOKING BACK...LOOKING AHEAD

The study of light follows the topics of waves and sound, although the wave nature of light has been mentioned only briefly. In many cases, especially those involving diagrams, the concepts in this chapter were more easily dealt with by treating light as rays.

Light is a form of energy that is transmitted through transparent media but is transmitted most easily through a vacuum. For most observations, light appears to travel in a straight line (rectilinear propagation), as verified by the formation of shadows, including solar and lunar eclipses. The speed of light in a vacuum or in air to three significant digits is 3.00×10^8 m/s.

Mirrors and lenses can be used to obtain images of objects. Real images can be produced only by converging mirrors and lenses for objects located at a distance greater than the focal length. Virtual images can be produced by all types of mirrors (plane, converging, and diverging) and both types of lenses (converging and diverging). Images can be described by three main characteristics, attitude, type, and location (including image distance), and the linear magnification of each mirror or lens can be found. Images in mirrors can be found in ray diagrams by applying the law of reflection ($\theta_i = \theta_r$) or by using rules of reflection for each type of mirror. They can also be determined by applying the mirror equation. Images in lenses can be found in ray diagrams by applying Snell's law of refraction or by using rules of refraction for each type of lens, or they can be determined by applying the thin-lens equation.

The use of lenses is based on the refraction of light. Refraction also results in the dispersion of light and internal reflection in media that have a higher index of refraction than their surroundings. Several applications of mirrors, lenses, refraction, dispersion, and total internal reflection were presented in this chapter.

We have hinted that there is more than one scientific model used to explain how light behaves. In Chapter 16 we will compare the wave model with the particle model.

The study of optical instruments, such as cameras, telescopes, microscopes, and the human eye, relies heavily on the topics studied in this chapter. These instruments are presented in Chapter 17.

CONCEPTS AND SKILLS

Having completed this chapter, you should be able to do the following:

- Describe evidence to support the notion that light is a form of energy.
- Use ray diagrams to draw and label shadows produced by solar and lunar eclipses.
- Calculate the linear magnification of a mirror or a lens using the ratio of heights or the ratio of distances.
- Draw and label ray diagrams to locate images in all types of mirrors, state the three main characteristics of each image found, and calculate the magnification of the mirror in each case.
- Apply the mirror equation (in terms of f, d_o, and d_i) to find any one of the variables given the other two, and calculate quantities such as image height and magnification.
- Do calculations involving the index of refraction and the speed of light in different media.
- Apply Snell's law of refraction to determine the index of refraction or angle of incidence or refraction for light travelling from one medium into another.

- Draw ray diagrams of the refraction of light through transparent blocks or prisms.
- Recognize the colours of the visible spectrum and describe the cause of the dispersion of white light as it passes through a prism.
- Describe the two main conditions under which total internal reflection occurs.
- Calculate the critical angle in a medium by applying Snell's law of refraction.
- Describe applications of total internal reflection.
- Draw and label ray diagrams to locate images in lenses, state the characteristics of each image, and calculate the magnification of the lens in each case.
- Apply the thin-lens equation (in terms of f, d_o, and d_i) to find any one of the variables in terms of the others, and calculate quantities such as image height and magnification.
- Compare lenses with curved mirrors.

KEY TERMS

You should be able to define each of the following words or phrases:

luminous object
incandescent source
gas discharge source
fluorescent source
light-emitting diode (LED)
phosphorescent material
chemiluminescent material
bioluminescent animal
rectilinear propagation of light
umbra
penumbra
solar eclipse
lunar eclipse
light year
specular (or regular) reflection
diffuse (or irregular) reflection
real image
virtual image
linear magnification
angle of incidence
angle of reflection
converging mirror
diverging mirror
focal point
focal length
reversibility of light rays
spherical aberration
refraction of light
angle of refraction
index of refraction
lateral displacement
dispersion
visible spectrum
angle of deviation
mirage
critical angle
total internal reflection
primary rainbow
secondary rainbow
fibre optics
lens
converging lens
diverging lens
primary focal point
secondary focal point
chromatic aberration

Chapter Review

MULTIPLE-CHOICE QUESTIONS

15-67 Which of the following statements describe the images formed by an object in a curved mirror?
1. Diverging mirrors always produce real images.
2. Virtual images are always upright relative to the object.
3. Real images are always inverted relative to the object.
4. A virtual image is always smaller than the object.

(a) 1, 2, 3; (b) 2 and 3; (c) 2, 3, and 4; (d) 1, 3, and 4; (e) 1, 2, 3, and 4

15-68 An object is located in front of a converging mirror and its image is located on the same side of the mirror but farther from the mirror. If d_o is the object distance and f is the focal length, then
(a) $d_o > 2f$ and the image is real
(b) $d_o < f$ and the image is real
(c) $f < d_o < 2f$ and the image is virtual
(d) $f < d_o < 2f$ and the image is real

15-69 For the fish in position B in Figure 15-66, the image seen by observer O is closest to which position? (a) A, (b) C, (c) D, (d) E, (e) F

15-70 In Figure 15-66, fish B can see fish F by
(a) looking directly at it only
(b) looking directly at it or viewing a virtual image reflected at the water's surface
(c) looking directly at it or viewing a real image reflected at the water's surface
(d) looking directly at it or viewing an image caused by the refraction of light in the water

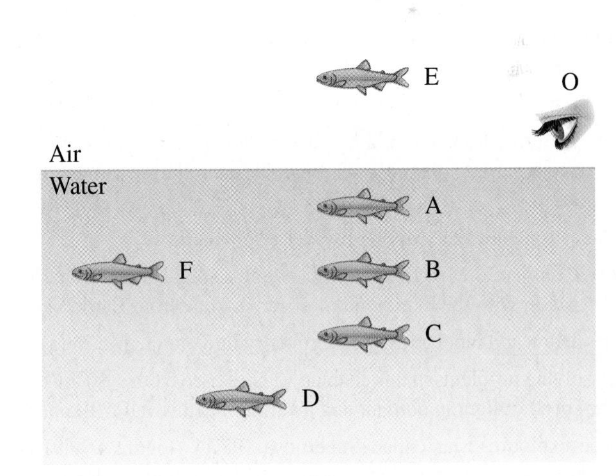

Figure 15-66 Questions 15-69 and 15-70.

15-71 A ray of white light approaches a rectangular transparent block at an angle of incidence of 70°. If the lateral displacement of the emerging red ray is LD_R and, and that of the emerging blue ray is LD_B, then
(a) $LD_R = LD_B$, (b) $LD_R < LD_B$, (c) $LD_R > LD_B$

15-72 A ray of blue light enters the prism shown in Figure 15-67. The ray that emerges is (a), (b), (c), (d), (e)

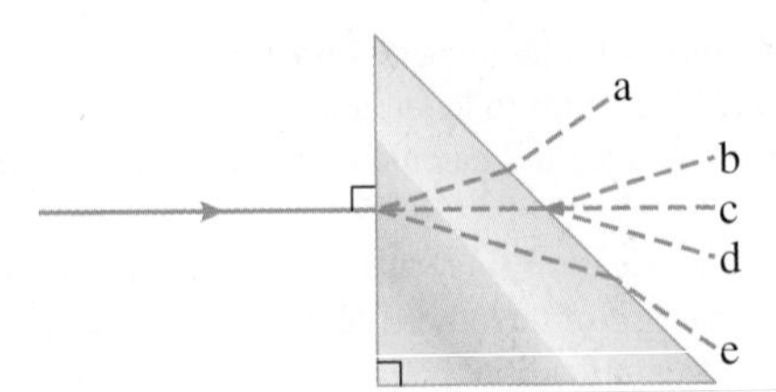

Figure 15-67 Question 15-72.

15-73 For a beam of sunlight entering a water droplet, let the indexes of refraction for red and green light be n_R and n_G respectively, and the speeds of the colours be v_R and v_G. In comparing the properties of the red and green light in the droplet,

(a) $n_R < n_G$; $v_R > v_G$
(b) $n_R < n_G$; $v_R < v_G$
(c) $n_R > n_G$; $v_R > v_G$
(d) $n_R > n_G$; $v_R \leq v_G$
(e) $n_R = n_G$; $v_R = v_G$

15-74 Objects O_1 and O_2 are placed at a distance f from a converging lens L_1 and a diverging lens L_2 respectively. The resulting images are I_1 and I_2 respectively. Which of the following statements about I_1 and I_2 and their magnifications are true?

(a) I_1 is nonexistent; I_2 is virtual and $|M| = 1$
(b) I_1 is virtual and $|M| = 1$; I_2 is virtual and $|M| < 1$
(c) I_1 is real and $|M| > 1$; I_2 is virtual and $|M| > 1$
(d) I_1 is nonexistent; I_2 is virtual and $|M| < 1$

15-75 Lenses L_1 and L_2 in Figure 15-68 are made of different materials having indexes of refraction n_1 and n_2 respectively. Light travels in L_1 and L_2 at speeds v_1 and v_2 respectively. Comparing the properties of the lenses,

(a) $n_1 > n_2$; $v_1 > v_2$
(b) $n_1 < n_2$; $v_1 > v_2$
(c) $n_1 > n_2$; $v_1 < v_2$
(d) $n_1 < n_2$; $v_1 < v_2$

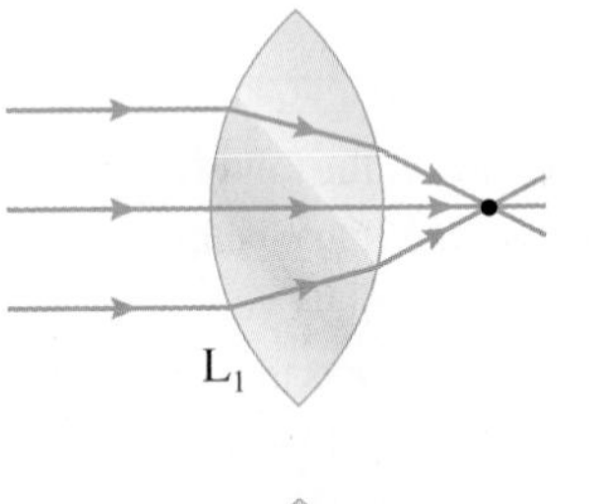

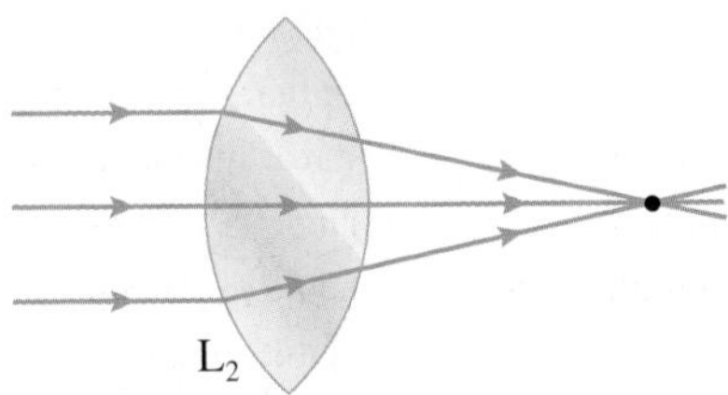

Figure 15-68 Question 15-75.

Review Questions and Problems

15-76 Describe how the sundial in Figure 1-3 (c) on page 3 relates to the concepts described in this chapter.

15-77 (a) What are the relative positions of the Sun, Moon, and Earth during a solar eclipse? a lunar eclipse? an annular eclipse?

(b) Describe how it would be possible for observers on Earth to see a partial solar eclipse; a partial lunar eclipse; an annular eclipse.

15-78 An astronaut on the Moon observes a solar eclipse.

(a) Compare the likelihood of observing a solar eclipse from the Moon with the likelihood of observing one from Earth.

(b) Draw and label a ray diagram verifying your answer in (a).

15-79 In solving problems in this chapter we have used 3.00×10^8 m/s as the speed of light in *both* air and a vacuum. Is this valid? Explain.

15-80 The exploding star called Supernova 1987A (Figure 15-69) is 1.68×10^5 light years away from the solar system. Express this distance in metres and gigametres.

NASA/AP Photo

Figure 15-69 Supernova 1987A (Question 15-80).

DID YOU KNOW?

As mentioned in Chapter 9, page 247, Supernova 1987A was discovered by Canadian Ian Shelton. Astronomers applied the designation "A" in the name as a convention just in case a second supernova was discovered in 1987. However, the chances of that occurring were very slight since the previous supernova visible to the naked eye was observed by Johannes Kepler in 1604.

15-81 The average distance from Earth to the Moon is 3.84×10^8 m.

(a) Express this distance in light years.

(b) How long does it take a beam of laser light to travel from Earth to the Moon and back again?

15-82 Which is safer for bike riding at night: clothing with specular or diffuse reflection? Explain.

15-83 Refer to Figure 15-70.

(a) What type of mirror is used to produce the images observed?

(b) How many mirrors are needed in this case?

(c) What is the angle between the mirrors? How can you tell?

Pavel L Photo and Video/ Shutterstock

Figure 15-70 A fun house set of mirrors (Question 15-83).

15-84 Why are angles of incidence, reflection, and refraction measured from a normal?

15-85 Use a ray diagram to locate the image of the object in front of the plane mirror shown in Figure 15-71. Describe the main characteristics of the image.

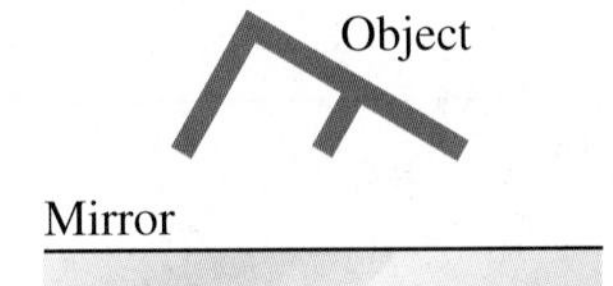

Figure 15-71 Question 15-85.

15-86 Describe what you would observe if you were to place a plane mirror along edges AB, BC, and CD in Figure 15-72. Try it.

Figure 15-72 Question 15-86.

15-87 Describe the differences between real and virtual images.

15-88 In ray diagrams involving either a single mirror or a single lens, how can real images be distinguished from virtual images? Can the same information be inferred by knowing the magnification of the mirror or lens, or by knowing the image distance? Explain.

15-89 For each situation described below, use equations to determine the image distance and linear magnification. State the type and attitude of each image.

(a) A person whose face is 22 cm long is looking into a cosmetic mirror from a distance of 28 cm. The focal length of the mirror is +72 cm.

(b) The mirror described in (a) above is used to place the image of a glowing bulb onto a screen. The bulb is 14 cm high and is located 1.2 m from the mirror.

(c) The person in (a) above is looking into a mirror of focal length −54 cm from a distance of 36 cm.

15-90 **Fermi Question:** Describe how you would determine a reasonable estimate of the focal length of the diverging mirror on the passenger side of a vehicle.

15-91 Figure 15-73 shows a mirror located at a blind corner. Explain the type of mirror used and its advantages.

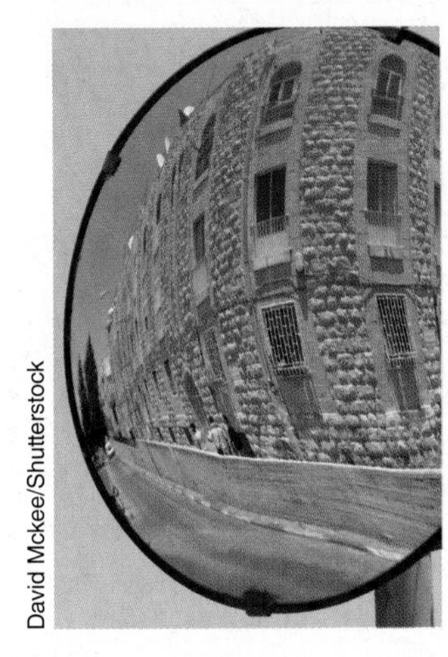

David Mckee/Shutterstock

Figure 15-73 Question 15-91.

15-92 Figure 15-74 shows three different scenes that relate to the concepts in this chapter. Describe the physics principles that explain each situation.

(a)

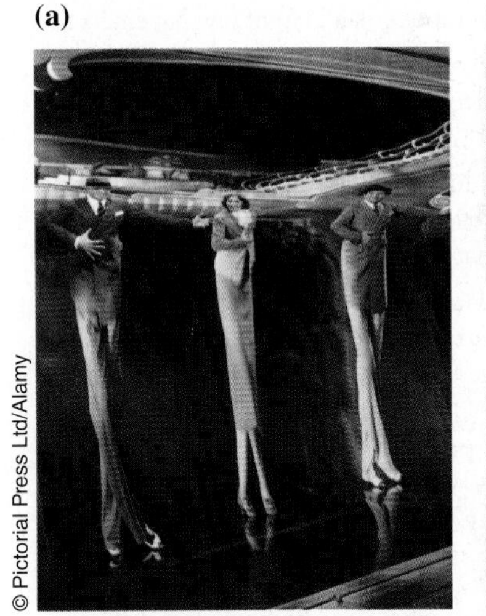

© Pictorial Press Ltd/Alamy

(b)

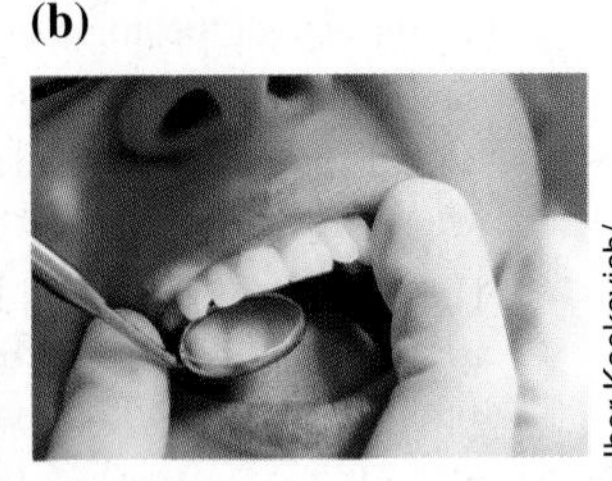

Ihar Kaskevich/ Shutterstock

(c)

© National Geographic Image Collection/Alamy

Figure 15-74 Question 15-92. **(a)** This dance routine featured Fred Astaire, Gracie Allen, and George Burns in the classic movie *A Damsel in Distress* (1937). **(b)** A dentist's mirror. **(c)** Shimmering "water" on a desert highway.

15-93 Determine whether each situation described below results in a real image, a virtual image, or no image.

(a) mirror: $f = +28$ cm; $d_o = +14$ cm

(b) mirror: $f = -18$ cm; $d_o = +18$ cm

(c) lens: $f = -38$ cm; $d_o = +20$ cm

(d) lens: $f = +11$ cm; $d_o = +11$ cm

(e) mirror: $M = -2.2$

(f) lens: $M = 0.28$

15-94 Compare spherical aberration in converging mirrors and converging lenses. How is the aberration corrected in each case?

15-95 The designer of a "house of mirrors" wants a curved mirror to produce a real image that is magnified by a factor of −1.2 times when the object (a person) is 1.1 m from the mirror. What is the required focal length of the mirror?

15-96 The speed of light in carbon disulfide is 1.84×10^8 m/s. A ray of light in air strikes the surface of the carbon disulfide at an angle of incidence of 37.1°.

(a) What is the index of refraction of this medium?

(b) What is the angle of refraction in the carbon disulfide?

(c) Determine the critical angle (relative to air) for light in the carbon disulfide.

15-97 The index of refraction of carbon tetrachloride is 1.46. Determine

(a) the speed of light in this medium

(b) the angle of refraction in air for a light ray in the carbon tetrachloride having an angle of incidence of 22.8°

15-98 State the relationship between the index of refraction of a medium and

(a) the speed of light in the medium

(b) the size of the critical angle of the medium (surrounded by air)

(c) the size of the angle of refraction of a light ray that enters the medium from air

15-99 What is the index of refraction of a medium (surrounded by air) whose critical angle is 36.8°?

15-100 The indexes of refraction for ethyl alcohol and sodium chloride are 1.361 and 1.544 respectively. What is the ratio of the speed of light in ethyl alcohol to that in sodium chloride?

15-101 White light in air is aimed toward crown glass at an angle of incidence of 32.8°. Use the data in Table 15-3 to determine the angle of refraction for (a) orange light and (b) violet light.

15-102 What is the critical angle for blue light in crown glass surrounded by air? (Reference: Table 15-3)

15-103 A certain converging lens made of crown glass has a focal length of 16.5 cm for green light. Does the focal length increase or decrease for these colours: (a) red, (b) yellow, (c) blue?

15-104 When white light passes through a clear glass window, why does it not disperse into its spectral colours?

15-105 In Figure 15-75, an object starts far from the mirror or lens and moves gradually closer. Each image corresponds to one, and only one, position of the object. Match each object to the best corresponding image. Describe what patterns you observe.

15-106 For each situation described below, use equations to determine the image distance and image height. State the attitude and type of each image.

(a) A person's face, which is 23 cm long, is located 32 cm from a fish-eye lens in a doorway. The focal length of the lens is −8.0 cm.

(b) A detective inspects a 14 mm long piece of broken glass through a magnifying lens of focal length 57 mm from a distance of 26 mm.

(c) A projector with a lens system of focal length 6.4 cm places the image of a drawing that is 24 mm high onto a screen. The image is in focus when the drawing is 23.4 cm from the lens.

15-107 Assume that the effective distance from the lens of your eye to the retina, where a real image is formed, is 19.0 mm. Determine the effective focal length of your eye when you are looking at an object that is at a distance from your lens of (a) 25.0 cm and (b) 55.5 cm. (The ability of your eye to adjust its focal length, called accommodation, is one of the concepts presented in Chapter 17 in more detail.)

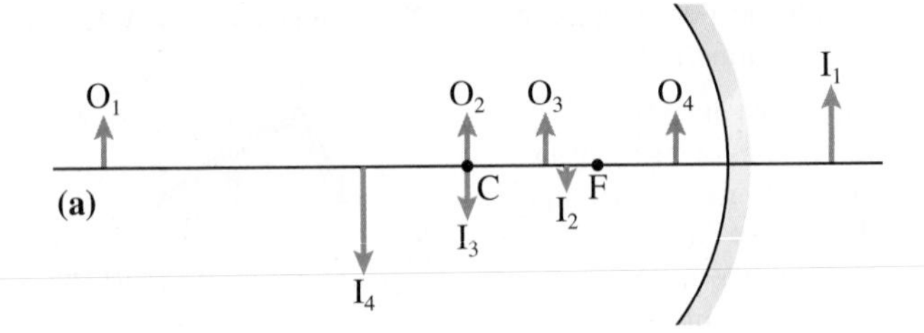

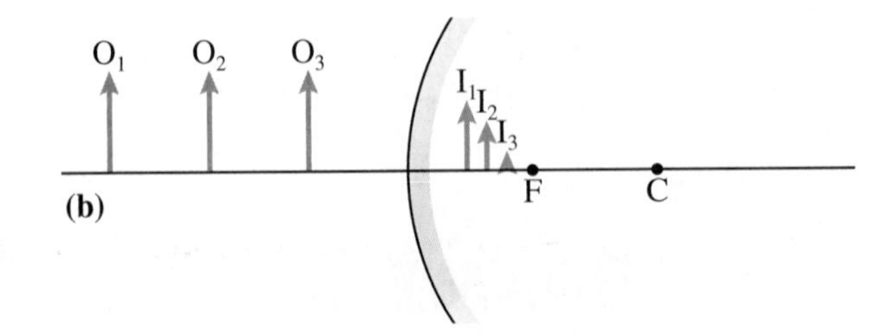

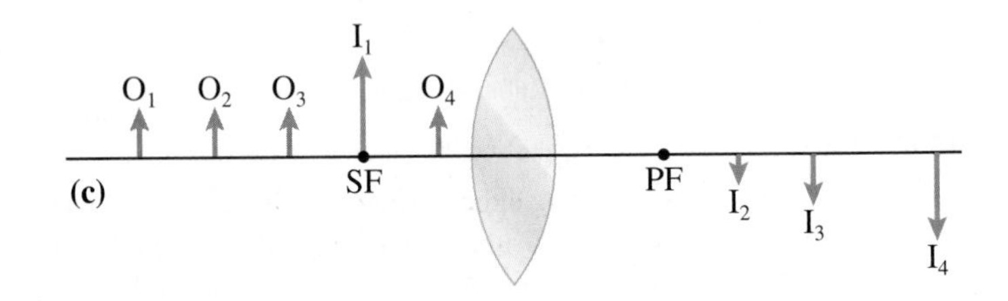

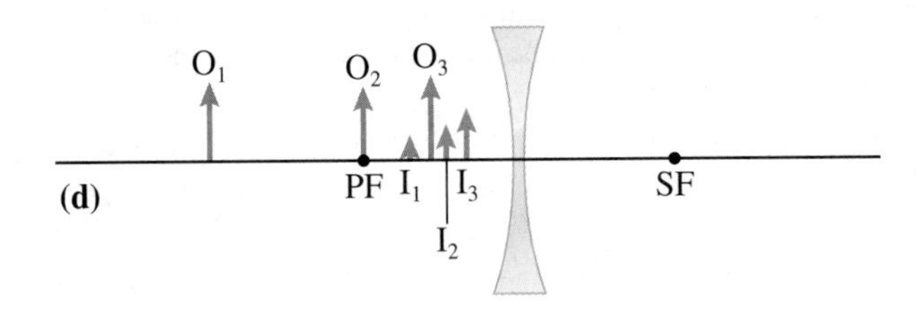

Figure 15-75 Question 15-105.

15-108 Why does chromatic aberration occur for lenses but not for curved mirrors?

Applying Your Knowledge

15-109 Fermi Question: Estimate how far light can travel in air during your average reaction time.

15-110 A patient must be 6.0 m from a certain eye-testing chart in order for the vision test to be valid. The chart is available with either normal or backward printing. How can a physician whose office is only 3.6 m long use the chart in an appropriate way? Draw a diagram to show a precise solution.

15-111 Fermi Question: Estimate the focal length of the optical device in each of the figures listed below. If there is no focal length, state why.

(a) a single mirror in Figure 15-1

(b) the group of mirrors in Figure 15-1

(c) the focusing lens in Figure 15-17 (b)

(d) the main part of the sculpture in Figure 15-20

(e) the concave reflectors in Figure 15-26 (b)

(f) the solar cooker in Figure 15-26 (c)

15-112 Light from air enters a set of transparent materials at an angle of incidence of 30.0°, as shown in Figure 15-76.

(a) Determine the angle of refraction in the sixth medium ($n_6 = 2.00$).

(b) What pattern have you observed in solving this problem?

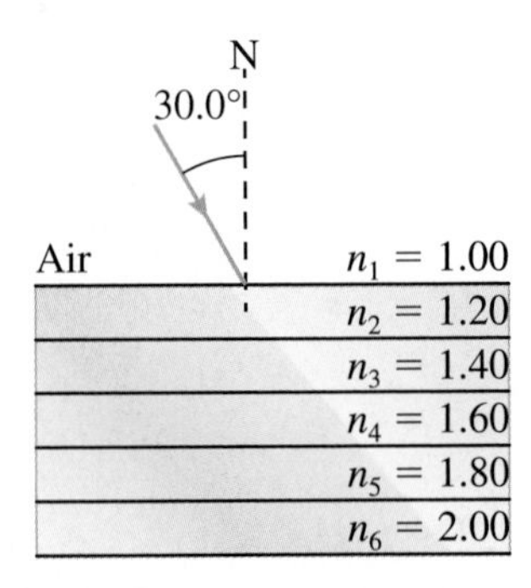

Figure 15-76 Question 15-112.

15-113 One of the ways to measure the speed of electromagnetic radiation, including light, is to set up a standing wave pattern using waves of known frequency. A movable transmitter/receiver sends signals that bounce off a reflector and interfere with the initial signal. When a standing wave resonance occurs, another movable receiver is used to determine the distance between nodes in the standing wave pattern. If a 9677 MHz signal produces a standing wave pattern with nodes separated by 15.50 cm, what is the speed of the electromagnetic radiation?

15-114 A cosmetic mirror has a shape that allows both real and virtual images to be observed. One such mirror is made so that a virtual image is 2.00 times the size of the object when the object is 18.2 cm from the mirror. Assuming the mirror is spherical, determine its radius of curvature.

15-115 Light from one last olive in a container filled with olive oil ($n = 1.47$) leaves the surface of the oil in the middle of the container and just passes over the edge of the container, as shown in Figure 15-77. Determine the height of the container.

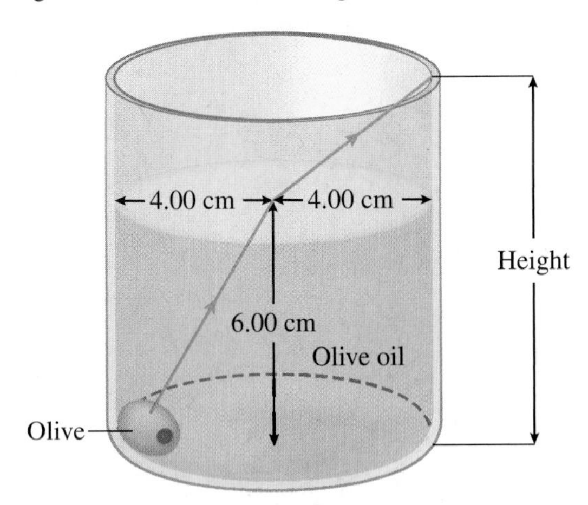

Figure 15-77 Question 15-115.

15-116 In an equilateral prism with a horizontal base, the dispersion of white light is most easily seen when the refracted light in the middle of the visible spectrum (yellow light) is parallel to the base of the prism. Determine the angle of incidence of yellow light in air that creates this situation in a prism made of crown glass. (Reference: Table 15-3)

15-117 A ray of light in air enters a rectangular glass block ($n = 1.52$) and emerges on the side of the block that is parallel to the incident side. The block is 42.4 mm wide. If the angle of incidence in the air is 26.8°, determine the lateral displacement of the ray that emerges back into the air.

15-118 For light going from air into a certain glass, the following data are observed: angle of incidence,

θ_i, (°) 10.0 20.0 30.0 40.0 50.0 60.0 70.0 80.0

angle of refraction, θ_R, (°) 6.65 13.2 19.5 25.4 30.7 35.3 38.8 41.0

(a) For each set of data find the ratio θ_i/θ_R.

(b) For each set of data find the ratio of $\sin\theta_i/\sin\theta_R$. What does this ratio represent?

(c) Within what range of values would Ptolemy's prediction about the ratio θ_i/θ_R reasonably apply?

15-119 Snell's law ($n_1 \sin\theta_1 = n_2 \sin\theta_2$) applies if the angles are measured from the normal. Would it also apply if the angles were measured from the interface between the two media? If so, verify that it does with an example. If not, offer a new form of the law.

15-120 Rather than travelling in a straight line, some of the light observed in Figure 15-78 is following a parabolic curve. Explain why this occurs.

Figure 15-78 Question 15-120.

15-121 Words and expressions new to a reader can often be understood by analyzing their components. Try doing this with a device called an *achromatic doublet*. What do you think the device is and what do you think it is used for?

15-122 The thin-lens equation expresses the lens focal length in terms of the object distance and image distance, but another equation, called the *lens-maker's equation*, expresses the focal length in terms of the index of refraction of the glass and the radii of curvature of the lens. This equation is

$$\frac{1}{f} = (n-1)\left(\frac{1}{R_1} + \frac{1}{R_2}\right)$$

where R_1 is the radius of curvature of the first (incident) surface of the lens, and R_2 is the radius of curvature of the second (emergent) surface of the lens. The radii are negative if the centre of curvature is on the incident side, and positive if the centre of curvature is on the emergent side. The radius of a plane surface is assumed to be infinity.

(a) Determine the focal length of a plano-convex lens having $R_1 = +28.2$ cm and made of glass with $n = 1.56$.

(b) A double concave lens ($n = 1.52$) has a radius of curvature on the incident side of -15.4 cm and a focal length of -48.1 cm. What is the radius of curvature of the emergent side?

Wave Optics 16

© Prisma Bildagentur AG/Alamy

Figure 16-1 In this night-time wintery scene in Austria, the way the light appears to spread out from bright sources can be explained by wave optics.

CHAPTER OUTLINE

If you had been at the scene shown in the photograph in Figure 16-1, your eyes would have observed normal, spherical light sources. The streams of light emanating in various directions from each source are observed only after the light passed through a special filter on the camera that took the photo. This is just one of many wave phenomena of light found in this chapter. Other effects include the colours of soap bubbles and oil slicks, and the polarization of light.

It is evident from the title of this chapter that we will explore concepts regarding the wave nature of light. This differs from the use of light rays to visualize the reflection and refraction of light presented in Chapter 15. But if light acts like a wave, what are the period, frequency, and wavelength of the waves? Does the universal wave equation apply to light? What evidence do we have that light undergoes diffraction and interference? Many of the important concepts addressed in this chapter were

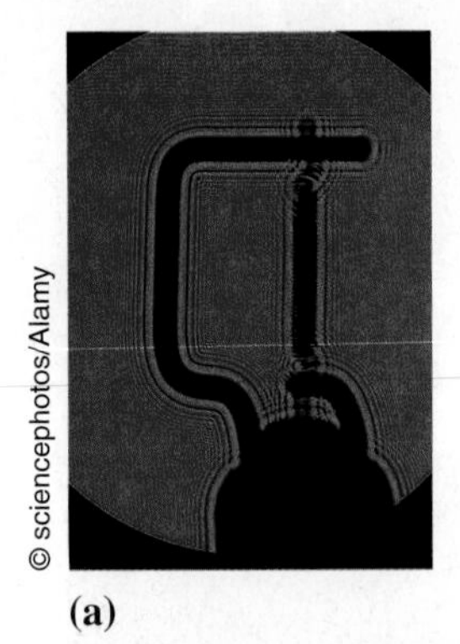

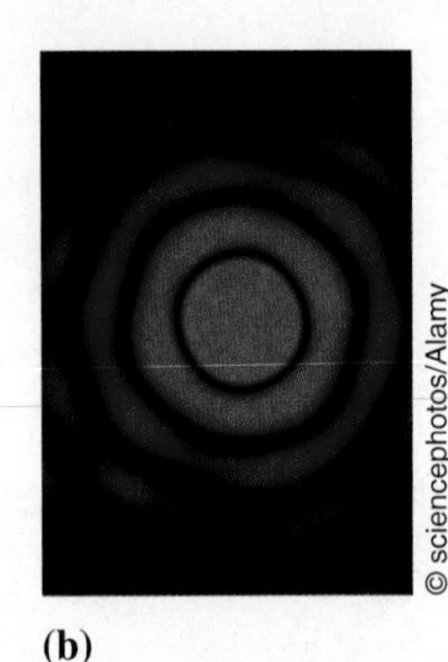

(a) (b)

Figure 16-2 A diffraction pattern is produced when red laser light is directed **(a)** past a light bulb filament or **(b)** through a tiny circular aperture.

presented in Chapter 13, where properties of mechanical oscillations and waves were first introduced and discussed. We revisit these ideas here, this time in the context of the physical nature of light.

16.1 Diffraction of Light and Huygens' Principle

Diffraction is the bending of a wave in a single medium around an obstacle or through an aperture. It was defined and shown to occur for water waves in Section 13.8 and for sound waves in Section 14.5. Does diffraction also occur for light? (Remember that diffraction occurs in a single medium; it differs greatly from refraction, which occurs when waves travel from one medium to another, as discussed in Sections 15.3 and 16.7.)

Under most circumstances, we do not observe the diffraction of light. However, if you look at a distant street light through a narrow gap between your fingers, or through the fibres of a sheer curtain or other thin piece of cloth, you will notice bright and dark regions. These regions are caused by diffraction. Examples of diffraction patterns of light around a light bulb filament and through a fine aperture are shown in Figure 16-2.

To explain how light diffracts around a sharp edge or through a small aperture, let us look at an old but useful theory about waves. The diffraction of light was first observed by an Italian university professor named Francesco Grimaldi (1618–1663), whose book describing his observation was published in 1665. That same year, the English scientist Robert Hooke began to describe his version of a wave model of light. Shortly after, in 1678, a Dutch physicist named Christian Huygens (1629–1695) proposed an idea regarding wave properties that is still used in simple ways today. Huygens' idea was that the present shape and location of a wavefront can be used to predict the future shape and location of the wavefront. **Huygens' principle** states

Huygens' principle every point on a wavefront acts as a source of a new wavelet, and the new wavefront is the forward envelope of the set of new wavelets

Every point on a wavefront acts as a source of a new wavelet, and the new wavefront is the forward envelope of the set of new wavelets.

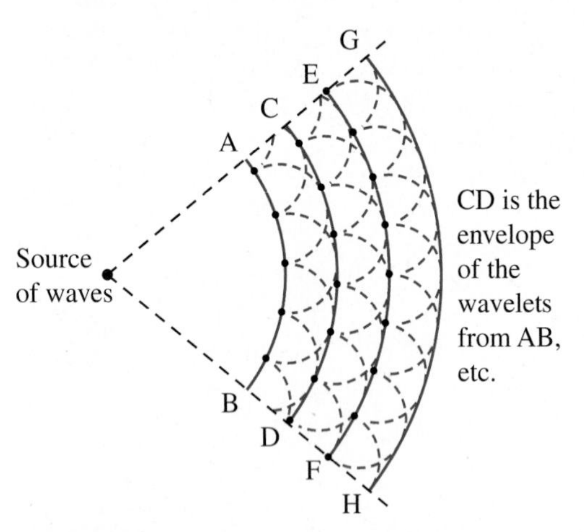

Figure 16-3 Applying Huygens' principle to wavefronts from a point source.

Figure 16-3 shows Huygens' principle applied to wavefronts moving outward from a point source and travelling at a constant speed in all directions. In three dimensions, the wavefronts would be spherical, but in the two-dimensional diagram shown here the wavefronts are circular. Only a segment of these wavefronts are shown for clarity. Consider several points along the segment of the wavefront AB. Each point acts as if it is the source of a new wavelet, and each wavelet advances the same distance in the same time. The wavelets overlap and, where the leading edges of all the new wavelets join together, an envelope is formed tangent to the wavelets. This envelope is the new segment of the wavefront CD. Similarly, points on CD act as sources of new wavelets, causing the next new segment of the wavefront, EF, and so on.

Let us now apply Huygens' principle to plane waves passing by the sharp edge of a barrier. As shown in Figure 16-4, propagating the wavelets forward at the sharp edge, in the direction of wave motion, results in the observed bending around this obstruction. This bending effect is what we observe as diffraction.

In order to explain the light and dark regions in the diffraction patterns shown in Figure 16-2, we must look more closely at how the propagating wavelets in Figure 16-4 interfere to cause the more complex pattern. This will be explored further in the quantitative analysis of the diffraction of light through a narrow aperture (Section 16.3).

Huygens' principle is helpful in understanding qualitatively the diffraction and interference of light. Huygens was a contemporary of Isaac Newton, but Newton had difficulty accepting the theory of wavelets, especially in trying to explain the straight line motion of light. Newton was aware of Grimaldi's discovery of diffraction, but

DID YOU KNOW?

Christian Huygens (pronounced hoi′ guns) lived at a time when emperors and kings commissioned scientists to enhance the image of their court. Huygens worked in the French court of King Louis XIV, and his career was noted for several inventions and discoveries. Based on Galileo's discovery of the regular period of a pendulum, Huygens was the first to build an accurate pendulum clock. In 1673, he proposed that a standard unit of length be based on the length of a pendulum that has a period of exactly 2.0 s. (This length is about the same as 1.0 m.) Using telescopes of his own design, he discovered the rings of Saturn. Today he is remembered mostly for his work on wave optics.

suggested that if light acted truly like a wave, it would diffract much more than was observed, something like sound waves around corners or through open windows. Neither Newton nor Huygens realized just how small the wavelengths of visible light were, which is an important factor in the degree to which the waves diffract when encountering everyday objects. Thus, Newton favoured the particle model of light, and his followers stubbornly adhered to this model, until a very important discovery, discussed in the next section, was made.

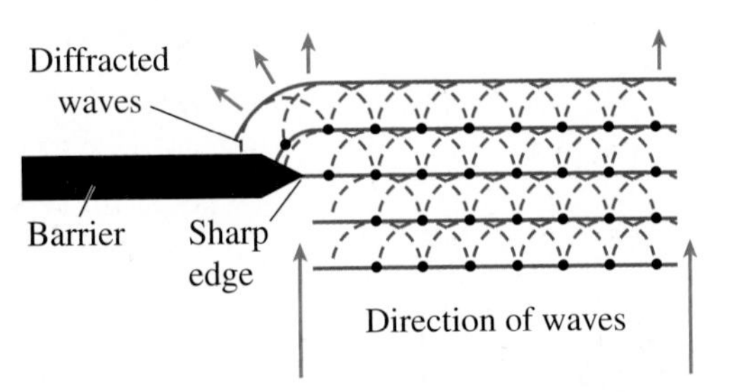

Figure 16-4 Applying Huygens' principle to plane waves passing a sharp edge.

TRY THIS!

"Illuminating" Animations of Huygens' Principle

Through a quick online search, explore some of the many animations available of Huygens' principle, especially the illustrations related to diffraction of waves.

EXERCISES

16-1 Under what condition(s) will the diffraction of Huygens' wavelets be at their maximum as the waves pass (a) around a barrier and (b) through an aperture? (**Note:** These concepts were presented for water waves in Section 13.8.)

16-2 Why do we not observe the diffraction of light through a window?

16-3 Like visible light, radio waves are part of the electromagnetic spectrum, which means they should display wave properties. Speculate on why radio waves undergo significant diffraction around most objects in our daily experience.[1]

16-4 Draw a sketch showing how Huygens' principle would be applied in each case:

(a) Waves of wavelength 1.5 cm are passing by a thin barrier that is 1.5 cm wide.

(b) Waves of wavelength 1.0 cm are passing through an aperture that is 1.0 cm wide.

16.2 Double-Slit Interference of Light

Interference is an important wave property, so if light has a wave nature, we should expect it to display interference patterns. (Interference of waves in one dimension was described in Section 13.7.) However, in your everyday experience, the chances are extremely slight that you will notice any interference of light. For example, imagine that you and a friend are looking at two different light sources in a room such that light from the sources must cross in order to reach your eyes. As illustrated in Figure 16-5, the paths of light may cross, but, as you know from experience, there is no observed interference: the light reaches your eyes completely undisturbed.

Despite this, it is simple nowadays to set up a demonstration of light interference. One way is to pass a laser beam through a set of two narrow slits separated by a tiny distance and observe the resulting interference pattern on a screen. The setup to view this "double-slit interference pattern" is illustrated in Figure 16-6 (a), and the resulting pattern of bright and dark regions, called fringes, is shown in (b).

Another technique for observing a double-slit interference pattern is to hold an opaque plate with two fine slits etched on it in front of one eye and look at an incandescent light source that has a straight filament. The slits and the filament should be parallel for maximum effect, as shown in Figure 16-7. The pattern has a bright fringe in the centre and, to the sides, bright fringes of spectral colours separated by dark fringes. The bright fringes are called maxima and the dark fringes are called minima.

The pattern described here is called *Young's double-slit interference pattern*, named after Thomas Young, an English physician and physicist. Between the years 1801 and

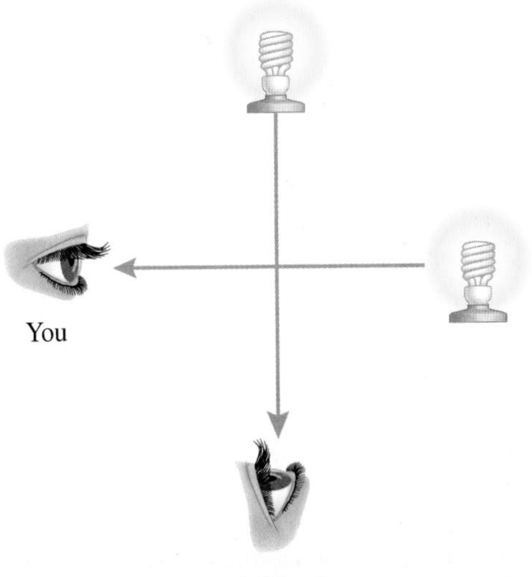

You

A friend

Figure 16-5 Although the paths of light cross, the observers see no effects of interference.

[1]The electromagnetic spectrum is presented in detail in Section 16.6.

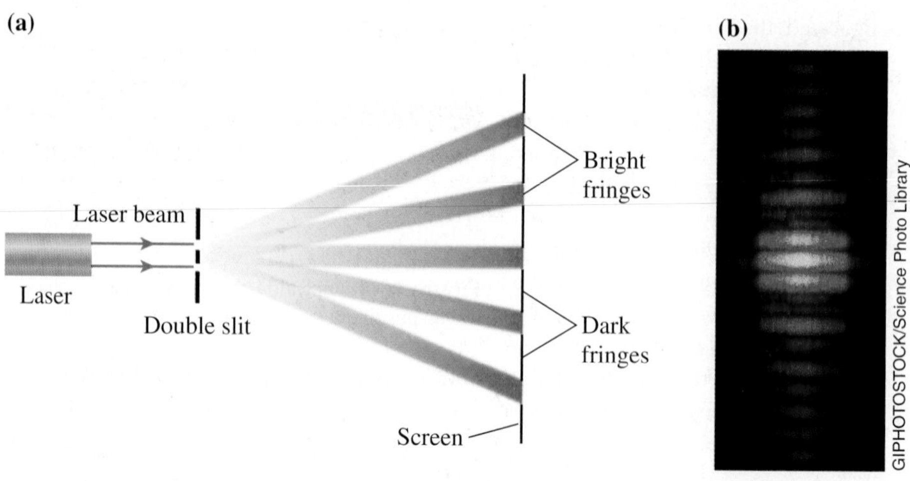

Figure 16-6 Viewing a double-slit interference pattern. **(a)** A representation of a laser beam passing through two close narrow slits, producing an interference pattern on a screen. **(b)** The double-slit interference pattern using laser light showing the bright and dark fringes on a screen.

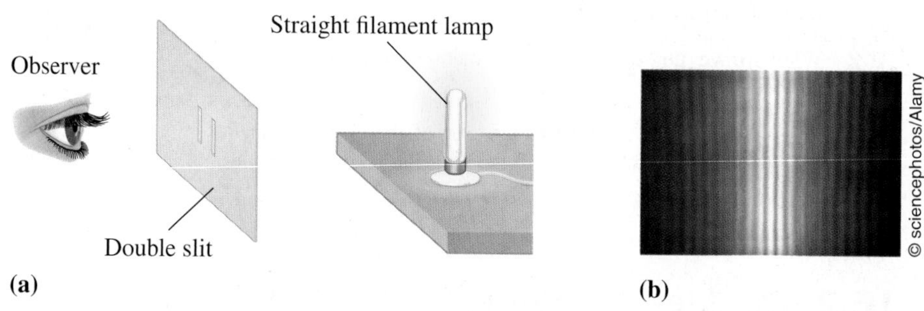

Figure 16-7 Using a straight filament to view a double-slit interference pattern. **(a)** The setup. **(b)** This pattern has a white central maximum and other coloured fringes.

1807, Young performed experiments using sunlight as the source (since it would be more than 70 years before the incandescent light bulb was invented). He had sunlight pass through a tiny pinhole that was in line with a set of two other pinholes that were extremely close together. Behind the double pinhole, he placed a screen on which he observed what occurred. If the light had been able to pass through the pinholes and continue in a straight line the way particles would be assumed to travel, the pattern on the screen would have been simply two bright spots. However, the pattern he observed was an interference pattern similar to a double-slit pattern, which is the pattern we will analyze mathematically.

PROFILES IN PHYSICS

Thomas Young (1773–1829): A Multi-Talented Scientist

From the start of his life, Thomas Young's achievements were most impressive (Figure 16-8). By age 4, he had read the Bible cover to cover twice. By age 14, he had learned eight languages, including the classical Latin and Greek, as well as Turkish, Hebrew, Persian, and Arabic. In university, he studied medicine, but he was able to devote much of his life to scientific research thanks to the financial security provided by an inheritance from an uncle.

Among his many scientific achievements are discoveries related to human vision. He discovered how the eye is able to focus on objects at different distances, the process called accommodation; he discovered the cause of the visual defect called astigmatism; and he proposed the theory of colour vision based on three types of receptors sensitive to red, green, and blue light. It was his study of human vision combined with his interest in the human voice that led to his experiment with the interference of light.

Young also researched the topics of heat, fluids, and mechanics. His discoveries regarding the elastic properties of materials are still recognized in the constant called Young's modulus, which is a characteristic property of materials, also known as the elastic modulus.

Figure 16-8 Thomas Young.

Sir Thomas Young MD, FRS, Briggs, Henry Perronet (1792–1844)/Royal Society, London, UK/Bridgeman Images

Young's interests went much beyond pure scientific research. He also found time to study Egyptian history, and he was among the first to decipher Egyptian hieroglyphics.

Young's discovery of the wave property of light, known as interference, was not immediately accepted in England, the country where Newton and his followers preferred the particle model. However, in continental Europe, especially in France, his work was soon hailed as an important turning point in the theory of light.

Observing the interference of water waves is helpful in understanding the interference of light waves. Figure 16-9 shows a simulation of the pattern created by two in-phase sources oscillating at a constant frequency in a tank with a shallow layer of water. Minima are produced wherever a crest from one source meets a trough from the other source. Maxima are produced wherever a crest from one source meets a crest from the other source, or where a trough from one source meets a trough from the other source.

To explain the creation of a double-slit interference pattern in terms of waves, consider Figure 16-10 (a). Light from a distant source (the Sun, for example) passes through the first slit, and Huygens' wavelets spread out from that slit toward the next slits. When a given wavelet arrives at the two slits, the wavelet's phase will be maintained as it passes through the slits. Thus, beyond the slits, there are now two new sets of Huygens' wavelets that can interfere with each other, causing maxima where the crests from one slit meet the crests from the other slit and minima between the maxima. From here on, the pattern resembles the two-source interference pattern observed in water (Figure 16-9), with maxima (the bright fringes) and minima or nodal lines (the dark fringes). Figure 16-10 (b) shows the pattern observed on the screen with the same labels for maxima ($M = 0, 1, 2, 3, \ldots$) and minima ($N = 1, 2, 3, \ldots$) used in the two-source interference pattern of water in Figure 16-9.

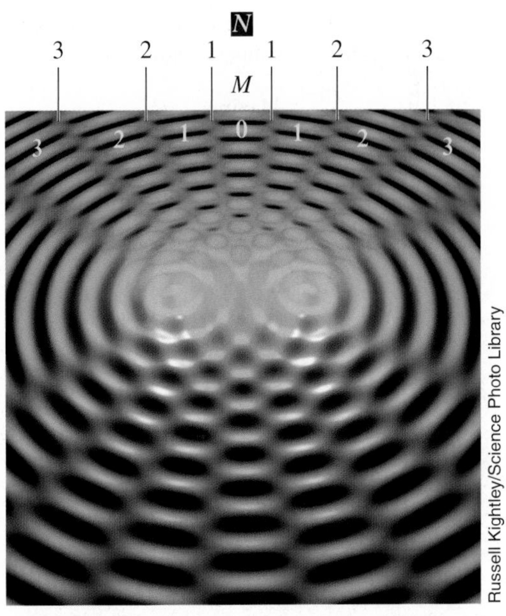

Russell Kightley/Science Photo Library

Figure 16-9 A two-source interference pattern of water waves helps to visualize how a two-source interference pattern of light is produced. The maxima, starting with the central maximum, $M = 0$, consist of very bright and very dark components, and the minima, starting with $N = 1$ on either side of the central maximum, consist of neutral components.

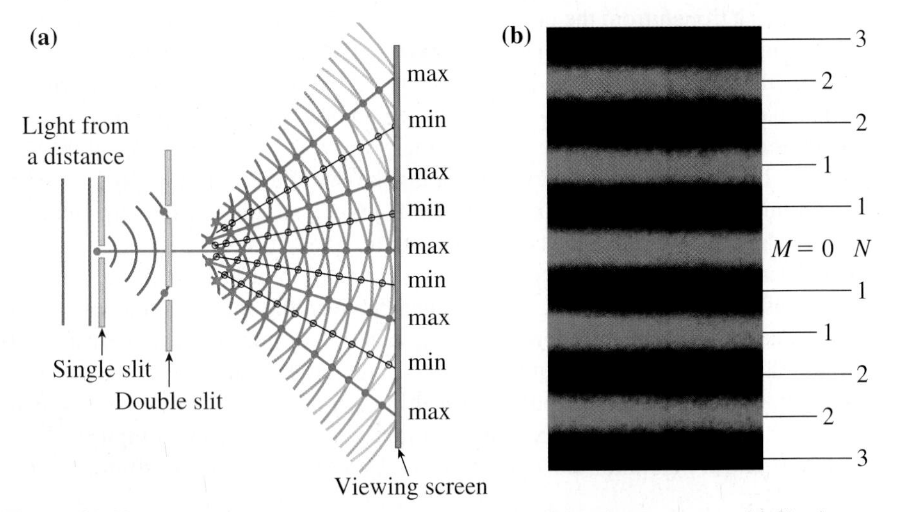

Figure 16-10 Using Huygens' wavelets to explain a double-slit interference pattern. **(a)** Circular wavelets originate at each slit. After the double slit, the wavelets interfere to produce maxima where the crests from one slit meet the crests from the other slit (the red dots) and minima between the maxima. **(b)** The interference pattern observed on a screen with maxima (M) and minima (N) labelled.

coherent waves waves that maintain their phase relationship and have the same wavelength

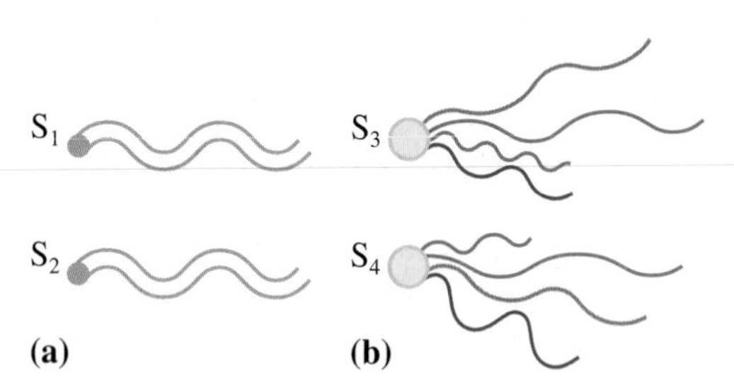

Figure 16-11 Comparing coherent and incoherent light sources. **(a)** Sources S_1 and S_2 are producing coherent light. **(b)** Sources S_3 and S_4 are producing incoherent light.

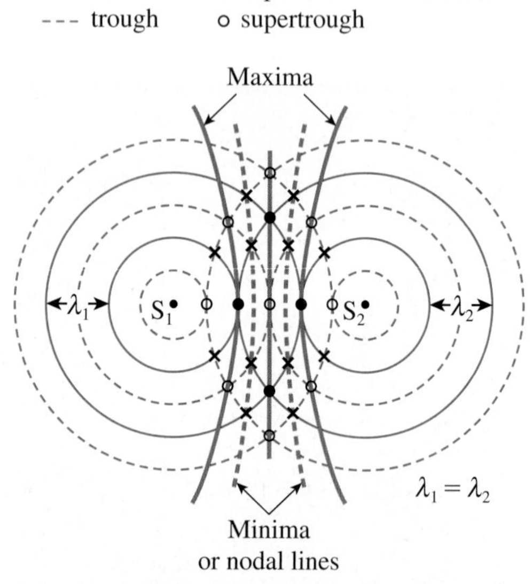

Figure 16-12 Examining how a two-source interference pattern in water is created.

Coherence of Light

We are now able to explain why the interference of light is normally not observed. When a two-source interference pattern is set up in water, the two sources have the same frequency and thus the same wavelength. Furthermore, they maintain the same phase relationship, which means that if they start out in phase they remain in phase, or if they start out 180° out of phase they remain that way. Waves that maintain their phase relationship and have the same wavelength are called **coherent waves**. All other waves are incoherent, and if incoherent waves are observed in water, there will be no distinguishable or constant pattern. See Figure 16-11.

All the interference patterns described in this section result from coherent light sources. Thomas Young achieved coherent light by passing light through a single pinhole. A double slit parallel to the straight filament of an incandescent lamp achieves the same results. And the unique light from a laser is coherent, a feature that makes laser light extremely useful.[2]

Light from ordinary sources, such as flames or light bulbs, is produced by the trillions of atoms and/or molecules that make up the source. The emission of light waves from such sources lasts for tiny fractions of a second, and there are so many millions of emissions per second that there is no way such light could be coherent. Thus, we normally do not observe the interference effects of light.

Analyzing Two-Source Interference Patterns

We begin the mathematical analysis of interference patterns by considering a two-source interference pattern in water. In Figure 16-12, the solid circles represent crests and the broken circles represent troughs, both of which are travelling away from the sources, S_1 and S_2. The wavelengths, λ_1 and λ_2, are equal because the frequencies of the sources are equal. Whenever a crest from S_1 meets a trough from S_2, destructive interference occurs and a *minimum* amplitude results. In this case, the minimum is a node because the amplitudes of the waves from the two sources are equal. A line of adjacent nodes in an interference pattern is called a **nodal line**. It is represented by the dashed green lines in Figure 16-12. Although nodal lines appear to be straight far away from the sources of waves, they are actually in the shape of a hyperbola.

nodal line a line joining adjacent nodes in a two-source interference pattern

Again considering Figure 16-12, whenever a crest from one source meets a crest from the other source, a supercrest occurs; in a water tank with light shining through the water, this type of maximum appears as a bright spot. Whenever a trough from one source meets a trough from the other source, a supertrough occurs; in water, this type of maximum is a dark spot. The maxima appear to travel away from the sources.

The mathematical analysis of a two-source interference pattern can be useful for finding unknown quantities, such as the wavelength of a wave. We begin this analysis by numbering the minimum lines, or nodal lines, $N_1, N_2, N_3, \ldots$ on both sides of the perpendicular bisector of an imaginary line joining the two sources (Figure 16-13). Similarly, the maximum lines are numbered $M_0, M_1, M_2, \ldots$, starting at the perpendicular bisector, which is also called the central maximum.

Consider the distance travelled by individual waves in order to produce destructive or constructive interference. The distance between the source and a position of interference is called the **path length**. At any position of interference, there are two path lengths, one from each source, and the magnitude of the difference between these path lengths is called the **path difference** (symbol *P.D.*). To discover the path difference along the various maximum and minimum lines, refer to Figure 16-14. Along the central maximum ($M = 0$), P_1 is a supercrest with a path difference of

path length the distance between the source and a position of a two-source interference pattern

path difference the magnitude of the difference between the path lengths from the two sources of an interference pattern; symbol *P.D.*

$$P.D. = |P_1S_1 - P_1S_2| = |3\lambda - 3\lambda| = 0$$

[2]Details about laser light are presented in Section 17.7.

Also, P_2 is a supertrough with $P.D. = |P_2S_1 - P_2S_2| = |4.5\lambda - 4.5\lambda| = 0$

In fact, for any position along the central maximum, the path difference is zero. Now, consider the next maximum line, $M = 1$, on both sides of the central maximum. At P_3, the path difference is

$$P.D. = |P_3S_1 - P_3S_2| = |4\lambda - 3\lambda| = 1\lambda$$

and at P_4,

$$P.D. = |P_4S_1 - P_4S_2| = |5\lambda - 4\lambda| = 1\lambda$$

Thus, along $M = 1$, the path difference is 1λ. Looking carefully at Figure 16-14, you should be able to prove to yourself that the following pattern exists.

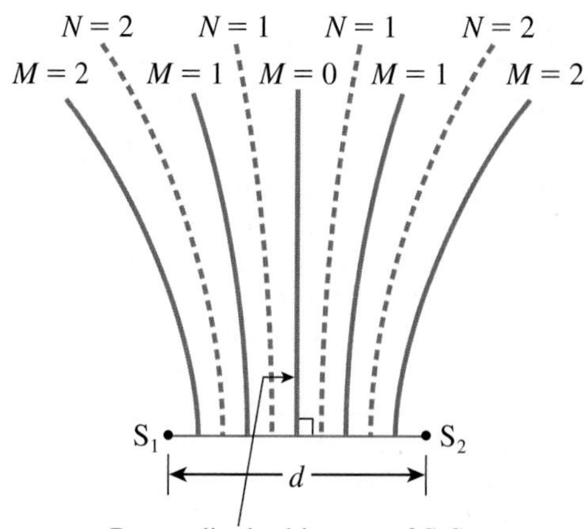

Figure 16-13 Minima and maxima in a two-source interference pattern. The central maximum is labelled $M = 0$, and the source separation is labelled d.

Maximum Line Number Path Difference

Maximum Line Number	Path Difference
$M = 0$	0
$M = 1$	1λ
$M = 2$	2λ
$M = 3$	3λ
↓	↓
$M = M$	$M\lambda$

Thus, we conclude that, for lines of constructive interference (maxima) in a two-source interference pattern, the path difference is

$$P.D. = M\lambda, \text{ where } M = 0, 1, 2, 3, \ldots \quad (16\text{-}1)$$

and λ is the wavelength of the waves producing the pattern.

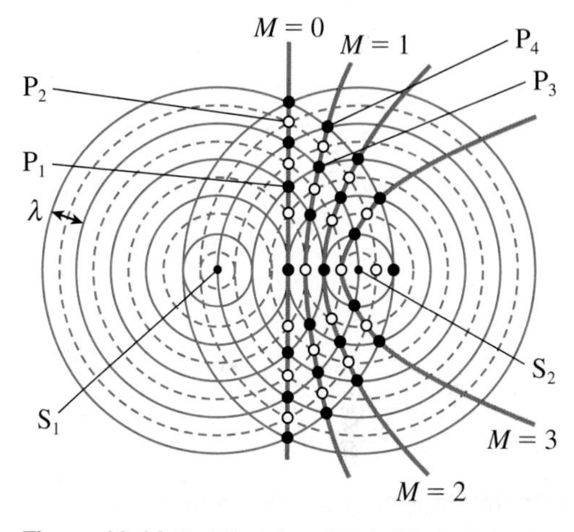

Figure 16-14 Path length and path difference in a two-source interference pattern. Only the central maximum and the maxima to the right are shown here.

Similar arguments can be used to determine the path difference along any of the lines of interference called minima, or nodal lines, $N = 1$, $N = 2$, etc. It is left as an exercise to prove that, *for lines of destructive interference (nodal lines or minima) in a two-source interference pattern, the path difference is*

$$P.D. = (N - ½)\lambda, \text{ where } N = 1, 2, 3, \ldots \quad (16\text{-}2)$$

SAMPLE PROBLEM 16-1

Point X is a nodal point along the second nodal line in a two-source interference pattern in a water tank. The path lengths from X to the sources are 28.5 cm and 31.3 cm. What is the wavelength of the water waves producing the pattern?

Solution

For a nodal line (Eqn. 16-2), $P.D. = (N - ½)\lambda$

$$\therefore \lambda = \frac{P.D.}{(N - ½)} = \frac{|28.5\,\text{cm} - 31.3\,\text{cm}|}{2 - ½} = 1.9\,\text{cm}$$

Thus, the wavelength is 1.9 cm.

Learning Tip

The diagrams and equations for the two-source interference patterns discussed in this section are for the special circumstance when the sources are in phase. Both the pattern and the equations that will be derived change if the frequencies are different or the sources are not oscillating in phase. For example, if the two sources have identical frequencies but are exactly opposite in phase, the perpendicular bisector of the line joining the sources is a minimum, *not* a maximum. Furthermore, if the sources have quite different frequencies, no distinguishable pattern is noticeable.

TRY THIS!

Young's Experiment on the Net

There are several equations and static diagrams in this section that are easier to understand and visualize if you view diagrams and animations on the Internet. Find sites related to *Young's double-slit experiment* or *Young's interference experiment*. Relate what you find to the diagrams and equation derivations in this section. Also, compare the patterns created using single-colour light (such as red laser light) and white light.

Two-Source Interference Far from the Sources

For an interference pattern in water, the path difference equations can be used to find the wavelength fairly accurately. However, the equations become less useful when studying other types of waves, such as light, where the separation of the sources becomes very small and the distance to the maximum or minimum points becomes relatively large. Then, the path difference equations are used to derive equations involving other quantities besides M, N, and λ, as shown next.

Consider sources S_1 and S_2 separated by a distance d, and an interference point P at a distance much greater than d from the sources (Figure 16-15). If we draw point A on the path line PS_1 so that $PA = PS_2$, then the path difference of the two paths is $|PS_1 - PS_2| = AS_1$, as seen in the diagram. Now, if PS_1 and $PS_2 \gg d$, then figure AS_1S_2 is essentially a right-angled triangle that contains the angle $\theta' = AS_2S_1$ such that $\sin\theta' = AS_1/S_1S_2$ or

$$\sin\theta' = \frac{\text{path difference}}{d}$$

Substituting the path difference equations (Eqns. 16-1 and 16-2) into this last equation, we get

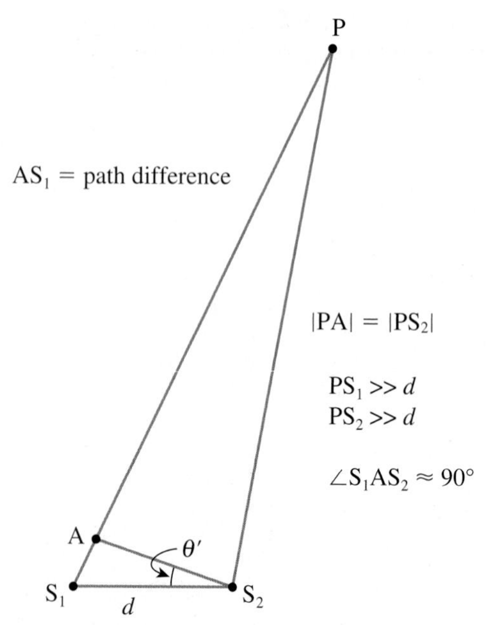

Figure 16-15 Another look at path difference in a two-source interference pattern.

$$\text{For maxima:} \quad \sin\theta' = \frac{M\lambda}{d}, \text{ where } M = 0, 1, 2, 3, \ldots \qquad \textbf{(16-3)}$$

$$\text{For minima:} \quad \sin\theta' = \frac{(N - \frac{1}{2})\lambda}{d}, \text{ where } N = 1, 2, 3, \ldots \qquad \textbf{(16-4)}$$

Let us look at the physical meaning of the equation $\sin\theta' = M\lambda/d$ by rearranging the equation to obtain $M = d\ \sin\theta'/\lambda$. This equation certainly is true when $\theta' = 0°$, in which case $\sin\theta' = 0$ and $M = 0$, so the maximum line is the central maximum. The maximum value of $\sin\theta'$ is 1 (when $\theta' = 90°$), in which case the value of M varies directly as the source separation d and inversely as the wavelength λ. This can be verified by observing a two-source interference pattern while varying d and λ, one at a time.

For large distances from the sources, the angle θ' in Figure 16-15 is so small that it would be impossible to measure it directly. Thus, to apply the equations involving $\sin\theta'$ we must find another expression for it that can be measured easily. Another diagram is drawn, this time with point P as far away as possible on the page (Figure 16-16). The perpendicular distance from P to the central maximum line is given the symbol x, and the distance from P to the midpoint of the line joining the sources is given the symbol L. The information in Figure 16-16 (a) proves that, for large L and small d values, the angle θ very closely equals the angle θ'. Therefore, $\sin\theta = x/L$, $\theta = \theta'$, and $\sin\theta' = M\lambda/d$, and we can write two important equations involving measurable quantities in a two-source interference pattern.

Math Tip

Angles in interference pattern equations are usually measured in degrees, so if your calculator is set on radians, change the setting to degrees for problems involving interference patterns.

Thus, for maxima:

$$\frac{x}{L} = \frac{M\lambda}{d}, \text{ where } M = 0, 1, 2, 3, \ldots \qquad \textbf{(16-5)}$$

and for minima or nodal lines,

$$\frac{x}{L} = \frac{(N - \frac{1}{2})\lambda}{d}, \text{ where } N = 1, 2, 3, \ldots \qquad \textbf{(16-6)}$$

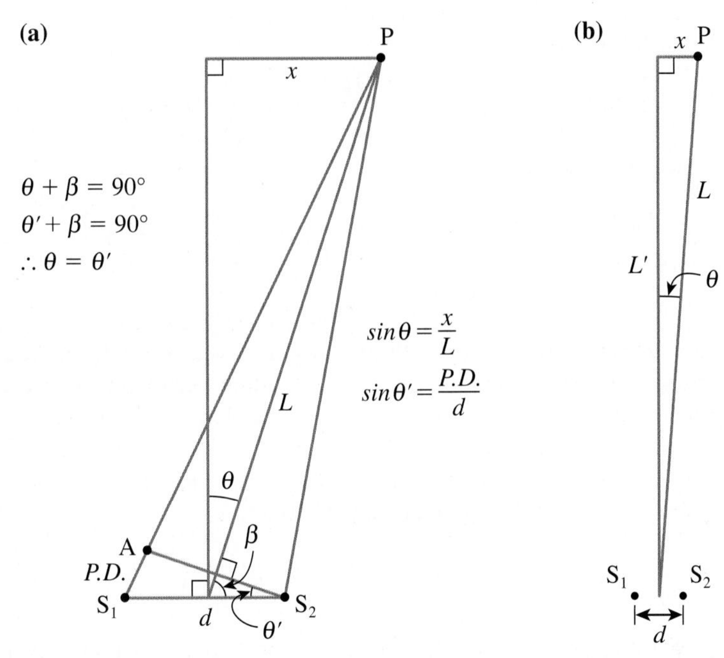

Figure 16-16 (a) Developing equations for a two-source interference pattern. **(b)** More realistic dimensions for the equations. Notice that for small angles of θ, the distance L used in the equation derivations is essentially equal to L', the distance from the sources to the screen.

Figure 16-16 (b) shows a more realistic diagram involving these equations, and the next sample problem reveals how the equations are applied to find the wavelength of the waves producing an interference pattern.

SAMPLE PROBLEM 16-2

Two in-phase sources of water waves, separated by 6.2 cm, produce an interference pattern. A student measures the distance from the point halfway between the sources to a position on the second maximum to be 34 cm. If the perpendicular distance from the same position to the central maximum line is 11 cm, what is the wavelength of the waves producing the pattern?

Solution

We are given that $d = 6.2$ cm, $L = 34$ cm, $x = 11$ cm, and $M = 2$.

$$\text{From} \quad \frac{x}{L} = \frac{M\lambda}{d},$$

$$\lambda = \frac{xd}{ML} = \frac{(11\,\text{cm})(6.2\,\text{cm})}{(2)(34\,\text{cm})} = 1.0\,\text{cm}$$

Thus, the wavelength is 1.0 cm.

Measuring the Wavelengths of Visible Light

We now can apply the equations derived for the two-source interference pattern of water waves to light. For convenience, let us summarize the important equations related to the variables shown in Figures 16-13 and 16-16.

Math Tip

Notice that the answer to Sample Problem 16-2 has two significant digits even though the M value has only one digit. Recall that M is a whole number and thus is not considered in our final determination of significant digits.

For maxima or bright fringes:

$$\sin\theta = \frac{M\lambda}{d}; \quad \sin\theta = \frac{x}{L}; \quad \frac{x}{L} = \frac{M\lambda}{d}, \text{ where } M = 0, 1, 2, \ldots \qquad \textbf{(16-3)}$$

For minima or dark fringes:

$$\sin\theta = \frac{(N - \frac{1}{2})\,\lambda}{d}; \quad \sin\theta = \frac{x}{L}; \quad \frac{x}{L} = \frac{(N - \frac{1}{2})\,\lambda}{d}, \text{ where } N = 1, 2, 3, \ldots \qquad \textbf{(16-4)}$$

In Sample Problem 16-3, we apply these concepts to see how Thomas Young succeeded in being the first person to perform the important measurements of the wavelengths of visible light. As you will discover, the wavelengths of visible light range from approximately 4×10^{-7} m to about 7×10^{-7} m. In most cases, we will convert these wavelengths to nanometres ($1.0 \text{ nm} = 1.0 \times 10^{-9}$ m), so the range of wavelengths is from approximately 400 nm to about 700 nm.

SAMPLE PROBLEM 16-3

Light passing through a double slit produces an interference pattern on a screen located 1.80 m from the slits. The slits are separated by a distance of 0.20 mm, and the distance from the middle of the central maximum to the middle of the first bright fringe is 5.1 mm (Figure 16-17). Determine the wavelength of the waves producing this part of the fringe. (This wavelength is the average wavelength of white light, as found originally by Young.)

$M = 1$ 5.1 mm $M = 0$ $M = 1$ $d = 0.20$ mm

Figure 16-17 Sample Problem 16-3.

Solution

We apply Eqn. 16-3, $\frac{x}{L} = \frac{M\lambda}{d}$, to isolate the wavelength, λ, and substitute using $M = 1$ and consistent units.

$$\lambda = \frac{xd}{ML} = \frac{5.1\,\text{mm} \times 0.20\,\text{mm}}{1 \times 1.80\,\text{m}} = \frac{5.1 \times 10^{-3}\,\text{m} \times 2.0 \times 10^{-4}\,\text{m}}{1.80\,\text{m}} = 5.7 \times 10^{-7}\,\text{m}$$

Therefore, the average wavelength of the white light is 5.7×10^{-7} m, or 5.7×10^{2} nm.

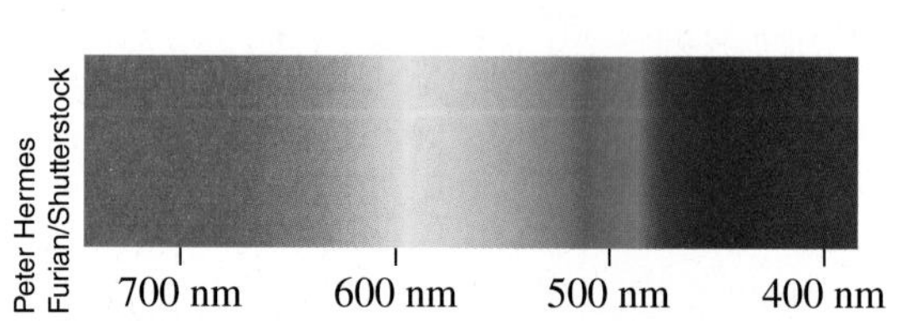

Figure 16-18 Colours of the visible spectrum.

Young's value of 570 nm for the average wavelength of white light is relatively close to the currently accepted value of 555 nm. From the observations found in the double-slit interference pattern, it is evident that different colours of light have different wavelengths. The range of light colours from red to violet visible to the human eye is called the *visible spectrum* (Figure 16-18). Since $x \propto \lambda$ and x is smaller for colours at the violet end of the spectrum, we conclude that violet light has short wavelengths (in the region of 400 nm) and red light has longer wavelengths (in the region of 700 nm). The wavelength ranges of the colours of the visible spectrum are given in Table 16-1.

Table 16-1

Wavelengths of the Visible Spectrum

Colour	Range of Wavelengths (nm)
Red	610 to 700
Orange	590 to 610
Yellow	570 to 590
Green	500 to 570
Blue	450 to 500
Violet	400 to 450

all values are approximate

The diameter of a human hair is approximately 100 times the average wavelength of visible light. Considering such short wavelengths, it is little wonder that early scientists had difficulty in finding evidence of the wave nature of light. Shortly after Young's discovery of the interference effects of light and his measurements of the wavelengths of light however, scientists who had favoured the particle model of light began to agree with scientists who favoured the wave model. But Young's discovery did not provide all the answers. Like so many other discoveries, it presented more questions: for example, if light is a wave, does it require a medium through which to travel, and is it a transverse or longitudinal wave? We will explore these questions later in the chapter.

EXERCISES

16-5 State how the first variable depends on the second variable.

(a) the *spacing* of the minima and maxima of a double-slit interference pattern and the *wavelength* of the waves producing the pattern

(b) the *spacing* of the minima and maxima and the *distance* from the slits to the screen

(c) the *sine of the angle of separation* of the minima and maxima and the *slit separation*

16-6 Use the two-source interference pattern in Figure 16-14 to prove that the path difference for minima (nodal lines) is $P.D. = (N - \frac{1}{2})\lambda$, where $N = 1, 2, 3, \ldots$

16-7 Two sources of circular waves in a water tank are producing waves with a wavelength of 2.8 cm.

(a) Determine on which maximum line the path difference is 8.4 cm.

(b) Determine on which nodal line the path difference is 7.0 cm.

16-8 Determine the unknown quantity in each case.

(a) $P.D. = 2.43 \times 10^{-6}$ m; $N = 5$; $\lambda = ?$

(b) $L = 4.55$ m; $N = 3$; $\lambda = 475$ nm; $d = 1.90 \times 10^{-4}$ m; $x = ?$

(c) $L = 3.50$ m; $M = 7$; $d = 2.00 \times 10^{-4}$ m; $x = 5.64$ cm; $\lambda = ?$

(d) $L = 3.85$ m; $x = 3.43$ cm; $\lambda = 547$ nm; $d = 0.215$ mm; $N = ?$

16-9 Light from a sodium lamp is used to produce a double-slit interference pattern on a screen that is 2.45 m from the slits. The fifth bright line of the sodium spectrum is located 3.00 cm to one side of the central maximum. The slit separation is 2.40×10^{-4} m. Determine

(a) the angle of separation between the central maximum and the fifth bright line

(b) the wavelength (in nanometres) of the light giving rise to this fifth bright line

(c) the colour of the light (Reference: Table 16-1)

16-10 White light passing through two slits separated by 0.180 mm produces an interference pattern on a screen that is 4.00 m away. Determine the width (in centimetres) of the spectrum from violet (4.00×10^2 nm) to red (7.00×10^2 nm) for the third maximum to one side of the central maximum.

16-11 (a) Derive an equation for the distance between adjacent maxima (Δx) in a double-slit interference pattern in terms of L, d, and λ.

(b) In an interference pattern produced by light of wavelength 660 nm passing through a double-slit of separation 0.21 mm, the distance from the middle of $M = 1$ to the middle of $M = 6$ is 9.5 cm. Determine the distance between the double slit and the screen.

16-12 A laser beam with light of wavelength 638.2 nm is used to check a manufacturer's claim that a double slit has a separation of 0.200 mm. The distance to the screen is 3.00 m and the measured distance from the central maximum to the middle of the fourth minimum is 33.2 mm. What is the measured slit separation?

16-13 Determine the greatest number of maximum lines in a two-source interference pattern when the ratio of the source separation to the wavelength is 4:1.

16.3 Single-Slit Diffraction and Resolution

In the double-slit interference of light, each slit causes waves to diffract outward as if originating from a point source of circular waves. In order to have a large amount of diffraction, the width of each slit, w, must be approximately equal to or less than the wavelength of the diffracting waves. Another way of saying this is $\frac{w}{\lambda} \leq 1$ or $w \leq \lambda$. In this section, we consider a different example of diffraction, namely, the diffraction that occurs when light passes through a single slit whose width is much larger than the wavelength of the light; that is, $\frac{w}{\lambda} \gg 1$ or $w \gg \lambda$. Although the width is much larger than the wavelength, it is still very small in terms of everyday measurements. For example, a slit of width 0.025 mm, which is less than the diameter of a human hair, is still about 50 times larger than the average wavelength of visible light. We will begin with single-slit diffraction of monochromatic (single colour) light, as shown in Figure 16-19.

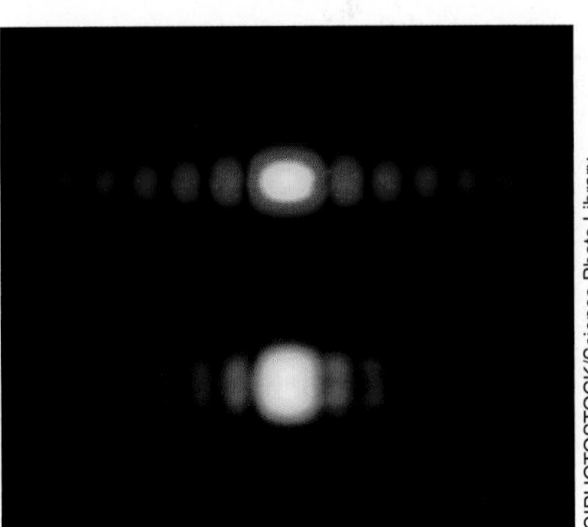
GIPHOTOSTOCK/Science Photo Library

Figure 16-19 Single-slit diffraction patterns created by red and green laser lights. How does the spacing of the green pattern compare to that of the red pattern? What does that indicate about the relative wavelengths of the red and green lights?

A Qualitative Look at Single-Slit Diffraction

Figure 16-20 shows several features related to single-slit diffraction. In (a), the pattern on the screen shows what we would expect to see if light travelled in a straight line through the slit, without experiencing diffraction. The illustrations in (b) and (c) show what actually happens when monochromatic light diffracts through a single slit.

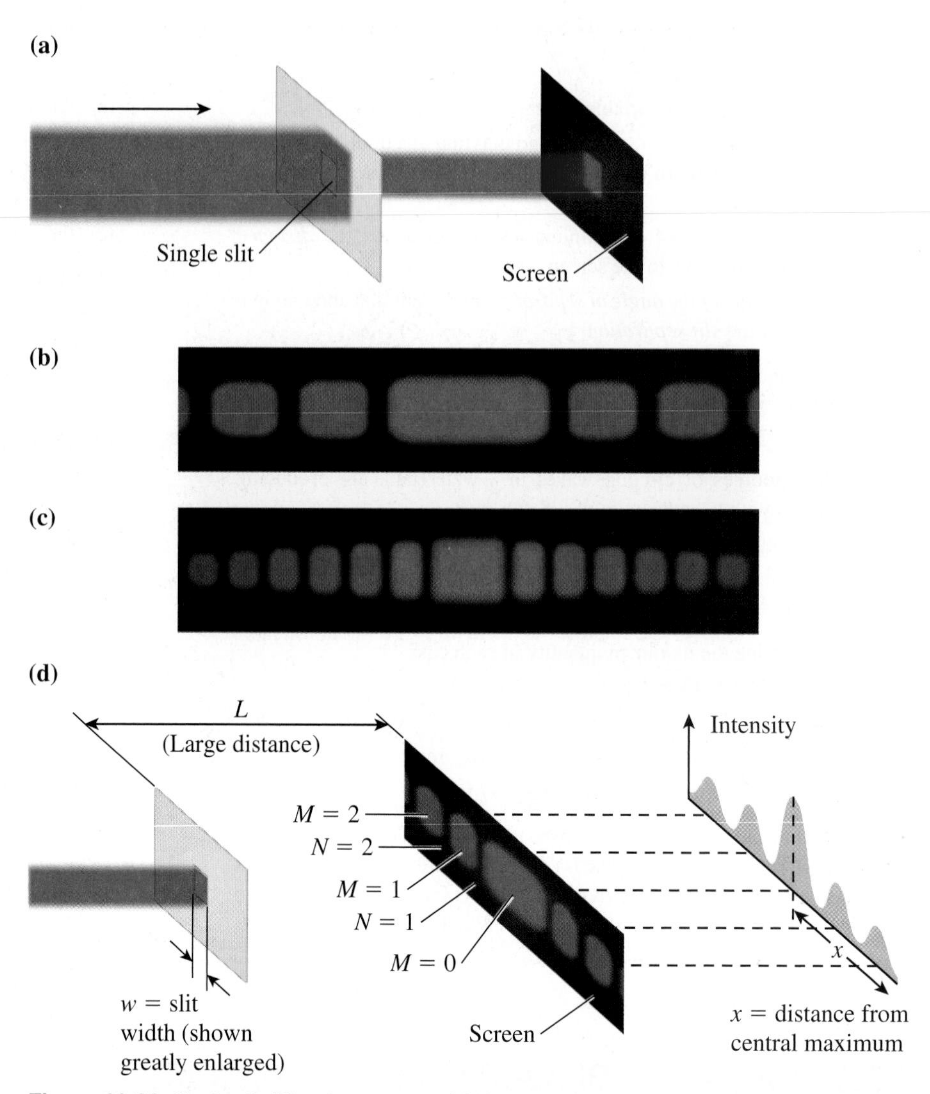

Figure 16-20 Single-slit diffraction patterns. **(a)** Without diffraction, the image on the screen would look like the slit. **(b)** One example of a single-slit diffraction pattern. **(c)** Another example created by a slit that is wider than in (b). **(d)** Single-slit diffraction terminology as well as the relationship between the intensity of the pattern and the distance from the middle of the central maximum.

A bright central maximum is surrounded by dark fringes, followed by alternating bright and dark fringes. Can you judge which of these two patterns was produced by the narrower slit? The diagram in (d) shows another diffraction pattern coupled to an illustration of the intensity of the pattern as a function of the distance from the middle of the central maximum. It also illustrates some of the terminology associated with single-slit diffraction. Refer also to Figure 13-55 on page 406, which shows the diffraction of water waves through a single aperture. By comparing the illustrations of diffraction through a single slit, it is evident that diffraction supports the wave theory of light.

Notice in Figure 16-20 (d) that the distance between the single slit and the screen is large. This means that the paths of the light wavelets that produce the central maximum are parallel (or nearly parallel) to each other. If the screen is closer, this parallelism can be achieved by using a converging lens between the slit and the screen. In either case, the type of diffraction in which the paths of the light wavelets are parallel is called **Fraunhofer diffraction**. It is named after the first person who explained the pattern, Josef Fraunhofer (1787–1826), a German optician. It is this type of diffraction that we

Fraunhofer diffraction the type of diffraction in which the paths of the Huygens' light wavelets are parallel

will analyze mathematically. (Another type of diffraction, called *Fresnel diffraction*, occurs when the paths of the light wavelets between the lens or slit and the screen or observer are not parallel. Fresnel diffraction is named after Augustin Fresnel (1788–1827), a French physicist. Fresnel diffraction is more complex mathematically than Fraunhofer diffraction.)

Quantitative Analysis of a Single-Slit Diffraction Pattern

A convenient way to analyze a single-slit diffraction pattern is to apply Huygens' principle to the light striking the slit. Consider a wavefront of monochromatic light just entering a slit of width w, as shown in Figure 16-21 (a). The wavefront can be replaced with a series of sources of new wavelets, all of which are in phase, as in (b). We will number these "Huygens' sources" from 1 to 8 (although any convenient number of sources may be used). If we follow the wavelets near the middle of the slit, we see that the wavefronts at the screen are at a maximum. The resulting bright fringe is a *central maximum*. Next, consider the diagram in (c), in which a path of the wavelets makes an angle θ to the initial path of light. The size of the angle is such that the path difference ($P.D.$) between source 4 and source 8 is half the wavelength of the light. This means that the wavelets from 4 and 8 will arrive at the screen out of phase, resulting in destructive interference. Similarly, wavelets 1 and 5 undergo destructive interference at the screen, as do 2 and 6, and 3 and 7. In this way, all the wavelets from one half of the slit cancel all the wavelets from the other half, producing a node or minimum fringe at the location shown.

The dimensions in Figure 16-21 (c) are greatly exaggerated; in reality, w is much smaller than L. Thus, the angles labelled θ in the diagram (c) are essentially equal, and from the small triangle near the slit, $\sin\theta = \dfrac{P.D.}{w/2}$

For the first node or minimum ($N = 1$), the path difference is $\lambda/2$. Therefore

$$\sin\theta = \frac{\lambda/2}{w/2}$$

$$\text{or} \quad \sin\theta = \frac{\lambda}{w}$$

for the first minimum fringe (for which $N = 1$) in a single-slit diffraction pattern. From this equation, we can see that $\sin\theta$, and thus θ, increases as the wavelength increases, and also increases as the width decreases. Therefore, as the wavelength becomes longer or the slit becomes narrower, the diffraction pattern spreads out more.

To locate the next node or minimum fringe in terms of the angle θ, let us consider the eight Huygens' sources as sets of two. As illustrated in Figure 16-22, 1 and 3 have a path difference of $\lambda/2$ and thus cancel; likewise, cancellation occurs for sources 6 and 8 (as shown), as well as other sets of two. Thus, for $N = 2$, $\sin\theta = \dfrac{P.D.}{w/4}$

For the second node or minimum ($N = 2$), the path difference is again $\lambda/2$. Therefore

$$\sin\theta = \frac{\lambda/2}{w/4}$$

$$\text{or} \quad \sin\theta = \frac{2\lambda}{w}$$

To locate the node $N = 3$, we can divide the wavefront at the slit into six sets of sources such that destructive interference occurs at the screen. In this case, $\sin\theta = \dfrac{3\lambda}{w}$.

Thus, in general, the minimum fringes of a single-slit diffraction pattern occur when

(a)

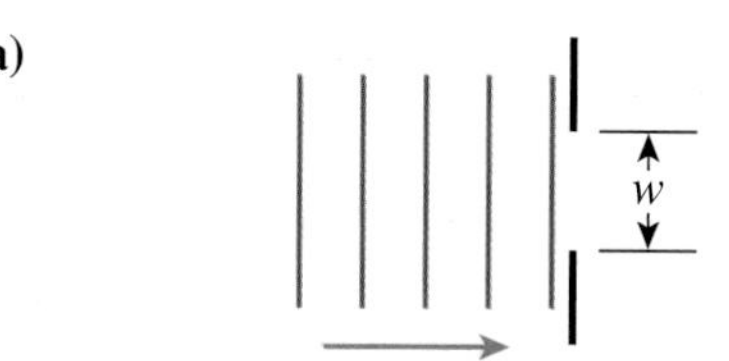

(b)

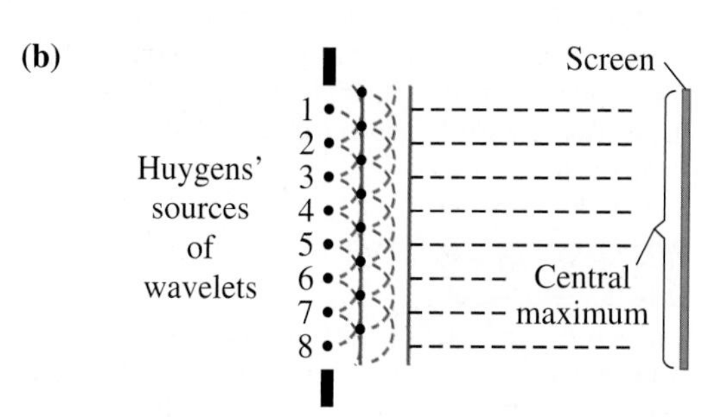

(c)

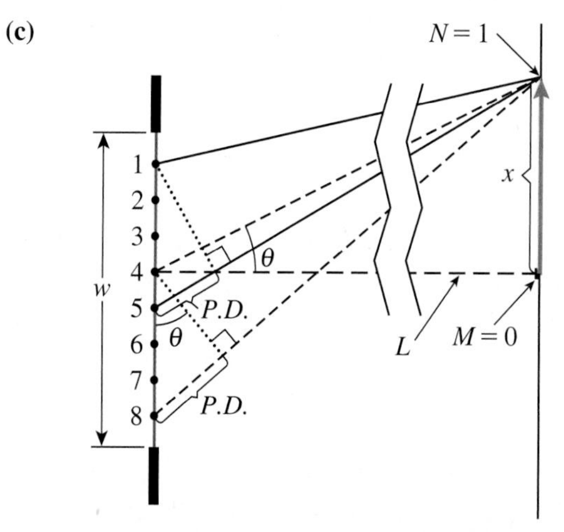

Figure 16-21 Using Huygens' wavelets to explain single-slit diffraction. **(a)** Monochromatic light enters a single slit. **(b)** The wavelets that travel straight reinforce each other and cause a maximum. (Diffracted wavelets are not shown here.) **(c)** The first minimum fringe ($N = 1$) occurs when pairs of wavelet sources separated by $w/2$ have a path difference ($P.D.$) of $\lambda/2$ and thus cancel. In actual situations, L is much larger than x or w.

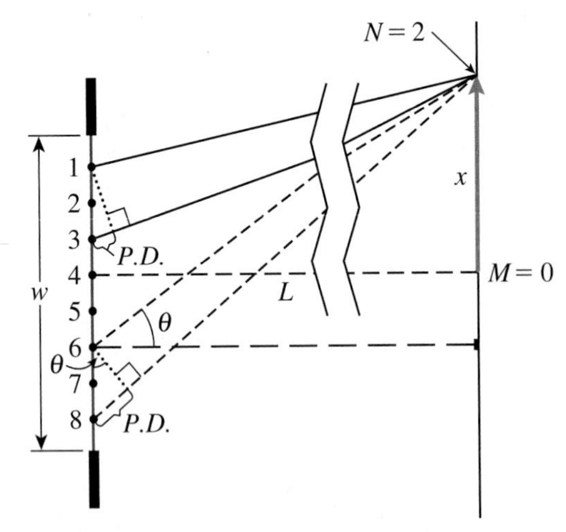

Figure 16-22 Locating the second node or minimum fringe. In realistic situations, the lines drawn from points 1, 3, 6, and 8 toward $N = 2$ are essentially parallel because L and x are so much larger than w.

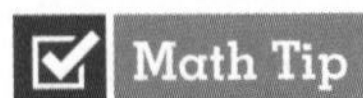
Math Tip

For very small angles, $\tan\theta$ and $\sin\theta$ are essentially equal. Use your calculator to verify this fact then draw a sketch of a right-angled triangle with one small angle to help you visualize why the equality exists.

$$\sin\theta = \frac{N\lambda}{w} \quad \text{where } N = 1, 2, 3, \ldots \qquad \textbf{(16-7)}$$

Finally, in Figure 16-22, we see that if x is the distance from the middle of the central maximum to a minimum fringe and L is the distance from the slit to the screen, then $\tan\theta = x/L$. But $x \ll L$, and therefore θ is small and $\tan\theta = \sin\theta$. Thus, the following relationship applies to the *minimum fringes* in a single-slit diffraction pattern:

$$\frac{x}{L} = \frac{N\lambda}{w}, \text{ where } N = 1, 2, 3, \ldots \qquad \textbf{(16-8)}$$

when $x \ll L$ or $L \gg x$.

Notice that we have derived an equation (Eqn. 16-8) for minima but not for maxima in a single-slit diffraction pattern. Experimentally, the minima are dark, narrow bands that are quite obvious. However, the maxima are relatively bright and wide. In each of these wide regions there is a point where the intensity is maximum, but it is not in the centre of the region, and the mathematics to determine the position of each maximum is complicated.

SAMPLE PROBLEM 16-4

Monochromatic light of wavelength 570 nm is aimed at a single slit that is 0.15 mm wide.

(a) Determine the width of the central maximum on a screen located 3.2 m away from the slit.

(b) Determine the angle at which the third minimum fringe occurs.

Solution

(a) The width of the central maximum can be determined by calculating the distance between the first minima on each side of the central maximum. From

$$\frac{x}{L} = \frac{N\lambda}{w},$$

$$x = \frac{LN\lambda}{w} = \frac{(3.2\,\text{m})(1)(5.7 \times 10^{-7}\,\text{m})}{1.5 \times 10^{-4}\,\text{m}} = 1.2 \times 10^{-2}\,\text{m}$$

Therefore, each of the minima on either side is 1.2×10^{-2} m from the centre of the diffraction pattern. Thus, the central maximum is 2.4×10^{-2} m wide, from one first minimum to the other.

(b) Using Eqn. 16-7,

$$\sin\theta = \frac{N\lambda}{w} = \frac{(3)(5.7 \times 10^{-7}\,\text{m})}{1.5 \times 10^{-4}\,\text{m}} = 0.0114$$

$$\therefore \quad \theta = \sin^{-1}(0.0114) = 0.65^\circ$$

Thus, the third minimum fringe occurs at an angle of 0.65°.

In the derivation of the equations involving single-slit diffraction, it was observed that the wavelets of light experienced interference. Diffraction and interference both result from the superposition of waves, but they are somewhat different properties of waves. However, when waves pass through apertures or around obstacles, both interference and diffraction occur simultaneously. In fact, a discussion of the double-slit interference pattern is more complete if it includes diffraction effects through each of the two slits. Figure 16-23 illustrates how diffraction affects the pattern observed in double-slit interference of monochromatic light.

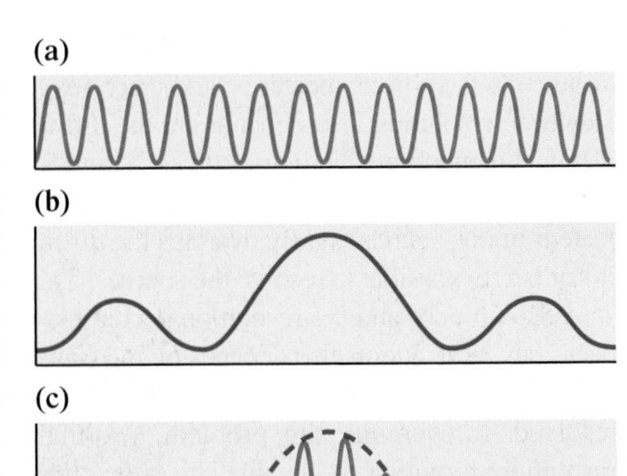

Figure 16-23 The intensity of the maxima in a double-slit interference pattern is affected by the diffraction of light through the individual slits. **(a)** Without diffraction the intensity of each maximum in the double-slit pattern would be the same. **(b)** The diffraction pattern produced by each slit. **(c)** The observed double-slit pattern is a combination of interference and diffraction.

Resolution

Imagine seeing a car with its headlights on approaching from a long distance away on an otherwise dark highway. From a far distance, the two headlights appear to be a single light source; they do not appear as separate, or "resolved," objects. At some much closer distance, they begin to appear to be separate light sources, and at closer distances, the sources are totally resolved. **Resolution** is the ability to distinguish two or more objects as separate entities (Figure 16-24).

(a)

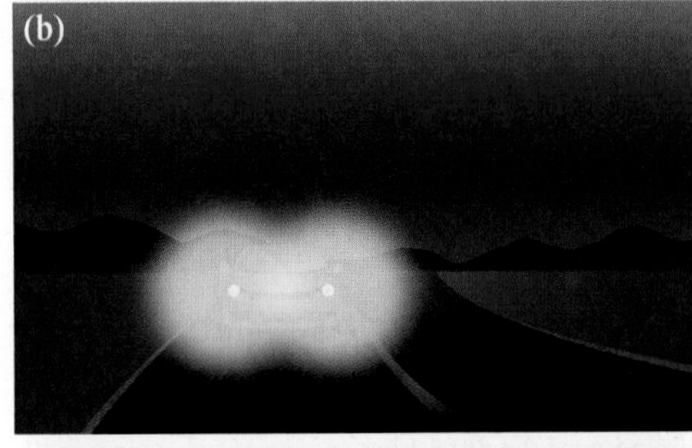

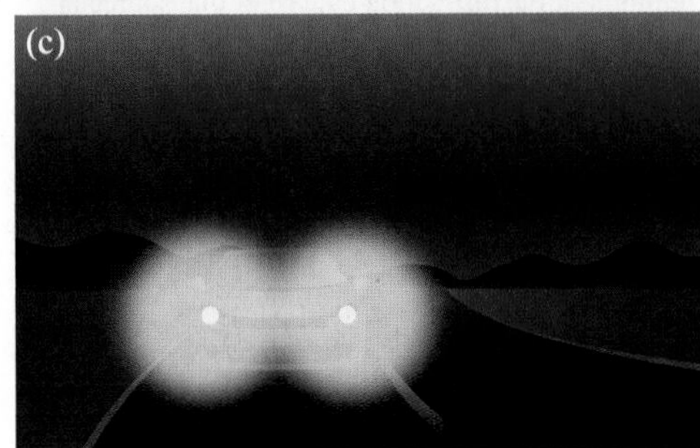

Figure 16-24 Resolving two headlights. **(a)** At a large distance, the two lights are unresolved, so they appear as one. **(b)** At some closer distance, the two lights are just resolved. **(c)** At an even closer distance, the two lights are distinctly separate.

resolution the ability to distinguish two or more objects as separate entities

Testing Resolution

Position your textbook vertically so you can view the pattern in Figure 16-25 from several metres away. Determine the maximum distance from the pattern at which you are just able to resolve the parallel bars using

- both eyes
- your left eye only
- your right eye only

If you wear eyeglasses or contact lenses, repeat the tests without them. Compare your resolving ability with that of your peers.

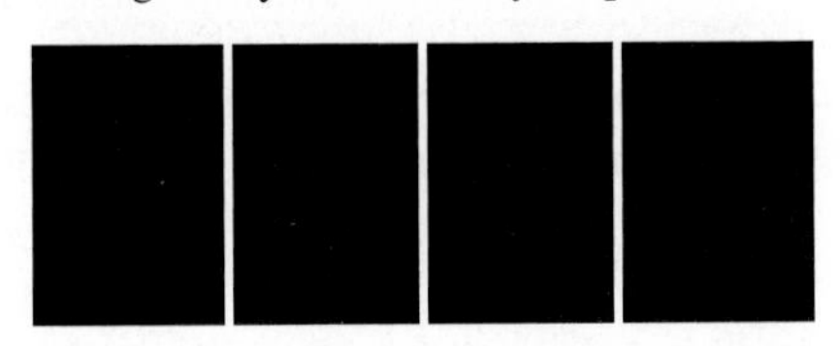

Figure 16-25 Pattern used to test your ability to resolve objects.

Resolution refers not only to emitted light from sources, but also reflected light from opaque objects. Consider, for example, some of the paintings of the French Impressionist movement. An artist brushed dabs of paint onto a canvas so that, from a close distance, the dabs appeared separate from each other; in other words, they were resolved. However, from a few metres or more away, depending on the separation of the dabs of paint, the dabs are not resolved, and an impression of a continuous colour pattern is seen (Figure 16-26).

Figure 16-26 The French impressionist artist Georges Seurat (1859–1891) used a "pointillism" style in which numerous dots on the screen were about 2 mm in diameter. When viewed from a distance, the individual dots cannot be resolved, allowing the viewer to see the intended image titled "Gray Weather, Grande Jatte."

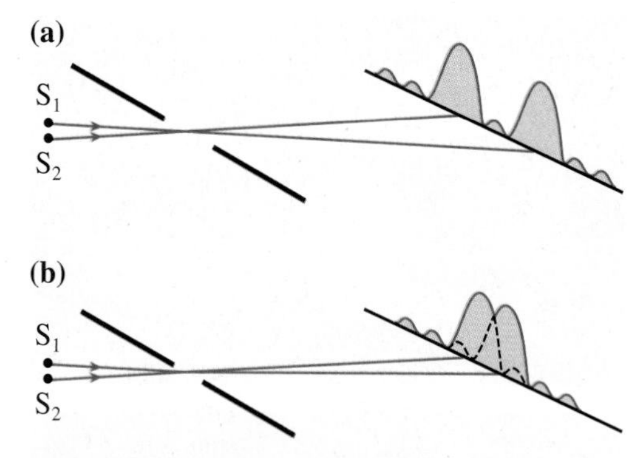

Figure 16-27 Observing how resolution depends on diffraction. **(a)** S_1 and S_2 can be resolved at the screen when the central maxima do not meet. **(b)** With a narrower slit, the diffraction is greater and S_1 and S_2 cannot be resolved at the screen because the central maxima overlap.

Rayleigh's criterion for light passing through a single slit, two sources are just resolved if the first minimum of the diffraction pattern of one source is in line with the middle of the central maximum of the other source

DID YOU KNOW?

Rayleigh's criterion is named in honour of Lord J.W.S. Rayleigh, an English physicist (1842–1919), who was one of the first people to be awarded the Nobel Prize in Physics. He was given the prize in 1904 for co-discovering argon.

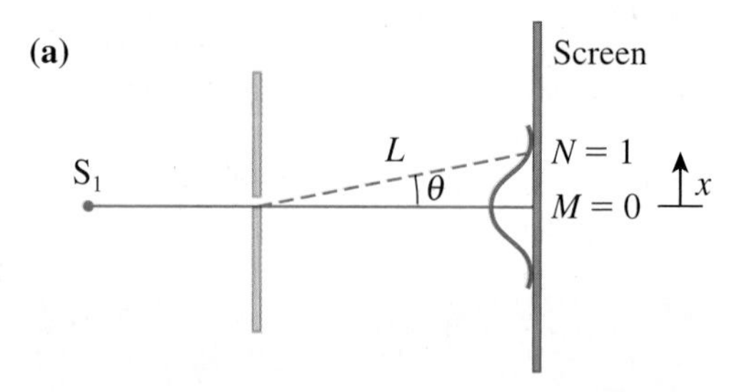

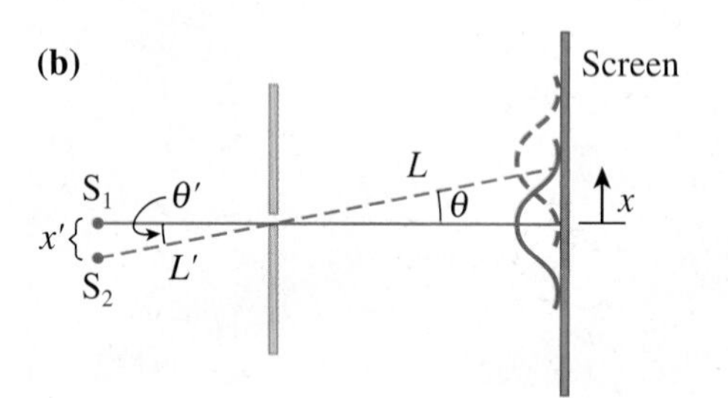

Figure 16-28 Rayleigh's criterion and angular separation. **(a)** The basic diffraction pattern of a single source, S_1. **(b)** Two sources, S_1 and S_2, are just resolved if the first minimum of one source coincides with the middle of the central maximum of the other source.

It is evident from the examples of the headlights and art that resolution decreases as the distance to the objects increases, and increases as the separation of the objects increases. However, resolution also depends on the diffraction of light. To see why, consider two identical monochromatic light sources, S_1 and S_2, whose light passes through a single slit and produces a diffraction pattern on a screen. As shown in Figure 16-27 (a), the diffraction pattern of one source slightly overlaps the diffraction pattern of the other source, making it just barely possible to resolve the sources. Then, as the slit width decreases, the pattern spreads out more and the resolution decreases. Thus, as diffraction increases, resolution decreases. (Resolution also depends on the wavelength of the light. Can you tell how?)

Any two observers may find it difficult to decide when two sources appear to be resolved. To overcome this problem, a formal rule has been defined for determining resolution through a single slit. This rule, known as **Rayleigh's criterion**, states

For light passing through a single slit, two sources are just resolved if the first minimum of the diffraction pattern of one source is in line with the middle of the central maximum of the other source.

We can extend Eqn. 16-7, $\sin\theta = \frac{N\lambda}{w}$, which becomes $\sin\theta = \frac{\lambda}{w}$ when $N = 1$, for the first minimum in a single-slit interference pattern. Figure 16-28 (a) shows the same situation depicted in Figure 16-20 (d) in a more simplified way. In this case, $\sin\theta$ also equals x/L. Thus $\sin\theta = \frac{\lambda}{w} = \frac{x}{L}$. Now in Figure 16-28 (b), light from two sources, S_1 and S_2, passes through the single slit such that $M = 0$ for S_2 and $N = 1$ for S_1. Thus, the angular separation of S_1 and S_2 at the slit, θ', is equal to the angle θ. Because the triangles shown are similar, we know that $x/L = x'/L'$, so we can write

$$\sin\theta = \frac{\lambda}{w} = \frac{x'}{L'} \qquad \textbf{(16-9)}$$

where θ is the angular separation of two sources, λ is the wavelength of the light from the sources, x' is the separation of the sources, and L' is the distance from the sources to the single slit.

In summary, for monochromatic light of wavelength λ, the minimum angular separation of the central maxima that allows resolution of the sources is equal to the angle that the two sources subtend with the slit, as shown in Figure 16-28. Thus, Eqn. 16-9 applies to the situation.

SAMPLE PROBLEM 16-5

Two laser beams, both having a wavelength of 620 nm, are aimed toward a single slit that is 3.2 m away. If the laser beams are 4.4 cm apart at the sources, how wide must the slit be to just resolve the beams?

Solution

The criterion is $\sin\theta = \frac{\lambda}{w} = \frac{x'}{L'}$,

$$\therefore \quad w = \frac{L'\lambda}{x'} = \frac{(3.2\,\text{m})(6.2 \times 10^{-7}\,\text{m})}{4.4 \times 10^{-2}\,\text{m}} = 4.5 \times 10^{-5}\,\text{m}, \quad \text{or} \quad 0.045\,\text{mm}$$

Thus, the slit must be 0.045 mm wide.

The diffraction of light is an important factor in determining how well our eyes can resolve objects, as well as the limitations of such devices as microscopes and telescopes. These optical devices have circular apertures, and the criterion for the minimum angular separation of two sources to allow resolution is somewhat different than for a single slit. (Details about the human eye and other optical instruments are found in Chapter 17.)

EXERCISES

16-14 Why is it said that a single-slit pattern results from both diffraction and interference?

16-15 A single narrow slit is obtained by supporting two razor blades so their edges are close and parallel. A laser beam is aimed through the slit, and the width of the slit is slowly made smaller. Describe what happens to the diffraction pattern observed.

16-16 For violet light of wavelength 420 nm, what is the angle between the initial path of light and the first nodal line of a single-slit diffraction pattern when the slit width is 350 times as large as the wavelength? Would your answer change if the only changes in the question were the colour and its wavelength? (Assume two significant digits in given data.)

16-17 Monochromatic light passes through a single slit of width 0.085 mm, and produces a diffraction pattern on a screen located 2.4 m from the slit. Determine the distance between the $N = 2$ minima if the wavelength of the light is 650 nm.

16-18 The slit in the previous question is adjustable. How wide (in millimetres) must it become to have a central maximum that is 22 mm wide with light of wavelength 450 nm? (Assume the screen location is the same.)

16-19 In Figure 16-19, use a ruler to help you determine an estimate for the ratio of the red light wavelength to the green light wavelength. Refer to Table 16-1 to check your estimate. (Use the average wavelengths of the red and green light.)

16-20 State whether the resolution of two objects seen through a single slit increases, decreases, or stay the same if the

(a) separation of the objects increases

(b) distance to the objects increases

(c) slit width increases

(d) wavelength of the light increases

16-21 Two red light point sources ($\lambda = 670$ nm) subtend an angle of 0.32° with the centre of a single slit. The sources are 6.4 cm apart and a screen is 1.8 m from the slit. The images of the two sources are just resolved on the screen.

(a) What is the width of the slit in millimetres?

(b) How far are the light sources from the slit?

(c) What is the distance in centimetres between the two maxima on the screen caused by the two light sources?

16.4 Diffraction Gratings and Spectroscopy

We have seen that when coherent light passes through a double slit, it undergoes both interference and diffraction. When coherent light passes through many closely spaced slits, all parallel to each other, again it undergoes interference and diffraction. Although the resulting pattern has similarities to a double-slit pattern, it is more distinct and more useful. Figure 16-29 shows two diffraction patterns caused when red laser light passes through sets of slits, one with twice the number of slits per centimetre as the other. (Compare these patterns with the double-slit interference pattern of laser light shown in Figure 16-6 (b)).

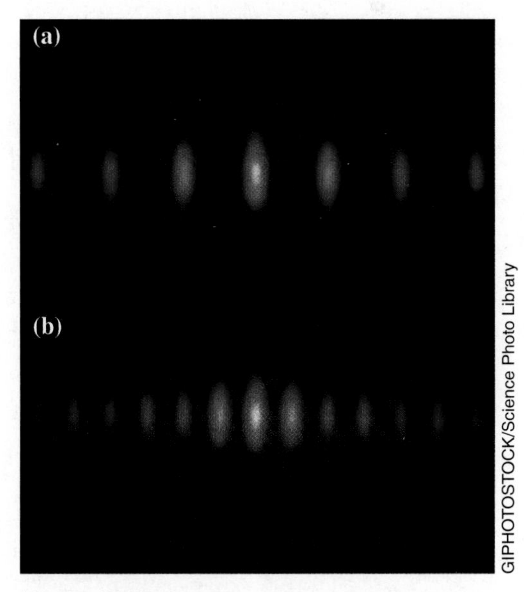

Figure 16-29 Diffraction grating patterns of red laser light. **(a)** The grating has 500 lines/cm, which causes the maxima to be spread out and very distinct. **(b)** The grating has 250 lines/cm, causing the maxima to be closer and less distinct than in (a).

An optical device made of many uniformly spaced, parallel slits or grooves is called a **diffraction grating**. (Such a device could also be called an interference grating, but the traditional name will be used here.) Diffraction gratings are of three main types, depending on how they are made. A **transmission grating** is made by ruling fine parallel grooves on a smooth glass surface with a diamond-tipped tool. The grooves are opaque to light while the spaces between the grooves act as transparent slits. A **reflection grating** is made by ruling fine parallel grooves on a smooth metal surface. As its name implies, this type of grating causes the interference of reflected light rather than transmitted light. Both the transmission and the reflection gratings require precision equipment and a relatively long time to manufacture, so they are expensive. An inexpensive type of grating, called a **replica grating**, is made by pouring molten plastic over a master grating.

diffraction grating an optical device made of many uniformly spaced, parallel slits or grooves

transmission grating a diffraction grating made by ruling fine parallel grooves on a smooth glass surface with a diamond-tipped tool

reflection grating a diffraction grating made by ruling fine parallel grooves on a smooth metal surface

replica grating a diffraction grating made by pouring molten plastic over a master grating

Terminology Tip

The maxima in a diffraction grating pattern tend to be distinct and bright, and the tradition is to call them an order number, such as zero-order, first-order, and so on, rather than the central maximum ($M = 0$), etc.

When the plastic solidifies, it is peeled off the master and fastened to a flat piece of glass or stiff plastic for support. Typical diffraction gratings may have between 2000 and 10 000 lines per centimetre.

The description of how an interference pattern is created in a double-slit can be applied and extended to a diffraction grating. Only the transmission grating will be analyzed mathematically here.

Consider Figure 16-30, which shows wavefronts of monochromatic light passing through a set of five slits. Like the double-slit pattern, the distance from the slits to the screen is very large compared to the spacing of the slits. In (a), the light coming through each slit is coherent, and the Huygens' wavelets that travel straight ahead interfere constructively, producing a maximum interference, or central maximum or zero-order maximum, labelled $M = 0$ in the diagram. In (b), we consider wavelets whose direction deviates by an angle θ with the original direction. At the screen, the path difference between each set of two slits is λ, so once again, constructive interference occurs, causing the first-order maximum $M = 1$. In (c), the second-order maximum, $M = 2$, is found when the path difference from each set of adjacent slits is 2λ. These maxima occur on both sides of the central maximum. Thus, in general, a maximum in a multiple-slit diffraction pattern occurs when the path difference, $P.D. = M\lambda$, where $M = 0, 1, 2, 3, \ldots$

Like the double-slit pattern, $\sin\theta = P.D./d$, so, in general,

$$\sin\theta = \frac{M\lambda}{d}, \text{ where } M = 0, 1, 2, 3, \ldots \quad \textbf{(16-10)}$$

(a)

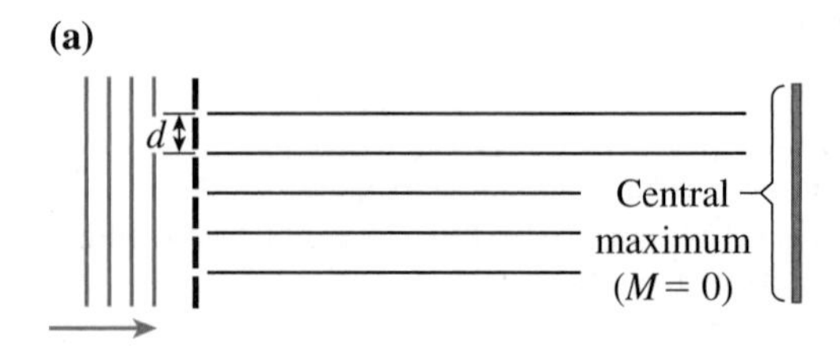

(b)

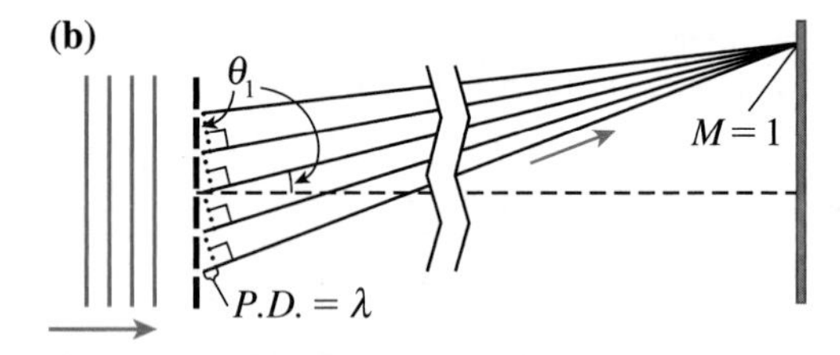

(c)

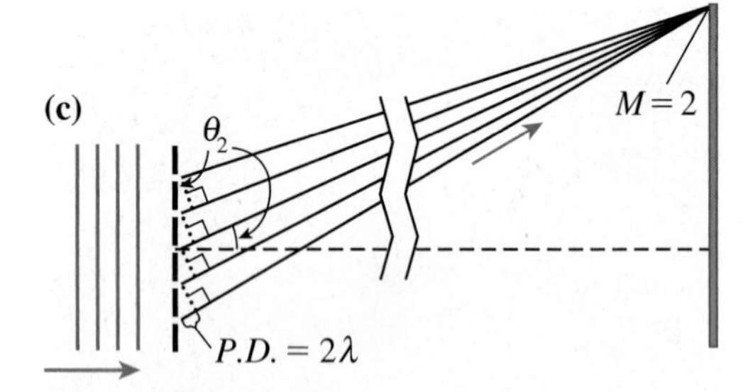

Figure 16-30 Deriving an equation for the positions of maxima produced by a diffraction grating. **(a)** When the path length is the same from all slits, the $P.D. = 0$ and the central maximum is observed. **(b)** When the $P.D.$ between the adjacent slits is 1λ, the maximum $M = 1$ is produced. **(c)** When the $P.D.$ is 2λ, the maximum $M = 2$ is produced.

TRY THIS!

Comparing CD and DVD Diffraction

Speculate on the maximum data capacity of a CD compared to a DVD: Which one contains more information? Based on your speculation and given that CDs and DVDs are made with numerous grooves, predict whether the CD diffraction pattern viewed in white light will be more spread out, less spread out, or the same spacing as that of a DVD. Test your prediction using a typical music CD and a typical movie DVD viewed at various angles in reflected light from a bright white source.

Mathematically, Eqn. 16-10 is similar to Eqn. 16-3 for double-slit maxima positions. However, several differences are observed when comparing a diffraction-grating pattern experimentally to a double-slit interference pattern. First, the angles of the lower-order maxima in the diffraction pattern are much larger than the angles of the lower-order maxima in a double-slit pattern. The large angles occur because of the very small values of slit separation. This means that the grating pattern is much more spread out than the double-slit pattern, and the total number of maxima is less. Also, the approximation that $\sin\theta = \tan\theta$, used in the double-slit pattern, cannot be used here, and if the equation $x/L = M\lambda/d$ is used, L must be measured from the grating to the maximum, not from the grating to the screen. Another difference is that the maxima in the grating pattern are much sharper and brighter than the maxima in the double-slit pattern, which in turn means that the minima are wide.

The advantages provided by the spread-out pattern and the sharp maxima in diffraction-grating patterns are applied in viewing and analyzing light spectra and studying the structure of matter using x-ray diffraction. Details of these applications are described later in this section.

SAMPLE PROBLEM 16-6

Light from a monochromatic source passes through a diffraction grating with 5000 lines/cm, and the second-order maxima (i.e., $M = 2$) on either side of the central maximum make an angle of 36.1° with the central maximum (Figure 16-31). Assume three significant digits.

(a) What is the wavelength of the light from the source?

(b) What colour is the light?

Solution

(a) First, we find the slit separation, d, which is the reciprocal of 5000 lines/cm:

$$(1/5000)\,\text{cm} = 2.00 \times 10^{-4}\,\text{cm}$$
$$= 2.00 \times 10^{-6}\,\text{m}$$

Now from Eqn. 16-10, $$\sin\theta = \frac{M\lambda}{d}$$

$$\therefore \lambda = \frac{d\sin\theta}{M} = \frac{(2.00 \times 10^{-6}\,\text{m})(\sin 36.1°)}{2} = 5.89 \times 10^{-7}\,\text{m}$$

Thus, the wavelength of the light is 5.89×10^{-7} m, or 589 nm.

(b) According to Table 16-1, the light with a wavelength of 589 nm is yellow.

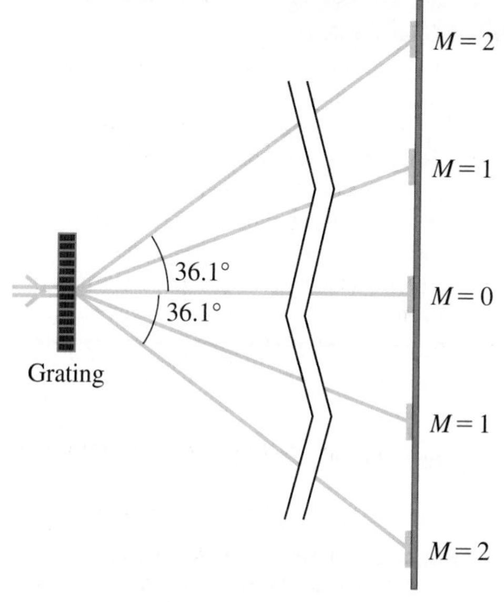

Figure 16-31 Sample Problem 16-6.

spectrum a distribution of energy emitted arranged in order of wavelength

continuous spectrum an unbroken band of emissions occurring at a wide range of wavelengths

line spectrum a set of emissions at multiple distinct wavelengths producing specific bright lines separated by dark spaces

Spectroscopy

In optics, we often discuss the light emission from a source in terms of its spectrum. The source's **spectrum** is the distribution of energy emitted arranged in order of wavelength. For the various monochromatic sources discussed so far in this chapter, their spectrum is simply all of its energy emitted at one wavelength. White light from an incandescent light bulb however, has a more complicated energy spectrum, with emissions occurring at a wide range of wavelengths; such emissions are called **continuous spectra**. In these spectra, the visible portion appears as an unbroken band of colours, as was shown in Figure 16-18. Gases at low pressure produce spectra with bright emissions at multiple distinct wavelengths, as if the source was an arrangement of several monochromatic emitters. Such a bright-light spectrum is called a **line spectrum**, examples of which are shown in Figure 16-32. These distinct wavelengths are unique to the atom or molecule that is emitting the light. **Spectroscopy** is the study of spectra, often undertaken to investigate the elemental or molecular composition of the unknown sample.

Gases can emit light when they are stimulated with heat or electricity. Each element or compound in the gaseous state emits its own unique set of wavelengths, just as each person has his or her own unique fingerprints. Various methods can be used to observe the spectra of such light sources. For example, a **spectrograph** is a photograph of a spectrum, and a **spectroscope** is a device that uses either a prism or a diffraction grating to obtain a spectrum that can be viewed by a person's eye. We will consider the grating spectroscope here.

Figure 16-33 illustrates the basic design of a grating spectroscope using, in this instance, the light emitted by hydrogen in a transparent tube called a discharge tube. The hydrogen is stimulated by electricity to emit light. The light becomes parallel when it passes through a set of lenses in the collimator. The parallel rays strike a diffraction grating, and the resulting interference pattern is viewed through a microscope. The microscope can be moved to various angular positions, and the measured angles can be used to determine the wavelengths of the light emitted from the hydrogen gas (using Eqn. 16-10, $\sin\theta = M\lambda/d$).

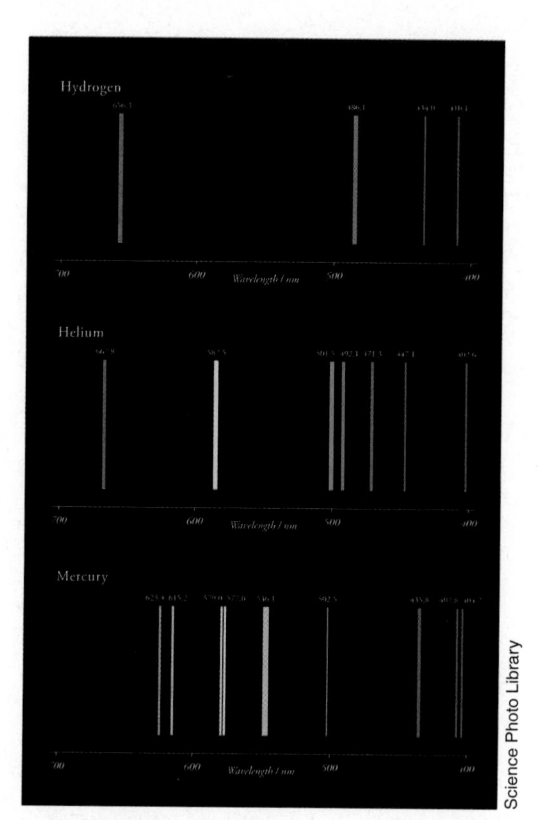

Figure 16-32 Line spectra for the gases hydrogen, helium, and mercury.

DID YOU KNOW?

Colour vision helps our eyes act as crude spectroscopes when we try to identify light sources by their colour. Neon lights used in advertising appear red; sodium lamps used on many lamp posts are orangish-yellow; and mercury lights used on some lamp posts are bluish. However, spectrometers and spectroscopes can be used to determine exactly which colours are emitted by such sources.

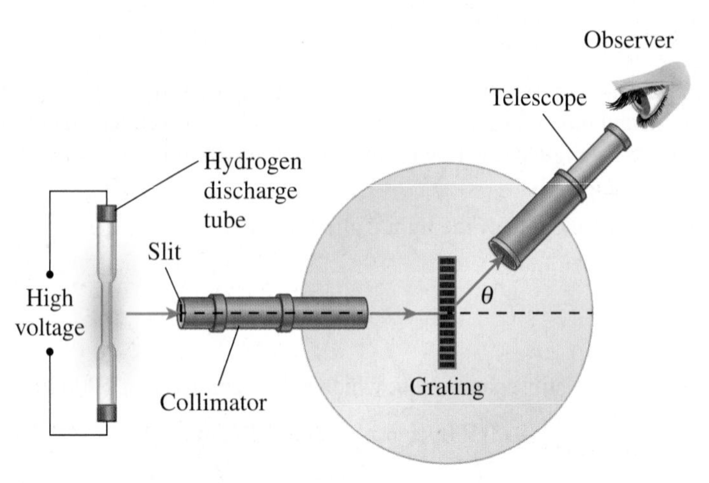

Figure 16-33 The design of a typical grating spectroscope.

spectroscopy the study of spectra

spectrograph a photograph of a spectrum

spectroscope a device that uses either a prism or a diffraction grating to obtain a spectrum

absorption spectrum a white-light spectrum that has specific wavelengths absorbed by elements or compounds, resulting in dark bands spaced throughout

We now consider the line spectrum of the light emitted by stimulated hydrogen atoms in greater detail. This spectrum consists of a red line, a blue line, and various violet lines, some of which are shown in Figure 16-34. Through a diffraction grating, the bright central maximum is the same colour as the light from the hydrogen in the tube. On either side of the central maximum however, are spectra of the colours emitted by the hydrogen, with the first-order spectrum ($M = 1$) being the brightest. The equation $\sin\theta = M\lambda/d$ can be applied to determine the wavelengths of the different colours of light of the hydrogen. (See Question 16-30.)

If the discharge tube with hydrogen is replaced with a discharge tube containing a different element or compound in the gaseous state, a completely different spectrum is observed. Thus, the spectrum of light emitted by substances can be used to determine the composition of the substance. Astronomers have applied this principle to discover the composition of stars and the atmosphere of distant planets, as well as the gases (mainly hydrogen) that exist in interstellar space.

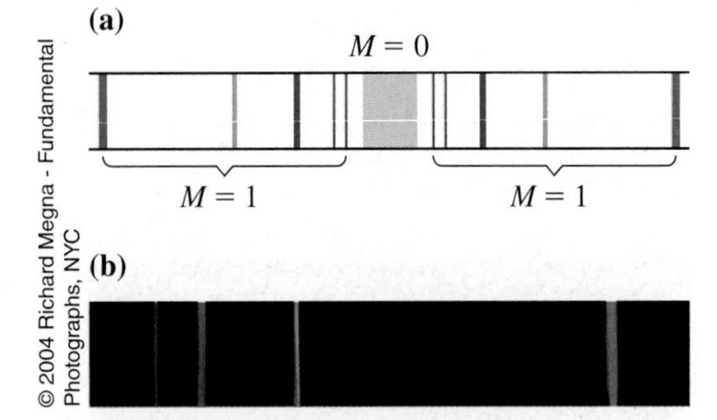

Figure 16-34 **(a)** Details of the spectra of atomic hydrogen produced by a diffraction grating. **(b)** Photograph of the main part of the spectrum to the right of the central maximum.

An element or compound that emits light of specific wavelengths also absorbs light of the same wavelengths. When a continuous white light spectrum is incident on a sample of elements or compounds that absorb some of the light, the resulting **absorption spectrum** has dark bands spaced throughout. For example, the white light emitted from the Sun must pass through the cooler gases at the outer surface of the Sun. These gases absorb some of the Sun's light, resulting in characteristic dark bands in the spectrum observed from the Sun. By analyzing this absorption spectrum, astronomers have found that approximately two-thirds of the known elements are present in the solar atmosphere. Absorption spectra can also be used to analyze pollutants in Earth's atmosphere and identify the composition of unknown substances. To accomplish this latter task, light with a known spectrum is passed through a solution of the unknown substance and then through a diffraction grating. The wavelengths of the dark bands in the resulting absorption spectrum can be used to determine the elements or compounds present in the substance (Figure 16-35).

Figure 16-35 Photo of the absorption spectrum of atomic hydrogen. Notice that the positions of the absorption lines correspond to the positions of the emission lines on the left side of the diagram in Figure 16-34(a).

Diffraction in Two and Three Dimensions

Our discussion of diffraction of light through single slits, double slits, and gratings has been restricted to one dimension only. What pattern do we observe for a two-dimensional array of slits? The photograph on the first page of this chapter shows an example of such a pattern. As you would expect, the maximum fringes form a two-dimensional pattern. The photograph was taken with a two-dimensional star-pattern diffraction filter placed over the lens. Each bright light in the photograph produces a central maximum and a set of spectral fringes. Similar patterns can be seen by viewing a distant light source, such as a street lamp, through a piece of sheer cloth.

Nature has provided us with three-dimensional diffraction gratings, namely the regular crystals of some solids, such as sodium chloride, that have regular arrays of atoms. The spacing of these atoms is so tiny however, that visible light cannot create a diffraction pattern through them. The atomic spacing is typically about 0.1 nm, which is thousands of times smaller than the wavelengths of visible light. Are there any waves that have a wavelength in the 0.1 nm range? In 1912, a German physicist named Max von Laue predicted that x rays, which are part of the same electromagnetic spectrum that visible light belongs to, could be used to create a diffraction pattern in solid crystals. (Section 16.6 provides more details of the electromagnetic spectrum.) This brilliant prediction turned out to be true. When x rays are passed through three-dimensional crystals, a diffraction pattern is formed due to the regular array of atoms in the crystal (Figure 16-36).

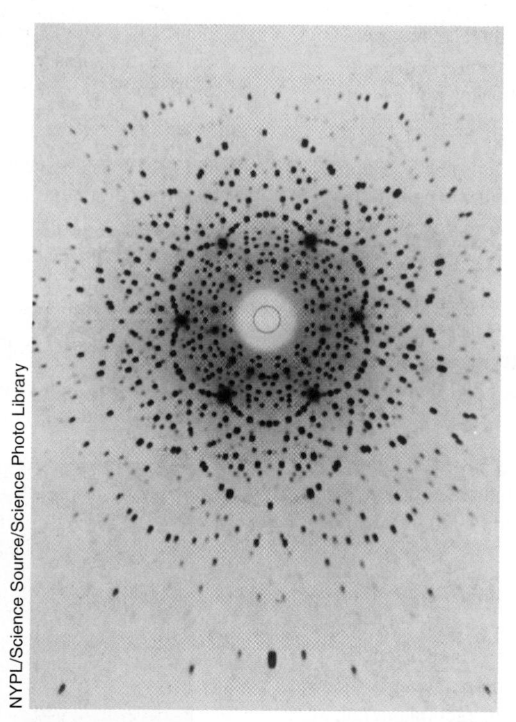

NYPL/Science Source/Science Photo Library

Figure 16-36 An x-ray diffraction pattern of the hexagonal crystals of beryl, a mineral composed of beryllium aluminum cyclosilicate.

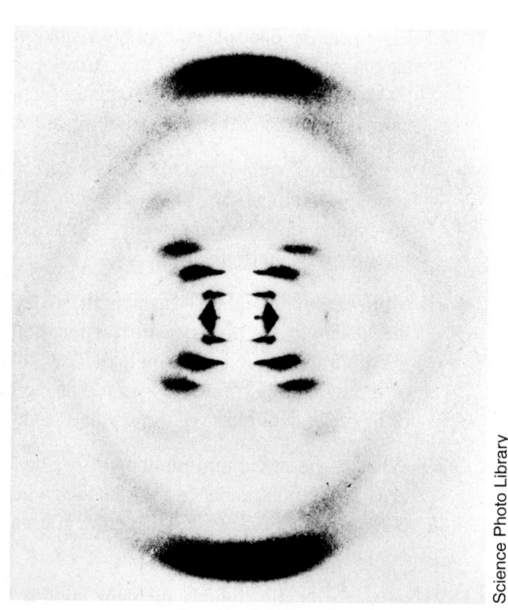

Science Photo Library

Figure 16-37 This historically important x-ray diffraction image of a DNA molecule was taken in 1953 by British scientist Rosalind Franklin, who used it to determine that a DNA molecule consisted of two interwoven strands. This discovery helped James Watson and Francis Crick develop an understanding of the double helix nature of DNA for which they received a Nobel Prize in 1962.

The study of x-ray diffraction patterns, called **x-ray crystallography**, is a branch of physics with applications that extend to other sciences. Crystals with known atomic spacings can be used to determine unknown wavelengths of x rays. Conversely, x rays of known wavelength can be used to determine the atomic structure of unknown crystals. In fact, x-ray diffraction was instrumental in the discovery of the double-helical structure of DNA molecules, the carriers of the genetic code (Figure 16-37).

x-ray crystallography the study of x-ray diffraction patterns in crystals

TRY THIS!

Franklin and Other Important Women in Science

Although Rosalind Franklin contributed greatly to the understanding of the structure and function of DNA, she did not share a Nobel Prize with Watson and Crick because she died of cancer four years before the prize was awarded. (The prize is awarded only to living recipients.) However, Franklin is recognized as one of an influential group of female scientists who made contributions of fundamental importance to the fields of physics, chemistry, biology, and medicine. Search online to find out who among the "Top 10 Women in Science" had the greatest influence in physics.[3]

DID YOU KNOW?

Max von Laue won the Nobel Prize in Physics in 1914 for his discovery of x-ray diffraction through solid crystals. Even today, the maximum fringes that occur when x rays reflect from crystals are called Laue spots. In 1915, the father-and-son team of William and Lawrence Bragg, two English physicists, won the Nobel Prize in Physics for studying the structure of crystals using x-ray diffraction.

EXERCISES

16-22 Two gratings are labelled 2500 lines/cm and 10 000 lines/cm, respectively. Compare the diffraction patterns produced when monochromatic light is passed through each of these gratings.

16-23 The maxima in a diffraction grating pattern are brighter and sharper than the maxima in a double-slit interference pattern. Assuming equal intensities of light are incident upon both the grating and the double slit, is the grating giving us something for nothing? Explain.

[3]See also the profile of Marie Curie in Chapter 24.

16-24 Light near the opposite ends of the visible spectrum, having wavelengths of 420 nm and 680 nm, strike a diffraction grating with 5000 lines/cm. Determine the angular deflection of both wavelengths at the second order maxima. Assume three significant digits.

16-25 White light, with wavelengths between 400 nm and 700 nm, strikes a grating with 4000 lines/cm. Determine whether the $M = 2$ and $M = 3$ fringes overlap. Show your reasoning. Assume three significant digits.

16-26 Monochromatic light of wavelength 515 nm is used to determine the spacing of the lines on a diffraction grating. The second order maximum is located at an angle of 32.8° to the original direction of the light. Determine the spacing of the lines *and* the number of lines per centimetre of the grating.

16-27 What is the maximum number of bright fringes, including the central maximum, that can be obtained when light of wavelength 600 nm passes through a diffraction grating with 8000 lines/cm? Assume three significant digits.

16-28 Monochromatic light is incident upon a grating with 2.50×10^3 lines/cm, and the diffraction pattern is viewed on a screen located 3.80 m from the grating. The first bright fringe is located 64.5 cm from the middle of the central maximum. What is the wavelength of the light? (Remember that for a diffraction grating, $\sin\theta$ does not equal $\tan\theta$.) Assume three significant digits.

16-29 A photographer without a diffraction filter improvises with a fine piece of sheer cloth placed over the camera lens. The cloth is placed so the threads that run horizontally have double the spacing of the threads that run vertically. The photographer takes a picture of a single bright street light. Draw a sketch of the resulting pattern.

16-30 A student uses a spectroscope to view the spectrum of hydrogen gas in a discharge tube. The spectroscope grating has 8.00×10^3 lines/cm (three significant digits). The angles of deviation of three of the bright lines in the first order maximum are found to be 20.3°, 22.9°, and 31.7°, respectively. Calculate the wavelengths that produce these bright lines, and state what colour each wavelength corresponds to.

16-31 Describe the features of the light and grating used to produce the pattern shown in Figure 16-38.

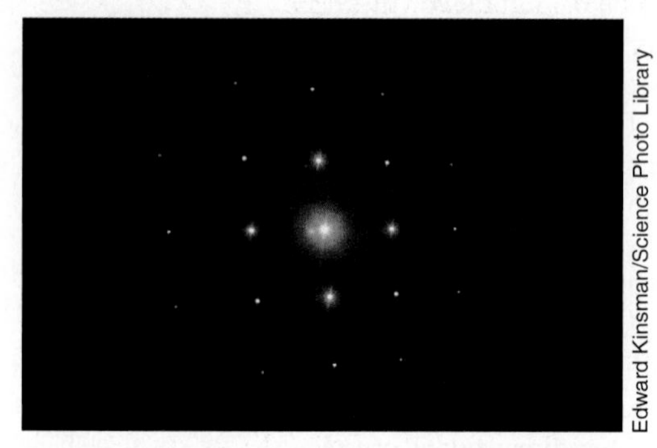

Edward Kinsman/Science Photo Library

Figure 16-38 Diffraction pattern (Question 16-31).

16.5 Thin-Film Interference

Oil and gasoline have rather drab colours, but under special conditions, they display beautiful colour fringes that appear to change when viewed from different angles. This occurs when white light strikes an extremely thin film of oil or gasoline that is spread out over a surface of water. Likely you have seen these coloured fringes in automobile service stations or truck stops. The colours of the light reflected from the thin film are an example of **iridescence**, which is caused by interference. Other examples are the colours seen in soap films, on fish scales and insect wings, and in the plumage of some birds, such as the head feathers of a male Mallard duck and some hummingbirds. See Figure 16-39.

iridescence the creation of colour fringes caused by thin-film interference of light

DID YOU KNOW?

Iridescence comes from the Greek words *iris* and *iridis*, which mean rainbow. In Greek mythology, Iris was the goddess of the rainbow.

The fringes in thin films are produced by the interference effects of light. The fringes are most easily observed when the light reflects from the thin film, but it can also be observed when the light is transmitted through the film. In white light, the interference fringes have various colours, but in monochromatic light, the fringes are bright and dark where constructive and destructive interference occur.

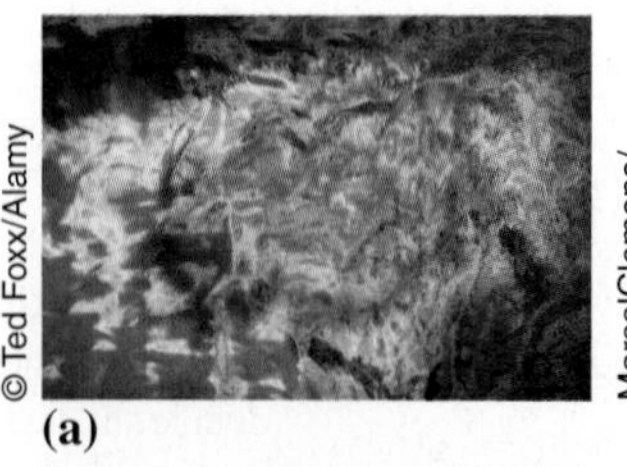

© Ted Foxx/Alamy

(a)

MarcelClemens/Shutterstock

(b)

MVPhoto/Shutterstock

(c)

Figure 16-39 Examples of colour fringes caused by thin-film interference of reflected light. **(a)** An oil slick on a wet surface. **(b)** Soap bubbles. **(c)** A male Mallard duck.

To explain how the interference of light can cause the fringes observed in thin films, we begin by reviewing what occurs when a wave travelling in one medium reaches a second medium. What was described for mechanical waves in Section 13.6 can be applied to light waves as well. The following statements summarize the three possible situations when light in one medium strikes a different medium.

- Light that is transmitted from one medium to another undergoes no phase change.
- Light reflected from the surface of a medium having a higher index of refraction, and hence a slower speed, undergoes a complete phase change. (In terms of transverse wave components, a crest becomes a trough.) This is called fixed-end reflection, or negative reflection.
- Light reflected from the surface of a medium having a lower index of refraction, and hence a faster speed, undergoes no phase change. This approximates an open-end reflection, or positive reflection.

Consider, for example, a coherent beam of monochromatic light in air striking a thin film of oil resting on water. The indexes of refraction of the three media are $n_{air} = 1.00$, $n_{oil} = 1.48$, and $n_{water} = 1.33$. Assume the beam strikes at a small angle of incidence, as shown in Figure 16-40. Part of the beam reflects off the air–oil interface and undergoes a phase change because the oil has a higher index of refraction than the air. Another part of the beam transfers with the same phase into the oil and reflects off the oil–water interface, this time without a phase change because the water has a lower index of refraction than the oil. This reflected light transfers with the same phase back into the air and interferes with the light reflected from the top surface. In (a), the thickness, t, of the oil film is much less than the wavelength, λ, of the light in the film, so the two reflected beams are opposite in phase and cancel. This causes a dark fringe to occur, as can be seen in the magnified view.

In (b), the thickness of the oil film is $\lambda/4$. Thus, the path difference of the two beams is $\lambda/2$. These two reflected beams are now in phase, causing constructive interference or a bright fringe. Again, the magnified view shows this more clearly. Using similar arguments for increasing thicknesses of the oil film, we find that destructive interference occurs when the path difference is $0\lambda, 1\lambda, 2\lambda, 3\lambda, \ldots$, and constructive

DID YOU KNOW?

Some birds have feathers that produce iridescence by the diffraction of reflected light rather than by thin-film interference. Diffraction is what gives pigeon necks the appearance of changing colours as their heads jerk back and forth when they walk.

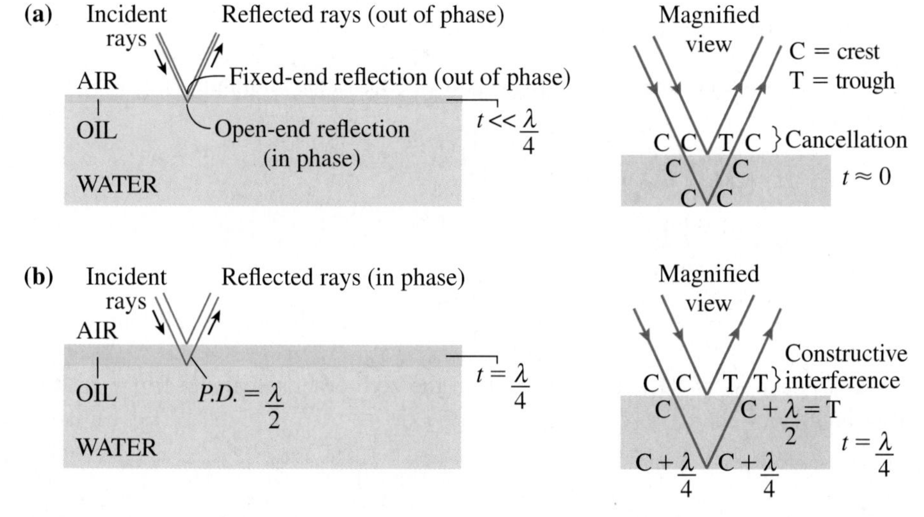

Figure 16-40 Thin-film interference of reflected light, with λ representing the wavelength of the light in the thin film. **(a)** When $t \ll \lambda/4$, the reflected beams are out of phase, causing destructive interference and thus a dark fringe. **(b)** When $t = \lambda/4$, the beam that reflects off the water travels an extra $\lambda/2$, so the two reflected beams are now in phase.

interference occurs when the path difference is $0.5\,\lambda, 1.5\,\lambda, 2.5\,\lambda, \ldots$, and so on. Since the path difference is approximately $2t$ (when the incident light is approximately normal to the film), we can write the relationships below.

Minima and maxima in reflection patterns from soap bubbles or oil slicks,

$$2t = N\lambda, \text{ where } N = 0, 1, 2, 3, \ldots \text{ represents the nodes or minimum fringes} \quad \textbf{(16-11)}$$

$$2t = (M + ½)\,\lambda, \text{ where } M = 0, 1, 2, 3, \ldots \text{ represents the maximum fringes} \quad \textbf{16-12)}$$

In these equations, t is the thickness of the film in which the wavelength of light is λ.

It is important to remember that these equations apply *only* if the thin film has a higher index of refraction than the media surrounding it on both sides. Thus, it applies not only to oil between air and water but also to a soap film surrounded by air. It is also important to remember that the wavelength is the wavelength of the light in the film. This can be found by knowing the wavelength of the light in air and the index of refraction, n, of the film and applying Eqn. 15-5 $n = \frac{c}{v} = \frac{f\lambda_{\text{air}}}{f\lambda} = \frac{\lambda_{\text{air}}}{\lambda}$ because the frequency remains constant as the light travels from the air into the film. Thus, with λ representing the wavelength of the light in the film,

$$n = \frac{\lambda_{\text{air}}}{\lambda} \quad \text{or} \quad \lambda = \frac{\lambda_{\text{air}}}{n}$$

SAMPLE PROBLEM 16-7

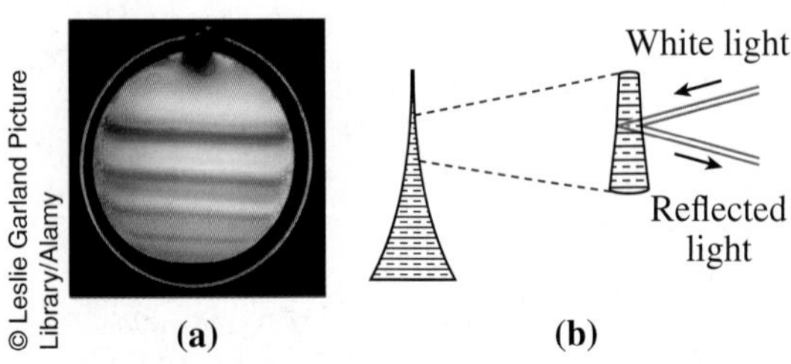

Figure 16-41 Sample Problem 16-7. **(a)** Loop with soap-film interference pattern observed in reflected white light. **(b)** Side view of the soap film.

A wire loop is dipped into soapy water and is then held vertically so the force of gravity causes the soap film to be thicker as it nears the bottom of the loop (Figure 16-41). The thin film of soap (with $n = 1.33$) reflects white light to an observer. What is the smallest thickness of the film for which the observer sees blue reflected light (wavelength 474 nm in air)?

Solution

The first maximum fringe is $M = 0$. This corresponds to the minimum thickness (from Eqn. 16-12):

$$2t = (M + ½)\,\lambda \text{ where } \lambda = \lambda_{\text{air}}/n$$

$$2t = \frac{(0 + ½)\,\lambda_{\text{air}}}{n}$$

$$\therefore \quad t = \frac{½\lambda_{air}}{2n} = \frac{\lambda_{air}}{4n} = \frac{4.74 \times 10^{-7}\,\text{m}}{(4)(1.33)} = 8.91 \times 10^{-6}\,\text{m}, \quad \text{or} \quad 89.1\,\text{nm}$$

Thus, a thickness of 89.1 nm will result in the zero-order maximum fringe for this wavelength of light.

It is evident from the above sample problem that each component wavelength will exhibit maxima upon reflection at a different thickness of the thin film. Thus, in white light, a film of gradually changing thickness gives rise to the spectral separation of colours upon reflection.

We have considered only incident light that is normal or nearly normal to the thin film. Interference can also occur at much larger angles of incidence, but the path difference of the interfering light beams is a more complicated calculation (Figure 16-42).

We have seen that constructive interference of reflected light occurs at a thickness of $\lambda/4$ if the film has an index of refraction greater than the surrounding media. What happens if the film of thickness $\lambda/4$ has an index of refraction between the indexes of the surrounding media? This can occur if a soap film remains on a transparent glass, and it has important applications in the manufacture of lenses and solar cells.

Usually when light strikes a glass lens, about 96% of the light is transmitted. The remaining light is reflected or absorbed. Since most expensive cameras and microscopes have between 6 and 8 lenses, a high portion of the incident light will be reflected, resulting in a dim image. Moreover, double reflection within each lens can cause a blurred image. To overcome these problems, each lens is coated with a thin film that will result in a minimum fringe upon reflection of certain wavelengths. For glass lenses, this film is usually made of magnesium fluoride (MgF_2), which has an index of refraction of 1.38. It can be shown (as you will be asked to do in Question 16-34) that, when a magnesium fluoride film is located between air ($n = 1.00$) and glass ($n \geq 1.50$), destructive interference of the reflected light will occur when the film thickness is $\lambda/4$. Such a film is called a **quarter-wavelength film**. In general,

Minima and maxima in reflection patterns from quarter-wavelength films,

$$2t = (N + \tfrac{1}{2})\lambda_{film}, \text{ where } N = 0, 1, 2, 3, \ldots \text{ for nodes or minimum fringes} \quad (16\text{-}13)$$

$$2t = M\lambda_{film}, \text{ where } M = 0, 1, 2, 3, \ldots \text{ for maximum fringes} \quad (16\text{-}14)$$

These equations apply *only* if the thin film has an index of refraction between the indexes of its surroundings.

SAMPLE PROBLEM 16-8

A quarter-wavelength film of magnesium fluoride ($n = 1.38$) is applied on a glass camera lens ($n = 1.56$). The film is 99.6 nm thick. Assuming that this is the smallest thickness possible for a minimum fringe upon reflection, what are the wavelength and the colour of the light for which this film is designed?

Solution

The smallest thickness occurs when $N = 0$ in Eqn. 16-13: $2t = (N + \tfrac{1}{2})\lambda$.

$$\text{Thus, } \lambda = \frac{2t}{(N + \tfrac{1}{2})} = \frac{2 \times 9.96 \times 10^{-8}\,\text{m}}{\tfrac{1}{2}} = 3.98 \times 10^{-7}\,\text{m}$$

$$\text{Now, from } \lambda_{air} = n\lambda = 1.38(3.98 \times 10^{-7}\,\text{m}) = 5.49 \times 10^{-7}\,\text{m}$$

Thus, the wavelength of the light for which a minimum fringe occurs upon reflection is 5.49×10^{-7} m, or about 550 nm, which is at the middle of the visible spectrum. From Table 16-1, this wavelength corresponds to green.

A lens with a coating thickness that has a minimum fringe upon reflection for green or yellow light will not result in a minimum fringe at the red and violet ends of the spectrum. Red and violet combine to produce a purple hue, and that is exactly what you will observe when white light reflects from most lenses in cameras, binoculars, telescopes, and other optical devices.

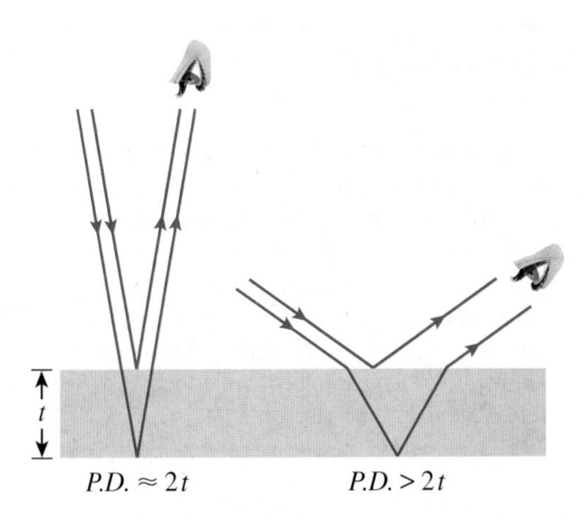

Figure 16-42 Thin-film interference of light at an angle of incidence much greater than 0° to the normal.

TRY THIS!

Soap Film Patterns

Use a fine metal wire about 20 cm long to make a round loop about 3 cm in diameter with the ends fashioned into a twisted handle. Add a soap bubble mixture to a mug. (If you don't have a commercial mixture, try liquid dish soap in water. In either case, if glycerin is available, a small amount will help the film last longer.) Dip the loop into the soap solution and hold the loop vertically (as in Figure 16-41 (a)) as you view the pattern in reflected white light.

- As the pattern changes, what colour(s) appear near the top of the loop?
- Does changing the angle of the loop help the film last longer before it bursts?
- What happens to the spacing of the colour fringes toward the bottom of the film? Why does this happen?

quarter-wavelength film a thin film with a thickness that is one-quarter the wavelength of the light in the film, often used to minimize reflections from air–glass boundaries

Lens Coatings

Locate a bright light source that can be used to view the light reflected off the lens surfaces of various optical devices. Choose a camera, binoculars, a telescope, a microscope, or a pair of eyeglasses with an anti-glare film. What do the colours of the light reflected from the lens coatings have in common? Do some lens coatings appear to be more effective at reflecting light than others?

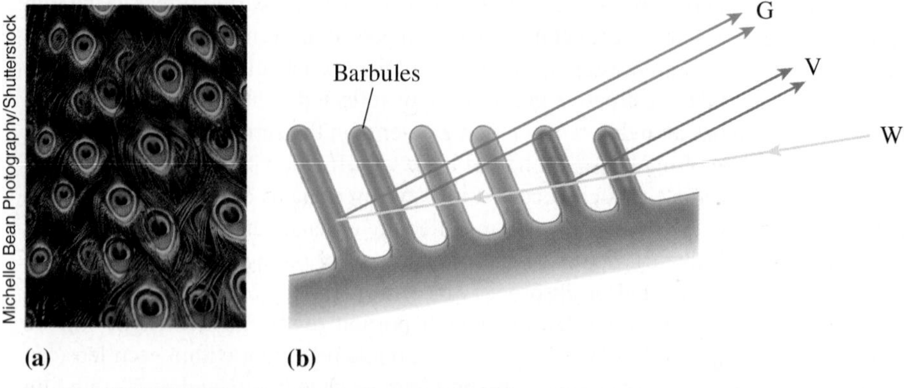

Figure 16-43 **(a)** The tail feathers of a peacock. **(b)** A simplified view of how interference occurs when light reflects from the feather barbules. White light (W) strikes the barbules and, when viewed from slightly different angles, various reflected wavelengths interfere constructively to produce colours such as violet (V) and green (G).

Some cars have an interior rear-view mirror that applies thin-film technology to reduce the reflective glare from bright lights of a following vehicle. The mirror has a thin coating of gel that is electrically conductive. Sensors reacting to bright light change the gel's properties, causing it to go from clear and reflective to opaque and non-reflective.

The iridescence of some bird feathers such as the head feathers of a male Mallard duck (seen in Figure 16-39 (c)) or the tail feathers of a peacock (Figure 16-43 (a)) results from interference of light similar to thin-film interference. The feathers are composed of barbules, tiny fibre-like nanostructures. When white light reflects from the barbules, the wavelength dependence of the resulting interference of the light results in maxima for different colours at different angles (Figure 16-43 (b)).

Air-Wedge Interference

air-wedge interference a type of thin-film interference in which there is a film of air between transparent plates

Another example of thin-film interference occurs when air is located between two glass surfaces. In this case, the air is the film and the interference is called **air-wedge interference**. As seen in Figure 16-44, destructive interference of reflected light occurs when the path difference is approximately 0λ, $\lambda/2$, $2\lambda/2$, $3\lambda/2$, ... Similarly, constructive interference of reflected light occurs when the path difference is $\lambda/4$, $3\lambda/4$, $5\lambda/4$, ... Thus,

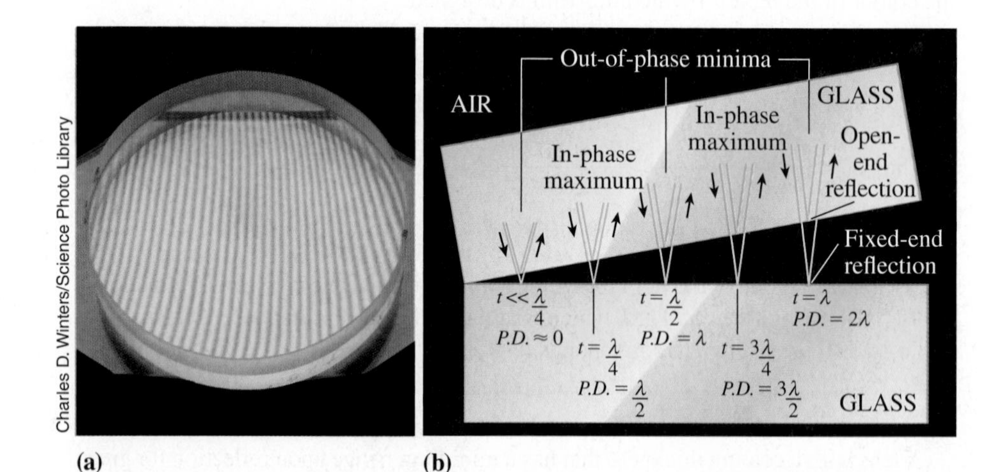

Figure 16-44 Interference of reflected light in an air wedge. **(a)** An example of an air wedge between two circular plates of glass. **(b)** An exaggerated view showing the reflected light beams. (In a real situation, the thickness of the air wedge is extremely small and the light comes through the top glass to the observer.)

the minima and maxima can be found using the same equations developed for a film between two media of lower index of refraction. Therefore, for air-wedge interference,

$$2t = N\lambda, \text{ where } N = 0, 1, 2, 3, \ldots \text{ for minima} \quad (16\text{-}15)$$

$$2t = (M + ½)\lambda, \text{ where } M = 0, 1, 2, 3, \ldots \text{ for maxima} \quad (16\text{-}16)$$

Air-wedge interference can be used to determine the precision with which a lens or glass plate has been made. For example, if a convex lens resting on an extremely flat glass plate is viewed in reflected monochromatic light, a series of circular bright and dark fringes is seen as the air wedge thickness increases. These fringes are called Newton's rings, because Isaac Newton first described them. If the lens is poorly ground, the rings will not be circular. Likewise, flat glass plates can be tested on a well-made plate called an optical flat. The surface of such a flat varies in height by no more than about 1/100 of the wavelength of light. A glass plate resting on an optical flat will produce parallel fringes if the plate is well made. The glass plate being tested in Figure 16-44 (a) is fairly well made because the interference fringes are close to parallel.

EXERCISES

16-32 Light travelling in medium A strikes the surface of medium B. In each case, state whether the portion of the light that reflects is in phase or out of phase with the incident light.

(a) $n_A > n_B$, where n is the index of refraction

(b) $v_A > v_B$, where v is the speed of light

16-33 Monochromatic yellow light ($\lambda = 583$ nm in air) is directed vertically downward and strikes a horizontal oil film ($n = 1.48$) on water ($n = 1.33$). Determine the three least possible thicknesses of the oil film that would give rise to maximum fringes upon reflection of this light.

16-34 Use the theory of fixed-end and open-end reflection to prove that a quarter-wavelength coating of magnesium fluoride ($n = 1.38$) over glass ($n \geq 1.50$) causes a minimum fringe upon reflection for light having a wavelength equal to four times the thickness of the coating.

16-35 In order to reduce the reflection of incident light of a particular wavelength, a silicon solar cell is manufactured with a thin layer of silicon monoxide ($n = 1.45$) over the actual silicon ($n = 3.50$). Determine the smallest thickness of silicon monoxide that results in a minimum fringe in the reflection of light of wavelength 5.50×10^{-7} m.

16-36 A soap film ($n = 1.33$) has a uniformly increasing thickness that ranges from 200 nm to 900 nm. Determine how many red fringes ($\lambda_{air} = 650$ nm) will be observed in the interference pattern of reflected light from this film.

16-37 A soap film ($n = 1.33$) has a thickness of $3\lambda_{film}/4$ for a certain monochromatic light. Is the reflected light maximum or minimum if the soap film is (a) surrounded by air and (b) between air and glass?

16-38 Photographs of Newton's rings (like the one in Figure 16-67 near the end of the chapter) show a dark circle in the middle of the pattern. This circle, produced when a convex glass surface sits on a flat glass surface, persists even if the convex surface is pressed tightly against the flat surface. Why is the reflected light a minimum at the centre?

16-39 Monochromatic light of wavelength λ strikes a convex lens resting on (a) an optical flat and (b) another convex lens that has the same size and shape, as illustrated in Figure 16-45. Compare the interference patterns viewed by reflected light. (A sketch will help.) Assume all surfaces are well made and the maximum air gaps are 1λ in (a) and 2λ in (b).

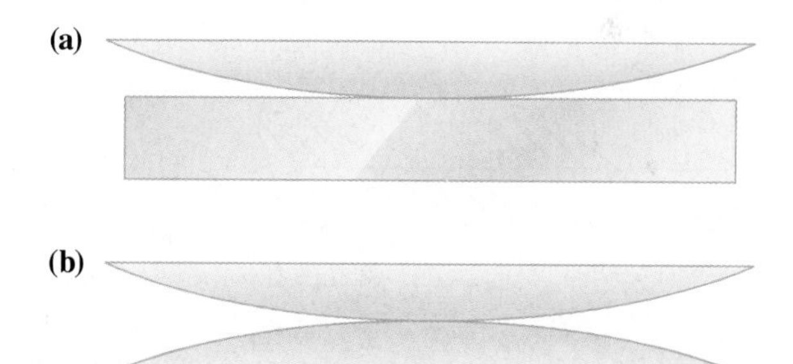

Figure 16-45 Question 16-39.

16-40 Two flat glass plates 8.0 cm long are touching at one end and are separated by a fine fibre at the other end. The air wedge between the plates is viewed in reflected monochromatic light ($\lambda_{air} = 5.0 \times 10^2$ nm) and parallel bright fringes are seen every 0.50 mm.

(a) By how much does the thickness of the air wedge increase with every new bright fringe?

(b) What is the thickness of the fibre?

16-41 **Fermi Question:** Assuming the glass plates have a diameter equal to your hand span, estimate the change in thickness (in millimetres) of the air wedge from the left side to the right side of the pattern shown in Figure 16-44 (a).

16.6 The Electromagnetic Spectrum

Having observed several situations in which light displays properties of waves, we are now ready to summarize our notions of the theories of light. After Thomas Young's double-slit experiment and measurement of the wavelengths of visible light at the beginning of the 19th century, an increasing number of scientists supported the wave theory of light. By the middle of that century, the speed of light in water had been determined, further strengthening the wave theory and weakening the particle theory.

During the same period of time, several discoveries were made about electricity and magnetism. When electrical currents were first being discovered, there was no reason to suspect that this had anything to do with magnetism or that electricity and magnetism had anything to do with light. But as we will discuss in further detail in Chapter 23, in 1831 a British physicist named Michael Faraday (1791–1867) demonstrated that a changing magnetic field can create an electrical current.[4] In other words, a magnetic field that changes in time gives rise to an electric field that also changes in time. (The concept of a field was introduced in Chapter 9, where a field was defined as any physical quantity that has a specific value at each point in space and time. In the same chapter, gravitational field was described in detail.)

In 1864, a Scottish mathematician and physicist named James Clerk Maxwell (1831–1879) proposed that a changing electric field also gives rise to a changing magnetic field, and modified the theory of the day that described mathematically the source of magnetic fields in the universe. With this modification, Maxwell demonstrated mathematically that changing electric and magnetic fields behave as waves in empty space—waves that travel at a speed of 3.00×10^8 m/s! He knew that this was the speed of light in a vacuum, so he hypothesized (correctly) that light must be a wave with electric and magnetic components. Such a wave is called an **electromagnetic wave**. Thus, in 1864, Maxwell proposed that *light travels through space as an electromagnetic wave*. Maxwell's modified theory was confirmed experimentally in 1888 by German physicist Heinrich Hertz (1857–1894), nine years after Maxwell's death.

electromagnetic wave a wave with electric and magnetic components that travels in a vacuum at a speed of 3.00×10^8 m/s

Maxwell summarized his theory in a set of equations, called Maxwell's equations. These are far beyond the scope of this book, but the consequences of his theory are what are important to our understanding of light. We will look at these consequences in terms of what is now known about the creation and propagation of light.

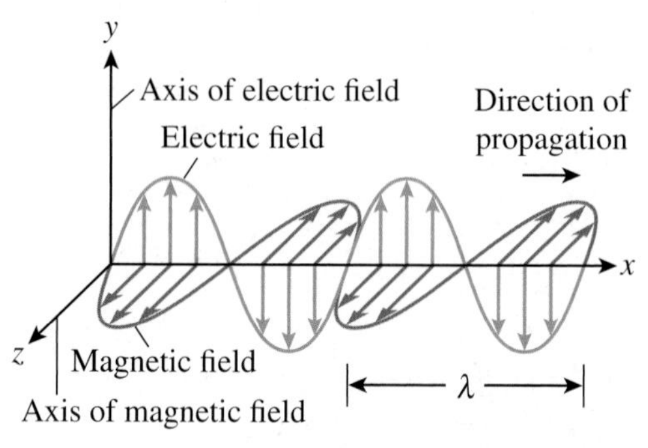

Figure 16-46 An electromagnetic wave at a given instant in time.

Electromagnetic waves, including visible light, can be created by the acceleration of electrically charged particles. The charged particles are often electrons, although other types of accelerated charged particles also produce electromagnetic waves. *Radio waves* are produced by the oscillations of electrons in the antennas of the radio stations; as the electrons oscillate, they undergo acceleration and radiate electromagnetic waves.

Light travels through space in the form of electromagnetic waves. Such a wave consists of two perpendicular components: an electric field and a magnetic field (thus the word "electro-magnetic"). The electric and magnetic fields are both perpendicular to the direction of propagation of the electromagnetic wave, as shown in Figure 16-46. (Be sure you understand this diagram before you read on.)

Electromagnetic waves are transverse. This is illustrated in Figure 16-46 which shows that the electric and magnetic components oscillate in planes that are perpendicular to the direction of propagation. The distance from one wave peak to the next is one wavelength, λ, which is shown in only one of several possible ways in the diagram. (You will discover further evidence of the transverse nature of light in the final section of this chapter.)

Electromagnetic waves produced by charged particles with regular periodic accelerations have the same frequency as the accelerations causing them. Since an electromagnetic wave has a frequency and a wavelength, and travels at a constant speed in a vacuum, it must obey the universal wave equation, $c = f\lambda$, where c is the speed of the

[4]See Section 23.5 for a thorough discussion of Faraday's law.

wave, f is the frequency of the wave, and λ is the wavelength. (If an electromagnetic wave is travelling in a vacuum, its speed is 3.00×10^8 m/s, and this speed is given the symbol c rather than v.)

Now, let us relate the consequences of Maxwell's theory to the wavelengths of visible light, discovered by Thomas Young. The visible colours range from red, with long wavelengths, to violet, with short wavelengths. The human eye receives the different wavelengths and interprets them as different colours.

SAMPLE PROBLEM 16-9

What is the range of frequencies of visible light, assuming the range of wavelengths in a vacuum is from 400 nm to 700 nm? (Use three significant digits.)

Solution

From $c = f\lambda$,

$$f_1 = \frac{c}{\lambda_1} = \frac{3.00 \times 10^8 \text{ m/s}}{4.00 \times 10^{-7} \text{ m}} = 7.50 \times 10^{14} \text{ Hz}$$

$$f_2 = \frac{c}{\lambda_2} = \frac{3.00 \times 10^8 \text{ m/s}}{7.00 \times 10^{-7} \text{ m}} = 4.29 \times 10^{14} \text{ Hz}$$

Thus, the range of frequencies is from 4.29×10^{14} Hz to 7.50×10^{14} Hz.

DID YOU KNOW?

James Clerk Maxwell was a brilliant mathematician whose theories of electromagnetic waves were based on the solutions to mathematical equations rather than on experimental results. Although he was a theoretical physicist and a mathematician, he was given the position of Professor of Experimental Physics at Cambridge University, England, in 1871 where he used a fund given to the university to set up the Cavendish Laboratory, a famous science laboratory that he directed until his death in 1879. Approximately eight years after his death, his prediction that accelerating charged particles cause electromagnetic waves was proven experimentally. This led to numerous applications, such as the transmission of telegraph, radio, and television signals.

The range of frequencies and wavelengths (in a vacuum) of the visible spectrum are illustrated in part of Figure 16-47. Notice that the visible spectrum forms only a small portion of the entire set of electromagnetic waves, called the **electromagnetic spectrum**. The waves in this spectrum all have common properties that the speed is constant and they obey the relation $v = f\lambda$, where $v = c$ in a vacuum. However, the waves have different energies associated with them. As shown in the diagram, the higher-frequency waves have higher energies than the lower-frequency waves. *Microwaves*, such as those used in microwave ovens, have frequencies and energies higher than radio waves. *Infrared waves*, which have various uses, including heat therapy, have lower frequencies and energies than visible light (Figure 16-48). (*Infra* means lower than.) *Ultraviolet waves*, the type that can cause sunburns and can also lead

electromagnetic spectrum the entire set of electromagnetic waves that obey the universal wave equation and can travel in a vacuum at 3.00×10^8 m/s

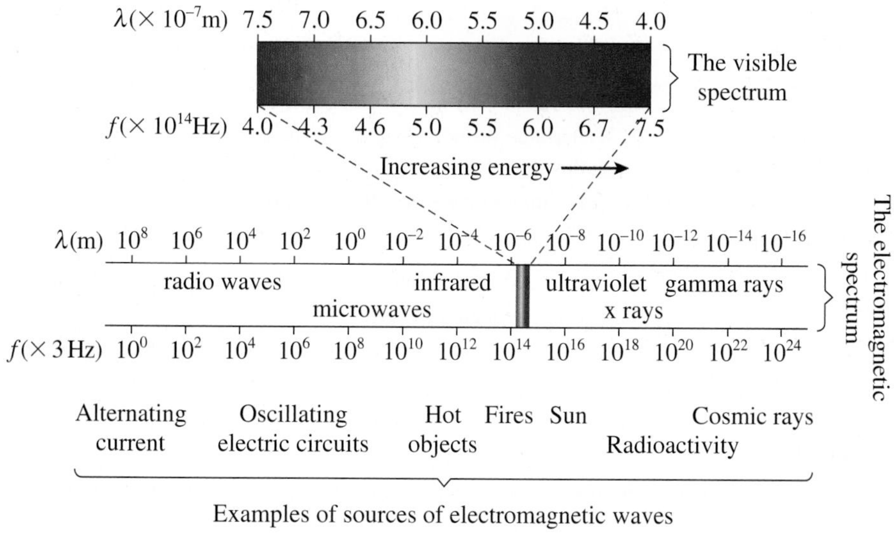

$\lambda(\times 10^{-7}$m) 7.5 7.0 6.5 6.0 5.5 5.0 4.5 4.0

The visible spectrum

$f(\times 10^{14}$Hz) 4.0 4.3 4.6 5.0 5.5 6.0 6.7 7.5

Increasing energy →

λ(m) 10^8 10^6 10^4 10^2 10^0 10^{-2} 10^{-4} 10^{-6} 10^{-8} 10^{-10} 10^{-12} 10^{-14} 10^{-16}

radio waves, microwaves, infrared, ultraviolet, x rays, gamma rays

The electromagnetic spectrum

$f(\times 3$Hz) 10^0 10^2 10^4 10^6 10^8 10^{10} 10^{12} 10^{14} 10^{16} 10^{18} 10^{20} 10^{22} 10^{24}

Alternating current, Oscillating electric circuits, Hot objects, Fires, Sun, Radioactivity, Cosmic rays

Examples of sources of electromagnetic waves

Figure 16-47 The visible spectrum and the electromagnetic spectrum.

Figure 16-48 A camera sensitive to infrared light captured this image of hot soup. The scale at the right indicates how the colours in the image relate to the temperatures of everything in the room, including the soup bowl and tabletop.

to skin cancer, have higher frequencies and energies than visible light. (*Ultra* means higher than.) *X rays* and *gamma rays* have still higher frequencies and energies. Notice that, in many cases, the classifications of waves overlap each other. For example, radio waves overlap with microwaves, etc.

The different energies associated with the different parts of the electromagnetic spectrum can be thought of in terms of the energies required to produce the waves: more energy is required to accelerate charges at higher frequencies, so the electromagnetic wave it emits will have a higher energy. The bottom part of Figure 16-47 lists some of the ways in which charged particles are made to accelerate to produce electromagnetic waves. Once the electromagnetic waves are produced, they can be used for a wide variety of applications. For example, a modern surgical device called a Gamma Knife uses radioactive cobalt-60 to produce gamma rays. These high-energy electromagnetic waves can bombard brain tumours in a precise way without scalpels or incisions.[5]

Like many other scientific theories, Maxwell's theory of electromagnetic waves opened the door to many more questions and discoveries. As the 20th century approached, the search for a greater understanding of the nature of light involved the question of what medium, if any, was required for the propagation of electromagnetic waves. The answers to this question and others led to the photon theory of light, in which light is treated as consisting of photons with specific amounts of energy, called quanta. Experiments demonstrate that light displays properties of particles in certain circumstances, and properties of waves in other circumstances. Details of the photon nature of light are discussed later in the text (Section 17.7) in the Optical Instruments chapter.

DID YOU KNOW?

Microwave ovens used in homes typically use 2.45 GHz electromagnetic waves to heat food. Polar molecules in the food, such as water molecules, oscillate at this high frequency to try to align with the oscillating electric field of the microwaves. This agitation results in an increased temperature within the food to a depth of about 3.8 cm. The principle of the microwave oven was discovered by accident in 1945 by Percy Spencer, who noticed that as he was experimenting with electromagnetic waves in the research laboratories at the Raytheon Corporation, a chocolate bar in his pocket started to melt!

TRY THIS!

Measuring the Speed of Light in the Kitchen

Remove the rotating tray from your microwave oven. Scatter chocolate chips uniformly over a clean piece of cardboard that fits inside your microwave oven snugly. Put the chocolate chips into the oven and heat them for a few seconds—just until the chocolate starts to melt. Remove the cardboard and chocolate carefully, without jostling, and notice that there are periodic lines of melted chips across the cardboard: these correspond to maxima in the standing-wave pattern of the electromagnetic waves inside the oven. You can use the distance between the maxima to calculate the wavelength of the electromagnetic waves. Look on the inside of the door for the frequency of your oven and use your data in the universal wave equation to verify the speed of light. (If you can't find the frequency, use the value stated in the Did You Know? box regarding microwave ovens.) Then enjoy some chocolate!

EXERCISES

16-42 A laser beam is travelling horizontally eastward and the magnetic component of the laser waves is in the horizontal plane. In what plane is the electric component?

16-43 In each case, which type of wave is associated with the higher energy?

(a) microwaves; x rays
(b) infrared light; ultraviolet light
(c) green light; blue light

16-44 A hydrogen gas emission tube emits light with a wavelength of 411 nm in a vacuum. What are the frequency and colour of this light?

16-45 Determine the wavelength (in air or a vacuum) of each of the following.

(a) a microwave with a frequency of 99.9 MHz
(b) a gamma ray with a frequency of 4.17×10^{22} Hz

16-46 For one AM radio station and one FM radio station, calculate

(a) the frequency in hertz
(b) the wavelength of the emitted waves (in air or a vacuum)

[5]This is discussed in more detail in Chapter 24, Figure 24-18.

16.7 Polarization of Light

Have you ever compared polarizing sunglasses with regular sunglasses by looking through them? Both types of sunglasses reduce the brightness of the light, but only the polarizing glasses reduce the glare off roads, water, cars, or other objects. How do polarizing sunglasses reduce glare? In what ways can light be polarized? What applications are made of the polarization of light? These and other questions are explored in this section.

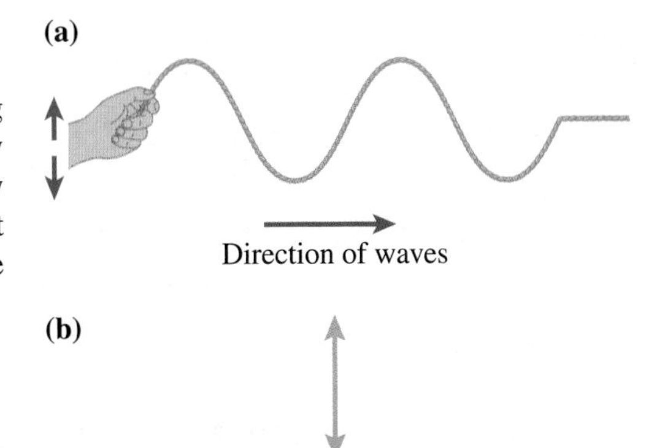

Figure 16-49 A plane-polarized wave. **(a)** The hand is oscillating in the vertical plane producing waves on the rope that are polarized vertically. **(b)** Symbol for a vertically polarized wave approaching the observer.

Polarization of Transverse Waves

In Sections 16.1 to 16.5, we investigated several instances in which light displayed wave properties, and in Section 16.6 we learned of the theory that light travels as a transverse electromagnetic wave. Evidence that light waves are transverse rather than longitudinal is found in the polarization of light.

Imagine transverse waves being generated on a rope in the vertical plane, as shown in Figure 16-49 (a). Such a wave is said to be **plane polarized**, or **linearly polarized**, which means the wave is oscillating in one plane only; in this case, the vertical plane. Figure 16-49 (b) shows a convenient symbol indicating plane polarization in the vertical plane; it resembles a view of the rope with the wave travelling toward the observer.

plane-polarized (or **linearly polarized**) **wave** the property of a wave in which the wave is oriented in one plane only

Transverse waves, for example on a rope, can be polarized in any plane that is perpendicular to the direction of propagation of the wave. Figure 16-50 shows a horizontally polarized wave with its symbol, as well as a randomly polarized set of waves, again with a convenient symbol. (Besides plane polarization, transverse waves can undergo circular polarization or elliptical polarization. We will restrict our discussion to plane polarization.)

Longitudinal waves cannot be polarized because they transfer by oscillations that are parallel to their direction of travel. Thus, since light waves can be polarized those waves must be transverse.

Ordinary light from such sources as the Sun or an incandescent light bulb is not polarized. A source of light, such as the filament of a light bulb, has millions of atoms, each of which can emit light in any direction. Thus, the light emitted has millions upon millions of waves that are oriented randomly.

If ordinary light is not polarized, how can it become polarized? The four methods described here are called selective absorption, reflection, scattering, and double refraction.

(a) A horizontally polarized wave and its symbol.

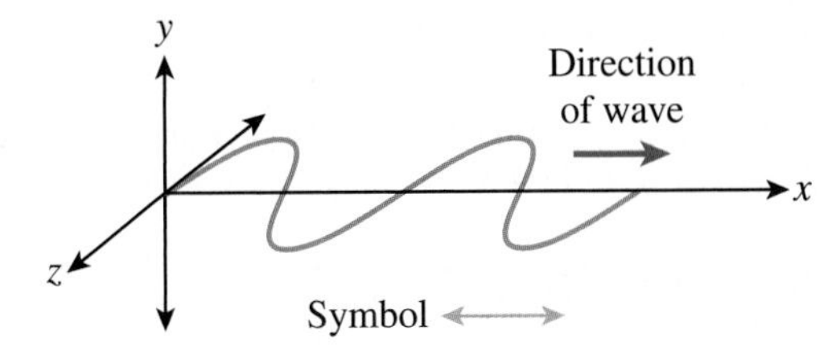

(b) Randomly polarized waves, with symbol.

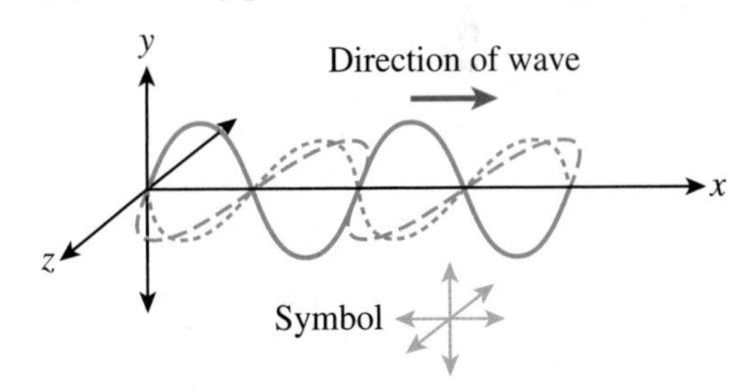

Figure 16-50 Other examples of plane-polarized waves. As in Figure 16-49, the symbols of the polarized waves are drawn from the perspective of the wave approaching the observer. **(a)** A horizontally polarized wave and its symbol. **(b)** Randomly polarized waves, with symbol.

Light Polarization by Selective Absorption

Certain crystals found in nature absorb the transverse light waves polarized in all planes except one; in other words, most of the electromagnetic waves have been selectively absorbed. Thus, the light transmitted through these crystals is plane polarized. Crystals or other materials that transmit light in one plane of polarization and absorb it in the other planes are called **dichroic** (Figure 16-51 (a)). The semi-precious stone called tourmaline has transparent varieties that are dichroic (Figure 16-51 (b)).

Dichroic materials can be used for manufacturing **polarizing filters**, which are materials that allow light to be transmitted in a single plane. The most common polarizing material in use today is called **Polaroid**, which is a plastic absorbing filter containing a dichroic substance; it was invented by an American scientist, Edwin Land (1909–1991). In one technique of manufacturing Polaroid, thin sheets of a substance composed of long-chain molecules of a hydrocarbon, such as polyvinyl alcohol, are stretched so that the molecules become parallel. Then the stretched sheets are dipped in iodine, which attaches to the long molecules and causes them to become electrical conductors. When light interacts with these conducting molecules, it is absorbed.

dichroic material a crystal or other material that transmits light in one plane and absorbs it in the other planes

polarizing filter a material that allows light to be transmitted in a single plane

Polaroid a plastic absorbing filter containing a dichroic substance

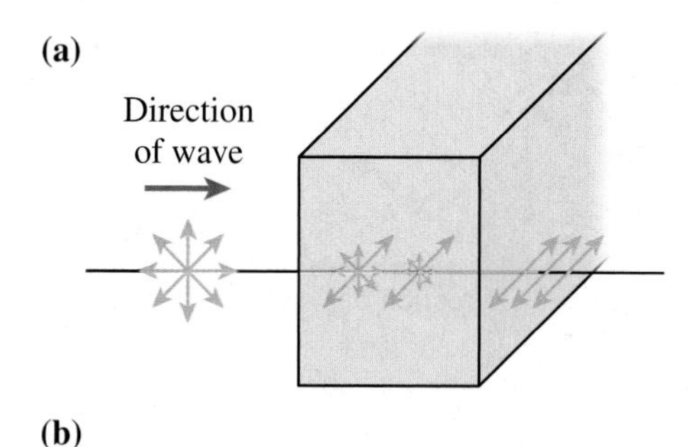

Imfoto/Shutterstock

Figure 16-51 **(a)** A diagram of a dichroic material. In this case, the material transmits light in the horizontal plane. **(b)** An uncut tourmaline gemstone.

axis of polarization the axis of a polarizing filter that allows the transmission of a plane polarized wave

DID YOU KNOW?

The dates given in reference books for Edwin Land's contribution of Polaroid filters vary from about 1928 to 1938. Land was born in 1909, and some books say he discovered the process of making polarizing filters at 19 years of age and developed the technology of mass production later. Of the over 500 inventions that he patented, one of the most famous is the instant camera, which he conceived when he took a photo of his three-year-old daughter and she asked why she could not see the photo right away. The instant camera was manufactured by the Polaroid-Land Corporation, so it is also called a Polaroid camera, after the company name. However, this name has nothing to do with the polarization of light.

If unpolarized light from a source, such as a light bulb, strikes a Polaroid filter, the transmitted light is plane polarized in a direction that is parallel to what is called the **axis of polarization** of the filter. (The note[6] at the bottom of the page explains why the polarization axis is perpendicular to the long-chain molecules in the filter.) Often, a little mark on the filter indicates the direction of this axis (Figure 16-52 (a)). If you look through such a filter at a light bulb, you will not detect any polarization. However, if you place a second Polaroid filter between the first one and your eyes and rotate either filter, you will discover that, when the two axes of polarization are perpendicular to each other, all or almost all the light is absorbed (Figure 16-52 (b)). (A small amount of very short wavelength light may still be transmitted.)

To help visualize polarization, consider vertically polarized waves on a rope travelling toward a set of barriers, each with a slit (Figure 16-53). When the slits of the barriers are both parallel to the plane in which the waves are oscillating, the waves will be transmitted. If, however, the second slit is perpendicular to the first slit, the wave energy will be absorbed. (This analogy is helpful to understand why the axes of the two polarizers must be perpendicular to absorb all the light, but it is opposite to the physics of what occurs when light interacts with the long-chain molecules in the polarizing material, as described in Footnote 6.)

To determine the amount of plane-polarized light transmitted through a polarizing material, consider a transverse wave of amplitude A_0 striking a polarizer that has a vertical polarizing axis. If the polarization axis of the wave is at an angle θ relative to the vertical (Figure 16-54), the amplitude of the transmitted light is the vertical component of the amplitude of the original wave:

$$A = A_0 \cos\theta \quad (16\text{-}17)$$

where A_0 is the original amplitude and A is the amplitude of the transmitted wave.

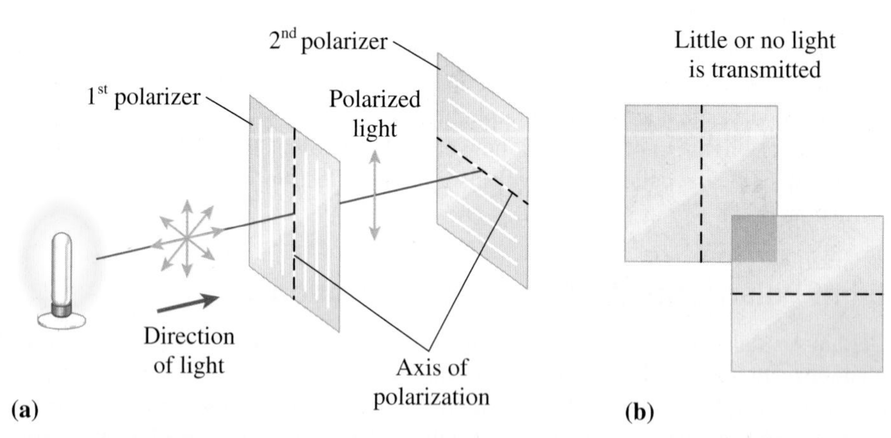

Figure 16-52 Polarizing filters. **(a)** Light that passes through the first polarizing filter becomes polarized in the vertical plane, and is then absorbed as it reaches the second polarizing filter. **(b)** When two polarizing filters are crossed at 90°, little or no light is transmitted.

[6] The explanation of how long-chain conducting molecules absorb light in all but one plane is based on the electromagnetic wave theory, described briefly in the previous section. Like other electromagnetic waves, light waves consist of an electric component and a magnetic component that are perpendicular to each other. The electric component is important in polarization because, when it is parallel to the stretched electrically conducting molecules, it sets up an electric current in the molecules and becomes absorbed. Electric components of the light wave perpendicular to the molecules do not interact with the molecules, and are thus transmitted. Waves with electric components at angles between these extremes are partially absorbed and partially transmitted. Thus, the axis of polarization is perpendicular to the alignment of the molecules of the type of polarizing filter described.

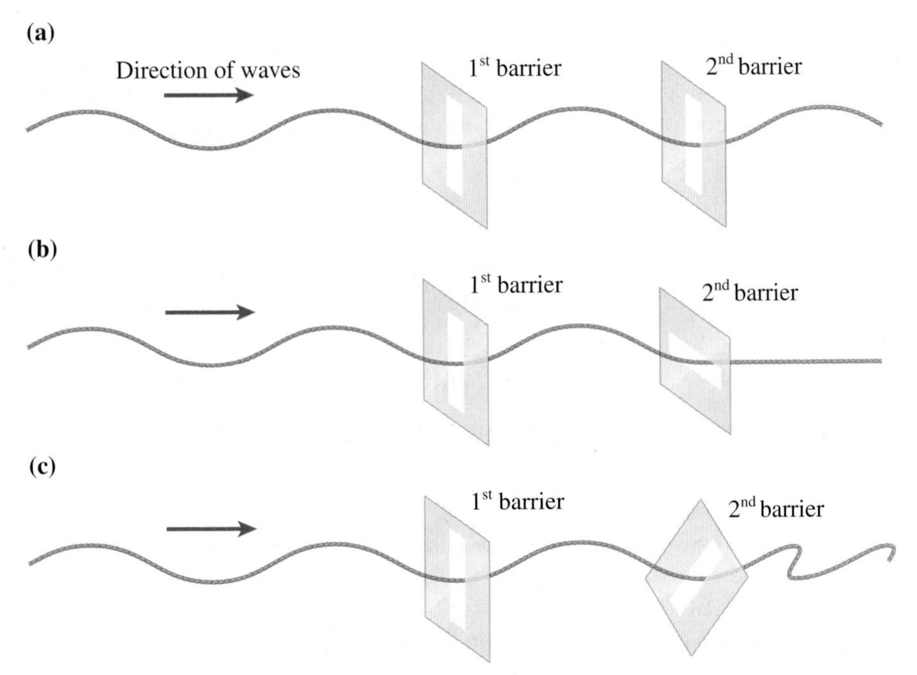

Figure 16-53 The transmission and absorption of plane polarized waves on a rope. **(a)** Maximum transmission and minimum absorption occur when the slits in the barriers are parallel. **(b)** Minimum transmission and maximum absorption occur when the slits in the barriers are perpendicular. **(c)** Some absorption and some transmission occur when the second slit is at an angle to the first slit.

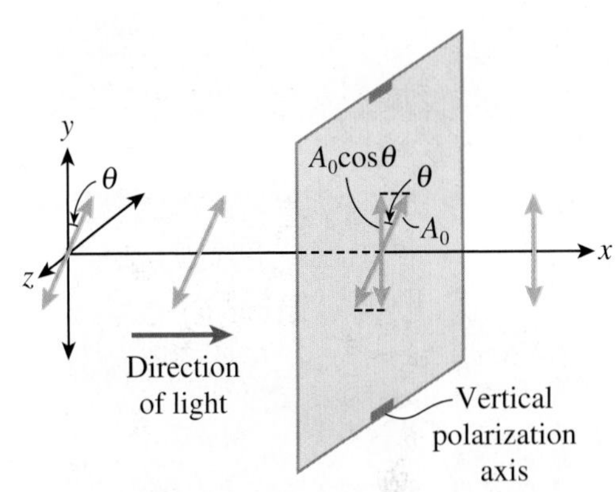

Figure 16-54 When a plane-polarized wave is polarized at an angle θ to the polarization axis of the filter, the amplitude of the transmitted wave is $A_0 \cos\theta$.

It is known that the intensity of light is proportional to the square of the amplitude, so we can write

$$I = I_0 \cos^2\theta \qquad \textbf{(16-18)}$$

where I_0 is the intensity of the incident plane-polarized light, I is the intensity of the transmitted light, and θ is the angle between the incoming plane-polarized light and the polarization axis of the filter.

Let us extend the situation of a wave in a single plane to unpolarized light, which has waves with polarization axes in all planes perpendicular to the direction of propagation. Figure 16-55 uses a small number of waves to represent all the possible number of waves striking a polarizing filter with a vertical polarizing axis. To determine the total intensity of the transmitted light, we add the components of the individual waves. It should be evident that 100% of the light in the vertical plane will be transmitted, none of the light in the horizontal plane will be transmitted, and 50% of the light in the plane at 45° will be transmitted (since $\cos 45° = 0.707$, and $\cos^2 45° = 0.500$). Considering these and the remaining components, it turns out that the average transmitted intensity is 50% of the incident light. This value is a maximum because some of the energy is absorbed even without polarization.

We can conclude that when unpolarized light of intensity I_0 is aimed toward a set of two polarizing filters, the intensity transmitted through the first filter is $I_1 = 0.5I_0$ and the intensity transmitted through the second filter is $I_2 = I_1 \cos^2\theta$ (Eqn. 16-18), where θ is the angle between the polarization axes of the two filters.

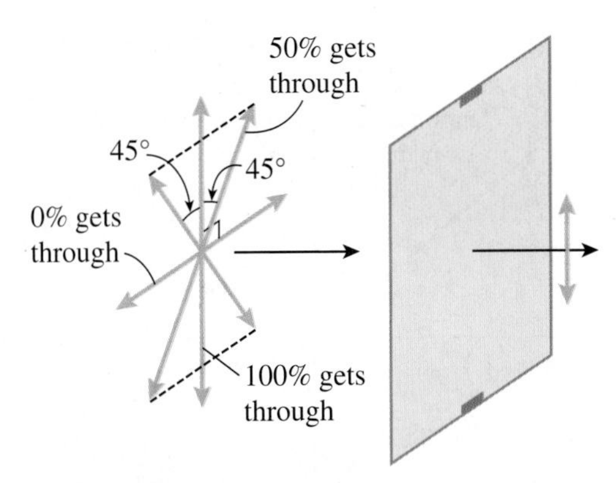

Figure 16-55 A maximum of 50% of the intensity of the original unpolarized light is transmitted through a polarizing filter.

(a)

(b)

Figure 16-56 Using a polarization filter to reduce glare. These two photos were taken at the same location. **(a)** shows the reflected glare from a restaurant patio barrier (a vertically polarized reflection), allowing you to see the blue bench, etc. **(b)** shows how the polarized reflected light can be absorbed by using a polarizing filter, allowing you to see the tables and chairs on the patio.

polarization angle the angle of incidence or reflection at which complete polarization of reflected light occurs; symbol, θ_p

SAMPLE PROBLEM 16-10

Unpolarized light strikes a set of two polarizing filters whose axes are at 33° to each other. What fraction of the intensity of the unpolarized light is observed after the second filter?

Solution

As was just shown, the maximum intensity transmitted through the first filter is 50%, or one-half the intensity of the unpolarized light, I_o. From Eqn. 16-18, the intensity transmitted through the second filter is

$$I_2 = I_1 \cos^2\theta = (0.50 I_0)\cos^2 33° = 0.35 I_0$$

Thus, 35% of the initial intensity is transmitted.

An interesting consequence of the transmission of polarized light is that two polarizing sheets can be crossed so no light is transmitted, but when a third sheet is placed between the other two at an angle of 45° to both of them, a fraction of the light gets through. You will be asked to verify this in Question 16-50.

Light Polarization by Reflection

When light strikes flat, non-metallic surfaces such as glass, water, pavement, or paint, it is partially reflected and partially absorbed or transmitted.[7] At most angles of incidence, the reflected portion is partially polarized. The direction of this polarization is parallel to the surface, so reflection from a horizontal surface produces light polarized horizontally. This polarized light is seen as a bright glare that can be eliminated by placing a polarizing filter between the reflecting surface and the observer. Polarizing sunglasses are worn with their axis of polarization in the vertical plane, which means they will absorb light waves in the horizontal plane. Polarizing sunglasses do not eliminate the glare from vertical surfaces such a glass windows (unless the wearer tilts the glasses by 90°) because such glare is caused by vertically polarized light. Polarizing filters used on cameras can be rotated in any direction to eliminate glare in any one plane at a time. (See Figure 16-56.)

The degree of polarization from a reflecting surface depends on the material as well as the angle of incidence of the light. The complete polarization of reflected light occurs at an angle of incidence or reflection called the **polarization angle**, θ_p. In 1812, a Scottish physicist named David Brewster discovered that the polarization angle for light travelling from one medium (n_1) into a second medium (n_2), is given by

$$\tan\theta_p = \frac{n_2}{n_1} \qquad (16\text{-}19)$$

where n_1 is the index of refraction of the first medium and n_2 is the index of refraction of the second medium (Figure 16-57).

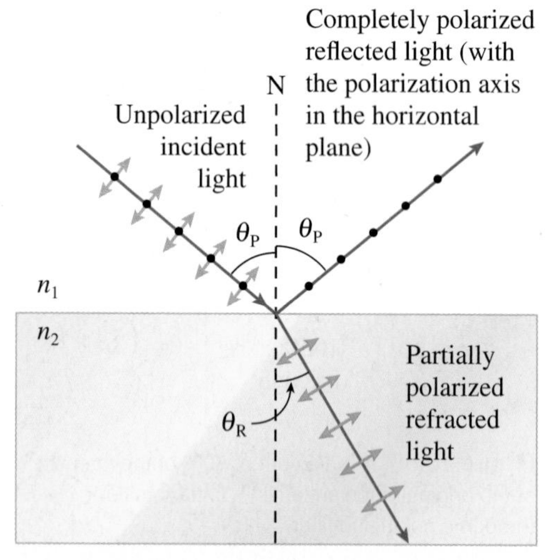

Figure 16-57 The polarization angle is the angle at which complete polarization of reflected light occurs.

SAMPLE PROBLEM 16-11

(a) Determine the polarization angle for light striking water ($n = 1.33$) from air.

(b) What is the angle between the reflected ray in the air and the refracted ray in the water?

[7] Notice that shiny metallic surfaces, such as polished aluminum or stainless steel, are not included in the list of surfaces because they do not absorb light the way non-metallic surfaces do.

Solution

(a) From $\tan\theta_p = \dfrac{n_2}{n_1}$

$$\therefore\ \theta_p = \tan^{-1}\left(\frac{n_2}{n_1}\right) = \tan^{-1}\left(\frac{1.33}{1.00}\right) = 53.06° \text{ or } 53.1°$$

(b) Figure 16-58 illustrates the situation. Using Snell's law,

$$n_1\sin\theta_p = n_2\sin\theta_R$$

$$\theta_R = \sin^{-1}\left(\frac{n_1\sin\theta_p}{n_2}\right) = \sin^{-1}\left(\frac{1.00\sin 53.06°}{1.33}\right) = 36.9°$$

Thus the angle of refraction is 36.9°. Notice that 36.9° + 53.1° = 90.0°. Therefore, the angle between the reflected ray and the refracted ray is 90.0°.

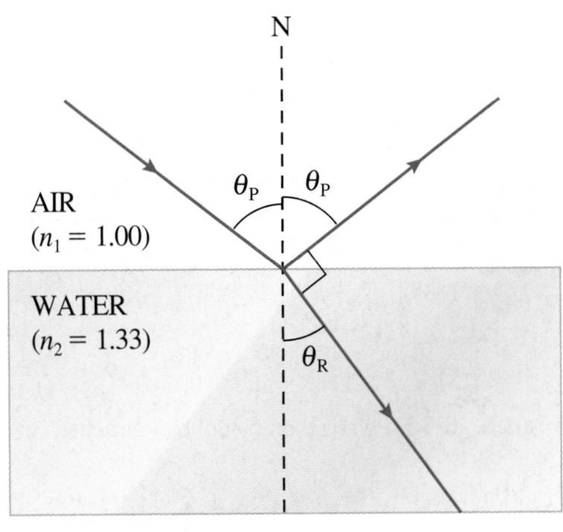

Figure 16-58 Sample Problem 16-11.

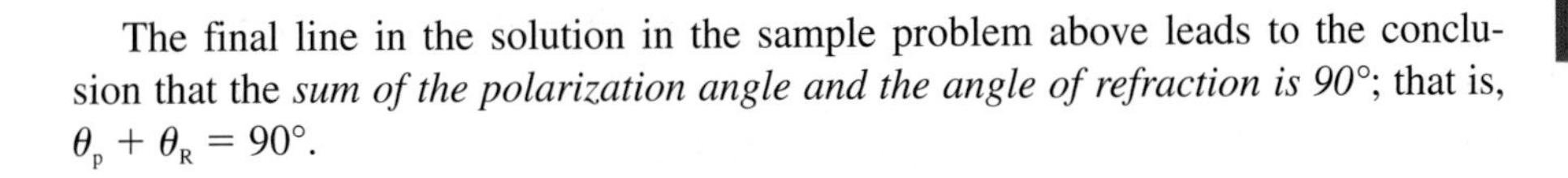

The final line in the solution in the sample problem above leads to the conclusion that the *sum of the polarization angle and the angle of refraction is 90°*; that is, $\theta_p + \theta_R = 90°$.

DID YOU KNOW?

The polarization angle is also known as Brewster's angle and Eqn. 16-19 is called Brewster's law. David Brewster (1781–1868) was the person who invented the kaleidoscope.

Light Polarization by Scattering

When light strikes a system of atoms and molecules, such as a gas, the electrons in the atoms and molecules can absorb the light and re-radiate part of it. The absorption and re-radiation of light by a substance is called **scattering**. A common example of the scattering of light occurs in the sky when sunlight is incident upon air molecules. The amount of scattering depends on the size of the wavelength of the light relative to the particles doing the scattering. White light from the Sun consists of many different wavelengths, but it is the blue waves that are scattered the most in Earth's atmosphere, so the sky appears to be blue. The longer wavelengths, from yellow to red, are transmitted through the atmosphere more readily, so the Sun appears to be yellow when it is viewed more directly and reddish when the light passes through more atmosphere (at sunrise and sunset).

scattering the absorption and re-radiation of light by a substance

Scattered light is polarized. To see why, refer to Figure 16-59, which shows unpolarized light from the Sun striking air molecules in Earth's atmosphere. The Sun, the observer, and the scattering molecules form an observation plane. To simplify our discussion, we will consider only the transverse light waves from the Sun that are

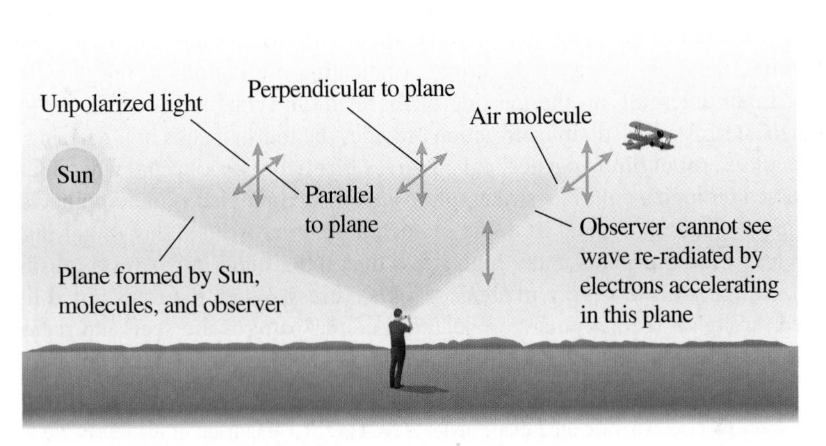

Figure 16-59 The polarization of scattered light in the sky.

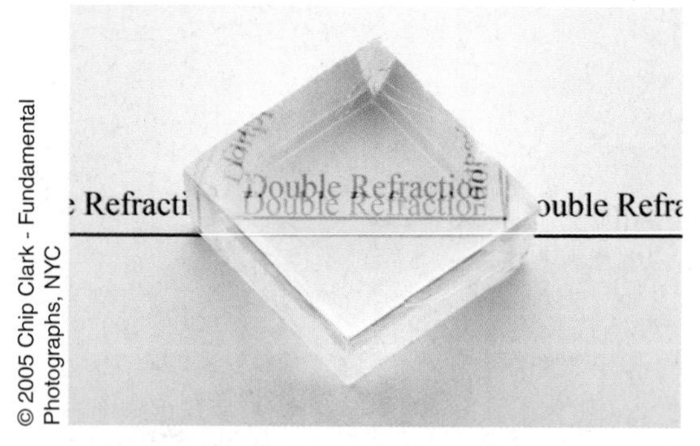

Figure 16-60 Polarization by double refraction.

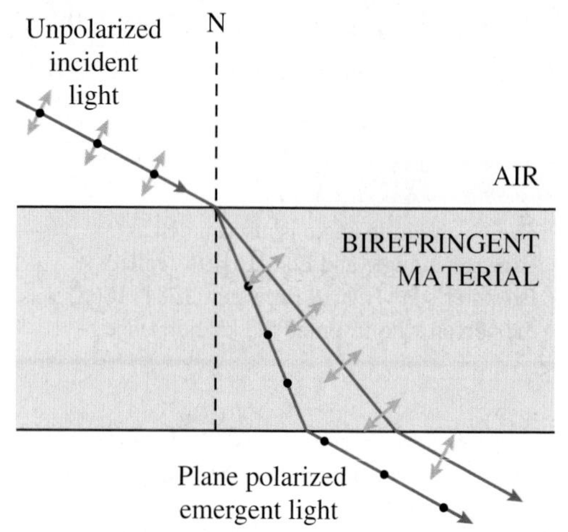

Figure 16-61 Polarization by double refraction.

birefringence the double refraction that occurs in a transparent crystalline substance in which the speed of light can differ in different directions

stress birefringence refraction of light in a transparent medium that depends on the mechanical stress applied

DID YOU KNOW?

There is evidence that the Vikings, who sailed the northern seas more than 1000 years ago, used a dichroic mineral called cordierite to help them navigate. Near the Arctic Circle in the summer the Sun dips below the horizon, so it cannot be used directly for navigation, but the sky remains bright so the stars commonly used for navigation cannot be seen. The Vikings could look at the scattered light of the sky through cordierite and use their observations to determine directions.

plane polarized either parallel or perpendicular to the observation plane.[8] Waves that are polarized parallel to the observation plane strike electrons in the air molecules and cause them to oscillate parallel to the observation plane. Any light re-radiated by these electrons cannot be in the direction of the oscillation, so the observer will not see any of this light. However, waves that are polarized perpendicular to the observation plane strike electrons in the molecules and cause them to oscillate perpendicular to the plane. The re-radiated light in this case will be in the observation plane, perpendicular to these oscillations. Thus, if the observation plane is horizontal, the scattered light is polarized vertically, a situation that can be observed at sunrise and sunset. Other directions of polarization occur at other times of the day, and the direction polarized depends on the particular observation plane based on where the observer looks in the sky. (The best way to observe polarized scattered light in the sky is to view the sky away from the Sun through a polarizing filter. Never look directly toward the Sun, even through a filter.)

The scattering of light has many applications. Photographers take into consideration the scattering of light in the sky as well as from the surroundings. Some insects, such as bees, and other animals, such as the horseshoe crab, have eyes that are sensitive to polarized light. Thus, they can navigate using the scattered light in the sky.

Light Polarization by Double Refraction

When light travels from air into another transparent medium, its speed changes and it refracts (unless the incident light is normal to the surface). In many substances, such as glass and water, the speed of light is the same in all directions, so the amount of refraction can be found by applying Snell's law ($n_1 \sin\theta_1 = n_2 \sin\theta_2$). In some crystalline substances however, the speed of light may be different in different directions, and light that enters can be split so that two portions refract different amounts. The double refraction that occurs in a transparent crystalline substance in which the speed of light can differ in different directions is called **birefringence**. Calcite (also called Iceland spar) and quartz are examples of birefringent media.

If a prism made of birefringent material is used to view an object, a double image of the object is seen. This observation has been known for hundreds of years because calcite was found in its natural state in Iceland by sailors who took samples of the crystal to other parts of the world (Figure 16-60).

How does light entering a birefringent medium "know" which portion should travel at which speed? The answer depends on the plane in which the electric field of the transverse light waves are oscillating. This explains why double refraction causes polarization of light. Figure 16-61 shows an unpolarized light beam incident upon a birefringent medium. Waves with a polarization axis parallel to the surface slow down more and thus refract more, as shown by the dots (indicating the direction that is out of the page). Think of these dots as electric fields oscillating into and out of the page. Waves polarized in the plane that is perpendicular to the surface slow down less and thus refract less, as shown by the arrows (indicating oscillations in the plane of the page). Light emerging on the far side of the medium remains split into two plane-polarized beams. Thus, double refraction provides the fourth means of polarizing light.

Another form of birefringence, called **stress birefringence**, occurs when light that has passed through a polarizer strikes a transparent medium that is undergoing various amounts of stress. A typical example of such a material is Plexiglas. Light interacts differently with molecules under high stress than those under low stress, causing the polarized light to rotate more in areas of higher stress. When the transmitted light is viewed through a second polarizer, beautiful colour patterns are seen and regions of high stress can be identified by the closeness of the colour bands. A similar method

[8]To be more precise, what we are talking about is the electric component of the transverse electromagnetic waves.

is applied to Plexiglas models of bridges, tools, and machine parts to determine the magnitude and direction of stress during the designing stages as well as in researching stresses on components of the human body (Figure 16-62).

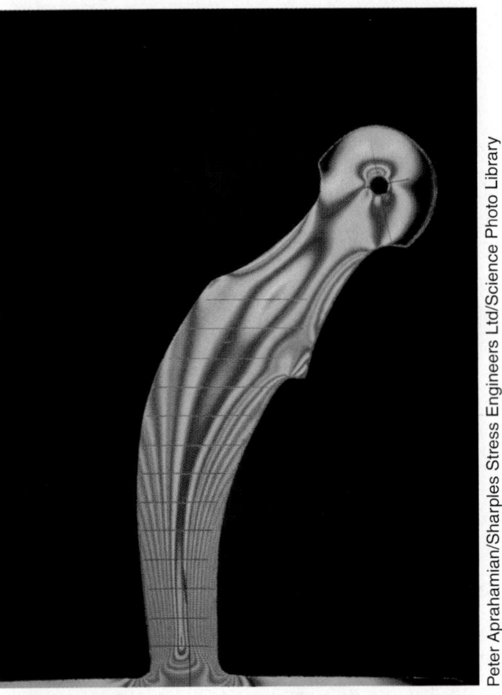

Figure 16-62 This Plexiglas model of a human hip joint placed between two polarizing plates is used to study the stresses on various parts of the joint. The greatest stresses are located where the interference patterns are closest together. Once the engineers are satisfied with the model, they use the design to create the real hip replacement or other component.

Experimenting with Polarizing Filters or Sunglasses

If polarizing filters are not available for this activity, you can use polarizing sunglasses instead. *Caution: Do not use the Sun as a "distant bright light source."*

Using One Polarizer

- Find examples of glare from a non-metallic horizontal surface (calm water, a car, smooth glass, or other shiny surface). Use the polarizer to try to eliminate the glare as much as possible. Estimate the polarizing angle (θ_P) for each situation.
- Find an example of glare from a vertical surface, such as a glass window. Describe how the polarizer must be held to reduce the glare to a minimum. Try to obtain results that resemble those in Figure 16-56. Estimate the polarizing angle (θ_P) in this case.
- View the light scattered from the sky. Try both clear and slightly cloudy regions of the sky. Remember that the polarization of scattered light is most easily observed if the Sun, the particles observed, and the observer form a 90° angle.

Using Two Polarizers

- As you look at a bright light source through a polarizer held horizontally, slowly rotate the second polarizer in your line of view. Under what condition(s) is the absorption maximum? Is there any part of the visible spectrum that you can see when the absorption is maximum?
- View parts of a piece of moulded plastic, such as a clear CD or DVD case or even a sealable sandwich bag, placed between two polarizers. Locate regions of high stress. Try twisting or stretching the plastic to see what happens to the pattern.

Using Three Polarizers

- While viewing a distant bright light source through two crossed polarizers, insert a third polarizer at an angle of 45° between the first two. What happens?

EXERCISES

16-47 Name two sources of unpolarized light. How is this light different from plane polarized light?

16-48 Can sound waves be polarized? Explain.

16-49 Unpolarized light is incident upon a pair of polarizing sheets. What maximum fraction of the intensity of this light is transmitted through the sheets if their axes are oriented at an angle of (a) 30°, (b) 60°, (c) 90°? Why is the word "maximum" used here?

16-50 Two polarizing sheets are placed with their axes crossed, then a third sheet is placed between them at an angle of 45° to both of them. What fraction of the initial intensity of unpolarized light is transmitted through this set of sheets?

16-51 A store has several pairs of sunglasses that are labelled "Polaroid." Describe how you would check to be sure the glasses are indeed polarizing using (a) two pairs of sunglasses and (b) a single pair of sunglasses.

16-52 You are given a Polaroid plate that has not had its polarizing axis marked. How would you determine the orientation of the polarization axis of the plate?

16-53 (a) What is the polarization angle (the angle of maximum polarization) for light striking glass ($n = 1.52$) from air?

(b) What is the angle of refraction in the glass for the situation in (a) above?

16-54 Determine the polarization angle for light that is travelling in water ($n = 1.33$) toward glass ($n = 1.56$).

16-55 A person is facing toward the Sun while fishing on a calm lake ($n = 1.33$).

(a) In what plane is the light reflected from the lake polarized?

(b) The person puts on a pair of polarizing sunglasses. In what plane should the polarization axis of the glasses be to reduce glare from the water?

(c) If the reflected light from the lake exhibits maximum polarization, at what angle is the Sun above the horizon?

16-56 What colour does the sky appear to be on the Moon where there is no atmosphere? Explain why.

16-57 When clouds are not in a dark shadow, they appear to be white in daylight. Explain why the clouds are white while the surrounding sky is blue. (**Hint:** Water droplets are different in size from air molecules.)

16-58 A prism made of calcite or some other birefringent material is placed on a printed page and a double image is seen. Predict what you would observe if you viewed the double image through a polarizing filter and rotated the filter through 360°. (If possible, try this.)

LOOKING BACK...LOOKING AHEAD

In this chapter, we discussed many light phenomena that can be explained if we consider light to be a wave. We compared the characteristics of light waves to the characteristics of mechanical waves (Chapter 13). Light undergoes diffraction and interference when it interacts with objects whose size is approximately the same as the wavelength of the light. The interference patterns that occur when light passes through single or double slits is used to determine the wavelengths of visible light. The ability to resolve two objects depends on the diffraction of light (as well as on other factors). The concepts and equations for double-slit interference were extended to multiple-slit diffraction gratings, which provide one means of studying various kinds of spectra.

Fringes observed when light reflects off (or is transmitted through) thin films or air wedges are also interference phenomena. These fringe patterns can be analyzed mathematically and the interference can be applied in practical ways.

Light is an electromagnetic wave that is produced by charged particles undergoing acceleration. Such a wave propagates through space at a constant speed of 3.00×10^8 m/s (in a vacuum). The visible spectrum is only a small part of the set of electromagnetic waves called the electromagnetic spectrum. The waves have different frequencies, wavelengths, and energies, as well as many different uses.

The transverse nature of electromagnetic waves is verified by the ability of light to be plane polarized. The four ways of polarizing light are selective absorption, reflection, scattering, and double refraction. Polarization has several applications.

In the next chapter, we explore applications of light in optical instruments, including cameras, microscopes, telescopes, lasers, and the human eye.

CONCEPTS AND SKILLS

Having completed this chapter you should now be able to do the following:

- Apply Huygens' principle in a diagram to illustrate why diffraction of light occurs.
- Manipulate the equations involving double-slit interference variables (θ, λ, x, L, d, N, or M) to determine unknown quantities.
- Manipulate the equations involving single-slit interference variables (θ, λ, x, L, w, N, or M) to determine unknown quantities.
- State what factors affect the resolution of two objects and describe how they do so.
- Apply Rayleigh's criterion to determine the angular separation for resolution through a single slit.
- Manipulate the equations involving diffraction grating variables (θ, λ, x, L, d, and M) to determine unknown quantities.
- Distinguish between a continuous spectrum and a line spectrum, and between an emission spectrum and an absorption spectrum.
- Describe applications of spectroscopy and diffraction in two and three dimensions.
- Manipulate the equations involving thin-film and air-wedge interference variables (t, λ, λ_{air}, n, N, and M) to determine unknown quantities.
- Describe applications of both thin-film and air-wedge interference.
- Describe how an electromagnetic wave is produced.
- Apply the universal wave equation to determine any one of the speed, frequency, or wavelength of an electromagnetic wave given the other two quantities.
- Describe how light can become plane polarized by selective absorption, reflection, scattering, and double refraction.
- Determine the portion of the intensity of light transmitted through a single polarizing filter, or through a set of two or more filters, knowing the angle between the axes of polarization of the filters.
- Determine the polarization angle for light reflecting off the surface of a medium of known index of refraction.
- Describe applications of the polarization of light.

KEY TERMS

You should be able to define or explain each of the following words or phrases:

Huygens' principle	path difference	diffraction grating	spectrum
coherent waves	Fraunhofer diffraction	transmission grating	continuous spectrum
nodal line	resolution	reflection grating	line spectrum
path length	Rayleigh's criterion	replica grating	spectroscopy

spectrograph
spectroscope
absorption spectrum
x-ray crystallography
iridescence
quarter-wavelength film
air-wedge interference
electromagnetic wave
electromagnetic spectrum
plane-polarized wave
linearly polarized wave
dichroic material
polarizing filter
Polaroid
axis of polarization
polarization angle
scattering
birefringence
stress birefringence

Chapter Review

MULTIPLE-CHOICE QUESTIONS

16-59 For incident light of wavelength λ in a double-slit experiment, the path difference from the two slits to the middle of the second-order maximum is

(a) $\lambda/4$
(b) $\lambda/2$
(c) λ
(d) 2λ
(e) 0

16-60 In a single-slit interference pattern, the light source is changed from red light to blue light. The corresponding interference pattern

(a) spreads out more because the wavelength increases
(b) becomes narrower because the frequency decreases
(c) spreads out more because the frequency increases
(d) becomes narrower because the wavelength decreases
(e) remains the same width because, as the wavelength decreases, the frequency increases

16-61 For the double-slit interference pattern of monochromatic light in Figure 16-63 to go from (a) to (b), which of the following adjustments or combination of adjustments could produce the change?

(i) The slit spacing decreases.
(ii) The slit spacing increases.
(iii) The wavelength of the light increases.
(iv) The wavelength of the light decreases.

(a) (i) and (iv) only
(b) (i) and (iii) only
(c) (ii) and (iv) only
(d) (i) only

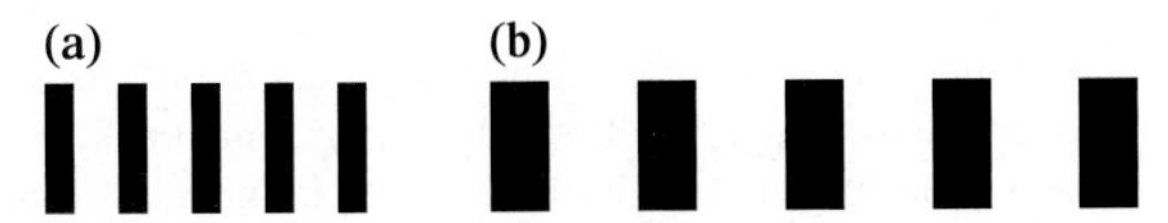

Figure 16-63 Question 16-61. The dark bars are minima and the white bars are maxima.

16-62 A diffraction grating diffracts violet light through an angle that is

(a) independent of the frequency of the light
(b) the same as the angle for red light
(c) less than the angle for red light
(d) greater than the angle for red light
(e) none of these

16-63 When a beam of light travels from air into water at an angle of incidence greater than 0°, it undergoes a change in

(a) speed and direction only
(b) speed, direction, and wavelength only
(c) speed, direction, wavelength, and frequency
(d) wavelength and frequency only
(e) direction, speed, and frequency only

16-64 Which statement is false?

(a) Both transverse and longitudinal waves can be polarized.
(b) Both transverse and longitudinal waves can diffract.
(c) Both transverse and longitudinal waves can interfere.
(d) Both transverse and longitudinal waves can be reflected.
(e) Both transverse and longitudinal waves can be refracted.

16-65 The polarization angle for light striking a transparent medium is 49°. If unpolarized light strikes the medium at an angle of incidence of 49°, which of the following statements is/are correct?

(a) The reflected light is completely polarized.
(b) Any light that refracts does so at an angle of 49°.
(c) Any light that refracts does so at an angle of 41°.
(d) Both (a) and (b) are true.
(e) Both (a) and (c) are true.

Review Questions and Problems

16-66 What is monochromatic light?

16-67 State what happens to the number of nodal lines in a two-source interference pattern if

(a) only the source separation is decreased
(b) only the wavelength of the sources is decreased
(c) only the amplitude of oscillation of the sources is decreased

16-68 Determine the wavelength of the waves producing a two-source interference pattern in each case.

(a) On the fourth nodal line, the path difference is 14 cm.
(b) On the third maximum (from the central maximum), the path difference is 66 mm.
(c) With a source separation of 3.2 cm, the second nodal line makes an angle of 2.8° with the central maximum.

16-69 In each case, one variable of a possible five for a two-source interference pattern is unknown. Solve for it.

(a) $M = 3$; $d = 4.2$ cm; $x = 16$ cm; $L = 86$ cm; $\lambda = ?$

(b) $M = 2$; $d = 2.0$ mm; $x = 4.2$ mm; $\lambda = 5.5 \times 10^{-7}$ m; $L = ?$

(c) $N = 3$; $\lambda = 2.0 \times 10^{-4}$ m; $x = 8.6 \times 10^{-2}$ m; $L = 2.2$ m; $d = ?$

16-70 A two-source interference pattern is set up using waves with a wavelength that is 0.25 times the source separation. At what angles from the central maximum are the first three nodal lines?

16-71 In a two-source interference pattern, the maximum line $M = 1$ makes an angle of 12° to the central maximum. If the source separation is 6.6 mm, what is the wavelength of the waves?

16-72 Figure 16-64 shows a double-slit interference pattern on a screen. Assume that C is along the perpendicular bisector of a line joining the two slits and assume the light is monochromatic.

(a) Using the symbols M and N, what labels (e.g., $M = 2$ or $N = 1$) should be assigned to each position from A to E?

(b) State the path difference in terms of the wavelength of the light at each position from A to E.

Figure 16-64 Question 16-72.

16-73 Light of wavelength 444 nm is used to compare a double-slit pattern and a single-slit pattern. The double slits are separated by 0.150 mm and the single slit is 0.150 mm wide. In both cases, the distance from the slit(s) to the screen is 3.65 m. Find the distance in millimetres from the middle of the central maximum to the first dark fringe for

(a) the double-slit pattern

(b) the single-slit pattern

(c) What is the ratio of these distances?

16-74 A point source emits light of wavelength 520 nm toward a single slit of width 0.085 mm. The light creates an interference pattern on a screen located 2.2 m from the slit.

(a) What is the width of the central maximum?

(b) What angle do the first-order dark fringes subtend with the slit?

16-75 The situation in the previous question is altered by adding a second point source of the same wavelength, with both sources located 3.6 m from the slit.

(a) Determine the distance (in centimetres) between the sources so they are just resolved on the screen.

(b) What angle do the sources subtend with the slit?

16-76 Assume that each rectangle in Figure 16-65 represents a bright light source. Describe how the resolution of the sources would compare in each of the following cases.

(a) viewed from a distance of 10 m and then 20 m

(b) viewed when the sources are lined up perpendicular to your line of sight and then at a 45° angle to your line of sight

Figure 16-65 Question 16-76.

16-77 A red laser beam of wavelength 650 nm is directed through a slit of width 0.62 mm and a diffraction pattern is observed on a screen 3.6 m from the slit.

(a) Determine the angular width of the central maximum found on the screen.

(b) What is the width in millimetres of the central maximum on the screen?

16-78 How does the resolution of two point sources of light depend on (a) the wavelength of the light and (b) the width of the aperture through which the point sources are viewed?

16-79 Does the spacing of the maxima in a diffraction grating pattern increase, decrease, or stay the same if

(a) the number of lines per centimetre of the grating increases?

(b) the wavelength of the light increases?

16-80 Monochromatic light is incident upon a diffraction grating with 4500 lines/cm. At a distance of 1.60 m from the grating, the diffraction pattern is observed on a screen, and the distance between the $M = 0$ and $M = 2$ maxima is found to be 76 cm. Determine the wavelength of the light. (Recall that because the angles involved in diffraction grating patterns are large, $x/L = \tan\theta \neq \sin\theta$.)

16-81 White light incident upon a diffraction grating with 5.0×10^3 lines/cm produces a pattern on a screen 1.8 m from the grating. The first blue and red maxima are located at angles from the central maximum of 7.9° and 11° respectively. What is the ratio of the red wavelength to the blue wavelength?

16-82 **Fermi Question:** Estimate how many wavelengths of yellow light would fit into a cross-section of a pane of window glass. Why is even thin glass not considered to be a "thin film"?

16-83 A thin film of plastic ($n = 1.52$) is placed over another material ($n = 1.36$), and light with a wavelength in air of 555 nm reflects from the plastic.

(a) What minimum thickness of plastic will result in a maximum fringe upon reflection?

(b) If the film were three times as thick as found in (a) above, would there be a maximum or minimum fringe observed upon reflection? Explain.

16-84 A quarter-wavelength coating of magnesium fluoride ($n = 1.38$) covers a lens ($n = 1.56$). The coating is 86.5 nm thick. What are the wavelength and colour of the light for which this coating is optimized?

16-85 Do the equations for air-wedge interference correspond to an air–oil–water situation or to a quarter-wavelength film of magnesium fluoride over glass? Explain.

16-86 Refer to the soap-bubble photograph, Figure 16-39 (b).

(a) What is the cause of the black area at the top surface of each bubble?

(b) **Fermi Question:** Estimate the thickness of the soap film ($n = 1.33$) at the outer edge of each black region.

16-87 Monochromatic light of wavelength 430 nm reflects off two parallel glass plates, each 12.0 cm long, separated by an air wedge. Assuming the plates touch at one end and are separated by a hair of diameter 0.064 mm at the other end, determine the number of bright fringes per centimetre observed in the pattern.

16-88 What does x-ray diffraction through crystals reveal about the structure of matter?

16-89 **Fermi Question:** Estimate or determine the diameter of a hair from your head, then state what waves in the electromagnetic spectrum would diffract most easily around the hair.

16-90 An electron oscillates back and forth in a radio transmission tower (Figure 16-66) at a frequency of 1.12 MHz. Determine the frequency and wavelength (in air) of the electromagnetic waves emitted by these oscillating charges.

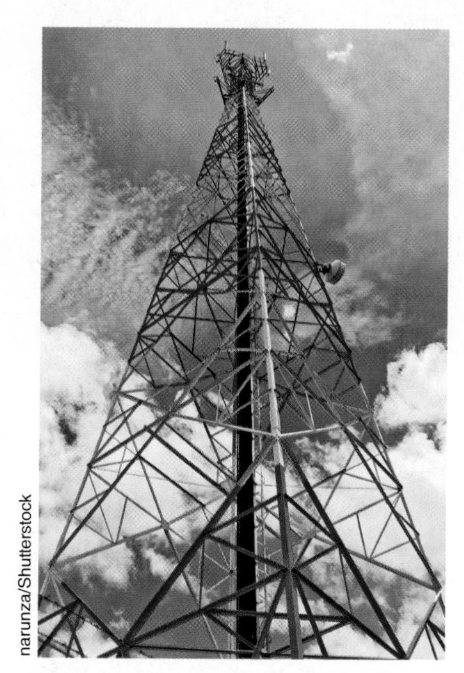

Figure 16-66 A radio transmission tower (Question 16-90).

16-91 (a) What properties does visible light have in common with other parts of the electromagnetic spectrum?

(b) What properties are different?

16-92 Assume that electromagnetic waves travelling in air reflect off a barrier so that a standing wave pattern is set up with 65 cm between the nodes.

(a) What is the frequency of these waves?

(b) To what part of the electromagnetic spectrum do these waves belong?

16-93 Compare the methods used for manufacturing polarizing filters and diffraction gratings.

16-94 What experimental evidence strongly suggests that light waves are transverse?

16-95 In certain types of Polaroid filter, the stretched molecules absorb light waves in which the electric fields oscillate parallel to the molecules. Compare the plane of the oscillating magnetic field of those absorbed waves with the axis of transmission of the filter.

16-96 Unpolarized light passes through two polarizing filters. What should be the angle between the polarizing axes if the transmitted intensity is (a) $\frac{1}{4}$ and (b) $\frac{1}{2}$ the original intensity?

16-97 The angle between the axes of two polarizing filters is 52°. What percentage of the initial intensity of unpolarized light is absorbed?

16-98 Light in air strikes a transparent medium at an angle of incidence equal to the polarization angle of the medium. For the light that enters the medium, the angle of refraction is 31.5°.

(a) Draw a diagram of this situation, showing the angles of incidence, reflection, and refraction.

(b) Determine the index of refraction of the medium.

Applying Your Knowledge

16-99 What would happen to the spacing of the maxima in a double-slit interference pattern if the entire setup were viewed under water? Why?

16-100 Monochromatic light passes through a double slit and creates an interference pattern on a screen 2.84 m from the slits. The slit separation is 0.160 mm and the distance between the first maxima on either side of the central maximum is 2.12 cm.

(a) What is the wavelength of the light?

(b) What is the angular separation between the central maximum and the adjacent bright fringes?

(c) Compare the sine and tangent of the angle found in (b) above. What is the significance of your answer in terms of the derivation of the double-slit interference equations?

16-101 Determine the maximum number of bright fringes, including the central maximum, that can be observed in a double-slit interference pattern using 685 nm light passing through slits separated by 6.20×10^{-6} m.

16-102 A single slit of adjustable width is used to view a diffraction pattern of 628 nm light.

(a) Assuming the slit width starts at about 6300 nm, what happens to the pattern as the slit width decreases?

(b) As the slit width decreases, at what width will there no longer be any dark fringes observed?

16-103 What are the advantages of a diffraction grating over a prism for use in a spectroscope?

16-104 Derive two equations for monochromatic light *transmitted* through a thin film of soap of variable thickness t surrounded by air. The wavelength of the light in the soap film is λ_{film}. One equation should be for maxima, showing the relation between t, λ_{film}, and $M = 0, 1, 2, \ldots$ The other equation should be for minima, showing the relation between t, λ_{film}, and $N = 0, 1, 2, 3, \ldots$ It will be useful to consider three film thicknesses: $t \ll \lambda_{film}/4$, $t = \lambda_{film}/4$, and $t = \lambda_{film}/2$. (**Hint:** Consider two rays of light, one being transmitted straight through the film and the other being reflected twice off the interior surface of the film.)

16-105 Unpolarized light is incident upon a set of four polarizing filters, each at an angle of 30° to the previous one. What percentage of the intensity of the original light is transmitted through the filters?

16-106 Figure 16-67 shows a photograph of the interference pattern viewed using a Newton's rings apparatus, which was mentioned in Section 16.5. Describe what can be inferred by analyzing this pattern.

Andrew Lambert Photography/Science Photo Library

Figure 16-67 Question 16-106.

16-107 Polarizing sunglasses can eliminate glare from non-metallic surfaces. Why are they unable to do the same with shiny metallic surfaces?

16-108 Figure 16-68 shows a photograph of a protractor placed between two polarizers. What does the pattern reveal about the structure of the protractor?

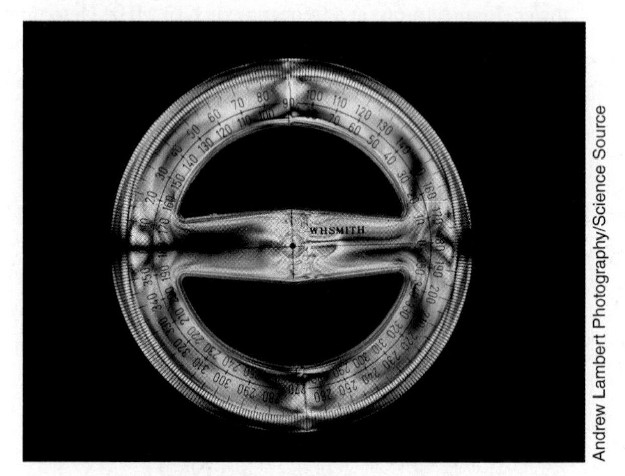

Andrew Lambert Photography/Science Source

Figure 16-68 Question 16-108.

Optical Instruments 17

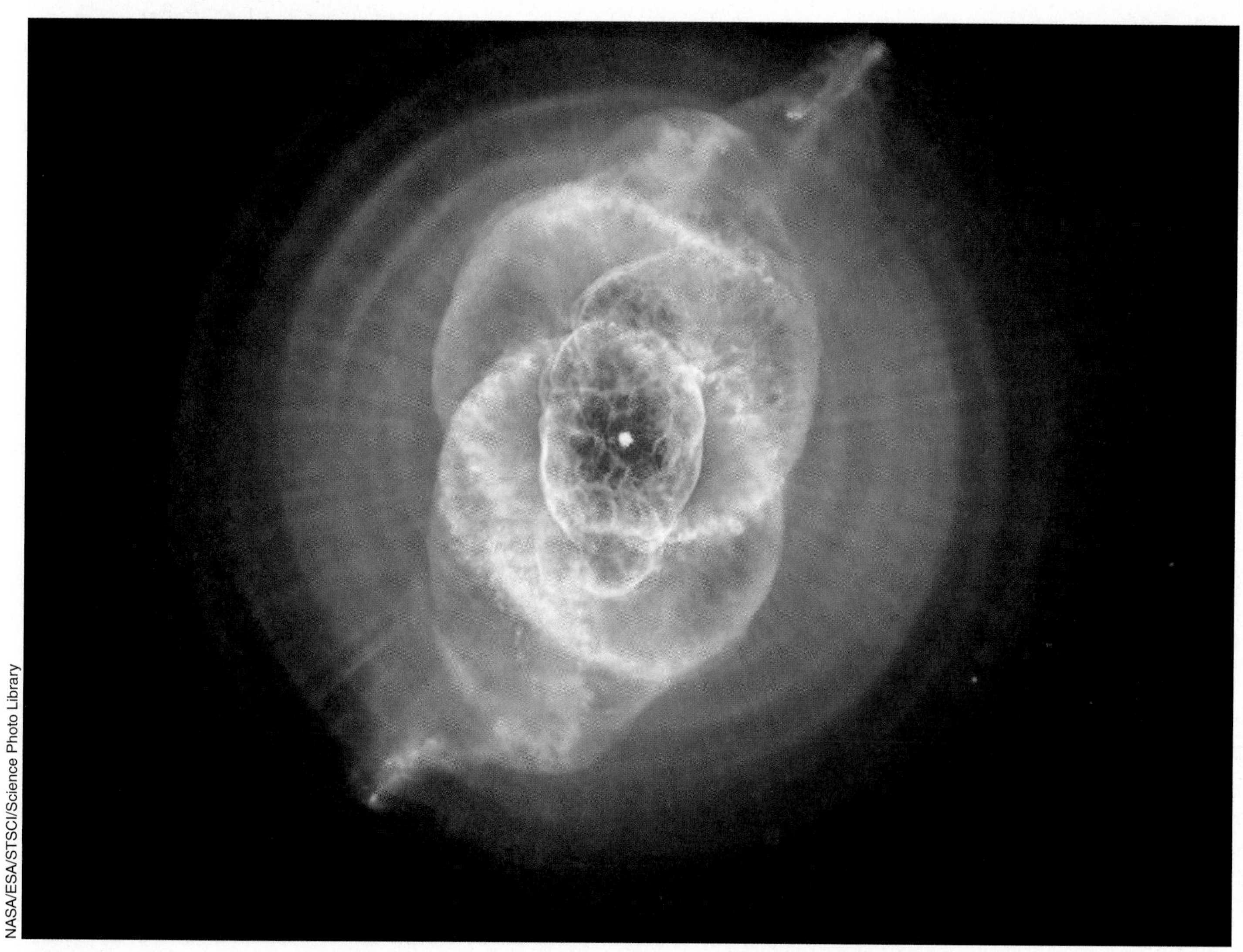

NASA/ESA/STSCI/Science Photo Library

Figure 17-1 This image of the Cat's Eye Nebula is one of numerous photographs of objects in space taken by the Hubble Space Telescope. The nebula is about 3300 light years from our solar system, so the view we see now actually occurred more than 3300 years ago.

When viewing the night sky, our eyes can see the Moon, some planets in the solar system, and numerous stars, including many in our Milky Way galaxy. But there are many more intriguing objects in the universe that become visible by using telescopes. Among the most beautiful objects in the universe are nebulas, which are collections of gases often resulting from the ejection of materials from a star that is either shedding layers or exploding violently as it nears the end of its existence. (*Nebula* comes from a Latin word meaning mist.) Figure 17-1 shows an image of the Cat's Eye Nebula taken by the Hubble Space Telescope, which is in a near-circular orbit above Earth's atmosphere. This image has better clarity than

CHAPTER OUTLINE

would be possible using any telescope on Earth's surface where the atmosphere reduces image quality. The outer rings of the nebula were caused by layers of material ejected at different times from the surface of a central dying star. The other complex patterns are composed of various gases. The specific images generated of the Cat's Eye Nebula depend on the wavelengths of the electromagnetic waves used to obtain them, from infrared, through the visible spectrum, and all the way to x rays. These images enable astronomers to analyze the structure, origin, and evolution of the universe. The Hubble Space Telescope is just one of many telescopes that not only increase our understanding, but also help bolster the public's appreciation of the vastness and beauty of the universe.

We begin this chapter with a discussion of the human eye, the most important and versatile optical instrument. Then the magnifier, microscopes, telescopes (including the Hubble Space Telescope), cameras, and lasers will be presented. Because these instruments involve a lens or a mirror, or sometimes both, the ray diagrams for lenses and mirrors in Chapter 15 will be utilized. Furthermore, the function of these instruments is influenced by the wave nature of light, so the concepts from Chapter 16 also will be applied.

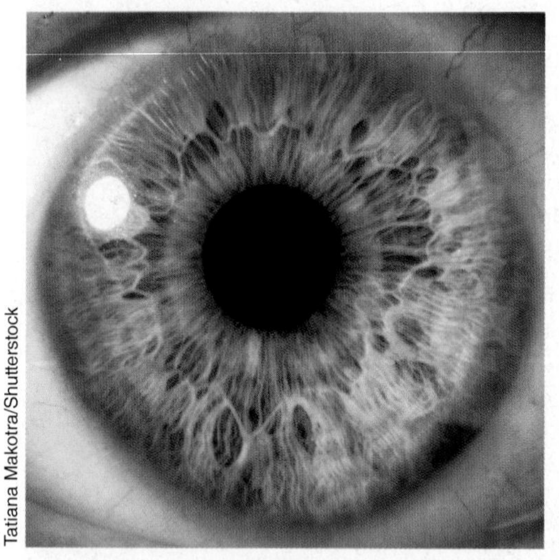
Tatiana Makotra/Shutterstock

Figure 17-2 This close-up view of a human eye reveals only the outer areas of a very complex, sensitive organ.

17.1 The Human Eye and Vision

The human eye (Figure 17-2) is a complex and fascinating part of our anatomy, with many features that are superior to other optical instruments. The eye can adjust to a range of intensities that span a factor of about 10^{10}; it has rapid automatic focusing abilities; it is self-cleaning and self-lubricating; it can repair some types of damage; it has a self-regulating pressure system; it can be aimed in many different directions through muscle control; it can perceive a wide range of colours; and a set of two eyes provides a field of view of over 180° as well as depth perception. The following exploration of the optics of the eye will show the complexity of this amazing instrument.

The Structure and Functions of the Eye

The basic structure of the eye is shown in Figure 17-3. The eye is nearly spherical, with an average diameter from 20 mm to 25 mm. It is protected by the tough, lightproof *sclera*, which is white and covers the entire eye except the area at the front of the eye. Surrounding the eye is a layer of fat that reduces the effect of sharp shocks. Inside the sclera is the *choroid*, a vascular or blood-vessel layer. The **cornea** is a fixed transparent layer at the front of the eye with an index of refraction of about 1.37. The cornea does approximately two-thirds of the focusing of the light rays that enter the eye. Most of the remaining focusing is done by the converging **lens**, which is composed of a fibrous substance with an average index of refraction of about 1.40. Two important jelly-like substances, the *aqueous humour* and the *vitreous humour*, help the eye maintain its shape. These substances have an index of refraction of about 1.33 (the same as water). The aqueous humour supplies nutrients to the cornea and lens, and also supplies cells to repair damage to those components. Most cells in the body that need repair get help from the blood system. However, having blood vessels spread throughout the cornea or lens would greatly impair our vision, so the role of the aqueous humour in supplying cells to repair the living cells of the cornea or lens is important.

cornea a fixed transparent layer at the front of the eye with an index of refraction of 1.37

lens (of the human eye) a fibrous substance behind the pupil of the eye with an average index of refraction of about 1.40

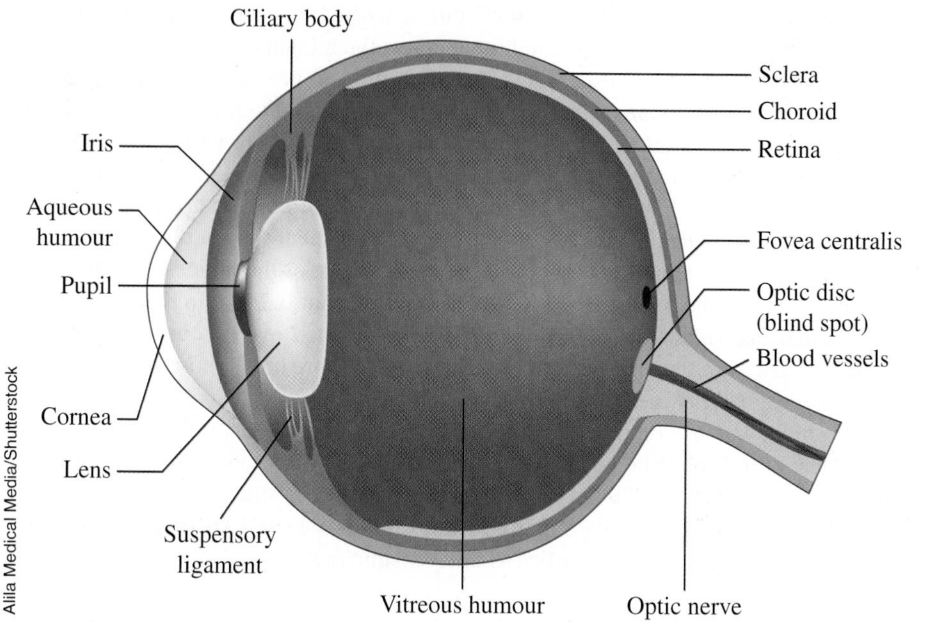

Figure 17-3 The basic structure of the human eye.

retina the rear layer of the eye that is light sensitive; it consists of numerous blood vessels, nerves, and receptors

far point the farthest distance at which objects appear in focus by an eye; for healthy eyes the far point is stated as infinity

near point the closest distance at which nearby objects can be focused by an eye; for healthy eyes the near point is usually considered to be 25 cm

accommodation the process of controlling the shape of the eye's lens by the ciliary muscles

After the light rays pass through the eye they strike the light-sensitive area at the rear of the eye called the **retina**, which consists of numerous blood vessels, receptors, and nerves. There, the light rays form a real, inverted image on the retina (described shortly), just as a converging glass lens can form a real image on a screen behind the lens.

The lens is somewhat flexible, and its shape is controlled by muscles in the *ciliary body*, which is attached to the *suspensory ligament*. When a person is viewing a distant object, these muscles are relaxed and the lens has a normal shape. When the person is viewing a nearby object, the ciliary muscles force the lens to become thicker so the image can remain in focus. The greatest distance at which objects are in focus on the retina when the ciliary muscles are relaxed is called the **far point**, while the closest distance at which nearby objects can be focused is called the **near point**. For normal, healthy eyes that can see Mars or the Moon clearly, the far point is stated as infinity and the near point is usually considered to be 25 cm. These values vary greatly for people with vision defects (to be discussed later). The process of changing the shape of the lens is called **accommodation** (Figure 17-4). If a person views nearby objects for an extended time, the ciliary muscles remain under tension, resulting in eyestrain. This problem is alleviated by looking at distant objects periodically.

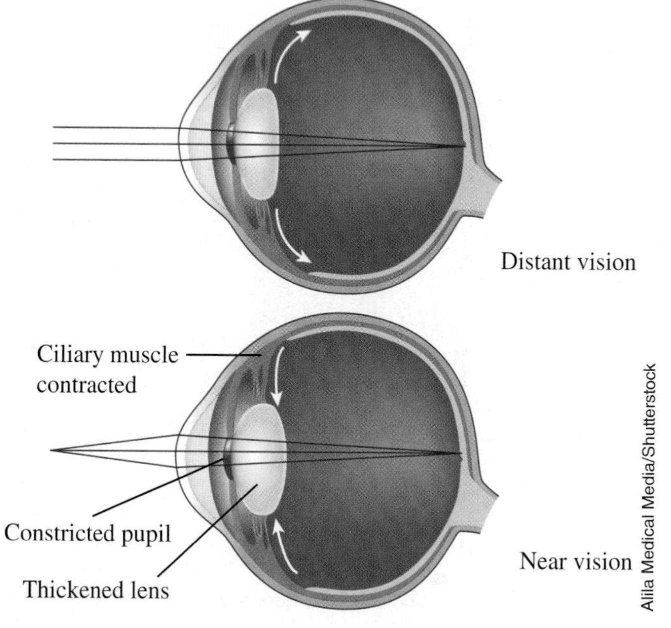

Figure 17-4 The ciliary muscles control the accommodation from the distant position (far point) to the near point. When a person looks at a distant object (the top diagram), the ciliary muscles are relaxed. When the person looks at a nearby object, the ciliary muscles exert a tension illustrated by the white arrows in the lower diagram, causing the lens to change shape.

DID YOU KNOW?

The eye loses its focusing ability under water because the index of refraction of the cornea is close to the index of refraction of water. Divers wearing face masks have an air space between the water and their eyes, allowing for normal focusing. Fish have a similar problem when they are out of water, although some fish that are frequently near the surface have a double set of eyes, one set for above water and a lower set for below the surface of the water. Dolphin's eyes have a special structure to help them see equally well in air as in water, and the Eurasian otter has eyes that become more convex underwater, enabling them to see fish better.

Your Near and Far Points

Determine the approximate near point and far point of each of your eyes, one at a time. If you wear corrective lenses, perform this activity both with and without the lenses.

pupil (of the eye) the opening of the eye through which light enters the lens

cones receptors in the eye responsible for day vision and sensitive to colours

rods receptors in the eye that react to grey and black shades and are responsible for night vision

The **pupil** of the eye is the "window" through which the light enters the lens. It appears black because most of the light that enters the human eye is absorbed inside. The pupil is surrounded by the *iris*, the coloured portion of the eye. The iris controls the size of the pupil, which ranges in diameter from about 2 mm to 8 mm. In bright light the pupil contracts, and in dim light the pupil dilates (enlarges) to let in more light.

The retina has two types of receptors, named **cones** and **rods** after their basic shapes (Figure 17-5). Each eye has more than 6 million cones that are sensitive to colour (discussed in detail shortly) and are concentrated near the middle of the retina. When the eye looks straight at an object the clearest part of the image is located in a tiny region called the *fovea centralis*, which is only about 0.3 mm in diameter. Each eye also has approximately 120 million rods concentrated mainly outside the fovea centralis. Cones are responsible for colour vision. Rods are much more sensitive than cones and respond better than cones when light intensity is low; thus, rods are responsible for night vision. Cones and rods send electrical signals to the brain in response to the amount of light absorbed by the pigment in their cell membranes, and these signals are changed into perceived images. The retina is sensitive to stimuli other than visible light. Astronauts in space have reported flashes in their eyes, later discovered to be caused by high-energy atomic particles called cosmic rays.

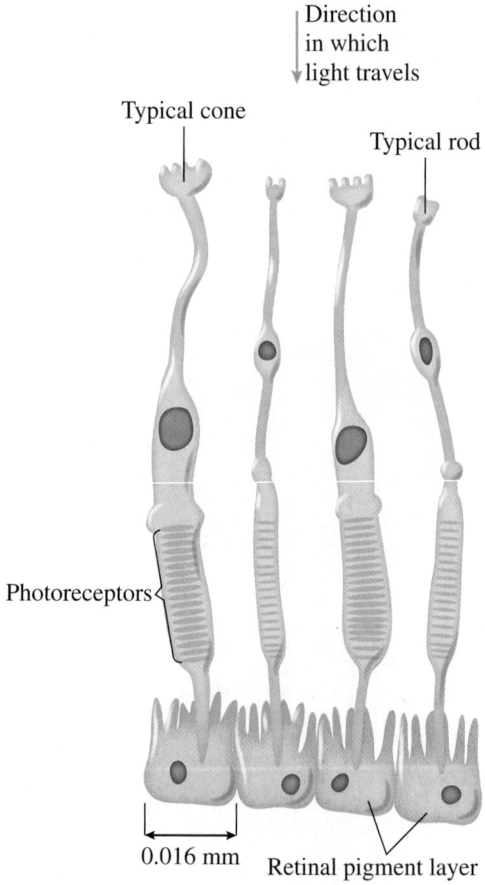

Direction in which light travels

Typical cone

Typical rod

Photoreceptors

0.016 mm

Retinal pigment layer

Figure 17-5 The rod and cone structure of the eye.

DID YOU KNOW?

Why do some animals, such as the fox featured in Figure 17-6, have eyes that appear to "glow in the dark"? The answer is that they have a reflective coating behind the rods to give the rods a second opportunity to absorb light. This allows for more sensitive low-light viewing for hunting at night, and causes the eyes to appear bright when a light shines into them.

Figure 17-6 A fox's eyes reflecting light.

Light energy received by the rods and cones initiates a chemical reaction, which in turn sets up electrical impulses that are transmitted via the retinal nerves to the larger **optic nerve**. This nerve transmits the signals to the brain, which interprets what we see. At the location in each eye where the retinal nerves join the optic nerve, the *optic disc* or *blind spot* occurs. This blind spot can be demonstrated only when one eye is being used.

TRY THIS!

Determining the Angle of Your Blind Spot

Hold the book at arm's length. Cover your left eye and stare directly at the letter **A** in Figure 17-7 with your right eye. Move the book toward you until the letter **B** disappears. This shows the location of your blind spot of your right eye.

- Determine the angle θ between the perpendicular line from A toward your right eye and the direction of your blind spot by applying $\tan\theta$.
- Repeat the measurement for your left eye by staring directly at the letter B and noting when A disappears.
- What happens if you move the book closer to or farther from you?

Figure 17-7 Determining the angle to the blind spot of each eye.

optic nerve the nerve that transmits the signals from the eye to the brain, which interprets what we see

primary light colours the three light colours, red, green, and blue, that when added together produce white light

secondary light colours the three light colours, yellow, magenta, and cyan, that are produced by combining two of the primary light colours

Our vision is enhanced by other mechanisms. For example, having two eyes (binocular vision) rather than just one allows us to judge distances more accurately. Viewing a nearby object forces our two eyes to aim closer to each other and, with experience, we then can judge the approximate distance to the object. Yet another example is the ability of the eye to retain an image for about 1/25 s after the object is removed. If a new image replaces the initial one after a short period the images appear to be in constant motion. This capacity of the eye allows us to enjoy movies. Each movie frame is shown for a fraction of a second, but its image remains in the eye until the image from the next frame arrives. For almost a century, the movie industry has shown movies at 24 frames per second (24 fps), but newer formats use higher rates, such as 48 fps, causing the movie to look smoother and more lifelike, but by some reports causing dizziness in certain viewers.

The images we see may also be influenced by our experience and by the surroundings of what we are viewing. Figure 17-8 illustrates examples of optical illusions that trick the eye.

(a)

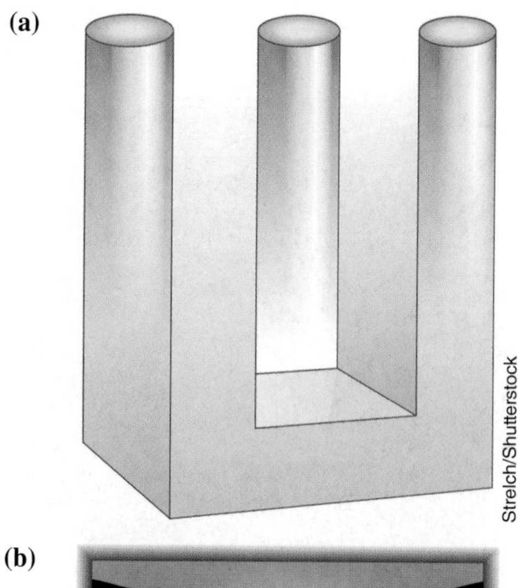

Strelch/Shutterstock

(b)

Peteri/Shutterstock

Figure 17-8 Optical illusions. **(a)** Our experience informs us that the figure shown has three prongs. However, on close inspection you would agree that the object would be impossible to build. **(b)** Does the image you see alternate between a vase and two people?

Colour Vision

As described earlier, the cones in the retina are sensitive to colours. The theory of colour vision is complex and not fully understood, so a simplified explanation of the "three-colour theory" will be described here. This theory proposes that for humans and other primates there are three types of cones. (Most mammals have two types of cones and birds have four types of cones.) In humans, the three types of cones are most sensitive to 570 nm (red) light, 534 nm (green) light, and 445 nm (blue) light.[1] These three colours are called the **primary light colours** because when they are all combined together, white light is produced. When any two primary colours are combined, the colour produced is a **secondary light colour**, yellow, magenta, or cyan. Figure 17-9 shows addition of the primary light colours to produce the secondary light colours and white light.

TACKLING MISCONCEPTIONS

Comparing Light Colours and Colour Pigments

Do not confuse the addition of light colours or energies with the addition of colour pigments. Pigments or paints absorb or subtract light energies rather than adding them. The primary pigment colours are the secondary light colours, and when mixed they theoretically absorb all incident light, producing black or nearly black.

The primary light colours can be added in various combinations and intensities to produce all the possible shades of colour. Similarly, the cones in the eye can respond in various combinations to allow us to view all the different shades of colours. For example, if the cones sensitive to green and red are stimulated equally, we see yellow.

A demonstration supporting the three-colour theory of cones involves staring at a coloured object until some of the cones in the retina become fatigued. To observe *retinal fatigue*, place a small, bright red object on a white background; stare at the centre of the object for about 40 s, then stare intently at one spot on a piece of white paper. The cones sensitive to red will have become fatigued (the chemical process cannot keep up with the stimulation), so when you stare at a white object, which is reflecting red, green, and blue light, only the cones sensitive to green and blue respond normally. Thus, you observe the colour cyan in the shape of the original object. (This applies to anyone with normal colour vision.)

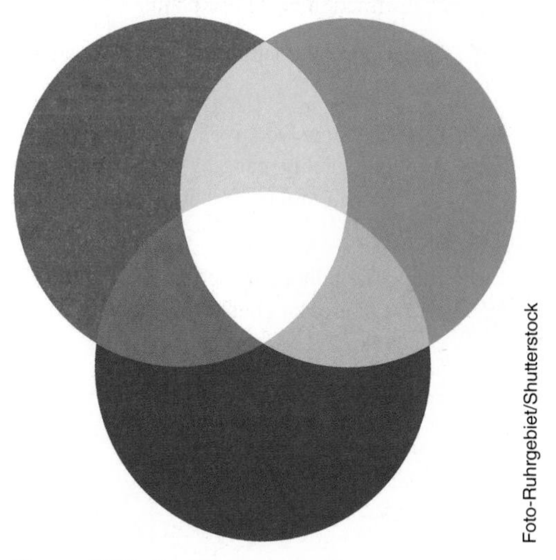

Foto-Ruhrgebiet/Shutterstock

Figure 17-9 If any two of the three bright primary colour light sources (red, green, and blue) overlap on a white background, they produce the secondary light colours (yellow, cyan, and magenta). If all three primary light sources overlap on a white background, white light results.

[1]The wavelengths listed are at the approximate middle of the range of wavelengths to which each type of receptor is sensitive.

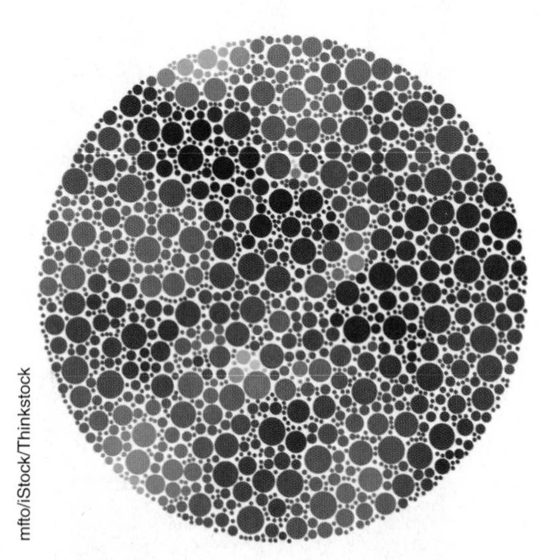

Figure 17-10 If you can see the number 2 in this image, you do not have red-green colour blindness.

TRY THIS!

Observing Retinal Fatigue

Use a small brightly coloured object (or set of objects) placed on a white background to demonstrate retinal fatigue of the cones as described in the text. (You can even try staring at the pattern in Figure 17-9.) Be sure you understand why you observe the coloured afterimages. Also, use a black and white object to try to observe retinal fatigue of the rods. How would you design a flag so that when you observe its afterimage you would see the true colours of your country's flag?

Colour blindness is a defect of the eye in which certain shades of colour are not clear. Its cause is that some cones do not respond to the light energies received. In one form of colour blindness, the cones needed to see red do not react, thus reducing the number of colours seen. In another type of colour blindness, the red and green cones respond simultaneously to all colours of the red-to-green portion of the visible spectrum, so most reds and greens appear to be yellow. Other types of colour blindness are usually less severe. About 8% of all males and 0.5% of all females suffer from some sort of colour blindness. Figure 17-10 shows one example of many possible tests for colour blindness.

EXERCISES

17-1 (a) State what you could do to initiate a change in the pupil of your eye and describe what that change would be.
(b) Repeat (a) for the lens of your eye.

17-2 State the function(s) of each of these parts of the eye: the cornea, sclera, aqueous humour, rods, cones, and optic nerve.

17-3 (a) In which part of the eye, the lens or the cornea, is the speed of light slower? Explain.
(b) In which of these two parts does the greater amount of refraction occur? Explain.

17-4 Under normal conditions, we do not notice the blind spot in our eyes. Why not?

17-5 Describe how you would make a booklet that when flipped through quickly page-by-page would appear to be a cartoon character in motion. What principle of vision is applied in your creation?

17-6 Rank the silhouette heights in Figure 17-11 from shortest to tallest, and then check your answer with a ruler. What do you conclude?

Figure 17-11 Question 17-6.

17-7 State which cones in the human eye must be activated in order to see (a) red, (b) cyan, (c) magenta, and (d) yellow.

17-8 If you stare at a bright yellow circle and then at a white piece of paper, what is the colour of the afterimage you see? Explain why this happens.

17.2 Vision Defects and Their Corrections

Normal, healthy eyes allow the formation of a clear image on the retina of the eye for objects at distances from about 25 cm (the near point) to infinity (the far point). However, more than half the population of North America suffers from vision defects resulting in a blurred image on the retina. The four general types of focusing problems are discussed below.

In **myopia**, or **nearsightedness**, the image in the eye comes to a focus in front of the retina. This problem is usually caused by an eye that is too long or a cornea that is too sharply curved. To a person with myopic vision, distant objects appear out of focus (blurry). One way to correct this fault uses a diverging lens, as Figure 17-12 and the next sample problem demonstrate. (Another method will be described shortly.) The thin-lens equation (Eqn. 15-8) developed in Section 15.5, $\frac{1}{f} = \frac{1}{d_o} + \frac{1}{d_i}$, can be applied to the corrective lenses. It is important to remember the sign conventions used in this equation.

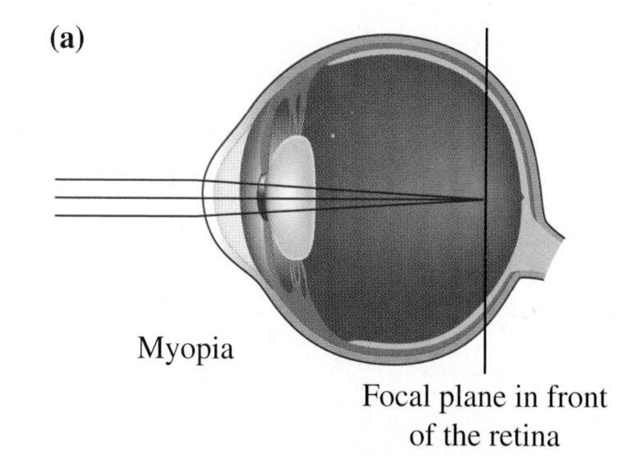

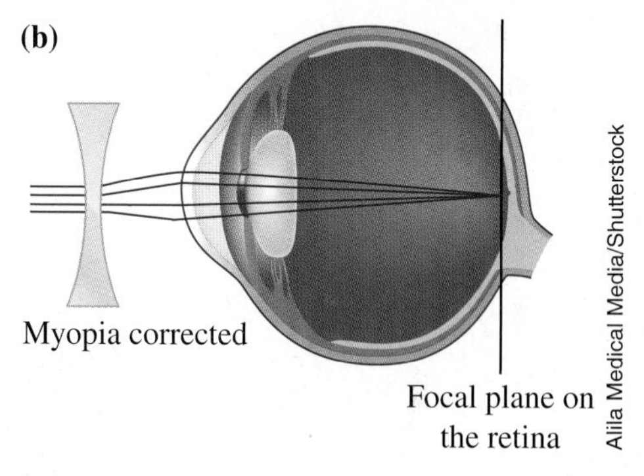

Figure 17-12 Myopia or nearsightedness. **(a)** The image comes to a focus in front of the retina. **(b)** A diverging lens can be used to correct the defect.

myopia (or nearsightedness) the condition of the eye in which the image comes to a focus in front of the retina; also called nearsightedness

SAMPLE PROBLEM 17-1

A person with myopic vision has a far point of 75 cm. Determine the focal length of a lens that will provide a far point of infinity. Assume the distance from the eye to the lens is negligible.

Solution

Figure 17-13 illustrates how to approach this problem. When the eye views a distant object ($d_o \approx \infty$), the image of that object produced by the corrective lens is upright and virtual, and located 75 cm from the lens. The eye looks at the image, and forms a final image on the retina. Thus, $d_i = -75$ cm, with the negative sign being used for the virtual image. Now,

$$\frac{1}{f} = \frac{1}{d_o} + \frac{1}{d_i} = \frac{1}{\infty} + \frac{1}{-75 \text{ cm}}$$

$$\therefore f = -75 \text{ cm}$$

Therefore the focal length of the lens is -0.75 m, with the negative sign indicating a diverging lens.

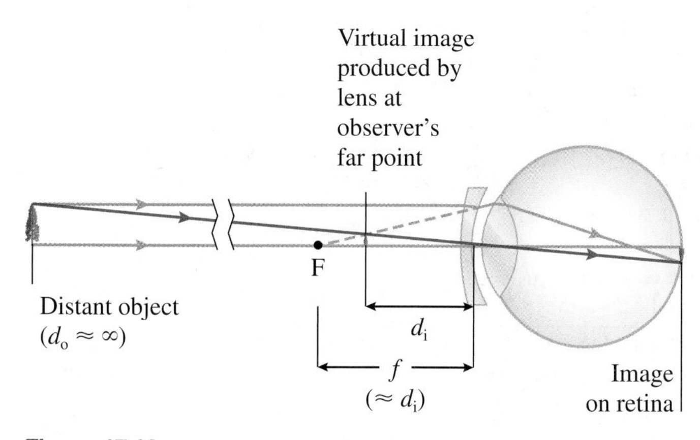

Figure 17-13 Sample Problem 17-1. (Distances shown do not have a common scale.)

Learning Tip

Although the eye acts as a converging lens, corrective lenses for eyes can be either converging or diverging. To review the sign conventions for lens variables, refer to Section 15.5.

In **hyperopia**, or **farsightedness**, the image comes to a focus behind the retina. The usual cause is a shortened eye. In fact, young children are often a little farsighted until their eyes grow to the appropriate size. To a person with hyperopic vision, objects close to the eye appear blurred. Figure 17-14 shows that this defect can be corrected using a converging lens. (Hyperopia is also called *hypermetropia*.)

hyperopia (or farsightedness) the condition of the eye in which image comes to a focus behind the retina; also called hypermetropia

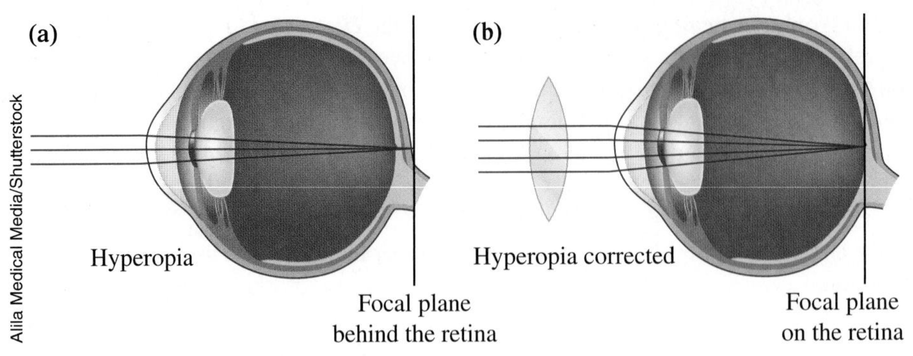

Figure 17-14 Hyperopia or farsightedness. **(a)** The image comes to a focus behind the retina. **(b)** A converging lens corrects the defect.

SAMPLE PROBLEM 17-2

A person with hyperopic vision has a near point of 120 cm. Determine the focal length of the corrective lens needed to bring the near point to 25 cm. Assume that the lens-to-eye distance is negligible.

Solution

Figure 17-15 illustrates that when the person views a nearby object ($d_o = 25$ cm), the image of that object produced by the corrective lens is upright, virtual, and located at a distance of 120 cm from the lens. The eye looks at this image and forms an image of it on the retina. Thus, $d_i = -120$ cm (for the virtual image). Now,

$$\frac{1}{f} = \frac{1}{d_o} + \frac{1}{d_i} = \frac{1}{25\text{ cm}} + \frac{1}{-120\text{ cm}}$$

$$\therefore f = 32\text{ cm}$$

Thus, the focal length is $+0.32$ m, with the positive sign indicating a converging lens.

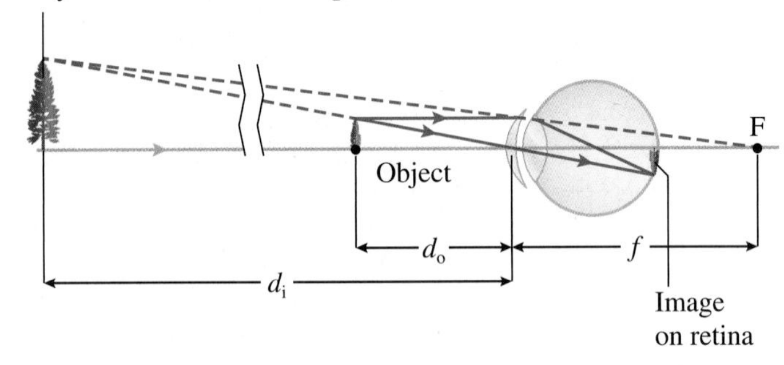

Figure 17-15 Sample Problem 17-2. (Distances shown do not have a common scale.)

In practice, the refractive power of a lens is indicated rather than its focal length.[2] As described in a "Did You Know?" box in Section 15.5, the power is the reciprocal of the focal length, $P = 1/f$, measured in dioptres, d, or m^{-1}. Thus, in the previous two sample problems, the power of the diverging lens is $1/(-0.75\text{ m}) = -1.3$ d, and the power of the converging lens is $1/(0.32\text{ m}) = 3.1$ d.

[2]The advantage of using power (P) rather than focal length is that if two or more thin lenses are adjacent, their combined power can be found by simply adding the individual powers.

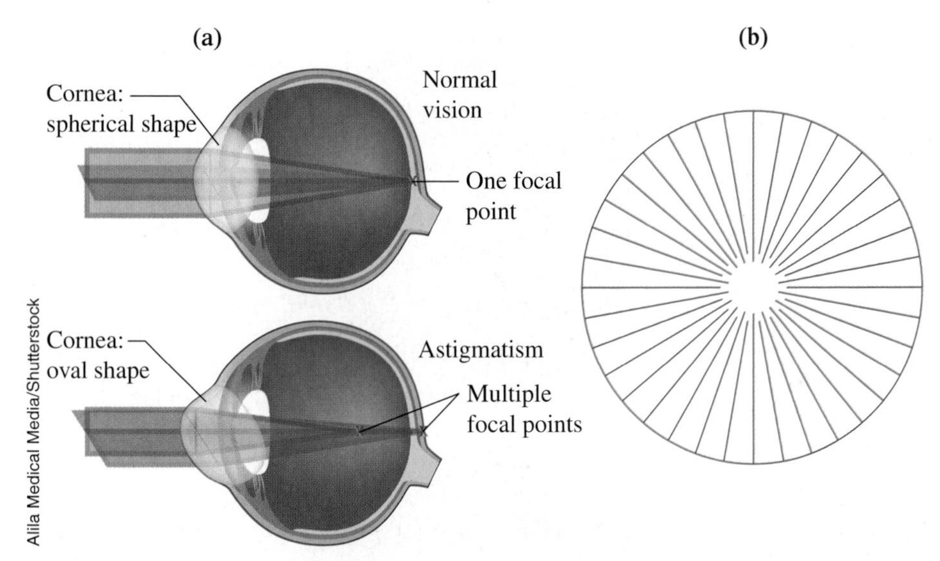

Figure 17-16 **(a)** Comparing astigmatism with normal vision. (The cornea shapes are exaggerated for clarity.) **(b)** To test for astigmatism, hold the book at arm's length and view the diagram with one eye at a time. If all the lines appear equally clear and thick to your unaided eyes, you do not have astigmatism.

DID YOU KNOW?

The World Health Organization estimates that there may be a billion people worldwide who do not have access to the corrective eyeglasses they need. To help alleviate this problem, one research group called *Adaptive Eyecare* manufactures framed eyeglasses with lenses that contain clear silicon oil. A tiny pump on the frame can be used to change the amount of oil in each lens, altering the lens shape and thus the focal length to suit the user. These adjustable glasses are longer-lasting and much less expensive than most prescription eyewear used in developed countries.

Astigmatism, or asymmetrical focusing, results when the cornea has an uneven surface which causes better focusing in one plane than in other planes (Figure 17-16 (a)). To a person with astigmatic vision, equal-sized lines in different directions appear to have different clarity or thicknesses. Figure 17-16 (b) gives a simple test for astigmatism. This defect can be corrected by using a cylindrical lens with a different focal length in different planes. It can also be corrected by wearing contact lenses, which cause a layer of tears to form between the contact lens and the cornea. This layer helps correct the astigmatism.

astigmatism a condition of the eye in which the focusing is asymmetrical; caused by a cornea with an uneven surface and resulting in better focusing in one plane than in other planes

Presbyopia is a focusing problem that occurs when a person's eyes can no longer accommodate to view nearby objects; in other words, the near point increases. It is normally associated with advancing age. The lens becomes less pliable with continued use and the ciliary muscles have ever-greater difficulty in focusing the lens. To a person with presbyopic vision, nearby objects, especially printed words, appear out of focus. The problem can be corrected by using reading glasses with converging lenses. If a person with presbyopia also has another defect, such as myopia, bifocal lenses can be worn (Figure 17-17). As its name implies, a bifocal lens has two focal lengths, one each for the upper and lower parts of the lens.

presbyopia a focusing problem that occurs when a person's eyes can no longer accommodate to view nearby objects

Figure 17-17 An example of the shape of a bifocal lens; the lower part of the lens is for reading.

A popular alternative to eyeglasses are contact lenses. Athletes and workers in some hazardous occupations who require corrective lenses often choose contact lenses for safety reasons, while many other people who wear them do so for cosmetic reasons. Contact lenses, which have been in use since the 1950s and currently are worn by millions of people around the world, are made of either rigid or soft plastic. The lens rests on a film of tears on the cornea. The original rigid lenses were made of glass, which caused problems because the glass did not allow oxygen to pass through to the cornea. Rigid lenses made of special plastic are gas permeable and are more useful for correcting astigmatism than soft plastic lenses. Currently, most contact lenses, including those used for cosmetic purposes, are made of silicon hydrogels that are highly permeable to oxygen and can be worn for extended periods, including overnight.

A common approach to a more permanent vision correction is *laser eye surgery*. Although the basic physics of the laser is described later in the chapter, all you need to understand here is that lasers applied to surgery can act like extremely sharp, accurate cutting or burning tools. One example of laser eye surgery is shown in Figure 17-18. In this procedure, a thin layer of the cornea is cut with one type of laser creating a pocket

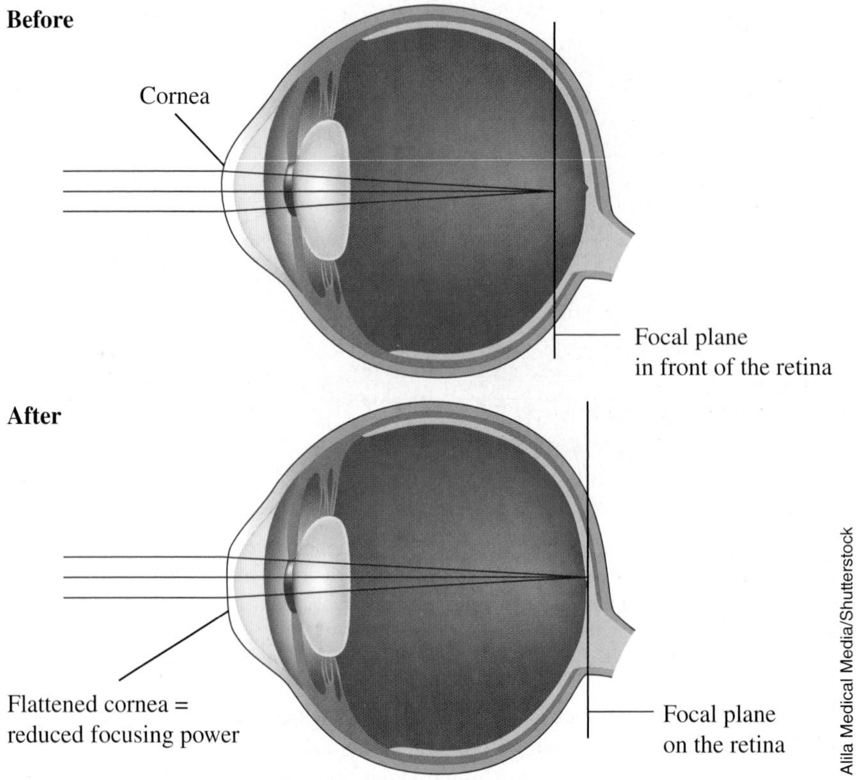

Alila Medical Media/Shutterstock

Figure 17-18 For an eye with myopia, laser surgery can remove part of the cornea to reduce the focusing power of the eye and cause the image to reach the retina.

DID YOU KNOW?

Problems other than those related to optical principles can affect many people. Of the thousands of people with epilepsy, about 5% may experience the photosensitive kind in which grand mal seizures are caused by flashing lights. The seizures can be initiated by strobe lights at concerts or movies or by festive flashing lights during the winter season. Reducing the flashing rate to less than 3 Hz helps to reduce the chances of a seizure.

TRY THIS!

Analyzing Vision Problems

Analyze the vision problems of a friend or family member who wears framed corrective lenses by holding the lenses close to a printed page and using the criteria listed below.

- A diverging lens held close to an object makes the object appear smaller.
- A converging lens held close to an object makes the object appear larger.
- A cylindrical lens, when rotated, causes the object viewed to appear to change shape.

that is folded back. Then a second type of laser is used carefully to reshape the cornea by eliminating selected underlying tissue.[3] Finally the cornea layer is replaced. The result is that the light rays entering the eye focus properly at the retina.

Lastly, research is being carried out to design systems to aid people with severe vision defects. For example, optoelectrical prosthetic devices are being designed that are directly implanted. They are useful for diseases such as macular degeneration where there is a problem in communication between the retina and the optic nerve. Another option involves "bionic lenses." One vision-aid device uses a small camera attached to an eyeglass frame and connected to a portable computer, which in turn is connected to a bone-conductive speaker. The system converts visual information into audio signals so the scene the user is facing can be interpreted. As well, bionic lenses have nano-electronic circuits and LEDs that produce virtual images similar to the heads-up displays on the windshields in aircraft cockpits. These bionic lenses can help the visually impaired, but can also provide applications for the general public.

EXERCISES

17-9 A person has a near point of 14 cm and a far point of 72 cm.
(a) What type of defect does the person have?
(b) Can the person see an object clearly at a distance of 12 cm? 20 cm? 85 cm?
(c) What type of lens will allow the person to see distant objects clearly?

17-10 State the type of lens used to correct (a) myopia, (b) astigmatism, (c) hyperopia, and (d) presbyopia.

17-11 Determine the focal length (in metres) and power (in dioptres) of a contact lens that will provide normal vision to a myopic person who has a far point of 125 cm.

[3] Section 17.7 discusses the specific details comparing the two types of lasers used in this surgery.

17-12 Determine the focal length (in metres) and power (in dioptres) of a contact lens that will provide normal vision to a hyperopic person who has a near point of 88 cm.

17-13 A person is wearing contact lenses with a focal length of +1.6 m, providing a near point of 28 cm.

(a) What type of defect does the person have?

(b) What is the power of the lenses (in dioptres)?

(c) Determine the person's near point without the lenses.

17.3 The Simple Magnifier

Likely you have used a simple magnifier (or magnifying glass) to closely inspect small printing or a biological specimen. Such a magnifier is a simple converging lens, with the object closer to the lens than the focal point to produce an enlarged, virtual image. The use of lenses to magnify images began in the late 16th century. Probably the best of the early magnifiers were made in Holland by Anton von Leeuwenhoek who designed small single lenses of high quality. The magnifier used by jewellers, called the loupe, is similar to Leeuwenhoek's lenses.

It is useful to develop equations for the magnification of a magnifier by considering not only the equations for lenses (from Chapter 15) but also the average near point of the human eye. For our discussion, we will use 25 cm as this near point. Although we could consider the linear magnification (M) of the magnifier, we will introduce another way of determining magnification using angles. Imagine viewing an object such as a lamp located several metres away and then viewing it again when you are closer, as shown in Figure 17-19. At a far distance, the lamp subtends (or makes) a small angle at your eye and it looks small, while at a closer distance the lamp subtends a larger angle at your eye and it looks bigger. The **angular size** of the lamp or any other object is the angle subtended by the object at the observer's eye. As shown in Figure 17-19, the angular size, φ, is found by $\tan\varphi = h_o/d_o$, where "o" represents the object. Thus,

angular size the angle subtended by an object at the observer's eye or other optical instrument

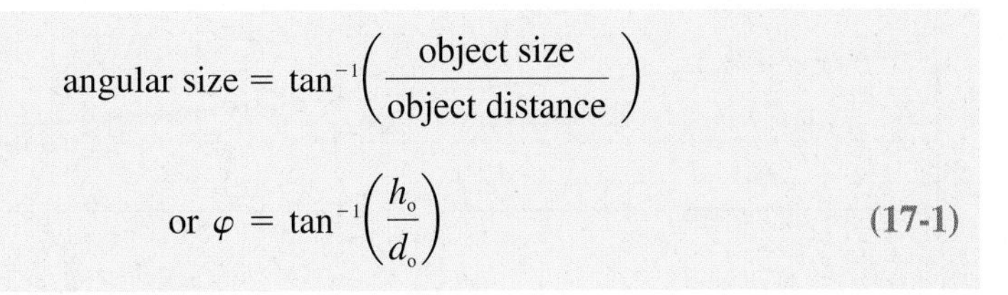

$$\text{angular size} = \tan^{-1}\left(\frac{\text{object size}}{\text{object distance}}\right)$$

$$\text{or } \varphi = \tan^{-1}\left(\frac{h_o}{d_o}\right) \qquad (17\text{-}1)$$

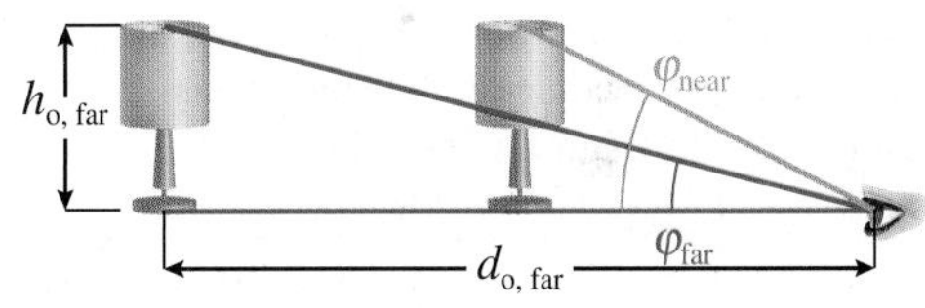

Figure 17-19 As the distance to an object becomes less, the angle it subtends at the eye increases, and the object looks bigger.

SAMPLE PROBLEM 17-3

A person views a hand-held device that is 15 cm by 4.5 cm by 1.3 cm from a distance of 25 cm. What range of angular sizes does the person see?

Solution

The largest angular size is seen when $h_o = 15$ cm, and the smallest angular size is seen when $h_o = 1.3$ cm. Applying Eqn. 17-1,

$$\varphi_{large} = \tan^{-1}\left(\frac{h_o}{d_o}\right) = \tan^{-1}\left(\frac{15\text{ cm}}{25\text{ cm}}\right) = 31°$$

$$\varphi_{small} = \tan^{-1}\left(\frac{h_o}{d_o}\right) = \tan^{-1}\left(\frac{1.3\text{ cm}}{25\text{ cm}}\right) = 3.0°$$

Thus, the angular sizes of the device viewed range from 31° down to 3.0°.

Math Tip

When an angle is less than about 10°, a handy relationship to use is $\theta \approx \tan\theta \approx \sin\theta$, where θ is the angle measured in radians. For example, in the second part of Sample Problem 17-3, we could have used

$$\varphi_{\text{small}} \approx \tan\varphi_{\text{small}} \approx \frac{1.3 \text{ cm}}{25 \text{ cm}} \approx 0.052 \text{ rad}$$

Converting to degrees,

$\varphi_{\text{small}} \approx 0.052 \text{ rad} \times \dfrac{360°}{2\pi \text{ rad}} \approx 3.0°$, which is the same answer (to two significant digits) as obtained before. This small-angle approximation will be applied shortly to optical instruments.

Logically, the maximum angular size of an object in focus viewed with the unaided eye occurs when the object is located at the observer's near point, as illustrated in Figure 17-20 (a). Taking the near point to be 25 cm and applying the relation $\tan\theta \approx \theta$ for small angles, the approximate maximum angular size of an object in focus is thus

$$\varphi_{\text{max}} \approx \frac{h_o}{25 \text{ cm}} \qquad \textbf{(17-2)}$$

where φ_{max} is measured in radians and h_o is expressed in centimetres

Now if a converging lens, that is, a simple magnifier, is placed in front of the eye to produce an enlarged image, the angular size of the image also changes. The largest angular size of an image in focus occurs when the object is located just inside the focal point of the lens and its virtual image is located at the observer's near point, as shown in Figure 17-20 (b). The angular size in radians of the image is $\varphi' \approx \dfrac{h_i}{25 \text{ cm}}$ (assuming a small angle); but by similar triangles, this ratio is equal to $\dfrac{h_o}{d_o}$. Thus, $\varphi' \approx \dfrac{h_o}{d_o}$.

angular magnification (or **magnifying power**) the ratio of the angular size of an image with the lens to the angular size without the lens

We can now define the **angular magnification** (also called the **magnifying power**), symbol m, of the lens. It is the ratio of the angular size with the lens to the angular size without the lens. Therefore,

$$m = \frac{\varphi'}{\varphi} \qquad \textbf{(17-3)}$$

TRY THIS!

Estimating the Size of the Moon

Outdoors on a clear night with the Moon visible, hold a quarter in your outstretched arm so it just hides the Moon. Given that the average Earth–Moon distance is 7.4×10^8 m, use the angular size of the quarter to determine an estimate of the diameter of the Moon.

(a)

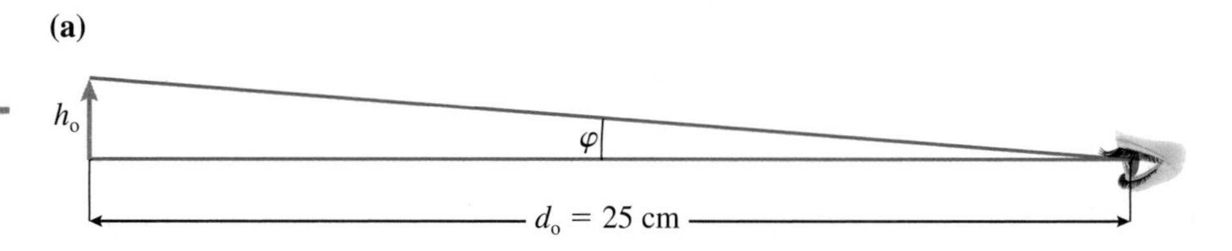

(b)

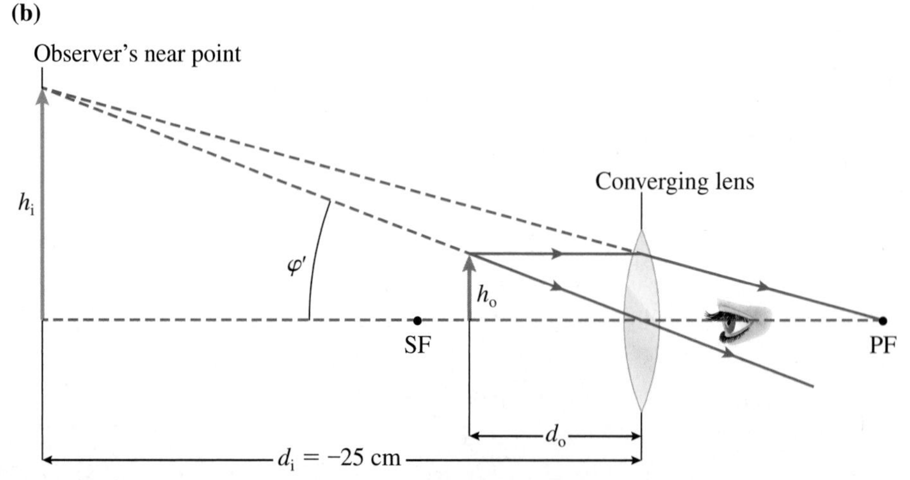

Figure 17-20 Using a converging lens to change the angular size of an image. **(a)** Assuming a near point of 25 cm, without a lens the maximum angular size in radians of an object is $\varphi_{\text{max}} \approx h_o/(25 \text{ cm})$. **(b)** With a converging lens in front of the eye, the image is located at the observer's near point, and the angular size in radians increases to $\varphi' \approx h_i/(25 \text{ cm})$.

We can use Eqn. 17-3 to determine the maximum angular magnification, m_{max}, of a magnifier in terms of the focal length of the lens. First, we substitute for the angular sizes:

$$m = \frac{\varphi'}{\varphi} \approx \frac{h_o/d_o}{h_o/25\text{ cm}} \approx \frac{25\text{ cm}}{d_o} \quad \text{where } d_o \text{ is in centimetres}$$

Then we combine this relationship with the thin-lens equation (Eqn. 15-8):

$$\frac{1}{f} = \frac{1}{d_o} + \frac{1}{d_i} \quad \text{or} \quad \frac{1}{d_o} = \frac{1}{f} - \frac{1}{d_i}$$

Thus, $m_{max} \approx 25\text{ cm}\left(\frac{1}{d_o}\right)$ becomes $m_{max} \approx 25\text{ cm}\left(\frac{1}{f} - \frac{1}{d_i}\right) \approx \frac{25\text{ cm}}{f} - \frac{25\text{ cm}}{d_i}$

But $d_i = -25$ cm (for maximum angular magnification), so

$$m_{max} \approx \frac{25\text{ cm}}{f} - \frac{25\text{ cm}}{-25\text{ cm}}$$

$$m_{max} \approx \frac{25\text{ cm}}{f} + 1 \qquad \textbf{(17-4)}$$

where f is the focal length expressed in centimetres.

In this derivation, the image was located at the observer's near point. Under normal conditions this situation would result in eyestrain, so the image is usually located somewhat farther away. As you will see in the following sample problem, the magnification changes only slightly when the image is farther from the observer, particularly for a lens of short focal length. Furthermore, as the image distance increases away from the near point, m_{max} decreases since it is related to $1/d_i$. Therefore, the calculation of magnification with Eqn. 17-4 truly is a maximum value for a lens of a given focal length.

SAMPLE PROBLEM 17-4

A lens of focal length +5.0 cm is used to magnify the details of an antique coin. The coin is located at or near the focal point of the lens. Determine the maximum angular magnification of the lens and the object distance when the image is

(a) at the near point

(b) located far from the lens

Solution

(a) Because the image is at the near point, we apply Eqn. 17-4. Using a near point of 25 cm,

$$m_{max} \approx \frac{25\text{ cm}}{f} + 1 \approx \frac{25\text{ cm}}{5.0\text{ cm}} + 1 \approx 5.0 + 1 \approx 6.0\text{ times}$$

To determine the object distance, we apply the equation $m_{max} \approx 25\text{ cm}\left(\frac{1}{d_o}\right)$:

$$d_o \approx 25\text{ cm}\left(\frac{1}{m_{max}}\right) = \frac{25\text{ cm}}{6.0} = 4.2\text{ cm}$$

Thus, when the image is at the near point, the maximum angular magnification is 6.0 and the object distance is 4.2 cm from the lens.

(b) With a large image distance d_i, the ratio $1/d_i$ approaches zero, so from the lens equation, $\frac{1}{f} = \frac{1}{d_o} + \frac{1}{d_i}$ becomes $\frac{1}{f} \approx \frac{1}{d_o}$, and d_o approaches f. This means that the object is located at the focal point. Substituting the given values,

$$m_{max} \approx 25 \text{ cm} \left(\frac{1}{f} - \frac{1}{d_i}\right) \approx 25 \text{ cm} \left(\frac{1}{f}\right) \approx \frac{25 \text{ cm}}{5.0 \text{ cm}} \approx 5.0 \text{ times}$$

Thus, the maximum magnification for this lens ranges from approximately 5.0 times (when the object is at the focal point) to 6.0 times (when the object is closer to the lens than the focal point).

We conclude from this example that the angular magnification provided by the lens does not change radically as the image viewed moves farther away. Thus, a person can view the image with a relaxed eye without a great loss of magnification.

Eqn. 17-4, derived for angular magnification in terms of focal length, reveals that magnification increases for lenses of shorter focal length. However, short focal length lenses must be relatively thick, so they create both spherical and chromatic aberration. Commonly, a single lens is made to have a magnification of only about 3 or 4 times, but when combined with at least one other lens to reduce aberrations, the magnifier can have a magnification of perhaps 10 to 20 times.

TACKLING MISCONCEPTIONS

Comparing Types of Magnification

It is important to distinguish linear magnification from angular magnification. When using a magnifier or other device with a lens, the closer the object is to the focal point of the lens, the farther away the image becomes, and the linear magnification calculation approaches infinity. Angular magnification for a magnifier turns out to be about 3 to 10 times, which corresponds more closely to what you observe when using the device.

EXERCISES

17-14 A 1.65 m person is jogging toward you. Determine the angular size in degrees of the person at a distance of (a) 155 m and (b) 15.0 m.

17-15 A rose is 12 cm in diameter. At what distance from an observer's eye would it subtend an angle of 1.5°?

17-16 **Fermi Question:** Choose some object, such as a doorway, on the far side of the room. Estimate the angle that the object subtends (makes) at your eye.

17-17 Determine the approximate focal length of a simple magnifier that provides an angular magnification of 8.0 with the image located

(a) at a near point of 25 cm

(b) at a very large distance from the lens

17-18 Compare the approximate relaxed eye magnification and the maximum angular magnification for a lens of focal length (a) 12.5 cm and (b) 2.5 cm.

17-19 A converging lens of focal length 6.0 cm is used as a simple magnifier. Determine the approximate maximum magnification produced by the lens for

(a) a nearsighted person whose near point is 18 cm (without corrective lenses)

(b) a farsighted person whose near point is 84 cm (without corrective lenses)

17.4 The Compound Light Microscope

The magnification limitation of a simple magnifier can be partially overcome using two converging lenses (or two sets of converging lenses) in an optical instrument called a **compound light microscope**. In such an instrument, the lens closest to the observer is called the **eyepiece lens** or simply the **eyepiece**. The lens closest to the object is called the **objective lens** or simply the **objective**. These lenses are positioned so that the image produced by the objective is magnified by the eyepiece (Figure 17-21). As will be shown shortly, a large magnification is obtained when the focal lengths of the lenses are small.

compound light microscope an optical instrument with two lenses or lens systems that magnify nearby objects

eyepiece lens (or eyepiece) the lens in an optical instrument closest to the observer

objective lens (or objective) the lens in an optical instrument closest to the object

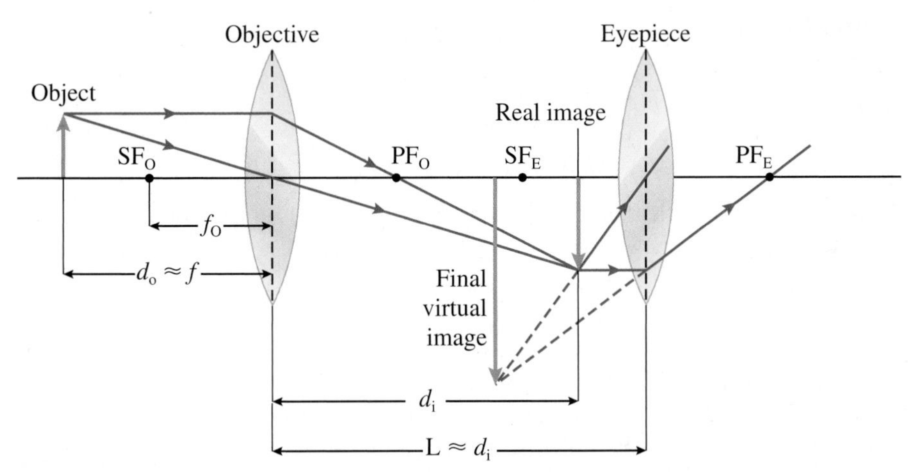

Figure 17-21 Ray diagram for a typical compound microscope. (This diagram is not drawn to scale; in an actual microscope the approximations shown are valid.) Notice that the real image formed by the objective becomes the "object" magnified by the eyepiece.

As you might have learned using microscopes, the total magnification, M_T, of the two lenses is the product of the objective's linear magnification, M_O, and the eyepiece's angular magnification, m_E. Thus,

$$M_T = M_O m_E \quad (17\text{-}5)$$

One way to determine the total magnification is to apply the equations for thin lenses (Eqns. 15-8 and 15-9) to the objective and then multiply by the angular magnification of the eyepiece. For the eyepiece calculations, the "object" is actually the image produced by the objective, as seen in Figure 17-21. Since the eyepiece acts as a magnifier with the "object" near the focal point, its maximum angular magnifications is $m_{max} \approx (25 \text{ cm})/(f_E)$.

Terminology Tips

There are some subtle differences in the symbols used in this topic. *M* is linear magnification; *m* is angular magnification (but could be mass in a different context); O represents the objective lens (e.g., M_O); E represents the eyepiece lens (e.g., m_E); o represents the object (e.g., h_o); i represents the image (e.g., d_i). Notice that the variables are in italics and the subscripts are not.

SAMPLE PROBLEM 17-5

A compound microscope has an objective with $f_O = 2.00$ cm and an eyepiece with $f_E = 2.5$ cm separated by 24 cm. Determine the total magnification of the microscope when it is used to view a specimen located 2.20 cm from the objective.

Solution

To find the linear magnification of the objective, we apply Eqn. 15-8 to find the image distance and then Eqn. 15-9 to find the linear magnification. Starting with the objective,

$$\frac{1}{f_o} = \frac{1}{d_o} + \frac{1}{d_i}$$

$$\frac{1}{d_i} = \frac{1}{f_o} - \frac{1}{d_o}$$

$$\frac{1}{d_i} = \frac{1}{2.00 \text{ cm}} - \frac{1}{2.20 \text{ cm}}$$

$$d_i = 22 \text{ cm}$$

$$M_O = -\frac{d_i}{d_o} = -\frac{22 \text{ cm}}{2.2 \text{ cm}} = -10$$

We next find the angular magnification of the eyepiece:

$$m \approx \frac{25\text{ cm}}{f_E} = \frac{25\text{ cm}}{2.5\text{ cm}} = 10$$

$$M_T = M_O m_E \approx (-10)(10) \approx -1.0 \times 10^2$$

Thus, the total magnification is approximately -1.0×10^2 times. The negative sign indicates an inverted image.

We can also determine the magnification of the objective in terms of its focal length. In Figure 17-21, we see that the object is just beyond the focal point of the objective; thus, $d_o \approx f_O$. Also, microscopes are designed so that the image distance is close to the distance between the objective and the eyepiece, which in turn is close to the length, L, of the microscope. Thus, $d_i \approx L$. Now from the lens magnification equation (Eqn. 15-9), $\frac{h_i}{h_o} = -\frac{d_i}{d_o}$,

$$M_O = \frac{-d_i}{d_o} \approx \frac{-L}{f_O}$$

where L is the length of the microscope and f_O is the focal length of the objective. These distances are both approximate, so we now have the following equation for the approximate magnification of the image in a compound microscope:

$$M_T = M_O m_E \approx \left(\frac{-L}{f_O}\right)\left(\frac{25\text{ cm}}{f_E}\right) \quad \textbf{(17-6)}$$

where all distances are measured in centimetres. The negative sign indicates an inverted image.

SAMPLE PROBLEM 17-6

A compound microscope, 28 cm in length, has an objective of focal length 1.4 cm and an eyepiece of focal length 2.5 cm. Determine the approximate total magnification of the microscope.

Solution

The approximate total magnification is given by Eqn. 17-6:

$$M_T = M_O m_E \approx \frac{-(L)\,(25\text{ cm})}{f_O f_E} \approx \frac{-(28\text{ cm})(25\text{ cm})}{(1.4\text{ cm})(2.5\text{ cm})} \approx -2.0 \times 10^2\text{ times}$$

Thus, the total magnification is approximately -2.0×10^2 times.

Most higher-quality compound light microscopes have a selection of objectives with varying magnifications (Figure 17-22). Thus, the total magnification can be varied. The problems of spherical and chromatic aberration are overcome by using lens combinations for the objective and the eyepiece rather than single lenses. Even with aberration reduced to a minimum, however, the magnification of a microscope is limited by the effects of the diffraction of light. As the magnification increases, the size of the diffraction pattern increases along with the image size. At very high

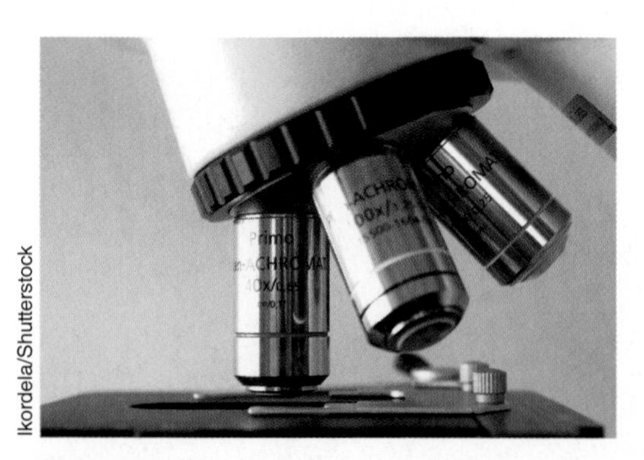
Ikordela/Shutterstock

Figure 17-22 A typical microscope with a choice of three objective lenses.

magnifications, the diffraction patterns begin to overlap, and the resolution of fine detail reaches a limit. As described in Section 16.3, the resolution increases with shorter wavelengths, so ultraviolet light or x rays can be used to obtain better resolution, and some modern microscopes use these shorter wavelength waves. A similar principle is applied in the design and operation of electron microscopes in which electrons display wave properties with wavelengths approximately 100 000 times smaller than visible light.

EXERCISES

17-20 A compound microscope has an objective of focal length 0.500 cm and an eyepiece of focal length 5.0 cm. A specimen is located 0.530 cm from the objective. Determine the approximate total magnification of the microscope for this situation.

17-21 The eyepiece and objective lenses of a compound microscope have focal lengths of 2.6 cm. If the distance from the objective to the eyepiece is 22 cm, what is the approximate total magnification of the microscope?

17-22 A microscope designer would like to obtain a total magnification of −400 times using an eyepiece ($m_E = 10$ times) in conjunction with an objective. For a microscope length of 25 cm, what focal length of the objective will achieve the desired total magnification? (Assume two significant digits.)

17.5 Telescopes

An **optical telescope** is an instrument that uses optical components to make distant objects appear closer. In this section we look at a variety of optical telescopes, such as refracting and reflecting telescopes, including ones placed in space. We also explore other types of telescopes that use invisible parts of the electromagnetic spectrum, such as radio waves, x rays, and gamma rays.

optical telescope an instrument that uses optical components to make distant objects appear closer

Refracting Telescopes

The idea of placing two lenses at opposite ends of a long tube to view distant objects appears to have been first tried in Holland by Hans Lippershey in 1608. However, it was Galileo, in Italy, who, after hearing of the device, made his own telescope and with it discovered previously unknown bodies in the sky, such as some of the moons of Jupiter. A telescope made with lenses is called a **refracting telescope**. Galileo offered his first telescopes to the officials of the powerful city of Venice, easily convincing them of their practical military use of seeing approaching ships a full two hours before the unaided eye could. The officials were duly impressed and soon doubled Galileo's salary.

refracting telescope a telescope made with lenses or lens systems

In general, refracting telescopes are classified as either astronomical or terrestrial. An **astronomical telescope** is one that produces an inverted image. A basic refracting astronomical telescope consists of two lenses, called the objective and the eyepiece, just as in microscopes. The main purpose of the telescope objective is to gather as much light as possible from distant stars, galaxies, and whatever else is in the sky. (This differs from the main purpose of a microscope objective, which is to give a magnified image.) To accomplish this, the objective has a relatively large aperture. Furthermore, it also has a fairly long focal length, f_O, and produces a real, inverted image of the object only slightly beyond the focal point. This real image then becomes the "object" of the eyepiece, which has a focal length f_E and is located so its focal point lies just beyond the real image. Thus, the eyepiece acts as a magnifier. As Figure 17-23 (a) shows, the two lenses are separated by a distance approximately equal to the sum of their focal lengths, $f_O + f_E$. The final image is virtual and inverted. The inversion is only a minor inconvenience when viewing stars or galaxies in the sky. Producing an upright image would require a third lens, resulting in a loss of light intensity.

astronomical telescope a telescope that produces an inverted image

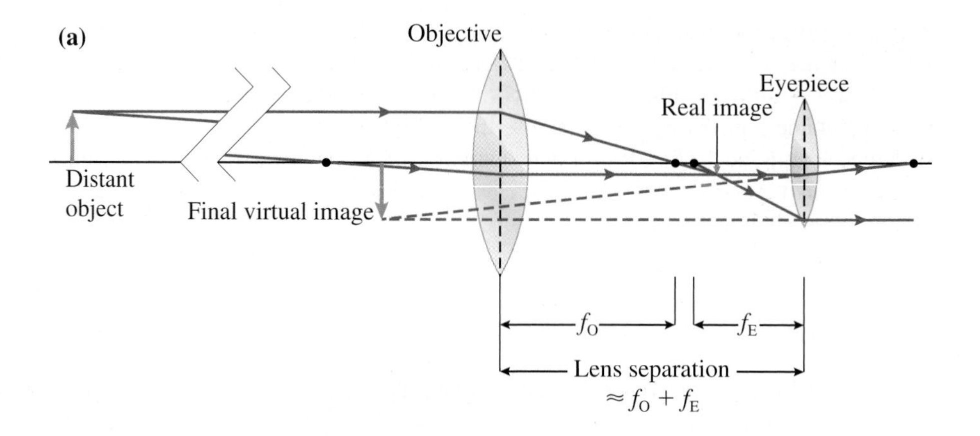

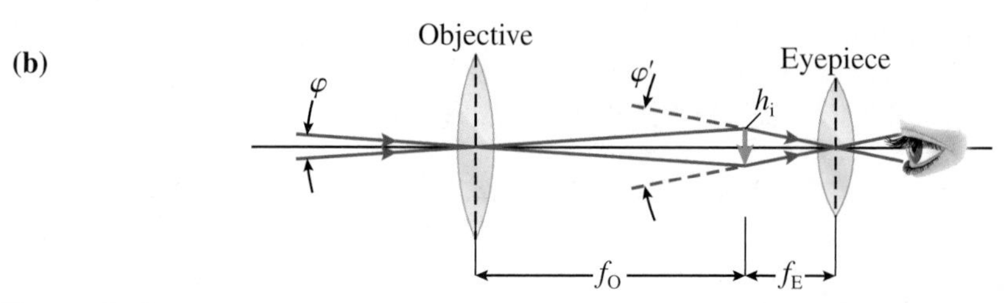

Figure 17-23 **(a)** A refracting astronomical telescope. **(b)** Determining angular magnification of an astronomical telescope. To simplify the diagram, the focal point of the objective coincides with the focal point of the eyepiece. Notice that the angle at the eyepiece (φ') is larger than the angle that would be viewed without the telescope (φ).

The angular magnification of the astronomical telescope (m) is the ratio of the angle subtended by the image seen in the eyepiece (φ') to the angle subtended by the object seen with the unaided eye (φ). Thus, $m = \frac{\varphi'}{\varphi}$. Because the object viewed is always extremely far away, the angle it subtends at the unaided eye is basically equal to the angle it subtends at the objective. This angle is illustrated in Figure 17-23 (b). Using this diagram and the fact that for very small angles measured in radians $\tan\varphi \approx \varphi$, the following equations for φ and φ' can be written:

$$\varphi \approx \frac{-h_i}{f_O} \qquad \varphi' \approx \frac{h_i}{f_E}$$

Now the angular magnification is

$$\frac{\varphi'}{\varphi} \approx \frac{h_i/f_E}{-h_i/f_O} \approx \frac{-f_O}{f_E}$$

$$\text{or } m \approx \frac{-f_O}{f_E} \qquad \textbf{(17-7)}$$

where m is the angular magnification of an astronomical telescope, f_O is the focal length of the objective, and f_E is the focal length of the eyepiece. The negative sign corresponds to an inverted image.

This equation is an approximation, but it shows that a large magnification is achieved by having a large focal length objective and a short focal length eyepiece.

SAMPLE PROBLEM 17-7

A refracting astronomical telescope has an objective of focal length 112 cm and an angular magnification of –28. Determine the focal length of the eyepiece and the approximate distance between the two lenses.

Solution

Applying Eqn. 17-7,

$$m \approx \frac{-f_{\mathrm{O}}}{f_{\mathrm{E}}}, \therefore f_{\mathrm{E}} \approx \frac{-f_{\mathrm{O}}}{m} \approx \frac{-112 \text{ cm}}{-28} \approx 4.0 \text{ cm}$$

Thus, the focal length of the eyepiece is 4.0 cm, and the approximate distance between the lenses is $f_{\mathrm{O}} + f_{\mathrm{E}} \approx 112 \text{ cm} + 4.0 \text{ cm} \approx 116 \text{ cm}$.

After the invention of the first telescopes, scientists built larger and larger refracting telescopes to extend their view farther into the sky. In 1897 the largest refracting telescope in the world, with a 1.0 m diameter objective, was built in Wisconsin, U.S.A. (Figure 17-24). One of the reasons larger refracting telescopes have not been built since 1897 is the difficulty in supporting a heavy glass converging lens that is thinner around the circumference. Other reasons will be described shortly.

When viewing distant objects on Earth, it is inconvenient to look at an inverted image. The problem is overcome by using a **terrestrial telescope**, which is a telescope that produces an upright image. Two examples of terrestrial telescopes are shown in Figure 17-25.

terrestrial telescope a telescope that produces an upright image

Emilio Segre Visual Archives/American Institute of Physics/Science Photo Library

Figure 17-24 Although it was built in 1897, the Yerkes Observatory in Wisconsin, U.S.A., is the world's largest refracting telescope. The telescope contributed greatly to improving our understanding of the universe and testing theories, including those of Albert Einstein who received the Nobel Prize in Physics in 1921, the same year this photograph was taken. (Einstein is seen to the right of centre.)

(a)

vvoe/Shutterstock

(b)

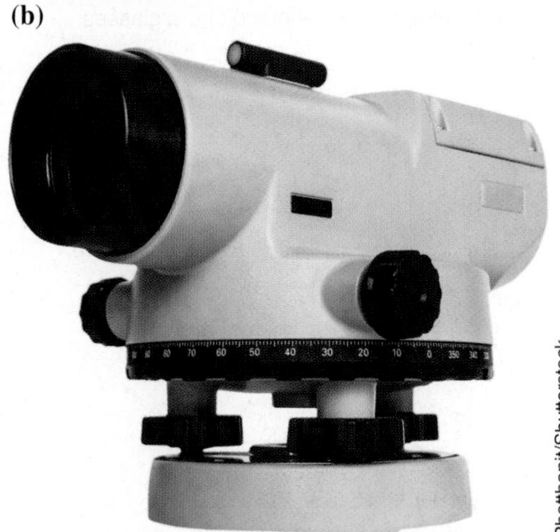

Phatthanit/Shutterstock

Figure 17-25 **(a)** A viewing scope is a simple terrestrial telescope. **(b)** A theodolite, an optical instrument used for surveying, can be adjusted in the horizontal and vertical planes to obtain highly precise measurements of angles. This theodolite has a finding scope on the top.

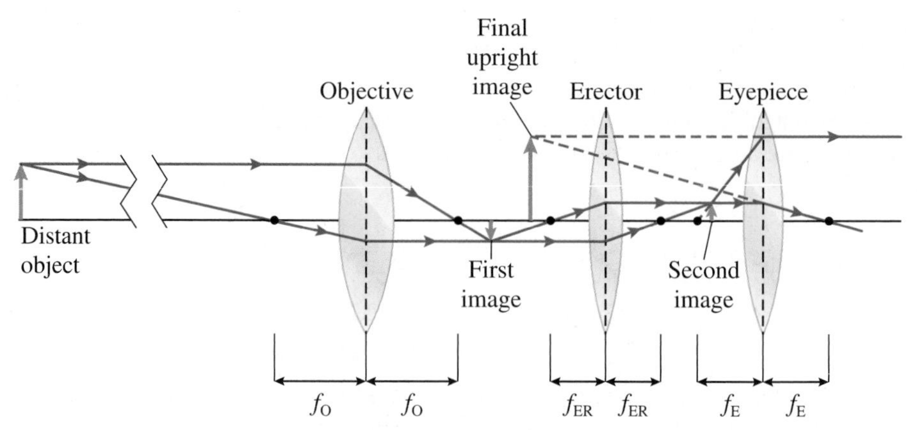

Figure 17-26 The erector lens allows the terrestrial telescope to obtain an upright image.

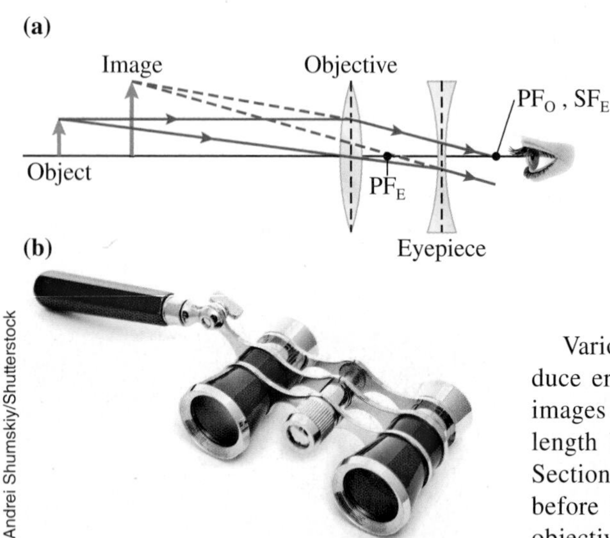

Figure 17-27 **(a)** The basic principle of the Galilean telescope. **(b)** Vintage opera glasses.

One way to obtain an upright or erect image is to place an erecting lens of focal length f_{ER} between the objective and the eyepiece of a refracting astronomical telescope. This lens must be located so it produces an upright, real image which can then be magnified by the eyepiece (Figure 17-26). To achieve a high magnification, such a telescope must become rather long, a problem sometimes overcome by constructing the telescope in cylindrical sections that can be pushed together when the telescope is not in use.

Various ways have been devised to reduce the length of telescopes that produce erect images. Binoculars are conveniently short, yet they produce upright images by using total internal reflection in prisms to obtain a large enough path length between the objective and the eyepiece. (Binoculars were described in Section 15.3.) Yet another idea is to intercept the image produced by the objective before it has become inverted, that is, before it reaches the focal point of the objective. This can be accomplished using a diverging lens, as illustrated in Figure 17-27 (a). Such an arrangement was invented by Galileo in 1609, so it is called a Galilean telescope. Opera glasses are an example of a Galilean telescope. Their magnification is typically only about 3 to 4 times (Figure 17-27 (b)).

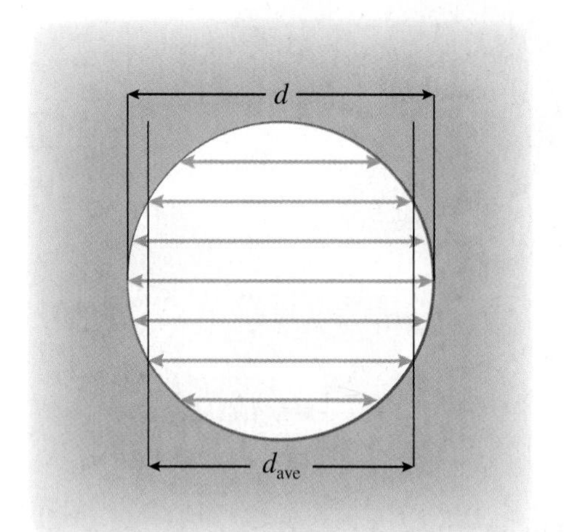

Figure 17-28 The average value of the distance across a circular aperture is less than the diameter d of the aperture by a factor of 1.22. This factor is used to modify the equation for diffraction in a single slit so it can be used for a lens.

Limiting Factors of a Telescope

When a telescope is used to view objects in space, it is limited in its ability to resolve any two adjacent stars or other objects. One of the limiting factors of resolution depends on the diffraction of light through a lens. You learned in Section 16.3 that, for a single slit, Rayleigh's criterion for single-slit diffraction is $\sin\theta = \frac{\lambda}{w}$ (Eqn. 16-9), where θ is the angular separation of two sources, λ is the wavelength of the light from the sources, and w is the slit width. If we replace a slit with a circular aperture or lens of diameter d, we can't simply replace w with d, because the light wavelets passing through the aperture "see" an effective diameter that is lower by a factor found to be 1.22, as illustrated in Figure 17-28. Thus, since the effective diameter is $d/1.22$, we rewrite Rayleigh's criterion as $\sin\theta = \frac{1.22\lambda}{d}$ for a lens or other circular aperture of diameter d. Figure 17-29 illustrates the variables involved for a circular aperture. Notice that $d \gg \lambda$, which means that $\sin\theta \approx \theta$ when θ is measured in radians. Furthermore, for small angles, $\theta \approx \frac{x}{L}$, where x is the separation of the objects viewed and

L is the distance from the objects to the lens. Thus, we can now write the equations for Raleigh's criterion for the minimum angular separation of objects for a lens:

$$\theta_{\text{min}} \approx \frac{1.22\lambda}{d} \approx \frac{x}{L} \tag{17-8}$$

where θ_{min} is the minimum angle in radians subtended by two separate objects that can be resolved by a lens, λ is the wavelength of the light, d is the diameter of the lens, x is the separation of the objects, and L is the distance from the sources to the lens.

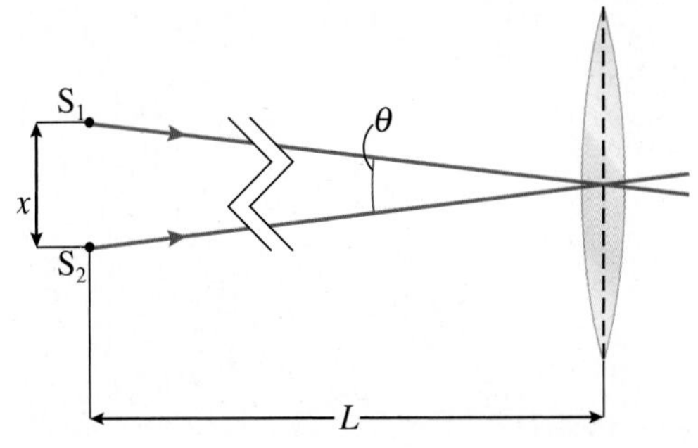

Figure 17-29 A simplified diagram using light rays from two sources that subtend an angle θ at the lens.

SAMPLE PROBLEM 17-8

Suppose a field biologist is using a telescope with an objective of diameter 4.0 cm to view two insects separated by 2.0 cm.

(a) What is the minimum angle of separation of the insects in white light to ensure their images are resolved?

(b) What is the distance at which the insects can be just resolved?

Solution

Given, $d = 4.0\text{ cm} = 4.0 \times 10^{-2}\text{ m}$; $x = 2.0\text{ cm} = 2.0 \times 10^{-2}\text{ m}$; $\lambda = 550\text{ nm} = 5.5 \times 10^{-7}\text{ m}$ for an average value for white light.

(a) To find θ we apply Eqn. 17-8:

$$\theta_{\text{min}} \approx \frac{1.22\lambda}{d} \approx \frac{1.22(5.5 \times 10^{-7}\text{ m})}{4.0 \times 10^{-2}\text{ m}} \approx 1.7 \times 10^{-5}\text{ rad}$$

Thus, the minimum angular separation of the objects for resolution for this telescope is 1.7×10^{-5} rad.

(b) To find the distance required we rearrange Eqn. 17-8, $\theta_{\text{min}} \approx \frac{x}{L}$, to solve for L.

$$L \approx \frac{x}{\theta_{\text{min}}} \approx \frac{2.0 \times 10^{-2}\text{ m}}{1.7 \times 10^{-5}\text{ rad}} \approx 1.2 \times 10^{3}\text{ m}$$

Thus, the telescope is able to just resolve two insects separated by 2.0 cm from a distance of 1.2×10^3 m. At any greater distance, the telescope will not resolve the insects.

It is evident from the discussion that larger-diameter lenses can resolve separate objects better (Eqn. 17-8), but large lenses have disadvantages (besides their large mass, mentioned earlier). Converging lenses are thin around their circumference, which makes them fragile where they must be supported. The effect of gravity on a heavy lens causes the lens to sag, resulting in slight distortion. Furthermore, spherical and chromatic aberration reduce the optical quality of the lens. But it is the large masses of moving air in Earth's atmosphere that set the main limits on resolution. (The "twinkling" of stars is an effect caused by light passing through air.) We now explore solutions to these limitations.

Reflecting Telescopes

The largest optical astronomical telescopes in the world are **reflecting telescopes** that use converging parabolic mirrors rather than converging lenses to gather distant light. Reflecting telescopes overcome several weaknesses of refracting telescopes. First, a

reflecting telescope a telescope that uses a converging parabolic mirror rather than a converging lens to gather distant light

Figure 17-30 The Canada–France–Hawaii Telescope is a joint observatory on the summit of Mauna Kea in Hawaii, above much of Earth's atmosphere at an elevation of 4200 m. The mirror has a diameter of 3.58 m and the telescope uses both visible and infrared wavelengths for observation, but, like many observatories, plans to add additional instruments. As of 2014, there are 12 telescopes on Mauna Kea.

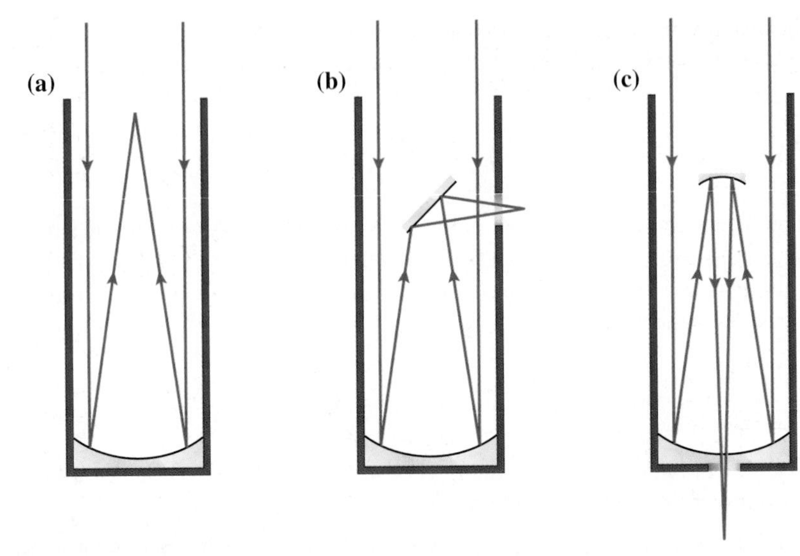

Figure 17-31 Common designs of reflecting telescopes. **(a)** The *prime focus* design can be used on large telescopes where instruments or even an observer are located at the focal point. **(b)** The *Newtonian focus*, designed by Isaac Newton, is relatively small and is popular with amateur astronomers. **(c)** The *Cassegrain focus* can be used on telescopes of all sizes, from small reflectors used by amateurs to the largest telescopes.

large mirror can be entirely supported from beneath, whereas a lens can be supported only around the outer circumference, the thinnest and weakest part of the lens. Supporting a heavy lens in this way causes distortion of the lens and thus distortion of the image. Furthermore, a mirror has but one surface that requires polishing, whereas a lens has two surfaces. Also, a mirror does not produce chromatic aberration as does a lens. Figure 17-30 shows an example of a desirable location for reflecting telescopes.

Reflecting astronomical telescopes are usually located on mountain tops to get as high above the majority of Earth's atmosphere as possible and away from the bright lights of cities. Some reflecting telescopes are so large that a person can stand in them to view the image at the focal point. Various ways can be used to direct the light gathered by the parabolic mirror to a convenient focus outside the main part of the telescope. Some of these ways are illustrated in Figure 17-31.

Modern reflecting telescopes have incorporated various major design changes. The single converging mirror is now replaced with smaller converging mirrors arranged in a circle or a mosaic pattern. Light gathered by these multiple-mirrored telescopes is combined using plane mirrors that bring the light to a focus. Then, rather than viewing the image or photographing it in a conventional way, the light is detected by a much more sensitive device, for example a charge-coupled device, or CCD. A CCD is made with a thin layer of silicon (the substance used in electronic circuit chips) divided into small squares. Light striking these squares generates an electric charge that can be recorded and analyzed by computer. CCDs are commonly used in digital instruments, including still and video cameras. Anyone who has taken videos with such a camera knows how sensitive a CCD is, even in dim light, yet it is not readily overexposed. Furthermore, modern optical telescopes are used to gather visible or nearly visible light. However, such light is only a small portion of the electromagnetic spectrum. Reflecting telescopes can be built to receive such waves as infrared, ultraviolet, x rays, gamma rays (Figure 17-32), and radio waves, described shortly.

Figure 17-32 This pair of multi-segmented reflecting telescopes is located on La Palma, one of the Canary Islands, at an elevation of 2200 m. The 17-m-diameter mirrors are supported on a metal structure and are used to analyze the annihilation of dark matter and collect indirect evidence of black holes as they suck in huge amounts of energy around them. (They are called the MAGIC or "Major Atmospheric Gamma-ray Imaging Cherenkov" Telescopes.)

Reflecting telescopes tend to be very expensive, but researchers have found ways to reduce the cost. One innovative technique is to build a rotating parabolic base into which a highly reflective liquid, such as liquid mercury, is placed. As the base rotates, the liquid moves up the wall and becomes a parabolic telescope that can view straight upward perpendicular to the ground. At a diameter of 6.0 m, the Zenith Telescope, operated by the University of British Columbia, is the largest of its kind in the world.

David Parker/Science Photo Library

Figure 17-33 The Arecibo Radio Telescope in Puerto Rico is built inside a natural depression. With a diameter of 305 m, the main collecting dish is the largest electromagnetic wave-gathering instrument in the world. Waves reflect from the main dish to the moveable receiver suspended by cables above the dish. Among its many discoveries, this radio telescope was the first to identify planets orbiting distant stars.

It rotates at 8.5 rpm and is used for long-term surveys of the sky, among various projects. In the future, a liquid-based telescope may be placed on the Moon.

Radio Telescopes

After the accidental discovery in 1931 of electromagnetic waves in the wavelength range of radio waves coming from space, astronomers began to build telescopes to gather radio waves. Such a telescope is called a **radio telescope**. It gathers waves with wavelengths that range from about 10 cm to 20 cm, almost 10^6 times longer than visible light wavelengths. Thus, the parabolic reflectors used in radio telescopes are large (Figure 17-33), but they have the advantage of being lightweight because they can be made with low-density metal or a wire mesh design. Signals from the parabolic reflector are focused on an antenna or other receiver and are sent to a control centre for recording and analysis.

radio telescope a telescope that gathers radio waves from space

Astronomers have found that radio waves from space have added greatly to our knowledge and understanding of the universe. Radio waves can penetrate Earth's atmosphere without undergoing much absorption. They can penetrate portions of space where visible light is absorbed by tiny interstellar particles. Furthermore, radio wavelength signals have allowed astronomers to discover objects previously unknown.

DID YOU KNOW?

A "satellite dish" receives signals from a communication satellite in a manner similar to the way in which a radio telescope receives signals from space.

One disadvantage of radio telescopes is that they have relatively poor resolution (recall from Rayleigh's criterion that $\sin\theta \propto \lambda$, which means that the diffraction is larger for these relatively long wavelengths). To overcome this problem, radio telescopes can be built to operate together in large arrays (Figure 17-34) or by combining information from parabolic reflectors located several thousand kilometres apart.

Space Telescopes

The greatest limitation of Earth-bound telescopes is caused by air masses moving in Earth's atmosphere. This problem is overcome by placing telescopes in space above the atmosphere. Such a venture is obviously expensive, but the cost is justified by the rewards. Within a short period of time after it is placed in space, a modern telescope can reveal an impressive amount of data about the structure and history of the universe.

Consider, for example, the Hubble Space Telescope (HST), the reflecting telescope mentioned in the chapter introduction. It is named after Edwin P. Hubble (1889–1953), an American astronomer who discovered receding galaxies in our expanding universe.

© Oleg Nekhaev/Shutterstock

Figure 17-34 A very large array (VLA) of radio telescopes in Siberia, Russia.

Stocktrek Images/Thinkstock

Figure 17-35 The size of the Hubble Space Telescope is evident when compared to the astronauts who were guided by a space shuttle's robotic arm as they performed repairs on the telescope. The HST has five cameras and spectral analysis devices, all of which have high resolution.

The HST (Figure 17-35) is more sophisticated and sensitive than any telescope built before it. It measures 13 m by 4.3 m, has a mass over 11 500 kg, and its primary mirror is 2.4 m in diameter. Lasers were used to check the polishing of the fine surface of the mirror in a process that took 28 months prior to launching the HST in 1990.

During its lifespan of more than 20 years, the HST has maintained a nearly circular orbit 560 km above Earth's surface, where the atmospheric interference is negligible. Here it can detect objects in the universe seven times farther away than Earth-bound telescopes, and can detect details of previously observed objects with much better resolution. Data from instruments aboard the HST are communicated to control centres on Earth where astronomers and astrophysicists analyze them.

One of the biggest technical problems of operating space telescopes is to keep them aimed at a particular star for up to 10 h while the telescope travels at a high speed (e.g., about 29 000 km/h) around Earth. Engineers have designed guidance systems to accomplish this task.

Other Telescopes

As discussed previously, waves from various parts of the electromagnetic spectrum can be detected from all directions in the universe. Modern telescopes and other receivers sensitive to chosen parts of the electromagnetic spectrum reveal information about quasars (quasi-stellar objects), black holes, neutron stars, supernova explosions, comets, and the evolution of stars and galaxies, to name just a few of their discoveries. Some of these telescopes are used on Earth; others are (or will be) placed in space. For example, a satellite called *Gaia* with two telescopes has been launched by the European Space Agency into an orbit on the side of the Sun opposite to Earth. It will gather enormous amounts of data from the Milky Way Galaxy, enabling scientists to create a detailed three-dimensional view of the galaxy. (Engineers state that this satellite could measure the diameter of a human hair accurately from a distance of 1000 km!) Another plan calls for placing a space telescope sensitive to long wavelengths in orbit on the side of the Moon opposite Earth where there will be no interference from radio waves we produce on Earth.

As new, more powerful telescopes are developed, scientists are excited about what we will learn about the universe and our place in it.

EXERCISES

17-23 (a) State the advantages and disadvantages of a refracting astronomical telescope compared with a refracting terrestrial telescope.

(b) Why are astronomical telescopes placed on mountaintops or in space?

(c) What are the advantages of radio telescopes? Why are the new radio telescopes built in arrays?

17-24 The largest refracting astronomical telescope in the world (Figure 17-24) has an objective of focal length 19.5 m and an eyepiece of focal length 10.0 cm. Determine the maximum angular magnification of this telescope.

17-25 Two lenses of focal length 2.6 cm and 39 cm are available for a refracting astronomical telescope.

(a) Which lens should be used for the objective?

(b) What maximum angular magnification is possible using these lenses?

(c) What is the approximate length of the telescope?

17-26 What focal length eyepiece must be used to obtain a magnification of −12.5 times in a refracting telescope that has an objective of focal length 23 cm?

17-27 The largest refracting telescope in the world is 1.0 m in diameter and the Keck II reflecting telescope on Mauna Kea, Hawaii, is 1.0×10^1 m in diameter.

(a) What is the ratio of the amount of light gathered by the larger telescope to that gathered by the smaller one?

(b) Describe the advantage(s) of the reflecting telescope over the refracting telescope.

17-28 What designs are used to increase the light gathering ability of optical reflecting telescopes?

17.6 Cameras and Photography

A lens camera is a light-proof box in which a converging lens forms a real image on a detector or sensor.[4] The principle of the camera was described as early as the 10th century by the Arabian scholar Alhazen, who used a pinhole camera to view a solar eclipse safely. The great Italian scientist and artist Leonardo da Vinci (1452–1519) described the *camera obscura*, a device used to produce an image on a screen from which an artist could copy a scene. (*Camera obscura* is Latin for "dark room.") In da Vinci's era, the camera obscura produced a faint image because only a small amount of light passed through the pinhole. By the middle of the 16th century, however, lenses were being used to produce a brighter and sharper image (Figure 17-36).

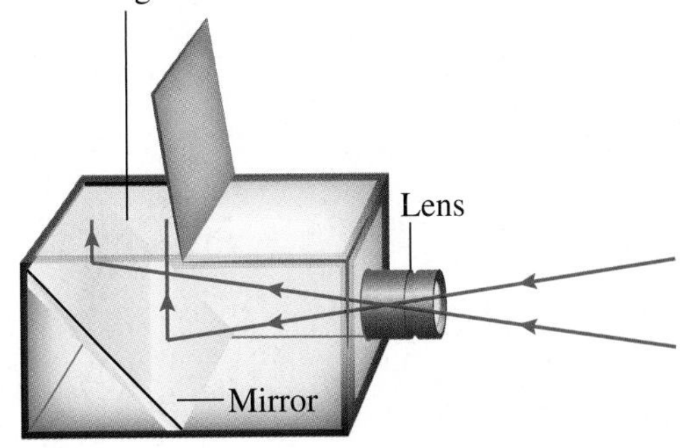

Figure 17-36 A camera obscura with a lens, in use about 300 years ago.

In the early 18th century, scientists became aware that light causes silver salts to turn dark. Within the next hundred years, this discovery made possible the invention of a photographic film that was sensitive to light. Thus, true photography began in the 19th century. During much of the 19th century, lens cameras used a single photographic plate that was both bulky and fragile. In the 1880s George Eastman, an American inventor (1854–1932), introduced roll film, and soon thereafter daylight-loading, roll-film cameras came into use. In 1935, colour film that required only one exposure and could be used in ordinary cameras was introduced. Since then, great improvements in convenience and specialized uses have made photography a major industry and important hobby in much of the world.

Today the field of photography is open to almost anyone who is interested. Easy-to-use digital equipment (Figure 17-37) is available in the form of still cameras, video cameras, and personal devices such as mobile phones, tablets, and webcams. Professional photographers use more sophisticated equipment in studios or for press and industrial photography. Specialized equipment is available for the movie and television industries as well as for such applications as high-speed, underwater, aerial, microscope, or infrared photography.

DID YOU KNOW?

Some historians surmise that certain "Old Master" painters, such as the 17th century Johannes Vermeer of Holland, used a *camera obscura* to project images onto a screen and then paint the scene in fine detail.

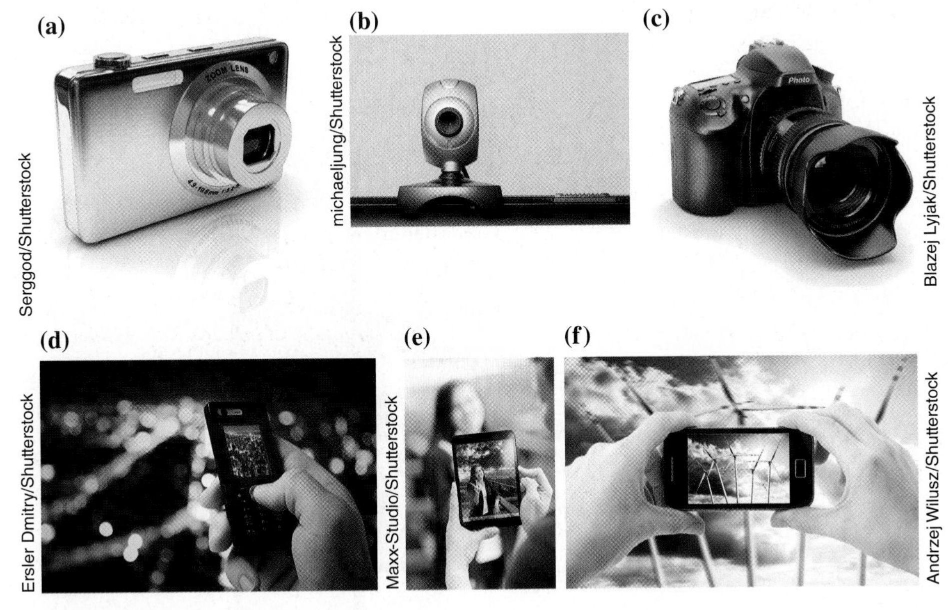

Figure 17-37 Examples of digital cameras.

[4]Most cameras today use CCDs to convert light energy to digital information. CCDs (charge-coupled devices) were described in conjunction with reflecting telescopes on page 572. Another type of detector is called a complementary-symmetrical metal oxide semiconductor, or CMOS.

Structure and Operation of the Camera

Modern lens cameras, whether simple or sophisticated, contain the following main parts:

- light-proof box, which supports the entire apparatus
- a converging lens or system of lenses, which gathers light and focuses it onto a detector
- a shutter, which controls the amount of light passing through the lens when the picture is taken
- the detector, which records the image on an electronic device
- a display screen or viewfinder, which allows the photographer to see what he or she is photographing

The simplest digital cameras use a lens or lens system that acts like a single converging lens. The focal length of this lens determines how large the camera must be to support it. Standard digital-camera lenses have a fixed focal length of about 40 mm, but small mobile devices may have a focal length down to 3 mm or even less. Figure 17-38 shows a ray diagram illustrating the image formation of a typical digital camera. To see how the variables labelled in the diagram relate to the size of a camera, consider once again the thin-lens equation (Eqn. 15-8): $\frac{1}{f} = \frac{1}{d_o} + \frac{1}{d_i}$. Since the object distance, d_o, is so much larger than the image distance, d_i, this equation can be reduced to $\frac{1}{f} \approx \frac{1}{d_i}$, from which we get $d_i \approx f$. Thus, the converging camera lens produces a real, inverted image of a distant object on the detector at a distance approximately equal to the focal length of the lens.

As with single lenses described in Chapter 15, linear magnification (M) can be found using $M = \frac{h_i}{h_o} = \frac{-d_i}{d_o}$, and this equation can be rearranged to obtain the equation for the image size, $h_i = \frac{-h_o d_i}{d_o}$.

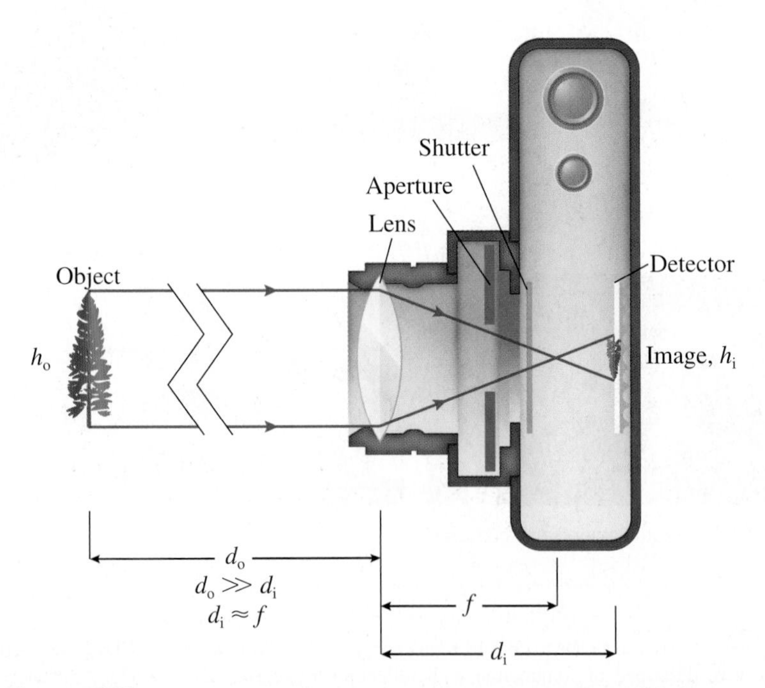

Figure 17-38 Ray diagram showing the formation of an image in a digital camera.

SAMPLE PROBLEM 17-9

The label on the lens cover of a small digital camera indicates a focal length of 4.3 mm.

(a) When taking a photo of a distant object, what is the approximate distance between the lens and the detector?

(b) When the camera is now focused at an object only 5.5 cm from the lens, what is the new distance between the lens and the detector?

(c) What can we conclude based on the calculations in (a) and (b)?

Solution

(a) For a distant object, d_o is much larger than d_i, so $1/d_o$ approaches zero and can be ignored. Thus, from the thin-lens equation,

$$\frac{1}{f} \approx \frac{1}{d_i} \text{ and } d_i \approx f = 4.3 \text{ mm}$$

Thus, the approximate distance between the lens and the detector is 4.3 mm.

(b) Using $d_o = 5.5 \text{ cm} = 55 \text{ mm}$,

$$\frac{1}{d_i} = \frac{1}{f} - \frac{1}{d_o}$$

$$\frac{1}{d_i} = \frac{1}{4.3 \text{ mm}} - \frac{1}{55 \text{ mm}}$$

$$\therefore d_i = 4.7 \text{ mm}$$

Therefore, when the object is only 5.5 cm from the lens, the image is 4.7 mm from the lens.

(c) We can conclude that when this camera changes its focus from a distant object to a very close one, the lens must move only a short distance (4.7 mm – 4.3 mm = 0.4 mm in this case) for the image on the detector to remain in focus.

If you experiment with a digital camera that has automatic (auto-) focus, similar to the one described in Sample Problem 17-9, listen carefully as you change the object distance of the camera's view. For example, if you hold your hand a short distance in front of the lens and then remove it, you can hear the lens shifting position to obtain a clear image of the more distant object.

Controlling Camera Variables

Fixed-focal-length cameras, including the one described in Sample Problem 17-9, are common and simple to use, and the quality of the photos they produce is continually improving. However, many cameras are available that allow users more versatility to control photographic variables. Of the numerous variables that can be adjusted, we will consider mainly those that relate to physics principles rather than subjective artistic principles.

One of the easiest camera variables to control is the focal length of the lens. In contrast to a fixed-focal-length lens, a variable-focal-length lens has a range of focal lengths that can be adjusted by using the zoom control or by interchanging lenses on more complex cameras with removable lenses.[5] For example, the lens of a small digital camera may be labelled 4.3 mm to 21.3 mm, which indicates the range of focal lengths.

[5]Most professional photographers use a digital single-lens reflex (DSLR) camera, often with interchangeable telephoto and zoom lenses.

DID YOU KNOW?

A "Quick Response Code" or QR code is a two-dimensional label (Figure 17-39) used for advertising, tracking, identification, and many other purposes. It can be detected by smart phone, tablet, or similar device, and then analyzed digitally. The code uses squares in three corners of the QR image and usually smaller squares near the fourth quarter to identify orientation and other features. The remaining pattern contains the information communicated. (Notice that this two-dimensional code differs from the barcode scanned with laser light, discussed in Section 17.7.)

Sergey Peterman/Shutterstock

Figure 17-39 Using a Quick Response (QR) code.

(a)

(b)

(c)

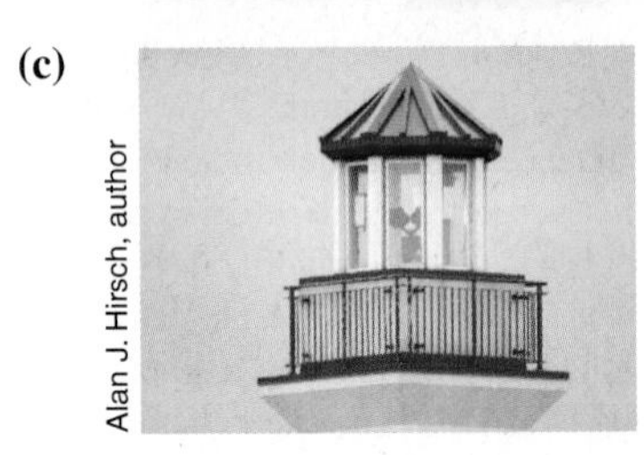

Figure 17-40 The effect of varying the focal length of a camera lens from short or wide-angle view in **(a)**, to medium in **(b)**, to long or close-up view in **(c)**.

Calculating the ratio of these values, (21.3 mm)/(4.3 mm) = 5, reveals that the longest focal length available magnifies the image 5 times as much as the shortest focal length available. This explains the label "5 ×" on the lens. The quantity 5 × is an example of an **optical zoom**, a camera feature in which the image magnification can be increased by increasing the focal length of the lens.[6] Figure 17-40 shows three photographs that illustrate the effects of controlling the optical zoom. Notice in Figure 17-40 that as the focal length increases the area of the scene decreases. In terms of the angular size described for other optical instruments, we see that the angular size is inversely proportional to the focal length of the lens; in other words, a short focal length yields a wide-angle view, and a long focal length provides a close-up, small-angle view.

The **aperture** (d) is the diameter to which the shutter opens. It is indicated by the ***f*-stop number**, which is the ratio of the focal length (f) of the lens to the diameter of the aperture. Thus,

$$f\text{-stop number} = \frac{f}{d} \quad (17\text{-}9)$$

where f is the focal length of the lens and d is the diameter of the aperture.

Typical *f*-stop numbers range from 1.2 to 32, depending on the lens, although inexpensive cameras have a much lower range. When the *f*-stop number is changed from one number to the next larger or smaller number, the diameter changes by $\sqrt{2}$, so the area changes by a factor of $(\sqrt{2})^2 = 2$. Thus, at $d = f/8$ there is approximately twice as much light entering the lens as at $d = f/11$. We can relate the aperture setting to **exposure time**, also called the shutter speed, which is the amount of time the shutter remains open when a photograph is taken. It is stated in seconds or fractions of a second, for instance 1/60 s and 1/1000 s. Figure 17-41 shows a typical leaf shutter with the *f*-stop adjusted from $d = f/11$ on the left to $d = f/8$ on the right. If a photograph is properly exposed at *f*/11 with an exposure time of 1/50 s, then at *f*/8 or about double the aperture area, the new exposure time should be 1/100 s.

optical zoom a camera feature in which the image magnification can be increased by increasing the focal length of the lens

aperture the diameter to which a camera shutter opens; symbol d

***f*-stop number** the ratio of the focal length (f) of a camera lens to the diameter of the aperture

exposure time the amount of time the shutter remains open while a photograph is taken (also called shutter speed)

SAMPLE PROBLEM 17-10

When a lens of focal length 50 mm is set at *f*/16, the correct exposure time for a particular scene is 1/100 s. Then the lens aperture is adjusted to *f*/8. (Assume two significant digits for the calculations.)

(a) Calculate the aperture (d) at the two settings.

(b) What is the new exposure time at *f*/8?

Solution

(a) At $f/16$, $d = \frac{f}{16} = \frac{50 \text{ mm}}{16} = 3.125$ mm, keeping two extra digits

At $f/8$, $d = \frac{f}{8} = \frac{50 \text{ mm}}{8} = 6.25$ mm, keeping one extra digit

Thus, the aperture is about 3.1 mm at *f*/16 and about 6.3 mm at *f*/8.

(b) At $f/8$ the diameter of the lens is 2 times as large as at $f/16$, so the area receiving light is 4 times as large. Thus, the exposure time should be 1/4 of the original value, or 1/400 s.

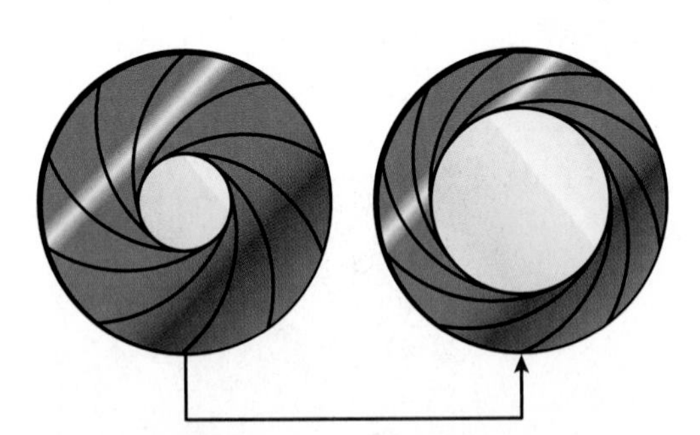

Figure 17-41 If the diameter of the aperture increases by a factor of $\sqrt{2}$ or 1.4, the area increases by a factor of $(\sqrt{2})^2$ or 2, etc.

[6]Another feature, called *digital zoom*, involves manipulating the digital image without altering the focal length of the lens. The quality of photos magnified this way tends to be poor, but with improved technology digital zoom capabilities may continue to improve.

Controlling the aperture makes special effects possible. At a small aperture, only the central portion of the lens is used, producing little aberration and allowing the scene photographed to be in sharp focus. Such a photograph has a larger **depth of field**, the range of object distances over which the image is clear. However, if the aperture becomes very small, the clarity of the photograph is affected by diffraction effects. At a large aperture (e.g., *f*/2), the outside portion of the lens is used, resulting in a lot of aberration. In this case the photograph has centred subjects in focus but a small depth of field (Figure 17-42).

depth of field the range of object distances over which an image in a camera (or other optical device) is clear

The use of computer systems in cameras makes photography easier for professional and amateur photographers alike. One use of such computers is in providing different exposure times for different segments of the sensor. The lens "views" the scene in several different segments. Then the computer compares the brightness of the segments with numerous pre-programmed examples, and adjusts the camera to give the best possible exposure in each of the segments. This type of lens is not easily "fooled" by bright regions surrounding dark subjects, or vice versa.

anmo/Shutterstock

Figure 17-42 The beauty of early spring flowers is emphasized with a shallow depth of field. In this case, the crocuses are in perfect focus, but the background and snowy foreground are out of focus and blurry.

TRY THIS!

Controlling Camera Variables

This activity allows you to apply physics principles to photography. Most steps can be done with any digital camera that has both automatic and manual controls. You may need the user's manual (often available online) to learn how to control features of your own camera.

- Familiarize yourself with the camera's main features, such as the automatic (auto) setting, various manual settings, shutter release, viewing screen, and focal length settings.
- View a specific scene through the viewfinder, and, if possible, observe the effect of manually adjusting the exposure time, the focal length, and the aperture.
- Once you understand how to control as many of the variables as possible, design a controlled experiment in which you vary the variables one at a time. View the resulting photographs and describe how you would apply what you learned to enhance your digital photography skills.

DID YOU KNOW?

Researchers at the Massachusetts Institute of Technology have created movies that show the motion of light by using a special camera that was able to take images with an incredibly short exposure time of 1.7×10^{-12} s!

EXERCISES

17-29 A camera lens of focal length 150 mm can be adjusted so that its distance from the sensor is between 150 mm and 156 mm. Determine the distance to the nearest and farthest objects on which this lens can focus. Assume three significant digits in given data.

17-30 The focal length of a zoom lens is adjusted from 80 mm to 200 mm.

(a) What happens to the magnification of the image on the film?

(b) Under what conditions does your answer hold?

17-31 A photographer wants to fill the 24.0-mm-high space of a sensor with the image of a person who is 1.50 m tall, located 8.50 m away. What focal length lens will accomplish this?

17-32 Compare the light-gathering ability of a lens that has its aperture changed in the following ways:

(a) from *f*/2 to *f*/4

(b) from *f*/5.6 to *f*/1.4

(c) from *f*/2.8 to *f*/22

17-33 A lens of 200 mm focal length is set at *f*/11 and 1/60 s. (Assume two significant digits.) To have the same exposure, what is the new exposure time if the aperture is changed to

(a) *f*/16?

(b) *f*/4.0?

17-34 A lens of focal length 40 mm is set at 1/250 s and *f*/5.6. (Assume two significant digits.)

(a) What is the diameter of the aperture?

(b) To have the same exposure, what is the new aperture setting if the exposure time is changed to 1/1000 s?

17-35 How does a photograph with a low depth of field compare to one with a large depth of field? How is a low depth of field achieved with a digital camera?

lightpoet/Shutterstock

Figure 17-43 Physicists conduct experiments using lasers in optics laboratories.

photon (or light quantum) a particle-like bundle or packet of light energy

laser Light Amplification by Stimulated Emission of Radiation

spontaneous emission the emission of photons from an atom that becomes excited by absorption of a photon or by collision with a particle such as an electron; the emission occurs very quickly after excitation

metastable state an excited state in which an atom can remain without undergoing spontaneous emission for a relatively long time

stimulated emission the emission of photons from an atom in the metastable state

DID YOU KNOW?

Many of today's applications in physics are possible because of the *photoelectric effect*, first explained in detail by Albert Einstein. If a photon of light with enough energy strikes a metal surface, an electron at the surface can gain enough of that photon's energy to break free of the forces binding it to the metal. It was Einstein's explanation of this effect that earned him the Nobel Prize in Physics in 1921. His revelations provided the background necessary to develop the means by which stimulated emissions can be used to control light in a laser.

17.7 Lasers and Holography

What is a laser? a device used at store checkouts? a medical tool? a fun "toy" in a light show? a laboratory tool (Figure 17-43)? or all of these? How does the origin of the word laser help us understand how it operates? When was the laser invented? Why was the laser once called "a solution looking for a problem?" We have many questions about lasers; now let us explore some answers.

In order to understand the special nature of light emitted by a laser, we first must explore the *quantum nature of light*. In 1905, Albert Einstein postulated (correctly) that light consists of particle-like bundles or packets of energy called **photons** or **light quanta** (singular, quantum). As featured in the diagram of the electromagnetic spectrum in Figure 16-47 on page 537, the energy of any part of the electromagnetic spectrum increases as the wavelength decreases (or the frequency increases). Thus, for instance, a photon of violet light has a higher energy than a photon of red light. The energy can be stated in joules (J), but it is common to use the electron-volt, eV, described in Section 7.6 where the equivalency is given as $1.00\ \text{eV} = 1.60 \times 10^{-19}\ \text{J}$. Consider, for example, the red light of a type of laser with a wavelength of 633 nm. A photon of this light has an energy of 1.96 eV, whereas a photon of violet light, with a wavelength of 445 nm, has 2.79 eV of energy.[7]

The word **laser** stands for Light Amplification by Stimulated Emission of Radiation. In a laser, one photon stimulates an excited atom to emit another photon of the same energy. (This process is discussed in more detail below.) Now there are two photons of the same energy and wavelength, and they are in phase and travelling in the same direction. Each of these photons can stimulate another atom to emit a photon, resulting in a total of four photons, which can then stimulate the emission of more photons, and so on. Thus, the original photon is amplified into many, many photons in a strong, unidirectional beam of light. The beam is monochromatic because there is only one wavelength and frequency. The beam is also coherent, which means that the waves are in phase and unidirectional. These properties are in contrast with the incoherent, polychromatic nature of ordinary white light.[8]

The physical principles involved in laser operation were first applied to microwaves in the development of *masers* (the *m* standing for microwave) in 1953, by the American physicist, Charles H. Townes. The first working laser (in the visible region of the spectrum) was developed in 1960 by Theodore Maiman at Hughes Research Laboratories in California. In 1964, Townes received the Nobel Prize in Physics, along with two Soviet laser pioneers, Aleksander Prokhorov and Nikolai Basov.

Atoms can emit photons by one of two possible processes. **Spontaneous emission** is the emission of photons after an atom is excited by the absorption of a photon or by collision with a particle such as an electron. The atom then emits a photon when it de-excites to a lower energy level. This spontaneous emission occurs very quickly after the original excitation—within about 10^{-8} s. The second possible process involves what is called a **metastable state**, an excited state in which the atom can remain without undergoing spontaneous emission for a long time (on the atomic scale, about 10^{-4} to 10^{-3} s). When in this metastable state, an atom will de-excite and emit a photon if stimulated by another photon of the "right" energy that happens by. Thus, **stimulated emission** is the emission of photons from atoms in the metastable state. The "right" energy has to correspond to the energy difference between the excited metastable state and a lower-energy state in the atom. If atoms are not

[7]The fact that the energy of a photon increases as the wavelength decreases or the frequency increases is evident in the equation for the energy of a photon, $E = hf$, where h is Planck's constant, 6.63×10^{-19} J. Since $c = f\lambda$, this equation can also be written $E = (hc)/\lambda$, where c is the speed of light.

[8]Coherent and incoherent light sources were discussed in Section 16.2 in connection with Young's double-slit experiment.

first excited to metastable states in the laser, they de-excite quickly via spontaneous emission, and stimulated emission occurs infrequently. For a laser to operate there must be more atoms excited into metastable states than there are in the lowest or "ground" state, a situation called **population inversion**. The initial excitation is often done with an electric discharge.

population inversion a situation in which more atoms are excited into metastable states than there are in the ground state

There are many different types of lasers. The first successful laser used a ruby rod, which emitted red light. Now there are liquids, gases, semiconductors, glasses, and crystals employed in lasers. Some lasers produce visible light, and others work in the invisible ultraviolet or infrared regions of the electromagnetic spectrum. Some lasers can be "tuned" to give monochromatic light over a wide range of possible wavelengths, while others operate at only one wavelength.

One of the most common inexpensive lasers has a gas mixture, about 85% helium and 15% neon, as the lasing material. You have probably seen helium–neon (He-Ne) lasers—they emit red light—at supermarket checkout counters. The operation of this type of laser depends on a lucky coincidence: one excited metastable state of helium, 20.61 eV above the ground state, has almost the same excitation energy (20.66 eV) as another metastable state in neon. An electric discharge excites helium atoms into the metastable state (E_1 in Figure 17-44). When an excited helium atom and a ground-state neon atom collide, the helium atom can de-excite and provide energy directly to excite the neon atom to its metastable state (E_2 in Figure 17-44). (The additional 0.05 eV of energy is provided from the kinetic energy of the atoms prior to the collision.) A neon atom in this state can readily participate in stimulated photon emission if a photon of energy 1.96 eV passes nearby. When stimulated, the neon atom drops to level E_3, emitting a visible "red" photon of energy 1.96 eV (wavelength 633 nm). The atom, still excited, then moves quickly to the ground state (E_0) via transitions that are not part of the laser action. You might wonder why the helium is used at all; why not just use the electric discharge to excite the neon atoms directly to their metastable states? Since neon has ten electrons and many different energy levels, the probability of exciting a neon atom to the right energy level for laser action is small. It is much easier to excite the helium since it has only two electrons, and then have helium collide with neon.

One of the useful features of laser light is that it is unidirectional. We mentioned that a photon emitted via stimulated emission has the same direction as the stimulating photon, and this property leads to the directionality. However, when a laser is turned on, the first few photons emitted come from spontaneous emission from the metastable states, and have random directions. These photons then act as the stimulators, so why does a laser not have beams in a number of different directions? The answer lies in the construction of a laser (Figure 17-45), which is normally a long, narrow tube with a mirror at each end. Photons that are emitted in any direction other than along the tube are absorbed by the tube walls. Photons emitted along the tube direction are reflected back and forth by the mirrors, stimulating the emission of many photons in this direction, and a strong beam builds up parallel to the length of the tube. One of the mirrors is essentially 100% reflecting, and the other reflects most of the light. The light transmitted through this "leaky" mirror forms the external laser beam.

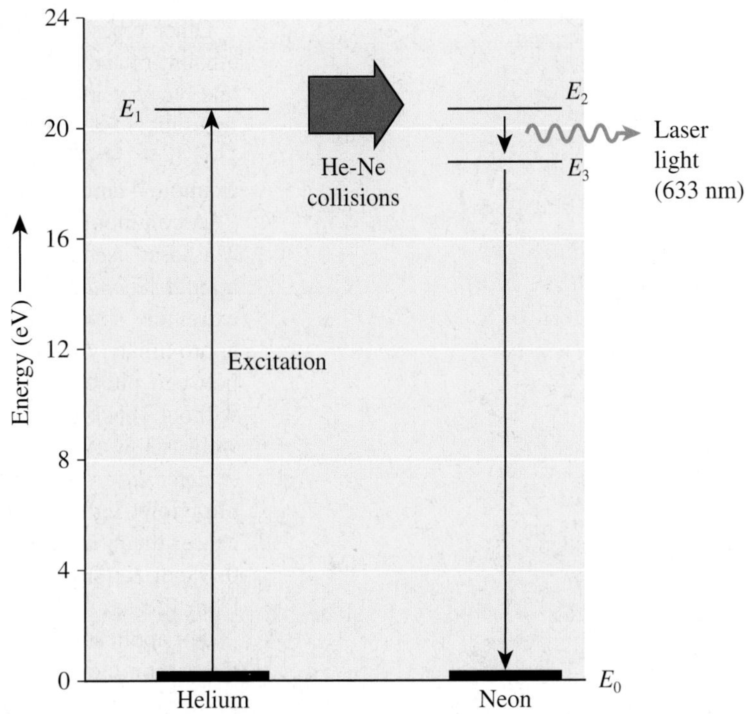

Figure 17-44 Important energy levels in a He-Ne laser.

Using Lasers

When lasers were first invented, people wondered what to do with them. What uses could there be for a bright, coherent, linear beam of light? Although lasers appeared initially to be a "solution looking for a problem," applications were

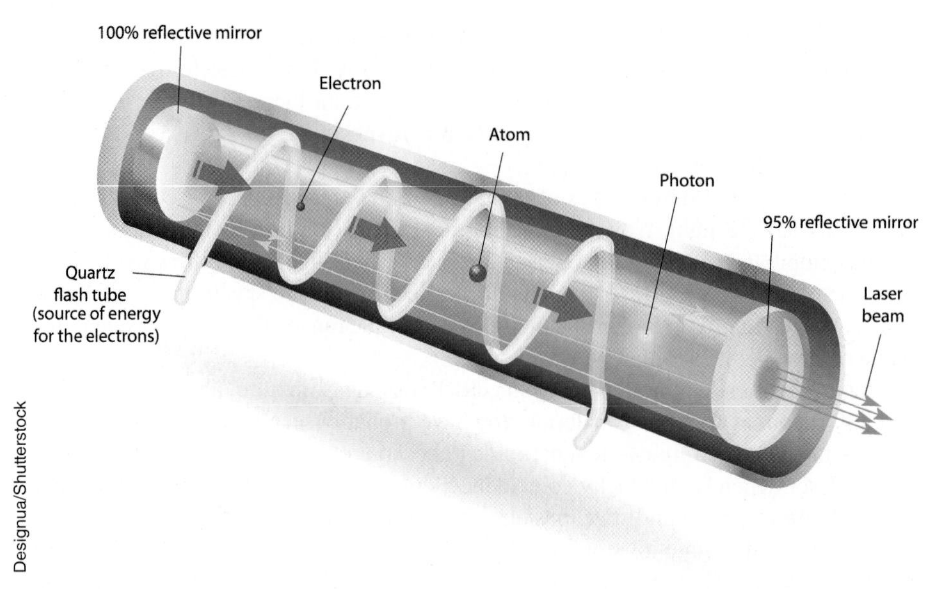

Figure 17-45 Mirrors at the ends of the laser tube ensure that only photons emitted along the direction of the tube contribute to the laser beam.

DID YOU KNOW?

Infrared laser beams with widths of only $\frac{1}{4}$ mm are routinely used in telecommunications to send telephone calls along fibre optics cables. One laser and one optical fibre can handle up to 32 000 calls simultaneously.

not long in coming, and lasers are now used in medicine, industry, commerce, communications, forensic science, etc. Some of the numerous uses of lasers are shown in Figure 17-46 and described below.

The helium–neon laser already described has found a number of applications. For example, it is used in supermarket checkout scanners, in which the reflected laser beam carries the product-identification information contained in the black and white stripes of the Universal Product Code (UPC). This information is relayed to a computer that retrieves the current price of the product and other information. Because a laser beam travels in a straight line, He-Ne lasers are used in guiding saws in lumber mills, in controlling machines that drill tunnels for railroads and highways, and in surveying.

Other types of lasers are in use in medicine. Because lasers concentrate a large amount of light energy onto a small area, they can be used to seal bleeding ulcers and blood vessels, vaporize tumours, and "weld" breaks in the retina. Lasers can be focused so finely that nearby cells are unaffected. By using optical fibres to carry the laser light, physicians can perform procedures in areas of the body—the stomach, for example—unreachable by other means except major surgery.

A common medical application is in laser eye surgery, described in Section 17.2. The laser used to make a surgical incision to create a flap in the cornea is called a *femtosecond laser*. The prefix femto means 10^{-15}, so this laser creates pulses of extremely short duration ($\approx 10^{-15}$ s). The pulses are directed toward the cornea at approximately every 6×10^{-5} s, so the pulse duration is much less than the time between pulses. In such a short time, the laser energy can perform an incision without shock or heat damage to nearby material. This enables high precision as well as low risk and fast recovery. After the cornea layer has been cut open, an *excimer laser* is used to reshape parts of the cornea. An excimer laser produces ultraviolet light that is easily absorbed by organic material. The absorbed energy causes the molecular bonds to break down and surface material to disappear (similar to vaporization) in a controlled operation. Wavelengths of excimer UV light range from 126 nm to 351 nm.

For applications requiring large amounts of energy, such as cutting steel saw blades, the carbon dioxide laser (which emits infrared light) is most often used. Lasers are proliferating in jobs such as welding, drilling, and heat treating, where precision and a large quantity of heat go hand in hand.

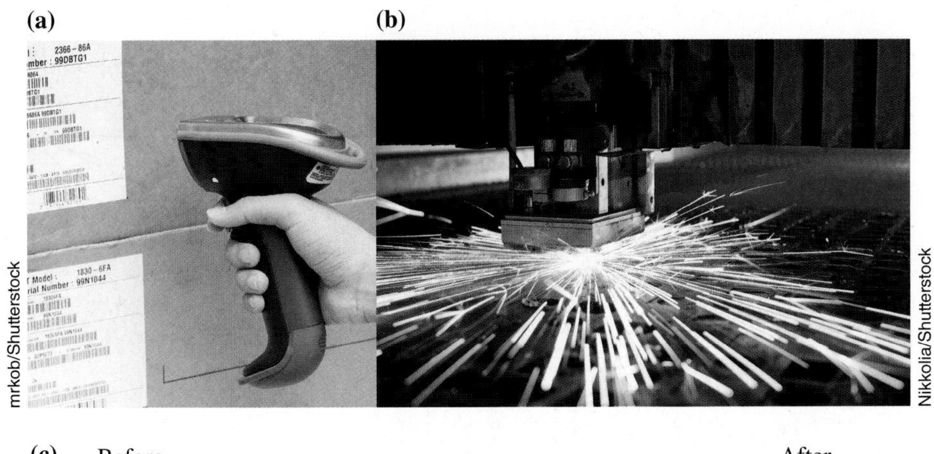

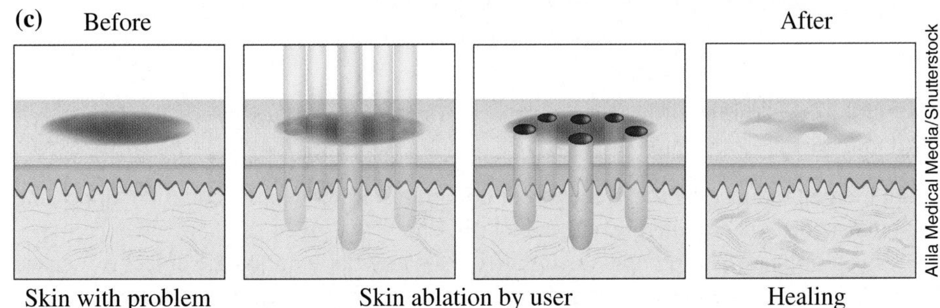

Figure 17-46 Applications of lasers: **(a)** Information on the UPC is read and transmitted quickly. **(b)** An industrial laser cutting tool. **(c)** Skin problems can be removed using a process known as "fractional laser skin resurfacing." (The removal is called ablation.)

Semiconductor lasers are small and inexpensive, and the infrared light emitted is used in the operation of compact disk (CD) players and in fibre optics communication systems. On CDs, the audio or video information is encoded in a pattern of tiny pits on the disk, and retrieved by reflection of the laser light.

Lasers are also used in

- monitoring small movements of Earth to give advance warning of earthquakes
- missile guidance systems
- high-precision fabrication in the semiconductor industry
- fusion research; it is hoped that intense laser beams can create the high temperatures and pressures required for nuclear fusion to occur[9]
- detecting fingerprints many years old on materials (such as leather) not suitable for standard fingerprint-detection techniques
- one method of 3-D printing (also called additive manufacturing) in which layers of material are laid down successively and fused into shapes by a computer-controlled laser; this versatile technology is being applied in the medical, dental, and biotech industries, as well as in jewellery, fashion, architecture, and many other fields.

Holography

Who would have thought when the laser was developed in the 1960s that one of its most fascinating uses would be the production of three-dimensional images in a process called **holography**? A few decades ago, most people had not seen a **hologram**,

holography the process in which laser light is used to produce a three-dimension image

hologram the three-dimensional image produced by holography using laser light

[9]More on the subject of fusion is presented in Chapter 24.

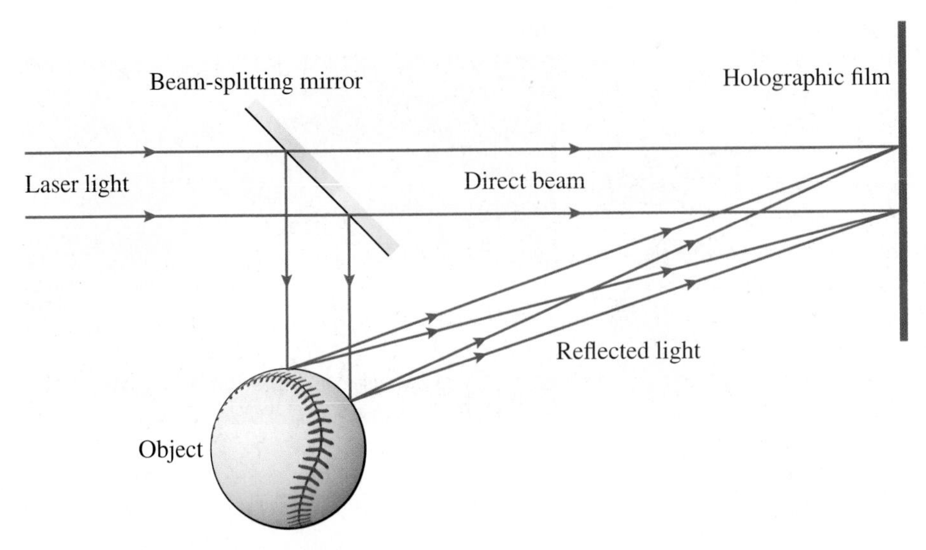

Figure 17-47 Making a hologram.

DID YOU KNOW?

Federal agencies are constantly trying to apply physics principles to reduce the success rate of counterfeiting paper currency. If you check out a modern polymer-based $20 Canadian bill, you will see and feel features such as holographic stripes, watermarks, raised printing, tactile marks, printing threads visible in UV light, transparent areas with detailed metallic images that change colours when the viewer tilts the bill, and hidden numbers in the metallic maple leaf (visible when held up to a bright white light). Forging this bill would be complex, to say the least.

the image produced by the process, but holograms are now quite common, appearing in magazines, on clothing, on currency bills, etc.

In an ordinary photograph, the light from a particular point on the object is focused to one point on the detector or sensor, but, in holography light from many points on the object travels to each point on the sensor. Figure 17-47 shows the traditional way a hologram is produced. Light from a laser is split into two parts by a half-silvered mirror. One part travels directly to the holographic sensor, and the other part reaches the sensor only by reflection from the object. On the sensor, the light from the two paths interferes constructively in some places and destructively in others, thus producing a speckled pattern of light and dark areas.

After the pattern is produced and is illuminated with a laser (Figure 17-48), a three-dimensional image of the object is produced. If the viewer's head moves from side to side or up and down, different views of the object result, just as if the original object were present. The production of a hologram depends on the fact that laser light is coherent. Because the rays in the original laser light are all in phase, they have specific phase differences when they combine on the film to produce the interference pattern. When the film is illuminated with coherent laser light, the transmitted light picks up

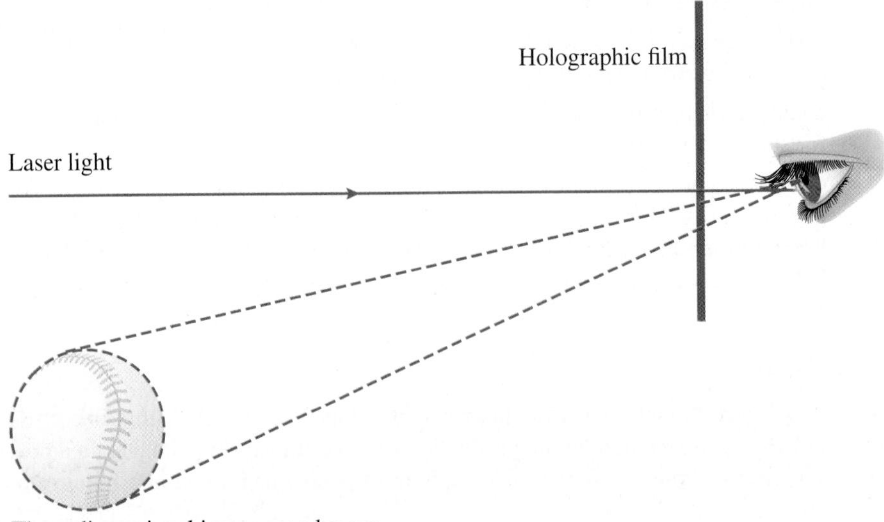

Figure 17-48 Viewing a hologram.

the phase difference information from the film and appears just as if it originated from the object itself. Cylindrical holograms can be made such that the viewer can walk completely around them and see a 360° image of the original object.

Some holograms can be viewed by reflection in white light. These are the common holograms that you have seen on credit cards, in magazines, etc. These *white-light holograms* normally give the impression of a three-dimensional image when viewed from side to side, but *not* up and down. This type of hologram is created with laser light in the same way as described in the previous paragraph, then the resulting hologram is "masked" in a certain way to remove the up-and-down information and produce a second hologram that can be viewed in white light.

Figures 17-47 and 17-48 illustrate what is referred to as the *geometric model* of holograms, that is, a model in which light travels in geometrically straight lines. However, holograms have properties that cannot be understood using this geometric model; *diffraction* effects must also be included. When a hologram is illuminated by a laser beam as in Figure 17-48, the beam diffracts around the various light and dark regions on the hologram to produce diffraction patterns which, superimposed, make up the three-dimensional image. We will not delve into the details of diffraction in holography, but you should be aware that diffraction is important for a complete understanding of this field.

Holography has a wide range of uses other than simply producing three-dimensional images. For example, it can be used in pattern recognition, such as fingerprint-matching or searching for occurrences of a particular word in a book. Microwave (radar) holography is applied in *side-looking radar* in airplanes to produce finely detailed images used in geological surveying, cartography, and military surveillance. Holography is also employed in detecting stresses in manufactured materials, and in a variety of other applications (Figure 17-49).

DID YOU KNOW?

Recent advances combining computing with lasers have generated holograms called volumetric displays by forming voxels, which are pixels in a three-dimensional grid.

(a)

Neftali/Shutterstock.com

(b)

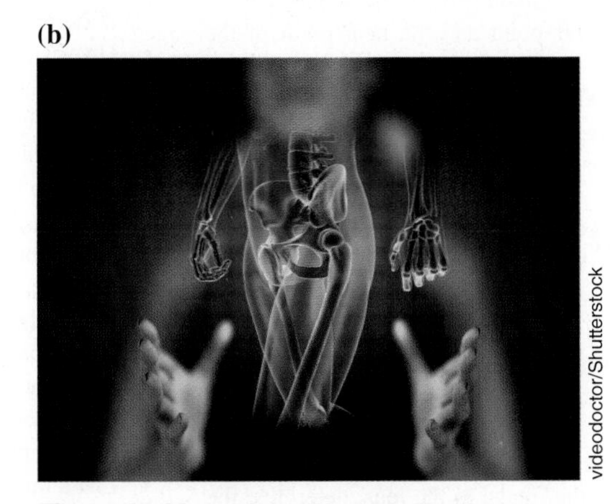

videodoctor/Shutterstock

Figure 17-49 **(a)** This 2001 United Kingdom stamp commemorates the 100th anniversary of the Nobel Prize in Physics with a hologram of a boron molecule. **(b)** A human radiography scan on a hologram provides a three-dimensional view.

EXERCISES

17-36 List three ways in which laser light differs from the light emitted by a normal light bulb.

17-37 What is the frequency of light emitted by a He-Ne laser? Its wavelength is 633 nm.

17-38 In each case below, state whether the first variable named is directly proportional to, inversely proportional to, or independent of the second variable.

(a) E_{photon}; λ_{photon} (b) E_{photon}; f_{photon} (c) f_{photon}; λ_{photon}

17-39 (a) What type of mirror is placed at each end of a typical laser tube?

(b) What is the function of each of these mirrors?

17-40 Why is laser light dangerous?

17-41 List four uses of lasers.

17-42 Does the process of holography support or refute the wave nature of light? Explain.

LOOKING BACK...LOOKING AHEAD

This chapter presented several optical instruments that are applications of the concepts found in Chapters 15 and 16. The equations and ray diagrams for lenses were especially useful in analyzing optical instruments. These instruments have limited resolution ability due to the wave nature of light.

Vision and vision defects were described, including the use of the thin-lens equation to determine the focal length of lenses needed to correct vision defects. Angular size was introduced and used to define the angular magnification of a simple magnifier. These concepts were applied to microscopes and telescopes, which were found to have distinctly different characteristics and equations for total magnification. Several different types of telescopes were described, including modern examples. Variables associated with photography (exposure time, focal length, and aperture) were analyzed mathematically. Finally, we used the quantum nature of atoms to discuss the operation of lasers, and presented a number of applications of these fascinating devices, including holography.

Topics in upcoming chapters include heat and thermodynamics, followed by electricity and magnetism.

CONCEPTS AND SKILLS

Having completed this chapter you should now be able to do the following:

- State what advantages we gain because our eyes have accommodation, binocular vision, and image retention.
- State the names and functions of the main parts of the human eye.
- Recognize the meaning of the far point and the near point of the human eye.
- Explain the role of cones and rods in human vision.
- Describe the three-colour theory of colour vision.
- Describe how retinal fatigue can be demonstrated, and explain its cause.
- Describe defects of the human eye and corrections for those defects.
- Apply the thin-lens equation to determine the focal length of a correcting lens for either myopic vision (given the far point) or hyperopic vision (given the near point).
- Given any two of the height of an object, the distance to the object, or the angular size of the object in a simple magnifier, determine the third quantity.
- Given any two of the angular magnification of a lens, the focal length of the lens, and the distance of the image to the lens, determine the third quantity.
- Given any three of the total magnification of a compound microscope, the length of the microscope, and the focal lengths of the objective and eyepiece, determine the fourth quantity.
- Distinguish between (a) refracting and reflecting telescopes, (b) astronomical and terrestrial telescopes, and (c) radio and optical telescopes.
- Given any two of the magnitude of the magnification of a telescope and the focal lengths of the objective and eyepiece, determine the third quantity.
- Describe the advantages of placing telescopes in space.
- Apply the lens equations (developed in Chapter 15, Eqns. 15-8 and 15-9) to camera lenses to determine any one of the related variables (f, d_i, d_o, M, h_i, and h_o).
- Given the *f*-stop number and the focal length of a lens, calculate the aperture.
- Given the correct *f*-stop number and exposure time for a particular scene, calculate other combinations of *f*-stop number and exposure time that will yield the proper exposure.
- Describe how a laser produces its light.
- State important properties of laser light.
- List a number of applications of laser light.
- Describe how a hologram is produced.

KEY TERMS

You should be able to define or explain each of the following words or phrases.

cornea	cones	hyperopia	compound light microscope
lens (of the human eye)	rods	farsightedness	eyepiece lens
retina	optic nerve	astigmatism	eyepiece
far point	primary light colours	presbyopia	objective lens
near point	secondary light colours	angular size	objective
accommodation	myopia	angular magnification	optical telescope
pupil	nearsightedness	magnifying power	refracting telescope

astronomical telescope
terrestrial telescope
reflecting telescope
radio telescope
optical zoom
aperture
f-stop number
exposure time
depth of field
photons
light quantum
laser
spontaneous emission
metastable state
stimulated emission
population inversion
holography
hologram

Chapter Review

MULTIPLE-CHOICE QUESTIONS

17-43 The human eye's accommodation allows us to
(a) view motion pictures that are shown at a minimum rate of 24 fps (frames per second)
(b) view with clarity objects that range in distance from our near point to our far point
(c) relax our eyes by viewing nearby objects
(d) view a range of light intensities that spans a factor of about 10^{10}
(e) perceive a wide range of colours

17-44 If you stare at a bright magenta circle for a minute and then look at a white background, the after image is
(a) red, (b) blue, (c) green, (d) yellow, (e) cyan

17-45 For a person with presbyopia, two possible corrections are
(a) a diverging lens or laser surgery to increase the eye's focal length
(b) a converging lens or laser surgery to increase the eye's focal length
(c) a diverging lens or laser surgery to decrease the eye's focal length
(d) a converging lens or laser surgery to decrease the eye's focal length

17-46 The power of an eyeglass corrective lens can be measured in
(a) only dioptres^{-1}
(b) only metres^{-1}
(c) dioptres^{-1} or metres^{-1}
(d) dioptres or metres^{-1}
(e) dioptres or metres

17-47 Simple magnifiers A and B have focal lengths of 4.0 cm and 8.0 cm respectively. Compared with A, B has
(a) higher magnification but lower spherical and chromatic aberration
(b) lower magnification but higher spherical and chromatic aberration
(c) lower magnification and lower spherical and chromatic aberration
(d) higher magnification and higher spherical and chromatic aberration

17-48 For a compound microscope with an eyepiece of focal length f_E and an objective of focal length f_O, a high magnification can be achieved by having
(a) high f_E and high f_O
(b) low f_E and low f_O
(c) low f_E and high f_O
(d) high f_E and low f_O

17-49 In an astronomical refracting telescope, chromatic aberration occurs because
(a) different colours of light have different indexes of refraction in the glass lenses
(b) different colours of light totally internally reflect at different angles off the inside surface of the objective lens
(c) the weight of the objective lens causes distortion around the outer part of the lens
(d) it is impossible to grind the surfaces of the glass lenses to be perfectly smooth

17-50 A zoom lens of a digital camera has its focal length adjusted from 100 mm to 300 mm. The resulting image compared to the original image has
(a) an increased magnification and angular size
(b) a decreased magnification and angular size
(c) an increased magnification and decreased angular size
(d) a decreased magnification and increased angular size

17-51 The *f*-stop number of a lens on a digital camera is changed from 8 to 2 while the focal length of the lens is kept constant. Compared to the first setting, the second setting has
(a) 4 times the diameter and 4 times the light entering
(b) ¼ the diameter and ⅛ the amount of light entering
(c) 8 times the diameter and 4 times as much light entering
(d) 4 times the diameter and 16 times the light entering
(e) ⅛ the diameter and ¼ the amount of light entering

17-52 To make a hologram requires
(a) light that is coherent and undergoes constructive interference only
(b) light that is incoherent and undergoes both constructive and destructive interference
(c) light that is coherent and undergoes destructive interference only
(d) light that is coherent and undergoes both constructive and destructive interference

Review Questions and Problems

17-53 State the main function of these parts of the human eye:
(a) pupil
(b) iris
(c) lens
(d) retina

17-54 If a lens inside an eye has a focal length of 25 mm, would that focal length remain the same or become larger or smaller if the lens were surrounded totally by air? Why?

17-55 How is the accommodation of the eye controlled?

17-56 If you want to see an object clearly in bright light, why is it best to look straight at the object?

17-57 What feature of the eye allows us to enjoy movies?

17-58 How many horizontal lines are parallel to the horizontal line at the bottom of the drawing in Figure 17-50? Use a ruler or other straight edge to check your answer. What do you conclude?

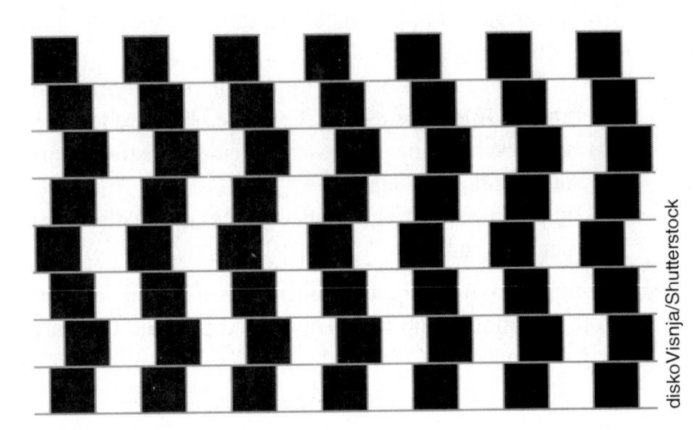

Figure 17-50 Question 17-58.

17-59 What changes occur in the human eye when it goes from looking at a brightly lit nearby book to viewing a dark, distant scene?

17-60 To what colours are the cones of the human eye most sensitive? What occurs in the retina to produce red? yellow? white?

17-61 **Fermi Question:** You are outdoors on a dark night viewing the sky with a telescope of diameter 25 cm. Estimate how many times more light the telescope can gather than your own viewing eye in the dark.

17-62 In each case below, the near point and far point, respectively, of a person's vision are given. State the type of vision (or defect) for each person, and name the type of lens, if any, needed to correct the defect.
(a) 14 cm; 11 m
(b) 25 cm; infinity
(c) 75 cm; infinity
(d) 65 cm; 7.5 m

17-63 A person with hyperopic vision wears correcting lenses of strength +3.25 d (dioptres). What is the person's near point without lenses?

17-64 What is astigmatism? How can it be corrected?

17-65 Compare the action of the human eye with that of a digital camera. Include a description of the parts of the eye that correspond to the aperture, lens, shutter, and detector of the camera.

17-66 State whether the image viewed in each of the following instruments is real or virtual.
(a) digital camera
(b) human eye
(c) simple magnifier
(d) microscope
(e) astronomical telescope
(f) terrestrial telescope
(g) Galilean telescope
(h) reflecting telescope

17-67 The diameter of an atom is in the order of 10^{-10} m. Why is it impossible to see an atom using even the best optical instrument?

17-68 A telescope with an objective of 14 cm diameter and 85.0 cm focal length is used to view the Moon, which is an average distance of 3.84×10^8 m away. For light of wavelength 485 nm,
(a) What is the minimum angle of resolution of the telescope?
(b) What minimum separation of features on the Moon can be resolved?

17-69 Toronto's CN Tower rises 555 m above the ground. Determine the angular size of the structure from a distance of (a) 8.00 km and (b) 123 km (the maximum distance from which the tower can be seen).

17-70 Anton van Leeuwenhoek, the first maker of high-quality lenses, discovered bacteria with a simple magnifier of focal length about 1.24 mm. Assuming he allowed his eye to be relaxed, what was the magnifying power (angular magnification) of the magnifier he used?

17-71 A certain jeweller's loupe produces a magnification of 12.9 when the image viewed is at the observer's near point, assumed to be 22 cm. What is the focal length of the loupe?

17-72 A compound microscope has an eyepiece of focal length 2.0 cm and an objective that produces a magnification of −20 times. Using two significant digits and assuming the image is viewed with a relaxed eye, what is the total magnification of the microscope?

17-73 A compound microscope, used with a relaxed eye, has an eyepiece and an objective, both with focal lengths of 22 mm. What length of the microscope would achieve a total magnification of −100 times? (Assume two significant digits.)

17-74 Describe the similarities and differences between a microscope and an astronomical refracting telescope.

17-75 A 9.0 mm object is located 44 mm from the objective of a two-lens optical instrument. The objective, with a focal length of 20 mm, is located 54 mm from the eyepiece, which has a focal length of 30 mm. Draw and label a ray diagram to locate the final image of the object. State the attitude and type of the image, and determine the magnification.

17-76 The parabolic mirror (the objective) of the reflecting telescope at Mt. Palomar in California has a focal length of 1.68 m. What focal length of the eyepiece will produce a magnification of −124 times?

17-77 What is the function of the third (middle) lens in a terrestrial telescope? Is the image it produces real or virtual? Explain.

17-78 Two of each of the lenses whose focal lengths are given are available for student experimentation: +55 cm; +22 cm; +4.0 cm; −16 cm; −4.0 cm. State which lenses you would use to make each of the instruments listed below. Explain each choice.
(a) a compound microscope
(b) a refracting astronomical telescope
(c) a Galilean telescope
(d) a terrestrial telescope

17-79 (a) What type of device is shown in Figure 17-51? How can you tell?

(b) What are the properties of the electromagnetic waves the device receives?

iurii/Shutterstock

Figure 17-51 Question 17-79.

17-80 State the function of these parts of a lens camera: (a) shutter, (b) zoom lens, and (c) aperture control.

17-81 A telephoto camera lens of focal length 200 mm can be adjusted so its distance from the sensor is between 200 mm and 212 mm. (Assume three significant digits.)

(a) Determine the distance to the nearest object this lens can focus on.

(b) At the distance found in (a), what size of object will completely fill the 24.0 mm space of the sensor?

17-82 By what factor does the light-gathering ability of a camera lens increase or decrease when its aperture changes in the following ways?

(a) from *f*/2.8 to *f*/16, (b) from *f*/11 to *f*/4.0

17-83 A camera lens of focal length 85 mm is set for proper exposure at *f*/11 and 1/15 s.

(a) What is the disadvantage of using an exposure time of 1/15 s rather than 1/60 s?

(b) If the exposure time is adjusted to 1/60 s, what is the correct aperture setting?

17-84 A lens of focal length 50 mm is set at 1/30 s and *f*/22 for the correct exposure of a scene. (Assume two significant digits.)

(a) If the aperture is now adjusted to *f*/5.6, what is the new exposure time?

(b) At *f*/5.6, is the depth of field larger than or smaller than the depth of field at *f*/22?

17-85 What does the acronym *laser* stand for?

17-86 What is meant by a metastable state in an atom?

17-87 A laser amplifies light. Does this mean that it violates conservation of energy? Explain briefly.

Applying Your Knowledge

17-88 A person with a near point of 78 cm and a far point of 180 cm requires bifocal lenses.

(a) Determine the focal length of the upper part of the lens (for distance vision). Assume two significant digits in given data.

(b) Determine the focal length of the lower part of the lens (for close-up vision), assuming a corrected near point of 25 cm.

17-89 The 102 cm diameter objective lens of the Yerkes Observatory telescope can resolve point objects separated by 1.00 cm from a distance of 14.8 km in certain light. What is the wavelength of this light?

17-90 The magnification of a refracting astronomical telescope is -32. If the objective and eyepiece are 82.5 cm apart, determine the focal length of each lens.

17-91 The Hubble Space Telescope can peer into space up to an estimated 14 billion light years (1.4×10^{10} l.y.). For light in the visible range, estimate the minimum separation (in light years and metres) of two stars that can just be resolved at the edge of the viewable universe by the telescope, which has a diameter of 2.4 m. Compare this distance to the distance between the Sun and its nearest neighbour, Proxima Centauri, which is 4.3 l.y.

17-92 Refer to the space telescopes shown in Figure 17-52.

(a) Describe features that are common to two or three of the telescopes.

(b) Describe any feature you see that is unique to each telescope.

(c) Which telescope do you think might be a replacement for the Hubble Space Telescope? Research the "James Webb Space Telescope" to check your answer.

(a) © edobric/Shutterstock

(b) Neo Edmund/Shutterstock

(c) Alex Mit/Shutterstock

Figure 17-52 Question 17-92.

17-93 Assume your responsibility is to design camera lenses for underwater photography. You want to obtain a lens that yields the same results as a 200 mm telephoto lens in air. Describe what you would do.

17-94 A telephoto lens of focal length 400 mm can be adjusted so the maximum image distance is 436 mm. (Assume three significant digits in given data.) Through what range of distances can this lens be focused?

17-95 A laser beam from, say, a He-Ne laser is normally not visible as it travels, but is visible where it strikes an object. How then is it possible to produce photographs that clearly show the paths of laser beams such as that seen in Figure 17-43?

17-96 A small He-Ne laser (wavelength 633 nm) has a power output of about 1 mW. How many photons are emitted per second?

17-97 Figure 17-53 shows the Horsehead Nebula along with several stars. Describe how an astrophysicist would attempt to identify what gases are found in the interstellar spaces.

Figure 17-53 The Horsehead Nebula (Question 17-97).

17-98 Figure 17-54 shows the medical condition called glaucoma, which was described briefly in the topic of pressure, Section 12.2.

(a) What causes the pressure to build up in the aqueous humour?

(b) Why does the final drawing show that the retina and optic nerve experience pressure when they are as far away as possible from the aqueous humour?

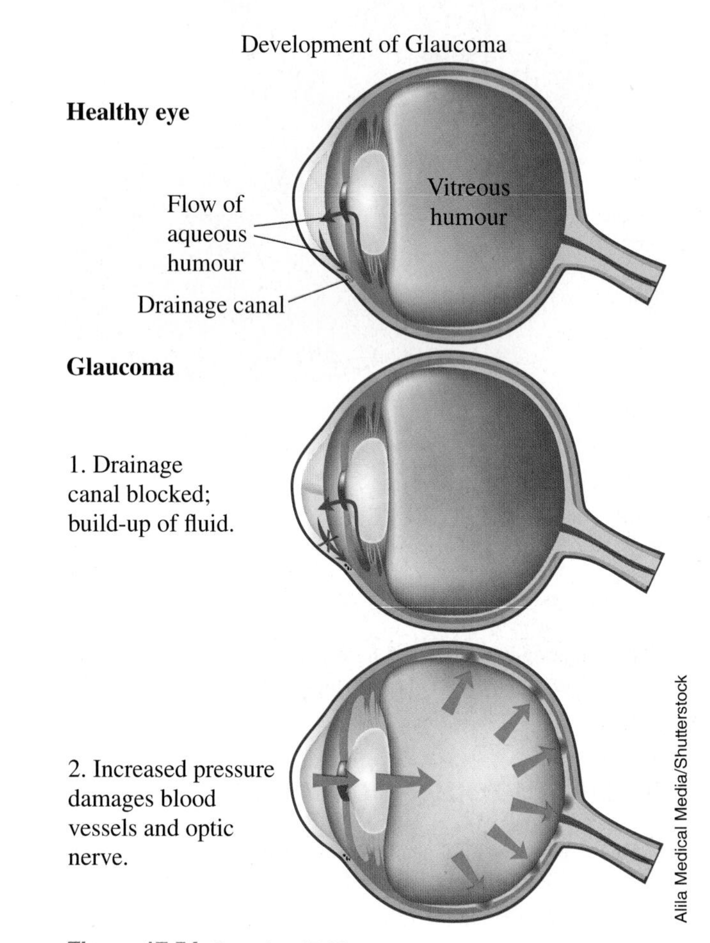

Figure 17-54 Question 17-98.

Heat and Thermodynamics 18

cowardlion/Shutterstock

Figure 18-1 In an automobile engine, heat from the burning of gasoline is used to generate useful kinetic energy. However, because of a property of nature known as the second law of thermodynamics (Section 18.5), there is an *unavoidable* loss of about 60% of the energy produced. Combined with a 20% energy loss due to engine friction, this leaves only about 20% left to run the car.

In our everyday lives we use devices such as automobiles (Figure 18-1), refrigerators, and air conditioners that operate according to physical laws involving temperature and heat. We rely also on electricity that has been generated by the burning of fossil fuels or by the heat generated in nuclear reactors. On cold winter days we wear bulky coats and sweaters to reduce heat loss from our bodies, and buildings have thermal insulation in the walls and ceilings to help keep indoor temperatures comfortable in all seasons. Without even thinking about it, in our daily way of life we are constantly using the concepts of temperature, heat, and other forms of energy such as electrical energy generated by heat.

CHAPTER **OUTLINE**

18.1 Temperature and Internal Energy

From an early age we have been accustomed to recognizing objects as being hot and cold, often by touching them. But what is temperature? Is it the same as heat? Often these two terms are used interchangeably in everyday conversation: we say "the heat is unbearable today" or instead "the temperature is very high." However, temperature and heat are actually different physical quantities. In this section we discuss temperature and the related concept of internal energy (or thermal energy); heat will be introduced in the next section.

Temperature

temperature a measure of the average translational kinetic energy of atoms and molecules

It is useful to think of the **temperature** of an object as being a measure of the average translational kinetic energy of the atoms and molecules in the object. As this kinetic energy increases, the temperature increases, and if kinetic energy decreases, temperature decreases. In many common situations, the temperature is in fact proportional to the kinetic energy, but this proportionality is not mathematically exact in all circumstances; its limitations will be discussed later in this section. When the translational kinetic energy of atoms and molecules increases, objects typically expand as a result, and when kinetic energy decreases, objects usually contract. Monitoring or measuring this expansion or contraction is often used as a way of measuring temperature.

thermometer a device that can measure temperature

A **thermometer** is any device that can measure temperature. You have probably used a thermometer containing a liquid such as mercury or coloured ethanol in a narrow capillary tube surrounded by transparent glass or plastic (Figure 18-2). As the temperature increases, the liquid expands and moves further up the capillary tube. Another style of thermometer uses a bimetallic strip, made by bonding two different metals that expand by different amounts for the same temperature increase (Figure 18-3). As temperature increases, the strip bends. Bimetallic strips are often formed into a thin spiral and used in thermometers that have a dial display (Figure 18-4). Another type of thermometer detects infrared radiation emitted by an object to establish the temperature. This last type will be discussed further in Section 18.4.

Thermometers such as these must be calibrated. One convenient calibration point is the temperature of an ice–water mixture (at atmospheric pressure), that is, the temperature at which water freezes, which is defined to be zero degrees Celsius (0°C). Another handy calibration point is the temperature at which water boils (at atmospheric pressure), defined to be 100°C. In order to calibrate a liquid-based thermometer such as shown in Figure 18-2, the positions of the liquid at these two calibration

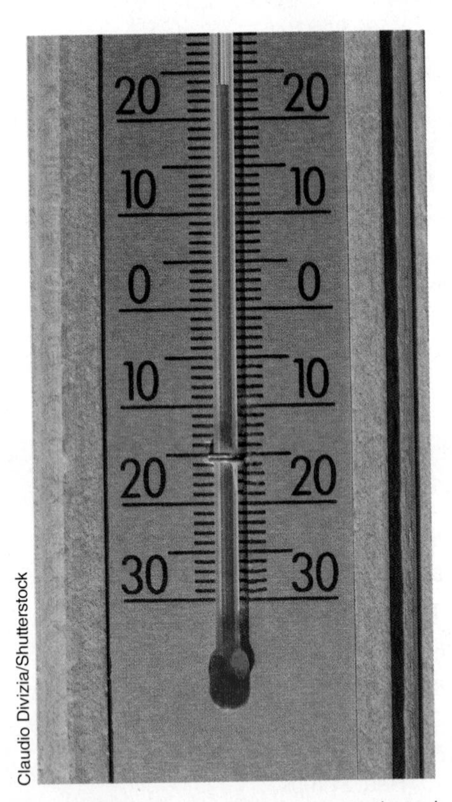

Figure 18-2 As temperature increases, the red liquid in this thermometer expands and moves further up the tube.

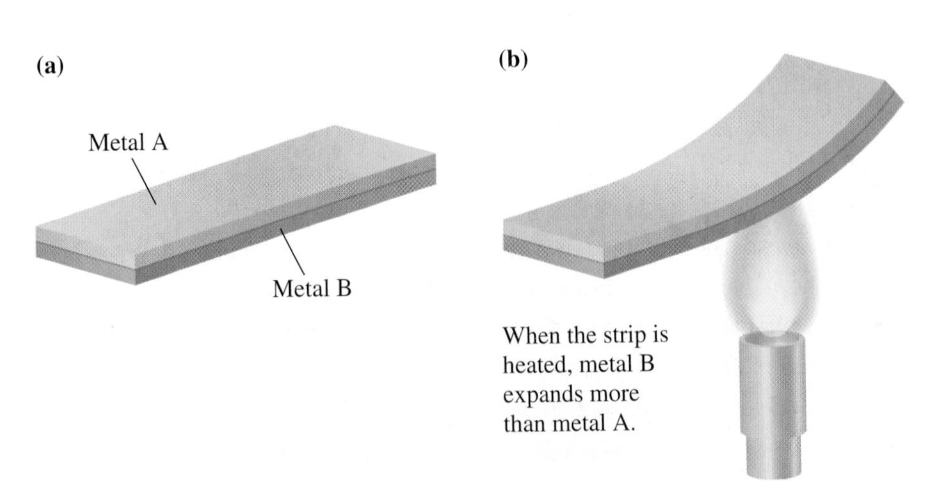

Figure 18-3 **(a)** A bimetallic strip. **(b)** As temperature increases, the strip bends.

temperatures are marked on the thermometer and the distance between these marks is divided into 100 equal units.

Different liquids have different thermal expansion properties, and although a mercury-based thermometer will agree with an ethanol-based thermometer at the calibration temperatures, there will be discrepancies between the readings at intermediate temperatures. A thermometer that bypasses this difficulty is the constant-volume gas thermometer, which is based on the ideal gas law, $PV = nRT$ (Eqn. 12-10, introduced in Section 12.2). An **ideal gas** is one for which this equation is valid at all temperatures and pressures. This is an idealization, but every gas regardless of its chemical composition obeys this law if the gas is at low density and at a temperature well above the point at which it liquefies. What is noteworthy about this law is that the temperature T of n moles of gas contained in a volume V at pressure P can be calculated from the law as $T = \frac{PV}{nR}$ (recall that R is the molar gas constant). A constant-volume gas thermometer consists of a bottle of known volume V containing a known quantity n of gas; a pressure gauge attached to the bottle shows the pressure P inside. The bottle is immersed in a liquid or other environment for which the temperature is to be measured, the pressure is recorded, and the temperature is calculated. Experiments show that temperatures determined by such a thermometer are independent of the type of gas.

The ideal gas law also indicates that for a given sample of gas at constant volume, the pressure P is proportional to the temperature T. Therefore, if P is measured at calibration temperatures T, a graph of P versus T will be linear with a slope of nR/V. Experimental results confirm this linearity. Of course, the numerical value of the slope differs from one gas sample to another, since n and V will generally be different. What is surprising is that for *any* gas sample, if the straight line on the graph of P versus T is extrapolated toward decreasing temperature (Figure 18-5), it intersects the temperature axis (i.e., at $P = 0$) at a temperature of –273.15°C, indicating that this temperature represents a lower bound of temperature for physical processes. Hence this temperature is defined as the **absolute zero** of temperature, and serves as the zero point of the Kelvin temperature scale.

Early gas thermometers used the freezing and boiling points of water for temperature calibration, but there were experimental difficulties because these temperatures depend on pressure. Because of these problems in calibration, in 1954 the International Committee on Weights and Measures adopted a new procedure based on two new points. One of these is absolute zero, and the other is the **triple point of water**, which is the single temperature and pressure at which water, ice, and water vapour can coexist in equilibrium. The triple point temperature and pressure are 0.01°C (or 273.16 K) and 4.58 mm of mercury. This point is easily reproducible experimentally. The SI unit of temperature, the **kelvin**, is *defined* as 1/273.16 of the temperature of the triple point of water.

Dejan Lazarevic/Shutterstock

Figure 18-4 This thermometer uses a bimetallic strip in the form of a thin spiral located behind the dial and attached to the centre of the needle.

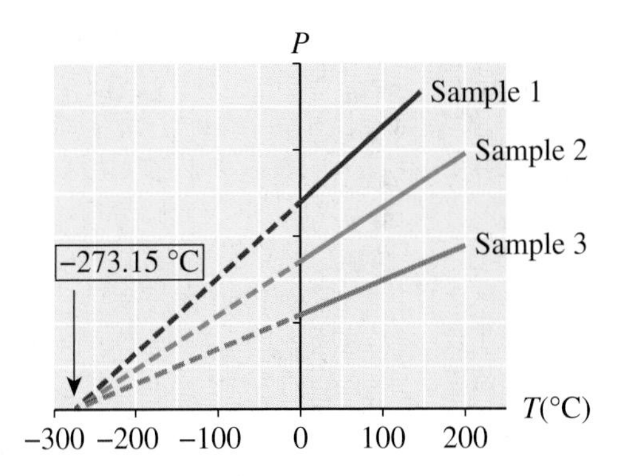

Figure 18-5 Pressure versus temperature graphs for three different gas samples. The straight lines for the samples have been extrapolated (dashed lines) toward decreasing temperatures and all intersect the temperature axis at −273.15°C.

ideal gas a gas for which the ideal gas law, $PV = nRT$, is valid at all temperatures and pressures

absolute zero a temperature of −273.15°C, or 0 K, representing a lower bound of temperature for physical processes

triple point of water the single temperature and pressure at which water, ice, and water vapour can coexist in equilibrium

kelvin 1/273.16 of the temperature of the triple point of water

Temperature and Kinetic Energy

It was stated earlier in this section that temperature and average translational kinetic energy are proportional to each other in many situations. The specific quantitative relation between the average translational kinetic energy of molecules (or atoms) and (absolute) temperature T is given by

$$\left(\frac{1}{2}mv^2\right)_{av} = \frac{3}{2}k_B T \qquad (18\text{-}1)$$

where m = mass of one molecule
k_B = Boltzmann's constant = 1.381×10^{-23} J/K (Section 12.2)
T = absolute temperature

This equation applies to gases, liquids, and solids at a temperature high enough that quantum effects are negligible—room temperature usually suffices. For solids, the atoms vibrate in place, but still have a kinetic energy associated with the vibration. For lower temperatures, quantum effects make the equation inaccurate to some extent. Eqn. 18-1 can be rearranged to give an expression for the molecular mean-square speed, and then taking the square root gives the root-mean-square (rms) speed:

$$\text{mean square speed } (v^2)_{av} = \frac{3k_BT}{m} \tag{18-2}$$

$$\text{rms speed } v_{rms} = \sqrt{\frac{3k_BT}{m}} \tag{18-3}$$

SAMPLE PROBLEM 18-1

Determine the rms speed in air at 25°C of **(a)** hydrogen molecules, and **(b)** oxygen molecules. Refer to the inside back cover for atomic masses of hydrogen and oxygen.

Solution

(a) Use Eqn. 18-3, $v_{rms} = \sqrt{\frac{3k_BT}{m}}$, with $T = (25 + 273)\text{ K} = 298\text{ K}$. Hydrogen is diatomic, with a mass per molecule given by

$$\frac{2.0\text{ g}}{\text{mol}} \times \frac{1\text{ kg}}{10^3\text{ g}} \times \frac{1\text{ mol}}{6.02 \times 10^{23}\text{ molecules}} = 3.32 \times 10^{-27}\text{ kg/molecule}$$

$$\text{Then, } v_{rms} = \sqrt{\frac{3k_BT}{m}} = \sqrt{\frac{3(1.38 \times 10^{-23}\text{ J/K})(298\text{ K})}{3.32 \times 10^{-27}\text{ kg}}} = 1.9 \times 10^3\text{ m/s}$$

Thus, the rms speed of the hydrogen molecules is 1.9×10^3 m/s.

(b) Oxygen molecules are also diatomic, with a mass per molecule of

$$\frac{32\text{ g}}{\text{mol}} \times \frac{1\text{ kg}}{10^3\text{ g}} \times \frac{1\text{ mol}}{6.02 \times 10^{23}\text{ molecules}} = 5.32 \times 10^{-26}\text{ kg/molecule}$$

$$v_{rms} = \sqrt{\frac{3k_BT}{m}} = \sqrt{\frac{3(1.38 \times 10^{-23}\text{ J/K})(298\text{ K})}{5.32 \times 10^{-26}\text{ kg}}} = 4.8 \times 10^2\text{ m/s}$$

The rms speed of the oxygen molecules is 4.8×10^2 m/s.

In Section 9.5 it was shown that the minimum speed required for an object to escape from Earth's gravitational field is 1.1×10^4 m/s. Since hydrogen molecules have an rms speed of 1.9×10^3 m/s, there is a high probability that random molecular collisions will at some time provide any given hydrogen molecule with a speed larger than the escape speed, and the molecule will be able to leave Earth's atmosphere forever. As a result, there is a negligible amount of hydrogen in Earth's atmosphere. The mass of a helium atom is only twice that of a hydrogen molecule, and helium also has a large enough rms speed that there are virtually no helium atoms left in the atmosphere. Fortunately, the rms speed of oxygen is sufficiently low that very few oxygen molecules leave the atmosphere.

DID YOU KNOW?

Helium is produced only as a result of the natural radioactive decay of certain elements such as uranium in Earth's crust, and therefore its supply is limited. Helium has many important uses: as a liquid (at a temperature of −269°C) it cools magnets in magnetic resonance imagery (MRI) machines in hospitals, and as a gas it is used in the manufacture of computer chips and as a protective atmosphere for arc welding. Since any leakage of helium to the atmosphere results in the eventual escape of this gas because of its high rms speed, there is concern about the frivolous use of helium in party balloons, for example.

Internal Energy

Eqn. 18-1 shows the quantitative connection between the translational kinetic energy of molecules and temperature, but molecules have other types of energy as well. For example, oxygen molecules rotate and vibrate, and therefore they have rotational and vibrational kinetic energies. In addition, they have electric potential energy (Section 20.2). The total energy of all the molecules in a system, including all the energy types just mentioned, is called the **internal energy** or the **thermal energy**, symbolized as U. The internal energy of a system is the product of the average energy of each molecule and the number of molecules.

internal energy (or thermal energy) total energy of all the molecules in a system, including translational, rotational, and vibrational kinetic energies, as well as electric potential energy

In a *monatomic* gas (helium, neon, argon, etc.), the atoms do not rotate or vibrate, and so the only type of energy is translational kinetic energy. Therefore, for a monatomic gas having N atoms each of mass m, the total internal energy U, using Eqn. 18-1, is

$$\text{monatomic gas} \quad U = N\left(\tfrac{1}{2}mv^2\right)_{\text{av}} = \tfrac{3}{2}Nk_{\text{B}}T \qquad \textbf{(18-4)}$$

Eqn. 18-4 can be rewritten in terms of the molar gas constant R (8.314 J·mol^{-1}·K^{-1}), since $k_{\text{B}} = \dfrac{R}{N_{\text{A}}}$ (Section 12.2). Making this substitution gives $U = \tfrac{3}{2}N\dfrac{R}{N_{\text{A}}}T = \tfrac{3}{2}\dfrac{N}{N_{\text{A}}}RT$. The quantity N/N_A is just the number of moles of gas, n, and so the total internal energy can be written

$$\text{monatomic gas} \quad U = N\left(\tfrac{1}{2}mv^2\right)_{\text{av}} = \tfrac{3}{2}nRT \qquad \textbf{(18-5)}$$

SAMPLE PROBLEM 18-2

What is the total internal energy of 2.0 mol of neon gas at a temperature of 15°C?

Solution

Since neon is a monatomic gas, use Eqn. 18-5:

$$U = \tfrac{3}{2}nRT = \tfrac{3}{2}(2.0\,\text{mol})(8.314\,\text{J}\cdot\text{mol}^{-1}\cdot\text{K}^{-1})[(273 + 15)\,\text{K}] = 7.2 \times 10^3\,\text{J}$$

Therefore, the total internal energy of 2.0 mol of neon gas at 15°C is 7.2×10^3 J.

EXERCISES

18-1 At what temperature (in kelvins) is the average translational kinetic energy of atoms and molecules equal to 7.00×10^{-21} J?

18-2 What is the rms speed of carbon dioxide molecules at 12°C? (Reference: inside back cover)

18-3 At what temperature (in degrees Celsius) is the rms speed of water molecules equal to 627 m/s? (Reference: inside back cover)

18-4 Determine the total internal energy of 8.7×10^{24} atoms of argon gas at 31°C.

18-5 How many moles of monatomic gas at 21°C would have a total internal energy of 6.2×10^5 J?

heat energy transferred between objects (or systems) as a result of a temperature difference between them

violetkaipa/Shutterstock

Figure 18-6 As heat is transferred from the hot coffee to the surrounding air, the temperature and internal energy of the coffee decrease, while the temperature and internal energy of the air increase.

18.2 Heat, Specific Heat, and Latent Heat

In the previous section, temperature and internal energy were discussed, and we now introduce the related concept of heat and its connection to changes in temperature of objects. **Heat** (symbolized as Q) is energy transferred between objects (or systems) as a result of a temperature difference between them. For example, as a hot cup of coffee cools, heat is transferred from the coffee to the surrounding air (Figure 18-6). As a result, the temperature and internal energy of the coffee decrease, and the temperature and internal energy of the air increase.

The amount of heat Q needed to increase the temperature of a mass m from T_1 to T_2 is proportional to m, and is also proportional to the temperature change $\Delta T = T_2 - T_1$. The amount of heat required also depends on the type of material. Raising the temperature of 1 kg of water by 1 K requires 4186 J of heat, but only 128 J is required to raise the temperature of 1 kg of lead by 1 K. Summarizing these relationships in the form of an equation,

$$Q = mC\Delta T \quad (18\text{-}6)$$

The quantity C depends on the type of material undergoing the temperature change, and is known as the **specific heat capacity** or, more commonly, the **specific heat**. For a given material, it is the amount of heat per unit mass and per unit temperature change needed to increase or decrease the temperature of the material. Its SI unit is joule·kilogram^{-1}·kelvin^{-1} (or J·kg^{-1}·K^{-1}). Note that a temperature *change* has the same numerical value in kelvins and in degrees Celsius. For example, if an object has an initial temperature of 20°C (293 K) and a final temperature of 26°C (299 K), then its change in temperature is 6°C (6 K). As a result, the unit of specific heat can be written as either J·kg^{-1}·K^{-1} or J·kg^{-1}·(°C)$^{-1}$.

specific heat (or specific heat capacity) amount of heat per unit mass and per unit temperature change needed to increase or decrease the temperature of the material

Terminology Tip

The term "specific heat" is rather an unfortunate one that is potentially confusing. It is not a "heat" (which would have units of joules), but rather heat per unit mass and per unit temperature (with units of joule·kilogram^{-1}·kelvin^{-1}).

Values of specific heats for a number of common substances are given in Table 18-1. The values tabulated were determined at room temperature (20°C) and one atmosphere of pressure, unless otherwise noted. (The specific heat of a material depends somewhat on both temperature and pressure.)

In Table 18-1 notice the large value of specific heat for water; this makes water an excellent coolant, since a relatively small mass of water can absorb (or release) a large quantity of heat without undergoing a large increase (decrease) in temperature. This explains why lakes do not become extremely hot in the summer, and why it takes so long for them to get cold in the autumn. Since cells and tissues of living systems consist largely of water, not surprisingly they have specific heats close to that of water. For example, the specific heat of an entire mouse is approximately 3450 J·kg^{-1}·K^{-1}.

Table 18-1

Specific Heats of Common Substances

Substance	Specific heat C (J·kg^{-1}·K^{-1} or J·kg^{-1}·(°C)$^{-1}$)
Lead	128
Mercury	139
Copper	387
Aluminum	895
Ice (−10°C)	2.22×10^3
Ethyl alcohol	2.43×10^3
Water	4186

SAMPLE PROBLEM 18-3

People lacking fireproof cooking pots often heated liquids by placing hot stones in a container holding a liquid. Indigenous peoples of North America concentrated maple sap to syrup using this method. If a hot rock with specific heat of 843 J·kg^{-1}·(°C)$^{-1}$, mass of 0.48 kg, and initial temperature of 223°C is placed in water having mass of 1.5 kg and initial temperature of 15°C, what will be the final temperature of the water and rock?

Solution

heat lost by rock = heat gained by water
Let the final temperature be T_F. Using Eqn. 18-6 ($Q = mC\Delta T$) with subscripts "R" for rock and "W" for water, and writing each ΔT so that it is a positive quantity,

$$m_R C_R(T_R - T_F) = m_W C_W(T_F - T_W)$$

Substituting known quantities, and using the specific heat of water from Table 18-1,

$$(0.48\,\text{kg})(843\,\text{J}\cdot\text{kg}^{-1}\cdot(°\text{C})^{-1})(223°\text{C} - T_F) = (1.5\,\text{kg})(4186\,\text{J}\cdot\text{kg}^{-1}\cdot(°\text{C})^{-1})(T_F - 15°\text{C})$$

$$9.02 \times 10^4\,\text{J} - (405\,\text{J}\cdot(°\text{C})^{-1})T_F = (6.28 \times 10^3\,\text{J}\cdot(°\text{C})^{-1})T_F - 9.42 \times 10^4\,\text{J}$$

$$(9.02 + 9.42) \times 10^4\,\text{J} = [(405 + 6.28 \times 10^3)\,\text{J}\cdot(°\text{C})^{-1}]T_F$$

$$T_F = 28°\text{C}.$$

Thus, the final temperature of the water and rock is 28°C.

Latent Heats

So far we have been discussing adding heat to a substance (or removing heat from it) with the assumption that the substance does not change its state; that is, it does not melt, solidify, vaporize, or condense. If the temperature of a solid is increased high enough, the solid will begin to melt, and the solid–liquid mixture will remain at this melting temperature (i.e., $\Delta T = 0$) until the solid has completely melted. If heat continues to be added, then the temperature of the liquid will increase. Similarly, if a liquid is cooled until it begins to solidify (or freeze), the liquid–solid mixture will stay at that freezing temperature (which is the same as the solid's melting temperature) until all the liquid has solidified. If heat continues to be withdrawn, the temperature of the solid will decrease.

Since there is no temperature change during the melting or solidification processes, Eqn. 18-6 cannot be used to determine the heat added or removed. Experiments show that for a solid of mass m to liquefy, the heat Q that must be added is given by

$$Q = mL_F \qquad (18\text{-}7)$$

where L_F is the **latent heat of fusion** of the material, that is, the heat per unit mass that is added as the material melts. The SI unit of L_F is joule per kilogram (J/kg, or J·kg^{-1}). Eqn. 18-7 also applies when heat Q is removed and the object solidifies.

latent heat of fusion heat per unit mass added (or withdrawn) as a particular material melts (or solidifies)

There is a similar result when a liquid is vaporized (boiled) or a gas condenses. There is no temperature change during these processes, and the heat added or taken away is

$$Q = mL_V \qquad (18\text{-}8)$$

where L_V is the **latent heat of vaporization** of the material, that is, the heat per unit mass that is added (or withdrawn) as the material vaporizes (or condenses). The SI unit of L_V is the same as that of L_F, joule per kilogram (J/kg, or J·kg^{-1}).

latent heat of vaporization heat per unit mass added (or withdrawn) as a particular material vaporizes (or condenses)

Table 18-2 lists values of melting and boiling temperatures, as well as latent heats of fusion and vaporization for several substances at a pressure of one standard atmosphere. Note that the latent heats of fusion and vaporization for water are quite large: it takes a lot of heat to melt ice and to vaporize (or boil) water. One result is that in spring it takes a long time to thaw the ice on lake surfaces. We sweat to cool ourselves during and after exercise (Figure 18-7) because evaporating the water from the skin removes a lot of heat from our bodies. Even when the skin is not wet, a typical person is vaporizing water from the skin and lungs, accounting for 16% and 8%, respectively,

Table 18-2

Latent Heats of Fusion and Vaporization

Substance	Melting Temperature (°C)	Latent Heat of Fusion (J/kg)	Boiling Temperature (°C)	Latent Heat of Vaporization (J/kg)
Mercury	−39	1.1×10^4	357	2.96×10^5
Lead	327	2.3×10^4	1744	8.58×10^5
Nitrogen	−210	2.6×10^4	−196	2.00×10^5
Hydrogen	−259	5.8×10^4	−253	4.55×10^5
Water	0.00	3.34×10^5	100	2.26×10^6
Aluminum	660	4.0×10^5	2467	1.05×10^7
Sulfur	119	3.81×10^4	445	3.26×10^5

Maridav/Shutterstock

Figure 18-7 Evaporating sweat from skin removes a great deal of heat from a person's body.

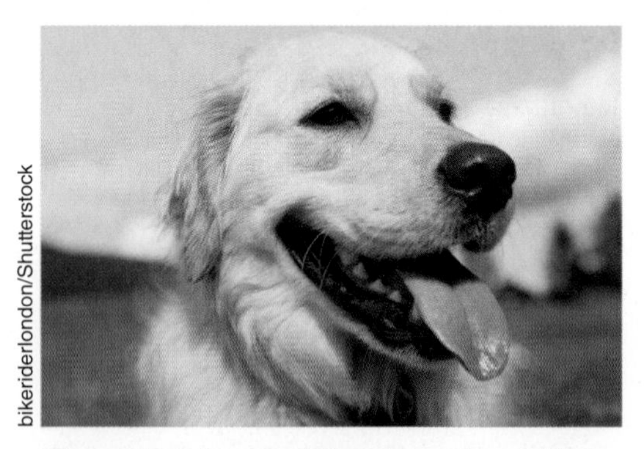

bikeriderlondon/Shutterstock

Figure 18-8 Dogs cool down by panting rapidly to evaporate water from their tongues.

of the body's normal heat loss. When a dog becomes hot (Figure 18-8), it pants to move air rapidly over the surface of its tongue to vaporize water and reduce its body temperature.

SAMPLE PROBLEM 18-4

A bartender is making a simple drink that is prepared by stirring an alcoholic beverage and an ice cube. The beverage has a mass of 34 g, a specific heat of 2.8×10^3 $J \cdot kg^{-1} \cdot (°C)^{-1}$, and an initial temperature of 20°C (two significant digits). The ice cube has an initial temperature of 0°C and a mass of 7.4 g.

(a) What is the temperature of the beverage after the stirring?

(b) If the drink is made with double the mass of beverage but still only one ice cube, what is the temperature after stirring?

Solution

(a) There is a heat exchange between the warm beverage and the cold ice. Heat is transferred from the beverage to the ice and, as a result, the beverage cools and the ice melts. However, there are three possible outcomes, depending on the relative amounts of beverage and ice:

scenario 1 (small amount of beverage, lots of ice): the beverage cools down to 0°C, *some* of the ice melts to produce water, and the final temperature of the mixture of beverage, water, and ice is 0°C

scenario 2 (large amount of beverage, small amount of ice): the beverage cools down somewhat, melting all the ice in the process and also warming the resulting water to a temperature greater than 0°C; the final temperature of the beverage and water is between 0°C and 20°C

scenario 3 (the beverage is the right amount to *just* melt all the ice): the beverage cools to 0°C, all the ice melts to produce water at 0°C, and the final temperature of the beverage and water is 0°C

To find out which scenario is applicable, we first calculate how much heat is required to melt all the ice, and also calculate how much heat would be given off by the beverage if it were to cool down all the way from 20°C to 0°C. We then compare these two values.

Using Eqn. 18-7 and the value of the latent heat of fusion of ice from Table 18-2 to determine how much heat is needed to melt all the ice (mass m_{ice}),

$$Q = m_{ice}L_F = \left(7.4\,g \times \frac{1\,kg}{10^3\,g}\right)(3.34 \times 10^5\,J/kg) = 2.5 \times 10^3\,J$$

Use Eqn. 18-6 with subscripts "B" for beverage to calculate the heat released from the beverage in cooling from 20°C to 0°C:

$$Q = m_B C_B \Delta T_B = \left(34\,g \times \frac{1\,kg}{10^3\,g}\right)(2.8 \times 10^3\,J\cdot kg^{-1}\cdot(°C)^{-1})(20°C) = 1.9 \times 10^3\,J$$

Since the heat released by the beverage is not enough to melt all the ice, this corresponds to "scenario 1," in which the beverage cools all the way down to 0°C, and some of the ice melts, producing water at 0°C.

Thus, the final temperature is 0°C.

(b) If the mass of the beverage is twice that in part (a), then the heat given off by the beverage if it were to cool to 0°C would be double the value calculated in (a), that is, $2 \times (1.9 \times 10^3\ J) = 3.8 \times 10^3$ J. This amount is more than the heat required to melt all the ice (2.5×10^3 J), and so we have "scenario 2" in which the final temperature will be between 0°C and 20°C. In terms of the heat transferred,

heat released from beverage in cooling from 20°C to final temperature T_F

= (heat to melt ice) + (heat to raise temperature of water from 0°C to T_F)

Using subscripts "B" for beverage and "W" for water,

$$m_B C_B \Delta T_B = m_{ice} L_F + m_W C_W \Delta T_W$$

The mass m_W of water is the same as the mass m_{ice} of the original ice, and so we will use the symbol $m_{ice,W}$ for both. Writing each ΔT so that it is a positive quantity,

$$m_B C_B(20°C - T_F) = m_{ice,W} L_F + m_{ice,W} C_W (T_F - 0°C)$$

$$(0.068\,kg)(2.8 \times 10^3\,J\cdot kg^{-1}\cdot(°C)^{-1})(20°C - T_F)$$

$$= (0.0074\,kg)(3.34 \times 10^5\,J/kg) + (0.0074\,kg)(4186\,J\cdot kg^{-1}\cdot(°C)^{-1})(T_F - 0°C)$$

$$3.81 \times 10^3\,J - (190\,J\cdot(°C)^{-1})T_F = 2.47 \times 10^3\,J + (31\,J\cdot(°C)^{-1})T_F$$

$$(3.81 - 2.47) \times 10^3\,J = [(190 + 31)\,J\cdot(°C)^{-1}]T_F$$

$$T_F = 6.1°C$$

Hence, the final temperature is 6.1°C if the mass of beverage is doubled.

DID YOU KNOW?

The Fahrenheit temperature scale, which is used by a few countries including the U.S.A., was put forward in 1724 by the German physicist Daniel Fahrenheit (1686–1736). He set the zero of the scale as the temperature at which a concentrated salt solution freezes, which is approximately −18°C. In many parts of Canada, salt is used on roads (Figure 18-9) and sidewalks in the winter to melt ice and snow, but if the temperature goes below −18°C, the resulting salt water freezes. A recent alternative to using salt alone is to utilize a mixture of beet juice and salt. Although this de-icer is more expensive than salt, it does not freeze until the temperature reaches −32°C (or −26°F).

Peter Gudella/Shutterstock

Figure 18-9 A truck distributing salt to a road. The freezing temperature of salt water is −18°C.

EXERCISES (REFERENCE: TABLES 18-1 AND 18-2)

18-6 Discuss how thermal energy (or internal energy) differs from heat.

18-7 How much heat must be applied to a piece of copper of mass 182 g to increase its temperature from 18°C to 32°C?

18-8 A person has a cup of very hot tea and adds a small amount of cold water to cool it. If the tea has a mass of 219 g and an initial temperature of 87°C, and the water has a temperature of 12°C and a mass of 27 g, what is the final temperature of the tea and water? Assume that the specific heat of tea is the same as that of water.

18-9 How much total heat is required to raise the temperature of 1.5 kg of water from 19°C to 100°C and then to boil half of it away?

18-10 Ice having a mass of 0.750 kg and a temperature of 0°C is added to ethyl alcohol at a temperature of 38°C. All the ice melts and the resulting water and alcohol have a final temperature of 11°C. What is the mass of the alcohol?

18.3 Thermal Expansion

When an object is heated, its internal atomic vibrations increase, and as a result the average distance between the atoms becomes larger and there is an increase in all dimensions of the object. Different objects expand by different amounts for the same temperature change; for example, an aluminum bar will expand twice as much as a steel bar of the same size. Experiments show that the change in length, ΔL, of an object is proportional to both the change in temperature, ΔT, and the original length L of the object. Expressing this in terms of an equation,

$$\Delta L = \alpha L \Delta T \tag{18-9}$$

coefficient of linear expansion a measure of how much the length of a substance expands per unit change in temperature

The constant of proportionality α is the **coefficient of linear expansion** of the substance being heated. This coefficient (unit K^{-1} or $(°C)^{-1}$) is a measure of how much the length of the substance expands per unit change in temperature. Eqn. 18-9 can also be used for contraction, as objects contract when cooled.[1] The coefficient α depends somewhat on the temperature T and on the temperature change ΔT, but for the purposes of this book we will treat it as essentially constant for a given substance.

Real objects are three dimensional and will expand in all dimensions, giving an increase in surface area and volume. Consider the area of a surface of an object that has been heated (Figure 18-10). The initial area was XY, but now the length X has increased to $X + \Delta X$, and similarly Y has increased to $Y + \Delta Y$. Therefore, the new area is

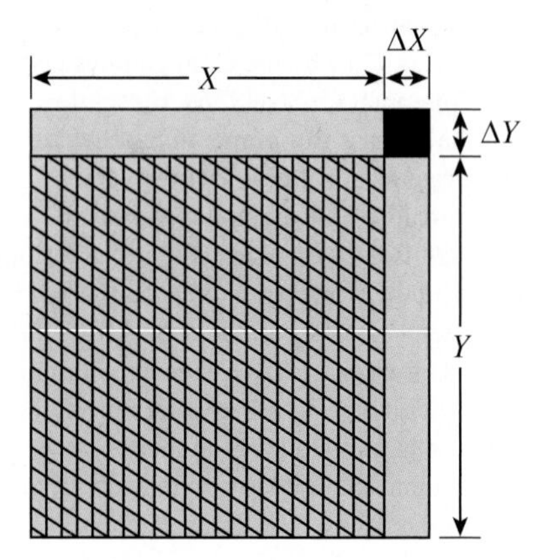

Figure 18-10 The increase in area of a heated surface.

$$(X + \Delta X)(Y + \Delta Y) = XY + X\Delta Y + Y\Delta X + \Delta X\Delta Y$$

Normally ΔX and ΔY are quite small, and hence the product $\Delta X\Delta Y$ is extremely small and can be neglected. Thus, we have

$$A_{new} = XY + X\Delta Y + Y\Delta X$$

Using Eqn. 18-9, $\Delta X = \alpha X \Delta T$ and $\Delta Y = \alpha Y \Delta T$, and hence

$$A_{new} = XY + X\alpha Y\Delta T + Y\alpha X\Delta T = XY + 2\alpha XY\Delta T$$

Thus, the *increase* in area ΔA is $2\alpha XY\Delta T$. Writing XY as the initial area A gives

$$\Delta A = 2\alpha A \Delta T \tag{18-10}$$

Eqn. 18-10 has the same form as Eqn. 18-9 for change in length, except that α has been replaced by 2α.

In a similar way, it can be shown that the change in volume, ΔV, is given by

$$\Delta V = 3\alpha V \Delta T = \beta V \Delta T \tag{18-11}$$

coefficient of volume expansion a measure of how much the volume of a substance expands per unit change in temperature

The quantity $\beta = 3\alpha$ is the **coefficient of volume expansion** (unit K^{-1} or $(°C)^{-1}$), which is a measure of how much the volume of a substance expands per unit change in temperature. As was the case for the coefficient of linear expansion α, the value of β depends to some extent on T and ΔT, but for any given substance we will treat β

[1]The anomalous expansion of water as it cools from 4°C to 0°C was discussed in Section 12.1.

Table 18-3

Coefficients of Linear Expansion α and Volume Expansion β Near Room Temperature

Substance	α (K^{-1} or (°C)$^{-1}$)	Substance	β (K^{-1} or (°C)$^{-1}$)
Aluminum	2.4×10^{-5}	Ethyl alcohol	1.1×10^{-4}
Brass	1.9×10^{-5}	Gasoline	9.5×10^{-4}
Concrete	1.2×10^{-5}	Glycerine	4.9×10^{-4}
Copper	1.7×10^{-5}	Mercury	1.8×10^{-4}
Glass	0.9×10^{-5}	Water	2.0×10^{-4}
Lead	2.9×10^{-5}	Air[2]	3.67×10^{-3}
Steel	1.1×10^{-5}	Helium[2]	3.665×10^{-3}

as effectively constant. Table 18-3 lists coefficients of linear expansion α and volume expansion β for a variety of materials. Notice that values of α are given only for solids, since liquids and gases cannot maintain a fixed shape (such as shown in Figure 18-10) on their own. However, a coefficient of volume expansion can still be defined for liquids and gases.

TACKLING MISCONCEPTIONS

Thermal Expansion of a Hole

Suppose that an object has a hole in it. If the object is heated, does the hole become larger or smaller? Many people think that the hole would be squeezed and become smaller, but that is not the case: the hole becomes larger. As the temperature increases, the spacing between the atoms lining the edge of the hole increases and therefore the hole's circumference increases (Figure 18-11). The hole increases to a size that would accommodate the amount of expanded solid material that would fill the space if the hole were not there.

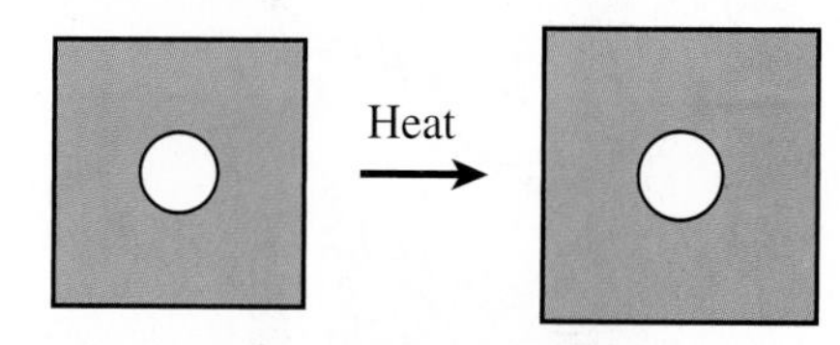

Figure 18-11 As an object with a hole is heated, there is an increase in the interatomic spacing of all the atoms, including those that line the edge of the hole. Hence, the hole's circumference increases, and the hole becomes larger.

SAMPLE PROBLEM 18-5

A circular aluminum washer with a hole 9.500 mm in diameter at 293 K (20°C) is to be heated until it will fit around a circular bolt 9.550 mm in diameter. To what temperature must the washer be heated?

Solution

Use Eqn. 18-9, with α for aluminum from Table 18-3:

$$\Delta L = \alpha L \Delta T, \text{which gives } \Delta T = \frac{\Delta L}{\alpha L} = \frac{(9.550 - 9.500)\,\text{mm}}{(2.4 \times 10^{-5}(°\text{C})^{-1})(9.500\,\text{mm})} = 219°\text{C}$$

The final temperature is (20 + 219)°C = 239°C.

Therefore, correct to two significant digits, the required temperature is 2.4×10^{2} °C.

SAMPLE PROBLEM 18-6

Because of increasing concentrations of carbon dioxide and other atmospheric greenhouse gases (Section 18.4), the temperature on Earth has been increasing for several decades and is expected to continue increasing. Determine the average increase in the depth of the oceans due to thermal expansion for an increase of 1°C in temperature. The average ocean depth is approximately 4.0×10^{3} m.

[2]The coefficient of volume expansion for gases depends on the type of process taken as heat is applied. The values given for air and helium are for an expansion at constant pressure at a temperature near 0°C.

TRY THIS!

Easy Jar Opening

The next time you want to open a sealed glass jar that has a metal lid (Figure 18-12), run hot water over the lid for about 15–20 s, and you will find that the lid is easier to remove. As seen in Table 18-3, the coefficient of linear expansion for any metal is larger than that of glass, and so the lid expands more than the glass as temperature increases, and fits more loosely. As well, in such a short time, only the lid is heated appreciably since the water cannot make its way to the glass that is directly under the lid.

Figure 18-12 Opening a glass jar that has a metal lid is easier if hot water is run over the lid first.

Solution

Table 18-3 gives the coefficient of volume expansion of water, $\beta = 2.0 \times 10^{-4}\,(°C)^{-1}$. From Eqn. 18-11, the coefficient of linear expansion is

$$\alpha = \frac{\beta}{3} = \frac{2.0 \times 10^{-4}(°C)^{-1}}{3} = 6.7 \times 10^{-5}(°C)^{-1}$$

Using Eqn. 18-9,

$$\Delta L = \alpha L \Delta T = (6.7 \times 10^{-5}(°C)^{-1})(4.0 \times 10^{3}\,\text{m})(1°C) \approx 0.3\ \text{m}$$

Therefore, the average depth of the oceans would increase by about 0.3 m if the water temperature increases by 1°C.

DID YOU KNOW?

The global average air temperature has increased by about 0.8°C since 1850, with 0.5°C of that increase occurring since 1980. Although an increase in ocean depth of 0.3 m (Sample Problem 18-6) might not seem large, it would have serious consequences for coastal areas such as the eastern seaboard of the U.S.A., alluvial-plain countries such as Bangladesh, and island nations such as the Maldives. Higher temperatures will also melt land-based ice sheets in Greenland and Antarctica, causing a rise in sea level greater than that due to thermal expansion. (Melting of ice already floating in the oceans will have no effect on sea levels.) Climate scientists estimate that Greenland and Antarctica combined have lost an average of about 200 km^3 each year since 1992.

EXERCISES (REFERENCE: TABLE 18-3)

18-11 A copper rod of length 2.0 m is heated from 15°C to 85°C. By how much does its length increase?

18-12 An aluminum duct in a factory has a small hole in it. When steam of temperature 182°C is flowing through the duct, the hole's radius is 9.00 mm. When the steam flow is turned off, the temperature falls to 2°C. What is the hole's radius at this temperature?

18-13 A steel baking sheet of dimensions 31 cm × 45 cm and temperature 21°C is placed in an oven and heated to a temperature of 232°C. By how much does the area of the sheet increase?

18-14 What temperature change is required to increase the volume of gasoline by 1.0%?

18-15 **Fermi Question:** An aluminum spoon at room temperature is accidentally dropped into a pot of boiling water. Estimate the change in the spoon's (a) length, and (b) volume. Use Table 18-3.

18.4 Energy Transfer

In Section 18.2 heat was defined as energy transferred between objects as a result of a temperature difference between them. We now explore in detail the three processes by which heat can be transferred into or out of an object: conduction, convection, and radiation.

Conduction

conduction heat transfer as a result of atomic and molecular collisions between two objects at different temperatures in physical contact, or between different-temperature regions of the same object

Conduction is heat transfer as a result of atomic and molecular collisions between two objects at different temperatures in physical contact, or between different-temperature regions of the same object. These objects could be solids, liquids, or gases.

Consider a solid object of thickness Δx and cross-sectional area A (Figure 18-13). One surface has a temperature T_1 that is higher than the temperature T_2 at the opposite surface. The molecules at the hot surface are vibrating more rapidly than those at the cold surface, and they collide with adjacent molecules, increasing their vibrational energy. These molecules in turn collide with other molecules closer to the cold surface, and the increase in vibrational energy gradually moves through the solid to the colder side. In other words, heat energy Q is being conducted from the hot surface to the cold surface.

The *rate* of heat transfer is the energy Q transferred divided by the time t. Recall that energy divided by time is power P, and so we write $P = Q/t$. This power is proportional to the temperature difference $\Delta T = T_1 - T_2$ and the area A, and inversely proportional to the thickness Δx. The power is proportional also to the substance's **thermal conductivity** (symbol k), which is a measure of how well the substance conducts heat. Combining all these parameters, the rate of heat transfer by conduction is given by

$$P = \frac{Q}{t} = \frac{kA\Delta T}{\Delta x} \qquad (18\text{-}12)$$

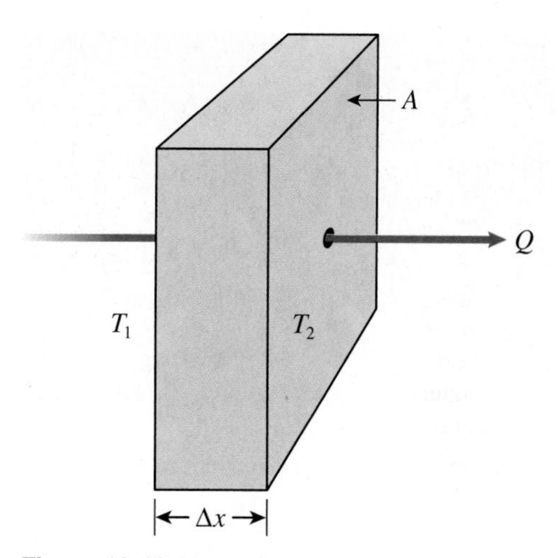

Figure 18-13 Heat Q is conducted from a region of high temperature T_1 to one of lower temperature T_2, through an object of thickness Δx and cross-sectional area A.

thermal conductivity a measure of how well a substance conducts heat

Numerical values of thermal conductivity are provided for several substances in Table 18-4. Notice that metals have high thermal conductivity. Some of the electrons in a metal are free to move quickly over long distances in the metal. It is these electrons, called conduction electrons, that account for the ability of metals to conduct heat and electrical current easily. In contrast, air has a very low thermal conductivity. Many items that are useful for thermal insulation, such as wool sweaters and fibreglass home insulation, have low thermal conductivity because of tiny air pockets in them. Notice that the thermal conductivity of argon gas is even lower than that of air. Some double-paned windows have argon instead of air between the panes of glass to decrease the thermal conductivity, and the small thermal conductivity of polyurethane foam is due to argon filling the small cells in the foam.

Light fluffy snow has a thermal conductivity of only 0.08 $W{\cdot}m^{-1}{\cdot}K^{-1}$ because of air trapped between the snowflakes. In one experiment, the temperature at ground level in a forest with a snow cover of 75 cm showed no variation during a period of nine days even though the air temperature above the snow fluctuated between $-4°C$ and $-33°C$

TRY THIS!

Feeling Thermal Conductivity

Try holding several room-temperature objects tightly, one at a time, and judge whether they feel warm or cold. Use objects made of different materials: metal, wood, glass, cloth, etc. Our perception of whether or not an object is cold or warm is often determined by the thermal conductivity of the object, rather than its actual temperature. If something has high thermal conductivity, it conducts heat quickly away from your hand and feels cold even though it is at room temperature. An object with low thermal conductivity feels warm.

Table 18-4

Thermal Conductivities

Substance	k ($W{\cdot}m^{-1}{\cdot}K^{-1}$) (or $W{\cdot}m^{-1}{\cdot}(°C)^{-1}$)	Substance	k ($W{\cdot}m^{-1}{\cdot}K^{-1}$) (or $W{\cdot}m^{-1}{\cdot}(°C)^{-1}$)
Aluminum	205	Concrete	1
Copper	385	Cork	0.04
Mercury	8.3	Fibreglass	0.04
Silver	427	Glass	0.80
Steel	50.2	Ice	1.6
Muscle	0.4	Wood	0.12–0.04
Skin	0.3	Air	0.024
Polyurethane foam	0.02	Argon	0.016

psynovec/Shutterstock

Figure 18-14 Light fluffy snow is an excellent thermal insulator because of entrapped air.

(Figure 18-14). Deep snow provides excellent protection for vegetation, small mammals, and hibernating or pupating insects.

SAMPLE PROBLEM 18-7

A plywood sheet has width 1.22 m, length 2.44 m, and thickness 5.0 mm. How much heat is transferred in one day through the sheet if the temperature is −9°C on one side and 21°C on the other? The thermal conductivity of the plywood is 0.12 W·m^{-1}·(°C)$^{-1}$.

Solution

Use Eqn. 18-12: $\frac{Q}{t} = \frac{kA\Delta T}{\Delta x}$, which gives $Q = \frac{kA\Delta T}{\Delta x}t$

$$Q = \frac{(0.12\,\text{W}\cdot\text{m}^{-1}\cdot(°\text{C})^{-1})(1.22\,\text{m} \times 2.44\,\text{m})[(21-(-9))°\text{C}]}{5.0 \times 10^{-3}\,\text{m}}[(24 \times 60 \times 60)\,\text{s}]$$

$$= 1.9 \times 10^{8}\,\text{J}$$

Thus, the heat transferred through the plywood in one day is 1.9×10^8 J.

DID YOU KNOW?

The highest known thermal conductivity at room temperature belongs to diamond, with $k \approx 2000$ W·m^{-1}·K^{-1}. The next highest is silver at 427 W·m^{-1}·K^{-1}. You might have heard that in the diamond business, diamonds are referred to as "ice." This is not because diamonds look like ice, but rather because they feel like ice since they conduct heat so rapidly. If someone offers to sell you a bag of diamonds, put your hand into the bag to find out if the contents feel "ice-cold" before you buy.

SAMPLE PROBLEM 18-8

A beverage cooler is made of plastic foam (thermal conductivity = 0.040 W·m^{-1}·(°C)$^{-1}$) of thickness 2.0 cm. The ends of the cooler have dimensions 30 cm × 30 cm, and the top, bottom, and sides are 50 cm × 30 cm. Ice at an initial temperature of 0°C is placed inside the cooler. Assume two significant digits in all given data.

(a) What is the rate of heat flow into the cooler if the outside temperature is 35°C?

(b) How much ice melts per hour?

Solution

(a) Eqn. 18-12 will be used: $P = \frac{Q}{t} = \frac{kA\Delta T}{\Delta x}$

The total area through which heat flows is

$$A = 2 \times (0.30\,\text{m})(0.30\,\text{m}) + 4 \times (0.50\,\text{m})(0.30\,\text{m}) = 0.78\,\text{m}^2$$

$$\therefore P = \frac{Q}{t} = \frac{kA\Delta T}{\Delta x} = \frac{(0.040\,\text{W}\cdot\text{m}^{-1}\cdot(°\text{C})^{-1})(0.78\,\text{m}^2)((35-0)°\text{C})}{0.020\,\text{m}} = 54.6\,\text{W}$$

Therefore, correct to two significant digits, the rate of heat flow is 55 W.

(b) From part (a), $\frac{Q}{t} = 54.6\,\text{W} = 54.6\,\frac{\text{J}}{\text{s}}$

In one hour (3600 s), the heat transferred is $Q = 54.6\,\frac{\text{J}}{\text{s}} \times 3600\,\text{s} = 1.97 \times 10^5\,\text{J}$.

Using Eqn. 18-7 ($Q = mL_F$), with $L_F = 3.34 \times 10^5$ J/kg for ice (Table 18-2),

$$m = \frac{Q}{L_F} = \frac{1.97 \times 10^5\,\text{J}}{3.34 \times 10^5\,\text{J/kg}} = 0.59\,\text{kg}$$

Hence, 0.59 kg of ice will melt per hour.

Thermal Resistance

A useful concept used for thermal insulation installed in walls and ceilings of buildings is the **thermal resistance,** or ***R*-value**, which is defined as the thickness of an insulating material divided by its thermal conductivity:

thermal resistance (*R*-value) thickness of an insulating material divided by its thermal conductivity

$$\text{thermal resistance} \quad R = \frac{\Delta x}{k} \qquad (18\text{-}13)$$

In terms of *R*-value, Eqn. 18-11 becomes

$$P = \frac{Q}{t} = \frac{A\Delta T}{R} \qquad (18\text{-}14)$$

Since R is proportional to Δx, doubling the thickness of an insulator doubles its *R*-value, and cuts the heat loss rate (P) in half. The *R*-value is inversely proportional to k, and therefore an insulator with a low thermal conductivity k will have a high *R*-value and a small heat loss rate.

The SI unit for R can be determined from Eqn. 18-13 ($R = \Delta x/k$):

$$\text{SI unit of } R = \frac{\text{m}}{\text{W}\cdot\text{m}^{-1}\cdot\text{K}^{-1}} = \text{W}^{-1}\cdot\text{m}^2\cdot\text{K (or W}^{-1}\cdot\text{m}^2\cdot{}^\circ\text{C)}$$

The value of R in SI units is often referred to as the RSI value. However, the numbers usually quoted commercially for the *R*-value of insulation (e.g., R-24) are unfortunately based on an awkward set of units: Q in British thermal units (Btu), t in hours, A in square feet, Δx in inches, and ΔT in degrees Fahrenheit. The *R*-value in these units is 5.67 times the RSI value. In this book we will use only RSI values for thermal resistance R; some representative values are given in Table 18-5.

An important advantage of using *R*-values instead of thermal conductivities is that if heat is being transferred through one material and then another (for example, through an exterior wall and then through an interior wall in a building), the *R*-values of the individual materials are simply added to give the total *R*-value of the combination, as illustrated in Sample Problem 18-9.

Table 18-5

Thermal Resistances *R* for Materials of Thickness 1 cm

Material	R ($\text{W}^{-1}\cdot\text{m}^2\cdot\text{K}$) (or $\text{W}^{-1}\cdot\text{m}^2\cdot{}^\circ\text{C}$)
Copper	2.6×10^{-5}
Plywood	8.6×10^{-2}
Glass	1.3×10^{-2}
Brick	7.7×10^{-3}
Concrete	1.0×10^{-2}
Gypsum wall board	6.3×10^{-2}
Fibreglass	0.25
Polyurethane foam	0.43
Air (non-convecting)	0.42

SAMPLE PROBLEM 18-9

A wall consists of an exterior layer of bricks of thickness 10.1 cm, an inner layer of fibreglass insulation of thickness 8.9 cm, and an interior layer of gypsum wallboard of thickness 1.3 cm (Figure 18-15). The dimensions of the wall are 2.4 m × 4.9 m. The exterior and interior temperatures are −13°C and 21°C respectively.

(a) How much heat flows through the wall in 1.0 h?

What is the temperature at the interface between

(b) the fibreglass and bricks?

(c) the fibreglass and wallboard?

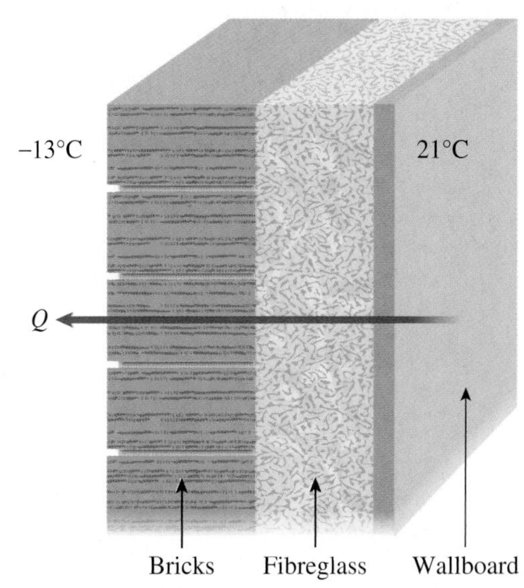

Figure 18-15 Sample Problem 18-9. Heat Q conducted through a wall consisting of bricks, fibreglass, and wall board.

Solution

(a) Start with Eqn. 18-14 and rearrange to solve for Q: $P = \frac{Q}{t} = \frac{A\Delta T}{R} \therefore Q = \frac{At\Delta T}{R}$

The total R-value is the sum of the R-values for the three materials, which can be calculated for the various thicknesses, by starting with the R-values for 1 cm thicknesses from Table 18-5:

$$\text{Bricks: } R_{BR} = \frac{7.7 \times 10^{-3}\ \text{W}^{-1}\cdot\text{m}^2\cdot{}^\circ\text{C}}{1\ \text{cm}} \times 10.1\ \text{cm} = 0.0778\ \text{W}^{-1}\cdot\text{m}^2\cdot{}^\circ\text{C}$$

$$\text{Fibreglass: } R_{FG} = \frac{0.25\ \text{W}^{-1}\cdot\text{m}^2\cdot{}^\circ\text{C}}{1\ \text{cm}} \times 8.9\ \text{cm} = 2.23\ \text{W}^{-1}\cdot\text{m}^2\cdot{}^\circ\text{C}$$

$$\text{Wallboard: } R_{WB} = \frac{6.3 \times 10^{-2}\ \text{W}^{-1}\cdot\text{m}^2\cdot{}^\circ\text{C}}{1\ \text{cm}} \times 1.3\ \text{cm} = 0.0819\ \text{W}^{-1}\cdot\text{m}^2\cdot{}^\circ\text{C}$$

$$\begin{aligned}\text{Total } R &= R_{BR} + R_{FG} + R_{WB} \\ &= (0.0778 + 2.23 + 0.0819)\ \text{W}^{-1}\cdot\text{m}^2\cdot{}^\circ\text{C} = 2.39\ \text{W}^{-1}\cdot\text{m}^2\cdot{}^\circ\text{C}\end{aligned}$$

Notice that the thermal resistance of the fibreglass is much greater than that of the bricks or wallboard. Now calculate the heat through the wall with the known total thermal resistance:

$$Q = \frac{At\Delta T}{R} = \frac{(2.4\ \text{m} \times 4.9\ \text{m})(3600\ \text{s})[(21 - (-13)^\circ\text{C}]}{2.39\ \text{W}^{-1}\cdot\text{m}^2\cdot{}^\circ\text{C}} = 6.02 \times 10^5\ \text{J}$$

Thus, the heat flowing through the wall in 1.0 h is 6.0×10^5 J (correct to two significant digits).

(b) An important key to finding the temperature at the interface between the bricks and fibreglass is to recognize that *the amount of heat transmitted through each of the materials (bricks, fibreglass, and wallboard) in the 1.0 h time interval must be the same (6.0×10^5 J)*. Suppose this was not the case; for example, suppose that the heat transferred outward through the fibreglass was larger than that transferred outward through the bricks. This would mean that more heat would be entering the fibreglass–bricks interface than would be leaving it. As a result, the temperature of this interface region would increase indefinitely; this would make no physical sense.[3]

To find the temperature at the fibreglass–bricks interface, use the equation $Q = \frac{At\Delta T}{R}$ from part (a), but consider the temperature change and thermal resistance only for the bricks, since we know the temperature on one side of the bricks already:

$$Q = \frac{At\Delta T_{BR}}{R_{BR}} \therefore \Delta T_{BR} = \frac{QR_{BR}}{At} = \frac{(6.02 \times 10^5\ \text{J})(0.0778\ \text{W}^{-1}\cdot\text{m}^2\cdot{}^\circ\text{C})}{(2.4\ \text{m} \times 4.9\ \text{m})(3600\ \text{s})} = 1.1^\circ\text{C}$$

Thus, the temperature change across the bricks is 1.1°C. Since the temperature on the outside surface of the bricks is −13°C, the temperature on the inside surface of the bricks is (−13 + 1.1)°C = −12°C.

Therefore, the temperature at the brick–fibreglass interface is −12°C.

(c) To find the temperature at the interface between the fibreglass and wallboard, we follow the same procedure as in part (b). We could consider the temperature change across either the fibreglass or the wallboard. Using the wallboard,

$$Q = \frac{At\Delta T_{WB}}{R_{WB}} \therefore \Delta T_{WB} = \frac{QR_{WB}}{At} = \frac{(6.02 \times 10^5\ \text{J})(0.0819\ \text{W}^{-1}\cdot\text{m}^2\cdot{}^\circ\text{C})}{(2.4\ \text{m} \times 4.9\ \text{m})(3600\ \text{s})} = 1.2^\circ\text{C}$$

[3]If either the exterior temperature or interior temperature were to change, then for a short time period the temperature at all points in between would change as the heat flow adjusted, but for the longer-term steady-state situation, the temperature at a given point (such as an interface) must remain constant.

The temperature at the inside surface of the wallboard is 21°C, and therefore the temperature at the fibreglass–wallboard interface is (21 − 1.2)°C = 20°C.

Thus, at the interface of the fibreglass and wallboard, the temperature is 20°C (correct to two significant digits).

Considering the total temperature change of 34°C across the wall, note that there is only 1°C across the bricks and 1°C across the wall board, and therefore 32°C across the highly insulating fibreglass.

Convection

Convection is a process that involves the physical movement of a fluid (liquid or gas) that transfers heat from one location (or object) to another. It can be *natural*, as in the upward motion of hot air above a campfire (Figure 18-16), or the movement of sea currents such as the Gulf Stream. Alternatively, convection can be *forced*, such a furnace fan moving heated air through a building, or the pumping of coolant liquid through a car engine and radiator.

convection a process that involves the physical movement of a fluid (liquid or gas) that transfers heat from one location (or object) to another

An important example of natural convection in cold climates is the effect of wind in removing heat from warm objects. Wind chill equivalent temperatures are regularly provided in weather reports when temperatures are low and winds are high. As seen in Table 18-6, if the actual air temperature is −10°C and the wind speed is 30 km/h, the wind chill equivalent temperature is −20°C. This means that heat will be transferred from a person's body at the same rate as in air at −20°C with only a very light wind.

The rate of heat transfer by convection depends on many factors. For example, heat loss from a hot water pipe depends on the pipe's orientation; the loss is greater from a horizontal pipe than from a vertical one. As a result, there is no single equation to

Alexander Ishchenko/Shutterstock

Figure 18-16 The upward convective movement of the heated air above a campfire is one of the mechanisms removing energy from the fire.

Table 18-6
Wind Chill Equivalent Temperatures

Wind Speed	Air Temperature (°C)				
(km/h)	**0**	**−10**	**−20**	**−30**	**−40**
10	−3	−15	−27	−39	−51
20	−5	−18	−30	−43	−56
30	−6	−20	−30	−46	−59
40	−7	−21	−34	−48	−61
50	−8	−22	−35	−49	−63
60	−9	−23	−36	−50	−64

Increasing risk of frostbite for most people within 30 min of exposure.
High risk for most people in 5 to 10 min of exposure.
High risk for most people in 2 to 5 min of exposure.
High risk for most people in 2 min of exposure.

Source: Environment Canada.

calculate heat transfer by convection. For any particular situation, experiments must be done to determine an equation and parameters that are appropriate for that scenario (see Question 18-81). The only common feature about convective heat transfer rate P is that it is proportional to the area A in contact with the fluid ($P \propto A$).

Radiation

thermal radiation (or **radiation**) heat transfer by the emission of electromagnetic energy from an object

The third mechanism of heat transfer is the process of **thermal radiation** (or **radiation**), in which energy is emitted by an object via electromagnetic waves.[4] In contrast with conduction and convection, which require a material medium, radiation can travel through either a medium or a vacuum. Any object at a temperature above absolute zero (0 K) emits radiation. The intensity I of the radiation, that is, the power P emitted per unit area A of the emitting object, is given by Stefan's law[5]:

$$I = \frac{P}{A} = \sigma T^4 \quad \textbf{(18-15)}$$

where σ = Stefan–Boltzmann constant = $5.670 \times 10^{-8}\ \text{W}\cdot\text{m}^{-2}\cdot\text{K}^{-4}$
T = absolute temperature (K)

Suppose that we use Eqn. 18-15 to estimate the rate P at which radiated heat is emitted from the skin of a person. Skin temperature is about 34°C (i.e., 307 K) and the surface area of an adult person is approximately 1 m^2, and hence

$$P_{\text{emitted}} = \sigma T_{\text{skin}}^4 A \approx (5.670 \times 10^{-8}\ \text{W}\cdot\text{m}^{-2}\cdot\text{K}^{-4})(307\ \text{K})^4(1\ \text{m}^2) \approx 5 \times 10^2\ \text{W}$$

The basal metabolic rate—that is, the rate at which body heat is generated—for a typical resting person is about 100 W, but the calculation above indicates that the radiated heat rate is roughly five times this value! In other words, although the person is generating energy at a rate of only 100 J/s, it is radiating away at a much faster rate, and the person's temperature should be falling rapidly. The resolution of this apparent problem is that the person also receives radiation from the environment: air, walls, furniture, etc. If the surrounding temperature is roughly 20°C (or 293 K), then the rate at which radiated energy is received by the person is

$$P_{\text{received}} = \sigma T_{\text{surroundings}}^4 A \approx (5.670 \times 10^{-8}\ \text{W}\cdot\text{m}^{-2}\cdot\text{K}^{-4})(293\ \text{K})^4(1\ \text{m}^2) \approx 4 \times 10^2\ \text{W}$$

Thus, the 400 W "shortfall" is made up by the radiated energy received from the surroundings.

In general, if an object is at absolute temperature T_1 and the surroundings are at an absolute temperature T_2, then the net intensity (I_{net}) of radiation that the object emits and receives is given by

$$I_{\text{net}} = \sigma(T_1^4 - T_2^4) \quad \textbf{(18-16)}$$

As well, since $I = \frac{P}{A}$, then

$$I_{\text{net}} = \frac{P_{\text{net}}}{A} \text{ or } P_{\text{net}} = I_{\text{net}}A \quad \textbf{(18-17)}$$

thermography the detection of thermal radiation emitted by buildings, people, etc., by infrared-sensitive cameras

thermogram images produced by thermography

atmospheric greenhouse effect (or **greenhouse effect**) absorption by atmospheric gases of some of the radiation emitted by Earth's surface, and the resulting increase in temperature of Earth's surface because of the radiation emitted by the gases back to the surface

The radiation emitted by objects has a wide range of wavelengths. For heat radiated by humans and other animals, as well as low-temperature objects such as buildings and

[4]Electromagnetic waves were discussed in Section 16.6.
[5]Josef Stefan (1835–1893), Austrian physicist.

plants in our environment, most of the radiation is in the invisible infrared portion of the electromagnetic spectrum. As temperature increases, the peak intensity gradually shifts to the red end of the visible spectrum, and at higher temperatures to the blue end (Figure 18-17).

The thermal radiation emitted by buildings, people, etc., can be detected by infrared-sensitive cameras in a technique called **thermography**. The images produced are referred to as **thermograms**. Figure 18-18 shows the thermogram of a house, indicating large heat loss from the attic and from some windows. Thermograms such as this show areas where heat loss should be reduced to decrease heating costs. Thermography is also used by firefighters to enable them to see through smoke, and has been helpful in locating people who have been lost in remote areas. Medical thermometers (Figure 18-19) that detect infrared radiation and convert this information to a temperature reading, usually of an eardrum, are common.

Shawn Hempel/Shutterstock

(a)

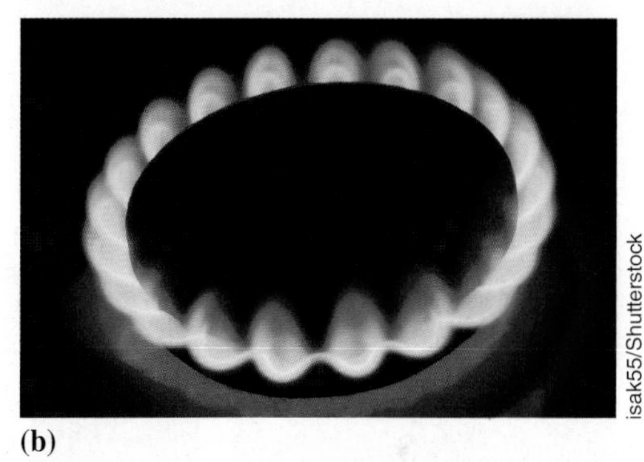
isak55/Shutterstock

(b)

Figure 18-17 As temperature increases from typical low temperatures in our everyday environment, the peak radiation intensity moves from the invisible infrared portion of the spectrum to **(a)** the red region of the visible spectrum, and at even higher temperatures to the **(b)** blue region of the visible spectrum.

Dario Sabljak/Shutterstock

Figure 18-18 The thermogram of this house shows high heat loss (red) from the attic and some windows.

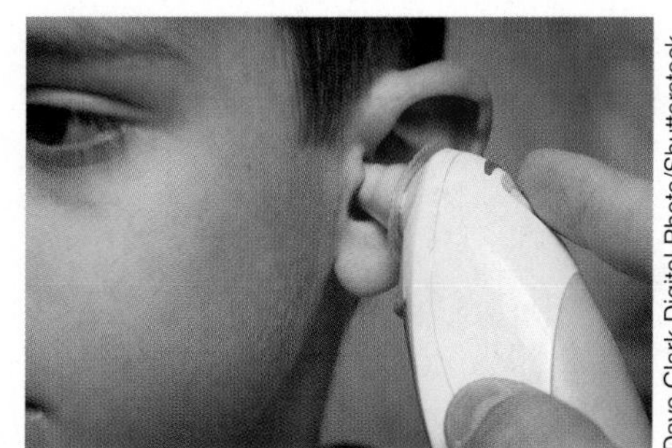
Dave Clark Digital Photo/Shutterstock

Figure 18-19 This thermal radiation thermometer detects infrared radiation emitted by the eardrum and provides a reading of the eardrum's temperature.

The Greenhouse Effect

You have probably heard the term "The Greenhouse Effect" in connection with global warming and climate change. What is this effect and how does it work? An important key to understanding this effect is a knowledge of thermal radiation, in particular Stefan's law, $I = P/A = \sigma T^4$ (Eqn. 18-15).

The Sun emits radiation, which has an intensity at the Sun's surface given by Stefan's law. As this radiation travels through space in all directions, its intensity decreases since the area over which the power is delivered is increasing. The solar radiation intensity at the top of Earth's atmosphere, averaged over the entire Earth, is 342 W/m^2. Clouds and atmospheric gases reflect away about 31% of the incoming energy, leaving 236 W/m^2 incident on Earth's surface. This radiation warms the surface to a temperature, T_{Earth}, which then also emits radiation according to Stefan's law ($I_{\text{emitted}} = \sigma T^4_{\text{Earth}}$). For Earth's surface to have, on average, a steady-state equilibrium temperature T_{Earth}, the incident intensity (236 W/m^2) and the emitted intensity must be equal. Setting $236\ \text{W/m}^2 = I_{\text{emitted}} = \sigma T^4_{\text{Earth}}$, and solving for T_{Earth} gives a value of 254 K or -19°C. However, the actual mean temperature of Earth's surface is 288 K, or 15°C, which is 34°C higher than our calculation predicted. The difference is due to the greenhouse effect, described in the next paragraph.

The **atmospheric greenhouse effect** (or **greenhouse effect**) refers to the absorption by atmospheric gases of some of the radiation emitted by Earth's surface. These gases then emit their own radiation equally back to Earth's surface and out into space. This additional radiation on Earth's surface increases its temperature to the observed value of 288 K. The gases that contribute most strongly to this greenhouse effect are carbon dioxide and methane. Carbon dioxide concentrations have been rising steadily from the pre-industrial revolution value of 280 parts per million (ppm) in the 1800s to 399 ppm in December 2014 (Source: Mauna Loa Observatory, Hawaii). As stated in the previous section, global average air temperature has been increasing. Climatologists and other scientists are predicting that as temperature continues to increase, there will be more severe storms, higher ocean levels, and worldwide climate change; there is evidence of these changes already.

Any planet with an atmosphere experiences a warming because of the greenhouse effect. Mars has a thin atmosphere and only a small effect, but Venus has a dense atmosphere of CO_2, and the greenhouse effect produces a huge temperature increase, giving a surface temperature of approximately 700 K.

You might be wondering why the name "greenhouse" effect is used for this phenomenon. The "trapping" of the infrared radiation from Earth by atmospheric gases is similar to what happens in greenhouses (as well as in cars parked in a sunny location). Visible sunlight passes through the glass of greenhouses, warming the interior. Objects in the warm interior then emit thermal infrared (IR) radiation. Glass is not very transparent to IR radiation, and most of the IR radiation simply reflects back from the glass and remains in the greenhouse. Thus, greenhouses become quite hot. Often, portions of the glass roofs and walls can be opened wide to let heat escape if needed.

DID YOU KNOW?

Studies have shown that when people tell a lie, they experience an increase in temperature around the nose and at the orbital muscle at the corner of each eye; this temperature change can be detected via thermography.

Biological Thermal Regulation

Many animals must maintain their body temperatures within a rather narrow range in order to function well, and so they need ways in which to release or retain heat as needed. We have already mentioned that humans sweat and dogs pant to release heat when required, but what do other animals do in hot weather? Many simply seek shade or a low-temperature burrow, or cool themselves in water or mud. Elephants use their trunks to spray themselves with water (Figure 18-20 (a)), and rabbits increase blood flow to their ears to release heat (Figure 18-20 (b)). Not surprisingly, rabbits in warm climates have larger ears for this purpose than their cold-climate cousins. Elephants also use their large ears, which have many blood vessels, to dissipate heat, and there is evidence that the *Stegosaurus* (Figure 18-20 (c)) used the large plates on its back for the same purpose. When hot, birds flatten their feathers and fur-bearing mammals compress their fur, to force out insulating air.

When weather turns cold, animals with fur develop a thicker winter coat that has many tiny air pockets for good insulation, and birds "fluff themselves up" to trap air in their feathers. Some mammals curl up to reduce their surface area and retain heat (Figure 18-21 (a)), and many birds and mammals huddle together (Figure 18-21 (b)). Some animals, such as mice, increase their metabolic rate at low temperatures to generate extra body heat, while others, such as lizards, decrease their metabolic rate to conserve energy. Of course there are also animals who spend the winter hibernating: bears, turtles, bats, hedgehogs, snails, and snakes, to name just a few.

Plants respond to temperature changes as well. Deciduous trees lose their leaves in the fall, and in hot dry weather leaves close their stomata to prevent water loss. Many desert plants such as cacti exist in hot climates thanks to their thick, water-storing stems. One of the most unusual temperature-sensitive plants is the *Rhododendron*, which is a flowering shrub. *Rhododendrons* do not lose their leaves in winter, but their leaves droop and curl (Figure 18-22), and if the temperature goes up or down by only a few degrees Celsius, the orientation and shape of the leaves change. Some people say that they can tell the outdoor temperature by looking at the leaves of *Rhododendrons*. The large temperature sensitivity of this phenomenon is still not well understood, but it has been suggested that the drooping protects the leaves from high solar irradiance during the day and cold temperatures at night, and that the curling prevents damage to cellular membranes during thawing, which often occurs in the early morning.

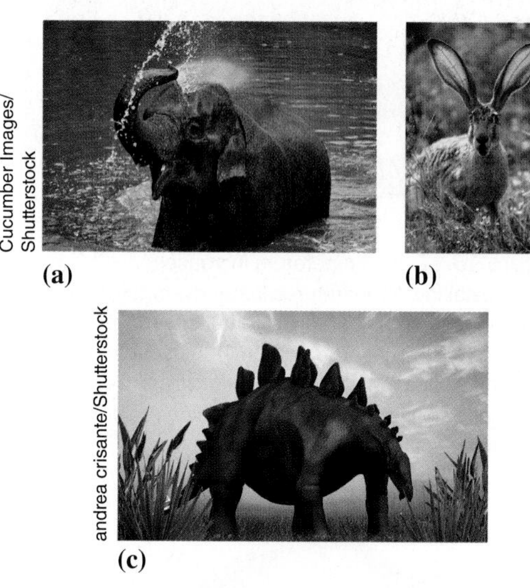
Cucumber Images/Shutterstock © Joe McDonald/Corbis andrea crisante/Shutterstock

(a) (b) (c)

Figure 18-20 **(a)** This elephant uses water for cooling, and **(b)** the rabbit relies on blood vessels in its large ears to dissipate heat. **(c)** It is believed that the *Stegosaurus* used blood flow in the large plates on top of its back for releasing excess heat.

Stefan Ataman/Shutterstock

(a)

© david tipling/Alamy

(b)

Figure 18-21 **(a)** This Husky dog curls up to reduce its surface area and retain heat; notice how it covers its nose, which has no fur. **(b)** Penguins huddle together to keep warm.

swa182/Shutterstock

Figure 18-22 The drooping and curling of *Rhododendron* leaves in winter is highly temperature dependent. Relatively small temperature changes produce noticeable changes in the appearance of the leaves.

EXERCISES

18-16 Describe how heat is transferred by conduction, convection, and radiation.

18-17 On a winter morning, you step in your bare feet from a carpet onto a tile floor, both of which are at the same temperature. Which one feels cooler? Why?

18-18 A small pond has a surface area of 153 m^2 and a constant depth of 2.5 m. The temperature of the pond water is 8°C at the bottom and 23°C at the top. The thermal conductivity of water is 0.58 W·m^{-1}·K^{-1}.

(a) At what rate is heat conducted from the top to the bottom of the pond?

(b) How much heat is transferred in 45 min?

18-19 The thermal resistance of a particular type of wood of thickness 2.2 cm is 0.24 W^{-1}·m^2·K. Determine the thermal conductivity of the wood.

18-20 The wall of a building consists of an exterior layer of bricks of thickness 0.10 m and interior plywood sheeting of thickness 2.0 cm. The dimensions of the wall are 2.4 m × 4.9 m. The outside and inside temperatures are −19°C and 22°C respectively. Using Table 18-5, determine

(a) the rate of heat transfer through the wall

(b) the temperature at the interface between the bricks and plywood

18-21 The filament of a 60 W incandescent light bulb has a temperature of 3300 K. What is the surface area of the filament in square millimetres? Assume two significant digits in given data and that 100% of the energy is radiated.

18-22 What is the net power radiated from a mouse having a temperature of 37°C and a surface area of 72 cm^2, in air at 25°C?

18.5 First and Second Laws of Thermodynamics

Many electric power plants burn fuel such as coal or natural gas to generate electricity, and car engines burn gasoline to provide kinetic energy, but these processes have efficiencies much less than 100%. In order to understand the limitations on these efficiencies, we need to consider the first and second laws of thermodynamics.

first law of thermodynamics the change in internal energy (ΔU) of a system equals the sum of the heat Q added to the system and the work W done on the system

The **first law of thermodynamics** is essentially a statement of conservation of energy of a system, with the internal energy being included explicitly. In words, the first law states that the change in internal energy (ΔU) of a system equals the sum of the heat Q added to the system and the positive work W done on the system; that is, $\Delta U = Q + W$. If the system does positive work W on its surroundings, then $\Delta U = Q - W$. For example, if a gas is heated and expands against a piston, the increase in internal energy of the gas equals the heat provided to the gas minus the work done by the gas on the piston.

If positive work W is done on a system, $\Delta U = Q + W$ (18-18a)

If a system does positive work W on its surroundings, $\Delta U = Q - W$ (18-18b)

As another example, consider the situation shown in Figure 18-23 (a), where water of mass m is at rest in a tank at the top of an incline. A valve is then opened at the bottom of the tank, allowing the water to flow down the incline into a lower tank Figure 18-23 (b), where the water comes to rest. The force of gravity does positive work W on the water as it flows, and this results in an increase in the internal energy U of the water; in other words, the temperature T of the water increases. From Section 18.2 we know that the energy required for a temperature change ΔT of a mass m is $mC\Delta T$, where C is the specific heat of the substance (water in this case). As discussed in Section 7.3, as the water moves downward through a vertical elevation y, the work done by gravity is equal to mgy. Thus, $\Delta U = Q + W$ (Eqn. 18-18a) gives $mC\Delta T = Q + mgy$. Recall that heat Q is energy transferred between objects as a result of a temperature difference; in this particular situation, no energy was provided because of a temperature difference, and therefore $Q = 0$. Thus, we have $mC\Delta T = mgy$. This equation is a statement of energy conservation, indicating that the initial gravitational potential energy mgy has been converted to internal energy.

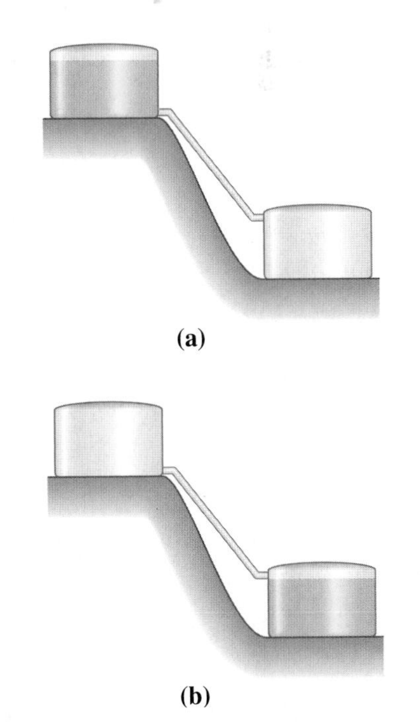

Figure 18-23 **(a)** Water is at rest in an elevated tank. **(b)** The water has flowed down the incline to the lower tank, with a resulting increase in the water's temperature.

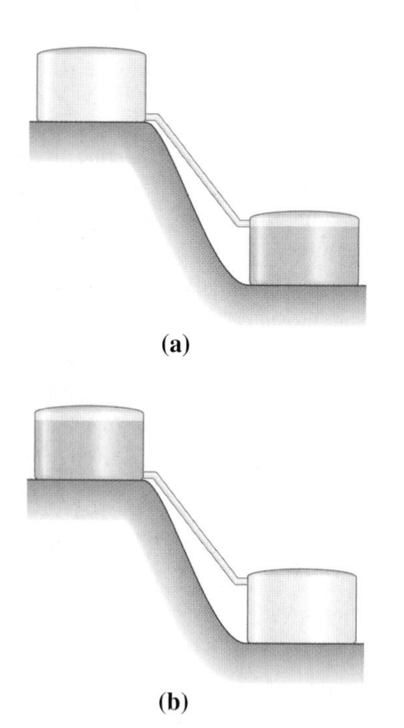

Figure 18-24 **(a)** Water is at rest in the lower tank. **(b)** Conservation of energy would allow the water to flow on its own to the higher tank, converting internal energy to gravitational potential energy and cooling the water.

second law of thermodynamics disorder is more probable than order

Consider now Figure 18-24 (a), which shows water at rest in the lower tank. Imagine now that this water flows up the incline to the upper tank on its own accord, converting internal energy into gravitational potential energy, and becoming cooler as it goes uphill Figure 18-24 (b). This uphill flow is, of course, impossible, but it is not conservation of energy that makes it so. In principle we could write $mC|\Delta T| = mgy$ for energy conservation for the uphill motion, but we know that nature does not work that way. In order to explain the non-reversibility of the water flow, another law of nature must be invoked: the second law of thermodynamics.

The Second Law of Thermodynamics

The **second law of thermodynamics** can be stated in many ways: probably the easiest is that disorder is more probable than order.[6] It is easy to create disorder from order, but difficult to generate order from disorder. The internal energy of the water in the lower tank in Figure 18-24 (a) is disordered energy; the molecules are moving randomly in all directions. It is essentially impossible that all the molecules would simultaneously move in the same direction and travel up the incline. On the other hand, gravitational potential energy is an example of ordered energy, since the vertical direction is a specific direction giving order to this energy. It is simple to convert this ordered energy to disordered internal energy as the water flows into the lower tank in Figure 18-23 (b).

We now apply these ideas to an important example of commercial energy production: the boiling of water to produce steam for the generation of electricity in a thermal electric power station. The heat to boil the water could come from the burning of fossil fuels, from nuclear reactors, or from solar energy. In a fossil-fuel electric plant, the fuel is burned to vaporize water in a boiler (Figure 18-25, label 1), producing steam at high pressure and temperature. The steam passes over the blades of a turbine (2), causing it to spin. The turbine turns an electric generator (3), which produces an electric current in the external wires. The steam loses energy in spinning the turbine, and its pressure and temperature are now lower (4). The low-temperature steam is condensed back to water (5) to begin the process over again. The condensation is often done by passing the steam through pipes cooled by water from a river or lake (Figure 18-26). The warm

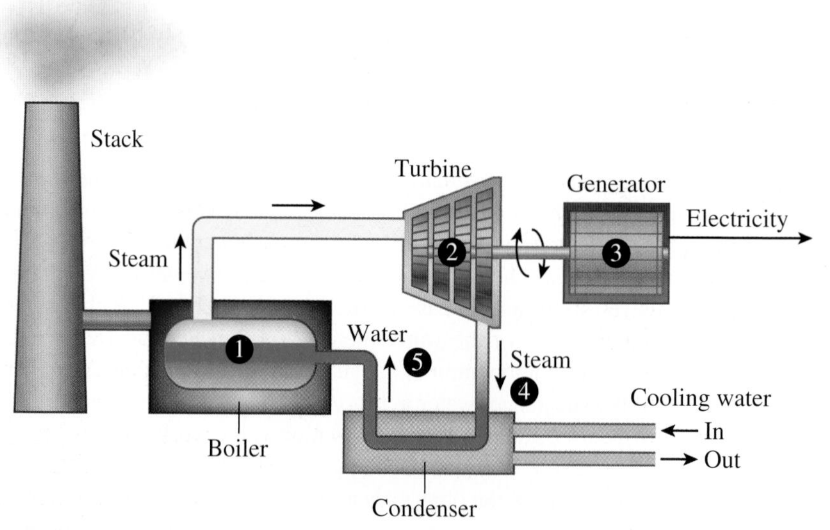

Figure 18-25 Schematic diagram of a fossil-fuel electric power plant.

[6]A quantitative statement will be given later in this section.

cooling water is returned to the river or lake. Alternatively, warm cooling water from the condenser can cool by falling through a cooling tower (Figure 7-35) and can then be recycled through the condenser.

The energy of the hot steam is disordered internal energy, but the spinning turbine and the electric current are ordered. The electric plant is performing a conversion of disordered energy into ordered energy, which is contrary to the natural tendency from order to disorder, and it is impossible to convert all the disordered internal energy into useful ordered energy. In this regard the condenser performs an important function. The heat discarded into the cooling water increases its internal energy and its disorder. If heat were not discarded to the cooling water (or elsewhere), the electric plant could not function: it cannot convert all the disordered energy of the steam into ordered energy. The disordered energy of the hot steam is converted partly into ordered energy of the turbine and electric current, and partly into disordered energy of the cooling water.

Miks Mihails Ignats/Shutterstock

Figure 18-26 A fossil-fuel thermal electric generating plant with a nearby source of cooling water.

We will see later in this section that the cooling in the condenser also improves the power-plant efficiency. In addition, from a practical point of view, if the steam leaving the turbine were not condensed to water, it would require all the energy output from the turbine to pump the waste steam from the turbine back to the boiler. This waste steam is at low pressure and the boiler is at high pressure, and the steam would not flow to the boiler without pumping. In order to move liquid water from the condenser to the boiler, a pump (not shown in Figure 18-25) is also required. However, the volume of the condensed water is roughly 1000 times smaller than the corresponding volume of steam, and hence the energy required to pump it to the boiler is about 1000 times smaller than to pump the same mass of steam. (The energy required can be expressed as energy = (difference in pressure) × volume.)

Heat Engines and Entropy

A thermal electric generating station is an example of a general type of machine called a **heat engine**, which is any device that produces mechanical work via a transfer of heat. The second law of thermodynamics can be stated specifically for heat engines: *no device can be constructed that, operating in a cycle (like an engine), accomplishes only the transfer of heat from some source resulting in its* ***complete*** *conversion to mechanical work.*

heat engine a device that produces mechanical work via a transfer of heat

Figure 18-27 shows the general structure of a heat engine. Heat Q_H is removed from a hot reservoir at absolute temperature T_H, some useful work W is done, and heat Q_C is discarded to a cold reservoir at absolute temperature T_C. The efficiency η at which a heat engine operates can be expressed in terms of W and Q_H. We begin with Eqn. 7-9b:

$$\text{efficiency } \eta = \frac{\text{useful } E_{\text{out}}}{\text{total } E_{\text{in}}} \quad (7\text{-}9b)$$

For a heat engine, the useful energy out is the work W, and the total energy in is the heat Q_H, and therefore

$$\text{heat engine } \eta = \frac{W}{Q_H} \quad (18\text{-}19)$$

This expression for efficiency can also be written in terms of the two heat energies Q_H and Q_C. We start with the first law of thermodynamics: $\Delta U = Q - W$ (Eqn. 18-18b). Since the engine operates in a cycle, its initial and final states are the same, and

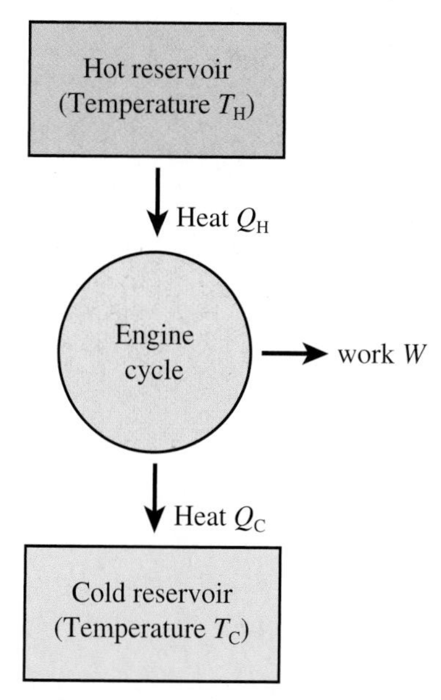

Figure 18-27 Schematic diagram of a heat engine.

therefore $\Delta U = 0$. Heat Q_H has been added to the engine and Q_C has been removed, giving a net heat added of $Q = Q_H - Q_C$. Thus, the first law equation becomes $0 = (Q_H - Q_C) - W$, or

$$\text{heat engine } W = Q_H - Q_C \qquad (18\text{-}20)$$

Substituting for W in Eqn. 18-19 gives

$$\text{heat engine } \eta = \frac{Q_H - Q_C}{Q_H} = 1 - \frac{Q_C}{Q_H} \qquad (18\text{-}21)$$

DID YOU KNOW?

Sometimes the first and second laws of thermodynamics are paraphrased as

First law: When it comes to energy, you don't get something for nothing.

Second law: You can't even break even.

entropy a measure of disorder

Notice from Eqn. 18-21 that the efficiency of a heat engine is always less than 1 (i.e., less than 100%), as expected from the second law of thermodynamics.

In the next few paragraphs, Eqn. 18-21 for efficiency will be converted into a more convenient one involving only the temperatures T_H and T_C. We have already mentioned disordered and ordered energy. **Entropy** is a measure of disorder, and another way to state the second law of thermodynamics is that *in any energy transfer or conversion within a closed system, the entropy of the system must increase* (or in rare cases, remain constant). In a thermal electric plant, if all the disordered thermal energy of the hot steam were to be converted solely to ordered electric energy, then there would be a decrease in the total entropy, in violation of the second law.

It can be shown in a derivation beyond the scope of this book that when heat Q is added to a system at *absolute* temperature T, then the change in entropy (ΔS) of the system is given by

$$\Delta S = \frac{Q}{T} \qquad (18\text{-}22)$$

If heat is removed from an object, then the change in entropy is negative: $\Delta S = -\frac{Q}{T}$. With Q in joules and T in kelvins, ΔS has units of joules per kelvin (J/K).

When a system receives heat, the heat is an input of disorganized energy, since it consists of random molecular motion. Hence, heat input increases entropy and, conversely, heat loss decreases entropy. But why is temperature in the denominator in Eqn. 18-22? In cold objects, there is not much molecular motion, and so a given quantity of heat will stir things up quite noticeably, increasing the disorder a great deal. But if the same amount of heat is added to a hot object (already having a large amount of disorder), the entropy increase is smaller.

Eqn. 18-22 will now be applied to the case of a heat engine in which heat Q_H is converted to work W and the remainder Q_C is discarded. The system under consideration is the engine plus the two thermal reservoirs, as shown in Figure 18-27. Since the engine runs in cycles, then at the end of each cycle it is in the same state at which it began. Therefore, $\Delta S = 0$ for the engine alone. The change in entropy as the heat Q_H flows out of the hot reservoir at temperature T_H is $\Delta S = -Q_H/T_H$, and the entropy change as the heat Q_C flows into the cold reservoir at temperature T_C is $\Delta S = Q_C/T_C$. There is no change in entropy associated with the organized mechanical work W, and thus the total entropy change is

$$\Delta S = \frac{-Q_H}{T_H} + \frac{Q_C}{T_C}$$

According to the second law of thermodynamics, the entropy can never decrease; that is, $\Delta S \geq 0$, which gives

$$\frac{-Q_H}{T_H} + \frac{Q_C}{T_C} \geq 0 \quad \text{or} \quad \frac{Q_C}{T_C} \geq \frac{Q_H}{T_H}$$

Rearranging, $\dfrac{Q_C}{Q_H} \geq \dfrac{T_C}{T_H}$

Returning to Eqn. 18-21 for the efficiency η of a heat engine, $\eta = 1 - \dfrac{Q_C}{Q_H}$,

and substituting $\dfrac{Q_C}{Q_H} \geq \dfrac{T_C}{T_H}$,

$$\text{heat engine } \eta \leq 1 - \frac{T_C}{T_H} \qquad \textbf{(18-23)}$$

Eqn. 18-23 is a handy compact expression for the efficiency in terms of the temperatures of the two thermal reservoirs. The equality sign in Eqn. 18-23 holds only in the special case of what is known as a reversible heat engine (for which $\Delta S = 0$), and gives the maximum possible efficiency of a heat engine:

$$\text{heat engine } \eta_{max} = 1 - \frac{T_C}{T_H} \qquad \textbf{(18-24)}$$

A reversible heat engine (also known as a Carnot[7] engine) is essentially an unachievable idealization. It can operate only if the temperature difference $(T_H - T_C)$ is infinitesimal, or in the case of a finite temperature difference, if the engine has an infinite amount of time in which to work. Nonetheless, Eqn. 18-24 serves as a useful upper limit for the efficiency of a heat engine in terms of easily measurable parameters, that is, the two temperatures. It is particularly convenient for a heat engine such as a thermal power plant where there are only two relevant temperatures. In an internal combustion engine (Figure 18-28), which is a heat engine, there are a number of different temperatures at various stages in the combustion cycle and Eqn. 18-24 cannot be used directly. However, Eqns. 18-19 to 18-21 are still valid for automobile engines.

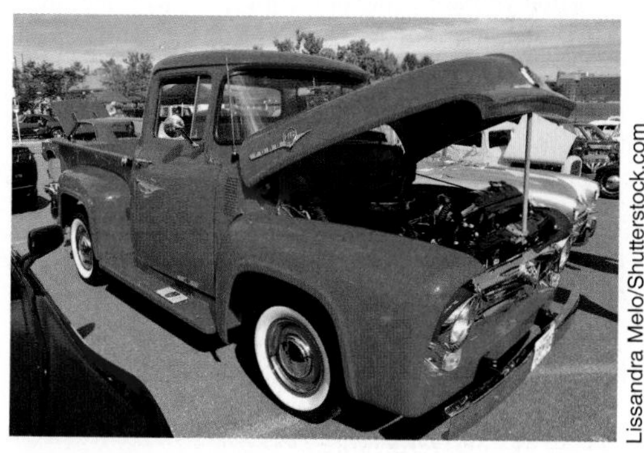
Lissandra Melo/Shutterstock.com

Figure 18-28 Internal combustion engines are heat engines with a rather low overall efficiency of only about 20%.

SAMPLE PROBLEM 18-10

A particular automobile engine receives 6.00×10^3 J of heat from the combustion of gasoline in each cycle, and delivers 1.14×10^3 J of useful mechanical work.

(a) What is the percentage efficiency of this engine?

(b) How much heat is discarded per cycle?

Solution

(a) Using Eqn. 18-19, $\eta = \dfrac{W}{Q_H} = \dfrac{1.14 \times 10^3\text{ J}}{6.00 \times 10^3\text{ J}} = 0.190 \times 100\% = 19.0\%$

Thus, the engine's efficiency is 19.0%.

(b) From the first law of thermodynamics for a heat engine (Eqn. 18-20),

$W = Q_H - Q_C$, which gives $Q_C = Q_H - W = (6.00 - 1.14) \times 10^3\text{ J} = 4.86 \times 10^3\text{ J}$.

Therefore, the amount of heat discarded is 4.86×10^3 J per cycle.

[7]Sadi Carnot (1797–1832), French physicist and engineer.

SAMPLE PROBLEM 18-11

The melting temperature of lead is 327°C, and its latent heat of fusion is 2.3×10^4 J/kg. What is the change in the entropy of 2.0 kg of lead as it melts?

Solution

The heat required to melt the lead is given by Eqn. 18-7:

$$Q = mL_F = (2.0\,\text{kg})(2.3 \times 10^4\,\text{J/kg}) = 4.6 \times 10^4\,\text{J}$$

Eqn. 18-22 gives the change in entropy: $\Delta S = \dfrac{Q}{T} = \dfrac{4.6 \times 10^4\ \text{J}}{(327 + 273)\,\text{K}} = 77\,\text{J/K}.$

Thus, the change in entropy as the lead melts is 77 J/K.

SAMPLE PROBLEM 18-12

In a fossil-fuel plant such as shown in Figure 18-25, the typical temperature of the steam is about 500°C and the cooling water from the lake or river has a temperature of roughly 20°C. Determine the maximum possible efficiency of such a plant. Express the answer as a percentage with two significant digits.

Solution

Use Eqn. 18-24: $\eta_{max} = 1 - \dfrac{T_C}{T_H} = 1 - \dfrac{(20 + 273)\,\text{K}}{(500 + 273)\,\text{K}} = 0.62 = 62\%$

Thus, the maximum possible efficiency is 62%.

Actual efficiencies for fossil-fuel electric plants are only about 30–40%, as a result of energy dissipated by friction, imperfect insulation, the irreversible transfer of heat from the hot reservoir to the cold reservoir, etc.

DID YOU KNOW?

When you are doing something to increase order in your environment, such as tidying your desk or arranging clothes in your closet, you are actually making things more disorganized and increasing entropy. How can this be? As you are performing your organizing tasks, you are moving the air around you and increasing its disorder, and you are also heating the nearby air and the objects that you are touching and again increasing disorder. The net result is that the amount of disorder generated outweighs the small amount of order you have created.

EXERCISES

18-23 A gas absorbs 4.0×10^3 J of heat energy while doing 1.5×10^3 J of work to push a piston. What is the change in internal energy of the gas?

18-24 Steam having a mass of 3.5 kg and a temperature of 100.0°C condenses to water at 100.0°C. What is the change in entropy of the steam during this condensation? (Reference: Table 18-2)

18-25 (a) The first law of thermodynamics for a heat engine can be written as $W = Q_H - Q_C$ (Eqn. 18-20). Describe in words what each of the three variables represents.

(b) Which of the three quantities is the useful energy? Which is the input energy?

(c) Why must the efficiency of a heat engine be less than 100%.

18-26 A paper mill (Figure 18-29) has an electrical generating station that produces electricity by burning scrap wood to boil water to create steam that passes through a turbine. The water is then condensed by cold river water at 5°C and returned to the boiler. If the temperature of the steam is 350°C, what is this plant's maximum possible efficiency (as a percentage with two significant digits)?

Igor Grochev/Shutterstock

Figure 18-29 Question 18-26.

18.6 Refrigerators, Air Conditioners, and Heat Pumps

We are all familiar with refrigerators (Figure 18-30), which are effectively heat engines operating in reverse. A heat engine generates heat Q_H in a hot thermal reservoir, uses it to do useful work W, and discards waste heat Q_C in a cold thermal reservoir. A refrigerator uses work W (usually provided by electricity) to extract heat Q_C from a cold thermal reservoir (the inside of the refrigerator), and expels heat Q_H to a hot thermal reservoir (the room in which the refrigerator is located), as shown in Figure 18-31. By the first law of thermodynamics, the sum of the external work provided and the heat extracted must equal the heat expelled:

Figure 18-30 A refrigerator operates by using external work to extract heat from the cold interior of the refrigerator and expelling it to the warmer surroundings.

$$\text{refrigerator } W + Q_C = Q_H \tag{18-25}$$

In the operation of a refrigerator, the useful energy is the heat Q_C removed from the interior of the refrigerator. The parameter that indicates how well this operation is performed is known as the **coefficient of performance (C.O.P.)**, defined as the ratio of the heat removed to the amount of work provided:

$$\text{refrigerator C.O.P.} = \frac{Q_C}{W} \tag{18-26}$$

Since Q_C and W are both energies and typically are expressed in the same energy unit, C.O.P. is a dimensionless quantity.

From Eqn. 18-25 the work is $W = Q_H - Q_C$, and substitution of this expression for W into Eqn. 18-26 gives

$$\text{refrigerator C.O.P.} = \frac{Q_C}{Q_H - Q_C} = \frac{1}{\frac{Q_H}{Q_C} - 1} \tag{18-27}$$

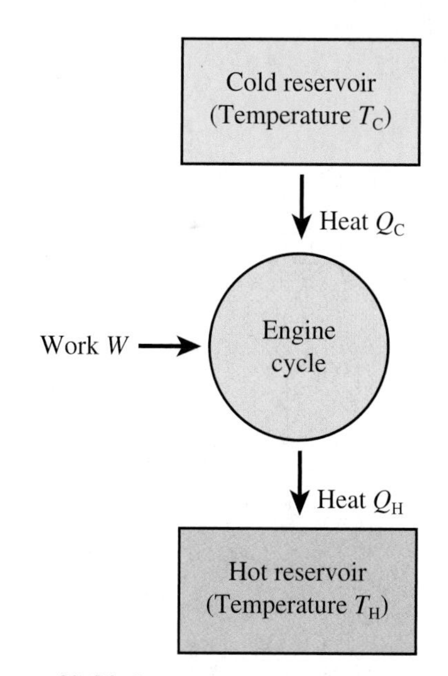

Figure 18-31 Schematic diagram of a refrigerator. Note the similarities and differences with Figure 18-27 for a heat engine.

By using an analysis involving entropy, similar to that used in Section 18.5 for heat engines, it can be shown (Question 18-89) that $Q_H/Q_C \geq T_H/T_C$, where T_H and T_C are absolute temperatures. Substituting this relation into Eqn. 18-27 gives

$$\text{refrigerator C.O.P.} \leq \frac{1}{\frac{T_H}{T_C} - 1} \tag{18-28}$$

Therefore, the maximum C.O.P. of a refrigerator is

$$\text{refrigerator C.O.P.}_{max} = \frac{1}{\frac{T_H}{T_C} - 1} \tag{18-29}$$

coefficient of performance (C.O.P.) for a refrigerator or air conditioner, the ratio of the heat removed from the cold reservoir to the amount of work provided

For a refrigerator with a temperature of $T_C = 263$ K (i.e., -10°C) in the freezer compartment and $T_H = 293$ K (20°C) in the surrounding room, substitution into Eqn. 18-29 gives C.O.P.$_{max}$ = 8.77. Recalling from Eqn. 18-26 that C.O.P. = Q_C/W, this means that for 1 J of external work applied, 8.77 J of heat can be removed from the interior of the refrigerator. This might seem to imply an "efficiency" of 877%, but what has been calculated is not the efficiency of an energy *conversion*, but rather a parameter (the C.O.P.) that indicates how well a refrigerator *moves* heat. A refrigerator is not converting disordered heat into organized work.

A good refrigerator will have an actual C.O.P. between about 3 and 5. This can be improved by using thicker insulation, installing tighter door seals, and increasing the efficiency of the mechanical components (motor and compressor).

SAMPLE PROBLEM 18-13

A refrigerator is operating with a temperature of −11°C in the freezer compartment in a room at 22°C. It is running at 55% of its maximum C.O.P. A glass casserole dish of mass 334 g and specific heat $8.4 \times 10^2\ \text{J}\cdot\text{kg}^{-1}\cdot(°\text{C})^{-1}$ is placed in the main refrigerator compartment, where the temperature is 3°C. The dish contains leftover stew of mass 445 g and specific heat $3.9 \times 10^3\ \text{J}\cdot\text{kg}^{-1}\cdot(°\text{C})^{-1}$. The dish and stew have a temperature of 31°C. How much electrical energy (in kilowatt·hours) must be used to reduce the temperature of the dish and stew to 3°C?

Solution

Start with Eqn. 18-29:

$$\text{C.O.P.}_{\text{max}} = \frac{1}{\dfrac{T_H}{T_C} - 1} = \frac{1}{\dfrac{(22 + 273)\,\text{K}}{(-11 + 273)\,\text{K}} - 1} = 7.94$$

Actual C.O.P. = 0.55 (C.O.P.$_{\text{max}}$) = 0.55 (7.94) = 4.37

Use Eqn. 18-6 ($Q = mC\Delta T$) to determine the heat Q_C to be removed from the stew and glass, writing each ΔT as a positive quantity:

$$\begin{aligned} Q_C &= m_{\text{stew}}C_{\text{stew}}\Delta T_{\text{stew}} + m_{\text{glass}}C_{\text{glass}}\Delta T_{\text{glass}} \\ &= (0.445\,\text{kg})(3.9 \times 10^3\,\text{J}\cdot\text{kg}^{-1}\cdot(°\text{C})^{-1})[(31-3)°\text{C}] \\ &\quad + (0.334\,\text{kg})(8.4 \times 10^2\,\text{J}\cdot\text{kg}^{-1}\cdot(°\text{C})^{-1})[(31-3)°\text{C}] \\ &= 5.64 \times 10^4\,\text{J} \end{aligned}$$

n. 18-26 gives C.O.P. in terms of Q_C and the electrical energy (or work) W needed:

$$\text{C.O.P.} = \frac{Q_C}{W}, \text{ from which } W = \frac{Q_C}{\text{C.O.P.}} = \frac{5.64 \times 10^4\,\text{J}}{4.37} = 1.29 \times 10^4\,\text{J}$$

Since $1\,\text{kW}\cdot\text{h} = 10^3\,\frac{\text{J}}{\text{s}} \times 3600\,\text{s} = 3.6 \times 10^6\,\text{J}$, then W in kilowatt·hours is

$$W = 1.29 \times 10^4\,\text{J} \times \frac{1\,\text{kW}\cdot\text{h}}{3.6 \times 10^6\,\text{J}} = 3.6 \times 10^{-3}\,\text{kW}\cdot\text{h}$$

Thus, the electrical energy required to cool the dish of stew is 3.6×10^{-3} kW·h.

How a Refrigerator Works

A refrigerator operates by alternately compressing a fluid and allowing it to expand, and between these processes the fluid either absorbs or gives off heat. Until recent years, a fluid commonly used in household refrigerators was Freon (CCl_2F_2), one of the chlorofluorocarbons (CFCs) responsible for the decline of ozone in the upper atmosphere. CFCs also contribute to global warming and climate change. Hydrogenated Freons are now being used that break down much more rapidly in the environment if they are accidentally released from a refrigerator. In industry, ammonia is frequently employed as the working fluid.

To explain the processes in a refrigerator cycle, we begin at the expansion valve in Figure 18-32. To the right of the valve, the fluid is a warm liquid at high pressure. When a sensor in the refrigerator indicates that cooling is required, the expansion valve is opened. The valve has only a very small opening, and liquid sprays slowly through it into the lower-pressure region to the left. Because of the pressure decrease, some of the liquid vaporizes, thus removing energy from the liquid, and the temperature falls. The liquid is now cold, with a small quantity of it having been vaporized. This cold liquid flows through coils in the refrigeration box, removing heat from the contents of the box. The liquid vaporizes during this process, and the coils that it is travelling through are called the evaporator. The fluid is now a warm, low-pressure vapour, which passes into the compressor. This device is run by an electric motor and compresses the vapour, heating it somewhat, and the fluid becomes a hot, high-pressure vapour. It passes through coils (usually at the rear of the refrigerator), giving up heat to the surroundings. In the process, it condenses (hence the name "condenser" in Figure 18-32) to form a warm liquid, which is ready to reenter the cycle.

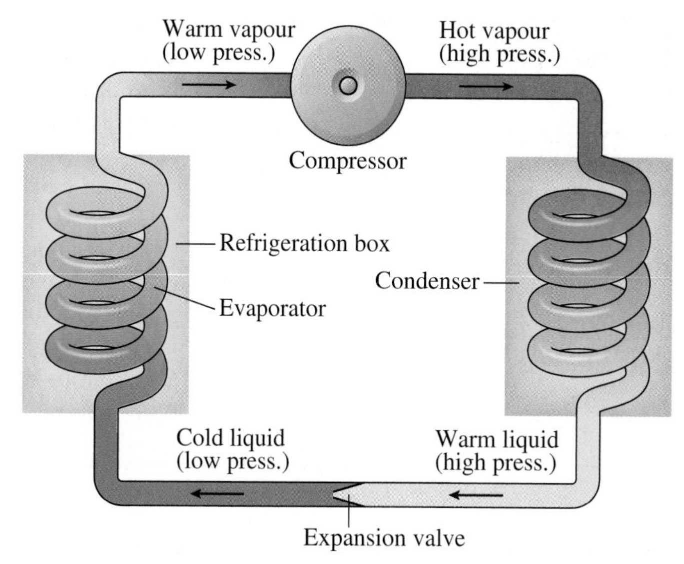

Figure 18-32 The refrigeration cycle.

DID YOU KNOW?

When a refrigerator is operating, the heat given off through the condenser coils is appreciable, and a little attention to these coils can make a refrigerator work more efficiently. If the coils are dusty, the efficiency of heat transfer from the condenser decreases, and so it is useful to clean the coils occasionally. As well, if the heat from the coils cannot easily escape from the area surrounding the refrigerator, some of it will be conducted back into the refrigeration box, requiring an increase in refrigeration cycles. Home refrigerators are sometimes placed in a recess in a wall, hampering heat escape; it is useful to have ample air circulation around the refrigerator, especially at the top, since hot air rises. Ideally, it would be best to allow the heat to vent through the rear wall, heating another room.

Air Conditioners and Heat Pumps

Air conditioners and heat pumps are essentially the same device as a refrigerator. An air conditioner (Figure 18-33) is a refrigerator with the evaporator (where cooling occurs) inside a house or car, and the condenser (where heating occurs) outdoors or perhaps in a window or hole in a wall so that the heat can easily be sent outside. The C.O.P. of an air conditioner has the same definition as that of a refrigerator:

$$\text{air conditioner C.O.P.} = \frac{Q_C}{W} \leq \frac{1}{\frac{T_H}{T_C} - 1} \quad \textbf{(18-30)}$$

Sometimes air conditioners are labelled with an energy efficiency ratio (EER). This is the rate at which heat is removed divided by the electrical power required. Unfortunately, the units are awkward, with the heat removal rate in British thermal units per hour, and the electrical power in watts. The numerical value of the EER in British thermal units per watt·hour is 3.4 times the C.O.P.

A heat pump used to heat a building is just an air conditioner with the locations of the evaporator and condenser reversed. The evaporator is placed outside the building to extract heat from the air, or from the ground or groundwater via a network of pipes laid underground. Since the heat pump is used for heating, the useful energy is the heat Q_H delivered to the interior of the building, and the C.O.P. is defined by C.O.P. = Q_H/W. This can be related (Question 18-92) to the absolute temperature T_H inside the building and the absolute temperature T_C of the external heat reservoir:

$$\text{heat pump C.O.P.} = \frac{Q_H}{W} \leq \frac{1}{1 - \frac{T_C}{T_H}} \quad \textbf{(18-31)}$$

GSPhotography/Shutterstock

Figure 18-33 This air conditioner has the evaporator inside the house, and the compressor and condenser outdoors.

Heat from a Refrigerator

Warm your hands by feeling the condenser coils at the back of a refrigerator when it is running.

Many heat pumps can serve as heaters in the winter and as air conditioners in the summer. In the winter, the working fluid flows in a direction such that the condenser is inside the building. In the summer, a simple flick of a switch causes the compressor to drive the fluid in the opposite direction. Thus, the evaporator coils become the condenser coils, and vice versa; the interior of the building is now cooled.

Suppose that a heat pump is operating as a heater with an exterior temperature of 0°C (i.e., $T_C = 273$ K), and an interior temperature of 20°C (i.e., $T_H = 293$ K). Substitution into Eqn. 18-31 gives a maximum possible C.O.P. of 15. The actual C.O.P. of a typical heat pump working between these temperatures is about 3. This means that for an input of 1 J of electric energy, the pump provides 3 J of heat energy to the building's interior. However, as the exterior temperature drops below about −10°C, the typical C.O.P. drops below 1, and it becomes more efficient simply to use an electric heater.

EXERCISES

18-27 Describe how refrigerators, air conditioners, and heat pumps are similar. How do they differ?

18-28 (a) Consider Eqn. 18-24 for the maximum efficiency of a heat engine. In order to have a large efficiency, should the temperatures T_C and T_H be as close as possible, or should T_H be as large as possible and T_C as small as possible?

(b) Repeat, using Eqn. 18-29 (or 18-30) for the maximum C.O.P. of a refrigerator or air conditioner.

(c) Repeat, using Eqn. 18-31 for a heat pump.

18-29 A refrigerator that has a C.O.P. of 1.3 consumes 1.6 MJ of electrical energy over a particular time period. During this time,

(a) how much heat (in megajoules) is removed from the refrigerator compartment?

(b) how much heat is dissipated by the condenser?

18-30 What is the maximum C.O.P. that is theoretically possible for an air conditioner operating between the interior of a house at 22.0°C and the outside air at 33.0°C?

18-31 A heat pump operating with a C.O.P. of 2.4 consumes 0.75 kW·h of electrical energy during a certain period of time. During this time,

(a) how much heat (in megajoules) is transferred into the high-temperature reservoir?

(b) how much heat (in megajoules) is transferred out of the low-temperature reservoir?

LOOKING BACK...LOOKING AHEAD

The three preceding chapters discussed optics: the geometric optics of mirrors and lenses, wave optics involving interference and diffraction, and optical instruments such as telescopes, cameras, and the human eye. The present chapter has focused on heat and thermodynamics. We introduced the important concepts of temperature, heat, and internal energy, and discussed how to determine the amount of heat required to change the temperature of a substance, or to melt or vaporize it. Expansion and contraction of objects as a result of temperature changes were examined, as well as heat transfer by conduction, convection, and radiation. The greenhouse effect in Earth's atmosphere was explained in terms of emission and absorption of thermal radiation. The first and second laws of thermodynamics were introduced and used to discuss the operation and efficiency of heat engines such as thermal electrical generating stations and automobile engines. Devices that move heat, such as air conditioners and heat pumps, were also described and their coefficients of performance defined.

The next chapter, covering the topic of electric charge and electric field, is the first of a three-chapter sequence focusing on electricity.

CONCEPTS AND SKILLS

Having completed this chapter, you should now be able to do the following:

- Distinguish between the concepts of temperature, internal energy, and heat.
- Describe how various types of thermometers operate.
- Perform calculations involving temperature and the mean square speed and root-mean-square speed of molecules and atoms.
- Solve numerical problems involving heat and changes in temperature and changes of state of substances.
- Use equations involving coefficients of linear and volume expansion to do calculations related to thermal expansion and contraction.
- Describe how heat is transferred by conduction, convection, and radiation.

- Solve numerical problems involving thermal conductivity, heat transfer, and thermal resistance, as well as problems related to intensity of thermal radiation.
- Discuss how emission and absorption of thermal radiation produces the greenhouse effect in Earth's atmosphere.
- Use the first and second laws of thermodynamics to describe the operation of a heat engine, and discuss why its efficiency must be less than 100%.
- Solve numerical problems about heat engines, involving efficiency, heat transferred, useful work done, and temperatures of thermal reservoirs.
- Describe how refrigerators, air conditioners, and heat pumps work, and perform related calculations involving coefficient of performance, heat transferred, etc.

KEY TERMS

You should be able to define or explain each of the following words or phrases:

temperature
thermometer
ideal gas
absolute zero
triple point of water
kelvin (unit)
internal energy
thermal energy
heat
specific heat
specific heat capacity
latent heat of fusion
latent heat of vaporization
coefficient of linear expansion
coefficient of volume expansion
conduction
thermal conductivity
thermal resistance
R-value
convection
thermal radiation
radiation
thermography
thermogram
atmospheric greenhouse effect
greenhouse effect
first law of thermodynamics
second law of thermodynamics
heat engine
entropy
coefficient of performance (C.O.P.)

Chapter Review

MULTIPLE-CHOICE QUESTIONS

18-32 The internal energy of krypton gas at room temperature consists of

(a) only translational kinetic energy
(b) only translational kinetic energy and rotational kinetic energy
(c) only translational kinetic energy, rotational kinetic energy, and vibrational kinetic energy
(d) translational kinetic energy, rotational kinetic energy, vibrational kinetic energy, and electric potential energy

18-33 The specific heat of 2 kg of water is

(a) twice the specific heat of 1 kg of water
(b) one-half the specific heat of 1 kg of water
(c) the same as the specific heat of 1 kg of water

18-34 The specific heat of material A is three times the specific heat of material B, and its mass is twice that of B. Both A and B have the same initial temperature, and equal amounts of heat are applied to both of them. If the change of temperature for A is ΔT, what change of temperature will B experience? (Assume that no change of state occurs.)

(a) ΔT
(b) $3\Delta T/2$
(c) $2\Delta T/3$
(d) $\Delta T/6$
(e) $6\Delta T$

18-35 Two steel bars are heated. One bar is twice the length of the other and has a temperature increase that is three times that of the shorter one. What is the ratio of the change in length of the longer bar to that of the shorter one?

(a) 1
(b) 6
(c) 2
(d) 3
(e) $\frac{1}{2}$

18-36 In Figure 18-34, drawing A shows two metal blocks, one at temperature T_1 and the other at temperature $T_2 < T_1$. Between them is a copper rod through which heat is conducted from one block to the other. Drawing B shows the same metal blocks at the same temperatures as in drawing A. Between them are two copper rods, each of which is half the length of the copper rod in A. The diameter of each copper rod in B is the same as that in A. The rate of heat conduction between the blocks in B is

(a) $\frac{1}{4}$ the rate of heat conduction in A
(b) $\frac{1}{2}$ the rate of heat conduction in A
(c) the same as the rate of heat conduction in A
(d) 2 times the rate of heat conduction in A
(e) 4 times the rate of heat conduction in A

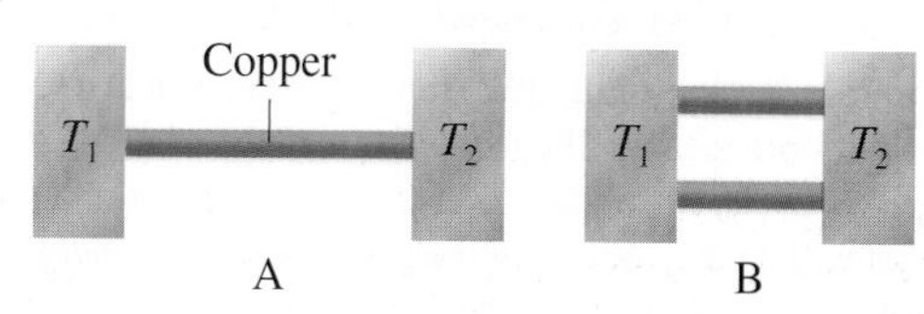

Figure 18-34 Question 18-36.

18-37 During a sunny spring day, the temperature of the outside wall of a house (Figure 18-35) increases from 10.0°C to 20.0°C. What is the ratio of the intensity of radiation emitted by the wall at 20.0°C to that emitted at 10.0°C?

(a) 1.04
(b) 1.15
(c) 2.00
(d) 16.0

Elena Elisseeva/Shutterstock

Figure 18-35 Question 18-37.

18-38 As heat is added to a quantity of air, the internal energy of the air increases by 6.5 J and the air does 1.0 J of work as it expands against a piston. How much heat was added to the air?

(a) 1.0 J
(b) 6.5 J
(c) 5.5 J
(d) 7.5 J

18-39 Ethyl alcohol is cooled to −114°C and solidifies at this temperature. During the solidification, the entropy of the ethyl alcohol

(a) increases
(b) decreases
(c) does not change

18-40 Liquid benzene is heated to 80.4°C and boils at this temperature. During the boiling, the entropy of the benzene

(a) increases
(b) decreases
(c) does not change

18-41 A refrigerator door is left wide open in a well-insulated apartment. The average temperature in the apartment will

(a) increase
(b) decrease
(c) not change

Review Questions and Problems

18-42 Describe how thermal expansion is involved in a thermometer containing (a) a column of liquid, and (b) a bimetallic strip.

18-43 Determine the average translational kinetic energy of water molecules at 19°C.

18-44 What is the internal energy of 1.8 mol of xenon gas at 25°C?

18-45 If heat is applied at a rate of 55 W to 87 g of water initially at 22°C, how long will it take for the water temperature to increase to 54°C?

18-46 Ice having a mass of 65.2 g and a temperature of 0°C is added to a liquid of mass 431 g at a temperature of 35°C. The liquid has a specific heat of 2.72×10^3 $J \cdot kg^{-1} \cdot K^{-1}$. Assuming no heat loss to the surroundings, what is the final temperature of this system?

18-47 **Fermi Question:** An uncooked roasting chicken is removed from a refrigerator and placed in an oven for cooking. Estimate how much heat is required to raise the temperature of the chicken to fully cooked (80°C).

18-48 A metal bar of length 8.024 m undergoes a temperature increase of 123°C. Its length increases to 8.044 m. Determine the metal's average coefficient of linear expansion during the temperature change.

18-49 A brass cube has a temperature of 15°C. At what temperature will its volume have increased by 0.21%?

18-50 Gasoline is being pumped from a tanker truck into an underground storage tank. The temperature of the gasoline in the tanker truck is 24°C. A volume of 49 237 L is pumped from the truck. What will be the volume of the gasoline in the underground tank, once it has cooled to 7°C there?

18-51 The temperature of 2.54 kg of mercury is increased by applying 2.3×10^4 J of heat to it. What is the resulting percentage increase in the volume of the mercury?

18-52 Ice on a frozen lake is 2.3 cm thick (Figure 18-36). The air temperature above the ice is −11°C, and the water temperature below the ice is 1°C. How much heat is transferred through 1.0 km^2 of ice in 2.0 h? (Reference: Table 18-4)

lumen-digital/Shutterstock

Figure 18-36 Question 18-52.

18-53 Ice at 0°C is placed inside a wooden box of thickness 1.5 cm. The ends of the box have dimensions of 32 cm × 32 cm, and the top, bottom, and sides are each 62 cm × 32 cm. The temperature outside the box is 25°C. In 1.5 h, 2.15 kg of the ice melts. What is the thermal conductivity of the wood?

18-54 A large window has dimensions 1.50 m × 2.50 m. It consists of two panes of glass, each of thickness 3.0 mm, separated by a layer of argon gas of thickness 1.4 mm. The air temperatures outside and inside the window are 33°C and 22°C respectively. What is the rate of heat conduction through the window? (Reference: Table 18-4)

18-55 A home has a concrete basement floor 15 m × 15 m and 9.0 cm thick. On a winter day the ground below is at 4.5°C while the basement room is at 20.0°C.

(a) Determine the rate of heat conduction through the floor. (Reference: Table 18-5)

(b) A do-it-yourself enthusiast glues a plywood floor of thickness 1.0 cm onto the concrete. What percentage of the heat loss is eliminated?

(c) What is the temperature at the plywood–concrete interface?

18-56 Figure 18-37 shows an outdoor wooden deck after the first autumn snowfall. Explain why there is no snow on the nail heads.

Ernie McFarland, author

Figure 18-37 Why is there no snow on the nail heads on this deck? (Question 18-56)

18-57 The intensity of radiation emitted by a particular star is 1.2×10^8 W/m^2. What is the surface temperature of the star?

18-58 During the course of a day, a thermal electric power plant generates 2.5×10^{14} J of heat, and discards 1.6×10^{14} J of heat to the cooling water. What is the percentage efficiency of this plant?

18-59 A nuclear power plant is generating 957 MW of electrical power with an efficiency of 34%.

(a) How much heat does it generate in 45 min?

(b) How much heat does it release to cooling water in 45 min?

18-60 If the maximum possible efficiency of a thermal electric generating station is 57.0% and the temperature at which it discards waste heat is 24°C, what is the temperature (in degrees Celsius) of the steam used to generate the electricity?

18-61 **Fermi Question:** Estimate the change in the entropy of water as it freezes at 0°C in the ice cube tray shown in Figure 18-38.

Brian Mueller/Shutterstock

Figure 18-38 Question 18-61.

18-62 The expression for the C.O.P. of a refrigerator or air conditioner is C.O.P. = Q_C/W, whereas for a heat pump the expression is C.O.P. = Q_H/W. Explain why these equations are different.

18-63 A refrigerator is operating at a C.O.P. of 1.08, which is 14% of the maximum C.O.P. theoretically possible. The temperature of the freezer compartment is −11°C. What is the temperature (to the nearest degree Celsius) of the surrounding room?

18-64 A window air conditioner (Figure 18-39) extracts 5.88 MJ of heat per hour from the room being cooled, and discards 9.84 MJ per hour to the outside air. What is the electrical power consumption (in kilowatts) of the air conditioner?

Joseph McCullar/Shutterstock

Figure 18-39 Question 18-64.

18-65 A heat pump is removing heat from several metres below the surface of the ground, where the temperature is 13°C, and delivering heat to the interior of a house at 21°C. What is the maximum C.O.P. that is theoretically possible for the heat pump?

Applying Your Knowledge

18-66 (a) The average temperature on Mars is −5°C. Assuming that Eqn. 18-3 is still approximately valid at this temperature, determine the rms speed for nitrogen gas (N_2) at this temperature.

(b) Would you expect to find nitrogen molecules in the atmosphere of Mars? (Reference: Table 9-5)

18-67 A person who is exercising vigorously generates about 3.0×10^2 W of power over and above the power generated when resting. Assume that all this extra power is removed by evaporation of water (i.e., sweat) from the skin. What mass of water would be evaporated in 15 min? Assume that the initial temperature of the water is 37°C (body temperature).

18-68 An electric current flowing through an electronic circuit device made of 35 mg of silicon delivers 6.2 mW of power to the device. If the cooling system for the circuit fails with the result that there is no heat transfer possible out of the circuit device, what will be the rate of temperature increase (in degrees Celsius per minute) for the device? The specific heat of silicon is 703 $J \cdot kg^{-1} \cdot (°C)^{-1}$.

18-69 Determine the amount of heat delivered to a person's skin if it comes in contact with (a) 22 g of hot water initially at 100.0°C, and (b) 22 g of steam initially at 100.0°C. Assume that the hot water is cooled to skin temperature (34°C), and that the steam condenses and is then cooled to skin temperature (34°C). Which situation gives the more severe burn?

18-70 A pail contains 4.00 kg of water at a temperature of 21°C. A person adds 0.25 kg of ice at 0°C, and then an aluminum frying pan of mass 1.6 kg and specific heat 895 $J \cdot kg^{-1} \cdot K^{-1}$. The resulting temperature of all the contents of the pail is again 21°C. Assuming no heat loss to the pail and the surroundings, what was the temperature of the frying pan before being placed in the water?

18-71 **Fermi Question:** Bridges expand and contract as the temperature changes, and in order to prevent cracks and warping, bridges have expansion joints (Figure 18-40). Estimate how much a 50 m length of bridge built of concrete and steel will expand between an average winter day and an average summer day. Express your answer in centimetres.

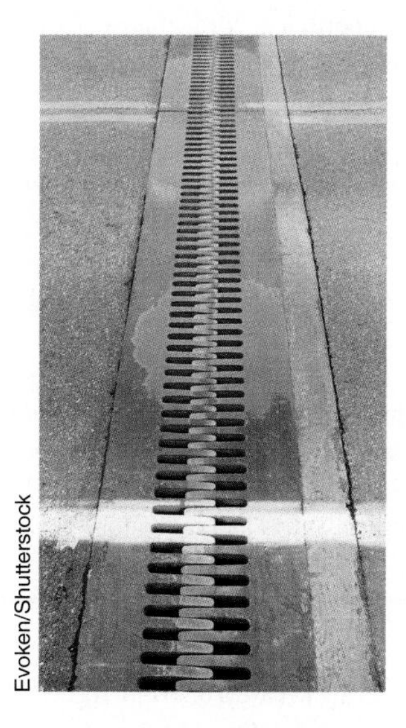

Evoken/Shutterstock

Figure 18-40 Expansion joints on a bridge allow for expansion and contraction with changing temperature (Question 18-71).

18-72 The density of lead at 0°C is 1.1342×10^4 kg/m^3. What is its density at 95°C? Assume that the value of the coefficient of thermal expansion given in Table 18-3 is valid over the range from 0°C to 95°C.

18-73 Figure 18-41 shows a copper washer with a gap in it. The width of the gap when the washer is at 135°C is 4.01 mm. What is the gap width at −25°C?

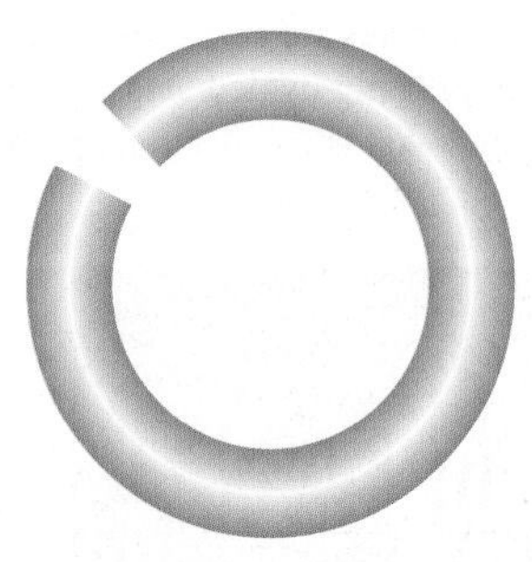

Figure 18-41 A copper washer with a gap (Question 18-73).

18-74 Airplane rivets are cooled to the temperature of solid CO_2 ("dry ice") before being inserted tightly into rivet holes of radius 2.250 mm. At the dry ice temperature of −78.5°C the aluminum rivets have the same radius as the holes in the airplane components, which are not cooled.

(a) What is the radius of each rivet at 22.0°C before being cooled?

(b) Why are the rivets cooled before being inserted into the holes?

18-75 If the density of a substance at temperatures T_1 and T_2 are ρ_1 and ρ_2 respectively, show that $\rho_2 = \dfrac{\rho_1}{1 + \beta\Delta T}$, where $\Delta T = T_2 - T_1$, and β is the coefficient of volume expansion of the substance.

18-76 Coffee having a volume of 272 mL completely fills a cup made of Pyrex glass, which has a coefficient of linear expansion of 3.2×10^{-6} $(°C)^{-1}$. The coffee and cup are at a temperature of 21°C, and are placed in a microwave oven and heated to 89°C. Assuming that the coefficient of volume expansion for coffee is the same as that for water, determine the volume of coffee that overflows from the cup.

DID YOU KNOW?

When a hot liquid is poured into a container of ordinary glass, the hot interior glass surface expands rapidly while the exterior surface remains briefly at room temperature. The glass may not be able to withstand the difference in expansion of the two surfaces and might break. Pyrex glass has a coefficient of linear expansion which is only about one-third that of ordinary glass, and therefore the difference in expansion between the surfaces is reduced and the glass is much less likely to break. Glass kitchen measuring cups and laboratory beakers are usually made of Pyrex glass.

18-77 Two copper plates, each 5.00 cm thick, have a 0.70 mm sheet of glass sandwiched between them. The outer surface of one copper plate is kept in contact with flowing ice water at 0°C, and the outer surface of the other is in contact with steam at 100°C (three significant digits). What power is conducted through a 9.0 cm × 9.0 cm area, and what are the temperatures at the two copper-glass interfaces?

18-78 A steel cooking pot is sitting on a burner on a stove. The bottom of the pot has a thickness of 5.0 mm and a radius of 9.0 cm. There is water at 100.0°C in the pot, and 0.100 kg of water evaporates every 2.0 min. What is the temperature of the lower surface of the pot, which is in contact with the burner?

18-79 The walls and the horizontal ceiling of a small hut are made of wood of thickness 3.8 cm with a thermal conductivity of $0.080\ \text{W·m}^{-1}\text{·K}^{-1}$, and have an interior layer of insulation 2.5 cm thick. The floor area is 3.5 m × 4.0 m, and the walls are 2.5 m high. In order to keep the interior temperature at 18°C when the exterior temperature is −2°C, it is found that a 1.0 kW electric heater must be used. Determine the thermal conductivity of the insulation. Neglect heat loss through the floor.

18-80 A circular steel steam pipe of outside diameter 30.0 cm contains steam at 152°C. It is insulated on the outside with 2.0 mm of insulation having thermal conductivity $0.040\ \text{W·m}^{-1}\text{·K}^{-1}$. The room temperature in which the pipe passes is 22°C. For a 3.0 m length of pipe, how much heat is lost per minute through the insulation? (Since the steel has a very high thermal conductivity compared to the insulation, its insulating effect can be neglected.)

18-81 An experiment is performed to determine the rate of heat loss (P) by *convection* from a vertical window when there is no wind. The experimental result is $P = 3.75A(\Delta T)^{1/4}$, where P is in watts, A is the surface area in square metres, and ΔT is the temperature difference in kelvins between the outside window surface and the outside air. If the coefficient of thermal conductivity of the glass is $0.84\ \text{W·m}^{-1}\text{·K}^{-1}$, what is the temperature difference across a window 3.0 mm thick, when the outside surface is at a temperature of 0°C and the outside air is at −20°C?

18-82 A warm-water radiator has a temperature of 42°C and a surface area of 4.5 m^2. The *net* radiation intensity that it is emitting is 129 W/m^2. What is the temperature of the surroundings in degrees Celsius?

18-83 Rigel is the brightest star in the constellation Orion (Figure 18-42) and is the seventh brightest star in the night sky. Rigel has a surface temperature of 1.1×10^4 K, and radiates energy at a rate of 2.7×10^{32} W.

(a) Determine the radius of Rigel in kilometres.

(b) What is the ratio of the radius of Rigel to the average radius of Earth (6.4×10^3 km)?

Santia/Shutterstock

Figure 18-42 Rigel in the constellation Orion is the bright star at the bottom right of the image (Question 18-83).

18-84 In Table 7-3, the efficiency of hydroelectric plants is given as 95%, whereas the efficiency of steam-electric plants is only 30–40%. Explain why there is such a large difference between these values.

18-85 While travelling 100 km, an automobile burns 8.0 L of gasoline having an energy density of 3.6×10^7 J/L.

(a) How much heat is provided by the gasoline?

(b) If the efficiency of the engine is 19%, how much heat is discarded to the environment?

18-86 **Fermi Question:** An oil-burning thermal electric power plant is producing 1000 MW of electrical power. The energy density of the oil is 3.7×10^7 J/L. Determine an approximate value for the volume of oil burned in one day.

18-87 During a time period of one month, a certain refrigerator extracts 362 MJ of heat from the refrigerator box and discards 453 MJ of heat to the condenser. If electricity costs 12¢/(kW·h), how much does it cost (to the nearest cent) to run the refrigerator for the month?

18-88 Water of mass 582 g and temperature 21°C is poured into ice cube trays and placed in the freezer compartment of a refrigerator that has a C.O.P. of 3.8. How much electrical energy (in megajoules) must be used to cool the water and convert it to ice at 0°C?

18-89 Use the second law of thermodynamics and the concept of entropy to show that, for a refrigerator, $Q_H/Q_C \geq T_H/T_C$. The symbols are as defined in Section 18.6.

18-90 A certain heat pump is able to remove 3.0×10^3 J of energy per second from groundwater. The pump uses 1.0 kW of electrical power. What is its C.O.P.?

18-91 Electrical power of 975 W is being used by a particular heat pump to provide heat to a building by extracting heat from the ground. The heat pump has a C.O.P. of 2.6. In a time period of 8.0 h, how much heat (in megajoules)

(a) is transferred to the building?

(b) is removed from the ground?

18-92 Starting with the definition of the C.O.P. of a heat pump (C.O.P. $= Q_H/W$), use the first and second laws of thermodynamics and the concept of entropy to show that C.O.P. $\leq \dfrac{1}{1 - \dfrac{T_C}{T_H}}$. The symbols are as defined in Section 18.6.

Electric Charge and Electric Field 19

Dave Massey/Shutterstock

Figure 19-1 Pollen grains are attracted to the hairs and legs of a bee by electric forces.

It is difficult to imagine life without electricity. Just stop and think of everything we use that is electrically powered—lights, computers, mobile devices, heating, air conditioning, hair dryers, televisions, toasters, and so on. In the natural world, many organisms hunt for food by detecting electric fields in their environment—bees even use electric charges and forces in their important role as pollinators (Figure 19-1)! In order to understand what electricity is and how it works, we start with a discussion of electric charge and electric field.

CHAPTER **OUTLINE**

19.1 Electric Charge and Atoms

When you comb your hair in dry weather, you often get "fly-away hair." You can achieve the same effect by removing a sweater, and if you shuffle across a carpet, you can sometimes cause a spark to jump when you touch a doorknob. These all are examples of objects becoming electrically charged through friction. Before describing how things become charged, we need to see where charges reside.

nucleus the dense core at the centre of the atom

proton a positively charged particle within the nucleus of the atom

neutron an electrically neutral particle within the nucleus of the atom

electron a negatively charged particle outside the nucleus of the atom

You already know that matter consists of atoms. At the core of each atom is a small dense **nucleus** consisting of **protons** and **neutrons**. The protons are electrically charged, and by convention, they are said to have *positive* charge. A neutron is electrically neutral, and has about the same mass as a proton. Outside the nucleus is a cloud of *negatively* charged **electrons**. The charge on an electron has the same magnitude as that on a proton, and this size of charge is fundamental—*all* charges found in nature are integer multiples of this basic charge.[1] Electrons are much lighter than a nucleus—each electron has a mass of only about 1/2000 of that of a proton or neutron. A typical nucleus is only of the order of 10^{-14} m across, and has a volume of only about 10^{-42} m^3. The distance from the nucleus to the orbital electrons is about 10^{-10} m, which is 10000 times the nuclear diameter. This means that an atom is mainly empty space!

positive ion an atom that has more protons than electrons

negative ion an atom that has more electrons than protons

In an electrically neutral atom, there are equal numbers of negatively charged electrons and positively charged protons, giving a total (or net) charge of zero. Most objects around you right now—this book, your clothes, your pet alligator—are electrically neutral (or nearly so) because the electrons and protons are equal in number. If an atom has more protons than electrons, the net charge is then positive and the atom is referred to as a **positive ion**. If an atom has more electrons than protons, there is an overall negative charge, resulting in a **negative ion**.

EXERCISES

19-1 State the sign of the charge on
(a) an electron
(b) a proton
(c) a nucleus
(d) an ion that has more electrons than protons
(e) an ion that has more protons than electrons

19-2 **Fermi Question:** Imagine that the nucleus of an atom is as large as an apple. Estimate the distance from the nucleus (apple) to the electrons on this scale.

19-3 Protons and neutrons are made of smaller particles called quarks, which have charges that are fractions of the charge on a single proton or electron. For example, the up quark and down quark have charges that are +2/3 and −1/3, respectively, of the proton charge.
(a) A single proton consists of three quarks, a combination of only up and down quarks. What combination must it be?
(b) Which three quarks (only ups and downs) make up a neutron?

19.2 Transfer of Electric Charge

The ancient Greeks knew as early as 600 BCE that when amber (fossilized tree resin) is rubbed with wool, it can pick up bits of straw and dust. If you pass a plastic comb through your hair, or rub a plastic ruler with a cloth, it will attract small bits of paper. What is happening is that electrons are being transferred from one object to the other

[1]As we will discuss in Question 19-3, the elementary particles known as quarks have fractions of this basic charge. However, quarks are never directly observed or found in isolation, but exist only in combinations. The particular combinations that they form, as found in nature, always end up with a total charge that is an integer multiple of *e*, the charge of an electron or proton.

as they rub together. Thus, the objects become electrically charged—one positively and one negatively—and can exert electric forces on each other and on other objects.

Why do electrons move from one object to another when they rub together? Different atoms and molecules have different attractions, or **affinities**, for electrons. When amber is rubbed with wool, the molecules in the amber and wool come into close contact, and since the amber molecules have a greater affinity for electrons, they remove electrons from the wool. The amber becomes negatively charged and the wool positively charged. Only a tiny fraction of the total charge in an object is transferred during rubbing—the negatively charged amber has only a slight excess of electrons over protons, and the positively charged wool has a slight deficiency of electrons. Materials can be placed in a list to show their relative electron affinities. In Table 19-1 you can see that, for example, amber has a greater affinity for electrons than does wool.

Table 19-1

Relative Electron Affinities of Common Materials

	Material
Increasing electron affinity (−) ↑	PVC (vinyl)
	Polyethylene
	Amber
	Rubber
	Copper
	Ebonite
	Paraffin wax
	Silk
	Lead
	Wool
	Glass
↓ Decreasing electron affinity (+)	Acetate
	Animal fur

The word "electron" comes from the Greek "elektron," which means amber (Figure 19-3). Electrons are transferred to amber when it is rubbed with wool (and most other materials). As we can see in Table 19-1, amber is said to have a high electron affinity.

When objects are rubbed together, new charge is not created. *The net electric charge created in any process is zero;* in other words, electric charge is conserved. This is the **law of conservation of electric charge**, discovered in the mid-1700s by several people independently. This law is as fundamental in nature as the laws of conservation of energy and momentum.

Electric Force

Suppose that two glass rods are each charged positively by rubbing with silk, and that one rod is then suspended from a thread. If the other rod is brought nearby, the

DID YOU KNOW?

The terms "positive" and "negative" in reference to electric charge were introduced by Benjamin Franklin, the American scientist, statesman, and inventor. Although he made many contributions to the study of electricity, he is most often remembered for experimenting with a kite in a thunderstorm (Figure 19-2), in order to study the electric nature of lightning. Although everyone has likely heard the story, there is much debate about whether he actually conducted this experiment or merely devised it as a "thought experiment" to explore the electrical nature of lightning from a theoretical perspective.

Fer Gregory/Shutterstock

Figure 19-2 The veracity of one of the most famous stories of Benjamin Franklin, in which he studied electricity by flying a kite in a thunderstorm, is the subject of debate among science historians.

electron affinity ease with which an atom or molecule attracts an electron

law of conservation of electric charge new charge is not created in any process, it is merely transferred from one object to the other or from one region to another

Galyna Andrushko/Shutterstock

Figure 19-3 Amber, fossilized tree resin shown here with a frog preserved inside, was studied by the ancient Greeks because of its high electron affinity.

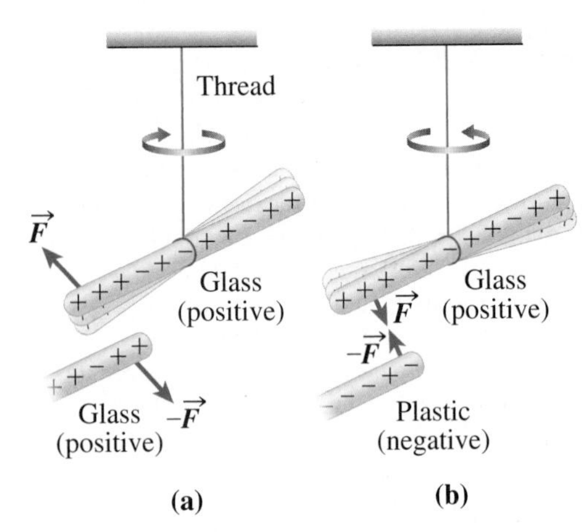

Figure 19-4 **(a)** Like charges repel. **(b)** Unlike charges attract.

suspended rod twists away (Figure 19-4 (a)). This simple experiment shows that *like charges repel*. Rubbing a plastic[2] rod with fur will charge it negatively, and if it is held close to the suspended glass rod, the glass rod twists toward the plastic (Figure 19-4 (b)), showing that *unlike charges attract*.

The Frenchman, Charles Dufay (1698–1739), discovered the nature of the **electric force**, which is an attractive force between unlike charges and a repulsive force between like charges. Dufay found also that there are only two kinds of charge. In fact, the electric force is what holds an atom together: there is an attraction between the negative electrons and the positive nucleus. In addition, the electric force—in various guises called hydrogen bonds, ionic bonds, etc.—causes atoms and ions to form molecules, solids, and liquids. In other words, electric forces are responsible for holding everything together, from the electrons and protons in a single atom to the hydrogen and oxygen atoms in every water molecule in the universe! Relative to the gravitational force, the electric force is extremely strong. The magnitude of the electric force between an electron and proton is about 2×10^{39} times their gravitational attraction. However, as strong as the electric force is, it is no match for the nuclear strong force (defined in Chapter 5) that holds the positively charged protons close together in the nucleus in spite of their electric repulsion.

electric force an attractive force between opposite charges and a repulsive force between like charges

Static Electricity

static electricity situations in which charges are at rest

electric polarization the spatial separation of positive and negative charges within a molecule or material

Static electricity refers to situations in which charges are at rest. You might have noticed that after combing your hair, the comb (which has become charged through rubbing with your hair) can attract something that is electrically *neutral*, such as a piece of paper. How is this possible? In each molecule in the paper, the electrons are repelled by the negative comb, and some of them move to the end of the molecule farthest from the comb (Figure 19-5). Thus, the ends of each molecule have opposite charges, one negative and the other positive. Such a molecule is said to be electrically polarized, or to have **electric polarization**. The negative comb is closer to the positive end of each molecule than to the negative end. The electric force depends on the distance between charges: the closer the charges, the stronger the force (Section 19.4). Thus, the attraction between the positive end of each molecule and the comb is greater than the repulsion between the negative end and the comb, and the paper moves toward the comb.

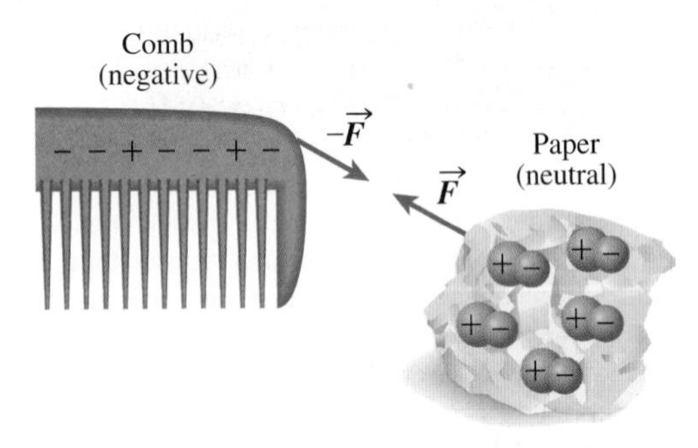

Figure 19-5 Because of electric polarization, a charged comb can attract an electrically neutral piece of paper.

polar molecules molecules such as water with a permanent electric polarization

Once a comb or other object is charged, it does not stay charged forever. Charged ions in the air can remove or add electrons from a charged object; airborne water molecules do this as well. Water molecules (Figure 19-6) have a permanent electric polarization, and hence are called **polar molecules**. The oxygen atom has a net negative charge and the hydrogen atoms a net positive charge. A negatively charged comb can lose electrons to hydrogen atoms in airborne water, and a positively charged glass rod can gain electrons from the oxygen atoms. On dry days, objects can retain a net charge much longer than on damp, humid days.

Static electricity is also responsible for clothes clinging to the body and lint sticking to clothing, especially in cold, dry climates. As well, it causes dust particles to adhere to newly polished tables. One way to overcome these problems is to use a humidifier to add more water molecules to the air so that the static charges can be neutralized more quickly. Another solution is to use an anti-static spray.

Dangerous explosions have occurred when large oil tankers were washed by powerful jets of water. The explosions were caused by spark discharges from the water jets that had become charged as the water came out of the nozzle. Providing a safe pathway for charge movement is essential here to avoid dangerous accumulations.

DID YOU KNOW?

Electric forces play a role in pollination. As bees fly around searching for flowers, they become charged by the air. The slight charge on the bee induces a polarization in the pollen grains which pulls the pollen toward the bee (Figure 19-1), just like bits of paper attracted to a charged comb.

[2]You will not find "plastic" as a single entry in Table 19-1 because different types of plastic (vinyl, acetate, etc., in the table) have different electron affinities.

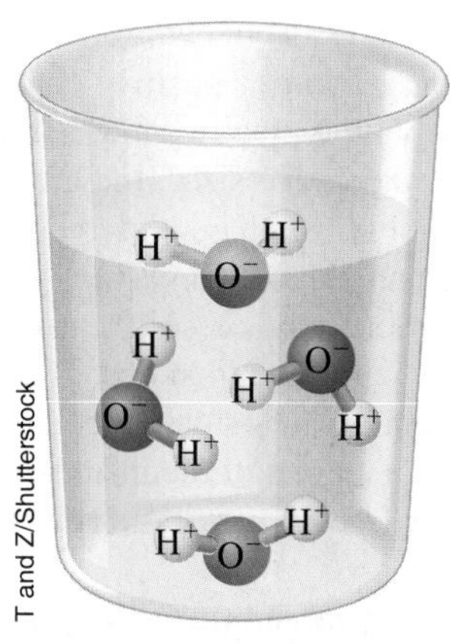

T and Z/Shutterstock

Figure 19-6 Water molecules have a permanent electric polarization and are important in dissipating the excess charge on charged objects.

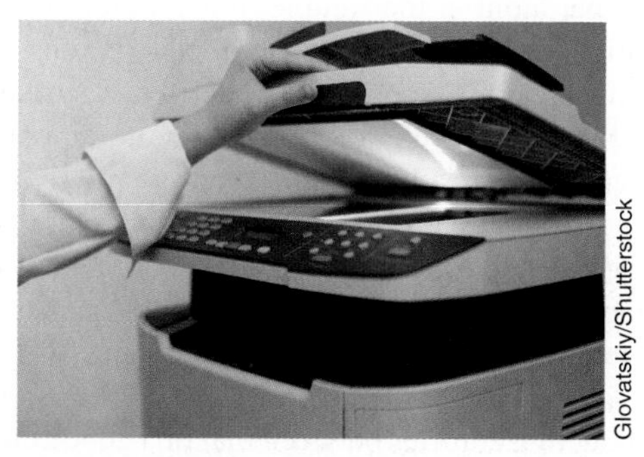

Glovatskiy/Shutterstock

Figure 19-7 Photocopiers and laser printers use principles from electrostatics in order to generate images on paper.

Despite these hazards, static electricity does have its benefits. Principles from static electricity are used to generate images on paper by laser printers and photocopiers (Figure 19-7). In the kitchen, static electricity forces help plastic wrap to cling to food or containers, thus sealing the food from the surrounding air and maintaining its freshness. An electronic air cleaner uses electric charging to purify the air circulating in a home with a central forced-air system. Millions of dust and pollen particles pass through the home's air ducts. The particles that are small enough to pass through a filter screen reach a cell where they are given a net positive charge. They are then attracted to negatively charged collectors, where they accumulate. The collectors must be washed periodically to maintain the air cleaner's efficiency. This approach to air filtration will be discussed in more detail later in the chapter, when we explore the operation of industrial-scale filters in Section 19.6.

DID YOU KNOW?

The stargazer fish (Figure 19-8) is one of several marine creatures that can generate an electric shock to defend itself against predators. The stargazer uses its fins to dig into the sandy seafloor where it lies in wait for prey, as shown here.

James van den Broek/Shutterstock

Figure 19-8 A stargazer fish, lying in wait for prey, is capable of generating an electric shock to defend itself from becoming prey to bigger fish.

EXERCISES

19-4 If paraffin wax is rubbed with wool,
- (a) what sign of charge results on each? (Refer to Table 19-1.)
- (b) what type of particles move during the transfer of charge?
- (c) do the particles named in (b) move onto, or off, the wool?

19-5 If a balloon is rubbed against a wall, it "sticks" there and can remain for several hours (Figure 19-25). Explain.

19-6 A positively charged glass rod attracts another object. What can you conclude about the charge on the other object?

19-7 If a plastic comb or ruler is charged by rubbing and then used to attract small bits of paper (Figure 19-9), often the paper will be attracted to the plastic, touch it, and then immediately fly away. Explain.

GIPhotoStock/Science Photo Library

Figure 19-9 Question 19-7.

19.3 Conductors and Insulators

You probably are aware that metals such as copper are used in electrical wiring, and that these wires are encased in plastic or rubber so that we cannot touch the metal and receive a shock (Figure 19-10). Why are metals good **electrical conductors**; that is, why do they allow charge to move through them easily? Why are other materials, such as plastic, good **electrical insulators**, which do not permit charge to flow easily?

electrical conductors materials, such as metals, in which charge moves easily

electrical insulators materials, such as plastic, in which charge does not move easily

demarcomedia/Shutterstock

Figure 19-10 Household electrical wiring is made typically of copper encased in plastic.

conduction electrons electrons that are very loosely bound to atoms in conducting materials such as metals

semiconductors materials with charge-conducting abilities between that of a good conductor and that of a good insulator

superconductors materials in which there is no resistance to the movement of charge below a certain critical operating temperature

In metals, some of the electrons are very loosely attached to the atoms. In fact, we generally think of metals as a highly ordered arrangement of positive charges surrounded by a sea or cloud of delocalized electrons, with typically one free electron per atom in the sample in most metals. Given their relative sizes, the thermal motion of the much heavier positive ions is negligible compared to the fast, random thermal motion of these free electrons. When an electric force is applied, the resulting acceleration gives rise to an overall movement of these **conduction electrons**. In insulators, the electrons are tightly bound to atoms and molecules, and it is much more difficult to get charge to move. If one end of an insulator acquires a charge, say through friction, this charge tends to stay at that end of the insulator. In contrast, a charge acquired at one end of a conductor is free to move, and typically becomes distributed all along the conductor.

There are also **semiconductors** such as silicon and germanium, with charge-conducting abilities between that of a good conductor and that of a good insulator. In the electronics industry, semiconductors are usually "doped" with selected impurity elements to acquire specific electrical properties; such materials are widely used in the manufacture of electronic devices. **Superconductors** are materials having absolutely no resistance to the movement of charge. For most superconducting materials, zero resistance occurs only at extremely low temperatures, making practical, everyday applications of these fascinating materials beyond our reach at present.

Charging by Contact (Conduction) and by Induction

charging by contact (charging by conduction) the transfer of charge from one object to another by direct contact between the objects

Since electrons can move fairly easily in conductors, it is possible to charge a conductor in a variety of ways. If a neutral conducting sphere on an insulated stand is touched with a positively charged object, such as a glass rod (Figure 19-11), some electrons will be attracted from the sphere to the positive rod. If the rod is then removed, the sphere will have a net positive charge; the positive charge on the rod will have been reduced somewhat. This method of charging a conductor is called **charging by contact** or **charging by conduction**. This technique works also for insulators to a small extent, but since electrons flow poorly in insulators, there is little transfer of charge. To charge a conductor *negatively* by contact (conduction), we touch it with a

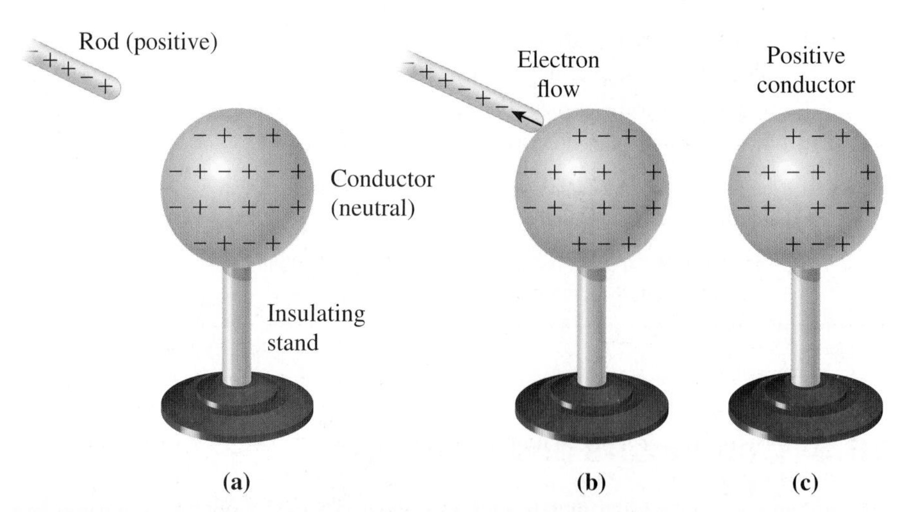

Figure 19-11 Charging by contact (conduction). **(a)** A charged rod is brought near a neutral conductor. **(b)** When the rod is in contact with the conductor, some electrons are attracted to the positive rod from the sphere. **(c)** When the rod is removed, the conductor has net positive charge as a result of this transfer of electrons to the rod.

negatively charged object. Some electrons are repelled from the negative object to the conductor, which acquires a net negative charge.

Another method of charging a conductor does not involve direct contact with a charged object. If a negatively charged plastic rod is held near a conductor, some electrons in the conductor move (still in the conductor) until they are far from the rod (Figure 19-12 (a)). Thus, the conductor has a net positive charge at one side, and a net negative charge at the other. The conductor is then connected by a wire to the ground, that is, to Earth (Figure 19-12 (b)), or simply is touched by a person, thus allowing charge flow through the person to ground. Earth is a conductor with a large amount of charge; when a charged object is connected to Earth through a conducting wire, large amounts of charge can move readily from the object to Earth or from Earth to the object. The process of conducting a charge to or from the ground is called **grounding**. The electrical connection to the ground is indicated on diagrams by the symbol shown in Figure 19-12 (b). During grounding, the negative rod repels electrons from the sphere to the ground. If the grounding connection and *then* the rod are removed, the conductor is left with a net positive charge (Figure 19-12 (c)). This method is **charging by induction**; a net charge is *induced* in the conductor by the charged rod. What about the increased number of electrons in Earth as a result of this procedure? Adding electrons to Earth can be compared to pouring water from a glass into the ocean—it makes a big difference to the glass, but little to the ocean.

grounding transferring excess charge from an object to Earth or from Earth to an object

charging by induction an excess charge is left on a conductor through the transfer of charge between the conductor and Earth in the presence of another charged object

leaf electroscope a sensitive instrument used to detect charge and to determine its sign

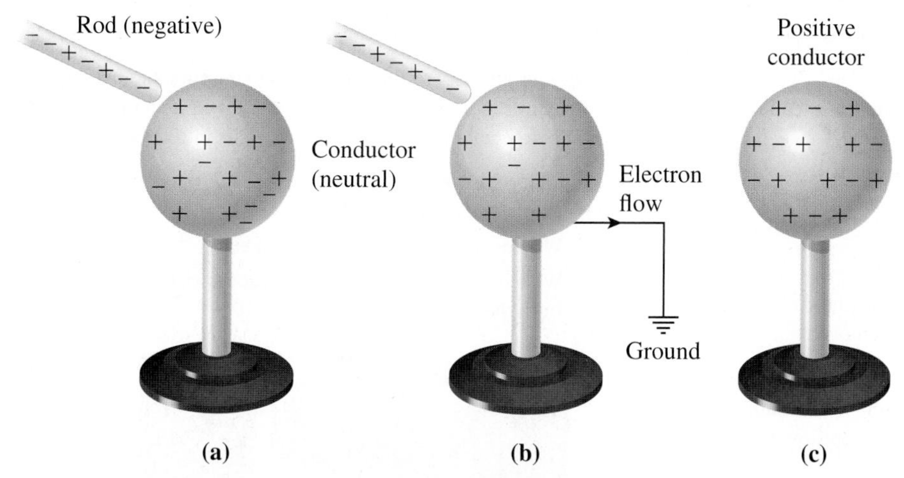

Figure 19-12 Charging by induction. **(a)** A charged rod is brought near a neutral conductor; the electrons in the conductor redistribute such that there is a net negative charge away from the rod. **(b)** When the conductor is then connected to ground, there is a path by which the repelled electrons can move further from the rod to ground. **(c)** When the grounding wire is removed and then the rod is removed, the conductor is left with a net positive charge as a result of this transfer of electrons to ground.

DID YOU KNOW?

A gecko is able to climb walls because each foot is covered in about half a million hairs (Figure 19-13). When these hairs are pressed against the wall, an electrical interaction results in a strong electrical attraction—strong enough to overcome gravity!

Figure 19-13 Each gecko foot is covered with about half a million hairs, which interact with the wall through electric forces.

A **leaf electroscope** (Figure 19-14) is a sensitive instrument used to detect charge and to determine its sign. It consists of two thin metal leaves attached to the bottom of a metal strip supported in an insulating stand. On top of the strip is a metal cap. One or both of the leaves are movable. The cap of the electroscope can be charged with a charge of known sign by conduction or induction. The charge distributes itself along the conducting strip and the metal leaves, which repel each other (Figure 19-14)). The separation of the leaves is an indication of the magnitude of the charge.

An electroscope with a charge of known sign can be used to determine the sign of an unknown charge. For example, suppose that an electroscope has an excess of electrons,

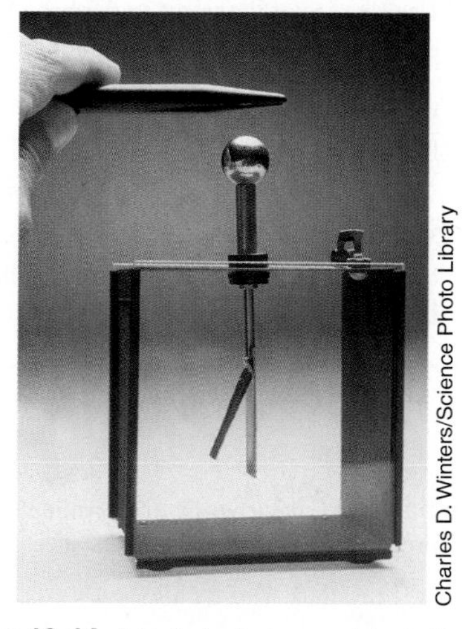

Figure 19-14 A leaf electroscope detects the presence of charge on nearby objects.

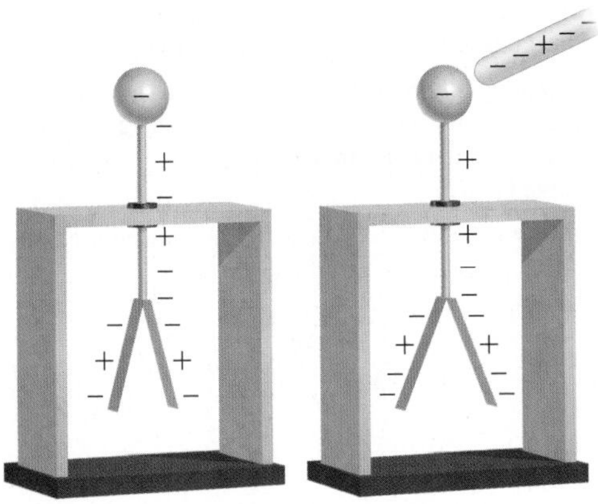

(a) (b)

Figure 19-15 **(a)** A negatively charged electroscope. **(b)** A negatively charged rod nearby pushes more electrons into the leaves, thus increasing their separation.

and thus a net negative charge (Figure 19-15 (a)). If an unknown charge is brought close to the electroscope cap and the leaves move farther apart (Figure 19-15 (b)), there must be more electrons moving into the leaves, thus increasing their mutual repulsion. This could happen only if the unknown charge is negative and is pushing electrons from the cap to the leaves. If the leaves move closer together, the unknown charge is positive. In either case, the amount of movement of the leaves is an indication of the magnitude of the unknown charge.

DID YOU KNOW?

If you are ever outside on a stormy day and your hair starts to stand on end, your life could be in imminent danger. Thunder clouds contain large amounts of charge, which get mirrored by opposite charges concentrating in the ground below. If these charges build up in the ground beneath you, they can travel upward through you, causing your hair to behave like the leaves of the electroscope in Figure 19-14. You need to get to shelter immediately before a massive discharge takes place in the form of a lightning strike!

EXERCISES

19-8 Silver is a slightly better electrical conductor than copper, but copper is more widely used. Why?

19-9 In charging a conductor by contact, does the conductor acquire a charge of the same sign as the other charged object?

19-10 Repeat the previous question for charging by induction.

19-11 Using a series of diagrams, describe how to give an insulated conductor a negative charge by induction.

19-12 In charging by induction, what would happen if the rod were removed before the grounding connection was removed?

19-13 When a conductor is being charged by conduction, electrons flow briefly until the conductor and other object reach some equilibrium charge concentration. Why do the electrons stop flowing?

19-14 Use a series of diagrams to describe how you could charge a leaf electroscope positively by (a) conduction and (b) induction.

19-15 A rod with an unknown charge is brought close to the cap of a positively charged electroscope and the leaves move farther apart. What is the sign of the unknown charge? Explain, using diagrams.

19-16 What happens if a negatively charged object is brought close to the cap of a neutral electroscope? What happens when the object is removed again? Explain, using diagrams.

19-17 Repeat the previous question for a positively charged object.

19-18 Use your answers to the previous two questions to explain why an electroscope is more useful if it is given a charge of known sign before an unknown charged object is introduced.

19-19 Children often get interesting hairstyles at the bottom of plastic playground slides (Figure 19-16).

(a) Why does the girl's hair stand on end?

(b) What happens when her dad reaches out to lift her down from the slide?

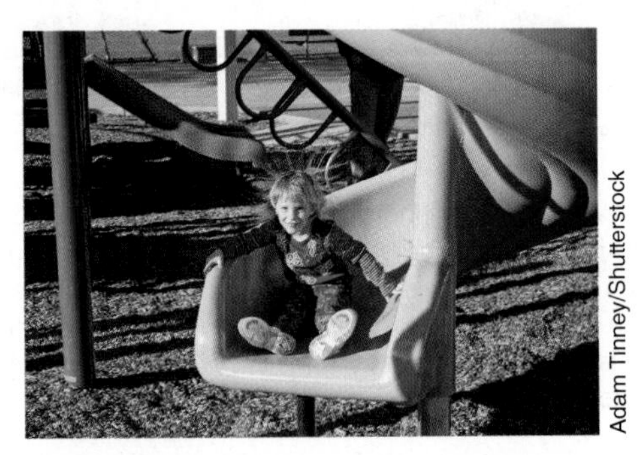

Adam Tinney/Shutterstock

Figure 19-16 Getting charged up in the playground (Question 19-19).

torsion balance device used to determine how electric force depends on charge and distance

Coulomb's law the force between two charges is proportional to the product of the charges and inversely proportional to the square of the distance between them

19.4 Coulomb's Law

Electrostatics experiments are tricky to perform. Charges leak from objects, a slight rubbing can charge something that should remain neutral, etc. As a result, the experimental foundations of static electricity proceeded more slowly than those of mechanics. In 1687, Newton published his law of universal gravitation, but it was not until 1788 that the corresponding force law between charges was established by Charles Augustin Coulomb (Figure 19-17) of France. Coulomb used a device called a **torsion balance** (Figure 19-18) to determine how electric force depends on charge and distance. Coulomb's torsion balance was similar to that used by Cavendish in 1798 for his determination of the universal gravitation constant (Section 9.1).

Coulomb's law of force between charges states that for charged objects that are much smaller than the distance between them, the magnitude of the electric force is proportional to the (absolute value of the) product of the charges, and inversely proportional to the square of the distance between them.

Figure 19-17 Charles Augustin Coulomb (1736–1806). After a career as a military engineer, he retired at age 53 to pursue scientific research. Using a torsion balance (which he invented), he discovered the force law between electric charges.

Mathematically, we write Coulomb's law as

$$F = \frac{k|q_1 q_2|}{r^2} \tag{19-1}$$

where F is the magnitude of the force, q_1 and q_2 are the two charges, r is the distance between the charges (Figure 19-19), and k is a constant of proportionality.

Note the similarity to the universal gravitation law, in which the force is proportional to the product of masses, and is inversely proportional to the square of the separation distance:

$$F \propto \frac{m_1 m_2}{r^2}$$

However, whereas gravity only attracts, the electric force can either attract or repel.

The SI unit of charge is the coulomb (C). For practical reasons related to accuracy of measurements, the coulomb is defined in terms of electric current; we will return to this definition in Chapter 20 once we have defined current. A net charge of 1 C is quite large; the net charge produced on an object by rubbing is typically of the order of a microcoulomb ($\mu C = 10^{-6}$ C) or less; the charge that flows in a lightning bolt is about 25 C; your body contains about 4×10^9 C of positive charge, and the same amount of negative charge.

The absolute value of the charge on an electron (or proton) is a fundamental constant represented by "e," and has the value

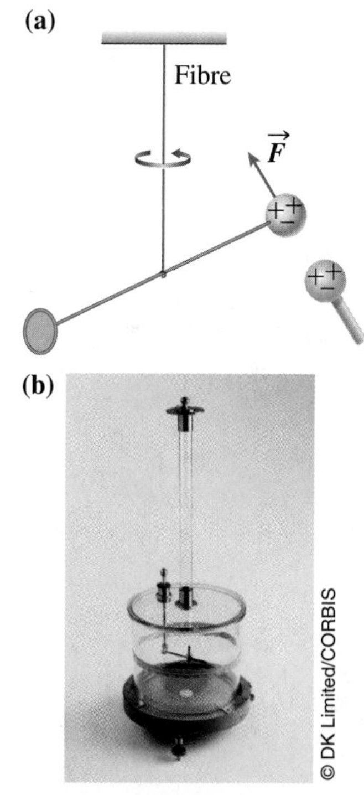

Figure 19-18 **(a)** Schematic diagram of Coulomb's apparatus. The forces between charges can be determined by measuring the twist of the fibre. **(b)** A torsion balance, such as the one designed and used by Coulomb.

$$e = 1.60 \times 10^{-19}\text{ C} \tag{19-2}$$

In SI units, the proportionality constant in Coulomb's law is

$$k = 8.99 \times 10^9\text{ N}\cdot\text{m}^2/\text{C}^2 \tag{19-3}$$

for charges separated by vacuum.

For charges separated by other materials, the electric force is reduced from its vacuum value because molecules in the materials "shield" the charges. For example, for charges in water, the electric force is reduced by a factor of 80 from the vacuum value. For charges in air (a poor shield), the value of k for vacuum is usually used; this introduces an error of only about one part in two thousand.

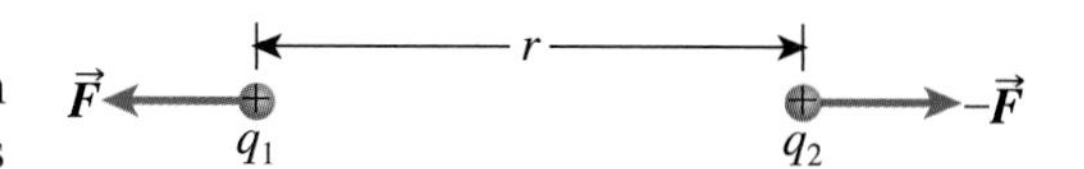

Figure 19-19 Coulomb's law. The directions of the forces shown are associated with interactions between like charges.

SAMPLE PROBLEM 19-1

What is the electric force (magnitude and direction) on the electron in a hydrogen atom? The electron is, on average, a distance of 5.3×10^{-11} m from the nucleus in hydrogen.

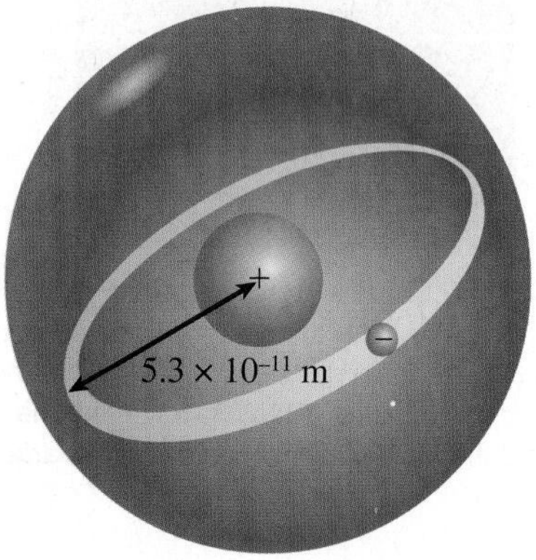

Solution

A hydrogen nucleus (Figure 19-20) is a proton, having a charge $q_1 = e = 1.60 \times 10^{-19}$ C, which also equals the absolute value of the charge on the electron (q_2). The magnitude of the force on the electron is given by Eqn. 19-1:

$$F = \frac{k|q_1 q_2|}{r^2}$$

Substituting values

$$F = \frac{(8.99 \times 10^9 \text{ N}\cdot\text{m}^2/\text{C}^2)|(1.60 \times 10^{-19}\text{ C})(-1.60 \times 10^{-19}\text{ C})|}{(5.3 \times 10^{-11}\text{ m})^2}$$

$$= 8.2 \times 10^{-8}\text{ N}$$

Thus, the magnitude of the force is 8.2×10^{-8} N. Since the nucleus is positive and the electron is negative, the force on the electron is attractive, that is, toward the nucleus.

5.3 x 10⁻¹¹ m

Figure 19-20 Sample Problem 19-1.

SAMPLE PROBLEM 19-2

Two charges separated by a distance d exert an electric force on each other. If the same charges are separated by a distance of $d/4$, by what factor will the magnitude of the electric force increase or decrease?

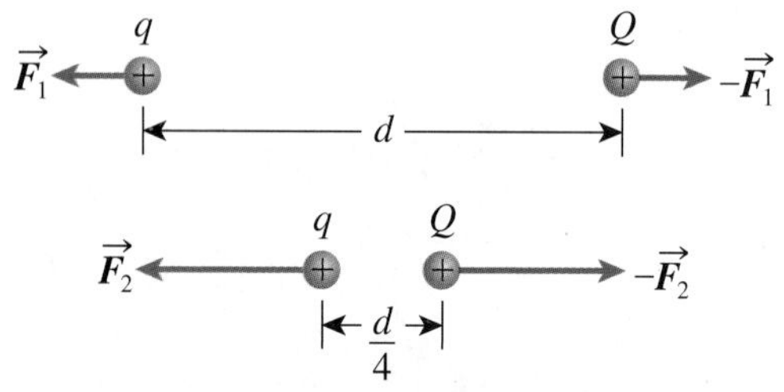

Solution

Let the two charges be q and Q, and the magnitude of the force be F_1 at separation d, and F_2 at separation $d/4$ (Figure 19-21). We are asked to find F_2 relative to F_1. When the separation is d, we have

$$F_1 = \frac{k|qQ|}{d^2} \qquad \text{Eqn. [1]}$$

When the separation is $d/4$,

$$F_2 = \frac{k|qQ|}{(d/4)^2} \qquad \text{Eqn. [2]}$$

Dividing Eqn. [2] by Eqn. [1]

$$\frac{F_2}{F_1} = \frac{k|qQ|}{(d/4)^2} \div \frac{k|qQ|}{d^2} = \frac{k|qQ|}{d^2/16} \times \frac{d^2}{k|qQ|} = 16$$

Hence, as the separation decreases from d to $d/4$, the force increases by a factor of 16.

$\vec{F}_1$ q Q $-\vec{F}_1$ d

$\vec{F}_2$ q Q $-\vec{F}_2$ $\frac{d}{4}$

Figure 19-21 Sample Problem 19-2.

SAMPLE PROBLEM 19-3

An alpha (α) particle (i.e., a helium nucleus, having charge $+2e$) is passing near a hydrogen nucleus (charge $+e$) and a carbon nucleus (charge $+6e$), as shown in Figure 19-22 (a). (Note the 77° and 36° angles.) The distances from the α particle to the hydrogen nucleus and carbon nucleus are 9.0×10^{-11} m and 1.5×10^{-10} m, respectively. What are the magnitude and direction of the resultant electric force on the α particle?

Solution

The resultant force on the α particle is the *vector* sum of the forces due to the other two nuclei. These forces are repulsive, since all the charges are positive. In Figure 19-22 (b) we label these forces $\vec{F}_C$ and $\vec{F}_H$ (due to the carbon and hydrogen nuclei respectively). We choose a convenient *x*-*y* coordinate system, and write *x*- and *y*-components of the forces in terms of given angles. The magnitude F_C is

$$F_C = \frac{k|q_1 q_2|}{r^2}$$

Since F_C is the magnitude of the force exerted on the α particle by the carbon nucleus, we use

$$q_1 = +2e = 3.20 \times 10^{-19}\ \text{C}\ (\alpha\ \text{particle})$$

$$q_2 = +6e = 9.60 \times 10^{-19}\ \text{C}\ (\text{carbon nucleus})$$

$$r = 1.5 \times 10^{-10}\ \text{m}$$

Thus,

$$F_C = \frac{(8.99 \times 10^9\ \text{N}\cdot\text{m}^2/\text{C}^2)(3.2 \times 10^{-19}\ \text{C})(9.60 \times 10^{-19}\ \text{C})}{(1.5 \times 10^{-10}\ \text{m})^2} = 1.23 \times 10^{-7}\ \text{N}$$

Repeating this calculation with $q_2 = +e$ (hydrogen nucleus) and $r = 9.0 \times 10^{-11}$ m gives $F_H = 5.68 \times 10^{-8}$ N. From Figure 19-22 (b), the *x*-component of the resultant force is

$$\Sigma F_x = F_H \cos 77° - F_C \cos 36°$$

$$= (5.68 \times 10^{-8}\ \text{N}) \cos 77° - (1.23 \times 10^{-7}\ \text{N}) \cos 36°$$

$$= -8.65 \times 10^{-8}\ \text{N}$$

Similarly, the *y*-component of the resultant force is

$$\Sigma F_y = F_H \sin 77° + F_C \sin 36° = 1.28 \times 10^{-7}\ \text{N}$$

These resultant *x*- and *y*-components are added in Figure 19-22 (c). The magnitude of the resultant force ($\vec{F}_R$) is given by

$$F_R = \sqrt{(-8.65 \times 10^{-8}\ \text{N})^2 + (1.28 \times 10^{-7}\ \text{N})^2} = 1.5 \times 10^{-7}\ \text{N}$$

The direction of $\vec{F}_R$ can be specified by angle θ in Figure 19-22 (c):

$$\tan\theta = \frac{1.28 \times 10^{-7}\ \text{N}}{8.66 \times 10^{-8}\ \text{N}} \quad \text{which gives } \theta = 56°$$

Thus, the resultant force on the α particle due to the presence of the other two charges is 1.5×10^{-7} N at an angle $\theta = 56°$ as shown in Figure 19-22 (c).

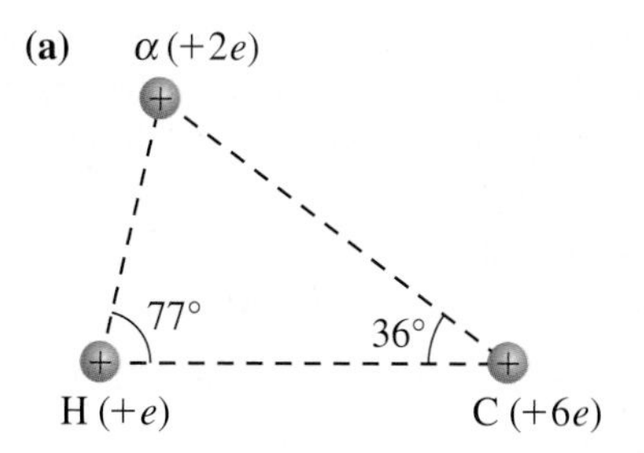

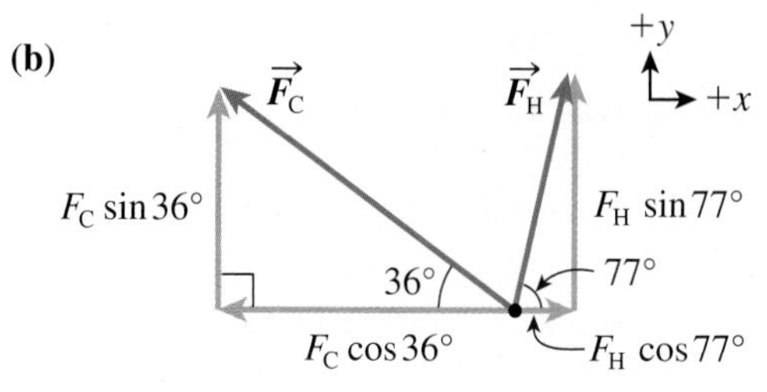

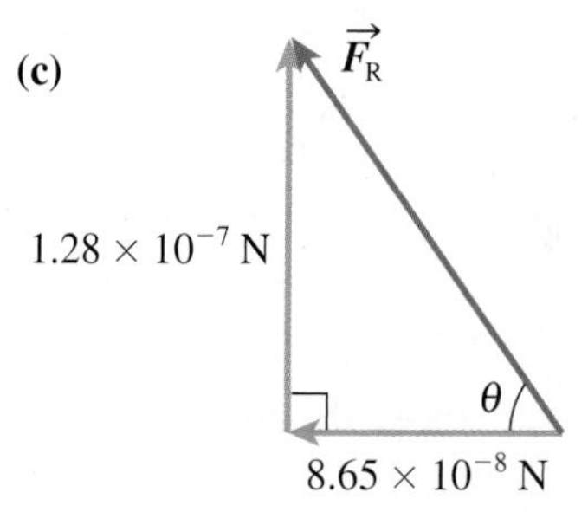

Figure 19-22 Sample Problem 19-3. **(a)** Positions of the charges. **(b)** Forces on α particle. **(c)** The resultant force.

TRY THIS!

Repulsive Balloons

With two balloons and some string, you can demonstrate the repulsive force between like charges. Suspend two balloons, equally inflated, from the top of a door frame by their strings and some tape. They should be about 2 or 3 cm apart and hanging at approximately the same height. Now charge each balloon (separately) by rubbing it with a woollen sweater. What happens? With a protractor app on a smart phone, estimate the angle that the strings make with the vertical and estimate the charge on each balloon (see Question 19-31). How does this compare with your answer to Question 19-32?

EXERCISES

19-20 An electron is 1.5×10^{-10} m from an oxygen nucleus (charge $+1.3 \times 10^{-18}$ C). What are the magnitude and direction of the force on (a) the electron? (b) the oxygen nucleus?

19-21 If two charged objects are moved apart so that the separation distance increases by a factor of 3, does the force between them increase or decrease? By what factor?

19-22 If two charges move closer together so that the force between them increases by a factor of 2.0, by what factor has their separation distance decreased? (Such factors are usually quoted as numbers greater than 1.)

19-23 Sketch the shape of a graph of the magnitude of the electric force versus separation distance between two point charges.

19-24 How many electrons make up a charge of magnitude 1.00 C?

19-25 At what distance would the force between two protons have a magnitude of 1.00 N?

19-26 What is the ratio of the magnitudes of the electric and gravitational forces between two electrons? (Refer to inside back cover for relevant data.)

19-27 Three charges are in a straight line. Charge #1 ($+4.5\ \mu C$) is separated by 4.3 cm from charge #2 ($-3.7\ \mu C$), which is between #1 and #3. Charge #2 is 3.7 cm from #3 ($-4.1\ \mu C$). What are the magnitude and direction of the force on (a) charge #2? (b) charge #1?

19-28 Three oxygen nuclei (charge on each $= +8e$) are arranged in an equilateral triangle with sides that are 1.5×10^{-10} m in length. Calculate the magnitude and direction of the resultant force on each nucleus.

19-29 Four equal charges of the same sign are placed at the corners of a square. What is the direction of the resultant force on each charge?

19-30 Figure 19-23 shows the positions of two α particles (charge on each is $+2e$) and an electron. What are the magnitude and direction of:

(a) the resultant force on the electron? (**Hint:** Choose the $+x$-axis along the line from the electron to the right-hand α particle.)

(b) the acceleration of the electron? (See inside back cover for relevant data.)

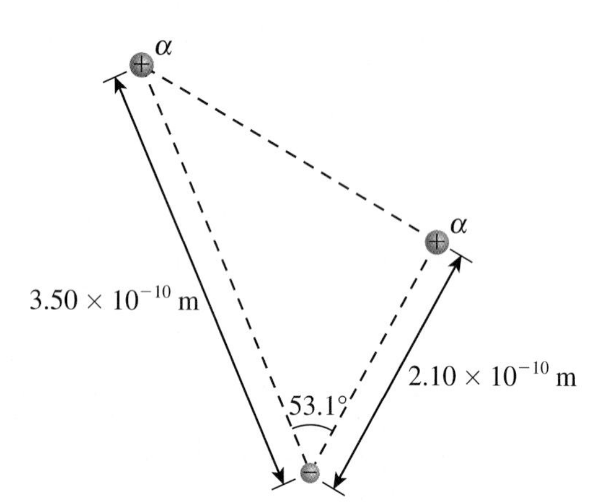

Figure 19-23 Question 19-30.

19-31 Two small spheres of equal mass $m = 3.2 \times 10^{-3}$ kg are suspended from a common point by threads each of length 31 cm (Figure 19-24). Equal positive charges are placed on the spheres, which move to equilibrium positions as shown.

(a) Draw a free-body diagram for either sphere at equilibrium.

(b) What is the charge on each sphere?

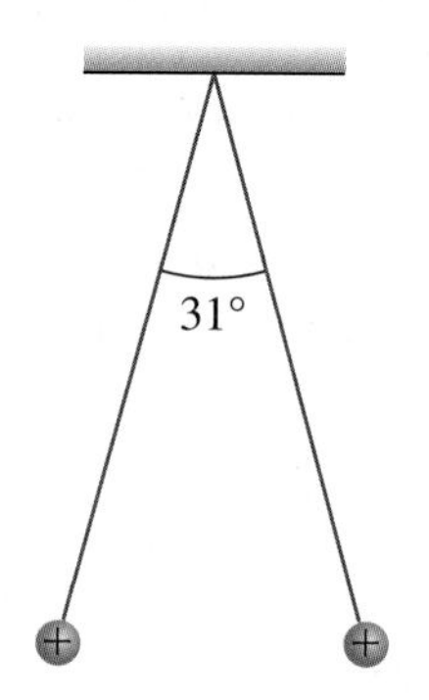

Figure 19-24 Question 19-31.

19-32 **Fermi Question:** Estimate the amount of charge required to keep a balloon stuck to a wall (Figure 19-25).

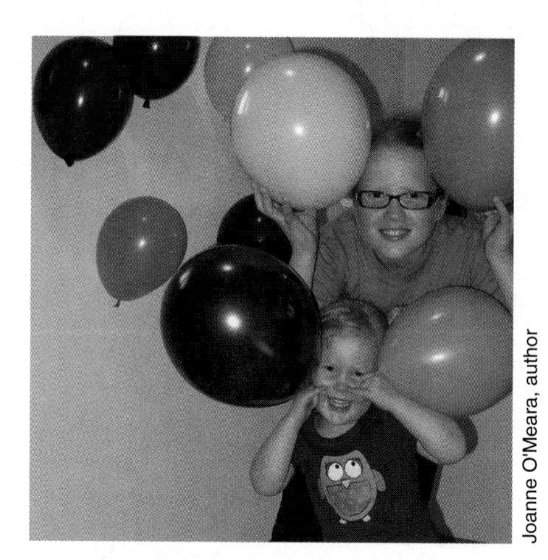

Figure 19-25 Question 19-32.

19.5 Electric Field

It is easy to understand how you can exert a force on this book when you are lifting it—you are touching it. But how does an electron in a hydrogen atom "know" that there is a proton in the nucleus exerting a force on it? Or how does Earth "know" that the Sun is attracting it? This difficulty with forces acting at a distance, in contrast with contact forces, has led to the concept of fields, as discussed in Chapter 9. Physicists commonly refer to gravitational fields (Chapter 9), electric fields, and magnetic fields.[3]

[3]Electric fields and magnetic fields were discussed in terms of their role in electromagnetic radiation in Section 16.6.

As we discussed with masses and gravitational fields, any charge has an associated electric field in the space around it, and when another charge encounters the field, this second charge experiences a force. The force is thus a two-step phenomenon: first the field is created, and then there is a force on any charge in the field. Electric field is more than just an abstraction; in advanced physics courses, students learn that the electric field can carry energy and momentum in the form of electromagnetic waves. The concept of electric (and magnetic) fields has led to the development of radio, television, radar, and a wide variety of other devices such as electric motors, generators, etc.

Imagine that you have charged a comb by passing it through your hair. An electric field exists in the space around the comb, and any charge (say q') in the vicinity will experience a force, $\vec{F}$. The **electric field**, $\vec{E}$, is defined as the ratio of this force to the charge:

electric field ($\vec{E}$) electric force per unit charge; unit N/C

$$\vec{E} = \frac{\vec{F}}{q'} \tag{19-4}$$

The SI unit of electric field is newton/coulomb (N/C). Notice that the definition of electric field is similar to that of gravitational field (Chapter 9), which is the ratio of gravitational force to mass.

The charge q' is not part of the charge on the comb creating the field; it is a charge that we imagine to be in the field created by the comb; this is often referred to as the **visiting charge** or **test charge**. We assume that this charge does not alter the charge distribution on the comb significantly. We only introduce this visiting charge as a means of determining the effect that the charged comb has on its surroundings.

visiting charge (test charge) a charge that is brought into a region of electric field in which it experiences an electric force

The electric field in a region is a vector quantity; its direction is the same as that of the force exerted on a positive charge. *Therefore, the electric field due to a single positive charge points away from the charge; for a single negative charge, the electric field points toward the charge.*

If the field due to a charge distribution is known, the force on a visiting charge q' located in the field is given by rearranging Eqn. 19-4 to give

$$\vec{F} = q'\vec{E} \tag{19-5}$$

Electric Field Due to a Single Charge

The magnitude of the electric field at a distance r away from a point charge q is a straightforward derivation. We start with the magnitude of the force on a charge q', using Eqn. 19-1

$$F = \frac{k|qq'|}{r^2}$$

where r is the distance separating the charges.[4] To determine the magnitude of the field due to q, we simply divide F by q', which gives

$$E = \frac{k|q|}{r^2} \tag{19-6}$$

[4]We assume here that the charges q and q' are distributed over regions of space that are small compared to the distance r (i.e., that the charges can be considered to be "point" charges), or that the charge distributions have spherical symmetry, as discussed in Chapter 9 for gravitational forces and fields.

? TACKLING MISCONCEPTIONS

The Electric Field Is Not Influenced by the Amount of Visiting Charge at That Location!

Be careful! The definition of electric field as described in Eqn. 19-4 often leads students to think that the visiting charge, q', somehow plays a role in the magnitude of the electric field at that location. Just as we discussed with masses and gravitational forces and fields in Chapter 9, there need to be two (or more) charges present in order for there to be an electric force acting. However, *one* single lonely point charge is all that is needed for there to be an associated electric field everywhere in space. The *electric force* describes the interaction between two or more specific charges, whereas the *electric field* merely describes the effect that one charge has on its surroundings in general.

DID YOU KNOW?

Sharks, as well as a variety of other creatures, can detect weak electric fields created by their prey (Figure 19-26); sharks can sense fields with magnitudes as low as 1×10^{-6} N/C! The duck-billed platypus can find food such as shrimp in extremely murky water because it can sense the electric field generated by tail flicks of the shrimp. In experiments investigating this phenomenon, a platypus attacked a small 1.5 V battery as if it were food.

Figure 19-26 A shark hunting for food can detect the electric fields generated by its prey.

Notice that the magnitude of the electric field due to a single point charge has the same inverse-square dependence on distance as the magnitude of the electric force. Notice also that the form of Eqn. 19-6 is very similar to Eqn. 9-4, the magnitude of the gravitational field (g) of mass M at a distance r from the mass $\left(g = \frac{GM}{r^2}\right)$.

SAMPLE PROBLEM 19-4

What is the electric field (magnitude and direction) at a distance of 1.5×10^{-10} m from a proton?

Solution

Since a proton has a positive charge ($+e$), the electric field is directed away from the proton. Using Eqn. 19-6, the magnitude of the field is

$$E = \frac{k|q|}{r^2} = \frac{(8.99 \times 10^9\ \text{N}\cdot\text{m}^2/\text{C}^2)(1.60 \times 10^{-19}\ \text{C})}{(1.5 \times 10^{-10}\ \text{m})^2} = 6.4 \times 10^{10}\ \text{N/C}$$

Thus, the electric field is 6.4×10^{10} N/C at this distance from the charge, directed away from the proton.

Electric Field Due to More than One Point Charge

The resultant electric field due to many charges is just the vector sum of the individual fields. For example, Figure 19-27 (a) shows the separate fields due to two point charges at a particular location in space (denoted by *), and Figure 19-27 (b) shows the resultant electric field at that location. As in any vector sum, each component of the resultant vector (Figure 19-27 (c)) is simply the sum of the corresponding components of the individual vectors:

$$E_{Rx} = E_{1x} + E_{2x} \text{ and } E_{Ry} = E_{1y} + E_{2y}$$

Adding the components of electric field vectors to give the resultant electric field is just the same as adding the components of any type of vector to give the resultant.

TRY THIS!

Electric Field Hockey

Search for "Electric Field Hockey PhET Simulation" online and play around with attractive and repulsive forces between point charges. Make sure that you try something harder than the practice screen that first appears; trying to get the charged puck in the net around obstacles using the attractive and repulsive forces of other charges is harder than it might first seem!

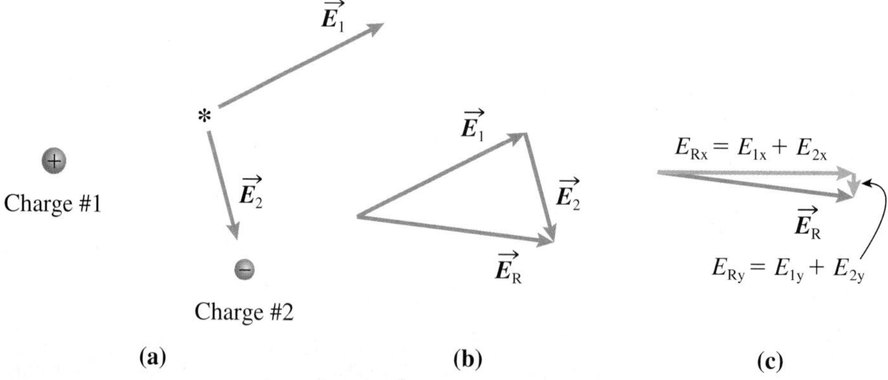

Figure 19-27 **(a)** Electric fields (at location indicated by *) due to the presence of two charges. **(b)** Resultant field $\vec{E}_R$ at that location. **(c)** Components of $\vec{E}_R$ at that location.

EXERCISES

19-33 What is the direction of the electric field at each of the following positions?

(a) near a proton

(b) near an electron

(c) near a carbon nucleus

(d) near a negatively charged chloride ion

(e) halfway between two electrons

19-34 The magnitude of the electric field at a point P is 2.3×10^7 N/C. What is the magnitude of the electric force that would be exerted at P on (a) an electron? (b) an oxygen nucleus (charge $+8e$)?

19-35 Determine the electric field (magnitude and direction) at a distance of 2.1×10^{-10} m from an α particle (charge $+2e$).

19-36 An electron experiences an acceleration of 1.3×10^6 m/s^2 northward because of an electric field. What are the magnitude and direction of the field at that location?

19-37 The magnitude of the electric field at a point P near an electron is 3.53×10^9 N/C. No other charges are present.

(a) What is the distance between P and the electron?

(b) What is the direction of the electric field at P?

19-38 Two charges have locations as shown in Figure 19-28. Charge q_1 is -3.6×10^{-8} C, and charge q_2 is -5.2×10^{-8} C. What is the electric field (magnitude and direction) at (a) point A? (b) point B?

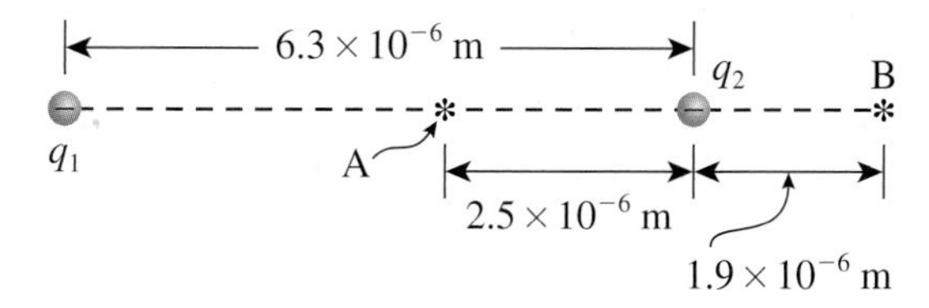

Figure 19-28 Questions 19-38 and 19-39.

19-39 For the two charges in the previous question, where is the electric field zero?

19-40 What is the electric field (magnitude and direction) at point D in Figure 19-29 due to the two charges? Charges q_1 and q_2 are -1.21 μC and $+3.40$ μC respectively.

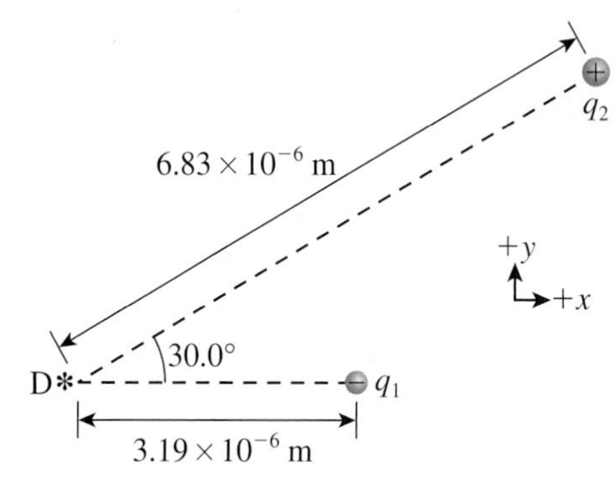

Figure 19-29 Question 19-40.

19-41 **Fermi Question:** If a (spherically symmetric) puffer fish (Figure 19-30) has a net charge on the order of 10 μC, from how far away would a shark be able to detect it? Neglect the effect of water here.

Figure 19-30 Question 19-41.

19.6 Electric Field Lines

electric field lines a means of visualizing how an electric field varies in space, in both magnitude and direction

Have you ever seen an electric field? Of course not. But it is helpful to visualize electric fields through **electric field lines**, which are lines whose orientation and spacing indicate the direction and relative magnitude of the field. Figure 19-31 shows the electric field lines around a negative point charge. The arrows on the lines point inward, showing the direction of the field. As the distance from the charge increases, the lines become farther apart, indicating a decrease in the magnitude of the field. The actual field lines form a *three-dimensional pattern;* we are looking at the lines only in a two-dimensional plane.

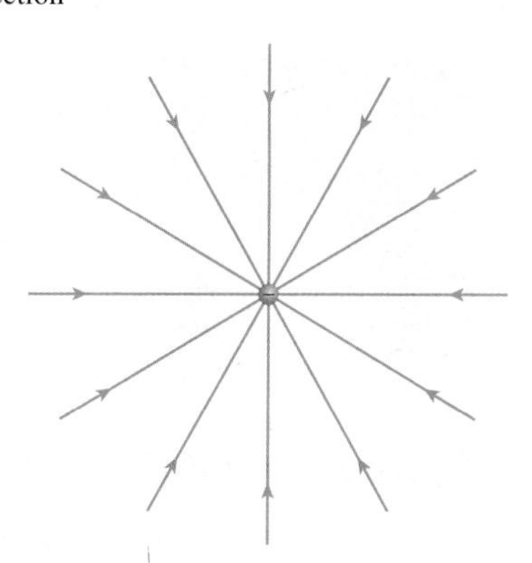

Figure 19-31 Electric field lines around a negative point charge.

Figure 19-32 shows the electric field lines around a positive and a negative point charge of equal magnitude. This pattern illustrates five key points about field lines:

- The tangent to the field line at any location gives the direction of the resultant electric field at that location. For example, at point P the resultant field ($\vec{E}_R$) is the vector sum of the fields due to the two charges and is in the direction of the tangent to the field line at that location, as shown in Figure 19-32.

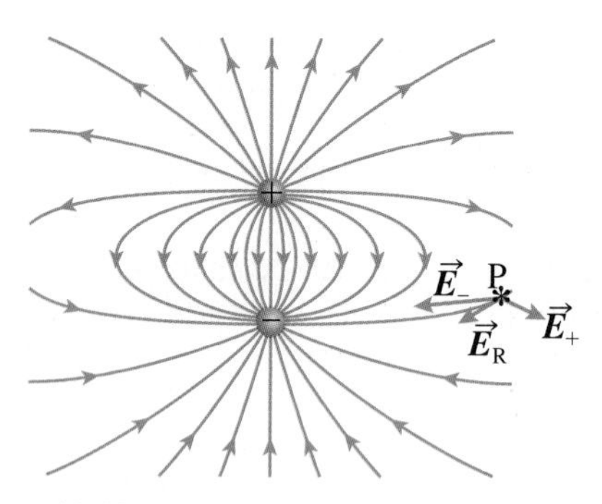

Figure 19-32 Electric field lines around a pair of positive and negative point charges of equal magnitude. The resultant field ($\vec{E}_R$) at P is the vector sum of the fields ($\vec{E}_+$ and $\vec{E}_-$) due to the two charges.

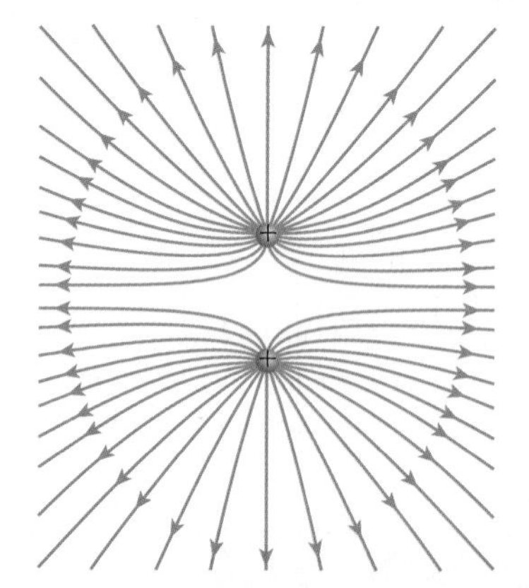

Figure 19-33 Electric field lines around two equal positive point charges.

- The arrows on electric field lines point from positive charges toward negative charges, in other words, the direction of the field line at any location is in the direction of the force that would be exerted on a positive test charge placed at that location.
- Where field lines are close together, the field is large; where lines are far apart, the field is small.
- Field lines originate at positive charges and terminate at negative charges.
- Field lines can never cross one another. If they did, this would mean that the electric field at the point of intersection had more than one direction at that point, which is not physically possible. In which direction would a positive charge experience a net electric force? This can't be ambiguous!

Figure 19-33 shows the field lines around two equal positive point charges. The field is weak between the charges, as shown by the small number of field lines there. This particular pattern provides a clear illustration of the repulsion of like charges, such as these two positive charges. How would the pattern change if the charges were both negative?

The electric field lines close to a section of a very large plate of uniformly distributed positive charge are shown in Figure 19-35 (a). The field is the vector sum of the fields due to all the positive charges and is uniform, that is, constant in magnitude and direction. Notice that the field lines are equally spaced and neither converge nor diverge. The field is perpendicular to the plate; we can justify this by considering the field at a point P close to the plate (Figure 19-35 (b)). The charge at point A on the plate produces a field $\vec{E}_A$ at P, and the charge at point B produces a field $\vec{E}_B$. Points A and B have been chosen to be symmetric relative to point P. Hence, when $\vec{E}_A$ and $\vec{E}_B$ are added to give the resultant field $\vec{E}_R$, this field is perpendicular to the plate, as shown. Another way of thinking about this is that field components parallel to the plate cancel out due to symmetry and the resultant field $\vec{E}_R$ is perpendicular to the plate. The field at point P due to *all* the charges on the plate can be determined by summing the fields due to charges taken in symmetrical pairs such as A and B. Since the field due to any symmetrical pair is perpendicular to the plate, the total field must be perpendicular to the plate.

DID YOU KNOW?

For people at serious risk of heart attack, their doctor may recommend an implantable cardioverter-defibrillator (ICD). Recent advances in electronics have led to the development of a sufficiently small device that can be implanted under the skin in order to monitor the electrical signals from the heart muscle, Figure 19-34. If irregularities are detected, the device can deliver a brief electrical impulse to the heart to return the organ to regular operation—all within the blink of an eye!

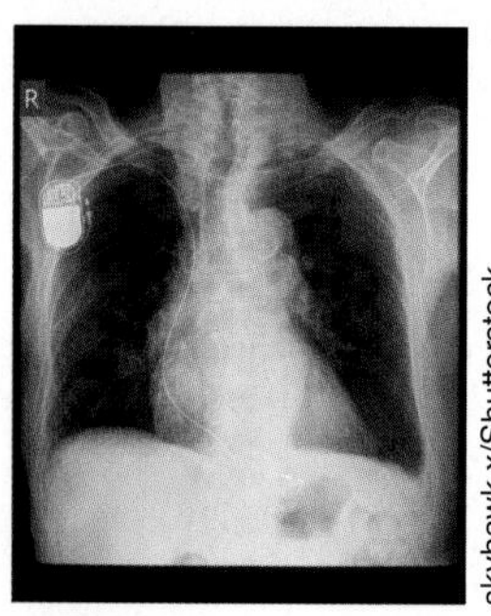

Figure 19-34 An x-ray image of the chest of a patient that has an ICD in place.

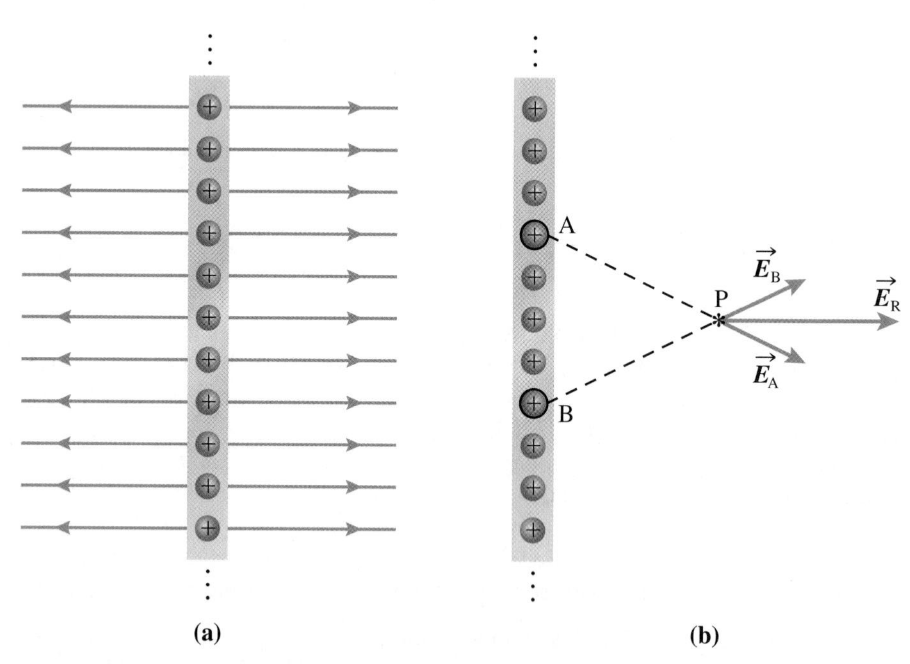

Figure 19-35 **(a)** Electric field lines close to a section of a large plate of uniformly distributed positive charge. **(b)** By symmetry, the resultant electric field near a large plate is perpendicular to that plate.

Two parallel plates of uniform charge, one positive and the other of equal magnitude but negative, are often used to produce a region of uniform electric field (Figure 19-36). The total field is the sum of the individual fields from the two plates. Since the field due to the positive plate points away from the positive plate and toward the negative plate, and the field due to the negative plate also points in this direction, the total field between the plates is twice the field due to one plate alone. Outside the plates, the fields due to the two plates point in opposite directions; therefore, the resultant field is essentially zero in this region. Note that the field is uniform near the middle of the plates, but non-uniform near the ends, where the cancellation of field components mentioned in the previous paragraph does not apply. (Why not?) When solving problems, we usually assume that the field is uniform throughout the entire region between the plates.

Oppositely charged plates are used in a number of devices to deflect moving charged particles. Inkjet printers (Figure 19-37) are common household devices that make use of such deflection plates, as we will explore in greater detail in Sample Problem 19-5.

Charged metal surfaces (but not in the shape of flat plates) are used in **electrostatic precipitators**, which remove smoke particles emitted from industrial chimneys (Figure 19-38). As smoke and dust particles move, they acquire electric charges by friction. In electrostatic precipitators, the particles pass through a metal tube containing a central metal rod. The rod and tube are given opposite charges and each charged particle is attracted to either the tube or the rod, where the particle's charge is neutralized. As the particles collect, they agglomerate and become large enough to settle out.

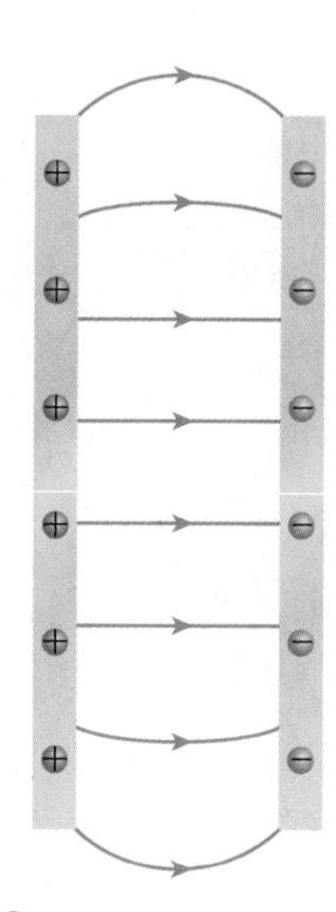

Figure 19-36 Electric field lines between oppositely charged plates.

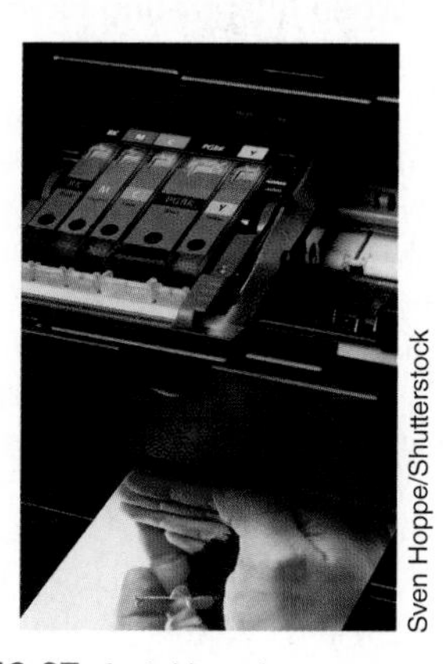

Figure 19-37 An inkjet printer uses oppositely charged plates to deflect the charged ink drops; see Sample Problem 19-5.

electrostatic precipitator a filter that uses electric fields to remove particulate from the emissions of industrial chimneys

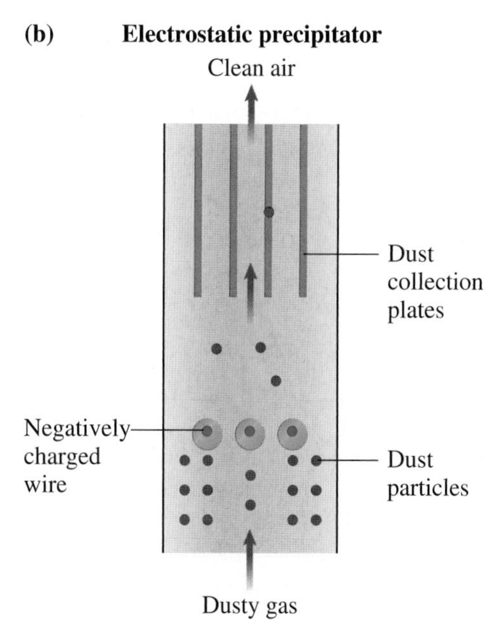

Figure 19-38 **(a)** Electrostatic precipitators use electric fields to remove particulate matter from the emissions of an industrial chimney. **(b)** The electric field from the wires charges the dust particles as they pass through. These charged dust particles are then attracted to the collection plates before the exhaust is released.

Discovery of the Elementary Charge—The Millikan Experiment

During the years 1910–1913, the American physicist, Robert A. Millikan, performed experiments in which he was able to establish the value of the **elementary charge**, *e*, which corresponds to the absolute value of the charge of one electron (or one proton);

elementary charge the smallest amount of charge observed in nature; all charges are integer multiples of this elementary charge; symbol *e*

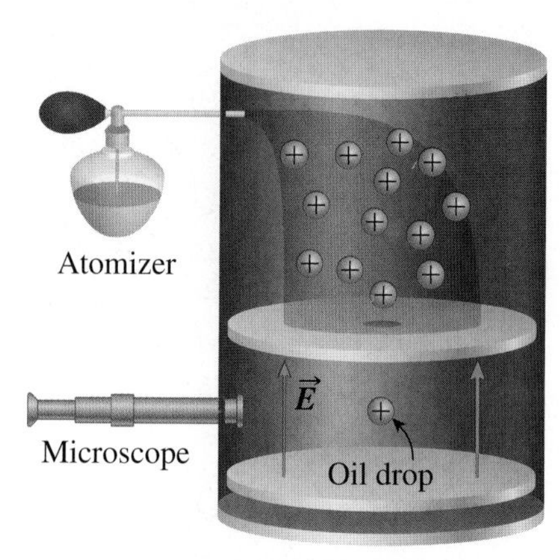

Figure 19-39 A schematic illustration of the Millikan apparatus.

 TACKLING MISCONCEPTIONS

Be Careful When Interpreting Field Line Diagrams!

At any point where there is no field line, it does not necessarily mean that there is zero field. The spaces between lines are there to indicate the relative magnitude of the field in different regions. In order to determine the direction of the electric field at a location where there is no field line illustrated, draw a tangent to the field lines on either side of the point in question.

see Eqn. 19-2. He sprayed tiny drops of oil from an atomizer into a chamber; in the process, the drops acquired small net charges. By observing an individual drop through a microscope (Figure 19-39), he was able to measure the terminal speed of the drop falling through the air under the influence of gravity (downward) and air friction (upward). In Chapter 12, we derived the relationship between terminal speed (v_t) for a small spherical object in a viscous fluid such as air and the radius (r) of the sphere (Eqn. 12-29):

$$v_t = \frac{2}{9}\frac{r^2 g}{\eta}(\rho - \rho_f)$$

where $\eta = 1.8 \times 10^{-5}$ Pa·s is the viscosity of air, $g = 9.8$ m/s^2 is acceleration due to gravity, and ρ and ρ_f are the densities of the oil and air respectively, which are both known values. So, by measuring v_t for each drop, Millikan was able to determine the radius, r, from Eqn. 12-29.

Once the radius of the drop has been determined, the mass is also known, since the oil has a given density (ρ). The uniform electric field is then turned on, causing an upward electric force on the drop that has a magnitude given by qE. The drop quickly reaches a new terminal speed, upward. By analyzing the forces acting on the drop and applying the same derivation that led to Eqn. 12-29, the charge on the drop, q, can be determined once the new terminal speed has been measured, provided that the strength of the electric field between the plates (E) is known.

Millikan and his colleagues measured the charges on thousands of drops, and found that they were always integer multiples of e; that is, charges such as $q = 2e$ and $q = -4e$ were found, but never ones such as $2.3e$ or $-4.5e$. He thus concluded that the charge e represented the elementary charge, and was awarded the 1923 Nobel Prize in Physics for his work.

Motion of a Charged Particle in a Uniform Electric Field

What does baseball have to do with ink jet printers? Plenty, as it turns out. A pitched or hit baseball follows a parabolic trajectory as it moves in a uniform gravitational field (i.e., as it experiences a constant acceleration due to gravity) when it starts with a velocity that has a component perpendicular to the gravitational field. Similarly, charged ink drops follow a parabolic trajectory between the deflecting plates as they move in a uniform electric field, provided that they start with a velocity that has a component perpendicular to the electric field as well. The acceleration of the baseball has a constant vertical component (9.80 m/s^2 downward) and no horizontal component. The acceleration of a positively charged particle in a uniform electric field has a constant component in the direction of the field, and no component perpendicular to the field. (A negatively charged particle has an acceleration opposite to the field direction.) Whether the acceleration arises from gravity or electricity, the object will experience the same parabolic path when the initial velocity has a component that is perpendicular to the acceleration.

SAMPLE PROBLEM 19-5

A charged drop of ink (mass $= 1.1 \times 10^{-10}$ kg, charge $= -8.0 \times 10^{-11}$ C) is projected between the deflecting plates in an inkjet printer with an initial velocity of 25 m/s parallel to the plates (Figure 19-40). The uniform electric field between the plates is 8.0×10^4 N/C upward.

(a) What is the acceleration (magnitude and direction) of the ink drop while between the plates?

(b) How long does it take to go through the plates?

(c) How far has it dropped or risen (specify which) when it leaves the plates?

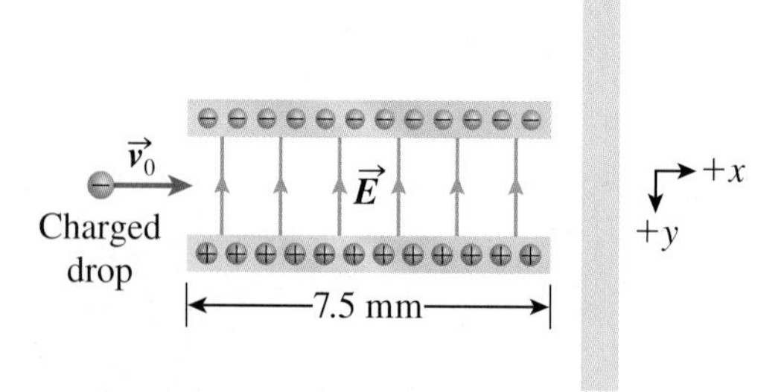

Figure 19-40 Sample Problem 19-5.

Solution

(a) To determine the acceleration, we will use Newton's second law ($\Sigma\vec{F} = m\vec{a}$), and therefore need to know the forces acting on the drop. The electric field is upward, and since the drop has a negative charge, the electric force on it is downward. The magnitude of the electric force is the product of the magnitude of the charge on the drop and the electric field:

$$F = qE = (8.0 \times 10^{-11}\ \text{C})\ (8.0 \times 10^{4}\ \text{N/C}) = 6.4 \times 10^{-6}\ \text{N}$$

This force acts downward, which we define to be the $+y$ direction (Figure 19-40). We use

$$\Sigma F_y = ma_y$$

Hence, $a_y = \dfrac{\Sigma F_y}{m} = \dfrac{6.4 \times 10^{-6}\ \text{N}}{1.1 \times 10^{-10}\ \text{kg}} = 5.8 \times 10^{4}\ \text{m/s}^2$

Thus, the acceleration is $5.8 \times 10^{4}\ \text{m/s}^2$ downward. (How does the acceleration due to gravity for the drop compare to this acceleration?)

(b) To determine the time taken to pass through the plates, we consider motion in the x-direction. There is no force in this direction, and therefore the x-component of acceleration is zero ($a_x = 0$), giving a constant x-component of velocity. Just as we did in projectile-motion problems in Chapter 4, we use Eqn. 4-10

$$x = x_0 + v_{0x}t$$

Letting $x_0 = 0$ as the drop enters the region between the deflecting plates, we have

$$x = v_{0x}t$$

$$t = \frac{x}{v_{0x}} = \frac{7.5 \times 10^{-3}\ \text{m}}{25\ \text{m/s}} = 3.0 \times 10^{-4}\ \text{s}$$

Thus, the drop takes 3.0×10^{-4} s to go through the plates.

(c) In the y-direction, the drop's initial velocity component is zero ($v_{0y} = 0$), and since the acceleration is downward, the drop will move downward. Its y-displacement can be calculated from Eqn. 4-13:

$$y = y_0 + v_{0y}t + \tfrac{1}{2}a_y t^2$$

Setting $y_0 = 0$ as the drop enters the space between the plates, and substituting $v_{0y} = 0$

$$\begin{aligned} y &= \tfrac{1}{2}a_y t^2 \\ &= \tfrac{1}{2}(5.82 \times 10^{4}\ \text{m/s}^2)\ (3.0 \times 10^{-4}\ \text{s})^2 \\ &= 2.6 \times 10^{-3}\ \text{m} \end{aligned}$$

Hence, the drop is now 2.6×10^{-3} m (or 2.6 mm) lower in the y-direction when it leaves the plates relative to its initial position.

Notice that this solution is exactly analogous to the solution of projectile-motion problems in Chapter 4. Furthermore, the degree to which the ink drop is deflected is directly related to how much charge it has. To generate an image on the paper, the printer only has to control how much charge it puts on each drop in order to determine where the drop is positioned on the page.

Moving Charges with Electric Fields

With a woollen sweater, charge a plastic comb via friction. With water running slowly out of a kitchen or bathroom tap, bring the comb close to the stream and observe the effect (Figure 19-41). What would happen if you had the opposite charge on the comb? How does the moisture in the air affect your results; that is, is it more dramatic to conduct this experiment in the summer or the winter? You can experiment with a number of variables here: the amount of charge on the comb, the separation between the comb and the stream, the speed of the water flow … how do these affect the amount of bend in the water?

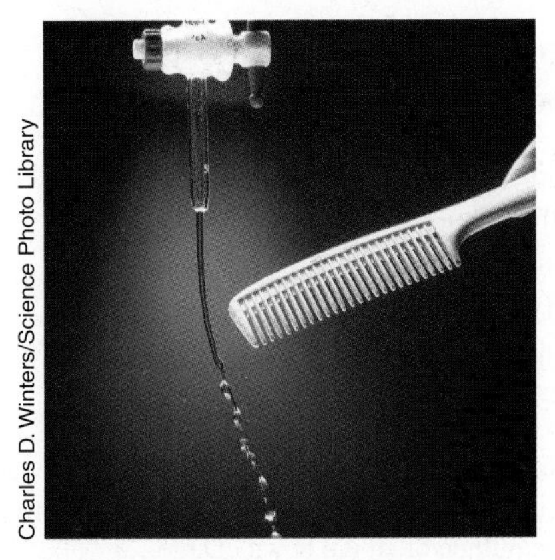

Charles D. Winters/Science Photo Library

Figure 19-41 An electrically charged comb will attract the polar molecules in a stream of water.

EXERCISES

19-42 Sketch the field line pattern around
(a) a single positive point charge;
(b) two equal negative point charges, separated in space.

19-43 A proton is moving from left to right. Describe its resulting motion if it enters a uniform electric field that is directed
(a) from left to right
(b) from right to left
(c) downward
(d) upward
You might wish to use diagrams in your descriptions.

19-44 Answer the previous question for an electron.

19-45 Suppose that in the Millikan experiment an electric field of magnitude 5.0×10^5 N/C is applied to a tiny oil drop of mass 2.45×10^{-14} kg, and that this electric field is just the right magnitude to hold the drop motionless against gravity. How many units of elementary charge are on the drop? Since the drop is not moving you do not need to consider the effects of air resistance.

19-46 A proton having an initial speed of 3.4×10^5 m/s enters a uniform electric field of magnitude 6.2×10^9 N/C having the same direction as the proton's velocity. After a time interval of 1.2×10^{-12} s, (a) what is the proton's speed? (b) How far has it travelled?

19-47 An electron with an initial speed of 5.5×10^6 m/s enters a uniform electric field that acts only to slow the motion of the electron. The field's magnitude is 3.7×10^6 N/C.
(a) Relative to the electron's velocity, in what direction is the field?
(b) How much time is required for the electron to come to rest (momentarily)?
(c) How far has the electron travelled in this time?

19-48 A proton is released from rest at the surface of a positively charged plate, and travels to a parallel negatively charged plate 2.1 cm away in a time of 1.6×10^{-8} s. What is the electric field (magnitude and direction) between the plates?

19-49 An electron travels between deflecting plates with an initial velocity of 2.0×10^7 m/s parallel to the plates, which lie in a horizontal plane. The electric field is 2.2×10^4 N/C downward, and the plates have a length of 4.0 cm. When the electron leaves the plates, (a) how far has it dropped or risen? (Specify which.)
(b) What is its velocity (magnitude and direction)?

LOOKING BACK...LOOKING AHEAD

We have just begun our study of electricity. In this chapter, we have discussed basic atomic structure, paying particular attention to electric charges in atoms. We have described how charges can be transferred from one object to another by friction. The properties of conductors and insulators were introduced, and we showed how conductors can become charged by contact and by induction. Coulomb's law, which is the basic law of electric forces, was presented, as was the fundamental concept of electric field. The use of electric field lines was discussed as a means of visualizing electric fields. Finally, we showed how the motion of charged particles in a uniform electric field can be analogous to projectile motion under the influence of gravity.

In the next chapter, we discuss some electrical concepts that you have almost certainly heard of in your daily activities: voltage and current. What exactly is a volt? What is an ampere? How is voltage (or electric potential difference) related to energy? Answers to these questions and more are coming up in Chapter 20.

CONCEPTS AND SKILLS

Having completed this chapter, you should now be able to do the following:

- Describe the basic structure of atoms and nuclei.
- Describe how objects can become charged.
- Discuss how a charged object can attract an electrically neutral object.
- Explain the term electric polarization.
- Discuss how charged objects "leak" charge.
- Explain the difference between electric conductors and insulators.
- Know how an electroscope works to determine the sign and relative magnitude of an unknown charge.
- State Coulomb's law and use it in solving problems.
- Define electric field and explain how the electric force on a charge arises from the electric field.
- Calculate the electric fields of point charges in order to solve related problems.
- Discuss the basic properties of electric field lines, and draw field lines around distributions of charge.
- Solve problems involving the motion of charged particles in a uniform electric field.

KEY TERMS

You should be able to define or explain each of the following words or phrases:

nucleus
proton
neutron
electron
positive ion
negative ion
electron affinity
law of conservation of electric charge
electric force
static electricity
electric polarization
polar molecules
electrical conductors
electrical insulators
conduction electrons
semiconductors
superconductors
charging by contact
charging by conduction
grounding
charging by induction
leaf electroscope
torsion balance
Coulomb's law
electric field ($\vec{E}$)
visiting charge
test charge
electric field lines
electrostatic precipitator
elementary charge

Chapter Review

MULTIPLE-CHOICE QUESTIONS

19-50 As two charges move apart, the force between them decreases by a factor of 27.0. By what factor has their separation distance increased?

(a) 27.0
(b) 729
(c) 5.20
(d) 3.00
(e) 2.28

19-51 Two charges $+q$ and $+2q$ experience a mutual force of magnitude F when separated by a distance d. What is the magnitude of the force between two charges $-2q$ and $-3q$ separated by $d/4$?

(a) 96 F
(b) 12 F
(c) 24 F
(d) 32 F
(e) 48 F

19-52 Four charges are placed at the corners of a square, one at each corner. The charges have the same magnitude, but not the same sign. Two of the charges are positive, and are at diagonally opposite corners of the square. The other two charges are negative, and are also at opposite corners. The resultant force on each charge

(a) is zero
(b) is directed toward the centre of the square
(c) is directed away from the centre of the square
(d) is directed along one side of the square
(e) has a direction not given in any of the previous answers

19-53 Two small spheres of equal mass are suspended from a common point by threads of equal length. A charge $+q$ is placed on one sphere, and $+3q$ on the other. Which of the drawings in Figure 19-42 best shows the resulting equilibrium positions of the spheres?

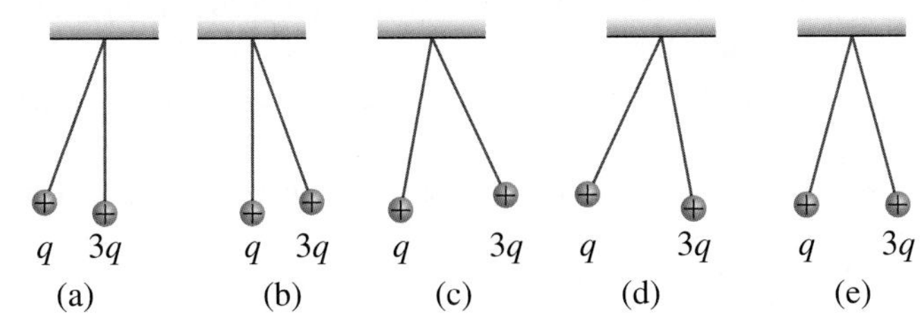

Figure 19-42 Question 19-53.

19-54 Which of the drawings in Figure 19-43 best shows the total electric field at point P, which is equidistant from two equal but opposite charges $+q$ and $-q$?

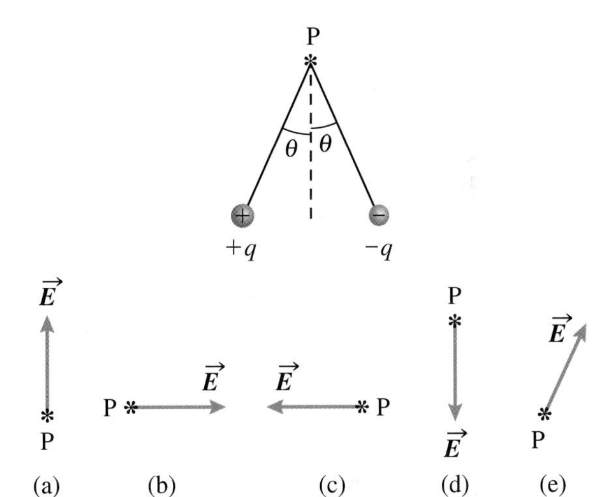

Figure 19-43 Questions 19-54 and 19-55.

19-55 Repeat the previous question if both charges are $+q$.

19-56 An electron with an initial velocity eastward enters a uniform electric field and is deflected northward. What is the direction of the electric field in the region?

(a) eastward
(b) westward
(c) southward
(d) northward
(e) north-eastward

19-57 Which of the drawings in Figure 19-44 best shows the resulting path of an electron (with initial velocity $\vec{v}$) that enters a region containing a uniform electric field directed from left to right?

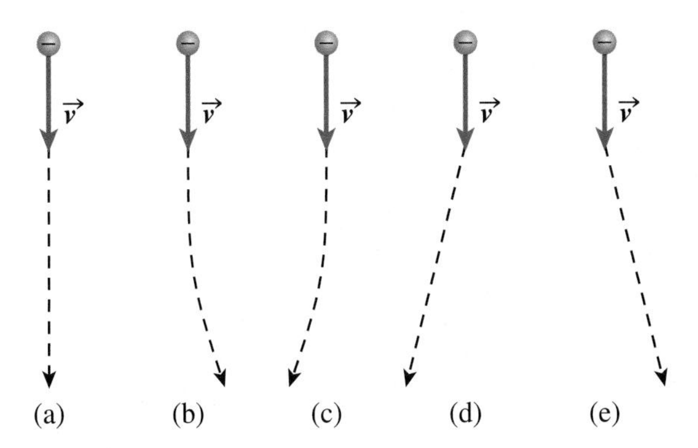

Figure 19-44 Question 19-57.

Review Questions and Problems

19-58 If you walk across a carpet, you can sometimes cause a spark when you touch a doorknob (or another person!). Explain why this happens and discuss how it could be prevented.

19-59 Explain how a charged object such as a comb can attract an electrically neutral object such as a piece of paper.

19-60 Why do static electricity experiments not work well on humid days?

19-61 If you hold a conducting sphere in your hand instead of placing it on an insulating stand and attempt to charge it by contact, what happens? Explain.

19-62 (a) Use diagrams to describe how to charge a conductor negatively by induction.

(b) If a negatively charged plastic rod is brought close to the cap of a negatively charged electroscope, what happens? Explain, using diagrams.

19-63 A positively charged glass rod is brought close to the cap of a charged electroscope. The leaves move closer together.

(a) What is the sign of the charge on the electroscope? Explain, using diagrams.

(b) If the rod moves closer to the electroscope cap, the leaves start to move farther apart. Explain.

19-64 If r represents the distance separating two charges, and F represents the magnitude of the electric force between them, sketch the shape of a graph of F versus $1/r^2$.

19-65 An α particle (charge $+2e$) experiences an electric force of magnitude 3.2×10^{-8} N near a proton.

(a) What is the magnitude of the force on the proton?

(b) Is the force repulsive or attractive?

(c) How far apart are the α particle and the proton?

19-66 Figure 19-45 shows the separations of a proton, an electron, and an α particle (charge $+2e$). What is the force (magnitude and direction) on the proton?

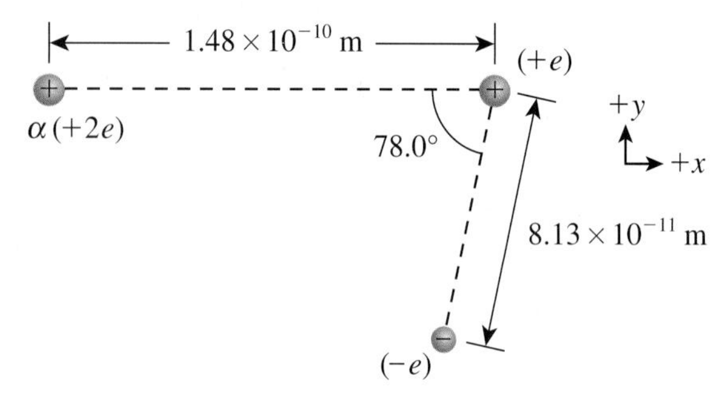

Figure 19-45 Question 19-66.

19-67 Earth has an electric field pointing toward its centre and having an average magnitude of about 1.5×10^2 N/C at its surface.

(a) What is the sign of the charge on Earth?

(b) What is the charge on Earth, assuming that all the charge is concentrated at the centre? This assumption is valid for a spherically symmetric object such as Earth with a surface charge. (Earth's radius = 6.37×10^6 m)

DID YOU KNOW?

The charge on Earth (Question 19-67) is the result of lightning, which transfers charge from clouds to Earth. Earth's charge gradually leaks into the atmosphere, but is continually replenished by lightning strikes by thunderstorms around the world (Figure 19-46).

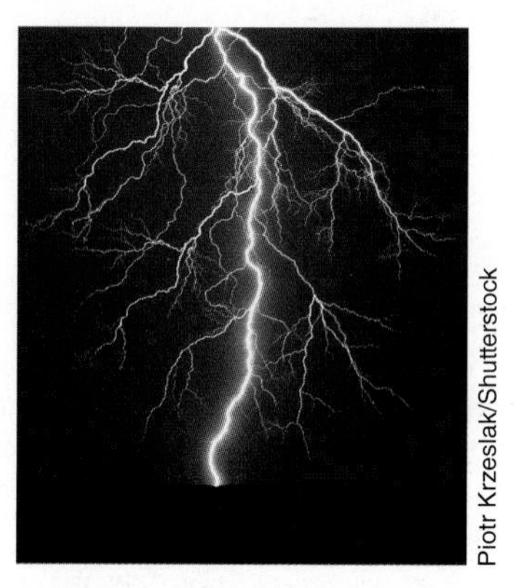

Piotr Krzeslak/Shutterstock

Figure 19-46 Lightning transfers large amounts of charge from the clouds to Earth (Question 19-67).

19-68 If the distance from a charge is increased by a factor of 6.0, does the magnitude of the electric field increase or decrease? by what factor?

19-69 What are the magnitude and direction of the electric field 5.5×10^{-10} m away from an electron?

19-70 What is the magnitude of the electric field at the surface of a uranium nucleus? The nucleus has a radius of 6.8 fm and a charge of $+92e$. Assume that the charge is distributed uniformly throughout the nucleus (assumed spherical), and therefore the electric field is the same as if the charge is concentrated at the centre.

19-71 If an electron is placed at each corner of an equilateral triangle that has sides with a length of 1.2×10^{-10} m, what are the magnitude and direction of the electric field at the midpoint of one of the sides?

19-72 A charged water droplet of mass 1.41×10^{-11} kg and charge $+5.22 \times 10^{-15}$ C enters a uniform electric field of magnitude 1.30×10^2 N/C with an initial velocity of 7.30 cm/s opposite to the field direction.

(a) What is the acceleration (magnitude and direction) of the droplet?

(b) After 2.00 s, what is its velocity (magnitude and direction)?

(c) How far has it moved?

Applying Your Knowledge

19-73 Many biological processes occur in an aqueous environment. Because water is a polar molecule, it shields the electric fields of charged particles such as ions that are in the water. Hence, the electric forces between charges in water are reduced relative to the forces if the charges were in air. Eqn. 19-6, which gives the magnitude of the electric field E at a distance r from a charge q, can be altered to account for the screening effect of water (or any other material) by including a dielectric constant, K, in the equation

$$E = \frac{k|q|}{K r^2}$$

The dielectric constant, which is dimensionless, is a measure of the screening effect of the material: the larger the value of K, the larger the screening effect and the smaller the electric field due to a charge in that medium. Values of K for a few materials are given in Table 19-2. The large value of K for water indicates that it strongly screens electric fields. Consider a sodium ion Na^+ and a chloride ion Cl^- separated by 5.0 nm in aqueous solution.

(a) What is the electric field (magnitude and direction) due to the sodium ion, at the position of the chloride ion?

(b) Determine the electric force (magnitude and direction) on the chloride ion due to the sodium ion.

Table 19-2

Dielectric Constants

Substance	Dielectric Constant K
Air	1.005
Palmitic acid*	2.3
Stearic acid*	2.3
Wood (approximate)	5.0
Ethanol	26.8
Water	80.4

*components of biological membranes

19-74 **Fermi Question:** Repeat Question 19-41, taking into account now that the shark and puffer fish are separated by water rather than air/vacuum. (See the previous question for guidance.)

19-75 You are given two identical, uncharged, metal spheres on insulating stands. Explain how you could give them equal but opposite charges using only a plastic rod rubbed with fur. You are not permitted to touch the rod to either sphere.

19-76 A charge of $+2.0\ \mu C$ is separated by 12 cm from a charge of $-3.2\ \mu C$. Where can a third charge be placed so that it experiences no resultant force?

19-77 Two particles of total charge 4.8×10^{-19} C experience a repulsive electric force of magnitude 3.8×10^{-8} N when they are 1.1×10^{-10} m apart. Determine the charge on each particle.

19-78 Two spheres of equal mass 4.5 g are suspended from a common point by threads of equal length (25 cm). The spheres are given charges of like sign, with one sphere receiving a charge 4.5 times that on the other. The spheres move to equilibrium positions such that the angle between the threads is 41° (Figure 19-47). What is the magnitude of the charge on each sphere?

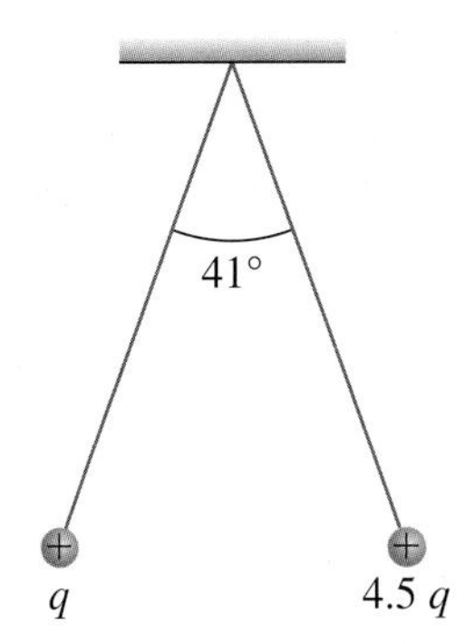

Figure 19-47 Question 19-78.

19-79 The electric field is zero at a point one-quarter of the way along the straight line joining two unknown point charges q_1 and q_2. What can you conclude about the ratio of these charges? What can you conclude about the signs of the charges?

19-80 A proton has an acceleration of magnitude 1.1×10^{14} m/s^2 in a uniform electric field.

(a) What is the magnitude of the field?

(b) How long would it take the proton to reach a speed of 3.0×10^5 m/s (1/1000 of the speed of light) if it starts from rest?

(c) How far would it travel in this time?

19-81 An electron enters a uniform electric field with an initial speed of 7.8×10^6 m/s. The field acts only to increase the electron's speed. After 3.4×10^{-8} s, the kinetic energy of the electron has increased by 55%. What are the magnitude and direction of the electric field?

19-82 Two parallel plates 4.50 cm apart are given equal but opposite charges so that an electric field of magnitude 5.41×10^4 N/C exists between them. An electron is released from rest at the negative plate at the same time that a proton is released from rest at the positive one. Neglect the force exerted between the particles themselves.

(a) How far are they from the positive plate when they pass each other?

(b) What is the kinetic energy of each particle when it strikes the opposite plate?

19-83 An α particle (charge $+2e$, mass 6.6×10^{-27} kg) enters a uniform electric field as shown in Figure 19-48. The initial speed of the particle is 4.3×10^5 m/s, and the magnitude of the field is 3.0×10^5 N/C.

(a) Sketch the resulting path of the α particle.

(b) Where on your sketch is the speed of the particle a minimum?

(c) What is this minimum speed?

(d) When the particle's y-component of displacement is zero, how much time has elapsed, and what is the total displacement?

19-84 **Fermi Question:** Determine an approximate value for the total amount of positive charge (or negative charge) in your body. Use an estimate of your body's mass and assume that your body is 100% water, which has a molar mass of 0.018 kg/mol.

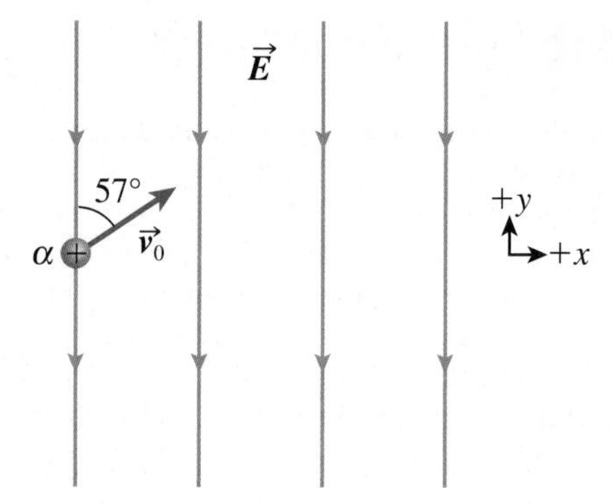

Figure 19-48 Question 19-83.

Electric Potential Energy, Electric Potential, and Current

20

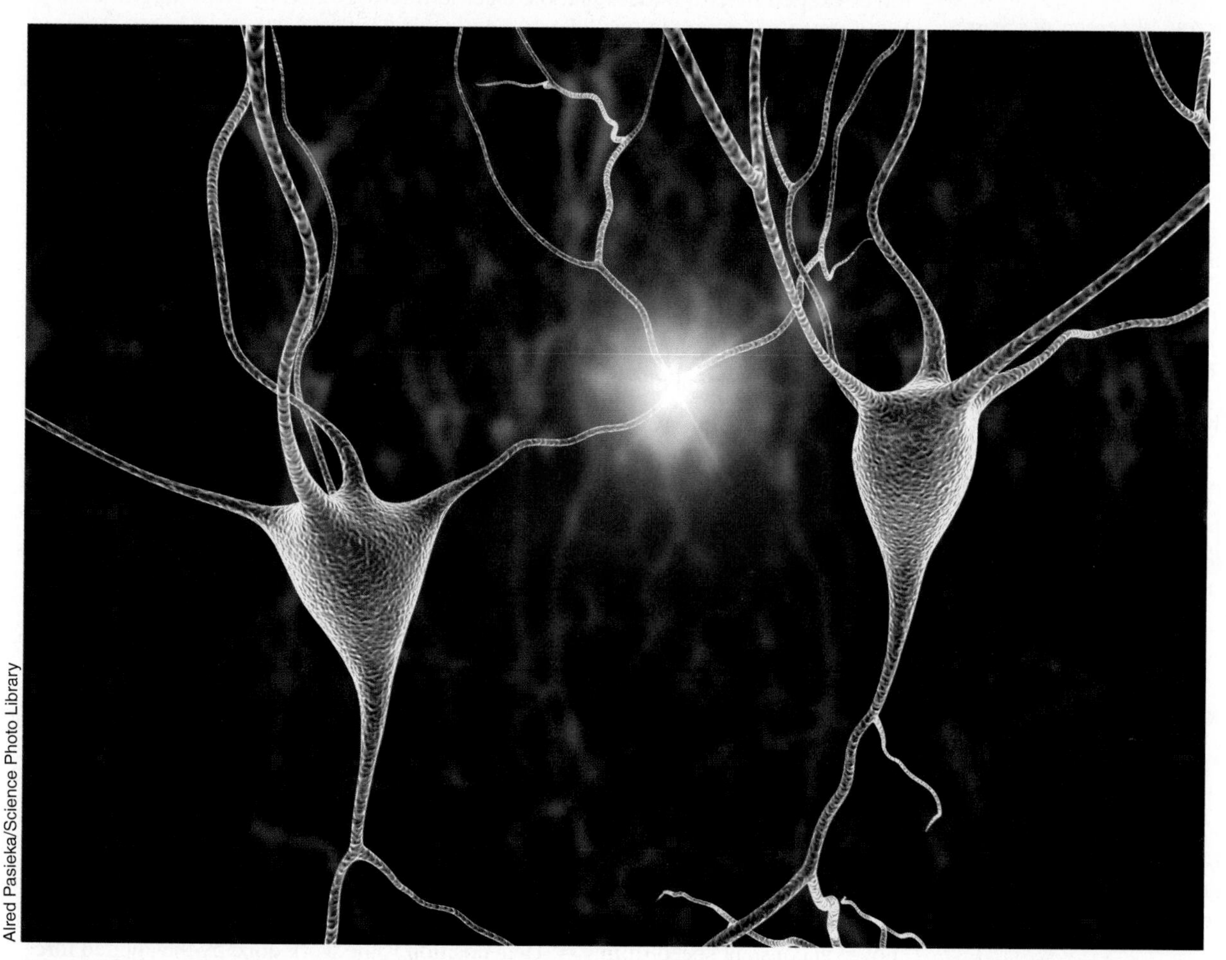

Alred Pasieka/Science Photo Library

Figure 20-1 A computer-generated image of neurons, the cells that provide the critical link between our brain and the rest of our body. All messages relayed from the brain via these neurons are carried as electric currents.

We live in a permanently electrified environment. The most spectacular evidence of this electrification is lightning, which strikes Earth 50 to 100 times each second and kills thousands of people worldwide every year. Lightning causes forest fires, electrical power interruptions, and damage to computers and communications equipment. However, it also contributes to the production of atmospheric fixed nitrogen, which is essential for the growth of plants. Even on a clear day, there is a downward electric field of about 100 to 200 N/C near Earth, resulting from a net negative charge on Earth. In order to understand more about lightning and many other electrical phenomena, including commercial electricity and nerve impulses in our own bodies (Figure 20-1), we need a thorough understanding of electric potential energy, electric potential, and current.

CHAPTER OUTLINE

20.1 Electric Potential Energy in a Uniform Electric Field

electric potential energy the energy of a charge due to its position in an electric field; symbol U

Physicists are always looking for similarities in different areas of physics. General conclusions or concepts developed in one area can often be applied in another. In the previous chapter, we discussed the interesting parallels between the gravitational forces exerted by two masses and the electric forces exerted by two charges. In both cases, the concept of field was introduced. It turns out that another concept introduced in studying gravity, namely potential energy, can play an important role in our understanding of electrical interactions as well.

Figure 20-2 (a) shows an apple of mass m close to Earth, and (b) shows a positive charge q near a large, negatively charged plate. Near Earth, the gravitational acceleration (or gravitational field) $\vec{g}$ is constant, and near the plate the electric field $\vec{E}$ is constant, or uniform. The force on the apple is $m\vec{g}$; the force on the positive charge is $q\vec{E}$. If the apple is moved from Earth ($y = 0$) to a position y above Earth, it gains gravitational potential energy, mgy. Similarly, if the positive charge is moved from the plate ($y = 0$) to position y (Figure 20-2 (b)), it gains electric potential energy, qEy. **Electric potential energy** is the energy of a charge due to its position in an electric field, and is often symbolized as U:

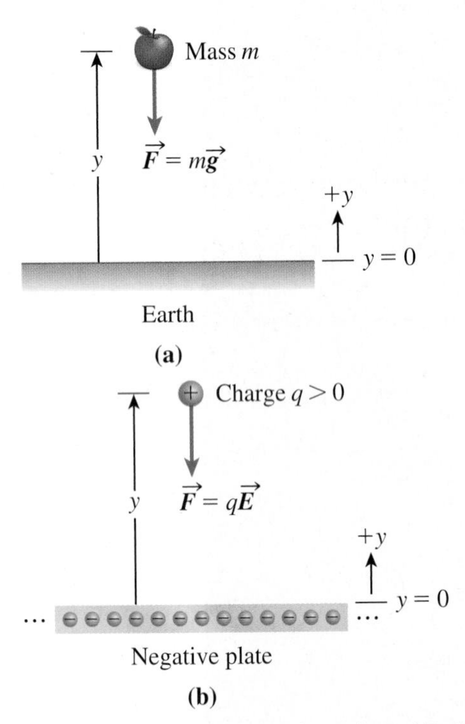

Figure 20-2 **(a)** An apple near the surface of Earth is analogous to **(b)** a positive charge near a large negatively charged plate. The apple has gravitational potential energy, mgy, and the charge has electric potential energy, qEy.

$$U = qEy \tag{20-1}$$

where charge q is present in a *uniform* electric field of magnitude E.

Another way to arrive at the same result is to consider the work done on each object. Suppose that the apple is lifted at constant speed from $y = 0$ to some final position, y. Since the acceleration is zero, the resultant force must be zero, and the upward lifting force must be equal in magnitude to the force of gravity, $m\vec{g}$, on the apple. Hence, the work, W, done on the apple by the lifting force is

$$W = F_y\Delta y = (mg)(y - 0) = mgy$$

This work represents a transfer of energy (due to the force applied in lifting the apple) into gravitational potential energy, mgy. In a similar way, if the positive charge is moved at constant speed from $y = 0$ to a position y, the work done by the applied force in order to counter the electric force is

$$W = F_y\Delta y = (qE)(y - 0) = qEy$$

The electric potential energy of the positive charge has been increased by an amount qEy.

If the apple is dropped from rest, it loses gravitational potential energy as it falls and gains an equivalent amount of kinetic energy. If the positive charge is released, it "falls" toward the negative plate as a result of the attractive electric force, losing electric potential energy and gaining an equal amount of kinetic energy. Notice that *a mass or charge released from rest always moves in the direction of decreasing potential energy.*

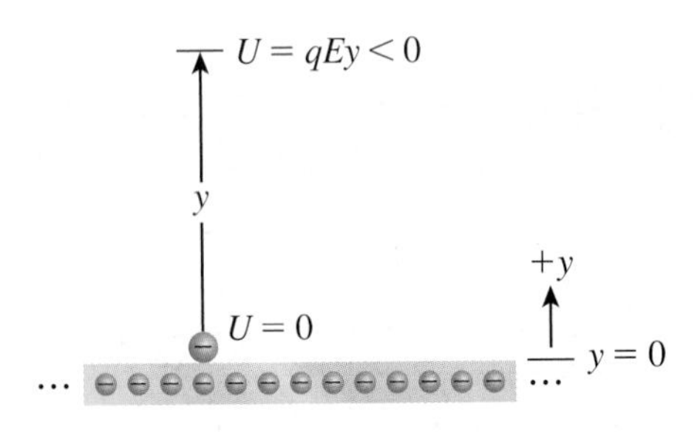

Figure 20-3 A negative charge released at the surface of the negative plate is repelled. It moves in the direction of decreasing electric potential energy (U), from zero potential energy at the plate to negative potential energy at position y. As the potential energy decreases, the kinetic energy increases by an equivalent amount.

You can see that gravitational potential energy and electric potential energy are analogous to each other. However, there is one important difference: mass can only be positive, whereas charge can be either positive or negative. What if the charge q is negative? We can still use Eqn. 20-1 to determine the electric potential energy in a uniform electric field, as it turns out. As shown in Figure 20-3, a negative charge located at the surface of the negatively charged plate ($y = 0$) has zero electric potential energy just as in the case of the positive charge at that location in Figure 20-2. However, at a position $y > 0$, a negative charge will have negative electric potential energy, based on Eqn. 20-1. A negative charge released from $y = 0$ is repelled by the negative plate and moves in the direction of decreasing potential energy, that is, it moves from zero potential energy at the plate to negative potential energy away from the plate. At the same time, the kinetic energy of the charge increases. A negative charge released from some position y away from the plate is repelled to even greater values of y, thus decreasing its potential energy even more.

When the equation $E_p = mgy$ is used for gravitational potential energy, the $+y$-direction must be upward, that is, opposite to the direction of the gravitational field and gravitational acceleration. Similarly, when using $U = qEy$ for determining the electric potential energy in a uniform electric field, *the $+y$-direction must be opposite to the electric field direction.*

The equation $E_p = mgy$ gives the gravitational potential energy *relative* to the position $y = 0$. Correspondingly, $U = qEy$ gives the electric potential energy relative to $y = 0$. When solving problems, you are free to choose the reference position of $y = 0$. In problems involving gravitational potential energy, it was suggested in Chapter 7 that you choose $y = 0$ at the lowest possible position so that all potential energies would be positive. However, as pointed out in Chapter 7, what is really important is the *change* in potential energy as an object moves from one position to another, and this change does not depend on your choice of $y = 0$ position. In problems involving electric potential energy in a uniform field, charges can be positive or negative, and it is often impossible to choose a position for $y = 0$ so that all electric potential energies in the problem are positive. Hence, it is usually most convenient simply to choose the location for $y = 0$ so that all y-values in a problem are positive.

DID YOU KNOW?

The energy emitted in the form of light by a fluorescent tube or neon sign comes from electric potential energy. Electrons are accelerated through an electric field inside the tube or sign, thereby decreasing their electric potential energy. This energy is converted into kinetic energy, which is absorbed by the gas molecules inside the device due to collisions with the electrons. The gas molecules then emit this absorbed energy in the form of UV and visible light. The chemical composition of the gas inside dictates the colour of light observed: neon emits red light (Figure 20-4), helium light is pale yellow, whereas argon and mercury vapour emit blue light.

Figure 20-4 The red glow of neon signs is due to the transformation of the electric potential energy of electrons into kinetic energy, which is then absorbed by the neon gas through collisions.

SAMPLE PROBLEM 20-1

A large, positively charged plate produces an electric field of magnitude 1.3×10^3 N/C. An electron is released from a point that is 1.2×10^{-2} m from the plate. Use energy methods to determine its speed just as it strikes the plate.

Solution

The field near a positively charged plate points away from the plate, and since the $+y$-direction must be opposite to this, $+y$ is toward the plate (Figure 20-5). We choose $y = 0$ as the initial position of the electron so that the electron's final position at the plate will be positive, and therefore slightly easier to deal with. (However, our final answer will not depend on our choice of $y = 0$ position.)

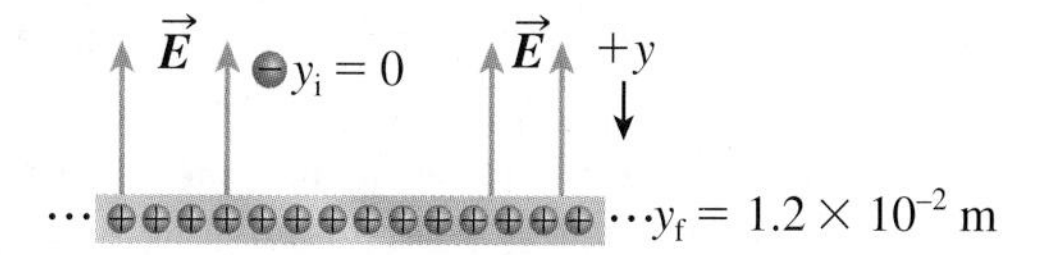

Figure 20-5 Sample Problem 20-1.

The initial electric potential energy ($U_i = qEy_i$) of the electron is zero because $y_i = 0$. Its final potential energy (U_f) at the surface of the plate is

$$\begin{aligned} U_f &= qEy_f \\ &= (-1.60 \times 10^{-19}\,\text{C})(1.3 \times 10^3\,\text{N/C})(1.2 \times 10^{-2}\,\text{m}) \\ &= -2.5 \times 10^{-18}\,\text{J} \end{aligned}$$

Watch signs when substituting into Eqn. 20-1. The charge q for an electron is negative, E is the *magnitude* of the electric field and hence must be positive, and y (positive in this case) is determined by our choice of $y = 0$ position.

By the law of conservation of energy, the sum of the electron's initial potential energy and its initial kinetic energy (E_{Ki}), both of which are zero, equals the sum of its final potential and kinetic energies:

$$U_i + E_{Ki} = U_f + E_{Kf}$$

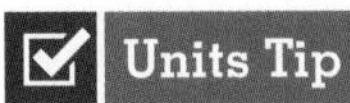

Units Tip

Notice the units in $U = qEy$. The left-hand side has units of joules whereas the right-hand side has units of: $C \cdot (N/C) \cdot m = N{\cdot}m = J$.

Therefore, $0 + 0 = U_f + \frac{1}{2}mv^2$

Solving for v, the speed of the electron as it reaches the plate is

$$v = \sqrt{\frac{-2U_f}{m}}$$

Do not be concerned by the negative sign inside the square root; remember that U_f is also negative.

Substituting $m = 9.11 \times 10^{-31}$ kg for the mass of the electron, and $U_f = -2.50 \times 10^{-18}$ J,

$$v = \sqrt{\frac{-2(-2.50 \times 10^{-18}\ \text{J})}{9.11 \times 10^{-31}\ \text{kg}}}$$

$$= 2.3 \times 10^{6}\ \text{m/s}$$

Learning Tip

When you need to know the mass of an electron or another physical constant, refer to the inside back cover.

Thus, the electron's speed just as it hits the plate is 2.3×10^6 m/s.

Note: Just as in Chapter 7 when working through problems involving gravitational potential energy, we could have also solved this problem by determining the force on the electron based on the field strength, and then used this to determine the acceleration of the charge. The last step in this approach would then be to calculate the final speed of the electron through the one-dimensional equations of motion with constant acceleration.

EXERCISES

20-1 Sketch the shape of a graph of electric potential energy versus height (y) for a proton above a large, negatively charged, horizontal plate. Choose $y = 0$ at the surface of the plate.

20-2 Repeat the previous question for an electron above the plate.

20-3 A typical fair-weather electric field near Earth is 1.5×10^2 N/C downward. Relative to the surface of Earth, what is the electric potential energy of (a) an electron 2.0 m above the surface? (b) a proton 2.0 m above the surface?

20-4 When thunderstorms occur, the electric field in the atmosphere near Earth is directed upward because of a large quantity of negative charge in the lower portions of the clouds. A typical electric field magnitude in this situation is 1.0×10^4 N/C. Relative to the surface of Earth, what are the electric potential energies of the two particles in the previous question? (**Hint:** Remember that the $+y$-direction must be opposite to the direction of the field.)

20-5 What are the gravitational potential energies of the two particles in the previous question, relative to the surface of Earth?

20-6 Each of the following objects is released from rest at point A. In which direction (toward B or away from B) will each object move?

(a) A proton has an electric potential energy of zero at A, $+3.5 \times 10^{-18}$ J at B.

(b) A baseball has a gravitational potential energy of 4.2 J at A, 2.1 J at B.

(c) A toboggan (Figure 20-6) has a gravitational potential energy of -2×10^4 J at A, -1×10^4 J at B.

(d) A positive ion has an electric potential energy of -3.4×10^{-17} J at A, $+2.1 \times 10^{-17}$ J at B.

Figure 20-6 The popular winter pastime of converting gravitational potential energy into kinetic energy, once you get up to the top of the hill that is! (Question 20-6 (c))

20-7 An α particle (charge $+2e$) is located 0.87 cm from a large, negatively charged plate. The field due to the plate has a magnitude of 3.56×10^3 N/C.

(a) What is the particle's electric potential energy relative to the surface of the plate?

(b) If the α particle is released from rest, what is its kinetic energy as it strikes the plate?

20-8 (a) When an electron is emitted inside a vertical fluorescent tube, will it fall because of gravity or rise because of Earth's downward electric field? Earth's average field strength is 1.5×10^2 N/C and the tube is off so there is no additional electric field to consider.

(b) Use energy methods to determine the electron's speed after moving 0.25 m in the tube, assuming that it starts from rest.

20-9 An electron is released from rest halfway between two parallel, oppositely charged plates. The field between the plates has a magnitude of 1.25×10^3 N/C.

(a) What is the sign of the charge on the plate toward which the electron moves?

(b) Just before it hits the plate, the electron has a speed of 2.36×10^6 m/s. Use energy methods to determine the separation of the plates.

20-10 A proton travelling at 3.54×10^5 m/s enters a region of uniform electric field of magnitude 2.18×10^5 N/C in the same direction as the proton's velocity. What is the proton's speed after travelling 1.20 cm? Use energy methods.

20-11 **Fermi Question:** Estimate the ratio of the gravitational potential energy of a skydiver as she leaves the airplane (Figure 20-7) to the electric potential energy of a charged key on a kite string in a thunderstorm. In both cases, use Earth as your reference position.

Figure 20-7 Question 20-11.

20.2 Electric Potential Energy of Point Charges

The electric field near the surface of Earth on a clear day is approximately constant, and the field near a large charged plate is also constant. However, any arbitrary distribution of charges gives rise to a non-constant field, and it is important to remember that Eqn. 20-1 given in Section 20.1 is applicable *only* for calculating the electric potential energy in the special case of a uniform (i.e., constant) electric field.

We now consider the electric potential energy associated with point charges, that is, charges that are each distributed over a small enough volume that they can be considered to be concentrated at single points. Any charge distribution can be thought of as a collection of point charges, and thus our discussion here could be applied to very complicated charge distributions. However, for now we limit ourselves to the electric potential energy of only two point charges, as shown in Figure 20-8. Since electric potential energy is related to the strength of electrical interaction between two charges, we would expect the potential energy to increase as the amount of charge involved (q_1 and q_2) increases, and decrease as the charges get further apart, that is, as r increases. Specifically,

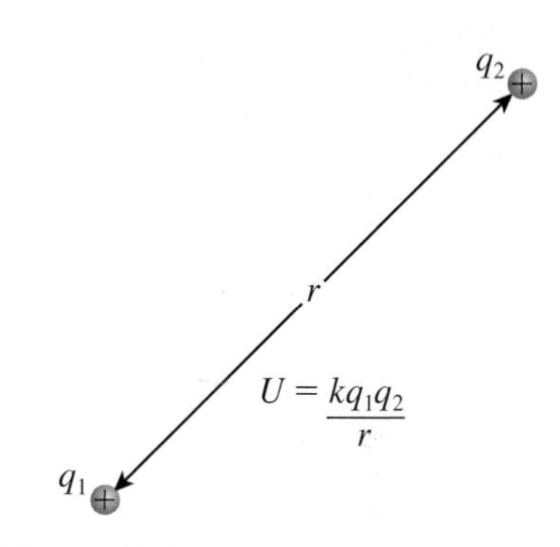

Figure 20-8 Electric potential energy, U, of two point charges.

The electric potential energy, U, of charges q_1 and q_2, separated by a distance r, is defined as

$$U = \frac{kq_1q_2}{r} \tag{20-2}$$

where k is the constant in Coulomb's law ($k = 8.99 \times 10^9$ N·m²/C²).

Although this is the potential energy of the two-charge system, often we think of one as the source charge and the other as the visiting charge, and refer to the potential energy of the visiting charge due to its location in the electric field of the source. This is very much like our discussions of charge interactions in Chapter 19 in which we often referred to the source charge and the test/visiting charge when discussing electric forces and fields.

The expression kq_1q_2/r is analogous to $-GmM/r$ for gravitational potential energy (Chapter 9). The negative sign included in gravitational potential energy reflects the attractive nature of the gravitational force. If q_1 and q_2 are of opposite sign, that is, if the electric force between them is attractive, the product q_1q_2 and the electric potential energy (U) are negative. When q_1 and q_2 have the same sign, U is positive.

TACKLING MISCONCEPTIONS

What Does Negative Potential Energy Mean?

The concept of an object having negative potential energy can seem somewhat strange at first glance. However, there is no physical reason for an object to be required to have positive potential energy, either gravitational or electrical. Negative values for potential energy merely arise based on the selection of the reference position at which the potential energy is zero. The most important feature of the potential energy of an object is what happens when the object changes position in the field, not the sign of the potential energy at a particular point in space. Unlike potential energy, *kinetic energy* (the energy of motion) will always be positive. What would you conclude about the mass or speed of an object if it had negative kinetic energy?

Units Tip

Notice the units for the right-hand side of Eqn. 20-2:

$$\frac{\text{N}\cdot\text{m}^2}{\text{C}^2} \times \frac{\text{C}\cdot\text{C}}{\text{m}} = \text{N}\cdot\text{m} = \text{J}$$

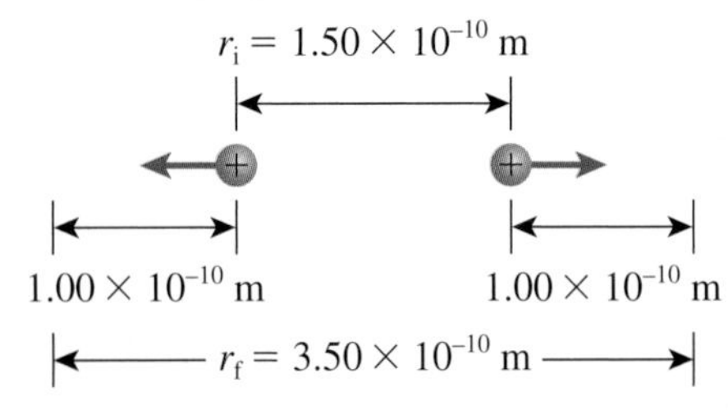

Figure 20-9 Sample Problem 20-2 (c).

Notice that as $r \rightarrow \infty$, $U \rightarrow 0$. Therefore, unlike the previous scenario of electric potential energy in uniform electric fields, we are *not* free to choose the position where electric potential energy is zero when we are dealing with systems of two or more point charges. It is implicit in Eqn. 20-2 that the electric potential energy of point charges approaches zero when the charges are separated by a very large distance; our reference position for $U = 0$ is *always* a charge separation of $r = \infty$ for a system of point charges.

SAMPLE PROBLEM 20-2

(a) What is the electric potential energy of two protons separated by 1.50×10^{-10} m?

(b) If the protons (1.50×10^{-10} m apart) were released from rest, in which direction would they move? Does the electric potential energy of the system increase or decrease as they move?

(c) What would be the speed of these protons after each one has moved 1.00×10^{-10} m?

Solution

(a) The electric potential energy of the protons can be determined by Eqn. 20-2:

$$U = \frac{kq_1q_2}{r}$$

$$= \frac{(8.99 \times 10^9\,\text{N}\cdot\text{m}^2/\text{C}^2)(1.60 \times 10^{-19}\,\text{C})(1.60 \times 10^{-19}\,\text{C})}{1.50 \times 10^{-10}\,\text{m}}$$

$$= 1.53 \times 10^{-18}\,\text{J}$$

Thus, the electric potential energy of the two-proton system is 1.53×10^{-18} J at this separation.

(b) If the protons were released from rest, their mutual electric repulsion would cause them to move away from each other. As the protons move apart, their separation (r in Eqn. 20-2) increases. Since the electric potential energy is inversely proportional to r, an increasing separation results in a decreasing electric potential energy for the two-proton system when released from rest.

(c) As the protons move apart (Figure 20-9), their electric potential energy decreases and their kinetic energy increases by an equal amount by the law of conservation of energy. The initial sum of potential energy (U_i) and kinetic energy (E_{Ki}) equals the final sum ($U_f + E_{Kf}$):

$$U_i + E_{Ki} = U_f + E_{Kf}$$

U_i is known from part (a), and $E_{Ki} = 0$ since the protons start from rest. To determine U_f, we use the same method as in part (a); the protons are now 3.50×10^{-10} m apart and we find from Eqn. 20-2 that U_f is 6.58×10^{-19} J. Therefore, the final kinetic energy of the system is

$$E_{Kf} = U_i + E_{Ki} - U_f$$

$$E_{Kf} = 1.53 \times 10^{-18}\,\text{J} + 0\,\text{J} - 0.658 \times 10^{-18}\,\text{J}$$

$$E_{Kf} = 0.87 \times 10^{-18}\,\text{J}$$

Therefore, the final kinetic energy of the two-proton system is 8.7×10^{-19} J. Since the protons are in identical situations, the increase in kinetic energy of the system due to the decrease in potential energy will be shared equally between the two charges:

$$E_{Kf} = \tfrac{1}{2}mv^2 + \tfrac{1}{2}mv^2$$

$$E_{Kf} = mv^2$$

$$v = \sqrt{\frac{E_{Kf}}{m}}$$

$$v = \sqrt{\frac{8.7 \times 10^{-19}\,\text{J}}{1.67 \times 10^{-27}\,\text{kg}}}$$

$$v = 2.3 \times 10^4\,\text{m/s}$$

Thus, the speed of each proton is 2.3×10^4 m/s. (This answer has only two digits because the subtraction of the potential energies to determine E_{Kf} results in a number with only two significant digits.)

SAMPLE PROBLEM 20-3

(a) What is the electric potential energy of a proton and an electron that are separated by 1.50×10^{-10} m?

(b) If the charges (1.50×10^{-10} m apart) are released from rest, in which direction do they move? Does the electric potential energy of the system increase or decrease as they move?

(c) What is the total kinetic energy of the system after the charges have moved to a separation of 9.50×10^{-11} m?

Solution

(a) The electric potential energy of the two-charge system can again be determined by Eqn. 20-2:

$$U = \frac{kq_1q_2}{r}$$

$$= \frac{(8.99 \times 10^9\,\text{N}\cdot\text{m}^2/\text{C}^2)(-1.60 \times 10^{-19}\,\text{C})(1.60 \times 10^{-19}\,\text{C})}{1.50 \times 10^{-10}\,\text{m}}$$

$$= -1.53 \times 10^{-18}\,\text{J}$$

Thus, the electric potential energy of this system is -1.53×10^{-18} J. Make sure that you include the sign of the charges involved when using Eqn. 20-2; this is very important in understanding charged-particle interactions, as we shall see in the rest of this solution.

(b) If the oppositely charged particles were released from rest, their mutual electric attraction would cause them to move toward each other. As the charges move closer, their separation (r in Eqn. 20-2) decreases. Since the electric potential energy is inversely proportional to r, a decreasing separation results in the negative electric potential energy becoming more negative, that is, a decreasing electric potential energy for the two-charge system when released from rest.

(c) As the opposite charges move closer, their electric potential energy decreases and their kinetic energy increases by an equal amount by the law of conservation of energy. The initial sum of potential energy (U_i) and kinetic energy (E_{Ki}) equals the final sum ($U_f + E_{Kf}$):

$$U_i + E_{Ki} = U_f + E_{Kf}$$

Problem-Solving Tip

Positive and Negative Changes in Electric Potential Energy

- The potential energy of the system decreases when opposite charges get closer.
- The potential energy of the system decreases when the same sign charges get further apart.
- The potential energy of the system increases when the charges are moved against the electric field.
- Due to the principle of conservation of energy, the kinetic energy of a charge increases when its potential energy decreases, and its kinetic energy decreases when its potential energy increases.

U_i is known from part (a), and $E_{Ki} = 0$ since the charges start from rest. To determine U_f, we use the same method as in part (a); the charges are now 9.50×10^{-11} m apart and we find from Eqn. 20-2 that U_f is -2.42×10^{-18} J. Therefore, the final kinetic energy of the system is

$$E_{Kf} = U_i + E_{Ki} - U_f$$

$$E_{Kf} = -1.53 \times 10^{-18}\text{ J} + 0\text{ J} - (-2.42 \times 10^{-18}\text{ J})$$

$$E_{Kf} = 2.42 \times 10^{-18}\text{ J} - 1.53 \times 10^{-18}\text{ J}$$

$$E_{Kf} = 0.89 \times 10^{-18}\text{ J}$$

Therefore, the final kinetic energy of the two-charge system is 8.9×10^{-19} J. The negative sign of the electric potential energy of two opposite charges was crucial to our understanding of how these charges interact: If we had used the absolute values of the charges involved, we would have ended up with a negative final kinetic energy at the end!

EXERCISES

20-12 Sketch a graph of the electric potential energy versus separation distance r for:

(a) two positive charges

(b) two negative charges

(c) one positive and one negative charge.

20-13 (a) What is the electric potential energy of two point charges separated by a very large distance?

(b) If the charges have different signs, does the electric potential energy increase or decrease as the separation decreases?

(c) Will the charges with different signs move toward each other or away from each other if released from rest? Is this in the direction of increasing or decreasing electric potential energy?

20-14 Two protons are separated by 6.0×10^{-15} m in a uranium nucleus. What is their electric potential energy?

20-15 In the ground state of the hydrogen atom, the electron is, on average, a distance of 5.3×10^{-11} m from the proton in the nucleus. What is the electric potential energy of such an atom?

20-16 **Fermi Question:** How much energy would be required to ionize all the hydrogen atoms in a glass of water (Figure 20-10)? (See the previous question for guidance.)

Efired/Shutterstock

Figure 20-10 Question 20-16.

20-17 The electric potential energy of an electron near a uranium nucleus (charge $+92e$) is -1.41×10^{-16} J. What is the distance between the electron and the nucleus?

20-18 Two α particles (charge $+2e$, mass 6.6×10^{-27} kg each) are released from rest, separated by 3.40×10^{-12} m.

(a) Will their separation increase or decrease? Is this increasing or decreasing electric potential energy?

(b) When the speed of each particle is 1.7×10^5 m/s, how far apart are they?

20.3 Electric Potential

electric potential electric potential energy per unit charge at a particular location in space within an electric field

volt SI unit for electric potential, potential difference, or voltage (V)

You have undoubtedly seen signs (Figure 20-11) such as: "Danger. High Voltage! Keep Out." You might already know that the voltage available from normal electrical outlets in North America is 120 volts, and you probably know that different batteries have different voltages: 1.5 volts, 6 volts, etc. What exactly is voltage?

To answer this question, we start by considering a charge placed at some point P where there is an electric field (Figure 20-12). This charge will have electric potential

Oliver Sved/Shutterstock

krivinis/Shutterstock

PH studio/Shutterstock

Nikolaj Kondratenko/Shutterstock

Figure 20-11 High voltage warning signs from around the world.

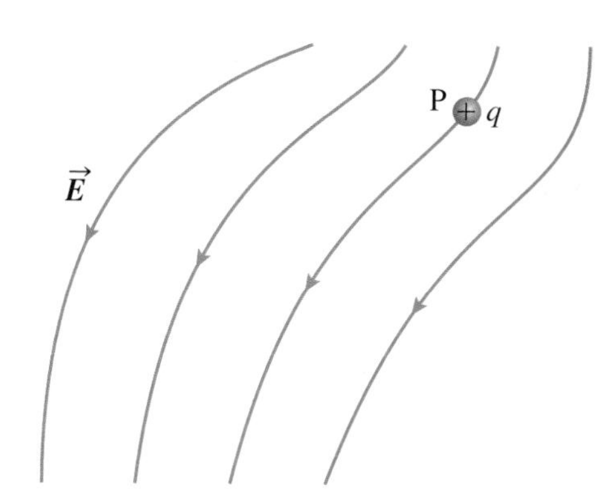

Figure 20-12 The electric potential at a point P in an electric field is the electric potential energy per unit charge q placed at that point.

TACKLING MISCONCEPTIONS

The Difference between Electric Potential and Electric Potential Energy

Be careful! The definition of electric potential, as described in Eqn. 20-3, often leads students to think that the visiting charge, q, somehow plays a role in the amount of electric potential found at that location. Just as we discussed with the difference between electric force and electric field in Chapter 19, there needs to be *two (or more)* charges present in order for there to be an *electric potential energy* associated with the system. However, *one* single lonely point charge is all that is needed for there to be an associated *electric potential* everywhere in space. *Electric potential energy* describes the interaction between two or more specific charges, whereas *electric potential* merely describes the effect that one charge has on its surroundings in general.

energy, which will depend on the amount of charge it has, as well as the electric field and the position of the charge within it. The larger the charge, the larger is the potential energy. The **electric potential** at point P is defined as the electric potential energy per unit charge. By dividing out the charge q, electric potential depends only on the electric field and the position within it. Using the customary symbol V for electric potential,

$$V = \frac{U}{q} \qquad (20\text{-}3)$$

Be careful: the terminology here can get a little confusing. Electric potential is not the same thing as electric potential energy, although, as we can see in Eqn. 20-3, they are closely related. Since both potential energy and charge are scalar quantities, electric potential is also a scalar.

Notice that we refer to the electric potential at a particular point in space. The electric potential is the result of the presence of source charges that produce the electric field at that point. The electric potential is not associated with any particular visiting charge *at* that point. If a charge happens to be placed at a point of known electric potential V, the charge's potential energy U can be determined by simply rearranging Eqn. 20-3 to give

$$U = qV \qquad (20\text{-}4)$$

Since electric potential is energy per unit charge, one possible SI unit is joule/coulomb (J/C); this is given the special name of **volt**, abbreviated V, where 1 volt (V) = 1 joule/coulomb (J/C). This unit is named in honour of the Italian physicist Alessandro Volta (Figure 20-13), credited with inventing the electric battery. Notice that the symbol for electric potential is written as V (italicized), and that the unit is also V (not italicized).

Since the electric potential energy is always relative to some position, such as our chosen $y = 0$ position for a uniform field or infinite separation for two point charges, the electric potential is also relative to this position. As is the case with

Salvatore Chiariello/Shutterstock

Figure 20-13 A monument in honour of Alessandro Volta (1745–1827), the Italian physicist credited with the invention of the battery. The monument is found in Volta's hometown of Como, Italy.

potential difference the change in electric potential from one location in an electric field to another

voltage see potential difference

potential energy, what is important is not the actual value of electric potential but rather the *change* in electric potential, or **potential difference** ΔV, from one point to another. This potential difference is commonly called **voltage**. When we say that a car battery provides 12 V, this means that the potential difference between one terminal and the other is 12 V. The battery terminal with the higher electric potential is referred to as the *positive* terminal, the lower-potential terminal being *negative*.

Very often, a quoted voltage is the potential difference between a point and Earth (or "ground"). For example, a 120 V electrical outlet provides a potential difference of 120 V between one of its wires and ground. If you read that the ionosphere[1] has a voltage of 3×10^5 V, it is implicitly assumed that this is the potential difference between the ionosphere and Earth.

The term "potential difference" is often used somewhat imprecisely. If a charge moves from a point where the electric potential is +5 V to a point where it is +12 V, the change in electric potential, that is, the potential difference, is +7 V. If the charge moves from +12 V to +5 V, the potential difference is now −7 V. However, it is common practice to say that the charge has moved through a potential difference of 7 V in either case. In other words, *potential difference often means the absolute value of the change in electric potential.* We need to be a little careful with signs when solving problems involving electric potential and potential difference.

Since we know that the electric potential energy of a charge q at a point where the electric potential is V is given by $U = qV$ (Eqn. 20-4), the *change* in the potential energy of the charge as it moves is given by

$$\Delta U = q\Delta V \quad \text{(20-5a)}$$

$$\text{or} \quad U_2 - U_1 = q(V_2 - V_1) \quad \text{(20-5b)}$$

where "1" and "2" represent two points in space (Figure 20-15).

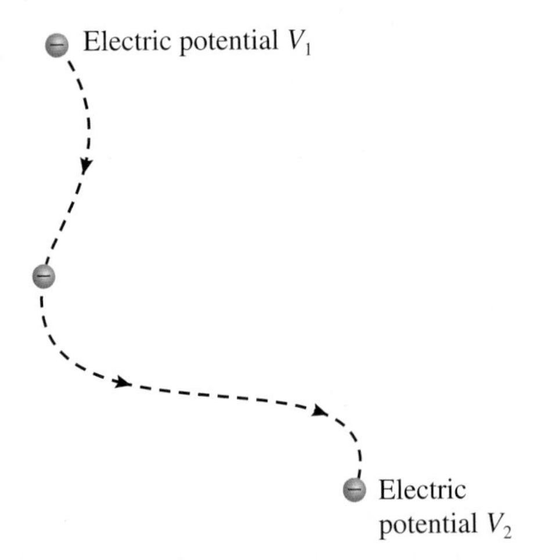

Figure 20-15 As the charge q moves from point 1 to 2, its electric potential energy changes by $\Delta U = q\Delta V$, that is, $U_2 - U_1 = q(V_2 - V_1)$.

DID YOU KNOW?

An electrocardiogram (ECG or EKG) is a recording of electric potential differences (as a function of time) on the skin as a result of the electrical activity of the heart (Figure 20-14). The potential differences detected are of the order of a few millivolts (mV). An electroencephalogram (EEG) is a recording of the electrical activity of the cerebral cortex, with potential differences in the range of a few tens of microvolts (μV).

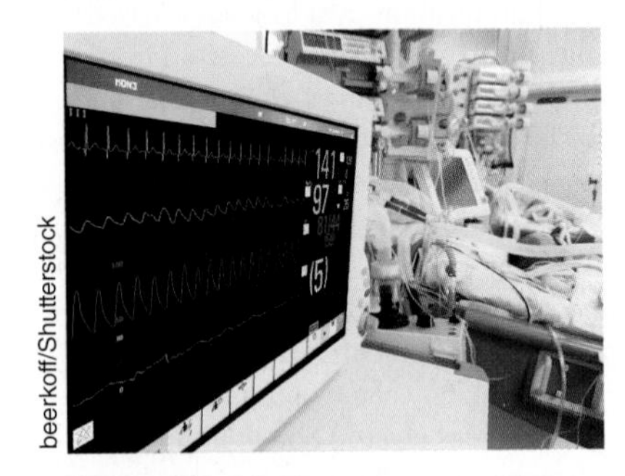

beerkoff/Shutterstock

Figure 20-14 The green trace across the top of the monitor is a display of the electrical activity of the heart detected through an ECG.

[1]The ionosphere is a region of the atmosphere, from about 80 to more than 600 km above Earth, where there are many ions produced by solar radiation and by energetic particles from the Sun.

A very useful property of Eqn. 20-5 is that it is valid regardless of the particular path followed by the charge from point 1 to point 2.

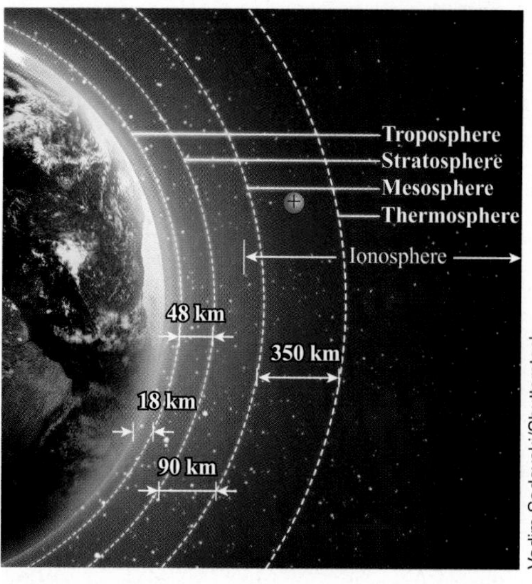

Figure 20-16 Sample Problem 20-4.

SAMPLE PROBLEM 20-4

(a) If the electric potential at a particular point in the ionosphere is $+3.0 \times 10^5$ V relative to Earth, what is the electric potential energy of a proton at that location (Figure 20-16)?

(b) Repeat for an electron.

Solution

(a) This problem is solved with a straightforward substitution into Eqn. 20-4:

$$U = qV = (1.60 \times 10^{-19}\ \text{C})(3.0 \times 10^5\ \text{V}) = 4.8 \times 10^{-14}\ \text{J}$$

Therefore, the electric potential energy of the proton is 4.8×10^{-14} J relative to Earth.

(b) For an electron, the only change is that the charge is negative. Hence, the electric potential energy of the electron is -4.8×10^{-14} J at this location in the ionosphere relative to Earth.

SAMPLE PROBLEM 20-5

When a car horn is sounded (Figure 20-17), charge moves from one terminal of the car battery, through the horn, and back to the other battery terminal. If the battery has a voltage of 12 V, the charge has been moved through a potential difference of 12 V. (As mentioned earlier, this commonly means that the absolute value of the potential difference is 12 V.) The charge moves in the direction of decreasing electric potential energy, and hence the charge loses potential energy, which is converted into sound energy in the horn (and some thermal energy). How much energy is converted, if the magnitude of charge that moves is 8.0 C?

Solution

We are not told what sign of charge is moving (to be discussed in Section 20.5), nor whether the change in electric potential is positive or negative. Therefore, we consider only the absolute value of quantities, using Eqn. 20-5

$$|\Delta U| = |q\Delta V| = (8.0\,\text{C})(12\,\text{V}) = 96\ \text{J}$$

Thus, 96 J of electric potential energy is converted to sound (and thermal energy).

Units Tip

In problems involving electric potential, it is often handy to remember that a volt is a joule per coulomb (1 V = 1 J/C). For example, when using the equation $U = qV$, charge q in coulombs times electric potential V in volts (i.e., joules per coulomb) gives electric potential energy in joules.

Figure 20-17 Sample Problem 20-5.

The Electron Volt—An Energy Unit

In atomic and subatomic physics, the joule is an enormous energy unit. Tiny particles such as electrons have very small energies when measured on the scale of joules as we have seen in the preceding Sample Problems in this chapter. For example, the kinetic energy of an electron moving with a speed of 1×10^6 m/s (one million metres per second!) is only 5×10^{-19} J. A more common energy unit on these scales is the electron volt (eV), and its multiples, keV (10^3 eV), MeV (10^6 eV), etc. (introduced in Section 7.6).

electron volt a unit of energy that corresponds to the change in electric potential energy experienced by an electron moving through a potential difference of 1 V (eV)

An **electron volt** is defined as the electric potential energy gained or lost by an electron in moving through a potential difference of one volt. If an electron moves to a point where the electric potential has increased by 1 V, its electric potential energy decreases by 1 eV (since $\Delta U = q\Delta V$, and $q < 0$ for an electron). If the electron moves so that the electric potential decreases by 1 V, it gains electric potential energy of 1 eV. In either case, it is common to say that the electron has moved through a potential difference of 1 V.

We can find the joule equivalent of 1 eV from $|\Delta U| = |q\Delta V|$. Using $\Delta V = 1$ V and, for an electron, $q = -e = -1.60 \times 10^{-19}$ C:

$$|\Delta U| = |q\Delta V| = (1.60 \times 10^{-19}\,\text{C})(1\,\text{V}) = 1.60 \times 10^{-19}\,\text{J}$$

Thus,

$$1\,\text{eV} = 1.60 \times 10^{-19}\,\text{J} \qquad \textbf{(20-6)}$$

As an example, converting the 5×10^{-19} J of kinetic energy of an electron moving at 1×10^6 m/s to electron volts gives

$$(5 \times 10^{-19}\,\text{J}) \times \left(\frac{1\,\text{eV}}{1.60 \times 10^{-19}\,\text{J}}\right) = 3\,\text{eV}$$

which is much more convenient to deal with than 5×10^{-19} J.

The electron volt is particularly easy to use when dealing with charged particles that are moving through potential differences. The electric potential energy gained or lost by an electron moving through 1 V is 1 eV, by definition. From the equation $|\Delta U| = |q\Delta V|$, we see that the potential energy change is proportional to the charge, and also proportional to the potential difference. Thus, if an electron moves through 2 V, the energy gained or lost is simply 2 eV; through 10 V, the energy is 10 eV, etc. Since a proton has the same magnitude of charge as an electron, the electric potential energy it gains or loses in moving through a potential difference of, say, 5 V is also 5 eV. An α particle has a charge of $+2e$; if it moves through 1 V, its energy gained or lost is 2 eV. If it moves through 3 V, it gains or loses 6 eV.

SAMPLE PROBLEM 20-6

An electron is accelerated from rest through a potential difference of 3.1×10^6 V in a particle accelerator designed for cancer therapy (Figure 20-18).

(a) As the electron is accelerated, does it move toward higher, or lower, electric potential?

(b) What is its final kinetic energy in electron volts?

Figure 20-18 Sample Problem 20-6.

Solution

(a) We know that particles always move toward decreasing potential energy when released from rest. Let us arbitrarily define the electron's initial potential energy (U_i) as zero, which means that the initial electric potential (V_i) is also zero, since $V_i = U_i/q$ from Eqn. 20-3. The electron will then move toward negative potential energy, as the electric field inside the accelerator acts on the charge to decrease its electric potential energy. Therefore, the final potential energy of the electron (U_f)

is less than zero, and, since the electron has a negative charge, V_f must be positive. In other words, the electron is moving toward *higher* electric potential, $\Delta V > 0$, as it moves to reduce its electric potential energy such that $\Delta U < 0$.

(b) By the law of conservation of energy, the sum of the electron's potential energy and kinetic energy must remain constant:

$$U_i + E_{Ki} = U_f + E_{Kf}$$

$$\text{or} \quad \Delta E_K = -\Delta U$$

In other words, the increase in kinetic energy equals the decrease in electric potential energy. Since the electron starts from rest, $E_{Ki} = 0$ and $\Delta E_K = E_{Kf}$. Therefore

$$E_{Kf} = -\Delta U$$

Eqn. 20-5 tells us that the change in electric potential energy of the electron (ΔU) is given by $q\Delta V$. Therefore, the final kinetic energy of the electron that starts from rest and moves through a potential difference of ΔV is given by

$$E_{Kf} = -q\Delta V$$

To arrive at an answer in electron volts, the charge q must be expressed in multiples of e (the absolute value of the elementary electric charge). For the electron, $q = -1\ e$. The potential difference ΔV must be expressed in volts: $\Delta V = +3.1 \times 10^6$ V in this case.

Thus, $E_{Kf} = -(-1\ e)(+3.1 \times 10^6\ \text{V}) = 3.1 \times 10^6$ eV (or 3.1 MeV)

Hence, the electron's final kinetic energy is 3.1×10^6 eV. As it accelerates through $+3.1 \times 10^6$ V, the electron loses 3.1×10^6 eV of electric potential energy, and gains 3.1×10^6 eV of kinetic energy.

Learning Tip

In general, if negative particles are released from rest, they move toward higher or increasing electric potential; thus, $\Delta V > 0$. Positive particles move toward lower or decreasing electric potential; hence, $\Delta V < 0$. *In either case, all charged particles move toward decreasing electric potential energy when released from rest.*

EXERCISES

20-19 A chloride ion with a single net negative charge (Cl^-) has an associated electric potential energy of -4.2×10^{-19} J at its present location. What is the electric potential at the position of the ion?

20-20 The membranes of nerve cells (or neurons) in their normal resting state have an electric potential difference across them (Figure 20-19). The electric potential in a cell's interior is typically –85 mV relative to the cell's exterior. If a sodium ion (Na^+) diffuses across a cell membrane,

(a) In which direction will it travel, starting from rest?

(b) What is the resulting change in electric potential energy of the ion as it crosses the membrane?

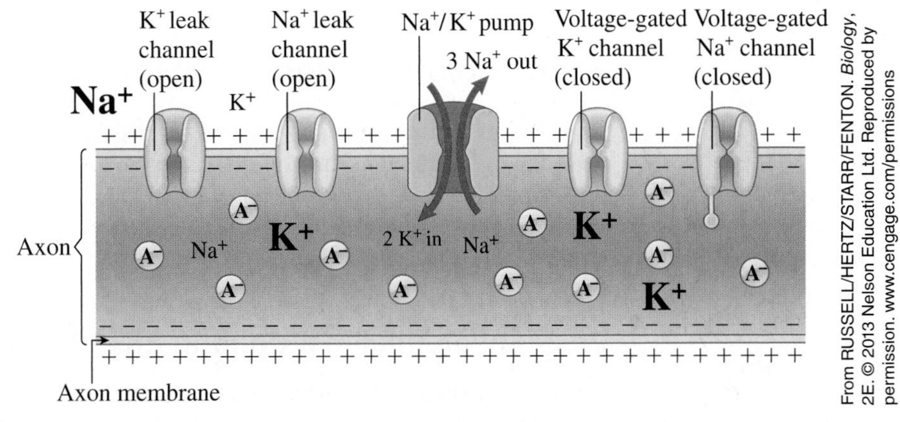

Figure 20-19 Na^+ and K^+ diffuse through the membrane while the Na^+/K^+ pumps work to move them back again to maintain the potential difference. The negatively charged proteins, amino acids, and other molecules cannot readily pass through the membrane (collectively labelled A^- here) (Question 20-20).

20-21 On a particular day, two points (A) 1.00 m and (B) 1.50 m above Earth have electric potentials of 152 V and 228 V respectively (both relative to ground).

(a) What is the electric potential energy in electron volts of an α particle (charge $+2e$) at each point, relative to ground?

(b) Will the α particle move from A to B, or B to A, when released from rest?

20-22 A flashlight (Figure 20-20) is powered by two 1.5 V batteries, providing a total potential difference of 3.0 V. When the flashlight is turned on for 4.0 s, the total charge that moves through the bulb is 12 C.

(a) What amount of energy has been provided by the batteries?

(b) What is the power supplied by the batteries?

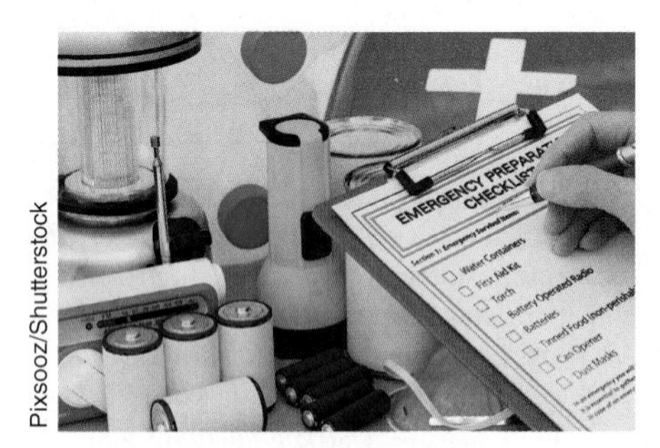

Pixsooz/Shutterstock

Figure 20-20 Question 20-22.

20-23 (a) How much energy is dissipated by a lightning bolt that transfers -22 C of charge to Earth through a potential difference of 3.2×10^7 V? Express the answer in megajoules.

(b) As the negative charge goes from the cloud to Earth, is it moving toward higher, or lower, electric potential?

20-24 Perform the following unit conversions:

(a) 3.74×10^{-18} J to electron volts

(b) 511 keV to joules

(c) 2.98×10^{-13} J to mega-electron volts

20-25 A proton is accelerated from rest through a potential difference of 4.2×10^6 V in a particle accelerator.

(a) What is the change in the proton's electric potential energy (in electron volts)?

(b) What is the change in the proton's kinetic energy (in electron volts)?

20-26 What potential difference is required to give a helium nucleus (charge $+2e$) an increase in kinetic energy of 8 MeV?

20-27 An electron is moving to the right from position A with a speed of 1.43×10^6 m/s, as shown in Figure 20-21. At position A there is an electric potential of $+2.53$ V. When the electron reaches position B, there is an electric potential of $+1.35$ V. What is the electron's speed as it reaches position B?

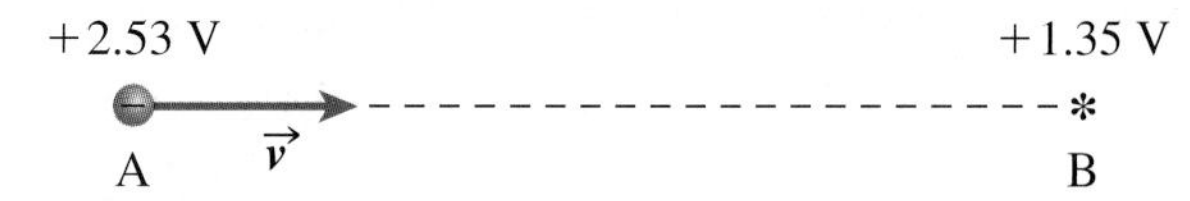

Figure 20-21 Question 20-27.

20-28 Why do you think birds can sit unharmed on high voltage transmission lines (Figure 20-22)? (**Hint:** Think about what ΔV is between their feet and hence the value of ΔU through their bodies.)

holstphoto/Shutterstock

Figure 20-22 Question 20-28.

20.4 Relating Electric Potential and Electric Field

As we discussed in the previous section, the electric potential at a particular location is related to the electric field arising from source charges. In this section we will explore the relationship between electric potential and the electric field in more detail for two specific scenarios: electric potential in a uniform electric field and electric potential in an electric field due to point source charges.

Electric Potential in a Uniform Electric Field

A number of situations in which a uniform electric field exists have been described: near Earth's surface, near a large charged plate, and between oppositely charged plates. A uniform electric field can also exist across the membrane of a biological cell as a result of charges on both the exterior and interior surfaces (Figure 20-19).

Figure 20-23 shows a large negatively charged plate that produces a uniform electric field $\vec{E}$ close to the plate. We will investigate how the electric potential, V, varies with distance d from the plate in such an electric field.

We start by considering the electric potential energies of a positive test charge q as it is moved against the electric field, starting at position 1 at the surface of the plate and ending at position 2 at a distance d away from the plate (Figure 20-23). As discussed in Section 20.1, Eqn. 20-1 can be used to calculate the electric potential energy in this situation. As shown, the $y = 0$ position is defined to be at the plate, and the $+y$-direction is opposite to the field direction. At position 2, $y = d$, and hence the electric potential energy of the charge is given by Eqn. 20-1:

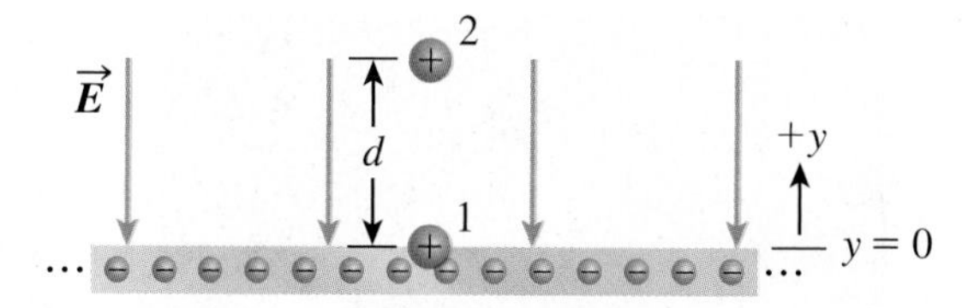

Figure 20-23 In a *uniform* electric field of magnitude E, the change in electric potential between positions 1 and 2 is $(V_2 - V_1) = Ed$, or $\Delta V = Ed$, where d is the distance between these locations measured along a straight line in the direction of the electric field.

$$U_2 = qEd$$

At position 1, $y = 0$, and hence the electric potential energy is zero:

$$U_1 = 0$$

The electric potential at positions 2 and 1 can be determined by using the defining equation for potential: $V = U/q$ (Eqn. 20-3). At position 2,

$$V_2 = \frac{U_2}{q} = \frac{qEd}{q} = Ed$$

At position 1,

$$V_1 = \frac{U_1}{q} = 0$$

We now subtract V_1 from V_2 to give the change in electric potential, ΔV, from position 1 to position 2:

$$\Delta V = V_2 - V_1 = Ed - 0$$

$$\text{Thus, } \Delta V = Ed \qquad \textbf{(20-7)}$$

where E is the magnitude of the *uniform* electric field in the region, and d is the distance between locations measured along a straight line in the direction of the electric field.

So, when dealing with a uniform electric field, we can calculate the potential difference between two positions in the field simply by knowing the magnitude of the field and the distance between the two positions. Note that d must be measured parallel to the field direction, as shown in Figure 20-23. If we now rearrange Eqn. 20-7,

$$E = \frac{\Delta V}{d} \qquad \textbf{(20-8)}$$

where E is the magnitude of the *uniform* electric field in the region.

Eqn. 20-8 shows us that the magnitude of the uniform electric field can be determined by the potential difference between two positions and the distance between them. Notice the unit of electric field implied by Eqn. 20-8. With ΔV in volts and d in metres, E has units of volts per metre (V/m). This is a more common unit for electric field than is newtons per coulomb (N/C), but the two are completely equivalent SI units: 1 V/m = 1 N/C.

The unit of volts per metre has a very useful physical interpretation: it tells us by how much the electric potential changes with distance. If the uniform electric field near Earth is 150 V/m downward, then for every metre moved downward the electric potential changes by 150 V. In particular, the electric potential decreases by 150 V

iodrakon/Shutterstock

Figure 20-24 Sparks arising from the conduction of electricity through air.

when moving 1 m in the direction of the electric field, and increases by 150 V when moving 1 m in the opposite direction.

Dry air can withstand a maximum electric field of approximately 3×10^6 V/m. If exposed to an electric field of about this magnitude, the air molecules become ionized and form a conducting path for the charges creating the field. The charges travel rapidly along this path, generating light and thermal energy in what we call a *spark* (Figure 20-24). You can easily estimate the potential difference giving rise to any sparks that you see. Suppose that you shuffle across the carpet in dry weather and generate a spark that is 2 cm (0.02 m) long when you touch a doorknob. As indicated above, the magnitude of the electric field required in dry air for a spark to occur is 3×10^6 V/m. Substituting these numbers into Eqn. 20-7 gives

$$\Delta V = Ed = (3 \times 10^6 \text{ V/m})(0.02 \text{ m}) = 6 \times 10^4 \text{ V}$$

Hence, a 2 cm spark results from a potential difference of about 6×10^4 V. Lightning, which is essentially a very large spark, is discussed in Section 20.5; such sparks arise from a very large potential difference between Earth and the storm clouds above.

As with our analysis of masses moving around in Earth's gravitational field (Chapters 5 and 7), we can solve problems involving the motion of charges in uniform electric fields via two complementary approaches: (1) by determining the acceleration of the charge through Newton's second law in conjunction with the electric force acting on the charge, or (2) by analyzing the change in kinetic energy resulting from the change in electric potential energy as the charge moves in the electric field. The first approach involves vectors, while the second approach only involves scalars, which is mathematically more straightforward. The approach you take for a given problem must also be influenced by the specifics of what is being asked, in addition to considering which method is more straightforward. Regardless of the method used, you will get the same result—one just might get you there a little more directly! Table 20-1 summarizes these complementary approaches to understanding the motion of a charge in a uniform electric field.

Table 20-1

Analyzing the Motion of Charges in Uniform Electric Fields

Dynamics/Kinematics (vector quantities)	Conservation of Energy (scalar quantities)
$\vec{E}$: arises from a particular configuration of source charges	V: arises from a particular configuration of source charges
$\vec{F} = q\vec{E}$ is then the force on a particular charge q placed in the electric field	$\Delta U = q\,\Delta V$ is the change in the electric potential energy of charge q moving through a potential difference ΔV
$\vec{a} = \frac{\vec{F}}{m} = \frac{q\vec{E}}{m}$ is the resulting acceleration of charge q with mass m in the electric field $\vec{E}$ (when no other forces need to be considered)	$\Delta E_K = -\Delta U = -q\,\Delta V$ is the resulting change in the kinetic energy of charge q moving through a potential difference ΔV

Electric Potential Due to Point Charges

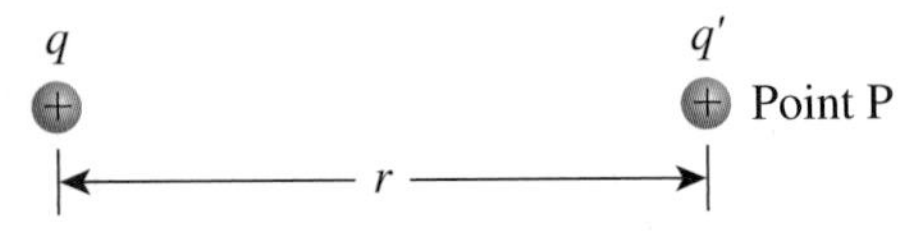

Figure 20-25 At point P, the electric potential, V, due to q is given by $V = kq/r$.

Figure 20-25 shows two point charges: q, which we will consider to be the source charge producing an electric field in space, and q', the test charge located at point P which is a distance r from the source charge. The electric potential energy of the two-charge system is $U = kqq'/r$ (Eqn. 20-2). Recalling that electric potential is electric potential energy per unit charge, we can easily determine the potential at point P by dividing U by q':

$$V = \frac{U}{q'} = \frac{kqq'}{r} \times \frac{1}{q'} = \frac{kq}{r}$$

Therefore,

$$V = \frac{kq}{r} \qquad (20\text{-}9)$$

is the electric potential at a distance r from a point charge q. As discussed in Section 20.2, the reference location where the electric potential is zero in this scenario is defined as $r \rightarrow \infty$.

Math Tip

The similarity between Eqn. 20-9, the electric potential near a point charge, and Eqn. 19-6, the electric field near a point charge, is not an accident. The electric potential, $V = kq/r$, and the magnitude of the electric field at that location, $E = kq/r^2$, only differ by a factor of $1/r$. For those of you with a background in calculus, in general we can express this relationship as $|\vec{E}| = \left|\frac{dV}{dr}\right|$, that is, the magnitude of the electric field equals the magnitude of the rate of change of the electric potential as a function of distance away from the source.

SAMPLE PROBLEM 20-7

(a) What is the electric potential at point A in Figure 20-26 due to the nearby electron and proton?

(b) Determine the potential difference ($V_B - V_A$) between points A and B.

(c) If a Cl^- ion is introduced into this system and moves from A to B, what is its change in electric potential energy?

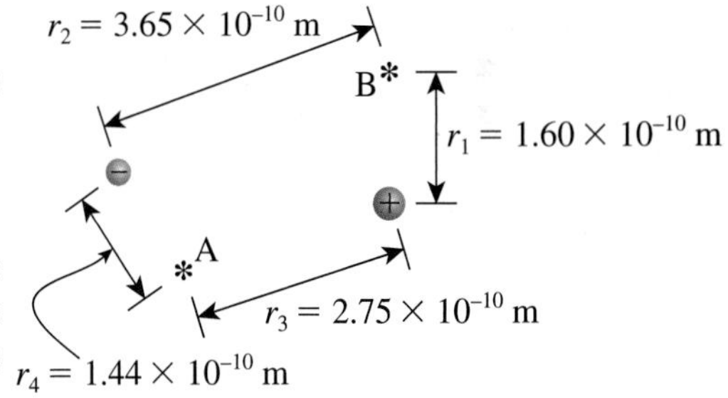

Figure 20-26 Sample Problem 20-7.

Solution

(a) The electric potential at A due to the electron and proton is just the sum of the separate electric potentials due to these two particles. Since electric potential is a scalar quantity, this is a straightforward addition. Thus, using Eqn. 20-9, the electric potential at A is

$$V_A = \frac{kq_p}{r_3} + \frac{kq_e}{r_4}$$
$$= k\left[\frac{q_p}{r_3} + \frac{q_e}{r_4}\right]$$

where q_p and q_e represent the charges on the proton and electron, respectively. Substituting numbers,

$$V_A = 8.99 \times 10^9\,\text{N}\cdot\text{m}^2/\text{C}^2\left[\frac{1.60 \times 10^{-19}\,\text{C}}{2.75 \times 10^{-10}\,\text{m}} + \frac{-1.60 \times 10^{-19}\,\text{C}}{1.44 \times 10^{-10}\,\text{m}}\right]$$
$$= -4.8\,\text{V}$$

Thus, the electric potential at point A is -4.8 V.

(b) The electric potential at point B can be calculated in a similar way:

$$V_B = k\left[\frac{q_p}{r_1} + \frac{q_e}{r_2}\right]$$
$$= 8.99 \times 10^9\,\text{N}\cdot\text{m}^2/\text{C}^2\left[\frac{1.60 \times 10^{-19}\,\text{C}}{1.60 \times 10^{-10}\,\text{m}} + \frac{-1.60 \times 10^{-19}\,\text{C}}{3.65 \times 10^{-10}\,\text{m}}\right]$$
$$= 5.0\,\text{V}$$

Notice that the electric potential at B is positive whereas the electric potential at A is negative. This is because we are closer to the negative source charge at A and closer to the positive source charge at B.

Given the values of the electric potential at these two locations, the potential difference is $V_B - V_A = (5.0 - (-4.8))\,\text{V} = 9.8\,\text{V}$.

Therefore, the potential difference in moving from location A to location B is 9.8 V.

Units Tip

When electric potential is calculated using $V = kq/r$, the SI units on the right-hand side are

$N{\cdot}m^2/C^2$ for k

C for q

m for r

This gives a unit for electric potential (V) of $(N{\cdot}m^2/C^2){\cdot}C/m$, which is $N{\cdot}m/C$. The product $N{\cdot}m$ is equivalent to a joule (J), and so the unit for electric potential is J/C, which is a volt (V).

(c) If the negatively charged Cl^- ion moves from A to B, its change in electric potential energy can be determined from Eqn. 20-5:

$$\Delta U = q\Delta V$$

that is, $U_B - U_A = q(V_B - V_A)$

Since $q = -e = -1.60 \times 10^{-19}$ C, and $\Delta V = V_B - V_A = 9.8$ V in this problem,

$$\Delta U = (-1.60 \times 10^{-19}\,\text{C})(9.8\,\text{V}) = -1.6 \times 10^{-18}\,\text{J}$$

Thus, the electric potential energy of the negatively charged Cl^- ion *decreases* by 1.6×10^{-18} J (or 9.8 eV).

As you may have noticed in the sample problems in this chapter, it can be very easy to run into trouble with the signs of charges. Table 20-2 provides a useful summary of the changes in electric potential energy and electric potential encountered by charges moving in electric fields to serve as an aid when working through such problems.

Table 20-2

Changes in Electric Potential Energy (*U*) and Electric Potential (*V*)

Sign of Charge	*U*	*V*	Scenario
+	decreases	decreases	when the charge moves as a result of the electric field
–	decreases	increases	
+	increases	increases	when the charge is moved against the electric field
–	increases	decreases	

EXERCISES

20-29 Points A and B in a uniform electric field have electric potentials of 5 V and 7 V respectively. The line joining A and B is parallel to the field direction. In which direction does the electric field point?

20-30 On a particular day, the uniform electric field near Earth is 232 V/m downward. If we define the electric potential to be zero at ground level, what is the electric potential 25 m above the ground? Be sure to specify whether this is positive or negative with respect to ground level.

20-31 After walking on a carpet on a dry day, you "zapped" a friend with a spark 1.0 cm long. What was the approximate potential difference between you and your friend?

20-32 A uniform electric field exists across the membrane of a neuron, a nerve cell such as those depicted in the opening image of this chapter (Figure 20-1). For a neuron having a membrane thickness of 7.2 nm and a potential difference of 87 mV across the membrane, what is the magnitude of the electric field?

20-33 Figure 20-27 shows two parallel, oppositely charged plates with a potential difference of 6.0 V between them. The plates are separated by 1.1 cm. Assume that the electric potential of the negative plate is zero.

(a) What is the electric potential of the positive plate? Be sure to specify whether this is positive or negative with respect to the negative plate.

(b) What is the electric potential at point A, which is 0.28 cm above the positive plate? Be sure to specify whether this is positive or negative with respect to the negative plate.

(c) What is the electric field (magnitude and direction) between the plates?

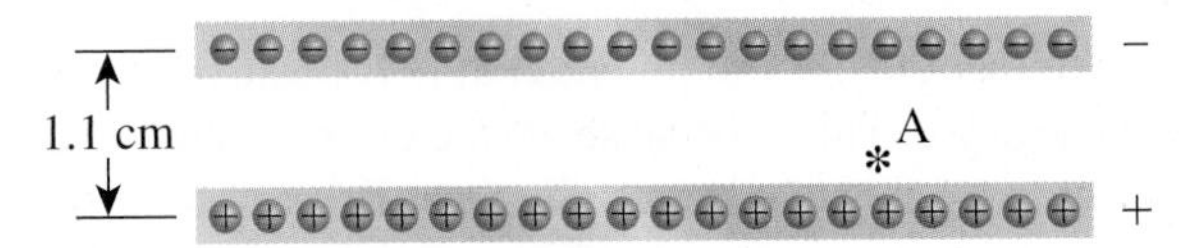

Figure 20-27 Question 20-33.

20-34 If the uniform electric field near Earth is 125 V/m downward, what is the electric potential energy (relative to Earth) of a dust particle having a charge of $-84e$, floating in the air 3.40 m above Earth?

20-35 A proton is accelerated from rest from a positively charged plate to a parallel negatively charged plate. The separation of the plates is 8.9 mm. If the potential difference between the plates is 75.3 V, what is the speed of the proton just as it hits the negative plate? (The mass of a proton is given in the inside back cover.)

20-36 What is the electric potential at a position that is 1.25×10^{-9} m away from (a) an electron? (b) a sodium ion Na^{+}?

20-37 At what distance from a chlorine nucleus (charge $+17e$) is the electric potential 1.0 V? Assume that the nucleus acts as a point charge.

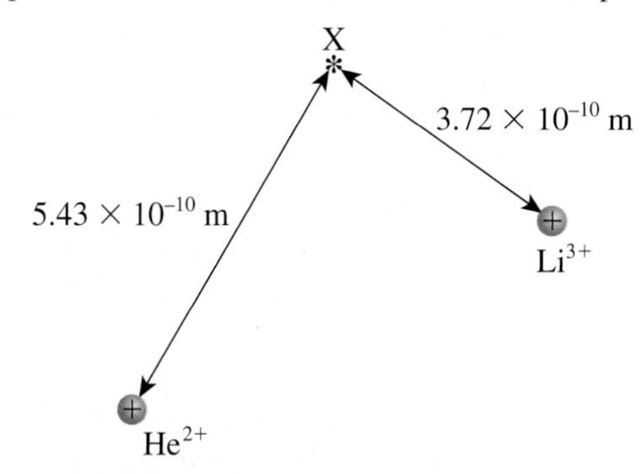

Figure 20-28 Questions 20-38 and 20-39.

20-38 What is the electric potential at point "X" in Figure 20-28?

20-39 (a) If a chloride ion Cl^{-} were at point "X" in the previous question, what would its electric potential energy be?

(b) Where is the reference position at which the electric potential energy is zero in this case? Explain the meaning of the sign of the electric potential energy of the chloride ion at location X.

20-40 Two point charges are positioned as shown in Figure 20-29. The potential difference $(V_T - V_R)$ between points T and R on the line joining the charges is 1.86 MV. How far is T from the positive charge?

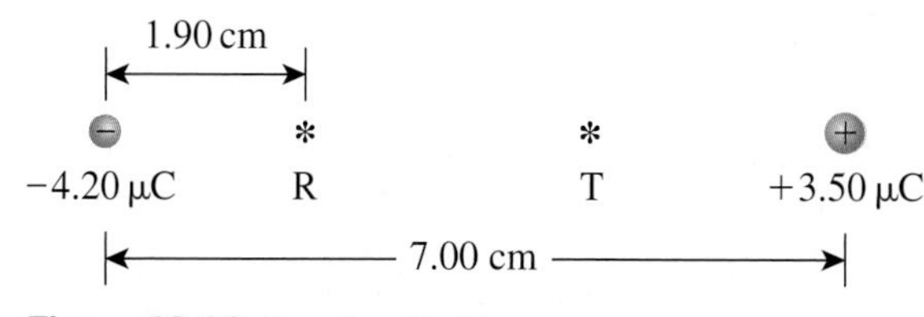

Figure 20-29 Question 20-40.

20-41 **Fermi Question:** Estimate the potential difference that gives rise to the spark shown in Figure 20-24.

20.5 Electric Current

electric current rate of movement of charge; symbol I; SI unit ampere (A)

ampere unit of electric current, equivalent to 1 coulomb per second; symbol A

As we mentioned at the beginning of Section 20.3, the voltage available from standard electrical outlets in North America is 120 V. Since electric potential (or voltage) is electric potential energy per unit charge, this means that for every coulomb of charge that moves, 120 J of electric potential energy is converted into other forms of energy. For example, in a toaster the electric potential energy is converted into thermal energy and some light. You are probably also aware that household circuits are protected by fuses or circuit breakers, which stop the movement of charge when the electric current exceeds a certain number of amperes, such as when you try to use the toaster, microwave, and coffee maker all at the same time in the morning. What *is* current, and what is an ampere?

Electric current is a measure of how much charge moves through a region per unit time. It is defined as the net charge moved divided by the time taken, and is usually given the symbol I. Writing charge as q and time as t,

$$I = \frac{q}{t} \quad \text{(20-10)}$$

Figure 20-30 André Marie Ampère (1775–1836), French physicist and mathematician. Ampère showed that currents travelling in two parallel wires exert magnetic forces on each other, and studied the relationship between current and magnetic field; these topics will be discussed in greater detail in Chapter 22.

From Eqn. 20-10, we see that one possible SI unit for current is coulomb/second (C/s); this is given the name of **ampere** (A) in honour of André Marie Ampère (Figure 20-30). If one coulomb of charge moves per second, the current is one ampere. In a typical household circuit, a fuse or circuit breaker limits the current to a maximum of 15 A.

When the concept of current was being established in the 1800s, the atomic nature of matter was not well understood, and a principle was established that *conventional*

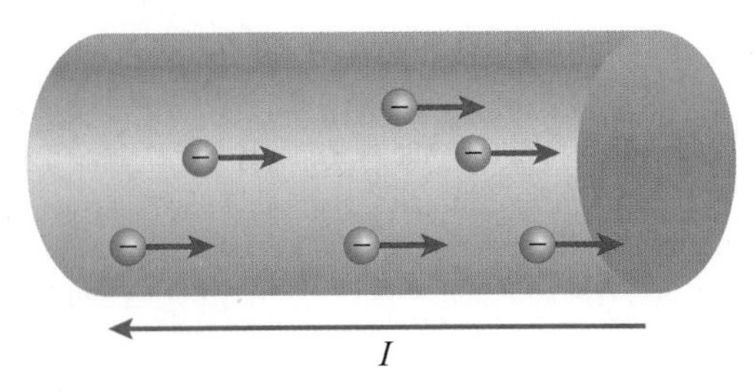

Figure 20-31 The movement of negative electrons from left to right is equivalent to a current from right to left.

current is in the direction of the movement of positive charge. This convention is still used today, even though we now know that most currents in conductors are due to the movement of the negatively charged electrons. The movement of electrons from left to right in a wire (Figure 20-31) is equivalent to a current (i.e., the movement of positive charge) in the opposite direction, from right to left.

TRY THIS!

Currents Used by Household Appliances

There are a number of appliances around the house that require fairly high currents to operate. If you examine them closely, you will often find specifications stamped directly on them. Compare the currents drawn by such everyday tools as your hairdryer, your clothes dryer, and your stove. **Note:** Frequently, the specifications given are the potential difference used (120 V in North America) and the power consumption in watts. Consult Sample Problem 20-8 to convert the power consumption into the current required for operation.

Electricity as a Hazard

Which is more dangerous, high voltage or high current? If harmless sparks between your finger and a doorknob result from a potential difference of about 10^4 V, then why is 120 V from a household outlet considered to be dangerous? In response, it is often said that "it is the current that kills." When you generate a spark between a finger and a doorknob, the potential difference and the instantaneous current are high, but the current is extremely short-lived and affects only a small part of your body. However, if you were to allow the potential difference from a standard 120 V outlet to continuously produce a current through your body, it would probably be lethal.

The current produced in the body depends on the potential difference and on the electrical resistance (Chapter 21) of the body; this resistance in turn depends on the points of contact on the body and on the skin condition (moist or dry). A current of approximately 3 to 8 mA through the body produces a mild tingling sensation. Above 10 mA, pain is felt, and 50 mA produces a severe shock. A current of about 100 to 200 mA (0.1 to 0.2 A) results in death. We will explore the electric current generated for a given potential difference in a material, as well as the distribution of current and potential difference around electric circuits in detail in Chapter 21.

DID YOU KNOW?

Electric fences are used in farming to keep livestock in a particular area or to keep predators, such as coyotes, out (Figure 20-32). These fences are designed to create a path for current to travel through the animal to ground should they come in contact with the wires. The amount of current generated in the animal depends in part on the potential difference between the wire and ground; in fencing used as a deterrent, this potential difference is adjusted such that the resulting current is uncomfortable for the animal rather than anything more dangerous.

Tony Campbell/iStock/Thinkstock

Figure 20-32 Electric fences are designed to deliver a small current to animals that come in contact in order to deter them from crossing the boundary.

SAMPLE PROBLEM 20-8

During a time interval of 5.0 s, 15 C of charge moves through a light bulb.

(a) How much current travels through the bulb?

(b) If the potential difference between the ends of the wires leading into the light bulb is 9.0 V, how much energy is dissipated in the light bulb in 5.0 s?

(c) How much power is consumed by the light bulb?

Solution

(a) Using Eqn. 20-10, current is charge divided by time:

$$I = \frac{q}{t} = \frac{15\,\text{C}}{5.0\,\text{s}} = 3.0\,\text{A}$$

Thus, the current is 3.0 A.

(b) This is similar to Sample Problem 20-5. The absolute value of the change in the electric potential energy is given by Eqn. 20-5:

$$|\Delta U| = |q\Delta V|$$

In 5.0 s, the charge q that moves is 15 C.

Thus, $|\Delta U| = |q\Delta V| = (15\text{ C})(9.0\text{ V}) = 1.4 \times 10^2\text{ J}$

Therefore, as the charge moves through the light bulb, 1.4×10^2 J of energy is dissipated. (This means that 1.4×10^2 J of electric potential energy is converted into 1.4×10^2 J of thermal energy and light energy.)

(c) Since part (b) tells us that 1.4×10^2 J of energy is dissipated in the bulb in 5.0 s, we can determine the rate of energy dissipated (or power) by

$$P = \frac{E}{t} = \frac{1.35 \times 10^2\,\text{J}}{5.0\,\text{s}} = 27\,\text{W}$$

Therefore, the bulb is consuming 27 W of power.

Applications of Electricity

Electricity is used in so many ways that it is impossible to imagine our lives today without it. Most people tend to take for granted the common uses in which electric potential energy is converted into some other form of energy. Light energy is produced in light bulbs; thermal energy in ovens, toasters, and hair dryers; mechanical energy (or energy of motion) in electric mixers, drills, and other tools; and sound energy in telephones and radios. Chemical energy is involved in such processes as the electroplating of silver onto less expensive nickel to make jewellery and silverware. There are countless common uses of electricity for domestic, commercial, and industrial applications, as well as for transportation and leisure activities.

Aside from these everyday uses, there are numerous unique and exciting specific applications that scientists and engineers have discovered or designed. Consider for example, some of the applications of electricity that relate to the human body. Because the body has its own electrical system, many medical uses of electric potential energy exist. Pacemakers stimulate the heart electrically and allow people with certain heart conditions to lead normal lives (Figure 20-33 (a)). Defibrillators send a large current through the body for a small fraction of a second to restore a normal heartbeat if the heart has been beating irregularly or to shock it back into beating if it has stopped (Figure 20-33 (b)). Other medical applications of electricity include the nerve conduction velocity test (NCV) as well as vagus nerve stimulation (VNS). The nerve conduction velocity test (Figure 20-34 (a)) is a diagnostic procedure in which the physician measures the time it takes for an electrical impulse to travel a given distance through the nervous system. An abnormal result from an NCV test may indicate nerve damage in the region. Vagus nerve stimulation is a therapeutic technique used for the treatment of epilepsy (Figure 20-34 (b)). A small device, much like a pacemaker, is implanted under the skin in the patient's chest, from where it sends a regular stream of electrical pulses to the vagus nerve at the base of the neck in an attempt to control seizures.

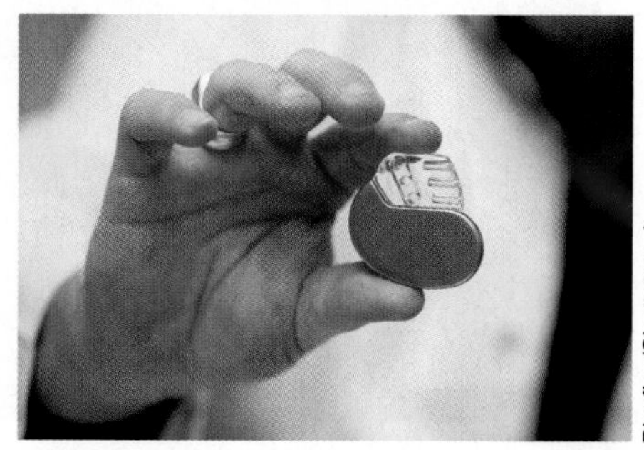

Picsfive/Shutterstock

(a)

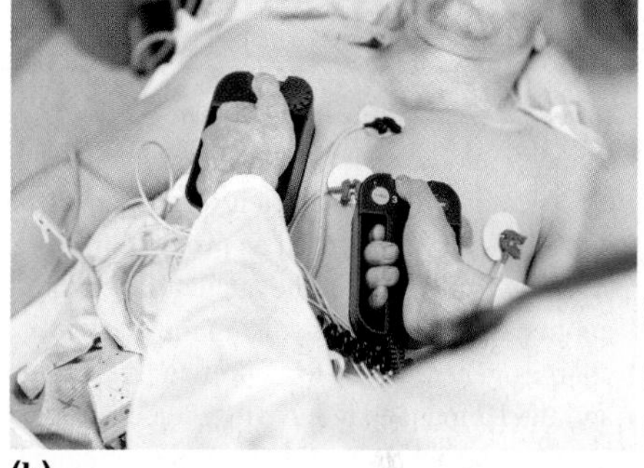

Tyler Olson/Shutterstock

(b)

Figure 20-33 **(a)** A pacemaker may be implanted under the skin of a patient with a history of heart disease in order to maintain the regular electrical functioning of the heart. **(b)** Defibrillators are used extensively in emergency rooms to rapidly restore a normal heartbeat.

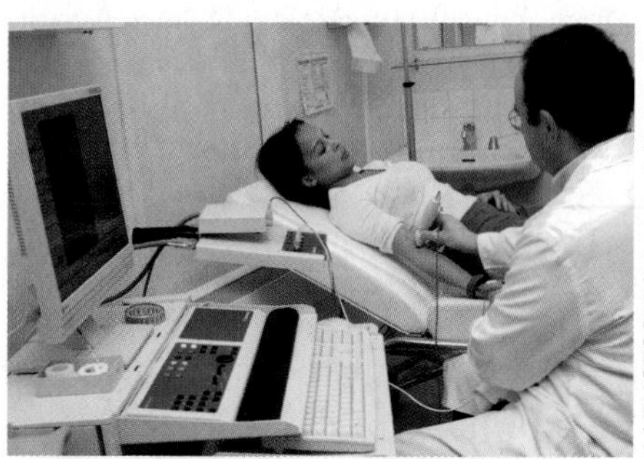

AJ PHOTO/HOP Americain/Science Photo Library

(a)

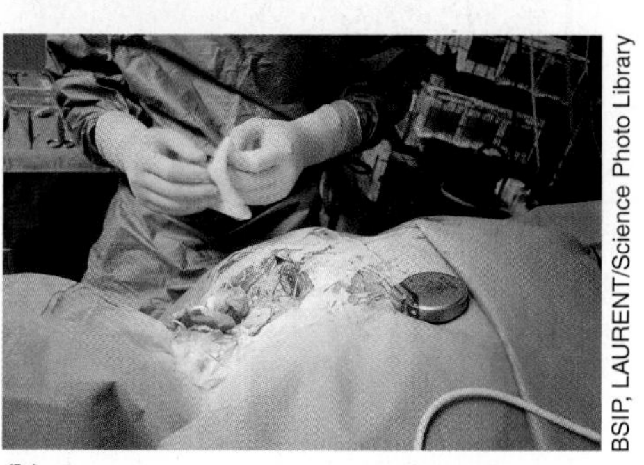

BSIP, LAURENT/Science Photo Library

(b)

Figure 20-34 **(a)** A patient undergoing a nerve conduction velocity test. **(b)** A surgeon implanting the pacemaker-like control unit for vagus nerve stimulation.

Lightning

Now that you have learned about electric potential and current, we can discuss one of nature's most fascinating displays, lightning (Figure 20-35). As we discussed in Chapter 19, the apocryphal story of Benjamin Franklin flying a kite in a thunderstorm in 1752 is one of the earliest connections between electricity in the laboratory and in nature. Whether he conducted this experiment or not, Franklin was one of the first scientists to make the link between lightning in nature and the charges involved in static electricity experiments in the laboratory.

There are roughly 2000 active thunderstorms around the world at any time. Lightning bolts carry negative charge from clouds to Earth, with the result that Earth has a permanent charge of about -5×10^5 C, and the upper atmosphere has an equivalent positive charge. In the absence of thunderstorms, these charges would gradually "leak" into each other, and Earth and the atmosphere would become electrically neutral. However, the constant activity of thunderstorms maintains the charge separation.

Piotr Krzeslak/Shutterstock

Figure 20-35 Electricity in the form of a lightning strike.

DID YOU KNOW?

Since the body's nervous system runs on electrical impulses, electricity has been used to try to relieve pain as well as promote healing. There is evidence that the ancient Egyptians applied this principle by using the electric charges generated by torpedo fish to ease pain, and the Romans may have used electric eels to treat headaches and arthritis. Today, medical practitioners use electrical nerve stimulation to treat certain types of pain. One approach is the use of transcutaneous electrical nerve stimulation, or TENS, which is very common in physiotherapy (Figure 20-36). A small current travels under the skin to the nerves beneath, with the purpose of stimulating the body's inherent pain response mechanisms. There have been contradicting research findings as to the benefits of TENS for pain relief: some studies have shown no benefit over the placebo effect whereas others have demonstrated a positive effect. The widespread technique, therefore, remains somewhat controversial.

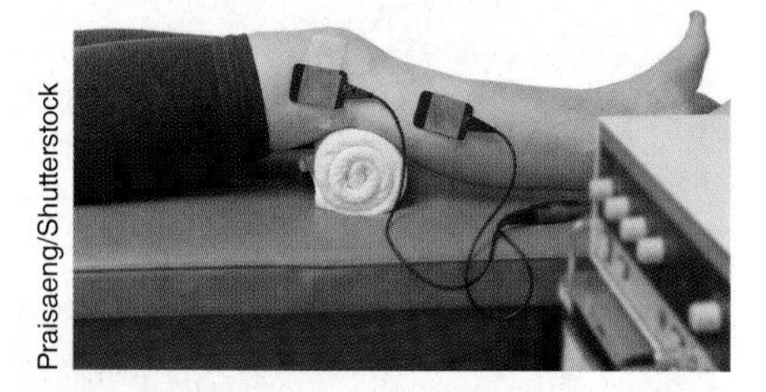

Praisaeng/Shutterstock

Figure 20-36 Physiotherapy for muscle and joint injury often makes use of small electric currents to relieve pain in a technique called transcutaneous electrical nerve stimulation (TENS).

A thundercloud is predominantly negatively charged at the bottom and positively charged at the top. The details of the charging mechanisms are still not well understood: different charging theories involve updrafts and downdrafts, collisions, friction, as well as melting and freezing. Most theories attribute the charge distribution to a difference in size between the positive and negative charge carriers (such as raindrops, ice crystals, ions, and hail). According to these theories, the positively charged particles tend toward the top of the cloud; the negatively charged particles are located predominately near the bottom.

In the absence of a thunderstorm, the electric field near Earth is downward because of Earth's overall negative charge; its magnitude is about 100 to 200 V/m. However, because of the large negative charge at the base of a thundercloud, the field between Earth and the cloud is upward, and has a typical magnitude of 10 000 V/m.

A lightning bolt is simply the rapid movement of electrons from the bottom of the cloud to the ground. (There is often lightning within the cloud as well.) Because of the large electric field between the ground and the cloud, the air molecules become ionized and form a conducting path (Figure 20-37). An electric field of about 3×10^6 V/m is needed to ionize dry air at atmospheric pressure, but the moist air and lower pressure associated with a thunderstorm produces ionization with lower magnitude electric fields. The electric field has to reach the ionizing value in only a small region near the cloud. Once the current begins, the large concentration of negative charge in the leading edge of the lightning bolt creates a large field nearby that causes the ionization and current to continue. Usually the initial ionization near the cloud is accompanied by ionization and charge movement near the ground as well, as large amounts of positive charge build up in the ground directly beneath the thundercloud. Thus, there are two channels of current, one starting at the cloud and one at the ground; these come together, typically about 50 m above the ground, to form a lightning bolt.

A typical flash of lightning consists of three to five current pulses, separated by times of about 40 ms. The peak current in the pulses is roughly 2×10^4 A, and the

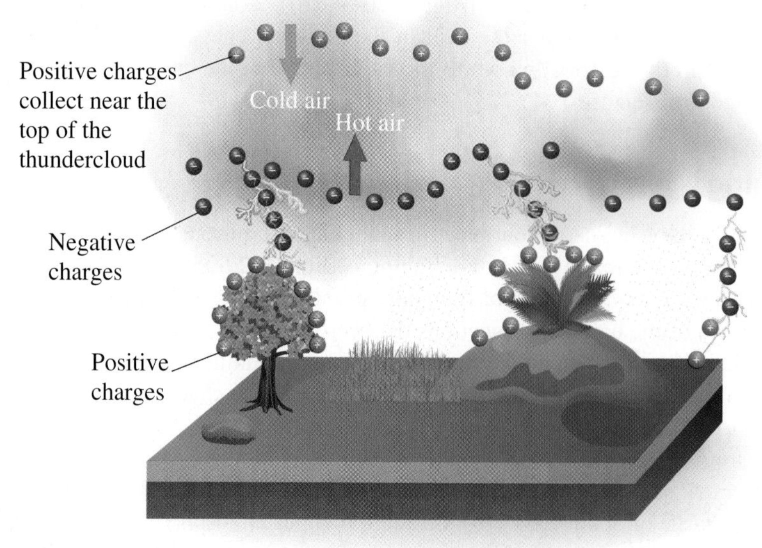

Designua/Shutterstock

Figure 20-37 Lightning strikes occur when the air molecules between the cloud and Earth become ionized, providing a pathway for the opposite charges to travel toward each other.

temperature in the current channel can reach 30000°C. The total charge transferred from the cloud to Earth is about −25 C (with wide variability), and since the potential difference between the cloud and ground is typically 5×10^8 V, the total energy released is of the order of 10^{10} J. Most of this energy goes into thermal energy, but some goes into light, sound, and radio waves.

Lightning tends to strike high objects such as tall buildings and trees, as these provide the shortest conducting path to ground. Every year people die because they stand under tall trees to try to keep dry during thunderstorms. Lightning rods are commonly used on buildings to provide a conducting path along which lightning can travel safely instead of passing through the building itself and causing damage. During a thunderstorm you should not stand near a tall tree or any other solitary object. If you are standing up or lying down and lightning strikes, the hazard is far greater than if you are squatting. The best choice is to seek shelter, for example in a car. If someone has been struck by lightning, it is worthwhile to administer proper first-aid revival techniques, even if the person appears lifeless. Reported cases verify that revival is possible.

We see lightning almost as soon as it strikes because the light travels at 3×10^8 m/s. However, the sound (thunder) travels at only about 340 m/s (or ⅓ km/s), and so we hear the thunder a few seconds afterward. You can estimate the distance (in kilometres) to a lightning strike by counting the number of seconds between seeing the lightning and hearing the thunder, and then simply dividing by three.

Household Electrical Circuits

By looking at the main fuse box or breaker box at home, determine the maximum current allowed in each circuit. Also determine the maximum total current permitted in the supply wires leading into your home.

DID YOU KNOW?

The CN Tower (Figure 20-38) in downtown Toronto opened in 1976. This communications and observation tower stands 553.33 m in height, which made it the tallest tower and free-standing structure in the world from its completion in 1976 until 2010. Given its record-setting height, lightning strikes the CN Tower an average of 50 to 75 times per year. Lightning rods in the form of long copper strips run down the exterior of the tower to 42 grounding rods buried deep below the surface to prevent damage. Scientists and engineers continue to study the charge conduction and dissipation associated with strikes to the tower today as a means of developing a greater understanding of lightning strikes to tall structures, and to aid in the design of protective measures for our increasingly sensitive electronic equipment.

Figure 20-38 The CN Tower in Toronto, Ontario, during a thunderstorm.

EXERCISES

20-42 If there is a movement of 3.4 C of charge in 1.7 s, what is the current?

20-43 The current through a flashlight bulb is 0.350 A. If it is used for 23 s, how much charge has moved through the bulb?

20-44 When a nerve cell is stimulated electrically, mechanically, or chemically, the permeability of the cell membrane changes and sodium ions (Na^+) can diffuse more readily into the cell. If 1.5×10^6 Na^+ ions move through the membrane in 5.0 ms, what is the corresponding electric current?

20-45 In a lightning bolt, negative charge moves from a cloud to the ground. What is the direction of the current?

20-46 A current of 2.35 A travels along a wire.

(a) How many coulombs of charge cross any cross-section of the wire per second?

(b) How many electrons cross any cross-section per second? (Assume that electrons are the only charges involved in generating this current.)

20-47 In a particular lightning bolt, −24 C of charge is transferred in 0.21 s through a potential difference of 5.5×10^8 V.

(a) What is the average current?

(b) How much electric potential energy is dissipated?

20-48 Suppose that, after seeing a lightning flash, you count six seconds until you hear the thunder. Approximately how far away was the flash?

20-49 When a large potential difference is applied to the terminals of a hydrogen gas-discharge tube, the hydrogen gas ionizes, producing free electrons and protons. The electrons move toward the positive terminal, and the protons toward the negative terminal. The resulting current is the sum of the currents due to the electrons and protons.

(a) What is the direction of the current?

(b) What is the total current if, in 1.00 s, 3.50×10^{18} electrons and 1.35×10^{18} protons move in opposite directions past any cross-section in the tube?

20-50 **Fermi Question:**

(a) Estimate the charge moved by your car battery on a cold morning when you are trying to start your vehicle.

(b) Car batteries have a potential difference of 12 V between the terminals. Estimate the power output at start-up on a cold morning.

LOOKING BACK...LOOKING AHEAD

In the previous chapter we began our exploration of electricity by discussing electric charges and how they interact, focusing our discussion on the forces between charges (Coulomb's law) as well as the somewhat abstract concept of electric fields associated with charged particles.

The focus of this chapter has been on understanding how charges interact by exploring the concepts of electric potential energy and electric potential. Electric potential energy and electric potential have been presented for the cases of uniform electric fields and point charges, and the electron volt has been introduced as a useful unit of energy. In addition, electric current has been defined as the movement of charge, and the natural phenomenon of lightning has been used to illustrate electric potential, charge, and current.

In the next chapter we will explore a fundamental relationship between potential difference, current, and a new concept, electrical resistance. Specifically, we will explore the current that is generated for a given potential difference between two locations. We will use this relationship, known as Ohm's law, to study direct-current (DC) circuits.

CONCEPTS AND SKILLS

Having completed this chapter, you should now be able to do the following:

- Use electric potential energy in problems involving uniform electric fields or point charges.
- Understand what happens (in terms of increasing or decreasing potential energy) when a mass or charge moves when released from rest.
- Understand the relation between volt, joule, and coulomb.
- Solve problems involving electric potential energy, electric potential, and potential difference in uniform electric fields.
- Solve problems involving electric potential energy, electric potential, and potential difference associated with point charges.
- Define an electron volt, and use it in determining changes in energy when charged particles are accelerated through potential differences.
- State the relation between the magnitude of a uniform electric field, potential difference between two points, and distance.
- State a common SI unit for electric field other than N/C.
- Define electric current and its SI unit, and solve related problems.
- State the direction of electric current and its relation to the direction of the motion of electrons.
- Discuss the dangers associated with electric current and voltage.
- Describe some of the numerous examples of medical and everyday applications of electricity discussed in this chapter.

KEY TERMS

You should be able to define or explain each of the following words or phrases:

electric potential energy
electric potential
volt
potential difference
voltage
electron volt
electric current
ampere

Chapter Review

MULTIPLE-CHOICE QUESTIONS

20-51 An electron is released from rest in an electric field. In which direction will it move?

(a) toward decreasing electric potential, which for an electron is toward decreasing potential energy
(b) toward increasing electric potential, which for an electron is toward decreasing potential energy
(c) toward increasing electric potential, which for an electron is toward increasing potential energy
(d) toward decreasing electric potential, which for an electron is toward increasing potential energy
(e) none of the above

20-52 A potassium ion (K^+) is at rest at the surface of the membrane of a neuron, where there is an electric field. In which direction will it move?

(a) toward decreasing electric potential, which for a K^+ ion is toward decreasing potential energy
(b) toward increasing electric potential, which for a K^+ ion is toward decreasing potential energy
(c) toward increasing electric potential, which for a K^+ ion is toward increasing potential energy
(d) toward decreasing electric potential, which for a K^+ ion is toward increasing potential energy
(e) none of the above

20-53 Electric potential is

(a) equal to electric potential energy
(b) the absolute value of electric potential energy
(c) the negative of electric potential energy
(d) electric potential energy per unit distance
(e) electric potential energy per unit charge

20-54 Scenario #1: A charge q is placed at a distance of r from a charge Q. Scenario #2: A charge $2q$ is placed at a distance of $2r$ from a charge Q. All the charges are positive. The electric potential at the location of charge q in Scenario #1 is

(a) the same as the electric potential at the location of charge $2q$ in Scenario #2
(b) twice the electric potential at the location of charge $2q$ in Scenario #2
(c) one-half of the electric potential at the location of charge $2q$ in Scenario #2
(d) four times the electric potential at the location of charge $2q$ in Scenario #2
(e) one-quarter of the electric potential at the location of charge $2q$ in Scenario #2

20-55 Scenario #1: A charge q is placed at a distance of r from a charge Q. Scenario #2: A charge $2q$ is placed at a distance of $2r$ from a charge Q. All the charges are positive. The electric potential energy of the system in Scenario #1 is

(a) the same as the electric potential energy of the system in Scenario #2
(b) twice the electric potential energy of the system in Scenario #2
(c) one-half the electric potential energy of the system in Scenario #2
(d) four times the electric potential energy of the system in Scenario #2
(e) one-quarter of the electric potential energy of the system in Scenario #2

20-56 Car batteries are often rated in ampere·hours (A·h) (Figure 20-39). An ampere·hour is a quantity of

(a) current
(b) time
(c) electric potential
(d) charge
(e) energy

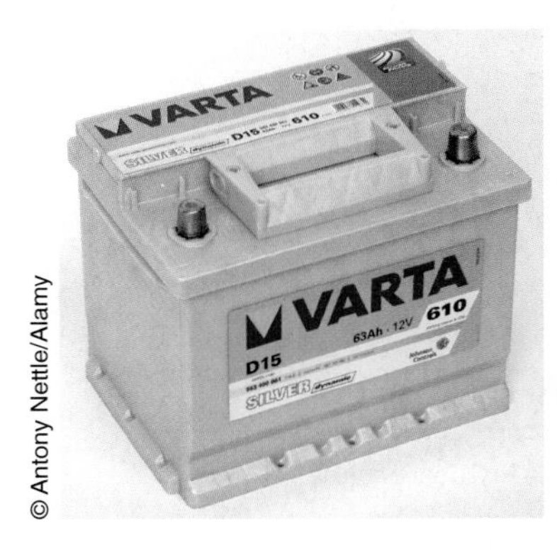

Figure 20-39 This battery is rated as 63 A·h, as shown on the front (Question 20-56).

Review Questions and Problems

20-57 A uniform electric field between two parallel, oppositely charged plates is directed from west to east.

(a) In which direction does an electron move if its electric potential energy is decreasing?

(b) If the electron's potential energy changes by 1.2×10^{-17} J as it moves 3.5 cm, what is the magnitude of the electric field?

20-58 Under a particular thundercloud there is a uniform electric field of 9.3×10^{3} V/m upward. An electron in the field, initially at rest, travels freely for 3.5×10^{-6} m before undergoing a collision with a gas molecule. Just before it hits the molecule, what is its speed? Use energy methods.

20-59 What is the electric potential energy of an electron and an α particle (charge $+2e$) separated by 6.9×10^{-12} m?

20-60 When a stimulus is applied to a nerve cell, sodium ions (Na^{+}) rush through the cell's membrane. If a Na^{+} ion moves through a potential difference of 42 mV in a time of 1.2 ms, what is the change, in electron volts, in the electric potential energy of the ion?

20-61 X-ray machines (Figure 20-40) used to generate anatomical images require a beam of electrons accelerated through a large potential difference in order to create the x rays. An electron in a mammography unit is accelerated from rest through a potential difference of 25 kV.

(a) Does the electron move toward higher, or lower, electric potential?

(b) What is the change in the electron's electric potential energy? Express your answer in kilo-electron volts and in joules.

(c) What is the change in the electron's kinetic energy?

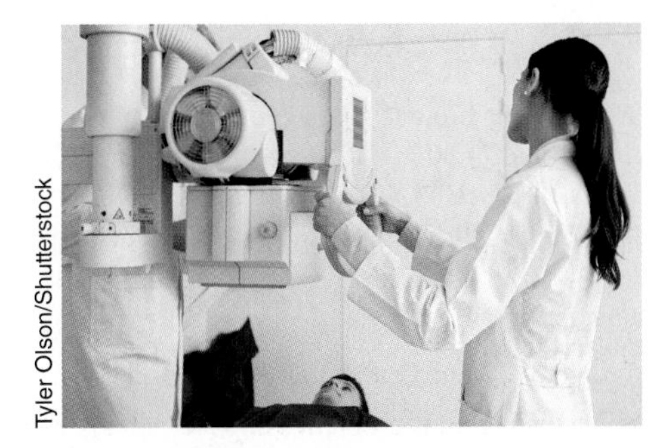

Figure 20-40 Question 20-61.

20-62 The electric field inside a straight piece of wire is approximately uniform and its direction is along its axis. A wire that is 1.50 m in length has a measured potential difference of 5.3×10^{-2} V between the ends. What is the magnitude of the electric field in the wire?

20-63 (a) If the electric potential increases from 101 V to 313 V in a vertical distance of 1.50 m near Earth, what is the magnitude of the uniform vertical electric field?

(b) If the electric potential increases with elevation, is the direction of this field upward or downward?

20-64 A pair of oppositely charged parallel plates is separated by 7.8 mm. If the resulting electric field between the plates has a magnitude of 6.41×10^{3} V/m, what is the electric potential difference between the plates?

20-65 If an electron moves from the negative plate to the positive plate in the previous question, what is the change in its (a) kinetic energy (in electron volts)? (b) electric potential energy (in electron volts)?

20-66 Sketch the shape of a graph of electric potential versus distance from

(a) a positive point charge

(b) a negative point charge

20-67 What is the electric potential at a distance of 1.5×10^{-10} m from a carbon nucleus ($q = +6e$)?

20-68 The electric potential at a point 3.43×10^{-12} m from an electron and at a distance d from an α particle (charge $+2e$) is zero. Determine d.

20-69 The current through a car radio (Figure 20-41) is 0.753 A, and the potential difference across it is 12 V.

(a) How much charge moves through the radio in 30.0 s?

(b) How much energy is dissipated in the radio in this time?

(c) What is the power used by the radio?

Syda Productions/Shutterstock

Figure 20-41 Question 20-69.

20-70 In 15.4 s, 13.5 C of charge moves through a flashlight bulb. What is the current?

20-71 In the previous question, if the potential difference across the bulb is 3.0 V, how much energy is dissipated in the bulb?

20-72 In an oscilloscope (Figure 20-42), a beam of electrons is accelerated at the back of the main tube, passes through deflecting plates, and hits the screen. What is the direction of the current in the beam?

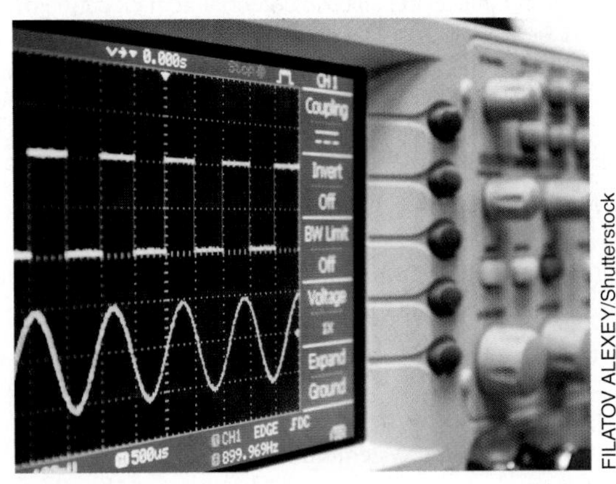

FILATOV ALEXEY/Shutterstock

Figure 20-42 Question 20-72.

Applying Your Knowledge

20-73 A particular proton in space experiences two fields: a gravitational field (1.30×10^5 N/kg toward planet X), and an electric field (2.47×10^{-4} N/C away from planet X).

(a) If the proton is released from rest, in which direction will it move?

(b) After moving 1.50 m, what will be its speed? Use energy methods and assume uniform fields.

20-74 Solve the previous problem using Newton's second law and constant-acceleration kinematics.

20-75 A proton is at rest 1.42×10^{-10} m from a uranium nucleus (charge $+92e$), which can be treated as a point charge since its size is much smaller than the proton–uranium distance. Assume that the uranium nucleus remains at rest.

(a) What is the initial electric potential energy of the system?

(b) In which direction does the proton move once it is released from rest?

(c) How far is the proton from the nucleus when its speed is 1.30×10^5 m/s?

20-76 A lithium nucleus ($q = +3e$) accelerates from rest through a potential difference of 4.6 kV.

(a) What is the final kinetic energy of the nucleus (in kilo-electron volts)?

(b) What is the work done (in joules) on the nucleus by the electric force during the acceleration? (**Hint:** Remember the work-energy theorem.)

20-77 A potassium ion (K^+) and a calcium ion (Ca^{2+}) are separated by 1.50×10^{-11} m. At what point on the line joining them is

(a) the electric field zero?

(b) the electric potential zero?

20-78 In an oscilloscope, an electron is accelerated from rest through a potential difference of 1.14×10^3 V. It then passes through deflecting plates that are 2.00 cm apart, across which there is a potential difference of 412 V (Figure 20-43). The electron enters the plate region with a velocity that is perpendicular to the electric field of the plates. When the electron leaves the plate region, what distance will it have moved toward the positive plate?

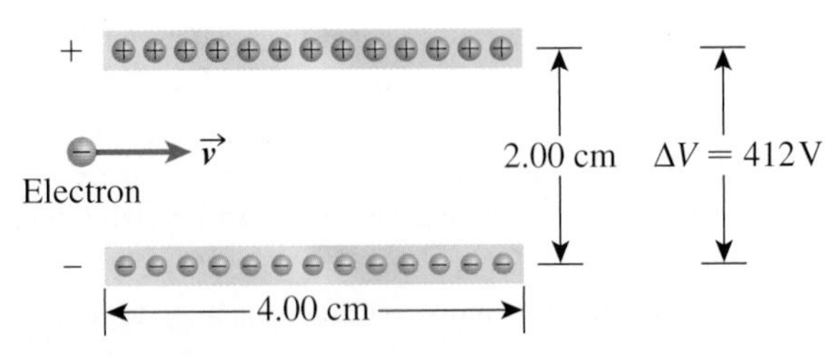

Figure 20-43 Question 20-78.

Electrical Resistance and Circuits

21

Anton Balazh/Shutterstock

Figure 21-1 A night-time view of North America from space. How fundamentally different would our world be without electricity and electrical circuits?

So many aspects of modern civilization—automobiles, televisions, lights (Figure 21-1), computers, communication devices, home appliances, heating systems, air conditioning, sound systems, etc.—involve the movement of electric current around specific loops called circuits. The amount of current moving in a circuit depends on the applied potential difference and also on the resistance of the circuit to the movement of charge. This chapter deals with the relationship between potential difference, resistance, and current, and also considers another important quantity in electrical circuits: the power consumed.

CHAPTER **OUTLINE**

21.1 Electrical Resistance and Ohm's Law

If you connect a garden hose to a faucet and turn it on, the amount of water passing through the hose per second depends on a number of factors: the water pressure at the faucet, as well as the hose's size, shape, and interior roughness. Similarly, when you turn on a flashlight, the amount of electric charge that moves per second (i.e., the current) depends on the potential difference between the terminals of the battery (analogous to the water pressure), and on various properties of the bulb filament, connecting wires, and contact points. These properties include materials, sizes, and (for contacts) degree of oxidation. All the factors that do not relate to the potential difference provided by the battery are taken into account in a single property: electrical resistance.

Suppose that we apply a potential difference (ΔV), across the ends of an object (such as a light bulb, wire, or nerve fibre) and measure the resulting current (I). The electrical **resistance** (R) of the object is *defined* as the ratio of potential difference to current:

resistance the ratio of the potential difference required to the resulting current produced in an object; symbol R; SI unit: ohm (Ω)

$$R = \frac{\Delta V}{I} \tag{21-1}$$

The ratio $(\Delta V)/I$ has units of volt/ampere (V/A), but this is given the special name of **ohm**, the SI unit of resistance, in honour of the German physicist Georg Simon Ohm (1787–1854). It is represented by the uppercase Greek letter omega (Ω). The resistance of an object in ohms tells us the potential difference required to produce a current of one ampere in the object; that is, 1 ohm (Ω) = 1 volt/ampere (V/A). The relation $R = (\Delta V)/I$ is often written in the form

ohm SI unit for resistance, equivalent to volts/ampere; symbol Ω

$$\Delta V = IR \tag{21-2}$$

and is referred to as **Ohm's law**. This is not really a law of nature in the same sense as Newton's laws; rather, it is just a rearrangement of the definition of resistance. Moreover, Ohm's law written in the form shown in Eqn. 21-2 is misleading, as it suggests that the current, I, is the independent variable and the potential difference, ΔV, is the dependent variable. In essentially all the cases we will be discussing in this book, the exact opposite is true: we apply a given potential difference to a circuit, through a battery or from the wall socket, and the current generated depends on this potential difference as well as the resistance of the circuit. The rearrangement of Eqn. 21-1, the definition of resistance, should therefore really be written as

Ohm's law an empirical relationship for some materials in which the current generated (I) is directly proportional to the potential difference applied across the object (ΔV)

$$I = \frac{\Delta V}{R} \tag{21-3}$$

If an object such as a metal wire is said to obey Ohm's law, or is said to be *ohmic,* this means that Eqn. 21-3 is valid (at constant temperature) regardless of the size of the applied potential difference. In other words, the resistance of the wire is constant at a given temperature. If various potential differences (ΔV) are applied to the wire, and a graph of the resulting current (I) is plotted versus ΔV, a straight line of slope $1/R$ results (Figure 21-2 (a)).

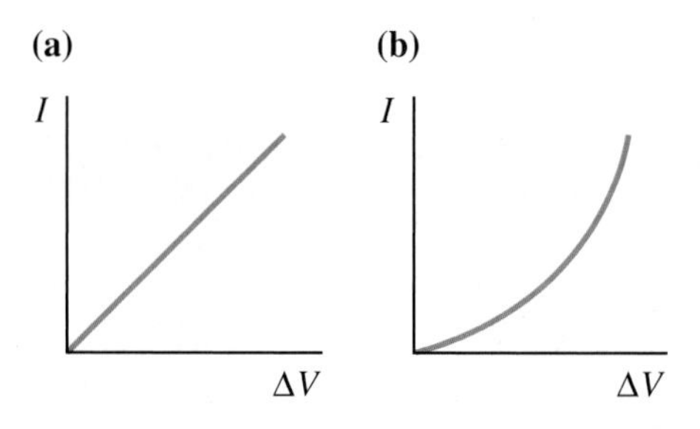

Figure 21-2 **(a)** Graph of I versus ΔV for a device (a metallic wire, for example) that obeys Ohm's law. **(b)** Representative graph of I versus ΔV for a device that does not obey Ohm's law.

There are many devices that do not have constant resistance at constant temperature; their resistance depends on the applied potential difference. Examples include semiconductors and transistors, and we say that these devices *do not obey Ohm's law* or that they are *non-ohmic*. A graph of I versus ΔV for such a device is non-linear (Figure 21-2 (b)). Nonetheless, we can still define the resistance of such a device as the ratio of $(\Delta V)/I$ in any given situation, but if the potential difference changes, this ratio (i.e., the resistance) also changes. In the remainder of this chapter, we assume that all devices are ohmic unless stated otherwise.

All conductors have some electrical resistance, but for normal electrical wiring, it is rather small. For example, a 30 cm length of copper wire of diameter 1.0 mm has a resistance of only 0.0065 Ω. In order to control the current in many electronic devices, standardized resistors (Figure 21-3) are used. These resistors come in a wide range of resistances, and are usually made of the semi-conductor carbon or a thin metal wire, surrounded by a plastic insulating casing. The resistance value is colour-coded on the outside of the plastic (Figure 21-4 and Table 21-1). For example, from Table 21-1, coloured bands in the order green-red-brown-silver form the code for 5, 2, 1, and ± 10%. The first two digits in this case give the number 52, the next digit indicates the power of ten, 10^1, and the last band gives the uncertainty in the resistance (± 10%). Therefore, the resistance is $52 \times 10^1\ \Omega \pm 10\%$ or $5.2 \times 10^2\ \Omega \pm 10\%$.

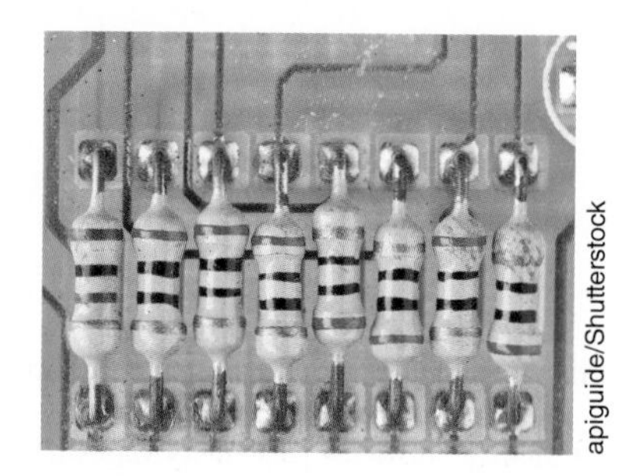

Figure 21-3 Standardized resistors use a universal colour code to indicate the resistance of the component.

Table 21-1

The Resistor Colour Code

Colour	First or Second Digit	Power of Ten	Uncertainty
Black	0	$10^0 = 1$	
Brown	1	10^1	
Red	2	10^2	
Orange	3	10^3	
Yellow	4	10^4	
Green	5	10^5	
Blue	6	10^6	
Violet	7	10^7	
Gray	8	10^8	
White	9	10^9	
Gold		10^{-1}	± 5%
Silver		10^{-2}	± 10%
No colour			± 20%

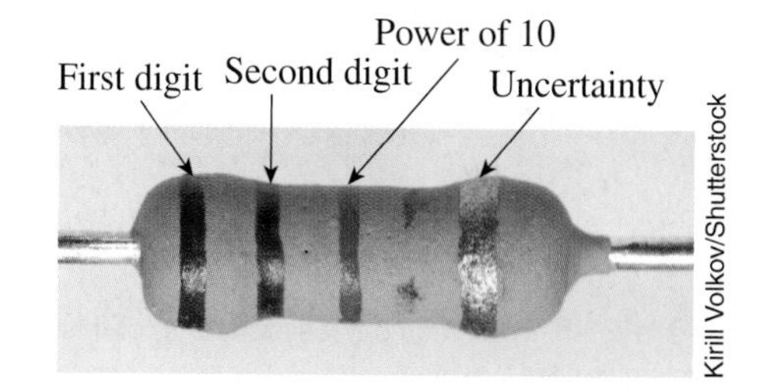

Figure 21-4 Interpreting the colour code of a resistor. From Table 21-1, this component has a resistance of $75 \times 10^3\ \Omega$, ± 5%.

SAMPLE PROBLEM 21-1

A potential difference is applied across the ends of a 150 Ω resistor, and a current of 0.12 A results. What is the applied potential difference?

Solution

Ohm's law in the form of Eqn. 21-2 is used:

$$\Delta V = IR = (0.12\ \text{A})(150\ \Omega) = 18\ \text{V}$$

Thus, the applied potential difference is 18 V across the ends of this resistor given the resulting current.

Electrical Resistivity

The electrical resistance of a copper wire is less than that of an aluminum wire of the same size. Why? The ability of a material to conduct current depends on the number of conduction electrons (Section 19.3) per unit volume and on the interaction of these moving electrons with the atoms of the material. Different materials have different numbers of conduction electrons and different degrees of electron–atom interaction. We say that different materials have different electrical **resistivities**, as defined below.

resistivity an inherent property of a medium that measures the difficulty with which it conducts electric current; in many materials, resistivity varies with temperature; symbol ρ; SI unit: Ω·m

The electrical resistance (R) of, for example, a wire depends not only on the material of which it is made, but also on its size. The resistance is proportional to the wire's length (L) and inversely proportional to its cross-sectional area (A):

$$R \propto \frac{L}{A}$$

This relation makes intuitive sense. We expect that a long wire will have a higher resistance than a short one because there are more obstacles to electron movement. A larger cross-sectional area means that there is more area across which the electrons can travel, and hence there is a lower resistance. The constant of proportionality in this relation is defined to be the electrical resistivity of the material, which is an intrinsic property of the material that reflects the difficulty with which it conducts electric current. The lower-case Greek letter rho (ρ) is used to represent electrical resistivity (Figure 21-5). Hence,

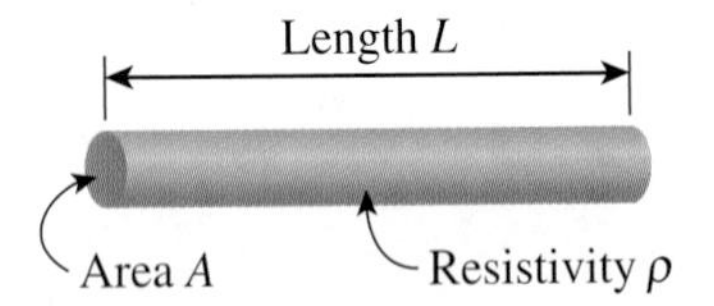

Figure 21-5 The electrical resistance, R, of this wire is given by $R = \rho L/A$.

$$R = \rho \frac{L}{A} \qquad (21\text{-}4)$$

To determine the SI unit of ρ, we rearrange Eqn. 21-4 to give $\rho = RA/L$. Substituting units on the right-hand side, we have: Ω·m²/m = Ω·m. Thus, the SI unit of resistivity is Ω·m (ohm·metre).

Table 21-2 gives approximate resistivities of some common materials at 20°C. The values depend to some extent on purity, heat treatment, and other factors. Copper is widely used in wiring because of its low resistivity and reasonable price. Notice that the resistivities of insulators are much, much larger than that of typical conductors.

The resistivity of any *metallic conductor* increases as temperature increases. At higher temperatures, the increased thermal vibration of the atoms in the conductor impedes the movement of electrons more strongly. Over a moderate range of temperatures, the resistivity of a conductor varies with temperature T as

$$\rho = \rho_0[1 + \alpha(T - T_0)] \qquad (21\text{-}5)$$

where ρ_0 is the resistivity at some reference temperature T_0 (typically 20°C), and α is called the temperature coefficient of resistivity. Some values of α are given in Table 21-2. As temperature decreases, the resistivity of a conductor typically decreases gradually. In the special case of materials known as *superconductors* (Section 19.3), the resistivity drops precipitously to zero below a certain critical temperature.

Table 21-2

Approximate Resistivities and Temperature Coefficients at 20°C

Material	Resistivity ($\Omega\cdot$m)	Temperature Coefficient α (°C)$^{-1}$
Conductors		
Silver	1.6×10^{-8}	3.8×10^{-3}
Copper	1.7×10^{-8}	3.9×10^{-3}
Gold	2.4×10^{-8}	8.3×10^{-3}
Aluminum	2.8×10^{-8}	3.9×10^{-3}
Tungsten	5.5×10^{-8}	4.5×10^{-3}
Iron	9.7×10^{-8}	6.5×10^{-3}
Mercury	96×10^{-8}	9×10^{-4}
Semiconductors		
Carbon	3.5×10^{-5}	-5×10^{-4}
Axoplasm	1.1×10^{0}	
Silicon (pure)	2.5×10^{3}	-7×10^{-2}
Nerve membrane	1.6×10^{7}	
Insulators		
Glass	10^{10} to 10^{14}	
Wood	10^{8} to 10^{11}	

SAMPLE PROBLEM 21-2

(a) What is the electrical resistance of 1.0 km of copper wire of diameter 2.2 mm at 20°C?

(b) If the temperature increases by 20°C, what is the new resistance?

Solution

(a) To determine the resistance of the wire, we need only to substitute values into Eqn. 21-4, being careful that L and A have units of metres and metres^2 respectively. The resistivity (ρ) of copper is $1.7 \times 10^{-8}\ \Omega\cdot$m at 20°C from Table 21-2, and the length (L) is 1.0 km, which equals 1.0×10^3 m. We must calculate the area (A) from the given diameter (2.2 mm). Since a diameter is given, we assume that the wire has a circular cross-section of area $A = \pi r^2$, where r is the radius (1.1 mm or 1.1×10^{-3} m). Hence,

$$R = \rho\frac{L}{A} = \rho\frac{L}{\pi r^2} = (1.7 \times 10^{-8}\ \Omega\cdot\text{m})\frac{1.0 \times 10^3\ \text{m}}{\pi(1.1 \times 10^{-3}\ \text{m})^2} = 4.5\ \Omega$$

Thus, the resistance of this wire is 4.5 Ω at 20°C.

(b) As the temperature increases, the wire's resistivity increases according to Eqn. 21-5:

$$\rho = \rho_0[1 + \alpha(T - T_0)]$$

Substituting values for ρ_0 and α from Table 21-2, and given that $T - T_0 = 20$°C,

$$\rho = (1.7 \times 10^{-8}\ \Omega\cdot\text{m})[1 + (3.9 \times 10^{-3}(°\text{C})^{-1}(20°\text{C}))] = 1.8 \times 10^{-8}\ \Omega\cdot\text{m}$$

Substituting this new value of ρ into $R = \rho L/A$, with L and A as in part (a), gives $R = 4.8\ \Omega$.

Therefore, the resistance of this wire is 4.8 Ω at 40°C.

DID YOU KNOW?

The fluids and tissues of the human body conduct electricity extremely well. It turns out that the overall resistance of the body in certain scenarios depends on the composition of the individual, specifically on the relative proportions of lean and fat soft tissue along the path. A diagnostic technique known as bioimpedance analysis (or BIA) has been used clinically to measure the composition of soft tissue in patients by determining the current generated when a small potential difference is applied between two points (Figure 21-6).

Figure 21-6 The measurement of resistance between the hands and the feet has been used to determine the relative proportions of fat and lean soft tissue in individuals through a technique called bioimpedance analysis.

EXERCISES

21-1 If a current of 2.5 A results when a potential difference of 23 V is applied to a resistor, what is the resistance?

21-2 (a) The resistance between two well-separated points on the human body is about $1.5 \times 10^3\ \Omega$. If a potential difference of 120 V is applied across these points, how much current is generated?

(b) If the skin is wet, the resistance drops to $5.0 \times 10^2\ \Omega$. What will the current be in this case for an applied potential difference of 120 V?

21-3 What does it mean if a circuit device obeys Ohm's law? if it does not obey Ohm's law?

21-4 What is the resistance (including uncertainty) of each of the following resistors?

(a) orange-orange-brown-silver

(b) red-yellow-silver-gold

(c) orange-white-yellow

21-5 A number of copper wires have the same length, but different cross-sectional areas, A. Sketch the shape of a graph of the resistances, R, of the wires (a) versus A (b) versus $1/A$.

21-6 What is the resistance of an aluminum wire of length 16 m and diameter 1.7 mm at 20°C?

21-7 The resistivity of axoplasm in a squid axon is 0.30 Ω·m. What is the length of a squid axon having a resistance of $1.5 \times 10^5\ \Omega$ and a radius of 0.025 cm? (An axon, which is a long projection of a neuron cell, carries nerve impulses as discussed in Chapter 20.)

21-8 An iron wire and a copper wire of the same length have the same potential difference applied to them. If the currents are the same, what is the ratio of

(a) the resistances of the wires?

(b) the cross-sectional areas of the wires?

(c) the radii of the wires?

(d) the diameters of the wires?

21-9 As temperature increases, does the resistivity of a metallic conductor increase or decrease? Why?

21-10 What is the resistivity of gold at 41°C?

21-11 A particular metal has a temperature coefficient of resistivity of $6.8 \times 10^{-3}\ (°C)^{-1}$. By what percentage does the resistivity of the metal change if the temperature increases by 5.0°C?

21-12 **Fermi Question:** Estimate the total resistance of all of the copper wiring in your home/apartment.

21.2 Batteries and Electrical Circuits

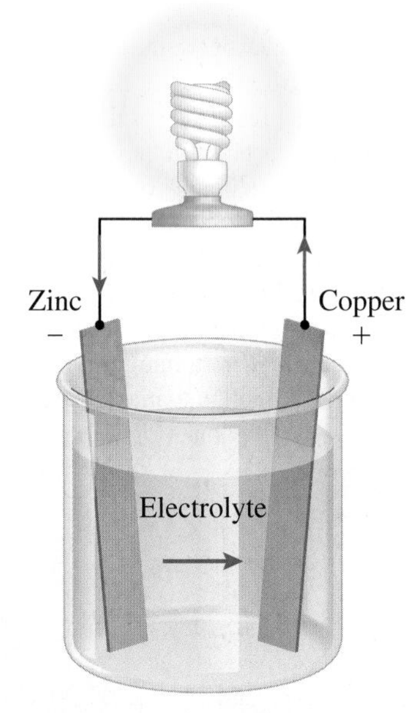

Figure 21-7 An elementary electric cell.

electric cell two different metals separated by a conducting liquid

electrolyte the conducting liquid used to separate two different metals in an electric cell

battery a combination of more than one electric cell, used as a source of energy

Everyone has used a battery—to provide energy for a flashlight, camera, cell phone, cordless power tool, remote control for various electronic devices, etc. But how do batteries work? How is the potential difference generated between the battery terminals? Who discovered batteries?

In the 1780s, Luigi Galvani (1737–1798), an anatomy professor in Italy, carried out a series of experiments involving the contraction of frogs' leg muscles due to electrical stimulation. In doing so, he was opening two separate areas of science: neurophysiology and current electricity. He was surprised to find that two dissimilar metals in contact with each other could induce the required stimulation. Galvani believed that the source of electricity was in the frog itself, and that the metals merely provided a conducting path. Alessandro Volta extended Galvani's experiments, and was able to show that the frog was acting merely as a detector of a potential difference created by the contact of the two metals. He found that some pairs of metals worked better than others, and he listed the metals in an electrochemical series showing their effectiveness. Metals far apart in the series provide a large potential difference.

This potential difference between two metals in contact led Volta to his greatest contribution to science: the invention of the electric (or voltaic) cell and the battery. An **electric cell** (Figure 21-7) consists essentially of two different metals (carbon can also be used) in electrical contact via a conducting liquid such as a dilute acid. This conducting liquid is called an **electrolyte**. A **battery** is a combination of several electric cells connected together. However, the term "battery" is commonly used even when there is only one cell.

The electric cell shown in Figure 21-7 operates in the following way. Some of the atoms in the zinc electrode dissolve in the electrolyte; they enter the solution as positive zinc ions, leaving behind electrons in the electrode. Thus, the zinc electrode becomes negatively charged. The same process occurs at the copper electrode but much more slowly, so that the zinc becomes negatively charged with respect to the copper. A potential difference is thus established between the electrodes. If the electrodes are then connected, say, to a flashlight bulb, electrons travel from the zinc, through the bulb, to the copper. An electrical circuit has been established. Eventually, one of the electrodes becomes so dissolved that a current cannot be maintained; the battery is "dead." Dry cell batteries used in flashlights, etc., consist of two different metals separated by an electrolytic paste rather than an acidic solution.

In circuit diagrams, a battery is indicated by the symbol shown at the top of Figure 21-8. The two horizontal lines having unequal lengths represent the battery terminals, and the two vertical lines leading away from the terminals represent connecting wires. *The longer of the two terminal lines represents the positive (high electric potential) terminal, and the shorter terminal line indicates the negative (low electric potential) terminal.* Strictly speaking, the symbol indicates only one electric cell, but it is commonly used to denote a battery. The proper symbol for a battery is a series of several alternating short and long lines to represent a number of cells connected together.

Battery
Switch (open)
Voltmeter
Conductor
Ground
Ammeter
Resistor
Light bulb
Alternating current generator

Figure 21-8 Conventional symbols used in circuit diagrams. In the symbol representing a battery, the longer of the two horizontal lines indicates the positive terminal, and the shorter horizontal line indicates the negative terminal.

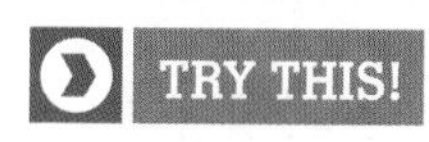

Lemon Power!

The slightly acidic juice of a lemon can provide the electrolyte you need to generate your very own battery at home. You will need some shiny, clean, copper pennies or copper strips, several fresh lemons, as well as some galvanized nails from the hardware store. Galvanized nails have been coated in zinc to prevent rusting. Insert a copper penny in one side of the lemon and the nail in the other; these are your two terminals of the lemon battery. Now connect the terminals to a small LED bulb via leads with alligator clips and observe the results. You can experiment with connecting several "batteries" in series (as shown in Figure 21-9 (a)) or in parallel, which will be discussed in more detail in Section 21.3. If you have access to a multimeter (see Section 21.3), you could measure the potential differences generated with your lemons, as shown in Figure 21-9 (b).

(a)

Martyn F. Chillmaid/
Science Photo Library

(b)

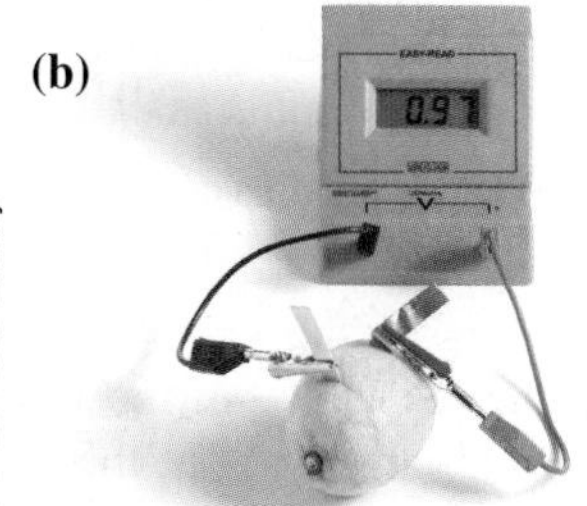

Science Photo Library

Figure 21-9 Turn lemons into electrical power!

Electrical Circuits

You have undoubtedly heard of electrical circuits—in your home, in computers, in automobiles, etc. An **electrical circuit** is simply a conducting path that forms a closed loop. For example, when you turn on a car headlight, current travels from the positive terminal of the car battery through the light to the negative terminal of the battery, and through the battery back to the starting point at the positive terminal. Note that any point in the circuit can be chosen as the starting point since charges begin moving almost instantly in all parts of the circuit when the battery is connected.

electrical circuit a conducting path that forms a closed loop

DID YOU KNOW?

Photovoltaic cells (Figure 21-10) provide another means of generating an electric current in a circuit. These cells are most commonly made of semiconductors such as silicon. When light from the Sun is absorbed by the photovoltaic cell, there is sufficient energy to free an electron from its associated atom. With an applied potential difference across the cell, the free electrons become a current which can be used to power a calculator, for example. Solar energy is also used with photovoltaic cells to power devices on a grander scale, such as entire households; however, efficiencies tend to be relatively low (~12 to 18% is typical). Most of the Sun's light does not have the right amount of energy to free the electrons from the photovoltaic cell, resulting in such low efficiencies.

Figure 21-10 Photovoltaic cells, such as the one seen in the top right corner of this solar-powered calculator, provide another means of generating an electric current.

open circuit an open switch or a broken wire in a circuit which creates an opening in the closed loop; current cannot travel through an open circuit

closed circuit a complete closed loop in which a current can readily travel

If current is not present in a circuit (because a wire has been broken or a switch has not been closed, for example), the circuit is an **open circuit**. If current is present, we have a **closed circuit**. Circuits are represented on paper by circuit diagrams, in which the various circuit elements or devices have specific symbols. Some common symbols are shown in Figure 21-8. A resistor is symbolized by a zigzag line, as shown. Connecting wires are normally considered to be conductors having negligible resistance (effectively, $R = 0$), and are shown by straight lines. Since the potential difference (ΔV) across a resistor can be written as $\Delta V = IR$, and since $R = 0$ for connecting wires, then *the potential difference from one end of a connecting wire to the other must be zero*. In other words, the electric potential along a connecting wire is constant.

SAMPLE PROBLEM 21-3

In the circuit represented in Figure 21-11,

(a) what is the direction of the resulting current?

(b) what are the following potential differences:

$V_B - V_C$? $V_A - V_D$? $V_B - V_A$? $V_C - V_D$?

(c) if we define the electric potential at the negative terminal of the battery to be zero, what are the electric potentials at points A, B, C, and D?

(d) what is the current in the circuit (assuming that the resistance is known to two significant digits)?

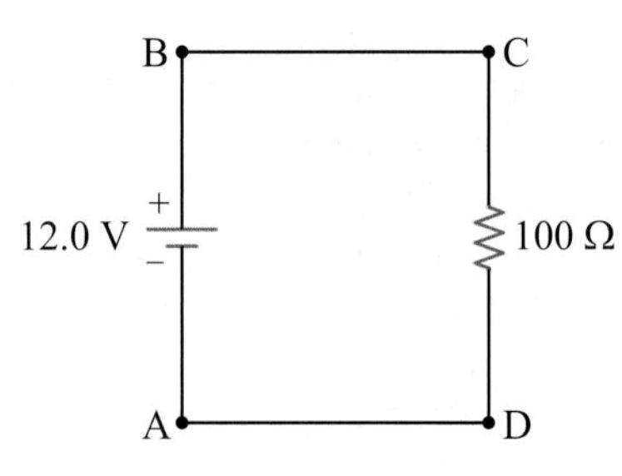

Figure 21-11 Sample Problem 21-3.

Solution

(a) The current is in the direction of the movement of positive charge. The positive terminal of the battery is indicated by the longer line (at the top of the battery symbol in Figure 21-11). Therefore, the current travels from this terminal to B, to C, through the resistor to D, to A, and back to the battery via the negative terminal. In short, the current is *clockwise* around the circuit.

(b) Points B and C are on a connecting wire of negligible resistance. Hence, $V_B - V_C = 0\text{ V}$. Similarly, $V_A - V_D = 0\text{ V}$.

Point B is at the same electric potential as the positive battery terminal, since the resistance of the wire connecting point B to this terminal is negligible. Similarly, A is at the same electric potential as the negative terminal. Therefore, $V_B - V_A$ equals the potential difference across the battery; that is, $V_B - V_A = 12.0\text{ V}$.

Point C has the same electric potential as B, and D has the same electric potential as A.

Therefore, $V_C - V_D = V_B - V_A = 12.0\text{ V}$.

(c) If the electric potential at the negative terminal of the battery is defined as zero, then $V_A = 0\text{ V}$.

Since $V_B - V_A = 12.0\text{ V}$ (from (b)), then $V_B = 12.0\text{ V}$.

Because B and C are at the same electric potential, $V_C = 12.0\text{ V}$.

Points A and D are at the same electric potential, and hence $V_D = 0\text{ V}$.

(d) We can calculate the current from Ohm's law in the form of Eqn. 21-4:

$$I = \frac{\Delta V}{R}$$

The potential difference (ΔV) across the 100 Ω resistor is 12.0 V, and thus

$$I = \frac{\Delta V}{R} = \frac{12.0\,\text{V}}{100\,\Omega} = 0.12\,\text{A}$$

The current through the resistor is 0.12 A. Therefore, by conservation of charge, the current everywhere in this circuit must be 0.12 A.

DID YOU KNOW?

Electrical circuits can be found in some surprising locations! Figure 21-12 (a) is an illustration of a retinal implant, an electrode array that is surgically embedded on the retina in order to restore partial vision to patients suffering from degenerative retinal diseases such as retinitis pigmentosa and age-related macular degeneration. This electronic prosthesis works by first sending visual information of the surroundings from a video camera mounted on special sunglasses to a pocket computer. The information is processed and projected through the patient's eye onto photovoltaic cells imbedded on the chip in the retina. Each cell converts the received light into a pulse of current with an intensity related to the intensity of light received from the sunglasses. These pulses of current are then relayed to the nearby neurons in the inner retina, bypassing the diseased retinal tissue entirely. The first such device was approved for use in the United States in 2013, consisting of an implanted 60-electrode array that is just 3 mm by 5 mm in size. Figure 21-12 (b) is a composite image of a patient with a prosthetic retina (left) and a simulation of what they would "see" (centre and right).

(a)

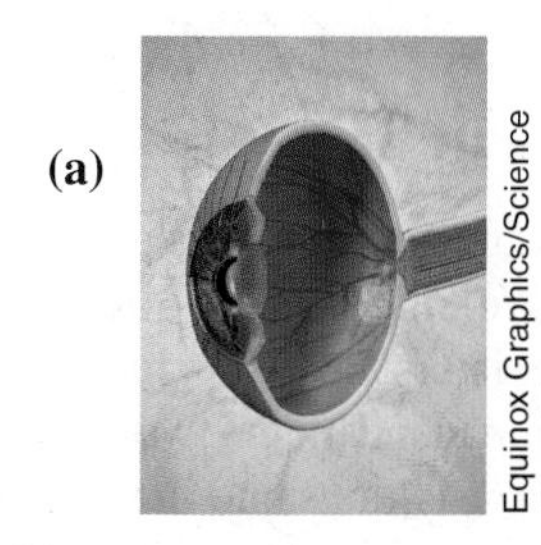

Equinox Graphics/Science Photo Library

(b)

Philippe Psaila/Science Photo Library

Figure 21-12 **(a)** An illustration of a retinal implant, an array of 60 electrodes on a 3 mm by 5 mm chip. **(b)** A composite image of a patient with a prosthetic retina (left) and a simulation of the regions of light and dark that they would "see" as a result (centre and right).

TRY THIS!

Build Your Own Simple Circuit

With a flashlight bulb, a C- or D-type battery, two pieces of wire, an elastic band, a clothespin, and some electrical tape you can create a simple circuit much like the one depicted in Figure 21-11. Wrap the exposed/stripped end of one wire a couple of times around the metal base of the flashlight bulb and then position it inside the clothespin. Wrap the elastic band a couple of times around the length of the battery; use the elastic to hold the metal base of your light bulb within the clothespin securely against the positive terminal of the battery. Take the second wire and use both the elastic band and some electrical tape to secure one end to the negative terminal of the battery. See what happens when you complete the circuit by touching the free ends of the wires together. Figure 21-13 is an illustration of your assembled circuit.

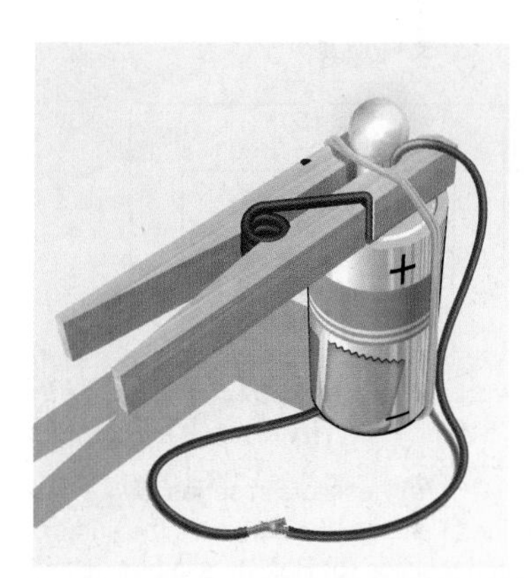

Figure 21-13 An illustration of your assembled circuit made up of everyday household items.

EXERCISES

21-13 Draw a circuit diagram consisting only of a 330 Ω resistor, a 15 V battery, and some wires, connected in a closed circuit such that current is generated. How large is the current?

21-14 For the circuit in Figure 21-14,

(a) what is the direction of the resulting current?

(b) what are the following potential differences?

$V_W - V_Z$? $V_X - V_W$? $V_X - V_Y$? $V_Z - V_Y$?

(c) if we define the electric potential at the negative terminal of the battery to be zero, what are the electric potentials at points W, X, Y, and Z?

(d) what is the current through the resistor? through wire XY? through the battery?

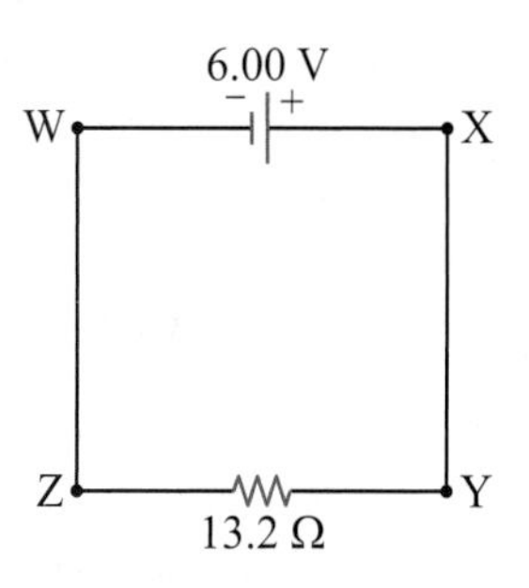

Figure 21-14 Question 21-14.

21-15 In the circuit in Figure 21-15, the resulting current is measured to be 2.38 A. What is

(a) the resistance R?

(b) the direction of the current through the resistor?

(c) the potential difference $V_A - V_C$?

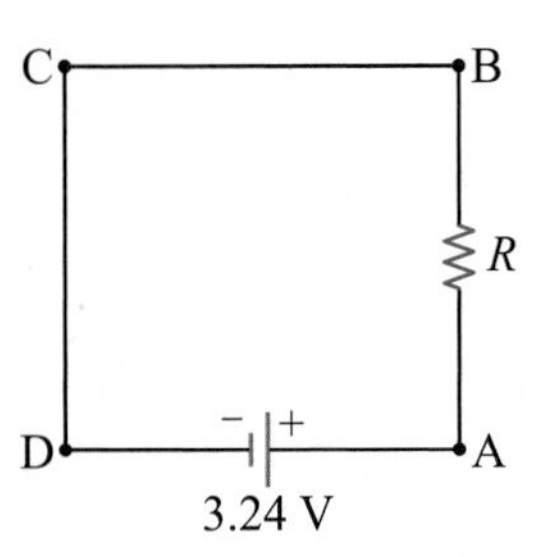

Figure 21-15 Questions 21-15 and 21-16.

21-16 For the circuit in Figure 21-15, if 9.7 C of charge moves through the battery:

(a) how long has the current been running?

(b) how much energy does the battery supply during this time?

21-17 **Fermi Question:** Estimate the current generated in the simple circuit constructed in the Try This! activity on page 685 (Figure 21-13).

21.3 Resistors in Series and Parallel

Commercial resistors commonly used in electronic devices have only certain values of resistance, as indicated by the colour code in Table 21-1. However, resistors can be connected together in various ways so that the total resistance of the combination can be virtually any value that might be required.

If resistors are connected so that the same current travels through each one (Figure 21-16 (a)), they are said to be connected in **series**. If they are connected so that the current splits into separate branches (Figure 21-16 (b)) with the same potential difference across each resistor, the resistors are said to be connected in **parallel**. The terms "series" and "parallel" are not restricted only to resistors; they apply to similar connections of any circuit components.

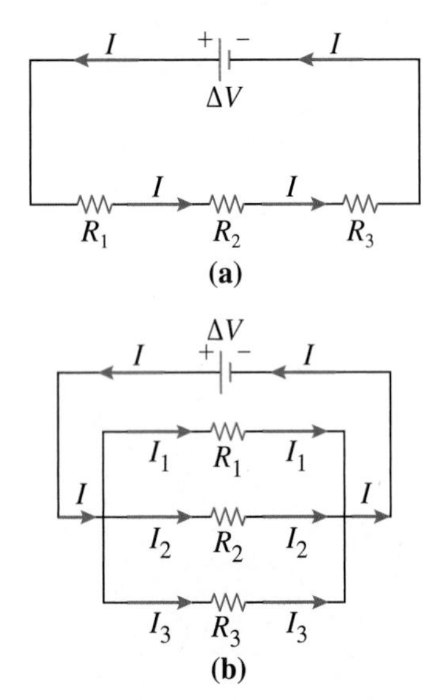

Figure 21-16 **(a)** Resistors in series. By conservation of charge, the current is the same through each resistor. **(b)** Resistors in parallel. By conservation of energy, the potential difference across each resistor is the same, and the total current provided by the battery is split among the resistors.

Equivalent Resistance of Resistors in Series

For resistors in series, we can determine an equivalent resistance, that is, a single resistance value for all the resistors together. We start with two resistors in series (Figure 21-17 (a)). In Figure 21-17 (b), the two resistors have been replaced by a single resistor having resistance R. If this resistor is "equivalent" to the two original resistors, this means that the current travelling in the circuit is the same in the two situations shown in Figures 21-17 (a) and (b). This will allow us to determine the relationship between R, R_1, and R_2.

In Figure 21-17 (b), the battery potential difference, ΔV, is applied across the single resistor, R, and hence, we can use Ohm's law:

$$\Delta V = IR \qquad \text{Eqn. [1]}$$

where I is the current in the circuit. Turning our attention now to Figure 21-17 (a), we can write the potential difference between A and B (i.e., the battery potential difference, ΔV) as the sum of two parts: ΔV_1 across R_1, and ΔV_2 across R_2.

$$\Delta V = \Delta V_1 + \Delta V_2 \qquad \text{Eqn. [2]}$$

Or, more explicitly,

$$\Delta V_1 + \Delta V_2 = (V_A - V_C) + (V_C - V_B) = V_A - V_B = \Delta V$$

By conservation of charge, the current through both R_1 and R_2 is the same: I. Hence, we can write $\Delta V_1 = IR_1$, and $\Delta V_2 = IR_2$, using Ohm's law. Thus, Eqn. [2] becomes

$$\Delta V = IR_1 + IR_2$$

$$\text{or} \quad \Delta V = I(R_1 + R_2) \qquad \text{Eqn. [3]}$$

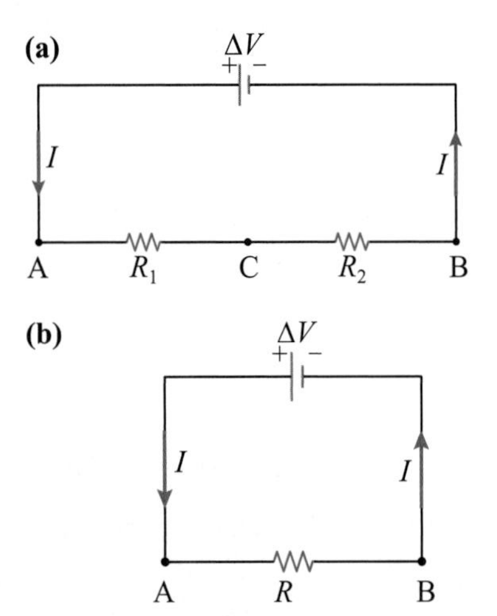

Figure 21-17 **(a)** Two resistors in series, R_1 and R_2. **(b)** The equivalent resistance, R, where $R = R_1 + R_2$.

We now compare Eqns. [1] and [3]. In order for the current I to be the same in the two situations, the single resistance R that is equivalent to the series combination of R_1 and R_2 must be

$$R = R_1 + R_2$$

For more than two resistors in series, we could use the same approach to show that the equivalent resistance is just the sum of all the single resistances:

$$R = \sum_{i=1}^{n} R_i, \text{ where } n \text{ is the number of resistors in series} \qquad (21\text{-}6)$$

Thus, for example, if we have three resistors in series, 100 Ω, 200 Ω, and 300 Ω, the total resistance is just the sum, 600 Ω.

series the current through each component in series is the same due to conservation of charge

parallel the potential difference across each component in parallel is the same due to conservation of energy

Equivalent Resistance of Resistors in Parallel

Figure 21-18 shows (a) two resistors in parallel and (b) the single equivalent resistance. As was the case with resistors in series, if this single resistor is "equivalent" to the two original resistors, this means that the total current travelling in the circuit is the same in the two situations shown in Figure 21-18.

Applying Ohm's law to the circuit in Figure 21-18 (b), the current due to the potential difference being applied across the equivalent resistance R is

$$I = \frac{\Delta V}{R} \qquad \text{Eqn. [4]}$$

Considering now Figure 21-18 (a), we see that the total current I splits at point X into two parts, I_1 and I_2. Since charge is conserved, the sum of these two currents must equal I; otherwise, charge would accumulate at the junction point X. *At any junction point in a circuit, the total current travelling in must equal the total current travelling out.*

$$I = I_1 + I_2 \qquad \text{Eqn. [5]}$$

The potential difference across *each* of the two resistors (R_1 and R_2) is the battery potential difference ΔV (or $V_X - V_Y$). Thus, for resistor 1, we can write the current as $I_1 = \Delta V/R_1$, and for resistor 2, $I_2 = \Delta V/R_2$. Substituting these expressions for I_1 and I_2 in Eqn. [5],

$$I = \frac{\Delta V}{R_1} + \frac{\Delta V}{R_2} \qquad \text{Eqn. [6]}$$

Now, whether there are two resistors in parallel as in Figure 21-18 (a), or the single *equivalent* resistor (Figure 21-18 (b)), the current, I, from the battery must be the same. Hence, we can replace I in Eqn. [6] with $(\Delta V)/R$ from Eqn. [4]:

$$\frac{\Delta V}{R} = \frac{\Delta V}{R_1} + \frac{\Delta V}{R_2}$$

Dividing each term by ΔV, we get the expression for the equivalent resistance R:

$$\frac{1}{R} = \frac{1}{R_1} + \frac{1}{R_2}$$

For more than two resistors in parallel, we just add more terms on the right-hand side:

$$\frac{1}{R} = \sum_{i=1}^{n} \frac{1}{R_i}, \text{ where } n \text{ is the number of resistors in parallel} \qquad (21\text{-}7)$$

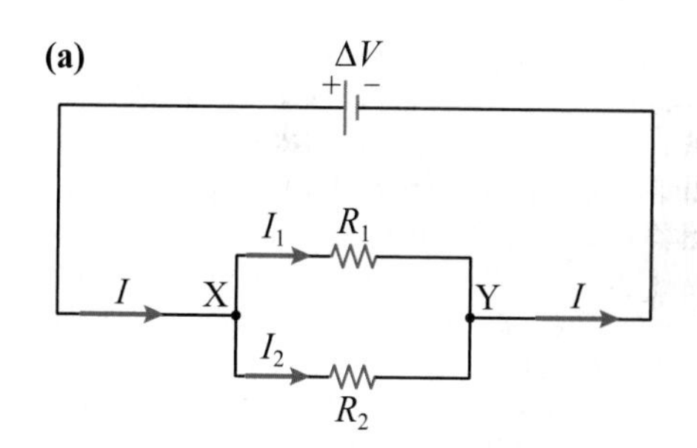

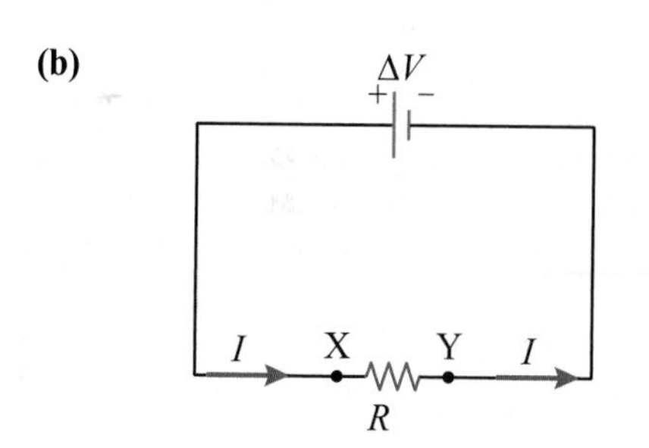

Figure 21-18 **(a)** Two resistors in parallel, R_1 and R_2. **(b)** The equivalent resistance, R, where $\frac{1}{R} = \frac{1}{R_1} + \frac{1}{R_2}$.

TACKLING MISCONCEPTIONS

The Movement of Charge in a Circuit

Current is not something that travels from a battery and gets smaller or becomes "used up" as it makes its way around a circuit. In a simple circuit such as the car headlight circuit, the current is the same in all parts of the circuit. This follows from the law of conservation of charge. For every ampere of current entering, say, a resistor, an ampere of current leaves the other end. Otherwise, a charge would build up on the resistor, which would violate the fundamental law of the conservation of charge.

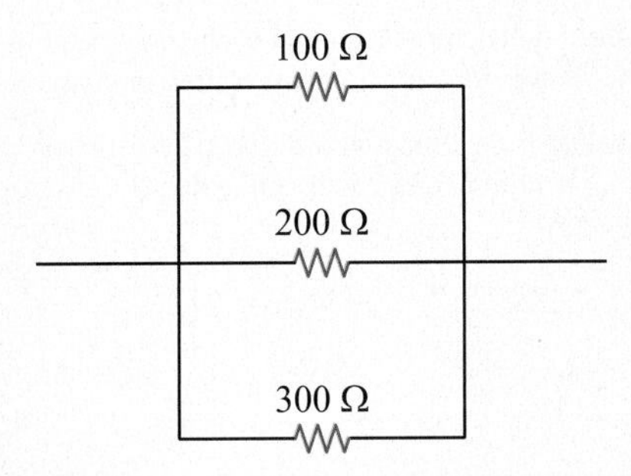

Figure 21-19 Three resistors in parallel. The equivalent resistance (54.5 Ω) is less than any of the single resistances.

If we have three resistors in parallel, say 100 Ω, 200 Ω, and 300 Ω (Figure 21-19), we can calculate the equivalent resistance from Eqn. 21-7:

$$\frac{1}{R} = \frac{1}{R_1} + \frac{1}{R_2} + \frac{1}{R_3}$$

$$= \frac{1}{100\ \Omega} + \frac{1}{200\ \Omega} + \frac{1}{300\ \Omega}$$

We use a common denominator of 600 Ω to add the fractions:

$$\frac{1}{R} = \frac{6}{600\ \Omega} + \frac{3}{600\ \Omega} + \frac{2}{600\ \Omega} = \frac{11}{600\ \Omega}$$

Therefore, $R = \frac{600\ \Omega}{11} = 54.5\ \Omega$ (assuming three significant digits).

Notice that *the equivalent resistance of resistors in parallel is less than any of the single resistances.*

TRY THIS!

Simulating Series and Parallel Resistors with Straws

First, you need to make yourself a nice thick milkshake—off to a good start so far, right? You will also need 6 equal length, equal diameter drinking straws. For your series "resistors," carefully tape 3 straws together in a line (Figure 21-20 (a)) to make a single straw that is three times the length of an individual one. For your parallel "resistors," tape 3 straws together side-by-side (Figure 21-20 (b)). Now try to drink the milkshake with each set of straws. In which configuration (series or parallel) does the milkshake flow more easily? Enjoy! **Note:** You can also explore this effect by testing how easy/hard it is to move a ping pong ball through a fixed distance by blowing through the series and parallel straw configurations if you prefer.

Figure 21-20 **(a)** Three drinking straws in series. **(b)** Three drinking straws in parallel. In which configuration does the milkshake flow more easily?

SAMPLE PROBLEM 21-4

In the circuit in Figure 21-21, what is

(a) the current through the battery?

(b) the potential difference across resistor 1?

(c) the current in resistor 2?

(Assume three significant figures for all the resistances.)

Solution

(a) To find the current through the battery, we need first to determine the equivalent resistance of the three resistors. Then Ohm's law can be used. Resistors 2 and 3 are in parallel with each other, and this two-resistor combination is in series with resistor 1. For the parallel combination, the equivalent resistance, which we will denote as R_{23}, can be calculated from Eqn. 21-7:

$$\frac{1}{R_{23}} = \frac{1}{R_2} + \frac{1}{R_3} = \frac{1}{200\ \Omega} + \frac{1}{300\ \Omega}$$

Rearranging and solving for R_{23}, we get: $R_{23} = 120\ \Omega$.

This 120 Ω resistance is in series with the 100 Ω resistor. Therefore, to determine the total resistance (R) of all the resistors, all we need to do is add these two series resistances:

$$R = R_{23} + R_1 = 120\ \Omega + 100\ \Omega = 220\ \Omega$$

The current in the circuit can now be easily determined from Ohm's law, $I = \Delta V/R$, since we know the potential difference and the resistance (Figure 21-22),

$$I = \frac{\Delta V}{R} = \frac{9.00\ \text{V}}{220\ \Omega} = 4.09 \times 10^{-2}\ \text{A}$$

Thus, the total current through the battery is 4.09×10^{-2} A (or 40.9 mA).

(b) To calculate the potential difference (ΔV_1) across resistor 1, we again use Ohm's law. The current is the same as that calculated in part (a), since this resistor is in series with the battery, by conservation of charge it must have the same current as the battery since there is no alternate route for the charge to take. Therefore

$$\Delta V_1 = IR_1 = (4.09 \times 10^{-2}\ \text{A})(100\ \Omega) = 4.09\ \text{V}$$

Hence, the potential difference across resistor 1 is 4.09 V.

(c) The current through resistor 2 is not the total current calculated in (a). In Figure 21-21, the total current passes through resistor 1 from c to b, and then branches into two parts from b to a, through resistors 2 and 3.

The easiest way to determine the current in resistor 2 is through Ohm's law, $\Delta V_2 = I_2R_2$. We know R_2, but do not know ΔV_2 or I_2. Therefore, to find I_2, we need first to calculate ΔV_2. The total potential difference from c to a is 9.00 V, and in part (b) we determined that the potential difference from c to b is 4.09 V. Thus, the potential difference from b to a—that is, the potential difference across each of resistors 2 and 3—is

$$\Delta V_2 = \Delta V_3 = (9.00 - 4.09)\text{ V} = 4.91\text{ V}$$

Now, from Ohm's law: $$I_2 = \frac{\Delta V_2}{R_2} = \frac{4.91\text{ V}}{200\ \Omega} = 2.45 \times 10^{-2}\text{ A}$$

Thus, the current through resistor 2 is 2.45×10^{-2} A (or 24.5 mA).

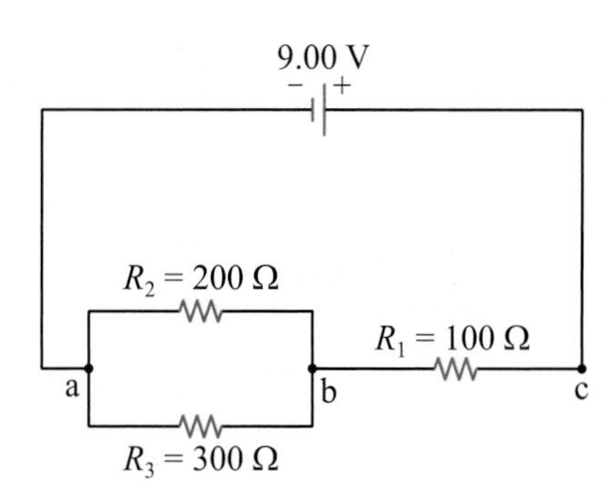

Figure 21-21 Sample Problem 21-4.

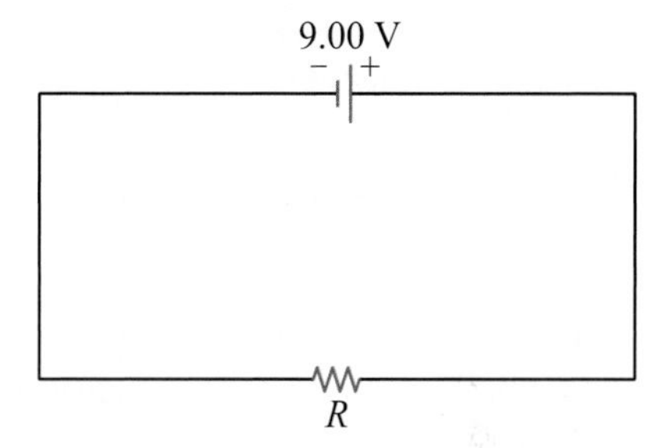

Figure 21-22 R is the equivalent resistance of R_1, R_2, and R_3 shown in Figure 21-21.

After working through Sample Problem 21-4, you may be wondering how much current goes through resistor 3. We could determine this in more than one way, but the most straightforward approach is to recognize that the total current through the battery (40.9 mA) gets split between the parallel branches. Therefore, 40.9 mA – 24.5 mA = 16.4 mA goes through resistor 3. Notice that the lower resistance ($R_2 = 200\ \Omega$) branch has the higher share of the total current. This is always the case, a result we often refer to as the current preferring the path of least resistance.

SAMPLE PROBLEM 21-5

In Figure 21-23 (a), the currents at M, N, and P, are 7.0 A, 4.0 A, and 1.0 A, respectively. What is the current through R_2?

Solution

Figure 21-23 (b) shows the currents at junction point Q. The polarity of the battery produces a counterclockwise current in the circuit. Therefore, the 7.0 A current travels from M into Q, the 4.0 A current travels from Q to N (and then through R_1), and the 1.0 A current travels from Q to P (and through R_3). By conservation of charge, we know that the total current entering the junction at Q must equal the total current leaving. Therefore, labelling the current through R_2 as I_2,

$$7.0\text{ A} = 4.0\text{ A} + I_2 + 1.0\text{ A}$$

Solving for I_2: $I_2 = (7.0 - 4.0 - 1.0)\text{ A} = 2.0\text{ A}$

Hence, the current through R_2 is 2.0 A.

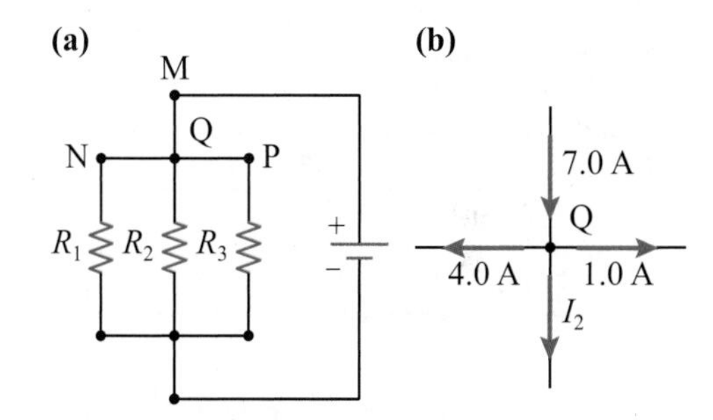

Figure 21-23 **(a)** Sample Problem 21-5. **(b)** Currents at junction point Q.

SAMPLE PROBLEM 21-6

Figure 21-24 shows a 9.0 V battery providing current to three resistors in parallel. Switch S is closed.

(a) What is the potential difference across each resistor?

(b) What is the current through each resistor?

(c) What is the current from B to C? from A to B?

(d) What is the current through the battery?

(e) If switch S is now opened, what are the answers to the above questions?

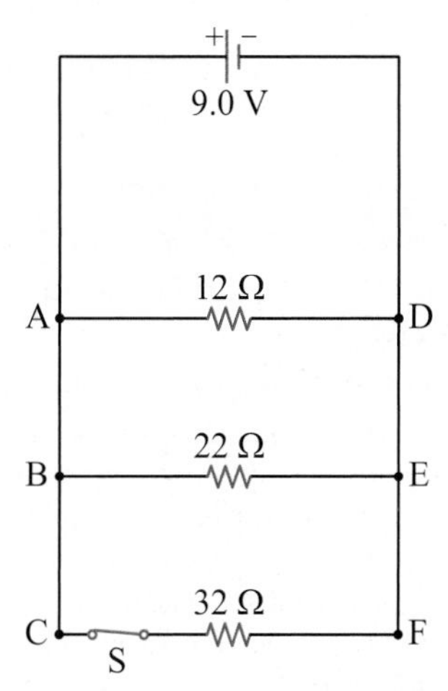

Figure 21-24 Sample Problem 21-6.

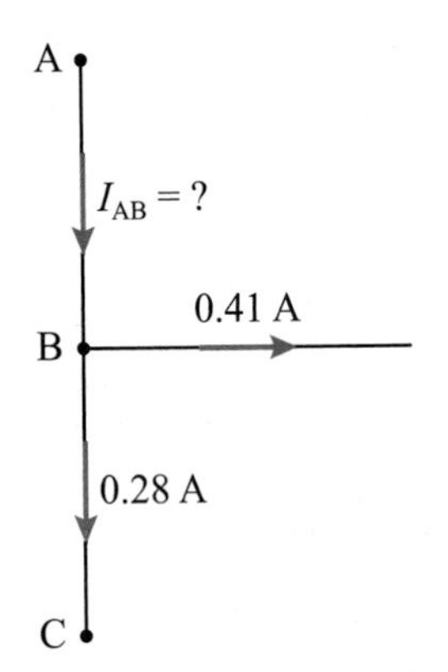

Figure 21-25 Currents entering and leaving point B (Sample Problem 21-6).

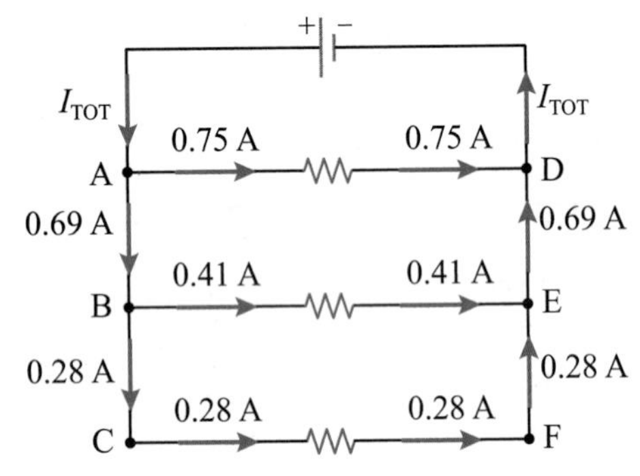

Figure 21-26 The currents in the circuit (Sample Problem 21-6).

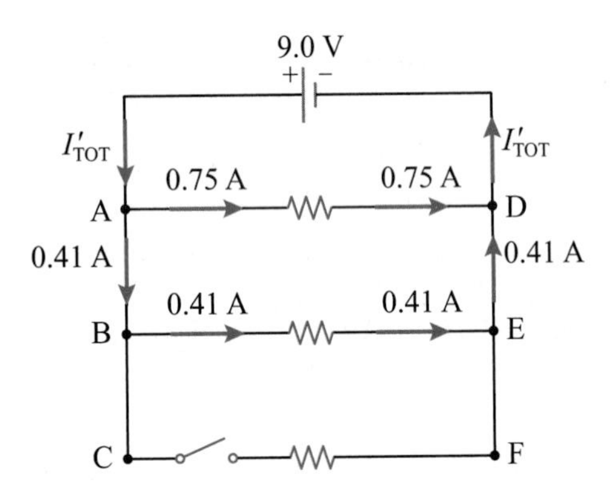

Figure 21-27 The circuit with the switch open between C and F (Sample Problem 21-6).

TACKLING MISCONCEPTIONS

Voltage, Not Current, Is a Fixed Quantity with DC Batteries

As we saw in Sample Problem 21-6, the current that is provided by a battery is not a fixed quantity. The potential difference across the terminals is a standard value, for example 1.5 V for a 'AA' type battery, but the current that results when the battery is connected to a circuit depends on the particular components involved and their configuration.

Solution

(a) Points A, B, and C are each connected by wires of negligible resistance to the positive terminal of the battery, and hence are at the same electric potential as this terminal. Similarly, points D, E, and F are at the same electric potential as the negative terminal of the battery. Therefore, the potential difference from A to D, from B to E, and from C to F is just the potential difference across the battery terminals, 9.0 V. Thus, the potential difference across each resistor is 9.0 V.

(b) Ohm's law can be used to determine the current through each resistor. Each resistance is known and the potential difference is 9.0 V across each resistor. As an example, for the 12 Ω resistor,

$$I = \frac{\Delta V}{R} = \frac{9.0\,\text{V}}{12\,\Omega} = 0.75\,\text{A}$$

Similarly, the currents through the 22 Ω and 32 Ω resistors are 0.41 A and 0.28 A respectively. Again, notice that the smallest resistance branch has the largest current and the largest resistance branch has the smallest current.

(c) The current that travels from B to C also travels through the 32 Ω resistor (and then from F to E). Hence, this current is the same as the current through the 32 Ω resistor, which we determined in part (b) to be 0.28 A.

To determine the current from A to B, consider Figure 21-25, which shows the currents entering and leaving point B. We are asked to find current I_{AB}. The current from B to C is 0.28 A, and the 0.41 A current travelling to the right is the current through the 22 Ω resistor (calculated in part (b)). Since current in equals current out at junction point B,

$$I_{AB} = 0.28\,\text{A} + 0.41\,\text{A} = 0.69\,\text{A}$$

Thus, the current from B to C is 0.28 A, and the current from A to B is 0.69 A.

(d) The current through the battery can be found from Figure 21-26, showing the various currents in the circuit. Notice how the total current, I_{TOT}, from the battery splits at point A and then splits further at B. The currents recombine at E and D to give a current of I_{TOT} returning to the battery. We can find I_{TOT} by considering either point A or D; we will use A. The current entering this point must equal the current leaving:

$$I_{TOT} = 0.69\,\text{A} + 0.75\,\text{A} = 1.44\,\text{A}$$

Hence, there is a current of 1.44 A through the battery.

(e) The circuit with switch S open (between C and F) is shown in Figure 21-27. The open switch prevents the movement of charge through the bottom resistor, and hence, no current travels from B to C, nor from F to E. For the other two resistors, there is still a potential difference of 9.0 V across them and unbroken paths for the movement of charge. Therefore the currents through these resistors are unchanged by the opening of the switch. (Compare the currents through these resistors in Figures 21-26 and 21-27.)

The current from A to B is now just the current through the middle resistor, that is, 0.41 A. The total current provided by the battery can be determined by noting that the current entering point A must equal the sum of the currents leaving point A. Designating the total current provided by the battery as I'_{TOT}, we have

$$I'_{TOT} = 0.75\,\text{A} + 0.41\,\text{A} = 1.16\,\text{A}$$

Thus, because the switch is open, the current through the battery has been reduced from 1.44 A to 1.16 A.

Redrawing Circuit Diagrams

In the circuit diagrams that have been shown so far, resistors in series have been drawn in the same straight-line portion of the circuit. However, the two resistors shown in Figure 21-28 (a) are also in series, since the same current travels through each of them. Figure 21-28 (b) illustrates the same circuit with the resistors in a straight line. When solving problems, it can be helpful to redraw the circuit to show series resistors in a straight line.

Similarly, if the same potential difference is applied across two or more resistors, the resistors are said to be in parallel, although they might not be drawn in portions of the circuit that are geometrically parallel (Figure 21-29 (a)). It can be useful to redraw circuit diagrams to show parallel resistors in geometrically parallel sections of the circuit for clarity (Figure 21-29 (b)), especially if the circuit consists of many components.

Math Tip

You can check the answer to Sample Problem 21-6 (d) for (i) the switch open and (ii) the switch closed configurations by determining the equivalent resistance (R) of (i) a 12 Ω and a 22 Ω resistor in parallel, and (ii) 12 Ω, 22 Ω, and 32 Ω resistors in parallel. Then $I = \Delta V/R$ gives the current through the battery in each case with $\Delta V = 9.0$ V.

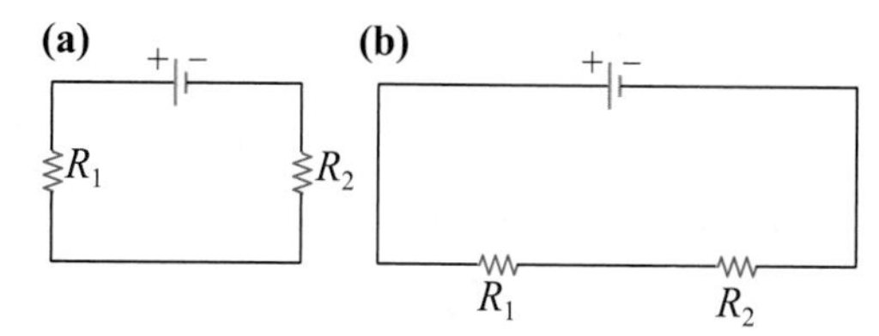

Figure 21-28 **(a)** The two resistors are in series, since the same current travels through each one. **(b)** The same circuit, redrawn to show clearly that the resistors are in series.

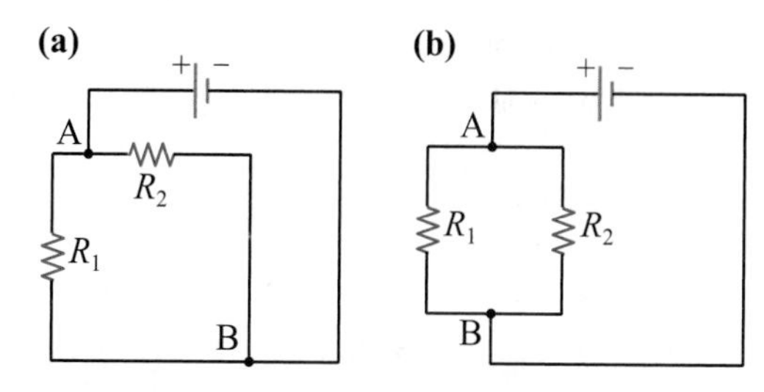

Figure 21-29 **(a)** The same potential difference is being applied across the two resistors, which are therefore in parallel. **(b)** The diagram has been redrawn to show clearly that the resistors are in parallel.

Measuring Potential Difference and Current

In order to measure potential difference and current in laboratories, electronics workshops, or in your home, **voltmeters** and **ammeters** are used. Many meters are **multimeters**—that is, multipurpose meters that can measure a wide range of potential differences and currents (as well as resistances). Dials or push-buttons on a multimeter allow the user to change easily from one function to another (Figure 21-30).

A voltmeter measures the potential difference (or voltage) *between* two points in a circuit. For example, Figure 21-31 illustrates the use of a voltmeter (indicated by a "V" inside a circle) to determine the potential difference between A and B, that is, the potential difference across resistor R_1. The two connecting wires from the meter are placed at points A and B at opposite ends of the resistor, thus placing the meter in *parallel* with the resistor. When a meter is used as a voltmeter, it has a large internal resistance, which results in only a small current travelling through the meter. Thus, using the meter has a negligible effect on the rest of the circuit. The current through the meter is proportional to the potential difference being applied, and the meter has been calibrated at the factory so that this potential difference can be read.

An ammeter measures the current *at* a particular point in a circuit. The circuit is opened at that point and the meter is inserted, so that the current must travel through the meter. Thus, the meter is placed in *series* in the circuit. In Figure 21-32, an ammeter (indicated by an "A" inside a circle) has been inserted to measure the current through point X. When a meter is used as an ammeter, it has a very small internal resistance, so that the current in the circuit is essentially unaffected by the insertion of the meter. Again, the meter has been pre-calibrated, in this case, so that current can be read.

voltmeter a device that measures the potential difference between two points in a circuit

ammeter a device that measures the current through a particular point in a circuit

multimeter a device that can be used to measure potential difference, current, as well as resistance

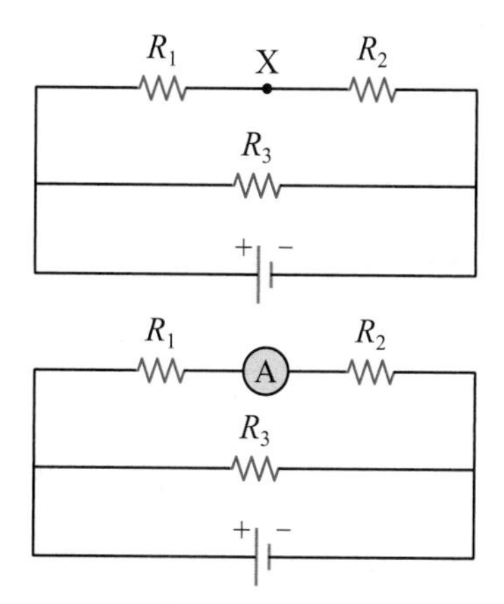

Figure 21-32 To measure the current at point X, the ammeter is inserted at this location. The ammeter will then measure the current through the resistors R_1 and R_2 with which it is in series.

Ekaterina Kondratova/Shutterstock

Figure 21-30 Digital multimeters can measure a wide range of potential differences, currents, as well as resistances.

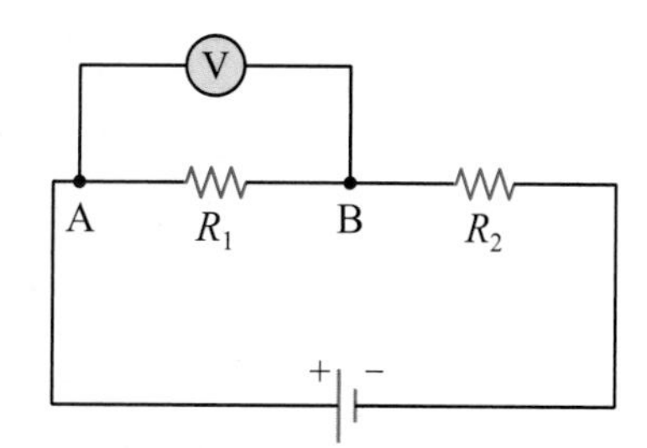

Figure 21-31 A voltmeter is placed in parallel with a resistor (R_1) in order to measure the potential difference across this component.

EXERCISES

21-18 What is the equivalent resistance of two 15 Ω resistors connected

(a) in series?

(b) in parallel?

21-19 What is the equivalent resistance of the resistor network in Figure 21-33? (Assume three significant digits.)

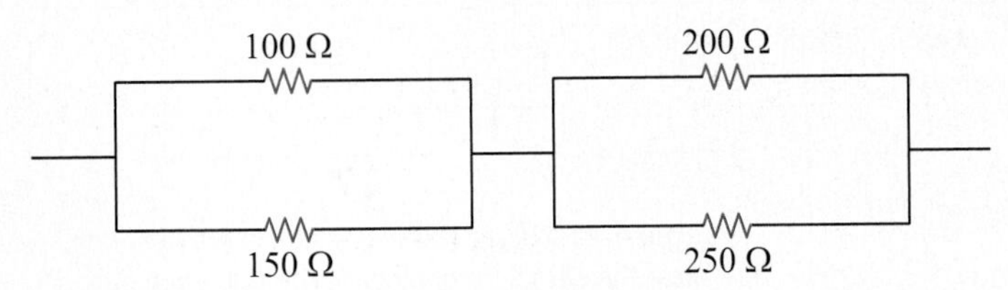

Figure 21-33 Question 21-19.

21-20 You are given a box of 100 Ω resistors. Draw diagrams to show how to connect these resistors to give a total resistance of

(a) 300 Ω

(b) 50 Ω

(c) 25 Ω

(d) 250 Ω.

21-21 In Figure 21-34, what is the current through the battery in the circuit?

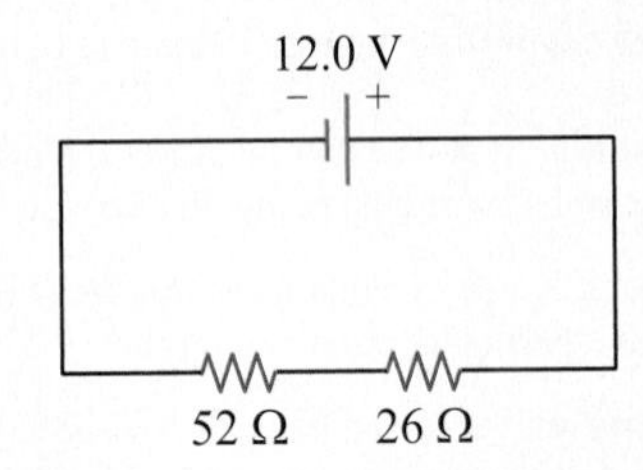

Figure 21-34 Questions 21-21 and 21-22.

21-22 Repeat the previous question if the two resistors are connected in parallel.

21-23 Figure 21-35 shows a portion of a circuit. What is the unknown current?

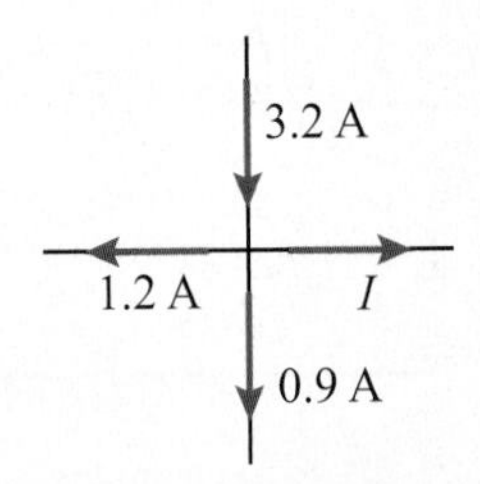

Figure 21-35 Question 21-23.

21-24 In Figure 21-36, what is

(a) the current through the 47 Ω resistor?

(b) the direction of the current through the 47 Ω resistor?

(c) the potential difference $V_A - V_B$? $V_B - V_A$?

(d) the potential difference across the 84 Ω resistor?

(e) the electric potential at point C, if we define the electric potential at the negative terminal of the battery to be zero?

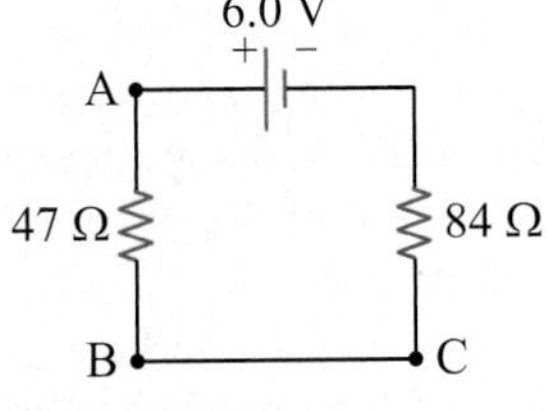

Figure 21-36 Question 21-24.

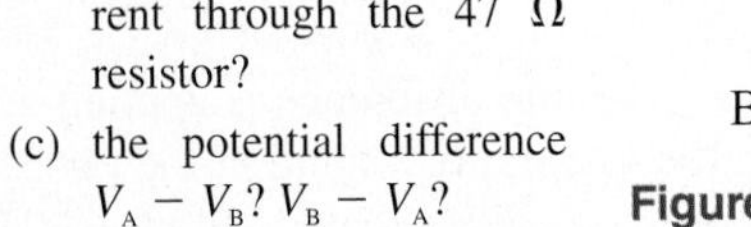

21-25 In Figure 21-37, what is

(a) the potential difference across each resistor?

(b) the current through each of the 150 Ω resistors?

(c) the current through the battery?

(d) the current entering point A through the 250 Ω resistor?

(e) the current leaving point A toward the negative terminal of the battery?

(f) the current at point B?

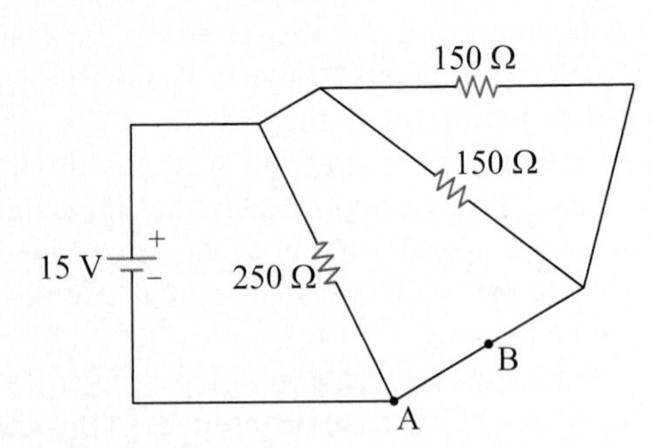

Figure 21-37 Question 21-25.

21-26 A circuit is constructed consisting of a 6.0 V battery, two resistors (75 Ω, 50 Ω) connected in series with the battery, and three resistors (200 Ω, 150 Ω, 170 Ω) connected in parallel with each other. This three-resistor combination is in series with the first two resistors and the battery.

(a) Draw a diagram of the circuit.

(b) What is the current through the battery in the circuit? (Assume two significant digits in each resistance value.)

(c) What is the current through the 150 Ω resistor?

(d) What is the potential difference across the 50 Ω resistor?

21-27 (a) For the circuit shown in Figure 21-38, what is the potential difference across each resistor? What is the current through each resistor? Switches S_1 and S_2 are closed.

(b) Repeat if switch S_2 is open, but S_1 remains closed.

(c) Repeat if both switches are open.

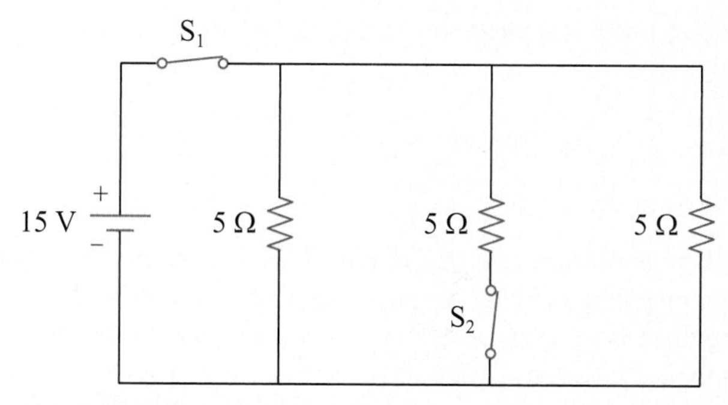

Figure 21-38 Question 21-27.

21-28 What are the readings of the voltmeters and ammeters in Figure 21-39? Assume two significant digits in the resistance values.

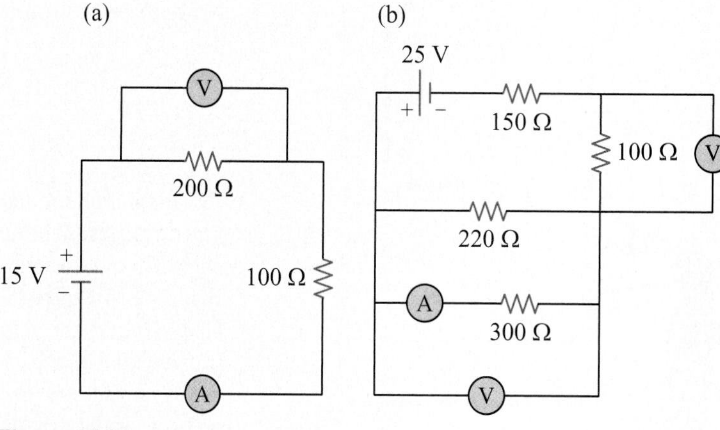

Figure 21-39 Question 21-28.

21.4 Electrical Energy and Power

If you look at almost any electrical appliance carefully, you will see a power rating in watts. For example, a hairdryer might be labelled as 2100 W, a radio as 21 W, and a television as 115 W. Since a watt is equivalent to a joule per second, this electrical power of an appliance is the amount of electric potential energy converted per second into other forms of energy. For instance, the 850 W toaster shown in Figure 21-40 converts 850 J of electric potential energy into thermal energy each second. But how is electrical power related to potential difference? to current? to resistance?

To answer these questions, we consider a simple circuit consisting only of a 100 Ω resistor and a 10 V battery (Figure 21-41). We can calculate the current in the circuit from Ohm's law:

$$I = \frac{\Delta V}{R} = \frac{10\text{ V}}{100\ \Omega} = 0.1\text{ A}$$

Therefore, 0.1 C of charge moves past any point in the circuit every second. Since current I is defined as the charge q divided by the time t, then the charge q that moves during a time t is just the product of I and t:

$$q = It$$

If we think of positive charge as travelling (i.e., considering conventional current), then in moving from A to B across the battery, the charge moves to higher electric potential and its electric potential energy increases by

$$\Delta U = q\Delta V \qquad \text{(Eqn. 20-5(a))}$$

To determine the power P provided by the battery, we divide the energy by the time:

$$P = \frac{E}{t} = \frac{\Delta U}{t} = \frac{q\Delta V}{t}$$

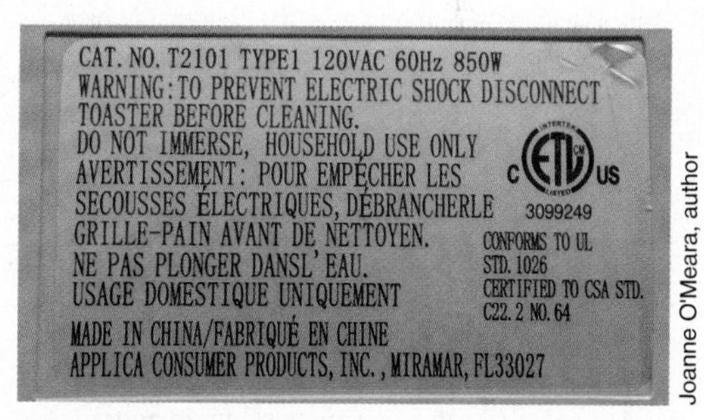

Joanne O'Meara, author

Figure 21-40 This label on a toaster indicates that it uses 850 W of power.

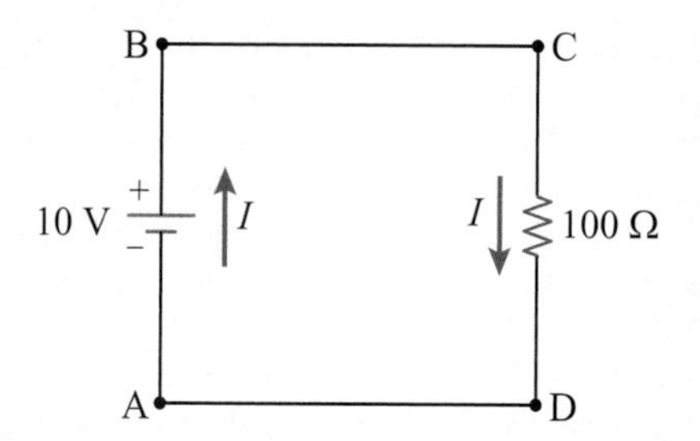

Figure 21-41 As charge moves from A to B through the battery, it gains electric potential energy. In moving from C to D through the resistor, it loses electric potential energy, which is converted to thermal energy by the resistor.

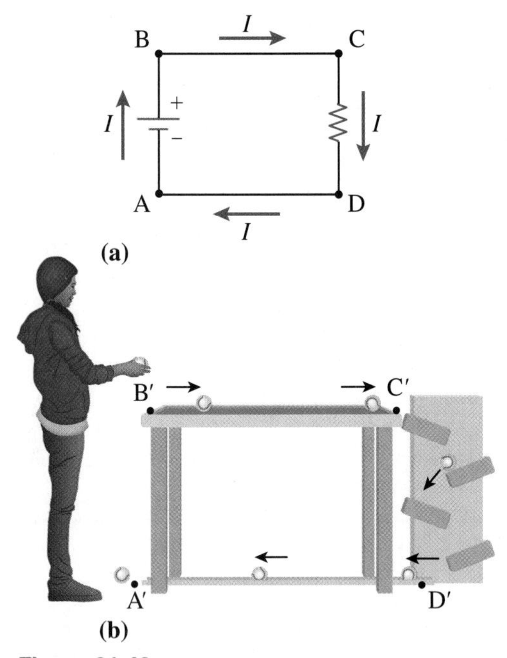

Figure 21-42 **(a)** The current through a battery and a resistor is analogous to **(b)** baseballs being lifted by a person, and then falling through soft rubber baffles.

TACKLING MISCONCEPTIONS

Does Power Consumed by a Resistor Depend on ΔV or ΔV^2?

The relationships $P = (\Delta V)I$, $P = I^2R$, and $P = (\Delta V)^2/R$ can be a source of confusion. For example, from $P = (\Delta V)I$, it appears that $P \propto \Delta V$. However, from $P = (\Delta V)^2/R$, it seems instead that $P \propto (\Delta V)^2$. How can P be proportional to both ΔV and $(\Delta V)^2$? The answer lies in how the other quantities, I and R, are changing (or not changing). In saying that the equation $P = (\Delta V)I$ implies $P \propto \Delta V$, we are assuming that I is constant; in other words, the constant of proportionality is I. Suppose that ΔV doubles, but with no change in I; P must double; that is, P is proportional to ΔV. How is it possible for ΔV to double without changing I? The resistance R must also be changing for this to occur. If ΔV doubles, and we double R at the same time, I will remain constant. Thus, to say that $P \propto \Delta V$ means that I is constant and R is changing. In saying that $P \propto (\Delta V)^2$ (from $P = (\Delta V)^2/R$), what assumptions are implicitly being made about R and I?

Now, q/t is the current I, and thus the electrical power supplied by the battery is

$$P = (\Delta V)I \tag{21-8}$$

This important and simple relationship, $P = (\Delta V)I$, tells us how much power is provided or consumed by a component in a circuit, across which there is a potential difference ΔV, and through which there is a current I. In the battery–resistor circuit shown in Figure 21-41, the battery provides an electrical power of $P = (10\text{ V})(0.1\text{ A}) = 1\text{ W}$. As charge moves through the resistor, it moves to lower electric potential, thus losing electric potential energy, which is converted into thermal energy in the resistor. (Resistors get hot as current moves through them.) The electrical power removed or dissipated by the resistor, or the thermal power generated in it, is also 1 W, since the potential difference across the resistor is 10 V and the current through it is 0.1 A. This is a direct consequence of conservation of energy: the rate of energy delivered to the circuit by the battery equals the rate of energy consumed by the various components to which it is connected.

It is useful to consider the gravitational analogue of this simple circuit. As the electric charges move from A to B (Figure 21-42 (a)), their electric potential energy is increased by the battery. A similar situation occurs if a person lifts baseballs at constant speed (Figure 21-42 (b)) from A′ to B′, thus increasing their gravitational potential energy. As the current moves from B to C, there is no change in electric potential, and thus no change in energy. As the balls roll from B′ to C′ along a horizontal (frictionless) surface, there is no energy change. As the current moves from C to D, electric potential energy is converted into thermal energy. As the balls roll down the soft rubber baffles from C′ to D′, gravitational potential energy is converted into thermal energy. (We assume that the baffles are designed so that the kinetic energy of the balls at entrance and exit is the same.) Finally, there is no change in energy as the current moves from D to A, or as the balls from roll from D′ to A′ again along a horizontal (frictionless) surface.

Power Dissipated in a Resistor

The relation $P = (\Delta V)I$ can be written in two other ways for the case where P represents the power dissipated in a resistor. We can substitute $\Delta V = IR$ into $P = (\Delta V)I$ to get $P = (IR)(I) = I^2R$. Therefore, we can also write the power dissipated in a resistor R through which a current I travels as

$$P = I^2R \tag{21-9}$$

Alternatively, we can get a relation for the power dissipated in the resistor that does not include current by substituting $I = (\Delta V)/R$ into $P = (\Delta V)I$, giving $P = \Delta V\left[\dfrac{\Delta V}{R}\right] = \dfrac{(\Delta V)^2}{R}$. Therefore, we can write the power dissipated in the resistor R with a potential difference ΔV across it as

$$P = \frac{(\Delta V)^2}{R} \tag{21-10}$$

In summary, there are three ways to calculate the power consumed by a resistor: Eqns. 21-8, 21-9, and 21-10. Which one is most convenient to use in a given situation depends on which quantities are known.

SAMPLE PROBLEM 21-7

In the circuit in Figure 21-43, determine the power

(a) dissipated by R_1 (assuming two significant digits in the value for R_1)

(b) dissipated by R_2

(c) provided by the battery.

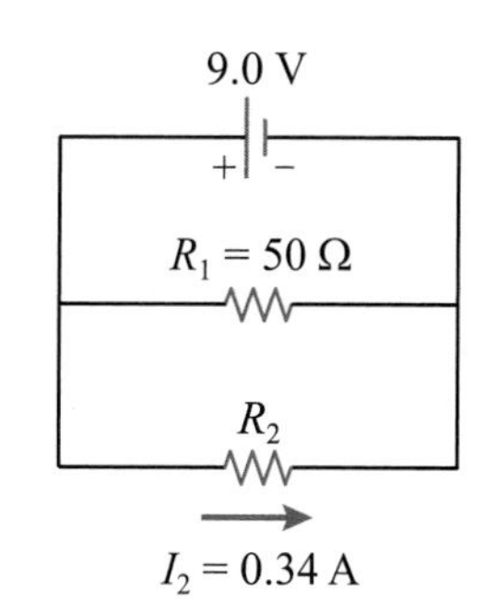

Figure 21-43 Sample Problem 21-7.

Solution

(a) The potential difference across each resistor is 9.0 V since the resistors are each in parallel with the 9.0 V battery.

For resistor 1, we know the resistance R_1 (50 Ω) and the potential difference (9.0 V). Therefore, the easiest way to calculate the power (P_1) dissipated is to use Eqn. 21-10:

$$P_1 = \frac{(\Delta V)^2}{R_1} = \frac{(9.0\,\text{V})^2}{50\,\Omega} = 1.6\,\text{W}$$

Thus, the power dissipated by R_1 is 1.6 W.

(b) For resistor 2, we know the potential difference and the current, but not the resistance. Therefore, to find the power, we use Eqn. 21-8:

$$P_2 = (\Delta V)I_2 = (9.0\,\text{V})(0.34\,\text{A}) = 3.1\,\text{W}$$

Hence, the power dissipated by R_2 is 3.1 W.

(c) There are a number of ways to find the power provided by the battery. As just one example, we could determine the current through R_1 using Ohm's law, add this to the current through R_2 to find the total current through the battery, and then use $P = (\Delta V)I$ to find the battery power.

However, the easiest way is simply to use conservation of energy. The energy provided by the battery equals the thermal energy produced in both resistors together. Since power is the rate of energy production or consumption, it is also true that the power provided by the battery equals the thermal power dissipated in both resistors. Thus, the battery power (P_B) is just the sum of the powers calculated in (a) and (b):

$$P_B = P_1 + P_2 = 1.6\,\text{W} + 3.1\,\text{W} = 4.7\,\text{W}$$

Hence, the battery provides 4.7 W of power in this circuit.

DID YOU KNOW?

Incandescent light bulbs have very low efficiency in converting electrical energy into light. As discussed in Chapter 7, these devices are only about 5% efficient, with 95% of the electrical energy being converted to heat instead of light. Fluorescent lights are better, with efficiencies of approximately 20%, while LEDs are better yet with efficiencies in the 30 to 50% range. People tend to prefer incandescent bulbs, however, as the light emitted is closer to sunlight in its emission frequencies. Research is ongoing into alternatives to incandescent lights that are cheaper to produce than fluorescent or LED devices, emit light that better mimics sunlight, and have improved efficiencies. One such approach is a polymer light source known as the field-induced polymer electroluminescent light, or FIPEL for short (Figure 21-44). The device is essentially a thin plastic foil with a patented mixture of nanoparticles embedded in a polymer matrix, manufactured at a fraction of the cost of fluorescent bulbs. When an electric current is applied to the film, the resulting excitation gives rise to visible light that is closer to sunlight than standard fluorescent bulbs.

Figure 21-44 Researchers at Wake Forest University working on FIPEL lighting technology.

EXERCISES

21-29 (a) For the relation $P = (\Delta V)^2/R$, sketch the shape of the graph of P versus ΔV, assuming constant R. (b) Repeat, plotting P versus $(\Delta V)^2$.

21-30 (a) For the relation $P = (\Delta V)^2/R$, sketch the shape of the graph of P versus R, assuming constant ΔV. (b) Repeat, plotting P versus $1/R$.

21-31 What is the electrical power provided by a 6.0 V battery that is supplying a current of 1.25 A?

21-32 Suppose that a person receives an electric shock as a result of touching a "live" wire in a household circuit. Assume that the person's skin is dry, and hence the body's resistance is 1.5×10^3 Ω. (The resistance would be less if the skin were wet.) What is the thermal power dissipated in the person's body if the current generated during the shock is 0.24 A?

21-33 A 12.0 V battery and a 150 Ω (two significant digits) resistor are connected to form a closed circuit. Determine (a) the electrical power provided by the battery and (b) the thermal power dissipated in the resistor.

21-34 Three small, identical light bulbs are connected in series with a 6.0 V battery. If the power dissipated in each bulb is 5.4 W, what is the (a) current through each bulb? (b) potential difference across each bulb? (c) resistance of each bulb's filament?

21-35 In the circuit in Figure 21-45,

(a) by how much does the electric potential energy of +2.0 C of charge increase or decrease (specify which) as the charge moves from A to B?

(b) Repeat, but as the charge moves from C to D.

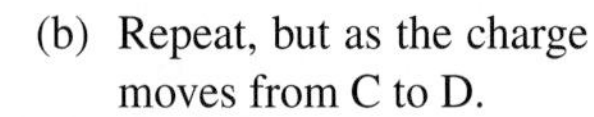

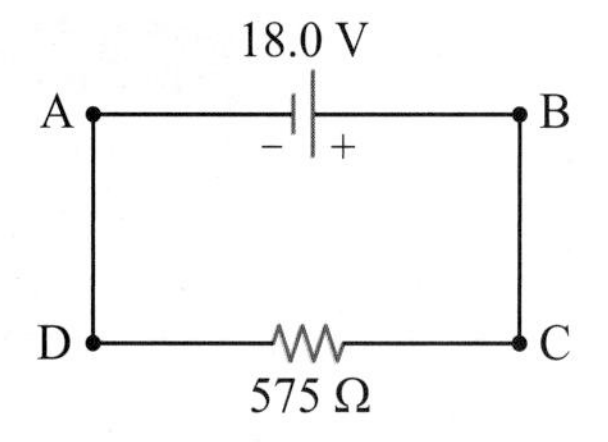

Figure 21-45 Question 21-35.

21-36 In the circuit in Figure 21-46, the current through the 300 Ω resistor is 194 mA.

(a) How much thermal power is dissipated in this resistor?

(b) What is the potential difference across the terminals of the battery?

(c) How much thermal power is dissipated in each of the other resistors?

(d) How much electrical power is provided by the battery?

(Assume two significant digits in the resistance values in this problem.)

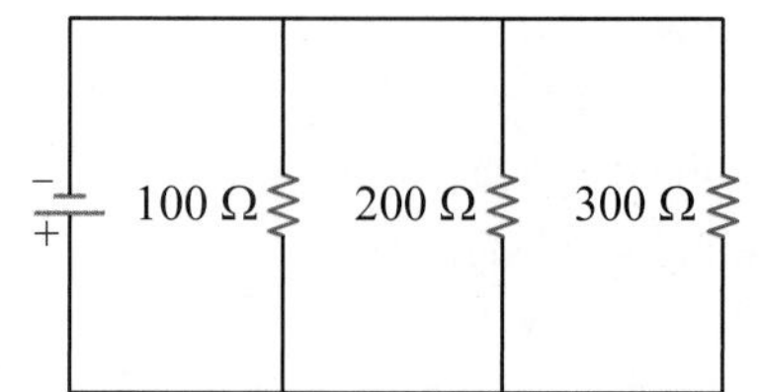

Figure 21-46 Question 21-36.

21-37 **Fermi Question:** Estimate the number of standard incandescent light bulbs that could be run from a standard rooftop solar panel assembly (Figure 21-47) on a sunny day.

manfredxy/Shutterstock

Figure 21-47 Question 21-37.

21.5 AC Circuits

You might have noticed that all the circuits that we have considered so far have involved batteries as an energy source. What about standard electrical circuits in your home? The energy is provided by falling water, nuclear fission, wind turbines, the burning of coal, oil, gas, etc., then converted to electrical energy and carried to your home through transmission lines.

direct current (DC) current travels in a circuit in only one direction, typically driven by a battery (or a collection of batteries)

alternating current (AC) current reverses direction in a circuit in a periodic fashion, such as the current generated by commercial power stations

There is a fundamental difference between a battery-powered circuit and a household circuit. In a battery circuit, the current travels in only one direction, and is called **direct current (DC)**. In the circuits that power your lights, television, refrigerator, etc., at home, the current changes direction 120 times each second, and is called **alternating current (AC)**.[1]

The alternating current is driven by a potential difference that changes sign 120 times per second. In a two-prong electrical outlet, one connector is said to be "neutral," that is, at ground potential. The other connector is at a higher electric potential than the neutral one for 1/120 s, then at a lower electric potential for the next 1/120 s, and so on.

A graph of this potential difference is shown in Figure 21-48 (a). You probably recognize it as a sine curve. Figure 21-48 (b) shows the current that results when this potential difference is applied to a 100 W light bulb.

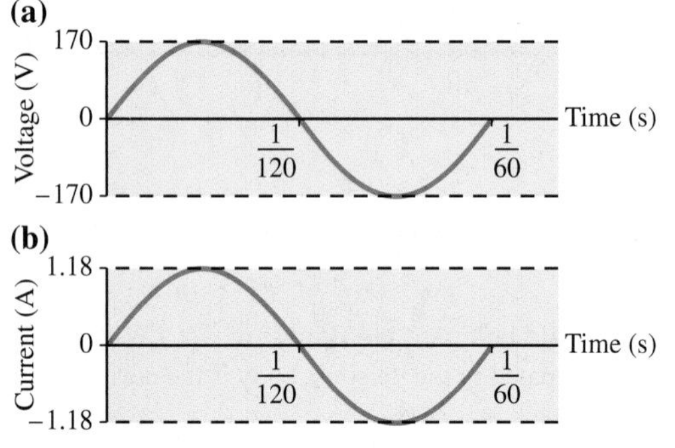

Figure 21-48 **(a)** Standard alternating potential difference. **(b)** The alternating current that results when the potential difference in (a) is applied to a 100 W light bulb.

You might know that the standard potential difference in North America is referred to as "120 V, 60 Hz", as shown on the bottom of the toaster in Figure 21-40. It is easy to see where the 60 Hz comes from: the potential difference goes through a complete cycle 60 times per second. But why is it 120 V, when the peak potential difference is clearly 170 V? The 120 V value is the effective potential difference, ΔV_{eff}, averaged over a complete cycle. It would be inappropriate to assign 170 V as the effective potential difference because this peak value is reached only twice per cycle (counting both +170 V and −170 V). When the potential difference varies

[1]Alternating current is used for commercial electricity because it can be transmitted at high potential difference more easily than direct current, producing less thermal energy loss in the transmission lines themselves. We will discuss this in more detail in Chapter 22 in our Profile on Nikola Tesla.

sinusoidally in time, as shown in Figure 21-48 (a), the effective value over a complete cycle is given by

$$\Delta V_{eff} = \frac{\Delta V_{peak}}{\sqrt{2}} \qquad (21\text{-}11)$$

which gives us an effective potential difference of 120 V when the peak value is 170 V.

Similarly, when we say that a sinusoidally varying AC current is, say 3 A, this is the effective current. The peak current is larger by a factor of $\sqrt{2}$. More formally, what we are calling the effective potential difference or current is actually the **root-mean-square (rms)** value, that is, the square root of the mean (average) of the square of the potential difference (or current), as discussed in Chapter 18 with respect to the speeds of molecules associated with thermal motion.

root-mean-square (rms) potential difference the effective potential difference over a cycle for AC circuits; also known as rms voltage

root-mean-square (rms) current the effective current over a cycle for AC circuits

Notice that we have not said that the average potential difference is 120 V. Because the potential difference is continually changing sign, the average potential difference is zero (over an integer number of cycles). However, its average effect is not zero. Consider a light bulb operating on AC. When the potential difference is positive, current travels in one direction through the bulb's filament, and electric potential energy is converted to light and thermal energy. When the potential difference is negative, current moves in the opposite direction, and light and thermal energy are again generated. Thus, even though the average potential difference is zero, energy is, nonetheless, delivered.

As is the case with DC circuits, the electrical power consumed is the product of potential difference and current. However, since the potential difference and current are varying, the (instantaneous) power also varies (Figure 21-49). This power goes from a value of zero (when $\Delta V = I = 0$), to a positive maximum (when $\Delta V > 0$ and $I > 0$), back to zero, to a positive maximum (when $\Delta V < 0$ and $I < 0$), etc. In Figure 21-49, the maximum instantaneous power consumption is 200 W; however, the effective power consumption over a cycle is 100 W.

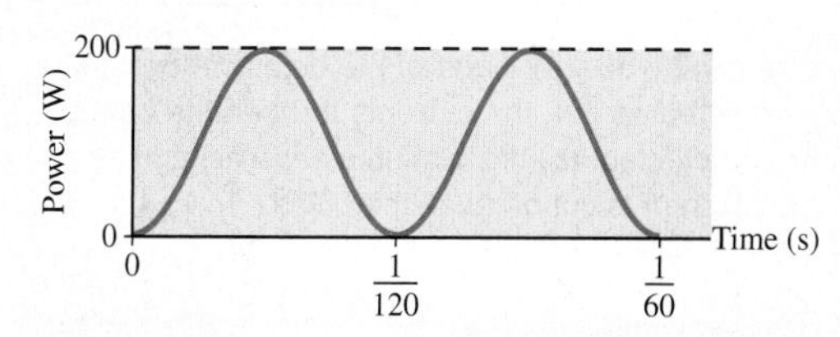

Figure 21-49 The product of potential difference and current gives the (instantaneous) power consumption of a 100 W light bulb. The maximum power is 200 W; the effective power consumed over a cycle is 100 W.

This **effective power** consumption is what is of interest to the electrical consumer. When we speak of power in an AC circuit, we do not usually mean the instantaneous power, but rather the effective power. The effective power is related in a straightforward way to the rms potential difference and current:

effective power the power supplied or consumed by a component in an AC circuit, averaged over a complete cycle

$$P_{eff} = \Delta V_{rms} I_{rms} \qquad (21\text{-}12)$$

Notice that Eqn. 21-12 conveniently has the same form as Eqn. 21-8 that we developed for DC power in the previous section. In Eqn. 21-12, the quantities P_{eff}, ΔV_{rms}, and I_{rms} are sometimes referred to simply as the power, potential difference, and current without the adjectives effective, or rms; however, we will be specific in this text.

The relations $P = (\Delta V)^2/R$ and $P = I^2R$ that we derived for DC can also be used for AC. In these equations, P becomes the effective power, and ΔV and I are the rms values of potential difference and current. The resistance R for an ohmic resistor is the same whether DC or AC is used. In summary, then, we have three equations again for determining the effective power supplied or consumed by a resistive component in an AC circuit:

$$P_{eff} = \Delta V_{rms} I_{rms} \qquad (21\text{-}12)$$

$$P_{eff} = \frac{(\Delta V_{rms})^2}{R} \qquad (21\text{-}13)$$

$$P_{eff} = I_{rms}^2 R \qquad (21\text{-}14)$$

where P_{eff} is the effective power supplied or consumed by the component, ΔV_{rms} is the rms potential difference across the component, I_{rms} is the rms current through the component, and R is the resistance.

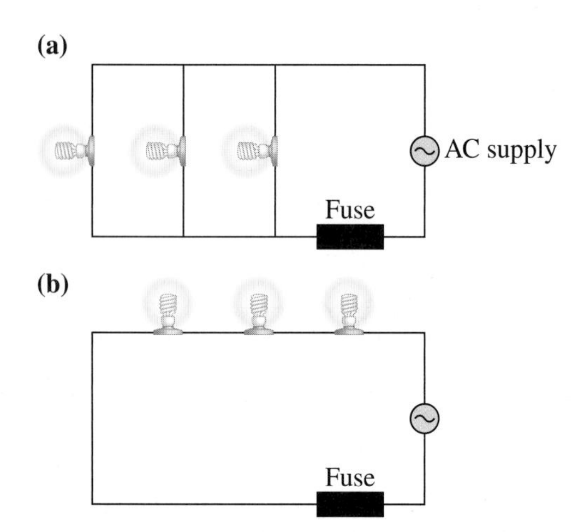

Figure 21-50 **(a)** If one of the bulbs connected in parallel burns out, the currents through the others are unaffected. **(b)** If a bulb in a series circuit burns out, current is cut off to all the bulbs.

DID YOU KNOW?

Christmas-tree lights (Figure 21-51) in common use a few decades ago were often wired in series: if one bulb burned out, the entire string of lights went out, and it was a time-consuming exercise to discover which bulb needed replacement. Modern Christmas mini-lights are connected in series, but when one bulb burns out, the rest of the lights continue to function. The bulbs have been cleverly designed so that, when the standard potential difference is applied to a string containing a burned-out bulb, a small spark is produced inside this bulb.[2] The spark melts the insulation off a tiny jumper wire in the bulb, resulting in an electrical connection across the interior of the bulb. This allows current to move through the rest of the circuit and bypass the filament of the burned-out bulb.

Figure 21-51 A clever re-design has resulted in Christmas-tree lights that do not result in the entire string going out when one bulb dies.

Strictly speaking, these three equations apply only in the analysis of AC circuits that are purely resistive; that is, there are only resistors and power supplies in the circuit. If we were to introduce other components, such as capacitors and inductors, the analysis of the effective power consumed as well as the relationship between current and potential difference becomes more complicated. In this text, we will limit ourselves to the analysis of resistive AC circuits only.

SAMPLE PROBLEM 21-8

The power rating of a toaster (Figure 21-40) is 850 W.

(a) What is the rms current in the toaster when connected to a household electrical outlet in North America, which provides an rms potential difference of 120 V?

(b) What is the peak current?

(c) What is the resistance of the toaster's filament?

Solution

(a) Since we know the power of the toaster as well as the rms potential difference across it, we can determine the rms current by rearranging Eqn. 21-12:

$$I_{rms} = \frac{P_{eff}}{\Delta V_{rms}} = \frac{850\,W}{120\,V} = 7.1\,A$$

Thus, the rms current through the toaster is 7.1 A.

(b) The peak current is $\sqrt{2}$ times the rms current:

$$I_{peak} = \sqrt{2}\,I_{rms} = \sqrt{2}(7.1\,A) = 1.0 \times 10^1\,A$$

Hence, the peak current is 1.0×10^1 A.

(c) To determine the resistance of the toaster's filament, we can rearrange Eqn. 21-13, which gives

$$R = \frac{(\Delta V_{rms})^2}{P_{eff}} = \frac{(120\,V)^2}{850\,W} = 17\,\Omega$$

(Because we know the rms current from part (a), we could also have used Eqn. 21-14, $P_{eff} = I_{rms}^2 R$, to calculate the resistance.)

Thus, the resistance of the filament is 17 Ω.

Practical AC Electricity

In a typical household circuit, the outlets, lights, etc., are connected in *parallel*. To understand why this is done, consider Figure 21-50 (a), which shows a 120 V AC supply connected to three light bulbs in parallel. In this way, the potential difference across each bulb is 120 V, regardless of how many of the bulbs are turned on. If one bulb is turned off, unscrewed, or burns out, the current through this bulb becomes zero, but this does not affect the current through the rest of the bulbs, each of which is still supplied with 120 V. In contrast, if the bulbs were connected in series, as shown in Figure 21-50 (b), then, if one bulb were to burn out, the current to the entire circuit is eliminated. Household circuits themselves are also connected in parallel with each other, so that, if current to one circuit is eliminated, other circuits are unaffected.

Notice the fuse in Figure 21-50; each household circuit has a fuse (or circuit breaker) as a safety device. The fuse is in series with the AC source so that the total

[2]There is no current initially in the string because there is no conducting path through the burned-out bulb, and therefore, the entire 120 V is applied across this bulb, producing a spark. When the string is functioning properly, the potential difference is divided equally across all the bulbs, and the small potential difference across any one bulb is not large enough to create a spark.

current in the circuit travels through the fuse. If too many appliances are turned on in a circuit, thus producing a very large total current, or if an appliance malfunctions and draws too much current, the fuse melts (or the circuit breaker "trips"), and no current can travel through the circuit. A typical household circuit has a fuse or breaker rated at 15 A; larger currents overheat the wires and constitute a fire hazard.

You are undoubtedly familiar with three-prong electrical outlets (Figure 21-52), with two slotted terminals and one rounded. One of the slotted terminals is connected to ground, and the other is "live"; that is, it provides the alternating potential difference relative to ground. The rounded terminal is also connected to ground and functions as a safety feature. An electrical device that typically requires a large current, such as a power saw, has its third wire connected to the casing of the saw. If the "live wire" inside the saw becomes loose and makes contact with the casing, a large current travels through the third wire to ground and the circuit fuse "blows." In the absence of the third wire, the body of anyone touching the casing of the saw could provide an electrical path to ground and the person could be electrocuted.

Figure 21-52 The third prong in this outlet operates as a safety feature to protect against electrical shocks.

Some appliances such as clothes dryers and stoves are connected to 240 V circuits. These devices need a large amount of power, and the larger potential difference permits a smaller current. (Considering $I_{rms} = P_{eff}/\Delta V_{rms}$, if ΔV_{rms} is doubled, then the same effective power will be consumed using half the current.) If the power were delivered at 120 V, the current would be so large that the wires would become hot enough to be a fire hazard. The resistance of wires is small, but not zero, and for a given wire of resistance R, an rms current I_{rms} produces an effective thermal power $P_{eff} = I_{rms}^2 R$. Hence, reducing the rms current in a wire by a factor of two decreases the heating effect by a factor of four.

A common unit used to express electrical *energy* consumption is the kilowatt·hour (kW·h), discussed in Chapter 7. The product of power in kilowatts and time in hours gives energy in kilowatt·hours. Beware: although kilowatt·hour contains "watt," it is an energy unit, not a power unit (1 kW·h = 3.60×10^6 J).

EXERCISES

21-38 Sketch the shape of the graph of potential difference versus time for the DC potential difference from a 12 V battery supplying current to a resistor.

21-39 A 60 W light bulb is in a lamp plugged into a standard 120 V outlet. Determine

(a) the rms current through the bulb (assuming two significant digits in the given data)

(b) the peak current

(c) the resistance of the bulb's filament.

21-40 A girl accidentally sticks a metal paperclip into the "live" connector of a 120 V outlet. If the resulting rms current that travels between her hand and her feet is 92 mA, what is the resistance between her hand and feet?

21-41 Figure 21-53 shows three identical 100 W light bulbs connected to an AC source that provides an rms potential difference of 120 V.

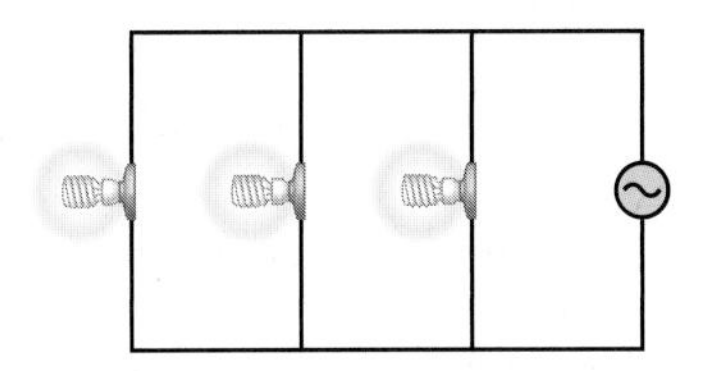

Figure 21-53 Question 21-41.

(a) If the bulbs are "on," what is the rms current through each bulb? (Assume two significant digits in the given data.)

(b) If the middle bulb is unscrewed, what is the rms current through the other two bulbs?

21-42 A small television uses 105 W of effective electrical power. If it is turned on for 3.00 h, what is its energy consumption in (a) kilowatt·hours and (b) joules? (c) What is the cost of this energy, if electrical energy costs 13¢/(kW·h)?

21-43 A 1.10 kW hairdryer and a 60.0 W light bulb are turned on in the same 120 V AC parallel circuit. Assuming three significant digits in the given potential difference, determine the rms current in the (a) hairdryer, (b) light bulb, and (c) fuse in the circuit. (d) If the fuse is rated at 15 A, will it blow?

21-44 The 60 Hz frequency of AC power remains almost constant, but the peak value of the potential difference does fluctuate, depending on demand and on equipment problems. (A "brownout" occurs when the potential difference drops so much that lights become noticeably dimmer.) If an electric baseboard heater uses a power of 1.20 kW when the rms potential difference is 120 V (three significant digits), what power will it use if the rms potential difference drops to 105 V? Assume that the resistance of the heater does not change at the different power (and temperature) levels.

LOOKING BACK...LOOKING AHEAD

In the previous two chapters on electricity, we discussed the basic concepts of electric charge, Coulomb's law, electric fields, potential energy, electric potential, and current.

This chapter has focused on resistance and circuits, both DC and AC. Electrical resistance has been defined and its relation to the resistivity and size of a conductor has been presented. We have discussed Ohm's law, that is, the relation between potential difference, current, and resistance, and have used it in calculating these quantities in simple circuits. The equivalent resistance of resistors in series and in parallel has been determined, and we have seen how to calculate electrical energy and power in circuits.

This chapter concludes our discussion of electricity. The next chapter is the first in our discussion of magnetism, a closely related area of study as it explores the interactions between charges when they are moving, that is, the interactions that occur between currents in wires.

CONCEPTS AND SKILLS

Having completed this chapter, you should now be able to do the following:

- Define resistance and state its SI unit.
- State Ohm's law and use it in determining potential difference, current, and resistance in simple circuits.
- Explain the difference between electrical resistivity and electrical resistance, and use resistivities in related problems.
- Describe how the resistivity of a metallic conductor changes with temperature, and use temperature coefficients of resistivity in problems.
- Describe the basic construction of a battery.
- Draw and use electrical circuit diagrams.
- Calculate the equivalent resistance of resistors in series and/or parallel.
- Solve circuit problems involving resistors in series and/or parallel.
- Explain how voltmeters and ammeters are used in circuits to measure potential difference and current.
- State the relation between electrical power, potential difference, and current, and use this relation (and similar ones involving resistance) in problems concerning electrical power.
- Describe the difference between DC and AC.
- Discuss what is meant by peak and rms potential difference and current.
- Solve problems related to resistive AC circuits.

KEY TERMS

You should be able to define or explain each of the following words or phrases:

resistance
ohm
Ohm's law
resistivity
electric cell
electrolyte
battery
electrical circuit
open circuit
closed circuit
series (electrical connections)
parallel (electrical connections)
voltmeter
ammeter
multimeter
direct current (DC)
alternating current (AC)
root-mean-square (rms) potential difference
root-mean-square (rms) current
effective power

Chapter Review

MULTIPLE-CHOICE QUESTIONS

21-45 At a given temperature, which piece of copper wire has the highest resistance?

(a) 2.0 m length, 1.0×10^{-6} m^2 cross-sectional area
(b) 4.0 m length, 1.0×10^{-6} m^2 cross-sectional area
(c) 2.0 m length, 3.0×10^{-6} m^2 cross-sectional area
(d) 4.0 m length, 3.0×10^{-6} m^2 cross-sectional area
(e) 1.0 m length, 3.0×10^{-6} m^2 cross-sectional area

21-46 In the circuit of Figure 21-54, which switch(es) must be closed to produce a current in conductor FG?

(a) 1 and 4 only
(b) 4 only
(c) 1, 2, and 3 only
(d) 2 and 3 only
(e) 1 only

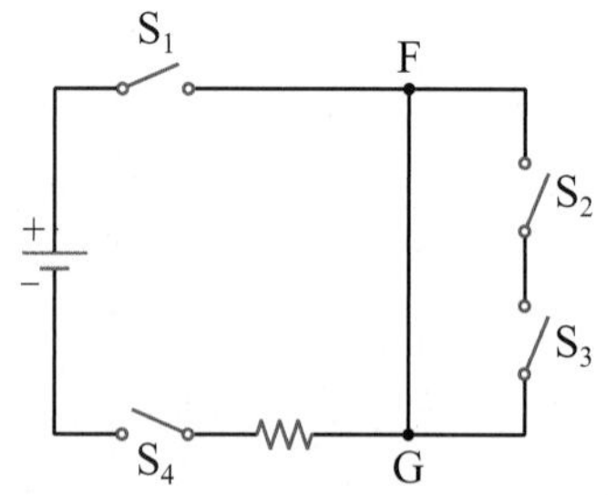

Figure 21-54 Question 21-46.

21-47 Which quantity must be the same for each component in a series circuit?

(a) potential difference
(b) power
(c) current
(d) resistance
(e) energy

21-48 Repeat the previous question for a parallel circuit.

21-49 Two resistors are connected in a circuit with a battery as shown in Figure 21-55. If resistor R_1 is then removed from the circuit, leaving a gap where it used to be, but leaving the connecting wires and resistor R_2 intact, which of the following is true?

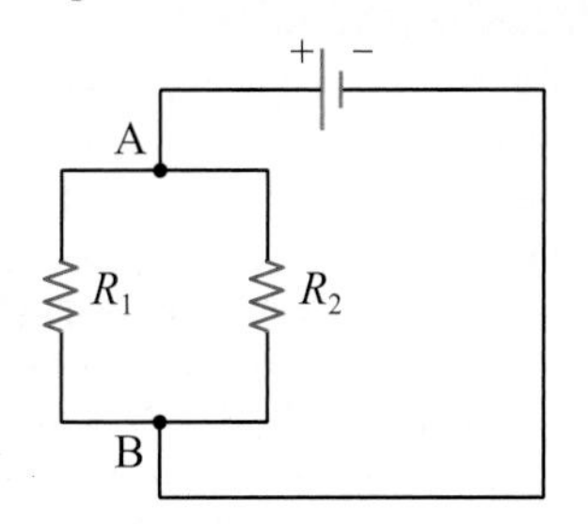

Figure 21-55 Question 21-49.

(a) The potential difference across R_2 increases, and the current through R_2 increases.
(b) The potential difference across R_2 decreases, and the current through R_2 decreases.
(c) The potential difference across R_2 increases, and the current through R_2 decreases.
(d) The potential difference across R_2 remains the same, and the current through R_2 increases.
(e) The potential difference across R_2 remains the same, and the current through R_2 remains the same.

Review Questions and Problems

21-50 A current of only about 25 μA may cause ventricular fibrillation during open-heart surgery. Determine the potential difference that would produce this current if the resistance of the heart is 240 Ω.

21-51 A number of copper wires of the same diameter have different lengths (L). Sketch the shape of a graph of the resistances of the wires versus L.

21-52 What is the diameter (in millimetres) of a 35 cm long aluminum wire if its resistance is 1.0 Ω?

21-53 An aluminum wire and a copper wire of the same length have the same potential difference applied to them. If the currents are the same, what is the ratio of the diameters of the wires?

21-54 As the temperature of a conductor is increased by 7.5°C, its resistivity increases by 3.2%. What is its temperature coefficient of resistivity?

21-55 Is the resistance of the filament of a light bulb greater when the bulb is on or off?

21-56 In Figure 21-56, follow the changes in electric potential as a positive charge moves around the circuit in the order M → N → O → P → M (i.e., following the conventional current direction). Assume that the electric potential at the negative battery terminal is defined to be zero. Specifically, what are the following changes in electric potential?

(a) $V_N - V_M$
(b) $V_O - V_N$
(c) $V_P - V_O$
(d) $V_M - V_P$
(e) Use your results to calculate the total change in electric potential in going around the circuit, starting and finishing at point M.
(f) What would be your answer to (e) if the path started and finished at N?

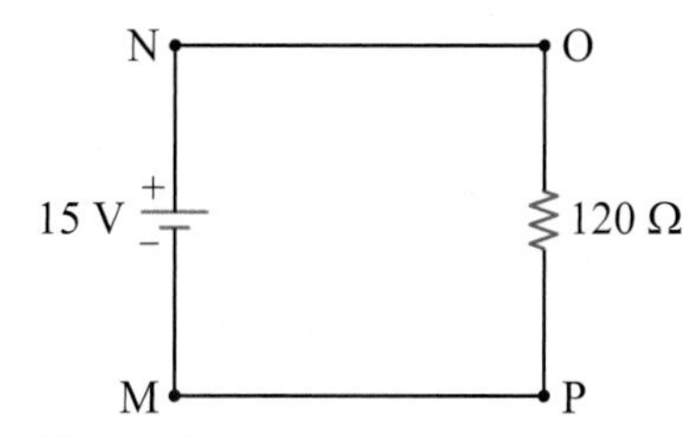

Figure 21-56 Question 21-56.

21-57 The equivalent resistance of the network shown in Figure 21-57 is 139 Ω. What is the resistance "X?"

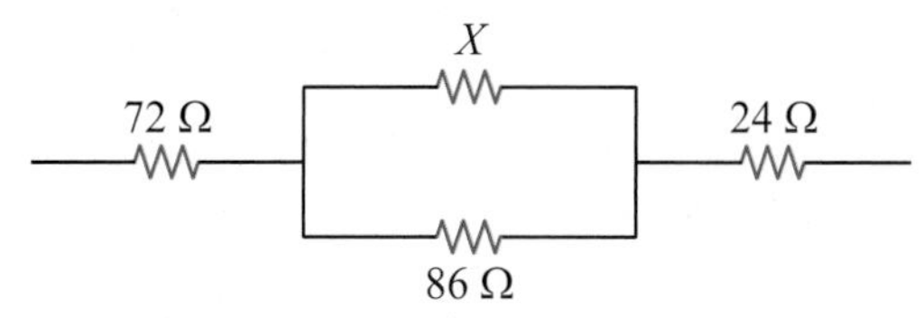

Figure 21-57 Question 21-57.

21-58 Figure 21-58 shows a portion of a circuit. What is the reading on the ammeter?

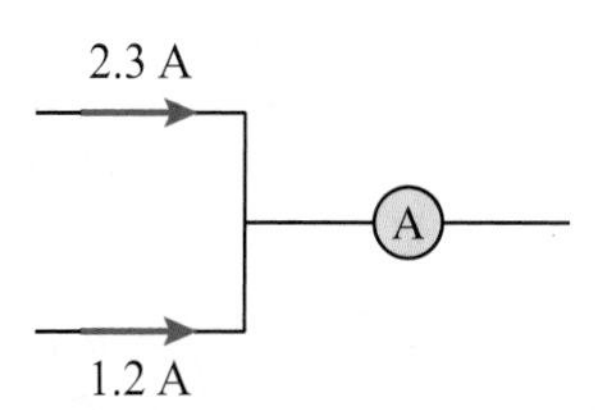

Figure 21-58 Question 21-58.

21-59 In Figure 21-59, what is the
(a) current through the 150 Ω resistor?
(b) potential difference across the 250 Ω resistor?
(c) current through the 350 Ω resistor?
(d) electric potential at point A (assuming zero electric potential at the negative terminal of the battery)?

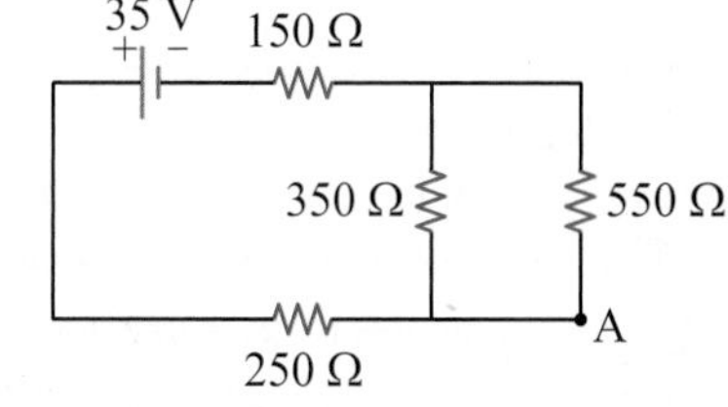

Figure 21-59 Questions 21-59 and 21-60.

21-60 In Figure 21-59, what would be the correct location of (a) a voltmeter to measure the potential difference across the 250 Ω resistor? (b) an ammeter to measure the current in the 350 Ω resistor?

21-61 Three resistors (1 Ω, 2 Ω, and 3 Ω) are connected together in series with a 6 V battery to form a closed circuit. What are the potential differences across the resistors?

21-62 (a) As current moves through a resistor, does it go from high to low or low to high electric potential? (b) Repeat, for the movement of electrons.

21-63 A light bulb is powered by a battery (of negligible internal resistance).

(a) A second identical bulb is then connected in series in the circuit. How do the brightnesses of the two bulbs compare with each other, and how do they compare with the brightness of the original single bulb? Explain.

(b) Repeat, if the second bulb is connected in parallel.

21-64 In the circuit shown in Figure 21-60, the battery maintains a constant potential difference across its terminals. The bulbs are all identical and are initially lit. In the parts that follow, in response to "What happens ...?" indicate whether the quantity increases, decreases, or is unchanged. Explain each answer briefly.

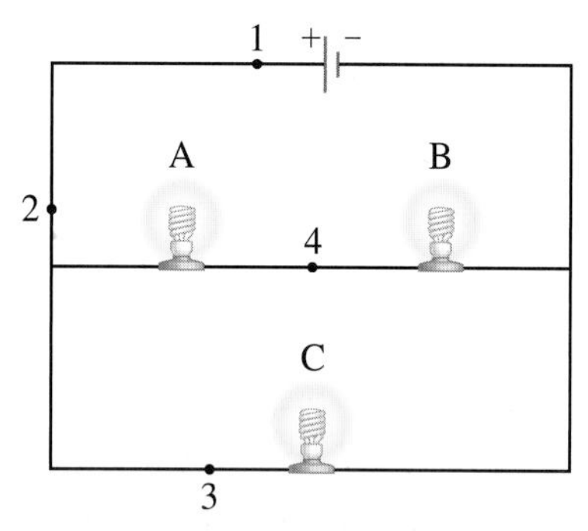

Figure 21-60 Question 21-64.

(a) Compare the brightnesses of the bulbs initially.

(b) If bulb C is now unscrewed from its socket, what happens simultaneously to the brightnesses of the three bulbs? (Answer for each bulb individually.) What happens simultaneously to the currents at points 1, 2, and 3?

(c) Bulb A is now unscrewed from its socket. (Bulb C remains unscrewed.) What happens simultaneously to the brightnesses of the bulbs, and to the currents at points 1, 2, and 3?

(d) With all three bulbs inserted in their sockets and lit, a wire of negligible resistance is connected from point 1 at the positive terminal of the battery to point 4. What happens simultaneously to the potential difference across bulb A? What happens simultaneously to the brightnesses of the three bulbs, and to the currents at points 1, 2, and 3?

21-65 Why are electrical circuits in automobiles connected together in parallel?

21-66 Using the relation $P = I^2R$, sketch the shape of a graph of (a) P versus I and (b) P versus I^2. Assume that R is constant.

21-67 If a number of extension cords are connected together to form an ultra-long cord, the rms potential difference available to an appliance plugged into the end is less than 120 V. Why?

21-68 As a light bulb "grows old," its filament becomes thinner. How does this affect its (a) resistance and (b) power? (Assume that the potential difference is constant.)

21-69 For how long must a 100 W light bulb be turned on to use 1.0 J of energy? (Assume two significant digits in the power rating.)

DID YOU KNOW?

Car battery technology is a key element in developing electric vehicles that have mass appeal. Electric vehicles that run purely on battery power currently on the market (Figure 21-61) typically have limited driving distances before needing to recharge as well as long recharging times required. In addition, such vehicles tend to be significantly more expensive than their fossil-fuel burning counterparts. MacAUTO, a research centre at McMaster University in Hamilton, ON, Canada, is working to improve all aspects of electric vehicle performance—from batteries to motors, as well as powertrains and charging stations—in order to accelerate the widespread adoption of these low-emission vehicles.

Tom Wang/Shutterstock

Figure 21-61 Advances in battery technology are being investigated in order to reduce costs and improve performance of electric vehicles.

21-70 If the rms potential difference available from an outlet changes from 120 V to 110 V (three significant digits), what is the new peak potential difference?

21-71 High-voltage DC power supplies are devices that plug into a standard AC outlet and produce a large DC potential difference between two terminals as output. They are often designed to have a large internal resistance in series with the output terminals. Why is this safer than having a small internal resistance and the same output potential difference?

21-72 In the circuit in Figure 21-62, what is the power (a) dissipated in each resistor and (b) supplied by the battery?

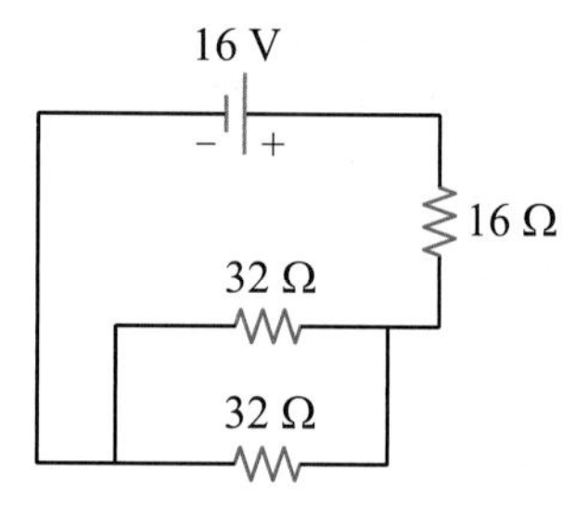

Figure 21-62 Question 21-72.

21-73 A heart defibrillator (Figure 21-63) provides a potential difference of 8.0×10^3 V through a patient's torso to reestablish normal beating of the heart. If the pathway of the current through the person has a resistance of 7.0×10^2 Ω, and the charge moves for 3.0 ms,

(a) how large is the current?

(b) how much charge moves?

(c) how much energy is deposited by the current?

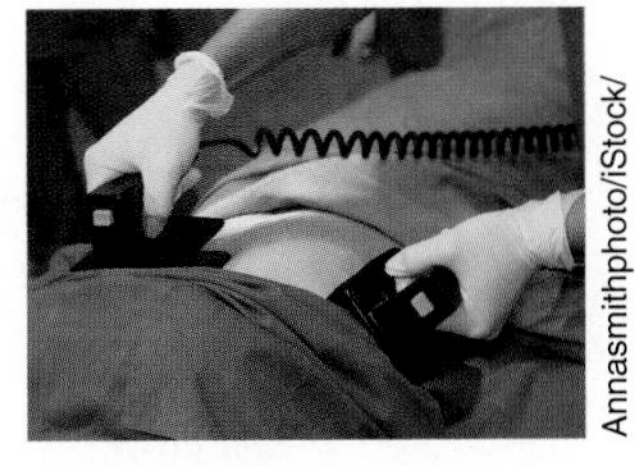
Annasmithphoto/iStock/Thinkstock

Figure 21-63 Question 21-73.

21-74 An electric clothes-dryer consumes 2.5 kW of power.

(a) What is the rms current if the power is provided at an rms potential difference of 120 V?

(b) What is the rms current if the power is provided at an rms potential difference of 240 V?

(c) What is the advantage of using 240 V for the dryer?

Applying Your Knowledge

21-75 In a single-loop circuit, the current is 1.50 A. When another resistor (1.2 kΩ) is added in series, the current is 1.30 A. What was the original resistance in the circuit?

21-76 In Figure 21-64, if we define the electric potential at the negative terminal of the battery to be zero, what is the electric potential at X? at Y?

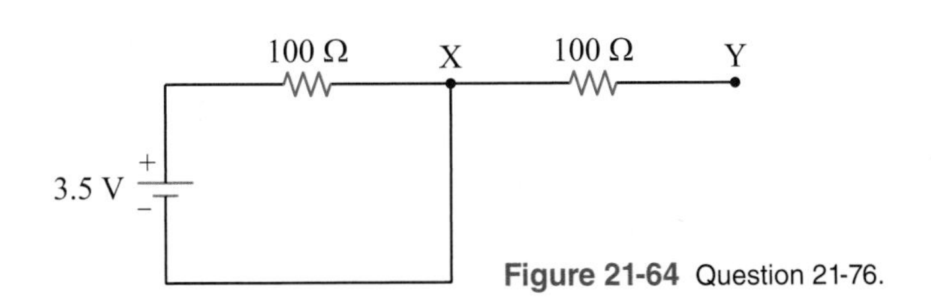

Figure 21-64 Question 21-76.

21-77 A piece of copper wire of resistance 3.2×10^{-2} Ω is melted and reformed so that its length increases by a factor of 4.0. What is its new resistance?

21-78 In Figure 21-65, if I_1 and I_2 represent the currents through resistors R_1 and R_2 respectively, at which lettered point(s) could an ammeter be connected to measure the sum $I_1 + I_2$?

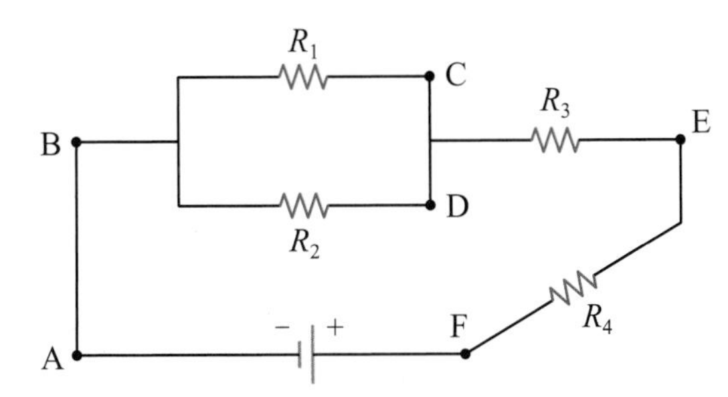

Figure 21-65 Question 21-78.

21-79 For two resistors R_1 and R_2 connected in parallel, show that the ratio of the currents through them is given by $I_1/I_2 = R_2/R_1$.

21-80 In Figure 21-66, what is the current through (a) the 200 Ω resistor and (b) the battery? Assume three significant digits in given data.

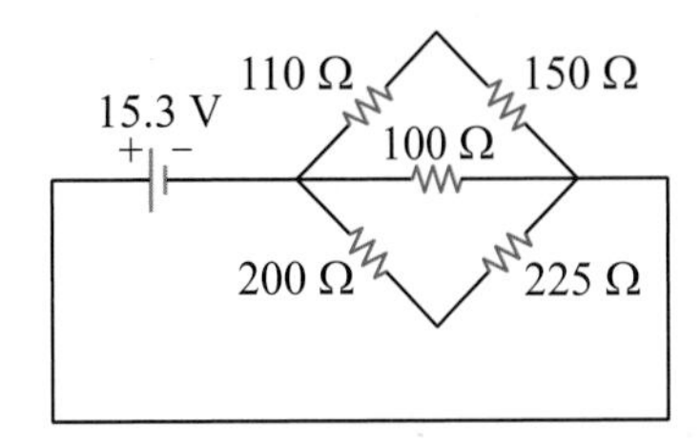

Figure 21-66 Question 21-80.

21-81 In Figure 21-67, a current of 59.5 mA travels through the 100 Ω resistor. What is the unknown resistance R? Assume three significant digits in the given resistances.

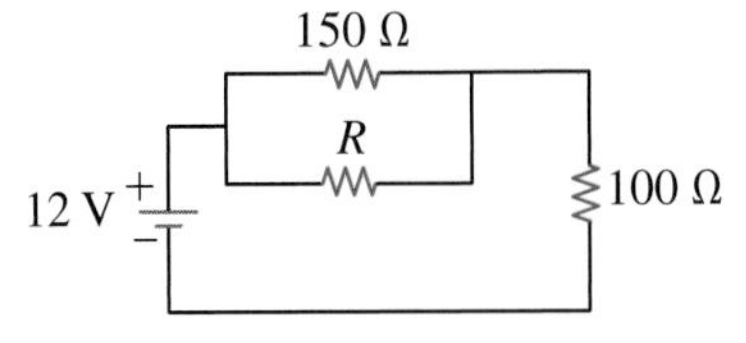

Figure 21-67 Question 21-81.

21-82 In Figure 21-68, the battery is providing 5.4 W of power. What is the resistance R?

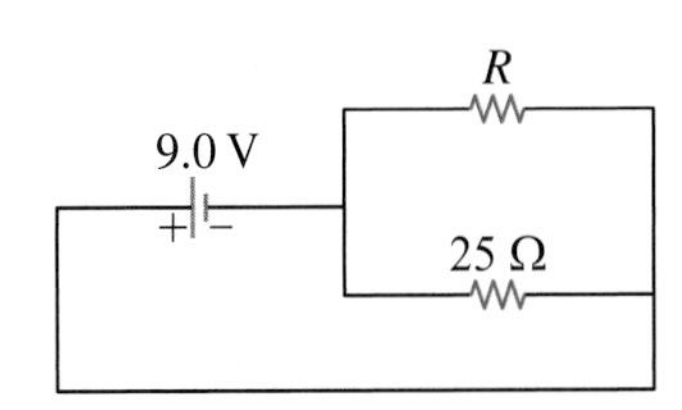

Figure 21-68 Question 21-82.

Magnetism 22

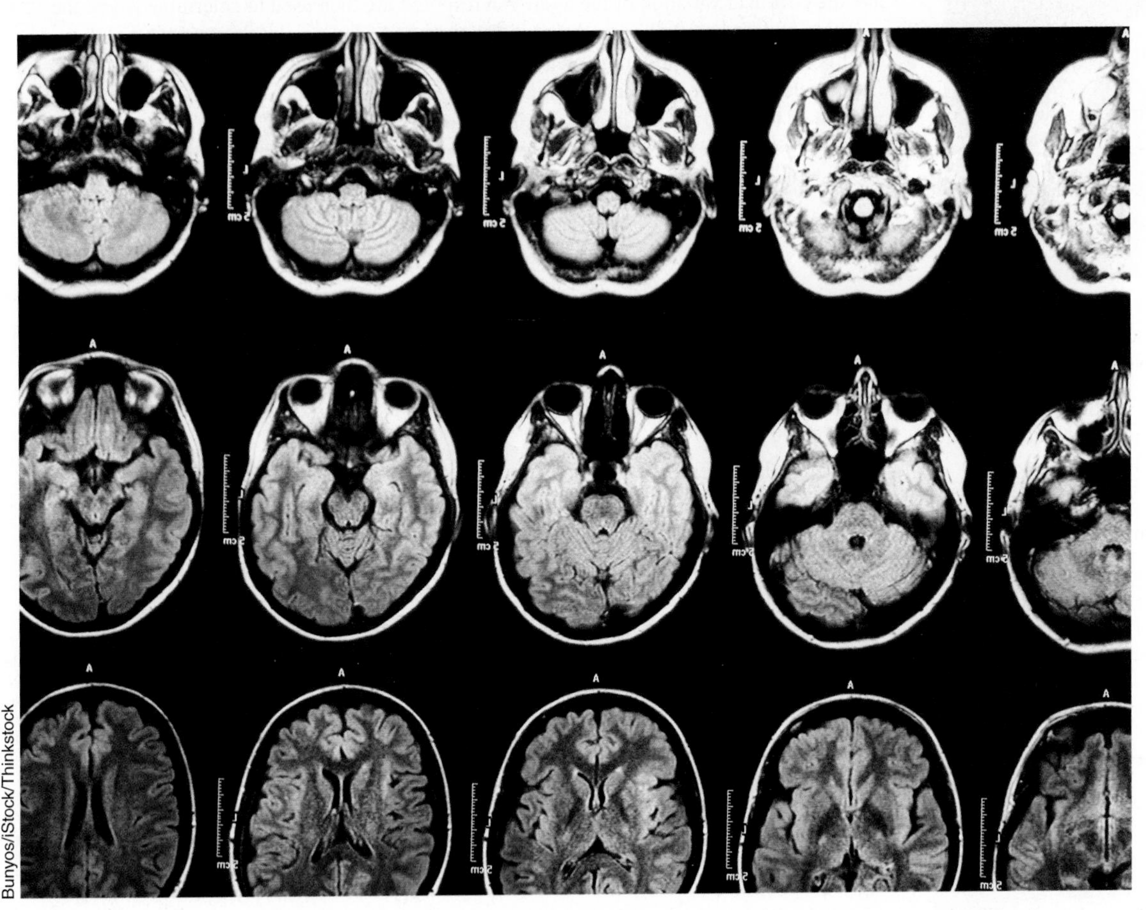

Bunyos/iStock/Thinkstock

Figure 22-1 The exquisite anatomical detail rendered by scanning the brain with magnetic fields in an MRI unit.

Magnetism plays a bigger role in your life than you probably realize. All forms of modern consumer electronics (cellphones, MP3 players, DVD players, personal computers, tablets, etc.) require magnetism to generate video and sound as well as to store and retrieve information. In fact, modern technology would not exist without magnetism. Magnets are employed in electric motors—whether in food processors, hairdryers, vacuum cleaners, or electric drills. Your car runs with the aid of several hundred magnets, both within the motor and in the various electronic sensors relaying information about vehicle operations. Magnets or magnetic materials are found in loudspeakers, bank cards, telephones, cupboard door latches, microwave ovens, and so on. Strong magnetic fields are used to generate incredibly detailed images inside our bodies through magnetic resonance imaging (MRI), as shown in Figure 22-1. Perhaps most importantly, magnets are used in electrical generating stations. Where would the world be without magnetism?

CHAPTER **OUTLINE**

DID YOU KNOW?

Magnetic Resonance Imaging (MRI) is a technique used extensively in medicine to generate anatomical images with exquisite detail (Figure 22-1). The patient is placed inside a large magnetic field which causes the hydrogen in the tissues to respond. The magnetic field applied within the patient varies spatially, and the hydrogen response depends on the chemical environment in which the atom finds itself. Both the spatial variation of the field and the chemical variation of the hydrogen response are then used to determine where the resulting hydrogen signal comes from in the body and what type of tissue is located there. MRI is often better at providing clear contrast between different types of soft tissue (e.g., fat, muscle, connective tissue) than other imaging techniques such as those based on x rays, which will be discussed in greater detail in Chapter 24.

Olivier Le Moal/Shutterstock

Figure 22-2 The pole of a compass (or any suspended magnet) that points toward the north is called its north pole.

22.1 Magnetic Fields

Almost everyone has played with magnets at one time or another. It is common knowledge that magnets attract iron, they have north and south poles, and that Earth itself is a magnet. But did you know that magnetism is caused by moving charged particles, and that the magnetic field of Earth has reversed its direction many times in its history?

The first discovery of magnetism is lost in antiquity, but we do know that by about 600 BCE the ancient Greeks had found rocks that attracted each other in a region called Magnesia. These rocks were called "magnets" after their place of discovery. By the second century CE, the Chinese also were aware of magnetism, and by the eleventh century they were using magnetic compasses in navigation at sea. A compass (Figure 22-2) is just a magnet supported at its centre so that it is free to rotate horizontally. The end of a compass (or any suspended magnet) that points toward the geographic north on Earth is called its **north pole**, and the other end is its **south pole**.[1]

north pole the end of a magnet that is attracted toward the geographic north on Earth

south pole the end of a magnet that is attracted toward the geographic south on Earth

ferromagnetic materials (or **magnetic materials**) a few substances found on Earth (such as iron, nickel, cobalt, and their alloys) that exert strong magnetic forces on other materials

A north pole of one magnet and the south pole of another exert an attractive force on each other. However, two north poles repel each other, and two south poles also repel (Figure 22-3). In other words, *opposite magnetic poles attract each other, and like magnetic poles repel.* This is similar to the behaviour of electric charges. Only a few substances, such as iron, nickel, cobalt, and their alloys, display strong magnetic properties. They are called **ferromagnetic materials**, or simply **magnetic materials**. Materials that show only weak magnetic effects are called non-magnetic materials.

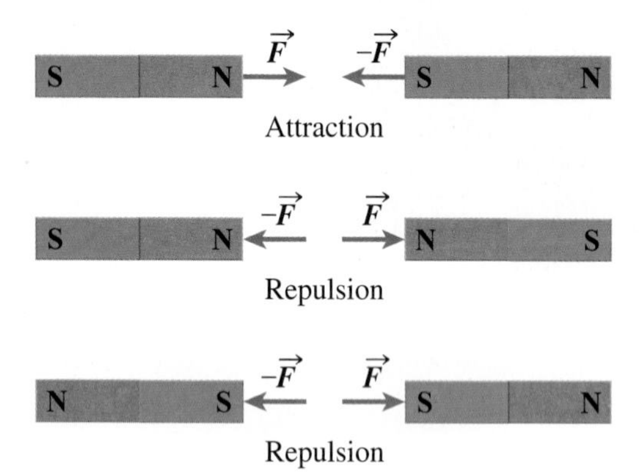

Figure 22-3 Opposite magnetic poles attract each other; like poles repel.

In Chapter 9 we introduced the physical concept of field when discussing gravitation. Remember that we described a field as any physical quantity that has a specific value at each point in space and time, and gravitational fields were a useful way of describing the effect that a mass has on any other mass nearby. Similarly, we use the concept of the magnetic field due to a magnet to understand the effect that the magnet has on any other magnet (or magnetic material) nearby. The magnitude of a magnetic field will be defined quantitatively in Section 22.2; we will start with understanding its direction.

As we discussed in Chapter 19, electric fields can be mapped out around the source of the field—that is, a charge or collection of charges—by determining the strength and direction of the force exerted on a visiting test charge at various locations. In magnetism, we can also map out the magnetic field around the source by determining the strength and direction of the force exerted on a visiting magnet. In practice this is readily done by moving a compass around near a magnet and observing the direction that the north pole of the compass points at various locations. Figure 22-4 shows the mapping of various magnetic field lines around a bar magnet using a compass,

TACKLING MISCONCEPTIONS

Magnetic versus Geographic North on Earth

Because of how we have defined north and south poles of magnets, the geographic north[1] on Earth is actually the location of a magnetic south pole, since magnetic north poles (in compasses, etc.) are attracted to it; see Figure 22-12 for an illustration.

[1]As discussed in greater detail shortly, the *geographic* north pole and the *magnetic* south pole are not located at exactly the same point on Earth; see Figure 22-18.

and Figure 22-5 illustrates the overall pattern of field lines observed in and around the magnet. Note that the field lines in Figures 22-4 and 22-5 form closed loops, and *outside* the magnet they go from the north pole to the south pole (remember this as N-S in alphabetical order). As with electric field line patterns, the magnetic field is strongest where the lines are closest together and individual field lines can never cross. At any point in Figure 22-5, the magnetic field is tangent to the field line at that point.

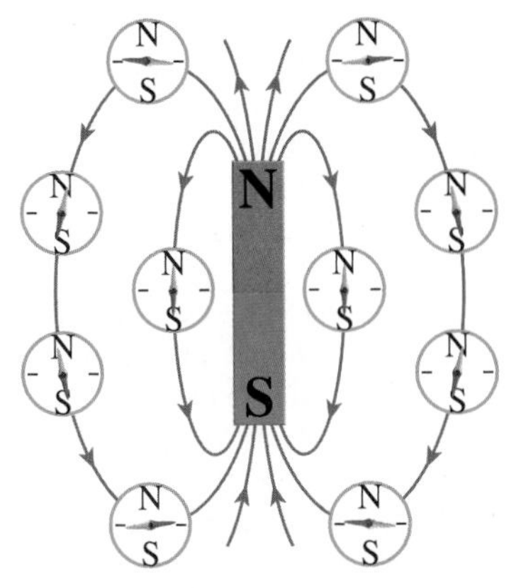

Figure 22-4 Mapping magnetic field lines near a bar magnet using a compass. The red arrowhead indicates the north pole of the compass, attracted to the south pole of the bar magnet.

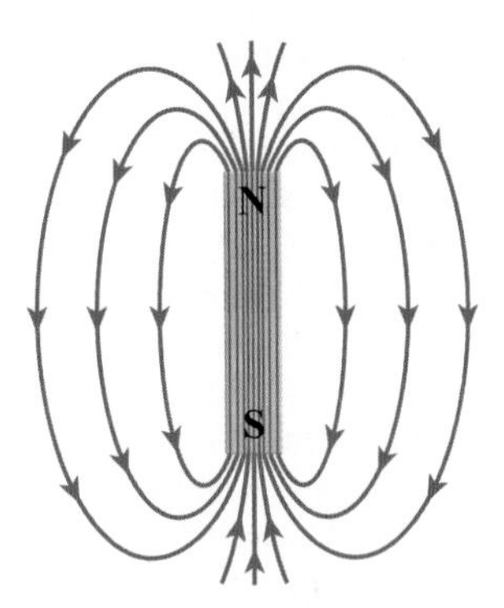

Figure 22-5 Magnetic field lines for a bar magnet.

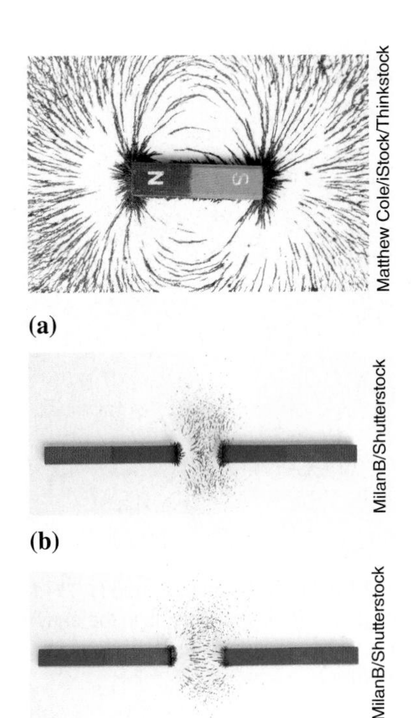

Figure 22-6 **(a)** Iron filings near a single magnet. **(b)** Iron filings showing the repulsive field lines when like poles are brought together. **(c)** Iron filings showing the attractive field lines when opposite poles are brought together.

Another technique that is often used in the classroom or lecture hall to explore magnetic field patterns involves the sprinkling of iron filings onto a piece of paper below or on top of a magnet. These iron filings become oriented along the field lines. Figure 22-6 shows the pattern produced by iron filings when close to one and two magnets.

To this point, we have discussed magnetism only with respect to individual magnets such as those played with by children or found in compass needles. Although magnetism has been used for navigation for millennia, it is only relatively recently that the phenomenon has become better understood. In 1820, Danish physicist Hans Christian Ørsted (Figure 22-8) first made the observation that an electric current in a wire produces a magnetic field, which he discovered accidentally during a lecture demonstration. His discovery was important in establishing a connection between electricity and magnetism; until his observations, electricity and magnetism had been considered by the scientific community to be two entirely separate concepts in physics. From his observations and many subsequent investigations by others, we now know that

> Any moving charge produces a magnetic field (as well as an electric field). When a moving charge encounters a magnetic field created by another moving charge, each charge experiences an equal and opposite magnetic force.

Of course the statement that a charge is moving depends on the frame of reference (Chapter 4). Two electrons moving side by side at the same velocity relative to you are at rest relative to each other. Relative to you, each electron produces both an electric and a magnetic field, and as a result the other electron experiences both an electric and a magnetic force. Relative to the electrons—that is, in a reference frame in which the electrons are stationary—each electron produces only an electric field, resulting in an electric force on the other electron. Thus, an electric force in one reference frame is equivalent to an electric force plus a magnetic force in another frame. Electric and magnetic forces are two different manifestations of the same fundamental interaction, often referred to as the electromagnetic interaction.

DID YOU KNOW?

Although moving a compass around or sprinkling iron filings near a magnet is a straightforward way to see the direction of the field, neither of these give quantitative information about the magnetic field strength in the region. Many modern-day devices for measuring magnetic fields are based on the Hall effect, which will be discussed in greater detail later in the chapter (Question 22-88 and Figure 22-71). Many smartphones and tablets contain such magnetic-field sensors (Figure 22-7).

Figure 22-7 You likely have a magnetic-field sensor built into your mobile device, enabling you to navigate by compass if necessary.

Figure 22-8 Hans Christian Ørsted (1777–1851) conducted important experiments in the early 1800s to gain a better understanding of the sources of magnetism.

Build Your Own Halbach Array

What is a Halbach array, you might ask, and why would I possibly want to build one? Although the technical term makes it sound terribly exotic, you probably have a refrigerator door covered in Halbach arrays. This ingenious design is why some of your everyday fridge magnets have a side that sticks to the fridge and a side that does not. You will need 10 (or more) small magnets, such as the neodymium or ceramic block magnets available at most hardware stores. You will also need a strong adhesive glue and a clamp or vise of some kind. With the magnets all stuck together in a stack, label the top surface of each with a dot from a permanent marker. This will allow you to keep track of the poles of each magnet as you assemble your array. We will build two different arrays with our magnets and compare the fields from each: an ordinary array as illustrated in Figure 22-9 (a) and a Halbach array as illustrated in Figure 22-9 (b).

To build the ordinary array, glue five magnets side by side such that the surfaces with the dots are all facing the same direction as shown in Figure 22-9 (a). Using block-shaped rather than disk-shaped magnets will help in your assembly, as will the clamp or vise to keep the magnets from trying to flip over and stick to their neighbours! Leave the array in the clamp or vise long enough to ensure the glue has dried.

To build the Halbach array, glue five magnets side by side with the marked surfaces arranged as follows (looking from above): left, bottom, right, top, left (Figure 22-9 (b)). Again, this will be harder to do if you have disk magnets or if you don't have some way of securing the magnets as you assemble them. The attractive force between the magnets is quite strong and they will try to flip around to stick together. Leave the array in the clamp or vise long enough to ensure the glue has dried.

Now you can compare the magnetic fields from the two different arrays:

- Try to stick each side of the two arrays to your fridge door. You should find that the Halbach array exhibits distinct asymmetry.
- Use a magnetic field monitoring app on a smartphone or tablet to measure quantitatively the field strength on each side of the two arrays.

Note: Not all refrigerator doors are magnetic. If this is the case in your house or apartment, try sticking your magnets to the sides of the fridge, as these are often magnetic even if the door is not. If that doesn't work, try to pick up paperclips with each side of the array and see what happens.

(a)

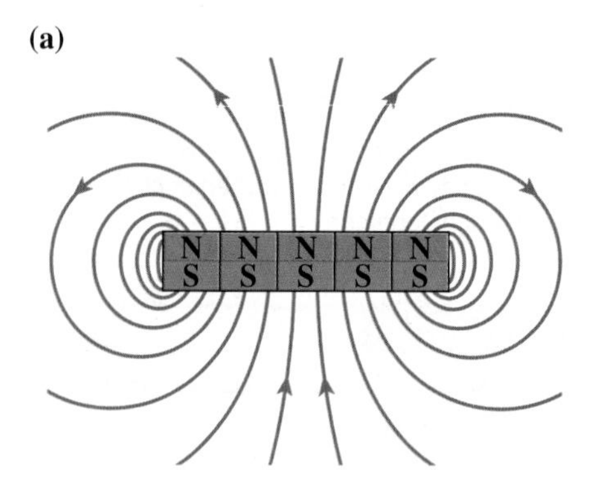

(b)

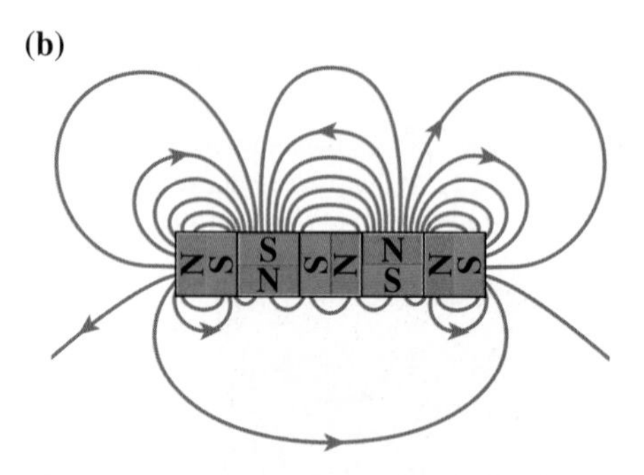

Figure 22-9 **(a)** An ordinary array of magnets with the north poles aligned results in a symmetric magnetic field. **(b)** In the case of a Halbach array, the north poles are intentionally arranged in such a way as to generate a strong magnetic field on one side and a weak field on the other.

Domains

Given that magnetic fields arise from moving charges, you are probably wondering about all those magnets stuck to your fridge—where are the moving charges there? Don't forget that those magnets are made up of many, many atoms with lots of electrons. These electrons are moving, which results in a small magnetic field. In most materials, the magnetic fields of the electrons are random in direction, and the total magnetic field of all the electrons adds (vectorially) to give essentially zero. However, in magnetic materials the magnetic fields of individual electrons can be aligned to give a large total magnetic field.

If you break a magnet in half, you do not get one magnet with only a north pole and one magnet with only a south pole. Instead, you produce a pair of magnets (Figure 22-10). If you break a magnet in several spots, you get several magnets. This phenomenon is related to the microscopic structure of magnetic materials, which consist of tiny regions called **domains**. In each domain, which has a maximum length or width of about 1 mm, the magnetic fields due to the electrons are very closely aligned. Thus, each domain acts like a small magnet with north and south poles at opposite ends. In a non-magnetized object the domains are randomly oriented (Figure 22-11 (a)), and the total magnetic field due to all the domains is close to zero. However, if the object is placed in an external magnetic field, the domains tend to line up with the external field

domains regions within a magnetic material in which the magnetic fields of the electrons are aligned; overall, each domain behaves much like a small magnet with a strong magnetic field as well as north and south poles

(Figures 22-11 (b) and (c)). Even when the external field is removed, the alignment of the domains often remains.

A magnet can remain magnetized for a long time. However, if it is dropped, some of the domains can be vibrated into random orientations, thus weakening the overall magnetic field. If a magnet is heated, the increase in thermal vibration can cause a complete loss of alignment. For each magnetic material there is a critical temperature, known as the **Curie temperature**, above which magnetic effects are lost. The Curie temperature for iron is 1043 K.

How does a magnet attract a non-magnetized object such as a paperclip, bottle cap, or coin? The field of the magnet causes the domains in the paperclip to tend to be aligned with their south poles facing the north pole of the magnet. Thus, the paperclip becomes a weak magnet itself that is then attracted to the stronger magnet.

Figure 22-10 Breaking a magnet does not produce isolated north and south poles, but rather more magnets.

Curie temperature the temperature above which magnetic effects are lost

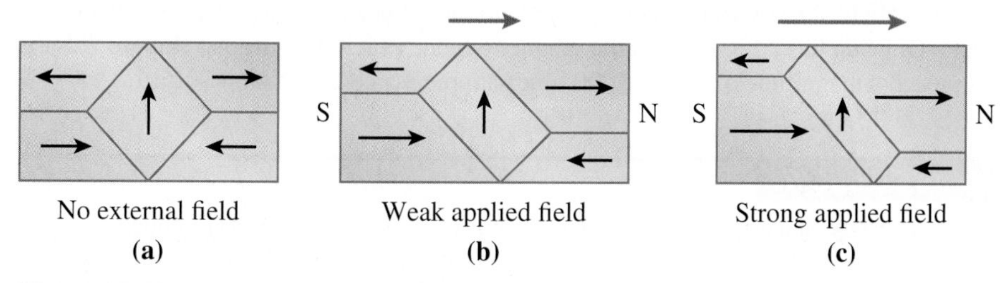

Figure 22-11 **(a)** Randomly arranged domains in a non-magnetized piece of magnetic material. Each domain acts as a small magnet with its north pole at the tip of the arrow. The length of each arrow indicates the relative magnitude of the magnetic field due to the domain. **(b)** When the material is placed in an external field, the domains tend to become aligned with that external field. In this case, the external field is directed to the right, as shown above the piece of material. By comparing (b) with (a), we see that the domains in this direction have grown at the expense of the other domains. **(c)** When a larger external field is applied, again to the right, the domains aligned in this direction become significantly larger while the others become smaller. There is an overall larger magnetic field associated with this object in (c) than there is in (b).

DID YOU KNOW?

Magnetars are a rare subset of pulsars (stars that emit pulses of radiation) with ultrahigh magnetic fields. In April, 2013, astronomers discovered a magnetar less than a light year (9.5×10^{15} m) from the black hole at the centre of our galaxy. While this might seem like a big distance, it is effectively at the edge of the region in which the black hole's gravity dominates physical processes. By studying the magnetic field of this celestial object, scientists will get a much better understanding of black hole dynamics and the theory of general relativity.

Magnetic Fields in Our Solar System

Figure 22-12 shows a sketch of Earth's magnetic field, which is similar to that of a large bar magnet. Notice that the geographic north is a magnetic south pole, as discussed in our Tackling Misconceptions box earlier in this section. Actually, the geographic north pole and the magnetic south pole are not located at exactly the same point on Earth. Therefore, the *direction* that a compass needle points (referred to as **magnetic north**) generally differs from true geographic north. At any given location on Earth, the angle between the geographic north direction and the magnetic north direction is called the **magnetic declination**. Typical magnetic declinations in North America range from about 20° W to 20° E, where 20° W means that a compass points 20° west of geographic north. For example, the magnetic declination is 11° W in Toronto, Ontario; 17° E in Vancouver, British Columbia; and 19° W in St. John's, Newfoundland.

The magnetic field at Earth's surface varies from place to place, being strongest at the poles and weakest at the equator. Its direction relative to horizontal also varies. At the poles the field is essentially vertical and at the equator it is horizontal. At any point on Earth's surface the angle that the field makes relative to the horizontal is called the **magnetic inclination**. This angle can be measured by a **dipping needle**, which is a compass free to rotate in a vertical plane.

The actual magnetic field of Earth is more complicated than Figure 22-12 suggests. For example, near the tips of Africa and South America there are regions where the field has the opposite direction to what we would expect. It is now generally acknowledged that Earth's magnetic field is due to the motion of free electrons in the molten

magnetic north the direction that a compass needle points on Earth

magnetic declination the angle between magnetic north and geographic north

magnetic inclination the angle of Earth's magnetic field relative to the horizontal

dipping needle a compass that rotates in the vertical plane to measure magnetic inclination

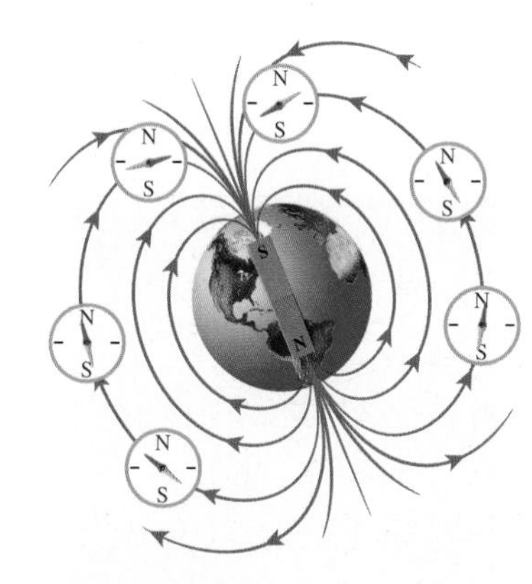

Figure 22-12 The magnetic field of Earth arises from moving charges in our molten core. Note that the *geographic* north is actually associated with the *magnetic* south pole of Earth's field.

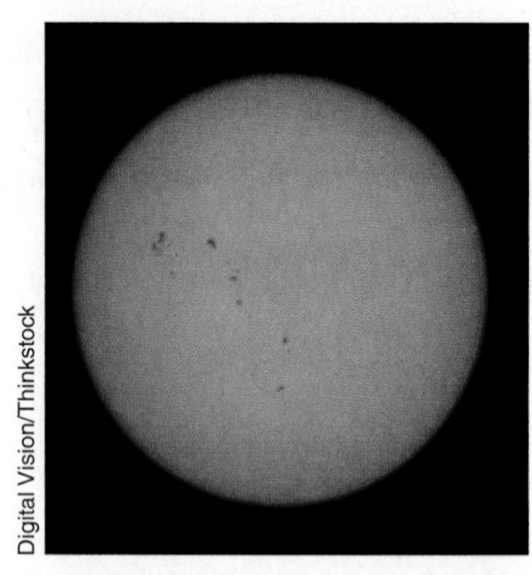

Figure 22-13 This image of our Sun clearly shows the presence of "sunspots," the dark regions in the image. The magnitude of the magnetic field in sunspots can be 100 to 5000 times the value found on Earth's surface.

outer core of Earth, about 3×10^3 km below the surface, but the details of the production of the field are still being investigated by scientists today. Detailed studies of the direction of magnetization of rocks indicate that the magnetic field of Earth reverses itself from time to time—nine times in the past 3.6 million years, most recently about 0.7 million years ago. For over a century geophysicists have observed a steady weakening of Earth's magnetic field. If this trend continues, the field would become zero in about 1.5×10^3 years, and then it could easily become reversed.

It is known that there are magnetic fields associated with other planets in our solar system: Mercury, Mars, Jupiter, Saturn, and Uranus. The absence of a field on Venus is a surprise, since our present understanding of the internal structure of Venus predicts a weak magnetic field. In addition, we would expect the field of Mars to have about the same magnitude as Earth's, but it is smaller by a factor of about 500. Mercury's magnetic field is approximately 1/100 of Earth's, but our present model of Mercury predicts no magnetic field. As you can see, our understanding of planetary magnetic fields is incomplete. Our Sun has a highly variable magnetic field, which in sunspots[2] (Figure 22-13) can reach a magnitude of 100 to 5000 times a typical value on Earth's surface.

DID YOU KNOW?

The constantly changing magnetic field generated deep inside the Sun has profound effects: It causes the temperature in the solar atmosphere to rise to millions of degrees and occasionally causes gigantic explosions that have effects throughout the solar system. To better understand solar magnetism, scientists are currently studying the observations made when two comets passed through the Sun's hot outer atmosphere in 2011, a fortuitous case of nature providing us with scientific probes of the solar magnetic field in a region that is very difficult to investigate otherwise.

magnetotactic describes an organism that uses Earth's magnetic field to orient itself in the environment

Biomagnetism

Imagine if people had their own built-in compasses; this could be of tremendous use to us in navigating around in Earth's magnetic field. A particular type of bacterium does have its own compass and uses it to orient itself in its environment. As a result this kind of bacterium is said to be **magnetotactic** (from the Latin *tactus* meaning a touch). An electron micrograph of one such bacterium is shown in Figure 22-14. Each particle in the chain inside the bacterium consists of a magnetized piece of magnetite (Fe_3O_4), and hence the chain acts as a long magnet.

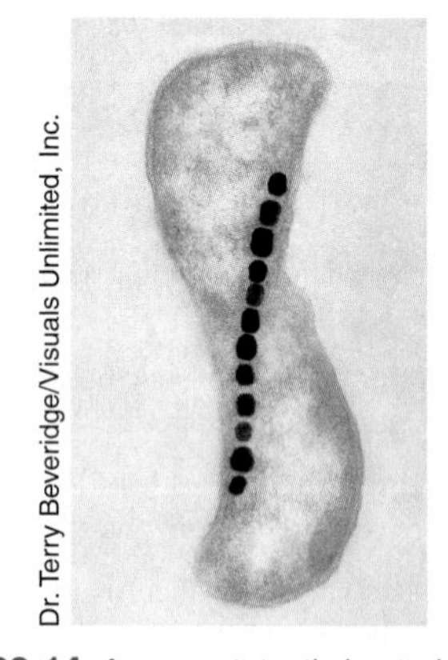

Figure 22-14 A magnetotactic bacterium. Each particle in the chain inside the bacterium is a magnetized piece of magnetite.

What advantage is it to these bacteria to have magnets? They live in mud at the bottom of swamps and ponds, and if the mud is disturbed so that the bacteria are stirred out of it, it is useful for them to be able to orient themselves to swim down to the mud again. In the northern hemisphere the local magnetic field has a downward component (Figure 22-12). The internal magnets of the bacteria allow them to detect the magnetic field so that they can swim down along the direction of a magnetic field line. Similar bacteria in the southern hemisphere have magnets of the opposite polarity, so that they too swim downward.

Although the mechanisms are not clearly understood, there is evidence that more complex organisms sense Earth's magnetic field as a means of navigation. For example, experiments with sea turtle hatchlings have demonstrated that their direction of migration is influenced by the external magnetic field in their environment (Figure 22-15). Scientists are also studying butterflies, beluga whales, homing pigeons, and foraging honeybees to better understand the role, if any, Earth's magnetic field plays in their interactions with the environment.

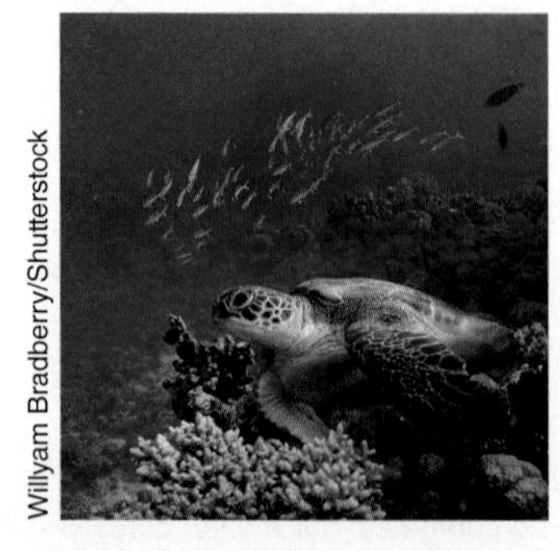

Figure 22-15 A sea turtle senses Earth's magnetic field during migration.

[2]Sunspots are regions of the Sun that have a lower temperature than the surrounding areas, and thus appear black. However, sunspots are still hot: if a sunspot could be viewed away from the rest of the Sun, it would glow red.

EXERCISES

22-1 Suppose that you are given a bar magnet and another piece of metal. Describe how you would determine whether the other piece of metal is (a) magnetic, (b) a magnet.

22-2 In Figure 22-16 each circle represents a compass. Redraw the images and show the direction of each compass needle.

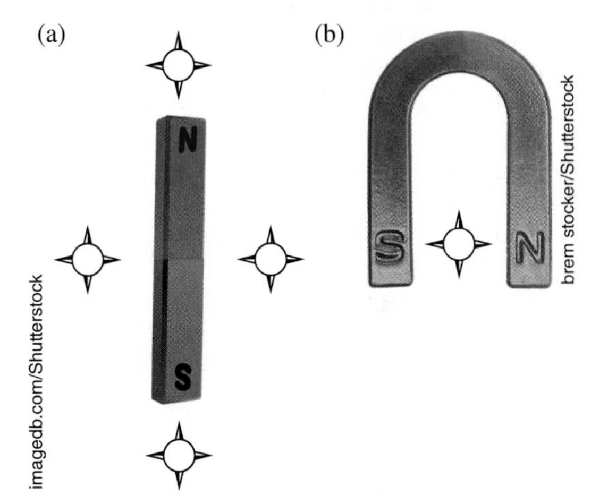

Figure 22-16 Mapping magnetic fields with a compass (Questions 22-2 and 22-3).

22-3 In Figure 22-16 (b), what is the direction of the magnetic field at the position of the compass?

22-4 What kind(s) of field(s) surround a moving charged particle?

22-5 Use the concept of domains to explain the following:

(a) When a magnetic material is being magnetized, it reaches a point (called saturation) where it can be magnetized no further.

(b) Hammering a magnet can demagnetize it.

22-6 Why do iron filings show the magnetic field direction near a magnet?

22-7 Explain how a magnet attracts a non-magnetized iron nail.

22-8 Give examples of magnetism in use in the kitchen (Figure 22-17).

Figure 22-17 Identifying some of the many uses of magnetism in our daily life (Question 22-8).

22-9 **Fermi Question:** How many atoms of iron are there in a sewing needle? How could a sewing needle be used to make a compass?

22-10 **Fermi Question:** A plot of the movement of Earth's magnetic pole in the northern hemisphere since 1600 is shown in Figure 22-18. Estimate the total distance that the pole has moved since 1600. What is the average speed of this magnetic drift?

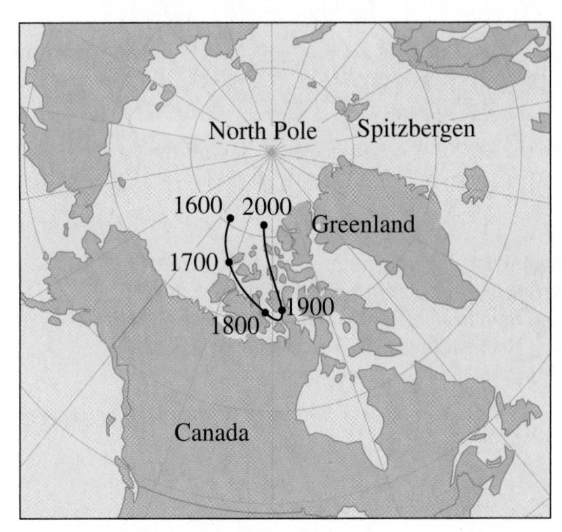

Figure 22-18 The position of Earth's magnetic pole in the northern hemisphere since 1600 (Question 22-10). Each grid line represents 10° of latitude or longitude.

22-11 **Fermi Question:** Based on the measurements of the field strength of Earth since 1900 (Figure 22-19),

(a) by what year will Earth's field have decreased to zero strength?

(b) what do you think will be the impact if Earth's magnetic field disappears?

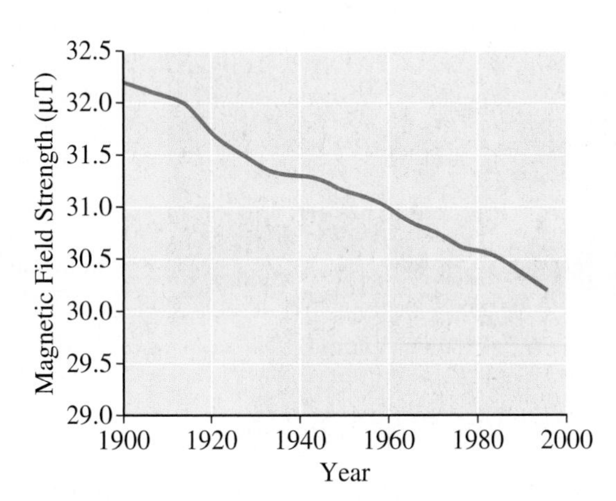

Figure 22-19 The strength of Earth's magnetic field since 1900 (Question 22-11).

22.2 Magnetic Fields Due to Currents

Figure 22-20 An industrial electromagnet can lift up to 2×10^4 kg of metal.

Have you ever wondered how cars and scrap metal can be picked up by electromagnets (Figure 22-20)? The fundamental physics involved here is simply the production of a magnetic field by an electric current.

Figure 22-21 shows a long, straight, current-carrying wire. A magnetic field is created around the wire such that each magnetic field line is a circle in a plane perpendicular to the wire, with its centre on the wire. The field direction can be determined using a **right-hand rule** (Figure 22-21): point the thumb of your right hand in the direction of the conventional current (positive charge flow), and your fingers will wrap around the wire in the direction of the associated magnetic field.

right-hand rule (magnetic field associated with a current-carrying wire) point the thumb of your right hand in the direction of the conventional current and your fingers will wrap around the wire in the direction of the associated magnetic field

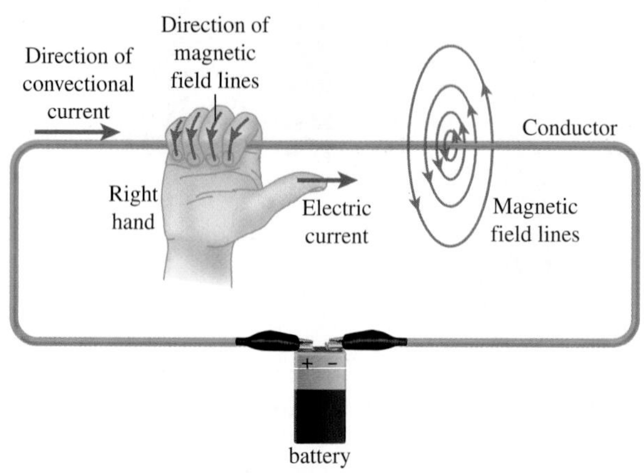

Figure 22-21 The magnetic field around a straight, current-carrying wire. The magnetic field forms circles around the wire, the orientation of which can be determined by the "right-hand rule."

tesla SI unit for the magnitude of a magnetic field; symbol T

Figure 22-22 A statue of Nikola Tesla (1856–1943) greets visitors to the observation deck at Niagara Falls, NY. Tesla made many contributions to high-voltage technology and designed the Niagara Falls power-generating station.

Experimentally it has been observed that the magnitude of the magnetic field (B) associated with a current (I) in a long straight wire is proportional to the current and inversely proportional to the distance (r) from the wire. This should make intuitive sense: The magnetic field is larger when closer to the source, and the magnetic field increases when the current that creates it increases. Mathematically, this relationship is written as

$$B = \frac{\mu_0 I}{2\pi r} \qquad (22\text{-}1)$$

where μ_0 is a constant (given below) and π is the familiar 3.14159...

Note that μ is the lower-case Greek letter mu, pronounced "mew."

The SI unit for the magnitude of the magnetic field (B) is the **tesla** (T), in honour of the Serbian-American scientist and inventor Nikola Tesla (Figure 22-22). With current (I) in amperes (A) and distance (r) in metres (m),

$$\mu_0 = 4\pi \times 10^{-7}\ \text{T}\cdot\text{m/A}$$

Strictly speaking, Eqn. 22-1 applies only to calculations of magnetic fields close to very long straight wires. We will restrict our discussions to such scenarios, as more complicated geometries are beyond the scope of this book.

The magnetic field of Earth has a magnitude of the order of 10^{-5} to 10^{-4} T at its surface, and the field of a strong bar magnet is about 0.1 T. Other representative magnetic field strengths are listed in Table 22-1. Another common unit for magnetic field is the **gauss** (G), named after the great German mathematician, physicist, and astronomer

gauss a common (non-SI) unit for the magnitude of a magnetic field; symbol G

Karl Friedrich Gauss (Figure 22-23). Instruments used to measure magnetic fields are often called **gaussmeters**. The approximate field strength at the surface of Earth is 1 G, which makes this unit more convenient than teslas in some scenarios.

$$1\text{ G} = 10^{-4}\text{ T}$$

Figure 22-23 Karl Friedrich Gauss (1777–1855). Gauss is one of history's greatest mathematicians, with celebrated contributions in number theory. He also did important work in astronomy, electricity, and magnetism, and is credited with inventing the electric telegraph.

gaussmeter an instrument used to measure magnetic fields, named after the German scientist Karl Friedrich Gauss

Table 22-1

Typical Values of Magnetic Field Strengths in the Universe

Object	Magnetic Field (T)	Magnetic Field (G)
Surface of Earth	10^{-5} to 10^{-4}	0.1 to 1
Strong bar magnet	0.1	1×10^{3}
Magnetic resonance imager	1.5 (most common)	1.5×10^{4}
Magnetic levitation trains	~ 5	$\sim 5 \times 10^{4}$
Largest steady-state magnet in laboratory setting	45	4.5×10^{5}
Surface of neutron star	10^{8}	10^{12}

SAMPLE PROBLEM 22-1

What is the magnetic field (magnitude and direction) at point P in Figure 22-24, 1.25×10^{-2} m from a long straight wire carrying a current of 2.75 A?

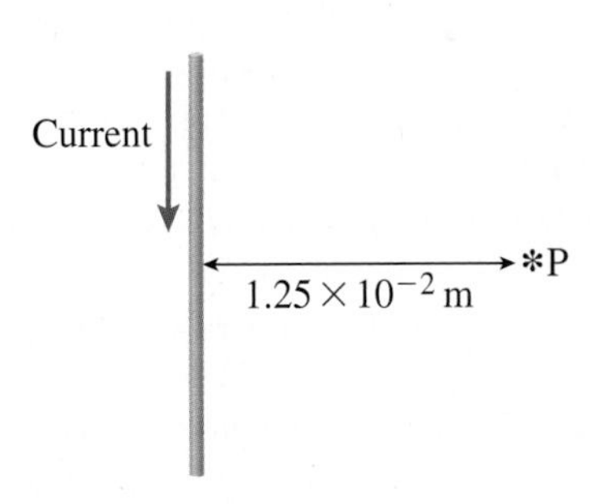

Figure 22-24 Sample Problem 22-1.

Solution

We use the right-hand rule to find the direction of the magnetic field at P. With your right thumb pointing in the direction of the current (toward the bottom of the page), your fingers curl around the wire such that they are directed out of the page at P. Thus, the magnetic field at P has a direction that is out of the page.

To calculate the magnitude of the field, we use Eqn. 22-1:

$$B = \frac{\mu_0 I}{2\pi r} = \frac{(4\pi \times 10^{-7}\text{ T}\cdot\text{m/A})(2.75\text{ A})}{2\pi(1.25 \times 10^{-2}\text{ m})}$$

Before proceeding further with the calculation, notice that we can easily divide $4\pi \times 10^{-7}$ in the numerator by 2π in the denominator to give 2×10^{-7}.

$$B = \frac{(2 \times 10^{-7}\text{T}\cdot\text{m/A})(2.75\text{ A})}{1.25 \times 10^{-2}\text{ m}} = 4.40 \times 10^{-5}\text{ T}$$

Hence, the magnetic field at P is 4.40×10^{-5} T directed out of the page.

Notice the units of each quantity:

$$\mu_0\text{: T}\cdot\text{m/A} \quad I\text{: A} \quad 2\pi\text{: unitless} \quad r\text{: m}$$

Therefore, the units of B work out as $\frac{(\text{T}\cdot\text{m/A}) \times (\text{A})}{\text{m}} = \text{T}$.

Problem-Solving Tip

This might seem obvious, but be careful when you are using the right-hand rule to determine the direction of the magnetic field. You might be surprised at how often students use their left hand, ending up with the wrong answer!

PROFILES IN PHYSICS

Nikola Tesla (1856–1943), Father of the Modern Electrical Age

Legend has it that Nikola Tesla (Figure 22-25) was born during a violent thunderstorm on July 10, 1856 in Smiljan, Lika, which was then part of the Austro-Hungarian Empire region of Croatia. While the midwife considered this to be a bad omen, his mother was said to have declared that "he will be a child of light." He began his career as an electrical engineer in Budapest before joining the Continental Edison Company in Paris in 1882. In 1884, Tesla moved to New York City, where he continued to work for Thomas Edison on direct current motors and generators. This turned out to be the start of a lifelong rivalry between these two great inventors, with Tesla lasting only two years at Edison's New York office before leaving to start Tesla Electric Light and Manufacturing.

Photograph by Professor Slaby in Himmel und Erde, volume 1 page 8/The Canadian Press

Figure 22-25 Nikola Tesla produces "artificial lightning" in his laboratory, circa 1900.

Their parting of ways in 1886 is largely attributed to the beginning of the "current wars," in which Tesla promoted the idea of using alternating current (AC) for widespread electrical power transmission, whereas Edison backed the use of direct current (DC). In 1887, Tesla filed for seven U.S. patents related to AC power, covering the complete system from generators, transformers, and transmission lines to electrical lighting and motors. This attracted the attention of the wealthy industrialist George Westinghouse, who negotiated a deal with Tesla in 1888 to license his AC patents. With the involvement of Westinghouse, a full-scale industrial war began between the AC and DC camps. There was a lot at stake, as Edison and his financial backers had invested considerably in DC infrastructure. Edison argued that DC systems required lower voltage and were therefore much safer for household use. However, there were significant issues with energy losses when transmitting over distances, which Tesla maintained could be solved by the use of AC. The public relations campaign was so fierce that the DC supporters resorted to electrocuting animals during public presentations to demonstrate the dangers of alternating current! However, after Westinghouse Corporation successfully illuminated the first all-electric World's Fair in Chicago in 1893 through AC power and, in collaboration with Tesla, demonstrated AC power generation and wide-spread transmission from Niagara Falls, it was clear that AC had won the "current wars."

Although such success should be more than sufficient to guarantee Tesla's place in history, he continued to invent in a number of fields, obtaining roughly 300 patents worldwide over his career, from wireless transmission of electrical energy and fluorescent lighting to vertical take-off technology for aircraft. When Guglielmo Marconi claimed success in the first trans-Atlantic radio transmission in 1901, Tesla reportedly said: "Marconi is a good fellow. Let him continue. He is using seventeen of my patents." That supportive attitude disappeared, however, when Marconi, backed by Edison, was issued a U.S. patent for the invention of the radio in 1904, followed by the Nobel Prize in 1909, after his earlier patent applications had been repeatedly rejected due to the priority of patents belonging to Tesla and other inventors. Tesla sued Marconi for patent infringement, finally winning his case when the U.S. Supreme Court ruled in Tesla's favour in 1943, several months after his death. This was too late for the history books, however, as Marconi is widely cited as the inventor of radio.

From these intense rivalries with peers such as Edison and Marconi, you might have concluded that Tesla was somewhat awkward in terms of personal interactions. Although he was a celebrity in his day, making the cover of *Time* magazine in 1931, he died in relative obscurity in 1943 and was almost destitute. Tesla was definitely an unusual character: physically striking at 6'2" and slim with piercing grey eyes, he had a photographic memory, a powerful imagination, excessive hygiene habits as well as a fixation on the number three, reportedly having to walk around the block three times before entering a building. Perhaps the creators of the popular television show "Big Bang Theory" used some of these characteristics as inspiration for Sheldon Cooper! More literal representations of Nikola Tesla have been seen on the big screen in recent years as well, with David Bowie playing the eccentric inventor in the 2006 Hollywood film *The Prestige* and a biopic slated for release in 2015 entitled *Tesla*. Regardless of his unusual personal characteristics, his role in bringing electrical distribution systems to the world deserves wide-spread recognition even now, more than 70 years after his death.

Magnetic Field of a Circular Loop and a Solenoid

Circular loops of wire are often used in devices such as electromagnets, electric motors, etc. As before, we can use a right-hand rule to determine the overall direction of the magnetic field inside and outside the loop (Figure 22-26 (a)). As shown, the thumb of the right hand is in the direction of the current and the fingers curl around the wire in the same sense as the resulting magnetic field lines. Figure 22-26 (b) shows the magnetic field pattern around a single circular loop. Notice that the magnetic field lines in Figure 22-26 (b) resemble those drawn in Figure 22-5 for a bar magnet with the north pole above the loop and the south pole below the loop.

Although the magnitude of the field at any arbitrary point near the loop is complicated to calculate, symmetry in the configuration makes the calculation of the magnetic field at the centre of the loop reasonably straightforward. The resulting expression is given by

$$B = \frac{\mu_0 I}{2R} \qquad (22\text{-}2)$$

where I is the current in the loop and R is the radius of the loop. The constant μ_0 is as defined in Eqn. 22-1.

Be very careful with Eqns. 22-1 and 22-2! They are very similar in form; the only difference is that Eqn. 22-1 has a π in the denominator. Eqn. 22-1 allows us to calculate the magnetic field at any point of interest that is a distance r from a long straight wire that is carrying a current I. In Eqn. 22-2, you are calculating the magnetic field at the centre of the circle, and R in this equation means the radius of that circle.

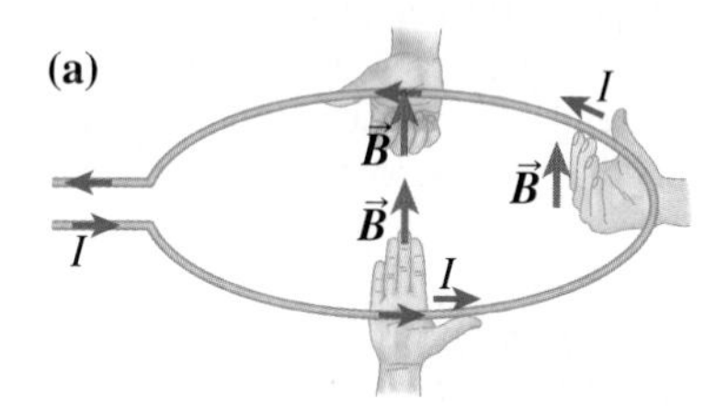

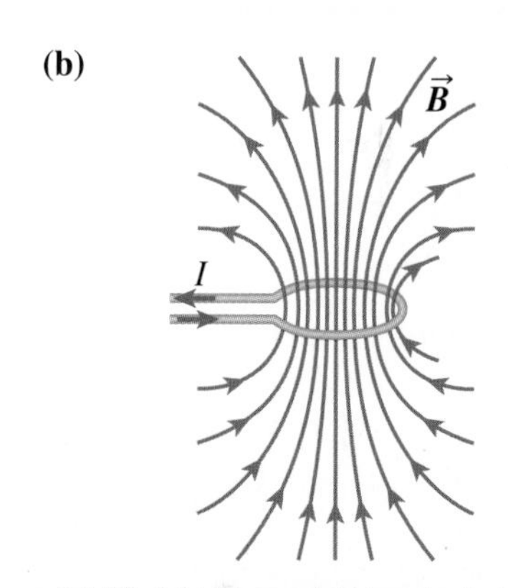

Figure 22-26 **(a)** Using a right-hand rule to determine the direction of the magnetic field inside and outside a circular loop of current-carrying wire. **(b)** The overall magnetic field pattern near a circular loop.

Often loops of wire are wound tightly into a helical coil called a **solenoid** (Figure 22-27 (a)). The magnetic field at any point near the solenoid is the sum of the fields due to the individual loops. Figure 22-27 (b) shows the magnetic field lines in a plane along the centre of the solenoid, as if the solenoid has been cut length-wise in half. In this depiction, the wires appear in cross-section as either ⊙ or ⊗ representing current coming out of, or into, the page, respectively. As we discussed with an individual loop of current, it is perhaps even more evident in Figure 22-27 (b) that the magnetic field lines are very similar to those of a bar magnet (Figure 22-5). In Figure 22-27 (b), can you tell which end of the solenoid is its north pole?

solenoid a device made of loops of wire wrapped into a helical coil inside which the magnetic field is relatively uniform

right-hand rule (magnetic field associated with a solenoid) wrap the fingers of your right hand around the solenoid in the direction of current flow and your extended thumb will be pointing in the direction of the associated magnetic field inside the solenoid

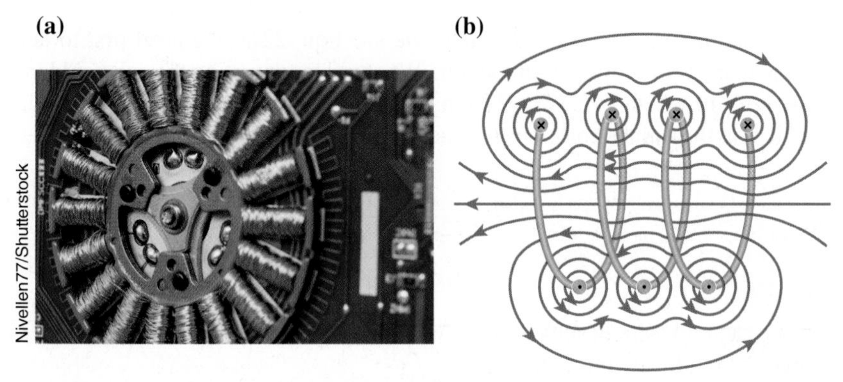

Figure 22-27 **(a)** An array of 18 solenoids, each one consisting of a series of tightly wound loops of copper wire forming a cylinder. The configuration of solenoids shown here is from the inside of a motor. **(b)** The magnetic field associated with a single solenoid. For simplicity, this illustration has only four loops of wire in the solenoid, but commercial devices contain many more loops as seen in (a).

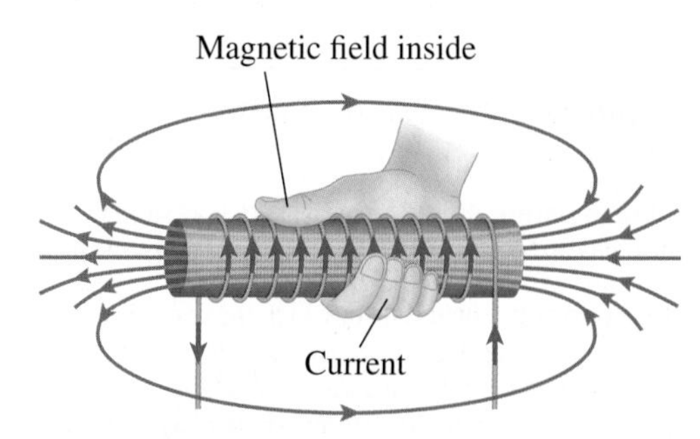

Figure 22-28 Using a right-hand rule to determine the magnetic field direction inside a solenoid.

We can use a **right-hand rule** to determine the magnetic field direction inside the solenoid. Wrap the fingers of the right hand around the solenoid in the direction of current flow and the thumb points in the direction of the field inside the solenoid (Figure 22-28).

DID YOU KNOW?

Magnetic Hill is a popular tourist destination in Moncton, New Brunswick, at one time the third most popular attraction in Canada after Niagara Falls and Banff National Park. At Magnetic Hill visitors drive their car to the bottom, put the transmission in neutral, release the brakes and enjoy the unusual sensation of their car appearing to roll uphill as if "pulled by a magnet"! In fact, there is no magnetism involved here at all; the effect is an optical illusion. The road slopes downhill at this point but the adjacent topography makes it appear to slope uphill. This optical illusion confused horse and cart drivers too before the era of motorized vehicles, as the locals were never quite sure when to snap the reins to get their horse to speed up and when to apply the brakes to the cart when travelling on what was then known as Fools' Hill.

Notice that the field lines in Figure 22-27 (b) are almost uniformly spaced inside the solenoid; they are quite uniformly spaced when the solenoid is made up of many more tightly wound loops than is shown here for simplicity. This indicates that the field is essentially constant in this region. By adding up the contributions to the magnetic field from each of the loops of wire through integral calculus, we can express the resulting field inside the solenoid quite simply as

$$B = \mu_0 n I \tag{22-3}$$

where n is the number of loops *per unit length* (i.e., the number of loops per metre along the solenoid length) and I is the current in the solenoid.

Strictly speaking, Eqn. 22-3 applies only to infinitely long solenoids. In practice, however, this expression is a reasonable approximation for the magnitude of the magnetic field inside a finite solenoid, as long as we are not trying to determine the field near the ends. As you can see in Eqn. 22-3, there is no spatial term in the expression such as r, x, y, or z, which means that the magnetic field is the same everywhere inside the solenoid as we had noted from the field lines in Figure 22-27 (b).

SAMPLE PROBLEM 22-2

What are the magnitude and direction of the magnetic field inside the solenoid of Figure 22-29? The solenoid has 26.0 loops of wire in its 2.72 cm length, and the current passing through it is 0.789 A.

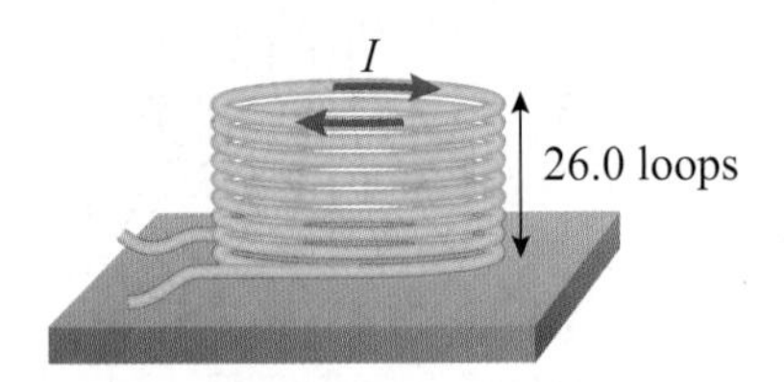

Figure 22-29 Sample Problem 22-2.

Solution

Using the right-hand rule, we wrap the fingers of the right hand in the given direction of the current around the solenoid, and the thumb points in the direction of the magnetic field inside the solenoid, that is, downward (toward the bottom of the page).

To calculate the magnitude of the field, we use Eqn. 22-3. We need first to determine n, the number of loops per unit length. We start by converting the solenoid length of 2.72 cm to m, which is 2.72×10^{-2} m. Since there are 26.0 loops in a length of 2.27×10^{-2} m, the number of loops per metre is

$$n = \frac{26.0}{2.72 \times 10^{-2}\ \text{m}}$$

$$= 956\ \text{m}^{-1}$$

Now we can make direct substitutions into Eqn. 22-3 to determine the magnitude of the magnetic field inside this solenoid as

$$B = \mu_0 n I$$

$$= (4\pi \times 10^{-7}\ \text{T}\cdot\text{m/A})\ (955.9\ \text{m}^{-1})\ (0.789\ \text{A})$$

$$= 9.48 \times 10^{-4}\ \text{T}$$

Hence, the magnetic field inside this solenoid is 9.48×10^{-4} T, downward.

electromagnet a coil or loop of wires with a magnetic core

relative permeability a measure of how strongly a material can be magnetized by an external magnetic field

diamagnetic materials in which the overall magnetic field is slightly reduced because there is a small magnetic field inside the material that is in the *opposite* direction to the external (much larger) field (e.g., copper, silver, graphite)

Electromagnets and Permeability

The magnetic field due to a solenoid can be increased greatly if a magnetic material is placed inside. The field due to the current passing through the solenoid magnetizes the material; the material's domains are aligned, and the material produces its own magnetic field in the same direction as the field of the solenoid. Therefore, the total magnetic field is the sum of that due to the current and that due to the material. A solenoid (or any similar coil of loops of wire) with a magnetic core is commonly referred to as an **electromagnet** (Figure 22-30).

With a magnetic core, the total magnetic field is increased by a factor of 10^3 to 10^4 relative to the field without a core; we describe this by saying that magnetic materials have a large **relative permeability**. Relative permeability is a measure of how strongly a material can be magnetized by an external magnetic field. With a *non-magnetic* core of say, aluminum, the total field is increased very little, and for a core of silver, the total field actually decreases by a small amount. Both aluminum and silver are said to have a small permeability. Table 22-2 lists the relative permeability of a variety of materials. Note that the values of less than 1 (for silver, etc.) indicate that the overall magnetic field is reduced in magnitude when such a material is present; these materials are called **diamagnetic**. For many materials, the values are quite close to 1. In these substances, the total magnetic field is largely unchanged by its presence. In the materials with high relative permeability (iron, cobalt, nickel), a range of values is given since the degree to which the domains align in such substances can vary greatly with the particular conditions.

Zigzag Mountain Art/Shutterstock

Figure 22-30 An electromagnet consists of a solenoid and an iron core. The magnetic field is many times larger than that due to the current in the solenoid alone.

Table 22-2

Relative Permeability of Various Materials

Material	Relative Permeability
Sodium	1.000 007 2
Aluminum	1.000 022
Sodium chloride	0.999 986
Copper	0.999 99
Silver	0.999 834
Graphite	0.999 984
Platinum	1.000 26
Uranium	1.000 40
Iron	150 to 150 000
Cobalt	70 to 250
Nickel	100 to 600

Build Your Own Electromagnet

You will need one *iron* nail that is 15 cm or more in length, some insulated copper wire, and one D battery. Wrap the copper wire as tightly as you can around the shaft of the nail; remember that Eqn. 22-3 tells us that the solenoid field will be bigger with more loops per unit length (n). Connect the battery to the two ends of the copper wire after you have stripped the insulation back a little, and voila! Test out the strength of your electromagnet by investigating how many staples or paperclips it can lift. You can explore a number of variables with this experiment: How does n influence the number of staples the electromagnet can lift? What happens if you replace the iron nail with aluminum foil or a plastic rod? How many staples can the electromagnet lift if you use more than one D battery (in series)? Many tablets and smartphones have a built-in magnetometer that measures magnetic fields up to 1 T, mainly for use in navigation. Coupled with specific magnetic field monitoring apps, you could make this a truly quantitative study, for example, by plotting the magnetic field strength as a function of the number of loops of wire per unit length. Have fun!

Electromagnets have a wide variety of applications: in loudspeakers and doorbells, in reading and writing data to the hard drive in computers, etc. Even the Large Hadron Collider, the massive particle physics experimental facility in Switzerland and France, makes extensive use of electromagnets in keeping its subatomic particles on track (Figure 22-31). The crane shown lifting scrap metal in Figure 22-20 uses an electromagnet that produces a magnetic field strong enough to lift a car. Electromagnets are also used in switches called relays. (See Applying Your Knowledge Question 22-79 at the end of this chapter.)

Photo by Rex Feature/The Canadian Press

Figure 22-31 The Compact Muon Solenoid (CMS) experiment is one of two large, general-purpose, particle physics detectors built for the Large Hadron Collider (LHC) at CERN in Switzerland and France. The magnetic field of the CMS is approximately 4 T.

Some other practical devices use a solenoid with an iron plunger inserted partway. When a current flows through the solenoid, the magnetic field that is produced attracts the plunger completely into the solenoid. In one common doorbell design

(a)

ueuaphoto/Shutterstock

(b)

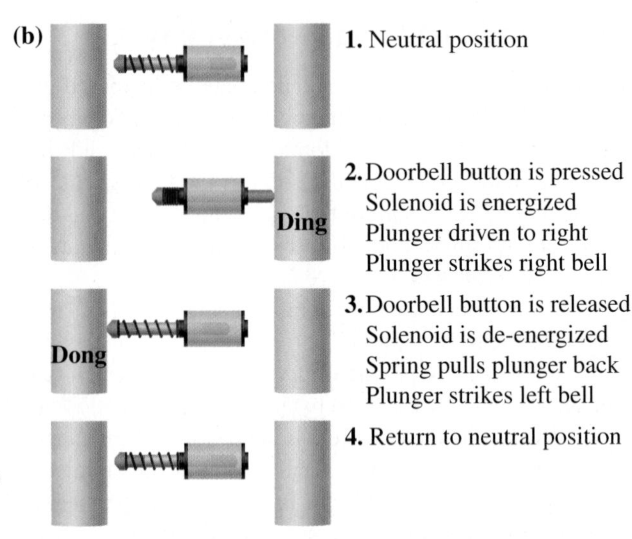

Figure 22-32 **(a)** When the button is pressed at the door, the circuit is closed and the solenoid is energized. **(b)** In this simple design, the plunger gets pulled into the solenoid once the circuit is complete, striking the first chime ("ding"). When the button is released, the plunger is pushed out by the compressed spring and hits the second chime ("dong").

Maglev magnetic levitation, used to reduce friction for high-speed commuter trains

(Figure 22-32) the plunger strikes a chime ("ding") when it moves into the solenoid as a result of current flowing when the switch is closed. The plunger then strikes the second chime ("dong") when the switch opens as the button at the door is released. Solenoids with a movable core are also used in dishwashers and washing machines to move plunger-type valves to control the water flow.

Strong electromagnets are also used for "**Maglev**" trains (Figure 22-33). Maglev is short for magnetic levitation, which means that the train carriage floats on a pocket of air above the track thereby significantly reducing frictional energy losses in the system. The magnetized coil in the track repels the large magnets under the train, allowing the carriage to hover as much as 1 to 10 cm above the track. The significant reduction in friction as well as advanced aerodynamic design allow these trains to reach speeds of more than 500 km/h, which is twice the top speed of a conventional commuter train! While Maglev transportation was first proposed more than a century ago, the first commercial high-speed Maglev train only began full operation in December of 2003. The Shanghai Transrapid line runs from the city centre to the Pudong airport, a distance of 30 km that requires less than 10 minutes to complete.

Hung Chung Chih/Shutterstock.com

Figure 22-33 The first commercially operating high-speed Maglev train runs from the centre of Shanghai, China, to the outlying airport with a top operational commercial speed of 431 km/h.

TRY THIS!

Magnetic Levitation

You can demonstrate the phenomenon of magnetic levitation (Maglev) with a few simple components. The difficulty with most attempts at levitation involving permanent magnets is stability: positioning one magnetic north pole above another gives rise to a repulsive force, but usually is followed by the top magnet flipping over and sticking to the bottom one rather than hovering in air. One way to provide stability is by confining the motion of the magnets: Figure 22-34 (a) shows two ring magnets with like poles directly above and below each other confined in their motion by being threaded onto a pencil. Another way to achieve stability is to tailor the spatial dependence of the magnetic field. Coupled with a diamagnetic material, such as graphite, you can demonstrate levitation at home with this technique. You will need nine 3 mm cube rare-earth magnets, some glue, and several pieces of 0.5 mm diameter pencil "leads," which these days are made of graphite (some brands work better than others). For a stable levitation field, the magnets should be glued together in a 3-by-3 array with the north poles of adjoining magnets alternately pointing up and down. Break a small piece of pencil graphite and balance it above the centre of the 3-by-3 array (Figure 22-34 (b)); the levitation height will not be huge but you should be able to slide a piece of paper between the graphite and the magnets below to show that it is hovering above the surface. Try blowing very gently through a straw on one end of the graphite to get it to spin in place. With the low friction of the system, the graphite will continue spinning in place for quite some time.

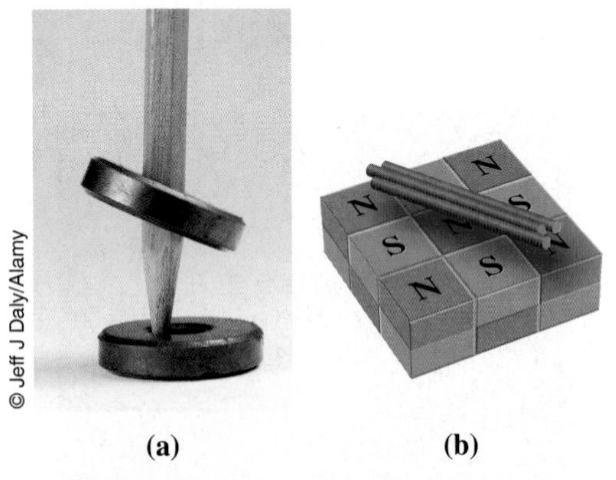

© Jeff J Daly/Alamy

(a) (b)

Figure 22-34 **(a)** One ring magnet levitates above the other with stability provided by the pencil through their centres. **(b)** An array of 9 cubic rare-earth magnets provides a stable levitation force for graphite from a pencil.

EXERCISES

22-12 A current of 2.50 A flows vertically upward in a long straight wire. What are the magnitude and direction of the magnetic field at a point 25.0 cm to the east of the wire in the horizontal plane?

22-13 Figure 22-35 shows magnetic field lines near a wire. What is the direction of the current?

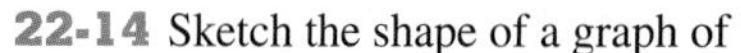
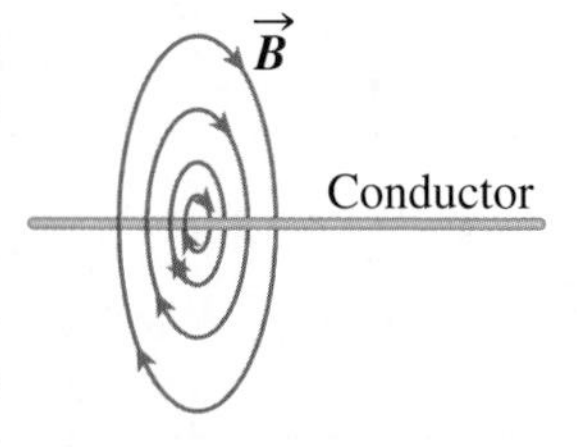

Figure 22-35 Question 22-13.

22-14 Sketch the shape of a graph of

(a) the magnitude of the magnetic field (B) versus distance (r) from a current-carrying, long, straight wire,

(b) the magnitude of the magnetic field (B) versus position (x) along the central axis of a solenoid. (Assume an infinite solenoid here.)

22-15 The magnetic field at the centre of a circular loop of wire of radius 1.50 cm is 3.48×10^{-4} T in magnitude.

(a) What is the current in the wire?

(b) If the loop is in a horizontal plane and the field at its centre is upward, is the current flowing clockwise, or counterclockwise, as viewed from above?

22-16 What is the magnitude of the magnetic field due to a current proportional to, whether it is flowing in a straight wire, a single loop of wire, or a solenoid?

22-17 A solenoid is placed so that its axis is vertical. When a current passes through the solenoid, the top of the solenoid becomes a magnetic south pole. As viewed from above, is the current flowing clockwise, or counterclockwise?

22-18 A solenoid 15.0 cm long has 157 loops of wire. If the magnetic field in the centre has a magnitude of 3.55×10^{-3} T, what is the current in the solenoid?

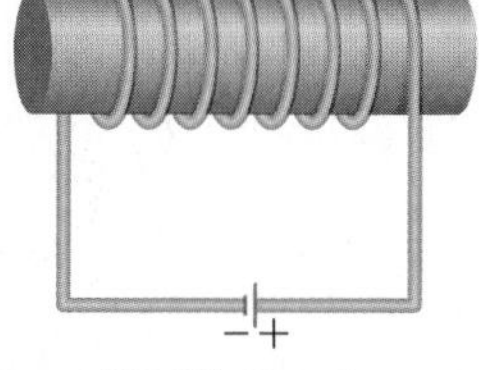

Figure 22-36 Questions 22-19 and 22-20.

22-19 Figure 22-36 shows an electromagnet. Redraw the image and sketch the resulting magnetic field lines. Be sure to indicate the field direction.

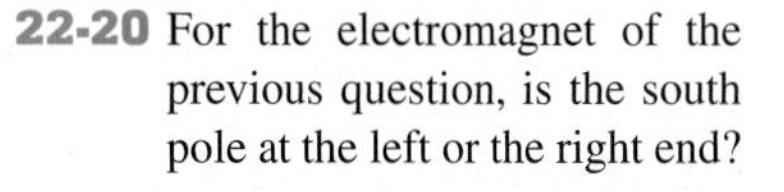

22-20 For the electromagnet of the previous question, is the south pole at the left or the right end?

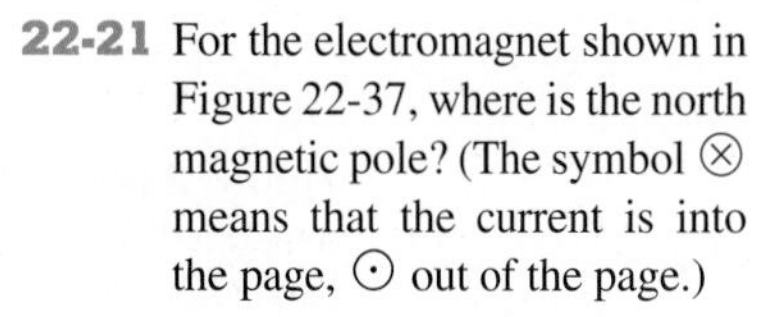

22-21 For the electromagnet shown in Figure 22-37, where is the north magnetic pole? (The symbol ⊗ means that the current is into the page, ⊙ out of the page.)

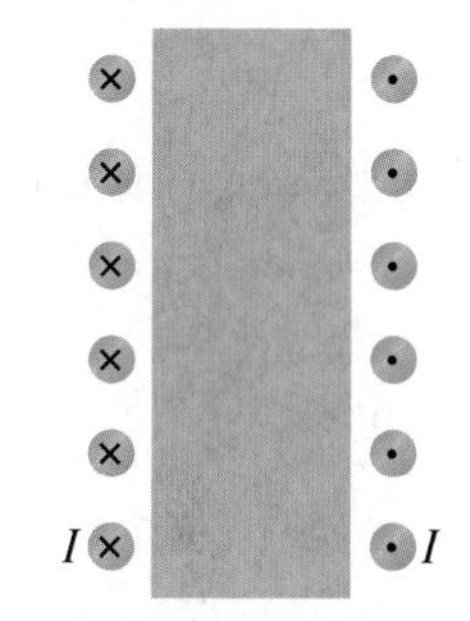

Figure 22-37 Question 22-21.

22-22 **Fermi Question:** There have been some experiments with DC power transmission. A large, high-voltage, DC transmission line may carry a total current of 1×10^4 A. Estimate the magnetic field that would be measured standing at the base of a tower holding such a line.

22-23 **Fermi Question:** Standing directly under a large, high-voltage, DC transmission line such as that described in the previous question, in which direction does your compass needle point if the current above you runs (horizontally) from south to north?

22.3 Magnetic Force on a Charged Particle

So far in this chapter we have discussed the source of magnetic fields such as magnets, currents in wires, solenoids, etc. We now turn our attention to the effect that magnetic fields have on charged particles. Charged particles that are *moving* experience a magnetic force when they are in a magnetic field. This force played a crucial role in the discovery of the electron, and is an important part of our understanding of the northern and southern lights (*aurora borealis* and *aurora australis*), as seen in Figure 22-38 and on the cover of this book.

Figure 22-38 The spectacular northern and southern lights at Earth's poles arise from magnetic forces acting on moving charged particles in the atmosphere.

Figure 22-39 (a) shows a positive charge q moving with velocity $\vec{v}$ in a magnetic field $\vec{B}$. The charge experiences a force that is perpendicular to both the velocity and the magnetic field. As shown in Figure 22-39 (a), when the positive charge is moving in the $-y$-direction in a magnetic field that is in the $+x$-direction, the resulting magnetic force is in the $+z$-direction. To determine the direction of the force, we use another **right-hand rule**: point the fingers of your right hand in the direction of the velocity, then curl your fingers toward the magnetic field vector (you might have to turn your hand upside down or sideways to do this), and your thumb will point in the direction of the force (Figure 22-39 (b)). There are three important things to remember when using this right-hand rule:

- The velocity and field vectors must be drawn so that they originate at the same point.

DID YOU KNOW?

The *aurora borealis* is named after the north wind, *Boreas*, in Greek mythology. The *aurora australis* is named after the south wind, *Auster*.

right-hand rule (magnetic force on a positive moving charge) align the fingers of your right hand with the direction of $\vec{v}$, curl them in the direction of $\vec{B}$. Your extended thumb is then aligned with the magnetic force $\vec{F}$ on a positive moving charge.

- The fingers must be curled from $\vec{v}$ toward $\vec{B}$ through the smaller of the two angles between these two vectors.
- The force on a negative charge is in the opposite direction to that determined above.

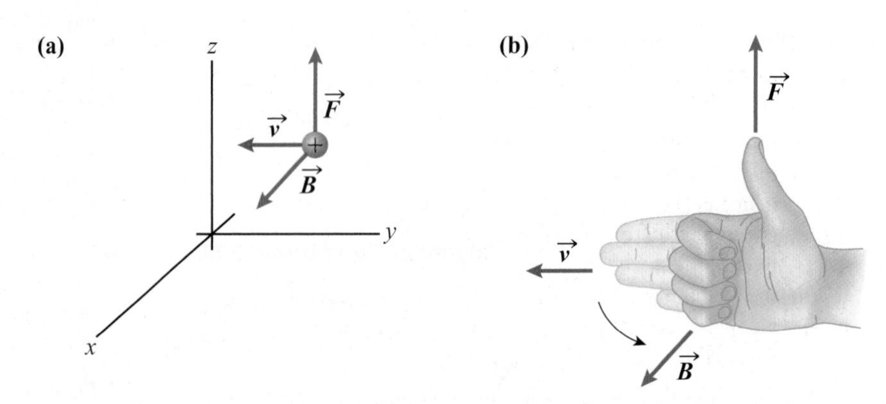

Figure 22-39 **(a)** A (positive) charge q moving with velocity $\vec{v}$ in a magnetic field $\vec{B}$ experiences a magnetic force $\vec{F}$, perpendicular to both $\vec{v}$ and $\vec{B}$. **(b)** Using a right-hand rule to determine the direction of $\vec{F}$.

Once we have determined the direction of the magnetic force through the right-hand rule, we can use Eqn. 22-4 to calculate the magnitude of that force.

The magnitude of the magnetic force (F) on a moving charge (q) is given by

$$F = |q|vB\sin\theta \tag{22-4}$$

where θ is the (smaller) angle between the velocity of the charge ($\vec{v}$) and the magnetic field in the region ($\vec{B}$).

Problem-Solving Tip

Use extra caution when determining the magnitude and direction of the magnetic force acting on a *negative* moving charge. When calculating the magnitude of the force with Eqn. 22-4, use the absolute value of the charge ($|q|$). Then use the right-hand rule as shown in Figure 22-39 (b) to determine the direction for a positive charge. The final answer is then the positive magnitude from Eqn. 22-4, and the *opposite* direction as that determined from the right-hand rule.

The force has a maximum magnitude if $\theta = 90°$ and is zero if $\theta = 0°$ or $180°$. Recall that v and B represent the magnitudes of $\vec{v}$ and $\vec{B}$, and hence are positive quantities. Eqn. 22-4 will always yield a positive answer, as expected when determining the magnitude of a vector quantity.

If you are familiar with vector cross-products, you will recognize that both the magnitude and the direction of the magnetic force can be expressed compactly by

$$\vec{F} = q\vec{v} \times \vec{B} \tag{22-5}$$

Since the magnetic force is always perpendicular to the velocity, it can never change the *speed* (i.e., the *magnitude* of the velocity) of a charged particle; it can change only the *direction* of its velocity. It is the ideal force to make a charged particle go in a circle, since it is always at right angles to the velocity.

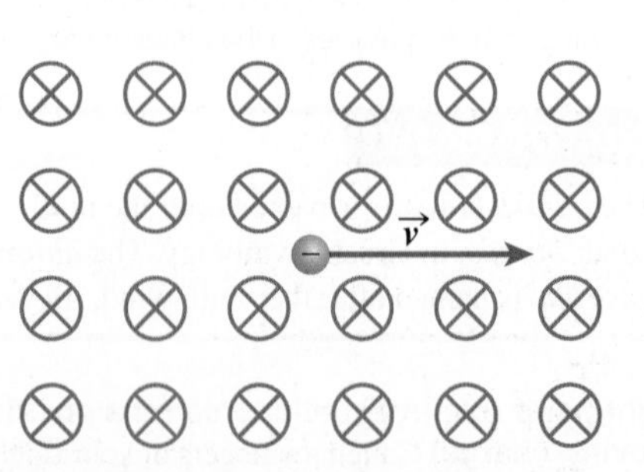

Figure 22-40 An electron moves to the right through a region of uniform magnetic field directed into the page (Sample Problem 22-3).

SAMPLE PROBLEM 22-3

An electron travelling at 2.32×10^6 m/s to the right encounters a uniform magnetic field of 4.35×10^{-4} T directed into the page, Figure 22-40. The electron's velocity and the magnetic field are perpendicular to each other.

(a) What is the direction of the force on the electron at the instant shown in Figure 22-40?

(b) What is the radius of the resulting circular orbit of the electron?

Solution

(a) To determine the direction of the magnetic force, we use the right-hand rule. We hold our open right hand such that the fingers are aligned to the right. We then rotate our hand so that the fingers can curl into the page, that is, in the direction of the magnetic field. Our extended thumb is pointing up, toward the top of the page, meaning that the magnetic force on a *positive* charge would be in this direction. Since an electron is negatively charged, the force on it is toward the bottom of the page at the instant shown in Figure 22-40.

(b) Since the magnetic force is perpendicular to the velocity, the electron follows a circular orbit (Figure 22-41). At the instant when the electron is at the top of the circle, we define the $+y$-direction to be toward the bottom of the page, that is, in the direction of the (centripetal) acceleration of the electron. We then use Newton's second law:

$$\Sigma F_y = ma_y$$

Since the electron moves in a circle, we can write the acceleration as $a_y = v^2/r$, where v is the electron's speed and r is the radius of the circle.

Thus, $\Sigma F_y = \dfrac{mv^2}{r}$

The only force that is acting is the magnetic force, which has a magnitude of $|q|vB \sin\theta$ based on Eqn. 22-4. We can therefore write

$$|q|vB \sin\theta = \frac{mv^2}{r}$$

Rearranging this expression for the unknown, r, we get

$$r = \frac{mv^2}{|q|vB \sin\theta}$$

$$r = \frac{mv}{|q|B \sin\theta}$$

The angle (θ) between $\vec{v}$ and $\vec{B}$ in this problem is 90°, and hence $\sin\theta = 1$. All the other variables are known, so we can substitute values and solve for r:

$$r = \frac{(9.11 \times 10^{-31}\ \text{kg})(2.32 \times 10^{6}\ \text{m/s})}{(1.60 \times 10^{-19}\ \text{C})(4.35 \times 10^{-4}\ \text{T})(1)}$$

$$r = 3.04 \times 10^{-2}\ \text{m}$$

Therefore, the radius of the circle is 3.04×10^{-2} m.

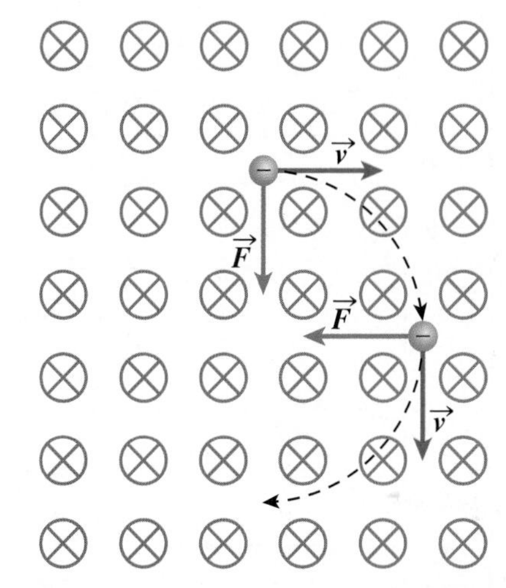

Figure 22-41 The resulting circular motion of an electron in a uniform magnetic field (Sample Problem 22-3).

A photograph of electrons in a circular orbit in a uniform magnetic field is shown in Figure 22-42. The electron path is visible because the electrons are travelling in a low-pressure gas; the gas atoms emit visible light when they lose energy acquired through collisions with the electrons. The field is perpendicular to the plane of the photograph. Based on the resulting counterclockwise circular path, is the field directed into or out of the page? (See Question 22-24 for the answer.)

Notice in Sample Problem 22-3 that the radius of the circular path of the electron is proportional to its mass-to-charge ratio, which enabled British scientist J. J. Thomson

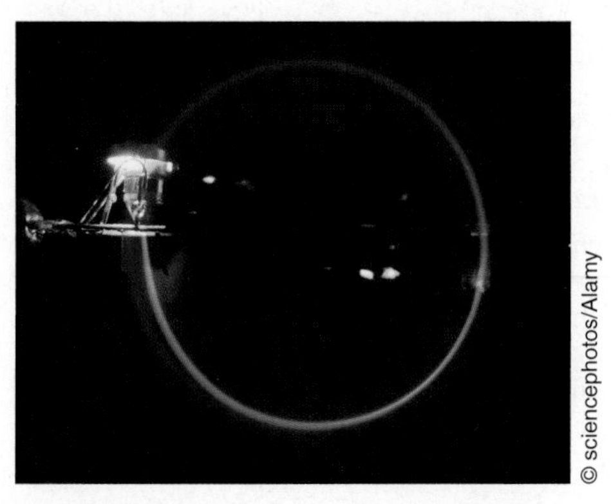

Figure 22-42 Electrons in a circular orbit in a low-pressure gas. There is a uniform magnetic field perpendicular to the plane of the photograph. The source of electrons is the illuminated nozzle on the left side of the image, with the beam travelling counterclockwise here.

(1856–1940) to measure this ratio of fundamental constants as early as 1897. It was the experimental work of Thomson that led scientists to realize that "cathode rays" were made up of subatomic charged particles—now known as electrons. Almost 15 years after Thomson first measured the m_e/e ratio, Millikan and his collaborators were able to measure the charge of the electron directly, as discussed in detail in Chapter 19. Between the two experimental approaches, the mass of the electron was therefore also determined. Thomson received the 1906 Nobel Prize in Physics for his discovery of the electron.

mass spectrometer a device that uses the dependence of the radius of the circular path on particle mass to separate atoms and molecules; used extensively in research and industrial laboratories

The dependence of the radius of the resulting circular path on the mass of the charge is also used in modern-day devices called **mass spectrometers**, in which atoms and molecules are sorted according to their masses by first being charged and then travelling through a magnetic field. Mass spectrometers are used, for example, to identify pollutants by measuring their molecular masses.

As we saw in Sample Problem 22-3, when a charged particle has a velocity that is perpendicular to the magnetic field, it undergoes circular motion. What happens when a charged particle is moving in a magnetic field with a velocity that is parallel to the magnetic field? What happens when the charge moves with a velocity that has components that are perpendicular and parallel to the field? To answer these questions we recall Eqn. 22-4, which gives the magnitude of the magnetic force on a charge moving in a magnetic field. The magnitude of the force depends on $\sin\theta$, where θ is the angle between the velocity vector ($\vec{v}$) and the magnetic field ($\vec{B}$). Therefore when a charge is moving in a direction that is parallel to the field, $\theta = 0°$ and the magnetic force on that charge is also zero. If there is no exerted force, the charge will continue moving in a straight line parallel to the field (Figure 22-43 (a)).

For the more general case in which a charged particle has a velocity that has a component parallel and a component perpendicular to the field, we can analyze the resulting motion by thinking about what happens in each direction. For the component of the velocity that is parallel to the field, $\theta = 0°$ and the magnetic force is also zero. Therefore the component of the velocity that is parallel to the field remains unchanged (Figure 22-43 (a)). For the component of the velocity that is perpendicular to the field, as discussed in Sample Problem 22-3, the particle undergoes circular motion in this plane (Figure 22-43 (b)). Now, in the general case in which a charge has both components of velocity, the combined effect is a helical path as shown in Figure 22-43 (c), where the long axis of the helix is parallel to the field and the circular motion occurs in the plane perpendicular to the field.

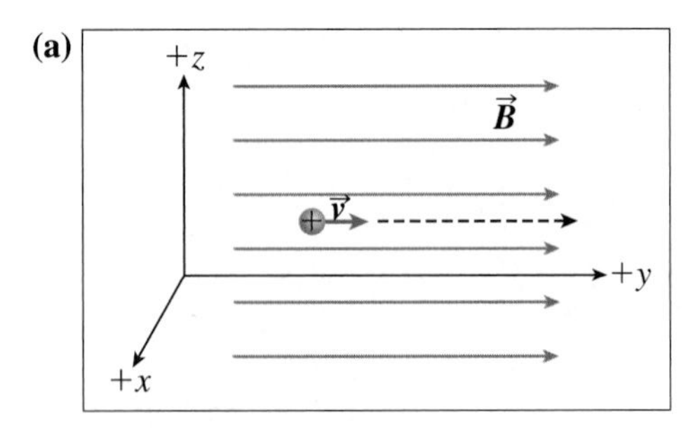

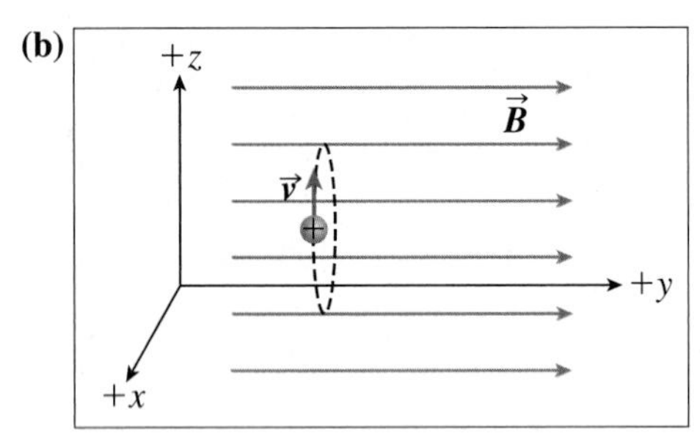

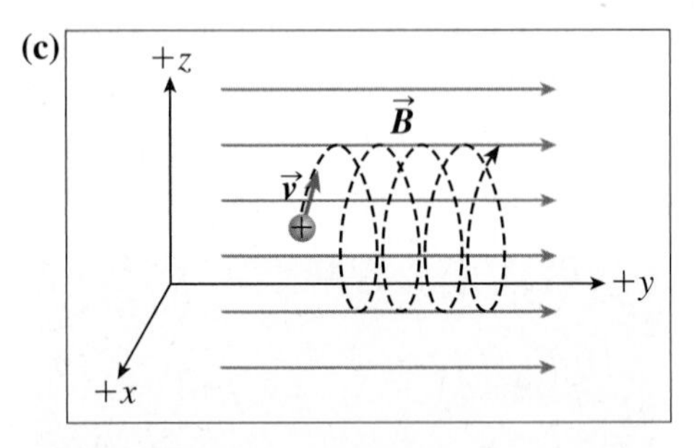

Figure 22-43 **(a)** The straight-line path of a charged particle that moves parallel (or anti-parallel) to the magnetic field. **(b)** The circular path of a charged particle that moves perpendicular to the magnetic field. **(c)** The helical path of a charged particle that moves at an angle (that is not 0°, 90°, or 180°) to the magnetic field.

The Earth's Magnetic Field and the Auroras

The *aurora borealis* (Figures 22-38 and 22-44 (b)) and *aurora australis* are beautiful, haunting, natural phenomena. The aurora often appears as dancing ribbons of light, mostly green in colour but occasionally with hints of red, purple, or pink. These spectacular displays are most often seen in the Arctic region of Earth, with a corresponding display in the Antarctic (witnessed mostly by penguins).

solar wind high-energy charged particles, mainly protons and electrons, emitted by the Sun

What causes the auroras? Why are they predominately seen at the poles? Earth is constantly being bombarded by a stream of high-energy charged particles, primarily protons and electrons, emitted by the Sun, known as the **solar wind**. When these particles collide with atoms and molecules in our atmosphere, the molecules absorb energy from the charged particles in the solar wind. This absorbed energy is then released by the atoms and molecules in the form of visible light, similar to the emission shown in Figure 22-42, with the different gases in the atmosphere giving rise to the various colours of light observed. The green light is emitted mainly by oxygen for example.

So the collisions between charged particles in the solar wind and molecules in our atmosphere give rise to the ethereal light show, but why do we see the light from these interactions only in the polar regions? In order to understand this better, remember that Earth is like a giant bar magnet (Figure 22-12). When the charged particles in the solar wind encounter the magnetic field of Earth, in general they will be travelling with velocities that have components both parallel and perpendicular to the field. Therefore, the solar wind follows a helical path as shown in Figure 22-44 (a) along the magnetic field lines. When the charged particles enter the polar regions they are travelling with a velocity that is essentially parallel (or anti-parallel) to the magnetic field, which means that they are not deflected by the field. This is why we see the auroras at the poles (Figure 22-44 (b)): the charged particles at the poles are able to reach the gas molecules in the atmosphere and collide. At lower latitudes, the solar wind is deflected by the magnetic field of Earth.

DID YOU KNOW?

The magnetic field of the Sun that powers the solar winds is quite dynamic: the north and south magnetic poles flip approximately every 11 years, with the most recent occurrence in November 2013. In comparison, the last time Earth's magnetic field flipped was almost 800 000 years ago.

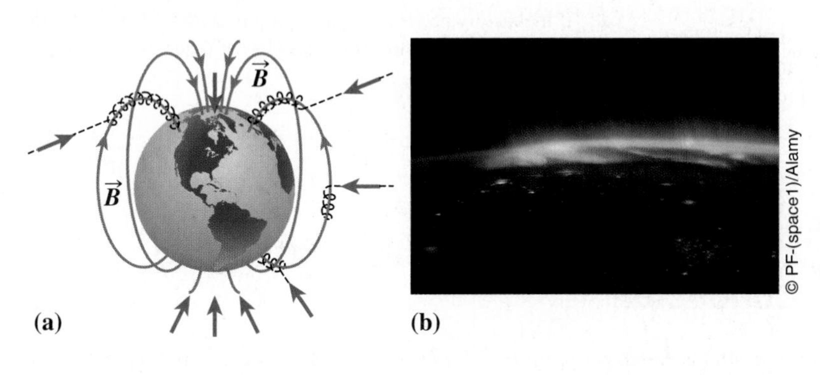

Figure 22-44 **(a)** Charged particles from the Sun travel in helical paths along the magnetic field lines of Earth. Particles arriving at the poles are not deflected, resulting in the observation of the auroras in these regions on Earth. **(b)** The aurora borealis at the north pole, as seen from the International Space Station.

EXERCISES

22-24 Based on the resulting counter-clockwise circular path of the electrons in Figure 22-42, is the magnetic field directed into or out of the page?

22-25 In Figure 22-45, what is the direction of the magnetic force on each charged particle moving with velocity $\vec{v}$ in the magnetic field $\vec{B}$ shown? (The symbol ⊗ means that the vector points into the page, ⊙ out of the page.)

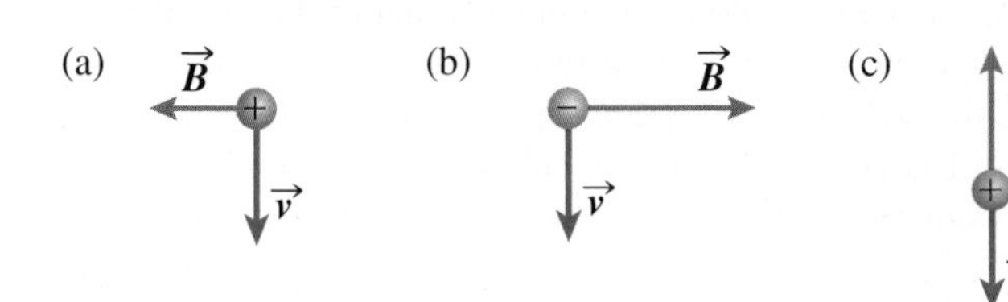

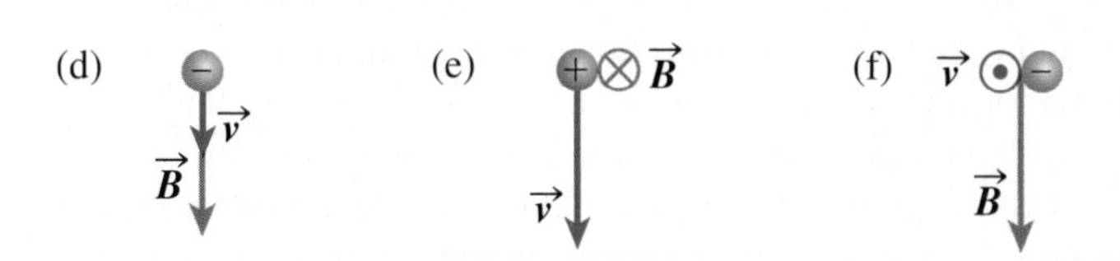

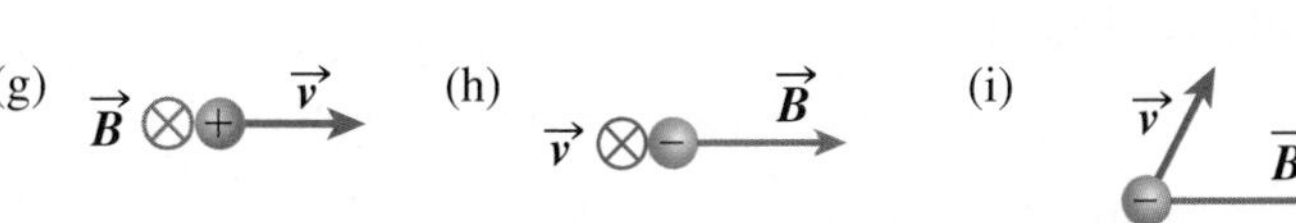

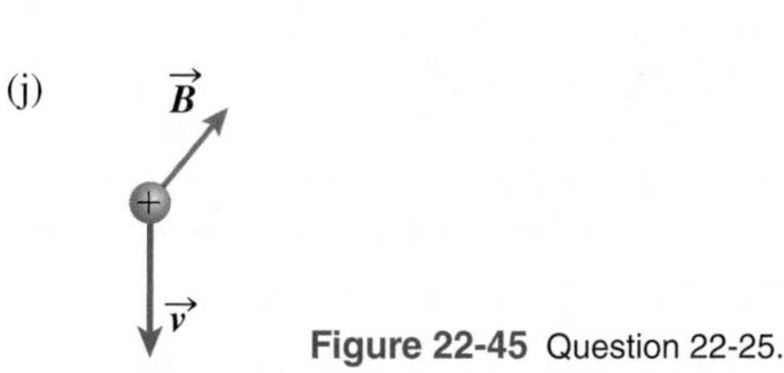

Figure 22-45 Question 22-25.

22-26 A proton with a velocity of 3.59×10^6 m/s north encounters a uniform magnetic field of 5.45×10^{-4} T south. Both the proton's velocity and the magnetic field are in the horizontal plane. What are the magnitude and direction of the magnetic force on the proton?

22-27 Repeat the previous question if the direction of the magnetic field is east.

22-28 An electron travelling west is deflected toward the north by a magnetic field. The electron is in the horizontal plane at all times. What is the direction of the field?

22-29 Use Eqn. 22-4 to express teslas in terms of kilograms, coulombs, and seconds.

22-30 Is it possible to accelerate a *stationary* electron with a magnetic field? an electric field? Explain.

22-31 An electron enters a uniform magnetic field of magnitude 7.2×10^{-2} T (Figure 22-46). If the electron's speed is 2.5×10^6 m/s, determine the acceleration (magnitude and direction) of the electron when it is at the position shown.

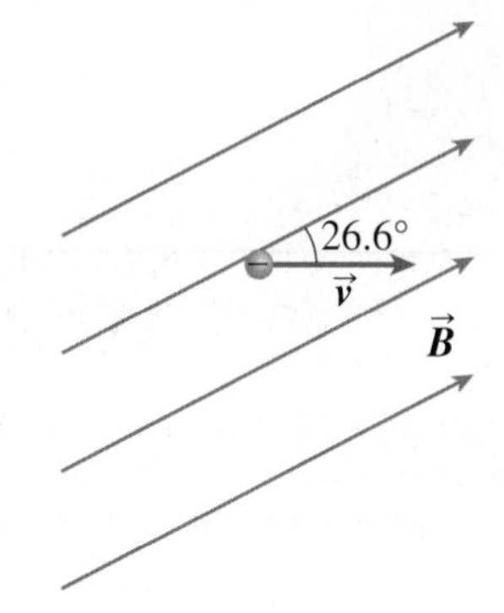

Figure 22-46 Question 22-31.

22-32 An α particle (charge $+2e$, mass 6.64×10^{-27} kg) travelling at 3.93×10^5 m/s enters a uniform magnetic field of magnitude 6.43×10^{-2} T. The field is perpendicular to the particle's velocity. What is the radius of the resulting circular orbit?

22-33 In the previous question, what is the period of revolution of the α particle?

22-34 An electron is accelerated from rest through a potential difference of 2.5 kV and then enters a uniform magnetic field that is perpendicular to the electron's velocity. The electron follows a circular path of radius 6.7 cm. What is the magnitude of the magnetic field?

22-35 Describe the subsequent motion of a proton that enters a uniform magnetic field, if the angle between the proton's velocity and the magnetic field is 35°.

22-36 A proton, travelling at 7.00×10^5 m/s east in the horizontal plane, enters a region of space where there is a uniform magnetic field of 4.56×10^{-3} T downward. In order for the proton to continue moving at constant velocity, what must be the magnitude and direction of a uniform electric field acting on the proton? (Neglect gravity in this case as it is orders of magnitude smaller than the electric and magnetic forces.)

22-37 What is the work done by a magnetic force on a charged particle always equal to? Explain your answer.

22-38 **Fermi Question:** In PET imaging (Section 24.6) it is ideal to have a small particle accelerator on site called a cyclotron. The cyclotron has a magnetic field that is perpendicular to the plane of the charged particles in order to keep the beam moving in a circular path. If the facility needs to produce protons with 4 MeV of kinetic energy, what is the diameter of the region of magnetic field inside the device?

22.4 Magnetic Force on a Current-Carrying Conductor

Figure 22-47 Your hairdryer is just one of the many appliances in your home that uses magnetic forces on currents to run electric motors.

Every time you use a hairdryer (Figure 22-47), vacuum cleaner, electric fan, or anything that has an electric motor, you are making use of the force that a current-carrying conductor experiences in a magnetic field. In electric motors, current passes through coils of wire situated between poles of a magnet and the resulting magnetic force causes the coils to rotate. It should not come as a surprise to learn that there is a magnetic force on a current-carrying conductor. In Section 22.3 we discussed the magnetic force on charges that are free to move anywhere in space. In the case of conductors, the charges can move, but only within the confines of the conductors.

Figure 22-48 shows a current-carrying wire in the magnetic field between the poles of a horseshoe magnet. Each moving charged particle in the wire experiences a force due to the magnetic field as expressed in Eqns. 22-4 and 22-5. Since each particle is confined within the wire, the magnetic force is effectively exerted on the wire itself and the direction of this force can be determined by the right-hand rule described in Section 22.3. We will follow the convention that current corresponds to the flow of positive charge; in Figure 22-48, each positive charge is moving out of the page along the wire. Pointing the fingers of your right hand in the direction of this velocity, and then curling your fingers upward in the direction of the magnetic field, your thumb points to the left to give the direction of the force on the charge, and hence on the wire.

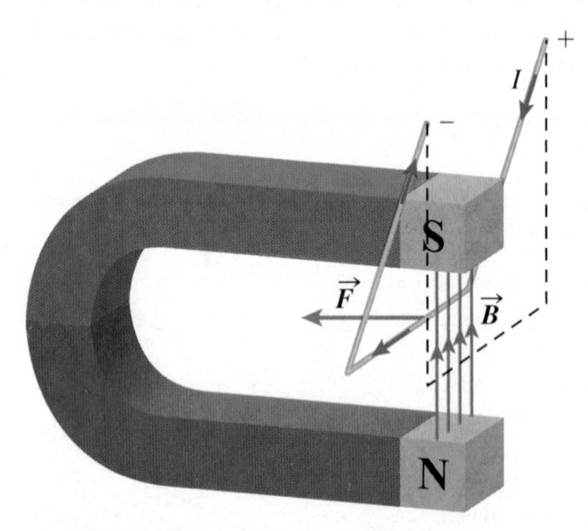

Figure 22-48 The current-carrying wire experiences an overall magnetic force to the left, deflecting it inward within the horseshoe magnet.

In dealing with magnetic forces on wires, we usually restate the **right-hand rule** as follows: point the fingers of your right hand in the direction of the current, curl your fingers toward the magnetic field vector, and your thumb will point in the direction of the force.

right-hand rule (magnetic force on a current-carrying wire) align your fingers with the direction of the current, curl them in the direction of $\vec{B}$. Your extended thumb is then aligned with the magnetic force $\vec{F}$ on that current.

Using the right-hand rule, we can readily determine the direction of the overall magnetic force on a current-carrying wire, but what is the magnitude of that force? For this we need to start with Eqn. 22-4, which tells us the magnitude of the magnetic force on an individual charged particle in the wire:

$$F = |q|\, vB \sin\theta \quad \text{(magnetic force on an individual charge } q\text{)}$$

where q is the charge, v is the speed of that moving charge, and θ is the angle between the direction of motion and the magnetic field in the region. If we consider all the charges together in the wire, we can represent the number of charges per unit volume as n, which has SI units of metre^{-3}. Therefore, the total number of charges in a wire of length L and cross-sectional area A is $L \times A \times n$, and the total amount of charge in the wire is $L \times A \times n \times |q|$. The total force on the wire will therefore be

$$F = LAn|q|vB \sin\theta \quad \text{(magnetic force on all of the charges in the wire)}$$

since the force on individual charges in the wire contributes to the overall magnitude (Figure 22-49).

Recall from Chapter 20 that current is simply a measure of charge traversing a region per unit time:

$$I = \frac{\Delta q}{\Delta t}$$

In this case, the charges are moving with an average speed of v along the length of the wire. As before, there is a total charge in the wire of $L \times A \times n \times |q|$. This amount of charge takes a time of Δt to travel the length L of the wire at an average speed of v, which means that $\Delta t = L/v$. Therefore, the current in the wire can be expressed as

$$I = \frac{\Delta q}{L} v$$

$$I = \frac{LAn|q|}{L} v$$

$$I = An|q|v$$

Since the total force on the wire is given by $F = L\,(A\ n\ |q|\ v)\ B\ \sin\theta$, we can now substitute our expression for current to simplify our expression for the overall force as

$$F = L(nA|q|v)B\sin\theta$$

$$F = LIB\sin\theta$$

Starting with the force on an individual charge q (Eqn. 22-4), we have now demonstrated that the magnitude of the total force on the wire, that is, the sum of the forces on all the moving charges, is given by

$$F = ILB\sin\theta \qquad (22\text{-}6)$$

where I is the current, L is the length of the wire, B is the magnitude of the uniform magnetic field, and θ is the angle between the straight wire and the magnetic field.

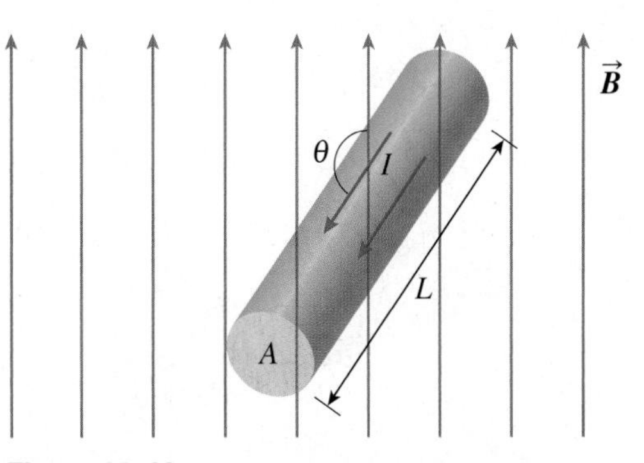

Figure 22-49 Calculating the overall magnetic force on a current-carrying wire of length L and cross-sectional area A in a magnetic field $\vec{B}$.

Equation 22-6 applies to a straight wire in a uniform magnetic field in which the angle between the current and the magnetic field is the same everywhere. We would need to use integral calculus to determine the magnetic force on current-carrying wires in other scenarios, which is beyond the scope of this book.

SAMPLE PROBLEM 22-4

Figure 22-50 shows a wire carrying a current that flows out of the page located between the poles of a horseshoe magnet. What is the direction of the magnetic force on the wire?

Solution

Recall that the magnetic field lines outside a magnet point from the north to the south pole. Thus, the magnetic field direction between the poles is to the left. Using the right-hand rule, we align the fingers of our right hand with the current (out of the page) and then curl them in the direction of the magnetic field, toward the left. Our extended thumb points down, toward the bottom of the page. Thus, the force on the wire is toward the bottom of the page in this case.

Figure 22-50 Sample Problem 22-4.

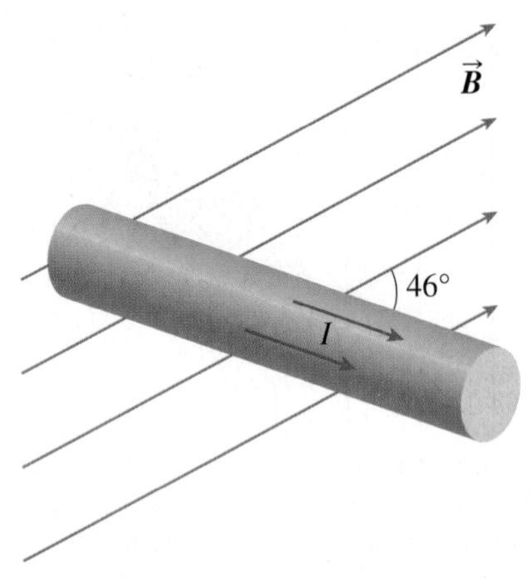

Figure 22-51 Sample Problem 22-5.

SAMPLE PROBLEM 22-5

What is the magnetic force (magnitude and direction) on the straight wire shown in Figure 22-51? The wire has a length of 7.5×10^{-2} m and carries a current of 0.93 A. The uniform magnetic field has a magnitude of 5.5×10^{-3} T. Both the wire and the magnetic field are in the plane of the page.

Solution

We will first determine the direction of the force. Using the right-hand rule, we align the fingers of our right hand in the direction of the current, and curl them in the direction of the magnetic field. Our extended thumb is then pointing in the direction of the magnetic force on the wire. Therefore, the magnetic force on the wire is directed out of the page.

We calculate the magnitude of the force from Eqn. 22-6:

$$F = ILB \sin\theta$$

$$F = (0.93 \text{ A})(7.5 \times 10^{-2} \text{ m})(5.5 \times 10^{-3} \text{ T}) \sin 46°$$

$$F = 2.8 \times 10^{-4} \text{ N}$$

Therefore, the total magnetic force on the wire is 2.8×10^{-4} N, out of the page.

Table 22-3

Right-Hand Rules Used in Understanding Magnetic Interactions

Scenario	Procedure
Magnetic field associated with a current-carrying wire	Direction of magnetic field lines; Right hand; Electric current
Magnetic field associated with a solenoid	Magnetic field inside; Current
Magnetic force on a positive moving charge	$\vec{F}$; $\vec{v}$; $\vec{B}$
Magnetic force on a current-carrying wire	$\vec{F}$; $\vec{B}$; I

There have now been four different right-hand rules presented in this chapter, which can get a little confusing! Table 22-3 provides a summary of these rules in one location to help you when reviewing.

Electric Motors

All electric motors, whether in food processors or battery-powered screwdrivers, operate using the same basic physics: The force on a current-carrying wire in a magnetic field is used to create motion. The basic elements of any electric motor are shown in Figure 22-53 (a): a current-carrying loop of wire and a magnet. The current is supplied by an external energy source such as a battery. Consider the force exerted on each segment of wire in the loop using Eqn. 22-6 and the right-hand rule. In segment QR, the wire and magnetic field are parallel; therefore, $\theta = 0°$, $\sin\theta = 0$, and there is zero force on segment QR. Similarly, there is no force exerted on segment SP. For segment PQ, $\theta = 90°$ and hence there is a magnetic force; using the right-hand rule, the direction is found to be into the page. (Remember that the magnetic field points from the north to the south pole.) Similarly, $\theta = 90°$ for segment RS and the direction of the force based on the right-hand rule is out of the page. The effect of these forces is illustrated in Figure 22-53 (b). The inward force on segment PQ and the outward force on RS produce a rotation of the loop in the direction shown. Thus, we have a rotating motor, which could be used to drive a device such as an electric drill.

The above description of an electric motor describes the basic principles involved, but there are problems with the design details. Once the loop has rotated past the vertical position shown in Figure 22-53 (b), the magnetic forces on it will begin to rotate the loop in the opposite direction since the current is now flowing down on the left side (RS) and up on the right side (PQ). Thus, the loop will not rotate continuously in one direction, as we want in a motor. This problem is solved through the use of a **commutator**, which is a metal ring split into two parts (Figure 22-53 (c)). Its purpose is to reverse the direction of the current in the loop every half revolution so that the magnetic forces always rotate the loop in the same direction. The commutator rotates with the loop and electrical contact with the external current source is provided by **brushes**, which are stationary metal pieces that touch the rotating commutator.

commutator a split metal ring used in electric motors to reverse the current direction with every half rotation

brushes stationary pieces of metal in electrical contact with rotating electrical components such as commutators

In real motors, the single loop is replaced by many loops wrapped around an iron core called an **armature**, and the permanent magnets, called field magnets, are often replaced by electromagnets. The loops are wound in several separate coils called windings. At any one time the current flows in only the one winding whose orientation results in the largest force.

armature the rotating iron core in an electric motor or generator

Build Your Own Electric Motor

With some simple equipment you can build your very own electric motor. You will need a C or a D battery, some insulated copper wire, two large-eyed long sewing needles, clay, electrical tape, and a small, strong circular magnet. Wrap the copper wire around a pen or marker 30 to 40 times. Slide this coil off the pen and secure the loops by wrapping each end of the wire around the coil a few times. Extend the ends of the wire away from the loop in opposite directions, as shown in Figure 22-52. Very carefully remove the top half of the insulation from each of the loose ends of the wire—make sure that you strip the insulation from the same side (top or bottom) on each end. (Why?) Use the clay to secure the battery on its side on a flat surface. Then use the electrical tape to connect the sewing needles to each end of the battery, standing vertically upright. The coil can then be placed in the eyes of the needles. Tape the small magnet to the battery, directly underneath the coil. Give your coil a spin and see what happens! What happens if you try to spin it the other way? Why? What would happen with a bigger magnet? What would happen with a second battery connected in series? With a little care, the motor you build will spin continuously until the battery is drained.

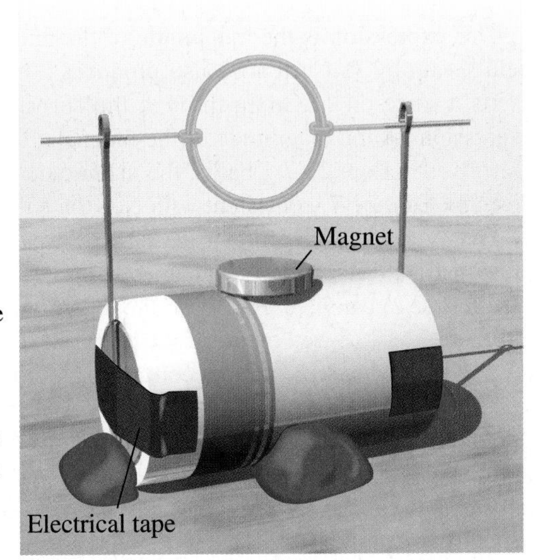

Figure 22-52 Your homemade electric motor: using magnetic forces to convert electrical energy stored in the battery to rotational kinetic energy.

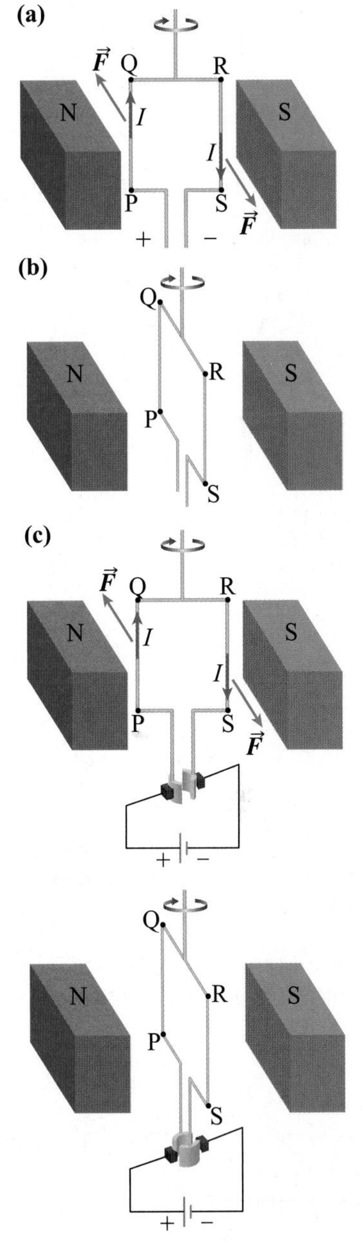

Figure 22-53 **(a)** Basic elements of a motor: a current-carrying loop of wire in a magnetic field. **(b)** The magnetic forces on segments PQ and RS produce a rotation of the loop, with the axis of rotation in the plane of the page. **(c)** In order to keep the loop rotating through the field, the current has to change direction with every half rotation. This is accomplished by connecting the loop to the battery through a commutator (split ring) and brushes.

Magnetic Force between Parallel Conductors

Since a current in a wire produces a magnetic field (Section 22.2), and a current-carrying wire in a magnetic field experiences a force, then two current-carrying wires should exert magnetic forces on each other. Indeed this is the case. Two parallel current-carrying wires, separated by a distance r, are shown in Figure 22-54, along with the magnetic field due *only* to the current I_1 for clarity. As given by Eqn. 22-1, the magnitude of the field at a distance r from I_1 is

$$B_1 = \frac{\mu_0 I_1}{2\pi r}$$

The other wire, carrying a current I_2, experiences a magnetic force due to this field. From Eqn. 22-6, a length L of wire with current I_2 will experience a magnetic force due to field B_1 that has a magnitude given by

$$F = I_2 L B_1 \sin\theta$$

From Figure 22-54 we can see that the angle θ between the current I_2 and the magnetic field B_1 is 90° when the two wires are parallel to each other. Therefore $\sin\theta = 1$ and,

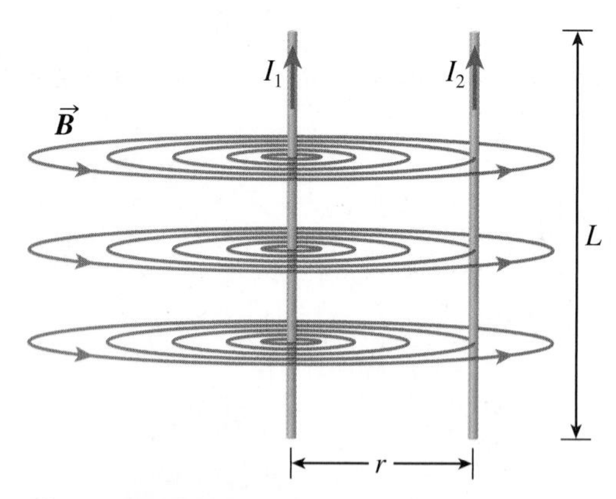

Figure 22-54 Two parallel wires of length L carrying currents I_1 and I_2 separated by a distance r. For clarity, only the magnetic field due to the current I_1 is shown.

substituting the expression for the magnitude of the magnetic field B_1 at the location of the second wire, we get the magnitude of the force acting on wire 2:

$$F = \frac{\mu_0 I_1 I_2 L}{2\pi r} \qquad (22\text{-}7)$$

gives the magnitude of the magnetic force acting on each parallel wire of length L, separated by a distance r, carrying currents of I_1 and I_2.

This expression is the magnitude of the magnetic force on I_2 due to the magnetic field set up by I_1. Current I_2 also produces a magnetic field, which correspondingly exerts a force on I_1. Going through the derivation again, we end up with the same expression for the magnitude of the magnetic force on I_1 due to the magnetic field of I_2 as given in Eqn. 22-7. That is, the magnitude of the magnetic force on *either* wire is given by Eqn. 22-7, consistent with Newton's third law.

The direction of the force on each wire can be determined by the right-hand rules previously introduced in this chapter. Consider first the force on I_2. A right-hand rule (Section 22.2) can be used to give the direction of the magnetic field due to I_1, shown in Figure 22-54. Based on this right-hand rule, the magnetic field due to I_1 will be into the page at the location of the second wire, I_2. Now we apply the second right-hand rule to determine the direction of the resulting force on I_2 as described in this section, with the fingers of our right hand directed up along I_2 and then curling them into the page in the direction of $\vec{B}_1$, our extended thumb is pointing to the left. Therefore, the force on I_2 due to the current in wire 1 is to the left, toward wire 1. Using the directions of the current I_1 and the field due to I_2 at the position of I_1, we can apply these right-hand rules again to determine that the magnetic force on I_1 due to the current in wire 2 is to the right, toward wire 2. In other words, *currents travelling in the same direction in parallel wires attract each other*. We can also show that *currents travelling in opposite directions in parallel wires repel each other,* by applying a similar analysis. In either scenario, the attractive or repulsive magnetic forces will have magnitudes on each wire given by Eqn. 22-7.

EXERCISES

22-39 For each drawing in Figure 22-55, what is the direction of the magnetic force on the current-carrying wire?

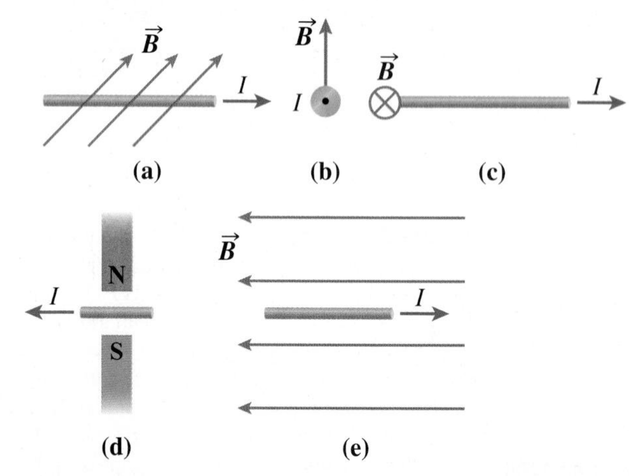

Figure 22-55 Question 22-39.

22-40 A wire carries a current vertically upward in a magnetic field that is perpendicular to the wire. If the magnetic force on the wire is northward, what is the direction of the field?

22-41 What is the magnitude of the magnetic force on a wire of length 7.5 cm, carrying a current of 1.5 A, when placed in a uniform magnetic field of magnitude 4.3×10^{-2} T? The angle between the wire and the field is 35°.

22-42 What is the magnetic force exerted on the wire shown in Figure 22-56? Assume that there is a uniform magnetic field of magnitude 3.6×10^{-3} T in the region directly between the poles of the magnets and that the field is

Figure 22-56 Question 22-42.

negligible elsewhere. The length AB is 2.5 cm, and the current in the wire is 0.92 A, running perpendicular to the magnet as shown.

22-43 A horizontal copper wire lies in the east–west direction. The direction of *electron* flow in the wire is eastward. Assuming that Earth's magnetic field is horizontal and exactly northward, what is the direction of the force on the wire?

22-44 In Figure 22-57, do the magnetic forces cause the loop to rotate clockwise, or counterclockwise? Explain.

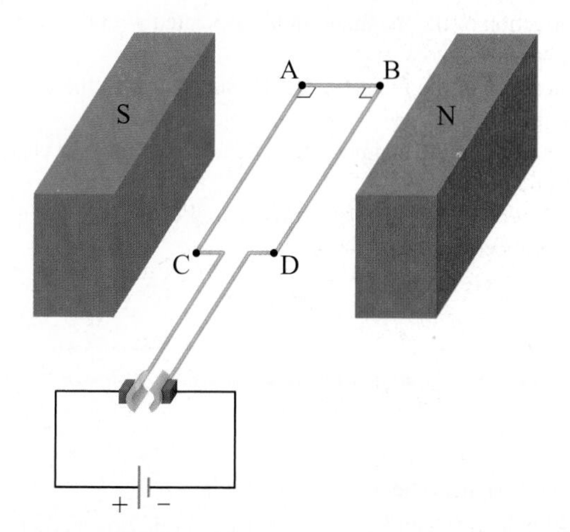

Figure 22-57 Questions 22-44, 22-45, and 22-46.

22-45 Describe two possible ways to make the loop in Figure 22-57 rotate in the opposite direction.

22-46 Determine the magnitude and direction of the magnetic force in Figure 22-57 on each of the wire segments AB, AC, and BD at the moment shown. Assume that there is a uniform magnetic field of magnitude 4.32×10^{-3} T between the poles. Length AB is 1.23 cm and AC = BD = 6.14 cm; there is a current of 0.445 A in the circuit.

22-47 Two long, straight, parallel wires carry currents.

(a) If the currents are in the same direction, do the wires attract or repel one another?

(b) If the currents are in opposite directions, do the wires attract or repel one another?

22-48 What is the magnitude of the force on 17 cm of a wire carrying a current of 0.75 A, situated 28 cm from a parallel wire with a current of 1.3 A in the same direction?

22-49 Draw two parallel wires carrying currents in opposite directions.

(a) Use a right-hand rule to determine the direction of the magnetic field at the wire on the right due to the wire on the left, and indicate this direction in your drawing.

(b) Use a right-hand rule to determine the direction of the force on the wire on the right due to the wire on the left, and show this force on your drawing.

22-50 Two current-carrying wires are moved farther apart, with no change in current in either wire. Sketch the shape of a graph of the magnitude of the force (F) on either wire versus the distance (r) apart.

22-51 The magnetic force *per unit length* on a current-carrying wire is 3.56×10^{-9} N/m in magnitude. The current in the wire is 0.550 A, and it is separated from a parallel wire by 0.750 m. What is the current in the other wire?

DID YOU KNOW?

A *ferrofluid* (Figure 22-58) is a liquid that becomes strongly magnetized in the presence of a magnetic field. Ferrofluids are typically made up of nanoscale magnetic particles suspended in a fluid, usually either an organic solvent or water. Each tiny particle is thoroughly coated with a surfactant to inhibit clumping. Ferrofluids can have exceptionally high relative permeability and are used in electronic devices to form liquid seals around sensitive components as well as in many mechanical engineering applications due to its friction-reducing capabilities.

Oliver Hoffmann/Shutterstock

Figure 22-58 Ferrofluid flowing from one magnet to another.

LOOKING BACK...LOOKING AHEAD

Various aspects of electricity were the topics in the three previous chapters: charge, current, electric field, potential, resistance, and so on.

The focus of this chapter has been magnetism, an area of science closely associated with electricity. We have seen that moving charges produce magnetic fields, and in turn that a moving charge experiences a force when in a magnetic field. The first two sections of the chapter dealt with the production of magnetic fields by magnets, Earth, various current configurations (straight wire, loop, solenoid), and electromagnets.

In the final two sections we have discussed the magnetic force exerted on moving charged particles, whether free or in a conductor. Electric motors were introduced as practical examples of using the force on a current-carrying conductor in a magnetic field.

In the next chapter we continue our discussion of magnetism with another facet closely linked to electricity: the production of a current in a conductor by a magnetic field. This phenomenon is of great importance in electrical generators and transformers.

CONCEPTS AND SKILLS

Having completed this chapter, you should now be able to do the following:

- Discuss what is meant by magnetic and non-magnetic materials.
- Discuss how magnetic fields are produced.
- Discuss the similarities and differences between electric and magnetic fields.
- Explain the domain theory of ferromagnetism.
- Discuss properties of Earth's magnetic field.
- Sketch the magnetic field lines around a bar magnet, straight wire, loop, and solenoid.
- Solve problems involving the magnitude of the magnetic field around a straight wire, at the centre of circular loops, and inside a solenoid.
- Explain how electromagnets work, and give some practical examples.
- Solve problems involving the magnetic force on a moving charged particle.
- Discuss some useful examples of magnetic forces on charged particles.
- Explain the natural phenomenon of the auroras.
- Solve problems involving the magnetic force on a current-carrying conductor.
- Explain the basic operation of an electric motor.
- Solve problems involving the magnetic force between parallel conductors.
- Use the appropriate right-hand rule to determine:
  - the direction of the magnetic field associated with the current in a straight wire
  - the direction of the magnetic field associated with the current in a solenoid
  - the direction of the magnetic force acting on a moving charge in a magnetic field
  - the direction of the magnetic force acting on a current in a wire in the presence of a magnetic field

KEY TERMS

You should be able to define or explain each of the following words or phrases:

north pole (of a magnet)
south pole (of a magnet)
ferromagnetic materials
magnetic materials
domains
Curie temperature
magnetic north
magnetic declination
magnetic inclination
dipping needle
magnetotactic
right-hand rule (magnetic field associated with a current-carrying wire)
tesla (SI unit)
gauss (unit)
gaussmeter
solenoid
right-hand rule (magnetic field associated with a solenoid)
electromagnet
relative permeability
diamagnetic
Maglev (trains)
right-hand rule (magnetic force on a positive moving charge)
mass spectrometer
solar wind
right-hand rule (magnetic force on a current-carrying wire)
commutator
brushes
armature

Chapter Review

MULTIPLE-CHOICE QUESTIONS

22-52 In Figure 22-59 which of the magnetic field configurations between the poles of the two bar magnets is/are correct?

(a) II, IV, and V
(b) I, II, and V
(c) III and IV
(d) II only
(e) IV only

Figure 22-59 Questions 22-52 and 22-59.

22-53 In order for a charged particle to follow a circular path in a uniform magnetic field, the particle's velocity

(a) must be parallel/anti-parallel to the magnetic field
(b) must be perpendicular to the magnetic field
(c) must have components that are both parallel and perpendicular to the magnetic field
(d) none of the above

22-54 In order for a charged particle to follow a helical path in a uniform magnetic field, the particle's velocity

(a) must be parallel/anti-parallel to the magnetic field
(b) must be perpendicular to the magnetic field
(c) must have components that are both parallel and perpendicular to the magnetic field
(d) none of the above

22-55 In order for a charged particle to follow a straight line path in a uniform magnetic field, the particle's velocity

(a) must be parallel/anti-parallel to the magnetic field
(b) must be perpendicular to the magnetic field
(c) must have components that are both parallel and perpendicular to the magnetic field
(d) none of the above

22-56 A proton experiences a magnetic force upward due to a northward magnetic field. What is the direction of the proton's velocity? (The magnetic field and the proton's velocity are in the horizontal plane.)

(a) north
(b) south
(c) east
(d) west

22-57 An electron experiences a magnetic force upward due to a northward magnetic field. What is the direction of the electron's velocity? (The magnetic field and the electron's velocity are in the horizontal plane.)

(a) north (c) east
(b) south (d) west

22-58 What is the direction of the current in wire segment PQ in Figure 22-60 if the magnetic forces are producing a clockwise rotation of the loop?

(a) from P to Q (b) from Q to P

Figure 22-60 Question 22-58.

Review Questions and Problems

22-59 If a compass was placed between the two poles shown in image (II) in Figure 22-59, which way would it point?

22-60 Discuss the similarities and differences between the electric field associated with a positive point charge and the magnetic field associated with a straight, current-carrying wire.

22-61 A variable current passes through a single loop of wire. Sketch a graph of the magnitude of the magnetic field at the loop's centre versus current.

22-62 If the magnitude of the magnetic field near a straight wire is to be no larger than 2.35×10^{-6} T at a distance of 0.300 m from the wire, what is the maximum current that the wire can carry?

22-63 What is the direction of the magnetic field at P in Figures 22-61 (a), (b), and (c)?

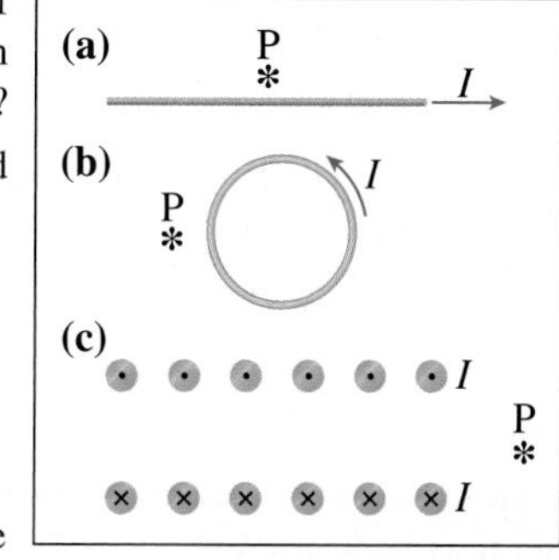

Figure 22-61 Question 22-63.

22-64 Sketch the magnetic field lines around

(a) a bar magnet
(b) a straight wire
(c) a single loop of wire
(d) a solenoid

In parts (b), (c), and (d), assume that a current is flowing.

22-65 What is the magnitude of the magnetic field at the centre of a solenoid having 15.0 loops per cm of length, when the current is 0.382 A?

22-66 Figure 22-62 shows a solenoid wound around a hollow cardboard tube. The current to the solenoid also flows through a variable resistor. In each part below, state whether the indicated change will cause the magnitude of the magnetic field inside the solenoid to increase, decrease, or remain the same.

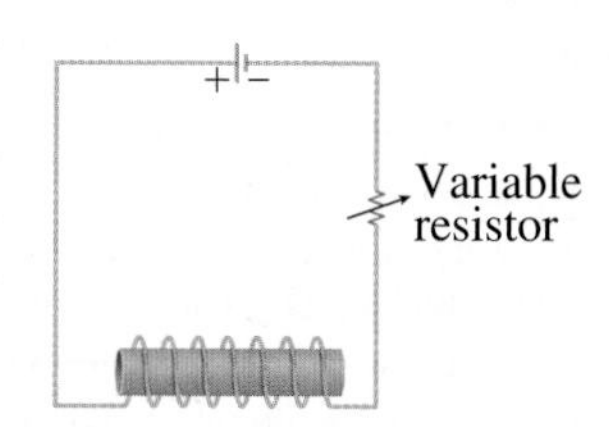

Figure 22-62 Question 22-66.

(a) The resistance of the variable resistor is decreased; no other variables are changed.

(b) An iron rod is placed inside the cardboard tube; no other variables are changed.

(c) The number of turns (per unit length) in the solenoid is decreased; no other variables are changed.

22-67 A proton enters a uniform magnetic field with a speed of 5.92×10^5 m/s as shown in Figure 22-63. The field in the region has a magnitude of 6.40×10^{-1} T. What is the force (magnitude and direction) on the proton?

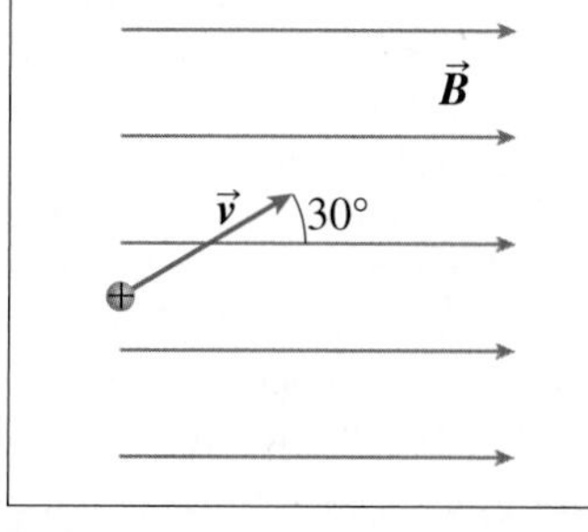

Figure 22-63 Question 22-67.

22-68 Repeat the previous question for an electron. All the other parameters of the problem remain the same.

22-69 If a charged particle travels in a straight line in a particular region, it is possible that the magnetic field there is zero. However, there are two other possibilities that involve a non-zero magnetic field. Describe them.

22-70 An electron travels along a circular path of radius 4.51 mm in a uniform magnetic field of magnitude 5.60×10^{-3} T. What is the speed of the electron?

22-71 What is the force (magnitude and direction) on the wire in Figure 22-64 if it carries a current of 0.78 A? The magnitude of the magnetic field is 3.4×10^{-5} T, and the length of the wire is 28 cm.

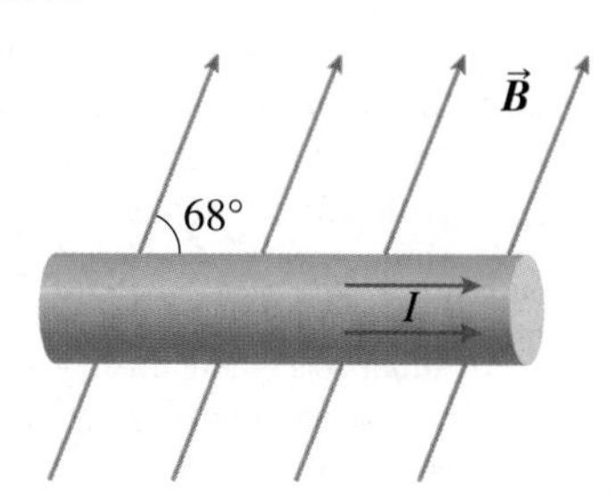

Figure 22-64 Question 22-71.

22-72 The angle (θ) between a current-carrying wire and a uniform magnetic field is varied from 0° to 90°. Sketch the shape of the graph of the magnitude of the magnetic force on the wire versus θ, if all other quantities remain constant.

22-73 Figure 22-65 shows a wire situated in a uniform magnetic field. If the force on the wire is in the direction shown, what is the direction of *electron flow* in the wire?

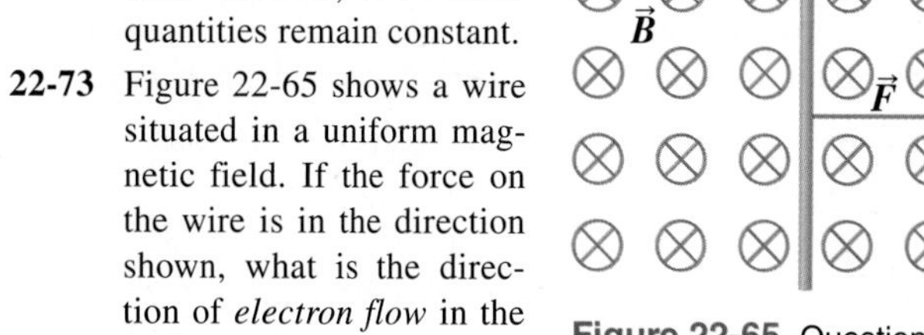

Figure 22-65 Question 22-73.

22-74 Suppose that you make a battery-powered motor from a coil of wire, two magnets, and a commutator, but find that the magnetic forces are weak. What changes could you make in your design to make the forces stronger?

22-75 Two long parallel wires carry currents in opposite directions.

(a) Do the wires attract, or repel, each other?

(b) How far apart are the wires if the force *per unit length* on each wire is 2.5×10^{-7} N/m in magnitude, when each wire carries a current of 1.5 A?

Applying Your Knowledge

22-76 In Section 22.2 we discussed the magnetic field at the centre of a single loop of wire, where the magnitude of this field can be determined by Eqn. 22-2. If we ran a current I through N loops of wire, all centred on the same point P as shown in Figure 22-66, what is the resulting magnitude of the magnetic field at P? The radius of all of the loops is considered to be the same (R), as shown.

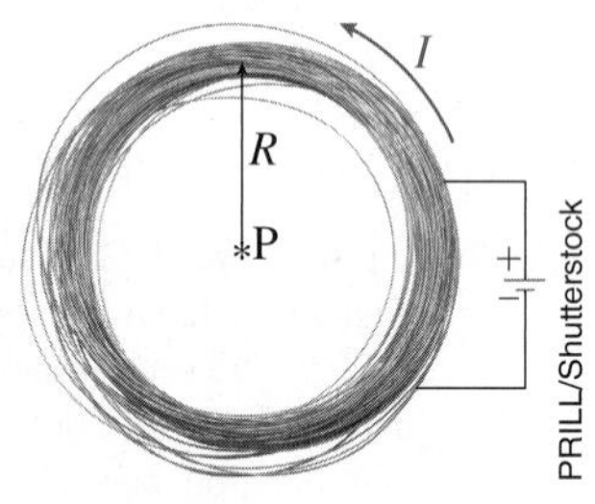

Figure 22-66 Questions 22-76, 22-77, and 22-78.

22-77 Using the result from Question 22-76, a current of 0.736 A flows through 10.0 circular loops of wire. The loops, each of radius 1.15 cm, are packed tightly together and have essentially a common centre. What is the magnitude of the field at the centre?

22-78 A tightly packed coil of wire, consisting of circular loops of radius 0.950 cm, is required to have a magnetic field magnitude no less than 1.00×10^{-4} T at its centre when a current of 1.20 A is flowing. Using the result from Question 22-76, what is the minimum number of loops necessary to generate this field strength?

22-79 A *relay* is a switch that uses a small current in an electromagnet to move a switch that closes a circuit with a much larger current flow. A relay is often used to turn on a large current in situations where this current is not to flow through the main switch. For example, the starter switch of a car (Figure 22-67) uses a relay so that the large starter current will not pass through the wires under the dashboard. A relay is simply an electromagnet, with a fixed iron core, that attracts a soft iron rod on a pivot when a small current flows through the electromagnet. When the rod moves, it closes the switch in the high-current circuit. Draw a diagram of a relay. Can you think of places other than automobiles where relays might be used?

Figure 22-67 The starter switch on a car uses a relay to protect the driver from large currents in the dashboard (Question 22-79).

22-80 Which end of the solenoid shown in Figure 22-68 is the north pole?

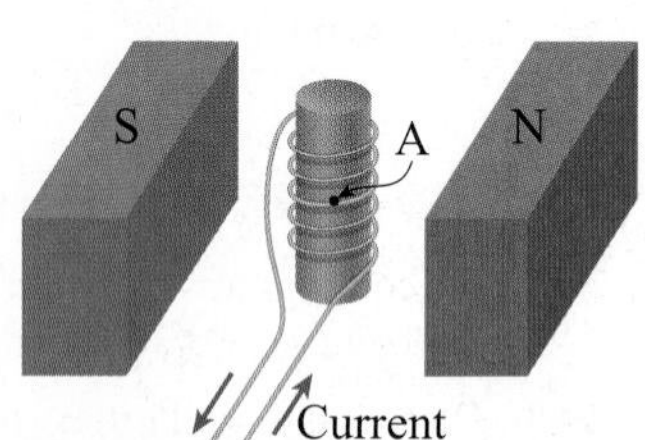

Figure 22-68 Questions 22-80 and 22-81.

22-81 Figure 22-68 shows a coil of wire wound around an iron armature that is free to rotate around an axis through A. Does the armature rotate clockwise, or counterclockwise, for the current direction shown? (**Hint:** Use your answer from Question 22-80 and think about the direction that the north pole of the solenoid will move relative to the external magnets.)

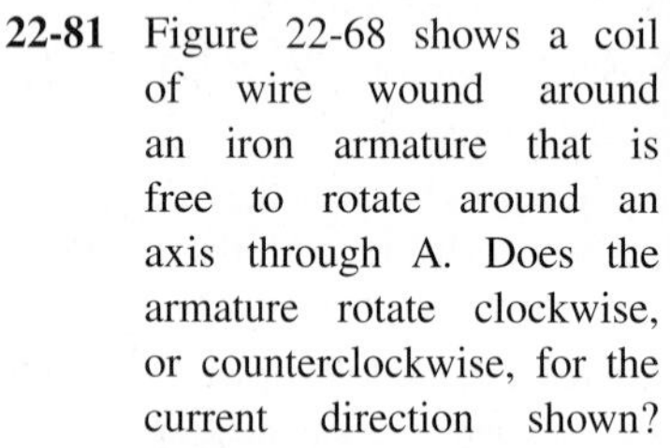

22-82 On Magnetic Island, Australia (Figure 22-69), the horizontal component of Earth's magnetic field is 3.27×10^{-5} T and the magnetic declination is 7.52° east (i.e., a compass points 7.52° east of geographic north). A solenoid having 125 loops in its 11.5 cm length is oriented with its axis east–west. A current of 25.5 mA flows through the solenoid in a counterclockwise direction as seen from the west, or down on the north side and up on the south side of the coil. If a horizontal compass is placed in the centre of the solenoid, in what direction will it point?

DID YOU KNOW?

Magnetic Island (Figure 22-69) is near the Great Barrier Reef off the coast of Queensland, Australia. Over two-thirds of the area on the island is protected as national park space. Magnetic Island was named by Captain Cook in 1770 as he believed that the island had an unusual magnetic field that was interfering with his compass readings. No one has been able to reproduce his observations or find reasons for an unusual magnetic field, but the name has stuck!

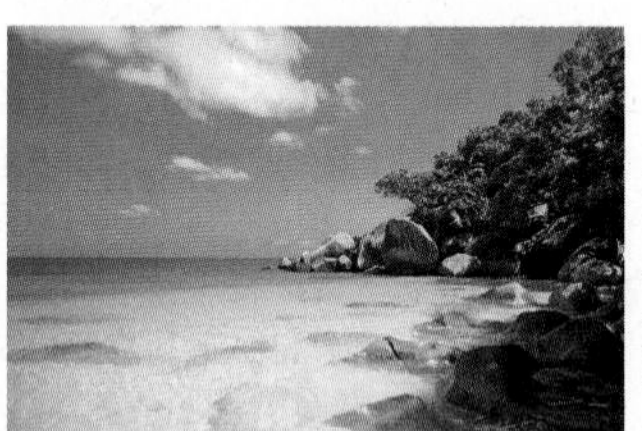

Figure 22-69 Magnetic Island, Australia (Question 22-82).

22-83 A proton with a kinetic energy of 23 keV follows a circular path of radius 24 cm in a uniform magnetic field. What is the magnitude of the field?

22-84 An α particle (charge $+2e$) has a circular orbit in a uniform magnetic field. Describe its subsequent motion if an electric field is also applied in the same direction as the magnetic field.

22-85 When currents of different magnitude travel in two long parallel wires that are 2.0 m apart, the *force per unit length* on each wire is 6.0×10^{-7} N/m in magnitude. When one current is increased by 1.0 A, and the other decreased by 1.0 A, with no other changes to the system, the magnitude of the force per unit length decreases to 4.0×10^{-7} N/m. What are the two original currents?

22-86 A common laboratory exercise for students is to measure the m_e/e ratio of an electron based on a similar experiment as that conducted by Thomson. The electrons are first accelerated from rest through a potential difference of ΔV, and then enter a region of magnetic field with a velocity that is perpendicular to the field. By varying the potential difference (ΔV) of the acceleration and the current that creates the magnetic field (I), students can measure the resulting radii of circular paths and calculate m_e/e from their data. (The magnetic field in the region is directly proportional to the current; i.e., $|\vec{B}| = cI$, where c is a constant of known value and appropriate units.) Derive an expression that relates the square of the radius (r^2) of the electron's path in the field to the m_e/e ratio for an electron, the accelerating potential difference (ΔV), and the current creating the magnetic field (I). How would you determine the m_e/e ratio if you plotted r^2 versus ΔV for a constant magnetic field? How would you determine the m_e/e ratio if you plotted r^2 versus $1/I^2$ for a constant accelerating potential difference?

22-87 Give examples of the use of magnetic fields in a modern vehicle (Figure 22-70).

Figure 22-70 Magnetism plays an important role in modern automotives (Question 22-87).

22-88 As mentioned earlier in the chapter, many modern devices use the Hall effect to measure the strength of a magnetic field. When a current-carrying conductor is placed in a magnetic field, the positive charge carriers will accumulate at one surface of the conductor due to the magnetic force (Eqn. 22-5) while the negative charges accumulate at the opposite surface, as shown in Figure 22-71. This charge separation across the conductor gives rise to an electric field and an electric force, which acts in the opposite direction to the magnetic force. An equilibrium is established when the electric force equals the magnetic force, resulting in a steady-state potential difference (ΔV) that can be measured across the conductor. Derive the relationship between the potential difference measured between the opposite surfaces at equilibrium for a given magnitude of magnetic field (B). Given that v is the speed of the charge carriers generating the current, express your relationship explicitly in terms of current (I) instead of v (see the derivation of Eqn. 22-6 on page 725). Your answer will contain geometry terms.

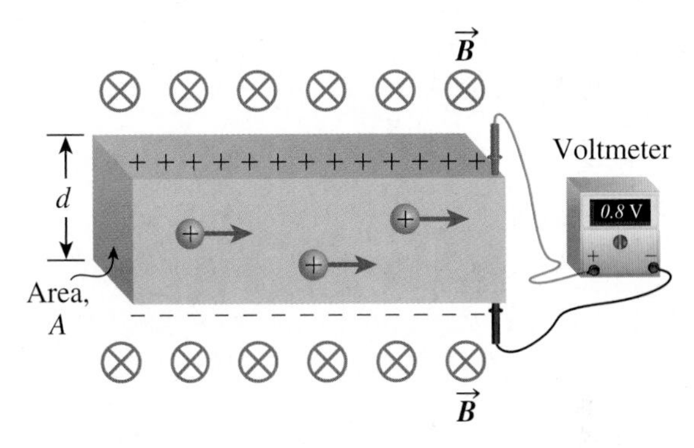

Figure 22-71 The Hall effect (Question 22-88).

DID YOU KNOW?

The Swedish car manufacturer Volvo has been experimenting with magnets embedded in the road as a means of guiding self-driving cars (Figure 22-72). The cars contain an array of magnetic-field sensors in order to detect the changes in field as they travel over the magnets. Such an approach would be costly to implement, but does have the advantage of being more resilient in challenging climates with snow, ice, and fog than self-driving cars that make use of GPS and optical sensors.

Figure 22-72 The use of magnets embedded in the road to guide self-driving cars.

Electromagnetic Induction 23

Jesus Keller/Shutterstock

Figure 23-1 Wind turbines cover the landscape in order to harness wind energy for electrical generation. The wind-blown turbine blades rotate coils of wire within a magnetic field in order to produce electric currents.

What do your electric toothbrush, an electric guitar, metal detectors at airports, and the brakes on a subway train have in common? In order to function, all these devices rely on the precise interplay of electric and magnetic fields, the subject of discussion in this chapter. In fact, virtually all the electricity that you use is generated by moving an electrical conductor in a magnetic field in a process called electromagnetic induction. In electrical power plants around the world, energy from a variety of sources is used to rotate huge coils of wire between poles of electromagnets to produce electric currents. The coils are rotated by falling water, high-pressure steam heated from the burning of fossil fuels (oil, gas, or coal) or from nuclear fission of uranium, or, more directly, by turbine blades turning in the wind (Figure 23-1). With all the innovations being explored in energy production around the world—from solar to geothermal, biomass to tidal—in all cases the technique still relies on the generation of electric currents in conductors moving through magnetic fields in the process known as electromagnetic induction. Furthermore, our whole electrical distribution system relies on transformers (Section 23.4) in order to minimize energy losses, which also operate through induction. In other words, the modern world would not exist without electromagnetic induction, as you will learn in more detail in this chapter.

CHAPTER **OUTLINE**

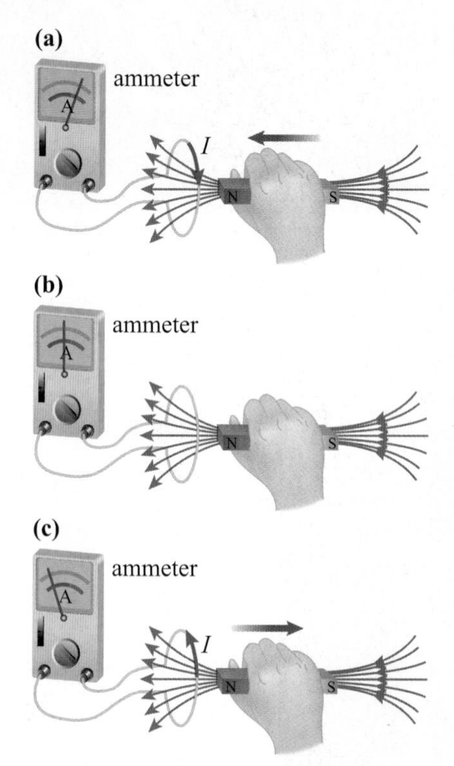

Figure 23-2 **(a)** Moving a magnet into a wire loop induces a current in the loop. **(b)** When the magnet and loop are stationary, there is no current in the loop. **(c)** Moving the magnet out of the loop results in a current in the opposite direction as that observed in (a).

electromagnetic induction the induction (creation) of an electric potential difference or voltage in a conductor by a changing magnetic field in the region

23.1 Producing an Electric Current with a Magnetic Field

After Ørsted's discovery in 1820 that an electric current produces a magnetic field, as discussed in Section 22.1, it was natural to ask if a magnetic field could produce an electric current. The answer—yes—was discovered independently in 1831 by Michael Faraday and Joseph Henry.

It is quite easy to demonstrate the creation of an electric current by a magnetic field. If you have access to the appropriate equipment, you can try some of the following demonstrations yourself. If a magnet is moved into a loop of wire (Figure 23-2 (a)), a current is induced (created) in the loop, as shown by the deflection of the pointer of the ammeter. If the same magnet is then moved out of the loop, a current is induced in the opposite direction (Figure 23-2 (c)). Note that the zero of the ammeter scale is straight up at the centre of the display, and that the deflections in Figures 23-2 (a) and (c) show opposite current directions. If the magnet and loop are both stationary in any position, there is zero current detected in the loop (Figure 23-2 (b)). It is important to note that it does not have to be the magnet that moves: a current is detected in the loop when it moves relative to the magnet as well (Figure 23-3).

The important feature of the above observations is that when the magnet is moving, the magnetic field due to the magnet is changing inside the loop. *Any time a magnetic field inside a closed circuit is changing, a current is induced in the circuit* (Figure 23-3).

What if the coil is not connected to an external circuit (Figure 23-4)? When the magnet moves, a current starts to flow along the coil itself with the result that one end of the coil becomes negatively charged and the other end positively charged. This current stops very quickly because of the electric repulsion of the negatively charged end and the negative electrons trying to flow toward it. Because the two ends of the coil have opposite net charges, there is an electric potential difference (i.e., a voltage) between them. This voltage would then have an associated current when the coil is connected to an external circuit (while the magnet or coil is moving). The changing magnetic field in the coil induces a voltage across the ends of the coil, and this in turn has an associated current if there is a closed circuit in which to flow. This process of inducing a voltage in a conductor (whatever its shape) whenever there is a changing magnetic field is called **electromagnetic induction**.

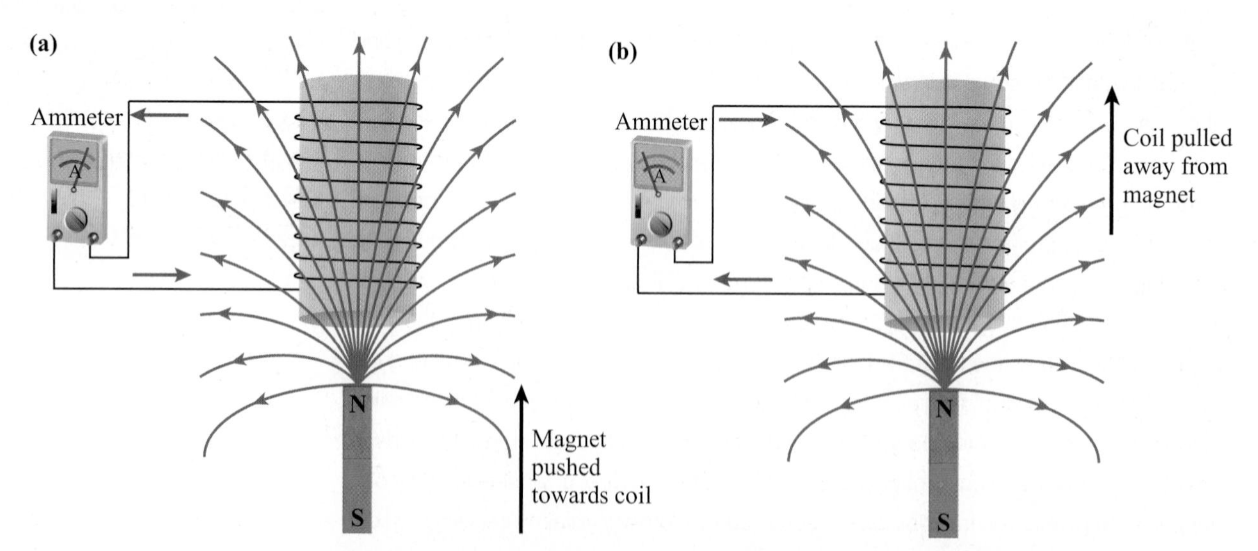

Figure 23-3 When there is relative movement between the bar magnet and the nearby coils, the magnetic field inside the coils changes. With this changing magnetic field, a current is induced in the coils. **(a)** When the magnet moves toward the coil, the direction of the resulting current is as shown. **(b)** When the coil moves away from the magnet, the direction of the resulting current is as shown.

Through extensive experimentation, Faraday discovered that the induced potential difference is directly proportional to three factors:

- the number of loops in the coil
- the rate at which the magnetic field inside the coil is changing
- the cross-sectional area of the coil (when the magnetic field is approximately uniform across the area)

Increasing any of these factors increases the induced potential difference: If the number of loops or the cross-sectional area is doubled, the voltage doubles. Similarly, if the rate at which the field is changing is doubled, the voltage doubles. The rate of changing field can be controlled through the strength of the magnet, the speed of the relative motion of the magnet or coil, and the relative positions of the magnet and the coil.

In all our discussions of induction thus far we have focused on a permanent magnet as the source of our changing magnetic field inside a coil or loop of wire. However, as we learned in Chapter 22, current-carrying wires are also a source of magnetic fields. Experimentally it has been observed that if we have a changing magnetic field due to a current-carrying wire, there will be an induced potential difference in a nearby coil or loop. For example, in Figure 23-5 the DC power supply is connected to a variable resistor; by changing the resistance of the circuit, we change the current in that circuit. This gives rise to a variable magnetic field inside the secondary coil on the right-hand side, thereby generating a current in this coil. However, if the current in the primary circuit is constant and there is no relative motion, there is no induced voltage in the nearby coil or loop of the secondary circuit.

Many microphones (Figure 23-7) use induction. In one model a small coil of wire is attached to a thin membrane and placed close to a small magnet. When sound waves hit the membrane, the membrane and coil vibrate, and the movement of the coil in the magnetic field induces a voltage across the ends of the coil. The frequency of the voltage matches the frequency of the sound waves, and thus the microphone produces an electrical signal corresponding to the sound.

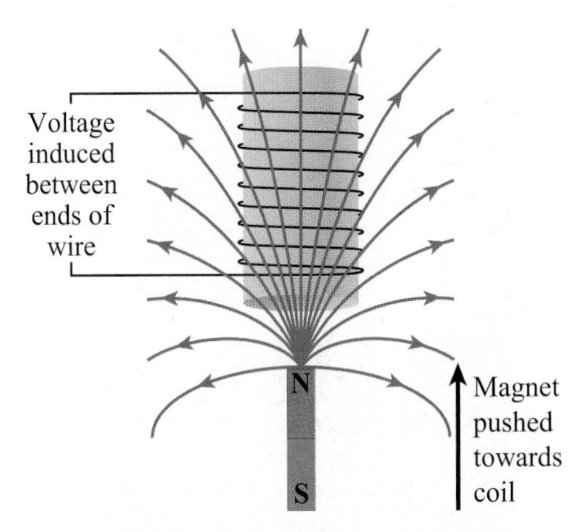

Figure 23-4 Changing the magnetic field inside a coil induces a potential difference (i.e., a voltage) between the ends of the coil.

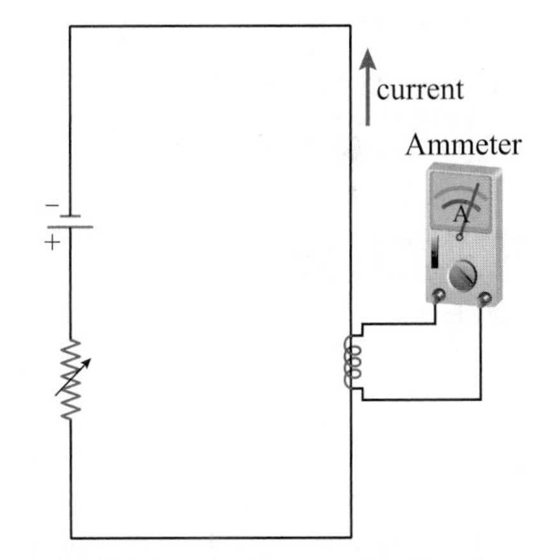

Figure 23-5 Changing the current in the primary circuit by changing the resistance results in a varying magnetic field inside the coil. This gives rise to an induced voltage and current in the secondary coil, as seen in the ammeter reading.

TRY THIS!

Use Electromagnetic Induction to Convert Your Energy into Electricity

An emergency radio or flashlight that uses a hand crank for power makes use of electromagnetic induction (Figure 23-6). The hand crank moves a coil inside a magnetic field, and the resulting induced current is used to power the device. You can experiment with such a radio by investigating the effect of different variables on the amount of play time you can generate: How long does the radio play for 10 turns versus 50 turns? How long does the radio play for 50 turns completed as quickly as you can versus 50 turns completed slowly? If you don't have access to a hand crank radio, you can also experiment with a hand crank flashlight or a "shake" flashlight. In all these devices your mechanical energy is converted into electrical energy by the movement of a coil through a magnetic field.

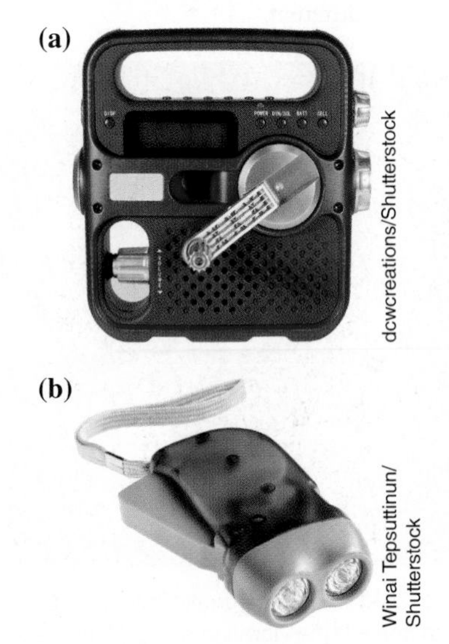

Figure 23-6 An emergency radio or a flashlight that use mechanical energy to generate power are examples of electromagnetic induction in action.

Electromagnetic Induction in a Uniform Magnetic Field

So far we have chosen our examples of electromagnetic induction to involve coils of wire because the effects can easily be observed. Although not as easy to demonstrate in the lecture hall or classroom, a potential difference can also be induced between the ends of a straight conducting wire or rod moving in a uniform magnetic field.

Paul Vinten/Shutterstock

Figure 23-7 Electromagnetic induction is used in many microphones.

slidewire generator a circuit in which a current is generated when a metallic rod or wire slides along stationary wires through a region of uniform magnetic field

This phenomenon can readily be understood in terms of the magnetic force on moving charges as we discussed in Chapter 22.

Figure 23-9 shows a conducting wire or rod moving to the right in a region in which there is a uniform magnetic field directed into the page. Each free (conduction) electron is moving from left to right across the magnetic field along with the rod and thus experiences a magnetic force. We can use a right-hand rule from Chapter 22 to determine that this force is toward the bottom end of the rod. The free electrons move in this direction until the accumulated negative charge at the end prevents further electron flow. Because of the flow of electrons along the rod, the ends acquire net charges: negative at the bottom, positive at the top. Thus, *a potential difference is established between the ends of the rod,* with the top end at a positive electric potential relative to the bottom end in this scenario, and we have another example of electromagnetic induction.

If the moving rod has sliding contacts with stationary wires in order to create a closed circuit (Figure 23-10), current flows in the circuit while the rod is moving. This configuration is called a **slidewire generator**. Since the magnetic force on a moving charge is proportional to the speed of the moving charge (Eqn. 22-4), it turns out that the induced potential difference is also proportional to the speed of the wire. Both the magnetic force and the induced potential difference are proportional to the magnitude of the magnetic field. As well, since a longer wire has more electrons to be affected by the magnetic force, the potential difference is proportional to the length of the sliding rod or wire. In equation form,

$$\Delta V = B v L \tag{23-1}$$

where ΔV is the potential difference between the ends of the rod, B is the magnitude of the uniform magnetic field in the region, and v and L are the speed and length of the rod that is moving perpendicular to the field.

Let's pause for a moment to examine Eqn. 23-1 with respect to the SI units on each side of the equation

$$[\text{V}] = [\text{T}][\text{m/s}][\text{m}]$$

Since 1 V = 1 J/C and 1 T = 1 N/(C·m/s) (from Eqn. 22-4), this becomes

$$[\text{J/C}] = [\text{N/(C}\cdot\text{m/s)}][\text{m/s}][\text{m}]$$

The metres per second cancel out on the right hand side, leaving

$$[\text{J/C}] = [\text{N/C}][\text{m}]$$

Since 1 J = 1 N·m, we have consistent units on the left and right side, as expected.

Remember that Eqn. 23-1 applies only to the slidewire generator scenario in which the sliding component moves perpendicular to the uniform magnetic field. As we discussed in Chapter 22, the magnitude of the magnetic force on a moving charge depends on the angle between the velocity and the field. The magnetic force (and therefore induced potential difference here) is greatest when that angle is 90° since $F = q\,v\,B\sin\theta$ (Eqn. 22-4). If the wire moves at some arbitrary angle θ relative to the magnetic field, instead of being perpendicular to it, the magnetic force on each free electron in the wire is reduced. Hence the induced voltage is smaller, decreasing to zero when the angle is 0° or 180°. Thus, when the sliding wire moves parallel (or anti-parallel) to the field lines, there is no magnetic force on the electrons, and no potential difference or current is induced.

DID YOU KNOW?

Coils of wire are positioned underneath each string of an electric guitar. The vibrating guitar string, which is made of a ferromagnetic material such as steel or nickel, induces a current in these coils with a frequency that matches that of the string (Figure 23-8). The oscillating current is then relayed to an amplifier to generate the corresponding sound. There are often two or three sets of such coils for each guitar, each with slightly different responses to the vibrating string. Depending on which set is connected to the amplifier, the sound coming out of the speakers can be quite different.

Kuzma/iStock/Thinkstock

Figure 23-8 Electromagnetic induction is used to convert the vibrating string of the electric guitar into an electrical signal for the amplifier.

Although the preceding discussion focuses on a straight wire or rod moving perpendicular to a uniform magnetic field, electromagnetic induction will also be observed with a variety of conductor geometries in regions of non-uniform magnetic field. The concept is the same: an induced voltage can be generated when any shape of conductor moves through a region of non-uniform magnetic field, depending on the orientation of the conductor relative to the field and the direction of motion relative to the field. The mathematics, however, get more complicated when we are not dealing with straight rods travelling at constant speed in a direction that is always perpendicular to a uniform magnetic field.

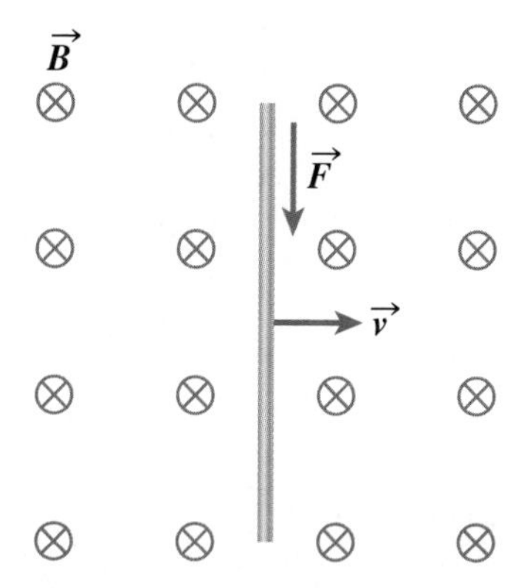

Figure 23-9 A straight wire or metallic rod moving through a region in which there is a uniform magnetic field. Each electron in the wire experiences a magnetic force toward the bottom end.

SAMPLE PROBLEM 23-1

A rod of length 0.15 m moves with a speed of 0.20 m/s in a uniform magnetic field of magnitude 3.6×10^{-3} T (Figure 23-11). The wire moves with a velocity perpendicular to the field, as shown.

(a) What is the voltage induced between the ends of the rod?

(b) Which end (X or Y) of the rod is at the higher electric potential?

(c) If the resistance R is 75 Ω, what is the current in the circuit?

(d) What would happen to the answers to (a), (b), and (c) if the velocity of the wire were reversed and the rod moved to the left through this region?

Solution

(a) The potential difference induced between the two ends of the rod is given by Eqn. 23-1:

$$\begin{aligned}\Delta V &= B\,v\,L \\ &= (3.6 \times 10^{-3}\,\text{T})(0.20\,\text{m/s})(0.15\,\text{m}) \\ &= 1.1 \times 10^{-4}\,\text{V}\end{aligned}$$

Thus, the induced voltage across the rod is 1.1×10^{-4} V while it moves at constant speed through the uniform magnetic field.

(b) To determine which end of the rod is at the higher electric potential, we need to find the direction of charge flow by using a right-hand rule (Chapter 22) to determine the direction of the magnetic force on charges in the moving rod. Since the charges are moving to the right along with the rod, we align the fingers of our right hand with their velocity and rotate our hand so that the fingers curl out of the page in the direction of the magnetic field. Our extended thumb, pointing down, now indicates the direction of force on a positive charge. This means that the bottom of the rod will become positively charged while the top becomes negatively charged. Therefore, point Y is at the higher electric potential.

(c) The current through the circuit can be determined from Ohm's law, $\Delta V = IR$. Thus,

$$\begin{aligned}I &= \frac{\Delta V}{R} \\ &= \frac{1.08 \times 10^{-4}\,\text{V}}{75\,\Omega} \\ &= 1.4 \times 10^{-6}\,\text{A}\end{aligned}$$

As a result of electromagnetic induction, the current through the circuit is 1.4×10^{-6} A while the rod moves through the magnetic field.

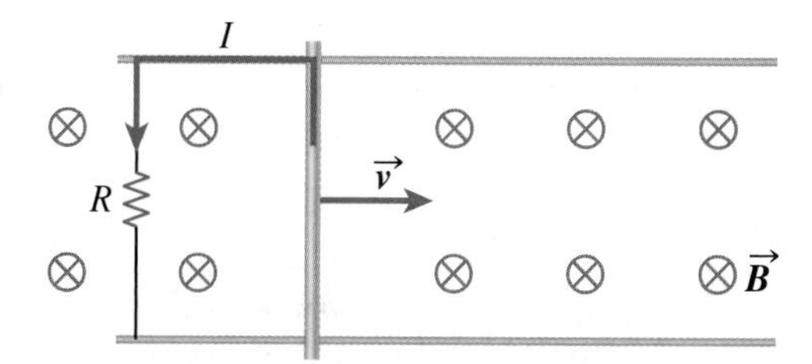

Figure 23-10 When the ends of the vertical moving rod are in contact with an external circuit, current flows while the rod is moving. This configuration is called a slidewire generator.

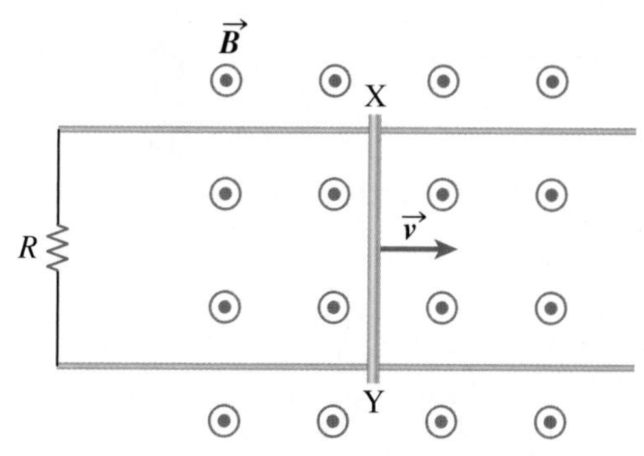

Figure 23-11 Sample Problem 23-1.

(d) If the wire's velocity were reversed, the magnetic force on the free electrons would be reversed (use a right-hand rule to check this), and therefore point X would be at a higher potential than Y. The size of the induced voltage and current would be the same as before, but the current would have the opposite direction. Therefore, the answers to (a) and (c) would not change, but the answer to (b) would be X instead of Y.

EXERCISES

23-1 A coil of wire forms a closed circuit with an ammeter (Figure 23-12). A magnet can be moved near one end (A) of the coil. When the north pole of the magnet is moved toward A, current flows in the circuit in a particular direction which is indicated on the ammeter as a positive value. Describe the current in the circuit when

(a) the south pole of the magnet is moved toward A
(b) the south pole is moved toward A, but faster than in (a)
(c) the north pole is held stationary near A
(d) the south pole of the magnet is moved away from A
(e) A is moved toward the south pole of the magnet
(f) the number of loops in the coil is increased, and the south pole of the magnet is moved toward A

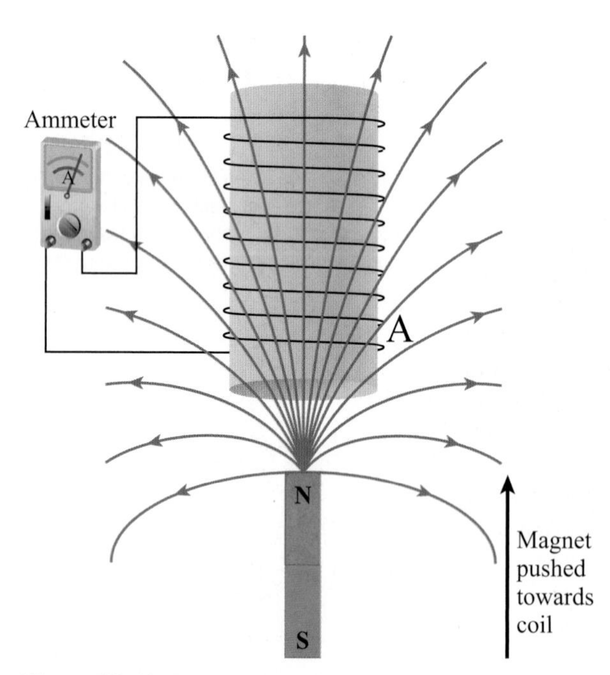

Figure 23-12 Questions 23-1 and 23-2.

23-2 A magnet is held close to one end of a coil (Figure 23-12). If it is moved away from the coil, a voltage is induced in the coil. If it is moved toward the coil at the same speed over the same distance, a voltage is again induced. Is there any difference between these two voltages?

23-3 A current-carrying wire passes through the centre of a coil (Figure 23-13). There is no relative motion of the current-carrying wire and the coil. Is there a voltage induced in the coil if

(a) the current is constant
(b) the current is decreasing

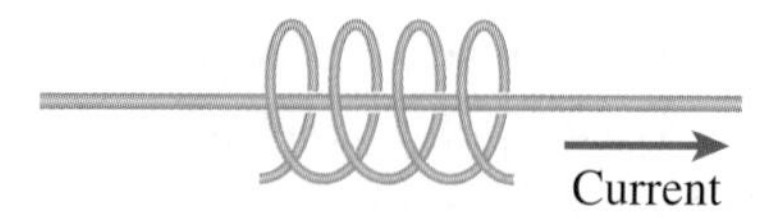

Figure 23-13 Question 23-3.

23-4 Figure 23-14 shows two circuits. In one, a coil is connected in series with a battery and a switch. In the second, a coil (very close to the first coil) is in series with a resistor. Does a current flow in this second coil when (a) the switch is being closed and (b) after the switch has been closed for some time? Explain your answers.

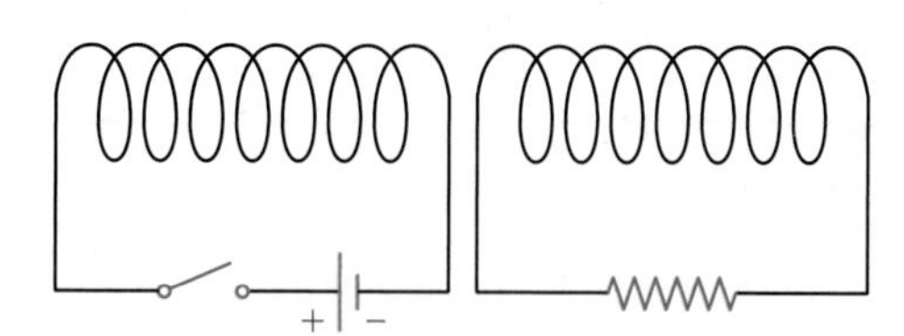

Figure 23-14 Question 23-4.

23-5 Figure 23-15 shows a wire (in cross-section) in a magnetic field. The field is directed to the right in the plane of the page and the wire is perpendicular to the page. Is there a potential difference between the ends of the wire if it moves

(a) toward W
(b) toward X
(c) toward Y

Explain your answers.

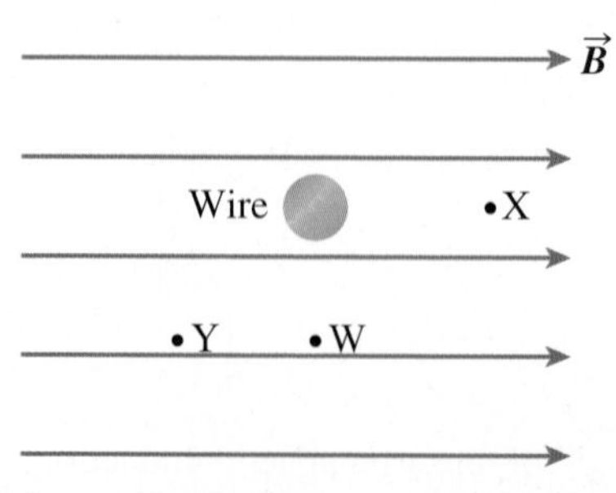

Figure 23-15 Questions 23-5 and 23-6.

23-6 In Figure 23-15 toward which point should the wire be moved to induce the maximum voltage?

23-7 A wire (in cross-section) between the poles of a horseshoe magnet is shown in Figure 23-16. Should the wire be moved toward A or B to induce a potential difference between the ends? Why?

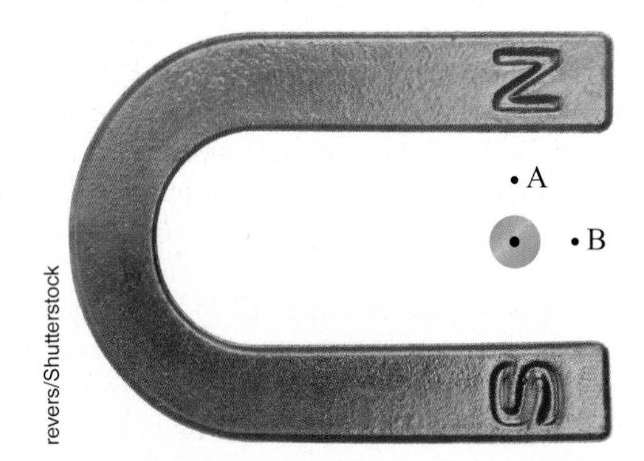

Figure 23-16 Question 23-7.

23-8 Figure 23-17 shows a moving rod MN that makes sliding electrical contact with stationary wires connected to a resistor. The rod, of length 3.5 cm, is moving perpendicular to a uniform magnetic field of magnitude 7.6×10^{-3} T.

(a) Which end (M or N) of the rod is at lower electric potential?

(b) In which direction does the current flow through the resistor?

(c) If the speed of the rod decreases, what happens to the current?

(d) If the current through the 52 Ω resistor is 2.3×10^{-6} A, what is the induced voltage in the wire?

(e) What is the speed of the wire?

(f) If the magnitude of the magnetic field were doubled, what would be the effect(s) on the system? Assume that the speed of the rod is unchanged.

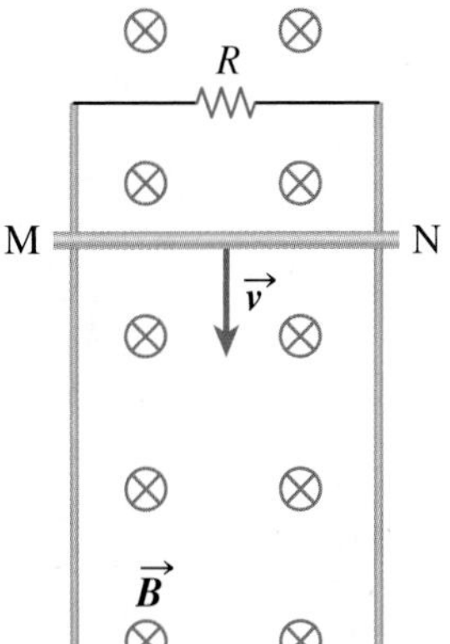

Figure 23-17 Questions 23-8 and 23-9.

23-9 If the magnetic field in the previous question were reversed in direction with all other variables remaining the same, what effect(s) would this have?

23-10 **Fermi Question:** How much power could be reasonably generated in a slidewire generator made out of a modified table tennis table (Figure 23-18)? How long would the power last? Where does this electrical power come from?

Figure 23-18 Question 23-10.

23.2 Lenz's Law

Lenz's law the direction of the current induced in a conductor will always oppose the change that produced this current

You might have noticed an interesting phenomenon if you performed the demonstrations with a magnet and coil mentioned in the previous section. If you move a magnet into a coil, you feel a force that resists the motion of the magnet. Similarly, when the magnet is pulled out of the coil, there is a force resisting the magnet's motion. These forces exist because the induced current in the coil sets up a magnetic field that opposes the motion of the magnet. Heinrich Lenz (Figure 23-19) summarized this concept in a statement now known as **Lenz's law**.

Lenz's Law: The direction of the current generated in a conductor due to electromagnetic induction will always oppose the *change* that produced that current.

No matter how many times you read that sentence, it is bound to sound a little confusing. It is easiest to understand Lenz's law by going through some examples. Let's start by returning to one of the scenarios we discussed in Section 23.1, that of moving the north pole of a bar magnet toward (or into) a coil (Figure 23-3 (a)). As we discussed, the ammeter demonstrates that a current is induced in the coil when the magnet moves toward it; therefore there is another magnetic field generated due to this

Figure 23-19 Heinrich Lenz (1804–1864).

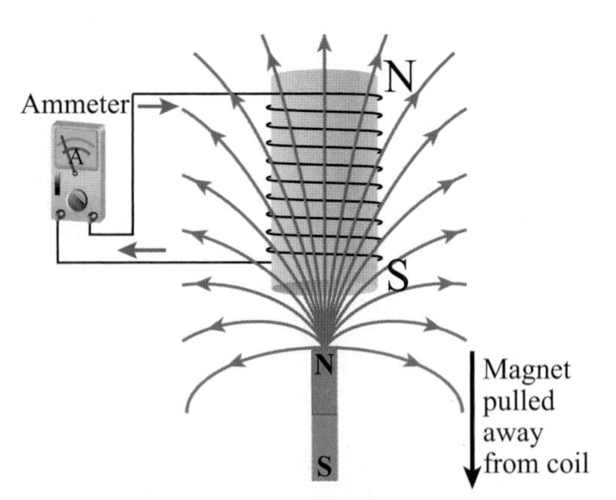

Figure 23-20 The induced current in the coil creates a south pole near the bar magnet's north pole; the resulting magnetic attraction opposes the motion of the magnet away from the coil.

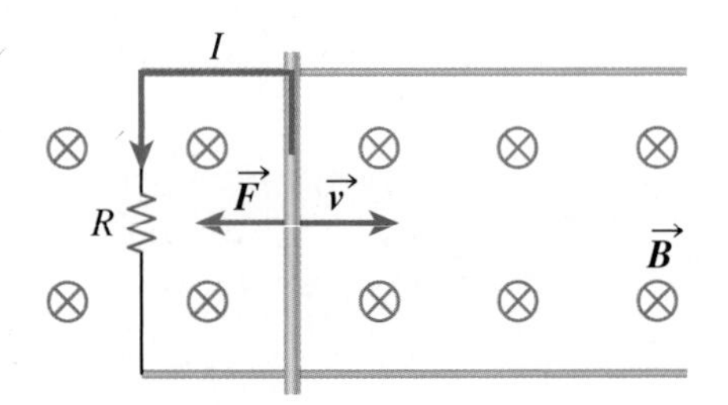

Figure 23-21 The magnetic force on the rod arising from the induced current acts to slow the moving rod down.

current. Given the direction of the induced current as indicated by the ammeter, with the appropriate right-hand rule from Chapter 22 we conclude that the magnetic field associated with the coil has a north pole at the bottom, near the north pole of the bar magnet. These like poles will exert repulsive forces on each other, opposing the motion of the bar magnet that is causing the induced current in the coil. That is, the induced current has a direction such that it opposes the change (i.e., the motion of the magnet) that created the current.

If the magnet's north pole is now moved away from the coil (Figure 23-20), the induced current in the coil is in the opposite direction of that observed in the preceding scenario. Therefore, based on the right-hand rule, the north pole of the coil is now at the top, and the south pole is close to the north pole of the bar magnet. The attraction between the south pole of the coil and the north pole of the bar magnet resists the magnet's motion away from the coil.

Let's also use Lenz's law to look at the case of a metallic rod moving in a uniform magnetic field (Figure 23-21) such as the slidewire generators discussed in Section 23.1. The rod makes electrical contact with stationary wires so that current can flow in a rectangular loop. We saw in Section 23.1 that if the rod is moved to the right in a magnetic field pointing into the page, a current is induced up through the rod and around the loop in a counterclockwise sense (Figure 23-21). This current itself is of course a flow of (positive) charge, and since there is an external magnetic field, there is now a magnetic force acting on the rod as it moves to the right. With the directions of current and magnetic field shown in Figure 23-21, the force on the moving rod (by a right-hand rule) is to the left, that is, in the direction opposite to the rod's velocity. In other words, the direction of the induced current is such that a magnetic force is developed that opposes the change (the motion of the rod) that created this current. In order to move the rod, an external force to the right must be continually provided. Work is done by this force as the wire moves, and hence energy is supplied, which appears as thermal energy in the circuit as current travels through the resistor.

Through this last example it is perhaps most clear that Lenz's law is a natural consequence of the important principle of conservation of energy. If the current in the metallic rod was generated in the opposite direction as a result of electromagnetic induction, there would be a magnetic force in the same direction as the rod's initial velocity. This would therefore accelerate the rod, which represents an increasing kinetic energy with no corresponding decrease in energy of another form—which would be great for solving the world's energy supply needs but is sadly not physically possible!

Lenz's Law in Action

You will need a steel/iron ball that is approximately 1.5 cm in diameter as well as a spherical or cylindrical neodymium magnet of a similar size. You will also need a piece of copper pipe that is approximately 30 cm in length with a diameter that is just a bit larger than the magnet and the steel ball. All these materials should be readily available at a local hardware store or online for minimal cost.

With the copper pipe secured vertically in a clamp or held vertically by a partner, drop the iron ball through the pipe. (Watch your toes!) Time how long it takes to come out at the bottom of the pipe. Now drop the spherical/cylindrical magnet through the pipe and time its fall through the same height. What is happening?

DID YOU KNOW?

The braking mechanism on free fall rides (Figure 23-22) such as "Drop Tower" at Canada's Wonderland is a large-scale version of the Try This! activity described here. If you conducted the experiment, you would have observed that without any external energy source required, the magnetic sphere/cylinder was significantly slowed in its descent. From this activity we can conclude that braking systems based on electromagnetic induction are incredibly safe: they will always work even if there is a total power failure—definitely an important design feature in an amusement park ride!

Figure 23-22 The braking systems on free fall rides are Lenz's law in action.

EXERCISES

23-11 Figure 23-23 shows a conducting coil wrapped around a hollow cardboard tube. The south pole of a magnet is pushed toward the coil.

(a) Based on Lenz's law, what is the polarity of the induced magnetic field associated with the current in the coil; that is, which end of the coil is the north pole and which end of the coil is the south pole?

(b) Does the current flow into or out of the coil at B?

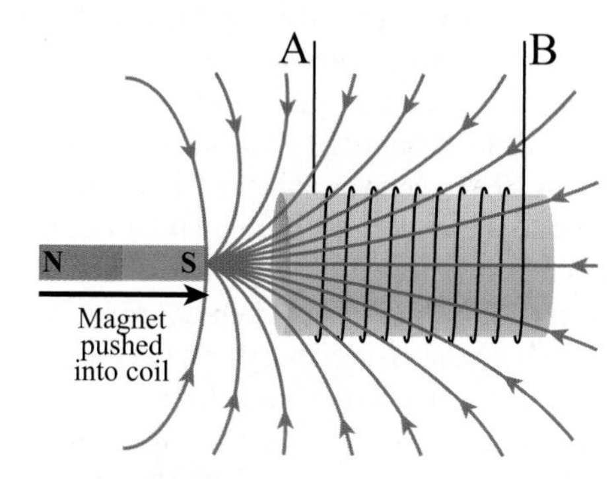

Figure 23-23 Question 23-11.

23-12 Repeat the previous question if the south pole of the magnet is pulled away from the coil with the same general configuration.

23-13 A coil whose axis is vertical has an induced current that is clockwise when viewed from the top. Describe two ways that this current could have been induced by moving a bar magnet near the top of the coil.

23-14 The north pole of a magnet is moved vertically down, toward the top end of a vertical coil. When viewed from the top, is the resulting *electron flow* clockwise or counterclockwise around the coil?

23-15 Figure 23-24 shows a rod moving in a uniform magnetic field.

(a) What is the direction of the induced current in the rod?

(b) What is the direction of the magnetic force on the rod due to this current?

(c) In what direction must an external force be applied to keep the rod moving at constant velocity?

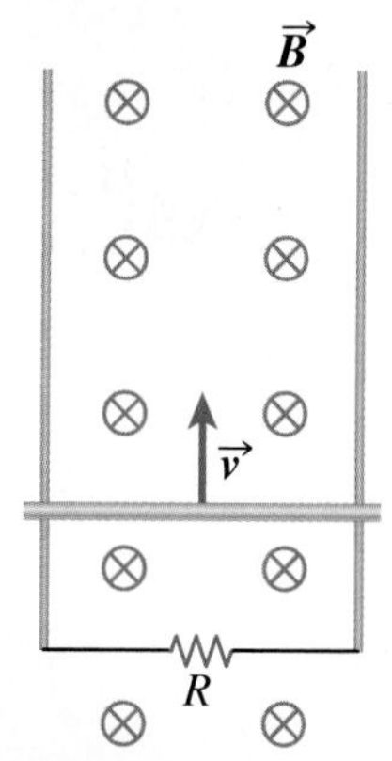

Figure 23-24 Questions 23-15 and 23-16.

23-16 In the previous question, the external force provided to keep the rod moving at constant speed does positive work on the wire, and therefore supplies energy to the system. Into what form(s) of energy does this go? Neglect friction. (We will investigate the work done by the external force in a slidewire generator quantitatively in Question 23-54.)

23-17 Figure 23-25 shows the current direction in a rod moving in a direction that is perpendicular to a uniform magnetic field in a slidewire generator.

(a) Is the wire moving toward the top or the bottom of the page?

(b) What is the direction of the magnetic force on the rod due to the induced current flow?

(c) In what direction must an external force be applied in order to maintain the motion of the rod at constant speed?

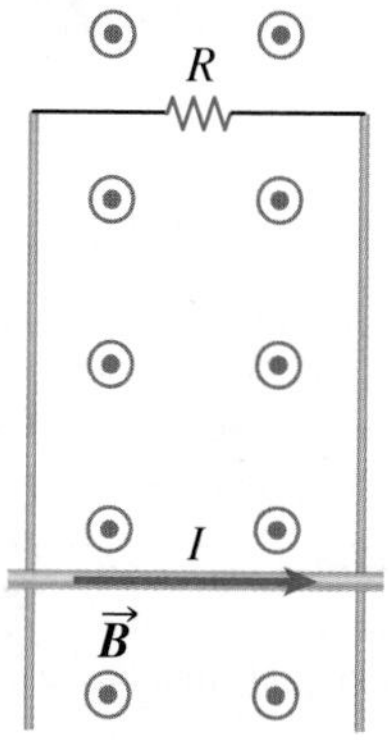

Figure 23-25 Questions 23-17 and 23-18.

23-18 (a) If you were asked to calculate the induced current in the previous question, which of the following quantities would you need to know?

i. the diameter of the rod
ii. the length of the rod
iii. the resistance of the resistor
iv. the speed of the rod
v. the magnitude of the magnetic field

(b) If you were asked to calculate the magnitude of the magnetic force acting on the rod due to the induced current in the previous question, which quantities would you need to know?

i. the diameter of the rod
ii. the length of the rod
iii. the resistance of the resistor
iv. the speed of the rod
v. the magnitude of the magnetic field

23.3 Electric Generators

In Chapter 22 we saw that an electric motor is constructed by making use of the force exerted on a current in an external magnetic field; this magnetic force is used to cause rotation. In this sense the electric motor is converting electrical energy (the current in the coil) into mechanical energy (the rotation of the coil) through a magnetic force. The construction of electric generators is physically very similar to that of motors, but the operation is essentially the reverse. In an **electric generator**, a conducting

electric generator a device that converts mechanical energy to electrical energy through the motion of a conducting coil through a magnetic field

Figure 23-26 The old-fashioned waterwheel is the predecessor to modern hydroelectric generators.

coil of wire is forced to rotate in a magnetic field, and a voltage is induced in the coil. Assuming that the coil is connected to an external circuit, an induced current is generated in the coil and circuit. In essence, mechanical energy (the rotation of the coil) is converted into electrical energy (the induced current in the coil) through electromagnetic induction. The generator coils in electrical power plants are rotated by a number of possible means, such as falling water (hydroelectric power), by high-pressure steam (coal, natural gas, nuclear power), wind turbines, etc. The waterwheel shown in Figure 23-26 is the pre-industrial revolution predecessor of hydroelectric generators; such a waterwheel used the potential energy of the falling water to physically turn mechanical industrial equipment, whereas a modern day hydroelectric station uses the falling water to turn generator coils, thereby providing electrical power for a vast array of industrial and household equipment.

We discuss the basic operation of a generator by considering a single rectangular loop being rotated in a magnetic field (Figure 23-28) by mechanical means. Each of the two long wires at the end of the loop is attached to a metal ring called a **slip ring** (R_1 and R_2). As the loop is rotated, the slip rings rotate as well. Stationary **metal brushes** (B_1 and B_2) provide electrical contact between the slip rings and an external circuit just as was described in Section 22.4 when discussing electric motors. In Figure 23-28, slip ring and brush R_1 and B_1 are always connected to the wire along BC, whereas R_2 and B_2 are always connected to the wire along EF.

In Figure 23-28 (a), the wire segment BC is moving upward because of the mechanical clockwise rotation of the loop about its axis. The magnetic field lines

slip ring the rotating metal ring providing electrical contact between the rotating coil in a motor or generator and the external circuit to which it is connected

metal brushes the stationary electrical contact between the slip ring of the motor or generator and the external circuit to which it is connected

DID YOU KNOW?

There is a generator under the hood of your car that operates in much the same fashion as described here. The alternator converts the mechanical energy of the movement of pistons into electrical energy in the form of an alternating current. The pistons are connected to the crankshaft, which converts the up-and-down motion into rotational motion, and the crankshaft spins the rotor shaft inside the alternator (Figure 23-27). The motion of the conducting loops inside the magnetic field of the alternator results in an induced current that is used to charge the battery, power the headlights, etc.

Figure 23-27 The generator inside your car engine, known as the alternator, converts the mechanical energy of the pistons into a current through electromagnetic induction.

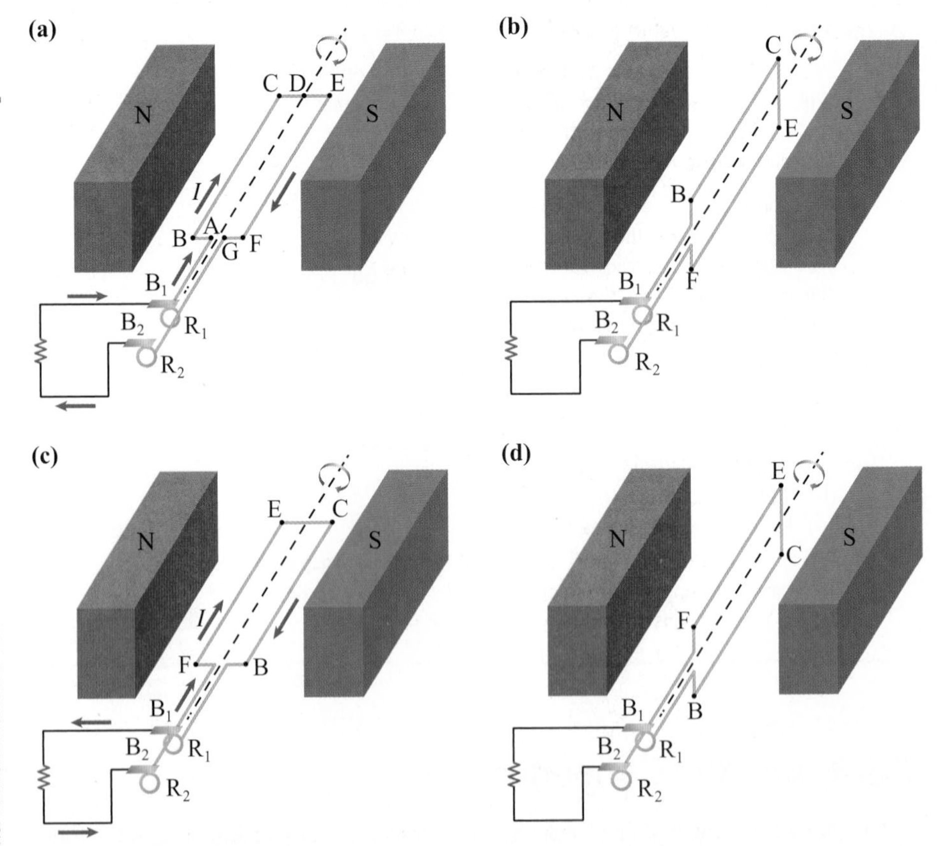

Figure 23-28 A single-loop electric generator. **(a)** As wire segment BC moves upward and EF moves downward, current is induced in the direction shown. **(b)** When the loop is vertical, the wires are moving parallel/anti-parallel to the field and, momentarily, there is no induced current. **(c)** When the loop has rotated through 180° from that shown in (a), the current is in the opposite direction in the external circuit. **(d)** After a further rotation of 90°, the loop is again vertical and, momentarily, there is no induced current.

point from left to right (north to south), and by a right-hand rule the magnetic force on the free electrons in BC is toward B. Thus, electrons flow toward B, and the conventional current is said to flow toward C. Similarly in segment EF, moving downward because of the rotation, the induced current is toward F (check this). Hence, the movement of segments BC and EF each induce a current in the direction shown in the loop and the external circuit. At this point in the rotation of the loop the current is at a maximum, since BC and EF are moving perpendicular to the field lines.

Figure 23-28 (b) shows the loop after rotation through 90°. In this vertical orientation, segments BC and EF are moving parallel or anti-parallel to the magnetic field lines. Therefore, as the loop moves through this point in its rotation, there is no induced voltage or current.

The position of the loop after rotation through a further 90° is shown in Figure 23-28 (c). This is essentially the same as Figure 23-28 (a) except that the positions of BC and EF have been reversed. Thus, the current is at a maximum again as the wires are moving perpendicular to the field at this moment, but now the current is travelling from F to E and from C to B, that is, in the direction opposite to that in Figure 23-28 (a). Therefore, the current in the external circuit has reversed its direction. Thus, we have an **alternating-current (AC) generator** that produces a current that reverses direction with each half-turn of the coil. Figure 23-28 (d) shows a further rotation of 90°, which results in the loop being vertical again with, momentarily, no current flowing.

alternating-current (AC) generator an electric generator in which the rotation of a coil through a magnetic field gives rise to a current in an external circuit that reverses direction with each half-turn of the coil

A graph of the induced current in the coil (and external circuit) versus time is shown in Figure 23-29. It alternates back and forth in direction as the coil rotates, passing through zero current when the loop is vertical and having maximum value when the loop is horizontal. If the loop is rotated at a constant rate, the graph has sinusoidal dependence, with the frequency of oscillation dependent on the rate of rotation of the coil. This is the physics underlying all commercial generation of electricity, such as hydroelectric power (Figure 23-30). Here the falling water provides the energy needed to turn the loops.

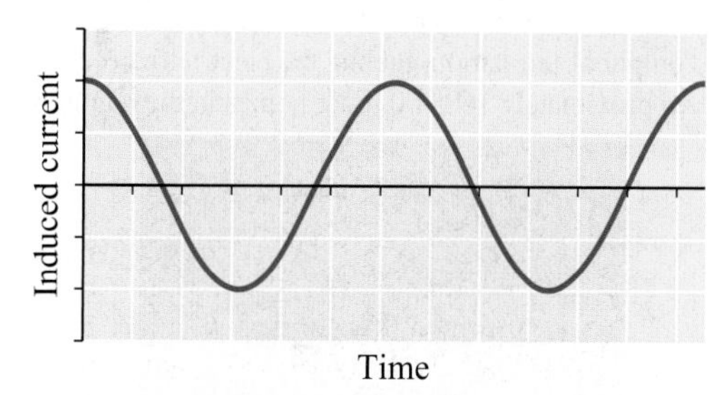

Figure 23-29 Sinusoidal variation of the induced current produced by an AC generator when the coil rotates at a constant rate.

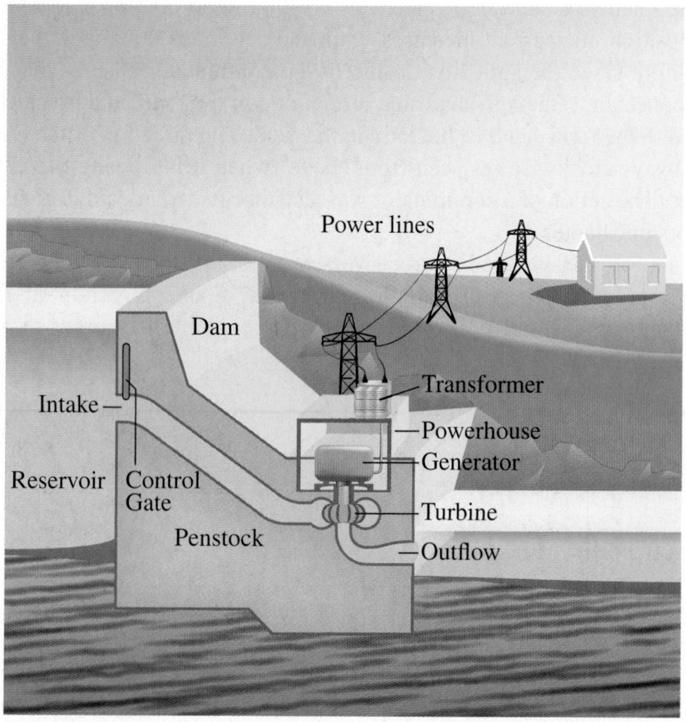

Figure 23-30 The basic design of a hydroelectric generating station.

TRY THIS!

Operate Your Own AC Generator

The PhET simulations developed at the University of Colorado have a great activity centred on all aspects of electromagnetic induction. Just search for "PhET generator simulation" and you will find an interactive animation that allows you to control (a) the strength of the magnetic field, (b) the mechanical rate of rotation, (c) the size and number of loops of the receiving coil in which you are generating the induced current, and (d) whether your indicator of current is a light bulb or an ammeter. There is also a tab here that allows you to play around with transformers ... more on that shortly!

Figure 23-31 A hydroelectric generating station in the Canadian Rockies.

direct-current (DC) generator an electric generator in which the rotation of a coil through a magnetic field gives rise to a current that travels in one direction only

DID YOU KNOW?

The site of the first commercial, large-scale, hydroelectric power plant, Niagara Falls, Ontario, is still a major supplier of electricity in the region. In 2013, the 10.2 km Niagara Tunnel Project was completed (Figure 23-32), a major engineering feat that funnels additional water to the generating station from the Niagara River at a rate of 500 m^3/s. This additional water flow increases the power output of the station, fueling the electricity needs of approximately 160 000 more homes in the region.

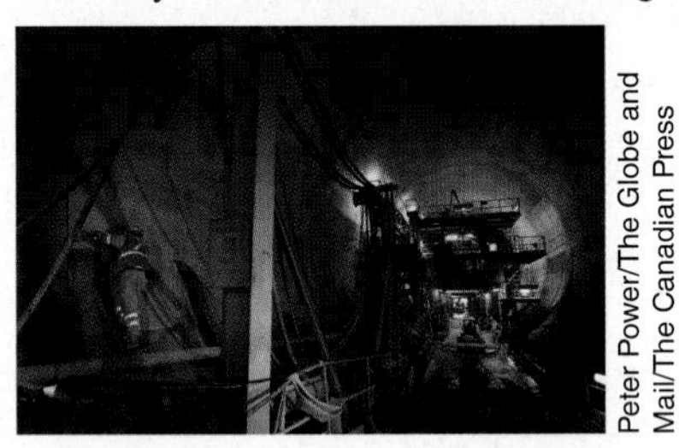

Figure 23-32 A construction worker involved in the final stages of the 10.2 km Niagara Tunnel Project.

Commercial Generation of Electricity

Commercial generators in electrical power plants use several sets of coils, each with a large number of windings, and they usually have electromagnets instead of permanent magnets. In North America the generators are rotated at a frequency of 60 Hz, but 50 Hz is used in many countries. New generators have been getting physically larger and their electrical power output has been increasing. In 1930, the largest generator had a power output of 200 MW (MW = megawatt = 10^6 watts), and the average was about 20 MW. Now there are generators with outputs in *excess* of 1000 MW.

The basic design of a hydroelectric station is illustrated in Figure 23-30 on page 745, and a photograph of such a station is Figure 23-31. Gravitational potential energy of the water at the top of the dam is converted to kinetic energy of the water in the tunnel, which in turn is changed to rotational kinetic energy of the generator, and finally to electrical energy. In a thermal-electric plant, high-pressure steam is produced by heating water which is then used to turn the generators. In a nuclear power plant, again water is changed to steam, in this case by the heat given off in the nuclear fission (breaking apart) of uranium; this will be discussed in greater detail in Chapter 24. Regardless of the type of plant, the production of electricity by the rotation of the generators is the same. The differences lie only in what is used to turn the generators.

DC Generators

In some circumstances, circuit components require current that travels in only one direction. With a slight modification in design, these AC generators will generate direct current instead. Specifically, if the slip rings of an AC generator are replaced by a split-ring commutator, we have a **direct-current (DC) generator**, one that produces current in only one direction. A single-loop DC generator is illustrated in Figure 23-33, and at the instant shown, the induced current is in the direction indicated with brush B_1 in contact with commutator segment C_1, and brush B_2 in contact with C_2. When the loop has rotated by 90° from the position shown, it is oriented vertically, the current drops to zero, and the brushes (B_1 and B_2) begin to make contact with the other commutator segments (C_2 and C_1 respectively). As the loop continues to rotate past the vertical position, the current increases again and continues flowing in the same direction as before since the split ring design of the commutator ensures that brush B_1 is always connected to the left-hand side of the loop in the field, and brush B_2 is always connected to the right-hand of the loop in the field. The direction of the current in the loop is always clockwise (as seen from above) when it is flowing based on a right-hand rule. (The action of a commutator was also discussed in detail in connection with DC motors in Chapter 22.)

A graph of the induced current versus time for a single-loop generator is given in Figure 23-34. Although the current is in one direction, it is certainly not constant (as it would be from a battery, for example). In order to generate an

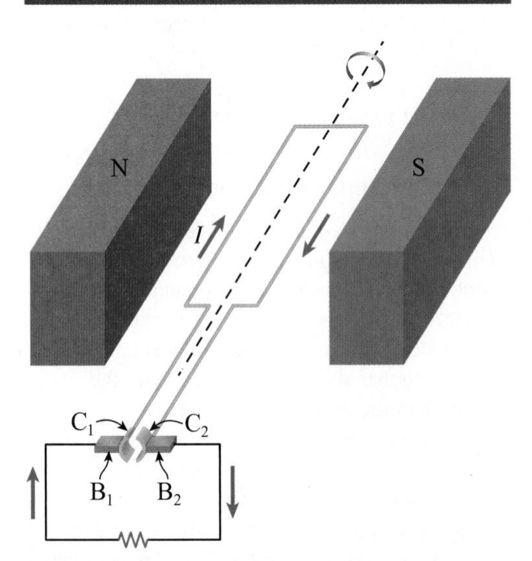

Figure 23-33 A single-loop DC generator. The design of the commutators (C_1 and C_2) produces a current in only one direction.

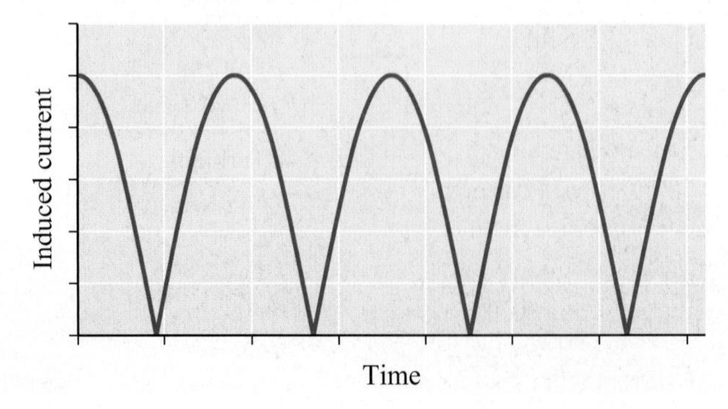

Figure 23-34 Induced current for a single-loop DC generator.

almost constant current, commercial DC generators have many sets of windings and corresponding commutator segments (Figure 23-35 (a)). At any given phase of the rotation, the brushes make contact with only those commutator segments connected to loops that can provide almost maximum current. Hence, the induced current is essentially constant, as illustrated in Figure 23-35 (b).

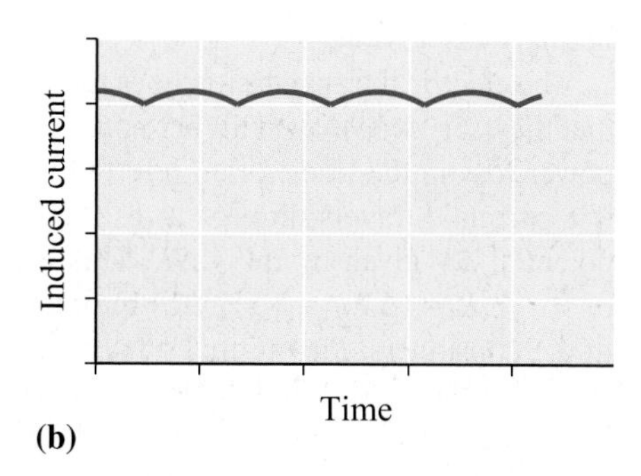

Figure 23-35 **(a)** A DC generator with many sets of windings and commutator segments. **(b)** The resulting induced current.

TRY THIS!

Convert Your DC Motor into a Generator

In Section 22.4 we included instructions on building your own simple electric motor to convert electrical energy into rotational motion. Given the similarities in construction between a motor and a generator, we can make a few modifications to your motor now to generate electricity from mechanical energy (Figure 23-36).

Replace the battery with a small inverted cardboard or plastic box, something that will provide structural support for the sewing needles and coil of wire. Secure the magnet to the top of the box and the needles to the sides for stability using modelling clay. Now connect the sewing needles to the leads of an LED light bulb with electrical tape. Your generator is ready to go, you just need to provide some mechanical energy to rotate the coil. You can do this in many ways; one idea is to use a straw to blow across the top side of the coil. Congratulations, you have just made your very own wind turbine! (**Note:** If you have difficulty getting the bulb to light up, make sure you have the leads connected properly: the longer lead of the LED bulb is the + terminal.)

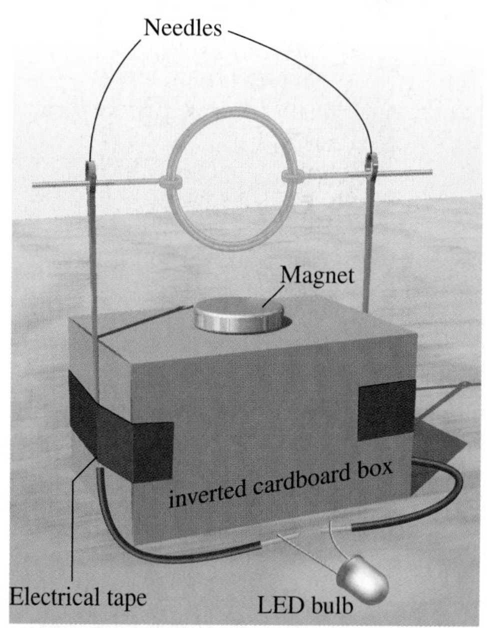

Figure 23-36 Your homemade DC generator using electromagnetic induction to convert mechanical energy into electrical energy.

EXERCISES

23-19 What is the direction of the induced current through the resistor in the external circuit due to the single-loop AC generator in Figure 23-37 at the instant shown?

23-20 What happens to the maximum value of the voltage produced by an AC generator if

(a) the number of loops in the coil is increased;

(b) the rotation rate of the coil is increased;

(c) a stronger magnet is used;

(d) a soft iron core is used in the coil instead of an air core?

You can make use of the PhET simulation referenced in the Try This! activity on page 745 to test out your answers to parts (a), (b), and (c).

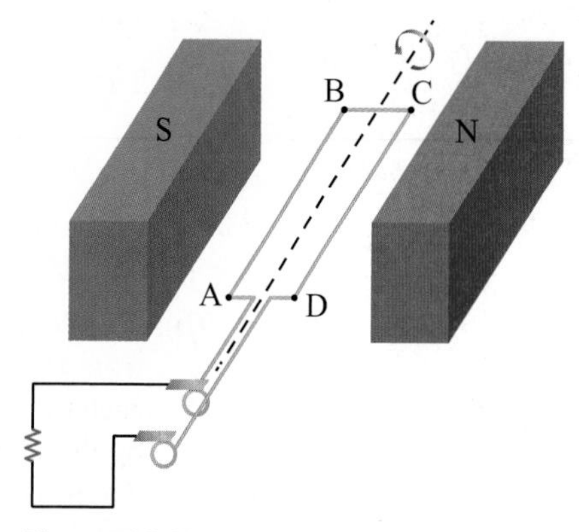

Figure 23-37 Question 23-19.

23.4 Transformers

transformer device that changes (transforms) voltage

You are probably aware of transformer stations (Figure 23-38 (a)) in your community, and transformers on utility poles (Figure 23-38 (b)) in local neighbourhoods. What are transformers? Why do we use them? The answer to the first question is easy: a **transformer** is a device that changes (transforms) voltage, and we will describe how it does this later in the section. The answer to the second question requires some preliminary discussion.

The electrical transmission lines that connect generating stations and the consumers that they serve have a small, but nonzero, resistance. Therefore, there is some thermal power loss in the lines, often referred to as "I^2R" loss. (Recall from Chapter 21 that if a current I travels through a resistance R, there is a resulting decrease in electric potential ΔV given by $\Delta V = IR$. This is converted to thermal energy (heat) at a rate of $P = I^2R = (\Delta V)I = (\Delta V)^2/R$.) In order that as much power as possible be delivered to the consumers, the thermal power lost in the transmission lines must be small. As shown in the sample problems that follow, this requires large transmission voltages. Transformers are used to increase voltage for long-distance transmission, and then to decrease voltage for safe distribution to consumers.

AlexKZ/Shutterstock

(a)

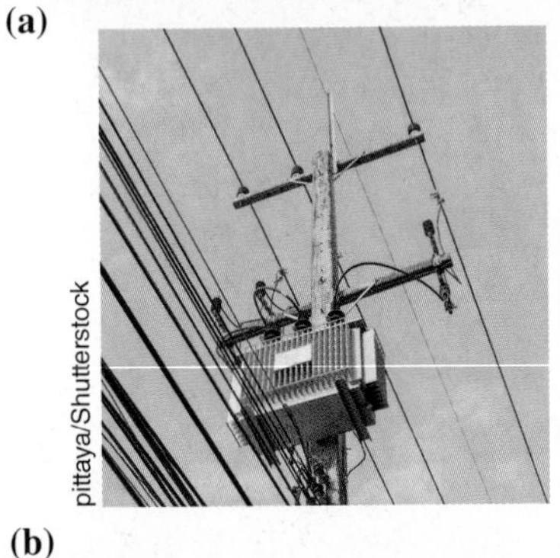
pittaya/Shutterstock

(b)

Figure 23-38 **(a)** A neighbourhood transformer station. **(b)** Transformer on a utility pole.

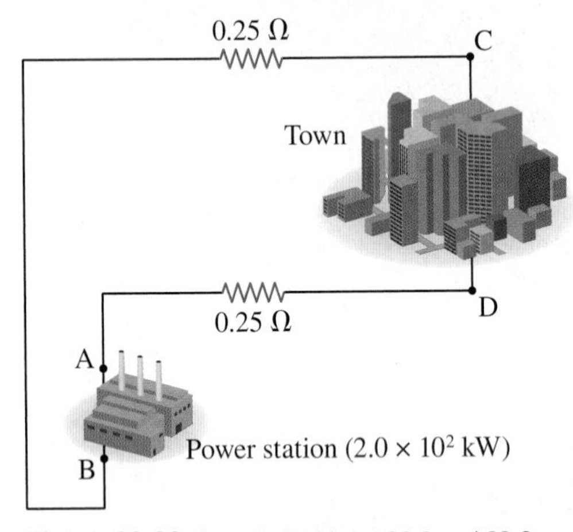

Figure 23-39 Sample Problems 23-2 and 23-3.

SAMPLE PROBLEM 23-2

Suppose that a transmission line connecting a small generating station to a town consists of a pair of wires, each with a resistance of 0.25 Ω (Figure 23-39). The power and voltage supplied by the electrical plant are 2.0×10^2 kW and 1.0×10^3 V (1.0 kV) respectively. This means that the potential difference between the "terminals" of the power plant (points A and B) is 1.0 kV. Determine:

(a) the current generated by the power plant

(b) the power dissipated (P_D) in the lines

(c) the percentage of the total power that is dissipated in the lines

Solution

(a) The current can be determined by a re-arrangement of $P = (\Delta V)I$, where P is the total power provided by the plant, ΔV is the voltage across the plant, and I is the resulting current generated by the plant.

$$I = \frac{P}{\Delta V}$$

When substituting numbers, we need to include power in watts rather than kilowatts: the power of the plant in watts is 2.0×10^5 W.

Then,
$$I = \frac{2.0 \times 10^5 \text{ W}}{1.0 \times 10^3 \text{ V}}$$
$$= 2.0 \times 10^2 \text{ A}$$

Thus, the current generated by the power station is 2.0×10^2 A when operating at 1.0 kV.

(b) To calculate the power dissipated in the lines, we will use one of the expressions from $P_D = (\Delta V)I = I^2R = (\Delta V)^2/R$, but we have to be careful to understand what each of ΔV, I, and R mean here. Since we want to calculate the power consumed by the transmission line, ΔV is the decrease in electric potential along the line, that is, between points A and D or B and C; R is the resistance of the line (0.25 Ω) and I is the current travelling along the line (2.0×10^2 A from (a)). Be careful! ΔV in this equation is NOT 1.0 kV—this is the potential difference between the terminals of the power station. We don't know what ΔV is for one of the transmission lines, but we don't need to know at this stage since we already have the other two variables, R and I. Therefore we use $P_D = I^2R$, knowing both I and R,

$$\begin{aligned} P_D &= I^2R \\ &= (2.0 \times 10^2 \text{A})^2 (0.25\ \Omega) \\ &= 1.0 \times 10^4 \text{W} \end{aligned}$$

Doubling this to take both lines into account, we get a total power dissipation of 2.0×10^4 W in the transmission lines when the power station operates at 1.0 kV.

(c) As a percentage of the total power, the power dissipated in the lines is

$$\begin{aligned} \frac{P_D}{P} \times 100\% &= \frac{2.0 \times 10^4 \text{W}}{2.0 \times 10^5 \text{W}} \times 100\% \\ &= 10\% \end{aligned}$$

Thus, 10% of the total power is lost in the lines when the power station operates at 1.0 kV. This is a lot of wasted energy!

DID YOU KNOW?

You may have a transformer sitting on your desk (or under it) right now! The power adapters used to connect laptops to a wall socket contain a transformer to reduce the voltage from the wall to the levels needed by your laptop circuitry.

SAMPLE PROBLEM 23-3

Repeat the previous problem for the same total power being supplied by the station, but this time at a voltage of 1.0×10^2 kV (1.0×10^5 V).

Solution

(a) The current generated by the station is now calculated as

$$\begin{aligned} I &= \frac{P}{\Delta V} \\ I &= \frac{2.0 \times 10^5 \text{W}}{1.0 \times 10^5 \text{V}} \\ &= 2.0 \text{A} \end{aligned}$$

Thus, the current generated by the power station is 2.0 A when operating at 1.0×10^2 kV, which is significantly lower than the current generated in Sample Problem 23-2.

(b) The power dissipated in one line is now

$$\begin{aligned} P_D &= I^2R \\ &= (2.0 \text{A})^2 (0.25\ \Omega) \\ &= 1.0 \text{W} \end{aligned}$$

Thus, the power dissipated in two lines is 2.0 W, which is significantly lower than that determined in the previous sample problem.

(c) The percentage of the total power that is dissipated in this case is

$$\begin{aligned} \frac{P_D}{P} \times 100\% &= \frac{2.0 \text{W}}{2.0 \times 10^5 \text{W}} \times 100\% \\ &= 0.0010\% \end{aligned}$$

Thus, by increasing the voltage of the power station output by a factor of 100, the percentage of the total power lost in the transmission lines decreases by a factor of 10 000!

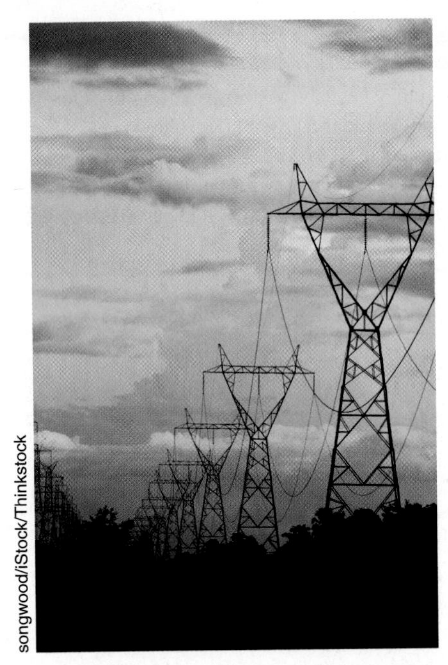

Figure 23-40 High-voltage transmission towers and lines are designed to protect people from accidentally coming in contact with them.

From the above calculations you can see why it is beneficial to transmit electrical energy at high voltage: a greater percentage of the power is delivered to the consumer. Long-distance transmission lines have voltages typically in excess of 500 kV. The voltage needs to be reduced when the energy is finally distributed to consumers because high voltages are very dangerous. High-voltage power lines are strung from huge towers (Figure 23-40) to prevent people from coming into contact with them, and to prevent sparking to the ground. The high voltage is decreased in steps at transformer stations until finally the utility pole transformer in your neighbourhood (Figure 23-38 (b)) lowers the voltage from about 4 kV to 120 V for home use in North America, and 240 V for home use in many other parts of the world.

Another advantage of high-voltage transmission is that a given line can carry more power: one 750 kV line, for instance, has the carrying capacity of about five 350 kV lines. (You can confirm this from $P = (\Delta V)^2/R$.) Thus using high voltage helps to reduce construction of transmission lines and, with increasing land costs, environmental and esthetic concerns, this is an important consideration.

In addition to being used in the large-scale transmission of electrical energy, transformers are used in situations where the standard 120 V from an outlet is inappropriate. For example, transformers are used in doorbell circuits so that the voltage across the switch (the doorbell button) is considerably less than 120 V. This reduces the possibility that the person pushing the button will receive an electric shock. For a similar reason transformers are used in providing power to children's electric trains. Transformers are also employed to increase voltages for the operation of devices such as neon lights.

How a Transformer Works

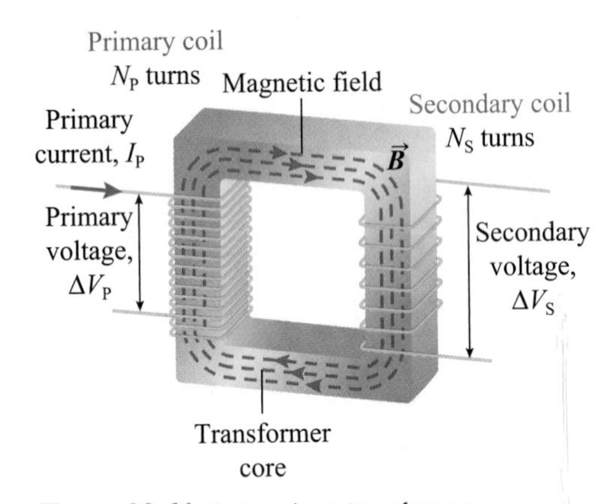

Figure 23-41 A step-down transformer.

We have seen *why* transformers are used in the transmission of electrical energy. Now we can discuss *how* they work.

A transformer consists of two coils wound around a common iron core (Figure 23-41). One coil is the primary, or input, coil, shown on the left; an AC voltage from an external source is applied across the ends of this coil. The other coil is the secondary, or output, coil, shown on the right. As the input current alternates in time, an alternating magnetic field is created inside the primary coil. The magnetic field lines tend to stay within the iron core, and thus the alternating magnetic field exists also within the part of the core inside the secondary coil. This changing field inside the secondary coil induces a voltage across its ends. The larger the number of loops (often called turns) in the secondary coil, the larger this output (secondary) voltage. The secondary voltage is also proportional to the primary voltage, and inversely proportional to the number of turns in the primary coil. We can summarize these statements as

$$\text{secondary voltage} = \frac{\text{number of secondary turns}}{\text{number of primary turns}} \cdot \text{primary voltage}$$

or, using the notation shown in Figure 23-41,

$$\Delta V_S = \frac{N_S}{N_P} \cdot \Delta V_P \qquad (23\text{-}2)$$

turns ratio the ratio of the number of turns in the secondary (output) coil to the number of turns in the primary (input) coil in a transformer

step-up transformer a device in which the output voltage is larger than the input voltage

Thus, for a given primary voltage, any desired secondary voltage can be obtained by changing the **turns ratio** (N_S/N_P). If the secondary voltage is larger than the primary voltage, the transformer is called a **step-up transformer**, since it "steps up" the voltage. If the secondary voltage is smaller than the primary, it is a **step-down transformer**. Since the number of turns in the secondary coil is less than that in the primary coil in Figure 23-41, this is a step-down transformer.

step-down transformer a device in which the output voltage is smaller than the input voltage

The electrical power supplied to the primary coil is often written as the product of the voltage ΔV_P and the current I_P through the coil:

$$P = \Delta V_P I_P$$

Similarly, the electrical power available from the secondary coil is

$$P = \Delta V_S I_S$$

where I_S is the current that is generated in the secondary coil when it is connected to a closed circuit. Notice that we have not subscripted these powers. Transformers are very efficient at transferring electrical energy from the primary circuit to the secondary circuit. Indeed, an efficiency of 99% is typical. Therefore, for most purposes we assume 100% efficiency and write

$$P = \Delta V_P I_P = \Delta V_S I_S \quad (23\text{-}3)$$

We can rearrange $\Delta V_P I_P = \Delta V_S I_S$ to express the secondary current generated as a result of the power input to the primary coil as

$$I_S = \frac{\Delta V_P}{\Delta V_S} \cdot I_P \quad (23\text{-}4)$$

Thus, in a step-up transformer, if the secondary coil has 3 times the number of turns as the primary coil, the secondary voltage is larger than the primary voltage by a factor of 3 based on Eqn. 23-2, and the secondary current is smaller than the primary current by a factor of 3. *Increasing the voltage through the transformer decreases the current.* Similarly, in a step-down transformer, the secondary current is larger than the primary current; *decreasing the voltage increases the current.*

DID YOU KNOW?

Since DC voltage applied to the primary coil of a transformer does not produce a changing magnetic field, transformers cannot be used in DC circuits. One of the major arguments in favour of AC in the "Current Wars" (discussed in the Profile on Nikola Tesla in Chapter 22) is that AC voltage can be easily changed by transformers, allowing for long-range transmission of electrical power with minimal loss, as well as safe operation in everyday use.

SAMPLE PROBLEM 23-4

A neon sign requires 12 kV for operation.

(a) What is the turns ratio of the transformer used to provide the voltage if a primary voltage of 120 V is supplied?

(b) If the sign uses 75 W of power, what are the primary and secondary currents?

Solution

(a) We are given the primary and secondary voltages ($\Delta V_P = 120\text{ V}$; $\Delta V_S = 1.2 \times 10^4\text{ V}$), and are asked to find the turns ratio (N_S/N_P). We simply use Eqn. 23-2 and rearrange:

$$\Delta V_S = \frac{N_S}{N_P} \cdot \Delta V_P$$

$$\frac{N_S}{N_P} = \frac{\Delta V_S}{\Delta V_P}$$

$$\frac{N_S}{N_P} = \frac{1.2 \times 10^4\text{ V}}{120\text{ V}}$$

$$\frac{N_S}{N_P} = 1.0 \times 10^2$$

Therefore, in order for the transformer to convert 120 V into 12 kV, the secondary coil must have 100 times the number of loops that exist in the primary coil.

(b) From Eqn. 23-3, $P = \Delta V_P I_P = \Delta V_S I_S$. Therefore, given that the sign uses 75 W of power, we can determine the secondary current as

$$I_S = \frac{P}{\Delta V_S} = \frac{75\ \text{W}}{1.2 \times 10^4\ \text{V}} = 6.3 \times 10^{-3}\ \text{A}$$

We could use the same equation again to determine the current in the primary coil, given the potential difference on this side. Or, given everything that we know, we could also use Eqn. 23-4 and rearrange:

$$I_S = \frac{\Delta V_P}{\Delta V_S} \cdot I_P$$

$$I_P = \frac{\Delta V_S}{\Delta V_P} \cdot I_S$$

$$I_P = \frac{1.2 \times 10^4\ \text{V}}{120\ \text{V}} \cdot (6.25 \times 10^{-3}\ \text{A})$$

$$I_P = 6.3 \times 10^{-1}\ \text{A}$$

Thus, the primary and secondary currents are 0.63 A and 6.3×10^{-3} A respectively. Notice that the ratio of primary current to secondary current is 1.0×10^2, that is, the inverse of the turns ratio.

TRY THIS!

Transformers—More Than Meets the Eye

Going back to the PhET simulation discussed in Section 23.3, click on the "Transformer" tab at the top. With the battery connected to the primary coil, what is the output in the secondary coil? Now select the AC power supply to connect to the primary coil and experiment with changing the number of primary turns, secondary turns, and the primary voltage. You can use either a light bulb for qualitative observations of the secondary voltage or a meter for more quantitative investigations. You can move the coils around relative to each other and see what happens when you increase the size of the secondary coil (called the "pickup coil" here).

EXERCISES

23-21 A transformer changes 120 V to 24 kV. If there are 5.0×10^3 turns in the secondary coil, how many turns are there in the primary?

23-22 A transformer for an electric train set (Figure 23-42) has 950 turns in the primary coil, and 63 in the secondary.

(a) What kind of transformer is it?

(b) By what factor does it change the voltage? (Factors are normally given as numbers larger than 1.)

(c) If the input voltage is 120 V, what is the output voltage?

Steve Mann/Shutterstock

Figure 23-42 Question 23-22.

23-23 What is the turns ratio of a transformer used in a local transformer station to reduce the voltage from 115 kV to 44 kV?

23-24 A transformer increases the voltage from 120 V to 985 V.

(a) Which is larger, the primary or the secondary current?

(b) What is the ratio of the primary current to the secondary current?

23-25 A power of 94 W at 11 kV is provided to the primary coil of a step-down transformer with a turns ratio of 0.35. Determine

(a) the primary current

(b) the secondary voltage

(c) the secondary current

(d) the electrical power supplied to the secondary circuit by the secondary coil.

23-26 Why do transformers not work in DC circuits?

23-27 A generating station is connected to a city by transmission lines with a total resistance of 2.5 Ω. The station provides 32 MW of electrical power at 75 kV. Calculate

(a) the current generated by the station

(b) the power lost in the lines

(c) the percentage of the total power delivered to the city

23-28 If the output voltage from the generating station in the previous question were increased to 100 kV, would your answers to (a), (b), and (c) increase, decrease, or remain the same?

23-29 Fermi Question: A child's toy "oven" can operate on either 2 AA batteries in series or connected to the wall outlet through a transformer. What is the turns ratio needed in this transformer (in North America)? Is this a step-up or step-down transformer?

23-30 Fermi Question: A British tourist visiting Calgary has brought along an adapter and converter in order to be able to plug in her hair dryer (Figure 23-43). What is the turns ratio inside this converter? Is this a step-up or step-down transformer?

Paul Michael Hughes/Shutterstock

Figure 23-43 Electrical outlets in Europe operate at a much higher voltage than their North American counterparts (Question 23-30).

23.5 Magnetic Flux

magnetic flux a measure of the strength of the magnetic field over a specific area

In previous sections of this chapter we have discussed electromagnetic induction and its role in the operation of generators and transformers. A useful (but somewhat abstract) concept that provides a more compact statement of the relationship between the changing magnetic field and the induced voltage is that of **magnetic flux**, and it also gives us an alternative view of the working of generators and transformers. In essence, magnetic flux is a measure of the strength of the magnetic field over a specific area. In what follows, the magnetic flux due to a *uniform* magnetic field is discussed in greater detail.

Most often we consider the magnetic flux through a loop; this flux is defined as the product of the loop's cross-sectional area and the component of the magnetic field that is perpendicular to the plane of the loop. Mathematically, we write

$$\phi = (A)(B\cos\theta)$$

$$\text{or}\quad \phi = BA\cos\theta \qquad (23\text{-}5)$$

where ϕ (the uppercase Greek letter phi) is magnetic flux, B is the magnitude of the uniform magnetic field, A is the area of the loop, and θ is the angle between the magnetic field and a line *perpendicular* to the plane of the loop (Figure 23-44).

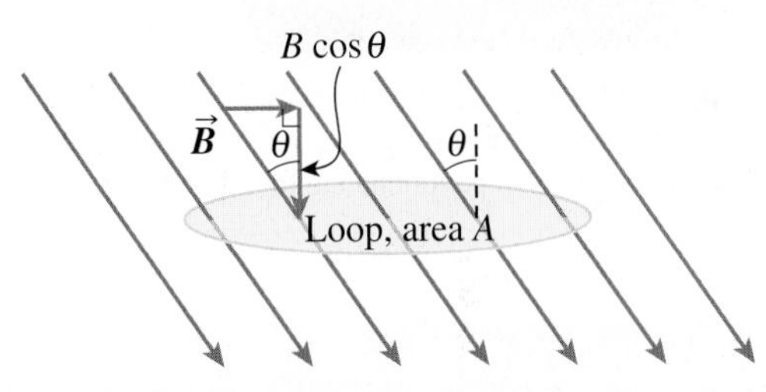

Figure 23-44 The magnetic flux through the loop is $\phi = BA\cos\theta$.

weber SI unit of magnetic flux; 1 weber = 1 Wb = 1 tesla·metre^2 = 1 T·m^2

Note that the area A used in Eqn. 23-5 is the full cross-sectional area of the loop, provided that the region containing the magnetic field is larger than the loop itself. This will always be the case in our discussions, unless otherwise explicitly noted. In SI units, magnetic flux is in T·m^2 (tesla·metre^2), which is given the name of **weber** (Wb), in honor of Wilhelm Eduard Weber (1804–1891), a German physicist who made many contributions to the study of magnetism.

When the magnetic field is perpendicular to the plane of the loop, $\theta = 0°$ and Eqn. 23-5 becomes $\phi = BA$ (Figure 23-45 (a)). When the field is parallel to the loop, $\theta = 90°$ and $\phi = 0$ (Figure 23-45 (b)). In a sense, we can think of the flux as being proportional to the number of field lines passing through the loop. This number depends on the relative orientation of the field and loop as shown in Figures 23-44 and 23-45. It will also depend on how big the area of the loop is, as well as the strength of the magnetic field in the region: in a strong field the field lines are close together and would therefore have more lines going through the loop than a weaker field.

Why is this concept of magnetic flux useful? With it, we can write **Faraday's law**, which summarizes the phenomenon of electromagnetic induction in a short statement that is applicable in a wide range of situations:

Faraday's Law: The voltage induced in a loop (or any closed conducting path) is equal to the absolute value of the rate of change of the magnetic flux through that loop.

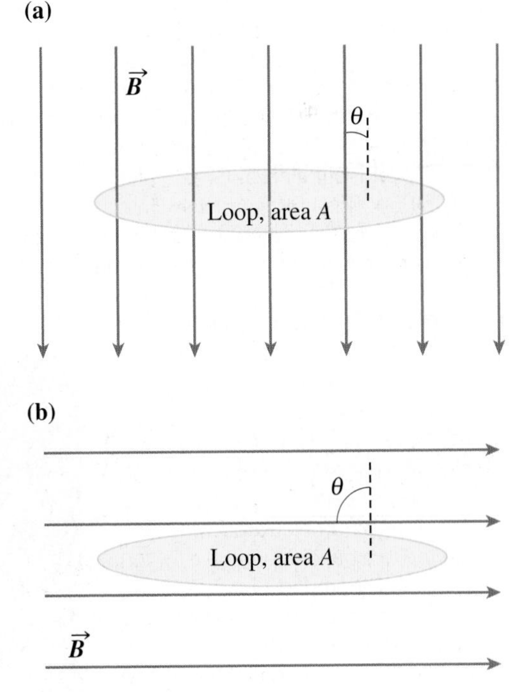

Figure 23-45 **(a)** When the magnetic field is perpendicular to the plane of the loop, $\theta = 0°$ and $\phi = BA$. **(b)** When the field is parallel to the loop, $\theta = 90°$ and $\phi = 0$.

Faraday's law the induced voltage in a closed conducting path is equal to the absolute value of the rate of change of the magnetic flux through that path

Since we are often dealing with more than one loop (in a coil, for example), this version of Faraday's law is usually written mathematically for N loops:

$$\Delta V = N\left|\frac{\Delta\phi}{\Delta t}\right| \tag{23-6}$$

where Δt is the time interval during which ϕ changes by an amount $\Delta\phi$.

loraks/Shutterstock

Figure 23-46 Induction stoves are much more efficient than conventional stoves as they use resistive heating rather than conduction to transfer energy.

DID YOU KNOW?

Figure 23-46 is a close up of an induction stove, in which electromagnetic induction is used to heat food. Conventional stoves use heat from the element to heat the pan or pot by conduction, which then heats the food inside. This is a highly inefficient way to cook your food as energy is lost at each stage. With an induction stove, the "element" controls a rapidly changing magnetic field. When a conducting pot is placed on top, there is an induced current in the metal due to the changing magnetic flux in the conductor, which heats the pot resistively rather than through conduction, which is much more efficient.

PROFILES IN PHYSICS

Michael Faraday (1791–1867): Experimentalist Extraordinaire!

Michael Faraday was born in 1791, one of four children in a poor London household in which food was often a scarcity. Faraday had no formal education: he learned to read through Sunday school at church and started working for a book-binding shop in London at the age of 13 in order to help provide for his family. His intense curiosity and thirst for knowledge led him to read as many books as he could while working there. Faraday got his "break" into science by being given the opportunity to attend four chemistry lectures by Sir Humphry Davy at the Royal Institution of Great Britain in 1812. Faraday followed up by applying for work with Davy: he sent the renowned professor a bound copy of the notes he had taken at the lectures. Faraday was hired by Davy at the next opening as a laboratory assistant.

After apprenticing with Davy for several years, Faraday struck out on his own, earning much acclaim as an analytical chemist. In the fall of 1831, Faraday demonstrated electromagnetic induction through the construction of a rudimentary transformer: an insulated wire connected to a battery through a switch was wound around one side of an iron ring. A second insulated wire was wound around the other side, connected to an ammeter. When he closed the switch, he observed a current in the second wire. When he opened the switch again, he observed a current again in the second wire, this time having the same magnitude but opposite direction as that observed in the first instance. Continued experiments with magnets and coils led to the conclusion that the potential difference induced was proportional to the rate of change of the magnetic flux, as summarized here in Eqn. 23-6.

Faraday's experimental work was hugely important to the understanding of electricity and magnetism and their inherent relationship. Mechanically inclined, he is credited with building the first electric motor, the first AC generator, and the first transformer. He also did extensive fundamental work in electrochemistry, coining terms such as ion, cathode, anode, and electrode. In recognition of his essential role in this field, the unit of electrical capacitance is named farad in his honour. His work was key in the subsequent development by James Clerk Maxwell of the theory of electromagnetism.

© GL Archive/Alamy

Figure 23-47 Michael Faraday delivering one of his famous Christmas lectures for the general public in London, England.

Perhaps because of how he got his start in science, Faraday was committed to providing the general public with opportunities to learn about cutting-edge developments. He was a hugely popular speaker: his annual Christmas Day public lectures at the Royal Institution in London were routinely full to capacity (Figure 23-47). Children and adults alike were delighted and amazed by his demonstrations and explanations. Who knows how many future scientists he inspired in these presentations!

We now illustrate how Faraday's law is useful in thinking about induced voltage in a loop. Figure 23-48 (a) shows a loop in a uniform magnetic field. In order to induce a voltage across the ends of the loop, we need to *change* the magnetic flux through the loop. This can be done in a variety of ways:

- by increasing or decreasing the magnitude of the magnetic field (Figure 23-48 (b))
- by changing the area of the loop (Figure 23-48 (c))
- by changing the orientation of the loop, most commonly by rotating it; this is considered further below

Regardless of the method used to change the flux, the faster the change is made, the larger the induced voltage generated.

A single-loop generator—that is, a loop being rotated in a magnetic field—is shown in Figure 23-49. In diagram (a) the plane of the loop is parallel to the field, and hence there is no magnetic flux through the loop. As the loop rotates from this position, the flux quickly becomes non-zero. Indeed, with the loop in this orientation, the rate of change in flux is maximum, and hence the maximum possible voltage is induced across the ends of the loop. In Figure 23-49 (b) the loop is perpendicular to the field, and the flux through the loop is as large as possible. As the loop rotates near this position, the flux changes only slowly from its maximum value. Thus, only a small voltage is induced across the ends of the loop. In fact, when the loop is perpendicular to the field, the voltage is instantaneously zero. As the loop rotates through its various positions, the induced voltage varies sinusoidally with time (if the rotation rate is constant), as discussed in Section 23.3.

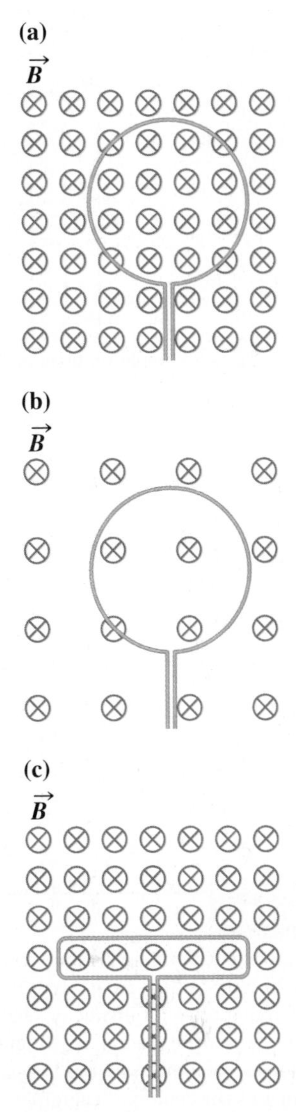

Figure 23-48 **(a)** A loop in a uniform magnetic field. **(b)** Decreasing the magnitude of the magnetic field decreases the magnetic flux through the loop. **(c)** Changing the loop's shape changes its area and the magnetic flux through the loop.

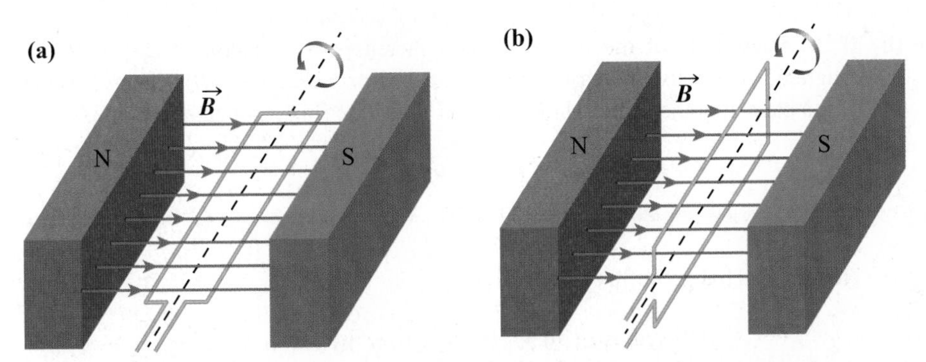

Figure 23-49 **(a)** When the plane of the loop is parallel to the magnetic field, the magnetic flux through the loop is zero. As the loop rotates from this position the flux changes quickly and the maximum voltage is induced in the loop. **(b)** When the loop and field are perpendicular, the flux through the loop is a maximum, but the rate of change of flux is zero. Hence, the induced voltage is zero.

We now show how the induced voltage in a straight wire moving in a uniform magnetic field can be explained in terms of magnetic flux. Figure 23-50 illustrates such a wire that makes contact with stationary conductors connecting the wire to a complete circuit. The loop that constitutes the circuit is WXZY. As the wire moves in the direction shown, the flux (think of the number of field lines) through the loop increases, and hence there is a voltage induced in the loop. The faster the wire moves, the faster the flux increases, and the larger the voltage. We could also cause the flux to increase more rapidly by increasing the magnetic field and/or the length of the wire.

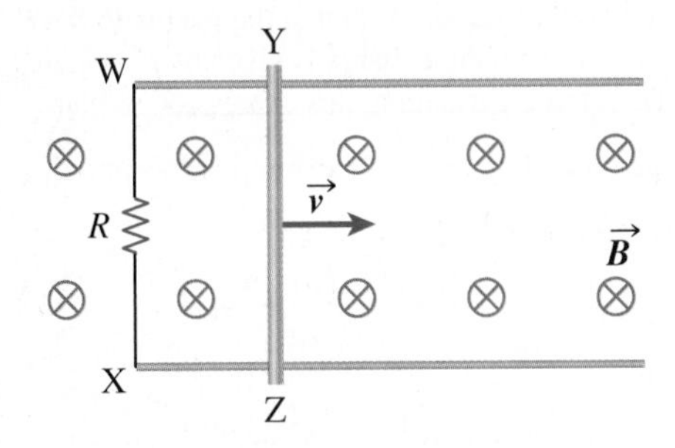

Figure 23-50 As the sliding wire moves, the magnetic flux through the loop WXZY increases, and hence a voltage is induced in the loop.

SAMPLE PROBLEM 23-5

A coil of wire with 2.0 loops, each of cross-sectional area 5.5 cm², is in a uniform magnetic field of magnitude 5.8×10^{-2} T (Figure 23-51). The field is perpendicular to the plane of the loops in the coil.

(a) What is the magnetic flux through one loop of the coil?

(b) If the magnitude of the magnetic field increases by a factor of 2.6 in a time of 1.5 s, what is the voltage induced in the coil?

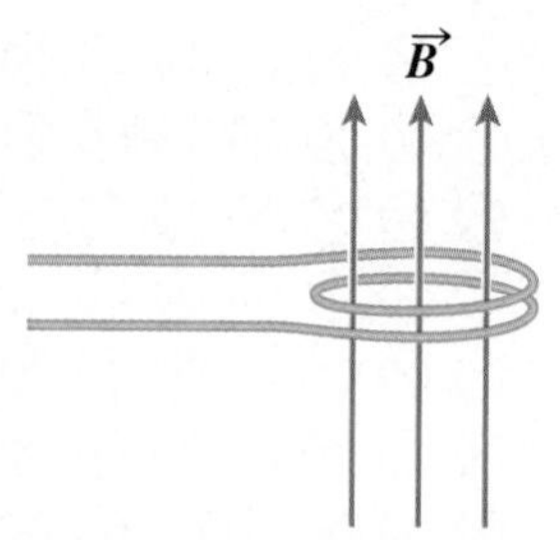

Figure 23-51 Sample Problem 23-5.

Solution

(a) To calculate the magnetic flux through *one* loop, we use $\phi = BA\cos\theta$. It is given that the field and the plane of the loops are perpendicular. Since θ is the angle between the field and a line perpendicular to the plane of the loops, $\theta = 0°$ and $\cos\theta = 1$. Thus in this configuration, $\phi = BA$, the maximum possible flux. Before substituting values of B and A, we convert the area in cm² to m²:

$$5.5\,\text{cm}^2 = 5.5\,\text{cm}^2 \times \frac{(1\,\text{m})^2}{(100\,\text{cm})^2}$$

$$= 5.5 \times 10^{-4}\,\text{m}^2$$

Then

$$\phi = BA$$

$$= (5.8 \times 10^{-2}\,\text{T})(5.5 \times 10^{-4}\,\text{m}^2)$$

$$= 3.2 \times 10^{-5}\,\text{Wb}$$

(b) If the magnitude of the magnetic field increases by a factor of 2.6, then the magnetic flux must increase by the same factor (since $\phi \propto B$) when all other parameters stay the same. Thus, the final magnetic flux through one loop is

$$\phi_f = 2.6(3.19 \times 10^{-5}\,\text{Wb})$$

$$= 8.29 \times 10^{-5}\,\text{Wb}$$

Hence, the change in flux through one loop is

$$\Delta\phi = (8.29 \times 10^{-5} - 3.19 \times 10^{-5})\,\text{Wb}$$

$$= 5.10 \times 10^{-5}\,\text{Wb}$$

The induced voltage is given by

$$\Delta V = N\left|\frac{\Delta\phi}{\Delta t}\right|$$

$$= (2.0) \times \frac{5.10 \times 10^{-5}\,\text{Wb}}{1.5\,\text{s}}$$

$$= 6.8 \times 10^{-5}\,\text{V}$$

Therefore, the induced potential difference between the ends of the coil is 6.8×10^{-5} V.

Units Tip

It is worthwhile examining the units in Sample Problem 23-5 (b) as they are likely somewhat unfamiliar to you. In Faraday's law, the left-hand side of the equation has SI units of volts or joules/coulomb, which is the same as newton·metres per coulomb. Therefore, we have

$$\text{V} = \frac{\text{Wb}}{\text{s}}$$

$$\frac{\text{J}}{\text{C}} = \frac{\text{T}\cdot\text{m}^2}{\text{s}}$$

$$\frac{\text{N}\cdot\text{m}}{\text{C}} = \frac{\text{T}\cdot\text{m}^2}{\text{s}}$$

In order to see how these two sides are dimensionally the same, recall that the magnetic force on a moving point charge is given by $F = qvB$. Therefore, rearranging this expression we can see that $1\,\text{T} = 1\frac{\text{N}\cdot\text{s}}{\text{C}\cdot\text{m}}$. Substituting this into the right-hand side above, we get

$$\frac{\text{N}\cdot\text{m}}{\text{C}} = \frac{\text{N}\cdot\text{s}}{\text{C}\cdot\text{m}}\,\frac{\text{m}^2}{\text{s}}$$

or

$$\frac{\text{N}\cdot\text{m}}{\text{C}} = \frac{\text{N}\cdot\text{m}}{\text{C}}$$

One more example of the use of electromagnetic induction in our daily lives is found in sheds and garages across the country: the bicycle speedometer (Figure 23-52). A small permanent magnet is mounted on one of the spokes of the front wheel with a receiver coil positioned on one of the front forks. As the permanent magnet on the wheel passes the receiver coil, a pulse of current is generated by the changing magnetic flux, which is relayed to the processor mounted on the handlebars. The rate of current pulses recorded is simply the frequency of revolution of the wheel, which therefore can be used to determine the speed as well as the distance travelled by the bicycle. In our final sample problem in this chapter we will use Faraday's law in a Fermi-style question to investigate the typical voltage induced in the receiver coil of a standard bicycle speedometer.

Dick Stada/Shutterstock

Figure 23-52 Electromagnetic induction can be used to tell you how fast and how far your bicycle travels.

SAMPLE PROBLEM 23-6

Fermi Question: Estimate the typical voltage induced in the receiver coil of a standard bicycle speedometer (Figure 23-52).

Solution

A strong permanent magnet will likely have a magnetic field strength on the order of 1 T at its surface. However, the receiver coil is separated from the magnet by a centimetre or two (otherwise the wheel wouldn't turn very well!), so we'll assume a magnetic field strength at the receiver coil of approximately 1×10^{-2} T. The receiver coil needs to be fairly small to be mounted on the fork of the bicycle, so let's estimate the area of the receiver to be approximately 0.1 cm^2, as it will be smaller than 1 cm^2 but bigger than 0.01 cm^2. Therefore, as the magnet passes by, the change in magnetic flux is

$$\Delta\phi \approx (1 \times 10^{-2}\,\text{T})\cdot(1 \times 10^{-5}\,\text{m}^2)$$
$$\approx 1 \times 10^{-7}\,\text{Wb}$$

Now we need to estimate the time it takes for the magnet to pass the receiver coil. A reasonably fit cyclist will be moving at approximately 40 km/h on a nice flat road, which is roughly 10 m/s. Since the magnet is mounted approximately halfway between the axel and the rim, the magnet will have a speed of about 5 m/s (recall that $v = \omega R$ from Chapter 10). If the magnet has a diameter on the order of 1 cm, it will pass completely in front of the receiver in

$$\Delta t = \frac{\Delta d}{v} \approx \frac{0.01\,\text{m}}{5\,\text{m/s}} \approx 0.002\,\text{s}$$

Substituting these estimates into Eqn. 23-6, we get

$$\Delta V = N\left|\frac{\Delta\phi}{\Delta t}\right|$$
$$\approx N \times \left(\frac{1 \times 10^{-7}\,\text{Wb}}{0.002\,\text{s}}\right)$$
$$\approx N(5 \times 10^{-5}\,\text{V})$$

Lastly, if we assume that the receiver coil has on the order of $N = 100$ loops, we conclude that a typical voltage generated in a standard bicycle speedometer is approximately 5 mV.

DID YOU KNOW?

Electromagnetic induction comes in handy when you are in a hurry in your car. At many traffic-light controlled intersections, the left-turn lane has a sensor imbedded in the concrete. This sensor is often a large coil that detects the presence of your iron-containing vehicle by induction as you drive over it. The sensor then signals to the control system that there is a vehicle waiting to turn left and your wait is shortened accordingly!

DID YOU KNOW?

If you have an electric toothbrush (Figure 23-53 (a)), you may have wondered how it can charge its battery without any obvious electrical contact between the toothbrush and the base that is plugged into the wall outlet. By now you might be able to guess the source of this charging current: electromagnetic induction! The toothbrush and base act as coupled coils when in close proximity: a time-varying current in the base coil results in an induced current in the toothbrush circuit to recharge the batteries, without any metal electrical contacts between the two pieces. This is especially important in a bathroom as it minimizes the possibility of water getting inside the toothbrush or shorting out the charger base. However, inductive charging, as it is called, is not reserved for the bathroom. Figure 23-53 (b) shows a wireless charger for mobile devices that works through induction, and Figure 23-53 (c) is a photograph of a wireless charging system designed to look like a manhole cover. This ingenious design is being deployed in a pilot project in New York City in 2014 in parking spots designated for the recharging of electric vehicles.

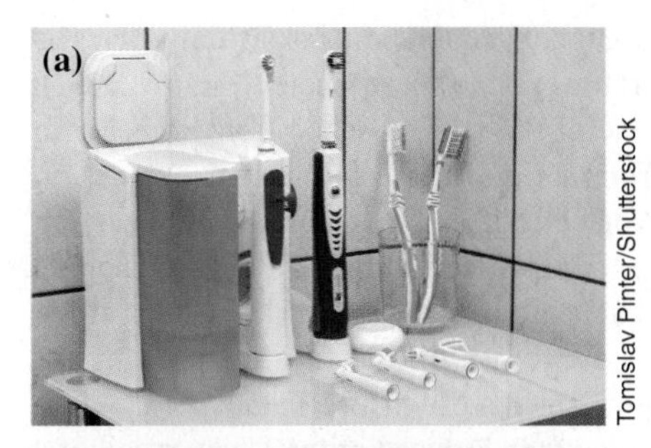

Tomislav Pinter/Shutterstock

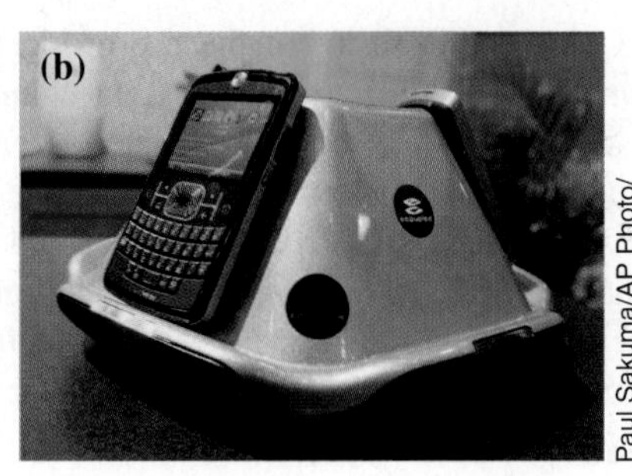

Paul Sakuma/AP Photo/ The Canadian Press

Rex Features/The Canadian Press

Figure 23-53 Three more examples of the use of electromagnetic induction in our daily lives. **(a)** A charger for an electric toothbrush. **(b)** A dock for charging a smartphone. **(c)** A manhole cover that charges electric cars.

EXERCISES

23-31 A magnetic field is perpendicular to a loop of conducting wire. In each of the following situations, will the magnetic flux through the loop increase, decrease, or remain the same?

(a) The magnitude of the magnetic field is increased.

(b) The loop's shape is changed so that its cross-sectional area is decreased.

(c) The loop is rotated so that the field makes a 35° angle with a line perpendicular to the plane of the loop.

23-32 A generator is constructed by having a coil of wire rotate in a uniform magnetic field.

(a) When is the magnetic flux through the coil at its maximum?

(b) When is the induced voltage in the coil at its maximum?

23-33 Explain how a transformer works in terms of magnetic flux.

23-34 A coil of conducting wire consists of 12 loops, each of cross-sectional area 10.5 cm^2. The coil is in a uniform magnetic field of magnitude 5.8×10^{-4} T. The field is perpendicular to the plane of the loops in the coil.

(a) What is the magnetic flux through one loop of the coil?

(b) If the coil is rotated so that there is a 67° angle between the field and a line perpendicular to the plane of the loops, what is the magnetic flux through one loop of the coil?

23-35 The magnetic flux through a loop increases from 4.5×10^{-3} Wb to 9.7×10^{-3} Wb in a time of 6.7×10^{-2} s. What is the induced voltage between the ends of the loop?

23-36 A coil of conducting wire is situated in a uniform magnetic field of magnitude 9.13×10^{-2} T. The coil consists of 55.0 loops, each of cross-sectional area 4.55×10^{-4} m^2, and the magnetic field is perpendicular to the plane of the loops in the coil.

(a) What is the magnetic flux through one loop of the coil?

(b) If the magnitude of the magnetic field decreases to 3.37×10^{-2} T in 4.52×10^{-3} s, what is the induced voltage between the ends of the coil?

LOOKING BACK...LOOKING AHEAD

The last several chapters discussed a wide range of topics in electricity and magnetism: electric field, electric potential, current, resistance, circuits, and the production and use of magnetic fields.

The focus of the present chapter has been on how magnetism can be used to produce a voltage between the ends of a conductor, in the process called electromagnetic induction. We have seen how a voltage can be induced across the ends of a coil by a changing magnetic field, and how moving a straight conductor in a magnetic field induces a voltage across the ends of the conductor. Lenz's law, which allows us to determine the direction of the induced current (in a closed circuit), was introduced. We discussed the important commercial applications of electromagnetic induction: electric generators and transformers that are used widely around the world in the production and transmission of electricity. Finally, the concept of magnetic flux and its relation to induction was introduced.

In the next, and final chapter of the book, we study nuclear physics. We will return to a number of familiar themes of the text, such as conservation of energy, and introduce several important additional concepts such as the mass–energy equivalence theory put forward originally by Albert Einstein in the famous equation $E = mc^2$. Our approach to the introduction to nuclear physics in the following chapter is to build on the foundational physics knowledge conveyed thus far in the text, with a strong emphasis on the role that physics plays in medical diagnosis and treatment.

CONCEPTS AND SKILLS

Having completed this chapter, you should now be able to do the following:

- Describe how to induce a voltage between the ends of a conducting coil.
- State the factors that determine the size of the induced voltage in a coil.
- Describe how to induce a voltage by moving a straight conductor in a magnetic field.
- State the factors that determine the size of the induced voltage in a moving straight conductor.
- Solve qualitative problems related to the voltage induced in a coil or straight conductor.
- State and use Lenz's law.
- Discuss the basic operation of an electric generator.
- Solve qualitative problems on single-loop generators.
- Describe the basic design of an electrical generating station.
- Discuss why transformers are used.
- Describe how step-up and step-down transformers work.
- Solve problems involving voltages and numbers of turns in transformers.
- Define and calculate magnetic flux through a loop in a uniform magnetic field.
- State Faraday's law of induction.
- Describe electromagnetic induction in a coil, a moving straight conductor, an electric generator, and a transformer in terms of changing magnetic flux.
- Solve quantitative problems regarding induced voltage in a coil by using rate of change of magnetic flux.

KEY TERMS

You should be able to define or explain each of the following words or phrases:

electromagnetic induction
slidewire generator
Lenz's law
electric generator
slip ring
metal brushes
alternating-current (AC) generator
direct-current (DC) generator
transformer
turns ratio
step-up transformer
step-down transformer
magnetic flux
weber (unit)
Faraday's law

Chapter Review

MULTIPLE-CHOICE QUESTIONS

23-37 A bar magnet moving into a coil of wire induces a voltage in the coil. Which of the following will *not* result in an increase in the induced voltage?

(a) a stronger bar magnet is used
(b) more loops are added to the coil
(c) the magnet is moved more quickly into the coil
(d) the cross-sectional area of the coil is increased
(e) none of the above

23-38 Assume that the magnetic field is uniform between the poles of the magnet shown in Figure 23-54. In which direction do we need to move the wire in order to generate an induced voltage between its ends?

(a) toward point A
(b) toward point B
(c) toward point C

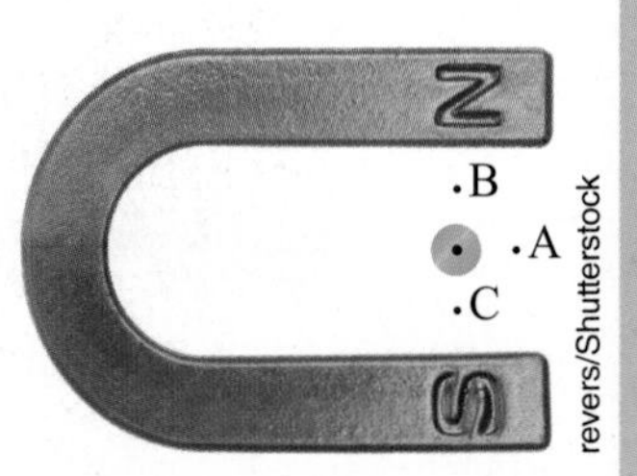

Figure 23-54 Question 23-38.

23-39 A vertical wire is situated in a uniform magnetic field directed northward in the horizontal plane. In order to induce the maximum voltage in the wire, in which direction(s) should it be moved?

(a) north
(b) south
(c) east

23-40 Figure 23-55 shows a wire moving in a uniform magnetic field. The wire makes contact with stationary wires so that current can flow around a rectangular loop. The induced current in the loop is

(a) clockwise
(b) counterclockwise

23-41 The direction of the magnetic force on the moving wire in Figure 23-55, due to the induced current in it, is toward

(a) the top of the page
(b) the bottom of the page
(c) the left
(d) the right

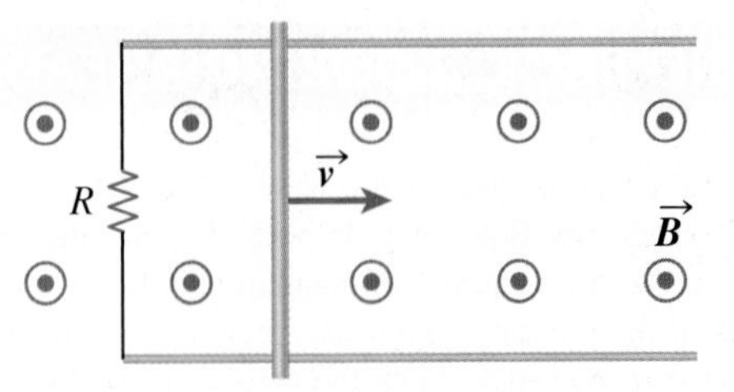

Figure 23-55 Questions 23-40 to 23-43.

23-42 The direction of the external force that must be applied to the wire in Figure 23-55 to maintain its motion is toward

(a) the top of the page
(b) the bottom of the page
(c) the left
(d) the right

Review Questions and Problems

23-43 (a) What is the induced voltage between the ends of the sliding wire in Figure 23-55 if the wire's speed is 55 cm/s, its length 17 cm, the magnitude of the magnetic field 3.4×10^{-3} T, and the resistance of the resistor 1.2 kΩ? The wire's velocity is perpendicular to the magnetic field.

(b) What is the induced current around the rectangular loop?

23-44 In Figures 23-56 (a) and (b), determine the direction of the induced current in the resistor and label the north and south poles of the coil.

(a)

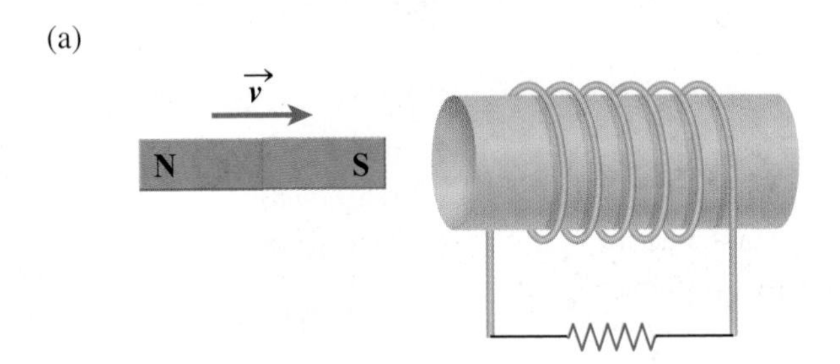

(b)

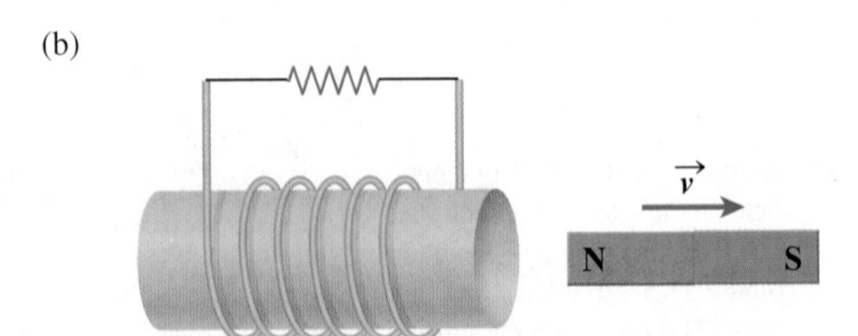

Figure 23-56 Question 23-44.

23-45 What is the direction of the rotation of the loop in the AC generator shown in Figure 23-57?

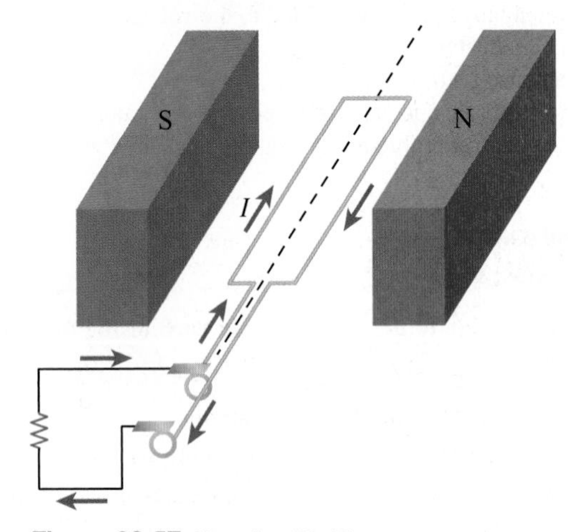

Figure 23-57 Question 23-45.

23-46 An electric doorbell uses a transformer to convert 120 V to 6.0 V. If there are 42 turns in the secondary coil, how many turns are there in the primary?

23-47 Why is high voltage used for long-distance transmission of electrical energy?

23-48 (a) What is the turns ratio of a transformer that converts 22 kV to 225 kV?

(b) If the input current is 115 A, what is the input power?

(c) What is the secondary current?

23-49 A small generating station produces 95 kW of electrical power that is transmitted by two wires each having a resistance of 0.065 Ω. How much power is saved by transmitting the power at 1.5 kV instead of 120 V?

23-50 A coil of conducting wire is situated in a uniform magnetic field. The number of loops in the coil and the coil's shape cannot be changed. List two ways that the magnetic flux through the coil might be increased.

23-51 Figure 23-58 shows a coil consisting of 3.0 loops of wire in a uniform magnetic field that makes a 70° angle with a plane *parallel* to the loops. If the cross-sectional area of each loop is 6.3 cm^2 and the magnitude of the magnetic field is 3.8×10^{-3} T, what is the magnetic flux through one loop of the coil?

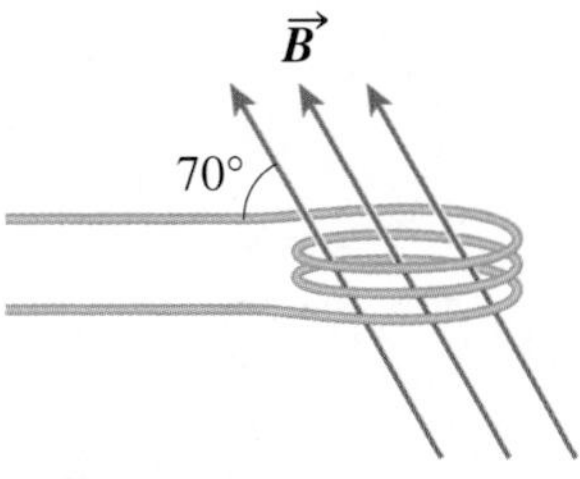

Figure 23-58 Question 23-51.

23-52 A coil of conducting wire is situated in a uniform magnetic field that is perpendicular to the plane of the loops in the coil (Figure 23-59). The coil has 4.0 loops of wire, each of cross-sectional area 4.3 cm^2, and the field has a magnitude of 0.56 T and a direction as shown. A voltage of 1.3 V is induced between the ends of the coil as the field is increased to 0.73 T.

(a) Over what time interval is the voltage increased?

(b) In what direction is the current induced in the resistor that is connected to the coil?

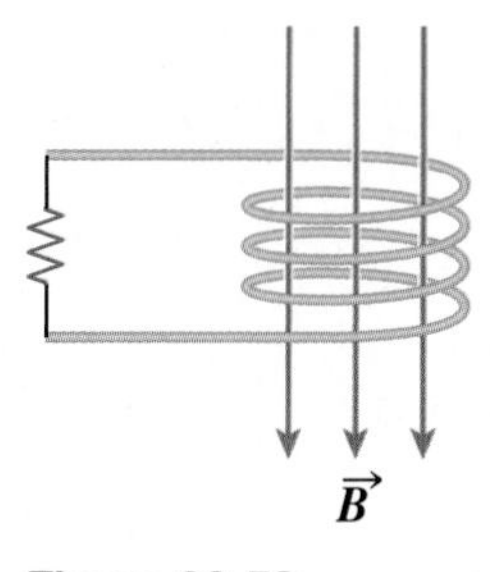

Figure 23-59 Question 23-52.

Applying Your Knowledge

23-53 (a) What is the current induced in the sliding wire in Figure 23-60? The uniform magnetic field has a magnitude of 8.7×10^{-3} T, the wire's speed is 0.45 m/s, its length is 0.22 m, and the resistance shown is 78 Ω. The wire's velocity is perpendicular to the magnetic field.

(b) What is the direction of the current through the sliding wire?

23-54 (a) In Figure 23-60, what is the external force required (magnitude and direction) to keep the wire moving at constant speed? Use the numerical values given in the previous question for your answer.

(b) What is the rate of work done by this force?

(c) What is the power dissipated by the resistance in the loop?

(d) How do your answers to (b) and (c) compare? What does this mean?

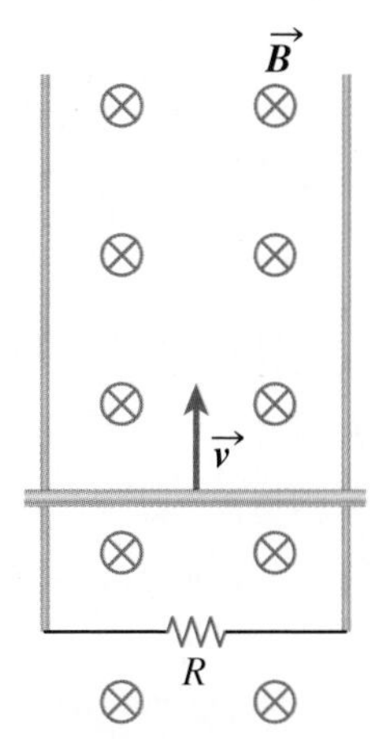

Figure 23-60 Questions 23-53 and 23-54.

23-55 Figure 23-61 shows two conducting loops and their respective circuits. One loop (#1) is in series with an ammeter, and the other (#2) with a resistor, a battery, and a switch. As the switch is closed, a current is induced in circuit #1 because of the changing magnetic field in loop #1 due to the increasing current in loop #2. This is quite similar to the experiment conducted by Faraday in 1831 (see Profiles in Physics on page 754).

(a) Is the direction of the current in loop #1 clockwise, or counterclockwise, as seen in Figure 23-61, as the switch in circuit #2 is closed?

(b) Describe the current in loop #1 as the switch remains closed.

(c) What happens to the current in loop #1 as the switch is opened?

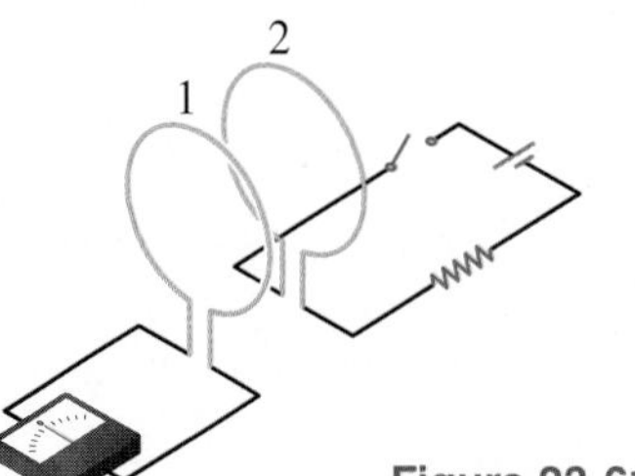

Figure 23-61 Question 23-55.

23-56 A coil of conducting wire is connected to a 74 Ω resistor to form a closed circuit. A uniform magnetic field is perpendicular to the 153 loops in the coil, each of cross-sectional area 2.5 cm^2. As the field magnitude is increased to 3.7×10^{-2} T in a time of 4.32×10^{-2} s, a current of 0.29 mA is induced in the circuit. What is the original magnitude of the magnetic field?

23-57 Use the steps described below to derive the relation $\Delta V = vBL$ given in Section 23.1 for the potential difference induced (ΔV) across the ends of a straight conducting wire of length L, moving with speed v in a direction perpendicular to a magnetic field of magnitude B.

(a) In terms of B, L, and d (see Figure 23-62), determine an expression for the magnetic flux through the conducting loop MNOP.

(b) In a time interval Δt, how much does the magnetic flux through the loop *increase*? Express your answer in terms of B, L, v, and Δt.

(c) Use Faraday's law to show that $\Delta V = vBL$.

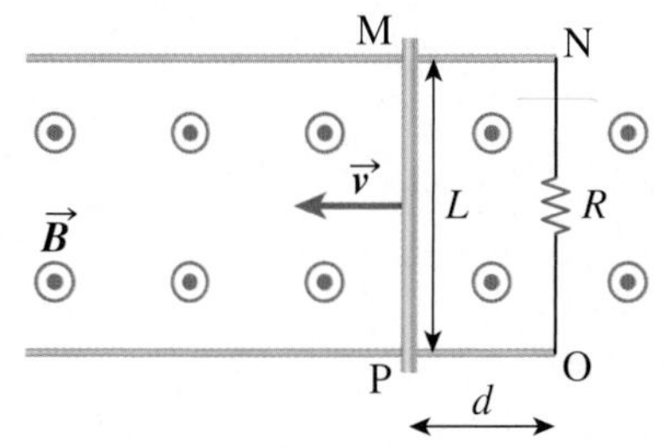

Figure 23-62 Question 23-57.

Nuclear Physics: Theory and Medical Applications 24

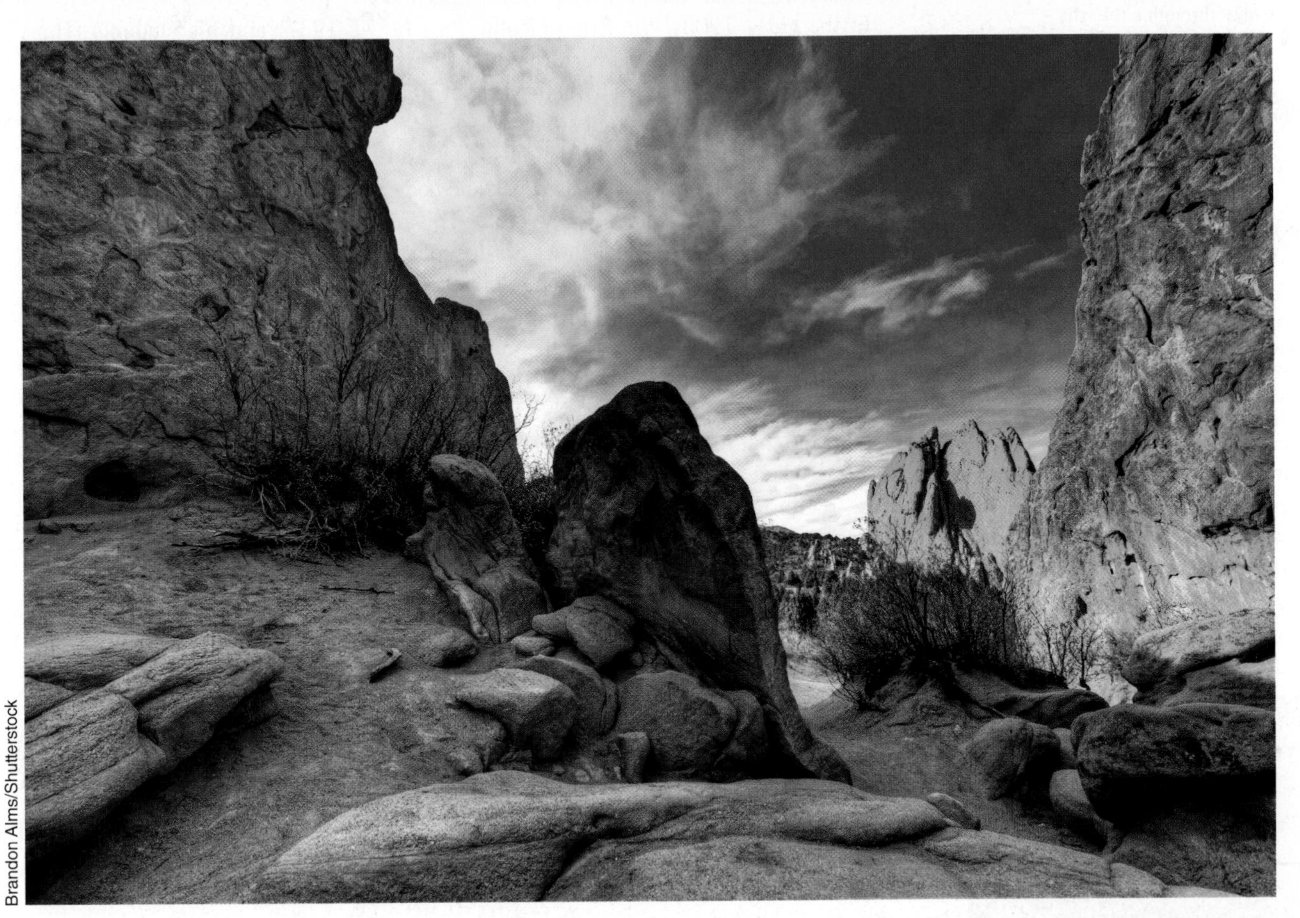

Brandon Alms/Shutterstock

Figure 24-1 Colorado Springs, Colorado, has a higher than average level of natural radioactivity due to its elevation and the local geology.

Radioactivity is all around us: in the air we breathe, the ground we walk on, the materials we build with, the foods we eat, and the water we drink. Since Earth formed and life developed, we have been immersed in this "background" radiation. In this chapter we will discuss nuclear structure and radioactivity, allowing us to learn about the many useful applications in modern society, as well as the health concerns associated with radiation. Specifically, we will learn how nuclear reactors work, how the Sun's energy is produced, how your own body is naturally radioactive, and how radiation is used in many ways to diagnose and treat disease. By the end of the chapter you will have a greater understanding of the causes for variations in background radiation, resulting in locations such as Colorado Springs (Figure 24-1) having higher than average levels.

CHAPTER **OUTLINE**

24.1 Structure of Nuclei

nucleus the dense core of an atom that contains most of its mass

It is hard to imagine the size of atoms; there are about 10^6 atoms across the period at the end of this sentence. And yet an atom is mostly empty space; most of its mass is concentrated in a central region (the **nucleus**) that takes up only about $1/10^{13}$ of the atom's volume. Since nuclei are so small, they cannot be seen directly, and we have to infer their existence and properties from experiments.

In the early 1900s, Sir Ernest Rutherford (1871–1937) and his students (Hans Geiger and Ernest Mardsen) performed landmark experiments that indicated the existence of nuclei. Positively charged particles called alpha (α) particles were used to bombard thin foils of gold and silver. Most of the α particles passed straight through the foils, but a few were deflected almost backward, as illustrated in Figure 24-2. Such a huge deflection indicated that an α particle had encountered a massive concentration of positive charge, the nucleus of an atom. This result was counter to the model of the atom that was popular at that time: Sir Joseph John Thomson, an English physicist, had proposed that the positive charge was spread out over the entire volume of the atom and that the negative electrons were imbedded in this sphere. (This is the so-called "plum pudding" model of the atom, with the electrons being the "plums" and the positive charge the "pudding.") The experiments conducted in Rutherford's laboratory gave indisputable evidence that the positive charge was concentrated in a small volume and that matter consists mainly of empty space.

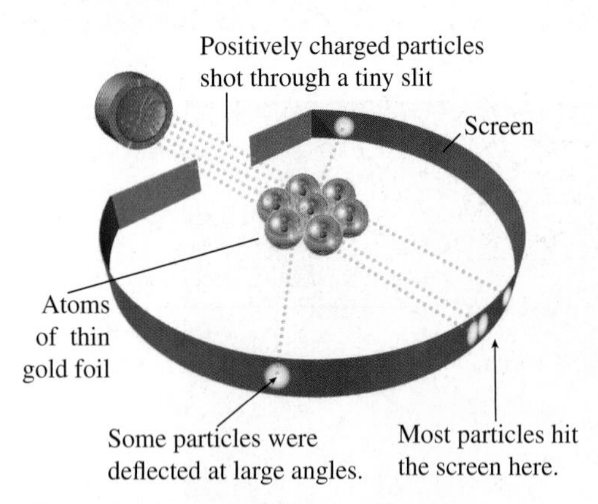

Figure 24-2 An illustration of the famous Rutherford scattering experiment conducted in the early 1900s that resulted in a better understanding of how mass is distributed within an atom.

nucleon a particle found within the nucleus of an atom, either a proton or a neutron

As depicted in Figure 24-3, we now know that the nucleus of an atom—the small, dense region at the atomic core—contains positively charged protons and electrically neutral neutrons. Protons and neutrons are sometimes referred to as **nucleons**, that is, particles in the nucleus. The nucleons attract each other through the nuclear strong force, which is much stronger than the electrical repulsion between the positively charged protons. A typical nucleus is of the order of 10^{-14} m in diameter, and the surrounding electrons are, on average, about 10^{-10} m from the centre. Protons and neutrons have roughly the same mass, with neutrons being slightly more massive: each nucleon has a mass that is about 2000 times bigger than that of an electron. The masses of a proton (m_p), a neutron (m_n), and an electron (m_e) are

$$m_p = 1.673 \times 10^{-27}\,\text{kg}$$
$$m_n = 1.675 \times 10^{-27}\,\text{kg}$$
$$m_e = 9.109 \times 10^{-31}\,\text{kg}$$

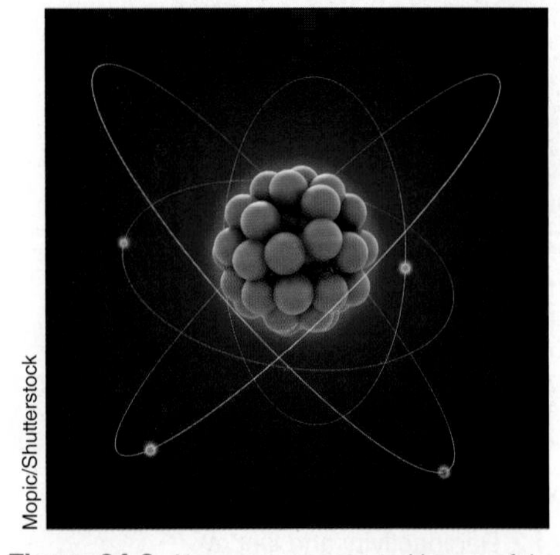

Figure 24-3 A computer-generated image of the distribution of mass in the atom. The protons are represented as red spheres packed tightly in the nucleus with the neutrons, represented as blue spheres. The orbits of the electrons are much further from the centre of the atom, not to scale in this image.

atomic mass unit a non-SI unit of mass that is often used in atomic and nuclear physics

As we can see, the masses of the atomic particles are very small in SI units. Therefore, atomic- and nuclear-scale masses are often expressed in the non-SI unit called the **atomic mass unit** (u). The mass of a carbon atom having 6 protons and 6 neutrons in the nucleus, and 6 bound electrons, is defined as exactly 12 u.

With this definition,

$$1\text{ u} = 1.660\,54 \times 10^{-27}\,\text{kg}$$

Thus, the mass of a proton, neutron, and electron in atomic mass units are

$$m_p = 1.007\,28\text{ u}$$
$$m_n = 1.008\,66\text{ u}$$
$$m_e = 5.486 \times 10^{-4}\text{ u}$$

Notice that the proton and neutron masses are each approximately 1 u while the mass of the electron is considerably smaller.

The number of protons in a nucleus is called the **atomic number**, symbolized by Z. This number determines the chemical element associated with the nucleus; for example, all nuclei with $Z = 6$ are carbon nuclei. The atomic number is sometimes referred to as the **proton number**. The **neutron number**, N, is the number of neutrons in the nucleus. The total number of protons plus neutrons is the **atomic mass number** (often called just **mass number**), and is designated by A, where $A = Z + N$. As an example, a carbon nucleus with 6 protons and 7 neutrons has $Z = 6$ and $A = 13$. The mass of an atom in u is approximately equal to the atomic mass number of its nucleus.

atomic number or **proton number** the total number of protons in a given nucleus; symbol Z

neutron number the total number of neutrons in a given nucleus; symbol N

atomic mass number or **mass number** the total number of nucleons (protons and neutrons) in a given nucleus; symbol A

nuclide a particular type of nucleus with a given atomic number (Z) and mass number (A)

All the different types of nuclei, as distinguished by their atomic numbers and mass numbers, are referred to as **nuclides**. Thus, we refer to a "chart of the nuclides," which is a chart that is similar in some ways to the chemical periodic table, listing all the types of nuclei and some of their properties, as illustrated in Figure 24-4. Note that, unlike the periodic table, the chart of the nuclides includes properties of thousands of known nuclides in the universe.

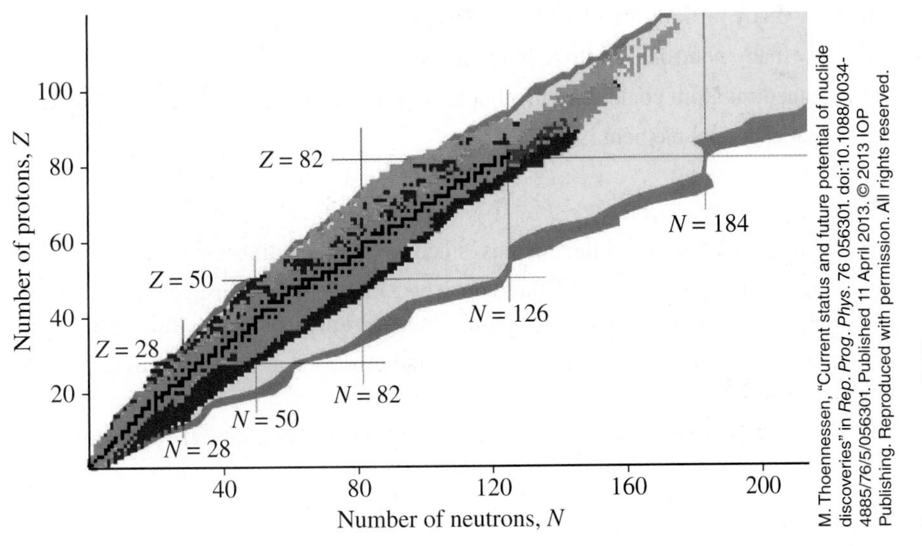

Figure 24-4 The chart of the nuclides. Similar in some ways to the periodic table, this much more extensive graphic summarizes important characteristics of all the known nuclides in the universe.

DID YOU KNOW?

Neutrons were not discovered until 1932 by James Chadwick, a student and colleague of Rutherford. Chadwick received the Nobel Prize for Physics in 1935 for this discovery.

To symbolize a given nuclide, we use the general form

$${}^{A}_{Z}\mathrm{X}$$

where X is the chemical element symbol, and Z and A are the atomic number and mass number respectively.

It should be noted that the subscript Z can also be interpreted as the charge of the nuclide in units of e. For example, a nucleus having 8 protons and 8 neutrons has $Z = 8$ and $A = 16$. This is a particular nuclide of oxygen (O): the charge of the nucleus is $+8e$ and this nuclide can be represented by the symbol ${}^{16}_{8}\mathrm{O}$. Notice that this notation is somewhat redundant since both the Z-value (8) and the chemical symbol (O) indicate the element oxygen. As a result, it is quite common to refer to this particular nuclide as simply ${}^{16}\mathrm{O}$.

Each chemical element has a variety of different possible nuclides, all with the same atomic number but different mass numbers. For example, carbon nuclei, which always have 6 protons, can have anywhere from 3 to 10 neutrons, thus giving mass numbers ranging from 9 to 16. In symbols, we have ${}^{9}_{6}\mathrm{C}$, ${}^{10}_{6}\mathrm{C}$, ..., ${}^{16}_{6}\mathrm{C}$. These nuclei are often referred to as carbon-9, carbon-10, etc. Nuclei that contain the same number of protons but different numbers of neutrons are **isotopes**, and thus the above carbon nuclei are called isotopes of

isotopes nuclei that contain the same number of protons but different numbers of neutrons

carbon. All isotopes of a given element are not equally common. For example, 99% of the stable carbon in the environment consists of ${}^{12}_{6}C$, and the remaining 1% is ${}^{13}_{6}C$. The standard chemical atomic mass for carbon, 12.01115 u/atom or 12.01115 g/mol, arises from a weighted average of atomic masses of ${}^{12}_{6}C$-atoms and ${}^{13}_{6}C$-atoms (including electrons), with the weighting factors being proportional to natural abundances. All the other carbon isotopes are **radioactive** (Section 24.2); that is, they decay or transform into nuclei of other elements after a period of time. For instance, a nucleus of ${}^{10}_{6}C$ has an average lifetime of only about 28 s, and then decays into a nucleus of boron. As we can see in Figure 24-4, only a very small portion of the chart of the nuclides consists of **stable nuclides** (represented as black squares), which can exist forever and do not naturally transform into other nuclei over time.

radioactive nuclei nuclei that spontaneously decay or transform into other nuclei over time

stable nuclei nuclei that can exist forever and do not naturally transform into other nuclei over time

SAMPLE PROBLEM 24-1

A nucleus has $Z = 13$ and $A = 27$.

(a) How many protons are there in the nucleus?

(b) How many neutrons are there in the nucleus?

(c) In a neutral atom containing this nucleus, how many bound electrons are there?

(d) What chemical element is this?

Solution

(a) There are 13 protons in the nucleus, since the atomic number Z is 13.

(b) Since the mass number is 27, there must be 27 nucleons in this nucleus. Given that there are 13 protons, there must be $27 - 13 = 14$ neutrons in the nucleus.

(c) In a neutral atom, there are as many bound electrons as there are protons in the nucleus; thus there are 13 electrons.

(d) From the periodic table of the elements (inside the back cover), we see that the element with atomic number 13 is aluminum (Al).

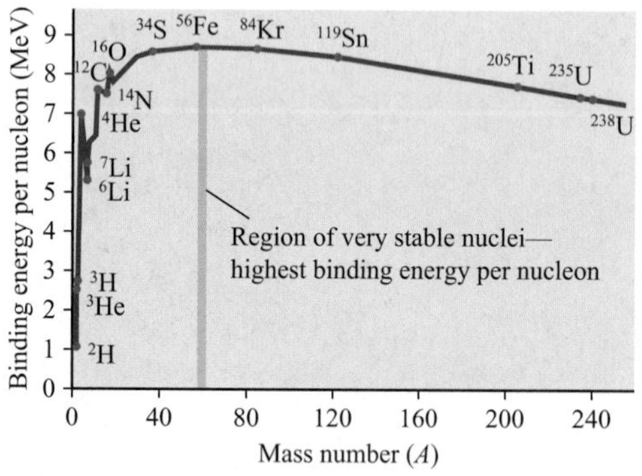

Imagine taking some protons and neutrons and putting them together to make a nucleus. Surprisingly, the mass of the nucleus is less than the mass of its separate protons and neutrons when not in the nucleus. By mass here, we mean **rest mass**, that is, the mass of stationary particles. As summarized in one of the most famous equations in all of science, $E = mc^2$, Einstein proposed in 1905 that mass and energy are equivalent, which means that the energy of a nucleus is less than the energy of its individual nucleons. Since energy is always conserved, the "missing" energy must be somewhere. It is released (as radiation, kinetic energy, etc.) when a nucleus is "assembled" from separate nucleons. This energy is called **binding energy**: the energy released when nucleons bind together to form a nucleus. This amount of energy would be required to break the nucleus apart into individual nucleons. Figure 24-5 illustrates the relationship between nuclear binding energy and mass number A for the stable nuclei in the universe. The curve reaches a maximum around $A = 60$. Based on these calculations, ${}^{56}Fe$ is the most stable nucleus in existence since it requires the greatest amount of energy to break apart.

Figure 24-5 Binding energies of various stable nuclei as a function of mass number A. ${}^{56}Fe$ has the highest binding energy per nucleon of all the stable nuclei, an important characteristic that we will return to in our discussion of fission and fusion in Section 24.4.

From HAWKES/IQBAL/MANSOUR. *Physics for Scientists and Engineers*, 1E. © 2014 Nelson Education Ltd. Reproduced by permission. www.cengage.com/permissions

rest mass the mass of stationary particles (Einstein's Theory of Special Relativity tells us that mass increases with speed)

binding energy the energy released when nucleons bind together to form a nucleus

In the following Sample Problem (24-2), we will determine the binding energy of a particular nucleus based on the difference in mass between the individual subatomic particles (protons and neutrons) and the mass of the assembled nucleus. We will then use Einstein's famous equation ($E = mc^2$) to determine the energy equivalence of this mass difference. In order to complete such a calculation you will need mass data, which are provided in Table 24-1.

Table 24-1

Atomic Masses for Selected Nuclides

Atomic Number (Z)	Element	Symbol	Mass Number (A)	Atomic Mass (u)
0	(neutron)	n	1	1.008 66
1	hydrogen	H	1	1.007 83
1	hydrogen (deuterium)	D	2	2.014 10
1	hydrogen (tritium)	T	3	3.016 05
2	helium	He	4	4.002 60
3	lithium	Li	7	7.016 01
5	boron	B	10	10.0129
6	carbon	C	12	12.000 00
8	oxygen	O	16	15.994 92
26	iron	Fe	56	55.934 94
29	copper	Cu	63	62.929 60
84	polonium	Po	218	218.008 97
86	radon	Rn	222	222.017 58
92	uranium	U	235	235.0439
92	uranium	U	238	238.050 79

SAMPLE PROBLEM 24-2

Using Table 24-1, determine the binding energy of a nucleus of $^{7}_{3}\text{Li}$.

Solution

The masses that are generally given in tables such as 24-1 are *atomic* masses, that is, masses of neutral atoms including the Z electrons. From Table 24-1, the mass of a $^{7}_{3}\text{Li}$ *atom* is 7.016 01 u. To calculate the binding energy we first determine the mass of the 10 *separate* particles in the atom: 3 protons, 4 neutrons, and 3 electrons. Then we subtract the atomic mass (7.016 01 u) from the total mass of all the constituents. Since the mass of the atomic electrons is included both before and after "assembly" of the atom, the difference in mass is due solely to the nuclear components. For the 3 protons and 3 electrons, we simplify the process by using the mass of 3 hydrogen *atoms* (each consisting of 1 proton and 1 bound electron). From Table 24-1, the mass of 3 hydrogen atoms is

$$3 \times 1.007\,83\ \text{u} = 3.023\,49\ \text{u}$$

The mass of 4 neutrons (from Table 24-1) is

$$4 \times 1.008\,66\ \text{u} = 4.034\,64\ \text{u}$$

Adding these, we get the total mass of the separate particles to be 7.058 13 u, which is greater than the 7.016 01 u mass of the assembled $^{7}_{3}\text{Li}$ atom. The difference in mass is 0.042 12 u. To change this to energy we convert the mass to kilograms, and then use $E = mc^2$ (where c is the speed of light, 3.00×10^8 m/s):

$$0.042\,12\,\text{u} = (0.042\,12\,\text{u}) \times \left(\frac{1.660\,54 \times 10^{-27}\,\text{kg}}{1\,\text{u}}\right)$$

$$= 6.994 \times 10^{-29}\,\text{kg}$$

Then,
$$E = mc^2$$
$$= (6.994 \times 10^{-29}\,\text{kg})(3.00 \times 10^8\,\text{m/s})^2$$
$$= 6.29 \times 10^{-12}\,\text{J}$$

Therefore, the energy associated with the binding of these particles together in the form of an atom of ${}^{7}_{3}\text{Li}$ is 6.29×10^{-12} J. This is a small value in the standard SI unit for energy, as is typically the case when discussing atomic- and nuclear-scale energies. Therefore, binding energies are often quoted in electron-volts (eV) or its common multiples, keV or MeV, where 1 eV = 1.60×10^{-19} J, as first introduced in Section 7.6.

Thus,
$$6.29 \times 10^{-12}\,\text{J} = 6.29 \times 10^{-12}\,\text{J} \times \frac{1\,\text{eV}}{1.60 \times 10^{-19}\,\text{J}}$$
$$= 3.93 \times 10^7\,\text{eV}$$

The binding energy of a ${}^{7}_{3}\text{Li}$ nucleus is 3.93×10^7 eV, or 39.3 MeV.

In order to compare our answer with the values plotted in Figure 24-5, we need to divide by the total number of nucleons associated with this nuclide, that is, 7. This gives 5.6 MeV per nucleon, which is consistent with the placement of the data point for ${}^{7}\text{Li}$ in Figure 24-5.

EXERCISES

24-1 A nucleus has 11 protons and 12 neutrons. What is its
(a) atomic number? (b) atomic mass number?

24-2 What is a nucleon?

24-3 What chemical element has $Z = 10$? (Consult the periodic table inside the back cover.)

24-4 A nucleus has $Z = 15$ and $A = 31$.
(a) How many protons and how many neutrons does it have?
(b) In a neutral atom containing this nucleus, how many electrons are there?
(c) What chemical element is this?

24-5 Carbon-14 and nitrogen-14 have the same mass number, but are different chemical elements. Explain.

24-6 A nucleus has the symbol ${}^{39}_{19}\text{K}$.
(a) What is its mass number?
(b) How many neutrons does it have?
(c) What is its chemical name?

24-7 Which of the following describe nuclei that are isotopes of the same element?
(a) 6 protons, 7 neutrons (c) $A = 14, Z = 6$
(b) $A = 13, Z = 7$ (d) ${}^{13}_{5}\text{B}$

24-8 What is the complete symbol for a nucleus that has 9 protons and 20 neutrons?

24-9 Use Table 24-1 to determine the binding energy (in joules) of a nucleus of ${}^{4}_{2}\text{He}$.

24-10 Using Table 24-1, calculate the binding energy (in MeV) of a nucleus of iron-56. Check that your answer is consistent with the data presented in Figure 24-5.

24-11 **Fermi Question:** How many hydrogen and oxygen nuclei are present in an Olympic-sized swimming pool?

DID YOU KNOW?

Carbon dating is a commonly used technique to determine the age of organic materials based on the content of one of the radioactive isotopes of carbon (${}^{14}_{6}\text{C}$). This technique is able to date objects up to about 40 000 years old, and was originally introduced to the scientific community by the American scientist Willard Libby in 1949, for which he received the Nobel Prize in Chemistry in 1960. Figure 24-6 shows a team of Italian scientists examining human remains found in a crypt of the cemetery of Porto Ercole. Carbon dating, in conjunction with many other tests, was employed by this team to try to solve the mystery surrounding the death of the Baroque artist Caravaggio. We will discuss dating techniques in more detail in Section 24.3.

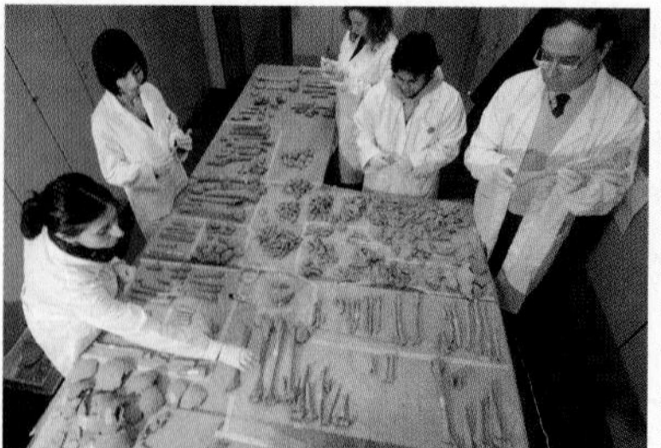
Eric Vandeville/ABACAPRESS.COM/ The Canadian Press

Figure 24-6 A team of Italian scientists uses carbon dating, among other techniques, in understanding the origins of human remains found in a crypt in Porto Ercole in Italy.

24.2 Radioactivity and Decay

When most people hear the term radioactivity, they think of dangerous substances with significant health concerns for those nearby. However, much of the radioactivity to which we are exposed occurs naturally—in soil, rocks, air, building materials, and even in our own bodies—without causing serious health issues. What is radioactivity? Why are some nuclei radioactive and others not? What kinds of radioactivity are there? In this section we explore the answers to these questions.

You might have noticed that in most of the nuclei mentioned in Section 24.1, the numbers of protons and neutrons were approximately equal. For example, in $^{12}_{6}C$ there are 6 protons and 6 neutrons. For relatively small nuclei with fewer than about 25 protons, having a roughly equal number of neutrons present results in stability, that is, the nuclei do not naturally transform into other nuclei.

Figure 24-7 provides a closer look at a specific region of the full chart of the nuclides shown in Figure 24-4. The nuclides are arranged vertically in increasing atomic number, with hydrogen ($Z = 1$) at the bottom and sulfur ($Z = 16$) at the top. The isotopes of a particular element are arranged horizontally in a row with increasing numbers of neutrons (N). As we can see in Figure 24-7, stable nuclei (blue rectangles) occur only for particular combinations of proton and neutron numbers. For low-Z nuclei, stable nuclei tend to occur when $N = Z$; for high-Z nuclei, $N > Z$ gives rise to stability. There are no stable nuclei with $Z = 43$ or 61, nor with $Z > 82$.

All unstable nuclei can be thought of as having either too many protons or too many neutrons; these nuclei are radioactive. Such nuclei **decay** or transform by emitting particles, and in the process change into different types of nuclei. **Radioactivity** is therefore simply the process of changing from one nuclide to another by the spontaneous emission of a particle. An unstable nucleus will always eventually change into a stable one, although a long series of decays into other unstable nuclei might first occur. The ultimate goal is to adjust the proton–neutron numbers to achieve stability.

decay transformation of one nuclide to another to achieve stability

radioactivity the process of changing from one nuclide to another by the spontaneous emission of a particle

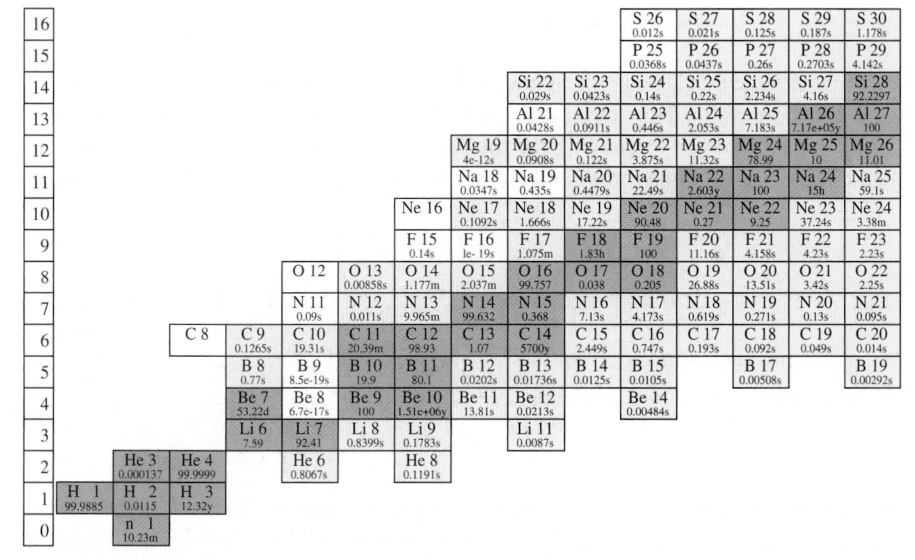

Figure 24-7 A closer look at a region of the chart of the nuclides shown in Figure 24-4. Vertically the nuclides are arranged in increasing atomic number, with hydrogen ($Z = 1$) at the bottom and sulfur ($Z = 16$) at the top. Horizontally, the isotopes of a particular element are arranged in a row with increasing numbers of neutrons (N). Stable nuclei (blue rectangles) occur only for particular combinations of proton and neutron numbers.

Since the goal is to achieve an appropriate number of protons and neutrons in the nucleus, you might guess that radioactive nuclei emit protons and neutrons to achieve this stability. It is not quite that straightforward, however. Instead, radioactive nuclei typically undergo one of two types of decay in order to achieve stability: alpha (α) or beta (β) emission. In addition, α or β decay is sometimes accompanied by the emission of gamma (γ) rays.

Alpha (α) Decay

alpha (α) particle a helium nucleus, a tightly bound collection of two protons and two neutrons

An **alpha (α) particle** is simply a helium nucleus ($^{4}_{2}He$), a tightly bound collection of two protons and two neutrons. It is called an α particle merely for historic reasons: scientists didn't know what these particles were when they were first being studied. Alpha decay tends to occur with more massive radioactive nuclei: the emission of an alpha particle results in the nucleus decreasing in both atomic number (by 2) and mass number (by 4). An example is the decay of radium-226 into radon-222, illustrated in Figure 24-8. This can be written as a nuclear decay equation:

$$^{226}_{88}Ra \rightarrow ^{222}_{86}Rn + ^{4}_{2}He$$

There are several important aspects to α decay illustrated in the process above:

- Electrical charge is conserved: the sum of the atomic numbers of the decay products (on the right hand side of the arrow) equals the atomic number of the original nucleus.
- Mass and energy are conserved: the sum of the mass numbers of the decay products equals the mass number of the original nucleus. The small difference in mass between the products and the original nucleus due to differences in binding energies is carried away as kinetic energy, mainly by the α particle.
- The original nucleus is called the **parent nucleus** (^{226}Ra); the resulting nucleus is called the **daughter nucleus** (^{222}Rn).

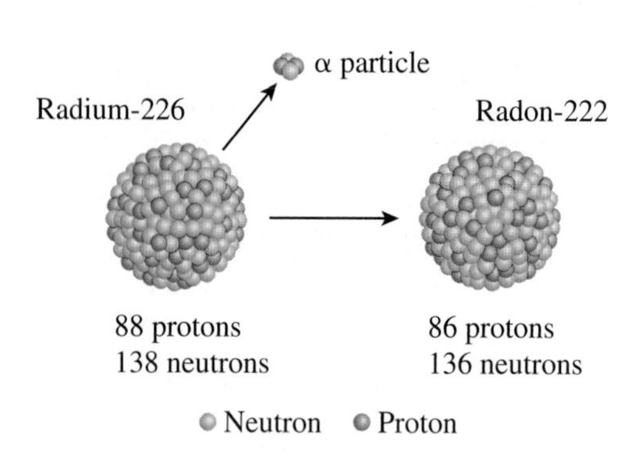

Figure 24-8 The α decay of radium-226.

parent nucleus the original nucleus that undergoes nuclear transformation

daughter nucleus the product nucleus that results from nuclear transformation

The α particle that is emitted has a well-defined energy, based on the energy released when the more stable daughter nucleus is formed from the less stable parent. By comparing the mass of the parent with the total mass of the products in the decay of ^{226}Ra, for example, we find that there is 4.87 MeV released as kinetic energy in this process.

SAMPLE PROBLEM 24-3

(a) Write the nuclear decay equation for the α decay of $^{222}_{86}Rn$.

(b) Determine the amount of energy released in this process.

Solution

(a) Given that this isotope of radon is an alpha emitter, we know that the atomic number of the daughter nucleus is $86 - 2 = 84$, which is polonium (from the periodic table inside the back cover). We also know that the mass number of the daughter nucleus is $222 - 4 = 218$. Therefore, the nuclear decay equation for this process is

$$^{222}_{86}Rn \rightarrow ^{218}_{84}Po + ^{4}_{2}He$$

(b) First we need to determine the mass difference between the products and the parent nucleus. From Table 24-1, the total mass of the products is (218.008 97 u + 4.002 60 u) or 222.011 57 u. Table 24-1 also tells us that the parent nucleus has an atomic mass of 222.017 58 u, which is greater than the mass of the products. This means that energy was released in the process, which we can calculate via $E = mc^2$. The mass difference is given by 222.017 58 u − 222.011 57 u = 0.006 01 u. In order to convert this to the equivalent energy, we need to express the mass difference in kilograms:

$$0.006\,01\text{ u} = (0.006\,01\text{ u}) \times \left(\frac{1.660\,54 \times 10^{-27}\text{ kg}}{1\text{ u}}\right)$$

$$= 9.980 \times 10^{-30}\text{ kg}$$

Then,

$$E = mc^2$$

$$= (9.980 \times 10^{-30}\text{ kg})(3.00 \times 10^{8}\text{ m/s})^2$$

$$= 8.98 \times 10^{-13}\text{ J}$$

Converting this energy difference to electron volts,

$$8.98 \times 10^{-13}\text{ J} = 8.98 \times 10^{-13}\text{ J} \times \frac{1\text{ eV}}{1.60 \times 10^{-19}\text{ J}}$$

$$= 5.61 \times 10^{6}\text{ eV}$$

Therefore, there is 5.61 MeV of energy released in the α decay of ^{222}Rn.

DID YOU KNOW?

You may have one or more radioactive α emitters in your house or apartment. One type of smoke detector uses americium-241, a source of α particles, to detect smoke particles in the air as they interact with the continuous emission of α particles from the source (Figure 24-9).

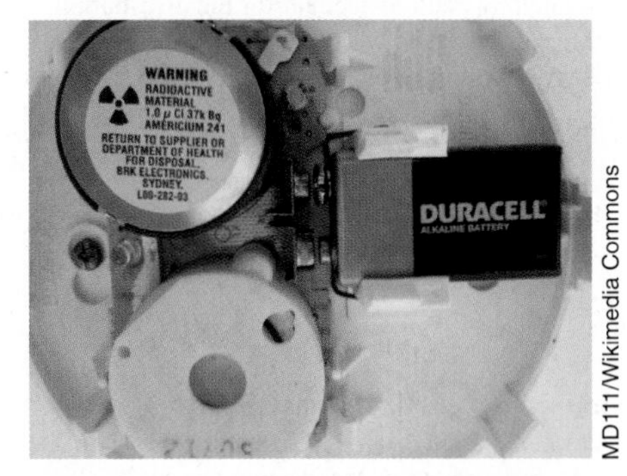

MD111/Wikimedia Commons

Figure 24-9 One type of household smoke detector uses an α-particle emitter to sense particulate matter in the air. If your smoke detector uses such a source there will be a radioactive symbol on the device such as the black and yellow sticker shown here.

Beta (β) Decay

β⁻ Emission

beta (β) decay a radioactive decay process in which the nucleus emits a beta (β) particle

beta (β⁻) particle an electron emitted in radioactive decay by nuclei with an excess of neutrons; represented in decay equations as $_{-1}^{0}\text{e}$

The vast majority of radioactive nuclei attempt to attain stability by a second process, **beta (β) decay**. Nuclei that lie to the right of the region of stability in Figure 24-7 have too many neutrons (not enough protons). In order to become stable, they need to gain positive charge or, equivalently, lose negative charge. Such nuclei often decay by the emission of an electron, called a **beta (β⁻) particle** in this scenario. This may seem odd, giving a fundamental particle like an electron another name, but this is merely due to history as it wasn't clear to scientists at the time that the newly discovered β⁻ particle was actually an electron. Since a β⁻ particle is simply an electron, it is represented in $_{Z}^{A}\text{X}$ notation as $_{-1}^{0}\text{e}$ given that the mass number is zero and the charge of the particle is $-1e$.

Essentially, in β⁻ decay, a neutron in the nucleus transforms into a proton and an electron; the proton remains in the nucleus and the electron is ejected. Thus, the atomic number of the nucleus increases by 1, but the atomic mass number is unchanged. As an example, consider the isotopes of carbon. Carbon-12 and carbon-13 are stable, but carbon-14, -15, and -16 are neutron-rich and undergo β⁻ decay. For carbon-15 the nuclear decay equation for this process can be written as

$$^{15}_{6}\text{C} \rightarrow {}^{15}_{7}\text{N} + {}_{-1}^{0}\text{e} + \bar{\nu}$$

The daughter nucleus, nitrogen-15, is stable in this case. The symbol $\bar{\nu}$ (an antineutrino) will be discussed further shortly. Notice again here that *the sum of the atomic numbers of the decay products equals the atomic number of the original nucleus* (conservation of charge). Similarly, *the sum of the resulting mass numbers equals the original mass number* (conservation of mass/energy). These "sum rules" for Z and A are valid for all types of decay and for all processes involving nuclei.

DID YOU KNOW?

Radiation is used in surprising places. Mantles used in propane lamps, such as those used by campers (Figure 24-10), contain radioactive thorium. This is why the instructions that come with such mantles always advise you to wash your hands after handling them.

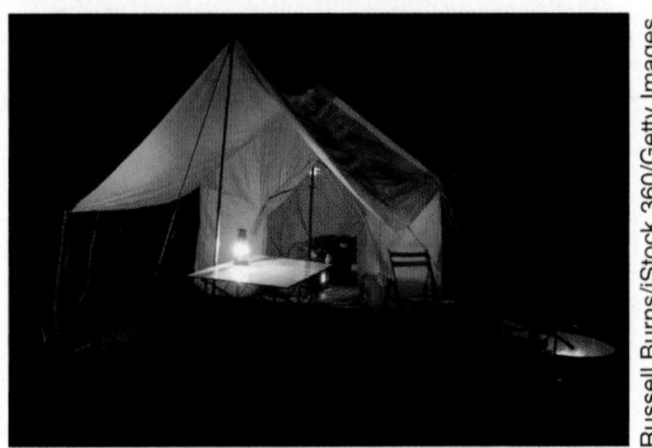

Russell Burns/iStock 360/Getty Images

Figure 24-10 The mantle inside a camping propane lamp contains radioactive thorium.

Unlike the particles emitted in α decay, it has been observed that the β^- particles from a specific transformation have a range of possible energies. This was a significant puzzle to the early researchers in the field of radiation physics: conservation of energy and mass require a β^- particle to have a single, well-defined energy related to the small difference in mass between the parent and daughter nuclei if it is the only other product of the decay. The emission of an additional particle, called the **anti-neutrino**, was postulated by Wolfgang Pauli in 1930 to explain this inconsistency between theory and experiment, but it took another 20+ years before the anti-neutrino was first detected experimentally. These curious particles have no charge (0 for the atomic number), essentially no mass on the scale of nucleon masses (0 for the mass number), and interact very weakly with matter. They are nevertheless an essential part of the β-decay process, as they carry energy away.

anti-neutrino a weakly interacting particle postulated by Pauli to account for the distribution of energies observed with β^- particles emitted in β decay, represented as $\bar{\upsilon}$ in decay equations

β^+ Emission

A nucleus to the left of the stable region in Figure 24-7 has an excess of protons. A common mode of decay for such a nucleus involves the emission of a **positron (β^+ particle)**, which is a particle having the same mass as an electron, but a charge of $+1e$. A positron is the **anti-particle** of an electron since it has the same mass but the opposite charge. Given the notation used to represent a β^- particle discussed previously, you may already have guessed that the positron is symbolized as ${}^{0}_{+1}\text{e}$, as this indicates a charge of $+1e$ and zero mass number.

positron (β^+ particle) the anti-particle of the electron, having the same mass but the opposite charge, represented in decay equations as ${}^{0}_{+1}\text{e}$

anti-particle a particle that has the same mass as its particle counterpart but the opposite charge; the positron and electron are anti-particles of each other

In positron emission, a proton in the nucleus transforms into a neutron that stays in the nucleus and a positron that is ejected. Thus, the atomic number of the nucleus decreases by 1 and the atomic mass number is unchanged. As examples, ${}^{9}_{6}\text{C}$, ${}^{10}_{6}\text{C}$, and ${}^{11}_{6}\text{C}$ are all positron emitters, and the decay of ${}^{10}_{6}\text{C}$ is summarized by

$${}^{10}_{6}\text{C} \rightarrow {}^{10}_{5}\text{B} + {}^{0}_{+1}\text{e} + \upsilon$$

In the case of positron emission, there is a corresponding **neutrino** (υ) released as well.

neutrino a weakly interacting particle postulated by Pauli to account for the distribution of energies observed with β^+ particles emitted in β decay, represented as υ in decay equations

When a daughter nucleus from either α or β decay is unstable, it too decays, creating yet another type of nucleus. A sequence of decays from one unstable nucleus to the next, eventually ending in a stable nucleus, is called a **decay series**. Figure 24-11

decay series a sequence of decays from one unstable nucleus to the next, eventually ending in a stable nucleus

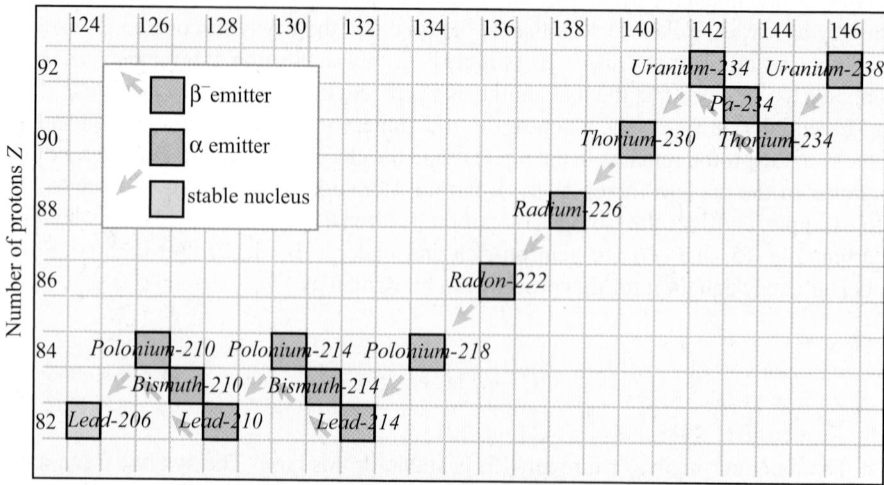

Figure 24-11 A decay series starting with uranium-238 and ending with lead-206 (stable). Alpha decay results in a decrease in proton and neutron numbers each by two, whereas β^- decay results in an increase by one in the proton number and a decrease by one in the neutron number. There are 14 transformations before stability is reached.

shows a decay series beginning with uranium-238 and ending with lead-206, which is stable. This series is one of three that occur naturally; the other two begin with thorium-232 and uranium-235. These three series account for much of the natural radiation to which people are exposed. This will be discussed in greater detail in Section 24.5.

SAMPLE PROBLEM 24-4

The helium isotope $^{6}_{2}He$ is unstable. **(a)** What kind of decay would it likely undergo? **(b)** What would be the daughter nucleus? **(c)** Write the nuclear decay equation for this process.

Solution

(a) This nucleus has 2 protons and 4 neutrons. Thus, there is an excess of neutrons, and we would expect it to undergo β^- decay, emitting an electron. (Indeed, this is its mode of decay.)

(b) Let the daughter nucleus be represented by $^{A}_{Z}X$. When a nucleus undergoes β^- decay, emitting an electron, the process is essentially the conversion of a neutron into a proton and an electron. Therefore, the mass number of the daughter is the same as the mass number of the parent; so $A = 6$. Also, the atomic number of the daughter is one more than that of the parent; so $Z = 3$. Based on the periodic table inside the back cover, $Z = 3$ corresponds to lithium (Li). Therefore, the daughter nucleus is $^{6}_{3}Li$.

(c) In general, the nuclear decay equation for β^- decay is

$$^{A}_{Z}X \rightarrow {}^{A}_{Z+1}Y + {}^{0}_{-1}e + \bar{\upsilon}$$

where X is the parent nucleus with atomic number Z and mass number A; Y is the daughter nucleus with correspondingly altered atomic number; $^{0}_{-1}e$ is the beta particle, and $\bar{\upsilon}$ is the anti-neutrino. In this case, $^{6}_{2}He$ is the parent nucleus and $^{6}_{3}Li$ is the daughter. Therefore, the nuclear decay equation for this process is

$$^{6}_{2}He \rightarrow {}^{6}_{3}Li + {}^{0}_{-1}e + \bar{\upsilon}$$

Gamma Rays

Often when a daughter nucleus is created via α or β decay, it has excess energy that is emitted in the form of gamma (γ) rays. **Gamma rays** are high-energy electromagnetic radiation, as discussed in Chapter 16. Since γ radiation has no electric charge and no (rest) mass, γ emission does not involve the change of atomic number or mass number of the nucleus; it is merely the release of excess energy from the nucleus with no inherent change to the nuclear composition. Although γ rays are typically associated with nuclear decay, they are also the high-energy photons generated in annihilation between a particle and its corresponding anti-particle.

Table 24-2 summarizes the general forms of the nuclear decay equations for all of the transformations discussed in this section. In the table we have used the following notation:

- X is the parent nucleus with atomic number Z and mass number A
- Y is the daughter nucleus with correspondingly altered atomic number and mass number

DID YOU KNOW?

When an electron and a positron, or any particle and anti-particle pair meet, they undergo **annihilation**, in which their mass and energy are converted into two high-energy photons. The energies and momenta of the two photons are dictated by the fundamental conservation principles discussed extensively throughout this book. An imaging technique called **Positron Emission Tomography (PET)** (Figure 24-12) is based on this annihilation process and will be discussed in greater detail in Section 24.6.

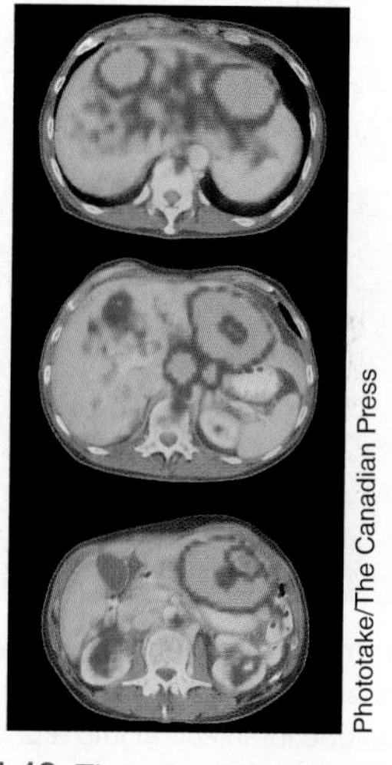

Phototake/The Canadian Press

Figure 24-12 Three images of the torso generated, in part, through the detection of the two high-energy photons emitted when electrons and positrons undergo annihilation during a PET scan.

annihilation an interaction that occurs when any particle/anti-particle pair meet: their mass and energy are converted into two high-energy photons in a process in which total energy and momentum are conserved

Positron Emission Tomography (PET) an imaging technique used to diagnose disease based on the annihilation of electrons and positrons

gamma rays high-energy electromagnetic radiation, often emitted by a nucleus to release excess energy without any inherent change to the nuclear composition

Table 24-2

General Forms of Nuclear Decay Equations

Process	Decay Equation
α decay	$^{A}_{Z}X \rightarrow {}^{A-4}_{Z-2}Y + {}^{4}_{2}He$
β^- decay	$^{A}_{Z}X \rightarrow {}^{A}_{Z+1}Y + {}^{0}_{-1}e + \bar{\upsilon}$
β^+ decay	$^{A}_{Z}X \rightarrow {}^{A}_{Z-1}Y + {}^{0}_{1}e + \upsilon$
γ-ray emission	$^{A}_{Z}X^* \rightarrow {}^{A}_{Z}X + \gamma$

DID YOU KNOW?

Some larger medical centres have a Department of Nuclear Medicine, which is responsible for preparing radioactive isotopes for use in diagnosing and treating disease. Figure 24-13 shows a staff member carefully preparing a sample for injection with the proper protective equipment in use. This sample may well be that of technetium-99, an isotope that emits a 140 keV γ ray. This isotope is considered to be the "workhorse" of Nuclear Medicine, used in tens of millions of procedures annually around the world, with approximately 30 to 40% of the world's supply of this isotope produced in Canada. We will discuss several examples of radioactive isotopes used medically in greater detail in Section 24.6.

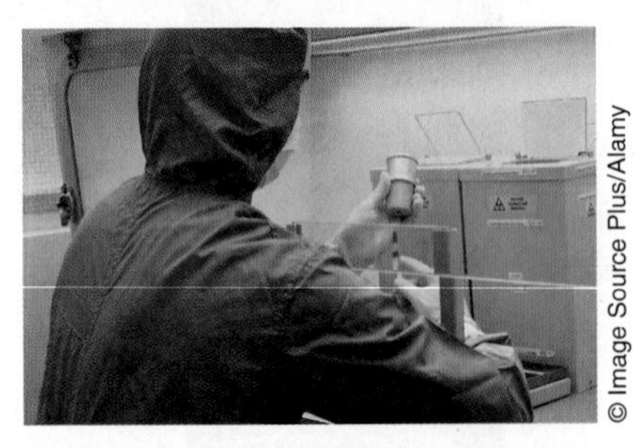

© Image Source Plus/Alamy

Figure 24-13 A staff member in the Department of Nuclear Medicine in a large medical centre preparing a radioactive isotope for injection into a patient.

- ${}_{-1}^{0}e$ is a β^- particle (electron)
- ${}_{1}^{0}e$ is a β^+ particle (positron)
- $\bar{v}$ is an anti-neutrino
- v is a neutrino
- X* refers to an isotope with excess energy

The Discovery of Radioactivity

There is a great deal written about the scientific method, but many of the greatest discoveries in science have been made accidentally. In 1896, a French physicist named Henri Becquerel (Figure 24-14) developed some photographic film on which uranium samples had been stored by chance in a closed drawer. He was surprised to find that the developed film showed the silhouettes of the uranium pieces. A less alert person might have assumed that the film exposure had been accidental, but Becquerel decided to investigate this observation further. Through his subsequent work, Becquerel is credited with discovering radioactivity, as he had detected radiation from the uranium samples with his exposed film.

Shortly after Becquerel's discovery, the married scientific team of Marie and Pierre Curie (Figure 24-15) became interested in radioactivity and performed a number of important experiments. In particular, they succeeded in separating two new elements from uranium ores. These elements, much more radioactive than uranium, were named polonium (after Marie Curie's home country, Poland) and radium. In 1903 the Curies and Becquerel shared the Nobel Prize in Physics. You can read more about Marie Sklodowska Curie in the following Profile.

As the early investigators of radioactivity probed the secrets of radiation, one of the ways they discovered to distinguish α, β, and γ radiation was the effect of a magnetic field on the radiation. Recall from Chapter 22 that a charged particle moving in a magnetic field experiences a force. Thus, α and β particles experience a magnetic force, whereas γ rays do not. Figure 24-16 shows the general behaviour of the various kinds of radiation in a magnetic field. Since α and β^- particles have charges of opposite sign, the forces on these particles are in opposite directions. Because α particles are much more massive than β particles, the acceleration of β particles is much larger and they move on circular arcs of a smaller radius.

Photos.com/Thinkstock

Figure 24-14 Portrait of Henri Becquerel (1852–1908), the French scientist credited with the discovery of radioactivity.

Photos.com/Thinkstock

Figure 24-15 Pierre Curie (1859–1906) and Marie Curie (1867–1934).

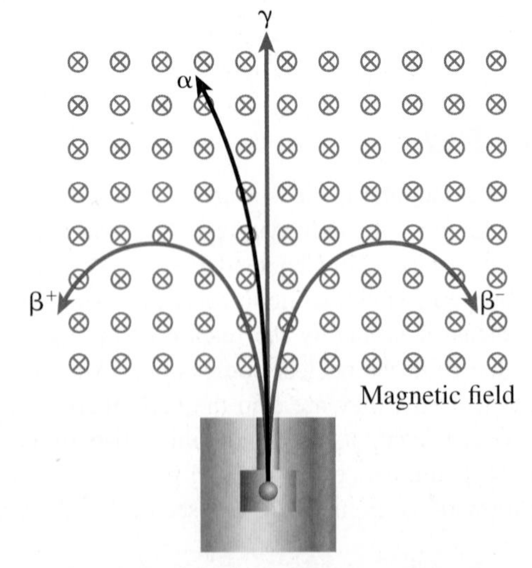

Figure 24-16 The effect of a magnetic field on various kinds of radiation. The magnetic field here is uniform and directed into the page, as indicated by the symbol ⊗.

Pioneering Female Scientist Marie Sklodowska Curie (1867–1934)

"All my life through, the new sights of Nature made me rejoice like a child."

MARIE CURIE, FROM HER 1923 BIOGRAPHY OF HER HUSBAND ENTITLED *PIERRE CURIE*.

Marie Sklodowska was born in Warsaw, Poland, in 1867. Marie was the youngest of five children, and having two teachers for parents meant that education was very important in their household. Marie was encouraged to pursue the physical sciences in particular by her cousin, Józef Boguski, who ran the Warsaw Museum of Industry, where Marie was allowed to conduct physics and chemistry experiments on weekends!

Marie began her university-level studies in secret in Poland, through the "flying university," so named because each session was held in a different private residence in a clandestine nature. Approximately 70% of the student body was female, with classes open to anyone regardless of gender or social status, unlike the public institutions of the day in Poland. In 1891, Marie followed her older sister Bronya to Paris to study at Sorbonne University. Although her preparation at the flying university was insufficient for the high expectations at Sorbonne, Marie worked hard and finished first in her master's degree physics course in 1893.

In 1894, through research collaborators, Marie was introduced to Pierre Curie, a physicist who was the laboratory chief at the Municipal School of Industrial Physics and Chemistry in Paris at the time. They married in July 1895 and welcomed their first child, Irène, to the family in 1897. In what must have been a most unusual arrangement for the time, Pierre's widowed father, a retired physician, became Irène's trusted caregiver while Marie continued her scientific work.

Marie chose to study the recently observed phenomenon of "uranium rays" reported by Henri Becquerel for her doctoral thesis. She conducted many important experiments, which informed a crucial hypothesis: The rays emitted by uranium compounds are an atomic property of the element uranium. This idea was revolutionary and led to a new era of scientific understanding.

In 1898, after further experiments with uranium and thorium, Marie coined the term "radioactivity" to describe this newly observed phenomenon. Pierre joined Marie in her studies of radioactive minerals. In July 1898, the Curies reported their discovery of a new element, which they named polonium in honour of Marie's homeland. In December 1898, they reported discovering a second new element, which they named radium from the Latin word for ray.

In 1903, based on her work on radioactivity, Marie became the first woman to earn a doctorate in France. Later in the same year, Marie was honoured with the joint awarding of a Nobel Prize in Physics, shared with Henri Becquerel and Pierre. In 1904, Pierre and Marie welcomed a second daughter to their family, Ève. This was followed shortly thereafter by a tragic accident: Pierre was killed while crossing a busy street in Paris in 1906.

After Pierre's unexpected death, Sorbonne University invited Marie to take up her husband's academic post, becoming the first female professor at this prestigious institution. Marie continued her work and in 1911 became the first scientist to be awarded a second Nobel Prize, this time in Chemistry. Marie dedicated most of the rest of her life to the newly created Radium Institute, as both a tribute to Pierre's memory and in service to the advancement of science.

Health effects of prolonged exposure to ionizing radiation began for Marie in 1920 when she was diagnosed with cataracts. She became more gravely ill in 1934, with an incurable blood disease "probably because it had been injured by a long accumulation of radiations." Marie Sklodowska Curie died on July 4, 1934. In 1935, Marie's legacy was recognized once again by the Nobel Committee, when her daughter Irène and son-in-law Frédéric Joliot were awarded the prize in Chemistry for their work in generating radioactive elements through nuclear reactions. With almost too many firsts to itemize here, Marie Sklodowska Curie, shown in her laboratory in Paris in Figure 24-17, was truly a pioneer of modern physics.

Photos.com/Thinkstock

Figure 24-17 Marie Curie in her laboratory in Paris in 1912.

The Production of Radioactive Nuclei

How are radioactive nuclei made? If they always decay eventually into stable nuclei, why are there any radioactive nuclei in existence at all? To answer the first question, we must first consider how nuclei are formed in general. In the process of nuclear fusion (Section 24.4) that occurs in stars, small nuclei such as hydrogen are fused together to make larger nuclei. Elements up to $Z = 26$ (iron) are believed to be produced this way. More massive nuclei are generated in supernova explosions, the death phase of very large stars. The nuclei that have been created by fusion during a star's lifetime, and by the supernova, are distributed far into space by these explosions. Many of the nuclei in your body right now were made in stars. (Isn't it nice to think that you are made of stardust?)

Some nuclei produced in stars and distributed in this way happen to be radioactive, and if they decay slowly enough, they exist for a very long time. The key to whether a radioactive nuclide occurs commonly is simply its rate of decay. If it decays quickly, not much remains (more about this in Section 24.3). Nuclei of the most common isotope of uranium, $^{236}_{92}\text{U}$, have an average lifetime of several billion years and thus $^{236}_{92}\text{U}$ is reasonably abundant on Earth. Since $^{236}_{92}\text{U}$ is the parent in a long decay series, all long-lived members in the series are also found to be relatively common. Another example of an abundant natural radioactive nuclide is potassium-40, which constitutes 0.1% of all natural potassium; nuclei of potassium-40 exist on average for over a billion years.

nuclear reaction a process in which two or more nuclei (or particles) interact to produce new nuclei

transmutation the changing of one type of nucleus into another; this can be achieved through a nuclear reaction or through nuclear decay

Not all radioactive nuclei are produced in stars; some are continually produced by the interaction of cosmic rays (primarily high-energy protons) with atoms in Earth's atmosphere. For example, radioactive carbon-14, which is useful in dating archaeological artifacts (Section 24.3), is produced from atmospheric nitrogen. Nuclei of nitrogen are struck by neutrons generated by collisions between cosmic rays and other nuclei, and carbon-14 is created along with a proton through the reaction

$$^{14}_{7}\text{N} + {}^{1}_{0}\text{n} \rightarrow {}^{14}_{6}\text{C} + {}^{1}_{1}\text{H}$$

The production of $^{14}_{6}\text{C}$ is one example of a **nuclear reaction**: a process in which a nucleus (such as $^{14}_{7}\text{N}$) interacts with another nucleus or particle (such as $^{1}_{0}\text{n}$) to produce a different nucleus (such as $^{14}_{6}\text{C}$). This is also an example of a **transmutation**, that is, the changing of one type of nucleus into another. All nuclear reactions and all α and β decays are transmutations. However, α and β decays are not nuclear reactions because they do not involve the interaction of two nuclei, or of one nucleus and another particle. In all transmutations (decays and reactions), atomic number and mass number are conserved.

Some radioactive nuclei on Earth today have been produced by human activities rather than cosmological processes. For example, there are radioactive nuclei in the atmosphere and soil due to historic nuclear weapons testing. In addition, as we will discuss in more detail in Section 24.6, many radioactive nuclei are made for the purposes of diagnosis or treatment of disease by bombarding stable nuclei with high-energy particles from a reactor or particle accelerator. For example, in the 1950s, Canadian scientist Dr. Harold E. Johns developed a cancer treatment using the beta-emitting isotope cobalt-60, produced by collisions of neutrons from a reactor and the stable nuclide cobalt-59:

$$^{59}_{27}\text{Co} + {}^{1}_{0}\text{n} \rightarrow {}^{60}_{27}\text{Co} \rightarrow {}^{60}_{28}\text{Ni} + {}^{0}_{-1}\text{e} + \bar{\upsilon} + \gamma$$

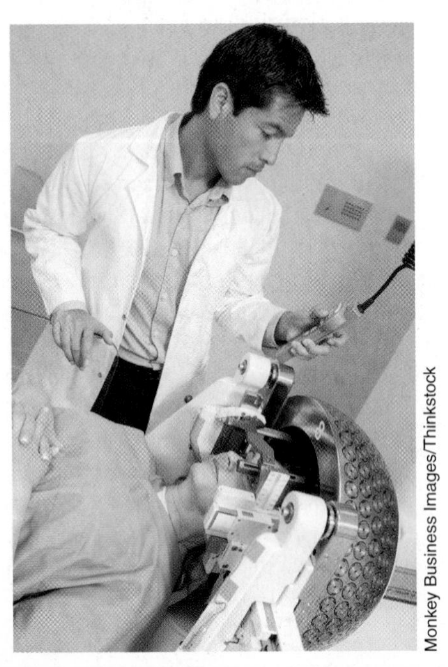

Monkey Business Images/Thinkstock

Figure 24-18 In a Gamma Knife treatment, specialized equipment directs ~200 beams of γ rays from ^{60}Co onto a tumour or other target. This treatment is typically used when a tumour or other abnormality in the brain is too hard to reach with standard neurosurgery.

The modern-day version of this treatment can be seen in Figure 24-18. The patient is being positioned inside a device called the Gamma Knife, which consists of approximately 200 cobalt-60 sources around the array. The healthcare team then delivers a carefully designed treatment to the patient that ensures the maximum effect at the site of the tumour or other abnormality, with minimal effect elsewhere.

EXERCISES

24-12 (a) How do the numbers of protons and neutrons compare for stable low-Z nuclei?

(b) How do the numbers of protons and neutrons compare for stable high-Z nuclei?

24-13 Determine the unknown quantities in each of the following nuclear decay equations:

(a) ${}^{38}_{20}\text{Ca} \rightarrow {}^{38}_{19}\text{K} + ?$

(b) ${}^{?}_{?}\text{Er} \rightarrow {}^{4}_{2}\text{He} + {}^{150}_{66}\text{Dy}$

(c) ${}^{298}_{92}\text{U} \rightarrow {}^{4}_{?}\text{He} + ?$

(d) ${}^{40}_{19}\text{K} \rightarrow {}^{0}_{-1}\text{e} + ?$

24-14 In the decay shown in part (a) of the previous question, which nucleus is the parent? daughter?

24-15 Which one of the following statements is true for β^- decay?

(a) The mass number increases by 1.

(b) The mass number decreases by 1.

(c) The atomic number increases by 1.

(d) The atomic number decreases by 1.

24-16 Repeat the previous question for β^+ decay.

24-17 The nuclide ${}^{18}_{10}\text{Ne}$ is unstable.

(a) What type of decay is it likely to undergo?

(b) Write the nuclear equation for the decay.

24-18 Repeat the previous question for the unstable nuclide ${}^{28}_{12}\text{Mg}$.

24-19 How does γ radiation differ from visible light?

24-20 When a nucleus emits a γ ray, how are its atomic number and mass number changed?

24-21 Which type of radiation is not affected by a magnetic field?

24-22 True or false? If false, explain your answer. All radioactive isotopes are generated by human activity.

24-23 Identify the unknown quantity in each of the following nuclear reactions:

(a) ${}^{14}_{7}\text{N} + {}^{1}_{0}\text{n} \rightarrow {}^{14}_{6}\text{C} + ?$

(b) $? + {}^{1}_{0}\text{n} \rightarrow {}^{60}_{27}\text{Co}$

(c) ${}^{6}_{3}\text{Li} + {}^{3}_{2}\text{He} \rightarrow {}^{8}_{5}\text{B} + ?$

24.3 Characteristics of Radioactive Decay

half-life the time required for half of the original sample of a nuclide to decay; symbol $T_{1/2}$

average lifetime the average time a particular nucleus of a given isotope will exist before decay; symbol T_{avg}

We mentioned in Section 24.2 that the relative abundance of radioactive nuclides depends in large part on how quickly they decay. Some types of nuclei decay very quickly, and others live for billions of years. The colours in Figure 24-7 are used to distinguish short-lived from long-lived isotopes. As mentioned previously, the blue rectangles are the stable isotopes. Arranged around this region of stability, we see isotopes that are relatively long lived (green), many that are quite short lived (yellow), and some nuclei that have an intermediate lifetime (red).

As an example, nuclei of ${}^{11}_{6}\text{C}$ live for 29 min on average; however, it is impossible to predict how long a given ${}^{11}_{6}\text{C}$ nucleus will exist before decaying. Some ${}^{11}_{6}\text{C}$ nuclei live for less than a second, and others live for months. The 29 min is just the *average* lifetime. We usually specify the typical lifetime for a radioactive nucleus by stating its **half-life** ($T_{1/2}$), defined as the time required for half of the parent nuclei in a given sample to decay. For example, the half-life of ${}^{3}_{1}\text{H}$ (tritium) is 12 y. Thus, if a sample of tritium today has 8×10^{23} nuclei (mass = 4 g), then in 12 y, 4×10^{23} nuclei will have decayed and 4×10^{23} will remain. After another 12 y, half this remaining number will decay, leaving 2×10^{23} nuclei of ${}^{3}_{1}\text{H}$ in the sample. After another 12 y, 1×10^{23} nuclei will remain, and so on. After n half-lives, the number of nuclei in the sample has decreased by a factor of 2^n from the original number. A graph of the number of ${}^{3}_{1}\text{H}$ nuclei as a function of time is shown in Figure 24-19.

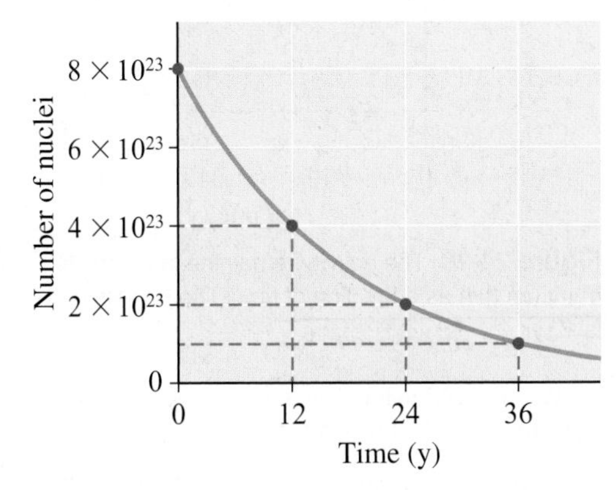

Figure 24-19 The number of nuclei remaining in a sample of tritium (${}^{3}_{1}\text{H}$) as a function of time. The half-life of ${}^{3}_{1}\text{H}$ is 12 y.

Notice in Figure 24-19 that half the original nuclei live longer than 12 y, some of them much longer. Therefore, the *average* lifetime is longer than the half-life. It can be shown (using calculus) that the **average lifetime** (T_{avg}) is related to the half-life ($T_{1/2}$) by

$$T_{avg} = 1.44\,T_{1/2} \qquad (24\text{-}1)$$

Hence, the average lifetime of a tritium nucleus is $1.44 \times 12\text{ y} = 17\text{ y}$.

activity the number of nuclei that decay per unit time; symbol A

becquerel the SI unit of activity, one decay per second; symbol Bq

The number of nuclei that decay per unit time is called the **activity** (A) of the sample. The SI unit of activity is the **becquerel** (Bq), which is one decay per second. The activity is a measure of the strength of the sample; a high-activity sample corresponds to many decays in a given period of time. The activity is directly proportional to the number of remaining nuclei in the sample; as the number of nuclei in a sample decreases with time, the activity decreases proportionately. The activity also depends on the half-life of the isotope: a shorter-lived nuclide with a short half-life will decay more quickly, and therefore has greater strength than a longer-lived isotope.

$$A \propto N \quad (24\text{-}2)$$

where N is the number of radioactive nuclei in the sample. Or

$$A = \lambda N \quad (24\text{-}3)$$

where λ is called the decay constant for the particular isotope.

decay constant a property of a given radioactive nuclide that describes the rate of decay, related to half-life; symbol λ

Notation Tip

Be careful with notation in this chapter, particularly when it comes to N and A. N can represent either the number of neutrons in a nuclide or the number of nuclei in a sample in the case of radioactive decay. Similarly, A can represent either the mass number of a nuclide or the activity of a sample in the case of radioactive decay. The meaning should be clear from the given scenario as well as the units provided.

The constant of proportionality in Eqn. 24-3 is called the **decay constant** (λ) of the isotope. When the decay constant has units of s^{-1}, activity will have units of decays per second (Bq). It can be demonstrated that the decay constant and the half-life are related by

$$T_{1/2} = \frac{0.693}{\lambda} \quad (24\text{-}4)$$

The longer the half-life, the smaller the value of the decay constant, which corresponds to a smaller activity as discussed above. When the half-life is in seconds, the decay constant will have units of s^{-1}.

Given the relationship in Eqn. 24-2, a graph of activity (A) versus time has the same general shape as that of the number of remaining nuclei (N) versus time. As before, after a period of one half-life, the activity of the sample is reduced by a factor of 2. Figure 24-20 shows a graph of activity versus time for our tritium sample that starts with 8.0×10^{23} nuclei. Given that the half-life of $^{3}_{1}H$ is 12 y, or 3.8×10^{8} s, we can use Eqn. 24-4 to determine the decay constant for this isotope: $\lambda = 0.693/T_{1/2} = 0.693/(3.78 \times 10^{8}\ s) = 1.8 \times 10^{-9}\ s^{-1}$. This is the scaling factor by which the y-axis of Figure 24-19 is converted from number of nuclei (N) to activity (A), via Eqn. 24-3. So at time $t = 0$, there are 8.0×10^{23} $^{3}_{1}H$ nuclei present, which corresponds to an initial activity of 1.5×10^{15} Bq. Because activity in becquerels often involves very large numbers, another (non-SI) unit is sometimes used: the **curie** (Ci), for which 1 Ci is equivalent to 3.7×10^{10} Bq.

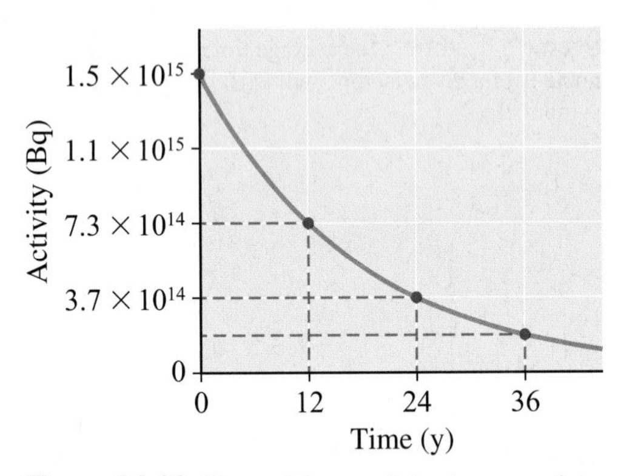

Figure 24-20 The activity remaining in a sample of tritium ($^{3}_{1}H$) as a function of time. The half-life of $^{3}_{1}H$ is 12 y.

curie a non-SI unit of activity, for which 1 Ci is equal to 3.7×10^{10} Bq; symbol Ci

SAMPLE PROBLEM 24-5

The half-life of uranium-238 is 4.5×10^{9} y.

(a) What is the decay constant of this isotope?

(b) How long would it take for the activity of a sample of uranium-238 to decrease by a factor of 16?

Solution

(a) From Eqn. 24-4, we can determine the decay constant given the half-life of ^{238}U. In order to determine the decay constant in s^{-1}, we first need to convert the half-life to seconds:

$$4.5 \times 10^9 \text{y} \times \frac{365 \text{ d}}{1 \text{ y}} \times \frac{24 \text{ h}}{1 \text{ d}} \times \frac{3600 \text{ s}}{1 \text{ h}} = 1.4 \times 10^{17} \text{s}$$

Now, Eqn. 24-4 gives $T_{½} = \frac{0.693}{\lambda}$ or $\lambda = \frac{0.693}{T_{½}}$. Therefore

$$\lambda = \frac{0.693}{1.4 \times 10^{17} \text{s}} = 4.9 \times 10^{-18} \text{s}^{-1}$$

Therefore, given the half-life of ^{238}U, the decay constant of this isotope is $4.9 \times 10^{-18} \text{s}^{-1}$.

(b) Since $16 = 2^4$, it would take 4 half-lives for the activity to decrease by a factor of 16. Thus, the time required would be

$$4(4.5 \times 10^9 \text{y}) = 1.8 \times 10^{10} \text{y}$$

DID YOU KNOW?

The size of the curie was selected because it represents the activity of 1 g of radium-226. (Radium was one of the new elements that the Curies isolated from uranium ores.)

So far we have explored the relationship between the number of nuclei present (N) and the activity (A) of the sample, and we have discussed some of the characteristics of decay in terms of the half-life and/or decay constant of the particular isotope. What we haven't yet discussed quantitatively is the time dependence of either the number of nuclei present or the activity, which are illustrated graphically in Figures 24-19 and 24-20. Perhaps you have already made an educated guess, based on the shapes of the curves shown in these two figures: both the number of nuclei present and the activity of the sample decrease exponentially in time. In particular, it can be shown (with calculus) that

$$N = N_0 e^{-\lambda t} \quad \textbf{(24-5)}$$

where N_0 is the initial number of nuclei present, N is the number of nuclei present at time t, and λ is the decay constant of the isotope.

Note that the presence of the decay constant in the exponential term means that isotopes with short half-lives and therefore large decay constants will see a rapid decrease in the number of nuclei present in a sample as it decays away quickly.

From Eqns. 24-5 and 24-3, we can readily determine the corresponding time dependence of the activity (A) of a sample.

Multiplying both sides of Eqn. 24-5 by λ,

$$\lambda N = \lambda N_0 e^{-\lambda t}$$

Substituting $A = \lambda N$ (Eqn. 24-3) for the left-hand side and $A_0 = \lambda N_0$ on the right-hand side, we get

$$A = A_0 e^{-\lambda t} \quad \textbf{(24-6)}$$

where A_0 is the initial activity of the sample, A is the activity of the sample at time t, and λ is the decay constant of the isotope.

TRY THIS!

Watch Alpha Emitters Decay!

The PhET simulations designed by researchers at the University of Colorado are a great source of interactive, research-based simulations of physical phenomena available online for free. Using your favourite search engine, look for the Alpha Decay PhET simulation. Through this simulation you can investigate the random nature of the decay of an individual radioactive nucleus as well as the overall exponential decrease in number of nuclei over time. Suggestions for using this simulation:

1. Through the "Multiple Atoms" tab, add 10 nuclei from the Bucket o' Polonium and watch the pattern of decayed nuclei across the top as a function of time. How many of the 10 nuclei decay before the half-life mark? How many of the 10 nuclei decay after the half-life mark? Repeat a few times to check for consistency.
2. Again through the Multiple Atoms tab, reset all (far right) to start from the beginning. Empty the Bucket o' Polonium by repeatedly adding 10 nuclei, click on "Reset All Nuclei" and then use the Pause button at the bottom to collect data every second on how many ^{211}Po nuclei remain as a function of time. Plot these data as a semi-log plot and fit a straight line. What is the slope?

SAMPLE PROBLEM 24-6

The initial activity of a sample of $^{9}_{6}$C, which has a half-life of 0.127 s, is 1.40×10^{12} Bq.

(a) What is the average lifetime of a nucleus of $^{9}_{6}$C?

(b) How many nuclei are present in the sample initially?

(c) What is the activity of the sample after 1.0 s?

Solution

(a) The average lifetime is determined by Eqn. 24-1:

$T_{avg} = 1.44\, T_{½} = (1.44)\,(0.127\text{ s}) = 0.183\text{ s}$

Thus, the average lifetime of a ^{9}C nucleus is 0.183 s.

(b) From Eqn. 24-3 we can determine the number of nuclei present given the initial activity. Based on the half-life of $^{9}_{6}$C, the decay constant of this isotope is

$$\lambda = \frac{0.693}{T_{½}}$$

or

$$\lambda = \frac{0.693}{0.127\text{ s}} = 5.457\text{ s}^{-1}$$

Substituting this into Eqn. 24-3 gives

$$N = \frac{A}{\lambda} = \frac{1.40 \times 10^{12}\text{ Bq}}{5.457\text{ s}^{-1}} = 2.57 \times 10^{11}\text{ nuclei}$$

There are 2.57×10^{11} nuclei present in the sample at time $t = 0$.

(c) Now that we know the decay constant, determining the activity after 1.0 s is simply a direct substitution into Eqn. 24-6:

$$A = A_0 e^{-\lambda t} = (1.40 \times 10^{12}\text{ Bq})e^{-(5.457\text{ s}^{-1})(1.0\text{ s})} = 5.97 \times 10^{9}\text{ Bq}.$$

Therefore, the activity after 1.0 s is 5.97×10^{9} Bq.

Radioisotope Dating

radioisotope dating the use of a radioactive isotope to determine the age of an object of historic or archeological interest

When an archaeologist or historian says that an ancient wooden bowl is say, 1500 y old, this age has often been determined by **radioisotope dating**, that is, the use of a radioactive substance to find an object's age. The most common isotope used is carbon-14 ($^{14}_{6}$C). This particular isotope makes up a small fraction of all carbon in the environment and is assimilated into plants and animals in the same way as the stable isotopes of carbon ($^{12}_{6}$C and $^{13}_{6}$C). Once a plant or animal dies, it is no longer exchanging carbon isotopes with the environment; therefore, the amount of the stable isotopes present ($^{12}_{6}$C and $^{13}_{6}$C) remains the same for all eternity. However, the amount of $^{14}_{6}$C present when the plant or animal dies will slowly decay over time, as described by Eqn. 24-6. Thus, the ratio of $^{14}_{6}$C nuclei to other carbon nuclei in the organic object gradually decreases after the organism dies. Since the half-life of $^{14}_{6}$C is 5.7×10^{3} y, this ratio decreases by a factor of two every 5.7×10^{3} y. In the case of say, a wooden bowl, its age can be determined by measuring this ratio, knowing that the steady-state ratio observed in living organisms is approximately one $^{14}_{6}$C atom present for every 10^{12} atoms of $^{12}_{6}$C.

SAMPLE PROBLEM 24-7

Human remains located in an archeological dig have been removed to a forensics laboratory for further examination (Figure 24-21). The $^{14}C/^{12}C$ ratio is measured to be 0.795 times that found in plants and animals living today. Estimate the age of the remains.

Solution

Since the relative amount of ^{14}C present in the sample is 0.795, this is the relative activity (A) at time t. The initial relative activity (A_0) is, by definition, 1.0, assuming that there has been no change in the proportions of carbon isotopes in the environment over time. With Eqn. 24-6, we can determine the unknown quantity t, the time since the person died and stopped exchanging carbon with the environment, if we know the decay constant λ of this isotope. Given that the half-life of ^{14}C is 5.7×10^3 y, we can determine the decay constant with Eqn. 24-4: $\lambda = \frac{0.693}{T_{1/2}}$ or $\lambda = \frac{0.693}{5.7 \times 10^3\,\text{y}} = 1.22 \times 10^{-4}\,\text{y}^{-1}$.
Now we can solve for the age of the remains, t:

$$A = A_0 e^{-\lambda t}$$

$$e^{-\lambda t} = \frac{A}{A_0}$$

$$\ln(e^{-\lambda t}) = \ln\left(\frac{A}{A_0}\right)$$

$$-\lambda t = \ln\left(\frac{A}{A_0}\right)$$

$$t = \frac{1}{\lambda}\ln\left(\frac{A_0}{A}\right)$$

$$t = \frac{1}{1.22 \times 10^{-4}\,\text{y}^{-1}}\ln\left(\frac{1.0}{0.795}\right)$$

$$t = 1.9 \times 10^3\,\text{y}$$

Therefore the age of these remains is approximately 1900 y.

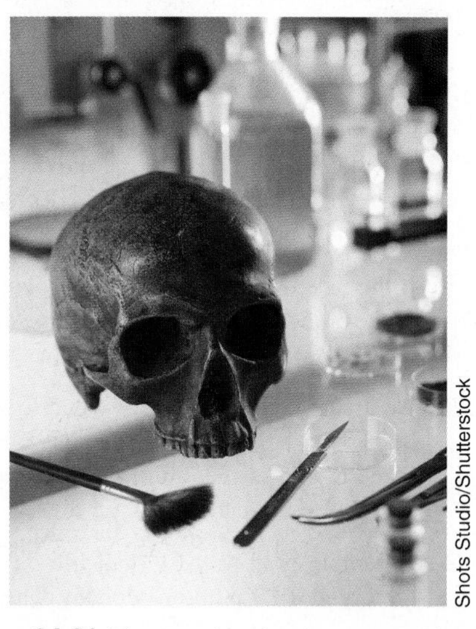

Figure 24-21 The age of ancient human remains can be assessed in a forensics laboratory with radioisotope dating techniques, such as carbon-14 dating.

Units Tip

Note that we did not convert half-life to seconds here; instead we determined the decay constant in units of y^{-1} rather than s^{-1}, resulting in our final answer of age having units of years rather than seconds. Depending on what is known and what is unknown in a problem, you may find it easier to work your answer in this way—just make sure you are consistent and carry your units through to make sure your final answer makes sense!

Carbon-14 dating has some limitations. For objects older than about 3×10^4 y, the amount of $^{14}_{6}C$ is very small and hard to detect. As well, the ratio of $^{14}_{6}C$ to stable carbon in the atmosphere has not been constant with time as we assumed in Sample Problem 24-7. More thorough dating with $^{14}_{6}C$ accounts for the variations in the ratio of $^{14}_{6}C$ to stable carbon over time, principally through calibration with tree rings.

Other isotopes can be used for dating. For example, $^{238}_{92}U$, with a very long half-life of 4.5×10^9 y, can be used in the dating of rocks. If the rocks have fossils in them, this dating can establish the age of the fossils.

DID YOU KNOW?

By 1967, nuclear weapons testing had increased the amount of radioactive ^{14}C in the atmosphere by about 60%. The radiation exposure to a person due to this additional ^{14}C is small; however, it has caused complications for carbon dating since it can no longer be assumed that the percentage of ^{14}C in the atmosphere has remained constant.

EXERCISES

24-24 The half-life of iodine-131 is 8.0 d. How long will it take for the number of nuclei in an iodine-131 sample to be reduced by a factor of 8.0?

24-25 After 8 half-lives what fraction of the original activity does a radioactive source have?

24-26 Carbon-14 has a half-life of 5.7×10^3 y. Assume that the initial number of nuclei present is 2.4×10^{16}. What is the activity of ^{14}C (in becquerels) present after 4 half-lives?

24-27 Sulfur-37 has a half-life of 5.1 min. What percentage of the initial activity of ^{37}S is present after (a) 3.0 min, (b) 12 min, and (c) 32 min?

24-28 Nuclei of $^{12}_{5}\text{B}$ have a half-life of 0.020 s. What is the average lifetime of these nuclei?

24-29 The average lifetime of fluorine-22 is 5.8 s. What is the half-life of this nuclide?

24-30 How long does it take for the activity of a sample of sodium-22, which has a half-life of 2.6 y, to decrease from 1.6×10^{12} Bq to 1.3×10^{11} Bq?

24-31 If the $^{14}_{6}\text{C}/^{12}_{6}\text{C}$ ratio in an ancient wooden bowl is 0.62 times that of the ratio in present day organic materials, how old is the bowl? The half-life of ^{14}C is 5.7×10^3 y.

24-32 An archeological artefact is dated to approximately 8100 y old by carbon dating. What is the $^{14}_{6}\text{C}/^{12}_{6}\text{C}$ ratio in this artefact compared to that of the ratio in present day organic materials? The half-life of ^{14}C is 5.7×10^3 y.

24-33 A sample of copper-61 (half-life 3.3 h) is delivered to a laboratory at 1:00 p.m. on Monday. The activity of the sample is measured to be 3.4×10^9 Bq at 8:30 pm on Tuesday (the next day). What was the sample's activity on delivery?

24-34 The half-life of $^{38}_{17}\text{Cl}$ is 37.0 min. How long does it take for the number of *decayed* $^{38}_{17}\text{Cl}$ nuclei in a sample to equal 87.5% of the original number of nuclei in the sample?

24-35 Argon-41 has a half-life of 1.8 h. If the initial activity of a sample of argon-41 is 3.5×10^{10} Bq, what will be its activity after 10.3 h?

24-36 How long will it take for the activity of an oxygen-15 sample to decrease from 9.8×10^{13} Bq to 2.2×10^{13} Bq? The half-life of oxygen-15 is 2.0 min.

24-37 The average human body contains about 0.2 g of ^{87}Rb, 0.02 g of ^{40}K, and 2×10^{-8} g of ^{14}C. All these isotopes are radioactive. What is the approximate total activity in the average human body due to these three radioactive isotopes? (The half-lives are 4.9×10^{10} y, 1.2×10^9 y, and 5.7×10^3 y, respectively.)

24.4 Nuclear Fission and Fusion

nuclear fission the breaking apart of a large nucleus into smaller nuclei

nuclear fusion the joining together (fusing) of small nuclei to form a larger one

Nuclear fission is the breaking apart of a large nucleus (such as uranium-235), whereas **nuclear fusion** is the joining (fusing) of small nuclei such as isotopes of hydrogen. In both processes the final (rest) mass is less than the initial (rest) mass. Since energy and mass are equivalent ($E = mc^2$), it seems that the final energy is less than the initial energy, in violation of the law of conservation of energy. However, the "missing" energy exists as kinetic energy of the product nuclei. Thus, both fission and fusion generate nuclear kinetic energy, or thermal energy. In other words, the nuclei get hotter. In electrical power reactors, thermal energy from fission is released in a controlled way, and heats water to produce steam for turning electrical generators (Figure 24-22). In a nuclear weapon, the thermal energy from fission or fusion is released rapidly as an explosion.

How is it possible for energy to be released when nuclei break apart (fission) and also when nuclei join together (fusion)? The answer lies in the size of the nuclei involved. Very small and very large nuclei are not as tightly bound together as mid-size nuclei, as illustrated in the graph of nuclear binding energies shown in Figure 24-5. As discussed previously, ^{56}Fe has the highest binding energy per nucleon of all known stable isotopes in the universe, and is therefore the most tightly bound. This means that binding energy will be released when mid-size nuclei ($50 < A < 60$) are produced, either by fission of very large nuclei or by fusion of very small ones.

Vaclav Volrab/Shutterstock

Figure 24-22 These large concrete structures are called cooling towers; they are designed to remove excess heat in the form of steam from the nuclear reactor at an electrical power plant.

Nuclear Fission

We now consider nuclear fission in some detail. We use as an example the fission of uranium-235 ($^{235}_{92}\text{U}$), which is commonly used in nuclear reactors. Uranium-235 constitutes only 0.7% of natural uranium, the more common isotope (99.3%) being uranium-238. Nuclei of $^{235}_{92}\text{U}$ are very stable, having a half-life of 7×10^8 y. However, a $^{235}_{92}\text{U}$ nucleus can be made to undergo fission if hit with a slowly moving neutron. The result is that two smaller, more tightly bound nuclei are produced, one with a mass number typically between 80 and 110, and the other between 120 and 145. In addition, a few neutrons (usually 2 or 3) are released. A typical nuclear equation involving fission of $^{235}_{92}\text{U}$ is

$$^{235}_{92}\text{U} + ^{1}_{0}\text{n} \rightarrow ^{140}_{56}\text{Ba} + ^{93}_{36}\text{Kr} + 3(^{1}_{0}\text{n})$$

Barium-140 and krypton-93 are only two examples of a wide variety of possible fission products.

If the neutrons released during fission are slowed down, they can then cause fission in other $^{235}_{92}U$ nuclei, which in turn produce more neutrons that cause more fission events, and so on. Thus, a **nuclear chain reaction** (Figure 24-23) is produced in which more and more nuclei undergo fission, and more and more energy is released.

nuclear chain reaction each fission event produces multiple neutrons that can then give rise to more fission events

DID YOU KNOW?

Nuclei of uranium-235 can spontaneously undergo fission without being struck by a neutron. However, the half-life for this rare type of decay is 2×10^{17} y, much longer than the half-life of 7×10^{8} y for the more common α decay of U-235.

Mpanchenko/Shutterstock

Figure 24-23 The instigation of a nuclear chain reaction. The release of 3 neutrons with each ^{235}U fission event provides an ongoing supply for more and more nuclei to undergo the process.

Nuclear Reactors

A **nuclear reactor** is a device in which a controlled fission reaction occurs. The fuel consists of uranium, either natural or **enriched uranium** which contains $^{235}_{92}U$ to a concentration of 3 to 4%, and is usually in the form of the chemical compound UO_2. This material is shaped into cylindrical pellets 9 to 10 mm in diameter and 10 to 15 mm long (about the size of the last joint in a finger). These pellets are placed in hollow metal tubes about 3 m long and 1 cm in diameter, and constitute the **fuel rods** (Figure 24-24). There are about 10^4 to 10^5 fuel rods in the core of a reactor. In order to slow down the neutrons so that a chain reaction can occur, a material called a **moderator** is used. Common moderators are water (H_2O), carbon in the form of graphite, and "heavy" water (deuterium oxide, D_2O, where $D = ^{2}_{1}H$).

If all the neutrons produced by fission then resulted in further fission events, there would be a runaway chain reaction and an explosion. Therefore, **control rods** consisting of a neutron-absorbing material such as boron or cadmium are inserted into the reactor. These control rods can be moved in and out of the fuel area by the reactor operators to adjust the rate of fission. Because a great deal of thermal energy is generated during fission, the fuel area must be cooled by passing water or another liquid or gas over it. In electrical power reactors this **coolant** is used to heat water, producing steam to turn generators. Many of the nuclei produced by fission are highly radioactive (such as ^{140}Ba and ^{93}Kr); hence, the core of the reactor must be shielded by high-density materials that absorb much of the radiation from these nuclei. The basic features of a power reactor are shown in Figure 25-25.

There are many types of reactors used around the world, with a variety of fuels, moderators, and coolants. Table 24-3 lists features of the most common types.

nuclear reactor a device in which a controlled fission reaction occurs

enriched uranium uranium sample in which the percentage of ^{235}U present is higher (~3%) than naturally found in the environment (0.7%)

fuel rods assembly containing uranium oxide pellets that serve as the fuel in the nuclear reactor

moderator material used to slow down the neutrons released by fission in order to instigate a nuclear chain reaction (e.g., graphite, water, heavy water)

control rods neutron-absorbing material (often boron or cadmium) used to prevent runaway chain reactions in reactor core

coolant liquid (often water) or gas that passes through the fuel area in order to remove the large amounts of thermal energy produced during fission

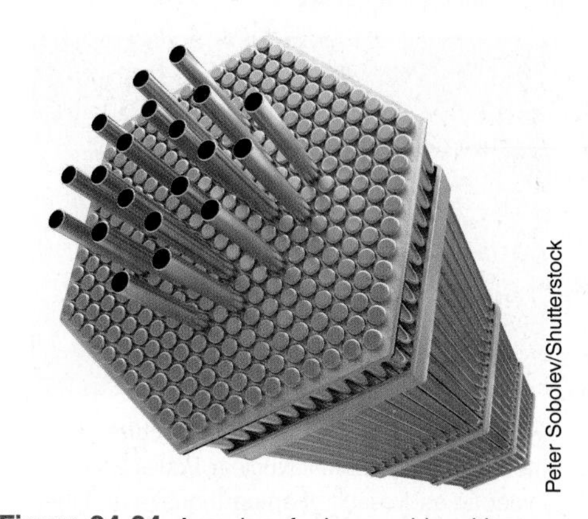

Figure 24-24 A nuclear fuel assembly, with many fuel rods mounted together in a closely packed array.

A cold water intake
B pump
C reactor
D fuel rods
E control rods
F steam generator
G steam line
H containment structure
I turbine
J generator
K transformer
L cooling water condenser
M cooling tower

Figure 24-25 The basic design of a typical nuclear power reactor.

DID YOU KNOW?

As the neutrons are slowed by the moderator, they are said to be *thermalized* and they become *thermal neutrons*. When they slow down, they have the same temperature as their surroundings. We say that they are then in thermal equilibrium with their environment.

DID YOU KNOW?

One of the most serious accidents at a nuclear power plant in history occurred on March 11, 2011, when the Fukushima Daiichi Plant in Japan was damaged by an earthquake and tsunami (Figure 24-26). Key safety systems failed and radioactive materials were released. A thorough clean-up and decommissioning is expected to take decades to complete. As serious as this accident was, the overall release of radiation was approximately 20 times less than the release associated with the Chernobyl nuclear reactor meltdown in April 1986, as safety systems have been continually improved in the intervening years.

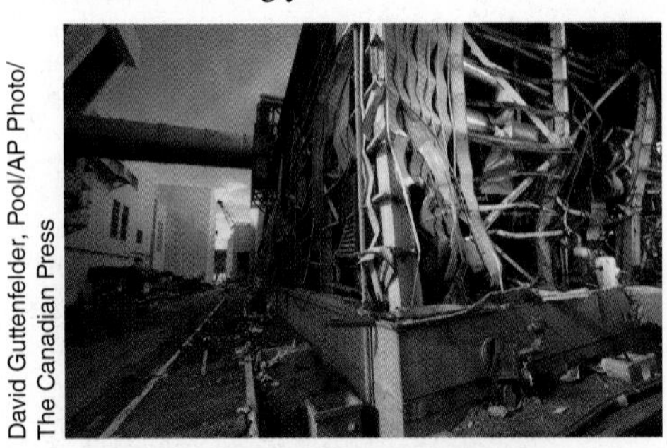

David Guttenfelder, Pool/AP Photo/ The Canadian Press

Figure 24-26 Damaged infrastructure at the Fukushima Daiichi Nuclear Power Plant in Japan as a result of an earthquake and the associated tsunami that hit on March 11, 2011.

Table 24-3

Features of Nuclear Reactors

Type[1]	Fuel	Moderator	Coolant	Comments
PWR	enriched U	H_2O	H_2O	most widely used
BWR	enriched U	H_2O	H_2O	The water that goes through the reactor also directly spins the turbines.
CANDU	natural U	D_2O	D_2O	Can be refuelled while operating (others cannot).
HTGR	natural U	graphite	helium or CO_2	high efficiency for generating electricity
LMFBR	natural U + plutonium (PuO_2)	None	liquid sodium	Creates the nuclear fuel $^{239}_{94}Pu$.

[1]PWR: pressurized water reactor
BWR: boiling water reactor
CANDU: CANada Deuterium Uranium
HTGR: high-temperature gas reactor
LMFBR: liquid metal, fast-neutron breeder reactor

Notice the term **breeder reactor** in this table: this refers to a reactor designed to convert much of the uranium-238 (the most common isotope of uranium) in the fuel to plutonium-239, which itself is fissionable and thus can be used as a fuel. In a breeder reactor, more fuel is generated in this way than is used in the reactor. In a conventional reactor, some plutonium-239 is created and its fission contributes about 40% of the total energy produced; however, less fuel is generated than is used.

breeder reactor converts the ^{238}U in the fuel to ^{239}Pu which is then used as fuel

The used reactor fuel is highly radioactive and must be stored for centuries before the nuclei in it have decayed to stable nuclei. This is the major disadvantage of nuclear energy, although the general public tends to worry more about the possibility of accidents at reactors. One major advantage of nuclear energy is that it does not produce air pollution that can contribute to climate change (a major problem in fossil fuel-based power plants).

Energy Released in Fission

The equivalence of energy and mass can be used to calculate the actual energy released in fission, similar to the calculations we performed in Section 24.1 of binding energies of nuclei. For a typical fission of $^{235}_{92}U$ we have

$$^{235}_{92}U + ^{1}_{0}n \rightarrow ^{140}_{56}Ba + ^{93}_{36}Kr + 3(^{1}_{0}n)$$

We can assume that both the neutron and the uranium nucleus are initially at rest (since they both would be moving slowly), and therefore the initial kinetic energy is zero. Thus, the total energy before the process is just the total energy associated with the rest mass of the uranium atom and the neutron. From Table 24-1 the atomic rest mass is 1.0087 u for a neutron, and 235.0439 u for a U-235 atom. Thus, the total rest mass before fission takes place is 236.0526 u.

The total rest mass (in u) after the fission process is as follows:

139.9106	($^{140}_{56}Ba$)
92.9299	($^{93}_{36}Kr$)
+ 3.0261	$3(^{1}_{0}n)$
235.8666	(Total)

Notice that the total rest mass of the system is lower after fission; the fission process results in more stable nuclei, as illustrated in the graph of nuclear binding energies shown in Figure 24-5. In order to determine the energy released as a result of the fission of one ^{235}U nucleus, we use $E = mc^2$ to convert this mass difference to energy. Specifically, the mass decreases by 236.0526 – 235.8666 = 0.1860 u as a result of the fission event. Converting this mass difference to units of kilograms,

$$0.1860\,u \times \frac{1.66054 \times 10^{-27}\,kg}{1\,u} = 3.089 \times 10^{-28}\,kg$$

In terms of energy, this corresponds to

$$E = mc^2 = (3.089 \times 10^{-28}\,kg)(2.998 \times 10^8\,m/s)^2 = 2.776 \times 10^{-11}\,J$$

Thus, the total *rest* energy after fission of one ^{235}U nucleus is 2.776×10^{-11} J (or 173.3 MeV) less than that of the system before the event. Where did this energy go? The total energy of a closed system is always constant. If the total rest energy has decreased, then there must be an increase in another form of energy. In this case, it is the kinetic energy associated with the movement of the resulting nuclei and neutrons (Figure 24-27). In other words:

Rest energy before fission = (rest energy + kinetic energy) after fission

TRY THIS!

Safely Operate Your Own Nuclear Reactor!

The PhET simulations designed by researchers at the University of Colorado are a great source of interactive, research-based simulations of physical phenomena available online for free. Using your favourite search engine, look for the Nuclear Fission PhET simulation. Through this simulation you can investigate the mechanism of neutron-induced fission of one nucleus of ^{235}U (first tab), chain reactions involving ^{235}U (second tab), as well as the effect of control rods in a reactor vessel (third tab). When investigating the nature of chain reactions, you can add ^{238}U nuclei to the sample. What happens when a neutron hits ^{238}U? (You can aim the neutrons in this part of the simulation.) Why do you think the fuel in this type of reactor needs to be enriched, that is, have the percentage of ^{235}U nuclei increased? The reactor simulation is a great visualization of the dynamic process of maintaining power output in the core while operating at a safe temperature. What effect do the control rods have in the core?

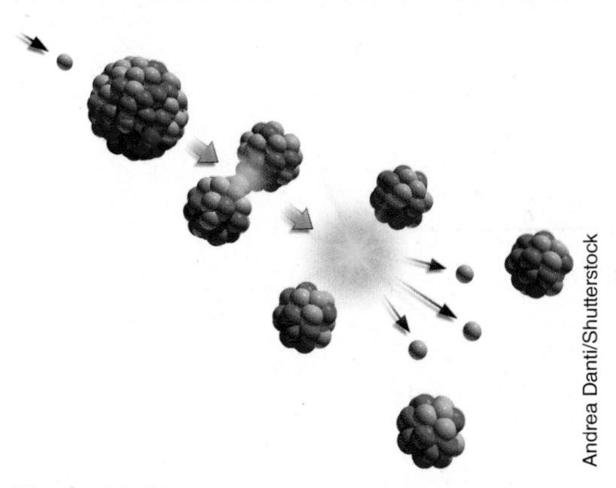

Figure 24-27 The fission of a large nucleus such as ^{235}U into two smaller nuclei along with the emission of neutrons results in a significant release of energy. The released energy results in fast-moving daughter nuclei and neutrons.

DID YOU KNOW?

One kilogram of natural uranium (Figure 24-28 (a)) will release approximately 500 000 MJ through *nuclear* fission, whereas 1 kg of coal (Figure 24-28 (b)) will release approximately 35 MJ through *chemical* combustion. This amount of coal will light a 100 W light bulb for about 4 days, whereas the same mass of natural uranium will light a 100 W light bulb for about 160 years!

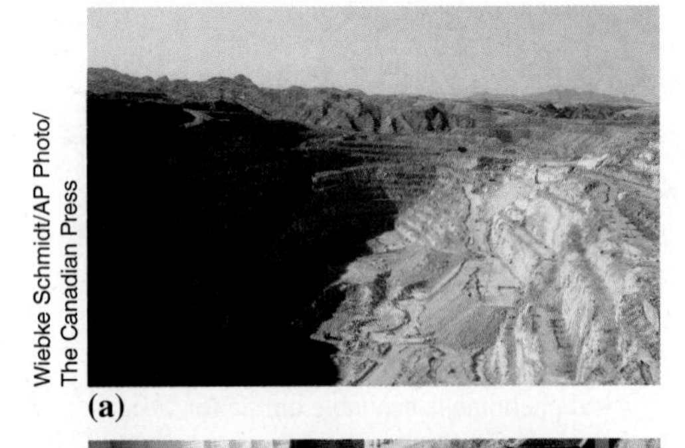

Wiebke Schmidt/AP Photo/ The Canadian Press

(a)

bissell/iStock 360/Getty Images

(b)

Figure 24-28 Coal and uranium are two natural resources that are mined and processed in order to power our technological world. **(a)** Rössing Uranium Mine in Namibia. This mine is the third largest source of uranium in the world, accounting for 7.8% of world production in 2006. **(b)** Vast amounts of coal are shipped by rail across the continent every year.

To be strictly correct, it must be mentioned that not quite all of the decrease in rest energy shows up as kinetic energy of the fission products. A small percentage of the energy is released by the daughter nuclei in the form of beta particles and/or gamma rays.

The percentage decrease in rest energy during a fission event is about 0.08%. In any process (nuclear, chemical, etc.) in which energy is released, the rest energy of the resulting particles is less than the original rest energy. In the chemical oxidation of carbon to form carbon dioxide, thermal energy (i.e., kinetic energy of the carbon dioxide molecules) is produced because the rest energy of the resulting carbon dioxide molecule is less than the rest energy of the initial carbon atom and oxygen molecule. However, in such *chemical* reactions, the percentage decrease in rest energy (and rest mass) is much smaller than in *nuclear* reactions such as fission or fusion. For example, in the oxidation of carbon, the decrease in rest energy is only $1 \times 10^{-8}\%$. In other words, the energy available in a kilogram of fissile material, such as uranium, is many, many orders of magnitude greater than the energy available in a kilogram of fossil fuel.

Nuclear Fusion

Nuclear fusion is the process of producing more stable nuclei by the combining or fusing of smaller nuclei. Figure 24-29 is an illustration of a chain of fusion reactions in which protons ultimately become a ^{4}He nucleus. This is the process taking place in the Sun, producing essentially all the energy in our solar system that allows life to exist on our planet. As with fission, the end products of a fusion reaction have lower rest mass—and therefore lower energy—than the smaller nuclei that existed before the event. Therefore, there is energy released in a fusion reaction in the form of kinetic energy of the reaction products, which could be harnessed for power generation. The fusion reaction that shows the greatest promise for success in power production is

$$^2_1\text{H} + ^3_1\text{H} \rightarrow ^4_2\text{He} + ^1_0\text{n}$$

This is one of the fusion reactions taking place in the Sun. Each fusion of two nuclei in this reaction produces 18 MeV of energy based on the rest masses given in Table 24-1. We leave the details of this calculation as an Exercise (Question 24-42).

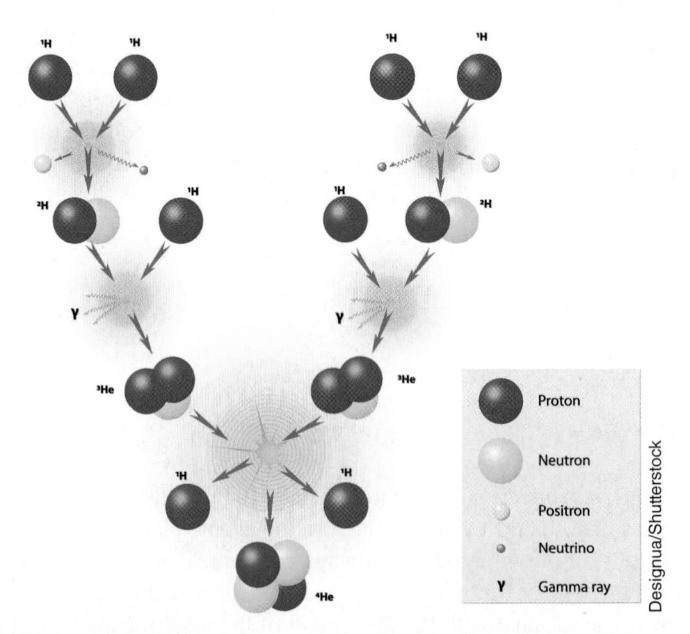

Designua/Shutterstock

Figure 24-29 Nuclear fusion is the process of producing more stable nuclei by the combining or fusing of smaller nuclei. In this illustration, protons ultimately become a ^{4}He nucleus through a chain of fusion reactions.

The products of nuclear fusion tend to be less radioactive than those of fission. Hence, there is developmental work in progress on fusion reactors, but such reactors are still many years away from commercial production. In order for nuclei to fuse, very high temperatures are required (10^7 to 10^8 °C). Thus, two of the major problems facing designers of fusion reactors are the production of high temperatures and the containment of the hot nuclei.

Two methods of containment being explored are **inertial confinement** and **magnetic confinement**. In the former, intense laser beams strike frozen fuel pellets—a combination of deuterium (${}^2_1H$) and tritium (${}^3_1H$)—and cause both a heating and an implosion of the fuel so that fusion occurs. The released energy, carried mainly by neutrons released in the fusion reaction, is absorbed by a coolant surrounding the fusion vessel (Figure 24-30 (a)). In magnetic confinement, magnetic fields are used to contain the fuel inside a closed region such as a torus, which is a doughnut-shaped ring (Figure 24-30 (b)). The fuel might be heated by inducing electric currents in it or by other means.

DID YOU KNOW?

A doughnut-shaped fusion device, such as the one shown in Figure 24-30 (b), is often called a *tokamak,* which comes from a Russian acronym for "toroidal magnetic chamber."

inertial confinement the use of frozen fuel pellets that are bombarded by laser beams to contain the fuel inside the core of a fusion reactor

magnetic confinement the use of magnetic fields to contain the fuel inside the core of a fusion reactor

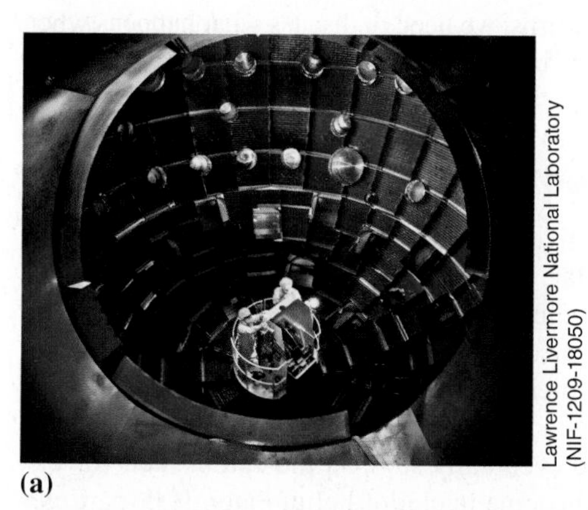

Lawrence Livermore National Laboratory (NIF-1209-18050)

(a)

Nataliya Hora/Shutterstock

(b)

Figure 24-30 **(a)** The target chamber at the National Ignition Facility (NIF) in Livermore, CA, USA. The NIF is a large, laser-based inertial confinement fusion research facility located at the Lawrence Livermore National Laboratory. **(b)** The magnetic confinement fusion device called COMPASS housed at the Institute of Plasma Physics in Prague, Czech Republic.

EXERCISES

24-38 Determine the unknown quantities in the following fission and fusion reactions. Processes (c), (d), and (e) occur in the Sun (Figure 24-31) and other stars.

(a) ${}^{235}_{92}U + {}^1_0n \rightarrow 3({}^?_0n) + {}^{95}_?Y + {}^?_{53}I$

(b) ${}^1_0n + {}^?_{94}Pu \rightarrow {}^{141}_?? + {}^{96}_{39}Y + 3({}^1_0n)$

(c) ${}^{12}_6C + {}^4_?? \rightarrow {}^?_8O$

(d) ${}^{12}_6C + {}^{12}? \rightarrow {}^?_{11}Na + {}^1_1?$

(e) ${}^8_?? + {}^?_2He \rightarrow {}^{12}_6C$

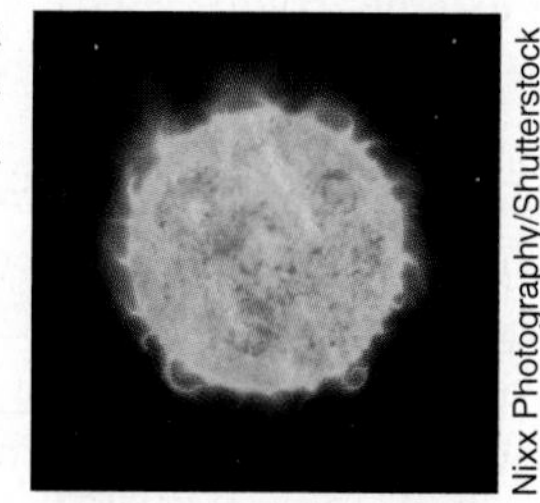

Nixx Photography/Shutterstock

Figure 24-31 Our Sun is a giant ball of gas with approximately 600 million tonnes of hydrogen fusing together to create helium every second.

24-39 State the function of each of the following components of a nuclear reactor: moderator, control rods, coolant.

24-40 Determine the energy released (in joules) in the following fission of ${}^{235}U$:

$${}^{235}_{92}U + {}^1_0n \rightarrow {}^{141}_{56}Ba + {}^{92}_{36}Kr + 3({}^1_0n)$$

The rest masses involved are

${}^{235}_{92}U$: 235.0439 u ${}^1_0n$: 1.0087 u

${}^{141}_{56}Ba$: 140.9141 u ${}^{92}_{36}Kr$: 91.9250 u

24-41 Convert your answer in the previous question to MeV.

24-42 Determine the kinetic energy (in MeV) produced by the fusion of one deuterium (${}^2_1H$) and one tritium (${}^3_1H$) nucleus in the following reaction:

$${}^2_1H + {}^3_1H \rightarrow {}^4_2He + {}^1_0n$$

The rest masses are

${}^2_1H$: 2.0141 u ${}^3_1H$: 3.0160 u

${}^4_2He$: 4.0026 u ${}^1_0n$: 1.0087 u

24-43 A commercial fusion reactor needs to produce about 3×10^3 MW of thermal power. Use your answer to the previous question to determine how many deuterium–tritium fusion reactions must occur per second in order to generate this power.

24-44 **Fermi Question:** What is the approximate energy produced by fission per kilogram of $^{235}_{92}U$? (Since the answer is only required to be approximate, assume that the mass of one $^{235}_{92}U$ nucleus is 235 u.)

24-45 **Fermi Question:** Approximately what mass of $^{235}_{92}U$ is required to operate a nuclear reactor for 1 year if the thermal power produced is 2×10^3 MW? Refer to the previous question for useful information.

24-46 **Fermi Question:** What is the approximate mass of $^{235}_{92}U$ that produces an energy release equivalent to about 20 kilotons of TNT? (1 kiloton of TNT releases about 5×10^{12} J.) This is the energy that was released when the plutonium-based fission bomb was dropped on Nagasaki on August 9, 1945.

24.5 Interactions between Radiation and Matter

As we alluded to in our profile of Marie Curie, working extensively with radiation without the appropriate safety precautions can cause serious health effects. How? In order to better understand the health concerns, we need to discuss what happens when radiation interacts with the human body.

electron–ion pair the products generated in material such as biological tissue when charged particles (α and β particles) travel through and cause localized ionizations

Interactions between Charged Particles and Matter

Since α and β particles have net positive or negative electric charge, their interaction with matter is somewhat similar. As they pass through a substance, they remove electrons from nearby atoms, thus producing positive ions in their wake. Hence, all along the path of these particles, **electron–ion pairs** are produced. These ionizations serve to transfer energy from the α and β particles to the medium through which they move. As a result, the α and β particles slow down as they travel through the material and eventually stop. At the end of their travels, α particles then attract electrons and become nuclei of helium atoms, β^- particles become orbiting electrons in atoms, and β^+ particles annihilate with local electrons to produce γ rays.

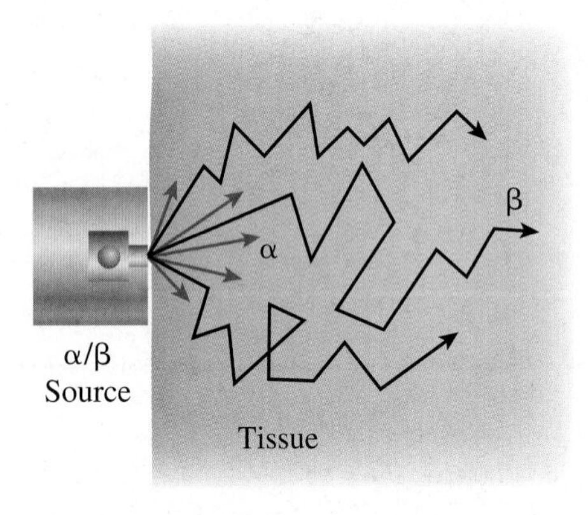

Figure 24-32 An illustration of the path of an α particle in tissue (red arrows): typically quite straight and short. The paths followed by β particles in tissue (black lines) are more circuitous and typically longer than those of α particles.

Alpha particles are much more massive, and for a given initial energy, have a much smaller speed than β particles. In addition, α particles have twice the charge ($+2e$) compared to that of β particles. Therefore, they produce many more electron–ion pairs in a given distance along their path and lose their energy rapidly; an α particle can be stopped by a piece of paper and will not penetrate the outer layer of skin when humans are exposed to an external source. Because of their large mass, α particles tend to "plow straight ahead," with a typical path being straight and short (Figure 24-32, red arrows).

Conversely, since a β particle has a relatively small mass, the electric forces exerted on it by surrounding electrons and nuclei can easily change its velocity vector, and thus its path is somewhat crooked (Figure 24-32, black lines). Relatively high-energy β particles can travel several millimetres in tissue before stopping. With both α and β particles, there is a definitive **range** over which the particles will travel before coming to a stop. This range depends on both the initial energy of the particle and the composition of the material in which it is travelling. For example, the range of a 4 MeV α particle in air is about 2.5 cm and about 20 μm in tissue. The corresponding values for a 4 MeV β particle are approximately 15 m in air and 2 cm in tissue. (This is an uncommonly high energy for a β particle, but not unusual for an α particle.) Table 24-4 provides some representative values of range for α and β particles in tissue as a function of initial particle energy.

range the finite distance travelled by a charged particle (α or β) in matter before coming to a stop; this value depends on the initial energy of the particle and the composition of the material in which it is travelling

To simplify determinations of the range of alpha and beta particles in tissue, empirical equations have been developed that relate range and particle energy. These equations eliminate the need to look up values in tables such as Table 24-4.

The following are the commonly used empirical expressions for the range of alpha and beta particles in tissue:

Alpha particle range (R_α) in tissue in μm

$$E_\alpha < 4\,\text{MeV} \quad R_\alpha = 5.6 \times E_\alpha \qquad (24\text{-}7)$$

$$4\,\text{MeV} < E_\alpha < 8\,\text{MeV} \quad R_\alpha = 12.4 \times E_\alpha - 26.2 \qquad (24\text{-}8)$$

Beta particle range (R_β) in tissue in cm

$$E_\beta < 3\,\text{MeV} \quad R_\beta = 0.412 \times (E_\beta)^n$$

$$\text{where } n = 1.27 - 0.0954 \times \ln(E_\beta) \qquad (24\text{-}9)$$

$$E_\beta > 3\,\text{MeV} \quad R_\beta = 0.530 \times E_\beta - 0.106 \qquad (24\text{-}10)$$

In all four equations, the energy of the alpha or beta particle used in the calculation *must* be in units of MeV.

Table 24-4

Ranges of α and β Particles in Tissue as a Function of Energy

Energy	α Range (cm)	β Range (cm)
50 keV	3×10^{-5}	0.004
100 keV	5×10^{-5}	0.01
500 keV	1×10^{-4}	0.2
1 MeV	5×10^{-4}	0.4
4 MeV	2×10^{-3}	2
7 MeV	6×10^{-3}	2.5
10 MeV	0.01	5
100 MeV	0.6	30

SAMPLE PROBLEM 24-8

Unlike β particles, α particles from many radionuclides have a range that is less than the average thickness of the human epidermis (~0.007 cm) and therefore are not typically cause for concern in terms of causing damage to underlying tissue. Determine the range in tissue of the α particle emitted in the decay of $^{214}_{84}\text{Po}$ to $^{210}_{82}\text{Pb}$ **(a)** using empirical Eqn. 24-7 or 24-8; and **(b)** using the data provided in Table 24-4. The following atomic rest mass data are provided:

Particle/Nucleus	Mass (u)
$^{214}_{84}\text{Po}$	213.9952
$^{210}_{82}\text{Pb}$	209.9842
$^{4}_{2}\text{He}$	4.002 60

Solution

In order to determine the range in tissue with either method, we need to determine the energy of the α particle emitted in this decay. The decay equation can be written as $^{214}_{84}\text{Po} \rightarrow {}^{210}_{82}\text{Pb} + {}^{4}_{2}\text{He}$, and the atomic mass data provided tell us the energy available from this decay for the α particle. The total rest mass before decay takes place is 213.9952 u. The total rest mass (in u) after decay is as follows:

$$\begin{array}{rl} 209.9842 & (^{210}_{82}\text{Pb}) \\ +\ 4.0026 & (^{4}_{2}\text{He}) \\ \hline 213.9868 & (\text{Total}) \end{array}$$

In order to determine the energy released as a result of the decay of one ^{214}Po nucleus, we use $E = mc^2$ to convert the mass difference to energy. Specifically, the mass decreases by (213.9952 – 213.9868) u = 0.0084 u as a result of α decay. Converting this mass difference to units of kg,

$$0.0084\,\text{u} \times \frac{1.660\,54 \times 10^{-27}\,\text{kg}}{1\,\text{u}} = 1.4 \times 10^{-29}\,\text{kg}$$

In terms of energy, this corresponds to

$$E = mc^2 = (1.4 \times 10^{-29}\,\text{kg})(3.00 \times 10^{8}\,\text{m/s})^2 = 1.3 \times 10^{-12}\,\text{J}$$

DID YOU KNOW?

In determining how much of the released energy is transferred to each decay product, it turns out that we simply need their relative masses. Since the parent nucleus in the decay is at rest before the event, the initial momentum of the system is zero. Therefore, the momenta of the products must sum to zero after the decay has occurred. In the case of the α decay of ^{214}Po discussed in Sample Problem 24-8, this means that the momentum of the α particle is equal and opposite to the momentum of the ^{210}Pb nucleus. Through this analysis, much like many calculations carried out in Chapter 8, we find that the α particle has $\frac{210}{4 + 210} = 0.98$ of the available energy, whereas the ^{210}Pb nucleus only has $\frac{4}{4 + 210} = 0.02$ of the available energy. The less massive object has much more kinetic energy for the same momentum, as expected.

Thus, the energy released in the α decay of one ^{214}Po nucleus is 1.3×10^{-12} J or 7.8 MeV. It should be noted that this energy is shared between the daughter nucleus (^{210}Pb) and the α particle, but the less massive decay product (^{4}He) will carry away the vast majority of this energy due to conservation of momentum. For our purposes, it is reasonable to estimate that the released α particle has approximately 7.8 MeV of kinetic energy. See the Did You Know? box on page 789 for how the energy is shared between the decay products in more detail.

(a) From Eqns. 24-7 and 24-8, we see that Eqn. 24-8 is the relevant one for an alpha particle of 7.8 MeV. Substituting into this equation then gives

$$R_\alpha = 12.4 \times E_\alpha - 26.2$$

$$R_\alpha = 12.4 \times (7.8\,\text{MeV}) - 26.2$$

$$R_\alpha = 71\ \mu\text{m}$$

Therefore the range of this alpha particle is 71 μm in tissue.

(b) From Table 24-4, we can interpolate the range values given for 7 MeV and 10 MeV α particles. We find that the range in tissue of an α particle emitted in the decay of ^{214}Po is approximately 7×10^{-3} cm, or 70 μm. Therefore, the α particles emitted by this isotope will not get through the epidermal layer of the skin to reach living cells beneath.

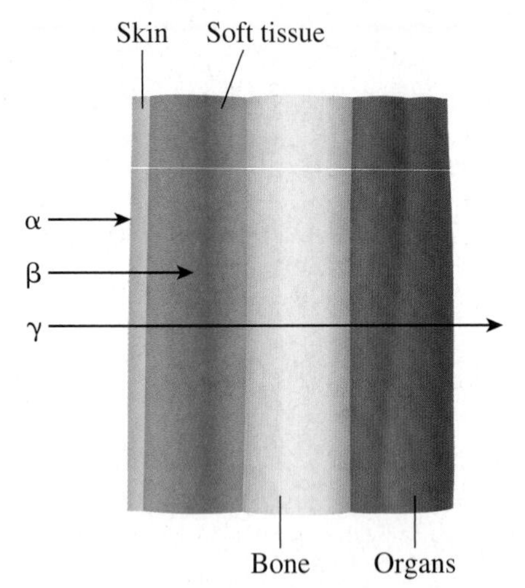

Figure 24-33 Relative penetration strength of α, β, and γ radiation in tissue. High energy γ rays can penetrate quite far in biological tissue.

Interactions between γ Rays and Matter

Whereas α and β particles produce electron–ion pairs all along their paths, γ rays (electromagnetic radiation) interact with matter via single infrequent events and tend to travel much further than charged particles, as illustrated in Figure 24-33. There are three main types of interaction that result in the transfer of energy from the γ ray to the material through which it travels: the **photoelectric effect**, **Compton scattering**, and electron–positron **pair production**. In the photoelectric effect, an atom absorbs an incident γ ray and emits an electron, thus producing an ion. In Compton scattering, the γ ray scatters as it interacts with an atom, resulting in a new direction of propagation for the photon as well as a reduction in energy—there is also an electron emitted in this process. In the case of pair production, which occurs only if a γ ray has an energy greater than twice the rest energy of an electron ($2m_ec^2 = 1.02$ MeV), it is possible for the γ ray to be converted into an electron and positron pair; this is the opposite process to annihilation. In order for energy and momentum to be conserved, this process must occur near a nucleus, which takes up some energy and momentum. The relative probabilities for the three types of interaction depend on the γ-ray energy and on the type of material in which the γ ray is travelling. In all three of these processes, the electromagnetic energy in the form of a γ ray is converted to energetic charged particles in the material; these in turn can cause further ionizations, as discussed when we explored the interactions of α and β particles with matter.

Since α, β, and γ radiation can produce ions, they are referred to as **ionizing radiation**. X rays and high-frequency ultraviolet light are also types of ionizing radiation. Microwave radiation, infrared, visible light, radio waves, and ultrasound are all forms of **non-ionizing radiation**; the health effects associated with non-ionizing radiation tend to be less serious than those from ionizing radiation, and typically centre on issues of heating.

As discussed above, α and β particles have finite ranges, determined by the particles' initial energy and the type of material through which they travel. However, given the infrequent and random nature in which γ rays interact with matter, it is impossible to assign finite ranges to γ rays. As γ rays go further and further into a

photoelectric effect the interaction between a γ ray and an atom that results in ionization of the atom; the γ ray is absorbed and an electron is ejected from the atom

Compton scattering a γ ray is scattered from an atom, resulting in a new direction of propagation for the photon as well as a reduction in energy; there is also an electron emitted in this process

pair production the opposite of an annihilation event: if a γ ray has greater than 1.02 MeV of energy, which is twice the rest mass energy of an electron, the photon can be converted into an electron and positron pair of particles

ionizing radiation forms of radiation that have sufficient energy to create electron–ion pairs in matter

non-ionizing radiation forms of radiation that do not have sufficient energy to create electron–ion pairs in matter

material, more and more of them are likely to undergo interactions, and collectively the number of γ rays remaining decreases (Figure 24-34). However, if we were to watch the progress of a particular γ ray in the material, it might travel only a short distance or it might go for a relatively long distance before interacting. There is even a finite probability that a particular γ ray would pass straight through; this is an inherently random process that is ruled by probabilities and likelihoods, not definitive features such as finite ranges. What we can say though is that γ rays with more initial energy will, on average, travel further into a given material than γ rays with less initial energy (Figure 24-34 (a)). We also know that γ rays of a specific energy will travel further in less dense materials such as air than they will in more dense materials such as lead (Figure 24-34 (b)).

We can characterize the probability of interacting in a given material for a given γ-ray energy with the **attenuation coefficient** (μ), which depends on photon energy and sample composition. It can be demonstrated that the number of photons remaining (N) after traversing a distance x of the material is given by

$$N = N_0 e^{-\mu x} \quad (24\text{-}11)$$

where N_0 is the initial number of photons hitting the surface of the material. The attenuation coefficient has units of m^{-1} or cm^{-1}. The probability of interaction can also be characterized in the form of the **mean free path** of the photon in this medium:

$$\text{mean free path} = \mu^{-1} \quad (24\text{-}12)$$

The mean free path, with units of metres or centimetres, is the average distance travelled in a given material by a γ ray of a specific energy before interacting in some way.

(a)

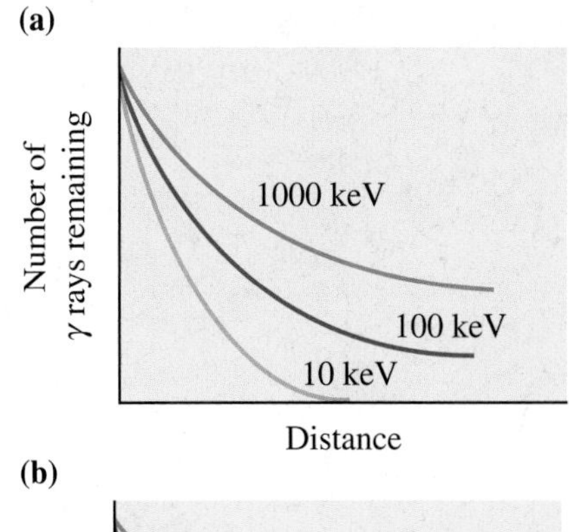

(b)

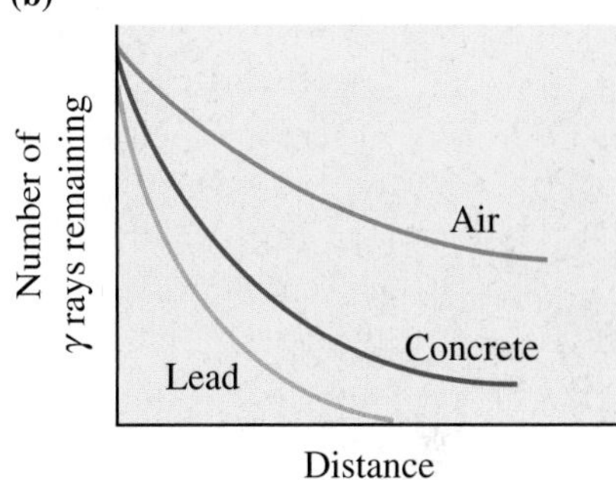

Figure 24-34 Gamma rays do not have a finite range. The number remaining decreases exponentially with distance into the material. The rate at which the number decreases depends on **(a)** the initial energy of the photon and **(b)** the material through which it travels.

attenuation coefficient a property of a given material that characterizes the probability of a photon of a specific energy interacting; symbol μ

mean free path the average distance travelled before interacting by a γ ray of a specific energy in a specific medium

SAMPLE PROBLEM 24-9

The radioisotope ^{137}Cs is a fission product that has been found to be useful in a number of industrial applications (Figure 24-35). This isotope has a half-life of 30 years and decays to ^{137}Ba through the emission of β^- particles. In 95% of decays, the daughter nucleus, ^{137}Ba, is in an excited state, which means that there is an associated γ ray emitted. In these decays, the range of the beta particle in tissue is approximately 0.2 cm and the γ ray has an energy of 662 keV. **(a)** Determine the approximate maximum energy of the beta particle emitted in these decays. **(b)** The attenuation coefficient in tissue for γ rays of energy 662 keV has been measured experimentally to be 0.0905 cm^{-1}. Determine the mean free path of this γ ray in tissue. **(c)** Determine the thickness of tissue over which the number of photons decreases by a factor of 10. This is sometimes referred to as the **tenth-value layer** of the material.

Courtesy of Seaman Nuclear Corporation, www.seamannuclear.com

Figure 24-35 A portable device that uses radiation emitted by ^{137}Cs to measure the thickness and/or moisture content of road surfaces.

tenth-value layer the thickness of material required to reduce the number of photons getting through by a factor of 10

Solution

(a) From Table 24-4 we can see that β particles having a range in tissue of 0.2 cm have an energy of 500 keV. Therefore, the maximum energy of the beta particles emitted in these decays is approximately 500 keV. (The actual value is 512 keV.)

(b) Since the mean free path is defined as the inverse of the attenuation coefficient (Eqn. 24-12), the average distance travelled by a 662 keV photon in tissue before interacting is mean free path $= \dfrac{1}{0.0905\,\text{cm}^{-1}} = 11.0\,\text{cm}$. A mean free path of 11.0 cm for these 662 keV photons is considerably larger than the range of the β particles emitted in these decays, and only represents the average distance between interactions. Many of the γ rays will travel considerably further than 11 cm before transferring all their energy to the tissue.

(c) Eqn. 24-11 tells us that the number of photons (N) travelling a distance x in the medium is given by the exponential function $N = N_0 e^{-\mu x}$. In order to determine the thickness (x) of tissue over which the number of photons decreases by a factor of 10, we isolate the equation for x, let $N_0 = 1$, $N = 0.1$, and substitute the given value of the attenuation coefficient as follows:

$$N = N_0 e^{-\mu x}$$

$$e^{-\mu x} = \frac{N}{N_0}$$

$$\ln(e^{-\mu x}) = \ln\left(\frac{N}{N_0}\right)$$

$$-\mu x = \ln\left(\frac{N}{N_0}\right)$$

$$x = \frac{1}{\mu}\ln\left(\frac{N_0}{N}\right), \text{ since } \ln\left(\frac{N_0}{N}\right) = -\ln\left(\frac{N}{N_0}\right)$$

$$x = \frac{1}{0.0905\,\text{cm}^{-1}}\ln\left(\frac{1}{0.1}\right)$$

$$x = 25\,\text{cm}$$

Therefore, a thickness of 25 cm of tissue will result in a reduction in the number of γ rays by a factor of 10.

DID YOU KNOW?

Although we concluded in Sample Problem 24-8 that the α particles from ^{214}Po decay do not pose a health hazard in cases of external exposure, inhaled dust containing this radioisotope can be deposited in the lungs. Once inside the lungs, the alpha particles have sufficient range to reach the basal cells of the epithelium. Exposure to α particles from inhaled radioisotopes such as ^{214}Po has been cited as the cause of increased rates of lung cancer among uranium miners.

Biological Effects of Ionizing Radiation

The ability of ionizing radiation to produce ions makes it a potential health hazard. Ionized biological molecules such as proteins, water, and DNA can react chemically with surrounding molecules, with the result that there is alteration or damage to the originally ionized molecules and to other molecules. If enough molecules are changed in a cell, its ability to function normally is impaired and it might even die. If enough cells die, the entire life of the organism is at risk. Even if a damaged cell remains alive, damage to its DNA can cause cancer or genetic alterations in offspring. In the early to mid-1900s, many people exposed to radiation in the workplace contracted cancer and other radiation-related illnesses as discussed in our profile on Marie Curie in this chapter. The need for protective measures for those occupationally exposed to radiation became evident and the field of **Health Physics**, which ensures the safety, continuous monitoring, and training for all those who deal with radiation in the workplace, was born. Today, working with ionizing radiation is a significantly safer undertaking as a result. There is a growing worldwide demand for health physicists. Career options range from industrial settings, nuclear power plants, medical facilities, government laboratories, and academia; any setting in which ionizing radiation is used routinely demands professionals who understand the potential hazards and their prevention.

Health Physics field in which the beneficial uses of ionizing radiation in medicine, industry, and research are balanced with the need to protect workers and the general public from potential health hazards'

Although much of our focus here has been on the negative effects of radiation exposure, there is growing scientific evidence for an effect called **hormesis**, in which exposure to low levels of radiation is beneficial (Figure 24-36). This is not a new phenomenon; after all, this is the mechanism by which immunization works and muscle is built through resistance training. In the past few decades, experiments have repeatedly demonstrated that low-level exposure to radiation can stimulate repair mechanisms and result in the activation of the immune system. Although the science in the laboratory is clear, the implications for regulatory bodies in terms of establishing guidelines for medical and occupational exposures remain controversial. In most of the developed world, regulatory bodies continue to assume that the risks of radiation effects increase linearly with exposure, with no acknowledgment of hormetic effects at low levels; this is the simplest and most conservative approach to adopt.

Figure 24-36 The water in the spa town of Bad Gastein, Austria, contains elevated levels of the radioactive gas radon. Every year, thousands of people expose themselves to low levels of radon for therapeutic/hormetic purposes in facilities ranging from rustic old mines to upscale spas and clinics.

hormesis an adaptive biological reaction to low doses of chemical toxins, radiation, or other forms of stress that is damaging, even fatal, in higher doses

Radioactivity in the Environment

Figure 24-37 gives a breakdown of the various sources of radiation to which we are typically exposed. Your own body is radioactive containing naturally occurring ^{40}K, ^{14}C, and other radioactive isotopes. A typical adult contains about 1.7×10^{-2} g of ^{40}K, which has an activity of about 4×10^3 Bq. As we can see in Figure 24-37, this internal radiation is approximately 5% of the total average annual radiation exposure.

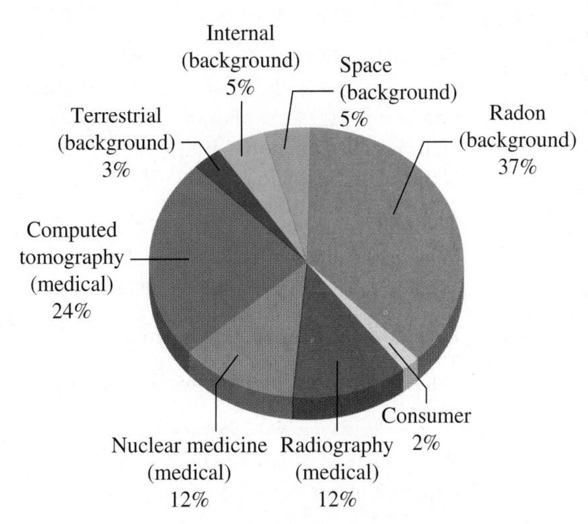

Figure 24-37 Sources of natural and manufactured radiation exposure in our environment.

background radiation radiation from environmental sources

Roughly 50% of the radiation to which we are exposed is natural radiation; this is often referred to by health physicists as **background radiation** as it is always there in the background. Some of this radiation is from internal sources within our bodies and some of it is external. The single biggest contribution is radon, as seen in Figure 24-37. Radon is a radioactive gas that is produced in the decay of uranium and thorium in the soil and rocks around us, as mentioned in Section 24.2. When radon gas escapes from the ground outdoors it gets diluted and does not pose a health risk. However, in some confined spaces, such as modern homes designed for energy efficiency, radon can accumulate to relatively high levels and become a health hazard. Because radon is a gas, it can be inhaled and become a source of radiation exposure inside the lungs. Long-term exposure to high levels of radon in the home may increase the risk of developing lung cancer. Radon exposure is believed to be the second leading cause of lung cancer in Canada after smoking.

Your exposure to natural radiation depends on where you live. Different kinds of soil and rock contain different levels of radioactive isotopes. As well, elevation plays a role: the higher the elevation, the greater the exposure to cosmic rays. As discussed in the

Test Your Radiation Knowledge

In your favourite search engine, type in "CNSC Fact or Fiction: Nuclear Life in Canada." This will take you to a 10-question true or false quiz about radiation in our natural and technological environments. Many of the answers should be clear once you've read through this chapter! (**Note:** CNSC stands for the Canadian Nuclear Safety Commission. This is the division of the federal government responsible for ensuring the safe use of radioactivity in all aspects of modern life.)

opening of the chapter, areas in Colorado have higher levels of natural radiation, as much as twice the average value in the United States.

The remaining 50% of our radiation exposure comes from manufactured sources, principally for medical and dental applications. X rays are a type of ionizing radiation and interact with matter in the same way as γ rays. We will discuss various applications of ionizing radiation in medicine in Section 24.6; the advantages of using radiation in diagnosing and treating disease must be weighed against potential health concerns arising from the exposure.

EXERCISES

24-47 Suppose that you have a radioactive source that produces both α and β particles. Would this source be a greater health hazard to you if you hold it in your hand for a lengthy period of time, or if you swallow it? Discuss.

24-48 List five types of ionizing radiation.

24-49 Determine the range in tissue of the α particle emitted by ^{239}Pu. The following atomic rest mass data are provided:

Particle/Nucleus	Mass (u)
$^{239}_{94}$Pu	239.0522
$^{235}_{92}$U	235.0439
$^{4}_{2}$He	4.002 60

24-50 Determine the range in tissue of the α particle emitted by ^{241}Am. The following atomic rest mass data are provided:

Particle/Nucleus	Mass (u)
$^{241}_{95}$Am	241.0568
$^{237}_{93}$Np	237.0482
$^{4}_{2}$He	4.002 60

24-51 Determine the range in tissue of the 606 keV β particle emitted by ^{131}I. This radioisotope is used medically to diagnose and treat various thyroid conditions; see Section 24.6 for more details.

24-52 Determine the range in tissue of the 1.7 MeV β particle emitted by ^{32}P. This is a commonly used radioisotope in biology research labs, often for a labelling technique with amino acids or nucleotides.

24-53 What are the three processes by which photons (γ rays, x rays) transfer energy to matter? Describe these processes.

24-54 The attenuation coefficient of lead for 100 keV photons is 63 cm^{-1}. What is the thickness of lead required to decrease the number of 100 keV photons getting through by a factor of 2? (This is called the **half-value layer** (HVL) of the material.) What is the mean free path of 100 keV photons in lead?

24-55 The half-value layer of lead for 500 keV photons is 3.8 mm. What is the attenuation coefficient of lead for 500 keV photons? What is the mean free path of 500 keV photons in lead?

24-56 How does ionizing radiation damage biological tissue?

24-57 What is hormesis? Give three examples of hormesis at work in biological systems.

24-58 What is background radiation? List three sources of background radiation in our everyday lives.

half-value layer thickness of material required to reduce the number of photons getting through by a factor of 2; symbol HVL

24.6 Medical Applications

Although radiation is potentially hazardous, it is used routinely in a number of beneficial ways: in industrial settings (quality control and automated processes, e.g., Figure 24-35), smoke detectors (Figure 24-9), radioactive dating (Figure 24-6), camping equipment (Figure 24-10), and in medical procedures for both the diagnosis and treatment of a variety of diseases. We will discuss several such applications here ranging from the use of x rays for diagnostic imaging to medical linear accelerators used in irradiating cancerous growths.

X-Ray Imaging (Radiography)

As discussed in Section 24.5, as photons travel through tissue they are absorbed or scattered and the number of photons getting through decreases exponentially with tissue thickness. The rate at which the number of photons decreases depends on the attenuation coefficient of the material, which is different for bone and soft tissue due to their differences in density. If a detection system, such as photosensitive film, is placed on the far side of the region of interest to be imaged, regions of lower attenuation coefficient will result in more photons reaching the detector. The resulting image consists of bright white regions of high density (bone) where the film was less exposed and darker regions of low density (soft tissue) where the film was more exposed. Figure 24-38 (a) depicts the process of generating an image using x-ray transmission. Figure 24-38 (b) is a typical diagnostic image generated in this way, in this case of a fractured shoulder.

Most of us have had an x-ray image taken at some point in our lives, perhaps many at this stage. In general, generating images with x rays is called **radiography**, and medical applications include diagnosing broken bones, assessing dental health, diagnosing breast cancer (x-ray imaging of breast tissue is called **mammography**), as well as identifying the presence and location of abnormalities such as cysts or tumours.

By replacing the photosensitive film with a digital video recorder, it is possible to acquire real-time moving images of the patient. This is a technique called **fluoroscopy**, and is used extensively to image the gastrointestinal (GI) tract and to guide surgeons in certain procedures. Fluoroscopy often involves much longer exposure time to the x rays than standard radiography, so the health risks must be balanced by the benefits in terms of diagnosis and treatment.

In many cases, **contrast agents** are used in radiography to enhance the image generated from a particular region of the body. For example, the highly attenuating and chemically inert element barium is often given to patients as a drink in order to deliver this contrast agent to the GI tract for imaging (Figure 24-39 (a)), whereas iodine-based contrast agents are regularly injected into the blood stream for x-ray images of the cardiovascular system (Figure 24-39 (b)). **Angiography** is the term used to refer to the generation of x-ray images of arteries and veins in the body.

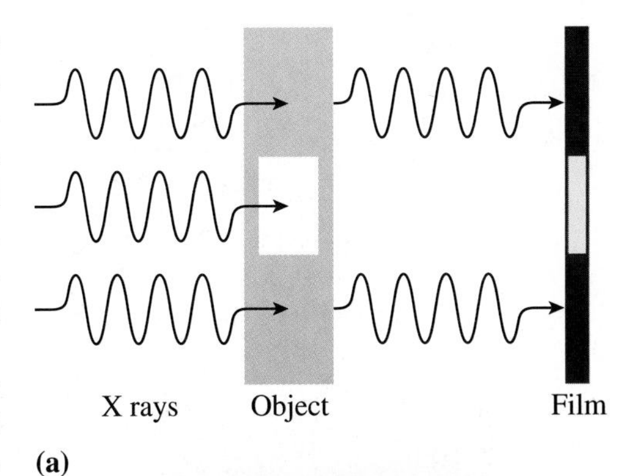

(a)

joloei/Shutterstock

(b)

Figure 24-38 **(a)** Generating an x-ray image by passing photons through an object containing regions of different attenuation coefficients. The bone (white) has a higher attenuation coefficient so the film remains unexposed behind it, whereas the soft tissue (beige) has more photons passing through to expose the film behind. **(b)** A typical x-ray image illustrating a shoulder fracture. Notice the dark region of the low density air inside the lung, as well as the bright white regions of the bones in the arm, shoulder, and chest.

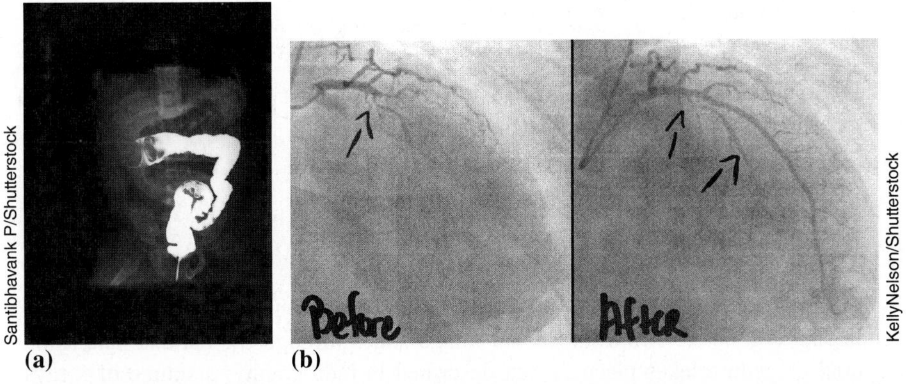

(a) (b)

Figure 24-39 **(a)** Generating an x-ray image with barium contrast agent. The highly attenuating barium in the intestines shows the anatomy of the organ quite clearly. **(b)** Generating an x-ray image with iodine contrast agent. The side-by-side images are taken before and after surgical intervention to reopen blocked cardiac arteries, a life-saving intervention for many patients.

radiography the use of x rays to generate an image

mammography the use of x rays to generate an image of breast tissue for the diagnosis of disease

fluoroscopy the use of a digital video recorder to capture real-time moving images of a patient with x rays

contrast agents highly attenuating inert materials used to enhance the image generated from a particular region of the body

angiography the generation of x-ray images of the cardiovascular system, often applied to imaging the blood supply in the brain and in the heart

SAMPLE PROBLEM 24-10

In order to appreciate the mechanism by which tissues are differentiated in an x-ray image, determine the percentage of 40 keV photons getting through to the film or digital video recorder in the case of

(a) 1 cm of muscle ($\mu = 0.28\ \text{cm}^{-1}$ at 40 keV)

(b) 1 cm of bone ($\mu = 1.2\ \text{cm}^{-1}$ at 40 keV)

(c) 1 cm of tissue containing barium contrast agent ($\mu = 24\ \text{cm}^{-1}$ at 40 keV).

Solution

Here we need to apply Eqn. 24-11 to determine the fractional transmission through each of the relevant materials.

(a)

$$N = N_0 e^{-\mu x}$$

$$\frac{N}{N_0} = e^{-\mu x}$$

$$\frac{N}{N_0} = e^{-(0.28\ \text{cm}^{-1})(1.0\ \text{cm})} = 0.76$$

Therefore, 76% of 40 keV photons will get through 1.0 cm of muscle to reach the film or digital video recorder.

(b)

$$N = N_0 e^{-\mu x}$$

$$\frac{N}{N_0} = e^{-\mu x}$$

$$\frac{N}{N_0} = e^{-(1.2\ \text{cm}^{-1})(1.0\ \text{cm})} = 0.30$$

Therefore, 30% of 40 keV photons will get through 1.0 cm of bone to reach the film or digital video recorder.

(c)

$$N = N_0 e^{-\mu x}$$

$$\frac{N}{N_0} = e^{-\mu x}$$

$$\frac{N}{N_0} = e^{-(24\ \text{cm}^{-1})(1.0\ \text{cm})} = 3.8 \times 10^{-11}$$

Therefore, only 3.8×10^{-9}% of 40 keV photons will get through 1.0 cm of tissue containing barium contrast agent!

CT (or CAT) Scans

Computed Tomography (CT) or Computed Axial Tomography (CAT) uses x rays and computers to generate two-dimensional image slices through the patient. The same essential procedure takes place as that described in radiography: a source of x rays irradiates the patient and the x rays travel through to the detection system on the far side. Regions with higher attenuation coefficient will result in fewer photons reaching the detector. The difference between CT and standard radiography is that the measurement is then repeated again and again, with the x-ray source and the detector rotating around the patient to get transmission measurements along many different paths. The system is coupled to a computer, which is able to combine the transmission data from each path through the patient to construct the final image of the internal organs at a particular position along the length of the subject.

Figure 24-40 (a) illustrates this process and Figure 24-40 (b) is an example of the type of anatomical detail provided to clinicians based on different attenuation properties of tissues, all without requiring a scalpel. As with radiography, contrast agents such as barium or iodine are sometimes used with CT to enhance the information from a particular organ or blood vessel. Because a CT scan is constructed from many x-ray transmission measurements, the radiation exposure associated with such a diagnostic

Rotating x-ray source
Direction of rotation
Rotating x-ray detectors
Motorized platform
(a)

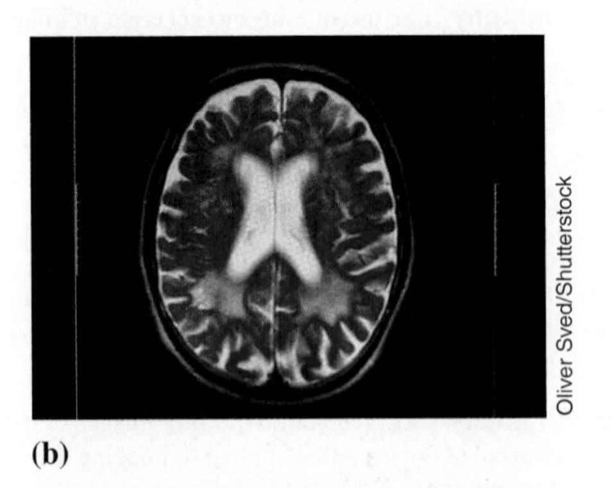

(b)

Figure 24-40 **(a)** A CT scan is obtained by making many x-ray transmission measurements through the patient from different angles around the subject. The data are then used to construct a "slice" through the patient through a complex computational process called backprojection. **(b)** CT images of the brain provide clinicians with incredibly detailed views without having to perform invasive procedures.

procedure is many times more than that associated with standard radiography techniques. CT scans are commonly acquired to identify abnormalities in the abdominal region, the lungs, and the brain.

Biological Tracers

Radioactive isotopes are often used as **tracers** within the field known as **nuclear medicine**, a medical specialization that uses radiation to diagnose and treat disease. A chemical compound containing a radioactive isotope is injected into a patient and the pathway that the compound follows can be traced by detecting the radiation emitted. This technique can be used to monitor the flow of blood, food, or water, and to detect blockages that prevent flow. It can also be used to discover where various elements become concentrated in biological systems.

tracer a chemical compound injected into a subject in order to track the pathway followed after injection

nuclear medicine a medical specialization that uses radioactive isotopes to diagnose and treat disease

The medical isotope ^{99m}Tc, used in approximately 80% of all nuclear medicine scans in Canada, is the workhorse of nuclear medicine. Here we use the "m" after the atomic mass to indicate that it is a meta-stable isotope, decaying via γ-ray emission to the ground state of ^{99}Tc with a half-life of about 6 hours. This isotope is an excellent choice as a tracer in the human body because it emits a γ ray of 140 keV, which easily escapes the body and is readily detected, but it does not emit any other forms of radiation such as α or β particles that would give rise to unnecessary patient exposure. It is used in tests on the heart, the circulatory system, and organs.

Approximately 50% of the ^{99m}Tc procedures done in Canada are for heart imaging. A widely used diagnostic procedure for heart health is the imaging of the blood infusion to the heart using a low dose of the radioactive tracer in the blood. A large radiation detector rotates about the chest of the patient, detecting the γ rays emitted by ^{99m}Tc. Image reconstruction via a computer allows for the display of an image that depicts where the radioisotope has accumulated within the field of view of the detection system, resulting in an image that shows the distribution of blood to the heart muscle. A series of images is often collected as a function of time after the tracer is injected. This is sometimes coupled with requiring the patient to exercise on a treadmill to assess blood flow to the heart at rest and during stress (Figure 24-41).

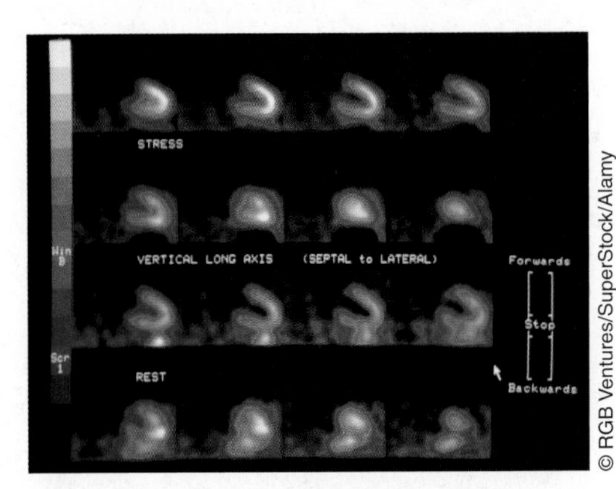

Figure 24-41 Blood infusion to the heart: a series of images of the heart muscle depicting the accumulation of the radioactive tracer ^{99m}Tc delivered by the blood supply. Such images are useful to clinicians in identifying abnormalities in blood delivery to this very important muscle.

SAMPLE PROBLEM 24-11

Measuring the rate of accumulation of iodine in the thyroid gland was the first application of radioactive tracers in medicine. A standard activity of ^{131}I is administered orally and the remaining activity is measured 24 h later. The percentage of ^{131}I that is taken up by the thyroid can then be calculated in order to identify abnormalities. When 3.7×10^5 Bq of ^{131}I is administered orally, it is determined that there is an activity of 1.5×10^5 Bq remaining in the thyroid after 24 h. If the half-life of ^{131}I is 8.0 d, how much of the initial activity was taken up by the thyroid in this patient?

Solution

Based on the half-life, we can determine what fraction of ^{131}I will remain after 24 h via Eqn. 24-6:

$$A = A_0 e^{-\lambda t}$$

$$\frac{A}{A_0} = e^{-\lambda t}$$

To determine λ:

$$\lambda = \frac{0.693}{T_{1/2}} = \frac{0.693}{8.0 \text{ d}}$$

$$\therefore \lambda = 0.0866 \text{ d}^{-1}$$

Now substituting in:

$$\frac{A}{A_0} = e^{-(0.0866\ \text{d}^{-1})(1.0\text{d})}$$

$$\frac{A}{A_0} = 0.917$$

Therefore, after 24 h (or 1.0 d), there will be 92% of the iodine originally taken up by the thyroid that has not decayed. Since we are told that 1.5×10^5 Bq remains in the thyroid after 24 h (A), we can determine A_0:

$A_0 = A/(0.917) = (1.5 \times 10^5\ \text{Bq})/(0.917) = 1.64 \times 10^5\ \text{Bq}$

Therefore, after the oral administration of 3.7×10^5 Bq of ^{131}I, only 1.6×10^5 Bq has accumulated in the thyroid. This represents a percentage uptake of

% uptake = $[(1.64 \times 10^5\ \text{Bq})/(3.7 \times 10^5\ \text{Bq})] \times 100\% = 44\%$

Positron Emission Tomography (PET) Scans

The x-ray imaging techniques that we have discussed here all provide medical personnel with anatomical information without the use of a scalpel. PET imaging provides information that is very different from the images acquired by CT because it gives insight into tissue *function,* not just its structure.

To create a PET scan, a radioactive tracer that emits positrons (β^+ particles) is injected into a patient and accumulates in the organ or region to be studied. The most commonly used radioisotope tracer with PET scans is ^{18}F attached to a glucose analogue called fluorodeoxyglucose (or FDG). After the ^{18}F-labelled FDG solution is injected into the patient, it will accumulate preferentially at actively metabolizing sites in the tissue. Any positron that is emitted will encounter an electron in a surrounding molecule and annihilate with it, as discussed in the Did You Know? box on page 773. Before annihilation, these positrons and electrons have velocities that are small in magnitude and randomly oriented, so the initial momentum of the positron–electron system is essentially zero. Since momentum is constant in the annihilation process, the total momentum of the two γ rays must also be zero. Therefore, the γ rays must move in opposite directions. The γ rays are detected in a ring-shaped detector (Figure 24-42 (a)) surrounding the patient and, with the aid of a computer, the source of the γ rays can be pinpointed and an image created showing where the radioactive material has accumulated (Figure 24-42 (b)).

PET scans are used to detect cancer and to create images during cancer therapy to determine if the therapy is effective. They can also be used to scan the heart to look for indications of coronary artery disease, or to evaluate damage from a heart attack. PET scans of the brain are useful in examining patients who have memory problems that might be the result of Alzheimer's disease, brain tumours, or other disorders.

The major obstacle to the wide-spread introduction of PET scanners is the relatively short half-life of most positron emitters. For example, the half-life of ^{18}F is 1.8 h. This means that the isotope needs to be produced shortly before injection into the patient, so any facility with a PET scanner will also have an in-house cyclotron accelerator in order to maintain the necessary supply of positron emitters. This is a costly and technologically complicated additional infrastructure that is not necessary when working with CT or MRI machines.

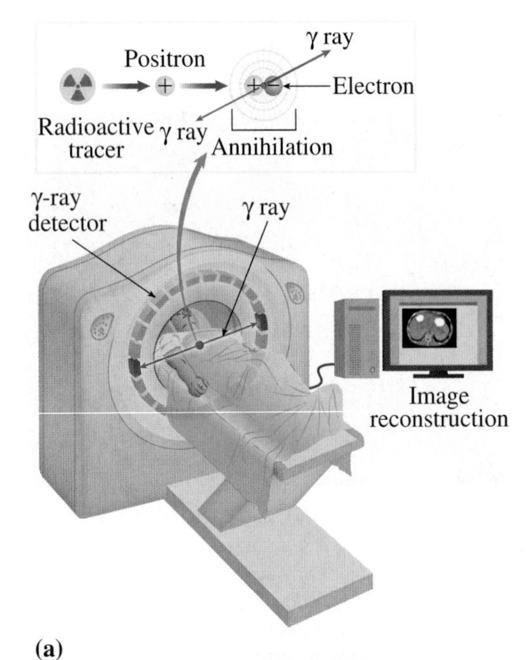

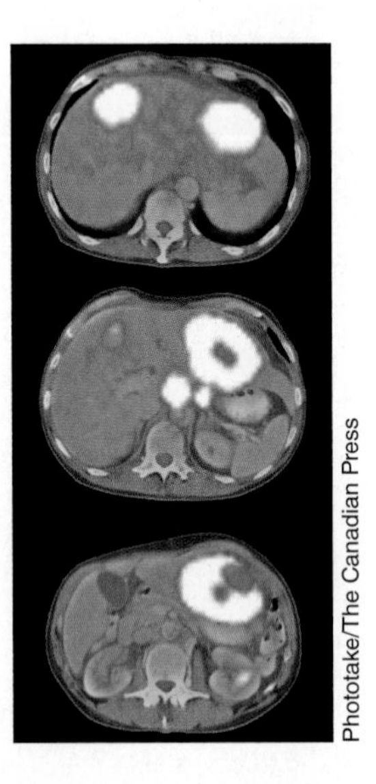

Figure 24-42 **(a)** Schematic illustration of how a PET image is formed. **(b)** Three cross-sectional images of the torso. The underlying grey-scale image is a CT scan depicting the anatomy, while the colour-scale image is a PET scan that illustrates the relative metabolic activity of the organs.

Radiation Therapy

In all the preceding examples of using radiation in medicine, we have discussed diagnostic techniques. In addition, ionizing radiation is used in treating disease, mainly focusing on tumour destruction in cancer patients. Given that ionizing radiation can cause tissue damage as discussed in Section 24.5, radiation therapy involves the very careful delivery of radiation to the tumour site, while minimizing the exposure of the surrounding healthy tissue. This is a delicate balance: all the cells in the tumour need to be destroyed to prevent the cancer from returning, with minimal damage done to the healthy tissue. Each tumour and each patient are unique; therefore, each treatment plan needs to be individualized.

External Beam Radiotherapy

Figure 24-43 summarizes the process that is undertaken for each cancer treatment involving the external delivery of radiation. In some cases, cancer is more effectively treated by the implantation of radioactive materials directly into the tumour, a treatment called **brachytherapy**, which will be discussed in more detail shortly.

brachytherapy the treatment of cancer by implantation of radioactive materials directly into the tumour

DID YOU KNOW?

The term brachytherapy comes from the Greek word *brachys* which means "short distance." By design, the radiation emitted by the implanted brachytherapy sources travels only a short distance, thereby confining the exposure to the target volume.

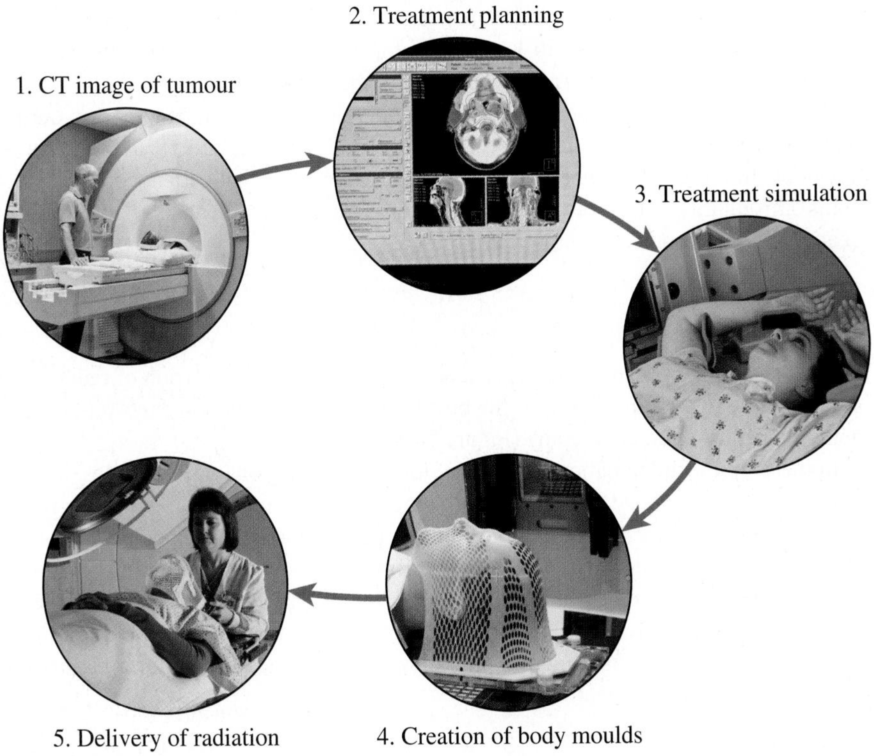

1. CT image of tumour

2. Treatment planning

3. Treatment simulation

4. Creation of body moulds

5. Delivery of radiation

EPSTOCK/Shutterstock, © Henry Westheim/Asia Photo Connection, Mark Kostich/iStock/Thinkstock, Mark Kostich/iStock/Thinkstock, © Luca DiCecco/Alamy

Figure 24-43 The process underlying every individualized radiation therapy treatment of cancer using an external source.

The first step on the journey is a CT or MRI image of the tumour site, to identify the boundaries of the tumour region to be irradiated as well as the location of any sensitive tissues in the vicinity. This information is then used to decide on various treatment options: external source or brachytherapy? If an external source treatment is selected, will this tumour be irradiated with x rays, electrons, or γ rays? The next stage, treatment planning, involves extensive computational work. The patient will be irradiated with x rays, electrons, or γ rays from many different directions and with different intensities in order to tailor the exposure to the particular anatomy of the tumour site. Figure 24-44 shows a simplified version of such a treatment plan.

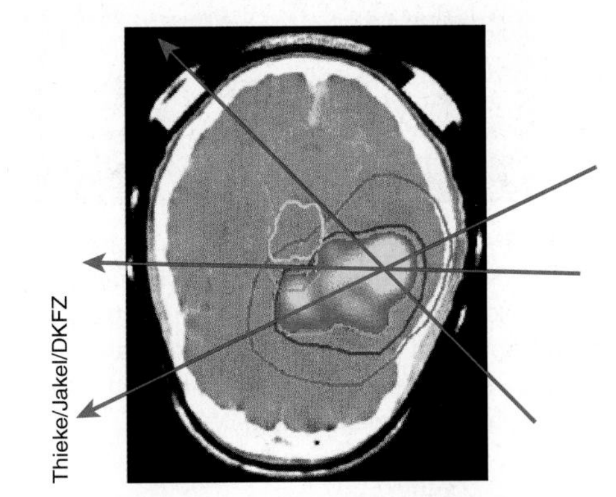

Thieke/Jakel/DKFZ

Figure 24-44 Treatment planning involves complicated decisions about different directions and intensities of radiation beams, as well as how many treatments will be delivered over an extended time period, all with the goal of maximizing the exposure to the tumour while minimizing damage to healthy tissue.

The image in Figure 24-44 identifies the **target volume**, which is the brain tumour to be irradiated. The image also identifies tissue that is very sensitive to radiation (outlined in yellow) and therefore the treatment must avoid exposure in this region as much as

target volume region of the body identified for high levels of radiation in treatment planning

multi-leaf collimator a device that provides clinicians the ability to shape the radiation beam with each delivery in order to better isolate the exposure to the tumour, thereby protecting the surrounding healthy tissue

Intensity Modulated Radiation Therapy (IMRT) high-precision radiation therapy that allows for the radiation exposure to conform more precisely to the shape of the tumour by modulating—or controlling—the intensity of the radiation beam in each delivery

possible. By delivering multiple beams of radiation from different angles from the side closest to the tumour, it is possible to deliver high exposure inside the tumour where the paths intersect, with significantly less exposure elsewhere. Treatment plans involve many angles of irradiation, not just the three shown in Figure 24-44.

A relatively recent development in external beam irradiation is the use of **multi-leaf collimators** in a treatment called **Intensity Modulated Radiation Therapy (IMRT)**. With such a device installed on the beam delivery system, the shape of the radiation beam with each delivery becomes part of the treatment planning process, which allows for even greater protection of the healthy tissue surrounding the tumour. Figure 24-45 (a) shows a typical multi-leaf collimator in use today. Figure 24-45 (b) illustrates the incredible advantage in radiation localization when a multi-leaf collimator is used in treatment planning. The upper image shows the high exposure region (red) for a treatment plan developed with conventional external beam irradiation for a Hodgkin lymphoma (HL) patient. The target volume, outlined in red, is well within this high exposure zone, but there are also sensitive organs such as the esophagus, spinal cord, heart, lungs, and breast tissue in the area. The lower image shows the capabilities of treatment planning when a multi-leaf collimator is added to the system: the red zone of high exposure is now quite nicely confined to the target volume.

With the precision of radiation localization required, patient movement during treatment can be a significant cause for concern. Therefore, great care is taken in patient positioning and immobilization, while also ensuring that patient comfort is not compromised; this is the motivation behind steps 3 and 4, treatment simulation and the creation of body moulds, identified in Figure 24-43.

(a)

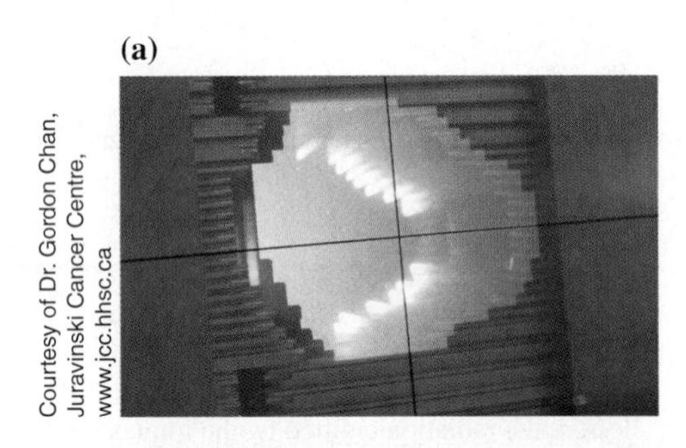

Courtesy of Dr. Gordon Chan, Juravinski Cancer Centre, www.jcc.hhsc.ca

(b)

Vitaliana De Sanctis, et al. "Intensity modulated radiotherapy in early stage Hodgkin lymphoma patients: Is it better than three dimensional conformal radiotherapy?" In *Radiation Oncology* 2012 Vol: 7(1):129. Figure 1. DOI: 10.1186/1748-717X-7-129. © 2012 De Sanctis et al.; licensee BioMed Central Ltd.

Figure 24-45 **(a)** A multi-leaf collimator is used to shape the radiation beam for better localization of radiation exposure to the tumour. **(b)** An illustration of the advantage of using a multi-leaf collimator. The upper image is the exposure with conventional treatment, the lower image is with IMRT.

Brachytherapy

In some cases it has been found experimentally that the direct insertion of radioactive isotopes into the tumour allows for the best localization of radiation exposure, a treatment referred to as brachytherapy (Figure 24-46).

In these cases, the radiation involved is low-energy photons or β particles, as they deposit their energy very close to the implanted sources. In addition, since the radiation source is imbedded in the target volume, patient movement does not influence where the high radiation exposure occurs. Brachytherapy is most commonly used in cases of prostrate and cervical cancer, with iodine-125 or palladium-103 "seeds" implanted in a tumour site in conjunction with careful treatment planning, as with external beam irradiations. Figure 24-47 shows an x ray acquired after a patient has had seeds implanted for treatment.

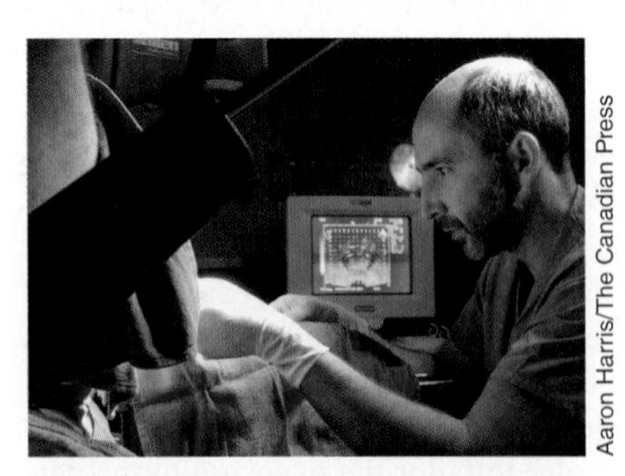

Aaron Harris/The Canadian Press

Figure 24-46 Radiation oncologist Dr. Gerard Morton inserts needles into the prostate of a patient during high dose-rate brachytherapy for prostate cancer at Toronto's Sunnybrook Regional Cancer Centre.

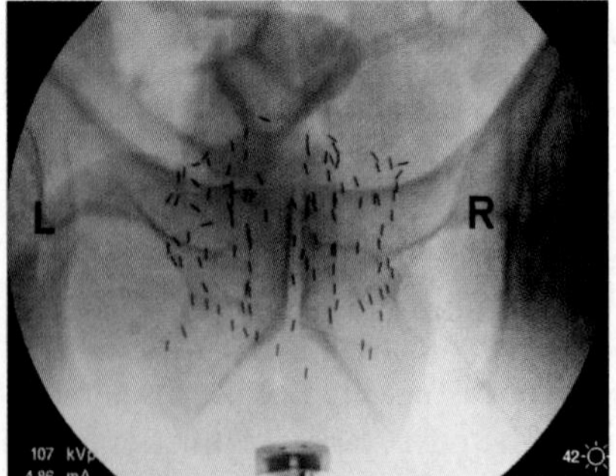

© phillip beron/Alamy

Figure 24-47 An x-ray image of the pelvis of a patient treated for prostate cancer through the implantation of radioactive iodine-125 seeds. The specific placement of the numerous seeds is determined by treatment planning calculations for this particular patient.

EXERCISES

24-59 Describe how an x-ray image is formed.

24-60 Explain why contrast agents are necessary in order to generate an x-ray image of blood flow.

24-61 Based on the data provided in Sample Problem 24-10, what is the half-value layer of (a) muscle, (b) bone, and (c) barium-infused tissue, for 40 keV photons?

24-62 When 3.7×10^5 Bq of ^{131}I is administered orally, it is determined that there is an activity of 8.0×10^4 Bq remaining in the thyroid after 24 hours. If the half-life of ^{131}I is 8.0 d, what percentage of the initial activity was taken up by the thyroid in this patient?

24-63 The overall iodine uptake in a patient administered 3.7×10^5 Bq of ^{131}I was determined to be 15%. What was the measured activity of ^{131}I in the thyroid after 24 hours? (The half-life of ^{131}I is 8.0 d.)

24-64 What is a major (a) advantage and (b) disadvantage of PET imaging?

24-65 Describe the process of treatment planning in radiation therapy.

24-66 How does IMRT result in better localization of radiation in the target volume?

24-67 What is a major (a) advantage and (b) disadvantage of brachytherapy?

LOOKING BACK...LOOKING AHEAD

The present chapter has focused on the dense object at the centre of an atom: the nucleus. We have discussed the structure of nuclei, and how radioactive nuclei decay by the emission of α and β particles and γ rays. The concept of half-life has been presented, as well as its application to radioactive dating. We have explored the processes of nuclear fission and fusion, and have seen how nuclear reactors work. The interaction of radiation with matter has been described, with an emphasis on interactions in biological tissue. A number of applications of radioactivity to medicine, biology, and industry have been discussed. Throughout this, our final chapter of the book, we have returned to familiar concepts explored elsewhere, such as conservation of energy, charge, and momentum. As with the preceding chapters, we have discussed a variety of real-world applications, as these provide meaningful context for the fundamental physics principles.

CONCEPTS AND SKILLS

Having completed this chapter, you should now be able to do the following:

- Describe the structure of nuclei.
- Calculate the binding energy of a nucleus.
- Describe α, β, and γ decay.
- Determine unknown quantities in nuclear equations that represent decays and nuclear reactions.
- Determine the energy released in a particular decay, given the masses involved.
- Discuss the synthesis of nuclei in stars and supernovae.
- Solve problems involving half-life and activity, or half-life and number of nuclei.
- Discuss radioactive dating.
- Describe the processes of nuclear fission and fusion, and how energy is released in these processes.
- Discuss the operation of a nuclear reactor, and state the function of components such as fuel rods, moderator, and control rods.
- Describe possible designs of fusion reactors.
- Calculate the energy released in fission and fusion reactions.
- Describe how various types of radiation interact with matter.
- Solve problems related to the range of α and β particles, and the attenuation coefficient of γ rays.
- State the biological effects of ionizing radiation.
- State the major sources of radiation to which we are exposed.
- Describe practical applications of radioactivity.
- Explain how radiographic, CT, and PET images are acquired.
- Discuss the use of radioactive isotopes as tracers in medicine.
- Describe the use of ionizing radiation in treating cancer, in terms of both external beam irradiation and brachytherapy.

KEY TERMS

You should be able to define or explain each of the following words or phrases:

nucleus	atomic number	atomic mass number	isotopes
nucleon	proton number	mass number	radioactive nuclei
atomic mass unit	neutron number	nuclide	stable nuclei

rest mass
binding energy
decay
radioactivity
alpha (α) particle
parent nucleus
daughter nucleus
beta (β) decay
beta (β^-) particle
anti-neutrino
positron (β^+ particle)
anti-particle
neutrino
decay series
annihilation
Positron Emission Tomography (PET)
gamma rays
nuclear reaction
transmutation
half-life
average lifetime
activity
becquerel (unit)
decay constant
curie (unit)
radioisotope dating
nuclear fission
nuclear fusion
nuclear chain reaction
nuclear reactor
enriched uranium
fuel rods
moderator
control rods
coolant
breeder reactor
inertial confinement
magnetic confinement
electron–ion pair
range
photoelectric effect
Compton scattering
pair production
ionizing radiation
non-ionizing radiation
attenuation coefficient
mean free path
tenth-value layer
Health Physics
hormesis
background radiation
half-value layer
radiography
mammography
fluoroscopy
contrast agents
angiography
tracer
nuclear medicine
brachytherapy
target volume
multi-leaf collimators
Intensity Modulated Radiation Therapy (IMRT)

Chapter Review

MULTIPLE-CHOICE QUESTIONS

24-68 Isotopes have the same

(a) atomic mass number
(b) atomic number
(c) number of neutrons

24-69 Which of the following statements is true for β^- decay?

(a) The mass number (*A*) increases by 1 and the atomic number (*Z*) is unchanged.
(b) *A* is unchanged and *Z* increases by 1.
(c) *A* and *Z* each decrease by 1.
(d) *A* is unchanged and *Z* decreases by 1.
(e) *A* decreases by 1 and *Z* increases by 1.

24-70 Which of the following statements is true for β^+ decay?

(a) The mass number (*A*) increases by 1 and the atomic number (*Z*) is unchanged.
(b) *A* is unchanged and *Z* increases by 1.
(c) *A* and *Z* each decrease by 1.
(d) *A* is unchanged and *Z* decreases by 1.
(e) *A* decreases by 1 and *Z* increases by 1.

24-71 Which of the following statements is true for α decay?

(a) The mass number (*A*) increases by 1 and the atomic number (*Z*) is unchanged.
(b) *A* is unchanged and *Z* increases by 1.
(c) *A* decreases by 1 and *Z* increases by 1.
(d) *A* is unchanged and *Z* decreases by 1.
(e) none of the above

24-72 Which of the following statements is true for γ decay?

(a) Both the mass number (*A*) and the atomic number (*Z*) are unchanged.
(b) *A* increases by 1 and *Z* is unchanged.
(c) *A* is unchanged and *Z* increases by 1.
(d) *A* decreases by 1 and *Z* increases by 1.
(e) *A* is unchanged and *Z* decreases by 1.

24-73 Which of the following is the correct daughter nucleus formed in the α decay of ${}^{238}_{92}\mathrm{U}$?

(a) ${}^{242}_{94}\mathrm{Pu}$
(b) ${}^{238}_{91}\mathrm{Pa}$
(c) ${}^{234}_{90}\mathrm{Th}$
(d) ${}^{238}_{93}\mathrm{Np}$
(e) ${}^{238}_{92}\mathrm{U}$

24-74 Which of the following is the correct daughter nucleus formed in the β^+ decay of ${}^{208}_{83}\mathrm{Bi}$?

(a) ${}^{212}_{85}\mathrm{At}$
(b) ${}^{208}_{82}\mathrm{Pb}$
(c) ${}^{204}_{81}\mathrm{Tl}$
(d) ${}^{208}_{84}\mathrm{Po}$
(e) ${}^{208}_{83}\mathrm{Bi}$

24-75 Which of the following is the correct additional reaction product produced in the following reaction: ${}^{16}_{8}\mathrm{O} + {}^{16}_{8}\mathrm{O} \rightarrow {}^{20}_{10}\mathrm{Ne} + ?$

(a) ${}^{16}_{8}\mathrm{O}$
(b) ${}^{18}_{9}\mathrm{F}$
(c) ${}^{20}_{10}\mathrm{Ne}$
(d) ${}^{14}_{7}\mathrm{N}$
(e) ${}^{12}_{6}\mathrm{C}$

24-76 A particular radioactive isotope has a half-life of 4.0 h. What fraction of the initial activity is remaining after 12 h?

(a) 1/2
(b) 1/3
(c) 1/8
(d) 1/16
(e) 1/32

24-77 Radioactive isotopes are used routinely in some research laboratories. At the beginning of the day on Monday (8 a.m.), a research assistant measures the activity of a particular radioactive isotope to be 1.6×10^7 Bq. At the beginning of the following day, 8 a.m. on Tuesday, the activity is now measured as 4.0×10^6 Bq. What is the half-life of this isotope?

(a) 6.0 h
(b) 12 h
(c) 18 h
(d) 24 h
(e) 48 h

24-78 Which of the following statements is true in nuclear fission?

(a) Two smaller nuclei combine together to make a larger nucleus.
(b) One large nucleus splits into two smaller nuclei.
(c) Mass number is not conserved in the process.
(d) Atomic number is not conserved in the process.
(e) Energy is not conserved in the process.

24-79 Which of the following statements is true in nuclear fusion?

(a) Two smaller nuclei combine together to make a larger nucleus.
(b) One large nucleus splits up into two smaller nuclei.
(c) Mass number is not conserved in the process.
(d) Atomic number is not conserved in the process.
(e) Energy is not conserved in the process.

24-80 When ^{235}U absorbs a slow-moving neutron, it then undergoes fission. Why does this lead to a chain reaction?

(a) The fission daughter nuclei are radioactive, resulting in subsequent α and β decay.
(b) Each fission releases multiple neutrons, which then go on to start more fission reactions with other ^{235}U nuclei in the sample.
(c) The fission daughter nuclei recombine through fusion to form ^{235}U again.
(d) all of the above

24-81 Which of the following is NOT a process by which photons (x rays and γ rays) transfer energy to the medium through which they are travelling?

(a) Compton scattering
(b) pair production
(c) photoelectric effect
(d) annihilation

24-82 Compared to the range of a 4 MeV α particle in tissue, the range of a 4 MeV β particle in tissue is

(a) smaller
(b) comparable
(c) larger

24-83 Compared to the mean free path of a 100 keV γ ray in tissue, the mean free path of a 100 keV γ ray in lead is

(a) smaller
(b) comparable
(c) larger

24-84 Compared to the mean free path of a 100 keV γ ray in tissue, the mean free path of a 1 MeV γ ray in tissue is

(a) smaller
(b) comparable
(c) larger

24-85 ^{131}Cs is occasionally used as a brachytherapy source. Which of the following statements is/are correct regarding this isotope?

(a) It emits 662 keV γ rays.
(b) It has a mean free path of more than 5 mm in lead.
(c) It has a short half-life of 9.7 days.
(d) Statements (a) and (b) are true.
(e) None of the above is true.

Review Questions and Problems

24-86 For a nucleus of $^{23}_{11}Na$, what are the

(a) atomic mass number,
(b) atomic number,
(c) number of neutrons, and
(d) chemical name?

24-87 What is the difference between a nucleon and a nuclide?

24-88 Why does it make no sense to ask which chemical element has $A = 16$?

24-89 What is the binding energy (in MeV) of a nucleus of ^{16}O? Use Table 24-1.

24-90 What is the binding energy (in MeV) of a nucleus of ^{63}Cu? Use Table 24-1.

24-91 Determine the unknown quantity in each of the following nuclear equations and name the process:

(a) $^{?}_{54}Xe \rightarrow {}^{0}_{?}e + {}^{118}_{53}I$
(b) $^{175}_{?}Pt \rightarrow {}^{?}_{2}He + {}^{171}_{76}Os$
(c) $^{12}_{6}C + {}^{1}_{1}H \rightarrow ? + {}^{1}_{0}n$

24-92 How much energy (in MeV) is released in the following α decay: $^{154}_{68}Er \rightarrow {}^{4}_{2}He + {}^{150}_{66}Dy$? The relevant atomic masses are as follows:

$^{154}_{68}Er$: 153.9328 u
$^{4}_{2}He$: 4.0026 u
$^{150}_{66}Dy$: 149.9256 u

24-93 It is likely that the trans-uranium elements (those with $Z > 92$) existed early in Earth's history. Explain how it is possible that these elements are not found now, although a large amount of uranium can be found.

24-94 After how many half-lives is the activity of a radioactive source reduced to 1/128 of the original activity?

24-95 The half-life of ^{65}Ni is 2.56 h.

(a) What is the average lifetime of this nuclide?
(b) What is the decay constant of this nuclide?

24-96 The activity of a sample of $^{55}_{24}Cr$ (half-life 3.6 min) is 4.5×10^{11} Bq at 1:15 p.m. How many nuclei of ^{55}Cr will remain at 1:51 p.m. on the same day?

24-97 It takes a sample of ^{222}Th 5.6×10^{10} y to decrease in activity by a factor of 16. What is the half-life of this isotope?

24-98 Chlorine-38 is produced by cosmic rays interacting with the atmosphere. If 7.5×10^6 nuclei of ^{38}Cl are produced at essentially the same time, what is the activity of ^{38}Cl remaining 5.5 h later? The half-life of ^{38}Cl is 37 min.

24-99 What is the meaning of each of the following terms?

(a) enriched uranium

(b) nuclear chain reaction

(c) breeder reactor

24-100 In what ways are nuclear power stations and fossil-fuel power stations (a) similar, and (b) different?

24-101 The following describes a fission reaction taking place with ^{239}Pu in a breeder reactor:

$$^{239}_{94}\text{Pu} + ^{1}_{0}\text{n} \rightarrow ^{100}_{40}\text{Zr} + ^{138}_{54}\text{Xe} + ?(^{1}_{0}\text{n})$$

(a) How many neutrons are produced in the process?

(b) What is the energy released in this reaction?

The relevant rest masses are

$^{239}_{94}$Pu: 239.0522 u

$^{1}_{0}$n: 1.0087 u

$^{100}_{40}$Zr: 99.9178 u

$^{138}_{54}$Xe: 137.9140 u

24-102 Calculate the energy released in the following fusion reaction, which occurs in the Sun and other stars.

$$^{3}_{2}\text{He} + ^{3}_{2}\text{He} \rightarrow ^{4}_{2}\text{He} + 2(^{1}_{1}\text{H})$$

The relevant rest masses are

$^{3}_{2}$He: 3.0160 u

$^{4}_{2}$He: 4.0026 u

$^{1}_{1}$H: 1.0078 u

24-103 Determine the range in tissue of α particles emitted in the decay of ^{222}Rn. The following atomic rest mass data are provided:

Particle/Nucleus	Mass (u)
$^{222}_{86}$Rn	222.0176
$^{218}_{84}$Po	218.0090
$^{4}_{2}$He	4.002 60

24-104 Determine the range in tissue of the 167 keV β particles emitted in the decay of ^{35}S. This is a commonly used radioisotope in biology research labs, often as a labelling technique with amino acids or nucleotides.

24-105 Determine the range in tissue of the 643 keV β particles emitted in the decay of ^{90}Y. This beta emitter is used in brachytherapy to treat liver cancer.

24-106 The half-value layer of tissue for the γ ray emitted by ^{99m}Tc is 4.6 cm. What is the tenth-value layer of tissue for this γ ray?

24-107 The tenth-value layer of steel for the γ ray emitted by ^{60}Co is 7.3 cm. What is the mean free path in steel for this γ ray?

24-108 What is the main difference between standard radiography and fluoroscopy?

24-109 Describe how a CT image is formed.

24-110 An overall iodine uptake in a patient administered 3.7×10^5 Bq of ^{131}I was determined to be 32%. What was the measured activity of ^{131}I in the thyroid 48 h after the iodine was administered? The half-life of ^{131}I is 8.0 d.

Applying Your Knowledge

24-111 It has been observed that nature favours symmetrical nuclear configurations. For example, nuclei in which there are an even number of protons and an even number of neutrons (even–even nuclei) are much more stable than nuclei with an odd number of protons and an odd number of neutrons (odd–odd nuclei). In fact, there are only five stable odd–odd nuclei to be found in the entire Chart of the Nuclides ($^{2}_{1}$H, $^{6}_{3}$Li, $^{10}_{5}$B, $^{14}_{7}$N, and $^{180}_{73}$Ta)! Calculate the binding energy *per nucleon* of the following odd–odd/even–even pairs of nuclei to see how they compare:

(a) $^{2}_{1}$H and $^{4}_{2}$He

(b) $^{10}_{5}$B and $^{12}_{6}$C

Note that in each case the even–even nucleus is formed from adding one more proton and one more neutron to the associated odd–odd nucleus.

24-112 The radius of a $^{12}_{6}$C nucleus is approximately 2.7 fm. What is the density of this nucleus?

24-113 $^{232}_{90}$Th (thorium) is the parent in a long decay series. It decays by α emission, its daughter by β^-, and the next daughter by β^-. What is the nuclide at this point in the series?

24-114 Radioactive ^{212}Pb decays to the stable ^{208}Pb through a three-step decay chain: β^- decay, β^- decay, followed by α decay. Write out the series of decay reactions fully.

24-115 Question 24-45 in Section 24.4 asks you to find the approximate mass of $^{235}_{92}$U required to operate a nuclear reactor for one year if the thermal power produced is 2×10^3 MW.

(a) What approximate mass of coal would be required to produce the same thermal power? Burning 1 kg of coal produces about 3×10^6 J of energy.

(b) If deuterium fusion is used to provide the power, what approximate mass of deuterium would be required? Assume that the fusion of two deuterium nuclei produces about 4 MeV of energy.

24-116 Figure 24-48 illustrates the attenuation coefficient of various biological tissues in the human body as a function of photon energy. Why is a mammography image (all "soft tissue") generated with incident photons that have much lower energy than those used in standard radiography imaging involving bones and soft tissue?

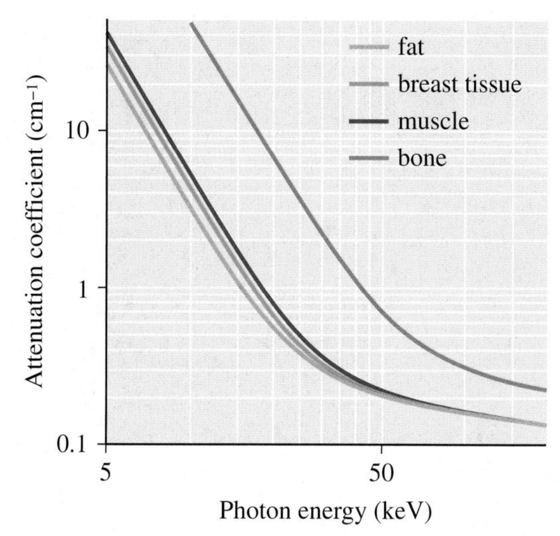

Figure 24-48 Attenuation coefficient data for a variety of biological tissues as a function of photon energy.

24-117 When a neutron is absorbed by a ^{10}B nucleus, the resulting reaction produces an energetic α particle and a ^{7}Li nucleus, as described in the equation

$$^{10}_{5}B + ^{1}_{0}n \rightarrow ^{4}_{2}He + ^{7}_{3}Li$$

Determine the range in tissue of the α particle produced in this reaction. **Note:** The masses of the products are quite similar here so make sure that you account for how the energy is divided between the α particle and the ^{7}Li nucleus. (See the Did You Know? box on page 789.)

The following atomic rest mass data are provided:

Particle/Nucleus	Mass (u)
$^{10}_{5}B$	10.0129
$^{1}_{0}n$	1.0087
$^{7}_{3}Li$	7.0160
$^{4}_{2}He$	4.002 60

DID YOU KNOW?

The nuclear reaction in Question 24-117 has been investigated for medical applications in a technique called boron neutron capture therapy. The idea is to get biologically inert ^{10}B into a tumour or other site of interest, and then bombard the region with neutrons that can react with the ^{10}B to generate α particles and ^{7}Li nuclei. The ionizing radiation then causes extensive highly localized damage, resulting in tissue damage within a few micrometres of the site of the nuclear reaction.

GLOSSARY

absolute pressure total pressure taking into account all sources of pressure p. 329

absolute zero a temperature of −273.15°C, or 0 K, representing a lower bound of temperature for physical processes p. 593

absorption spectrum a white-light spectrum that has specific wavelengths absorbed by elements or compounds, resulting in dark bands spaced throughout p. 528

acceleration the time rate of change of velocity p. 32

acceleration due to gravity the rate of increase of velocity with time for an object falling with negligible air resistance p. 44

accommodation the process of controlling the shape of the eye's lens by the ciliary muscles p. 553

acoustics the qualities of a room that determine how well sound is heard p. 459

action-at-a-distance forces forces such as gravity that do not require contact between objects in order to be exerted p. 107

activity the number of nuclei that decay per unit time; symbol A p. 778

adhesive forces forces exerted between unlike molecules p. 334

air-wedge interference a type of thin-film interference in which there is a film of air between transparent plates p. 534

alpha (α) particle a helium nucleus, a tightly bound collection of two protons and two neutrons p. 770

alternating current (AC) current reverses direction in a circuit in a periodic fashion, such as the current generated by commercial power stations p. 696

alternating-current (AC) generator an electric generator in which the rotation of a coil through a magnetic field gives rise to a current in an external circuit that reverses direction with each half-turn of the coil p. 745

ammeter a device that measures the current through a particular point in a circuit p. 691

ampere unit of electric current, equivalent to 1 coulomb per second; symbol A p. 669

amplitude (of oscillation) the maximum displacement of an oscillating object from its equilibrium position; symbol A p. 375

angiography the generation of x-ray images of the cardiovascular system, often applied to imaging the blood supply in the brain and in the heart p. 795

angle of deviation the angle between the incident ray and the refracted ray beyond the prism for a prism with non-parallel sides p. 489

angle of incidence the angle between the incident ray and the normal; symbol θ_i p. 476

angle of reflection the angle between the reflected ray and the normal; symbol θ_r p. 476

angle of refraction the angle between a refracted ray and the normal; symbol θ_R p. 485

angular acceleration see instantaneous angular acceleration

angular displacement change in angular position of a rotating object p. 257

angular magnification the ratio of the angular size of an image with the lens to the angular size without the lens (also called magnifying power) p. 562

angular momentum the product of moment of inertia and angular velocity (for an object rotating about a fixed axis) p. 277

angular position the position of a point on a rotating object, measured as an angle between an arbitrary axis (drawn outward from the centre of rotation) and the line from the centre to the chosen point p. 256

angular size the angle subtended by an object at the observer's eye or other optical instrument p. 561

angular velocity instantaneous rotation rate of a rotating object; the limit as $\Delta t \rightarrow 0$ of angular displacement $\Delta\theta$ divided by the time interval Δt taken (also called instantaneous angular velocity) p. 257

annihilation an interaction that occurs when any particle/anti-particle pair meet: their mass and energy are converted into two high-energy photons in a process in which total energy and momentum are conserved p. 773

anti-neutrino a weakly interacting particle postulated by Pauli to account for the distribution of energies observed with β particles emitted in β decay, represented as $\bar{v}$ in decay equations p. 772

antinode a position of maximum displacement in an interference pattern p. 403

anti-particle a particle that has the same mass as its particle counterpart but the opposite charge; the positron and electron are anti-particles of each other p. 772

aperture the diameter to which a camera shutter opens; symbol d p. 578

apparent weight difference between the force of gravity and the buoyant force on an object p. 347

Archimedes' principle the upward buoyant force on an object immersed wholly or partially in a fluid is equal in magnitude to the weight of the fluid displaced by the object p. 347

armature the rotating iron core in an electric motor or generator p. 727

astigmatism a condition of the eye in which the focusing is asymmetrical; caused by a cornea with an uneven surface and resulting in better focusing in one plane than in other planes p. 559

astronomical telescope a telescope that produces an inverted image p. 567

atmospheric greenhouse effect absorption by atmospheric gases of some of the radiation emitted by Earth's surface, and the resulting increase in temperature of Earth's surface because of the radiation emitted by the gases back to the surface (also called greenhouse effect) p. 608

atomic mass number/mass number the total number of nucleons (protons and neutrons) in a given nucleus; symbol A p. 765

atomic mass unit a non-SI unit of mass that is often used in atomic and nuclear physics p. 764

atomic number the total number of protons in a given nucleus (also called proton number); symbol Z p. 765

attenuation coefficient a property of a given material that characterizes the probability of a photon of a specific energy interacting; symbol μ p. 791

average acceleration the ratio of the change in velocity to the elapsed time p. 32

average angular acceleration a measure of how rapidly, on average, the rotation rate of a rotating object is changing during a time interval; change in angular velocity ($\Delta\omega$) divided by the time interval (Δt) taken p. 257

average angular velocity average rotation rate of a rotating object during a time interval; angular displacement $\Delta\theta$ divided by the time interval Δt taken p. 257

average lifetime the average time a particular nucleus of a given isotope will exist before decay; symbol T_{avg} p. 777

average speed the ratio of the total distance travelled to the elapsed time p. 22

average velocity the ratio of the displacement to the elapsed time for that displacement p. 26

axis of equilibrium the line along which a medium lies when not disturbed by a wave p. 397

axis of polarization the axis of a polarizing filter that allows the transmission of a plane polarized wave p. 540

background radiation radiation from environmental sources p. 793

barometer an instrument that measures atmospheric pressure p. 340

barometric equation the relationship expressing the exponential decrease of atmospheric pressure as a function of elevation above sea level p. 337

base unit a standard unit of measurement from which other units may be derived (also called fundamental unit) p. 5

battery a combination of more than one electric cell, used as a source of energy p. 682

beat frequency the number of beats per second; symbol f_{beat} p. 440

beats the interference pattern of periodic sounds of increasing and decreasing loudness caused by sounds of nearly equal frequency p. 439

becquerel the SI unit of activity, one decay per second; symbol Bq p. 778

Bernoulli's equation at any point in the flow of an ideal fluid, the sum of the pressure, P, the kinetic energy per unit volume, $\frac{1}{2}\rho v^2$, and the gravitational potential energy per unit volume, ρgy, is constant. p. 352

Bernoulli's principle fluid pressure decreases (increases) when fluid speed increases (decreases) p. 355

beta (β) decay a radioactive decay process in which the nucleus emits a beta (β) particle p. 771

beta (β^-) particle an electron emitted in radioactive decay by nuclei with an excess of neutrons; represented in decay equations as ${}^{0}_{-1}e$ p. 771

beta (β^+) particle see positron

binding energy the energy released when nucleons bind together to form a nucleus p. 766

bioluminescent animal an animal that can emit light energy because of chemical reactions p. 469

birefringence the double refraction that occurs in a transparent crystalline substance in which the speed of light can differ in different directions p. 544

black hole a region around a collapsed star where gravity is so strong that not even light can escape p. 247

body waves waves that move through Earth's interior (also called seismic waves); consist of primary and secondary waves p. 396

bow wave a two-dimensional overlapping of water waves created where the speed of a boat exceeds the speed of the waves p. 437

brachytherapy the treatment of cancer by implantation of radioactive materials directly into the tumour p. 799

breaking point the point on the stress–strain graph for a given material at which the material breaks (also called fracture point) p. 308

breeder reactor converts the ${}^{238}U$ in the fuel to ${}^{239}Pu$, which is then used as fuel p. 785

brittle the adjective used to describe a material that stretches very little beyond the elastic limit before breaking p. 308

brushes stationary pieces of metal in electrical contact with rotating electrical components such as commutators p. 726

bulk modulus ratio of bulk stress to bulk strain p. 312

bulk strain for an object subject to bulk stress, the ratio of change in volume to original volume (also called volume strain) p. 312

bulk stress the change in pressure (force per unit area) exerted on all surfaces of an object (also called volume stress) p. 312

buoyant force an upward force exerted on an object in a fluid p. 346

centre of gravity (CG) the point in an object where the force of gravity can be assumed to act; same as the centre of mass in most practical situations p. 291

centre of mass (CM) point located at the average position of the mass of an object p. 288

centrifugal "force" a fictitious force that is needed to explain the motion of an object relative to a rotating noninertial frame of reference; this "force" is directed outward from the centre of rotation p. 153

centripetal acceleration the acceleration that occurs in uniform circular motion; it is an instantaneous acceleration toward the centre of the circle p. 90

charging by conduction see charging by contact

charging by contact the transfer of charge from one object to another by direct contact between the objects (also called charging by conduction) p. 632

charging by induction an excess charge is left on a conductor through the transfer of charge between the conductor and Earth in the presence of another charged object p. 633

chemiluminescent material a material that produces light through a chemical reaction without a noticeable increase in temperature p. 469

chromatic aberration a problem with lenses in which colours caused by the dispersion of light occur around the perimeter of the lens p. 501

closed circuit a complete closed loop in which a current can readily travel p. 684

coefficient of kinetic friction the constant of proportionality (which depends on the types of materials in contact) between the magnitude of kinetic friction and the magnitude of the normal force p. 141

coefficient of linear expansion a measure of how much the length of a substance expands per unit change in temperature p. 600

coefficient of performance (C.O.P.) for a refrigerator or air conditioner, the ratio of the heat removed from the cold reservoir to the amount of work provided p. 617

coefficient of static friction the constant of proportionality (which depends on the types of materials in contact) between the maximum magnitude of static friction and the magnitude of the normal force p. 140

coefficient of volume expansion a measure of how much the volume of a substance expands per unit change in temperature p. 600

coherent waves waves that maintain their phase relationship and have the same wavelength p. 514

cohesive forces forces exerted between like molecules p. 344

commutator a split metal ring used in electric motors to reverse the current direction with every half rotation p. 726

completely inelastic collision the total final kinetic energy is less than the total initial kinetic energy. After such a collision, the objects stick together and move with the same velocity. The decrease in the total kinetic energy is the maximum that is possible. p. 207

component of a vector a projection of a vector along an axis, usually of a rectangular coordinate system p. 67

compound light microscope an optical instrument with two lenses or lens systems that magnify nearby objects p. 564

compressibility reciprocal of bulk modulus p. 312

compression (of object) the shortening of an object as a result of forces pushing inward on it p. 305

compression (of wave) the portion of a longitudinal wave where the particles are compressed p. 394

compression wave see longitudinal wave

compressive strain for an object that has been compressed, the ratio of the magnitude of the decrease in length, ΔL, to the original unstretched length, L_0 p. 305

compressive stress ratio of the magnitude of the external force $\vec{F}$ compressing an object to the cross-sectional area A of the object p. 305

Compton scattering a γ ray is scattered from an atom, resulting in a new direction of propagation for the photon as well as a reduction in energy; there is also an electron emitted in this process p. 790

conduction heat transfer as a result of atomic and molecular collisions between two objects at different temperatures in physical contact, or between different-temperature regions of the same object p. 602

conduction electrons electrons that are very loosely bound to atoms in conducting materials such as metals p. 632

cones receptors in the eye responsible for day vision and sensitive to colours p. 554

consonance an indication of pleasant, harmonious pairs of sounds p. 444

constructive interference an interference of waves in which the resulting displacement is larger than the displacements of the individual waves p. 402

contact angle for a liquid on a solid, the angle between the solid surface under the liquid and the edge of the liquid surface where it touches the solid p. 344

contact forces forces such as friction and normal force that, in order to be exerted, require contact between objects p. 107

continuous spectrum an unbroken band of emissions occurring at a wide range of wavelengths p. 527

contrast agents highly attenuating inert materials used to enhance the image generated from a particular region of the body p. 795

control rods neutron-absorbing material (often boron or cadmium) used to prevent runaway chain reactions in a reactor core p. 783

convection a process that involves the physical movement of a fluid (liquid or gas) that transfers heat from one location (or object) to another p. 607

converging lens a lens that causes light rays from a distant source to converge to a focal point p. 495

converging mirror a mirror with a concave surface that causes parallel light rays to converge p. 479

coolant liquid (often water) or gas that passes through the fuel area in order to remove the large amounts of thermal energy produced during fission p. 783

Coriolis "force" a fictitious force that is needed to explain the motion of an object relative to a rotating noninertial frame of reference; this "force" has a direction perpendicular to the object's velocity relative to the rotating frame p. 153

cornea a fixed transparent layer at the front of the eye with an index of refraction of 1.37 p. 552

Coulomb's law the force between two charges is proportional to the product of the charges and inversely proportional to the square of the distance between them p. 634

crest the upper part of a transverse wave p. 394

critical angle the angle of incidence in a medium of greater index of refraction that allows the light in a medium of lower index of refraction to disappear; symbol θ_c p. 491

curie a non-SI unit of activity, for which 1 Ci is equal to 3.7×10^{10} Bq; symbol Ci p. 778

Curie temperature the temperature above which magnetic effects are lost p. 709

damped harmonic motion periodic or repeated motion in which the amplitude of oscillation and the energy decrease with time p. 389

daughter nucleus the product nucleus that results from nuclear transformation p. 770

decay transformation of one nuclide to another to achieve stability p. 769

decay constant a property of a given radioactive nuclide that describes the rate of decay, related to half-life; symbol λ p. 778

decay series a sequence of decays from one unstable nucleus to the next, eventually ending in a stable nucleus p. 772

density mass per unit volume of a substance; symbol ρ p. 326; the inertial property of gases, liquids, and solids; symbol ρ (the Greek letter rho) p. 418

depth of field the range of object distances over which an image in a camera (or other optical device) is clear p. 579

derived unit a measurement unit that can be stated in terms of one or more of the base (SI) units p. 7

destructive interference an interference of waves in which the resulting displacement is smaller than the displacements of the individual waves p. 402

diamagnetic materials in which the overall magnetic field is slightly reduced because there is a small magnetic field inside the material that is in the *opposite* direction to the external (much larger) field; e.g., copper, silver, graphite p. 716

diastolic pressure minimum blood pressure exerted by the heart during a person's cardiac cycle p. 334

dichroic material a crystal or other material that transmits light in one plane and absorbs it in the other planes p. 539

diffraction the bending of a wave in a single medium as it passes by a barrier or through an opening p. 406

diffraction grating an optical device made of many uniformly spaced, parallel slits or grooves p. 526

diffuse reflection reflection of light off an irregular surface (also called irregular reflection) p. 475

dimensional analysis the process of using dimensions, such as length, mass, and time, to analyze a problem or an equation p. 10

dipping needle a compass that is free to rotate in the vertical plane; used to measure the magnetic inclination at a particular location p. 709

direct current (DC) current travels in a circuit in only one direction, typically driven by a battery (or a collection of batteries) p. 696

direct-current (DC) generator an electric generator in which the rotation of a coil through a magnetic field gives rise to a current that travels in one direction only p. 746

dispersion the splitting of white light into its spectral colours as it passes through a transparent medium p. 489

displacement the change of position from the initial point to the final point p. 25

dissonance an indication of unpleasant, inharmonious pairs of sounds p. 444

diverging lens a lens that causes light rays from a distant source to diverge as if they originated from a virtual focal point p. 495

diverging mirror a mirror with a convex shape that causes parallel light rays to diverge p. 479

domains regions within a magnetic material in which the magnetic fields of the electrons are aligned; overall each domain behaves much like a small magnet with a strong magnetic field as well as north and south poles p. 708

Doppler effect the observed changing frequency of waves due to the relative motion of the source of waves and the receiver p. 433

drag force for an object moving in a fluid, the force exerted on the object by the fluid, in a direction opposite to the object's velocity p. 363

ductile the adjective used to describe a material that stretches a lot beyond the elastic limit before breaking p. 308

dynamic equilibrium the state of a moving object on which the sum of forces and sum of torques are both zero p. 292

dynamics the study of forces and their effects on the motion of objects p. 106

echolocation the finding of objects using reflected ultrasonic sounds p. 428

effective power the power supplied or consumed by a component in an AC circuit, averaged over a complete cycle p. 697

efficiency ratio of useful energy out to total energy in, during an energy conversion p. 183

elastic collision the total final kinetic energy is equal to the total initial kinetic energy in such a collision p. 207

elastic limit for a given material, the maximum stress at which the material exhibits elasticity p. 307

elastic modulus the constant of proportionality between stress and strain (if the stress is sufficiently small) p. 303

elastic range the region of the stress–strain graph in which a given material exhibits elasticity p. 307

elasticity the ability of a material, having been stressed, to return to its original shape once the stress is removed p. 307

electric cell two different metals separated by a conducting liquid p. 682

electric current rate of movement of charge; symbol I; SI unit ampere (A) p. 669

electric field ($\vec{E}$) electric force per unit charge; unit N/C p. 639

electric field lines a means of visualizing how an electric field varies in space, in both magnitude and direction p. 641

electric force an attractive force between opposite charges and a repulsive force between like charges p. 630

electric generator a device that converts mechanical energy to electrical energy through the motion of a conducting coil through a magnetic field p. 743

electric polarization the spatial separation of positive and negative charges within a molecule p. 630

electric potential electric potential energy per unit charge at a particular location in space within an electric field p. 658

electric potential energy the energy of a charge due to its position in an electric field; symbol U p. 652

electrical circuit a conducting path that forms a closed loop p. 683

electrical conductors materials, such as metals, in which charge moves easily p. 631

electrical force a force that exists between charged objects or particles, and is responsible for holding together the atoms and molecules in all objects p. 132

electrical insulators materials, such as plastic, in which charge does not move easily p. 631

electrolyte the conducting liquid used to separate two different metals in an electric cell p. 682

electromagnet a coil or loop of wires with a magnetic core p. 763

electromagnetic induction the induction (creation) of an electric potential difference or voltage in a conductor by a changing magnetic field in the region p. 736

electromagnetic interaction a name given to the combination of electrical and magnetic forces, which are very closely related p. 132

electromagnetic spectrum the entire set of electromagnetic waves that obey the universal wave equation and can travel in a vacuum at 3.00×10^8 m/s p. 537

electromagnetic wave a wave with electric and magnetic components that travels in a vacuum at a speed of 3.00×10^8 m/s p. 536

electron a negatively charged particle outside the nucleus of the atom p. 628

electron affinity ease with which an atom or molecule attracts an electron p. 629

electron–ion pair the products generated in material such as biological tissue when charged particles (α and β particles) travel through and cause localized ionizations p. 788

electron volt a unit of energy that corresponds to the change in electric potential energy experienced by an electron moving through a potential difference of 1 V; symbol eV p. 662

electrostatic precipitator a filter that uses electric fields to remove particulate from the emissions of industrial chimneys p. 643

elementary charge the smallest amount of charge observed in nature; all charges are integer multiples of this elementary charge; symbol e p. 643

energy return on investment (EROI) the energy obtained from a source per unit of energy required (or spent) to obtain it p. 186

enriched uranium uranium sample in which the percentage of ^{235}U present is higher (~3%) than naturally found in the environment (0.7%) p. 783

entropy a measure of disorder p. 614

equation of continuity for an ideal fluid flowing in a tube the product of cross-sectional area and speed is constant, i.e., volume flow rate is constant p. 350

equitempered scale the musical scale commonly used for musical compositions, with a standard frequency of 440 Hz and 12 intervals per octave p. 445

escape speed minimum speed needed to escape from the gravitational pull of a celestial object p. 246

exposure time the amount of time the shutter remains open when a photograph is taken (also called shutter speed) p. 578

eyepiece lens (eyepiece) the lens in an optical instrument closest to the observer p. 564

far point the farthest distance at which objects appear in focus by an eye; for healthy eyes the far point is stated as infinity p. 553

Faraday's law the induced voltage in a closed conducting path is equal to the absolute value of the rate of change of the magnetic flux through that path p. 754

farsightedness see hyperopia

Fermi question see order-of-magnitude estimation

ferromagnetic materials a few substances found on Earth (such as iron, nickel, cobalt, and their alloys) that exert strong magnetic forces on other materials (also called magnetic materials) p. 706

fibre optics the applications of the transmission of light in transparent fibres p. 493

fictitious force a force that must be invented to explain the motion of an object relative to a noninertial frame of reference (also called pseudoforce or inertial force) p. 152

field a physical quantity that has a specific value at each point in space and time p. 232

first harmonic see fundamental mode

first law of thermodynamics the change in internal energy (ΔU) of a system equals the sum of the heat Q added to the system and the work W done on the system p. 611

first mode see fundamental mode

flow rate see volume flow rate

fluid a substance that flows and takes the shape of its container p. 326

fluid dynamics the study of fluids in motion and objects moving in fluids p. 349

fluid statics the study of fluids at rest p. 326

fluorescent source an object that emits light when struck by high-energy waves or particles p. 468

fluoroscopy the use of a digital video recorder to capture real-time moving images of a patient with x rays p. 795

focal length (of a curved mirror) the distance from the focal point of a curved mirror to the vertex; symbol f p. 479

focal point the position where the incident rays parallel and close to the principal axis of an optical device meet, or appear to meet, after they strike the device (also called the principal focus); symbol F p. 479

force diagram see free-body diagram (FBD)

force of friction the force parallel to surfaces in contact with each other, due to attractive electrical forces between molecules in the surfaces p. 106

forced oscillation a vibration in which the driving frequency is different from the resonant frequency, resulting in a small amplitude p. 391

fracture point see breaking point

frame of reference a coordinate system fixed to an object, such as Earth, relative to which a motion can be observed p. 94; any object (Earth, bus, particle, etc.) relative to which the positions, velocities, etc., of other objects can be measured p. 151

Fraunhofer diffraction the type of diffraction in which the paths of the Huygens' light wavelets are parallel p. 520

free-body diagram (FBD) a diagram in which only the forces acting on a particular object are shown (also called force diagram) p. 107

free fall the action of any object falling with negligible air resistance p. 44

frequency of revolution the number of revolutions per second of an object in uniform circular motion; symbol f p. 92

friction factor the constant of proportionality between the magnitude of the drag force exerted by a fluid on a small object moving in the fluid, and the speed of the object p. 363

frictional heating the production of thermal energy as the result of kinetic friction acting between two sliding objects p. 142

***f*-stop number** the ratio of the focal length (f) of a camera lens to the diameter of the aperture p. 578

fuel rods assembly containing uranium oxide pellets that serve as the fuel in the nuclear reactor p. 783

fundamental frequency the lowest resonant frequency in a standing wave pattern; corresponds to the fundamental mode p. 404

fundamental mode the mode with the longest wavelength on a standing wave (also called the first mode or the first harmonic) p. 404

fundamental unit see base unit

$\vec{g}$ the symbol for the acceleration due to gravity p. 46

gamma rays high-energy electromagnetic radiation, often emitted by a nucleus to release excess energy without any inherent change to the nuclear composition p. 773

gas discharge source a gas contained in a glass tube that becomes luminous when an electric current passes through it p. 468

gauge pressure the difference between the pressure in one location and the pressure in another location (usually the atmosphere) p. 329

gauss a common (non-SI) unit for the magnitude of a magnetic field; symbol G p. 712

gaussmeter an instrument used to measure magnetic fields, named after the German scientist Karl Friedrich Gauss p. 713

gravitational field the gravitational force per unit mass at a particular location relative to a massive object such as Earth, the Sun, etc. p. 232

gravitational potential energy energy due to elevation; $E_p = mgy$ close to Earth's surface p. 170

gravity an attractive force between all objects p. 106

greenhouse effect warming of Earth's atmosphere due to the absorption of radiation (especially infrared radiation) by atmospheric gases such as carbon dioxide; see also atmospheric greenhouse effect p. 186

grounding transferring excess charge from an object to Earth or from Earth to an object p. 633

half-life the time required for half the original sample of a nuclide to decay; symbol $T_{½}$ p. 777

half-value layer thickness of material required to reduce the number of photons getting through by a factor of 2; symbol HVL p. 794

harmonic motion see periodic motion

harmonics the collection of frequencies of a musical tone p. 442

Health Physics field in which the beneficial uses of ionizing radiation in medicine, industry, and research are balanced with the need to protect workers and the general public from potential health hazards p. 792

heat energy transferred between objects (or systems) as a result of a temperature difference between them pp. 176, 596

heat engine a device that produces mechanical work via a transfer of heat p. 613

hologram the three-dimensional image produced by holography using laser light p. 583

holography the process in which laser light is used to produce a three-dimension image p. 583

Hooke's law the proportional relation between the amount of elongation and the force causing the elongation; written as $F_x = \pm kx$ p. 303

horizontal range the horizontal distance that a projectile travels from launch to landing p. 83

hormesis an adaptive biological reaction to low doses of chemical toxins, radiation, or other forms of stress that is damaging, even fatal, in higher doses p. 793

human audible range the range of frequencies that a human ear can detect p. 423

Huygens' principle every point on a wavefront acts as a source of a new wavelet, and the new wavefront is the forward envelope of the set of new wavelets p. 510

hydraulic lift a mechanical device to lift objects, operated by a fluid under pressure (also called hydraulic press) p. 335

hydraulic press see hydraulic lift

hyperopia the condition of the eye in which image comes to a focus behind the retina (also called farsightedness) p. 552

ideal fluids incompressible fluids in which flow is streamline and viscosity is negligible p. 349

ideal gas a gas for which the ideal gas law, $PV = nRT$, is valid at all temperatures and pressures p. 593

ideal gas law the relationship $PV = nRT$, where P is the pressure exerted on a gas, V is the volume of the gas, n is the number of moles of gas, R is the molar gas constant, and T is the temperature in kelvins p. 337

impulse the change in an object's momentum ($\Delta\vec{p}$) p. 197

incandescent source an object that emits light due to a high temperature p. 468

index of refraction the ratio of the speed of light in a vacuum to the speed of light in another medium p. 486

inelastic collision the total final kinetic energy is not equal to the total initial kinetic energy in such a collision p. 207

inertia the tendency of objects to continue moving with constant velocity or to stay at rest (if there is no resultant force) p. 114

inertial confinement the use of frozen fuel pellets that are bombarded by laser beams to contain the fuel inside the core of a fusion reactor p. 787

inertial force see fictitious force

inertial frame of reference a frame of reference in which the law of inertia and other laws of physics are obeyed p. 152

infrasonic sounds sound frequencies below 25 Hz, the low end of the human audible range (also called infrasounds) p. 428

infrasounds see infrasonic sounds

instantaneous acceleration the acceleration at any instant p. 36

instantaneous angular acceleration a measure of how rapidly the rotation rate of a rotating object is changing at a given instant; the limit as $\Delta t \rightarrow 0$ of change in angular velocity $\Delta\omega$ divided by the time interval Δt taken (also called angular acceleration) p. 257

instantaneous angular velocity see angular velocity

instantaneous speed the speed at any given instant p. 24

instantaneous velocity the velocity at any particular instant of time p. 27

intensity level the logarithm to the base 10 of the ratio of the intensity of a sound to the reference intensity, 1.0×10^{-12} W/m^2; symbol β; measured in bels (B) or decibels (dB) p. 424

Intensity Modulated Radiation Therapy (IMRT) high-precision radiation therapy that allows for the radiation exposure to conform more precisely to the shape of the tumour by modulating—or controlling—the intensity of the radiation beam in each delivery p. 800

internal energy total energy of all the molecules in a system, including translational, rotational, and vibrational kinetic energies, as well as electric potential energy (also called thermal energy) p. 595

ionizing radiation forms of radiation that have sufficient energy to create electron–ion pairs in matter p. 790

iridescence the creation of colour fringes caused by thin-film interference of light p. 530

irregular reflection see diffuse reflection

isotopes nuclei that contain the same number of protons but different numbers of neutrons p. 765

joule the SI unit of work and energy, equivalent to a newton·metre (N·m) or kilogram·metre^2/second^2 (kg·m^2/s^2) p. 162

kelvin 1/273.16 of the temperature of the triple point of water p. 593

Kepler's first law the orbit of each planet is an ellipse with the Sun at one focus p. 238

Kepler's second law the line joining the Sun and the planet sweeps out equal areas in equal time intervals p. 238

Kepler's third law the square of the time (T^2) in which a planet completes one revolution about the Sun is proportional to the cube of its average distance (r^3) from the Sun p. 238

kinematics the study of motion p. 22

kinetic energy energy of motion, equal to $\frac{1}{2}mv^2$, where m and v represent an object's mass and speed respectively p. 167

kinetic friction the friction force exerted between surfaces moving relative to each other p. 106

laminar flow see streamline flow

laser Light Amplification by Stimulated Emission of Radiation p. 580

latent heat of fusion heat per unit mass added (or withdrawn) as a particular material melts (or solidifies) p. 597

latent heat of vaporization heat per unit mass added (or withdrawn) as a particular material vaporizes (or condenses) p. 597

lateral displacement the sideways displacement of a light ray that strikes a transparent rectangular block (with parallel sides) at a non-zero angle of incidence p. 488

law of conservation of angular momentum if the net torque acting on an object (or system of objects) is zero, then the angular momentum of the object (or system) is conserved p. 277

law of conservation of electric charge new charge is not created in any process, it is merely transferred from one object to the other or from one region to another p. 629

law of conservation of energy energy can be converted into different forms, but it cannot be created or destroyed p. 173

law of conservation of momentum if the resultant force acting on a system is zero, then the momentum of the system is conserved p. 201

law of inertia another name for Newton's first law of motion p. 114

law of universal gravitation two particles are attracted to each other by a force directed along the line between them; the magnitude of the force is proportional to the product of the particles' masses and inversely proportional to the square of the distance between the particles. p. 225

leaf electroscope a sensitive instrument used to detect charge and to determine its sign p. 633

lens a transparent device that produces images by altering the direction of light p. 495

lens (of the human eye) a fibrous substance behind the pupil of the eye with an average index of refraction of about 1.40 p. 552

Lenz's law the direction of the current induced in a conductor will always oppose the change that produced this current p. 741

light quantum see photon

light year the distance light travels in a vacuum in one Earth year p. 474

light-emitting diode (LED) an efficient source that emits light when an electric current travels through a semiconductor p. 468

line spectrum a set of emissions at multiple distinct wavelengths producing specific bright lines separated by dark spaces p. 527

linear density the mass per unit length or inertial property of a string; symbol μ (the Greek letter mu) p. 417

linear magnification the ratio of the image height to the object height in an optical device p. 476

linearly polarized wave see plane-polarized wave

longitudinal oscillation a vibration in which the motion is parallel to the longitudinal axis p. 374

longitudinal wave a wave in which the particles of the medium move parallel to the direction of wave propagation (also called compression wave) p. 394

luminous object any object that is a source of light energy p. 468

lunar eclipse a situation in which light from the Sun toward the Moon is blocked by Earth p. 471

Mach number the ratio of the speed of the source to the speed of sound in air at that location p. 437

Maglev magnetic levitation, used to reduce friction for high-speed commuter trains p. 718

magnetic confinement the use of magnetic fields to contain the fuel inside the core of a fusion reactor p. 787

magnetic declination the angle between magnetic north and geographic north p. 709

magnetic flux a measure of the strength of the magnetic field over a specific area p. 753

magnetic inclination the angle of Earth's magnetic field relative to the horizontal; this angle varies with location on Earth p. 709

magnetic materials see ferromagnetic materials

magnetic north the direction that a compass needle points on Earth p. 709

magnetotactic describes an organism that uses Earth's magnetic field to orient itself in the environment p. 710

magnifying power see angular magnification

mammography the use of x rays to generate an image of breast tissue for the diagnosis of disease p. 795

manometer a U-shaped transparent tube partly filled with a liquid, used to measure gauge pressure p. 332

mass number see atomic mass number

mass spectrometer a device that uses the dependence of the radius of the circular path on particle mass to separate atoms and molecules; used extensively in research and industrial laboratories p. 722

mean free path the average distance travelled by a γ ray before interacting in a specific medium p. 791

measurement a numerical quantity found by measuring with a device or an instrument p. 2

mechanical energy the sum of kinetic energy, gravitational potential energy, and elastic potential energy p. 175

metal brushes the stationary electrical contact between the slip ring of the motor or generator and the external circuit to which it is connected p. 744

metastable state an excited state in which an atom can remain without undergoing spontaneous emission for a relatively long time p. 580

mirage an optical illusion caused by the refraction of light in layers of air of varying density p. 489

moderator material used to slow down the neutrons released by fission in order to instigate a nuclear chain reaction (e.g., graphite, water, heavy water) p. 783

modulus the elastic property of gases, liquids, and solids; symbol B p. 418

moment arm perpendicular distance from an axis of rotation to the line of action of a force p. 273

moment of inertia rotational analogue of mass; for an object rotating about a fixed axis, the moment of inertia is the sum, for all particles on the object, of the product of each particle's mass and the square of its distance from the rotation axis (also called rotational inertia) p. 268

momentum the product of the object's mass and velocity, usually represented by the symbol $\vec{p}$ p. 196

multi-leaf collimator a device that provides clinicians with the ability to shape the radiation beam with each delivery in order to better isolate the exposure to the tumour, thereby protecting the surrounding healthy tissue p. 800

multimeter a device that can be used to measure potential difference, current, as well as resistance p. 691

myopia the condition of the eye in which the image comes to a focus in front of the retina (also called nearsightedness) p. 557

nanotechnology the application of the study of the properties of extremely tiny particles, some as small as an atom p. 4

near point the closest distance at which nearby objects can be focused by an eye; for healthy eyes the near point is usually considered to be 25 cm p. 553

nearsightedness see myopia

negative ion an atom that has more electrons than protons p. 668

net force see resultant force

net vector see resultant

neutrino a weakly interacting particle postulated by Pauli to account for the distribution of energies observed with β^+ particles emitted in β decay, represented as υ in decay equations (neutrinos and anti-neutrinos are anti-particles of each other) p. 772

neutron an electrically neutral particle within the nucleus of the atom p. 628

neutron number the total number of neutrons in a given nucleus; symbol N p. 765

newton (N) the magnitude of resultant force that, if applied to an object of mass 1 kg, will produce an acceleration of magnitude 1 m/s^2; symbol N p. 111

Newtonian fluid a fluid that has a viscosity that varies only with temperature p. 359

Newton's first law of motion if the resultant force on an object is zero, the object's velocity is constant; that is, the object travels in a straight line at constant speed (which could be zero) p. 114

Newton's second law for rotation the resultant torque acting on an object equals the product of the object's moment of inertia and the object's angular acceleration p. 275

Newton's second law of motion If the resultant force on an object is not zero, the object experiences an acceleration in the direction of the resultant force. The magnitude of the acceleration is proportional to the magnitude of the resultant force, and inversely proportional to the mass of the object. p. 110

Newton's third law of motion if object A exerts a force on object B, an equal but opposite force is simultaneously exerted by B on A p. 128

nodal line a line joining adjacent nodes in a two-source interference pattern p. 514

node the resultant when a crest meets a trough of equal amplitude and length during destructive interference of transverse pulses or waves; a position of zero amplitude in an interference pattern p. 403

noninertial frame of reference a frame of reference in which the law of inertia and other laws of physics do not appear to be obeyed p. 152

non-ionizing radiation forms of radiation that do not have sufficient energy to create electron–ion pairs in matter p. 790

non-Newtonian fluid a fluid that has a viscosity that depends on parameters other than temperature, such as flow speed, applied force, etc. p. 359

non-uniform motion motion with a changing velocity, in other words a change in speed (the magnitude of velocity), or a change in direction, or both p. 29

normal force the force perpendicular to surfaces in contact with each other p. 106

north pole the end of a magnet that is attracted toward the geographic north on Earth p. 706

nuclear chain reaction each fission event produces multiple neutrons that can then give rise to more fission events p. 783

nuclear fission the breaking apart of a large nucleus into smaller nuclei p. 782

nuclear fusion the joining together (fusing) of small nuclei to form a larger one p. 782

nuclear medicine a medical specialization that uses radioactive isotopes to diagnose and treat disease p. 797

nuclear reaction a process in which two or more nuclei (or particles) interact to produce new nuclei p. 776

nuclear reactor a device in which a controlled fission reaction occurs p. 783

nuclear strong force the attractive force that holds together the protons and neutrons in nuclei p. 132

nuclear weak force a nuclear force related to processes such as beta-decay, and which has been shown to be essentially the same as the electromagnetic interaction p. 132

nucleon a particle found within the nucleus of an atom, either a proton or a neutron p. 764

nucleus the dense core at the centre of the atom; the dense core of an atom that contains most of its mass pp. 628, 764

nuclide a particular nucleus with a given atomic number (Z) and mass number (A) p. 765

objective lens (objective) the lens in an optical instrument closest to the object p. 564

octave a sound interval with one sound having double the frequency of another sound p. 445

ohm SI unit for resistance, equivalent to volts/ampere; symbol Ω p. 678

Ohm's law an empirical relationship for some materials in which the current generated (I) is directly proportional to the potential difference applied across the object (ΔV) p. 678

open circuit an open switch or a broken wire in a circuit which creates an opening in the closed loop; current cannot travel through an open circuit p. 684

optic nerve the nerve that transmits the signals from the eye to the brain, which interprets what we see p. 554

optical telescope an instrument that uses optical components to make distant objects appear closer p. 567

optical zoom a camera feature in which the image magnification can be increased by increasing the focal length of the lens p. 578

order-of-magnitude estimation a calculation based on reasonable assumptions to obtain a value expressed to a power of 10 (also called a Fermi question after Enrico Fermi) p. 14

oscillation the periodic or harmonic motion of a particle or mechanical system (also called vibration) p. 374

pair production the opposite of an annihilation event: if a photon/γ ray has greater than 1.02 MeV of energy, which is twice the rest mass energy of an electron, the photon can be converted into an electron/positron pair p. 790

parallel the potential difference across each component in parallel is the same due to conservation of energy p. 687

parent nucleus the original nucleus that undergoes nuclear transformation p. 770

pascal SI unit of pressure, equivalent to newton per square metre; symbol Pa p. 328

Pascal's principle a change in pressure applied to an enclosed fluid at rest is transferred undiminished to all parts of the fluid and to its container p. 335

path difference the magnitude of the difference between the path lengths from the two sources of an interference pattern; symbol *P.D.* p. 514

path length the distance between the source and a position of a two-source interference pattern p. 514

penumbra the lighter part of a shadow where some light falls p. 471

percent error the difference between the measured and accepted values of a measurement expressed as a percentage p. 12

percussion the striking of one object against another p. 455

period of revolution the time for one revolution of an object in uniform circular motion; symbol T p. 92

periodic motion motion that is repeated at regular intervals of time (also called harmonic motion) p. 374

periodic wave a travelling wave produced by a source oscillating at some constant frequency p. 396

phase the part of a cycle at which an object with SHM is found p. 376

phosphorescent material a material that becomes luminous when struck by high-energy waves or particles, and remains luminous for a period of time p. 468

photoelectric effect the interaction between a γ ray and an atom that results in ionization of the atom; the γ ray is absorbed and an electron is ejected from the atom p. 790

photon a particle-like bundle or packet of light energy (also called light quantum) p. 580

plane-polarized wave the property of a wave in which the wave is oriented in one plane only (also called linearly polarized wave) p. 539

plastic range the region of the stress–strain graph in which a given material exhibits plasticity p. 307

plasticity the failure of a material to return to its initial size and shape when stress is removed p. 307

Poiseuille's law the relationship between volume flow rate, pressure change, viscosity of a fluid, and flow tube length and radius p. 360

polar molecules molecules such as water with a permanent electric polarization p. 630

polarization angle the angle of incidence or reflection at which complete polarization of reflected light occurs; symbol θ_p p. 542

polarizing filter a material that allows light to be transmitted in a single plane p. 539

Polaroid a plastic absorbing filter containing a dichroic substance p. 539

population inversion a situation in which more atoms are excited into metastable states than there are in the ground state p. 581

positive ion an atom that has more protons than electrons p. 628

positron (β^+ particle) the anti-particle of the electron, having the same mass but the opposite charge, represented in decay equations as ${}^{0}_{+1}e$ p. 772

Positron Emission Tomography (PET) an imaging technique used to diagnose disease based on the annihilation of electrons and positrons p. 773

possible error see uncertainty

potential difference the change in electric potential from one location in an electric field to another p. 660

power rate at which energy is produced or used, i.e., energy divided by time p. 181

presbyopia a focusing problem that occurs when a person's eyes can no longer accommodate to view nearby objects p. 559

pressure the ratio of force magnitude F to area A, when a force of magnitude F is applied in a direction perpendicular to a surface of area A; symbol P p. 328

pressure gradient the ratio of pressure change to flow tube length in the flow of a viscous fluid p. 360

primary focal point the position where rays of light parallel to the principal axis of a lens come together (for a converging lens) or appear to spread out from (for a diverging lens); symbol PF p. 496

primary light colours the three light colours, red, green, and blue, that when added together produce white light p. 554

primary rainbow a rainbow caused by the single internal reflections of sunlight in raindrops, resulting in the colours red, orange, yellow, green, blue, and violet p. 492

primary (P) waves longitudinal or compression waves that travel through both solid and liquid parts of Earth's interior at a speed of about 4–8.0 km/s p. 396

principal focus see focal point

principle of superposition the resulting displacement of two interfering pulses at a point is the algebraic sum of the displacements of the individual pulses at that point p. 404

projectile an object moving through the air without any propulsion system and following a curved path p. 81

projectile motion the curved motion of a projectile p. 81

projectile motion (formal definition) motion with a constant horizontal velocity combined with constant vertical acceleration caused by gravity, resulting in a parabolic path p. 82

proportional limit for a given material, the maximum stress at which the stress and strain are proportional p. 307

proton a positively charged particle within the nucleus of the atom p. 628

proton number see atomic number

pseudoforce see fictitious force

pupil (of the eye) the opening of the eye through which light enters the lens p. 554

qualitative description a statement that indicates the quality of some object, event, or idea p. 2

quality (of a musical tone) a subjective measure of how pleasing a tone is to the human ear (also called tonal quality or timbre) p. 442

quantitative description a statement that indicates a quantity p. 2

quarter-wavelength film a thin film with a thickness that is one-quarter the wavelength of the light in the film, often used to minimize reflections from air–glass boundaries p. 533

radian a unit of angular measure; one radian (approximately 57.3°) is the central angle in a circle that subtends an arc of length one radius p. 290

radiation see thermal radiation

radio telescope a telescope that gathers radio waves from space p. 573

radioactive nuclei nuclei that spontaneously decay or transform into other nuclei over time p. 766

radioactivity the process of changing from one nuclide to another by the spontaneous emission of a particle p. 769

radiography the use of x rays to generate an image p. 795

radioisotope dating the use of a radioactive isotope to determine the age of an object of historical or archeological interest p. 780

range the finite distance travelled by a charged particle (α or β) in matter before coming to a stop; this value depends on the initial energy of the particle and the composition of the material in which it is travelling p. 788

rarefaction the portion of a longitudinal wave where the particles are spread apart or rarefied p. 394

Rayleigh's criterion for light passing through a single slit, two sources are just resolved if the first minimum of the diffraction pattern of one source is in line with the middle of the central maximum of the other source p. 524

real image an image that can be placed onto a screen so the observer does not have to look at the optical device to see the image; for a mirror, it is in front of the mirror p. 475

rectilinear propagation of light the observation that light appears to travel in a straight line in a vacuum or in a single medium p. 471

reflecting telescope a telescope that uses converging parabolic mirrors rather than converging lenses to gather distant light p. 571

reflection grating a diffraction grating made by ruling fine parallel grooves on a smooth metal surface p. 526

refracting telescope a telescope made with lenses or lens systems p. 567

refraction of light the bending of light as it travels from one transparent medium to another at an angle p. 484

regular reflection see specular reflection

relative permeability a measure of how strongly a material can be magnetized by an external magnetic field p. 716

relative velocity the velocity of an object relative to a specific frame of reference p. 94

replica grating a diffraction grating made by pouring molten plastic over a master grating p. 526

resistance the ratio of the potential difference required to the resulting current produced in an object; symbol R; unit ohm (Ω) p. 678

resistivity an inherent property of a medium that measures the difficulty with which it conducts electric current; in many materials, resistivity varies with temperature; symbol ρ; SI unit Ω m p. 680

resolution the ability to distinguish two or more objects as separate entities p. 523

resonance a natural oscillation caused by an input force at the resonant frequency that results in a large amplitude p. 391

resonant frequency the natural frequency at which an oscillation occurs most easily p. 391

rest mass the mass of stationary particles (Einstein's Theory of Special Relativity tells us that mass increases with speed) p. 766

resultant the result of a vector addition (or subtraction) (also called a resultant vector, a net vector, or a vector sum) p. 61

resultant displacement the vector addition of the individual displacements that an object has undergone p. 75

resultant force total of the forces acting on an object (also called net force, total force, or sum of the forces) p. 108

resultant vector see resultant

retina the rear layer of the eye that is light-sensitive; it consists of numerous blood vessels, nerves, and receptors p. 553

reverberation time the amount of time for a sound in a room to decrease in intensity by a factor of one million from the maximum value p. 459

reversibility of light rays the property of light that allows a path of light to be traced by reversing the direction of a known path p. 480

Reynolds number a dimensionless parameter involving a fluid's density, average speed, and viscosity, as well as the diameter of the tube in which the fluid is flowing; if Reynolds number is greater than about 2000, the flow is turbulent p. 362

right-hand rule (magnetic field associated with a current-carrying wire) point the thumb of your right hand in the direction of the conventional current and your fingers will wrap around the wire in the direction of the associated magnetic field p. 712

right-hand rule (magnetic field associated with a solenoid) wrap the fingers of your right hand around the solenoid in the direction of current flow and your extended thumb will be pointing in the direction of the associated magnetic field inside the solenoid p. 715

right-hand rule (magnetic force on a current-carrying wire) align your fingers with the direction of the current, curl them in the direction of $\vec{B}$. Your extended thumb is then aligned with the magnetic force $\vec{F}$ on that current p. 724

right-hand rule (magnetic force on a positive moving charge) align the fingers of your right hand with the direction of $\vec{v}$, curl them in the direction of $\vec{B}$. Your extended thumb is then aligned with the magnetic force $\vec{F}$ on a positive moving charge p. 719

rms voltage see root-mean-square (rms) potential difference

rods receptors in the eye that react to grey and black shades and are responsible for night vision p. 554

root-mean-square (rms) current the effective current over a cycle for AC circuits p. 697

root-mean-square (rms) potential difference the effective potential difference over a cycle for AC circuits (also known as rms voltage) p. 697

rotation axis a line about which an object rotates; relative to this axis, each point on the rotating object travels on a circle with its centre on the axis p. 255

rotational inertia see moment of inertia

rotational kinetic energy the kinetic energy associated with an object's rotation about an axis p. 268

***R*-value** see thermal resistance

safety factor the ratio between the ultimate strength and the maximum allowable stress for safety; a typical value is about 10 p. 300

scalar quantity a quantity with magnitude (or size) but no direction (also called a scalar) p. 16

scattering the absorption and re-radiation of light by a substance p. 543

scientific notation the method of expressing very large or small numbers using a non-zero digit before the decimal point and other digits after it, then multiplying by the appropriate power of 10 (also called standard form) p. 6

second law of thermodynamics disorder is more probable than order; in any energy transfer or conversion within a closed system, the entropy of the system must increase (or in rare cases, remain constant) pp. 612, 654

secondary focal point the position on the side of a lens opposite to the primary focal point and equidistant from the lens; symbol SF p. 496

secondary light colours the three light colours, yellow, magenta, and cyan, that are produced by combining two of the primary light colours p. 559

secondary rainbow a rainbow caused by two internal reflections of sunlight in raindrops, observed higher in the sky than the primary rainbow, resulting in the colours violet, blue, green, yellow, orange, and red p. 492

secondary (S) waves transverse waves that travel only through the solid part of Earth's interior at a speed of about 2–5 km/s p. 396

seismic waves see body waves

semiconductors materials with charge-conducting abilities between that of a good conductor and that of a good insulator p. 632

series the current through each component in series is the same due to conservation of charge p. 687

shear modulus ratio of shear stress to shear strain, for shear strains small enough that shear strain and shear stress are proportional p. 310

shear strain for an object subject to a shear stress, the ratio of the distance Δx that one surface has moved (relative to the opposite surface) to the distance L between the surfaces p. 310

shear stress ratio of magnitude of force F to area A, when a force $\vec{F}$ is applied parallel to a surface of area A on an object p. 310

shear wave see transverse wave

shock wave cone an overlapping series of compressions in the air where an aircraft is travelling at a supersonic speed p. 437

shutter speed see exposure time

SI the international metric system, Système International d'Unités p. 5

significant digits the digits in any measurement that are reliably known p. 12

simple harmonic motion a periodic vibratory motion such that the force (and hence the acceleration) is directly proportional to the displacement; symbol SHM p. 375

slidewire generator a circuit in which a current is generated when a metallic rod or wire slides along stationary wires through a region of uniform magnetic field p. 738

slip ring the rotating metal ring providing electrical contact between the rotating coil in a motor or generator and the external circuit to which it is connected p. 744

solar eclipse a situation in which light from the Sun toward Earth is blocked by the Moon p. 471

solar wind high-energy charged particles, mainly protons and electrons, emitted by the Sun p. 722

solenoid a device made of loops of wire wrapped into a helical coil inside which the magnetic field is relatively uniform p. 715

sonar an acronym for sound navigation and ranging p. 428

sonic boom a very loud noise created near Earth's surface as a supersonic aircraft above creates a shock wave cone p. 438

sound a wave disturbance that originates with an oscillation, travels through a material medium, and can be detected by a listener or receiver p. 414

sound barrier a high pressure region in the air as an object approaches the speed of sound p. 437

sound intensity a measure of the power of a sound per unit area, measured in watts per square metre, or W/m^2 p. 423

south pole the end of a magnet that is attracted toward the geographic south on Earth p. 706

specific gravity (s.g.) ratio of the density of an object to that of water at 4°C p. 327

specific heat amount of heat per unit mass and per unit temperature change needed to increase or decrease the temperature of the material (also called specific heat capacity) p. 596

specific heat capacity see specific heat

spectrograph a photograph of a spectrum p. 528

spectroscope a device that uses either a prism or a diffraction grating to obtain a spectrum p. 528

spectroscopy the study of spectra p. 528

spectrum a distribution of energy emitted arranged in order of wavelength p. 527

specular reflection reflection of light off a smooth, regular surface (also called regular reflection) p. 475

speed the time rate of change of distance p. 22

spherical aberration a problem with spherical curved mirrors in which the reflected rays from an object do not meet at the same location, resulting in a blurred image p. 481

sphygmomanometer a device used to measure blood pressure, consisting of an inflatable cuff and a mercury manometer or other pressure meter p. 333

spontaneous emission the emission of photons from an atom that becomes excited by absorption of a photon or by collision with a particle such as an electron; the emission occurs very quickly after excitation p. 580

spring constant force required per unit elongation of a spring; a measure of stiffness of a spring; symbol k, SI unit newton per metre (N/m) p. 303

stable equilibrium a situation in which an object will return to equilibrium after being disturbed p. 300

stable nuclei nuclei that can exist forever and do not naturally transform into other nuclei over time p. 766

standard atmosphere see standard atmospheric pressure

standard atmospheric pressure 1 atm = 101.3 kPa, typical pressure exerted by atmosphere at ground level (also called standard atmosphere) p. 329

standard form see scientific notation

standard unit a unit of measurement that can be reproduced according to its definition and does not change p. 5

standing wave the interference pattern that results when periodic waves of equal wavelength and amplitude travelling in opposite directions in the same medium meet p. 404

static electricity situations in which charges are at rest p. 630

static equilibrium the state of a stationary object on which the sum of forces and sum of torques are both zero p. 292

static friction the friction force exerted between surfaces that are not moving relative to each other p. 106

statics study of forces and torques acting on structures at rest p. 288

step-down transformer a device in which the output voltage is smaller than the input voltage p. 751

step-up transformer a device in which the output voltage is larger than the input voltage p. 750

stimulated emission the emission of photons from an atom in the metastable state p. 580

Stokes's law the relationship between the magnitude of the drag force on a small spherical object moving in a fluid, the object's radius and speed, and the coefficient of viscosity of the fluid p. 363

strain a measure of the amount of deformation that occurs as a result of a stress; see also tensile strain; bulk strain; shear strain; compressive strain p. 303

streamline the smooth path followed by a particle in streamline flow p. 349

streamline flow fluid flow that is smooth and steady; each fluid particle passing through a particular point always follows the same smooth path (also called laminar flow) p. 349

stress a quantity related to the force causing a deformation, on a force-per-unit-area basis; see also tensile stress; bulk stress; shear stress; compressive stress p. 303

stress birefringence refraction of light in a transparent medium that depends on the mechanical stress applied p. 544

sum of the forces see resultant force

superconductors materials in which there is no resistance to the movement of charge below a certain critical operating temperature p. 632

surface tension magnitude of intermolecular force per unit length in a direction perpendicular to a line of molecules on a liquid's surface; symbol γ p. 342

surfactants molecules that lower the surface tension of liquids p. 343

systolic pressure maximum blood pressure exerted by the contraction of the heart in a person's cardiac cycle p. 334

tangent to a curve a straight line parallel to a curved line at a particular point p. 29

tangential acceleration for any point on an object rotating about a fixed axis, the tangential acceleration for the point is an acceleration component that increases or decreases the speed at that point; this component is tangential to the circle on which the point travels, and is perpendicular to the centripetal component of acceleration, which is directed toward the centre of the circle p. 263

target volume region of the body identified for high levels of radiation in treatment planning p. 799

temperature a measure of the average translational kinetic energy of atoms and molecules p. 592

tensile strain for an object that has been stretched, the ratio of the increase in length ΔL to the original unstretched length L_0 p. 304

tensile stress ratio of the magnitude of the external force $\vec{F}$ stretching an object to the cross-sectional area A of the object p. 304

tension force exerted by ropes, strings, cables, tendons, etc. p. 106

tenth-value layer the thickness of material required to reduce the number of photons getting through by a factor of 10 p. 791

terminal speed the maximum speed reached by an object falling in a gas or liquid; constant downward speed as a particle settles in a fluid pp. 49, 363

terrestrial telescope a telescope that produces an upright image p. 569

tesla SI unit for the magnitude of a magnetic field; symbol T p. 712

test charge see visiting charge

thermal conductivity a measure of how well a substance conducts heat p. 603

thermal energy the energy associated with the haphazard motion (including vibrations and rotations) of atoms and molecules; when kinetic friction acts on an object, thermal energy is produced (also called internal energy) p. 176

thermal radiation heat transfer by the emission of electromagnetic energy from an object (also called radiation) p. 608

thermal resistance (R-value) thickness of an insulating material divided by its thermal conductivity p. 605

thermogram image produced by thermography p. 608

thermography the detection of thermal radiation emitted by buildings, people, etc. by infrared-sensitive cameras p. 608

thermometer a device that can measure temperature p. 592

threshold of hearing the lowest sound intensity that a human can hear; estimated to be about 1.0×10^{-12} W/m^2 for a young child p. 424

threshold of pain the minimum sound intensity at which a human begins to feel pain; estimated to be 1.0 W/m^2 at 1000 Hz p. 424

timbre see quality (of a musical tone)

tonal quality see quality (of a musical tone)

tone a sound produced by a musical instrument or a singing voice p. 443

torque rotational effect of a force; torque equals ± the product of the magnitude of the force and the moment arm (perpendicular distance from the rotation axis to the line of action of the force) p. 273

torsion twisting of an object due to an applied torque p. 310

torsion balance device used to determine how electric force depends on charge and distance p. 634

torsional oscillation a twisting around a rotational axis p. 375

total force see resultant force

total internal reflection the situation in which light in a medium of higher index of refraction strikes the interface with a medium of lower index of refraction at an angle greater than the critical angle, causing the light to reflect p. 491

tracer a chemical compound injected into a subject in order to track the pathway followed after injection p. 797

transducer a device that converts electrical signals into oscillations and then sound oscillations p. 428

transformer device that changes (transforms) voltage p. 748

translational kinetic energy an object's kinetic energy associated with the motion of the centre (of mass) of an object p. 268

transmission grating a diffraction grating made by ruling fine parallel grooves on a smooth glass surface with a diamond-tipped tool p. 526

transmutation the changing of one type of nucleus into another; this can be achieved through a nuclear reaction or through nuclear decay p. 776

transverse oscillation a vibration in which the basic motion is perpendicular to the rest axis p. 374

transverse wave a wave in which the particles of the medium move perpendicular to the direction of wave propagation (also called shear wave) p. 434

trigonometry the branch of mathematics that deals with the relationships between the angles and sides of triangles and the calculations based on them p. 67

triple point of water the single temperature and pressure at which water, ice, and water vapour can coexist in equilibrium p. 593

trough the lower part of a transverse wave p. 394

truss any structure built with triangular shapes p. 315

turbulent flow irregular and chaotic fluid flow p. 349

turns ratio the ratio of the number of turns in the secondary (output) coil to the number of turns in the primary (input) coil in a transformer p. 750

ultimate strength the maximum stress that a material can withstand without breaking p. 307

ultrasonic sounds sound frequencies above 20 kHz, the high end of the human audible range (also called ultrasounds) p. 428

ultrasounds see ultrasonic sounds

umbra the part of a shadow where light is completely blocked p. 471

uncertainty in measurement, the range of values in which the true value is expected to lie (also called possible error) p. 12

uniform circular motion motion at a constant speed in a circle (or arc) of constant radius p. 90

uniform motion motion at a constant speed in a straight line (one dimension) p. 24

universal gravitation constant given the symbol G, a constant used in the law of universal gravitation having a value of 6.67×10^{-11} N·m^2/kg^2 p. 225

universal wave equation $v = f\lambda$ where v is the speed of a periodic wave and f and λ are the frequency and wavelength, respectively, of the periodic wave p. 399

unstable equilibrium a situation in which an object in static equilibrium tends to fall or keep moving away from the equilibrium position after being disturbed p. 300

vector quantity a quantity with both magnitude and direction (also called a vector) p. 16

vector sum see resultant

velocity the time rate of change of position p. 26

vibration see oscillation

virtual image an image that cannot be placed onto a screen so the observer can see the image only by looking at the optical device; for a mirror, it is behind the mirror p. 475

viscosity internal friction between adjacent layers of moving fluid p. 349

visible spectrum the components of white light visible to the human eye p. 489

visiting charge/test charge a charge that is brought into a region of electric field in which it experiences an electric force p. 639

volt SI unit for electric potential, potential difference, or voltage; symbol V p. 658

voltage see potential difference

voltmeter a device that measures the potential difference between two points in a circuit p. 691

volume flow rate volume of fluid flowing per unit time, or product of cross-sectional area and speed (also called flow rate) p. 350

volume strain see bulk strain

volume stress see bulk stress

watt SI unit of power, abbreviated as W; equivalent to joule per second (J/s) p. 181

wave a disturbance that transmits energy and/or information, but not matter p. 394

wavelength (of the wave) the shortest distance between any two in-phase points on a wave; symbol λ p. 397

weber SI unit of magnetic flux; 1 weber = 1 Wb = 1 tesla · metre^2 = 1 T·m^2 p. 753

weight a term sometimes used to refer to the gravitational force on an object p. 113

weightlessness the sensation associated with being in free fall, in which an object is in motion with only the force of gravity acting p. 240

work when a constant force $\vec{F}$ is applied to an object while the object undergoes a displacement $\Delta\vec{r}$, the work done on the object is given by $W = F\Delta r\cos\theta$, where θ is the angle between $\vec{F}$ and $\Delta\vec{r}$ p. 162

work-energy theorem the total work done on an object by all the forces acting on it equals the change in the object's kinetic energy p. 167

x-ray crystallography the study of x-ray diffraction patterns in crystals p. 529

Young's modulus ratio of tensile stress to tensile strain, for strains small enough that strain and stress are proportional p. 304

Chapter 1 Measurement and Types of Quantities

Exercises

1-1 Answers will vary.

(a) Food can be sweet, salty, bitter, sour, or umami (savory).

(b) An odour can be musty, smoky, fruity, etc.

1-2 Answers will vary. Examples of qualitative descriptions are friendly, fun-loving, honest; examples of quantitative descriptions can be height, mass, shoe size.

1-3 (a) quantitative (b) qualitative (c) qualitative (d) quantitative (e) quantitative (f) quantitative

1-4 Answers will vary.
time: time to recharge a battery; time available between classes
length: height adjustment of a bicycle seat; distance from home to work or school
mass: mass of frozen food defrosting in a microwave oven; mass of a parcel sent by courier
volume: volume of books that a knapsack can hold; volume of water needed to keep hydrated during a long-distance run

1-5 Answers will vary: electrical voltage: 6.0 V; temperature: 100°C; power: 60 W

1-6 Some of the disadvantages are

- Most celestial bodies that are visible at night are not visible during the daytime, and vice versa.
- Cloudy conditions interfere with observation of any celestial body.
- Accuracy is difficult to achieve.
- Convenience is minimal.

1-7 Theoretical aspects of physics involve posing questions, creating ideas to research answers to those questions, experimenting, measuring, analyzing, and collaborating, which leads to theories and more questions. The theoretical research and discovery leads to applications that, in most cases, help to improve our lives. One example is the discovery of current electricity, which has led to countless, very useful, electrical devices.

1-8 The original metre was defined in terms of the distance from the equator to the North Pole, a distance that could only be assumed because it was impossible to measure. The original second was defined in terms of a mean solar day, a quantity that is not constant because Earth's rotation is very gradually slowing down.

1-9 (a) 2×10^{41} (b) 1.7×10^{42} (c) 1×10^{83}. Mass has by far the greatest range.

1-10 1×10^{9} s, or 31 years

1-11 4×10^{11} stars

1-12 1×10^{57} atoms

1-13 (a) 8.4×10^{-15} (b) 8×10^{-36} (c) 8.0×10^{8} (d) 1.94×10^{5} m/s

1-14 A base unit is a standard unit of measurement from which other units may be derived. In the SI, examples are the metre (m), kilogram (kg), and second (s). A derived unit is a measurement unit stated in terms of one or more base units. Examples are a unit for speed (m/s), a unit for surface area (m^2), and a unit for solid volume (m^3).

1-15 Four examples are: watt ($W = kg{\cdot}m^2{\cdot}s^{-3}$); pascal ($Pa = kg{\cdot}s^{-2}$); volt ($V = kg{\cdot}m^2{\cdot}s^{-3}{\cdot}A^{-1}$); becquerel ($Bq = s^{-1}$)

1-16 Some of the patterns are the prefixes from 10^3 to 10^{-3} change by a factor 10^1; the remaining prefixes change by a factor of 10^3; the symbols for the large numbers (from mega upward) are capital letters, and all the other symbols are lower case; the origins of the prefixes are all non-English words; some original meanings relate to the power of 10 (e.g., Greek *femten*, or 15, is used for 10^{-15}), while some others relate to a power of 10^3 (e.g., Italian *setta* or 7 is $(10^3)^7$ or 10^{21}).

1-17 (a) 1.3 m (b) 3×10^{-8} m (c) 1.23×10^{7} m (d) 1.486×10^{9} m

1-18 (a) 2×10^{-2} s (b) 8.6×10^{4} μm (c) 3.28×10^{-6} Mg (d) 1.05×10^{5} kHz (e) 2.4×10^{6} mW/m² (f) 9.8×10^{-12} m/μs² (g) 4.7×10^{3} kg/m³ (h) 0.53 people/ha

1-19 (a) 65 cm (b) 8.17 m

1-20 62.1 mi/h

1-21 (a) L/T (b) L/T (c) L/T^2 (d) M/L^3

1-22 (a) speed (b) length (c) density (d) acceleration

1-23 (a) $M{\cdot}L/T^2$ (b) $M{\cdot}L^2/T^3$ (c) T^{-1}

1-24 The dimensions are both L.

1-25 L/T^3

1-26 Both sides of the equation have dimensions L.

1-27 To compare the uncertainties, begin by expressing them in the same unit. Thus, $\pm 100\,g \times \dfrac{1\,kg}{1000\,g} = \pm 0.1\,kg$. The most expensive scale is likely the first one, 42.40 kg ± 0.005 kg, which provides the greatest number of significant digits and the smallest uncertainty. The least expensive scale is likely the last one, which has the largest uncertainty, ± 0.1 kg.

1-28 (a) 1 (b) 5 (c) 3 (d) 4

1-29 (a) 3.85×10^{4} Gm (b) 9.40×10^{-4} MW (c) 5.51×10^{1} dam (d) 8.77×10^{2} kL (e) 7.66×10^{-2} μg

1-30 8%

1-31 3.12 m, 0.446 m²

1-32 (a) 2.306×10^{-30} kg (b) $1.671\,710 \times 10^{-27}$ kg

1-33 $1.499\,832 \times 10^{11}$ m, $1.492\,144 \times 10^{11}$ m

1-34 499 s

1-35 (a) about 10^{14} cells (b) about 10^{9} to 10^{10} kg (c) about 10^{4} rotations (d) about 5 populations

1-36 Answers will vary. Some examples of scalar quantities are area, volume, speed, density, energy, power, and frequency.

1-37 Answers will vary. Two examples of displacement are a ball tossed 10 m west and a walk of 100 m south from the bus stop to the residence. Two examples of velocity are a motorbike travelling at 50 km/h north and a jogger running along a path at 5 m/s southeast.

Chapter Review

Multiple-Choice Questions

1-38 (c)

1-39 (e)

1-40 (a)

1-41 (c)

1-42 (c)

1-43 (c)

Review Questions and Problems

1-44 Measurement is important

(a) in society in order to have efficient communication in numerous aspects of our lives, including manufacturing, building infrastructure, selling, and buying.

(b) in physics in order to design and perform experiments and develop theories and applications resulting from experiments and discoveries.

1-45 The base units are the metre (m) for length, the second (s) for time, and the kilogram (kg) for mass.

1-46 It isn't necessary to have a base unit for area because area can be expressed in terms of the base unit for length.

1-47 about 9 dam, or 90 m

1-48 (a) 8.85×10^3 m, 8.85×10^4 dm, 8.85×10^5 cm

(b) 1.90×10^5 kg, 1.90×10^8 g, 1.90×10^{10} cg

(c) 6.9×10^{15} s, 6.9×10^{21} μs, 6.9×10^{-3} Es

1-49 For a student of mass 60 kg, the ratios would be

(a) $3 \times 10^2/1$

(b) $4 \times 10^{-3}/1$

(c) 9/1

(d) 10/1

(e) $5 \times 10^{-4}/1$

1-50 2.6×10^5 kg, 2.6×10^2 Mg

1-51 (a) 1.97×10^2 km (b) 2.2×10^{-2} m/s^2 (c) 1.1×10^{-7} m/s^2

1-52 (a) r is the radius; b is the base; h is the height

(b) The dimension of πr^2 is L^2, and the dimension of $bh/2$ is L^2. So we conclude that area has the dimension L^2.

1-53 (a) The symbols are dimensions: M for mass, L for length, and T for time.

(b) speed (v): $[v] = L/T$

acceleration (a): $[a] = L/T^2$

area (A): $[A] = L^2$

(c) $[P] = M{\cdot}L^2/T^3$, $[p] = M/LT^2$

(d) $[E] = M{\cdot}[v]^2$

(e) $[P] = M{\cdot}[v]{\cdot}[a]$

1-54 It is possible for the measurements to be multiplied (e.g., area × length = volume, or $L^2 \times L = L^3$), but it is not possible to add them (e.g., you can't add area and length).

1-55 Wood contracts as it dries, so it must be fully dried (cured) and totally contracted before being marked to ensure accurate scale divisions.

1-56 2%

1-57 496 m

1-58 3.06 m

1-59 17.9 m^2

1-60 1.3×10^3 cm^3

1-61 (a) approx. 2×10^3 to 3×10^3 paces (b) approx. 10^4 to 10^5 kernels (c) approx. 10^4 L (d) approx. 10^8 to 10^9 beats (e) about 2×10^9 people

1-62 (a) vector (b) scalar (c) scalar (d) scalar (e) vector (f) scalar

Applying Your Knowledge

1-63 The meanings are not exactly the same, but they are close. The SI is based on multiples of 10, whereas computers and data storage are based on powers of 2. For example, in the SI, kilo means 10^3, or 1000, and mega means 10^6, but a kilobyte is 2^{10} bytes, or 1024 bytes, and a megabyte is 2^{20} bytes, or 1.05×10^6 bytes.

1-64 about 10^9 to 10^{10} cells

1-65 about 10^8 L

1-66 about 1 s

1-67 One microcentury is 52.6 min, which is close to the average 50 min lecture.

1-68 $V^{1/3}$, $V^{2/3}$

1-69 (a) 27.8 m/s (b) 3.5×10^2 km/h (c) To change m/s to km/h, multiply by 3.6; to change km/h to m/s, divide by 3.6.

1-70 (a) 30 ms (b) 6×10^6 years

1-71 360°/24 h = 15°/h; approx. 2 h

1-72 about 6×10^4 mm or 60 m

Chapter 2 One-Dimensional Kinematics

Exercises

2-1 (a) 9.26 m/s (b) 1.58×10^3 m/s (c) 1.02×10^3 m/s (d) The range of speeds is from 2.1×10^{-4} m/s to 6.9×10^{-5} m/s.

2-2 (a) 5.00×10^2 s (b) 2.56 s

2-3 (a) $d = \text{speed}_{av}\,(\Delta t)$ (b) The equation is dimensionally correct because both sides of the equation have dimension [L].

2-4 (a) 12.0 m (b) 1.04×10^2 km, 1.04×10^5 m (c) 884 km (d) 7.19×10^2 km

2-5 (a) uniform motion (b) The object is speeding up. (c) The object is slowing down.

2-6 about 1 h to 2 h

2-7 vector quantity; the velocity (both the speed and the direction) of the wind

2-8 (a) if motion is in one direction with no reversals of direction of motion

(b) yes, if an object reverses its direction of motion

(c) No; at most, the displacement's magnitude equals the distance (as in (a)).

2-9 (a) Speed (instantaneous) indicates how fast an object is moving; it is a scalar quantity. Velocity is a vector quantity; it is similar to speed in that at any time the magnitude of velocity equals the speed, but velocity also includes direction.

(b) The motion must be in one direction with no reversals of direction of motion.

2-10 (a) 3.0×10^1 km/h (b) 0.0 km/h (c) The scalar speed involves total distance (24 km) but the vector velocity involves displacement (0 km).

2-11 (a) 0.57 m/s (b) −0.26 m (c) −0.10 m/s

2-12 (a) 1.0×10^1 m/s, 5.0 m/s (b) 2.5 m/s west, 13 m/s east, 0 m/s (c) 0 m/s, 2.5 m/s (d) 13 m/s east

2-13 5.6 m forward

2-14 2.0×10^2 s

2-15 (a) The object is moving west away from the origin ($x = 0$) with constant velocity, then the velocity increases during a very short time interval and the object continues west with a larger constant velocity. Finally, the velocity is reduced to zero (during a very short time interval) and the object stays at one position, west of the initial starting position.

(b) The object begins at a position that is east of the origin ($x = 0$) and travels toward the origin (i.e., westward) with a velocity that has a continually decreasing magnitude; the object eventually reaches the origin.

(c) The object begins at a position south of the origin ($x = 0$) and travels northward at a constant velocity, eventually passing and going beyond the origin.

2-16 0.75 m northward = displacement

2-17 approx. 6.9 cm/s south, approx. 15 cm/s south, 0 cm/s

2-18 (a), (b), and (d)

2-19 (a) constant southward velocity (b) increasing southward velocity (c) decreasing southward velocity

2-20 Yes, a car that is travelling west and slowing down has a westward velocity but an eastward acceleration.

2-21 (a) 0.585 m/s² (b) 36.1 m/s²

2-22 1.37×10^3 m/s² west

2-23 (a) 1.54 (km/h)/s west (b) 0.427 m/s² west

2-24 The first slope is $m = a = 40$ cm/s².

2-25 (a) The v-t graph is shown below.

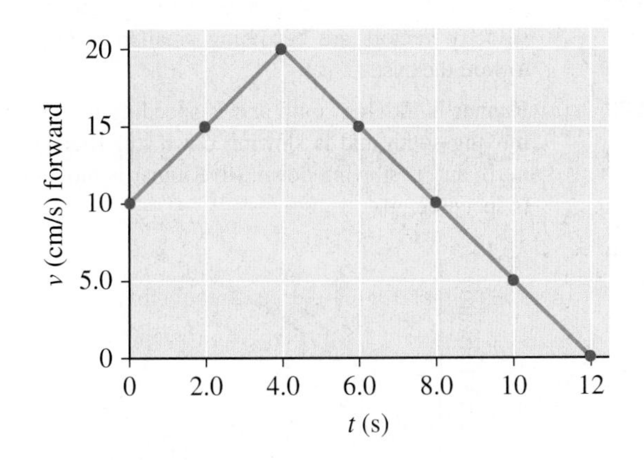

(b) The slopes are 2.5 cm/s² forward and −2.5 cm/s² forward.

(c) The required a-t graph is shown below.

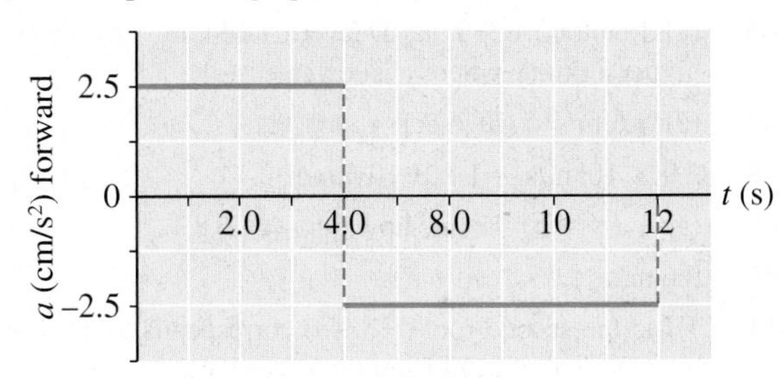

2-26 The final point on the v-t graph is 0.5 m/s west at 3.0 s. The resulting velocity–time graph is shown below.

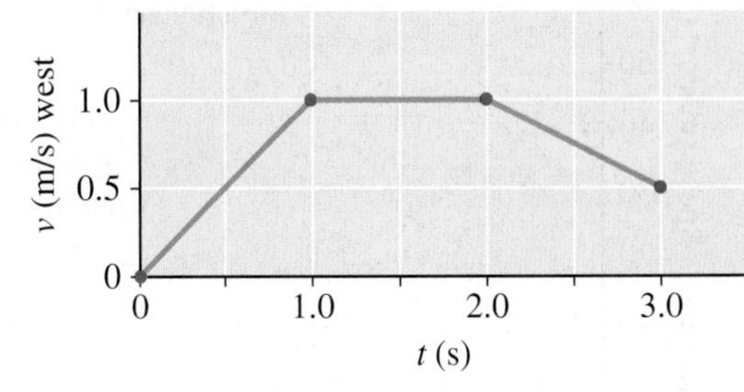

2-27 during constant acceleration

2-28 (a) The object speeds up from rest with a continually decreasing acceleration; after reaching a maximum velocity, the object slows down with constant acceleration until it comes to rest.

(b) The object undergoes constant positive acceleration, followed by constant negative acceleration (of smaller magnitude but for a greater time period than the positive acceleration).

(c) The object undergoes increasing positive acceleration, followed by constant positive acceleration for a greater amount of time than the increasing acceleration.

2-29 Eqn. 2.9.

2-30 The equation is dimensionally correct because both sides are L^2/T^2.

2-31 $t = \dfrac{2\Delta x}{v_0 + v}$

2-32 Add the areas of the rectangle and the triangle, which is equivalent to Eqn. 2-8.

2-33 8.4 m/s

2-34 4.3×10^4 m/s², in opposite direction to ball's initial velocity

2-35 (a) 13 m (b) 7.9 m/s

2-36 (a) 1.0×10^{-2} h (b) 0.70 km (c) 1.7 (km/h)/s forward

2-37 533 m/s

2-38 about 1 m/s² in a direction opposite to the initial motion

2-39 13 m/s

2-40 (a) 1.06×10^{-8} s (b) 4.10×10^{15} m/s² east

2-41 (a) same magnitude, opposite direction (b) same (c) zero (d) 9.8 m/s² down at all points

2-42 (a) 33 m/s (b) 37 m/s (c) 24 m/s

2-43 From 1.0 s to 2.0 s, the distance fallen is 19.6 m − 4.9 m = 14.7 m, which is 3 × 4.9 m.

2-44 (a) 21 m/s upward (b) 22 m

2-45 (a) 63 m (b) 35 m/s

2-46 (a) London: 1.330 m; Denver: 1.334 m (b) Denver is at a higher altitude, where g is lower.

2-47 (a) 1.6 m/s^2 (b) 6.2:1

2-48 2.9×10^3 m/s = 1.1×10^4 km/h

2-49 (a) 1.3 s (b) 13 m/s upward (c) 6.8 s

2-50 9.6 m/s

2-51 1.1 s; the second root (−3.9 s) corresponds to an object having been thrown upward from the water level.

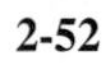

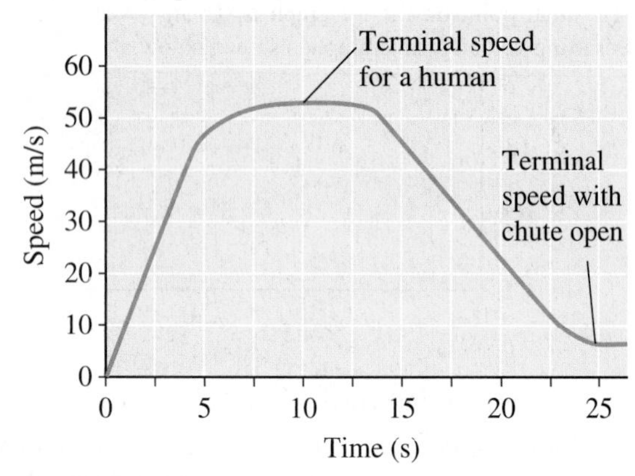

Chapter Review

Multiple-Choice Questions

2-53 (d)

2-54 (a)

2-55 (b)

2-56 (d)

2-57 (b)

Review Questions and Problems

2-58 52.426 m/s

2-59 (a) 5.0×10^1 m (b) approx. 10 to 17 car lengths

2-60 2.6 min

2-61 No; jogger 1 is travelling at a constant speed that is larger than the (constant) speed of jogger 2.

2-62 (a) 1.30×10^2 km (b) 65 km/h

2-63 (a) 1.3×10^2 s (b) 21 m/s

2-64 0.87 s

2-65 17 m

2-66 (a) 3.3×10^4 m/s = 1.2×10^5 km/h (b) 2.1×10^4 m/s (c) zero

2-67 (a) They are equal. (b) They are equal. (c) The magnitude of the instantaneous velocity equals the instantaneous speed.

2-68 (a) 1.7×10^2 m/s (b) 2.0×10^2 m/s (c) 75 m/s east

2-69 (a) 18 m (b) 47 m/s (c) 31 m/s

2-70 The motion begins from a position that is opposite in direction to the motion; for 2 time units there is uniform motion in the positive direction; for the next time unit there is uniform motion at a slow speed in the negative direction; for the final time unit there is uniform motion at a high speed in the negative direction, ending at the position that had occurred at 1 time unit.

2-71 The slope of a line on a position–time graph indicates the velocity. If the line is curved, the slope of the tangent to the curve at a particular instant indicates the velocity at that instant.

2-72 (a) The area under the line (or curve) on a velocity–time graph indicates the displacement.

(b) The slope of the line on a velocity–time graph indicates the acceleration.

2-73

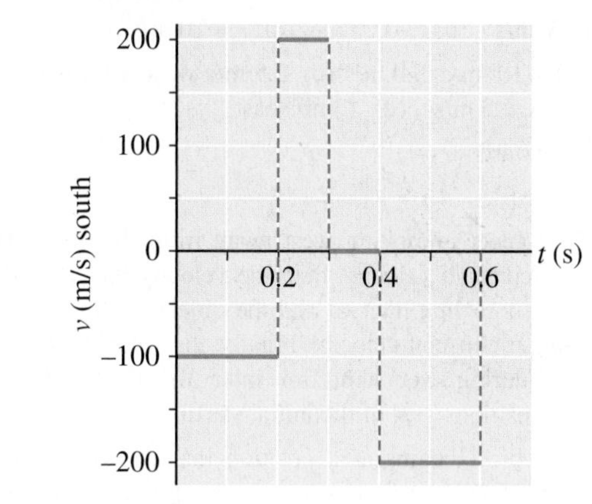

2-74 Uniform motion:

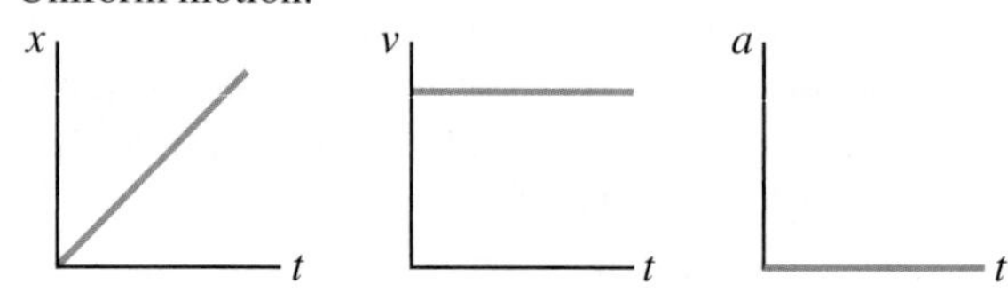

Constant acceleration:

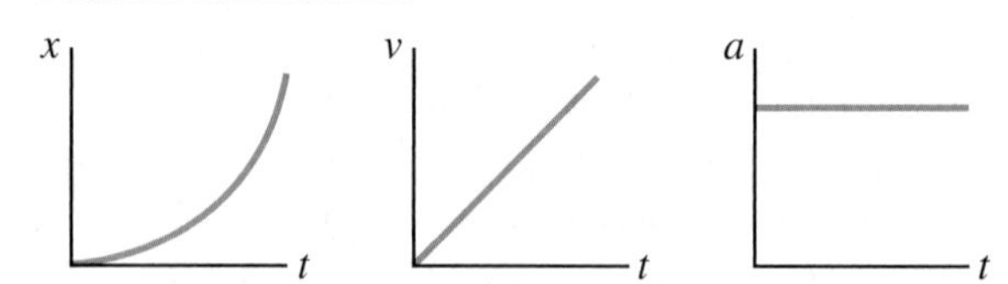

2-75 (a) 30 s, 65 s, 100 s (b) positive: 0–30 s, 65–100 s; negative: 30–65 s, 100–120 s (c) approx. $+6.9 \times 10^{-2}$ m/s, approx. -1.3×10^{-1} m/s

2-76 At the top of the path of a ball thrown vertically upward, the ball is instantaneously at rest (i.e., has zero speed), but it has non-zero acceleration (9.8 m/s^2 downward).

2-77 (c) The acceleration is shown as westerly, and the eastward velocity vectors are becoming smaller as the dog moves toward the east.

2-78 (a) Runner is moving south and is speeding up. (b) Runner is moving south and is slowing down. (c) Runner is moving north and is slowing down. (d) Runner is moving north and is speeding up.

2-79

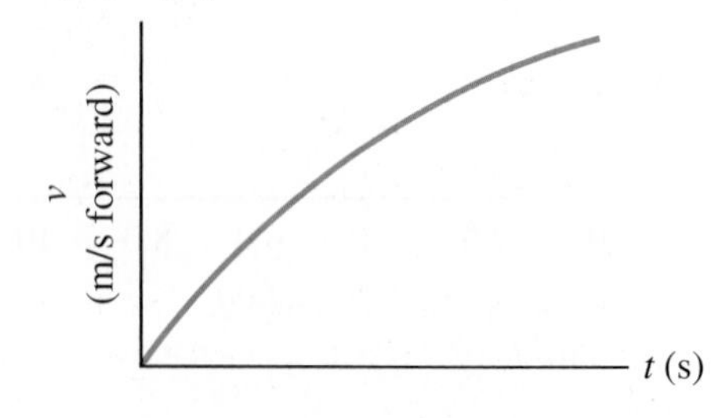

2-80

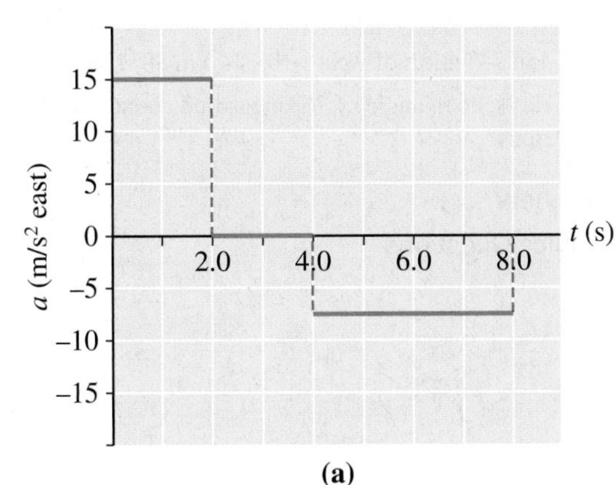

(a)

150
120
90
60
30
0
0
2.0
4.0
6.0
8.0
t (s)
x (m east)

(b)

2-81 (a)

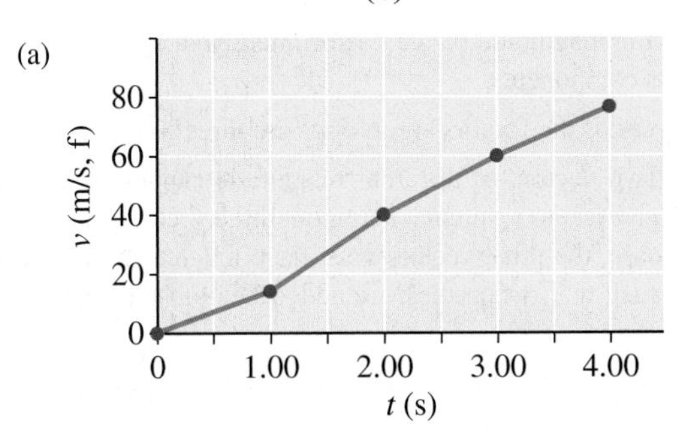

(b) 1.00–2.00 s, 0.00–1.00 s, 24.4 m/s^2 forward, 14.4 m/s^2 forward (c) 18.6 m/s^2

2-82 (a)

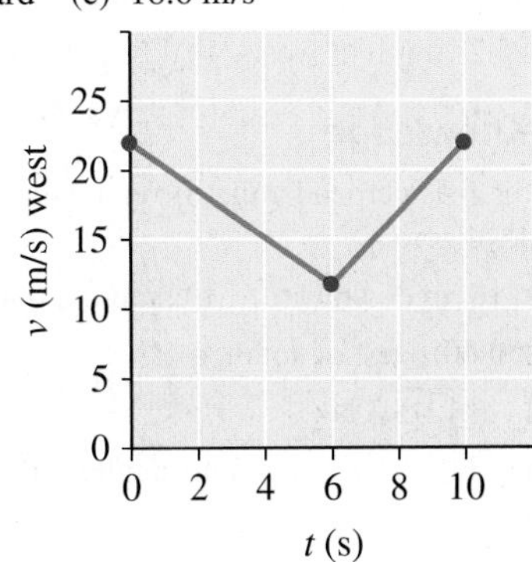

(b) 1.7×10^2 m west

2-83 (a) 0.0 s, 6.0 s, 12 s (b) 4.0 s, 8.0 s, 14 s (c) 4.0–8.0 s, 14–16 s (d) approx. 35 m/s^2 north, approx. 12 m/s^2 south

2-84 (a) 21 m/s^2 (b) 25 m/s (c) $2.1g$

2-85 When a ball is tossed vertically upward, at the top of the flight it reverses directions but its acceleration remains 9.8 m/s^2 downward.

2-86 1.7×10^2 m

2-87 (a) starting from rest, a positive constant acceleration of relatively high magnitude; uniform motion; negative constant acceleration with low magnitude; negative constant acceleration with high magnitude

(b) starting with a fairly high forward velocity; negative constant acceleration eventually coming to a stop and then reversing direction with the same constant acceleration

(c) constant positive acceleration followed by uniformly diminishing positive acceleration

2-88 (a) 1.6×10^2 m/s^2 forward (b) 0.16 s

2-89 (a) 16 s (b) 12 s

2-90 Only the first runner has to accelerate from rest; the remaining three runners are able to receive the baton transferred in the relay while running at nearly full speed. Practice helps improve a smooth, efficient baton transfer at a high speed.

2-91 for 120 km/h and 5 bottles: 99 m

2-92 6.1 m/s upward

2-93 28 h

2-94 The final velocities of the stones are equal.

2-95 Refer to the graphs below, where S represents the steel ball and T represents the tennis ball.

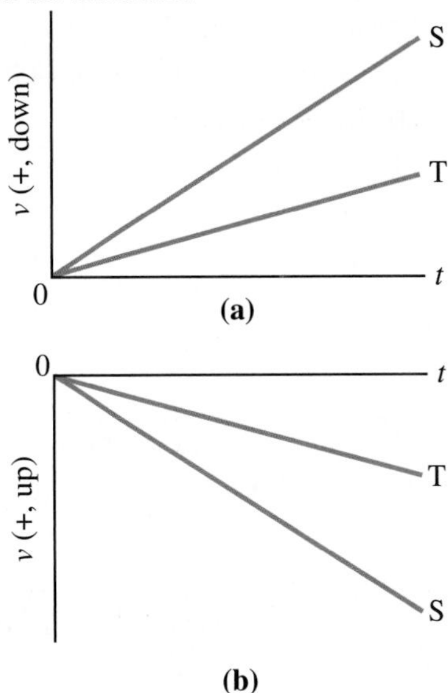

2-96 (C); neglecting air resistance, the motion of a ball going up and then down under the influence of gravity is symmetric.

2-97 0.34 s, 3.1 s

Applying Your Knowledge

2-98 At an estimated acceleration of magnitude about $3g$, the patient is advised not to play tennis at this stage.

2-99 95 km/h

2-100 62 m/s

2-101 (a) 0.0 s and 45 s (b) 75 s (c) 9.0×10^2 m

2-102 (a) 2.9×10^2 m (b) 16 s after the fish passes the barracuda

2-103 (a) 0.32 s (b) 4.8×10^{-2} s (c) The vaulter takes almost (0.32 s/0.048 s) = 7 times as long to travel first 50 cm as the final 50 cm, thus appearing to be in "slow motion" at the top of the vault.

2-104 (a) 7.9 to 22g (b) 1.6 cm (c) Teens may still be growing so their muscles and bones are not as strong as they will be later. Also, it is possible safety equipment is not as protective as the (potentially more expensive) equipment used by professional athletes.

2-105 3.13 m/s^2

2-106 (a) 91 m (b) The motion is not constant acceleration at 9.8 m/s^2 downward because the ball experiences air resistance. The graphs are shown below.

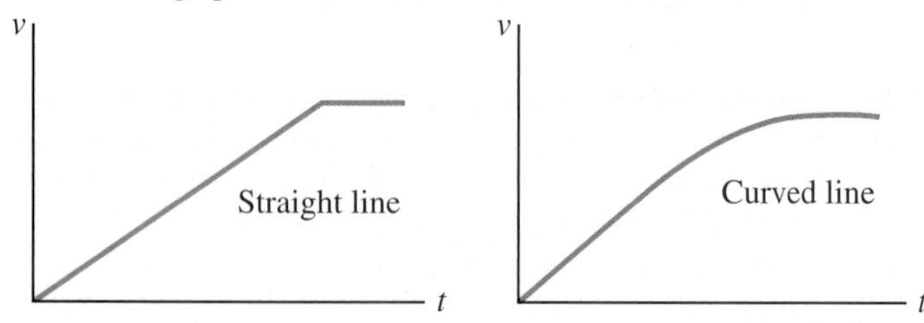

Chapter 3 Vectors and Trigonometry

Exercises

3-1 (a) & (b) 22 m/s

3-2 $\vec{A}$ = 21 m/s 35° west of north, $\vec{B}$ = 35 m/s 45° south of east

3-3 Possible scales for the vectors are

(a) 1 cm = 1 × 10^3 km

(b) 1 cm = 0.01 m/s

(c) 1 cm = 10 N

(d) 1 cm = 50 N

(e) 1 cm = 30 N

3-4 3.0 m/s^2 upward

3-5 about 1 × 10^3 km (in the direction chosen)

3-6 (a) 40 km 77° north of east (b) 3.9 × 10^2 N 69° south of west

3-7 (a) 7.9 units 3° east of north

(b) 7.9 units 3° west of south

(c) $(\vec{Y} - \vec{X})$ is equal in magnitude but opposite in direction to $(\vec{X} - \vec{Y})$.

(d) If the vectors are equal (in magnitude and direction), then both subtractions are zero.

3-8 40 m 65° north of east

3-9 (a) 1.7 m (b) 57.8 mm (c) 0.76 km (d) 53.1 cm

3-10 (a) 2.0 × 10^2 paces (b) 1.5 × 10^2 m

3-11 (a) 5.3 cm (b) 20.1° (c) 58° (d) 15.0 m

3-12 47 cm

3-13 (a) $x = (5.5 \text{ cm}) \cos 16° = 5.3$ cm; $y = (5.5 \text{ cm}) \sin 16° = 1.5$ cm

(b) $x = (-44 \text{ km}) \cos 33° = -37$ km; $y = (44 \text{ km}) \sin 33° = 24$ km

(c) $x = (-42 \text{ cm/s}) \cos 41° = -32$ cm/s; $y = (-42 \text{ cm/s}) \sin 41° = -28$ cm/s

3-14 $g_x = 2.7$ m/s^2, $g_y = 9.4$ m/s^2

3-15 5.0 × 10^1 km 13° west of north

3-16 (a) 36 km 34° north of west (b) 14 km 38° south of east

3-17 5.8 × 10^2 N, at an angle of 76° measured counterclockwise from the $-x$-axis

Chapter Review

Multiple-Choice Questions

3-18 (c)

3-19 (a)

3-20 (c)

3-21 (d)

3-22 (e)

3-23 (b)

Review Questions and Problems

3-24 15.6°

3-25 19 m

3-26 7.8 × 10^2 m

3-27 No; the component of a vector is its projection along an axis, so the component's magnitude may equal the vector's magnitude if the vector and axis are parallel, but the component's magnitude is less than the vector's magnitude in other situations.

3-28 $(A^2 + B^2)^{1/2}$

3-29 No; the magnitude of the vector must be at least as large as one of its components.

3-30 (a) yes; if the vectors are in opposite directions

(b) Two vectors of different magnitude cannot be combined to give a zero resultant vector, but three vectors can. In the latter case, the three vectors must have magnitudes and directions such that, when they are added head-to-tail, the resultant is zero.

3-31 17 km west, 17 km west, 89 km east, 89 km west

3-32 4.7 × 10^2 N vertically upward

3-33 3.1 × 10^2 m 7.5° north of east

3-34 1.2 × 10^2 m 83° counterclockwise from the $+x$-axis

3-35 (a) 7.0 km 5.2° north of west (b) 11 km 3.3° north of west

Applying Your Knowledge

3-36 Either vector $\vec{B}$ is perpendicular to vector $\vec{A}$, or $\vec{B}$ and/or $\vec{A}$ has a magnitude of zero.

3-37 (a) approx. 10^6 m downward (b) approx. 0 m

3-38 1.0 × 10^3 km 60° west of south

3-39 (a) 18.8 N (b) 45.3 N

3-40 The length is 3.0 × 10^1 cm and the angles are 69°, 69°, and 42°.

3-41 1.4 × 10^9 km

3-42 (a) $\theta = \tan^{-1} \dfrac{\text{rise}}{\text{run}}$

(b), (c) Answers will vary. An example is a rise of 20 cm and a run of 30 cm, with an angle of 34°.

(d) Answers will depend on the student's preference. See (b), (c) above for an example.

Chapter 4 Two-Dimensional Kinematics

Exercises

4-1 (a) 78 m 61° west of north (b) 8.6×10^2 m 29° south of east (c) 3.3 km 41° north of east

4-2 (a) 77.0 km (b) 18 km 29° south of west (c) 61.6 km/h (d) 15 km/h 29° south of west

4-3 (a) 1.5 cm/s (b) 1.5 cm/s horizontally to the left; 1.5 cm/s up and to the right, 60° above horizontal (c) 1.2 cm/s vertically downward

4-4 (a) 1.3×10^3 m 42° north of east (b) 5.6 m/s, 5.2 m/s 42° north of east

4-5 (a) A to the left, B toward the bottom of the page, C to the right, D up and to the right
(b) A down and to the left, B down and to the right, C up and to the right, D to the right (perhaps slightly downward)

4-6 (a) 17 km 32° west of south (b) $v_{av,x} = -38$ km/h, $v_{av,y} = -61$ km/h

4-7 (a) A rough sketch would yield an estimate of about 2 m/s² at an angle of 45° west of south.
(b) 2.4 m/s² 45° west of south

4-8 0.85 m/s² 50° south of west

4-9 (a) 2.52×10^{-2} m/s² (b) 1.63×10^{-7} m/s² (c) 7.04 m/s²

4-10 15 m/s 65° south of east

4-11 17 m/s at an angle of 8.8° above the horizontal

4-12 $a_{av,x} = 9.0 \times 10^{-3}$ m/s²; $a_{av,y} = -2.5 \times 10^{-2}$ m/s²

4-13 $a_V = 9.8$ m/s² downward; $a_h = 0$

4-14 The airplane has a propulsion system.

4-15 (a) 0.42 s (b) 0.71 m horizontally from the edge of the table

4-16 (a) 0.714 s (b) 17.1 m (c) 25.0 m/s at an angle of 16.3° below the horizontal (d) 0.38 m above the net

4-17 One value is given here. At 2.0 s, the position components are $x = 20$ m and $y = 19.6$ m lower, and the velocity is 20 m/s 76° below the horizontal.

The table below shows the results of the vertical drop calculations as well as the horizontal position at the times indicated:

t (s)	0	0.50	1.0	1.5	2.0
y (m)	0	1.225	4.9	11.0	19.6
x (m)	0	5.0	10	15	20

The results are shown in the table below:

t (s)	0	0.50	1.0	1.5	2.0
$\vec{v}$ (m/s; ° below the horizontal)	5.0; 0	7.0; 44	11; 63	16; 71	20; 76

4-18 43 m/s

4-19 (a) $t_{RISE} = t_{FALL}$ (b) 9.8 m/s² downward

4-20 (a) 1.2×10^3 m (b) 16 s (c) 2.5×10^3 m

4-21 8.9 m above

4-22 (a) 0.53 s (b) 2.8 m (c) 6.0 m/s at an angle of 29° below the horizontal

4-23 (a) 0.87 s (b) 2.3 m (c) 11 m/s at an angle of 76° below the horizontal

4-24 (a) The table below summarizes the values for the different angles:

angle (°)	20	40	45	50	60
horizontal range (m)	4.7×10^2	7.3×10^2	7.4×10^2	7.3×10^2	4.7×10^2

(b) The horizontal ranges for any two angles equidistant from 45° are equal, and the maximum horizontal range occurs when the launch angle is 45°.

4-25 (a) 4.9×10^3 m (b) 2.6×10^4 m

4-26 The estimated distance is about 10 m.

4-27 The speed of an object in uniform circular motion remains constant.

4-28 (a) west, north, south (b) 2.0×10^1 m/s²

4-29 (a) 5.2 m/s² (b) 37 m/s²

4-30 The direction is still toward the centre of the circle.

4-31 (a) a_c decreases by a factor of 2. (b) a_c increases by a factor of 4.

4-32 0.10 m/s²

4-33 1.2×10^2 m/s²

4-34 (a) 4.9×10^{12} m/s² (b) 5.0×10^{11}

4-35 0.4 to 0.5 m/s²

4-36 (a) 3×10^{-5} m/s² (b) 3×10^{-4}%

4-37 The correct equations are shown:
(a) $\vec{v}_{AC} = \vec{v}_{AB} + \vec{v}_{BC}$
(b) $\vec{v}_{DE} = \vec{v}_{DL} + \vec{v}_{LE}$
(c) $\vec{v}_{NM} = \vec{v}_{NT} + \vec{v}_{TM}$; $\vec{v}_{MN} = \vec{v}_{MT} + \vec{v}_{TN}$
(d) $\vec{v}_{LP} = \vec{v}_{LM} + \vec{v}_{MN} + \vec{v}_{NO} + \vec{v}_{OP}$

4-38 (a) 3.9 m/s forward (b) 1.7 m/s forward (c) 3.0 m/s 21° to the left of the forward direction

4-39 7.3×10^2 km 41° south of west from Vancouver

4-40 (a) 43 m/s 40° north of west (b) 5.0×10^1 m/s 56° north of west (c) 72 m/s 57° north of west

4-41 (a) 1.0 m/s 53° to the shore (b) 9.0×10^1 m downstream from the point across from the starting position (c) 49° upstream from a direct line across the river

4-42 The canoeist who aims straight across the river takes less time to cross the river and reaches the other side first.

4-43 (a) 6.2×10^2 km/s 15° north of east (b) 6.0×10^2 km/h 22° north of east

Chapter Review

Multiple-Choice Questions

4-44 (b)

4-45 (e)

4-46 (a)

4-47 (c)

4-48 (d)

Review Questions and Problems

4-49 They are equal.

4-50 No; it is only as large if $\vec{A}$ and $\vec{B}$ have the same direction.

4-51 67 m, 17 m

4-52 (a) 0.71 m/s, 0.79 m/s (b) for each, 0.50 m/s at an angle of 45° up from the bottom edge of the square

4-53 (a) 11.3 m (b) 2.3 m/s (c) 8.2 m 33° below the horizontal (d) 1.6 m/s 33° below the horizontal

4-54 (a) 17 m 29° east of south (b) 2.7 m/s 29° east of south

4-55 at A, 100 km/h 24° north of east; at B, 100 km/h east; at C, 100 km/h 46° south of east

4-56 35 cm 207° counterclockwise from the $+x$-axis

4-57 gas pedal, brake pedal, steering wheel

4-58 0.90 m/s^2 39° west of south

4-59 (a) 13 m/s (b) 13 m/s east, 13 m/s north, 13 m/s 45° north of west (c) 8.6 m/s north (d) 2.7 m/s^2 45° south of west, 1.9 m/s^2 west

4-60 27 m/s 32° south of east

4-61 54 m/s east

4-62 at the top of the curved path; twice: just after being released and just prior to landing

4-63 zero and 9.8 m/s^2 downward

4-64 (a) 0.50 s (b) 0.3 m (c) 19 m/s 15° below the horizontal

4-65 29 m/s horizontally

4-66 (a) 1.6×10^2 s (b) 1.2×10^2 km (c) 31 km

4-67 (a) 1.1×10^2 m (b) 24 s (c) 6.2×10^2 m

4-68 The angles are 52°, 75°, and 0.5°, respectively.

4-69 instantaneous acceleration

4-70 no (**Hint:** Apply the equation $a_c = \frac{4\pi^2 r}{T^2}$.)

4-71 1.4×10^2 m

4-72 About 1 Hz

4-73 0.202 cm/s^2, 5.45×10^{-5} cm/s^2, 2.75×10^{-7} cm/s^2

4-74 5.4 s

4-75 (a) 73 m/s^2 (b) 7.4g (c) The caption in Figure 4-52(b) reveals several differences and advantages.

4-76 (a) The ball appears to be falling vertically downward toward the floor.

(b) Relative to Earth or a person standing on the ground outside the train, the ball has a velocity forward and a constant acceleration downward, so its path is a parabolic curve typical of projectile motion.

4-77 (a) The instantaneous velocity is south and the instantaneous acceleration is west (toward the centre of the circle).

(b) The instantaneous velocity is west (to the right) and downward at an angle to the horizontal. The instantaneous acceleration is the acceleration due to gravity, and is thus straight downward.

4-78 2.6×10^2 km/h 29° south of east

4-79 (a) 0.50 m/s (b) 0.94 m/s 58° to the shore (c) 39° upstream from a line directly across the river

4-80 3.5×10^2 km/h 56° south of east

Applying Your Knowledge

4-81 40°, 49°; the Nepal side is steeper.

4-82 (a) 83° south of east (b) 15 s

4-83 **Hint:** The equation to apply involves x_{MAX}, v_0^2, and g.

4-84 (a) 28 m/s (b) about 4 s (c) about 20 m

4-85 no; as can be shown by the vector subtraction of two velocity vectors, where the final velocity is greater in magnitude than the initial velocity for a particle moving in a circle of constant radius

4-86 no; short by 1.3 m

4-87 2.2×10^{-2} Hz, 45 s, 7.0×10^1 m/s

4-88 (a) 0.60 m (b) 6.0 m/s 35° above the horizontal

4-89 (a) 6.3 m/s^2 (b) about 1 m/s^2

4-90 17 m

4-91 (a) 19° north of west (b) 62 km/h

4-92 87 km/h

Chapter 5 Newton's Laws of Motion

Exercises

5-1 $\vec{F}_G$ downward, $\vec{T}$ upward

5-2 $\vec{F}_G$ downward, $\vec{F}_N$ upward

5-3 $\vec{F}_G$ downward, $\vec{F}_N$ upward

5-4 $\vec{F}_G$ downward, $\vec{F}_{buoyant}$ and $\vec{F}_f$ upward

5-5 $\vec{F}_G$ downward, $\vec{T}$ upward

5-6 $\vec{F}_G$ downward, $\vec{F}_N$ upward, $\vec{F}_f$ horizontally opposite to direction of motion

5-7 $\vec{F}_G$ downward

5-8 $\vec{F}_G$ downward, $\vec{F}_N$ upward, $\vec{F}_f$ horizontally opposite to direction of motion, $\vec{T}$ at 30° above horizontal

5-9 $\vec{F}_G$ downward, $\vec{F}_N$ perpendicular to ramp, $\vec{F}_f$ parallel to ramp and opposite to direction of motion, $\vec{T}$ parallel to ramp (at 20° above horizontal)

5-10 (a) 0 N (b) 19 N downward

5-11 (a) 94 N downward (b) 0 N

5-12 3 N horizontally forward

5-13 0.59 N at 53° up from horizontal

5-14 6.15×10^3 N at 23.9° above horizontal

5-15 678 N at 17.1° east of north

5-16 162 N

5-17 31 N at 30° south of east

5-18 7 N (considering significant digits)

5-19 4.8 N

5-20 3.51×10^{15} m/s^2

5-21 (a) downward (b) 9.8 m/s^2 (c) downward (d) downward, 9.8 m/s^2, downward (e) downward, 9.8 m/s^2, downward

5-22 Same answers as for Question 5-21. (a) downward (b) 9.8 m/s^2 (c) downward (d) downward, 9.8 m/s^2, downward (e) downward, 9.8 m/s^2, downward

5-23 (a)

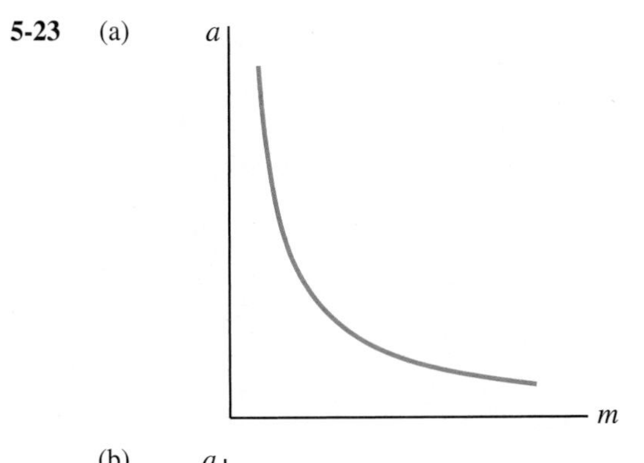

(b)

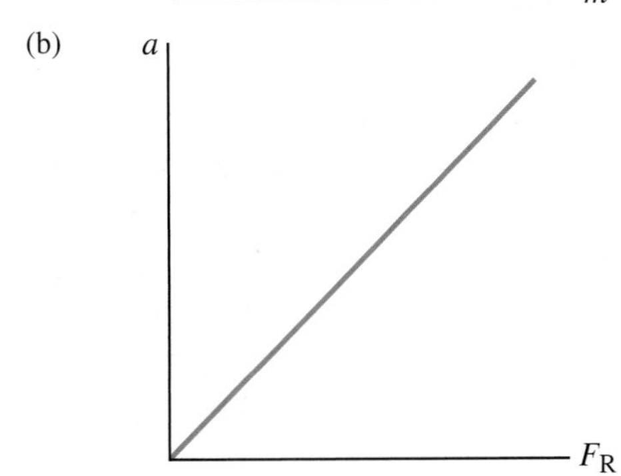

5-24 24.0 N

5-25 12.0 g

5-26 8.8×10^{-19} N

5-27 3.7 m/s^2

5-28 9% (decrease)

5-29 (a) 1.1×10^4 m/s^2 (b) 6.1×10^{-4} s (c) 1.1×10^3

5-30 approx. 3×10^3 N to 10×10^3 N

5-31 yes

5-32 yes

5-33 no

5-34 yes

5-35 yes

5-36 no

5-37 no

5-38 yes

5-39 35 N, opposite to boy's push

5-40 (a) 2.62×10^3 kg (b) 2.57×10^4 N

5-41 (a) 708 N (b) 708 N

5-42 (a) 11 N (b) 26 N (c) 47 N

5-43 1.5×10^{-13} N

5-44 (a) 133 N (b) 55.4 N (c) 1.40×10^2 N, 59.3 N

5-45 1.3 m/s^2 downward

5-46 (a) 2.07×10^3 N upward (b) normal force

5-47 (a) 3.01×10^3 N (b) 8 kg

5-48 602 N

5-49 (a) 602 N (b) 482 N (c) 0 N (d) No (e) 542 N, 542 N

5-50 approx. 3×10^{12} N

5-51 (a) 4.3×10^3 m/s^2 upward (b) 2.8×10^3 N upward, brick (c) 5.1

5-52 (a) 2.1×10^{-4} N (b) ground (c) 2.1×10^{-6} N downward (d) 0.073 m

5-53 (a) 85 N, 447 N (b) 455 N at 79° above horizontal

5-54 (a) 2.5 N (b) 1.8 N, 6.7 N (c) 3.1 N at $\theta = 54°$

5-55 $T_1 = 49.9$ N; $T_2 = 40.8$ N

5-56 (a) 0.98 m/s^2 (b) 13 N (c) 66 N

5-57 (a) 0.744 m/s^2 (b) 12.0 N (last rope), 21.8 N (middle rope)

5-58 (a) 0.81 kg (b) 7.6 N

5-59 (a) 5.7 m/s^2 (b) 15 N

5-60 (a) 4.8 m/s^2 (b) 19 N

5-61 (a) 2.7×10^2 N (b) 28 N

5-62 (a) 32.8 N upward (b) 6.10 N in opposite direction to sleigh's motion

5-63 (a) 37.8 N upward (b) 7.96 N in opposite direction to sleigh's motion (c) 38.6 N at 11.9° from vertical

5-64 (a) 25.0 kg (b) 86 N

5-65 (a) 1.51 m/s^2 (b) 385 N (c) 1.70 m

5-66 (a) $mg - F \sin\theta$ (b) $F \cos\theta$

5-67 5.8 m/s^2

5-68 8.8°

5-69 $g \sin\phi$

5-70 (a) gravitational force toward Jupiter, exerted by Jupiter on the Sun

(b) downward normal force, exerted by the book on the table

(c) force exerted by the pan on the chef, directed toward the oven

(d) forward force exerted by the water on the canoeist (and paddle)

(e) upward gravitational force exerted by the apple on Earth

(f) downward force (friction) exerted by the hailstone on the air

5-71 (a) 1.0×10^4 N forward (b) 1.0×10^4 N backward (c) 2.0×10^4 N forward

5-72 The flyboard (and the two hand attachments) force water downward as water jets, and by Newton's third law of motion, the water exerts an upward force on the board and the person.

5-73 (a) 1.8 kg (b) 0.37 N

5-74 (a) 1.0 m/s^2 (b) 56 N (c) 56 N

5-75 (a) electrical force (b) gravity (c) electrical force (d) gravity (e) electrical force (f) electrical force (g) nuclear strong force

Chapter Review

Multiple-Choice Questions

5-76 (a)

5-77 (b)

5-78 (e)

5-79 (c)

5-80 (d)

5-81 (d)

5-82 (c)

5-83 (b)

Review Questions and Problems

5-84 3.5 N at 72° east of north

5-85 $m = 3.6$ kg

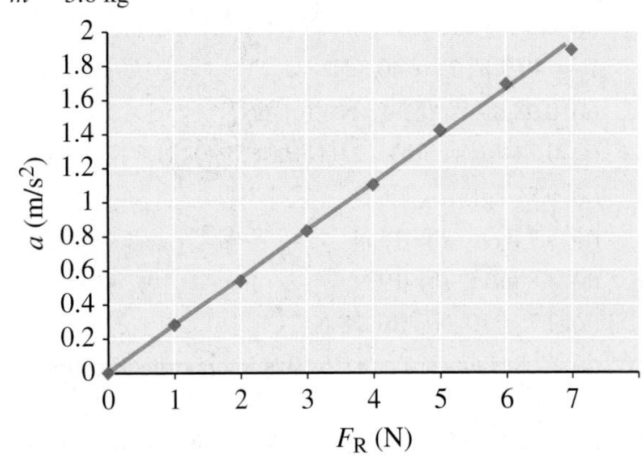

5-86 59 m/s^2

5-87 2.5×10^2 N

5-88 27 N

5-89 0.87 m/s^2; the length of the string and the mass of the washer are not needed.

5-90 0 N

5-91 2.0×10^3 N at 63° below the $+x$-axis

5-92 0.92 m/s^2

5-93 2.0×10^2 N

5-94 (a) 0.213 m/s^2 (b) 2.41×10^{-2} N

5-95 electrical force

Applying Your Knowledge

5-96 2.9×10^3 N

5-97 approx. (3 to 8) $\times 10^2$ N

5-98 (a) 1.55 m/s^2 (b) 295 N

5-99 29 N upward and to the right, at 52° above horizontal

5-100 2.0 kg

5-101 1.2 kg

5-102 (a)

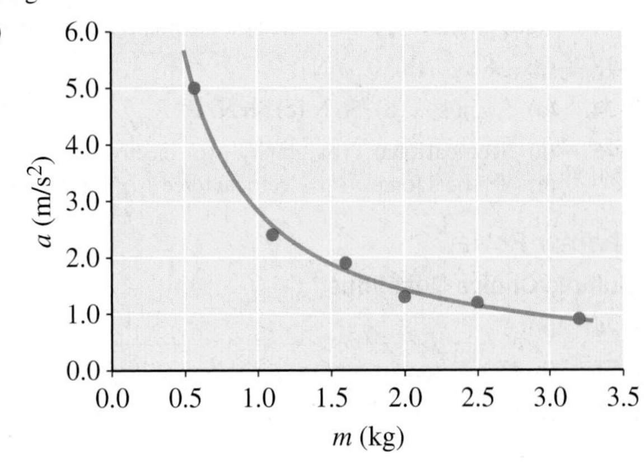

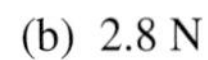
(b) 2.8 N

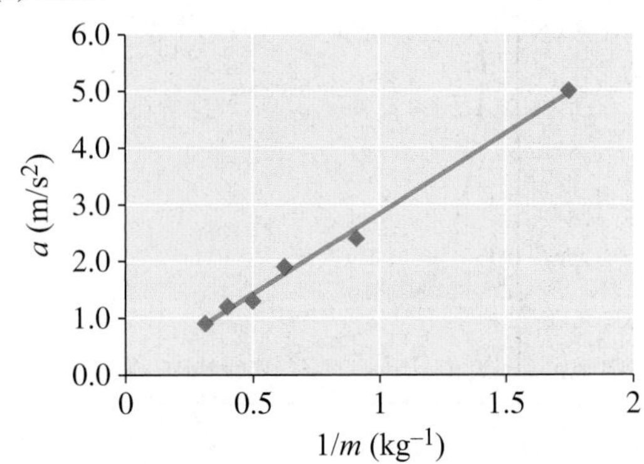

5-103 (a) 4.2 m/s (b) 2.6×10^3 N

5-104 (a) 1.4 m/s^2 (b) $T_{12} = 2.9 \times 10^2$ N; $T_{23} = 3.4 \times 10^2$ N

5-105 1.4×10^3 N

5-106 5.4 m/s^2 downward

5-107

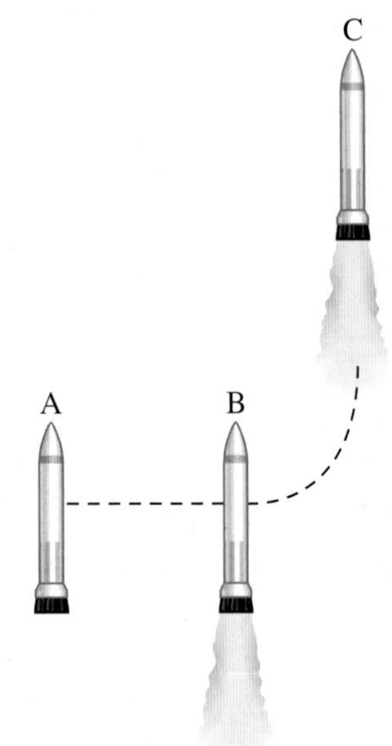

5-108 304 N, upward and to the right at 18.6° above horizontal

5-109 7.9 m/s

5-110 23°

5-111 (a) up (b) 0.27 m/s^2 (c) 30.7 N

5-112 (a) 18.3 N (b) 70.1 N

5-113

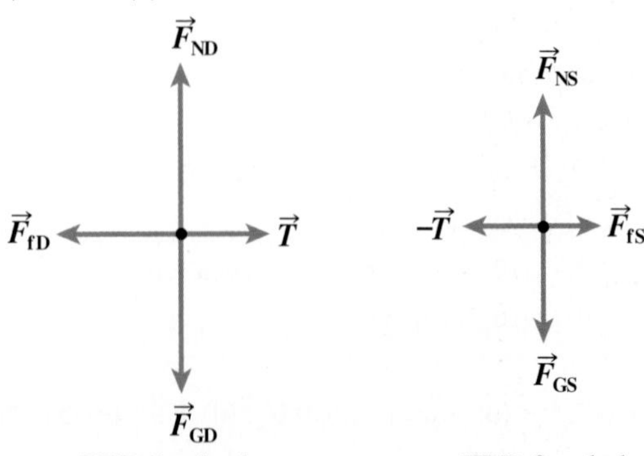

FBD for donkey FBD for sled

5-114 (a) 0.92 m/s (b) 73 N east (c) 73 N west

5-115 (a) 0.54 N (b) 0.54 N (c) 0.54 N

5-116 (a) 97 m^2/s^2 (b) 24 N (c) 23 N (horizontal, backward), 8.6 N (downward) (d) 5.5×10^2 N at 88° below horizontal

5-117 3.02×10^{-2} m

5-118 (a) 3.6 kg (b) 0.96 m/s^2

5-119 (a) 2 cm (b) 2 (c) 3.3 m/s^2

Chapter 6 Applying Newton's Laws

Exercises

6-1 If friction did not exist, a person's feet would simply slip on the surface underneath their feet, and walking would be impossible.

6-2 For a plate sliding eastward across a table, the friction force is westward and kinetic.

6-3 (a) 55 N, 0 m/s^2 (b) 51 N, 0.94 m/s^2

6-4 (a) 4.5×10^2 N (b) 4.1×10^2 N

6-5 Yes; she slides.

6-6 0.51

6-7 (a) 1.8 m/s^2 (b) 0.19

6-8 (a) gravity $m_{\text{small}}\vec{g}$ downward, normal force $\vec{F}_{\text{N,small}}$ upward and horizontal static friction $\vec{F}_S$ exerted by large book (b) static friction (c) 0.26

6-9 $m_{\text{large}}\vec{g}$ downward, $\vec{F}_{\text{N,large}}$ upward, horizontal $\vec{F}_{\text{applied}}$, horizontal $\vec{F}_K$, vertical $-\vec{F}_{\text{N,small}}$ and horizontal $-\vec{F}_S$ exerted by small book

6-10 59 N

6-11 3.3 m/s^2

6-12 (a) $\mu_S = \tan\phi$ (b) Adjust incline to angle θ so that block slides at constant velocity. $\mu_K = \tan\theta$.

6-13 approx. 0.5 to 0.6

6-14 (a) $\vec{F}_G$ toward centre of Earth (b) $\vec{F}_G$ toward Sun (c) vector sum of $\vec{F}_G$ down and $\vec{F}_N$ up (d) horizontal static friction $\vec{F}_S$ exerted by road (e) horizontal component of normal force $\vec{F}_N$ exerted by road (f) electrical force exerted by nucleus

6-15 (a) vertical inside wall of interior cylindrical chamber (b) horizontal normal force (c) 14 m (d) 14 m/s (e) 1.3×10^3 N

6-16 (a) 5.5×10^3 m/s (b) 5.1×10^9 s, 1.6×10^2 yr

6-17 (a) electrical force (b) 8.3×10^{-8} N, toward nucleus

6-18 (a) at A, gravity and tension down; at B, gravity down and tension toward right; at C, gravity down and tension up; at D, gravity down and tension toward left

(b) at A, sum of gravity down and tension down; at B, tension toward right; at C, vector sum of tension up and gravity down, of magnitude $T - mg$; at D, tension toward left

6-19 306 N

6-20 (a) static friction (b) 0.26 (c) increase (d) no change

6-21 23 m/s

6-22 (a) gravity down and tension down (b) 3.1 N

6-23 (a) $T = m(v^2/r - g)$ (b) T decreases (c) $T \rightarrow 0$ (d) $v_{\text{min}} = \sqrt{gr}$ (e) 2.9 m/s, yes

6-24 (a) Relative to the train, the ball accelerates toward the front of the train.

(b) The acceleration of the ball is toward the front of the train.

(c) The force causing this acceleration is not a real one.

6-25 (a) A (b) forces on ball are $\vec{F}_G$ and $\vec{T}$; ball accelerates because of horizontal component of $\vec{T}$ (c) forces are $\vec{F}_G$, $\vec{T}$, and $\vec{F}_{\text{fictitious}}$ (toward rear of bus); ball is not accelerating; magnitude of $\vec{F}_{\text{fictitious}}$ toward rear equals magnitude of horizontal component of $\vec{T}$ toward front

6-26 Relative to Earth, you tend to continue to travel in a straight line (neglecting forces exerted by friction and the seat belt), while the car moves to the right. Relative to the car, you tend to move to the left because of the fictitious centrifugal "force."

6-27 away from C

Chapter Review

Multiple-Choice Questions

6-28 (c)

6-29 (b)

6-30 (c)

6-31 (d)

6-32 (d)

Review Questions and Problems

6-33

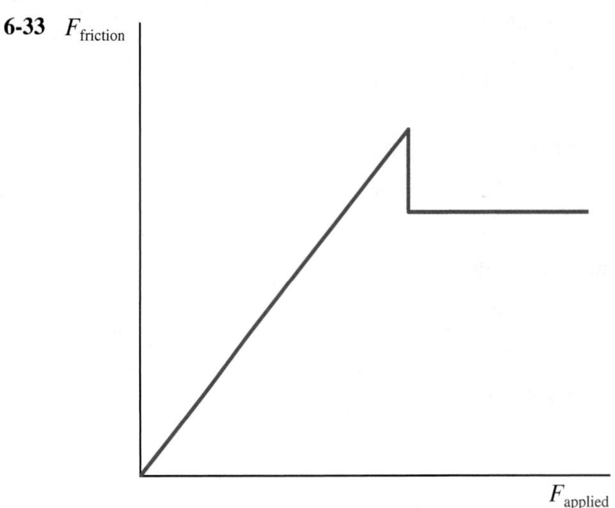

6-34 The refrigerator does not move.

6-35 96 N

6-36 0.97 s

6-37 (a) the air (b) 2.3 N

6-38 (a) 0.032 m (b) 3.1×10^{-8} s

6-39 Since the girl is moving in a circle, her velocity is changing and there is an acceleration, which means that there is a resultant force.

Applying Your Knowledge

6-40 0.35

6-41 (a) static friction (b) 5.9 m/s^2

6-42 (a) 24 N (b) 2.0×10^1

6-43 (a) 1.3×10^2 N (b) 11 N (c) 73°

6-44 0.47

6-45 4.4 kg

6-46 (a) 3.8 m/s^2 (b) 0.16

6-47 13 m

6-48 3.6 m/s^2; downward

6-49 (a) 2.7×10^2 N (b) 43 N

6-50 1.3 m/s^2

6-51 (a) forces acting: $m\vec{g}$ downward and $\vec{T}$ upward and to right (b) horizontal component of string tension (c) 2.13 m/s (d) 1.82 s

6-52 7.49 m/s

6-53 (a) forces acting: $m\vec{g}$ downward and $\vec{F}_N$ downward (b) gravity and normal force (c) $F_N \rightarrow 0$ (d) $v_{min} = \sqrt{gr}$ (e) approx. 8 m/s for $r \approx 6$ m

6-54 4.0×10^2 N

6-55 approx. 2 N

6-56 (a) forces acting: $m\vec{g}$ downward, $\vec{F}_N$ upward, $\vec{F}_S$ in direction of acceleration

(b) forces acting: $m\vec{g}$ downward, $\vec{F}_N$ upward, $\vec{F}_S$ in direction of acceleration, $\vec{F}_{fictitious}$ opposite to direction of acceleration

(c) 4.6 m/s^2

6-57 (a) forces acting: $m\vec{g}$ down, $\vec{F}_{N,floor}$ up, $\vec{F}_{N,wall}$ horizontally toward centre; $\vec{F}_{N,wall}$ provides $\vec{a}_c$

(b) forces acting: $m\vec{g}$ down, $\vec{F}_{N,floor}$ up, $\vec{F}_{N,wall}$ horizontally toward centre, $\vec{F}_{centrifugal}$ away from centre. Since $m\vec{g} = -\vec{F}_{N,floor}$ and $\vec{F}_{centrifugal} = -\vec{F}_{N,wall}$, then $\Sigma\vec{F} = 0$ and the person appears to be stationary.

Chapter 7 Work, Energy, and Power

Exercises

7-1 no

7-2 no

7-3 yes, positive

7-4 yes, negative

7-5 yes, positive

7-6 yes, negative

7-7 no

7-8 no

7-9 no

7-10 $\vec{F} \perp \Delta\vec{r}$

7-11 $W = 0$

7-12 zero

7-13 kg·m^2/s^2

7-14 (a) −49.1 J (b) 49.1 J (c) 0 J

7-15 up, −49.1 J; down, 49.1 J

7-16 6.3 m

7-17 (a) 0 J (b) 0 J

7-18 0 J

7-19 (a) 1.13×10^3 N (b) 7.2×10^2 J, -7.2×10^2 J

7-20 32°

7-21 1.29×10^3 J

7-22 1.29×10^3 J; the work required to pull an object up an incline is independent of the angle of the incline.

7-23 8.14×10^3 J

7-24 Kinetic energy is a scalar quantity.

7-25 (a) 4 (b) 9

7-26 2.43×10^5 J

7-27 (a) 3.21×10^5 J (b) 32% (c) 7.8×10^4 J

7-28 11 m/s

7-29 1.2×10^{-13} kg

7-30 (a) 3.09 J (b) 4.18 m/s

7-31 (a) −0.66 J (b) 4.1 m/s

7-32 8.6×10^{-2} N

7-33 0.45 m/s

7-34 approx. 2×10^3 J to 3×10^3 J

7-35 kg·m^2/s^2, kg·m^2/s^2, yes, joule (J)

7-36 1.8×10^3 J

7-37 (a) 4.29 J, 0 J (b) 0 J, −4.29 J (c) −4.29 J, −4.29 J

7-38 15 m

7-39 (a) -9.3×10^2 J (b) 9.3×10^2 J (c) 9.3×10^2 J

7-40 A rollercoaster always starts by going up a hill so that the coaster gains gravitational potential energy, which is converted to kinetic energy as the coaster goes downward.

7-41 (a) 0.056 J (b) 0 J (c) −0.056 J (d) 0.056 J, 1.5 m/s

7-42 (a) 17 m/s (b) decrease

7-43 same speeds

7-44 (a) 2.1 m/s (b) 0.43 m (c) 2.1 m/s

7-45 11 m/s

7-46 (a) 6.10×10^6 J (b) 13.6 m/s

7-47 (a) 29 m/s (b) 29 m/s

7-48 11 cm

7-49 joule

7-50 (a) thermal energy (b) thermal energy and kinetic energy

7-51 (a) -2.3×10^2 J (b) 2.3×10^2 J

7-52 7.4 m

7-53 (a) 0.099 J, 0 J (b) thermal energy (c) 0.18 N

7-54 1.1 m/s

7-55 1.1 m/s

7-56 (a) 3.6×10^5 J (b) kinetic energy (c) 1.3×10^3 kg (d) 0.59

7-57 1.0×10^{-7} J

7-58 (a) 2.18×10^{-18} J (b) 2.1×10^{-13} J (c) 9.6 J (d) 6.7×10^{-4} Cal (e) 3.5×10^{-3} Btu (f) 1.81×10^{-2} Btu

7-59 3.6×10^2 kJ

7-60 (a) 28 W (b) 3.90×10^5 J (c) 0.10 kW·h (d) 4.0 kW

7-61 83 W

7-62 44 min

7-63 1×10^{34} J

7-64 23¢

7-65 5.6×10^4 W

7-66 1.84 m/s

7-67 2.2×10^4 J

7-68 0.042 Cal

7-69 9

7-70 approx. (2 to 4) $\times 10^3$ W

7-71 18 MJ

7-72 (a) kinetic energy (b) 2.6×10^6 J (c) 6.3×10^6 J

7-73 28%

7-74 6%

Chapter Review

Multiple-choice Questions

7-75 (e)

7-76 (a)

7-77 (b)

7-78 (a)

7-79 (b)

7-80 (d)

7-81 (c)

7-82 (a)

7-83 (b)

7-84 (c)

Review Questions and Problems

7-85 (a) Work is a scalar. (b) Kinetic energy is a scalar. (c) Force is a vector. (d) Velocity is a vector. (e) Speed is a scalar. (f) Gravitational potential energy is a scalar. (g) Distance is a scalar. (h) Displacement is a vector. (i) Thermal energy is a scalar.

7-86

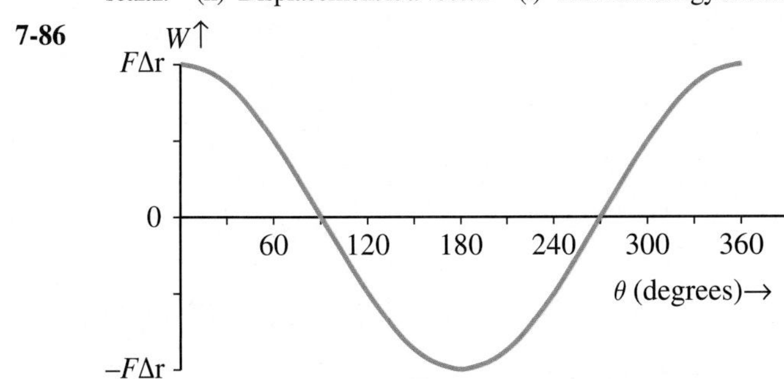

7-87 6.1×10^{-20} J

7-88 Earth's speed would increase.

7-89

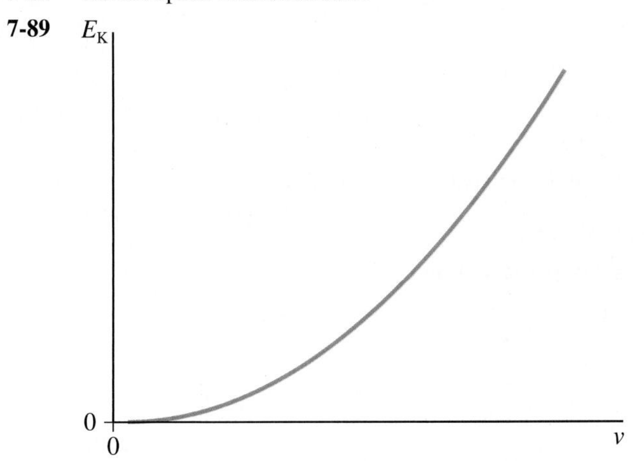

7-90 25.7 m

7-91 51 N

7-92 0.94 m

7-93 (a) 1.2 J (b) 0.22 N (c) −0.11 J, 0.11 J (d) 3.5 m/s

7-94 1.5 m/s

7-95 8.7×10^4 MJ

7-96 1.4×10^7 m/s

7-97 (a) 0.090 kW·h, 3.2×10^5 J (b) 2×10^4 J

7-98 (a) 2.2×10^6 J (b) 3.1×10^7 J

7-99 1.2×10^6 J

7-100 3.0×10^6 J, 7.1×10^2 Cal

7-101 3.5¢

7-102 (a) 1.1 kW·h (b) 10¢

Applying Your Knowledge

7-103 gravity, 0 J; normal force, 0 J; applied force, 21 J; friction, −18 J

7-104 gravity, 0 J; normal force, 0 J; applied force, 21 J; friction, −18 J

7-105 (a) −23 J (b) 0 J

7-106 0.14 J

7-107 See Solutions Manual for proof.

7-108 5.6 m

7-109 4.7 m/s

7-110 8.9 m/s

7-111 (a) 4.9 m/s (b) back to original vertical level

7-112 12.9°

7-113 2.5 m/s

7-114 (a) 2.3×10^2 N, 1.3×10^2 N (b) 1.4 m/s (c) 2.0×10^2 J

7-115 (a) 6.4×10^2 N, 1.2×10^2 N (b) 6.1 m/s (c) 3.1×10^3 J

7-116 28 MW

7-117 (a) approx. 10^{10} J for a 20-yr-old person (b) approx. 10^3 d

7-118 4.67×10^{10} J

7-119 (a) −58 J (b) 58 J (c) 58 W (d) 36 N, -4.6×10^2 J, 4.6×10^2 W, 8.0, 3

7-120 (a) 1.8×10^2 J (b) 73 W (c) 2.9×10^2 W

7-121 (a) 2.3×10^5 kg/s (b) 8.9×10^2 W (c) 1.1×10^5

7-122 1.2×10^3 kg

7-123 0.019 J

Chapter 8 Momentum and Collisions

Exercises

8-1 (a) kg·m/s (b) kg·m/s for both

8-2 (a) 6.4×10^2 kg·m/s east (b) 5.0×10^{-15} kg·m/s downward (c) 1.2×10^2 kg·m/s west (d) 84 kg·m/s downward

8-3 Typical person at top speed – about 3×10^2 kg·m/s; car on a highway – about 3×10^4 kg·m/s; the momentum of the car is about 1×10^2 larger than that of the person

8-4

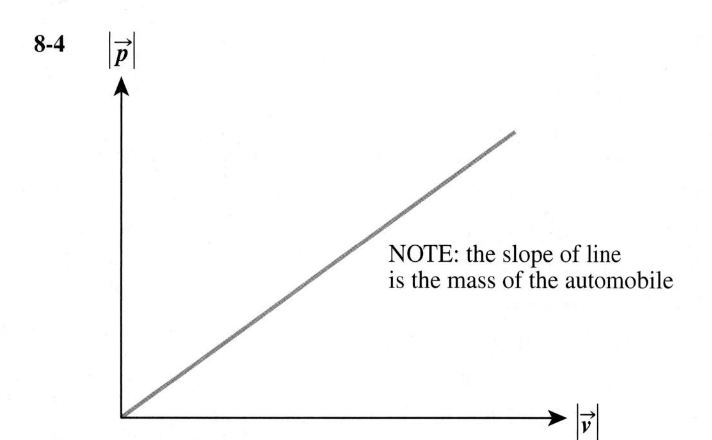

8-5 32 m/s

8-6 Impulse is the same as the change in momentum.

8-7 (a) 0.25 kg·m/s downward (b) zero (c) 0.25 kg·m/s upward

8-8 3.3×10^4 kg·m/s

8-9 12 ms

8-10 (a) 0.42 kg·m/s downward (b) 0.37 kg·m/s upward (c) 0.79 kg·m/s upward (d) 1.9×10^2 N upward

8-11 The magnitude of the force of gravity is small compared to the force exerted by the floor.

8-12 2.48 kg·m/s, 0.514 kg·m/s

8-13 (a)

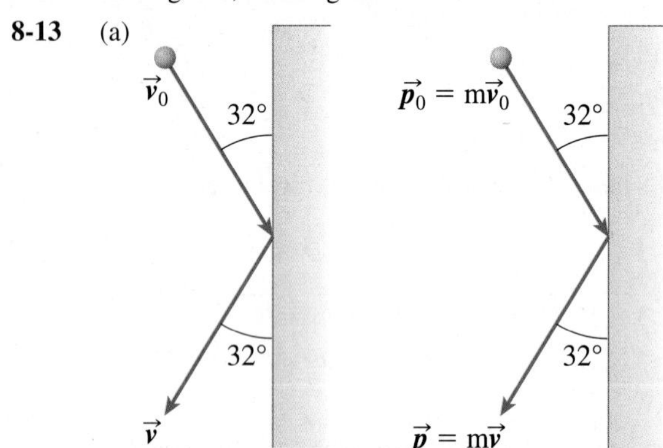

(b) 2.5×10^{-23} kg·m/s to left (c) 1.3×10^{-10} N to left

8-14 (a) 3.1×10^3 N at 19° above the $+x$-axis (b) 3.1×10^3 N at 19° below the $-x$-axis

8-15 (a), (b), (d), (e), (f) and (g) yes; (c) and (h) no

8-16 2.6 m/s in opposite direction to boy's velocity

8-17 50.6 kg

8-18 44 m/s

8-19 (a) downward (b) upward (c) equal in magnitude (d) zero (e) conserved (f) upward

8-20 1.9 m/s in original direction of cart's velocity

8-21 (a) 4.95 m/s east (b) 2.34×10^4 kg·m/s west, 2.34×10^4 kg·m/s east (c) equal in magnitude, opposite in direction (d) zero

8-22 48 kg

8-23 16.2 km/h

8-24 36 m/s

8-25 0 m/s

8-26 (a) completely inelastic (b) equal to (c) completely inelastic (d) inelastic

8-27 0 m/s

8-28 0 m/s; 815 m/s in direction of initial velocity

8-29 (a) 85 km/h north (b) 4.1×10^6 J, 4.0×10^6 J, 1×10^5 J

8-30 (a) 2.4×10^2 g (b) inelastic

8-31 $\vec{v}_1' = 0.4$ m/s in the opposite direction to its initial motion; $\vec{v}_2' = 3.1$ m/s in the same direction that m_1 was originally moving

8-32 515 m/s

8-33 $m/2$

8-34 (a)

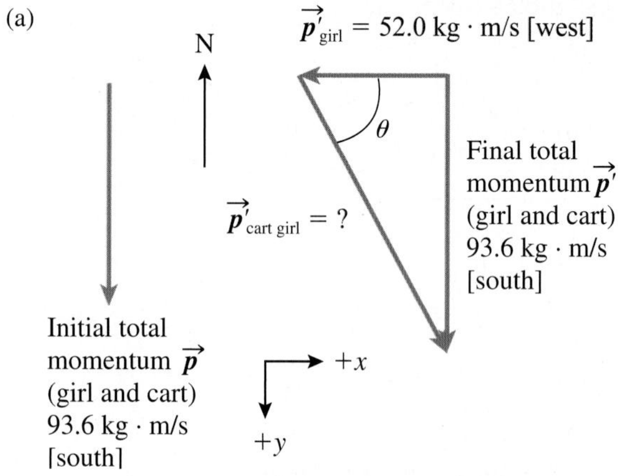

(b) 4.1 m/s at 61° south of east

8-35 3.0×10^1 km/h at 51° south of east

8-36 (a) and (b) 3.4×10^3 km/h

8-37 (a)

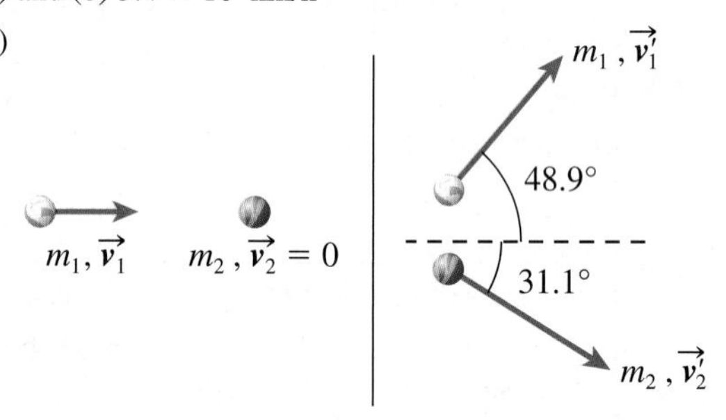

(b) 1.18 m/s at 48.9°, 1.72 m/s at 31.1°

(c)

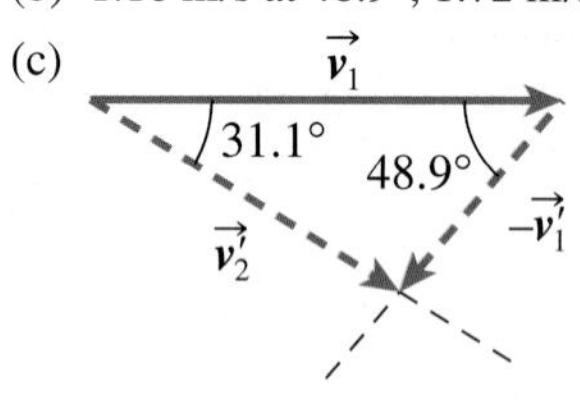

(d) no

8-38 66° from initial direction of neutron's velocity

Chapter Review

Multiple-Choice Questions

8-39 (d)

8-40 (e)

8-41 (c)

8-42 (d)

8-43 (a)

8-44 (c)

8-45 (a)

Review Questions and Problems

8-46

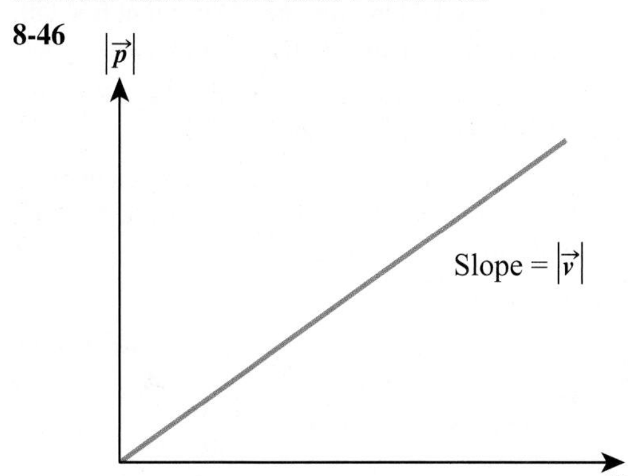

8-47 25 m/s

8-48 3.2×10^5 N east

8-49 (a) 28 m/s horizontally (b) The speed of the ball leaving the racquet increases with greater follow-through since the contact time increases, which results in an increased change in momentum for the ball.

8-50 31 N

8-51 15 kg

8-52 1.00 m/s

8-53 Answers will vary.

8-54 0.619 km/s

8-55 2.2 m/s in opposite direction to its original velocity

8-56 1.90×10^2 m/s toward Jupiter

8-57 (a) completely inelastic (b) inelastic (c) elastic

8-58 small mass and large speed

8-59 7.79×10^{-2} m/s for 253 g puck, 1.88 m/s for 232 g puck; Both pucks move in same direction as initially moving puck.

8-60 1.6 kg

8-61 (a) approximately 40 kg (b) 35 kg

8-62 (a)

Initial

Final

Puck 1
$\vec{v}_1 = 5.4$ m/s

Puck 2
$\vec{v}_2 = 0$

1

46°

33°

2

$+y$

$+x$

(b) 3.0 m/s; 4.0 m/s (c) approximately 2.7 m/s and 4.0 m/s

8-63 (a) 9.1 m/s at 26° north of west (b) 31%

Applying Your Knowledge

8-64 1.50 m/s

8-65 5.8 m/s in opposite direction to velocity of man and go-kart

8-66 11 m/s

8-67 7.4 m

8-68 23 m/s, 13° below the $-x$-axis, or 13° below the horizontal to the left

8-69 8.0×10^1 kg/s

8-70 (a) 7.1 m/s (b) The bowling ball is deflected by 6.0° after its elastic collision with the pin.

8-71 (a) 0.99 m/s, 49° to the initial direction of the moving stone (b) yes, since the opponent's stone slides a distance of 3.0 m before stopping, which is greater than the radius of the 'house'

8-72 55 kg

8-73 381 m/s

8-74 $2m/3$

8-75 1.78

8-76 Since $p = mv$ is the same for all objects, then $v = p/m$, that is, $v \propto 1/m$, and a graph of v versus m is:

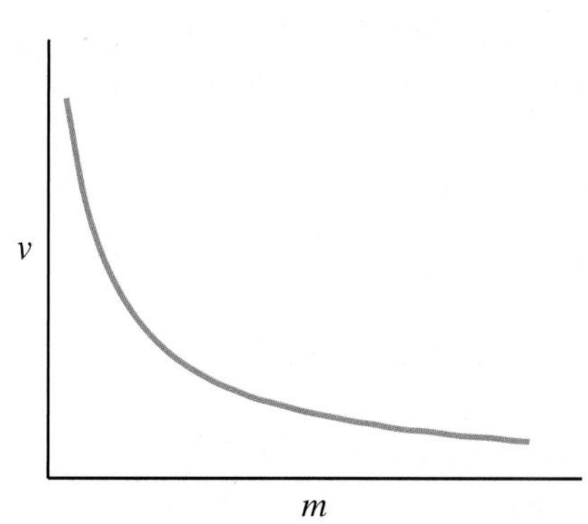

8-77 rubber ball

8-78 bounces off

8-79 (a) 8.6 kg·m/s, 4.8 kg·m/s (b) 1.0×10^3 N, 5.8×10^2 N (c) 1.2×10^3 N at 29° above horizontal

8-80 3.00 m/s west; Did you pay attention to significant digits in the north-south direction?

8-81 (a) 2.3 m/s (b) 2.5 m/s

8-82 (b) $mv/(m + M)$ (d) $h = m^2v^2/(2g(m + M)^2)$
(e) $v = ((m + M)/m)\sqrt{2gh}$ (f) 6.6×10^2 m/s

8-83 (a) 0.80 m/s (b) 7.7 N

8-84 4.9 m/s at 12° west of north

8-85 (a) $v'_{1x} = (m_1 - m_2)\, v_{1i\,x}/(m_1 + m_2)$; $v'_{2x} = 2m_1 v_{1i\,x}/(m_1 + m_2)$
(b) $v'_{1x} = 0$; $v'_{2x} = v_{1i\,x}$
(c) $v'_{1x} \approx v_{1i\,x}$; $v'_{2x} \approx 2v_{1i\,x}$
(d) $v'_{1x} \approx -v_{1i\,x}$; $v'_{2x} \approx 2m_1 v_{1i\,x}/m_2$

8-86 approx. 10^8 kg·m/s, 5×10^5 N

8-87 approx. 10^{12} kg·m/s; about 10^4 times greater than momentum of the cruise ship in previous problem

8-88 approx. 1–2 minutes; Lucas (little brother) reaches the shore before Cameron (big brother) does.

Chapter 9 Gravitation

Exercises

9-1 (a) 2.0×10^{20} N toward centre of Earth (b) 2.0×10^{20} N toward centre of Moon

9-2

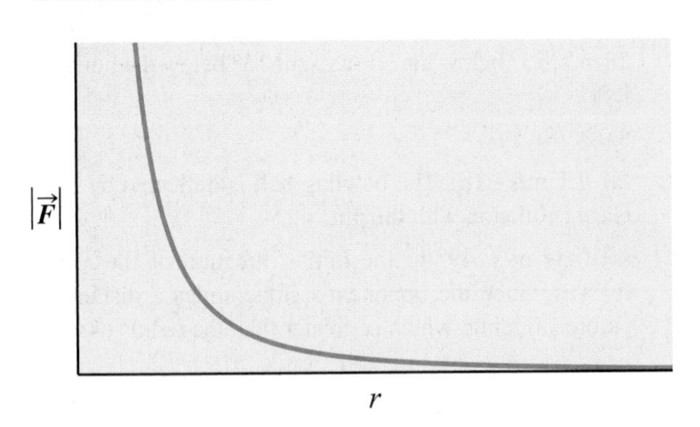

9-3 (a) yes (b) no (c) yes (d) yes (e) no

9-4 4

9-5 3

9-6 $5r_E$

9-7 (a) 1.14×10^{-11} N (b) 1.25×10^{11} times smaller than the strength of the force of gravity due to Earth on one orange

9-8 approx. 1×10^{-10} N

9-9 7.3×10^{-2} m/s² toward the centre of Earth

9-10 (a) $g/4$ (b) $g/9$ (c) $g/100$

9-11 8.08 m/s²

9-12 (a) 3.0×10^6 m (b) 2.5×10^3 N

9-13 $4g$

9-14 (a) 3.7 m/s² (b) approx. 3% difference

9-15 (a) 7.3×10^{22} kg (b) 1.3×10^2 N for a mass of 8.0×10^1 kg

9-16 (a) increase (b) decrease

9-17 approx. −0.2%

9-18 5.90×10^{-3} N/kg toward the centre of the Sun

9-19 (a) 5.42×10^{-9} N (b) 1.08×10^{-8} N

9-20 (a) 2.6×10^3 km (b) 0.24 N

9-21 1.3×10^{-10} N/kg toward the centre of the ball

9-22 5.3×10^{-3} N/kg toward the centre of the comet

9-23 approx. 7×10^{-6} N/kg, assuming that the orbit of Neptune is the outer boundary of the solar system

9-24 (a) 6.0 m, 2.0×10^1 m (b) 1.0×10^1 m, 2.0×10^1 m

9-25 The speed of Earth is fast enough to keep it in orbit about the Sun.

9-26 The gravitational force acts perpendicular to the direction of motion of the satellite.

9-27 (a) 7.61×10^3 m/s (b) 5.69×10^3 s (or 1.58 h)

9-28 (a) 1.3×10^4 km (b) 5.6×10^3 m/s

9-29 1.68×10^3 m/s

9-30 (a) 8.78 m/s² (b) 90% of g (c) 7.69 km/s (d) 5.50×10^3 s, or 1.53 h

9-31 Most rapidly January 4th, least rapidly July 5th

9-32 Each value is between 0.97 and 1.01 yr²/AU³; yes

9-33 (a) 2.94×10^{-19} s²/m³ (b) 2.01×10^{30} kg

9-34 6.48×10^{23} kg

9-35 The astronaut is not weightless because he/she is not in free fall; more forces are acting on the astronaut than just gravity.

9-36 (a) The astronaut cannot grab onto anything on the walls, ceiling, etc., to help her move around.

(b) In order to move, she could throw something, for example, a glove or a pen, and she would recoil in the opposite direction (using Newton's third law of motion).

9-37 (a) no (b) with a straw

9-38 1.3×10^2 km/h; this is larger than the actual top speed since some of the ride involves slowing down!

9-39 8×10^6 s; 20 Earth-years corresponds to roughly 80 Mercury-years!

9-40 (a) kg·m²/s² (b) kg·m²/s² (c) yes (d) joule (J), which is equivalent to kg·m²/s²

9-41 (a) -5.29×10^{33} J (b) 0 J

9-42 (a) 2.30×10^9 J (b) 2.2×10^9 J (c) 2.2×10^9 J

9-43 2.1×10^3 m/s

9-44 (a) $\Delta E_P = -\Delta E_K$ (b) 6.6 km/s

9-45 2.88×10^3 m/s

9-46 2.1×10^4 m/s

9-47 (a) 1.6×10^8 m/s (b) 54%

9-48 1.1×10^{26} kg

9-49 1.0×10^1 km

9-50 (a) 1.2×10^{31} kg (b) 6.1 solar masses

9-51 6×10^5 m/s

Chapter Review

Multiple-Choice Questions

9-52 (a)

9-53 (d)

9-54 (c)

9-55 (a)

9-56 (e)

9-57 (a)

9-58 (b)

9-59 (c)

9-60 (d)

9-61 (b)

9-62 (b)

9-63 (c)

Review Questions and Problems

9-64 1.1×10^{-2} m/s² toward centre of Sun

9-65 4.1×10^{23} N

9-66 (a) 4.45 m/s^2 toward centre of Earth (b) 5.56 × 10^4 N toward centre of Earth

9-67 0.09 m

9-68 2.3 × 10^{-3} m/s^2 at 5.1° from probe–Earth line

9-69 1.11 × 10^{26} kg

9-70 0.69 N toward centre of Earth

9-71 2.0 × 10^{-1} N/kg

9-72 5.1 × 10^4 km

9-73 The satellite is moving sufficiently fast that, as it falls toward Earth, the spherical shape of the planet curves away from the satellite.

9-74 Weightlessness means that an object is in free fall; no forces other than that of gravity are acting on the object, including normal forces from the object's surroundings.

9-75 (a) 9.68 × 10^6 m (b) 2.63 h

9-76 It is at its closest distance to the Sun. Kepler's second law

9-77 1.90 × 10^{27} kg

9-78 (a)

(b)

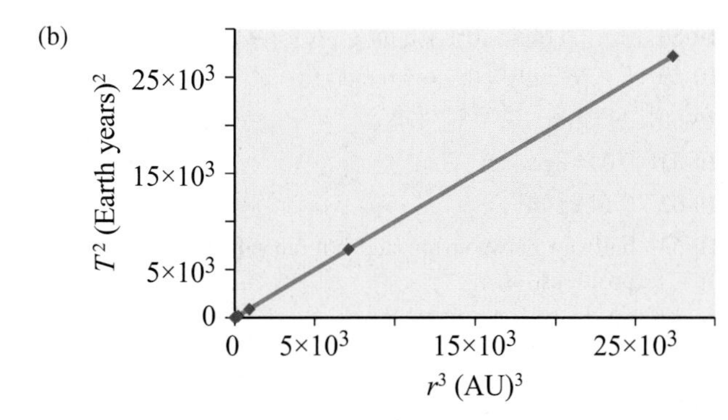

The data from Table 9-4 are shown with solid squares. It is apparent that these data satisfy Kepler's third law: $T^2 \propto r^3$.

9-79 3.1 × 10^8 m, 1.8 × 10^5 s

9-80 7.9 × 10^{11} m

9-81 1.6 × 10^2 m/s

9-82 Your direction does not matter.

9-83 1.2 × 10^7 m

9-84 (a) 1.7 × 10^2 km/s (b) no

9-85 13 km

Applying Your Knowledge

9-86 4.77 × 10^{20} N at 24.6° from line to Sun

9-87 $0.7g_E$

9-88 (a) 0.61 m/s^2 toward the centre of Earth (b) 2.2 × 10^2 N toward the centre of Earth

9-89 (a) 17 m (b) 6.3 m/s

9-90 0.75 r_E

9-91 3.46 × 10^8 m; closer to the Moon since it has a weaker gravitational field

9-92 The total momentum remains constant at zero.

9-93 0.82

9-94 (a) 6.56 m (b) 6.59 m

9-95 1.53 s

9-96 (a) 3.6 × 10^7 m (b) 3.1 × 10^3 m/s

9-97 See Solutions Manual for proof.

9-98 (a) 6.7 × 10^{-15} s^2/m^3 (b) 8.8 × 10^{25} kg

9-99 (a) 9.83 × 10^{-14} s^2/m^3 (b) 2.0 × 10^7 m

9-100 (a) at the equator (b) 0.464 km/s (c) east, in direction of rotation of Earth (d) near the equator and has a large ocean to the east instead of a heavily populated land mass in case of emergencies at liftoff

9-101 2.7 × 10^6 m/s

9-102 Using the values given in the previous question for mass and radius of the neutron star, we get a difference in gravitational acceleration of ~2 × 10^8 m/s^2 for a person that is approximately 1 m tall; you would be ripped apart with such differences in gravitational acceleration.

9-103 $f_{max} = \sqrt{\dfrac{G\rho}{3\pi}}$

9-104 ~3 years

Chapter 10 Rotational Motion

Exercises

10-1 (a) 1.00 rad (b) 2.00 rad (c) 6.28 rad

10-2 (a) 57.3°, 0.159 rev (b) 115°, 0.318 rev (c) 360°, 1.00 rev

10-3 (a) 183 rad/min (b) 175 deg/s (c) 29.1 rev/min

10-4 7.27 × 10^{-5} rad/s

10-5 (a) 42.1 rad/min^2 (b) 0.670 deg/s^2 (c) 1.17 × 10^{-2} rad/s^2

10-6 14 rad/s

10-7 191 rad/s^2

10-8 (a) 4.0 × 10^2 rev (b) 3.1 × 10^3 rad (c) −32 rad/s^2

10-9 0.394 rad/s, 0.738 rad

10-10 (a) 25 rad/s^2 (b) 78 rev

10-11 (a) −3.8 rad/s^2 (b) 0.45 s (c) 0.48 s

10-12 14 rev

10-13 approx. 4 to 8 rad/s

10-14 2.9 m/s

10-15 1.4 × 10^3 rpm

10-16 9.4 m/s^2

10-17 0.82 m

10-18 (a) toward the centre of rotation (b) $a_t = 0$ (c) $\alpha > 0$

10-19 (a) 27.2 m/s^2 (b) 3.90 m/s^2 (c) 27.5 m/s^2

10-20 (a) 12 rad/s (b) 12 rev (c) 3.6 m/s (d) 43 m/s^2, −0.36 m/s^2

10-21 92 rad/s

10-22 0.12 m/s^2

10-23 (a) 99 m/s (b) 2.2×10^2 m (c) 65 rad/s^2 (d) 2.9×10^2 rad/s (e) 1.0×10^2

10-24 3.17 m

10-25 kg·m^2/s^2 = joule

10-26 1.8×10^7 kg·m^2

10-27 2.8 J

10-28 24 J

10-29 $\omega = \frac{\sqrt{gy}}{R}$

10-30 (a) 2.2×10^2 N·m (b) 2.8×10^2 N

10-31 (a) 72 N·m, clockwise (b) 0 N·m, no rotation (c) 41 N·m, counterclockwise (d) 0 N·m, no rotation (e) 65 N·m, counterclockwise

10-32 A large steering wheel provides a large moment arm to produce the considerable torque needed to turn the big front wheel assembly.

10-33 33 N·m

10-34 93 rad/s^2

10-35 0.725 kg·m^2

10-36 4.6 N

10-37 3.87 rad/s^2

10-38 0.13 kg·m^2/s

10-39 0.169 kg·m^2/s

10-40 (a) increase (b) 1.8

10-41 (a) 41 J (b) 4.5 rad/s (c) 19 J

Chapter Review

Multiple-Choice Questions

10-42 (c)

10-43 (b)

10-44 (e)

10-45 (b)

10-46 (c)

10-47 (a)

Review Questions and Problems

10-48 (a) 1.50 rad (b) 85.7°, 0.238 rev

10-49 (a) 29 rpm (b) 3.0 rad/s

10-50 2.1 s

10-51 (a) 1.3×10^2 rad/s (b) 1.3×10^3 rpm

10-52 (a) 8.8×10^3 rev/min^2 (b) 1.80×10^2 rev (c) 2.84×10^3 rpm

10-53 (a) 2.51 s (b) -1.50×10^2 rad/s^2

10-54 (a) 5.3 rad/s, 14.2 rad/s (b) decreased by factor of 2.7

10-55

v
slope = ω
0
0
r

10-56

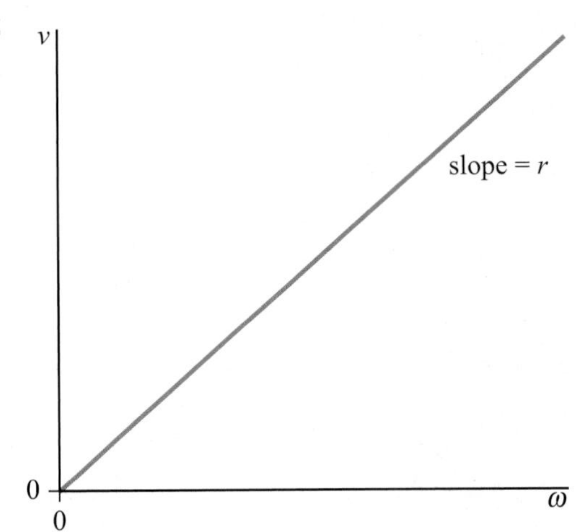

10-57 6.1 rpm

10-58 (a) 2.5 m/s^2 (b) 2.3 m/s^2 (c) 3.4 m/s^2

10-59 (a) 76 rad/s (b) −7.6 rad/s^2

10-60 29 kg·m^2

10-61 0.044 kg

10-62 0.14 kg·m^2

10-63 halfway between top and bottom positions; at top and bottom positions

10-64 (a) $|\tau_{rope}| = 2.3 \times 10^2$ N·m; $|\tau_{gravity}| = 2.7 \times 10^2$ N·m (b) clockwise

10-65 0.35 N·m

10-66 (a) 96 kg·m^2/s (b) 5.0 rad/s (c) 7.5×10^2 J, 2.4×10^2 J (d) Some kinetic energy is absorbed as thermal energy in muscles. No.

Applying Your Knowledge

10-67 419 rad/s, 503 rad/s

10-68 (a) 3.0×10^1 rev/min (b) -2.4×10^2 rev/min^2

10-69 203 rev

10-70 (a) 1.59 s (b) 157 rad/s (c) for A 26.6 rev, for B 19.8 rev

10-71 (a) 31 rad/s (b) 3.6×10^2 rad/s^2

10-72 (a) 5.0 cm (b) 5.6 rad/s (c) 5.0 s

10-73 (a) $2\omega R$ (b) ωR (c) 0 (d) $\sqrt{2}\,\omega R$

10-74 52.3 J

10-75 approx. 0.03 to 0.1 kg·m^2

10-76 4.4 rad/s

10-77 21 rad/s

10-78 (a) -103 rad/s^2 (b) 25.5 s (c) 5.31×10^3 rev

10-79 1.9×10^2 N·m

10-80 (a) counterclockwise (b) 1.2 rad/s^2

10-81 (a) $\dfrac{3g}{2\sqrt{2}L}, \dfrac{3g}{4\sqrt{2}}$ (b) No, because the moment arm changes.

10-82 Moment of inertia depends on the *square* of the distance from the mass to the rotation axis. Although the arms have a small mass, increasing their distance from the rotation axis has a large effect.

10-83 3.6 rad/s^2

10-84 (a) 1.9×10^2 rad/s (b) 1.1×10^{38} kg·m^2 (c) 1.9×10^6 m/s (d) 3.6×10^8 m/s^2

10-85 64 J

10-86 approx. 3 to 4

Chapter 11 Statics, Stability, and Elasticity

Exercises

11-1 (a) $x_{CM} = 2.2$ cm, $y_{CM} = -4.1$ cm (b) 4.6 cm

11-2 $x_{CM} = 0.40$ m, $y_{CM} = 0.047$ m

11-3 (a) $x_{CM,table} = 0.420$ m, $y_{CM,table} = 0.340$ m (b) $x_{CM,box} = 0.27$ m, $y_{CM,box} = 0.38$ m

11-4 (a) and (b) The CM of object A follows a parabolic projectile-motion trajectory.

11-5 In the high jump, kinetic energy is converted to gravitational potential energy, and the height that the CM reaches above the ground is determined by the gravitational potential energy ($E_P = mgy_{CM}$). If the jumper's CM stays above the bar during the jump, as with pre-Fosbury methods, then the bar height reached will not be as high as the jumper's CM. If the jumper can go over the bar with the CM actually below the bar, then the bar height reached will be higher than the CM height.

11-6 No; if only one force acts on an object, the object will undergo an acceleration in the direction of the force.

11-7 A person exerts a force on a box initially at rest and causes the CM of the box to accelerate, but the force is directed toward the CM and does not provide a torque to rotate the box.

11-8 When an airplane propeller starts to rotate from rest, the net torque is not zero, but the net force is zero (since the CM of the propeller does not move).

11-9 2.1 m from pivot

11-10 approx. (1 to 2) $\times 10^2$ N·m

11-11 $F_1 = 31$ N, $F_2 = 58$ N

11-12 (a) 1.2×10^2 N (b) 1.1×10^2 N toward right and downward, at 23° below horizontal (c) 1.1×10^2 N toward left and upward, at 23° above horizontal

11-13 (a) 41 N horizontally away from wall (b) 1.5×10^2 N upward and toward wall, at 75° above horizontal

11-14 (a) 2.3×10^2 N horizontally away from wall (b) 7.3×10^2 N upward and toward wall, at 72° above horizontal

11-15 0.33

11-16 1.0×10^2 N

11-17 to ensure that the CM of the crane plus its load is always above the base of the crane

11-18 D, C, B, A

11-19 (a) lean forward (b) lean backward (c) lean to right

11-20 drooping, since it lowers the CM of the walker-pole system, thus increasing stability

11-21 1.3×10^2 N

11-22 (a) pascal (or newton per square metre) (b) pascal (c) unitless (d) pascal (e) unitless

11-23 5.8×10^6 Pa

11-24 0.022 m

11-25 1.8 kg

11-26 (a) 0.0048 = 0.48% (b) 4.4×10^5 Pa (c) 9.1×10^7 Pa

11-27 Large blood vessel walls consist of elastin and collagen. For small pressure increases, it is mainly elastin with its small Young's modulus that contributes to the stress–strain curve, and the strain increases substantially. The collagen with its large Young's modulus is slack. For larger stresses, the collagen tightens up and is the main contributor to the shape of the stress–strain curve, and the rate of increase of strain is lower.

11-28 (a) 5×10^3 N (b) 0.02 cm

11-29 (a) pascal (or newton per square metre) (b) pascal (c) unitless (d) pascal (e) pascal (f) pascal (g) unitless

11-30 (a) #2 (b) #1 (c) #3

11-31 4×10^2 N

11-32 2.7×10^{-4}

11-33 3.0×10^{10} Pa

11-34 0.59°

11-35 9.9×10^2 N·m

11-36 -1.2 cm^3

11-37 2.1×10^9 Pa

11-38 $B = 1.4 \times 10^{11}$ Pa; $Y = 1.1 \times 10^{11}$ Pa; $S = 4.5 \times 10^{10}$ Pa; copper

11-39 A horizontal slab (of any material) supported only at its ends sags under its own weight and becomes compressed along the top, but experiences tension on the bottom. If a material is weak under tension, the tension along the bottom can result in cracking or breaking of the slab.

Chapter Review

Multiple-Choice Questions

11-40 (b)

11-41 (c)

11-42 (a)

11-43 (d)

11-44 (c)

11-45 (e)

11-46 (a)

11-47 (b)

11-48 (c)

11-49 (a)

11-50 (d)

11-51 (e)

Review Questions and Problems

11-52 (a) 4.66×10^3 km (b) 1.71×10^3 km below Earth's surface

11-53 $x_{CM} = 0.964$ m; $y_{CM} = 0.0790$ m

11-54 48°

11-55 1.1×10^2 N

11-56 (a) 27 cm (b) 1.27 N

11-57 $\vec{F}_A = 3.5 \times 10^3$ N down; $\vec{F}_B = 4.3 \times 10^3$ N up

11-58 1.01 m

11-59 $\vec{F}_{upper} = 49$ N to left; $\vec{F}_{lower} = 69$ N up and to right, at 45° to horizontal

11-60 $F_B = 1.64 \times 10^3$ N; $\vec{F}_H = 1.54 \times 10^3$ N down and to the right at an angle of 63.1° below the horizontal

11-61 The support base enclosed by four feet is much larger than the base enclosed by two feet.

11-62 (a) 7.8×10^{-3} m (b) 1.1 m

11-63 (a) 1.0×10^7 Pa, 4.1% (b) 1.8×10^8 Pa, 72%

11-64 22 cm^2

11-65 (a) 1.6×10^{-4} (b) 5.2×10^{-5} m

11-66 0.12 m

11-67 1.1

11-68 −0.067%

Applying Your Knowledge

11-69 $x_{CM} = -0.062$ m; $y_{CM} = 0.044$ m

11-70 From the bottom. For each bottom pair (or whatever number) of hanging objects, the CM of the pair can be determined easily, and the single string supporting the pair of objects is tied above that CM. As the mobile is then constructed upward, each pair of objects can be treated as a single hanging object, and the procedure of determining the position of various vertical strings can be done fairly easily. Starting from the top, a great deal of trial and error is needed.

11-71 (a) 8.0×10^2 kg (b) 3.2×10^4 N upward and to right, at 51° above horizontal

11-72 (a) 7.4×10^2 N (b) 4.2×10^2 N downward and to right, at 51° below horizontal

11-73 1.1×10^4 N

11-74 (a) 3.0×10^3 N (b) 2.7×10^3 N, up and to right, at 42° above horizontal

11-75 163 N

11-76 $\theta_{min} = \tan^{-1}\left(\frac{1}{2\mu_S}\right)$

11-77 (a) 1.1×10^4 N (b) 19

11-78 341 N, up and to right, at 32.2° above horizontal

11-79 (a) 2.4×10^4 Pa (b) 2.7×10^{10} Pa

11-80 (a) 7.9×10^2 N (b) 7.3×10^2 N, down and to left, at 85° below horizontal

11-81 approx. 5×10^3 N

11-82 (a) rear: 5.9×10^2 N; front: 2.9×10^2 N (b) 18°

11-83 27 N

11-84 copper 3.6 mm, aluminum 11 mm

11-85 2.4 mm

11-86 (a) 2.2 cm (b) 1.2

11-87 $\frac{5}{6}\ell$ from wire A

11-88 $\frac{9}{14}\ell$ from wire A

11-89 (a) 7.7 m/s (b) 2.6×10^2 m/s^2 (c) 9.6×10^3 N (d) 3.3×10^7 Pa (e) 22%

Chapter 12 Fluid Statics and Dynamics

Exercises

12-1 (a) 7.4×10^3 kg/m^3 (b) 1.26×10^3 kg/m^3

12-2 23.5 kg

12-3 Use a bathroom scale to measure mass, and immerse yourself "completely" in water in the bathtub for an estimation of volume; expect a value slightly less than 1.00×10^3 kg/m^3.

12-4 (a) 11.3 (b) 0.917

12-5 (a) The same force will be applied to a smaller area with small gloves, and therefore the pressure will be greater.

(b) The same force will be applied to a smaller contact area between the two walnuts, and therefore the pressure will be greater.

12-6 6.9×10^3 kPa

12-7 For a 70 kg person: (a) approx. 2×10^1 kPa (b) approx. 1×10^2 kPa (c) approx. 1×10^2 kPa

12-8 (a) 6.2×10^6 N (b) 121 kPa

12-9 514 kPa

12-10 87 kPa, 188 kPa

12-11 459 mm Hg

12-12 9.3×10^2 kPa

12-13 148 kPa

12-14 43 mm

12-15 approx. 1 m

12-16 75 kPa

12-17 2.4 km

12-18 When using a straw, a person reduces the pressure in the lungs, mouth, etc., and the atmospheric pressure pushes the water up the straw into the mouth.

12-19 (a) 101 kPa (b) 10.3 m, which is an unreasonably large height

12-20 102 kPa

12-21 0.097 N/m

12-22 The paper clip on water would be more likely to remain on the surface because the surface tension of water is much greater than that of ethyl alcohol.

12-23 123°

12-24 Since the surface tension of lung surfactant is extremely small, any additives will cause the surface tension to increase, thereby making breathing more difficult.

12-25 11%

12-26 (a) 0.659 (b) 188 N (c) 0 N

12-27 (a) 6.4×10^4 N (b) 6.5×10^3 kg

12-28 approx. 1 N

12-29 61 kg

12-30 2.4

12-31 (a) 0.075 m^3/s (b) 7.9 m/s

12-32 0.58

12-33 (a) The spoiler results in a downward force on the rear of the car. The shape of the spoiler deflects air upward, and by Newton's third law of motion, the air exerts a downward force on the car.

(b) This downward force results in a larger normal force exerted on the rear tires, and therefore a larger friction force, which means more traction.

12-34 (a) Between the buildings there is a small cross-sectional area in which the wind must move, and according to the equation of continuity the wind speeds up.

(b) From Bernoulli's principle, the barometric pressure in the high-speed wind will be smaller than the pressure in the lower-speed wind in the countryside.

(c) Because the high-speed wind has low pressure, and the interior of a building has normal (higher) pressure, the pressure difference across the window results in a net force outward that can blow a window out of its frame.

12-35 (a) The water in the U-shaped trap prevents sewer gases from coming up into the sink and the building.

(b) Plumbing design (b) is better. In design (a), if sewage were flowing at high speed and low pressure, the higher atmospheric pressure in the sink could force the water in the trap into the sewer, thereby letting sewer gases enter the building. In design (b), the vent to the atmosphere ensures that the sewage is at atmospheric pressure.

12-36 7.8 m/s

12-37 8.4 m/s

12-38 0.040 N

12-39 8.6×10^2 Pa/m

12-40 Because of the viscosity of the gas, the pressure falls as the gas flows along the pipeline (according to Poiseuille's law), and therefore the pressure needs to be increased periodically by compressors.

12-41 (a) Because of lower air density, the aerodynamic lift is reduced. (b) The lower air density gives a smaller drag force.

12-42 (a) 4.8 m/s (b) 0.020 m/s

12-43 8.4 mm

12-44 (a) 3.6×10^{-6} m/s (b) 1.7×10^{-13} N

12-45 33 m/s

12-46 1.1×10^4

Chapter Review

Multiple-Choice Questions

12-47 (d)

12-48 (e)

12-49 (c)

12-50 (a)

12-51 (b)

12-52 (a)

12-53 (c)

12-54 (a)

12-55 (c)

12-56 (b)

12-57 (d)

12-58 (a)

12-59 (d)

12-60 (a)

12-61 (e)

12-62 (c)

Review Questions and Problems

12-63 0.11 m^2

12-64 18.1 kPa

12-65 10.3 m

12-66 8.8 atm

12-67 1.3×10^3 kg/m^3

12-68 (a) 95 mm Hg (b) 13 kPa

12-69 2.2×10^2 kg

12-70 (a) 0.43 kg/m^3 (b) The air pressure is so low that it is easier for water molecules to escape from the liquid water with less thermal energy than at lower elevations (and higher air pressure).

12-71

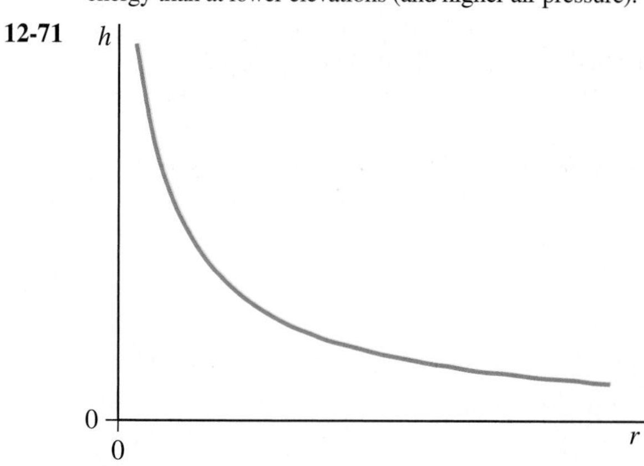

12-72 0.052 mm

12-73 (a) 19.0 N (b) 1.9 m/s^2 upward

12-74 2.9×10^3 kg

12-75 65 N

12-76 3.9×10^2 m^3

12-77 1.9×10^{-3} m/s

12-78 (a) 1.2 m/s (b) 24 mm^2

12-79 (a) 2.6 m/s (b) 1.2×10^4 Pa

12-80 (a) 2.4×10^{-4} m^3/s (b) 1.2 m/s (c) 9.6×10^4 Pa

12-81 2.0 mm

12-82 8

12-83 3.5×10^2, no

12-84 (a) 3.1×10^{-6} m/s (b) 54 min

Applying Your Knowledge

12-85 approx. (1 to 3) $\times 10^5$ Pa

12-86 12 kPa, 113 kPa

12-87 0.64 P_0

12-88 9.77 N/kg

12-89 6.5×10^{-3} N/m

12-90 0.029 g

12-91 $\rho = \rho_f \dfrac{W_a}{W_a - W_f}$. See Solutions Manual for proof.

12-92 (a) 9.2% (b) 1.5×10^2 m

12-93 (a) 0.059 m (b) 0.81 m/s^2

12-94 (a) 1.00×10^{-2} kg (b) 781 kg/m^3

12-95 29%

12-96 (a) 0.064 kg (b) 0.159 kg

12-97 1.6×10^4 N

12-98 1.4%

12-99 (a) 1.3×10^3 Pa (b) $v_1 = 3.5$ m/s, $v_2 = 45$ m/s

12-100 7.4 N

12-101 15.0 cm

12-102 0.76 m

12-103 3.4 cm^3/min

12-104 2.5×10^6 g/mol

Chapter 13 Oscillations and Waves

Exercises

13-1 (a) transverse (b) longitudinal (c) longitudinal

13-2 (a) 4.0 s (b) 0.29 s (c) 4.0×10^{-2} s to 5.0×10^{-5} s

13-3 (a) 20.1 Hz (b) 12 Hz (c) 1.5 Hz

13-4 (a) 1.7×10^2 cm (b) ½ cycle difference (opposite phase) (c) 0.27 s

13-5 Both sides of the equation have the dimension of time.

13-6 Let y be the displacement from the equilibrium position: for (a), (c), and (d), at $y = +A$ and $-A$; and for (b) and (e), at $y = 0$.

13-7 25 N/m

13-8 2.6 kg

13-9 Lengthen the pendulum.

13-10 (a) 96 cm (b) 2.5 s, 0.41 Hz

13-11 2.1 m

13-12 3.71 m/s^2

13-13 −2.7 cm

13-14 (a) 12.7 cm (b) -1.00×10^2 cm/s (c) -7.85×10^2 cm/s^2

13-15 The graph, shown below, is sinusoidal starting and ending at $x = +18.0$ cm when $t = 0.000$ s and 0.800 s respectively.

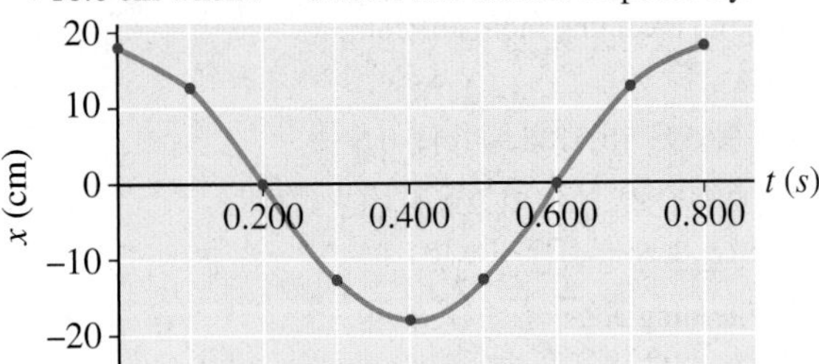

13-16 2.50 cm

13-17 ±4.5 cm

13-18 The slopes of the curve on the velocity graph at several times equal the accelerations at those times.

13-19 (a) 12 cm, 38 cm (b) 25 cm (c) 13 cm

13-20 The dimensions of the expression are L/T.

13-21 (a) 91 N/m (b) 0.40 J

13-22 (a) 0.215 m (b) 9.39 m/s (c) 6.51 m/s

13-23 (a) 0.58 J, 4.5 m/s (b) 0.25 m (c) 6.3 m/s

13-24 2×10^1 N/m

13-25 (a) A, B, and C, respectively (b) 0.100 m (c) 1.0×10^3 N/m (d) 9.1 m/s

13-26 (a) $v = 2\pi f \sqrt{(A^2 - x^2)}$ (b) 1.5 m/s

13-27 (a) medium (b) slow (c) fast (d) medium (e) fast

13-28 The resonant frequency increases as the length decreases. When walking, the frequency of your pace is low and your extended arms tend to swing at or near that low frequency. When running, your pace frequency increases, making it more natural to oscillate your arms at a higher frequency. This occurs with your arms bent at the elbow, which is equivalent to a shorter length of pendulum.

13-29 (a) Jump up and down in phase with the system to maximize the amplitude of oscillation.

(b) Jump up and down out of phase with the trampoline to minimize the amplitude of oscillation.

13-30 (a) Particles in longitudinal waves move parallel to the direction of propagation, and particles in transverse waves move perpendicular to the direction of propagation.

(b) Longitudinal waves can travel through solids, liquids, and gases, and transverse waves can travel most easily in solids.

(c) Longitudinal waves consist of compressions and rarefactions, and transverse waves consist of crests and troughs.

13-31 A tsunami is caused by an underwater earthquake or a volcanic eruption, not by tidal forces.

13-32 27 min

13-33 (a) The particles oscillate perpendicular to the direction of the wave in symmetry around the equilibrium position.

(b) The particles oscillate back and forth parallel to the direction of the wave in symmetry around the equilibrium position.

13-34 In both cases, the energy is transmitted in the same direction the wave travels.

13-35 B down, C up, D down, E up

13-36 8.80×10^2 Hz

13-37 (a) 11 cm (b) 44 cm (c) 3.0 dm (d) 2.0×10^2 cm/s (e) 14 dm/s

13-38 (a) 0.30 Hz (b) 18 m

13-39 (a) 1.5×10^3 m/s (b) 73 m

13-40 1.11×10^{-5} s

13-41 (a) 0.800 MHz (b) 1.91×10^{-3} m

13-42 (a) negative (b) positive

13-43 The leading edge of the transmitted pulse is 2.0 cm beyond the interface; it is the same shape and phase as the incident pulse, but only half as long and smaller in amplitude. The leading edge of the reflected pulse, which is travelling to the left from the interface, is negative (i.e., on the opposite side of the axis of equilibrium) and it is 4.0 cm from the interface, with the same wavelength and smaller amplitude because some energy went into the transmitted pulse. Refer to the diagram.

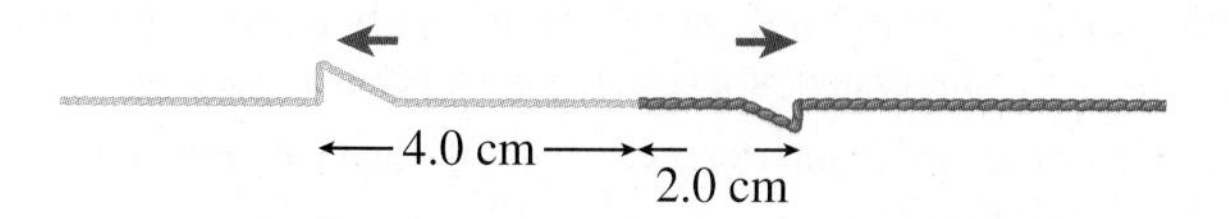

13-44 The transmitted pulse is positive (i.e., on the same side of the axis of equilibrium), twice as long in wavelength, and a little smaller in amplitude. Its leading edge is 4.0 cm beyond the interface. The reflected pulse is positive with the leading edge 2.0 cm from the interface, and is the same wavelength as the incident pulse, but with smaller amplitude. Refer to the diagram.

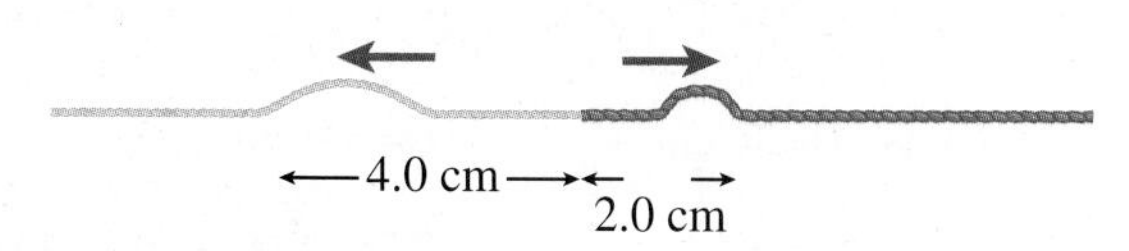

13-45 (a) constructive (b) destructive (c) constructive (d) destructive

13-46 The two waves must have the same size (i.e., wavelength and amplitude) and as they approach each other, their corresponding components must be opposite in phase.

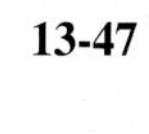

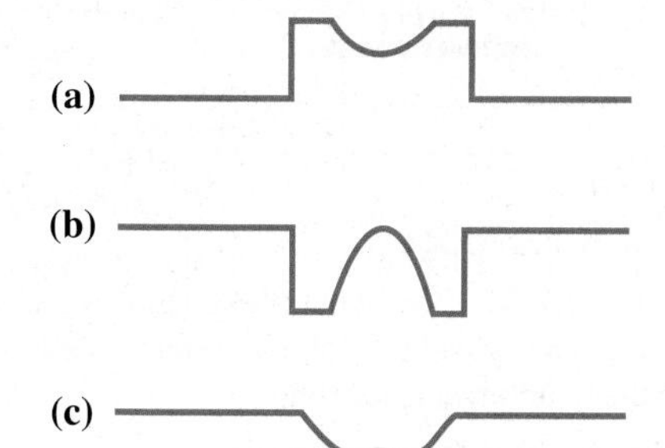

13-48 No, a standing wave can be set up only at specific frequencies such that there are n antinodes between the ends of the rope, where $n = 1, 2, 3, \ldots$

13-49 The wavelength is 3.0 m.

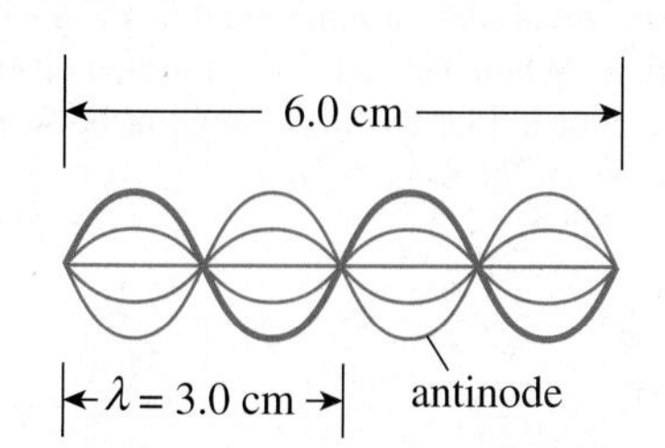

13-50 (a) 0.61 Hz (b) 1.2 Hz (c) 1.8 Hz

13-51 (a) 7.6 m, 9.1 m/s (b) 4.8 Hz, 6.0 Hz

13-52 (a) 32 cm (b) 96 cm (c) 1.6 m/s (d) 5.0 Hz (e) 6th

13-53 The size of the shadow decreases.

13-54 Maximum diffraction occurs when $\lambda \geq w$.

13-55 Maximum diffraction occurs when $\lambda \geq w$.

13-56 Light waves have a very small wavelength, so diffraction would be observed only under very special circumstances.

Chapter Review

Multiple-Choice Questions

13-57 (c)

13-58 (d)

13-59 (a)

13-60 (d)

13-61 (c)

13-62 (e)

Review Questions and Problems

13-63 (a) $T = \frac{1}{f}$

(b) $a \propto x$

(c) $T \propto 1/\sqrt{k}$

(d) $f \propto 1/\sqrt{m}$

(e) $T \propto \sqrt{L}$

(f) $f \propto \sqrt{g}$

(g) hertz = cycles/s

(h) $v_{max} \propto A$

(i) $E_{S,\,max} = E_{K,max}$

(j) $t_{damping}$ decreases as friction increases.

(k) $\lambda \propto \frac{1}{f}$

(l) The amount of diffraction is small if $\lambda \ll w$ and maximum if $\lambda/w \geq 1$.

13-64 1.1×10^{-2} s, 9.0×10^1 Hz

13-65 H stands for harmonic, which means repeated at regular intervals of time.

13-66 6.9 Hz

13-67 (a) 9.8×10^3 N/m (b) The period would increase because $T \propto \sqrt{m}$.

13-68 The frequency increases because the pendulum system is shorter.

13-69 The clock will lose time, so the pendulum should be shortened because $T \propto \frac{1}{\sqrt{g}}$.

13-70 (a) $4A$ (b) $x = 0, x = \pm A$ (c) $x = \pm A, x = 0$

13-71 8.5 s, 0.12 Hz

13-72 9.822 m/s^2

13-73 4 times

13-74 0.220 Hz

13-75 (a) 0.25 mm (b) -2.1×10^3 mm/s (c) -1.8×10^6 mm/s^2

13-76 (a) 0.449 m (b) 4.04 m/s

13-77 (a) 5.3×10^{-2} J (b) 0.50 m/s (c) 0.33 m/s (d) 5.3×10^{-2} J

13-78 0.21 kg

13-79 Answers will vary because of the large number of possibilities.

(a) Useful damping occurs in springs and shock absorbers for all types of road transportation vehicles, for weigh scales, and for our eardrums.

(b) Damping that is not useful occurs in devices meant to have large oscillations while exercising (e.g., on a trampoline) or having fun at a playground (e.g., on a swing). It also occurs when friction must be overcome when operating a pendulum clock or similar device.

13-80 Any isolated system undergoing SHM will experience damping and then will eventually come to rest. To keep the motion going, the system would require a periodic input of energy, so it would no longer be isolated.

13-81

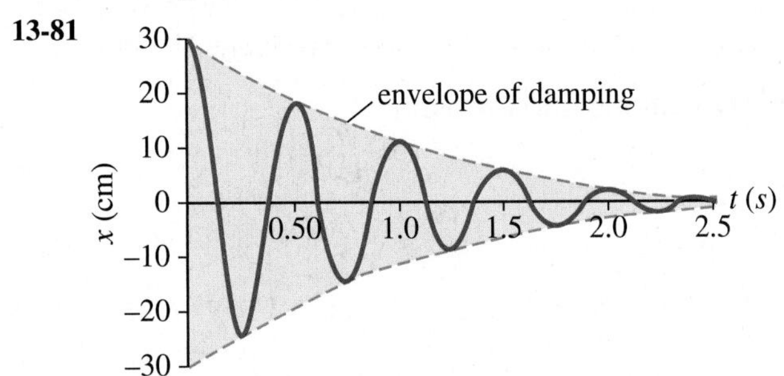

13-82 If the car is pushed back and forth at the resonant frequency of the system, the amplitude of motion can be built up enough to (possibly) free the car.

13-83 ½ f, f, ¼ f (because any frequency above the resonant frequency would involve a forced oscillation, resulting in a lower amplitude)

13-84 Stirring at the resonant frequency of the eggs in the bowl would cause a build-up of amplitude, possibly spilling the contents over the edge, so stirring at a different frequency is advised.

13-85 The main function of a wave is to transfer energy.

13-86 (a) The energy would travel faster when the rod is struck with a blow parallel to the rod, causing a longitudinal wave. A transverse wave, caused when the blow is perpendicular to the rod, travels more slowly than the longitudinal wave.

(b) P waves are longitudinal and travel more quickly than S waves, which are transverse.

13-87 7×10^2 s later

13-88 0.0665 m

13-89 108 MHz to 87.5 MHz

13-90 566 m to 186 m

13-91

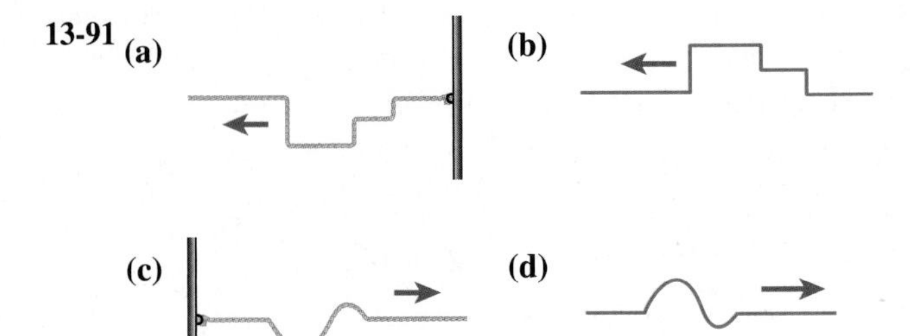

13-92 (a) 2λ (b) positive (c) As the wave goes from X (a denser material) to Y, the reflected wave is positive because the reflection is an open-end reflection. The wavelength of the reflected wave in X is the same as that of the incident wave in X, but the amplitude is smaller because some of the energy is transferred to Y.

13-93 (a) Constructive interference occurs when a compression meets a compression or a rarefaction meets a rarefaction.

(b) Destructive interference occurs when a compression meets a rarefaction.

13-94 (a) 10 mm, 2.0 mm (b) 10 mm, 16.0 mm

13-95 (a) 18 cm (b) 69 Hz

(c)

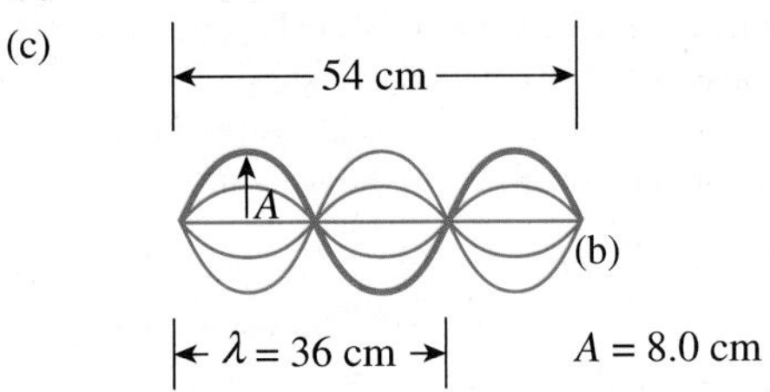

13-96 220 Hz, 880 Hz

13-97 (a) (b)

Applying Your Knowledge

13-98 (a) 13 Hz, 6.7 Hz (b) about 3 Hz and 1 to 1.5 Hz (c) The human circulatory and respiratory systems would not be able to handle the requirements for flying.

13-99 (a) 3.7 cm (b) 2.6 Hz (c) 9.6 m

13-100 (a) 5.7×10^2 N/m (b) 63 kg

13-101 (a) When viewed from a direction perpendicular to the plane of Callisto's orbit, the motion would look like the lower part of Figure 13-13, with Jupiter at the centre of the circle and Callisto revolving around Jupiter in a counterclockwise direction. When viewed from Earth, however, the motion would appear like the mass moving back and forth in the upper part of the same figure.

(b)

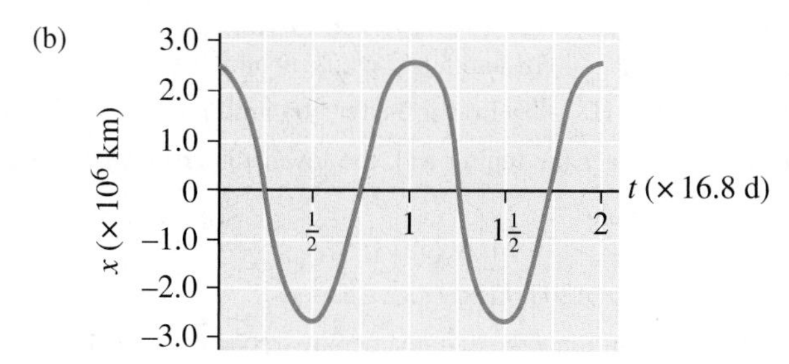

13-102 5.1 m/s, 8.3×10^3 m/s^2

13-103 All values are approximate: (a) about 1 m (b) 7×10^2 N/m (c) 0.5 Hz (d) 3 m/s, 1×10^1 m/s^2.

13-104 Answers will vary. Some possible features to help reduce oscillations are

- shock absorbers beneath the seat
- springs beneath the seat and smaller sets of springs within the seat cushions
- strong back and side supports for travelling over rough terrain
- adjustable seat height to improve comfort

13-105 Had the troops "broken step" so they all would be taking steps at random frequencies, the amplitude of oscillation of the bridge would not have been able to build up as it had with resonance.

13-106 (a) The height of the tides, in other words, the vertical amplitude of the water, can be maximized when resonance occurs. So when a large amplitude caused by resonance coincides with a large amplitude caused by tidal action, a very high tide results.

(b) A dam across the mouth of the bay would change the effective length of the resonating cavity, thus changing the resonant frequency. This would mean a lower likelihood of a very high tide because the frequency (or period) due to tidal forces would no longer coincide with the frequency of the water in the bay.

13-107 The two acrobats have to be nearly completely out of phase in their oscillations to achieve the positions shown in the photograph. This would be easier to accomplish if the two pendulums have the same length and thus the same resonant frequency. At the beginning of the motion on the pendulums, the catcher would move right as the jumper moves left, both acrobats building up amplitude with each cycle of oscillation. When the amplitude is large enough (as judged by lots of practice), the jumper releases herself from her pendulum just as the catcher is approaching his maximum amplitude at a speed of zero.

13-108 There are many possible answers, so the answers will vary. One choice would be to design various structures for mooring ships and boats in a harbour and testing the structures in the tank. Should the support pillars be rectangular, circular, elliptical, or some other shape? What materials would be best to use to provide the highest strength? Other choices could be to test the design of headwaters, industrial buildings, homes, roads, and other infrastructure, and floating devices for humans and pets.

13-109 (a) A strong wind would be expected to push against the cable, the pendulums, and Wallenda himself, causing an oscillation whose amplitude could build up to a dangerous size. The pendulums attached tightly to the horizontal cable are all different in length, so they have different resonant frequencies. As the wind causes any one particular pendulum to begin swinging, the cable will try to twist (a torsional oscillation) with the same frequency as the pendulum. However, that twisting will be counteracted by the other nearby pendulums which are trying to oscillate at a different frequency. The pendulums are absorbing much of the input energy of the wind.

(b) It would be very difficult for anyone, even someone as experienced and careful as Wallenda, to walk along a cable 5.1 cm in diameter while varying the frequency of his steps. As discussed in the chapter, with a constant stepping frequency on a bridge or a cable, oscillations can build up. These oscillations, combined with the pendulum oscillations, became severe enough twice during Wallenda's crossing that he knelt down with one knee on the cable to reduce the amplitude of the oscillation.

13-110 0.86 m

Chapter 14 Sound and Music

Exercises

14-1 (a) the piano strings (b) the mosquito's wings (c) air molecules

14-2 The air molecules become compressed (increased density) and then rarefied (decreased density) at the frequency of the longitudinal sound wave that passes through the air.

14-3 At lower temperatures the air molecules move more slowly so they collide more slowly as the longitudinal sound wave passes along.

14-4 (a) longitudinal wave (b) 0.783 m

14-5 (a) about 1 km (b) For every 3 s after a bolt of lightning is seen until the sound of the resulting thunder is heard, the lightning occurred approximately 1 km away.

14-6 0.588 s

14-7 15.3 s

14-8 16.2°C

14-9 increase by a factor of 4

14-10 The speed of sound is fastest in solids, where the particles are close together.

14-11 The equation is dimensionally correct, with both sides having dimensions of length/time.

14-12 1.47×10^3 m/s

14-13 3.1×10^3 m/s

14-14 8.3×10^2 kg/m^3

14-15 4.9×10^{10} N/m^2

14-16 7.0×10^{10} N/m^2

14-17 sound energy → mechanical energy → electrical energy

14-18 (a) 10^3 (b) 10^2

14-19 70 dB

14-20 (a) 40 dB (b) 10 dB

14-21 79.8 dB

14-22 88 dB

14-23 (a) 1.0×10^{-2} W/m^2 (b) 3.2×10^{-7} W/m^2

14-24 (a) 6.0 dB (b) 16 dB (c) 26 dB The pattern is, for every 10-fold increase in sound intensity, the intensity level increases by 10 dB.

14-25 93 dB

14-26 55 dB

14-27 9.0×10^1 m

14-28 82.0 ms

14-29 8.0×10^4 Hz

14-30 A transducer converts electrical signals into mechanical oscillations and then into sound oscillations.

14-31 1.5 mm to 0.075 mm

14-32 Ultrasonics is used in diagnostic medicine to test abnormalities of the eye or brain, and in therapeutic medicine to provide pain relief or remove brain tumors.

14-33 Briefly, $f' = \frac{v}{\lambda'} = \frac{v}{(v + v_S)/f} = f\left(\frac{v}{v + v_S}\right)$

14-34 Briefly,

$$f' = f - \frac{v_L}{\lambda}$$

$$= f - \frac{v_L}{v/f} \text{ (because } \lambda = v/f\text{)}$$

$$= f - \frac{fv_L}{v}$$

$$= f\left(1 - \frac{v_L}{v}\right)$$

$$f' = f\left(\frac{v - v_L}{v}\right)$$

14-35 (a) 996 Hz (b) 858 Hz

14-36 21.9 m/s

14-37 5.029×10^4 Hz

14-38 943 Hz, 722 Hz

14-39 (a) 9.2×10^3 m/s (b) 9.3×10^2 m/s (c) 28 m/s

14-40 324 m/s

14-41 The sound barrier is a region of high air pressure as the aircraft pushes the sound compressions closer and closer together. Breaking through this high-pressure region causes oscillations that regular aircraft are not built to handle.

14-42 No; at subsonic speeds the aircraft does not catch up to its own sound waves, so does not produce a shock cone.

14-43 $\theta = \sin^{-1}\frac{v}{v_S}$, which becomes smaller as v_S increases. The sharpness increases.

14-44 3.20

14-45 Wherever there are people, animals, and buildings, sonic booms from supersonic aircraft can cause both physical and psychological damage. Over an ocean, sonic booms are much less likely to cause problems, especially if the aircraft flies at high altitudes.

14-46 Sound displays several properties of waves, including diffraction around corners or through openings, interference (e.g., in the production of beats), resonance (e.g., in musical instruments), and the transmission of energy (e.g., with sympathetic oscillations).

14-47 (a) 1.32 m, 0.331 m (b) about 1.3 m and 0.33 m

14-48 Sounds from low-range speakers have long wavelengths that diffract more than the waves from high-range speakers. Thus, the high-range speakers are more directional.

14-49 1 Hz, 4 Hz, 5 Hz, 14 Hz, 18 Hz, 19 Hz. The lower beat frequencies (1 Hz, 4 Hz, and 5 Hz) would sound relatively pleasant.

14-50 (a) 444 Hz (b) Lower the tension until no beats are heard.

14-51 The longer the tuning fork the lower the resonant frequency of the fork.

14-52 784.0 Hz, 1176.0 Hz, 2352.0 Hz

14-53 784.8 Hz, 1308.0 Hz, 1831.2 Hz

14-54 The addition of the fundamental frequency and the third harmonic yields a more complex and pleasing sound than the fundamental frequency by itself.

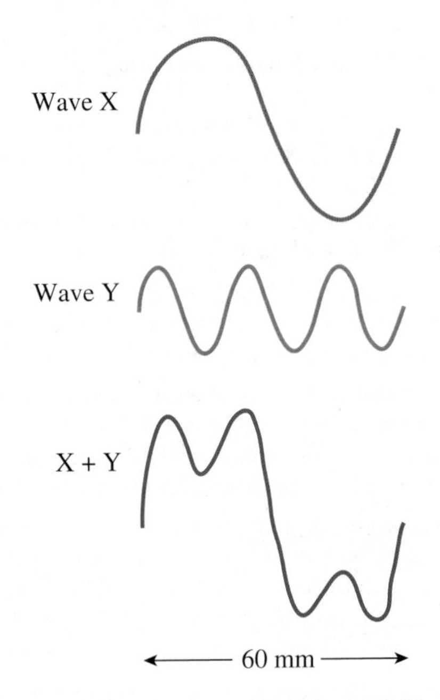

14-55 (a) 196.0 Hz, 784.0 Hz (b) 643.0 Hz, 2572 Hz

14-56 (a) 1152.0 Hz (b) 1365.2 Hz

14-57 (a) 3/2 (high) (b) 3/1 (high) (c) 2/1 (high) (d) 41/40 (low)

14-58 1396.9 Hz, 164.8 Hz

14-59 1046.51 Hz, 932.33 Hz

14-60 600.00 Hz, 673.48 Hz, 755.95 Hz, 848.53 Hz, 952.44 Hz, 1069.08 Hz, 1200.00 Hz, 1346.95 Hz, 1511.91 Hz, 1697.06 Hz, 1904.88 Hz, 2138.16 Hz, 2400.00 Hz

14-61 The 130 g/m string has a lower frequency.

14-62 Harmonics up to the 4th one can be heard by an average, young, healthy human ear.

14-63 200 Hz, 400 Hz

14-64 (a) 4.0×10^2 m/s (b) 66 N

14-65 1.4×10^2 m/s

14-66 8.9×10^2 Hz

14-67 0.98 m

14-68 The speed in X is twice as fast as in Y.

14-69 (a) transverse standing wave, ½ λ between nodes

(b) longitudinal standing wave, ½ λ between nodes

14-70 (a) 382 Hz (b) The speed and frequency become higher. (c) 392 Hz, flat

14-71 (a) 0.335 m (b) 0.168 m

14-72 65.3 Hz, 196 Hz, 327 Hz

14-73 Sawtooth and square waves are shown in Figure 14-56 in the text. See diagrams below.

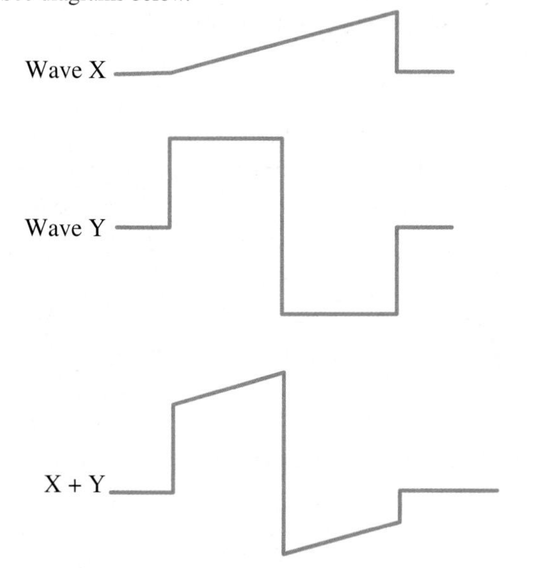

14-74 Answers will vary and should include a discussion of features that reflect or redirect sound, as well as those that absorb some sound.

14-75 25 dB

14-76 The coats will tend to absorb sounds quite well, so the reverberation time will be reduced.

14-77 The chamber walls have a design that would reduce the reflection of sound waves that normally occurs off smooth, flat surfaces. Also, likely the wall material is textured in order to absorb sound waves.

14-78 17 m

Chapter Review

Multiple-Choice Questions

14-79 (d)

14-80 (b)

14-81 (c)

14-82 (b)

14-83 (a)

14-84 (b)

14-85 (b)

14-86 (c)

Review Questions and Problems

14-87 1.44×10^3 m

14-88 1.11×10^3 s

14-89 The sound in the pipe, by 0.308 s

14-90 (a) $I \propto 1/r^2$ (b) $\beta = 10 \log \frac{I}{I_0}$

14-91 (a) $\frac{1}{4} I_1$ (b) $1/16\, I_1$ (c) $4I_1$

14-92 5.01×10^{-2} W/m^2

14-93 93 dB

14-94 (a) 98.8 dB (b) 114 dB

14-95 (a) 1.66×10^3 m (b) 2.26×10^9 N/m^2

14-96 (a) 1.25×10^3 Hz (b) 1.07×10^3 Hz (c) 1.28×10^3 Hz

14-97 The airplane (or other source of sound) could move from a low altitude to a high altitude.

14-98 (a) 1.98×10^3 km/h (b) 67.5°

14-99 3.55×10^3 km/h south

14-100 (a) A major factor that determines the amount of diffraction is the wavelength of the sound waves relative to the size of the opening or barrier. The maximum diffraction occurs when the wavelength is approximately equal to the size of the opening or barrier.

(b) about 2 kHz

14-101 436 Hz

14-102 880 Hz, 1760 Hz, 3520 Hz

14-103 (a) The sound is a rather boring, pure sound.

(b) The sound is a pure sound about the same frequency as the sound in (a), somewhat more harsh and quite boring.

(c) This is a higher quality sound because it has a fundamental frequency combined with overtones.

14-104 (a) 1318.4 Hz (or 1318.5 Hz) (b) 164.8 Hz

14-105 Starting with the simplest ratio, the order is 2:1, 3:2, 5:4, 17:16, and 97:96.

14-106 392 Hz, 784 Hz, 1176 Hz

14-107 1.4×10^2 N

14-108 (a) 0.220 m, 0.439 m (b) 0.110 m, 0.329 m

14-109 The auditorium's reverberation time gives a good indication of the main use the room is designed to accommodate. For example, a reverberation time of about 3 s is good for orchestral performances. To increase the reverberation time, more reflection (and/or less absorption) of sound would be needed. This can be accomplished by adding reflecting baffles and/or placing pieces of more reflecting materials at several appropriate places in the room. To decrease the reverberation time, more absorption of sound would be needed. This could be accomplished by adding sound absorbing materials spaced appropriately on the room surfaces.

14-110 The reading must drop by 60 dB, so it is 96 dB − 60 dB = 36 dB. A reverberation time of 0.95 s is rather short, so the acoustics of the hall would be more appropriate for speech or theatrical performances than for concerts.

Applying Your Knowledge

14-111 (a) 0.88 m (b) 353 m/s, 4.0×10^2 Hz (c) 0.88 m (d) 1.1×10^3 Hz

14-112 (a) 25 mm or 0.025 m (b) 3.5×10^3 Hz

(c) The lowest point in the graph in Figure 14-13 occurs at a frequency of about 3500 Hz. This means that the length of the auditory canal helps determine the ear's sensitivity.

14-113 The calculated tension is 1.28×10^3 N, and the meter reading is 1.20×10^3 N, so the meter reading has a fairly high error (approx. 6%).

14-114 (a) 6.3×10^5 times (b) 2.5 times

14-115 4.0×10^1 MHz

14-116 The tip's speed would have to be greater than the speed of sound in air, i.e., about 350 m/s or faster to create a sonic boom.

14-117 Answers will vary.

(a) The hall in the photograph has surfaces on the ceiling and walls that are not smooth and flat, so sounds will be unlikely to reflect as much as they would off flat, harsh surfaces. Even the seats have a soft, cushiony appearance designed to absorb sound, thus helping to improve acoustics.

(b) Designs should enhance sound absorption and reduce sound reflection.

14-118 2.6×10^2 m

Chapter 15 Reflection, Refraction, and Dispersion of Light

Exercises

15-1 With air in the jar, the ringing bell can be heard as the gong strikes the metal plate. When the air is removed from the jar, the oscillating gong remains visible, but the sound it produces can no longer be heard.

15-2 **(a)** Lunar eclipse.

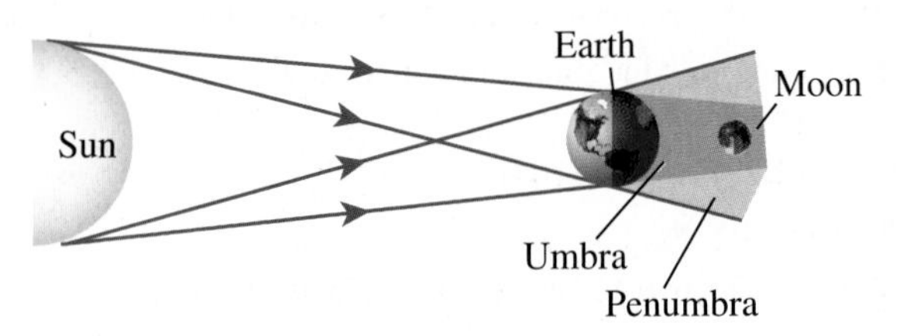

(b) Annular eclipse.

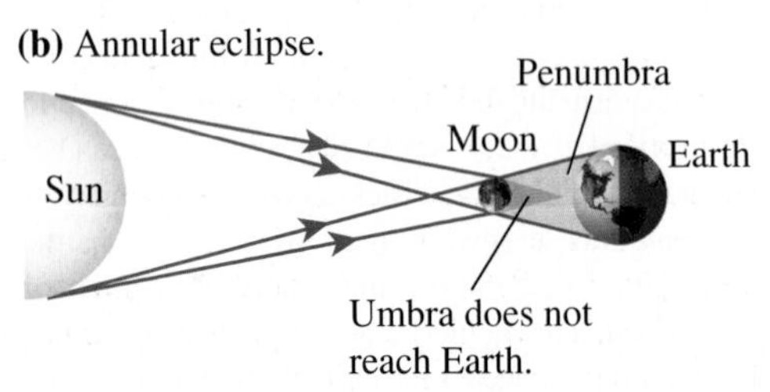

15-3 3×10^{-5} s; human reaction time alone would make such a short time interval impossible to measure.

15-4 23%

15-5 3.00×10^{11} m

15-6 (a) 3.13×10^8 m/s

(b) At half the frequency, the reflected light passing between the teeth would be observable only half as often, so the light would be only half as bright.

15-7 about 8 times/s

15-8 3.00×10^8 m/s

15-9 At double the frequency, the mirror would have reflected the light after 1/4 of a revolution instead of after 1/8 of a revolution, and Michelson would still have seen the light source.

15-10 (a) 1.89×10^{22} m (b) 1.5×10^{10} light years

15-11 6.48×10^2 s or 10.8 min

15-12 Diffuse reflection is the reflection of light off an irregular surface, whereas specular reflection is reflection of light off a smooth, regular surface.

15-13 (a) The image is upright and virtual.

(b) 45 cm

(c) 45 cm + 45 cm = 90 cm

(d) 5 cm/s + 5 cm/s = 10 cm/s

15-14 (a) 17° (b) Both angles are 0°.

15-15 (a) (b)

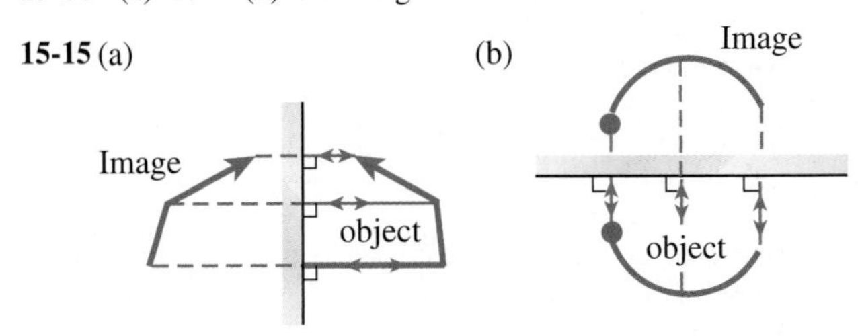

15-16 There are three images, as shown in the diagram below.

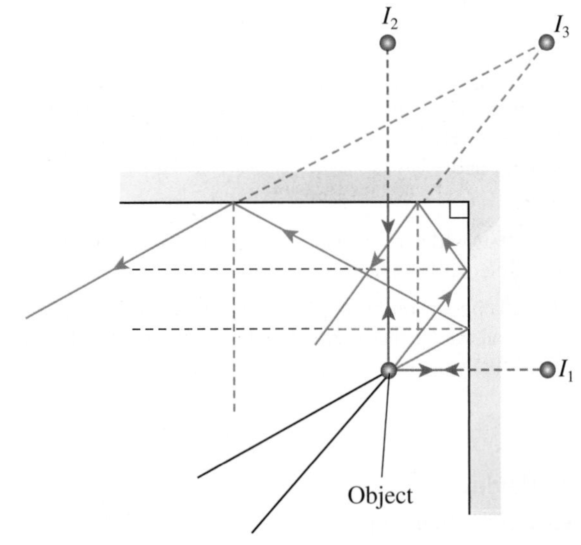

15-17 (a) 3.8 cm, −0.50 (b) 7.7 cm, −2.1 (c) −2.3 cm, +1.9 (d) −1.9 cm, +0.25

15-18 In all three cases the results are close to the calculated results in Exercise 15-17, but they are not exact.

(a) The image is inverted and real, with $d_i \approx 3.8$ cm and $M \approx -0.5$.

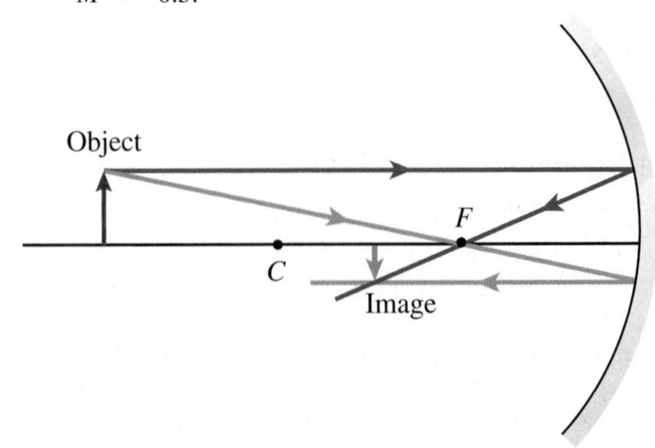

(b) The image is inverted and real, with $d_i \approx 7.3$ cm and $M \approx -1.9$.

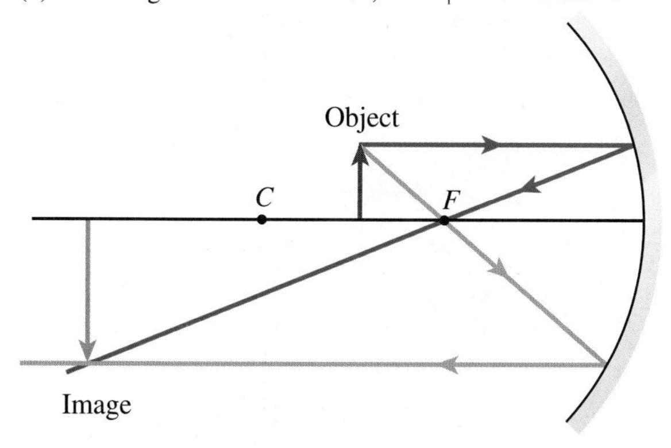

(c) The image is upright and virtual, with $d_i \approx -1.6$ cm and $M \approx +1.9$.

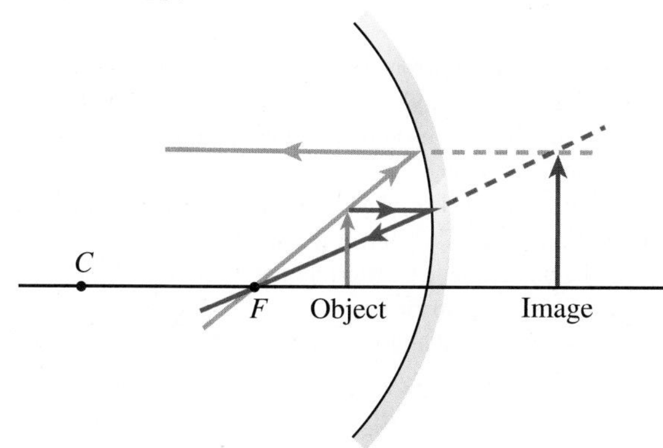

(d) The image is upright and virtual, with $d_i \approx -1.9$ cm and $M \approx +0.2$.

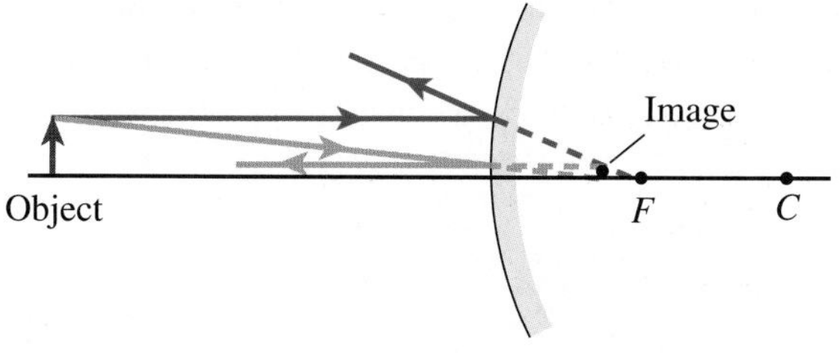

15-19 2.0 cm, 0.50

15-20 $d_i = 8.5$ cm, which is the focal length of the mirror.

15-21 The theoretical focal length of a plane mirror is ∞, so $d_i = -d_o$.

15-22 (a) $f = 15$ cm, converging mirror (b) $f = 60$ cm, converging mirror (c) $f = -20$ cm, diverging mirror (d) $f = \infty$, plane mirror

15-23 $d_i = -60$ cm, $M = +2.5$, $h_i = 58$ cm, upright

15-24 The image is 24 cm from the mirror, 4.8 cm high, upright, and has a magnification of +0.030.

15-25 The window should be slanted outward at the top so the light from the Sun or a streetlight reflects more toward the ground than outward toward a passerby.

15-26 1.35

15-27 2.19×10^8 m/s

15-28 4.43×10^{-10} s

15-29 20.8 m

15-30 (a) toward N (b) toward N (c) away from N

15-31 64.0°

15-32 The gemstone has an index of refraction of 1.65, but zircon has an index of refraction of 1.93, so this gemstone is not zircon; according to Table 15.2, it is heavy flint.

15-33 With the liquid in a transparent, rectangular container, aim an incident ray of light toward the container and measure the angle of incidence in the air and the angle of refraction in the liquid. Calculate the index of refraction and apply the equation $n = c/v$ to find the speed, v.

15-34 68.0°

15-35 $b = f = h = 30.0°$ and $c = d = e = g = 18.4°$

15-36 35.0°, 39.7°

15-37 41.8°

15-38 Blue refracts more. The angle of refraction is 0.2° less for the blue than for the red.

15-39 This type of mirage, called a superior mirage, occurs when air at Earth's surface is colder (and thus more dense) and air higher up is warmer (and thus less dense), causing the light from the object to be refracted down toward the observer.

15-40 First, the light must be travelling within a medium of higher index of refraction than its surroundings. Second, the angle of incidence of the light within the medium must be greater than the critical angle for the situation.

15-41 31.2°

15-42 The index of refraction is 1.65 and the material is heavy flint.

15-43 (a) B (b) A (c) B

15-44 42.8°

15-45 2.15×10^8 m/s

15-46 51.0°

15-47 (a) 37.3° (b) 39.3°; the heavy flint glass would sparkle slightly more because of its lower critical angle.

15-48 The cladding should have a lower index of refraction than the core fibres in order for total internal reflection to occur within the core fibres.

15-49

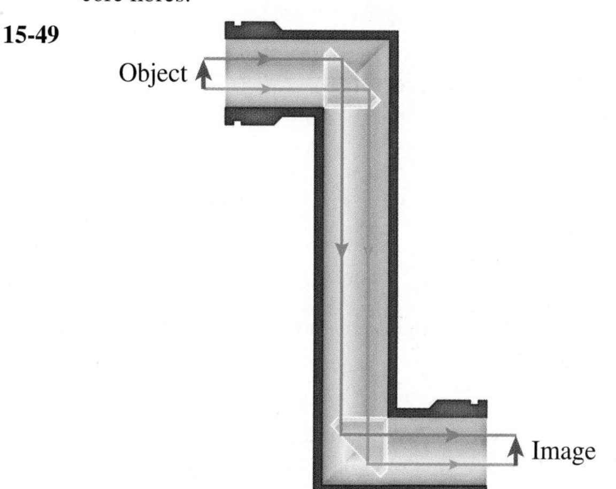

15-50 Total internal reflection occurs within a material with an index of refraction greater than its surroundings. For crown glass, $n = 1.52$, which is larger than that of air (1.00) as well as that of water (1.33), so the periscope would function properly.

15-51 In air, diamond has a critical angle of 24.4° but in water its critical angle is found to be 33.3°. Thus, a diamond in water would sparkle much less that a diamond in air because its critical angle is greater in water.

15-52

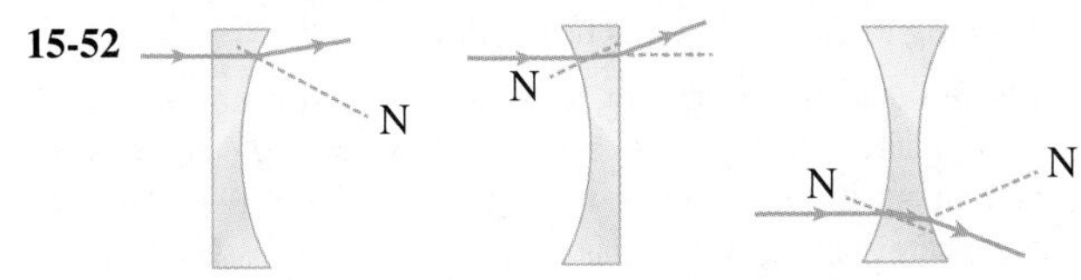

15-53 (a) The first lens has a greater focal length because it is thinner, and the light rays parallel to the principal axis will not refract as much as in the thicker lens.

(b) The first lens has a greater focal length because it is thinner, and the light rays parallel to the principal axis will not refract as much as in the thicker lens.

(c) The first lens has a greater focal length because the light rays parallel to the principal axis will refract less in the lens with the lower index of refraction.

15-54 Hold the lens so that light from a distant source (such as the Sun or a distant bright scene) is focused onto a sheet of paper and measure the distance from the middle of the lens to the paper.

15-55 If light from a distant source can be brought to a focus on a screen, wall, or piece of paper, the lens is converging. If no image of a distant source can be located on the screen, the lens is a diverging lens.

15-56 A double concave lens is a diverging lens, so it cannot be used to place an image onto a screen.

15-57 An air bubble surrounded by glass causes the opposite effect of glass surrounded by air, so a bubble (which is convex) acts like a diverging lens.

15-58 (a) +9.0 cm, −2.0 (b) −3.0 cm, +2.0 (c) −1.5 cm, +0.50

15-59 In all three cases, the results are the same as those found by calculations.

(a) The image is inverted and real, with d_i = 9.0 cm and $M = -2.0$.

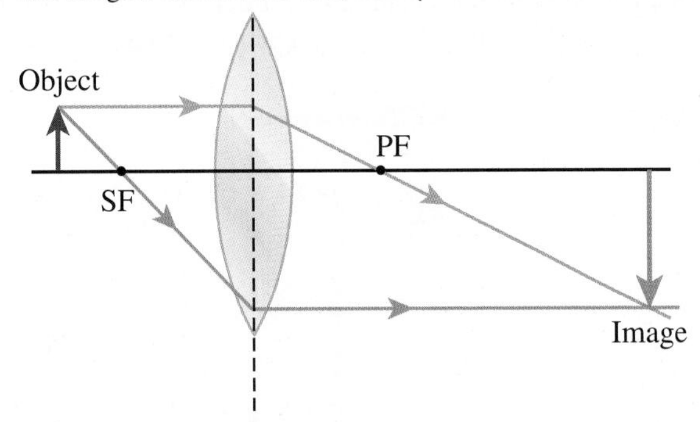

(b) The image is upright and virtual, with d_i = −3.0 cm and $M = +2.0$.

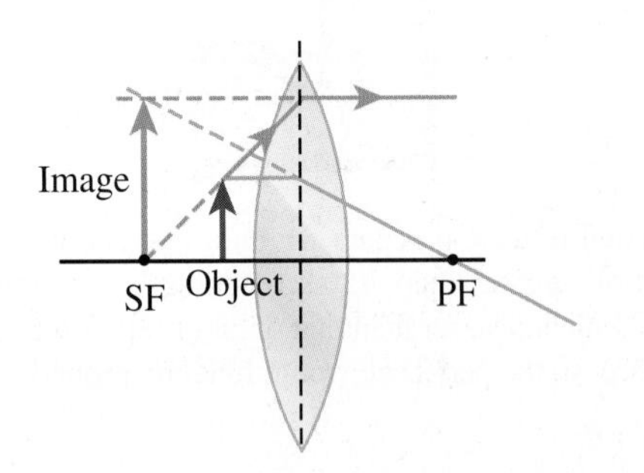

(c) The image is upright and virtual, with d_i = −1.5 cm and M = +0.50.

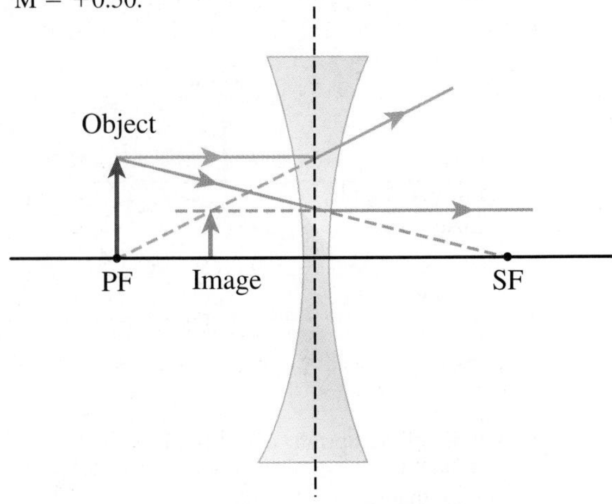

15-60 For converging mirrors and lenses,

- The image of any object beyond the focal point is inverted and real, with $|M|$ becoming larger as the object moves closer to the focal point.
- No image occurs when the object is located at the focal point.
- The image of an object located between the focal point and the optical device is upright and virtual, with $|M|>1$.

For diverging mirrors and lenses, the image of any object is always upright and virtual, with $|M|$ always <1.

15-61 The screen should be located a distance d_i = +0.154 m from the lens.

15-62 This question applies the reversibility of light rays. (a) converging lens (b) 1.0*f*

15-63 The mold should be located 33 cm from the lens.

15-64 The object should be located 30 cm from the lens.

15-65 (a) converging lens (b) +32 cm (c) 31 cm

15-66

High *n*

Low *n*

Light Source

Chapter Review

Multiple-Choice Questions

15-67 (b)

15-68 (d)

15-69 (a)

15-70 (b)

15-71 (b)

15-72 (d)

15-73 (a)

15-74 (d)

15-75 (c)

Review Questions and Problems

15-76 The shadow cast by the opaque rod of the sundial occurs because of the rectilinear propagation of light. The rod blocks light from the Sun, causing an umbra.

15-77 (a) For a solar eclipse, the Moon is between the Sun and Earth. For a lunar eclipse, Earth is between the Moon and the Sun. For an annular eclipse, the Moon is between the Sun and Earth but slightly farther from Earth than in a regular solar eclipse.

(b) For a partial solar eclipse, the observer must be in the penumbra of the Moon's shadow. For a partial lunar eclipse, the Moon must be in the penumbra of Earth's shadow. For an annular eclipse, the observer must be located beyond the umbra of the Moon's shadow.

15-78 (a) Earth's shadow is larger than the Moon's shadow, so the astronaut on the Moon is much more likely to observe a solar eclipse than when on Earth.

(b) As shown in the diagram below, the Moon would be in Earth's umbra for a fairly long time.

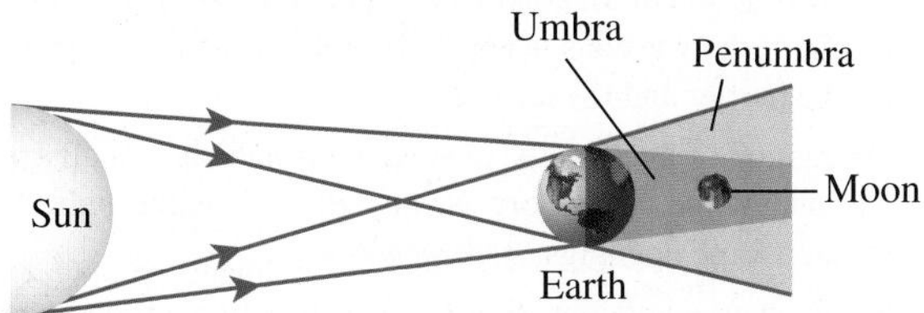

15-79 In most situations in this chapter, we have used three significant digits, in which case the speed of light is 3.00×10^8 m/s in both air and a vacuum. If four or more significant digits are required, the speeds differ, as stated in Section 15.3, where the speed of light in air was given as $2.997\ 05 \times 10^8$ m/s and in a vacuum it was given as $2.997\ 92 \times 10^8$ m/s.

15-80 1.59×10^{21} m, 1.59×10^{12} Gm

15-81 (a) 4.06×10^{-8} l.y. (b) 2.56 s

15-82 Clothing with specular reflection is safer than diffuse reflection because more light reflects from the bike rider to a vehicle's driver.

15-83 (a) plane mirrors (b) two mirrors (c) 60°

15-84 In many cases, the surface of the object that a light ray strikes is curved, and it would be difficult or impossible to measure the angle between the curved surface and the light ray. So a normal is drawn and the angle between it and the light ray is measured. For consistency, that normal is drawn even for non-curved surfaces, such as rectangular blocks or plane mirrors.

15-85 The image is upright and virtual, with a magnification of 1.0 and an image distance equal to the object distance.

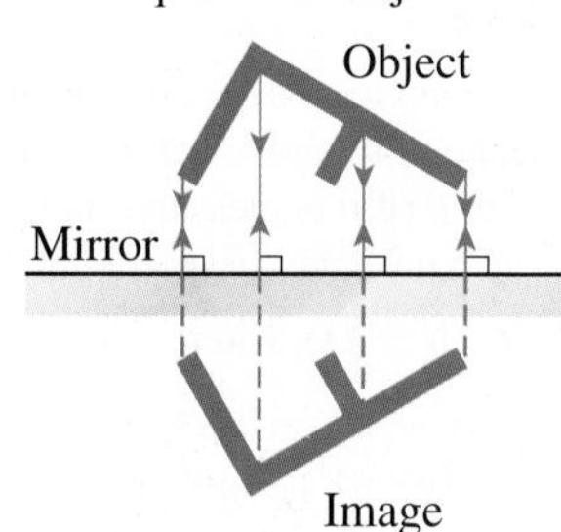

15-86 Along AB and CD, the image is upright and the reflection reads IMAGE. Along BC, the image is inverted and remains reversed.

15-87 Real images can be placed onto a screen and they are inverted (relative to the initial object). Virtual images cannot be placed onto a screen, so the observer must look at the optical device to see the images. They are upright (relative to the initial object).

15-88 Relative to the object, real images are inverted, their magnification is negative, and d_i is positive, whereas virtual images are upright, M is positive, and d_i is negative.

15-89 (a) −46 cm, 1.6, virtual, upright

(b) 1.8 m, −1.5, real, inverted

(c) −22 cm, 0.60, virtual, upright

15-90 about 1×10^2 cm

15-91 The mirror is a diverging mirror, which gives a wider angle of view than a plane mirror and allows observers from various directions to see around a corner.

15-92 (a) An elongated, irregular diverging mirror, similar to a "fun-house" mirror, was used to create the effect in the dance scene shown.

(b) The dentist's mirror reflects light into an area difficult to illuminate and also reflects light from that area to the dentist.

(c) An inferior mirage occurs when light refracts through hot, low-density air and the observer in a moving vehicle sees a shiny image that continually recedes.

15-93 (a) The object is closer than the focal length of a converging mirror, so the image is virtual.

(b) The mirror is diverging, so the image is virtual.

(c) The lens is diverging, so the image is virtual.

(d) The object is at the focal point of a converging lens, so there is no image.

(e) The image is real.

(f) The image is virtual.

15-94 In both cases, light rays parallel to the principal axis and far from it do not reflect or refract through the focal point, so the image is either blurry or less distinct near the outer edge of the device. For a converging mirror, the aberration is corrected using a parabolic rather than a spherical mirror. For a converging lens, the aberration is corrected by covering or otherwise not using the outer area of the lens.

15-95 0.60 m

15-96 (a) 1.63 (b) 21.7° (c) 37.8°

15-97 (a) 2.05×10^8 m/s (b) 34.5°

15-98 (a) $v_X \propto \frac{1}{n_X}$ (b) $\sin\theta_c = \sin^{-1}\left(\frac{1}{n_X}\right)$ (c) $\theta_X \propto \sin^{-1}\left(\frac{1}{n_X}\right)$

15-99 1.67

15-100 1.134

15-101 (a) orange light, 20.9° (b) violet light, 20.7°

15-102 41.0°

15-103 The focal length becomes shorter as the index of refraction becomes larger. Thus (a) $f_R > f_G$ (b) $f_Y > f_G$ (c) $f_B < f_G$

15-104 The spectral colours may be dispersed within the glass, but they undergo only a very small amount of lateral displacement in the thin glass, and we do not observe dispersion. (Recall that dispersion is more readily observed using a triangular prism.)

15-105 (a) $O_1 \rightarrow I_2$; $O_2 \rightarrow I_3$; $O_3 \rightarrow I_4$; $O_4 \rightarrow I_1$. As the object approaches the converging mirror, the image becomes larger and farther from the focal point and is inverted and real, but then becomes upright and virtual when the object is inside the focal point.

(b) $O_1 \rightarrow I_3$; $O_2 \rightarrow I_2$; $O_3 \rightarrow I_1$. As the object approaches the diverging mirror, the image is always upright, virtual, and smaller than the object, but gradually becomes larger.

(c) $O_1 \rightarrow I_2$; $O_2 \rightarrow I_3$; $O_3 \rightarrow I_4$; $O_4 \rightarrow I_1$. As the object approaches the converging lens, the image becomes larger and moves farther from the lens, is inverted and real, but then becomes upright and virtual when the object is inside the secondary focal point.

(d) $O_1 \rightarrow I_1$; $O_2 \rightarrow I_2$; $O_3 \rightarrow I_3$. As the object approaches the diverging lens, the image is always upright, virtual, and smaller than the object, but gradually becomes larger.

15-106 (a) $d_i = -6.4$ cm; $h_i = 4.6$ cm; virtual and upright

(b) $d_i = -48$ mm; $h_i = 26$ mm; virtual and upright

(c) $d_i = 8.8$ cm; $h_i = -0.90$ cm; real and inverted

15-107 (a) +1.77 cm (b) +1.84 cm

15-108 Dispersion of light, caused when the components of white light travel at different speeds in the outer parts of a lens, can occur in lenses but not in mirrors.

Applying Your Knowledge

15-109 3×10^7 m

15-110 With a backward eye-testing chart on one wall and a plane mirror on the opposite wall, the patient should look at the mirror from a distance of 2.4 m. Refer to the diagram.

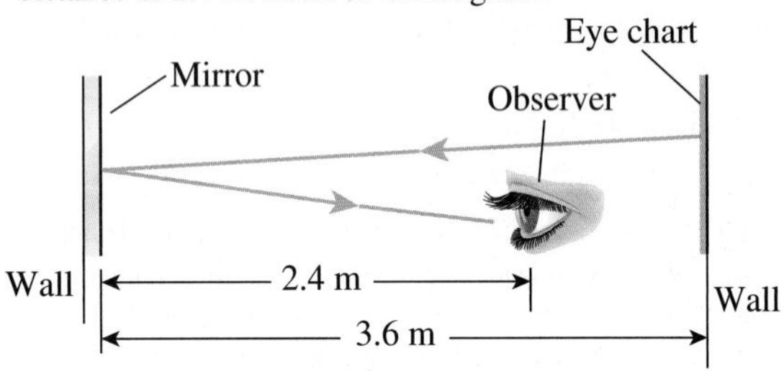

15-111 (a) no focal length for a plane mirror (b) about 10^2 m (c) about 0.1 m (d) a little less than 10^1 m (e) about 1 m (f) about 1 m

15-112 (a) 14.5°

(b) Snell's law can be applied either step-by-step or by equating the data for any pair of materials.

15-113 3.000×10^8 m/s

15-114 72.8 cm

15-115 2.84 cm

15-116 49.3°

15-117 7.8 mm

15-118 (a), (b) Using 30° as an example, $\theta_i/\theta_R = 1.54$ and $\sin\theta_i/\sin\theta_R = 1.50$. The ratio $\sin\theta_i/\sin\theta_R$ represents the index of refraction of the glass.

(c) Ptolemy's prediction applies only to small angles, up to about 20°.

15-119 $n_1 \cos\theta_1 = n_2 \cos\theta_2$

15-120 There are two principles involved in the operation of the fountain: First, the motion of the jets of water is influenced by gravity and each jet follows a parabolic path. (Refer to projectile motion in Chapter 4.) Second, bright light beams at the origin of each jet of water travel within the water and undergo total internal reflection wherever the light beam strikes the inside surface of the stream.

15-121 "Chroma" relates to colour, so achromatic means eliminating colour. Doublet means two lenses, so an achromatic doublet is an arrangement of two lenses: the main converging lens that has chromatic aberration, and a second lens that is diverging and reduces or eliminates the aberration.

15-122 (a) 5.0×10^1 cm (b) 4.0×10^1 cm

Chapter 16 Wave Optics

Exercises

16-1 The diffraction of the wavelets will be maximum when the wavelength of the wavelets is approximately the same as the width of (a) the barrier and (b) the aperture.

16-2 Diffraction of light can be observed through very small apertures because of the very short wavelengths of visible light, but not through large openings, such as windows.

16-3 Radio waves undergo significant diffraction around regularly sized objects because they have a much longer wavelength than the waves of visible light.

16-4 (a)

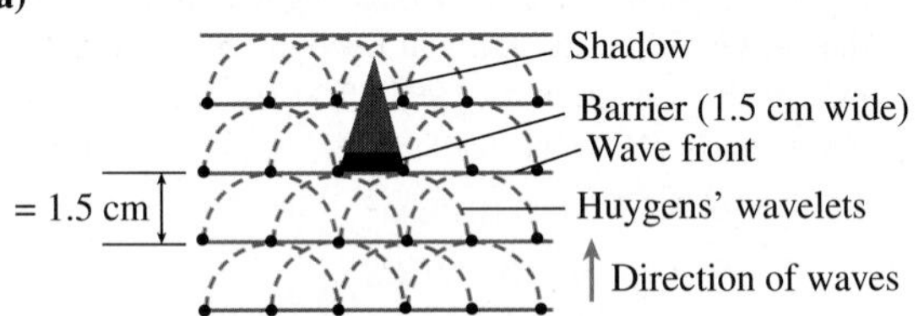

(b)

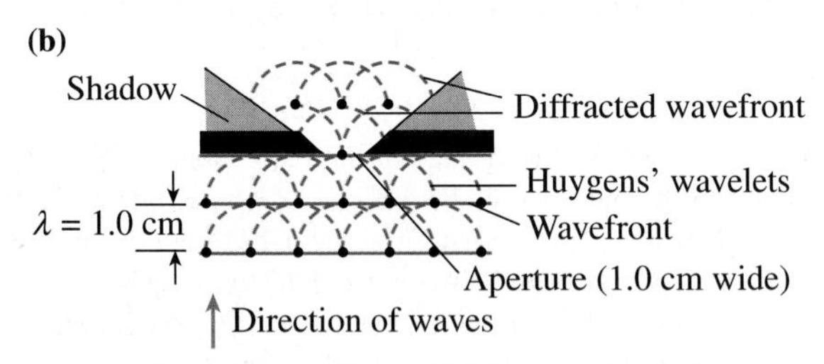

16-5 (a) The spacing is directly proportional to the wavelength.

(b) The spacing is directly proportional to the distance from the slits to the screen.

(c) The sine of the angle of separation is inversely proportional to the slit separation, and since θ is small, $\sin\theta$ is approximately equal to θ (if θ is measured in radians), and therefore the angle is also inversely proportional the slit separation.

16-6 In general, $P.D. = (N - \frac{1}{2}\lambda)$, where $N = 1, 2, 3, \ldots$

16-7 (a) $M = 3$ (b) $N = 3$

16-8 (a) 5.4×10^{-7} m (b) 2.84×10^{-2} m (c) 4.60×10^{-7} m (d) 4

16-9 (a) 0.702° (b) 588 nm (c) yellow

16-10 2.00 cm

16-11 (a) $\Delta x = L\lambda/d$ (b) 6.0 m

16-12 0.202 mm

16-13 9 maxima

16-14 When light is aimed at a narrow slit, the light is observed to spread out, a result of diffraction, but it is also observed to have light and dark fringes, a result of interference.

16-15 As the slit width decreases, the diffraction pattern spreads out more. (As Eqn. 16-8 shows, x is inversely proportional to w.)

16-16 0.16°; by applying Eqn. 16-7, with $N = 1$, you can see that the answer would be the same because $\sin\theta$ will remain 1/350 even if the wavelength changes.

16-17 7.3 cm

16-18 0.098 mm

16-19 To two significant digits, both ratios are 1.2:1.0.

16-20 (a) increases (b) decreases (c) increases (d) decreases

16-21 (a) 0.12 mm (b) 11 m (c) 1.0 cm

16-22 The grating labelled 10 000 lines/cm has smaller spacings than the other grating, so the pattern will be more spread out and will have a lower number of maxima.

16-23 Nature never gives us something for nothing. The brighter, sharper maxima are a result of a greater concentration of light at the maxima. This contrasts with a double-slit pattern in which the maxima are more spread out.

16-24 24.8°, 42.8°

16-25 By applying Eqn. 16-3 ($\sin\theta = M\lambda/d$) to the given data to three significant digits, we see that the $M = 2$ spectrum is spread out from 18.7° to 34.1°, while the $M = 3$ spectrum is spread out from 28.7° to 57.1°. Thus, the patterns overlap between 28.7° (the violet end of the $M = 3$ spectrum) and 34.1° (the red end of the $M = 2$ spectrum).

16-26 1.90×10^{-4} cm/line, 5.26×10^{3} lines/cm

16-27 5 maxima

16-28 6.69×10^{-7} m

16-29 The diagram below shows two overlapping diffraction patterns, one vertical and one horizontal, with the vertical pattern spread out farther than the horizontal one because the vertical spacing is smaller and $\sin\theta$ is inversely proportional to the spacing. (Note that only the spacings are illustrated here, not the faint spectral colours that would be possible to see.)

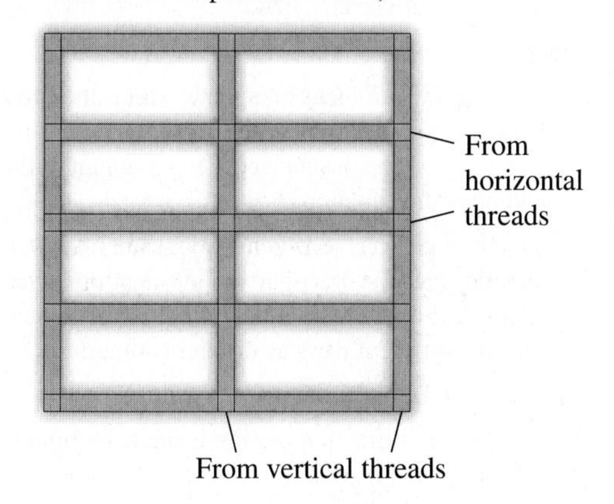

16-30 The wavelengths are 4.34×10^{-7} m (violet), 4.86×10^{-7} m (blue), and 6.57×10^{-7} m (red).

16-31 The light is in the green range and the grating consists of two perpendicular sets of very closely spaced lines. The line spacings are equal in the vertical and horizontal directions.

16-32 (a) in phase (open-end reflection) (b) out of phase (fixed-end reflection)

16-33 98.5 nm, 295 nm, 492 nm

16-34 Consider two in-phase incident waves striking the film, one reflecting out of phase at the air–film interface due to fixed-end reflection, and the other passing into the film and then reflecting out of phase from the glass. It then goes back to the air–film interface (having travelled a total distance of ½λ in the film) where it is now out of phase with the first wave, causing a minimum.

16-35 94.8 nm

16-36 3 red fringes

16-37 (a) maximum (b) minimum

16-38 The minimum occurs because the light reflecting off the inside surface of the convex surface undergoes open-end reflection, while the light that reflects nearby off the flat surface undergoes fixed-end reflection, causing the two reflections to be out of phase.

16-39 When the reflected light is viewed, concentric circular patterns of minima and maxima will be observed in both cases, with the very centre of each pattern dark (a minimum). However, the pattern in (a) will be more spread out than that in (b). (Recall that adjacent maxima or adjacent minima are created when the air–wedge thickness changes by one-half the wavelength of the light.)

16-40 (a) 2.5×10^{2} nm (b) 4.0×10^{-5} m

16-41 about 1×10^{-2} mm

16-42 the vertical plane

16-43 (a) x rays (b) ultraviolet light (c) blue light

16-44 7.30×10^{14} Hz, violet

16-45 (a) 3.00 m (b) 7.19×10^{-15} m

16-46 (a) Examples are AM 710 kHz = 7.10×10^{5} Hz, and FM 96.3 MHz = 9.63×10^{7} Hz.

(b) for the examples chosen in (a), 423 m and 3.12 m

16-47 Light from the Sun and from light bulbs is unpolarized, which means that the waves from the sources oscillate randomly in various orientations. Plane-polarized light has waves oriented in a single plane.

16-48 Sound waves cannot be polarized because they are longitudinal.

16-49 The values are "maximum" because some of the light's energy is absorbed by the filter. (a) 3/8 (b) 1/8 (c) 0

16-50 1/8

16-51 (a) Cross one lens of a pair of sunglasses with one lens of the second pair, and if no light is transmitted through the crossed lenses, they must be polarizing lenses.

(b) Find an example in the room of glare off a horizontal surface, such as the floor, then view the glare though the sunglasses; if the glare is eliminated (or at least substantially reduced), the glasses must have polarizing lenses.

16-52 One way is to view the polarized light reflected off a non-metallic surface through the Polaroid plate. If the surface is horizontal, the reflected light is polarized horizontally, so if you orient the Polaroid plate so the glare is removed from the reflection, you can conclude that you are holding it such that the polarization axis is vertical.

16-53 (a) 56.7° (b) 33.3°

16-54 49.6°

16-55 (a) The light reflected from a (non-metallic) horizontal surface is polarized in the horizontal plane.

(b) The axis of polarization should be in the vertical plane so that the electric components of the waves become absorbed by the filter, thus reducing the glare.

(c) 36.9° above the horizon

16-56 The sky appears black on the Moon because there are no atmospheric particles to scatter the light.

16-57 Air molecules are relatively small and tend to scatter light waves with shorter wavelengths, such as blue light. Water droplets are larger and have a variety of sizes, allowing many more wavelengths, including long wavelengths, to scatter from them, creating white or off-white clouds.

16-58 Assume for this description that the double image consists of two letter *a*'s, one of which is slightly above and to the left of the other. If you view the double image through a polarizing filter such that you see only the upper-left *a*, as you rotate the filter through 90°, that *a* will disappear and the lower-right *a* will appear. Rotating the filter a further 90° will produce the first *a*, then rotating through another 90° (a total rotation so far of 270°) will produce the second *a*.

Chapter Review

Multiple-Choice Questions

16-59 (d)

16-60 (d)

16-61 (b)

16-62 (c)

16-63 (b)

16-64 (a)

16-65 (e)

Review Questions and Problems

16-66 Monochromatic light is light of a single wavelength and thus a single colour.

16-67 Use Eqn. 16-4 with $\theta = 90°$.

(a) N is reduced. (b) N is increased. (c) N is the same.

16-68 (a) 4.0 cm (b) 22 mm (c) 0.10 cm

16-69 (a) 0.26 cm (b) 7.6 m (c) 0.013 m

16-70 7.2°; 22°; 39°

16-71 1.4 mm

16-72 A: $M = 3$, $P.D. = 3\lambda$; B: $N = 2$, $P.D. = 1.5\lambda$; C: $M = 0$, $P.D. = 0$; D: $M = 1$, $P.D. = \lambda$; E: $N = 3$, $P.D. = 2.5\lambda$

16-73 (a) 5.50 mm (b) 10.8 mm (c) 0.500

16-74 (a) 2.7 cm (b) 0.70°

16-75 (a) 2.2 cm (b) 0.35°

16-76 (a) Resolution is better at 10 m.

(b) At a 45° angle, the effective distance between the sources is lower, so there is lower resolution.

16-77 (a) 0.12° (b) 7.5 mm

16-78 (a) As the wavelength increases, diffraction also increases, so resolution decreases.

(b) As the aperture width increases, the resolution also increases.

16-79 (a) increases (b) increases

16-80 4.8×10^{-7} m

16-81 1.4:1

16-82 A window pane's thickness is in the order of 10^4 wavelengths of yellow light. This is huge compared to thin films, which are generally less than one wavelength in thickness.

16-83 (a) 91.3 nm (b) The reflected light would again be a maximum. In (a), the plastic film is $\lambda_{film}/4$ thick. So in (c) it is $3\lambda_{film}/4$ thick. In the latter case, the path length of the reflected light is one wavelength longer than in (a), so the phase of the reflected wave is the same when it emerges from the thin film.

16-84 477 nm

16-85 The reflected light from a one-quarter wavelength air wedge is maximum, which would also be the case with an air–oil–water situation. (This can be shown using basic theory of fixed-end and open-end reflections combined with path length of the waves.) The quarter wavelength film of magnesium fluoride over glass produces a minimum reflection, not a maximum.

16-86 (a) The force of gravity causes the rounded top surface of the bubble to become very thin, and the light that reflects off the surface interferes destructively as the light striking the top surface reflects out of phase and the light striking the inside film surface reflects in phase, causing the dark surface.

(b) Answers will vary: about 8×10^{-8} m.

16-87 25 fringes/cm

16-88 X rays have extremely short wavelengths (thousands of times smaller than visible light) and they diffract through crystals that have regular spacings of atoms. The x-ray diffraction patterns thus reveal the atomic structure of crystals.

16-89 Assume it takes about 10 hair diameters to equal 1 mm, so the diameter of each hair is 0.1 mm, or 1×10^{-4} m. The electromagnetic wave that would diffract most easily around a hair would have a similar wavelength, which from Figure 16-47 is in the infrared range of wavelengths.

16-90 268 m

16-91 (a) All parts of the electromagnetic spectrum consist of oscillating electric and magnetic waves that can travel in a vacuum, have a constant speed in a vacuum (3.00×10^8 m/s), and obey the equation $c = f\lambda$.

(b) Visible light has different frequencies, wavelengths, and energies than the other parts of the electromagnetic spectrum. Also, visible light can be detected by the human eye, which sees the different parts as different colours.

16-92 (a) 2.3×10^8 Hz (b) the microwave region

16-93 A transmission diffraction grating is made by ruling fine parallel grooves on a smooth glass surface with a diamond-tip tool. A much cheaper replica diffraction grating is made by pouring molten plastic over a master grating and allowing it to cool before being pulled off. Diffraction gratings can also be made by using holography. Polarizing filters can be made by stretching long-chain molecules on sheets so they are parallel, then dipping the sheets in iodine to cause the molecules to become electric conductors.

16-94 Light can be polarized, and since longitudinal waves cannot be polarized, light must consist of transverse waves.

16-95 The direction of the magnetic field is parallel to the axis of transmission.

16-96 (a) 45° (b) 0°

16-97 81%

16-98 (a)

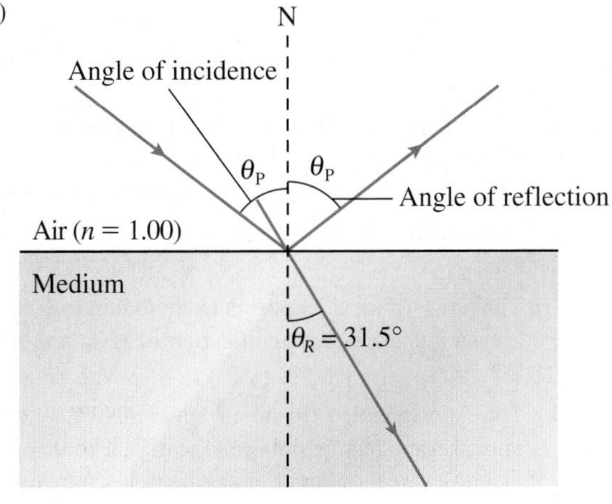

(b) 1.63

Applying Your Knowledge

16-99 The spacing (Δx) of the maxima in a double-slit interference pattern is directly proportional to the wavelength of the light, and under water the wavelength of the light is lower than that in air, so the spacing is closer.

16-100 (a) 5.97×10^{-7} m (b) 0.214° (c) The sine and tangent are very close to each other. This means that, in deriving the interference equations, the perpendicular distance from the slits to the screen is basically equal to the distance from the slits to the $M = 1$ maxima.

16-101 9

16-102 (a) At the beginning, the slit width is about 10 times the wavelength, so the pattern would be spread out slightly, like that in Figure 16-20(c). Then, as the width decreases, the pattern spreads out more and more, like the pattern illustrated in Figure 16-20(b).

(b) When the slit width decreases to about the size of the wavelength (628 nm), the diffraction pattern spreads to its maximum amount, so there will no longer be any dark fringes.

16-103 A diffraction grating is less cumbersome than a prism; it can spread the spectral colours out more and, with a high number of lines on the grating, the spectra can be very bright.

16-104 For maxima, $2t = M\lambda_{film}$, where $M = 0,1,2, \ldots$, and for minima $2t = (N - ½)\lambda_{film}$, where $N = 1,2,3, \ldots$.

16-105 21%

16-106 The interference pattern is quite irregular, which means that the two surfaces in contact are not very regular or smooth. The pattern is viewed in reflected white light, as is evident by the spectral colours of the fringes. The air wedge is narrower toward the bottom of the pattern where the fringes are closer together. It is possible that the screws holding the apparatus together are tighter at the bottom than at the top.

16-107 The electric component of the electromagnetic wave interacts with the various atoms and molecules in a non-metallic surface and the light is absorbed. With a shiny metal surface, the free electrons in the metal oscillate because of the incident electric field and reradiate (reflect) the light back into the medium where the light came from.

16-108 The pattern reveals areas of high stress wherever the spectral colours are closest to one another (e.g., at the very centre of the protractor).

Chapter 17 Optical Instruments

Exercises

17-1 (a) Move from very bright to dark surroundings. The pupil's diameter will increase about 8 mm.

(b) The lens can change shape as you change your view from distant to nearby.

17-2 The functions are
cornea: performs about 2/3 of the focusing of the light rays that enter the eye
sclera: provides protection for the eye (except for the front of the eye)
aqueous humour: supplies nutrients to the cornea and lens, and also supplies cells to repair damage to those components
rods: react to low-intensity light
cones: responsible for colour vision
optic nerve: transmits signals from the retina to the brain

17-3 The index of refraction of the lens (n = 1.40) is slightly higher than the index of refraction of the cornea (n = 1.37), so in the lens, the speed of light is slower and the refraction is greater.

17-4 Normally, we look at our surroundings with both eyes, in which case we would not be able to notice the blind spot. To demonstrate the blind spot in one eye, the other eye must be closed or covered.

17-5 The booklet would contain drawings of the cartoon character on several pages with slight changes in the character's position as the pages progress. As you flip through the booklet quickly page-by-page, the cartoon character will appear to be moving. This result is based on the eye's ability to retain an image for about 1/25 s after the object is removed.

17-6 Although the silhouettes appear progressively taller from left to right, measuring them reveals that they are all the same height. The illusion is caused by the lines surrounding the silhouettes, which fool our perspective view.

17-7 (a) red (b) green and blue (c) red and blue (d) red and green

17-8 The white paper reflects the three primary light colours, red, green, and blue, but when the cones sensitive to red and green are fatigued by having stared at a yellow circle, the colour observed is blue.

17-9 (a) The person has myopia.

(b) The person can see an object clearly at a distance of 20 cm but not at 12 cm or 85 cm.

(c) A diverging lens can be used to provide better vision.

17-10 (a) diverging lens

(b) cylindrical lens with a different focal length in different planes

(c) converging lens

(d) converging lens (in a pair of reading glasses) or a bifocal lens

17-11 $f = -0.125$ m and $P = -8.00$ d, with the negative sign indicating a diverging lens.

17-12 $f = +0.35$ m and $P = +2.9$ d, with the positive sign indicating a converging lens.

17-13 (a) Because the focal length of the lens is positive, the lens is a converging lens and the defect is hyperopia.

(b) +0.63 d

(c) 0.34 m

17-14 (a) 0.610° (b) 6.28°

17-15 4.6 m

17-16 Answers will vary, depending on the height of the door and the distance from it. For example, if you estimate the height of the door to be about 2.5 m and the distance from the door to be about 7.5 m, then the angle it subtends at the eye can be estimated using $\theta \approx \tan^{-1}\left(\frac{2.5\text{ m}}{7.5\text{ m}}\right) \approx 18°$. (If sine had been used, the angle in this case would be almost the same, at 19°.)

17-17 (a) about 3.6 cm (b) about 3.1 cm

17-18 (a) Relaxed eye magnification is about 2.0; maximum angular magnification is about 3.0.

(b) Relaxed eye magnification is about 1.0×10^1; maximum angular magnification is about 11.

17-19 (a) about 4.0 (b) about 15

17-20 about −83 times; the negative sign indicates an inverted image.

17-21 about −81 times

17-22 approx. 0.63 cm

17-23 (a) A refracting astronomical telescope has one less set of lenses (the erector set) than a refracting terrestrial telescope, so it absorbs less light and thus can provide a brighter image. However, it produces an inverted image, so it is not convenient for viewing objects on Earth.

(b) Viewing celestial objects through Earth's atmosphere results in absorption of light as well as distortion, so astronomical telescopes are more useful if they are above all or most of the atmosphere.

(c) Radio telescopes receive radio waves, a part of the electromagnetic spectrum that can penetrate regions of space that visible light cannot and pass through Earth's atmosphere with less absorption than visible light. New radio telescopes are built in arrays to improve the resolution of the telescopes.

17-24 about −195 times

17-25 (a) The longer focal length lens (39 cm) should be used for the objective lens.

(b) about −15 times

(c) about 42 cm

17-26 about 1.8 cm

17-27 (a) The light-gathering ability of the Mauna Kea telescope is 1.0×10^2 times as great as that of the Keck telescope.

(b) A reflecting telescope can have a large mirror supported from beneath, whereas a lens can be supported only around the outer circumference, which is thin and weak. A mirror has only one surface that requires polishing, whereas a lens has two surfaces. Also, a mirror does not produce chromatic aberration, but a lens does.

17-28 The light gathering ability of optical reflecting telescopes can be increased by
- using multi-segmented mirrors to bring the light to a focus
- using CCDs rather than film
- using a variety of waves of the electromagnetic spectrum

17-29 The object can be from 3.9 m to a great distance (infinity) from the lens.

17-30 (a) The range of focal lengths is from 80 mm to 200 mm, which allows a change of magnification of (200 mm)/(80 mm) = 2.5 times.

(b) The adjustment to the zoom lens must be a change to the optical zoom feature only, not the digital zoom option, which magnifies images but produces poorer picture quality.

17-31 13.4 cm

17-32 (a) decreased by 4 times (b) increased by 16 times (c) decreased by about 64 times

17-33 (a) about 1/30 s (b) about 1/450 s

17-34 (a) 7.1 mm (b) *f*/2.8

17-35 In a photograph with a low depth of field, the main subject is in focus but the surroundings at smaller or greater distances from the camera are out of focus. In a photograph with a high depth of field, the entire scene is in clear focus. A low depth of field can be achieved using a long focal length of the camera's lens and focusing on a subject that is as close as possible and that looks clear or in focus. The surroundings closer or farther away than the subject should then be somewhat out of focus or blurry.

17-36 Laser light consists of light waves that are in phase, unidirectional, and monochromatic (single wavelength).

17-37 4.74×10^{14} Hz

17-38 (a) $E_{\text{photon}} \propto \frac{1}{\lambda_{\text{photon}}}$ (b) $E_{\text{photon}} \propto f$ (c) $f_{\text{photon}} \propto \frac{1}{\lambda_{\text{photon}}}$

17-39 (a) The mirror at each end of a typical laser is flat and highly reflective, with all the light reflected at one end and most of the light reflected at the other end.

(b) These mirrors help cause the buildup of many photons moving parallel to the tube, with the mirror that is not 100% reflective allowing the coherent laser light to pass through and form the laser beam.

17-40 Laser light concentrates a high amount of energy in a small area, which can cause high heat and a dangerous situation if not controlled and used properly.

17-41 Answers may vary because there are so many uses. Examples are reading barcodes, surveying, eye surgery, and fibre optics communications systems.

17-42 The process of holography supports the wave nature of light because it is based on the interference and diffraction of light, both of which are wave properties.

Chapter Review

Multiple-Choice Questions

17-43 (b)

17-44 (c)

17-45 (d)

17-46 (d)

17-47 (c)

17-48 (b)

17-49 (a)

17-50 (c)

17-51 (d)

17-52 (d)

Review Questions and Problems

17-53 (a) The pupil allows light to enter the eye.

(b) The iris controls the size of the pupil in bright and dim light.

(c) The lens does about 1/3 of the focusing of the light rays that enter the eye.

(d) The retina receives the light energy and transforms it into electrical signals that are sent to the brain for interpretation.

17-54 The eye's lens is surrounded by substances that have an index of refraction nearly as high as the lens, so when the light rays travel through the eye, they do not refract as much as they would if the lens were surrounded by air. The increased refraction if the lens were surrounded by air means that the light will be bent more and the focal length will be smaller.

17-55 The eye's accommodation is controlled by the ciliary muscles that change the shape of the lens from normal (when viewing a distant object) to a short focal length (when viewing a nearby object).

17-56 If you look straight at an object, the object's image will focus on the *fovea centralis*, which is the most sensitive area in the retina and contains cones.

17-57 The eye is able to retain an image for about 1/25 s, so when viewing a movie that is played at about 24 frames per second, new images replace previous ones quickly enough for the images to appear to be moving.

17-58 At first glance, none of the lines appear to be parallel to the bottom horizontal line. However, when a straight edge is used to check the lines, it turns out they are all parallel to the bottom line. As in previously viewed optical illusions, our eyes can be fooled by the surroundings of what we look at.

17-59 Two changes occur: First the eye accommodates by changing its focal length from a short to a relaxed focal length. Second, the pupil aperture increases as the eye becomes adapted to the darker scene.

17-60 The colours are the primary light colours, red, green, and blue. To produce red, the retina's red cones must react. To produce yellow, the retina's red and green cones must react. To produce white, the retina's red, green, and blue cones must react.

17-61 about 1×10^3 times

17-62 (a) myopia, or short sightedness; diverging lens

(b) normal vision; no corrective lens needed

(c) hyperopia, or far-sightedness; converging lens

(d) presbyopia as well as myopia; bifocal lens

17-63 1.3 m

17-64 Astigmatism is a condition of the eye in which the focusing is asymmetrical; in other words, the focusing is different in different planes. It can be corrected using a cylindrical lens with a different focal length in different planes, or using contact lenses.

17-65 Light enters the eye or a camera through an aperture and a lens. The focal length of the eye's lens can be adjusted somewhat to allow clear viewing of both nearby and distant objects. The focal length of most cameras can be adjusted. The eye's iris controls the pupil size to allow the eye to adjust to both dark and bright conditions. The camera's shutter can be controlled to allow different amounts of light to enter the camera under different lighting conditions. The retina of the eye and the detector of a camera detect the light.

17-66 (a) real (b) real (c) virtual (d) virtual (e) virtual (f) virtual (g) virtual (h) real

17-67 Optical instruments use visible light with wavelengths in the region of about 5×10^{-7} m, which is about 5000 times as large as the diameter of an atom (at 10^{-10} m). Thus, it would be impossible to use an optical instrument to view an atom.

17-68 (a) about 4.2×10^{-6} rad

(b) The telescope can resolve features separated by a distance of about 1.6×10^3 m.

17-69 (a) 3.97° (b) 0.259°

17-70 about 2.0×10^2 times

17-71 about 1.8 cm

17-72 about -2.5×10^2 times

17-73 approx. 19 cm

17-74 Both the microscope and the astronomical telescope have two lenses (or two main sets of lenses), the objective and the eyepiece. In both cases the final image is inverted and virtual, with a magnification greater than one. The purpose of the objective lens in a microscope is to give a magnified image, the focal lengths of the objective and eyepiece are both fairly small, the object is located just beyond the focal point of the objective lens, and the total magnification can be found using the product of the magnification of the two lenses. In an astronomical telescope, the purpose of the objective lens is to gather as much light as possible, so it has a large aperture. It also has a large focal length compared to the eyepiece. The angular magnification of the telescope is usually found using $m = \frac{-f_O}{f_E}$.

17-75 The attitude of the final image is inverted compared to the original object, and the magnification is about −2.3 times.

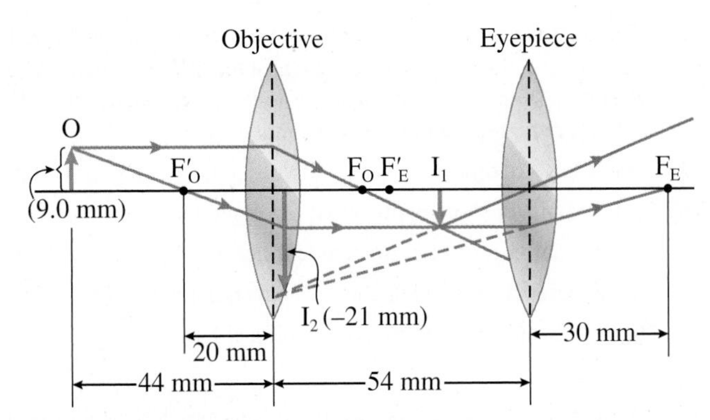

17-76 about 1.35 cm

17-77 The middle lens in a terrestrial telescope ensures that the final image viewed in the telescope is upright. The final image is virtual because it is created by the converging eyepiece with the "object" located just inside its focal point.

17-78 One example of a possible choice is given for each case:

(a) two +4.0 cm lenses

(b) a +55 cm lens for the objective and a +4.0 cm lens for the eyepiece

(c) a +22 cm lens for the objective and a −16 cm lens for the eyepiece

(d) a +55 cm lens for the objective, a +22 cm lens for the erector, and a +22 cm lens for the eyepiece

17-79 (a) The device is a radio telescope, which is built with a metal or wire mesh design in a parabolic shape.

(b) The waves received by this telescope are long-wavelength radio waves, which are part of the electromagnetic spectrum.

17-80 (a) The shutter controls the amount of light passing through the lens when the picture is taken.

(b) The zoom lens is a variable-focal length lens that gathers light and focuses it onto a detector.

(c) The aperture control sets the diameter of the lens to which the shutter opens.

17-81 (a) 3.5 m (b) 0.40 m

17-82 (a) decreased by 33 times (b) increased by 7.6 times

17-83 (a) At 1/15 s, the chances of shaking the camera while the shutter is opened is greater than at 1/60 s, so blurriness is more likely to occur.

(b) *f*/5.5 or *f*/5.6

17-84 (a) 1/460 s (or commonly, 1/500 s)

(b) At $d = f/5.6$, the aperture is much larger than at $d = f/22$, so the outside of the lens is used, causing more blurriness and aberration; thus, the depth of field is smaller than at $d = f/22$.

17-85 The acronym *laser* stands for "light amplification by stimulated emission of radiation."

17-86 A metabolic state in an atom is an excited state in which the atom can remain without undergoing spontaneous emission for a long time (on the atomic scale, about 10^{-4} to 10^{-3} s). When in this metastable state, an atom will de-excite and emit a photon if stimulated by another photon of the "right" energy that happens by, leading to stimulated emission of photons.

17-87 The fact that a laser amplifies light does not violate conservation of energy because the amount of input energy required to produce the laser light exceeds the output energy of the laser light.

Applying Your Knowledge

17-88 (a) −0.18 m, with the negative sign indicating a diverging lens

(b) +0.37 m, with the positive sign indicating a converging lens

17-89 about 5.65×10^{-7} m

17-90 objective: approx. 8.0×10^{1} cm; eyepiece: about 2.5 cm

17-91 The distance between the two stars is about 3.7×10^{18} m, or 3.9×10^{3} l.y., which is 9.1×10^{2} times as far as the star nearest our solar system.

17-92 (a) All the space telescopes need a method of converting solar energy into electrical energy, but the solar panels in (a) are not obvious, whereas those in (b) and (c) are. All three telescopes are reflecting telescopes, with (a) and (c) having segmented mirrors so the converging mirror can be as large as possible.

(b) In (b), the arrangement has two telescopes, not just one as in (a) and (c). The second, smaller telescope could be for aligning the main mirror or for gathering a different part of the electromagnetic spectrum than the main mirror. Also in (b), it looks as if the mirrors are more protected from possible collisions with particles in space, whereas the other two telescopes are not as protected. The telescopes in (a) and (c) look very futuristic, while the one in (b) looks traditional. It is not possible to see how the electromagnetic radiation is collected in (b), but in the other two telescopes, we can see that the radiation reflected from the segmented mirrors is focused to another mirror that reflects the radiation back to the very centre of the main mirror arrangement, where the signals are gathered and, presumably, sent to Earth.

(c) Without researching the Internet, it would be hard to guess which of the modern telescopes is the James Webb one, but it turns out to be the first one. It is a huge, multi-nation project that will provide an exciting replacement to the Hubble Space Telescope. (Note that the photo shown in the text shows only the side that faces away from the Sun, whereas the photo galleries on the Net show several different viewpoints.)

17-93 We know that the index of refraction of water differs from the index of refraction of air by a ratio of $\frac{n_W}{n_A} = \frac{1.33}{1.00} = 1.33$. This means that the light travelling from the water to the camera's glass lens (with an index of refraction of about 1.6) will refract less than if the camera were in air. Thus, to achieve a focal length of 200 mm in water, the telephoto lens would need a focal length in air of considerably less than 200 mm in order to provide the required refractive power. (In fact, the focal length would need to be roughly 90 mm.)

17-94 The lens can focus from 4.8 m to infinity.

17-95 The laser beam must reflect off something in order for us to see it or take a photograph of it. The easiest way to see the beam is to add mist from a water-spray container to the air. (Of course, it would be possible to see a laser beam if it were aimed toward an eye, but that would be extremely dangerous and foolish!)

17-96 about 3×10^{15} photons/s

17-97 Stars and galaxies emit light that can be analyzed using spectroscopy, which was described in Section 16.4. Gases between the stars or galaxies and Earth absorb some of the light, causing an absorption spectrum. Astronomers can analyze the absorption spectrum to determine the composition of gases in interstellar space.

17-98 (a) Fluids slowly but continuously enter the aqueous humour, but with the condition of glaucoma the liquids are unable to drain away, thus causing the liquid pressure to build up.

(b) According to Pascal's principle (Section 12.2), a change in pressure applied to an enclosed fluid at rest is transferred undiminished to all parts of the fluid and to its container. So any increase in pressure at the front of the eye is distributed throughout the entire interior of the eye.

Chapter 18 Heat and Thermodynamics

Exercises

18-1 338 K

18-2 4.0×10^2 m/s

18-3 11°C

18-4 5.5×10^4 J

18-5 1.7×10^2 mol

18-6 Thermal energy is the total energy of all the molecules in a system, including translational, rotational, and vibrational kinetic energies, and as well as electric potential energy. Heat is energy transferred between objects (or systems) as a result of a temperature difference between them.

18-7 9.9×10^2 J

18-8 79°C

18-9 2.2×10^6 J

18-10 4.3 kg

18-11 2.4×10^{-3} m

18-12 8.96 mm

18-13 6.5 cm^2

18-14 11°C

18-15 (a) approx. 0.03 cm (b) approx. 0.01 cm^3

18-16 conduction: atomic and molecular collisions transfer heat from hot regions to cold regions
convection: the physical movement of a fluid transfers heat from one region to another
radiation: the emission of electromagnetic radiation removes energy from an object

18-17 The tile floor feels cooler because its thermal conductivity is larger.

18-18 (a) 5.3×10^2 W (b) 1.4×10^6 J

18-19 0.092 W·m^{-1}·K^{-1}

18-20 (a) 1.9×10^3 W (b) −6°C

18-21 8.9 mm^2

18-22 0.55 W

18-23 2.5×10^3 J

18-24 -2.1×10^4 J/K

18-25 (a) W is the useful work done by the engine, Q_H is the heat extracted from a hot reservoir, and Q_C is the heat discarded to a cold reservoir.

(b) The useful energy is the work W, and the input energy is the heat Q_H.

(c) By the second law of thermodynamics, it is impossible to convert the disordered internal energy in the hot reservoir to ordered work with 100% efficiency.

18-26 55%

18-27 Refrigerators, air conditioners, and heat pumps are all essentially the same device. In each case, a warm vapour is compressed to produce a hot vapour that is then condensed to create a warm liquid. This liquid passes through an expansion valve to produce a cold liquid that then absorbs heat (in a unit called the evaporator) to produce a warm vapour that is returned to the compressor. The three types of appliances differ in the locations of the various components. In a refrigerator, the evaporator is inside the refrigeration box, and the condenser is outside the refrigerator box; that is, it is in contact with air in the surrounding room. In an air conditioner, the evaporator is inside a building or car, and the condenser is outdoors. In a heat pump, the evaporator is in the outdoor air or ground, and the condenser is inside a building.

18-28 (a) T_H large and T_C small (b) close (c) close

18-29 (a) 2.1 MJ (b) 3.7 MJ

18-30 27

18-31 (a) 6.5 MJ (b) 3.8 MJ

Chapter Review

Multiple-Choice Questions

18-32 (a)

18-33 (c)

18-34 (e)

18-35 (b)

18-36 (e)

18-37 (b)

18-38 (d)

18-39 (b)

18-40 (a)

18-41 (a)

Review Questions and Problems

18-42 (a) As temperature increases (decreases), the liquid expands (contracts) and moves upward (downward) in a thin tube.

(b) A bimetallic strip consists of two metals that have different coefficients of thermal expansion. As the temperature changes, the strip bends because of the different amounts of expansion of the two metals.

18-43 6.0×10^{-21} J

18-44 6.7×10^3 J

18-45 2.1×10^2 s

18-46 13°C

18-47 approx. $(2 \text{ to } 6) \times 10^5$ J

18-48 2.0×10^{-5} (°C)$^{-1}$

18-49 52°C

18-50 4.844×10^4 L

18-51 1.2%

18-52 6.0×10^{12} J

18-53 0.080 W·m^{-1}·K^{-1}

18-54 4.3×10^2 W

18-55 (a) 3.9×10^4 W (b) 49% (c) 12.4°C

18-56 The metal nails have a higher thermal conductivity than the wooden deck, and conduct heat quickly to the snow on the nail heads, melting the snow there.

18-57 6.8×10^3 K

18-58 36%

18-59 (a) 7.6×10^{12} J (b) 5.0×10^{12} J

18-60 418°C

18-61 approx. $(-2$ to $-3) \times 10^2$ J/K

18-62 For a refrigerator or an air conditioner, the useful energy is the energy Q_C removed from the cold reservoir. For a heat pump, the useful energy is the energy Q_H moved into the hot reservoir.

18-63 23°C

18-64 1.10 kW

18-65 37

Applying Your Knowledge

18-66 (a) approx. 4.4×10^2 m/s (b) No; the rms speed is close enough to the escape speed (5.0×10^3 m/s) that a significant concentration of nitrogen would not be expected.

18-67 0.11 kg

18-68 15°C/min

18-69 (a) 6.1×10^3 J (b) 5.6×10^4 J; steam gives the more severe burn.

18-70 95°C

18-71 approx. 2 to 3 cm

18-72 1.1249×10^4 kg/m^3

18-73 4.00 mm

18-74 (a) 2.255 mm (b) As the rivets expand when their temperature returns to normal, a very snug fit is ensured.

18-75 given in question

18-76 3.5 mL

18-77 6.9×10^2 W, 11°C and 89°C

18-78 107.4°C

18-79 0.05 W·m^{-1}·(°C)$^{-1}$

18-80 4.4×10^5 J

18-81 0.028°C

18-82 22°C

18-83 (a) 1.6×10^8 km (b) 2.5×10^4

18-84 In a hydroelectric plant, the energy conversion is from ordered energy (gravitational potential energy of water, and then kinetic energy of the falling water) to ordered energy (electricity). In a steam-electric plant, the energy conversion is from disordered energy (internal energy of hot steam) to ordered energy (electricity), and from the second law of thermodynamics this type of conversion cannot be done with high efficiency.

18-85 (a) 2.9×10^8 J (b) 2.3×10^8 J

18-86 approx. (6 to 8) $\times 10^6$ L

18-87 $3.03

18-88 0.065 MJ

18-89 given in question

18-90 4.0

18-91 (a) 73 MJ (b) 45 MJ

18-92 given in question

Chapter 19 Electric Charge and Electric Field

Exercises

19-1 (a) negative (b) positive (c) positive (d) negative (e) positive

19-2 approx. 1 km

19-3 (a) up, up, and down (b) up, down, and down

19-4 (a) wax negative, wool positive (b) electrons (c) off the wool

19-5 Charge is transferred from the wall to the balloon: one becomes + and one becomes −; the opposite charges attract, with a sufficiently strong force to overcome the downward force of gravity. Eventually, the excess charge on the balloon 'leaks' away and the balloon falls to the ground.

19-6 The other object could have a negative charge, or it could be electrically neutral.

19-7 Upon contact, the paper gains the same sign of charge as that on the comb; these like charges then repel each other.

19-8 Silver is more expensive than copper.

19-9 yes

19-10 no

19-11 Step 1

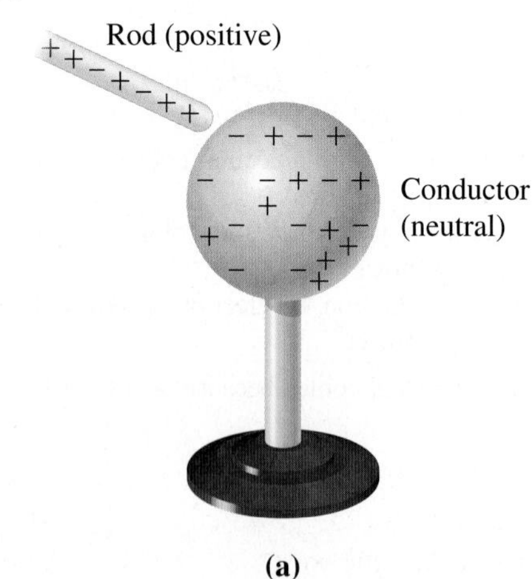

(a)

Step 2

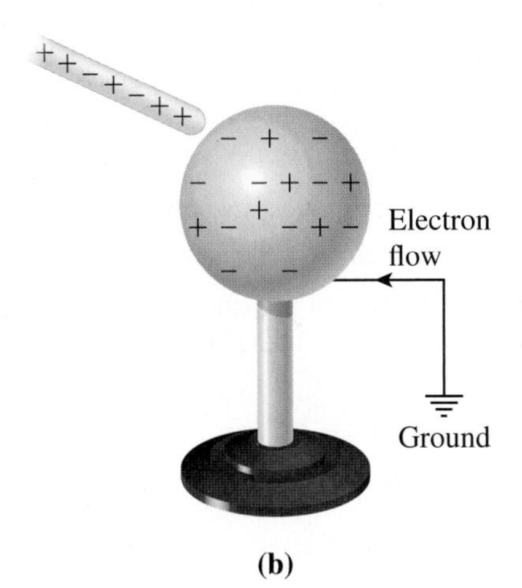

(b)

Step 3

(c)

19-12 As the rod is removed, electrons would flow through the ground connection and the final charge on the insulated conductor would be neutral.

19-13 The electrons stop flowing because the concentrations of charge on the conductor and the other object have become equal.

19-14 (a) Charging by conduction:

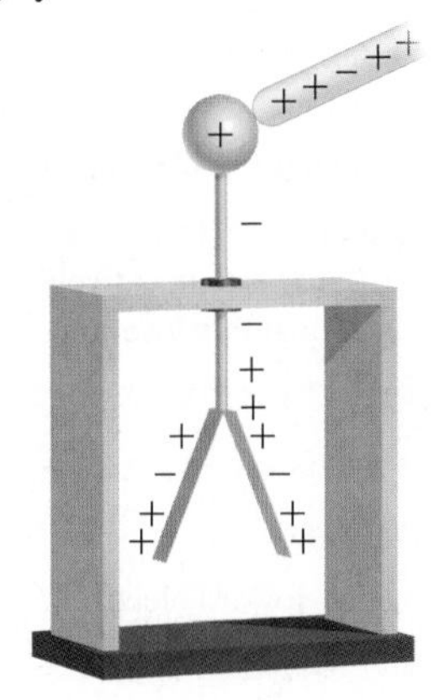

(b) Charging by induction:

Step 1

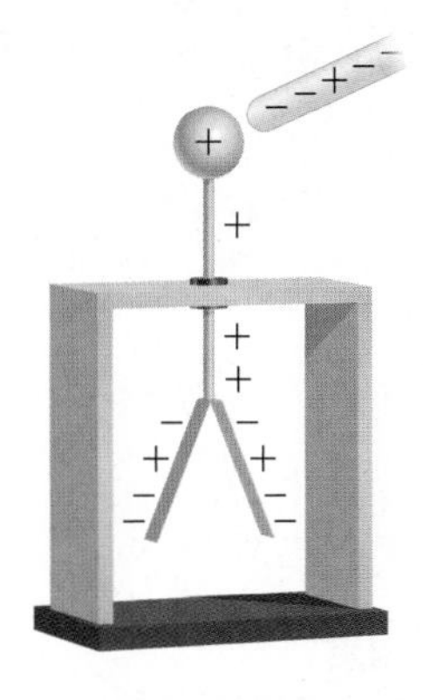

Step 2

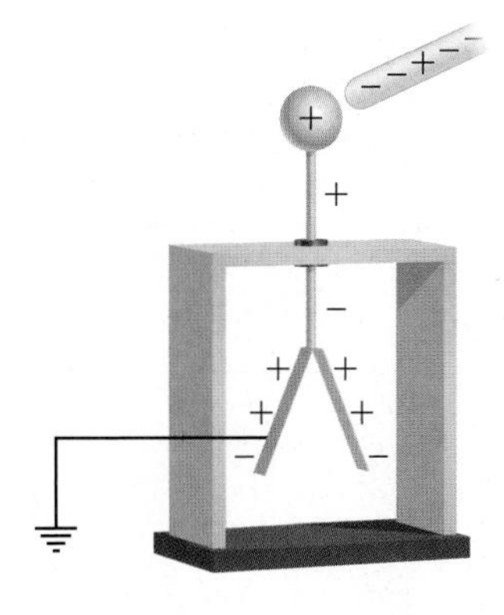

Step 3

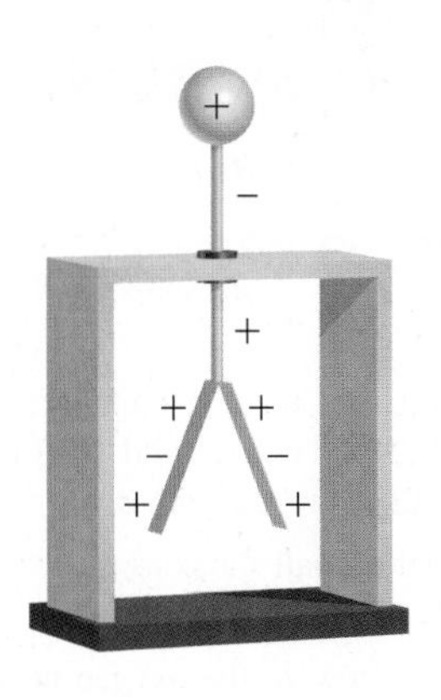

19-15 positive

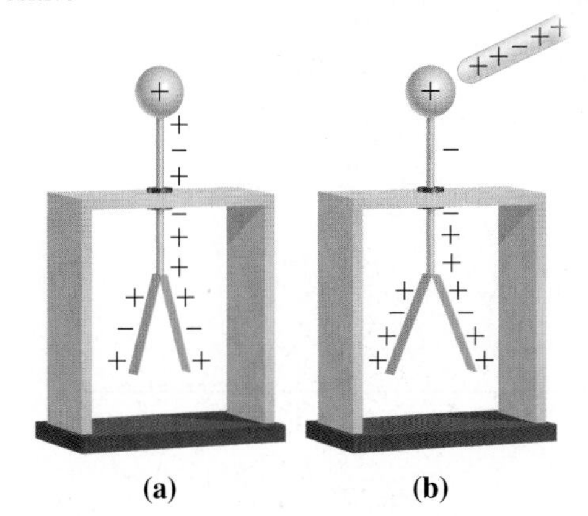

(a) **(b)**

19-16 The leaves move apart while the object is present, and return to their original position when the object is removed.

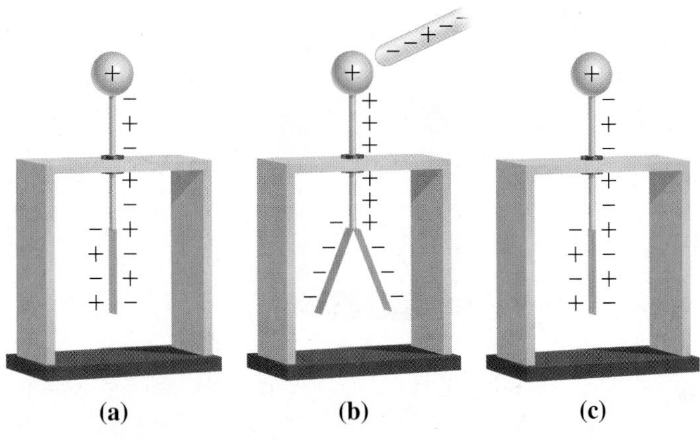

19-17 The leaves move apart while the object is present, and return to their original position when the object is removed.

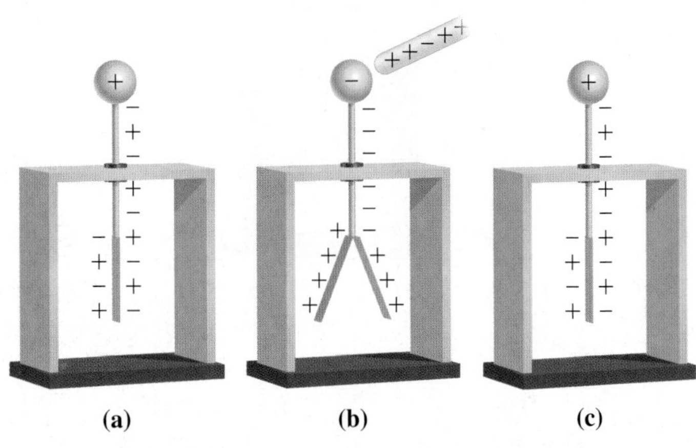

19-18 The electroscope can only tell us the sign of the charge on the unknown object if the electroscope starts off with a charge itself of known sign.

19-19 (a) The friction between the child and the slide results in a transfer of charge to the child. This excess charge serves to repel the hairs from each other.

(b) Dad may get a small shock as the excess charge on the child finds a path to ground through Dad!

19-20 (a) 8.3×10^{-8} N toward the oxygen nucleus (b) 8.3×10^{-8} N toward the electron

19-21 decreases by factor of 9

19-22 1.4

19-23

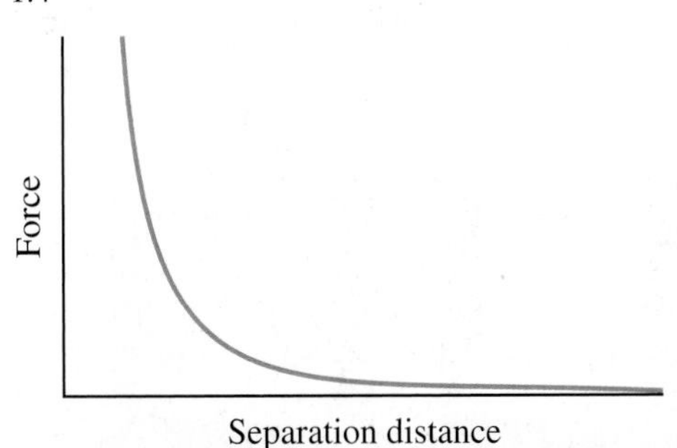

19-24 6.25×10^{18}

19-25 1.52×10^{-14} m

19-26 4.16×10^{42}

19-27 (a) 1.8×10^{2} N toward #1 (b) 1.1×10^{2} N toward #2

19-28 1.1×10^{-6} N away from centre of triangle

19-29 away from the centre of the square

19-30 (a) 1.30×10^{-8} N at 13.3° to the left of the line from the electron to the closer α (b) 1.43×10^{22} m/s² in the same direction as in (a)

19-31 (a)

31°

74.5°

$\vec{T}$ $\vec{F}_E$ $\vec{F}_G$

(b) 1.6×10^{-7} C

19-32

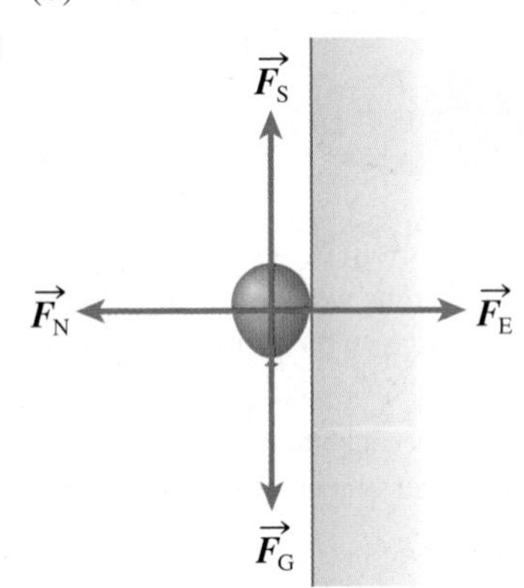

approximately 5×10^{-7} C, using mass of balloon ~10 g, radius of balloon ~10 cm, and the coeff. of static friction ~0.5

19-33 (a) away from proton (b) toward electron (c) away from carbon nucleus (d) toward negatively charged chloride ion (e) Electric field is zero at this location.

19-34 (a) 3.7×10^{-12} N (b) 2.9×10^{-11} N

19-35 6.5×10^{10} N/C away from α

19-36 7.4×10^{-6} N/C southward

19-37 (a) 6.38×10^{-10} m (b) toward electron

19-38 (a) 5.2×10^{13} N/C toward q_2 (b) 1.3×10^{14} N/C toward q_2 (and q_1)

19-39 2.9×10^{-6} m to right of q_1

19-40 5.99×10^{14} N/C at 33.2° below the $+x$-axis

19-41 approximately 10 km

19-42 (a)

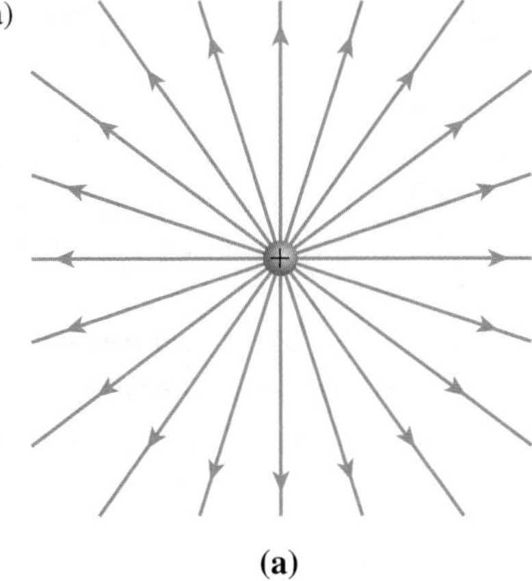

(a)

(b)

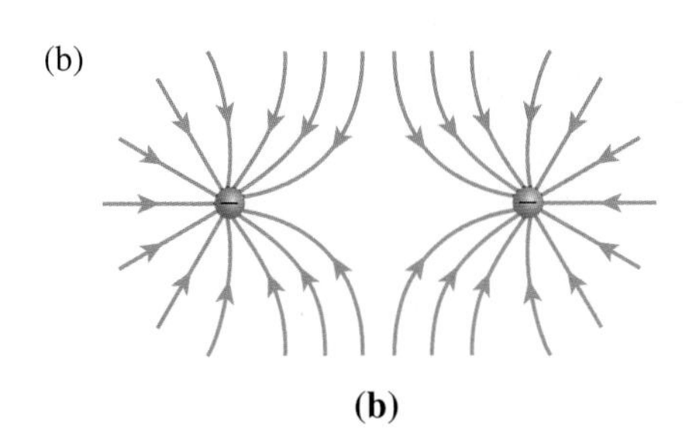

(b)

19-43 (a) The proton continues moving from left to right, and speeds up.

(b) The proton continues moving from left to right, but slows down. If the field is not turned off, the proton will eventually stop and reverse direction, and will speed up in the opposite direction.

(c) The proton continues moving from left to right with a constant velocity, but also develops an ever-increasing downward component of velocity.

(d) The proton continues moving from left to right with a constant velocity, but also develops an ever-increasing upward component of velocity.

19-44 (a) The electron continues moving from left to right, but slows down. If the field is not turned off, the electron will eventually stop and reverse direction, and will speed up in the opposite direction.

(b) The electron continues moving from left to right, and speeds up.

(c) The electron continues moving from left to right with a constant velocity, but also develops an ever-increasing upward component of velocity.

(d) The electron continues moving from left to right with a constant velocity, but also develops an ever-increasing downward component of velocity.

19-45 3

19-46 (a) 1.1×10^6 m/s (b) 8.4×10^{-7} m

19-47 (a) parallel (b) 8.5×10^{-12} s (c) 2.3×10^{-5} m

19-48 1.7×10^6 N/C from positive to negative plate

19-49 (a) risen, by 7.7×10^{-3} m (b) 2.1×10^7 m/s, 21° above horizontal

Chapter Review

Multiple-Choice Questions

19-50 (c)

19-51 (e)

19-52 (b)

19-53 (e)

19-54 (b)

19-55 (a)

19-56 (c)

19-57 (c)

Review Questions and Problems

19-58 Friction between the carpet and you results in a buildup of charge on you; when you touch the doorknob you discharge and this results in a spark. Avoid wearing shoes that have good insulation so that you don't build up charge in the first place (i.e., better connection to ground).

19-59 The charged object results in an electrical polarization in the neutral object. Opposite charges are attracted and like charges are repelled. The opposite charges are closer, and therefore the overall force is attractive.

19-60 Water moisture in the air is a good conductor; therefore, it is hard to build up a lot of charge on objects in humid air.

19-61 The sphere cannot be charged because any charge that is placed on it will immediately be neutralized by your connection to Earth.

19-62 (a)

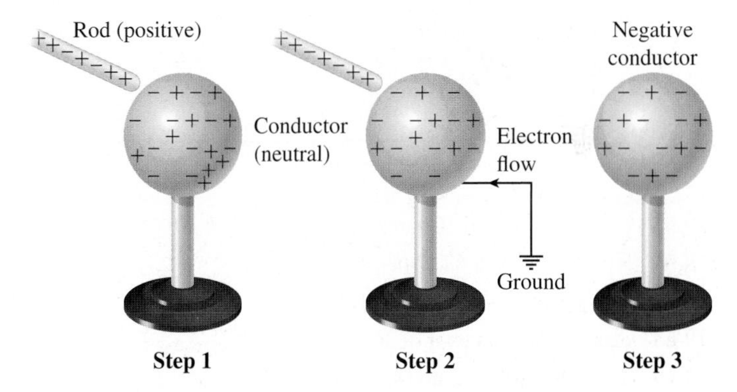

(b)

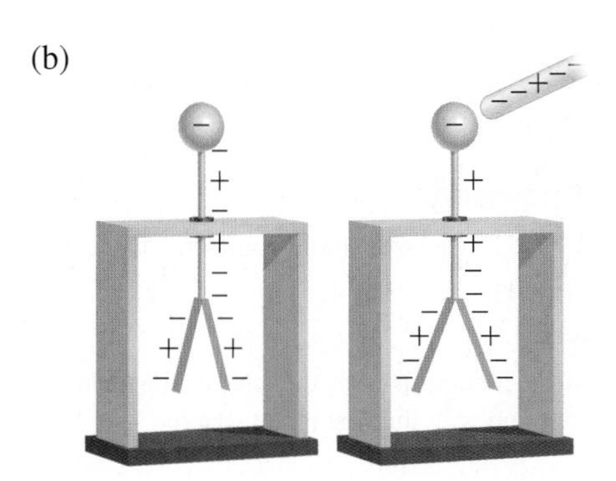

The leaves move further apart when the negatively charged rod is near.

19-63 (a) negative

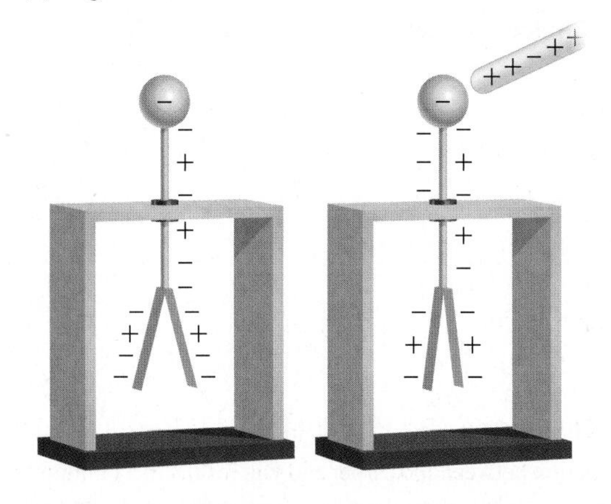

(b) Eventually, enough of the negative charges move up to the cap that the leaves are left with a net positive charge, at which point they repel each other.

19-64 The graph will be linear, will pass through the origin, and will have a positive slope of $k|q_1q_2|$.

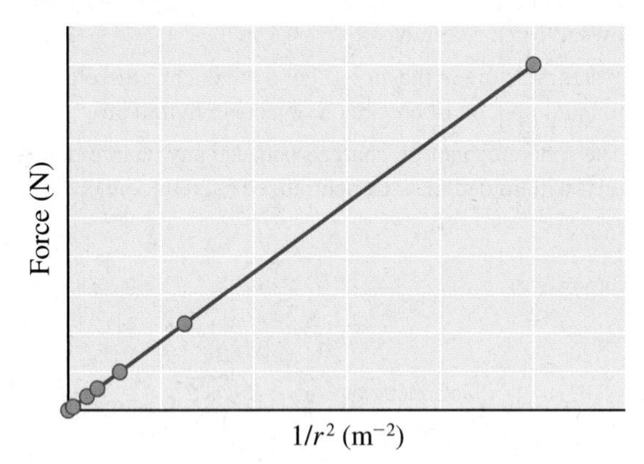

19-65 (a) 3.2×10^{-8} N (b) repulsive (c) 1.2×10^{-10} m

19-66 3.67×10^{-8} N at 68.0° below the $+x$-axis

19-67 (a) negative (b) -6.8×10^5 C

19-68 decreases, by a factor of 36

19-69 4.8×10^9 N/C, toward the electron

19-70 2.9×10^{21} N/C

19-71 1.3×10^{11} N/C, toward the centre of the triangle

19-72 (a) 4.81×10^{-2} m/s², opposite to the initial velocity (b) 2.33 cm/s, opposite to initial velocity (c) 4.97 cm

Applying Your Knowledge

19-73 (a) 7.2×10^5 N/C, away from the sodium ion (b) 1.1×10^{-13} N, toward the sodium ion

19-74 approximately 1 km

19-75 Charge one by induction with the plastic rod then charge the other by induction with the first metal sphere.

19-76 on the side opposite the negative charge, 45 cm from the positive charge

19-77 3.2×10^{-19} C; 1.6×10^{-19} C

19-78 1.1×10^{-7} C; 5.0×10^{-7} C

19-79 If the point at which $E = 0$ is closer to q_1, then $q_1 = q_2/9$; otherwise, $q_2 = q_1/9$. The charges have the same sign (either both positive or both negative).

19-80 (a) 1.1×10^6 N/C (b) 2.7×10^{-9} s (c) 4.1×10^{-4} m

19-81 3.2×10^2 N/C, opposite to the electron's velocity

19-82 (a) 2.45×10^{-3} cm (b) 3.90×10^{-16} J for each particle

19-83 (a)

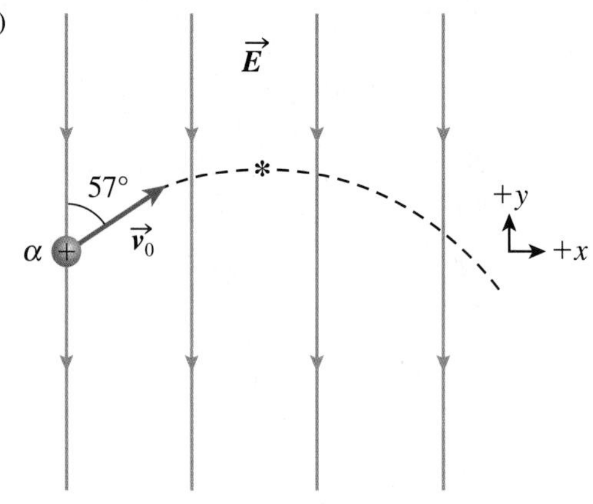

(b) At the location indicated by the asterisk in the diagram for part (a).

(c) 3.6×10^5 m/s

(d) 3.2×10^{-8} s; 1.2×10^{-2} m in the $+x$-direction

19-84 approximately 10^9 C

Chapter 20 Electric Potential Energy, Electric Potential, and Current

Exercises

20-1

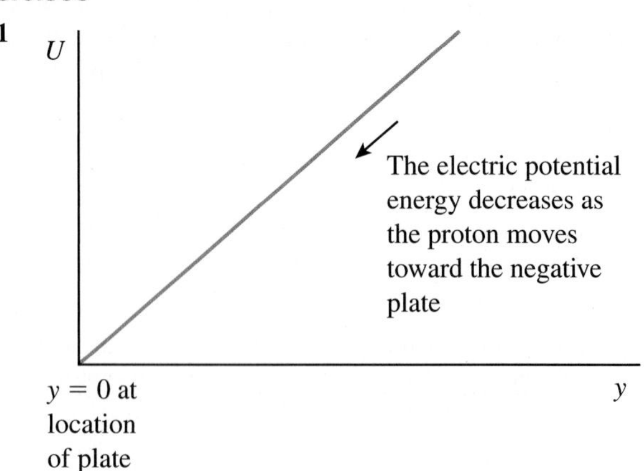

20-2 $y = 0$ at location of plate

y

The electric potential energy decreases as the electron moves away from the negative plate

U

20-3 (a) -4.8×10^{-17} J (b) 4.8×10^{-17} J

20-4 (a) 3.2×10^{-15} J (b) -3.2×10^{-15} J

20-5 (a) 1.8×10^{-29} J (b) 3.3×10^{-26} J

20-6 (a) away from B (b) toward B (c) away from B (d) away from B

20-7 (a) 9.9×10^{-18} J (b) 9.9×10^{-18} J

20-8 (a) rise (b) 3.6×10^{6} m/s

20-9 (a) positive (b) 2.54×10^{-2} m

20-10 7.91×10^{5} m/s

20-11 approximately 2×10^{2}. Assumptions: skydiver mass ~50 kg and height ~4000 m; key on a kite: charge ~1 mC, electric field in a storm ~10^{4} N/C, height ~10 m

20-12 (a)

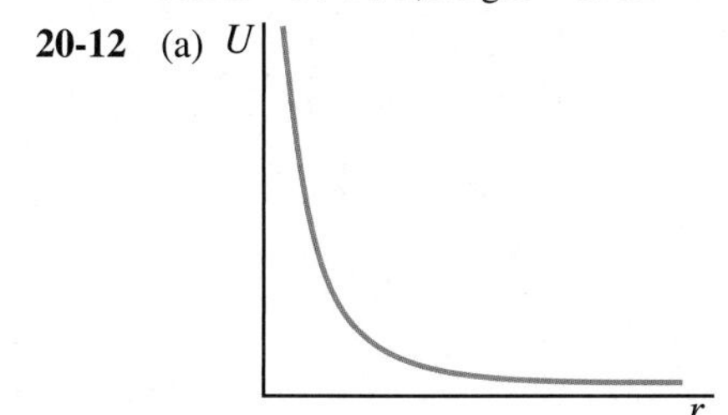

(b)

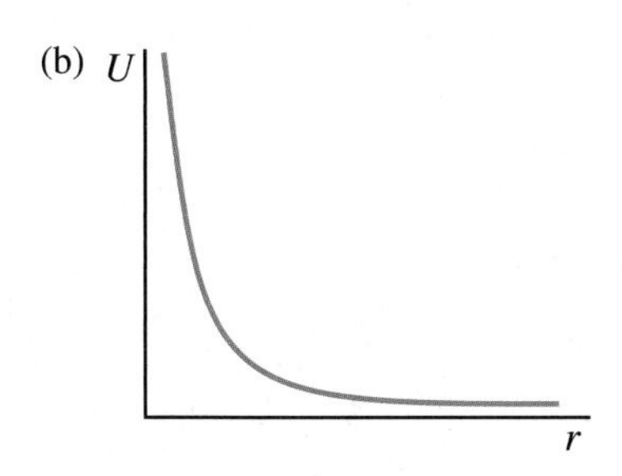

(c)

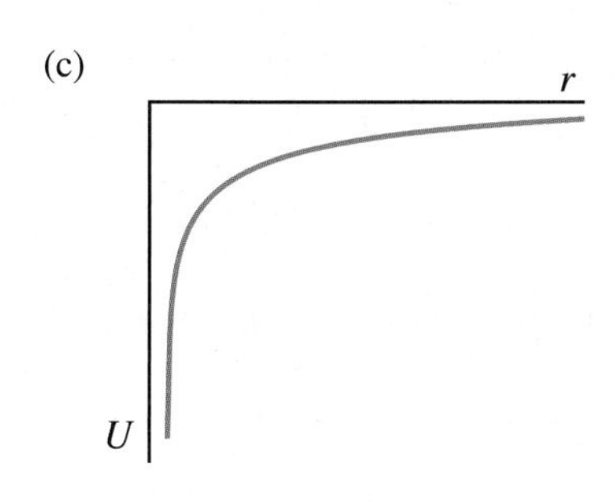

20-13 (a) zero (b) decreases (c) toward each other, in the direction of decreasing U

20-14 3.8×10^{-14} J

20-15 -4.3×10^{-18} J

20-16 approximately 4×10^{7} J based on the electron and proton in each atom having an average separation of ~1×10^{-10} m, and a glass of water (250 mL, or 250 g) containing ~8×10^{24} molecules of water or ~16×10^{24} atoms of H

20-17 1.50×10^{-10} m

20-18 (a) separation increases, electric potential energy decreases (b) 1.2×10^{-11} m

20-19 2.6 V

20-20 (a) from high to low electric potential, outside to inside; (b) -1.4×10^{-20} J

20-21 (a) 304 eV at A and 456 eV at B (b) from B to A in the direction of decreasing electric potential energy

20-22 (a) 36 J (b) 9.0 W

20-23 (a) 7.0×10^{2} MJ (b) higher

20-24 (a) 23.4 eV (b) 8.18×10^{-14} J (c) 1.86 MeV

20-25 (a) -4.2×10^{6} eV (b) $+4.2 \times 10^{6}$ eV

20-26 -4×10^{6} V or -4 mv

20-27 1.28×10^{6} m/s

20-28 Each foot is at the same electric potential; therefore $\Delta V \sim 0$ and therefore $\Delta U \sim 0$.

20-29 from B toward A

20-30 $+5.8 \times 10^{3}$ V

20-31 approximately 3×10^{4} V

20-32 1.2×10^{7} V/m

20-33 (a) +6.0 V (b) +4.5 V (c) 5.5×10^{2} V/m, from the positive plate to the negative plate

20-34 -5.71×10^{-15} J

20-35 1.2×10^{5} m/s

20-36 (a) -1.15 V (b) $+1.15$ V

20-37 2.4×10^{-8} m

20-38 16.9 V

20-39 (a) -2.70×10^{-18} J

(b) The reference position is at a point infinitely far away. The negative electric potential energy means that the chloride ion is in a more stable configuration (lower potential energy) at this location than infinitely far away, which is consistent with the attractive force between the negative chloride ion and the positive source charges.

20-40 2.4 cm

20-41 The spark looks like it is slightly longer than the width of the man's hand, so approximately 15 cm in length; therefore, the potential difference is approximately 4.5×10^{5} V.

20-42 2.0 A

20-43 8.1 C

20-44 4.8×10^{-11} A

20-45 The direction of the current is from the ground to the cloud.

20-46 (a) 2.35 C (b) 1.47×10^{19}

20-47 (a) 1.1×10^2 A (b) 1.3×10^{10} J

20-48 2 km

20-49 (a) The direction of the current is toward the negative terminal. (b) 0.776 A

20-50 (a) max. time turning the key ~5 s; current ~100 A; therefore, charge moved is 500 C
(b) 500 C moved through 12 V gives a change in electric potential energy of ~6000 J; therefore, the power output is approximately 1000 W in 5 s

Chapter Review

Multiple-Choice Questions

20-51 (b)

20-52 (a)

20-53 (e)

20-54 (b)

20-55 (a)

20-56 (d)

Review Questions and Problems

20-57 (a) from east to west (b) 2.1×10^3 V/m

20-58 1.1×10^5 m/s

20-59 -6.7×10^{-17} J

20-60 4.2×10^{-2} eV

20-61 (a) toward higher electric potential
(b) -25 keV $= -4.0 \times 10^{-15}$ J (c) $+4.0 \times 10^{-15}$ J

20-62 3.5×10^{-2} V/m

20-63 (a) 141 V/m (b) The field is downward.

20-64 5.0×10^1 V

20-65 (a) 5.0×10^1 eV (b) -5.0×10^1 eV

20-66 (a)

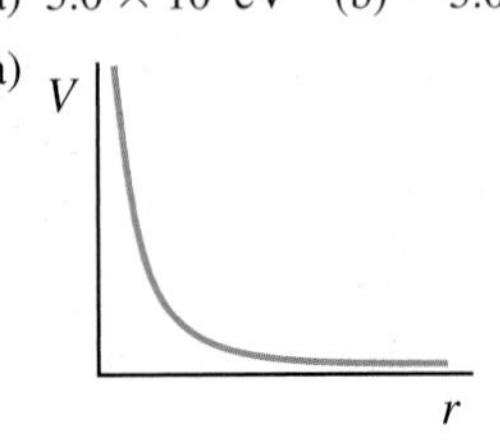

(b)

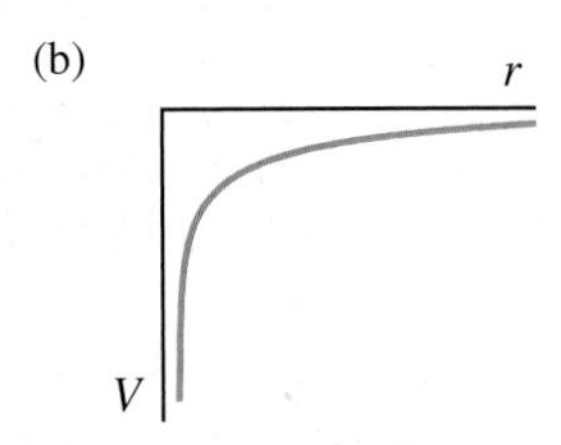

20-67 58 V

20-68 6.86×10^{-12} m

20-69 (a) 22.6 C (b) 2.7×10^2 J (c) 9.0 W

20-70 0.877 A

20-71 41 J

20-72 The direction of the current is from the screen toward the back of the main tube.

Applying Your Knowledge

20-73 (a) toward planet X (b) 565 m/s

20-74 565 m/s

20-75 (a) 1.49×10^{-16} J (b) The proton moves away from the uranium nucleus when released from rest. (c) 1.57×10^{-10} m

20-76 (a) 14 keV (b) 2.2×10^{-15} J

20-77 (a) 6.21×10^{-12} m from the K^+ ion (b) nowhere

20-78 7.23×10^{-3} m

Chapter 21 Electrical Resistance and Circuits

Exercises

21-1 9.2 Ω

21-2 (a) 0.080 A (b) 0.24 A

21-3 If a circuit device obeys Ohm's law, the current through the device changes proportionally to changes in the potential difference across it. If a circuit device does not obey Ohm's law, the current varies non-linearly when the potential difference is varied.

21-4 (a) 3.3×10^2 Ω ± 10% (b) 0.24 Ω ± 5% (c) 3.9×10^5 Ω ± 20%

21-5 (a)

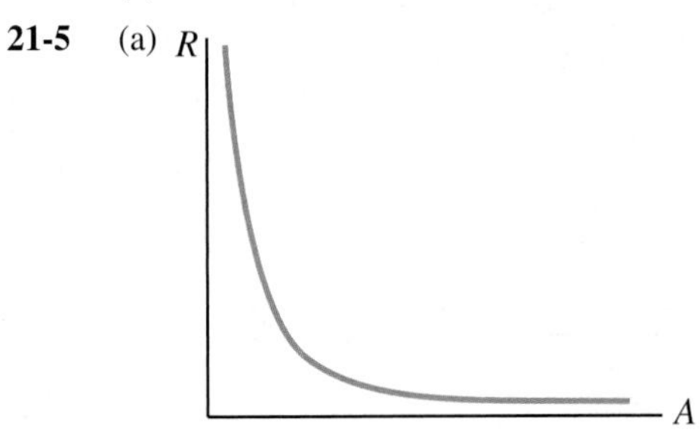

(b)

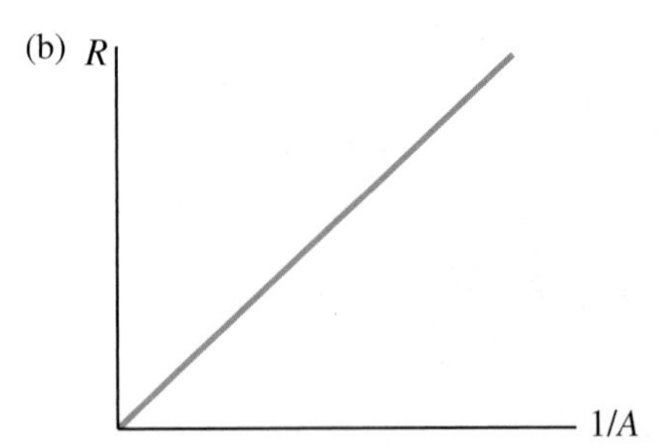

21-6 0.20 Ω

21-7 0.098 m

21-8 (a) $R_{Fe}/R_{Cu} = 1.0$ (b) $A_{Fe}/A_{Cu} = 5.7$ (c) $r_{Fe}/r_{Cu} = 2.4$ (d) $d_{Fe}/d_{Cu} = 2.4$

21-9 Resistivity increases with temperature because the increased thermal energy of the atoms in the conductor results in more collisions with the conduction electrons.

21-10 $2.8 \times 10^{-8}\ \Omega\cdot$m

21-11 3.4%

21-12 Using $\rho = 1.7 \times 10^{-8}\ \Omega\cdot$m, $L \sim 500$ m, and $A \sim 1 \times 10^{-8}$ m^2 gives $R \sim 1000\ \Omega$.

21-13

$R = 330\ \Omega$

15 V

$I = 0.045$ A

21-14 (a) clockwise (b) 0 V, 6.00 V, 0 V, −6.00 V (c) 0 V, 6.00 V, 6.00 V, 0 V (d) 0.455 A in each case

21-15 (a) 1.36 Ω (b) toward top of diagram (c) 3.24 V

21-16 (a) 4.1 s (b) 31 J

21-17 Assuming that the bulb has a resistance of approximately 10 Ω and is connected to a single 1.5 V battery, the current is ~0.15 A.

21-18 (a) $3.0 \times 10^{1}\ \Omega$ (b) 7.5 Ω

21-19 171 Ω

21-20 (a) resistor network with equivalent resistance of 300 Ω, where each resistor R is 100 Ω

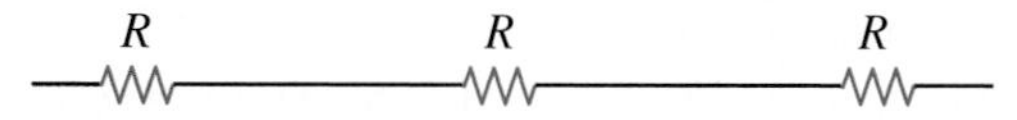

(b) resistor network with equivalent resistance of 50 Ω, where each resistor R is 100 Ω

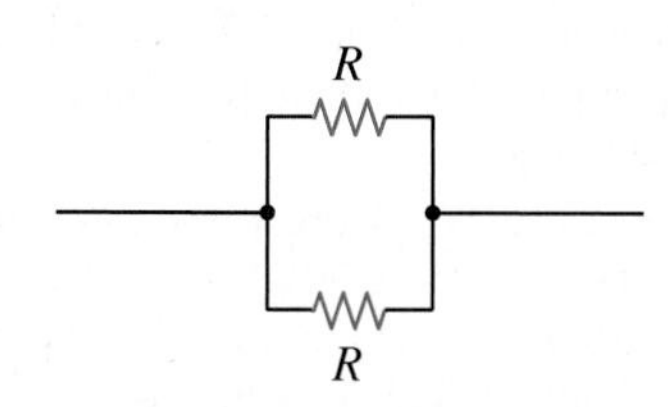

(c) resistor network with equivalent resistance of 25 Ω, where each resistor R is 100 Ω

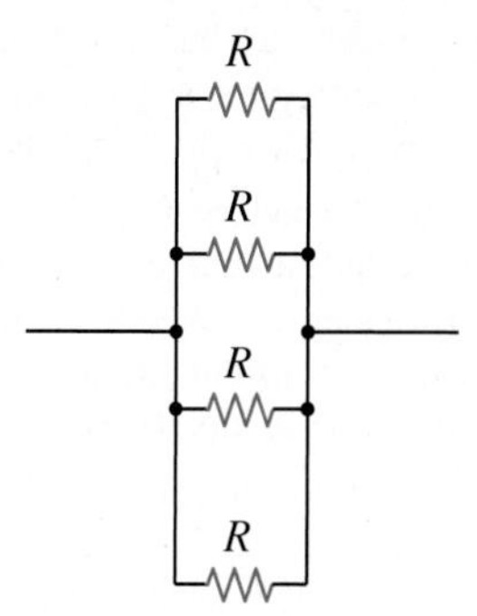

(d) resistor network with equivalent resistance of 250 Ω, where each resistor R is 100 Ω

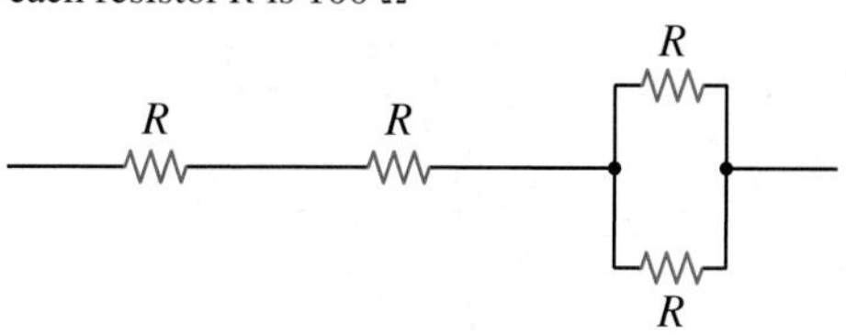

21-21 0.15 A

21-22 0.69 A

21-23 1.1 A

21-24 (a) 0.046 A (b) from A to B (c) +2.2 V, −2.2 V (d) 3.8 V (e) +3.8 V

21-25 (a) 15 V for each resistor (b) 0.10 A for each 150 Ω resistor (c) 0.26 A (d) 0.060 A (e) 0.26 A (f) 0.20 A

21-26 (a) A possible sketch for the circuit is

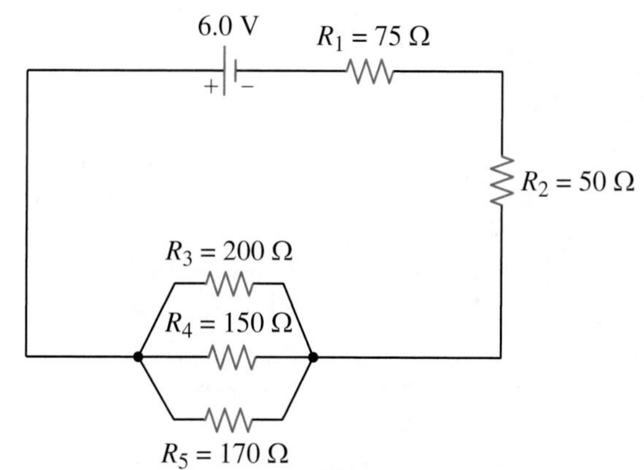

(b) 0.033 A (c) 0.013 A (d) 1.6 V

21-27 (a) 15 V and 3 A for each resistor (b) 0 V and 0 A for the middle resistor, 15 V and 3 A for the other resistors (c) 0 V and 0 A for all resistors

21-28 (a) 1.0×10^{1} V, 0.050 A (b) 8.4 V across the 300 Ω resistor, 6.6 V across the 100 Ω resistor, 0.028 A

21-29

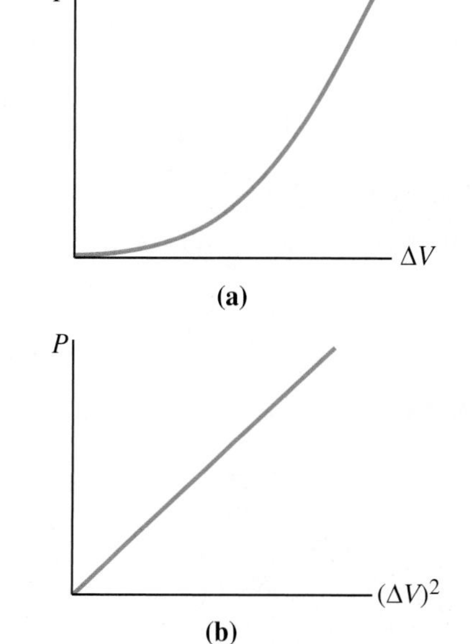

21-30 P vs R

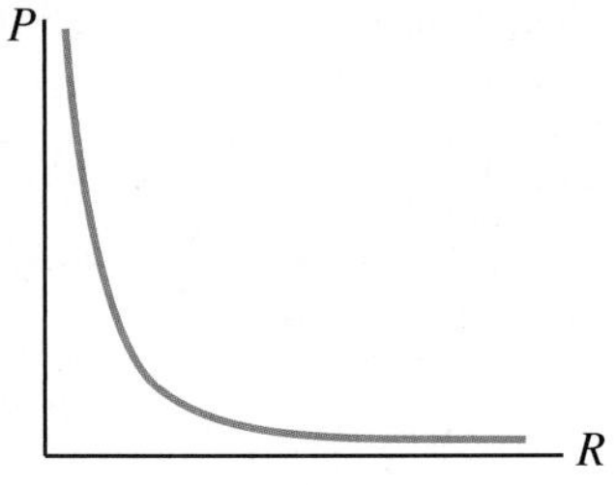

(a)

P vs $1/R$

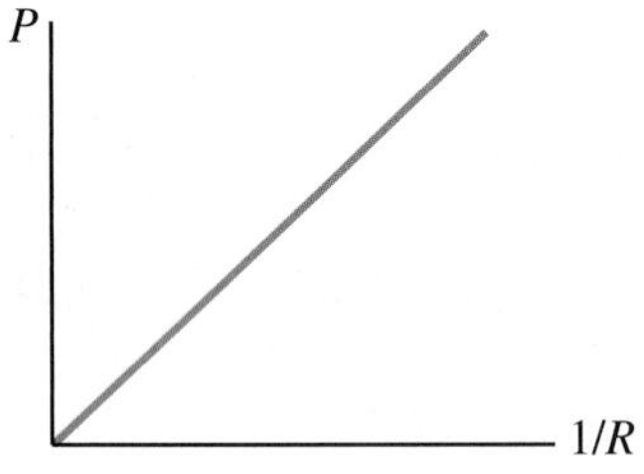

(b)

21-31 7.5 W

21-32 86 W

21-33 (a) 0.96 W (b) 0.96 W

21-34 (a) 2.7 A through each bulb (b) 2.0 V across each bulb (c) 0.74 Ω

21-35 (a) increases by 36 J (b) decreases by 36 J

21-36 (a) 11 W (b) 58 V (c) $P_{100} = 34$ W; $P_{200} = 17$ W (d) 62 W

21-37 Assume approx. 500 W/m^2 of sunlight on a sunny day. Rooftop panel area is about 50 m^2, which gives 25 000 W received by panels. Assuming about 15% efficiency, about 3750 W of energy is generated, or enough for approx. 40 light bulbs (100 W bulbs).

21-38 ΔV

12 V

Time

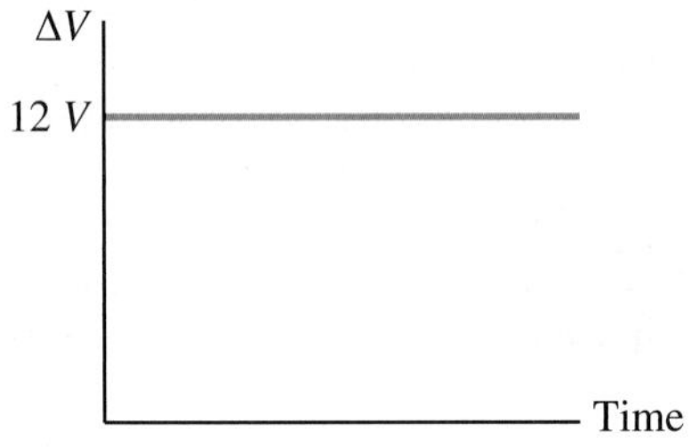

21-39 (a) 0.50 A (b) 0.71 A (c) 2.4×10^2 Ω

21-40 1.3×10^3 Ω

21-41 (a) 0.83 A through each bulb (b) 0.83 A through each of the other two bulbs

21-42 (a) 0.315 kW·h (b) 1.13×10^6 J (c) 4.1¢

21-43 (a) 9.17 A (b) 0.500 A (c) 9.67 A (d) no

21-44 919 W

Chapter Review

Multiple-Choice Questions

21-45 (b)

21-46 (a)

21-47 (c)

21-48 (a)

21-49 (e)

Review Questions and Problems

21-50 6.0×10^{-3} V, or 6.0 mV

21-51 R vs L

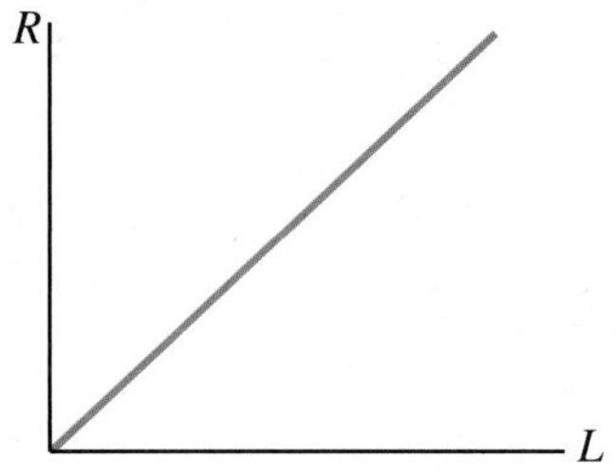

21-52 0.11 mm

21-53 $d_{Al}/d_{Cu} = 1.3$

21-54 4.3×10^{-3} (°C)$^{-1}$

21-55 on

21-56 (a) 15 V (b) 0 V (c) −15 V (d) 0 V (e) 0 V (f) 0 V

21-57 86 Ω

21-58 3.5 A

21-59 (a) 0.057 A (b) 14 V (c) 0.035 A (d) 21 V

21-60 (a) The voltmeter would be in parallel with the 250 Ω resistor, with one lead of the voltmeter connected just to the left of the resistor and the other lead connected just to the right of the resistor.

(b) The ammeter would be connected in series with the 350 Ω resistor, either just above the resistor on the circuit diagram or just below it.

21-61 $\Delta V_1 = 1$ V; $\Delta V_2 = 2$ V; $\Delta V_3 = 3$ V

21-62 (a) from high to low electric potential (b) from low to high electric potential

21-63 (a) The two bulbs are equal in brightness, and this brightness is less than the brightness of the original single bulb. When the second bulb is added in series, the total resistance in the circuit is increased, and the current decreases (from Ohm's law, $I = \Delta V/R$). Since the current has decreased, the brightness is less than in the case of the single bulb. Since the two resistors have the same (reduced) current through them, the brightness of each is equal.

(b) The two bulbs are equal in brightness, and this brightness is equal to the brightness of the original single bulb. When the second bulb is added in parallel, the same potential difference (i.e., the battery potential difference) is applied to each bulb, and the same amount of current passes through each bulb, giving equal brightness. Since the battery potential difference is applied to each bulb, the current through each bulb is the same as if the other bulb didn't even exist. Hence, the brightness of each bulb is the same as the brightness of the original single bulb.

21-64 (a) In brightness, C > A = B. The battery potential difference is being applied across bulb C, and also across the series combination of A & B. The resistance of A & B together is greater than that of C alone; hence, the current through A & B is less than that through C, and A & B are not as bright as C.

(b) The brightness of C decreases to zero; A & B are unchanged (since the battery potential difference is still applied across A & B together). Current at 3 has decreased to zero; current at 1 and 2 has decreased since the battery is no longer providing current to bulb C.

(c) There is now no current generated anywhere in the circuit; none of the bulbs are lit.

(d) The left side of bulb A and the right side of bulb A are now at the same electric potential, and hence the potential difference across A is zero. Therefore, no current travels through A, and its brightness decreases to zero. The battery potential difference is now being applied across bulbs B and C in parallel. The brightness of C remains unchanged, and the brightness of B increases until it is the same as C. The current at point 3, that is, the current through C, is unchanged. The current at point 2 decreases; it is now only the current through C, whereas previously it was the sum of the current through C and the current through A & B. The current at point 1 is the total current in the circuit; the current through C is unchanged, but the current through B has increased, and therefore the current at 1 increases.

21-65 Electrical circuits in automobiles are connected in parallel so that all the circuits will be provided with the same potential difference (typically 12 V) regardless of how many circuits are being used.

21-66 P

I

(a)

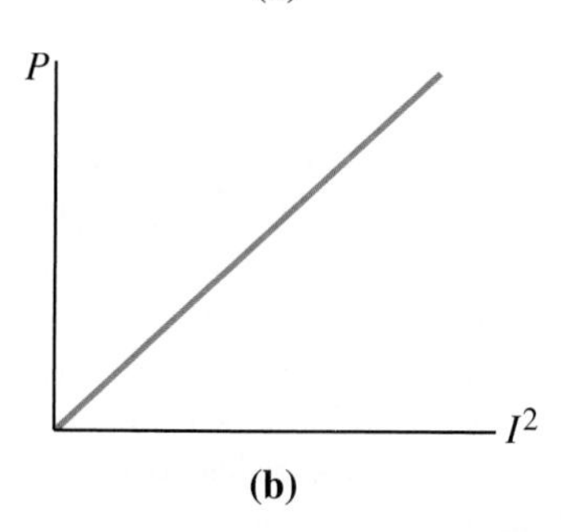

(b)

21-67 Although the resistance of an extension cord is small, it is not zero. Since resistance is proportional to length, the resistance of an ultra-long extension cord can be large enough that there is an appreciable decrease in electric potential along its length. Therefore, the rms potential difference available to an appliance plugged into the end of it will be less than 120 V.

21-68 (a) Resistance is inversely proportional to cross-sectional area, and therefore a thinner filament has a higher resistance.

(b) Since the potential difference ΔV is constant and the resistance R is increasing, the power P must be decreasing: from the relation $P = (\Delta V)^2/R$.

21-69 1.0×10^{-2} s

21-70 156 V

21-71 It is safest to have a large internal resistance within the high voltage power supplies because if a person were to accidentally touch the high voltage terminals, the large internal resistance would prevent a large current from travelling and causing serious injury to the person.

21-72 (a) $P_{16} = 4.0$ W; $P_{32} = 2.0$ W each (b) 8.0 W

21-73 (a) 11 A (b) 0.034 C (c) 2.7×10^2 J

21-74 (a) 21 A (b) 1.0×10^1 A (c) lower current required, therefore safer

Applying Your Knowledge

21-75

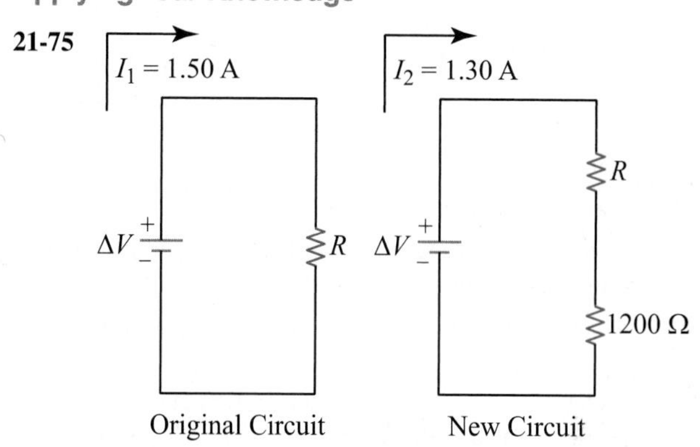

Original Circuit New Circuit

$R = 7.8$ kΩ

21-76 0 V; 0 V

21-77 0.51 Ω

21-78 A, B, E, or F

21-79 See Solutions Manual for proof.

21-80 (a) 0.0360 A (b) 0.248 A

21-81 316 Ω

21-82 38 Ω

Chapter 22 Magnetism

Exercises

22-1 (a) magnetic if it is attracted to both poles (b) a magnet if it is attracted to one pole and repelled by the other

22-2

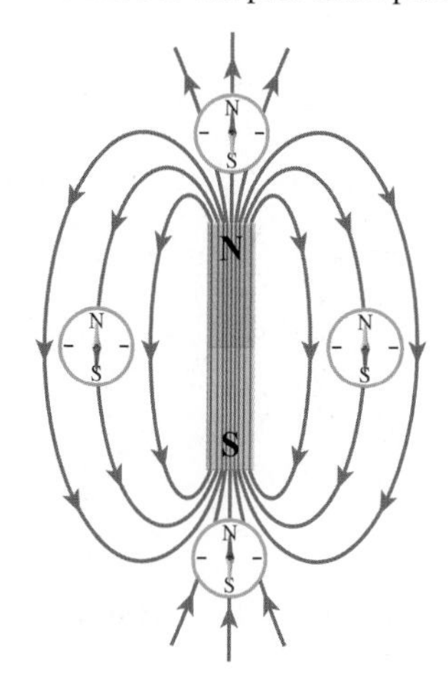

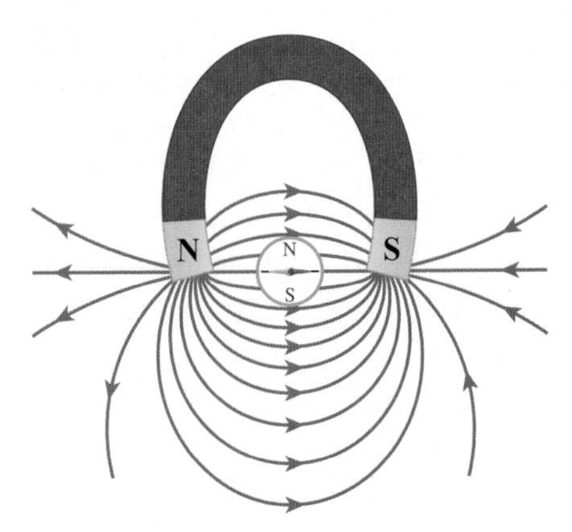

22-3 to the right

22-4 electric and magnetic fields

22-5 (a) All domains are aligned. (b) The force of the hammer blows physically reorients the domains.

22-6 The domains in the iron filings align with the external field.

22-7 The magnetic field of the magnet causes the domains in the nail to tend to be aligned with the magnetic field. As such, the domains are oriented so that their south poles are facing the north pole of the magnet. In this way, the nail itself becomes a weak magnet that is then attracted to the stronger magnet.

22-8 microwave oven, magnetic strip on wall to hold utensils, magnetic closures for drawers and cupboard doors, induction stovetop, magnets on fridges ...

22-9 Approximately 10^{23} atoms; magnetize the needle in the field of a strong permanent magnet, float the needle on a cork in a dish with water and give it a spin. The needle will align with Earth's magnetic field.

22-10 ~2500 km, ~6 km/year

22-11 (a) circa 3500 (b) no more magnetic storms, more cosmic rays entering the atmosphere causing heating, some animals may have navigational issues

22-12 2.00×10^{-6} T, north

22-13 to the left

22-14 (a)

B

r

(b)

B

x

22-15 (a) 8.31 A (b) counterclockwise as viewed from above

22-16 current

22-17 clockwise

22-18 2.70 A

22-19

$\vec{B}$

I

− +

22-20 right

22-21 below the magnet

22-22 approx. 50 μT

22-23 approx. NW

22-24 out of the page

22-25 (a) into page (b) into page (c) zero force (d) zero force (e) right (f) left (g) up (h) up (i) out of the page (j) out of the page

22-26 zero force

22-27 3.13×10^{-16} N, vertically down

22-28 vertically down

22-29 $T = kg/(C \cdot s)$

22-30 It is not possible with a magnetic field because the magnetic force cannot change the speed of a charge, only its direction; it is possible with an electric field.

22-31 1.4×10^{16} m/s^2, into the page

22-32 0.127 m

22-33 2.03×10^{-6} s

22-34 2.5×10^{-3} T

22-35 helical motion with the axis of the helix along the direction of the field and the circular component in the plane perpendicular to the field

22-36 3.19×10^{3} N/C, south

22-37 Zero; the force always acts perpendicular to the velocity. The magnetic force cannot change the speed of the charge, only its direction of motion because of this perpendicularity.

22-38 approx. 60 cm in diameter

22-39 (a) out of the page (b) to the left (c) up, toward the top of the page (d) out of the page (e) zero force

22-40 east

22-41 2.8×10^{-3} N

22-42 8.3×10^{-5} N (up, toward the top of the page)

22-43 vertically down

22-44 Clockwise; the force on the left side of the loop is up, and the force on the right side of the loop is down.

22-45 (1) Flip the magnet so that the north pole is on the left and the south pole is on the right. (2) Change the terminals of the battery so that the left side of the loop is connected to the negative terminal and the right side of the loop is connected to the positive terminal.

22-46 $\vec{F}_{AB}$ = zero; $\vec{F}_{AC} = 1.18 \times 10^{-4}$ N (up); $\vec{F}_{BD} = 1.18 \times 10^{-4}$ N (down)

22-47 (a) attract (b) repel

22-48 1.2×10^{-7} N, toward the other wire

22-49

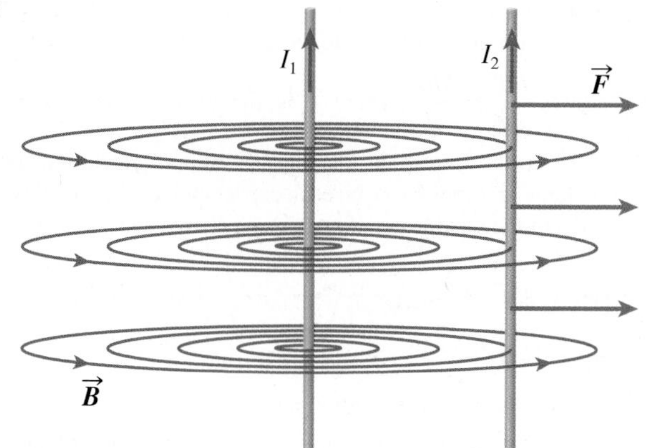

22-50

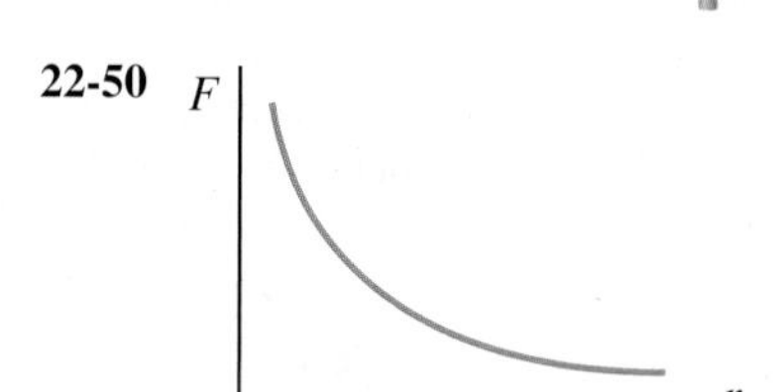

22-51 2.43×10^{-2} A

Chapter Review

Multiple-Choice Questions

22-52 (b)

22-53 (b)

22-54 (c)

22-55 (a)

22-56 (c)

22-57 (d)

22-58 (a)

Review Questions and Problems

22-59 to the right

22-60 electric field: associated with any charged particle, electric force is parallel (or antiparallel) to the electric field, electric field acts on any nearby charges; magnetic field: associated with moving charges only, magnetic force is perpendicular to the magnetic field, magnetic field only acts on nearby moving charges

22-61

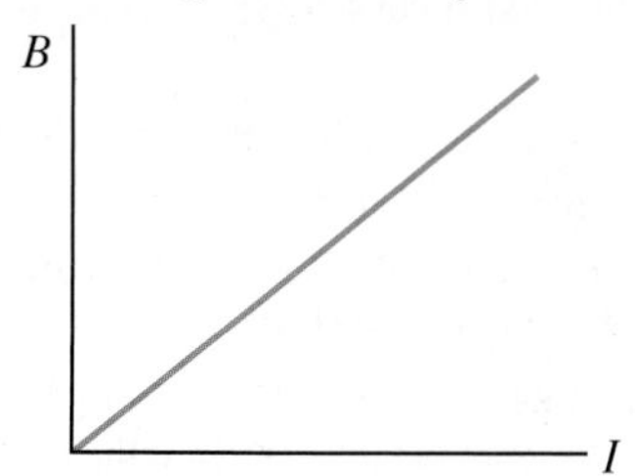

22-62 3.52 A

22-63 (a) out of the page (b) into the page (c) to the right

22-64 (a)

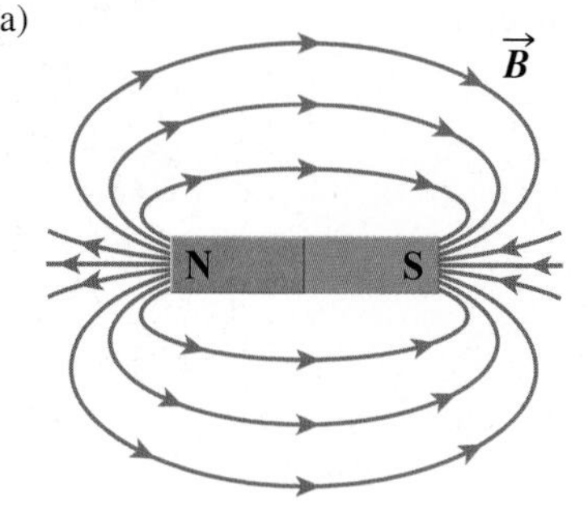

(b)

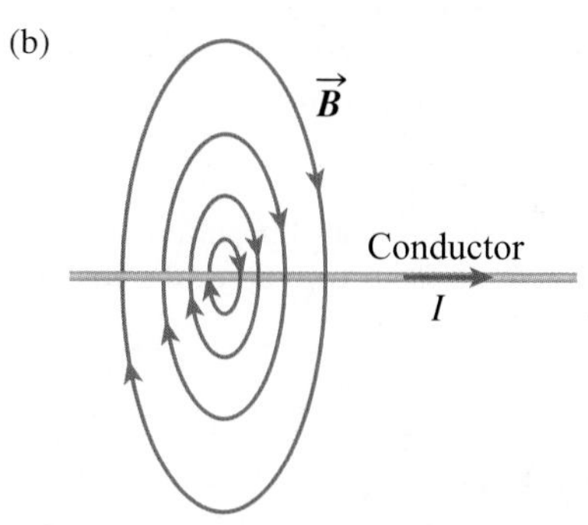

(c)

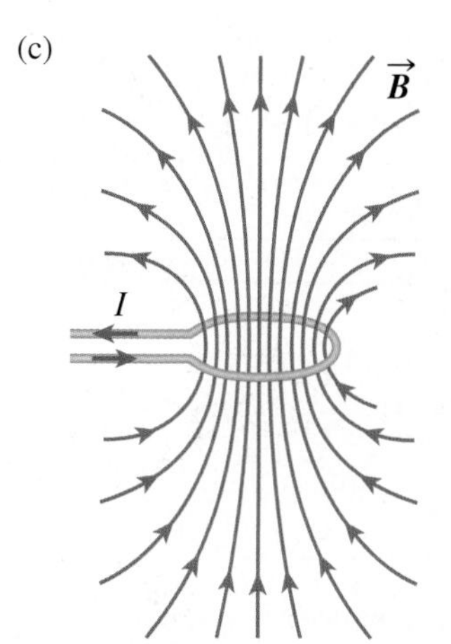

(d)

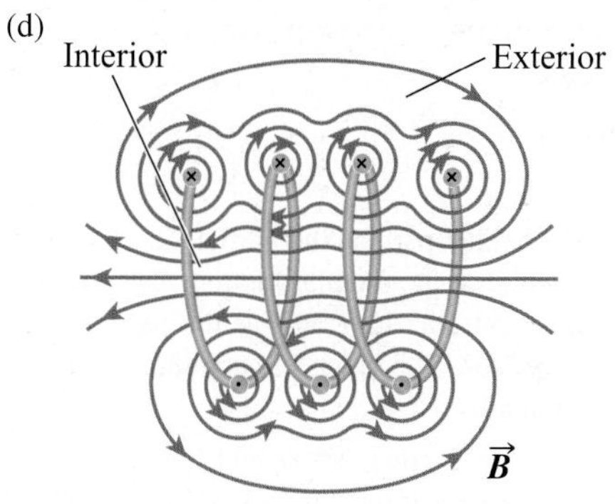

22-65 7.20×10^{-4} T

22-66 (a) increase (b) increase (c) decrease

22-67 3.03×10^{-14} N, into the page

22-68 3.03×10^{-14} N, out of the page

22-69 The particle travels either parallel or anti-parallel to the magnetic field.

22-70 4.44×10^{6} m/s

22-71 6.9×10^{-6} N, out of the page

22-72 F

0 10 20 30 40 50 60 70 80 90

Angle (degrees)

22-73 up, toward the top of the page

22-74 (1) increase the battery potential difference (2) more coils of wire, (3) insert a ferromagnetic core inside the coil (4) stronger magnets

22-75 (a) repel (b) 1.8 m

Applying Your Knowledge

22-76 $B = \dfrac{\mu_0 NI}{2R}$

22-77 4.02×10^{-4} T

22-78 $N \geq 2$

22-79

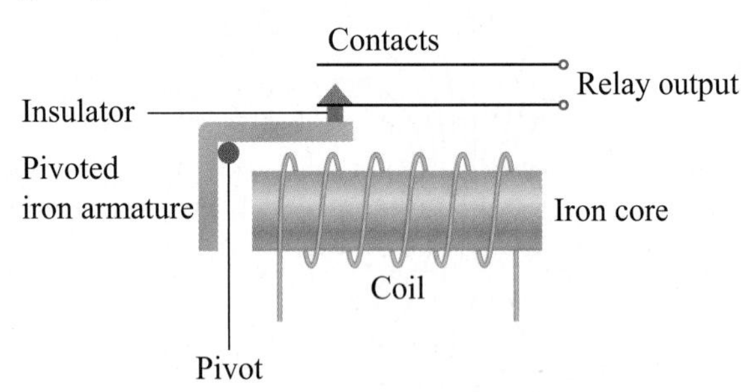

amplifiers, irons, hot water heaters, oven/stove

22-80 The top of the solenoid is the north pole.

22-81 counterclockwise

22-82 46.7° north of west

22-83 9.1×10^{-2} T

22-84 It will follow a helical path as the particle is accelerated in the plane perpendicular to the circular motion. The coil of the helix will stretch out along the long axis as the particle increases in speed in this direction. The radius of curvature will also decrease as the speed increases.

22-85 Originally one wire carried 2.0 A and the other wire carried 3.0 A. The 2.0 A current is then decreased by 1.0 A and the 3.0 A current is increased by 1.0 A.

22-86 $r^2 = \dfrac{2}{c^2}\dfrac{m_e}{e}\dfrac{\Delta V}{I^2}$; the slope of r^2 vs. ΔV gives m_e/e by: $\dfrac{m_e}{e} = \text{slope}\cdot\dfrac{c^2 I^2}{2}$;

the slope of r^2 vs. $1/I^2$ gives m_e/e by: $\dfrac{m_e}{e} = \text{slope}\cdot\dfrac{c^2}{2\Delta V}$

22-87 Sound-system speakers, starter motor, alternator, fan motors (interior and engine), engine speed sensors, door locks, magnetic door closures, side-view mirror adjuster motors, windshield wiper motor, DVD player, seat adjuster motor, window motor, etc.

22-88 $\Delta V = \dfrac{IdB}{nqA} = \dfrac{IdB}{\rho A}$, where d is the distance between top and bottom surfaces, n is the number of charge carriers per unit volume, q is the charge on each carrier and A is the cross-sectional area of the conductor; $\rho = nq$ is the charge density (C/m^3)

Chapter 23 Electromagnetic Induction

Exercises

23-1 (a) negative current reading on ammeter
(b) negative current reading on ammeter, but larger than in (a)
(c) no current
(d) positive current reading on ammeter
(e) negative current reading on ammeter
(f) negative current reading on ammeter, but larger than in (a)

23-2 Polarity is reversed, but it is the same magnitude.

23-3 (a) no (b) yes

23-4 (a) Yes; the magnetic field changes in the second circuit as the switch is closed in the first.
(b) No; once the switch has been closed for a bit, the current in the first circuit is constant, so there is no changing magnetic field in the second circuit and therefore no induced current.

23-5 (a) Yes; the wire moves perpendicular to the field. (b) No; the wire moves parallel to the field, therefore no voltage. (c) Yes; the wire moves in a direction that is not parallel to the field, so there is a voltage.

23-6 toward W because the wire moves perpendicular to the field here

23-7 toward B as this is perpendicular to the magnetic field lines

23-8 (a) M is lower potential. (b) from right to left (c) Current decreases. (d) 1.2×10^{-4} V (e) 0.45 m/s (f) Induced voltage and induced current double.

23-9 N would be lower potential and the current would go through the resistor from left to right. All quantitative answers about the speed of the rod and the induced voltage remain the same.

23-10 Approximately 2 W; approximately 3 s; the mechanical power needed to move the sliding rod at constant speed is the source of the power.

23-11 (a) south pole at left end, north pole at right end (b) current flows into the coil at B, out of the coil at A

23-12 (a) north pole at left end, south pole at right end (b) current flows out of the coil at B, into the coil at A

23-13 The north pole of the bar magnet is moved away from the coil, or the south pole of the bar magnet is moved toward the coil.

23-14 clockwise

23-15 (a) right to left (b) toward the bottom of the page (c) toward the top of the page

23-16 thermal energy in the heating of the wires and resistor as the current flows through the circuit

23-17 (a) moving toward top of page (b) toward bottom of page (c) toward top of page

23-18 (a) ii, iii, iv, and v (b) ii, iii, iv, and v

23-19 up through resistor

23-20 (a) increases (b) increases (c) increases (d) increases

23-21 25 turns

23-22 (a) step-down transformer (b) decreases voltage by a factor of 15 (c) 8.0 V

23-23 $N_S/N_P = 0.38$

23-24 (a) Primary current is larger in a step-up transformer.
(b) $I_P/I_S = 8.2$

23-25 (a) 8.6×10^{-3} A (b) 3.9 kV (c) 2.4×10^{-2} A (d) 94 W

23-26 There needs to be a changing current in order to generate a changing magnetic field for electromagnetic induction.

23-27 (a) 4.3×10^{2} A (b) 4.6×10^{5} W (c) 99%

23-28 (a) decrease (b) decrease (c) increase

23-29 Approximately 1/40; this is a step-down transformer.

23-30 Approximately 2; this is a step-up transformer.

23-31 (a) increase (b) decrease (c) decrease

23-32 (a) when the coil is perpendicular to the field (b) when the coil is parallel to the field

23-33 When there is an AC current in the primary coil, there is a time-varying magnetic field in the core. This time-varying magnetic field means there is a time-varying magnetic flux through the secondary coil, which means there is an induced voltage.

23-34 (a) 6.1×10^{-7} Wb (b) 2.4×10^{-7} Wb

23-35 7.8×10^{-2} V

23-36 (a) 4.15×10^{-5} Wb (b) 0.319 V

Chapter Review

Multiple-Choice Questions

23-37 (e)

23-38 (a)

23-39 (c)

23-40 (a)

23-41 (c)

23-42 (d)

Review Questions and Problems

23-43 (a) 3.2×10^{-4} V (b) 2.6×10^{-7} A

23-44 (a) Current goes from right to left through the resistor; south pole on left-hand side, north pole on right. (b) Current goes from left to right through the resistor; south pole on right-hand side, north pole on left.

23-45 counterclockwise

23-46 840 turns

23-47 reduces the I^2R power loss in the lines

23-48 (a) 10.2 (b) 2.53×10^{6} W (c) 11.2 A

23-49 81 kW saved

23-50 arranging the coil so that it is perpendicular to the field; increasing the magnetic field strength

23-51 2.2×10^{-6} Wb

23-52 (a) 2.2×10^{-4} s (b) from top to bottom through the resistor

Applying Your Knowledge

23-53 (a) 1.1×10^{-5} A (b) from right to left through the sliding wire

23-54 (a) 2.1×10^{-8} N, up (b) 9.5×10^{-9} W (c) 9.5×10^{-9} W (d) The rate of work done by the external force is equal to the rate of energy dissipated by the resistance in the loop; conservation of energy.

23-55 (a) counterclockwise (b) Once current reaches steady state in loop #2, there is no current in loop #1, as the magnetic flux is no longer changing. (c) Current in loop #1 will appear again when the switch is opened; same magnitude as observed in (a) but now in the clockwise direction.

23-56 1.3×10^{-2} T

23-57 (a) $\Phi = B\,L\,d$ (b) $\Delta\Phi = B\,L\,v\,\Delta t$ (c) $\Delta V = B\,L\,v$

Chapter 24 Nuclear Physics: Theory and Medical Applications

Exercises

24-1 (a) 11 (b) 23

24-2 A particle (proton or neutron) in the nucleus of an atom is called a nucleon.

24-3 neon

24-4 (a) 15 protons, 16 neutrons (b) 15 electrons (c) phosphorus

24-5 Carbon has atomic number $(Z) = 6$, whereas nitrogen has atomic number $(Z) = 7$. They are different chemical elements because they have different numbers of protons in their nuclei.

24-6 (a) 39 (b) 20 neutrons (c) potassium

24-7 (A) and (C) are both isotopes of carbon

24-8 $^{29}_{9}\text{F}$

24-9 4.54×10^{-12} J

24-10 4.94×10^{2} MeV; 8.8 MeV per nucleon

24-11 approx. 1×10^{32} nuclei in total given that there are approximately 4×10^{31} water molecules present

24-12 (a) $N \approx Z$; (b) $N > Z$

24-13 (a) $^{38}_{20}\text{Ca} \rightarrow {}^{38}_{19}\text{K} + {}^{0}_{1}\text{e} + \upsilon$
(b) $^{154}_{68}\text{Er} \rightarrow {}^{4}_{2}\text{He} + {}^{150}_{66}\text{Dy}$
(c) $^{298}_{92}\text{U} \rightarrow {}^{4}_{2}\text{He} + {}^{294}_{90}\text{Th}$
(d) $^{40}_{19}\text{K} \rightarrow {}^{0}_{-1}\text{e} + {}^{40}_{20}\text{Ca} + \bar{\upsilon}$

24-14 parent: $^{38}_{20}\text{Ca}$; daughter: $^{38}_{19}\text{K}$

24-15 C

24-16 D

24-17 (a) β^{+} decay (b) $^{18}_{10}\text{Ne} \rightarrow {}^{18}_{9}\text{F} + {}^{0}_{1}\text{e} + \upsilon$

24-18 (a) β^{-} decay (b) $^{28}_{12}\text{Mg} \rightarrow {}^{28}_{13}\text{Al} + {}^{0}_{-1}\text{e} + \bar{\upsilon}$

24-19 higher energy (higher frequency or lower wavelength)

24-20 Atomic number and mass number are unchanged.

24-21 γ rays, because they have no charge

24-22 This is false because most of the radioactivity is generated by naturally found materials and cosmological processes.

24-23 (a) ${}^{14}_{7}N + {}^{1}_{0}n \rightarrow {}^{14}_{6}C + {}^{1}_{1}H$

(b) ${}^{59}_{27}Co + {}^{1}_{0}n \rightarrow {}^{60}_{27}Co$

(c) ${}^{6}_{3}Li + {}^{3}_{2}He \rightarrow {}^{8}_{5}B + {}^{1}_{0}n$

24-24 24 d

24-25 1/256

24-26 5.8×10^{3} Bq

24-27 (a) 67% (b) 20% (c) 1.3%

24-28 0.029 s

24-29 4.0 s

24-30 9.4 y

24-31 approx. 3900 y

24-32 The ratio in the artefact is approximately 0.37 times that of the present ${}^{14}_{6}C/{}^{12}_{6}C$ ratio.

24-33 2.5×10^{12} Bq

24-34 111 min

24-35 6.6×10^{8} Bq

24-36 4.3 min

24-37 approx. 9×10^{3} Bq

24-38 (a) ${}^{235}_{92}U + {}^{1}_{0}n \rightarrow 3({}^{1}_{0}n) + {}^{95}_{39}Y + {}^{138}_{53}I$

(b) ${}^{1}_{0}n + {}^{239}_{94}Pu \rightarrow {}^{141}_{55}Cs + {}^{96}_{39}Y + 3({}^{1}_{0}n)$

(c) ${}^{12}_{6}C + {}^{4}_{2}He \rightarrow {}^{16}_{8}O$

(d) ${}^{12}_{6}C + {}^{12}_{6}C \rightarrow {}^{23}_{11}Na + {}^{1}_{1}H$

(e) ${}^{8}_{4}Be + {}^{4}_{2}He \rightarrow {}^{12}_{6}C$

24-39 moderator: slows down the neutrons released in fission so that they can undergo another fission reaction

control rods: made of material that absorbs neutrons well; used to control the rate of fission in the core

coolant: liquid or gas used to remove heat from the core released by the fission reactions

24-40 2.8×10^{-11} J

24-41 180 MeV

24-42 18 MeV

24-43 Approximately 1×10^{21} reactions per second are required.

24-44 approx. 8×10^{13} J/kg

24-45 approx. 800 kg

24-46 approx. 1 kg

24-47 When holding it in your hand, the health concern arises from the β particles only, since the α particles will likely have insufficient range to penetrate the dead layer of skin on the surface. However, when swallowed, both the α and β particles will contribute to the health effects.

24-48 x rays, γ rays, α particles, β particles, high-frequency UV light

24-49 energy released in reaction: 5.3 MeV; 5.2 MeV given to the α particle; therefore, a range of 38 μm in tissue

24-50 energy released in reaction: 5.6 MeV; 5.5 MeV given to the α particle; therefore, a range of 42 μm in tissue

24-51 0.21 cm

24-52 0.79 cm

24-53 (1) photoelectric effect: a photon is absorbed and an electron is emitted

(2) Compton scatter: a photon changes direction and transfers partial energy to the atom, and an electron is emitted

(3) pair production: a photon with energy greater than 1.02 MeV is converted into an electron–positron pair

24-54 half-value layer: 110 μm; mean free path: 160 μm

24-55 attenuation coefficient: 1.8 cm^{-1}; mean free path: 0.55 cm

24-56 Electron–ion pairs interact with biological molecules, causing localized molecular damage or alteration; damaged cells impair biological function, leading to mutation or cellular death.

24-57 an adaptive response by an organism to a stress of some kind; (1) injection of an inactive virus (vaccination), (2) resistance training for muscle development, (3) low level radiation exposure to trigger repair mechanisms

24-58 the radiation to which we are exposed in our natural and technological world; (1) cosmic rays, (2) radon, (3) medical procedures, (4) internal sources such as ${}^{40}K$, etc.

24-59 Photons are differentially attenuated through the subject. Regions of high attenuation result in less exposed film on the far side, while regions of low attenuation result in more exposed film.

24-60 Blood and surrounding tissue have essentially identical attenuation properties so introducing a contrast agent that is highly attenuating will allow the vessels to be seen on the image.

24-61 (a) 2.5 cm (b) 0.58 cm (c) 290 μm

24-62 24%

24-63 5.1×10^{4} Bq

24-64 (a) provides functional information rather than anatomical (b) requires an on-site cyclotron accelerator to make the short-lived positron emitters

24-65 (1) generate an image of the tumour with CT (2) determine the particular details of the irradiation based on the anatomy (3) simulate the treatment to make sure the exposure is distributed as desired (4) create moulds to immobilize the patient (5) deliver treatment

24-66 Multi-leaf collimators allow for the radiation beam to be shaped with each beam delivery.

24-67 (a) good radiation localization with patient movement issues (b) more invasive procedure than external irradiation

Chapter Review

Multiple-Choice Questions

24-68 (b)

24-69 (b)

24-70 (d)

24-71 (e)

24-72 (a)

24-73 (c)

24-74 (b)

24-75 (e)

24-76 (c)

24-77 (b)

24-78 (b)

24-79 (a)

24-80 (b)

24-81 (d)

24-82 (c)

24-83 (a)

24-84 (c)

24-85 (c)

Review Questions and Problems

24-86 (a) 23 (b) 11 (c) 12 (d) sodium

24-87 A nucleon (neutron or proton) is merely a particle that is found within the nucleus. A nuclide is a nucleus with a specific number of protons and neutrons, such as ^{16}O or ^{238}U. A nuclide is made up of a specific mix of nucleons.

24-88 Many chemical elements have isotopes with mass number 16 (e.g., ^{16}C, ^{16}N, ^{16}O, ^{16}F). Only the atomic number (Z) is unique for a given chemical element; therefore, it would make sense to ask which chemical element has $Z = 16$, to which the answer is sulfur.

24-89 130 MeV

24-90 550 MeV

24-91 (a) $^{118}_{54}Xe \rightarrow {}^{0}_{1}e + {}^{118}_{53}I$; β^+ decay
(b) $^{175}_{78}Pt \rightarrow {}^{4}_{2}He + {}^{171}_{76}Os$; α decay
(c) $^{12}_{6}C + {}^{1}_{1}H \rightarrow {}^{12}_{7}N + {}^{1}_{0}n$; nuclear reaction

24-92 4.30 MeV

24-93 They have relatively short half-lives compared to the age of Earth, whereas the half-lives of the isotopes of uranium are quite long.

24-94 7 half-lives

24-95 (a) 3.69 h (b) 0.271 h^{-1}

24-96 1.4×10^{11} nuclei remain.

24-97 1.4×10^{10} y

24-98 4.8 Bq

24-99 (a) uranium sample in which the percentage of ^{235}U present is higher than naturally found in the environment (b) fission event in which the reaction produces multiple neutrons that can then each cause subsequent fission events (c) converts the ^{238}U to ^{239}Pu, which is then used as fuel

24-100 (a) Both use thermal energy to create steam, which then drives turbines.
(b) Nuclear power generates thermal energy from nuclear processes, whereas fossil-fuel stations generate thermal energy from chemical processes (combustion of fossil fuel). In addition, nuclear power does not create greenhouse gases, but does result in radioactive waste that must be carefully sequestered.

24-101 (a) 2 (b) 198 MeV

24-102 13 MeV

24-103 energy released in the decay: 5.6 MeV; 5.5 MeV given to the α particle; therefore, a range of 42 μm in tissue is expected

24-104 310 μm

24-105 0.23 cm

24-106 15 cm

24-107 3.2 cm

24-108 The film in radiography is replaced with a camera in fluoroscopy to obtain real-time moving images of the anatomy.

24-109 An x-ray source irradiates a patient, and the degree of attenuation is determined by a detector on the far side. The source–detector assembly is then rotated around the patient, collecting attenuation data along many paths through the patient. A computer uses a process called back-projection to form the resulting 2D map of attenuation properties of the patient imaged.

24-110 1.0×10^5 Bq

Applying Your Knowledge

24-111 (a) $^{2}_{1}H$ and $^{4}_{2}He$: 1.1 MeV/nucleon and 7.1 MeV/nucleon respectively (b) $^{10}_{5}B$ and $^{12}_{6}C$: 6.5 MeV/nucleon and 7.7 MeV/nucleon respectively

24-112 2.4×10^{17} kg/m^3

24-113 $^{228}_{90}Th$

24-114 $^{212}_{82}Pb \rightarrow {}^{212}_{83}Bi + {}^{0}_{-1}e + \bar{\upsilon}$
$^{212}_{83}Bi \rightarrow {}^{212}_{84}Po + {}^{0}_{-1}e + \bar{\upsilon}$
$^{212}_{84}Po \rightarrow {}^{208}_{82}Pb + {}^{4}_{2}He$

24-115 (a) 2.1×10^{10} kg of coal (b) 660 kg

24-116 The attenuation coefficient of the various 'soft tissues' (i.e., breast tissue, fat, muscle) are more differentiated at lower photon energies. Standard radiographic imaging is not done at these lower energies, however, because the photons do not travel as far through tissue; thereby making it much harder to image deep into the body, which is not as much of a concern in mammography.

24-117 Energy released in the reaction: 2.80 MeV; 1.78 MeV is given to the α particle; therefore, a range of 9.99 μm in tissue is expected

INDEX

Note: Page numbers with "f" indicate figures; those with "t" indicate tables, and those with "n" indicate footnotes

E

F

G

H

I

M

N

O

Q

R

S

T

U

X

Y

Z

NUMERICAL CONSTANTS

FUNDAMENTAL PHYSICAL CONSTANTS

Name	Symbol	Value
Atomic mass unit	u	$1.660\ 539 \times 10^{-27}$ kg
Avogadro's number	N_A	6.022×10^{23} mol^{-1}
Boltzmann's constant	k_B	1.381×10^{-23} J·K^{-1}
Fundamental unit of charge	e	1.602×10^{-19} C
Coulomb's law constant	k	8.99×10^{9} N·m^2·C^{-2}
Permeability of free space	μ_0	$4\pi \times 10^{-7}$ T·m·A^{-1}
Mass of electron	m_e	$9.109\,383 \times 10^{-31}$ kg $= 5.485\ 799 \times 10^{-4}$ u
Mass of neutron	m_n	$1.674\,927 \times 10^{-27}$ kg $= 1.008\ 665$ u
Mass of proton	m_p	$1.672\,622 \times 10^{-27}$ kg $= 1.007\ 276$ u
Molar gas constant	R	8.314 J·mol^{-1}·K^{-1}
Planck's constant	h	6.626×10^{-34} J·s
Speed of light in vacuum	c	2.998×10^{8} m/s
Stefan–Boltzmann constant	σ	5.670×10^{-8} W·m^{-2}·K^{-4}
Universal gravitation constant	G	6.67×10^{-11} N·m^2·kg^{-2}

DATA FOR THE SOLAR SYSTEM

Object	Mass (kg)	Radius (m)	Mean density (kg/m^3)	Surface gravity (m/s^2)	Period of rotation (days)	Mean distance from Earth (m)
Moon	7.35×10^{22}	1.74×10^{6}	3.34×10^{3}	1.62	27.3	3.84×10^{8}
Sun	1.99×10^{30}	6.96×10^{8}	1.41×10^{3}	274	≈ 26	1.50×10^{11}

Planet	Mean distance from Sun (10^6 km)	Period of revolution (Earth years)	Mass (kg)	Equatorial radius (10^3 km)	Surface gravity (Earth g = 1)	Period of rotation[1] (Earth days)
Mercury	58	0.241	3.30×10^{23}	2.44	0.39	58.6
Venus	108	0.615	4.87×10^{24}	6.05	0.90	243
Earth	150	1.00	5.98×10^{24}	6.38	1.00	0.997
Mars	228	1.88	6.42×10^{23}	3.40	0.38	1.03
Jupiter	778	11.9	1.90×10^{27}	71.4	2.58	0.41
Saturn	1.43×10^{3}	29.5	5.67×10^{26}	60.3	1.11	0.43
Uranus	2.87×10^{3}	84.0	8.70×10^{25}	25.3	1.07	0.65
Neptune	4.50×10^{3}	165	1.03×10^{26}	24.6	1.40	0.77
Pluto[2]	5.91×10^{3}	248	6.6×10^{23}	1.20	0.059	6.4

[1] relative to the stars

[2] In 2006, the International Astronomical Union demoted Pluto from a major planet to a dwarf planet, which is a planet that has too small a gravitational field to have cleared its orbit of other objects. As of 2014 there are four other dwarf planets that have been named: Ceres, Haumea, Makemake, and Eris. The demotion Pluto does not take away from the amazing story of how Pluto was discovered in 1930 after painstaking research by a 24-year-old lab assistant, Clyde W. ugh, working in the Lowell Observatory in Flagstaff, Arizona.